**Solar System Data**

| Body | Mass (kg) | Mean Radius (m) | Period (s) | Distance from Sun (m) |
|---|---|---|---|---|
| Mercury | $3.18 \times 10^{23}$ | $2.43 \times 10^{6}$ | $7.60 \times 10^{6}$ | $5.79 \times 10^{10}$ |
| Venus | $4.88 \times 10^{24}$ | $6.06 \times 10^{6}$ | $1.94 \times 10^{7}$ | $1.08 \times 10^{11}$ |
| Earth | $5.98 \times 10^{24}$ | $6.37 \times 10^{6}$ | $3.156 \times 10^{7}$ | $1.496 \times 10^{11}$ |
| Mars | $6.42 \times 10^{23}$ | $3.37 \times 10^{6}$ | $5.94 \times 10^{7}$ | $2.28 \times 10^{11}$ |
| Jupiter | $1.90 \times 10^{27}$ | $6.99 \times 10^{7}$ | $3.74 \times 10^{8}$ | $7.78 \times 10^{11}$ |
| Saturn | $5.68 \times 10^{26}$ | $5.85 \times 10^{7}$ | $9.35 \times 10^{8}$ | $1.43 \times 10^{12}$ |
| Uranus | $8.68 \times 10^{25}$ | $2.33 \times 10^{7}$ | $2.64 \times 10^{9}$ | $2.87 \times 10^{12}$ |
| Neptune | $1.03 \times 10^{26}$ | $2.21 \times 10^{7}$ | $5.22 \times 10^{9}$ | $4.50 \times 10^{12}$ |
| Pluto | $\approx 1.4 \times 10^{22}$ | $\approx 1.5 \times 10^{6}$ | $7.82 \times 10^{9}$ | $5.91 \times 10^{12}$ |
| Moon | $7.36 \times 10^{22}$ | $1.74 \times 10^{6}$ | — | — |
| Sun | $1.991 \times 10^{30}$ | $6.96 \times 10^{8}$ | — | — |

**Physical Data Often Used[a]**

| | |
|---|---|
| Acceleration due to gravity | $9.80\ \mathrm{m/s^2}$ |
| Average earth-moon distance | $3.84 \times 10^{8}$ m |
| Average earth-sun distance | $1.496 \times 10^{11}$ m |
| Average radius of the earth | $6.37 \times 10^{6}$ m |
| Density of air (20°C and 1 atm) | $1.20\ \mathrm{kg/m^3}$ |
| Density of water (20°C and 1 atm) | $1.00 \times 10^{3}\ \mathrm{kg/m^3}$ |
| Mass of the earth | $5.98 \times 10^{24}$ kg |
| Mass of the moon | $7.36 \times 10^{22}$ kg |
| Mass of the sun | $1.99 \times 10^{30}$ kg |
| Standard atmospheric pressure | $1.013 \times 10^{5}$ Pa |

[a] These are the values of the constants as used in the text.

**Some Prefixes for Powers of Ten**

| Power | Prefix | Abbreviation | Power | Prefix | Abbreviation |
|---|---|---|---|---|---|
| $10^{-18}$ | atto | a | $10^{1}$ | deka | da |
| $10^{-15}$ | femto | f | $10^{2}$ | hecto | h |
| $10^{-12}$ | pico | p | $10^{3}$ | kilo | k |
| $10^{-9}$ | nano | n | $10^{6}$ | mega | M |
| $10^{-6}$ | micro | $\mu$ | $10^{9}$ | giga | G |
| $10^{-3}$ | milli | m | $10^{12}$ | tera | T |
| $10^{-2}$ | centi | c | $10^{15}$ | peta | P |
| $10^{-1}$ | deci | d | $10^{18}$ | exa | E |

# PHYSICS

## For Scientists & Engineers

Third edition

*Updated Version*

**VOLUME II**

*Isaac Newton's manuscripts on the aberration of light and the prism he used for his experiments. (Erich Lessing/ MAGNUM)*

# PHYSICS

## For Scientists & Engineers

Third edition

*Updated Version*

VOLUME II

Raymond A. Serway
*James Madison University*

**SAUNDERS GOLDEN SUNBURST SERIES**
**Saunders College Publishing**
**Philadelphia Fort Worth Chicago**
**San Francisco Montreal Toronto**
**London Sydney Tokyo**

Text Typeface: Caledonia
Compositor: Progressive Typographers, Inc.
Acquisitions Editor: John Vondeling
Developmental Editor: Ellen Newman
Managing Editor: Carol Field
Senior Project Manager: Sally Kusch
Copy Editor: Will Eaton
Manager of Art and Design: Carol Bleistine
Art and Design Coordinator: Caroline McGowan
Text Designer: Edward A. Butler
Cover Designer: Lawrence R. Didona
Text Artwork: Rolin Graphics
Layout Artist: Tracy Baldwin
Production Manager: Jay Lichty
Director of EDP: Tim Frelick
Marketing Manager: Marjorie Waldron

Cover: Beams of red, green, and blue light strike a D-shaped plastic lens and are bent toward a focus. Photography by S. Schwartzenberg/© The Exploratorium.

Printed in the United States of America

Physics for Scientists & Engineers, 3/e, VOLUME II, Updated Printing

0-03-096029-0

Library of Congress Catalog Card Number: 89-43326

345 032 98765432

# Preface

This two-volume textbook is intended for a course in introductory physics for students majoring in science or engineering. The book is an extended version of *Physics for Scientists and Engineers* in that Volume II includes seven additional chapters covering selected topics in modern physics. This material on modern physics has been added to meet the needs of those universities which choose to cover the basic concepts of quantum physics and its application to atomic, molecular, solid state, and nuclear physics as part of their curriculum.

The entire contents of the text could be covered in a three-semester course, but it is possible to use the material in shorter sequences with the omission of selected chapters and sections. The mathematical background of the student taking this course should ideally include one semester of calculus. If that is not possible, the student should be enrolled in a concurrent course in introduction to calculus.

*Star trails around the south celestial pole. (© Anglo-Australian Telescope Board 1980)*

## OBJECTIVES

The main objectives of this introductory physics textbook are twofold: to provide the student with a clear and logical presentation of the basic concepts and principles of physics, and to strengthen an understanding of the concepts and principles through a broad range of interesting applications to the real world. In order to meet these objectives, emphasis is placed on sound physical arguments. At the same time, I have attempted to motivate the student through practical examples that demonstrate the role of physics in other disciplines.

## COVERAGE

The material covered in this book is concerned with fundamental topics in classical physics and modern physics. The book is divided into six parts. In the first volume, Part I (Chapters 1 – 15) deals with the fundamentals of Newtonian mechanics and the physics of fluids; Part II (Chapter 16 – 18) covers wave motion and sound; Part III (Chapters 19 – 22) is concerned with heat and thermodynamics. In this volume, Part IV (Chapters 23 – 34) treats electricity and magnetism, Part V (Chapters 35 – 38) covers light and optics, and Part VI (Chapters 39 – 47) deals with relativity and modern physics.

**Features** Most instructors would agree that the textbook selected for a course should be the student's major "guide" for understanding and learning the subject matter. Furthermore, a textbook should be easily accessible and should be styled and written for ease in instruction. With these points in mind, I have included many pedagogic features in the textbook which are intended to enhance its usefulness to both the student and instructor. These are as follows:

**Organization** The book is divided into six parts: mechanics, wave motion and sound, heat and thermodynamics, electricity and magnetism, light and optics, and modern physics. Each part includes an overview of the subject matter to be covered in that part and some historical perspectives.

**Style** As an aid for rapid comprehension, I have attempted to write the book in a style that is clear, logical, and succinct. The writing style is somewhat informal and relaxed, which I hope students will find appealing and enjoyable to read. New terms are carefully defined, and I have tried to avoid jargon.

**Previews** Most chapters begin with a chapter preview, which includes a brief discussion of chapter objectives and content.

**Important Statements and Equations** Most important statements and definitions are set in blue print for added emphasis and ease of review. Important equations are highlighted with a tan screen for review or reference.

**Marginal Notes** Comments and marginal notes set in blue print are used to locate important statements, equations, and concepts in the text.

*Experiencing circular motion on a roller coaster. (© Zerschling, Photo Researchers)*

**Illustrations** The readability and effectiveness of the text material and worked examples are enhanced by the large number of figures, diagrams, photographs, and tables. Full color is used to add clarity to the artwork and to make it as realistic as possible. For example, vectors are color-coded, and curves in *xy*-plots are drawn in color. Three-dimensional effects are produced with the use of color airbrushed areas, where appropriate. The color photographs have been carefully selected, and their accompanying captions have been written to serve as an added instructional tool. Several chapter-opening photographs, particularly in the chapters on mechanics, include color-coded vector overlays that illustrate and present physical principles more clearly and apply them to real-world situations. A complete description of the pedagogical use of color appears on p. xvii.

**Mathematical Level** Calculus is introduced gradually, keeping in mind that a course in calculus is often taken concurrently. Most steps are shown when basic equations are developed, and reference is often made to mathematical appendices at the end of the text. Vector products are introduced later in the text where they are needed in physical applications. The dot product is introduced in Chapter 7, "Work and Energy." The cross product is introduced in Chapter 11, which deals with rotational dynamics.

**Worked Examples** A large number of worked examples of varying difficulty are presented as an aid in understanding concepts. In many cases, these examples will serve as models for solving the end-of-the-chapter problems. The examples are set off in a blue box, and the solution answers are highlighted with a tan screen. Most examples are given titles to describe their content.

**Worked Example Exercises** Many of the worked examples are followed immediately by exercises with answers. These exercises are intended to make the textbook more interactive with the student, and to test immediately the student's understanding of problem-solving techniques. The exercises represent extensions of the worked examples and are numbered in case the instructor wishes to assign them.

**Units** The international system of units (SI) is used throughout the text. The British engineering system of units (conventional system) is used only to a limited extent in the chapters on mechanics, heat, and thermodynamics.

**Biographical Sketches** Throughout the text I have added short biographies of important scientists to give the third edition a more historical emphasis.

**Problem-Solving Strategies and Hints** As an added feature of the third edition, I have included general strategies for solving the types of problems featured in both the examples and in the end-of-chapter problems. It is my hope that this feature will help the students identify the steps in solving a problem and eliminate any uncertainty they might have. This feature is highlighted by a light blue screen for emphasis and ease of location.

**Summaries** Each chapter contains a summary which reviews the important concepts and equations discussed in that chapter. The summaries are highlighted with a tan screen.

**Thought Questions** A list of questions requiring verbal answers is given at the end of each chapter. Some questions provide the student with a means of self-testing the concepts presented in the chapter. Others could serve as a basis for initiating classroom discussions. Answers to most questions are included in the Student Study Guide that accompanies the text.

**Problems** An extensive set of problems is included at the end of each chapter. Answers to odd-numbered problems are given at the end of the book; these pages have colored edges for ease of location. This Updated Version contains approximately 700 new problems, most of which are at the intermediate level. Some of these are laboratory problems, marked with a △, that allow the student to write solutions based on real data. For the convenience of both the student and the instructor, about two thirds of the problems are keyed to specific sections of the chapter. The remaining problems, labeled "Additional Problems," are not keyed to specific sections. In my opinion, assignments should consist mainly of the keyed problems to help build self-confidence in students.

In general, the problems within a given section are presented so that the straightforward problems are first, followed by problems of increasing difficulty. For ease in identifying the intermediate-level problems, the problem number is printed in blue. I have also included a small number of challenging problems, which are indicated by a problem number printed in magenta.

**Laboratory Problems** Selected problems throughout the text use real data to challenge students' problem-solving skills. These problems are marked with a △ for identification.

**Spreadsheet Problems and Examples** Selected problems and examples throughout the text have been keyed with a color box to indicate that there are spreadsheets on a separate disk that accompany these problems. This is an option; these problems can also be solved analytically without using the software. The spreadsheets will help the student work some of the difficult problems; instructions for use of the spreadsheets are found in Appendix F. The data disk is available free to instructors who adopt the text and can be used with a variety of programs (e.g., Lotus 1-2-3) that the instructor may already have.

**Calculator/Computer Problems** Numerical problems that can best be solved with the use of programmable calculators or a computer are given in a selected number of chapters. These will be useful in those courses where the instructor wishes to put programming skills to practice.

**Guest Essays** I have included 17 essays, written by guest authors, on topics of current interest to scientists and engineers. Three of the essays in the Updated Version are entirely new. The essays are intended as supplemental readings for the student and are marked with a blue bar at the edge of the page so they can be located easily.

*Hurricane Elena photographed from space. (NASA)*

**Special Topics** Many chapters include special topic sections which are intended to expose the student to various practical and interesting applications of physical principles. Most of these are considered optional, and as such are labeled with an asterisk (*).

**Appendices and Endpapers** Several appendices are provided at the end of the text, including the new appendix with instructions for problem-solving with spreadsheets. Most of the appendix material represents a review of mathematical techniques used in the text, including scientific notation, algebra, geometry, trigonometry, differential calculus, and integral calculus. Reference to these appendices is made throughout the text. Most mathematical review sections include worked examples and exercises with answers. In addition to the mathematical reviews, the appendices contain tables of physical data, conversion factors, atomic masses, and the SI units of physical quantities, as well as a periodic chart. Other useful information, including fundamental constants and physical data, planetary data, a list of standard prefixes, mathematical symbols, the Greek alphabet, and standard abbreviation of units appears on the endpapers.

## CHANGES IN THE UPDATED VERSION

A number of changes and improvements have been made in preparing the Updated Version of this text. Many of these changes are in response to comments and suggestions offered by users of the text and reviewers of the manuscript. The following represent the major changes in the Updated Version:

1. Approximately 700 new problems, including many at the intermediate level. Some of these are Laboratory Problems, marked with a $\Delta$, which allow students to write solutions based on real data. *All of the problems have been thoroughly reviewed for accuracy.*
2. Some of the photographs in the mechanics section include color-coded vector overlays to illustrate and present physical principles more clearly.
3. Three entirely new guest essays on dark matter, chaos, and the third law of thermodynamics.
4. New expanded ancillary package featuring a new Instructor's Manual with solutions to all problems, including the 700 new problems in the Updated Version. *All solutions have been reviewed for accuracy.* Other items in the expanded package include Physics Videodisc, Demonstration Videotape, Selected Solutions Transparency Masters, and Serway Physics Problem Set for *Interactive Physics.* The package is described in more detail below.
5. Careful attention to accuracy, as always, marks the Updated Version. The text, artwork, and problems have all been rigorously reviewed, and we believe the book is as close to error-free as possible. If you believe you have found an error, please send your comments to Chiara Puffer, Editorial Assistant, Physics, Saunders College Publishing, Public Ledger Building, 620 Chestnut St., Suite 560, Philadelphia, PA 19106-3399. We will make every effort to correct the error in the next printing.

*Computer simulation of the pattern of air flow around a space shuttle. (NASA)*

## ANCILLARIES

The Updated Version has a new expanded package:

**Instructor's Manual with Solutions** This manual consists of complete, worked-out solutions to all the problems in the text and answers to even-numbered problems.

The solutions to the 700 new problems in the Updated Version are included; they are marked so the instructor can identify the new problems. *All solutions have been carefully reviewed for accuracy.*

**Saunders Physics Videodisc** contains 70 physics demonstrations and 500 still images, 200 that correlate directly with illustrations from Serway's *Physics for Scientists and Engineers, third edition* and a wide variety of other sources.

**Physics Demonstration Videotape** by J. C. Sprott of the University of Wisconsin, Madison, is a unique two-hour video-cassette divided into 12 primary topics. Each topic contains between four and nine demonstrations for a total of 70 physics demonstrations.

**Selected Solutions Transparency Masters** include selected worked-out solutions that can be used in the classroom when transferred to acetates.

**Printed Test Bank** This printed test bank contains 2300 multiple choice questions from the software disk. It is provided as another source of test questions and is helpful for the instructor who does not have access to a computer.

**Computerized Test Bank** Available for the IBM PC, Apple II, and Macintosh computers, this test bank contains over 2300 multiple choice questions, representing every chapter of the text. The IBM and Macintosh versions have been revised in the Updated Version. The Macintosh Test Bank includes a new, user-friendly program. The test bank enables the instructor to create many unique tests and permits the editing of questions as well as addition of new questions. The software program solves all problems and prints each answer on a separate grading key. All questions have been reviewed for accuracy.

**Serway Physics Problem Set for *Interactive Physics*** contains approximately 100 problems from the text on disk for use with *Interactive Physics* 2.0 for the Macintosh by Knowledge Revolution.

**So You Want to Learn Physics: A Preparation for Scientists and Engineers**, by Rodney Cole, University of California, Davis, is a preparatory physics text or supplement that offers content material, worked examples, and a review of techniques learned in class to students who need additional preparation for calculus-based physics. The friendly, straight-forward tone encourages students to experiment with mathematics as it is used in physics.

The following ancillaries accompanied the third edition and are still available:

**Student Study Guide** The Study Guide contains chapter objectives, a skills section that reviews mathematical techniques, and suggested approaches to problem-solving methodology; notes from selected chapter sections, which include a glossary of important terms, theorems, and concepts; answers to selected end-of-chapter questions; and solutions to selected end-of-chapter problems.

**Student Study Guide with Computer Exercises (IBM or Macintosh)** The Study Guide also includes the option of using a select group of computer programs (presented in special computer modules) that are interactive in nature. The student's input will have direct and immediate effect on the output. This feature will enable students to work through many challenging numerical problems and experience the power of the computer in scientific computations. The courseware disk is available for the IBM or Macintosh computers and is packaged with the Study Guide.

**Spreadsheet Data Disk** The Spreadsheet Data Disk (IBM) contains problems and examples keyed to the text. The Spreadsheet Disk is a data disk that can be used with a variety of spreadsheet programs (e.g., Lotus 1-2-3) that the instructor may already have. Use of the Spreadsheet Disk is optional.

**Overhead Transparency Acetates** This collection of transparencies consists of 300 full-color figures from the text and features large print for easy viewing in the classroom.

**Physics Laboratory Manual** To supplement the learning of basic physical principles while introducing laboratory procedures and equipment, each chapter of the laboratory manual includes a pre-laboratory assignment, objectives, an equipment list, the theory behind the experiment, experimental procedure, calculations, graphs, and questions. In addition, a laboratory report is provided for each experiment so the student can record data, calculations, and experimental results.

**Instructor's Manual for Physics Laboratory Manual** Each chapter contains a discussion of the experiment, teaching hints, answers to selected questions, and a post-laboratory quiz with short answer and essay questions. We have also included a list of the suppliers of scientific equipment and a summary of the equipment needed for all the laboratory experiments in the manual.

## TEACHING OPTIONS

This book is structured in the following sequence of topics: Volume I includes classical mechanics, matter waves, and heat and thermodynamics; Volume II includes electricity and magnetism, light waves, optics, relativity, and modern physics. This presentation is a more traditional sequence, with the subject of matter waves presented before electricity and magnetism. Some instructors may prefer to cover this material after completing electricity and magnetism (after Chapter 34). The chapter on relativity was placed near the end of the text because this topic is often treated as an introduction to the era of "modern physics." If time permits, instructors may choose to cover Chapter 39 in Volume II after completing Chapter 14, which concludes the material on Newtonian mechanics.

For those instructors teaching a two-semester sequence, some sections and chapters could be deleted without any loss in continuity. I have labeled these with asterisks (*) in the Table of Contents and in the appropriate sections of the text. For student enrichment, some of these sections or chapters could be given as extra reading assignments. The guest essays could also serve the same purpose.

*Vertical-axis wind generator. (U.S. Department of Energy)*

## ACKNOWLEDGMENTS

I thank the following people for their suggestions and assistance during the preparation of the Updated Version of the third edition: Stephen Baker, Rice University; Kenneth Brownstein, University of Maine; C.H. Chan, The University of Alabama in Huntsville; Clifton Bob Clark, University of North Carolina at Greensboro; Walter C. Connolly, Appalachian State University; James L. DuBard, Binghamton–Southern College; Joe L. Ferguson, Mississippi State University; R.H. Garstang, University of Colorado at Boulder; Jerome W. Hosken, City College of San Francisco; Francis A. Liuima, Boston College; Ralph V. McGrew, Broome Community College; David Murdock, Tennessee Technological University; C.W. Scherr, University of Texas at Austin; Eric Sheldon, University of Massachusetts–Lowell; Richard R. Sommerfield, Foothill College; Herman Trivilino, San Jacinto College North; Steve van Wyk, Chapman College; Joseph Veit, Western Washington University; T.S. Venkataraman, Drexel University; and Noboru Wada, Colorado School of Mines.

The third edition of this textbook was prepared with the guidance and assistance of many professors who reviewed part or all of the manuscript. I wish to acknowledge the following scholars and express my sincere appreciation for their suggestions, criticisms, and encouragement:

George Alexandrakis, University of Miami; Bo Casserberg, University of Minnesota; Soumya Chakravarti, California State Polytechnic University; Edward Chang, University of Massachusetts, Amherst; Hans Courant, University of Minnesota; F. Paul Esposito, University of Cincinnati; Clark D. Hamilton, National Bureau of Standards; Mark Heald, Swarthmore College; Paul Holoday, Henry Ford Community College;

Larry Kirkpatrick, Montana State University; Barry Kunz, Michigan Technological University; Douglas A. Kurtze, Clarkson University; Robert Long, Worcester Polytechnic Institute; Nolen G. Massey, University of Texas at Arlington; Charles E. McFarland, University of Missouri at Rolla; James Monroe, The Pennsylvania State University, Beaver Campus; Fred A. Otter, University of Connecticut; Eric Peterson, Highland Community College; Jill Rugare, DeVry Institute of Technology; Charles Scherr, University of Texas at Austin; John Shelton, College of Lake County; Kervork Spartalian, University of Vermont; Robert W. Stewart, University of Victoria; James Stith, United States Military Academy; Carl T. Tomizuka, University of Arizona; Som Tyagi, Drexel University; James Walker, Washington State University; George Williams, University of Utah; and Edward Zimmerman, University of Nebraska, Lincoln.

Special thanks go to the many people who provided me with useful comments and suggestions for improvement during the development of this third edition. These include Albert A. Bartlett, David R. Currot, Chelcie Liu, Howard C. McAllister, A. J. Slavin, J. C. Sprott, and William W. Wood.

I would also like to thank the following professors for their suggestions during the development of the prior editions of this textbook:

Elmer E. Anderson, University of Alabama; Wallace Arthur, Fairleigh Dickinson University; Duane Aston, California State University at Sacramento; Richard Barnes, Iowa State University; Marvin Blecher, Virginia Polytechnic Institute and State University; William A. Butler, Eastern Illinois University; Don Chodrow, James Madison University; Clifton Bob Clark, University of North Carolina at Greensboro; Lance E. De Long, University of Kentucky; Jerry S. Faughn, Eastern Kentucky University; James B. Gerhart, University of Washington; John R. Gordon, James Madison University; Herb Helbig, Clarkson University; Howard Herzog, Broome Community College; Larry Hmurcik, University of Bridgeport; William Ingham, James Madison University; Mario Iona, University of Denver; Karen L. Johnston, North Carolina State University; Brij M. Khorana, Rose-Hulman Institute of Technology; Carl Kocher, Oregon State University; Robert E. Kribel, Jacksonville State University; Fred Lipschultz, University of Connecticut; Francis A. Liuima, Boston College; Charles E. McFarland, University of Missouri, Rolla; Clem Moses, Utica College; Curt Moyer, Clarkson University; Bruce Morgan, U.S. Naval Academy; A. Wilson Nolle, The University of Texas at Austin; Thomas L. O'Kuma, San Jacinto College North; George Parker, North Carolina State University; William F. Parks, University of Missouri, Rolla; Philip B. Peters, Virginia Military Institute; Joseph W. Rudmin, James Madison University; James H. Smith, University of Illinois at Urbana-Champaign; Edward W. Thomas, Georgia Institute of Technology; Gary Williams, University of California, Los Angeles; George A. Williams, University of Utah; and Earl Zwicker, Illinois Institute of Technology.

I would like to thank the following people for contributing many interesting new problems and questions to the text: Ron Canterna, University of Wyoming; Paul Feldker, Florissant Valley Community College; Roger Ludin, California Polytechnic State University; Richard Reimann, Boise State University; Jill Rugare, DeVry Institute of Technology; Stan Shepard, The Pennsylvania State University; Som Tyagi, Drexel University; Steve van Wyk, Chapman College; and James Walker, Washington State University.

I would also like to thank the following people for writing guest essays: Isaac D. Abella, University of Chicago; Albert A. Bartlett, University of Colorado at Boulder; Gordon Batson, Clarkson University; Leon Blitzer, University of Arizona; Steven Csorna, Vanderbilt University; Roger A. Freedman and Paul K. Hansma, University of California, Santa Barbara; Robert G. Fuller, University of Nebraska; Clark D. Hamilton, National Bureau of Standards; Edward Lacy; Samson A. Marshall, Michigan Technological Institute; John D. Meakin, University of Delaware; Philip Morrison, Massachusetts Institute of Technology; Brian B. Schwartz, Brooklyn College, C.U.N.Y., and the American Physical Society; J. Clint Sprott, University of Wisconsin; Virginia Trimble, University of California, Irvine, and University of Maryland, College Park; Hans Christian von Baeyer, College of William and Mary; Clifford Will, Washington Univer-

*A surfer "riding the pipe" on a wave. (© Doug Peebles/Index Stock International)*

sity; Dean A. Zollman, Kansas State University; and Alma C. Zook, Pomona College. I appreciate the assistance of Carl T. Tomizuka in coordinating the essays.

I am especially grateful to the following people for their careful accuracy reviews of all the problems and examples in the text: Stanley Bashkin, University of Arizona; Jeffrey J. Braun, University of Evansville; Louis H. Cadwell, Providence College, Ralph V. McGrew, Broome Community College; Charles D. Teague, Eastern Kentucky University; and Steve van Wyk, Chapman College.

I appreciate the assistance of Jeffrey J. Braun, Charles Teague and Steve van Wyk in reorganizing the problem sets. My grateful thanks also go to Steve van Wyk and Louis H. Cadwell for the preparation of the Instructor's Manual that accompanies the text. I am indebted to my colleague and friend John R. Gordon for his many contributions during the development of this text, his continued encouragement and support, and for his expertise in revising the Student Study Guide. I am grateful to David Oliver for developing the computer software that accompanies the Student Study Guide and to David Stetser for developing the Spreadsheet Data Disks that accompany the text and the Spreadsheet Appendix in the back of this text. Support for this work has been provided by Miami University — Middletown and the Department of Physics and Astronomy, Center for Advanced Studies, University of New Mexico, Albuquerque, New Mexico. I thank David Loyd for preparing the Physics Laboratory Manual and accompanying Instructor's Manual that can be used with this text. I appreciate the assistance of Louis H. Cadwell in preparing the answers that appear at the end of the text, and in preparing some of the test questions for the Computerized Test Bank and Printed Test Bank. I also thank the staff of the Physics Department at Georgia Tech for providing many of the questions for this test bank. I am grateful to Mario Iona for making many excellent suggestions for improving the figures in the text. I thank Sarah Evans, Ellen Newman, Henry Leap, and Jim Lehman for locating and/or providing many excellent photographs. I thank my son Mark for writing many of the biographical sketches included in this edition. I thank Agatha Brabon, Linda Delosh, Mary Thomas, Georgina Valverde, and Linda Miller for an excellent job in typing various stages of the original manuscript. During the development of this textbook, I have benefited from valuable discussions with many people including Subash Antani, Gabe Anton, Randall Caton, Don Chodrow, Jerry Faughn, John R. Gordon, William Ingham, David Kaup, Len Ketelsen, Henry Leap, H. Kent Moore, Charles McFarland, Frank Moore, Clem Moses, William Parks, Dorn Peterson, Joe Rudmin, Joe Scaturro, Alex Serway, John Serway, Georgio Vianson, and Harold Zimmerman. Special recognition is due to my mentor and friend, Sam Marshall, a gifted teacher and scientist who helped me sharpen my writing skills while I was a graduate student.

Special thanks and recognition go to the professional staff at Saunders College Publishing for their fine work during the development and production of this text, especially Ellen Newman, Senior Developmental Editor; Sally Kusch, Senior Project Manager; and Carol Bleistine, Manager of Art and Design. I thank John Vondeling, Associate Publisher, for his great enthusiasm for the project, his friendship, and his confidence in me as an author. I am most appreciative of the excellent artwork by Tom Mallon and the excellent design work by Edward A. Butler.

A special note of appreciation goes to the hundreds of students at Clarkson University who used the first edition of this text in manuscript form during its development. I am most grateful for the supportive environment provided by James Madison University. I also wish to thank the many users of the second edition who submitted suggestions and pointed out errors. With the help of such cooperative efforts, I hope to have achieved my main objective; that is, to provide an effective textbook for the student.

And last, I thank my wonderful family for their continued patience and understanding. The completion of this enormous task would not have been possible without their endless love and faith in me.

**Raymond A. Serway**
*James Madison University*
*Harrisonburg, Virginia*

# To the Student

I feel it is appropriate to offer some words of advice which should be of benefit to you, the student. Before doing so, I will assume that you have read the preface, which describes the various features of the text that will help you through the course.

*Computer simulation of gaseous flow around a post in a rocket engine. (U.S. Department of Energy)*

## HOW TO STUDY

Very often instructors are asked "How should I study physics and prepare for examinations?" There is no simple answer to this question, but I would like to offer some suggestions based on my own experiences in learning and teaching over the years.

First and foremost, maintain a positive attitude towards the subject matter, keeping in mind that physics is the most fundamental of all natural sciences. Other science courses that follow will use the same physical principles, so it is important that you understand and be able to apply the various concepts and theories discussed in the text.

## CONCEPTS AND PRINCIPLES

It is essential that you understand the basic concepts and principles *before* attempting to solve assigned problems. This is best accomplished through a careful reading of the textbook before attending your lecture on that material. In the process, it is useful to jot down certain points which are not clear to you. Take careful notes in class, and then ask questions pertaining to those ideas that require clarification. Keep in mind that few people are able to absorb the full meaning of scientific material after one reading. Several readings of the text and notes may be necessary. Your lectures and laboratory work should supplement the text and clarify some of the more difficult material. You should reduce memorization of material to a minimum. Memorizing passages from a text, equations, and derivations does not necessarily mean you understand the material. Your understanding of the material will be enhanced through a combination of efficient study habits, discussions with other students and instructors, and your ability to solve the problems in the text. Ask questions whenever you feel it is necessary.

## STUDY SCHEDULE

It is important to set up a regular study schedule, preferably on a daily basis. Make sure to read the syllabus for the course and adhere to the schedule set by your instructor. The lectures will be much more meaningful if you read the corresponding textual material before attending the lecture. As a general rule, you should devote about two hours of study time for every hour in class. If you are having trouble with the course, seek the advice of the instructor or students who have taken the course. You may find it necessary to seek further instruction from experienced students. Very often, instructors will offer review sessions in addition to regular class periods. It is important that you avoid the practice of delaying study until a day or two before an exam. More often than not, this will lead to disastrous results. Rather than an all-night study session, it is better to briefly review the basic concepts and equations, followed by a good night's rest. If you feel in need of additional help in understanding the concepts, preparing for exams, or in problem-solving, we suggest that you acquire a copy of the Student Study Guide which accompanies the text, which should be available at your college bookstore.

## USE THE FEATURES

You should make full use of the various features of the text discussed in the preface. For example, marginal notes are useful for locating and describing important equations and concepts, while important statements and definitions are highlighted in color. Many useful tables are contained in appendices, but most are incorporated in the text where they are used most often. Appendix B is a convenient review of mathematical techniques. Answers to odd-numbered problems are given at the end of the text, and answers to most end-of-chapter questions are provided in the study guide. Exercises (with answers), which follow some worked examples, represent extensions of those examples, and in most cases you are expected to perform a simple calculation. Their purpose is to test your problem-solving skills as you read through the text. Problem-Solving Strategies and Hints are included in selected chapters throughout the text to give you additional information to help you solve problems. An overview of the entire text is given in the table of contents, while the index will enable you to locate specific material quickly. Footnotes are sometimes used to supplement the discussion or to cite other references on the subject. Many chapters include problems that require the use of programmable calculators or computers. Problems and examples with boxes can be solved either analytically or with the use of spreadsheets available from your instructor. These are intended for those courses that place some emphasis on numerical methods. You may want to develop appropriate programs for some of these problems even if they are not assigned by your instructor.

After reading a chapter, you should be able to define any new quantities introduced in that chapter, and discuss the principles and assumptions that were used to arrive at certain key relations. The chapter summaries and the review sections of the study guide should help you in this regard. In some cases, it will be necessary to refer to the index of the text to locate certain topics. You should be able to correctly associate with each physical quantity a symbol used to represent that quantity and the unit in which the quantity is specified. Furthermore, you should be able to express each important relation in a concise and accurate prose statement.

*A model steam engine. (Courtesy of CENCO)*

## THE IMPORTANCE OF PROBLEM SOLVING

R.P. Feynman, Nobel laureate in physics, once said, "You do not know anything until you have practiced." In keeping with this statement, I strongly advise that you develop the skills necessary to solve a wide range of problems. Your ability to solve problems will be one of the main tests of your knowledge of physics, and therefore you should try to solve as many problems as possible. It is essential that you understand basic concepts and principles before attempting to solve problems. It is good practice to try to find alternate solutions to the same problem. For example, problems in mechanics can be solved using Newton's laws, but very often an alternative method using energy considerations is more direct. You should not deceive yourself into thinking you understand the problem after seeing its solution in class. You must be able to solve the problem and similar problems on your own.

The method of solving problems should be carefully planned. A systematic plan is especially important when a problem involves several concepts. First, read the problem several times until you are confident you understand what is being asked. Look for any key words that will help you interpret the problem, and perhaps allow you to make certain assumptions. Your ability to interpret the question properly is an integral part of problem solving. You should acquire the habit of writing down the information given in a problem, and decide what quantities need to be found. You might want to construct a table listing quantities given, and quantities to be found. This procedure is sometimes used in the worked examples of the text. After you have decided on the method you feel is appropriate for the situation, proceed with your solution. General problem-solving strategies of this type are included in the text and are highlighted by a light blue screen.

I often find that students fail to recognize the limitations of certain formulas or physical laws in a particular situation. It is very important that you understand and remember the assumptions which underlie a particular theory or formalism. For example, certain equations in kinematics apply only to a particle moving with constant acceleration. These equations are not valid for situations in which the acceleration is not constant, such as the motion of an object connected to a spring, or the motion of an object through a fluid.

## GENERAL PROBLEM-SOLVING STRATEGY

Most courses in general physics require the student to learn the skills of problem solving, and examinations are largely composed of problems that test such skills. This brief section describes some useful ideas which will enable you to increase your accuracy in solving problems, enhance your understanding of physical concepts, eliminate initial panic or lack of direction in approaching a problem, and organize your work. One way to help accomplish these goals is to adopt a problem-solving strategy. Many chapters in this text will include a section labeled "Problem-Solving Strategies and Hints" which should help you through the "rough spots."

In developing problem-solving strategies, five basic steps are commonly used.

1. Draw a suitable diagram with appropriate labels and coordinate axes if needed.
2. As you examine what is being asked in the problem, identify the basic physical principle (or principles) that are involved, listing the knowns and unknowns.
3. Select a basic relationship or derive an equation that can be used to find the unknown, and solve the equation for the unknown symbolically.
4. Substitute the given values along with the appropriate units into the equation.
5. Obtain a numerical value for the unknown. The problem is verified and receives a check mark if the following questions can be properly answered: Do the units match? Is the answer reasonable? Is the plus or minus sign proper or meaningful?

Diagram

Given Data

Basic Equation

Working Equation

Evaluation and Check

A menu for problem solving.

One of the purposes of this strategy is to promote accuracy. Properly drawn diagrams can eliminate many sign errors. Diagrams also help to isolate the physical principles of the problem. Symbolic solutions and carefully labeled knowns and unknowns will help eliminate other careless errors. The use of symbolic solutions should help you think in terms of the physics of the problem. A check of units at the end of the problem can indicate a possible algebraic error. The physical layout and organization of your problem will make the final product more understandable and easier to follow. Once you have developed an organized system for examining problems and extracting relevant information, you will become a more confident problem solver.

---

EXAMPLE A person driving in a car at a speed of 20 m/s applies the brakes and stops in a distance of 100 m. What was the acceleration of the car?

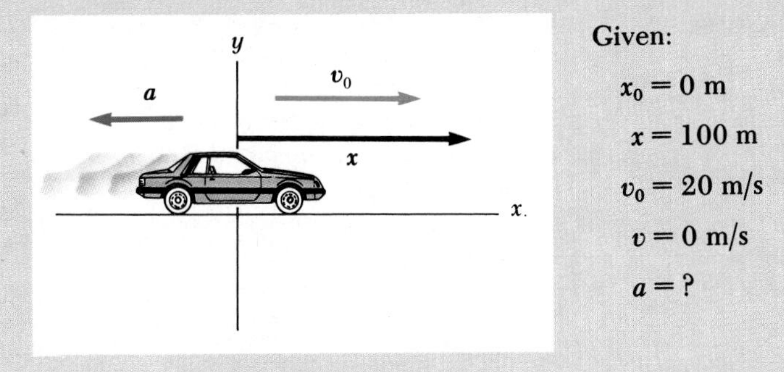

$$v^2 = v_0^2 + 2a(x - x_0)$$

$$v^2 - v_0^2 = 2a(x - x_0)$$

$$a = \frac{v^2 - v_0^2}{2(x - x_0)}$$

$$a = \frac{(0\ \text{m/s})^2 - (20\ \text{m/s})^2}{2(100\ \text{m})} = -2\ \text{m/s}^2$$

$$\frac{\text{m}^2/\text{s}^2}{\text{m}} = \frac{\text{m}}{\text{s}^2}$$

## EXPERIMENTS

Physics is a science based upon experimental observations. In view of this fact, I recommend that you try to supplement the text through various type of "hands-on" experiments, either at home or in the laboratory. These can be used to test ideas and models discussed in class or in the text. For example, the common "Slinky" toy is excellent for studying traveling waves; a ball swinging on the end of a long string can be used to investigate pendulum motion; various masses attached to the end of a vertical spring or rubber band can be used to determine their elastic nature; an old pair of Polaroid sunglasses and some discarded lenses and magnifying glass are the components of various experiments in optics; you can get an approximate measure of the acceleration of gravity by dropping a ball from a known height by simply measuring the time of its fall with a stopwatch. The list is endless. When physical models are not available, be imaginative and try to develop models of your own.

## AN INVITATION TO PHYSICS

It is my sincere hope that you too will find physics an exciting and enjoyable experience, and that you will profit from this experience, regardless of your chosen profession. Welcome to the exciting world of physics.

The scientist does not study nature because it is useful; he studies it because he delights in it, and he delights in it because it is beautiful. If nature were not beautiful, it would not be worth knowing, and if nature were not worth knowing, life would not be worth living.

HENRI POINCARÉ

# Pedagogical Use of Color

The various colors that you will see in the illustrations of this text are used to improve clarity and understanding. Many figures with three-dimensional representations are air-brushed in various colors to make them as realistic as possible.

Color coding has been used in various parts of the book to identify specific physical quantities. The following schemes should be noted.

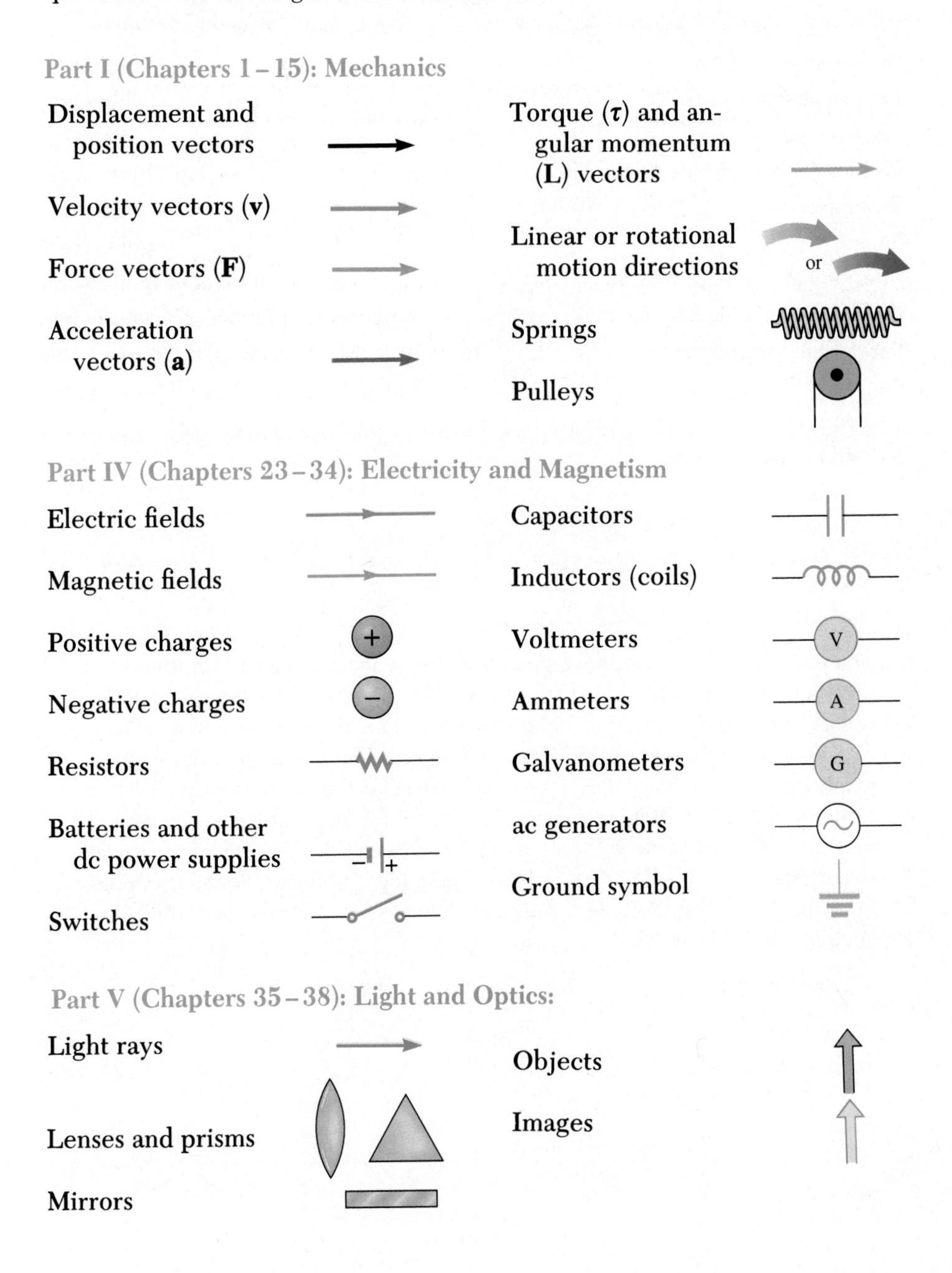

Most graphs are presented with curves plotted in either red or blue, and coordinate axes are in black. Several colors are used in those graphs where many physical quantities are plotted simultaneously, or in those cases where different processes may be occurring and need to be distinguished.

Finally, motional shading effects have been incorporated in many figures to remind the readers that they are dealing with a dynamic system rather than a static system. These figures will appear to be somewhat like a "multiflash" photograph of a moving system, with faint images of the "past history" of the system's path. In some figures, a broad, colored arrow is used to indicate the direction of motion of the system.

In addition to the use of color in the figures, the pedagogy in the text has been enhanced with color as well. We have used the following color-coded system:

| | |
|---|---|
| Important equations | tan screen |
| Important statements | blue type |
| Marginal notes | blue type |
| Problem-Solving Strategy and Hints | light blue screen |
| Examples | blue box |
| Summaries | gold screen |
| Intermediate-level problems | problem number in blue type |
| Challenging problems | problem number in magenta type |
| Spreadsheet problems | black or color box around problem number |
| Laboratory problems | black triangle next to problem number |

The Guest Essays are marked with a blue bar on the edge of the page so they can be located easily.

The publisher and author have gone to extreme measures in attempting to ensure the publication of an error-free text. The manuscript, galleys, and page proofs have been carefully checked by the author, the editors, and a battery of reviewers. While we realize that a 100% error-free text may not be humanly possible, Serway's *Physics for Scientists and Engineers* is very close. Confirmed in this belief, we are offering $5.00 for any first-time error you may find. (Note that we will only pay for each error the first time it is brought to our attention.) Please write to John Vondeling, Publisher, Saunders College Publishing, The Public Ledger Building, 620 Chestnut Street, Suite 560, Philadelphia, PA, 19106-3477.

# Introducing the Book

5

## The Laws of Motion

*A sailboat with its spinnaker billowing runs before the wind. What force or forces are causing the forward motion? Does the water exert a force on the boat? (© Doug Peebles/ Index Stock International)*

In the previous two chapters on kinematics, we described the motion of particles based on the definition of displacement, velocity, and acceleration. However, we would like to be able to answer specific questions related to the causes of motion, such as "What mechanism causes motion?" and "Why do some objects accelerate at a higher rate than others?" In this chapter, we shall describe the change in motion of particles using the concepts of force, mass, and momentum. We shall then discuss the three basic laws of motion, which are based on experimental observations and were formulated nearly three centuries ago by Sir Isaac Newton.

5.1 INTRODUCTION TO CLASSICAL MECHANICS

*The purpose of classical mechanics is to provide a connection between the motion of a body and the forces acting on it.* Keep in mind that classical mechanics deals with objects that are large compared with the dimensions of atoms

95

**PHOTOGRAPHS**

**Chapter opening photographs have been carefully selected to relate to the principles discussed in the chapter. The captions serve as an added instructional tool. Some photographs, particularly in the chapters on mechanics, have color-coded vector overlays.**

**CHAPTER PREVIEWS**

**Include a brief discussion of chapter objectives and content.**

COLOR-CODED LINE ART

**Color is used to add clarity and three-dimensional effects. A complete description of the color system used appears on page xvii.**

106 CHAPTER 5 THE LAWS OF MOTION

$$\sum F_x = T = ma_x \qquad \text{or} \qquad a_x = \frac{T}{m}$$

In this situation, there is no acceleration in the $y$ direction. Applying $\Sigma F_y = ma_y$ with $a_y = 0$ gives

$$N - W = 0 \qquad \text{or} \qquad N = W$$

That is, the normal force is equal to and opposite the weight.

If $T$ is a *constant* force, then the acceleration, $a_x = T/m$, is also a constant. Hence, the equations of kinematics from Chapter 3 can be used to obtain the displacement, $\Delta x$, and velocity, $v$, as functions of time. Since $a_x = T/m =$ constant, these expressions can be written

$$\Delta x = v_0 t + \tfrac{1}{2}\left(\frac{T}{m}\right)t^2$$

$$v = v_0 + \left(\frac{T}{m}\right)t$$

where $v_0$ is the velocity at $t = 0$.

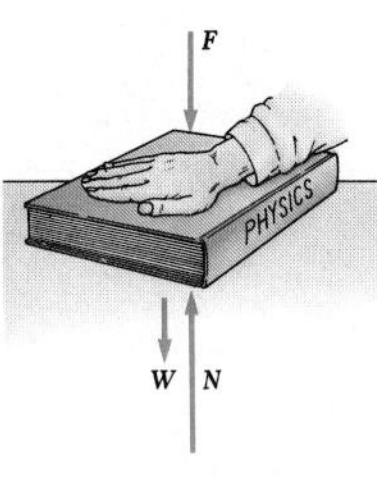


Figure 5.8 When one pushes downward on an object with a force $F$, the normal force $N$ is greater than the weight. That is, $N = W + F$.

In the example just presented, we found that the normal force $N$ was equal in magnitude and opposite the weight $W$. *This is not always the case.* For example, suppose you were to push down on a book with a force $F$ as in Figure 5.8. In this case, $\Sigma F_y = 0$ gives $N - W - F = 0$, or $N = W + F$. Other examples in which $N = W$ will be presented later.

Consider a lamp of weight $W$ suspended from a chain of negligible weight fastened to the ceiling, as in Figure 5.9a. The free-body diagram for the lamp is shown in Figure 5.9b, where the forces on it are the weight, $W$, acting downward, and the force of the chain on the lamp, $T$, acting upward. The force $T$ is the constraint force in this case. (If we cut the chain, $T = 0$ and the body executes free fall.)

If we apply the first law to the lamp, noting that $a = 0$, we see that since there are no forces in the $x$ direction, the equation $\Sigma F_x = 0$ provides no helpful information. The condition $\Sigma F_y = 0$ gives

$$\sum F_y = T - W = 0 \qquad \text{or} \qquad T = W$$

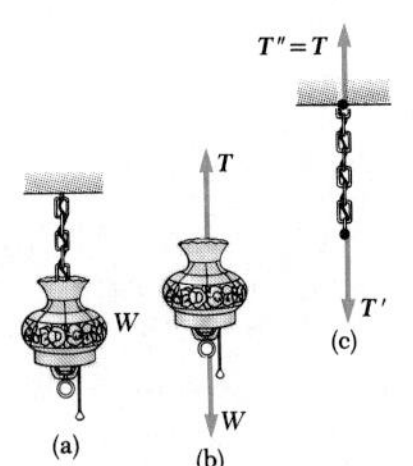


Figure 5.9 (a) A lamp of weight $W$ suspended by a light chain from a ceiling. (b) The forces acting on the lamp are the force of gravity, $W$, and the tension in the chain, $T$. (c) The forces acting on the chain are $T'$, that exerted by the lamp, and $T''$, that exerted by the ceiling.

Note that $T$ and $W$ are *not* action-reaction pairs. The reaction to $T$ is $T'$, the force exerted on the chain by the lamp, as in Figure 5.9c. The force $T'$ acts downward and is transmitted to the ceiling. That is, the force of the chain on the ceiling, $T'$, is *downward* and equal to $W$ in magnitude. The ceiling exerts an equal and opposite force, $T'' = T$, on the chain, as in Figure 5.9c.

**PROBLEM-SOLVING STRATEGY**

The following procedure is recommended when dealing with problems involving the application of Newton's laws:

1. Draw a simple, neat diagram of the system.
2. Isolate the object of interest whose motion is being analyzed. Draw a free-body diagram for this object, that is, a diagram showing *all external forces acting on the object.* For systems containing more than one object, draw *separate* diagrams for each object. *Do not* include forces that the object exerts on its surroundings.

PROBLEM-SOLVING STRATEGIES AND HINTS

**Outlines general strategies for solving the types of problems featured in both the examples and the end-of-chapter problems. These steps are highlighted with a light blue screen so they can be easily located.**

3. Establish convenient coordinate axes for each body and find the components of the forces along these axes. Now, apply Newton's second law, $\Sigma \boldsymbol{F} = m\boldsymbol{a}$, in *component* form. Check your dimensions to make sure that all terms have units of force.
4. Solve the component equations for the unknowns. Remember that you must have as many independent equations as you have unknowns in order to obtain a complete solution.
5. It is a good idea to check the predictions of your solutions for extreme values of the variables. You can often detect errors in your results by doing so.

EXAMPLE 5.2 A Traffic Light at Rest

A traffic light weighing 100 N hangs from a cable tied to two other cables fastened to a support, as in Figure 5.10a. The upper cables make angles of 37° and 53° with the horizontal. Find the tension in the three cables.

*Solution* First we construct a free-body diagram for the traffic light, as in Figure 5.10b. The tension in the vertical cable, $T_3$, supports the light, and so we see that $T_3 = W = 100$ N. Now we construct a free-body diagram for the knot that holds the three cables together, as in Figure 5.10c. This is a convenient point to choose because all forces in question act at this point. We choose the coordinate axes as shown in Figure 5.10c and resolve the forces into their $x$ and $y$ components:

| Force | $x$ component | $y$ component |
|---|---|---|
| $T_1$ | $-T_1 \cos 37°$ | $T_1 \sin 37°$ |
| $T_2$ | $T_2 \cos 53°$ | $T_2 \sin 53°$ |
| $T_3$ | 0 | $-100$ N |

The condition for equilibrium $\Sigma \boldsymbol{F} = 0$ gives us the equations

$$(1) \quad \sum F_x = T_2 \cos 53° - T_1 \text{ os } 37° = 0$$

$$(2) \quad \sum F_y = T_1 \sin 37° + T_2 \sin 53° - 100 \text{ N} = 0$$

From (1) we see that the horizontal components of $\boldsymbol{T}_1$ and $\boldsymbol{T}_2$ must be equal in magnitude, and from (2) we see that the sum of the vertical components of $\boldsymbol{T}_1$ and $\boldsymbol{T}_2$ must balance the weight of the light. We can solve (1) for $T_2$ in terms of $T_1$ to give

$$T_2 = T_1\left(\frac{\cos 37°}{\cos 53°}\right) = 1.33T_1$$

This value for $T_2$ can be substituted into (2) to give

$$T_1 \sin 37° + (1.33T_1)(\sin 53°) - 100 \text{ N} = 0$$

$$T_1 = 60.0 \text{ N}$$

$$T_2 = 1.33T_1 = 79.8 \text{ N}$$

Exercise 2 In what situation will $T_1 = T_2$?
Answer When the supporting cables make equal angles with the horizontal support.

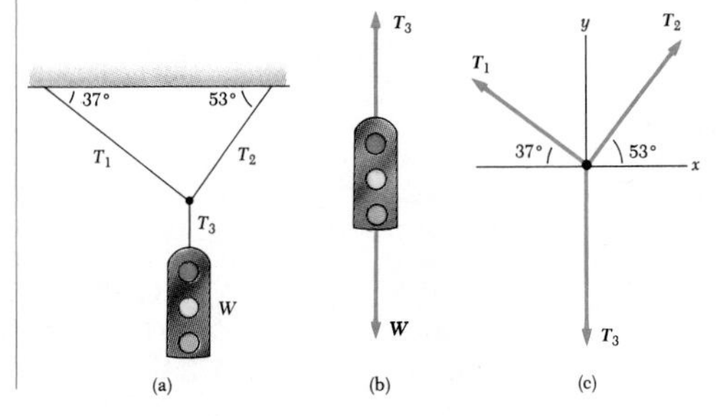


Figure 5.10 (Example 5.2) (a) A traffic light suspended by cables. (b) Free-body diagram for the traffic light. (c) Free-body diagram for the knot.

## WORKED EXAMPLE EXERCISES

**Many of the worked examples are followed immediately by exercises with answers to make the textbook more interactive and to test the student's understanding of problem-solving techniques.**

## WORKED EXAMPLES

**Presented as an aid in understanding concepts, these examples serve as models for solving the end-of-chapter problems. The examples are set off in a blue box, and the solution answer is highlighted with a tan screen.**

## SUMMARY

Newton's first law — **Newton's first law** states that a body at rest will remain at rest or a body in uniform motion in a straight line will maintain that motion unless an external resultant force acts on the body.

Newton's second law — **Newton's second law** states that the time rate of change of momentum of a body is equal to the resultant force acting on the body. If the mass of the body is constant, the net force equals the product of the mass and its acceleration, or $\Sigma \mathbf{F} = m\mathbf{a}$.

Inertial frame — Newton's first and second laws are valid in an inertial frame of reference. An **inertial frame** is one in which an object, subject to no net external force, moves with constant velocity including the special case of $v = 0$.

**Mass** is a scalar quantity. The mass that appears in Newton's second law is called **inertial mass.**

Weight — The **weight** of a body is equal to the product of its mass and the acceleration of gravity, or $\mathbf{W} = m\mathbf{g}$.

Newton's third law — **Newton's third law** states that if two bodies interact, the force exerted on body 1 by body 2 is equal to and opposite the force exerted on body 2 by body 1. Thus, an isolated force cannot exist in nature.

Forces of friction — The r... surface is ... mum forc... $f_s \leq \mu_s N$, w... When a b... opposite t... nitude of ... *friction.* U...

**More on I...**

As we hav... ing Newt... recognize... construct... the free-b... of mechan... free-body... to constru... lems. Whe... you const...

As us... notes a no...

## SUMMARY

**Reviews important concepts and equations discussed in the chapter.**

## MARGINAL NOTES

**Comments and notes in blue print are used to locate important statements, equations, and concepts in the chapter.**

## PROBLEMS KEYED TO SECTIONS

**Allows students to refer back to the sections to look at a worked example for help in solving problems.**



its speed increases to $7.0 \times 10^5$ m/s in a distance of 5.0 cm. Assuming its acceleration is constant, (a) determine the force on the electron and (b) compare this force with the weight of the electron, which we neglected.

25. A 15-lb block rests on the floor. (a) What force does the floor exert on the block? (b) If a rope is tied to the block and run vertically over a pulley and the other end attached to a free-hanging 10-lb weight, what is the force of the floor on the 15-lb block? (c) If we replace the 10-lb weight in (b) by a 20-lb weight, what is the force of the floor on the 15-lb block?

Section 5.8 Some Applications of Newton's Laws

26. Find the tension in each cord for the systems described in Figure 5.22. (Neglect the mass of the cords.)

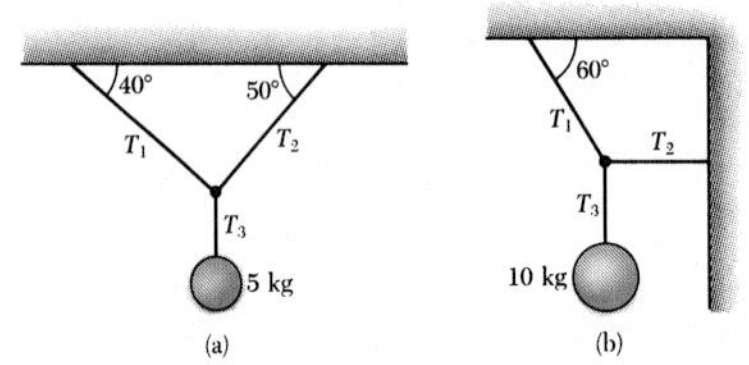


**Figure 5.22** (Problem 26).

27. Three 10-kg objects are suspended as shown (Fig. 5.23). Determine each of the labeled tension forces.

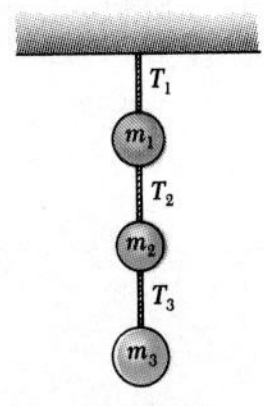


**Figure 5.23** (Problem 27).

28. A 200-N weight is tied to the middle of a strong rope, and two people pull at opposite ends of the rope in an attempt to lift the weight. (a) What force $\mathbf{F}$ must each person apply to suspend the weight as shown in Figure 5.24? (b) Can they pull in such a way as to make the rope horizontal? Explain.

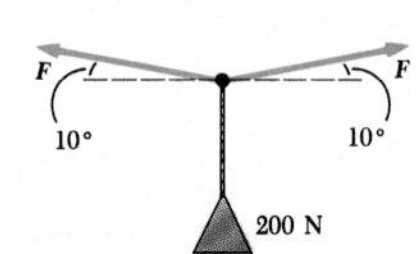


**Figure 5.24** (Problem 28).

29. A 50-kg mass hangs from a rope 5 m in length, which is fastened to the ceiling. What horizontal force applied to the mass will deflect it 1 m sideways from the vertical and maintain it in that position?

30. The systems shown in Figure 5.25 are in equilibrium. If the spring scales are calibrated in N, what do they read in each case? (Neglect the mass of the pulleys and strings, and assume the incline is smooth.)

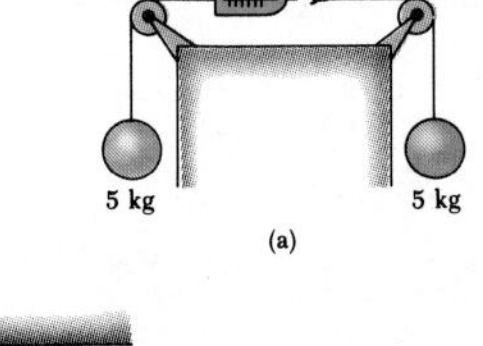


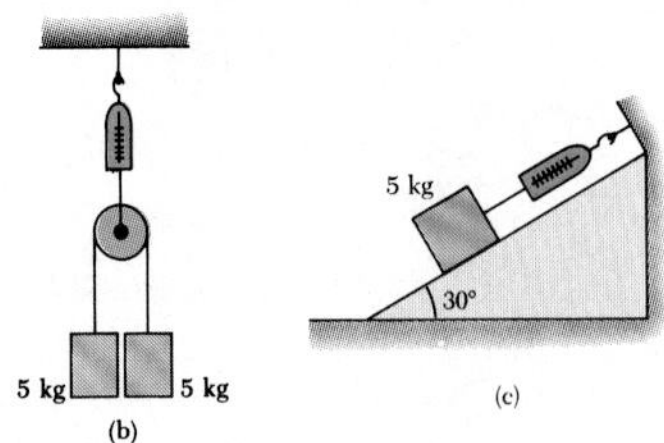


**Figure 5.25** (Problem 30).

31. A bag of cement hangs from three wires as shown (Fig. 5.26). Two of the wires make angles $\theta_1$ and $\theta_2$ with the horizontal. If the system is in equilibrium, (a) show that

$$T_1 = \frac{W \cos(\theta_2)}{\sin(\theta_1 + \theta_2)}$$

(b) Given that $W = 200$ N, $\theta_1 = 10°$ and $\theta_2 = 25°$, find the tensions $T_1$, $T_2$, and $T_3$ in the wires.

32. A woman at an airport is pulling her 20-kg suitcase at constant speed by pulling on a strap at an angle $\theta$

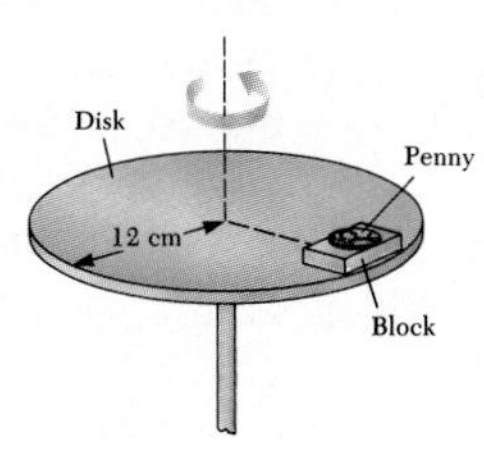


**Figure 6.24** (Problem 50).

through its center, as shown in Figure 6.25. The period of revolution of the hoop is $T$. The bead has mass $m$ and is at rest relative to the hoop at the angle $\theta$. Answer the following in terms of the given parameters and the acceleration due to gravity $g$. (a) Find the normal force $\boldsymbol{N}$ exerted on the bead by the wire. (b) Find the angle between the normal force and the axis of rotation. (c) Find the centripetal force exerted on the bead.

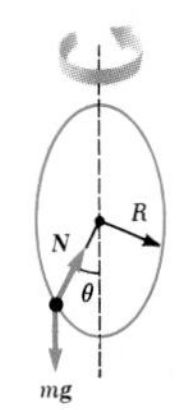


**Figure 6.25** (Problem 51).

52. An amusement park ride consists of a rotating circular platform 8 m in diameter from which "bucket seats" are suspended at the end of 2.5-m chains. (For simplicity only one chain is shown [Figure 6.26], though for stability at least two would be used.) When the system rotates the chains holding the seats make an angle $\theta = 28°$ with the vertical. (a) What is the speed of the seat? (b) If a child of mass 40 kg sits in the 10-kg seat, what is the tension in the chain?

53. The following experiment is performed in a spacecraft that is at rest where the net gravitational field is zero. A small sphere is injected into a viscous medium with initial velocity $v_0$. The sphere experiences a resistive force $\boldsymbol{R} = -b\boldsymbol{v}$. Find the velocity of the sphere as a function of time. (*Hint:* Apply Newton's second law, write $a$ as $dv/dt$, separate the variables, and integrate the equation.)

**CALCULATOR/COMPUTER PROBLEMS**

54. A hailstone of mass $4.8 \times 10^{-4}$ kg and radius 0.5 cm falls through the atmosphere and experiences a net force given by Equation 6.9. This expression can be written in the form

$$m\frac{dv}{dt} = mg - Kv^2$$

where $K = \frac{1}{2}C\rho A$, $\rho = 1.29$ kg/m³ and $C = 0.5$. (a) What is the terminal velocity of the hailstone? (b) Use a method of numerical integration to find the velocity and position of the hailstone at 1-s intervals, taking $v_0 = 0$. Continue your calculation until terminal velocity is reached.

55. A 0.5-kg block slides down a 30° incline of length 1 m. The coefficient of kinetic friction between the block and the incline varies with the block's velocity according to the expression

$$\mu = 0.3 + 1.2\sqrt{v}$$

where $v$ is in m/s. (a) Use a numerical method to find the velocity of the block at intervals of 10 cm during its motion. (b) If the length of the plane is extended to several km, will the block reach terminal velocity? If so, what is its terminal velocity, and at what point does it occur on the incline?

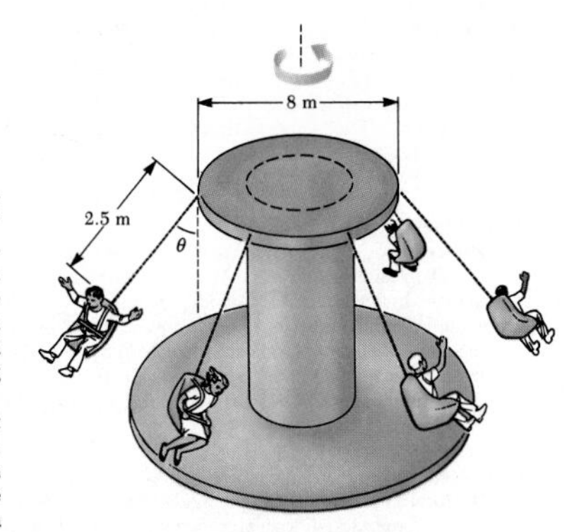


**Figure 6.26** (Problem 52).

## SPREADSHEET PROBLEMS

**Spreadsheets on a separate disk available from your instructor accompany these problems. These problems can also be solved analytically without using the software.**

## PROBLEMS GRADED BY LEVEL OF DIFFICULTY

**Intermediate-level problems are indicated by printing the problem number in blue print; challenging problems have a magenta problem number.**

ESSAY SCALING—THE PHYSICS OF LILLIPUT 19

# ESSAY

## Scaling—the Physics of Lilliput*

Philip Morrison
*Massachusetts Institute of Technology*

The fictional traveler Lemuel Gulliver spent a busy time in a kingdom called Lilliput, where all living things—men, cattle, trees, grass—were exactly similar to our world, except that they were all built on the scale of one inch to the foot. Lilliputians were a little under 6 inches high, on the average, and built proportionately just as we are. Gulliver also visited Brobdingnag, the country of the giants, who were exactly like men but 12 times as tall. As Swift described it, daily life in both kingdoms was about like ours (in the 18th century). His commentary on human behavior is still worth reading, but we shall see that people of such sizes just could not have been as he described them.

Long before Swift lived, Galileo understood why very small or very large models of man could not be like us, but apparently Dean Swift had never read what Galileo wrote. One character in Galileo's "Two New Sciences" says, "Now since . . . in geometry, . . . mere size cuts no figure, I do not see that the properties of circles, triangles, cylinders, cones, and other solid figures will change with their size. . . ." But his physicist friend replies, "The common opinion is here absolutely wrong." Let us see why.

We start with the strength of a rope. It is easy to see that if one man who pulls with a certain strength can almost break a certain rope, two such ropes will just withstand the pull of two men. A single large rope with the same total area of cross-section as the two smaller ropes combined will contain just double the number of fibers of one of the small ropes, and it will also do the job. In
or rope is proportional to its area of cro
Experience and theory agree in this co
holds, not only for ropes or cables supp
supporting a thrust. The thrust which a c
a given material, is also proportional to
Now the body of a man or an animal i
skeleton—supported by various braces
But the weight of the body which must b
flesh and bone present, that is, to the vo
Let us now compare Gulliver with th
Since the giant is exactly like Gulliver in
sions is 12 times the corresponding one
columns and braces is proportional to the
of their linear dimension (strength $\propto L^2$),
as Gulliver's. Because his weight is propo
$12^3$ or 1728 times as great as Gulliver's.
ratio a dozen times smaller than ours. Just
much trouble as we should have in carry
In reality, of course, Lilliput and Br
effects of a difference in scale if we com
The smaller ones are not scale models o
sponding leg bones of two closely rela
gazelle, the other a bison. Notice that th
geometrically to that of the smaller. It is
ing the scale change, which would make
Galileo wrote very clearly on this ve
dingnag, or of any normal-looking giant
giant the same proportion of limb as that
a harder and stronger material for makin
strength in comparison with men of me

*Adapted from PSSC PHYSICS, 2nd edition,
Development Center, Inc., Newton, MA.

## GUEST ESSAYS

**Intended as supplemental readings for the student. The blue bar at the edges of the pages help students locate them easily. The essays have been placed at the ends of chapters to avoid interrupting the main textual material.**

ESSAY SCALING—THE PHYSICS OF LILLIPUT 23

very biggest things we can make which have some roundness, which are fully three-dimensional, are buildings and great ships. These lack a good deal of being a thousand times larger than men in their linear dimensions.

Within our present technology our scaling arguments are important. If we design a new large object on the basis of a small one, we are warned that new effects too small to detect on our scale may enter and even become the most important things to consider. We cannot just scale up and down blindly, geometrically, but by scaling in the light of physical reasoning, we can sometimes foresee what changes will occur. In this way we can employ scaling in intelligent airplane design, for example, and not arrive at a jet transport that looks like a bee—and won't fly.

### Suggested Readings


Galilei, Galileo, "Dialogues Concerning Two New Sciences," trans. by Henry Crew and Alphonso De Salvio, Evanston, Northwestern University Press, 1946, pp. 1–6, 125–128.

Haldane, J. B. S., "On Being the Right Size." *World of Mathematics*, Vol. II, edited by James R. Newman. New York, Simon & Schuster, 1956.

Holcomb, Donald F. and Philip Morrison, *My Father's Watch—Aspects of the Physical World*, Englewood Cliffs, NJ, Prentice Hall, 1974, pp. 68–83.

Smith, Cyril S., "The Shape of Things," *Scientific American*, January, 1954, p. 58.

Thompson, D'Arcy W., "On Magnitude," in *On Growth and Form*, Cambridge University Press, 1952 and 1961.


### Essay Questions

1. The leg bones of one animal are twice as strong as those of another closely related animal of similar shape. (a) What would you expect to be the ratio of these animals' heights? (b) What would you expect to be the ratio of their weights?
2. A hummingbird must eat very frequently and even then must have a highly concentrated form of food such as sugar. What does the concept of scaling tell you about the size of a hummingbird?
3. About how many Lilliputians would it take to equal the mass of one citizen of Brobdingnag?

### Essay Problems

1. The total surface area of a rectangular solid is the sum of the areas of the six faces. If each dimension of a given rectangular solid is doubled, what effect does this have on the total surface area?
2. A hollow metal sphere has a wall thickness of 2 cm. If you increase both the diameter and thickness of this sphere so that the overall volume is three times the original overall volume, how thick will the shell of the new sphere be?
3. If your height and all your other dimensions were doubled, by what factor would this change (a) your weight? (b) the ability of your leg bones to support your weight?
4. According to the zoo, an elephant of mass $4.0 \times 10^3$ kg consumes $3.4 \times 10^2$ times as much food as a guinea pig of mass 0.70 kg. They are both warm-blooded, plant-eating, similarly shaped animals. Find the ratio of their surface areas, which is approximately the ratio of their heat losses, and compare it with the known ratio of food consumed.
5. A rectangular water tank is supported above the ground by four pillars 5 m long whose diameters are 20 cm. If the tank were made 10 times longer, wider, and deeper, what diameter pillars would be needed? How much more water would the tank hold?
6. How many state maps of scale 1 : 1 000 000 would you need to cover the state with those maps?

## SUGGESTED READINGS, ESSAY QUESTIONS, AND PROBLEMS

**Most of the essays contain suggested readings, questions, and problems for added flexibility in covering these topics.**

# Contents Overview

*Thermogram produced with an infrared scanner. (VANSCAN® Thermogram by Daedalus Enterprises, Inc.)*

# Contents

* These sections are optional.



*The first integrated circuit. (Courtesy of Texas Instruments)*

*Aurora Borealis seen from inside a dome on the western shore of Hudson Bay. (© David Hiser/Photographers Aspen)*

*Scientist checking a laser-cutting device. (Philippe Plailly, Science Photo Library/Photo Researchers, Inc.)*

*This dramatic one-minute exposure captures multiple lightning bolts illuminating Kitt Peak National Observatory in Arizona, illustrating electrical breakdown in the atmosphere. (© Gary Ladd 1972)*

# PART IV
# Electricity and Magnetism

*For the sake of persons of . . . different types, scientific truth should be presented in different forms, and should be regarded as equally scientific, whether it appears in the robust form and the vivid coloring of a physical illustration, or in the tenuity and paleness of a symbolic expression.*

JAMES CLERK MAXWELL

We now begin the study of that branch of physics which is concerned with electric and magnetic phenomena. The laws of electricity and magnetism play a central role in the operation of various devices such as radios, televisions, electric motors, computers, high-energy accelerators, and a host of electronic devices used in medicine. However, more fundamentally, we now know that the interatomic and intermolecular forces that are responsible for the formation of solids and liquids are electric in origin. Furthermore, such forces as the pushes and pulls between objects and the elastic force in a spring arise from electric forces at the atomic level.

Evidence in Chinese documents suggests that magnetism was known as early as around 2000 B.C. The ancient Greeks observed electric and magnetic phenomena possibly as early as 700 B.C. They found that a piece of amber, when rubbed, becomes electrified and attracts pieces of straw or feathers. The existence of magnetic forces was known from observations that pieces of a naturally occurring stone called *magnetite* ($Fe_3O_4$) are attracted to iron. (The word *electric* comes from the Greek word for amber, *elecktron.* The work *magnetic* comes from the name of a northern central district of Greece where magnetite was found, *Magnesia.*)

In 1600, William Gilbert discovered that electrification was not limited to amber but is a general phenomenon. Scientists went on to electrify a variety of objects, including chickens and people! Experiments by Charles Coulomb in 1785 confirmed the inverse-square force law for electricity.

It was not until the early part of the 19th century that scientists established that electricity and magnetism are, in fact, related phenomena. In 1820, Hans Oersted discovered that a compass needle is deflected when placed near a circuit carrying an electric current. In 1831, Michael Faraday, and almost simultaneously, Joseph Henry, showed that, when a wire is moved near a magnet (or, equivalently, when a magnet is moved near a wire), an electric current is observed in the wire. In 1873, James Clerk Maxwell used these observations and other experimental facts as a basis for formulating the laws of electromagnetism as we know them today. (*Electromagnetism* is a name given to the combined fields of electricity and magnetism.) Shortly thereafter (around 1888), Heinrich Hertz verified Maxwell's predictions by producing electromagnetic waves in the laboratory. This was followed by such practical developments as radio and television.

Maxwell's contributions to the science of electromagnetism were especially significant because the laws he formulated are basic to *all* forms of electromagnetic phenomena. His work is comparable in importance to Newton's discovery of the laws of motion and the theory of gravitation.

# 23
# Electric Fields

*Photograph of a carbon filament incandescent lamp. Thomas Edison's first light bulb also used a piece of carbonized cotton thread as its filament. The carbon filament lamp emits a different light spectrum than that produced by a tungsten lamp because of its composition and because it operates at a lower temperature. (Courtesy of CENCO)*

The electromagnetic force between charged particles is one of the fundamental forces of nature. In this chapter, we begin by describing some of the basic properties of electrostatic forces. We then discuss Coulomb's law, which is the fundamental law of force between any two charged particles. The concept of an electric field associated with a charge distribution is then introduced, and its effect on other charged particles is described. The method for calculating electric fields of a given charge distribution from Coulomb's law is discussed, and several examples are given. Then the motion of a charged particle in a uniform electric field is discussed. We conclude the chapter with a brief discussion of the oscilloscope.

## 23.1 PROPERTIES OF ELECTRIC CHARGES

A number of simple experiments can be performed to demonstrate the existence of electrical forces and charges. For example, after running a comb through your hair, you will find that the comb will attract bits of paper. The attractive force is often strong enough to suspend the pieces of paper. The same effect occurs with other rubbed materials, such as glass or rubber.

Biographical Sketch

**Charles Coulomb**
*(1736–1806)*

Charles Coulomb, the great French physicist after whom the unit of electric charge called the *coulomb* was named, was born in Angoulême in 1736. He was educated at the École du Génie in Mézieres, graduating in 1761 as a military engineer with a rank of First Lieutenant. Coulomb served in the West Indies for nine years, where he supervised the building of fortifications in Martinique.

In 1774, Coulomb became a correspondent to the Paris Academy of Science. There he shared the Academy's first prize for his paper on magnetic compasses and also received first prize for his classic work on friction, a study that was unsurpassed for 150 years. During the next 25 years, he presented 25 papers to the Academy on electricity, magnetism, torsion, and applications to the torsion balance, as well as several hundred committee reports on engineering and civil projects.

Coulomb took full advantage of the various positions he held during his lifetime. For example, his experience as an engineer led him to investigate the strengths of materials and determine the forces that affect objects on beams, thereby contributing to the field of structural mechanics. He also contributed to the field of ergonomics. His research provided a fundamental understanding of the ways in which people and animals can best do work and greatly influenced the subsequent research of Gaspard Coriolis (1792–1843).

Coulomb's major contribution to science was in the field of electrostatics and magnetism, in which he made use of the torsion balance he developed (see Fig. 23.2). The paper describing this invention also contained a design for a compass using the principle of torsion suspension. His next paper gave proof of the inverse square law for the electrostatic force between two charges.

Coulomb died in 1806, five years after becoming president of the Institut de France (formerly the Paris Academy of Science). His research on electricity and magnetism brought this area of physics out of traditional natural philosophy and made it an exact science.

(Photograph courtesy of AIP Niels Bohr Library, E. Scott Barr Collection)

Another simple experiment is to rub an inflated balloon with wool. The balloon will then adhere to the wall or the ceiling of a room, often for hours. When materials behave in this way, they are said to be *electrified,* or to have become **electrically charged.** You can easily electrify your body by vigorously rubbing your shoes on a wool rug. The charge on your body can be sensed and removed by lightly touching (and startling) a friend. Under the right conditions, a visible spark is seen when you touch one another, and a slight tingle will be felt by both parties. (Experiments such as these work best on a dry day, since an excessive amount of moisture can lead to a leakage of charge from the electrified body to the earth by various conducting paths.)

In a systematic series of rather simple experiments, one finds that there are two kinds of electric charges, which were given the names **positive** and **negative** by Benjamin Franklin (1706–1790). To demonstrate this fact, consider a hard rubber rod that has been rubbed with fur and then suspended by a nonmetallic thread as in Figure 23.1. When a glass rod that has been rubbed with silk is brought near the rubber rod, the rubber rod is attracted toward the glass rod. On the other hand, if two charged rubber rods (or two charged glass rods) are brought near each other, as in Figure 23.1b, the force between them will be repulsive. This observation shows that the rubber and glass are in two different states of electrification. On the basis of these observations, we conclude that *like charges repel one another and unlike charges attract one another.*

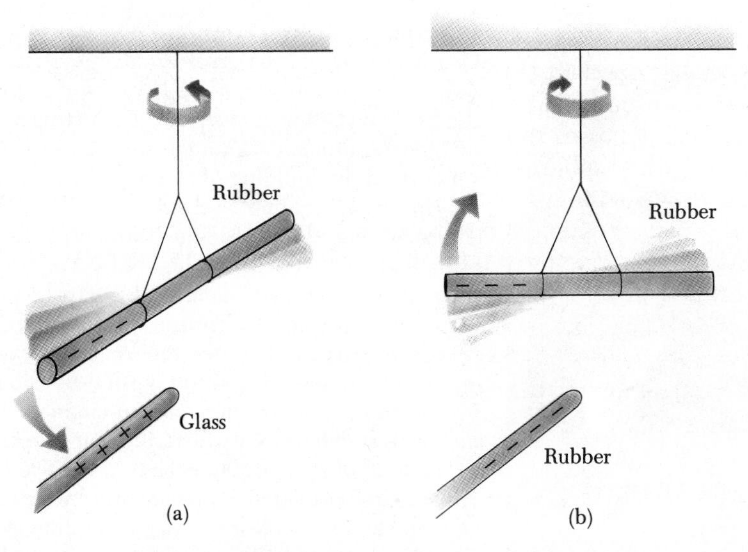


**Figure 23.1** (a) A negatively charged rubber rod, suspended by a thread, is attracted to a positively charged glass rod. (b) A negatively charged rubber rod is repelled by another negatively charged rubber rod.

Using the convention suggested by Franklin, the electric charge on the glass rod is called *positive,* and that on the rubber rod is called *negative.* Therefore any charged body that is attracted to a charged rubber rod (or repelled by a charged glass rod) must have a positive charge. Conversely, any charged body that is repelled by a charged rubber rod (or attracted to a charged glass rod) has a negative charge on it.

Charge is conserved

Another important aspect of Franklin's model of electricity is the implication that *electric charge is always conserved.* That is, when one body is rubbed against another, charge is not created in the process. The electrified state is due to a *transfer* of charge from one body to the other. Therefore, one body gains some amount of negative charge while the other gains an equal amount of positive charge. For example, when a glass rod is rubbed with silk, the silk obtains a negative charge that is equal in magnitude to the positive charge on the glass rod. We now know from our understanding of atomic structure *that it is the negatively charged electrons that are transferred* from the glass to the silk in the rubbing process. Likewise, when rubber is rubbed with fur, electrons are transferred from the fur to the rubber, giving the rubber a net negative charge and the fur a net positive charge. This is consistent with the fact that neutral, uncharged matter contains as many positive charges (protons within atomic nuclei) as negative charges (electrons).

Charge is quantized

In 1909, Robert Millikan (1868–1953) discovered that electric charge always occurs as some integral multiple of some fundamental unit of charge, $e$. In modern terms, the charge $q$ is said to be **quantized.** That is, electric charge exists as discrete "packets." Thus, we can write $q = Ne$, where $N$ is some integer. Other experiments in the same period showed that the electron has a charge $-e$ and the proton has an equal and opposite charge, $+e$. Some elementary particles, such as the neutron, have no charge. A neutral atom must contain as many protons as electrons.

Electric forces between charged objects were measured quantitatively by Coulomb using the torsion balance, which he invented (Fig. 23.2). Using this apparatus, Coulomb confirmed that the electric force between two small charged spheres is proportional to the inverse square of their separation, that is, $F \propto 1/r^2$. The operating principle of the torsion balance is the same as that of the apparatus used by Cavendish to measure the gravitational constant (Section 14.2), with masses replaced by charged spheres. The electric force between the charged spheres produces a twist in the suspended fiber. Since the restoring torque of the twisted fiber is proportional to the angle through which it rotates, a measurement of this angle provides a quantitative measure of the electric force of attraction or repulsion. If the spheres are charged by rubbing, the electrical force between the spheres is very large compared with the gravitational attraction; hence the gravitational force can be neglected.

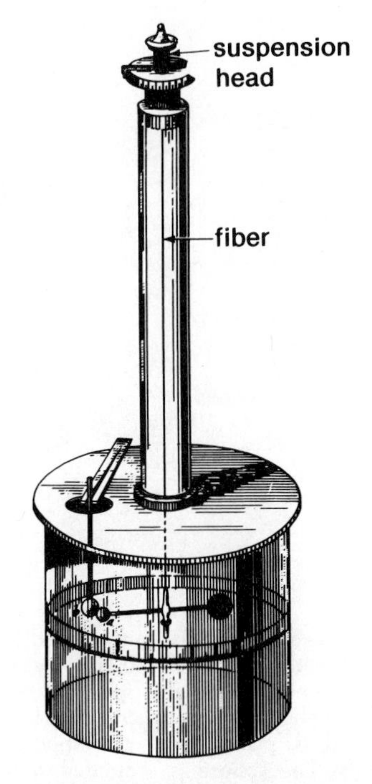


**Figure 23.2** Coulomb's torsion balance, which was used to establish the inverse-square law for the electrostatic force between two charges. (Taken from Coulomb's 1785 memoirs to the French Academy of Sciences.)

From our discussion thus far, we conclude that electric charge has the following important properties:

1. There are two kinds of charges in nature, with the property that unlike charges attract one another and like charges repel one another.
2. The force between charges varies as the inverse square of their separation.
3. Charge is conserved.
4. Charge is quantized.

## 23.2 INSULATORS AND CONDUCTORS

It is convenient to classify substances in terms of their ability to conduct electrical charge.

> **Conductors** are materials in which electric charges move quite freely, whereas **insulators** are materials that do not readily transport charge.

Materials such as glass, rubber, and lucite fall into the category of insulators. When such materials are charged by rubbing, only the area that is rubbed becomes charged and the charge is unable to move to other regions of the material.

Metals are good conductors

In contrast, materials such as copper, aluminum, and silver are good conductors. When such materials are charged in some small region, the charge readily distributes itself over the entire surface of the conductor. If you hold a copper rod in your hand and rub it with wool or fur, it will not attract a small piece of paper. This might suggest that a metal cannot be charged. On the other hand, if you hold the copper rod by a lucite handle and then rub, the rod will remain charged and attract the piece of paper. This is explained by noting that in the first case, the electric charges produced by rubbing will readily move from copper through your body and finally to earth. In the second case, the insulating lucite handle prevents the flow of charge to earth.

*Semiconductors* are a third class of materials, and their electrical properties are somewhere between those of insulators and conductors. Silicon and germanium are well-known examples of semiconductors commonly used in the fabrication of a variety of electronic devices. The electrical properties of semiconductors can be changed over many orders of magnitude by adding controlled amounts of certain foreign atoms to the materials.

Charging by induction

When a conductor is connected to earth by means of a conducting wire or copper pipe, it is said to be **grounded.** The earth can then be considered an infinite "sink" to which electrons can easily migrate. With this in mind, we can understand how to charge a conductor by a process known as **induction.**

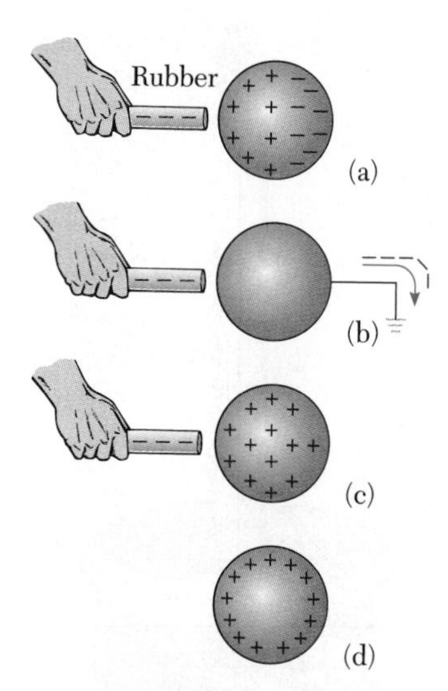


**Figure 23.3** Charging a metallic object by induction. (a) The charge on a neutral metallic sphere is redistributed when a charged rubber rod is placed near the sphere. (b) The sphere is grounded, and some of the electrons leave the conductor. (c) The ground connection is removed, and the sphere has a nonuniform positive charge. (d) When the rubber rod is removed, the sphere becomes uniformly charged.

To understand induction, consider a negatively charged rubber rod brought near a neutral (uncharged) conducting sphere insulated from ground. That is, there is no conducting path to ground (Fig. 23.3a). The region of the sphere nearest the negatively charged rod will obtain an excess of positive charge, while the region of the sphere farthest from the rod will obtain an equal excess of negative charge. (That is, electrons in the part of the sphere nearest the rod migrate to the opposite side of the sphere.) If the same experiment is performed with a conducting wire connected from the sphere to ground (Fig. 23.3b), some of the electrons in the conductor will be repelled to earth. If the wire to ground is then removed (Fig. 23.3c), the conducting sphere will contain an excess of *induced* positive charge. Finally, when the rubber rod is removed from the vicinity of the sphere (Fig. 23.3d), the induced positive charge remains on the ungrounded sphere. Note that the charge remaining on the sphere is uniformly distributed over its surface because of the repulsive forces among the like charges. In the process, the electrified rubber rod loses none of its negative charge.

Thus, we see that charging an object by induction requires no contact with the body inducing the charge. This is in contrast to charging an object by rubbing (that is, charging by *conduction*), which does require contact between the two objects.

A process which is very similar to that of charging by induction in conductors also takes place in insulators. In most neutral atoms or molecules the center of positive charge coincides with the center of negative charge. However, in the presence of a charged object, these centers may shift slightly, resulting in more positive charge on one side of the molecule than on the other. This effect, known as **polarization,** will be discussed more completely in Chapter 26. This realignment of charge within individual molecules produces an induced charge on the surface of the insulator as shown in Figure 23.4. With these ideas, you should be able to explain why a comb that has been rubbed through hair will attract bits of neutral paper, or why a balloon that has been rubbed against your clothing is able to stick to a neutral wall.

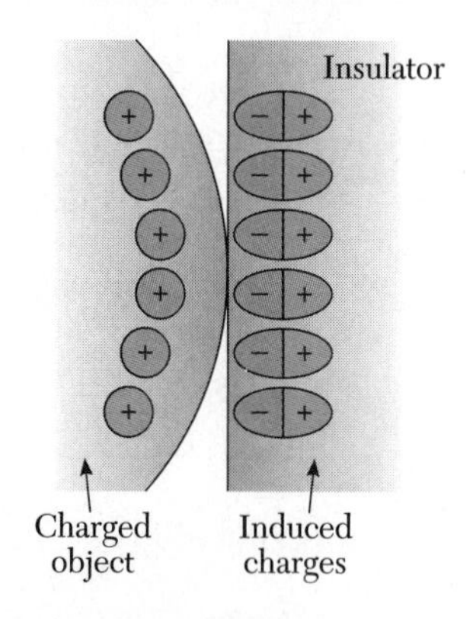


**Figure 23.4** The charged object on the left induces charges on the surface of an insulator at the right.

## 23.3 COULOMB'S LAW

In 1785, Coulomb established the fundamental law of electric force between two stationary, charged particles. Experiments show that an **electric force** has the following properties: (1) The force is inversely proportional to the square of the separation, $r$, between the two particles and is directed along the line joining the particles. (2) The force is proportional to the product of the charges $q_1$ and $q_2$ on the two particles. (3) The force is attractive if the charges are of opposite sign and repulsive if the charges have the same sign. From these observations, we can express the magnitude of the electric force between the two charges as

Coulomb's law

$$F = k\frac{|q_1||q_2|}{r^2} \qquad (23.1)$$

where $k$ is a constant called the *Coulomb constant.* In his experiments, Coulomb was able to show that the exponent of $r$ was 2 to within an uncertainty of a few percent. Modern experiments have shown that the exponent is 2 to a precision of a few parts in $10^9$.

The constant $k$ in Equation 23.1 has a value that depends on the choice of units. The unit of charge in SI units is the coulomb (C). The coulomb is defined in terms of a unit current called the *ampere* (A), where current equals the rate of flow of charge. (The ampere will be defined in Chapter 27.) When the current in a wire is 1 A, the amount of charge that flows past a given point in the wire in 1 s is 1 C. The Coulomb constant $k$ in SI units has the value

$$k = 8.9875 \times 10^9 \ \mathrm{N \cdot m^2/C^2} \tag{23.2}$$

Coulomb constant

To simplify our calculations, we shall use the approximate value

$$k \cong 9.0 \times 10^9 \ \mathrm{N \cdot m^2/C^2} \tag{23.3}$$

The constant $k$ is also written

$$k = \frac{1}{4\pi\epsilon_0}$$

where the constant $\epsilon_0$ is known as the *permittivity of free space* and has the value

$$\epsilon_0 = 8.8542 \times 10^{-12} \ \mathrm{C^2/N \cdot m^2} \tag{23.4}$$

The smallest unit of charge known in nature is the charge on an electron or proton.[1] The charge of an electron or proton has a magnitude

$$|e| = 1.60219 \times 10^{-19} \ \mathrm{C} \tag{23.5}$$

Charge on an electron or proton

Therefore, 1 C of charge is equal to the charge of $6.3 \times 10^{18}$ electrons (that is, $1/e$). This can be compared with the number of free electrons in 1 $\mathrm{cm^3}$ of copper,[2] which is of the order of $10^{23}$. Note that 1 C is a substantial amount of charge. In typical electrostatic experiments, where a rubber or glass rod is charged by friction, a net charge of the order of $10^{-6}$ C ($= 1\ \mu$C) is obtained. In other words, only a very small fraction of the total available charge is transferred between the rod and the rubbing material.

The charges and masses of the electron, proton, and neutron are given in Table 23.1.

When dealing with Coulomb's force law, you must remember that force is a *vector* quantity and must be treated accordingly. Furthermore, note that *Coulomb's law applies exactly only to point charges or particles.* The electric

[1] No unit of charge smaller than $e$ has been detected as a free charge; however, some recent theories have proposed the existence of particles called *quarks* having charges $e/3$ and $2e/3$. Although there is experimental evidence for such particles inside nuclear matter, *free* quarks have never been detected. We shall discuss other properties of quarks in Chapter 47 of the extended version of this text.

[2] A metal atom, such as copper, contains one or more outer electrons, which are weakly bound to the nucleus. When many atoms combine to form a metal, the so-called free electrons are these outer electrons, which are not bound to any one atom. These electrons move about the metal in a manner similar to gas molecules moving in a container.

**TABLE 23.1 Charge and Mass of the Electron, Proton, and Neutron**

| Particle | Charge (C) | Mass (kg) |
|---|---|---|
| Electron (e) | $-1.6021917 \times 10^{-19}$ | $9.1095 \times 10^{-31}$ |
| Proton (p) | $+1.6021917 \times 10^{-19}$ | $1.67261 \times 10^{-27}$ |
| Neutron (n) | 0 | $1.67492 \times 10^{-27}$ |

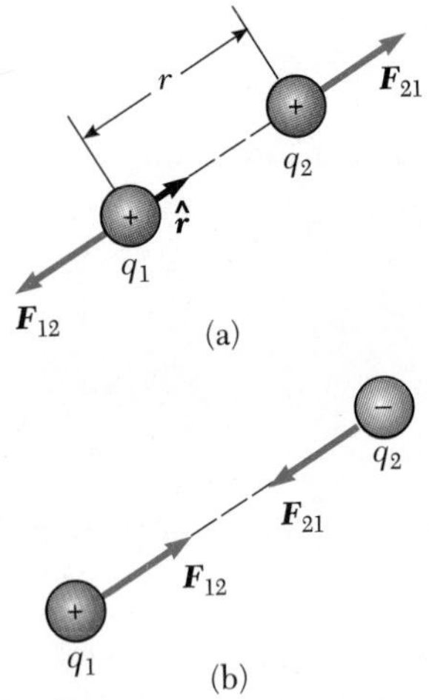

force on $q_2$ due to $q_1$, written $\mathbf{F}_{21}$, can be expressed in vector form as

$$\mathbf{F}_{21} = k\,\frac{q_1 q_2}{r^2}\,\hat{\mathbf{r}} \tag{23.6}$$

where $\hat{\mathbf{r}}$ is a unit vector directed from $q_1$ to $q_2$ as in Figure 23.5a. Since Coulomb's law obeys Newton's third law, the electric force on $q_2$ due to $q_1$ is equal in magnitude to the force on $q_2$ due to $q_1$ and in the opposite direction, that is, $\mathbf{F}_{12} = -\mathbf{F}_{21}$. Finally, from Equation 23.6 we see that if $q_1$ and $q_2$ have the same sign, the product $q_1q_2$ is positive and the force is repulsive, as in Figure 23.5a. On the other hand, if $q_1$ and $q_2$ are of opposite sign, as in Figure 23.5b, the product $q_1q_2$ is negative and the force is attractive.

When more than two charges are present, the force between any pair of charges is given by Equation 23.6. Therefore, the resultant force on any one of them equals the *vector* sum of the forces due to the various individual charges. This principle of *superposition* as applied to electrostatic forces is an experimentally observed fact. For example, if there are four charges, then the resultant force on particle 1 due to particles 2, 3, and 4 is given by

$$\mathbf{F}_1 = \mathbf{F}_{12} + \mathbf{F}_{13} + \mathbf{F}_{14}$$

**Figure 23.5** Two point charges separated by a distance $r$ exert a force on each other given by Coulomb's law. Note that the force on $q_1$ is equal to and opposite the force on $q_2$. (a) When the charges are of the same sign, the force is repulsive. (b) When the charges are of the opposite sign, the force is attractive.

**EXAMPLE 23.1 Find the Resultant Force**

Consider three point charges located at the corners of a triangle, as in Figure 23.6, where $q_1 = q_3 = 5\ \mu\text{C}$, $q_2 = -2\ \mu\text{C}$ ($1\ \mu\text{C} = 10^{-6}$ C), and $a = 0.1$ m. Find the resultant force on $q_3$.

*Solution* First, note the direction of the individual forces on $q_3$ due to $q_1$ and $q_2$. The force on $q_3$ due to $q_2$ is attractive since $q_2$ and $q_3$ have opposite signs. The force on $q_3$ due to $q_1$ is repulsive since they are both positive.

Now let us calculate the magnitude of the forces on $q_3$. The magnitude of the force on $q_3$ due to $q_2$ is given by

$$F_{32} = k\,\frac{|q_3||q_2|}{a^2}$$

$$= \left(9.0 \times 10^9\ \frac{\text{N}\cdot\text{m}^2}{\text{C}^2}\right)\frac{(5 \times 10^{-6}\ \text{C})(2 \times 10^{-6}\ \text{C})}{(0.1\ \text{m})^2}$$

$$= 9.0\ \text{N}$$

Note that since $q_3$ and $q_2$ are opposite in sign, $\mathbf{F}_{32}$ is to the left as shown in Figure 23.6.

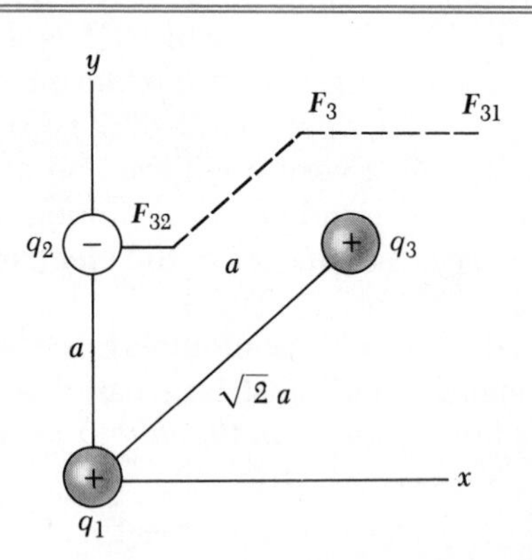


**Figure 23.6** (Example 23.1) The force on $q_3$ due to $q_1$ is $\mathbf{F}_{31}$. The force on $q_3$ due to $q_2$ is $\mathbf{F}_{32}$. The *total* force, $\mathbf{F}_3$, on $q_3$ is the *vector* sum $\mathbf{F}_{31} + \mathbf{F}_{32}$.

The magnitude of the force on $q_3$ due to $q_1$ is given by

$$F_{31} = k\frac{|q_3||q_1|}{(\sqrt{2}a)^2}$$

$$= \left(9.0 \times 10^9 \frac{\text{N}\cdot\text{m}^2}{\text{C}^2}\right) \frac{(5 \times 10^{-6}\text{ C})(5 \times 10^{-6}\text{ C})}{2(0.1\text{ m})^2}$$

$$= 11\text{ N}$$

The force $\mathbf{F}_{31}$ is repulsive and makes an angle of 45° with the $x$ axis. Therefore, the $x$ and $y$ components of $\mathbf{F}_{31}$ are equal, with magnitude given by $F_{31}\cos 45° = 7.9$ N. The force $\mathbf{F}_{32}$ is in the negative $x$ direction. Hence, the $x$ and $y$ components of the resultant force on $q_3$ are given by

$$F_x = F_{31x} + F_{32} = 7.9\text{ N} - 9.0\text{ N} = -1.1\text{ N}$$

$$F_y = F_{31y} = 7.9\text{ N}$$

We can also express the resultant force on $q_3$ in unit-vector form as $\mathbf{F}_3 = (-1.1\mathbf{i} + 7.9\mathbf{j})$ N.

Exercise 1 Find the magnitude and direction of the resultant force on $q_3$.
Answer 8.0 N at an angle of 98° with the $x$ axis.

**EXAMPLE 23.2 Where is the Resultant Force Zero?**
Three charges lie along the $x$ axis as in Figure 23.7. The positive charge $q_1 = 15\ \mu$C is at $x = 2$ m, and the positive charge $q_2 = 6\ \mu$C is at the origin. Where must a *negative* charge $q_3$ be placed on the $x$ axis such that the resultant force on it is zero?

*Solution* Since $q_3$ is negative and both $q_1$ and $q_2$ are positive, the forces $\mathbf{F}_{31}$ and $\mathbf{F}_{32}$ are both attractive, as indicated in Figure 23.7. If we let $x$ be the coordinate of $q_3$, then the forces $F_{31}$ and $F_{32}$ have magnitudes given by

$$F_{31} = k\frac{|q_3||q_1|}{(2-x)^2} \quad \text{and} \quad F_{32} = k\frac{|q_3||q_2|}{x^2}$$

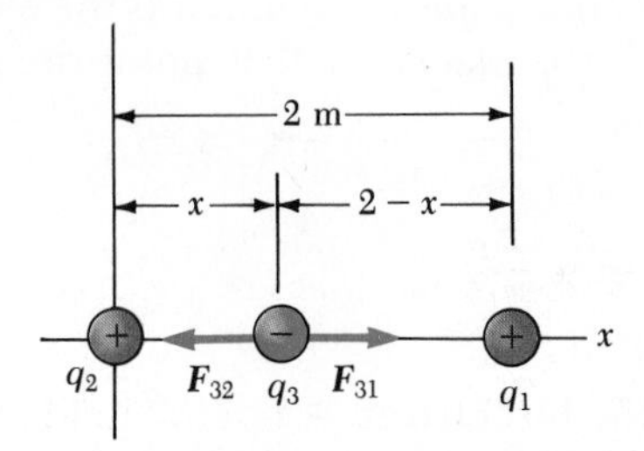


**Figure 23.7** (Example 23.2) Three point charges are placed along the $x$ axis. The charge $q_3$ is negative, whereas $q_1$ and $q_2$ are positive. If the net force on $q_3$ is zero, then the force on $q_3$ due to $q_1$ must be equal and opposite to the force on $q_3$ due to $q_2$.

If the resultant force on $q_3$ is zero, then $\mathbf{F}_{32}$ must be equal to and opposite $\mathbf{F}_{31}$, or

$$k\frac{|q_3||q_2|}{x^2} = k\frac{|q_3||q_1|}{(2-x)^2}$$

Since $k$ and $q_3$ are common to both sides, we solve for $x$ and find that

$$(2-x)^2|q_2| = x^2|q_1|$$

$$(4 - 4x + x^2)(6 \times 10^{-6}\text{ C}) = x^2(15 \times 10^{-6}\text{ C})$$

Solving this quadratic equation for $x$, we find that $x = 0.775$ m. Why is the negative root not acceptable?

**EXAMPLE 23.3 The Hydrogen Atom**
The electron and proton of a hydrogen atom are separated (on the average) by a distance of approximately $5.3 \times 10^{-11}$ m. Find the magnitude of the electrical force and the gravitational force between the two particles.

*Solution* From Coulomb's law, we find that the attractive electrical force has the magnitude

$$F_e = k\frac{|e|^2}{r^2} = 9.0 \times 10^9 \frac{\text{N}\cdot\text{m}^2}{\text{C}^2} \frac{(1.6 \times 10^{-19}\text{ C})^2}{(5.3 \times 10^{-11}\text{ m})^2}$$

$$= 8.2 \times 10^{-8}\text{ N}$$

Using Newton's universal law of gravity and Table 23.1, we find that the gravitational force has the magnitude

$$F_g = G\frac{m_e m_p}{r^2}$$

$$= \left(6.7 \times 10^{-11}\frac{\text{N}\cdot\text{m}^2}{\text{kg}^2}\right)$$

$$\times \frac{(9.11 \times 10^{-31}\text{ kg})(1.67 \times 10^{-27}\text{ kg})}{(5.3 \times 10^{-11}\text{ m})^2}$$

$$= 3.6 \times 10^{-47}\text{ N}$$

The ratio $F_e/F_g \approx 3 \times 10^{39}$. Thus the gravitational force between charged atomic particles is negligible compared with the electrical force.

**EXAMPLE 23.4 Find the Charge on the Spheres**
Two identical small charged spheres, each having a mass of $3 \times 10^{-2}$ kg, hang in equilibrium as shown in Figure 23.8a. If the length of each string is 0.15 m and the angle $\theta = 5°$, find the magnitude of the charge on each sphere, assuming the spheres have identical charges.

*Solution* From the right triangle in Figure 23.8a, we see that $\sin\theta = a/L$. From the known length of the string and

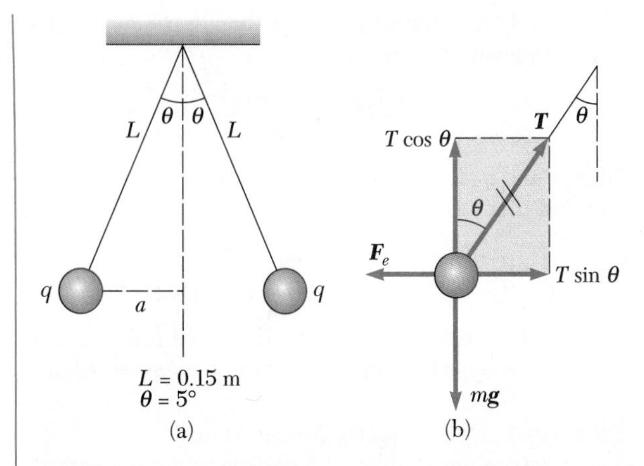


**Figure 23.8** (Example 23.4) (a) Two identical spheres, each with the same charge $q$, suspended in equilibrium by strings. (b) The free-body diagram for the charged spheres on the left side.

the angle the string makes with the vertical, the distance $a$ is calculated to be

$$a = L \sin\theta = (0.15 \text{ m}) \sin 5° = 0.013 \text{ m}$$

Therefore, the separation of the spheres is $2a = 0.026$ m.

The forces acting on one of the spheres are shown in Figure 23.8b. Because the sphere is in equilibrium, the resultants of the forces in the horizontal and vertical directions must separately add up to zero:

$$(1) \qquad \sum F_x = T \sin\theta - F_e = 0$$

$$(2) \qquad \sum F_y = T \cos\theta - mg = 0$$

From (2), we see that $T = mg/\cos\theta$, and so $T$ can be eliminated from (1) if we make this substitution. This gives a value for the electric force, $F_e$:

$$(3) \qquad F_e = mg \tan\theta$$

$$= (3 \times 10^{-2} \text{ kg})(9.80 \text{ m/s}^2)\tan(5°)$$

$$= 2.57 \times 10^{-2} \text{ N}$$

From Coulomb's law (Eq. 23.1), the electric force between the charges has magnitude given by

$$F_e = k\frac{|q|^2}{r^2}$$

where $r = 2a = 0.026$ m and $|q|$ is the magnitude of the charge on each sphere. Note that the term $|q|^2$ arises here because we have assumed that the charge is the same on both spheres. This equation can be solved for $|q|^2$ to give the charge as follows:

$$|q|^2 = \frac{F_e r^2}{k} = \frac{(2.57 \times 10^{-2} \text{ N})(0.026 \text{ m})^2}{9 \times 10^9 \text{ N} \cdot \text{m}^2/\text{C}^2}$$

$$|q| = \boxed{4.4 \times 10^{-8} \text{ C}}$$

Exercise 2 If the charge on the spheres is negative, how many electrons had to be added to the spheres to give a net charge of $-4.4 \times 10^{-8}$ C?
Answer $2.7 \times 10^{11}$ electrons.

## 23.4 THE ELECTRIC FIELD

The gravitational field **g** at a point in space was defined in Chapter 14 to be equal to the gravitational force ***F*** acting on a test mass $m_0$ divided by the test mass. That is, $\mathbf{g} = \mathbf{F}/m_0$. In similar manner, an electric field at a point in space can be defined in terms of the electric force acting on a test charge $q_0$ placed at that point. To be more precise,

> the electric field vector ***E*** at a point in space is defined as the electric force ***F*** acting on a positive test charge placed at that point divided by the magnitude of the test charge $q_0$:

Definition of electric field

$$\mathbf{E} \equiv \frac{\mathbf{F}}{q_0} \qquad (23.7)$$

Note that ***E*** is the field *external* to the test charge—not the field produced by the test charge. The vector ***E*** has the SI units of newtons per coulomb (N/C). The direction of ***E*** is in the direction of ***F*** since we have assumed that ***F*** acts on a positive test charge. Thus, we can say that *an electric field exists at a point if a test charge at rest placed at that point experiences an electrical force.* Once the

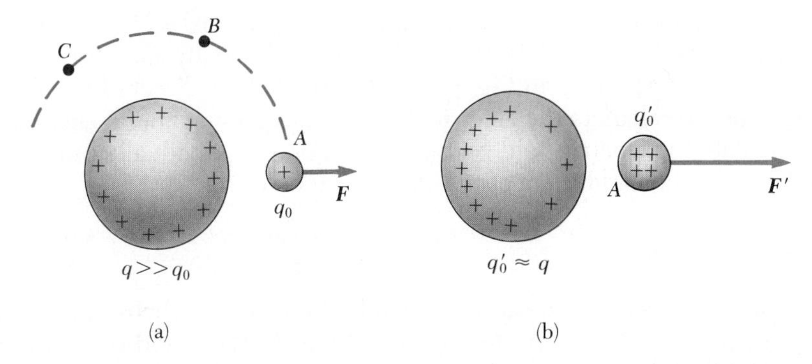


**Figure 23.9** (a) When a small test charge $q_0$ is placed near a conducting sphere of charge $q$ (where $q \gg q_0$), the charge on the conducting sphere remains uniform. (b) If the test charge $q_0'$ is of the order of the charge on the sphere, the charge on the sphere is nonuniform.

electric field is known at some point, the force on *any* charged particle placed at that point can be calculated from Equation 23.7. Furthermore, the electric field is said to exist at some point (even empty space) regardless of whether or not a test charge is located at that point.

When Equation 23.7 is applied, we must assume that the test charge $q_0$ is small enough such that it does not disturb the charge distribution responsible for the electric field.[3] For instance, if a vanishingly small test charge $q_0$ is placed near a uniformly charged metallic sphere as in Figure 23.9a, the charge on the metallic sphere, which produces the electric field, will remain uniformly distributed. Furthermore, the force $\boldsymbol{F}$ on the test charge will have the same magnitude at points $A$, $B$, and $C$, which are equidistant from the sphere. If the test charge is large enough ($q_0' \gg q_0$) as in Figure 23.9b, the charge on the metallic sphere will be redistributed and the ratio of the force to the test charge at point $A$ will be different: ($F'/q_0' \neq F/q_0$). That is, because of this redistribution of charge on the metallic sphere, the electric field at point $A$ set up by the sphere in Figure 23.9b must be different from that of the field at point $A$ in Figure 23.9a. Furthermore, the distribution of charge on the sphere will change as the smaller charge is moved from point $A$ to point $B$ or $C$.

Consider a point charge $q$ located a distance $r$ from a test charge $q_0$. According to Coulomb's law, the force on the test charge is given by

$$\boldsymbol{F} = k\frac{qq_0}{r^2}\hat{\boldsymbol{r}}$$

Since the electric field at the position of the test charge is defined by $\boldsymbol{E} = \boldsymbol{F}/q_0$, we find that the electric field *at the position of $q_0$ due to the charge $q$* is given by

$$\boldsymbol{E} = k\frac{q}{r^2}\hat{\boldsymbol{r}} \qquad (23.8)$$

[3] To be more precise, the test charge $q_0$ should be infinitesimally small to ensure that its presence does not affect the original charge distribution. Therefore, strictly speaking, we should replace Equation 23.7 by the expression

$$\boldsymbol{E} = \lim_{q_0 \to 0} \frac{\boldsymbol{F}}{q_0}$$

It is impossible to follow this prescription strictly in any experiment since no charges smaller in magnitude than $e$ are known to exist. However, as a practical matter, it is almost always possible to select a sufficiently small test charge to obtain any desired degree of accuracy.

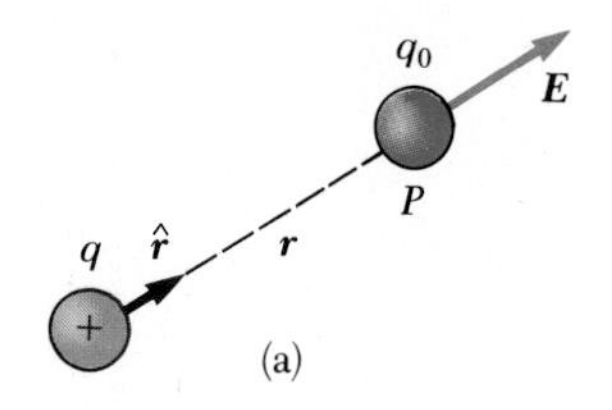


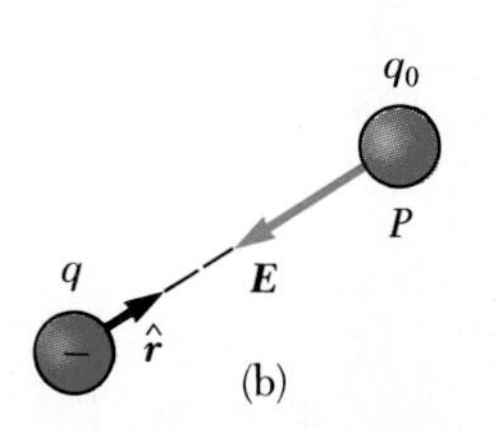


**Figure 23.10** A test charge $q_0$ at the point $P$ is at a distance $r$ from a point charge $q$. (a) If $q$ is positive, the electric field at $P$ points radially *outward* from $q$. (b) If $q$ is negative, the electric field at $P$ points radially *inward* toward $q$.

where $\hat{r}$ is a unit vector that is directed away from $q$ toward $q_0$ (Fig. 23.10). If $q$ is *positive*, as in Figure 23.10a, the field is directed radially *outward* from this charge. If $q$ is *negative*, as in Figure 23.10b, the field is directed *toward* $q$.

In order to calculate the electric field due to a group of point charges, we first calculate the electric field vectors at the point $P$ individually using Equation 23.8 and then add them *vectorially*. In other words,

the total electric field due to a group of charges equals the vector sum of the electric fields of all the charges.

This **superposition principle** applied to fields follows directly from the superposition property of electric forces. Thus, the electric field of a group of charges (excluding the test charge $q_0$) can be expressed as

$$\mathbf{E} = k \sum_i \frac{q_i}{r_i^2} \hat{\mathbf{r}}_i \tag{23.9}$$

where $r_i$ is the distance from the $i$th charge, $q_i$, to the point $P$ (the location of the test charge) and $\hat{r}_i$ is a unit vector directed from $q_i$ toward $P$.

### EXAMPLE 23.5 Electric Force on a Proton

Find the electric force on a proton placed in an electric field of $2 \times 10^4$ N/C directed along the positive $x$ axis.

*Solution* Since the charge on a proton is

$$+e = +1.6 \times 10^{-19}\ \text{C},$$

the electric force on it is

$$\mathbf{F} = e\mathbf{E} = (1.6 \times 10^{-19}\ \text{C})(2 \times 10^4 \mathbf{i}\ \text{N/C})$$
$$= 3.2 \times 10^{-15} \mathbf{i}\ \text{N}$$

where $\mathbf{i}$ is a unit vector in the positive $x$ direction. The weight of the proton is calculated to be equal to $mg = (1.67 \times 10^{-27}\ \text{kg})(9.8\ \text{m/s}^2) = 1.6 \times 10^{-26}$ N. Hence, we see that the magnitude of the gravitational force in this case is negligible compared with the electric force.

### EXAMPLE 23.6 Electric Field Due to Two Charges

A charge $q_1 = 7\ \mu$C is located at the origin, and a second charge $q_2 = -5\ \mu$C is located on the $x$ axis 0.3 m from the origin (Figure 23.11). Find the electric field at the point $P$ with coordinates (0, 0.4) m.

*Solution* First, let us find the magnitudes of the electric fields due to each charge. The fields $\mathbf{E}_1$ due to the 7-$\mu$C charge and $\mathbf{E}_2$ due to the $-5$-$\mu$C charge at $P$ are shown in Figure 23.11. Their magnitudes are given by

$$E_1 = k\frac{|q_1|}{r_1^2} = \left(9.0 \times 10^9 \frac{\text{N} \cdot \text{m}^2}{\text{C}^2}\right) \frac{(7 \times 10^{-6}\ \text{C})}{(0.4\ \text{m})^2}$$

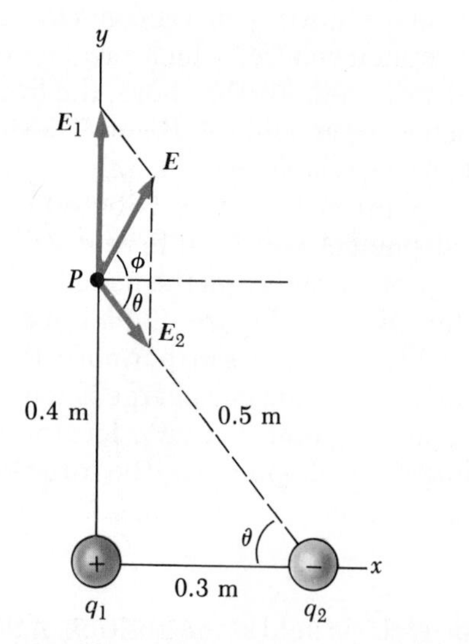


**Figure 23.11** (Example 23.6) The total electric field $\mathbf{E}$ at $P$ equals the vector sum $\mathbf{E}_1 + \mathbf{E}_2$, where $\mathbf{E}_1$ is the field due to the positive charge $q_1$ and $\mathbf{E}_2$ is the field due to the negative charge $q_2$.

$$= 3.94 \times 10^5\ \text{N/C}$$

$$E_2 = k\frac{|q_2|}{r_2^2} = \left(9.0 \times 10^9 \frac{\text{N} \cdot \text{m}^2}{\text{C}^2}\right) \frac{(5 \times 10^{-6}\ \text{C})}{(0.5\ \text{m})^2}$$
$$= 1.8 \times 10^5\ \text{N/C}$$

The vector $\mathbf{E}_1$ has only a $y$ component. The vector $\mathbf{E}_2$ has an $x$ component given by $E_2 \cos\theta = \frac{3}{5}E_2$ and a negative $y$

component given by $-E_2 \sin\theta = -\frac{4}{5}E_2$. Hence, we can express the vectors as

$$\mathbf{E}_1 = 3.94 \times 10^5 \mathbf{j} \text{ N/C}$$
$$\mathbf{E}_2 = (1.1 \times 10^5 \mathbf{i} - 1.4 \times 10^5 \mathbf{j}) \text{ N/C}$$

The resultant field $\mathbf{E}$ at $P$ is the superposition of $\mathbf{E}_1$ and $\mathbf{E}_2$:

$$\mathbf{E} = \mathbf{E}_1 + \mathbf{E}_2 = (1.1 \times 10^5 \mathbf{i} + 2.5 \times 10^5 \mathbf{j}) \text{ N/C}$$

From this result, we find that $\mathbf{E}$ has a magnitude of $2.7 \times 10^5$ N/C and makes an angle $\phi$ of 66° with the positive $x$ axis.

Exercise 3 Find the electric force on a test charge of $2 \times 10^{-8}$ C placed at $P$.
Answer $5.4 \times 10^{-3}$ N in the same direction as $\mathbf{E}$.

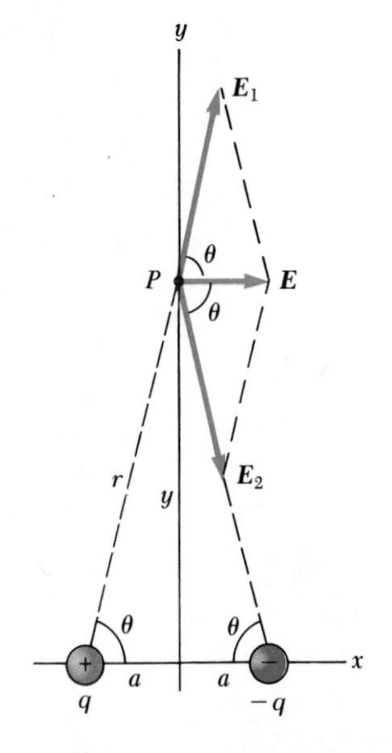


Figure 23.12 (Example 23.7) The total electric field $\mathbf{E}$ at $P$ due to two equal and opposite charges (an electric dipole) equals the vector sum $\mathbf{E}_1 + \mathbf{E}_2$. The field $\mathbf{E}_1$ is due to the positive charge $q$, and $\mathbf{E}_2$ is the field due to the negative charge $-q$.

### EXAMPLE 23.7 Electric Field of a Dipole

An **electric dipole** consists of a positive charge $q$ and a negative charge $-q$ separated by a distance $2a$, as in Figure 23.12. Find the electric field $\mathbf{E}$ due to these charges along the $y$ axis at the point $P$, which is a distance $y$ from the origin. Assume that $y \gg a$.

*Solution* At $P$, the fields $\mathbf{E}_1$ and $\mathbf{E}_2$ due to the two charges are equal in magnitude, since $P$ is equidistant from the two equal and opposite charges. The total field $\mathbf{E} = \mathbf{E}_1 + \mathbf{E}_2$, where the magnitudes of $\mathbf{E}_1$ and $\mathbf{E}_2$ are given by

$$E_1 = E_2 = k\frac{q}{r^2} = k\frac{q}{y^2 + a^2}$$

The $y$ components of $\mathbf{E}_1$ and $\mathbf{E}_2$ cancel each other. The $x$ components are equal since they are both along the $x$ axis. Therefore, $\mathbf{E}$ lies along the $x$ axis and has a magnitude equal to $2E_1 \cos\theta$. From Figure 23.12 we see that $\cos\theta = a/r = a/(y^2 + a^2)^{1/2}$. Therefore,

$$E = 2E_1 \cos\theta = 2k\frac{q}{(y^2 + a^2)}\frac{a}{(y^2 + a^2)^{1/2}}$$
$$= k\frac{2qa}{(y^2 + a^2)^{3/2}}$$

Using the approximation $y \gg a$, we can neglect $a^2$ in the denominator and write

$$E \approx k\frac{2qa}{y^3} \qquad (23.10)$$

Thus we see that along the $y$ axis the field of a *dipole* at a distant point varies as $1/r^3$, whereas the more slowly varying field of a *point charge* goes as $1/r^2$. This is because at distant points, the fields of the two equal and opposite charges almost cancel each other. The $1/r^3$ variation in $E$ for the dipole is also obtained for a distant point along the $x$ axis (Problem 61) and for a general distant point. The dipole is a good model of many molecules, such as HCl.

As we shall see in later chapters, neutral atoms and molecules behave as dipoles when placed in an external electric field. Furthermore, many molecules, such as HCl, are permanent dipoles. (HCl is essentially an $H^+$ ion combined with a $Cl^-$ ion.) The effect of such dipoles on the behavior of materials subjected to electric fields will be discussed in Chapter 26.

## 23.5 ELECTRIC FIELD OF A CONTINUOUS CHARGE DISTRIBUTION

In the previous section, we showed how to calculate the electric field of a point charge using Coulomb's law. The total field of a group of point charges was obtained by taking the vector sum of the individual fields due to all the charges. *This procedure makes use of the superposition principle as applied to the electrostatic field.*

Very often the charges of interest are close together compared with their distances to points of interest. In such situations, the system of charges can be considered to be *continuous*. That is, we imagine that the system of closely spaced charges is equivalent to a total charge that is continuously distributed through a volume or over some surface.

A continuous charge distribution

To evaluate the electric field of a continuous charge distribution, the following procedure is used. First, we divide the charge distribution into small elements each of which contains a small charge $\Delta q$, as in Figure 23.13. Next, we use Coulomb's law to calculate the electric field due to one of these elements at a point $P$. Finally, we evaluate the total field at $P$ due to the charge distribution by summing the contributions of all the charge elements (that is, by applying the superposition principle).

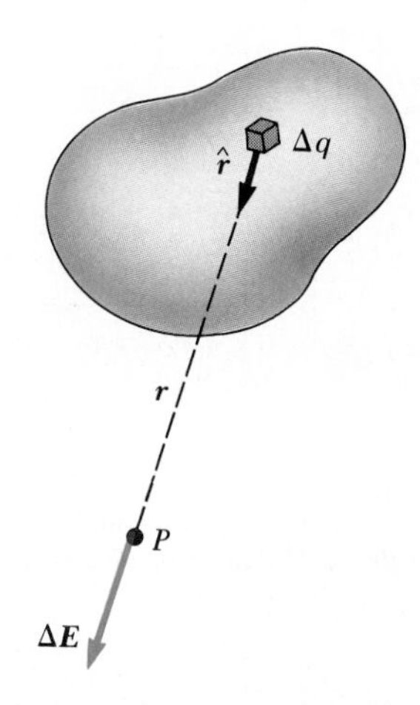


**Figure 23.13** The electric field at $P$ due to a continuous charge distribution is the vector sum of the fields due to all the elements $\Delta q$ of the charge distribution.

The electric field at $P$ due to one element of charge $\Delta q$ is given by

$$\Delta \mathbf{E} = k \frac{\Delta q}{r^2} \hat{\mathbf{r}}$$

where $r$ is the distance from the element to point $P$ and $\hat{\mathbf{r}}$ is a unit vector directed from the charge element toward $P$. The total electric field at $P$ due to all elements in the charge distribution is approximately given by

$$\mathbf{E} \approx k \sum_i \frac{\Delta q_i}{r_i^2} \hat{\mathbf{r}}_i$$

where the index $i$ refers to the $i$th element in the distribution. If the separation between elements in the charge distribution is small compared with the distance to $P$, the charge distribution can be approximated to be continuous. Therefore, the total field at $P$ in the limit $\Delta q_i \rightarrow 0$ becomes

Electric field of a continuous charge distribution

$$\mathbf{E} = k \lim_{\Delta q_i \rightarrow 0} \sum_i \frac{\Delta q_i}{r_i^2} \hat{\mathbf{r}}_i = k \int \frac{dq}{r^2} \hat{\mathbf{r}} \tag{23.11}$$

where the integration is a *vector* operation and must be treated with caution. We shall illustrate this type of calculation with several examples. In these examples, we shall assume that the charge is *uniformly* distributed on a line or a surface or throughout some volume. When performing such calculations, it is convenient to use the concept of a charge density along with the following notations:

If a charge $Q$ is uniformly distributed throughout a volume $V$, the *charge per unit volume*, $\rho$, is defined by

Volume charge density

$$\rho \equiv \frac{Q}{V} \tag{23.12}$$

where $\rho$ has units of $C/m^3$.

If a charge $Q$ is uniformly distributed on a surface of area $A$, the *surface charge density*, $\sigma$, is defined by

Surface charge density

$$\sigma \equiv \frac{Q}{A} \tag{23.13}$$

where $\sigma$ has units of $C/m^2$.

Finally, if a charge $Q$ is uniformly distributed along a line of length $\ell$, the *linear charge density*, $\lambda$, is defined by

$$\lambda \equiv \frac{Q}{\ell} \qquad (23.14)$$

Linear charge density

where $\lambda$ has units of C/m.

If the charge is *nonuniformly* distributed over a volume, surface, or line, we would have to express the charge densities as

$$\rho = \frac{dQ}{dV} \qquad \sigma = \frac{dQ}{dA} \qquad \lambda = \frac{dQ}{d\ell}$$

where $dQ$ is the amount of charge in a small volume, surface, or length element.

**EXAMPLE 23.8 The Electric Field Due to a Charged Rod**

A rod of length $\ell$ has a uniform positive charge per unit length $\lambda$ and a total charge $Q$. Calculate the electric field at a point $P$ along the axis of the rod, a distance $d$ from one end (Fig. 23.14).

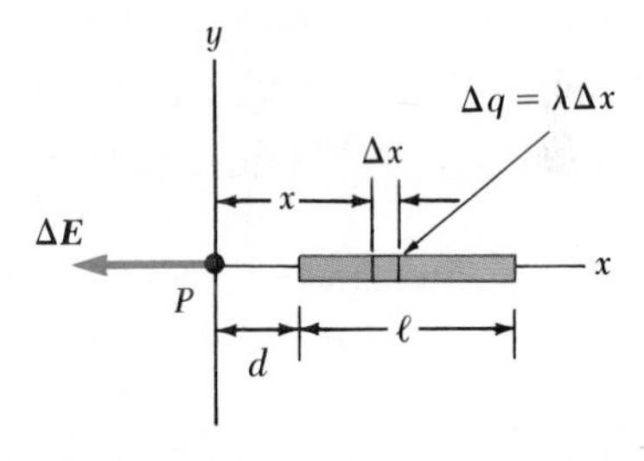


**Figure 23.14** (Example 23.8) The electric field at $P$ due to a uniformly charged rod lying along the $x$ axis. The field at $P$ due to the segment of charge $\Delta q$ is given by $k\ \Delta q/x^2$. The total field at $P$ is the vector sum over all segments of the rod.

*Solution* For this calculation, the rod is taken to be along the $x$ axis. The ratio of $\Delta q$, the charge on the segment to $\Delta x$, the length of the segment, is equal to the ratio of the total charge to the total length of the rod. That is, $\Delta q/\Delta x = Q/\ell = \lambda$. Therefore, the charge $\Delta q$ on the small segment is given by $\Delta q = \lambda\ \Delta x$.

The field $\Delta E$ due to this segment at the point $P$ is in the negative $x$ direction, and its magnitude is given by [4]

$$\Delta E = k\frac{\Delta q}{x^2} = k\frac{\lambda\ \Delta x}{x^2}$$

Note that each element produces a field in the negative $x$ direction, and so the problem of summing their contributions is particularly simple in this case. The total field at $P$ due to all segments of the rod, which are at different distances from $P$, is given by Equation 23.11, which in this case becomes

$$E = \int_d^{\ell+d} k\lambda \frac{dx}{x^2}$$

where the limits on the integral extend from one end of the rod ($x = d$) to the other ($x = \ell + d$). Since $k$ and $\lambda$ are constants, they can be removed from the integral. Thus, we find that

$$E = k\lambda \int_d^{\ell+d} \frac{dx}{x^2} = k\lambda \left[-\frac{1}{x}\right]_d^{\ell+d}$$

$$= k\lambda\left(\frac{1}{d} - \frac{1}{\ell + d}\right)$$

$$= \frac{kQ}{d(\ell + d)} \qquad (23.15)$$

where we have used the fact that the total charge $Q = \lambda\ell$. From this result we see that if the point $P$ is *far* from the rod ($d \gg \ell$), then $\ell$ in the denominator can be neglected, and $E \approx kQ/d^2$. This is just the form you would expect for a point charge. Therefore, at large distances from the rod, the charge distribution appears to be a point charge of magnitude $Q$. The use of the limiting technique ($d \to \infty$) is often a good method for checking a theoretical formula.

[4] It is important that you understand the procedure being used to carry out integrations such as this. First, choose an element whose parts are all equidistant from the point at which the field is being calculated. Next, express the charge element $\Delta q$ in terms of the other variables within the integral (in this example, there is one variable, $x$.) In examples that have spherical or cylindrical symmetry, the variable will be a radial coordinate.

### EXAMPLE 23.9 The Electric Field of a Uniform Ring of Charge

A ring of radius $a$ has a uniform positive charge per unit length, with a total charge $Q$. Calculate the electric field along the axis of the ring at a point $P$ lying a distance $x$ from the center of the ring (Fig. 23.15a).

*Solution* The magnitude of the electric field at $P$ due to the segment of charge $\Delta q$ is

$$\Delta E = k\frac{\Delta q}{r^2}$$

This field has an $x$ component $\Delta E_x = \Delta E \cos\theta$ along the axis of the ring and a component $\Delta E_\perp$ perpendicular to the axis. But as we see in Figure 23.15b, the resultant field at $P$ must lie along the $x$ axis since the perpendicular components sum up to zero. That is, the perpendicular component of any element is canceled by the perpendicular component of an element on the opposite side of the ring. Since $r = (x^2 + a^2)^{1/2}$ and $\cos\theta = x/r$, we find that

$$\Delta E_x = \Delta E \cos\theta = \left(k\frac{\Delta q}{r^2}\right)\frac{x}{r} = \frac{kx}{(x^2 + a^2)^{3/2}}\Delta q$$

In this case, all segments of the ring give the *same* contribution to the field at $P$ since they are all equidistant from this point. Thus, we can easily sum over all segments to get the total field at $P$:

$$E_x = \sum \frac{kx}{(x^2 + a^2)^{3/2}}\Delta q = \frac{kx}{(x^2 + a^2)^{3/2}}Q \qquad (23.16)$$

This result shows that the field is zero at $x = 0$. Does this surprise you?

Exercise 4 Show that at large distances from the ring ($x \gg a$) the electric field along the axis approaches that of a point charge of magnitude $Q$.

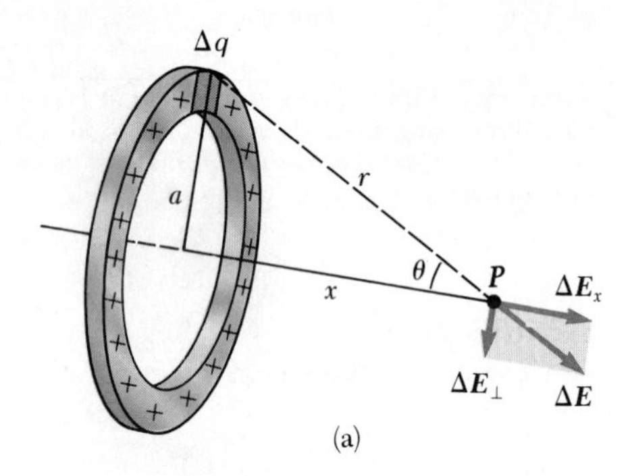


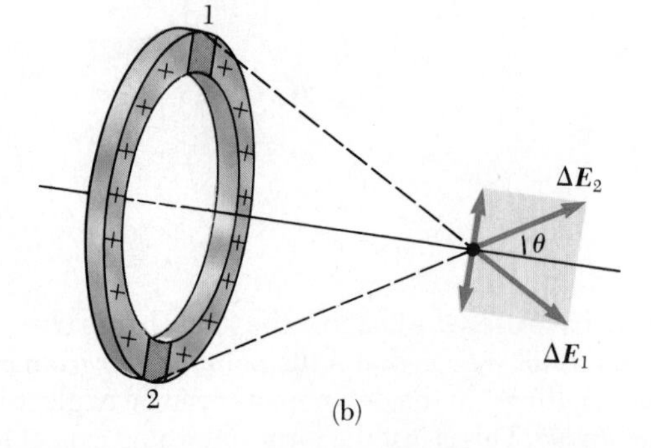


**Figure 23.15** (Example 23.9) A uniformly charged ring of radius $a$. (a) The field at $P$ on the $x$ axis due to an element of charge $\Delta q$. (b) The total electric field at $P$ is along the $x$ axis. Note that the perpendicular component of the electric field at $P$ due to segment 1 is canceled by the perpendicular component due to segment 2, which is opposite segment 1.

### EXAMPLE 23.10 The Electric Field of a Uniformly Charged Disk □

A disk of radius $R$ has a uniform charge per unit area $\sigma$. Calculate the electric field along the axis of the disk, a distance $x$ from its center (Fig. 23.16).

*Solution* The solution to this problem is straightforward if we consider the disk as a set of concentric rings. We can then make use of Example 23.9, which gives the field of a given ring of radius $r$, and sum up contributions of all rings making up the disk. By symmetry, the field on an axial point must be parallel to this axis.

The ring of radius $r$ and width $dr$ has an area equal to $2\pi r\, dr$ (Fig. 23.16). The charge $dq$ on this ring is equal to the area of the ring multiplied by the charge per unit area, or $dq = 2\pi\sigma r\, dr$. Using this result in Equation 23.16 (with $a$ replaced by $r$) gives for the field due to the ring the expression

$$dE = \frac{kx}{(x^2 + r^2)^{3/2}}(2\pi\sigma r\, dr)$$

To get the total field at $P$, we integrate this expression over the limits $r = 0$ to $r = R$, noting that $x$ is a constant, which gives

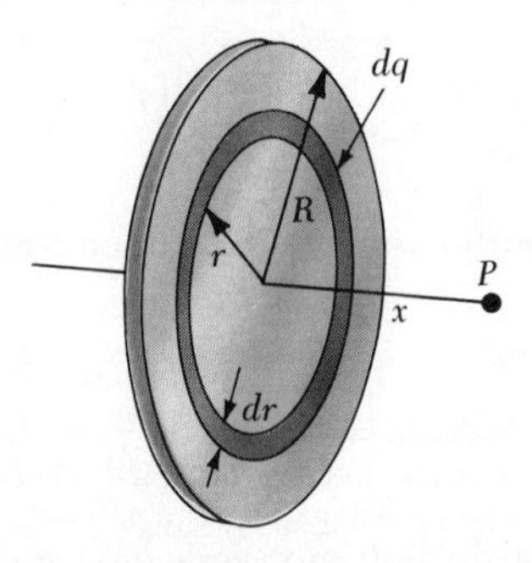


**Figure 23.16** (Example 23.10) A uniformly charged disk of radius $R$. The electric field at an axial point $P$ is directed along this axis, perpendicular to the plane of the disk.

$$E = kx\pi\sigma \int_0^R \frac{2r\,dr}{(x^2 + r^2)^{3/2}}$$

$$= kx\pi\sigma \int_0^R (x^2 + r^2)^{-3/2}\,d(r^2)$$

$$= kx\pi\sigma \left[\frac{(x^2 + r^2)^{-1/2}}{-1/2}\right]_0^R$$

$$= 2\pi k\sigma \left(\frac{x}{|x|} - \frac{x}{(x^2 + R^2)^{1/2}}\right) \qquad (23.17)$$

The result is valid for all values of $x$. The field close to the disk along an axial point can also be obtained from Equation 23.17 by letting $x \to 0$ (or $R \to \infty$). This gives

$$E = 2\pi k\sigma = \frac{\sigma}{2\epsilon_0} \qquad (23.18)$$

where $\epsilon_0$ is the permittivity of free space, given by Equation 23.4. As we shall find in the next chapter, the same result is obtained for the field of a uniformly charged infinite sheet.

## 23.6 ELECTRIC FIELD LINES

A convenient aid for visualizing electric field patterns is to draw lines pointing in the same direction as the electric field vector at any point. These lines, called **electric field lines,** are related to the electric field in any region of space in the following manner:

1. The electric field vector $\boldsymbol{E}$ is *tangent* to the electric field line at each point.
2. The number of lines per unit area through a surface perpendicular to the lines is proportional to the strength of the electric field in that region. Thus $\boldsymbol{E}$ is large when the field lines are close together and small when they are far apart.

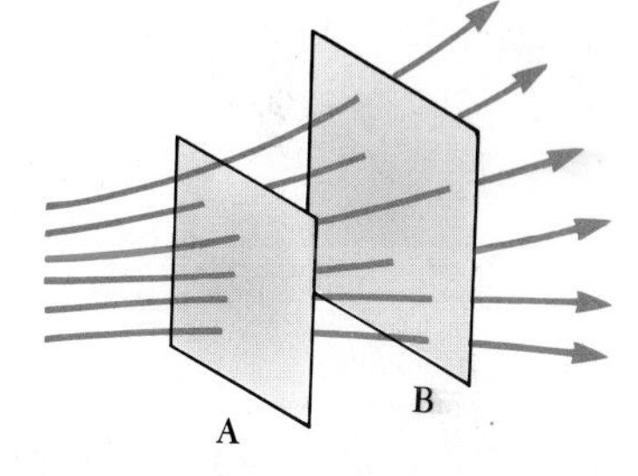


**Figure 23.17** Electric field lines penetrating two surfaces. The magnitude of the field is greater on surface A than on surface B.

These properties are illustrated in Figure 23.17. The density of lines through surface A is greater than the density of lines through surface B. Therefore, the electric field is more intense on surface A than on surface B. Furthermore, the field drawn in Figure 23.17 is nonuniform since the lines at different locations point in different directions.

Some representative electric field lines for a single positive point charge are shown in Figure 23.18a. Note that in this two-dimensional drawing we

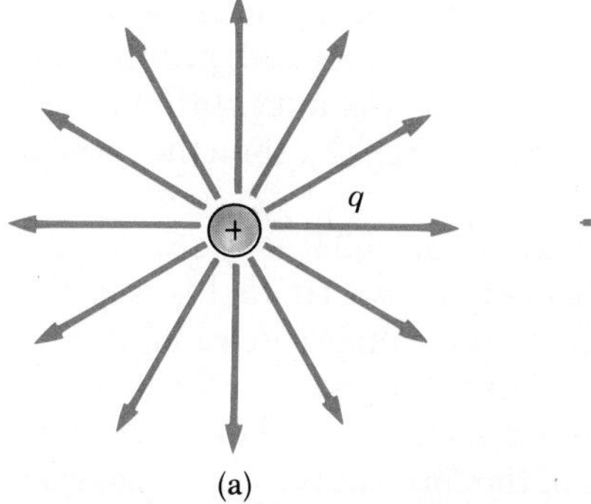


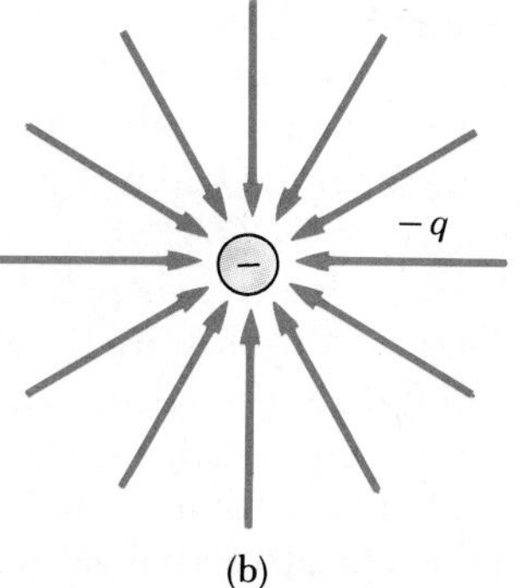


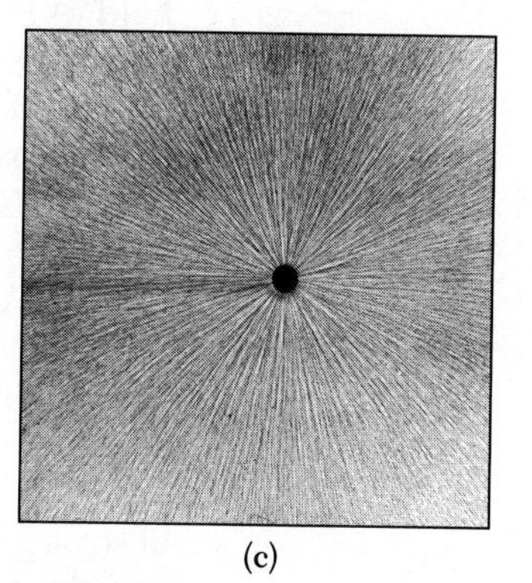


**Figure 23.18** The electric field lines for a point charge. (a) For a positive point charge, the lines are radially outward. (b) For a negative point charge, the lines are radially inward. Note that the figures show only those field lines that lie in the plane containing the charge. (c) The dark areas are small pieces of thread suspended in oil, which align with the electric field produced by a small charged conductor at the center. (Photo courtesy of Harold M. Waage, Princeton University)

show only the field lines that lie in the plane containing the point charge. The lines are actually directed radially outward from the charge in *all* directions, somewhat like the needles of a porcupine. Since a positive test charge placed in this field would be repelled by the charge $q$, the lines are directed radially away from the positive charge. Similarly, the electric field lines for a single negative point charge are directed toward the charge (Fig. 23.18b). In either case, the lines are along the radial direction and extend all the way to infinity. Note that the lines are closer together as they get near the charge, indicating that the strength of the field is increasing.

The rules for drawing electric field lines for any charge distribution are as follows:

**Rules for drawing electric field lines**

1. The lines must begin on positive charges and terminate on negative charges, or at infinity in the case of an excess of charge.
2. The number of lines drawn leaving a positive charge or approaching a negative charge is proportional to the magnitude of the charge.
3. No two field lines can cross.

Is this visualization of the electric field in terms of field lines consistent with Coulomb's law? To answer this question, consider an imaginary spherical surface of radius $r$ concentric with the charge. From symmetry, we see that the magnitude of the electric field is the same everywhere on the surface of the sphere. The number of lines, $N$, that emerge from the charge is equal to the number that penetrate the spherical surface. Hence, the number of lines per unit area on the sphere is $N/4\pi r^2$ (where the surface area of the sphere is $4\pi r^2$). Since $E$ is proportional to the number of lines per unit area, we see that $E$ varies as $1/r^2$. This is consistent with the result obtained from Coulomb's law, that is, $E = kq/r^2$.

It is important to note that electric field lines are not material objects. They are used only to provide us with a qualitative description of the electric field. One problem with this model is the fact that one always draws a finite number of lines from each charge, which makes it appear as if the field were quantized and acted only in a certain direction. The field, in fact, is continuous—existing at every point. Another problem with this model is the danger of getting the wrong impression from a two-dimensional drawing of field lines being used to describe a three-dimensional situation.

Since charge is quantized, the number of lines leaving any material object must be $0, \pm C'e, \pm 2C'e, \ldots$, where $C'$ is an arbitrary (but fixed) proportionality constant. Once $C'$ is chosen, the number of lines is not arbitrary. For example, if object 1 has charge $Q_1$ and object 2 has charge $Q_2$, then the ratio of number of lines is $N_2/N_1 = Q_2/Q_1$.

The electric field lines for two point charges of equal magnitude, but opposite signs (the electric dipole), are shown in Figure 23.19. In this case, the number of lines that begin at the positive charge must equal the number that terminate at the negative charge. At points very near the charges, the lines are nearly radial. The high density of lines between the charges indicates a region of strong electric field. The attractive nature of the force between the charges can also be seen from Figure 23.19.

Figure 23.20 shows the electric field lines in the vicinity of two equal positive point charges. Again, the lines are nearly radial at points close to either charge. The same number of lines emerge from each charge since the charges are equal in magnitude. At large distances from the charges, the field

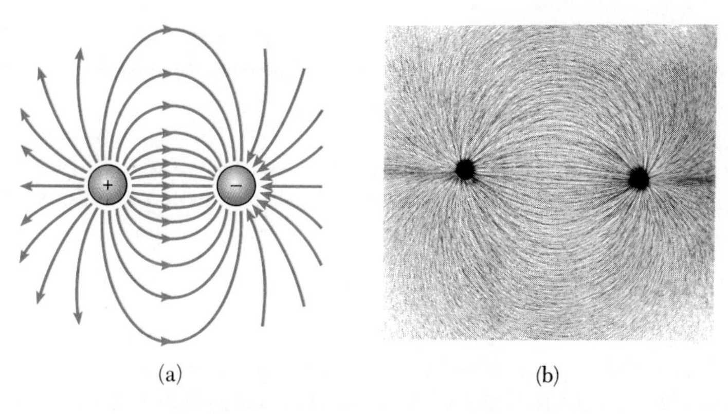


**Figure 23.19** (a) The electric field lines for two equal and opposite point charges (an electric dipole). Note that the number of lines leaving the positive charge equals the number terminating at the negative charge. (b) The photograph was taken using small pieces of thread suspended in oil, which align with the electric field. (Photo courtesy of Harold M. Waage, Princeton University)

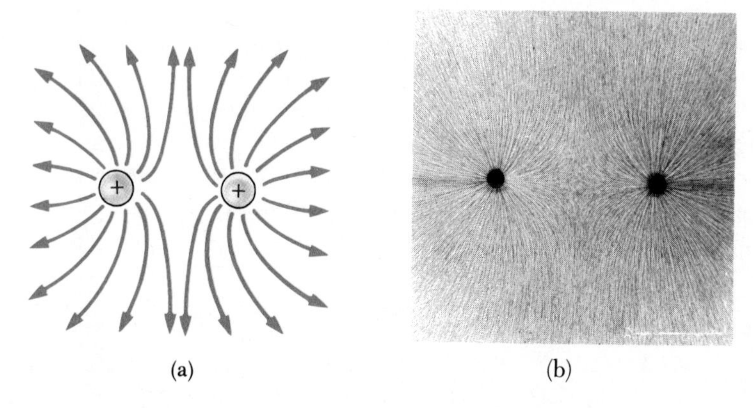


**Figure 23.20** (a) The electric field lines for two positive point charges. (b) The photograph was taken using small pieces of thread suspended in oil, which align with the electric field. (Photo courtesy of Harold M. Waage, Princeton University)

is approximately equal to that of a single point charge of magnitude $2q$. The bulging out of the electric field lines between the charges indicates the repulsive nature of the electric force between like charges.

Finally, in Figure 23.21 we sketch the electric field lines associated with a positive charge $+2q$ and a negative charge $-q$. In this case, we see that the number of lines leaving the charge $+2q$ is twice the number entering the charge $-q$. Hence only half of the lines that leave the positive charge enter the negative charge. The remaining half terminate on a negative charge we assume to be located at infinity. At large distances from the charges (large compared with the charge separation), the electric field lines are equivalent to those of a single charge $+q$.

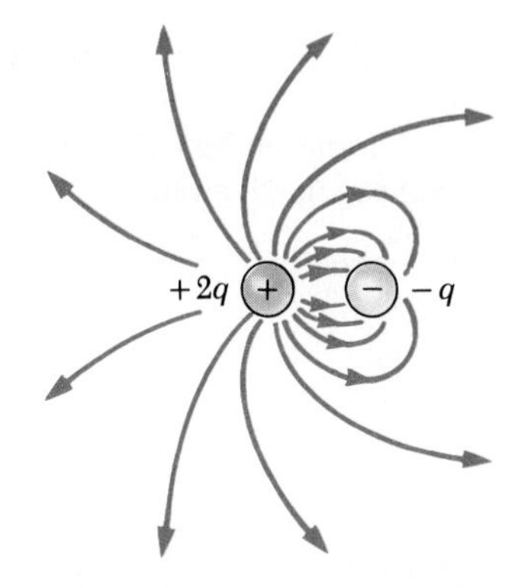


**Figure 23.21** The electric field lines for a point charge $+2q$ and a second point charge $-q$. Note that two lines leave the charge $+2q$ for every one that terminates on $-q$.

## 23.7 MOTION OF CHARGED PARTICLES IN A UNIFORM ELECTRIC FIELD

In this section we describe the motion of a charged particle in a uniform electric field. As we shall see, the motion is equivalent to that of a projectile moving in a uniform gravitational field. When a particle of charge $q$ is placed in an electric field $\mathbf{E}$, the electric force on the charge is $q\mathbf{E}$. If this is the only force exerted on the charge, then Newton's second law applied to the charge gives

$$\mathbf{F} = q\mathbf{E} = m\mathbf{a}$$

where $m$ is the mass of the charge and we assume that the speed is small compared with the speed of light. The acceleration of the particle is therefore given by

$$\mathbf{a} = \frac{q\mathbf{E}}{m} \tag{23.19}$$

If $\mathbf{E}$ is uniform (that is, constant in magnitude and direction), we see that the acceleration is a constant of the motion. If the charge is positive, the acceleration will be in the direction of the electric field. If the charge is negative, the acceleration will be in the direction *opposite* the electric field.

EXAMPLE 23.11 An Accelerating Positive Charge

A positive point charge $q$ of mass $m$ is released from rest in a uniform electric field $\mathbf{E}$ directed along the $x$ axis as in Figure 23.22. Describe its motion.

*Solution* The acceleration of the charge is constant and given by $qE/m$. The motion is simple linear motion along the $x$ axis. Therefore, we can apply the equations of kinematics in one dimension (from Chapter 3):

$$x - x_0 = v_0t + \tfrac{1}{2}at^2 \qquad v = v_0 + at$$

$$v^2 = v_0^2 + 2a(x - x_0)$$

Taking $x_0 = 0$ and $v_0 = 0$ gives

$$x = \tfrac{1}{2}at^2 = \frac{qE}{2m}\,t^2$$

$$v = at = \frac{qE}{m}\,t$$

$$v^2 = 2ax = \left(\frac{2qE}{m}\right)x$$

The kinetic energy of the charge after it has moved a distance $x$ is given by

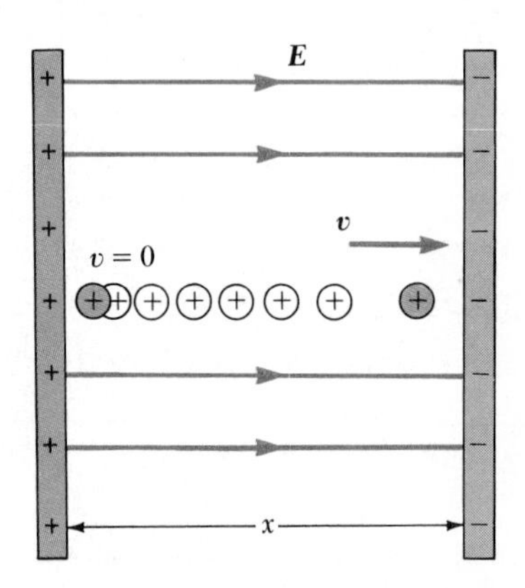


**Figure 23.22** (Example 23.11) A positive point charge $q$ in a uniform electric field $\mathbf{E}$ undergoes constant acceleration in the direction of the field.

$$K = \tfrac{1}{2}mv^2 = \tfrac{1}{2}m\left(\frac{2qE}{m}\right)x = qEx$$

This result can also be obtained from the work-energy theorem, since the work done by the electric force is $F_ex = qEx$ and $W = \Delta K$.

The electric field in the region between two oppositely charged flat metal plates is approximately uniform (Fig. 23.23). Suppose an electron of charge $-e$ is projected horizontally into this field with an initial velocity $v_0\mathbf{i}$. Since the electric field $\mathbf{E}$ is in the positive $y$ direction, the acceleration of the electron is in the negative $y$ direction. That is,

$$\mathbf{a} = -\frac{eE}{m}\,\mathbf{j} \tag{23.20}$$

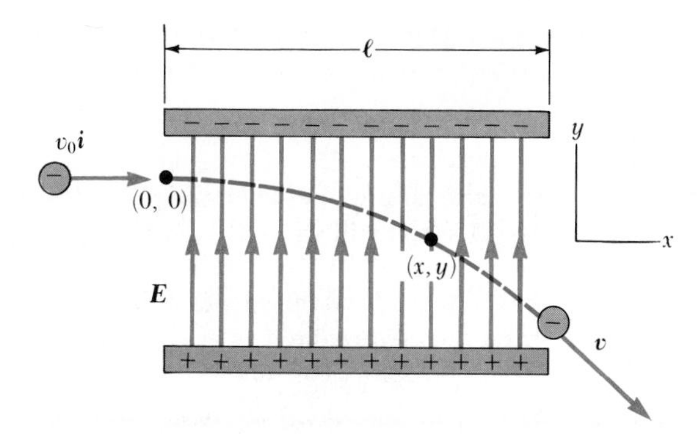


**Figure 23.23** An electron is projected horizontally into a uniform electric field produced by two charged plates. The electron undergoes a downward acceleration (opposite ***E***), and its motion is parabolic.

Because the acceleration is constant, we can apply the equations of kinematics in two dimensions (from Chapter 4) with $v_{x0} = v_0$ and $v_{y0} = 0$. The components of velocity of the electron after it has been in the electric field a time $t$ are given by

$$v_x = v_0 = \text{constant} \tag{23.21}$$

$$v_y = at = -\frac{eE}{m}t \tag{23.22}$$

Likewise, the coordinates of the electron after a time $t$ in the electric field are given by

$$x = v_0 t \tag{23.23}$$

$$y = \tfrac{1}{2}at^2 = -\tfrac{1}{2}\frac{eE}{m}t^2 \tag{23.24}$$

Substituting the value $t = x/v_0$ from Equation 23.23 into Equation 23.24, we see that $y$ is proportional to $x^2$. Hence, the trajectory is a parabola. After the electron leaves the region of uniform electric field, it continues to move in a straight line with a speed $v > v_0$.

Note that we have neglected the gravitational force on the electron. This is a good approximation when dealing with atomic particles. For an electric field of $10^4$ N/C, the ratio of the electric force, $eE$, to the gravitational force, $mg$, for the electron is of the order of $10^{14}$. The corresponding ratio for a proton is of the order of $10^{11}$.

EXAMPLE 23.12 An Accelerated Electron

An electron enters the region of a uniform electric field as in Figure 23.23, with $v_0 = 3 \times 10^6$ m/s and $E = 200$ N/C. The width of the plates is $\ell = 0.1$ m. (a) Find the acceleration of the electron while in the electric field.

Since the charge on the electron has a magnitude of $1.60 \times 10^{-19}$ C and $m = 9.11 \times 10^{-31}$ kg, Equation 23.20 gives

$$\boldsymbol{a} = -\frac{eE}{m}\boldsymbol{j} = -\frac{(1.6 \times 10^{-19}\ \text{C})(200\ \text{N/C})}{9.11 \times 10^{-31}\ \text{kg}}\boldsymbol{j}$$

$$= -3.51 \times 10^{13}\boldsymbol{j}\ \text{m/s}^2$$

(b) Find the time it takes the electron to travel through the region of the electric field.

The horizontal distance traveled by the electron while in the electric field is $\ell = 0.1$ m. Using Equation

23.23 with $x = \ell$, we find that the time spent in the electric field is given by

$$t = \frac{\ell}{v_0} = \frac{0.1 \text{ m}}{3 \times 10^6 \text{ m/s}} = 3.33 \times 10^{-8} \text{ s}$$

(c) What is the vertical displacement $y$ of the electron while it is in the electric field?

Using Equation 23.24 and the results from (a) and (b), we find that

$$y = \tfrac{1}{2}at^2 = -\tfrac{1}{2}(3.51 \times 10^{13} \text{ m/s}^2)(3.33 \times 10^{-8} \text{ s})^2$$

$$= -0.0195 \text{ m} = -1.95 \text{ cm}$$

If the separation between the plates is smaller than this, the electron will strike the positive plate.

Exercise 5 Find the speed of the electron as it emerges from the electric field.
Answer $3.22 \times 10^6$ m/s.

## *23.8 THE OSCILLOSCOPE

The oscilloscope is an electronic instrument widely used in making electrical measurements. The main component of the oscilloscope is the cathode ray tube (CRT), shown in Figure 23.24. This tube is commonly used to obtain a visual display of electronic information for other applications, including radar systems, television receivers, and computers. The CRT is a vacuum tube in which electrons are accelerated and deflected under the influence of electric fields.

The electron beam is produced by an assembly called an *electron gun,* located in the neck of the tube. The assembly shown in Figure 23.24 consists of a heater (H), a cathode (C), and a positively charged anode (A). An electric current maintained in the heater causes its temperature to rise, which in turn heats the cathode. The cathode reaches temperatures high enough to cause electrons to be "boiled off." Although they are not shown in the figure, the electron gun also includes an element that focuses the electron beam and one that controls the number of electrons reaching the anode (that is, a brightness control). The anode has a hole in its center that allows the electrons to pass through without striking the anode. These electrons, if left undisturbed, travel in a straight-line path until they strike the face of the CRT. The screen at the

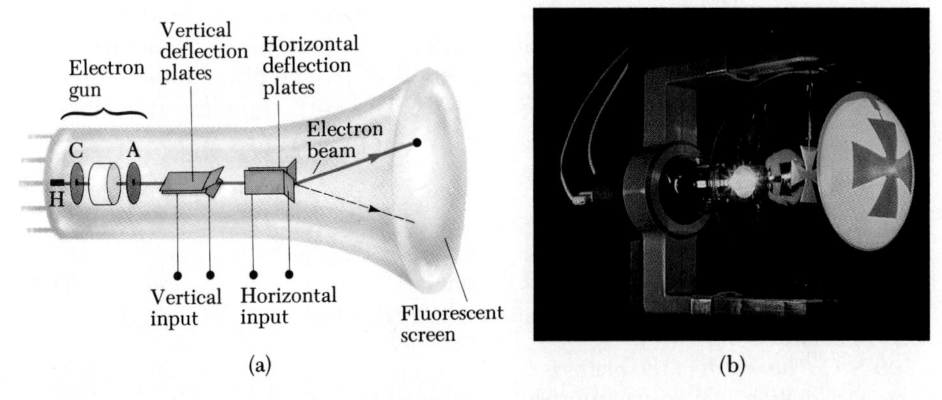


**Figure 23.24** (a) Schematic diagram of a cathode ray tube. Electrons leaving the hot cathode C are accelerated to the anode A. The electron gun is also used to focus the beam, and the plates deflect the beam. (b) Photograph of a "Maltese Cross" tube showing the shadow of a beam of cathode rays falling on the tube's luminescent screen. The hot filament also produces a beam of light and a second shadow of the cross. (Courtesy of CENCO)

front of the tube is coated with a material that emits visible light when bombarded with electrons. This results in a visible spot of light on the screen of the CRT.

The electrons are deflected in various directions by two sets of plates placed at right angles to each other in the neck of the tube. In order to understand how the deflection plates operate, first consider the horizontal deflection plates in Figure 23.24a. External electric circuits are used to control and change the amount of charge present on these plates, with positive charge being placed on one plate and negative on the other. (In Chapter 25 we shall see that this can be accomplished by applying a voltage across the plates.) This increasing charge creates an increasing electric field between the plates, which causes the electron beam to be deflected from its straight-line path. The tube face is slightly phosphorescent and therefore glows briefly after the electron beam moves from one point to another on the screen. Slowly increasing the charge on the horizontal plates causes the electron beam to move gradually from the center toward the side of the screen. Because of the phosphorescence, however, one sees a horizontal line extending across the screen instead of the simple movement of the dot. The horizontal line can be maintained on the screen by rapid, repetitive tracing.

The vertical deflection plates act in exactly the same way as the horizontal plates, except that changing the charge in them with external controlling circuits causes a vertical line on the tube face. In practice, the horizontal and vertical deflection plates are used simultaneously. To see how the oscilloscope can display visual information, let us examine how we could observe the sound wave from a tuning fork on the screen. For this purpose, the charge on the horizontal plates changes in such a manner that the beam sweeps across the face of the tube at a constant rate. The tuning fork is then sounded into a microphone, which changes the sound signal to an electric signal that is applied to the vertical plates. The combined effect of the horizontal and vertical plates causes the beam to sweep the tube horizontally and up and down at the same time, with the vertical motion corresponding to the tuning fork signal. A pattern such as that shown in Figure 23.25 is seen on the screen.

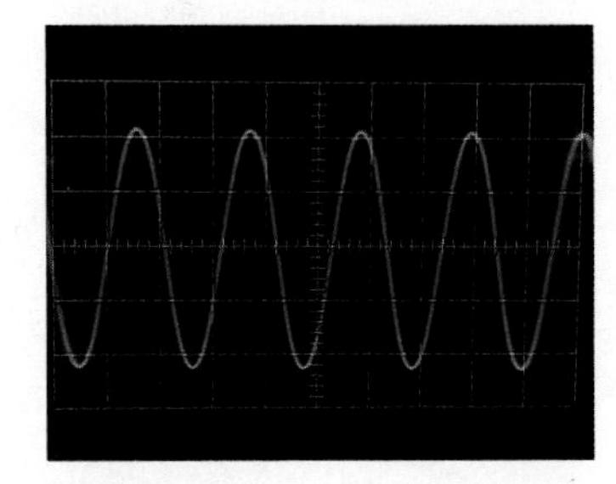

**Figure 23.25** A sinusoidal wave produced by a wave generator and displayed on the oscilloscope. (Courtesy of Henry Leap and Jim Lehman)

## SUMMARY

Properties of electric charges

**Electric charges** have the following important properties:

1. Unlike charges attract one another and like charges repel one another.
2. Electric charge is always conserved.
3. Charge is quantized, that is, it exists in discrete packets that are some integral multiple of the electronic charge.
4. The force between charged particles varies as the inverse square of their separation.

**Conductors** are materials in which charges move quite freely. Some examples of good conductors are copper, aluminum, and silver. **Insulators** are materials that do not readily transport charge. Some examples are glass, rubber, and wood.

**Coulomb's law** states that the electrostatic force between two stationary, charged particles separated by a distance $r$ has a magnitude given by

Coulomb's law

$$F = k\frac{|q_1||q_2|}{r^2} \tag{23.1}$$

where the constant $k$ has the value

Coulomb constant

$$k = 8.9875 \times 10^9\ \text{N}\cdot\text{m}^2/\text{C}^2 \tag{23.2}$$

The smallest unit of charge known to exist in nature is the charge on an electron or proton. The magnitude of this charge $e$ is given by

Charge on an electron or proton

$$|e| = 1.60219 \times 10^{-19}\ \text{C} \tag{23.5}$$

The **electric field *E*** at some point in space is defined as the electric force $\boldsymbol{F}$ that acts on a small positive test charge placed at that point divided by the magnitude of the test charge $q_0$:

Definition of electric field

$$\boldsymbol{E} = \frac{\boldsymbol{F}}{q_0} \tag{23.7}$$

The electric field due to a point charge $q$ at a distance $r$ from the charge is given by

Electric field of a point charge $q$

$$\boldsymbol{E} = k\frac{q}{r^2}\hat{\boldsymbol{r}} \tag{23.8}$$

where $\hat{\boldsymbol{r}}$ is a unit vector directed from the charge to the point in question. The electric field is directed radially outward from a positive charge and is directed *toward* a negative charge.

The *electric field* due to a group of charges can be obtained using the **superposition principle.** That is, the total electric field equals the *vector sum* of the electric fields of all the charges at some point:

Electric field of a group of charges

$$\boldsymbol{E} = k\sum_i \frac{q_i}{r_i^2}\hat{\boldsymbol{r}}_i \tag{23.9}$$

Similarly, the electric field of a continuous charge distribution at some point is given by

Electric field of a continuous charge distribution

$$\boldsymbol{E} = k\int \frac{dq}{r^2}\hat{\boldsymbol{r}} \tag{23.11}$$

where $dq$ is the charge on one element of the charge distribution and $r$ is the distance from the element to the point in question.

**Electric field lines** are useful for describing the electric field in any region of space. The electric field vector $\boldsymbol{E}$ is always tangent to the electric field lines at every point. Furthermore, the number of lines per unit area through a surface perpendicular to the lines is proportional to the magnitude of $\boldsymbol{E}$ in that region.

A charged particle of mass $m$ and charge $q$ moving in an electric field $\boldsymbol{E}$ has an acceleration $\boldsymbol{a}$ given by

Acceleration of a charge in an electric field

$$\boldsymbol{a} = \frac{q\boldsymbol{E}}{m} \tag{23.19}$$

If the electric field is uniform, the acceleration is constant and the motion of the charge is similar to that of a projectile moving in a uniform gravitational field.

### PROBLEM-SOLVING STRATEGY AND HINTS

1. **Units:** When performing calculations that involve the use of the Coulomb constant $k(=1/4\pi\epsilon_0)$ which appears in Coulomb's law, charges must be in coulombs, and distances in meters. If they appear in other units, you must convert them.
2. **Applying Coulomb's law to point charges:** It is important to remember to use the superposition principle properly when dealing with a collection of interacting charges. When several charges are present, the resultant force on any one of the charges is the *vector sum* of the forces due to the individual forces. You must be very careful in the algebraic manipulation of vector quantities. It may be useful to review the material on vector addition in Chapter 2.
3. **Calculating the electric field of point charges:** Remember that the superposition principle can also be applied to electric fields, which are also vector quantities. To find the total electric field at a given point, first calculate the electric field at the point due to each individual charge. The resultant field at the point is the vector sum of the fields due to the individual charges.
4. **Continuous charge distributions:** When you are confronted with problems that involve a continuous distribution of charge, the vector sums for evaluating the total electric field at some point must be replaced by vector integrals. The charge distribution is divided into infinitesimal pieces, and the vector sum is carried out by integrating over the entire charge distribution. You should review Examples 8, 9, and 10, which demonstrate such procedures.
5. **Symmetry:** Whenever dealing with either a distribution of point charges or a continuous charge distribution, you should take advantage of any symmetry in the system to simplify your calculations.
6. **Motion of charged particles in uniform electric fields:** The motion of a charged particle in a uniform electric field (one that is constant in magnitude and direction) can be described by using the equations of projectile motion developed in Chapter 4. In this case, the uniform acceleration is given by $\mathbf{a} = q\mathbf{E}/m$.

## QUESTIONS

1. Sparks are often observed (or heard) on a dry day when clothes are removed in the dark. Explain.
2. Explain from an atomic viewpoint why charge is usually transferred by electrons.
3. A balloon is negatively charged by rubbing and then clings to a wall. Does this mean that the wall is positively charged? Why does the balloon eventually fall?
4. A light, uncharged metal sphere suspended from a thread is attracted to a charged rubber rod. After touching the rod, the sphere is repelled by the rod. Explain.
5. Explain what we mean by a neutral atom.
6. If a suspended object A is attracted to object B, which is charged, can we conclude that object A is charged? Explain.
7. A charged comb will often attract small bits of dry paper that fly away when they touch the comb. Explain.
8. Why do some clothes cling together and to your body after being removed from a dryer?
9. A large metal sphere insulated from ground is charged with an electrostatic generator while a person standing on an insulating stool holds the sphere while it is being charged. Why is it safe to do this? Why wouldn't it be safe for another person to touch the sphere after it has been charged?
10. What is the difference between charging an object by induction and charging by conduction?

11. What are the similarities and differences between Newton's Universal Law of Gravitation, $F = \frac{Gm_1m_2}{r^2}$, and Coulomb's Law, $F = \frac{kQ_1Q_2}{r^2}$?
12. Assume that someone proposes a theory that says people are bound to the earth by electric forces rather than by gravity. How could you prove this theory wrong?
13. Would life be different if the electron were positively charged and the proton were negatively charged? Does the choice of signs have any bearing on physical and chemical interactions? Explain.
14. When defining the electric field, why is it necessary to specify that the magnitude of the test charge be very small (i.e., take the limit as $q \to 0$)?
15. Two charged spheres each of radius $a$ are separated by a distance $r > 2a$. Is the force on either sphere given by Coulomb's law? Explain. (*Hint:* Refer back to Chapter 14 on gravitation.)
16. When is it valid to approximate a charge distribution by a "point charge"?
17. Is it possible for an electric field to exist in empty space? Explain.
18. Explain why electric field lines do not form closed loops.
19. Explain why electric field lines never cross. (*Hint:* **E** must have a unique direction at all points.)
20. A "free" electron and "free" proton are placed in an identical electric field. Compare the electric forces on each particle. Compare their accelerations.
21. Explain what happens to the magnitude of the electric field of a point charge as $r$ approaches zero.
22. A negative charge is placed in a region of space where the electric field is directed vertically upward. What is the direction of the electric force experienced by this charge?
23. A charge $4q$ is at a distance $r$ from a charge $-q$. Compare the number of electric field lines leaving the charge $4q$ with the number entering the charge $-q$.
24. In Figure 23.21, where do the extra lines leaving the charge $+2q$ end?
25. Consider two equal point charges separated by some distance $d$. At what point (other than $\infty$) would a third test charge experience no net force?
26. An uncharged, metallic coated, Ping–Pong ball is placed in the region between two horizontal parallel metal plates. If the two plates are charged, one positive and one negative, describe the motion the Ping–Pong ball will undergo.
27. A negative point charge $-q$ is placed at the point $P$ near the positively charged ring shown in Figure 23.15 of Example 23.9. If $x \ll a$, describe the motion of the point charge if it is released from rest.
28. Explain the differences between linear, surface, and volume charge densities, and give examples of when each would be used.
29. If the electron in Figure 23.23 is projected into the electric field with an arbitrary velocity $\boldsymbol{v}_0$ (at an angle to $\boldsymbol{E}$), will its trajectory still be parabolic? Explain.
30. If a metal object receives a positive charge, does its mass increase, decrease, or stay the same? What happens to the mass if the object is given a negative charge?
31. It has been reported that in some instances people near where a lightning bolt strikes the earth have had their clothes thrown off. Explain why this might happen.
32. Why should a ground wire be connected to the metal support rod for a television antenna?
33. Are the occupants of a steel-frame building safer than those in a wood-frame house during an electrical storm or vice versa? Explain.
34. A light piece of aluminum foil is draped over a wooden rod. When a rod with a positive charge is brought close to the foil, two parts of the foil stand apart. Why? What kind of charge is on the foil?
35. Why is it more difficult to charge an object by friction on a humid day than on a dry day?
36. How would you experimentally distinguish an electric field from a gravitational field?

## PROBLEMS

### Section 23.3 Coulomb's Law

1. Suppose that 1 g of hydrogen is separated into electrons and protons. Suppose also that the protons are placed at the Earth's north pole and the electrons are placed at the south pole. What is the resulting compressional force on the Earth?
2. (a) Calculate the number of electrons in a small silver pin, electrically neutral, with a mass of 10 g. Silver has 47 electrons per atom. The atomic weight of silver is 107.87. (b) Electrons are added to the pin until the net charge is 1 mC. How many electrons are added for every $10^9$ electrons already present?
3. Two protons in a molecule are separated by a distance of $3.8 \times 10^{-10}$ m. Find the electrostatic force exerted by one proton on the other.
4. A 6.7-$\mu$C charge is located 5.0 m from a $-8.4$-$\mu$C charge. Find the electrostatic force exerted by one charge on the other.
5. A 1.3-$\mu$C charge is located on the $x$ axis at $x = -0.5$ m, a 3.2-$\mu$C charge is located on the $x$ axis at $x = 1.5$ m, and a 2.5-$\mu$C charge is located at the origin. Find the net force on the 2.5-$\mu$C charge. All charges are positive.
6. Two identical small metal spheres attract each other with a force of 0.0853 N. The distance between the

spheres is 1.19 m. The spheres are brought into electrical contact with each other so that the net charge is shared equally. When returned to a separation of 1.19 m, the spheres repel each other with a force of 0.0196 N. Find the charge originally on each sphere.

7. Three point charges of 2 $\mu$C, 7 $\mu$C, and $-4$ $\mu$C are located at the corners of an equilateral triangle as in Figure 23.26. Calculate the net electric force on the 7-$\mu$C charge.

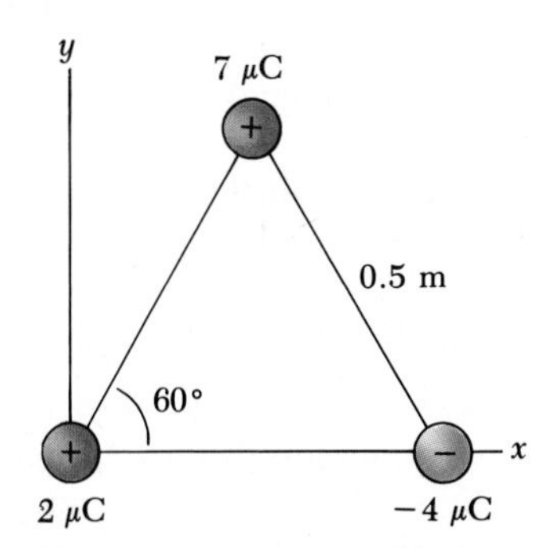


**Figure 23.26** (Problems 7 and 24).

8. Four point charges are situated at the corners of a square of sides $a$ as in Figure 23.27. Find the resultant force on the positive charge $q$.

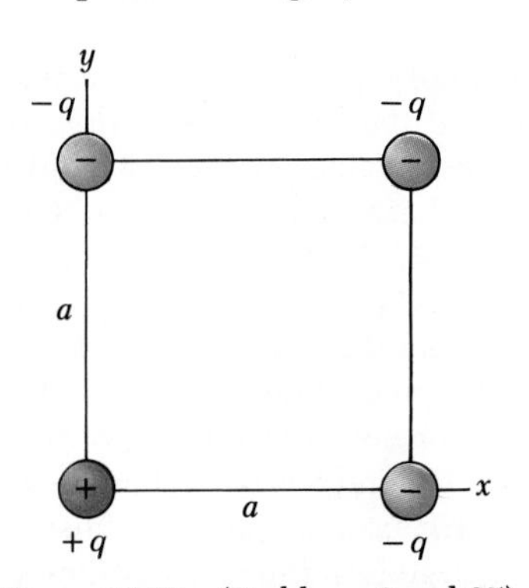


**Figure 23.27** (Problems 8 and 25).

9. Four identical point charges ($q = +10$ $\mu$C) are located on the corners of a rectangle as shown in Figure 23.28. The dimensions of the rectangle are $L = 60$ cm and $W = 15$ cm. Calculate the magnitude and direction of the net electrostatic force exerted on the charge at the lower left corner of the rectangle by the other three charges.

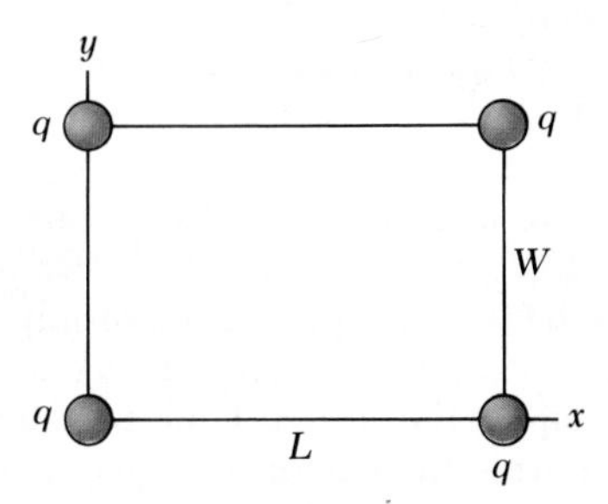


**Figure 23.28** (Problems 9 and 20).

10. Four arbitrary pointlike charges are located at the corners of a rigid square plate. Prove that the net torque about the center of the plate is zero.

11. Two small silver spheres, each with a mass of 100 g, are separated by a distance of 1 m. Calculate the fraction of the electrons in one sphere that must be transferred to the other in order to produce an attractive force of $10^4$ N (about a ton) between the spheres. (The number of electrons per atom of silver is 47, and the number of atoms per gram is Avogadro's number divided by the atomic weight of silver, 107.87.)

12. Richard Feynman once said that if two persons stood at arm's length from each other and each person had 1% more electrons than protons, the force of repulsion between the two people would be enough to lift a "weight" equal to that of the entire earth. Carry out an order-of-magnitude calculation to substantiate this assertion.

13. In a thundercloud there may be an electric charge of $+40$ C near the top of the cloud and $-40$ C near the bottom of the cloud. These charges are separated by about 2 km. What is the electric force between these two sets of charges?

### Section 23.4 The Electric Field

14. An airplane is flying through a thundercloud at a height of 2000 m. (This is a very dangerous thing to do because of updrafts, turbulence, and the possibility of electric discharge.) If there is a charge concentration of $+40$ C at height 3000 m within the cloud and $-40$ C at height 1000 m, what is the electric field $E$ at the aircraft?

15. What are the magnitude and direction of the electric field that will balance the weight of (a) an electron and (b) a proton? (Use the data in Table 23.1.)

16. An object having a net charge of 24 $\mu$C is placed in a uniform electric field of 610 N/C directed vertically. What is the mass of this object if it "floats" in this electric field?

17. A point charge of $-5.2$ $\mu$C is located at the origin. Find the electric field (a) on the $x$ axis at $x = 3$ m, (b) on the $y$ axis at $y = -4$ m, (c) at the point with coordinates $x = 2$ m, $y = 2$ m.

18. Find the total electric field along the line of the two charges shown in Figure 23.29 at the point midway between them.

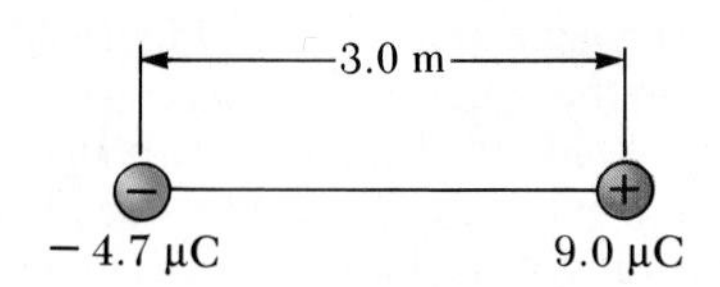


**Figure 23.29** (Problems 18 and 27).

19. Two equal point charges each of magnitude 2.0 $\mu$C are located on the $x$ axis. One is at $x = 1.0$ m, and the other is at $x = -1.0$ m. (a) Determine the electric field on the $y$ axis at $y = 0.5$ m. (b) Calculate the electric force on a third charge, of $-3.0$ $\mu$C, placed on the $y$ axis at $y = 0.5$ m.
20. Four identical point charges ($q = +6$ $\mu$C) are located on a rectangle as shown in Figure 23.28, with $L =$ 80 cm and $W = 20$ cm. Calculate the resultant electric field at the center of the rectangle.
21. Five equal, negative point charges $-q$ are placed symmetrically around a circle of radius $R$ as in Figure 23.30. Calculate the electric field $E$ at the center of the circle.

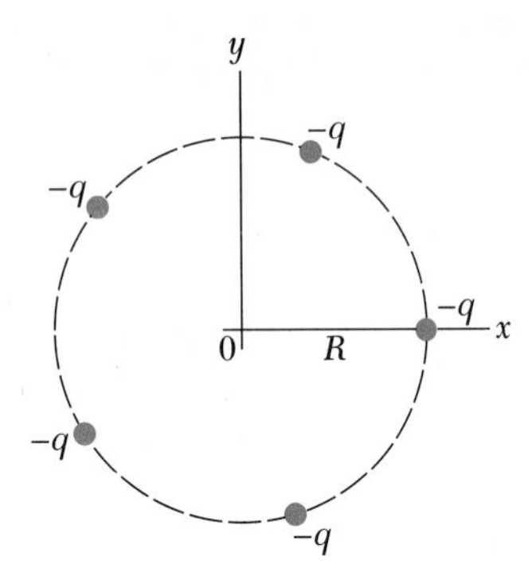


**Figure 23.30** (Problem 21).

22. Three identical point charges ($q = +2.7$ $\mu$C) are placed on the corners of an equilateral triangle whose sides have a length of 35 cm (see Figure 23.31). What is the magnitude of the resultant electric field at the center of the triangle?

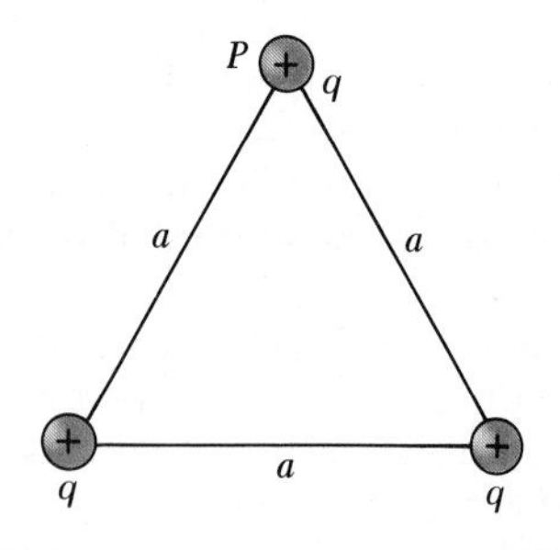


**Figure 23.31** (Problems 22 and 23).

23. Three equal positive charges $q$ are at the corners of an equilateral triangle of sides $a$ as in Figure 23.31. (a) At what point in the plane of the charges (other than $\infty$) is the electric field zero? (b) What are the magnitude and direction of the electric field at the point $P$ due to the two charges at the base of the triangle?
24. Three charges are at the corners of an equilateral triangle as in Figure 23.26. Calculate the electric field intensity at the position of the 2-$\mu$C charge due to the 7-$\mu$C and $-4$-$\mu$C charges.
25. Four charges are at the corners of a square as in Figure 23.27. (a) Find the magnitude and direction of the electric field at the position of the charge $-q$, the coordinates of which are $x = a$, $y = a$. (b) What is the electric force on this charge?
26. A charge of $-4$ $\mu$C is located at the origin, and a charge of $-5$ $\mu$C is located along the $y$ axis at $y =$ 2.0 m. At what point along the $y$ axis is the electric field zero?
27. In Figure 23.29, determine the point (other than $\infty$) at which the total electric field is zero.

### Section 23.5 Electric Field of a Continuous Charge Distribution

28. A rod 14 cm long is uniformly charged and has a total charge of $-22$ $\mu$C. Determine the magnitude and direction of the electric field along the axis of the rod, at a point 36 cm from its center.
29. A continuous line of charge lies along the $x$-axis, extending from $x = +x_0$ to positive infinity. The line carries a uniform linear charge density $\lambda_0$. What are the magnitude and direction of the electric field at the origin?
30. A line of charge starts at $x = +x_0$ and extends to positive infinity. If the linear charge density is given by $\lambda = \lambda_0 x_0/x$, determine the electric field at the origin.
31. A uniformly charged ring of radius 10 cm has a total charge of 75 $\mu$C. Find the electric field on the *axis* of the ring at (a) 1 cm, (b) 5 cm, (c) 30 cm, and (d) 100 cm from the center of the ring.
32. Show that the maximum field strength $E_m$ along the axis of a uniformly charged ring occurs at $x = a/\sqrt{2}$ (see Fig. 23.15) and has the value $Q/(6\sqrt{3}\pi\epsilon_0 a^2)$.
33. A sphere of radius 4 cm has a net charge of $+39$ $\mu$C. (a) If this charge is uniformly distributed throughout the volume of the sphere, what is the volume charge density? (b) If this charge is uniformly distributed on the sphere's surface, what is the surface charge density?
34. A uniformly charged disk of radius 35 cm carries a charge density of $7.9 \times 10^{-3}$ C/m$^2$. Calculate the electric field on the *axis* of the disk at (a) 5 cm, (b) 10 cm, (c) 50 cm, and (d) 200 cm from the center of the disk.
35. Example 23.10 derives the exact expression for the electric field at a point on the axis of a uniformly charged disk (see Equation 23.17). Consider a disk of radius $R = 3$ cm, having a uniformly distributed charge of $+5.2$ $\mu$C. (a) Using the result of Example 23.10, compute the electric field at a point on the axis and 3 mm from the center. Compare this answer to

the field computed from the near-field approximation (Equation 23.18). (b) Using the result of Example 23.10, compute the electric field at a point on the axis and 30 cm from the center of the disk. Compare this to the electric field obtained by treating the disk as a $+5.2\ \mu C$ point charge at a distance of 30 cm.

36. The electric field along the axis of a uniformly charged disk of radius $R$ and total charge $Q$ was calculated in Example 23.10. Show that the electric field at distances $x$ that are large compared with $R$ approaches that of a point charge $Q = \sigma\pi R^2$. (*Hint:* First show that $x/(x^2 + R^2)^{1/2} = (1 + R^2/x^2)^{-1/2}$ and use the binomial expansion $(1 + \delta)^n \approx 1 + n\delta$ when $\delta \ll 1$.)

37. A uniformly charged ring and a uniformly charged disk each have a charge of $+25\ \mu C$ and a radius of 3 cm. For each of these charged objects, determine the electric field at a point along the axis which is 4 cm from the center of the object.

38. A 10 gram piece of Styrofoam carries a net charge of $-0.7\ \mu C$ and "floats" above the center of a very large horizontal sheet of plastic which has a uniform charge density on its surface. What is the charge per unit area on the plastic sheet?

39. A uniformly charged insulating rod of length 14 cm is bent into the shape of a semicircle as in Figure 23.32. If the rod has a total charge of $-7.5\ \mu C$, find the magnitude and direction of the electric field at $O$, the center of the semicircle.

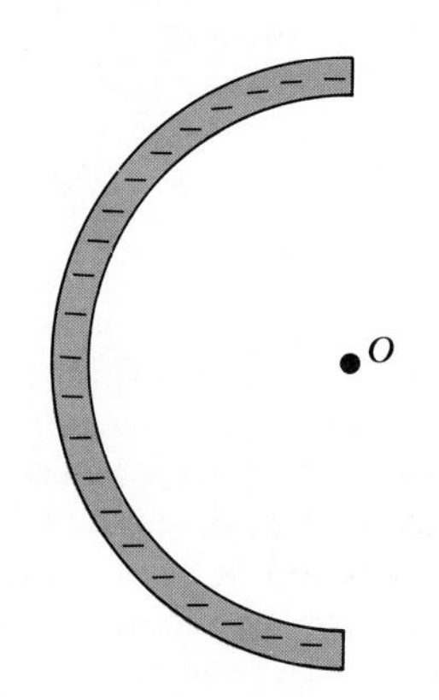


**Figure 23.32** (Problem 39).

## Section 23.6 Electric Field Lines

40. A positively charged disk has a uniform charge per unit area as described in Example 23.10. Sketch the electric field lines in a plane perpendicular to the plane of the disk passing through its center.

41. A negatively charged rod of finite length has a uniform charge per unit length. Sketch the electric field lines in a plane containing the rod.

42. Four equal positive point charges are at the corners of a square. Sketch the electric field lines in the plane of the square.

43. Figure 23.33 shows the electric field lines for two point charges separated by a small distance. (a) Determine the ratio $q_1/q_2$. (b) What are the signs of $q_1$ and $q_2$?

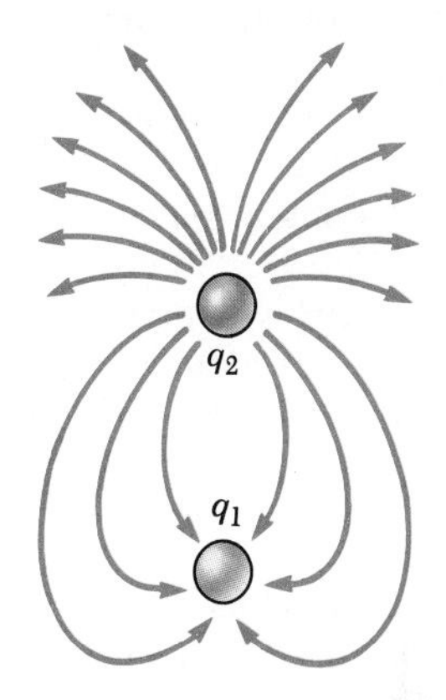


**Figure 23.33** (Problem 43).

## Section 23.7 Motion of Charged Particles in a Uniform Electric Field

44. An electron and a proton are each placed at rest in an external electric field of 520 N/C. Calculate the speed of each particle after 48 nanoseconds.

45. A proton accelerates from rest in a uniform electric field of 640 N/C. At some later time, its speed is $1.20 \times 10^6$ m/s (nonrelativistic since $v$ is much less than the speed of light). (a) Find the acceleration of the proton. (b) How long does it take the proton to reach this velocity? (c) How far has it moved in this time? (d) What is its kinetic energy at this time?

46. An electron with a speed of $3 \times 10^6$ m/s moves into a uniform electric field of 1000 N/C. The field is parallel to the electron's velocity and acts to decelerate the electron. How far does the electron travel before it is brought to rest?

47. The electrons in a particle beam each have a kinetic energy of $1.6 \times 10^{-17}$ J. What are the magnitude and direction of the electric field that will stop these electrons in a distance of 10 cm?

48. An electron traveling with an initial velocity equal to $8.6 \times 10^5 \mathbf{i}$ m/s enters a region of a uniform electric field given by $\mathbf{E} = 4.1 \times 10^3 \mathbf{i}$ N/C. (a) Find the acceleration of the electron. (b) Determine the time it takes for the electron to come to rest after it enters the field. (c) How far does the electron move in the electric field before coming to rest?

49. A proton is projected in the positive $x$ direction into a region of a uniform electric field $\mathbf{E} = -6 \times 10^5 \mathbf{i}$ N/C.

The proton travels 7 cm before coming to rest. Determine (a) the acceleration of the proton, (b) its initial speed, and (c) the time it takes the proton to come to rest.

**50.** A proton and an electron both start from rest and from the same point in a uniform electric field of 370 N/C. How far apart are they after 1 $\mu s$? (Ignore the attraction between the electron and the proton. If you like, you might imagine the experiment to be tried with the proton only, and then repeated with the electron only.)

51. A proton has an initial velocity of $4.50 \times 10^5$ m/s in the horizontal direction. It enters a uniform electric field of $9.60 \times 10^3$ N/C directed vertically. Ignore any gravitational effects and (a) find the time it takes the proton to travel 5.0 cm horizontally, (b) the vertical displacement of the proton after it has traveled 5.0 cm horizontally, and (c) the horizontal and vertical components of the proton's velocity after it has traveled 5.0 cm horizontally.

52. An electron is projected at an angle of 30° above the horizontal at a speed of $8.2 \times 10^5$ m/s, in a region of an electric field $\boldsymbol{E} = 390\boldsymbol{j}$ N/C. Neglect gravity and find: (a) the time it takes the electron to return to its initial height, (b) the maximum height reached by the electron, and (c) its horizontal displacement when it reaches its maximum height.

53. Protons are projected with an initial speed, given by $v_0 = 9.55 \times 10^3$ m/s, into a region where a uniform electric field, $\boldsymbol{E} = -720\boldsymbol{j}$ N/C, is present as shown in Figure 23.34. The protons are to hit a target that lies at a horizontal distance of 1.27 mm from the point where the protons are launched. Find (a) the two projection angles $\theta$ that will result in a hit, and (b) the total time of flight for each of these two trajectories.

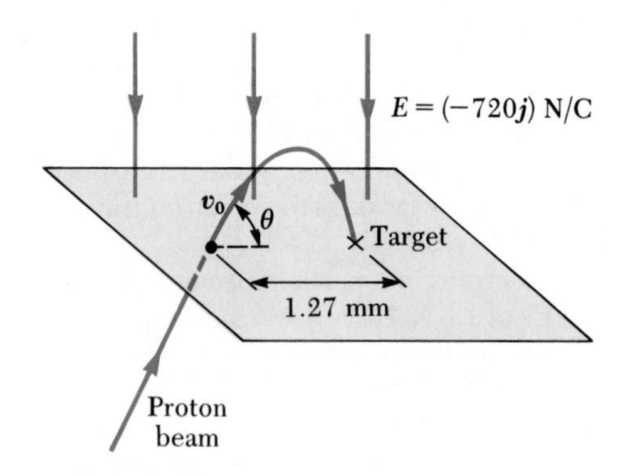


**Figure 23.34** (Problem 53).

## ADDITIONAL PROBLEMS

54. A small 2-g plastic ball is suspended by a 20-cm long string in a uniform electric field as shown in Figure 23.35. If the ball is in equilibrium when the string makes a 15° angle with the vertical as indicated, what is the net charge on the ball?

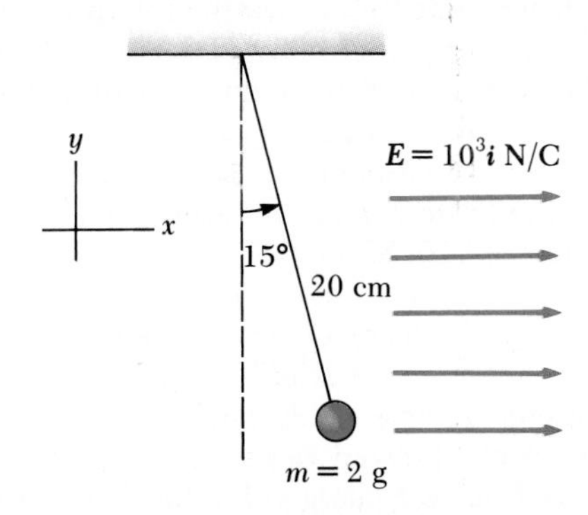


**Figure 23.35** (Problem 54).

55. A charged cork ball of mass 1 g is suspended on a light string in the presence of a uniform electric field as in Figure 23.36. When $\boldsymbol{E} = (3\boldsymbol{i} + 5\boldsymbol{j}) \times 10^5$ N/C, the ball is in equilibrium at $\theta = 37°$. Find (a) the charge on the ball and (b) the tension in the string.

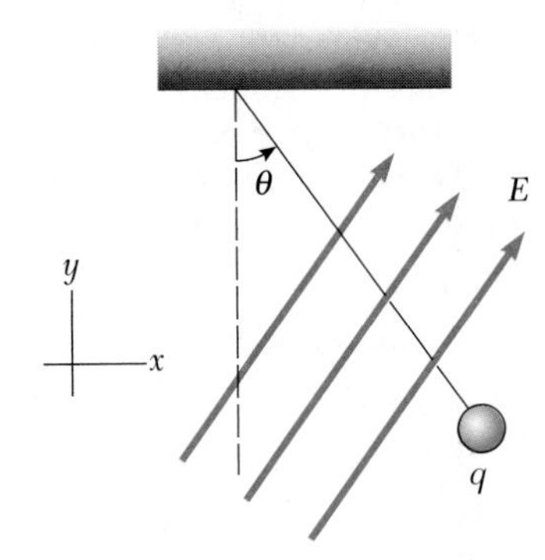


**Figure 23.36** (Problem 55).

56. Two small spheres each of mass 2 g are suspended by light strings 10 cm in length (Fig. 23.37). A uniform electric field is applied in the $x$ direction. If the spheres have charges equal to $-5 \times 10^{-8}$ C and $+5 \times 10^{-8}$ C, determine the electric field intensity

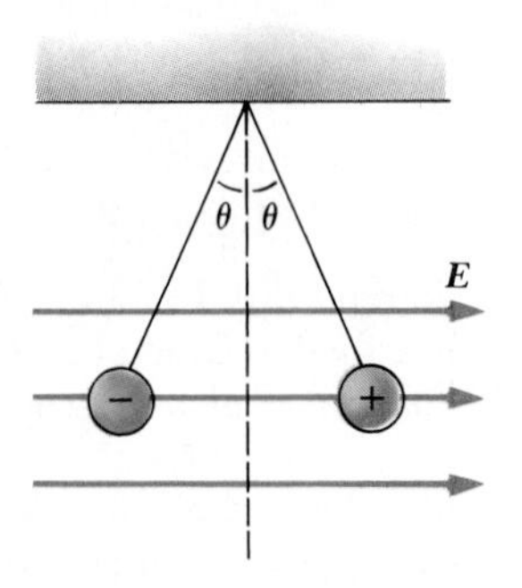


**Figure 23.37** (Problem 56).

that enables the spheres to be in equilibrium at an angle of $\theta = 10°$.

57. Two small spheres of mass $m$ are suspended from strings of length $\ell$ that are connected at a common point. One sphere has charge $Q$; the other has charge $2Q$. Assume the angles, $\theta_1$ and $\theta_2$, that the strings make with the vertical are small. (a) How are $\theta_1$ and $\theta_2$ related? (b) Show that the distance $r$ between the spheres is

$$r \simeq \left(\frac{4kQ^2\ell}{mg}\right)^{1/3}$$

58. Three charges of equal magnitude $q$ are fixed in position at the vertices of an equilateral triangle (Figure 23.38). The charge located at the origin in Figure 23.38 is negative; the other two are positive. A fourth charge $Q$ is free to move along the positive $x$ axis under the influence of the forces exerted by the three fixed charges. Locate an equilibrium position for $Q$.

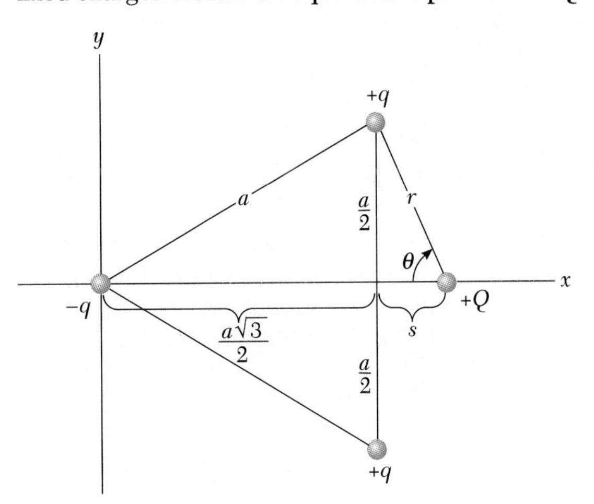


**Figure 23.38** (Problem 58).

59. Three identical small Styrofoam balls ($m = 2$ g) are suspended from a fixed point by three nonconducting threads, each with a length of 50 cm and with negligible mass. At equilibrium the three balls form an equilateral triangle with sides of 30 cm. What is the common charge $q$ that is carried by each ball?

60. Two small spheres of charge $Q$ are suspended from strings of length $\ell$ that are connected at a common point. One sphere has mass $m$; the other has mass $2m$. Assume the angles, $\theta_1$ and $\theta_2$, that the strings make with the vertical are small. (a) How are $\theta_1$ and $\theta_2$ related? (b) Show that the distance $r$ between the spheres is

$$r \simeq \left(\frac{3kQ^2\ell}{2mg}\right)^{1/3}$$

61. Consider the electric dipole shown in Figure 23.39. Show that the electric field at a *distant* point along the $x$ axis is given by $E_x \cong 2kp/x^3$, where $p = 2qa$ is the dipole moment.

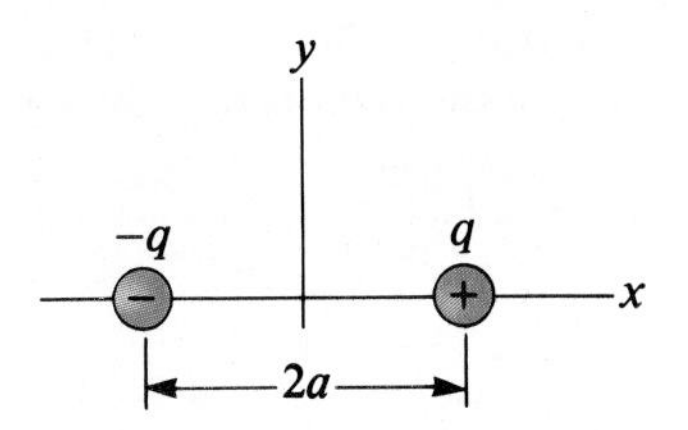


**Figure 23.39** (Problem 61).

62. Many molecules possess an electric dipole moment because the center of distribution of the positive charge (protons) does not exactly coincide with that of the negative charge (electrons). The electric dipole moment of a water molecule in its gaseous state is $6.24 \times 10^{-30}$ C·m. (a) If a water molecule is placed in an electric field of $10^4$ N/C, calculate the maximum torque that the field can exert on the molecule. (b) Find the range of the potential energies that the molecule may have in this field.

63. Three charges of equal magnitude $q$ reside at the corners of an equilateral triangle of side length $a$. Two of the charges are negative, and the other is positive, as shown in Figure 23.40. (a) Find the magnitude and direction of the electric field at point $P$, midway between the negative charges, in terms of $k$, $q$, and $a$. (b) Where must a $-4q$ charge be placed so that any charge located at point $P$ will experience no net electrostatic force ($\mathbf{F}_e = 0$)? In (b) let the distance between the $+q$ charge and point $P$ be one meter.

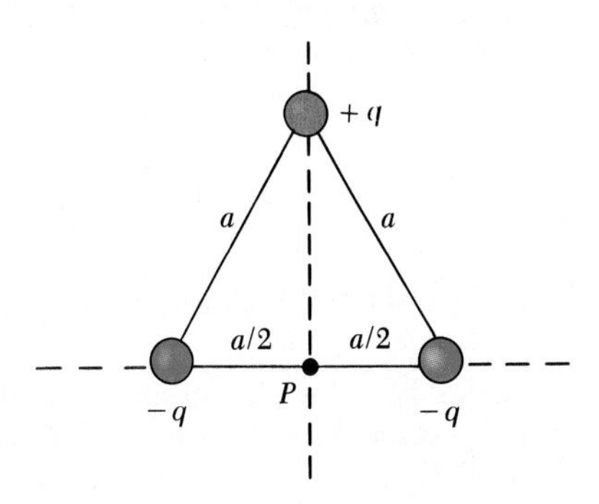


**Figure 23.40** (Problem 63).

64. Two rings each of radius $R$ have their axes oriented along the same line and are separated by a distance $2R$. One ring has uniform linear charge density $+\lambda$ and the other ring has uniform linear charge density $-\lambda$. Find (a) the electric field at point $P_1$ midway between the two rings on the common axis and (b) the electric field at point $P_2$ on the axis a distance $R$ outside the negatively charged ring.

65. Identical thin rods of length $2a$ carry equal charges $+Q$ uniformly distributed along their lengths. The rods lie along the $x$ axis with their centers separated by a distance $b > 2a$ (Figure 23.41). Show that the force exerted on the right rod is given by

$$F = \left(\frac{kQ^2}{4a^2}\right) \ln\left(\frac{b^2}{b^2 - 4a^2}\right)$$

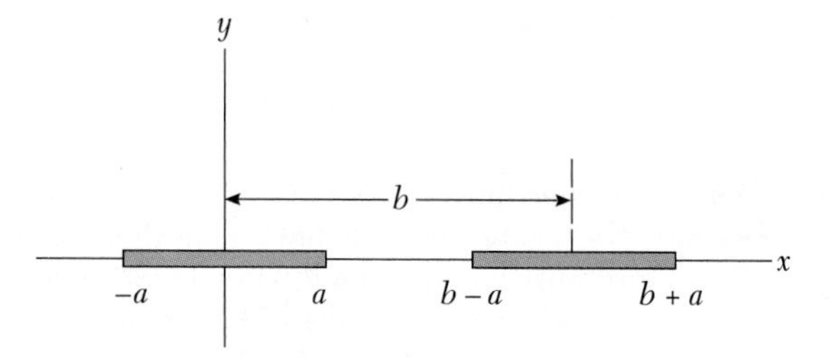


**Figure 23.41** (Problem 65).

66. A line of positive charge is formed into a semicircle of radius $R = 60$ cm as shown in Figure 23.42. The charge per unit length along the semicircle is described by the expression

$$\lambda = \lambda_0 \cos\theta$$

The total charge on the semicircle is 12 $\mu$C. Calculate the total force on a charge of 3 $\mu$C placed at the center of curvature of the semicircle.

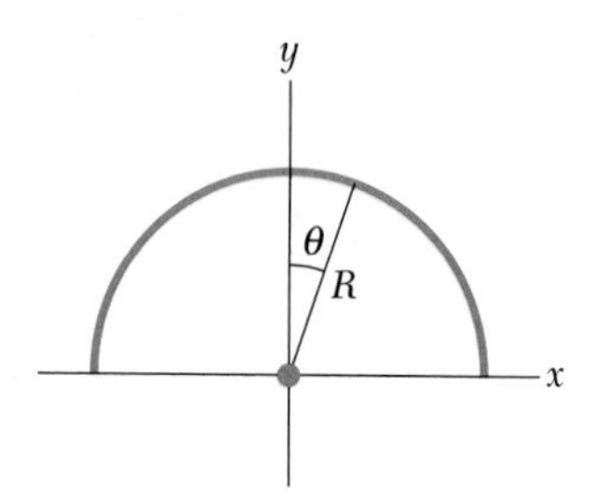


**Figure 23.42** (Problem 66).

67. Air will break down (lose its insulating quality) and sparking will result if the field strength is increased to about $3 \times 10^6$ N/C. (This field strength is also expressed as $3 \times 10^6$ V/m.) What acceleration will an electron experience in such a field? If the electron starts from rest, in what distance will it acquire a speed equal to 10 percent of the speed of light?

68. A line charge of length $\ell$ and oriented along the $x$ axis as in Figure 23.14 has a charge per unit length $\lambda$, which varies with $x$ as $\lambda = \lambda_0(x - d)/d$, where $d$ is the distance of the rod from the origin (point $P$ in the figure) and $\lambda_0$ is a constant. Find the electric field at the origin. (*Hint:* An infinitesimal element has a charge $dq = \lambda\, dx$, but note that $\lambda$ is *not* a constant.)

69. A thin rod of length $\ell$ and uniform charge per unit length $\lambda$ lies along the $x$ axis as shown in Figure 23.43. (a) Show that the electric field at the point $P$, a distance $y$ from the rod, along the perpendicular bisector has no $x$ component and is given by $E = 2k\lambda \sin\theta_0/y$. (b) Using your result to (a), show that the field of a rod of *infinite* length is given by $E = 2k\lambda/y$. (*Hint:* First calculate the field at $P$ due to an element of length $dx$, which has a charge $\lambda\, dx$. Then change variables from $x$ to $\theta$ using the facts that $x = y \tan\theta$ and $dx = y \sec^2\theta\, d\theta$ and integrate over $\theta$.)

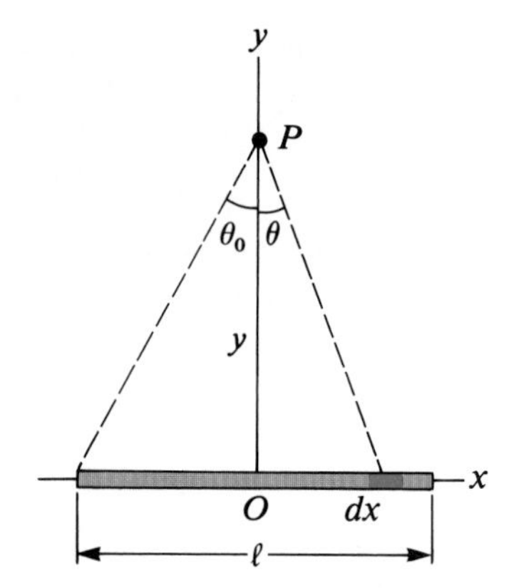


**Figure 23.43** (Problem 69).

70. A positive charge $+q$ of mass $M$ is free to move along the $x$ axis. It is in equilibrium at the origin, midway between a pair of identical point charges, $+q$, located on the $x$ axis at $x = +a$ and $x = -a$. The charge at the origin is displaced a small distance $x \ll a$ and released. Show that it can undergo simple harmonic motion with an angular frequency

$$\omega = \left(\frac{4kq^2}{Ma^3}\right)^{1/2}$$

71. A set of eight point charges, each of magnitude $+q$, is located on the corners of a cube of side $s$ as shown in Figure 23.44. (a) Determine $x$, $y$, and $z$ components of

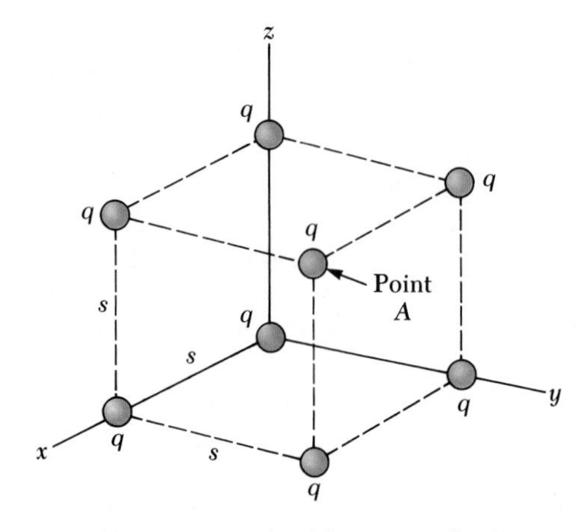


**Figure 23.44** (Problems 71 and 72).

the resultant force exerted on the charge located at point $A$ by the other charges. (b) What are the magnitude and direction of this resultant force?

72. Consider the charge distribution shown in Figure 23.44. (a) Show that the magnitude of the electric field at the center of any face of the cube has a value of $2.18\ kq/s^2$. (b) What is the direction of the electric field at the center of the top face of the cube?

73. Three point charges $q$, $-2q$, and $q$ are located along the $x$ axis as in Figure 23.45. Show that the electric field at the distant point $P$ ($y \gg a$) along the $y$ axis is given by

$$\mathbf{E} = -k\frac{3qa^2}{y^4}\mathbf{j}$$

This charge distribution, which is essentially that of two electric dipoles, is called an *electric quadrupole.* Note that $\mathbf{E}$ varies as $r^{-4}$ for the quadrupole, compared with variations of $r^{-3}$ for the dipole and $r^{-2}$ for the monopole (a single charge).

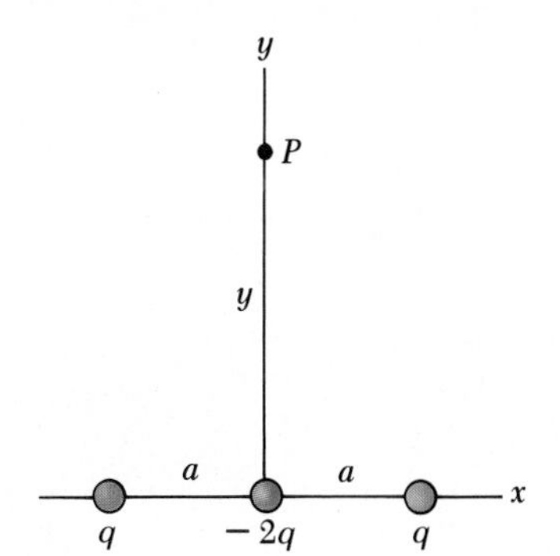


**Figure 23.45** (Problem 73).

74. An electric dipole in a uniform electric field is displaced slightly from its equilibrium position, as in Figure 23.46, where $\theta$ is small. The dipole moment is $p = 2qa$ and the moment of inertia of the dipole is $I$. If the dipole is released from this position, show that it exhibits simple harmonic motion with a frequency given by

$$f = \frac{1}{2\pi}\sqrt{\frac{pE}{I}}$$

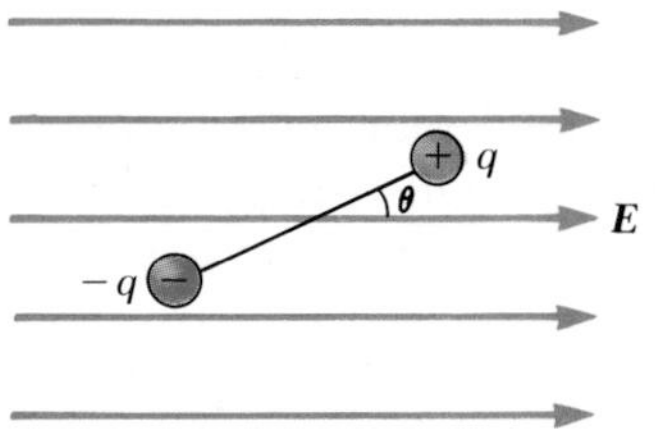

**Figure 23.46** (Problem 74).

75. A *negatively* charged particle $-q$ is placed at the center of a uniformly charged ring, where the ring has a total positive charge $Q$ as in Example 23.9. The particle, confined to move along the $x$ axis, is displaced a *small* distance $x$ along the axis (where $x \ll a$) and released. Show that the particle oscillates with simple harmonic motion along the $x$ axis with a frequency given by

$$f = \frac{1}{2\pi}\left(\frac{kqQ}{ma^3}\right)^{1/2}$$

76. In the Millikan oil drop experiment (see previous problem), the droplets are so tiny that they appear only as points of light in the microscope used to observe them. In order to find the radius (and hence the mass) of each droplet, we allow them to fall freely under gravity. The retarding force $\mathbf{F}$ exerted by the viscous air on a sphere of radius $r$ moving with speed $v$ through air is given by Stokes' law, $F = 6\pi\eta r v$, where $\eta$ is the coefficient of viscosity. (a) Find the SI units for $\eta$. (b) Show that when a falling droplet achieves a constant "terminal" velocity (signifying that the total force on the droplet is zero), the following relation is true, thus allowing the radius of the droplet to be determined:

$$v = \frac{2gr^2}{9\eta}(\rho_0 - \rho_a)$$

Here, $\rho_0$ and $\rho_a$ are the respective densities of the oil and air.

## CALCULATOR/COMPUTER PROBLEMS

77. A continuous charge is distributed along a rod lying along the $x$ axis as in Figure 23.14. The total charge on the rod is $Q = +16 \times 10^{-10}$ C, $d = 1.0$ m, and $\ell = 2.0$ m. Estimate the electric field at $x = 0$ by approximating the rod to be (a) a point charge at $x = 2.0$ m, (b) two point charges (each of charge $8 \times 10^{-10}$ C) at $x = 1.5$ m and $x = 2.5$ m, and (c) four point charges (each of charge $4 \times 10^{-10}$ C) at $x = 1.25$ m, $x = 1.75$ m, $x = 2.25$ m, and $x = 2.75$ m. (d) Write a program that will enable you to extend your calculations to 256 equally spaced point charges, and compare your result with that given by the exact expression, Equation 23.15.

78. Consider a uniform ring of charge located in the $yz$ plane as in Figure 23.15a, where $Q = +16 \times 10^{-10}$ C, and the radius $a = 1$ m. Estimate the electric field along the $x$ axis at $x = 3$ m by approximating the ring to be (a) a point charge at $x = 0$, (b) two point charges (each of charge $8 \times 10^{-10}$ C) diametrically opposite each other on the ring, and (c) four point charges (each of charge $4 \times 10^{-10}$ C) symmetrically spaced on the ring. (d) Write a program that will enable you to extend your calculations to 64 point charges equally spaced on the ring, and compare your result with that given by the exact expression, Equation 23.16.

# 24
# Gauss' Law

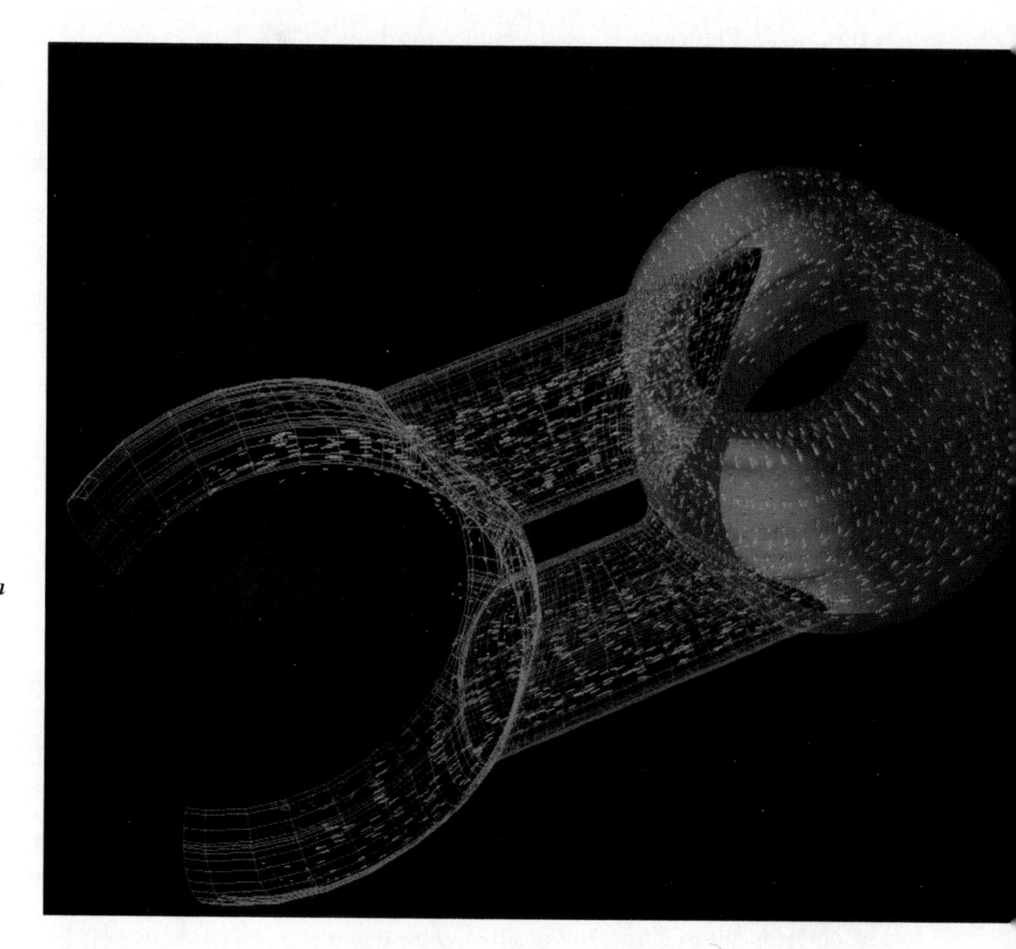

*This computer-generated picture shows the surfaces of a manifold in a space shuttle's main engine — just one of the many kinds of surfaces that crop up in scientific and mathematical problems. Gaussian surfaces, described in this chapter, are hypothetical closed surfaces, useful for solving many problems in electrostatics. (© Dale E. Boyer, Science Source/Photo Researchers)*

In the preceding chapter we showed how to calculate the electric field of a given charge distribution from Coulomb's law. This chapter describes an alternative procedure for calculating electric fields known as *Gauss' law*. This formulation is based on the fact that the fundamental electrostatic force between point charges is an inverse-square law. Although Gauss' law is a consequence of Coulomb's law, it is much more convenient for calculating the electric field of highly symmetric charge distributions. Furthermore, Gauss' law serves as a guide for understanding more complicated problems.

## 24.1 ELECTRIC FLUX

The concept of electric field lines was described qualitatively in the previous chapter. We shall now use the concept of electric flux to put this idea on a quantitative basis. *Electric flux is a measure of the number of electric field lines penetrating some surface.* When the surface being penetrated encloses some net charge, the net number of lines that go through the surface is proportional to the net charge within the surface. The number of lines counted is independent of the shape of the surface enclosing the charge. This is essentially a statement of Gauss' law, which we describe in the next section.

First consider an electric field that is uniform in both magnitude and direction, as in Figure 24.1. The electric field lines penetrate a rectangular surface of area $A$, which is perpendicular to the field. Recall that the number of lines per unit area is proportional to the magnitude of the electric field. Therefore, the number of lines penetrating the surface of area $A$ is proportional to the product $EA$. The product of the electric field strength, $E$, and a surface area $A$ perpendicular to the field is called the **electric flux, Φ**:

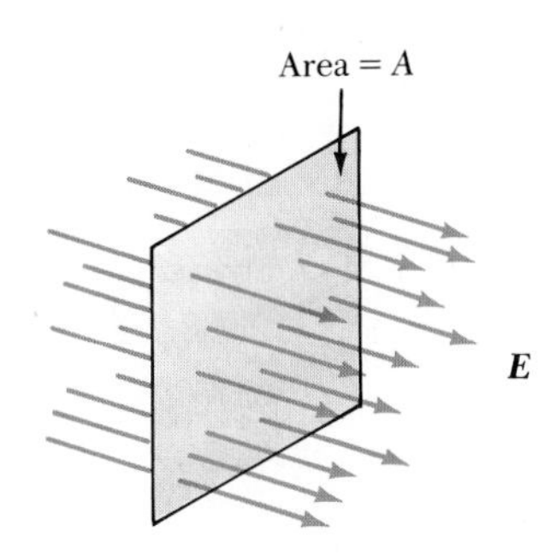


**Figure 24.1** Field lines of a uniform electric field penetrating a plane of area $A$ perpendicular to the field. The electric flux, $\Phi$, through this area is equal to $EA$.

$$\Phi = EA \tag{24.1}$$

From the SI units of $E$ and $A$, we see that electric flux has the units of $\mathrm{N \cdot m^2/C}$.

If the surface under consideration is not perpendicular to the field, the number of lines (or the flux) through it must be less than that given by Equation 24.1. This can be easily understood by considering Figure 24.2, where the normal to the surface of area $A$ is at an angle $\theta$ to the uniform electric field. Note that the number of lines that cross this area is equal to the number that cross the projected area $A'$, which is perpendicular to the field. From Figure 24.2 we see that the two areas are related by $A' = A \cos \theta$. Since the flux through the area $A$ equals the flux through $A'$, we conclude that the desired flux is given by

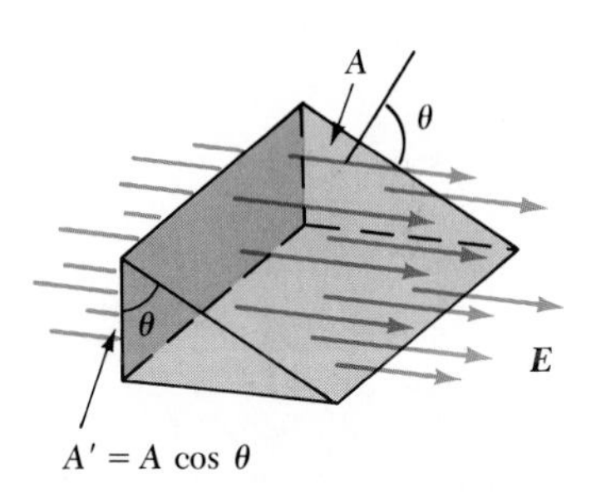


**Figure 24.2** Field lines for a uniform electric field through an area $A$ that is at an angle $\theta$ to the field. Since the number of lines that go through the shaded area $A'$ is the same as the number that go through $A$, we conclude that the flux through $A'$ is equal to the flux through $A$ and is given by $\Phi = EA \cos \theta$.

$$\Phi = EA \cos \theta \tag{24.2}$$

From this result, we see that the flux through a surface of fixed area has the maximum value, $EA$, when the surface is perpendicular to the field (or when the *normal* to the surface is parallel to the field, that is, $\theta = 0°$); the flux is zero when the surface is parallel to the field (or when the normal to the surface is perpendicular to the field, that is, $\theta = 90°$).

In more general situations, the electric field may vary over the surface in question. Therefore, our definition of flux given by Equation 24.2 has meaning only over a small element of area. Consider a general surface divided up into a large number of small elements, each of area $\Delta A$. The variation in the electric field over the element can be neglected if the element is small enough. It is convenient to define a vector $\Delta \boldsymbol{A}_i$ whose magnitude represents the area of the $i$th element and whose direction is *defined to be perpendicular* to the surface, as in Figure 24.3. The electric flux $\Delta\Phi_i$ through this small element is given by

$$\Delta\Phi_i = E_i\, \Delta A_i \cos \theta = \boldsymbol{E}_i \cdot \Delta \boldsymbol{A}_i$$

where we have used the definition of the scalar product of two vectors ($\boldsymbol{A} \cdot \boldsymbol{B} = AB \cos \theta$). By summing the contributions of all elements, we obtain the total flux through the surface.[1] If we let the area of each element approach

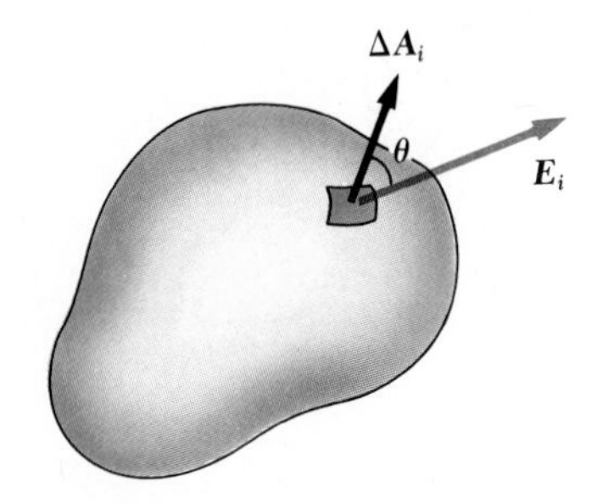


**Figure 24.3** A small element of a surface of area $\Delta A_i$. The electric field makes an angle $\theta$ with the normal to the surface (the direction of $\Delta \boldsymbol{A}_i$), and the flux through the element is equal to $E_i\, \Delta A_i \cos \theta$.

[1] It is important to note that drawings with field lines have their inaccuracies, since a small area (depending on its location) may happen to have too many or too few penetrating lines. At any rate, it is stressed that the basic definition of electric flux is $\int \boldsymbol{E} \cdot d\boldsymbol{A}$. The use of lines is only an aid for visualizing the concept.

zero, then the number of elements approaches infinity and the sum is replaced by an integral. Therefore *the general definition of electric flux is*

Definition of electric flux

$$\Phi \equiv \lim_{\Delta A_i \to 0} \sum \boldsymbol{E}_i \cdot \Delta \boldsymbol{A}_i = \int_{\text{surface}} \boldsymbol{E} \cdot d\boldsymbol{A} \tag{24.3}$$

Equation 24.3 is a surface integral, which must be evaluated over the hypothetical surface in question. In general, the value of $\Phi$ depends both on the field pattern and on the specified surface.

We shall usually be interested in evaluating the flux through a *closed surface.* (A closed surface is defined as a surface which divides space into an inside and an outside region, so that one cannot move from one region to the other without crossing the surface. The surface of a sphere, for example, is a closed surface.) Consider the closed surface in Figure 24.4. Note that the vectors $\Delta \boldsymbol{A}_i$ point in different directions for the various surface elements. At each point, these vectors are *normal* to the surface and, by convention, always point *outward.* At the elements labeled ① and ②, $\boldsymbol{E}$ is outward and $\theta < 90°$; hence the flux $\Delta\Phi = \boldsymbol{E} \cdot \Delta \boldsymbol{A}$ through these elements is positive. On the other hand, for elements such as ③, where the field lines are directed into the surface, $\theta > 90°$ and the flux becomes negative with $\cos\theta$. The total, or net, flux through the surface is proportional to the net number of lines penetrating the surface (where the net number means *the number leaving the volume surrounding the surface minus the number entering the surface*). If there are more lines leaving the surface than entering, the net flux is positive. If more lines enter than leave the surface, the net flux is negative. Using the symbol $\oint$ to represent an *integral over a closed surface,* we can write the net flux, $\Phi_c$, through a closed surface

$$\Phi_c = \oint \boldsymbol{E} \cdot d\boldsymbol{A} = \oint E_n \, dA \tag{24.4}$$

where $E_n$ represents the component of the electric field perpendicular, or normal, to the surface and the subscript c denotes a closed surface. Evaluating the net flux through a closed surface could be very cumbersome. However, if the field is normal to the surface at each point and constant in magnitude, the calculation is straightforward. The following example illustrates this point.

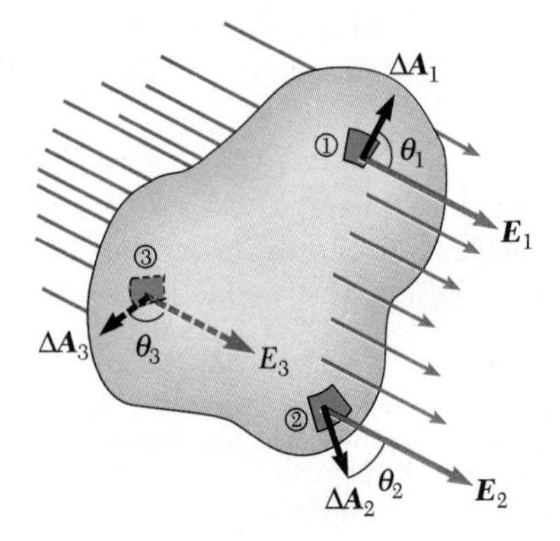


**Figure 24.4** A closed surface in an electric field. The area vectors $\Delta \boldsymbol{A}_i$ are, by convention, normal to the surface and point outward. The flux through an area element can be positive (elements ① and ②) or negative (element ③).

**EXAMPLE 24.1 Flux Through a Cube**
Consider a uniform electric field $\mathbf{E}$ oriented in the $x$ direction. Find the net electric flux through the surface of a cube of edges $\ell$ oriented as shown in Figure 24.5.

*Solution* The net flux can be evaluated by summing up the fluxes through each face of the cube. First, note that the flux through *four* of the faces is zero, since $\mathbf{E}$ is perpendicular to $d\mathbf{A}$ on these faces. In particular, the orientation of $d\mathbf{A}$ is perpendicular to $\mathbf{E}$ for the two faces labeled ③ and ④ in Figure 24.5. Therefore, $\theta = 90°$, so that $\mathbf{E}\cdot d\mathbf{A} = E\,dA\cos 90° = 0$. The fluxes through the planes parallel to the $yx$ plane are also zero for the same reason.

Now consider the faces labeled ① and ②. The net flux through these faces is given by

$$\Phi_c = \int_1 \mathbf{E}\cdot d\mathbf{A} + \int_2 \mathbf{E}\cdot d\mathbf{A}$$

For the face labeled ①, $\mathbf{E}$ is constant and inward while $d\mathbf{A}$ is outward ($\theta = 180°$), so that we find that the flux through this face is

$$\int_1 \mathbf{E}\cdot d\mathbf{A} = \int_1 E\,dA\cos 180° = -E\int_1 dA$$

$$= -EA = -E\ell^2$$

since the area of each face is $A = \ell^2$.

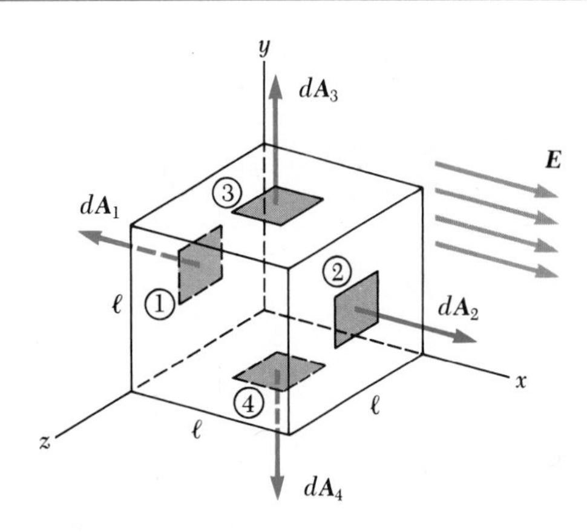


**Figure 24.5** (Example 24.1) A hypothetical surface in the shape of a cube in a uniform electric field parallel to the $x$ axis. The net flux through the surface is zero.

Likewise, for the face labeled ②, $\mathbf{E}$ is constant and outward and in the same direction as $d\mathbf{A}$ ($\theta = 0°$), so that the flux through this face is

$$\int_2 \mathbf{E}\cdot d\mathbf{A} = \int_2 E\,dA\cos 0° = E\int_2 dA = +EA = E\ell^2$$

Hence, the net flux over all faces is zero, since

$$\Phi_c = -E\ell^2 + E\ell^2 = 0$$

## 24.2 GAUSS' LAW

In this section we describe a general relation between the net electric flux through a closed surface (often called a *gaussian surface*) and the charge *enclosed* by the surface. This relation, known as *Gauss' law*, is of fundamental importance in the study of electrostatic fields.

First, let us consider a positive point charge $q$ located at the center of a sphere of radius $r$ as in Figure 24.6. From Coulomb's law we know that the magnitude of the electric field everywhere on the surface of the sphere is $E = kq/r^2$. Furthermore, the field lines are radial outward, and hence are perpendicular (or normal) to the surface at each point. That is, at each point $\mathbf{E}$ is parallel to the vector $\Delta\mathbf{A}_i$ representing the local element of area $\Delta A_i$. Therefore

$$\mathbf{E}\cdot\Delta\mathbf{A}_i = E_n\,\Delta A_i = E\,\Delta A_i$$

and from Equation 24.4 we find that the net flux through the gaussian surface is given by

$$\Phi_c = \oint E_n\,dA = \oint E\,dA = E\oint dA$$

since $E$ is constant over the surface and given by $E = kq/r^2$. Furthermore, for a spherical gaussian surface, $\oint dA = A = 4\pi r^2$ (the surface area of a sphere).

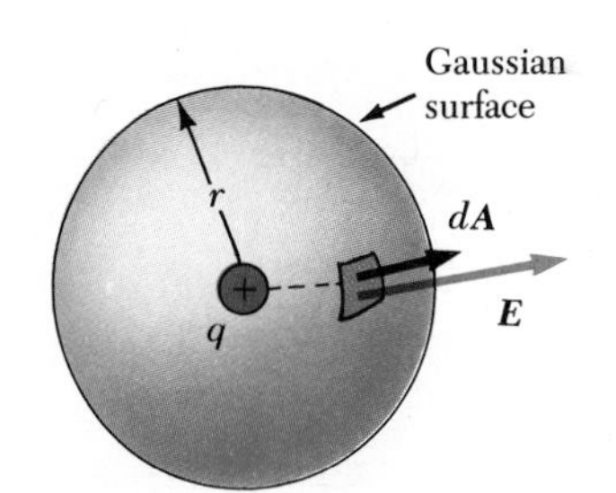


**Figure 24.6** A spherical surface of radius $r$ surrounding a point charge $q$. When the charge is at the center of the sphere, the electric field is normal to the surface and constant in magnitude everywhere on the surface.

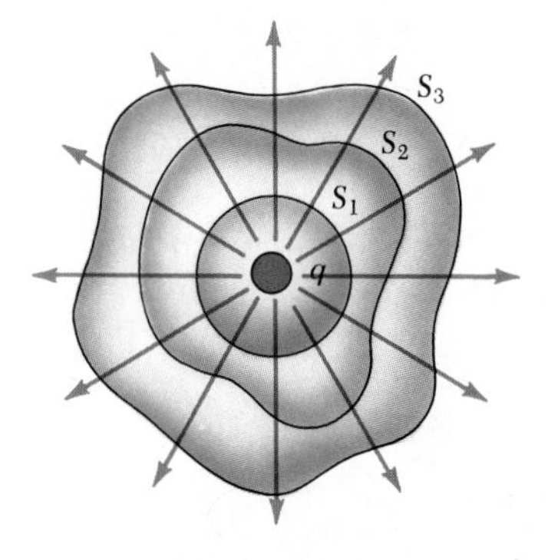


Figure 24.7 Closed surfaces of various shapes surrounding a charge $q$. Note that the net electric flux through each surface is the same.

Hence the net flux through the gaussian surface is

$$\Phi_c = \frac{kq}{r^2}(4\pi r^2) = 4\pi kq$$

Recalling that $k = 1/4\pi\epsilon_0$, we can write this in the form

$$\Phi_c = \frac{q}{\epsilon_0} \tag{24.5}$$

Note that this result, which is independent of $r$, says that the net flux through a spherical gaussian surface is proportional to the charge $q$ *inside* the surface. The fact that the flux is independent of the radius is a consequence of the inverse-square dependence of the electric field given by Coulomb's law. That is, $E$ varies as $1/r^2$, but the area of the sphere varies as $r^2$. Their combined effect produces a flux that is independent of $r$.

Now consider several closed surfaces surrounding a charge $q$ as in Figure 24.7. Surface $S_1$ is spherical, whereas surfaces $S_2$ and $S_3$ are nonspherical. The flux that passes through surface $S_1$ has the value $q/\epsilon_0$. As we discussed in the previous section, the flux is proportional to the number of electric field lines passing through that surface. The construction in Figure 24.7 shows that the number of electric field lines through the spherical surface $S_1$ is equal to the number of electric field lines through the nonspherical surfaces $S_2$ and $S_3$. Therefore, it is reasonable to conclude that the net flux through any closed surface is independent of the shape of that surface. (One can prove that this is the case if $E \propto 1/r^2$.) In fact, *the net flux through any closed surface surrounding a point charge $q$ is given by $q/\epsilon_0$.*

The net flux through a closed surface is zero if there is no charge inside

Now consider a point charge located *outside* a closed surface of arbitrary shape, as in Figure 24.8. As you can see from this construction, some electric field lines enter the surface, and others leave the surface. However, *the number of electric field lines entering the surface equals the number leaving the surface.* Therefore, we conclude that *the net electric flux through a closed surface that surrounds no charge is zero.* If we apply this result to Example 24.1, we can easily see that the net flux through the cube is zero, since it was assumed there was no charge inside the cube.

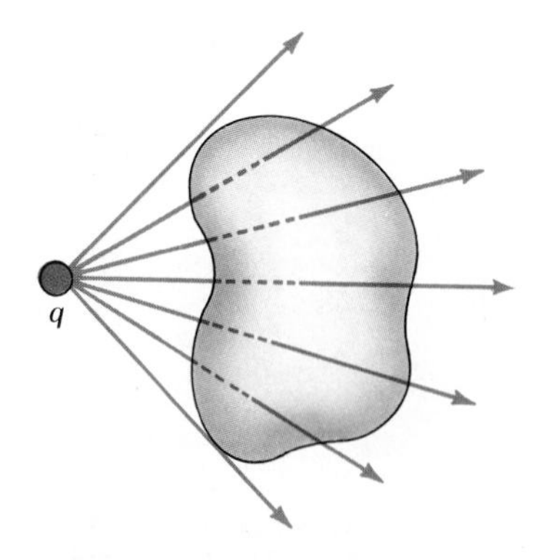


Figure 24.8 A point charge located *outside* a closed surface. In this case, note that the number of lines entering the surface equals the number leaving the surface.

Let us extend these arguments to the generalized case of many point charges, or a continuous distribution of charge. We shall make use of the **superposition principle,** which says that *the electric field due to many charges is the vector sum of the electric fields produced by the individual charges.* That is, we can express the flux through any closed surface as

$$\oint \mathbf{E} \cdot d\mathbf{A} = \oint (\mathbf{E}_1 + \mathbf{E}_2 + \mathbf{E}_3) \cdot d\mathbf{A}$$

where $\mathbf{E}$ is the total electric field at any point on the surface and $\mathbf{E}_1$, $\mathbf{E}_2$, and $\mathbf{E}_3$ are the fields produced by the individual charges at that point. Consider the system of charges shown in Figure 24.9. The surface S surrounds only one charge, $q_1$; hence the net flux through S is $q_1/\epsilon_0$. The flux through S due to the charges outside it is zero since each electric field line that enters S at one point leaves it at another. The surface S′ surrounds charges $q_2$ and $q_3$; hence the net flux through S′ is $(q_2 + q_3)/\epsilon_0$. Finally, the net flux through surface S″ is zero

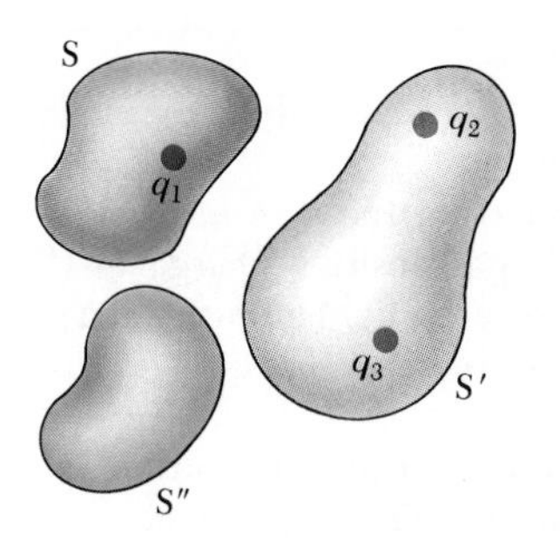


**Figure 24.9** The net electric flux through any closed surface depends only on the charge *inside* that surface. The net flux through surface S is $q_1/\epsilon_0$, the net flux through surface S′ is $(q_2 + q_3)/\epsilon_0$, and the net flux through surface S″ is zero.

since there is no charge inside this surface. That is, *all* lines that enter S″ at one point leave S″ at another.

**Gauss' law,** which is a generalization of the above discussion, states that the net flux through *any* closed surface is given by

$$\Phi_c = \oint \mathbf{E} \cdot d\mathbf{A} = \frac{q_{in}}{\epsilon_0} \tag{24.6}$$

where $q_{in}$ represents the *net charge inside* the gaussian surface and $\mathbf{E}$ represents the electric field at any point on the gaussian surface. In words,

Gauss' law states that the net electric flux through any closed gaussian surface is equal to the net charge inside the surface divided by $\epsilon_0$.

Gauss' law

A formal proof of Gauss' law is presented in Section 24.6. When using Equation 24.6, you should note that although the charge $q_{in}$ is the net charge inside the gaussian surface, the $\mathbf{E}$ that appears in Gauss' law represents the *total electric field,* which includes contributions from charges both inside and outside the gaussian surface. This point is often neglected or misunderstood.

In principle, Gauss' law can always be used to calculate the electric field of a system of charges or a continuous distribution of charge. However, in practice, *the technique is useful only in a limited number of situations where there is a high degree of symmetry.* As we shall see in the next section, *Gauss' law can be used to evaluate the electric field for charge distributions that have spherical, cylindrical, or plane symmetry.* If one carefully chooses the gaussian surface surrounding the charge distribution, the integral in Equation 24.6 will be easy to evaluate. You should also note that a gaussian surface is a mathematical surface and need not coincide with any real physical surface.

## 24.3 APPLICATION OF GAUSS' LAW TO CHARGED INSULATORS

In this section we give some examples of how to use Gauss' law to calculate $\mathbf{E}$ for a given charge distribution. It is important to recognize that *Gauss' law is useful when there is a high degree of symmetry in the charge distribution, as in the case of uniformly charged spheres, long cylinders, and plane sheets.* In such cases, it is possible to find a simple gaussian surface over which the surface integral given by Equation 24.6 is easily evaluated.

The surface should always be chosen such that it has the same symmetry as that of the charge distribution.

Gauss' law is useful for evaluating $E$ when the charge distribution has symmetry

The following examples should clarify this procedure.

**EXAMPLE 24.2 The Electric Field Due to a Point Charge**

Starting with Gauss' law, calculate the electric field due to an isolated point charge $q$ and show that Coulomb's law follows from this result.

*Solution* For this situation we choose a spherical gaussian surface of radius $r$ and centered on the point charge, as in Figure 24.10. The electric field of a positive point charge is radial outward by symmetry, and is therefore normal to the surface at every point. That is, $\mathbf{E}$ is parallel to $d\mathbf{A}$ at each point, and so $\mathbf{E}\cdot d\mathbf{A} = E\,dA$ and Gauss' law gives

$$\Phi_c = \oint \mathbf{E}\cdot d\mathbf{A} = \oint E\,dA = \frac{q}{\epsilon_0}$$

By symmetry, $E$ is constant everywhere on the surface, and so it can be removed from the integral. Therefore,

$$\oint E\,dA = E\oint dA = E(4\pi r^2) = \frac{q}{\epsilon_0}$$

where we have used the fact that the surface area of a sphere is $4\pi r^2$. Hence, the magnitude of the field at a distance $r$ from the charge $q$ is

$$E = \frac{q}{4\pi\epsilon_0 r^2} = k\frac{q}{r^2}$$

If a second point charge $q_0$ is placed at a point where the field is $E$, the electrostatic force on this charge has a magnitude given by

$$F = q_0 E = k\frac{qq_0}{r^2}$$

This, of course, is Coulomb's law. Note that this example is logically circular. It does, however, demonstrate the equivalence of Coulomb's law and Gauss' law.

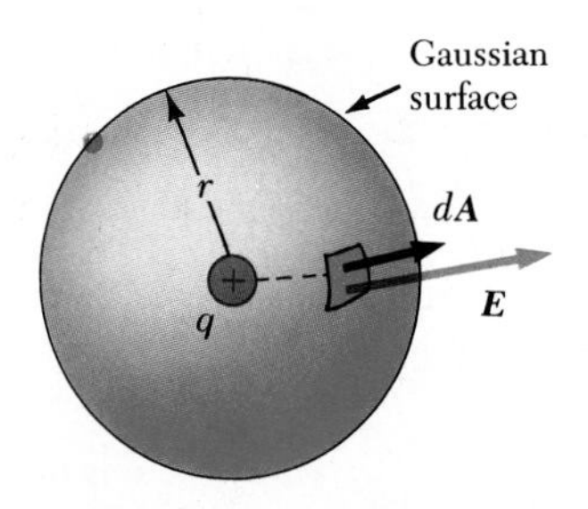


**Figure 24.10** (Example 24.2) The point charge $q$ is at the center of the spherical gaussian surface, and $\mathbf{E}$ is parallel to $d\mathbf{A}$ at every point on the surface.

**EXAMPLE 24.3 A Spherically Symmetric Charge Distribution** □

An insulating sphere of radius $a$ has a uniform charge density $\rho$ and a total positive charge $Q$ (Fig. 24.11). (a) Calculate the electric field intensity at a point *outside* the sphere, that is, for $r > a$.

*Solution* Since the charge distribution is spherically symmetric, we again select a spherical gaussian surface of radius $r$, concentric with the sphere, as in Figure 24.11a. Following the line of reasoning given in Example 24.2, we find that

$$E = k\frac{Q}{r^2} \qquad \text{(for } r > a\text{)} \qquad (24.7)$$

Note that this result is identical to that obtained for a point charge. Therefore, we conclude that, for a uniformly charged sphere, the field in the region external to the sphere is *equivalent* to that of a point charge located at the center of the sphere.

(b) Find the electric field intensity at a point *inside* the sphere, that is, for $r < a$.

*Solution* In this case we select a spherical gaussian surface with radius $r < a$, concentric with the charge distribution (Fig. 24.11b). To apply Gauss' law in this situation, it is important to recognize that the charge $q_{in}$ *within* the gaussian surface of volume $V'$ is a quantity *less* than the total charge $Q$. To calculate the charge $q_{in}$, we use the fact that $q_{in} = \rho V'$, where $\rho$ is the charge per unit volume and $V'$ is the volume enclosed by the gaussian surface, given by $V' = \frac{4}{3}\pi r^3$ for a sphere. Therefore,

$$q_{in} = \rho V' = \rho\left(\tfrac{4}{3}\pi r^3\right)$$

As in Example 24.2, the electric field is constant in magnitude everywhere on the spherical gaussian surface and

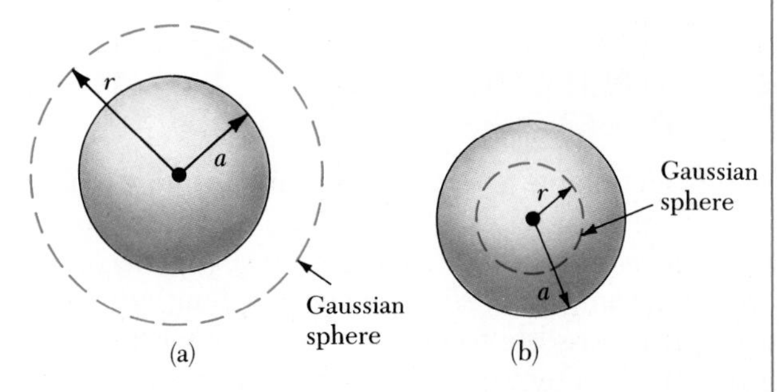


**Figure 24.11** (Example 24.3) A uniformly charged insulating sphere of radius $a$ and total charge $Q$. (a) The field at a point exterior to the sphere is $kQ/r^2$. (b) The field inside the sphere is due only to the charge *within* the gaussian surface and is given by $(kQ/a^3)r$.

is normal to the surface at each point. Therefore, Gauss' law in the region $r < a$ gives

$$\oint E\,dA = E\oint dA = E(4\pi r^2) = \frac{q_{in}}{\epsilon_0}$$

Solving for $E$ gives

$$E = \frac{q_{in}}{4\pi\epsilon_0 r^2} = \frac{\rho\,\frac{4}{3}\pi r^3}{4\pi\epsilon_0 r^2} = \frac{\rho}{3\epsilon_0}\,r$$

Since by definition $\rho = Q/\frac{4}{3}\pi a^3$, this can be written

$$E = \frac{Qr}{4\pi\epsilon_0 a^3} = \frac{kQ}{a^3}\,r \qquad \text{(for } r < a) \qquad (24.8)$$

Note that this result for $E$ differs from that obtained in (a). It shows that $E \rightarrow 0$ as $r \rightarrow 0$, as you might have guessed based on the spherical symmetry of the charge distribution. Therefore, the result fortunately eliminates the singularity that would exist at $r = 0$ if $E$ varied as $1/r^2$ inside the sphere. That is, if $E \propto 1/r^2$, the field would be infinite at $r = 0$, which is clearly a physically impossible situation. A plot of $E$ versus $r$ is shown in Figure 24.12.

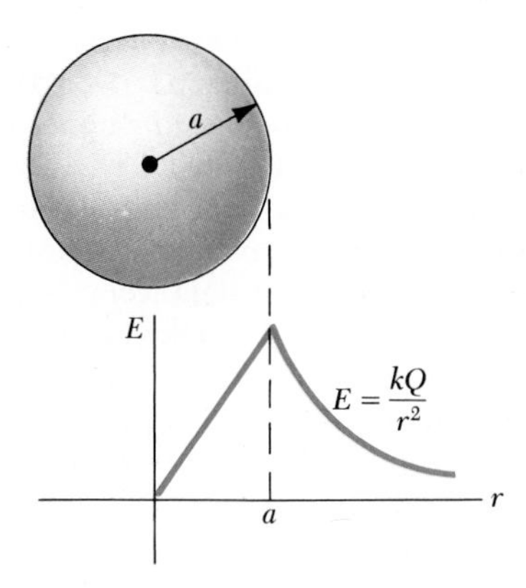


**Figure 24.12** (Example 24.3) A plot of $E$ versus $r$ for a uniformly charged insulating sphere. The field inside the sphere ($r < a$) varies linearly with $r$. The field outside the sphere ($r > a$) is the same as that of a point charge $Q$ located at the origin.

### EXAMPLE 24.4 The *E* Field of a Thin Spherical Shell

A thin *spherical shell* of radius $a$ has a total charge $Q$ distributed uniformly over its surface (Fig. 24.13). Find the electric field at points inside and outside the shell.

*Solution* The calculation of the field outside the shell is identical to that already carried out for the solid sphere in Example 24.3a. If we construct a spherical gaussian surface of radius $r > a$, concentric with the shell, then the charge inside this surface is $Q$. Therefore, the field at a point outside the shell is equivalent to that of a point charge $Q$ at the center:

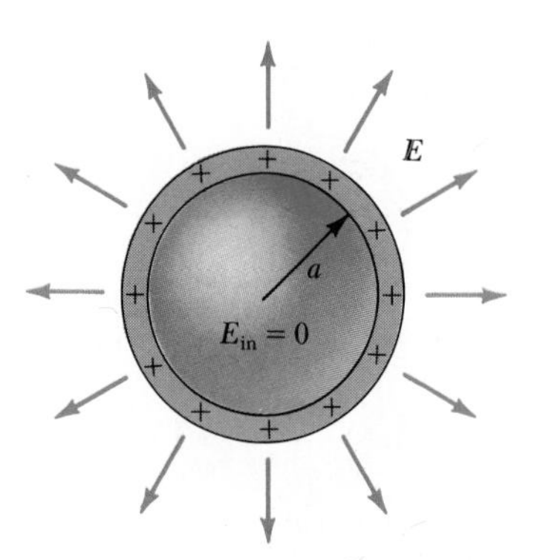


**Figure 24.13** (Example 24.4) The electric field inside a uniformly charged spherical shell is *zero*. The field outside is the same as that of a point charge having a total charge $Q$ located at the center of the shell.

$$E = k\frac{Q}{r^2} \qquad \text{(for } r > a)$$

*The electric field inside the spherical shell is zero.* This also follows from Gauss' law applied to a spherical surface of radius $r < a$. Since the net charge inside the surface is zero, and because of the spherical symmetry of the charge distribution, application of Gauss' law shows that $E = 0$ in the region $r < a$.

The same results can be obtained using Coulomb's law and integrating over the charge distribution. This calculation is rather complicated and will be omitted.

### EXAMPLE 24.5 A Cylindrically Symmetric Charge Distribution

Find the electric field at a distance $r$ from a uniform positive line charge of infinite length whose charge per unit length is $\lambda$ = constant (Fig. 24.14).

*Solution* The symmetry of the charge distribution shows that $\mathbf{E}$ must be perpendicular to the line charge and directed outward as in Figure 24.14a. The end view of the line charge shown in Figure 24.14b should help visualize the directions of the electric field lines. In this situation, we select a cylindrical gaussian surface of radius $r$ and length $\ell$ that is coaxial with the line charge. For the curved part of this surface, $\mathbf{E}$ is constant in magnitude and perpendicular to the surface at each point. Furthermore, the flux through the *ends* of the gaussian cylinder is *zero* since $\mathbf{E}$ is *parallel* to these surfaces.

The total charge inside our gaussian surface is $\lambda\ell$, where $\lambda$ is the charge per unit length and $\ell$ is the length of the cylinder. Applying Gauss' law and noting that $\mathbf{E}$ is

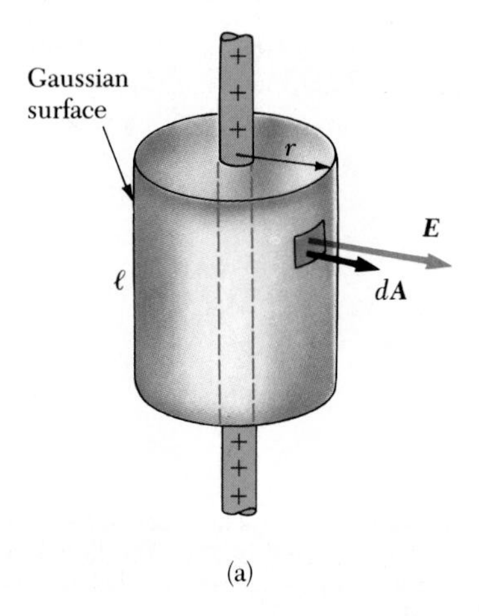


(a)

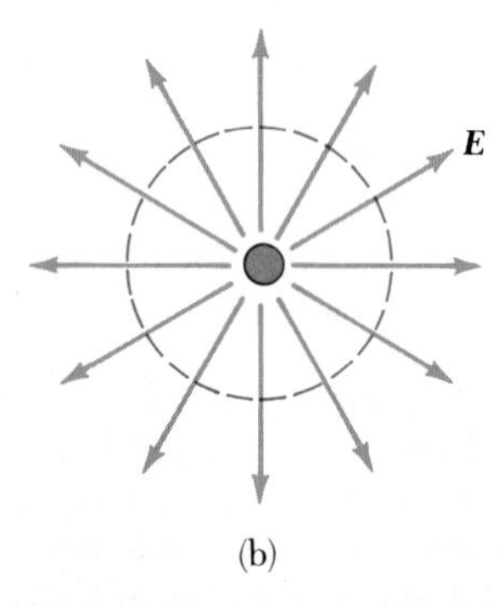


(b)

**Figure 24.14** (Example 24.5) (a) An infinite line of charge surrounded by a cylindrical gaussian surface concentric with the line charge. (b) The field on the cylindrical surface is constant in magnitude and perpendicular to the surface.

parallel to $d\mathbf{A}$ everywhere on the cylindrical surface, we find that

$$\Phi_c = \oint \mathbf{E} \cdot d\mathbf{A} = E \oint dA = \frac{q_{in}}{\epsilon_0} = \frac{\lambda \ell}{\epsilon_0}$$

But the area of the curved surface is $A = 2\pi r \ell$; therefore

$$E(2\pi r \ell) = \frac{\lambda \ell}{\epsilon_0}$$

$$E = \frac{\lambda}{2\pi \epsilon_0 r} = 2k \frac{\lambda}{r} \qquad (24.9)$$

Thus, we see that the field of a cylindrically symmetric charge distribution varies as $1/r$, whereas the field external to a spherically symmetric charge distribution varies as $1/r^2$. Equation 24.9 can also be obtained using Coulomb's law and integration; however, the mathematical techniques necessary for this calculation are more cumbersome.

If the line charge has a finite length, the result for $E$ is *not* the same as that given by Equation 24.9. For points close to the line charge and far from the ends, Equation 24.9 gives a good approximation of the actual value of the field. It turns out that Gauss' law is *not useful* for calculating $E$ for a finite line charge. This is because the electric field is no longer constant in magnitude over the surface of the gaussian cylinder. Furthermore, $\mathbf{E}$ is not perpendicular to the cylindrical surface at all points. When there is little symmetry in the charge distribution, as in this situation, it is necessary to calculate $\mathbf{E}$ using Coulomb's law.

It is left as a problem (Problem 35) to show that the $\mathbf{E}$ field *inside* a uniformly charged rod of finite thickness is proportional to $r$.

### EXAMPLE 24.6 A Nonconducting Plane Sheet of Charge

Find the electric field due to a nonconducting, infinite plane with uniform charge per unit area $\sigma$.

*Solution* The symmetry of the situation shows that $\mathbf{E}$ must be perpendicular to the plane and that the direction of $\mathbf{E}$ on one side of the plane must be opposite its direction on the other side, as in Figure 24.15. It is convenient to choose for our gaussian surface a small cylinder whose axis is perpendicular to the plane and whose ends each have an area $A$ and are equidistant from the plane. Here we see that since $\mathbf{E}$ is parallel to the cylindrical surface, there is no flux through this surface. The flux out of *each* end of the cylinder is $EA$ (since $\mathbf{E}$ is perpendicular to the ends); hence the *total* flux through our gaussian surface is

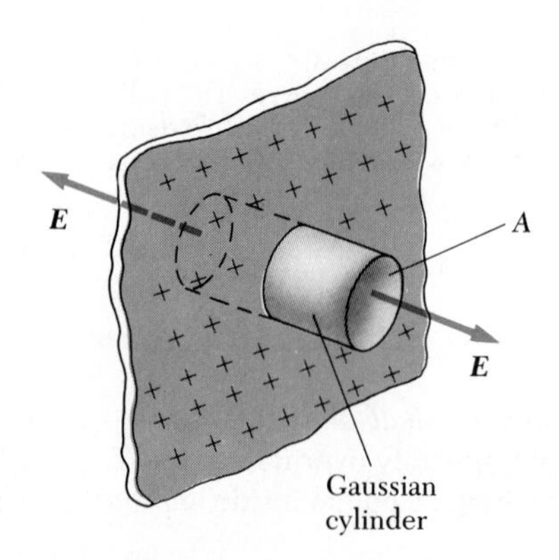


**Figure 24.15** (Example 24.6) A cylindrical gaussian surface penetrating an infinite sheet of charge. The flux through each end of the gaussian surface is *EA*. There is no flux through the cylinder's surface.

$2EA$. Noting that the total charge *inside* the surface is $\sigma A$, we use Gauss' law to get

$$\Phi_c = 2EA = \frac{q_{in}}{\epsilon_0} = \frac{\sigma A}{\epsilon_0}$$

$$E = \frac{\sigma}{2\epsilon_0} \qquad (24.10)$$

Since the distance of the surfaces from the plane does not appear in Equation 24.10, we conclude that $E = \sigma/2\epsilon_0$ at *any* distance from the plane. That is, the field is *uniform* everywhere.

An important configuration related to this example is the case of two parallel sheets of charge, with charge densities $\sigma$ and $-\sigma$, respectively (Problem 58). In this situation, the electric field is $\sigma/\epsilon_0$ *between* the sheets and approximately zero elsewhere.

## 24.4 CONDUCTORS IN ELECTROSTATIC EQUILIBRIUM

A good electrical conductor, such as copper, contains charges (electrons) that are not bound to any atom and are free to move about within the material. When there is no *net* motion of charge within the conductor, the conductor is in **electrostatic equilibrium.** As we shall see, *a conductor in electrostatic equilibrium* has the following properties:

Properties of a conductor in electrostatic equilibrium

1. The electric field is zero everywhere inside the conductor.
2. Any excess charge on an isolated conductor must reside entirely on its surface.
3. The electric field just outside a charged conductor is perpendicular to the conductor's surface and has a magnitude $\sigma/\epsilon_0$, where $\sigma$ is the charge per unit area at that point.
4. On an irregularly shaped conductor, charge tends to accumulate at locations where the radius of curvature of the surface is the smallest, that is, at sharp points.

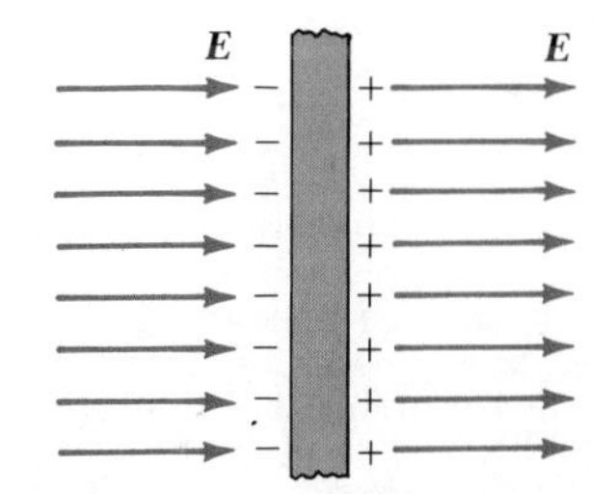


**Figure 24.16** A conducting slab in an external electric field $\boldsymbol{E}$. The charges induced on the surfaces of the slab produce an electric field which opposes the external field, giving a resultant field of zero in the conductor.

The first property can be understood by considering a conducting slab placed in an external field $\boldsymbol{E}$ (Fig. 24.16). In electrostatic equilibrium, the electric field *inside* the conductor must be zero. If this were not the case, the free charges would accelerate under the action of an electric field. Before the external field is applied, the electrons are uniformly distributed throughout the conductor. When the external field is applied, the free electrons accelerate to the left, causing a buildup of negative charge on the left surface (excess electrons) and of positive charge on the right (where electrons have been removed). These charges create their own electric field, which *opposes* the external field. The surface charge density increases until the magnitude of the electric field set up by these charges equals that of the external field, giving a net field of zero *inside* the conductor. In a good conductor, the time it takes the conductor to reach equilibrium is of the order of $10^{-16}$ s, which for most purposes can be considered instantaneous.

We can use Gauss' law to verify the second and third properties of a conductor in electrostatic equilibrium. Figure 24.17 shows an arbitrarily shaped insulated conductor. A gaussian surface is drawn inside the conductor as close to the surface as we wish. As we have just shown, the electric field everywhere inside the conductor is zero when it is in electrostatic equilibrium. Since the electric field is also zero at *every* point on the gaussian surface, we see that the net flux through this surface is zero. From this result and Gauss' law, we conclude that the net charge inside the gaussian surface is zero. Since

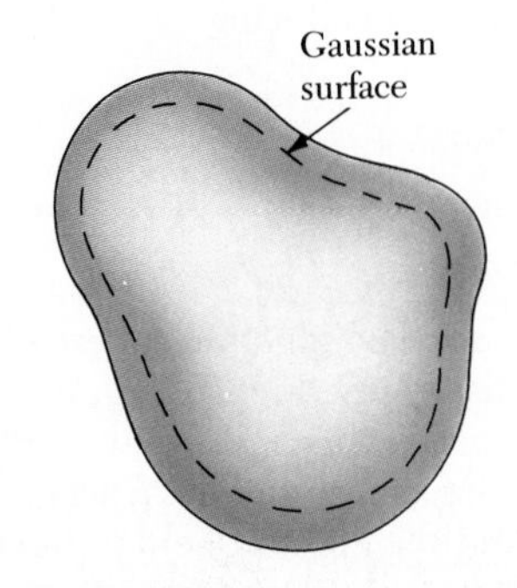


**Figure 24.17** An insulated conductor of arbitrary shape. The broken line represents a gaussian surface just inside the conductor.

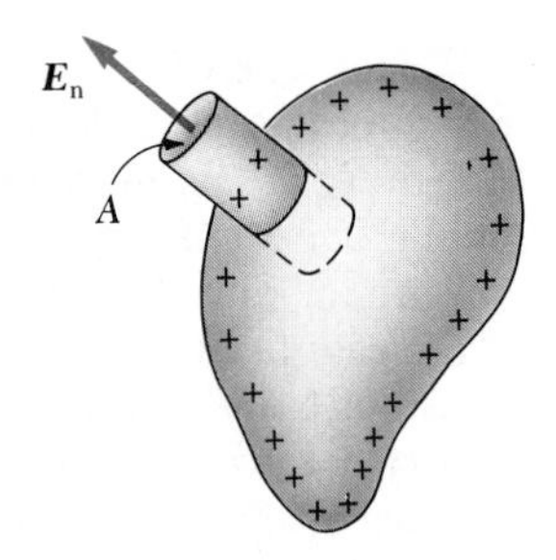


**Figure 24.18** A gaussian surface in the shape of a small cylinder is used to calculate the electric field just outside a charged conductor. The flux through the gaussian surface is $E_n A$. Note that $\boldsymbol{E}$ is zero inside the conductor.

Electric field pattern of a charged conducting plate near an oppositely charged conducting cylinder. Small pieces of thread suspended in oil align with the electric field lines. Note that (1) the electric field lines are perpendicular to the conductors and (2) there are no lines inside the cylinder ($E = 0$). (Courtesy of Harold M. Waage, Princeton University)

there can be no net charge inside the gaussian surface (which is arbitrarily close to the conductor's surface), *any net charge on the conductor must reside on its surface.* Gauss' law does *not* tell us how this excess charge is distributed on the surface. In Section 25.6 we shall prove the fourth property of a conductor in electrostatic equilibrium.

We can use Gauss' law to relate the electric field just outside the surface of a charged conductor in equilibrium to the charge distribution on the conductor. To do this, it is convenient to draw a gaussian surface in the shape of a small cylinder with end faces parallel to the surface (Fig. 24.18). Part of the cylinder is just outside the conductor, and part is inside. There is no flux through the face on the inside of the cylinder since $\boldsymbol{E} = 0$ inside the conductor. Furthermore, the field is normal to the surface. If $\boldsymbol{E}$ had a tangential component, the free charges would move along the surface creating surface currents, and the conductor would not be in equilibrium. There is no flux through the cylindrical face of the gaussian surface since $\boldsymbol{E}$ is tangent to this surface. Hence, the net flux through the gaussian surface is $E_n A$, where $E_n$ is the electric field just outside the conductor. Applying Gauss' law to this surface gives

$$\Phi_c = \oint E_n \, dA = E_n A = \frac{q_{in}}{\epsilon_0} = \frac{\sigma A}{\epsilon_0}$$

We have used the fact that the charge inside the gaussian surface is $q_{in} = \sigma A$, where $A$ is the area of the cylinder's face and $\sigma$ is the (local) charge per unit area. Solving for $E_n$ gives

Electric field just outside a charged conductor

$$E_n = \frac{\sigma}{\epsilon_0} \qquad (24.11)$$

EXAMPLE 24.7 A Sphere Inside a Spherical Shell

A solid conducting sphere of radius $a$ has a net positive charge $2Q$ (Fig. 24.19). A conducting spherical shell of inner radius $b$ and outer radius $c$ is concentric with the solid sphere and has a *net* charge $-Q$. Using Gauss' law, find the electric field in the regions labeled ①, ②, ③, and ④ and the charge distribution on the spherical shell.

*Solution* First note that the charge distribution on both spheres has spherical symmetry, since they are concentric. To determine the electric field at various distances $r$ from the center, we construct spherical gaussian surfaces of radius $r$.

To find $E$ inside the solid sphere of radius $a$ (region ①), we construct a gaussian surface of radius $r < a$. Since there can be no charge inside a conductor in electrostatic equilibrium, we see that $q_{in} = 0$, and so from Gauss' law $E_1 = 0$ for $r < a$. Thus we conclude that the net charge $2Q$ on the solid sphere is distributed on its outer surface.

In region ② between the spheres, where $a < r < b$, we again construct a spherical gaussian surface of radius $r$ and note that the charge inside this surface is $+2Q$ (the

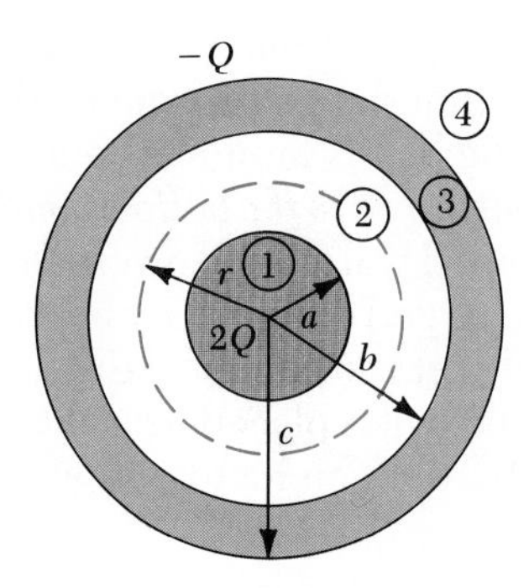


**Figure 24.19** (Example 24.7) A solid conducting sphere of radius $a$ and charge $2Q$ surrounded by a conducting spherical shell of charge $-Q$.

charge on the inner sphere). Because of the spherical symmetry, the electric field lines must be radial outward and constant in magnitude on the gaussian surface. Following Example 24.2 and using Gauss' law, we find that

$$E_2 A = E_2(4\pi r^2) = \frac{q_{in}}{\epsilon_0} = \frac{2Q}{\epsilon_0}$$

$$E_2 = \frac{2Q}{4\pi\epsilon_0 r^2} = \frac{2kQ}{r^2} \qquad (\text{for } a < r < b)$$

In region ④ outside both spheres, where $r > c$, the spherical gaussian surface surrounds a *total* charge of $q_{in} = 2Q + (-Q) = Q$. Therefore, Gauss' law applied to this surface gives

$$E_4 = \frac{kQ}{r^2} \qquad (\text{for } r > c)$$

Finally, consider region ③, where $b < r < c$. The electric field must be *zero* in this region since the spherical shell is also a conductor in equilibrium. If we construct a Gaussian surface of this radius, we see that $q_{in}$ must be zero since $E_3 = 0$. From this argument, we conclude that the charge on the *inner surface* of the *spherical shell* must be $-2Q$ to cancel the charge $+2Q$ on the solid sphere. (The charge $-2Q$ is induced by the charge $+2Q$ on the solid sphere.) Furthermore, since the net charge on the spherical shell is $-Q$, we conclude that the outer surface of the shell must have a charge equal to $+Q$.

## *24.5 EXPERIMENTAL PROOF OF GAUSS' LAW AND COULOMB'S LAW

When a net charge is placed on a conductor, the charge distributes itself on the surface in such a way that the electric field inside is zero. Since $\mathbf{E} = 0$ inside a conductor in electrostatic equilibrium, Gauss' law shows that there can be no net charge inside the conductor. We have seen that Gauss' law is a consequence of Coulomb's law (Example 24.2). Hence, it should be possible to test the validity of the inverse-square law of force by attempting to detect a net charge inside a conductor. If a net charge is detected anywhere but on the conductor's surface, Coulomb's law, and hence Gauss' law, is invalid. Many experiments, including early work by Faraday, Cavendish, and Maxwell, have been performed to show that the net charge on a conductor resides on its surface. In all reported cases, no electric field could be detected in a closed conductor. The most recent and precise experiments by Williams, Faller, and Hill in 1971 showed that the exponent of $r$ in Coulomb's law is $(2 + \delta)$, where $\delta = (2.7 \pm 3.1) \times 10^{-16}$!

The following experiment can be performed to verify that the net charge on a conductor resides on its surface. A positively charged metal ball at the end of a silk thread is lowered into an uncharged, hollow conductor through a small opening[2] (Fig. 24.20a). The hollow conductor is insulated from ground. The charged ball induces a negative charge on the inner wall of the hollow conductor, leaving an equal positive charge on the outer wall (Fig. 24.20b). The presence of positive charge on the outer wall is indicated by the deflection of an electrometer (a device used to measure charge). The deflection of the

[2] The experiment is often referred to as *Faraday's ice-pail experiment*, since it was first performed by Faraday using an ice pail for the hollow conductor.

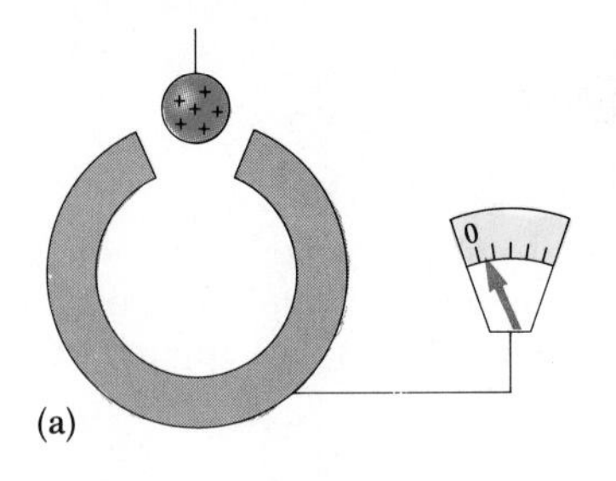


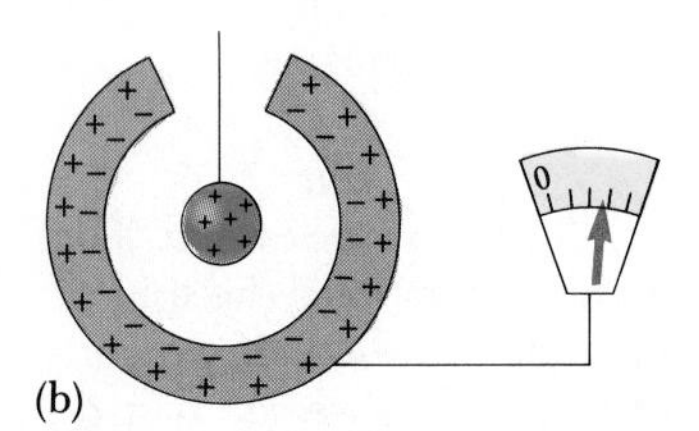


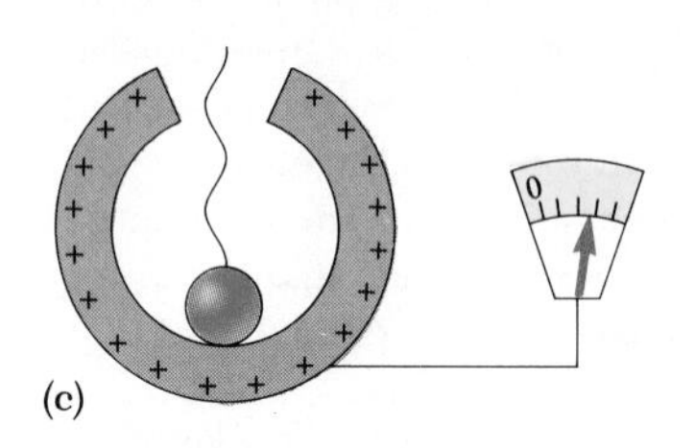


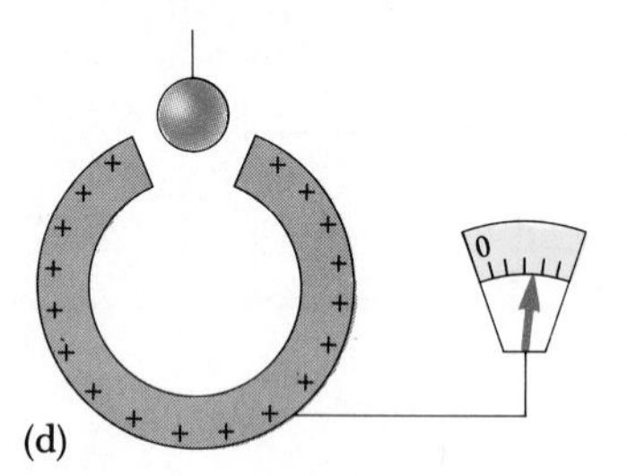


**Figure 24.20** An experiment showing that any charge transferred to a conductor resides on its surface in electrostatic equilibrium. The hollow conductor is insulated from ground, and the small metal ball is supported by an insulating thread.

electrometer remains unchanged when the ball touches the inner surface of the hollow conductor (Fig. 24.20c). When the ball is removed, the electrometer reading remains the same and the ball is found to be uncharged (Fig. 24.20d). This shows that *the charge transferred to the hollow conductor resides on its outer surface.* If a small charged metal ball is now lowered into the *center* of the charged hollow conductor, the charged ball will not be attracted to the hollow conductor. This shows that $\mathbf{E} = 0$ at the center of the hollow conductor. On the other hand, if a small charged ball is placed near the outside of the conductor, the ball will be repelled by the conductor, showing that $\mathbf{E} \neq 0$ outside the conductor.

## *24.6 DERIVATION OF GAUSS' LAW

One method that can be used to derive Gauss' law involves the concept of the *solid angle.* Consider a spherical surface of radius $r$ containing an area element $\Delta A$. The solid angle $\Delta\Omega$ subtended by this element at the center of the sphere is defined to be

$$\Delta\Omega \equiv \frac{\Delta A}{r^2}$$

From this expression, we see that $\Delta\Omega$ has no dimensions, since $\Delta A$ and $r^2$ both have the dimension of $L^2$. The unit of a solid angle is called the **steradian.** Since the total surface area of a sphere is $4\pi r^2$, the total solid angle subtended by the sphere at the center is given by

$$\Omega = \frac{4\pi r^2}{r^2} = 4\pi \text{ steradians}$$

Now consider a point charge $q$ surrounded by a closed surface of arbitrary shape (Fig. 24.21). The total flux through this surface can be obtained by evaluating $\mathbf{E} \cdot \Delta \mathbf{A}$ for each element of area and summing over all elements of the surface. The flux through the element of area $\Delta A$ is

$$\Delta\Phi = \mathbf{E} \cdot \Delta\mathbf{A} = E \cos\theta\, \Delta A = kq \frac{\Delta A \cos\theta}{r^2}$$

where we have used the fact that $E = kQ/r^2$ for a point charge. But the quantity $\Delta A \cos\theta/r^2$ is equal to the solid angle $\Delta\Omega$ subtended at the charge $q$ by the surface element $\Delta A$. From Figure 24.22 we see that $\Delta\Omega$ is equal to the solid angle subtended by the element of a spherical surface of radius $r$. Since the

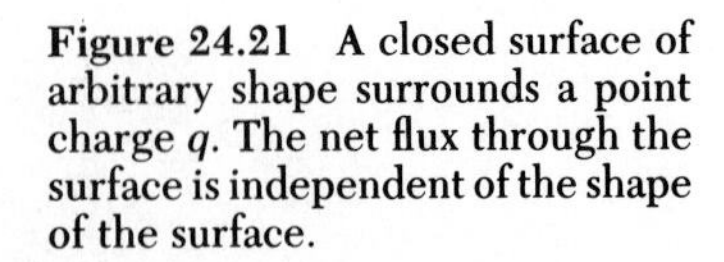

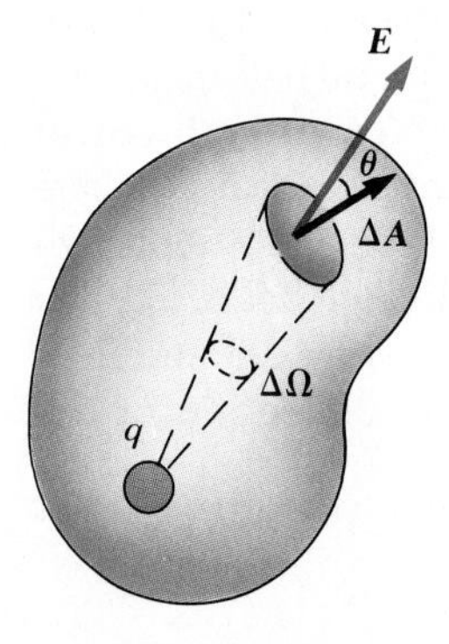


**Figure 24.21** A closed surface of arbitrary shape surrounds a point charge $q$. The net flux through the surface is independent of the shape of the surface.

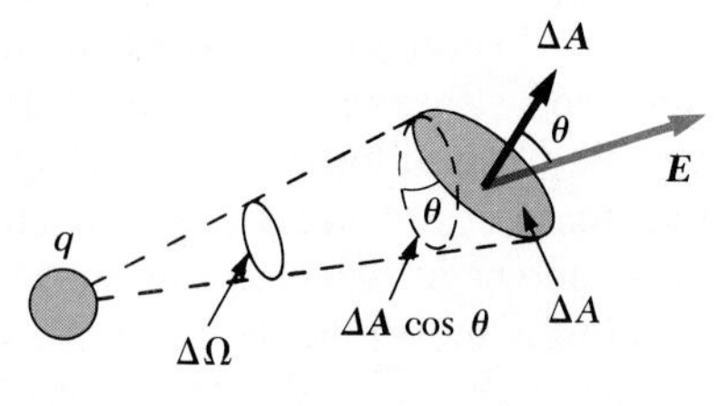


**Figure 24.22** The area element $\Delta A$ subtends a solid angle $\Delta\Omega = (\Delta A \cos\theta)/r^2$ at the charge $q$.

total solid angle at a point is $4\pi$ steradians, we see that the total flux through the closed surface is

$$\Phi_c = kq \oint \frac{dA \cos\theta}{r^2} = kq \oint d\Omega = 4\pi kq = \frac{q}{\epsilon_0}$$

Thus we have derived Gauss' law, Equation 24.6. Note that this result is independent of the shape of the closed surface and independent of the position of the charge within the surface.

## SUMMARY

**Electric flux** is a measure of the number of electric field lines that penetrate a surface. If the electric field is uniform and makes an angle $\theta$ with the normal to the surface, the electric flux through the surface is

Flux through a surface in a uniform electric field

$$\Phi = EA\cos\theta \tag{24.2}$$

In general, the electric flux through a surface is defined by the expression

Definition of electric flux

$$\Phi = \int_{\text{surface}} \mathbf{E} \cdot d\mathbf{A} \tag{24.3}$$

**Gauss' law** says that the net electric flux, $\Phi_c$, through any closed gaussian surface is equal to the *net* charge *inside* the surface divided by $\epsilon_0$:

Gauss' law

$$\Phi_c = \oint \mathbf{E} \cdot d\mathbf{A} = \frac{q_{\text{in}}}{\epsilon_0} \tag{24.6}$$

Using Gauss' law, one can calculate the electric field due to various symmetric charge distributions. Table 24.1 lists some typical results.

**TABLE 24.1 Typical Electric Field Calculations Using Gauss' Law**

| Charge Distribution | Electric Field | Location |
|---|---|---|
| Insulating sphere of radius $R$, uniform charge density, and total charge $Q$ | $k\frac{Q}{r^2}$ | $r > R$ |
| | $k\frac{Q}{R^3}r$ | $r < R$ |
| Thin spherical shell of radius $R$ and total charge $Q$ | $k\frac{Q}{r^2}$ | $r > R$ |
| | $0$ | $r < R$ |
| Line charge of infinite length and charge per unit length $\lambda$ | $2k\frac{\lambda}{r}$ | Outside the line charge |
| Nonconducting, infinite charged plane with charge per unit area $\sigma$ | $\frac{\sigma}{2\epsilon_0}$ | Everywhere outside the plane |
| Conductor of surface charge per unit area $\sigma$ | $\frac{\sigma}{\epsilon_0}$ | Just outside the conductor |
| | $0$ | Inside the conductor |

Properties of a conductor in electrostatic equilibrium

**A conductor in electrostatic equilibrium** has the following properties:

1. The electric field is zero everywhere inside the conductor.
2. Any excess charge on an isolated conductor must reside entirely on its surface.
3. The electric field just outside the conductor is perpendicular to its surface and has a magnitude $\sigma/\epsilon_0$, where $\sigma$ is the charge per unit area at that point.
4. On an irregularly shaped conductor, charge tends to accumulate where the radius of curvature of the surface is the smallest, that is, at sharp points.

### PROBLEM-SOLVING STRATEGY AND HINTS

Gauss' law may seem mysterious to you, and it is usually one of the most difficult concepts to understand in introductory physics. However, as we have seen, Gauss' law is very powerful in solving problems having a high degree of symmetry. In this chapter, you will only encounter problems with three kinds of symmetry: plane symmetry, cylindrical symmetry, and spherical symmetry. It is important to review Examples 2 through 7 and to use the following procedure:

1. First, select a gaussian surface which *has the same symmetry as the charge distribution.* For point charges or spherically symmetric charge distributions, the gaussian surface should be a sphere centered on the charge as in Examples 2, 3, 4 and 7. For uniform line charges or uniformly charged cylinders, your choice of a gaussian surface should be a cylindrical surface that is coaxial with the line charge or cylinder as in Example 5. For sheets of charge having plane symmetry, the gaussian surface should be a "pillbox" that straddles the sheet as in Example 6. Note that in all cases, the gaussian surface is selected such that the electric field has the same magnitude everywhere on the surface, and is directed perpendicular to the surface. This enables you to easily evaluate the surface integral that appears on the left side of Gauss' law, which represents the total electric flux through that surface.
2. Now evaluate the right side of Gauss' law, which amounts to calculating the total electric charge, $q_{in}$, *inside* the gaussian surface. If the charge density is uniform as is usually the case (that is, if $\lambda$, $\sigma$, or $\rho$ is constant), simply multiply that charge density by the length, area, or volume enclosed by the gaussian surface. However, if the charge distribution is *nonuniform,* you must integrate the charge density over the region enclosed by the gaussian surface. For example, if the charge is distributed along a line, you would integrate the expression $dq = \lambda\, dx$, where $dq$ is the charge on an infinitesimal element $dx$ and $\lambda$ is the charge per unit length. For a plane of charge, you would integrate $dq = \sigma\, dA$, where $\sigma$ is the charge per unit area and $dA$ is an infinitesimal element of area. Finally, for a volume of charge you would integrate $dq = \rho\, dV$, where $\rho$ is the charge per unit volume and $dV$ is an infinitesimal element of volume.
3. Once the left and right sides of Gauss's law have been evaluated, you can proceed to calculate the electric field on the gaussian surface assuming the charge distribution is given in the problem. Conversely, if the electric field is known, you can calculate the charge distribution that produces the field.

## QUESTIONS

1. If the net flux through a gaussian surface is zero, which of the following statements are true? (a) There are no charges inside the surface. (b) The net charge inside the surface is zero. (c) The electric field is zero everywhere on the surface. (d) The number of electric field lines entering the surface equals the number leaving the surface.
2. If the electric field in a region of space is zero, can you conclude there are no electric charges in that region? Explain.
3. A spherical gaussian surface surrounds a point charge $q$. Describe what happens to the flux through the surface if (a) the charge is tripled, (b) the volume of the sphere is doubled, (c) the shape of the surface is changed to that of a cube, and (d) the charge is moved to another position inside the surface.
4. If there are more electric field lines leaving a gaussian surface than there are entering the surface, what can you conclude about the *net* charge enclosed by that surface?
5. A uniform electric field exists in a region of space in which there are no charges. What can you conclude about the *net* electric flux through a gaussian surface placed in this region of space?
6. Explain why Gauss' law cannot be used to calculate the electric field of (a) an electric dipole, (b) a charged disk, (c) a charged ring, and (d) three point charges at the corners of a triangle.
7. If the total charge inside a closed surface is known but the distribution of the charge is unspecified, can you use Gauss' law to find the electric field? Explain.
8. Explain why the electric flux through a closed surface with a given enclosed charge is independent of the size or shape of the surface.
9. Consider the electric field due to a nonconducting infinite plane with a uniform charge density. Explain why the electric field does not depend on the distance from the plane in terms of the spacing of the electric field lines.
10. Use Gauss' law to explain why electric field lines must begin and end on electric charges. (*Hint:* Change the size of the gaussian surface.)
11. A point charge is placed at the center of an uncharged metallic spherical shell insulated from ground. As the point charge is moved off center, describe what happens to (a) the total induced charge on the shell and (b) the distribution of charge on the interior and exterior surfaces of the shell.
12. Explain why excess charge on a isolated conductor must reside on its surface, using the repulsive nature of the force between like charges and the freedom of motion of charge within the conductor.
13. A person is placed in a large hollow metallic sphere that is insulated from ground. If a large charge is placed on the sphere, will the person be harmed upon touching the inside of the sphere? Explain what will happen if the person also has an initial charge whose sign is opposite to that of the charge on the sphere.
14. How would the observations described in Figure 24.20 differ if the hollow conductor were grounded? How would they differ if the small charged ball were an insulator rather than a conductor?
15. What other experiment might be performed on the ball in Figure 24.20 to show that its charge was transferred to the hollow conductor?
16. What would happen to the electrometer reading if the charged ball in Figure 24.20 touched the inner wall of the conductor? the outer wall?
17. Two solid spheres, both of radius $R$, carry identical total charges, $Q$. One sphere is a good conductor while the other is an insulator. If the charge on the insulating sphere is uniformly distributed throughout its interior volume, how do the electric fields outside these two spheres compare? Are the fields identical inside the two spheres?

## PROBLEMS

### Section 24.1 Electric Flux

1. A flat surface having an area of 3.2 $m^2$ is rotated in a uniform electric field of intensity $E = 6.2 \times 10^5$ N/C. Calculate the electric flux through this area when the electric field is (a) perpendicular to the surface, (b) parallel to the surface, and (c) makes an angle of 75° with the plane of the surface.
2. An electric field of intensity $3.5 \times 10^3$ N/C is applied along the $x$ axis. Calculate the electric flux through a rectangular plane 0.35 m wide and 0.70 m long if (a) the plane is parallel to the $yz$ plane, (b) the plane is parallel to the $xy$ plane, and (c) the plane contains the $y$ axis and its normal makes an angle of 40° with the $x$ axis.
3. A uniform electric field $a\mathbf{i} + b\mathbf{j}$ intersects a surface of area $A$. What is the flux through this area if the surface lies (a) in the $yz$ plane? (b) in the $xz$ plane? (c) in the $xy$ plane?
4. Consider a closed triangular box resting within a horizontal electric field $E = 7.8 \times 10^4$ N/C as shown in Figure 24.23. Calculate the electric flux through

(a) the left-hand vertical surface ($A'$), (b) the slanted surface ($A$), and (c) the entire surface of the box.

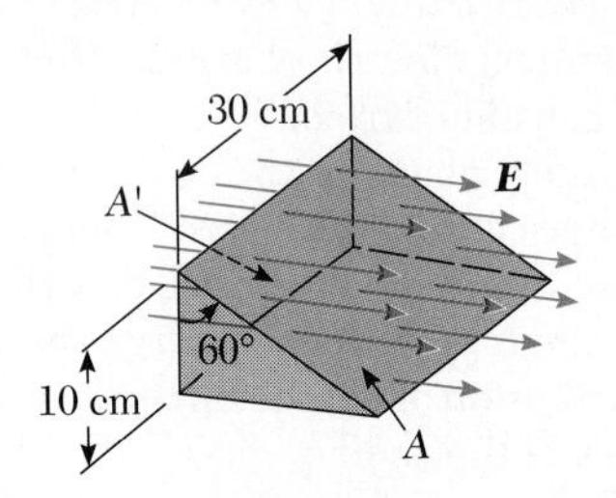


**Figure 24.23** (Problem 4).

5. A 40-cm diameter loop is rotated in a uniform electric field until the position of maximum electric flux is found. The flux in this position is measured to be $5.2 \times 10^5$ Nm²/C. What is the electric field strength?
6. A nonuniform electric field is given by the expression $\mathbf{E} = ay\mathbf{i} + bz\mathbf{j} + cx\mathbf{k}$, where $a$, $b$, and $c$ are constants. Determine the electric flux through a rectangular surface in the $xy$ plane, extending from $x = 0$ to $x = w$ and from $y = 0$ to $y = h$.
7. An electric field is given by $\mathbf{E} = az\mathbf{i} + bx\mathbf{k}$, where $a$ and $b$ are constants. Determine the electric flux through the triangular surface shown in Figure 24.24.

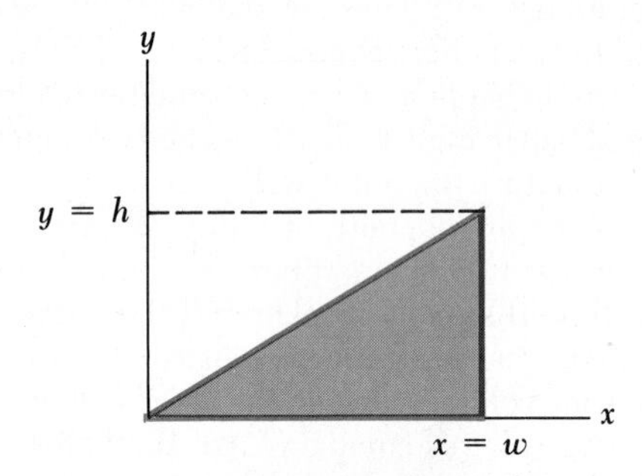


**Figure 24.24** (Problem 7).

8. A cone with a circular base of radius $R$ stands upright so that its axis is vertical. A uniform electric field $\mathbf{E}$ is applied in the vertical direction. Show that the flux through the cone's surface (not counting its base) is given by $\pi R^2 E$.
9. A pyramid with a 6-m square base and height of 4 m is placed in a vertical electric field of 52 N/C. Calculate the total electric flux through the pyramid's four slanted surfaces.

## Section 24.2 Gauss' Law

10. Four closed surfaces, $S_1$ through $S_4$, together with the charges $-2Q$, $+Q$, and $-Q$ are sketched in Figure 24.25. Find the electric flux through each surface.

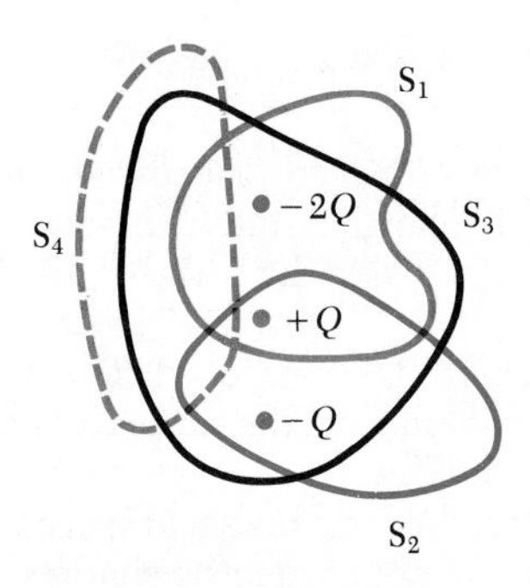


**Figure 24.25** (Problem 10).

11. A point charge of $+5\ \mu$C is located at the center of a sphere with a radius of 12 cm. What is the electric flux through the surface of this sphere?
12. The electric field in the earth's atmosphere is $E = 100$ N/C, pointing downward. Determine the electric charge on the earth.
13. A point charge of 12 $\mu$C is placed at the center of a *spherical* shell of radius 22 cm. What is the total electric flux through (a) the entire surface of the shell and (b) any hemispherical surface of the shell? (c) Do the results depend on the radius? Explain.
14. A charge of 12 $\mu$C is at the geometric center of a cube. What is the electric flux through one of the cube's faces?
15. The following charges are located inside a submarine: $+5\ \mu$C, $-9\ \mu$C, $+27\ \mu$C, and $-84\ \mu$C. Calculate the net electric flux through the submarine. Compare the number of electric field lines leaving the submarine with the number entering it.
16. A point charge of 0.0462 $\mu$C is inside a pyramid. Determine the total electric flux through the surface of the pyramid.
17. Five charges are placed in a closed box. Each charge (except the first) has a magnitude which is twice that of the previous one placed in the box. If all charges have the same sign and if (after all charges have been placed in the box) the net electric flux through the box is $4.8 \times 10^7$ Nm²/C, what is the magnitude of the smallest charge in the box? Does the answer depend on the size of the box?
18. The electric field everywhere on the surface of a hollow sphere of radius 0.75 m is measured to be equal to $8.90 \times 10^2$ N/C and points radially toward the center of the sphere. (a) What is the net charge within the sphere's surface? (b) What can you conclude about the nature and distribution of the charge inside the sphere?
19. A 10-$\mu$C charge located at the origin of a cartesian coordinate system is surrounded by a nonconducting hollow sphere of radius 10 cm. A drill with a radius of 1 mm is aligned along the $z$ axis, and a hole is drilled in

the sphere. Calculate the electric flux through the hole.

**20.** A charge of 170 $\mu$C is at the center of a cube of sides 80 cm. (a) Find the total flux through each face of the cube. (b) Find the flux through the whole surface of the cube. (c) Would your answers to (a) or (b) change if the charge were not at the center? Explain.

**21.** The total electric flux through a closed surface in the shape of a cylinder is $8.60 \times 10^4$ N·m²/C. (a) What is the net charge within the cylinder? (b) From the information given, what can you say about the charge within the cylinder? (c) How would your answers to (a) and (b) change if the net flux were $-8.60 \times 10^4$ N·m²/C?

**22.** A cube of sides 10 cm is centered at the origin. A point charge of 2 $\mu$C is located on the $y$ axis at $y = 20$ cm. (a) Sketch the electric lines for the point charge. (b) What is the net flux through the surface of the cube? (c) Repeat (a) and (b) if a second point charge of 4 $\mu$C is located at the center of the cube. (Neglect the lines that go through the edges and corners.)

**23.** A point charge $Q$ is located just above the center of the flat face of a hemisphere of radius $R$ as shown in Figure 24.26. (a) What is the electric flux through the curved surface of this hemisphere? (b) What is the electric flux through the flat face of this hemisphere?

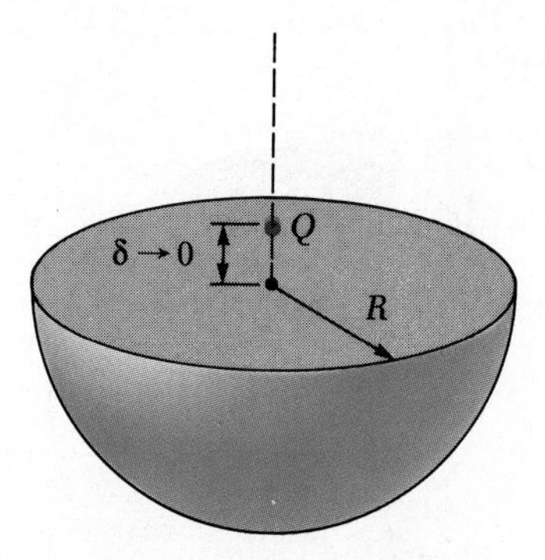


**Figure 24.26** (Problem 23).

## Section 24.3 Application of Gauss' Law to Charged Insulators

**24.** On a clear, sunny day, there is a vertical electrical field of about 130 N/C pointing down over flat ground or water. (The field can vary considerably in magnitude and may be reversed if clouds are overhead.) What is the surface charge density on the ground for these conditions?

**25.** Consider a thin spherical shell of radius 14 cm with a total charge of 32 $\mu$C distributed uniformly on its surface. Find the electric field for the following distances from the center of the charge distribution: (a) $r =$ 10 cm and (b) $r = 20$ cm.

**26.** An inflated balloon in the shape of a sphere of radius 12 cm has a total charge of 7 $\mu$C uniformly distributed on its surface. Calculate the electric field intensity at the following distances from the center of the balloon: (a) 10 cm, (b) 12.5 cm, (c) 30 cm.

**27.** An insulating sphere is 8 cm in diameter, and carries a $+5.7$ $\mu$C charge uniformly distributed throughout its interior volume. Calculate the charge enclosed by a concentric spherical surface with the following radii: (a) $r = 2$ cm and (b) $r = 6$ cm.

**28.** An insulating sphere of radius 10 mm has a uniform charge density $6 \times 10^{-3}$ C/m³. Calculate the electric flux through a concentric spherical surface with the following radii: (a) $r = 5$ mm, (b) $r = 10$ mm, and (c) $r = 25$ mm.

**29.** A solid sphere of radius 40 cm has a total positive charge of 26 $\mu$C uniformly distributed throughout its volume. Calculate the electric field intensity at the following distances from the center of the sphere: (a) 0 cm, (b) 10 cm, (c) 40 cm, (d) 60 cm.

**30.** An insulating sphere of 10 cm radius has a uniform charge density throughout its volume. If the magnitude of the electric field at a distance of 5 cm from the center is $8.6 \times 10^4$ N/C, what is the magnitude of the electric field at 15 cm from the center?

**31.** A spherically symmetric charge distribution has a charge density given by $\rho = a/r$ where $a$ is constant. Find the electric field as a function of $r$. [See the note in Problem 54.]

**32.** The charge per unit length on a *long*, straight filament is $-90$ $\mu$C/m. Find the electric field at the following distances from the filament: (a) 10 cm, (b) 20 cm, (c) 100 cm.

**33.** A uniformly charged, straight filament 7 m in length has a total positive charge of 2 $\mu$C. An uncharged cardboard cylinder 2 cm in length and 10 cm in radius surrounds the filament at its center, with the filament as the axis of the cylinder. Using any reasonable approximations, find (a) the electric field at the surface of the cylinder and (b) the total electric flux through the cylinder.

**34.** A cylindrical shell of radius 7 cm and length 240 cm has its charge uniformly distributed on its surface. The electric field intensity at a point 19 cm radially outward from its axis (measured from the midpoint of the shell) is $3.6 \times 10^4$ N/C. Use approximate relations to find (a) the net charge on the shell and (b) the electric field at a point 4 cm from the axis, measured from the midpoint.

**35.** Consider a long cylindrical charge distribution of radius $R$ with a uniform charge density $\rho$. Find the electric field at distance $r$ from the axis where $r < R$.

**36.** A nonconducting wall carries a uniform charge density of 8.6 $\mu C/cm^2$. What is the electric field at a distance of 7 cm from the wall? Does your result change as the distance from the wall is varied?

**37.** A large plane sheet of charge has a charge per unit area of 9.0 $\mu C/m^2$. Find the electric field intensity *just above the surface* of the sheet, measured from the sheet's midpoint.

## Section 24.4 Conductors in Electrostatic Equilibrium

**38.** A conducting spherical shell of radius 15 cm carries a net charge of $-6.4\ \mu C$ uniformly distributed on its surface. Find the electric field at points (a) just outside the shell and (b) inside the shell.

**39.** A long, straight metal rod has a radius of 5 cm and a charge per unit length of 30 nC/m. Find the electric field at the following distances from the axis of the rod: (a) 3 cm, (b) 10 cm, (c) 100 cm.

**40.** A square plate of copper of sides 50 cm is placed in an extended electric field of $8 \times 10^4$ N/C directed *perpendicular* to the plate. Find (a) the charge density of each face of the plate and (b) the total charge on each face.

**41.** A thin conducting plate 50 cm on a side lies in the $xy$ plane. If a total charge of $4 \times 10^{-8}$ C is placed on the plate, find (a) the charge density on the plate, (b) the electric field just above the plate, and (c) the electric field just below the plate.

**42.** A very large, thin, flat plate of aluminum of area $A$ has a total charge $Q$ uniformly distributed over its surfaces. If the same charge is spread uniformly over the *upper* surface of an otherwise identical glass plate, compare the electric fields just above the center of the upper surface of each plate.

**43.** A solid copper sphere 15 cm in radius has a total charge of 40 nC. Find the electric field at the following distances measured from the center of the sphere: (a) 12 cm, (b) 17 cm, (c) 75 cm. (d) How would your answers change if the sphere were hollow?

**44.** A hollow conducting sphere is surrounded by a larger concentric, spherical, conducting shell. The inner sphere has a net negative charge of $-Q$, and the outer sphere has a net positive charge of $+3Q$. The charges are in electrostatic equilibrium. Using Gauss's law, find the charges and the electric fields everywhere.

**45.** A solid conducting sphere of radius 2 cm has a positive charge of $+8\ \mu C$. A conducting spherical shell of inner radius 4 cm and outer radius 5 cm is concentric with the solid sphere and has a net charge of $-4\ \mu C$. Find the electric field at the following distances from the center of this charge configuration: (a) $r = 1$ cm, (b) $r = 3$ cm, (c) $r = 4.5$ cm, and (d) $r = 7$ cm.

**46.** Consider the data given in Problem 45. Calculate the net charge enclosed by a concentric spherical gaussian surface with the following radii: (a) $r = 1$ cm, (b) 3 cm, (c) $r = 4.5$ cm, and (d) $r = 7$ cm.

**47.** A *long*, straight wire is surrounded by a hollow metallic cylinder whose axis coincides with that of the wire. The solid wire has a charge per unit length of $+\lambda$, and the hollow cylinder has a *net* charge per unit length of $+2\lambda$. From this information, use Gauss' law to find (a) the charge per unit length on the inner and outer surfaces of the hollow cylinder and (b) the electric field outside the hollow cylinder, a distance $r$ from the axis.

**48.** The electric field on the surface of an irregularly shaped conductor varies from $5.6 \times 10^4$ N/C to $2.8 \times 10^4$ N/C. Calculate the local surface charge density at the point on the surface where the radius of curvature of the surface is (a) greatest and (b) smallest.

## Section 24.6 Derivation of Gauss' Law

**49.** A sphere of radius $R$ surrounds a point charge $Q$, located at its center. (a) Show that the electric flux through a circular cap of half-angle $\theta$ (see Figure 24.27) is given by

$$\Phi = \frac{Q}{2\epsilon_0}(1 - \cos\theta)$$

(b) What is the flux for $\theta = 90°$? (c) What is the flux for $\theta = 180°$?

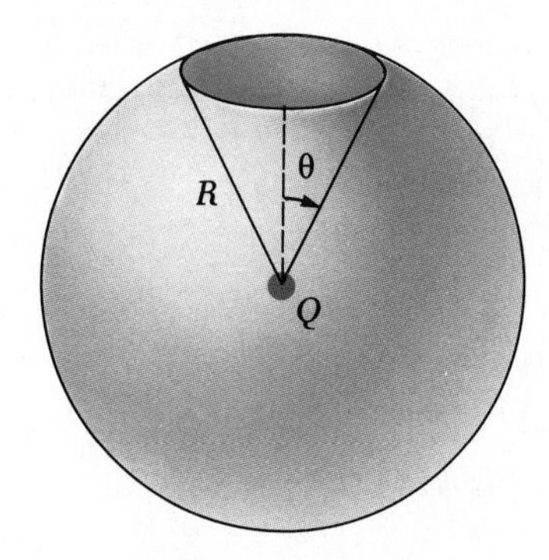


**Figure 24.27** (Problem 49).

## ADDITIONAL PROBLEMS

**50.** For the configuration shown in Figure 24.28, suppose that $a = 5$ cm, $b = 20$ cm, and $c = 25$ cm. Furthermore, suppose that the electric field at a point 10 cm from the center is measured to be $3.6 \times 10^3$ N/C radially *inward* while the electric field at a point 50 cm from the center is $2.0 \times 10^2$ N/C radially *outward*. From this information, find (a) the charge on the insulating sphere, (b) the net charge on the hollow conducting sphere, and (c) the total charge on the inner and outer *surfaces*, respectively, of the hollow conducting sphere.

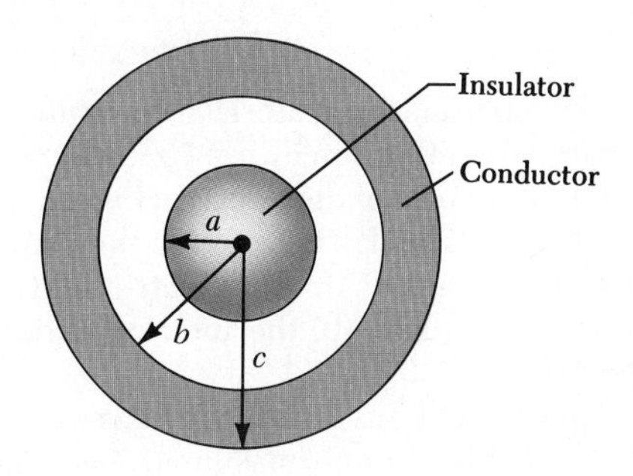


**Figure 24.28** (Problems 50 and 51).

51. A solid *insulating* sphere of radius $a$ has a uniform charge density $\rho$ and a total charge $Q$. Concentric with this sphere is an *uncharged, conducting* hollow sphere whose inner and outer radii are $b$ and $c$, as in Figure 24.28. (a) Find the electric field intensity in the regions $r < a$, $a < r < b$, $b < r < c$, and $r > c$. (b) Determine the induced charge per unit area on the inner and outer surfaces of the hollow sphere.
52. Consider an insulating sphere of radius $R$ and having a *uniform* volume charge density $\rho$. Plot the magnitude of the electric field, $E$, as a function of the distance from the center of the sphere, $r$. Let $r$ range over the interval $0 < r < 3R$ and plot $E$ in units of $\rho R/\epsilon_0$.
53. An early (incorrect) model of the hydrogen atom, suggested by J. J. Thomson, proposed that a positive cloud of charge $+e$ was uniformly distributed throughout the volume of a sphere of radius $R$, with the electron an equal-magnitude negative point charge $-e$ at the center. (a) Using Gauss' law, show that the electron would be in equilibrium at the center and, if displaced from the center a distance $r < R$, would experience a restoring force of the form $F = -Kr$ where $K$ is a constant. (b) Show that the force constant $K = ke^2/R^3$. (c) Find an expression for the frequency $f$ of simple harmonic oscillations that an electron would undergo if displaced a short distance ($<R$) from the center and released. (d) Calculate a numerical value for $R$ that would result in a frequency of $2.47 \times 10^{15}$ Hz, the most intense line in the hydrogen spectrum.
54. Consider a solid insulating sphere of radius $b$ with nonuniform charge density $\rho = Cr$. Find the charge contained within the radius when (a) $r < b$ and (b) $r > b$. (*Note:* The volume element $dV$ for a spherical shell of radius $r$ and thickness $dr$ is equal to $4\pi r^2\,dr$.)
55. A solid insulating sphere of radius $R$ has a *nonuniform* charge density that varies with $r$ according to the expression $\rho = Ar^2$, where $A$ is a constant and $r < R$ is measured from the center of the sphere. (a) Show that the electric field *outside* ($r > R$) the sphere is given by the expression $E = AR^5/5\epsilon_0 r^2$. (b) Show that the electric field *inside* ($r < R$) the sphere is given by $E = Ar^3/5\epsilon_0$. (*Hint:* Note that the total charge $Q$ on the sphere is equal to the integral of $\rho\,dV$, where $r$ extends from 0 to $R$; also note that the charge $q$ within a radius $r < R$ is *less* than $Q$. To evaluate the integrals, note that the volume element $dV$ for a spherical shell of radius $r$ and thickness $dr$ is equal to $4\pi r^2\,dr$.)
56. An infinitely long insulating cylinder of radius $R$ has a volume charge density that varies with the radius as

$$\rho = \rho_0\left(a - \frac{r}{b}\right),$$

where $\rho_0$, $a$, and $b$ are positive constants and $r$ is the distance from the axis of the cylinder. Use Gauss' law to determine the magnitude of the electric field at radial distances (a) $r < R$ and (b) $r > R$.
57. The flux of *any* vector field **V** through a surface can be defined as

$$\Phi_{\mathbf{V}} = \int_{\mathbf{S}} \mathbf{V}\cdot d\mathbf{A}$$

Radial fields whose magnitudes vary inversely as the square of the distance, such as the gravitational field

$$\mathbf{g} = \frac{\mathbf{F_g}}{m}$$

also obey Gauss' Law. Using Gauss' (gravitational) law, find the gravitational field **g** at a point distance $r$ from the center of the earth where $r < R_e$. Pretend that the earth's density is uniform.
58. Two infinite, nonconducting sheets of charge are parallel to each other as in Figure 24.29. The sheet on the left has a uniform surface charge density $\sigma$, and the one on the right has a uniform charge density $-\sigma$. Calculate the value of the electric field at points (a) to

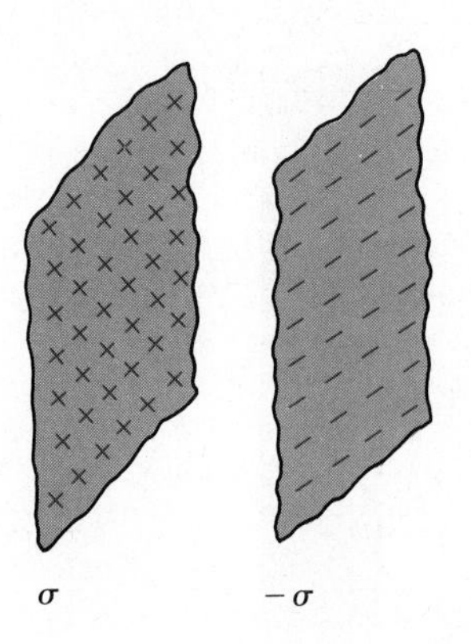


**Figure 24.29** (Problems 58 and 59).

the left of, (b) in between, and (c) to the right of the two sheets. (*Hint:* See Example 24.6.)

59. Repeat the calculations for Problem 58 when both sheets have *positive* uniform charge densities of value $\sigma$.

60. A closed surface with dimensions $a = b = 0.4$ m and $c = 0.6$ m is located as shown in Figure 24.30. The electric field throughout the region is *nonuniform* and given by

$$\mathbf{E} = (3 + 2x^2)\mathbf{i} \text{ N/C}$$

where $x$ is in meters. Calculate the net electric flux leaving the closed surface. What net charge is enclosed by the surface?

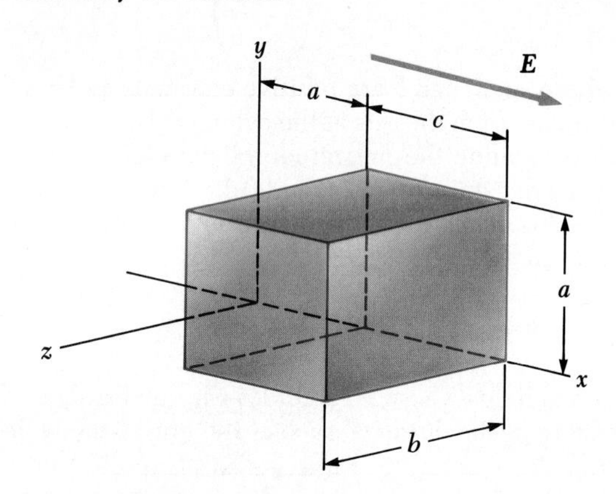


**Figure 24.30** (Problem 60).

61. A slab of insulating material (infinite in two of its three dimensions) has a uniform positive charge density $\rho$ as in the edge view of Figure 24.31. (a) Show that the electric field a distance $x$ from its center and inside the slab is $E = \rho x/\epsilon_0$. (b) Suppose an electron of charge $-e$ and mass $m$ is placed inside the slab. If it is released from rest at a distance $x$ from the center, show that the electron exhibits simple harmonic motion with a frequency given by

$$f = \frac{1}{2\pi}\sqrt{\frac{\rho e}{m\epsilon_0}}$$

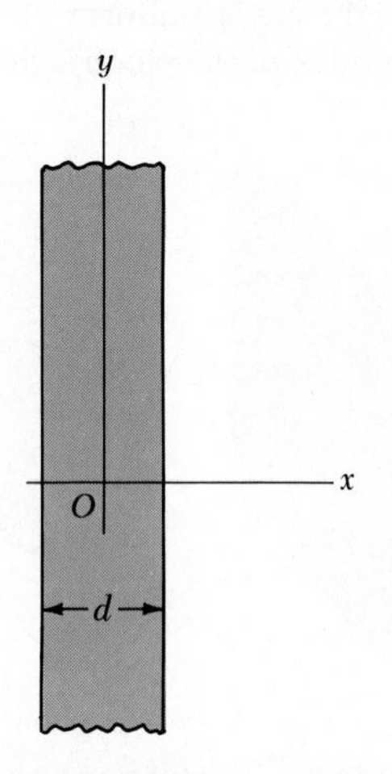


**Figure 24.31** (Problems 61 and 62).

62. A slab of insulating material has a *nonuniform* positive charge density given by $\rho = Cx^2$, where $x$ is measured from the center of the slab as in Figure 24.31, and $C$ is a constant. The slab is infinite in the $y$ and $z$ directions. Derive expressions for the electric field in (a) the exterior regions and (b) the interior region of the slab $(-d/2 < x < d/2)$.

63. A sphere of radius $2a$ is made of a nonconducting material that has a uniform volume charge density $\rho$. (Assume that the material does not affect the electric field.) A spherical cavity of radius $a$ is now removed from the sphere as shown in Figure 24.32. Show that the electric field within the cavity is uniform and is given by $E_x = 0$ and $E_y = \rho a/3\epsilon_0$. (*Hint:* the field within the cavity is the superposition of the field due to the original uncut sphere, plus the field due to a sphere the size of the cavity with a uniform negative charge density $-\rho$.)

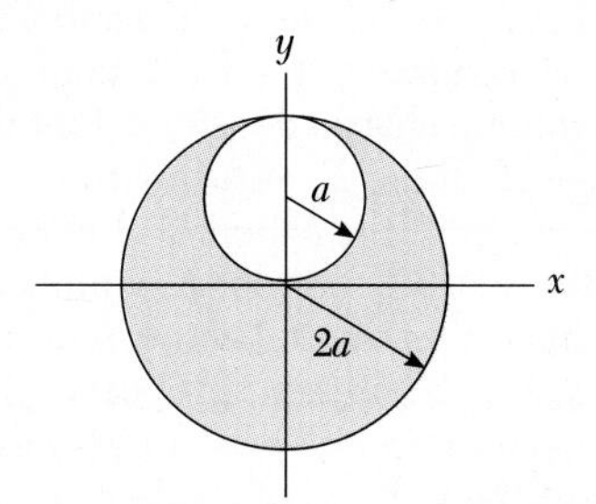


**Figure 24.32** (Problem 63).

64. A point charge $Q$ is located on the axis of a disk of radius $R$ at a distance $b$ from the plane of the disk (Fig. 24.33). Show that if one-fourth of the electric flux from the charge threads the disk, then $R = \sqrt{3}b$.

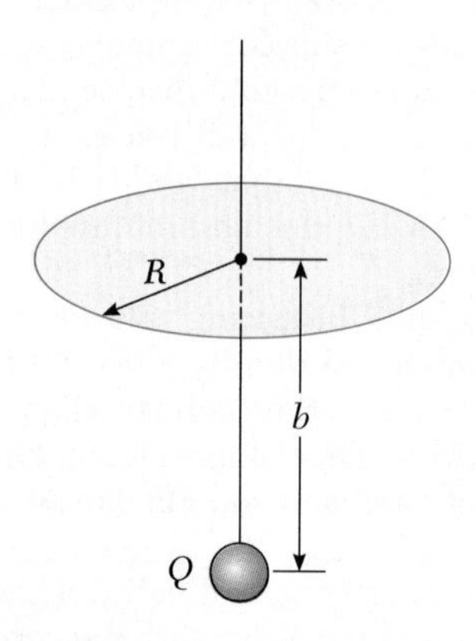


**Figure 24.33** (Problem 64).

# 25
# Electric Potential

*Jennifer is holding onto a charged sphere that reaches a potential of about 100,000 volts. The device that generates this high potential is called a Van de Graaff generator. Why do you suppose Jennifer's hair stands on end like the needles of a porcupine? Why is it important that she stand on a pedestal insulated from ground? (Courtesy of Henry Leap and Jim Lehman)*

The concept of potential energy was first introduced in Chapter 8 in connection with such conservative forces as the force of gravity and the elastic force of a spring. By using the law of energy conservation, we were often able to avoid working directly with forces when solving various mechanical problems. In this chapter we shall see that the energy concept is also of great value in the study of electricity. Since the electrostatic force given by Coulomb's law is conservative, one can conveniently describe electrostatic phenomena in terms of an electrical potential energy. This idea enables us to define a scalar quantity called *electric potential*. Because the potential is a scalar function of position, it offers a simpler way of describing electrostatic phenomena than does the electric field. In later chapters we shall see that the concept of the electric potential is of great practical value. In fact, the measured voltage between any two points in an electrical circuit is simply the difference in electric potential between the points.

## 25.1 POTENTIAL DIFFERENCE AND ELECTRIC POTENTIAL

In Chapter 14, we showed that the gravitational force is conservative. Since the electrostatic force, given by Coulomb's law, is of the same form as the universal law of gravity, it follows that the electrostatic force is also conservative. Therefore, it is possible to define a potential energy function associated with this force.

When a test charge $q_o$ is placed in an electrostatic field $\mathbf{E}$, the electric force on the test charge is $q_o\mathbf{E}$. The force $q_o\mathbf{E}$ is the vector sum of the individual forces exerted on $q_o$ by the various charges producing the field $\mathbf{E}$. It follows that the force $q_o\mathbf{E}$ is conservative, since the individual forces governed by Coulomb's law are conservative. The work done by the force $q_o\mathbf{E}$ is equal to the negative of the work done by an external agent. Furthermore, the work done by the electric force $q_o\mathbf{E}$ on the test charge for an infinitesimal displacement $d\mathbf{s}$ is given by

$$dW = \mathbf{F}\cdot d\mathbf{s} = q_o\mathbf{E}\cdot d\mathbf{s} \tag{25.1}$$

By definition, the work done by a conservative force equals the negative of the change in potential energy, $dU$; therefore, we see that

$$dU = -q_o\mathbf{E}\cdot d\mathbf{s} \tag{25.2}$$

For a finite displacement of the test charge between points $A$ and $B$, the **change in the potential energy** is given by

Change in potential energy

$$\Delta U = U_B - U_A = -q_o \int_A^B \mathbf{E}\cdot d\mathbf{s} \tag{25.3}$$

The integral in Equation 25.3 is performed along the path by which $q_o$ moves from $A$ to $B$ and is called a *path integral,* or *line integral.* Since the force $q_o\mathbf{E}$ is conservative, *this integral does not depend on the path taken between A and B.*

The **potential difference,** $V_B - V_A$, between the points $A$ and $B$ is defined as the change in potential energy divided by the test charge $q_o$:

Potential difference

$$V_B - V_A = \frac{U_B - U_A}{q_o} = -\int_A^B \mathbf{E}\cdot d\mathbf{s} \tag{25.4}$$

*Potential difference should not be confused with potential energy.* The potential difference is *proportional* to the potential energy, and we see from Equation 25.4 that the two are related by $\Delta U = q_o\,\Delta V$. Because potential energy is a scalar, electric potential is also a scalar quantity. Note that the change in the potential energy of the charge is the negative of the work done by the electric force. Hence, we see that

the potential difference $V_B - V_A$ equals the work per unit charge that an external agent must perform to move a test charge from $A$ to $B$ without a change in kinetic energy.

Equation 25.4 defines potential differences only. That is, only *differences* in $V$ are meaningful. The electric potential function is often taken to be zero at some convenient point. We shall usually choose the potential to be zero for a point at infinity (that is, a point infinitely remote from the charges producing the electric field). With this choice, we can say that the *electric potential at an arbitrary point equals the work required per unit charge to bring a positive test charge from infinity to that point.* Thus, if we take $V_A = 0$ at infinity in Equation 25.4, then the potential at any point $P$ is given by

$$V_P = -\int_{\infty}^{P} \mathbf{E} \cdot d\mathbf{s} \qquad (25.5)$$

In reality, $V_P$ represents the potential difference between the point $P$ and a point at infinity. (Equation 25.5 is a special case of Eq. 25.4.)

Since potential difference is a measure of energy per unit charge, the SI unit of potential is joules per coulomb, defined to be equal to a unit called the **volt** (V):

Definition of a volt

$$1 \text{ V} \equiv 1 \text{ J/C}$$

That is, 1 J of work must be done to take a 1-C charge through a potential difference of 1 V. Equation 25.4 shows that the potential difference also has units of electric field times distance. From this, it follows that the SI unit of electric field (N/C) can also be expressed as volts per meter:

$$1 \text{ N/C} = 1 \text{ V/m}$$

A unit of energy commonly used in atomic and nuclear physics is the **electron volt,** which is defined as *the energy that an electron (or proton) gains when moving through a potential difference of magnitude 1 V.* Since 1 V = 1 J/C and since the fundamental charge is equal to $1.6 \times 10^{-19}$ C, we see that the electron volt (eV) is related to the joule through the relation

The electron volt

$$1 \text{ eV} = 1.6 \times 10^{-19} \text{ C} \cdot \text{V} = 1.6 \times 10^{-19} \text{ J} \qquad (25.6)$$

For instance, an electron in the beam of a typical TV picture tube (or cathode ray tube) has a speed of $5 \times 10^7$ m/s. This corresponds to a kinetic energy of $1.1 \times 10^{-15}$ J, which is equivalent to $7.1 \times 10^3$ eV. Such an electron has to be accelerated from rest through a potential difference of 7.1 kV to reach this speed.

## 25.2 POTENTIAL DIFFERENCES IN A UNIFORM ELECTRIC FIELD

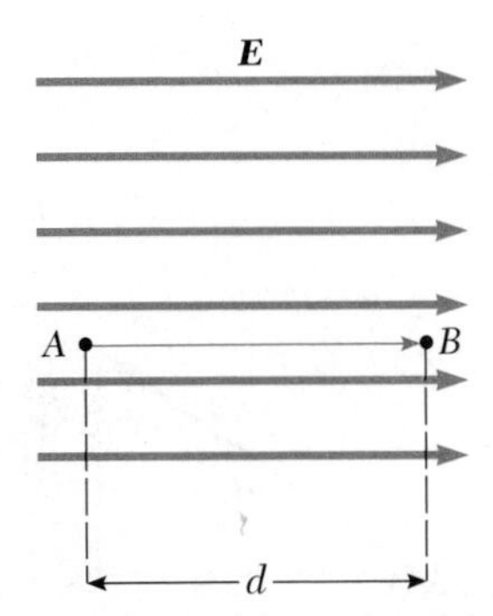

**Figure 25.1** The displacement of a charged particle from $A$ to $B$ in the presence of a uniform electric field $\mathbf{E}$.

In this section, we shall describe the potential difference between any two points in a *uniform* electric field. The potential difference is independent of the path between these two points; that is, the work done in taking a test charge from point $A$ to point $B$ is the same along all paths. This confirms that a static, uniform electric field is conservative. By definition, a force is conservative if it has this property (see Section 8.1).

First, consider a uniform electric field directed along the $x$ axis, as in Figure 25.1. Let us calculate the potential difference between two points,

$A$ and $B$, separated by a distance $d$, where $d$ is measured parallel to the field lines. If we apply Equation 25.4 to this situation, we get

$$V_B - V_A = \Delta V = -\int_A^B \mathbf{E}\cdot d\mathbf{s} = -\int_A^B E\cos 0°\ ds = -\int_A^B E\ ds$$

Since $E$ is constant, it can be removed from the integral sign, giving

Potential difference in a uniform $E$ field

$$\Delta V = -E\int_A^B ds = -Ed \tag{25.7}$$

The minus sign results from the fact that point $B$ is at a lower potential than point $A$; that is, $V_B < V_A$.

Now suppose that a test charge $q_o$ moves from $A$ to $B$. The change in its potential energy can be found from Equations 25.4 and 25.7:

$$\Delta U = q_o\,\Delta V = -q_oEd \tag{25.8}$$

From this result, we see that if $q_o$ is positive, $\Delta U$ is negative. This means that *a positive charge will lose electric potential energy when it moves in the direction of the electric field.* This is analogous to a mass losing gravitational potential energy when it moves to lower elevations in the presence of gravity. If a positive test charge is released from rest in this electric field, it experiences an electric force $q_o\mathbf{E}$ in the direction of $\mathbf{E}$ (to the right in Fig. 25.1). Therefore, it accelerates to the right, gaining kinetic energy. *As it gains kinetic energy, it loses an equal amount of potential energy.*

On the other hand, if the test charge $q_o$ is negative, then $\Delta U$ is positive and the situation is reversed. *A negative charge gains electric potential energy when it moves in the direction of the electric field.* If a negative charge is released from rest in the field $\mathbf{E}$, it accelerates in a direction opposite the electric field.[1]

Now consider the more general case of a charged particle moving between any two points in a uniform electric field directed along the $x$ axis, as in Figure 25.2. If $\mathbf{d}$ represents the displacement vector between points $A$ and $B$, Equation 25.4 gives

$$\Delta V = -\int_A^B \mathbf{E}\cdot d\mathbf{s} = -\mathbf{E}\cdot\int_A^B d\mathbf{s} = -\mathbf{E}\cdot\mathbf{d} \tag{25.9}$$

where again, we are able to remove $\mathbf{E}$ from the integral since it is constant. Further, the change in potential energy of the charge is

$$\Delta U = q_o\,\Delta V = -q_o\,\mathbf{E}\cdot\mathbf{d} \tag{25.10}$$

Finally, our results show that all points in a plane *perpendicular* to a uniform electric field are at the same potential. This can be seen in Figure 25.2, where the potential difference $V_B - V_A$ is *equal* to $V_C - V_A$. Therefore, $V_B = V_C$.

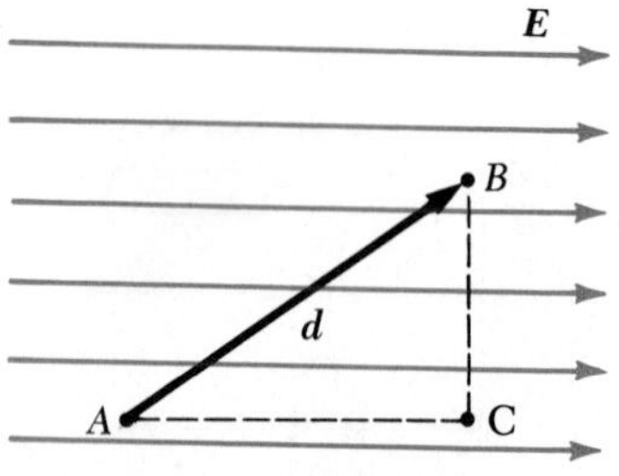

Figure 25.2 A uniform electric field directed along the positive $x$ axis. Point $B$ is at a lower potential than point $A$. Points $B$ and $C$ are at the *same* potential.

[1] Note that when a charged particle accelerates, it actually loses energy by radiating electromagnetic waves.

The name **equipotential surface** is given to any surface consisting of a continuous distribution of points having the same potential.

**An equipotential surface**

Note that since $\Delta U = q_0 \Delta V$, *no* work is done in moving a test charge between any two points on an equipotential surface. The equipotential surfaces of a uniform electric field consist of a family of planes all *perpendicular* to the field (Fig. 25.2). Equipotential surfaces for fields with other symmetries will be described in later sections.

### EXAMPLE 25.1 The Field Between Two Parallel Plates of Opposite Charge

A 12-V battery is connected between two parallel plates as in Figure 25.3. The separation between the plates is 0.3 cm, and the electric field is assumed to be uniform. (This assumption is reasonable if the plate separation is small compared to the plate size and if we do not consider points near the edges of the plates.) Find the electric field between the plates.

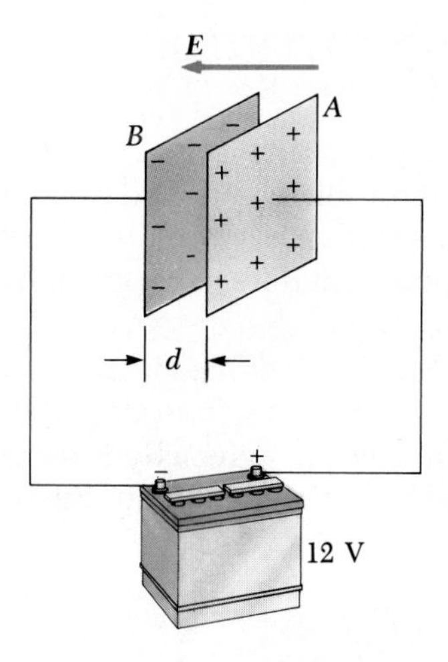


**Figure 25.3** (Example 25.1) A 12-V battery connected to two parallel plates. The electric field between the plates has a magnitude given by the potential difference divided by the plate separation $d$.

*Solution* The electric field is directed from the positive plate toward the negative plate. We see that the positive plate (at the right) is at a *higher* potential than the negative plate. Note that the potential difference between plates $B$ and $A$ must equal the potential difference between the battery terminals. This can be understood by noting that all points on a conductor in equilibrium are at the same potential,[2] and hence there is no potential difference between a terminal of the battery and any portion of the plate to which it is connected. Therefore, the magnitude of the electric field between the plates is

$$E = \frac{|V_B - V_A|}{d} = \frac{12\ \text{V}}{0.3 \times 10^{-2}\ \text{m}} = 4.0 \times 10^3\ \text{V/m}$$

This configuration, which is called a *parallel-plate capacitor*, will be examined in more detail in the next chapter.

[2] The electric field vanishes within a conductor in electrostatic equilibrium, and so the path integral $\int \mathbf{E} \cdot d\mathbf{s}$ between any two points within the conductor must be zero. A fuller discussion of this point is given in Section 25.6.

### EXAMPLE 25.2 Motion of a Proton in a Uniform Electric Field □

A proton is released from *rest* in a uniform electric field of $8 \times 10^4$ V/m directed along the positive $x$ axis (Fig. 25.4). The proton undergoes a displacement of 0.5 m in the direction of $\mathbf{E}$. (a) Find the *change* in the electric potential between the points $A$ and $B$.

Using Equation 25.4 and noting that the displacement is *in the direction* of the field, we have

$$\Delta V = V_B - V_A = -\int \mathbf{E} \cdot d\mathbf{s} = -\int_0^d E\,dx = -E\int_0^d dx$$

$$= -Ed = -\left(8 \times 10^4 \frac{\text{V}}{\text{m}}\right)(0.5\ \text{m})$$

$$= -4 \times 10^4\ \text{V}$$

Thus, the electric potential of the proton *decreases* as it moves from $A$ to $B$.

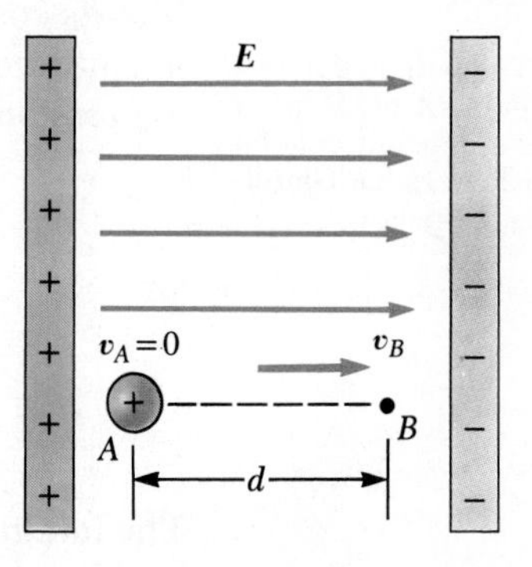


**Figure 25.4** (Example 25.2) A proton accelerates from $A$ to $B$ in the direction of the electric field.

(b) Find the change in potential energy of the proton for this displacement.

$$\Delta U = q_0\,\Delta V = e\,\Delta V$$
$$= (1.6 \times 10^{-19}\text{ C})(-4 \times 10^4\text{ V})$$
$$= \boxed{-6.4 \times 10^{-15}\text{ J}}$$

The negative sign here means that the potential energy of the proton decreases as it moves in the direction of **E**. This makes sense since as the proton *accelerates* in the direction of **E**, it gains kinetic energy and at the same time loses electrical potential energy (the total energy is conserved).

(c) Find the speed of the proton after it has been displaced from rest by 0.5 m.

If there are no forces acting on the proton other than the conservative electric force, we can apply the principle of conservation of mechanical energy in the form $\Delta K + \Delta U = 0$; that is, *the decrease in potential energy must be accompanied by an equal increase in kinetic energy*. Because the mass of the proton is given by the value $m_p = 1.67 \times 10^{-27}$ kg, we find

$$\Delta K + \Delta U = (\tfrac{1}{2}m_p v^2 - 0) - 6.4 \times 10^{-15}\text{ J} = 0$$

$$v^2 = \frac{2(6.4 \times 10^{-15})\text{ J}}{1.67 \times 10^{-27}\text{ kg}} = 7.66 \times 10^{12}\text{ m}^2/\text{s}^2$$

$$v = \boxed{2.77 \times 10^6\text{ m/s}}$$

If an electron were accelerated under the same circumstances, its speed would approach the speed of light and the problem would have to be treated by relativistic mechanics (Chapter 39).

## 25.3 ELECTRIC POTENTIAL AND POTENTIAL ENERGY DUE TO POINT CHARGES

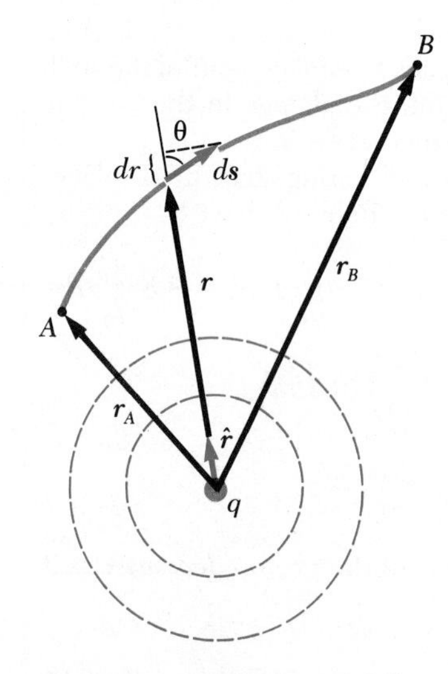


**Figure 25.5** The potential difference between points $A$ and $B$ due to a point charge $q$ depends *only* on the initial and final radial coordinates, $r_A$ and $r_B$, respectively.

Consider an isolated positive point charge $q$ as in Figure 25.5. Recall that such a charge produces an electric field that is radially outward from the charge. In order to find the electric potential at a distance $r$ from the charge, we begin with the general expression for the potential difference given by

$$V_B - V_A = -\int_A^B \mathbf{E}\cdot d\mathbf{s}$$

Since the electric field due to the point charge is given by $\mathbf{E} = kq\hat{\mathbf{r}}/r^2$, where $\hat{\mathbf{r}}$ is a unit vector directed from the charge to the field point, the quantity $\mathbf{E}\cdot d\mathbf{s}$ can be expressed as

$$\mathbf{E}\cdot d\mathbf{s} = k\frac{q}{r^2}\,\hat{\mathbf{r}}\cdot d\mathbf{s}$$

The dot product $\hat{\mathbf{r}}\cdot d\mathbf{s} = ds\cos\theta$, where $\theta$ is the angle between $\hat{\mathbf{r}}$ and $d\mathbf{s}$ as in Figure 25.5. Furthermore, note that $ds\cos\theta$ is the projection of $d\mathbf{s}$ onto $\mathbf{r}$, so that $ds\cos\theta = dr$. That is, any displacement $d\mathbf{s}$ produces a change $dr$ in the magnitude of $r$. With these substitutions, we find that $\mathbf{E}\cdot d\mathbf{s} = (kq/r^2)\,dr$, so the expression for the potential difference becomes

$$V_B - V_A = -\int E_r\,dr = -kq\int_{r_A}^{r_B}\frac{dr}{r^2} = \frac{kq}{r}\Bigg]_{r_A}^{r_B}$$

$$\boxed{V_B - V_A = kq\left[\frac{1}{r_B} - \frac{1}{r_A}\right]} \qquad (25.11)$$

The integral $-\int_A^B \mathbf{E}\cdot d\mathbf{s}$ is *independent* of the path between $A$ and $B$, as it must be. (We had already concluded that the electric field of a point charge is a conservative field, by analogy with the gravitational field of a point mass.)

Furthermore, Equation 25.11 expresses the important result that the potential difference between any two points $A$ and $B$ depends *only* on the *radial* coordinates $r_A$ and $r_B$. It is customary to choose the reference of potential to be zero at $r_A = \infty$. (This is quite natural since $V \propto 1/r_A$ and as $r_A \to \infty$, $V \to 0$.) With this choice, the electric potential due to a point charge at any distance $r$ from the charge is given by

$$V = k\frac{q}{r} \qquad (25.12)$$

Potential of a point charge

From this we see that $V$ is constant on a spherical surface of radius $r$. Hence, we conclude that *the equipotential surfaces (surfaces on which V remains constant) for an isolated point charge consist of a family of spheres concentric with the charge,* as shown in Figure 25.5. Note that the equipotential surfaces are perpendicular to the lines of electric force, as was the case for a uniform electric field.

The electric potential of two or more point charges is obtained by applying the superposition principle. That is, the total potential at some point $P$ due to several point charges is the sum of the potentials due to the individual charges. For a group of charges, we can write the total potential at $P$ in the form

$$V = k\sum_i \frac{q_i}{r_i} \qquad (25.13)$$

The potential of several point charges

where the potential is again taken to be zero at infinity and $r_i$ is the distance from the point $P$ to the charge $q_i$. Note that the sum in Equation 25.13 is an *algebraic sum* of scalars rather than a vector sum (which is used to calculate the electric field of a group of charges). Thus, it is much easier to evaluate $V$ than to evaluate $E$.

We now consider the potential energy of interaction of a system of charged particles. If $V_1$ is the electric potential due to charge $q_1$ at a point $P$, then the work required to bring a second charge, $q_2$, from infinity to the point $P$ without acceleration is given by $q_2V_1$. By definition, this work equals the potential energy, $U$, of the two-particle system when the particles are separated by a distance $r_{12}$ (Fig. 25.6).

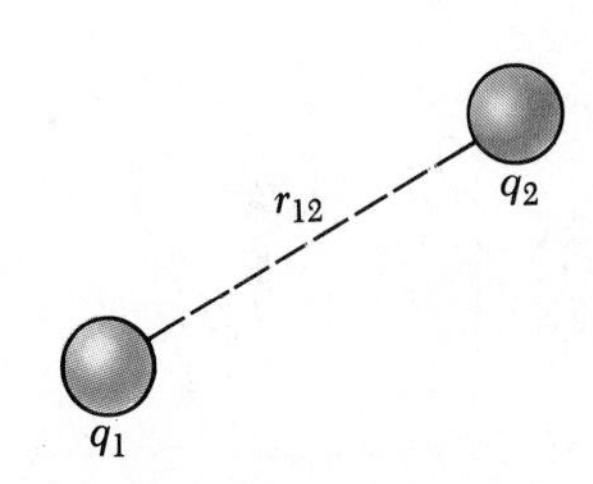


**Figure 25.6** If two point charges are separated by a distance $r_{12}$, the potential energy of the pair of charges is given by $kq_1q_2/r_{12}$.

Therefore, we can express the potential energy as

$$U = q_2V_1 = k\frac{q_1q_2}{r_{12}} \qquad (25.14)$$

Electric potential energy of two charges

Note that if the charges are of the same sign, $U$ is positive.[3] This is consistent with the fact that like charges repel, and so positive work must be done *on* the system to bring the two charges near one another. Conversely, if the charges are of opposite sign, the force is attractive and $U$ is negative. This means that negative work must be done to bring the unlike charges near one another.

[3] The expression for the electric potential energy for two point charges, Equation 25.14, is of the *same* form as the gravitational potential energy of two point masses given by $Gm_1m_2/r$ (Chapter 14). The similarity is not surprising in view of the fact that both are derived from an inverse-square force law.

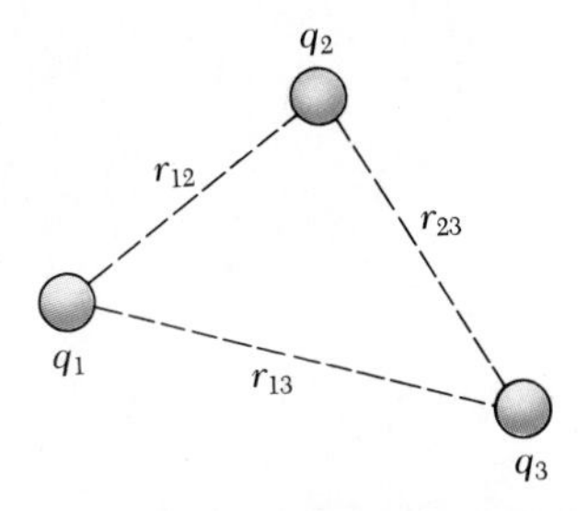


**Figure 25.7** Three point charges are fixed at the positions shown. The potential energy of this system of charges is given by Equation 25.15.

If there are more than two charged particles in the system, the total potential energy can be obtained by calculating $U$ for every *pair* of charges and summing the terms algebraically. As an example, the total potential energy of the three charges shown in Figure 25.7 is given by

$$U = k\left(\frac{q_1q_2}{r_{12}} + \frac{q_1q_3}{r_{13}} + \frac{q_2q_3}{r_{23}}\right) \tag{25.15}$$

Physically, we can interpret this as follows: Imagine that $q_1$ is fixed at the position shown in Figure 25.7, but $q_2$ and $q_3$ are at infinity. The work required to bring $q_2$ from infinity to its position near $q_1$ is $kq_1q_2/r_{12}$, which is the first term in Equation 25.15. The last two terms in Equation 25.15 represent the work required to bring $q_3$ from infinity to its position near $q_1$ and $q_2$. (You should show that the result is independent of the order in which the charges are transported.)

### EXAMPLE 25.3 The Potential Due to Two Point Charges

A 5-$\mu$C point charge is located at the origin, and a second point charge of $-2\ \mu$C is located on the $x$ axis at the position (3, 0) m, as in Figure 25.8a. (a) If the potential is taken to be zero at infinity, find the total electric potential due to these charges at the point $P$, whose coordinates are (0, 4) m.

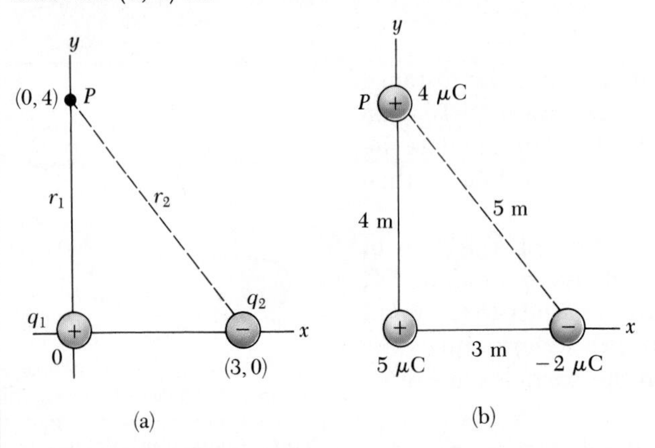


**Figure 25.8** (Example 25.3) The electric potential at the point $P$ due to the two point charges $q_1$ and $q_2$ is the algebraic sum of the potentials due to the individual charges.

The total potential at $P$ due to the two charges is given by

$$V_P = k\left(\frac{q_1}{r_1} + \frac{q_2}{r_2}\right)$$

Since $r_1 = 4$ m and $r_2 = 5$ m, we get

$$V_P = 9 \times 10^9 \frac{\text{N}\cdot\text{m}^2}{\text{C}^2}\left(\frac{5 \times 10^{-6}\ \text{C}}{4\ \text{m}} - \frac{2 \times 10^{-6}\ \text{C}}{5\ \text{m}}\right)$$

$$= \boxed{7.65 \times 10^3\ \text{V}}$$

(b) How much work is required to bring a third point charge of 4 $\mu$C from infinity to the point $P$?

$$W = q_3V_P = (4 \times 10^{-6}\ \text{C})(7.65 \times 10^3\ \text{V})$$

Since 1 V = 1 J/C, $W$ reduces to

$$W = \boxed{3.06 \times 10^{-2}\ \text{J}}$$

Exercise 1 Find the *total* potential energy of the system of three charges in the configuration shown in Figure 25.8b.
Answer $6.0 \times 10^{-4}$ J.

## 25.4 ELECTRIC POTENTIAL DUE TO CONTINUOUS CHARGE DISTRIBUTIONS

The electric potential due to a continuous charge distribution can be calculated in two ways. If the charge distribution is known, we can start with Equation 25.12 for the potential of a point charge. We then consider the potential due to a small charge element $dq$, treating this element as a point charge (Figure 25.9). The potential $dV$ at some point $P$ due to the charge element $dq$ is given by

$$dV = k\frac{dq}{r} \tag{25.16}$$

where $r$ is the distance from the charge element to the point $P$. To get the total potential at $P$, we integrate Equation 25.16 to include contributions from all elements of the charge distribution. Since each element is, in general, at a different distance from $P$ and since $k$ is a constant, we can express $V$ as

$$V = k\int \frac{dq}{r} \tag{25.17}$$

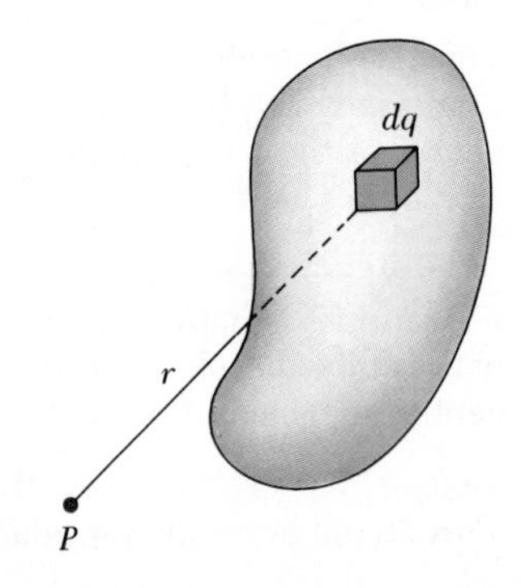


**Figure 25.9** The electric potential at the point $P$ due to a continuous charge distribution can be calculated by dividing the charged body into segments of charge $dq$ and summing the potential contributions over all segments.

In effect, we have replaced the sum in Equation 25.13 by an integral. Note that this expression for $V$ uses a particular choice of reference: the potential is taken to be zero for point $P$ located infinitely far from the charge distribution.

The second method for calculating the potential of a continuous charge distribution makes use of Equation 25.4. This procedure is useful when the electric field is already known from other considerations, such as Gauss' law. If the charge distribution is highly symmetric, we first evaluate $\boldsymbol{E}$ at any point using Gauss' law and then substitute the value obtained into Equation 25.4 to determine the potential difference between any two points. We then choose $V$ to be zero at *any* convenient point. Let us illustrate both methods with several examples.

### EXAMPLE 25.4 Potential Due to a Uniformly Charged Ring

Find the electric potential at a point $P$ located on the axis of a uniformly charged ring of radius $a$ and total charge $Q$. The plane of the ring is chosen perpendicular to the $x$ axis (Fig. 25.10).

*Solution* Let us take the point $P$ to be at a distance $x$ from the center of the ring, as in Figure 25.10. The charge element $dq$ is at a distance equal to $\sqrt{x^2+a^2}$ from the point $P$. Hence, we can express $V$ as

$$V = k\int \frac{dq}{r} = k\int \frac{dq}{\sqrt{x^2+a^2}}$$

In this case, *each* element $dq$ is at the *same distance* from the point $P$. Therefore, the term $\sqrt{x^2+a^2}$ can be removed from the integral and $V$ reduces to

$$V = \frac{k}{\sqrt{x^2+a^2}}\int dq = \frac{kQ}{\sqrt{x^2+a^2}} \tag{25.18}$$

The only variable that appears in this expression for $V$ is $x$. This is not surprising, since our calculation is valid only for points along the $x$ axis, where $y$ and $z$ are both zero. From the symmetry of the situation, we see that along the $x$ axis $\boldsymbol{E}$ can have only an $x$ component. Therefore, we can use the expression $E_x = -dV/dx$, which we shall derive in Section 25.5, to find the electric field at $P$:

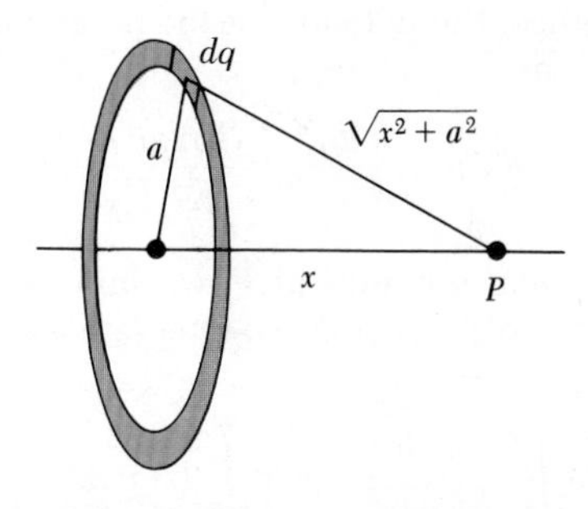


**Figure 25.10** (Example 25.4) A uniformly charged ring of radius $a$, whose plane is perpendicular to the $x$ axis. All segments of the ring are at the same distance from any axial point $P$.

$$\begin{aligned} E_x &= -\frac{dV}{dx} = -kQ\frac{d}{dx}(x^2+a^2)^{-1/2} \\ &= -kQ(-\tfrac{1}{2})(x^2+a^2)^{-3/2}(2x) \\ &= \frac{kQx}{(x^2+a^2)^{3/2}} \end{aligned} \tag{25.19}$$

This result agrees with that obtained by direct integration (see Example 23.9). Note that $E_x = 0$ at $x = 0$ (the center of the ring). Could you have guessed this from Coulomb's law?

Exercise 2 What is the electric potential at the center of the uniformly charged ring? What does the field at the center imply about this result?

Answer $V = kQ/a$ at $x = 0$. Because $E = 0$, $V$ must have a maximum or minimum value; it is in fact a maximum.

### EXAMPLE 25.5 Potential of a Uniformly Charged Disk

Find the electric potential along the axis of a uniformly charged disk of radius $a$ and charge per unit area $\sigma$ (Fig. 25.11).

*Solution* Again we choose the point $P$ to be at a distance $x$ from the center of the disk and take the plane of the disk perpendicular to the $x$ axis. The problem is simplified by dividing the disk into a series of charged rings. The potential of each ring is given by Equation 25.18 in Example 25.4. Consider one such ring of radius $r$ and width $dr$, as indicated in Figure 25.11. The area of the ring is $dA = 2\pi r\,dr$ (the circumference multiplied by the width), and the charge on the ring is $dq = \sigma\,dA = \sigma 2\pi r\,dr$. Hence, the potential at the point $P$ due to this ring is given by

$$dV = \frac{k\,dq}{\sqrt{r^2 + x^2}} = \frac{k\sigma 2\pi r\,dr}{\sqrt{r^2 + x^2}}$$

To find the *total* potential at $P$, we sum over all rings making up the disk. That is, we integrate $dV$ from $r = 0$ to $r = a$:

$$V = \pi k\sigma \int_0^a \frac{2r\,dr}{\sqrt{r^2 + x^2}} = \pi k\sigma \int_0^a (r^2 + x^2)^{-1/2} 2r\,dr$$

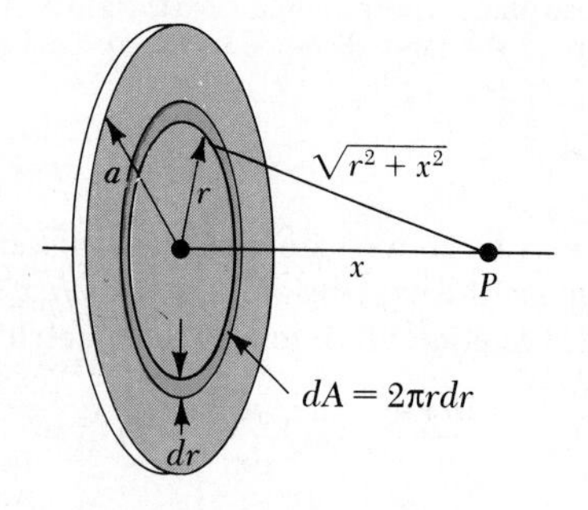


**Figure 25.11** (Example 25.5) A uniformly charged disk of radius $a$, whose plane is perpendicular to the $x$ axis. The calculation of the potential at an axial point $P$ is simplified by dividing the disk into rings of area $2\pi r\,dr$.

This integral is of the form $u^n\,du$ and has the value $u^{n+1}/(n+1)$, where $n = -\frac{1}{2}$ and $u = r^2 + x^2$. This gives the result

$$V = 2\pi k\sigma[(x^2 + a^2)^{1/2} - x] \qquad (25.20)$$

As in Example 25.4, we can find the electric field at any axial point by taking the negative of the derivative of $V$ with respect to $x$. This gives

$$E_x = -\frac{dV}{dx} = 2\pi k\sigma\left(1 - \frac{x}{\sqrt{x^2 + a^2}}\right) \qquad (25.21)$$

The calculation of $V$ and $\mathbf{E}$ for an arbitrary point off the axis is more difficult to perform.

### EXAMPLE 25.6 Potential of a Finite Line Charge

A rod of length $\ell$ located along the $x$ axis has a uniform charge per unit length and a total charge $Q$. Find the electric potential at a point $P$ along the $y$ axis at a distance $d$ from the origin (Figure 25.12).

*Solution* The element of length $dx$ has a charge $dq$ given by $\lambda\,dx$, where $\lambda$ is the charge per unit length, $Q/\ell$. Since this element is at a distance $r = \sqrt{x^2 + d^2}$ from the point $P$, we can express the potential at $P$ due to this element as

$$dV = k\frac{dq}{r} = k\frac{\lambda\,dx}{\sqrt{x^2 + d^2}}$$

To get the total potential at $P$, we integrate this expression over the limits $x = 0$ to $x = \ell$. Noting that $k$, $\lambda$ and $d$ are constants, we find that

$$V = k\lambda \int_0^\ell \frac{dx}{\sqrt{x^2 + d^2}} = k\frac{Q}{\ell}\int_0^\ell \frac{dx}{\sqrt{x^2 + d^2}}$$

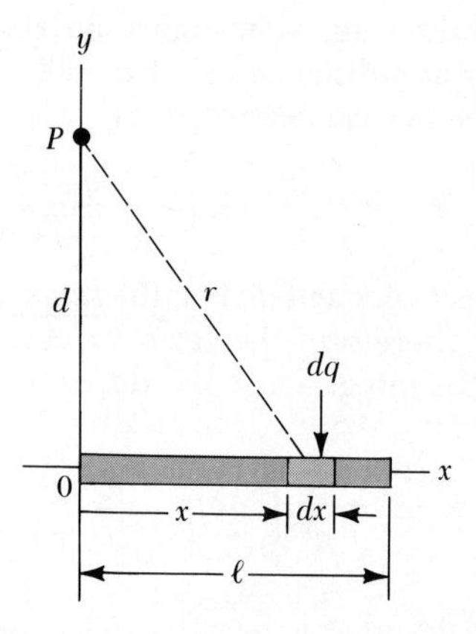


**Figure 25.12** (Example 25.6) A uniform line charge of length $\ell$ located along the $x$ axis. To calculate the potential at $P$, the line charge is divided into segments each of length $dx$, having a charge $dq = \lambda\,dx$.

This integral, found in most integral tables, has the value

$$\int \frac{dx}{\sqrt{x^2+d^2}} = \ln(x+\sqrt{x^2+d^2})$$

Evaluating $V$, we find that

$$V = \frac{kQ}{\ell}\ln\left(\frac{\ell+\sqrt{\ell^2+d^2}}{d}\right) \qquad (25.22)$$

### EXAMPLE 25.7 Potential of a Uniformly Charged Sphere

An insulating solid sphere of radius $R$ has a uniform positive charge density with total charge $Q$ (Fig. 25.13). (a) Find the electric potential at a point *outside* the sphere, that is, for $r > R$. Take the potential to be zero at $r = \infty$.

*Solution* In Example 24.3, we found from Gauss' law that the magnitude of the electric field *outside* a uniformly charged sphere is given by

$$E_r = k\frac{Q}{r^2} \qquad (\text{for } r > R)$$

where the field is directed radially outward when $Q$ is positive. To obtain the potential at an exterior point, such as $B$ in Figure 25.13, we substitute this expression for $E$ into Equation 25.5. Since $\mathbf{E}\cdot d\mathbf{s} = E_r\,dr$ in this case, we get

$$V_B = -\int_\infty^r E_r\,dr = -kQ\int_\infty^r \frac{dr}{r^2}$$

$$V_B = k\frac{Q}{r} \qquad (\text{for } r > R)$$

Note that the result is identical to that for the electric potential due to a point charge. Since the potential must be continuous at $r = R$, we can use this expression to obtain the potential at the surface of the sphere. That is, the potential at a point such as $C$ in Figure 25.13 is given by

$$V_C = k\frac{Q}{R} \qquad (\text{for } r = R)$$

(b) Find the potential at a point *inside* the charged sphere, that is, for $r < R$.

*Solution* In Example 24.3 we found that the electric field inside a uniformly charged sphere is given by

$$E_r = \frac{kQ}{R^3}r \qquad (\text{for } r < R)$$

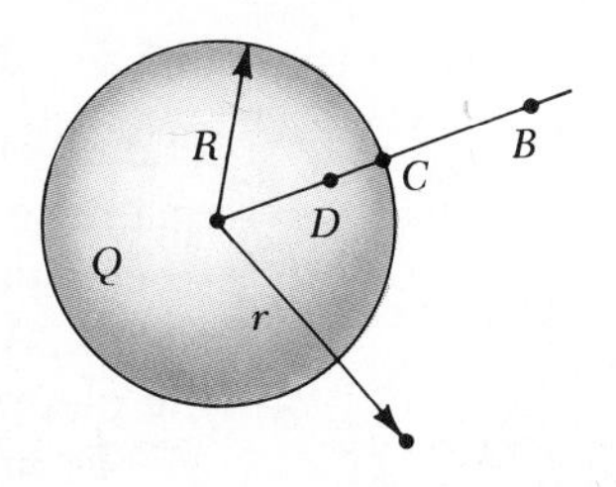


**Figure 25.13** (Example 25.7) A uniformly charged insulating sphere of radius $R$ and total charge $Q$. The electric potential at points $B$ and $C$ is equivalent to that of a point charge $Q$ located at the center of the sphere.

We can use this result and Equation 25.4 to evaluate the potential difference $V_D - V_C$, where $D$ is an interior point:

$$V_D - V_C = -\int_R^r E_r\,dr = -\frac{kQ}{R^3}\int_R^r r\,dr = \frac{kQ}{2R^3}(R^2 - r^2)$$

Substituting $V_C = kQ/R$ into this expression and solving for $V_D$, we get

$$V_D = \frac{kQ}{2R}\left(3 - \frac{r^2}{R^2}\right) \qquad (\text{for } r < R) \quad (25.23)$$

At $r = R$, this expression gives a result that agrees with that for the potential at the surface, that is, $V_C$. A plot of $V$ versus $r$ for this charge distribution is given in Figure 25.14.

**Exercise 3** What is the electric field at the center of a uniformly charged sphere? What is the electric potential at this point?
**Answer** At $r = 0$, $E = 0$ and $V_0 = 3kQ/2R$.

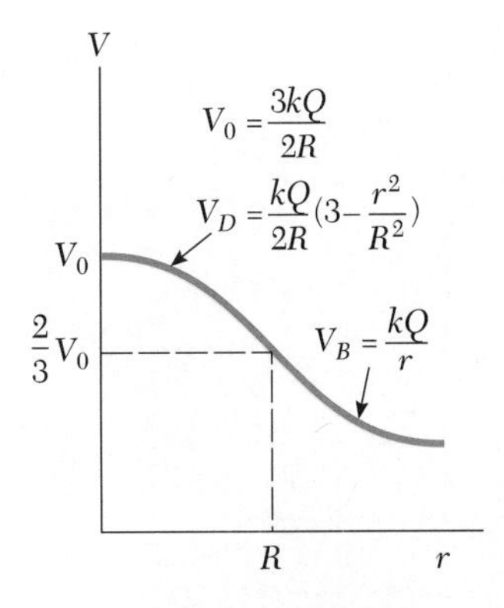


**Figure 25.14** (Example 25.7) A plot of the electric potential $V$ versus the distance $r$ from the center of a uniformly charged, insulating sphere of radius $R$. The curve for $V_D$ inside the sphere is parabolic and joins smoothly with the curve for $V_B$ outside the sphere, which is a hyperbola. The potential has a maximum value $V_0$ at the center of the sphere.

## *25.5 OBTAINING $E$ FROM THE ELECTRIC POTENTIAL

The electric field $\mathbf{E}$ and the potential $V$ are related by Equation 25.4. Both quantities are determined by a specific charge distribution. We now show how to calculate the electric field if the electric potential is known in a certain region. As we shall see, the electric field is simply the negative derivative of the electric potential.

From Equation 25.4 we can express the potential difference $dV$ between two points a distance $ds$ apart as

$$dV = -\mathbf{E} \cdot d\mathbf{s} \tag{25.24}$$

If the electric field has only *one* component, $E_x$, then $\mathbf{E} \cdot d\mathbf{s} = E_x\, dx$. Therefore, Equation 25.24 becomes $dV = -E_x\, dx$, or

$$E_x = -\frac{dV}{dx} \tag{25.25}$$

That is,

> the electric field is equal to the negative of the derivative of the potential with respect to some coordinate.

Note that the potential change is zero for any displacement perpendicular to the electric field. This is consistent with the notion of equipotential surfaces being perpendicular to the field, as in Figure 25.15a.

If the charge distribution has *spherical symmetry, where the charge density depends only on the radial distance r*, then the electric field is radial. In this case, $\mathbf{E} \cdot d\mathbf{s} = E_r\, dr$, and so we can express $dV$ in the form $dV = -E_r\, dr$. Therefore,

Equipotential surfaces are always perpendicular to the electric field lines

$$E_r = -\frac{dV}{dr} \tag{25.26}$$

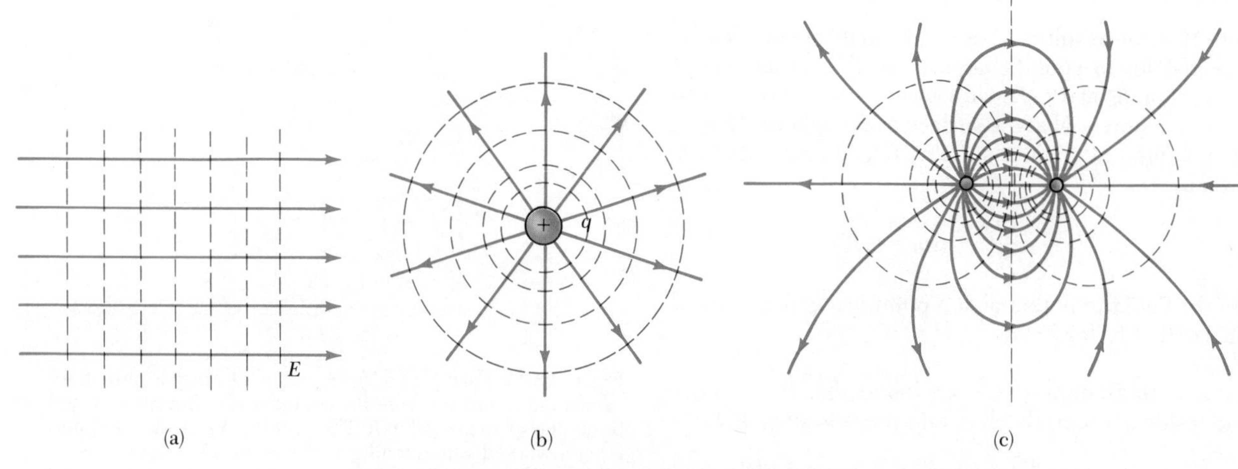


**Figure 25.15** Equipotential surfaces (dashed blue lines) and electric field lines (red lines) for (a) a uniform electric field produced by an infinite sheet of charge, (b) a point charge, and (c) an electric dipole. In all cases, the equipotential surfaces are *perpendicular* to the electric field lines at every point.

Note that the potential changes only in the radial direction, not in a direction perpendicular to $r$. Thus $V$ (like $E_r$) is a function only of $r$. Again, this is consistent with the idea that *equipotential surfaces are perpendicular to field lines.* In this case the equipotential surfaces are a family of spheres concentric with the spherically symmetric charge distribution (Figure 25.15b). The equipotential surfaces for the electric dipole are sketched in Figure 25.15c.

When a test charge is displaced by a vector $d\mathbf{s}$ that lies *within* any equipotential surface, then by definition $dV = -\mathbf{E}\cdot d\mathbf{s} = 0$. This shows that the equipotential surfaces must *always* be *perpendicular* to the electric field lines.

In general, the electric potential is a function of all three spatial coordinates. If $V(r)$ is given in terms of the rectangular coordinates, the electric field components $E_x$, $E_y$, and $E_z$ can readily be found from $V(x, y, z)$. The field components are given by

$$E_x = -\frac{\partial V}{\partial x} \qquad E_y = -\frac{\partial V}{\partial y} \qquad E_z = -\frac{\partial V}{\partial z}$$

In these expressions, the derivatives are called *partial derivatives.* This means that in the operation $\partial V/\partial x$, *one takes a derivative with respect to x while y and z are held constant.* For example, if $V = 3x^2y + y^2 + yz$, then

$$\frac{\partial V}{\partial x} = \frac{\partial}{\partial x}(3x^2y + y^2 + yz) = \frac{\partial}{\partial x}(3x^2y) = 3y\frac{d}{dx}(x^2) = 6xy$$

In vector notation, $\mathbf{E}$ is often written $\mathbf{E} = -\nabla V = -\left(\mathbf{i}\frac{\partial}{\partial x} + \mathbf{j}\frac{\partial}{\partial y} + \mathbf{k}\frac{\partial}{\partial z}\right)V$, where $\nabla$ is called the *gradient operator.*

**EXAMPLE 25.8 The Point Charge Revisited**
Let us use the potential function for a point charge $q$ to derive the electric field at a distance $r$ from the charge.

*Solution* The potential of a point charge is given by Equation 25.12:

$$V = k\frac{q}{r}$$

Since the potential is a function of $r$ only, it has spherical symmetry and we can apply Equation 25.26 directly to obtain the electric field:

$$E_r = -\frac{dV}{dr} = -\frac{d}{dr}\left(k\frac{q}{r}\right) = -kq\frac{d}{dr}\left(\frac{1}{r}\right)$$

$$E_r = \frac{kq}{r^2}$$

Thus, the electric field is radial and the result agrees with that obtained using Gauss' law.

**EXAMPLE 25.9 The Electric Potential of a Dipole**
An electric dipole consists of two equal and opposite charges separated by a distance $2a$, as in Figure 25.16. Calculate the electric potential and the electric field at the point $P$ on the $x$ axis and located a distance $x$ from the center of the dipole.

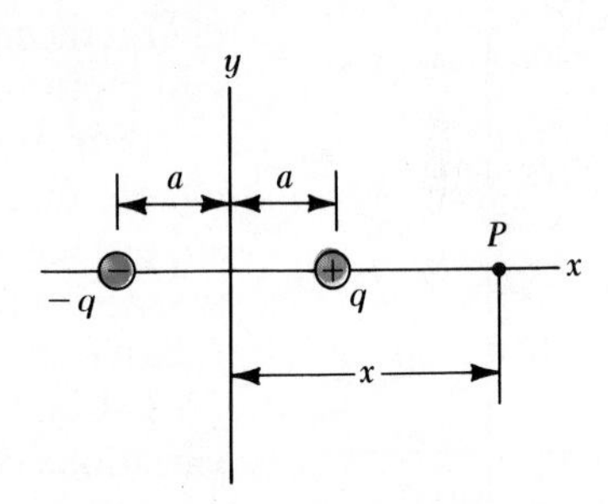


**Figure 25.16** (Example 25.9) An electric dipole located on the $x$ axis.

*Solution*

$$V = k\sum\frac{q_i}{r_i} = k\left(\frac{q}{x-a} - \frac{q}{x+a}\right) = \frac{2kqa}{x^2 - a^2}$$

If the point $P$ is far from the dipole, so that $x \gg a$, then $a^2$ can be neglected in the term $x^2 - a^2$ and $V$ becomes

$$V \approx \frac{2kqa}{x^2} \qquad (x \gg a)$$

Using Equation 25.25 and this result, the electric field at $P$ is given by

$$E = -\frac{dV}{dx} = \frac{4kqa}{x^3} \qquad \text{for } x \gg a$$

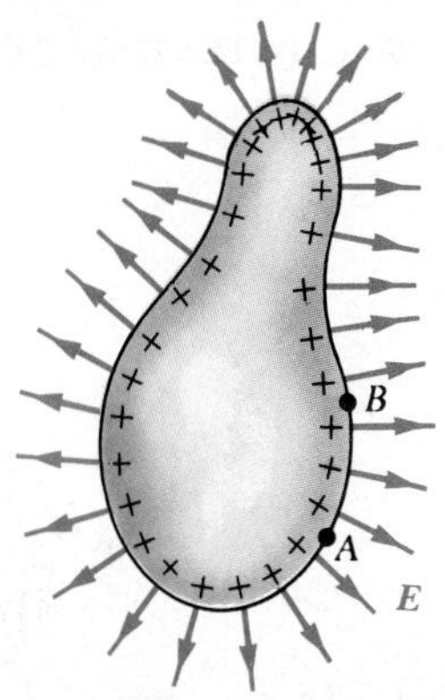

**Figure 25.17** An arbitrarily shaped conductor with an excess positive charge. When the conductor is in electrostatic equilibrium, all of the charge resides at the surface, $E = 0$ inside the conductor, and the electric field just outside the conductor is perpendicular to the surface. The potential is constant inside the conductor and is equal to the potential at the surface. The surface charge density is nonuniform.

## 25.6 POTENTIAL OF A CHARGED CONDUCTOR

In the previous chapter we found that when a solid conductor in equilibrium carries a net charge, the charge resides on the outer surface of the conductor. Furthermore, we showed that the electric field just outside the surface of a conductor in equilibrium is perpendicular to the surface while the field *inside* the conductor is zero. If the electric field had a component parallel to the surface, this would cause surface charges to move, creating a current and a nonequilibrium situation.

We shall now show that *every point on the surface of a charged conductor in equilibrium is at the same potential.* Consider two points $A$ and $B$ on the surface of a charged conductor, as in Figure 25.17. Along a surface path connecting these points, $\mathbf{E}$ is always perpendicular to the displacement $d\mathbf{s}$; therefore $\mathbf{E} \cdot d\mathbf{s} = 0$. Using this result and Equation 25.4, we conclude that the potential difference between $A$ and $B$ is necessarily zero. That is,

$$V_B - V_A = -\int_A^B \mathbf{E} \cdot d\mathbf{s} = 0$$

This result applies to *any* two points on the surface. Therefore, $V$ is constant everywhere on the surface of a charged conductor in equilibrium. That is,

> the surface of any charged conductor in equilibrium is an equipotential surface. Furthermore, since the electric field is zero inside the conductor, we conclude that the potential is constant everywhere inside the conductor and equal to its value at the surface.

Therefore, no work is required to move a test charge from the interior of a charged conductor to its surface. (Note that the potential is *not zero* inside the conductor even though the electric field is zero.)

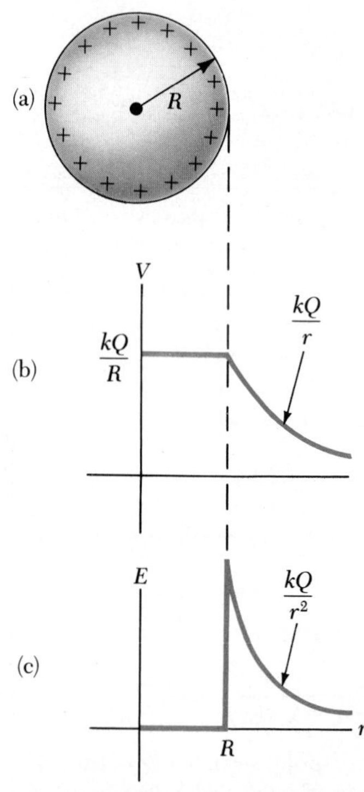

**Figure 25.18** (a) The excess charge on a conducting sphere of radius $R$ is uniformly distributed on its surface. (b) The electric potential versus the distance $r$ from the center of the charged conducting sphere. (c) The electric field intensity versus the distance $r$ from the center of the charged conducting sphere.

For example, consider a solid metal sphere of radius $R$ and total positive charge $Q$, as shown in Figure 25.18a. The electric field outside the charged sphere is given by $kQ/r^2$ and points radially outward. Following Example 25.7, we see that the potential at the interior and surface of the sphere must be $kQ/R$ relative to infinity. The potential outside the sphere is given by $kQ/r$. Figure 25.18b is a plot of the potential as a function of $r$, and Figure 25.18c shows the variations of the electric field with $r$.

When a net charge is placed on a spherical conductor, the surface charge density is uniform, as indicated in Figure 25.18a. However, if the conductor is nonspherical, as in Figure 25.17, the surface charge density is high where the radius of curvature is small and convex and low where the radius of curvature is small and concave. Since the electric field just outside a charged conductor is proportional to the surface charge density, $\sigma$, we see that *the electric field is large near points having a small convex radius of curvature and reaches very high values at sharp points.*

Figure 25.19 shows the electric field lines around two spherical conductors, one with a net charge $Q$ and one with zero net charge. In this case, the surface charge density is *not* uniform on either conductor. The larger sphere (on the right), with zero net charge, has negative charges induced on its side that faces the charged sphere and positive charge on its side opposite the charged sphere. The blue lines in Figure 25.19 represent the boundaries of the equipotential surfaces for this charge configuration. Again, you should notice that the field lines are perpendicular to the conducting surfaces. Fur-

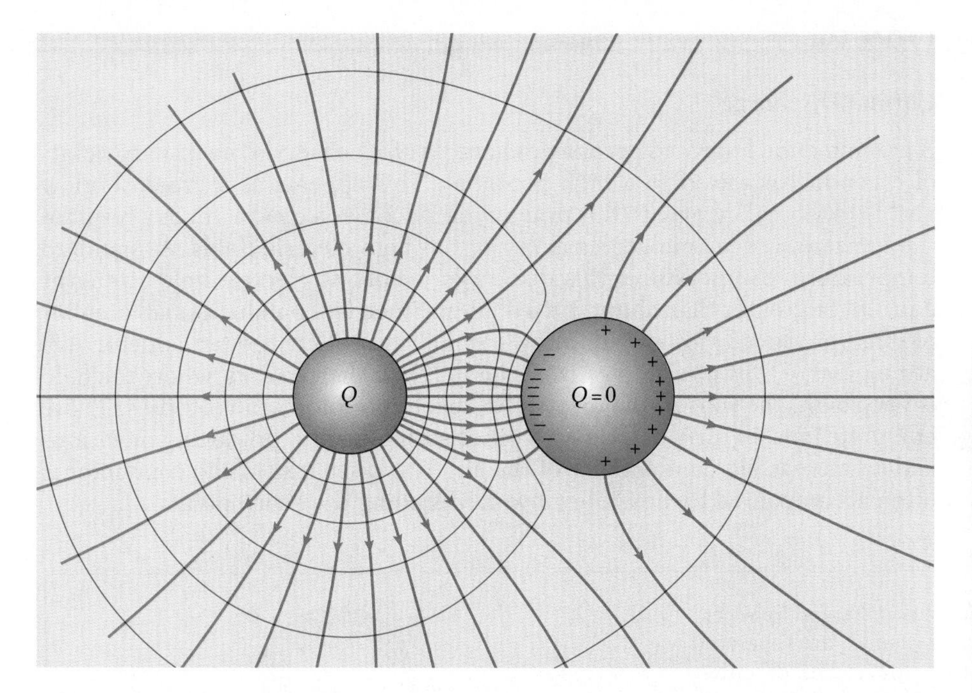


**Figure 25.19** The electric field lines (in red) around two spherical conductors. The smaller sphere on the left has a net charge $Q$, and the sphere on the right has zero net charge. The blue lines represent the edges of the equipotential surfaces. (From E. Purcell, *Electricity and Magnetism,* New York, McGraw-Hill, 1965, with permission of the Education Development Center, Inc.)

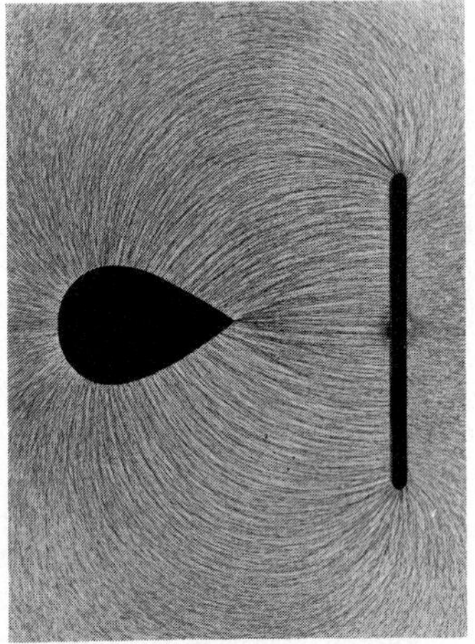

Electric field pattern of a charged conducting plate near an oppositely charged pointed conductor. Small pieces of thread suspended in oil align with the electric field lines. Note that the electric field is most intense near the pointed part of the conductor and at other points where the radius of curvature is small. (Courtesy of Harold M. Waage, Princeton University)

thermore, the equipotential surfaces are perpendicular to the field lines at the boundaries of the conductor and everywhere else in space.

### A Cavity Within a Conductor

Now consider a conductor of arbitrary shape containing a cavity as in Figure 25.20. Let us assume there are no charges *inside* the cavity. We shall show that *the electric field inside the cavity must be zero,* regardless of the charge distribution on the *outside* surface of the conductor. Furthermore, the field in the cavity is zero even if an electric field exists outside the conductor.

In order to prove this point, we shall use the fact that every point on the conductor is at the same potential, and therefore any two points $A$ and $B$ on the surface of the cavity must be at the same potential. Now *imagine* that a field $\mathbf{E}$ exists in the cavity, and evaluate the potential difference $V_B - V_A$ defined by the expression

$$V_B - V_A = -\int_A^B \mathbf{E} \cdot d\mathbf{s}$$

If $\mathbf{E}$ is non-zero, we can always find a path between $A$ and $B$ for which $\mathbf{E} \cdot d\mathbf{s}$ is always a positive number, and so the integral must be positive. However, since $V_B - V_A = 0$, the integral must also be zero. This contradiction can be reconciled only if $\mathbf{E} = 0$ inside the cavity. Thus, we conclude that a cavity surrounded by conducting walls is a field-free region as long as there are no charges inside the cavity.

This result has some interesting applications. For example, it is possible to shield an electronic circuit or even an entire laboratory from external fields by surrounding it with conducting walls. Shielding is often necessary when performing highly sensitive electrical measurements.

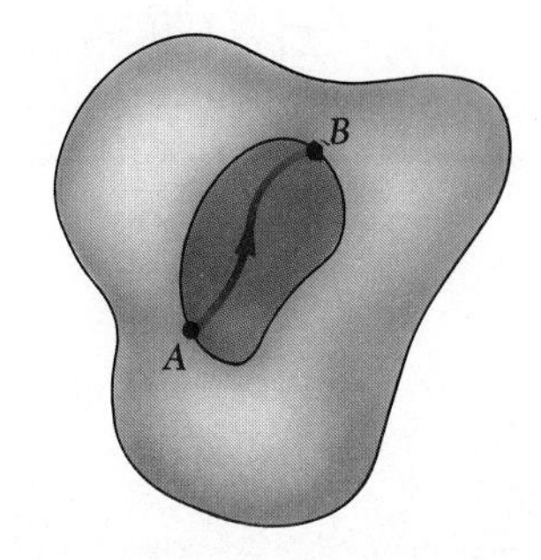


**Figure 25.20** A conductor in electrostatic equilibrium containing an empty cavity. The electric field in the cavity is *zero,* regardless of the charge on the conductor.

### Corona Discharge

A phenomenon known as **corona discharge** is often observed near sharp points of a conductor raised to a high potential. This appears as a greenish glow visible to the naked eye. In this process, air becomes a conductor as a result of the ionization of air molecules in regions of high electric fields. At standard temperature and pressure, this discharge occurs at electric field strengths equal to or greater than about $3 \times 10^6$ V/m. Since air contains a small number of ions (produced, for example, by cosmic rays), a charged conductor will attract ions of the opposite sign from the air. Near sharp points, where the field is very high, the ions in the air will be accelerated to high velocities. These energetic ions, in turn, collide with other air molecules, producing more ions and an increase in conductivity of the air. The discharge of the conductor is often accompanied by a visible glow surrounding the sharp points.

**EXAMPLE 25.10 Two Connected Charged Spheres**
Two spherical conductors of radii $r_1$ and $r_2$ are separated by a distance much larger than the radius of either sphere. The spheres are connected by a conducting wire as in Figure 25.21. If the charges on the spheres in equilibrium are $q_1$ and $q_2$, respectively, find the ratio of the field strengths at the surfaces of the spheres.

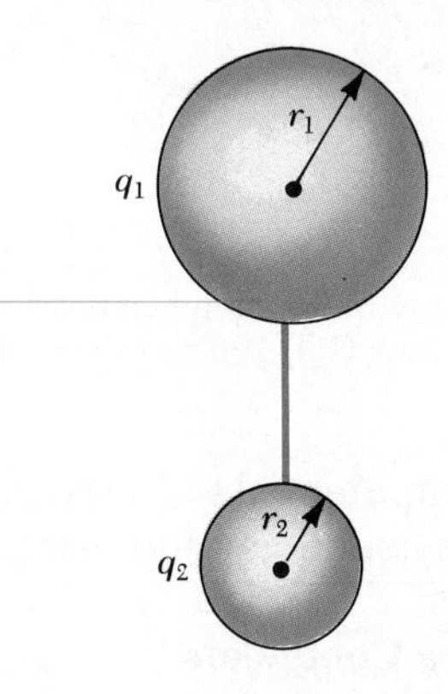


**Figure 25.21** (Example 25.10) Two charged spherical conductors connected by a conducting wire. The spheres are at the *same* potential, $V$.

*Solution* Since the spheres are connected by a conducting wire, they must both be at the *same* potential $V$, given by

$$V = k\frac{q_1}{r_1} = k\frac{q_2}{r_2}$$

Therefore, the ratio of charges is

$$(1) \qquad \frac{q_1}{q_2} = \frac{r_1}{r_2}$$

Since the spheres are very far apart, their surfaces are uniformly charged and we can express the electric fields at their surfaces as

$$E_1 = k\frac{q_1}{{r_1}^2} \quad \text{and} \quad E_2 = k\frac{q_2}{{r_2}^2}$$

Taking the ratio of these two fields and making use of (1), we find that

$$(2) \qquad \frac{E_1}{E_2} = \frac{r_2}{r_1}$$

Hence, the field is more intense in the vicinity of the smaller sphere.

## *25.7 THE MILLIKAN OIL-DROP EXPERIMENT

During the period 1909 to 1913, Robert Andrews Millikan (1868–1953) performed a brilliant set of experiments at the University of Chicago in which he measured the elementary charge on an electron, $e$, and demonstrated the quantized nature of the electronic charge. The apparatus used by Millikan, diagrammed in Figure 25.22, contains two parallel metal plates. Oil droplets charged by friction in an atomizer are allowed to pass through a small hole in the upper plate. A light beam directed horizontally is used to illuminate the oil droplets, which are veiwed by a telescope whose axis is at right angles to the

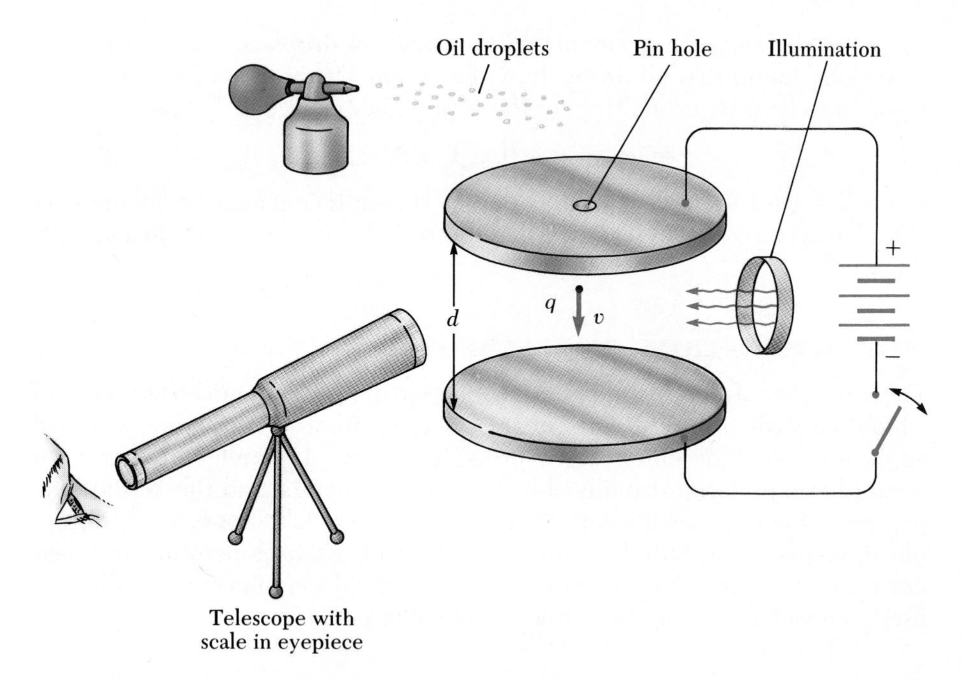


**Figure 25.22** A schematic view of the Millikan oil-drop apparatus.

beam. When the droplets are viewed in this manner, they appear as shining stars against a dark background, and the rate of fall of individual drops may be determined.[4]

Let us assume a single drop having a mass $m$ and carrying a charge $q$ is being viewed, and that its charge is negative. If there is no electric field present between the plates, the two forces acting on the charge are its weight, $mg$, acting downward, and an upward viscous drag force $D$, as indicated in Figure 25.23a. The drag force is proportional to the speed of the drop. When the drop reaches its terminal speed $v$, the two forces balance each other ($mg = D$).

Now suppose that an electric field is set up between the plates by connecting a battery such that the upper plate is at the higher potential. In this case, a third force $q\mathbf{E}$ acts on the charged drop. Since $q$ is negative and $\mathbf{E}$ is downward, the electric force is *upward* as in Figure 25.23b. If this force is large enough, the drop will move upward and the drag force $D'$ will act downward. When the upward electric force, $qE$, balances the sum of the weight and the drag force both acting downward, the drop reaches a new terminal speed $v'$.

With the field turned on, a drop moves slowly upward, typically at rates of *hundredths* of a centimeter per second. The rate of fall in the absence of a field is comparable. Hence, a single droplet with constant mass and radius may be followed for hours, alternately rising and falling, by simply turning the electric field on and off.

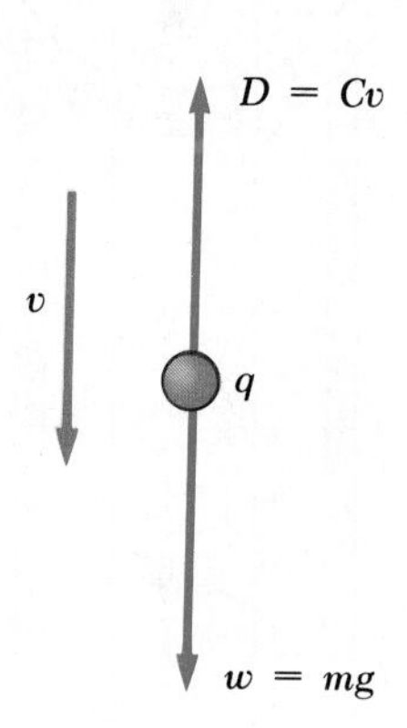


(a) Field off

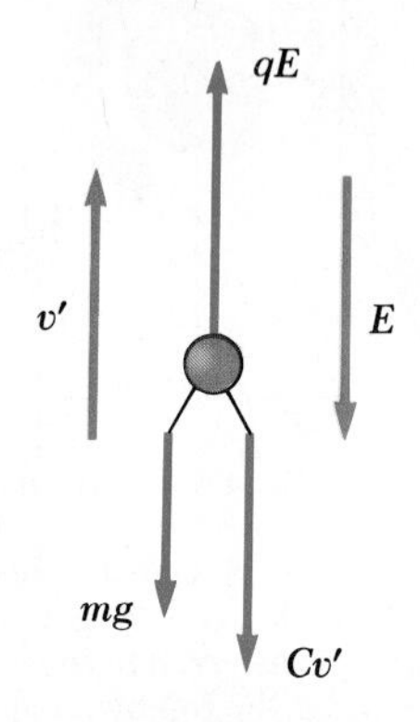


(b) Field on

**Figure 25.23** The forces on a charged oil droplet in the Millikan experiment.

[4] At one time, the oil droplets were termed "Millikan's Shining Stars." Perhaps this description has lost its popularity because of the generations of physics students who have experienced hallucinations, near blindness, migraine headaches, etc., while repeating his experiment!

After making measurements on thousands of droplets, Millikan and his coworkers found that all drops, to within about 1% precision, had a charge equal to some integer multiple of the elementary charge $e$. That is,

$$q = ne \qquad n = 0, \pm 1, \pm 2, \pm 3, \ldots \tag{25.27}$$

where $e = 1.60 \times 10^{-19}$ C. Millikan's experiment is conclusive evidence that charge is quantized. He was awarded the Nobel Prize in physics in 1923 for this work.

## *25.8 APPLICATIONS OF ELECTROSTATICS

The principles of electrostatics have been used in various applications, a few of which we shall briefly discuss in this section. Some of the more practical applications include electrostatic precipitators, used to reduce the level of atmospheric pollution from coal-burning power plants, and the xerography process, which has revolutionized imaging process technology. Scientific applications of electrostatic principles include electrostatic generators for accelerating elementary charged particles and the field-ion microscope, which is used to image atoms on the surface of metallic samples.

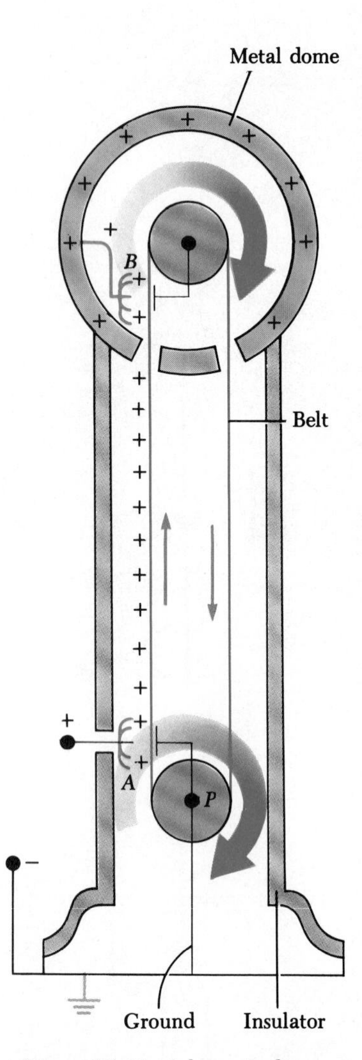

**Figure 25.24** Schematic diagram of a Van de Graaff generator. Charge is transferred to the hollow conductor at the top by means of a moving belt. The charge is deposited on the belt at point $A$ and is transferred to the hollow conductor at point $B$.

### The Van de Graaff Generator

In the previous chapter we described an experiment that demonstrates a method for transferring charge to a hollow conductor (the Faraday ice-pail experiment). When a charged conductor is placed in contact with the inside of a hollow conductor, all of the charge of the first conductor is transferred to the hollow conductor. In principle, the charge on the hollow conductor and its potential can be increased without limit by repeating the process.

In 1929 Robert J. Van de Graaff used this principle to design and build an electrostatic generator. This type of generator is used extensively in nuclear physics research. The basic idea of the Van de Graaff generator is described in Figure 25.24. Charge is delivered continuously to a high-voltage electrode on a moving belt of insulating material. The high-voltage electrode is a hollow conductor mounted on an insulating column. The belt is charged at $A$ by means of a corona discharge between comb-like metallic needles and a grounded grid. The needles are maintained at a positive potential of typically $10^4$ V. The positive charge on the moving belt is transferred to the high-voltage electrode by a second comb of needles at $B$. Since the electric field inside the hollow conductor is negligible, the positive charge on the belt easily transfers to the high-voltage electrode, regardless of its potential. In practice, it is possible to increase the potential of the high-voltage electrode until electrical discharge occurs through the air. Since the "breakdown" voltage of air is equal to about $3 \times 10^6$ V/m, a sphere 1 m in radius can be raised to a maximum potential of $3 \times 10^6$ V. The potential can be increased further by increasing the radius of the hollow conductor and by placing the entire system in a container filled with high-pressure gas.

Van de Graaff generators can produce potential differences as high as 20 million volts. Protons accelerated through such potential differences receive enough energy to initiate nuclear reactions between the protons and various target nuclei.

### The Electrostatic Precipitator

One important application of electrical discharge in gases is a device called an *electrostatic precipitator*. This device is used to remove particulate matter from combustion gases, thereby reducing air pollution. They are especially useful in coal-burning power plants and in industrial operations that generate large quantities of smoke. Current systems are able to eliminate more than 99% of the ash and dust (by weight) from the smoke. Figure 25.25 shows the basic idea of the electrostatic precipitator. A high voltage (typically 40 kV to 100 kV) is maintained between a wire running down the center of a duct and the outer wall, which is grounded. The wire is maintained at a negative potential with respect to the walls, and so the electric field is directed toward the wire. The electric field near the wire reaches high enough values to cause a corona discharge around the wire and the formation of positive ions, electrons, and negative ions, such as $O_2^-$. As the electrons and negative ions are accelerated toward the outer wall by the nonuniform electric field, the dirt particles in the streaming gas become charged by collisions and ion capture. Since most of the charged dirt particles are negative, they are also drawn to the outer wall by the electric field. By periodically shaking the duct, the particles fall loose and are collected at the bottom.

In addition to reducing the level of particulate matter in the atmosphere, the electrostatic precipitator also recovers valuable materials from the stack in the form of metal oxides.

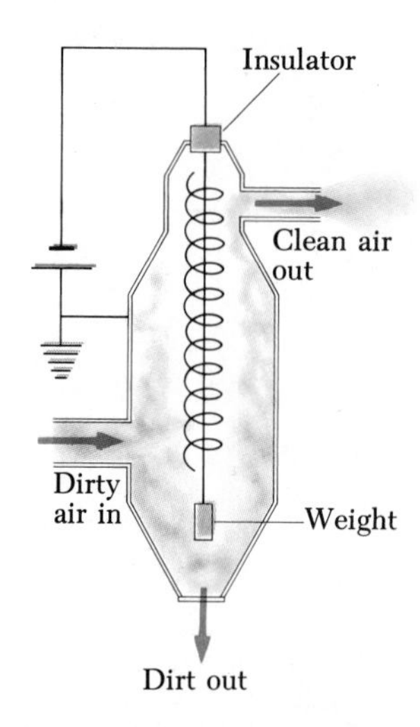


**Figure 25.25** Schematic diagram of an electrostatic precipitator. The high negative voltage maintained on the central wire creates an electrical discharge in the vicinity of the wire.

### Xerography

The process of xerography is widely used for making photocopies of letters, documents, and other printed materials. The basic idea for the process was developed by Chester Carlson, for which he was granted a patent in 1940. In 1947, the Xerox Corporation launched a full-scale program to develop automated duplicating machines using this process. The huge success of this development is quite evident; today, practically all modern offices and libraries have one or more duplicating machines, and the capabilities of modern machines are on the increase.

Some features of the xerographic process involve simple concepts from electrostatics and optics. However, the one idea that makes the process unique is the use of a photoconductive material to form an image. (A photoconductor is a material that is a poor conductor in the dark but becomes a good electrical conductor when exposed to light.)

The sequence of steps used in the xerographic process is illustrated in Figure 25.26. First, the surface of a plate or drum is coated with a thin film of the photoconductive material (usually selenium or some compound of selenium), and the photoconductive surface is given a positive electrostatic charge in the dark. The page to be copied is then projected onto the charged surface. The photoconducting surface becomes conducting only in areas where light strikes. In these areas, the light produces charge carriers in the photoconductor, which neutralize the positively charged surface. However, the charges remain on those areas of the photoconductor not exposed to light, leaving a latent (hidden) image of the object in the form of a positive surface charge distribution.

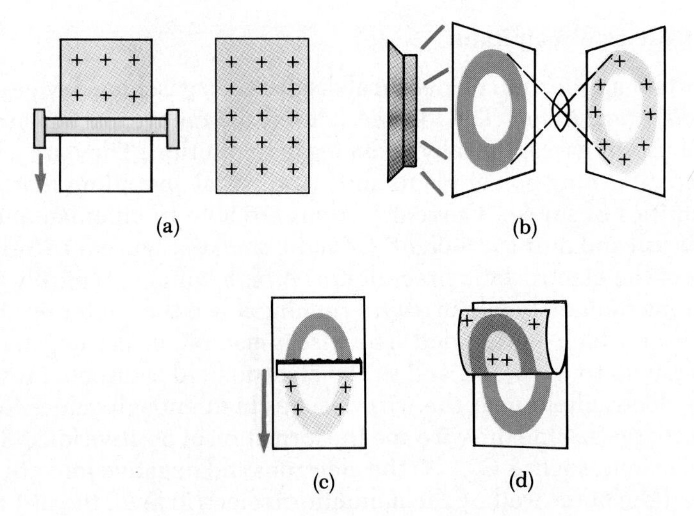


**Figure 25.26** The xerographic process: (a) The photoconductive surface is positively charged. (b) Through the use of a light source and lens, an image is formed on the surface in the form of hidden positive charges. (c) The surface containing the image is covered with a charged powder, which adheres only to the image area. (d) A piece of paper is placed over the surface and given a charge. This transfers the visible image to the paper, which is finally heat-treated to "fix" the powder to the paper.

Next, a negatively charged powder called a *toner* is dusted onto the photoconducting surface. The charged powder adheres only to those areas of the surface that contain the positively charged image. At this point, the image becomes visible. The image is then transferred to the surface of a sheet of positively charged paper.

Finally, the toner material is "fixed" to the surface of the paper through the application of heat. This results in a permanent copy of the original.

## The Field-Ion Microscope

In Section 25.6 we pointed out that the electric field intensity can be very high in the vicinity of a sharp point on a charged conductor. A device that makes use of this intense field is the *field-ion microscope,* which was invented in 1956 by E. W. Mueller of the Pennsylvania State University.

The basic construction of the field-ion microscope is shown in Figure 25.27. A specimen to be studied is fabricated from a fine wire, and a sharp tip is formed, usually by etching the wire in an acid. Typically, the diameter of the tip is about 0.1 $\mu$m (= 100 nm). The specimen is placed at the center of an evacuated glass tube containing a fluorescent screen. Next, a small amount of helium is introduced into the vessel. A very high potential difference is applied between the needle and the screen, producing a very intense electric field near the tip of the needle. It is important to cool the tip to at least the temperature of liquid nitrogen to obtain stable pictures. The helium atoms in the vicinity of this high-field region are ionized by the loss of an electron, which leaves the helium positively charged. The positively charged $He^+$ ions then accelerate to the negatively charged fluorescent screen. This results in a pattern on the screen that represents an image of the tip of the specimen.

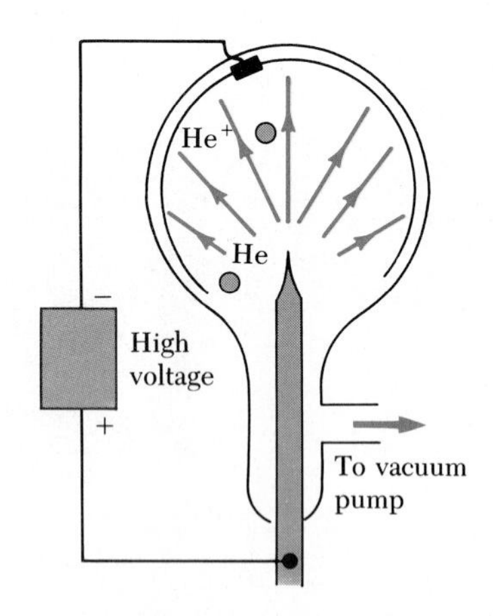


**Figure 25.27** Schematic diagram of a field-ion microscope. The electric field is very intense at the tip of the needle-shaped specimen.

Under the proper conditions (low specimen temperature and high vacuum), the images of the individual atoms on the surface of the sample are

**Figure 25.28** Field ion microscope image of the surface of a platinum crystal with a magnification of 1 000 000×. Individual atoms can be seen on surface layers using this technique. (Courtesy of Prof. T.T. Tsong, The Pennsylvania State University)

visible, and the atomic arrangement on the surface can be studied. Unfortunately, the high electric fields also set up large mechanical stresses near the tip of the specimen, which limits the application of the technique to strong metallic elements, such as tungsten and rhenium. Figure 25.28 represents a typical field-ion microscope pattern of a platinum crystal.

## SUMMARY

When a positive test charge $q_o$ is moved between points $A$ and $B$ in an electrostatic field $\mathbf{E}$, the **change in the potential energy** is given by

Change in potential energy

$$\Delta U = -q_o \int_A^B \mathbf{E} \cdot d\mathbf{s} \qquad (25.3)$$

The **potential difference** $\Delta V$ between points $A$ and $B$ in an electrostatic field $\mathbf{E}$ is defined as the change in potential energy divided by the test charge $q_o$:

Potential difference

$$\Delta V = \frac{\Delta U}{q_o} = -\int_A^B \mathbf{E} \cdot d\mathbf{s} \qquad (25.4)$$

where the electric potential $V$ is a scalar and has the units of J/C, defined to be 1 volt (V).

The potential difference between two points $A$ and $B$ in a *uniform* electric field $\mathbf{E}$ is given by

$$\Delta V = -Ed \tag{25.7}$$

where $d$ is the displacement in the direction *parallel* to $\mathbf{E}$.

**Equipotential surfaces** are surfaces on which the electric potential remains constant. Equipotential surfaces are *perpendicular* to the electric field lines.

The potential due to a point charge $q$ at any distance $r$ from the charge is given by

Potential of a point charge

$$V = k\frac{q}{r} \tag{25.12}$$

The potential due to a group of point charges is obtained by summing the potentials due to the individual charges. Since $V$ is a scalar, the sum is a simple algebraic operation.

The **potential energy of a pair of point charges** separated by a distance $r_{12}$ is given by

Electric potential energy of two charges

$$U = k\frac{q_1 q_2}{r_{12}} \tag{25.14}$$

This represents the work required to bring the charges from an infinite separation to the separation $r_{12}$. The potential energy of a distribution of point charges is obtained by summing terms like Equation 25.14 over all *pairs* of particles.

**TABLE 25.1 Potentials Due to Various Charge Distributions**

| Charge Distribution | Electric Potential | Location |
|---|---|---|
| Uniformly charged ring of radius $a$ | $V = k\frac{Q}{\sqrt{x^2 + a^2}}$ | Along the axis of the ring, a distance $x$ from its center |
| Uniformly charged disk of radius $a$ | $V = 2\pi k\sigma[(x^2 + a^2)^{1/2} - x]$ | Along the axis of the disk, a distance $x$ from its center |
| Uniformly charged, *insulating* solid sphere of radius $R$ and total charge $Q$ | $V = k\frac{Q}{r}$ | $r \geq R$ |
| | $V = \frac{kQ}{2R}\left(3 - \frac{r^2}{R^2}\right)$ | $r < R$ |
| Isolated *conducting* sphere of total charge $Q$ and radius $R$ | $V = k\frac{Q}{R}$ | $r \leq R$ |
| | $V = k\frac{Q}{r}$ | $r > R$ |

The **electric potential due to a continuous charge distribution** is given by

$$V = k \int \frac{dq}{r} \tag{25.17}$$

Electric potential due to a continuous charge distribution

If the electric potential is known as a function of coordinates $x$, $y$, $z$ the components of the electric field can be obtained by taking the negative derivative of the potential with respect to the coordinates. For example, the $x$ component of the electric field is given by

$$E_x = -\frac{dV}{dx} \tag{25.25}$$

Every point on the surface of a charged conductor in electrostatic equilibrium is at the same potential. Furthermore, the potential is constant everywhere inside the conductor and equal to its value at the surface. Table 25.1 lists potentials due to several charge distributions.

## PROBLEM-SOLVING STRATEGY AND HINTS

1. When working problems involving electric potential, remember that potential is a *scalar quantity* (rather than a vector quantity like the electric field), so there are no components to worry about. Therefore, when using the superposition principle to evaluate the electric potential at a point due to a system of point charges, you simply take the algebraic sum of the potentials due to each charge. However, you must keep track of signs. The potential for each positive charge ($V = kq/r$) is positive, while the potential for each negative charge is negative.
2. Just as in mechanics, only *changes* in potential are significant, hence the point where you choose the potential to be zero is arbitrary. When dealing with point charges or a finite-sized charge distribution, we usually define $V = 0$ to be at a point infinitely far from the charges. However, if the charge distribution itself extends to infinity, some other nearby point must be selected as the reference point.
3. The electric potential at some point $P$ due to a continuous distribution of charge can be evaluated by dividing the charge distribution into infinitesimal elements of charge $dq$ located at a distance $r$ from the point $P$. You then treat this element as a point charge, so that the potential at $P$ due to the element is $dV = k\, dq/r$. The total potential at $P$ is obtained by integrating $dV$ over the entire charge distribution. In performing the integration for most problems, it is necessary to express $dq$ and $r$ in terms of a single variable. In order to simplify the integration, it is important to give careful consideration of the geometry involved in the problem. You should review Examples 25.4 through 25.6 as guides for using this method.
4. Another method that can be used to obtain the potential due to a finite continuous charge distribution is to start with the definition of the potential difference given by Equation 25.4. If $\mathbf{E}$ is known or can be obtained easily (say from Gauss' law), then the line integral of $\mathbf{E} \cdot d\mathbf{s}$ can be evaluated. An example of this method is given in Example 25.7.
5. Once you know the electric potential at a point, it is possible to obtain the electric field at that point by remembering that *the electric field is equal to the negative of the derivative of the potential with respect to some coordinate.* Examples 25.4 and 25.5 illustrate how to use this procedure.

## QUESTIONS

1. In your own words, distinguish between electric potential and electrical potential energy.
2. A negative charge moves in the direction of a uniform electric field. Does its potential energy increase or decrease? Does the electric potential increase or decrease?
3. If a proton is released from rest in a uniform electric field, does its electric potential increase or decrease? What about its potential energy?
4. Give a physical explanation of the fact that the potential energy of a pair of like charges is positive whereas the potential energy of a pair of unlike charges is negative.
5. A uniform electric field is parallel to the $x$ axis. In what direction can a charge be displaced in this field without any external work being done on the charge?
6. Explain why equipotential surfaces are always perpendicular to electric field lines.
7. Describe the equipotential surfaces for (a) an infinite line of charge and (b) a uniformly charged sphere.
8. Explain why, under static conditions, all points in a conductor must be at the same electric potential.
9. If the electric potential at some point is zero, can you conclude that there are no charges in the vicinity of that point? Explain.
10. If the potential is constant in a certain region, what is the electric field in that region?
11. The electric field inside a hollow, uniformly charged sphere is zero. Does this imply that the potential is zero inside the sphere? Explain.
12. The potential of a point charge is defined to be zero at an infinite distance. Why can we not define the potential of an infinite line of charge to be zero at $r = \infty$?
13. Two charged conducting spheres of different radii are connected by a conducting wire as in Figure 25.21. Which sphere has the greater charge density?
14. What determines the maximum potential to which the dome of a Van de Graaff generator can be raised?
15. In what type of weather would a car battery be more likely to discharge and why?
16. Explain the origin of the glow that is sometimes observed around the cables of a high-voltage power line.
17. Why is it important to avoid sharp edges, or points, on conductors used in high-voltage equipment?
18. How would you shield an electronic circuit or laboratory from stray electric fields? Why does this work?
19. Why is it relatively safe to stay in an automobile with a metal body during a severe thunderstorm?
20. Walking across a carpet and then touching someone can result in a shock. Explain why this occurs.

## PROBLEMS

### Section 25.1 Potential Difference and Electric Potential

1. Concentric spherical surfaces surrounding a point charge at their center are *equipotential surfaces*. The intersections of these surfaces with a plane through their common center are *equipotential lines*. How much work is done in moving a charge $q$ a distance $s$ along an arc of an equipotential of circular shape and of radius $R$?
2. What change in potential energy does a 12-$\mu$C charge experience when it is moved between two points for which the potential difference is 65 V? Express the answer in eV.
3. (a) Calculate the speed of a proton that is accelerated from rest through a potential difference of 120 V. (b) Calculate the speed of an electron that is accelerated through the same potential difference.
4. Through what potential difference would one need to accelerate an electron in order for it to achieve a velocity of 40% of the velocity of light, starting from rest? ($c = 3.0 \times 10^8$ m/s.)
5. A deuteron (a nucleus which consists of one proton and one neutron) is accelerated through a 2.7-kV potential difference. (a) How much energy does it gain? (b) How fast would it be going if it started from rest?
6. What potential difference is needed to stop an electron with an initial speed of $4.2 \times 10^5$ m/s?
7. An ion accelerated through a potential difference of 115 V experiences an increase in potential energy of $7.37 \times 10^{-17}$ J. Calculate the charge on the ion.
8. In a tandem Van de Graaff accelerator a proton is accelerated through a potential difference of $14 \times 10^6$ V. Assuming that the proton starts from rest, calculate its (a) final kinetic energy in joules, (b) final kinetic energy in MeV, and (c) final speed.
9. A positron, when accelerated from rest between two points at a fixed potential difference, acquires a speed of 30% of the speed of light. What speed will be achieved by a *proton* if accelerated from rest between the same two points?

### Section 25.2 Potential Differences in a Uniform Electric Field

10. Consider two points in an electric field. The potential at point $P_1$ is $V_1 = -30$ V, and the potential at point $P_2$ is $V_2 = +150$ V. How much work is done by an external force in moving a charge $q = -4.7$ $\mu$C from $P_2$ to $P_1$?
11. How much work is done (by a battery, generator, or other source of electrical energy) in moving Avoga-

dro's number of electrons from an initial point where the electric potential is 9 V to a point where the potential is $-5$ V? (The potential in each case is measured relative to a common reference point.)

12. A capacitor consists of two parallel plates separated by a distance of 0.3 mm. If a 20-V potential difference is maintained between those plates, calculate the electric field strength in the region between the plates.
13. The electric field between two charged parallel plates separated by a distance of 1.8 cm has a uniform value of $2.4 \times 10^4$ N/C. Find the potential difference between the two plates. How much kinetic energy would be gained by a deuteron in accelerating from the positive to the negative plate?
14. Suppose an electron is released from rest in a uniform electric field whose strength is $5.9 \times 10^3$ V/m. (a) Through what potential difference will it have passed after moving 1 cm? (b) How fast will the electron be moving after it has traveled 1 cm?
15. An electron moving parallel to the $x$ axis has an initial velocity of $3.7 \times 10^6$ m/s at the origin. The velocity of the electron is reduced to $1.4 \times 10^5$ m/s at the point $x = 2$ cm. Calculate the potential difference between the origin and the point $x = 2$ cm. Which point is at the higher potential?
16. A positron has the same charge as a proton, but the same mass as an electron. Suppose a positron moves 5.2 cm in the direction of a uniform 480 V/m electric field. (a) How much potential energy does it gain or lose? (b) How much kinetic energy does it gain or lose?
17. A proton moves in a region of a uniform electric field. The proton experiences an increase in kinetic energy of $5 \times 10^{-18}$ J after being displaced 2 cm in a direction parallel to the field. What is the magnitude of the electric field?
18. A uniform electric field of magnitude 325 V/m is directed in the *negative y* direction in Figure 25.29. The coordinates of point $A$ are $(-0.2, -0.3)$ m, and those of point $B$ are (0.4, 0.5) m. Calculate the potential $V_B - V_A$ using the blue path.

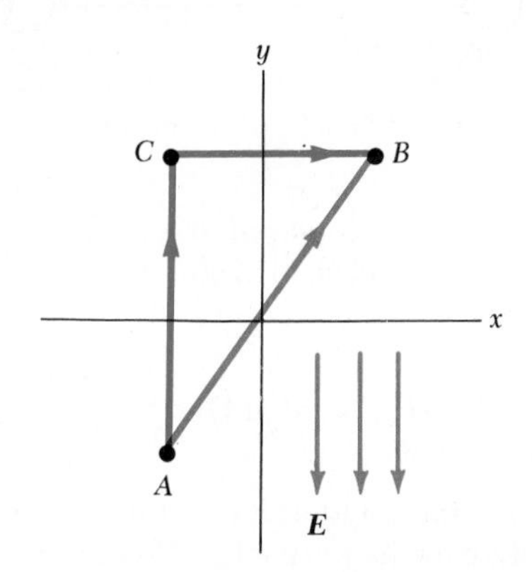


Figure 25.29 (Problems 18 and 19).

19. For the situation described in Problem 18, calculate the change in electric potential while going from point $A$ to point $B$ along the direct red path $AB$. Which point is at the higher potential?
20. A uniform electric field of magnitude 250 V/m is directed in the positive $x$ direction. (a) Suppose a $+12\ \mu$C charge moves from the origin to the point $(x, y) = (20$ cm, $50$ cm$)$. Through what potential difference did it move? (b) What was the change in its potential energy?

### Section 25.3 Electric Potential and Potential Energy Due to Point Charges

*Note: Assume a reference level for potential as $V = 0$ at $r = \infty$ unless the statement of the problem requires otherwise.*

21. At what distance from a point charge of 8 $\mu$C would the potential equal $3.6 \times 10^4$ V?
22. A small spherical object carries a charge of 8 nC. At what distance from the center of the object is the potential equal to 100 V? 50 V? 25 V? Is the spacing of the equipotentials proportional to the change in $V$?
23. At a distance $r$ away from a point charge $q$, the electrical potential is $V = 400$ V and the magnitude of the electric field is $E = 150$ N/C. Determine the value of $q$ and $r$.
24. Given two 2-$\mu$C charges, as shown in Figure 25.30 and a positive test charge $q = 1.28 \times 10^{-18}$ C at the origin, (a) what is the net force exerted on $q$ by the two 2-$\mu$C charges? (b) What field **E** do the two 2-$\mu$C charges produce at the origin? (c) What is the potential $V$ produced by the two 2-$\mu$C charges at the origin?

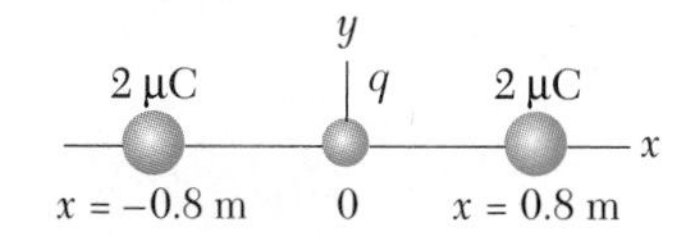


Figure 25.30 (Problem 24).

25. A $+2.8\ \mu$C charge is located on the $y$ axis at $y = +1.6$ m and a $-4.6\ \mu$C charge is located at the origin. Calculate the net electric potential at the point (0.4 m, 0).
26. A charge $+q$ is at the origin. A charge $-2q$ is at $x = 2.0$ m on the $x$ axis. (a) For what finite value(s) of $x$ is the electric field zero? (b) For what finite value(s) of $x$ is the electric potential zero?
27. The three charges shown in Figure 25.31 are at the vertices of an isosceles triangle. Calculate the electric potential at the *midpoint of the base*, taking $q = 7\ \mu$C.

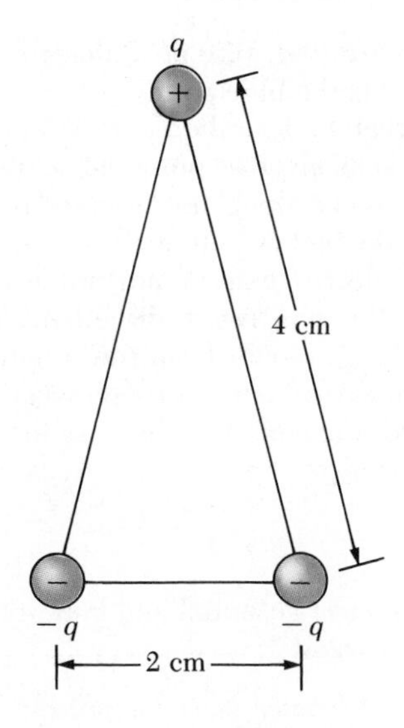


**Figure 25.31** (Problem 27).

28. Calculate the value of the electric potential at point $P$ due to the charge configuration shown in Figure 25.32. Use the values $q_1 = 5\ \mu C$, $q_2 = -10\ \mu C$, $a = 0.4$ m, and $b = 0.50$ m.

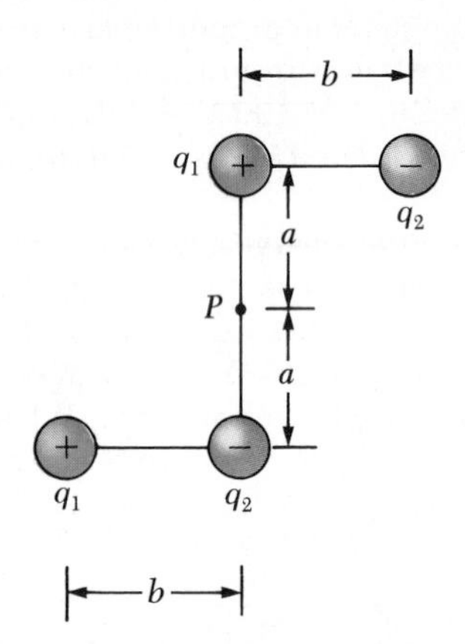


**Figure 25.32** (Problem 28).

29. Two point charges, $Q_1 = +5$ nC and $Q_2 = -3$ nC, are separated by 35 cm. (a) What is the potential energy of the pair? What is the significance of the algebraic sign of your answer? (b) What is the electric potential at a point midway between the charges?
30. A charge $q_1 = -9\ \mu C$ is located at the origin, and a second charge $q_2 = -1\ \mu C$ is located on the $x$ axis at $x = 0.7$ m. Calculate the electric potential energy of this pair of charges.
31. The Bohr model of the hydrogen atom states that the electron can only exist in certain allowed orbits. The radius of each Bohr orbit is given by the expression $r = n^2(0.0529\ \text{nm})$ where $n = 1, 2, 3, \ldots$ . Calculate the electric potential energy of a hydrogen atom when the electron is in the (a) first allowed orbit, $n = 1$, (b) second allowed orbit, $n = 2$, and (c) when the electron has escaped from the atom, $r = \infty$. Express your answers in electron volts.
32. Calculate the energy required to assemble the array of charges shown in Figure 25.33, where $a = 0.20$ m, $b = 0.40$ m, and $q = 6\ \mu C$.

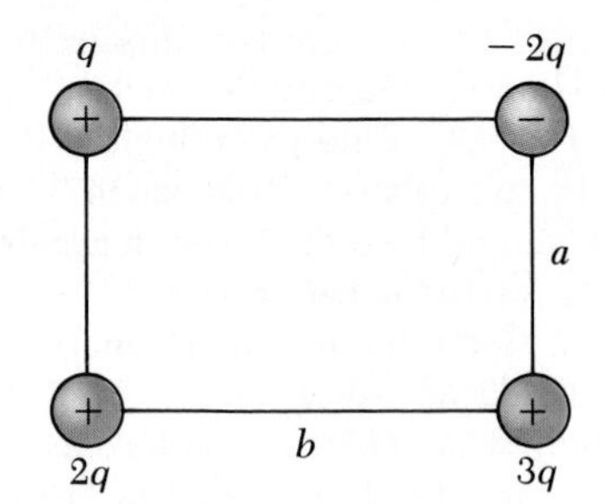


**Figure 25.33** (Problem 32).

33. Show that the amount of work required to assemble four identical point charges of magnitude $Q$ at the corners of a square of side $s$ is given by $5.41kQ^2/s$.
34. Four equal point charges of charge $q = +5\ \mu C$ are located at the corners of a 30 cm by 40 cm rectangle. Calculate the electric potential energy stored in this charge configuration.
35. Four charges are located at the corners of a rectangle as in Figure 25.34. How much energy would be expended in removing the two 4-$\mu$C charges to infinity?

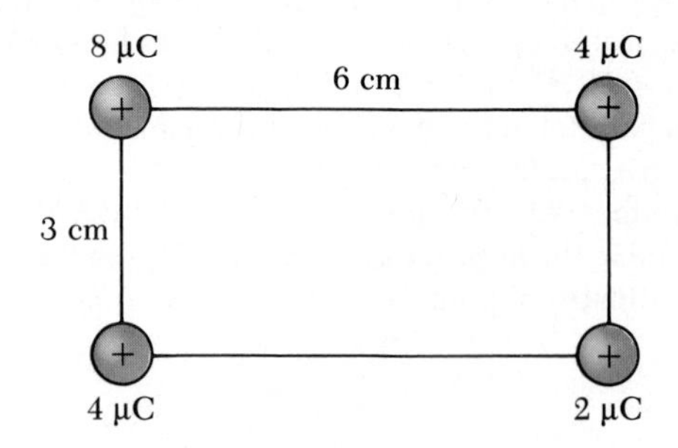


**Figure 25.34** (Problem 35).

36. How much work is required to assemble eight identical point charges, each of magnitude $q$, at the corners of a cube of side $s$?

### Section 25.4 Electric Potential Due to Continuous Charge Distributions

37. Consider a ring of radius $R$ with total charge $Q$ spread uniformly over its perimeter. What is the potential difference between the point at the center of the ring

and a point on the axis of the ring at a distance $2R$ from the center of the ring?

38. Consider a Helmholtz pair consisting of two coaxial rings of 30 cm radius, separated by a distance of 30 cm. (a) Calculate the electric potential at a point on their common axis midway between the two rings, assuming that each ring carries a uniformly distributed charge of $+5\ \mu C$. (b) What is the potential at this point if the two rings carry equal and opposite charges?

39. A rod of length $L$ (Fig. 25.35) lies along the $x$ axis with its left end at the origin and has a *nonuniform* charge density $\lambda = \alpha x$ (where $\alpha$ is a positive constant). (a) What are the units of the constant $\alpha$? (b) Calculate the electric potential at point $A$, a distance $d$ from the left end of the rod.

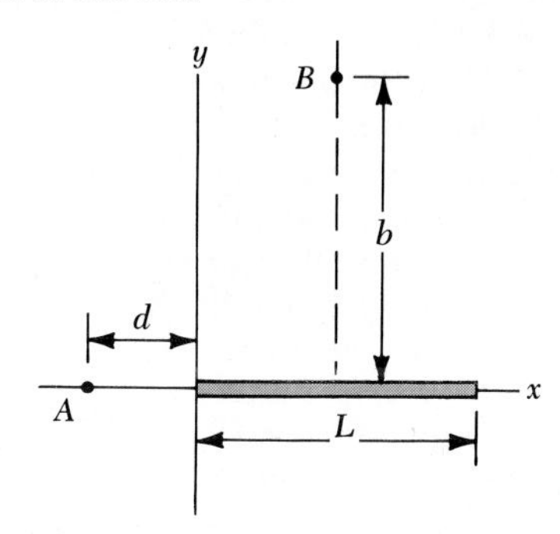


**Figure 25.35** (Problems 39 and 40).

40. For the arrangement described in the previous problem, calculate the electric potential at point $B$ on the perpendicular bisector of the rod a distance $b$ above the $x$ axis. Note that the rod has a *nonuniform* charge density $\lambda = \alpha x$.

41. Calculate the electric potential at point $P$ on the axis of the annulus shown in Figure 25.36, which has a uniform charge density $\sigma$ and inner and outer radii $a$ and $b$, respectively.

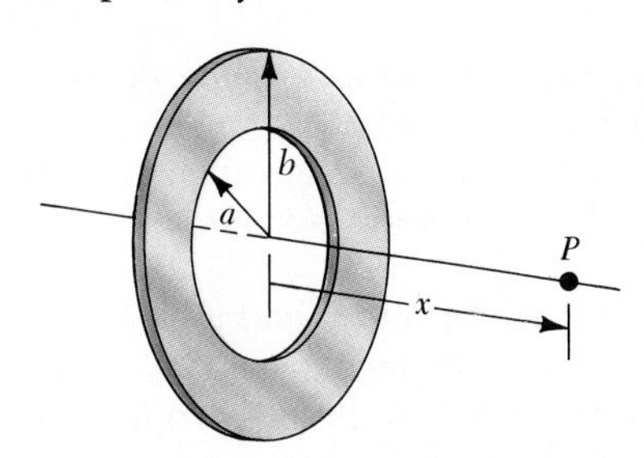


**Figure 25.36** (Problem 41).

## *Section 25.5 Obtaining *E* from the Electric Potential

42. The electric potential over a certain region of space is given by $V = 3x^2y - 4xz - 5y^2$ volts. Find (a) the electric potential and (b) the components of the electric field at the point $(+1, 0, +2)$ where all distances are in meters.

43. Over a certain region of space, the electric potential is given by $V = 5x - 3x^2y + 2yz^2$. Find the expressions for the $x$, $y$, and $z$ components of the electric field over this region. What is the magnitude of the field at the point $P$, which has coordinates (in meters) $(1, 0, -2)$?

44. The electric potential in a certain region is given by $V = 4xz - 5y + 3z^2$ volts. Find the magnitude of the electric field at the point $(+2, -1, +3)$ where all distances are in meters.

45. The potential in a region between $x = 0$ and $x = 6$ m is given by:

$$V = a + bx$$

where $a = 10$ V and $b = -7$ V/m. Determine (a) the potential at $x = 0$, 3 m, and 6 m and (b) the magnitude and direction of the electric field at $x = 0$, 3 m, and 6 m.

46. The electric potential in a certain region is given by

$$V = ax^2 + bx + c$$

$$a = 12\ \text{V/m}^2 \quad b = -10\ \text{V/m} \quad c = 62\ \text{V}$$

Determine (a) the magnitude and direction of the electric field at $x = +2$ m and (b) the position where the electric field is zero.

47. The electric potential inside a charged spherical conductor of radius $R$ is given by $V = kQ/R$ and outside the potential is given by $V = kQ/r$. Using $E_r = -\dfrac{dV}{dr}$, derive the electric field both (a) inside ($r < R$) and (b) outside ($r > R$) this charge distribution.

48. The electric potential inside a uniformly charged spherical insulator of radius $R$ is given by

$$V = \frac{kQ}{2R}\left(3 - \frac{r^2}{R^2}\right)$$

and outside by

$$V = \frac{kQ}{r}$$

Use $E_r = -\dfrac{dV}{dr}$ to derive the electric field both (a) inside ($r < R$) and (b) outside ($r > R$) this charge distribution.

49. Use the exact result from Example 25.9 to evaluate $V$ at $x = 3a$ for the described dipole when $a = 2$ mm and $q = 3\ \mu C$. Compare this answer to that which you would obtain if you used the approximate formula valid when $x \gg a$.

50. Use the exact result for the dipole potential from Example 25.9 to find the electric field at any point where $x > a$ on the axis of the dipole. Evaluate $E$ at $x = 3a$ if $a = 2$ mm and $q = 3\ \mu C$.

## Section 25.6 Potential of a Charged Conductor

51. How many electrons should be removed from an initially uncharged spherical conductor of radius 0.3 m to produce a potential of 7.5 kV at the surface?
52. Calculate the surface charge density, $\sigma$ (in $C/m^2$), for a solid spherical conductor of radius $R = 0.25$ m if the potential at a distance 0.5 m from the center of the sphere is 1300 V.
53. A spherical conductor has a radius of 14 cm and has a charge of $+26\ \mu C$. Calculate the electric field and the electric potential at the following distances from the center of this conductor: (a) $r = 10$ cm, (b) $r = 20$ cm, and (c) $r = 14$ cm.
54. Two spherical conductors of radii $r_1$ and $r_2$ are connected by a conducting wire as shown in Figure 25.21. If $r_1 = 0.94$ m, $r_2 = 0.47$ m, and the field at the surface of the smaller sphere is 890 N/C, calculate the total charge on the larger sphere assuming it is initially uncharged. Assume that the separation of the spheres is large compared to their radii.
55. Two charged spherical conductors are connected by a long conducting wire. A total charge of $+20\ \mu C$ is placed on this combination of two spheres. (a) If one has a radius of 4 cm and the other has a radius of 6 cm, what is the electric field near the surface of each sphere? (b) What is the electrical potential of each sphere?
56. An egg-shaped conductor has a charge of +43 nC placed on its surface. It has a total surface area of 38 $cm^2$. (a) What is the average surface charge density? (b) What is the electric field inside the conductor? (c) What is the (average) electric field just outside the conductor?

## *Section 25.8 Applications of Electrostatics

57. Consider the Van de Graaff generator with a 30-cm diameter dome operating in dry air. (a) What is the maximum potential of the dome? (b) What is the maximum charge on the dome?
58. (a) Calculate the *largest* amount of charge possible on the surface of a Van de Graaff generator with a 40-cm diameter dome surrounded by air. (b) What is the potential of this spherical dome when it has the charge calculated above?
59. What charge would have to be placed on the surface of a Van de Graaff generator, whose dome has a radius of 15 cm, to produce a spark across a 10-cm air gap?
60. What power in watts must a Van de Graaff generator deliver if it produces a 100-$\mu$A beam of protons at an energy of 12 MeV?

## ADDITIONAL PROBLEMS

61. At a certain distance from a point charge the field intensity is 500 V/m and the potential is $-3000$ V. (a) What is the distance to the charge? (b) What is the magnitude of the charge?
62. All of the corners, except one, of a 1-m cube are occupied by charges of $+1\ \mu C$. What is the electric potential at the empty corner?
63. Three point charges of magnitude $+8\ \mu C$, $-3\ \mu C$, and $+5\ \mu C$ are located on the corners of a triangle whose sides are each 9 cm long. Calculate the electric potential at the center of this triangle.
64. The electric potential just outside a charged conducting sphere is 200 V, and 10 cm farther from the center of the sphere the potential is 150 V. Find (a) the radius of the sphere and (b) the charge on the sphere.
65. Equal charges ($q = +2\ \mu C$) are placed at 30° intervals around the equator of a sphere with a radius of 1.2 m. What is the electrical potential (a) at the center of the sphere and (b) at the north pole of the sphere?
66. The charge distribution shown in Figure 25.37 is referred to as a linear quadrupole. (a) Find the exact expression for the potential at a point on the $x$ axis where $x > d$. (b) Show that the expression obtained in (a) reduces to

$$V = \frac{2kQd^2}{x^3}$$

when $x \gg d$.

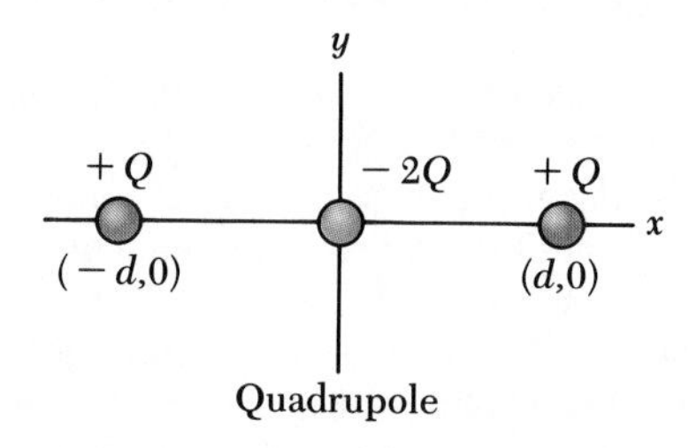


**Figure 25.37** (Problems 66, 67, and 75).

67. Use the exact result from Problem 66 to evaluate the potential for the linear quadrupole at $x = 3d$ if $d =$ 2 mm and $q = 3\ \mu C$. Compare this answer to what you obtain when you use the approximate result valid when $x \gg d$.
68. (a) Use the exact result from Problem 66 to find the electric field at any point along the axis of the linear quadrupole for $x > d$. (b) Evaluate $E$ at $x = 3d$ if $d =$ 2 mm and $Q = 3\ \mu C$.
69. Two point charges of equal magnitude are located along the $y$ axis at equal distances above and below the $x$ axis, as shown in Figure 25.38. (a) Plot a graph of the potential at points along the $x$ axis over the interval $-3a < x < 3a$. You should plot the potential in units of $kQ/a$, where $k$ is the Coulomb constant. (b) Let the charge located at $-a$ be *negative* and plot the potential along the $y$ axis over the interval $-4a < y < 4a$.

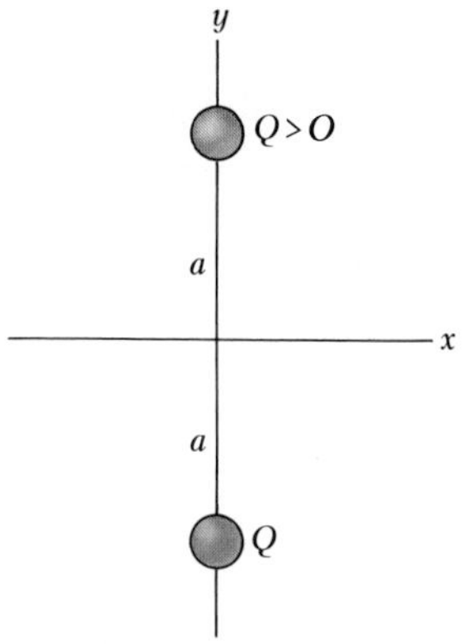

**Figure 25.38** (Problem 69).

70. The liquid-drop model of the nucleus suggests that high-energy oscillations of certain nuclei can split the nucleus into two unequal fragments plus a few neutrons. The fragments acquire kinetic energy from their mutual Coulombic repulsion. Calculate the Coulomb potential energy (in MeV) of two spherical fragments from a uranium nucleus having the following charges and radii: $+38e$ and radius $5.5 \times 10^{-15}$ m; $+54e$ and radius $6.2 \times 10^{-15}$ m, respectively. Assume that the charge is distributed uniformly throughout the volume of each spherical fragment and that their surfaces are initially in contact at rest. (The electrons surrounding the nucleus can be neglected.)
71. Two identical raindrops, each carrying surplus electrons on its surface to make a net charge $-q$ on each, collide and form a single drop of larger size. Before the collision, the characteristics of each drop are the following: (a) surface charge density $\sigma_0$, (b) electric field $\mathbf{E}_0$ at the surface, (c) electric potential $V_0$ at the surface (where $V \equiv 0$ at $r = \infty$). For the combined drop, find these three quantities in terms of their original values.
72. A Van de Graaff generator is operating so that the potential difference between the high-voltage electrode and the charging needles (points $B$ and $A$ in Figure 25.24) is $1.5 \times 10^4$ V. Calculate the power required to drive the belt (against electrical forces) at an instant when the effective current delivered to the high-voltage electrode is 500 $\mu$A.
73. Calculate the work that must be done to charge a spherical shell of radius $R$ to a total charge $Q$.
74. A point charge $+q$ is located at $x = -R$ and a point charge $-2q$ is located at the origin. Prove that the equipotential surface that has zero potential is a sphere centered at $(-4R/3, 0, 0)$ whose radius $r$ is given by $= 2R/3$.
75. From Gauss' law, the electric field set up by a uniform straight-line charge is

$$\mathbf{E} = \left(\frac{\lambda}{2\pi\epsilon_0 r}\right)\hat{\mathbf{r}}$$

where $\hat{\mathbf{r}}$ is a unit vector pointing radially away from the line, and $\lambda$ is the charge per meter along the line. Derive an expression for the potential difference between $r = r_1$ and $r = r_2$.
76. Consider two thin, conducting, spherical shells as in Figure 25.39. The inner shell has a radius $r_1 = 15$ cm and a charge of $+10$ nC. The outer shell has a radius $r_2 = 30$ cm and a charge of $-15$ nC. Find (a) the electric field $\mathbf{E}$ and (b) the electric potential $V$ in these regions, with $V \equiv 0$ at $r = \infty$:

| | |
|---|---|
| Region $A$: | inside the inner shell $(r < r_1)$ |
| Region $B$: | between the shells $(r_1 < r < r_2)$ |
| Region $C$: | outside the outer shell $(r > r_2)$ |

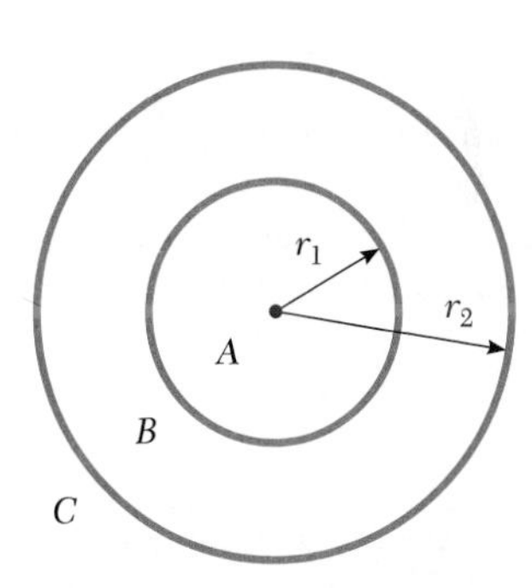


**Figure 25.39** (Problem 76).

77. The $x$ axis is the symmetry axis of a uniformly charged ring of radius $R$ and charge $Q$ (Fig. 25.40). A point charge $Q$ of mass $M$ is located at the center of the ring. When it is displaced slightly, the point charge accelerates along the $x$ axis to infinity. Show that the ultimate speed of the point charge is

$$v = \left(\frac{2kQ^2}{MR}\right)^{1/2}$$

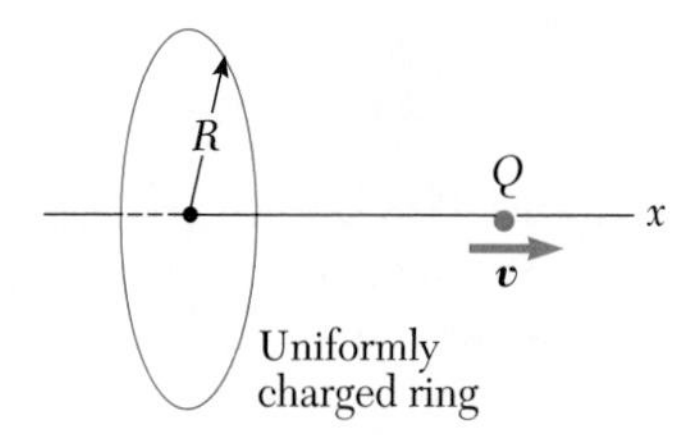


**Figure 25.40** (Problem 77).

78. The thin, uniformly charged rod shown in Figure 25.41 has a length $L$ and a linear charge density $\lambda$. Find an expression for the electric potential at point $P$, a distance $b$ along the positive $y$ axis.

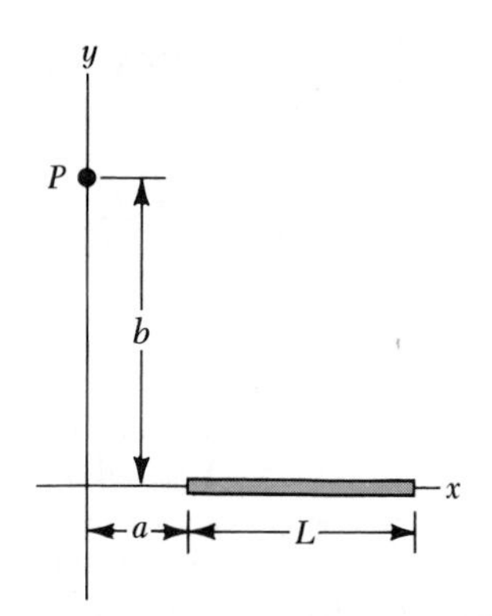


Figure 25.41 (Problem 78).

79. It is shown in Example 25.6 that the potential at a point $P$ a distance $d$ above one end of a uniformly charged rod of length $\ell$ lying along the $x$ axis is given by

$$V = \frac{kQ}{\ell} \ln\left(\frac{\ell + \sqrt{\ell^2 + d^2}}{d}\right)$$

Use this result to derive an expression for the $y$ component of the electric field at the point $P$. (*Hint:* Replace $d$ with $y$.)

80. Figure 25.42 shows several equipotential lines each labeled by its potential in volts. The distance between the lines of the square grid represents 1 cm. (a) Is the magnitude of the **E** field bigger at $A$ or $B$? Why? (b) What is **E** at $B$? (c) Represent what the **E** field looks like by drawing at least 8 field lines.

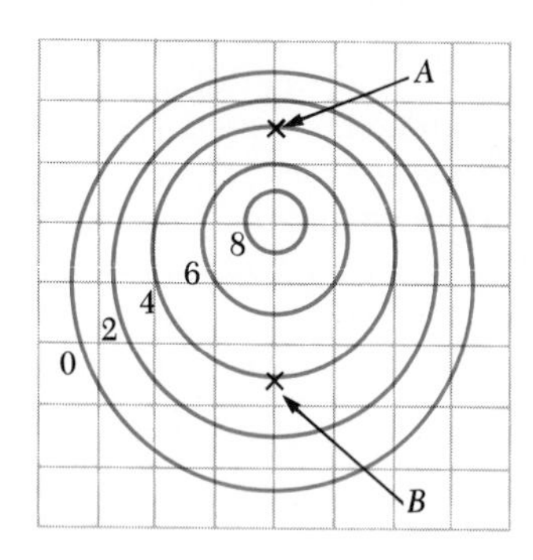


Figure 25.42 (Problem 80).

81. A dipole is located along the $y$ axis as in Figure 25.43. (a) At a point $P$, which is far from the dipole ($r \gg a$), the electric potential is given by

$$V = k\frac{p \cos\theta}{r^2}$$

where $p = 2qa$. Calculate the radial component of the associated electric field, $E_r$, and the perpendicular component, $E_\theta$. Note that $E_\theta = \frac{1}{r}\left(\frac{\partial V}{\partial \theta}\right)$. Do these results seem reasonable for $\theta = 90°$ and $0°$? for $r = 0$? (b) For the dipole arrangement shown, express $V$ in terms of rectangular coordinates using $r = (x^2 + y^2)^{1/2}$ and

$$\cos\theta = \frac{y}{(x^2 + y^2)^{1/2}}$$

Using these results and taking $r \gg a$, calculate the field components $E_x$ and $E_y$.

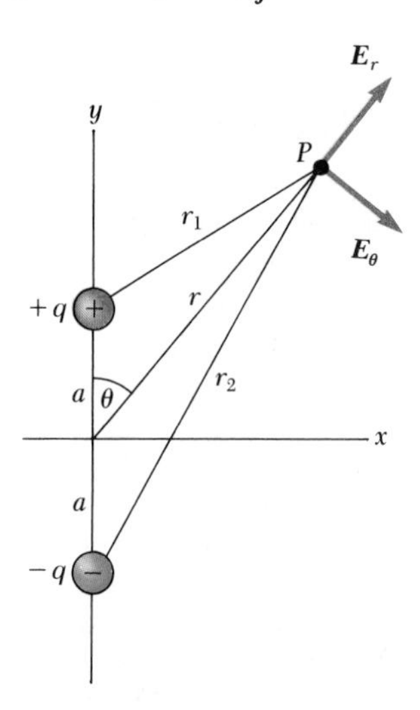


Figure 25.43 (Problem 81).

82. A disk of radius $R$ has a nonuniform surface charge density $\sigma = Cr$, where $C$ is a constant and $r$ is measured from the center of the disk (Fig. 25.44). Find (by direct integration) the potential at an axial point $P$ a distance $x$ from the disk.

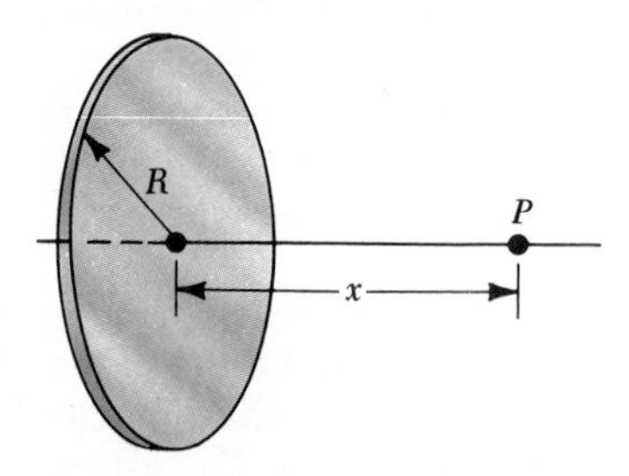


Figure 25.44 (Problem 82).

83. A solid sphere of radius $R$ has a *uniform* charge density $\rho$ and *total* charge $Q$. Derive an expression for the total electric potential energy of the charged sphere. (*Hint:* Imagine that the sphere is constructed by adding successive layers of concentric shells of charge $dq = (4\pi r^2\, dr)\rho$ and use $dU = V\, dq$.)

84. A Geiger-Müller counter is a type of radiation detector that essentially consists of a hollow cylinder (the cathode) of inner radius $r_a$ and a coaxial cylindrical

wire (the anode) of radius $r_b$ (Fig. 25.45). The charge per unit length on the anode is $\lambda$, while the charge per unit length on the cathode is $-\lambda$. (a) Show that the potential difference between the wire and the cylinder in the sensitive region of the detector is given by

$$V = 2k\lambda \ln\left(\frac{r_a}{r_b}\right)$$

(b) Show that the magnitude of the electric field over that region is given by

$$E = \frac{V}{\ln(r_a/r_b)}\left(\frac{1}{r}\right)$$

where $r$ is the distance from the center of the anode to the point where the field is to be calculated.

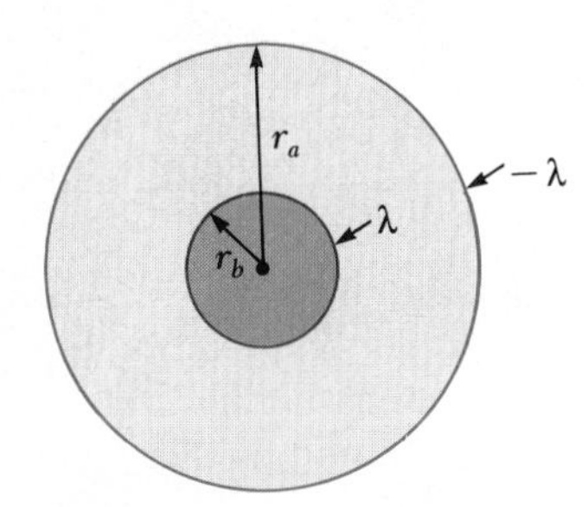


**Figure 25.45** (Problem 84).

## CALCULATOR/COMPUTER PROBLEMS

85. A uniformly charged rod is located along the $x$ axis as in Figure 23.14. The total charge on the rod is $+16 \times 10^{-10}$ C, $d = 1.0$ m, and $\ell = 2.0$ m. Estimate the electrical potential at $x = 0$ by approximating the rod to be (a) a point charge at $x = 2.0$ m, (b) two point charges (each of charge $+8 \times 10^{-10}$ C) at $x = 1.5$ m and $x = 2.5$ m, and (c) four point charges (each of charge $+4 \times 10^{-10}$ C) at $x = 1.25$ m, $x = 1.75$ m, $x = 2.25$ m, and $x = 2.75$ m. (d) Write a program that will enable you to extend your calculations to 256 equally spaced point charges, and compare your result with that given by the exact expression

$$V = k\frac{Q}{\ell}\ln\left(\frac{\ell + d}{d}\right)$$

86. A ring of radius 1 m has a uniform charge per unit length and a total charge of $+16 \times 10^{-10}$ C. The ring lies in the $yz$ plane, and its center is at $x = 0$, as in Figure 25.10. Estimate the electric potential along the $x$ axis at $x = 2$ m by approximating the ring to be (a) a point charge located at the origin, (b) two point charges (each of charge $+8 \times 10^{-10}$ C) diametrically opposite each other on the ring, and (c) four point charges (each of charge $+4 \times 10^{-10}$ C) symmetrically spaced on the ring. (d) Write a program that will enable you to extend your calculations to 64 point charges equally spaced on the ring, and compare your result with that given by the exact expression, Equation 25.18.

87. Three identical charges lie at the vertices of an equilateral triangle having sides 2 m long (Fig. 25.46). Locate positions of electrostatic equilibrium within the triangle.

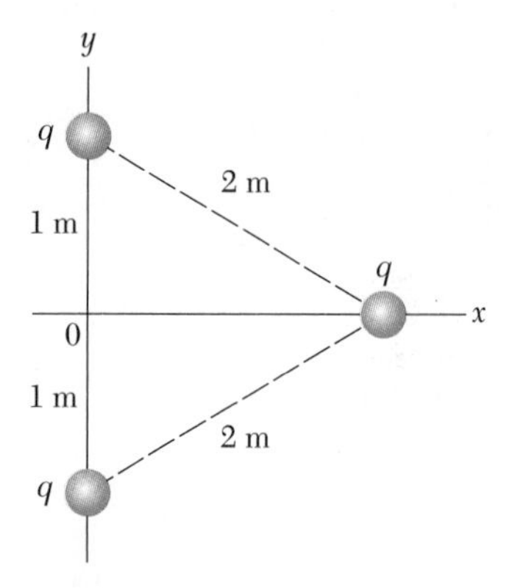


**Figure 25.46** (Problem 87).

# 26

# Capacitance and Dielectrics

*Very large capacitors are used as components in high-voltage power generators. In this photograph, a technician installs a safety switch in the "oil section" that contains the Marx generators seen at the right of the picture. This is the main high-voltage mechanical switch linking the Marx generators to accelerator modules. (Courtesy of Sandia National Laboratories)*

This chapter is concerned with the properties of capacitors, devices that store charge. Capacitors are commonly used in a variety of electrical circuits. For instance, they are used (1) to tune the frequency of radio receivers, (2) as filters in power supplies, (3) to eliminate sparking in automobile ignition systems, and (4) as energy-storing devices in electronic flash units.

A capacitor basically consists of two conductors separated by an insulator. We shall see that the capacitance of a given device depends on its geometry and on the material separating the charged conductors, called a *dielectric*. A dielectric is an insulating material having distinctive electrical properties that can best be understood as a consequence of the properties of atoms.

## 26.1 DEFINITION OF CAPACITANCE

Consider two conductors having a potential difference $V$ between them. Let us assume that the conductors have equal and opposite charges as in Figure 26.1. This can be accomplished by connecting the two uncharged conductors to the terminals of a battery. Such a combination of two conductors is called a *capacitor*. The potential difference $V$ is found to be proportional to the magnitude of the charge $Q$ on the capacitor.[1]

The **capacitance,** $C$, of a capacitor is defined as the ratio of the magnitude of the charge on either conductor to the magnitude of the potential difference between them:

$$C \equiv \frac{Q}{V} \tag{26.1}$$

Definition of capacitance

Note that by definition *capacitance is always a positive quantity.* Furthermore, since the potential difference increases as the stored charge increases, the ratio $Q/V$ is constant for a given capacitor. Therefore, the capacitance of a device is a measure of its ability to store charge and electrical potential energy.

From Equation 26.1, we see that capacitance has SI units of coulombs per volt. The SI unit of capacitance is the **farad** (F), in honor of Michael Faraday. That is,

$$[\text{Capacitance}] = 1\ \text{F} = 1\ \text{C/V}$$

The farad is a very large unit of capacitance. In practice, typical devices have capacitances ranging from microfarads ($1\ \mu\text{F} = 10^{-6}$ F) to picofarads ($1\ \text{pF} = 10^{-12}$ F). As a practical note, capacitors are often labeled mF for microfarads and mmF for micromicrofarads (picofarads).

As we shall show in the next section, the capacitance of a device depends on the geometrical arrangement of the conductors. To illustrate this point, let us calculate the capacitance of an isolated spherical conductor of radius $R$ and charge $Q$. (The second conductor can be taken as a concentric hollow conducting sphere of infinite radius.) Since the potential of the sphere is simply $kQ/R$ (where $V = 0$ at infinity), its capacitance is given by

$$C = \frac{Q}{V} = \frac{Q}{kQ/R} = \frac{R}{k} = 4\pi\epsilon_0 R \tag{26.2}$$

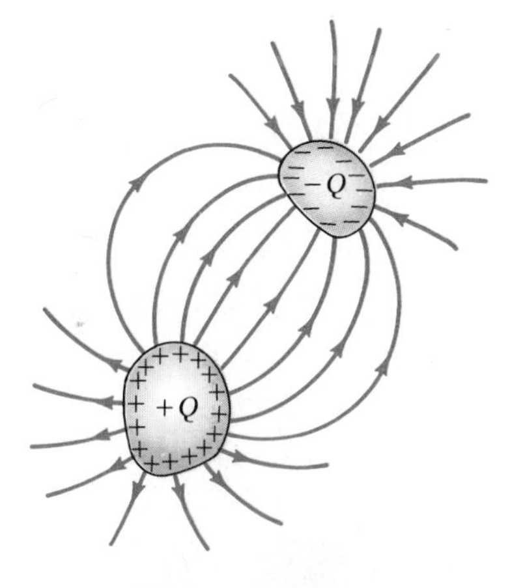


**Figure 26.1** A capacitor consists of two conductors isolated from each other and their surroundings. Once the capacitor is charged, the two conductors carry equal but opposite charges.

This shows that the capacitance of an isolated charged sphere is proportional to its radius and is independent of both the charge and the potential difference. For example, an isolated metallic sphere of radius 0.15 m has a capacitance of

$$C = 4\pi\epsilon_0 R = 4\pi(8.85 \times 10^{-12}\ \text{C}^2/\text{N}\cdot\text{m}^2)(0.15\ \text{m}) = 17\ \text{pF}$$

## 26.2 CALCULATION OF CAPACITANCE

The capacitance of a pair of oppositely charged conductors can be calculated in the following manner. A convenient charge of magnitude $Q$ is assumed, and the potential difference is calculated using the techniques described in the previous chapter. One then simply uses $C = Q/V$ to evaluate the capacitance.

[1] The proportionality between the potential difference and charge on the conductors can be proved from Coulomb's law or by experiment.

As you might expect, the calculation is relatively easy to perform if the geometry of the capacitor is simple.

Let us illustrate this with three geometries that we are all familiar with, namely, two parallel plates, two concentric cylinders, and two concentric spheres. In these examples, we shall assume that the charged conductors are separated by a vacuum. The effect of a dielectric material between the conductors will be treated in Section 26.5.

### The Parallel-Plate Capacitor

Two parallel plates of equal area $A$ are separated by a distance $d$ as in Figure 26.2. One plate has a charge $+Q$, the other, charge $-Q$. The charge per unit area on either plate is $\sigma = Q/A$. If the plates are very close together (compared with their length and width), we can neglect end effects and assume that the electric field is uniform between the plates and zero elsewhere. According to Example 24.6, the electric field between the plates is given by

$$E = \frac{\sigma}{\epsilon_0} = \frac{Q}{\epsilon_0 A}$$

The potential difference between the plates equals $Ed$; therefore

$$V = Ed = \frac{Qd}{\epsilon_0 A}$$

Substituting this result into Equation 26.1, we find that the capacitance is given by

$$C = \frac{Q}{V} = \frac{Q}{Qd/\epsilon_0 A}$$

$$C = \frac{\epsilon_0 A}{d} \tag{26.3}$$

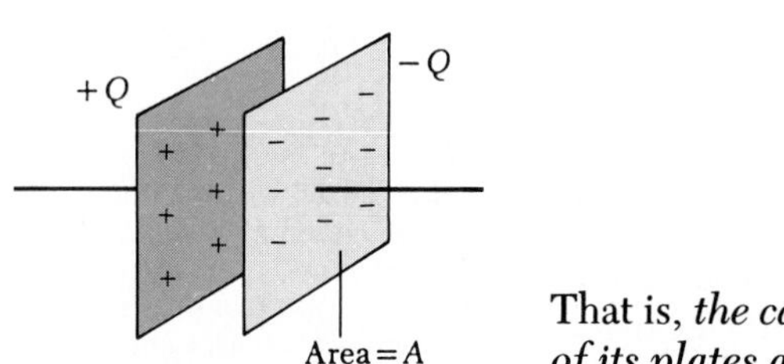


Figure 26.2 A parallel-plate capacitor consists of two parallel plates each of area $A$, separated by a distance $d$. The plates carry equal and opposite charges.

That is, *the capacitance of a parallel-plate capacitor is proportional to the area of its plates and inversely proportional to the plate separation.*

As you can see from the definition of capacitance, $C = Q/V$, the amount of charge a given capacitor is able to store for a given potential difference across its plates increases as the capacitance increases. Therefore, it seems reasonable that a capacitor constructed from plates having a large area should be able to store a large charge. The amount of charge needed to produce a given potential difference increases with decreasing plate separation.

A careful inspection of the electric field lines for a parallel-plate capacitor reveals that the field is uniform in the central region between the plates as in Figure 26.3a. However, the field is nonuniform at the edges of the plates. Figure 26.3b is a photograph of the electric field pattern of a parallel-plate capacitor showing the nonuniform field lines at its edges.

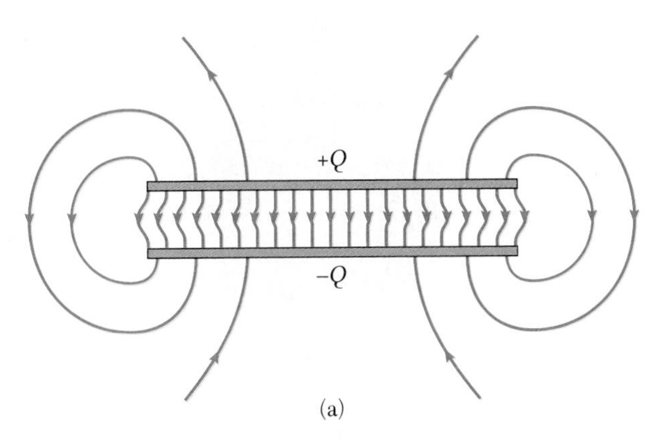


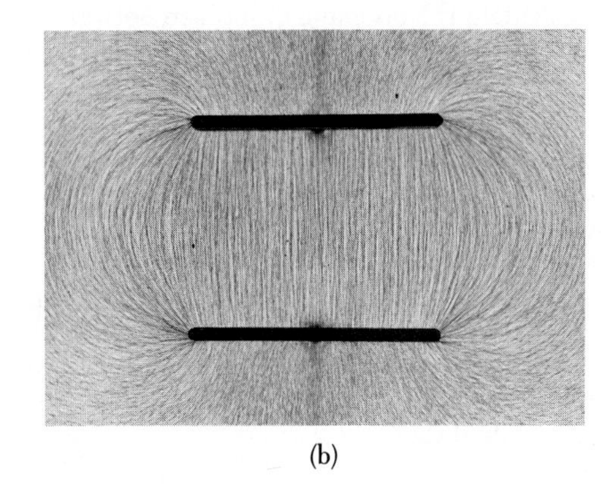


**Figure 26.3** (a) The electric fields between the plates of a parallel-plate capacitor is uniform near its center, but is nonuniform near its edges. (b) Electric field pattern of two oppositely charged conducting parallel plates. Small pieces of thread on an oil surface align with the electric field. Note the nonuniform nature of the electric field at the ends of the plates. Such end effects can be neglected if the plate separation is small compared to the length of the plates. (Courtesy of Harold M. Waage, Princeton University)

### EXAMPLE 26.1 Parallel-Plate Capacitor

**A parallel-plate capacitor has an area of $A = 2\ \text{cm}^2 = 2 \times 10^{-4}\ \text{m}^2$ and a plate separation of $d = 1\ \text{mm} = 10^{-3}\ \text{m}$. Find its capacitance.**

*Solution* From Equation 26.3, we find

$$C = \epsilon_0 \frac{A}{d} = \left(8.85 \times 10^{-12}\ \frac{\text{C}^2}{\text{N}\cdot\text{m}^2}\right)\left(\frac{2 \times 10^{-4}\ \text{m}^2}{1 \times 10^{-3}\ \text{m}}\right)$$

$$= 1.77 \times 10^{-12}\ \text{F} = \boxed{1.77\ \text{pF}}$$

*Exercise 1* If the plate separation of this capacitor is increased to 3 mm, find its capacitance. Answer: 0.59 pF.

### EXAMPLE 26.2 The Cylindrical Capacitor

**A cylindrical conductor of radius $a$ and charge $+Q$ is concentric with a larger cylindrical shell of radius $b$ and charge $-Q$ (Fig. 26.4a). Find the capacitance of this cylindrical capacitor if its length is $\ell$.**

*Solution* If we assume that $\ell$ is long compared with $a$ and $b$, we can neglect end effects. In this case, the field is perpendicular to the axis of the cylinders and is confined to the region between them (Fig. 26.4b). We must

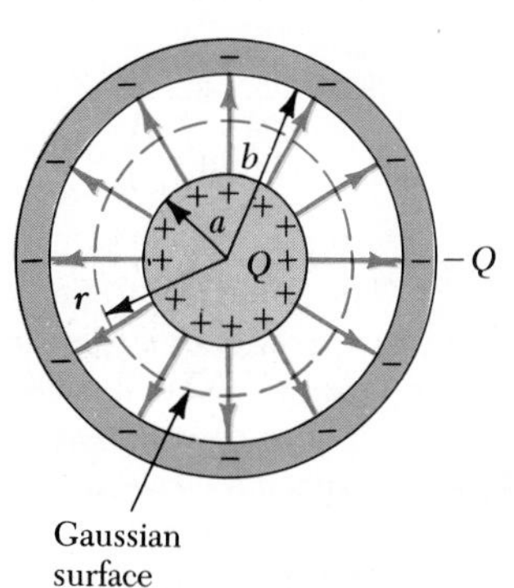


**Figure 26.4** (Example 26.2) (a) A cylindrical capacitor consists of a cylindrical conductor of radius $a$ and length $\ell$ surrounded by a coaxial cylindrical shell of radius $b$. (b) The end view of a cylindrical capacitor. The blue broken line represents the end of the cylindrical gaussian surface of radius $r$ and length $\ell$.

first calculate the potential difference between the two cylinders, which is given in general by

$$V_b - V_a = -\int_a^b \mathbf{E} \cdot d\mathbf{s}$$

where $\mathbf{E}$ is the electric field in the region $a < r < b$. In Chapter 24, we showed using Gauss' law that the electric field of a cylinder of charge per unit length $\lambda$ is given by $2k\lambda/r$. The same result applies here, since the outer cylinder does not contribute to the electric field inside it. Using this result and noting that $\mathbf{E}$ is along $r$ in Figure 26.4b, we find that

$$V_b - V_a = -\int_a^b E_r\, dr = -2k\lambda \int_a^b \frac{dr}{r} = -2k\lambda \ln\left(\frac{b}{a}\right)$$

Substituting this into Equation 26.1 and using the fact that $\lambda = Q/\ell$, we get

$$C = \frac{Q}{V} = \frac{Q}{\dfrac{2kQ}{\ell}\ln\left(\dfrac{b}{a}\right)} = \frac{\ell}{2k\ln\left(\dfrac{b}{a}\right)} \qquad (26.4)$$

$V$ is the magnitude of the potential difference given by $2k\lambda \ln(b/a)$, a *positive* quantity. That is, $V = V_a - V_b$ is *positive* since the inner cylinder is at the higher potential.

Our result for $C$ makes sense since it shows that the capacitance is proportional to the length of the cylinders. As you might expect, the capacitance also depends on the radii of the two cylindrical conductors. As an example, a coaxial cable consists of two concentric cylindrical conductors of radii $a$ and $b$ separated by an insulator. The cable carries currents in opposite directions in the inner and outer conductors. Such a geometry is especially useful for shielding an electrical signal from external influences. From Equation 26.4, we see that the capacitance per unit length of a coaxial cable is given by

$$\frac{C}{\ell} = \frac{1}{2k\ln\left(\dfrac{b}{a}\right)}$$

**EXAMPLE 26.3 The Spherical Capacitor**

A spherical capacitor consists of a spherical conducting shell of radius $b$ and charge $-Q$ that is concentric with a smaller conducting sphere of radius $a$ and charge $+Q$ (Fig. 26.5). Find its capacitance.

*Solution* As we showed in Chapter 24, the field outside a spherically symmetric charge distribution is radial and given by $kQ/r^2$. In this case, this corresponds to the field between the spheres ($a < r < b$). (The field is zero elsewhere.) From Gauss' law we see that only the inner sphere contributes to this field. Thus, the potential difference between the spheres is given by

$$V_b - V_a = -\int_a^b E_r\, dr = -kQ\int_a^b \frac{dr}{r^2} = kQ\left[\frac{1}{r}\right]_a^b$$

$$= kQ\left(\frac{1}{b} - \frac{1}{a}\right)$$

The magnitude of the potential difference is given by

$$V = V_a - V_b = kQ\frac{(b-a)}{ab}$$

Substituting this into Equation 26.1, we get

$$C = \frac{Q}{V} = \frac{ab}{k(b-a)} \qquad (26.5)$$

Exercise 2 Show that as the radius $b$ of the outer sphere approaches infinity, the capacitance approaches the value $a/k = 4\pi\epsilon_0 a$. This is consistent with the result obtained earlier (Eq. 26.2).

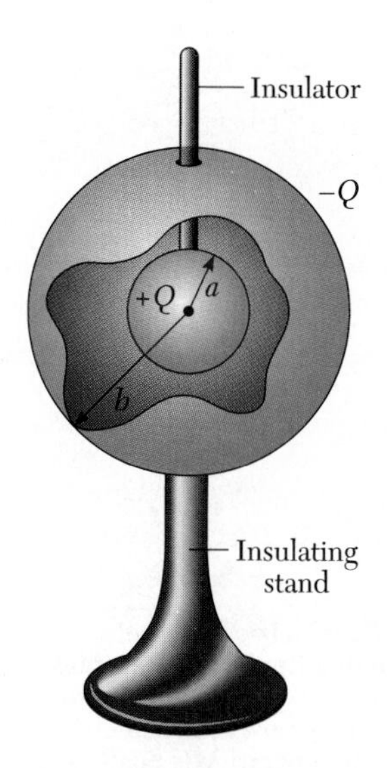


**Figure 26.5** (Example 26.3) A spherical capacitor consists of an inner sphere of radius $a$ surrounded by a concentric spherical shell of radius $b$. The electric field between the spheres is radial outward if the inner sphere is positively charged.

## 26.3 COMBINATIONS OF CAPACITORS

Two or more capacitors are often combined in circuits in several ways. The equivalent capacitance of certain combinations can be calculated using methods described in this section. The circuit symbols for capacitors and batteries, together with their color codes, are given in Figure 26.6. The positive terminal of the battery is at the higher potential and is represented by the longer vertical line in the battery symbol.

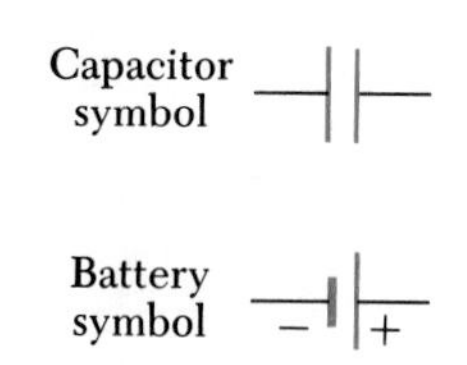


**Figure 26.6** Circuit symbols for capacitors and batteries. Note that capacitors are in blue, while batteries are in red.

### Parallel Combination

Two capacitors connected as shown in Figure 26.7a are known as a *parallel combination* of capacitors. The left plates of the capacitors are connected by a conducting wire to the positive terminal of the battery and are therefore at the same potential. Likewise, the right plates are connected to the negative terminal of the battery. When the capacitors are first connected in the circuit, electrons are transferred through the battery from the left plates to the right plates, leaving the left plates positively charged and the right plates negatively charged. The energy source for this charge transfer is the internal chemical energy stored in the battery, which is converted to electrical energy. The flow of charge ceases when the voltage across the capacitors is equal to that of the battery. The capacitors reach their maximum charge when the flow of charge ceases. Let us call the maximum charges on the two capacitors $Q_1$ and $Q_2$. Then the *total charge*, $Q$, stored by the two capacitors is

$$Q = Q_1 + Q_2 \qquad (26.6)$$

Suppose we wish to replace these two capacitors by one equivalent capacitor having a capacitance $C_{eq}$. This equivalent capacitor must have exactly the

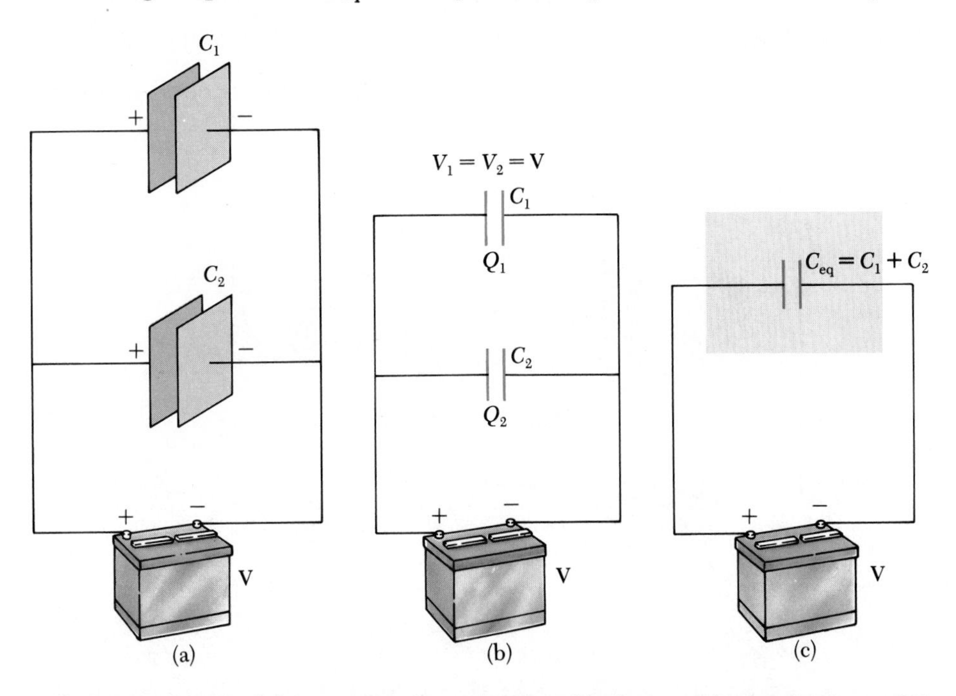


**Figure 26.7** (a) A parallel connection of two capacitors. (b) The circuit diagram for the parallel combination. (c) The potential difference is the same across each capacitor, and the equivalent capacitance is $C_{eq} = C_1 + C_2$.

same external effect on the circuit as the original two. That is, it must store $Q$ units of charge. We also see from Figure 26.7b that

> the potential difference across each capacitor in the parallel circuit is the same and is equal to the voltage of the battery, $V$.

From Figure 26.7c, we see that the voltage across the equivalent capacitor is also $V$. Thus, we have

$$Q_1 = C_1 V \qquad Q_2 = C_2 V$$

and, for the equivalent capacitor,

$$Q = C_{eq} V$$

Substituting these relations into Equation 26.6 gives

$$C_{eq} V = C_1 V + C_2 V$$

or

$$C_{eq} = C_1 + C_2 \qquad \begin{pmatrix}\text{parallel} \\ \text{combination}\end{pmatrix} \tag{26.7}$$

If we extend this treatment to three or more capacitors connected in parallel, the equivalent capacitance is found to be

$$C_{eq} = C_1 + C_2 + C_3 + \cdots \qquad \begin{pmatrix}\text{parallel} \\ \text{combination}\end{pmatrix} \tag{26.8}$$

Thus we see that *the equivalent capacitance of a parallel combination of capacitors is larger than any of the individual capacitances.*

### Series Combination

Now consider two capacitors connected in *series*, as illustrated in Figure 26.8a.

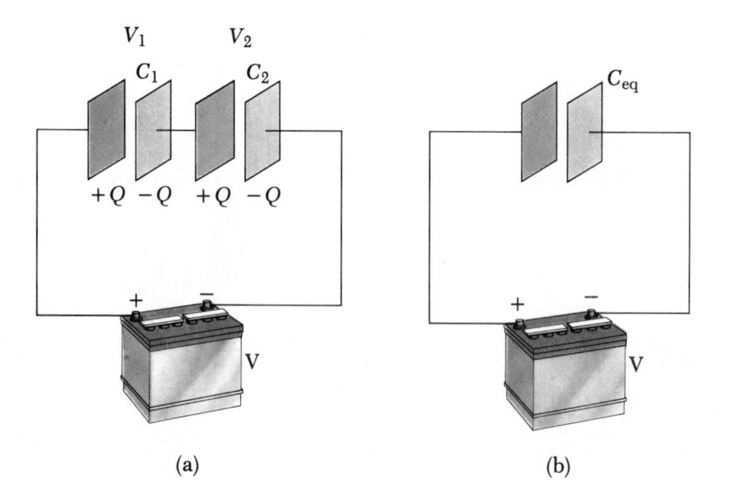


**Figure 26.8** A series connection of two capacitors. The charge on each capacitor is the same, and the equivalent capacitance can be calculated from the relation $\frac{1}{C_{eq}} = \frac{1}{C_1} + \frac{1}{C_2}$.

For this series combination of capacitors, the magnitude of the charge must be the same on all the plates.

To see why this must be true, let us consider the charge transfer process in some detail. We start with uncharged capacitors and follow what happens just after a battery is connected to the circuit. When the battery is connected, electrons are transferred from the left plate of $C_1$ to the right plate of $C_2$ through the battery. As this negative charge accumulates on the right plate of $C_2$, an equivalent amount of negative charge is forced off the left plate of $C_2$, leaving it with an excess positive charge. The negative charge leaving the left plate of $C_2$ accumulates on the right plate of $C_1$, where again an equivalent amount of negative charge leaves the left plate. The result of this is that *all of the right plates gain a charge of* $-Q$ *while all of the left plates have a charge of* $+Q$.

Suppose an equivalent capacitor performs the same function as the series combination. After it is fully charged, *the equivalent capacitor must end up with a charge of* $-Q$ *on its right plate and* $+Q$ *on its left plate.* By applying the definition of capacitance to the circuit shown in Figure 26.8b, we have

$$V = \frac{Q}{C_{eq}}$$

where $V$ is the potential difference between the terminals of the battery and $C_{eq}$ is the equivalent capacitance. From Figure 26.8a, we see that

$$V = V_1 + V_2 \tag{26.9}$$

where $V_1$ and $V_2$ are the potential differences across capacitors $C_1$ and $C_2$. In general, the potential difference across any number of capacitors in series is equal to the sum of the potential differences across the individual capacitors. Since $Q = CV$ can be applied to each capacitor, the potential difference across each is given by

$$V_1 = \frac{Q}{C_1} \qquad V_2 = \frac{Q}{C_2}$$

Substituting these expressions into Equation 26.9, and noting that $V = Q/C_{eq}$, we have

$$\frac{Q}{C_{eq}} = \frac{Q}{C_1} + \frac{Q}{C_2}$$

Cancelling $Q$, we arrive at the relationship

$$\frac{1}{C_{eq}} = \frac{1}{C_1} + \frac{1}{C_2} \qquad \begin{pmatrix}\text{series}\\ \text{combination}\end{pmatrix} \tag{26.10}$$

If this analysis is applied to three or more capacitors connected in series, the equivalent capacitance is found to be

$$\frac{1}{C_{eq}} = \frac{1}{C_1} + \frac{1}{C_2} + \frac{1}{C_3} + \cdots \qquad \begin{pmatrix}\text{series}\\ \text{combination}\end{pmatrix} \tag{26.11}$$

This shows that *the equivalent capacitance of a series combination is always less than any individual capacitance in the combination.*

**EXAMPLE 26.4 Equivalent Capacitance**
Find the equivalent capacitance between $a$ and $b$ for the combination of capacitors shown in Figure 26.9a. All capacitances are in $\mu$F.

*Solution* Using Equations 26.8 and 26.11, we reduce the combination step by step as indicated in the figure. The 1-$\mu$F and 3-$\mu$F capacitors are in *parallel* and combine according to $C_{eq} = C_1 + C_2$. Their equivalent capacitance is 4 $\mu$F. Likewise, the 2-$\mu$F and 6-$\mu$F capacitors are also in *parallel* and have an equivalent capacitance of 8 $\mu$F. The upper branch in Figure 26.9b now consists of two 4-$\mu$F capacitors in *series,* which combine according to

$$\frac{1}{C_{eq}} = \frac{1}{C_1} + \frac{1}{C_2} = \frac{1}{4\ \mu\text{F}} + \frac{1}{4\ \mu\text{F}} = \frac{1}{2\ \mu\text{F}}$$

$$C_{eq} = 2\ \mu\text{F}$$

Likewise, the lower branch in Figure 26.9b consists of two 8-$\mu$F capacitors in *series,* which give an equivalent of 4 $\mu$F. Finally, the 2-$\mu$F and 4-$\mu$F capacitors in Figure 26.9c are in *parallel* and have an equivalent capacitance of 6 $\mu$F. Hence, the equivalent capacitance of the circuit is 6 $\mu$F.

Exercise 3 Consider three capacitors having capacitances of 3 $\mu$F, 6 $\mu$F, and 12 $\mu$F. Find their equivalent capacitance if they are connected (a) in parallel, (b) in series.
Answer (a) 21 $\mu$F, (b) 1.71 $\mu$F.

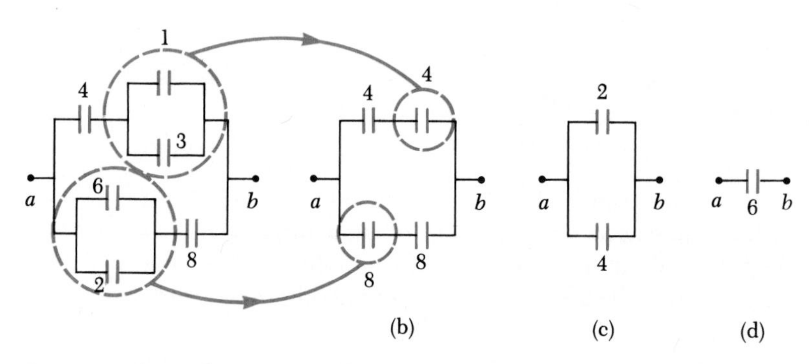


**Figure 26.9** (Example 26.4) To find the equivalent combination of the capacitors in (a), the various combinations are reduced in steps as indicated in (b), (c), and (d), using the series and parallel rules described in the text.

## 26.4 ENERGY STORED IN A CHARGED CAPACITOR

Almost everyone that works with electronic equipment has at some time verified that a capacitor is able to store energy. If the plates of a charged capacitor are connected together by a conductor, such as a wire, charge will transfer from one plate to the other until the two are uncharged. The discharge can often be observed as a visible spark. If you should accidentally touch the opposite plates of a charged capacitor, your fingers would act as a pathway by which the capacitor could discharge, which would result in an electric shock. The degree of shock you would receive depends on the capacitance and voltage applied to the capacitor. Such a shock could be fatal where high voltages are present, such as in the power supply of a television set.

Consider a parallel-plate capacitor that is initially uncharged, so that the initial potential difference across the plates is zero. Now imagine that the capacitor is connected to a battery and develops a maximum charge $Q$. We shall assume that the capacitor is charged *slowly* so that the problem can be considered as an electrostatic system. The final potential difference across the capacitor is $V = Q/C$. Since the initial potential difference is zero, the *average* potential difference during the charging process is $V/2 = Q/2C$. From this we might conclude that the work needed to charge the capacitor is given by $W = QV/2 = Q^2/2C$. Although this result is correct, a more detailed proof is desirable and is now given.

Suppose that $q$ is the charge on the capacitor at some instant during the charging process. At the same instant, the potential difference across the capacitor is $V = q/C$. The work necessary[2] to transfer an increment of charge $dq$ from the plate of charge $-q$ to the plate of charge $q$ (which is at the higher potential) is given by

$$dW = V\,dq = \frac{q}{C}\,dq$$

Thus, the total work required to charge the capacitor from $q = 0$ to some final charge $q = Q$ is given by

$$W = \int_0^Q \frac{q}{C}\,dq = \frac{Q^2}{2C}$$

But the work done in charging the capacitor can be considered as potential energy $U$ stored in the capacitor. Using $Q = CV$, we can express the electrostatic energy stored in a charged capacitor in the following alternative forms:

$$U = \frac{Q^2}{2C} = \tfrac{1}{2}QV = \tfrac{1}{2}CV^2 \qquad (26.12)$$

**Energy stored in a charged capacitor**

This result applies to *any* capacitor, regardless of its geometry. We see that the stored energy increases as $C$ increases and as the potential difference increases. In practice, there is a limit to the maximum energy (or charge) that can be stored. This is because electrical discharge will ultimately occur between the plates of the capacitor at a sufficiently large value of $V$. For this reason, capacitors are usually labeled with a maximum operating voltage.

The energy stored in a capacitor can be considered as being stored in the electric field created between the plates as the capacitor is charged. This description is reasonable in view of the fact that the electric field is proportional to the charge on the capacitor. For a parallel-plate capacitor, the potential difference is related to the electric field through the relationship $V = Ed$. Furthermore, its capacitance is given by $C = \epsilon_0 A/d$. Substituting these expressions into Equation 26.12 gives

$$U = \tfrac{1}{2}\frac{\epsilon_0 A}{d}(E^2 d^2) = \tfrac{1}{2}(\epsilon_0 A d)E^2 \qquad (26.13)$$

**Energy stored in a parallel-plate capacitor**

Since the volume of a parallel-plate capacitor that is occupied by the electric field is $Ad$, the *energy per unit volume* $u = U/Ad$, called the *energy density*, is

$$u = \tfrac{1}{2}\epsilon_0 E^2 \qquad (26.14)$$

**Energy density in an electric field**

[2] One mechanical analog of this process is the work required to raise a mass through some vertical distance in the presence of gravity.

Although Equation 26.14 was derived for a parallel-plate capacitor, the expression is generally valid. That is, the *energy density in any electrostatic field is proportional to the square of the electric field intensity at a given point.* (A formal proof of this statement is given in intermediate and advanced courses in electricity and magnetism.)

EXAMPLE 26.5 Rewiring Two Charged Capacitors □

Two capacitors $C_1$ and $C_2$ (where $C_1 > C_2$) are charged to the same potential difference $V_0$, but with opposite polarity. The charged capacitors are removed from the battery, and their plates are connected as shown in Figure 26.10a. The switches $S_1$ and $S_2$ are then closed as in Figure 26.10b. (a) Find the final potential difference between $a$ and $b$ after the switches are closed.

*Solution* The charges on the left-hand plates of the capacitors *before* the switches are closed are given by

$$Q_1 = C_1V_0 \qquad \text{and} \qquad Q_2 = -C_2V_0$$

The negative sign for $Q_2$ is necessary since this capacitor's polarity is *opposite* that of capacitor $C_1$. After the switches are closed, the charges on the plates redistribute until the total charge $Q$ shared by both capacitors is

$$Q = Q_1 + Q_2 = (C_1 - C_2)V_0$$

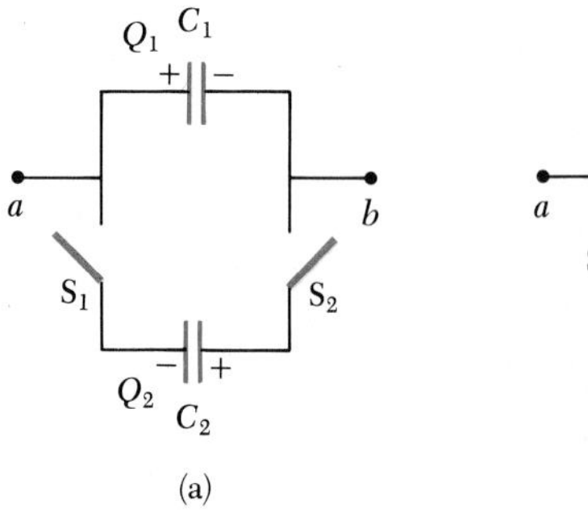


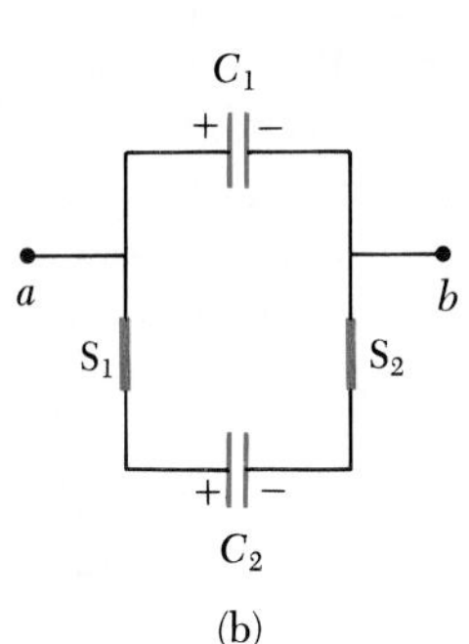


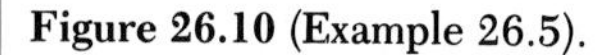

**Figure 26.10** (Example 26.5).

The two capacitors are now in *parallel*, so the final potential difference across each is the *same* and given by

$$V = \frac{Q}{C_1 + C_2} = \left(\frac{C_1 - C_2}{C_1 + C_2}\right)V_0$$

(b) Find the total energy stored in the capacitors before and after the switches are closed.

*Solution* Before the switches are closed, the total energy stored in the capacitors is given by

$$U_i = \tfrac{1}{2}C_1V_0^2 + \tfrac{1}{2}C_2V_0^2 = \tfrac{1}{2}(C_1 + C_2)V_0^2$$

After the switches are closed and the capacitors have reached an equilibrium charge, the total energy stored in the capacitors is given by

$$U_f = \tfrac{1}{2}C_1V^2 + \tfrac{1}{2}C_2V^2 = \tfrac{1}{2}(C_1 + C_2)V^2$$

$$= \tfrac{1}{2}(C_1 + C_2)\left(\frac{C_1 - C_2}{C_1 + C_2}\right)^2 V_0^2 = \left(\frac{C_1 - C_2}{C_1 + C_2}\right)^2 U_i$$

Therefore, the ratio of the final to the initial energy stored is

$$\frac{U_f}{U_i} = \left(\frac{C_1 - C_2}{C_1 + C_2}\right)^2$$

This shows that the final energy is *less* than the initial energy. At first, you might think that energy conservation has been violated, but this is not the case since we have assumed that the circuit is ideal. Part of the missing energy appears as heat energy in the connecting wires, which have resistance, and part of the energy is radiated away in the form of electromagnetic waves (Chapter 34).

## 26.5 CAPACITORS WITH DIELECTRICS

A *dielectric* is a nonconducting material, such as rubber, glass, or waxed paper. When a dielectric material is inserted between the plates of a capacitor, the capacitance increases. If the dielectric completely fills the space between the plates, the capacitance increases by a dimensionless factor $\kappa$, called the **dielectric constant.**

The following experiment can be performed to illustrate the effect of a dielectric in a capacitor. Consider a parallel-plate capacitor of charge $Q_0$ and capacitance $C_0$ in the absence of a dielectric. The potential difference across the capacitor as measured by a voltmeter is $V_0 = Q_0/C_0$ (Fig. 26.11a). Notice

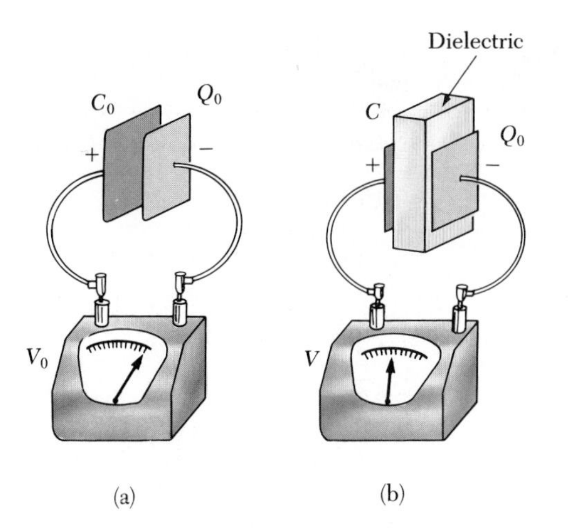


**Figure 26.11** When a dielectric is inserted between the plates of a charged capacitor, the charge on the plates remains unchanged, but the potential difference as recorded by an electrostatic voltmeter is reduced from $V_0$ to $V = V_0/\kappa$. Thus, the capacitance *increases* in the process by the factor $\kappa$.

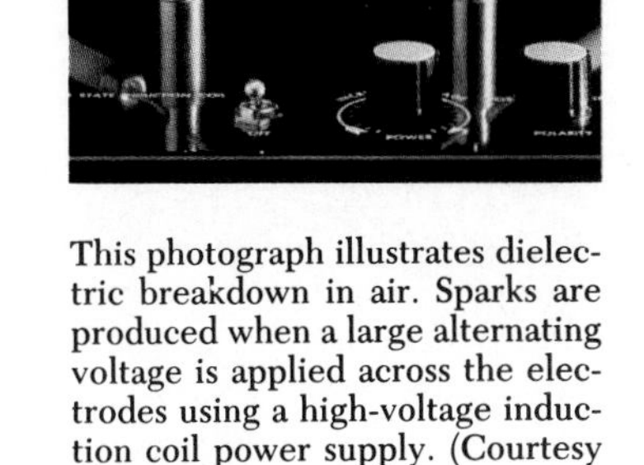

This photograph illustrates dielectric breakdown in air. Sparks are produced when a large alternating voltage is applied across the electrodes using a high-voltage induction coil power supply. (Courtesy of CENCO)

that the capacitor circuit is *open*, that is, the plates of the capacitor are *not* connected to a battery and charge cannot flow through an ideal voltmeter. (We shall discuss the voltmeter further in Chapter 28.) Hence, there is *no* path by which charge can flow and alter the charge on the capacitor. If a dielectric is now inserted between the plates as in Figure 26.11b, it is found that the voltmeter reading *decreases* by a factor $\kappa$ to a value $V$, where

$$V = \frac{V_0}{\kappa}$$

Since $V < V_0$, we see that $\kappa > 1$.

Since the charge $Q_0$ on the capacitor *does not change*, we conclude that the capacitance must change to the value

$$C = \frac{Q_0}{V} = \frac{Q_0}{V_0/\kappa} = \kappa\frac{Q_0}{V_0}$$

$$C = \kappa C_0 \tag{26.15}$$

where $C_0$ is the capacitance in the absence of the dielectric. That is, the capacitance *increases* by the factor $\kappa$ when the dielectric completely fills the region between the plates.[3] For a parallel-plate capacitor, where $C_0 = \epsilon_0 A/d$, we can express the capacitance when the capacitor is filled with a dielectric as

$$C = \kappa\frac{\epsilon_0 A}{d} \tag{26.16}$$

The capacitance of a filled capacitor is greater than that of an empty one by a factor $\kappa$.

From Equations 26.3 and 26.16, it would appear that the capacitance could be made very large by decreasing $d$, the distance between the plates. In

[3] If another experiment is performed in which the dielectric is introduced while the potential difference remains constant by means of a battery, the charge increases to a value $Q = \kappa Q_0$. The additional charge is supplied by the battery and the capacitance still increases by the factor $\kappa$.

**TABLE 26.1 Dielectric Constants and Dielectric Strengths of Various Materials at Room Temperature**

| Material | Dielectric Constant $\kappa$ | Dielectric Strength[a] (V/m) |
|---|---|---|
| Vacuum | 1.00000 | — |
| Air (dry) | 1.00059 | $3 \times 10^6$ |
| Bakelite | 4.9 | $24 \times 10^6$ |
| Fused quartz | 3.78 | $8 \times 10^6$ |
| Pyrex glass | 5.6 | $14 \times 10^6$ |
| Polystyrene | 2.56 | $24 \times 10^6$ |
| Teflon | 2.1 | $60 \times 10^6$ |
| Neoprene rubber | 6.7 | $12 \times 10^6$ |
| Nylon | 3.4 | $14 \times 10^6$ |
| Paper | 3.7 | $16 \times 10^6$ |
| Strontium titanate | 233 | $8 \times 10^6$ |
| Water | 80 | — |
| Silicone oil | 2.5 | $15 \times 10^6$ |

[a] The dielectric strength equals the maximum electric field that can exist in a dielectric without electrical breakdown.

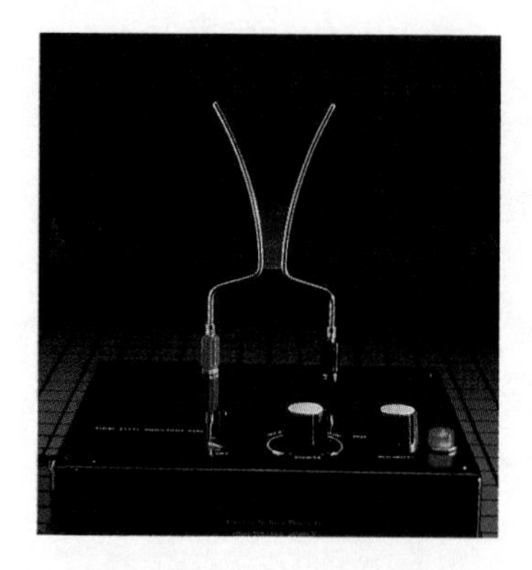

A continuous electrical discharge is produced between two electrodes when the applied voltage produces an electric field that exceeds the dielectric strength of air. What happens to the air in the vicinity of this discharge? As the discharge continues, the "sparks" rise to the top. Can you explain this behavior? (Courtesy CENCO)

practice, the lowest value of $d$ is limited by the electrical discharge that could occur through the dielectric medium separating the plates. For any given separation $d$, the maximum voltage that can be applied to a capacitor without causing a discharge depends on the *dielectric strength* (maximum electric field intensity) of the dielectric, which for air is equal to $3 \times 10^6$ V/m. If the field strength in the medium exceeds the dielectric strength, the insulating properties will break down and the medium will begin to conduct. Most insulating materials have dielectric strengths and dielectric constants greater than that of air, as Table 26.1 indicates. Thus, we see that a dielectric provides the following advantages:

1. A dielectric increases the capacitance of a capacitor.
2. A dielectric increases the maximum operating voltage of a capacitor.
3. A dielectric may provide mechanical support between the conducting plates.

### Types of Capacitors

Commercial capacitors are often made using metal foil interlaced with thin sheets of paraffin-impregnated paper or mylar, which serves as the dielectric material. These alternate layers of metal foil and dielectric are then rolled into the shape of a cylinder to form a small package (Fig. 26.12a). High-voltage capacitors commonly consist of a number of interwoven metal plates immersed in silicone oil (Fig. 26.12b). Small capacitors are often constructed from ceramic materials. Variable capacitors (typically 10 to 500 pF) usually consist of two interwoven sets of metal plates, one fixed and the other movable, with air as the dielectric.

An electrolytic capacitor is often used to store large amounts of charge at relatively low voltages. This device, shown in Figure 26.12c, consists of a metal foil in contact with an electrolyte — a solution that conducts electricity by virtue of the motion of ions contained in the solution. When a voltage is applied between the foil and the electrolyte, a thin layer of metal oxide (an

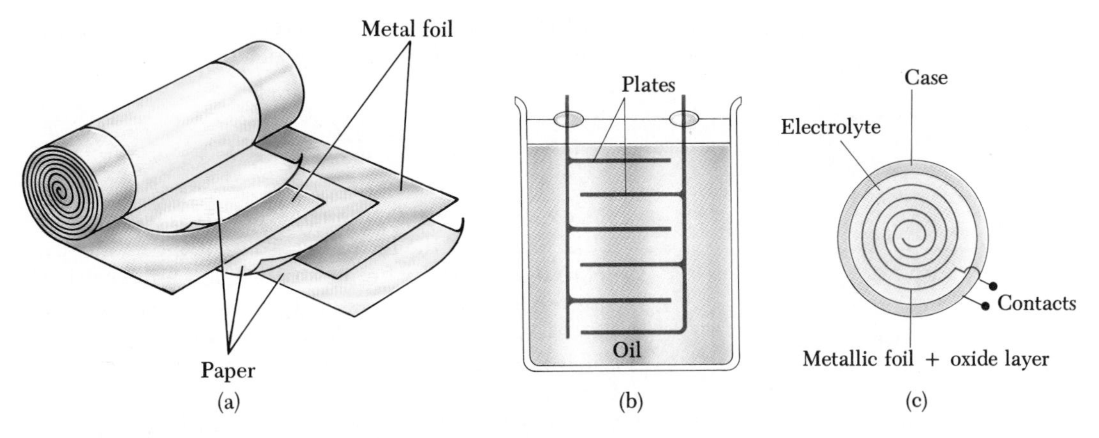


**Figure 26.12** Three commercial capacitor designs. (a) A tubular capacitor whose plates are separated by paper and then rolled into a cylinder. (b) A high-voltage capacit · consists of many parallel plates separated by insulating oil. (c) An electrolytic capacitor.

insulator) is formed on the foil, and this layer serves as the dielectric. Very large values of capacitance can be obtained because the dielectric layer is very thin.

When electrolytic capacitors are used in circuits, the polarity (the plus and minus signs on the device) must be installed properly. If the polarity of the applied voltage is opposite to what is intended, the oxide layer will be removed and the capacitor will conduct electricity rather than store charge.

**EXAMPLE 26.6 A Paper-Filled Capacitor**
A parallel-plate capacitor has plates of dimensions 2 cm × 3 cm. The plates are separated by a 1-mm thickness of paper. (a) Find the capacitance of this device.

*Solution* Since $\kappa = 3.7$ for paper (Table 26.1), we get

$$C = \kappa \frac{\epsilon_0 A}{d} = 3.7\left(8.85 \times 10^{-12} \frac{\text{C}^2}{\text{N}\cdot\text{m}^2}\right)\left(\frac{6 \times 10^{-4}\ \text{m}^2}{1 \times 10^{-3}\ \text{m}}\right)$$

$$= 19.6 \times 10^{-12}\ \text{F} = \boxed{19.6\ \text{pF}}$$

(b) What is the maximum charge that can be placed on the capacitor?

*Solution* From Table 26.1 we see that the dielectric strength of paper is $16 \times 10^6$ V/m. Since the thickness of the paper is 1 mm, the maximum voltage that can be applied before breakdown occurs is

$$V_{\text{max}} = E_{\text{max}} d = \left(16 \times 10^6 \frac{\text{V}}{\text{m}}\right)(1 \times 10^{-3}\ \text{m})$$

$$= 16 \times 10^3\ \text{V}$$

Hence, the maximum charge is given by

$$Q_{\text{max}} = CV_{\text{max}} = (19.6 \times 10^{-12}\ \text{F})(16 \times 10^3\ \text{V})$$

$$= \boxed{0.31\ \mu\text{C}}$$

*Exercise 4* What is the maximum energy that can be stored in the capacitor?
*Answer* $2.5 \times 10^{-3}$ J.

**EXAMPLE 26.7 Energy Stored Before and After**
A parallel-plate capacitor is charged with a battery to a charge $Q_0$, as in Figure 26.13a. The battery is then removed, and a slab of dielectric cor·tant $\kappa$ ·s inserted between the plates, as in Figure 26.1· ·. Find the energy stored in the capacitor before and after the dielectric is inserted.

*Solution* The energy stored in the capacitor in the absence of the dielectric is

$$U_0 = \tfrac{1}{2} C_0 V_0^2$$

Since $V_0 = Q_0/C_0$, this can be expressed as

$$U_0 = \frac{Q_0^2}{2C_0}$$

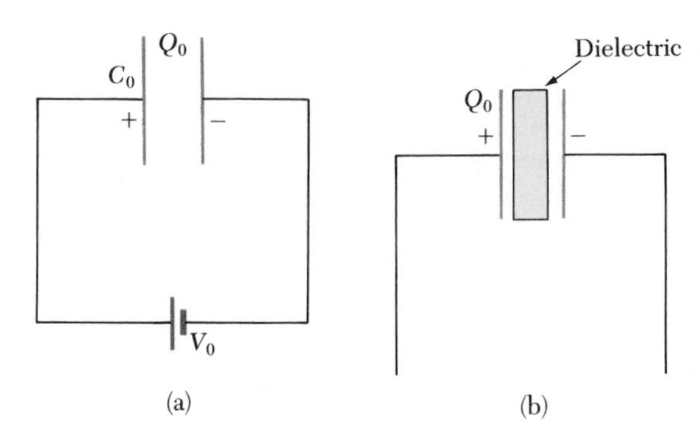


**Figure 26.13** (Example 26.7).

After the battery is removed and the dielectric is inserted between the plates, the *charge on the capacitor remains the same.* Hence, the energy stored in the presence of the dielectric is given by

$$U = \frac{Q_0^2}{2C}$$

But the capacitance in the presence of the dielectric is given by $C = \kappa C_0$, and so $U$ becomes

$$U = \frac{Q_0^2}{2\kappa C_0} = \frac{U_0}{\kappa}$$

Since $\kappa > 1$, we see that the final energy is *less* than the initial energy by the factor $1/\kappa$. This missing energy can be accounted for by noting that when the dielectric is inserted into the capacitor, it gets pulled into the device. The external agent must do negative work to keep the slab from accelerating. This work is simply the difference $U - U_0$. (Alternatively, the positive work done by the system on the external agent is given by $U_0 - U$.)

Exercise 5 Suppose that the capacitance in the absence of a dielectric is 8.50 pF, and the capacitor is charged to a potential difference of 12.0 V. If the battery is disconnected, and a slab of polystyrene ($\kappa = 2.56$) is inserted between the plates, calculate the energy difference $U - U_0$.
Answer 373 pJ

As we have seen, the energy of a capacitor is lowered when a dielectric is inserted between the plates, which means that work is done on the dielectric. This, in turn, implies that a force must act on the dielectric which draws it into the capacitor. This force originates from the nonuniform nature of the electric field of the capacitor near its edges as indicated in Figure 26.14. The horizontal component of this fringe field acts on the induced charges on the surface of the dielectric, producing a net horizontal force directed into the capacitor.

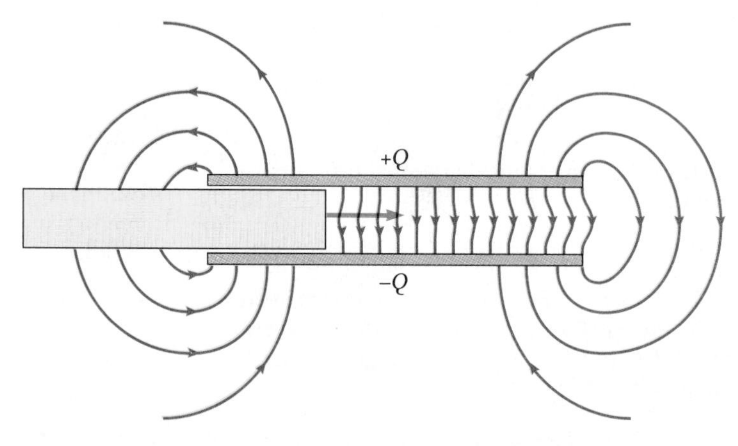


**Figure 26.14** The nonuniform electric field near the edges of a parallel-plate capacitor causes a dielectric to be pulled into the capacitor. Note that the field acts on the induced surface charges on the dielectric which are nonuniformly distributed.

## *26.6 ELECTRIC DIPOLE IN AN EXTERNAL ELECTRIC FIELD

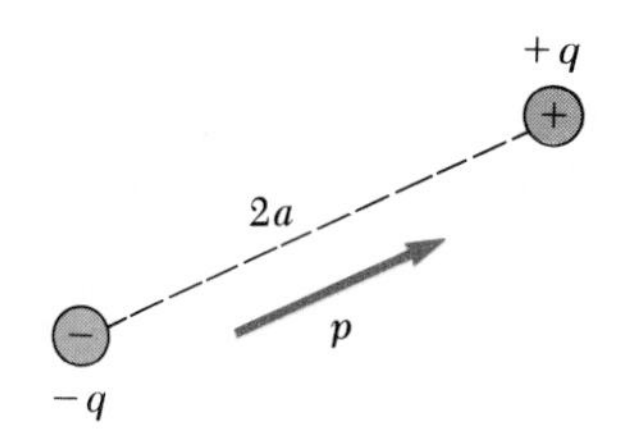


**Figure 26.15** An electric dipole consists of two equal and opposite charges separated by a distance $2a$.

The electric dipole, discussed briefly in Example 23.7, consists of two equal and opposite charges separated by a distance $2a$, as in Figure 26.15. Let us define the **electric dipole moment** of this configuration as the vector $\boldsymbol{p}$ whose magnitude is $2aq$ (that is, the separation $2a$ multiplied by the charge $q$).

$$p \equiv 2aq \tag{26.17}$$

Now suppose an electric dipole is placed in a uniform *external* electric field $\boldsymbol{E}$ as in Figure 26.16, where the dipole moment makes an angle $\theta$ with the field. The forces on the two charges are equal and opposite as shown, each having a magnitude of

$$F = qE$$

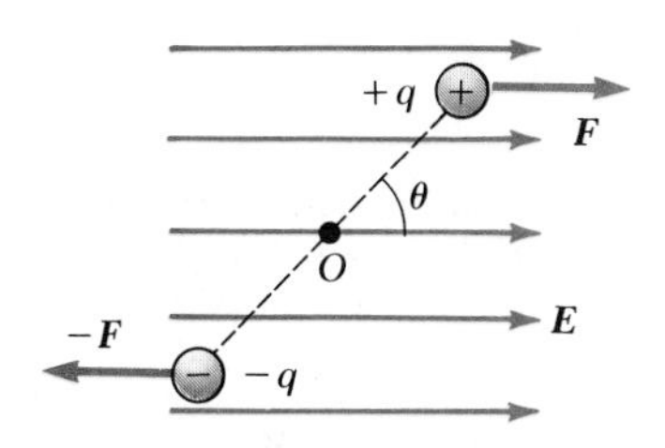


**Figure 26.16** An electric dipole in a uniform electric field. The dipole moment $\boldsymbol{p}$ is at an angle $\theta$ with the field, and the dipole experiences a torque.

Thus, we see that the net force on the dipole is *zero*. However, the two forces produce a net torque on the dipole, and the dipole tends to rotate such that its axis is aligned with the field. The torque due to the force on the positive charge about an axis through $O$ is given by $Fa \sin\theta$, where $a \sin\theta$ is the moment arm of $F$ about $O$. In Figure 26.16, this force tends to produce a clockwise rotation. The torque on the negative charge about $O$ is also $Fa \sin\theta$, so the net torque about $O$ is given by

$$\tau = 2Fa \sin\theta$$

Because $F = qE$ and $p = 2aq$, we can express $\tau$ as

$$\tau = 2aqE \sin\theta = pE \sin\theta \tag{26.18}$$

It is convenient to express the torque in vector form as the cross product of the vectors $\boldsymbol{p}$ and $\boldsymbol{E}$:

Torque on an electric dipole in an extended electric field

$$\boldsymbol{\tau} = \boldsymbol{p} \times \boldsymbol{E} \tag{26.19}$$

We can also determine the potential energy of an electric dipole as a function of its orientation with respect to the external electric field. In order to do this, you should recognize that work must be done by an external agent to rotate the dipole through a given angle in the field. The work done is then stored as potential energy in the system, that is, the dipole and the external field. The work $dW$ required to rotate the dipole through an angle $d\theta$ is given by $dW = \tau\, d\theta$ (Chapter 10). Because $\tau = pE \sin\theta$, and because the work is transformed into potential energy $U$, we find that for a rotation from $\theta_0$ to $\theta$, the change in potential energy is

$$U - U_0 = \int_{\theta_0}^{\theta} \tau\, d\theta = \int_{\theta_0}^{\theta} pE \sin\theta\, d\theta = pE \int_{\theta_0}^{\theta} \sin\theta\, d\theta$$

$$U - U_0 = pE[-\cos\theta]_{\theta_0}^{\theta} = pE(\cos\theta_0 - \cos\theta)$$

The term involving $\cos\theta_0$ is a constant that depends on the initial orientation of the dipole. It is convenient to choose $\theta_0 = 90°$, so that $\cos\theta_0 = \cos 90° = 0$. Furthermore, let us choose $U_0 = 0$ at $\theta_0 = 90°$ as our reference of potential energy. Hence, we can express $U$ as

$$U = -pE \cos\theta \tag{26.20}$$

This can be written as the dot product of the vectors $\boldsymbol{p}$ and $\boldsymbol{E}$:

Potential energy of an electric dipole in an external electric field

$$U = -\boldsymbol{p} \cdot \boldsymbol{E} \tag{26.21}$$

Molecules are said to be polarized when there is a separation between the "center of gravity" of the negative charges and that of the positive charges on the molecule. In some molecules, such as water, this condition is always present. This can be understood by inspecting the geometry of the water molecule. The molecule is arranged so that the oxygen atom is bonded to the hydrogen atoms with an angle of 105° between the two bonds (Fig. 26.17). The center of negative charge is near the oxygen atom, and the center of positive charge lies at a point midway along the line joining the hydrogen atoms (point $x$ in the diagram). Materials composed of molecules that are permanently polarized in this fashion have large dielectric constants. For example, the dielectric constant of water is quite large ($\kappa = 80$).

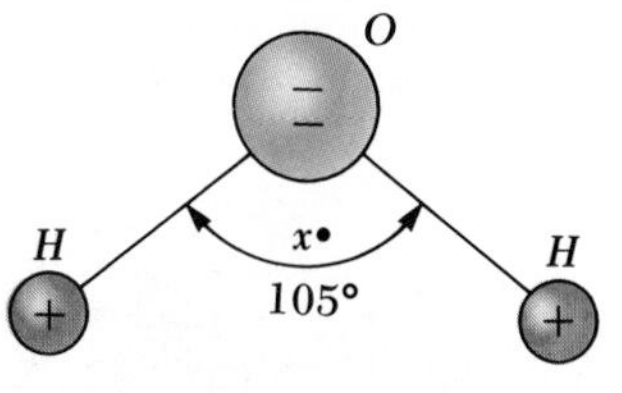

O
H
H
x
105°

**Figure 26.17** The water molecule, $H_2O$, has a permanent polarization resulting from its bent geometry.

A symmetrical molecule might have no permanent polarization, but a polarization can be induced by an external electric field. For example, if a linear molecule lies along the $x$ axis, an external electric field in the positive $x$ direction would cause the center of positive charge to shift to the right from its initial position and the center of negative charge to shift to the left. This *induced polarization* is the effect that predominates in most materials used as dielectrics in capacitors.

---

**EXAMPLE 26.8 The $H_2O$ Molecule**

The $H_2O$ molecule has a dipole moment of $6.3 \times 10^{-30}$ C · m. A sample contains $10^{21}$ such molecules, whose dipole moments are all oriented in the direction of an electric field of $2.5 \times 10^5$ N/C. How much work is required to rotate the dipoles from this orientation ($\theta = 0°$) to one in which all of the moments are perpendicular to the field ($\theta = 90°$)?

*Solution* The work required to rotate *one* molecule by 90° is equal to the difference in potential energy between the 90° orientation and the 0° orientation. Using Equation 26.20 gives

$$\begin{aligned} W = U_{90} - U_0 &= (-pE \cos 90°) - (-pE \cos 0°) \\ &= pE = (6.3 \times 10^{-30}\ \text{C} \cdot \text{m})(2.5 \times 10^5\ \text{N/C}) \\ &= 1.6 \times 10^{-24}\ \text{J} \end{aligned}$$

Since there are $10^{21}$ molecules in the sample, the *total* work required is given by

$$W_{\text{total}} = (10^{21})(1.6 \times 10^{-24}\ \text{J}) = 1.6 \times 10^{-3}\ \text{J}$$

---

## *26.7 AN ATOMIC DESCRIPTION OF DIELECTRICS

In Section 26.5 we found that the potential difference between the plates of a capacitor is reduced by the factor $\kappa$ when a dielectric is introduced. Since the potential difference between the plates equals the product of the electric field and the separation $d$, the electric field is also reduced by the factor $\kappa$. Thus, if $\boldsymbol{E}_0$ is the electric field without the dielectric, the field in the presence of a dielectric is

$$\boldsymbol{E} = \frac{\boldsymbol{E}_0}{\kappa} \tag{26.22}$$

This can be understood by noting that a dielectric can be polarized. At the atomic level, a polarized material is one in which the positive and negative charges are slightly separated. If the molecules of the dielectric possess permanent electric dipole moments in the absence of an electric field, they are called *polar molecules* (water is an example). The dipoles are randomly oriented in the absence of an electric field, as shown in Figure 26.18a. When an external field is applied, a torque is exerted on the dipoles, causing them to be partially aligned with the field, as in Figure 26.18b. The degree of alignment depends on temperature and on the magnitude of the applied field. In general, the alignment increases with decreasing temperature and with increasing electric field strength. The partially aligned dipoles produce an internal electric field that *opposes* the external field, thereby causing a reduction of the original field.

If the molecules of the dielectric do not possess a permanent dipole moment, they are called **nonpolar molecules.** In this case, an external electric field produces some charge separation, and the resulting dipole moments are said to be *induced.* These induced dipole moments tend to align with the external field, causing a reduction in the internal electric field.

With these ideas in mind, consider a slab of dielectric material in a uniform electric field $\mathbf{E}_0$ as in Figure 26.19a. Positive portions of the molecules are shifted in the direction of the electric field, and negative portions are shifted in the opposite direction. Hence, the applied electric field polarizes the dielectric. The net effect on the dielectric is the formation of an "induced" positive surface charge density $\sigma_i$ on the right face and an equal negative surface charge density on the left face, as shown in Figure 26.19b. These induced surface charges on the dielectric give rise to an induced electric field $\mathbf{E}_i$, which *opposes* the external field $\mathbf{E}_0$. Therefore, the net electric field $\mathbf{E}$ in the dielectric has a magnitude given by

$$E = E_0 - E_i \tag{26.23}$$

In the parallel-plate capacitor shown in Figure 26.20, the external field $E_0$ is related to the free charge density $\sigma$ on the plates through the relation $E_0 = \sigma/\epsilon_0$. The induced electric field in the dielectric is related to the induced

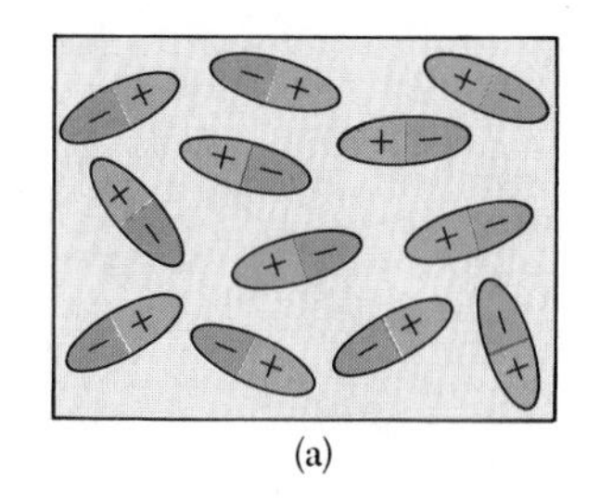


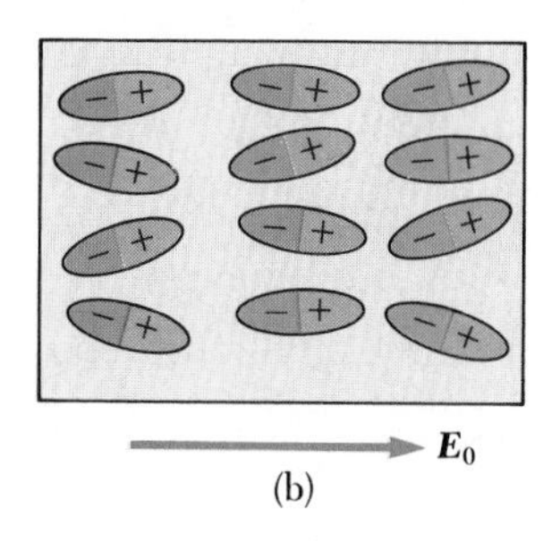


**Figure 26.18** (a) Molecules with a permanent dipole moment are randomly oriented in the absence of an external electric field. (b) When an external field is applied, the dipoles are partially aligned with the field.

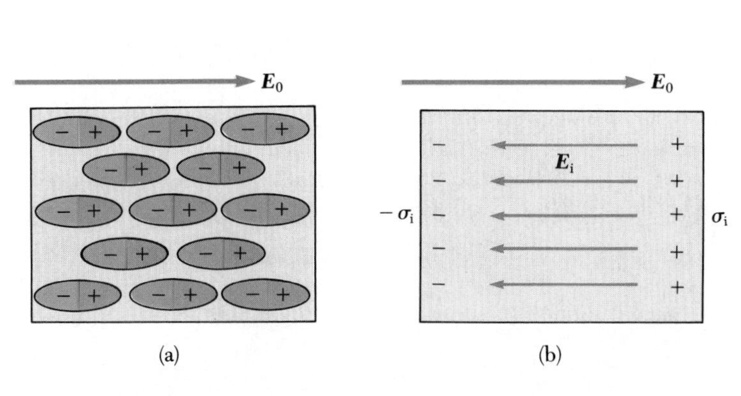


**Figure 26.19** (a) When a dielectric is polarized, the molecular dipole moments in the dielectric are aligned with the external field $\mathbf{E}_0$. (b) This polarization causes an induced negative surface charge on one side of the dielectric and an equal positive surface charge on the opposite side. This results in a reduction in the electric field within the dielectric.

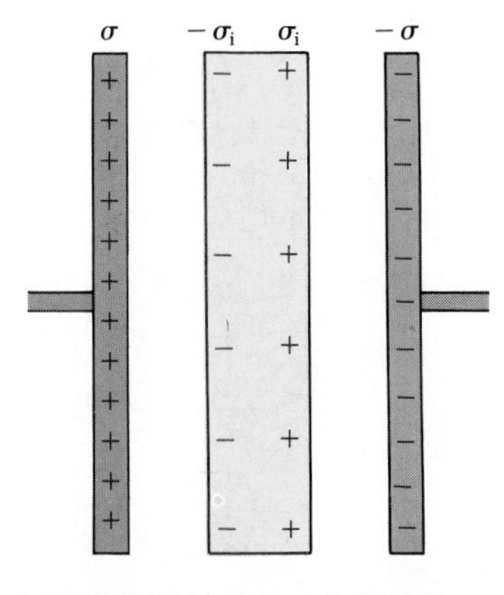


**Figure 26.20** Induced charge on a dielectric placed between the plates of a charged capacitor. Note that the induced charge density on the dielectric is *less* than the free charge density on the plates.

charge density $\sigma_i$ through the relation $E_i = \sigma_i/\epsilon_o$. Since $E = E_0/\kappa = \sigma/\kappa\epsilon_o$, substitution into Equation 26.23 gives

$$\frac{\sigma}{\kappa\epsilon_o} = \frac{\sigma}{\epsilon_o} - \frac{\sigma_i}{\epsilon_0}$$

$$\sigma_i = \left(\frac{\kappa - 1}{\kappa}\right)\sigma \qquad (26.24)$$

Because $\kappa > 1$, this shows that the charge density $\sigma_i$ induced on the dielectric is *less* than the free charge density $\sigma$ on the plates. For instance, if $\kappa = 3$, we see that the induced charge density on the dielectric is two thirds the free charge density on the plates. If there is no dielectric present, $\kappa = 1$ and $\sigma_i = 0$ as expected. However, if the dielectric is replaced by a *conductor*, for which $E = 0$, then Equation 26.23 shows that $E_0 = E_i$, corresponding to $\sigma_i = \sigma$. That is, the surface charge induced on the conductor will be equal to and opposite that on the plates, resulting in a net field of *zero* in the conductor.

**EXAMPLE 26.9 A Partially Filled Capacitor**

A parallel-plate capacitor has a capacitance $C_0$ in the absence of a dielectric. A slab of dielectric material of dielectric constant $\kappa$ and thickness $\frac{1}{3}d$ is inserted between the plates (Fig. 26.21a). What is the new capacitance when the dielectric is present?

*Solution* This capacitor is equivalent to two parallel-plate capacitors of the same area $A$ connected in series, one with a plate separation $d/3$ (dielectric filled) and the other with a plate separation $2d/3$ (Fig. 26.21b). (This step is permissible since there is no potential difference between the lower plate of $C_1$ and the upper plate of $C_2$.)[4]

From Equations 26.3 and 26.15, the two capacitances are given by

$$C_1 = \frac{\kappa\epsilon_o A}{d/3} \quad \text{and} \quad C_2 = \frac{\epsilon_o A}{2d/3}$$

Using Equation 26.10 for two capacitors combined in series, we get

$$\frac{1}{C} = \frac{1}{C_1} + \frac{1}{C_2} = \frac{d/3}{\kappa\epsilon_o A} + \frac{2d/3}{\epsilon_o A}$$

$$\frac{1}{C} = \frac{d}{3\epsilon_o A}\left(\frac{1}{\kappa} + 2\right) = \frac{d}{3\epsilon_o A}\left(\frac{1 + 2\kappa}{\kappa}\right)$$

$$C = \left(\frac{3\kappa}{2\kappa + 1}\right)\frac{\epsilon_o A}{d}$$

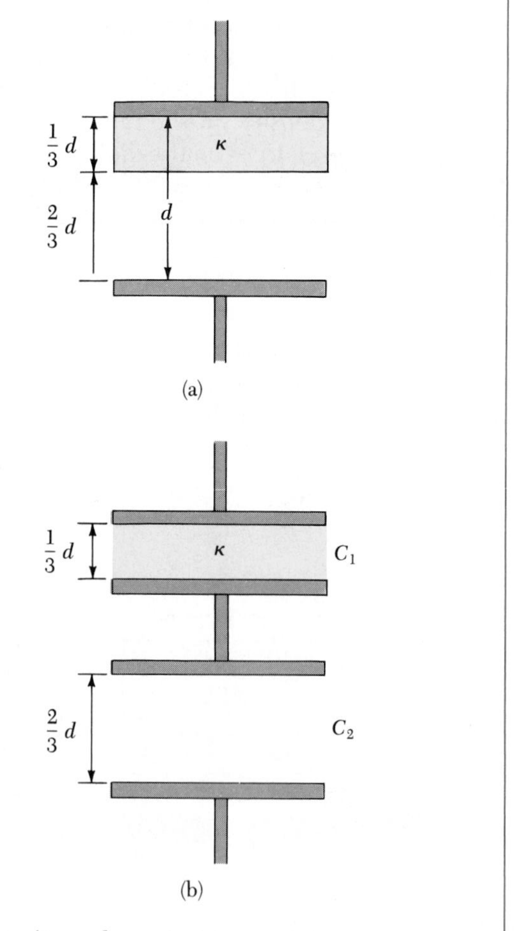


**Figure 26.21** (Example 26.9) (a) A parallel-plate capacitor of plate separation $d$ partially filled with a dielectric of thickness $d/3$. (b) The equivalent circuit of the capacitor consists of two capacitors connected in series.

[4] You could also imagine placing two thin metallic plates (with a coiled-up conducting wire between them) at the lower surface of the dielectric in Figure 26.21a and then pulling the assembly out until it becomes like Figure 26.21b.

Since the capacitance *without* the dielectric is given by $C_0 = \epsilon_0 A/d$, we see that

$$C = \left(\frac{3\kappa}{2\kappa + 1}\right)C_0$$

**EXAMPLE 26.10 Effect of a Metal Slab**

A parallel-plate capacitor has a plate separation $d$ and plate area $A$. An uncharged *metal* slab of thickness $a$ is inserted midway between the plates, as shown in Figure 26.22a. Find the capacitance of the device.

*Solution* This problem can be solved by noting that whatever charge appears on one plate of the capacitor must induce an *equal* and *opposite* charge on the metal slab, as shown in Figure 26.22a. Consequently, the net charge on the metal slab remains zero, and the field inside the slab is zero. Hence, the capacitor is equivalent to two capacitors in *series*, each having a plate separation $(d - a)/2$ as shown in Figure 26.22b. Using the rule for adding two capacitors in series we get

$$\frac{1}{C} = \frac{1}{C_1} + \frac{1}{C_2} = \frac{1}{\dfrac{\epsilon_0 A}{(d-a)/2}} + \frac{1}{\dfrac{\epsilon_0 A}{(d-a)/2}}$$

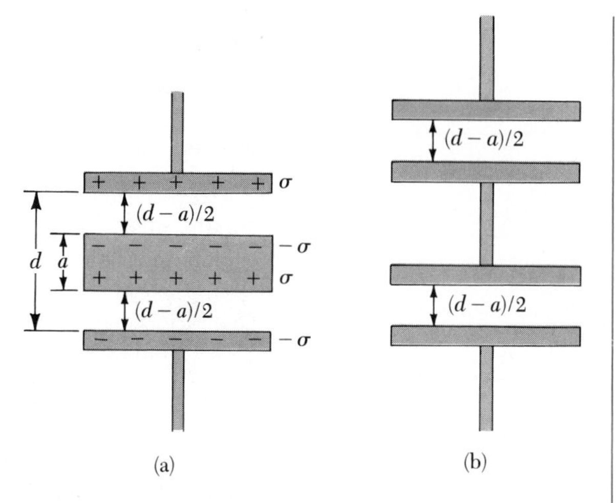


**Figure 26.22** (Example 26.10) (a) A parallel-plate capacitor of plate separation $d$ partially filled with a metal slab of thickness $a$. (b) The equivalent circuit of the device in (a) consists of two capacitors in series, each with a plate separation $(d - a)/2$.

Solving for $C$ gives

$$C = \frac{\epsilon_0 A}{d - a}$$

Note that $C$ approaches infinity as $a$ approaches $d$. Why?

## SUMMARY

A *capacitor* consists of two equal and oppositely charged conductors spaced very close together compared to their size with a potential difference $V$ between them. The **capacitance** $C$ of any capacitor is defined to be the ratio of the magnitude of the charge $Q$ on either conductor to the magnitude of the potential difference $V$:

$$C \equiv \frac{Q}{V} \qquad (26.1)$$

Definition of capacitance

The SI unit of capacitance is coulomb per volt, or farad (F), and 1 F = 1 C/V.

The capacitance of several capacitors is summarized in Table 26.2. The formulas apply when the charged conductors are separated by a vacuum.

If two or more capacitors are connected in *parallel*, the potential difference across them must be the same. The equivalent capacitance of a parallel combination of capacitors is given by

$$C_{eq} = C_1 + C_2 + C_3 + \cdots \qquad (26.8)$$

Parallel combinations

**TABLE 26.2 Capacitance and Geometry**

| Geometry | Capacitance | Equation |
|---|---|---|
| Isolated charged sphere of radius $R$ | $C = 4\pi\epsilon_0 R$ | (26.2) |
| Parallel-plate capacitor of plate area $A$ and plate separation $d$ | $C = \epsilon_0 \frac{A}{d}$ | (26.3) |
| Cylindrical capacitor of length $\ell$ and inner and outer radii $a$ and $b$, respectively | $C = \frac{\ell}{2k \ln\left(\frac{b}{a}\right)}$ | (26.4) |
| Spherical capacitor with inner and outer radii $a$ and $b$, respectively | $C = \frac{ab}{k(b-a)}$ | (26.5) |

If two or more capacitors are connected in *series*, the charge on them is the same, and the equivalent capacitance of the series combination is given by

Series combination

$$\frac{1}{C_{eq}} = \frac{1}{C_1} + \frac{1}{C_2} + \frac{1}{C_3} + \cdots \tag{26.11}$$

Work is required to charge a capacitor, since the charging process consists of transferring charges from one conductor at a lower potential to another conductor at a higher potential. The work done in charging the capacitor to a charge $Q$ equals the electrostatic potential energy $U$ stored in the capacitor, where

Energy stored in a charged capacitor

$$U = \frac{Q^2}{2C} = \tfrac{1}{2}QV = \tfrac{1}{2}CV^2 \tag{26.12}$$

When a dielectric material is inserted between the plates of a capacitor, the capacitance generally increases by a dimensionless factor $\kappa$ called the **dielectric constant.** That is,

$$C = \kappa C_0 \tag{26.15}$$

where $C_0$ is the capacitance in the absence of the dielectric. The increase in capacitance is due to a decrease in the electric field in the presence of the dielectric and to a corresponding decrease in the potential difference between the plates—assuming the charging battery is removed from the circuit before the dielectric is inserted. The decrease in $\boldsymbol{E}$ arises from an internal electric field produced by aligned dipoles in the dielectric. This internal field produced by the dipoles opposes the original applied field, and this results in a reduction in the net electric field.

An *electric dipole* consists of two equal and opposite charges separated by a distance $2a$. The **electric dipole moment $\boldsymbol{p}$** of this configuration has a magnitude given by

Electric dipole moment

$$p \equiv 2aq \tag{26.17}$$

The **torque** acting on an electric dipole in a uniform electric field $\mathbf{E}$ is given by

$$\tau = \mathbf{p} \times \mathbf{E} \quad (26.19)$$

Torque on an electric dipole in an extended electric field

The **potential energy** of an electric dipole in a uniform external electric field $\mathbf{E}$ is given by

$$U = -\mathbf{p} \cdot \mathbf{E} \quad (26.21)$$

Potential energy of an electric dipole in an external electric field

### PROBLEM-SOLVING STRATEGY AND HINTS

1. Be careful with your choice of units. To calculate capacitance in farads, make sure that distances are in meters and use the SI value of $\epsilon_0$. When checking consistency of units, remember that the units for electric fields can be either N/C or V/m.
2. When two or more unequal capacitors are connected in *series,* they carry the same charge, but the potential differences are not the same. Their capacitances add as reciprocals, and the equivalent capacitance of the combination is always *less* than the smallest individual capacitor.
3. When two or more capacitors are connected in *parallel,* the potential difference across each is the same. The charge on each capacitor is proportional to its capacitance, hence the capacitances add directly to give the equivalent capacitance of the parallel combination.
4. The effect of a dielectric on a capacitor is to *increase* its capacitance by a factor $\kappa$ (the dielectric constant) over its empty capacitance. The reason for this is that induced surface charges on the dielectric reduce the electric field inside the material from $E$ to $E/\kappa$.
5. Be careful about problems in which you may be connecting or disconnecting a battery to a capacitor. It is important to note whether modifications to the capacitor are being made while the capacitor is connected to the battery or after it is disconnected. If the capacitor remains connected to the battery, the *voltage* across the capacitor necessarily remains the *same* (equal to the battery voltage), and the charge will be proportional to the capacitance *however it may be modified* (say by inserting a dielectric). On the other hand, if you disconnect the capacitor from the battery *before* making any modifications to the capacitor, then its *charge* remains the same. In this case, as you vary the capacitance, the voltage across the plates will change in an inverse proportion to capacitance according to $V = Q/C$.

## QUESTIONS

1. What happens to the charge on a capacitor if the potential difference between the conductors is doubled?
2. The plates of a capacitor are connected to a battery. What happens to the charge on the plates if the connecting wires are removed from the battery? What happens to the charge if the wires are removed from the battery and connected to each other?
3. A farad is a very large unit of capacitance. Calculate the length of one side of a square, air-filled capacitor with a plate separation of 1 meter. Assume it has a capacitance of 1 farad.
4. A pair of capacitors are connected in parallel while an identical pair are connected in series. Which pair would be more dangerous to handle after being connected to the same voltage source? Explain.
5. If you are given 3 different capacitors $C_1$, $C_2$, $C_3$, how many different combinations of capacitance can you produce?
6. What advantage might there be in using 2 identical capacitors in parallel connected in series with another identical parallel pair, rather than using a single capacitor by itself?

7. Is it always possible to reduce a combination of capacitors to one equivalent capacitor with the rules we have just developed? Explain your answer.
8. Since the net charge in a capacitor is always zero, what does a capacitor store?
9. Since the charges on the plates of a parallel-plate capacitor are equal and opposite, they attract each other. Hence, it would take positive work to increase the plate separation. What happens to the external work done in this process?
10. Explain why the work needed to move a charge, $Q$, through a potential, $V$, is given by $W = QV$ whereas the energy stored in a charged capacitor is $U = \frac{1}{2}QV$. Where does the $\frac{1}{2}$ factor come from?
11. If the potential difference across a capacitor is doubled, by what factor does the energy stored change?
12. Why is it dangerous to touch the terminals of a high-voltage capacitor even after the applied voltage has been turned off? What can be done to make the capacitor safe to handle after the voltage source has been removed?
13. If you want to increase the maximum operating voltage of a parallel-plate capacitor, describe how you can do this for a fixed plate separation.
14. An air-filled capacitor is charged, then disconnected from the power supply, and finally connected to a voltmeter. Explain how and why the voltage reading changes when a dielectric is inserted between the plates of the capacitor.
15. Using the polar molecule description of a dielectric, explain how a dielectric affects the electric field inside a capacitor.
16. Explain why a dielectric increases the maximum operating voltage of a capacitor although the physical size of the capacitor does not change.
17. What is the difference between dielectric strength and the dielectric constant?
18. Where in a coaxial cable will electrical breakdown first occur if the cable is connected to an excessive potential difference?
19. Explain why a water molecule is permanently polarized. What type of molecule has no permanent polarization?
20. If a dielectric-filled capacitor is heated, how will its capacitance change? (Neglect thermal expansion and assume that the dipole orientations are temperature-dependent.)
21. In terms of induced charges, explain why a charged comb attracts small bits of paper.
22. If you were asked to design a capacitor where small size and large capacitance were required, what factors would be important in your design?

## PROBLEMS

### Section 26.1 Definition of Capacitance

1. The excess charge on each conductor of a parallel-plate capacitor is 53 $\mu$C. What is the potential difference between the conductors if the capacitance of the system is $4 \times 10^{-3}$ $\mu$F?
2. Show that the units $C^2/N \cdot m$ equal 1 F.
3. Two parallel wires are suspended in a vacuum. When the potential difference between the two wires is 52 V, each wire has a charge of 73 pC (the two charges are of opposite sign). Calculate the capacitance of the parallel-wire system.
4. Two conductors insulated from each other are charged by transferring electrons from one conductor to the other. After $1.6 \times 10^{12}$ electrons have been transferred, the potential difference between the conductors is found to be 14 V. What is the capacitance of the system?
5. A parallel-plate capacitor has a capacitance of 19 $\mu$F. What charge on each plate will produce a potential difference of 36 V between the plates of the capacitor?
6. An isolated conducting sphere can be considered as one element of a capacitor (the other element being a concentric sphere of infinite radius). (a) If the capacitance of this system is $9.1 \times 10^{-11}$ F, what is the radius of the sphere? (b) If the potential at the surface of the sphere is $2.8 \times 10^4$ V, what is the corresponding surface charge density?
7. An isolated charged conducting sphere of radius 12 cm creates an electric field of $4.9 \times 10^4$ N/C at a distance of 21 cm from its center. (a) What is its surface charge density? (b) What is its capacitance?
8. Two conducting spheres with diameters of 0.40 m and 1.0 m are separated by a distance that is large compared with the diameters. The spheres are connected by a thin wire and are charged to 7 $\mu$C. (a) How is this total charge shared between the spheres? (Neglect any charge on the wire.) (b) What is the potential of the system of spheres relative to $V = 0$ at $r = \infty$?
9. Two spherical conductors with radii $R_1$ and $R_2$ are separated by a sufficiently large distance that induction effects are negligible. The spheres are connected by a thin conducting wire and are brought to the same potential $V$ relative to $V = 0$ at $r = \infty$. (a) Determine the capacitance $C$ of the system, where $C = (Q_1 + Q_2)/V$. (b) What is the charge ratio $Q_1/Q_2$?

## Section 26.2 Calculation of Capacitance

**10.** An air-filled parallel plate capacitor is to have a capacitance of 1 F. If the distance between the plates is 1 mm, calculate the required surface area of each plate. Convert your answer to square miles.

**11.** A parallel-plate capacitor has a plate area of 12 $cm^2$ and a capacitance of 7 pF. What is the plate separation?

**12.** The plates of a parallel-plate capacitor are separated by 0.2 mm. If the space between the plates is air, what plate area is required to provide a capacitance of 9 pF?

**13.** When a potential difference of 150 V is applied to the plates of a parallel-plate capacitor, the plates carry a surface charge density of 30 nC/$cm^2$. What is the spacing between the plates?

14. A small object with a mass of 350 mg carries a charge of 30 nC and is suspended by a thread between the vertical plates of a parallel-plate capacitor. The plates are separated by 4 cm. If the thread makes an angle of 15° with the vertical, what is the potential difference between the plates?

15. An air-filled capacitor consists of two parallel plates, each with an area of 7.6 $cm^2$, separated by a distance of 1.8 mm. If a 20-V potential difference is applied to these plates, calculate (a) the electric field between the plates, (b) the surface charge density, (c) the capacitance, and (d) the charge on each plate.

**16.** A 1-megabit computer memory chip contains many 60-fF capacitors. Each capacitor has a plate area of 21 $\mu m^2$ ($21 \times 10^{-12}$ $m^2$). Determine the plate separation of such a capacitor (assume a parallel-plate configuration). The characteristic atomic diameter is $10^{-10}$ m = 1 Å. Express the plate separation in Å.

17. A circular parallel-plate capacitor with a spacing $d$ = 3 mm is charged to produce an electric field strength of $3 \times 10^6$ V/m. What plate radius $R$ is required if the stored charge $Q$ is 1 $\mu$C?

18. A capacitor is constructed of interlocking plates as shown in Figure 26.23 (a cross-sectional view). The separation between adjacent plates is 0.8 mm, and the total effective area of adjacent plates is 7 $cm^2$. Ignoring side effects, calculate the capacitance of the unit.

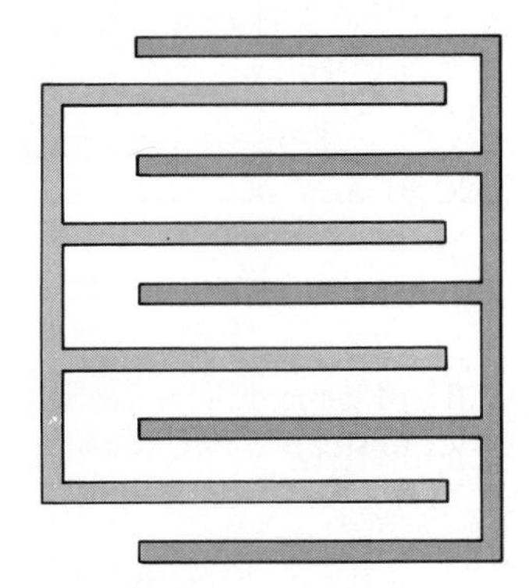

**Figure 26.23** (Problems 18 and 87).

19. One plate of a parallel-plate capacitor (with area $A$ = 100 $cm^2$) is grounded. The ungrounded plate carries a fixed charge $Q$. When the ungrounded plate is moved 0.5 cm farther away from the grounded plate, the potential difference between the plates increases by 200 V. Determine the magnitude of the charge $Q$.

20. An air-filled *cylindrical* capacitor has a capacitance of 10 pF and is 6 cm in length. If the radius of the outside conductor is 1.5 cm, what is the required radius of the inner conductor?

21. A 50-m length of coaxial cable has an inner conductor with a diameter of 2.58 mm and a charge of +8.1 $\mu$C. The surrounding conductor has an inner diameter of 7.27 mm and a charge of −8.1 $\mu$C. (a) What is the capacitance of this cable? (b) What is the potential difference between the two conductors?

22. A cylindrical capacitor has outer and inner conductors whose radii are in the ratio of $b/a = 4/1$. The inner conductor is to be replaced by a wire whose radius is one half of the original inner conductor. By what factor should the length be increased in order to obtain a capacitance equal to that of the original capacitor?

**23.** An air-filled spherical capacitor is constructed with inner and outer shell radii of 7 and 14 cm, respectively. (a) Calculate the capacitance of the device. (b) What potential difference between the spheres will result in a charge of 4 $\mu$C on each conductor?

24. Find the capacitance of the earth. (*Hint:* The outer conductor of the "spherical capacitor" may be considered as a conducting sphere at infinity where $V \equiv 0$.)

25. A spherical capacitor consists of a conducting ball with a diameter of 10 cm that is centered inside a grounded conducting spherical shell with an inner diameter of 12 cm. What capacitor charge is required to achieve a potential of 1000 V on the ball?

**26.** Estimate the maximum voltage to which a smooth, metallic sphere 10 cm in diameter can be charged without exceeding the dielectric strength of the dry air around the sphere.

## Section 26.3 Combinations of Capacitors

**27.** Two capacitors, $C_1 = 2\ \mu F$ and $C_2 = 16\ \mu F$, are connected in parallel. What is the value of the equivalent capacitance of the combination?

**28.** Calculate the equivalent capacitance of the two capacitors in the previous exercise if they are connected in series.

29. (a) Determine the equivalent capacitance for the capacitor network shown in Figure 26.24. (b) If the network is connected to a 12-V battery, calculate the potential difference across each capacitor and the charge on each capacitor.

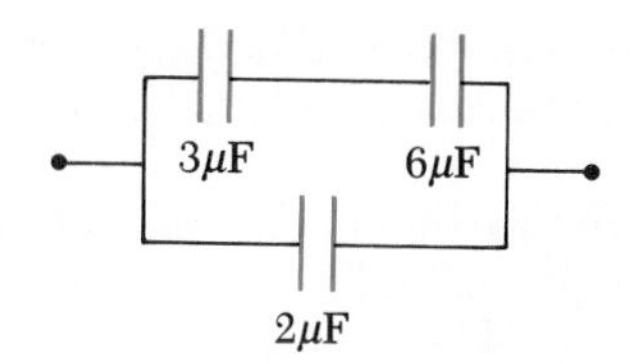


**Figure 26.24** (Problem 29).

30. Evaluate the effective capacitance of the configuration shown in Figure 26.25. Each of the capacitors is identical and has capacitance $C$.

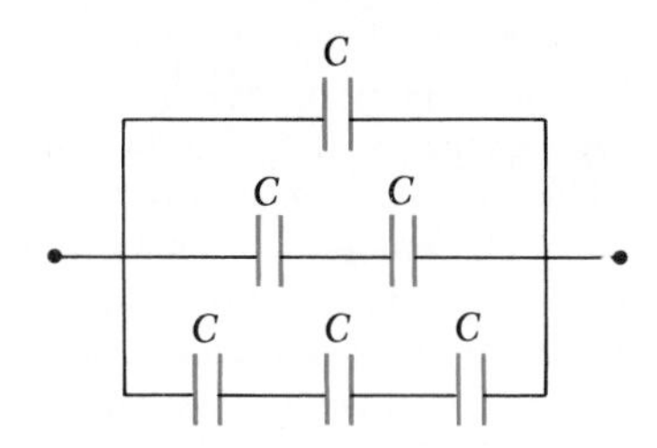


**Figure 26.25** (Problem 30).

31. Four capacitors are connected as shown in Figure 26.26. (a) Find the equivalent capacitance between points $a$ and $b$. (b) Calculate the charge on each capacitor if $V_{ab} = 15$ V.

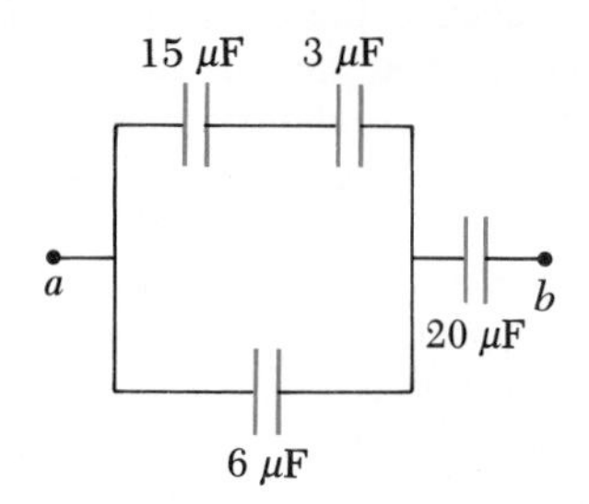


**Figure 26.26** (Problems 31 and 44).

32. (a) Figure 26.27 shows a network of capacitors between the terminals $a$ and $b$. Reduce this network to a single equivalent capacitor. (b) Determine the charge on the 4-$\mu$F and 8-$\mu$F capacitors when the capacitors are fully charged by a 12-V battery connected to the terminals. (c) Determine the potential difference across each capacitor.

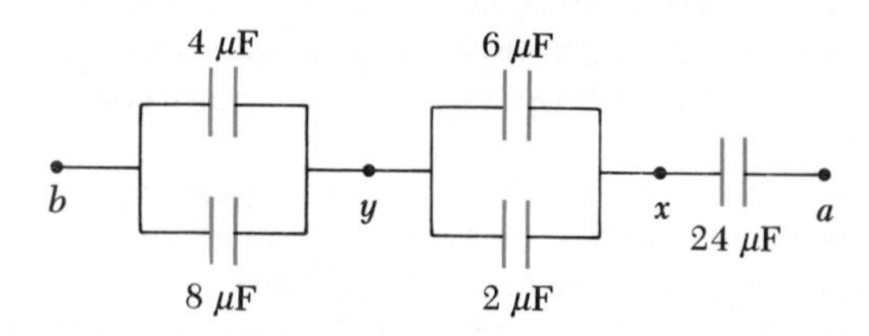


**Figure 26.27** (Problem 32).

33. Consider the circuit shown in Figure 26.28, where $C_1 = 6\ \mu$F, $C_2 = 3\ \mu$F, and $V = 20$ V. $C_1$ is first charged by the closing of switch $S_1$. Switch $S_1$ is then opened, and the charged capacitor is connected to the uncharged capacitor by the closing of $S_2$. Calculate the initial charge acquired by $C_1$ and the final charge on each of the two capacitors.

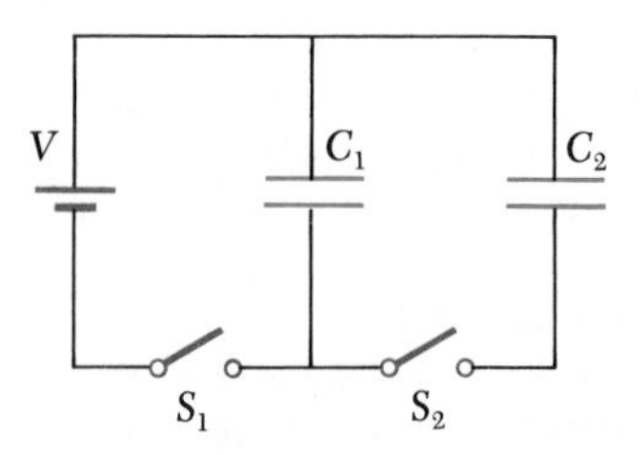


**Figure 26.28** (Problem 33).

34. Consider the group of capacitors shown in Figure 26.29. (a) Find the equivalent capacitance between points $a$ and $b$. (b) Determine the charge on each capacitor when the potential difference between $a$ and $b$ is 12 V.

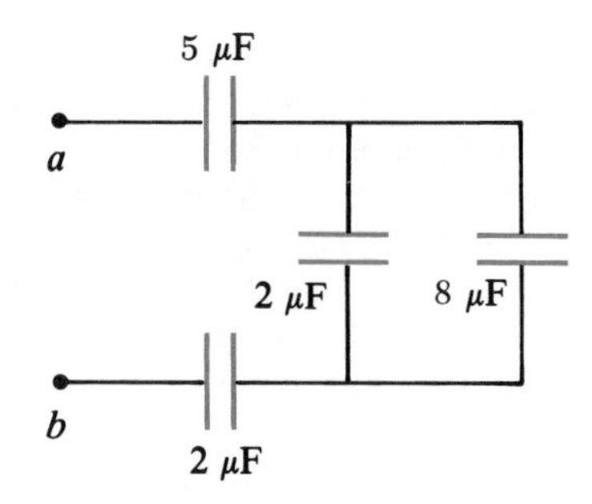


**Figure 26.29** (Problem 34).

35. Consider the combination of capacitors shown in Figure 26.30. (a) What is the equivalent capacitance between points $a$ and $b$? (b) Determine the charge on each capacitor if $V_{ab} = 4.8$ V.

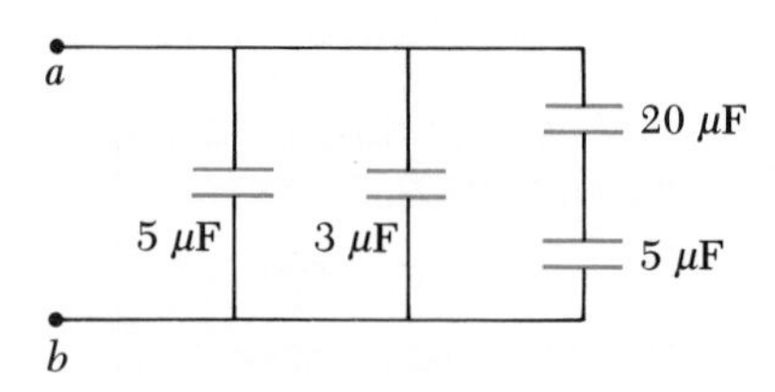


**Figure 26.30** (Problem 35).

36. How many 0.25-pF capacitors must be connected in parallel in order to store 1.2 $\mu$C of charge when connected to a battery providing a potential difference of 10 V?

37. A group of identical capacitors is connected first in series and then in parallel. The combined capacitance

in parallel is 100 times larger than for the series connection. How many capacitors are in the group?

38. Find the equivalent capacitance between points $a$ and $b$ for the group of capacitors connected as shown in Figure 26.31 if $C_1 = 5\ \mu F$, $C_2 = 10\ \mu F$, and $C_3 = 2\ \mu F$.

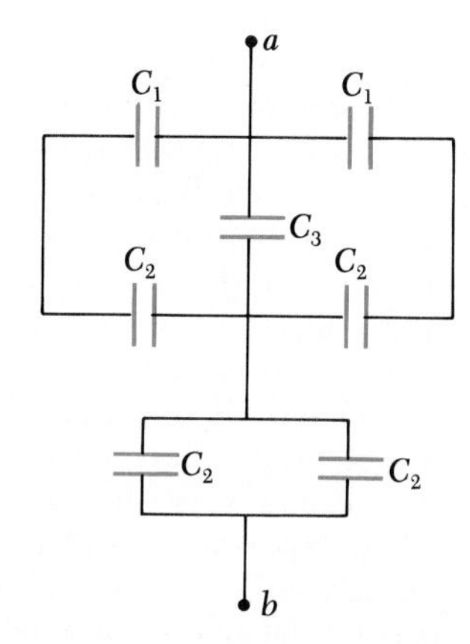


**Figure 26.31** (Problems 38 and 39).

39. For the network described in the previous exercise, if the potential between points $a$ and $b$ is 60 V, what charge is stored on the capacitor $C_3$?

40. Find the equivalent capacitance between points $a$ and $b$ in the capacitor network shown in Figure 26.32.

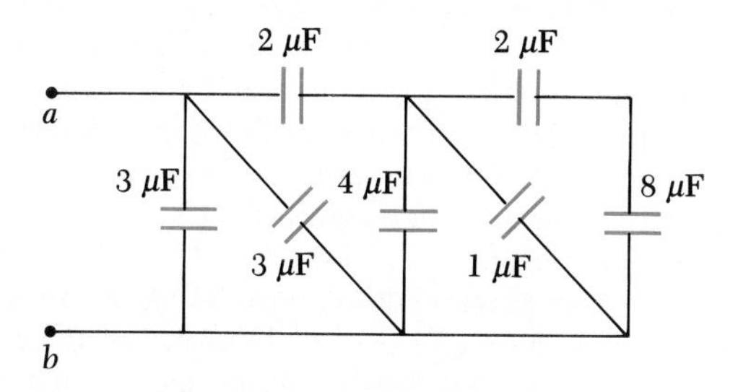


**Figure 26.32** (Problem 40).

41. Find the equivalent capacitance between points $a$ and $b$ in the combination of capacitors shown in Figure 26.33.

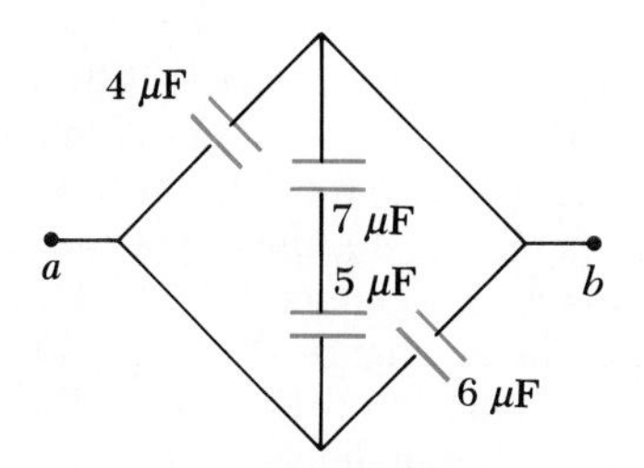


**Figure 26.33** (Problem 41).

42. How should four 2-$\mu$F capacitors be connected to have a total capacitance of (a) 8 $\mu$F, (b) 2 $\mu$F, (c) 1.5 $\mu$F, and (d) 0.5 $\mu$F?

43. A conducting slab with a thickness $d$ and area $A$ is inserted into the space between the plates of a parallel-plate capacitor with spacing $s$ and area $A$, as shown in Figure 26.34. What is the value of the capacitance of the system?

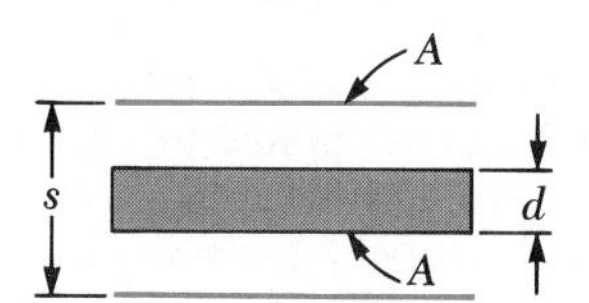


**Figure 26.34** (Problem 43).

## Section 26.4 Energy Stored in a Charged Capacitor

44. What total energy is stored in the group of capacitors shown in Figure 26.26 if $V_{ab} = 15$ V?

45. Calculate the energy stored in a 18-$\mu$F capacitor when it is charged to a potential of 100 V.

46. The energy stored in a particular capacitor is increased fourfold. What is the accompanying change in the (a) charge and (b) potential difference across the capacitor?

47. The energy stored in a 12-$\mu$F capacitor is 130 $\mu$J. Determine (a) the charge on the capacitor and (b) the potential difference across the capacitor.

48. Two capacitors, $C_1 = 25\ \mu F$ and $C_2 = 5\ \mu F$, are connected in parallel and charged with a 100-V power supply. (a) Calculate the total energy stored in the two capacitors. (b) What potential difference would be required across the same two capacitors connected in series in order that the combination store the same energy as in (a)?

49. A 16-pF parallel-plate capacitor is charged by a 10-V battery. If each plate of the capacitor has an area of 5 cm$^2$, what is the energy stored in the capacitor? What is the energy density (energy per unit volume) in the electric field of the capacitor if the plates are separated by air?

50. The energy density in a parallel-plate capacitor is given as $2.1 \times 10^{-9}$ J/m$^3$. What is the value of the electric field in the region between the plates?

51. A parallel-plate capacitor has a charge $Q$ and plates of area $A$. Show that the force exerted on each plate by the other is given by $F = Q^2/2\epsilon_0 A$. (*Hint:* Let $C = \epsilon_0 A/x$ for an arbitrary plate separation $x$; then require that the work done in separating the two charged plates be $W = \int F\,dx$.)

52. A uniform electric field $E = 3000$ V/m exists within a certain region. What volume of space would contain an energy equal to $10^{-7}$ J? Express your answer in cubic meters and in liters.

53. Show that the energy associated with a conducting sphere of radius $R$ and charge $Q$ surrounded by a vacuum is given by $U = kQ^2/2R$.
54. Use Equation 26.14 to make an explicit calculation of the energy stored in the field of a simple spherical capacitor. Show that $U = Q^2/2C$.

*Section 26.5 Capacitors with Dielectrics and *Section 26.7 An Atomic Description of Dielectrics

55. Determine (a) the capacitance and (b) the maximum voltage which can be applied to a Teflon-filled parallel-plate capacitor having a plate area of 175 cm$^2$ and insulation thickness of 0.04 mm. (See Table 26.1 for other dielectric properties.)
56. A parallel-plate capacitor is to be constructed using paper as a dielectric. If a maximum voltage before breakdown of 2500 V is desired, what thickness of dielectric is needed? (See Table 26.1 for other dielectric properties.)
57. A parallel-plate capacitor has a plate area of 0.64 cm$^2$. When the plates are in a vacuum, the capacitance of the device is 4.9 pF. (a) Calculate the value of the capacitance if the space between the plates is filled with nylon. (b) What is the maximum potential difference that can be applied to the plates without causing dielectric breakdown, or discharge?
58. A capacitor is constructed from two square metal plates of side length $L$ and separated by a distance $d$ (Fig. 26.35). One half of the space between the plates (top to bottom) is filled with polystyrene ($\kappa = 2.56$), and the other half is filled with neoprene rubber ($\kappa = 6.7$). Calculate the capacitance of the device, taking $L = 2$ cm and $d = 0.75$ mm. (*Hint:* The capacitor can be considered as two capacitors connected in parallel.)

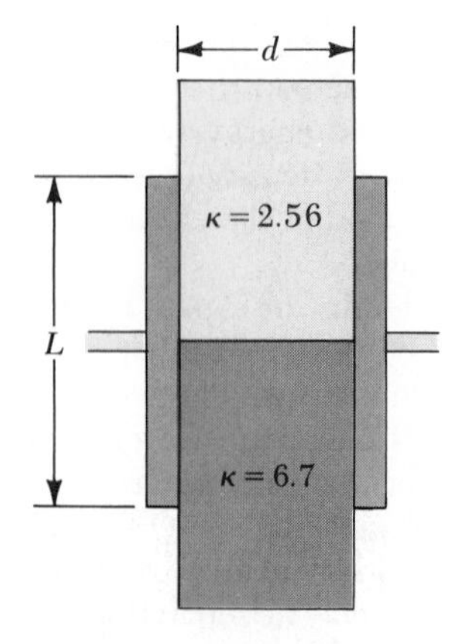


**Figure 26.35** (Problem 58).

59. A commercial capacitor is constructed as shown in Figure 26.12a. This particular capacitor is "rolled" from two strips of aluminum separated by two strips of paraffin-coated paper. Each strip of foil and paper is 7 cm wide. The foil is 0.004 mm thick, and the paper is 0.025 mm thick and has a dielectric constant of 3.7. What length should the strips be if a capacitor of $9.5 \times 10^{-8}$ F is desired? (Use the parallel-plate formula.)
60. A detector of radiation called a Geiger tube consists of a closed, hollow, conducting cylinder with a fine wire along its axis. Suppose that the internal diameter of the cylinder is 2.5 cm and that the wire along the axis has a diameter of 0.2 mm. If the dielectric strength of the gas between the central wire and cylinder is $1.2 \times 10^6$ V/m, calculate the maximum voltage $V_{max}$ that can be applied between the wire and the cylinder before breakdown occurs in the gas.
61. The plates of an isolated, charged capacitor are 1 mm apart and the potential difference across them is $V_0$. The plates are now separated to 4 mm (while the charge on them is preserved) and a slab of dielectric material is inserted, filling the space between the plates. The potential difference across the capacitor is now $V_0/2$. Find the dielectric constant of the material.
62. (a) What is the capacitance of a square parallel-plate capacitor measuring 5 cm on a side with a 0.2-mm gap between the plates if this gap is filled with Teflon? (b) What maximum voltage can this capacitor withstand? (c) What maximum energy can this capacitor store?
63. A parallel-plate capacitor having air between its plates is charged to 31.5 V. The capacitor is then isolated from the charging source and the volume between the plates is filled with Plexiglas. Determine the new potential difference across the capacitor. The dielectric constant of Plexiglas is 3.12.
64. Let $Q_0$ be the greatest charge that can be placed on the plates of an air-filled parallel-plate capacitor without causing electrical breakdown. Let $Q$ be the greatest charge that can be placed on the plates of this same capacitor when the gap between the plates is filled with neoprene rubber. What is the ratio $Q/Q_0$?
**65.** A sheet of 0.1-mm thick paper is inserted between the plates of a 340-pF air-filled capacitor with a plate separation of 0.4 mm. Calculate the new capacitance.
66. A wafer of titanium dioxide ($\kappa = 173$) has an area of 1 cm$^2$ and a thickness of 0.10 mm. Aluminum is evaporated on the parallel faces to form a parallel-plate capacitor. (a) Calculate the capacitance. (b) When the capacitor is charged with a 12-V battery, what is the magnitude of charge delivered to each plate? (c) For the situation in (b), what are the free and induced surface charge densities? (d) What is the electric field strength $E$?

## ADDITIONAL PROBLEMS

67. When two capacitors are connected in parallel, the equivalent capacitance is 4 $\mu$F. If the same capacitors are reconnected in series, the equivalent capacitance is one fourth the capacitance of one of the two capacitors. Determine the two capacitances.
68. For the system of capacitors shown in Figure 26.36, find (a) the equivalent capacitance of the system, (b) the potential across each capacitor, (c) the charge on each capacitor, and (d) the total energy stored by the group.

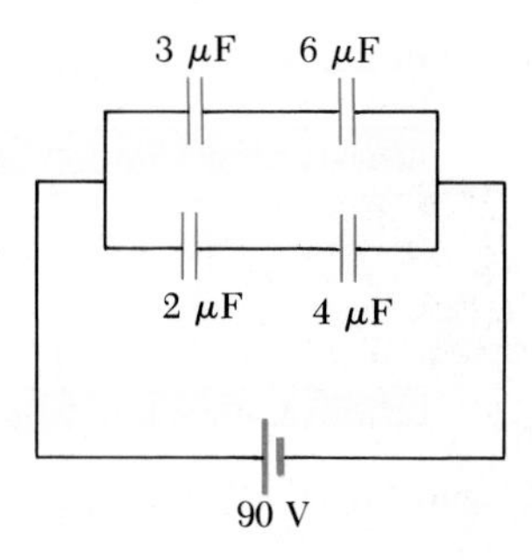


**Figure 26.36** (Problem 68).

69. When the voltage applied to a capacitor increases from 80 V to 110 V, the charge on the capacitor increases by $9.0 \times 10^{-5}$ C. Determine the capacitance.
70. A parallel-plate capacitor with air between its plates has a capacitance $C_0$. A slab of dielectric material with a dielectric constant $\kappa$ and a thickness equal to a fraction $f$ of the separation of the plates is inserted between the plates in contact with one plate. Find the capacitance $C$ in terms of $f$, $\kappa$, and $C_0$. Check your result by first letting $f$ approach zero and then letting it approach one.
71. When a certain air-filled parallel-plate capacitor is connected across a battery, it acquires a charge (on each plate) of 150 $\mu$C. While the battery connection is maintained, a dielectric slab is inserted into and fills the region between the plates. This results in the accumulation of an *additional* charge of 200 $\mu$C on each plate. What is the dielectric constant of the dielectric slab?
72. The energy stored in a 52-$\mu$F capacitor is used to melt a 6-mg sample of lead. To what voltage must the capacitor be initially charged, assuming the initial temperature of the lead is 20°C?
73. Three capacitors of 8 $\mu$F, 10 $\mu$F, and 14 $\mu$F are connected to the terminals of a 12-volt battery. How much energy does the battery supply if the capacitors are connected (a) in series and (b) in parallel?
74. When considering the energy supply for an automobile, the energy per unit mass of the supply is an important parameter. Using the following data, compare the energy per unit mass (J/kg) for gasoline, lead-acid batteries, and capacitors.

    *Gasoline:* 126,000 Btu/gal; density = 670 kg/m$^3$.

    *Lead-acid battery:* 12 V; 100 A·h; mass = 16 kg.

    *Capacitor:* potential difference at full charge = 12 V; capacitance = 0.1 F; mass = 0.1 kg.
75. An isolated capacitor of unknown capacitance has been charged to a potential difference of 100 V. When the charged capacitor is then connected in parallel to an uncharged 10-$\mu$F capacitor, the voltage across the combination is 30 V. Calculate the unknown capacitance.
76. A certain electronic circuit calls for a capacitor having a capacitance of 1.2 pF and a breakdown potential of 1000 V. If you have a supply of 6-pF capacitors each having a breakdown potential of 200 V, how could you meet this circuit requirement?
77. A 2-$\mu$F capacitor and a 3-$\mu$F capacitor have the same maximum voltage rating $V_{max}$. Due to this voltage limitation, the maximum potential difference that can be applied to a series combination of these capacitors is 800 V. Calculate the maximum voltage rating of the individual capacitors.
78. A 2-nF parallel-plate capacitor is charged to an initial potential difference $V_i = 100$ V and then isolated. The dielectric material between the plates is mica ($\kappa = 5$). (a) How much work is required to withdraw the mica sheet? (b) What is the potential difference of the capacitor after the mica is withdrawn?
79. A parallel-plate capacitor is constructed using a dielectric material whose dielectric constant is 3 and whose dielectric strength is $2 \times 10^8$ V/m. The desired capacitance is 0.25 $\mu$F, and the capacitor must withstand a maximum potential difference of 4000 V. Find the minimum area of the capacitor plates.
80. A parallel-plate capacitor is constructed using three different dielectric materials, as shown in Figure 26.37. (a) Find an expression for the capacitance of the device in terms of the plate area $A$ and $d$, $\kappa_1$, $\kappa_2$, and $\kappa_3$. (b) Calculate the capacitance using the values $A = 1$ cm$^2$, $d = 2$ mm, $\kappa_1 = 4.9$, $\kappa_2 = 5.6$, and $\kappa_3 = 2.1$.

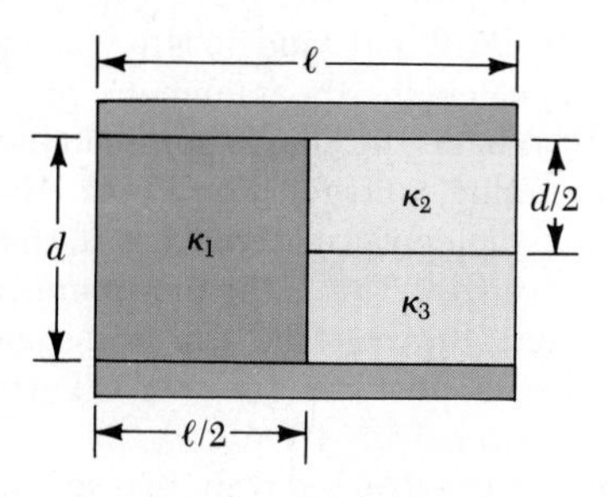


**Figure 26.37** (Problem 80).

81. In the arrangement shown in Figure 26.38, a potential $V$ is applied, and $C_1$ is adjusted so that the electrostatic voltmeter between points $b$ and $d$ reads zero. This "balance" occurs when $C_1 = 4\ \mu F$. If $C_3 = 9\ \mu F$ and $C_4 = 12\ \mu F$, calculate the value of $C_2$.

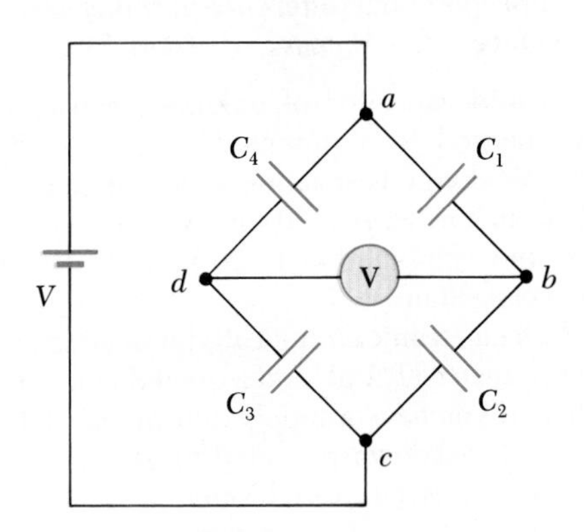


**Figure 26.38** (Problem 81).

82. It is possible to obtain large potential differences by first charging a group of capacitors connected in parallel and then activating a switch arrangement that in effect disconnects the capacitors from the charging source and from each other and reconnects them in a *series* arrangement. The group of charged capacitors is then discharged in *series*. What is the maximum potential difference that can be obtained in this manner by using ten capacitors each of 500 $\mu$F and a charging source of 800 V?

83. A parallel-plate capacitor of plate separation $d$ is charged to a potential difference $V_0$. A dielectric slab of thickness $d$ and dielectric constant $\kappa$ is introduced between the plates *while the battery remains connected to the plates.* (a) Show that the ratio of energy stored after the dielectric is introduced to the energy stored in the empty capacitor is given by $U/U_0 = \kappa$. Give a physical explanation for this increase in stored energy. (b) What happens to the charge on the capacitor? (Note that this situation is not the same as Example 26.7, in which the battery was removed from the circuit before introducing the dielectric.)

84. A parallel-plate capacitor is to be constructed using Pyrex glass as a dielectric. If the capacitance of the device is to be 0.2 $\mu$F and it is to be operated at 6000 V, (a) calculate the minimum plate area required. (b) What is the energy stored in the capacitor at the operating voltage? For Pyrex, use $\kappa = 5.6$. (*Note:* Each dielectric material has a characteristic dielectric strength. This is the maximum voltage per unit thickness the material can withstand without electrical breakdown or rupture. For Pyrex, the dielectric strength is $14 \times 10^6$ V/m.)

85. A capacitor is constructed from two square plates of sides $\ell$ and separation $d$. A material of dielectric constant $\kappa$ is inserted a distance $x$ into the capacitor, as in Figure 26.39. (a) Find the equivalent capacitance of the device. (b) Calculate the energy stored in the capacitor if the potential difference is $V$. (c) Find the direction and magnitude of the force exerted on the dielectric, assuming a constant potential difference $V$. Neglect friction and edge effects. (d) Obtain a numerical value for the force assuming that $\ell = 5$ cm, $V = 2000$ V, $d = 2$ mm, and the dielectric is glass ($\kappa = 4.5$). (*Hint:* The system can be considered as two capacitors connected in *parallel.*)

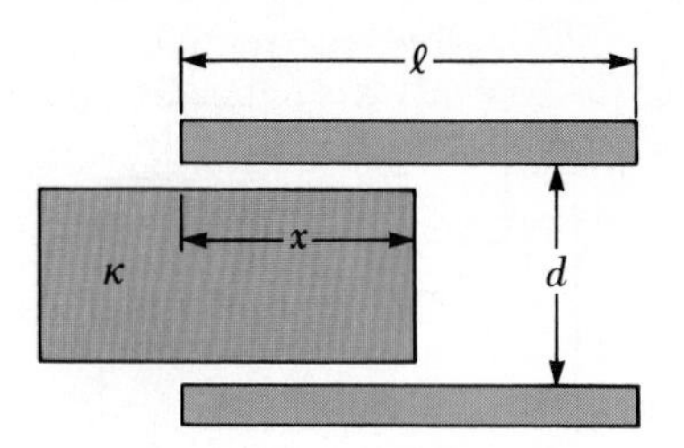


**Figure 26.39** (Problem 85).

86. Capacitors $C_1 = 6\ \mu F$ and $C_2 = 2\ \mu F$ are charged as a parallel combination across a 250-V battery. The capacitors are disconnected from the battery and from each other. They are then connected positive plate to negative plate and negative plate to positive plate. Calculate the resulting charge on each capacitor.

87. A stack of $N$ plates has alternate plates connected to form a capacitor similar to Figure 26.23. Adjacent plates are separated by a dielectric of thickness $d$. The dielectric constant is $\kappa$ and the area of overlap of adjacent plates is $A$. Show that the capacitance of this stack of plates is $C = \dfrac{\kappa\epsilon_0 A}{d}(N-1)$.

88. A coulomb balance is constructed of two parallel plates, each being 10 cm square. The upper plate is movable. A 25-mg mass is placed on the upper plate and the plate is observed to lower and the mass is then removed. When a potential difference is applied to the plates, it is found that the applied voltage must be 375 V to cause the upper plate to lower the same amount as it lowered when the mass was on it. If the force exerted on each plate by the other is given by $F = Q^2/2\epsilon_0 A$, calculate the following, assuming an applied voltage of 375 V: (a) The charge on the plates. (b) The electric field between the plates. (c) The separation distance of the plates. (d) The capacitance of this capacitor.

89. The inner conductor of a coaxial cable has a radius of 0.8 mm and the outer conductor's inside radius is 3.0 mm. The space between the conductors is filled with polyethylene, which has a dielectric constant of

2.3 and a dielectric strength of $18 \times 10^6$ V/m. What is the maximum potential difference that this cable can withstand?

**90.** You are optimizing coaxial cable design for a major manufacturer. Show that for a given outer conductor radius $b$, maximum potential difference capability is attained when the radius of the inner conductor is given by $a = b/e$ where $e$ is the base of natural logarithms.

**91.** Calculate the equivalent capacitance between the points $a$ and $b$ in Figure 26.40. Note that this is not a simple series or parallel combination. (*Hint:* Assume a potential difference $V$ between points $a$ and $b$. Write expressions for $V_{ab}$ in terms of the charges and capacitances for the various possible pathways from $a$ to $b$, and require conservation of charge for those capacitor plates that are connected to each other.)

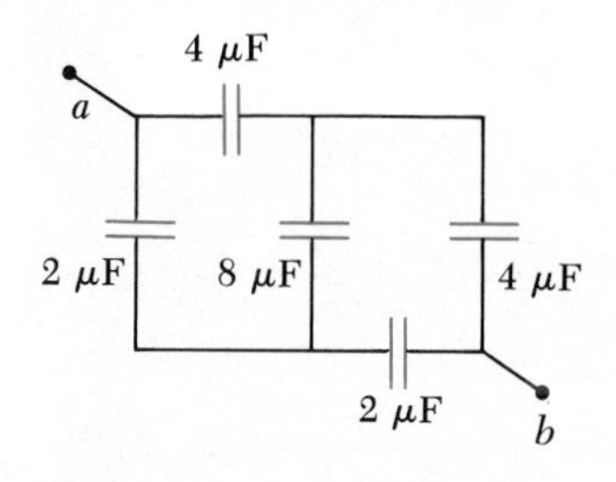


**Figure 26.40** (Problem 91).

**92.** Determine the effective capacitance of the combination shown in Figure 26.41. (*Hint:* Consider the symmetry involved!)

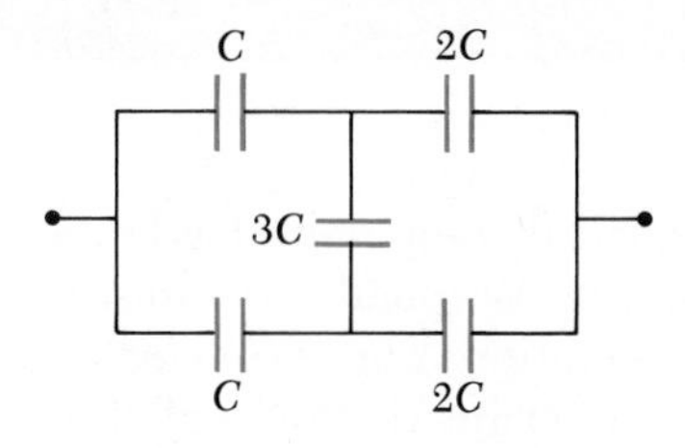


**Figure 26.41** (Problem 92).

**93.** Consider two *long*, parallel, and oppositely charged wires of radius $d$ with their centers separated by a distance $D$. Assuming the charge is distributed uniformly on the surface of each wire, show that the capacitance per unit length of this pair of wires is given by the following expression:

$$\frac{C}{\ell} = \frac{\pi\epsilon_0}{\ln\left(\frac{D-d}{d}\right)}.$$

"But we just don't have the technology to carry it out."

# 27
# Current and Resistance

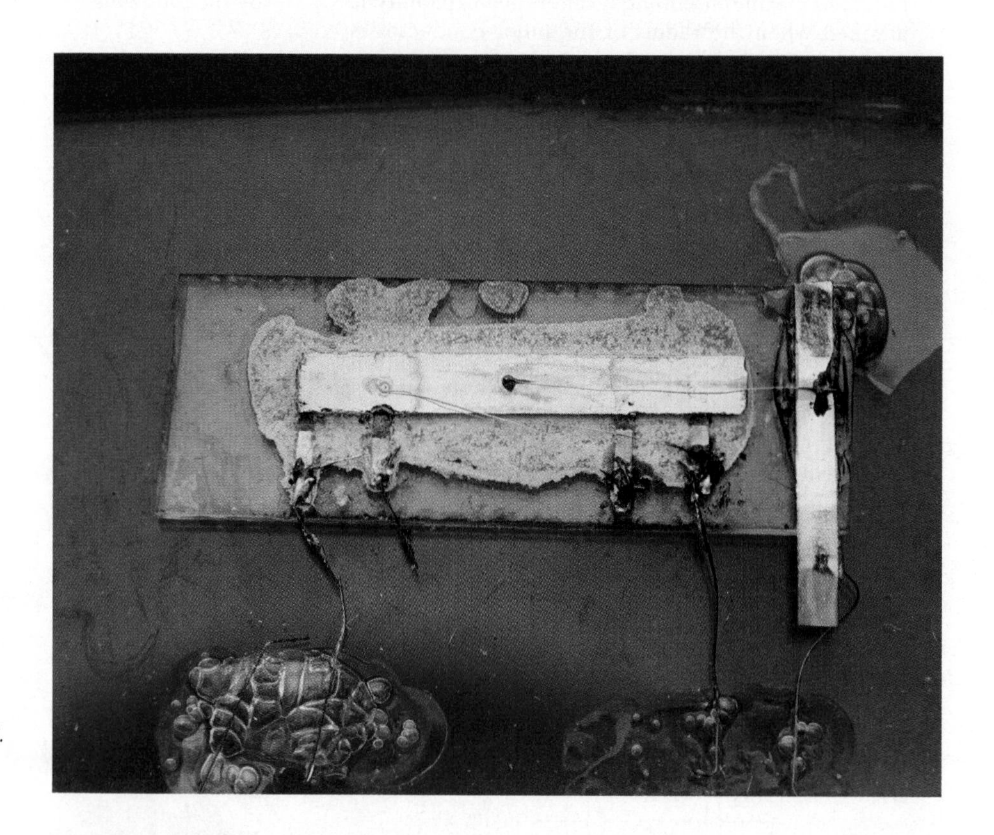

*The first integrated circuit, tested on September 12, 1958. (Courtesy of Texas Instruments)*

Thus far our discussion of electrical phenomena has been confined to charges at rest, or electrostatics. We shall now consider situations involving electric charges in motion. The term *electric current,* or simply *current,* is used to describe the rate of flow of charge through some region of space. Most practical applications of electricity deal with electric currents. For example, the battery of a flashlight supplies current to the filament of the bulb when the switch is turned on. A variety of home appliances operate on alternating current. In these common situations, the flow of charge takes place in a conductor, such as a copper wire. However, it is possible for currents to exist outside of a conductor. For instance, a beam of electrons in a TV picture tube constitutes a current.

In this chapter we shall first discuss the battery, one source of continuous current, followed by a definition of current and current density. A microscopic description of current will be given, and some of the factors that contribute to the resistance to the flow of charge in conductors will be discussed. Mechanisms responsible for the electrical resistance of various materials depend on the composition of the material and on temperature. A classical model is used to describe electrical conduction in metals, and some of the limitations of this model are pointed out.

## 27.1 THE BATTERY

Although electrical phenomena were known before 1800, electrical machines of that era were limited to devices that could produce static charge and large potential differences by means of friction. Such machines were capable of producing large sparks, but were of little practical value.

The electric battery, invented in 1800 by Alessandro Volta (1745–1827), was one of the most important practical discoveries in science. This invention represented the basis for a wide range of subsequent developments in electrical technology.

It is interesting to describe briefly some important events that led to Volta's invention. In 1786, Luigi Galvani (1737–1798) found that when a copper hook was inserted into the spinal cord of a frog, which in turn was hung from an iron railing, the leg muscles contracted. Galvani observed the same effect when other dissimilar metals were used. In reporting this unusual phenomenon, he proposed that the source of the charge was the muscle or nerve of the frog. Hence, he termed the source "animal electricity."

After hearing of Galvani's results, Volta proceeded to confirm and expand these experiments. He then offered the idea that the source of the charge was not the animal, but the contact between the two dissimilar metals, iron and copper. During his investigations, Volta recognized that the contact between the two metals required a moist conductor (such as the frog's muscle) to obtain a sizable effect. He eventually proved his point conclusively by showing that the effect occurred (although weakly) when the frog muscle was replaced by an inorganic substance. Further, he showed that certain pairs of metals produced a larger effect than others.

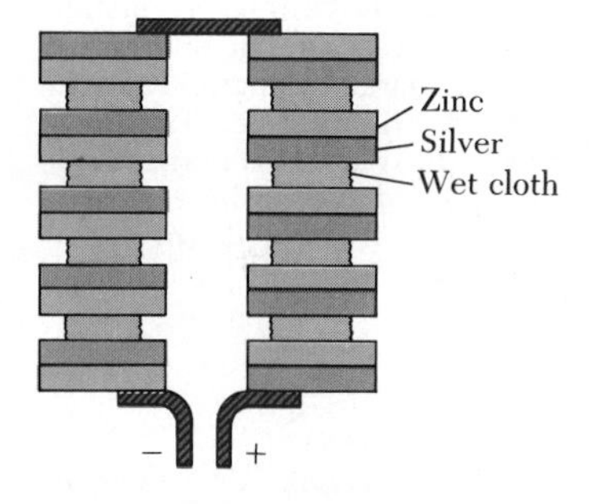


**Figure 27.1** Diagram of Volta's original pile. The cloth separating the plates is soaked in a salt solution. A potential difference is produced between the two end plates.

Volta then proceeded to invent a continuous source of electricity, the first battery. His original device, called the Voltaic pile, consisted of alternate disks of silver and zinc, as in Figure 27.1. Adjacent layers were separated by a cloth that had been soaked in a salt solution or dilute acid. The layered structure provided a continuous potential difference between the two ends, with an excess of positive charge at the silver end and an equal amount of negative charge at the zinc end. In effect, the pile was an energy converter, where internal chemical energy was converted into electric potential energy. Although this battery produced small potential differences compared to those produced by friction machines, it was able to provide a large electric charge, and hence proved to be of great practical importance. These early sources were very important for experiments because they provided a nearly constant potential difference.

There are many different kinds of batteries in use today. One of the most common types is the ordinary flashlight battery. These batteries are produced in a variety of shapes and sizes, but they all work in basically the same way. Figure 27.2 is a diagram of the interior of such a battery. In this particular battery, often referred to as a dry cell, the zinc case serves as the negative terminal, while the carbon rod down its center serves as the positive terminal. The space between the two terminals contains a paste-like mixture of manganese dioxide, ammonium chloride, and carbon.

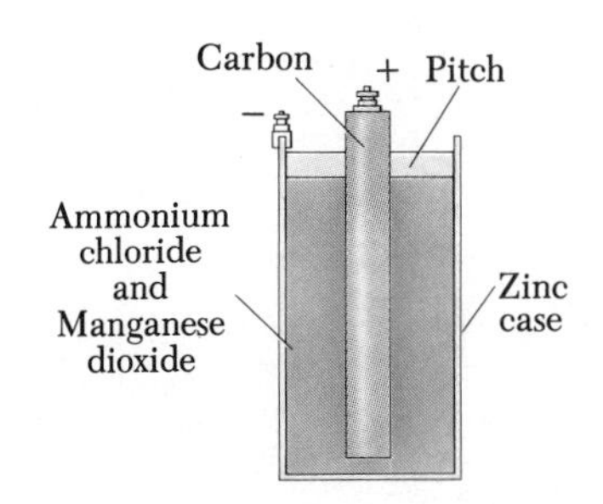


**Figure 27.2** Cross-sectional view of a dry cell.

When these materials are assembled in this fashion, two chemical reactions take place; one occurs at the zinc case, the other at the manganese dioxide layer surrounding the carbon rod. Positive charged zinc ions ($Zn^{2+}$) leave the case and enter the ammonium chloride paste, where they combine with chloride ions ($Cl^-$). (The chloride ions are present because a small percentage of the ammonium chloride dissociates, leaving some free chloride ions in the solution.) As each zinc ion is removed from the case, it leaves behind two electrons. As additional zinc ions leave the case, more electrons accumulate, leaving the zinc case with a net negative charge.

When a chloride ion breaks free from the ammonium chloride molecule, the remnant portion of the molecule becomes singly ionized. This positively charged ion is neutralized by the manganese dioxide, which supplies the needed electrons. As a result, the carbon rod surrounded by its manganese dioxide layer ends up with a net positive charge.

These chemical reactions and thus the charge separation do not continue without limit. The zinc case ultimately achieves such a strong negative charge that the zinc ions can no longer escape. A similar charge saturation occurs at the carbon rod.

## 27.2 ELECTRIC CURRENT

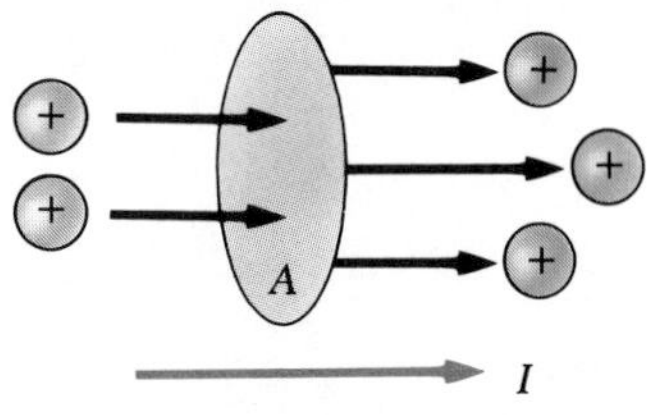

Whenever electric charges of like sign move, a *current* is said to exist. To define current more precisely, suppose the charges are moving perpendicular to a surface of area $A$ as in Figure 27.3. This area could be the cross-sectional area of a wire, for example. The **current** is *the rate at which charge flows through this surface.* If $\Delta Q$ is the amount of charge that passes through this area in a time interval $\Delta t$, the **average current, $I_{av}$,** is equal to the ratio of the charge to the time interval:

**Figure 27.3** Charges in motion through an area $A$. The time rate of flow of charge through the area is defined as the current $I$. The direction of the current is in the direction in which positive charge would flow if free to do so.

$$I_{av} = \frac{\Delta Q}{\Delta t} \tag{27.1}$$

If the rate at which charge flows varies in time, the current also varies in time and we define the **instantaneous current, $I$,** as the differential limit of the expression above:

Electric current

$$I \equiv \frac{dQ}{dt} \tag{27.2}$$

The SI unit of current is the **ampere** (A), where

The direction of the current

$$1\text{ A} = 1\text{ C/s} \tag{27.3}$$

That is, 1 A of current is equivalent to 1 C of charge passing through the surface in 1 s. In practice, smaller units of current are often used, such as the milliampere (1 mA $= 10^{-3}$ A) and the microampere (1 $\mu$A $= 10^{-6}$ A).

When charges flow through the surface in Figure 27.3, they can be positive, negative, or both. *It is conventional to choose the direction of the current to be in the direction of flow of positive charge.* In a conductor such as copper, the current is due to the motion of the negatively charged electrons. Therefore, when we speak of current in an ordinary conductor, such as a copper wire, *the direction of the current will be opposite the direction of flow of electrons.* On the other hand, if one considers a beam of positively charged protons in an accelerator, the current is in the direction of motion of the protons. In some cases, the current is the result of the flow of both positive and negative charges. This occurs, for example, in semiconductors and electrolytes. It is common to refer to a moving charge (whether it is positive or negative) as a mobile *charge carrier*. For example, the charge carriers in a metal are electrons.

It is instructive to relate current to the motion of the charged particles. To illustrate this point, consider the current in a conductor of cross-sectional area $A$ (Fig. 27.4). The volume of an element of the conductor of length $\Delta x$ (the shaded region in Fig. 27.4) is $A\,\Delta x$. If $n$ represents the number of mobile charge carriers per unit volume, then the number of mobile charge carriers in the volume element is given by $nA\,\Delta x$. Therefore, the charge $\Delta Q$ in this element is given by

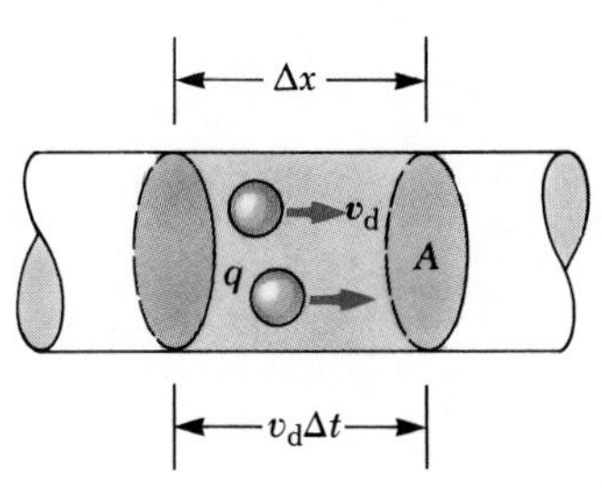


**Figure 27.4** A section of a uniform conductor of cross-sectional area $A$. The charge carriers move with a speed $v_d$, and the distance they travel in a time $\Delta t$ is given by $\Delta x = v_d\,\Delta t$. The number of mobile charge carriers in the section of length $\Delta x$ is given by $nAv_d\,\Delta t$, where $n$ is the number of mobile carriers per unit volume.

$$\Delta Q = \text{number of charges} \times \text{charge per particle} = (nA\,\Delta x)q$$

where $q$ is the charge on each particle. If the charge carriers move with a speed $v_d$, the distance they move in a time $\Delta t$ is given by $\Delta x = v_d\,\Delta t$. Therefore, we can write $\Delta Q$ in the form

$$\Delta Q = (nAv_d\,\Delta t)q$$

If we divide both sides of this equation by $\Delta t$, we see that the current in the conductor is given by

$$I = \frac{\Delta Q}{\Delta t} = nqv_dA \tag{27.4}$$

Current in a conductor

The velocity of the charge carriers, $v_d$, is actually an average velocity and is called the **drift velocity.** To understand the meaning of drift velocity, consider a conductor in which the charge carriers are free electrons. In an isolated conductor, these electrons undergo random motion similar to that of gas molecules. When a potential difference is applied across the conductor (say, by means of a battery), an electric field is set up in the conductor, which creates an electric force on the electrons and hence a current. In reality, the electrons do not simply move in straight lines along the conductor. Instead, they undergo repeated collisions with the metal atoms, which results in a complicated zigzag motion (Fig. 27.5). The energy transferred from the electrons to the metal atoms causes an increase in the vibrational energy of the atoms and a corresponding increase in the temperature of the conductor. However, despite the collisions, the electrons move slowly along the conductor (in a direction opposite $\boldsymbol{E}$) with an average velocity called the drift velocity, $\boldsymbol{v}_d$. The field does work on the electrons that exceeds the average loss due to collisions, which results in a net current. As we shall see in an example that follows, drift velocities are *much* smaller than the average speed between collisions. We shall discuss this model in more detail in Section 27.6. One can think of the collisions of the electrons within a conductor as being an effective internal friction (or drag force), similar to that experienced by the molecules of a liquid flowing through a pipe stuffed with steel wool.

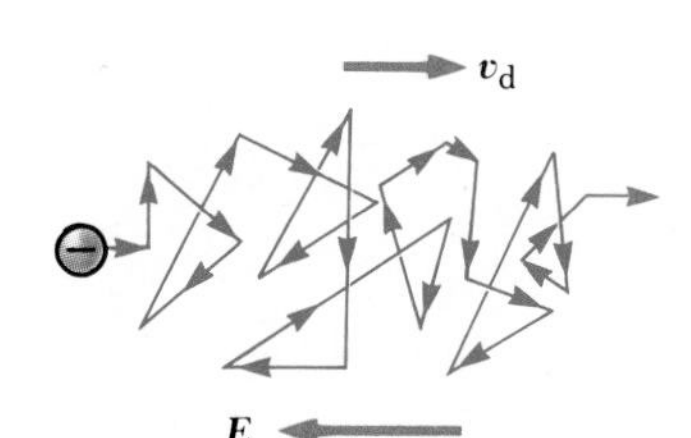


**Figure 27.5** A schematic representation of the zigzag motion of a charge carrier in a conductor. The changes in direction are due to collisions with atoms in the conductor. Note that the net motion of electrons is opposite the direction of the electric field. The zig-zag paths are actually parabolic segments.

The following quotation is an interesting and amusing description by W.F.G. Swann of electronic conduction in telephone cables.[1]

> Think of the cables which carry the telephone current in the form of electrons. In the absence of the current the electrons are moving in all directions. As many are

[1] W.F.G. Swann, *Physics Today*, June 1951, p. 9.

moving from left to right as are moving from right to left; and the nothingness which is there is composed of two equal and opposite halves, about a million million amperes per square centimeter in one direction, and a million million amperes per square centimeter in the other direction. The telephone current constitutes an upsetting of the balance to the extent of one hundredth of a millionth of an ampere per square centimeter, or about one part in a hundred million million million. Then if this one part in a hundred million million million is at fault by one part in a thousand, we ring up the telephone company and complain that the quality of the speech is faulty.

**EXAMPLE 27.1 The Drift Velocity in a Copper Wire**
A copper wire of cross-sectional area $3 \times 10^{-6}\ \text{m}^2$ carries a current of 10 A. Find the drift velocity of the electrons in this wire. The density of copper is $8.95\ \text{g/cm}^3$.

*Solution* From the periodic table of the elements, we find that the atomic weight of copper is 63.5 g/mole. Recall that one atomic mass of any substance contains Avogadro's number of atoms, $6.02 \times 10^{23}$ atoms. Knowing the density of copper enables us to calculate the volume occupied by 63.5 g of copper:

$$V = \frac{m}{\rho} = \frac{63.5\ \text{g}}{8.95\ \text{g/cm}^3} = 7.09\ \text{cm}^3$$

If we now assume that each copper atom contributes one free electron to the body of the material, we have

$$n = \frac{6.02 \times 10^{23}\ \text{electrons}}{7.09\ \text{cm}^3}$$

$$= 8.48 \times 10^{22}\ \text{electrons/cm}^3$$

$$= \left(8.48 \times 10^{22}\ \frac{\text{electrons}}{\text{cm}^3}\right)\left(10^6\ \frac{\text{cm}^3}{\text{m}^3}\right)$$

$$= 8.48 \times 10^{28}\ \text{electrons/m}^3$$

From Equation 27.4, we find that the drift velocity is

$$v_d = \frac{I}{nqA}$$

$$= \frac{10\ \text{C/s}}{(8.48 \times 10^{28}\ \text{m}^{-3})(1.6 \times 10^{-19}\ \text{C})(3 \times 10^{-6}\ \text{m}^2)}$$

$$= 2.46 \times 10^{-4}\ \text{m/s}$$

Example 27.1 shows that typical drift velocities are very small. In fact, the drift velocity is much smaller than the average velocity between collisions. For instance, electrons traveling with this velocity would take about 68 min to travel 1 m! In view of this low speed, you might wonder why a light turns on almost instantaneously when a switch is thrown. This can be explained by considering the flow of water through a pipe. If a drop of water is forced in one end of a pipe that is already filled with water, a drop must be pushed out the other end of the pipe. While it may take individual drops of water a long time to make it through the pipe, a flow initiated at one end produces a similar flow at the other end very quickly. In a conductor, the electric field that drives the free electrons travels through the conductor with a speed close to that of light. Thus, when you flip a light switch, the message for the electrons to start moving through the wire (the electric field) reaches them at a speed of the order of $10^8$ m/s.

## 27.3 RESISTANCE AND OHM'S LAW

Earlier, we found that there can be no electric field inside a conductor. However, this statement is true *only* if the conductor is in static equilibrium. The purpose of this section is to describe what happens when the charges are allowed to move in the conductor.

Charges move in a conductor to produce a current under the action of an electric field inside the conductor. An electric field can exist in the conductor in this case since we are dealing with charges in motion, a *nonelectrostatic* situation. This is in contrast with the situation in which a conductor in *electrostatic equilibrium* (where the charges are at rest) can have no electric field inside.

Consider a conductor of cross-sectional area $A$ carrying a current $I$. The **current density** $J$ in the conductor is defined to be the current per unit area. Since $I = nqv_dA$, the current density is given by

Current density

$$J \equiv \frac{I}{A} = nqv_d \tag{27.5}$$

where $J$ has SI units of $A/m^2$. This expression is valid only if the current density is uniform and the surface is perpendicular to the direction of the current. In general, the current density is a *vector quantity*. That is,

$$\boldsymbol{J} = nq\boldsymbol{v}_d \tag{27.6}$$

From this definition, we see once again that the current density, like the current, is in the direction of motion of the charges for positive charge carriers and opposite the direction of motion for negative charge carriers.

*A current density* ***J*** *and an electric field* **E** *are established in a conductor when a potential difference is maintained across the conductor.* If the potential difference is constant, the current in the conductor will also be constant. Very often, the current density in a conductor is proportional to the electric field in the conductor. That is,

Ohm's law

$$\boldsymbol{J} = \sigma \boldsymbol{E} \tag{27.7}$$

where the constant of proportionality $\sigma$ is called the **conductivity** of the conductor.[2] Materials that obey Equation 27.7 are said to follow Ohm's law, named after Georg Simon Ohm (1787–1854). More specifically,

> Ohm's law states that for many materials (including most metals), the ratio of the current density and electric field is a constant, $\sigma$, which is independent of the electric field producing the current.

Materials that obey Ohm's law, and hence demonstrate this linear behavior between **E** and ***J***, are said to be *ohmic*. The electrical behavior of most materials is quite linear for *very small changes* in the current. Experimentally, one finds that not all materials have this property. Materials that do not obey Ohm's law are said to be *nonohmic*. Ohm's law is *not* a fundamental law of nature, but an empirical relationship valid only for certain materials.

A form of Ohm's law that is more directly useful in practical applications can be obtained by considering a segment of a straight wire of cross-sectional area $A$ and length $\ell$, as in Figure 27.6. A potential difference $V_b - V_a$ is maintained across the wire, creating an electric field in the wire and a current. If the electric field in the wire is assumed to be uniform, the potential difference $V = V_b - V_a$ is related to the electric field through the relationship[3]

$$V = E\ell$$

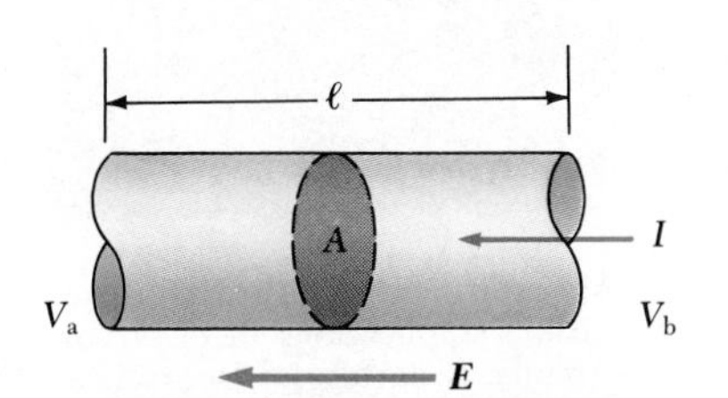


**Figure 27.6** A uniform conductor of length $\ell$ and cross-sectional area A. A potential difference $V_b - V_a$ maintained across the conductor sets up an electric field **E** in the conductor, and this field produces a current $I$.

[2] Do not confuse the conductivity $\sigma$ with the surface charge density, for which the same symbol is used.

[3] This result follows from the definition of potential difference:

$$V_b - V_a = -\int_a^b \mathbf{E} \cdot d\mathbf{s} = E\int_0^\ell dx = E\ell$$

Therefore, we can express the magnitude of the current density in the wire as

$$J = \sigma E = \sigma \frac{V}{\ell}$$

Since $J = I/A$, the potential difference can be written

$$V = \frac{\ell}{\sigma} J = \left(\frac{\ell}{\sigma A}\right) I$$

The quantity $\ell/\sigma A$ is called the **resistance** $R$ of the conductor:

Resistance of a conductor

$$R = \frac{\ell}{\sigma A} = \frac{V}{I} \tag{27.8}$$

From this result we see that resistance has SI units of volts per ampere. One volt per ampere is defined to be one ohm (Ω):

$$1\ \Omega \equiv 1\ \text{V/A}$$

That is, if a potential difference of 1 V across a conductor causes a current of 1 A, the resistance of the conductor is 1 Ω. For example, if an electrical appliance connected to a 120-V source carries a current of 6A, its resistance is 20 Ω.

The inverse of the conductivity of a material is called the **resistivity** $\rho$:

Resistivity

$$\rho \equiv \frac{1}{\sigma} \tag{27.9}$$

Using this definition and Equation 27.8, the resistance can be expressed as

Resistance of a uniform conductor

$$R = \rho \frac{\ell}{A} \tag{27.10}$$

where $\rho$ has the units ohm-meters (Ω·m). (The symbol $\rho$ for resistivity should not be confused with the same symbol used earlier in the text for mass density or charge density.) Every ohmic material has a characteristic resistivity, a parameter that depends on the properties of the material and on temperature. On the other hand, as you can see from Equation 27.10, the resistance of a substance depends on simple geometry as well as on the resistivity of the substance. Good electrical conductors have very low resistivity (or high conductivity), and good insulators have very high resistivity (low conductivity). Table 27.1 gives the resistivities of a variety of materials at 20°C.

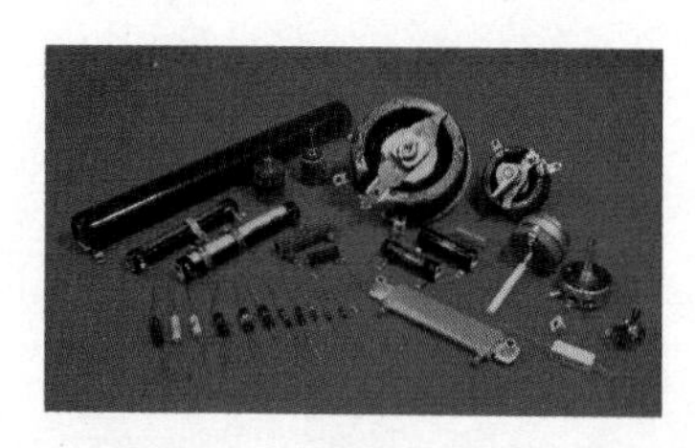

An assortment of resistors used for various applications in electronic circuits. (Courtesy of Henry Leap and Jim Lehman)

Equation 27.10 shows that the resistance of a given cylindrical conductor is proportional to its length and inversely proportional to its cross-sectional area. Therefore, if the length of a wire is doubled, its resistance doubles. Furthermore, if its cross-sectional area is doubled, its resistance drops by one-half. The situation is analogous to the flow of a liquid through a pipe. As the length of the pipe is increased, the resistance to liquid flow increases. As its cross-sectional area is increased, the pipe can more readily transport liquid.

All electric appliances such as toasters, heaters, and light bulbs have a fixed resistance. Most electric circuits make use of devices called **resistors** to control the current level in the various parts of the circuit. Two common types of resistors are the "composition" resistor containing carbon, which is a semi-

**TABLE 27.1 Resistivities and Temperature Coefficients of Resistivity for Various Materials**

| Material | Resistivity[a] ($\Omega \cdot m$) | Temperature Coefficient $\alpha$ [$(C°)^{-1}$] |
|---|---|---|
| Silver | $1.59 \times 10^{-8}$ | $3.8 \times 10^{-3}$ |
| Copper | $1.7 \times 10^{-8}$ | $3.9 \times 10^{-3}$ |
| Gold | $2.44 \times 10^{-8}$ | $3.4 \times 10^{-3}$ |
| Aluminum | $2.82 \times 10^{-8}$ | $3.9 \times 10^{-3}$ |
| Tungsten | $5.6 \times 10^{-8}$ | $4.5 \times 10^{-3}$ |
| Iron | $10 \times 10^{-8}$ | $5.0 \times 10^{-3}$ |
| Platinum | $11 \times 10^{-8}$ | $3.92 \times 10^{-3}$ |
| Lead | $22 \times 10^{-8}$ | $3.9 \times 10^{-3}$ |
| Nichrome[b] | $1.50 \times 10^{-6}$ | $0.4 \times 10^{-3}$ |
| Carbon | $3.5 \times 10^{-5}$ | $-0.5 \times 10^{-3}$ |
| Germanium | 0.46 | $-48 \times 10^{-3}$ |
| Silicon | 640 | $-75 \times 10^{-3}$ |
| Glass | $10^{10} - 10^{14}$ | |
| Hard rubber | $\approx 10^{13}$ | |
| Sulfur | $10^{15}$ | |
| Quartz (fused) | $75 \times 10^{16}$ | |

[a] All values at 20°C.
[b] A nickel-chromium alloy commonly used in heating elements.

conductor, and the "wire-wound" resistor, which consists of a coil of wire. Resistors are normally color-coded to give their values in ohms, as shown in Figure 27.7. Table 27.2 will enable you to translate from the color code to a specific value of resistance.

Ohmic materials, such as copper, have a linear current-voltage relationship over a large range of applied voltage (Fig. 27.8a). The slope of the $I$ versus $V$ curve in the linear region yields a value for $R$. Nonohmic materials have a nonlinear current-voltage relationship. One common semiconducting device that has nonlinear $I$ versus $V$ characteristics is the diode (Fig. 27.8b). The

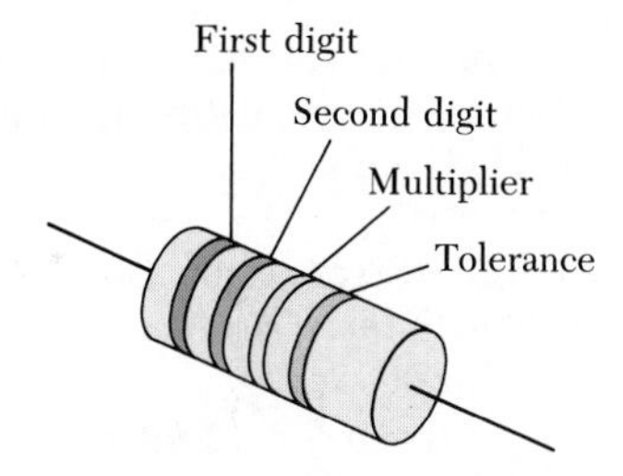


**Figure 27.7** The colored bands on a resistor represent a code for determining the value of its resistance. The first two colors give the first two digits in the resistance value. The third color represents the power of ten for the multiplier of the resistance value. The last color is the tolerance of the resistance value. As an example, if the four colors are orange, blue, yellow, and gold, the resistance value is $36 \times 10^4\ \Omega$ or 360 k$\Omega$, with a tolerance value of 18 k$\Omega$ (5%).

**TABLE 27.2 Color Code for Resistors**

| Color | Number | Multiplier | Tolerance (%) |
|---|---|---|---|
| Black | 0 | 1 | |
| Brown | 1 | $10^1$ | |
| Red | 2 | $10^2$ | |
| Orange | 3 | $10^3$ | |
| Yellow | 4 | $10^4$ | |
| Green | 5 | $10^5$ | |
| Blue | 6 | $10^6$ | |
| Violet | 7 | $10^7$ | |
| Gray | 8 | $10^8$ | |
| White | 9 | $10^9$ | |
| Gold | | $10^{-1}$ | 5% |
| Silver | | $10^{-2}$ | 10% |
| Colorless | | | 20% |

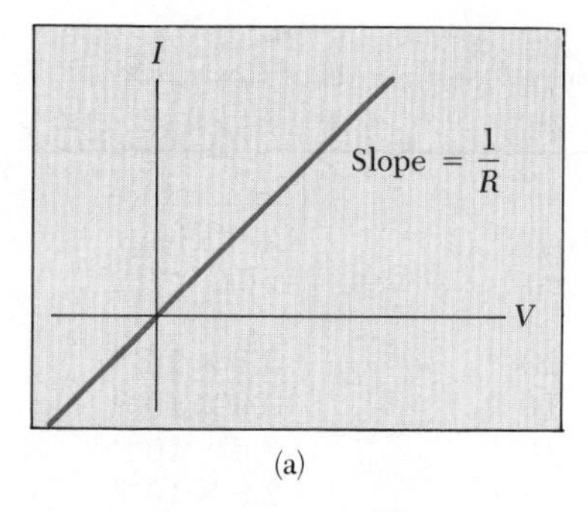


(a)

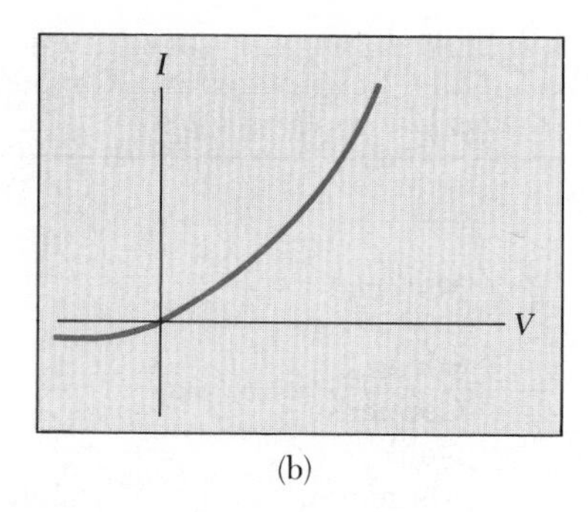


(b)

**Figure 27.8** (a) The current-voltage curve for an ohmic material. The curve is linear, and the slope gives the resistance of the conductor. (b) A nonlinear current-voltage curve for a semiconducting diode. This device does not obey Ohm's law.

effective resistance of this device (inversely proportional to the slope of its $I$ versus $V$ curve) is small for currents in one direction (positive $V$) and large for currents in the reverse direction (negative $V$). In fact, most modern electronic devices, such as transistors, have nonlinear current-voltage relationships; their proper operation depends on the particular way in which they violate Ohm's law.

**EXAMPLE 27.2 The Resistance of a Conductor**
Calculate the resistance of a piece of aluminum that is 10 cm long and has a cross-sectional area of $10^{-4}$ m². Repeat the calculation for a piece of glass of resistivity $10^{10}\ \Omega \cdot$ m.

*Solution* From Equation 27.10 and Table 27.1, we can calculate the resistance of the aluminum bar:

$$R = \rho \frac{L}{A} = (2.82 \times 10^{-8}\ \Omega \cdot \text{m})\left(\frac{0.1\ \text{m}}{10^{-4}\ \text{m}^2}\right)$$

$$= 2.82 \times 10^{-5}\ \Omega$$

Similarly, for glass we find

$$R = \rho \frac{L}{A} = (10^{10}\ \Omega \cdot \text{m})\left(\frac{0.1\ \text{m}}{10^{-4}\ \text{m}^2}\right) = 10^{13}\ \Omega$$

As you might expect, aluminum has a much lower resistance than glass. This is the reason that aluminum is a good conductor and glass is a poor conductor.

**EXAMPLE 27.3 The Resistance of Nichrome Wire**
(a) Calculate the resistance per unit length of a 22-gauge nichrome wire of radius 0.321 mm.

*Solution* The cross-sectional area of this wire is

$$A = \pi r^2 = \pi(0.321 \times 10^{-3}\ \text{m})^2 = 3.24 \times 10^{-7}\ \text{m}^2$$

The resistivity of nichrome is $1.5 \times 10^{-6}\ \Omega \cdot$ m (Table 27.1). Thus, we can use Equation 27.10 to find the resistance per unit length:

$$\frac{R}{\ell} = \frac{\rho}{A} = \frac{1.5 \times 10^{-6}\ \Omega \cdot \text{m}}{3.24 \times 10^{-7}\ \text{m}^2} = 4.6\ \Omega/\text{m}$$

(b) If a potential difference of 10 V is maintained across a 1-m length of the nichrome wire, what is the current in the wire?

*Solution* Since a 1-m length of this wire has a resistance of 4.6 Ω, Ohm's law gives

$$I = \frac{V}{R} = \frac{10\ \text{V}}{4.6\ \Omega} = 2.2\ \text{A}$$

Note that the resistance of the nichrome wire is about 100 times larger than that of the copper wire. Therefore, a copper wire of the same radius would have a resistance per unit length of only 0.052 Ω/m. A 1-m length of copper wire of the same radius would carry the same current (2.2 A) with an applied voltage of only 0.11 V.

Because of its high resistivity and its resistance to oxidation, nichrome is often used for heating elements in toasters, irons, and electric heaters.

*Exercise 1* What is the resistance of a 6-m length of 22-gauge nichrome wire? How much current does it carry when connected to a 120-V source?
*Answer* 28 Ω, 4.3 A.

Exercise 2 Calculate the current density and electric field in the wire assuming that it carries a current of 2.2 A.
Answer $6.7 \times 10^6$ A/m$^2$; 10 N/C.

EXAMPLE 27.4 The Resistance of a Coaxial Tube □
The gap between a pair of coaxial tubes is completely filled with silicon as in Figure 27.9a. The inner radius of the tube is $a = 0.500$ cm, the outer radius, $b = 1.75$ cm, and its length, $L = 15.0$ cm. Calculate the total resistance of the silicon when measured between the inner and outer tubes.

*Solution* The resistivity of copper is small compared with that of silicon, so its resistance can be neglected. In this type of problem, we must divide the conductor into elements of infinitesimal thickness over which the area may be considered constant. We can start by using the differential form of Equation 27.10, which is $dR = \rho\, d\ell/A$, where $dR$ is the resistance of a section of the conductor of thickness $d\ell$ and area $A$. In this example, we take as our element a hollow cylinder of thickness $dr$ and length $L$ as in Figure 27.9b. Any current that passes between the inner and outer tubes must pass radially through such elements, and the area through which it passes is $A = 2\pi rL$. (This would be the surface area of our hollow cylinder neglecting the area of its ends.) Hence, we can write the resistance of our hollow cylinder as

$$dR = \frac{\rho}{2\pi rL}\, dr$$

Since we wish to know the total resistance of the silicon, we must integrate this expression over $dr$ from $r = a$ to $r = b$. This gives

$$R = \int_a^b dR = \frac{\rho}{2\pi L}\int_a^b \frac{dr}{r} = \frac{\rho}{2\pi L}\ln\left(\frac{b}{a}\right)$$

Substituting in the values given, and using $\rho = 640\ \Omega\cdot\text{m}$ for silicon gives

$$R = \frac{640\ \Omega\cdot\text{m}}{2\pi(0.150\ \text{m})}\ln\left(\frac{1.75\ \text{cm}}{0.500\ \text{cm}}\right) = 851\ \Omega$$

Exercise 3 If a potential difference of 12.0 V is applied between the inner and outer copper tube, calculate the total current that passes between them.
Answer 14.1 mA

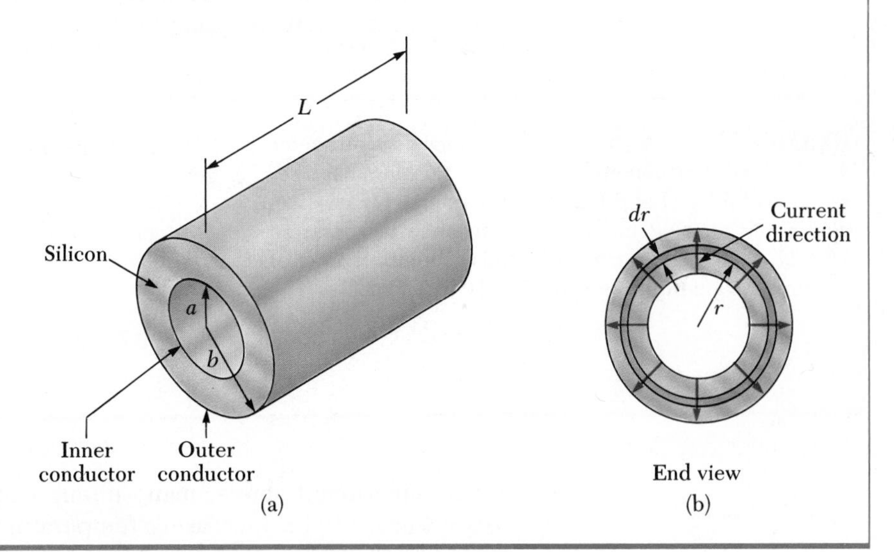


**Figure 27.9** (Example 27.4).

## 27.4 THE RESISTIVITY OF DIFFERENT CONDUCTORS

The resistivity of a conductor depends on a number of factors, one of which is temperature. For most metals, resistivity increases with increasing temperature. The resistivity of a conductor varies in an approximately linear fashion with temperature over a limited temperature range according to the expression

$$\rho = \rho_o[1 + \alpha(T - T_o)] \qquad (27.11)$$

Variation of $\rho$ with temperature

where $\rho$ is the resistivity at some temperature $T$ (in °C), $\rho_o$ is the resistivity at some reference temperature $T_o$ (usually taken to be 20°C), and $\alpha$ is called the **temperature coefficient of resistivity.** From Equation 27.11, we see that the temperature coefficient of resistivity can also be expressed as

Temperature coefficient of resistivity

$$\alpha = \frac{1}{\rho_o} \frac{\Delta\rho}{\Delta T} \tag{27.12}$$

where $\Delta\rho = \rho - \rho_o$ is the change in resistivity in the temperature interval $\Delta T = T - T_o$.

The resistivities and temperature coefficients for various materials are given in Table 27.1. Note the enormous range in resistivities, from very low values for good conductors, such as copper and silver, to very high values for good insulators, such as glass and rubber. An ideal, or "perfect," conductor would have zero resistivity, and an ideal insulator would have infinite resistivity.

Since the resistance of a conductor is proportional to the resistivity according to Equation 27.10, the temperature variation of the resistance can be written

$$R = R_o[1 + \alpha(T - T_o)] \tag{27.13}$$

Precise temperature measurements are often made using this property, as shown in the following example.

**EXAMPLE 27.5 A Platinum Resistance Thermometer**
A resistance thermometer made from platinum has a resistance of 50.0 Ω at 20°C. When immersed in a vessel containing melting indium, its resistance increases to 76.8 Ω. From this information, find the melting point of indium. For platinum, $\alpha = 3.92 \times 10^{-3}\ (\text{C}°)^{-1}$.

*Solution* Using Equation 27.13 and solving for $\Delta T$, we get

$$\Delta T = \frac{R - R_o}{\alpha R_o} = \frac{76.8\ \Omega - 50.0\ \Omega}{[3.92 \times 10^{-3}\ (\text{C}°)^{-1}](50.0\ \Omega)}$$

$$= 137\ \text{C}°$$

Since $\Delta T = T - T_o$ and $T_o = 20°\text{C}$, we find that $T = 157°\text{C}$.

As mentioned above, many ohmic materials have resistivities that increase linearly with increasing temperature, as shown in Figure 27.10. In reality, however, there is always a nonlinear region at very low temperatures, and the resistivity usually approaches some finite value near absolute zero (see magnified insert in Fig. 27.10). This residual resistivity near absolute zero is due primarily to collisions of electrons with impurities and imperfections in the metal. In contrast, the high-temperature resistivity (the linear region) is dominated by collisions of electrons with the metal atoms. We shall describe this process in more detail in Section 27.6.

Semiconductors, such as silicon and germanium, have intermediate values of resistivity. The resistivity of semiconductors generally decreases with increasing temperature, corresponding to a negative temperature coefficient of resistivity (Fig. 27.11). This is due to the increase in the density of charge carriers at the higher temperatures. Since the charge carriers in a

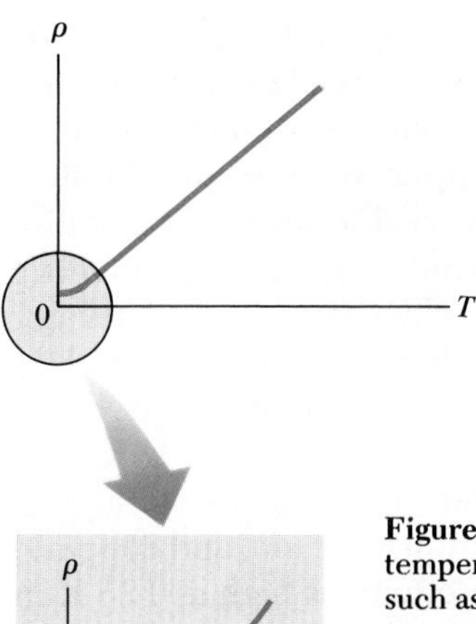


**Figure 27.10** Resistivity versus temperature for a normal metal, such as copper. The curve is linear over a wide range of temperatures, and $\rho$ increases with increasing temperature. As $T$ approaches absolute zero (insert), the resistivity approaches a finite value $\rho_0$.

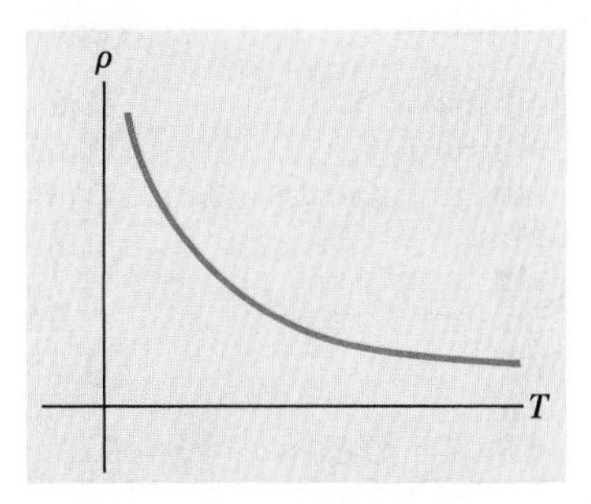


**Figure 27.11** Resistivity versus temperature for a pure semiconductor, such as silicon or germanium.

semiconductor are often associated with impurity atoms, the resistivity is very sensitive to the type and concentration of such impurities. The **thermistor** is a semiconducting thermometer that makes use of the large changes in its resistivity with temperature. We shall return to the study of semiconductors in the extended version of this text, Chapter 43.

## 27.5 SUPERCONDUCTORS

There is a class of metals and compounds whose resistance goes virtually to *zero* below a certain temperature, $T_c$, called the *critical temperature.* These materials are known as **superconductors.** The resistance-temperature graph for a superconductor follows that of a normal metal at temperatures above $T_c$ (Fig. 27.12). When the temperature is at or below $T_c$, the resistivity drops suddenly to zero. This phenomenon was discovered in 1911 by the Dutch physicist H. Kamerlingh Onnes when he was working with mercury, which is a superconductor below 4.2 K. Recent measurements have shown that the resistivities of superconductors below $T_c$ are less than $4 \times 10^{-25}\ \Omega \cdot \mathrm{m}$, which is around $10^{17}$ times smaller than the resistivity of copper and considered to be *zero* in practice.

Today there are thousands of known superconductors. Such common metals as aluminum, tin, lead, zinc, and indium are superconductors. Table 27.3 lists the critical temperatures of several superconductors. The value of $T_c$ is sensitive to chemical composition, pressure, and crystalline structure. It is interesting to note that copper, silver, and gold, which are excellent conductors, do not exhibit superconductivity.

One of the truly remarkable features of superconductors is the fact that once a current is set up in them, it will persist *without any applied voltage* (since $R = 0$). In fact, steady currents have been observed to persist in superconducting loops for several years with no apparent decay!

One of the most important recent developments in physics that has created much excitement in the scientific community has been the discovery of high temperature copper-oxide-based superconductors. The excitement

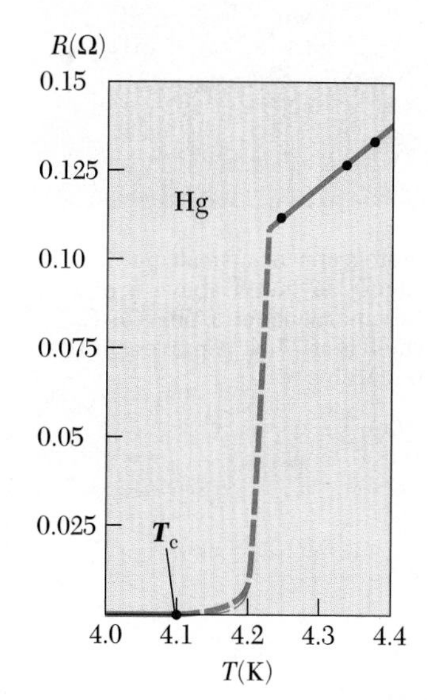


**Figure 27.12** Resistance versus temperature for mercury. The graph follows that of a normal metal above the critical temperature, $T_c$. The resistance drops to zero at the critical temperature, which is 4.15 K for mercury.

**TABLE 27.3 Critical Temperatures for Various Superconductors**

| Material | $T_c$ (K) |
|---|---|
| $Nb_3Ge$ | 23.2 |
| $Nb_3Sn$ | 18.05 |
| Nb | 9.46 |
| Pb | 7.18 |
| Hg | 4.15 |
| Sn | 3.72 |
| Al | 1.19 |
| Zn | 0.88 |
| $YBa_2Cu_3O_{7-\delta}$ | 92 |
| Bi-Sr-Ca-Cu-O | 105 |
| Tl-Ba-Ca-Cu-O | 125 |

Photograph of a small permanent magnet levitated above a disk of the superconductor $YBa_2Cu_3O_{7-\delta}$, which is at 77 K. (Courtesy of IBM Research)

began with a 1986 publication by Georg Bednorz and K. Alex Müller, two scientists working at the IBM Zurich Research Laboratory in Switzerland, who reported evidence for superconductivity at a temperature near 30 K in an oxide of barium, lanthanum, and copper. Bednorz and Müller were awarded the Nobel Prize in 1987 for their remarkable and important discovery. Shortly thereafter, a new family of compounds was open for investigation, and research activity in the field of superconductivity proceeded vigorously. In early 1987, groups at the University of Alabama at Huntsville and the University of Houston announced the discovery of superconductivity at about 92 K in an oxide of yttrium, barium, and copper ($YBa_2Cu_3O_7$). Late in 1987, teams of scientists from Japan and the United States reported superconductivity at 105 K in an oxide of bismuth, strontium, calcium, and copper. Most recently, scientists have reported superconductivity as high as 125 K in an oxide containing thallium. At this point, one cannot rule out the possibility of room temperature superconductivity, and the search for novel superconducting materials continues. These developments are very exciting and important both for scientific reasons and because practical applications become more probable and widespread as the critical temperature is raised.

An important and useful application of superconductivity has been the construction of superconducting magnets in which the magnetic field strengths are about ten times greater than those of the best normal electromagnets. Such superconducting magnets are being considered as a means of storing energy. The idea of using superconducting power lines for transmitting power efficiently is also receiving some consideration. Modern superconducting electronic devices consisting of two thin-film superconductors separated by a thin insulator have been constructed. These devices include magnetometers (a magnetic-field measuring device) and various microwave devices.

We shall return to the subject of superconductivity in more depth in Chapter 44 of the extended version of this text. This same material is also available as a supplement to the standard version of the text.

## 27.6 A MODEL FOR ELECTRICAL CONDUCTION

In this section we describe a classical model of electrical conduction in metals. This model leads to Ohm's law and shows that resistivity can be related to the motion of electrons in metals.

Consider a conductor as a regular array of atoms containing free electrons (sometimes called *conduction* electrons). These electrons are free to move through the conductor and are approximately equal in number to the number of atoms. In the absence of an electric field, the free electrons move in random directions through the conductor with average speeds of the order of $10^6$ m/s. (These speeds can be properly calculated only if a quantum mechanical description is used.) The situation is similar to the motion of gas molecules confined in a vessel. In fact, some scientists refer to conduction electrons in a metal as an *electron gas.* The conduction electrons are not totally "free" since they are confined to the interior of the conductor and undergo frequent collisions with the array of atoms. These collisions are the predominant mechanism for the resistivity of a metal at normal temperatures. Note that there is no current through a conductor in the absence of an electric field since the

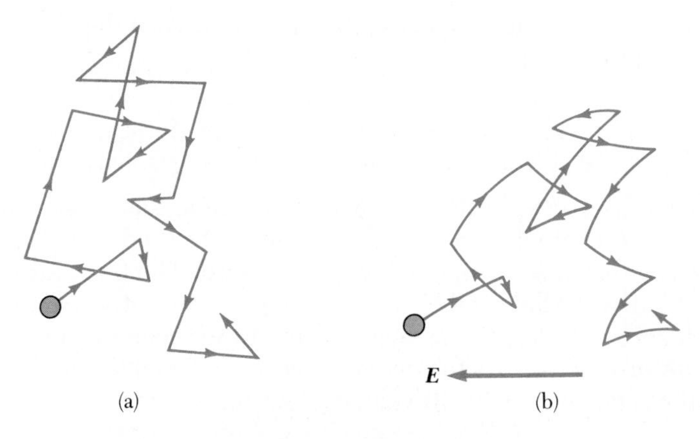


**Figure 27.13** (a) A schematic diagram of the random motion of a charge carrier in a conductor in the absence of an electric field. Note that the drift velocity is zero. (b) The motion of a charge carrier in a conductor in the presence of an electric field. Note that the random motion is modified by the field, and the charge carrier has a drift velocity.

*average velocity* of the free electrons is zero. That is, on the average, just as many electrons move in one direction as in the opposite direction, and so there is no net flow of charge.

The situation is modified when an electric field is applied to the metal. In addition to the random thermal motion just described, the free electrons drift slowly in a direction opposite that of the electric field, with an average drift speed $v_d$, which is much smaller (typically $10^{-4}$ m/s) than the average speed between collisions (typically $10^6$ m/s). Figure 27.13 provides a crude description of the motion of free electrons in a conductor. In the absence of an electric field, there is no net displacement after many collisions (Fig. 27.13a). An electric field $\boldsymbol{E}$ modifies the random motion and causes the electrons to drift in a direction opposite that of $\boldsymbol{E}$ (Fig. 27.13b). The slight curvature in the paths in Figure 27.13b results from the acceleration of the electrons between collisions, caused by the applied field. One mechanical system somewhat analogous to this situation is a ball rolling down a slightly inclined plane through an array of closely spaced, fixed pegs (Fig. 27.14). The ball represents a conduction electron, the pegs represent defects in the crystal lattice, and the component of the gravitational force along the incline represents the electric force $e\boldsymbol{E}$.

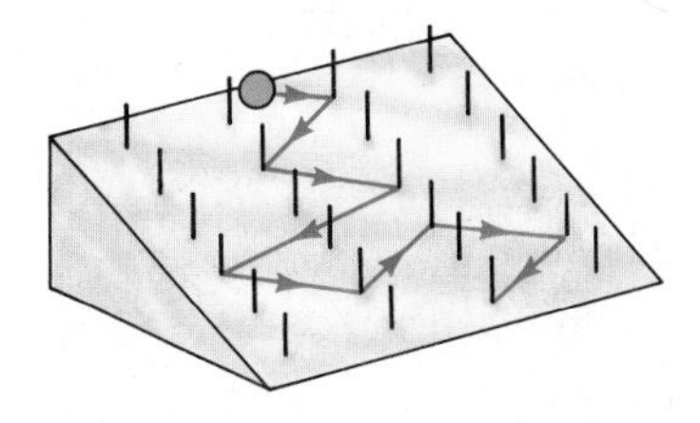

**Figure 27.14** A mechanical system somewhat analogous to the motion of charge carriers in the presence of an electric field. The collisions of the ball with the pegs represent the resistance to the ball's motion down the incline.

In our model, we shall assume that the excess energy acquired by the electrons in the electric field is lost to the conductor in the collision process. The energy given up to the atoms in the collisions increases the vibrational energy of the atoms, causing the conductor to heat up. The model also assumes that the motion of an electron after a collision is independent of its motion before the collision.

We are now in a position to obtain an expression for the drift velocity. When a mobile charged particle of mass $m$ and charge $q$ is subjected to an electric field $\boldsymbol{E}$, it experiences a force $q\boldsymbol{E}$. Since $\boldsymbol{F} = m\boldsymbol{a}$, we conclude that the acceleration of the particle is given by

$$\boldsymbol{a} = \frac{q\boldsymbol{E}}{m} \qquad (27.14)$$

This acceleration, which occurs for only a short time between collisions, enables the electron to acquire a small drift velocity. If $t$ is the time since the last collision and $\mathbf{v}_0$ is the initial velocity, then the velocity of the electron after a time $t$ is given by

$$\mathbf{v} = \mathbf{v}_0 + \mathbf{a}t = \mathbf{v}_0 + \frac{q\mathbf{E}}{m}t \tag{27.15}$$

We now take the average value of $\mathbf{v}$ over all possible times $t$ and all possible values of $\mathbf{v}_0$. If the initial velocities are assumed to be randomly distributed in space, we see that the average value of $\mathbf{v}_0$ is zero. The term $(q\mathbf{E}/m)t$ is the velocity added by the field at the end of one trip between atoms. If the electron starts with zero velocity, the average value of the second term of Equation 27.15 is $(q\mathbf{E}/m)\tau$, where $\tau$ is the *average time between collisions*. Because the average of $\mathbf{v}$ is equal to the drift velocity,[4] we have

**Drift velocity**

$$\mathbf{v}_d = \frac{q\mathbf{E}}{m}\tau \tag{27.16}$$

Substituting this result into Equation 27.6, we find that the magnitude of the current density is given by

**Current density**

$$J = nqv_d = \frac{nq^2E}{m}\tau \tag{27.17}$$

Comparing this expression with Ohm's law, $J = \sigma E$, we obtain the following relationships for the conductivity and resistivity:

**Conductivity**

$$\sigma = \frac{nq^2\tau}{m} \tag{27.18}$$

**Resistivity**

$$\rho = \frac{1}{\sigma} = \frac{m}{nq^2\tau} \tag{27.19}$$

The average time between collisions is related to the average distance between collisions $\ell$ (the mean free path, see Section 21.8) and the average thermal speed $\bar{v}$ through the expression[5]

$$\tau = \frac{\ell}{\bar{v}} \tag{27.20}$$

According to this classical model, the conductivity and resistivity do not depend on the electric field. This feature is characteristic of a conductor obeying Ohm's law. The model shows that the conductivity can be calculated from a knowledge of the density of the charge carriers, their charge and mass, and the average time between collisions.

[4] Since the collision process is random, each collision event is *independent* of what happened earlier. This is analogous to the random process of throwing a die. The probability of rolling a particular number on one throw is independent of the result of the previous throw. On the average, it would take six throws to come up with that number, starting at any arbitrary time.

[5] Recall that the thermal speed is the speed a particle has as a consequence of the temperature of its surroundings (Chapter 20).

**EXAMPLE 27.6 Electron Collisions in Copper**
(a) Using the data and results from Example 27.1 and the classical model of electron conduction, estimate the average time between collisions for electrons in copper at 20°C.

From Equation 27.19 we see that

$$\tau = \frac{m}{nq^2\rho}$$

where $\rho = 1.7 \times 10^{-8}\ \Omega \cdot \text{m}$ for copper and the carrier density $n = 8.48 \times 10^{28}$ electrons/m$^3$ for the wire described in Example 27.1. Substitution of these values into the expression above gives

$$\tau = \frac{(9.11 \times 10^{-31}\ \text{kg})}{(8.48 \times 10^{28}\ \text{m}^{-3})(1.6 \times 10^{-19}\ \text{C})^2(1.7 \times 10^{-8}\ \Omega \cdot \text{m})}$$

$$= \boxed{2.5 \times 10^{-14}\ \text{s}}$$

(b) Assuming the mean thermal speed for free electrons in copper to be $1.6 \times 10^6$ m/s and using the result from (a), calculate the mean free path for electrons in copper.

$$\ell = \bar{v}\tau = (1.6 \times 10^6\ \text{m/s})(2.5 \times 10^{-14}\ \text{s})$$

$$= \boxed{4.0 \times 10^{-8}\ \text{m}}$$

which is equivalent to 40 nm (compared with atomic spacings of about 0.2 nm). Thus, although the time between collisions is very short, the electrons travel about 200 atomic distances before colliding with an atom.

Although this classical model of conduction is consistent with Ohm's law, it is not satisfactory for explaining some important phenomena. For example, classical calculations for $\bar{v}$ using the ideal-gas model are about a factor of 10 smaller than the true values. Furthermore, according to Equations 27.19 and 27.20, the temperature variation of the resistivity is predicted to vary as $\bar{v}$, which, according to an ideal-gas model (Chapter 21), is proportional to $\sqrt{T}$. This is in disagreement with the linear dependence of resistivity with temperature for pure metals (Fig. 27.8). It is possible to account for such observations only by using a quantum mechanical model, which we shall describe briefly.

According to quantum mechanics, electrons have wavelike properties. If the array of atoms is regularly spaced (that is, periodic), the wavelike character of the electrons makes it possible for them to move freely through the conductor, and a collision with an atom is unlikely. For an idealized conductor, there would be no collisions, the mean free path would be infinite, and the resistivity would be zero. Electron waves are scattered only if the atomic arrangement is irregular (not periodic) as a result of, for example, structural defects or impurities. At low temperatures, the resistivity of metals is dominated by scattering caused by collisions between the electrons and impurities. At high temperatures, the resistivity is dominated by scattering caused by collisions between the electrons and the atoms of the conductor, which are continuously displaced as a result of thermal agitation. The thermal motion of the atoms causes the structure to be irregular (compared with an atomic array at rest), thereby reducing the electron's mean free path.

## 27.7 ELECTRICAL ENERGY AND POWER

If a battery is used to establish an electric current in a conductor, there is a continuous transformation of chemical energy stored in the battery to kinetic energy of the charge carriers. This kinetic energy is quickly lost as a result of collisions between the charge carriers and the lattice ions, resulting in an increase in the temperature of the conductor. Therefore, we see that the chemical energy stored in the battery is continuously transformed into thermal energy.

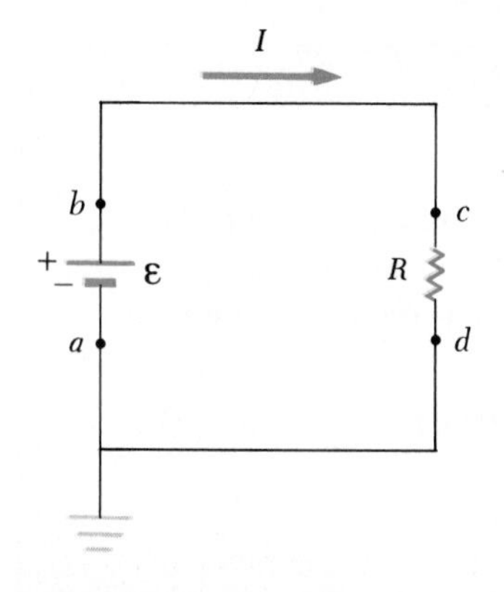


**Figure 27.15** A circuit consisting of a battery of emf $\mathcal{E}$ and resistance $R$. Positive charge flows in the clockwise direction, from the negative to the positive terminal of the battery. Points $a$ and $d$ are grounded.

Consider a simple circuit consisting of a battery whose terminals are connected to a resistor $R$, as shown in Figure 27.15. The symbol ⊣⊢ is used to designate a battery (or any other direct current source), and resistors are designated by the symbol —ʌʌʌ—. The positive terminal of the battery (the longer plate) is at the higher potential, while the negative terminal (the shorter plate) is at the lower potential. Now imagine following a positive quantity of charge $\Delta Q$ moving around the circuit from point $a$ through the battery and resistor and back to $a$. Point $a$ is a reference point that is grounded (ground symbol ⏚), and its potential is taken to be zero. As the charge moves from $a$ to $b$ through the battery, its electrical potential energy *increases* by an amount $V\,\Delta Q$ (where $V$ is the potential at $b$) while the chemical potential energy in the battery *decreases* by the same amount. (Recall from Chapter 25 that $\Delta U = q\,\Delta V$.) However, as the charge moves from $c$ to $d$ through the resistor, it *loses* this electrical potential energy as it undergoes collisions with atoms in the resistor, thereby producing thermal energy. Note that if we neglect the resistance of the interconnecting wires there is no loss in energy for paths $bc$ and $da$. When the charge returns to point $a$, it must have the same potential energy (zero) as it had at the start.[6]

The rate at which the charge $\Delta Q$ *loses* potential energy in going through the resistor is given by

$$\frac{\Delta U}{\Delta t} = \frac{\Delta Q}{\Delta t} V = IV$$

where $I$ is the current in the circuit. Of course, the charge regains this energy when it passes through the battery. Since the rate at which the charge loses energy equals the power $P$ lost in the resistor, we have

Power

$$P = IV \tag{27.21}$$

In this case, the power is supplied to a resistor by a battery. However, Equation 27.21 can be used to determine the power transferred to *any* device carrying a current $I$ and having a potential difference $V$ between its terminals.

Using Equation 27.21 and the fact that $V = IR$ for a resistor, we can express the power dissipated in the alternative forms

Power loss in a conductor

$$P = I^2R = \frac{V^2}{R} \tag{27.22}$$

When $I$ is in amperes, $V$ in volts, and $R$ in ohms, the SI unit of power is the watt (W). The power lost as heat in a conductor of resistance $R$ is called *joule heating*[7]; however, it is often referred to as an *$I^2R$ loss.*

A battery or any other device that provides electrical energy is called an *electromotive force,* usually referred to as an *emf.* The concept of emf will be discussed in more detail in Chapter 28. (The phrase *electromotive force* is an unfortunate one, since it does not describe a force but actually refers to a potential difference in volts.) *Neglecting the internal resistance of the battery, the potential difference between points a and b is equal to the emf $\mathcal{E}$ of the*

[6] Note that when the current reaches its steady-state value, there is *no* change with time in the kinetic energy associated with the current.

[7] It is called *joule heating* even though its dimensions are *energy per unit time,* which are dimensions of power.

*battery*. That is, $V = V_b - V_a = \mathcal{E}$, and the current in the circuit is given by $I = V/R = \mathcal{E}/R$. Since $V = \mathcal{E}$, the power supplied by the emf can be expressed as $P = I\mathcal{E}$, which, of course, equals the power lost in the resistor, $I^2R$.

**EXAMPLE 27.7 Power in an Electric Heater**
An electric heater is constructed by applying a potential difference of 110 V to a nichrome wire of total resistance 8 Ω. Find the current carried by the wire and the power rating of the heater.

*Solution* Since $V = IR$, we have

$$I = \frac{V}{R} = \frac{110\ \text{V}}{8\ \Omega} = 13.8\ \text{A}$$

We can find the power rating using $P = I^2R$:

$$P = I^2R = (13.8\ \text{A})^2(8\ \Omega) = 1.52\ \text{kW}$$

If we were to double the applied voltage, the current would double but the power would quadruple.

**EXAMPLE 27.8 Electrical Rating of a Lightbulb**
A light bulb is rated at 120 V/75 W. That is, its operating voltage is 120 V and it has a power rating of 75 W. The bulb is powered by a 120-V direct-current power supply. Find the current in the bulb and its resistance.

*Solution* Since the power rating of the bulb is 75 W and the operating voltage is 120 V, we can use $P = IV$ to find the current:

$$I = \frac{P}{V} = \frac{75\ \text{W}}{120\ \text{V}} = 0.625\ \text{A}$$

Using Ohm's law, $V = IR$, the resistance is calculated to be

$$R = \frac{V}{I} = \frac{120\ \text{V}}{0.625\ \text{A}} = 192\ \Omega$$

*Exercise 4* What would the resistance be in a lamp rated at 120 V and 100 W?
*Answer* 144 Ω.

## *27.8 ENERGY CONVERSION IN HOUSEHOLD CIRCUITS

The heat generated when current passes through a resistive material is used in many common devices. A cross-sectional view of the spiral heating element of an electric range is shown in Figure 27.16a. The material through which the current passes is surrounded by an insulating substance in order to prevent the current from flowing through the cook to the earth when he or she touches the pan. A material that is a good conductor of heat surrounds the insulator.

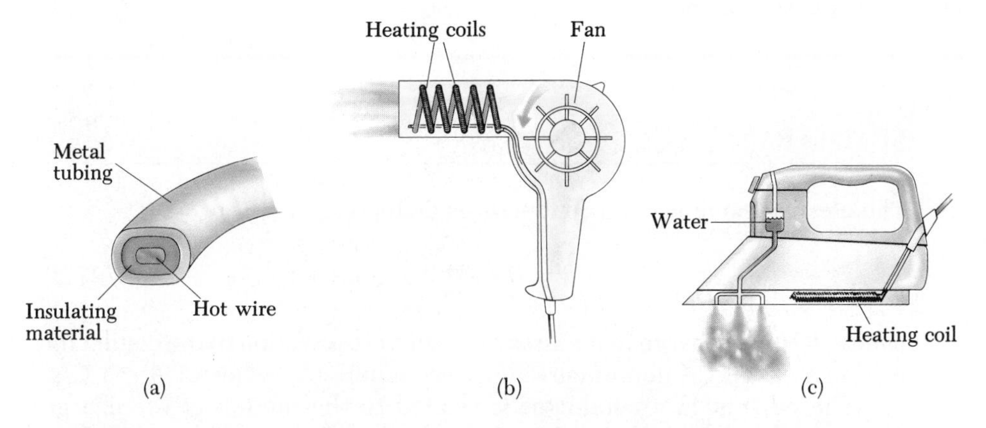


**Figure 27.16** (a) The cross-section of a heating element used in an electric range. (b) In a hair dryer, warm air is produced by blowing air from a fan past the heating coils. (c) In a steam iron, water is turned into steam by heat from a heating coil.

Figure 27.16b shows a common hair dryer, in which a fan blows air past heating coils. In this case the warm air can be used to dry hair, but on a broader scale this same principle is used to dry clothes and to heat buildings.

A final example of a household appliance that uses the heating effect of electric currents is the steam iron shown in Figure 27.16c. A heating coil warms the bottom of the iron and simultaneously turns water to steam, which is sprayed from jets located in the bottom of the iron.

The unit of energy the electric company uses to calculate energy consumption, the **kilowatt-hour,** is defined in terms of the unit of power. One kilowatt-hour (kWh) is the energy converted or consumed in 1 h at the constant rate of 1 kW. The numerical value of 1 kWh is

$$1\text{ kWh} = (10^3\text{ W})(3600\text{ s}) = 3.6 \times 10^6\text{ J} \tag{27.23}$$

On your electric bill, the amount of electricity used is usually stated in multiples of kWh.

EXAMPLE 27.9 The Cost of Operating a Lightbulb
How much does it cost to burn a 100-W lightbulb for 24 h if electricity costs eight cents per kilowatt-hour?

*Solution* A 100-W lightbulb is equivalent to a 0.1-kW bulb. Since the energy consumed equals power × time, the amount of energy you must pay for, expressed in kWh, is

$$\text{Energy} = (0.10\text{ kW})(24\text{ h}) = 2.4\text{ kWh}$$

If energy is purchased at eight cents per kWh, the cost is

$$\text{Cost} = (2.4\text{ kWh})(\$0.08/\text{kWh}) = \$0.19$$

That is, it will cost 19 cents to operate the lightbulb for one day. This is a small amount, but when larger and more complex devices are being used, the costs go up rapidly.

Demands on our energy supplies have made it necessary to be aware of the energy requirements of our electric devices. This is true not only because they are becoming more expensive to operate but also because, with the dwindling of the coal and oil resources that ultimately supply us with electrical energy, increased awareness of conservation becomes necessary. On every electric appliance is a label that contains the information you need to calculate the power requirements of the appliance. The power consumption in watts is often stated directly, as on a lightbulb. In other cases, the amount of current used by the device and the voltage at which it operates are given. This information and Equation 27.21 are sufficient to calculate the operating cost of any electric device.

Exercise 5 If electricity costs eight cents per kilowatt-hour, what does it cost to operate an electric oven, which operates at 20 A and 220 V, for 5 h?
Answer $1.76.

## SUMMARY

Electric current

The **electric current** $I$ in a conductor is defined as

$$I \equiv \frac{dQ}{dt} \tag{27.2}$$

where $dQ$ is the charge that passes through a cross section of the conductor in a time $dt$. The SI unit of current is the ampere (A), where 1 A = 1 C/s.

Current in a conductor

The current in a conductor is related to the motion of the charge carriers through the relationship

$$I = nqv_dA \tag{27.4}$$

where $n$ is the density of charge carriers, $q$ is their charge, $v_d$ is the drift velocity, and $A$ is the cross-sectional area of the conductor.

The **current density $\boldsymbol{J}$** in a conductor is defined as the current per unit area:

$$\boldsymbol{J} = nq\boldsymbol{v}_d \tag{27.6}$$

The current density in a conductor is proportional to the electric field according to the expression

Ohm's law

$$\boldsymbol{J} = \sigma \boldsymbol{E} \tag{27.7}$$

The constant $\sigma$ is called the **conductivity** of the material. The inverse of $\sigma$ is called the **resistivity**, $\rho$. That is, $\rho = 1/\sigma$.

A material is said to obey Ohm's law if its conductivity is independent of the applied field.

The **resistance** $R$ of a conductor is defined as the ratio of the potential difference across the conductor to the current:

Resistance of a conductor

$$R \equiv \frac{V}{I} \tag{27.8}$$

If the resistance is independent of the applied voltage, the conductor obeys Ohm's law.

If the conductor has a uniform cross-sectional area $A$ and a length $\ell$, its resistance is given by

Resistance of a uniform conductor

$$R = \frac{\ell}{\sigma A} = \rho \frac{\ell}{A} \tag{27.10}$$

The SI unit of resistance is volt per ampere, which is defined to be 1 ohm ($\Omega$). That is, $1\ \Omega = 1$ V/A.

The resistivity of a conductor varies with temperature in an approximately linear fashion, that is

Variation of $\rho$ with temperature

$$\rho = \rho_o[1 + \alpha(T - T_o)] \tag{27.11}$$

where $\alpha$ is the temperature coefficient of resistivity and $\rho_o$ is the resistivity at some reference temperature $T_o$.

In a classical model of electronic conduction in a metal, the electrons are treated as molecules of a gas. In the absence of an electric field, the average velocity of the electrons is zero. When an electric field is applied, the electrons move (on the average) with a **drift velocity $\boldsymbol{v}_d$**, which is opposite the electric field. The drift velocity is given by

Drift velocity

$$\boldsymbol{v}_d = \frac{q\boldsymbol{E}}{m}\tau \tag{27.16}$$

where $\tau$ is the average time between collisions with the atoms of the metal. The resistivity of the material according to this model is given by

Resistivity

$$\rho = \frac{m}{nq^2\tau} \tag{27.19}$$

where $n$ is the number of free electrons per unit volume.

Power

If a potential difference $V$ is maintained across a resistor, the **power,** or rate at which energy is supplied to the resistor, is given by

$$P = IV \tag{27.21}$$

Since the potential difference across a resistor is given by $V = IR$, we can express the power dissipated in a resistor in the form

Power loss in a resistor

$$P = I^2R = \frac{V^2}{R} \tag{27.22}$$

The electrical energy supplied to a resistor appears in the form of internal energy (thermal energy) in the resistor.

## QUESTIONS

1. Explain the chemistry involved in the operation of a "dry cell" battery.
2. In an analogy between traffic flow and electrical current, what would correspond to the charge $Q$? What would correspond to the current $I$?
3. What factors affect the resistance of a conductor?
4. What is the difference between resistance and resistivity?
5. We have seen that an electric field must exist inside a conductor that carries a current. How is this possible in view of the fact that in *electrostatics,* we concluded that $E$ must be zero inside a conductor?
6. Two wires A and B of circular cross-section are made of the same metal and have equal lengths, but the resistance of wire A is three times greater than that of wire B. What is the ratio of their cross-sectional areas? How do their radii compare?
7. What is required in order to maintain a steady current in a conductor?
8. Do all conductors obey Ohm's law? Give examples to justify your answer.
9. When the voltage across a certain conductor is doubled, the current is observed to increase by a factor of 3. What can you conclude about the conductor?
10. In the water analogy of an electric circuit, what corresponds to the power supply, resistor, charge, and potential difference?
11. Why might a "good" electrical conductor also be a "good" thermal conductor?
12. Use the atomic theory of matter to explain why the resistance of a material should increase as its temperature increases.
13. How does the resistance change with temperature for copper and silicon? Why are they different?
14. Explain how a current can persist in a superconductor without any applied voltage.
15. What single experimental requirement makes superconducting devices expensive to operate? In principle, can this limitation be overcome?
16. What would happen to the drift velocity of the electrons in a wire and to the current in the wire if the electrons could move freely without resistance through the wire?
17. If charges flow very slowly through a metal, why does it not require several hours for a light to come on when you throw a switch?
18. In a conductor, the electric field that drives the electrons through the conductor propagates with a speed close to the speed of light, although the drift velocity of the electrons is very small. Explain how these can both be true. Does the same electron move from one end of the conductor to the other?
19. Two conductors of the same length and radius are connected across the same potential difference. One conductor has twice the resistance of the other. Which conductor will dissipate more power?
20. When incandescent lamps burn out, they usually do so just after they are switched on. Why?
21. If you were to design an electric heater using nichrome wire as the heating element, what parameters of the wire could you vary to meet a specific power output, such as 1000 W?
22. Two light bulbs both operate from 110 V, but one has a power rating of 25 W and the other of 100 W. Which bulb has the higher resistance? Which bulb carries the greater current?
23. A typical monthly utility rate structure might go something like this: $1.60 for the first 16 kWh, 7.05 cents/kWh for the next 34 kWh used, 5.02 cents/kWh for the next 50 kWh, 3.25 cents/kWh for the next 100 kWh, 2.95 cents/kWh for the next 200 kWh, 2.35 cents/kWh for all in excess of 400 kWh. Based on these rates, what would be the charge for 327 kWh? From the standpoint of encouraging conservation of energy, what is wrong with this pricing method?

## PROBLEMS

### Section 27.2 Electric Current

1. Calculate the current in the case for which $3 \times 10^{12}$ electrons pass a given cross section of a conductor each second.
2. In a particular cathode ray tube, the measured beam current is 30 $\mu$A. How many electrons strike the tube screen every 40 s?
3. A small sphere that carries a charge of 8 nC is whirled in a circle at the end of an insulating string. The rotation frequency is $100\pi$ rad/s. What average current does this rotating charge represent?
4. The quantity of charge $q$ (in C) passing through a surface of area 2 cm$^2$ varies with time as $q = 4t^3 + 5t + 6$, where $t$ is in s. (a) What is the instantaneous current through the surface at $t = 1.0$ s? (b) What is the value of the current density?
5. The current $I$ (in A) in a conductor depends on time as $I = 2t^2 - 3t + 7$, where $t$ is in s. What quantity of charge moves across a section through the conductor during the interval $t = 2$ s to $t = 4$ s?
6. Suppose that the current through a conductor decreases exponentially with time according to

$$I(t) = I_0 e^{-t/\tau}$$

where $I_0$ is the intial current (at $t = 0$), and $\tau$ is a constant having dimensions of time. Consider a fixed observation point within the conductor. (a) How much charge passes this point between $t = 0$ and $t = \tau$? (b) How much charge passes this point between $t = 0$ and $t = 10\tau$? (c) How much charge passes this point between $t = 0$ and $t = \infty$?
7. Calculate the number of free electrons per cubic meter for gold, assuming one free electron per atom.
8. Calculate the average drift speed of electrons traveling through a copper wire with a cross-sectional area of 1 mm$^2$ when carrying a current of 1 A (values similar to those for the electric wire to your study lamp). It is known that about one electron per atom of copper contributes to the current. The atomic weight of copper is 63.54, and its density is 8.92 g/cm$^3$.
9. A copper bus bar has a cross section of 5 cm $\times$ 15 cm and carries a current with a density of 2000 A/cm$^2$. (a) What is the total current in the bus bar? (b) What amount of charge passes a given point in the bar per hour?
10. Figure 27.17 represents a section of a circular conductor of nonuniform diameter carrying a current of 5 A. The radius of cross section $A_1$ is 0.4 cm. (a) What is the magnitude of the current density across $A_1$? (b) If the current density across $A_2$ is one fourth the value across $A_1$, what is the radius of the conductor at $A_2$?

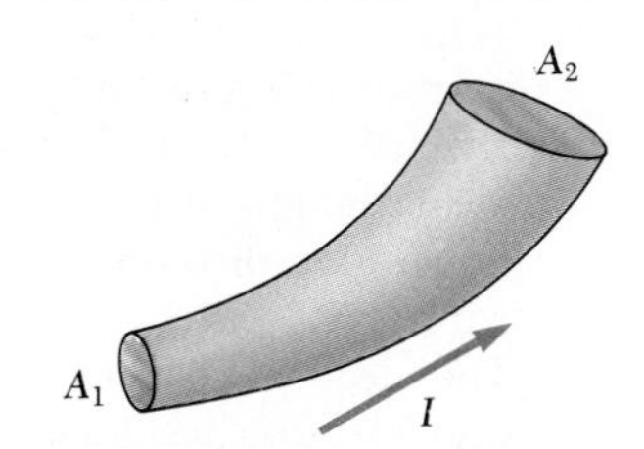


**Figure 27.17** (Problem 10).

11. A coaxial conductor with a length of 20 m consists of an inner cylinder with a radius of 3.0 mm and a concentric outer cylindrical tube with an inside radius of 9.0 mm. A uniformly distributed leakage current of 10 $\mu$A flows between the two conductors. Determine the leakage current density (in A/m$^2$) through a cylindrical surface (concentric with the conductors) that has a radius of 6.0 mm.

### Section 27.3 Resistance and Ohm's Law

12. A conductor of uniform radius 1.2 cm carries a current of 3 A produced by an electric field of 120 V/m. What is the resistivity of the material?
13. An electric field of 2100 V/m is applied to a section of silver of uniform cross section. Calculate the resulting current density if the specimen is at a temperature of 20°C.
14. A solid cube of silver (specific gravity = 10.50) has a mass of 90 g. (a) What is the resistance between opposite faces of the cube? (b) If there is one conduction electron for each silver atom, find the average drift speed of electrons when a potential difference of $10^{-5}$ V is applied to opposite faces. The atomic number of silver is 47, and its atomic mass is 107.87.
15. Calculate the resistance at 20°C of a 40-m length of silver wire having a cross-sectional area of 0.4 mm$^2$.
16. Eighteen-gauge wire has a diameter of 1.024 mm. Calculate the resistance of 15 m of 18-gauge copper wire at 20°C.
17. A 2.4-m length of wire that is 0.031 cm$^2$ in cross section has a measured resistance of 0.24 $\Omega$. Calculate the conductivity of the material.
18. What is the resistance of a tungsten filament 15 cm in length and 0.002 cm in diameter at 20°C?
19. What diameter copper wire has a resistance per unit length of $3.28 \times 10^{-3}$ $\Omega$/m at 20°C?
20. Determine the resistance at 20°C of a 1.5-m length of platinum wire that has a diameter of 0.10 mm.
21. A wire with a resistance $R$ is lengthened to 1.25 times its original length by pulling it through a small hole. Find the resistance of the wire after it is stretched.
22. Aluminum and copper wires of equal length are found to have the same resistance. What is the ratio of their radii?
23. Suppose that you wish to fabricate a uniform wire out of 1 g of copper. If the wire is to have a resistance of

$R = 0.5\ \Omega$, and all of the copper is to be used, what will be (a) the length and (b) the diameter of this wire?

24. What is the resistance of a device that operates with a current of 7 A when the applied voltage is 110 V?
25. A 0.9-V potential difference is maintained across a 1.5-m length of tungsten wire that has a cross-sectional area of 0.6 mm$^2$. What is the current in the wire?
26. The electron beam emerging from a certain high-energy electron accelerator has a circular cross-section of radius 1 mm. (a) If the beam current is 8 $\mu$A, find the current density in the beam, assuming that it is uniform throughout. (b) The speed of the electrons is so close to the speed of light that their speed can be taken as $c = 3 \times 10^8$ m/s with negligible error. Find the electron density in the beam. (c) How long does it take for an Avogadro's number of electrons to emerge from the accelerator?
27. A resistor is constructed of a carbon rod that has a uniform cross-sectional area of 5 mm$^2$. When a potential difference of 15 V is applied across the ends of the rod, there is a current of $4 \times 10^{-3}$ A in the rod. Find (a) the resistance of the rod and (b) the rod's length.
28. A current density of $6 \times 10^{-13}$ A/m$^2$ exists in the atmosphere where the electric field (due to charged thunderclouds in the vicinity) is 100 V/m. Calculate the electrical conductivity of the earth's atmosphere in this region.

### Section 27.4 The Resistivity of Different Conductors

29. An aluminum wire with a diameter of 0.1 mm has a uniform electric field of 0.2 V/m imposed along its entire length. The temperature of the wire is 50°C. Assume one free electron per atom. (a) Use the information in Table 27.1 and determine the resistivity. (b) What is the current density in the wire? (c) What is the total current in the wire? (d) What is the drift speed of the conduction electrons? (e) What potential difference must exist between the ends of a 2-m length of the wire to produce the stated electric field strength?
30. Calculate the percentage change in the resistance of a carbon filament when it is heated from room temperature to 160°C.
31. What is the fractional change in the resistance of an iron filament when its temperature changes from 25°C to 50°C?
32. The resistance of a platinum wire is to be calibrated for low-temperature measurements. A platinum wire with resistance 1 Ω at 20°C is immersed in liquid nitrogen at 77 K (−196°C). If the temperature response of the platinum wire is linear, what is the expected resistance of the platinum wire at −196°C? ($\alpha_{\text{platinum}} = 3.92 \times 10^{-3}/°\text{C}$)
33. If a copper wire has a resistance of 18 Ω at 20°C, what resistance will it have at 60°C? (Neglect any change in length or cross-sectional area due to the change in temperature.)
34. An aluminum rod has a resistance of 1.234 Ω at 20°C. Calculate the resistance of the rod at 120°C by accounting for the change in both the resistivity and dimensions of the rod.
35. At what temperature will tungsten have a resistivity four times that of copper? (Assume that the copper is at 20°C.)
36. A segment of nichrome wire is initially at 20°C. Using the data from Table 27.1, calculate the temperature to which the wire must be heated to double its resistance.
37. At 45°C, the resistance of a segment of gold wire is 85 Ω. When the wire is placed in a liquid bath, the resistance decreases to 80 Ω. What is the temperature of the bath?
38. A 500-W heating coil designed to operate from 110 V is made of nichrome wire 0.5 mm in diameter. (a) Assuming that the resistivity of the nichrome remains constant at its 20°C value, find the length of wire used. (b) Now consider the variation of resistivity with temperature. What power will the coil of part (a) actually deliver when it is heated to 1200°C?

### Section 27.6 A Model for Electrical Conduction

39. Calculate the current density in a gold wire in which an electric field of 0.74 V/m exists.
40. If the drift velocity of free electrons in a copper wire is $7.84 \times 10^{-4}$ m/s, calculate the electric field in the conductor.
41. Use data from Example 27.6 to calculate the collision mean free path of electrons in copper if the average thermal speed of conduction electrons is $8.6 \times 10^5$ m/s.
42. If the current through a given conductor is doubled, what happens to the (a) charge carrier density? (b) current density? (c) electron drift velocity? (d) average time between collisions?

### Section 27.7 Electrical Energy and Power

43. A 10-V battery is connected to a 120-Ω resistor. Neglecting the internal resistance of the battery, calculate the power dissipated in the resistor.
44. How much current is being supplied by a 200-V generator delivering 100 kW of power?
45. Suppose that a voltage surge produces 140 V for a moment. By what percentage will the output of a 120-V, 100-W light bulb increase, assuming its resistance does not change?
46. A particular type of automobile storage battery is characterized as "360-ampere-hour, 12 V." What total energy can the battery deliver?

47. If a 55-Ω resistor is rated at 125 W (the maximum allowed power), what is the maximum allowed operating voltage?

48. In a hydroelectric installation, a turbine delivers 1500 hp to a generator, which in turn converts 80% of the mechanical energy into electrical energy. Under these conditions, what current will the generator deliver at a terminal potential difference of 2000 V?

49. Suppose that you want to install a heating coil that will convert electric energy to heat at a rate of 300 W for a current of 1.5 A. (a) Determine the resistance of the coil. (b) The resistivity of the coil wire is $10^{-6}\ \Omega\cdot\text{m}$, and its diameter is 0.3 mm. Determine its length.

50. An electric heater with a resistance of 20 Ω requires 100 V across its terminals. A built-in switching circuit repetitively turns the heater on for 1 s and off for 4 s. (a) How much energy is produced by the heater in 1 h? (b) What is the average power delivered by the heater over a period of one cycle?

### Section 27.8 Energy Conservation in Household Circuits

51. What is the required resistance of an immersion heater that will increase the temperature of 1.5 kg of water from 10°C to 50°C in 10 min while operating at 110 V?

52. The heating element of a coffee maker operates at 120 V and carries a current of 2 A. Assuming that all of the heat generated is absorbed by the water, how long does it take to heat 0.5 kg of water from room temperature (23°C) to the boiling point?

53. Compute the cost per day of operating a lamp that draws 1.7 A from a 110-V line if the cost of electrical energy is 6 cents/kWh.

54. An electric heater operating at full power draws a current of 8 A from a 110-V circuit. (a) What is the resistance of the heater? (b) Assuming the resistance $R$ remains constant, at what voltage and current would the heater dissipate 750 W?

55. A certain toaster has a heating element made of nichrome resistance wire. When first connected to a 120-V voltage source (and the wire is at a temperature of 20°C) the initial current is 1.8 A, but the current begins to decrease as the resistive element heats up. When the toaster has reached its final operating temperature, the current has dropped to 1.53 A. (a) Find the power the toaster consumes when it is at its operating temperature. (b) What is the final temperature of the heating element?

## ADDITIONAL PROBLEMS

56. An electric utility company supplies a customer's house from the main power lines (120 V) with two copper wires, each 50 m long and having a resistance of 0.108 Ω per 300 m. (a) Find the voltage at the customer's house for a load current of 110 A. For this load current, find (b) the power the customer is receiving and (c) the power dissipated in the copper wires.

57. The potential difference across the filament of a lamp is maintained at a constant level while equilibrium temperature is being reached. It is observed that the steady-state current in the lamp is only one tenth of the current drawn by the lamp when it is first turned on. If the temperature coefficient of resistivity for the lamp at 20°C is $0.0045\ (\text{C}°)^{-1}$, and if the resistance increases linearly with increasing temperature, what is the final operating temperature of the filament?

58. The current in a resistor decreases by 3 A when the voltage applied across the resistor decreases from 12 V to 6 V. Find the resistance of the resistor.

59. An electric car is designed to run off a bank of 12-V batteries with total energy storage of $2 \times 10^7$ J. (a) If the electric motor draws 8 kW, what is the current delivered to the motor? (b) If the electric motor draws 8 kW as the car moves at a steady speed of 20 m/s, how far will the car travel before it is "out of juice"?

60. (a) A sheet of copper ($\rho = 1.7 \times 10^{-8}\ \Omega\cdot\text{m}$) is 2 mm thick and has surface dimensions of 8 cm × 24 cm. If the long edges are joined to form a hollow tube 24 cm in length, what is the resistance between the ends? (b) What mass of copper would be required to manufacture a spool of copper cable 1500 m in length and having a total resistance of 4.5 Ω?

61. A Wheatstone bridge can be used to measure the strain ($\Delta L/L_0$) of a wire (see Section 12.4), where $L_0$ is the length before stretching, $L$ is the length after stretching, and $\Delta L = L - L_0$. Let $\alpha = \Delta L/L_0$. Show that the resistance is $R = R_0\ (1 + 2\alpha + \alpha^2)$ for any length where $R_0 = \dfrac{\rho L_0}{A_0}$. Assume that the resistivity and volume of the wire stay constant.

62. The current in a wire decreases with time according to the relation $I = 2.5e^{-at}$ mA where $a = 0.833\ \text{s}^{-1}$. Determine the total charge that has flowed through the wire by the time the current has diminished to zero.

63. A resistor is constructed by forming a material of resistivity $\rho$ into the shape of a hollow cylinder of length $L$ and inner and outer radii $r_a$ and $r_b$, respectively (Fig. 27.18). In use, a potential difference is applied be-

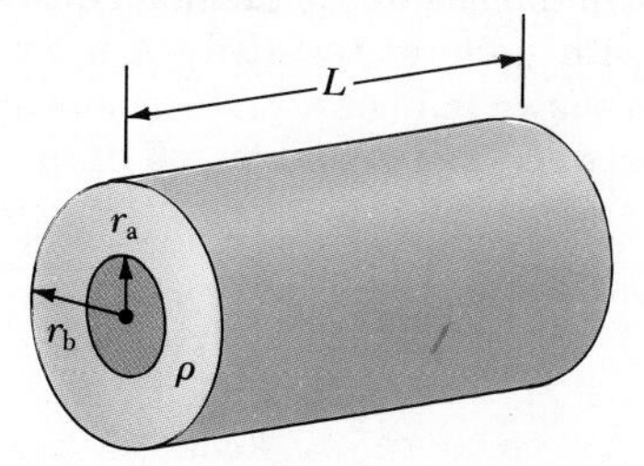


**Figure 27.18** (Problems 63 and 64).

tween the ends of the cylinder, producing a current parallel to the axis. (a) Find a general expression for the resistance of such a device in terms of $L$, $\rho$, $r_a$, and $r_b$. (b) Obtain a numerical value for $R$ when $L = 4$ cm, $r_a = 0.5$ cm, $r_b = 1.2$ cm, and the resistivity $\rho = 3.5 \times 10^5\ \Omega \cdot$ m.

64. Consider the device described in Problem 63. Suppose now that the potential difference is applied between the inner and outer surfaces so that the resulting current flows radially outward. (a) Find a general expression for the resistance of the device in terms of $L$, $\rho$, $r_a$, and $r_b$. (b) Calculate the value of $R$ using the parameter values given in (b) of Problem 63.

65. A more general definition of the temperature coefficient of resistivity is

$$\alpha = \frac{1}{\rho}\frac{d\rho}{dT}$$

where $\rho$ is the resistivity at temperature $T$. (a) Assuming that $\alpha$ is constant, show that

$$\rho = \rho_0 e^{\alpha(T-T_0)}$$

where $\rho_0$ is the resistivity at temperature $T_0$. (b) Using the series expansion ($e^x \approx 1 + x$; $x \ll 1$), show that the resistivity is given approximately by the expression $\rho = \rho_0[1 + \alpha(T - T_0)]$ for $\alpha(T - T_0) \ll 1$.

**66.** There is a close analogy between the flow of heat because of a temperature difference (Section 20.7) and the flow of electrical charge because of a potential difference. The thermal energy $dQ$ and the electrical charge $dq$ are both transported by free electrons in the conducting material. Consequently, a good electrical conductor is usually also a good heat conductor. Consider a thin conducting slab of thickness $dx$, area $A$, and electrical conductivity $\sigma$, with a potential difference $dV$ between opposite faces. Show that the current $I = dq/dt$ is given by

| Charge conduction | Analogous heat conduction (Eq. 20.15) |
|---|---|
| $\frac{dq}{dt} = -\sigma A \frac{dV}{dx}$ | $\frac{dQ}{dt} = -kA\frac{dT}{dx}$ |

In the analogous heat conduction equation, the rate of heat flow $dQ/dt$ (in SI units of joules per second) is due to a temperature gradient $dT/dx$, in a material of thermal conductivity $k$. What is the origin of the minus sign in the charge conduction equation?

67. Material with uniform resistivity $\rho$ is formed into a wedge as shown in Figure 27.19. Show that the resistance between face $A$ and face $B$ of this wedge is given by

$$R = \rho\frac{L}{w(y_2 - y_1)}\ln\left(\frac{y_2}{y_1}\right)$$

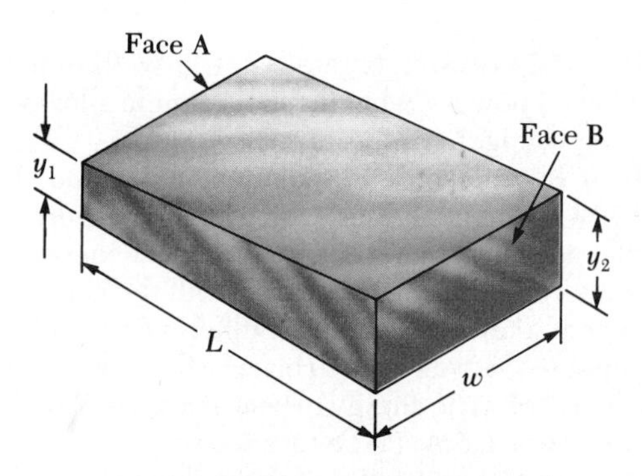


**Figure 27.19** (Problem 67).

68. A material of resistivity $\rho$ is formed into the shape of a truncated cone of altitude $h$ as in Figure 27.20. The bottom end has a radius $b$ and the top end has a radius $a$. Assuming a uniform current density through any circular cross section of the cone, show that the resistance between the two ends is

$$R = \frac{\rho}{\pi}\left(\frac{h}{ab}\right)$$

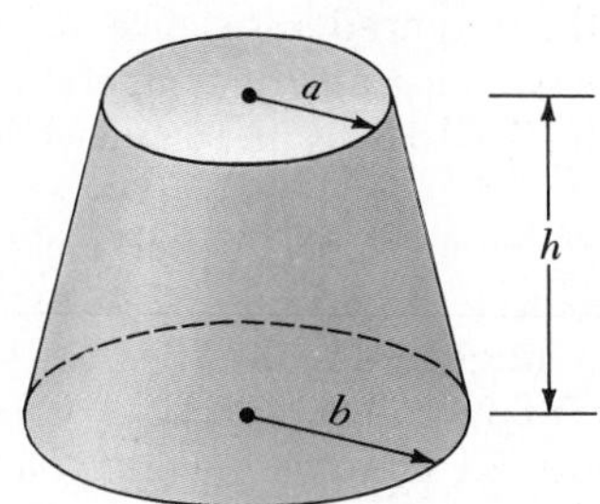


**Figure 27.20** (Problem 68).

Δ69. An experiment is conducted to measure the electrical resistivity of nichrome in the form of wires with different lengths and cross-sectional areas. For one set of measurements, a student uses #30 gauge wire, which has a cross sectional area of $7.3 \times 10^{-8}$ m$^2$. The voltage across the wire and the current in the wire are measured with a voltmeter and ammeter, respectively. For each of the measurements given in the table below taken on wires of three different lengths, calculate the resistance of the wires and the corresponding values of the resistivity. What is the average value of the resistivity, and how does it compare with the value given in Table 27.1?

| $L$(m) | $V$(V) | $I$(A) | $R(\Omega)$ | $\rho(\Omega\cdot$m$)$ |
|---|---|---|---|---|
| 0.54 | 5.22 | 0.500 | | |
| 1.028 | 5.82 | 0.276 | | |
| 1.543 | 5.94 | 0.187 | | |

# 28
# Direct Current Circuits

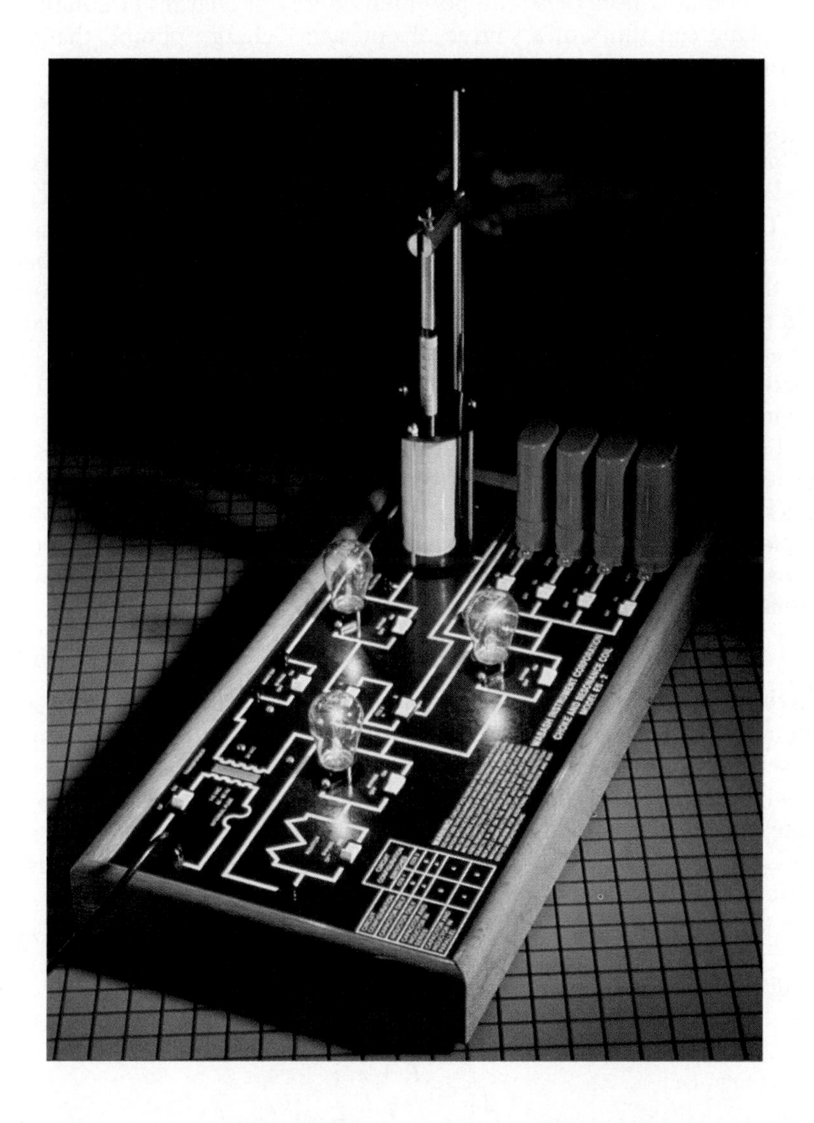

*This versatile circuit enables the experimenter to examine the properties of circuit elements such as capacitors and resistors and their effect on circuit behavior. (Courtesy of CENCO)*

This chapter is concerned with the analysis of some simple circuits whose elements include batteries, resistors, and capacitors in various combinations. The analysis of these circuits is simplified by the use of two rules known as *Kirchhoff's rules.* These rules follow from the laws of conservation of energy and conservation of charge. Most of the circuits analyzed are assumed to be in *steady state,* where the currents are constant in magnitude and direction. In one section we discuss circuits containing resistors and capacitors, for which the current varies with time. Finally, a number of common electrical devices and techniques are described for measuring current, potential differences, resistance, and emfs.

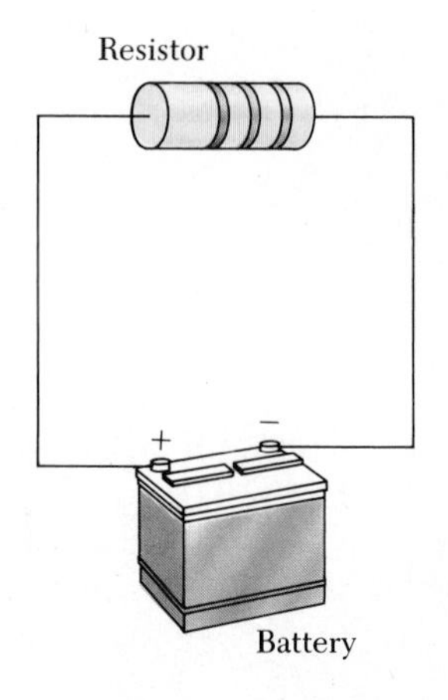


Figure 28.1 A circuit consisting of a resistor connected to the terminals of a battery.

## 28.1 ELECTROMOTIVE FORCE

In the previous chapter we found that a constant current can be maintained in a closed circuit through the use of a source of energy, called an **electromotive force** (abbreviated *emf*). A source of emf is any device (such as a battery or generator) that will *increase* the potential energy of charges circulating in a circuit. One can think of a source of emf as a "charge pump" that forces electrons to move in a direction opposite the electrostatic force on these negative charges inside the source. The emf, $\mathcal{E}$, of a source describes the work done per unit charge, and hence the SI unit of emf is the volt.

Consider the circuit shown in Figure 28.1, consisting of a battery connected to a resistor. We shall assume that the connecting wires have no resistance. The positive terminal of the battery is at a higher potential than the negative terminal. If we were to neglect the internal resistance of the battery itself, then the potential difference across the battery (the terminal voltage) would equal the emf of the battery. However, because a real battery always has some internal resistance $r$, the terminal voltage is not equal to the emf of the battery. The circuit shown in Figure 28.1 can be described by the circuit diagram in Figure 28.2a. The battery within the dotted rectangle is represented by a source of emf, $\mathcal{E}$, in series with the internal resistance, $r$. Now imagine a positive charge moving from $a$ to $b$ in Figure 28.2a. As the charge passes from the negative to the positive terminal of the battery, its potential *increases* by $\mathcal{E}$. However, as it moves through the resistance $r$, its potential *decreases* by an amount $Ir$, where $I$ is the current in the circuit. Thus, the terminal voltage of the battery, $V = V_b - V_a$, is given by[1]

$$V = \mathcal{E} - Ir \tag{28.1}$$

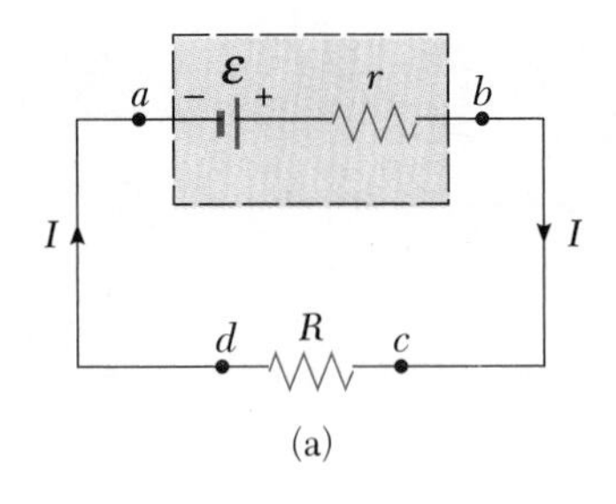


(a)

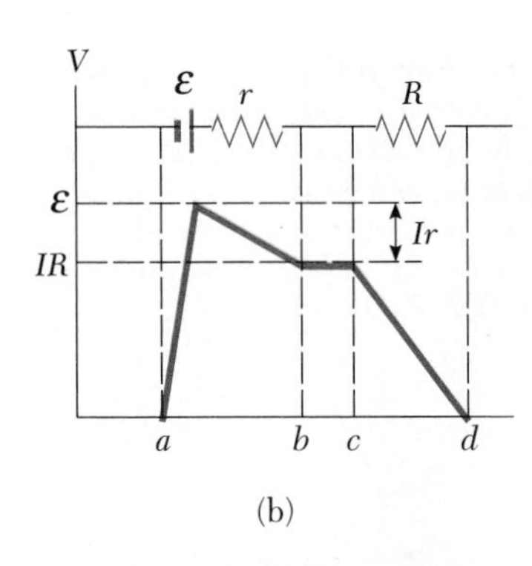


(b)

Figure 28.2 (a) Circuit diagram of a source of emf $\mathcal{E}$ of internal resistance $r$ connected to an external resistor $R$. (b) Graphical representation showing how the potential changes as the series circuit in (a) is traversed clockwise.

From this expression, note that $\mathcal{E}$ is equivalent to the **open-circuit voltage,** that is, the *terminal voltage when the current is zero.* Figure 28.2b is a graphical representation of the changes in potential as the circuit is traversed in the clockwise direction. By inspecting Figure 28.2a we see that the terminal voltage $V$ must also equal the potential difference across the external resistance $R$, often called the **load resistance.** That is, $V = IR$. Combining this with Equation 28.1, we see that

$$\mathcal{E} = IR + Ir \tag{28.2}$$

Solving for the current gives

$$I = \frac{\mathcal{E}}{R + r} \tag{28.3}$$

This shows that the current in this simple circuit depends on both the resistance external to the battery and the internal resistance. If the load resistance $R$ is much greater than the internal resistance $r$, we can neglect $r$ in this analysis. In many circuits we shall ignore this internal resistance.

If we multiply Equation 28.2 by the current $I$, the following expression is obtained:

$$I\mathcal{E} = I^2R + I^2r \tag{28.4}$$

[1] The terminal voltage in this case is less than the emf by an amount $Ir$. In some situations, the terminal voltage may *exceed* the emf by an amount $Ir$. This happens when the direction of the current is *opposite* that of the emf, as in the case of charging a battery with another source of emf.

This equation tells us that the total power output of the source of emf, $I\mathcal{E}$, is converted into power dissipated as joule heat in the load resistance, $I^2R$, *plus* power dissipated in the internal resistance, $I^2r$. Again, if $r \ll R$, then most of the power delivered by the battery is transferred to the load resistance.

**EXAMPLE 28.1 Terminal Voltage of a Battery**

A battery has an emf of 12 V and an internal resistance of 0.05 Ω. Its terminals are connected to a load resistance of 3 Ω. (a) Find the current in the circuit and the terminal voltage of the battery.

Using Equations 28.1 and 28.3, we get

$$I = \frac{\mathcal{E}}{R + r} = \frac{12\ \text{V}}{3.05\ \Omega} = 3.93\ \text{A}$$

$$V = \mathcal{E} - Ir = 12\ \text{V} - (3.93\ \text{A})(0.05\ \Omega) = 11.8\ \text{V}$$

As a check of this result, we can calculate the voltage drop across the load resistance $R$. This gives

$$V = IR = (3.93\ \text{A})(3\ \Omega) = 11.8\ \text{V}$$

(b) Calculate the power dissipated in the load resistor, the power dissipated by the internal resistance of the battery, and the power delivered by the battery.

The power dissipated by the load resistor is

$$P_R = I^2R = (3.93\ \text{A})^2(3\ \Omega) = 46.3\ \text{W}$$

The power dissipated by the internal resistance is

$$P_r = I^2r = (3.93\ \text{A})^2(0.05\ \Omega) = 0.772\ \text{W}$$

Hence, the power delivered by the battery is the sum of these quantities, or 47.1 W. This can be checked using the expression $P = I\mathcal{E}$.

**EXAMPLE 28.2 Matching the Load**

Show that the *maximum* power lost in the load resistance $R$ in Figure 28.2a occurs when $R = r$, that is, when the load resistance *matches* the internal resistance.

*Solution* The power dissipated in the load resistance is equal to $I^2R$, where $I$ is given by Equation 28.3:

$$P = I^2R = \frac{\mathcal{E}^2R}{(R + r)^2}$$

When $P$ is plotted versus $R$ as in Figure 28.3, we find that $P$ reaches a *maximum* value of $\mathcal{E}^2/4r$ at $R = r$. This can also be proved by differentiating $P$ with respect to $R$, setting the result equal to zero, and solving for $R$. The details are left as a problem (Problem 79).

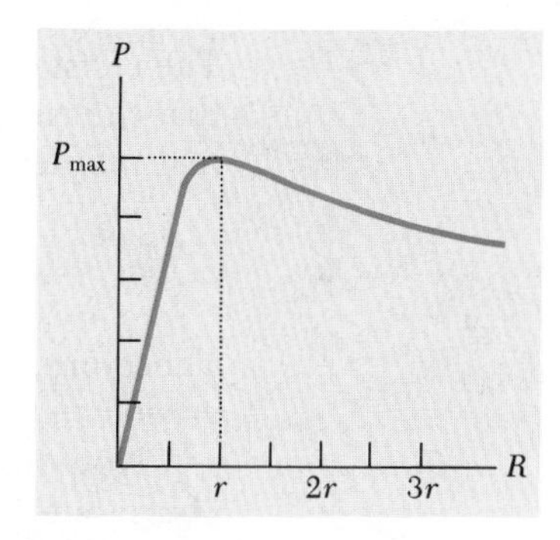


**Figure 28.3** Graph of the power $P$ delivered to a load resistor as a function of $R$. Note that the power into $R$ is a maximum when $R$ equals $r$, the internal resistance of the battery.

## 28.2 RESISTORS IN SERIES AND IN PARALLEL

When two or more resistors are connected together such that they have only one common point per pair, they are said to be in *series*. Figure 28.4 shows two resistors connected in series. Note that

For a series connection of resistors, the current is the same in each resistor

> the current is the same through each resistor since any charge that flows through $R_1$ must equal the charge that flows through $R_2$.

Since the potential drop from $a$ to $b$ in Figure 28.4b equals $IR_1$ and the potential drop from $b$ to $c$ equals $IR_2$, the potential drop from $a$ to $c$ is given by

$$V = IR_1 + IR_2 = I(R_1 + R_2)$$

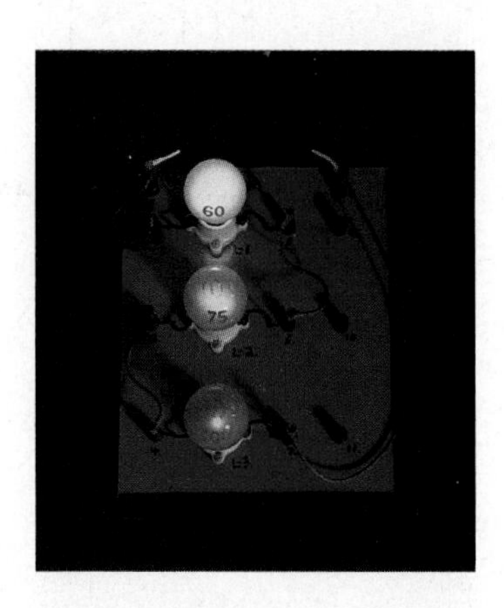


A series connection of three lamps, all rated at 120 V, with power ratings of 60 W, 75 W, and 200 W. Why are the intensities of the lamps different? Which lamp has the greatest resistance? How would their relative intensities differ if they were connected in parallel? (Courtesy of Henry Leap and Jim Lehman)

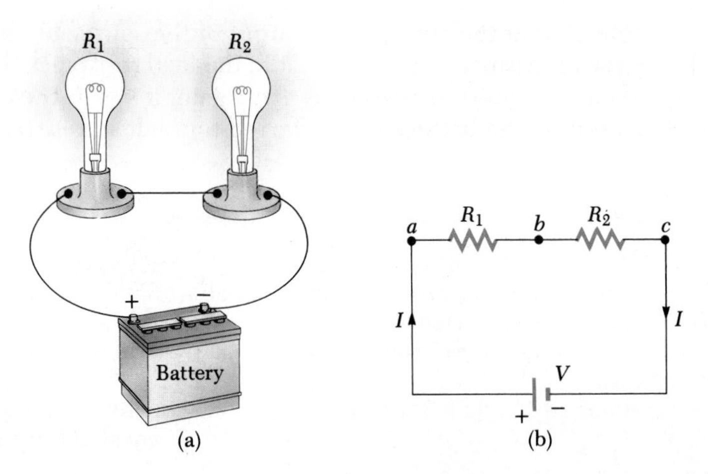


**Figure 28.4** Series connection of two resistors, $R_1$ and $R_2$. The current in each resistor is the same.

Therefore, we can replace the two resistors in series by a single *equivalent resistance* $R_{eq}$ whose value is the *sum* of the individual resistances:

$$R_{eq} = R_1 + R_2 \tag{28.5}$$

The resistance $R_{eq}$ is equivalent to the series combination $R_1 + R_2$ in the sense that the circuit current is unchanged when $R_{eq}$ replaces $R_1 + R_2$. The equivalent resistance of three or more resistors connected in series is simply

$$R_{eq} = R_1 + R_2 + R_3 + \cdots \tag{28.6}$$

Therefore, *the equivalent resistance of a series connection of resistors is always greater than any individual resistance.*

Note that if the filament of one light bulb in Figure 28.4 were to break, or "burn out," the circuit would no longer be complete (an open-circuit condition) and the second bulb would also go out. Some Christmas tree light sets (especially older ones) are connected in this way, and the agonizing experience of determining which bulb is burned out is a familiar one. Frustrating experiences such as this illustrate how inconvenient it would be to have all appliances in a house connected in series. In many circuits, fuses are used in series with other circuit elements for safety purposes. The conductor in the fuse is designed to melt and open the circuit at some maximum current, the value of which depends on the nature of the circuit. If a fuse is not used, excessive currents could damage circuit elements, overheat wires, and perhaps cause a fire. In modern home construction, circuit breakers are used in place of fuses. When the current in a circuit exceeds some value (typically 15 A), the circuit breaker acts as a switch and opens the circuit.

Now consider two resistors connected in *parallel* as shown in Figure 28.5.

In this case, there is an equal potential difference across each resistor.

However, the current in each resistor is in general not the same. When the current $I$ reaches point $a$ (called a *junction*), it splits into two parts, $I_1$ going

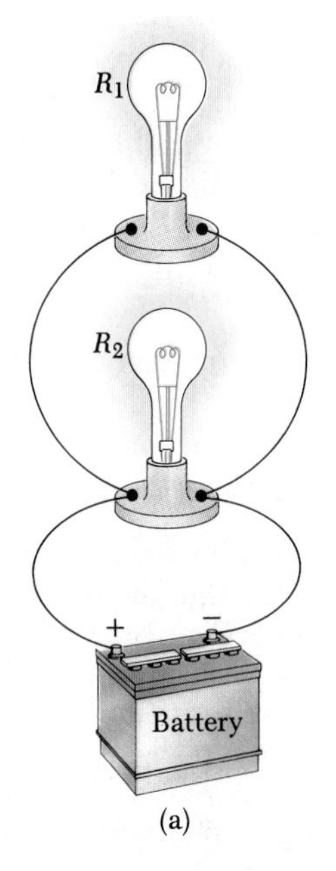


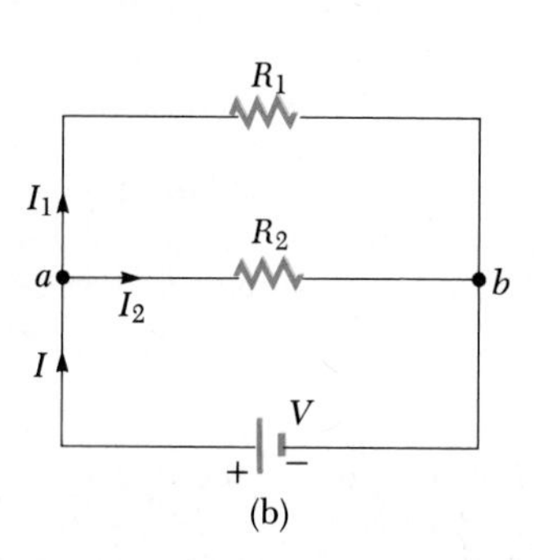


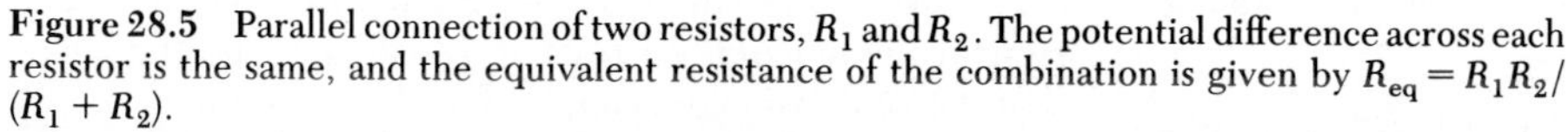

**Figure 28.5** Parallel connection of two resistors, $R_1$ and $R_2$. The potential difference across each resistor is the same, and the equivalent resistance of the combination is given by $R_{eq} = R_1R_2/(R_1 + R_2)$.

Three incandescent lamps with power ratings of 25 W, 75 W, and 150 W, connected in parallel to a voltage source of about 100 V. All lamps are rated at the same voltage. Why do the intensities of the lamps differ? Which lamp draws the most current? Which has the least resistance? (Courtesy of Henry Leap and Jim Lehman)

through $R_1$ and $I_2$ going through $R_2$. If $R_1$ is greater than $R_2$, then $I_1$ will be less than $I_2$. That is, the charge will tend to take the path of least resistance. Clearly, since charge must be conserved, the current $I$ that enters point $a$ must equal the total current leaving this point, $I_1 + I_2$:

$$I = I_1 + I_2$$

Since the potential drop across each resistor must be the *same*, Ohm's law gives

$$I = I_1 + I_2 = \frac{V}{R_1} + \frac{V}{R_2} = V\left(\frac{1}{R_1} + \frac{1}{R_2}\right) = \frac{V}{R_{eq}}$$

From this result, we see that the equivalent resistance of two resistors in parallel is given by

$$\frac{1}{R_{eq}} = \frac{1}{R_1} + \frac{1}{R_2} \tag{28.7}$$

This can be rearranged to give

$$R_{eq} = \frac{R_1R_2}{R_1 + R_2}$$

Georg Simon Ohm (1787–1854), German physicist. (Courtesy of AIP Niels Bohr Library, E. Scott Barr Collection)

An extension of this analysis to three or more resistors in parallel gives the following general expression:

Several resistors in parallel

$$\frac{1}{R_{eq}} = \frac{1}{R_1} + \frac{1}{R_2} + \frac{1}{R_3} + \cdots \tag{28.8}$$

It can be seen from this expression that the equivalent resistance of two or more resistors connected in parallel is always *less* than the smallest resistance in the group.

Household circuits are always wired such that the light bulbs (or appliances, etc.) are connected in parallel, as in Figure 28.5a. In this manner, each device operates independently of the others, so that if one is switched off, the others remain on. Equally important, each device operates on the same voltage.

Finally, it is interesting to note that parallel resistors combine in the same way that series capacitors combine, and vice versa.

**EXAMPLE 28.3 Find the Equivalent Resistance**
Four resistors are connected as shown in Figure 28.6a. (a) Find the equivalent resistance between $a$ and $c$.

The circuit can be reduced in steps as shown in Figure 28.6. The 8-Ω and 4-Ω resistors are in series, and so the equivalent resistance between $a$ and $b$ is 12 Ω (Eq. 28.5). The 6-Ω and 3-Ω resistors are in parallel, and so from Equation 28.7 we find that the equivalent resistance from $b$ to $c$ is 2 Ω. Hence, the equivalent resistance from $a$ to $c$ is 14 Ω.

(b) What is the current in each resistor if a potential difference of 42 V is maintained between $a$ and $c$?

The current $I$ in the 8-Ω and 4-Ω resistors is the same since they are in series. Using Ohm's law and the results from (a), we get

$$I = \frac{V_{ac}}{R_{eq}} = \frac{42\text{ V}}{14\ \Omega} = 3\text{ A}$$

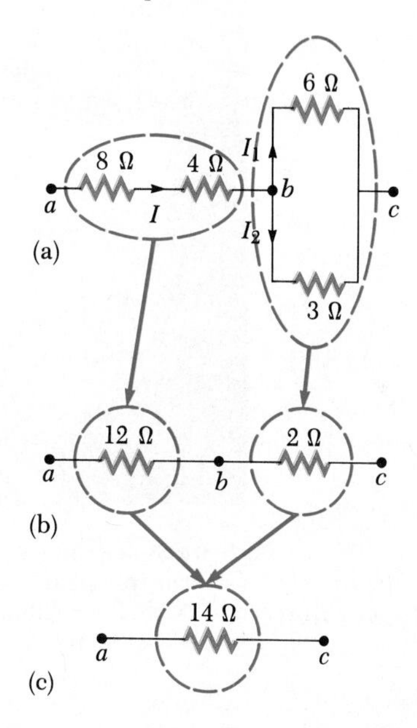


**Figure 28.6** (Example 28.3) The equivalent resistance of the four resistors shown in (a) can be reduced in steps to an equivalent 14-Ω resistor.

When this current enters the junction at $b$, it splits and part of the current passes through the 6-Ω resistor ($I_1$) and part goes through the 3-Ω resistor ($I_2$). Since the potential difference across these resistors, $V_{bc}$, is the *same* (they are in parallel), we see that $6I_1 = 3I_2$, or $I_2 = 2I_1$. Using this result and the fact that $I_1 + I_2 = 3$ A, we find that $I_1 = 1$ A and $I_2 = 2$ A. We could have guessed this from the start by noting that the current through the 3-Ω resistor has to be twice the current through the 6-Ω resistor in view of their relative resistances and the fact that the same voltage is applied to each of them.

As a final check, note that $V_{bc} = 6I_1 = 3I_2 = 6$ V and $V_{ab} = 12I = 36$ V; therefore, $V_{ac} = V_{ab} + V_{bc} = 42$ V, as it must.

**EXAMPLE 28.4 Three Resistors in Parallel**
Three resistors are connected in parallel as in Figure 28.7. A potential difference of 18 V is maintained between points $a$ and $b$. (a) Find the current in each resistor.

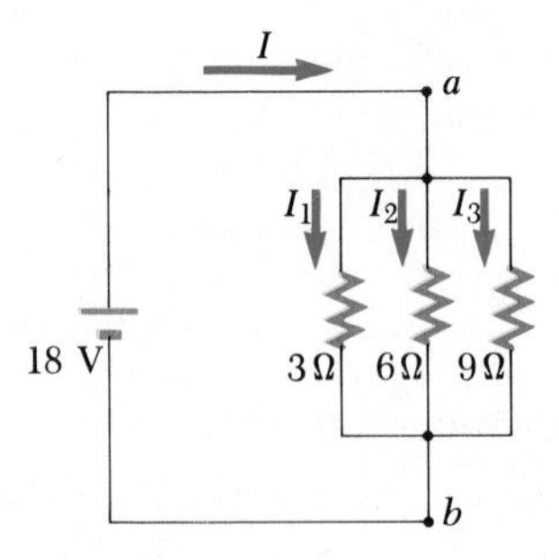


**Figure 28.7** (Example 28.4) Three resistors connected in parallel. The voltage across each resistor is 18 V.

The resistors are in parallel, and the potential difference across each is 18 V. Applying $V = IR$ to each resistor gives

$$I_1 = \frac{V}{R_1} = \frac{18\text{ V}}{3\ \Omega} = \boxed{6\text{ A}}$$

$$I_2 = \frac{V}{R_2} = \frac{18\text{ V}}{6\ \Omega} = \boxed{3\text{ A}}$$

$$I_3 = \frac{V}{R_3} = \frac{18\text{ V}}{9\ \Omega} = \boxed{2\text{ A}}$$

(b) Calculate the power dissipated by each resistor and the total power dissipated by the three resistors.

Applying $P = I^2R$ to each resistor gives

3-Ω: $P_1 = I_1{}^2R_1 = (6\text{ A})^2(3\ \Omega) = \boxed{108\text{ W}}$

6-Ω: $P_2 = I_2{}^2R_2 = (3\text{ A})^2(6\ \Omega) = \boxed{54\text{ W}}$

9-Ω: $P_3 = I_3{}^2R_3 = (2\text{ A})^2(9\ \Omega) = \boxed{36\text{ W}}$

This shows that the smallest resistor dissipates the most power since it carries the most current. (Note that you can also use $P = V^2/R$ to find the power dissipated by each resistor.) Summing the three quantities gives a total power of 198 W.

(c) Calculate the equivalent resistance of the three resistors.

We can use Equation 28.7 to find $R_{eq}$:

$$\frac{1}{R_{eq}} = \frac{1}{3} + \frac{1}{6} + \frac{1}{9}$$

$$R_{eq} = \boxed{\frac{18}{11}\ \Omega}$$

Exercise 1 Use the result for $R_{eq}$ to calculate the total power dissipated in the circuit.
Answer 198 W.

**EXAMPLE 28.5 Finding $R_{eq}$ by Symmetry Arguments**
Consider the five resistors connected as shown in Figure 28.8a. Find the equivalent resistance of the combination of resistors between points *a* and *b*.

*Solution* In this type of problem, it is convenient to assume a current entering junction *a* and then apply symmetry arguments. Because of the symmetry in the circuit (all 1-Ω resistors in the outside loop), the currents in branches *ac* and *ad* must be equal; hence, the potentials at points *c* and *d* must be equal. Since $V_c = V_d$, points *c* and *d* may be connected together without affecting the circuit, as in Figure 28.8b. Thus, the 5-Ω resistor may be removed from the circuit, and the circuit may be reduced as shown in Figures 28.8c and 28.8d. From this reduction, we see that the equivalent resistance of the combination is 1 Ω. Note that the result is 1 Ω regardless of what resistor is connected between *c* and *d*.

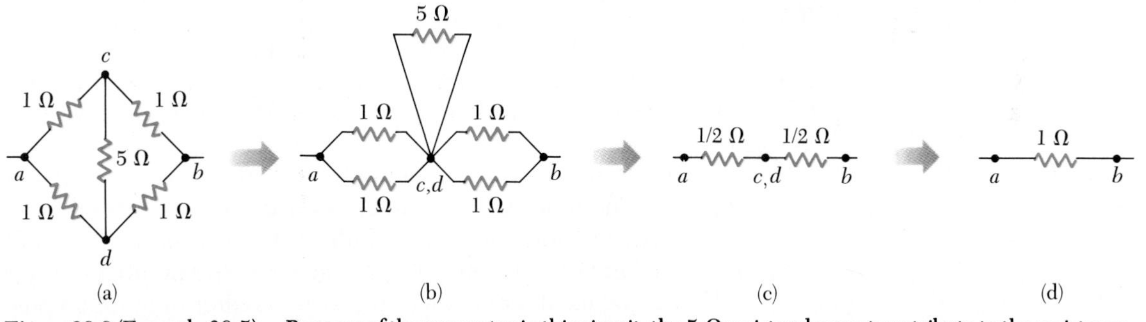


**Figure 28.8** (Example 28.5). Because of the symmetry in this circuit, the 5-Ω resistor does not contribute to the resistance between points *a* and *b* and can be disregarded.

## 28.3 KIRCHHOFF'S RULES

As we saw in the previous section, simple circuits can be analyzed using Ohm's law and the rules for series and parallel combinations of resistors. Very often it is not possible to reduce a circuit to a single loop. The procedure for analyzing more complex circuits is greatly simplified by the use of two simple rules called **Kirchhoff's rules:**

1. The sum of the currents entering any junction must equal the sum of the currents leaving that junction. (A **junction** is any point in the circuit where a current can split.)

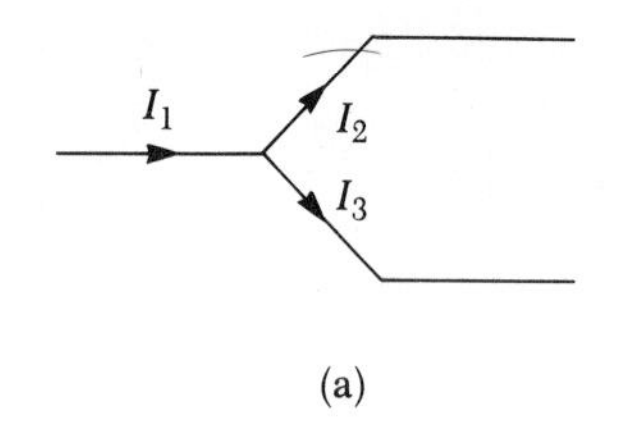


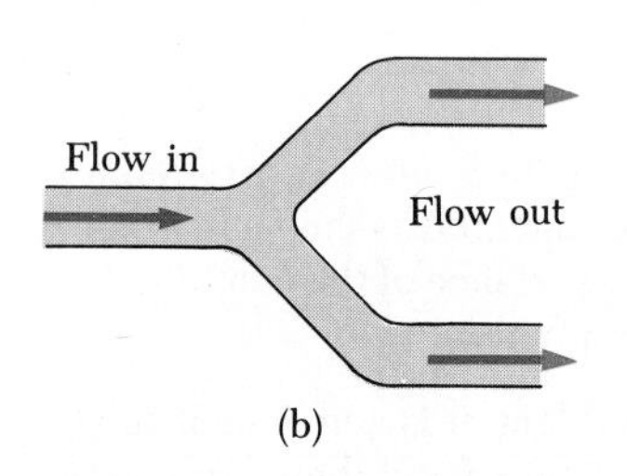


**Figure 28.9** (a) A schematic diagram illustrating Kirchhoff's junction rule. Conservation of charge requires that whatever current enters a junction must leave that junction. Therefore, in this case, $I_1 = I_2 + I_3$. (b) A mechanical analog of the junction rule: the flow out must equal the flow in.

2. The algebraic sum of the changes in potential across all of the elements around any closed circuit loop must be *zero.*

The first rule is a statement of **conservation of charge.** That is, whatever current enters a given point in a circuit must leave that point, since charge cannot build up at a point. If we apply this rule to the junction shown in Figure 28.9a, we get

$$I_1 = I_2 + I_3$$

Figure 28.9b represents a mechanical analog to this situation, in which water flows through a branched pipe with no leaks. The flow rate into the pipe equals the total flow rate out of the two branches.

The second rule follows from **conservation of energy.** That is, any charge that moves around *any* closed loop in a circuit (it starts and ends at the same point) must gain as much energy as it loses. Its energy may decrease in the form of a potential drop, $-IR$, across a resistor or as the result of having the charge go the reverse direction through a source of emf. In a practical application of the latter case, electrical energy is converted into chemical energy when a battery is charged; similarly, electrical energy may be converted into mechanical energy for operating a motor.

As an aid in applying the second rule, the following calculational tools should be noted. These points are summarized in Fig. 28.10.

1. If a resistor is traversed in the direction of the current, the change in potential across the resistor is $-IR$ (Fig. 28.10a).
2. If a resistor is traversed in the direction *opposite* the current, the change in potential across the resistor is $+IR$ (Figure 28.10b).
3. If a source of emf is traversed in the direction of the emf (from $-$ to $+$ on the terminals), the change in potential is $+\mathcal{E}$ (Fig. 28.10c).
4. If a source of emf is traversed in the direction opposite the emf (from $+$ to $-$ on the terminals), the change in potential is $-\mathcal{E}$ (Fig. 28.10d).

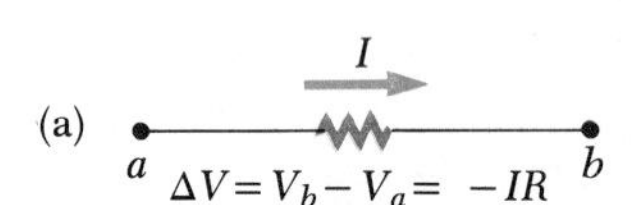


**Figure 28.10** Rules for determining the potential changes across a resistor and a battery, assuming the battery has no internal resistance.

There are limitations on the number of times you can use the junction rule and the loop rule. The junction rule can be used as often as needed so long as each time you write an equation, you include in it a current that has not been used in a previous junction rule equation. In general, the number of times the junction rule must be used is one fewer than the number of junction points in the circuit. The loop rule can be used as often as needed so long as a new circuit element (resistor or battery) or a new current appears in each new equation. In general, *the number of independent equations you need must at least equal the number of unknowns in order to solve a particular circuit problem.*

Complex networks with many loops and junctions generate large numbers of independent, linear equations and a corresponding large number of unknowns. Such situations can be handled formally using matrix algebra. Computer programs can also be written to solve for the unknowns.

The following examples illustrate the use of Kirchhoff's rules in analyzing circuits. In all cases, it is assumed that the circuits have reached steady-state conditions, that is, the currents in the various branches are constant. If a capacitor is included as an element in one of the branches, *it acts as an open circuit,* that is, the current in the branch containing the capacitor will be zero under steady-state conditions.

## PROBLEM-SOLVING STRATEGY AND HINTS: KIRCHHOFF'S RULES

1. First, draw the circuit diagram and assign labels and symbols to all the known and unknown quantities. You must assign a *direction* to the currents in each part of the circuit. Do not be alarmed if you guess the direction of a current incorrectly; the result will have a negative value, but *its magnitude will be correct.* Although the assignment of current directions is arbitrary, you must adhere *rigorously* to the assigned directions when applying Kirchhoff's rules.
2. Apply the junction rule (Kirchhoff's first rule) to any junction in the circuit which provides a relation between the various currents. (This step is easy!)
3. Now apply Kirchhoff's second rule to as many loops in the circuit as are needed to solve for the unknowns. In order to apply this rule, you must correctly identify the change in potential as you cross each element in traversing the closed loop (either clockwise or counterclockwise). Watch out for signs! We suggest that you follow the four "rules-of-thumb" listed above and summarized in Figure 28.10.
4. Finally, you must solve the equations simultaneously for the unknown quantities. Be careful in your algebraic steps, and check your numerical answers for consistency.

### EXAMPLE 28.6 A Single-Loop Circuit

A single-loop circuit contains two external resistors and two sources of emf as shown in Figure 28.11. The internal resistances of the batteries have been neglected. (a) Find the current in the circuit.

There are no junctions in this single-loop circuit, and so the current is the same in all elements. Let us assume that the current is in the clockwise direction as shown in Figure 28.11. Traversing the circuit in the clockwise direction, starting at point $a$, we see that $a \rightarrow b$ represents a potential increase of $+\mathcal{E}_1$, $b \rightarrow c$ represents a potential decrease of $-IR_1$, $c \rightarrow d$ represents a potential decrease of $-\mathcal{E}_2$, and $d \rightarrow a$ represents a potential decrease of $-IR_2$. Applying Kirchhoff's second rule gives

$$\sum_i \Delta V_i = 0$$

$$\mathcal{E}_1 - IR_1 - \mathcal{E}_2 - IR_2 = 0$$

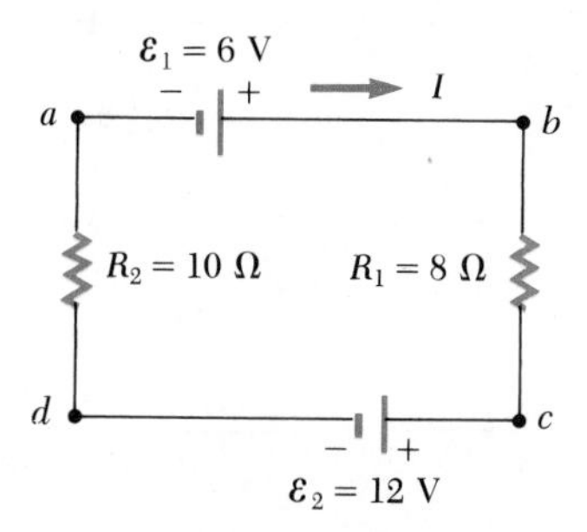


**Figure 28.11** (Example 28.6) A series circuit containing two batteries and two resistors, where the polarities of the batteries are in opposition to each other.

Solving for $I$ and using the values given in Figure 28.11, we get

$$I = \frac{\mathcal{E}_1 - \mathcal{E}_2}{R_1 + R_2} = \frac{6\text{ V} - 12\text{ V}}{8\ \Omega + 10\ \Omega} = -\frac{1}{3}\text{ A}$$

The negative sign for $I$ indicates that the direction of the current is *opposite* the assumed direction, or *counterclockwise.*

(b) What is the power lost in each resistor?

$$P_1 = I^2R_1 = (\tfrac{1}{3}\text{A})^2(8\ \Omega) = \frac{8}{9}\text{ W}$$

$$P_2 = I^2R_2 = (\tfrac{1}{3}\text{A})^2(10\ \Omega) = \frac{10}{9}\text{ W}$$

Hence, the total power lost is $P_1 + P_2 = 2$ W. Note that the 12-V battery delivers power $I\mathcal{E}_2 = 4$ W. Half of this power is delivered to the external resistors. The other half is delivered to the 6-V battery, which is being charged by the 12-V battery. If we had included the internal resistances of the batteries, some of the power would be dissipated as heat in the batteries, so that *less* power would be delivered to the 6-V battery.

### EXAMPLE 28.7 Applying Kirchhoff's Rules

Find the currents $I_1$, $I_2$, and $I_3$ in the circuit shown in Figure 28.12.

We shall choose the directions of the currents as shown in Figure 28.12. Applying Kirchhoff's first rule to junction $c$ gives

$$\text{(1)} \qquad I_1 + I_2 = I_3$$

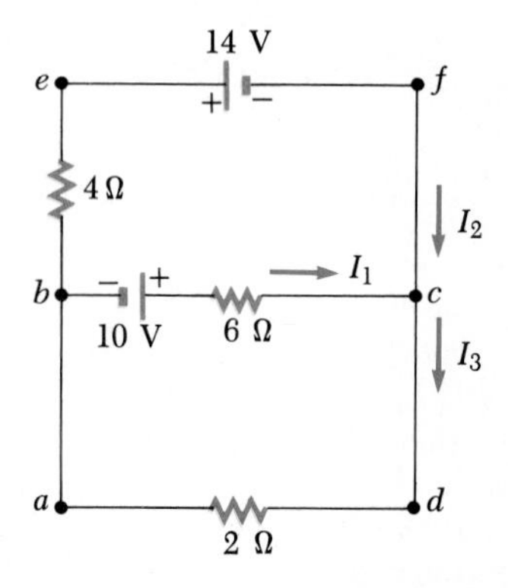


**Figure 28.12** (Example 28.7) A circuit containing three loops.

There are *three* loops in the circuit, *abcda, befcb,* and *aefda* (the outer loop). We need only *two* loop equations to determine the unknown currents. The third loop equation would give no new information. Applying Kirchhoff's second rule to loops *abcda* and *befcb* and traversing these loops in the clockwise direction, we obtain the following expressions:

(2) Loop *abcda:* $10\text{ V} - (6\ \Omega)I_1 - (2\ \Omega)I_3 = 0$

(3) Loop *befcb:* $-14\text{ V} - 10\text{ V} + (6\ \Omega)I_1 - (4\ \Omega)I_2 = 0$

Note that in loop *befcb,* a positive sign is obtained when traversing the 6-Ω resistor since the direction of the path is opposite the direction of the current $I_1$. A third loop equation for *aefda* gives $-14 = 2I_3 + 4\ I_2$, which is just the sum of (2) and (3). Expressions (1), (2), and (3) represent three linear, independent equations with three unknowns. We can solve the problem as follows: Substituting (1) into (2) gives

$$10 - 6I_1 - 2(I_1 + I_2) = 0$$

(4) $$10 = 8I_1 + 2I_2$$

Dividing each term in (3) by 2 and rearranging the equation gives

(5) $$-12 = -3I_1 + 2I_2$$

Subtracting (5) from (4) eliminates $I_2$, giving

$$22 = 11I_1$$

$$I_1 = 2\text{ A}$$

Using this value of $I_1$ in (5) gives a value for $I_2$:

$$2I_2 = 3I_1 - 12 = 3(2) - 12 = -6$$

$$I_2 = -3\text{ A}$$

Finally, $I_3 = I_1 + I_2 = -1$ A. Hence, the currents have the values

$$I_1 = 2\text{ A} \qquad I_2 = -3\text{ A} \qquad I_3 = -1\text{ A}$$

The fact that $I_2$ and $I_3$ are both negative indicates only that we chose the *wrong* direction for these currents. However, the numerical values are correct.

Exercise 2 Find the potential difference between points *b* and *c*.
Answer $V_b - V_c = 2$ V.

### EXAMPLE 28.8 A Multi-Loop Circuit

The multiloop circuit in Figure 28.13 contains three resistors, three batteries, and one capacitor. (a) Under steady-state conditions, find the unknown currents.

First note that *the capacitor represents an open circuit, and hence there is no current along path ghab under steady-state conditions.* Therefore, $I_{gf} = I_1$. Labeling the currents as shown in Figure 28.13 and applying Kirchhoff's first rule to junction *c*, we get

(1) $$I_1 + I_2 = I_3$$

Kirchhoff's second rule applied to loops *defcd* and *cfgbc* gives

(2) Loop *defcd:* $4\text{ V} - (3\ \Omega)I_2 - (5\ \Omega)I_3 = 0$

(3) Loop *cfgbc:* $8\text{ V} - (5\ \Omega)I_1 + (3\ \Omega)I_2 = 0$

From (1) we see that $I_1 = I_3 - I_2$, which when substituted into (3) gives

(4) $$8\text{ V} - (5\ \Omega)I_3 + (8\ \Omega)I_2 = 0$$

Subtracting (4) from (2), we eliminate $I_3$ and find

$$I_2 = -\tfrac{4}{11}\text{ A} = -0.364\text{ A}$$

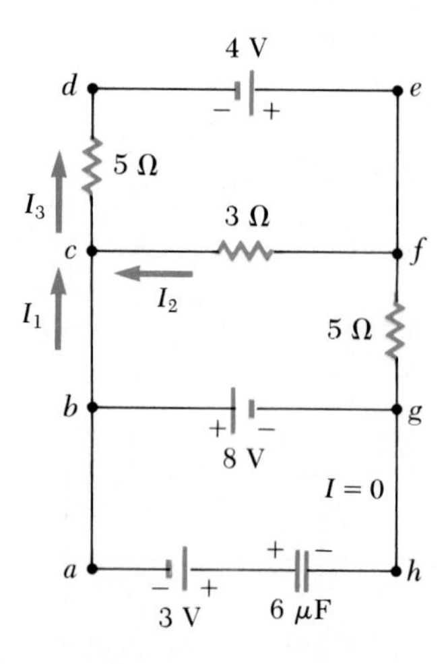


**Figure 28.13** (Example 28.8) A multiloop circuit. Note that Kirchhoff's loop equation can be applied to *any* closed loop, including one containing the capacitor.

Since $I_2$ is negative, we conclude that $I_2$ is from $c$ to $f$ through the 3-Ω resistor. Using this value of $I_2$ in (3) and (1) gives the following values for $I_1$ and $I_3$:

$$I_1 = 1.38 \text{ A} \qquad I_3 = 1.02 \text{ A}$$

Under state-steady conditions, the capacitor represents an *open* circuit, and so there is no current in the branch *ghab*.

(b) What is the charge on the capacitor?

We can apply Kirchhoff's second rule to loop *abgha* (or any loop that contains the capacitor) to find the potential difference $V_c$ across the capacitor:

$$-8 \text{ V} + V_c - 3 \text{ V} = 0$$

$$V_c = 11.0 \text{ V}$$

Since $Q = CV_c$, we find that the charge on the capacitor is equal to

$$Q = (6\ \mu\text{F})(11.0 \text{ V}) = 66.0\ \mu\text{C}$$

Why is the left side of the capacitor positively charged?

**Exercise 3** Find the voltage across the capacitor by traversing any other loop, such as the outside loop. **Answer** 11.0 V.

## 28.4 *RC* CIRCUITS

So far we have been concerned with circuits with constant currents, or so-called *steady-state circuits.* We shall now consider circuits containing capacitors, in which the currents may vary in time. When a potential difference is first applied across a capacitor, the rate at which it charges depends on its capacitance and on the resistance in the circuit.

### Charging a Capacitor

Consider the series circuit shown in Figure 28.14. Let us assume that the capacitor is initially uncharged. There is no current when the switch S is open (Fig. 28.14b). If the switch is closed at $t = 0$, charges will begin to flow, setting up a current in the circuit, and the capacitor will begin to charge (Fig. 28.14c). Note that during the charging process, charges do not jump across the plates of the capacitor since the gap between the plates represents an open circuit. Instead, charge is transferred from one plate to the other through the resistor, switch, and battery until the capacitor is fully charged. The value of the maximum charge depends on the emf of the battery. Once the maximum charge is reached, the current in the circuit is zero.

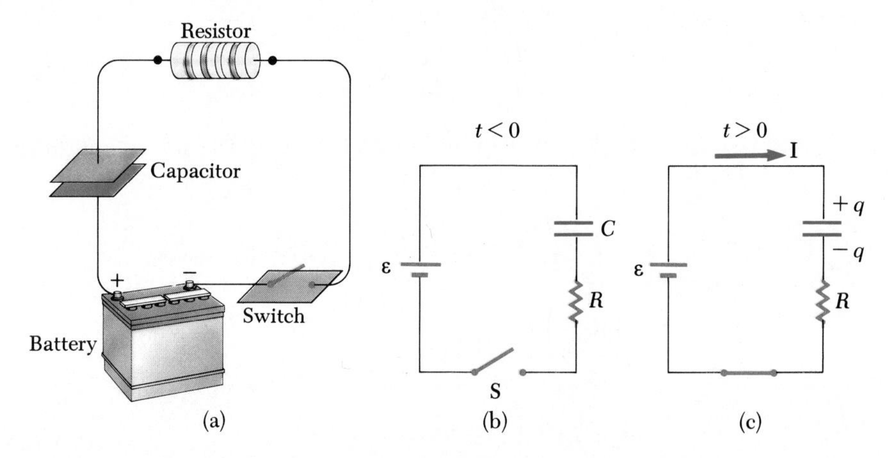


**Figure 28.14** (a) A capacitor in series with a resistor, battery, and switch. (b) Circuit diagram representing this system before the switch is closed, $t < 0$. (c) Circuit diagram after the switch is closed, $t > 0$.

To put this discussion on a quantitative basis, let us apply Kirchhoff's second rule to the circuit *after* the switch is closed. This gives

$$\mathcal{E} - IR - \frac{q}{C} = 0 \tag{28.9}$$

where $IR$ is the potential drop across the resistor and $q/C$ is the potential drop across the capacitor. Note that $q$ and $I$ are *instantaneous* values of the charge and current, respectively, as the capacitor is being charged.

We can use Equation 28.9 to find the initial current in the circuit and the maximum charge on the capacitor. At $t = 0$, when the switch is closed, the charge on the capacitor is zero, and from Equation 28.9 we find that the initial current in the circuit, $I_0$, is a maximum and equal to

Maximum current

$$I_0 = \frac{\mathcal{E}}{R} \qquad \text{(current at } t = 0\text{)} \tag{28.10}$$

At this time, *the potential drop is entirely across the resistor.* Later, when the capacitor is charged to its maximum value $Q$, charges cease to flow, the current in the circuit is zero, and *the potential drop is entirely across the capacitor.* Substituting $I = 0$ into Equation 28.9 gives the following expression for $Q$:

Maximum charge on the capacitor

$$Q = C\mathcal{E} \qquad \text{(maximum charge)} \tag{28.11}$$

To determine analytical expressions for the time dependence of the charge and current, we must solve Equation 28.9, a single equation containing two variables, $q$ and $I$. In order to do this, let us differentiate Equation 28.9 with respect to time. Since $\mathcal{E}$ is a constant, $d\mathcal{E}/dt = 0$ and we get

$$\frac{d}{dt}\left(\mathcal{E} - \frac{q}{C} - IR\right) = 0 - \frac{1}{C}\frac{dq}{dt} - R\frac{dI}{dt} = 0$$

Recalling that $I = dq/dt$, we can express this equation in the form

$$R\frac{dI}{dt} + \frac{I}{C} = 0$$

$$\frac{dI}{I} = -\frac{1}{RC}\,dt \tag{28.12}$$

Since $R$ and $C$ are constants, this can be integrated using the initial condition that at $t = 0$, $I = I_0$:

$$\int_{I_0}^{I} \frac{dI}{I} = -\frac{1}{RC}\int_0^t dt$$

$$\ln\left(\frac{I}{I_0}\right) = -\frac{t}{RC}$$

Current versus time

$$I(t) = I_0\, e^{-t/RC} = \frac{\mathcal{E}}{R}\, e^{-t/RC} \tag{28.13}$$

where $e$ is the base of the natural logarithm and $I_0 = \mathcal{E}/R$ is the initial current.

In order to find the charge on the capacitor as a function of time, we can substitute $I = dq/dt$ into Equation 28.13 and integrate once more:

$$\frac{dq}{dt} = \frac{\mathcal{E}}{R} e^{-t/RC}$$

$$dq = \frac{\mathcal{E}}{R} e^{-t/RC}\, dt$$

We can integrate this expression using the condition that $q = 0$ at $t = 0$:

$$\int_0^q dq = \frac{\mathcal{E}}{R}\int_0^t e^{-t/RC}\, dt$$

In order to integrate the right side of this expression, we use the fact that $\int e^{-ax}\, dx = -\frac{1}{a} e^{-ax}$. The result of the integration gives

$$q(t) = C\mathcal{E}[1 - e^{-t/RC}] = Q[1 - e^{-t/RC}] \qquad (28.14)$$

**Charge versus time for a capacitor being charged**

where $Q = C\mathcal{E}$ is the *maximum* charge on the capacitor.

Plots of Equations 28.13 and 28.14 are shown in Figure 28.15. Note that the charge is zero at $t = 0$ and approaches the maximum value of $C\mathcal{E}$ as $t \to \infty$ (Fig. 28.15a). Furthermore, the current has its maximum value of $I_0 = \mathcal{E}/R$ at $t = 0$ and decays exponentially to zero as $t \to \infty$ (Fig. 28.15b). The quantity $RC$, which appears in the exponential of Equations 28.13 and 28.14, is called the **time constant,** $\tau$, of the circuit. It represents the time it takes the current to decrease to $1/e$ of its initial value; that is, in a time $\tau$, $I = e^{-1}I_0 = 0.37I_0$. In a time $2\tau$, $I = e^{-2}I_0 = 0.135I_0$, and so forth. Likewise, in a time $\tau$ the charge will increase from zero to $C\mathcal{E}[1 - e^{-1}] = 0.63C\mathcal{E}$.

The following dimensional analysis shows that $\tau$ has the unit of time:

$$[\tau] = [RC] = \left[\frac{V}{I} \times \frac{Q}{V}\right] = \left[\frac{Q}{Q/T}\right] = [T]$$

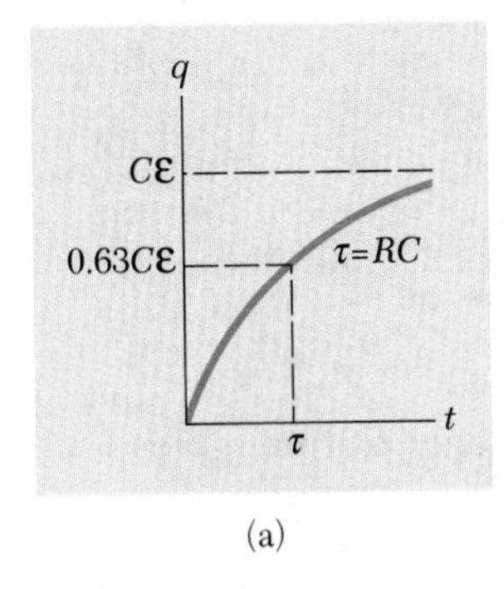


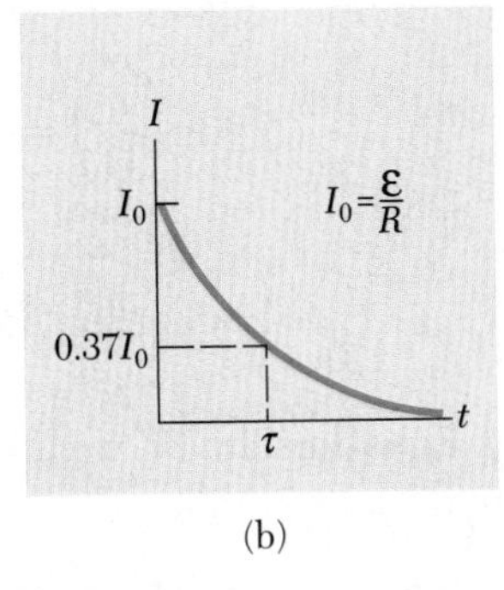


**Figure 28.15** (a) Plot of capacitor charge versus time for the circuit shown in Figure 28.14. After one time constant, $\tau$, the charge is 63% of the maximum value, $C\mathcal{E}$. The charge approaches its maximum value as $t$ approaches infinity. (b) Plot of current versus time for the $RC$ circuit shown in Figure 28.14. The current has its maximum value, $I_0 = \mathcal{E}/R$, at $t = 0$ and decays to zero exponentially as $t$ approaches infinity. After one time constant, $\tau$, the current decreases to 37% of its initial value.

The work done by the battery during the charging process is $Q\mathcal{E} = C\mathcal{E}^2$. After the capacitor is fully charged, the energy stored in the capacitor is $\frac{1}{2}Q\mathcal{E} = \frac{1}{2}C\mathcal{E}^2$, which is just half the work done by the battery. It is left as a problem to show that the remaining half of the energy supplied by the battery goes into joule heat in the resistor (Problem 82).

### Discharging a Capacitor

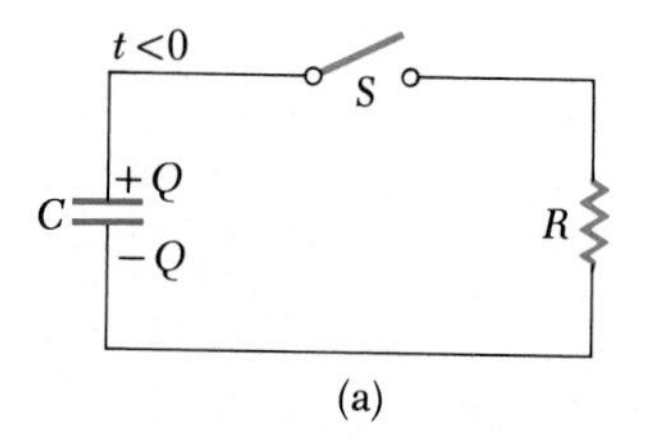


(b)

**Figure 28.16** (a) A charged capacitor connected to a resistor and a switch, which is open at $t < 0$. (b) After the switch is closed, a nonsteady current is set up in the direction shown and the charge on the capacitor decreases exponentially with time.

Now consider the circuit in Figure 28.16, consisting of a capacitor with an initial charge $Q$, a resistor, and a switch. When the switch is open (Fig. 28.16a), there is a potential difference of $Q/C$ across the capacitor and zero potential difference across the resistor since $I = 0$. If the switch is closed at $t = 0$, the capacitor begins to discharge through the resistor. At some time during the discharge, the current in the circuit is $I$ and the charge on the capacitor is $q$ (Fig. 28.16b). From Kirchhoff's second rule, we see that the potential drop across the resistor, $IR$, must equal the potential difference across the capacitor, $q/C$:

$$IR = \frac{q}{C} \tag{28.15}$$

However, the current in the circuit must equal the rate of *decrease* of charge on the capacitor. That is, $I = -dq/dt$, and so Equation 28.15 becomes

$$-R\frac{dq}{dt} = \frac{q}{C}$$

$$\frac{dq}{q} = -\frac{1}{RC}\,dt \tag{28.16}$$

Integrating this expression using the fact that $q = Q$ at $t = 0$ gives

$$\int_Q^q \frac{dq}{q} = -\frac{1}{RC}\int_0^t dt$$

$$\ln\left(\frac{q}{Q}\right) = -\frac{t}{RC}$$

**Charge versus time for a discharging capacitor**

$$q(t) = Q\,e^{-t/RC} \tag{28.17}$$

Differentiating Equation 28.17 with respect to time gives the current as a function of time:

$$I(t) = -\frac{dq}{dt} = \frac{Q}{RC}\,e^{-t/RC} = I_0\,e^{-t/RC} \tag{28.18}$$

where the initial current $I_0 = Q/RC$. Therefore, we see that both the charge on the capacitor and the current decay exponentially at a rate characterized by the time constant $\tau = RC$.

**EXAMPLE 28.9 Charging a Capacitor in an *RC* Circuit**

An uncharged capacitor and a resistor are connected in series to a battery as in Figure 28.17. If $\mathcal{E} = 12$ V, $C = 5\ \mu$F, and $R = 8 \times 10^5\ \Omega$, find the time constant of the circuit, the maximum charge on the capacitor, the maximum current in the circuit, and the charge and current as a function of time.

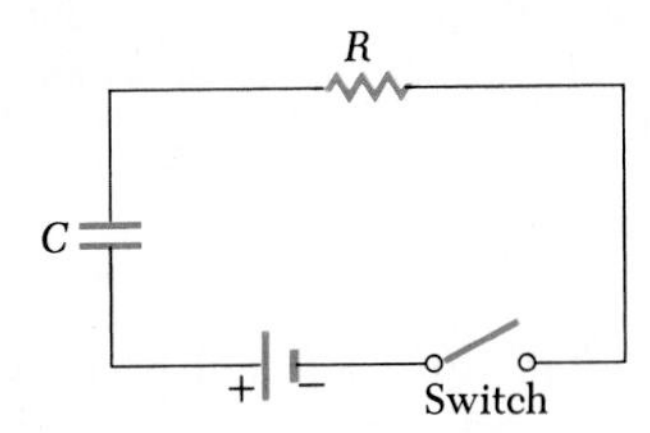


**Figure 28.17** (Example 28.9) The switch of this series *RC* circuit is closed at $t = 0$.

*Solution* The time constant of the circuit is $\tau = RC = (8 \times 10^5\ \Omega)(5 \times 10^{-6}\ \text{F}) = 4$ s. The maximum charge on the capacitor is $Q = C\mathcal{E} = (5 \times 10^{-6}\ \text{F})(12\ \text{V}) = 60\ \mu$C. The maximum current in the circuit is $I_0 = \mathcal{E}/R = (12\ \text{V})/(8 \times 10^5\ \Omega) = 15\ \mu$A. Using these values and Equations 28.13 and 28.14, we find that

$$q(t) = 60[1 - e^{-t/4}]\ \mu\text{C}$$

$$I(t) = 15\ e^{-t/4}\ \mu\text{A}$$

Graphs of these functions are given in Figure 28.18.

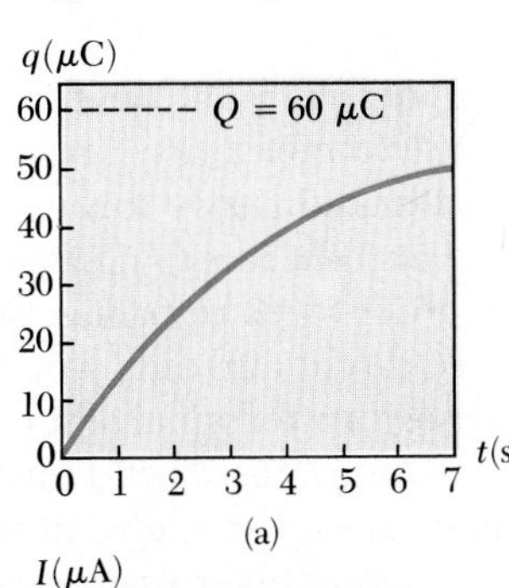


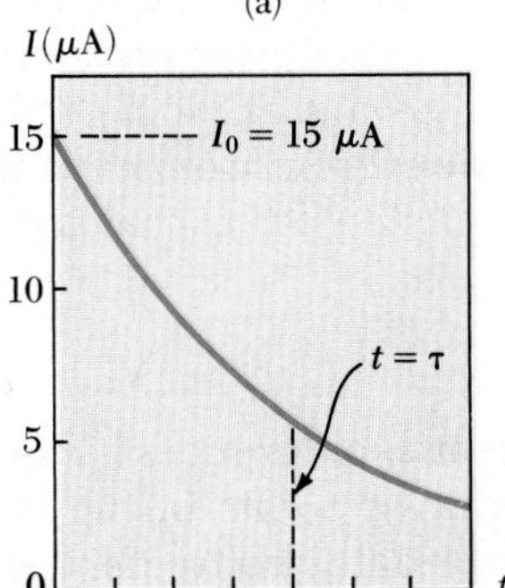


**Figure 28.18** (Example 28.9) Plots of (a) charge versus time and (b) current versus time for the *RC* circuit shown in Figure 28.17, with $\mathcal{E} = 12$ V, $R = 8 \times 10^5\ \Omega$, and $C = 5\ \mu$F.

Exercise 4 Calculate the charge on the capacitor and the current in the circuit after one time constant has elapsed.

Answer 37.9 $\mu$C, 5.52 $\mu$A.

**EXAMPLE 28.10 Discharging a Capacitor in an *RC* Circuit** □

Consider a capacitor $C$ being discharged through a resistor $R$ as in Figure 28.16. (a) After how many time constants will the charge on the capacitor drop to one-fourth of its initial value?

*Solution* The charge on the capacitor varies with time according to Equation 28.17,

$$q(t) = Qe^{-t/RC}$$

where $Q$ is the initial charge on the capacitor. To find the time it takes the charge $q$ to drop to one-fourth of its initial value, we substitute $q(t) = Q/4$ into this expression and solve for $t$:

$$\tfrac{1}{4}Q = Qe^{-t/RC}$$

or

$$\tfrac{1}{4} = e^{-t/RC}$$

Taking logarithms of both sides, we find

$$-\ln 4 = -\frac{t}{RC}$$

or

$$t = RC \ln 4 = 1.39RC$$

(b) The energy stored in the capacitor decreases with time as it discharges. After how many time constants will this stored energy reduce to one-fourth of its initial value?

*Solution* Using Equations 26.12 and 28.17, we can express the energy stored in the capacitor at any time $t$ as

$$U = \frac{q^2}{2C} = \frac{Q^2}{2C} e^{-2t/RC} = U_0 e^{-2t/RC}$$

where $U_0$ is the initial energy stored in the capacitor. As in part (a), we now set $U = U_0/4$ and solve for $t$:

$$\tfrac{1}{4}U_0 = U_0 e^{-2t/RC}$$

$$\tfrac{1}{4} = e^{-2t/RC}$$

Again, taking logarithms of both sides and solving for $t$ gives

$$t = \tfrac{1}{2}RC \ln 4 = 0.693RC$$

**Exercise 5** After how many time constants will the current in the $RC$ circuit drop to one-half of its initial value?
**Answer** $0.693RC$

**EXAMPLE 28.11 Heat Loss in a Resistor**

A 5-$\mu$F capacitor is charged to a potential difference of 800 V and is then discharged through a 25-k$\Omega$ resistor, as in Figure 28.16. What is the total energy which has been lost as joule heating in the resistor when the capacitor is fully discharged?

*Solution* We shall solve this problem in *two* ways. The first method, which is actually quite simple, is to note that the initial energy in the system equals the energy stored in the capacitor, given by $C\mathcal{E}^2/2$. Once the capacitor is fully discharged, the energy stored in it is zero. Since energy is conserved, the initial energy stored in the capacitor is transformed into thermal energy dissipated in the resistor. Using the given values of $C$ and $\mathcal{E}$, we find

$$\text{Energy} = \tfrac{1}{2}C\mathcal{E}^2 = \tfrac{1}{2}(5 \times 10^{-6}\text{ F})(800\text{ V})^2 = 1.60\text{ J}$$

The second method, which is more difficult but perhaps more instructive, is to note that as the capacitor discharges through the resistor, the rate at which heat is generated in the resistor (or the power loss) is given by $RI^2$, where $I$ is the instantaneous current given by Equation 28.18. Since the power is defined as the rate of change of energy, we conclude that the energy lost in the resistor in the form of heat must equal the time integral of $RI^2\,dt$. That is

$$\text{Energy} = \int_0^\infty RI^2\,dt = \int_0^\infty R(I_0 e^{-t/RC})^2\,dt$$

To evaluate this integral, we note that the initial current $I_0 = \mathcal{E}/R$, and all parameters are constants except for $t$. Thus, we find

$$\text{Energy} = \frac{\mathcal{E}^2}{R}\int_0^\infty e^{-2t/RC}\,dt$$

The integral in the expression above has a value of $RC/2$, so we find

$$\text{Energy} = \tfrac{1}{2}C\mathcal{E}^2$$

which agrees with the simpler approach, as it must. Note that this second approach can be used to find the energy lost as heat at *any* time after the switch is closed by simply replacing the upper limit in the integral by that specific value of $t$.

**Exercise 6** Show that the integral given in this example has the value of $RC/2$.

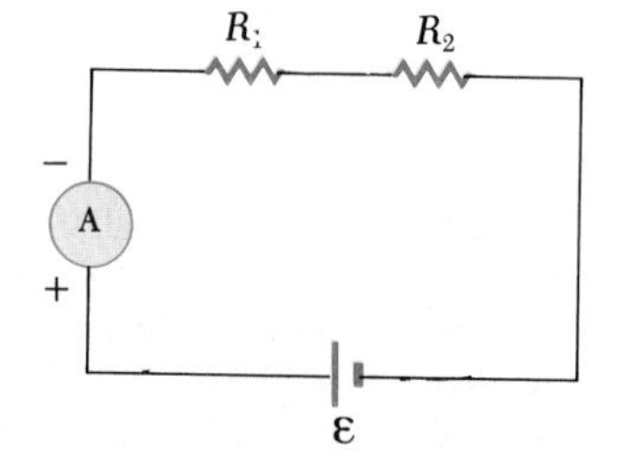


**Figure 28.19** The current in a circuit can be measured with an ammeter connected in series with the resistor and battery. An ideal ammeter has zero resistance.

## *28.5 ELECTRICAL INSTRUMENTS

### The Ammeter

Current is one of the most important quantities that one would like to measure in an electric circuit. A device that measures current is called an **ammeter.** The current to be measured must pass directly through the ammeter, so that the ammeter is in series with the current it is to measure, as in Figure 28.19. The wires usually must be cut in order to make connections to the ammeter. When using an ammeter to measure direct currents, you must be sure to connect it such that current enters the positive terminal of the instrument and exits at the negative terminal. **Ideally, an ammeter should have zero resistance so as not to alter the current being measured.** In the circuit shown in Figure 28.19, this condition requires that the ammeter's resistance be small compared to $R_1 + R_2$. Since any ammeter always has some resistance, its presence in the circuit will slightly reduce the current from its value when the ammeter is not present.

### The Voltmeter

A device that measures potential differences is called a **voltmeter.** The potential difference between any two points in the circuit can be measured by simply attaching the terminals of the voltmeter between these points without breaking the circuit, as in Figure 28.20. The potential difference across resistor $R_2$ is measured by connecting the voltmeter in parallel with $R_2$. Again, it is necessary to observe the polarity of the instrument. The positive terminal of the voltmeter must be connected to the end of the resistor at the higher potential, and the negative terminal to the low-potential end of the resistor. **An ideal voltmeter has infinite resistance so that no current will pass through it.** In Figure 28.20, this condition requires that the voltmeter must have a resistance that is very large compared to $R_2$. In practice, if this condition is not met, one should make corrections for the known resistance of the voltmeter.

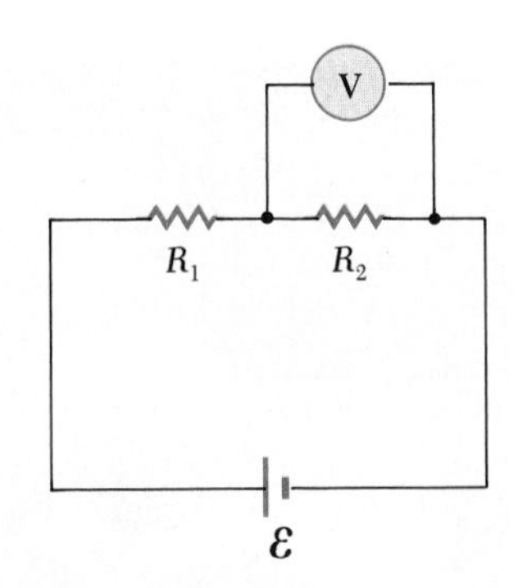


**Figure 28.20** The potential difference across a resistor can be measured with a voltmeter connected in parallel with the resistor. An ideal voltmeter has infinite resistance and does not affect the circuit.

### The Galvanometer

The **galvanometer** is the main component used in the construction of ammeters and voltmeters. The essential features of a common type, called the *D'Arsonval galvanometer,* are shown in Figure 28.21. It consists of a coil of wire mounted such that it is free to rotate on a pivot in a magnetic field provided by a permanent magnet. The basic operation of the galvanometer makes use of the fact that a torque acts on a current loop in the presence of a magnetic field. (The reason for this is discussed in detail in Chapter 29.) The torque experienced by the coil is proportional to the current through it. This means that the larger the current, the larger the torque and the more the coil will rotate before the spring tightens enough to stop the rotation. Hence, the amount of deflection is proportional to the current. Once the instrument is properly calibrated, it can be used in conjunction with other circuit elements to measure either currents or potential differences.

A typical off-the-shelf galvanometer is often not suitable for use as an ammeter. One of the main reasons for this is that a typical galvanometer has a resistance of about 60 Ω. An ammeter resistance this large would considerably alter the current in the circuit in which it is placed. This can easily be understood by considering the following example. Suppose you were to construct a simple series circuit containing a 3-V battery and a 3-Ω resistor. The current in such a circuit is 1 A. However, if you insert a 60-Ω galvanometer in the circuit to measure the current, the total resistance of the circuit would now be 63 Ω, and the current would be reduced to 0.048 A.

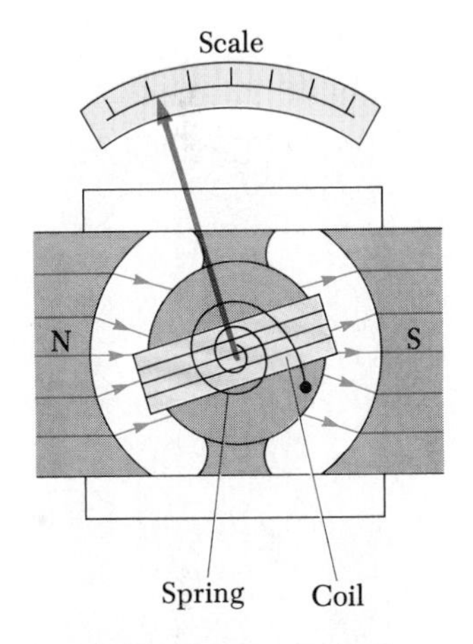


**Figure 28.21** The principal components of a D'Arsonval galvanometer. When current passes through the coil, situated in a magnetic field, the magnetic torque causes the coil to twist. The angle through which the coil rotates is proportional to the current through it because of the spring's torque.

A second factor that limits the use of a galvanometer as an ammeter is the fact that a typical galvanometer will give a full-scale deflection for very low currents, of the order of 1 mA or less. Consequently, such a galvanometer cannot be used directly to measure currents greater than this. However, one can convert a galvanometer into an ammeter by simply placing a resistor, $R_p$, in *parallel* with the galvanometer as in Figure 28.22a. The value of $R_p$, some-

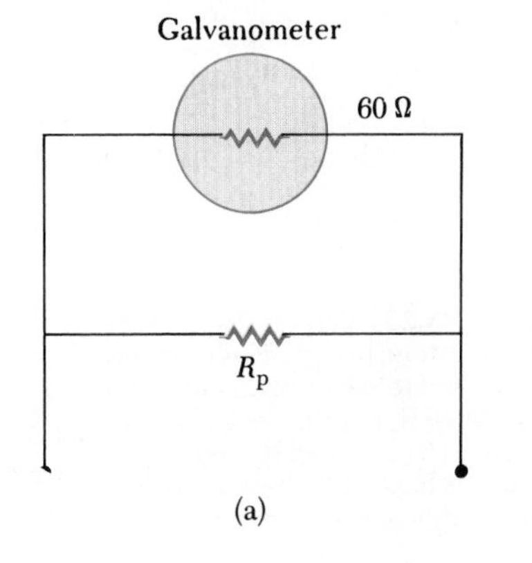


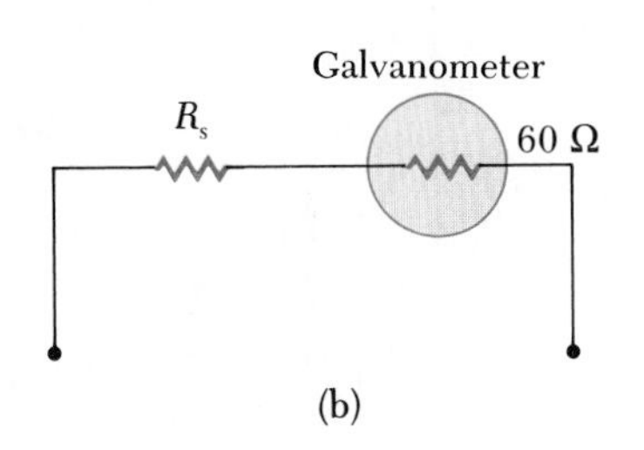


**Figure 28.22** (a) When a galvanometer is to be used as an ammeter, a resistor, $R_p$, is connected in parallel with the galvanometer. (b) When the galvanometer is used as a voltmeter, a resistor, $R_s$, is connected in series with the galvanometer.

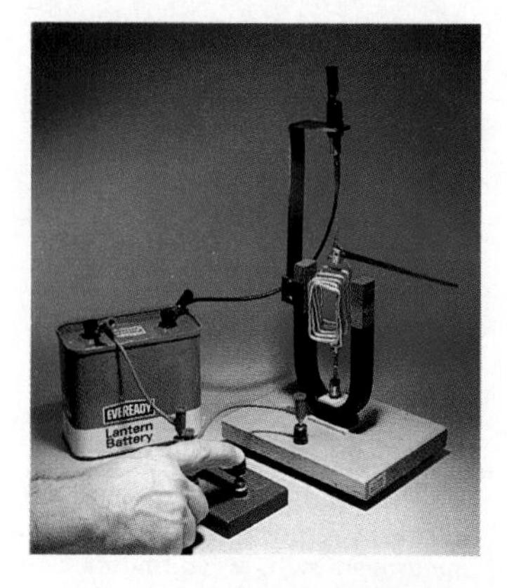


Large-scale model of a galvanometer movement. Why does the coil rotate about the vertical axis after the switch is closed? (Courtesy of Henry Leap and Jim Lehman)

times called the *shunt resistor*, must be very small compared to the resistance of the galvanometer so that most of the current to be measured passes through the shunt resistor. For example, if you wish to measure a current of 2 A with a galvanometer whose resistance is 60 Ω, the shunt resistance should have a value of about 0.03 Ω.

A galvanometer can also be used as a voltmeter by adding an external resistor, $R_s$, in series with it, as in Figure 28.22b. In this case, the external resistor must have a value which is very large compared to the resistance of the galvanometer. This will insure that the galvanometer will not significantly alter the voltage to be measured. For example, if you wish to measure a maximum voltage of 100 V with a galvanometer whose resistance is 60 Ω, the external series resistor should have a value of about $10^5$ Ω.

When a voltmeter is constructed with several available ranges, one selects various values of $R_s$ by using a switch that can be connected to a preselected set of resistors. The required value of $R_s$ increases as the maximum voltage to be measured increases.

## *28.6 THE WHEATSTONE BRIDGE

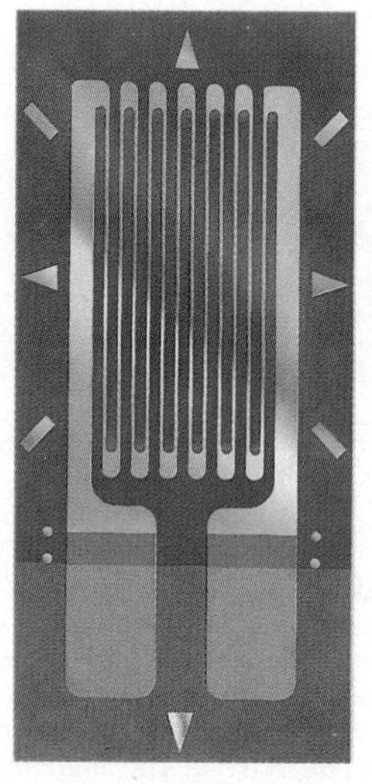

The strain gage, a device used for experimental stress analysis, consists of a thin coiled wire bonded to a flexible plastic backing. Stresses are measured by detecting changes in resistance of the coil as the strip bends. Resistance measurements are made with the gage as one element of a Wheatstone bridge. These devices are commonly used in modern electronic balances to measure the mass of an object.

Unknown resistances can be accurately measured using a circuit known as a **Wheatstone bridge** (Fig. 28.23). This circuit consists of the unknown resistance, $R_x$, three known resistors, $R_1$, $R_2$, and $R_3$ (where $R_1$ is a calibrated variable resistor), a galvanometer, and a source of emf. The principle of its operation is quite simple. The known resistor $R_1$ is varied until the galvanometer reading is zero, that is, until there is no current from $a$ to $b$. Under this condition the bridge is said to be balanced. Since the potential at point $a$ must equal the potential at point $b$ when the bridge is balanced, the potential difference across $R_1$ must equal the potential difference across $R_2$. Likewise, the potential difference across $R_3$ must equal the potential difference across $R_x$. From these considerations, we see that

$$(1) \qquad I_1R_1 = I_2R_2$$

$$(2) \qquad I_1R_3 = I_2R_x$$

Dividing (1) by (2) eliminates the currents, and solving for $R_x$ we find

$$R_x = \frac{R_2R_3}{R_1} \tag{28.19}$$

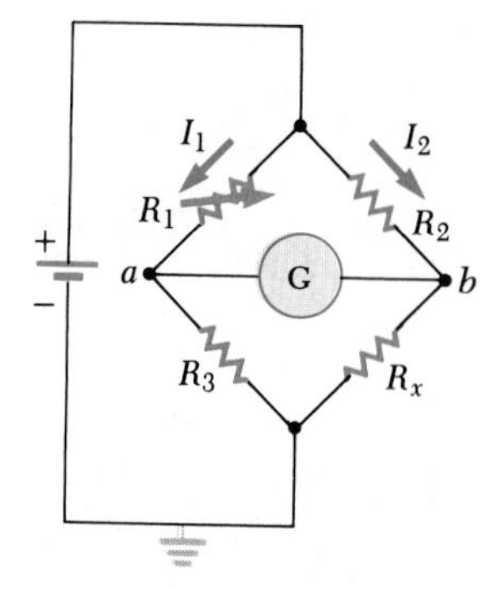


**Figure 28.23** Circuit diagram for a Wheatstone bridge. This circuit is often used to measure an unknown resistance $R_x$ in terms of known resistances $R_1$, $R_2$, and $R_3$. When the bridge is balanced, there is no current in the galvanometer.

Since $R_1$, $R_2$, and $R_3$ are known quantities, $R_x$ can be calculated. There are a number of similar devices that use the null measurement, such as a capacitance bridge (used to measure unknown capacitances). These devices do not require the use of calibrated meters and can be used with any source of emf.

When very high resistances are to be measured (above $10^5\ \Omega$), the Wheatstone bridge method becomes difficult for technical reasons. As a result of recent advances in the technology of such solid state devices as the field-effect transistor, modern electronic instruments are capable of measuring resistances as high as $10^{12}\ \Omega$. Such instruments are designed to have an extremely high effective resistance between their input terminals. For example, input resistances of $10^{10}\ \Omega$ are common in most digital multimeters.

Voltages, currents, and resistances are frequently measured by digital multimeters like the one shown in this photograph. (Courtesy of Henry Leap and Jim Lehman)

## *28.7 THE POTENTIOMETER

A **potentiometer** is a circuit that is used to measure an unknown emf, $\mathcal{E}_x$, by comparison with a known emf. Figure 28.24 shows the essential components of the potentiometer. Point $d$ represents a sliding contact used to vary the resistance (and hence the potential difference) between points $a$ and $d$. In a common version of the potentiometer, called a **slide-wire potentiometer,** the variable resistor is a wire with the contact point $d$ at some position on the wire. The other required components in this circuit are a galvanometer, a power source with emf $\mathcal{E}_0$, a standard reference battery, and the unknown emf, $\mathcal{E}_x$.

With the currents in the directions shown in Fig. 28.24, we see from Kirchhoff's first rule that the current through the resistor $R_x$ is $I - I_x$, where $I$ is the current in the lower branch (through the battery of emf $\mathcal{E}_0$) and $I_x$ is the current in the upper branch. Kirchhoff's second rule applied to loop $abcd$ gives

$$-\mathcal{E}_x + (I - I_x)R_x = 0$$

where $R_x$ is the resistance between points $a$ and $d$. The sliding contact at $d$ is now adjusted until the galvanometer reads zero (a balanced circuit). Under this condition, the current in the galvanometer and in the unknown cell is *zero* ($I_x = 0$), and the potential difference between $a$ and $d$ equals the unknown emf, $\mathcal{E}_x$. That is,

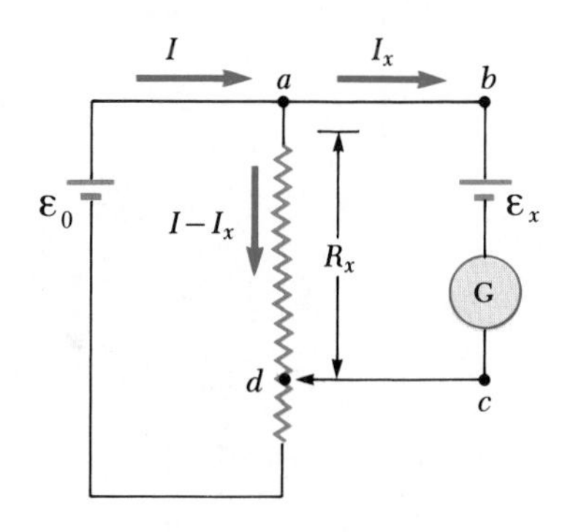


**Figure 28.24** Circuit diagram for a potentiometer. The circuit is used to measure an unknown emf $\mathcal{E}_x$ in terms of a known emf $\mathcal{E}_s$, provided by a standard cell.

$$\mathcal{E}_x = IR_x$$

Next, the cell of unknown emf is replaced by a standard cell of known emf, $\mathcal{E}_s$, and the above procedure is repeated. That is, the moving contact at $d$ is varied until a balance is obtained. If $R_s$ is the resistance between $a$ and $d$ when balance is achieved, then

$$\mathcal{E}_s = IR_s$$

where it is assumed that $I$ remains the same.

Combining this expression with the previous equation, $\mathcal{E}_x = IR_x$, we see that

$$\mathcal{E}_x = \frac{R_x}{R_s}\mathcal{E}_s \qquad (28.20)$$

This result shows that the unknown emf can be determined from a knowledge of the standard-cell emf and the ratio of the two resistances.

If the resistor is a wire of resistivity $\rho$, its resistance can be varied using sliding contacts to vary the length of the circuit. With the substitutions $R_s = \rho L_s/A$ and $R_x = \rho L_x/A$, Equation 28.20 reduces to

$$\mathcal{E}_x = \frac{L_x}{L_s}\mathcal{E}_s \tag{28.21}$$

According to this result, the unknown emf can be obtained from a measurement of the two wire lengths and the magnitude of the standard emf.

## *28.8 HOUSEHOLD WIRING AND ELECTRICAL SAFETY

Household circuits represent a very practical application of some of the ideas we have presented in this chapter concerning circuit analysis. In our world of electrical appliances, it is useful to understand the power requirements and limitations of conventional electrical systems and the safety measures that should be practiced to prevent accidents.

In a conventional installation, the utilities company distributes electrical power to individual homes with a pair of power lines. Each user is connected in parallel to these lines, as shown in Figure 28.25. The potential difference between these wires is about 120 V. The voltage alternates in time, but for the present discussion we shall assume a steady direct current (dc) voltage. (Alternating voltages and currents will be discussed in Chapter 33.) One of the wires is connected to ground, and the potential on the "live" wire oscillates relative to ground.[2]

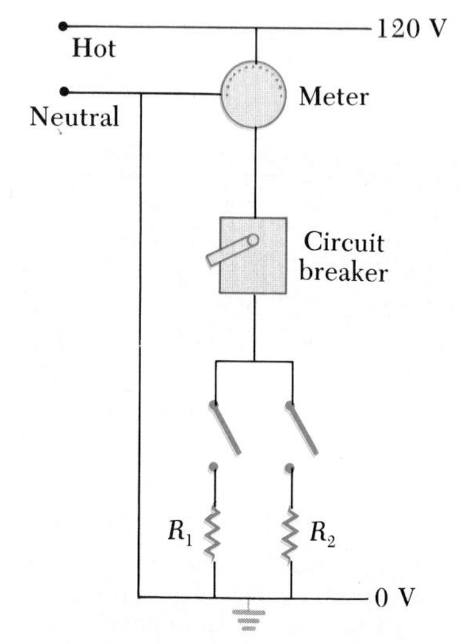


**Figure 28.25** Wiring diagram for a household circuit. The resistances $R_1$ and $R_2$ represent appliances or other electric devices which operate with an applied voltage of 120 V.

A meter and circuit breaker (or in older installations, a fuse) are connected in series with the wire entering the house as indicated in Figure 28.25. The circuit breaker is a device that protects against too large a current, which can cause overheating and fires. When the current exceeds some safe value (typically 15 A or 30 A), the circuit breaker disconnects the voltage source from the load. Some circuit breakers make use of the principle of the bimetallic strip discussed in Chapter 19.

The wire and circuit breaker are carefully selected to meet the current demands for that circuit. If a circuit is to carry currents as large as 30 A, a heavy wire and appropriate circuit breaker must be selected to handle this current. Other individual household circuits, which are normally used to power lamps and small appliances, often require only 15 A. Therefore, each circuit has its own circuit breaker to accommodate various load conditions.

As an example, consider a circuit in which a toaster, a microwave oven, and a heater are in the same circuit (corresponding to $R_1, R_2, \ldots$ in Fig. 28.25). We can calculate the current through each appliance using the expression $P = IV$. The toaster, rated at 1000 W, would draw a current of 1000/120 = 8.33 A. The microwave oven, rated at 800 W, would draw a current of 6.67 A, and the electric heater, rated at 1300 W, would draw a current of 10.8 A. If the three appliances are operated simultaneously, they will draw a total current of 25.8 A. Therefore, the circuit should be wired to handle at least this much current. In order to accommodate a small additional load, such as a 100-W lamp, a 30-A circuit should be installed. Alternatively, one could

[2] The phrase *live wire* is common jargon for a conductor whose potential is above or below ground.

operate the toaster and microwave oven on one 20-A circuit and the heater on a separate 20-A circuit.

Many heavy-duty appliances, such as electric ranges and clothes dryers, require 240 V for their operation. The power company supplies this voltage by providing a third live wire, which is 120 V *below* ground potential (Fig. 28.26). Therefore, the potential difference between this wire and the other live wire (which is 120 V above ground potential) is 240 V. An appliance that operates from a 240-V line requires half the current of one operating from a 120-V line; therefore smaller wires can be used in the higher-voltage circuit without overheating becoming a problem.

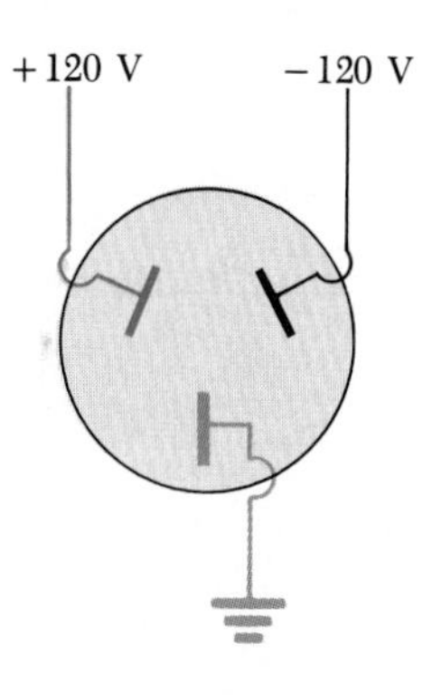


**Figure 28.26** Power connections for a 240-V appliance.

### Electrical Safety

When the live wire of an electrical outlet is connected directly to ground, the circuit is completed and a short-circuit condition exists. When this happens accidentally, a properly operating circuit breaker will "break" the circuit. On the other hand, a person can be electrocuted by touching the live wire of a frayed cord or other exposed conductors while in contact with ground. An exceptionally good ground contact might be made either by the person's touching a water pipe (which is normally at ground potential) or by standing on ground with wet feet, since water is not a good insulator. Such situations should be avoided at all costs.

Electrical shock can result in fatal burns, or it can cause the muscles of vital organs, such as the heart, to malfunction. The degree of damage to the body depends on the magnitude of the current, the length of time it acts, the location of the contact, and the part of the body through which the current passes. Currents of 5 mA or less can cause a sensation of shock, but ordinarily do little or no damage. If the current is larger than about 10 mA, the hand muscles contract and the person may be unable to release the live wire. If a current of about 100 mA passes through the body for only a few seconds, the result could be fatal. Such large currents will paralyze the respiratory muscles and prevent breathing. In some cases, currents of about 1 A through the body can produce serious (and sometimes fatal) burns. In practice, no contact with live wires should be regarded as safe.

Many 120-V outlets are designed to take a three-pronged power cord. (This feature is required in all new electrical installations.) One of these prongs is the live wire and two are common with ground. The additional ground connection is provided as a safety feature. Many appliances contain a three-pronged 120-V power cord with one of the ground wires connected directly to the casing of the appliance. If the live wire is accidentally shorted to the casing (which often occurs when the wire insulation wears off), the current will take the low-resistance path through the appliance to ground. In contrast, if the casing of the appliance is not properly grounded and a short occurs, anyone in contact with the appliance will experience an electric shock since his or her body will provide a low-resistance path to ground.

Special power outlets called ground-fault interrupters (GFI's) are now being used in kitchens, bathrooms, basements, and other hazardous areas of new homes. These devices are designed to protect persons from electrical shock by sensing small currents ($\approx 5$ mA) leaking to ground. When an excessive leakage current is detected, the current is shut off (interrupted) in less than one millisecond.

## SUMMARY

The **emf** of a battery is equal to the voltage across its terminals when the current is zero. That is, the emf is equivalent to the open-circuit voltage of the battery.

The **equivalent resistance** of a set of resistors connected in **series** is given by

Resistors in series

$$R_{eq} = R_1 + R_2 + R_3 + \cdots \tag{28.6}$$

The **equivalent resistance** of a set of resistors connected in *parallel* is given by

Resistors in parallel

$$\frac{1}{R_{eq}} = \frac{1}{R_1} + \frac{1}{R_2} + \frac{1}{R_3} + \cdots \tag{28.8}$$

Kirchhoff's rules

Complex circuits involving more than one loop are conveniently analyzed using two simple rules called **Kirchhoff's rules:**

1. The sum of the currents entering any junction must equal the sum of the currents leaving that junction.
2. The sum of the potential differences across each element around any closed-circuit loop must be *zero.*

The first rule is a statement of **conservation of charge.** The second rule is equivalent to a statement of **conservation of energy.**

When a resistor is traversed in the direction of the current, the change in potential, $\Delta V$, across the resistor is $-IR$. If a resistor is traversed in the direction opposite the current, $\Delta V = +IR$.

If a source of emf is traversed in the direction of the emf (negative to positive) the change in potential is $+\mathcal{E}$. If it is traversed opposite the emf (positive to negative), the change in potential is $-\mathcal{E}$.

If a capacitor is charged with a battery of emf $\mathcal{E}$ through a resistance $R$, the current in the circuit and charge on the capacitor vary in time according to the expressions

Current versus time

$$I(t) = \frac{\mathcal{E}}{R} e^{-t/RC} \tag{28.13}$$

Charge versus time

$$q(t) = Q[1 - e^{-t/RC}] \tag{28.14}$$

where $Q = C\mathcal{E}$ is the *maximum* charge on the capacitor. The product $RC$ is called the **time constant** of the circuit.

If a charged capacitor is discharged through a resistance $R$, the charge and current decrease exponentially in time according to the expressions

$$q(t) = Q\, e^{-t/RC} \tag{28.17}$$

$$I(t) = I_0\, e^{-t/RC} \tag{28.18}$$

where $I_0 = Q/RC$ is the initial current in the circuit and $Q$ is the initial charge on the capacitor.

A **Wheatstone bridge** is a particular circuit that can be used to measure an unknown resistance.

A **potentiometer** is a circuit that can be used to measure an unknown emf.

## QUESTIONS

1. Explain the difference between load resistance and internal resistance for a battery.
2. Under what condition does the potential difference across the terminals of a battery equal its emf? Can the terminal voltage ever exceed the emf? Explain.
3. Is the direction of current through a battery always from negative to positive on the terminals? Explain.
4. Two different sets of Christmas-tree lights are available. For set A, when one bulb is removed (or burns out), the remaining bulbs remain illuminated. For set B, when one bulb is removed, the remaining bulbs will not operate. Explain the difference in wiring for the two sets of lights.
5. How would you connect resistors in order for the equivalent resistance to be larger than the individual resistances? Give an example.
6. How would you connect resistors in order for the equivalent resistance to be smaller than the individual resistances? Give an example.
7. Given three lightbulbs and a battery, sketch as many different electric circuits as you can.
8. When resistors are connected in series, which of the following will be the same, for each resistor: potential difference, current, power?
9. When resistors are connected in parallel, which of the following will be the same for each resistor: potential difference, current, power?
10. What advantage might there be in using two identical resistors in parallel connected in series with another identical parallel pair, rather than just using a single resistor?
11. Are the two headlights on a car wired in series or in parallel? How can you tell?
12. An incandescent lamp connected to a 120-V source with a short extension cord will provide more illumination than if it were connected to the same source with a very long extension cord. Explain.
13. Embodied in Kirchhoff's rules are two conservation laws. What are they?
14. When can the potential difference across a resistor be positive?
15. With reference to Figure 28.13, suppose the wire between points $g$ and $h$ is replaced by a 10-Ω resistor. Explain why this change will *not* affect the currents calculated in Example 28.8.
16. With reference to Figure 28.27, describe what happens to the light bulb after the switch is closed. Assume the capacitor has a large capacitance and is initially uncharged, and assume that the light will illuminate when connected directly across the battery terminals.
17. What would be the internal resistance of an ideal ammeter and voltmeter? Why would these become ideal meters?

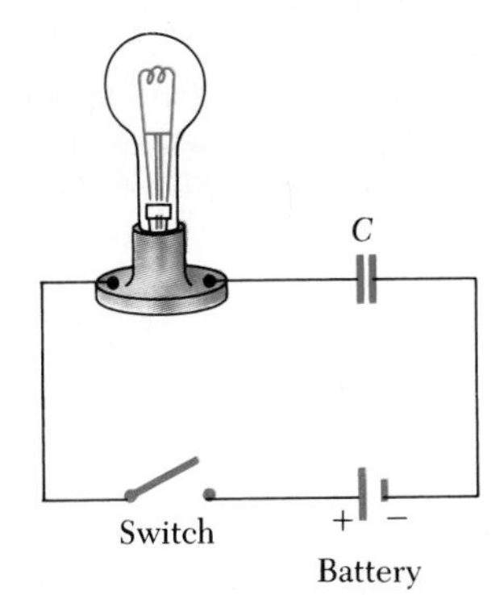


**Figure 28.27** (Question 16).

18. What are a Wheatstone bridge and potentiometer used to measure?
19. Although the internal resistance of the unknown and known emfs was neglected in the treatment of the potentiometer (Section 28.7), it is really not necessary to make this assumption. Explain why the internal resistances play no role in this measurement.
20. Why is it dangerous to turn on a light when you are in the bathtub?
21. Why is it possible for a bird to sit on a high-voltage wire without being electrocuted?
22. Suppose you fall from a building and on the way down grab a high-voltage wire. Assuming that the wire holds you, will you be electrocuted? If the wire then breaks, should you continue to hold onto an end of the wire as you fall?
23. Would a fuse work successfully if it were placed in parallel with the device it is supposed to protect?
24. What advantage does 120-V operation offer over 240 V? What disadvantages?
25. When electricians work with potentially live wires, they often use the backs of their hands or fingers to move wires. Why do you suppose they use this technique?
26. What procedure would you use to try to save a person who is "frozen" to a live high-voltage wire without endangering your own life?
27. At what levels of current do you experience (a) the sensation of shock, (b) involuntary muscle contractions, (c) paralysis of respiratory muscle, and (d) serious and possibly fatal burns? In practice what current level is regarded as safe?
28. If it is the current flowing through the body that determines how serious a shock will be, why do we see warnings of high voltage rather than high current near electric equipment?
29. Suppose you are flying a kite when it strikes a high-voltage wire. What factors determine how great a shock you receive?

30. A series circuit consists of three identical lamps connected to a battery as in the circuit shown at the right. When the switch *S* is closed, (a) what happens to the intensities of lamps *A* and *B*? (b) What happens to the intensity of lamp *C*? (c) What happens to the current in the circuit? (d) What happens to the voltage drop across the three lamps? (e) Does the power dissipated in the circuit increase, decrease, or remain the same?

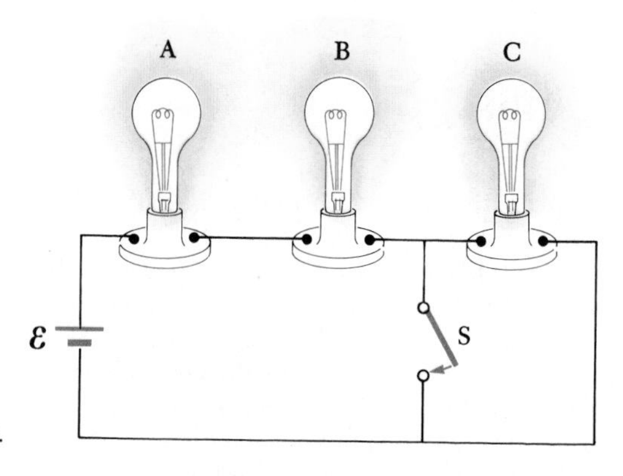


(Question 30).

## PROBLEMS

### Section 28.1 Electromotive Force

1. A battery with an emf of 12 V and internal resistance of 0.9 Ω is connected across a load resistor *R*. If the current in the circuit is 1.4 A, what is the value of *R*?
2. What power is dissipated in the internal resistance of the battery in the circuit described in Problem 1?
3. (a) What is the current in a 5.6-Ω resistor connected to a battery with an 0.2-Ω internal resistance if the terminal voltage of the battery is 10 V? (b) What is the emf of the battery?
4. If the emf of a battery is 15 V and a current of 60 A is measured when the battery is shorted, what is the internal resistance of the battery?
5. The current in a loop circuit that has a resistance of $R_1$ is 2 A. The current is reduced to 1.6 A when an additional resistor $R_2 = 3\ \Omega$ is added in series with $R_1$. What is the value of $R_1$?
6. A typical fresh AA dry cell has an emf of 1.50 V and an internal resistance of 0.311 Ω. (a) Find the terminal voltage of the battery when it supplies 58 mA to a circuit. (b) What is the resistance *R* of the external circuit?
7. A battery has an emf of 15 V. The terminal voltage of the battery is 11.6 V when it is delivering 20 W of power to an external load resistor *R*. (a) What is the value of *R*? (b) What is the internal resistance of the battery?
8. What potential difference will be measured across a 18-Ω load resistor when it is connected across a battery of emf 5 V and internal resistance 0.45 Ω?
9. Two 1.50-V batteries—with their positive terminals in the same direction—are inserted in series into the barrel of a flashlight. One battery has an internal resistance of 0.255 Ω, the other an internal resistance of 0.153 Ω. When the switch is closed, a current of 0.6 A occurs in the lamp. (a) What is the lamp's resistance? (b) What fraction of the power dissipated is dissipated in the batteries?

### Section 28.2 Resistors in Series and in Parallel

10. Two circuit elements with fixed resistances $R_1$ and $R_2$ are connected in *series* with a 6-V battery and a switch. The battery has an internal resistance of 5 Ω, $R_1 = 132\ \Omega$, and $R_2 = 56\ \Omega$. (a) What is the current through $R_1$ when the switch is closed? (b) What is the voltage across $R_2$ when the switch is closed?
11. Using only three resistors—2 Ω, 3 Ω, and 4 Ω—find all 17 different resistance values that may be obtained by various combinations of one or more resistors. Tabulate the values in order of increasing resistance.
12. A television repairman needs a 100-Ω resistor to repair a malfunctioning set. He is temporarily out of resistors of this value. All he has in his tool box is a 500-Ω resistor and two 250-Ω resistors. How can the desired resistance be obtained from the resistors on hand?
13. The current in a circuit is tripled by connecting a 500-Ω resistor in parallel with the resistance of the circuit. Determine the resistance of the circuit in the absence of the 500-Ω resistor.
14. Find the equivalent resistance between points *a* and *b* in Figure 28.28.

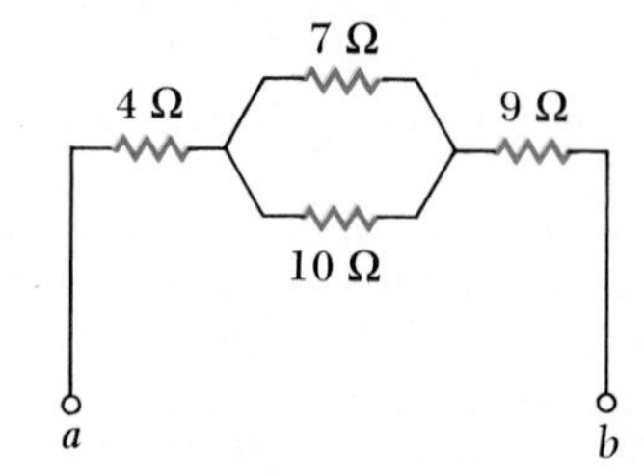


**Figure 28.28** (Problems 14 and 15).

15. A potential difference of 34 V is applied between points *a* and *b* in Figure 28.28. Calculate the current in each resistor.
16. Find the equivalent resistance between points *a* and *b* in Figure 28.29.

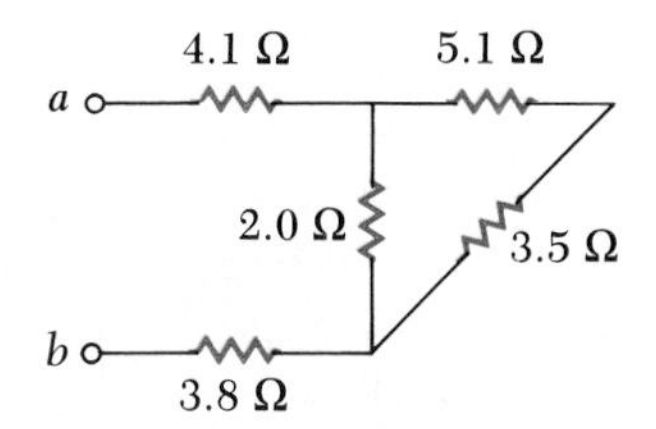


**Figure 28.29** (Problem 16).

17. Evaluate the effective resistance of the network of identical resistors, each having resistance $R$, shown in Figure 28.30.

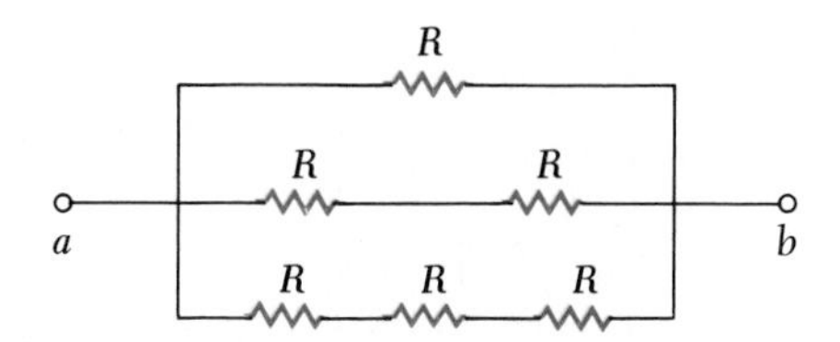


**Figure 28.30** (Problem 17).

18. In Figures 28.4 and 28.5, let $R_1 = 11\ \Omega$, $R_2 = 22\ \Omega$, and the battery have a terminal voltage of 33 V. (a) In the parallel circuit shown in Figure 28.5, which resistor uses more power? (b) Verify that the sum of the power ($I^2R$) used by each resistor equals the power supplied by the battery ($IV$). (c) In the series circuit, Figure 28.4, which resistor uses more power? (d) Verify that the sum of the power ($I^2R$) used by each resistor equals the power supplied by the battery ($P = IV$). (e) Which circuit configuration uses more power?
19. Calculate the power dissipated in each resistor in the circuit of Figure 28.31.

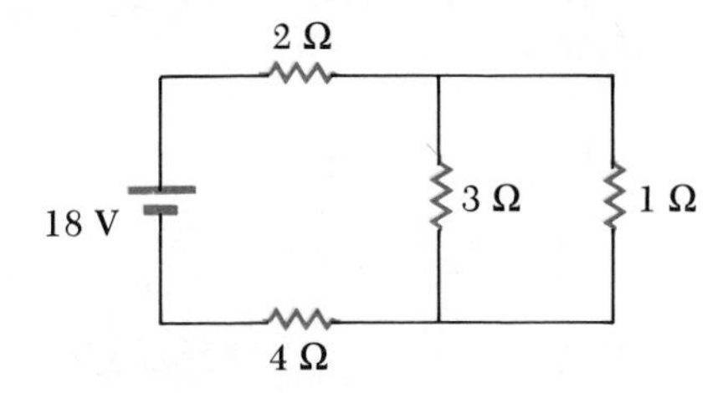


**Figure 28.31** (Problem 19).

20. Determine the equivalent resistance between the terminals $a$ and $b$ for the network illustrated in Figure 28.32.
21. Consider the circuit shown in Figure 28.33. Find (a) the current in the 20-Ω resistor and (b) the potential difference between points $a$ and $b$.

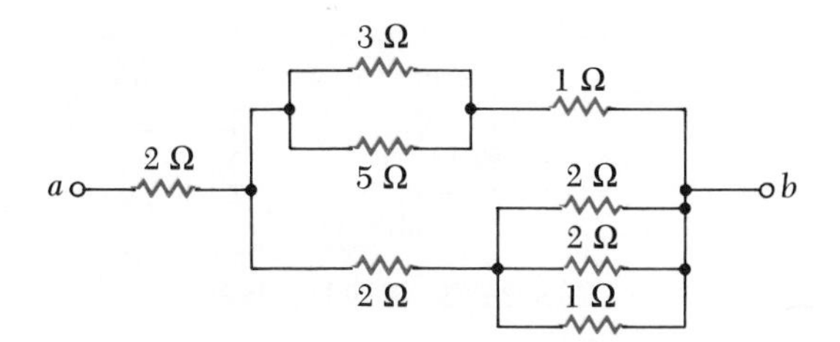


**Figure 28.32** (Problem 20).

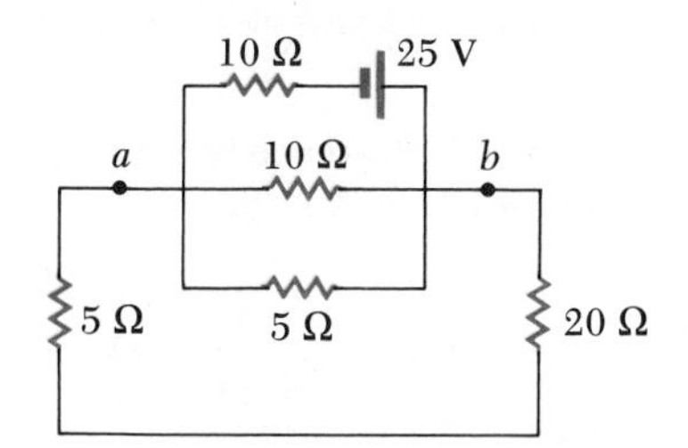


**Figure 28.33** (Problem 21).

22. In Figure 28.34, each resistor has a resistance of 1 Ω. Suppose that a given current $I$ enters at $a$ and comes out at $b$. By utilizing arguments based upon the symmetry of the network, show that the equivalent resistance $R_{eq}$ of the network from $a$ to $b$ is $\frac{2}{3}$ Ω. (*Hint:* What would the resistance be if the vertical resistors were absent?)

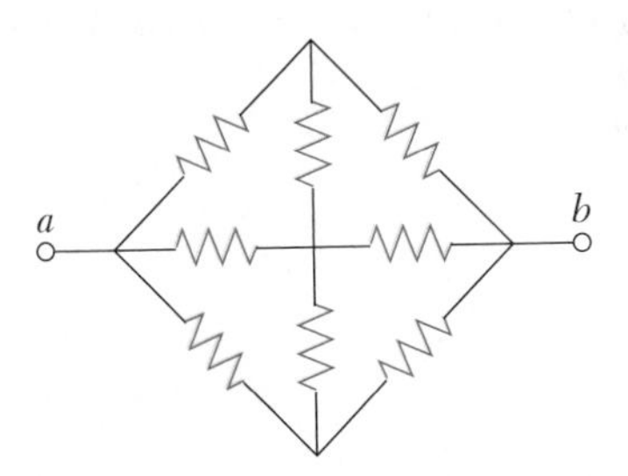


**Figure 28.34** (Problem 22).

23. Two resistors connected in series have an equivalent (combined) resistance of 690 Ω. When they are connected in parallel, their equivalent resistance is 150 Ω. Find the resistance of each of the resistors.
24. Three 100-Ω resistors are connected as shown in Figure 28.35. The maximum power that can be dissipated in any one of the resistors is 25 W. (a) What is the maximum voltage that can be applied to the terminals $a$ and $b$? (b) For the voltage determined in (a), what is the power dissipation in each resistor? What is the total power dissipation?

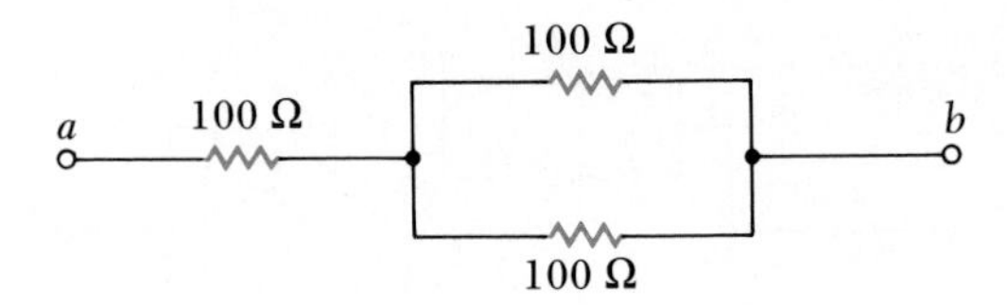


**Figure 28.35** (Problem 24).

Section 28.3 Kirchhoff's Rules

*The currents are not necessarily in the direction shown for some circuits.*

**25.** Find the potential difference between points *a* and *b* in the circuit in Figure 28.36.

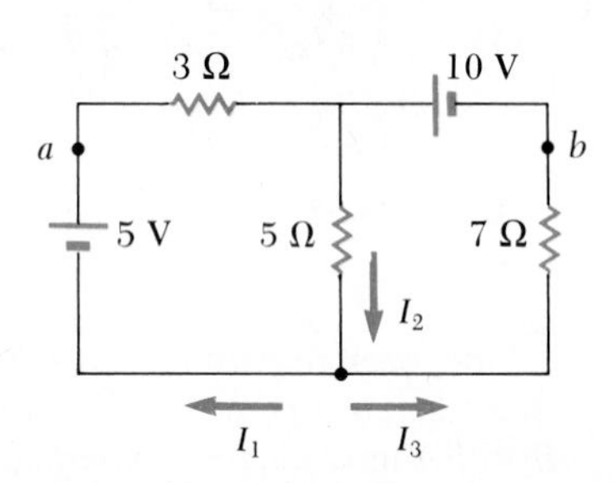


**Figure 28.36** (Problems 25 and 26).

**26.** Find the currents $I_1$, $I_2$, and $I_3$ in the circuit shown in Figure 28.36.

27. Determine the current in each of the branches of the circuit shown in Figure 28.37.

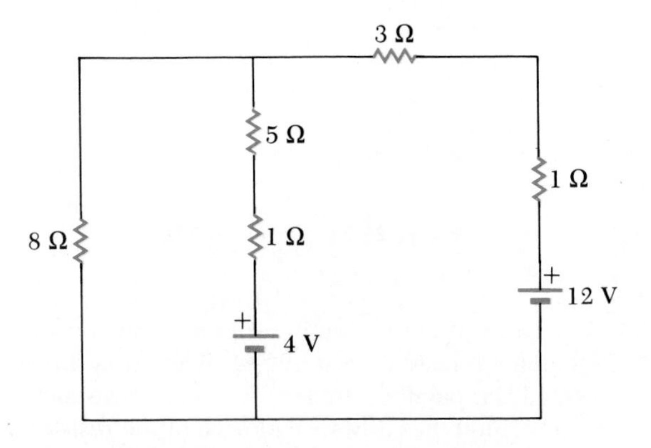


**Figure 28.37** (Problem 27).

**28.** Two batteries and two resistors are connected in the single loop shown in Figure 28.38. Given that the potential at point *d* equals zero, determine the potentials at points (a) *a*, (b) *b*, (c) *c*.

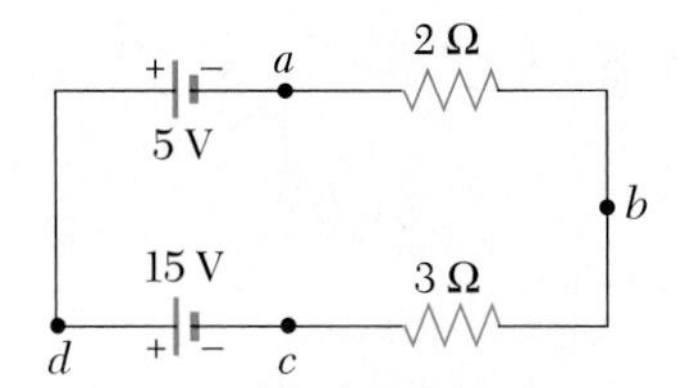


**Figure 28.38** (Problem 28).

**29.** A dead battery is "charged" by connecting it to the live battery of another car (Fig. 28.39). Determine the current in the starter and in the dead battery.

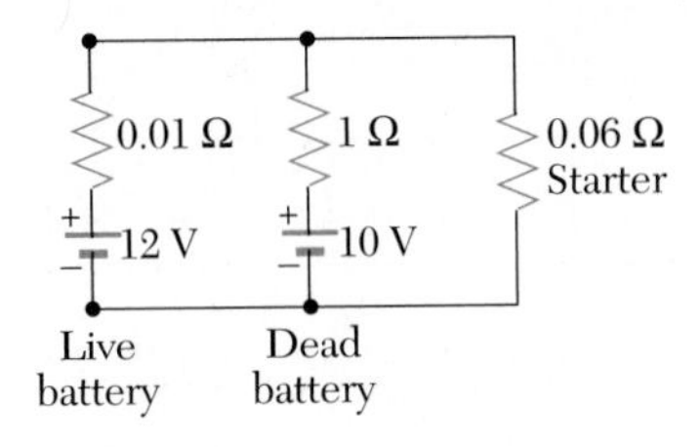


**Figure 28.39** (Problem 29).

30. For the network shown in Figure 28.40, show that the resistance $R_{ab} = (27/17)\ \Omega$.

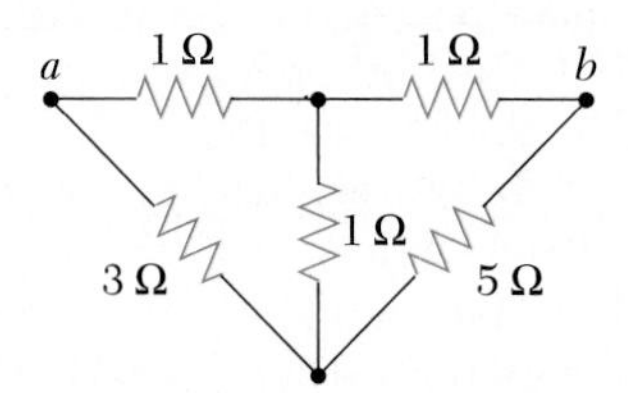


**Figure 28.40** (Problem 30).

31. Calculate each of the unknown currents $I_1$, $I_2$, and $I_3$ for the circuit of Figure 28.41.

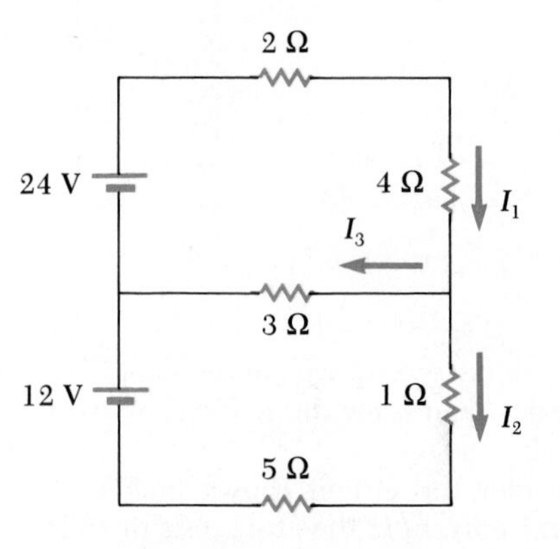


**Figure 28.41** (Problem 31).

**32.** The ammeter in the circuit shown in Figure 28.42 reads 2 A. Find the currents $I_1$ and $I_2$ and the value of $\mathcal{E}$.

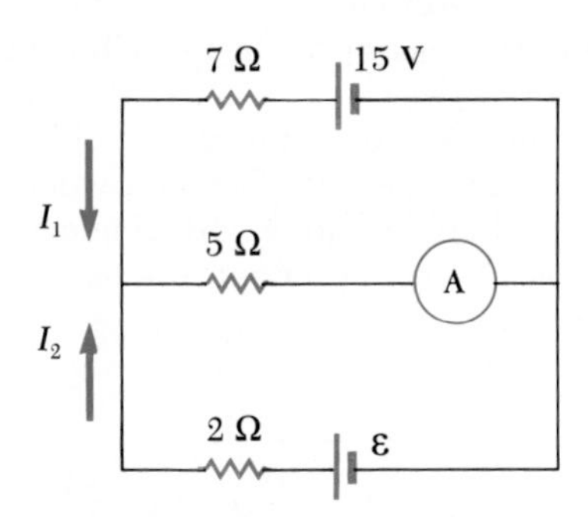


**Figure 28.42** (Problem 32).

**33.** Using Kirchhoff's rules, (a) find the current in each of the resistors in the circuit shown in Figure 28.43. (b) Find the potential difference between points $c$ and $f$. Which is at the higher potential?

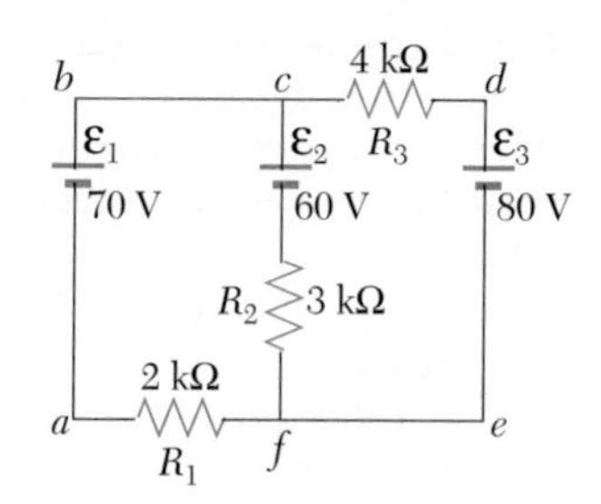


**Figure 28.43** (Problem 33).

**34.** Consider the circuit shown in Figure 28.44. Find the value of $I_1$, $I_2$, and $I_3$.

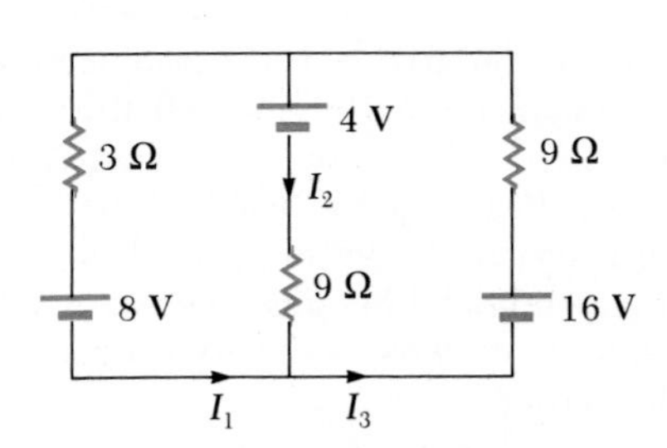


**Figure 28.44** (Problems 34 and 35).

**35.** (a) Find the value of $I_1$ and $I_3$ in the circuit of Figure 28.44 if the 4-V battery is replaced by a 5-$\mu$F capacitor. (b) Determine the charge on the 5-$\mu$F capacitor.

**36.** In Figure 28.45, calculate (a) the equivalent resistance of the network outside the battery, (b) the current through the battery, and (c) the current in the 6-Ω resistor.

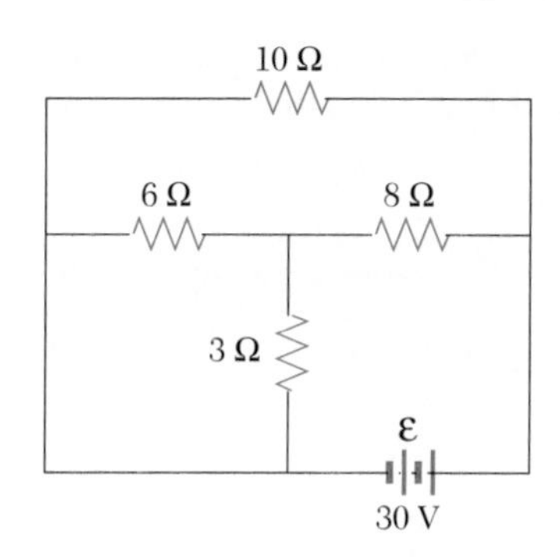


**Figure 28.45** (Problem 36).

**37.** For the circuit shown in Figure 28.46, calculate (a) the current in the 2-Ω resistor and (b) the potential difference between points $a$ and $b$.

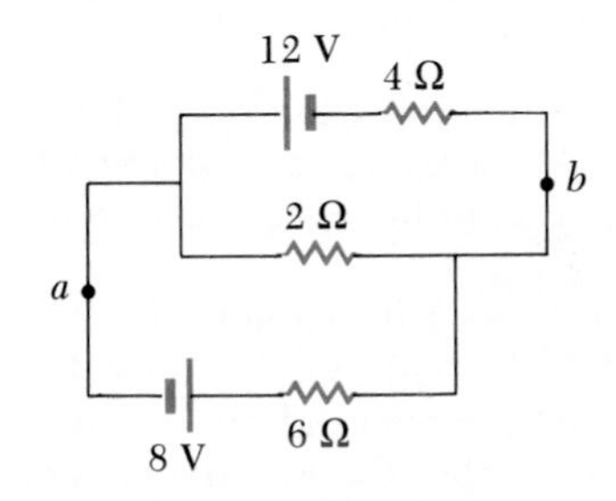


**Figure 28.46** (Problem 37).

**38.** Consider the circuit shown in Figure 28.47. Find the current in each of the resistors using Kirchhoff's rules.

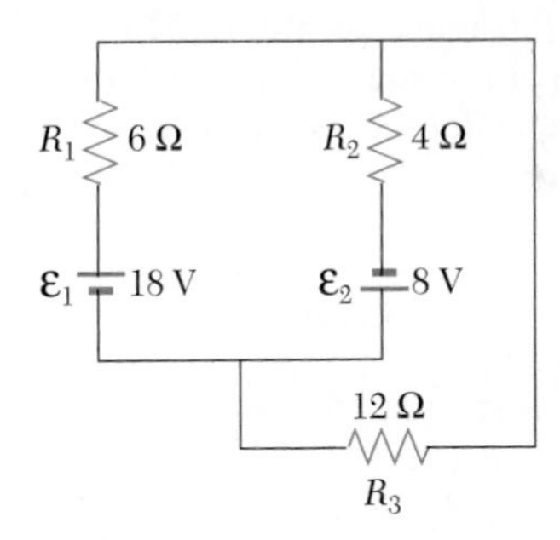


**Figure 28.47** (Problem 38).

**39.** Calculate the power dissipated in each resistor in the circuit shown in Figure 28.48.

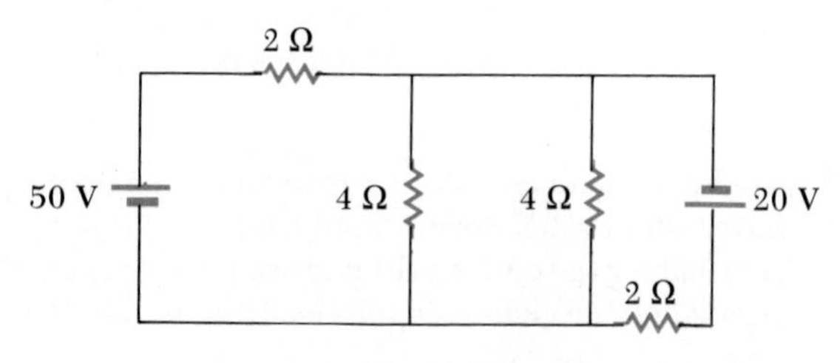


**Figure 28.48** (Problem 39).

40. Calculate the power dissipated in each resistor in the circuit of Figure 28.49.

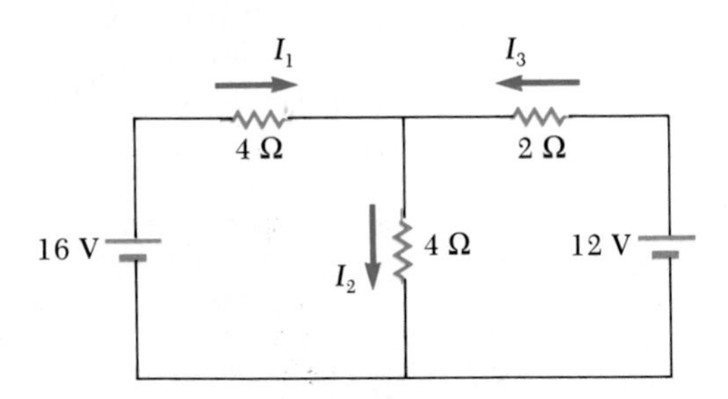


**Figure 28.49** (Problem 40).

## Section 28.4 *RC* Circuits

41. A fully-charged capacitor has 12 J of energy stored. How much energy remains stored when its charge has decreased to half its original value during a discharge?

42. Consider a series *RC* circuit (Fig. 28.14) for which $R = 1\ \text{M}\Omega$, $C = 5\ \mu\text{F}$, and $\mathcal{E} = 30$ V. Find (a) the time constant of the circuit and (b) the *maximum* charge on the capacitor after the switch is closed.

43. The switch in the *RC* circuit described in Problem 42 is closed at $t = 0$. Find the current in the resistor $R$ at a time 10 s after the switch is closed.

44. At $t = 0$, an uncharged capacitor of capacitance $C$ is connected through a resistance $R$ to a battery of constant emf $\mathcal{E}$ (Figure 28.50). (a) How long does it take for the capacitor to reach one half of its final charge? (b) How long does it take for the capacitor to become fully charged?

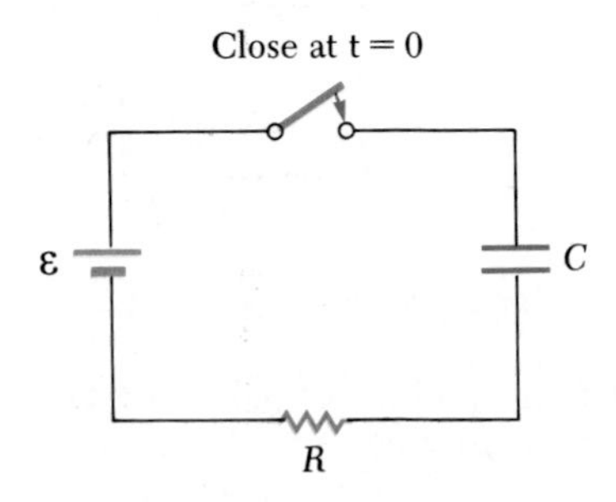


**Figure 28.50** (Problem 44).

45. A 4-MΩ resistor and a 3-μF capacitor are connected in series with a 12-V power supply. (a) What is the time constant for the circuit? (b) Express the current in the circuit and the charge on the capacitor as functions of time.

46. A 750-pF capacitor has an initial charge of 6 μC. It is then connected to a 150-MΩ resistor and allowed to discharge through the resistor. (a) What is the time constant for the circuit? (b) Express the current in the circuit and the charge on the capacitor as functions of time.

47. The circuit has been connected as shown in Figure 28.51 a "long" time. (a) What is the voltage across the capacitor? (b) If the battery is disconnected, how long does it take for the capacitor to discharge to 1/10 of its initial voltage?

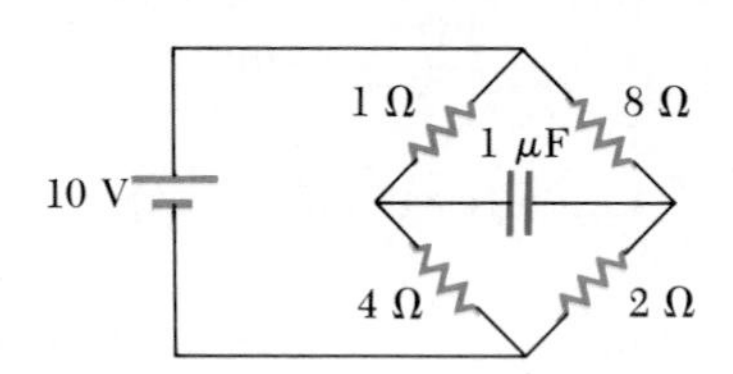


**Figure 28.51** (Problem 47).

48. A $2 \times 10^{-3}$-μF capacitor with an initial charge of 5.1 μC is discharged through a 1300-Ω resistor. (a) Calculate the current through the resistor 9 μs after the resistor is connected across the terminals of the capacitor. (b) What charge remains on the capacitor after 8 μs? (c) What is the maximum current through the resistor?

49. Consider the capacitor-resistor combination described in Problem 48. (a) How much energy is stored initially in the charged capacitor? (b) If the capacitor is completely discharged through the resistor, how much energy will be dissipated as heat in the resistor?

50. A capacitor in an *RC* circuit is charged to 60% of its maximum value in 0.9 s. What is the time constant of the circuit?

51. Dielectric materials used in the manufacture of capacitors are characterized by conductivities that are small but not zero. Therefore, a charged capacitor will slowly lose its charge by "leaking" across the dielectric. If a certain 3.6-μF capacitor leaks charge such that the potential difference decreases to half its initial value in 4 s, what is the equivalent resistance of the dielectric?

## °Section 28.5 Electrical Instruments

52. A typical galvanometer, which requires a current of 1.5 mA for full-scale deflection and has a resistance of 75 Ω, may be used to measure currents of much larger values. To enable the measuring of large currents without damage to the sensitive meter, a relatively small shunt resistor is wired in parallel with the meter movement similar to Figure 28.22a. Most of the cur-

rent will then flow through the shunt resistor. Calculate the value of the shunt resistor that enables the meter to be used to measure a current of 1 A at full-scale deflection. (*Hint:* Use Kirchhoff's laws.)

53. The same galvanometer movement as used in the previous problem may be used to measure voltages. In this case a large resistor is wired in series with the meter movement similar to Figure 28.22b, which in effect limits the current that flows through the movement when large voltages are applied. Most of the potential drop occurs across the resistor placed in series. Calculate the value of the resistor that enables the movement to measure an applied voltage of 25 V at full-scale deflection.

54. Consider a galvanometer with an internal resistance of 60 Ω. If the galvanometer deflects full scale when it carries a current of 0.5 mA, what value series resistance must be connected to the galvanometer if this combination is to be used as a voltmeter having a full-scale deflection for a potential difference of 1.0 V?

55. Assume that a galvanometer has an internal resistance of 60 Ω and requires a current of 0.5 mA to produce full-scale deflection. What value of resistance must be connected in parallel with the galvanometer if the combination is to serve as an ammeter with a full-scale deflection for a current of 0.1 A?

56. An ammeter is constructed with a galvanometer (see Fig. 28.22a) that requires a potential difference of 50 mV across the meter movement and a current of 1 mA through the movement to cause a full-scale deflection. Find the shunt resistance $R$ that will produce a full-scale deflection when a current of 5 A enters the ammeter.

57. A galvanometer with a full-scale sensitivity of 1 mA requires a 900-Ω series resistor to make a voltmeter reading full scale when 1 V is measured across the terminals (see Fig. 28.22b). What series resistor is required to make the same galvanometer into a 50-V (full-scale) voltmeter?

58. A current of 2.5 mA causes a given galvanometer movement to deflect full scale. The resistance of the movement is 200 Ω. (a) Show by means of a circuit diagram, using two resistors and three external jacks, how the meter movement may be made into a dual-range voltmeter. (b) Determine the values of the resistors needed to make the high range 0–200 V and the low range 0–20 V. Indicate these values on the diagram.

59. The same meter movement is given as in the previous problem. (a) Show by means of a circuit diagram, using two resistors and three external jacks, how the meter movement may be made into a dual-range ammeter. (b) Determine the values of the resistors needed to make the high range 0–10 A and the low range 0–1 A. Indicate these values on the diagram.

### *Section 28.6 The Wheatstone Bridge

60. A Wheatstone bridge of the type shown in Figure 28.23 is used to make a precise measurement of the resistance of a wire connector. The resistor shown in the circuit as $R_3$ is 1 kΩ. If the bridge is balanced by adjusting $R_1$ such that $R_1 = 2.5R_2$, what is the resistance of the wire connector, $R_x$?

61. Consider the case when the Wheatstone bridge shown in Figure 28.23 is *unbalanced.* Calculate the current through the galvanometer when $R_x = R_3 = 7\ \Omega$, $R_2 = 21\ \Omega$, and $R_1 = 14\ \Omega$. Assume the voltage across the bridge is 70 V, and neglect the galvanometer's resistance.

62. Consider the Wheatstone bridge shown in Figure 28.23. When the Wheatstone bridge is balanced the voltage drop across $R_x$ is 3.2 V and $I_1 = 200\ \mu A$. If the total current drawn from the power supply is 500 μA, what is the resistance, $R_x$?

63. The Wheatstone bridge in Figure 28.23 is balanced when $R_1 = 10\ \Omega$, $R_2 = 20\ \Omega$, and $R_3 = 30\ \Omega$. Calculate the value of $R_x$.

### *Section 28.7 The Potentiometer

64. Consider the potentiometer circuit shown in Figure 28.24. When a standard cell of emf 1.0186 V is used in the circuit, and the resistance between $a$ and $d$ is 36 Ω, the galvanometer reads zero. When the standard cell is replaced by an unknown emf, the galvanometer reads zero when the resistance is adjusted to 48 Ω. What is the value of the unknown emf?

### *Section 28.8 Household Wiring and Electrical Safety

65. An electric heater is rated at 1500 W, a toaster is rated at 750 W, and an electric grill is rated at 1000 W. The three appliances are connected to a common 120-V circuit. (a) How much current does each appliance draw? (b) Is a 25-A circuit sufficient in this situation? Explain.

66. A 1000-W toaster, an 800-W microwave oven, and a 500-W coffee pot are all plugged into the same 120-V outlet. If the circuit is protected by a 20-A fuse, will the fuse blow if all these appliances are used at once?

67. An 8-foot extension cord has two 18-gauge copper wires, each having a diameter of 1.024 mm. (a) How much power does this cord dissipate when carrying a current of 1 A? (b) How much power does this cord dissipate when carrying a current of 10 A?

68. Sometimes aluminum wiring is used instead of copper for economic reasons. According to the National Electrical Code, the maximum allowable current for 12-gauge copper wire with rubber insulation is 20 A. What should be the maximum allowable current in a 12-gauge aluminum wire if it is to dissipate the same power per unit length as the copper wire?

69. A 4-kW heater is wired for 240-V operation with nichrome wire having a total mass $M$. (a) How much current does the heater require? (b) How much current would a 120-V, 4-kW heater require? (c) If a 240-V, 4-kW heater and a 120-V, 4-kW heater have the same length resistance wires in them, how does the mass of the resistance wire in the 120-V heater compare to the mass of the resistance wire in the 240-V heater?

## ADDITIONAL PROBLEMS

70. Calculate the potential difference between the points $a$ and $b$ for the circuit shown in Figure 28.52 and identify which point is at the higher potential.

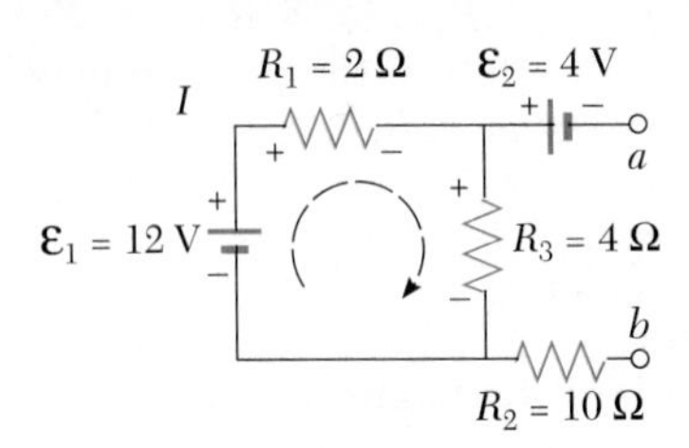


**Figure 28.52** (Problem 70).

71. When two unknown resistors are connected in series with a battery, 225 W are dissipated with a total current of 5 A. For the same total current, 50 W are dissipated when the resistors are connected in parallel. Determine the values of the two resistors.

72. Before the switch is closed in the circuit in Figure 28.53 there is no charge stored by the capacitor. Determine the currents in $R_1$, $R_2$, and $C$ (a) at the instant the switch is closed (that is, $t = 0$), and (b) after the switch is closed for a long period of time (that is, as $t \to \infty$).

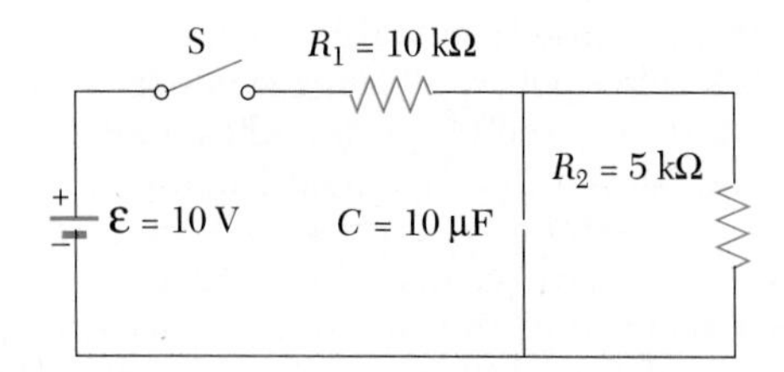


**Figure 28.53** (Problem 72).

73. Three resistors, each of value 3 Ω, are arranged in two different arrangements as shown in Figure 28.54. If the maximum allowable power for each individual resistor is 48 W, calculate the maximum power that can be dissipated by (a) the circuit shown in Figure 28.54a and (b) the circuit shown in Figure 28.54b.

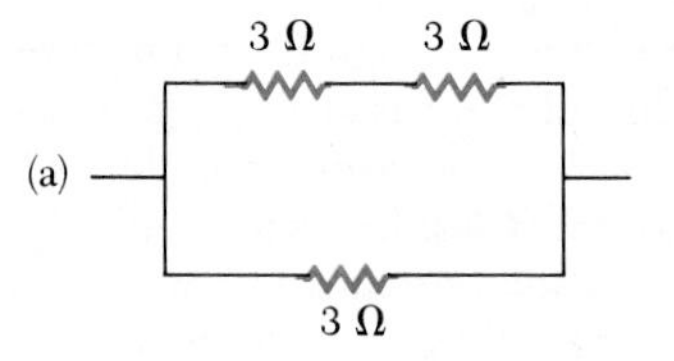


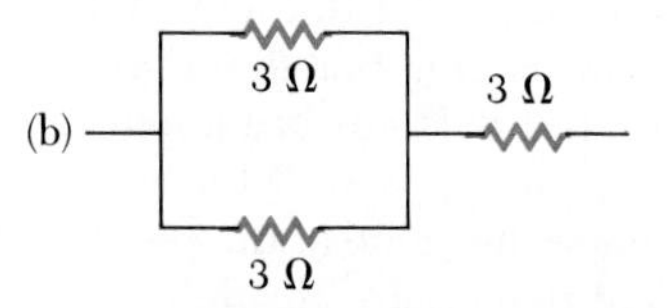


**Figure 28.54** (Problem 73).

74. (a) Calculate the current through the 6-V battery in Figure 28.55. (b) Determine the potential difference between points $a$ and $b$.

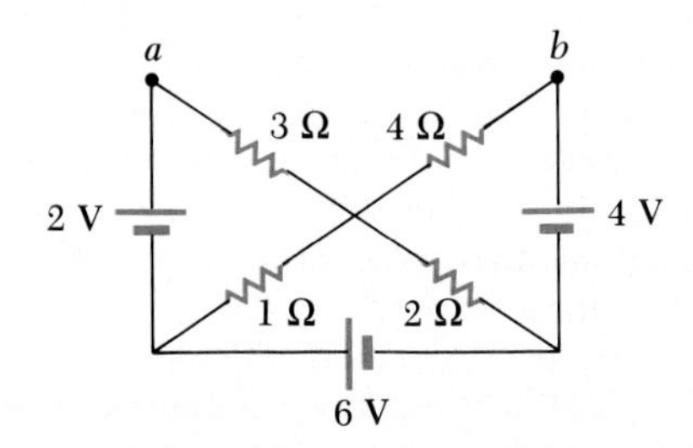


**Figure 28.55** (Problem 74).

75. Three 60-W, 120-V light bulbs are connected across a 120-V power source, as shown in Figure 28.56. Find (a) the total power dissipation in the three light bulbs and (b) the voltage across each of the bulbs. Assume that the resistance of each bulb conforms to Ohm's law (even though in reality the resistance increases markedly with current).

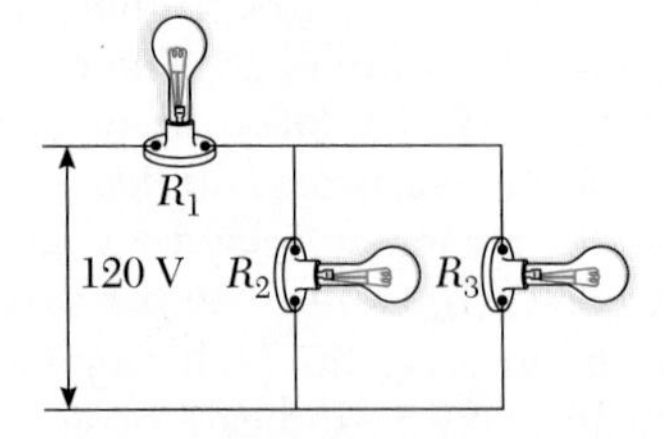


**Figure 28.56** (Problem 75).

76. Consider the circuit shown in Figure 28.57. Calculate (a) the current in the 4-Ω resistor, (b) the potential difference between points $a$ and $b$, (c) the terminal potential difference of the 4-V battery, and (d) the thermal energy expended in the 3-Ω resistor during 10 min of operation of the circuit.

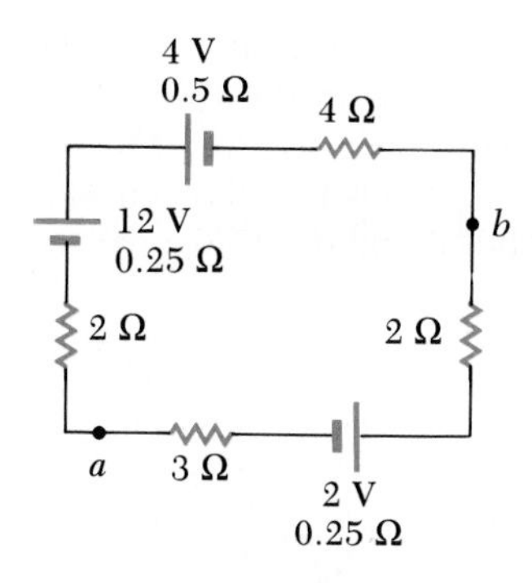


**Figure 28.57** (Problem 76).

77. The value of a resistor $R$ is to be determined using the ammeter-voltmeter setup shown in Figure 28.58. The ammeter has a resistance of 0.5 Ω, and the voltmeter has a resistance of 20 000 Ω. Within what range of actual values of $R$ will the measured values be correct to within 5% if the measurement is made using the circuit shown in (a) Figure 28.58a and (b) Figure 28.58b?

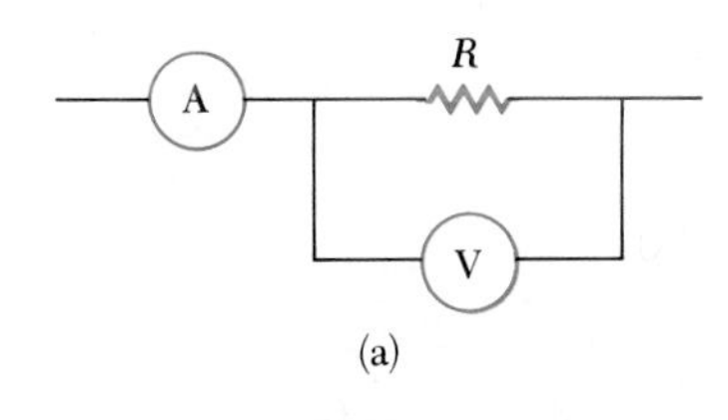


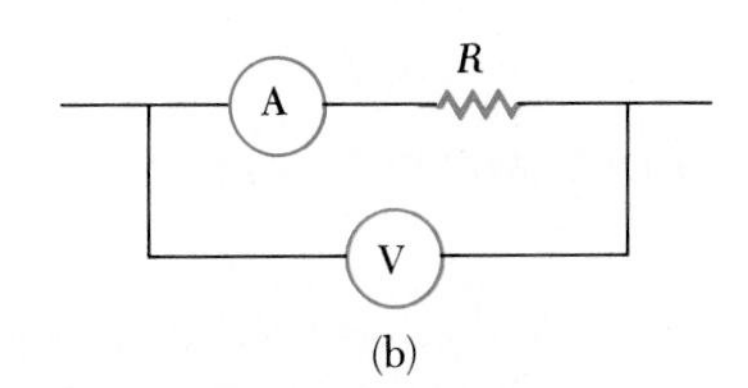


**Figure 28.58** (Problem 77).

78. A dc power supply has an open-circuit emf of 40 V and an internal resistance of 2 Ω. It is used to charge two storage batteries connected in series, each having an emf of 6 V and internal resistance of 0.3 Ω. If the charging current is to be 4 A, (a) what additional resistance should be added in series? (b) Find the power lost in the supply, the batteries, and the added series resistance. (c) How much power is converted to chemical energy in the batteries?
79. A battery has an emf $\mathcal{E}$ and internal resistance $r$. A variable resistor $R$ is connected across the terminals of the battery. Find the value of $R$ such that (a) the potential difference across the terminals is a maximum, (b) the current in the circuit is a maximum, (c) the power delivered to the resistor is a maximum.
80. Consider the circuit shown in Figure 28.59. (a) Calculate the current in the 5-Ω resistor. (b) What power is dissipated by the entire circuit? (c) Determine the potential difference between points $a$ and $b$. Which point is at the higher potential?

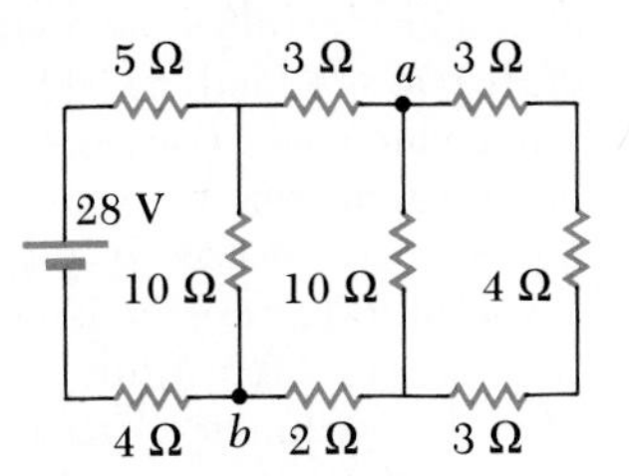


**Figure 28.59** (Problem 80).

81. The values of the components in a simple $RC$ circuit (Fig. 28.14) are as follows: $C = 1\ \mu\text{F}$, $R = 2 \times 10^6\ \Omega$, and $\mathcal{E} = 10$ V. At the instant 10 s after the switch in the circuit is closed, calculate (a) the charge on the capacitor, (b) the current in the resistor, (c) the rate at which energy is being stored in the capacitor, and (d) the rate at which energy is being delivered by the battery.
82. A battery is used to charge a capacitor through a resistor, as in Figure 28.14. Show that in the process of charging the capacitor, half of the energy supplied by the battery is dissipated as heat in the resistor and half is stored in the capacitor.
83. The switch in the circuit shown in Figure 28.60a closes when $V_c \geq 2V/3$ and opens when $V_c \leq V/3$. The voltmeter will show a voltage as plotted in Figure 28.60b. What is the period, $T$, of the waveform in terms of $R_A$, $R_B$, and $C$?

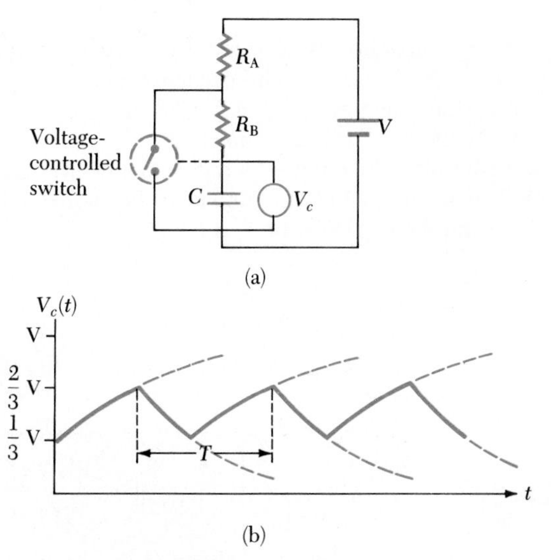


**Figure 28.60** (Problem 83).

84. Design a multirange dc voltmeter that is capable of a full-scale deflection for the following divisions of voltage: (a) 20 V, (b) 50 V, and (c) 100 V. Assume a meter movement which has a coil resistance of 60 Ω and gives a full-scale deflection for a current of 1 mA.

85. Design a multirange dc ammeter that is capable of a full-scale deflection for the following divisions of current: (a) 25 mA, (b) 50 mA, and (c) 100 mA. Assume a meter movement which has a coil resistance of 25 Ω and gives a full-scale deflection for a current of 1 mA.

86. A particular galvanometer serves as a 2-V full-scale voltmeter when a 2500-Ω resistor is connected in series with it. It serves as a 0.5-A full-scale ammeter when a 0.22-Ω resistor is connected in parallel with it. Determine the internal resistance of the galvanometer and the current required to produce full-scale deflection.

87. In Figure 28.61, suppose that the switch has been closed sufficiently long for the capacitor to become fully charged. Find (a) the steady-state current through each resistor and (b) the charge $Q$ on the capacitor. (c) The switch is now opened at $t = 0$. Write an equation for the current $i_{R_2}$ through $R_2$ as a function of time and (d) find the time that it takes for the charge on the capacitor to fall to $\frac{1}{5}$ of its initial value.

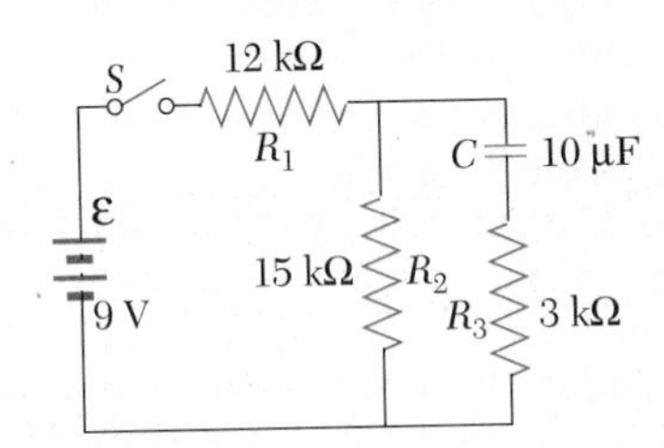


**Figure 28.61** (Problem 87).

88. A 10-$\mu$F capacitor is charged by a 10-V battery through a resistance $R$. The capacitor reaches a potential difference of 4 V in a period of 3 s after the charging began. Find the value of $R$.

89. (a) Determine the charge on the capacitor in Figure 28.62 when $R = 10\ \Omega$. (b) For what value of $R$ will the charge on the capacitor be zero?

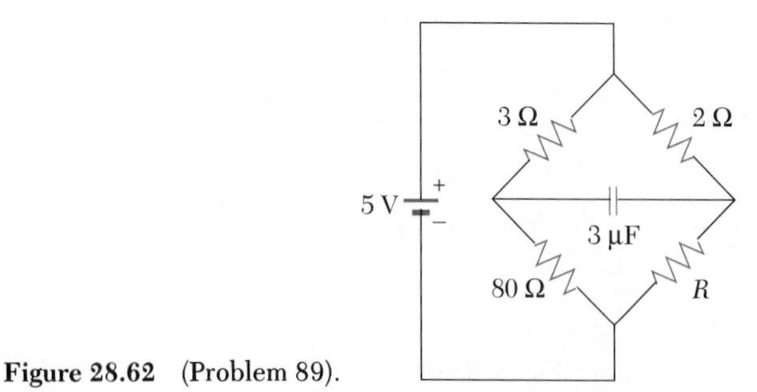


**Figure 28.62** (Problem 89).

90. (a) Using symmetry arguments, show that the current through any resistor in the configuration of Figure 28.63 is either $I/3$ or $I/6$. All resistors have the same resistance $r$. (b) Show that the equivalent resistance between points $A$ and $B$ is $(5/6)r$.

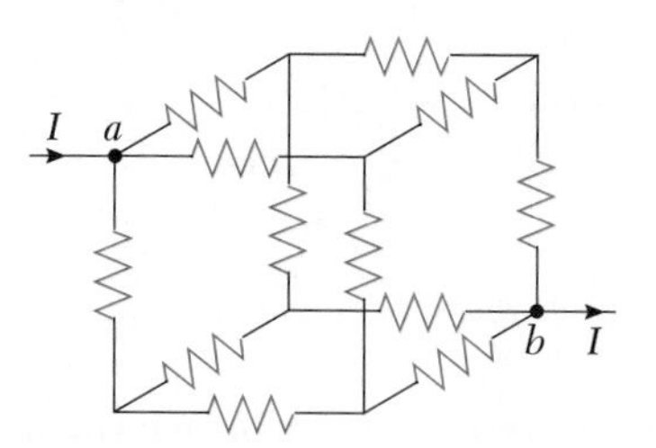


**Figure 28.63** (Problem 90).

Δ91. The circuit shown in Figure 28.64 is set up in the laboratory to measure an unknown capacitance $C$ using a voltmeter of resistance $R = 10\ \text{M}\Omega$ and a battery whose emf is 6.19 V. The data given in the table below is the measured voltage across the capacitor as a function of time, where $t = 0$ represents the time the switch is open. (a) Construct a graph of $\ln(\mathcal{E}/V)$ versus $t$, and perform a linear least squares fit to the data. (b) From the slope of your graph obtain a value for the time constant of the circuit, and a value for the capacitance.

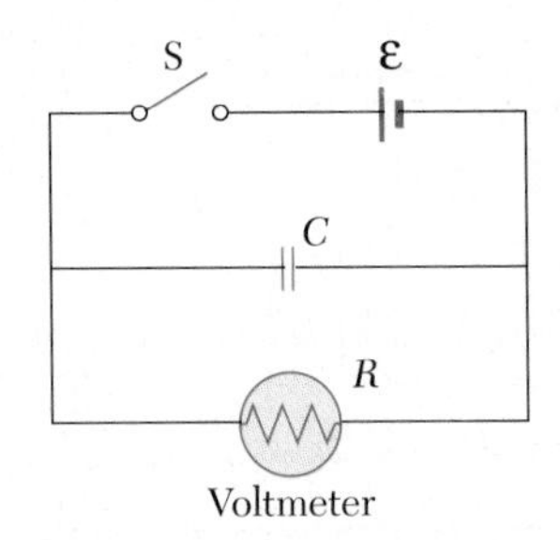


**Figure 28.64** (Problem 91).

| $V$ (V) | $t$ (s) | $\ln(\mathcal{E}/V)$ |
|---|---|---|
| 6.19 | 0 | |
| 5.55 | 4.87 | |
| 4.93 | 11.1 | |
| 4.34 | 19.4 | |
| 3.72 | 30.8 | |
| 3.09 | 46.6 | |
| 2.47 | 67.3 | |
| 1.83 | 102.2 | |

# ESSAY

## Exponential Growth

**Albert A. Bartlett**
*University of Colorado Boulder*

If something is growing at a constant rate, such as $P = 5\%$ per year, we refer to the growth as steady growth. It is also called **exponential growth** because the size, $N$, of the growing quantity at some time $t$ in the future is related to its present size, $N_0$ (at time $t = 0$), by the exponential function

$$N = N_0 e^{kt} \tag{E.1}$$

where $e = 2.718\ .\ .\ .$ is the base of the natural logarithims and $k$ is the annual percent growth rate $P$ divided by 100:

$$k = \frac{P}{100} \tag{E.2}$$

Note that if $k$ is positive, $N$ increases exponentially with time (exponential growth). If $k$ is negative, $N$ decreases exponentially with time (exponential decay). Some exam-

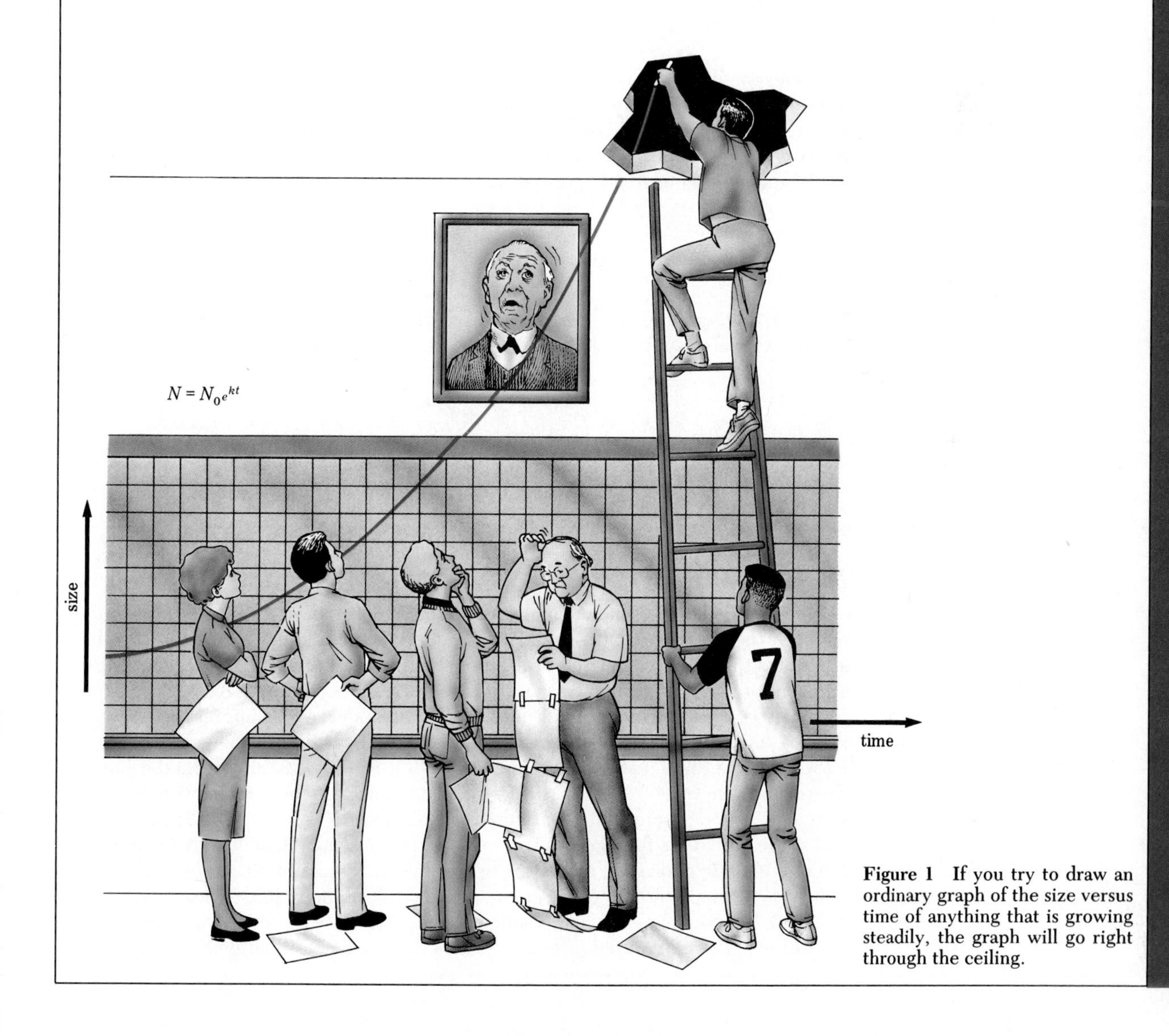


**Figure 1** If you try to draw an ordinary graph of the size versus time of anything that is growing steadily, the graph will go right through the ceiling.

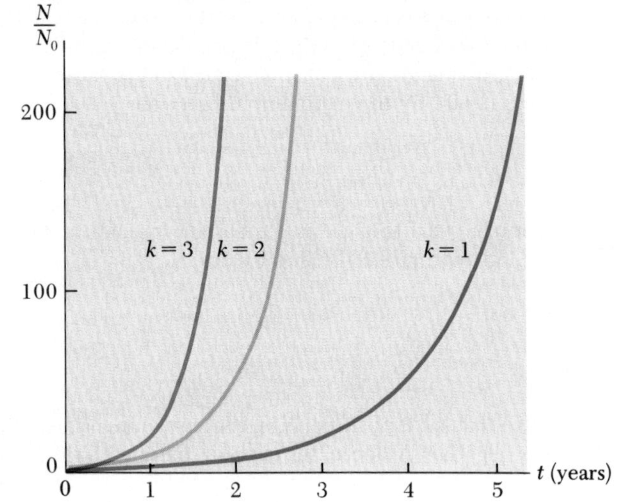


**Figure 2** Exponential growth curve $N/N_0 = e^{kt}$. Note, $N/N_0$ has the value 1 at $t = 0$.

ples of exponential decay are radioactive decay (Section 45.4) and the decay of the quantity of charge on the plates of a capacitor as the capacitor is discharging through a resistor (Section 28.4). Figures 1 and 2 are representative graphs of exponential growth. This essay describes several examples of exponential growth.

The condition of steady growth is represented by the simple differential equation

$$\frac{dN}{dt} = kN \tag{E.3}$$

In words, this equation says that the rate of change of the quantity $N$ is proportional to $N$. We can rearrange Equation E.3 to give

$$\frac{1}{N}\frac{dN}{dt} = k$$

In this form we see that the fractional change of $N$ per unit time is constant. An example would be the case where the quantity $N$ is growing 6% per year. In this case $P = 6\%$ per year and $k = 0.06\%$ per year.

The solution of the Equation E.3 is Equation E.1. If we can show that a quantity $N$ changes with time according to Equation E.3 it follows automatically that $N$ will obey Equation E.1.

For example, the fundamental concept of compound interest on a savings account in the bank is that the interest $\Delta N$ added to the number $N$ of dollars in the account in the time interval $\Delta t$ is proportional to the number of dollars in the account. The constant of proportionality is the interest rate. If $N = \$250$, $P = 8\%$ per year, and $\Delta t = 1$ year, then we have simple compounding once a year. The interest in that year is $\Delta N = 0.08 \times \$250 = \$20$ so that the value of $N$ at the end of the year would be $\$250 + \$20 = \$270$. In the next year $\Delta N = 0.08 \times \$270 = \$21.60$, and at the end of the second year, $N = \$270 + \$21.60 = \$291.60$. Suppose the interest were 8% per year, compounded semiannually. In the first half year $\Delta N = (0.08/2) \times \$250 = \$10$, and at the end of the first half-year, $N = \$260$. In the second half year, $\Delta N = 0.04 \times \$260 = \$10.40$, and at the end of the first full year, $N = \$270.40$. At the end of the first year, compounding once gave $N = \$270$. Compounding twice gave $N = \$270.40$. This suggests a very fundamental fact. The more frequently we compound the interest, the more rapid is the increase in the size of $N$. Equation E.3 represents the limiting case where $\Delta t$ approaches zero and the interest is *compounded continuously*. In this case at the end of one year we use Equation E.1:

$$N = \$250e^{0.08\times 1} = \$250 \times 1.08329 = \$270.82$$

Many banks calculate interest by compounding continuously. This leads to newpaper ads such as one that says "9.54% annual yield" which corresponds to a "9.11% annual rate." What this means is that the rate of 9.11% *compounded continuously* for one year gives the same result (yield) as a rate of 9.54% compounded once. In other words

$$e^{0.0911\times 1} = 1.0954 \tag{E.4}$$

It has been shown that the number of miles of highway in the United States obeys Equation E.3 so that the number of miles of highway will grow exponentially according to Equation E.1.

In steady growth, it takes a fixed length of time for a quantity to grow by a fixed fraction such as 5%. From this it follows that it takes a fixed longer length of time for that quantity to grow by 100%. Let us calculate the time required for the quantity $N$ to double in value, which is called the **doubling time,** $T_2$. We can obtain an expression for $T_2$ by writing Equation E.1 as $N/N_0 = e^{kt}$ and taking the natural logarithm of each side:

$$\ln\left(\frac{N}{N_0}\right) = kt$$

If we set $N = 2N_0$ (that is, we double $N_0$), then $T_2$ (which is the time $t$ when $N = 2N_0$) is

$$T_2 = \frac{\ln(2N_0/N_0)}{k} = \frac{\ln 2}{k} = \frac{0.693}{k}$$

Since $k = P/100$, this becomes

$$T_2 \approx \frac{70}{P} \tag{E.5}$$

Likewise, if we wanted the time for $N$ to *triple* in size, we would use the natural logarithm of 3, to find

$$T_3 = \frac{100 \ln 3}{P} \approx \frac{110}{P}$$

**EXAMPLE E.1 Compound Interest—The Eighth Wonder**
Suppose you put \$15 in a savings account at 9% annual interest to be compounded continuously. How large a sum of money would be in the account at the end of 200 years?

*Solution* $N_0 = \$15.$; $k = 9/100 = 0.09$ per year and $t = 200$ years. Therefore, from Equation E.1 we have

$$N = \$15 \times e^{(0.09\times 200)} = \$15 \times e^{18} = \$15 \times 6.57 \times 10^7 = \$985 \text{ million!}$$

Now you can see why a famous financier once said that he could not name the seven wonders of the ancient world but surely the eighth wonder would have to be compound interest!

*(Continued)*

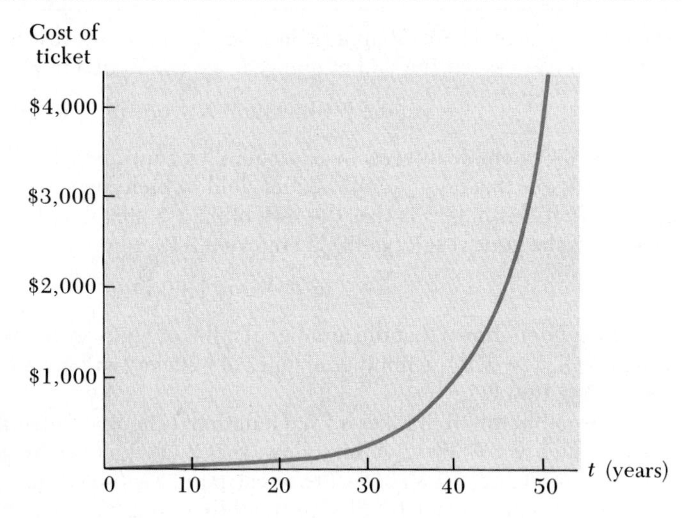


**Figure 3** Price of a $4 ticket with 14% annual inflation. The cost at $t = 0$ is $N_0 = \$4$.

**EXAMPLE E.2 The Consequences of Inflation**
Let us use the doubling time instead of the quantity $e$ to estimate the consequences of an annual inflation rate of 14% that continued for 50 years.

*Solution* First, we calculate $T_2$ from Equation E.5

$$T_2 = \frac{70}{14} = 5 \text{ years}$$

This inflation rate will cause prices to double every five years!

In the next step, we calculate the number of doubling times in 50 years:

$$\text{Number of doublings} = \frac{50 \text{ years}}{5 \text{ years/doubling}} = 10$$

Finally, we count up the consequences of each doubling by making use of a table, which shows us that, in 10 doubling times, prices will increase by a factor of 1024, which is approximately 1000. Thus, in 50 years of 14% annual inflation, the cost of a $4 ticket to the movies would increase to roughly $4000! (See Fig. 3.)

It is very convenient to remember that 10 doublings give an increase by a factor of approximately $10^3$, that 20 doublings give an increase of a factor of approximately $10^6$, that 30 doublings give an increase by a factor of approximately $10^9$, and so on.

| No. of Doublings | Price Increase Factor |
|---|---|
| 1 | $2 = 2^1$ |
| 2 | $4 = 2^2$ |
| 3 | $8 = 2^3$ |
| 4 | $16 = 2^4$ |
| 5 | $32 = 2^5$ |
| 6 | $64 = 2^6$ |
| 7 | $128 = 2^7$ |
| 8 | $256 = 2^8$ |
| 9 | $512 = 2^9$ |
| 10 | $1024 = 2^{10}$ |

**EXAMPLE E.3 The Increasing Rate of Energy Consumption**
For many years before 1975, consumption of electrical energy in the United States grew steadily at a rate of about 7% per year. By what factor would consumption increase if this growth rate continued for 40 years?

*Solution* In this case, $P = 7$, and so the doubling time from Equation E.5 is

$$T_2 = \frac{70}{7} = 10 \text{ years}$$

and the number of doublings in 40 years is

$$\text{Number of doubings} = \frac{40 \text{ years}}{10 \text{ years/doubling}} = 4$$

Therefore, in 40 years the amount of power consumed would be $2^4 = 16$ times the amount used today. That is, 40 years from now we would need 16 times as many electric generating plants as we have at the present. Furthermore, if those additional plants are similar to today's, then each day they would consume 16 times as much fuel as our present plants use, and there would be 16 times as much pollution and waste heat to contend with!

**EXAMPLE E.4 Annual Increase in World Population**

Populations tend to grow steadily. In July of 1987 we saw reports that the population of the earth had reached $5 \times 10^9$ people. The world birth rate was estimated to be 28 per 1000 each year while the annual death rate was estimated to be 11 per 1000. Thus, for every 1000 people, the population increase each year is $28 - 11 = 17$. For this growth rate we find

$$k = \frac{17}{1000} = 0.017 \text{ per year}$$

$$P = 100\,k = 1.7\% \text{ per year}$$

This growth rate seems so small that many people regard it as trivial and inconsequential. A proper perspective of this rate appears only when we calculate the doubling time:

$$T_2 = \frac{70}{1.7} = 41 \text{ years}$$

This simple calculation indicates that it is most likely that the world population will double within the life expectancy of today's students! At the most elemental level, this means that we have approximately 41 years to double world food production.

What is the annual increase in the earth's population? Since for one year $\Delta N \ll N$ we can get a good answer from Equation E.3:

$$\frac{\Delta N}{\Delta t} = 0.017 \times 5 \times 10^9 = 85 \text{ million per year}$$

*This annual increase in the world population is roughly one third of the population of the United States.*

Some illuminating calculations can be made based on the assumption that this rate of growth has been constant and will remain constant. These calculations will demonstrate that the growth rate has not been constant at this value in the past and cannot remain this high for very long.

**EXAMPLE E.5 When did Adam and Eve Live?**

When we use Equation E.1, setting $N = 5 \times 10^9$, $N_0 = 2$ (Adam and Eve) and $k = 0.017$ (from Example E.4), we have

$$5 \times 10^9 = 2e^{0.017t}$$

This gives $t = 1273$ years ago, or about 714 A.D.! This result proves that through essentially all of human history the population growth rate was very much smaller than it is today. It must have been near zero through most of human history.

**EXAMPLE E.6 Growth of Population Density**

The land area of the continents (excluding Antarctica) is $1.24 \times 10^{14}$ m$^2$. If this modest annual growth rate of 1.7% were to continue steadily in the future, how long

*(Continued)*

would it take for the population to reach a density of one person per square meter on the continents?

$$1.24 \times 10^{14} = 5 \times 10^{9} e^{0.017t}$$

Solving, we find $t$ is slightly less than 600 years.

**EXAMPLE E.7 Growth of the Mass of People**

If this very low rate of growth continued, how long would it take for the mass of people to equal the mass of the earth ($5.98 \times 10^{24}$ kg)? (Assume that the mass of an average person is 65 kg.)

$$5.98 \times 10^{24} = (5 \times 10^{9} \times 65) e^{0.017t}$$

This gives a value of $t$ of about 1800 years! We have assumed that the mass of a person is 65 kg.

The last two examples prove that the growth rate of world population cannot stay as high as it presently is for any extended period of time. Although world agricultural production has been just barely keeping pace with world population growth, millions are malnourished and many people are starving. However, we will not have to double food production in 41 years if we can lower the worldwide birth rate. If we fail to double world food production in 41 years, then the death rate will rise. Dramatic increases in world food production in recent decades are due almost exclusively to the rapid growth of the use of petroleum for powering machinery and for manufacturing fertilizers and insecticides. Indeed, it has been noted that "modern agriculture is the use of land to convert petroleum into food." The student must wonder how much longer we can continue the long history of approximately steady population growth when our food supplies are tied so closely to dwindling supplies of petroleum.

This brief introduction to the arithmetic of steady growth enables us to understand that, in all biological systems, the normal condition is the steady-state condition, where the birth rate is equal to the death rate. Growth is a short-term transient phenomenon that can never continue for more than a short period of time. Yet in the United States, business and government leaders at all levels, from local communities to Washington, D.C., would have us believe that steady growth forever is a goal we can achieve. They would have us believe that we should continue our population growth (the U.S. population increases by about 2 million people per year) and the growth in our rates of consumption of natural resources. We now hear about "sustainable growth" as though the addition of the adjective "sustainable" would render inoperable the laws of nature.

In contrast to all this optimism, please remember that someone once noted that "The greatest shortcoming of the human race is our inability to understand the exponential function."

### Suggested Readings

Bartlett, A. A., *Civil Engineering*, December 1969, pp. 71–72.

Bartlett, A. A., "The Exponential Function," *The Physics Teacher*, October 1976 to January 1979.

Bartlett, A. A., "The Forgotten Fundamentals of the Energy Crisis," *Am. J. Physics* 46:876, 1978.

Kerr, R. A., "Another Oil Resource Warning," *Science*, January 27, 1984, p. 382.

### Essay Problems

1. In the year 1626 Manhattan Island was purchased for $24. Assuming a continually compounded interest rate of 4.4%, calculate the current land valuation of the island.
2. The following "mystery" was taken from Deborah Hughes-Hallett: *Elementary Functions*, W. W. Norton, 1980, p. 264.

   The police were baffled by what seemed to be the perfect murder of a girl who had been found, apparently suffocated, in her kitchen. Finally, Sherlock Holmes was called in. With the aid of Dr. Watson's knowledge of botany, the mystery was solved and the following story told. The girl had been making bread in her kitchen, whose dimensions were 6 ft by 10 ft by 10 ft. She had formed the dough into a ball of volume 1/6 cubic feet and turned away to wash some dishes. At that moment Holmes' enemy, Professor Moriarty, had added a particularly virulent strain of yeast to the bread. As a result, the bread immediately started to rise, tripling in volume every 4 minutes. Before long, the dough filled the room, stopping the clock at 3:48 and squashing the girl to death against the wall. By the time Inspector Lestrade of Scotland Yard reached the scene the next day, the yeast had worked itself out and the dough returned to its original size.

At what time did Professor Moriarty add the yeast?

# 29
# Magnetic Fields

*The blue arc in this photograph indicates the circular path followed by an electron beam moving in a magnetic field. The vessel contains gas at very low pressure, and the beam is made visible as the electrons collide with the gas atoms, which in turn emit visible light. The magnetic field is produced by two coils (not shown). The apparatus can be used to measure the ratio of e/m for the electron.* (Courtesy of CENCO)

The behavior of bar magnets is well known to anyone who has studied science. Permanent magnets, which are usually made of alloys containing iron, will attract or repel other magnets. Furthermore, they will attract other bits of iron, which in turn can become magnetized. The list of important technological applications of magnetism is extensive. For instance, large electromagnets are used to pick up heavy loads. Magnets are also used in such devices as meters, transformers, motors, particle accelerators, and loudspeakers. Magnetic tapes are routinely used in sound recording, TV recording, and computer memories. Intense magnetic fields generated by superconducting magnets are currently being used as a means of containing the plasmas (heated to temperatures of the order of $10^8$ K) used in controlled nuclear fusion research.

## 29.1 INTRODUCTION

The phenomenon of magnetism was known to the Greeks as early as around 800 B.C. They discovered that certain stones, now called *magnetite* ($Fe_3O_4$), attract pieces of iron. Legend ascribes the name *magnetite* to the shepherd Magnes, "the nails of whose shoes and the tip of whose staff stuck fast in a magnetic field while he pastured his flocks." In 1269 Pierre de Maricourt, using a spherical natural magnet, mapped out the directions taken by a needle when placed at various points on the surface of the sphere. He found that the directions formed lines that encircle the sphere passing through two points diametrically opposite each other, which he called the *poles* of the magnet. Subsequent experiments showed that every magnet, regardless of its shape, has two poles, called *north* and *south poles,* which exhibit forces on each other in a manner analogous to electrical charges. That is, like poles repel each other and unlike poles attract each other.

An assortment of commercially available magnets. The four red magnets and the large black magnet on the left are made of an alloy of iron, aluminum, and cobalt. The six horseshoe magnets on the right are made of different nickel alloy steels. The rectangular magnets on the lower right are ceramics made of iron, nickel, and beryllium oxides. (Courtesy of CENCO)

In 1600 William Gilbert extended these experiments to a variety of materials. Using the fact that a compass needle orients in preferred directions, he suggested that the earth itself is a large permanent magnet. In 1750 John Michell (1724–1793) used a torsion balance to show that magnetic poles exert attractive or repulsive forces on each other and that these forces vary as the inverse square of their separation. Although the force between two magnetic poles is similar to the force between two electric charges, there is an important difference. Electric charges can be isolated (witness the electron or proton), whereas *magnetic poles cannot be isolated.* That is, *magnetic poles are always found in pairs.* All attempts thus far to detect an isolated magnetic monopole have been unsuccessful. No matter how many times a permanent magnet is cut, each piece will always have a north and a south pole.

The relationship between magnetism and electricity was discovered in 1819 when, during a lecture demonstration, the Danish scientist Hans Oersted found that an electric current in a wire deflected a nearby compass

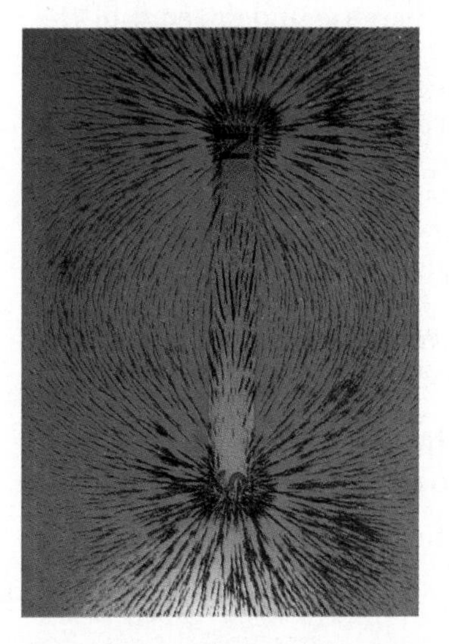

Magnetic field pattern of a bar magnet as displayed by iron filings on a sheet of paper. (Courtesy of Henry Leap and Jim Lehman)

Magnetic field patterns surrounding two bar magnets as displayed with iron filings. This demonstrates the magnetic field pattern between unlike poles. (Courtesy of Henry Leap and Jim Lehman)

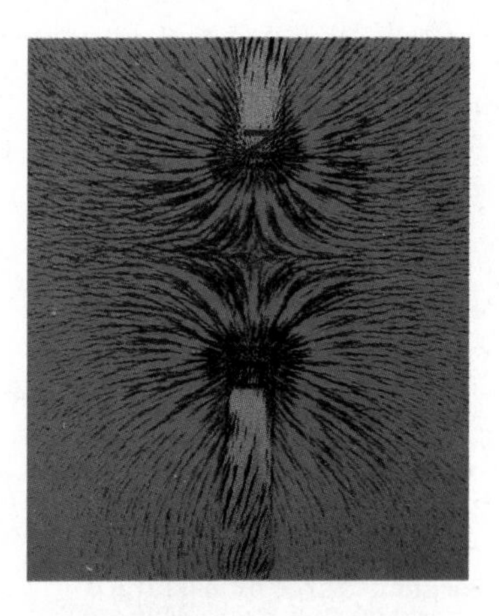

This demonstrates the magnetic field pattern between two like poles. (Courtesy of Henry Leap and Jim Lehman)

Hans Christian Oersted (1777–1851), Danish physicist. (The Bettmann Archive)

needle.[1] Shortly thereafter, André Ampère (1775–1836) obtained quantitative laws of magnetic force between current-carrying conductors. He also suggested that electric current loops of molecular size are responsible for *all* magnetic phenomena. This idea is the basis for the modern theory of magnetism.

In the 1820's, further connections between electricity and magnetism were demonstrated by Faraday and independently by Joseph Henry (1797–1878). They showed that an electric current could be produced in a circuit either by moving a magnet near the circuit or by changing the current in another, nearby circuit. These observations demonstrate that a changing magnetic field produces an electric field. Years later, theoretical work by Maxwell showed that a changing electric field gives rise to a magnetic field.

This chapter examines forces on moving charges and on current-carrying wires in the presence of a magnetic field. The source of the magnetic field itself will be described in Chapter 30.

## 29.2 DEFINITION AND PROPERTIES OF THE MAGNETIC FIELD

The electric field $\boldsymbol{E}$ at a point in space has been defined as the electric force per unit charge acting on a test charge placed at that point. Similarly, the gravitational field $\boldsymbol{g}$ at a point in space is the gravitational force per unit mass acting on a test mass.

We now define a magnetic field vector $\boldsymbol{B}$ (sometimes called the *magnetic induction* or *magnetic flux density*) at some point in space in terms of a magnetic force that would be exerted on an appropriate test object. Our test object is taken to be a charged particle moving with a velocity $\boldsymbol{v}$. For the time being, let us assume that there are no electric or gravitational fields present in the region of the charge. Experiments on the motion of various charged particles moving in a magnetic field give the following results:

Properties of the magnetic force on a charge moving in a $\boldsymbol{B}$ field

1. The magnetic force is proportional to the charge $q$ and speed $v$ of the particle.
2. The magnitude and direction of the magnetic force depend on the velocity of the particle and on the magnitude and direction of the magnetic field.
3. When a charged particle moves in a direction *parallel* to the magnetic field vector, the magnetic force $\boldsymbol{F}$ on the charge is *zero*.
4. When the velocity vector makes an angle $\theta$ with the magnetic field, the magnetic force acts in a direction perpendicular to both $\boldsymbol{v}$ and $\boldsymbol{B}$; that is, $\boldsymbol{F}$ is perpendicular to the plane formed by $\boldsymbol{v}$ and $\boldsymbol{B}$ (Fig. 29.1a).
5. The magnetic force on a positive charge is in the direction opposite the direction of the force on a negative charge moving in the same direction (Fig. 29.1b).
6. If the velocity vector makes an angle $\theta$ with the magnetic field, the magnitude of the magnetic force is proportional to $\sin\theta$.

These observations can be summarized by writing the magnetic force in the form

Magnetic force on a charged particle moving in a magnetic field

$$\boldsymbol{F} = q\boldsymbol{v} \times \boldsymbol{B} \tag{29.1}$$

[1] It is interesting to note that the same discovery was reported in 1802 by an Italian jurist, Gian Dominico Romognosi, but was overlooked, probably because it was published in a newspaper, *Gazetta de Trentino*, rather than in a scholarly journal.

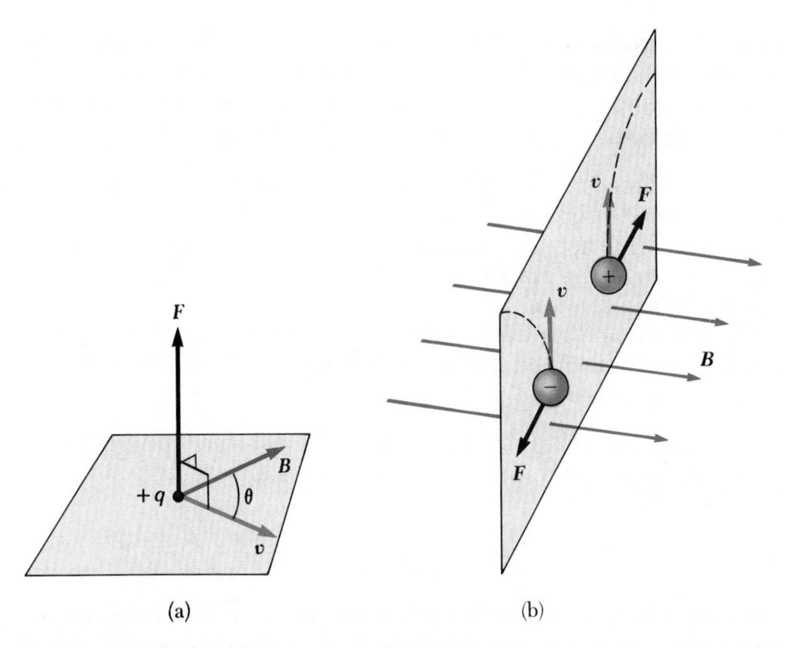


**Figure 29.1** The direction of the magnetic force on a charged particle moving with a velocity $\boldsymbol{v}$ in the presence of a magnetic field. (a) When $\boldsymbol{v}$ is at an angle $\theta$ to $\boldsymbol{B}$, the magnetic force is perpendicular to both $\boldsymbol{v}$ and $\boldsymbol{B}$. (b) In the presence of a magnetic field, the moving charged particles are deflected as indicated by the dotted lines.

where the direction of the magnetic force is in the direction of $\boldsymbol{v} \times \boldsymbol{B}$, which, by definition of the cross product, is perpendicular to both $\boldsymbol{v}$ and $\boldsymbol{B}$.

Figure 29.2 gives a brief review of the right-hand rule for determining the direction of the cross product $\boldsymbol{v} \times \boldsymbol{B}$. You point the four fingers of your right hand along the direction of $\boldsymbol{v}$, and then turn them until they point along the direction of $\boldsymbol{B}$. The thumb then points in the direction of $\boldsymbol{v} \times \boldsymbol{B}$. Since $\boldsymbol{F} = q\boldsymbol{v} \times \boldsymbol{B}$, $\boldsymbol{F}$ is in the direction of $\boldsymbol{v} \times \boldsymbol{B}$ if $q$ is positive (Fig. 29.2a) and *opposite* the direction of $\boldsymbol{v} \times \boldsymbol{B}$ if $q$ is negative (Fig. 29.2b). The magnitude of the magnetic force has the value

$$F = qvB \sin \theta \tag{29.2}$$

where $\theta$ is the angle between $\boldsymbol{v}$ and $\boldsymbol{B}$. From this expression, we see that $F$ is *zero* when $\boldsymbol{v}$ is parallel to $\boldsymbol{B}$ ($\theta = 0$ or $180°$). Furthermore, the force has its *maximum* value, $F = qvB$, when $\boldsymbol{v}$ is perpendicular to $\boldsymbol{B}$ ($\theta = 90°$).

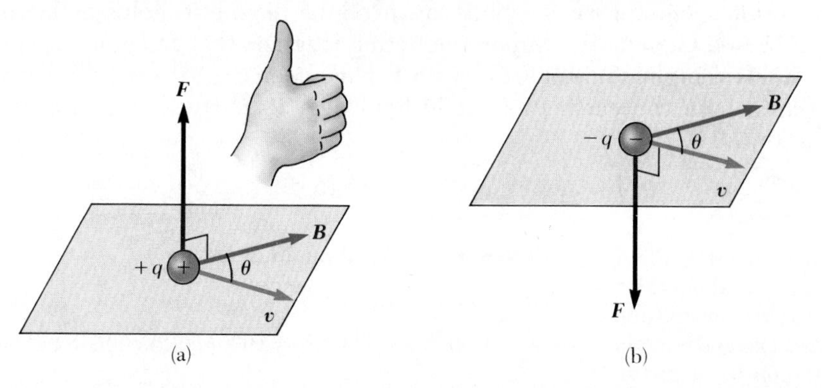


**Figure 29.2** The right-hand rule for determining the direction of the magnetic force $\boldsymbol{F}$ acting on a charge $q$ moving with a velocity $\boldsymbol{v}$ in a magnetic field $\boldsymbol{B}$. If $q$ is positive, $\boldsymbol{F}$ is upward in the direction of the thumb. If $q$ is negative, $\boldsymbol{F}$ is downward.

We can regard Equation 29.1 as an operational definition of the magnetic field at a point in space. That is, the magnetic field is defined in terms of a *sideways* force acting on a moving charged particle. There are several important differences between electric and magnetic forces:

Differences between electric and magnetic fields

1. The electric force is always in the direction of the electric field, whereas the magnetic force is perpendicular to the magnetic field.
2. The electric force acts on a charged particle independent of the particle's velocity, whereas the magnetic force acts on a charged particle only when the particle is in motion.
3. The electric force does work in displacing a charged particle, whereas the magnetic force associated with a steady magnetic field does *no* work when a particle is displaced.

This last statement is a consequence of the fact that when a charge moves in a steady magnetic field, the magnetic force is always *perpendicular* to the displacement. That is,

$$\mathbf{F}\cdot d\mathbf{s} = (\mathbf{F}\cdot \mathbf{v})\, dt = 0$$

since the magnetic force is a vector perpendicular to $\mathbf{v}$. From this property and the work-energy theorem, we conclude that the kinetic energy of a charged particle *cannot* be altered by a magnetic field alone. In other words,

A magnetic field cannot change the speed of a particle

when a charge moves with a velocity $\mathbf{v}$, an applied magnetic field can alter the direction of the velocity vector, but it cannot change the speed of the particle.

The SI unit of the magnetic field is the **weber per square meter** ($\mathrm{Wb/m^2}$), also called the **tesla** (T). This unit can be related to the fundamental units by using Equation 29.1: a 1-coulomb charge moving through a field of 1 tesla with a velocity of 1 m/s perpendicular to the field experiences a force of 1 newton:

$$[\mathrm{B}] = \mathrm{T} = \frac{\mathrm{Wb}}{\mathrm{m^2}} = \frac{\mathrm{N}}{\mathrm{C\cdot m/s}} = \frac{\mathrm{N}}{\mathrm{A\cdot m}} \qquad (29.3)$$

In practice, the cgs unit for magnetic field, called the **gauss** (G), is often used. The gauss is related to the tesla through the conversion

$$1\ \mathrm{T} = 10^4\ \mathrm{G} \qquad (29.4)$$

Conventional laboratory magnets can produce magnetic fields as large as about 25 000 G, or 2.5 T. Superconducting magnets that can generate magnetic fields as high as 250 000 G, or 25 T, have been constructed. This can be compared with the earth's magnetic field near its surface, which is about 0.5 G, or $0.5 \times 10^{-4}$ T.

EXAMPLE 29.1 A Proton Moving in a Magnetic Field
A proton moves with a speed of $8 \times 10^6$ m/s along the $x$ axis. It enters a region where there is a field of magnitude 2.5 T, directed at an angle of 60° to the $x$ axis and lying in the $xy$ plane (Fig. 29.3). Calculate the initial magnetic force and acceleration of the proton.

*Solution* From Equation 29.2, we get

$$F = qvB \sin\theta$$

$$= (1.6 \times 10^{-19}\ \mathrm{C})(8 \times 10^6\ \mathrm{m/s})(2.5\ \mathrm{T})(\sin 60^\circ)$$

$$= 2.77 \times 10^{-12}\ \mathrm{N}$$

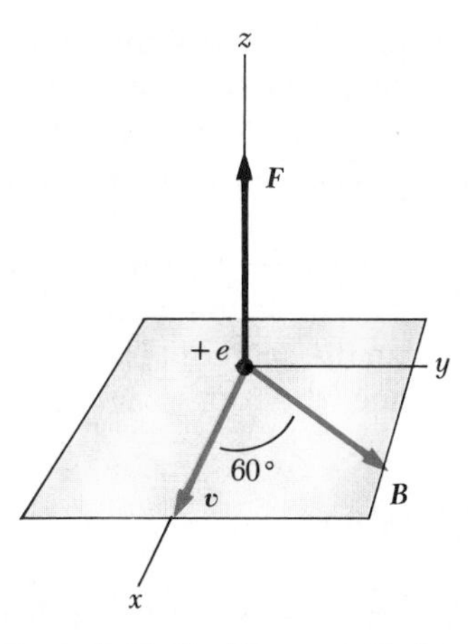


**Figure 29.3** (Example 29.1) The magnetic force $\boldsymbol{F}$ on a proton is in the positive $z$ direction when $\boldsymbol{v}$ and $\boldsymbol{B}$ lie in the $xy$ plane.

Since $\boldsymbol{v} \times \boldsymbol{B}$ is in the positive $z$ direction and since the charge is positive, the force $\boldsymbol{F}$ is in the positive $z$ direction.

Since the mass of the proton is $1.67 \times 10^{-27}$ kg, its initial acceleration is

$$a = \frac{F}{m} = \frac{2.77 \times 10^{-12}\ \text{N}}{1.67 \times 10^{-27}\ \text{kg}} = 1.66 \times 10^{15}\ \text{m/s}^2$$

in the positive $z$ direction.

Exercise 1 Verify that the units of $\boldsymbol{F}$ in the above calculation for the magnetic force reduce to newtons.

## 29.3 MAGNETIC FORCE ON A CURRENT-CARRYING CONDUCTOR

If a force is exerted on a single charged particle when it moves through a magnetic field, it should not surprise you that a current-carrying wire also experiences a force when placed in a magnetic field. This follows from the fact that the current represents a collection of many charged particles in motion; hence, the resultant force on the wire is due to the sum of the individual forces on the charged particles.

The force on a current-carrying conductor can be demonstrated by hanging a wire between the faces of a magnet as in Figure 29.4. In this figure, the magnetic field is directed into the page and covers the region within the shaded circle. When the current in the wire is zero, the wire remains vertical as in Figure 29.4a. However, when a current is set up in the wire directed upwards as in Figure 29.4b, the wire deflects to the left. If we reverse the current, as in Figure 29.4c, the wire deflects to the right.

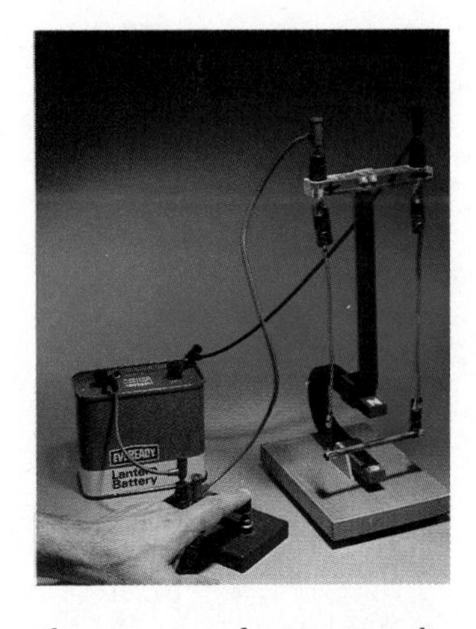


This apparatus demonstrates the force on a current-carrying conductor in an external magnetic field. Why does the bar swing *into* the magnet after the switch is closed? (Courtesy of Henry Leap and Jim Lehman)

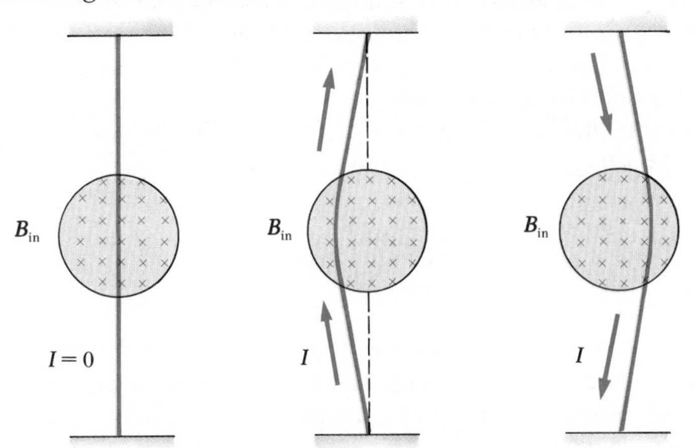


**Figure 29.4** A flexible vertical wire which is partially stretched between the faces of a magnet with the field (blue crosses) directed into the paper. (a) When there is no current in the wire, it remains vertical. (b) When the current is upwards, the wire deflects to the left. (c) When the current is downwards, the wire deflects to the right.

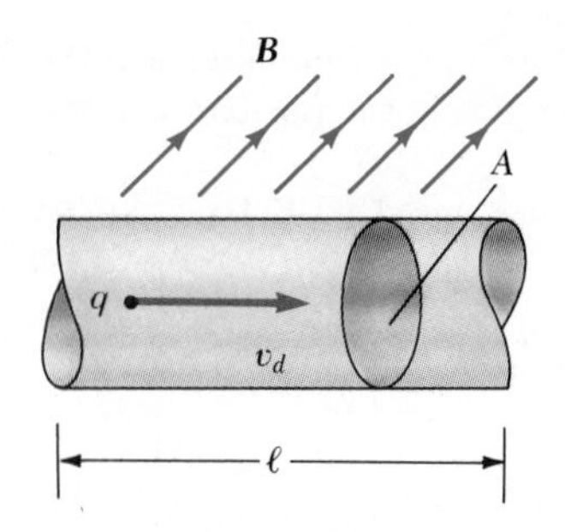


Figure 29.5 A section of a wire containing moving charges in an external magnetic field $\boldsymbol{B}$. The magnetic force on each charge is $q\boldsymbol{v_d} \times \boldsymbol{B}$, and the net force on a straight element is $I\boldsymbol{\ell} \times \boldsymbol{B}$.

Let us quantify this discussion by considering a straight segment of wire of length $\ell$ and cross-sectional area $A$, carrying a current $I$ in a uniform *external* magnetic field $\boldsymbol{B}$ as in Figure 29.5. The magnetic force on a charge $q$ moving with a drift velocity $\boldsymbol{v}_d$ is given by $q\boldsymbol{v_d} \times \boldsymbol{B}$. The force on the charge carriers is transmitted to the "bulk" of the wire through collisions with the atoms making up the wire. To find the total force on the wire, we multiply the force on one charge, $q\boldsymbol{v_d} \times \boldsymbol{B}$, by the number of charges in the segment. Since the volume of the segment is $A\ell$, the number of charges in the segment is $nA\ell$, where $n$ is the number of charges per unit volume. Hence, the total magnetic force on the wire of length $\ell$ is

$$\boldsymbol{F} = (q\boldsymbol{v_d} \times \boldsymbol{B})nA\ell$$

This can be written in a more convenient form by noting that, from Equation 27.4, the current in the wire is given by $I = nqv_dA$. Therefore, $\boldsymbol{F}$ can be expressed as

$$\boldsymbol{F} = I\boldsymbol{\ell} \times \boldsymbol{B} \tag{29.5}$$

where $\boldsymbol{\ell}$ is a vector in the direction of the current $I$; the magnitude of $\boldsymbol{\ell}$ equals the length $\ell$ of the segment. Note that this expression applies only to a straight segment of wire in a uniform external magnetic field. Furthermore, we have neglected the field produced by the current itself. (In fact, the wire cannot produce a force on itself.)

Now consider an arbitrarily shaped wire of uniform cross section in an external magnetic field, as in Figure 29.6. It follows from Equation 29.5 that the magnetic force on a very small segment $d\boldsymbol{s}$ in the presence of a field $\boldsymbol{B}$ is given by

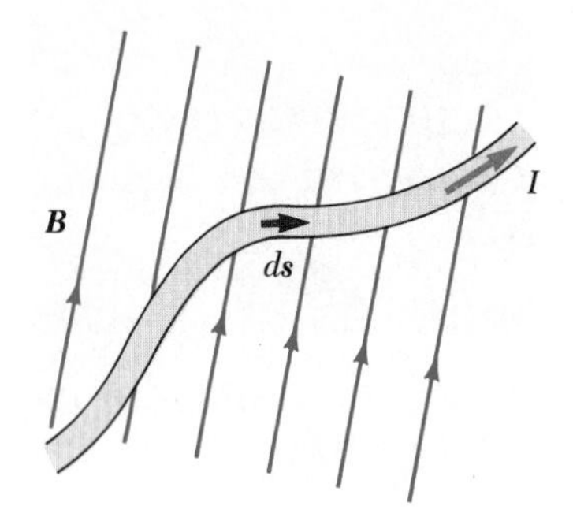


Figure 29.6 A wire of arbitrary shape carrying a current $I$ in an external magnetic field $\boldsymbol{B}$ experiences a magnetic force. The force on any segment $d\boldsymbol{s}$ is given by $I\,d\boldsymbol{s} \times \boldsymbol{B}$ and is directed *out* of the page.

$$d\boldsymbol{F} = I\,d\boldsymbol{s} \times \boldsymbol{B} \tag{29.6}$$

where $d\boldsymbol{F}$ is directed out of the page for the directions assumed in Figure 29.6. We can consider Equation 29.6 as an alternative definition of $\boldsymbol{B}$. That is, the field $\boldsymbol{B}$ can be defined in terms of a measurable force on a current element, where the force is a maximum when $\boldsymbol{B}$ is perpendicular to the element and zero when $\boldsymbol{B}$ is parallel to the element.

To get the total force $\boldsymbol{F}$ on the wire, we integrate Equation 29.6 over the length of the wire:

$$\boldsymbol{F} = I\int_a^b d\boldsymbol{s} \times \boldsymbol{B} \tag{29.7}$$

In this expression, $a$ and $b$ represent the end points of the wire. When this integration is carried out, the magnitude of the magnetic field and the direction the field makes with the vector $d\boldsymbol{s}$ (that is, the element orientation) may vary at each point.

Now let us consider two special cases involving the application of Equation 29.7. In both cases, the external magnetic field is taken to be constant in magnitude and direction.

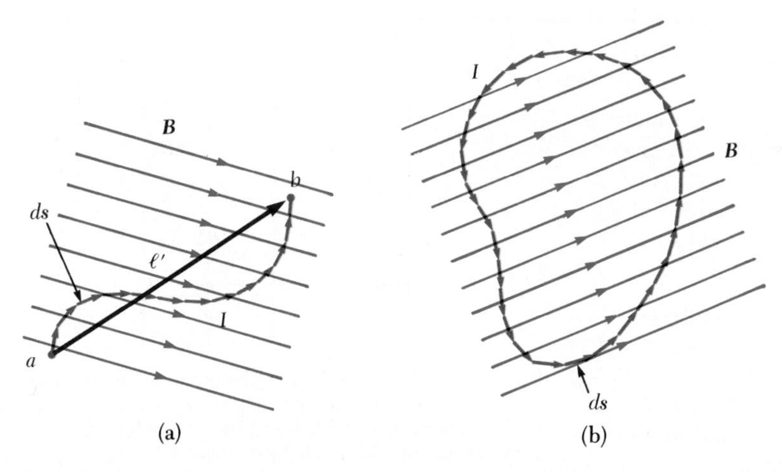


**Figure 29.7** (a) A curved conductor carrying a current $I$ in a uniform magnetic field. The magnetic force on the conductor is equivalent to the force on a straight segment of length $\ell'$ running between the ends of the wire, $a$ and $b$. (b) A current-carrying loop of arbitrary shape in a uniform magnetic field. The net magnetic force on the loop is 0.

### Case I

Consider a curved wire carrying a current $I$; the wire is located in a uniform external magnetic field $\mathbf{B}$ as in Figure 29.7a. Since the field is assumed to be uniform (that is, $\mathbf{B}$ has the same value over the region of the conductor), $\mathbf{B}$ can be taken outside the integral in Equation 29.7, and we get

$$\mathbf{F} = I\left(\int_a^b d\mathbf{s}\right) \times \mathbf{B} \tag{29.8}$$

But the quantity $\int_a^b d\mathbf{s}$ represents the *vector sum* of all the displacement elements from $a$ to $b$ as described in Figure 29.6. From the law of addition of many vectors (Chapter 2), the sum equals the vector $\boldsymbol{\ell}'$, which is directed from $a$ to $b$. Therefore, Equation 29.8 reduces to

$$\mathbf{F} = I\boldsymbol{\ell}' \times \mathbf{B} \tag{29.9}$$

Force on a wire in a uniform field

### Case II

An arbitrarily shaped, closed loop carrying a current $I$ is placed in a uniform external magnetic field $\mathbf{B}$ as in Figure 29.7b. Again, we can express the force in the form of Equation 29.8. In this case, the vector sum of the displacement vectors must be taken over the closed loop. That is,

$$\mathbf{F} = I\left(\oint d\mathbf{s}\right) \times \mathbf{B}$$

Since the set of displacement vectors forms a *closed polygon* (Fig. 29.7b), the vector sum must be *zero*. This follows from the graphical procedure of adding vectors by the *polygon method* (Chapter 2). Since $\oint d\mathbf{s} = 0$, we conclude that

$$\mathbf{F} = 0 \tag{29.10}$$

That is,

> the total magnetic force on any closed current loop in a uniform magnetic field is zero.

**EXAMPLE 29.2 Force on a Semicircular Conductor** □

A wire bent into the shape of a semicircle of radius $R$ forms a closed circuit and carries a current $I$. The circuit lies in the $xy$ plane, and a uniform magnetic field is present along the positive $y$ axis as in Figure 29.8. Find the magnetic forces on the straight portion of the wire and on the curved portion.

*Solution* The force on the straight portion of the wire has a magnitude given by $F_1 = I\ell B = 2IRB$, since $\ell = 2R$ and the wire is perpendicular to $\mathbf{B}$. The direction of $\mathbf{F}_1$ is *out* of the paper since $\boldsymbol{\ell} \times \mathbf{B}$ is outward. (That is, $\boldsymbol{\ell}$ is to the right in the direction of the current, and so by the rule of cross products, $\boldsymbol{\ell} \times \mathbf{B}$ is outward.)

To find the force on the curved part, we must first write an expression for the force $d\mathbf{F}_2$ on the element $d\mathbf{s}$. If $\theta$ is the angle between $\mathbf{B}$ and $d\mathbf{s}$ in Figure 29.8, then the magnitude of $d\mathbf{F}_2$ is given by

$$dF_2 = I|d\mathbf{s} \times \mathbf{B}| = IB \sin\theta\, ds$$

where $ds$ is the length of the small element measured along the circular arc. In order to integrate this expression, we must express $ds$ in terms of the variable $\theta$. Since $s = R\theta$, $ds = R\, d\theta$, and the expression for $dF_2$ can be written

$$dF_2 = IRB \sin\theta\, d\theta$$

To get the *total* force $F_2$ on the curved portion, we can integrate this expression to account for contributions from *all* elements. Note that the direction of the force on every element is the same: *into* the paper (since $d\mathbf{s} \times \mathbf{B}$ is inward). Therefore, the resultant force $\mathbf{F}_2$ on the curved

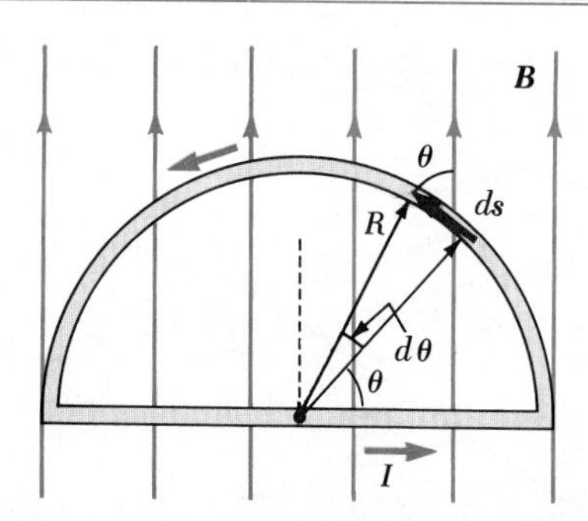


Figure 29.8 (Example 29.2) The net force on a closed current loop in a uniform magnetic field is *zero*. In this case, the force on the straight portion is $2IRB$ and outward, while the force on the curved portion is also $2IRB$ and inward.

wire must also be *into* the paper. Integrating $dF_2$ over the limits $\theta = 0$ to $\theta = \pi$ (that is, the entire semicircle) gives

$$F_2 = IRB \int_0^{\pi} \sin\theta\, d\theta = IRB[-\cos\theta]_0^{\pi}$$

$$= -IRB(\cos\pi - \cos 0) = -IRB(-1 - 1) = 2IRB$$

Since $F_2 = 2IRB$ and is directed *into* the paper while the force on the straight wire $F_1 = 2IRB$ is *out* of the paper, we see that the *net* force on the closed loop is *zero*. This result is consistent with Case II as described above and given by Equation 29.10.

Exercise 2 Find the force $\mathbf{F}_2$ on the semicircular part of the wire by making use of the more direct method, Case I, discussed in the text (Eq. 29.9).

## 29.4 TORQUE ON A CURRENT LOOP IN A UNIFORM MAGNETIC FIELD

In the previous section we showed how a force is exerted on a current-carrying conductor when the conductor is placed in an external magnetic field. With this as a starting point, we shall show that a torque is exerted on a current loop placed in a magnetic field. The results of this analysis will be of great practical value when we discuss motors in a future chapter.

Consider a rectangular loop carrying a current $I$ in the presence of a uniform magnetic field *in the plane of the loop,* as in Figure 29.9a. The forces on the sides of length $a$ are zero since these wires are parallel to the field; hence $d\mathbf{s} \times \mathbf{B} = 0$ for these sides. The magnitude of the forces on the sides of length $b$, however, is given by

$$F_1 = F_2 = IbB$$

The direction of $\mathbf{F}_1$, the force on the left side of the loop, is out of the paper and that of $\mathbf{F}_2$, the force on the right side of the loop, is into the paper. If we were to view the loop from an end view, as in Figure 29.9b, we would see the forces directed as shown. If we assume that the loop is pivoted so that it can rotate about point $O$, we see that these two forces produce a torque about $O$ that rotates the loop clockwise. The magnitude of this torque, $\tau_{max}$, is

$$\tau_{max} = F_1 \frac{a}{2} + F_2 \frac{a}{2} = (IbB)\frac{a}{2} + (IbB)\frac{a}{2} = IabB$$

where the moment arm about $O$ is $a/2$ for each force. Since the area of the loop is $A = ab$, the torque can be expressed as

$$\tau = IAB \tag{29.11}$$

Remember that this result is valid only when the field $\mathbf{B}$ is parallel to the plane of the loop. The sense of the rotation is clockwise when viewed from the bottom end, as indicated in Figure 29.9b. If the current were reversed, the forces would reverse their directions and the rotational tendency would be counterclockwise.

Now suppose the uniform magnetic field makes an angle $\theta$ with respect to a line perpendicular to the plane of the loop, as in Figure 29.10a. For convenience, we shall assume that the field $\mathbf{B}$ is perpendicular to the sides of length $b$. In this case, the magnetic forces $\mathbf{F}_3$ and $\mathbf{F}_4$ on the sides of length $a$ cancel each other and produce no torque since they pass through a common origin. However, the forces $\mathbf{F}_1$ and $\mathbf{F}_2$ acting on the sides of length $b$ form a couple and hence produce a torque about *any point*. Referring to the end view shown in Figure 29.10b, we note that the moment arm of the force $\mathbf{F}_1$ about the point $O$ is equal to $(a/2)\sin\theta$. Likewise, the moment arm of $\mathbf{F}_2$ about $O$ is also $(a/2)\sin\theta$. Since $F_1 = F_2 = IbB$, the net torque about $O$ has a magnitude given by

$$\tau = F_1 \frac{a}{2}\sin\theta + F_2 \frac{a}{2}\sin\theta$$

$$= IbB\left(\frac{a}{2}\sin\theta\right) + IbB\left(\frac{a}{2}\sin\theta\right) = IabB\sin\theta$$

$$= IAB\sin\theta$$

where $A = ab$ is the area of the loop. This result shows that the torque has the *maximum* value $IAB$ when the field is parallel to the plane of the loop ($\theta = 90°$)

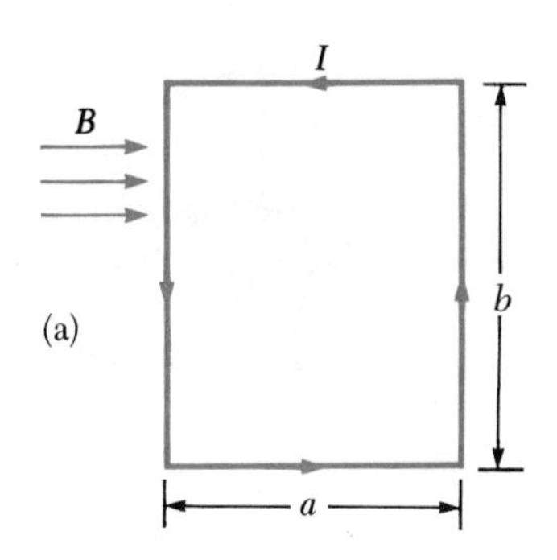


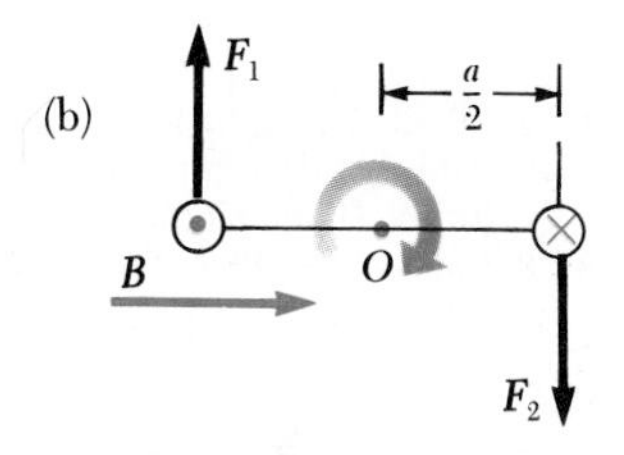


**Figure 29.9** (a) Front view of a rectangular loop in a uniform magnetic field. There are no forces on the sides of width $a$ parallel to $\mathbf{B}$, but there are forces acting on the sides of length $b$. (b) Bottom view of the rectangular loop shows that the forces $\mathbf{F}_1$ and $\mathbf{F}_2$ on the sides of length $b$ create a torque that tends to twist the loop clockwise as shown.

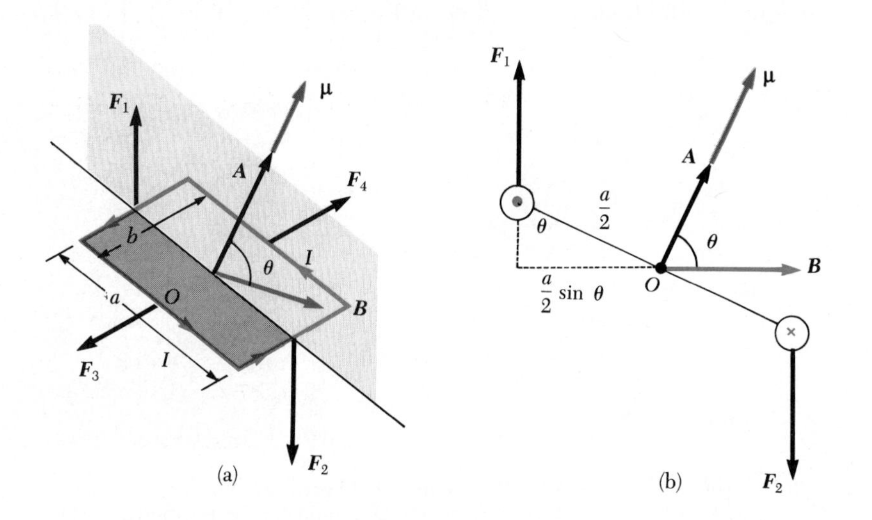


**Figure 29.10** (a) A rectangular current loop whose normal makes an angle $\theta$ with a uniform magnetic field. The forces on the sides of length $a$ cancel while the forces on the sides of width $b$ create a torque on the loop. (b) An end view of the loop. The magnetic moment $\boldsymbol{\mu}$ is in the direction normal to the plane of the loop.

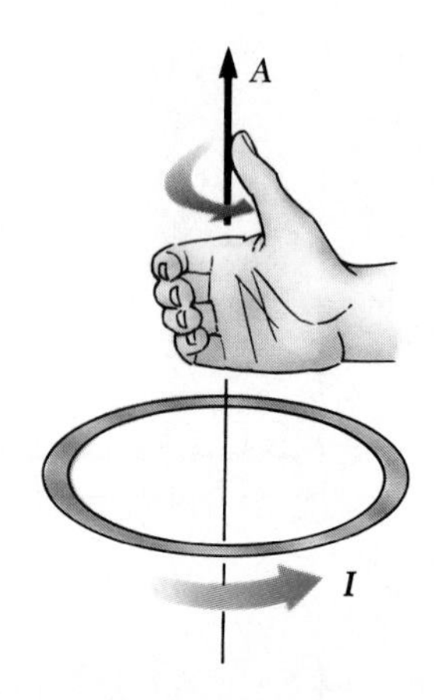


Figure 29.11 A right-hand rule for determining the direction of the vector **A**. The magnetic moment $\mu$ is also in the direction of **A**.

and is *zero* when the field is perpendicular to the plane of the loop ($\theta = 0$). As we see in Figure 29.10, the loop tends to rotate to smaller values of $\theta$ (that is, such that the normal to the plane of the loop rotates toward the direction of the magnetic field).

A convenient vector expression for the torque is the following cross-product relationship:

$$\boldsymbol{\tau} = I\mathbf{A} \times \mathbf{B} \tag{29.12}$$

where **A**, a vector perpendicular to the plane of the loop, has a magnitude equal to the area of the loop. The sense of **A** is determined by the right-hand rule as described in Figure 29.11. By rotating the four fingers of the right hand in the direction of the current in the loop, the thumb points in the direction of **A**. The product $I\mathbf{A}$ is defined to be the **magnetic moment** $\boldsymbol{\mu}$ of the loop. That is,

$$\boldsymbol{\mu} = I\mathbf{A} \tag{29.13}$$

The SI unit of magnetic moment is ampere-meter² (A · m²). Using this definition, the torque can be expressed as

Torque on a current loop

$$\boldsymbol{\tau} = \boldsymbol{\mu} \times \mathbf{B} \tag{29.14}$$

Note that this result is analogous to the torque acting on an electric dipole moment $\mathbf{p}$ in the presence of an external electric field $\mathbf{E}$, where $\boldsymbol{\tau} = \mathbf{p} \times \mathbf{E}$ (Section 26.6). If a coil has $N$ turns all of the same dimensions, the magnetic moment and the torque on the coil will clearly be $N$ times greater than in a single loop.

Although the torque was obtained for a particular orientation of $\mathbf{B}$ with respect to the loop, the equation $\boldsymbol{\tau} = \boldsymbol{\mu} \times \mathbf{B}$ is valid for any orientation. Furthermore, although the torque expression was derived for a rectangular loop, the result is valid for a loop of *any* shape.

It is interesting to note the similarity between the tendency for rotation of a current loop in an external magnetic field and the motion of a compass needle (or pivoted bar magnet) in such a field. Like the current loop, the compass needle and bar magnet can be regarded as magnetic dipoles. The similarity in their magnetic field lines is described in Figure 29.12. Note that one face of

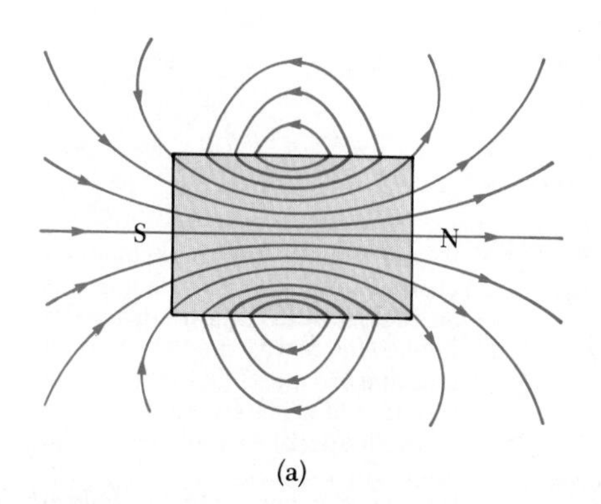


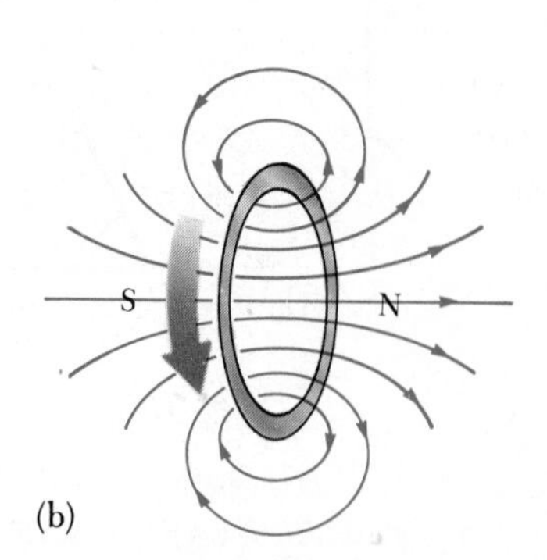


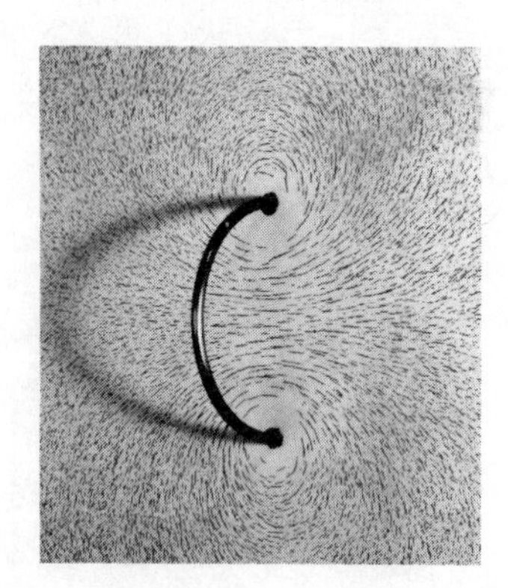


Figure 29.12 The similarity between the magnetic field patterns of (a) a bar magnet and (b) a current loop. (c) Photograph courtesy of Education Development Center, Newton, MA.

the current loop behaves as the north pole of a bar magnet while the opposite face behaves as the south pole. The field lines shown in Figure 29.12 are the patterns due to the bar magnet (Fig. 29.12a) and the current loop (Fig. 29.12b). There is *no* external field present in these diagrams. Furthermore, the diagrams are a simplified, two-dimensional description of the field lines.

**EXAMPLE 29.3 The Magnetic Moment of a Coil**
A rectangular coil of dimensions 5.40 cm × 8.50 cm consists of 25 turns of wire. The coil carries a current of 15 mA. (a) Calculate the magnitude of the magnetic moment of the coil.

*Solution* The magnitude of the magnetic moment of a current loop is given by $\mu = IA$ (see Eq. 29.13), where $A$ is the area of the loop. In this case, $A = (0.0540 \text{ m})(0.0850 \text{ m}) = 4.59 \times 10^{-3} \text{ m}^2$. Since the coil has 25 turns, and assuming that each turn has the same area $A$, we have

$$\mu_{\text{coil}} = \text{NIA} = (25)(15 \times 10^{-3} \text{ A})(4.59 \times 10^{-3} \text{ m}^2)$$

$$= 1.72 \times 10^{-3} \text{ A}\cdot\text{m}^2 = 1.72 \times 10^{-3} \text{ J/T}$$

(b) Suppose a magnetic field of magnitude 0.350 T is applied parallel to the plane of the loop. What is the magnitude of the torque acting on the loop?

*Solution* In general, the torque is given by $\boldsymbol{\tau} = \boldsymbol{\mu} \times \boldsymbol{B}$, where the vector $\boldsymbol{\mu}$ is directed perpendicular to the plane of the loop. In this case, $\boldsymbol{B}$ is *perpendicular* to $\boldsymbol{\mu}_{\text{coil}}$, so that

$$\tau = \mu_{\text{coil}} B = (1.72 \times 10^{-3} \text{ J/T})(0.350 \text{ T})$$

$$= 6.02 \times 10^{-4} \text{ N}\cdot\text{m}$$

Note that this is the basic principle behind the operation of a galvanometer coil discussed in Chapter 28.

**Exercise 3** Calculate the magnitude of the torque on the coil when the 0.350 T magnetic field makes angles of (a) 60° and (b) 0° with $\boldsymbol{\mu}$.
**Answer** (a) $5.21 \times 10^{-4}$ J (b) zero

## 29.5 MOTION OF A CHARGED PARTICLE IN A MAGNETIC FIELD

In Section 29.2 we found that the magnetic force acting on a charged particle moving in a magnetic field is always perpendicular to the velocity of the particle. From this property, it follows that

> the work done by the magnetic force is zero since the displacement of the charge is always perpendicular to the magnetic force. Therefore, a static magnetic field changes the direction of the velocity but does not affect the speed or kinetic energy of the charged particle.

Consider the special case of a positively charged particle moving in a uniform external magnetic field with its initial velocity vector *perpendicular* to the field. Let us assume that the magnetic field is *into* the page (this is indicated by the crosses in Fig. 29.13). The crosses are used to represent the *tail* of $\boldsymbol{B}$, since $\boldsymbol{B}$ is directed *into* the page. Later, we shall use dots to represent the *tip* of a vector directed *out* of the page. Figure 29.13 shows that the

> charged particle moves in a circle whose plane is perpendicular to the magnetic field.

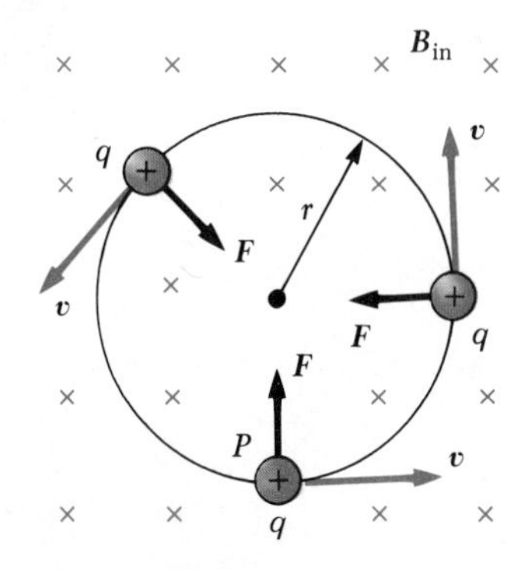

**Figure 29.13** When the velocity of a charged particle is perpendicular to a uniform magnetic field, the particle moves in a circular path whose plane is perpendicular to $\boldsymbol{B}$, which is directed into the page (the blue crosses represent the tail of the vector). The magnetic force, $\boldsymbol{F}$, on the charge is always directed toward the center of the circle.

This is because the magnetic force $\boldsymbol{F}$ is at right angles to $\boldsymbol{v}$ and $\boldsymbol{B}$ and has a constant magnitude equal to $qvB$. As the force $\boldsymbol{F}$ deflects the particle, the directions of $\boldsymbol{v}$ and $\boldsymbol{F}$ change continuously, as shown in Figure 29.13. Therefore the force $\boldsymbol{F}$ is a *centripetal force*, which changes only the direction of $\boldsymbol{v}$ while the speed remains constant. The sense of the rotation, as shown in

Figure 29.13, is counterclockwise for a positive charge. If $q$ were negative, the sense of the rotation would be reversed, or clockwise. Since the resultant force $\boldsymbol{F}$ in the radial direction has a magnitude of $qvB$, we can equate this to the required centripetal force, which is the mass $m$ multiplied by the centripetal acceleration $v^2/r$. From Newton's second law, we find that

$$F = qvB = \frac{mv^2}{r}$$

$$r = \frac{mv}{qB} \tag{29.15}$$

That is, the radius of the path is proportional to the momentum $mv$ of the particle and is inversely proportional to the magnetic field. The angular frequency of the rotating charged particle is given by

$$\omega = \frac{v}{r} = \frac{qB}{m} \tag{29.16}$$

The period of its motion (the time for one revolution) is equal to the circumference of the circle divided by the speed of the particle:

$$T = \frac{2\pi r}{v} = \frac{2\pi}{\omega} = \frac{2\pi m}{qB} \tag{29.17}$$

These results show that the angular frequency and period of the circular motion do not depend on the speed of the particle or the radius of the orbit. The angular frequency $\omega$ is often referred to as the **cyclotron frequency** since charged particles circulate at this frequency in one type of accelerator called a *cyclotron*, which will be discussed in Section 29.6.

If a charged particle moves in a uniform magnetic field with its velocity at some arbitrary angle to $\boldsymbol{B}$, its path is a helix. For example, if the field is in the $x$ direction as in Figure 29.14, there is no component of force in the $x$ direction, and hence $a_x = 0$ and the $x$ component of velocity, $v_x$, remains constant. On the other hand, the magnetic force $q\boldsymbol{v} \times \boldsymbol{B}$ causes the components $v_y$ and $v_z$ to

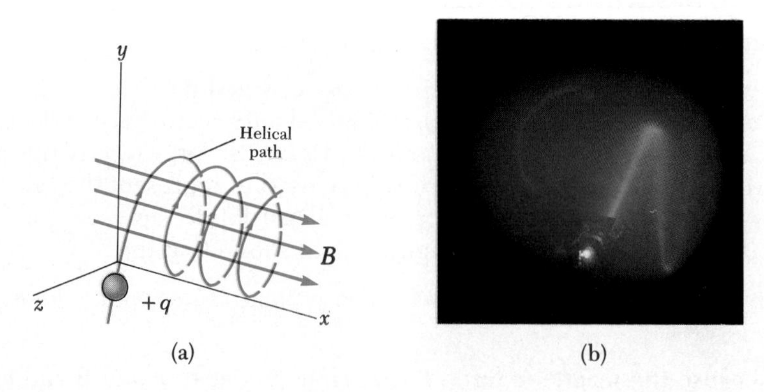


**Figure 29.14** (a) A charged particle having a velocity vector that has a component parallel to a uniform magnetic field moves in a helical path. (b) Photograph of the helical path followed by an electron beam when the beam is directed at an arbitrary angle to the magnetic field. (Photo courtesy of Henry Leap and Jim Lehman)

change in time, and the resulting motion is a helix having its axis parallel to the $\boldsymbol{B}$ field. The projection of the path onto the $yz$ plane (viewed along the $x$ axis) is a circle. (The projections of the path onto the $xy$ and $xz$ planes are sinusoids!) Equations 29.15 to 29.17 still apply, provided that $v$ is replaced by $v_\perp = \sqrt{v_y^2 + v_z^2}$.

**EXAMPLE 29.4 A Proton Moving Perpendicular to a Uniform Magnetic Field**

A proton is moving in a circular orbit of radius 14 cm in a uniform magnetic field of magnitude 0.35 T directed perpendicular to the velocity of the proton. Find the orbital speed of the proton.

*Solution* From Equation 29.15, we get

$$v = \frac{qBr}{m} = \frac{(1.60 \times 10^{-19}\ \text{C})(0.35\ \text{T})(14 \times 10^{-2}\ \text{m})}{1.67 \times 10^{-27}\ \text{kg}}$$

$$= 4.69 \times 10^6\ \text{m/s}$$

*Exercise 4* If an electron moves perpendicular to the same magnetic field with this speed, what is the radius of its circular orbit?
*Answer* $7.63 \times 10^{-5}$ m.

**EXAMPLE 29.5 The Bending of an Electron Beam**

In an experiment designed to measure the strength of a uniform magnetic field produced by a set of coils, the electrons are accelerated from rest through a potential difference of 350 V, and the beam associated with the electrons is measured to have a radius of 7.5 cm as in Figure 29.15. Assuming the magnetic field is perpendicular to the beam, (a) what is the magnitude of the magnetic field?

*Solution* First, we must calculate the speed of electrons using the fact that the increase in kinetic energy of the electrons must equal the change in their potential energy, $|e|V$ (because of conservation of energy). Since $K_i = 0$ and $K_f = mv^2/2$, we have

$$\tfrac{1}{2}mv^2 = |e|V$$

$$v = \sqrt{\frac{2|e|V}{m}} = \sqrt{\frac{2(1.60 \times 10^{-19}\ \text{C})(350\ \text{V})}{9.11 \times 10^{-31}\ \text{kg}}}$$

$$= 1.11 \times 10^7\ \text{m/s}$$

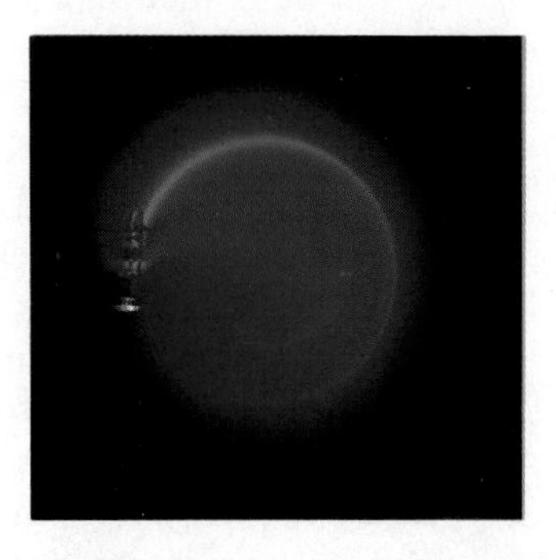

**Figure 29.15** The bending of an electron beam in an external magnetic field. The tube contains gas at very low pressure, and the beam is made visible as the electrons collide with the gas atoms, which in turn emit visible light. The apparatus used to take this photograph is part of a system used to measure the ratio $e/m$. (Courtesy of Henry Leap and Jim Lehman)

We can now use Equation 29.15 to find the strength of the magnetic field:

$$B = \frac{mv}{|e|r} = \frac{(9.11 \times 10^{-31}\ \text{kg})\ (1.11 \times 10^7\ \text{m/s})}{(1.60 \times 10^{-19}\ \text{C})(0.075\ \text{m})}$$

$$= 8.43 \times 10^{-4}\ \text{T}$$

(b) What is the angular frequency of revolution of the electrons?

*Solution* Using Equation 29.16, we find

$$\omega = \frac{v}{r} = \frac{1.11 \times 10^7\ \text{m/s}}{0.075\ \text{m}}$$

$$= 1.48 \times 10^8\ \text{rad/s}$$

*Exercise 5* What is the period of revolution of the electrons?
*Answer* $T = 42.5$ ns.

When charged particles move in a nonuniform magnetic field, the motion is rather complex. For example, in a magnetic field that is strong at the ends and weak in the middle, as in Figure 29.16, the particles can oscillate back and forth between the end points. Such a field can be produced by two current

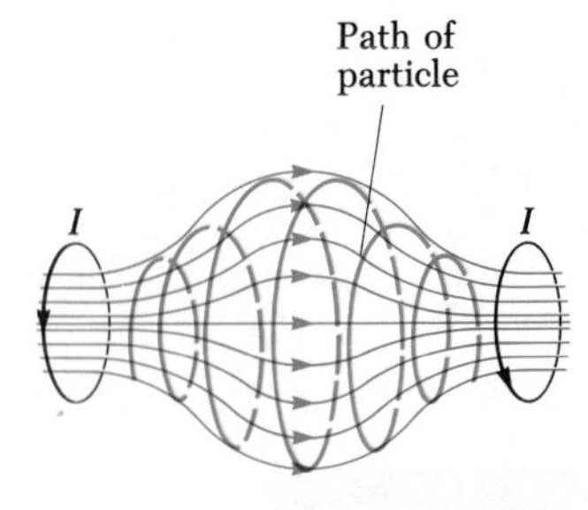


Figure 29.16 A charged particle moving in a nonuniform magnetic field represented by the blue lines (a magnetic bottle) spirals about the field (red path) and oscillates between the end points.

loops as in Figure 29.16. In this case, a charged particle starting at one end will spiral along the field lines until it reaches the other end, where it reverses its path and spirals back. This configuration is known as a *magnetic bottle* because charged particles can be trapped in it. This concept has been used to confine very hot gases ($T$ greater than $10^6$ K) consisting of electrons and positive ions, known as **plasmas.** Such a plasma-confinement scheme could play a crucial role in achieving a controlled nuclear fusion process, which could supply us with an almost endless source of energy. Unfortunately, the magnetic bottle has its problems. If a large number of particles is trapped, collisions between the particles cause them to eventually "leak" from the system.

The Van Allen radiation belts consist of charged particles (mostly electrons and protons) surrounding the earth in doughnut-shaped regions (Fig. 29.17a). These radiation belts were discovered in 1958 by a team of researchers under the direction of James Van Allen, using data gathered by instrumentation aboard the Explorer I satellite. The charged particles, trapped by the earth's nonuniform magnetic field, spiral around the earth's field lines from pole to pole. These particles originate mainly from the sun, but some come from stars and other heavenly objects. For this reason, these particles are given the name *cosmic rays*. Most cosmic rays are deflected by the earth's magnetic field and never reach the earth. However, some become trapped, and these make up the Van Allen belts. When these charged particles are in the earth's atmosphere over the poles, they often collide with other atoms, causing them to emit visible light. This is the origin of the beautiful Aurora Borealis, or Northern Lights (Fig. 29.17b). A similar phenomenon seen in the southern hemisphere is called the Aurora Australis.

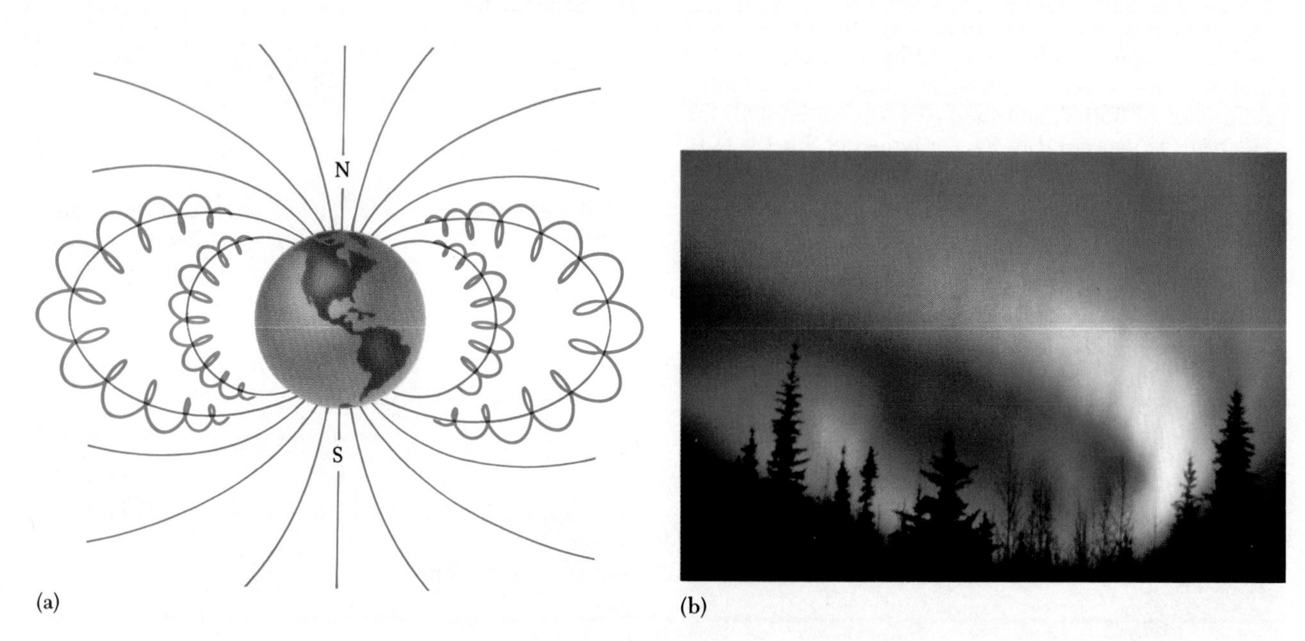


Figure 29.17 (a) The Van Allen belts are made up of charged particles (electrons and protons) trapped by the earth's nonuniform magnetic field. The field lines are in blue and the particle paths in red. (b) Aurora borealis, the Northern Lights, photographed near Fairbanks, Alaska. Auroras occur when cosmic rays—electrically charged particles originating mainly from the sun—become trapped in the earth's atmosphere over earth's magnetic poles and collide with other atoms, resulting in the emission of visible light. (Jack Finch/Science Photo Library)

## *29.6 APPLICATIONS OF THE MOTION OF CHARGED PARTICLES IN A MAGNETIC FIELD

In this section we describe some important devices that involve the motion of charged particles in uniform magnetic fields. For many situations, the charge under consideration will be moving with a velocity $\boldsymbol{v}$ in the presence of both an electric field $\boldsymbol{E}$ and a magnetic field $\boldsymbol{B}$. Therefore, the charge will experience both an electric force $q\boldsymbol{E}$ and a magnetic force $q\boldsymbol{v} \times \boldsymbol{B}$, and so the total force on the charge will be given by

Lorentz force

$$\boldsymbol{F} = q\boldsymbol{E} + q\boldsymbol{v} \times \boldsymbol{B} \qquad (29.18)$$

The force described by Equation 29.18 is known as the **Lorentz force.**

### Velocity Selector

In many experiments involving the motion of charged particles, it is important to have a source of particles that move with essentially the same velocity. This can be achieved by applying a combination of an electric field and a magnetic field oriented as shown in Figure 29.18. A uniform electric field vertically downward is provided by a pair of charged parallel plates, while a uniform magnetic field is applied perpendicular to the page (indicated by the crosses). Assuming that $q$ is positive, we see that the magnetic force $q\boldsymbol{v} \times \boldsymbol{B}$ is upward and the electric force $q\boldsymbol{E}$ is downward. If the fields are chosen such that the electric force balances the magnetic force, the particle will move in a straight horizontal line and emerge from the slit at the right. If we equate the upward magnetic force $qvB$ to the downward electric force $qE$, we find $qvB = qE$, from which we get

$$v = \frac{E}{B} \qquad (29.19)$$

Note that only those particles having this velocity will pass undeflected through the perpendicular electric and magnetic fields. In practice, $\boldsymbol{E}$ and $\boldsymbol{B}$ are adjusted to provide this specific velocity. The magnetic force acting on particles with velocities greater than this will be stronger than the electric force, and these particles will be deflected upward. Those with velocities less than this will be deflected downward.

Bubble chamber photograph. The spiral tracks at the bottom of the photograph are an electron-positron pair (left and right, respectively) formed by a gamma ray interacting with a hydrogen nucleus. An applied magnetic field causes the electron and the positron to be deflected in opposite directions. The track leaving from the cusp between the two spirals is an additional electron knocked out of a hydrogen atom during this interaction. (G. Holton, F.J. Rutherford, F.G. Watson, *Project Physics*, New York: HRW, 1981.)

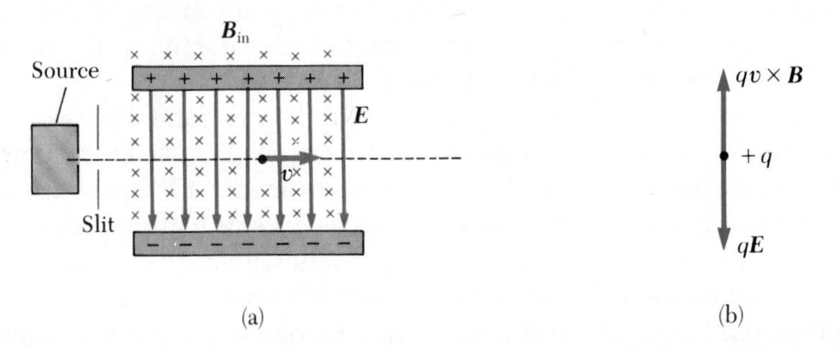


**Figure 29.18** (a) A velocity selector. When a positively charged particle is in the presence of both an inward magnetic field as indicated by the blue crosses, and a downward electric field indicated by the red arrows, it experiences both an electric force $q\boldsymbol{E}$ downward and a magnetic force $q\boldsymbol{v} \times \boldsymbol{B}$ upward. (b) When these forces balance each other as shown here, the particle moves in a horizontal line through the fields.

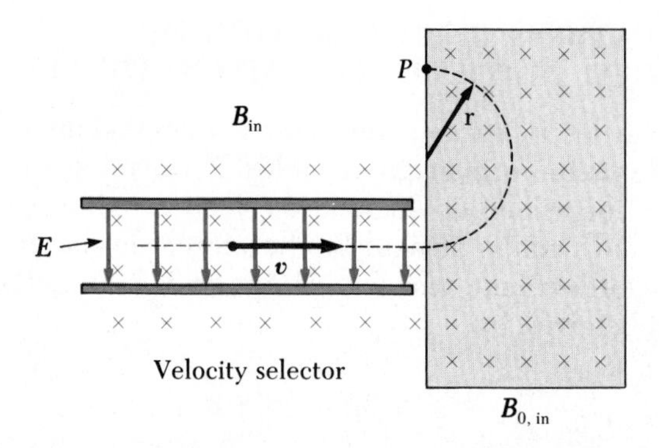


**Figure 29.19** A mass spectrometer. Charged particles are first sent through a velocity selector. They then enter a region where the magnetic field $\boldsymbol{B}_0$ (inward) causes positive ions to move in a semicircular path and strike a photographic film at $P$.

## The Mass Spectrometer

The **mass spectrometer** is an instrument that separates atomic and molecular ions according to their mass-to-charge ratio. In one version, known as the *Bainbridge mass spectrometer,* a beam of ions first passes through a velocity selector and then enters a uniform magnetic field $\boldsymbol{B}_0$ directed into the paper (Fig. 29.19). Upon entering the magnetic field $\boldsymbol{B}_0$, the ions move in a semicircle of radius $r$ before striking a photographic plate at $P$. From Equation 29.15, we can express the ratio $m/q$ as

$$\frac{m}{q} = \frac{rB_0}{v} \tag{29.20}$$

Assuming that the magnitude of the magnetic field in the region of the velocity selector is $B$ and using Equation 29.19, which gives the speed of the particle, we find that

$$\frac{m}{q} = \frac{rB_0B}{E} \tag{29.21}$$

Therefore, one can determine $m/q$ by measuring the radius of curvature and knowing the fields $B$, $B_0$, and $E$. In practice, one usually measures the masses of various isotopes of a given ion with the same charge $q$. Hence, the mass ratios can be determined even if $q$ is unknown.

A variation of this technique was used by Joseph John Thomson (1856–1940) in 1897 to measure the ratio $e/m$ for electrons. Figure 29.20a shows the basic apparatus used by Thomson in his measurements. Electrons are accelerated from the cathode to the anodes, collimated by slits in the anodes, and then allowed to drift into a region of crossed (perpendicular) electric and magnetic fields. The simultaneously applied $\boldsymbol{E}$ and $\boldsymbol{B}$ fields are first adjusted to produce an undeflected beam. If the $\boldsymbol{B}$ field is then turned off, the $\boldsymbol{E}$ field alone produces a measurable beam deflection on the phosphorescent screen. From the size of the deflection and the measured values of $\boldsymbol{E}$ and $\boldsymbol{B}$, the charge to mass ratio, $e/m$, may be determined. The results of this crucial experiment represent the discovery of the electron as a fundamental particle of nature.

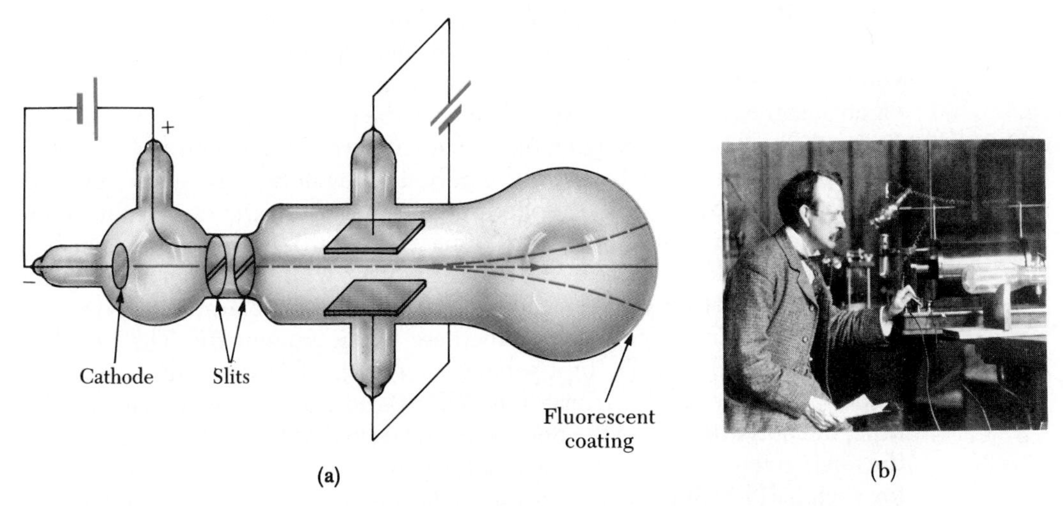


**Figure 29.20** (a) Thomson's apparatus for measuring $q/m$. Electrons are accelerated from the cathode, pass through two slits, and are deflected by both an electric field and a magnetic field (not shown, but directed into the paper). The deflected beam then strikes a phosphorescent screen. (b) J.J. Thomson in the Cavendish Laboratory, University of Cambridge.

### The Cyclotron

The **cyclotron,** invented in 1934 by E. O. Lawrence and M. S. Livingston, is a machine that can accelerate charged particles to very high velocities. Both electric and magnetic forces play a key role in the operation of the cyclotron. The energetic particles that emerge from the cyclotron are used to bombard other nuclei; this bombardment in turn produces nuclear reactions of interest to researchers. A number of hospitals use cyclotron facilities to produce radioactive substances that can be used in diagnosis and treatment.

A schematic drawing of a cyclotron is shown in Figure 29.21. Motion of the charges occurs in two semicircular containers, $D_1$ and $D_2$, referred to as

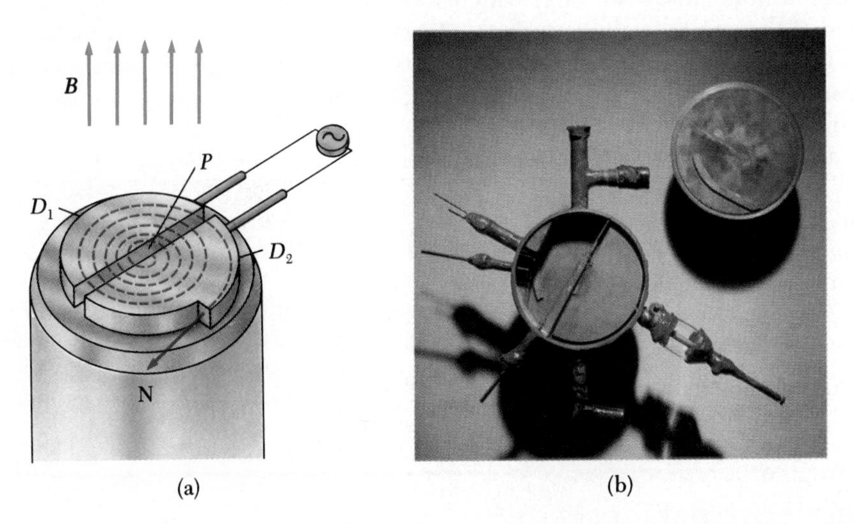


**Figure 29.21** (a) The cyclotron consists of an ion source, two dees across which an alternating voltage is applied, and a uniform magnetic field provided by an electromagnet. (The south pole of the magnet is not shown.) (b) The first cyclotron, invented by E.O. Lawrence and M.S. Livingston in 1934. (Courtesy of Lawrence Berkeley Laboratory, University of California)

*dees*. The dees are evacuated in order to minimize energy losses resulting from collisions between the ions and air molecules. A high-frequency alternating voltage is applied to the dees, and a uniform magnetic field provided by an electromagnet is directed perpendicular to the dees. Positive ions released at $P$ near the center of the magnet move in a semicircular path and arrive back at the gap in a time $T/2$, where $T$ is the period of revolution, given by Equation 29.17. The frequency of the applied voltage $V$ is adjusted such that the polarity of the dees is reversed in the same time it takes the ions to complete one half of a revolution. If the phase of the applied voltage is adjusted such that $D_2$ is at a *lower* potential than $D_1$ by an amount $V$, the ion will accelerate across the gap to $D_2$ and its kinetic energy will increase by an amount $qV$. The ion then continues to move in $D_2$ in a semicircular path of larger radius (since its velocity has increased). After a time $T/2$, it again arrives at the gap. By this time, the potential across the dees is reversed (so that $D_1$ is now negative) and the ion is given another "kick" across the gap. The motion continues such that for each half revolution, the ion gains additional kinetic energy equal to $qV$. When the radius of its orbit is nearly that of the dees, the energetic ions leave the system through an exit slit as shown in Figure 29.21.

It is important to note that the operation of the cyclotron is based on the fact that the time for one revolution is *independent* of the speed (or radius) of the ion.

We can obtain the maximum kinetic energy of the ion when it exits from the cyclotron in terms of the radius $R$ of the dees. From Equation 29.15 we find that $v = qBR/m$. Hence, the kinetic energy is given by

$$K = \tfrac{1}{2}mv^2 = \frac{q^2B^2R^2}{2m} \qquad (29.22)$$

When the energy of the ions exceeds about 20 MeV, relativistic effects come into play and the masses of the ions no longer remain constant. (Such effects will be discussed in Chapter 39.) For this reason, the period of the orbit increases and the rotating ions do not remain in phase with the applied voltage. Accelerators have been built which solve this problem by modifying the period of the applied voltage such that it remains in phase with the rotating ion. In 1977, protons were accelerated to 400 GeV (1 GeV = $10^9$ eV) in an accelerator in Batavia, Illinois. The system incorporates 954 magnets and has a circumference of 6.3 km (4.1 miles)!

**EXAMPLE 29.6 A Proton Accelerator**
Calculate the maximum kinetic energy of protons in a cyclotron of radius 0.50 m in a magnetic field of 0.35 T.

*Solution* Using Equation 29.22, we find that

$$K = \frac{q^2B^2R^2}{2m} = \frac{(1.6 \times 10^{-19}\ \text{C})^2(0.35\ \text{T})^2(0.50\ \text{m})^2}{2(1.67 \times 10^{-27}\ \text{kg})}$$

$$K = 2.34 \times 10^{-13}\ \text{J} = 1.46\ \text{MeV}$$

In this calculation, we have used the conversions 1 eV = $1.6 \times 10^{-19}$ J and 1 MeV = $10^6$ eV. The kinetic energy acquired by the protons is equivalent to the energy they would gain if they were accelerated through a potential difference of 1.46 MV!

## *29.7 THE HALL EFFECT

In 1879 Edwin Hall discovered that when a current-carrying conductor is placed in a magnetic field, a voltage is generated in a direction perpendicular to both the current and the magnetic field. This observation, known as the *Hall*

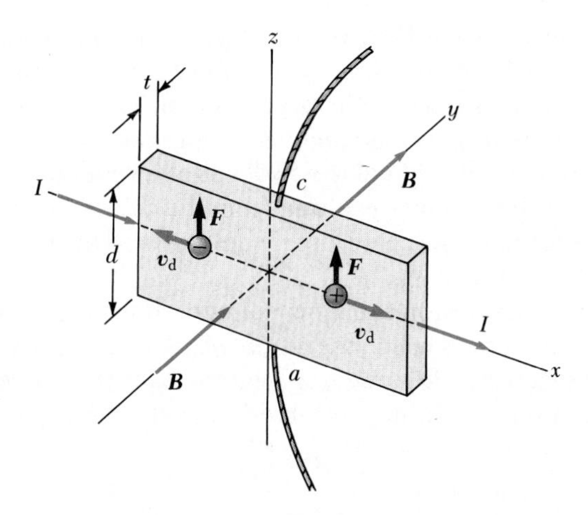


**Figure 29.22** To observe the Hall effect, a magnetic field is applied to a current-carrying conductor. When $I$ is in the $x$ direction and $\mathbf{B}$ in the $y$ direction as shown, both positive and negative charge carriers are deflected upward in the magnetic field. The Hall voltage is measured between points $a$ and $c$.

*effect,* arises from the deflection of charge carriers to one side of the conductor as a result of the magnetic force experienced by the charge carriers. A proper analysis of experimental data gives information regarding the sign of the charge carriers and their density. The effect also provides a convenient technique for measuring magnetic fields.

The arrangement for observing the Hall effect consists of a conductor in the form of a flat strip carrying a current $I$ in the $x$ direction as in Figure 29.22. A uniform magnetic field $\mathbf{B}$ is applied in the $y$ direction. If the charge carriers are electrons moving in the negative $x$ direction with a velocity $\mathbf{v}_d$, they will experience an *upward* magnetic force $\mathbf{F}$. Hence, the electrons will be deflected upward, accumulating at the upper edge and leaving an excess positive charge at the lower edge (Fig. 29.23a). This accumulation of charge at the edges will continue until the electrostatic field set up by this charge separation balances the magnetic force on the charge carriers. When this equilibrium condition is reached, the electrons will no longer be deflected upward. A sensitive voltmeter or potentiometer connected across the sample as shown in Figure 29.23 can be used to measure the potential difference generated across

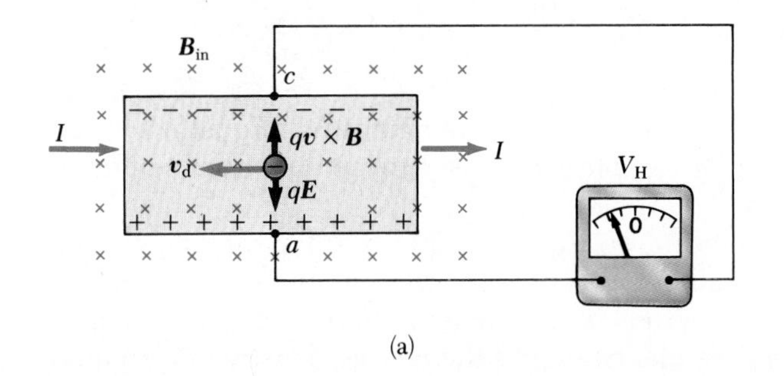


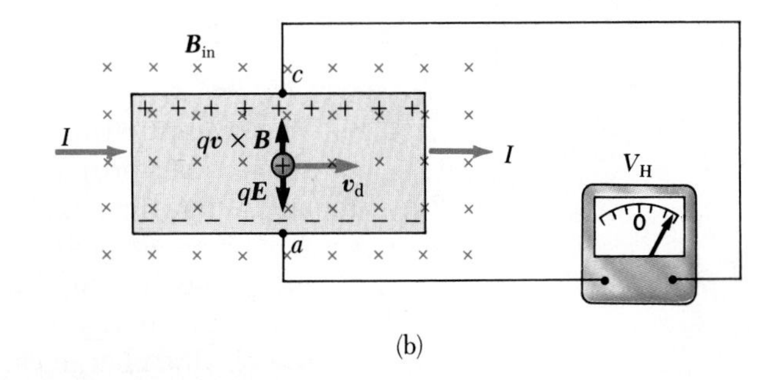


**Figure 29.23** (a) When the charge carriers are negative, the upper edge becomes negatively charged, and $c$ is at a lower potential than $a$. (b) When the charge carriers are positive, the upper edge becomes positively charged, and $c$ is at a higher potential than $a$. In either case, the charge carriers are no longer deflected when the edges become fully charged, that is, when there is a balance between the electrostatic force trying to combine the charges and the magnetic deflection force.

the conductor, known as the **Hall voltage** $V_H$. If the charge carriers are positive, and hence move in the positive $x$ direction as in Figure 29.23b, they will also experience an *upward* magnetic force $q\mathbf{v}_d \times \mathbf{B}$. This produces a buildup of positive charge on the upper edge and leaves an excess of negative charge on the lower edge. Hence, the sign of the Hall voltage generated in the sample is opposite the sign of the voltage resulting from the deflection of electrons. The sign of the charge carriers can therefore be determined from a measurement of the polarity of the Hall voltage.

To find an expression for the Hall voltage, first note that the magnetic force on the charge carriers has a magnitude $qv_dB$. In equilibrium, this force is balanced by the electrostatic force $qE_H$, where $E_H$ is the electric field due to the charge separation (sometimes referred to as the *Hall field*). Therefore,

$$qv_dB = qE_H$$

$$E_H = v_dB$$

If $d$ is taken to be the width of the conductor, then the Hall voltage $V_H$ measured by the potentiometer is equal to $E_Hd$, or

$$V_H = E_Hd = v_dBd \tag{29.23}$$

Thus, we see that the measured Hall voltage gives a value for the drift velocity of the charge carriers if $d$ and $B$ are known.

The number of charge carriers per unit volume (or charge density), $n$, can be obtained by measuring the current in the sample. From Equation 27.4, the drift velocity can be expressed as

$$v_d = \frac{I}{nqA} \tag{29.24}$$

where $A$ is the cross-sectional area of the conductor. Substituting Equation 29.24 into Equation 29.23 we obtain

The Hall voltage

$$V_H = \frac{IBd}{nqA} \tag{29.25}$$

Since $A = td$, where $t$ is the thickness of the sample, we can also express Equation 29.25 as

$$V_H = \frac{IB}{nqt} \tag{29.26}$$

The quantity $1/nq$ is referred to as the **Hall coefficient** $R_H$. Equation 29.26 shows that a properly calibrated sample can be used to measure the strength of an unknown magnetic field.

Since all quantities appearing in Equation 29.26 other than $nq$ can be measured, a value for the Hall coefficient is readily obtained. The sign and magnitude of $R_H$ give the sign of the charge carriers and their density. In most metals, the charge carriers are electrons and the charge density determined from Hall effect measurements is in good agreement with calculated values for monovalent metals, such as Li, Na, Cu, and Ag, where $n$ is approximately equal to the number of valence electrons per unit volume. However, this classical model is not valid for metals such as Fe, Bi, and Cd or for semiconductors such as silicon and germanium. These discrepancies can be explained only by using a model based on the quantum nature of solids.

**EXAMPLE 29.7 The Hall Effect for Copper**
A rectangular copper strip 1.5 cm wide and 0.1 cm thick carries a current of 5 A. A 1.2-T magnetic field is applied perpendicular to the strip as in Figure 29.23. Find the resulting Hall voltage.

*Solution* If we assume there is one electron per atom available for conduction, then we can take the charge density to be $n = 8.48 \times 10^{28}$ electrons/m³ (Example 27.1). Substituting this value and the given data into Equation 29.26 gives

$$V_H = \frac{IB}{nqt}$$

$$= \frac{(5\ \text{A})(1.2\ \text{T})}{(8.48 \times 10^{28}\ \text{m}^{-3})(1.6 \times 10^{-19}\ \text{C})(0.1 \times 10^{-2}\ \text{m})}$$

$$V_H = 0.442\ \mu\text{V}$$

Hence, the Hall voltage is quite small in good conductors. Note that the width of this sample is not needed in this calculation.

In semiconductors, where $n$ is much smaller than in monovalent metals, one finds a larger Hall voltage since $V_H$ varies as the inverse of $n$. Current levels of the order of 1 mA are generally used for such materials. Consider a piece of silicon with the same dimensions as the copper strip, with $n = 10^{20}$ electrons/m³. Taking $B = 1.2$ T and $I = 0.1$ mA, we find that $V_H = 7.5$ mV. Such a voltage is readily measured with a potentiometer.

## *29.8 THE QUANTUM HALL EFFECT

Although the Hall effect was discovered over one hundred years ago, it continues to be one of the most valuable techniques for helping scientists understand the electronic properties of metals and semiconductors. For example, in 1980, scientists reported that at low temperatures and very strong magnetic fields, a two-dimensional system of electrons in a semiconductor exhibits a conductivity given by $\sigma = i(e^2/h)$, where $i$ is a small integer, $e$ is the electronic charge, and $h$ is an atomic constant called Planck's constant. This behavior manifests itself as a series of plateaus in the Hall voltage as the applied magnetic field is varied. The quantized nature of this two-dimensional conductivity (or resistivity) was totally unanticipated. As its discoverer, Klaus von Klitzing stated, "It is quite astonishing that it is the total macroscopic conductance of the Hall device which is quantized rather than some idealized microscopic conductivity."

One of the important consequences of the quantum Hall effect is the ability to measure the ratio of fundamental constants, $e^2/h$, to an accuracy of at least one part in $10^5$. This provides a very accurate measure of the dimensionless fine structure constant, given by $\alpha = e^2/hc \approx 1/137$, since $c$ is an *exactly* defined quantity (the speed of light). In addition, the quantum Hall effect provides scientists with a new and convenient standard of resistance. The 1985 Nobel Prize in physics was awarded to von Klitzing for this fundamental discovery.

Another great surprise occurred in 1982 when scientists announced that in some nearly ideal samples at very low temperatures, the Hall conductivity could take on both integer values of $e^2/h$ and *fractional* values of $e^2/h$. Undoubtedly, future discoveries in this and related areas of science will continue to improve our understanding of the nature of matter.

## SUMMARY

The **magnetic force** that acts on a charge $q$ moving with a velocity $\boldsymbol{v}$ in an external magnetic field $\boldsymbol{B}$ is given by

Magnetic force on a charged particle moving in a magnetic field

$$\boldsymbol{F} = q\boldsymbol{v} \times \boldsymbol{B} \tag{29.1}$$

That is, the magnetic force is in a direction perpendicular both to the velocity of the particle and to the field. The *magnitude* of the magnetic force is given by

$$F = qvB \sin \theta \tag{29.2}$$

where $\theta$ is the angle between $\boldsymbol{v}$ and $\boldsymbol{B}$. From this expression, we see that $F = 0$ when $\boldsymbol{v}$ is parallel to (or opposite) $\boldsymbol{B}$. Furthermore, $F = qvB$ when $\boldsymbol{v}$ is perpendicular to $\boldsymbol{B}$.

The SI unit of $\boldsymbol{B}$ is the **weber per square meter** (Wb/m²), also called the **tesla** (T), where

$$[\mathrm{B}] = \mathrm{T} = \frac{\mathrm{Wb}}{\mathrm{m}^2} = \frac{\mathrm{N}}{\mathrm{A} \cdot \mathrm{m}} \tag{29.3}$$

If a straight conductor of length $\ell$ carries a current $I$, the force on that conductor when placed in a uniform *external* magnetic field $\boldsymbol{B}$ is given by

Force on a straight wire carrying a current

$$\boldsymbol{F} = I\boldsymbol{\ell} \times \boldsymbol{B} \tag{29.5}$$

where the direction of $\boldsymbol{\ell}$ is in the direction of the current and $|\boldsymbol{\ell}| = \ell$.

If an arbitrarily shaped wire carrying a current $I$ is placed in an *external* magnetic field, the force on a very small segment $d\boldsymbol{s}$ is given by

Force on a current element

$$d\boldsymbol{F} = I\, d\boldsymbol{s} \times \boldsymbol{B} \tag{29.6}$$

To determine the total force on the wire, one has to integrate Equation 29.6, keeping in mind that both $\boldsymbol{B}$ and $d\boldsymbol{s}$ may vary at each point.

The net magnetic force on any *closed* loop carrying a current in a uniform *external* magnetic field is *zero.*

The force on a current-carrying conductor of arbitrary shape in a uniform magnetic field is given by

Force on a wire in a uniform field

$$\boldsymbol{F} = I\boldsymbol{\ell}' \times \boldsymbol{B} \tag{29.9}$$

where $\boldsymbol{\ell}'$ is a vector directed from one end of the conductor to the opposite end.

The **magnetic moment** $\boldsymbol{\mu}$ of a current loop carrying a current $I$ is

Magnetic moment of a current loop

$$\boldsymbol{\mu} = I\boldsymbol{A} \tag{29.13}$$

where $\boldsymbol{A}$ is perpendicular to the plane of the loop and $|\boldsymbol{A}|$ is equal to the area of the loop. The SI unit of $\boldsymbol{\mu}$ is A · m².

The torque $\boldsymbol{\tau}$ on a current loop when the loop is placed in a uniform *external* magnetic field $\boldsymbol{B}$ is given by

Torque on a current loop

$$\boldsymbol{\tau} = \boldsymbol{\mu} \times \boldsymbol{B} \tag{29.14}$$

When a charged particle moves in an external magnetic field, the work done by the magnetic force on the particle is *zero* since the displacement is always *perpendicular* to the direction of the magnetic force. The external magnetic field can alter the direction of the velocity vector, but it cannot change the speed of the particle.

If a charged particle moves in a uniform external magnetic field such that its initial velocity is *perpendicular* to the field, the particle will move in a circle whose plane is *perpendicular* to the magnetic field. The radius $r$ of the circular path is given by

$$r = \frac{mv}{qB} \tag{29.15}$$

where $m$ is the mass of the particle and $q$ is its charge. The angular frequency (cyclotron frequency) of the rotating charged particle is given by

Cyclotron frequency

$$\omega = \frac{qB}{m} \tag{29.16}$$

If a charged particle is moving in the presence of both a magnetic field and an electric field, the total force on the charge is given by the **Lorentz force**,

Lorentz force

$$\boldsymbol{F} = q\boldsymbol{E} + q\boldsymbol{v} \times \boldsymbol{B} \tag{29.18}$$

That is, the charge experiences both an electric force $q\boldsymbol{E}$ and a magnetic force $q\boldsymbol{v} \times \boldsymbol{B}$.

## QUESTIONS

1. At a given instant, a proton moves in the positive $x$ direction in a region where there is a magnetic field in the negative $z$ direction. What is the direction of the magnetic force? Will the proton continue to move in the positive $x$ direction? Explain.
2. Two charged particles are projected into a region where there is a magnetic field perpendicular to their velocities. If the charges are deflected in opposite directions, what can you say about them?
3. If a charged particle moves in a straight line through some region of space, can you say that the magnetic field in that region is zero?
4. Suppose an electron is chasing a proton up this page when suddenly a magnetic field is formed perpendicular to the page. What will happen to the particles?
5. Why does the picture on a TV screen become distorted when a magnet is brought near the screen?
6. How can the motion of a moving charged particle be used to distinguish between a magnetic field and an electric field? Give a specific example to justify your argument.
7. List several similarities and differences in electric and magnetic forces.
8. Justify the following statement: "It is impossible for a constant (i.e., time independent) magnetic field to alter the speed of a charged particle."
9. In view of the above statement, what is the role of a magnetic field in a cyclotron?
10. A current-carrying conductor experiences no magnetic force when placed in a certain manner in a uniform magnetic field. Explain.
11. Is it possible to orient a current loop in a uniform magnetic field such that the loop will not tend to rotate? Explain.
12. How can a current loop be used to determine the presence of a magnetic field in a given region of space?
13. What is the *net* force on a compass needle in a uniform magnetic field?
14. What type of magnetic field is required to exert a resultant force on a magnetic dipole? What will be the direction of the resultant force?
15. A proton moving horizontally enters a region where there is a uniform magnetic field perpendicular to the proton's velocity, as shown in Figure 29.24. Describe

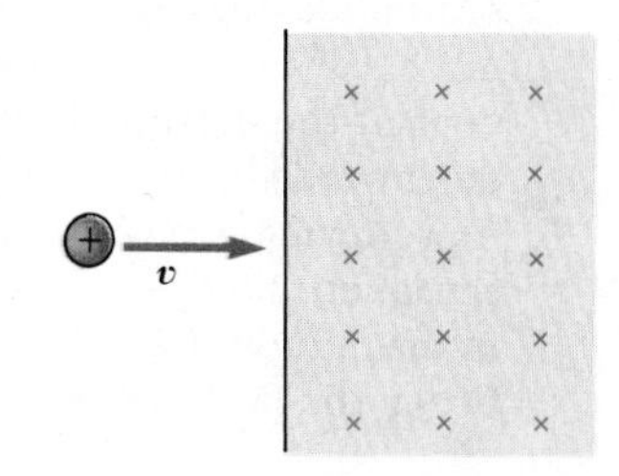


**Figure 29.24** (Question 15).

its subsequent motion. How would an electron behave under the same circumstances?

16. In a magnetic bottle, what reverses the direction of the velocity of the confined charged particles at the ends of the bottle? (*Hint:* Find the direction of the magnetic force on these particles in a region where the field becomes stronger and the field lines converge.)
17. In the cyclotron, why do particles of differing velocities take the same amount of time to complete one half of a revolution?
18. The *bubble chamber* is a device used for observing tracks of particles that pass through the chamber, which is immersed in a magnetic field. If some of the tracks are spirals and others are straight lines, what can you say about the particles?
19. Can a magnetic field set a resting electron into motion? If so, how?
20. You are designing a magnetic probe that uses the Hall effect to measure magnetic fields. Assume that you are restricted to using a given material and that you have already made the probe as thin as possible. What, if anything, can be done to increase the Hall voltage produced for a given magnetic field strength?
21. The electron beam in the photograph below is projected to the right. The beam deflects downward in the presence of a magnetic field produced by a pair of current-carrying coils. (a) What is the direction of the magnetic field? (b) What would happen to the beam if the current in the coils were reversed?

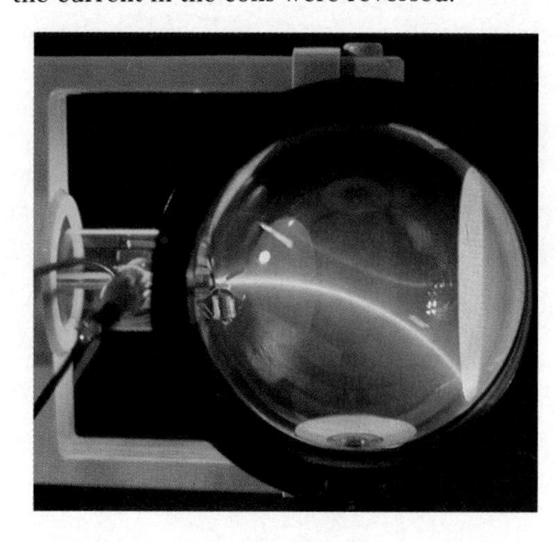

(Courtesy of CENCO)

## PROBLEMS

### Section 29.2 Definition and Properties of the Magnetic Field

1. Consider an electron near the magnetic equator. In which direction will it tend to be deflected if its velocity is directed (a) downward, (b) northward, (c) westward, or (d) southeastward?
2. An electron moving along the positive $x$ axis perpendicular to a magnetic field experiences a magnetic deflection in the negative $y$ direction. What is the direction of the magnetic field over this region?
3. An alpha particle (which is the nucleus of a helium atom) is moving with a northward velocity of $3.8 \times 10^5$ m/s in a region where the magnetic field is 1.9 T and points horizontally to the east. What are the magnitude and direction of the magnetic force on this alpha particle?
4. What force of magnetic origin is experienced by a proton moving north to south with a speed equal to $4.8 \times 10^6$ m/s at a location where the vertical component of the earth's magnetic field is 75 $\mu$T directed downward? In what direction is the proton deflected?
5. A proton moving with a speed of $4 \times 10^6$ m/s through a magnetic field of 1.7 T experiences a magnetic force of magnitude $8.2 \times 10^{-13}$ N. What is the angle between the proton's velocity and the field?
6. An electron is accelerated through 2400 V and then enters a region where there is a uniform 1.7-T magnetic field. What are (a) the maximum and (b) the minimum values of the magnetic force this charge can experience?
7. The magnetic field over a certain region is given by $\boldsymbol{B} = (4\boldsymbol{i} - 11\boldsymbol{j})$ T. An electron moves in the field with a velocity $\boldsymbol{v} = (-2\boldsymbol{i} + 3\boldsymbol{j} - 7\boldsymbol{k})$ m/s. Write out in unit-vector notation the force exerted on the electron by the magnetic field.
8. An electron is projected into a uniform magnetic field given by $\boldsymbol{B} = (1.4\boldsymbol{i} + 2.1\boldsymbol{j})$ T. Find the vector expression for the force on the electron when its velocity is $\boldsymbol{v} = 3.7 \times 10^5\boldsymbol{j}$ m/s.
9. A proton moves with a velocity of $\boldsymbol{v} = (2\boldsymbol{i} - 4\boldsymbol{j} + \boldsymbol{k})$ m/s in a region in which the magnetic field is given by $\boldsymbol{B} = (\boldsymbol{i} + 2\boldsymbol{j} - 3\boldsymbol{k})$ T. What is the magnitude of the magnetic force this charge experiences?
10. A proton moves perpendicular to a uniform magnetic field $\boldsymbol{B}$ with a speed of $10^7$ m/s and experiences an acceleration of $2 \times 10^{13}$ m/s$^2$ in the $+x$ direction when its velocity is in the $+z$ direction. Determine the magnitude and direction of the field.
11. Show that the work done by the magnetic force on a charged particle moving in a magnetic field is zero for any displacement of the particle.

### Section 29.3 Magnetic Force on a Current-Carrying Conductor

12. Calculate the magnitude of the force per unit length exerted on a conductor carrying a current of 22 A in a region where a uniform magnetic field has a magnitude of 0.77 T and is directed perpendicular to the conductor.
13. A wire carries a steady current of 2.4 A. A straight section of the wire, with a length of 0.75 m along the $x$ axis, lies within a uniform magnetic field, $\mathbf{B} = (1.6\,\mathbf{k})$T. If the current flows in the $+x$ direction, what is the magnetic force on the section of wire?
14. A conductor suspended by two flexible wires as in Figure 29.25 has a mass per unit length of 0.04 kg/m. What current must exist in the conductor in order for the tension in the supporting wires to be zero if the magnetic field over the region is 3.6 T into the page? What is the required direction for the current?

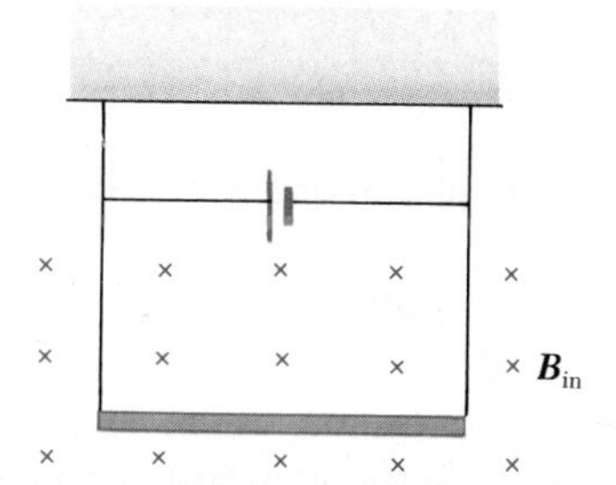

**Figure 29.25** (Problem 14).

15. A wire with a mass of 0.5 g/cm carries a 2-A current horizontally to the south. What are the direction and magnitude of the minimum magnetic field needed to lift this wire vertically upward?
16. A rectangular loop with dimensions 10 cm × 20 cm is suspended by a string, and the lower horizontal section of the loop is immersed in a magnetic field confined to a circular region (Fig. 29.26). If a current of 3 A is maintained in the loop in the direction shown, what are the direction and magnitude of the magnetic field required to produce a tension of $4 \times 10^{-2}$ N in the supporting string? (Neglect the mass of the loop.)

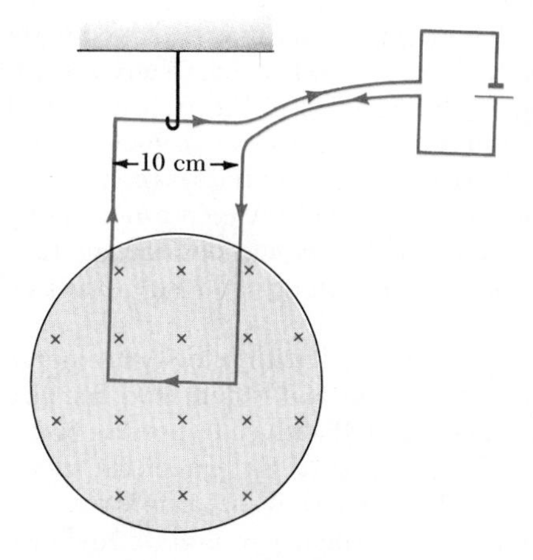

**Figure 29.26** (Problem 16).

17. A wire 2.8 m in length carries a current of 5 A in a region where a uniform magnetic field has a magnitude of 0.39 T. Calculate the magnitude of the magnetic force on the wire if the angle between the magnetic field and the direction of the current in the wire is (a) 60°, (b) 90°, (c) 120°.
18. In Figure 29.27, the cube is 40 cm on each edge. Four straight segments of wire—*ab*, *bc*, *cd*, and *da*—form a closed loop that carries a current $I = 5$ A as shown. A uniform magnetic field $\mathbf{B} = 0.02$ T is in the positive $y$ direction. Make a table showing the magnitude and direction of the force on each segment, listing them in the above order.

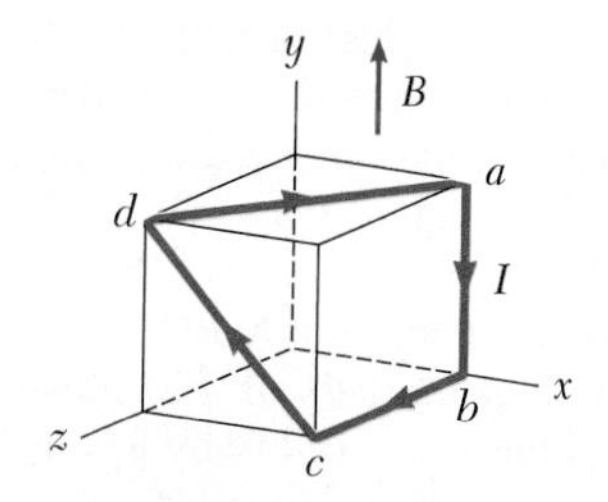

**Figure 29.27** (Problem 18).

19. A current $I = 15$ A is directed along the positive $x$ axis in a wire perpendicular to a magnetic field. The current experiences a magnetic force per unit length of 0.63 N/m in the negative $y$ direction. Calculate the magnitude and direction of the magnetic field in the region through which the current passes.
20. The Earth has a magnetic field of $0.6 \times 10^{-4}$ T, pointing 75° below the horizontal in a north-south plane. A 10-m-long straight wire carries a 15-A current. (a) If the current is directed horizontally toward the east, what are the magnitude and direction of the magnetic force on the wire? (b) What are the magnitude and direction of the force if the current is directed vertically upward?
21. A strong magnet is placed under a horizontal conducting ring of radius $r$ which carries a current $I$ as shown in Figure 29.28. If the magnetic lines of force make an angle $\theta$ with the vertical at the ring's location, what are the magnitude and direction of the resultant force on the ring?

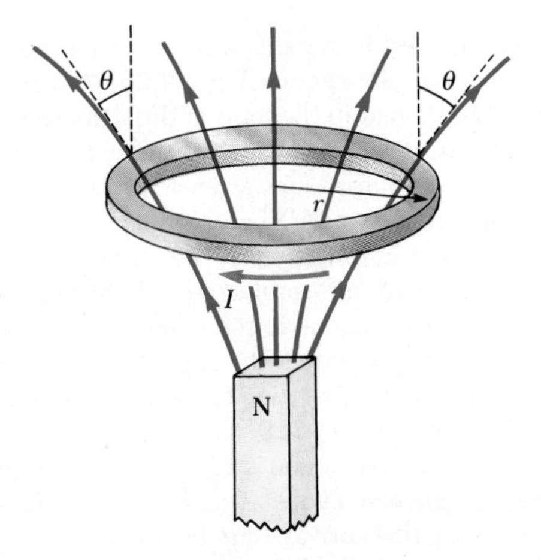


Figure 29.28 (Problem 21).

### Section 29.4 Torque on a Current Loop in a Uniform Magnetic Field

**22.** A current of 17 mA is maintained in a single circular loop of 2 m circumference. An external magnetic field of 0.8 T is directed parallel to the plane of the loop. (a) Calculate the magnetic moment of the current loop. (b) What is the magnitude of the torque exerted on the loop by the magnetic field?

**23.** A rectangular loop consists of 100 closely wrapped turns and has dimensions 0.4 m by 0.3 m. The loop is hinged along the $y$ axis, and the plane of the coil makes an angle of 30° with the $x$ axis (Fig. 29.29). What is the magnitude of the torque exerted on the loop by a uniform magnetic field of 0.8 T directed along the $x$ axis when the current in the windings has a value of 1.2 A in the direction shown? What is the expected direction of rotation of the loop?

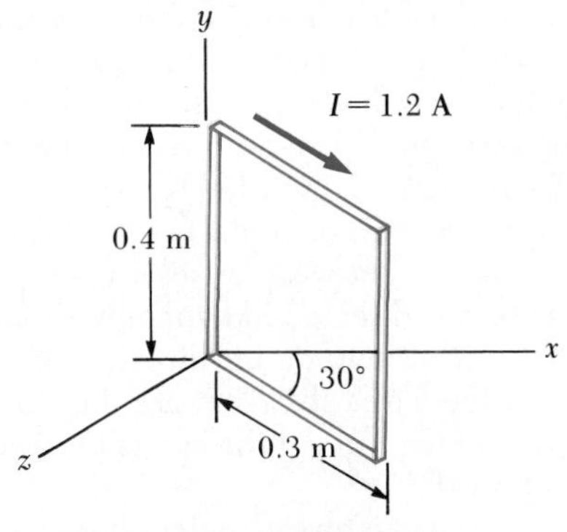


Figure 29.29 (Problem 23).

**24.** A small bar magnet is suspended in a uniform 0.25-T magnetic field. The maximum torque experienced by the bar magnet is $4.6 \times 10^{-3}$ N · m. Calculate the magnetic moment of the bar magnet.

**25.** A rectangular coil of 225 turns and area 0.45 m$^2$ is in a uniform magnetic field of 0.21 T. Measurements indicate that the maximum torque exerted on the loop by the field is $8 \times 10^{-3}$ N · m. (a) Calculate the current in the coil. (b) Would the value found for the required current be different if the 225 turns of wire were used to form a single-turn coil with the same shape of larger area? Explain.

**26.** A wire is formed into a circle with a diameter of 10 cm and placed in a uniform magnetic field of $3 \times 10^{-3}$ T. A current of 5 A passes through the wire. Find (a) the maximum torque that can be experienced by the current-carrying loop and (b) the range of potential energy the loop possesses for different orientations.

**27.** A circular coil of 100 turns has a radius of 0.025 m and carries a current of 0.1 A while in a uniform external magnetic field of 1.5 T. How much work must be done to rotate the coil from a position where the magnetic moment is parallel to the field to a position where the magnetic moment is opposite the field?

### Section 29.5 Motion of a Charged Particle in a Magnetic Field

**28.** The magnetic field of the earth at a certain location is directed vertically downward and has a magnitude of $0.5 \times 10^{-4}$ T. A proton is moving horizontally towards the west in this field with a velocity of $6.2 \times 10^6$ m/s. (a) What are the direction and magnitude of the magnetic force the field exerts on this charge? (b) What is the radius of the circular arc followed by this proton?

**29.** A singly charged positive ion has a mass of $3.2 \times 10^{-26}$ kg. After being accelerated through a potential difference of 833 V, the ion enters a magnetic field of 0.92 T along a direction perpendicular to the direction of the field. Calculate the radius of the path of the ion in the field.

**30.** A 2-keV electron moving perpendicular to the earth's magnetic field of 50 $\mu$T has a circular trajectory. Determine (a) the radius of the trajectory and (b) the time required for the electron to complete one circle. (c) Show that your answer to (b) is consistent with the cyclotron frequency of the electron.

**31.** What magnetic field would be required to constrain an electron whose energy is 725 eV to a circular path of radius 0.5 m?

**32.** A beam of protons (all with velocity $\mathbf{v}$) emerges from a particle accelerator and is deflected in a circular arc with a radius of 0.45 m by a transverse uniform magnetic field of 0.80 T. (a) Determine the speed $v$ of the protons in the beam. (b) What time is required for the deflection of a particular proton through an angle of 90°? (c) What is the energy of the particles in the beam?

33. A proton, a deuteron, and an alpha particle ($^4_2$He nucleus) are accelerated through a common potential difference $V$. The particles enter a uniform magnetic field $\mathbf{B}$ along a direction perpendicular to $\mathbf{B}$. The proton moves in a circular path of radius $r_p$. Find the value of the radii of the orbits of the deuteron, $r_d$, and the alpha particle, $r_\alpha$, in terms of $r_p$.

**34.** Calculate the cyclotron frequency of a proton in a magnetic field of 5.2 T.

35. A cosmic-ray proton in interstellar space has an energy of 10 MeV and executes a circular orbit with a radius equal to that of Mercury's orbit around the Sun ($5.8 \times 10^{10}$ m). What is the galactic magnetic field in that region of space?

36. A singly charged ion of mass $m$ is accelerated from rest by a potential difference $V$. It is then deflected by a uniform magnetic field (perpendicular to the ion's velocity) into a semicircle of radius $R$. Now a doubly-charged ion of mass $m'$ is accelerated through the same potential difference and deflected by the same magnetic field into a semicircle of radius $R' = 2R$. What is the ratio of the ions' masses?

37. A singly charged positive ion moving with a speed of $4.6 \times 10^5$ m/s leaves a spiral track of radius 7.94 mm in a photograph along a direction perpendicular to the magnetic field of a bubble chamber. The magnetic field applied for the photograph has a magnitude of 1.8 T. Compute the mass (in atomic mass units) of this particle, and, from that, identify the particle.

38. The accelerating voltage that is applied to an electron gun is 15 kV, and the horizontal distance from the gun to a viewing screen is 35 cm. What is the deflection caused by the vertical component of the Earth's magnetic field ($4 \times 10^{-5}$ T), assuming that any change in the horizontal component of the beam velocity is negligible?

### °Section 29.6 Applications of the Motion of Charged Particles in a Magnetic Field

**39.** A crossed-field velocity selector has a magnetic field of $10^{-2}$ T. What electric field strength is required if 10-keV electrons are to pass through undeflected?

40. A velocity filter consists of magnetic and electric fields described by $\mathbf{E} = E\mathbf{k}$ and $\mathbf{B} = B\mathbf{j}$. If $B = 0.015$ T, find the value of $E$ such that a 750-eV electron moving along the positive $x$ axis will be undeflected.

41. At the equator, near the surface of the earth, the magnetic field is approximately 50 $\mu$T northward, and the electric field is about 100 N/C downward. Find the gravitational, electric, and magnetic forces on a 100-eV electron moving eastward in a straight line in this environment.

42. Singly charged uranium ions are accelerated through a potential difference of 2 kV and enter a uniform magnetic field of 1.2 T directed perpendicular to their velocities. Determine the radius of the circular path followed by these ions assuming that they are (a) $U^{238}$ ions and (b) $U^{235}$ ions. How does the ratio of these path radii depend on the accelerating voltage and the magnetic field strength?

43. Consider the mass spectrometer shown schematically in Figure 29.19. The electric field between the plates of the velocity selector is 2500 V/m, and the magnetic field in both the velocity selector and the deflection chamber has a magnitude of 0.035 T. Calculate the radius of the path in the system for a singly charged ion with a mass $m = 2.18 \times 10^{-26}$ kg.

44. What is the required radius of a cyclotron designed to accelerate protons to energies of 34 MeV using a magnetic field of 5.2 T?

45. What is the minimum size of a cyclotron designed to accelerate protons to an energy of 18 MeV with a cyclotron frequency of $3 \times 10^7$ Hz?

**46.** At the Fermilab accelerator in Weston, Illinois, protons with momentum $4.8 \times 10^{-16}$ kg·m/s are held in a circular orbit of radius 1 km by an upward magnetic field. What upward magnetic field must be used to maintain the protons in this orbit?

47. A cyclotron designed to accelerate protons is provided with a magnetic field of 0.45 T and has a radius of 1.2 m. (a) What is the cyclotron frequency? (b) What is the maximum speed acquired by the protons?

48. (a) What must be the magnetic field strength within a 60-inch diameter cyclotron if that cyclotron is to accelerate protons to a maximum kinetic energy of 10.5 MeV? (b) At what frequency must the oscillator in the cyclotron operate? (c) If the frequency of the oscillator is maintained at the value found in (b), to what value must the magnetic field strength be altered if the cyclotron is to accelerate deuterons? (Note: A deuteron is a deuterium nucleus, consisting of a proton and a neutron bound together.) (d) After the cyclotron is adjusted to accelerate deuterons, what is the maximum kinetic energy of the deuterons produced?

49. The picture tube in a television uses magnetic deflection coils rather than electric deflection plates. Suppose an electron beam is accelerated through a 50-kV potential difference and then passes through a uniform magnetic field produced by these coils for 1 cm. The screen is located 10 cm from the center of the coils and is 50 cm wide. When the field is turned off, the electron beam hits the center of the screen. What field strength is necessary to deflect the beam to the side of the screen?

### °Section 29.7 The Hall Effect

50. A flat ribbon of silver with a thickness $t = 0.20$ mm is used for a Hall-effect measurement of a uniform magnetic field that is perpendicular to the ribbon, as

shown in Figure 29.30. The Hall coefficient for silver is $R_H = 0.84 \times 10^{-10}$ m³/C. (a) What is the effective density of charge carriers, $n$, in silver? (b) If a current $I = 20$ A produces a Hall voltage $V_H = 15$ $\mu$V, what is the magnitude of the applied magnetic field?

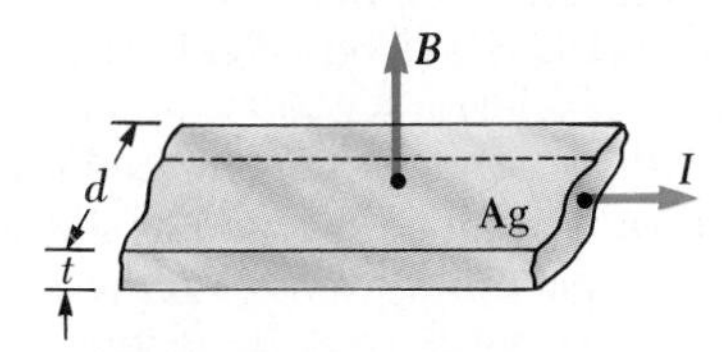


**Figure 29.30** (Problem 50).

51. A section of conductor 0.4 cm in thickness is used as the experimental specimen in a Hall effect measurement. If a Hall voltage of 35 $\mu$V is measured for a current of 21 A in a magnetic field of 1.8 T, calculate the Hall coefficient for the conductor.
52. Suppose the conductor shown in Figure 29.22 is copper and is carrying a current of 10 A in a magnetic field of 0.5 T. The width of the conductor $d$ is 1 cm and the thickness is 1 mm. Find the Hall potential across the width of the conductor.
53. In an experiment designed to measure the earth's magnetic field using the Hall effect, a copper bar 0.5 cm in thickness is positioned along an east-west direction. If a current of 8 A in the conductor results in a measured Hall voltage of $5.1 \times 10^{-12}$ V, what is the calculated value of the earth's magnetic field? (Assume that $n = 8.48 \times 10^{28}$ electrons/m³ and that the plane of the bar is rotated to be perpendicular to the direction of $\boldsymbol{B}$.)
54. A flat copper ribbon with a thickness of $\frac{1}{3}$ mm carries a steady current of 50 A and is located within a uniform magnetic field of 1.3 T that is directed perpendicular to the plane of the ribbon. If a Hall voltage of 9.6 $\mu$V is measured across the ribbon, what is the charge density of free electrons in the copper ribbon? What effective number of free electrons per atom does this result indicate?
55. The Hall effect can be used to measure the number of conduction electrons per unit volume $n$ for an unknown sample. The sample is 15 mm thick, and when placed in a 1.8-T magnetic field produces a Hall voltage of 0.122 $\mu$V while carrying a 12-A current. What is the value of $n$?
56. A Hall-effect probe for measuring magnetic fields is designed to operate with a 120-mA current in the probe. When the probe is placed in a uniform field of 0.08 T, it produces a Hall voltage of 0.7 $\mu$V. (a) When it is measuring an unknown field, the Hall voltage is 0.33 $\mu$V. What is the unknown field strength? (b) If the thickness of the probe in the direction of $\boldsymbol{B}$ is 2 mm, find the charge-carrier density (each of charge $e$).

## ADDITIONAL PROBLEMS

57. A wire with a mass of 1 g/cm is placed on a horizontal surface with a coefficient of friction of 0.2. The wire carries a current of 1.5 A toward the east, and moves horizontally to the north. What are the magnitude and the direction of the *smallest* magnetic field that enables the wire to move in this fashion?
58. Indicate the initial direction of the deflection of the charged particles as they enter the magnetic fields as shown in Figure 29.31.

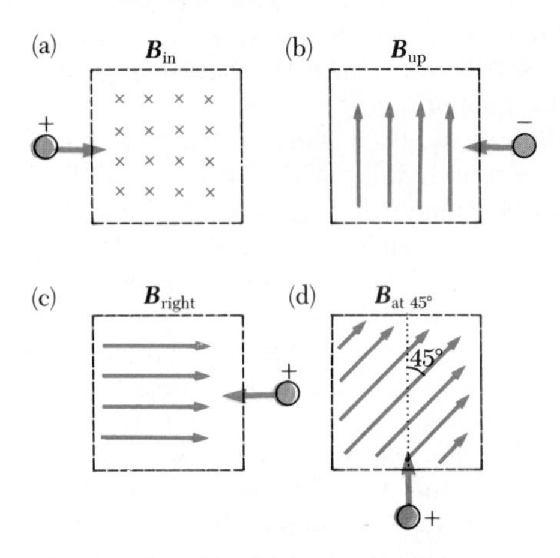


**Figure 29.31** (Problem 58).

59. A positive charge $q = 3.2 \times 10^{-19}$ C moves with a velocity $\boldsymbol{v} = (2\boldsymbol{i} + 3\boldsymbol{j} - \boldsymbol{k})$ m/s through a region where both a uniform magnetic field and a uniform electric field exist. (a) Calculate the total force on the moving charge (in unit-vector notation) if $\boldsymbol{B} = (2\boldsymbol{i} + 4\boldsymbol{j} + \boldsymbol{k})$ T and $\boldsymbol{E} = (4\boldsymbol{i} - \boldsymbol{j} - 2\boldsymbol{k})$ V/m. (b) What angle does the force vector make relative to the positive $x$ axis?
60. A cosmic-ray proton traveling at half the speed of light is heading directly toward the center of the earth in the plane of the earth's equator. Will it hit the earth? As an estimate, assume that the earth's magnetic field is $5 \times 10^{-5}$ T and extends out one earth diameter, or $1.3 \times 10^{7}$ m. Calculate the radius of curvature of the proton in this magnetic field.
61. A straight wire of mass 10 g and length 5 cm is suspended from two identical springs which, in turn, form a closed circuit (Fig. 29.32). The springs stretch a distance of 0.5 cm under the weight of the wire. The circuit has a *total* resistance of 12 Ω. When a magnetic field is turned on, directed *out* of the page (indicated by the dots in Fig. 29.32), the springs are observed to stretch an *additional* 0.3 cm. What is the strength of the magnetic field? (The upper portion of the circuit is fixed.)
62. A proton having velocity $\boldsymbol{v} = 2 \times 10^{8}\boldsymbol{i}$ m/s enters a uniform magnetic field having a strength of 1 T. The proton leaves the field with a velocity $\boldsymbol{v} = -2 \times 10^{8}\boldsymbol{j}$ m/s. Determine (a) the direction of the magnetic

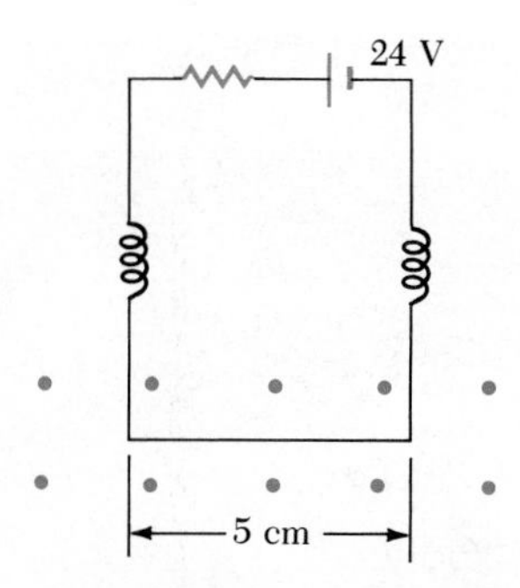


Figure 29.32 (Problem 61).

field, (b) the radius of curvature of the path of the proton while in the magnetic field, (c) the distance traveled by the proton while in the magnetic field, and (d) the time spent by the proton in the magnetic field.

63. Sodium melts at 210°F. Liquid sodium, an excellent thermal conductor, is used in some nuclear reactors to remove thermal energy from the reactor core. The liquid sodium can be moved through pipes by pumps that exploit the force on a moving charge in a magnetic field. The principle is as follows: imagine the liquid metal to be in a pipe having a rectangular cross section of width $w$ and height $h$. A uniform magnetic field perpendicular to the pipe affects a section of length $L$ (Fig. 29.33). An electric current directed perpendicular to the pipe and to the magnetic field produces a current density $J$. (a) Explain why this arrangement produces a force on the liquid that is directed along the length of the pipe. (b) Show that the section of liquid in the magnetic field experiences a pressure increase equal to $JLB$. (c) Calculate the current density required to produce a pressure increase equal to $JLB$.

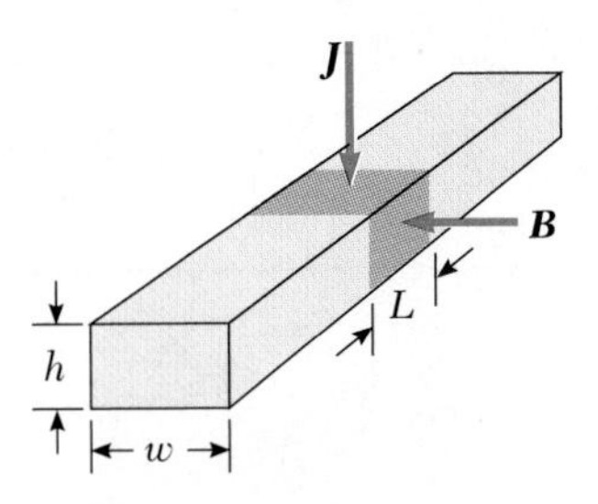


Figure 29.33 (Problem 63).

64. A thin rod of length $\ell$ is made of a nonconducting material and carries a uniform charge per unit length $\lambda$. The rod is rotated with angular velocity $\omega$ about an axis through its center, perpendicular to the length of the rod. Show that the magnetic dipole moment is $\omega\lambda\ell^3/24$. (*Hint:* consider the charge $dq$ located within the element $dx$ a distance $x$ from the axis.)

65. A mass spectrometer of the Bainbridge type is used to examine the isotopes of uranium. Ions in the beam emerge from the velocity selector with a speed equal to $3 \times 10^5$ m/s and enter a uniform magnetic field of 0.6 T directed perpendicular to the velocity of the ions. What is the distance between the impact points formed on the photographic plate by singly charged ions of $^{235}$U and $^{238}$U?

66. A cyclotron designed to accelerate deuterons has a magnetic field with a uniform intensity of 1.5 T over a region of radius 0.45 m. If the alternating potential applied between the dees of the cyclotron has a maximum value of 15 kV, what time is required for the deuterons to acquire maximum attainable energy?

67. A singly charged heavy ion is observed to complete five revolutions in a uniform magnetic field of magnitude $5 \times 10^{-2}$ T in 1.50 ms. Calculate the (approximate) mass of the ion in kg.

68. A uniform magnetic field of 0.15 T is directed along the positive $x$ axis. A positron moving with a speed of $5 \times 10^6$ m/s enters the field along a direction that makes an angle of 85° with the $x$ axis (Fig. 29.34). The motion of the particle is expected to be a helix, as described in Section 29.5. Calculate (a) the pitch $p$ and (b) the radius $r$ of the trajectory.

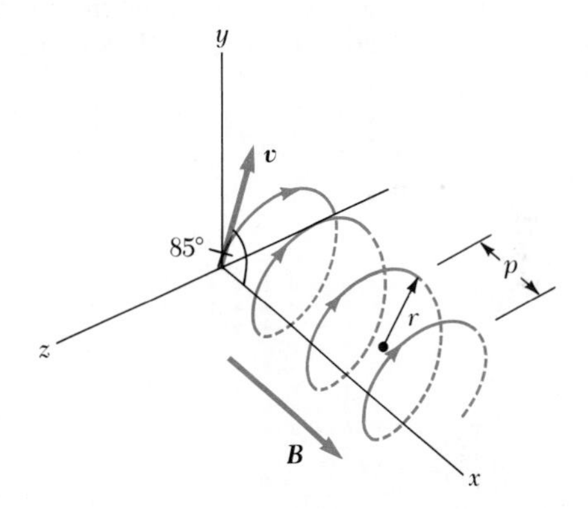


Figure 29.34 (Problem 68).

69. Consider an electron orbiting a proton and maintained in a fixed circular path of radius equal to $R = 5.29 \times 10^{-11}$ m by the Coulomb force of mutual attraction. Treating the orbiting charge as a current loop, calculate the resulting torque when the system is in an external magnetic field of 0.4 T directed perpendicular to the magnetic moment of the orbiting electron.

70. A proton moving in the plane of the page has kinetic energy of 6 MeV. It enters a magnetic field $B = 1$ T (into the page) at an angle $\theta = 45°$ to the linear boundary of the field as shown in Figure 29.35. (a) Find $x$, the distance from the point of entry to where the proton will leave the field. (b) Determine

$\theta'$, the angle between the boundary and the proton's velocity vector as it leaves the field.

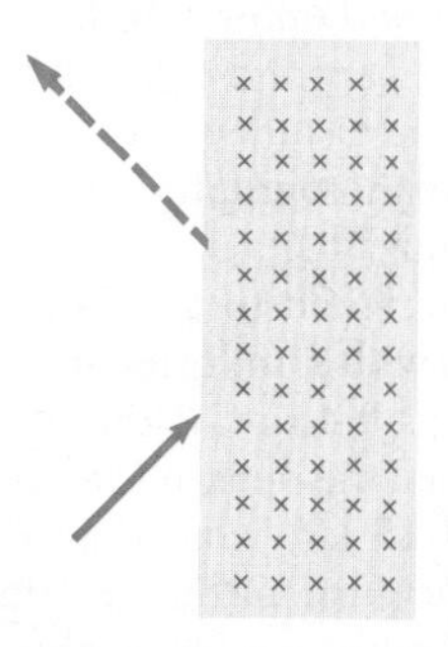

**Figure 29.35** (Problem 70).

71. Protons with kinetic energy of 5 MeV are moving in the positive $x$ direction and enter a magnetic field $\mathbf{B} = (0.5\ \text{T})\mathbf{k}$ directed out of the plane of the page and extending from $x = 0$ to $x = 1$ m as shown in Figure 29.36. (a) Calculate the $y$ component of the protons' momentum as they leave the magnetic field at $x = 1$ m. (b) Find the angle $\alpha$ between the initial velocity vector of the proton beam and the velocity vector after the beam emerges from the field. (*Hint:* Neglect relativistic effects and note that $1\ \text{eV} = 1.60 \times 10^{-19}$ J.)

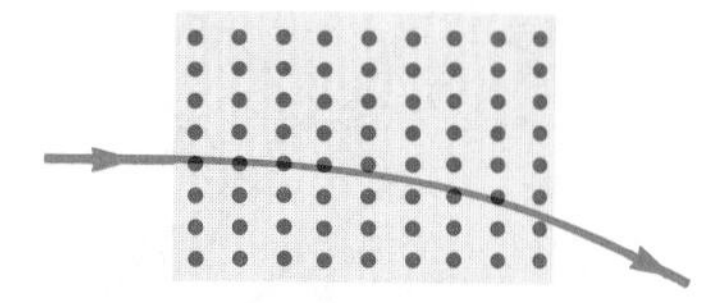

**Figure 29.36** (Problem 71).

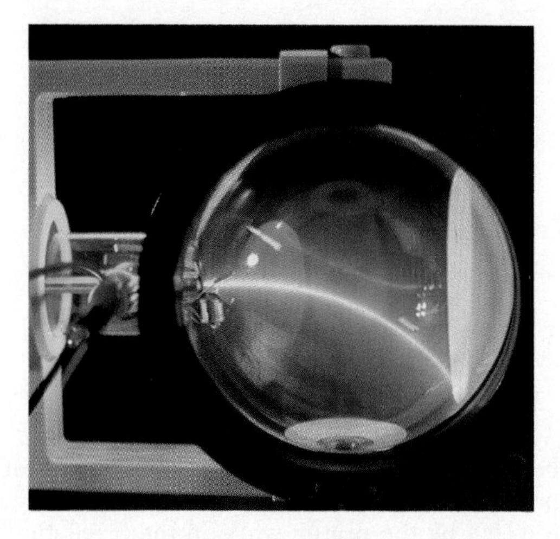

(Courtesy of CENCO)

△72. Table 29.1 shows measurements of a Hall voltage and corresponding magnetic field for a probe used to measure magnetic fields. (a) Make a plot of these data, and deduce a relationship between the Hall voltage and magnetic field. (b) If the measurements were taken with a current of 0.2 A and the sample is made from a material having a charge-carrier density of $10^{26}/\text{m}^3$, what is the thickness of the sample?

**TABLE 29.1**

| $V_H$ ($\mu$V) | $B$ (T) |
|---|---|
| 0 | 0 |
| 11 | 0.1 |
| 19 | 0.2 |
| 28 | 0.3 |
| 42 | 0.4 |
| 50 | 0.5 |
| 61 | 0.6 |
| 68 | 0.7 |
| 79 | 0.8 |
| 90 | 0.9 |
| 102 | 1.0 |

# 30

# Sources of the Magnetic Field

*An assortment of commercially available magnets. The four red magnets and the large black magnet on the left are made of an alloy of iron, aluminum, and cobalt. The six horseshoe magnets on the right are made of different nickel alloy steels. The rectangular magnets on the lower right are ceramics made of iron, nickel, and beryllium oxides. (Courtesy of CENCO)*

The preceding chapter treated a class of problems involving the magnetic force on a charged particle moving in a magnetic field. To complete the description of the magnetic interaction, this chapter deals with the origin of the magnetic field, namely, moving charges or electric currents. We begin by showing how to use the law of Biot and Savart to calculate the magnetic field produced at a point by a current element. Using this formalism and the superposition principle, we then calculate the total magnetic field due to a distribution of currents for several geometries. Next, we show how to determine the force between two current-carrying conductors, which leads to the definition of the ampere. We shall also introduce Ampère's law, which is very useful for calculating the magnetic field of highly symmetric configurations carrying steady currents. We apply Ampère's law to determine the magnetic field for several current configurations, including that of a solenoid.

This chapter is also concerned with some aspects of the complex processes that occur in magnetic materials. All magnetic effects in matter can be explained on the basis of effective current loops associated with atomic magnetic dipole moments. These atomic magnetic moments can arise both from the orbital motion of the electrons and from an intrinsic, or "built-in," property of the electrons known as *spin*. Our description of magnetism in matter will be based in part on the experimental fact that the presence of bulk matter generally modifies the magnetic field produced by currents. For example, when a material is placed inside a current-carrying solenoid, the material sets up its own magnetic field, which adds (vectorially) to the field previously present.

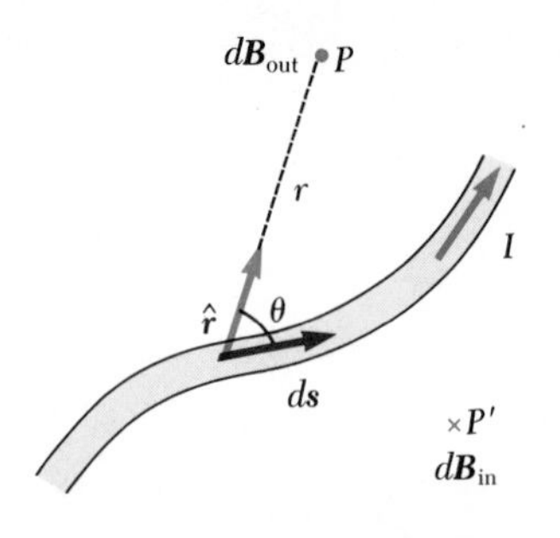


**Figure 30.1** The magnetic field $\boldsymbol{dB}$ at a point $P$ due to a current element $\boldsymbol{ds}$ is given by the Biot-Savart law, Equation 30.1. The field is out of the paper at $P$.

## 30.1 THE BIOT-SAVART LAW

Shortly after Oersted's discovery in 1819 that a compass needle is deflected by a current-carrying conductor, Jean Baptiste Biot and Felix Savart reported that a conductor carrying a steady current produces a force on a magnet. From their experimental results, Biot and Savart were able to arrive at an expression that gives the magnetic field at some point in space in terms of the current that produces the field. The *Biot-Savart law* says that if a wire carries a steady current $I$, the magnetic field $\boldsymbol{dB}$ at a point $P$ associated with an element $\boldsymbol{ds}$ (Fig. 30.1) has the following properties:

Properties of the magnetic field due to a current element

1. The vector $\boldsymbol{dB}$ is perpendicular both to $\boldsymbol{ds}$ (which is in the direction of the current) and to the unit vector $\hat{\boldsymbol{r}}$ directed from the element to the point $P$.
2. The magnitude of $\boldsymbol{dB}$ is inversely proportional to $r^2$, where $r$ is the distance from the element to the point $P$.
3. The magnitude of $\boldsymbol{dB}$ is proportional to the current and to the length $ds$ of the element.
4. The magnitude of $\boldsymbol{dB}$ is proportional to $\sin\theta$, where $\theta$ is the angle between the vectors $\boldsymbol{ds}$ and $\hat{\boldsymbol{r}}$.

The **Biot-Savart law** can be summarized in the following convenient form:

Biot-Savart law

$$\boldsymbol{dB} = k_m \frac{I\,\boldsymbol{ds} \times \hat{\boldsymbol{r}}}{r^2} \tag{30.1}$$

where $k_m$ is a constant that in SI units is exactly $10^{-7}$ Wb/A·m. The constant $k_m$ is usually written $\mu_0/4\pi$, where $\mu_0$ is another constant, called the **permeability of free space.** That is,

$$\frac{\mu_0}{4\pi} = k_m = 10^{-7}\ \text{Wb/A}\cdot\text{m} \tag{30.2}$$

Permeability of free space

$$\mu_0 = 4\pi k_m = 4\pi \times 10^{-7}\ \text{Wb/A}\cdot\text{m} \tag{30.3}$$

Hence, the Biot-Savart Law, Equation 30.1, can also be written

Biot-Savart law

$$\boldsymbol{dB} = \frac{\mu_0}{4\pi}\frac{I\,\boldsymbol{ds} \times \hat{\boldsymbol{r}}}{r^2} \tag{30.4}$$

It is important to note that the Biot-Savart law gives the magnetic field at a point only for a small element of the conductor. To find the *total* magnetic field $\boldsymbol{B}$ at some point due to a conductor of finite size, we must sum up contributions from all current elements making up the conductor. That is, we must evaluate $\boldsymbol{B}$ by integrating Equation 30.4:

$$\boldsymbol{B} = \frac{\mu_0 I}{4\pi}\int \frac{\boldsymbol{ds} \times \hat{\boldsymbol{r}}}{r^2} \tag{30.5}$$

where the integral is taken over the entire conductor. This expression must be handled with special care since the integrand is a vector quantity.

There are interesting similarities between the Biot-Savart law of magnetism and Coulomb's law of electrostatics. That is, the current element $I\,\boldsymbol{ds}$

produces a magnetic field, whereas a point charge $q$ produces an electric field. Furthermore, *the magnitude of the magnetic field varies as the inverse square of the distance from the current element,* as does the electric field due to a point charge.

However, the directions of the two fields are quite different. The electric field due to a point charge is radial. In the case of a positive point charge, $\mathbf{E}$ is directed from the charge to the field point. On the other hand, the magnetic field due to a current element is perpendicular to both the current element and the radius vector. Hence, if the conductor lies in the plane of the paper, as in Figure 30.1, $d\mathbf{B}$ points *out* of the paper at the point $P$ and into the paper at $P'$.

The examples that follow illustrate how to use the Biot-Savart law for calculating the magnetic induction of several important geometric arrangements. It is important that you recognize that the magnetic field described in these calculations is *the field due to a given current-carrying conductor.* This is not to be confused with any *external* field that may be applied to the conductor.

### EXAMPLE 30.1 Magnetic Field of a Thin Straight Conductor

Consider a thin, straight wire carrying a constant current $I$ and placed along the $x$ axis as in Figure 30.2. Let us calculate the total magnetic field at the point $P$ located at a distance $a$ from the wire.

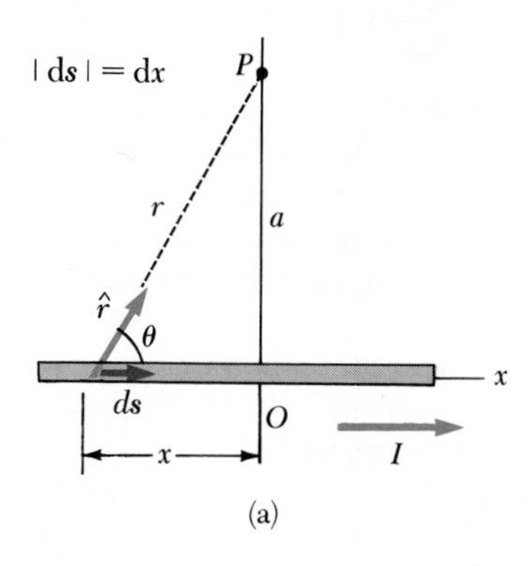


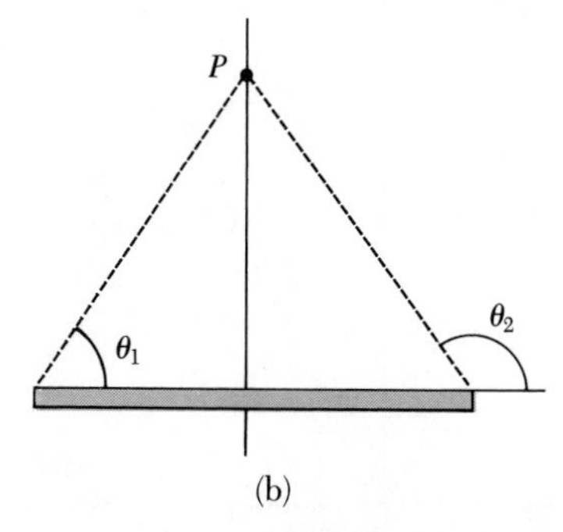


**Figure 30.2** (Example 30.1) (a) A straight wire segment carrying a current $I$. The magnetic field at $P$ due to each element $ds$ is out of the paper, and so the net field is also out of the paper. (b) The limiting angles $\theta_1$ and $\theta_2$ for this geometry.

*Solution* An element $d\mathbf{s}$ is at a distance $r$ from $P$. The direction of the field at $P$ due to this element is out of the paper, since $d\mathbf{s} \times \hat{\mathbf{r}}$ is out of the paper. In fact, *all* elements give a contribution directly out of the paper at $P$. Therefore, we have only to determine the magnitude of the field at $P$. In fact, taking the origin at $O$ and letting $P$ be along the *positive* $y$ axis, with $\mathbf{k}$ being a unit vector pointing *out* of the paper, we see that

$$d\mathbf{s} \times \hat{\mathbf{r}} = \mathbf{k}|d\mathbf{s} \times \hat{\mathbf{r}}| = \mathbf{k}(dx \sin \theta)$$

Substitution into Equation 30.4 gives $d\mathbf{B} = \mathbf{k}\, dB$, with

$$(1) \qquad dB = \frac{\mu_0 I}{4\pi} \frac{dx \sin \theta}{r^2}$$

In order to integrate this expression, we must relate the variables $\theta$, $x$, and $r$. One approach is to express $x$ and $r$ in terms of $\theta$. From the geometry in Figure 30.2a and some simple differentiation, we obtain the following relationship:

$$(2) \qquad r = \frac{a}{\sin \theta} = a \csc \theta$$

Since $\tan \theta = -a/x$ from the right triangle in Figure 30.2a,

$$x = -a \cot \theta$$

$$(3) \qquad dx = a \csc^2 \theta \, d\theta$$

Substitution of (2) and (3) into (1) gives

$$(4) \qquad dB = \frac{\mu_0 I}{4\pi} \frac{a \csc^2 \theta \sin \theta \, d\theta}{a^2 \csc^2 \theta} = \frac{\mu_0 I}{4\pi a} \sin \theta \, d\theta$$

Thus, we have reduced the expression to one involving only the variable $\theta$. We can now obtain the total field at $P$ by integrating (4) over all elements subtending angles ranging from $\theta_1$ to $\theta_2$ as defined in Figure 30.2b. This gives

$$B = \frac{\mu_0 I}{4\pi a}\int_{\theta_1}^{\theta_2} \sin\theta\, d\theta = \frac{\mu_0 I}{4\pi a}(\cos\theta_1 - \cos\theta_2) \qquad (30.6)$$

We can apply this result to find the magnetic field of any straight wire if we know the geometry and hence the angles $\theta_1$ and $\theta_2$.

Consider the special case of an infinitely long, straight wire. In this case, $\theta_1 = 0$ and $\theta_2 = \pi$, as can be seen from Figure 30.2b, for segments ranging from $x = -\infty$ to $x = +\infty$. Since $(\cos\theta_1 - \cos\theta_2) = (\cos 0 - \cos\pi) = 2$, Equation 30.6 becomes

$$B = \frac{\mu_0 I}{2\pi a} \qquad (30.7)$$

A three-dimensional view of the direction of $\boldsymbol{B}$ for a long, straight wire is shown in Figure 30.3. *The field lines are circles concentric with the wire and are in a plane perpendicular to the wire.* The magnitude of $\boldsymbol{B}$ is constant on any circle of radius $a$ and is given by Equation 30.7. A convenient rule for determining the direction of $\boldsymbol{B}$ is to grasp the wire with the right hand, with the thumb along the direction of the current. The four fingers wrap in the direction of the magnetic field.

Our result shows that the magnitude of the magnetic field is proportional to the current and decreases as the distance from the wire increases, as one might intuitively expect. Notice that Equation 30.7 has the same mathematical form as the expression for the magnitude of the electric field due to a long charged wire (Eq. 24.9).

**Exercise 1** Calculate the magnetic field of a long, straight wire carrying a current of 5 A, at a distance of 4 cm from the wire.
**Answer** $2.5 \times 10^{-5}$ T.

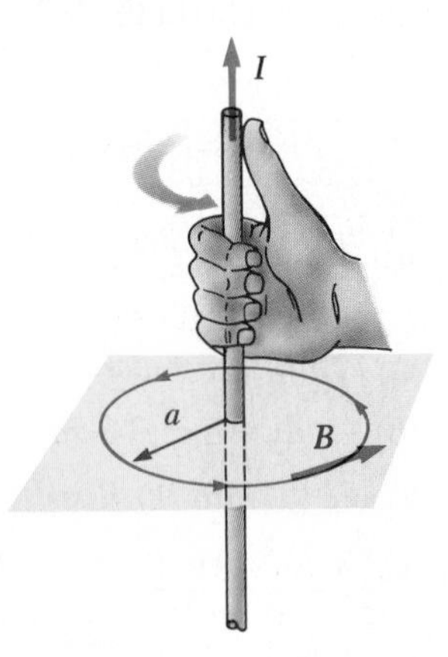


**Figure 30.3** The right-hand rule for determining the direction of the magnetic field due to a long, straight wire. Note that the magnetic field lines form circles around the wire.

### EXAMPLE 30.2 Field of a Current Loop

Calculate the magnetic field at the point $O$ for the current loop shown in Figure 30.4. The loop consists of two straight portions and a circular arc of radius $R$, which subtends an angle $\theta$ at the center of the arc. We shall ignore the contribution of the current in the short arcs near $O$.

*Solution* First, note that the magnetic field at $O$ due to the straight segments $OA$ and $OC$ is identically *zero*, since $d\boldsymbol{s}$ is parallel to $\hat{\boldsymbol{r}}$ along these paths and therefore $d\boldsymbol{s} \times \hat{\boldsymbol{r}} = 0$. This simplifies the problem because now we need to be concerned only with the magnetic field at $O$ due to the curved portion $AC$. Note that each element along the path $AC$ is at the same distance $R$ from $O$, and each gives a contribution $d\boldsymbol{B}$, which is directed into the paper at $O$. Furthermore, at every point on the path $AC$, we see that $d\boldsymbol{s}$ is perpendicular to $\hat{\boldsymbol{r}}$, so that $|d\boldsymbol{s} \times \hat{\boldsymbol{r}}| = ds$. Using this information and Equation 30.4, we get the following expression for the field at $O$ due to the segment $ds$:

$$dB = \frac{\mu_0 I}{4\pi}\frac{ds}{R^2}$$

Since $I$ and $R$ are constants, we can easily integrate this expression, which gives

$$B = \frac{\mu_0 I}{4\pi R^2}\int ds = \frac{\mu_0 I}{4\pi R^2}s = \frac{\mu_0 I}{4\pi R}\theta \qquad (30.8)$$

where we have used the fact that $s = R\theta$, where $\theta$ is measured in *radians*. The direction of $\boldsymbol{B}$ is *into* the paper at $O$ since $d\boldsymbol{s} \times \hat{\boldsymbol{r}}$ is into the paper for every segment.

For example, if an arc subtends an angle $\theta = \pi/2$ rad, we find from Equation 30.8 that $B = \mu_0 I/8R$.

**Exercise 2** A current loop in the form of a full circle of radius $R$ carries a current $I$. What is the magnitude of the magnetic field at its center?
**Answer** $\mu_0 I/2R$.

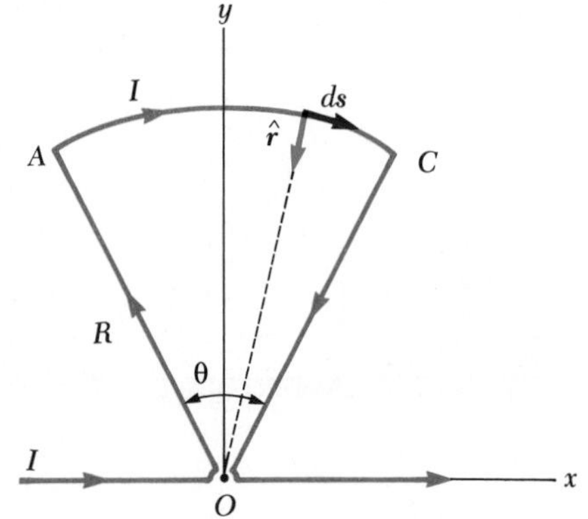


**Figure 30.4** (Example 30.2) The magnetic field at $O$ due to the current loop is into the paper. Note that the contribution to the field at $O$ due to the straight segments $OA$ and $OC$ is zero.

**EXAMPLE 30.3 Magnetic Field on the Axis of a Circular Current Loop**
Consider a circular loop of wire of radius $R$ located in the $yz$ plane and carrying a steady current $I$, as in Figure 30.5. Let us calculate the magnetic field at an axial point $P$ a distance $x$ from the center of the loop.

*Solution* In this situation, note that any element $\boldsymbol{ds}$ is perpendicular to $\hat{\boldsymbol{r}}$. Furthermore, all elements around the loop are at the same distance $r$ from $P$, where $r^2 = x^2 + R^2$. Hence, the *magnitude* of $\boldsymbol{dB}$ due to the element $\boldsymbol{ds}$ is given by

$$dB = \frac{\mu_0 I}{4\pi}\frac{|\boldsymbol{ds} \times \hat{\boldsymbol{r}}|}{r^2} = \frac{\mu_0 I}{4\pi}\frac{ds}{(x^2 + R^2)}$$

The direction of the magnetic field $\boldsymbol{dB}$ due to the element $\boldsymbol{ds}$ is perpendicular to the plane formed by $\hat{\boldsymbol{r}}$ and $\boldsymbol{ds}$, as shown in Figure 30.5. The vector $\boldsymbol{dB}$ can be resolved into a component $dB_x$, along the $x$ axis, and a component $dB_y$, which is perpendicular to the $x$ axis. When the components perpendicular to the $x$ axis are summed over the whole loop, the result is *zero*. That is, by symmetry any element on one side of the loop will set up a perpendicular component that cancels the component set up by an element diametrically opposite it. Therefore, we see that *the resultant field at P must be along the x axis* and can be found by integrating the components $dB_x = dB \cos\theta$, where this expression is obtained from resolving the vector $\boldsymbol{dB}$ into its components as shown in Figure 30.5. That is, $\boldsymbol{B} = \boldsymbol{i}B_x$, where $B_x$ is given by

$$B_x = \oint dB \cos\theta = \frac{\mu_0 I}{4\pi}\oint \frac{ds \cos\theta}{x^2 + R^2}$$

where the integral must be taken over the entire loop.

Since $\theta$, $x$, and $R$ are constants for all elements of the loop and since $\cos\theta = R/(x^2 + R^2)^{1/2}$, we get

$$B_x = \frac{\mu_0 IR}{4\pi(x^2 + R^2)^{3/2}}\oint ds = \frac{\mu_0 IR^2}{2(x^2 + R^2)^{3/2}} \qquad (30.9)$$

where we have used the fact that $\oint ds = 2\pi R$ (the circumference of the loop).

To find the magnetic field at the *center* of the loop, we set $x = 0$ in Equation 30.9. At this special point, this gives

$$B = \frac{\mu_0 I}{2R} \qquad (\text{at } x = 0) \qquad (30.10)$$

It is also interesting to determine the behavior of the magnetic field at large distances from the loop, that is, when $x$ is large compared with $R$. In this case, we can neglect the term $R^2$ in the denominator of Equation 30.9 and get

$$B \approx \frac{\mu_0 IR^2}{2x^3} \qquad (\text{for } x \gg R) \qquad (30.11)$$

Since the magnitude of the magnetic dipole moment $\boldsymbol{\mu}$ of the loop is defined as the product of the current and the area (Eq. 29.13), $\mu = I(\pi R^2)$ and we can express Equation 30.11 in the form

$$B = \frac{\mu_0}{2\pi}\frac{\mu}{x^3} \qquad (30.12)$$

This result is similar in form to the expression for the electric field due to an electric dipole, $E = kp/y^3$ (Eq. 23.10), where $p$ is the electric dipole moment. The pattern of the magnetic field lines for a circular loop is shown in Figure 30.6. For clarity, the lines are drawn only for one plane which contains the axis of the loop. The field pattern is axially symmetric.

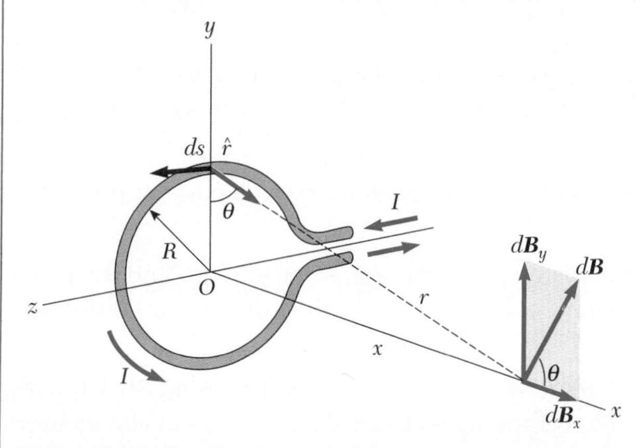


**Figure 30.5** (Example 30.3) The geometry for calculating the magnetic field at an axial point $P$ for a current loop. Note that by symmetry the total field $\boldsymbol{B}$ is along the $x$ axis.

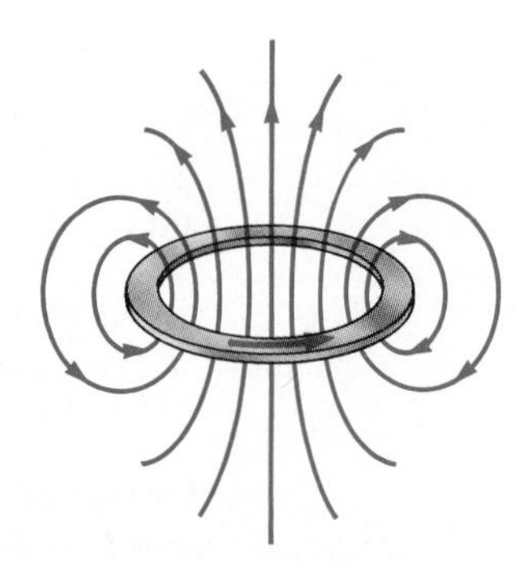

**Figure 30.6** Magnetic field lines for a current loop. Far from the loop, the field lines are identical in form to those of an electric dipole.

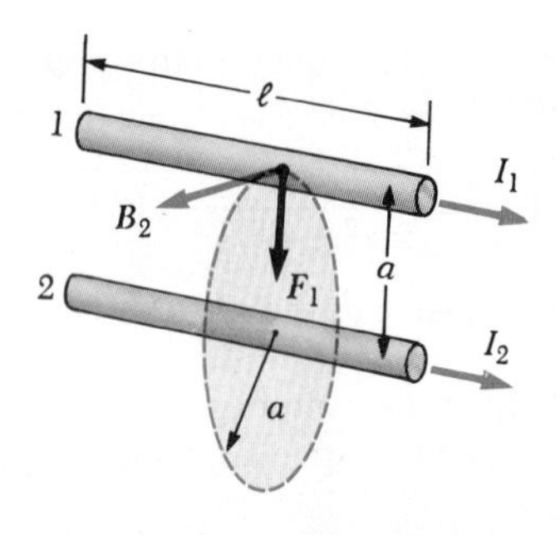


**Figure 30.7** Two parallel wires each carrying a steady current exert a force on each other. The field $\boldsymbol{B}_2$ at wire 1 due to wire 2 produces a force on wire 1 given by $F_1 = I_1 \ell B_2$. The force is attractive if the currents are parallel as shown and repulsive if the currents are antiparallel.

## 30.2 THE MAGNETIC FORCE BETWEEN TWO PARALLEL CONDUCTORS

In the previous chapter we described the magnetic force that acts on a current-carrying conductor when the conductor is placed in an external magnetic field. Since a current in a conductor sets up its own magnetic field, it is easy to understand that two current-carrying conductors will exert magnetic forces upon each other. As we shall see, such forces can be used as the basis for defining the ampere and the coulomb. Consider two long, straight, parallel wires separated by a distance $a$ and carrying currents $I_1$ and $I_2$ in the same direction, as in Figure 30.7. We can easily determine the force on one wire due to a magnetic field set up by the other wire. Wire 2, which carries a current $I_2$, sets up a magnetic field $\boldsymbol{B}_2$ at the position of wire 1. The direction of $\boldsymbol{B}_2$ is *perpendicular* to the wire, as shown in Figure 30.7. According to Equation 29.5, the magnetic force on a length $\ell$ of wire 1 is $\boldsymbol{F}_1 = I_1 \boldsymbol{\ell} \times \boldsymbol{B}_2$. Since $\boldsymbol{\ell}$ is perpendicular to $\boldsymbol{B}_2$, the magnitude of $\boldsymbol{F}_1$ is given by $F_1 = I_1 \ell B_2$. Since the field due to wire 2 is given by Equation 30.7,

$$B_2 = \frac{\mu_0 I_2}{2\pi a}$$

we see that

$$F_1 = I_1 \ell B_2 = I_1 \ell \left( \frac{\mu_0 I_2}{2\pi a} \right) = \frac{\ell \mu_0 I_1 I_2}{2\pi a}$$

We can rewrite this in terms of the force per unit length as

$$\frac{F_1}{\ell} = \frac{\mu_0 I_1 I_2}{2\pi a} \tag{30.13}$$

The direction of $\boldsymbol{F}_1$ is downward, toward wire 2, since $\boldsymbol{\ell} \times \boldsymbol{B}_2$ is downward. If one considers the field set up at wire 2 due to wire 1, the force $\boldsymbol{F}_2$ on wire 2 is found to be equal to and opposite $\boldsymbol{F}_1$. This is what one would expect, because Newton's third law of action-reaction must be obeyed.[1] When the currents are in opposite directions, the forces are reversed and the wires repel each other. Hence, we find that

> parallel conductors carrying currents in the same direction attract each other, whereas parallel conductors carrying currents in opposite directions repel each other.

The force between two parallel wires each carrying a current is used to define the **ampere** as follows:

> If two long, parallel wires 1 m apart carry the same current and the force per unit length on each wire is $2 \times 10^{-7}$ N/m, then the current is defined to be 1 A.

The numerical value of $2 \times 10^{-7}$ N/m is obtained from Equation 30.13, with $I_1 = I_2 = 1$ A and $a = 1$ m. Therefore, a mechanical measurement can be used

[1] Although the total force on wire 1 is equal to and opposite the total force on wire 2, Newton's third law does not apply when one considers two small elements of the wires that are not opposite each other. This apparent violation of Newton's third law and of conservation of momentum is described in more advanced treatments on electricity and magnetism.

to standardize the ampere. For instance, the National Bureau of Standards uses an instrument called a *current balance* for primary current measurements. These results are then used to standardize other, more conventional instruments, such as ammeters.

The SI unit of charge, the **coulomb,** can now be defined in terms of the ampere as follows:

If a conductor carries a steady current of 1 A, then the quantity of charge that flows through a cross section of the conductor in 1 s is 1 C.

## 30.3 AMPÈRE'S LAW

A simple experiment first carried out by Oersted in 1820 clearly demonstrates the fact that a current-carrying conductor produces a magnetic field. In this experiment, several compass needles are placed in a horizontal plane near a long vertical wire, as in Figure 30.8a. When there is no current in the wire, all compasses in the loop point in the same direction (that of the earth's field), as one would expect. However, when the wire carries a strong, steady current, the compass needles will all deflect in a direction tangent to the circle, as in Figure 30.8b. These observations show that the direction of $\boldsymbol{B}$ is consistent with the right-hand rule described in Section 30.1.

If the wire is grasped in the right hand with the thumb in the direction of the current, the fingers will wrap (or curl) in the direction of $\boldsymbol{B}$.

When the current is reversed, the compass needles in Figure 30.8b will also reverse.

Since the compass needles point in the direction of $\boldsymbol{B}$, we conclude that the lines of $\boldsymbol{B}$ form circles about the wire, as we discussed in the previous section. By symmetry, the magnitude of $\boldsymbol{B}$ is the same everywhere on a circular path that is centered on the wire and lying in a plane that is perpendicular to the wire. By varying the current and distance $r$ from the wire, one finds that $\boldsymbol{B}$ is proportional to the current and inversely proportional to the distance from the wire.

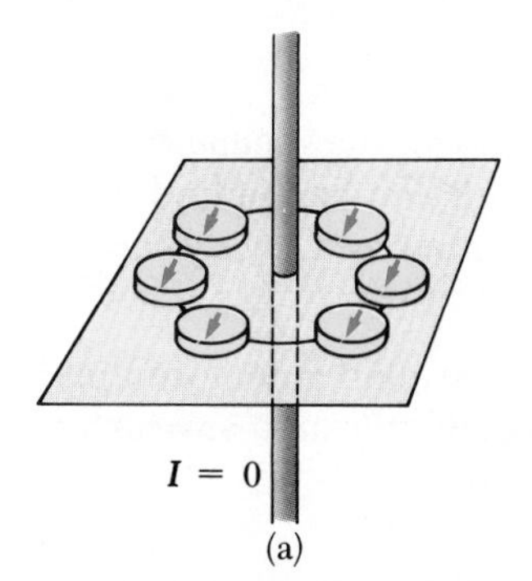


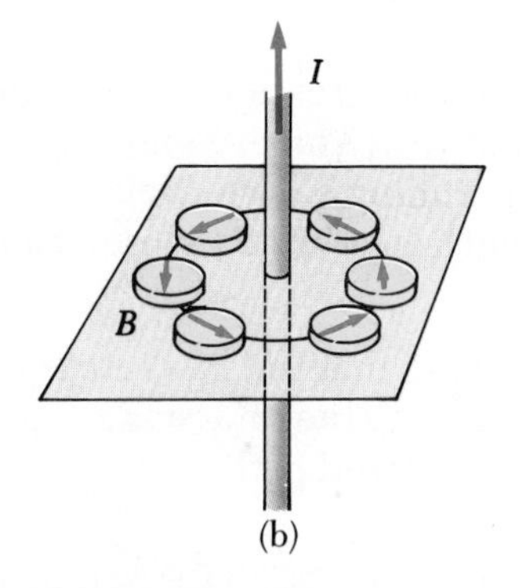


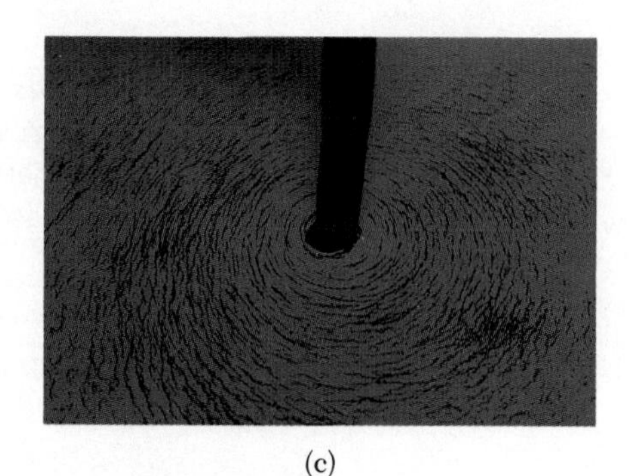


**Figure 30.8** (a) When there is no current in the vertical wire, all compass needles point in the same direction. (b) When the wire carries a strong current, the compass needles deflect in a direction tangent to the circle, which is the direction of $\boldsymbol{B}$ due to the current. (c) Circular magnetic field line surrounding a current-carrying conductor as displayed with iron filings. The photograph was taken using 30 parallel wires each carrying a current of $\frac{1}{2}$ A. (Photo courtesy of Henry Leap and Jim Lehman)

Biographical Sketch

**André-Marie Ampère**
*(1775–1836)*

André-Marie Ampère was a French mathematician, chemist, and philosopher who founded the science of electrodynamics. The unit of measure for electric current was named in his honor.

Ampère's genius, particularly in mathematics, became evident early in his life: he had mastered advanced mathematics by the age of 12. In his first publication, *Considerations on the Mathematical Theory of Games,* an early contribution to the theory of probability, he proposed the inevitability of a player's losing a game of chance to a player with greater financial resources.

Ampère is credited with the discovery of electromagnetism—the relationship between electric current and magnetic fields. His work in this field was influenced by the findings of Danish physicist Hans Christian Oersted. Ampère presented a series of papers expounding the theory and basic laws of electromagnetism, which he called electrodynamics, to differentiate it from the study of stationary electric forces, which he called electrostatics.

The culmination of Ampère's studies came in 1827 when he published his *Mathematical Theory of Electrodynamic Phenomena Deduced Solely from Experiment,* in which he derived precise mathematical formulations of electromagnetism, notably Ampère's law.

Many stories are told of Ampère's absent-mindedness, a trait he shared with Newton. In one instance, he forgot to honor an invitation to dine with the Emperor Napoleon.

Ampère's personal life was filled with tragedy. His father, a wealthy city official, was guillotined during the French Revolution, and his wife's death in 1803 was a major blow. Ampère died at the age of 63 of pneumonia. His judgment of his life is clear from the epitaph he chose for his gravestone: *Tandem felix* (Happy at last).

(Photo courtesy of AIP Niels Bohr Library)

Now let us evaluate the product $\boldsymbol{B} \cdot d\boldsymbol{s}$ and sum these products over the closed circular path centered on the wire. Along this path, the vectors $d\boldsymbol{s}$ and $\boldsymbol{B}$ are parallel at each point (Fig. 30.8b), so that $\boldsymbol{B} \cdot d\boldsymbol{s} = B\,ds$. Furthermore, $\boldsymbol{B}$ is constant in magnitude on this circle and given by Equation 30.7. Therefore, the sum of the products $B\,ds$ over the closed path, which is equivalent to the line integral of $\boldsymbol{B} \cdot d\boldsymbol{s}$, is given by

$$\oint \boldsymbol{B} \cdot d\boldsymbol{s} = B \oint ds = \frac{\mu_0 I}{2\pi r}(2\pi r) = \mu_0 I \tag{30.14}$$

where $\oint ds = 2\pi r$ is the circumference of the circle.

This result, known as **Ampère's law,** was calculated for the special case of a circular path surrounding a wire. However, the result can be applied in the general case in which an arbitrary closed path is threaded by a *steady current.* That is,

Ampère's law says that the line integral of $\boldsymbol{B} \cdot d\boldsymbol{s}$ around any closed path equals $\mu_0 I$, where $I$ is the total steady current passing through any surface bounded by the closed path.

Ampère's law

$$\oint \boldsymbol{B} \cdot d\boldsymbol{s} = \mu_0 I \tag{30.15}$$

*Ampère's law is valid only for steady currents.* Furthermore, *Ampère's law is useful only for calculating the magnetic field of current configurations with a high degree of symmetry,* just as Gauss' law is useful only for calculating the electric field of highly symmetric charge distributions. The following examples illustrate some symmetric current configurations for which Ampère's law is useful.

EXAMPLE 30.4 The $B$ Field of a Long Wire

A long, straight wire of radius $R$ carries a steady current $I_0$ that is uniformly distributed through the cross section of the wire (Figure 30.9). Calculate the magnetic field at a distance $r$ from the center of the wire in the regions $r \geq R$ and $r < R$.

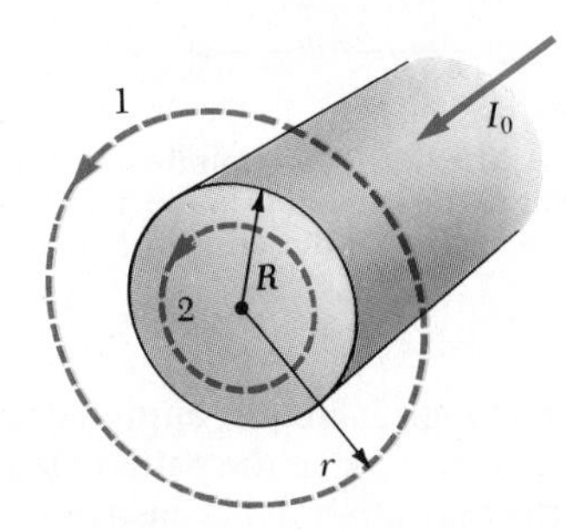


**Figure 30.9** (Example 30.4) A long, straight wire of radius $R$ carrying a steady current $I_0$ uniformly distributed across the wire. The magnetic field at any point can be calculated from Ampère's law using a circular path of radius $r$, concentric with the wire.

*Solution* In region 1, where $r \geq R$, let us choose a circular path of radius $r$ centered at the wire. From symmetry, we see that $\mathbf{B}$ must be constant in magnitude and parallel to $d\mathbf{s}$ at every point on the path. Since the total current passing through by path 1 is $I_0$, Ampère's law applied to the path gives

$$\oint \mathbf{B} \cdot d\mathbf{s} = B \oint ds = B(2\pi r) = \mu_0 I_0$$

$$B = \frac{\mu_0 I_0}{2\pi r} \quad \text{(for } r \geq R\text{)} \qquad (30.16)$$

which is identical in meaning to Equation 30.7.

Now consider the interior of the wire, that is, region 2, where $r < R$. In this case, note that the current $I$ enclosed by the path is *less* than the total current, $I_0$. Since the current is assumed to be uniform over the cross section of the wire, we see that the fraction of the current enclosed by the path of radius $r < R$ must equal the ratio of the area $\pi r^2$ enclosed by path 2 and the cross-sectional area $\pi R^2$ of the wire.[2] That is,

[2] Alternatively, the current linked by path 2 must equal the product of the current density, $J = I_0/\pi R^2$, and the area $\pi r^2$ enclosed by path 2.

$$\frac{I}{I_0} = \frac{\pi r^2}{\pi R^2}$$

$$I = \frac{r^2}{R^2} I_0$$

Following the same procedure as for path 1, we can now apply Ampère's law to path 2. This gives

$$\oint \mathbf{B} \cdot d\mathbf{s} = B(2\pi r) = \mu_0 I = \mu_0 \left( \frac{r^2}{R^2} I_0 \right)$$

$$B = \left( \frac{\mu_0 I_0}{2\pi R^2} \right) r \quad \text{(for } r < R\text{)} \qquad (30.17)$$

The magnetic field versus $r$ for this configuration is sketched in Figure 30.10. Note that inside the wire, $B \rightarrow 0$ as $r \rightarrow 0$. This result is similar in form to that of the electric field inside a uniformly charged rod.

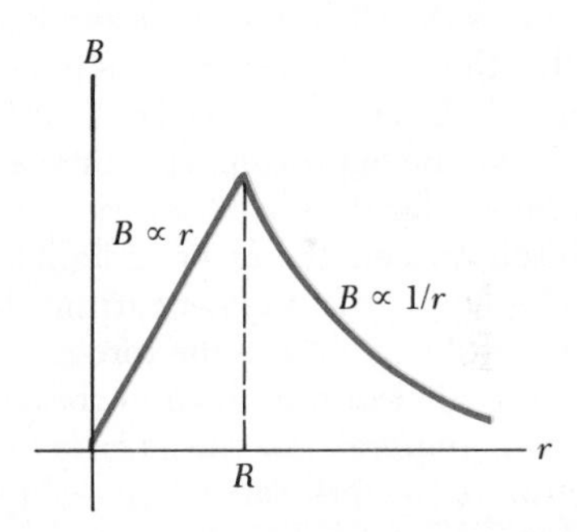


**Figure 30.10** A sketch of the magnetic field versus $r$ for the wire described in Example 30.4. The field is proportional to $r$ inside the wire and varies as $1/r$ outside the wire.

EXAMPLE 30.5 The Magnetic Field of a Toroidal Coil

The *toroidal coil* consists of $N$ turns of wire wrapped around a doughnut-shaped structure as in Figure 30.11. Assuming that the turns are closely spaced, calculate the magnetic field *inside* the coil, a distance $r$ from the center.

*Solution* To calculate the field inside the coil, we evaluate the line integral of $\mathbf{B} \cdot d\mathbf{s}$ over a circle of radius $r$. By symmetry, we see that the magnetic field is constant in magnitude on this path and tangent to it, so that $\mathbf{B} \cdot d\mathbf{s} = B\, ds$. Furthermore, note that the closed path threads $N$

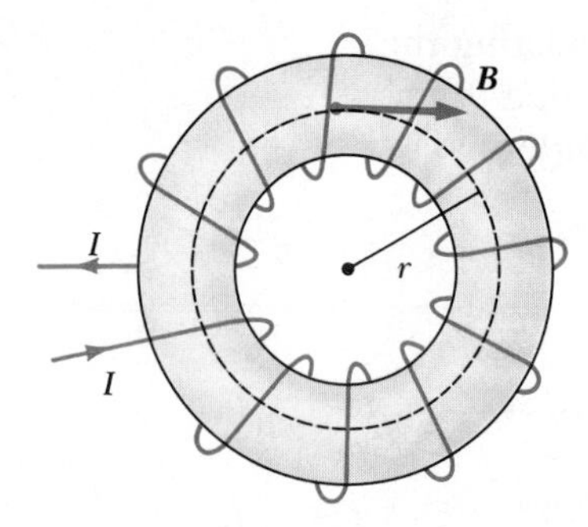


**Figure 30.11** (Example 30.6) A toroidal coil consists of many turns of wire wrapped around a doughnut-shaped structure (torus). If the coils are closely spaced, the field inside the toroidal coil is tangent to the dashed circular path and varies as $1/r$, and the exterior field is zero.

loops of wire, each of which carries a current $I$. Therefore, the right side of Ampère's law, Equation 30.15, is $\mu_0 NI$ in this case. Ampère's law applied to this path then gives

$$\oint \mathbf{B} \cdot d\mathbf{s} = B\oint ds = B(2\pi r) = \mu_0 NI$$

$$B = \frac{\mu_0 NI}{2\pi r} \qquad (30.18)$$

This result shows that $B$ varies as $1/r$ and hence is nonuniform within the coil. However, if $r$ is large compared with $a$, where $a$ is the cross-sectional radius of the toroid, then the field will be approximately uniform inside the coil. Furthermore, for the ideal toroidal coil, where the turns are closely spaced, the external field is *zero*. This can be seen by noting that the *net* current threaded by any circular path lying outside the toroidal coil is zero (including the region of the "hole in the doughnut"). Therefore, from Ampère's law one finds that $\mathbf{B} = 0$ in the regions exterior to the toroidal coil. In reality, the turns of a toroidal coil forms a helix rather than circular loops (the ideal case). As a result, there is always a small field external to the coil.

**EXAMPLE 30.6 Magnetic Field of an Infinite Current Sheet**

An infinite sheet lying in the $yz$ plane carries a surface current of density $\mathbf{J}_s$. The current is in the $y$ direction, and $J_s$ represents the current per unit length measured along the $z$ axis. Find the magnetic field near the sheet.

*Solution* To evaluate the line integral in Ampère's law, let us take a rectangular path around the sheet as in Figure 30.12. The rectangle has dimensions $\ell$ and $w$, where the sides of length $\ell$ are parallel to the surface. The *net* current through the loop is $J_s\ell$ (that is, the net current equals the current per unit length multiplied by the

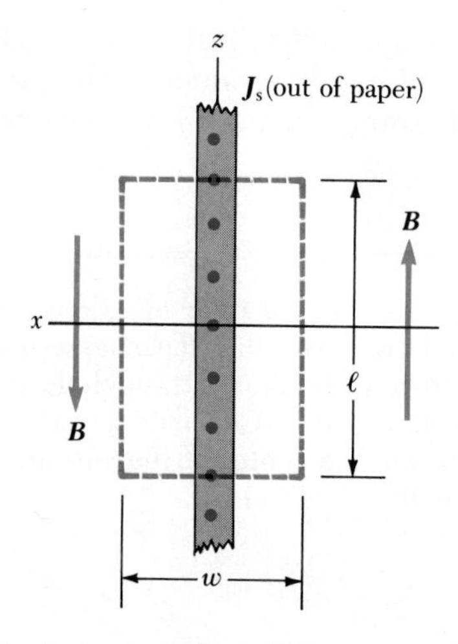


**Figure 30.12** A top view of an infinite current sheet lying in the $yz$ plane, where the current is in the $y$ direction (out of the paper). This view shows the direction of $\mathbf{B}$ on both sides of the sheet.

length of the rectangle). Hence, applying Ampère's law over the loop and noting that the paths of length $w$ do not contribute to the line integral (because the component of $\mathbf{B}$ along the direction of these paths is zero), we get

$$\oint \mathbf{B} \cdot d\mathbf{s} = \mu_0 I = \mu_0 J_s \ell$$

$$2B\ell = \mu_0 J_s \ell$$

$$B = \mu_0 \frac{J_s}{2} \qquad (30.19)$$

The result shows that *the magnetic field is independent of the distance from the current sheet.* In fact, the magnetic field is uniform and is everywhere parallel to the plane of the sheet. This is reasonable since we are dealing with an *infinite* sheet of current. The result is analogous to the uniform electric field associated with an infinite sheet of charge. (Example 24.6.)

**EXAMPLE 30.7 The Magnetic Force on a Current Segment**

A long straight wire oriented along the $y$ axis carries a steady current $I_1$ as in Figure 30.13. A rectangular circuit located to the right of the wire carries a current $I_2$. Find the magnetic force on the *upper horizontal segment* of the circuit that runs from $x = a$ to $x = a + b$.

*Solution* In this problem, you may be tempted to use Equation 30.13 to obtain the force. However, this result applies *only* to two *parallel* wires, and cannot be used here. The correct approach is to start with the force on a small segment of the conductor given by $d\mathbf{F} = I\, d\mathbf{s} \times \mathbf{B}$

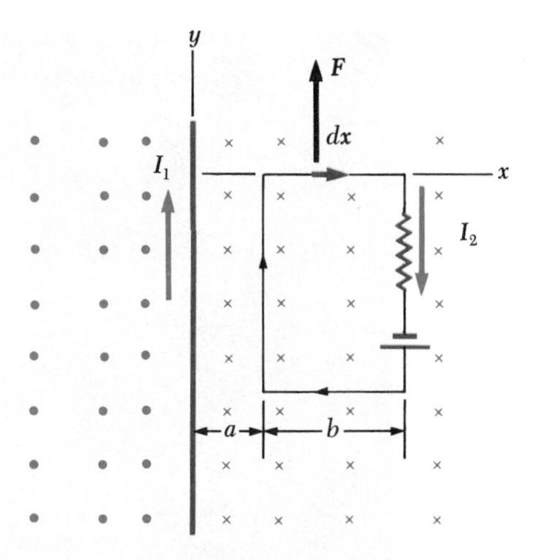


Figure 30.13 Example 30.7.

(Eq. 29.6), where in this case, $I = I_2$ and $\boldsymbol{B}$ is the magnetic field due to the long straight wire at the position of the segment of length $ds$. From Ampere's law, the field at a distance $x$ from the straight wire is given by

$$\boldsymbol{B} = \frac{\mu_0 I_1}{2\pi x}(-\boldsymbol{k})$$

where the field points *into* the page as indicated by the unit vector notation $(-\boldsymbol{k})$. Taking the length of our segment as $d\boldsymbol{s} = dx\,\boldsymbol{i}$, we find

$$d\boldsymbol{F} = \frac{\mu_0 I_1 I_2}{2\pi x}[\boldsymbol{i} \times (-\boldsymbol{k})]\,dx = \frac{\mu_0 I_1 I_2}{2\pi}\frac{dx}{x}\boldsymbol{j}$$

Integrating this equation over the limits $x = a$ to $x = a + b$ gives

$$\boldsymbol{F} = \frac{\mu_0 I_1 I_2}{2\pi}\ln x\Big]_a^{a+b}\boldsymbol{j} = \frac{\mu_0 I_1 I_2}{2\pi}\ln\left(1 + \frac{b}{a}\right)\boldsymbol{j}$$

The force points *upward* as indicated by the notation $\boldsymbol{j}$, and as shown in Figure 30.13.

Exercise 3 What is the force on the lower horizontal segment of the circuit?
Answer The force has the same magnitude as the force on the upper horizontal segment, but is directed *downward*.

## 30.4 THE MAGNETIC FIELD OF A SOLENOID

**A solenoid** is a long wire wound in the form of a helix. With this configuration, one can produce a reasonably uniform magnetic field within a small volume of the solenoid's interior region if the consecutive turns are closely spaced. When the turns are closely spaced, each can be regarded as a circular loop, and the net magnetic field is the vector sum of the fields due to all the turns.

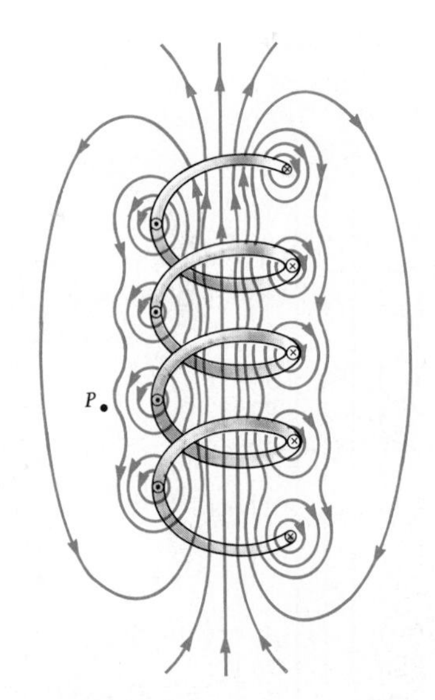


Figure 30.14 The magnetic field lines for a loosely wound solenoid. Adapted from D. Halliday and R. Resnick, *Physics*, New York, Wiley, 1978.

Figure 30.14 shows the magnetic field lines of a loosely wound solenoid. Note that the field lines inside the coil are nearly parallel, uniformly distributed, and close together. This indicates that the field inside the solenoid is uniform. The field lines between the turns tend to cancel each other. The field outside the solenoid is both nonuniform and weak. The field at exterior points, such as $P$, is weak since the field due to current elements on the upper portions tends to cancel the field due to current elements on the lower portions.

If the turns are closely spaced and the solenoid is of finite length, the field lines are as shown in Figure 30.15. In this case, the field lines diverge from one end and converge at the opposite end. An inspection of this field distribution exterior to the solenoid shows a similarity with the field of a bar magnet. Hence, one end of the solenoid behaves like the north pole of a magnet while the opposite end behaves like the south pole. As the length of the solenoid increases, the field within it becomes more and more uniform. One approaches the case of an *ideal solenoid* when the turns are closely spaced and the length is long compared with the radius. In this case, the field outside the solenoid is weak compared with the field inside the solenoid, and the field inside is uniform over a large volume.

We can use Ampère's law to obtain an expression for the magnetic field inside an ideal solenoid. A longitudinal cross section of part of our ideal sole-

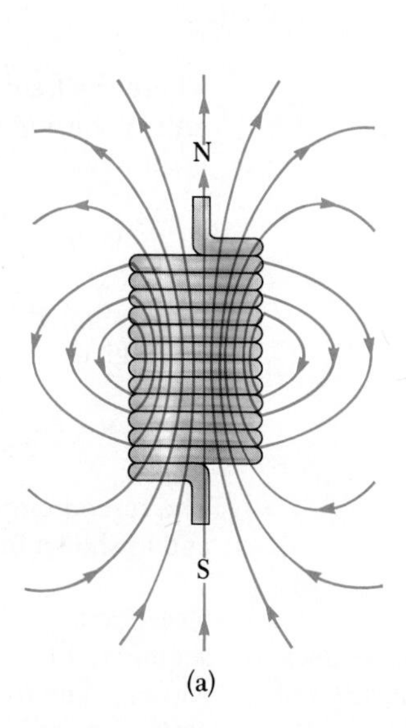


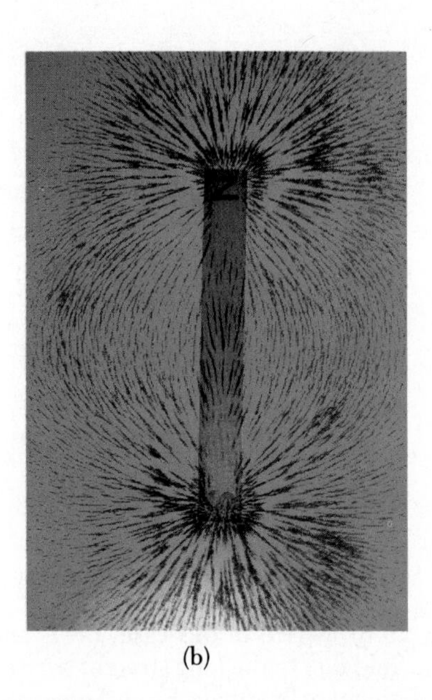


**Figure 30.15** (a) Magnetic field lines for a tightly wound solenoid of finite length carrying a steady current. The field inside the solenoid is nearly uniform and strong. Note that the field lines resemble those of a bar magnet, so that the solenoid effectively has north and south poles. (b) Magnetic field pattern of a bar magnet, as displayed by small iron filings on a sheet of paper. (Courtesy of Henry Leap and Jim Lehman)

noid (Fig. 30.16) carries a current $I$. For the ideal solenoid, $\mathbf{B}$ inside the solenoid is uniform and parallel to the axis and $\mathbf{B}$ outside is zero. Consider a rectangular path of length $\ell$ and width $w$ as shown in Figure 30.16. We can apply Ampère's law to this path by evaluating the integral of $\mathbf{B}\cdot d\mathbf{s}$ over each of the four sides of the rectangle. The contribution along side 3 is clearly zero, since $\mathbf{B} = 0$ in this region. The contributions from sides 2 and 4 are both zero since $\mathbf{B}$ is perpendicular to $d\mathbf{s}$ along these paths. Side 1, whose length is $\ell$, gives a contribution $B\ell$ to the integral since $\mathbf{B}$ is uniform and parallel to $d\mathbf{s}$ along this path. Therefore, the integral over the closed rectangular path has the value

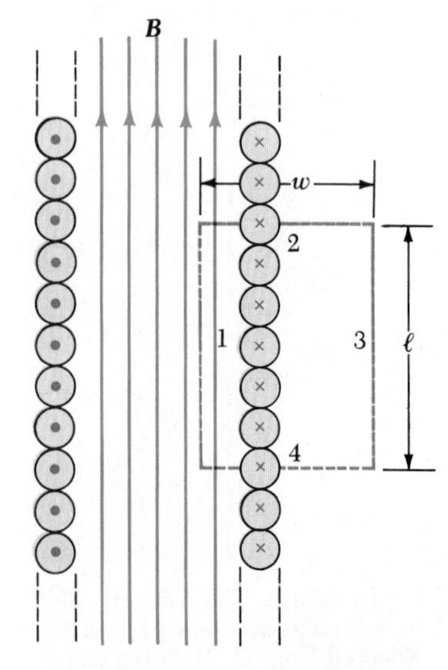


**Figure 30.16** A cross-sectional view of a tightly wound solenoid. If the solenoid is long relative to its radius, we can assume that the field inside is uniform and the field outside is zero. Ampère's law applied to the red dashed rectangular path can then be used to calculate the field inside the solenoid.

$$\oint \mathbf{B}\cdot d\mathbf{s} = \int_{\text{path 1}} \mathbf{B}\cdot d\mathbf{s} = B\int_{\text{path 1}} ds = B\ell$$

The right side of Ampère's law involves the *total* current that passes through the area bound by the path of integration. In our case, the total current through the rectangular path equals the current through each turn multiplied by the number of turns. If $N$ is the number of turns in the length $\ell$, then the total current through the rectangle equals $NI$. Therefore, Ampère's law applied to this path gives

$$\oint \mathbf{B}\cdot d\mathbf{s} = B\ell = \mu_0 NI$$

$$B = \mu_0 \frac{N}{\ell} I = \mu_0 nI \qquad (30.20)$$

where $n = N/\ell$ is the number of turns *per unit length* (not to be confused with $N$).

We also could obtain this result in a simpler manner by reconsidering the magnetic field of a toroidal coil (Example 30.5). If the radius $r$ of the toroidal coil containing $N$ turns is large compared with its cross-sectional radius $a$, then a short section of the toroidal coil approximates a solenoid with $n = N/2\pi r$. In this limit, we see that Equation 30.18 derived for the toroidal coil agrees with Equation 30.20.

Equation 30.20 is valid only for points near the center of a very long solenoid. As you might expect, the field near each end is smaller than the value given by Equation 30.20. At the very end of a long solenoid, the magnitude of the field is about one half that of the field at the center. The field at arbitrary axial points of the solenoid is derived in Section 30.5.

## *30.5 THE MAGNETIC FIELD ALONG THE AXIS OF A SOLENOID

Consider a solenoid of length $\ell$ and radius $R$ containing $N$ closely spaced turns and carrying a steady current $I$. Let us determine an expression for the magnetic field at an axial point $P$ inside the solenoid, as indicated in Figure 30.17.

Perhaps the simplest way to obtain the desired result is to consider the solenoid as a distribution of current loops. The field of any one loop along the axis is given by Equation 30.9. Hence, the net field in the solenoid is the superposition of fields from all loops. The number of turns in a length $dx$ of the solenoid is $(N/\ell)\ dx$; therefore the total current in a width $dx$ is given by $I(N/\ell)\ dx$. Then, using Equation 30.9, we find that the field at $P$ due to the section $dx$ is given by

$$dB = \frac{\mu_0 R^2}{2(x^2 + R^2)^{3/2}}\, I\left(\frac{N}{\ell}\right) dx \qquad (30.21)$$

This expression contains the variable $x$, which can be expressed in terms of the variable $\phi$, defined in Figure 30.17. That is, $x = R \tan \phi$, so that we have $dx = R \sec^2 \phi\ d\phi$. Substituting these expressions into Equation 30.21 and integrating from $\phi_1$ to $\phi_2$, we get

$$B = \frac{\mu_0 NI}{2\ell}\int_{\phi_1}^{\phi_2} \cos \phi\ d\phi = \frac{\mu_0 NI}{2\ell}(\sin \phi_2 - \sin \phi_1) \qquad (30.22)$$

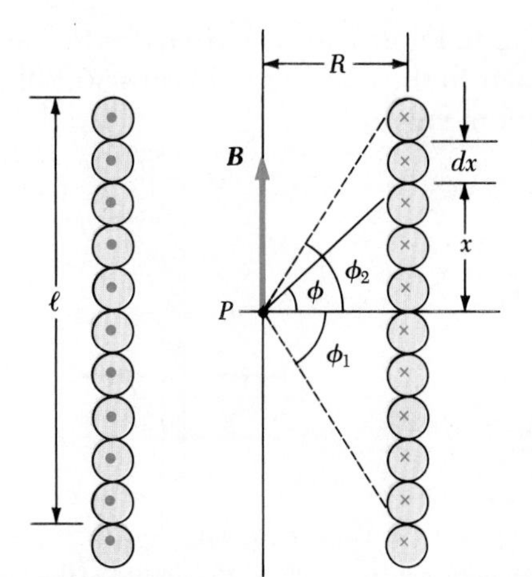


**Figure 30.17** The geometry for calculating the magnetic field at an axial point $P$ inside a tightly wound solenoid.

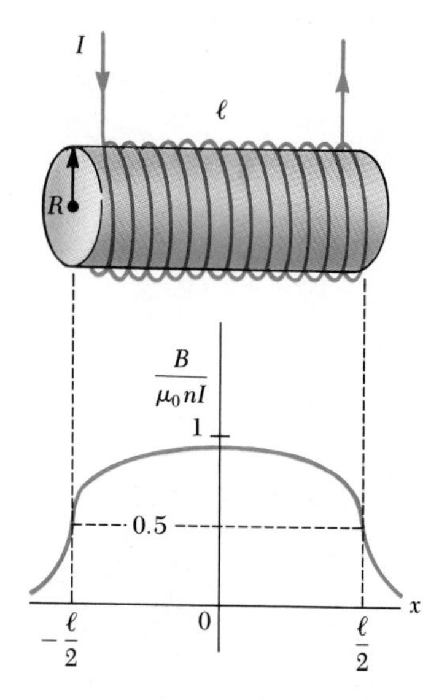


**Figure 30.18** A sketch of the magnetic field along the axis versus $x$ for a long, tightly wound solenoid. Note that the magnitude of the field at the ends is about one half the value of the center.

If $P$ is at the *midpoint* of the solenoid and if we assume that the solenoid is long compared with $R$, then $\phi_2 \approx 90°$ and $\phi_1 \approx -90°$; therefore

$$B \approx \frac{\mu_0 NI}{2\ell}(1+1) = \frac{\mu_0 NI}{\ell} = \mu_0 nI \qquad \text{(at the center)}$$

which is in agreement with our previous result, Equation 30.20.

If $P$ is a point at the end of a long solenoid (say, the bottom), then $\phi_1 \approx 0°$, $\phi_2 \approx 90°$, and

$$B \approx \frac{\mu_0 NI}{2\ell}(1+0) = \tfrac{1}{2}\mu_0 nI \qquad \text{(at the ends)}$$

This shows that the field at each end of a solenoid approaches *one half* the value at the solenoid's center as the length $\ell$ approaches infinity.

A sketch of the field at axial points versus $x$ for a solenoid is shown in Figure 30.18. If the length $\ell$ is large compared with $R$, the axial field will be quite uniform over most of the solenoid and the curve will be quite flat except at points near the ends. On the other hand, if $\ell$ is comparable to $R$, then the field will have a value somewhat less than $\mu_0 nI$ at the middle and will be uniform only over a small region of the solenoid.

## 30.6 MAGNETIC FLUX

The flux associated with a magnetic field is defined in a manner similar to that used to define the electric flux. Consider an element of area $dA$ on an arbitrarily shaped surface, as in Figure 30.19. If the magnetic field at this element is $\mathbf{B}$, then the magnetic flux through the element is $\mathbf{B}\cdot d\mathbf{A}$, where $d\mathbf{A}$ is a vector perpendicular to the surface whose magnitude equals the area $dA$. Hence, the total magnetic flux $\Phi_m$ through the surface is given by

Magnetic flux

$$\Phi_m = \int \mathbf{B}\cdot d\mathbf{A} \tag{30.23}$$

Consider the special case of a plane of area $A$ and a uniform field $\mathbf{B}$, which makes an angle $\theta$ with the vector $d\mathbf{A}$. The magnetic flux through the plane in this case is given by

$$\Phi_m = BA\cos\theta \tag{30.24}$$

If the magnetic field lies in the plane as in Figure 30.20a, then $\theta = 90°$ and the flux is zero. If the field is perpendicular to the plane as in Figure 30.20b, then $\theta = 0°$ and the flux is $BA$ (the maximum value).

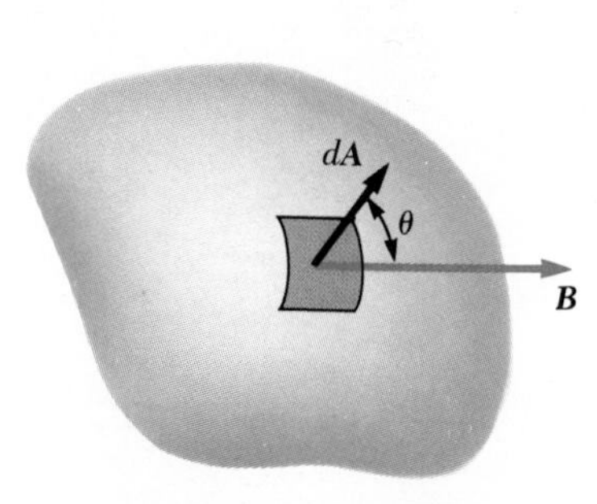


**Figure 30.19** The magnetic flux through an area element $dA$ is given by $\mathbf{B}\cdot d\mathbf{A} = BdA\cos\theta$. Note that $d\mathbf{A}$ is perpendicular to the surface.

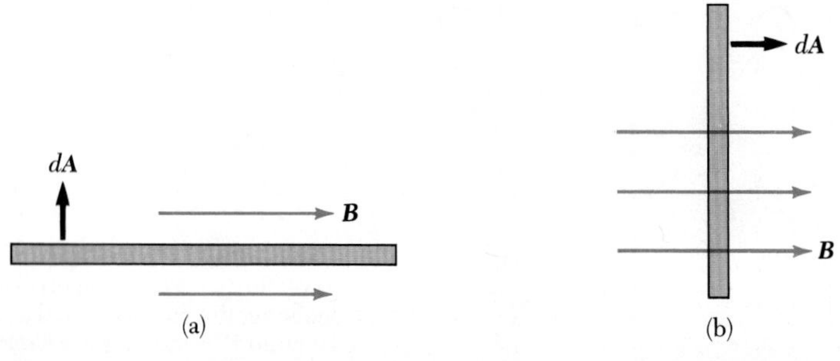


**Figure 30.20** (a) The flux is zero when the magnetic field is parallel to the surface of the plane (an edge view). (b) The flux is a maximum when the magnetic field is perpendicular to the plane.

Since $B$ has units of Wb/m², or T, the unit of flux is the weber (Wb), where 1 Wb = 1 T·m².

**EXAMPLE 30.8 Flux Through a Rectangular Loop** A rectangular loop of width $a$ and length $b$ is located a distance $c$ from a long wire carrying a current $I$ (Fig. 30.21). The wire is parallel to the long side of the loop. Find the total magnetic flux through the loop.

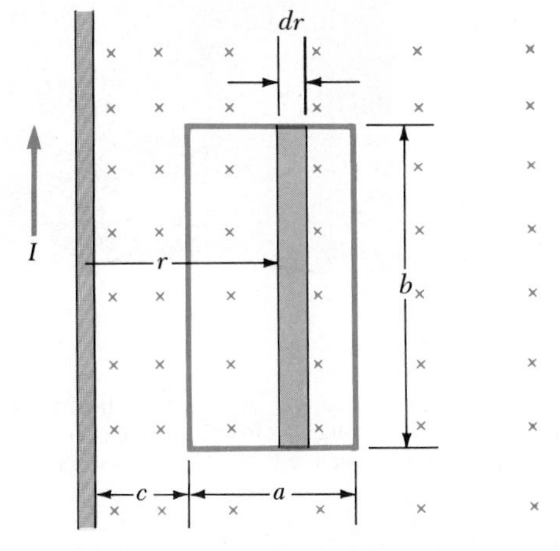


**Figure 30.21** (Example 30.8) The magnetic field due to the wire carrying a current $I$ is *not* uniform over the rectangular loop.

*Solution* From Ampère's law, we found that the magnetic field due to the wire at a distance $r$ from the wire is given by

$$B = \frac{\mu_0 I}{2\pi r}$$

That is, the field *varies* over the loop and is directed *into* the page as shown in Figure 30.21. Since $\mathbf{B}$ is parallel to $d\mathbf{A}$, we can express the magnetic flux through an area element $dA$ as

$$\Phi_m = \int B\, dA = \int \frac{\mu_0 I}{2\pi r}\, dA$$

*Note that since* $\mathbf{B}$ *is not uniform, but depends on* $r$, *it cannot be removed from the integral.* In order to carry out the integration, we first express the area element (the blue region in Fig. 30.21) as $dA = b\, dr$. Since $r$ is the only variable that now appears in the integral, the expression for $\Phi_m$ becomes

$$\Phi_m = \frac{\mu_0 I}{2\pi} b \int_c^{a+c} \frac{dr}{r} = \frac{\mu_0 I b}{2\pi} \ln r \Big]_c^{a+c}$$

$$= \frac{\mu_0 I b}{2\pi} \ln\left(\frac{a+c}{c}\right)$$

## 30.7 GAUSS' LAW IN MAGNETISM

In Chapter 24 we found that the flux of the electric field through a closed surface surrounding a net charge is proportional to that charge (Gauss' law). In other words, the number of electric field lines leaving the surface depends only on the net charge within it. This property is based in part on the fact that electric field lines originate on electric charges.

The situation is quite different for magnetic fields, which are continuous and form closed loops. Magnetic field lines due to currents do not begin or end at any point. The magnetic field lines of the bar magnet in Figure 30.22 illustrate this point. Note that for any closed surface, the number of lines entering that surface equals the number leaving that surface, and so the net magnetic flux is *zero*. This is in contrast to the case of a surface surrounding one charge of an electric dipole (Fig. 30.23), where the net electric flux is not zero.

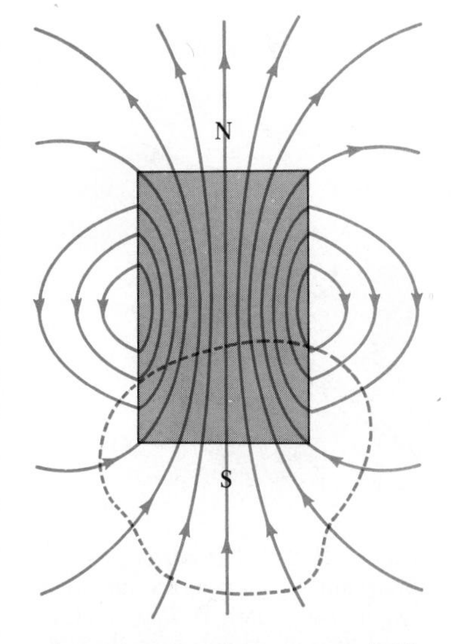


**Figure 30.22** The magnetic field lines of a bar magnet form closed loops. Note that the net flux through the closed surface surrounding one of the poles (or any other closed surface) is zero.

Gauss' law in magnetism states that the net magnetic flux through any closed surface is always zero:

$$\oint \mathbf{B} \cdot d\mathbf{A} = 0 \qquad (30.25)$$

This statement is based on the experimental fact that *isolated magnetic poles (or monopoles) have not been detected, and perhaps do not even exist.* The only

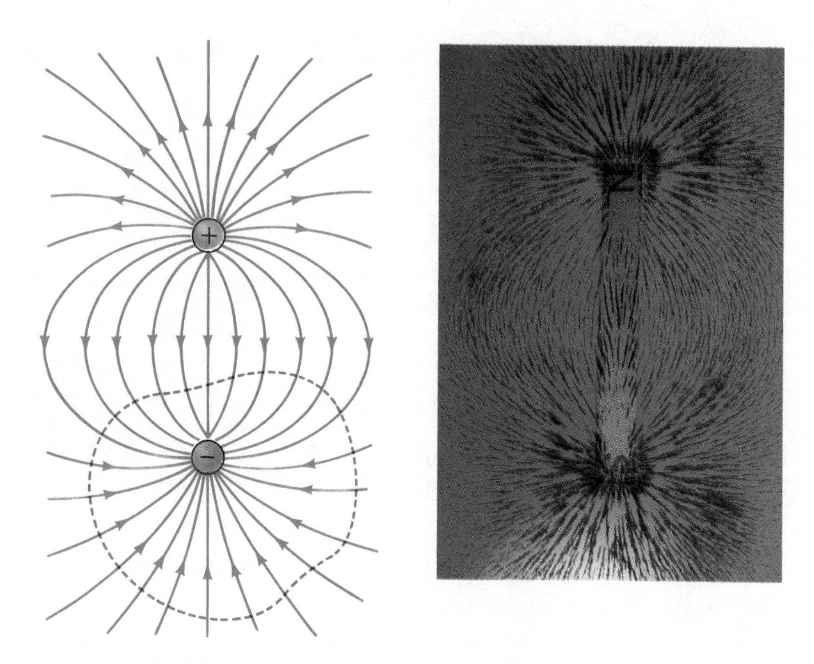


**Figure 30.23** (a) The electric field lines of an electric dipole begin on the positive charge and terminate on the negative charge. The electric flux through a closed surface surrounding one of the charges is *not* zero. (b) Magnetic field pattern of a bar magnet. (Courtesy of Henry Leap and Jim Lehman)

known sources of magnetic fields are magnetic dipoles (current loops), even in magnetic materials. In fact, all magnetic effects in matter can be explained in terms of magnetic dipole moments (effective current loops) associated with electrons and nuclei. This will be discussed further in Section 30.9.

## 30.8 DISPLACEMENT CURRENT AND THE GENERALIZED AMPÈRE'S LAW

We have seen that charges in motion, or currents, produce magnetic fields. When a current-carrying conductor has high symmetry, we can calculate the magnetic field using Ampère's law, given by Equation 30.15:

$$\oint \boldsymbol{B} \cdot d\boldsymbol{s} = \mu_0 I$$

where the line integral is over *any closed path through which the conduction current passes.* If $Q$ is the charge on the capacitor at any instant, the conduction current is defined by

$$I \equiv \frac{dQ}{dt}$$

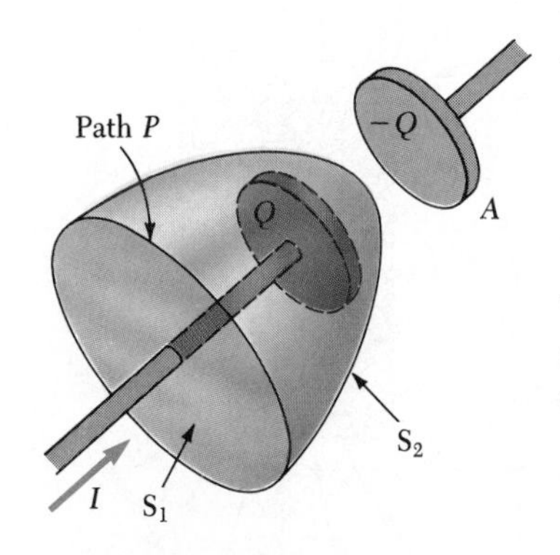


**Figure 30.24** The surfaces $S_1$ (in blue) and $S_2$ (in red) are bounded by the same path P. The conduction current passes only through $S_1$. This leads to a contradictory situation in Ampère's law which is resolved only if one postulates a displacement current through $S_2$.

We shall now show that *Ampère's law in this form is valid only if the conduction is constant in time.* Maxwell recognized this limitation and modified Ampère's law to include all possible situations.

We can understand this problem by considering a capacitor being charged as in Figure 30.24. The argument given here is equivalent to Maxwell's original reasoning. When the current $I$ changes with time (for example, when an ac voltage source is used), the charge on the plate changes, but *no conduction current passes between the plates.* Now consider the two surfaces $S_1$ and $S_2$ bounded by the same path P. Ampère's law says that the line integral of $\boldsymbol{B}\cdot d\boldsymbol{s}$ around this path must equal $\mu_0 I$, where $I$ is the total current through any surface bounded by the path P.

When the path P is considered to bound $S_1$, the result of the integral is $\mu_0 I$ since the current passes through $S_1$. However when the path bounds $S_2$, the result is *zero* since no conduction current passes through $S_2$. Thus, we have a contradictory situation which arises from the discontinuity of the current! Maxwell solved this problem by postulating an additional term on the right side of Equation 30.15, called the **displacement current,** $I_d$, defined as

$$I_d \equiv \epsilon_0 \frac{d\Phi_e}{dt} \tag{30.26}$$

Displacement current

Recall that $\Phi_e$ is the flux of the electric field, defined as $\Phi_e = \int \boldsymbol{E}\cdot d\boldsymbol{A}$.

As the capacitor is being charged (or discharged), the *changing* electric field between the plates may be thought of as a sort of current that bridges the discontinuity in the conduction current. When this expression for the current (Eq. 30.26) is added to the right side of Ampère's law, the difficulty represented by Figure 30.24 is resolved. No matter what surface bounded by the path P is chosen, some combination of conduction and displacement current will pass through it. With this new term $I_d$, we can express the generalized form of Ampère's law (sometimes called the **Ampère-Maxwell law**) as[3]

$$\oint \boldsymbol{B}\cdot d\boldsymbol{s} = \mu_0(I + I_d) = \mu_0 I + \mu_0\epsilon_0 \frac{d\Phi_e}{dt} \tag{30.27}$$

Ampère-Maxwell law

The meaning of this expression can be understood by referring to Figure 30.25. The electric flux through $S_2$ is $\Phi_e = \int \boldsymbol{E}\cdot d\boldsymbol{A} = EA$, where $A$ is the area of the plates and $E$ is the uniform electric field strength between the plates. If $Q$ is the charge on the plates at any instant, then one finds that $E = Q/\epsilon_0 A$ (Section 26.2). Therefore, the electric flux through $S_2$ is simply

$$\Phi_e = EA = \frac{Q}{\epsilon_0}$$

Hence, the displacement current $I_d$ through $S_2$ is

$$I_d = \epsilon_0 \frac{d\Phi_e}{dt} = \frac{dQ}{dt} \tag{30.28}$$

That is, the displacement current is precisely equal to the conduction current $I$ passing through $S_1$!

The central point of this formalism is the fact that

magnetic fields are produced both by conduction currents and by changing electric fields.

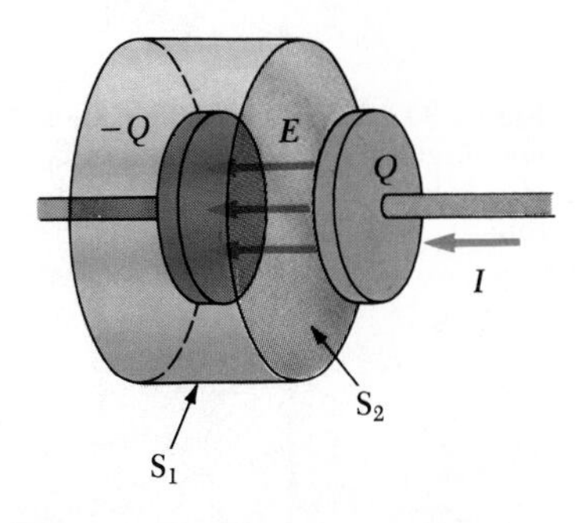


**Figure 30.25** The conduction current $I = dQ/dt$ passes through $S_1$. The displacement current $I_d = \epsilon_0\, d\Phi_e/dt$ passes through $S_2$. The two currents must be equal for continuity. In general, the total current through any surface bounded by some path is $I + I_d$.

[3] Strictly speaking, this expression is valid only in a vacuum. If a magnetic material is present, one must also include a magnetizing current $I_m$ on the right side of Equation 30.27 to make Ampère's law fully general. On a microscopic scale, $I_m$ is a current that is as real as the conduction current $I$.

**EXAMPLE 30.9 Displacement Current in a Capacitor**
An ac voltage is applied directly across an 8-$\mu$F capacitor. The frequency of the source is 3 kHz, and the voltage amplitude is 30 V. Find the displacement current between the plates of the capacitor.

*Solution* The angular frequency of the source is given by $\omega = 2\pi f = 2\pi(3 \times 10^3 \text{ Hz}) = 6\pi \times 10^3 \text{ s}^{-1}$. Hence, the voltage across the capacitor in terms of $t$ is

$$V = V_m \sin \omega t = (30 \text{ V}) \sin(6\pi \times 10^3 t)$$

We can make use of Equation 30.28 and of the fact that the charge on the capacitor is given by $Q = CV$ to find the displacement current:

$$I_d = \frac{dQ}{dt} = \frac{d}{dt}(CV) = C\frac{dV}{dt}$$

$$= (8 \times 10^{-6})\frac{d}{dt}[30 \sin(6\pi \times 10^3 t)]$$

$$= (4.52 \text{ A}) \cos(6\pi \times 10^3 t)$$

Hence, the displacement current varies sinusoidally with time and has a *maximum* value of 4.52 A.

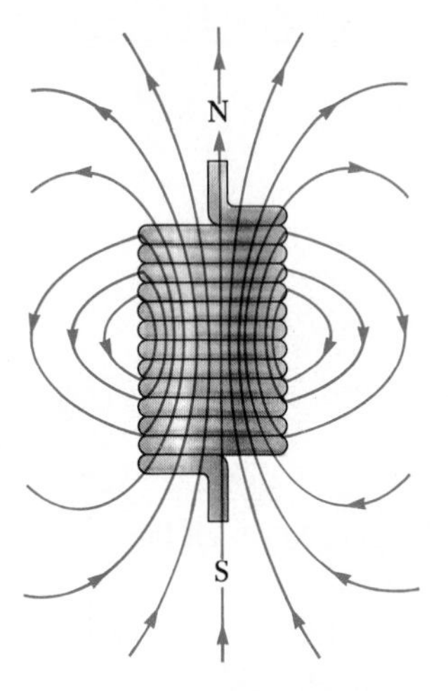

**Figure 30.26** Magnetic field lines for a tightly wound solenoid of finite length carrying a steady current.

## °30.9 MAGNETISM IN MATTER

The magnetic field produced by a current in a coil of wire gives us a hint as to what might cause certain materials to exhibit strong magnetic properties. Earlier we found that a coil like that shown in Figure 30.26 has a north and a south pole. In general, any current loop has a magnetic field and a corresponding magnetic moment. Similarly, the magnetic moments in a magnetized substance are associated with internal atomic currents. One can view these currents as arising from electrons orbiting around the nucleus and protons orbiting about each other inside the nucleus.

We shall begin this section with a brief discussion of the magnetic moments due to electrons. As we shall see, the net magnetic moment of an electron is due to a combination of its orbital motion and an intrinsic property called *spin*. The mutual forces between these magnetic dipole moments and their interaction with an external magnetic field are of fundamental importance in understanding the behavior of magnetic materials. We shall describe three categories of materials, paramagnetic, ferromagnetic, and diamagnetic. **Paramagnetic** and **ferromagnetic** materials are those that have atoms with permanent magnetic dipole moments. **Diamagnetic** materials are those whose atoms have no permanent magnetic dipole moments. For materials whose atoms have permanent magnetic moments, the diamagnetic contribution to the magnetism is usually overshadowed by paramagnetic or ferromagnetic effects.

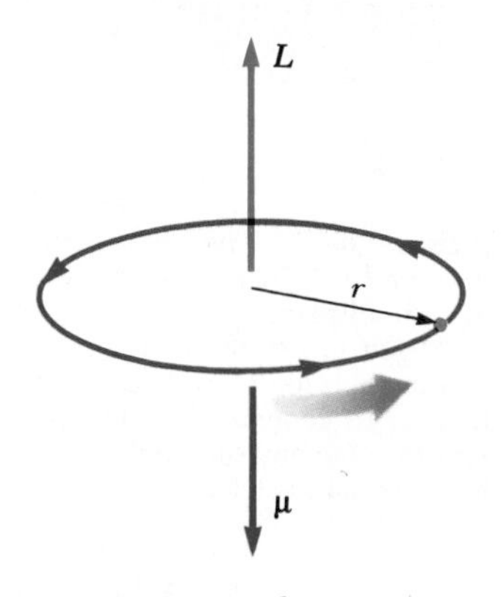

**Figure 30.27** An electron moving in a circular orbit of radius $r$ has an angular momentum $L$ and a magnetic moment $\mu$ which are in *opposite* directions.

### The Magnetic Moments of Atoms

It is instructive to begin our discussion with a classical model of the atom in which electrons are assumed to move in circular orbits about the much more massive nucleus. In this model, an orbiting electron is viewed as a tiny current loop, and the atomic magnetic moment is associated with this orbital motion. Although this model has many deficiencies, its predictions are in good agreement with the correct theory from quantum physics.

Consider an electron moving with constant speed $v$ in a circular orbit of radius $r$ about the nucleus, as in Figure 30.27. Since the electron travels a distance of $2\pi r$ (the circumference of the circle) in a time $T$, where $T$ is the time for one revolution, the orbital speed of the electron is $v = 2\pi r/T$. The effective

Oxygen, a paramagnetic substance, is attracted to a magnetic field. The liquid oxygen in this photograph is suspended between the poles of the magnet. (Courtesy of Leon Lewandowski)

current associated with this orbiting electron equals its charge divided by the time for one revolution. Using $T = 2\pi/\omega$ and $\omega = v/r$ we have

$$I = \frac{e}{T} = \frac{e\omega}{2\pi} = \frac{ev}{2\pi r}$$

The magnetic moment associated with this effective current loop is given by $\mu = IA$, where $A = \pi r^2$ is the area of the orbit. Therefore,

$$\mu = IA = \left(\frac{ev}{2\pi r}\right)\pi r^2 = \tfrac{1}{2}\, evr \qquad (30.29)$$

Orbital magnetic moment

Since the magnitude of the orbital angular momentum of the electron is given by $L = mvr$, the magnetic moment can be written as

$$\mu = \left(\frac{e}{2m}\right) L \qquad (30.30)$$

This result says that *the magnetic moment of the electron is proportional to its orbital angular momentum.* Note that since the electron is negatively charged, the vectors $\boldsymbol{\mu}$ and $\mathbf{L}$ point in *opposite* directions. Both vectors are perpendicular to the plane of the orbit as indicated in Figure 30.27.

A fundamental outcome of quantum physics is that the orbital angular momentum must be *quantized,* and is always some integer multiple of $\hbar = h/2\pi = 1.06 \times 10^{-34}$ J·s, where $h$ is Planck's constant. That is

Angular momentum is quantized

$$L = 0, \hbar, 2\hbar, 3\hbar, \ldots$$

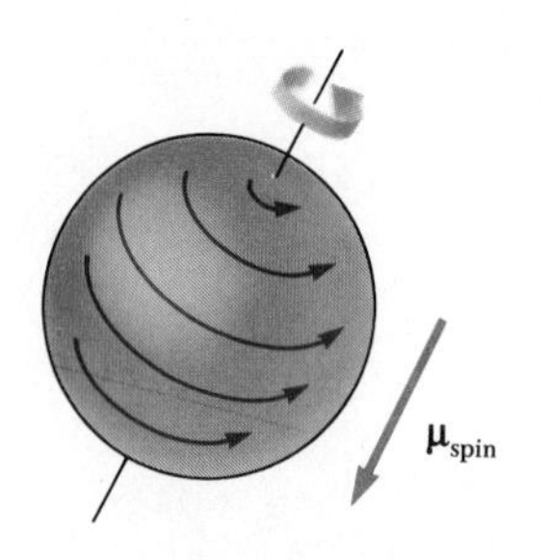


**Figure 30.28** Model of a spinning electron. The magnetic moment $\mu_{spin}$ can be viewed as arising from the effective current loops associated with a spinning charged sphere.

Hence, the smallest nonzero value of the magnetic moment is

$$\mu = \frac{e}{2m}\hbar \tag{30.31}$$

Since all substances contain electrons, you may wonder why all substances are not magnetic. The main reason is that in most substances, the magnetic moment of one electron in an atom is canceled by the moment of another electron in the atom orbiting in the opposite direction. The net result is that *the magnetic effect produced by the orbital motion of the electrons is either zero or very small for most materials.*

So far we have only considered the contribution to the magnetic moment of an atom from the orbital motion of the electron. However, an electron has another intrinsic property called *spin* which also contributes to the magnetic moment. In this regard, one can view the electron as a sphere of charge spinning about its axis as it orbits the nucleus, as in Figure 30.28. (This classical description of a spinning electron should not be taken literally. The property of spin can be understood only through a quantum mechanical model.) This spinning motion produces an effective current loop and hence a magnetic moment which is of the same order of magnitude as that due to the orbital motion. The magnitude of the spin angular momentum predicted by quantum theory is

Spin angular momentum

$$S = \frac{\hbar}{2} = 5.2729 \times 10^{-35}\ \text{J}\cdot\text{s}$$

The intrinsic magnetic moment associated with the spin of an electron has the value

Bohr magneton

$$\mu_B = \frac{e}{2m}\hbar = 9.27 \times 10^{-24}\ \text{J/T} \tag{30.32}$$

which is called the **Bohr magneton.**

In atoms or ions containing many electrons, the electrons usually pair up with their spins *opposite* each other, which results in a cancellation of the spin magnetic moments. However, atoms with an *odd* number of electrons must have at least one "unpaired" electron and a corresponding spin magnetic moment. The total magnetic moment of an atom is the vector sum of the orbital and spin magnetic moments. The magnetic moments of a few atoms and ions are given in Table 30.1. Note that some atoms such as helium and neon have zero moments because their individual moments cancel.

The nucleus of an atom also has a magnetic moment associated with its constituent protons and neutrons. However, the magnetic moment of a proton or neutron is small compared to the magnetic moment of an electron and can usually be neglected. This can be understood by inspecting Equation 30.32. Since the masses of the proton and neutron are much greater than that of the electron, their magnetic moments are much smaller by a factor of about $10^3$.

**TABLE 30.1 Magnetic Moments of Some Atoms and Ions**

| Atom (or ion) | Magnetic moment ($10^{-24}$ J/T) |
|---|---|
| H | 9.27 |
| He | 0 |
| Ne | 0 |
| $Ce^{3+}$ | 19.8 |
| $Yb^{3+}$ | 37.1 |

### Magnetization and Magnetic Field Strength

Magnetization

The magnetic state of a substance is described by a quantity called the **magnetization vector, *M*.** *The magnitude of the magnetization vector is equal to the magnetic moment per unit volume of the substance.* As you might expect, the total magnetic field in a substance depends on both the applied (external) field and the magnetization of the substance.

Consider a region where there exists a magnetic field $\mathbf{B}_0$ produced by a current-carrying conductor, such as the interior of a toroidal winding. If we now fill that region with a magnetic substance, the *total* field $\mathbf{B}$ in that region will be given by $\mathbf{B} = \mathbf{B}_0 + \mathbf{B}_m$ where $\mathbf{B}_m$ is the field produced by the magnetic substance. This contribution can be expressed in terms of the magnetization vector as $\mathbf{B}_m = \mu_0\mathbf{M}$; hence the total field in the substance becomes

$$\mathbf{B} = \mathbf{B}_0 + \mu_0\mathbf{M} \qquad (30.33)$$

Magnetic field strength

It is convenient to introduce another field quantity $\mathbf{H}$, called the **magnetic field strength.** This vector quantity is defined by the relation $\mathbf{H} = (\mathbf{B}/\mu_0) - \mathbf{M}$, or

$$\mathbf{B} = \mu_0(\mathbf{H} + \mathbf{M}) \qquad (30.34)$$

In SI units, the dimensions of both $\mathbf{H}$ and $\mathbf{M}$ are A/m.

To better understand these expressions, consider the region inside a toroidal coil which carries a current $I$. If the interior region is a vacuum, then $\mathbf{M} = 0$, and $\mathbf{B} = \mathbf{B}_0 = \mu_0\mathbf{H}$. Since $B_0 = \mu_0 nI$ inside a toroid, where $n$ is the number of turns per unit length in its windings, then $H = B_0/\mu_0 = \mu_0 nI/\mu_0$, or

$$H = nI \qquad (30.35)$$

That is, the magnetic field strength inside the toroid is due to the current in the its windings.

If the toroid core is now filled with some substance, and the current $I$ is kept constant, then $\mathbf{H}$ inside the substance will remain *unchanged,* with a magnitude $nI$. This is because the magnetic field strength $\mathbf{H}$ is due *solely* to the current in the coil. The total field $\mathbf{B}$, however, changes when the substance is introduced. From Equation 30.34, we see that part of $\mathbf{B}$ arises from the term $\mu_0\mathbf{H}$ associated with the toroidal current; the second contribution to $\mathbf{B}$ is the term $\mu_0\mathbf{M}$ due to the magnetization of the substance.

For a large class of substances, specifically paramagnetic and diamagnetic substances, the magnetization $\mathbf{M}$ is proportional to the magnetic field strength $\mathbf{H}$. In these linear substances, we can write

$$\mathbf{M} = \chi\mathbf{H} \qquad (30.36)$$

Magnetic susceptibility

where $\chi$ is a dimensionless factor called the **magnetic susceptibility.** If the sample is paramagnetic, $\chi$ is positive, in which case $\mathbf{M}$ is in the same direction as $\mathbf{H}$. If the substance is diamagnetic, $\chi$ is negative, and $\mathbf{M}$ is opposite $\mathbf{H}$. It is important to note that *this linear relationship does not apply to ferromagnetic substances.* The susceptibilities of some substances are given in Table 30.2.

**TABLE 30.2 Magnetic Susceptibilities of Some Paramagnetic and Diamagnetic Substances at 300 K**

| Paramagnetic Substance | $\chi$ | Diamagnetic Substance | $\chi$ |
|---|---|---|---|
| Aluminum | $2.3 \times 10^{-5}$ | Bismuth | $-1.66 \times 10^{-5}$ |
| Calcium | $1.9 \times 10^{-5}$ | Copper | $-9.8 \times 10^{-6}$ |
| Chromium | $2.7 \times 10^{-4}$ | Diamond | $-2.2 \times 10^{-5}$ |
| Lithium | $2.1 \times 10^{-5}$ | Gold | $-3.6 \times 10^{-5}$ |
| Magnesium | $1.2 \times 10^{-5}$ | Lead | $-1.7 \times 10^{-5}$ |
| Niobium | $2.6 \times 10^{-4}$ | Mercury | $-2.9 \times 10^{-5}$ |
| Oxygen (STP) | $2.1 \times 10^{-6}$ | Nitrogen (STP) | $-5.0 \times 10^{-9}$ |
| Platinum | $2.9 \times 10^{-4}$ | Silver | $-2.6 \times 10^{-5}$ |
| Tungsten | $6.8 \times 10^{-5}$ | Silicon | $-4.2 \times 10^{-6}$ |

Substituting Equation 30.36 for $\boldsymbol{M}$ into Equation 30.34 gives

$$\boldsymbol{B} = \mu_0(\boldsymbol{H} + \boldsymbol{M}) = \mu_0(\boldsymbol{H} + \chi\boldsymbol{H}) = \mu_0(1 + \chi)\boldsymbol{H}$$

or

$$\boldsymbol{B} = \kappa_m \boldsymbol{H} \tag{30.37}$$

where the constant $\kappa_m$ is called the **permeability** of the substance and has the value[4]

Permeability

$$\kappa_m = \mu_0(1 + \chi) \tag{30.38}$$

Substances may also be classified in terms of how their permeability $\kappa_m$ compares to $\mu_0$ (the permeability of free space) as follows

| | |
|---|---|
| Paramagnetic | $\kappa_m > \mu_0$ |
| Diamagnetic | $\kappa_m < \mu_0$ |
| Ferromagnetic | $\kappa_m \gg \mu_0$ |

Since $\chi$ is very small for paramagnetic and diamagnetic substances (see Table 30.2), $\kappa_m$ is nearly equal to $\mu_0$ in these cases. For ferromagnetic substances, however, $\kappa_m$ is typically several thousand times larger than $\mu_0$. Although Equation 30.37 provides a simple relation between $\boldsymbol{B}$ and $\boldsymbol{H}$, it must be interpreted with care when dealing with ferromagnetic substances. As mentioned earlier, $\boldsymbol{M}$ is not a linear function of $\boldsymbol{H}$ for ferromagnetic substances such as iron, nickel, and cobalt. This is because the value of $\kappa_m$ *is not a characteristic of the substance,* but depends on the previous state and treatment of the sample.

[4] The symbol $\mu$ is often used for permeability, but we have already used this same symbol for magnetic moment. For this reason, we use the symbol $\kappa_m$ for permeability.

**EXAMPLE 30.10 An Iron-Filled Toroid**

A toroidal winding carrying a current of 5 A is wound with 60 turns/m of wire. The core is iron, which has a magnetic permeability of $5000\mu_0$ under the given conditions. Find $H$ and $B$ inside the iron core.

*Solution* Using Equations 30.35 and 30.37, we get

$$H = nI = \left(60 \frac{\text{turns}}{\text{m}}\right)(5 \text{ A}) = 300 \frac{\text{A}\cdot\text{turns}}{\text{m}}$$

$$B = \kappa_m H = 5000\mu_0 H$$
$$= 5000\left(4\pi \times 10^{-7}\,\frac{\text{Wb}}{\text{A}\cdot\text{m}}\right)\left(300\,\frac{\text{A}\cdot\text{turns}}{\text{m}}\right)$$
$$= 1.88\ \text{T}$$

This value of $B$ is 5000 times larger than the field in the absence of iron!

Exercise 4 Determine the magnitude and direction of the magnetization inside the iron core.
Answer $M = 1.5 \times 10^6$ A/m; $\boldsymbol{M}$ is in the direction of $\boldsymbol{H}$.

### Ferromagnetism

Iron, cobalt, nickel, gadolinium, and dysprosium are strongly magnetic materials and are said to be ferromagnetic. Ferromagnetic substances are used to fabricate permanent magnets. Such substances contain atomic magnetic moments that tend to align parallel to each other even in a weak external magnetic field. Once the moments are aligned, the substance will remain magnetized after the external field is removed. This permanent alignment is due to a strong coupling between neighboring moments, which can only be understood in quantum mechanical terms.

All ferromagnetic materials contain microscopic regions called **domains,** within which all magnetic moments are aligned. These domains have volumes of about $10^{-12}$ to $10^{-8}$ m$^3$ and contain $10^{17}$ to $10^{21}$ atoms. The boundaries between the various domains having different orientations are called **domain walls.** In an unmagnetized sample, the domains are randomly oriented such that the net magnetic moment is zero as shown in Figure 30.29a. When the sample is placed in an external magnetic field, the domains tend to align with the field by rotating slightly, which results in a magnetized sample, as in Figure 30.29b. Observations show that domains initially oriented along the external field will grow in size at the expense of the less favorably oriented domains. When the external field is removed, the sample may retain a net magnetization in the direction of the original field.[5] At ordinary temperatures, thermal agitation is not sufficiently high to disrupt this preferred orientation of magnetic moments.

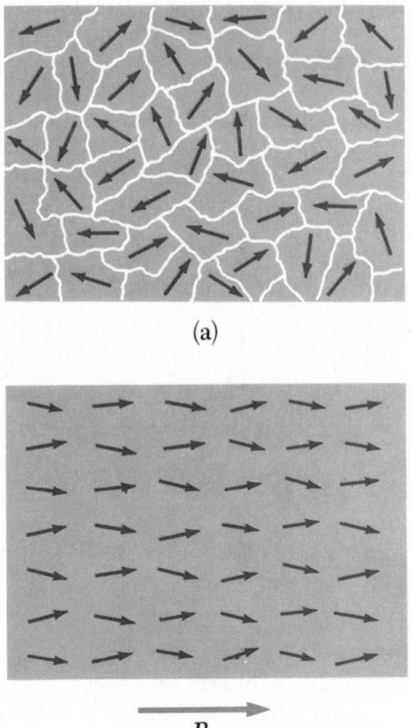

A typical experimental arrangement used to measure the magnetic properties of a ferromagnetic material consists of a toroid-shaped sample wound with $N$ turns of wire, as in Figure 30.30. This configuration is sometimes referred to as the **Rowland ring.** A secondary coil connected to a galvanometer is used to measure the magnetic flux. The magnetic field $\boldsymbol{B}$ within the core of the toroid is measured by increasing the current in the toroid coil from zero to $I$. As the current changes, the magnetic flux through the secondary coil changes by $BA$, where $A$ is the cross-sectional area of the toroid. Because of this changing flux, an emf is induced in the secondary coil that is proportional to the rate of change in magnetic flux. If the galvanometer in the secondary circuit is properly calibrated, one can obtain a value for $\boldsymbol{B}$ corresponding to any value of the current in the toroidal coil. The magnetic field $\boldsymbol{B}$ is measured first in the empty coil and then with the same coil filled with the magnetic substance. The magnetic properties of the substance are then obtained from a comparison of the two measurements.

(a)

$B_0$
(b)

**Figure 30.29** (a) Random orientation of atomic magnetic dipoles in an unmagnetized substance. (b) When an external field $\boldsymbol{B}_0$ is applied, the atomic magnetic dipoles tend to align with the field, giving the sample a net magnetization $\boldsymbol{M}$.

[5] It is possible to observe the domain walls directly and follow their motion under a microscope. In this technique, a liquid suspension of finely powdered ferromagnetic substance is applied to the sample. The fine particles tend to accumulate at the domain walls and shift with them.

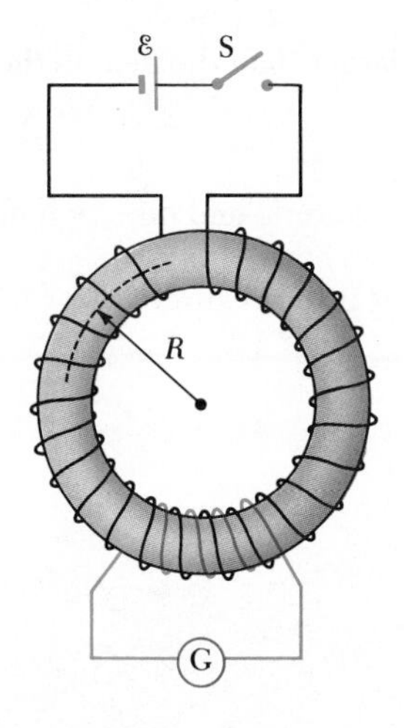


**Figure 30.30** Cross section of a toroidal winding arrangement used to measure the magnetic properties of a substance. The material under study fills the core of the toroid, and the secondary circuit containing the galvanometer measures the magnetic flux.

Now consider a toroidal coil whose core consists of unmagnetized iron. If the current in the windings is increased from zero to some value $I$, the field intensity $H$ increases linearly with $I$ according to the expression $H = nI$. Furthermore, the total field $B$ also increases with increasing current as shown in Figure 30.31. At point $O$, the domains are randomly oriented, corresponding to $B_m = 0$. As the external field increases, the domains become more aligned until all are nearly aligned at point $a$. At this point, the iron core is approaching saturation. (The condition of saturation corresponds to the case where all domains are aligned in the same direction.) Next, suppose the current is reduced to zero, thereby eliminating the external field. The $B$ versus $H$ curve, called a **magnetization curve,** now follows the path $ab$ shown in Figure 30.31. Note that at point $b$, the field $\mathbf{B}$ is not zero, although the external field is $\mathbf{B}_0 = 0$. This is explained by the fact that the iron core is now magnetized due to the alignment of a large number of domains (that is, $\mathbf{B} = \mathbf{B}_m$). At this point, the iron is said to have a *remanent magnetization*. If the external field is reversed in direction and increased in strength by reversing the current, the domains reorient until the sample is again unmagnetized at point $c$, where $\mathbf{B} = 0$. A further increase in the reverse current causes the iron to be magnetized in the opposite direction, approaching saturation at point $d$. A similar sequence of events occurs as the current is reduced to zero and then increased in the original (positive) direction. In this case, the magnetization curve follows the path $def$. If the current is increased sufficiently, the magnetization curve returns to point $a$, where the sample again has its maximum magnetization.

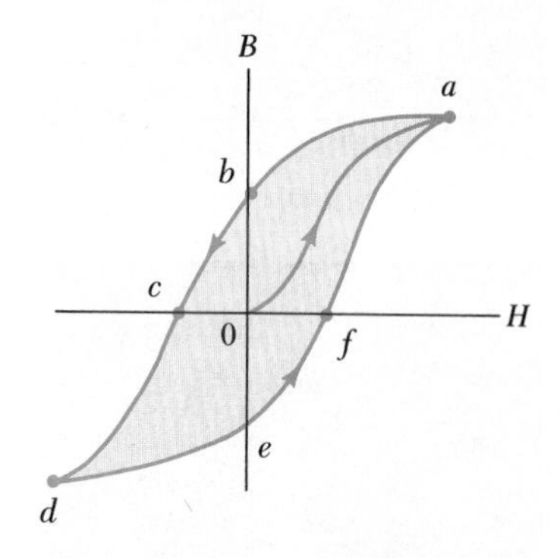


**Figure 30.31** Hysteresis curve for a ferromagnetic material.

The effect just described, called **magnetic hysteresis,** shows that the magnetization of a ferromagnetic substance depends on the history of the substance as well as the strength of the applied field. (The word *hysteresis* literally means to "lag behind.") One often says that a ferromagnetic substance has a "memory" since it remains magnetized after the external field is removed. The closed loop in Figure 30.31 is referred to as a *hysteresis loop*. Its shape and size depend on the properties of the ferromagnetic substance and on the strength of the maximum applied field. The hysteresis loop for "hard" ferromagnetic materials (used in permanent magnets) is characteristically wide as in Figure 30.32a, corresponding to a large remanent magnetization. Such materials cannot be easily demagnetized by an external field. This is in contrast to "soft" ferromagnetic materials, such as iron, that have a very narrow

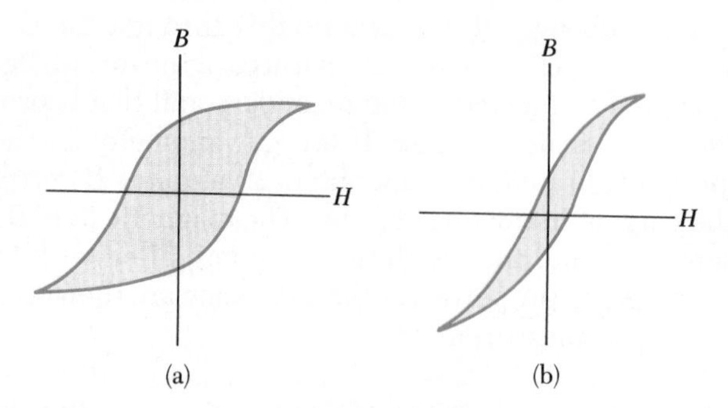


**Figure 30.32** Hysteresis curves for (a) a hard ferromagnetic material and (b) a soft ferromagnetic material.

hysteresis loop and a small remanent magnetization (Fig. 30.32b.) Such materials are easily magnetized and demagnetized. An ideal soft ferromagnet would exhibit no hysteresis and hence would have no remanent magnetization. A ferromagnetic substance can be demagnetized by carrying the substance through successive hysteresis loops, gradually decreasing the applied field as in Figure 30.33.

The magnetization curve is useful for another reason. *The area enclosed by the magnetization curve represents the work required to take the material through the hysteresis cycle.* The energy acquired by the sample in the magnetization process originates from the source of the external field, that is, the emf in the circuit of the toroidal coil. When the magnetization cycle is repeated, dissipative processes within the material due to realignment of the domains result in a transformation of magnetic energy into internal thermal energy, which raises the temperature of the substance. For this reason, devices subjected to alternating fields (such as transformers) use cores made of soft ferromagnetic substances, which have narrow hysteresis loops and a correspondingly small energy loss per cycle.

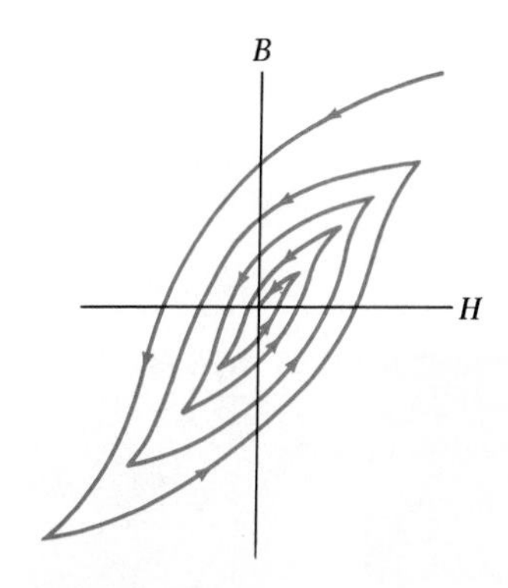


**Figure 30.33** Demagnetizing a ferromagnetic material by carrying it through successive hysteresis loops.

### Paramagnetism

Paramagnetic substances have a positive but small susceptibility ($0 < \chi \ll 1$), which is due to the presence of atoms (or ions) with *permanent* magnetic dipole moments. These dipoles interact only weakly with each other and are randomly oriented in the absence of an external magnetic field. When the substance is placed in an external magnetic field, its atomic dipoles tend to line up with the field. However, this alignment process must compete with the effects of thermal motion, which tends to randomize the dipole orientations.

Experimentally, one finds that the magnetization of a paramagnetic substance is proportional to the applied field and inversely proportional to the absolute temperature under a wide range of conditions. That is,

$$M = C\frac{B}{T} \qquad (30.39)$$

This is known as **Curie's law** after its discoverer Pierre Curie (1859–1906), and the constant $C$ is called **Curie's constant.** This shows that the magnetization increases with increasing applied field and with decreasing temperature. When $B = 0$, the magnetization is zero, corresponding to a random orientation of dipoles. At very high fields or very low temperatures, the magnetization approaches its maximum, or saturation, value corresponding to a complete alignment of its dipoles and Equation 30.39 is no longer valid.

It is interesting to note when the temperature of a ferromagnetic substance reaches or exceeds a critical temperature, called the **Curie temperature,** the substance loses its spontaneous magnetization and becomes paramagnetic (see Fig. 30.34). Below the Curie temperature, the magnetic moments are aligned and the substance is ferromagnetic. Above the Curie temperature, the thermal energy is large enough to cause a random orientation of dipoles, hence the substance becomes paramagnetic. For example, the Curie temperature for iron is 1043 K. A list of Curie temperatures for several ferromagnetic substances is given in Table 30.3.

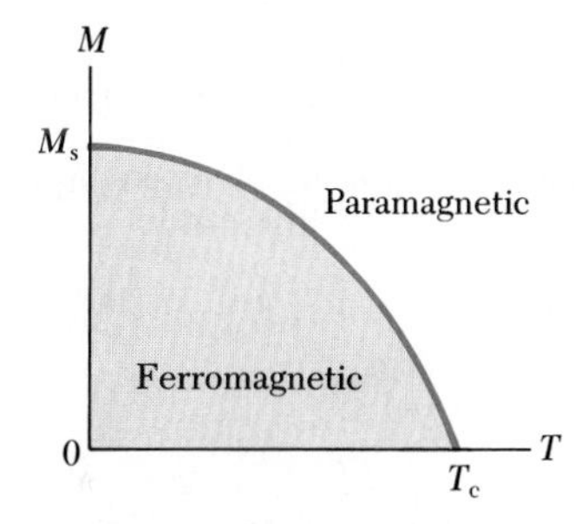


**Figure 30.34** Plot of the magnetization versus absolute temperature for a ferromagnetic substance. The magnetic moments are aligned (ordered) below the Curie temperature $T_c$, where the substance is ferromagnetic. The substance becomes paramagnetic (disordered) above $T_c$.

**TABLE 30.3 Curie Temperature for Several Ferromagnetic Substances**

| Substance | $T_c$ (K) |
|---|---|
| Iron | 1043 |
| Cobalt | 1394 |
| Nickel | 631 |
| Gadolinium | 317 |
| $Fe_2O_3$ | 893 |

### Diamagnetism

A diamagnetic substance is one whose atoms have no permanent magnetic dipole moment. When an external magnetic field is applied to a diamagnetic substance such as bismuth or silver, a weak magnetic dipole moment is *induced* in the direction opposite the applied field. Although the effect of diamagnetism is present in all matter, it is weak compared to paramagnetism or ferromagnetism.

We can obtain some understanding of diamagnetism by considering two electrons of an atom orbiting the nucleus in opposite directions but with the same speed. The electrons remain in these circular orbits because of the attractive electrostatic force (the centripetal force) of the positively charged nucleus. Because the magnetic moments of the two electrons are equal in magnitude and opposite in direction, they cancel each other and the dipole moment of the atom is zero. When an external magnetic field is applied, the electrons experience an additional force $q\mathbf{v} \times \mathbf{B}$. This added force modifies the centripetal force so as to increase the orbital speed of the electron whose magnetic moment is antiparallel to the field and decreases the speed of the electron whose magnetic moment is parallel to the field. As a result, the magnetic moments of the electrons no longer cancel, and the substance acquires a net dipole moment that opposes the applied field.

A small permanent magnet levitated above a disk of the superconductor $YBa_2Cu_3O_7$ cooled to liquid nitrogen temperature. (Courtesy of IBM Research)

As you recall from Chapter 27, superconductors are substances whose dc resistance is *zero* below some critical temperature characteristic of the substance. Certain types of superconductors also exhibit *perfect diamagnetism* in the superconducting state. As a result, an applied magnetic field is *expelled* by the superconductor so that the field is *zero* in its interior. This phenomenon of flux expulsion is known as the **Meissner effect.** If a permanent magnet is brought near a superconductor, the two substances will *repel* each other. This is illustrated in the photograph, which shows a small permanent magnet levitated above a superconductor maintained at 77 K. A more detailed description of the unusual properties of superconductors is presented in Chapter 44 of the extended version of this text.

**EXAMPLE 30.11 Saturation Magnetization**
Estimate the *maximum magnetization* in a long cylinder of iron, assuming there is one unpaired electron spin per atom.

*Solution* The maximum magnetization, called the *saturation magnetization,* is obtained when all the magnetic moments in the sample are aligned. If the sample contains $n$ atoms per unit volume, then the saturation magnetization $M_s$ has the value

$$M_s = n\mu$$

where $\mu$ is the magnetic moment per atom. Since the molecular weight of iron is 55 g/mole and its density is 7.9 g/cm$^3$, the value of $n$ is $8.5 \times 10^{28}$ atoms/m$^3$. Assuming each atom contributes one Bohr magneton (due to one unpaired spin) to the magnetic moment, we get

$$M_s = \left(8.5 \times 10^{28} \frac{\text{atoms}}{\text{m}^3}\right)\left(9.27 \times 10^{-24} \frac{\text{A}\cdot\text{m}^2}{\text{atom}}\right)$$

$$= 7.9 \times 10^5 \text{ A/m}$$

This is about one half the experimentally determined saturation magnetization for annealed iron, which indicates that there are actually *two* unpaired electron spins per atom.

## *30.10 MAGNETIC FIELD OF THE EARTH

When we speak of a small bar magnet's having a north and a south pole, we should more properly say that it has a "north-seeking" and a "south-seeking" pole. By this we mean that if such a magnet is used as a compass, one end will seek, or point to, the north geographic pole of the earth. Thus, we conclude that *a north magnetic pole is located near the south geographic pole, and a south magnetic pole is located near the north geographic pole.* In fact, the configuration of the earth's magnetic field, pictured in Figure 30.35, is very much like that which would be achieved by burying a bar magnet deep in the interior of the earth.

If a compass needle is suspended in bearings that allow it to rotate in the vertical plane as well as in the horizontal plane, the needle is horizontal with respect to the earth's surface only near the equator. As the device is moved northward, the needle rotates such that it points more and more toward the surface of the earth. Finally, at a point just north of Hudson Bay in Canada, the north pole of the needle would point directly downward. This location, first found in 1832, is considered to be the location of the south-seeking magnetic pole of the earth. This site is approximately 1300 mi from the earth's geographic north pole and varies with time. Similarly, the north-seeking magnetic pole of the earth is about 1200 miles away from the earth's geographic south pole. Thus, it is only approximately correct to say that a compass needle points north. The difference between true north, defined as the geographic north pole, and north indicated by a compass varies from point to point on the earth, and the difference is referred to as *magnetic declination.* For example, along a line through Florida and the Great Lakes, a compass indicates true north, whereas in Washington state, it aligns 25° east of true north.

Although the magnetic field pattern of the earth is similar to that which would be set up by a bar magnet deep within the earth, it is easy to understand why the source of the earth's field cannot be large masses of permanently magnetized material. The earth does have large deposits of iron ore deep beneath its surface, but the high temperatures in the earth's core prevent the iron from retaining any permanent magnetization. It is considered more likely that the true source is charge-carrying convection currents in the earth's core.

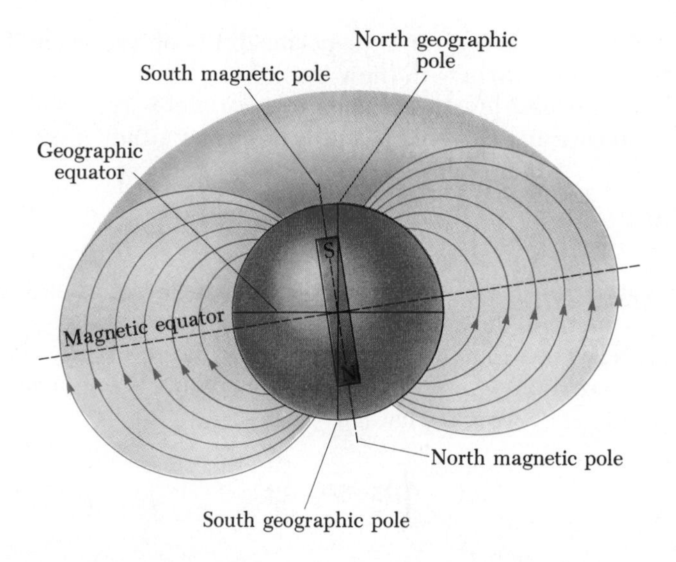


**Figure 30.35** The earth's magnetic field lines. Note that a magnetic south pole is at the north geographic pole and a magnetic north pole is at the south geographic pole.

Charged ions or electrons circling in the liquid interior could produce a magnetic field, just as a current in a loop of wire produces a magnetic field. There is also strong evidence to indicate that the strength of a planet's field is related to the planet's rate of rotation. For example, Jupiter rotates faster than the earth, and recent space probes indicate that Jupiter's magnetic field is stronger than ours. Venus, on the other hand, rotates more slowly than the earth, and its magnetic field is found to be weaker. Investigation into the cause of the earth's magnetism remains open.

There is an interesting sidelight concerning the earth's magnetic field. It has been found that the direction of the field has been reversed several times during the last million years. Evidence for this is provided by basalt (a type of rock that contains iron) that is spewed forth by volcanic activity on the ocean floor. As the lava cools, it solidifies and retains a picture of the earth's magnetic field direction. The rocks can be dated by other means to provide the evidence for these periodic reversals of the magnetic field.

## SUMMARY

The **Biot-Savart law** says that the magnetic field $d\mathbf{B}$ at a point $P$ due to a current element $d\mathbf{s}$ carrying a steady current $I$ is

Biot-Savart law

$$d\mathbf{B} = k_m \frac{I\, d\mathbf{s} \times \hat{\mathbf{r}}}{r^2} \qquad (30.1)$$

where $k_m = 10^{-7}$ Wb/A·m and $r$ is the distance from the element to the point $P$. To find the total field at $P$ due to a current-carrying conductor, we must integrate this vector expression over the entire conductor.

The **magnetic field** at a distance $a$ from a long, straight wire carrying a current $I$ is given by

Magnetic field of an infinitely long wire

$$B = \frac{\mu_0 I}{2\pi a} \qquad (30.7)$$

where $\mu_0 = 4\pi \times 10^{-7}$ Wb/A·m is the **permeability of free space.** The field lines are circles concentric with the wire.

The force per unit length between two parallel wires separated by a distance $a$ and carrying currents $I_1$ and $I_2$ has a magnitude given by

Force per unit length between two wires

$$\frac{F}{\ell} = \frac{\mu_0 I_1 I_2}{2\pi a} \qquad (30.13)$$

The force is attractive if the currents are in the same direction and repulsive if they are in opposite directions.

**Ampère's law** says that the line integral of $\mathbf{B} \cdot d\mathbf{s}$ around any closed path equals $\mu_0 I$, where $I$ is the total steady current passing through any surface bounded by the closed path. That is,

Ampère's law

$$\oint \mathbf{B} \cdot d\mathbf{s} = \mu_0 I \qquad (30.15)$$

Using Ampère's law, one finds that the fields inside a toroid and solenoid are given by

$$B = \frac{\mu_0 NI}{2\pi r} \tag{30.18}$$

Magnetic field inside a toroid

$$B = \mu_0 \frac{N}{\ell} I = \mu_0 nI \tag{30.20}$$

Magnetic field inside a solenoid

where $N$ is the total number of turns.

The **magnetic flux** $\Phi_m$ through a surface is defined by the surface integral

$$\Phi_m \equiv \int \boldsymbol{B} \cdot d\boldsymbol{A} \tag{30.23}$$

Magnetic flux

**Gauss' law of magnetism** states that the net magnetic flux through any closed surface is zero. That is, isolated magnetic poles (or magnetic monopoles) do not exist.

**A displacement current** $I_d$ arises from a time-varying electric flux and is defined by

$$I_d \equiv \epsilon_0 \frac{d\Phi_e}{dt} \tag{30.26}$$

Displacement current

The **generalized form of Ampère's law,** which includes the displacement current, is given by

$$\oint \boldsymbol{B} \cdot d\boldsymbol{s} = \mu_0 I + \mu_0 \epsilon_0 \frac{d\Phi_e}{dt} \tag{30.27}$$

Ampère-Maxwell law

This law describes the fact that magnetic fields are produced both by conduction currents and by changing electric fields.

The fundamental sources of all magnetic fields are the magnetic dipole moments associated with atoms. The atomic dipole moments can arise both from the orbital motions of the electrons and from an intrinsic property of electrons known as *spin.*

The magnetic properties of substances can be described in terms of their response to an external field. In a broad sense, materials can be described as being *ferromagnetic, paramagnetic,* or *diamagnetic.* The atoms of **ferromagnetic** and **paramagnetic** materials have permanent magnetic moments. **Diamagnetic** materials consist of atoms with no permanent magnetic moments.

When a paramagnetic or ferromagnetic material is placed in an external magnetic field, its dipoles tend to align parallel to the field, and this aligning in turn increases the net field. The increase in the field is quite small in the case of paramagnetic substances. This is because the magnetic dipoles in paramagnetic materials are randomly oriented in the absence of a magnetic field. The dipoles are partially aligned in the presence of an applied field.

## QUESTIONS

1. Is the magnetic field due to a current loop uniform? Explain.
2. A current in a conductor produces a magnetic field which can be calculated using the Biot-Savart law. Since current is defined as the rate of flow of charge, what can you conclude about the magnetic field due to stationary charges? What about moving charges?
3. Two parallel wires carry currents in opposite directions. Describe the nature of the resultant magnetic field due to the two wires at points (a) between the wires and (b) outside the wires in a plane containing the wires.
4. Explain why two parallel wires carrying currents in opposite directions repel each other.
5. Two wires carrying equal and opposite currents are twisted together in the construction of a circuit. Why does this technique reduce stray magnetic fields?
6. Is Ampère's law valid for all closed paths surrounding a conductor? Why is it not useful for calculating $\boldsymbol{B}$ for all such paths?
7. Compare Ampère's law with the Biot-Savart law. Which is the more general method for calculating $\boldsymbol{B}$ for a current-carrying conductor?
8. Is the magnetic field inside a toroidal coil uniform? Explain.
9. Describe the similarities between Ampère's law in magnetism and Gauss' law in electrostatics.
10. A hollow copper tube carries a current. Why is $\boldsymbol{B} = 0$ inside the tube? Is $\boldsymbol{B}$ nonzero outside the tube?
11. Why is $\boldsymbol{B}$ nonzero outside a solenoid? Why is $\boldsymbol{B} = 0$ outside a toroid? (The lines of $\boldsymbol{B}$ must form closed paths.)
12. Describe the change in the magnetic field inside a solenoid carrying a steady current $I$ if (a) the length of the solenoid is doubled, but the number of turns remains the same and (b) the number of turns is doubled, but the length remains the same.
13. A plane conducting loop is located in a uniform magnetic field that is directed along the $x$ axis. For what orientation of the loop is the flux through it a maximum? For what orientation is the flux a minimum?
14. What new concept did Maxwell's generalized form of Ampère's circuital law include?
15. A magnet attracts a piece of iron. The iron can then attract another piece of iron. On the basis of alignment of the domains, explain what happens in each piece of iron.
16. You are an astronaut stranded on a planet with no test equipment or minerals around. The planet does not even have a magnetic field. You have two bars of iron in your possession; one is magnetized, one is not. How could you determine which is magnetized?
17. Why will hitting a magnet with a hammer cause its magnetism to be reduced?
18. Will a nail be attracted to either pole of a magnet? Explain what is happening inside the nail.
19. The north-seeking pole of a magnet is attracted toward the geographic north pole of the earth. Yet, like poles repel. What is the way out of this dilemma?
20. A Hindu ruler once suggested that he be entombed in a magnetic coffin with the polarity arranged such that he would be forever suspended between heaven and earth. Is such magnetic levitation possible? Discuss.
21. Why is $\boldsymbol{M} = 0$ in a vacuum? What is the relationship between $\boldsymbol{B}$ and $\boldsymbol{H}$ in a vacuum?
22. Explain why some atoms have permanent magnetic dipole moments and others do not.
23. What factors can contribute to the total magnetic dipole moment of an atom?
24. Why is the susceptibility of a diamagnetic substance negative?
25. Why can the effect of diamagnetism be neglected in a paramagnetic substance?
26. Explain the significance of the Curie temperature for a ferromagnetic substance.
27. Discuss the difference between ferromagnetic, paramagnetic, and diamagnetic substances.
28. What is the difference between hard and soft ferromagnetic materials?
29. Should the surface of a computer disk be make from a "hard" or a "soft" ferromagnetic substance?
30. Explain why it is desirable to use hard ferromagnetic materials to make permanent magnets.
31. Why is an ordinary, unmagnetized steel nail attracted to a permanent magnet?
32. Would you expect the tape from a tape recorder to be attracted to a magnet? (Try it, but not with a recording you wish to save.)
33. Given only a strong magnet and a screwdriver, (a) how would you magnetize the screwdriver, and (b) how would you then demagnetize the screwdriver?
34. The photograph below shows two permanent magnets with holes through their centers. Note that the

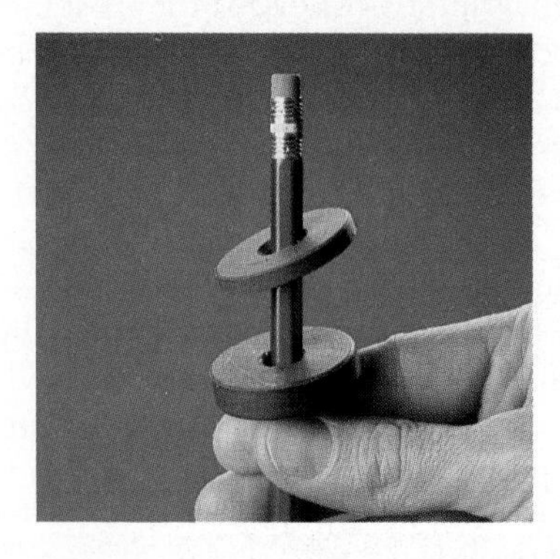

Question 34. Magnetic levitation using two ceramic magnets. (Courtesy of CENCO)

upper magnet is levitated above the lower magnet. (a) How does this occur? (b) What purpose does the pencil serve? (c) What can you say about the poles of the magnets from this observation? (d) If the upper magnet were inverted, what do you suppose would happen?

## PROBLEMS

### Section 30.1 The Biot-Savart Law

1. Calculate the magnitude of the magnetic field at a point 100 cm from a long, thin conductor carrying a current of 1 A.
2. A long, thin conductor carries a current of 10 A. At what distance from the conductor is the magnitude of the resulting magnetic field equal to $10^{-4}$ T?
3. A wire in which there is a current of 5 A is to be formed into a circular loop of one turn. If the required value of the magnetic field at the *center* of the loop is 10 $\mu$T, what is the required radius of the loop?
4. In Neils Bohr's 1913 model of the hydrogen atom, an electron circles the proton at a distance of $5.3 \times 10^{-11}$ m with a speed of $2.2 \times 10^{6}$ m/s. Compute the magnetic field strength produced by the electron's motion at the location of the proton.
5. A conductor in the shape of a square of edge length $\ell = 0.4$ m carries a current $I = 10$ A (Fig. 30.36). Calculate the magnitude and direction of the magnetic field produced at the *center* of the square.

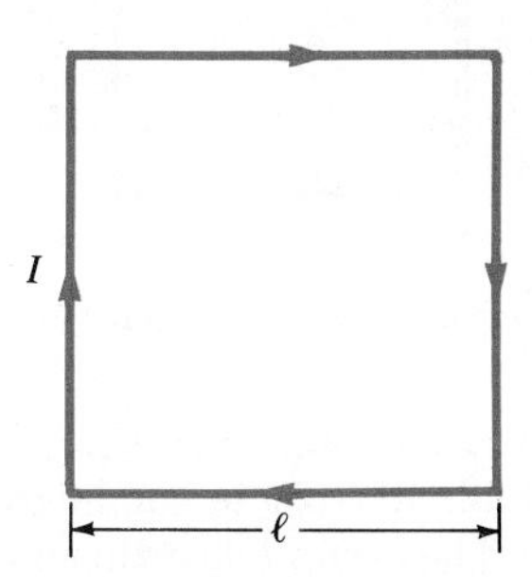


**Figure 30.36** (Problems 5 and 7).

6. A 12 cm $\times$ 16 cm rectangular loop of superconducting wire carries a current of 30 A. What is the magnetic field at the center of the loop?
7. If the total length of the conductor in Problem 5 is formed into a single *circular* turn with the *same* current, what is the value of the magnetic field at the center of the turn?
8. How many turns should be in a flat circular coil of radius 0.1 m in order for a current of 10 A to produce a magnetic field of $3 \times 10^{-3}$ T at its center?
9. Determine the magnetic field at a point $P$ that is a distance $x$ from the corner of an infinitely long wire that is bent at a right angle, as shown in Figure 30.37. The wire carries a steady current $I$.

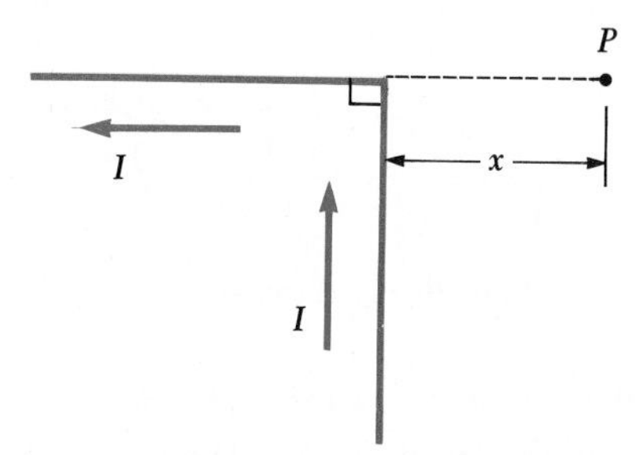


**Figure 30.37** (Problem 9).

10. A segment of wire of total length $4r$ is formed into a shape as shown in Figure 30.38 and carries a current $I = 6$ A. Find the magnitude and direction of the magnetic field at point $P$ when $r = 2\pi$ cm.

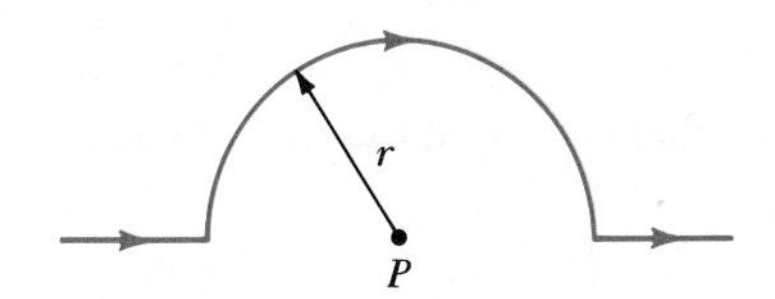


**Figure 30.38** (Problem 10).

11. A current path shaped as shown in Figure 30.39 produces a magnetic field at $P$, the center of the arc. If the arc subtends an angle of 30° and the total length of wire in the "pie-shaped" part of the path is 1.2 m, what are the magnitude and direction of the field produced at $P$ if the current is 3 A? Ignore the contribution to the field due to the current in the *small* arcs near $P$.

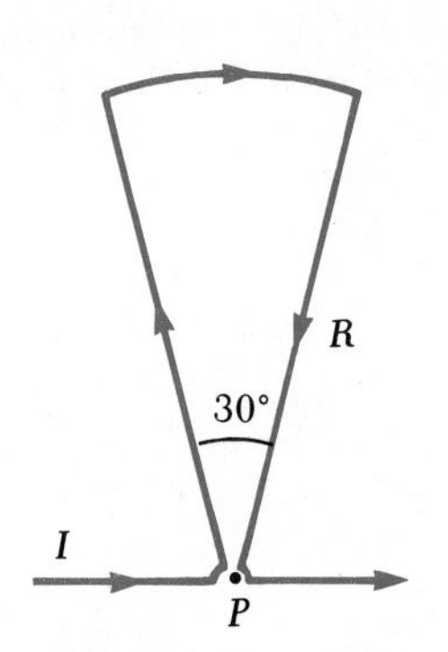


**Figure 30.39** (Problem 11).

12. Consider the current-carrying loop shown in Figure 30.40, formed of radial lines and segments of circles whose centers are at point $P$. Find the magnitude and direction of the magnetic field $\boldsymbol{B}$ at $P$.

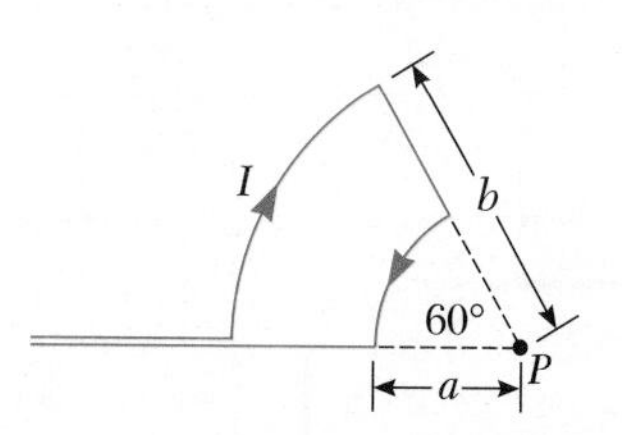


**Figure 30.40** (Problem 12).

13. A circular coil of wire whose radius is 10 cm has 100 turns. The current in the coil is 0.3 A. (a) Calculate the magnetic field at the center of the coil. (b) How far above or below the center of the coil are the points where the magnetic field has dropped to half its value at the center of the coil?

14. Recalling that the current density $J = nqv_d$ (Eq. 27.6), show that the Biot-Savart law can be written

$$d\boldsymbol{B} = \frac{\mu_0}{4\pi} \frac{q\boldsymbol{v}_d \times \hat{\boldsymbol{r}}}{r^2} n\, dV$$

where $dV$ is the volume element of the conductor and the drift velocity $\boldsymbol{v}_d$ is as defined in Chapter 27.

## Section 30.2 The Magnetic Force Between Two Parallel Conductors

15. Two long parallel conductors, separated by a distance $a = 10$ cm, carry currents in the same direction. If $I_1 = 5$ A and $I_2 = 8$ A, what is the force per unit length exerted on each conductor by the other?

16. Two long parallel wires, each having a mass per unit length of 40 g/m, are supported in a horizontal plane by strings 6 cm long as shown in Figure 30.41. Each wire carries the same current $I$, causing the wires to repel each other so that the angle $\theta$ between the supporting strings is 16°. (a) Are the currents in the same or opposite directions? (b) Find the magnitude of each current.

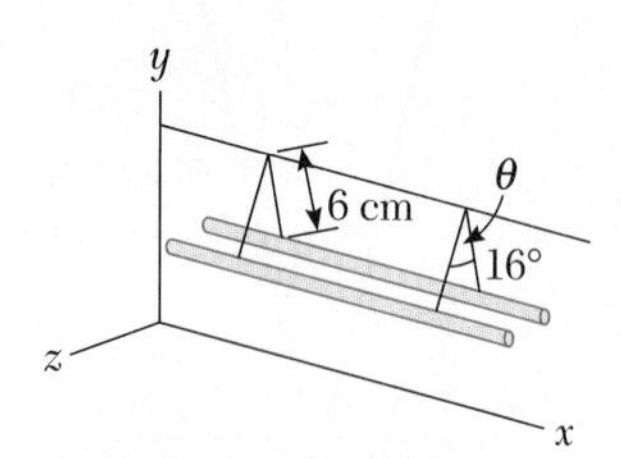


**Figure 30.41** (Problem 16).

17. Two parallel copper rods are 1 cm apart. Lightning sends a 10 000-ampere pulse of current along each conductor. Calculate the force per unit length on one conductor. Is the force attractive or repulsive?

18. Compute the magnetic force per unit length between two adjacent windings of a solenoid if each carries a current $I = 100$ A, and the center-to-center distance between the wires is 4 mm.

19. For the arrangement shown in Figure 30.42, the current in the long, straight conductor has the value $I_1 = 5$ A and lies in the plane of the rectangular loop, which carries a current $I_2 = 10$ A. The dimensions are $c = 0.1$ m, $a = 0.15$ m, and $\ell = 0.45$ m. Find the magnitude and direction of the *net force* exerted on the rectangle by the magnetic field of the straight current-carrying conductor.

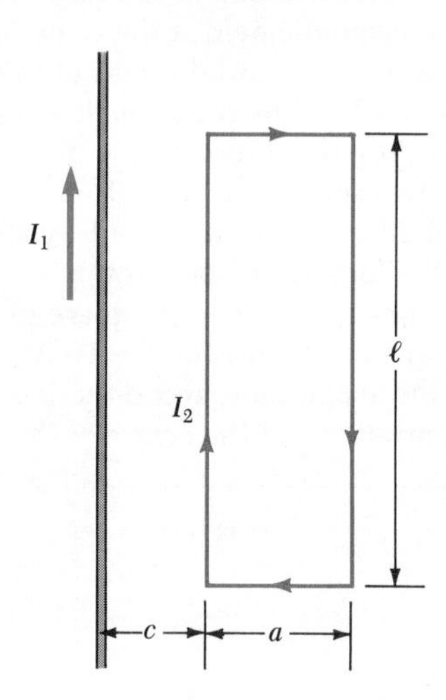


**Figure 30.42** (Problem 19).

20. Four long, parallel conductors carry equal currents $I = 5$ A. An end view of the conductors is shown in Figure 30.43. The current direction is out of the page at points $A$ and $B$ (indicated by the dots) and into the page at points $C$ and $D$ (indicated by the crosses). Calculate the magnitude and direction of the magnetic field at point $P$, located at the center of the square of edge length 0.2 m.

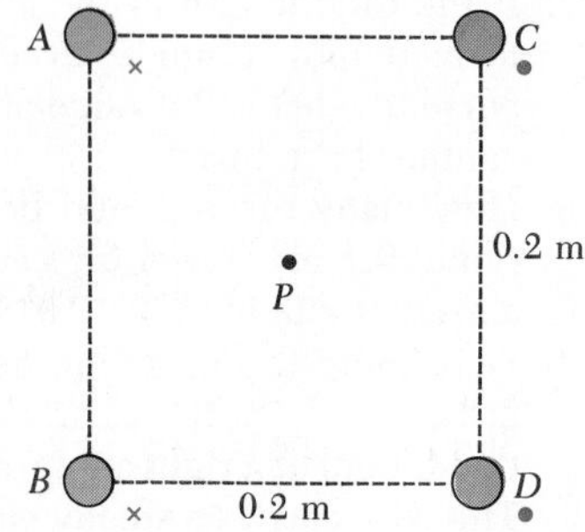


**Figure 30.43** (Problem 20).

21. Two long, parallel conductors carry currents $I_1 = 3$ A and $I_2 = 3$ A, both directed into the page in Figure 30.44. The conductors are separated by a distance of 13 cm. Determine the magnitude and direction of the resultant magnetic field at point $P$, located 5 cm from $I_1$ and 12 cm from $I_2$.

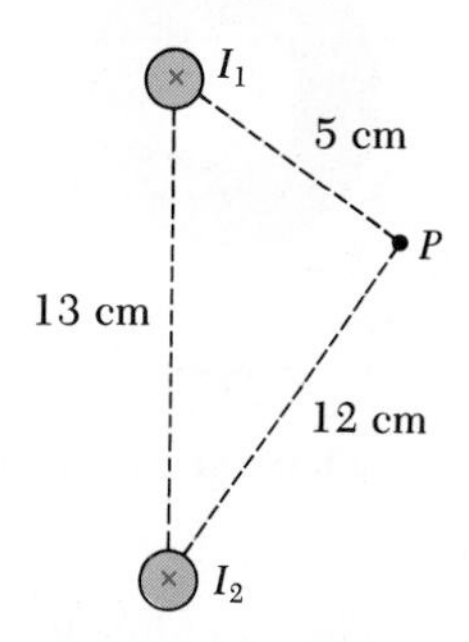


**Figure 30.44** (Problem 21).

## Section 30.3 Ampère's Law and Section 30.4 The Magnetic Field of a Solenoid

22. A closely wound long solenoid of overall length 30 cm has a magnetic field $B = 5 \times 10^{-4}$ T at its center due to a current $I = 1$ A. How many turns of wire are on the solenoid?
23. A superconducting solenoid is to be designed to generate a magnetic field of 10 T. (a) If the solenoid winding has 2000 turns/meter, what is the required current? (b) What force per unit length is exerted on the solenoid windings by this magnetic field?
24. What current is required in the windings of a long solenoid that has 1000 turns uniformly distributed over a length of 0.4 m in order to produce a magnetic field of magnitude $1.0 \times 10^{-4}$ T at the center of the solenoid?
25. Some superconducting alloys at very low temperatures can carry very high currents. For example, $Nb_3Sn$ wire at 10 K can carry $10^3$ A and maintain its superconductivity. Determine the maximum $B$ field which can be achieved in a solenoid of length 25 cm if 1000 turns of $Nb_3Sn$ wire are wrapped on the outside surface.
26. A toroidal winding (Fig. 30.11) has a total of 400 turns on a core with inner radius $a = 4$ cm and outer radius $b = 6$ cm. Calculate the magnitude of the magnetic field at a point midway between the inner and outer walls of the core when there is a current of 0.5 A maintained in the windings.
27. The magnetic coils of a Tokamak fusion reactor are in the shape of a toroid having an inner radius of 0.7 m and outer radius of 1.3 m. Inside the toroid is the plasma. If the toroid has 900 turns of large diameter wire, each of which carries a current of 14 000 A, find the magnetic field strength along (a) the inner radius of the toroid and (b) the outer radius of the toroid.
28. A cylindrical conductor of radius $R = 2.5$ cm carries a current $I = 2.5$ A along its length; this current is uniformly distributed throughout the cross section of the conductor. Calculate the magnetic field midway along the radius of the wire (that is, at $r = R/2$).
29. For the conductor described in Problem 28, find the distance beyond the surface of the conductor at which the magnitude of the magnetic field has the same value as the magnitude of the field at $r = R/2$.
30. Niobium metal becomes a superconductor (with electrical resistance equal to zero) when cooled below 9 K. If superconductivity is destroyed when the surface magnetic field exceeds 0.1 T, determine the maximum current a 2-mm diameter niobium wire can carry and remain superconducting.
31. A *packed bundle* of 100 long straight insulated wires forms a cylinder of radius $R = 0.5$ cm. (a) If each wire carries a 2 A current, what are the magnitude and direction of the magnetic force per unit length acting on a wire located 0.2 cm from the center of the bundle? (b) Would a wire on the outer edge of the bundle experience a greater or a smaller force compared to the wire 0.2 cm from the center?
32. In Figure 30.45, assume that both currents are in the negative $x$ direction. (a) Sketch the magnetic field pattern in the $yz$ plane. (b) At what distance $d$ along the $z$ axis is the magnetic field a maximum?

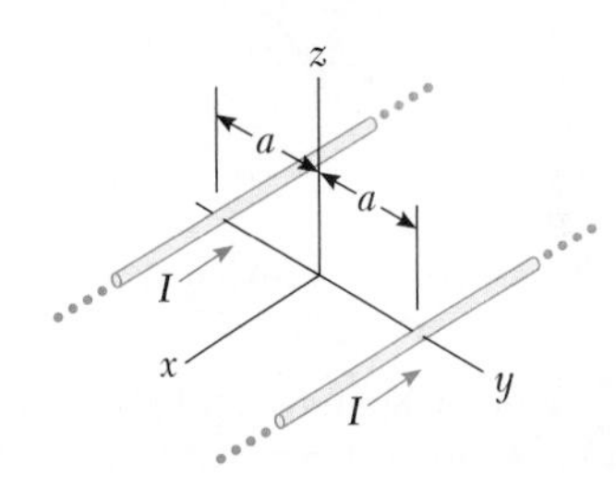


**Figure 30.45** (Problem 32).

## *Section 30.5 The Magnetic Field Along the Axis of a Solenoid

33. A short solenoid, with a length of 10 cm and a radius of 5 cm, consists of 200 turns of fine wire that carries a current of 15 A. What is the magnetic field strength $B$ at the center of the solenoid? For the same number of turns per unit length, what value of $B$ would result for $\ell \rightarrow \infty$?
34. A solenoid has 900 turns, carries a current of 3 A, has a length of 80 cm, and a radius of 2.5 cm. Calculate the magnetic field along its axis at (a) its center and (b) a point near the end.

35. A solenoid has 500 turns, a length of 50 cm, a radius of 5 cm, and carries a current of 4 A. Calculate the magnetic field at an axial point, a distance of 15 cm from the center (that is, 10 cm from one end).

### Section 30.6 Magnetic Flux

36. A toroid is constructed from $N$ *rectangular* turns of wire. Each turn has height $h$. The toroid has an inner radius $a$ and outer radius $b$. (a) If the toroid carries a current $I$, show that the total magnetic flux through the turns of the toroid is proportional to $\ln(b/a)$. (b) Evaluate this flux if $N = 200$ turns, $h = 1.5$ cm, $a = 2$ cm, $b = 5$ cm, and $I = 2$ A.
37. A cube of edge length $\ell = 2.5$ cm is positioned as shown in Figure 30.46. There is a uniform magnetic field throughout the region given by the expression $\mathbf{B} = (5\mathbf{i} + 4\mathbf{j} + 3\mathbf{k})$ T. (a) Calculate the flux through the shaded face of the cube. (b) What is the total flux through the six faces of the cube?

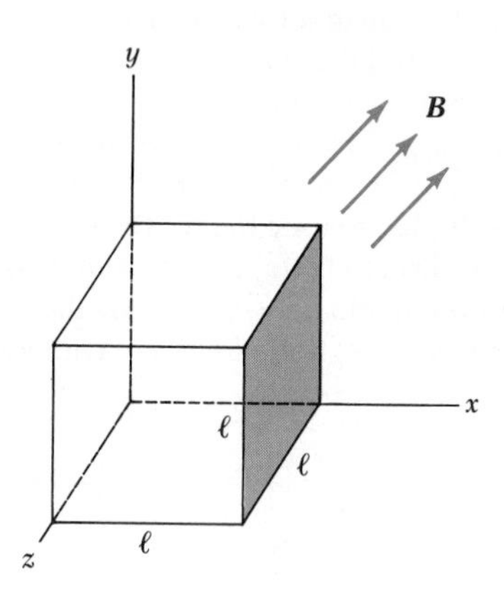


**Figure 30.46** (Problem 37).

38. A solenoid 2.5 cm in diameter and 30 cm in length has 300 turns and carries a current of 12 A. Calculate the flux through the surface of a disk of 5-cm radius that is positioned perpendicular to and centered on the axis of the solenoid, as in Figure 30.47.

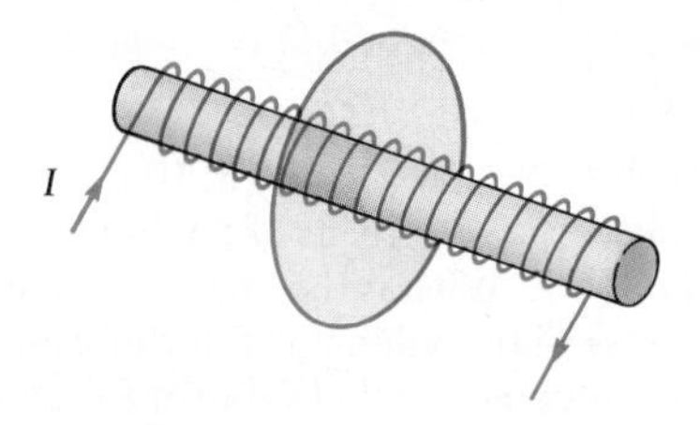


**Figure 30.47** (Problem 38).

39. Figure 30.48 shows an enlarged *end* view of the solenoid described in Problem 38. Calculate the flux through the blue area defined by an annulus with an inner radius of 0.4 cm and outer radius of 0.8 cm.

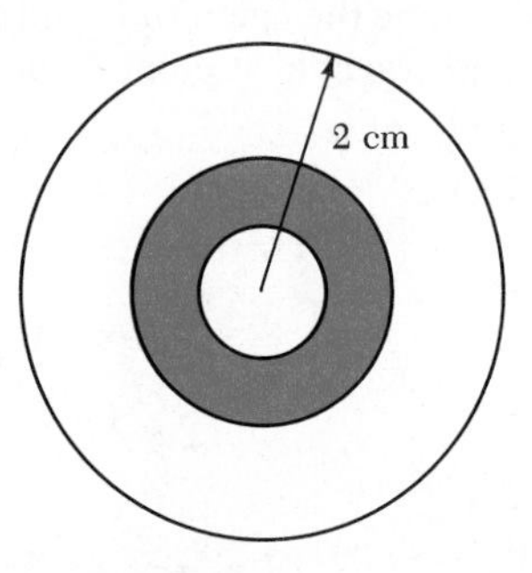


**Figure 30.48** (Problem 39).

40. A circular loop of wire of radius $R$ is placed in a uniform magnetic field $\mathbf{B}$ and is then spun at constant angular velocity $\omega$ about an axis through its diameter. Determine the magnetic flux through the loop as a function of time if the axis of rotation is (a) perpendicular to $\mathbf{B}$, and (b) parallel to $\mathbf{B}$.

### Section 30.8 Displacement Current and the Generalized Ampère's Law

41. The applied voltage across the plates of a 4-$\mu$F capacitor varies in time according to the expression

$$V_{\text{app}} = (8\text{ V})(1 - e^{-t/4})$$

where $t$ is in s. Calculate (a) the displacement current as a function of time and (b) the value of the current at $t = 4$ s.
42. A capacitor of capacitance $C$ has a charge $Q$ at $t = 0$. At that time, a resistor of resistance $R$ is connected to the plates of the charged capacitor. (a) Find the displacement current in the dielectric between the plates of the capacitor as a function of time. (b) Evaluate this displacement current at time $t = 0.1$ s if $C = 2.0\ \mu$F, $Q = 20\ \mu$C, and $R = 500$ k$\Omega$. (c) At what rate is the electric flux between the capacitor plates changing at $t = 0.1$ s?
43. A 0.1 A current is charging a capacitor with square plates, 5 cm on a side. If the plate separation is 4 mm, find (a) the rate of change of electric flux $d\Phi_E/dt$ between the plates and (b) the displacement current $I_d$ between the plates.
44. A 0.2-A current is charging a capacitor with circular plates, 10 cm in radius. If the plate separation is 4 mm, (a) what is the rate of increase of electric field $dE/dt$ between the plates? (b) What is the magnetic field between the plates at a radius of 5 cm from the center?

### Section 30.9 Magnetism In Matter

**45.** What is the relative permeability of a material that has a magnetic susceptibility of $10^{-4}$?

**46.** An iron-core toroid is wrapped with 250 turns of wire per meter of its length. The current in the winding is 8 A. Taking the magnetic permeability of iron to be $\kappa_m = 5000\mu_0$, calculate (a) the magnetic field strength, $\boldsymbol{H}$ and (b) the magnetic flux density, $\boldsymbol{B}$.

47. A toroidal winding with a mean radius of 20 cm and 630 turns (as in Figure 30.30) is filled with powdered steel whose magnetic susceptibility $\chi$ is 100. If the current in the windings is 3 A, find $B$ (assumed uniform) inside the toroid.

48. A toroid has an average radius of 9 cm. The current in the coil is 0.5 A. How many turns are required to produce a magnetic field strength of 700 A·turns/m within the toroid?

49. A magnetic field of flux density 1.3 T is to be set up in an iron-core toroid. The toroid has a mean radius of 10 cm, and magnetic permeability of $5000\mu_0$. What current is required if there are 470 turns of wire in the winding?

50. A toroidal solenoid has an average radius of 10 cm and a cross-sectional area of 1 cm$^2$. There are 400 turns of wire on the soft iron core, which has a permeability of $800\mu_0$. Calculate the current necessary to produce a magnetic flux of $5 \times 10^{-4}$ Wb through a cross section of the core.

51. A coil of 500 turns is wound on an iron ring ($\kappa_m = 750\mu_0$) of 20 cm mean radius and 8 cm$^2$ cross-sectional area. Calculate the magnetic flux $\Phi$ in this Rowland ring when the current in the coil is 0.5 A.

**52.** Show that the product of magnetic field strength $H$ and magnetic flux density $B$ has SI units of J/m$^3$.

**53.** In the text, we found that an alternative description for magnetic field $\boldsymbol{B}$ in terms of magnetic field strength $\boldsymbol{H}$ and magnetization $\boldsymbol{M}$ is $\boldsymbol{B} = \mu_0\boldsymbol{H} + \mu_0\boldsymbol{M}$. Relate the magnetic susceptibility $\chi$ to $|\boldsymbol{H}|$ and $|\boldsymbol{M}|$ for paramagnetic or diamagnetic materials.

54. Calculate the magnetic field strength $H$ of a magnetized substance characterized by a magnetization of $0.88 \times 10^6$ A·turns/m and a magnetic field of flux density 4.4 T. (*Hint:* See Problem 53.)

55. A magnetized cylinder of iron has a magnetic field $B = 0.04$ T in its interior. The magnet is 3 cm in diameter and 20 cm long. If the same magnetic field is to be produced by a 5-A current carried by an air-core solenoid having the same dimensions as the cylindrical magnet, how many turns of wire must be on the solenoid?

56. In Bohr's 1913 model of the hydrogen atom, the electron is in a circular orbit of radius $5.3 \times 10^{-11}$ m, and its speed is $2.2 \times 10^6$ m/s. (a) What is the magnitude of the magnetic moment due to the electron's motion? (b) If the electron orbits counterclockwise in a horizontal circle, what is the direction of this magnetic moment vector?

57. At saturation, the alignment of spins in iron can contribute as much as 2 Tesla to the total magnetic field $B$. If each electron contributes a magnetic moment of $9.27 \times 10^{-24}$ A·m$^2$ (one Bohr magneton), how many electrons per atom contribute to the saturated field of iron? (*Hint:* There are $8.5 \times 10^{28}$ iron atoms/m$^3$.)

### °Section 30.10 Magnetic Field of the Earth

58. A circular coil of 5 turns and a diameter of 30 cm is oriented in a vertical plane with its axis perpendicular to the horizontal component of the earth's magnetic field. A horizontal compass placed at the center of the coil is made to deflect 45° from magnetic North by a current of 0.60 A in the coil. (a) What is the horizontal component of the earth's magnetic field? (b) If a compass "dip" needle oriented in a vertical north-south plane makes an angle of 13° from the vertical, what is the total strength of the earth's magnetic field at this location?

59. The magnetic moment of the earth is approximately $8.7 \times 10^{22}$ A·m$^2$. (a) If this were caused by the complete magnetization of a huge iron deposit, how many unpaired electrons would this correspond to? (b) At 2 unpaired electrons per iron atom, how many kilograms of iron would this correspond to? (The density of iron is 7900 kg/m$^3$, and there are approximately $8.5 \times 10^{28}$ iron atoms/m$^3$.)

## ADDITIONAL PROBLEMS

**60.** A lightning bolt may carry a current of $10^4$ A for a short period of time. What is the resulting magnetic field at a point 100 m from the bolt?

**61.** Measurements of the magnetic field of a large tornado were made at the Geophysical Observatory in Tulsa, Oklahoma in 1962. If the tornado's field was $B = 1.5 \times 10^{-8}$ T pointing north when the tornado was 9 km east of the observatory, what current was carried up/down the funnel of the tornado?

62. The core of a solenoid having 250 turns per meter is filled with a material of unknown composition. When the current in the solenoid is 2 A, measurements reveal that the magnetic field within the core is 0.13 T. Determine the magnetic susceptibility of the material and classify it magnetically.

63. Two long, parallel conductors are carrying currents in the same direction as in Figure 30.49. Conductor A carries a current of 150 A and is held firmly in position. Conductor B carries a current $I_B$ and is allowed to slide freely up and down (parallel to A) between a set of nonconducting guides. If the linear density of conductor B is 0.10 g/cm, what value of current $I_B$ will

result in equilibrium when the distance between the two conductors is 2.5 cm?

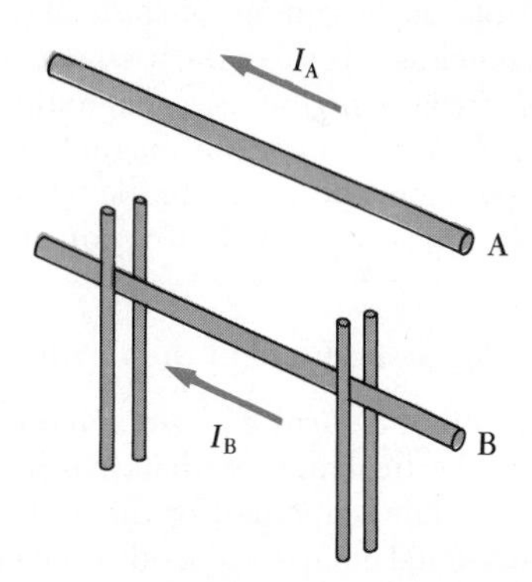


**Figure 30.49** (Problem 63).

64. Two parallel conductors carry current in opposite directions as shown in Figure 30.50. One conductor carries a current of 10 A. Point $A$ is at the *midpoint* between the wires and point $C$ is a distance $d/2$ to the right of the 10-A current. If $d = 18$ cm and $I$ is adjusted so that the magnetic field at $C$ is zero, find (a) the value of the current $I$ and (b) the value of the magnetic field at $A$.

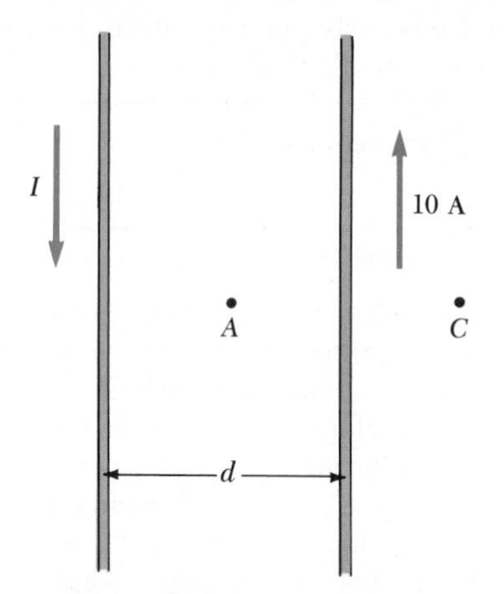


**Figure 30.50** (Problem 64).

65. A very long, thin strip of metal of width $w$ carries a current $I$ along its length as in Figure 30.51. Find the magnetic field in the *plane* of the strip (at an external point $P$) a distance $b$ from one edge.

66. A large nonconducting belt with a uniform surface charge density $\sigma$ moves with a speed $v$ on a set of rollers as shown in Figure 30.52. Consider a point *just above* the surface of the moving belt. (a) Find an expression for the magnitude of the magnetic field $\mathbf{B}$ at this point. (b) If the belt is positively charged, what is the direction of $\mathbf{B}$? (Note that the belt may be considered as an infinite sheet.)

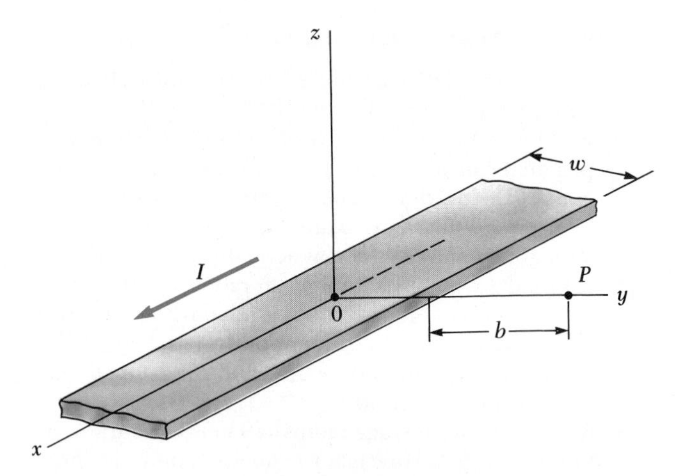


**Figure 30.51** (Problem 65).

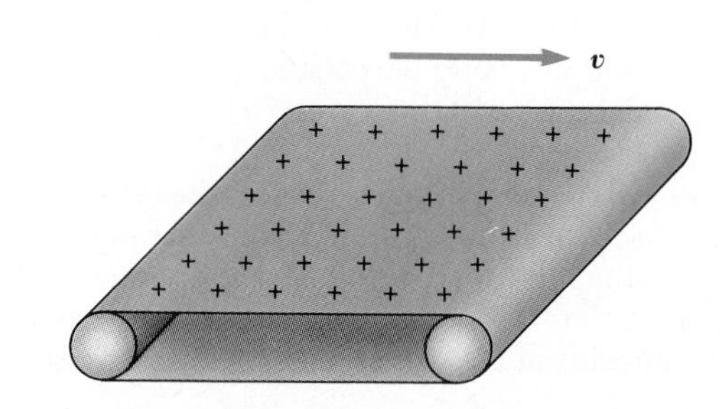


**Figure 30.52** (Problem 66).

67. A straight wire located at the equator is oriented parallel to the earth along the east-west direction. The earth's magnetic field at this point is horizontal and has a magnitude of $3.3 \times 10^{-5}$ T. If the mass per unit length of the wire is $2 \times 10^{-3}$ kg/m, what current must the wire carry in order that the magnetic force balance the weight of the wire?

**68.** The earth's magnetic field at either pole is about $0.7\ \text{G} = 7 \times 10^{-5}$ T. Using a model in which you assume that this field is produced by a current loop around the equator, determine the current that would generate such a field. ($R_e = 6.37 \times 10^6$ m.)

69. A nonconducting ring of radius $R$ is uniformly charged with a total positive charge $q$. The ring rotates at a constant angular velocity $\omega$ about an axis through its center, perpendicular to the plane of the ring. If $R = 0.1$ m, $q = 10\ \mu$C, and $\omega = 20$ rad/s, what is the resulting magnetic field on the axis of the ring a distance of 0.05 m from the center?

70. Consider a thin disk of radius $R$ mounted to rotate about the $x$ axis in the $yz$ plane. The disk has a positive uniform surface charge density $\sigma$ and angular velocity $\omega$. Show that the magnetic field at the center of the disk is given by $B = \frac{1}{2}\mu_0\sigma\omega R$.

71. Two circular coils of radius $R$ are each perpendicular to a common axis. The coil centers are a distance $R$ apart and a steady current $I$ flows in the same direction around each coil as shown in Figure 30.53. (a) Show that the magnetic field on the axis at a distance $x$ from the center of one coil is

$$B = \frac{\mu_0 I R^2}{2}\left[\frac{1}{(R^2 + x^2)^{3/2}} + \frac{1}{(2R^2 + x^2 - 2Rx)^{3/2}}\right]$$

(b) Show that $\frac{dB}{dx}$ and $\frac{d^2B}{dx^2}$ are both zero at a point *midway* between the coils. This means the magnetic field in the region midway between the coils is *uniform*. Coils in this configuration are called **Helmholtz coils.**

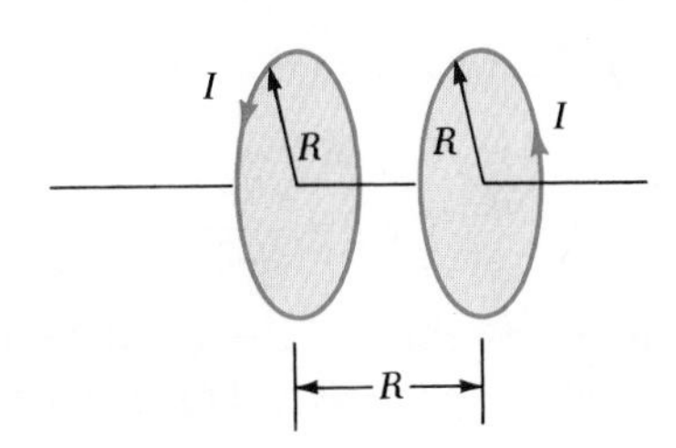


**Figure 30.53** (Problem 71).

72. Two identical, flat, circular coils of wire each have 100 turns and a radius of 0.50 m. If these coils are arranged as a set of Helmholtz coils and each coil carries a current of 10 A, determine the magnitude of the magnetic field at a point half way between the coils and on the axis of the coils. (See Figure 30.53.)

73. A long cylindrical conductor of radius $R$ carries a current $I$ as in Figure 30.54. The current density $J$, however, is *not* uniform over the cross section of the conductor but is a function of the radius according to $J = br$, where $b$ is a constant. Find an expression for the magnetic field $B$ (a) at a distance $r_1 < R$ and (b) at a distance $r_2 > R$, measured from the axis.

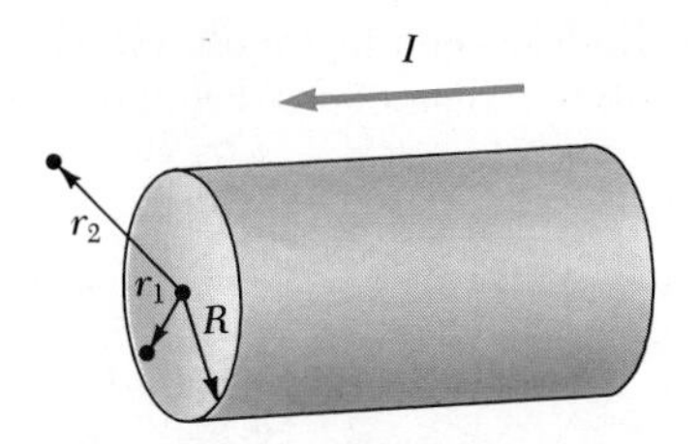


**Figure 30.54** (Problem 73).

74. A bar magnet (mass = 39.4 g, magnetic moment = 7.65 J/T, length = 10 cm) is connected to the ceiling by a string. A uniform external magnetic field is applied horizontally, as shown in Figure 30.55. The magnet is in equilibrium, making an angle $\theta$ with the horizontal. If $\theta = 5°$, determine the strength of the applied magnetic field.

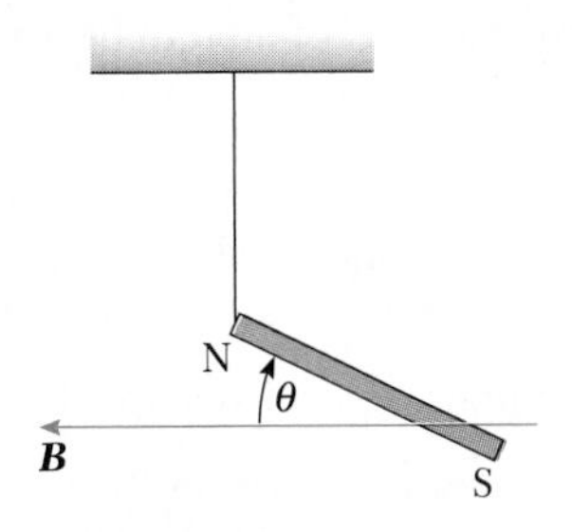


**Figure 30.55** (Problem 74).

75. Two circular loops are parallel, coaxial, and almost in contact, 1 mm apart (Fig. 30.56). Each loop is 10 cm in radius. The top loop carries a current of 140 A clockwise. The bottom loop carries 140 A counterclockwise. (a) Calculate the magnetic force that the bottom loop exerts on the top loop. (b) The upper loop has a mass of 0.021 kg. Calculate its acceleration, assuming that the only forces acting on it are the force in part (a) and its weight.

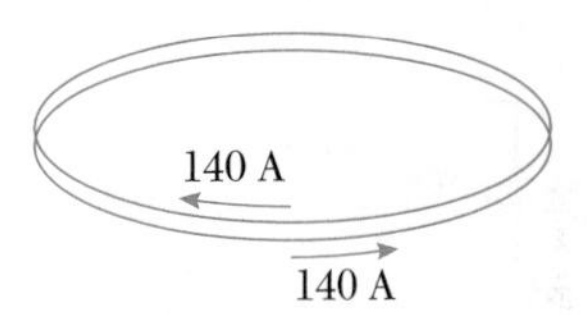


**Figure 30.56** (Problem 75).

76. For a research project, a student needs a solenoid that produces an interior magnetic field of 0.03 T. She decides to use a current of 1.0 A and a wire 0.50 mm in diameter. She winds the solenoid as layers on an insulating form 1.0 cm in diameter and 10.0 cm long. Determine the number of layers of wire needed and the total length of the wire.

77. A toroidal winding filled with a magnetic substance carries a steady current of 2 A. The coil contains a total of 1505 turns, has an average radius of 4 cm, and the core has a cross-sectional area of 1.21 cm$^2$. The total magnetic flux through a cross-section for the toroid is measured as $3 \times 10^{-5}$ Wb. Assume the flux density is constant. (a) What is the magnetic field strength $H$ within the core? (b) Determine the permeability of the core material.

78. A paramagnetic substance achieves 10% of its saturation magnetization when placed in a magnetic field of

5.0 T at a temperature of 4.0 K. The density of magnetic atoms in the sample is $8 \times 10^{27}$ atoms/m$^3$, and the magnetic moment per atom is 5 Bohr magnetons. Calculate the Curie constant for this substance.

79. The density of a specimen of a suspected new element is determined to be 4.15 g/cm$^3$. The saturation magnetism of the material is found to be $7.6 \times 10^4$ A/m, and the measured atomic magnetic moment is 1.2 Bohr magnetons. Calculate the expected value of the atomic weight of the element based on these values.

80. The force on a magnetic dipole $M$ aligned with a nonuniform magnetic field in the $x$ direction is given by $F_x = M\dfrac{dB}{dx}$. Suppose that 2 flat loops of wire each have radius $R$ and carry current $I$. (a) If the loops are arranged coaxially and separated by a large variable distance $x$, show that the magnetic force between them varies as $1/x^4$. (b) Evaluate the magnitude of this force if $I = 10$ A, $R = 0.5$ cm, and $x = 5.0$ cm.

81. A wire is formed into the shape of a square of edge length $L$ (Fig. 30.57). Show that when the current in the loop is $I$, the magnetic field at point $P$ a distance $x$ from the center of the square along its axis is given by

$$B = \frac{\mu_0 I L^2}{2\pi\left(x^2 + \dfrac{L^2}{4}\right)\sqrt{x^2 + \dfrac{L^2}{2}}}$$

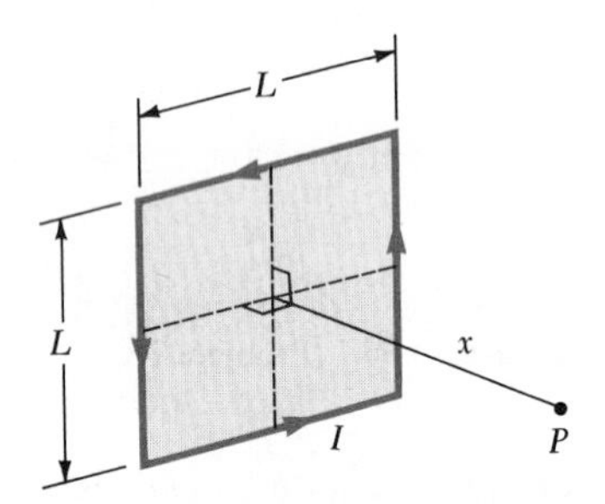


**Figure 30.57** (Problem 81).

82. A wire is bent into the shape shown in Figure 30.58a, and the magnetic field is measured at $P_1$ when the current in the wire is $I$. The same wire is then formed into the shape shown in Figure 30.58b, and the magnetic field measured at point $P_2$ when the current is again $I$. If the *total* length of wire is the same in each case, what is the ratio of $B_1/B_2$?

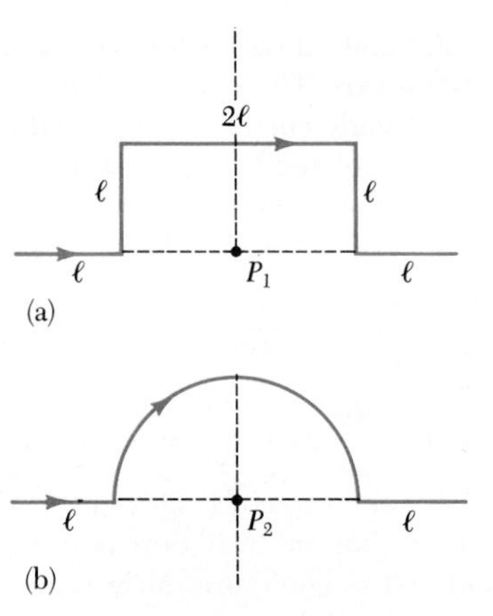


**Figure 30.58** (Problem 82).

83. A wire carrying a current $I$ is bent into the shape of an exponential spiral, $r = e^\theta$, from $\theta = 0$ to $\theta = 2\pi$ as in Figure 30.59. To complete a loop, the ends of the spiral are connected by a straight wire along the $x$ axis. Find the magnitude and direction of $\boldsymbol{B}$ at the origin. (*Hints:* Use the Biot-Savart Law. The angle $\beta$ between a radial line and its tangent line at any point on the curve $r = f(\theta)$ is related to the function in the following way:

$$\tan \beta = \frac{r}{dr/d\theta}$$

In this case $r = e^\theta$, thus $\tan \beta = 1$ and $\beta = \pi/4$. Therefore, the angle between $d\boldsymbol{s}$ and $\hat{\boldsymbol{r}}$ is $\pi - \beta = 3\pi/4$. Also

$$ds = \frac{dr}{\sin \pi/4} = \sqrt{2}\, dr$$

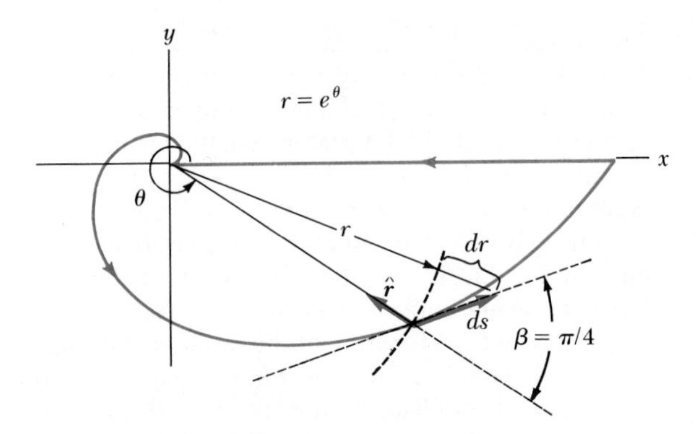


**Figure 30.59** (Problem 83).

84. **A long cylindrical conductor of radius $a$ has two cylindrical cavities of diameter $a$ through its entire length,**

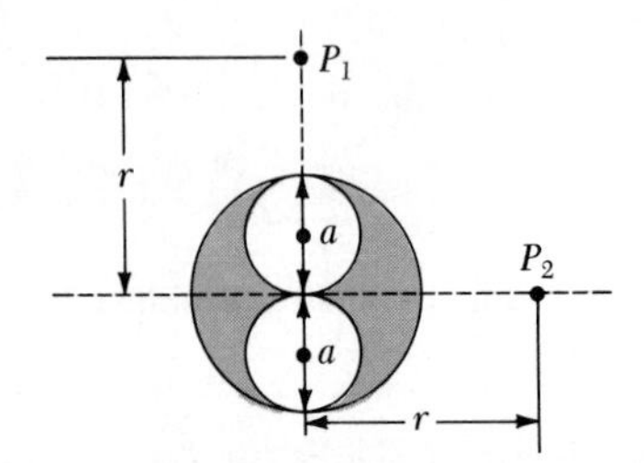


**Figure 30.60** (Problems 84 and 85).

as shown in Figure 30.60. A current, $I$, is directed out of the page and is uniform through a cross-section of the conductor. Find the magnitude and direction of the magnetic field at point $P_1$ in terms of $\mu_0$, $I$, $r$, and $a$.

85. Given the same conductor as described in Problem 84, find the magnitude and direction of the magnetic field at point $P_2$ as shown in Figure 30.60 in terms of $\mu_0$, $I$, $r$, and $a$.

86. Consider a flat circular current loop of radius $R$ carrying current $I$. Choose the $x$ axis to be along the axis of the loop with the origin at the center of the loop. Plot a graph of the ratio of the magnitude of the magnetic field at coordinate $x$ to that at the origin for $x = 0$ to $x = 5R$. It may be useful to use a programmable calculator or small computer to solve this problem.

87. A sphere of radius $R$ has a constant volume charge density $\rho$. Determine the magnetic field at the center of the sphere when it rotates as a rigid body with angular velocity $\omega$ about an axis through its center (Fig. 30.61).

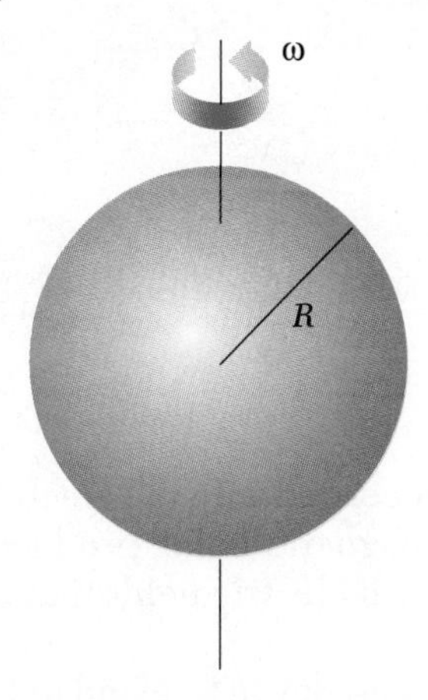


**Figure 30.61** (Problems 87 and 89).

△88. Table 30.4 is data taken on a ferromagnetic material. (a) Construct a magnetization curve from the data. Remember that $\boldsymbol{B} = \boldsymbol{B}_0 + \mu_0 \boldsymbol{M}$. (b) Determine the ratio $B/B_0$ for each pair of values of $B$ and $B_0$, and construct a graph of $B/B_0$ versus $B_0$. ($B/B_0$ is called the relative permeability, and it is a measure of the induced magnetic field.)

**TABLE 30.4**

| $B$(T) | $B_0$(T) |
|---|---|
| 0.2 | $4.8 \times 10^{-5}$ |
| 0.4 | $7.0 \times 10^{-5}$ |
| 0.6 | $8.8 \times 10^{-5}$ |
| 0.8 | $1.2 \times 10^{-4}$ |
| 1.0 | $1.8 \times 10^{-4}$ |
| 1.2 | $3.1 \times 10^{-4}$ |
| 1.4 | $8.7 \times 10^{-4}$ |
| 1.6 | $3.4 \times 10^{-3}$ |
| 1.8 | $1.2 \times 10^{-1}$ |

89. A sphere of radius $R$ has a constant volume charge density $\rho$. Determine the magnetic dipole moment of the sphere when it rotates as a rigid body with angular velocity $\omega$ about an axis through its center (Fig. 30.61).

# 31
# Faraday's Law

*Michael Faraday in his laboratory at the Royal Institute, London, in 1860. (Courtesy of The Bettmann Archive)*

Our studies so far have been concerned with the electric fields due to stationary charges and the magnetic fields produced by moving charges. This chapter deals with electric fields that originate from changing magnetic fields.

Experiments conducted by Michael Faraday in England in 1831 and independently by Joseph Henry in the United States that same year showed that an electric current could be induced in a circuit by a changing magnetic field. The results of these experiments led to a very basic and important law of electromagnetism known as *Faraday's law of induction.* This law says that the magnitude of the emf induced in a circuit equals the time rate of change of the magnetic flux through the circuit.

As we shall see, an induced emf can be produced in many ways. For instance, an induced emf and an induced current can be produced in a closed loop of wire when the wire moves into a magnetic field. We shall describe such experiments along with a number of important applications that make use of the phenomenon of electromagnetic induction.

With the treatment of Faraday's law, we complete our introduction to the fundamental laws of electromagnetism. These laws can be summarized in a set of four equations called *Maxwell's equations.* Together with the Lorentz force law, which we shall discuss briefly, they represent a complete theory for describing the interaction of charged objects. Maxwell's equations relate electric and magnetic fields to each other and to their ultimate source, namely, electric charges.

Biographical Sketch

**Joseph Henry**
*(1797–1878)*

Joseph Henry, an American physicist who carried out early experiments in electrical induction, was born in Albany, New York, in 1797. The son of a laborer, Henry had little schooling and was forced to go to work at a very young age. After working his way through Albany Academy to study medicine, then engineering, Henry became professor of mathematics and physics in 1826. He later became professor of natural philosophy at New Jersey College (now Princeton University).

In 1848, Henry became the first director of the Smithsonian Institute, where he introduced a weather-forecasting system based on meteorological information received by the electric telegraph. He was also the first president of The Academy of Natural Science, a position he held until his death in 1878.

Many of Henry's early experiments were with electromagnetism. He improved the electromagnet of William Sturgeon and made one of the first electromagnetic motors. By 1830, Henry had made powerful electromagnets by using many turns of fine insulated wire wound around iron cores. He discovered the phenomenon of self-induction but failed to publish his findings; as a result, credit was given to Michael Faraday.

Henry's contribution to science was ultimately recognized: in 1893 the unit of inductance was named the henry.

(Courtesy of AIP Niels Bohr Library, E. Scott Barr Collection)

## 31.1 FARADAY'S LAW OF INDUCTION

We begin by describing two simple experiments that demonstrate that a current can be produced by a changing magnetic field. First, consider a loop of wire connected to a galvanometer as in Figure 31.1. If a magnet is moved toward the loop, the galvanometer needle will deflect in one direction, as in Figure 31.1a. If the magnet is moved away from the loop, the galvanometer needle will deflect in the opposite direction, as in Figure 31.1b. If the magnet is held stationary relative to the loop, no deflection is observed. Finally, if the magnet is held stationary and the coil is moved either toward or away from the magnet, the needle will also deflect. From these observations, one concludes

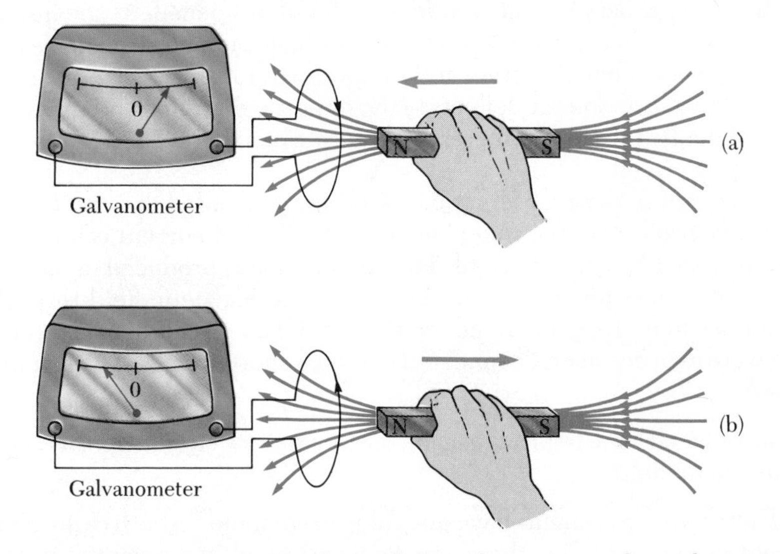


**Figure 31.1** (a) When a magnet is moved toward a loop of wire connected to a galvanometer, the galvanometer deflects as shown. This shows that a current is induced in the loop. (b) When the magnet is moved away from the loop, the galvanometer deflects in the opposite direction, indicating that the induced current is opposite that shown in (a).

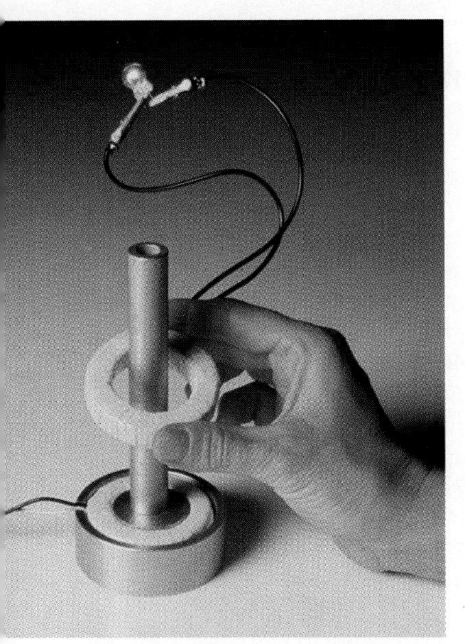

A demonstration of electromagnetic induction. An ac voltage is applied to the lower coil. A voltage is induced in the upper coil as indicated by the illuminated lamp connected to the upper coil. What do you think happens to the lamp's intensity as the upper coil is moved over the vertical tube? (Courtesy of CENCO)

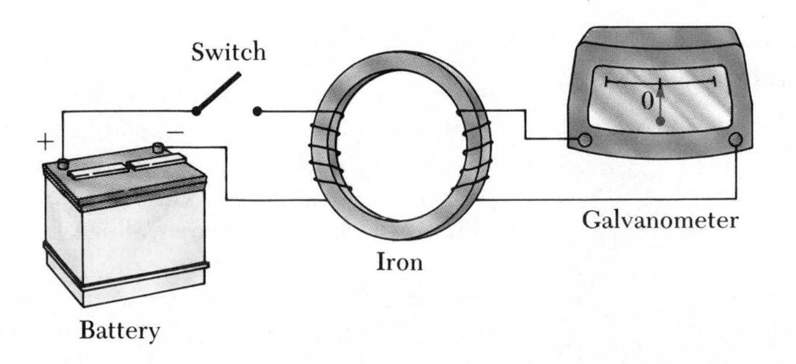


**Figure 31.2** Faraday's experiment. When the switch in the primary circuit at the left is closed, the galvanometer in the secondary circuit at the right deflects momentarily. The emf induced in the secondary circuit is caused by the changing magnetic field through the coil in this circuit.

that *a current is set up in the circuit as long as there is relative motion between the magnet and the coil.*[1]

These results are quite remarkable in view of the fact that *a current is set up in the circuit even though there are no batteries in the circuit!* We call such a current an *induced current,* which is produced by an *induced emf.*

Now let us describe an experiment, first conducted by Faraday, that is illustrated in Figure 31.2. Part of the apparatus consists of a coil connected to a switch and a battery. We shall refer to this coil as the *primary coil* and to the corresponding circuit as the primary circuit. The coil is wrapped around an iron ring to intensify the magnetic field produced by the current through the coil. A second coil, at the right, is also wrapped around the iron ring and is connected to a galvanometer. We shall refer to this as the *secondary coil* and to the corresponding circuit as the secondary circuit. There is no battery in the secondary circuit and the secondary coil is not connected to the primary coil. The only purpose of this circuit is to detect any current that might be produced by a change in the magnetic field.

At first sight, you might guess that no current would ever be detected in the secondary circuit. However, something quite amazing happens when the switch in the primary circuit is suddenly closed or opened. At the instant the switch in the primary circuit is closed, the galvanometer in the secondary circuit deflects in one direction and then returns to zero. When the switch is opened, the galvanometer deflects in the opposite direction and again returns to zero. Finally, the galvanometer reads zero when there is a steady current in the primary circuit.

As a result of these observations, Faraday concluded that *an electric current can be produced by a changing magnetic field.* A current cannot be produced by a steady magnetic field. The current that is produced in the secondary circuit occurs for only an instant while the magnetic field through the secondary coil is changing. In effect, the secondary circuit behaves as though there were a source of emf connected to it for a short instant. It is customary to say that

> an induced emf is produced in the secondary circuit by the changing magnetic field.

These two experiments have one thing in common. In both cases, an emf is induced in a circuit when the *magnetic flux* through the circuit *changes with*

[1] The exact magnitude of the current depends on the particular resistance of the circuit, but the existence of the current (or the algebraic sign) does *not.*

*time.* In fact, a general statement that summarizes such experiments involving induced currents and emfs is as follows:

The emf induced in a circuit is directly proportional to the time rate of change of magnetic flux through the circuit.

This statement, known as **Faraday's law of induction,** can be written

$$\mathcal{E} = -\frac{d\Phi_m}{dt} \tag{31.1}$$

Faraday's law

where $\Phi_m$ is the magnetic flux threading the circuit (Section 30.6), which can be expressed as

$$\Phi_m = \int \mathbf{B} \cdot d\mathbf{A} \tag{31.2}$$

The integral given by Equation 31.2 is taken over the area bounded by the circuit. The meaning of the negative sign in Equation 31.1 is a consequence of Lenz's law and will be discussed in Section 31.3. If the circuit is a coil consisting of $N$ loops all of the same area and if the flux threads all loops, the induced emf is given by

$$\mathcal{E} = -N\frac{d\Phi_m}{dt} \tag{31.3}$$

Suppose the magnetic field is uniform over a loop of area $A$ lying in a plane as in Figure 31.3. In this case, the flux through the loop is equal to $BA \cos \theta$; hence the induced emf can be expressed as

$$\mathcal{E} = -\frac{d}{dt}(BA \cos \theta) \tag{31.4}$$

From this expression, we see that an emf can be induced in the circuit in several ways: (1) The magnitude of $\mathbf{B}$ can vary with time; (2) the area of the circuit can change with time; (3) the angle $\theta$ between $\mathbf{B}$ and the normal to the plane can change with time; and (4) any combination of these can occur.

The following examples illustrate cases where an emf is induced in a circuit as a result of a time variation of the magnetic field.

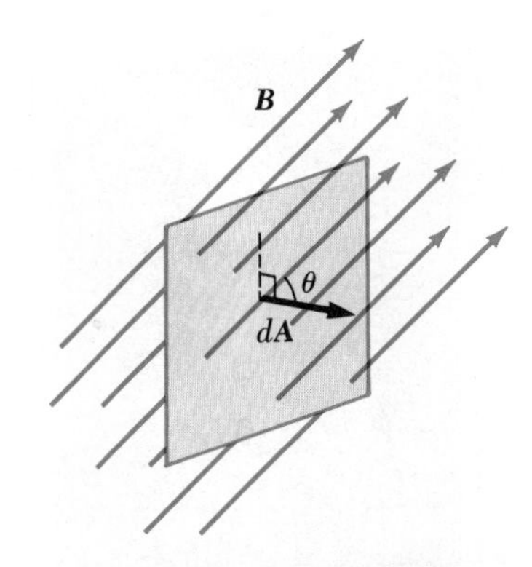


**Figure 31.3** A conducting loop of area $A$ in the presence of a uniform magnetic field $\mathbf{B}$, which is at an angle $\theta$ with the normal to the loop.

EXAMPLE 31.1 Application of Faraday's Law

A coil is wrapped with 200 turns of wire on the perimeter of a square frame of sides 18 cm. Each turn has the same area, equal to that of the frame, and the total resistance of the coil is 2 Ω. A uniform magnetic field is turned on perpendicular to the plane of the coil. If the field changes *linearly* from 0 to 0.5 Wb/m² in a time of 0.8 s, find the magnitude of the induced emf in the coil while the field is changing.

*Solution* The area of the loop is $(0.18 \text{ m})^2 = 0.0324 \text{ m}^2$. The magnetic flux through the loop at $t = 0$ is zero since $B = 0$. At $t = 0.8$ s, the magnetic flux through the loop is $\Phi_m = BA = (0.5 \text{ Wb/m}^2)(0.0324 \text{ m}^2) = 0.0162$ Wb. Therefore, the magnitude of the induced emf is

$$|\mathcal{E}| = \frac{N\,\Delta\Phi_m}{\Delta t} = \frac{200(0.0162 \text{ Wb} - 0 \text{ Wb})}{0.8 \text{ s}} = 4.05 \text{ V}$$

(Note that 1 Wb = 1 V·s.)

Exercise 1 What is the magnitude of the induced current in the coil while the field is changing?
Answer 2.03 A.

**EXAMPLE 31.2 An Exponentially Decaying *B* Field**
A plane loop of wire of area $A$ is placed in a region where the magnetic field is *perpendicular* to the plane. The magnitude of $\mathbf{B}$ varies in time according to the expression $B = B_0 e^{-at}$. That is, at $t = 0$ the field is $B_0$, and for $t > 0$, the field decreases exponentially in time (Fig. 31.4). Find the induced emf in the loop as a function of time.

*Solution* Since $\mathbf{B}$ is perpendicular to the plane of the loop, the magnetic flux through the loop at time $t > 0$ is given by

$$\Phi_m = BA = AB_0 e^{-at}$$

Also, since the coefficient $AB_0$ and the parameter $a$ are constants, the induced emf can be calculated from Equation 31.1:

$$\mathcal{E} = -\frac{d\Phi_m}{dt} = -AB_0 \frac{d}{dt} e^{-at} = aAB_0 e^{-at}$$

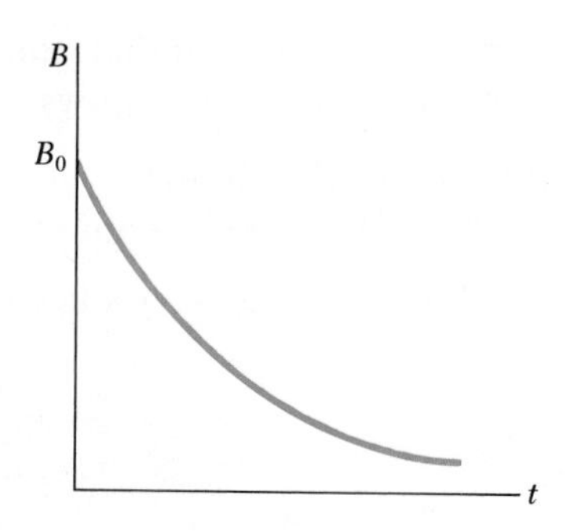


Figure 31.4 (Example 31.2) Exponential decrease of the magnetic field with time. The induced emf and induced current have similar time variations.

That is, the induced emf decays exponentially in time. Note that the maximum emf occurs at $t = 0$, where $\mathcal{E}_{max} = aAB_0$. Why is this true? The plot of $\mathcal{E}$ versus $t$ is similar to the $B$ versus $t$ curve shown in Figure 31.4.

Biographical Sketch

**Michael Faraday**
*(1791–1867)*

Michael Faraday was a British physicist and chemist who is often regarded as the greatest experimental scientist of the 1800s. His many contributions to the study of electricity include the invention of the electric motor, electric generator, and transformer, as well as the discovery of electromagnetic induction, the laws of electrolysis, the discovery of benzene, and the theory that the plane of polarization of light is rotated in an electric field.

Faraday was born in 1791 in rural England, but his family moved to London shortly thereafter. One of ten children and the son of a blacksmith, Faraday received a minimal education and became apprenticed to a bookbinder at age 14. He was fascinated by articles on electricity and chemistry and was fortunate to have an employer who allowed him to read books and attend scientific lectures. He received some education in science from the City Philosophical Society.

When Faraday finished his apprenticeship in 1812, he expected to devote himself to bookbinding rather than to science. That same year, Faraday attended a lecture by Humphry Davy, who made many contributions in the field of heat and thermodynamics. Faraday sent 386 pages of notes, bound in leather, to Davy; Davy was impressed and appointed Faraday his permanent assistant at the Royal Institution. Faraday toured France and Italy from 1813 to 1815 with Davy, visiting leading scientists of the time such as Volta and Vauquelin.

Despite his limited mathematical ability, Faraday succeeded in making the basic discoveries on which virtually all our uses of electricity depend. He conceived the fundamental nature of magnetism and, to a degree, that of electricity and light.

A modest man who was content to serve science as best he could, Faraday declined a knighthood and an offer to become president of the Royal Society. He was also a moral man; he refused to take part in the preparation of poison gas for use in the Crimean War.

Faraday died in 1867. His many achievements are recognized by the use of his name. The Faraday constant is the quantity of electricity required to deliver a standard amount of substance in electrolysis, and the SI unit of capacitance is the farad.

(Courtesy of AIP Niels Bohr Library)

## 31.2 MOTIONAL EMF

In Examples 31.1 and 31.2, we considered cases in which an emf is produced in a circuit when the magnetic field changes with time. In this section we describe the so-called **motional emf,** which is the emf induced in a conductor moving through a magnetic field.

First, consider a straight conductor of length $\ell$ moving with constant velocity through a uniform magnetic field directed into the paper as in Figure 31.5. For simplicity, we shall assume that the conductor is moving perpendicular to the field. The electrons in the conductor will experience a force along the conductor given by $\boldsymbol{F} = q\boldsymbol{v} \times \boldsymbol{B}$. Under the influence of this force, the electrons will move to the *lower* end and accumulate there, leaving a net positive charge at the upper end. An electric field is therefore produced within the conductor as a result of this charge separation. The charge at the ends builds up until the magnetic force $qvB$ is balanced by the electric force $qE$. At this point, charge stops flowing and the condition for equilibrium requires that

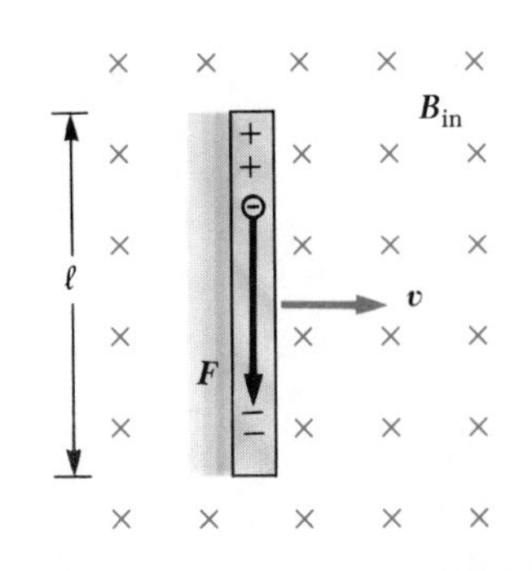

**Figure 31.5** A straight conducting bar of length $\ell$ moving with a velocity $\boldsymbol{v}$ through a uniform magnetic field $\boldsymbol{B}$ directed perpendicular to $\boldsymbol{v}$. An emf equal to $B\ell v$ is induced between the ends of the bar.

$$qE = qvB \qquad \text{or} \qquad E = vB$$

Since the electric field is constant, the electric field produced in the conductor is related to the potential difference across the ends according to the relation $V = E\ell$. Thus,

$$V = E\ell = B\ell v$$

where the upper end is at a higher potential than the lower end. Thus, *a potential difference is maintained as long as there is motion through the field. If the motion is reversed, the polarity of V is also reversed.*

A more interesting situation occurs if we now consider what happens when the moving conductor is part of a closed conducting path. This situation is particularly useful for illustrating how a changing magnetic flux can cause an induced current in a closed circuit. Consider a circuit consisting of a conducting bar of length $\ell$ sliding along two fixed parallel conducting rails as in Figure 31.6a. For simplicity, we assume that the moving bar has zero resistance and that the stationary part of the circuit has a resistance $R$. A uniform and constant magnetic field $\boldsymbol{B}$ is applied perpendicular to the plane of the circuit. As the bar is pulled to the right with a velocity $\boldsymbol{v}$, under the influence of an applied force $\boldsymbol{F}_{app}$, free charges in the bar experience a magnetic force along the length of the bar. This force, in turn, sets up an induced current since the charges are free to move in a closed conducting path. In this case, the rate of change of magnetic flux through the loop and the corresponding induced emf across the moving bar are proportional to the change in area of the loop as the bar moves through the magnetic field. As we shall see, if the bar is pulled to the right with a constant velocity, the work done by the applied force is dissipated in the form of joule heating in the circuit's resistive element.

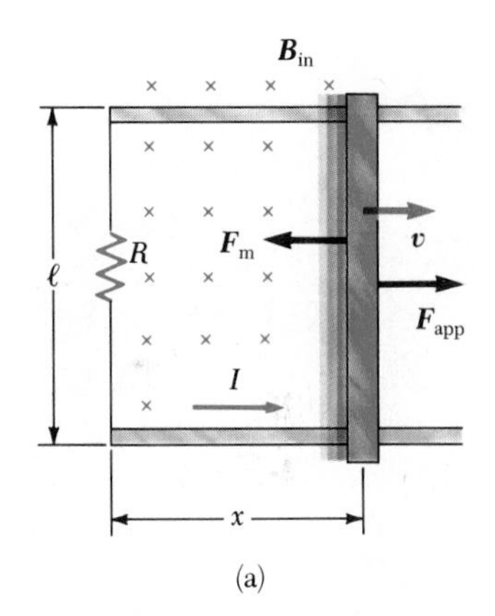


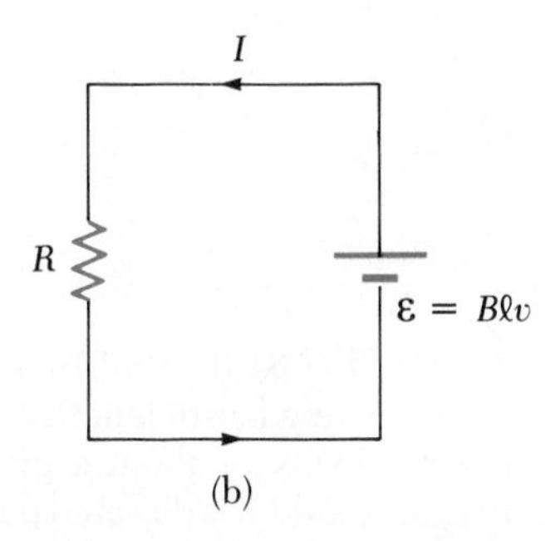

**Figure 31.6** (a) A conducting bar sliding with a velocity $\boldsymbol{v}$ along two conducting rails under the action of an applied force $\boldsymbol{F}_{app}$. The magnetic force $\boldsymbol{F}_m$ opposes the motion, and a counterclockwise current is induced in the loop. (b) The equivalent circuit of (a).

Since the area of the circuit at any instant is $\ell x$, the external magnetic flux through the circuit is given by

$$\Phi_m = B\ell x$$

where $x$ is the width of the circuit, which changes with time. Using Faraday's law, we find that the induced emf is

$$\mathcal{E} = -\frac{d\Phi_m}{dt} = -\frac{d}{dt}(B\ell x) = -B\ell\frac{dx}{dt}$$

$$\mathcal{E} = -B\ell v \tag{31.5}$$

If the resistance of the circuit is $R$, the magnitude of the induced current is given by

$$I = \frac{|\mathcal{E}|}{R} = \frac{B\ell v}{R} \tag{31.6}$$

The equivalent circuit diagram for this example is shown in Figure 31.6b.

Let us examine the system using energy considerations. Since there is no real battery in the circuit, one might wonder about the origin of the induced current and the electrical energy in the system. We can understand this by noting that the external force does work on the conductor, thereby moving charges through a magnetic field. This causes the charges to move along the conductor with some average drift velocity, and hence a current is established. From the viewpoint of energy conservation, the total work done by the applied force during some time interval should equal the electrical energy that the induced emf supplied in that same period. Furthermore, if the bar moves with constant speed, the work done must equal the energy dissipated as heat in the resistor in this time interval.

As the conductor of length $\ell$ moves through the uniform magnetic field $\mathbf{B}$, it experiences a magnetic force $\mathbf{F}_m$ of magnitude $I\ell B$ (Section 29.3). The direction of this force is opposite the motion of the bar, or to the left in Figure 31.6a.

If the bar is to move with a *constant* velocity, the applied force must be equal to and opposite the magnetic force, or to the right in Figure 31.6a. If the magnetic force acted in the direction of motion, it would cause the bar to accelerate once it was in motion, thereby increasing its velocity. This state of affairs would represent a violation of the principle of energy conservation. Using Equation 31.6 and the fact that $F_{app} = I\ell B$, we find that the power delivered by the applied force is

$$P = F_{app}v = (I\ell B)v = \frac{B^2\ell^2v^2}{R} = \frac{V^2}{R} \tag{31.7}$$

This power is equal to the rate at which energy is dissipated in the resistor, $I^2R$, as we would expect. It is also equal to the power $I\mathcal{E}$ supplied by the induced emf. This example is a clear demonstration of the conversion of mechanical energy into electrical energy and finally into thermal energy (joule heating).

EXAMPLE 31.3 Emf Induced in a Rotating Bar

A conducting bar of length $\ell$ rotates with a constant angular velocity $\omega$ about a pivot at one end. A uniform magnetic field $\mathbf{B}$ is directed perpendicular to the plane of rotation, as in Figure 31.7. Find the emf induced between the ends of the bar.

*Solution* Consider a segment of the bar of length $dr$, whose velocity is $\mathbf{v}$. According to Equation 31.5, the emf induced in a conductor of this length moving perpendicular to a field $\mathbf{B}$ is given by

$$(1) \qquad d\mathcal{E} = Bv\,dr$$

Each segment of the bar is moving perpendicular to $\mathbf{B}$, so there is an emf generated across each segment; the value of this emf is given by (1). Summing up the emfs induced across all elements, which are in series, gives the total emf between the ends of the bar. That is,

$$\mathcal{E} = \int Bv\,dr$$

In order to integrate this expression, note that the linear

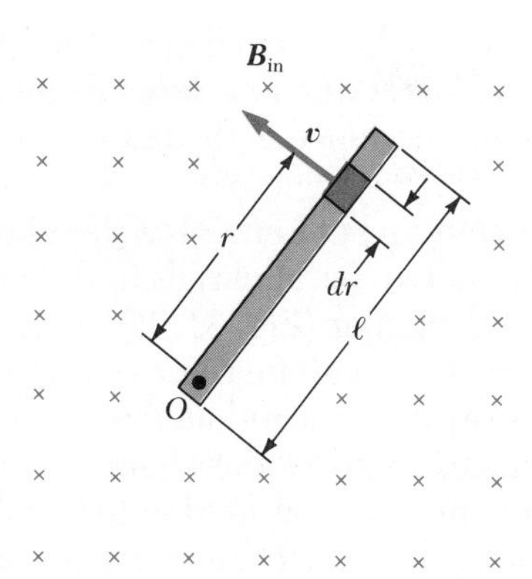


Figure 31.7 (Example 31.3) A conducting bar rotating about a pivot at one end in a uniform magnetic field that is perpendicular to the plane of rotation. An emf is induced across the ends of the bar.

speed of an element is related to the angular speed $\omega$ through the relationship $v = r\omega$. Therefore, since $B$ and $\omega$ are constants, we find that

$$\mathcal{E} = B\int v\,dr = B\omega\int_0^\ell r\,dr = \tfrac{1}{2}B\omega\ell^2$$

EXAMPLE 31.4 Magnetic Force on a Sliding Bar □

A bar of mass $m$ and length $\ell$ moves on two frictionless parallel rails in the presence of a uniform magnetic field directed into the paper (Fig. 31.8). The bar is given an initial velocity $\mathbf{v}_0$ to the right and is released. Find the velocity of the bar as a function of time.

*Solution* First note that the induced current is counterclockwise and the magnetic force is $F_m = -I\ell B$, where the negative sign denotes that the force is to the left and *retards* the motion. This is the *only* horizontal force acting on the bar, and hence Newton's second law applied to motion in the horizontal direction gives

$$F_x = ma = m\frac{dv}{dt} = -I\ell B$$

Since the induced current is given by Equation 31.6, $I = B\ell v/R$, we can write this expression as

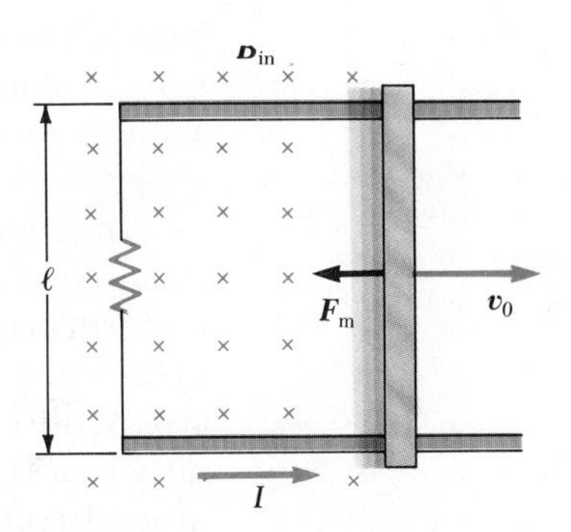


Figure 31.8 (Example 31.4) A conducting bar of length $\ell$ sliding on two fixed conducting rails is given an initial velocity $\mathbf{v}_0$ to the right.

$$m\frac{dv}{dt} = -\frac{B^2\ell^2}{R}v$$

$$\frac{dv}{v} = -\left(\frac{B^2\ell^2}{mR}\right)dt$$

Integrating this last equation using the initial condition that $v = v_0$ at $t = 0$, we find that

$$\int_{v_0}^{v}\frac{dv}{v} = \frac{-B^2\ell^2}{mR}\int_0^t dt$$

$$\ln\left(\frac{v}{v_0}\right) = -\left(\frac{B^2\ell^2}{mR}\right)t = -\frac{t}{\tau}$$

where the constant $\tau = mR/B^2\ell^2$. From this, we see that the velocity can be expressed in the exponential form

$$v = v_0e^{-t/\tau}$$

Therefore, the velocity of the bar decreases exponentially with time under the action of the magnetic retarding force. Furthermore, if we substitute this result into Equations 31.5 and 31.6, we find that the induced emf and induced current also decrease exponentially with time. That is,

$$I = \frac{B\ell v}{R} = \frac{B\ell v_0}{R}e^{-t/\tau}$$

$$\mathcal{E} = IR = B\ell v_0e^{-t/\tau}$$

## 31.3 LENZ'S LAW

The direction of the induced emf and induced current can be found from Lenz's law,[2] which can be stated as follows:

> The polarity of the induced emf is such that it tends to produce a current that will create a magnetic flux to oppose the change in magnetic flux through the loop.

[2] Developed by the German physicist Heinrich Lenz (1804–1865).

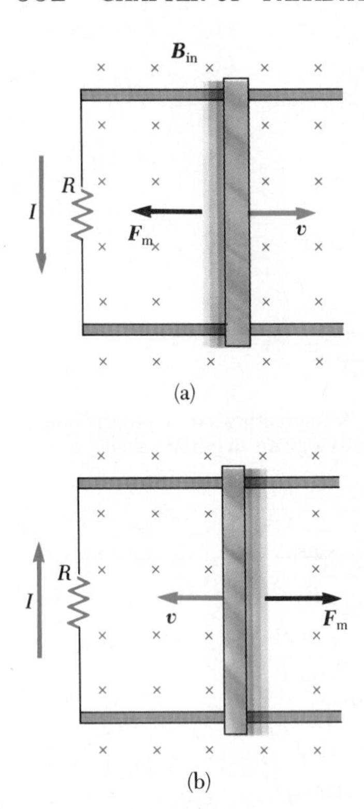


**Figure 31.9** (a) As the conducting bar slides on the two fixed conducting rails, the magnetic flux through the loop increases in time. By Lenz's law, the induced current must be *counterclockwise* so as to produce a counteracting flux *out* of the paper. (b) When the bar moves to the left, the induced current must be *clockwise*. Why?

That is, the induced current tends to keep the original flux through the circuit from changing. The interpretation of this statement depends on the circumstances. As we shall see, this law is a consequence of the law of conservation of energy.

In order to obtain a better understanding of Lenz's law, let us return to the example of a bar moving to the right on two parallel rails in the presence of a uniform magnetic field directed into the paper (Fig. 31.9a). As the bar moves to the right, the magnetic flux through the circuit increases with time since the area of the loop increases. Lenz's law says that the induced current must be in a direction such that the flux *it* produces opposes the change in the external magnetic flux. Since the flux due to the external field is increasing *into* the paper, the induced current, if it is to oppose the change, must produce a flux *out* of the paper. Hence, the induced current must be counterclockwise when the bar moves to the right to give a counteracting flux out of the paper in the region *inside* the loop. (Use the right-hand rule to verify this direction.) On the other hand, if the bar is moving to the left, as in Figure 31.9b, the magnet flux through the loop decreases with time. Since the flux is into the paper, the induced current has to be clockwise to produce a flux into the paper inside the loop. In either case, the induced current tends to maintain the original flux through the circuit.

Let us look at this situation from the viewpoint of energy considerations. Suppose that the bar is given a slight push to the right. In the above analysis, we found that this motion leads to a counterclockwise current in the loop. Let us see what happens if we assume that the current is clockwise. For a clockwise current $I$, the direction of the magnetic force on the sliding bar would be to the right. This force would accelerate the rod and increase its velocity. This, in turn, would cause the area of the loop to increase more rapidly, thus increasing the induced current, which would increase the force, which would increase the current, which would. . . . In effect, the system would acquire energy with no additional input energy. This is clearly inconsistent with all experience and with the law of conservation of energy. Thus, we are forced to conclude that the current must be counterclockwise.

Consider another situation, one in which a bar magnet is moved to the right toward a stationary loop of wire, as in Figure 31.10a. As the magnet moves to the right toward the loop, the magnetic flux through the loop in-

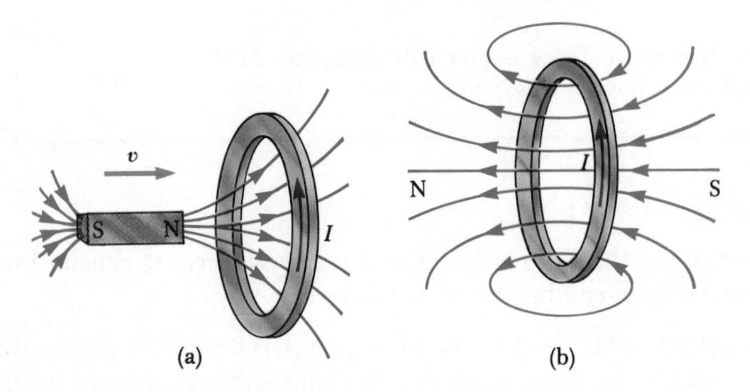


**Figure 31.10** (a) When the magnet is moved toward the stationary conducting loop, a current is induced in the direction shown. (b) This induced current produces its own flux to the left to counteract the increasing external flux to the right.

creases with time. To counteract this increase in flux to the right, the induced current produces a flux to the left, as in Figure 31.10b; hence the induced current is in the direction shown. Note that the magnetic field lines associated with the induced current oppose the motion of the magnet. Therefore, the left face of the current loop is a north pole and the right face is a south pole.

On the other hand, if the magnet were moving to the left, its flux through the loop, which is toward the right, would decrease in time. Under these circumstances, the induced current in the loop would be in a direction such as to set up a field through the loop directed from left to right in an effort to maintain a constant number of flux lines. Hence, the induced current in the loop would be opposite that shown in Figure 31.10b. In this case, the left face of the loop would be a south pole and the right face would be a north pole.

### EXAMPLE 31.5 Application of Lenz's Law

A coil of wire is placed near an electromagnet as shown in Figure 31.11a. Find the direction of the induced current in the coil (a) at the instant the switch is closed, (b) after the switch has been closed for several seconds, and (c) when the switch is opened.

*Solution* (a) When the switch is closed, the situation changes from a condition in which no lines of flux pass through the coil to one in which lines of flux pass through in the direction shown in Figure 31.11b. To counteract this change in the number of lines, the coil must set up a field from left to right in the figure. This requires a current directed as shown in Figure 31.11b.

(b) After the switch has been closed for several seconds, there is no change in the number of lines through the loop; hence the induced current is zero.

(c) Opening the switch causes the magnetic field to change from a condition in which flux lines thread through the coil from right to left to a condition of zero flux. The induced current must then be as shown in Figure 31.11c, so as to set up its own field from right to left.

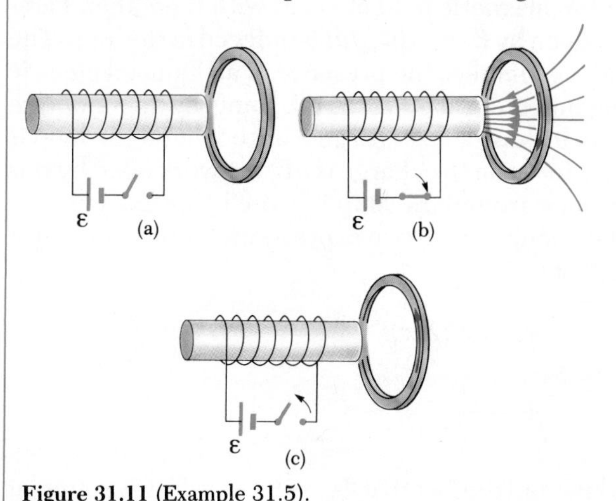


**Figure 31.11** (Example 31.5).

### EXAMPLE 31.6 A Loop Moving Through a *B* Field

A rectangular loop of dimensions $\ell$ and $w$ and resistance $R$ moves with constant speed $v$ to the right, as in Figure 31.12a. It continues to move with this speed through a region containing a uniform magnetic field $\boldsymbol{B}$ directed into the paper and extending a distance $3w$. Plot the flux, the induced emf, and the external force acting on the loop as a function of the position of the loop in the field.

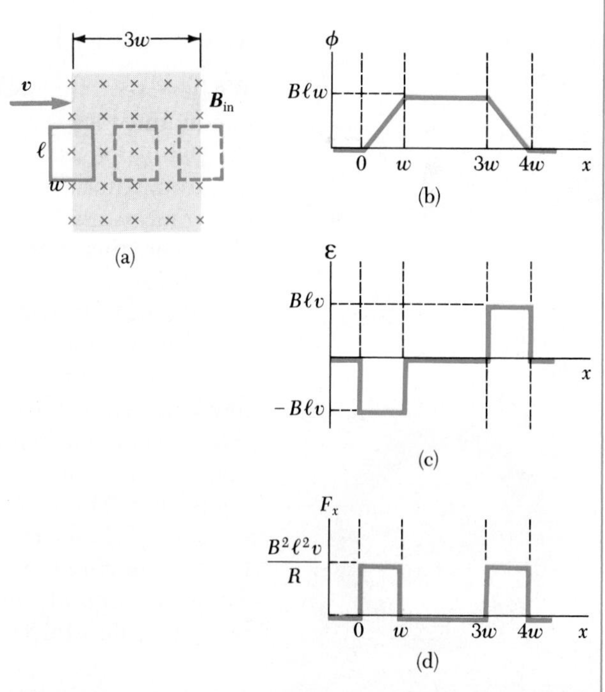


**Figure 31.12** (Example 31.6) (a) A conducting rectangular loop of width $w$ and length $\ell$ moving with a velocity $\boldsymbol{v}$ through a uniform magnetic field extending a distance $3w$. (b) A plot of the flux as a function of the position of the loop. (c) A plot of the induced emf versus the position of the leading edge. (d) A plot of the force versus position such that the velocity of the loop remains constant.

*Solution* Figure 31.12b shows the flux through the loop as a function of loop position. Before the loop enters the field, the flux is zero. As it enters the field, the flux increases linearly with position. Finally, the flux decreases linearly to zero as the loop leaves the field.

Before the loop enters the field, there is no induced emf since there is no field present (Fig. 31.12c). As the right side of the loop enters the field, the flux inward begins to increase. Hence, according to Lenz's law, the induced current is counterclockwise and the induced emf is given by $-B\ell v$. This motional emf arises from the magnetic force experienced by charges in the right side of the loop. When the loop is entirely in the field, the *change* in flux is zero, and hence the induced emf vanishes.

From another point of view, the right and left sides of the loop experience magnetic forces that tend to set up currents that cancel one another. As the right side of the loop leaves the field, the flux inward begins to decrease, a clockwise current is induced, and the induced emf is $B\ell v$. As soon as the left side leaves the field, the emf drops to zero.

The external force that must act on the loop to maintain this motion is plotted in Figure 31.12d. When the loop is not in the field, there is no magnetic force on it; hence the external force on it must be zero if $v$ is constant. When the right side of the loop enters the field, the external force necessary to maintain constant speed must be equal to and opposite the magnetic force on that side, given by $F_m = -I\ell B = -B^2\ell^2 v/R$. When the loop is entirely in the field, the flux through the loop is not changing with time. Hence, the net emf induced in the loop is zero, and the current is also zero. Therefore, no external force is needed to maintain the motion of the loop. (From another point of view, the right and left sides of the loop experience equal and opposite forces; hence, the net force is zero.) Finally, as the right side leaves the field, the external force must be equal to and opposite the magnetic force on the left side of the loop. From this analysis, we conclude that power is supplied only when the loop is either entering or leaving the field. Furthermore, this example shows that the induced emf in the loop can be zero even when there is motion through the field! Again, it is emphasized that an emf is induced in the loop *only* when the magnetic flux through the loop *changes* in time.

## 31.4 INDUCED EMFS AND ELECTRIC FIELDS

We have seen that a changing magnetic flux induces an emf and a current in a conducting loop. We therefore must conclude that *an electric field is created in the conductor as a result of the changing magnetic flux.* In fact, the law of electromagnetic induction shows that *an electric field is always generated by a changing magnetic flux*, even in free space where no charges are present. However, this induced electric field has properties that are quite different from those of an electrostatic field *produced by stationary charges.*

We can illustrate this point by considering a conducting loop of radius $r$ situated in a uniform magnetic field that is perpendicular to the plane of the loop, as in Figure 31.13. If the magnetic field changes with time, then Faraday's law tells us that an emf given by $\mathcal{E} = -d\Phi_m/dt$ is induced in the loop. The induced current that is produced implies the presence of an induced electric field $\mathbf{E}$, which must be tangent to the loop since all points on the loop are equivalent. The work done in moving a test charge $q$ once around the loop is equal to $q\mathcal{E}$. Since the electric force on the charge is $q\mathbf{E}$, the work done by this force in moving the charge once around the loop is given by $qE(2\pi r)$, where $2\pi r$ is the circumference of the loop. These two expressions for the work must be equal; therefore we see that

$$q\mathcal{E} = qE(2\pi r)$$

$$E = \frac{\mathcal{E}}{2\pi r}$$

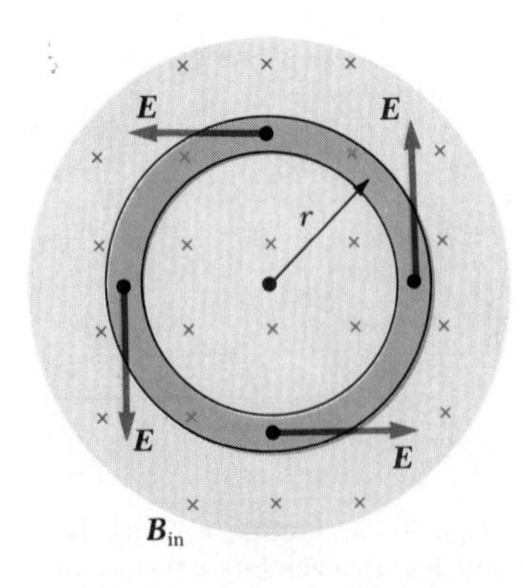


**Figure 31.13** A loop of radius $r$ in a uniform magnetic field perpendicular to the plane of the loop. If $\mathbf{B}$ changes in time, an electric field is induced in a direction tangent to the loop.

Using this result, Faraday's law, and the fact that $\Phi_m = BA = \pi r^2 B$ for a circular loop, we find that the induced electric field can be expressed as

$$E = -\frac{1}{2\pi r}\frac{d\Phi_m}{dt} = -\frac{r}{2}\frac{dB}{dt} \tag{31.8}$$

If the time variation of the magnetic field is specified, the induced electric field can easily be calculated from Equation 31.8. The negative sign indicates that the induced electric field **E** *opposes* the change in the magnetic field. It is important to understand that *this result is also valid in the absence of a conductor.* That is, a free charge placed in a changing magnetic field will also experience the same electric field.

The emf for any closed path can be expressed as the line integral of $\mathbf{E}\cdot d\mathbf{s}$ over that path. In more general cases, **E** may not be constant, and the path may not be a circle. Hence, Faraday's law of induction, $\mathcal{E} = -d\Phi m/dt$, can be written as

$$\oint \mathbf{E}\cdot d\mathbf{s} = -\frac{d\Phi_m}{dt} \tag{31.9}$$

Faraday's law

It is important to recognize that *the induced electric field* **E** *that appears in Equation 31.9 is a nonconservative, time-varying field that is generated by a changing magnetic field.* The field **E** that satisfies Equation 31.9 could not possibly be an electrostatic field for the following reason. If the field were electrostatic, and hence conservative, the line integral of $\mathbf{E}\cdot d\mathbf{s}$ over a closed loop would be zero, contrary to Equation 31.9.

EXAMPLE 31.7 Electric Field Due to a Solenoid

A long solenoid of radius $R$ has $n$ turns per unit length and carries a time-varying current that varies sinusoidally as $I = I_0 \cos \omega t$, where $I_0$ is the maximum current and $\omega$ is the angular frequency of the current source (Fig. 31.14). (a) Determine the electric field outside the solenoid, a distance $r$ from its axis.

First, let us consider an external point and take the path for our line integral to be a circle centered on the solenoid, as in Figure 31.14. By symmetry we see that the magnitude of **E** is constant on this path and tangent to it. The magnetic flux through this path is given by $BA = B(\pi R^2)$, and hence Equation 31.9 gives

$$\oint \mathbf{E}\cdot d\mathbf{s} = -\frac{d}{dt}[B(\pi R^2)] = -\pi R^2 \frac{dB}{dt}$$

$$E(2\pi r) = -\pi R^2 \frac{dB}{dt}$$

Since the magnetic field inside a long solenoid is given by Equation 30.20, $B = \mu_0 nI$, and $I = I_0 \cos \omega t$, we find that

$$E(2\pi r) = -\pi R^2 \mu_0 n I_0 \frac{d}{dt}(\cos \omega t) = \pi R^2 \mu_0 n I_0 \omega \sin \omega t$$

$$E = \frac{\mu_0 n I_0 \omega R^2}{2r} \sin \omega t \qquad (\text{for } r > R)$$

Hence, the electric field varies sinusoidally with time, and its amplitude falls off as $1/r$ outside the solenoid.

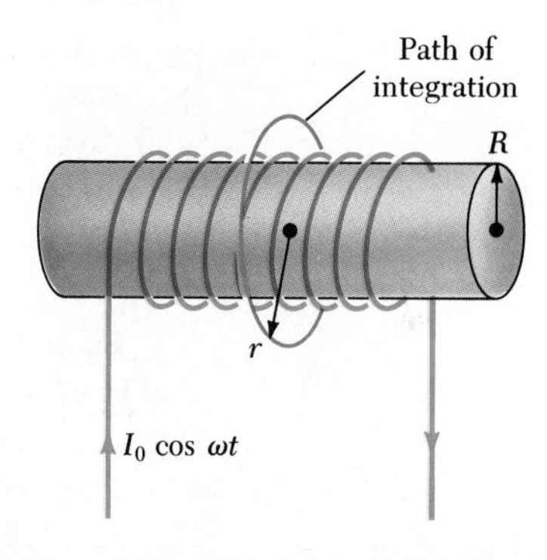


**Figure 31.14** (Example 31.7) A long solenoid carrying a time-varying current given by $I = I_0 \cos \omega t$. An electric field is induced both inside and outside the solenoid.

(b) What is the electric field inside the solenoid, a distance $r$ from its axis?

For an interior point ($r < R$), the flux threading an integration loop is given by $B(\pi r^2)$. Using the same procedure as in (a), we find that

$$E(2\pi r) = -\pi r^2 \frac{dB}{dt} = \pi r^2 \mu_0 n I_0 \omega \sin \omega t$$

$$E = \frac{\mu_0 n I_0 \omega}{2} r \sin \omega t \qquad (\text{for } r < R)$$

This shows that the amplitude of the electric field *inside* the solenoid increases linearly with $r$ and varies sinusoidally with time.

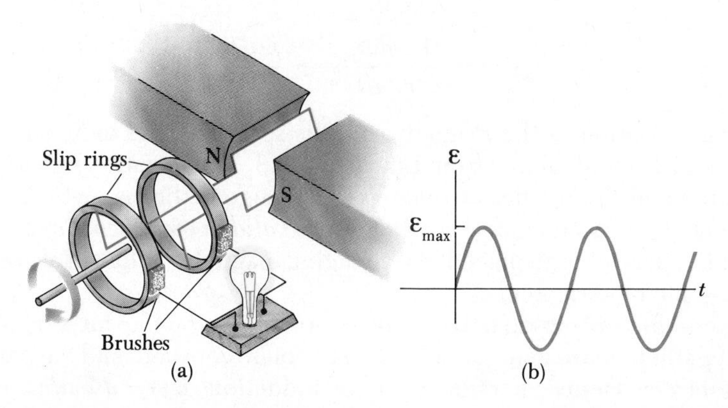


**Figure 31.15** (a) Schematic diagram of an ac generator. An emf is induced in a coil that rotates by some external means in a magnetic field. (b) The alternating emf induced in the loop plotted versus time.

## *31.5 GENERATORS AND MOTORS

Generators and motors are important devices that operate on the principle of electromagnetic induction. First, let us consider the **alternating current generator** (or ac generator), a device that converts mechanical energy to electrical energy. In its simplest form, the ac generator consists of a loop of wire rotated by some external means in a magnetic field (Fig. 31.15a). In commercial power plants, the energy required to rotate the loop can be derived from a variety of sources. For example, in a hydroelectric plant, falling water directed against the blades of a turbine produces the rotary motion; in a coal-fired plant, the heat produced by burning coal is used to convert water to steam and this steam is directed against the turbine blades. As the loop rotates, the magnetic flux through it changes with time, inducing an emf and a current in an external circuit. The ends of the loop are connected to slip rings that rotate with the loop. Connections to the external circuit are made by stationary brushes in contact with the slip rings.

To put our discussion of the generator on a quantitative basis, suppose that the loop has $N$ turns (a more practical situation), all of the same area $A$, and suppose that the loop rotates with a constant angular velocity $\omega$. If $\theta$ is the angle between the magnetic field and the normal to the plane of the loop as in Figure 31.16, then the magnetic flux through the loop at any time $t$ is given by

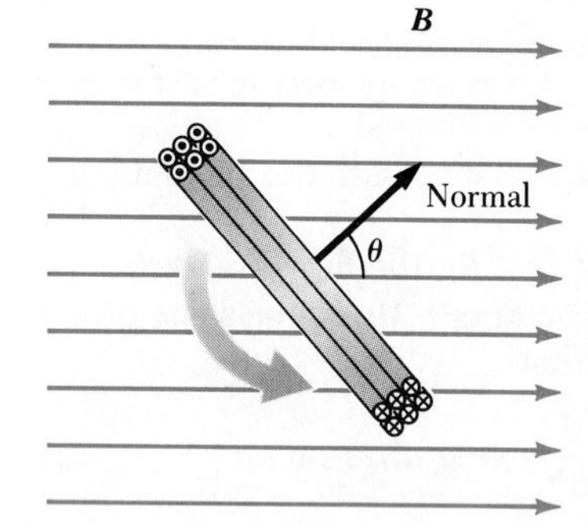


**Figure 31.16** A loop of area $A$ containing $N$ turns, rotating with constant angular velocity $\omega$ in the presence of a magnetic field. The emf induced in the loop varies sinusoidally in time.

$$\Phi_m = BA \cos \theta = BA \cos \omega t$$

where we have used the relationship between angular displacement and angular velocity, $\theta = \omega t$. (We have set the clock so that $t = 0$ when $\theta = 0$.) Hence, the induced emf in the coil is given by

$$\mathcal{E} = -N \frac{d\Phi_m}{dt} = -NAB \frac{d}{dt} (\cos \omega t) = NAB\omega \sin \omega t \qquad (31.10)$$

This result shows that the emf varies sinusoidally with time, as plotted in Figure 31.15b. From Equation 31.10 we see that the maximum emf has the value

$$\mathcal{E}_{max} = NAB\omega \qquad (31.11)$$

which occurs when $\omega t = 90°$ or $270°$. In other words, $\mathcal{E} = \mathcal{E}_{max}$ when the magnetic field is in the plane of the coil, and the time rate of change of flux is a maximum. Furthermore, the emf is *zero* when $\omega t = 0$ or $180°$, that is, when $\boldsymbol{B}$ is perpendicular to the plane of the coil, and the time rate of change of flux is zero. The frequency for commercial generators in the United States and Canada is 60 Hz, whereas in some European countries, 50 Hz is used. (Recall that $\omega = 2\pi f$, where $f$ is the frequency in hertz.)

**EXAMPLE 31.8 Emf Induced in a Generator**

An ac generator consists of 8 turns of wire each of area $A = 0.09\ \text{m}^2$ and total resistance 12 Ω. The loop rotates in a magnetic field $B = 0.5$ T at a constant frequency of 60 Hz. (a) Find the maximum induced emf.

First note that $\omega = 2\pi f = 2\pi(60\ \text{Hz}) = 377\ \text{s}^{-1}$. Using Equation 31.11 with the appropriate numerical values gives

$$\mathcal{E}_{max} = NAB\omega = 8(0.09\ \text{m}^2)(0.5\ \text{T})(377\ \text{s}^{-1}) = 136\ \text{V}$$

(b) What is the maximum induced current?

From Ohm's law and the results to (a), we find that the maximum induced current is

$$I_{max} = \frac{\mathcal{E}_{max}}{R} = \frac{136\ \text{V}}{12\ \Omega} = 11.3\ \text{A}$$

Exercise 2 Determine the time variation of the induced emf and induced current when the output terminals are connected by a low-resistance conductor.

Answers:

$$\mathcal{E} = \mathcal{E}_{max} \sin \omega t = (136\ \text{V}) \sin 377t$$

$$I = I_{max} \sin \omega t = (11.3\ \text{A}) \sin 377t$$

The **direct current** (dc) **generator** is illustrated in Figure 31.17a. Such generators are used, for instance, to charge storage batteries used in older style cars. The components are essentially the same as those of the ac generator, except that the contacts to the rotating loop are made using a split ring, or commutator.

In this configuration, the output voltage always has the same polarity and the current is a pulsating direct current as in Figure 31.17b. The reason for this can be understood by noting that the contacts to the split ring reverse their roles every half cycle. At the same time, the polarity of the induced emf reverses; hence the polarity of the split ring (which is the same as the polarity of the output voltage) remains the same.

A pulsating dc current is not suitable for most applications. To obtain a more steady dc current, commercial dc generators use many armature coils

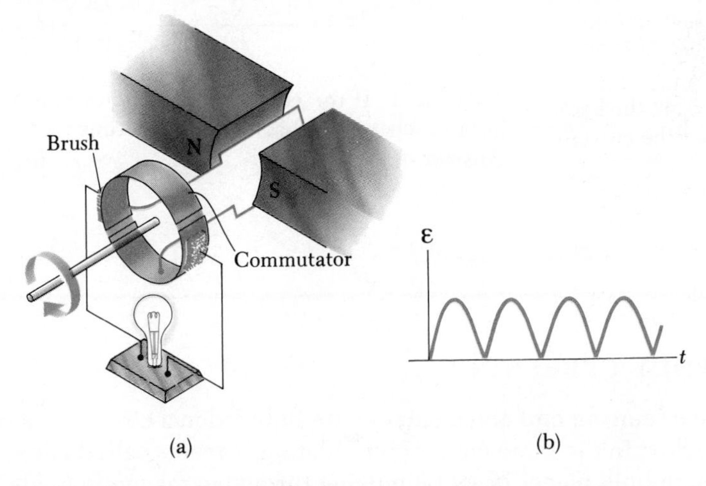


Figure 31.17 (a) Schematic diagram of a dc generator. (b) The emf versus time fluctuates in magnitude but always has the same polarity.

and commutators distributed so that the sinusoidal pulses from the various coils are out of phase. When these pulses are superimposed, the dc output is almost free of fluctuations.

**Motors** are devices that convert electrical energy into mechanical energy. Essentially, *a motor is a generator operating in reverse.* Instead of generating a current by rotating a loop, a current is supplied to the loop by a battery and the torque acting on the current-carrying loop causes it to rotate.

Useful mechanical work can be done by attaching the rotating armature to some external device. However, as the loop rotates, the changing magnetic flux induces an emf in the loop; this induced emf *always* acts to reduce the current in the loop. If this were not the case, Lenz's law would be violated. The back emf increases in magnitude as the rotational speed of the armature increases. (The phrase *back emf* is used to indicate an emf that tends to reduce the supplied current.) Since the voltage available to supply current equals the difference between the supply voltage and the back emf, the current through the armature coil is limited by the back emf.

When a motor is first turned on, there is initially no back emf and the current is very large because it is limited only by the resistance of the coil. As the coils begin to rotate, the induced back emf opposes the applied voltage and the current in the coils is reduced. If the mechanical load increases, the motor will slow down, which causes the back emf to decrease. This reduction in the back emf increases the current in the coils and therefore also increases the power needed from the external voltage source. For this reason, the power requirements are greater for starting a motor and for running it under heavy loads. If the motor is allowed to run under no mechanical load, the back emf reduces the current to a value just large enough to overcome energy losses due to heat and friction.

**EXAMPLE 31.9 The Induced Current in a Motor**
Assume that a motor having coils with a resistance of 10 Ω is supplied by a voltage of 120 V. When the motor is running at its maxiumum speed, the back emf is 70 V. Find the current in the coils (a) when the motor is first turned on and (b) when the motor has reached maximum speed.

*Solution* (a) When the motor is first turned on, the back emf is zero. (The coils are motionless.) Thus the current in the coils is a maximum and equal to

$$I = \frac{\mathcal{E}}{R} = \frac{120\ \text{V}}{10\ \Omega} = 12\ \text{A}$$

(b) At the maximum speed, the back emf has its maximum value. Thus, the effective supply voltage is now that of the external source minus the back emf. Hence, the current is reduced to

$$I = \frac{\mathcal{E} - \mathcal{E}_{\text{back}}}{R} = \frac{120\ \text{V} - 70\ \text{V}}{10\ \Omega} = \frac{50\ \text{V}}{10\ \Omega} = 5\ \text{A}$$

Exercise 3 If the current in the motor is 8 A at some instant, what is the back emf at this time?
Answer 40 V.

## *31.6 EDDY CURRENTS

As we have seen, an emf and a current are induced in a circuit by a changing magnetic flux. In the same manner, circulating currents called **eddy currents** are set up in bulk pieces of metal moving through a magnetic field. This can easily be demonstrated by allowing a flat metal plate at the end of a rigid bar to swing as a pendulum through a magnetic field (Fig. 31.18). The metal should be a material such as aluminum or copper. As the plate enters the field, the changing flux creates an induced emf in the plate, which in turn causes the free electrons in the metal to move, producing the swirling eddy currents. According to Lenz's law, the direction of the eddy currents must oppose the change that causes them. For this reason, the eddy currents must produce effective magnetic poles on the plate, which are repelled by the poles of the magnet, thus giving rise to a repulsive force that opposes the motion of the pendulum. (If the opposite were true, the pendulum would accelerate and its energy would increase after each swing, in violation of the law of energy conservation.) Alternatively, the retarding force can be "felt" by pulling a metal sheet through the field of a strong magnet.

As indicated in Figure 31.19, with $\boldsymbol{B}$ into the paper the eddy current is counterclockwise as the swinging plate *enters* the field in position 1. This is because the external flux into the paper is increasing, and hence by Lenz's law the induced current must provide a flux out of the paper. The opposite is true as the plate leaves the field in position 2, where the current is clockwise. Since the induced eddy current always produces a retarding force $\boldsymbol{F}$ when the plate enters or leaves the field, the swinging plate eventually comes to rest.

If slots are cut in the metal plate as in Figure 31.20, the eddy currents and the corresponding retarding force are greatly reduced. This can be understood since the cuts in the plate result in open circuits for any large current loops that might otherwise be formed.

The braking systems on many subway and rapid transit cars make use of electromagnetic induction and eddy currents. An electromagnet, which can be energized with a current, is positioned near the steel rails. The braking action occurs when a large current is passed through the electromagnet. The relative motion of the magnet and rails induces eddy currents in the rails, and the direction of these currents produces a drag force on the moving vehicle. The loss in mechanical energy of the vehicle is transformed into joule heat. Since the eddy currents decrease steadily in magnitude as the vehicle slows down, the braking effect is quite smooth. Eddy current brakes are also used in some mechanical balances and in various machines.

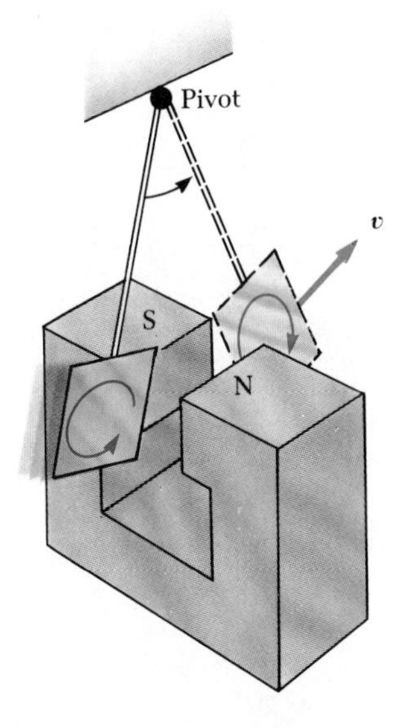


**Figure 31.18** An apparatus that demonstrates the formation of eddy currents in a conductor moving through a magnetic field. As the plate enters or leaves the field, the changing flux sets up an induced emf, which causes the eddy currents.

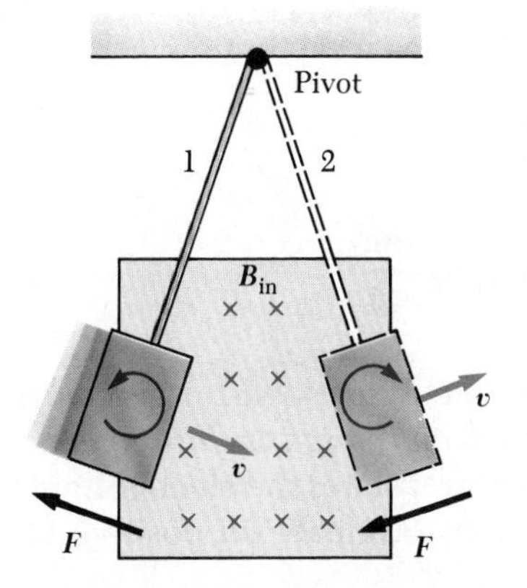


**Figure 31.19** As the conducting plate enters the field in position 1, the eddy currents are counterclockwise. However, in position 2, the currents are clockwise. In either case, the plate is repelled by the magnet and eventually comes to rest.

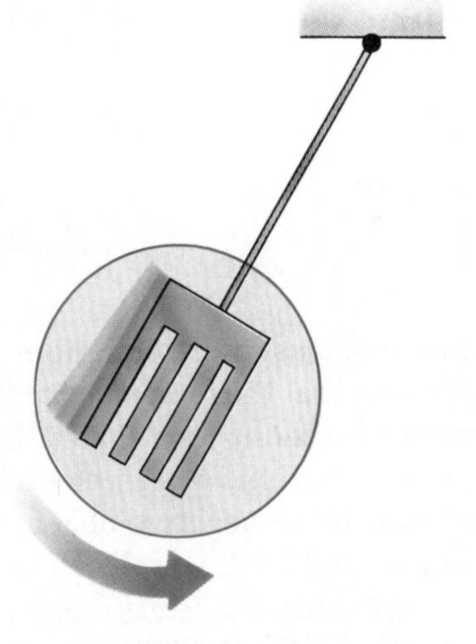

**Figure 31.20** When slots are cut in the conducting plate, the eddy currents are reduced and the plate swings more freely through the magnetic field.

Eddy currents are often undesirable since they dissipate energy in the form of heat. To reduce this energy loss, the moving conducting parts are often laminated, that is, built up in thin layers separated by a nonconducting material such as lacquer or a metal oxide. This layered structure increases the resistance of the possible paths of the eddy currents and effectively confines the currents to individual layers. Such a laminated structure is used in the cores of transformers and motors to minimize eddy currents and thereby increase the efficiency of these devices.

## 31.7 MAXWELL'S WONDERFUL EQUATIONS

We conclude this chapter by presenting four equations that can be regarded as the basis of all electrical and magnetic phenomena. These equations, known as Maxwell's equations, after James Clerk Maxwell, are as fundamental to electromagnetic phenomena as Newton's laws are to the study of mechanical phenomena. In fact, the theory developed by Maxwell was more far-reaching than even he imagined at the time, because it turned out to be in agreement with the special theory of relativity, as Einstein showed in 1905. As we shall see, Maxwell's equations represent laws of electricity and magnetism that have already been discussed. However, the equations have additional important consequences. In Chapter 34 we shall show that these equations predict the existence of electromagnetic waves (traveling patterns of electric and magnetic fields), which travel with a speed $c = 1/\sqrt{\mu_0 \epsilon_0} \approx 3 \times 10^8$ m/s, the speed of light. Furthermore, the theory shows that such waves are radiated by accelerating charges.

For simplicity, we present **Maxwell's equations** as applied to *free space*, that is, in the absence of any dielectric or magnetic material. The four equations are:

Gauss' law

$$\oint \mathbf{E} \cdot d\mathbf{A} = \frac{Q}{\epsilon_0} \tag{31.12}$$

Gauss' law in magnetism

$$\oint \mathbf{B} \cdot d\mathbf{A} = 0 \tag{31.13}$$

Faraday's law

$$\oint \mathbf{E} \cdot d\mathbf{s} = -\frac{d\Phi_m}{dt} \tag{31.14}$$

Ampère-Maxwell law

$$\oint \mathbf{B} \cdot d\mathbf{s} = \mu_0 I + \epsilon_0 \mu_0 \frac{d\Phi_e}{dt} \tag{31.15}$$

Let us discuss these equations one at a time. Equation 31.12 is *Gauss' law*, which states that the *total electric flux through any closed surface equals the net charge inside that surface divided by* $\epsilon_0$. This law relates the electric field to the charge distribution, where electric field lines originate on positive charges and terminate on negative charges.

Equation 31.13, which can be considered *Gauss' law in magnetism*, says that *the net magnetic flux through a closed surface is zero*. That is, the number of magnetic field lines that enter a closed volume must equal the number that leave that volume. This implies that magnetic field lines cannot begin or end at any point. If they did, this would mean that isolated magnetic monopoles existed at those points. The fact that isolated magnetic monopoles have not been observed in nature can be taken as a confirmation of Equation 31.13.

Equation 31.14 is *Faraday's law of induction,* which describes the relationship between an electric field and a changing magnetic flux. This law states that *the line integral of the electric field around any closed path (which equals the emf) equals the rate of change of magnetic flux through any surface area bounded by that path.* One consequence of Faraday's law is the current induced in a conducting loop placed in a time-varying magnetic field.

Equation 31.15 is the generalized form of Ampère's law, which describes a relationship between magnetic and electric fields and electric currents. That is, *the line integral of the magnetic field around any closed path is determined by the sum of the net conduction current through that path and the rate of change of electric flux through any surface bounded by that path.*

Once the electric and magnetic fields are known at some point in space, the force on a particle of charge $q$ can be calculated from the expression

$$\mathbf{F} = q\mathbf{E} + q\mathbf{v} \times \mathbf{B} \tag{31.16}$$

The Lorentz force

This is called the **Lorentz force.** Maxwell's equations, together with this force law, give a complete description of all classical electromagnetic interactions.

It is interesting to note the symmetry of Maxwell's equations. Equations 31.12 and 31.13 are symmetric, apart from the absence of a magnetic monopole term in Equation 31.13. Furthermore, Equations 31.14 and 31.15 are symmetric in that the line integrals of $\mathbf{E}$ and $\mathbf{B}$ around a closed path are related to the rate of change of magnetic flux and electric flux, respectively. "Maxwell's wonderful equations," as they were called by John R. Pierce,[3] are of fundamental importance not only to electronics but to all of science. Heinrich Hertz once wrote, "One cannot escape the feeling that these mathematical formulas have an independent existence and an intelligence of their own, that they are wiser than we are, wiser even than their discoverers, that we get more out of them than we put into them."

[3] John R. Pierce, *Electrons and Waves,* New York, Doubleday Science Study Series, 1964. Chapter 6 of this interesting book is recommended as supplemental reading.

## SUMMARY

**Faraday's law of induction** states that the emf induced in a circuit is directly proportional to the time rate of change of magnetic flux through the circuit. That is,

$$\mathcal{E} = -\frac{d\Phi_m}{dt} \tag{31.1}$$

Faraday's law

where $\Phi_m$ is the magnetic flux, given by

$$\Phi_m = \int \mathbf{B} \cdot d\mathbf{A}$$

When a conducting bar of length $\ell$ moves through a magnetic field $\boldsymbol{B}$ with a speed $v$ such that $\boldsymbol{B}$ is perpendicular to the bar, the emf induced in the bar (the so-called **motional emf**) is given by

Motional emf

$$\mathcal{E} = -B\ell v \tag{31.5}$$

**Lenz's law states that the induced current and induced emf in a conductor are in such a direction as to oppose the change that produced them.**

A general form of **Faraday's law of induction** is

Faraday's law in general form

$$\mathcal{E} = \oint \boldsymbol{E} \cdot d\boldsymbol{s} = -\frac{d\Phi_{\mathrm{m}}}{dt} \tag{31.9}$$

where $\boldsymbol{E}$ is a nonconservative, time-varying electric field that is produced by the changing magnetic flux.

When used with the Lorentz force law, $\boldsymbol{F} = q\boldsymbol{E} + q\boldsymbol{v} \times \boldsymbol{B}$, **Maxwell's equations,** given below in integral form, describe *all* electromagnetic phenomena:

Gauss' law (electricity)

$$\oint \boldsymbol{E} \cdot d\boldsymbol{A} = \frac{Q}{\epsilon_0} \tag{31.12}$$

Gauss' law (magnetism)

$$\oint \boldsymbol{B} \cdot d\boldsymbol{A} = 0 \tag{31.13}$$

Faraday's law

$$\oint \boldsymbol{E} \cdot d\boldsymbol{s} = -\frac{d\Phi_{\mathrm{m}}}{dt} \tag{31.14}$$

Ampère-Maxwell law

$$\oint \boldsymbol{B} \cdot d\boldsymbol{s} = \mu_0 I + \mu_0 \epsilon_0 \frac{d\Phi_{\mathrm{e}}}{dt} \tag{31.15}$$

The last two equations are of particular importance for the material discussed in this chapter. Faraday's law describes how an electric field can be induced by a changing magnetic flux. Similarly, the Ampère-Maxwell law describes how a magnetic field can be produced by both a conduction current and a changing electric flux.

## QUESTIONS

1. What is the difference between magnetic flux and magnetic field?
2. A circular loop is located in a uniform and constant magnetic field. Describe how an emf can be induced in the loop in this situation.
3. A loop of wire is placed in a uniform magnetic field. For what orientation of the loop is the magnetic flux a maximum? For what orientation is the flux zero?
4. As the conducting bar in Figure 31.21 moves to the right, an electric field is set up directed downward. If the bar were moving to the left, explain why the electric field would be upward.
5. As the bar in Figure 31.21 moves perpendicular to the field, is an external force required to keep it moving with constant velocity?

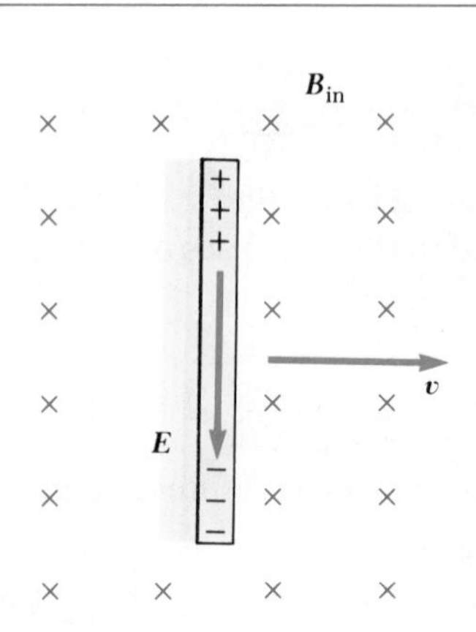


**Figure 31.21** (Questions 4 and 5).

6. The bar in Figure 31.22 moves on rails to the right with a velocity $\boldsymbol{v}$, and the uniform, constant magnetic field is *outward*. Why is the induced current clockwise? If the bar were moving to the left, what would be the direction of the induced current?

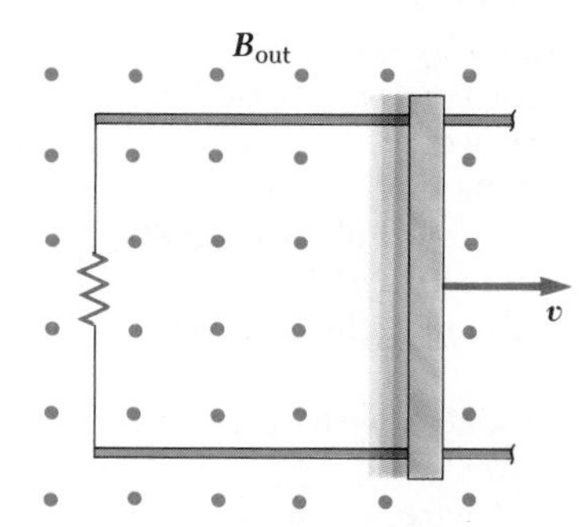


**Figure 31.22** (Questions 6 and 7).

7. Explain why an external force is necessary to keep the bar in Figure 31.22 moving with a constant velocity.
8. A large circular loop of wire lies in the horizontal plane. A bar magnet is dropped through the loop. If the axis of the magnet remains horizontal as it falls, describe the emf induced in the loop. How is the situation altered if the axis of the magnet remains vertical as it falls?
9. When a small magnet is moved toward a solenoid, an emf is induced in the coil. However, if the magnet is moved around inside a toroidal coil, there is *no* induced emf. Explain.
10. Will dropping a magnet down a long copper tube produce a current in the tube? Explain.
11. How is electrical energy produced in dams (that is, how is the energy of motion of the water converted into ac electricity)?
12. In a beam balance scale, an aluminum plate is sometimes used to slow the oscillations of the beam near equilibrium. The plate is mounted at the end of the beam, and moves between the poles of a small horseshoe magnet, attached to the frame. Why are the oscillations of the beam strongly damped near equilibrium?
13. What happens when the coil of a generator is rotated at a faster rate?
14. Could a current be induced in a coil by rotating a magnet inside the coil? If so, how?
15. When the switch in the circuit shown in Figure 31.23a is closed, a current is set up in the coil and the metal ring springs upward (see Fig. 31.23b). Explain this behavior.

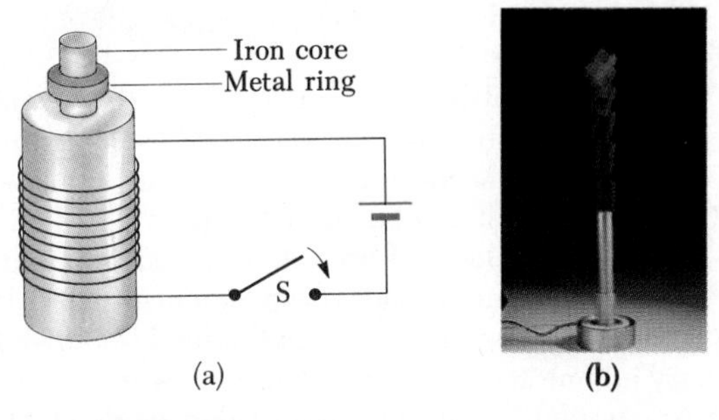


**Figure 31.23** (Questions 15 and 16). (Courtesy of CENCO)

16. Assume that the battery in Figure 31.23a is replaced by an alternating current source and the switch S is held closed. If the metal ring on top of the solenoid is held down, it will become *hot*. Why?
17. Identify the individual generally associated with each of Maxwell's four equations.
18. Do Maxwell's equations (Section 31.7) allow for the existence of magnetic "charges," that is, isolated N or S poles?

## PROBLEMS

### Section 31.1 Faraday's Law of Induction

1. A 50-turn rectangular coil of dimensions 5 cm × 10 cm is "dropped" from a position where $B = 0$ to a new position where $B = 0.5$ T and is directed perpendicular to the plane of the coil. Calculate the resulting average emf induced in the coil if the displacement occurs in 0.25 s.
2. A plane loop of wire consisting of a single turn of cross-sectional area 8.0 cm$^2$ is perpendicular to a magnetic field that increases uniformly in magnitude from 0.5 T to 2.5 T in a time of 1.0 s. What is the resulting induced current if the coil has a total resistance of 2 Ω?
3. A powerful electromagnet has a field of 1.6 T and a cross sectional area of 0.2 m$^2$. If we place a coil of 200 turns with a total resistance of 20 Ω around the electromagnet, and then turn off the power to the electromagnet in 0.02 s, what will be the induced current in the coil?
4. A square, single-turn coil 0.20 m on a side is placed with its plane perpendicular to a constant magnetic field. An emf of 18 mV is induced in the winding when the area of the coil decreases at a rate of 0.1 m$^2$/s. What is the magnitude of the magnetic field?
5. The plane of a rectangular coil of dimensions 5 cm by 8 cm is perpendicular to the direction of a magnetic field $\boldsymbol{B}$. If the coil has 75 turns and a total resistance of 8 Ω, at what rate must the magnitude of $\boldsymbol{B}$ change in order to induce a current of 0.1 A in the windings of the coil?

6. A tightly wound circular coil has 50 turns, each of radius 0.1 m. A uniform magnetic field is turned on along a direction perpendicular to the plane of the coil. If the field increases linearly from 0 to 0.6 T in a time of 0.2 s, what emf is induced in the windings of the coil?
7. A 30-turn circular coil of radius 4 cm and resistance 1 Ω is placed in a magnetic field directed perpendicular to the plane of the coil. The magnitude of the magnetic field varies in time according to the expression $B = 0.01t + 0.04t^2$, where $t$ is in s and $B$ is in T. Calculate the induced emf in the coil at $t = 5$ s.
8. A plane loop of wire of 10 turns, each of area 14 cm$^2$, is perpendicular to a magnetic field whose magnitude changes in time according to $B = (0.5\text{ T})\sin(60\pi t)$. What is the induced emf in the loop as a function of time?
9. A plane loop of wire of area 14 cm$^2$ with 2 turns is perpendicular to a magnetic field whose magnitude decays in time according to $B = (0.5\text{ T})e^{-t/7}$. What is the induced emf as a function of time?
10. A rectangular loop of area $A$ is placed in a region where the magnetic field is perpendicular to the plane of the loop. The magnitude of the field is allowed to vary in time according to $B = B_0 e^{-t/\tau}$, where $B_0$ and $\tau$ are constants. The field has a value of $B_0$ at $t \leq 0$. (a) Use Faraday's law to show that the emf induced in the loop is given by

$$\mathcal{E} = \frac{AB_0}{\tau} e^{-t/\tau}$$

(b) Obtain a numerical value for $\mathcal{E}$ at $t = 4$ s when $A = 0.16$ m$^2$, $B_0 = 0.35$ T, and $\tau = 2$ s. (c) For the values of $A$, $B_0$, and $\tau$ given in (b), what is the *maximum* value of $\mathcal{E}$?
11. A long solenoid has $n$ turns per meter and carries a current $I = I_0(1 - e^{-\alpha t})$, with $I_0 = 30$ A and $\alpha = 1.6\text{ s}^{-1}$. Inside the solenoid and coaxial with it is a loop that has a radius $R = 6$ cm and consists of a total of $N$ turns of fine wire. What emf is induced in the loop by the changing current? Take $n = 400$ turns/m and $N = 250$ turns. (See Fig. 31.24.)
12. A magnetic field of 0.2 T exists within a solenoid of 500 turns and a diameter of 10 cm. How rapidly (that is, within what period of time) must the field be reduced to zero magnitude if the average magnitude of the induced emf within the coil during this time interval is to be 10 kV?
13. A coil, formed by wrapping 50 turns of wire in the shape of a square, is positioned in a magnetic field so that the normal to the plane of the coil makes an angle of 30° with the direction of the field. It is observed that if the magnitude of the magnetic field is increased uniformly from 200 μT to 600 μT in 0.4 s, an emf of 80 mV is induced in the coil. What is the total length of the wire?
14. A long straight wire carries a current $I = I_0 \sin(\omega t + \delta)$ and lies in the plane of a rectangular loop of $N$ turns of wire, as shown in Figure 31.25. The quantities $I_0$, $\omega$, and $\delta$ are all constants. Determine the emf induced in the loop by the magnetic field due to the current in the straight wire. Assume $I_0 = 50$ A, $\omega = 200\pi\text{ s}^{-1}$, $N = 100$, $a = b = 5$ cm, and $\ell = 20$ cm.

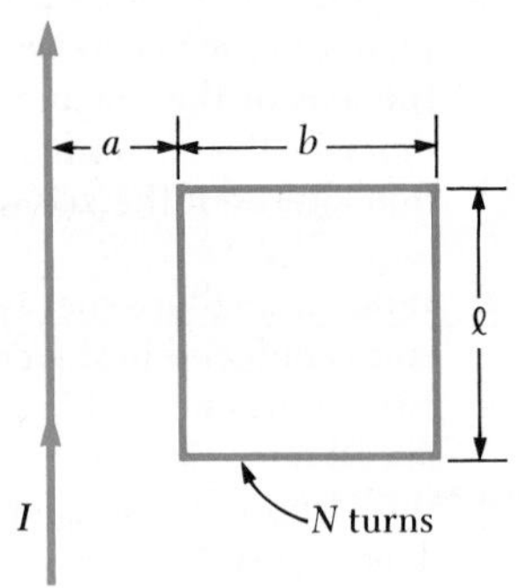


**Figure 31.25** (Problem 14).

15. A circular loop with a radius $R$ consists of $N$ tight turns of wire and is penetrated by an external magnetic field directed perpendicular to the plane of the loop. The magnitude of the field in the plane of the loop is $B = B_0(1 - r/2R)\cos\omega t$, where $R$ is the radius of the loop, and where $r$ is measured from the center of the loop, as shown in Figure 31.26. Determine the induced emf in the loop.

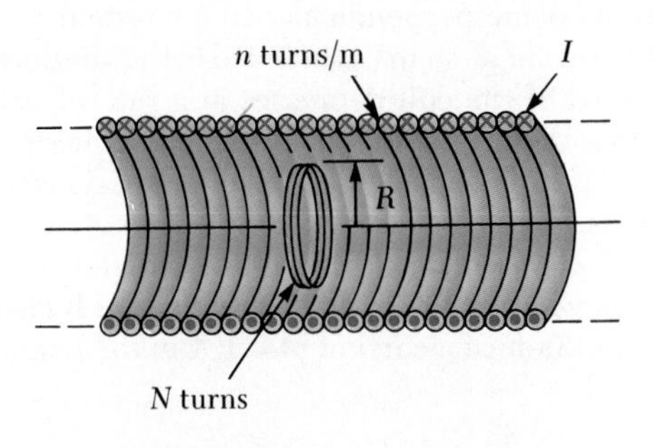


**Figure 31.24** (Problem 11).

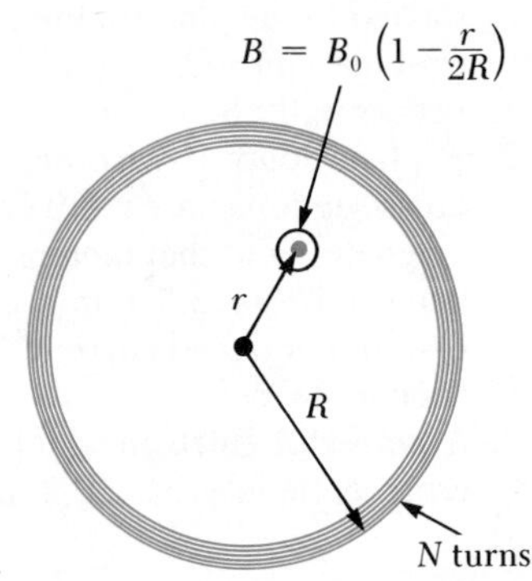


**Figure 31.26** (Problem 15).

16. A toroid with a rectangular cross section ($a = 2$ cm by $b = 3$ cm) and inner radius $R = 4$ cm consists of 500 turns of wire that carries a current $I = I_0 \sin \omega t$, with $I_0 = 50$ A and a frequency $f = \omega/2\pi = 60$ Hz. A loop that consists of 20 turns of wire links the toroid, as shown in Figure 31.27. Determine the emf induced in the loop by the changing current $I$.

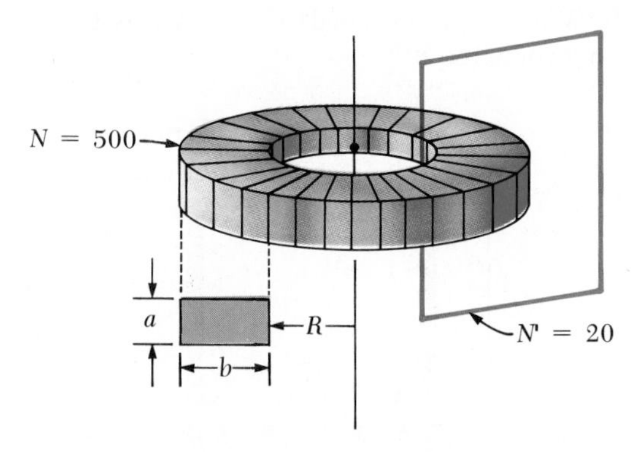


**Figure 31.27** (Problem 16).

## Section 31.2 Motional emf and Section 31.3 Lenz's Law

17. A measured average emf of 30 mV is induced in a small circular coil of 600 turns and 4 cm diameter under the following condition: It is rotated in a uniform magnetic field in 0.05 s from a position where the plane of the coil is parallel to the field to a position where the plane of the coil is at an angle of 45° to the field. What is the value of $B$ within the region where the measurement is made?

18. Consider the arrangement shown in Figure 31.28. Assume that $R = 6\ \Omega$, $\ell = 1.2$ m, and that a uniform 2.5-T magnetic field is directed *into* the page. At what speed should the bar be moved to produce a current of 0.5 A in the resistor?

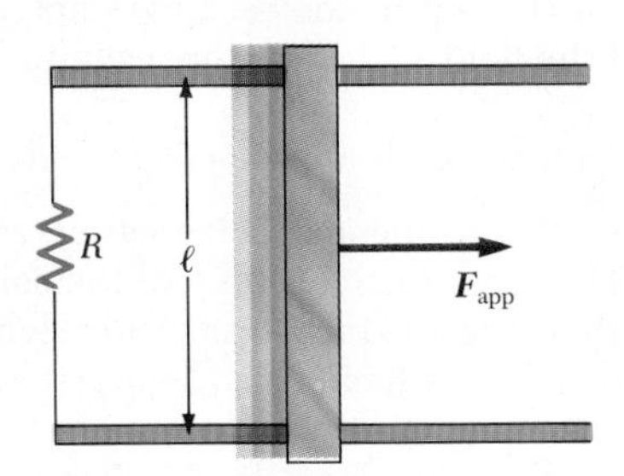


**Figure 31.28** (Problems 18, 19, and 20).

19. In the arrangement shown in Figure 31.28, a conducting bar moves to the right along parallel, frictionless conducting rails connected on one end by a 6-Ω resistor. A 2.5-T magnetic field is directed *into* the paper. Let $\ell = 1.2$ m and neglect the mass of the bar. (a) Calculate the applied force required to move the bar to the right at a *constant* speed of 2 m/s. (b) At what rate is energy dissipated in the resistor?

20. A conducting rod of length $\ell$ moves on two horizontal frictionless rails as shown in Figure 31.28. If a constant force of 1 N moves the bar at 2 m/s through a magnetic field $\boldsymbol{B}$ which is into the paper, (a) what is the current through an 8-Ω resistor $R$? (b) What is the rate of energy dissipation in the resistor? (c) What is the mechanical power delivered by the force $\boldsymbol{F}$?

21. Over a region where the *vertical* component of the earth's magnetic field is 40 $\mu$T directed downward, a 5-m length of wire is held along an east-west direction and moved horizontally to the north with a speed of 10 m/s. Calculate the potential difference between the ends of the wire and determine which end is positive.

22. A small airplane with a wing span of 14 m is flying due north at a speed of 70 m/s over a region where the vertical component of the earth's magnetic field is 1.2 $\mu$T. (a) What potential difference is developed between the wing tips? (b) How would the answer to (a) change if the plane were flying due east?

23. A helicopter has blades of length 3 m, rotating at 2 rev/s about a central hub. If the vertical component of the earth's magnetic field is $0.5 \times 10^{-4}$ T, what is the emf induced between the blade tip and the center hub?

24. A metal blade spins at a constant rate in the magnetic field of the earth as in Figure 31.7. The rotation occurs in a region where the component of the earth's magnetic field perpendicular to the plane of rotation is $3.3 \times 10^{-5}$ T. If the blade is 1.0 m in length and its angular velocity is $5\pi$ rad/s, what potential difference is developed between its ends?

25. A rigid thin conducting rod of length $L$ is mechanically rotated at constant angular velocity $\omega$ about an axis perpendicular to the rod and *through its center*. If a uniform magnetic field exists *parallel* to the rotation axis of the rod, (a) show that the induced emf between the center of the rod and one of its ends is proportional to $L^2$. (b) Evaluate the magnitude of this emf for $L = 0.2$ m, $\omega = 60$ rad/s, and $|\boldsymbol{B}| = 1.2$ T.

26. A 200-turn circular coil of radius 10 cm is located in a uniform magnetic field of 0.8 T such that the plane of the coil is perpendicular to the direction of the field. The coil is rotated at a constant rate (uniform angular velocity) through 90° in a time of 1.5 s, so that the plane of the coil is finally parallel to the direction of the field. (a) Calculate the *average* emf induced in the coil as a result of the rotation. (b) What is the instantaneous value of the emf in the coil at the moment the plane of the coil makes an angle of 45° with the magnetic field?

27. Use Lenz's law to answer the following questions concerning the direction of induced currents. (a) What is the direction of the induced current in resistor $R$ in

Figure 31.29a when the bar magnet is moved to the left? (b) What is the direction of the current induced in the resistor $R$ right after the switch S in the circuit of Figure 31.29b is closed? (c) What is the direction of the induced current in $R$ when the current $I$ in Figure 31.29c decreases rapidly to zero? (d) A copper bar is moved to the right while its axis is maintained perpendicular to a magnetic field, as in Figure 31.29d. If the top of the bar becomes positive relative to the bottom, what is the direction of the magnetic field?

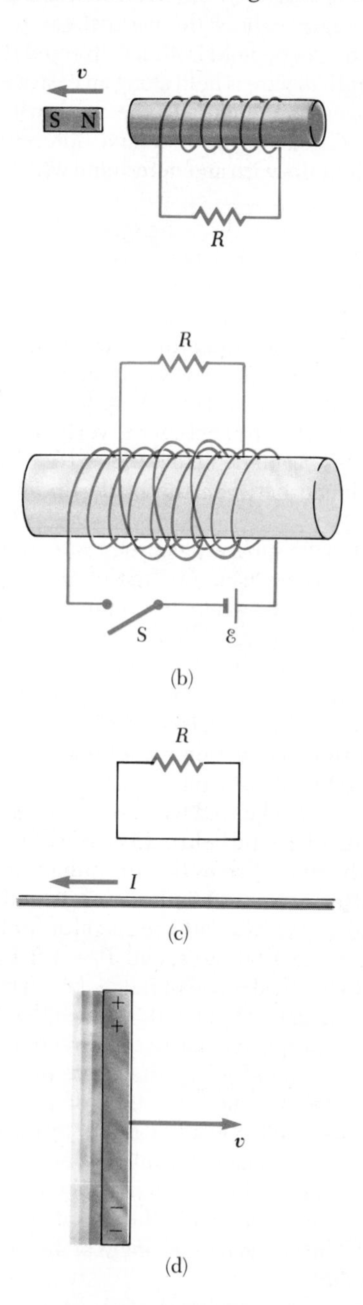


**Figure 31.29** (Problem 27).

28. A conducting rectangular loop of mass $M$, resistance $R$, and dimensions of $w$ wide by $\ell$ long falls from rest into a magnetic field $\mathbf{B}$ as shown in Figure 31.30. The loop accelerates until it reaches terminal speed, $v_t$. (a) Show that

$$v_t = \frac{MgR}{B^2w^2}$$

(b) Why is $v_t$ proportional to $R$? (c) Why is it inversely proportional to $B^2$?

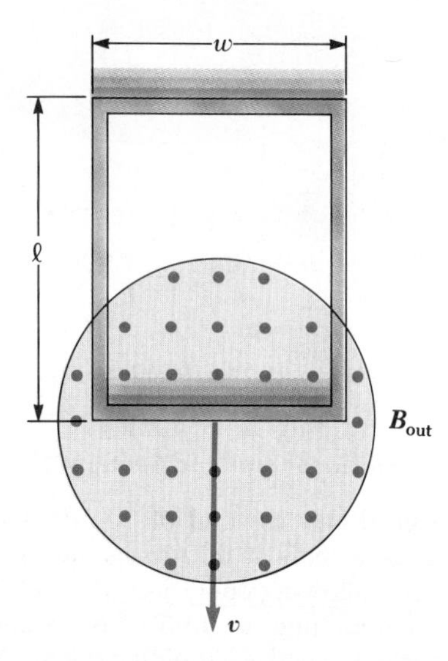


**Figure 31.30** (Problems 28 and 29).

29. A 0.15-kg wire in the shape of a closed rectangle 1 m wide and 1.5 m long has a total resistance of 0.75 Ω. The rectangle is allowed to fall through a magnetic field directed perpendicular to the direction of motion of the wire (Fig. 31.30). The rectangle accelerates downward until it acquires a *constant* speed of 2 m/s with the top of the rectangle not yet in that region of the field. Calculate the magnitude of $B$.

### Section 31.4 Induced emfs and Electric Fields

**30.** The current in a solenoid is increasing at a rate of 10 A/s. The cross-sectional area of the solenoid is $\pi$ cm$^2$ and there are 300 turns on its 15-cm length. What is the induced emf which acts to oppose the increasing current?

**31.** A single-turn, circular loop of radius $R$ is coaxial with a long solenoid of radius $r$ and length $\ell$ and having $N$ turns (Fig. 31.31). The variable resistor is changed so that the solenoid current decreases linearly from 7.2 A to 2.4 A in 0.3 s. If $r = 0.03$ m, $\ell = 0.75$ m, and $N = 1500$ turns, calculate the induced emf in the circular loop.

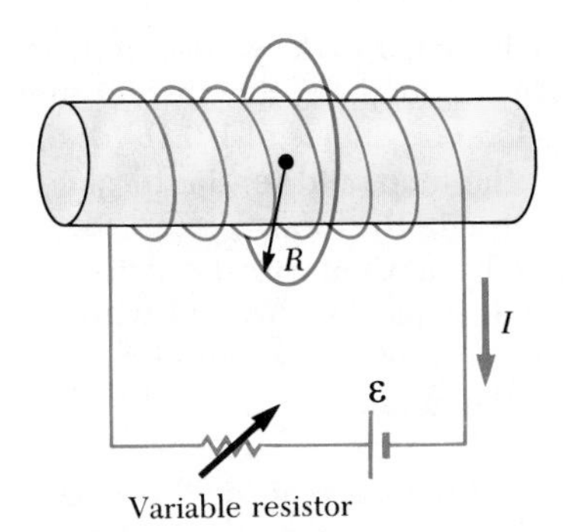


**Figure 31.31** (Problem 31).

32. A coil of 15 turns and radius 10 cm surrounds a long solenoid of radius 2 cm and $10^3$ turns/meter. If the current in the inner solenoid changes as $I = (5\text{ A})\sin(120t)$, what is the induced emf in the 15-turn coil?
33. A magnetic field directed into the page changes with time according to $B = (0.03t^2 + 1.4)$ T, where $t$ is in s. The field has a circular cross-section of radius $R =$ 2.5 cm (Fig. 31.32). What are the magnitude and direction of the electric field at point $P_1$ when $t = 3$ s and $r_1 = 0.02$ m?

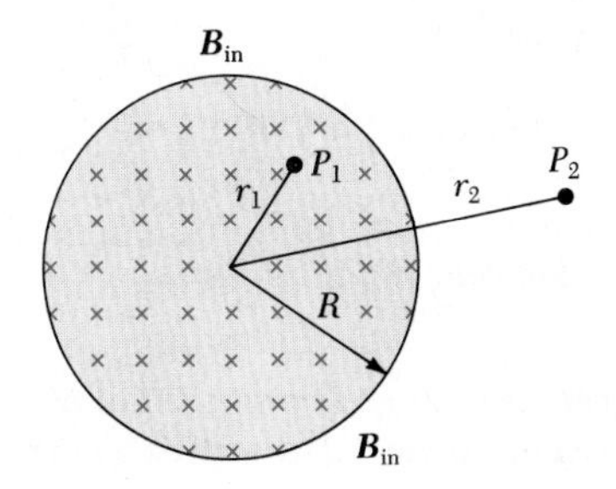


**Figure 31.32** (Problems 33 and 34).

34. For the situation described in Figure 31.32, the magnetic field varies as $B = (2t^3 - 4t^2 + 0.8)$ T, and $r_2 = 2R = 5$ cm. (a) Calculate the magnitude and direction of the force exerted on an electron located at point $P_2$ when $t = 2$ s. (b) At what time is the force equal to zero?
35. A long solenoid with 1000 turns/meter and radius 2 cm carries an oscillating current given by the expression $I = (5\text{ A})\sin(100\pi t)$. What is the electric field induced at a radius $r = 1$ cm from the axis of the solenoid? What is the direction of this electric field when the current is *increasing* counterclockwise in the coil?
36. A solenoid has a radius of 2 cm and has 1000 turns/m. The current varies with time according to the expression $I = 3e^{0.2t}$, where $I$ is in A and $t$ is in s. Calculate the electric field at a distance of 5 cm from the axis of the solenoid at $t = 10$ s.
37. An aluminum ring of radius 5 cm and resistance $3 \times 10^{-4}\ \Omega$ is placed on top of a long air-core solenoid with 1000 turns per meter and radius 3 cm as shown in Figure 31.33. At the location of the ring, the magnetic field due to the current in the solenoid is one-half that at the center of the solenoid. If the current in the solenoid is *increasing* at a rate of 270 A/s, (a) what is the induced current in the ring? (b) At the center of the ring, what is the magnetic field produced by the induced current in the ring? (c) What is the direction of the field in (b)?

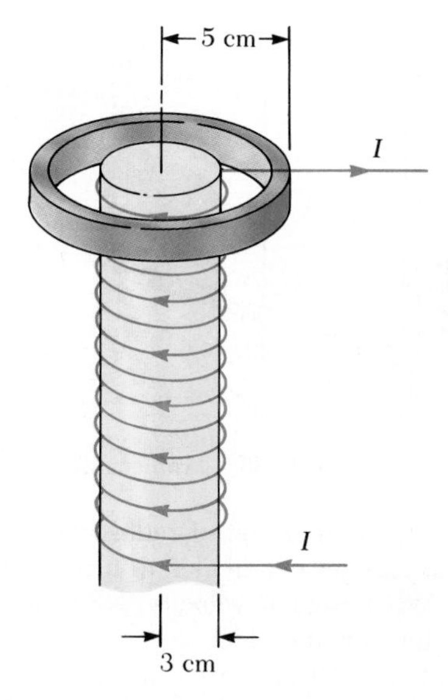


**Figure 31.33** (Problem 37).

38. A circular coil, enclosing an area of 100 cm², is made of 200 turns of copper wire as shown in Figure 31.34. Initially, a 1.1-T uniform magnetic field points perpendicularly *upward* through the plane of the coil. The direction of the field then reverses so that the final magnetic field has a magnitude of 1.1 T pointing *downward* through the coil. During the time the field is changing its direction, how much charge flows through the coil if the coil is connected to a 5-Ω resistor as shown?

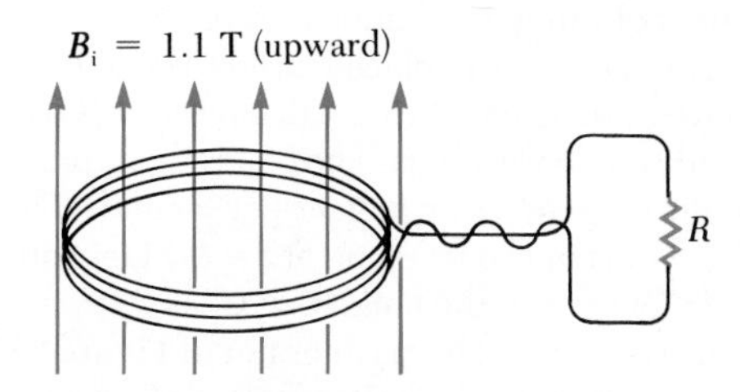


**Figure 31.34** (Problem 38).

°Section 31.5 Generators and Motors

**39.** A square coil (20 cm × 20 cm) that consists of 100 turns of wire rotates about a vertical axis at 1500 rpm, as indicated in Figure 31.35. The horizontal component of the earth's magnetic field at the location of the loop is $2 \times 10^{-5}$ T. Calculate the maximum emf induced in the coil by the earth's field.

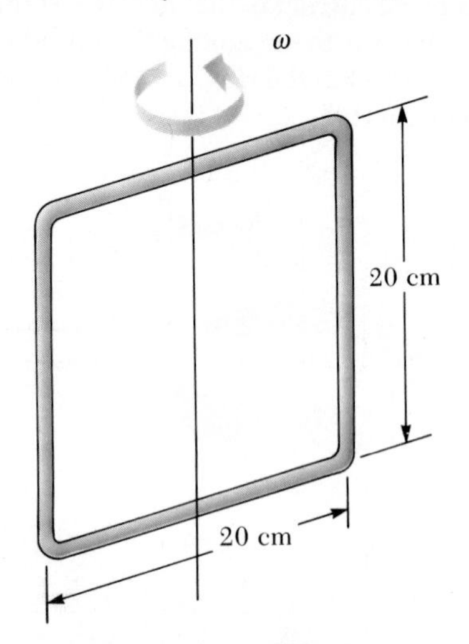


**Figure 31.35** (Problem 39).

**40.** A 400-turn circular coil of radius 15 cm is rotating about an axis perpendicular to a magnetic field of 0.02 T. What angular velocity will produce a maximum induced emf of 4 V?

41. A loop of area 0.1 m² is rotating at 60 rev/s with the axis of rotation perpendicular to a 0.2-T magnetic field. (a) If there are 1000 turns on the loop, what is the maximum voltage induced in the loop? (b) When the maximum induced voltage occurs, what is the orientation of the loop with respect to the magnetic field?

**42.** The coil of a simple ac generator develops a sinusoidal emf of maximum value 90.4 V and frequency 60 Hz. If the coil has dimensions of 10 cm by 20 cm and rotates in a magnetic field of 1.0 T, how many turns are in the winding?

43. A long solenoid, whose axis coincides with the $x$ axis, consists of 200 turns per meter of wire that carries a steady current of 15 A. A coil is formed by wrapping 30 turns of thin wire around a frame that has a radius of 8 cm. The coil is placed inside the solenoid and mounted on an axis that is a diameter of the coil and coincides with the $y$ axis. The coil is then rotated with an angular speed of $4\pi$ radians per second. (The plane of the coil is in the $yz$ plane at $t = 0$.) Determine the emf developed in the coil.

44. Let the rectangular loop generator of Figure 31.15 be a square 10 cm on a side. It is rotated at a frequency of 60 Hz in a uniform field of 0.80 T. Calculate (a) the flux through the loop as a function of time, (b) the emf induced in the loop, (c) the current induced in the loop for a loop resistance of 1.0 Ω, (d) the power dissipated in the loop, and (e) the torque that must be exerted to rotate the loop.

45. (a) What is the maximum torque delivered by an electric motor if it has 80 turns of wire wrapped on a rectangular coil, of dimensions 2.5 cm by 4 cm? Assume that the motor uses 10 A of current and that a uniform 0.8-T magnetic field exists within the motor. (b) If the motor rotates at 3600 rev/min, what is the (peak) power produced by the motor?

46. A semicircular conductor of radius $R = 0.25$ m is rotated about the axis $AC$ at a constant rate of 120 revolutions per minute (Fig. 31.36). A uniform magnetic field in all of the lower half of the figure is directed *out* of the plane of rotation and has a magnitude of 1.3 T. (a) Calculate the *maximum* value of the emf induced in the conductor. (b) What is the value of the *average* induced emf for each complete rotation? (c) How would the answers to (a) and (b) change if the uniform field $\boldsymbol{B}$ were allowed to extend a distance $R$ above the axis of rotation? (d) Sketch the emf versus time in each case.

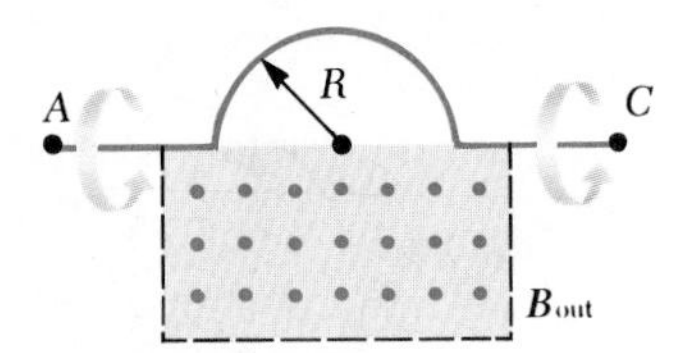


**Figure 31.36** (Problem 46).

47. A small rectangular coil composed of 50 turns of wire has an area of 30 cm², and carries a current of 1.5 A. When the plane of the coil makes an angle of 30° with a uniform magnetic field, the torque on the coil is 0.1 N · m. What is the strength of the magnetic field?

48. A bar magnet is spun at constant angular velocity $\omega$ about an axis as shown in Figure 31.37. A flat rectan-

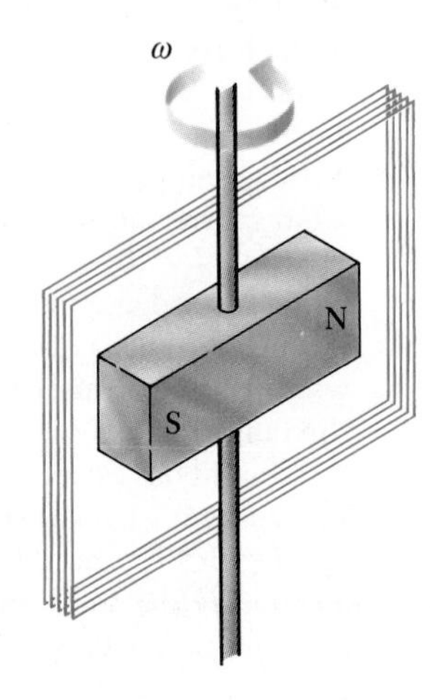


**Figure 31.37** (Problem 48).

gular conducting loop surrounds the magnet, and at $t = 0$, the magnet is oriented as shown. Sketch the induced current in the loop as a function of time, plotting counterclockwise currents as positive and clockwise currents as negative.

### *Section 31.6 Eddy Currents

49. A rectangular loop with resistance $R$ has $N$ turns, each of length $\ell$ and width $w$ as shown in Figure 31.38. The loop moves into a uniform magnetic field $\boldsymbol{B}$ with velocity $\boldsymbol{v}$. What is the magnitude and direction of the resultant force on the loop (a) as it enters the magnetic field? (b) as it moves within the magnetic field? (c) as it leaves the magnetic field?

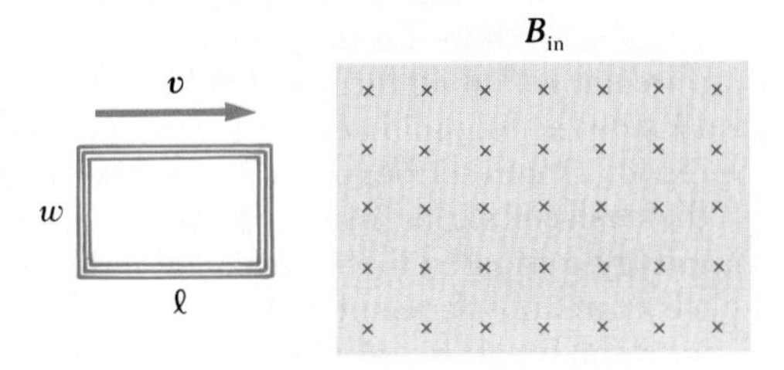


**Figure 31.38** (Problem 49).

50. A dime is suspended from a thread and hung between the poles of a strong horseshoe magnet as shown in Figure 31.39. The dime rotates at constant angular speed $\omega$ about a vertical axis. Let $\theta$ be the angle between the direction of $\boldsymbol{B}$ and the normal to the face of the dime, and sketch a graph of the torque due to induced currents as a function of $\theta$ for $0 \le \theta \le 2\pi$.

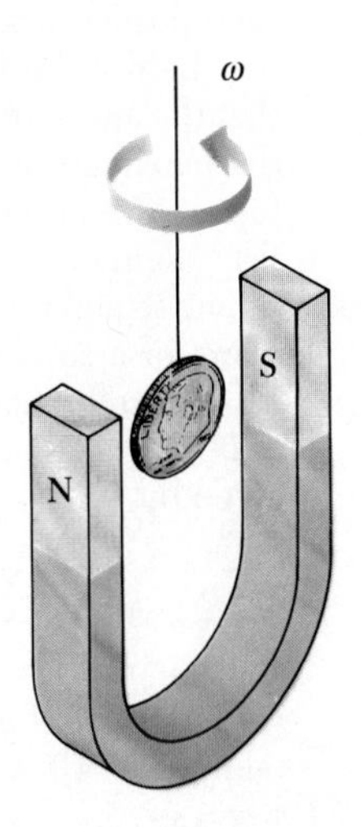


**Figure 31.39** (Problem 50).

### Section 31.7 Maxwell's Wonderful Equations

51. A proton moves through a uniform electric field given by $\boldsymbol{E} = 50\boldsymbol{j}$ V/m and a uniform magnetic field $\boldsymbol{B} = (0.2\boldsymbol{i} + 0.3\boldsymbol{j} + 0.4\boldsymbol{k})$ T. Determine the acceleration of the proton when it has a velocity of $\boldsymbol{v} = 200\boldsymbol{i}$ m/s.
52. An electron moves through a uniform electric field $\boldsymbol{E} = (2.5\boldsymbol{i} + 5.0\boldsymbol{j})$ V/m and a uniform magnetic field $\boldsymbol{B} = 0.4\boldsymbol{k}$ T. Determine the acceleration of the electron when it has a velocity of $\boldsymbol{v} = 10\boldsymbol{i}$ m/s.

## ADDITIONAL PROBLEMS

53. A conducting rod moves with a constant velocity $\boldsymbol{v}$ perpendicular to a long, straight wire carrying a current $I$ as in Figure 31.40. Show that the emf generated between the ends of the rod is given by

$$|\mathcal{E}| = \frac{\mu_0 v I}{2\pi r} \ell$$

In this case, note that the emf decreases with increasing $r$, as you might expect.

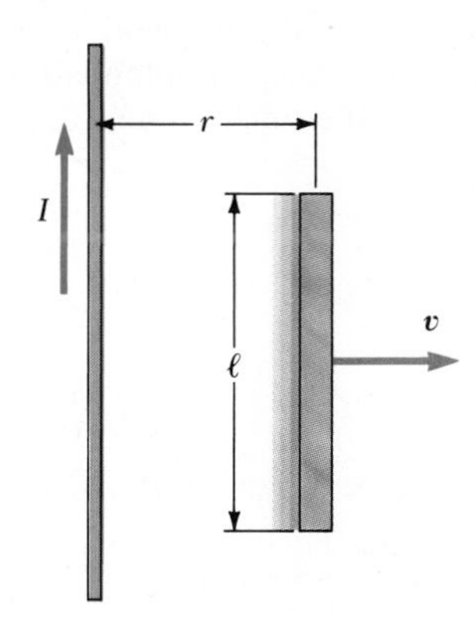


**Figure 31.40** (Problem 53).

54. A circular loop of wire 5 cm in radius is in a spatially uniform magnetic field, with the plane of the circular loop perpendicular to the direction of the field (Fig. 31.41). The magnetic field varies with time:

$$B(t) = a + bt \qquad a = 0.20 \text{ T} \qquad b = 0.32 \text{ T/s}$$

(a) Calculate the magnetic flux through the loop at $t = 0$. (b) Calculate the emf induced in the loop. (c) If the resistance of the loop is 1.2 Ω, what is the induced current? (d) At what rate is electric energy being dissipated in the loop?

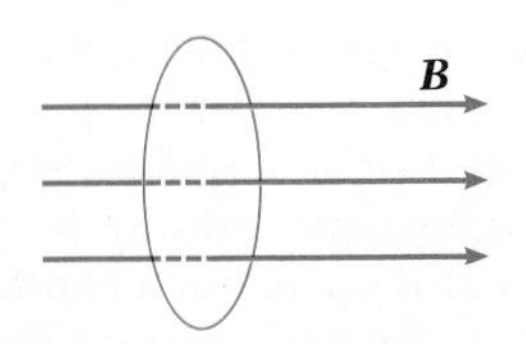


**Figure 31.41** (Problem 54).

55. Consider a long solenoid of length $\ell$ containing a core of permeability $\mu$. The core material is magnetized by increasing the current in the coil, so as to produce a changing field $dB/dt$. (a) Show that the rate at which work done against the induced emf in the coil is given by

$$\frac{dW}{dt} = I\mathcal{E} = HA\ell \frac{dB}{dt}$$

where $A$ is the area of the solenoid. (*Hint:* Use Faraday's law to find $\mathcal{E}$ and make use of Equation 30.34.) (b) Use the result of part (a) to show that the total work done in a complete hysteresis cycle equals the area enclosed by the $B$ versus $H$ curve (Fig. 30.31).

56. In Figure 31.42, a uniform magnetic field decreases at a constant rate $dB/dt = -K$, where $K$ is a positive constant. A circular loop of wire of radius $a$ containing a resistance $R$ and a capacitance $C$ is placed with its plane normal to the field. (a) Find the charge $Q$ on the capacitor when it is fully charged. (b) Which plate of the capacitor is at the higher potential? (c) Discuss the force that causes the separation of charges.

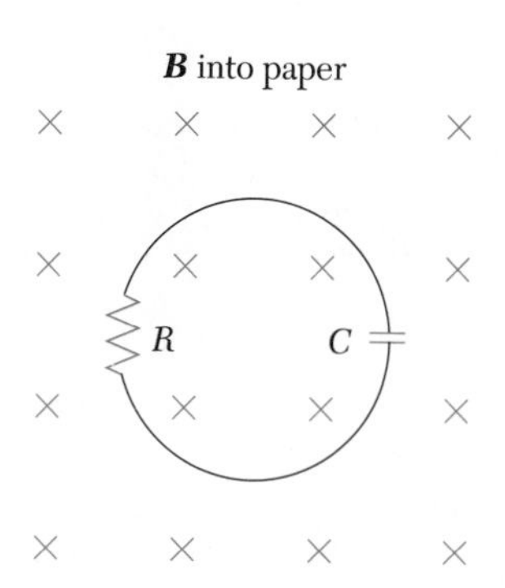


**Figure 31.42** (Problem 56).

57. A rectangular coil of $N$ turns, dimensions $\ell$ and $w$, and total resistance $R$ rotates with angular velocity $\omega$ about the $y$ axis in a region where a uniform magnetic field $\mathbf{B}$ is directed along the $x$ axis (Fig. 31.15a). The rotation is initiated so that the plane of the coil is perpendicular to the direction of $\mathbf{B}$ at $t = 0$. Let $\omega = 30$ rad/s, $N = 60$ turns, $\ell = 0.10$ m, $w = 0.20$ m, $R = 10\ \Omega$, and $B = 1.0$ T. Calculate (a) the maximum induced emf in the coil, (b) the maximum rate of change of magnetic flux through the coil, (c) the value of the induced emf at $t = 0.05$ s, and (d) the torque exerted on the loop by the magnetic field at the instant when the emf is a maximum.

58. A wire of mass $m$, length $d$, and resistance $R$ slides without friction on parallel rails as shown in Figure 31.43. A battery that maintains a constant emf $\mathcal{E}$ is connected between the rails, and a constant magnetic field $\mathbf{B}$ is directed perpendicular to the plane of the page. If the wire starts from rest, show that at time $t$ it moves with a speed

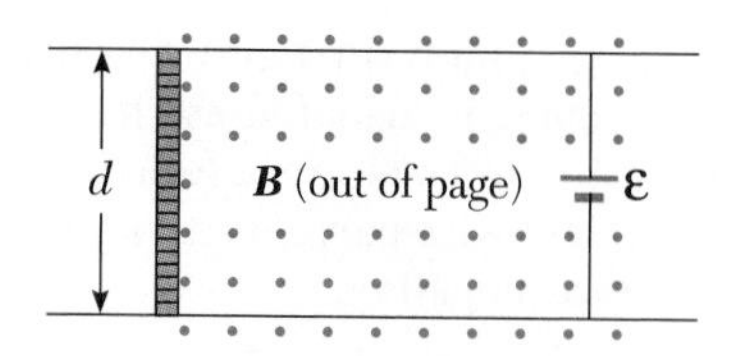


**Figure 31.43** (Problem 58).

$$v = \frac{\mathcal{E}}{Bd}(1 - e^{-B^2d^2t/mR})$$

59. A solenoid wound with 2000 turns/m is supplied with current that varies in time according to $I = 4\sin(120\pi t)$, where $I$ is in A and $t$ is in s. A small coaxial circular coil of 40 turns and radius $r = 5$ cm is located inside the solenoid near its center. (a) Derive an expression that describes the manner in which the emf in the small coil varies in time. (b) At what average rate is energy dissipated in the small coil if the windings have a total resistance of 8 Ω?

60. To monitor the breathing of a hospital patient, a thin belt is girded about the patient's chest. The belt is a 200-turn coil. During inhalation, the area within the coil increases by 39 cm$^2$. The earth's magnetic field is 50 $\mu$T and makes an angle of 28° with the plane of the coil. If a patient takes 1.80 s to inhale, find the average induced emf in the coil while the patient is inhaling.

61. An automobile has a vertical radio antenna 1.2 m long. The automobile travels at 65 km/h on a horizontal road where the earth's magnetic field is 50 $\mu$T directed downward (toward the north) at an angle of 65° below the horizontal. (a) Specify the direction that the automobile should move in order to generate the maximum motional emf in the antenna, with the top of the antenna positive relative to the bottom. (b) Calculate the magnitude of this induced emf.

62. A long straight wire is parallel to one edge and is in the plane of a single turn rectangular loop as in Figure 31.44. (a) If the current in the long wire varies in time as $I = I_0e^{-t/\tau}$, show that the induced emf in the loop is given by

$$\mathcal{E} = \frac{\mu_0 b}{2\pi}\frac{I}{\tau}\ln\left(1 + \frac{a}{d}\right)$$

(b) Calculate the value for the induced emf at $t = 5$ s taking $I_0 = 10$ A, $d = 3$ cm, $a = 6$ cm, $b = 15$ cm, and $\tau = 5$ s.

63. A conducting rod of length $\ell$ moves with velocity $\mathbf{v}$ along a direction parallel to a long wire carrying a steady current $I$. The axis of the rod is maintained perpendicular to the wire with the near end a distance

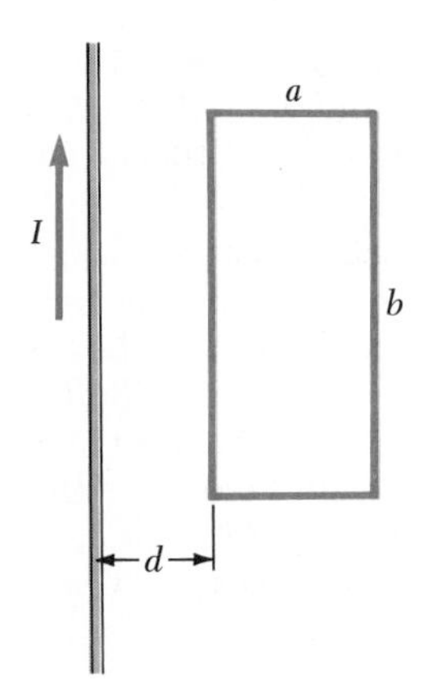


Figure 31.44 (Problem 62).

$r$ away, as shown in Figure 31.45. Show that the emf induced in the rod is given by

$$|\mathcal{E}| = \frac{\mu_0 I}{2\pi} v \ln\left(1 + \frac{\ell}{r}\right)$$

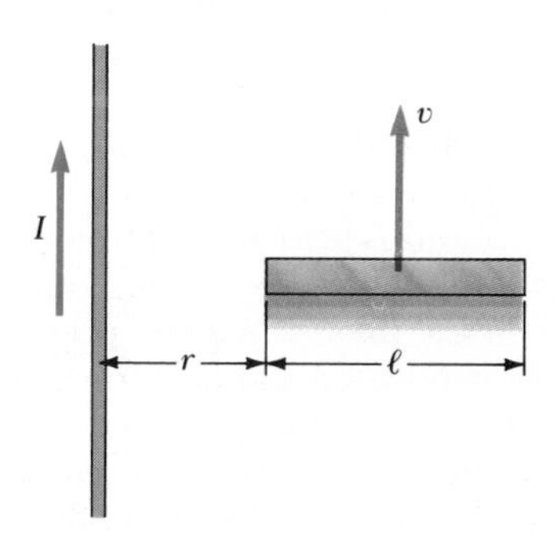


Figure 31.45 (Problem 63).

64. A rectangular loop of dimensions $\ell$ and $w$ moves with a constant velocity $\mathbf{v}$ away from a long wire that carries a current $I$ in the plane of the loop (Fig. 31.46). The total resistance of the loop is $R$. Derive an expression that gives the current in the loop at the instant the near side is a distance $r$ from the wire.

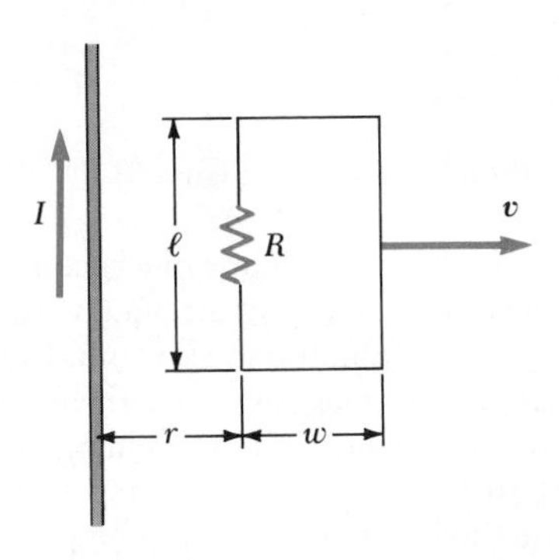


Figure 31.46 (Problem 64).

65. A square loop of wire with edge length $a = 0.2$ m is perpendicular to the earth's magnetic field at a point where $B = 15\ \mu$T, as in Figure 31.47. The total resistance of the loop and the wires connecting the loop to the galvanometer is 0.5 Ω. If the loop is suddenly collapsed by horizontal forces as shown, what total charge will pass through the galvanometer?

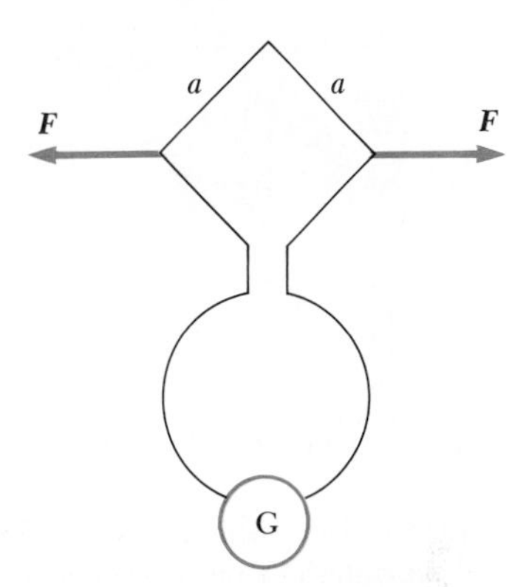


Figure 31.47 (Problem 65).

66. Magnetic field values are often determined by using a device known as a *search coil*. This technique depends on the measurement of the total charge passing through a coil in a time interval during which the magnetic flux linking the windings changes either because of the motion of the coil or because of a change in the value of $B$. (a) Show that if the flux through the coil changes from $\Phi_1$ to $\Phi_2$, the charge transferred through the coil between $t_1$ and $t_2$ will be given by $Q = N(\Phi_2 - \Phi_1)/R$ where $R$ is the resistance of the coil and associated circuitry (galvanometer). (b) As a specific example, calculate $B$ when a 100-turn coil of resistance 200 Ω and cross-sectional area 40 cm$^2$ produces the following results: A total charge of $5 \times 10^{-4}$ C passes through the coil when it is rotated in a uniform field from a position where the plane of the coil is perpendicular to the field into a position where the coil's plane is parallel to the field.

67. The magnetic flux threading a metal ring varies with time $t$ according to

$$\Phi_m = 3(at^3 - bt^2)\ \text{T}\cdot\text{m}^2 \qquad a = 2\ \text{s}^{-3} \qquad b = 6\ \text{s}^{-2}$$

The resistance of the ring is 3 Ω. Determine the *maximum current* induced in the ring during the interval from $t = 0$ to $t = 2$ s.

68. Figure 31.48 illustrates an arrangement similar to that discussed in Example 31.4, except in this case the bar is pulled horizontally across the set of parallel rails by a string (assumed massless) that passes over an ideal pulley and is attached to a freely suspended mass $M$. The uniform magnetic field has a magnitude $B$, the sliding bar has mass $m$, and the distance between the

rails is $\ell$. The rails are connected at one end by a load resistor $R$. Derive an expression that gives the value of the horizontal speed of the bar as a function of time, assuming that the suspended mass was released with the bar at rest at $t = 0$. Assume no friction between the rails and the bar.

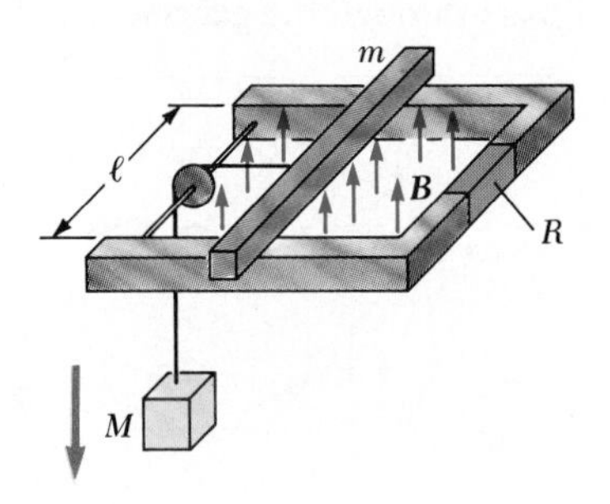


**Figure 31.48** (Problem 68).

69. In Figure 31.49, the rolling axle, 1.5 m long, is pushed along horizontal rails at a constant speed $v = 3$ m/s. A resistor $R = 0.4\ \Omega$ is connected to the rails at points $a$ and $b$, directly opposite each other. (The wheels make good electrical contact with the rails, so the axle, rails, and $R$ form a complete, closed-loop circuit. The only significant resistance in the circuit is $R$.) There is a uniform magnetic field $B = 0.08$ T vertically downward. (a) Find the induced current $I$ in the resistor. (b) What horizontal force $\boldsymbol{F}$ is required to keep the axle rolling at constant speed? (c) Which end of the resistor, $a$ or $b$, is at the higher electric potential? (d) After the axle rolls past the resistor, does the current in $R$ reverse direction?

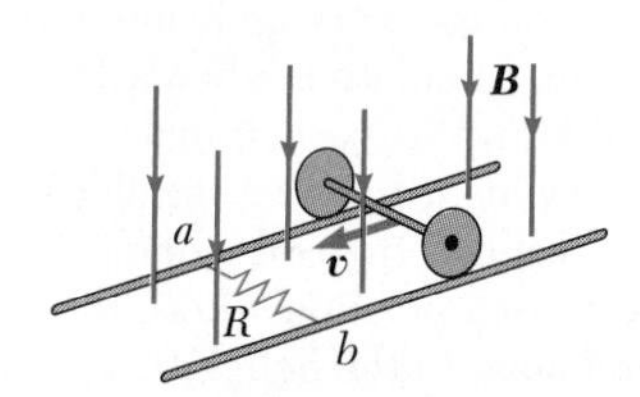


**Figure 31.49** (Problem 69).

70. Two infinitely long solenoids (seen in cross section) thread the circuit as shown in Figure 31.50. The solenoids have radii of 0.1 m and 0.15 m, respectively. The magnitude of $\boldsymbol{B}$ inside each is the same and is increasing at the rate of 100 T/s. What is the current in each resistor?

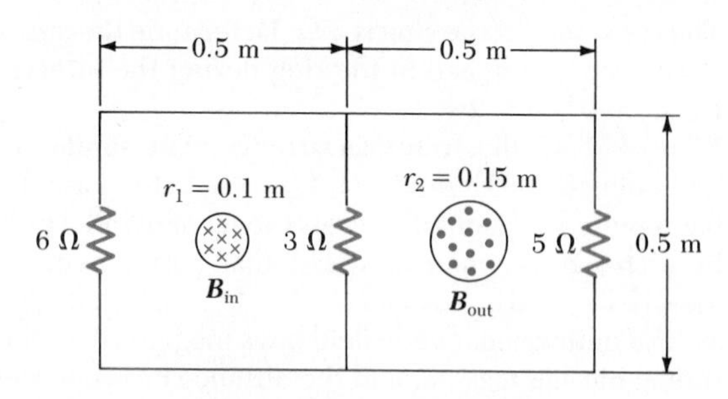


**Figure 31.50** (Problem 70).

71. Figure 31.51 shows a circular loop of radius $r$ that has a resistance $R$ spread uniformly throughout its length. The loop's plane is normal to the magnetic field $\boldsymbol{B}$ that decreases at a constant rate: $dB/dt = -K$, where $K$ is a positive constant. (a) What is the direction of the induced current? (b) Find the value of the induced current. (c) Which point, $a$ or $b$, is at the higher potential? Explain. (d) Discuss what force causes the current in the loop.

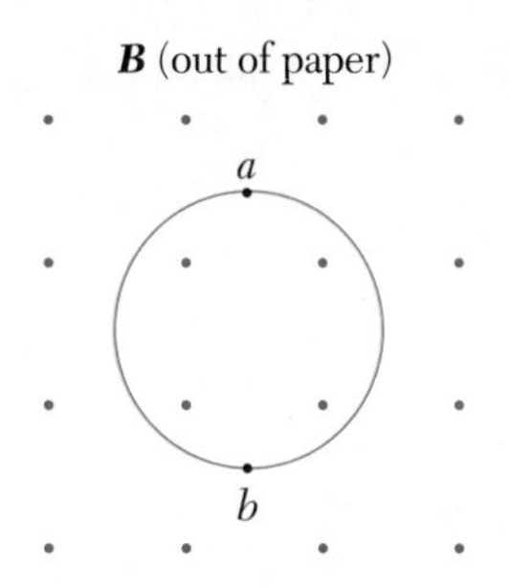


**Figure 31.51** (Problem 71).

72. A wire 30 cm long is held parallel to and at a distance of 80 cm above a long wire carrying current of 200 A that rests on the floor of a room (Fig. 31.52). The 30-cm wire is released and falls, remaining parallel with the current-carrying wire as it falls. Assume that the falling wire accelerates at a constant rate of $9.80\ \text{m/s}^2$ and derive an equation for the emf induced in the falling wire. Express your result as a function of the time $t$ after the wire was dropped. What is the induced emf 0.30 s after the wire is released?

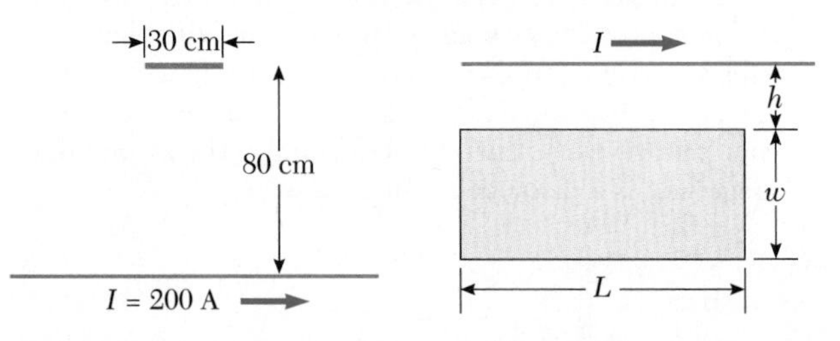


**Figure 31.52** (Problem 72). **Figure 31.53** (Problem 73).

73. (a) A loop of wire in the shape of a rectangle of width $w$ and length $L$ and a long straight wire carrying a current $I$ lie on a tabletop as shown in Figure 31.53. (a) Determine the magnetic flux through the loop. (b) Suppose that the current is changing with time according to $I = a + bt$, where $a$ and $b$ are constant. Determine the induced emf in the loop if $b = 10$ A/s, $h = 1$ cm, $w = 10$ cm, and $L = 100$ cm.

# 32

# Inductance

*Hydroelectric power is generated using the mechanical energy of water and electromagnetic induction. (© Tomas D. Friedmann 1971/Photo Researchers, Inc.)*

In the previous chapter, we saw that currents and emfs are induced in a circuit when the magnetic flux through the circuit changes with time. This phenomenon of electromagnetic induction has some practical consequences, which we shall describe in this chapter. First, we shall describe an effect known as *self-induction,* in which a time-varying current in a conductor induces an emf in the conductor that opposes the external emf that set up the current. This phenomenon is the basis of the element known as the *inductor,* which plays an important role in circuits with time-varying currents. We shall discuss the concepts of the energy stored in the magnetic field of an inductor and the energy density associated with a magnetic field.

Next, we shall study how an emf can be induced in a circuit as a result of a changing flux produced by an external circuit, which is the basic principle of *mutual induction.* Finally, we shall examine the characteristics of circuits containing inductors, resistors, and capacitors in various combinations. For example, we shall find that in a circuit containing only an inductor and a capacitor, the charge and current both oscillate in simple harmonic fashion. These oscillations correspond to a continuous transfer of energy between the electric field of the charged capacitor and the magnetic field of the current-carrying inductor.

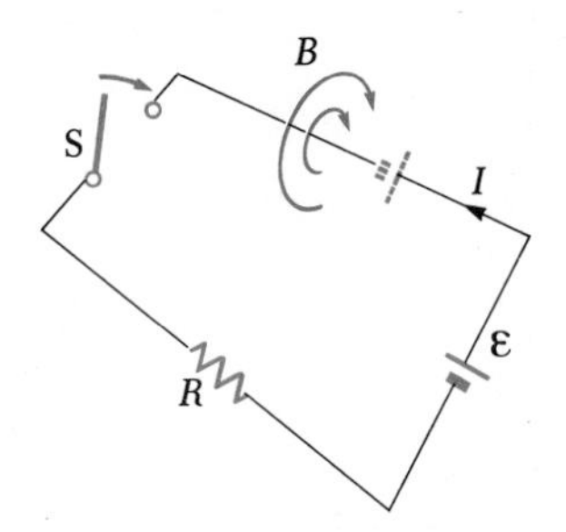


**Figure 32.1** After the switch in the circuit is closed, the current produces a magnetic flux through the loop. As the current increases toward its equilibrium value, the flux changes in time and induces an emf in the loop. The unlabelled battery is a symbol for the self-induced emf.

## 32.1 SELF-INDUCTANCE

Consider an isolated circuit consisting of a switch, resistor, and source of emf, as in Figure 32.1. When the switch is closed, the current doesn't immediately jump from zero to its maximum value, $\mathcal{E}/R$. The law of electromagnetic induction (Faraday's law) prevents this from occurring. What happens is the following: As the current increases with time, the magnetic flux through the loop due to this current also increases with time. This increasing flux induces an emf in the circuit that opposes the change in the net magnetic flux through the loop. By Lenz's law, the induced electric field in the wires must therefore be opposite the direction of the conventional current, and this opposing emf results in a *gradual* increase in the current. This effect is called *self-induction* since the changing flux through the circuit arises from the circuit itself. The emf that is set up in this case is called a **self-induced emf.** Later, in Section 32.4, we shall describe a related effect called *mutual induction* in which an emf is induced in one circuit as a result of a changing magnetic flux set up by another circuit.

To obtain a quantitative description of self-induction, we recall from Faraday's law that the induced emf is given by the negative time rate of change of the magnetic flux. The magnetic flux is proportional to the magnetic field, which in turn is proportional to the current in the circuit. Therefore, *the self-induced emf is always proportional to the time rate of change of the current.* For a closely spaced coil of $N$ turns of fixed geometry (a toroidal coil or the ideal solenoid), we find that

Induced emf

$$\mathcal{E} = -N\frac{d\Phi_m}{dt} = -L\frac{dI}{dt} \qquad (32.1)$$

where $L$ is a proportionality constant, called the **inductance** of the device, that depends on the geometric features of the circuit and other physical characteristics. From this expression, we see that the inductance of a coil containing $N$ turns is given by

Inductance of an $N$-turn coil

$$L = \frac{N\Phi_m}{I} \qquad (32.2)$$

where it is assumed that the same flux passes through each turn. Later we shall use this equation to calculate the inductance of some special current geometries.

From Equation 32.1, we can also write the inductance as the ratio

Inductance

$$L = -\frac{\mathcal{E}}{dI/dt} \qquad (32.3)$$

This is usually taken to be the defining equation for the inductance of any coil, regardless of its shape, size, or material characteristics. Just as resistance is a measure of the opposition to current, inductance is a measure of the opposition to the *change* in current.

The SI unit of inductance is the **henry** (H), which, from Equation 32.3, is seen to be equal to 1 volt-second per ampere:

$$1\ \text{H} = 1\ \frac{\text{V}\cdot\text{s}}{\text{A}}$$

As we shall see, *the inductance of a device depends on its geometry.* However, the calculation of a device's inductance can be quite difficult for complicated geometries. The examples below involve rather simple situations for which inductances are easily evaluated.

EXAMPLE 32.1 Inductance of a Solenoid
Find the inductance of a uniformly wound solenoid with $N$ turns and length $\ell$. Assume that $\ell$ is long compared with the radius and that the core of the solenoid is air.

*Solution* In this case, we can take the interior field to be uniform and given by Equation 30.20:

$$B = \mu_0 nI = \mu_0 \frac{N}{\ell} I$$

where $n$ is the number of turns per unit length, $N/\ell$. The flux through each turn is given by

$$\Phi_m = BA = \mu_0 \frac{NA}{\ell} I$$

where $A$ is the cross-sectional area of the solenoid. Using this expression and Equation 32.2 we find that

$$L = \frac{N\Phi_m}{I} = \frac{\mu_0 N^2 A}{\ell} \qquad (32.4)$$

This shows that $L$ depends on geometric factors and is proportional to the square of the number of turns. Since $N = n\ell$, we can also express the result in the form

$$L = \mu_0 \frac{(n\ell)^2}{\ell} A = \mu_0 n^2 A\ell = \mu_0 n^2(\text{volume}) \qquad (32.5)$$

where $A\ell$ is the volume of the solenoid.

EXAMPLE 32.2 Calculating Inductance and Emf
(a) Calculate the inductance of a solenoid containing 300 turns if the length of the solenoid is 25 cm and its cross-sectional area is 4 cm$^2$ = $4 \times 10^{-4}$ m$^2$.

*Solution* Using Equation 32.4 we get

$$L = \frac{\mu_0 N^2 A}{\ell} = (4\pi \times 10^{-7}\ \text{Wb/A}\cdot\text{m}) \frac{(300)^2(4 \times 10^{-4}\ \text{m}^2)}{25 \times 10^{-2}\ \text{m}}$$

$$= 1.81 \times 10^{-4}\ \text{Wb/A} = \boxed{0.181\ \text{mH}}$$

(b) Calculate the self-induced emf in the solenoid described in (a) if the current through it is *decreasing* at the rate of 50 A/s.

*Solution* Using Equation 32.1 and given that $dI/dt =$ 50 A/s, we get

$$\mathcal{E} = -L\frac{dI}{dt} = -(1.81 \times 10^{-4}\ \text{H})(-50\ \text{A/s})$$

$$= \boxed{9.05\ \text{mV}}$$

## 32.2 *RL* CIRCUITS

A circuit that contains a coil, such as a solenoid, has a self-inductance that prevents the current from increasing or decreasing instantaneously. A circuit element that has a large inductance is called an **inductor.** The circuit symbol for an inductor is ꝏꝏ. We shall always assume that the self-inductance of the remainder of the circuit is negligible compared with that of the inductor.

Consider the circuit consisting of a resistor, inductor, and battery shown in Figure 32.2. The internal resistance of the battery will be neglected. Suppose the switch S is closed at $t = 0$. The current will begin to increase, and due to the increasing current, the inductor will produce an emf (sometimes referred to as a *back emf*) that opposes the increasing current. In other words, the inductor acts like a battery whose polarity is opposite that of the real battery in the circuit. The back emf produced by the inductor is given by

$$\mathcal{E}_L = -L\frac{dI}{dt}$$

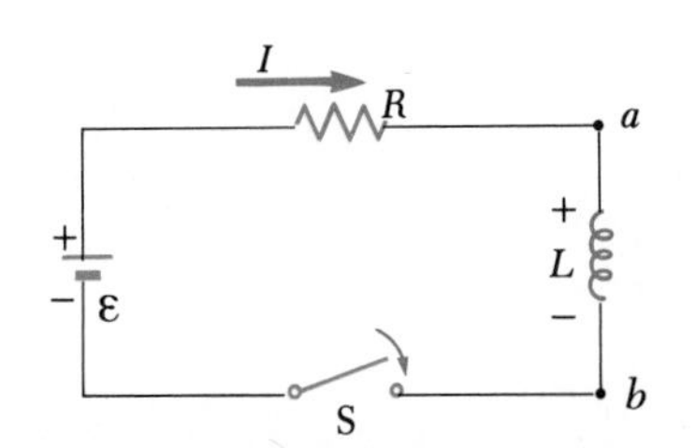


**Figure 32.2** A series *RL* circuit. As the current increases toward its maximum value, the inductor produces an emf that opposes the increasing current.

Since the current is increasing, $dI/dt$ is positive; therefore $\mathcal{E}_L$ is negative. This corresponds to the fact that there is a potential drop in going from $a$ to $b$ across the inductor. For this reason, point $a$ is at a higher potential than point $b$, as illustrated in Figure 32.2.

With this in mind, we can apply Kirchhoff's loop equation to this circuit:

$$\mathcal{E} - IR - L\frac{dI}{dt} = 0 \tag{32.6}$$

where $IR$ is the voltage drop across the resistor. We must now look for a solution to this differential equation, which is seen to be similar in form to that of the $RC$ circuit (Section 28.4).

To obtain a mathematical solution of Equation 32.6, it is convenient to change variables by letting $x = \frac{\mathcal{E}}{R} - I$, so that $dx = -dI$. With these substitutions, Equation 32.6 can be written

$$x + \frac{L}{R}\frac{dx}{dt} = 0$$

$$\frac{dx}{x} = -\frac{R}{L}\,dt$$

Integrating this last expression gives

$$\ln\frac{x}{x_0} = -\frac{R}{L}\,t$$

where the integrating constant is taken to be $-\ln x_0$. Taking the antilog of this result gives

$$x = x_0 e^{-Rt/L}$$

Since at $t = 0$, $I = 0$, we note that $x_0 = \mathcal{E}/R$. Hence, the last expression is equivalent to

$$\frac{\mathcal{E}}{R} - I = \frac{\mathcal{E}}{R}\,e^{-Rt/L}$$

$$I = \frac{\mathcal{E}}{R}\,(1 - e^{-Rt/L})$$

which represents the solution of Equation 32.6.

This mathematical solution of Equation 32.6, which represents the current as a function of time, can also be written:

$$I(t) = \frac{\mathcal{E}}{R}\,(1 - e^{-Rt/L}) = \frac{\mathcal{E}}{R}\,(1 - e^{-t/\tau}) \tag{32.7}$$

where the constant $\tau$ is the **time constant** of the $RL$ circuit:

Time constant of the *RL* circuit

$$\tau = L/R \tag{32.8}$$

It is left as an exercise to show that the dimension of $\tau$ is time. Physically, $\tau$ is the time it takes the current to reach $(1 - e^{-1}) = 0.63$ of its final value, $\mathcal{E}/R$.

Figure 32.3 represents a graph of the current versus time, where $I = 0$ at $t = 0$. Note that the final equilibrium value of the current, which occurs at $t = \infty$, is given by $\mathcal{E}/R$. This can be seen by setting $dI/dt$ equal to zero in Equation 32.6 (at equilibrium, the change in the current is zero) and solving for the current. Thus, we see that the current rises very fast initially and then gradually approaches the equilibrium value $\mathcal{E}/R$ as $t \rightarrow \infty$.

One can show that Equation 32.7 is a solution of Equation 32.6 by computing the derivative $dI/dt$ and noting that $I = 0$ at $t = 0$. Taking the first time derivative of Equation 32.7, we get

$$\frac{dI}{dt} = \frac{\mathcal{E}}{L} e^{-t/\tau} \qquad (32.9)$$

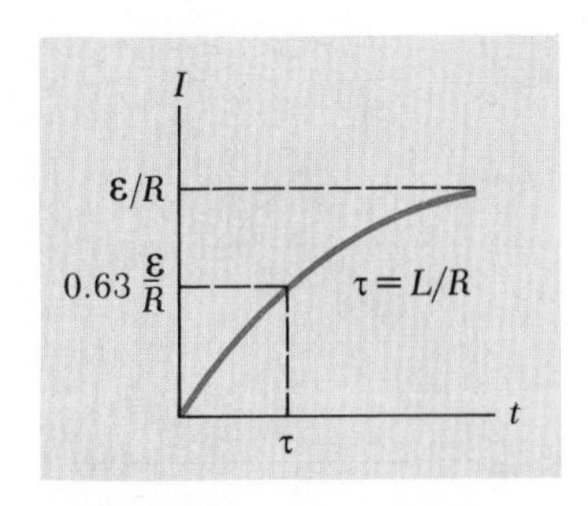


**Figure 32.3** Plot of the current versus time for the *RL* circuit shown in Figure 32.2. The switch is closed at $t = 0$, and the current increases toward its maximum value, $\mathcal{E}/R$. The time constant $\tau$ is the time it takes $I$ to reach 63% of its maximum value.

Substitution of this result together with Equation 32.7 will indeed verify that our solution satisfies Equation 32.6. That is,

$$\mathcal{E} - IR - L\frac{dI}{dt} = 0$$

$$\mathcal{E} - \frac{\mathcal{E}}{R}(1 - e^{-t/\tau})R - L\left(\frac{\mathcal{E}}{L}e^{-t/\tau}\right) = 0$$

$$\cancel{\mathcal{E}} - \cancel{\mathcal{E}} + \cancel{\mathcal{E}e^{-t/\tau}} - \cancel{\mathcal{E}e^{-t/\tau}} = 0$$

and the solution is verified.

From Equation 32.9 we see that the rate of increase of current, $dI/dt$, is a *maximum* (equal to $\mathcal{E}/L$) at $t = 0$ and falls off exponentially to zero as $t \rightarrow \infty$ (Fig. 32.4).

Now consider the *RL* circuit arranged as shown in Figure 32.5. The circuit contains two switches that operate such that when one is closed, the other is opened. Now suppose that $S_1$ is closed for a long enough time to allow the current to reach its equilibrium value, $\mathcal{E}/R$. If $S_1$ is now opened and $S_2$ is closed at $t = 0$, we have a circuit with no battery ($\mathcal{E} = 0$). If we apply Kirchhoff's circuit law to the upper loop containing the resistor and inductor, we obtain the expression

$$IR + L\frac{dI}{dt} = 0 \qquad (32.10)$$

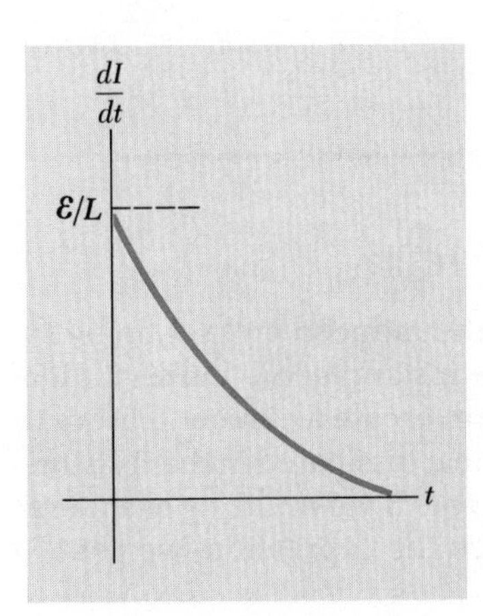


**Figure 32.4** Plot of $dI/dt$ versus time for the *RL* circuit shown in Figure 32.2. The rate of change of current is a maximum at $t = 0$ when the switch is closed. The rate $dI/dt$ decreases exponentially with time as $I$ increases toward its maximum value.

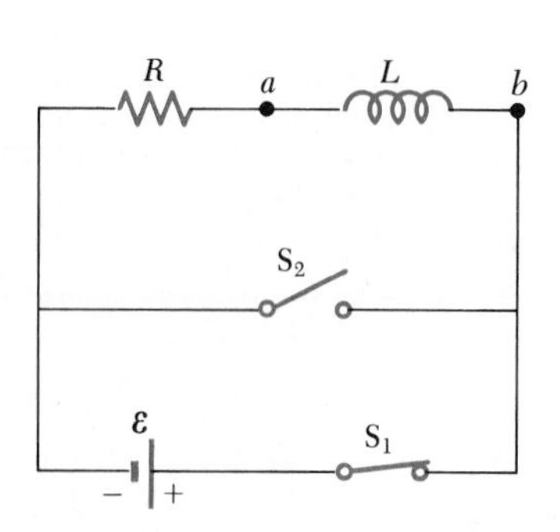


**Figure 32.5** An *RL* circuit containing two switches. When $S_1$ is closed and $S_2$ is open as shown, the battery is in the circuit. At the instant $S_2$ is closed, $S_1$ is opened and the battery is removed from the circuit.

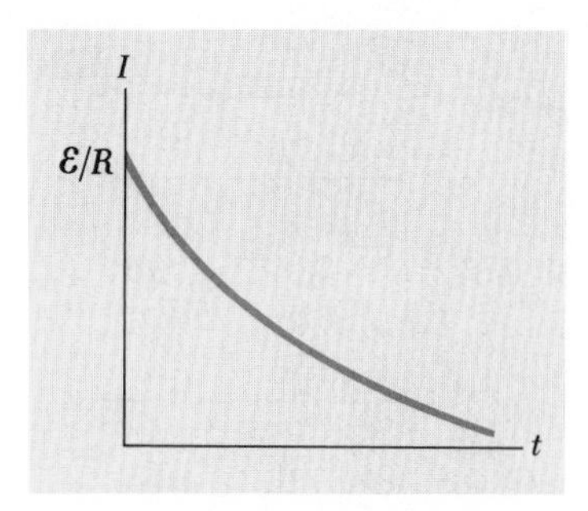


Figure 32.6 Current versus time for the circuit shown in Figure 32.5. At $t < 0$, $S_1$ is closed and $S_2$ is open. At $t = 0$, $S_2$ is closed, $S_1$ is open, and the current has its maximum value $\mathcal{E}/R$.

It is left as a problem (Problem 18) to show that the solution of this differential equation is

$$I(t) = \frac{\mathcal{E}}{R} e^{-t/\tau} = I_0 e^{-t/\tau} \tag{32.11}$$

where the current at $t = 0$ is given by $I_0 = \mathcal{E}/R$ and $\tau = L/R$.

The graph of the current versus time (Fig. 32.6) shows that the current is continuously decreasing with time, as one would expect. Furthermore, note that the slope, $dI/dt$, is always negative and has its maximum value at $t = 0$. The negative slope signifies that $\mathcal{E}_L = -L\,(dI/dt)$ is now *positive;* that is, point $a$ is at a lower potential than point $b$ in Figure 32.5.

EXAMPLE 32.3 Time Constant of an *RL* Circuit □
The circuit shown in Figure 32.7a consists of a 30-mH inductor, a 6-Ω resistor, and a 12-V battery. The switch is closed at $t = 0$. (a) Find the time constant of the circuit.

*Solution* The time constant is given by Equation 32.8

$$\tau = \frac{L}{R} = \frac{30 \times 10^{-3}\ \text{H}}{6\ \Omega} = 5.00\ \text{ms}$$

(b) Calculate the current in the circuit at $t = 2$ ms.

*Solution* Using Equation 32.7 for the current as a function of time (with $t$ and $\tau$ in ms), we find that at $t = 2$ ms

$$I = \frac{\mathcal{E}}{R}(1 - e^{-t/\tau}) = \frac{12\ \text{V}}{6\ \Omega}(1 - e^{-0.4}) = 0.659\ \text{A}$$

A plot of Equation 32.7 for this circuit is given in Figure 32.7b.

Exercise 1 Calculate the current in the circuit and the voltage across the resistor after one time constant has elapsed.
Answer 1.26 A, 7.56 V.

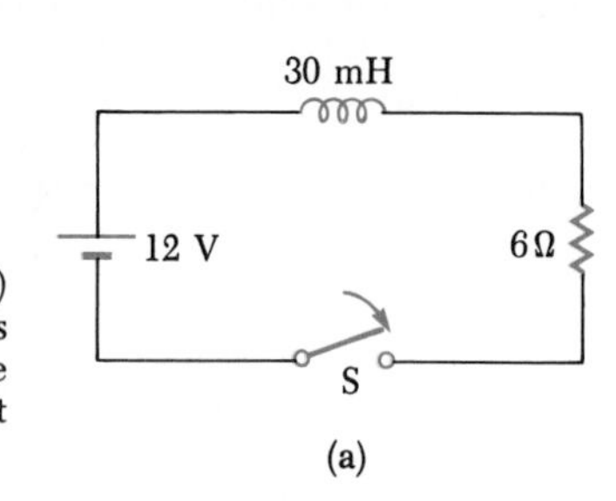


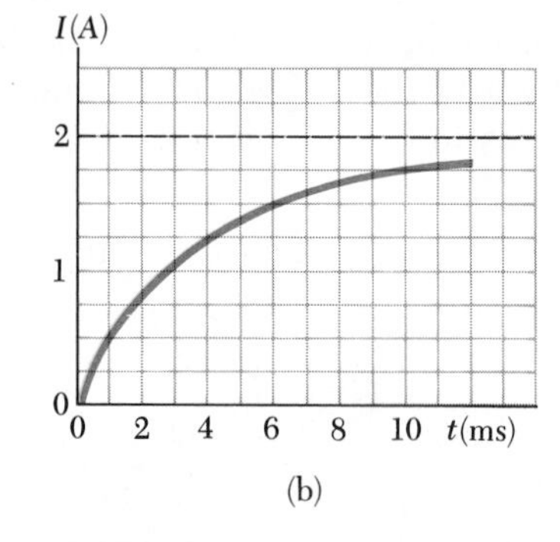


Figure 32.7 (Example 32.3) (a) The switch in this *RL* circuit is closed at $t = 0$. (b) A graph of the current versus time for the circuit in (a).

## 32.3 ENERGY IN A MAGNETIC FIELD

In the previous section we found that the induced emf set up by an inductor prevents a battery from establishing an instantaneous current. Hence, a battery has to do work against an inductor to create a current. Part of the energy supplied by the battery goes into joule heat dissipated in the resistor, while the remaining energy is stored in the inductor. If we multiply each term in Equation 32.6 by the current $I$ and rearrange the expression, we get

$$I\mathcal{E} = I^2R + LI\frac{dI}{dt} \tag{32.12}$$

This expression tells us that the rate at which energy is supplied by the battery, $I\mathcal{E}$, equals the sum of the rate at which joule heat is dissipated in the resistor, $I^2R$, and the rate at which energy is stored in the inductor, $LI\,(dI/dt)$. Thus, Equation 32.12 is simply an expression of energy conservation. If we let $U_m$ denote the energy stored in the inductor at any time, then the rate $dU_m/dt$ at which energy is stored in the inductor can be written

$$\frac{dU_m}{dt} = LI\frac{dI}{dt}$$

To find the total energy stored in the inductor, we can rewrite this expression as $dU_m = LI\,dI$ and integrate:

$$U_m = \int_0^{U_m} dU_m = \int_0^I LI\,dI$$

$$U_m = \tfrac{1}{2}LI^2 \tag{32.13}$$

Energy stored in an inductor

where $L$ is constant and has been removed from the integral. Equation 32.13 represents the energy stored as magnetic energy in the field of the inductor when the current is $I$. Note that it is similar in form to the equation for the energy stored in the electric field of a capacitor, $Q^2/2C$. In either case, we see that it takes work to establish a field.

We can also determine the energy per unit volume, or energy density, stored in a magnetic field. For simplicity, consider a solenoid whose inductance is given by Equation 32.5:

$$L = \mu_0 n^2 A\ell$$

The magnetic field of a solenoid is given by

$$B = \mu_0 nI$$

Substituting the expression for $L$ and $I = B/\mu_0 n$ into Equation 32.13 gives

$$U_m = \tfrac{1}{2}LI^2 = \tfrac{1}{2}\mu_0 n^2 A\ell\left(\frac{B}{\mu_0 n}\right)^2 = \frac{B^2}{2\mu_0}(A\ell) \tag{32.14}$$

Because $A\ell$ is the volume of the solenoid, the energy stored per unit volume in a magnetic field is given by

$$u_m = \frac{U_m}{A\ell} = \frac{B^2}{2\mu_0} \tag{32.15}$$

Magnetic energy density

Although Equation 32.15 was derived for the special case of a solenoid, *it is valid for any region of space in which a magnetic field exists.* Note that Equation 32.15 is similar in form to the equation for the energy per unit volume stored in an electric field, given by $\frac{1}{2}\epsilon_0 E^2$. In both cases, the energy density is proportional to the *square* of the field strength.

**EXAMPLE 32.4 What Happens to the Energy in the Inductor?**

Consider once again the *RL* circuit shown in Figure 32.5, in which switch $S_2$ is closed at the instant $S_1$ is opened (at $t = 0$). Recall that the current in the upper loop decays exponentially with time according to the expression $I = I_0 e^{-t/\tau}$, where $I_0 = \mathcal{E}/R$ is the initial current in the circuit and $\tau = L/R$ is the time constant. The energy stored in the magnetic field of the inductor gradually dissipates as thermal energy in the resistor. Let us show explicitly that all the energy stored in the inductor gets dissipated as heat in the resistor.

*Solution* The rate at which energy is dissipated in the resistor, $dU/dt$, (or the power) is equal to $I^2R$, where $I$ is the instantaneous current. That is,

$$\frac{dU}{dt} = I^2R = (I_0 e^{-Rt/L})^2 R = I_0^2 R e^{-2Rt/L} \qquad (1)$$

To find the total energy dissipated in the resistor, we integrate this expression over the limits $t = 0$ to $t = \infty$. (The upper limit of $\infty$ is used because it takes an infinite time for the current to reach zero.) Hence,

$$U = \int_0^\infty I_0^2 R e^{-2Rt/L}\, dt = I_0^2 R \int_0^\infty e^{-2Rt/L}\, dt \qquad (2)$$

The value of the definite integral is $L/2R$, so $U$ becomes

$$U = I_0^2 R \left(\frac{L}{2R}\right) = \frac{1}{2} L I_0^2$$

Note that this is equal to the initial energy stored in the magnetic field of the inductor, given by Equation 32.13, as we set out to prove.

*Exercise 2* Evaluate the integral on the right hand side of Equation (2) and show that it has the value $L/2R$.

**EXAMPLE 32.5 The Coaxial Cable**

A long coaxial cable consists of two concentric cylindrical conductors of radii $a$ and $b$ and length $\ell$, as in Figure 32.8. The inner conductor is assumed to be a thin cylindrical shell. Each conductor carries a current $I$ (the outer one being a return path). (a) Calculate the self-inductance $L$ of this cable.

*Solution* To obtain $L$, we must know the magnetic flux through any cross section between the two conductors. From Ampère's law (Section 30.3), it is easy to see that the magnetic field *between* the conductors is given by $B = \mu_0 I/2\pi r$. Furthermore, the field is zero outside the conductors and zero inside the inner hollow conductor. The field is zero outside since the *net* current through a circular path surrounding both wires is zero, and hence from Ampère's law, $\oint \mathbf{B} \cdot d\mathbf{s} = 0$. The field is zero inside the inner conductor since it is hollow and there is no current within a radius $r < a$.

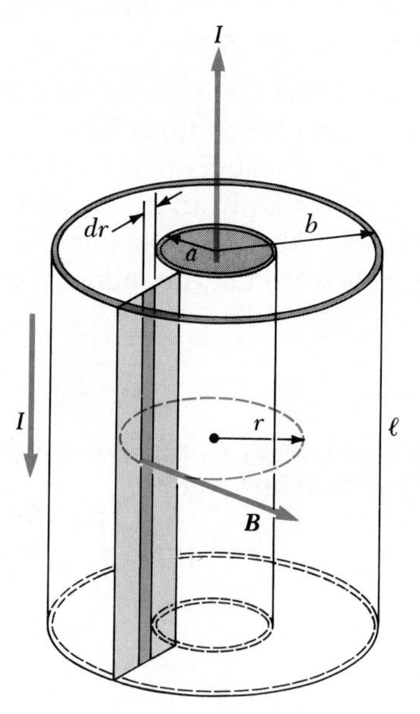


Figure 32.8 (Example 32.5) Section of a long coaxial cable. The inner and outer conductors carry equal and opposite currents.

The magnetic field is *perpendicular* to the shaded rectangular strip of length $\ell$ and width $(b - a)$. This is the cross section of interest. Dividing this rectangle into strips of width $dr$, we see that the area of each strip is $\ell dr$ and the flux through each strip is $B\, dA = B\ell\, dr$. Hence, the *total* flux through any cross section is

$$\Phi_m = \int B\, dA = \int_a^b \frac{\mu_0 I}{2\pi r} \ell\, dr = \frac{\mu_0 I \ell}{2\pi} \int_a^b \frac{dr}{r} = \frac{\mu_0 I \ell}{2\pi} \ln\left(\frac{b}{a}\right)$$

Using this result, we find that the self-inductance of the cable is

$$L = \frac{\Phi_m}{I} = \frac{\mu_0 \ell}{2\pi} \ln\left(\frac{b}{a}\right)$$

Furthermore, the self-inductance per unit length is given by $(\mu_0/2\pi)\ln(b/a)$.

(b) Calculate the total energy stored in the magnetic field of the cable.

*Solution* Using Equation 32.14 and the results to (a), we get

$$U_m = \tfrac{1}{2} L I^2 = \frac{\mu_0 \ell I^2}{4\pi} \ln\left(\frac{b}{a}\right)$$

## *32.4 MUTUAL INDUCTANCE

Very often the magnetic flux through a circuit varies with time because of varying currents in nearby circuits. This gives rise to an induced emf through a process known as **mutual induction,** so called because it depends on the interaction of two circuits.

Consider two closely wound coils as shown in the cross-sectional view of Figure 32.9. The current $I_1$ in coil 1, which has $N_1$ turns, creates magnetic field lines, some of which pass through coil 2, which has $N_2$ turns. The corresponding flux through coil 2 produced by coil 1 is represented by $\Phi_{21}$. We define the **mutual inductance** $M_{21}$ of coil 2 with respect to coil 1 as the ratio of $N_2\Phi_{21}$ and the current $I_1$:

$$M_{21} \equiv \frac{N_2\Phi_{21}}{I_1} \qquad (32.16)$$

Definition of mutual inductance

$$\Phi_{21} = \frac{M_{21}}{N_2} I_1$$

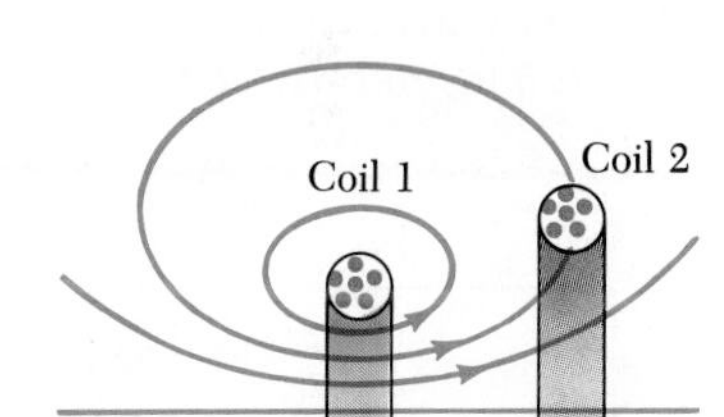


**Figure 32.9** A cross-sectional view of two adjacent coils. A current in coil 1 sets up a flux, part of which passes through coil 2.

The mutual inductance depends on the geometry of both circuits and on their orientation with respect to one another. Clearly, as the circuit separation increases, the mutual inductance decreases since the flux linking the circuits decreases.

If the current $I_1$ varies with time, we see from Faraday's law and Equation 32.16 that the emf induced in coil 2 by coil 1 is given by

$$\mathcal{E}_2 = -N_2 \frac{d\Phi_{21}}{dt} = -M_{21}\frac{dI_1}{dt} \qquad (32.17)$$

Similarly, if the current $I_2$ varies with time, the induced emf in coil 1 due to coil 2 is given by

$$\mathcal{E}_1 = -M_{12}\frac{dI_2}{dt} \qquad (32.18)$$

These results are similar in form to the expression for the self-induced emf $\mathcal{E} = -L\,(dI/dt)$. *The emf induced by mutual induction in one coil is always proportional to the rate of current change in the other coil.* If the rates at which the currents change with time are equal (that is, if $dI_1/dt = dI_2/dt$), then one finds that $\mathcal{E}_1 = \mathcal{E}_2$. Although the proportionality constants $M_{12}$ and $M_{21}$ appear to be different, one can show that they are equal. Thus, taking $M_{12} = M_{21} = M$, Equations 32.17 and 32.18 become

$$\mathcal{E}_2 = -M\frac{dI_1}{dt} \quad \text{and} \quad \mathcal{E}_1 = -M\frac{dI_2}{dt} \qquad (32.19)$$

The unit of mutual inductance is also the henry.

**EXAMPLE 32.6 Mutual Inductance of Two Solenoids**

A long solenoid of length $\ell$ has $N_1$ turns, carries a current $I$, and has a cross-sectional area $A$. A second coil containing $N_2$ turns is wound around the center of the first coil, as in Figure 32.10. Find the mutual inductance of the system.

*Solution* If the solenoid carries a current $I_1$, the magnetic field at its center is given by

$$B = \frac{\mu_0 N_1 I_1}{\ell}$$

Since the flux $\Phi_{21}$ through coil 2 due to coil 1 is $BA$, the mutual inductance is

$$M = \frac{N_2 \Phi_{21}}{I_1} = \frac{N_2 BA}{I_1} = \mu_0 \frac{N_1 N_2 A}{\ell}$$

For example, if $N_1 = 500$ turns, $A = 3 \times 10^{-3}\ \text{m}^2$, $\ell = 0.5$ m, and $N_2 = 8$ turns, we get

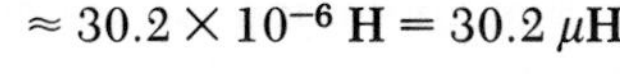

$$M = \frac{(4\pi \times 10^{-7}\ \text{Wb/A}\cdot\text{m})(500)(8)(3 \times 10^{-3}\ \text{m}^2)}{0.5\ \text{m}}$$

$$\approx 30.2 \times 10^{-6}\ \text{H} = 30.2\ \mu\text{H}$$

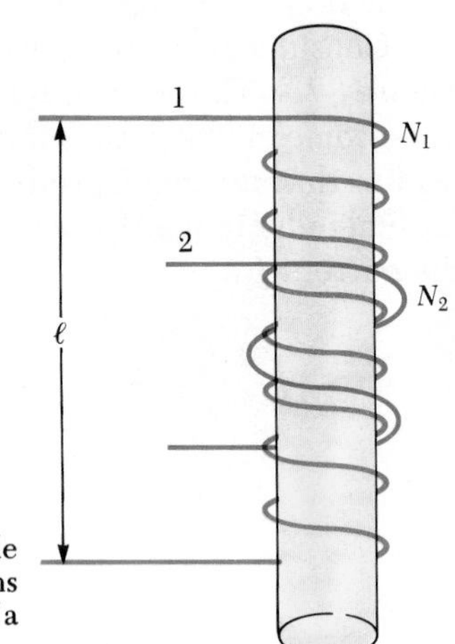


**Figure 32.10** (Example 32.6) A small coil of $N_2$ turns wrapped around the center of a long solenoid of $N_1$ turns.

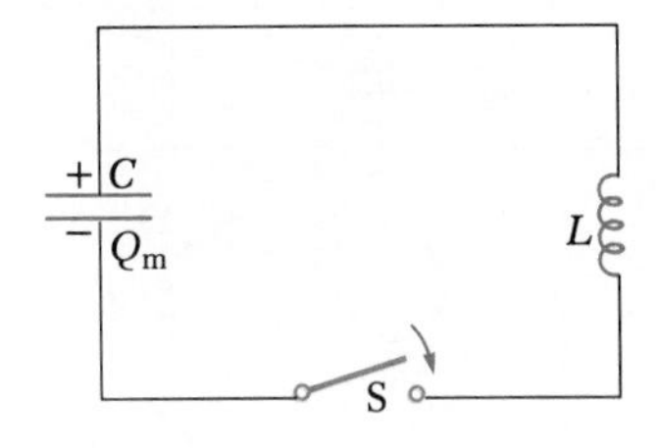


**Figure 32.11** A simple *LC* circuit. The capacitor has an initial charge $Q_m$, and the switch is closed at $t = 0$.

## 32.5 OSCILLATIONS IN AN *LC* CIRCUIT

When a *charged* capacitor is connected to an inductor as in Figure 32.11 and the switch is then closed, oscillations will occur in the current and charge on the capacitor. If the resistance of the circuit is zero, no energy is dissipated as joule heat and the oscillations will persist. In this section we shall neglect the resistance in the circuit.

In the following analysis, let us assume that the capacitor has an initial charge $Q_m$ and that the switch is closed at $t = 0$. It is convenient to describe what happens from an energy viewpoint.

When the capacitor is fully charged, the total energy $U$ in the circuit is stored in the electric field of the capacitor and is equal to $Q_m^2/2C$. At this time, the current is zero and so there is no energy stored in the inductor. As the capacitor begins to discharge, the energy stored in its electric field decreases. At the same time, the current increases and some energy is now stored in the magnetic field of the inductor. Thus, we see that energy is transferred from the electric field of the capacitor to the magnetic field of the inductor. When the capacitor is fully discharged, it stores no energy. At this time, the current reaches its maximum value and all of the energy is now stored in the inductor. The process then repeats in the reverse direction. The energy continues to transfer between the inductor and capacitor indefinitely, corresponding to oscillations in the current and charge.

A graphical description of this energy transfer is shown in Figure 32.12. The circuit behavior is analogous to the oscillating mass-spring system studied in Chapter 13. The potential energy stored in a stretched spring, $\frac{1}{2}kx^2$, is analogous to the potential energy stored in the capacitor, $Q_m^2/2C$. The kinetic energy of the moving mass, $\frac{1}{2}mv^2$, is analogous to the energy stored in the inductor, $\frac{1}{2}LI^2$, which requires the presence of moving charges. In Figure 32.12a, all of the energy is stored as potential energy in the capacitor at $t = 0$ (since $I = 0$). In Figure 32.12b, all of the energy is stored as "kinetic" energy

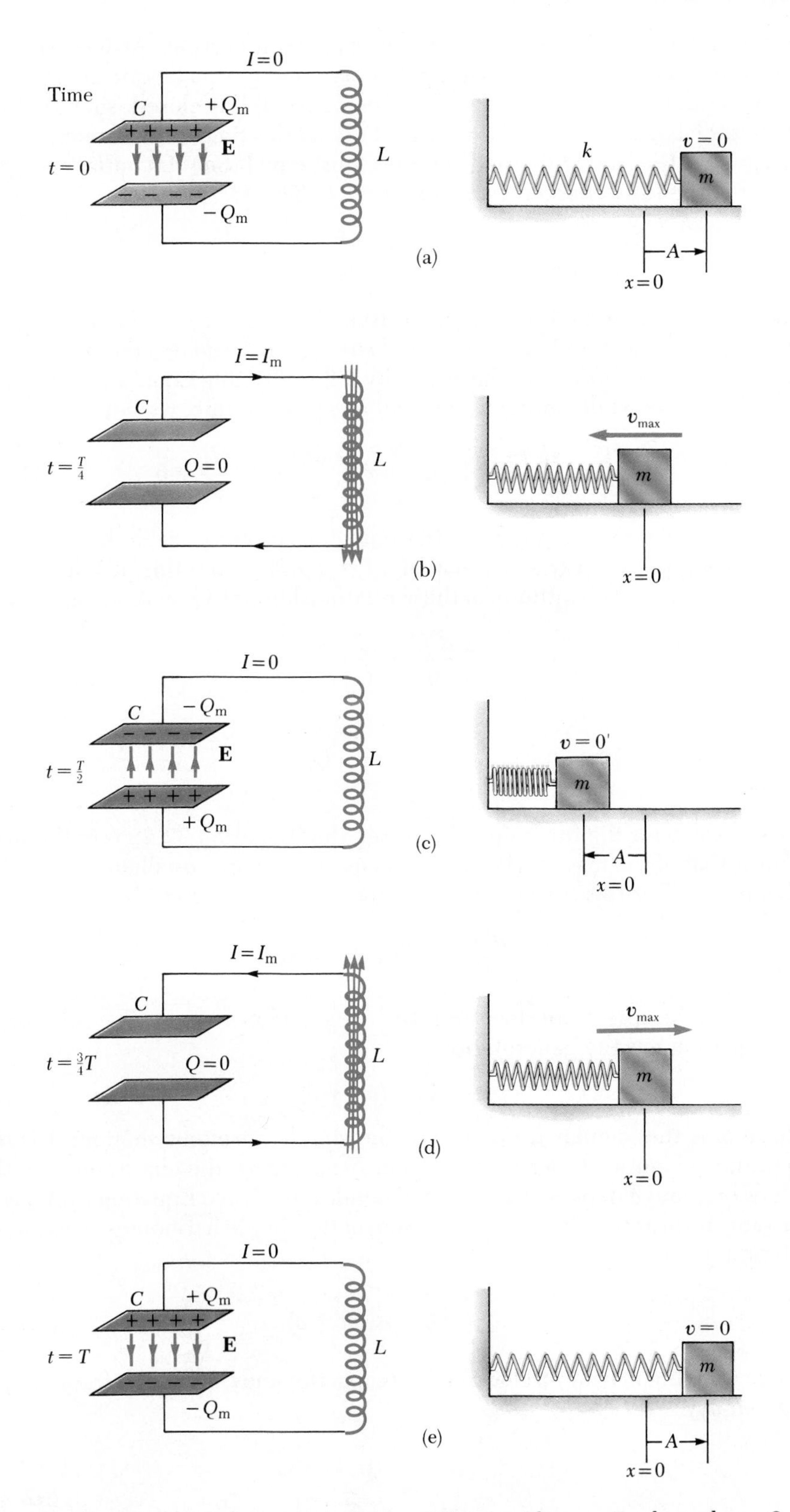


**Figure 32.12** Energy transfer in a resistanceless $LC$ circuit. The capacitor has a charge $Q_m$ at $t = 0$ when the switch is closed. The mechanical analog of this circuit, the mass-spring system, is shown at the right.

in the inductor, $\frac{1}{2}LI_m^2$, where $I_m$ is the maximum current. At intermediate points, part of the energy is potential energy and part is kinetic energy.

Consider some arbitrary time $t$ after the switch is closed, such that the capacitor has a charge $Q$ and the current is $I$. At this time, both elements store energy, but the sum of the two energies must equal the total initial energy $U$ stored in the fully charged capacitor at $t = 0$. That is,

Total energy stored in the $LC$ circuit

$$U = U_C + U_L = \frac{Q^2}{2C} + \frac{1}{2}LI^2 \tag{32.20}$$

Since we have assumed the circuit resistance to be zero, no energy is dissipated as joule heat and hence *the total energy must remain constant in time.* This means that $dU/dt = 0$. Therefore, by differentiating Equation 32.20 with respect to time while noting that $Q$ and $I$ vary with time, we get

The total energy in an $LC$ circuit remains constant; therefore, $dU/dt = 0$

$$\frac{dU}{dt} = \frac{d}{dt}\left(\frac{Q^2}{2C} + \frac{1}{2}LI^2\right) = \frac{Q}{C}\frac{dQ}{dt} + LI\frac{dI}{dt} = 0 \tag{32.21}$$

We can reduce this to a differential equation in one variable by using the relationship between $Q$ and $I$, namely, $I = dQ/dt$. From this, it follows that $dI/dt = d^2Q/dt^2$. Substitution of these relationships into Equation 32.21 gives

$$L\frac{d^2Q}{dt^2} + \frac{Q}{C} = 0$$

$$\frac{d^2Q}{dt^2} = -\frac{1}{LC}Q \tag{32.22}$$

We can solve for the function $Q$ by noting that Equation 32.22 is of the *same* form as that of the mass-spring system (simple harmonic oscillator) studied in Chapter 13. For this system, the equation of motion is given by

$$\frac{d^2x}{dt^2} = -\frac{k}{m}x = -\omega^2 x$$

where $k$ is the spring constant, $m$ is the mass, and $\omega = \sqrt{k/m}$. The solution of this equation has the general form

$$x = A\cos(\omega t + \delta)$$

where $\omega$ is the angular frequency of the simple harmonic motion, $A$ is the amplitude of motion (the maximum value of $x$), and $\delta$ is the phase constant; the values of $A$ and $\delta$ depend on the initial conditions. Since Equation 32.22 is of the same form as the differential equation of the simple harmonic oscillator, its solution is

Charge versus time for the $LC$ circuit

$$Q = Q_m\cos(\omega t + \delta) \tag{32.23}$$

where $Q_m$ is the maximum charge of the capacitor and the angular frequency $\omega$ is given by

Angular frequency of oscillation

$$\omega = \frac{1}{\sqrt{LC}} \tag{32.24}$$

Note that *the angular frequency of the oscillations depends solely on the inductance and capacitance of the circuit.*

Since $Q$ varies periodically, the current also varies periodically. This is easily shown by differentiating Equation 32.23 with respect to time, which gives

$$I = \frac{dQ}{dt} = -\omega Q_m \sin(\omega t + \delta) \tag{32.25}$$

Current versus time for the $LC$ current

To determine the value of the phase angle $\delta$, we examine the initial conditions, which in our situation require that at $t = 0$, $I = 0$ and $Q = Q_m$. Setting $I = 0$ at $t = 0$ in Equation 32.25 gives

$$0 = -\omega Q_m \sin \delta$$

which shows that $\delta = 0$. This value for $\delta$ is also consistent with Equation 32.23 and the second condition that $Q = Q_m$ at $t = 0$. Therefore, in our case, the time variation of $Q$ and that of $I$ are given by

$$Q = Q_m \cos \omega t \tag{32.26}$$

$$I = -\omega Q_m \sin \omega t = -I_m \sin \omega t \tag{32.27}$$

where $I_m = \omega Q_m$ is the *maximum* current in the circuit.

Graphs of $Q$ versus $t$ and $I$ versus $t$ are shown in Figure 32.13. Note that the charge on the capacitor oscillates between the extreme values $Q_m$ and $-Q_m$, and the current oscillates between $I_m$ and $-I_m$. Furthermore, the current is 90° out of phase with the charge. That is, when the charge reaches an extreme value, the current is zero, and when the charge is zero, the current has an extreme value.

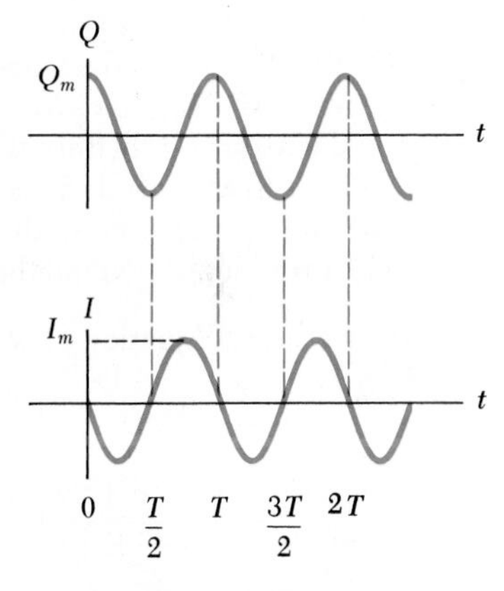


**Figure 32.13** Graphs of charge versus time and current versus time for a resistanceless $LC$ circuit. Note that $Q$ and $I$ are 90° out of phase with each other.

Let us return to the energy of the $LC$ circuit. Substituting Equations 32.26 and 32.27 in Equation 32.20, we find that the total energy is given by

$$U = U_C + U_L = \frac{Q_m^2}{2C}\cos^2 \omega t + \frac{LI_m^2}{2}\sin^2 \omega t \tag{32.28}$$

This expression contains all of the features that were described qualitatively at the beginning of this section. It shows that the energy of the system continuously oscillates between energy stored in the electric field of the capacitor and energy stored in the magnetic field of the inductor. When the energy stored in the capacitor has its maximum value, $Q_m^2/2C$, the energy stored in the inductor is zero. When the energy stored in the inductor has its maximum value, $\frac{1}{2}LI_m^2$, the energy stored in the capacitor is zero.

Plots of the time variations of $U_C$ and $U_L$ are shown in Figure 32.14. Note that the sum $U_C + U_L$ is a constant and equal to the total energy, $Q_m^2/2C$. An analytical proof of this is straightforward. Since the maximum energy stored in the capacitor (when $I = 0$) must equal the maximum energy stored in the inductor (when $Q = 0$),

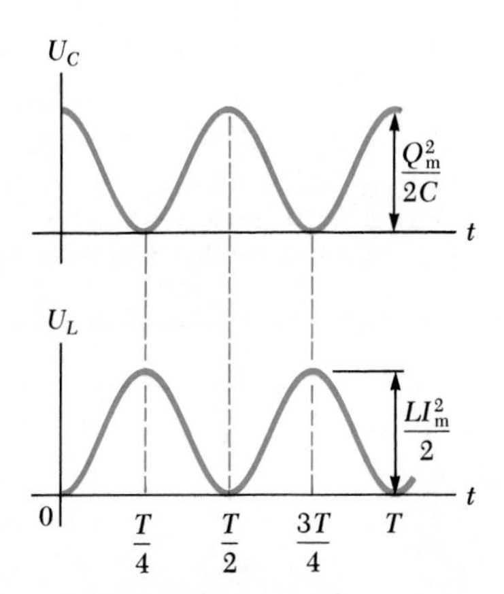


**Figure 32.14** Plots of $U_C$ versus $t$ and $U_L$ versus $t$ for a resistanceless $LC$ circuit. The sum of the two curves is a constant and equal to the total energy stored in the circuit.

$$\frac{Q_m^2}{2C} = \frac{1}{2}LI_m^2 \tag{32.29}$$

Substitution of this into Equation 32.28 for the total energy gives

$$U = \frac{Q_m^{\;2}}{2C}(\cos^2 \omega t + \sin^2 \omega t) = \frac{Q_m^{\;2}}{2C} \tag{32.30}$$

because $\cos^2 \omega t + \sin^2 \omega t = 1$.

You should note that the total energy $U$ remains constant *only* if energy losses are neglected. In actual circuits, there will always be some resistance and so energy will be lost in the form of heat. (In fact, even when the energy losses due to wire resistance are neglected, energy will also be lost in the form of electromagnetic waves radiated by the circuit.) In our idealized situation, the oscillations in the circuit persist indefinitely.

**EXAMPLE 32.7 An Oscillatory *LC* Circuit**
An $LC$ circuit has an inductance of 2.81 mH and a capacitance of 9 pF (Fig. 32.15). The capacitor is initially charged with a 12-V battery when the switch $S_1$ is open and switch $S_2$ is closed. $S_1$ is then closed at the same instant that $S_2$ is opened so that the capacitor is shorted across the inductor. (a) Find the frequency of oscillation.

*Solution* Using Equation 32.24 gives for the frequency

$$f = \frac{\omega}{2\pi} = \frac{1}{2\pi\sqrt{LC}}$$

$$= \frac{1}{2\pi[(2.81 \times 10^{-3}\ \text{H})(9 \times 10^{-12}\ \text{F})]^{1/2}} = 10^6\ \text{Hz}$$

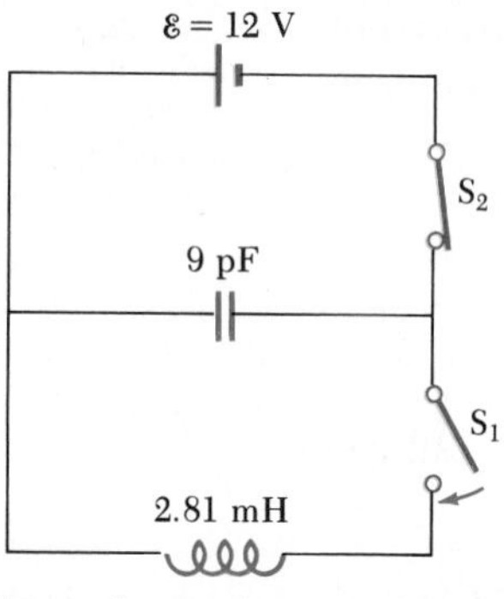

**Figure 32.15** (Example 32.7) First the capacitor is fully charged with the switch $S_1$ open and $S_2$ closed. Then, $S_1$ is closed at the same time that $S_2$ is being opened.

(b) What are the maximum values of charge on the capacitor and current in the circuit?

*Solution* The initial charge on the capacitor equals the maximum charge, and since $C = Q/V$, we get

$$Q_m = CV = (9 \times 10^{-12}\ \text{F})(12\ \text{V}) = 1.08 \times 10^{-10}\ \text{C}$$

From Equation 32.27, we see that the maximum current is related to the maximum charge:

$$I_m = \omega Q_m = 2\pi f Q_m$$
$$= (2\pi \times 10^6\ \text{s}^{-1})(1.08 \times 10^{-10}\ \text{C})$$
$$= 6.79 \times 10^{-4}\ \text{A}$$

(c) Determine the charge and current as functions of time.

*Solution* Equations 32.26 and 32.27 give the following expressions for the time variation of $Q$ and $I$:

$$Q = Q_m \cos \omega t = (1.08 \times 10^{-10}\ \text{C}) \cos \omega t$$
$$I = -I_m \sin \omega t = (-6.79 \times 10^{-4}\ \text{A}) \sin \omega t$$

where

$$\omega = 2\pi f = 2\pi \times 10^6\ \text{rad/s}$$

*Exercise 3* What is the total energy stored in the circuit?
*Answer* $6.48 \times 10^{-10}$ J.

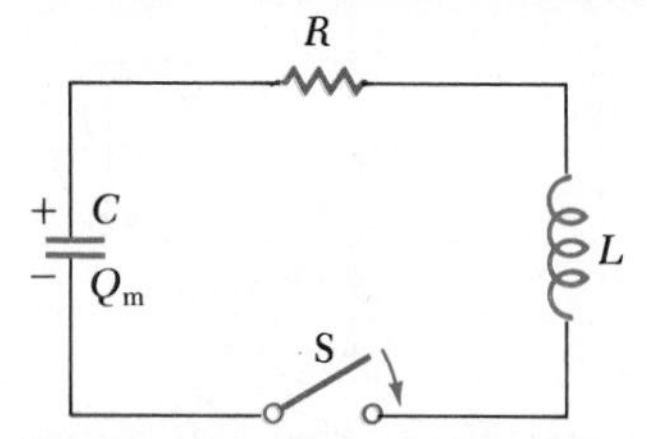


**Figure 32.16** A series *RLC* circuit. The capacitor has a charge $Q_m$ at $t = 0$ when the switch is being closed.

## *32.6 THE *RLC* CIRCUIT

We now turn our attention to a more realistic circuit consisting of an inductor, a capacitor, and a resistor connected in series, as in Figure 32.16. We shall assume that the capacitor has an initial charge $Q_m$ before the switch is closed. Once the switch is closed and a current is established, the total energy stored in the circuit at any time is given, as before, by Equation 32.20. That is, the energy stored in the capacitor is $Q^2/2C$, and the energy stored in the inductor is $\frac{1}{2}LI^2$. However, the total energy is no longer constant, as it was in the *LC* circuit, because of the presence of a resistor, which dissipates energy as heat. Since the rate of energy dissipation through a resistor is $I^2R$, we have

$$\frac{dU}{dt} = -I^2R \tag{32.31}$$

where the negative sign signifies that $U$ is *decreasing* in time. Substituting this result into the time derivative of Equation 32.20 gives

$$LI\frac{dI}{dt} + \frac{Q}{C}\frac{dQ}{dt} = -I^2R \tag{32.32}$$

Using the fact that $I = dQ/dt$ and $dI/dt = d^2Q/dt^2$, and dividing Equation 32.32 by $I$, we get

$$L\frac{d^2Q}{dt^2} + R\frac{dQ}{dt} + \frac{Q}{C} = 0 \tag{32.33}$$

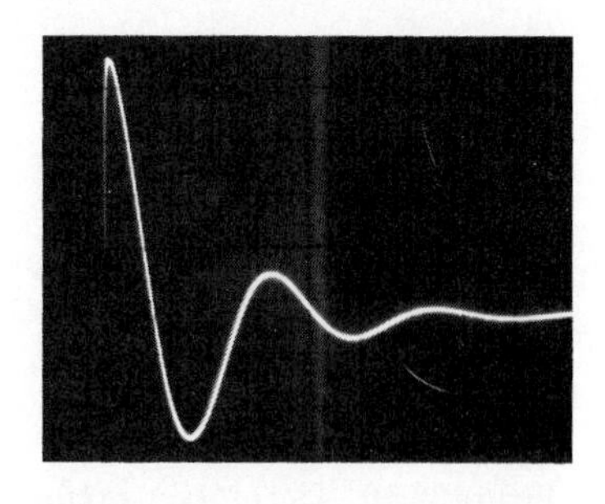

Oscilloscope pattern showing the decay in the oscillations of an *RLC* circuit. The parameters used were $R = 75\ \Omega$, $L = 10$ mH, $C = 0.19\ \mu$F, and $f = 300$ Hz. (Courtesy of J. Rudmin)

Note that the *RLC* circuit is analogous to the damped harmonic oscillator discussed in Section 13.6 and illustrated in Figure 32.17. The equation of motion for this mechanical system is

$$m\frac{d^2x}{dt^2} + b\frac{dx}{dt} + kx = 0 \tag{32.34}$$

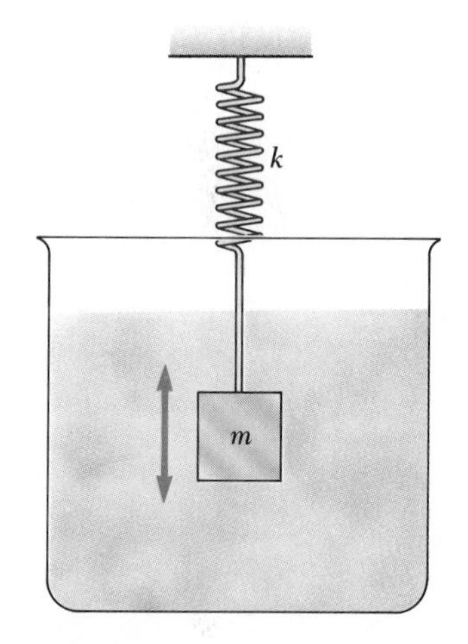


**Figure 32.17** A mass-spring system moving in a viscous medium with damped harmonic motion is analogous to an *RLC* circuit.

Comparing Equations 32.33 and 32.34, we see that $Q$ corresponds to $x$, $L$ corresponds to $m$, $R$ corresponds to the damping constant $b$, and $C$ corresponds to $1/k$, where $k$ is the force constant of the spring.

The analytical solution of Equation 32.33 is rather cumbersome and is usually covered in courses dealing with differential equations. Therefore, we shall give only a qualitative description of the circuit behavior.

In the simplest case, when $R = 0$, Equation 32.33 reduces to that of a simple *LC* circuit, as expected, and the charge and the current oscillate sinusoidally in time.

Next consider the situation where $R$ is reasonably small. In this case, the solution of Equation 32.33 is given by

$$Q = Q_m e^{-Rt/2L} \cos \omega_d t \tag{32.35}$$

where

$$\omega_d = \left[\frac{1}{LC} - \left(\frac{R}{2L}\right)^2\right]^{1/2} \tag{32.36}$$

That is, the charge will oscillate with *damped harmonic motion* in analogy with a mass-spring system moving in a viscous medium. From Equation 32.35, we see that when $R \ll \sqrt{4L/C}$ the frequency $\omega_d$ of the damped oscillator will be close to that of the undamped oscillator, $1/\sqrt{LC}$. Since $I = dQ/dt$, it follows that the current will also undergo damped harmonic motion. A plot of the charge versus time for the damped oscillator is shown in Figure 32.18. Note that the maximum value of $Q$ decreases after each oscillation, just as the amplitude of a damped harmonic oscillator decreases in time.

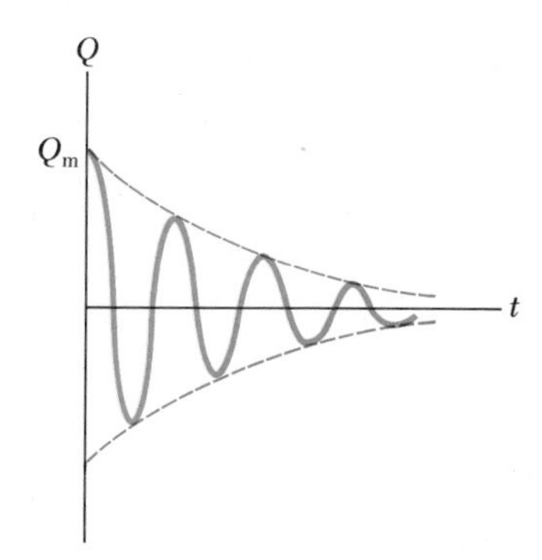


**Figure 32.18** Charge versus time for a damped *RLC* circuit. This occurs for $R \ll \sqrt{4L/C}$. The $Q$ versus $t$ curve represents a plot of Equation 32.35.

When we consider larger values of $R$, we find that the oscillations damp out more rapidly; in fact, there exists a critical resistance value $R_c$ above which

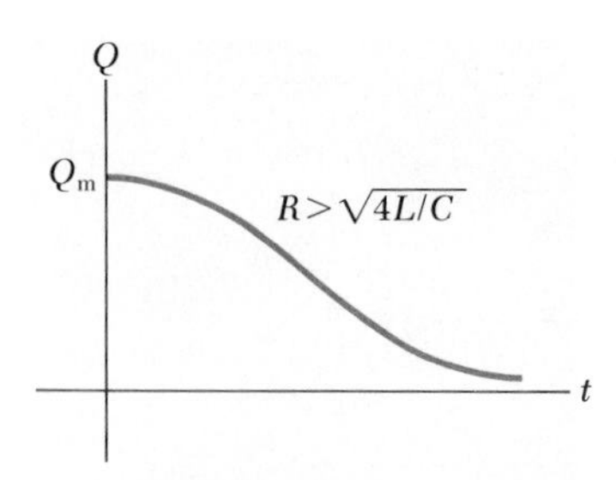


**Figure 32.19** Plot of $Q$ versus $t$ for an overdamped $RLC$ circuit, which occurs for values of $R > \sqrt{4L/C}$.

*no* oscillations occur. The critical value is given by $R_c = \sqrt{4L/C}$. A system with $R = R_c$ is said to be *critically damped.* When $R$ exceeds $R_c$, the system is said to be *overdamped* (Fig. 32.19).

## SUMMARY

When the current in a coil changes with time, an emf is induced in the coil according to Faraday's law. The **self-induced emf** is defined by the expression

Induced emf

$$\mathcal{E} = -L\frac{dI}{dt} \tag{32.1}$$

where $L$ is the *inductance* of the coil. Inductance is a measure of the opposition of a device to a *change* in current. Inductance has the SI unit of **henry** (H), where 1 H = 1 V·s/A.

The **inductance** *of any coil,* such as a solenoid or toroid, is given by the expression

Inductance of an $N$-turn coil

$$L = \frac{N\Phi_m}{I} \tag{32.2}$$

where $\Phi_m$ is the magnetic flux through the coil and $N$ is the total number of turns.

The inductance of a device depends on its geometry. For example, the *inductance of a solenoid* (whose core is a vacuum), as calculated from Equation 32.2, is given by

Inductance of a solenoid

$$L = \frac{\mu_0 N^2 A}{\ell} \tag{32.4}$$

where $N$ is the number of turns, $A$ is the cross-sectional area, and $\ell$ is the length of the solenoid.

If a resistor and inductor are connected in series to a battery of emf $\mathcal{E}$ as shown in Figure 32.2 and a switch in the circuit is closed at $t = 0$, the current in the circuit varies in time according to the expression

Current in an $RL$ circuit

$$I(t) = \frac{\mathcal{E}}{R}(1 - e^{-t/\tau}) \tag{32.7}$$

where $\tau = L/R$ is the *time constant* of the $RL$ circuit. That is, the current rises to an equilibrium value of $\mathcal{E}/R$ after a time that is long compared to $\tau$.

If the battery is removed from an $RL$ circuit, as in Figure 32.5 with $S_1$ open and $S_2$ closed, the current decays exponentially with time according to the expression

$$I(t) = \frac{\mathcal{E}}{R}e^{-t/\tau} \tag{32.11}$$

where $\mathcal{E}/R$ is the initial current in the circuit.

The **energy** *stored in the magnetic field of an inductor* carrying a current $I$ is given by

Energy stored in an inductor

$$U_m = \tfrac{1}{2}LI^2 \tag{32.13}$$

This result is obtained by applying the principle of energy conservation to the *RL* circuit.

The **energy per unit volume** (or energy density) at a point where the magnetic field is $B$ is given by

$$u_m = \frac{B^2}{2\mu_0} \tag{32.15}$$

Magnetic energy density

That is, the energy density is proportional to the square of the field at that point.

If two coils are close to each other, a changing current in one coil can induce an emf in the other coil. If $dI_1/dt$ is rate of change of current in the first coil, the emf induced in the second is given by

$$\mathcal{E}_2 = -M\frac{dI_1}{dt} \tag{32.19}$$

where $M$ is a constant called the **mutual inductance** of one coil with respect to the other.

If $\Phi_{21}$ is the magnetic flux through coil 2 due to the current $I_1$ in coil 1 and $N_2$ is the number of turns in coil 2, then the **mutual inductance** of coil 2 is given by

$$M_{21} = \frac{N_2\Phi_{21}}{I_1} \tag{32.16}$$

Mutual inductance

In an *LC* circuit with zero resistance, the charge on the capacitor and the current in the circuit vary in time according to the expressions

$$Q = Q_m \cos(\omega t + \delta) \tag{32.23}$$

$$I = \frac{dQ}{dt} = -\omega Q_m \sin(\omega t + \delta) \tag{32.25}$$

Charge and current versus time in an *LC* circuit

where $Q_m$ is the maximum charge on the capacitor, $\delta$ is a phase constant, and $\omega$ is the angular frequency of oscillation, given by

$$\omega = \frac{1}{\sqrt{LC}} \tag{32.24}$$

Frequency of oscillation in an *LC* circuit

The energy in an *LC* circuit continuously transfers between energy stored in the capacitor and energy stored in the inductor. The **total energy** of the *LC* circuit at any time $t$ is given by

$$U = U_C + U_L = \frac{Q_m{}^2}{2C}\cos^2 \omega t + \frac{LI_m{}^2}{2}\sin^2 \omega t \tag{32.28}$$

Energy of an *LC* circuit

where $I_m$ is the maximum current in the circuit. At $t = 0$, all of the energy is stored in the electric field of the capacitor ($U = Q_m{}^2/2C$). Eventually, all of this energy is transferred to the inductor ($U = LI_m{}^2/2$). However, the *total* energy remains constant since the energy losses are neglected in the ideal *LC* circuit.

The charge and current in an *RLC* circuit exhibit a damped harmonic behavior for small values of $R$. This is analogous to the damped harmonic motion of a mass-spring system in which friction is present.

## QUESTIONS

1. Why is the induced emf that appears in an inductor called a "counter" or "back" emf?
2. A circuit containing a coil, resistor, and battery is in steady state, that is, the current has reached a constant value. Does the coil have an inductance? Does the coil affect the value of the current in the circuit?
3. Does the inductance of a coil depend on the current in the coil? What parameters affect the inductance of a coil?
4. How can a long piece of wire be wound on a spool so that it has a negligible self-inductance?
5. A long fine wire is wound as a solenoid with a self-inductance $L$. If this is connected directly across the terminals of a battery, how does the maximum current depend on $L$?
6. For the series $RL$ circuit shown in Figure 32.20, can the back emf ever be greater than the battery emf? Explain.

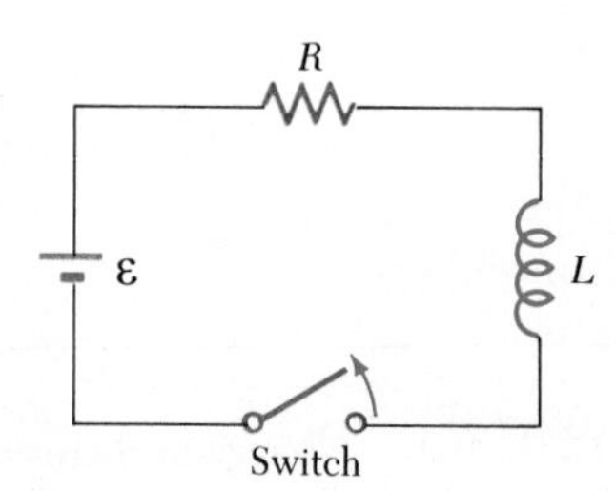


**Figure 32.20** (Questions 6 and 7).

7. Suppose the switch in the $RL$ circuit in Figure 32.20 has been closed for a long time and is suddenly opened. Does the current instantaneously drop to zero? Why does a spark tend to appear at the switch contacts when the switch is opened?
8. If the current in an inductor is doubled, by what factor does the stored energy change?
9. Discuss the similarities between the energy stored in the electric field of a charged capacitor and the energy stored in the magnetic field of a current-carrying coil.
10. What is the effective inductance of two isolated inductors, connected in series?
11. Discuss how the mutual inductance arises between the primary and secondary coils in a transformer.
12. The centers of two circular loops are separated by a fixed distance. For what relative orientation of the loops will their mutual inductance be a maximum? For what orientation will it be a minimum?
13. Two solenoids are connected in series such that each carries the same current at any instant. Is mutual induction present? Explain.
14. In the $LC$ circuit shown in Figure 32.12, the charge on the capacitor is sometimes zero, even though there is current in the circuit. How is this possible?
15. If the resistance of the wires in an $LC$ circuit were not zero, would the oscillations persist? Explain.
16. How can you tell whether an $RLC$ circuit will be over- or underdamped?
17. What is the significance of "critical damping" in an $RLC$ circuit?

## PROBLEMS

### Section 32.1 Self-Inductance

1. A 2-H inductor carries a steady current of 0.5 A. When the switch in the circuit is opened, the current disappears in 0.01 s. What is the induced emf that appears in the inductor during this time?
2. A "Slinky toy" spring has a radius of 4 cm and an inductance of 125 $\mu$H when extended to a length of 2 m. What is the total number of turns in the spring?
3. What is the inductance of a 510-turn solenoid that has a radius of 8 cm and an overall length of 1.4 m?
4. A small air-core solenoid has a length of 4 cm and a radius of 0.25 cm. If the inductance is to be 0.06 mH, how many turns per cm are required?
5. Show that the two expressions for inductance given by

$$L = \frac{N\Phi_m}{I} \quad \text{and} \quad L = -\frac{\mathcal{E}}{dI/dt}$$

have the same units.

6. Calculate the magnetic flux through a 300-turn, 7.2-mH coil when the current in the coil is 10 mA.
7. A 40-mA current is carried by a uniformly wound air-core solenoid with 450 turns, a 15-mm diameter, and 12-cm length. Compute (a) the magnetic field inside the solenoid, (b) the magnetic flux through each turn, and (c) the inductance of the solenoid. (d) Which of these quantities depends on the current?
8. A 0.388-mH inductor has a length that is four times its diameter. If it is wound with 22 turns per centimeter, what is its length?
9. An emf of 36 mV is induced in a 400-turn coil at an instant when the current has a value of 2.8 A and is changing at a rate of 12 A/s. What is the total magnetic flux through the coil?
10. The current in a 90-mH inductor changes with time as $I = t^2 - 6t$ (in SI units). Find the magnitude of the induced emf at (a) $t = 1$ s and (b) $t = 4$ s. (c) At what time is the emf zero?

11. A current $I = I_0 \sin \omega t$, with $I_0 = 5$ A and $\omega/2\pi = 60$ Hz, flows through an inductor whose inductance is 10 mH. What is the back emf as a function of time?

**12.** Three solenoidal windings of 300, 200, and 100 turns are wrapped at well-spaced positions along a cardboard tube of radius 1 cm. Each winding extends for 5 cm along the cylindrical surface. What is the equivalent inductance of the 600 turns when the three sets of windings are connected in series?

13. Two coils, A and B, are wound using *equal lengths* of wire. Each coil has the same number of turns per unit length, but coil A has twice as many turns as coil B. What is the ratio of the self-inductance of A to the self-inductance of B? (*Note:* The radii of the two coils are not equal.)

14. A toroid has a major radius $R$ and a minor radius $r$, and is tightly wound with $N$ turns of wire, as shown in Figure 32.21. If $R \gg r$, the magnetic field inside the toroid is essentially that of a long solenoid that has been bent into a large circle of radius $R$. Using the uniform field of a long solenoid, show that the self-inductance of such a toroid is given (approximately) by

$$L \cong \frac{\mu_0 N^2 A}{2\pi R}$$

(An exact expression for the inductance of a toroid with a rectangular cross-section is derived in Problem 78.)

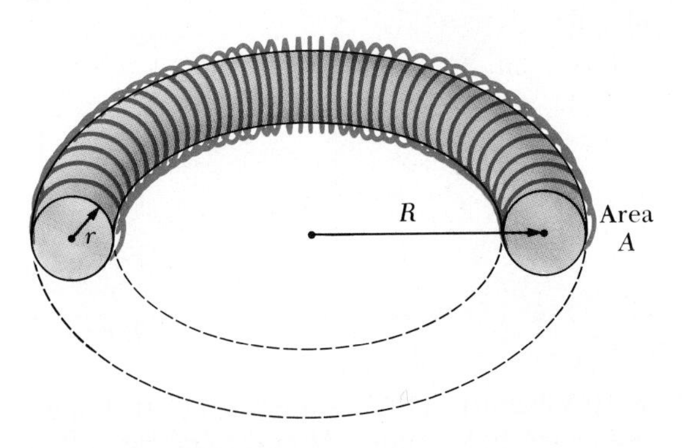


**Figure 32.21** (Problem 14).

**15.** A solenoid has 120 turns uniformly wrapped around a wooden core, which has a diameter of 10 mm and a length of 9 cm. (a) Calculate the inductance of this solenoid. (b) The wooden core is replaced with a soft iron rod with the same dimensions, but a magnetic permeability $\kappa_m = 800\mu_0$. What is the new inductance?

16. A solenoidal inductor contains 420 turns, is 16 cm in length, and has a cross-sectional area of 3 cm$^2$. What uniform rate of decrease of current through the inductor will produce an induced emf of 175 $\mu$V?

Section 32.2 *RL* Circuits

17. Verify by direct substitution that the expression for current given in Equation 32.7 is a solution of Kirchhoff's loop equation for the *RL* circuit as given by Equation 32.6.

18. Show that $I = I_0 e^{-t/\tau}$ is a solution of the differential equation

$$IR + L\frac{dI}{dt} = 0,$$

where $\tau = L/R$ and $I_0 = \mathcal{E}/R$ is the value of the current at $t = 0$.

19. Calculate the inductance in an *RL* circuit in which $R = 0.5\ \Omega$ and the current increases to one fourth its final value in 1.5 s.

**20.** A 12-V battery is connected in series with a resistor and an inductor. The circuit has a time constant of 500 $\mu$s, and the maximum current is 200 mA. What is the value of the inductance?

**21.** Show that the inductive time constant $\tau$ has SI units of seconds.

22. An inductor with an inductance of 15 H and resistance of 30 $\Omega$ is connected across a 100-V battery. (a) What is the *initial* rate of increase of current in the circuit? (b) At what rate is the current changing at $t = 1.5$ s?

**23.** A 12-V battery is about to be connected to a series circuit containing a 10-$\Omega$ resistor and a 2-H inductor. (a) How long will it take the current to reach 50% of its final value? (b) How long will it take to reach 90% of its final value?

24. Consider the circuit shown in Figure 32.22, taking $\mathcal{E} = 6$ V, $L = 8$ mH, and $R = 4\ \Omega$. (a) What is the inductive time constant of the circuit? (b) Calculate the current in the circuit at a time 250 $\mu$s after the switch $S_1$ is closed. (c) What is the value of the final steady-state current? (d) How long does it take the current to reach 80% of its maximum value?

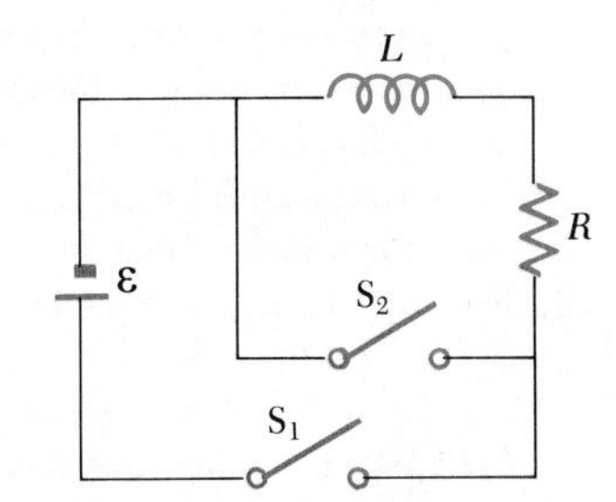


**Figure 32.22** (Problems 24, 27, and 28).

25. A 140-mH inductor and a 4.9-$\Omega$ resistor are connected with a switch to a 6-V battery as shown in Figure 32.23. (a) If the switch is thrown to the left (connecting the battery), how much time elapses before the current reaches 220 mA? (b) What is the current through the inductor 10 s after the switch is

closed? (c) Now the switch is quickly thrown from A to B. How much time elapses before the current falls to 160 mA?

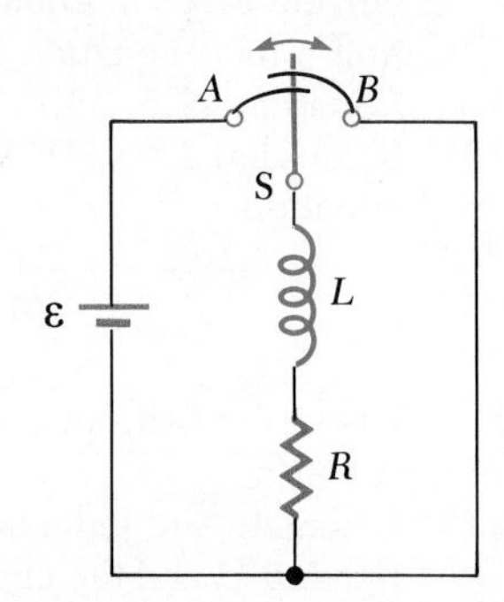


Figure 32.23 (Problem 25).

26. When the switch in Figure 32.24 is closed, the current takes 3.0 ms to reach 98% of its final value. If $R = 10\ \Omega$, what is the inductance $L$?

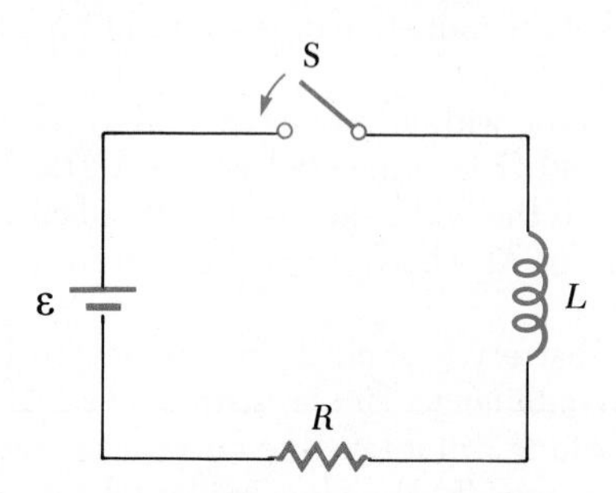


Figure 32.24 (Problems 26, 29, 30, 35).

27. Let the following values be assigned to the components in the circuit shown in Figure 32.22. $\mathcal{E} = 6$ V, $L = 24$ mH, and $R = 10\ \Omega$. (a) Calculate the current in the circuit at a time 0.5 ms after switch $S_1$ is closed. (b) What is the maximum value of the current in the circuit?
28. Assume that switch $S_1$ in the circuit of Figure 32.22 has been closed long enough for the current to reach its *maximum* value. If switch $S_1$ is now opened and switch $S_2$ closed at $t = 0$, after what time interval will the current in $R$ be 25% of the maximum value? (Use the numerical values given in Problem 27.)
29. For the $RL$ circuit shown in Figure 32.24, let $L = 3$ H, $R = 8\ \Omega$, and $\mathcal{E} = 36$ V. (a) Calculate the ratio of the potential difference across the resistor to that across the inductor when $I = 2$ A. (b) Calculate the voltage across the inductor when $I = 4.5$ A.
30. In the circuit shown in Figure 32.24 let $L = 7$ H, $R = 9\ \Omega$, and $\mathcal{E} = 120$ V. What is the self-induced emf 0.2 s after the switch is closed?
31. One application of an $RL$ circuit is the generation of high-voltage transients from a low-voltage dc source, as shown in Figure 32.25. (a) What is the current in the circuit a long time after the switch has been in position A? (b) Now the switch is thrown quickly from A to B. Compute the initial voltage across each resistor and the inductor. (c) How much time elapses before the voltage across the inductor drops to 12 V?

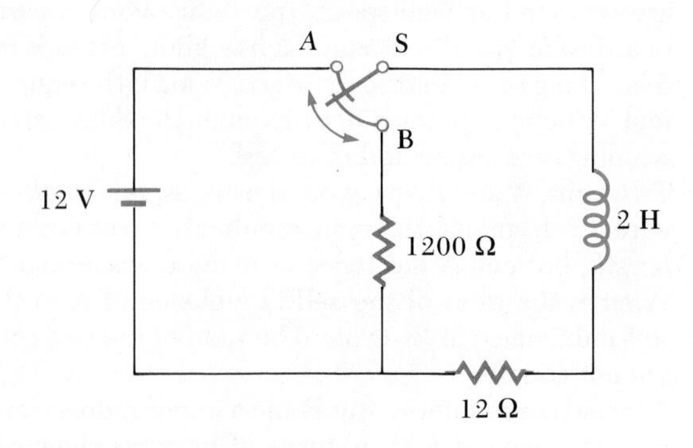


Figure 32.25 (Problem 31).

## Section 32.3 Energy in a Magnetic Field

32. Calculate the energy associated with the magnetic field of a 200-turn solenoid in which a current of 1.75 A produces a flux of $3.7 \times 10^{-4}$ Wb in each turn.
33. An air-core solenoid with 68 turns is 8 cm long and has a diameter of 1.2 cm. How much energy is stored in its magnetic field when it carries a current of 0.77 A?
34. Consider the circuit shown in Figure 32.26. What energy is stored in the inductor when the current reaches its final equilibrium value after the switch is closed?

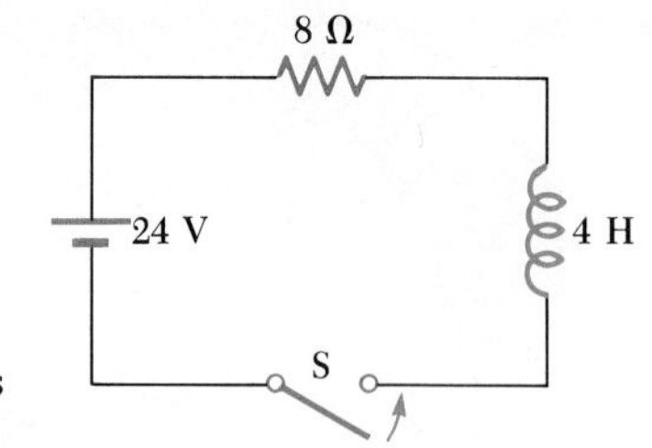


Figure 32.26 (Problems 34 and 42).

35. A 10-V battery, a 5-Ω resistor, and a 10-H inductor are connected in series. After the current in the circuit has reached its maximum value, calculate (a) the power supplied to the circuit by the battery, (b) the power dissipated in the resistor, (c) the power dissipated in the inductor, and (d) the energy stored in the magnetic field of the inductor.
36. At $t = 0$, a source of emf, $\mathcal{E} = 500$ V, is applied to a coil that has an inductance of 0.80 H and a resistance of 30 Ω. (a) Find the energy stored in the magnetic field when the current reaches half its maximum value. (b) How long after the emf is connected does it take for the current to reach this value?
37. The magnetic field inside a superconducting solenoid is 4.5 T. The solenoid has an inner diameter of 6.2 cm and a length of 26 cm. (a) Determine the magnetic

energy density in the field. (b) Determine the magnetic energy stored in the magnetic field within the solenoid.

**38.** A uniform *electric* field of magnitude $6.8 \times 10^5$ V/m throughout a cylindrical volume results in a total energy due to the electric field of 3.4 $\mu$J. What *magnetic* field over this same region will store the same total energy?

**39.** On a clear day, there is a vertical electric field near the earth's surface with a magnitude about 100 V/m. At the same time, the earth's magnetic field has a magnitude approximately $0.5 \times 10^{-4}$ T. Compute the energy density of the two fields.

**40.** A battery for which $\mathcal{E} = 15$ V is connected to an *RL* circuit for which $L = 0.6$ H and $R = 7\ \Omega$. When the current has reached one half of its final value, what is the total magnetic energy stored in the inductor?

**41.** Two inductors ($L_1 = 85\ \mu$H, and $L_2 = 200\ \mu$H) are connected in series with an 850-mA dc power supply. Calculate the energy stored in each inductor.

42. The switch in the circuit of Figure 32.26 is closed at $t = 0$. (a) Calculate the *rate* at which energy is being stored in the inductor after an elapsed time equal to the time constant of the circuit. (b) At what rate is energy being dissipated as joule heat in the resistor at this time? (c) What is the total energy stored in the inductor at this time?

43. An *RL* circuit in which $L = 4$ H and $R = 5\ \Omega$ is connected to a battery with $\mathcal{E} = 22$ V at time $t = 0$. (a) What energy is stored in the inductor when the current in the circuit is 0.5 A? (b) At what rate is energy being stored in the inductor when $I = 1$ A? (c) What power is being delivered to the circuit by the battery when $I = 0.5$ A?

44. The magnitude of the magnetic field outside a sphere of radius $R$ is given by $B = B_0(R/r)^2$, where $B_0$ is a constant. Determine the total energy stored in the magnetic field outside the sphere and evaluate your result for $B_0 = 5 \times 10^{-5}$ T and $R = 6 \times 10^6$ m, values appropriate for the earth's magnetic field.

### *Section 32.4 Mutual Inductance

**45.** Two nearby coils, A and B, have a mutual inductance $M = 28$ mH. What is the emf induced in coil A as a function of time when the current in coil B is given by $I = 3t^2 - 4t + 5$, where $I$ is in A when $t$ is in s?

**46.** Two coils, held in fixed positions, have a mutual inductance of 100 $\mu$H. What is the peak voltage in one of the coils when a sinusoidal current is given by $I(t) = (10\text{ A}) \sin(1000t)$ flows in the other coil?

**47.** An emf of 96 mV is induced in the windings of a coil when the current in a nearby coil is increasing at the rate of 1.2 A/s. What is the mutual inductance of the two coils?

48. A long solenoid consists of $N_1$ turns with a radius $R_1$. A second solenoid, with $N_2$ turns of radius $R_2$, has the same length as the first and lies entirely within the first solenoid, with their axes parallel. (a) Assume solenoid 1 carries a current $I$ and compute their mutual inductance. (b) Now assume that solenoid 2 carries the *same* current $I$ (and solenoid 1 carries no current) and compute their mutual inductance. Do you obtain the same result?

**49.** A coil of 50 turns is wound on a long solenoid as shown in Figure 32.10. The solenoid has a cross-sectional area of $8.8 \times 10^{-3}$ m$^2$ and is wrapped uniformly with 1000 turns per meter of length. Calculate the mutual inductance of the two windings.

50. A 70-turn solenoid is 5 cm long, 1 cm in diameter, and carries a 2-A current. A single loop of wire, 3 cm in diameter, is held perpendicular to the axis of the solenoid. What is the mutual inductance of the two if the plane of the loop passes through the center of the solenoid?

51. Two nearby solenoids, A and B, sharing the same cylindrical axis, have 400 and 700 turns, respectively. A current of 3.5 A in coil A produces a flux of 300 $\mu$Wb at the center of A and a flux of 90 $\mu$Wb at the center of B. (a) Calculate the mutual inductance of the two solenoids. (b) What is the self-inductance of coil A? (c) What emf will be induced in coil B when the current in coil A increases at the rate of 0.5 A/s?

52. Two single-turn circular loops of wire have radii $R$ and $r$, with $R \gg r$. The loops lie in the same plane and are concentric. (a) Show that the mutual inductance of the pair is $M = \mu_0 \pi r^2/2R$. (*Hint:* Assume that the larger loop carries a current $I$ and compute the resulting flux through the smaller loop.) (b) Evaluate $M$ for $r = 2$ cm and $R = 20$ cm.

### Section 32.5 Oscillations in an *LC* Circuit

53. A 1.0-$\mu$F capacitor is charged by a 40-V dc power supply. The fully-charged capacitor is then discharged through a 10-mH inductor. Find the *maximum* current that occurs in the resulting oscillations.

54. An *LC* circuit consists of a 20-mH inductor and a 0.5-$\mu$F capacitor. If the maximum instantaneous current in this circuit is 0.1 A, what is the greatest potential difference that appears across the capacitor?

55. An *LC* circuit of the type shown in Figure 32.11 has an inductance of 0.57 mH and a capacitance of 15 pF. The capacitor is charged to its maximum value by a 32-V battery. The battery is then removed from the circuit and the capacitor discharged through the inductor. (a) If all resistance in the circuit is neglected, determine the maximum value of the current in the oscillating circuit. (b) At what frequency does the circuit oscillate? (c) What is the maximum energy stored in the magnetic field of the inductor?

**56.** Calculate the inductance of an *LC* circuit that oscillates at a frequency of 120 Hz when the capacitance is 8 $\mu$F.

57. A fixed inductance $L = 1.05\ \mu H$ is used in series with a variable capacitor in the tuning section of a radio. What capacitance will tune the circuit into the signal from a station broadcasting at a frequency of 96.3 MHz?
58. An *LC* circuit (shown in Figure 32.11) contains an 82-mH inductor and a 17-$\mu$F capacitor, which initially carries a 180-$\mu$C charge. The switch is closed at $t = 0$. (a) Find the frequency (in Hz) of the resulting oscillations. At the instant $t = 1.0$ ms, find (b) the charge on the capacitor and (c) the current in the circuit.
59. (a) What capacitance must be combined with a 45-mH inductor in order to achieve a resonant frequency of 125 Hz? (b) What time interval elapses between accumulations of maximum charge of the *same* sign on a given plate of the capacitor?
60. An *LC* circuit carries a current that is oscillating with a period $T$. If the charge on the capacitor is at a maximum at $t = 0$, when will the energy stored in the electric field of the capacitor equal the energy stored in the magnetic field of the inductor? (Express your answer as a fraction of $T$.)
61. An *LC* circuit consists of a 3.3-H inductor and an 840-pF capacitor, initially carrying a 105-$\mu$C charge. At $t = 0$ the switch in Figure 32.11 is closed. Compute the following quantities at the instant $t = 2.0$ ms: (a) the energy stored in the capacitor; (b) the energy stored in the inductor; (c) the total energy in the circuit.
62. A 6-V battery is used to charge a 50-$\mu$F capacitor. The capacitor is then discharged through a 0.34-mH inductor. (a) Find the maximum charge on the capacitor. (b) Compute the maximum current in the circuit. (c) Compute the maximum energy stored in each of these components.

## *Section 32.6 The *RLC* Circuit

63. Consider the circuit shown in Figure 32.16. Let $R = 7.6\ \Omega$, $L = 2.2$ mH, and $C = 1.8\ \mu F$. (a) Calculate the frequency of the damped oscillation of the circuit. (b) What is the value of the critical resistance in the circuit?
64. Consider a series *LC* circuit ($L = 2.18$ H, $C = 6$ nF). What is the maximum value of a resistor that, if inserted in series with $L$ and $C$, will allow the circuit to continue to oscillate?
65. Consider an *LC* circuit with $L = 500$ mH and $C = 0.1\ \mu F$. (a) What is the resonant frequency ($\omega_0$) of this circuit? (b) If a resistance of 1000 $\Omega$ is introduced into this circuit, what would be the frequency of the (damped) oscillations? (c) What is the percent difference between the two frequencies?
66. Electrical oscillations are initiated in a series circuit with a capacitance $C$, inductance $L$, and resistance $R$. (a) If $R \ll \sqrt{4L/C}$ (weak damping), how much time elapses before the current amplitude in the circuit falls off to 50% of its initial value? (b) How long does it take for the energy in the circuit to decrease to 50% of its initial value?
67. (a) Show that the ratio of the oscillation frequency of a damped *LC* oscillator ($\omega_d$) to that of an undamped oscillator ($\omega_0$) may be expressed as

$$\frac{\omega_d}{\omega_0} = \sqrt{1 - \frac{R^2C}{4L}}$$

(b) What happens to this ratio if $L < 4R^2C$?
68. Consider an *RLC* series circuit consisting of a charged 500-$\mu$F capacitor connected to a 32-mH inductor and a resistor $R$. Calculate the frequency of the oscillations (in Hz) that result for the following values of $R$: (a) $R = 0$ (no damping); (b) $R = 16\ \Omega$ (critical damping: $R = \sqrt{4L/C}$); (c) $R = 4\ \Omega$ (underdamped: $R < \sqrt{4L/C}$); (d) $R = 64\ \Omega$ (overdamped: $R > \sqrt{4L/C}$).

## ADDITIONAL PROBLEMS

69. An inductor that has a resistance of 0.5 $\Omega$ is connected to a 5-V battery. One second after the switch is closed, the current through the circuit is 4 A. Calculate the inductance.
70. A soft iron rod ($\mu = 800\mu_0$) is used as the core of a solenoid. The rod has a diameter of 24 mm and is 10 cm long. A 10-m piece of 22-gauge copper wire (diameter = 0.644 mm) is wrapped around the iron rod in a single uniform layer, except for a 10-cm length at each end, to be used for connections. (a) How many turns of this wire can be wrapped around the rod? (*Hint:* The diameter of the wire adds to the diameter of the rod in determining the circumference of each turn. Also, the wire spirals diagonally along the surface of the rod.) (b) What is the resistance of this inductor? (c) What is its inductance?
71. A time-varying current $I$ is applied to an inductance of 5 H, as shown in Figure 32.27. Make a quantitative graph of the potential at point $a$ relative to that at point $b$. The current arrow indicates the direction of conventional current.

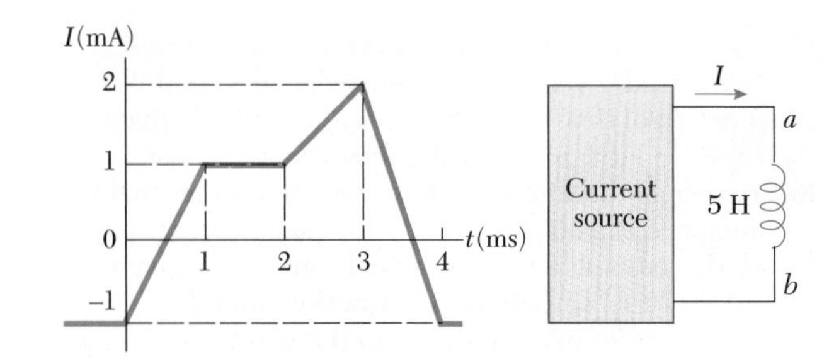


**Figure 32.27** (Problem 71).

72. A series *LR* circuit has $L = 0.1$ H and $R = 6\ \Omega$. At a certain moment the current is 5 A. (a) What is the

potential difference across the inductor? (b) Calculate the time rate of change of current.

73. The inductor in the circuit in Figure 32.28 has negligible resistance. When the switch is opened after having been closed for a long time, the current in the inductor drops to 0.25 A in 0.15 s. What is the inductance of the inductor?

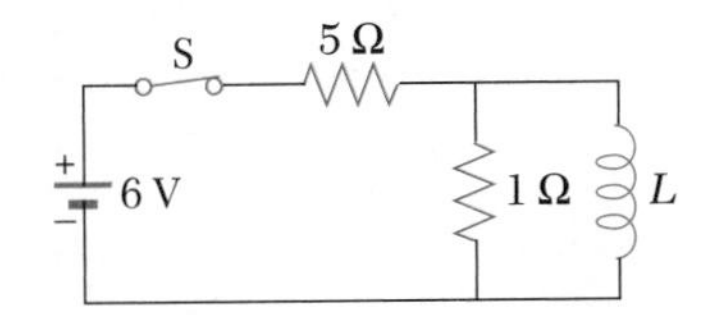


**Figure 32.28** (Problem 73).

74. A platinum wire 2.5 mm in diameter is connected in series to a 100-$\mu$F capacitor and a $1.2 \times 10^{-3}$-$\mu$H inductor to form an *RLC* circuit. The resistivity of platinum is $11 \times 10^{-8}\ \Omega \cdot$ m. Calculate the *maximum* length of wire for which the current in the circuit will oscillate.

75. Assume that the switch in the circuit shown in Figure 32.29 is initially in position 1. Show that if the switch is thrown from position 1 to position 2, all the energy stored in the magnetic field of the inductor will be dissipated as thermal energy in the resistor.

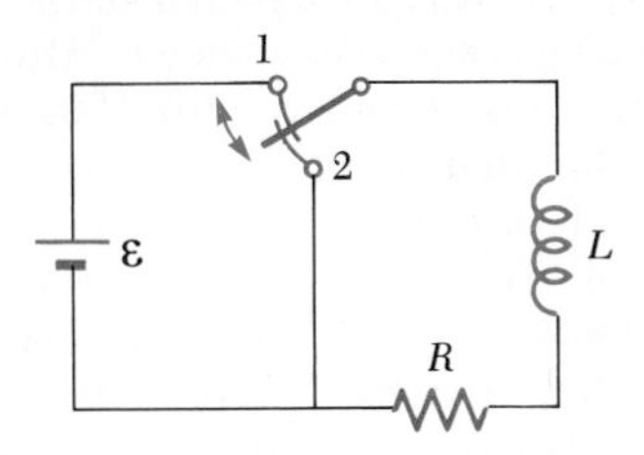


**Figure 32.29** (Problem 75).

76. The lead-in wires from a TV antenna are often constructed in the form of two parallel wires (Fig. 32.30).

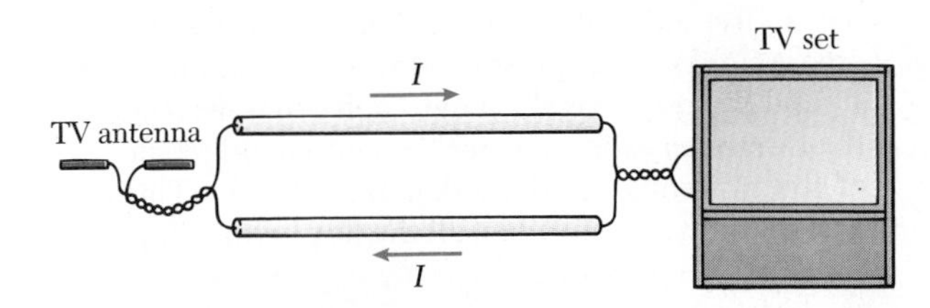


**Figure 32.30** (Problem 76).

(a) Why does this configuration of conductors have an inductance? (b) What constitutes the flux loop for this configuration? (c) Neglecting any magnetic flux inside the wires, show that the inductance of a length $x$ of this type of lead-in is

$$L = \frac{\mu_0 x}{\pi} \ln \left( \frac{w - a}{a} \right)$$

where $a$ is the radius of the wires and $w$ is the center-to-center separation of the wires.

77. At $t = 0$, the switch in Figure 32.31 is closed. By using Kirchhoff's laws for the instantaneous currents and voltages in this two-loop circuit, show that the current through the inductor is given by

$$I(t) = \frac{\mathcal{E}}{R_1} [1 - e^{-(R'/L)t}]$$

where $R' = R_1 R_2/(R_1 + R_2)$.

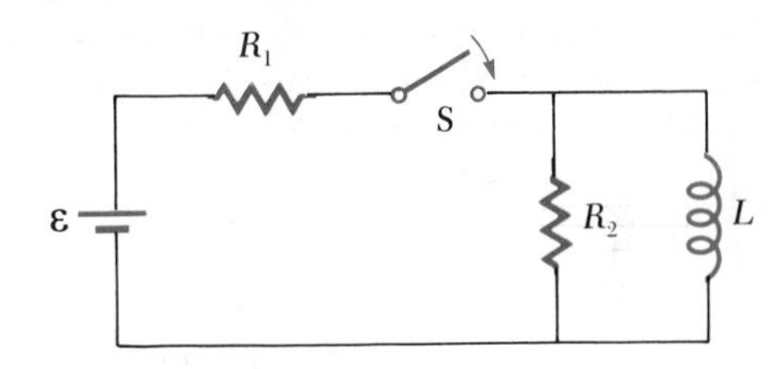


**Figure 32.31** (Problem 77).

78. The toroidal coil shown in Figure 32.32 consists of $N$ turns and has a rectangular cross section. Its inner and outer radii are $a$ and $b$, respectively. (a) Show that the self inductance of the coil is given by

$$L = \frac{\mu_0 N^2 h \ \ \ln}{2\pi \ \ \ b/a}$$

(b) Using this result, compute the self-inductance of a 500-turn toroid with $a = 10$ cm, $b = 12$ cm, and $h = 1$ cm. (c) In Problem 14, an approximate formula for the inductance of a toroid with $R \gg r$ was derived. To get a feel for the accuracy of this result, use the expression in Problem 14 to compute the (approximate) inductance of the toroid described in part (b).

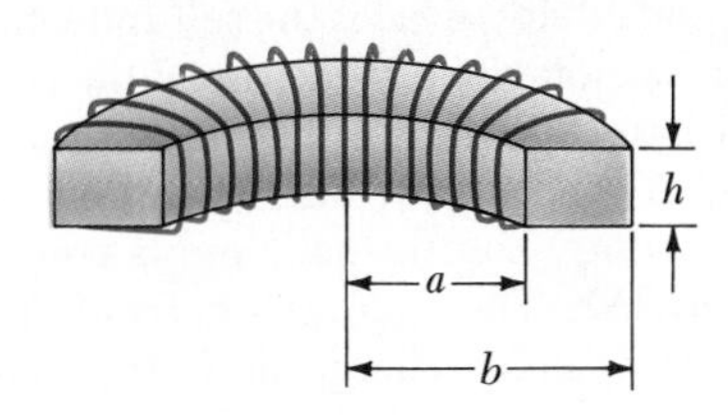


**Figure 32.32** (Problem 78).

79. In Figure 32.33, the switch is closed, and steady-state conditions are established in the circuit. The switch is now opened at $t = 0$. (a) Find the initial voltage $\mathcal{E}_0$ across $L$ just after $t = 0$. Which end of the coil is at the higher potential: $a$ or $b$? (b) Make freehand graphs of the currents in $R_1$ and in $R_2$ vs. $t$, treating the steady-state directions as positive. Show values before and

after $t = 0$. (c) How long after $t = 0$ does the magnitude of the current in $R_2$ drop exponentially to 2 mA?

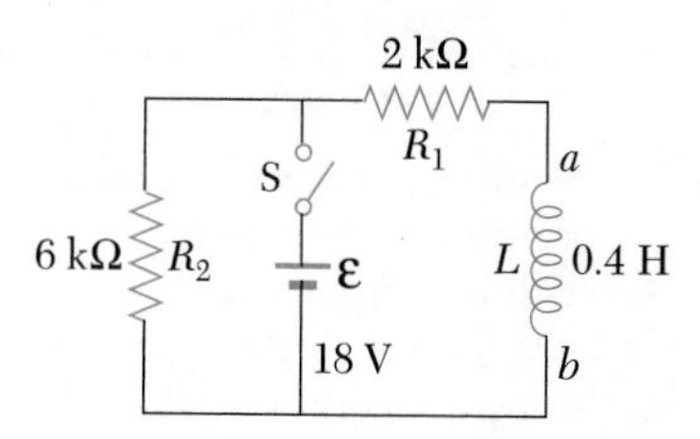


**Figure 32.33** (Problem 79).

80. The switch in the circuit in Figure 32.34 is closed at $t = 0$. Before the switch is closed, the capacitor is uncharged, and all currents are zero. Determine the currents in $L$, $C$, and $R$ and the potential differences across $L$, $C$, and $R$ (a) the instant after closing the switch, (b) long after the switch is closed.

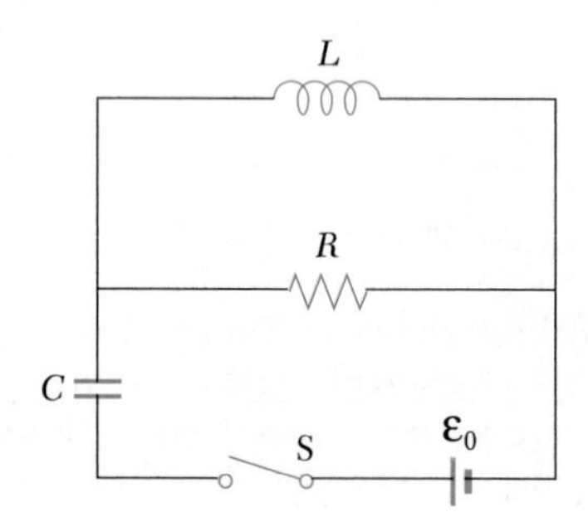


**Figure 32.34** (Problem 80).

81. Two long parallel wires, each of radius $a$, have their centers a distance $d$ apart and carry *equal* currents in *opposite* directions. Neglecting the flux within the wires themselves, calculate the inductance per unit length of such a pair of wires.
82. An air-core solenoid 0.5 m in length contains 1000 turns and has a cross-sectional area of 1 cm$^2$. (a) Neglecting end effects, what is the self-inductance? (b) A secondary winding wrapped around the center of the solenoid has 100 turns. What is the mutual inductance? (c) A constant current of 1 A flows in the secondary winding, and the solenoid is connected to a load of $10^3$ Ω. The constant current is suddenly stopped. How much charge flows through the load resistor?
83. To prevent damage from arcing in an electric motor, a discharge resistor is sometimes placed in parallel with the armature. If the motor is suddenly unplugged while running, this resistor limits the voltage that appears across the armature coils. Consider a 12-V dc motor that has an armature with a resistance of 7.5 Ω and an inductance of 450 mH. Assume the counter emf in the armature coils is 10 V when the motor is running at normal speed. (The equivalent circuit for the armature is shown in Figure 32.35.) Calculate the maximum resistance $R$ that will limit the voltage across the armature to 80 V when the motor is unplugged.

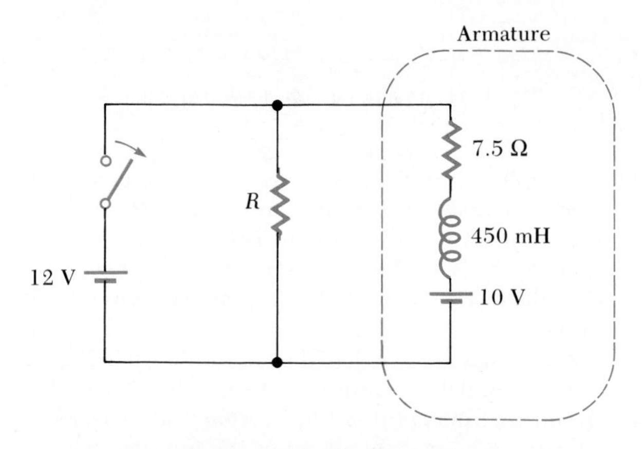


**Figure 32.35** (Problem 83).

84. A battery is in series with a switch and a 2-H inductor whose windings have a resistance $R$. After the switch is closed, the current rises to 80% of its final value in 0.4 s. Find the value of $R$.
85. A flat coil of wire has an inductance of 2 H and a resistance of 40 Ω. At $t = 0$, a battery of emf, $\mathcal{E} = 60$ V, is connected to the coil. Consider the state of affairs one time constant later. At this instant, find (a) the power delivered by the battery, (b) the Joule power developed in the resistance of the windings, and (c) the instantaneous rate at which energy is being stored in the magnetic field.
86. A toroidal solenoid has two separate sets of windings that are each spread uniformly around the toroid, with total turns $N_1$ and $N_2$, respectively. The toroid has a circumferential length $\ell$ and a cross-sectional area $A$. (a) Write expressions for the self-inductances $L_1$ and $L_2$, respectively, when each coil is used alone. (b) Derive an expression for the mutual inductance $M$ of the two coils. (c) Show that $M^2 = L_1L_2$. (This expression is true only when all the flux linking one coil also links the other coil.)

# 33
# Alternating Current Circuits

*High-voltage transmission lines at a power station in Oxfordshire, England. (© E. Nagele, FPG International)*

In this chapter, we shall describe the basic principles of simple alternating current (ac) circuits. We shall investigate the characteristics of circuits containing familiar elements and driven by a sinusoidal voltage. Our discussion will be limited to analyzing simple series circuits containing a resistor, inductor, and capacitor, both individually and in combination with each other. We shall make use of the fact that these elements respond linearly; that is, the ac current through each element is proportional to the instantaneous ac voltage across the element. We shall find that when the applied voltage of the generator is sinusoidal, the current in each element is also sinusoidal, but not necessarily in phase with the applied voltage. We conclude the chapter with two sections concerning the characteristics of *RC* filters, transformers, and power transmission.

## 33.1 AC SOURCES AND PHASORS

An ac circuit consists of combinations of circuit elements and a generator which provides the alternating current. The basic principles of the ac generator were described in Section 31.5. By rotating a coil in a magnetic field with constant angular velocity $\omega$, a sinusoidal voltage (emf) is induced in the coil. This instantaneous voltage, $v$, is given by

$$v = V_m \sin \omega t \tag{33.1}$$

where $V_m$ is the *peak voltage of the ac generator,* or the **voltage amplitude.** The angular frequency, $\omega$, is given by

$$\omega = 2\pi f = \frac{2\pi}{T}$$

where $f$ is the frequency of the source and $T$ is the period. Commercial electric-power plants in the United States use a frequency of $f = 60$ Hz (cycles per second), which corresponds to an angular frequency of $\omega = 377$ rad/s.

The primary aim of this chapter can be summarized as follows: Consider an ac generator connected to a series circuit containing $R$, $L$, and $C$ elements. If the voltage amplitude and frequency of the generator are given, together with the values of $R$, $L$, and $C$, find the resulting current as specified by its amplitude and its phase constant. In order to simplify our analysis of more complex circuits containing two or more elements, we shall use graphical constructions called *phasor diagrams.* In these constructions, alternating quantities, such as current and voltage, are represented by rotating vectors called **phasors.** The length of the phasor represents the amplitude (maximum value) of the quantity, while the projection of the phasor onto the vertical axis represents the instantaneous value of that quantity. Phasors rotate counterclockwise. As we shall see, the method of combining several sinusoidally varying currents or voltages with different phases is greatly simplified using this procedure. We will also use phasor diagrams in Chapters 37 and 38, which concern interference and diffraction of light waves.

## 33.2 RESISTORS IN AN AC CIRCUIT

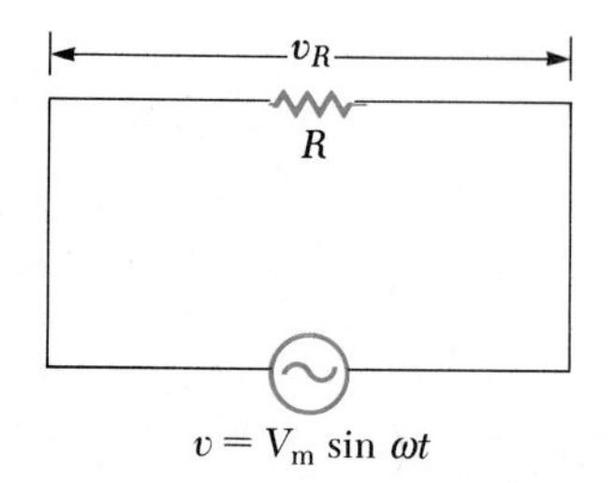


**Figure 33.1** A circuit consisting of a resistor $R$ connected to an ac generator, designated by the symbol —⊚—.

Consider a simple ac circuit consisting of a resistor and an ac generator (designated by the symbol —⊚—), as in Figure 33.1. At any instant, the algebraic sum of the potential increases and decreases around a closed loop in a circuit must be zero (Kirchhoff's loop equation). Therefore, $v - v_R = 0$, or

$$v = v_R = V_m \sin \omega t \tag{33.2}$$

where $v_R$ is the *instantaneous voltage drop across the resistor.* Therefore, the instantaneous current is equal to

$$i_R = \frac{v}{R} = \frac{V_m}{R} \sin \omega t = I_m \sin \omega t \tag{33.3}$$

where $I_m$ is the *peak current,* given by

Maximum current in a resistor

$$I_m = \frac{V_m}{R} \tag{33.4}$$

From Equations 33.2 and 33.3, we see that the instantaneous voltage drop across the resistor is

$$v_R = I_m R \sin \omega t \tag{33.5}$$

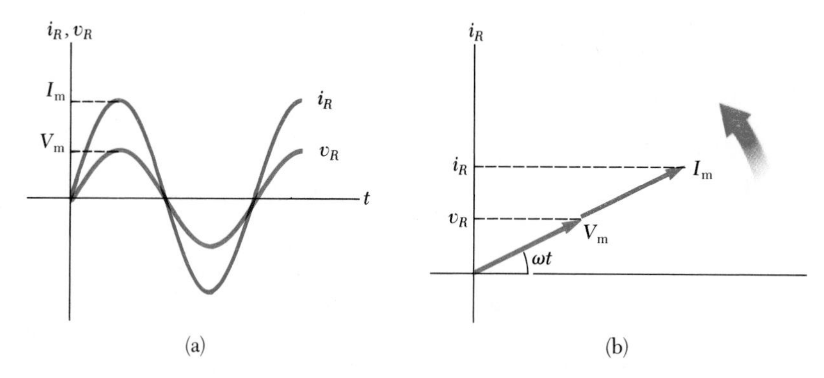


**Figure 33.2** (a) Plots of the current and voltage across a resistor as functions of time. The current is in phase with the voltage. (b) A phasor diagram for the resistive circuit, showing that the current is *in phase* with the voltage. The projections of the rotating arrows (phasors) onto the vertical axis represent the instantaneous values $v_R$ and $i_R$.

Since $i_R$ and $v_R$ both vary as sin $\omega t$ and reach their peak values at the *same time*, they are said to be *in phase*. Graphs of the voltage and current as functions of time (Fig. 33.2a) show that they each reach their peak and zero values at the same instant.

A phasor diagram may be used to represent the phase relationship between current and voltage. The lengths of the arrows correspond to $V_m$ and $L_m$. The projections of the arrows onto the vertical axis give $v_R$ and $i_R$. In the case of the single-loop resistive circuit, the current and voltage phasors lie along the same line, as in Figure 33.2b, since $i_R$ and $v_R$ are *in phase with each other*.

The current is in phase with the voltage for a resistor

Note that *the average value of the current over one cycle is zero*. That is, the current is maintained in one direction (the positive direction) for the same amount of time and at the same magnitude as in the opposite direction (the negative direction). However, the direction of the current has no effect on the behavior of the resistor in the circuit. This can be understood by realizing that collisions between electrons and the fixed atoms of the resistor result in an increase in the temperature of the resistor. Although this temperature increase depends on the magnitude of the current, it is independent of the direction of the current.

This discussion can be made quantitative by recalling that the rate at which electrical energy is converted to heat in a resistor, which is the power $P$, is given by

Power

$$P = i^2R$$

where $i$ is the instantaneous current in the resistor. Since the heating effect of a current is proportional to the *square* of the current, it makes no difference whether the current is direct or alternating, that is, whether the sign associated with the current is positive or negative. However, the heating effect produced by an alternating current having a maximum value of $I_m$ *is not the same* as that produced by a direct current of the same value. This is because the alternating current is at this maximum value for only a very brief instant of time during a cycle. What is of importance in an ac circuit is an average value of current referred to as the rms current. The term **rms** refers to *root mean square*, which simply means that one takes the square root of the average value of the square of the current. $I^2$ varies as $\sin^2 \omega t$, and one can show[1] that the average value of $i^2$ is $\frac{1}{2}I_m^{\,2}$ (Fig. 33.3). Therefore, the rms current, $I_{rms}$, is related to the peak value of the alternating current, $I_m$, as

rms current

$$I_{rms} = \frac{I_m}{\sqrt{2}} = 0.707 I_m \tag{33.6}$$

This equation says that an alternating current whose maximum value is 2A will produce the same heating effect in a resistor as a direct current of $(0.707)(2) = 1.414$ A. Thus, we can say that the average power dissipated in a resistor that carries an alternating current is $P_{av} = I_{rms}^{\,2}R$.

Alternating voltages are also best discussed in terms of rms voltages, and the relationship here is identical to the above, that is, the rms voltage, $V_{rms}$, is related to the peak value of the alternating voltage, $V_m$, as

rms voltage

$$V_{rms} = \frac{V_m}{\sqrt{2}} = 0.707 V_m \tag{33.7}$$

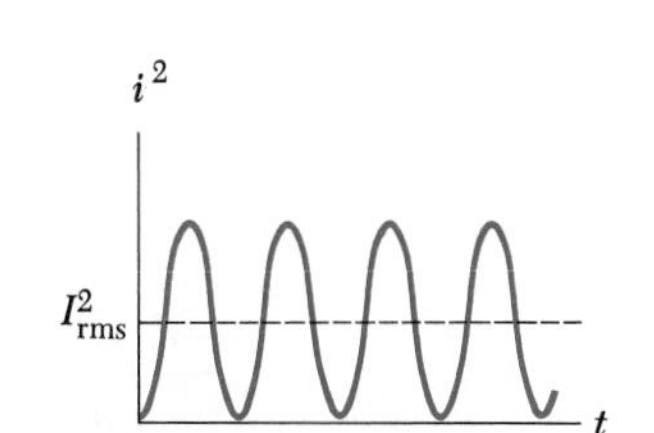


**Figure 33.3** Plot of the square of the current in a resistor versus time. The rms current is the square root of the average of the square of the current.

When one speaks of measuring an ac voltage of 120 V from an electric outlet, one is really referring to an rms voltage of 120 V. A quick calculation using Equation 33.7 shows that such an ac voltage actually has a peak value of about 170 V. In this chapter we shall use rms values when discussing alternating currents and voltages. One reason for this is that ac ammeters and voltmeters are designed to read rms values. Furthermore, we shall find that if we use rms values, many of the equations we use will have the same form as those used in the study of direct current (dc) circuits. Table 33.1 summarizes the notation that will be used in this chapter.

**TABLE 33.1 Notation Used in This Chapter**

| | Voltage | Current |
|---|---|---|
| Instantaneous value | $v$ | $i$ |
| Peak value | $V_m$ | $I_m$ |
| rms value | $V_{rms}$ | $I_{rms}$ |

[1] The fact that the square root of the average value of the square of the current is equal to $I_m/\sqrt{2}$ can be shown as follows. The current in the circuit varies with time according to the expression $i = I_m \sin \omega t$, so that $i^2 = I_m^{\,2} \sin^2 \omega t$. Therefore we can find the average value of $i^2$ by calculating the average value of $\sin^2 \omega t$. Note that a graph of $\cos^2 \omega t$ versus time is identical to a graph of $\sin^2 \omega t$ versus time, except that the points are shifted on the time axis. Thus, the time average of $\sin^2 \omega t$ is equal to the time average of $\cos^2 \omega t$ when taken over one or more complete cycles. That is,

$$(\sin^2 \omega t)_{av} = (\cos^2 \omega t)_{av}$$

With this fact and the trigonometric identity $\sin^2 \theta + \cos^2 \theta = 1$, we get

$$(\sin^2 \omega t)_{av} + (\cos^2 \omega t)_{av} = 2(\sin^2 \omega t)_{av} = 1$$

or

$$(\sin^2 \omega t)_{av} = \tfrac{1}{2}$$

When this result is substituted in the expression $i^2 = I_m^{\,2} \sin^2 \omega t$, we get $(i^2)_{av} = I_{rms}^{\,2} = I_m^{\,2}/2$, or $I_{rms} = I_m/\sqrt{2}$, where $I_{rms}$ is the rms current. The factor of $1/\sqrt{2}$ is only valid for sinusoidally varying currents. Other waveforms such as sawtooth variations have different factors.

**EXAMPLE 33.1 What is the rms Current?**
An ac voltage source has an output given by the expression $v = 200 \sin \omega t$. This source is connected to a 100-$\Omega$ resistor as in Figure 33.1. Find the rms current in the circuit.

*Solution* Compare the expression for the voltage output given above with the general form, $v = V_m \sin \omega t$. We see that the peak output voltage of the device is 200 V. Thus, the rms voltage output of the source is

$$V_{rms} = \frac{V_m}{\sqrt{2}} = \frac{200 \text{ V}}{\sqrt{2}} = 141 \text{ V}$$

Ohm's law can be used in resistive ac circuits as well as in dc circuits. The calculated rms voltage can be used with Ohm's law to find the rms current in the circuit:

$$I_{rms} = \frac{V_{rms}}{R} = \frac{141 \text{ V}}{100 \ \Omega} = 1.41 \text{ A}$$

Exercise 1 Find the peak current in the circuit.
Answer 2 A.

## 33.3 INDUCTORS IN AN AC CIRCUIT

Now consider an ac circuit consisting only of an inductor connected to the terminals of an ac generator as in Figure 33.4. If $v_L$ is the *instantaneous voltage drop across the inductor*, then Kirchhoff's loop rule applied to this circuit gives $v + v_L = 0$, or

$$v - L\frac{di}{dt} = 0$$

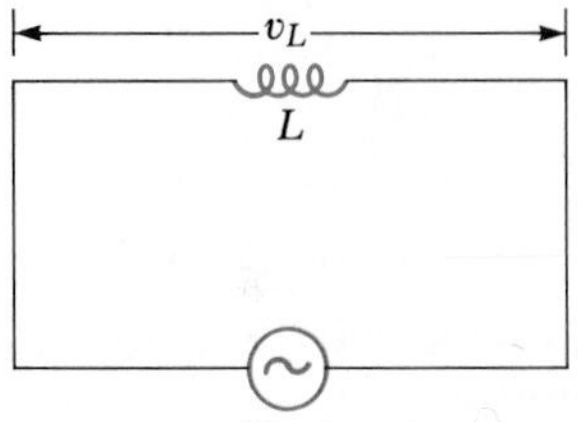

When we rearrange this equation and substitute $v = V_m \sin \omega t$, we get

$$L\frac{di}{dt} = V_m \sin \omega t \qquad (33.8)$$

$v_L$
$L$
$v = V_m \sin \omega t$

**Figure 33.4** A circuit consisting of an inductor $L$ connected to an ac generator.

Integrating this expression[2] gives the current as a function of time:

$$i_L = \frac{V_m}{L}\int \sin \omega t \, dt = -\frac{V_m}{\omega L}\cos \omega t \qquad (33.9)$$

Using the trigonometric identity $\cos \omega t = -\sin(\omega t - \pi/2)$, Equation 33.9 can also be expressed as

$$i_L = \frac{V_m}{\omega L}\sin\left(\omega t - \frac{\pi}{2}\right) \qquad (33.10)$$

Comparing this result with Equation 33.8 clearly shows that the current is out of phase with the voltage by $\pi/2$ rad, or 90°. A plot of the voltage and current versus time is given in Figure 33.5a. The voltage reaches its peak value at a time that is one quarter of the oscillation period *before* the current reaches its peak value. The corresponding phasor diagram for this circuit is shown in Figure 33.5b. Thus we see that

> for a sinusoidal applied voltage, the current always lags behind the voltage across an inductor by 90°.

The current in an inductor lags the voltage by 90°

This can be understood by noting that since the voltage across the inductor is proportional to $di/dt$, the value of $v_L$ is largest when the current is changing most rapidly. Since $i$ versus $t$ is a sinusoidal curve, $di/dt$ (the slope) is maximum when the curve goes through zero. This shows that $v_L$ reaches its maximum value when the current is zero.

[2] The constant of integration is neglected here since it depends on the initial conditions, which are not important for this situation.

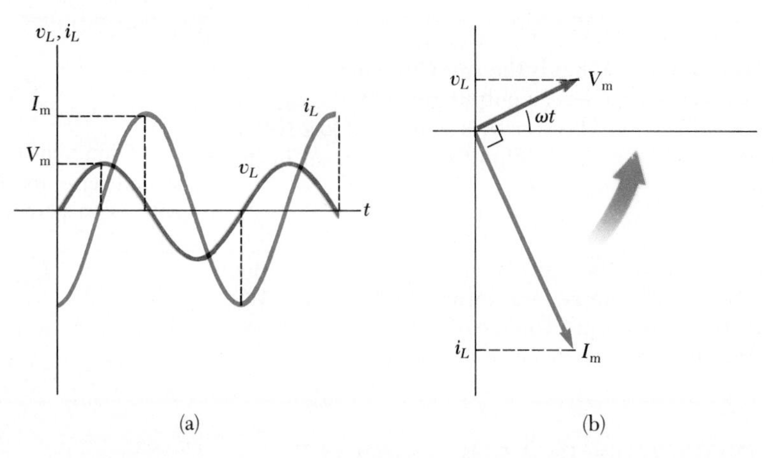


**Figure 33.5** (a) Plots of the current and voltage across the inductor as functions of time. The voltage *leads* the current by 90°. (b) The phasor diagram for the inductive circuit. Projections of the phasors onto the vertical axis give the instantaneous values $v_L$ and $i_L$.

From Equation 33.9 we see that the peak current $I_m$ is

Maximum current in an inductor

$$I_m = \frac{V_m}{\omega L} = \frac{V_m}{X_L} \tag{33.11}$$

where the quantity $X_L$, called the **inductive reactance,** is given by

Inductive reactance

$$X_L = \omega L \tag{33.12}$$

The rms current is given by an expression similar to Equation 33.11, with $V_m$ replaced by $V_{rms}$.

The term *reactance* is used so that it is not confused with *resistance.* Reactance is distinguished from resistance by the phase difference between $v$ and $i$. Recall that $i$ and $v$ are always in phase in a purely resistive circuit, whereas $i$ lags behind $v$ by 90° in a purely inductive circuit.

Using Equations 33.8 and 33.11, we find that the instantaneous voltage drop across the inductor can be expressed as

$$v_L = V_m \sin \omega t = I_m X_L \sin \omega t \tag{33.13}$$

We can think of Equation 33.13 as Ohm's law for an inductive circuit. It is left as a problem (Problem 8) to show that $X_L$ has the SI unit of ohm.

Note that the reactance of an inductor increases with increasing frequency. This is because at higher frequencies, the current must change more rapidly, which in turn causes an increase in the induced emf associated with a given peak current.

**EXAMPLE 33.2 A Purely Inductive AC Circuit**
In a purely inductive ac circuit (Fig. 33.4), $L = 25$ mH and the rms voltage is 150 V. Find the inductive reactance and rms current in the circuit if the frequency is 60 Hz.

*Solution* First, note that $\omega = 2\pi f = 2\pi(60) = 377\ \text{s}^{-1}$. Equation 33.12 then gives

$$X_L = \omega L = (377\ \text{s}^{-1})(25 \times 10^{-3}\ \text{H}) = 9.43\ \Omega$$

The rms current is given by

$$I_{rms} = \frac{V_L}{X_L} = \frac{150\text{ V}}{9.43\ \Omega} = 15.9\text{ A}$$

Exercise 2 Calculate the inductive reactance and rms current in the circuit if the frequency is 6 kHz.
Answers $X_L = 943\ \Omega$, $I_{rms} = 0.159$ A.

## 33.4 CAPACITORS IN AN AC CIRCUIT

Figure 33.6 shows an ac circuit consisting of a capacitor connected across the terminals of an ac generator. Kirchhoff's loop rule applied to this circuit gives $v - v_C = 0$, or

$$v = v_C = V_m \sin \omega t \tag{33.14}$$

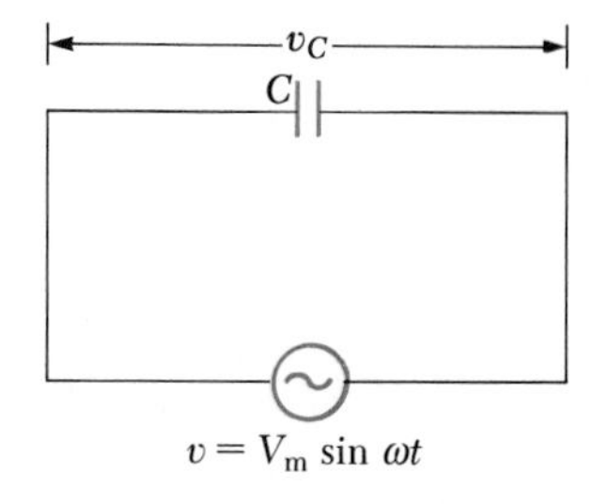


**Figure 33.6** A circuit consisting of a capacitor $C$ connected to an ac generator.

where $v_C$ is the *instantaneous voltage drop across the capacitor*. But from the definition of capacitance, $v_C = Q/C$, which when substituted into Equation 33.14 gives

$$Q = CV_m \sin \omega t \tag{33.15}$$

Since $i = dQ/dt$, differentiating Equation 33.15 gives the instantaneous current:

$$i_C = \frac{dQ}{dt} = \omega CV_m \cos \omega t \tag{33.16}$$

Again, we see that the current is not in phase with the voltage drop across the capacitor, given by Equation 33.14. Using the trigonometric identity $\cos \omega t = \sin\left(\omega t + \frac{\pi}{2}\right)$, we can express Equation 33.16 in the alternative form

$$i_C = \omega CV_m \sin\left(\omega t + \frac{\pi}{2}\right) \tag{33.17}$$

Comparing this expression with Equation 33.14, we see that the current is 90° out of phase with the voltage across the capacitor. A plot of the current and voltage versus time (Fig. 33.7a) shows that the current reaches its peak value

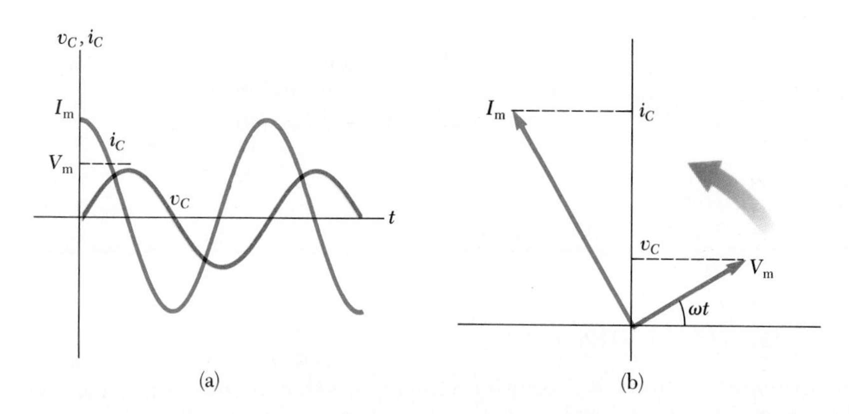


**Figure 33.7** (a) Plots of the current and voltage across the capacitor as functions of time. The voltage *lags behind* the current by 90°. (b) Phasor diagram for the purely capacitive circuit. Projections of the phasors onto the vertical axis gives the instantaneous values $v_C$ and $i_C$.

one quarter of a cycle sooner than the voltage reaches it peak value. The corresponding phasor diagram in Figure 33.7b also shows that

The current leads the voltage across the capacitor by 90°

for a sinusoidally applied emf, the current always leads the voltage across a capacitor by 90°.

From Equation 33.17, we see that the peak current in the circuit is

$$I_m = \omega C V_m = \frac{V_m}{X_C} \tag{33.18}$$

where $X_C$ is called the **capacitive reactance.**

Capacitive reactance

$$X_C = \frac{1}{\omega C} \tag{33.19}$$

The rms current is given by an expression similar to Equation 33.18, with $V_m$ replaced by $V_{rms}$.

Combining Equations 33.14 and 33.18, we can express the instantaneous voltage drop across the capacitor as

$$v_C = V_m \sin \omega t = I_m X_C \sin \omega t \tag{33.20}$$

The SI unit of $X_C$ is also the ohm. As the frequency of the circuit increases, the current increases but the reactance decreases. For a given maximum applied voltage $V_m$, the current increases as the frequency increases. As the frequency approaches zero, the capacitive reactance approaches infinity. Therefore, the current approaches zero. This makes sense since the circuit approaches dc conditions as $\omega \to 0$. Of course, a capacitor passes no current under steady-state dc conditions.

EXAMPLE 33.3 A Purely Capacitive AC Circuit
An 8-$\mu$F capacitor is connected to the terminals of an ac generator whose rms voltage is 150 V and whose frequency is 60 Hz. Find the capacitive reactance and the rms current in the circuit.

*Solution* Using Equation 33.19 and the fact that $\omega = 2\pi f = 377\ s^{-1}$ gives

$$X_C = \frac{1}{\omega C} = \frac{1}{(377\ s^{-1})(8 \times 10^{-6}\ F)} = 332\ \Omega$$

Hence, the rms current is

$$I_{rms} = \frac{V_{rms}}{X_C} = \frac{150\ V}{332\ \Omega} = 0.452\ A$$

Exercise 3 If the frequency is doubled, what happens to the capacitive reactance and the current?
Answer $X_C$ is halved, and $I$ is doubled.

## 33.5 THE *RLC* SERIES CIRCUIT

In the previous sections, we examined the effects that an inductor, a capacitor, and a resistor have when placed separately across an ac-voltage source. We shall now consider what happens when combinations of these devices are used.

Figure 33.8a shows a circuit containing a resistor, an inductor, and a capacitor connected in series across an ac-voltage source. As before, we assume that the applied voltage varies sinusoidally with time. It is convenient to assume that the applied voltage is given by

$$v = V_{\mathrm{m}} \sin \omega t$$

while the current varies as

$$i = I_{\mathrm{m}} \sin(\omega t - \phi)$$

The quantity $\phi$ is called the **phase angle** between the current and the applied voltage. Our aim is to determine $\phi$ and $I_{\mathrm{m}}$. Figure 33.8b shows the voltage versus time across each element in the circuit and their phase relations.

In order to solve this problem, we must construct and analyze the phasor diagram for this circuit. First, note that since the elements are in series, the current everywhere in the circuit must be the same at any instant. That is, *the ac current at all points in a series ac circuit has the same amplitude and phase.* Therefore, as we found in the previous sections, the voltage across each element will have *different* amplitudes and phases, as summarized in Figure 33.9. In particular, the voltage across the resistor is in phase with the current (Figure 33.9a), the voltage across the inductor leads the current by 90° (Fig. 33.9b), and finally, the voltage across the capacitor lags behind the current by 90° (Fig. 33.9c). Using these phase relationships, we can express the *instantaneous* voltage drops across the three elements as

$$v_R = I_{\mathrm{m}}R \sin \omega t = V_R \sin \omega t \tag{33.21}$$

$$v_L = I_{\mathrm{m}}X_L \sin\left(\omega t + \frac{\pi}{2}\right) = V_L \cos \omega t \tag{33.22}$$

$$v_C = I_{\mathrm{m}}X_C \sin\left(\omega t - \frac{\pi}{2}\right) = -V_C \cos \omega t \tag{33.23}$$

where $V_R$, $V_L$, and $V_C$ are the *peak* voltages across each element, given by

$$V_R = I_{\mathrm{m}}R \tag{33.24}$$

$$V_L = I_{\mathrm{m}}X_L \tag{33.25}$$

$$V_C = I_{\mathrm{m}}X_C \tag{33.26}$$

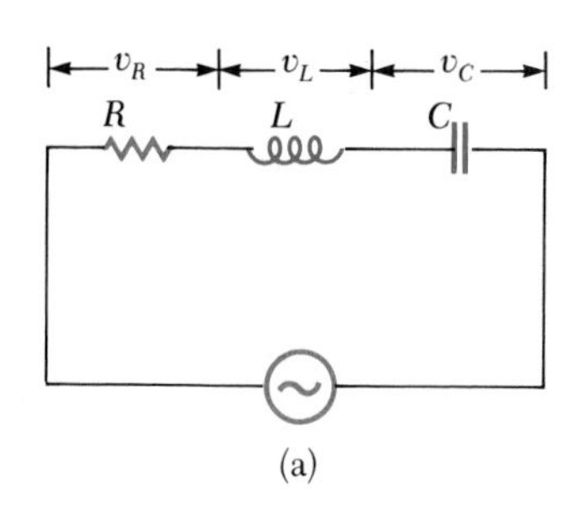


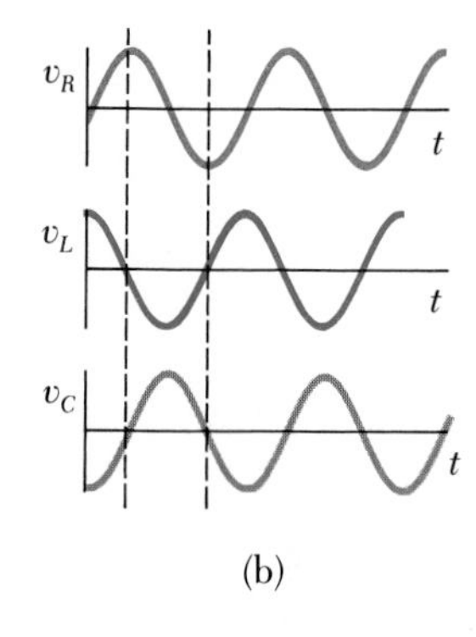


**Figure 33.8** (a) A series circuit consisting of a resistor, an inductor, and a capacitor connected to an ac generator. (b) Phase relations in the series *RLC* circuit shown in part (a).

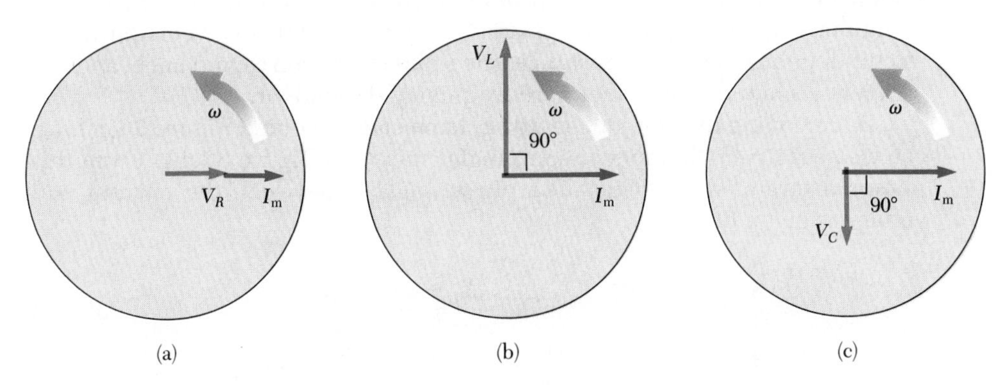


**Figure 33.9** Phase relationships between the peak voltage and current phasors for (a) a resistor, (b) an inductor, and (c) a capacitor.

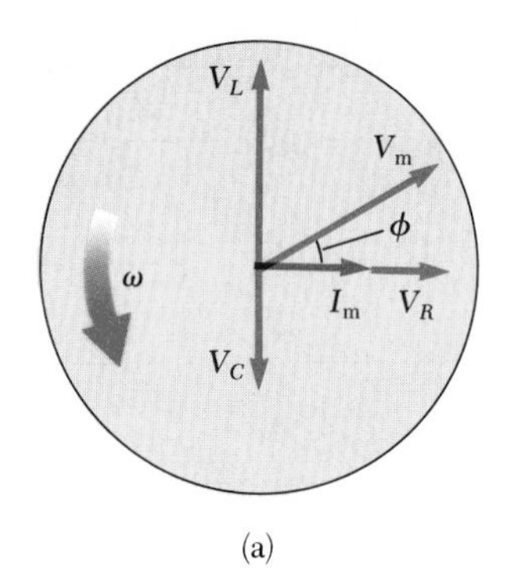


(a)

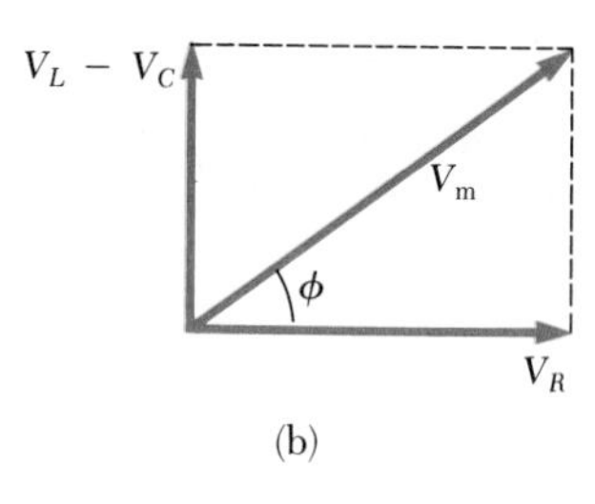


(b)

**Figure 33.10** (a) The phasor diagram for the series *RLC* circuit shown in Figure 33.8. Note that the phasor $V_R$ is *in phase* with the current phasor $I_m$, the phasor $V_L$ *leads* the phasor $I_m$ by 90°, and the phasor $V_C$ *lags behind* the phasor $I_m$ by 90°. The total voltage $V_m$ makes an angle $\phi$ with the current phasor $I_m$. (b) Simplified version of the phasor diagram shown in (a).

At this point, we could proceed by noting that the instantaneous voltage $v$ across the three elements equals the sum

$$v = v_R + v_L + v_C \tag{33.27}$$

Although this analytical approach is correct, it is simpler to obtain the sum by examining the phasor diagram.

Because the current in each element is the same at any instant, we can obtain the resulting phasor diagram by combining the three phasors shown in Figure 33.9. This gives the diagram shown in Figure 33.10a, where a single phasor $I_m$ is used to represent the current in each element. To obtain the vector sum of these voltages, it is convenient to redraw the phasor diagram as in Figure 33.10b. From this diagram, we see that the *vector* sum of the voltage amplitudes $V_R$, $V_L$, and $V_C$ equals a phasor whose length is the peak applied voltage, $V_m$, where the phasor $V_m$ makes an angle $\phi$ with the current phasor, $I_m$. Note that the voltage phasors $V_L$ and $V_C$ are in opposite directions along the same line, and hence we are able to construct the difference phasor $V_L - V_C$, which is perpendicular to the phasor $V_R$. From the right triangle in Figure 33.10b, we see that

$$V_m = \sqrt{V_R^2 + (V_L - V_C)^2} = \sqrt{(I_mR)^2 + (I_mX_L - I_mX_C)^2}$$

$$\mathbf{V_m = I_m \sqrt{R^2 + (X_L - X_C)^2}} \tag{33.28}$$

where $X_L = \omega L$ and $X_C = 1/\omega C$. Therefore, we can express the maximum current as

$$I_m = \frac{V_m}{\sqrt{R^2 + (X_L - X_C)^2}}$$

The **impedance** $Z$ of the circuit is defined to be

$$\mathbf{Z \equiv \sqrt{R^2 + (X_L - X_C)^2}} \tag{33.29}$$

Therefore, we can write Equation 33.28 in the form

$$\mathbf{V_m = I_m Z} \tag{33.30}$$

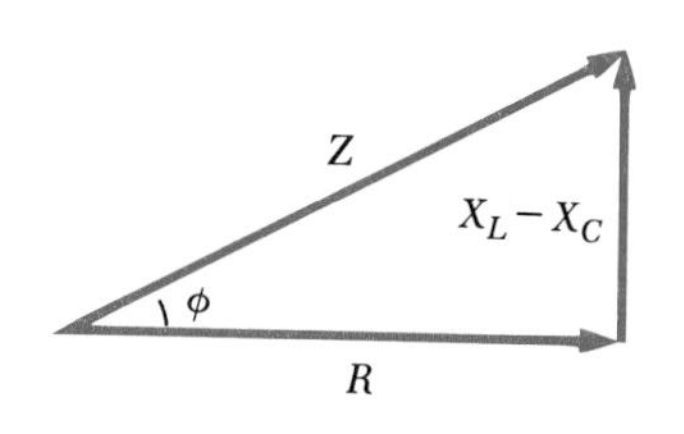


**Figure 33.11** The impedance triangle for a series *RLC* circuit which gives the relationship $Z = \sqrt{R^2 + (X_L - X_C)^2}$.

Impedance also has the SI unit of ohm. We can regard Equation 33.30 as a generalized Ohm's law applied to an ac circuit. Note that the current in the circuit depends upon the resistance, the inductance, the capacitance, and the frequency since the reactances are frequency dependent.

By removing the common factor $I_m$ from each phasor in Figure 33.10, we can also construct an impedance triangle, shown in Figure 33.11. From this phasor diagram, we find that the phase angle $\phi$ between the current and voltage is given by

Phase angle $\phi$

$$\tan \phi = \frac{X_L - X_C}{R} \tag{33.31}$$

For example, when $X_L > X_C$ (which occurs at high frequencies), the phase angle is positive, signifying that the current lags behind the applied voltage, as

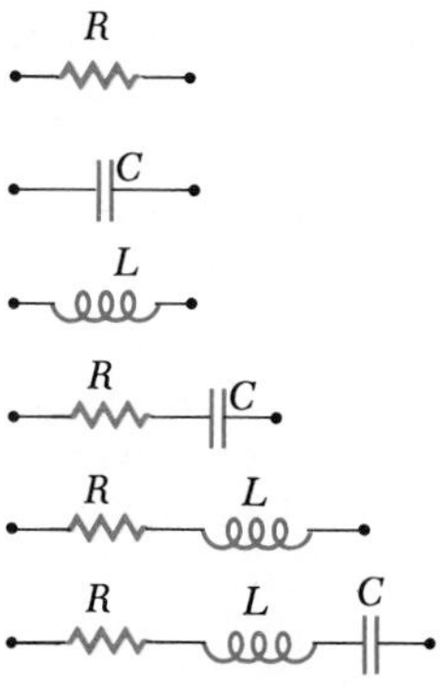

| Circuit Elements | Impedance, Z | Phase angle, $\phi$ |
|---|---|---|
| $R$ | $R$ | $0°$ |
| $C$ | $X_C$ | $-90°$ |
| $L$ | $X_L$ | $+90°$ |
| $R$ $C$ | $\sqrt{R^2+X_C^2}$ | Negative, between $-90°$ and $0°$ |
| $R$ $L$ | $\sqrt{R^2+X_L^2}$ | Positive, between $0°$ and $90°$ |
| $R$ $L$ $C$ | $\sqrt{R^2+(X_L-X_C)^2}$ | Negative if $X_C > X_L$<br>Positive if $X_C < X_L$ |

**Figure 33.12** The impedance values and phase angles for various circuit element combinations. In each case, an ac voltage (not shown) is applied across the combination of elements (that is, across the dots).

in Figure 33.10. On the other hand, if $X_L < X_C$, the phase angle is negative, signifying that the current leads the applied voltage. Finally, when $X_L = X_C$, the phase angle is zero. In this case, the ac impedance equals the resistance and the current has its *peak* value, given by $V_m/R$. The frequency at which this occurs is called the *resonance frequency,* which will be described further in Section 33.7.

Figure 33.12 gives impedance values and phase angles for various series circuits containing different combinations of circuit elements.

EXAMPLE 33.4 Analyzing a Series *RLC* Circuit

Analyze a series *RLC* ac circuit for which $R = 250\ \Omega$, $L = 0.6$ H, $C = 3.5\ \mu$F, $\omega = 377\ \text{s}^{-1}$, and $V_m = 150$ V.

*Solution* The reactances are given by $X_L = \omega L = 226\ \Omega$ and $X_C = 1/\omega C = 758\ \Omega$. Therefore, the impedance is equal to

$$Z = \sqrt{R^2 + (X_L - X_C)^2}$$

$$= \sqrt{(250\ \Omega)^2 + (226\ \Omega - 758\ \Omega)^2} = 588\ \Omega$$

The maximum current is given by

$$I_m = \frac{V_m}{Z} = \frac{150\text{ V}}{588\ \Omega} = 0.255\text{ A}$$

The phase angle between the current and voltage is

$$\phi = \tan^{-1}\left(\frac{X_L - X_C}{R}\right) = \tan^{-1}\left(\frac{226 - 758}{250}\right)$$

$$= -64.8°$$

Since the circuit is more capacitive than inductive, $\phi$ is negative and the current leads the applied voltage.

The *peak* voltages across each element are given by

$$V_R = I_m R = (0.255\text{ A})(250\ \Omega) = 63.8\text{ V}$$

$$V_L = I_m X_L = (0.255\text{ A})(226\ \Omega) = 57.6\text{ V}$$

$$V_C = I_m X_C = (0.255\text{ A})(758\ \Omega) = 193\text{ V}$$

Using Equations 33.21, 33.22, and 33.23, we find that the *instantaneous* voltages across the three elements can be written

$$v_R = (63.8\text{ V}) \sin 377t$$

$$v_L = (57.6\text{ V}) \cos 377t$$

$$v_C = (-193\text{ V}) \cos 377t$$

and the applied voltage is $v = 150 \sin(\omega t - 64.8°)$. The sum of the *peak* voltages is $V_R + V_L + V_C = 314$ V, which is much larger than the maximum voltage of the generator, 150 V. The former is a meaningless quantity. This is because when harmonically varying quantities are added, *both their amplitudes and their phases* must be taken into account and we know that the peak voltages across the different circuit elements occur at *different* times. That is, the voltages must be added in a way that takes account of the different phases. When this is done, Equation 33.28 is satisfied. You should verify this result.

## 33.6 POWER IN AN AC CIRCUIT

As we shall see in this section, *there are no power losses associated with pure capacitors and pure inductors in an ac circuit.* (A pure inductor is defined as one with no resistance or capacitance.) First, let us analyze the power dissipated in an ac circuit containing only a generator and a capacitor.

When the current begins to increase in one direction in an ac circuit, charge begins to accumulate on the capacitor and a voltage drop appears across it. When the voltage drop across the capacitor reaches its peak value, the energy stored in the capacitor is $\frac{1}{2}CV_m^2$. However, this energy storage is only momentary. The capacitor is charged and discharged twice during each cycle. In this process, charge is delivered to the capacitor during two quarters of the cycle, and is returned to the voltage source during the remaining two quarters. Therefore, *the average power supplied by the source is zero.* In other words, *a capacitor in an ac circuit does not dissipate energy.*

Similarly, the source must do work against the back emf of the inductor, which carries a current. When the current reaches its peak value, the energy stored in the inductor is a maximum and is given by $\frac{1}{2}LI_m^2$. When the current begins to decrease in the circuit, this stored energy is returned to the source as the inductor attempts to maintain the current in the circuit.

When we studied dc circuits in Chapter 27, we found that the power delivered by a battery to an external circuit is equal to the product of the current and the emf of the battery. Likewise, the instantaneous power delivered by an ac generator to any circuit is the product of the generator current and the applied voltage. For the *RLC* circuit shown in Figure 33.8, we can express the instantaneous power $P$ as

$$P = iv = I_m \sin(\omega t - \phi)[V_m \sin \omega t]$$
$$= I_m V_m \sin \omega t \sin(\omega t - \phi) \qquad (33.32)$$

Clearly this result is a complicated function of time and, in itself, is not very useful from a practical viewpoint. What is generally of interest is the average power over one or more cycles. Such an average can be computed by first using the trigonometric identity $\sin(\omega t - \phi) = \sin \omega t \cos \phi - \cos \omega t \sin \phi$. Substituting this into Equation 33.32 gives

$$P = I_m V_m \sin^2 \omega t \cos \phi - I_m V_m \sin \omega t \cos \omega t \sin \phi \qquad (33.33)$$

We now take the time average of $P$ over one or more cycles, noting that $I_m$, $V_m$, $\phi$, and $\omega$ are all constants. The time average of the first term on the right of Equation 33.33 involves the average value of $\sin^2 \omega t$, which is $\frac{1}{2}$, as shown in footnote 1. The time average of the second term on the right of Equation 33.33 is identically zero because $\sin \omega t \cos \omega t = \frac{1}{2} \sin 2\omega t$, whose average value is zero.

Therefore, we can express the **average power** $P_{av}$ as

$$P_{av} = \tfrac{1}{2} I_m V_m \cos \phi \qquad (33.34)$$

It is convenient to express the average power in terms of the rms current and rms voltage defined by Equations 33.6 and 33.7. Using these defined quantities, the average power becomes

Average power

$$P_{av} = I_{rms} V_{rms} \cos \phi \qquad (33.35)$$

where the quantity $\cos\phi$ is called the **power factor.** By inspecting Figure 33.10, we see that the maximum voltage drop across the resistor is given by $V_R = V_m \cos\phi = I_m R$. Using Equations 33.6 and 33.7 and the fact that $\cos\phi = I_m R/V_m$, we find that $P_{av}$ can be expressed as

$$P_{av} = I_{rms} V_{rms} \cos\phi = I_{rms}\left(\frac{V_m}{\sqrt{2}}\right)\frac{I_m R}{V_m} = I_{rms}\frac{I_m R}{\sqrt{2}}$$

$$P_{av} = I_{rms}^2 R \qquad (33.36)$$

In other words, the *average power delivered by the generator is dissipated as heat in the resistor,* just as in the case of a dc circuit. *There is no power loss in an ideal inductor or capacitor.* When the load is purely resistive, then $\phi = 0$, $\cos\phi = 1$, and from Equation 33.35 we see that $P_{av} = I_{rms} V_{rms}$.

**EXAMPLE 33.5 Average Power in a *RLC* Series Circuit**

Calculate the average power delivered to the series *RLC* circuit described in Example 33.4.

*Solution* First, let us calculate the rms voltage and rms current:

$$V_{rms} = \frac{V_m}{\sqrt{2}} = \frac{150\text{ V}}{\sqrt{2}} = 106\text{ V}$$

$$I_{rms} = \frac{I_m}{\sqrt{2}} = \frac{V_m/Z}{\sqrt{2}} = \frac{0.255\text{ A}}{\sqrt{2}} = 0.180\text{ A}$$

Since $\phi = -64.8°$, the power factor, $\cos\phi$, is 0.426, and hence the average power is

$$P_{av} = I_{rms} V_{rms} \cos\phi = (0.180\text{ A})(106\text{ V})(0.426)$$

$$= 8.13\text{ W}$$

The same result can be obtained using Equation 33.36.

## 33.7 RESONANCE IN A SERIES *RLC* CIRCUIT

A series *RLC* circuit is said to be in *resonance* when the current has its peak value. In general, the rms current can be written

$$I_{rms} = \frac{V_{rms}}{Z} \qquad (33.37)$$

where Z is the impedance. Substituting Equation 33.29 into 33.37 gives the relationship

$$I_{rms} = \frac{V_{rms}}{\sqrt{R^2 + (X_L - X_C)^2}} \qquad (33.38)$$

Because the impedance depends on the frequency of the source, we see that the current in the *RLC* circuit will also depend on the frequency. Note that the current reaches its peak when $X_L = X_C$, corresponding to $Z = R$. The frequency $\omega_0$ at which this occurs is called the **resonance frequency** of the circuit. To find $\omega_0$, we use the condition $X_L = X_C$, from which we get

$$\omega_0 L = \frac{1}{\omega_0 C}$$

$$\omega_0 = \frac{1}{\sqrt{LC}} \qquad (33.39)$$

Resonance frequency

Note that this frequency also corresponds to the natural frequency of oscillation of an *LC* circuit (Section 32.5). Therefore, the current in a series *RLC* circuit reaches its *peak* value when the frequency of the applied voltage matches the natural oscillator frequency, which depends only on *L* and *C*. Furthermore, at this frequency the current is in phase with the applied voltage.

A plot of the rms current versus frequency for a series *RLC* circuit is shown in Figure 33.13a. The data that are plotted assume a constant rms voltage of 5 mV, $L = 5\ \mu$H, and $C = 2$ nF. The three curves correspond to three different values of *R*. Note that in each case, the current reaches its peak value at the resonance frequency, $\omega_0$. Furthermore, the curves become narrower and taller as the resistance decreases.

By inspecting Equation 33.38, one must conclude that the current would become infinite at resonance when $R = 0$. Although the equation predicts this, real circuits always have some resistance, which limits the value of the current. Mechanical systems can also exhibit resonances. For example, when an undamped mass-spring system is driven at its natural frequency of oscillation, its amplitude increases with time, as we discussed in Chapter 13. Large-amplitude mechanical vibrations can be disastrous, as in the case of the Tacoma Narrows Bridge collapse. (See the essay in Chapter 13.)

It is also interesting to calculate the average power as a function of frequency for a series *RLC* circuit. Using Equation 33.36 together with Equation 33.37, we find that

$$P_{\text{av}} = I_{\text{rms}}^2 R = \frac{V_{\text{rms}}^2}{Z^2} R = \frac{V_{\text{rms}}^2 R}{R^2 + (X_L - X_C)^2} \qquad (33.40)$$

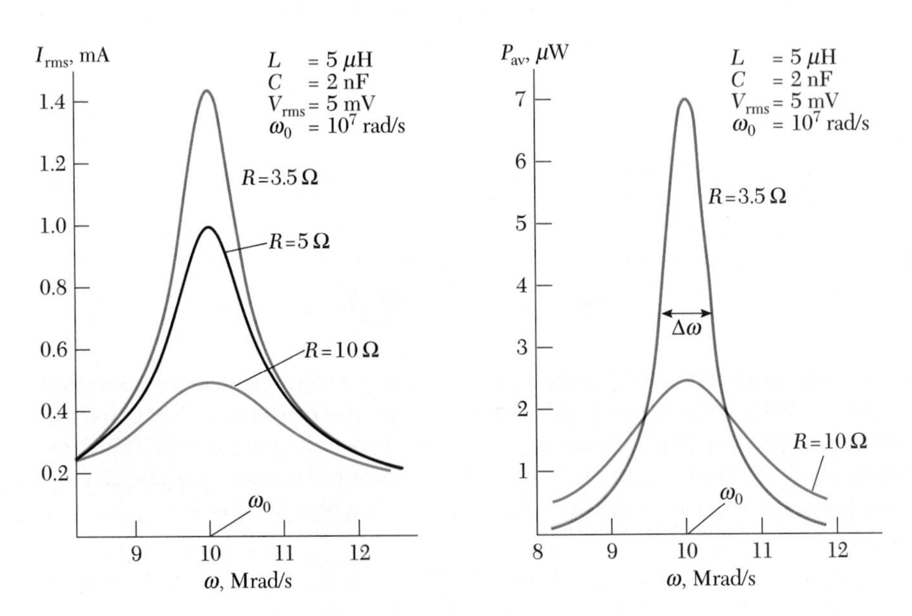


**Figure 33.13** (a) Plots of the rms current versus frequency for a series *RLC* circuit for three different values of *R*. Note that the current reaches its peak value at the resonance frequency $\omega_0$. (b) Plots of the average power versus frequency for the series *RLC* circuit for two different values of *R*.

Since $X_L = \omega L$, $X_C = \dfrac{1}{\omega C}$, and $\omega_0^2 = 1/LC$, the factor $(X_L - X_C)^2$ can be expressed as

$$(X_L - X_C)^2 = \left(\omega L - \frac{1}{\omega C}\right)^2 = \frac{L^2}{\omega^2}(\omega^2 - \omega_0^2)^2$$

Using this result in Equation 33.40 gives

$$P_{av} = \frac{V_{rms}^2 R\omega^2}{R^2\omega^2 + L^2(\omega^2 - \omega_0^2)^2} \qquad (33.41)$$

Power in an *RLC* circuit

This expression shows that at resonance, when $\omega = \omega_0$, the *average power is a maximum* and has the value $V_{rms}^2/R$. A plot of the average power versus the frequency $\omega$ of the applied voltage is shown in Figure 33.13b for the series *RLC* circuit described in Figure 33.13a, taking $R = 3.5\ \Omega$ and $R = 10\ \Omega$. As the resistance is made smaller, the curve becomes sharper in the vicinity of the resonance. The sharpness of the curve is usually described by a dimensionless parameter known as the **quality factor,** denoted by $Q_0$ (not to be confused with the symbol for charge), which is given by the ratio[3]

$$Q_0 = \frac{\omega_0}{\Delta\omega} \qquad (33.42)$$

where $\Delta\omega$ is the width of the curve measured between the two values of $\omega$ for which $P_{av}$ has *half* its maximum value (half-power points, see Figure 33.13b). It is left as a problem (Problem 87) to show that the width at the half-power points has the value $\Delta\omega = R/L$, so that

$$Q_0 = \frac{\omega_0 L}{R} \qquad (33.43)$$

Quality factor

That is, $Q_0$ is equal to the ratio of the inductive reactance to the resistance evaluated at the resonance frequency, $\omega_0$. Note that $Q_0$ is a dimensionless quantity.

The curves plotted in Figure 33.14 show that a high-$Q_0$ circuit responds to a very narrow range of frequencies, whereas a low-$Q_0$ circuit responds to a much broader range of frequencies. Typical values of $Q_0$ in electronic circuits range from 10 to 100. For example, $Q_0 = 14.3$ for the circuit described in Figure 33.13 when $R = 3.5\ \Omega$.

The receiving circuit of a radio is an important application of a resonant circuit. The radio is tuned to a particular station (which transmits a specific radio frequency signal) by varying a capacitor, which changes the resonant frequency of the receiving circuit. When the resonance frequency of the circuit matches that of the incoming radio wave, the current in the receiving circuit increases. This signal is then amplified and fed to a speaker. Since many signals are often present over a range of frequencies, it is important to design a high-$Q_0$ circuit in order to eliminate unwanted signals. In this manner, stations whose frequencies are near but not at the resonance frequency will give

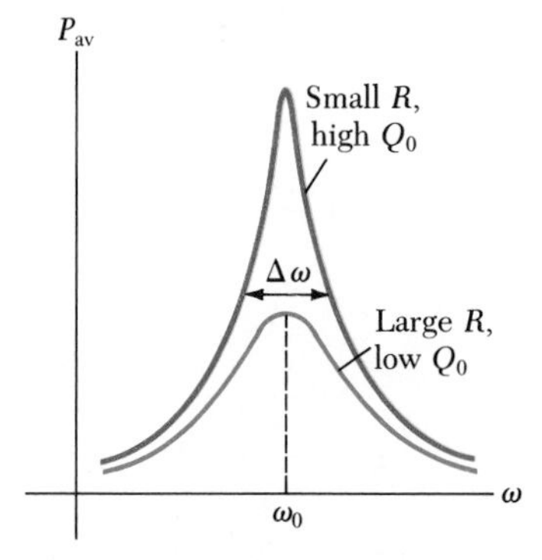


**Figure 33.14** Plots of the average power versus frequency for a series *RLC* circuit (see Eq. 33.41). The upper, narrow curve is for a small value of *B*, and the lower, broad curve is for a large value of *R*. The width $\Delta\omega$ of each curve is measured between points where the power is half its maximum value. The power is a maximum at the resonance frequency, $\omega_0$.

[3] The quality factor is also defined as the ratio $2\pi E/\Delta E$, where $E$ is the energy stored in the oscillating system and $\Delta E$ is the energy lost per cycle of oscillation. One can also define the quality factor for a mechanical system such as a damped oscillator.

negligibly small signals at the receiver relative to the one that matches the resonance frequency.

**EXAMPLE 33.6 A Resonating Series *RLC* Circuit**
Consider a series *RLC* circuit for which $R = 150\ \Omega$, $L = 20$ mH, $V_{rms} = 20$ V, and $\omega = 5000\ s^{-1}$. Determine the value of the capacitance for which the current has its peak value.

*Solution* The current has its peak value at the resonance frequency $\omega_0$, which should be made to match the "driving" frequency of 5000 $s^{-1}$ in this problem:

$$\omega_0 = 5 \times 10^3\ s^{-1} = \frac{1}{\sqrt{LC}}$$

$$C = \frac{1}{(25 \times 10^6\ s^{-2})L}$$

$$= \frac{1}{(25 \times 10^6\ s^{-2})(20 \times 10^{-3}\ H)} = 2.00\ \mu F$$

**Exercise 4** Calculate the maximum value of the rms current in the circuit.
**Answer** 0.133 A.

## *33.8 FILTER CIRCUITS

In this section, we give a brief description of *RC* filters, which are commonly used in ac circuits to modify the characteristics of a time-varying signal. A filter circuit can be used to smooth out or eliminate a time-varying voltage. For example, radios are usually powered by a 60-Hz ac voltage. The ac voltage is converted to dc using a *rectifier circuit.* After rectification, however, the voltage will still contain a small ac component at 60 Hz (sometimes called *ripple*), which must be filtered. This 60-Hz ripple must be reduced to a value much smaller than the audio signal to be amplified. Without filtering, the resulting audio signal includes an annoying hum at 60 Hz.

First, consider the simple series *RC* circuit shown in Figure 33.15a. The input voltage is across the two elements and is represented by $V_m \sin \omega t$. Since we shall be interested only in peak values, we can use Equation 33.30, which shows that the peak input voltage is related to the peak current by

$$V_{in} = I_m Z = I_m \sqrt{R^2 + \left(\frac{1}{\omega C}\right)^2}$$

If the voltage across the resistor is considered to be the output voltage, $V_{out}$, then from Ohm's law the peak output voltage is given by

$$V_{out} = I_m R$$

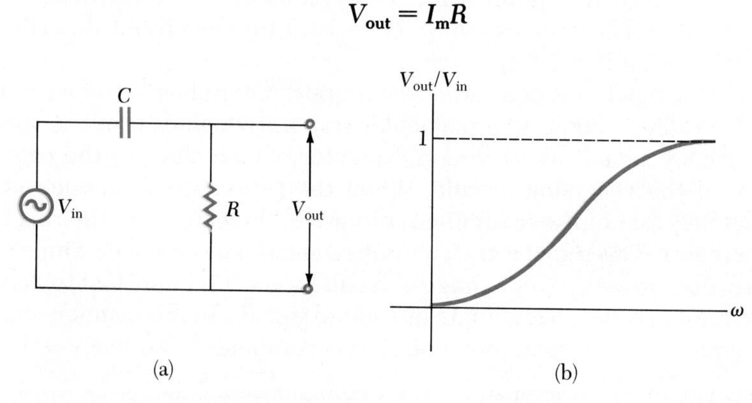


**Figure 33.15** (a) A simple *RC* high-pass filter. (b) Ratio of the output voltage to the input voltage for an *RC* high-pass filter.

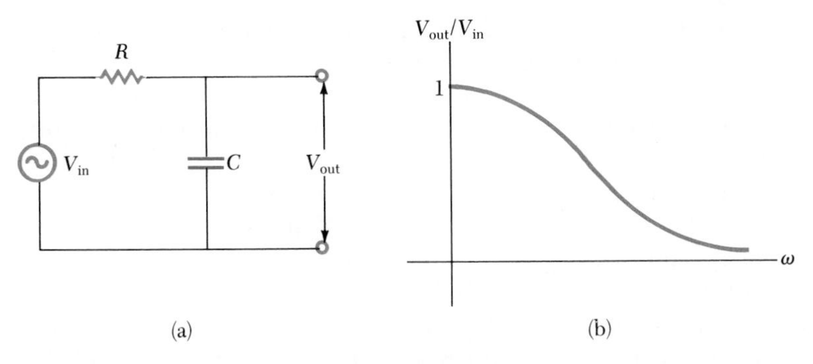


**Figure 33.16** (a) A simple *RC* low-pass filter. (b) Ratio of the output voltage to the input voltage for an *RC* low-pass filter.

Therefore, the ratio of the output voltage to the input voltage is given by

$$\frac{V_{out}}{V_{in}} = \frac{R}{\sqrt{R^2 + \left(\frac{1}{\omega C}\right)^2}} \tag{33.44}$$

High-pass filter

A plot of Equation 33.34, given in Figure 33.15b, shows that at low frequencies, $V_{out}$ is small compared with $V_{in}$, whereas at high frequencies the two voltages are equal. Since the circuit preferentially passes signals of higher frequency while low frequencies are filtered (or attenuated), the circuit is called an *RC high-pass filter.* Physically, the high-pass filter is a result of the "blocking action" of the capacitor to direct current or low frequencies.

Now consider the *RC* series circuit shown in Figure 33.16a, where the output voltage is taken across the capacitor. In this case, the peak voltage equals the voltage across the capacitor. Since the impedance across the capacitor is $X_C = 1/\omega C$,

$$V_{out} = I_m X_C = \frac{I_m}{\omega C}$$

Therefore, the ratio of the output voltage to the input voltage is given by

$$\frac{V_{out}}{V_{in}} = \frac{1/\omega C}{\sqrt{R^2 + \left(\frac{1}{\omega C}\right)^2}} \tag{33.45}$$

Low-pass filter

This ratio, plotted in Figure 33.16b, shows that in this case the circuit preferentially passes signals of low frequency. Hence, the circuit is called an *RC low-pass filter.*

We have considered only two simple filters. One can also use a series *RL* circuit as a high-pass or low-pass filter. It is also possible to design filters, called *band-pass filters,* that pass only a narrow range of frequencies.

## *33.9 THE TRANSFORMER AND POWER TRANSMISSION

When electrical power is transmitted over large distances, it is economical to use a high voltage and low current to minimize the $I^2R$ heating loss in the transmission lines. For this reason, 350-kV lines are common, and in many areas even higher-voltage (765 kV) lines are under construction. Such high-voltage transmission systems have met with considerable public resistance because of the potential safety and environmental problems they pose. At the receiving end of such lines, the consumer requires power at a low voltage and high current (for safety and efficiency in design) to operate such things as appliances and motor-driven machines. Therefore, a device is required that will increase (or decrease) the ac voltage $V$ and current $I$ without causing appreciable changes in the product $IV$. The *ac transformer* is the device used for this purpose.

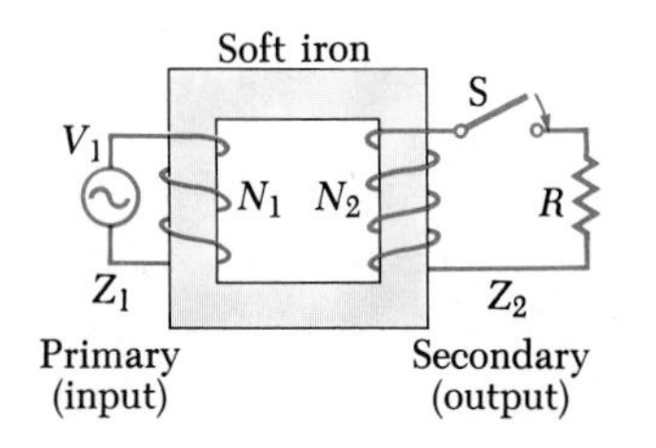


**Figure 33.17** An ideal transformer consists of two coils wound on the same soft iron core. An ac voltage $V_1$ is applied to the primary coil, and the output voltage $V_2$ is across the load resistance $R$.

In its simplest form, the ac transformer consists of two coils of wire wound around a core of soft iron as in Figure 33.17. The coil on the left, which is connected to the input ac voltage source and has $N_1$ turns, is called the *primary* winding (or primary). The coil on the right, consisting of $N_2$ turns and connected to a load resistor $R$, is called the *secondary*. The purpose of the common iron core is to increase the magnetic flux and to provide a medium in which nearly all the flux through one coil passes through the other coil. Eddy current losses are reduced by using a laminated iron core.[4] Soft iron is used as the core material to reduce hysteresis losses. Joule heat losses due to the finite resistance of the coil wires are usually quite small. Typical transformers have power efficiencies ranging from 90% to 99%. In what follows, we shall assume an ideal transformer, for which there are no power losses.

First, let us consider what happens in the primary circuit when the switch in the secondary circuit of Figure 33.17 is open. If we assume that the resistance of the primary coil is negligible relative to its inductive reactance, then the primary circuit is equivalent to a simple circuit consisting of an inductor connected to an ac generator (described in Section 33.3). Since the current is 90° out of phase with the voltage, the power factor, $\cos\phi$, is zero, and hence the average power delivered from the generator to the primary circuit is zero. Faraday's law tells us that the voltage $V_1$ across the primary coil is given by

$$V_1 = -N_1 \frac{d\Phi_m}{dt} \tag{33.46}$$

where $\Phi_m$ is the magnetic flux through each turn. If we assume that no flux leaks out of the iron core, then the flux through each turn of the primary equals the flux through each turn of the secondary. Hence, the voltage across the secondary coil is given by

$$V_2 = -N_2 \frac{d\Phi_m}{dt} \tag{33.47}$$

Since $d\Phi_m/dt$ is common to Equations 33.46 and 33.47, we find that

$$V_2 = \frac{N_2}{N_1} V_1 \tag{33.48}$$

When $N_2$ is greater than $N_1$, the output voltage $V_2$ exceeds the input voltage $V_1$. This is referred to as a *step-up transformer*. When $N_2$ is less than $N_1$,

[4] Losses in the core are present even under the condition of no load, that is, when the secondary circuit is open. Most of the power loss in this case is in the form of hysteresis losses as the core is magnetized cyclically.

the output voltage is less than the input voltage, and we speak of a *step-down transformer*.

When the switch in the secondary circuit is closed, a current $I_2$ is induced in the secondary. If the load in the secondary circuit is a pure resistance, $R_L$, the induced current will be in phase with the induced voltage. The power supplied to the secondary circuit must be provided by the ac generator that is connected to the primary circuit, as in Figure 33.18. An *ideal transformer* with a resistive load is one in which the energy losses in the transformer windings and core can be neglected. In an ideal transformer, the power supplied by the generator, $I_1V_1$, is equal to the power in the secondary circuit, $I_2V_2$. That is,

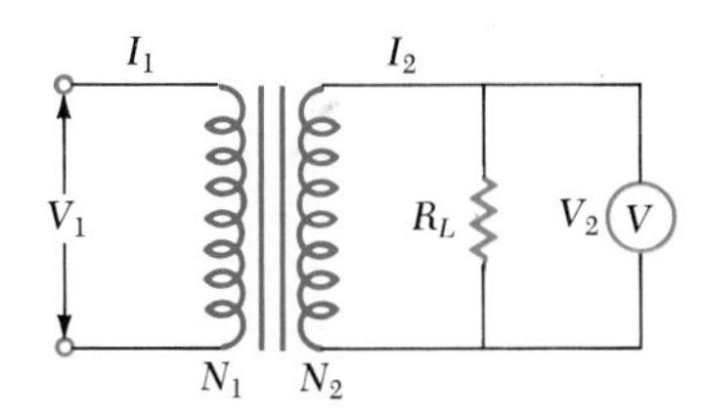

**Figure 33.18** Conventional circuit diagram for a transformer.

$$I_1V_1 = I_2V_2 \tag{33.49}$$

Clearly, the value of the load resistance $R$ determines the value of the secondary current, since $I_2 = V_2/R$. Furthermore, the current in the primary is $I_1 = V_1/R_{eq}$, where $R_{eq}$ is the equivalent resistance of the load resistance $R$ when viewed from the primary side, given by

$$R_{eq} = \left(\frac{N_1}{N_2}\right)^2 R \tag{33.50}$$

From this analysis, we see that a transformer may be used to match resistances between the primary circuit and the load. In this manner, one can achieve *maximum* power transfer between a given power source and the load resistance.

In real transformers, the power in the secondary is typically between 90% and 99% of the primary power. The energy losses are due mainly to hysteresis losses in the transformer core, and thermal energy losses from currents induced in the core and the coil windings themselves.

We can now understand why transformers are useful for transmitting power over long distances. By stepping up the generator voltage, the current in the transmission line is reduced, thereby reducing $I^2R$ losses. In practice, the voltage is stepped up to around 230 000 V at the generating station, then stepped down to around 20 000 V at a distributing station, and finally stepped down to 110–220 V at the customer's utility poles. The power is supplied by a three-wire cable. In the United States, two of these wires are "hot," with voltages of 110 V with respect to a common ground wire. Most home appliances operating on 110 V are connected in parallel between one of the hot wires and ground. Larger appliances, such as electric stoves and clothes dryers, require 220 V. This is obtained across the two hot wires, which are 180° out of phase so that the voltage difference between them is 220 V.

There is a practical upper limit to the voltages one can use in transmission lines. Excessive voltages could ionize the air surrounding the transmission lines, which could result in a conducting path to ground or to other objects in the vicinity. This, of course, would present a serious hazard to any living creatures. For this reason, a long string of insulators is used to keep high-voltage wires away from their supporting metal towers. Other insulators are used to maintain separation between wires.

**EXAMPLE 33.7 A Step-up Transformer**
A generator produces 10 A (rms) of current at 400 V. The voltage is stepped up to 4500 V by an ideal transformer and transmitted a long distance through a power line of total resistance 30 Ω. (a) Determine the percentage of power lost when the voltage is stepped up.

*Solution* From Equation 33.49, we find that the current in the transmission line is

$$I_2 = \frac{I_1 V_1}{V_2} = \frac{(10\text{ A})(400\text{ V})}{4500\text{ V}} = 0.89\text{ A}$$

Hence, the power lost in the transmission line is

$$P_{\text{lost}} = I_2{}^2 R = (0.89\text{ A})^2(30\ \Omega) = 24\text{ W}$$

Since the output power of the generator is $P = IV = (10\text{ A})(400\text{ V}) = 4000\text{ W}$, we find that the percentage of power lost is

$$\%\text{ power lost} = \left(\frac{24}{4000}\right) \times 100 = 0.6\%$$

(b) What percentage of the original power would be lost in the transmission line if the voltage were not stepped up?

*Solution* If the voltage were not stepped up, the current in the transmission line would be 10 A and the power lost in the line would be $I^2R = (10\text{ A})^2(30\ \Omega) = 3000\text{ W}$. Hence, the percentage of power lost would be

$$\%\text{ power lost} = \frac{3000}{4000} \times 100 = 75\%$$

This example illustrates the advantage of high-voltage transmission lines.

**Exercise 5** If the transmission line is cooled so that the resistance is reduced to 5 Ω, how much power will be lost in the line if it carries a current of 0.89 A?
Answer 4 W.

## SUMMARY

If an ac circuit consists of a generator and a resistor, the current in the circuit is in phase with the voltage. That is, the current and voltage reach their peak values at the same time.

The **rms current** and **rms voltage** in an ac circuit in which the voltages and current vary sinusoidally are given by the relations

$$I_{\text{rms}} = \frac{I_m}{\sqrt{2}} = 0.707 I_m \tag{33.6}$$

$$V_{\text{rms}} = \frac{V_m}{\sqrt{2}} = 0.707 V_m \tag{33.7}$$

where $I_m$ and $V_m$ are the peak values of the current and voltage, respectively.

If an ac circuit consists of a generator and an inductor, the current *lags behind* the voltage by 90°. That is, the voltage reaches its peak value one quarter of a period before the current reaches its peak value.

If an ac circuit consists of a generator and a capacitor, the current *leads* the voltage by 90°. That is, the current reaches its peak value one quarter of a period before the voltage reaches its peak value.

In ac circuits that contain inductors and capacitors, it is useful to define the **inductive reactance** $X_L$ and **capacitive reactance** $X_C$ as

Inductive reactance

$$X_L = \omega L \tag{33.12}$$

Capacitive reactance

$$X_C = \frac{1}{\omega C} \tag{33.19}$$

where $\omega$ is the angular frequency of the ac generator. The SI unit of reactance is the ohm.

The rms current in a series *RLC* circuit is

$$I_{\text{rms}} = \frac{V_{\text{rms}}}{\sqrt{R^2 + (X_L - X_C)^2}} \tag{33.38}$$

where $V_{\text{rms}}$ is the rms value of the applied voltage.

The quantity in the denominator of Equation 33.38 is defined as the **impedance** Z of the circuit, which also has the unit of ohm:

$$Z \equiv \sqrt{R^2 + (X_L - X_C)^2} \tag{33.29}$$

Impedance

In an *RLC* series ac circuit, the applied voltage and current are out of phase. The **phase angle** $\phi$ between the current and voltage is given by

$$\tan \phi = \frac{X_L - X_C}{R} \tag{33.31}$$

Phase angle

The sign of $\phi$ can be positive or negative, depending on whether $X_L$ is greater or less than $X_C$. The phase angle is zero when $X_L = X_C$.

The **average power** delivered by the generator in an *RLC* ac circuit is given by

$$P_{\text{av}} = I_{\text{rms}} V_{\text{rms}} \cos \phi \tag{33.35}$$

Average power

An equivalent expression for the average power is

$$P_{\text{av}} = I_{\text{rms}}^2 R \tag{33.36}$$

Average power

The average power delivered by the generator is dissipated as heat in the resistor. There is no power loss in an ideal inductor or capacitor.

A series *RLC* circuit is in resonance when the inductive reactance equals the capacitive reactance. When this condition is met, the current given by Equation 33.38 reaches its peak value. Setting $X_L = X_C$, one finds that the **resonance frequency** $\omega_0$ of the circuit has the value

$$\omega_0 = 1/\sqrt{LC} \tag{33.39}$$

Resonance frequency

The current in a series *RLC* circuit reaches its peak value when the frequency of the generator equals $\omega_0$, that is, when the "driving" frequency matches the resonance frequency.

A transformer is a device designed to raise or lower an ac voltage and current without causing an appreciable change in the product *IV*. In its simplest form, it consists of a primary coil of $N_1$ turns and a secondary coil of $N_2$ turns, both wound on a common soft iron core. When a voltage $V_1$ is applied across the primary, the voltage $V_2$ across the secondary is given by

$$V_2 = \frac{N_2}{N_1} V_1 \tag{33.48}$$

In an ideal transformer, the power delivered by the generator must equal the power dissipated in the load. If a load resistor *R* is connected across the secondary coil, this means that

$$I_1 V_1 = I_2 V_2 = \frac{{V_2}^2}{R} \tag{33.49}$$

## QUESTIONS

1. What is meant by the statement "the voltage across an inductor leads the current by 90°"?
2. Explain why the reactance of a capacitor decreases with increasing frequency, whereas the reactance of an inductor increases with increasing frequency.
3. Why does a capacitor act as a short circuit at high frequencies? Why does it act as an open circuit at low frequencies?
4. Explain how the acronym "ELI the ICE man" can be used to recall whether current leads voltage or voltage leads current in *RLC* circuits.
5. Why is the sum of the peak voltages across each of the elements in a series *RLC* circuit usually greater than the peak applied voltage? Doesn't this violate Kirchhoff's voltage law?
6. Does the phase angle depend on frequency? What is the phase angle when the inductive reactance equals the capacitive reactance?
7. In a series *RLC* circuit, what is the possible range of values for the phase angle?
8. If the frequency is doubled in a series *RLC* circuit, what happens to the resistance, the inductive reactance, and the capacitive reactance?
9. Energy is delivered to a series *RLC* circuit by a generator. This energy is dissipated as heat in the resistor. What is the source of this energy?
10. Explain why the average power delivered to an *RLC* circuit by the generator depends on the phase between the current and applied voltage.
11. A particular experiment requires a beam of light of very stable intensity. Why would an ac voltage be unsuitable for powering the light source?
12. What is the impedance of an *RLC* circuit at the resonance frequency?
13. Consider a series *RLC* circuit in which *R* is an incandescent lamp, *C* is some fixed capacitor, and *L* is a *variable* inductance. The source is 110 V ac. Explain why the lamp glows brightly for some values of *L* and does not glow at all for other values.
14. What is the advantage of transmitting power at high voltages?
15. What determines the peak voltage that can be used on a transmission line?
16. Why do power lines carry electrical energy at several thousand volts potential, but it is always stepped down to 240 V or 120 V as it enters your home?
17. Will a transformer operate if a battery is used for the input voltage across the primary? Explain.
18. How can the average value of a current be zero and yet the square root of the average squared current not be zero?
19. What is the time average of a sinusoidal potential with amplitude $V_m$? What is its rms voltage?
20. What is the time average of the "square-wave" potential shown in Figure 33.19? What is its rms voltage?

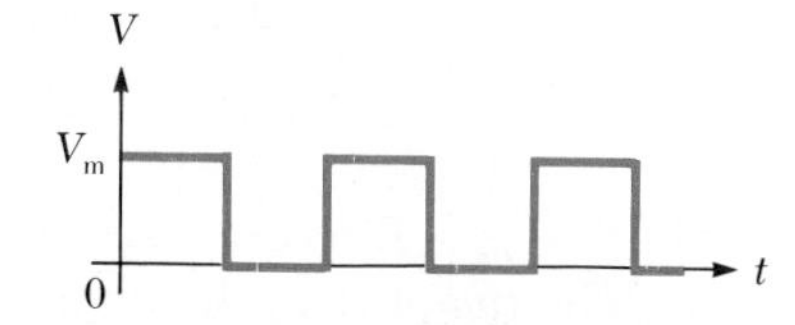


**Figure 33.19** (Question 20).

21. Do ac ammeters and voltmeters read peak, rms, or average values?
22. Is the voltage applied to a circuit always in phase with the current through a resistor in the circuit?
23. Would an inductor and a capacitor used together in an ac circuit dissipate any power?
24. Show that the peak current in an *RLC* circuit occurs when the circuit is in resonance.
25. Explain how the quality factor is related to the response characteristics of a receiver. Which variable most strongly determines the quality factor?
26. List some applications for a filter circuit.
27. The approximate efficiency of an incandescent lamp for converting electrical energy into heat is (a) 30%, (b) 60%, (c) 100%, or (d) 10%.
28. A night-watchman is fired by his boss for being wasteful and keeping all the lights on in the building. The night-watchman defends himself by claiming that the building is electrically heated, so his boss's claim is unfounded. Who should win the argument if this were to end up in a court of law?
29. Why are the primary and secondary coils of a transformer wrapped on an iron core that passes through both coils?
30. With reference to Figure 33.20, explain why the capacitor prevents a dc voltage from passing between A and B, yet allows an ac signal to pass from A to B. (The circuits are said to be capacitively coupled.)

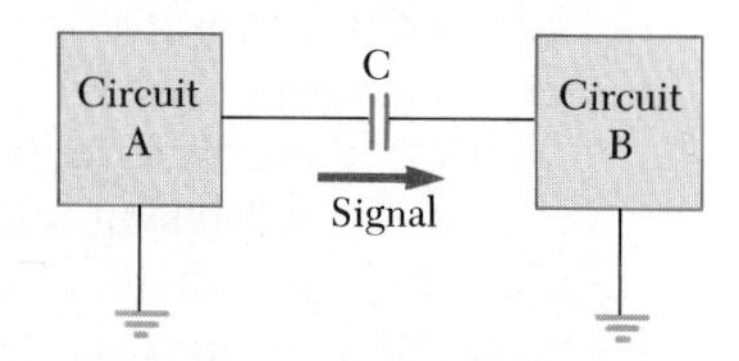


**Figure 33.20** (Question 30).

31. With reference to Figure 33.21, one finds that if *C* is made sufficiently large, an ac signal passes from A to ground rather than into B. Hence, the capacitor acts as a filter. Explain.

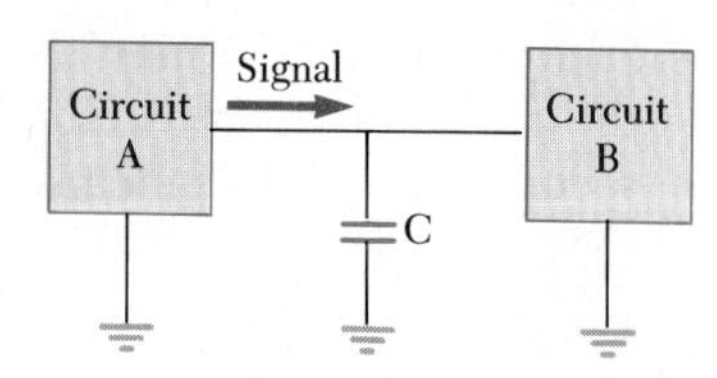


**Figure 33.21** (Question 31).

## PROBLEMS

*Assume all AC voltages and currents are sinusoidal, unless stated otherwise.*

### Section 33.2 Resistors in an AC Circuit

1. Show that the rms value for the sawtooth voltage shown in Figure 33.22 is given by $V_m/\sqrt{3}$.
2. (a) What is the resistance of a lightbulb that uses an average power of 75 W when connected to a 60-Hz power source with a peak voltage of 170 V? (b) What is the resistance of a 100-W bulb?

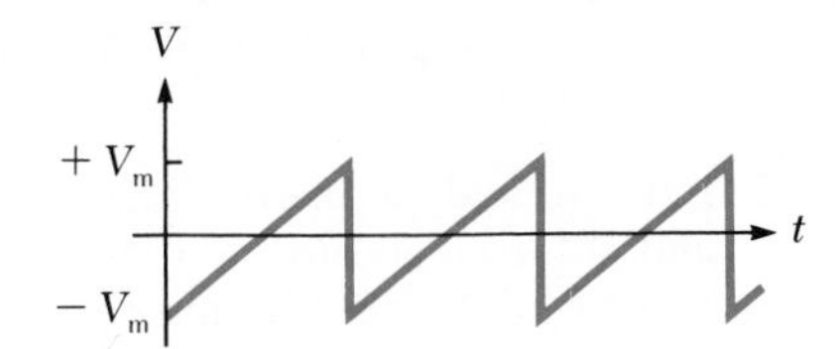


Figure 33.22 (Problem 1).

3. An ac power supply produces a peak voltage $V_m = 100$ V. This power supply is connected to a 24-Ω resistor, and the current and resistor voltage are measured with an ideal ac ammeter and voltmeter, as shown in Figure 33.23. What does each meter read?

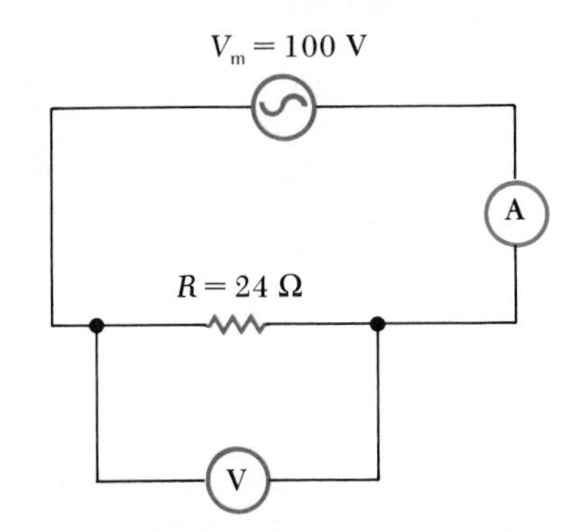


Figure 33.23 (Problem 3).

4. In the simple ac circuit shown in Figure 33.1, let $R = 60\ \Omega$, $V_m = 100$ V, and the frequency of the generator $f = 50$ Hz. Assume that the voltage across the resistor $V_R = 0$ when $t = 0$. Calculate (a) the peak current in the resistor and (b) the angular frequency of the generator.
5. Use the values given in Problem 4 for the circuit of Figure 33.1 to calculate the current through the resistor at (a) $t = \frac{1}{75}$ s and (b) $t = \frac{1}{150}$ s.
6. In the simple ac circuit shown in Figure 33.1, $R = 70\ \Omega$. (a) If $V_R = 0.25V_m$ at $t = 0.01$ s, what is the angular frequency of the generator? (b) What is the next value of $t$ for which $V_R$ will be $0.25V_m$?
7. The current in the circuit shown in Figure 33.1 equals 60% of the peak current at $t = 0.007$ s. What is the smallest frequency of the generator that gives this current?

### Section 33.3 Inductors in an AC Circuit

8. Show that the inductive reactance $X_L$ has the SI unit of ohm.
9. In a purely inductive ac circuit, as in Figure 33.4, $V_m = 100$ V. (a) If the peak current is 7.5 A at a frequency of 50 Hz, calculate the inductance $L$. (b) At what angular frequency $\omega$ will the maximum current be reduced to 2.5 A?
10. When a particular inductor is connected to a sinusoidal voltage with a 110-V amplitude, a peak current of 3 A appears in the inductor. (a) What will be the peak current if the frequency of the applied voltage is doubled? (b) What is the inductive reactance at these two frequencies?
11. An inductor is connected to a 20-Hz power supply that produces a 50-V rms voltage. What inductance is needed to keep the instantaneous current in the circuit below 80 mA?
12. An inductor has a 54-Ω reactance at 60 Hz. What will be the peak current if this inductor is connected to a 50-Hz source that produces a 100-V rms voltage?
13. For the circuit shown in Figure 33.4, $V_m = 80$ V, $\omega = 65\pi$ rad/s, and $L = 70$ mH. Calculate the current in the inductor at $t = 0.0155$ s.
14. (a) If $L = 310$ mH and $V_m = 130$ V in the circuit of Figure 33.4, at what frequency will the inductive reactance equal 40 Ω? (b) Calculate the peak value of the current in the circuit at this frequency.
15. What is the inductance of a coil that has an inductive reactance of 63 Ω at an angular frequency of 820 rad/s?

### Section 33.4 Capacitors in an AC Circuit

16. Show that the SI unit of capacitive reactance is the ohm.
17. (a) For what linear frequencies does a 22-μF capacitor have a reactance below 175 Ω? (b) Over this same frequency range, what would be the reactance of a 44-μF capacitor?
18. Calculate the capacitive reactance of a 10-μF capacitor when connected to an ac generator having an angular frequency of 95π rad/s.
19. A 98-pF capacitor is connected to a 60-Hz power supply that produces a 20-V rms voltage. What is the maximum charge that appears on either of the capacitor plates?
20. A sinusoidal voltage $v(t) = V_m \cos \omega t$ is applied to a capacitor as shown in Figure 33.24. (a) Write an expression for the instantaneous charge on the capacitor in terms of $V_m$, $C$, $t$, and $\omega$. (b) What is the instantaneous current in the circuit?

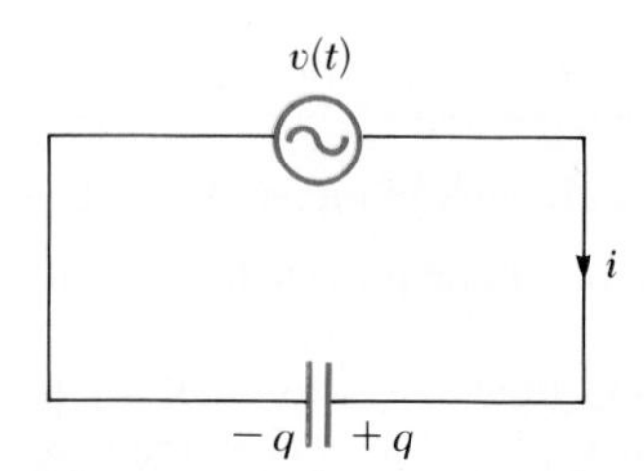


**Figure 33.24** (Problem 20).

**21.** What peak current will be delivered by an ac generator with $V_m = 48$ V and $f = 90$ Hz when connected across a 3.7-$\mu$F capacitor?

**22.** A variable-frequency ac generator with $V_m = 18$ V is connected across a $9.4 \times 10^{-8}$-F capacitor. At what frequency should the generator be operated to provide a peak current of 5 A?

**23.** The generator in a purely capacitive ac circuit (Fig. 33.6) has an angular frequency of $100\pi$ rad/s and $V_m = 220$ V. If $C = 20$ $\mu$F, what is the current in the circuit at $t = 0.004$ s?

## Section 33.5 The *RLC* Series Circuit

**24.** At what frequency will the inductive reactance of a 57-$\mu$H inductance equal the capacitive reactance of a 57-$\mu$F capacitor?

**25.** A series ac circuit contains the following components: $R = 150$ $\Omega$, $L = 250$ mH, $C = 2$ $\mu$F and a generator with $V_m = 210$ V operating at 50 Hz. Calculate the (a) inductive reactance, (b) capacitive reactance, (c) impedance, (d) peak current, and (e) phase angle.

**26.** A sinusoidal voltage $v(t) = (40\text{ V})\sin(100t)$ is applied to a series *RLC* circuit with $L = 160$ mH, $C = 99$ $\mu$F, and $R = 68$ $\Omega$. (a) What is the impedance of the circuit? (b) What is the current amplitude? (c) Determine the numerical values for $I_m$, $\omega$, and $\phi$ in the equation $i(t) = I_m \sin(\omega t - \phi)$.

**27.** An *RLC* circuit consists of a 150-$\Omega$ resistor, a 21-$\mu$F capacitor, and a 460-mH inductor, connected in series with a 120-V, 60-Hz power supply. (a) What is the phase angle between the current and the applied voltage? (b) Does the current or voltage reach its peak earlier?

**28.** A resistor ($R = 900$ $\Omega$), a capacitor ($C = 0.25$ $\mu$F), and an inductor ($L = 2.5$ H) are connected in series across a 240-Hz ac source for which $V_m = 140$ V. Calculate the (a) impedance of the circuit, (b) peak current delivered by the source, and (c) phase angle between the current and voltage. (d) Is the current leading or lagging behind the voltage?

**29.** A coil with an inductance of 18.1 mH and a resistance of 7 $\Omega$ is connected to a *variable*-frequency ac generator. At what frequency will the voltage across the coil lead the current by 45°?

**30.** A 400-$\Omega$ resistor, an inductor, and a capacitor are in series with a generator. When the frequency is adjusted to $600/\pi$ Hz, the inductive reactance is 700 $\Omega$. What is the *minimum* value of capacitance that will result in a circuit impedance of 910 $\Omega$?

**31.** An ac source with $V_m = 150$ V and $f = 50$ Hz is connected between points $a$ and $d$ in Figure 33.25. Calculate the *peak* voltages between points (a) $a$ and $b$, (b) $b$ and $c$, (c) $c$ and $d$, (d) $b$ and $d$.

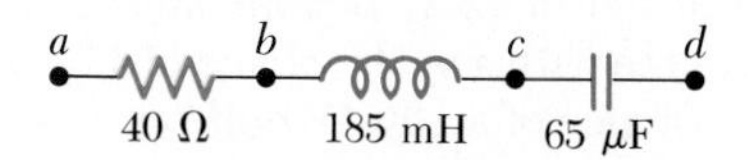


**Figure 33.25** (Problem 31).

**32.** Draw to scale a phasor diagram showing $Z$, $X_L$, $X_C$, and $\phi$ for an ac series circuit for which $R = 300$ $\Omega$, $C = 11$ $\mu$F, $L = 0.2$ H, and $f = 500/\pi$ Hz.

**33.** An inductor ($L = 400$ mH), a capacitor ($C = 4.43$ $\mu$F), and a resistor ($R = 500$ $\Omega$) are connected in series. A 50-Hz ac generator produces a peak current of 250 mA in the circuit. (a) Calculate the required peak voltage $V_m$. (b) Determine the angle by which the current in the circuit leads or lags behind the applied voltage.

**34.** A series *RLC* circuit with $R = 1500$ $\Omega$ and $C = 15$ nF is connected to an ac generator whose frequency can be varied. When the frequency is adjusted to 50.5 kHz, the rms current in the circuit reaches a maximum at 0.14 A. Determine (a) the inductance and (b) the rms value of the generator voltage.

## Section 33.6 Power in an AC Circuit

**35.** Calculate the average power delivered to the series *RLC* circuit described in Problem 25.

**36.** Consider the circuit described in Problem 28. (a) What is the power factor of the circuit? (b) What is the rms current in the circuit? (c) What average power is delivered by the source?

**37.** An ac voltage of the form (in SI units)

$$v = 100 \sin(1000t)$$

is applied to a series *RLC* circuit. If $R = 400$ $\Omega$, $C = 5.0$ $\mu$F, and $L = 0.50$ H, find the average power dissipated in the circuit.

**38.** An ac voltage with an amplitude of 100 V is applied to a series combination of a 200-$\mu$F capacitor, a 100-mH inductor, and a 20-$\Omega$ resistor. Calculate the power dissipation and the power factor for a frequency of (a) 60 Hz and (b) 50 Hz.

**39.** The rms terminal voltage of an ac generator is 200 V. The operating frequency is 100 Hz. Write the equation giving the output voltage as a function of time.

40. The average power in a circuit for which the rms current is 5 A is 450 W. Calculate the resistance of the circuit.
41. In a certain series *RLC* circuit, $I_{rms} = 9$ A, $V_{rms} = 180$ V, and the current leads the voltage by 37°. (a) What is the total resistance of the circuit? (b) Calculate the reactance of the circuit $(X_L - X_C)$.
42. A series *RLC* circuit has a resistance of 45 Ω and an impedance of 75 Ω. What average power will be delivered to this circuit when $V_{rms} = 210$ V?

## Section 33.7 Resonance in a Series *RLC* Circuit

43. Calculate the resonance frequency of a series *RLC* circuit for which $C = 8.4\ \mu$F and $L = 120$ mH.
44. (a) Compute the quality factor for each of the circuits described in Problems 26 and 27. (b) Which of these two circuits has the sharper resonance?
45. An *RLC* circuit is used in a radio to tune into an FM station broadcasting at 99.7 MHz. The resistance in the circuit is 12 Ω and the inductance is 1.40 $\mu$H. What capacitance should be used?
46. The tuning circuit of an AM radio is a parallel *LC* combination that has 1-Ω resistance. The inductance is 0.2 mH and the capacitor is variable, so that the circuit can resonate at frequencies between 550 kHz and 1650 kHz. Find the range of values for *C*.
47. A coil of resistance 35 Ω and inductance 20.5 H is in series with a capacitor and a 200-V (rms), 100-Hz source. The current in the circuit is 4 A (rms). (a) Calculate the capacitance in the circuit. (b) What is $V_{rms}$ across the coil?
48. A series *RLC* circuit has the following values: $L = 20$ mH, $C = 100$ nF, $R = 20$ Ω, and $V = 100$ V, with $v = V \sin \omega t$. Find (a) the resonant frequency, (b) the amplitude of the current at the resonant frequency, (c) the *Q* of the circuit, and (d) the amplitude of the voltage across the inductor at resonance.
49. Consider a series combination of a 10-mH inductor, a 100-$\mu$F capacitor, and a 10-Ω resistor. A 50-V (rms) sinusoidal voltage is applied to the combination. Calculate the rms current for (a) the resonant frequency, (b) half the resonant frequency, and (c) double the resonant frequency.

## °Section 33.8 Filter Circuits

50. Consider the circuit shown in Figure 33.15, with $R = 800$ Ω and $C = 0.09\ \mu$F. Calculate the ratio $V_{out}/V_{in}$ for (a) $\omega = 300\ s^{-1}$ and (b) $\omega = 7 \times 10^5\ s^{-1}$.
51. The *RC* high-pass filter shown in Figure 33.15 has a resistance $R = 0.50$ Ω. (a) What capacitance will give an output signal with one-half the amplitude of a 300-Hz input signal? (b) What is the gain $(V_{out}/V_{in})$ for a 600-Hz signal?
52. The *RC* low-pass filter shown in Figure 33.16 has a resistance $R = 90$ Ω and a capacitance $C = 8000$ pF. Calculate the gain $(V_{out}/V_{in})$ for an input frequency (a) $f = 600$ Hz, and (b) $f = 600$ kHz.
53. Assign the values of *R* and *C* given in Problem 50 to the circuit shown in Figure 33.16 and calculate $V_{out}/V_{in}$ for (a) $\omega = 300\ s^{-1}$ and (b) $\omega = 7 \times 10^5\ s^{-1}$.
54. (a) For the circuit shown in Figure 33.26, show that the maximum possible value of the ratio $V_{out}/V_{in}$ is unity. (b) At what frequency (expressed in terms of *R*, *L*, and *C*) does this occur?

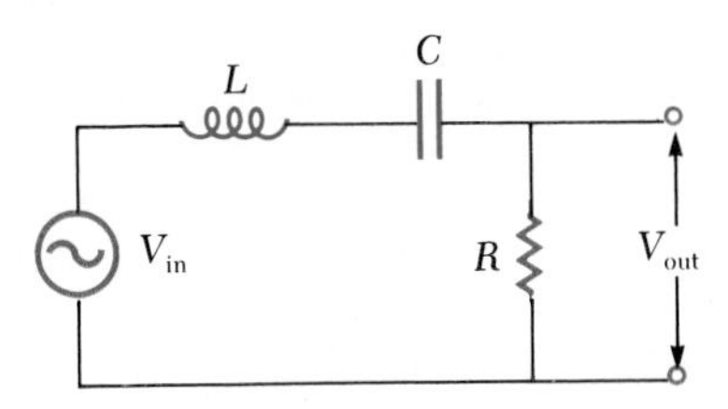


**Figure 33.26** (Problems 54 and 55).

55. The circuit shown in Figure 33.26 can be used as a filter to pass signals that lie in a certain frequency band. (a) Show that the gain $(V_{out}/V_{in})$ for an input voltage of frequency $\omega$ is given by

$$\frac{V_{out}}{V_{in}} = \frac{1}{\sqrt{1 + \left[\frac{(\omega^2/\omega_0{}^2) - 1}{\omega RC}\right]^2}}$$

(b) Let $R = 100$ Ω, $C = 0.050\ \mu$F, and $L = 0.127$ H. Compute the gain of this circuit for input frequencies $f_1 = 1.5$ kHz, $f_2 = 2.0$ kHz, and $f_3 = 2.5$ kHz.
56. Show that two successive high-pass filters with the same values of *R* and *C* give a combined gain

$$\frac{V_{out}}{V_{in}} = \frac{1}{1 + (1/\omega RC)^2}$$

57. Consider a low-pass filter followed by a high-pass filter, as shown in Figure 33.27. If $R = 1000$ Ω and $C = 0.050\ \mu$F, determine $V_{out}/V_{in}$ for a 2.0-kHz input frequency.

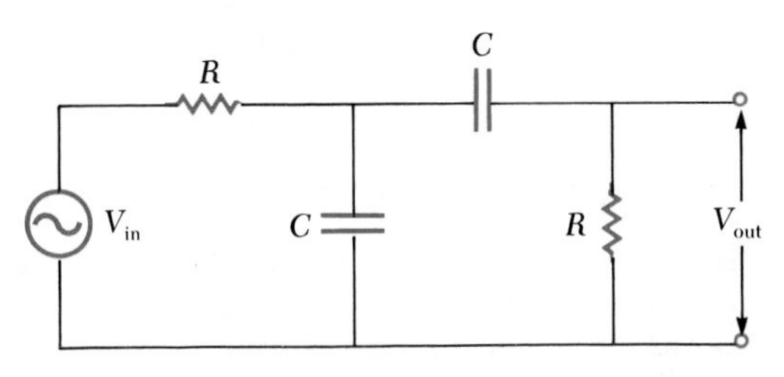


**Figure 33.27** (Problem 57).

°Section 33.9 The Transformer and Power Transmission

58. The primary winding of an electric train transformer has 400 turns, and the secondary has 50. If the input voltage is 120 V (rms) what is the output voltage?
59. A transformer has $N_1 = 350$ turns and $N_2 = 2000$ turns. If the input voltage is $v(t) = (170\text{ V})\cos\omega t$, what rms voltage is developed across the secondary coil?
60. Consider an ideal transformer with $N_1$ primary and $N_2$ secondary windings. Show that a step-up transformer (one with $N_2 > N_1$) actually reduces the current in the output by a factor of $N_1/N_2$.
61. A particular transformer is 95% efficient and has twice as many secondary windings as primary windings. If the primary windings carry a 5 A current at an rms voltage of 120 V, what are the secondary current and rms voltage?
62. A step-up transformer is designed to have an output voltage of 2200 V (rms) when the primary is connected across a 110-V (rms) source. (a) If there are 80 turns on the primary winding, how many turns are required on the secondary? (b) If a load resistor across the secondary draws a current of 1.5 A, what is the current in the primary, assuming ideal conditions?
63. If the transformer in Problem 62 has an efficiency of 95%, what is the current in the primary when the secondary current is 1.2 A?
64. The primary current of an ideal transformer is 8.5 A when the primary voltage is 77 V. Calculate the voltage across the secondary when a current of 1.4 A is delivered to a load resistor.

## ADDITIONAL PROBLEMS

65. A series *RLC* circuit consists of an 8-Ω resistor, a 5-μF capacitor, and a 50-mH inductor. A variable frequency source of 400 V (rms) is applied across the combination. Determine the power delivered to the circuit when the frequency is equal to one half of the resonance frequency.
66. A series *RLC* circuit has $R = 10\ \Omega$, $L = 2$ mH, and $C = 4\ \mu$F. Determine (a) the impedance at a frequency of 60 Hz, (b) the resonant frequency in Hz, (c) the impedance at resonance, and (d) the impedance at a frequency equal to one-half the resonant frequency.
67. In a series ac circuit, $R = 21\ \Omega$, $L = 25$ mH, $C = 17\ \mu$F, $V_m = 150$ V, and $\omega = \dfrac{2000}{\pi}\ \text{s}^{-1}$. (a) Calculate the peak current in the circuit. (b) Determine the peak voltage across each of the three elements. (c) What is the power factor for the circuit? (d) Show $X_L$, $X_C$, $R$, and $\phi$ in a phasor diagram for the circuit.
68. An *RL* series combination consisting of a 1.5-Ω resistor and a 2.5-mH inductor is connected to a 12.5-V (rms), 400-Hz generator. Determine (a) the impedance of the circuit, (b) the rms current, (c) the rms voltage across the resistor, and (d) the rms voltage across the inductor.
69. As a way of determining the inductance of a coil used in a research project, a student first connects the coil to a 12-V battery and measures a current of 0.63 A. The student then connects the coil to a 24-V (rms), 60-Hz generator and measures an rms current of 0.57 A. What value does the student calculate for the inductance?
70. A 2.5-V (rms), 100-Hz generator is connected in series with a 100-nF capacitor and a 2500-Ω resistor. Determine (a) the impedance of the circuit and (b) the rms current in the circuit.
71. A transmission line with a resistance per unit length of $4.5 \times 10^{-4}\ \Omega$/m is to be used to transmit 5000 kW of power over a distance of 400 miles ($6.44 \times 10^5$ m). The terminal voltage of the generator is 4500 V. (a) What is the line loss if a transformer is used to step up the voltage to 500 kV? (b) What fraction of the input power is lost to the line under these circumstances? (c) What difficulties would be encountered on attempting to transmit the 5000 kW of power at the generator voltage of 4500 V?
72. A transformer operating from 120 V (rms) supplies a 12-V lighting system for a garden. Eight lights, each rated 40 W, are installed in parallel. (a) Find the equivalent resistance of the total lighting system. (b) What current is in the secondary circuit? (c) What single resistance, connected across the 120 V supply, would consume the same power as when the transformer is used? Show that this equals the answer to part (a) times the square of the turns ratio.
73. *LC* filters are used as both high- and low-pass filters as were the *RC* filters in Section 33.8. However, all real inductors have resistance, as indicated in Figure 33.28, which must be taken into account. (a) Determine which circuit in Figure 33.28 is the high-pass filter and which is the low-pass filter. (b) Derive the

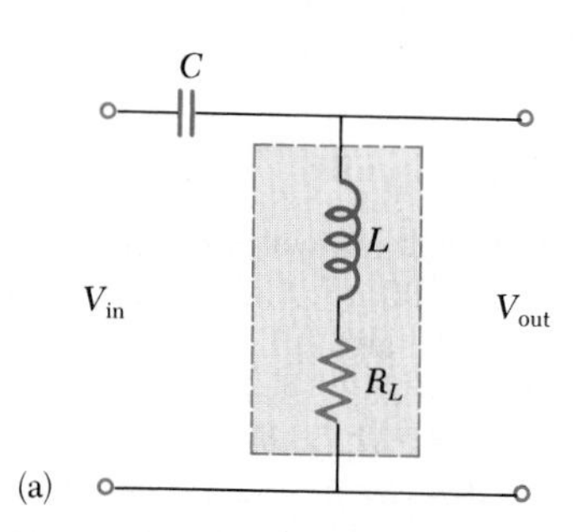


**Figure 33.28a** (Problem 73).

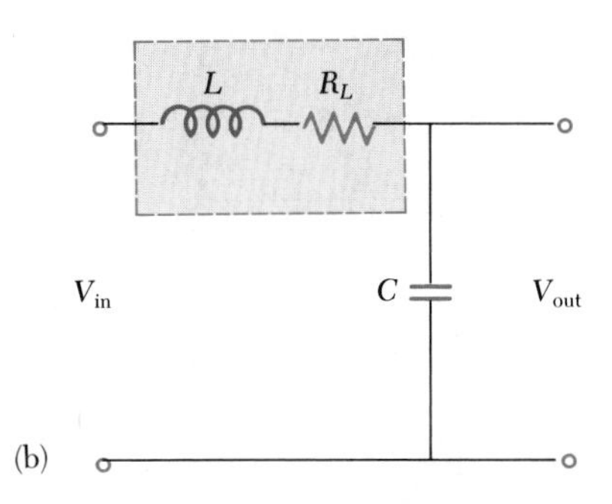


**Figure 33.28b** (Problem 73).

output/input formulas for each circuit following the procedure used for the *RC* filters in Section 33.8.

74. A resistor of 80 Ω and a 200-mH inductor are connected in *parallel* across a 100-V (rms), 60-Hz source. (a) What is the rms current in the resistor? (b) By what angle does the total current lead or lag behind the voltage?

75. An inductor is in series with an 80-Ω resistor and the combination is placed across a 110-V (rms), 60-Hz power source. If the resistor dissipates 50 W of power, find the inductance of the inductor.

76. The average power delivered to a series *RLC* circuit at frequency $\omega$ (Section 33.7) is given by Equation 33.41. (a) Show that the peak current can be written

$$I_m = \omega V_m[L^2(\omega_0{}^2 - \omega^2)^2 + (\omega R)^2]^{-1/2}$$

where $\omega$ is the operating frequency of the circuit and $\omega_0$ is the resonance frequency. (b) Show that the phase angle can be expressed as

$$\phi = \tan^{-1}\left[\frac{L}{R}\left(\frac{\omega_0{}^2 - \omega^2}{\omega}\right)\right]$$

77. Consider a series *RLC* circuit with the following circuit parameters: $R = 200\ \Omega$, $L = 663$ mH, and $C = 26.5\ \mu$F. The applied voltage has an amplitude of 50 V and a frequency of 60 Hz. Find the following amplitudes: (a) The current $i$, including its phase constant $\phi$ relative to the applied voltage $v$; (b) the voltage $V_R$ across the resistor and its phase relative to the current; (c) the voltage $V_C$ across the capacitor and its phase relative to the current; and (d) the voltage $V_L$ across the inductor and its phase relative to the current.

78. A voltage $v = 100 \sin \omega t$ (in SI units) is applied across a series combination of a 2-H inductor, a 10-$\mu$F capacitor, and a 10-Ω resistor. (a) Determine the angular frequency $\omega_0$ at which the power dissipated in the resistor is a maximum. (b) Calculate the power dissipated at that frequency. (c) Determine the two angular frequencies $\omega_1$ and $\omega_2$ at which the power dissipated is one-half the maximum value. [The $Q$ of the circuit is approximately $\omega_0/(\omega_2 - \omega_1)$.]

79. *Impedance matching:* A transformer may be used to provide maximum power transfer between two ac circuits that have different impedances. (a) Show that the ratio of turns $N_1/N_2$ needed to meet this condition is given by

$$\frac{N_1}{N_2} = \sqrt{\frac{Z_1}{Z_2}}$$

(b) Suppose you want to use a transformer as an impedance-matching device between an audio amplifier that has an output impedance of 8000 Ω and a speaker that has an input impedance of 8 Ω. What should be the ratio of primary to secondary turns on the transformer?

80. An ac source has an internal resistance of 3200 Ω. In order for the maximum power to be transferred to an 8-Ω resistive load $R_2$, a transformer is used between the source and the load. Assuming an ideal transformer, (a) find the appropriate turns ratio of the transformer. If the output voltage of the source is 80 V (rms), determine (b) the rms voltage across the load resistor and (c) the rms current in the load resistor. (d) Calculate the power dissipated in the load. (e) Verify that the ratio of currents is inversely proportional to the turns ratio.

81. Figure 33.29a shows a parallel *RLC* circuit, and the corresponding phasor diagram is given in Figure 33.29b. The instantaneous voltage (and rms voltage) across each of the three circuit elements is the same, and each is in phase with the current through the

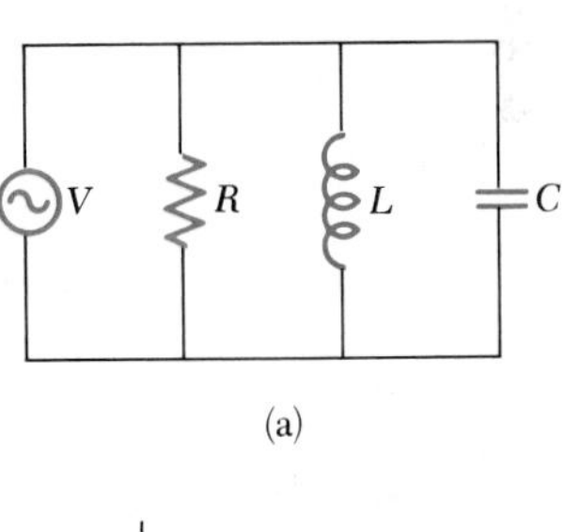


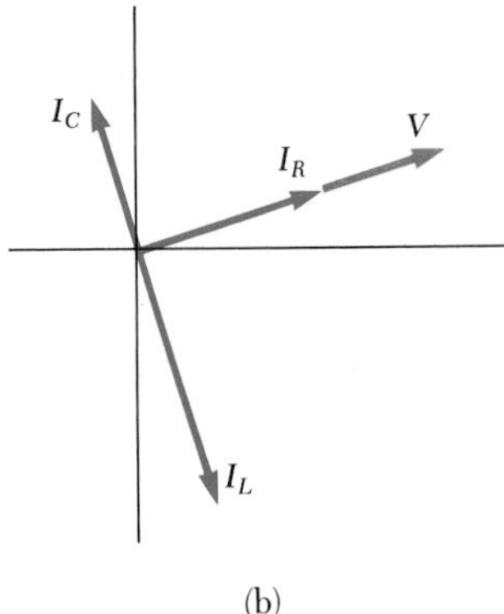


**Figure 33.29** (Problem 81).

resistor. The currents in $C$ and $L$ lead (or lag behind) the current in the resistor, as shown in Figure 33.29b. (a) Show that the rms current delivered by the source is given by

$$I_{\text{rms}} = V_{\text{rms}} \left[ \frac{1}{R^2} + \left( \omega C - \frac{1}{\omega L} \right)^2 \right]^{1/2}$$

(b) Show that the phase angle $\phi$ between $V_{\text{rms}}$ and $I_{\text{rms}}$ is given by

$$\tan \phi = R \left( \frac{1}{X_C} - \frac{1}{X_L} \right)$$

82. An 80-Ω resistor, a 200-mH inductor, and a 0.15-μF capacitor are connected in parallel across a 120-V (rms) source operating at an angular frequency of 374 rad/s. (a) What is the resonant frequency of the circuit? (b) Calculate the rms current in the resistor, inductor, and capacitor. (c) What is the rms current delivered by the source? (d) Is the current leading or lagging behind the voltage? By what angle?

83. Consider the phase-shifter circuit shown in Figure 33.30. The input voltage is described by the expression $v = 10 \sin 200t$ (in SI units). If $L = 500$ mH, find (a) the value of $R$ such that the output voltage $v_0$ lags the input voltage by 30° and (b) the amplitude of the output voltage.

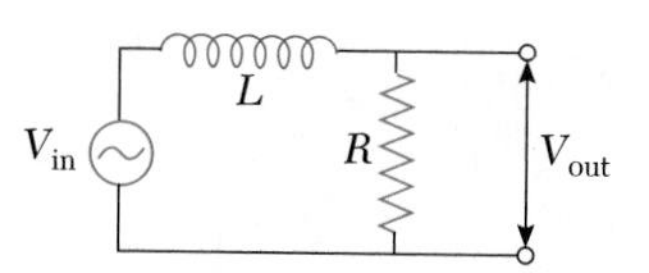


**Figure 33.30** (Problem 83).

## CALCULATOR/COMPUTER PROBLEMS

84. A series $RLC$ circuit is operating at 2000 Hz. At this frequency, $X_L = X_C = 1884$ Ω. The resistance of the circuit is 40 Ω. (a) Prepare a table showing the values of $X_L$, $X_C$, and $Z$ for $f = 300$, 600, 800, 1000, 1500, 2000, 3000, 4000, 6000, and 10 000 Hz. (b) Plot on the same set of axes $X_L$, $X_C$, and $Z$ as a function of $\ln f$.

85. Suppose the high-pass filter shown in Figure 33.15 has $R = 1000$ Ω and $C = 0.050$ μF. (a) At what frequency does $V_{\text{out}}/V_{\text{in}} = \frac{1}{2}$? (b) Plot $\log_{10}(V_{\text{out}}/V_{\text{in}})$ versus $\log_{10}(f)$ over the frequency range from 1 Hz to 1 MHz. (This log–log plot of gain versus frequency is known as a **Bode plot.**)

86. Suppose the low-pass filter shown in Figure 33.16 has $R = 1000$ Ω and $C = 0.050$ μF. (a) At what frequency does $V_{\text{out}}/V_{\text{in}} = \frac{1}{2}$? (b) Plot $\log_{10}(V_{\text{out}}/V_{\text{in}})$ versus $\log_{10}(f)$ over the frequency range from 1 Hz to 1 MHz.

87. A series $RLC$ circuit in which $R = 1$ Ω, $L = 1$ mH, and $C = 1$ nF is connected to an ac generator delivering 1 V (rms). Use a small computer to make a careful plot of the power delivered to the circuit as a function of the frequency and verify that the half-width of the resonance peak is $R/2\pi L$.

# 34
# Electromagnetic Waves

*An old-style rancher in St. Augustine Plains, New Mexico, rides past one of the 27 radio telescopes that comprise the Very Large Array (VLA). Arranged in a Y-shaped configuration on a system of railroad tracks, the radio telescopes of the VLA capture and focus electromagnetic waves from space. (© Danny Lehman)*

The waves we have described in Chapters 16, 17, and 18 are mechanical waves. Such waves correspond to the disturbance of a medium. By definition, mechanical disturbances such as sound waves, water waves, and waves on a string require the presence of a medium. This chapter is concerned with the properties of electromagnetic waves that (unlike mechanical waves) can propagate through empty space.

In Section 31.7 we gave a brief description of Maxwell's equations, which form the theoretical basis of all electromagnetic phenomena. The consequences of Maxwell's equations are far reaching and very dramatic for the history of physics. One of Maxwell's equations, the Ampère-Maxwell law, predicts that a time-varying electric field produces a magnetic field just as a time-varying magnetic field produces an electric field (Faraday's law). From this generalization, Maxwell introduced the concept of displacement current, a new source of a magnetic field. Thus, Maxwell's theory provided the final important link between electric and magnetic fields.

Astonishingly, Maxwell's formalism also predicts the existence of electromagnetic waves that propagate through space with the speed of light. This prediction was confirmed experimentally by Hertz, who first generated and

Biographical Sketch

**James Clerk Maxwell**
*(1831–1879)*

James Clerk Maxwell is generally regarded as the greatest theoretical physicist of the 19th century. Born in Edinburgh to a well-known Scottish family, he entered the University of Edinburgh at age 15, around the time that he discovered an original method for drawing a perfect oval. Maxwell was appointed to his first professorship in 1856 at Aberdeen. This was the beginning of a career during which Maxwell would develop the electromagnetic theory of light, the kinetic theory of gases, and explanations of the nature of Saturn's rings and of color vision.

Maxwell's development of the electromagnetic theory of light took many years and began with the paper "On Faraday's Lines of Force," in which Maxwell expanded upon Faraday's theory that electric and magnetic effects result from fields of lines of force surrounding conductors and magnets. His next publication, "On Physical Lines of Force," included a series of papers on the nature of electromagnetism. By considering how the motion of the vortices and cells could produce magnetic and electric effects, Maxwell was successful in explaining all the known effects of electromagnetism. He effectively showed that the lines of force behaved in a similar way.

Maxwell's other important contributions to theoretical physics were made in the area of the kinetic theory of gases. Here, he furthered the work of Rudolf Clausius, who in 1858 had shown that a gas must consist of molecules in constant motion colliding with one another and the walls of the container. This resulted in Maxwell's distribution of molecular velocities in addition to important applications of the theory to viscosity, conduction of heat, and diffusion of gases.

Maxwell's successful interpretation of Faraday's concept of the electromagnetic field resulted in the field equation bearing Maxwell's name. Formidable mathematical ability combined with great insight enabled Maxwell to lead the way in the study of the two most important areas of physics at that time. Maxwell died of cancer before he was 50.

(Photo courtesy of AIP Niels Bohr Library)

detected electromagnetic waves. This discovery has led to many practical communication systems, including radio, television, and radar. On a conceptual level, Maxwell unified the subjects of light and electromagnetism by developing the idea that light is a form of electromagnetic radiation.

Electromagnetic waves are generated by accelerating electric charges. The radiated waves consist of oscillating electric and magnetic fields, which are *at right angles to each other* and also *at right angles to the direction of wave propagation.* Thus, electromagnetic waves are transverse in nature. Maxwell's theory shows that the electric and magnetic field amplitudes, $E$ and $B$, in an electromagnetic wave are related by $E = cB$. At large distances from the source of the waves, the amplitudes of the oscillating fields diminish with distance, in proportion to $1/r$. The radiated waves can be detected at great distances from the oscillating charges. Furthermore, electromagnetic waves carry energy and momentum and hence exert pressure on a surface.

Electromagnetic waves cover a wide range of frequencies. For example, radio waves (frequencies of about $10^7$ Hz) are electromagnetic waves produced by oscillating currents in a radio tower's transmitting antenna. Light waves are a high-frequency form of electromagnetic radiation (about $10^{14}$ Hz) produced by oscillating electrons within atomic systems.

## 34.1 MAXWELL'S EQUATIONS AND HERTZ'S DISCOVERIES

The fundamental laws governing the behavior of electric and magnetic fields are Maxwell's equations, which were discussed in Section 31.7.[1] In this unified theory of electromagnetism, Maxwell showed that electromagnetic waves are a natural consequence of these fundamental laws. Recall that *Maxwell's equations* in free space are given by

$$\oint \mathbf{E} \cdot d\mathbf{A} = \frac{Q}{\epsilon_0} \tag{34.1}$$

$$\oint \mathbf{B} \cdot d\mathbf{A} = 0 \tag{34.2}$$

$$\oint \mathbf{E} \cdot d\mathbf{s} = -\frac{d\Phi_m}{dt} \tag{34.3}$$

$$\oint \mathbf{B} \cdot d\mathbf{s} = \mu_0 I + \mu_0 \epsilon_0 \frac{d\Phi_e}{dt} \tag{34.4}$$

As we shall see in the next section, one can combine Equations 34.3 and 34.4 and obtain a wave equation for both the electric and the magnetic field. In empty space ($Q = 0, I = 0$), these equations permit a wavelike solution, where the *wave velocity* $(\mu_0\epsilon_0)^{-1/2}$ *equals the measured speed of light.* This result led Maxwell to the prediction that light waves are, in fact, a form of electromagnetic radiation.

Electromagnetic waves were first generated and detected in 1887 by Hertz, using electrical sources. His experimental apparatus is shown schematically in Figure 34.1. An induction coil is connected to two spherical electrodes with a narrow gap between them (the transmitter). The coil provides short voltage surges to the spheres, making one positive, the other negative. A spark is generated between the spheres when the voltage between them reaches the breakdown voltage for air. As the air in the gap is ionized, it conducts more readily and the discharge between the spheres becomes oscillatory. From an electrical circuit viewpoint, this is equivalent to an $LC$ circuit, where the inductance is that of the loop and the capacitance is due to the spherical electrodes.

Since $L$ and $C$ are quite small, the frequency of oscillation is very high, $\approx 100$ MHz. (Recall that $\omega = 1/\sqrt{LC}$ for an $LC$ circuit.) Electromagnetic waves are radiated at this frequency as a result of the oscillation (and hence acceleration) of free charges in the loop. Hertz was able to detect these waves using a single loop of wire with its own spark gap (the receiver). This loop, placed several meters from the transmitter, has its own effective inductance, capacitance, and natural frequency of oscillation. Sparks were induced across the gap of the receiving electrodes when the frequency of the receiver was adjusted to match that of the transmitter. Thus, Hertz demonstrated that the oscillating current induced in the receiver was produced by electromagnetic waves radiated by the transmitter. Hertz's experiment is analogous to the mechanical phenomenon in which a tuning fork picks up the vibrations from another, identical oscillating tuning fork.

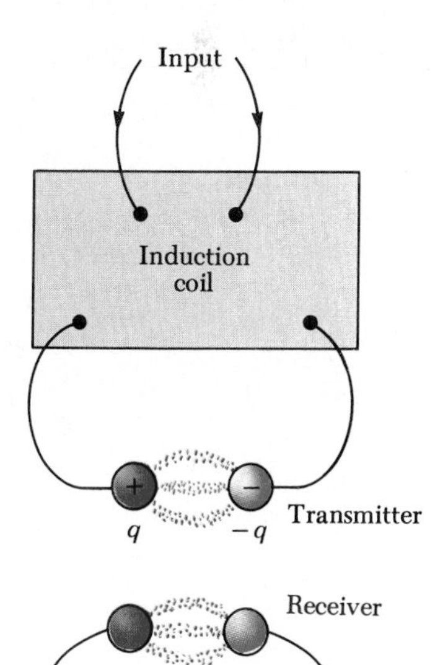


**Figure 34.1** Schematic diagram of Hertz's apparatus for generating and detecting electromagnetic waves. The transmitter consists of two spherical electrodes connected to an induction coil, which provides short voltage surges to the spheres, setting up oscillations in the discharge. The receiver is a nearby loop containing a second spark gap.

[1] The reader should review Section 31.7 as a background for the material in this chapter.

Biographical Sketch

**Heinrich Rudolf Hertz**
*(1857–1894)*

Heinrich Hertz was born in 1857 in Hamburg, Germany. He studied physics under Helmholtz and Kirchhoff at the University of Berlin. In 1885, Hertz accepted the position of Professor of Physics at Karlsruhe; it was here that he discovered radio waves in 1888, his most important accomplishment.

In 1889 Hertz succeeded Rudolf Clausius as Professor of Physics at the University of Bonn, where his experiments involving the cathode ray's penetration through certain metal films led him to the conclusion that cathode rays were waves rather than particles.

Discovering radio waves, demonstrating their generation, and determining their velocity are among Hertz's many achievements. After finding that the velocity of a radio wave was the same as that of light, Hertz showed that radio waves, like light waves, could be reflected, refracted, and diffracted.

Hertz died of blood poisoning at the age of 36. During his short life, he made many contributions to science. The hertz, equal to one complete vibration or cycle per second, is named after him.

In a series of experiments, Hertz also showed that the radiation generated by his spark-gap device exhibited the wave properties of interference, diffraction, reflection, refraction, and polarization, all of which are properties exhibited by light. Thus, it became evident that the radio-frequency waves had properties similar to light waves and differed only in frequency and wavelength.

Perhaps the most convincing experiment performed by Hertz was the measurement of the velocity of the radio-frequency waves. Radio-frequency waves of known frequency were reflected from a metal sheet and created an interference pattern whose nodal points (where $\boldsymbol{E}$ was zero) could be detected. The measured distance between the nodal points allowed determination of the wavelength $\lambda$. Using the relation $v = \lambda f$, Hertz found that $v$ was close to $3 \times 10^8$ m/s, the known speed of visible light.

Large oscillator as well as circular and square resonators used by Heinrich Hertz, 1886–88. (Photo Deutsches Museum Munich)

## 34.2 PLANE ELECTROMAGNETIC WAVES

The properties of electromagnetic waves can be deduced from Maxwell's equations. One approach that can be used to derive such properties would be to solve the second-order differential equation that can be obtained from Maxwell's third and fourth equations. A rigorous mathematical treatment of this sort is beyond the scope of this text. To circumvent this problem, we shall assume that the electric and magnetic vectors have a specific space-time behavior that is consistent with Maxwell's equations.

First, we shall assume that the electromagnetic wave is a *plane wave*, that is, one that travels in one direction. The plane wave we are describing has the following properties. The wave travels in the $x$ direction (the direction of propagation), the electric field $\mathbf{E}$ is in the $y$ direction, and the magnetic field $\mathbf{B}$ is in the $z$ direction, as in Figure 34.2. Waves in which the electric and magnetic fields are restricted to being parallel to certain lines in the $yz$ plane are said to be **linearly polarized waves.**[2] Furthermore, we assume that $E$ and $B$ at any point $P$ depend upon $x$ and $t$ and not upon the $y$ or $z$ coordinates of the point $P$.

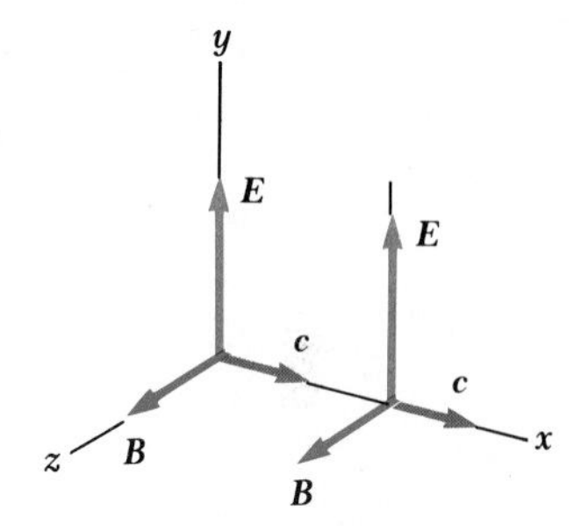


**Figure 34.2** A plane polarized electromagnetic wave traveling in the positive $x$ direction. The electric field is along the $y$ direction, and the magnetic field is along the $z$ direction. These fields depend only on $x$ and $t$.

We can relate $E$ and $B$ to each other by using Maxwell's third and fourth equations (Eqs. 34.3 and 34.4). In empty space, where $Q = 0$ and $I = 0$, these equations become

$$\oint \mathbf{E} \cdot d\mathbf{s} = -\frac{d\Phi_{\mathrm{m}}}{dt} \tag{34.5}$$

$$\oint \mathbf{B} \cdot d\mathbf{s} = \epsilon_0 \mu_0 \frac{d\Phi_{\mathrm{e}}}{dt} \tag{34.6}$$

Using these expressions and the plane wave assumption, one obtains the following differential equations relating $E$ and $B$. For simplicity of notation, we have dropped the subscripts on the components $E_y$ and $B_z$:

$$\frac{\partial E}{\partial x} = -\frac{\partial B}{\partial t} \tag{34.7}$$

$$\frac{\partial B}{\partial x} = -\mu_0 \epsilon_0 \frac{\partial E}{\partial t} \tag{34.8}$$

Note that the derivatives here are partial derivatives. For example, when $\partial E/\partial x$ is evaluated, we assume that $t$ is constant. Likewise, when evaluating $\partial B/\partial t$, $x$ is held constant. We shall derive these expressions from Maxwell's equations later in this section. Taking the derivative of Equation 34.7 and combining this with Equation 34.8 we get

$$\frac{\partial^2 E}{\partial x^2} = -\frac{\partial}{\partial x}\left(\frac{\partial B}{\partial t}\right) = -\frac{\partial}{\partial t}\left(\frac{\partial B}{\partial x}\right) = -\frac{\partial}{\partial t}\left(\frac{-\mu_0 \epsilon_0 \partial E}{\partial t}\right) \tag{34.9}$$

$$\frac{\partial^2 E}{\partial x^2} = \mu_0 \epsilon_0 \frac{\partial^2 E}{\partial t^2} \tag{34.10}$$

Wave equations for electromagnetic waves in free space

[2] Waves with other particular patterns of vibrations of $\mathbf{E}$ and $\mathbf{B}$ include *circularly polarized waves*. The most general polarization pattern is *elliptical*.

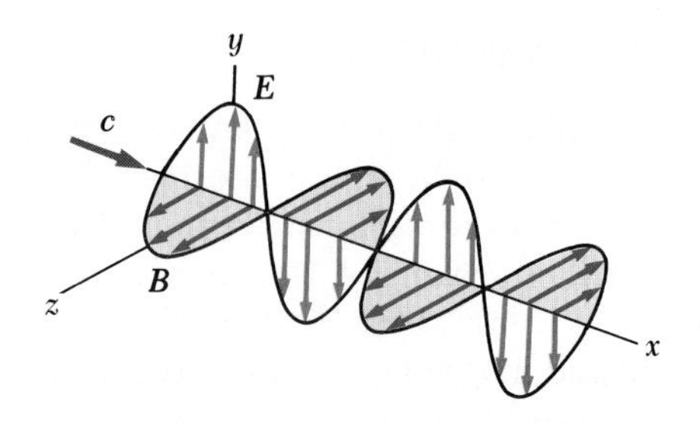


**Figure 34.3** Representation of a sinusoidal, plane polarized electromagnetic wave moving in the positive $x$ direction with a speed $c$. The drawing represents a snapshot, that is, the wave at some instant. Note the sinusoidal variations of $E$ and $B$ with $x$.

In the same manner, taking a derivative of Equation 34.8 and combining this with Equation 34.10, we get

$$\frac{\partial^2 B}{\partial x^2} = \mu_0 \epsilon_0 \frac{\partial^2 B}{\partial t^2} \tag{34.11}$$

Equations 34.10 and 34.11 both have the form of the general wave equation,[3] with a speed $c$ given by

$$c = \frac{1}{\sqrt{\mu_0 \epsilon_0}} \tag{34.12}$$

Taking $\mu_0 = 4\pi \times 10^{-7}$ Wb/A·m and $\epsilon_0 = 8.85418 \times 10^{-12}$ C²/N·m² in Equation 34.12, we find that

The speed of electromagnetic waves

$$c = 2.99792 \times 10^8 \text{ m/s} \tag{34.13}$$

Since this speed is precisely the same as the speed of light in empty space,[4] one is led to believe (correctly) that light is an electromagnetic wave.

The simplest plane wave solution is a sinusoidal wave, for which the field amplitudes $E$ and $B$ vary with $x$ and $t$ according to the expressions

Sinusoidal electric and magnetic fields

$$E = E_m \cos(kx - \omega t) \tag{34.14}$$

$$B = B_m \cos(kx - \omega t) \tag{34.15}$$

where $E_m$ and $B_m$ are the *maximum* values of the fields. The constant $k = 2\pi/\lambda$, where $\lambda$ is the wavelength, and the angular frequency $\omega = 2\pi f$, where $f$ is the number of cycles per second. The ratio $\omega/k$ equals the speed $c$, since

$$\frac{\omega}{k} = \frac{2\pi f}{2\pi/\lambda} = \lambda f = c$$

Figure 34.3 is a pictorial representation at one instant of a sinusoidal, linearly polarized plane wave moving in the positive $x$ direction.

[3] The general wave equation is of the form $(\partial^2 f/\partial x^2) = (1/v^2)(\partial^2 f/\partial t^2)$, where $v$ is the speed of the wave and $f$ is the wave amplitude. The wave equation was first introduced in Chapter 16, and it would be useful for the reader to review this material.

[4] Because of the redefinition of the meter in 1983, the speed of light is now a *defined* quantity with an *exact* value of $c = 2.99792458 \times 10^8$ m/s.

Taking partial derivatives of Equations 34.14 and 34.15, we find that

$$\frac{\partial E}{\partial x} = -kE_m \sin(kx - \omega t)$$

$$-\frac{\partial B}{\partial t} = -\omega B_m \sin(kx - \omega t)$$

Since these must be equal, according to Equation 34.7, we find that at any instant

$$kE_m = \omega B_m$$

$$\frac{E_m}{B_m} = \frac{\omega}{k} = c$$

The minus sign is ignored here since we are interested only in comparing the amplitudes. Using these results together with Equations 34.14 and 34.15, we see that

$$\frac{E_m}{B_m} = \frac{E}{B} = c \tag{34.16}$$

That is, *at every instant the ratio of the electric field to the magnetic field of an electromagnetic wave equals the speed of light.*

Finally, one should note that electromagnetic waves obey the *superposition principle,* since the differential equations involving $E$ and $B$ are *linear* equations. For example, two waves traveling in opposite directions with the same frequency could be added by simply adding the wave fields algebraically. Furthermore, we now have a theoretical value for $c$, given by the relation $c = 1/\sqrt{\mu_0\epsilon_0}$.

Let us summarize the properties of electromagnetic waves as we have described them:

Properties of electromagnetic waves

1. The solutions of Maxwell's third and fourth equations are wavelike, where both $E$ and $B$ satisfy the same wave equation.
2. Electromagnetic waves travel through empty space with the speed of light, $c = 1/\sqrt{\epsilon_0\mu_0}$.
3. The electric and magnetic field components of plane electromagnetic waves are perpendicular to each other and also perpendicular to the direction of wave propagation. The latter property can be summarized by saying that electromagnetic waves are transverse waves.
4. The relative magnitudes of $\mathbf{E}$ and $\mathbf{B}$ in empty space are related by $E/B = c$.
5. Electromagnetic waves obey the principle of superposition.

EXAMPLE 34.1 An Electromagnetic Wave

A plane electromagnetic sinusoidal wave of frequency 40 MHz travels in free space in the $x$ direction, as in Figure 34.4. At some point and at some instant, the electric field $\mathbf{E}$ has its *maximum* value of 750 N/C and is along the $y$ axis. (a) Determine the wavelength and period of the wave.

Since $c = \lambda f$ and $f = 40 \text{ MHz} = 4 \times 10^7 \text{ s}^{-1}$, we get

$$\lambda = \frac{c}{f} = \frac{3 \times 10^8 \text{ m/s}}{4 \times 10^7 \text{ s}^{-1}} = 7.50 \text{ m}$$

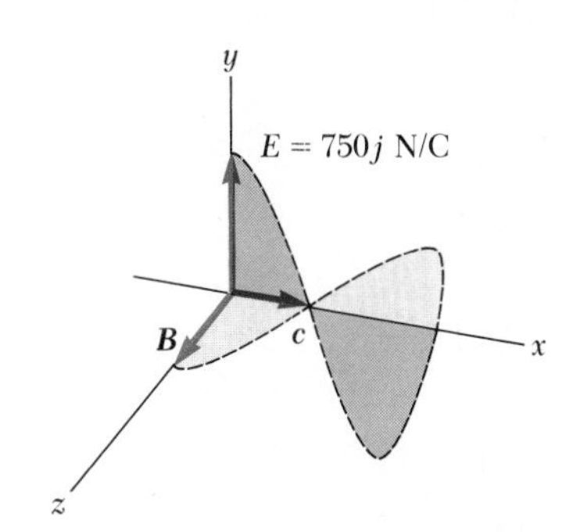


**Figure 34.4** (Example 34.1) At some instant, a plane electromagnetic wave moving in the $x$ direction has a maximum electric field of 750 N/C in the positive $y$ direction. The corresponding magnetic field at that point has a magnitude $E/c$ and is in the $z$ direction.

The period of the wave $T$ equals the inverse of the frequency, and so

$$T = \frac{1}{f} = \frac{1}{4 \times 10^7\ \text{s}^{-1}} = 2.5 \times 10^{-8}\ \text{s}$$

(b) Calculate the magnitude and direction of the magnetic field $\boldsymbol{B}$ when $\boldsymbol{E} = 750\boldsymbol{j}$ N/C.

From Equation 34.16 we see that

$$B_{\text{m}} = \frac{E_{\text{m}}}{c} = \frac{750\ \text{N/C}}{3 \times 10^8\ \text{m/s}} = 2.50 \times 10^{-6}\ \text{T}$$

Since $\boldsymbol{E}$ and $\boldsymbol{B}$ must be perpendicular to each other and both must be perpendicular to the direction of wave propagation ($x$ in this case), we conclude that $\boldsymbol{B}$ is in the $z$ direction.

(c) Write expressions for the space-time variation of the electric and magnetic field components for this plane wave.

We can apply Equations 34.14 and 34.15 directly:

$$E = E_{\text{m}} \cos(kx - \omega t) = (750\ \text{N/C}) \cos(kx - \omega t)$$

$$B = B_{\text{m}} \cos(kx - \omega t)$$
$$= (2.50 \times 10^{-6}) \cos(kx - \omega t)$$

where

$$\omega = 2\pi f = 2\pi(4 \times 10^7\ \text{s}^{-1}) = 8\pi \times 10^7\ \text{rad/s}$$

$$k = \frac{2\pi}{\lambda} = \frac{2\pi}{7.5\ \text{m}} = 0.838\ \text{m}^{-1}$$

We shall now give derivations for Equations 34.7 and 34.8. To derive Equation 34.7, we start with Faraday's law, that is, Equation 34.5:

$$\oint \boldsymbol{E} \cdot d\boldsymbol{s} = -\frac{d\Phi_{\text{m}}}{dt}$$

Again, let us assume that the electromagnetic plane wave travels in the $x$ direction with the electric field $\boldsymbol{E}$ in the positive $y$ direction and the magnetic field $\boldsymbol{B}$ in the positive $z$ direction.

Consider a thin rectangle lying in the $xy$ plane. The dimensions of the rectangle are width $dx$ and height $\ell$, as in Figure 34.5. To apply Equation 34.5, we must first evaluate the line integral of $\boldsymbol{E} \cdot d\boldsymbol{s}$ around this rectangle. The contributions from the top and bottom of this rectangle are zero since $\boldsymbol{E}$ is perpendicular to $d\boldsymbol{s}$ for these paths. We can express the electric field on the right side of the rectangle as

$$E(x + dx, t) \approx E(x, t) + \left.\frac{dE}{dx}\right]_{t\ \text{constant}} dx = E(x, t) + \frac{\partial E}{\partial x}\, dx$$

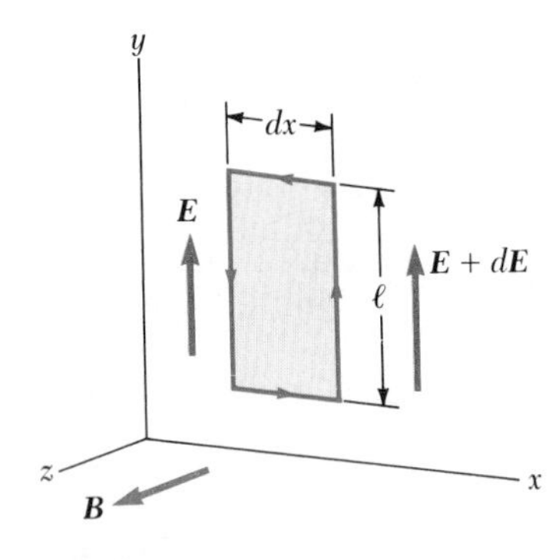


**Figure 34.5** As a plane wave passes through a rectangular path of width $dx$ lying in the $xy$ plane, the electric field in the $y$ direction varies from $\boldsymbol{E}$ to $\boldsymbol{E} + d\boldsymbol{E}$. This spatial variation in $\boldsymbol{E}$ gives rise to a time-varying magnetic field along the $z$ direction, according to Equation 34.19.

while the field on the left side is simply $E(x, t)$. Therefore, the line integral over this rectangle becomes approximately[5]

$$\oint \boldsymbol{E} \cdot d\boldsymbol{s} = E(x + dx, t) \cdot \ell - E(x, t) \cdot \ell \approx (\partial E/\partial x)\, dx \cdot \ell \qquad (34.17)$$

[5] Since $dE/dx$ means the change in $E$ with $x$ at a given instant $t$, $dE/dx$ is equivalent to the partial derivative $\partial E/\partial x$. Likewise, $dB/dt$ means the change in $B$ with time at a particular position $x$, and so we can replace $dB/dt$ by $\partial B/\partial t$.

Since the magnetic field is in the $z$ direction, the magnetic flux through the rectangle of area $\ell\ dx$ is approximately

$$\Phi_{\text{m}} = B\ell dx$$

(This assumes that $dx$ is small compared with the wavelength of the wave.) Taking the time derivative of the flux gives

$$\frac{d\Phi_{\text{m}}}{dt} = \ell\ dx \left.\frac{dB}{dt}\right]_{x\text{ constant}} = \ell\ dx \frac{\partial B}{\partial t} \tag{34.18}$$

Substituting Equations 34.17 and 34.18 into Equation 34.5 gives

$$\left(\frac{\partial E}{\partial x}\right) dx \cdot \ell = -\ell\ dx \frac{\partial B}{\partial t}$$

$$\frac{\partial E}{\partial x} = -\frac{\partial B}{\partial t} \tag{34.19}$$

Thus, we see that Equation 34.19 is equivalent to Equation 34.7.

In a similar manner, we can verify Equation 34.8 by starting with Maxwell's fourth equation in empty space (Eq. 34.6):

$$\oint \boldsymbol{B} \cdot d\boldsymbol{s} = \mu_0 \epsilon_0 \frac{d\Phi_{\text{e}}}{dt}$$

In this case, we evaluate the line integral of $\boldsymbol{B} \cdot d\boldsymbol{s}$ around a rectangle lying in the $yz$ plane and having width $dx$ and length $\ell$, as in Figure 34.6, where the magnetic field is in the $z$ direction. Using the sense of the integration shown, and noting that the magnetic field changes from $B(x, t)$ to $B(x + dx, t)$ over the width $dx$, we get

$$\oint \boldsymbol{B} \cdot d\boldsymbol{s} = B(x, t) \cdot \ell - B(x + dx, t) \cdot \ell = -(\partial B/\partial x)\ dx \cdot \ell \tag{34.20}$$

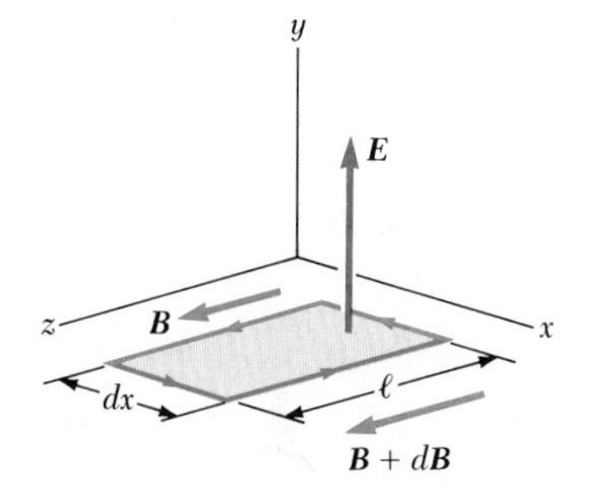


**Figure 34.6** As a plane wave passes through a rectangular curve of width $dx$ lying in the $xz$ plane, the magnetic field along $z$ varies from $\boldsymbol{B}$ to $\boldsymbol{B} + d\boldsymbol{B}$. This spatial variation in $\boldsymbol{B}$ gives rise to a time-varying electric field along the $y$ direction, according to Equation 34.22.

The electric flux through the rectangle is

$$\Phi_{\text{e}} = E\ell\ dx$$

which when differentiated with respect to time gives

$$\frac{\partial \Phi_{\text{e}}}{\partial t} = \ell\ dx \frac{\partial E}{\partial t} \tag{34.21}$$

Substituting Equations 34.20 and 34.21 into Equation 34.6 gives

$$-(\partial B/\partial x) dx \cdot \ell = \mu_0 \epsilon_0 \ell\ dx (\partial E/\partial t)$$

$$\frac{\partial B}{\partial x} = -\mu_0 \epsilon_0 \frac{\partial E}{\partial t} \tag{34.22}$$

which is equivalent to Equation 34.8.

## 34.3 ENERGY CARRIED BY ELECTROMAGNETIC WAVES

Electromagnetic waves carry energy, and as they propagate through space they can transfer energy to objects placed in their path. The rate of flow of energy in an electromagnetic wave is described by a vector **S**, called the **Poynting vector,** defined by the expression

Poynting vector

$$\mathbf{S} \equiv \frac{1}{\mu_0} \mathbf{E} \times \mathbf{B} \tag{34.23}$$

The magnitude of the Poynting vector represents the rate at which energy flows through a unit surface area perpendicular to the flow.

The direction of **S** is along the direction of wave propagation (Fig. 34.7). The SI units of the Poynting vector are $\text{J/s} \cdot \text{m}^2 = \text{W/m}^2$. (These are the units **S** must have since it represents the power per unit area, where the unit area is oriented at right angles to the direction of wave propagation.)

As an example, let us evaluate the magnitude of **S** for a plane electromagnetic wave where $|\mathbf{E} \times \mathbf{B}| = EB$. In this case

Poynting vector for a plane wave

$$S = \frac{EB}{\mu_0} \tag{34.24}$$

Since $B = E/c$, we can also express this as

$$S = \frac{E^2}{\mu_0 c} = \frac{c}{\mu_0} B^2 \tag{34.25}$$

*These equations for S apply at any instant of time.*

What is of more interest for a sinusoidal plane electromagnetic wave is the time average of S taken over one or more cycles, called the *wave intensity, I.* When this average is taken, one obtains an expression involving the time average of $\cos^2(kx - \omega t)$, which equals $\frac{1}{2}$. Hence, the average value of S (or the intensity of the wave) is

Wave intensity

$$I = S_{\text{av}} = \frac{E_{\text{m}} B_{\text{m}}}{2\mu_0} = \frac{E_{\text{m}}^2}{2\mu_0 c} = \frac{c}{2\mu_0} B_{\text{m}}^2 \tag{34.26}$$

where it is important to note that $E_{\text{m}}$ and $B_{\text{m}}$ represent *maximum* values of the fields. The constant $\mu_0 c$, called the **impedance of free space**, has the SI units of ohms and has the value

Impedance of free space

$$\mu_0 c = \sqrt{\frac{\mu_0}{\epsilon_0}} = 120\pi\ \Omega \approx 377\ \Omega$$

Recall that the energy per unit volume $u_e$, the instantaneous energy density associated with an electric field (Section 26.4), is given by

$$u_e = \tfrac{1}{2}\epsilon_0 E^2 \tag{26.14}$$

and that the instantaneous energy density $u_{\text{m}}$ associated with a magnetic field (Section 32.3) is given by

$$u_{\text{m}} = \frac{B^2}{2\mu_0} \tag{32.15}$$

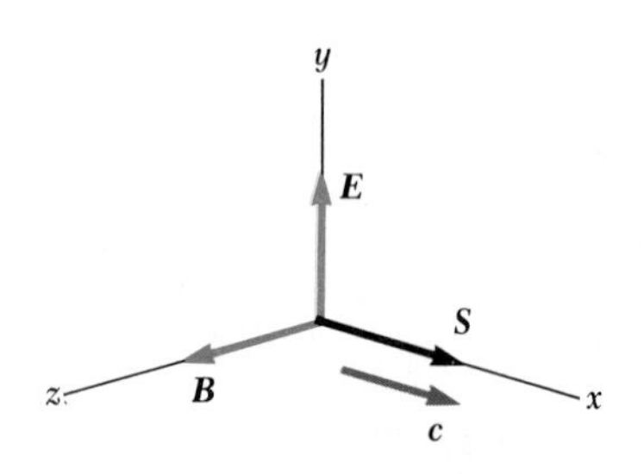


**Figure 34.7** The Poynting vector **S** for a plane electromagnetic wave moving in the *x* direction is along the direction of propagation.

Because *E* and *B* vary with time for an electromagnetic wave, we see that the energy densities also vary with time. Using the relationships $B = E/c$ and $c = 1/\sqrt{\epsilon_0 \mu_0}$, Equation 32.15 becomes

$$u_m = \frac{(E/c)^2}{2\mu_0} = \frac{\epsilon_0\mu_0}{2\mu_0} E^2 = \tfrac{1}{2}\epsilon_0 E^2$$

Comparing this result with Equation 32.15, we see that

$$u_m = u_e = \tfrac{1}{2}\epsilon_0 E^2 = \frac{B^2}{2\mu_0} \quad (34.27)$$

That is, *for an electromagnetic wave the instantaneous energy density associated with the magnetic field equals the instantaneous energy density associated with the electric field.* Hence, in a given volume the energy is equally shared by the two fields.

The **total instantaneous energy density** $u$ is equal to the sum of the energy densities associated with the electric and magnetic fields:

$$u = u_e + u_m = \epsilon_0 E^2 = \frac{B^2}{\mu_0} \quad (34.28)$$

Total energy density

When this is averaged over one or more cycles of an electromagnetic wave, we again get a factor of $\frac{1}{2}$. Hence, the total *average* energy per unit volume of an electromagnetic wave is given by

$$u_{av} = \epsilon_0 (E^2)_{av} = \tfrac{1}{2}\epsilon_0 E_m{}^2 = \frac{B_m{}^2}{2\mu_0} \quad (34.29)$$

Average energy density of an electromagnetic wave

Comparing this result with Equation 34.26 for the average value of $S$, we see that

$$I = S_{av} = cu_{av} \quad (34.30)$$

In other words, *the intensity of an electromagnetic wave equals the average energy density multiplied by the speed of light.*

**EXAMPLE 34.2 Fields Due to a Point Source**

A point source of electromagnetic radiation has an average power output of 800 W. Calculate the *maximum* values of the electric and magnetic fields at a point 3.50 m from the source.

*Solution* Recall from Chapter 17 that the wave intensity, $I$, at a distance $r$ from a point source is given by

$$I = \frac{P_{av}}{4\pi r^2}$$

where $P_{av}$ is the average power output of the source and $4\pi r^2$ is the area of a sphere of radius $r$ centered on the source. Since the intensity of an electromagnetic wave is also given by Equation 34.26, we have

$$I = \frac{P_{av}}{4\pi r^2} = \frac{E_m{}^2}{2\mu_0 c}$$

Solving for the maximum electric field, $E_m$, gives

$$E_m = \sqrt{\frac{\mu_0 c P_{av}}{2\pi r^2}}$$

$$= \sqrt{\frac{(4\pi \times 10^{-7}\ \text{N/A}^2)(3.00 \times 10^8\ \text{m/s})(800\ \text{W})}{2\pi(3.50\ \text{m})^2}}$$

$$= 62.6\ \text{V/m}$$

We can easily calculate the maximum value of the magnetic field using the result above and the relation $B_m = E_m/c$ (Eq. 34.16):

$$B_m = \frac{E_m}{c} = \frac{62.6\ \text{V/m}}{3.00 \times 10^8\ \text{m/s}} = 2.09 \times 10^{-7}\ \text{T}$$

Exercise 1 Calculate the value of the energy density at the point 3.50 m from the point source.
Answer $1.73 \times 10^{-8}\ \text{J/m}^3$

## 34.4 MOMENTUM AND RADIATION PRESSURE

*Electromagnetic waves transport linear momentum as well as energy.* Hence, it follows that pressure (radiation pressure) is exerted on a surface when an electromagnetic wave impinges on it. In what follows, we shall assume that the electromagnetic wave transports a total energy $U$ to a surface in a time $t$. If the surface *absorbs all* the incident energy $U$ in this time, Maxwell showed that the total momentum $\mathbf{p}$ delivered to this surface has a magnitude given by

Momentum delivered to an absorbing surface

$$p = \frac{U}{c} \quad \text{(complete absorption)} \tag{34.31}$$

Furthermore, if the Poynting vector of the wave is $\mathbf{S}$, the *radiation pressure* $P$ (force per unit area) exerted on the perfect absorbing surface is given by

Radiation pressure exerted on a perfect absorbing surface

$$P = \frac{S}{c} \tag{34.32}$$

We can apply these results to a perfect black body, where *all* of the incident energy is absorbed (none is reflected).

On the other hand, if the surface is a perfect reflector (for example, a mirror with a 100% reflecting surface), then the momentum delivered in a time $t$ for normal incidence is *twice* that given by Equation 34.31, or $2U/c$. That is, a momentum equal to $U/c$ is delivered by the incident wave and $U/c$ is delivered by the reflected wave, in analogy with a ball colliding elastically with a wall. Therefore,

$$p = \frac{2U}{c} \quad \text{(complete reflection)} \tag{34.33}$$

The momentum delivered to an arbitrary surface has a value between $U/c$ and $2U/c$, depending on the properties of the surface. Finally, the radiation pressure exerted on a perfect reflecting surface for normal incidence of the wave is given by[6]

$$P = \frac{2S}{c} \tag{34.34}$$

Although radiation pressures are very small (about $5 \times 10^{-6}\ \text{N/m}^2$ for direct sunlight), they have been measured using torsion balances such as the one shown in Figure 34.8. Light is allowed to strike either a mirror or a black disk, both of which are suspended from a fine fiber. Light striking the black disk is completely absorbed, and so all of its momentum is transferred to the disk. Light striking the mirror (normal incidence) is totally reflected, hence the momentum transfer is twice as great as that transferred to the disk. The radiation pressure is determined by measuring the angle through which the horizontal portion rotates. The apparatus must be placed in a high vacuum to eliminate the effects of air currents.

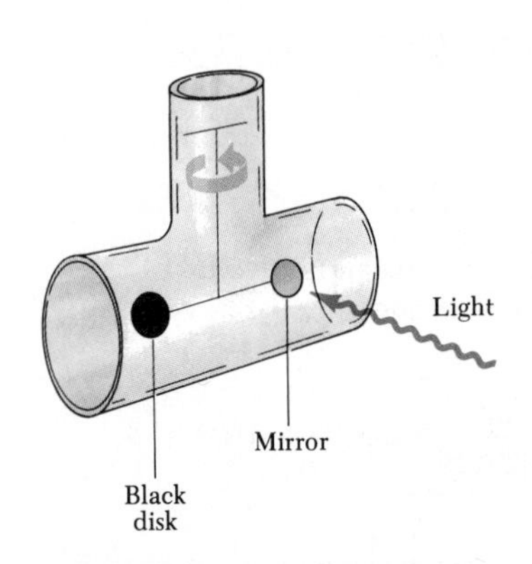


**Figure 34.8** An apparatus for measuring the pressure of light. In practice, the system is contained in a high vacuum.

[6] For *oblique* incidence, the momentum transferred is $2U \cos\theta/c$ and the pressure is given by $P = 2S\cos\theta/c$, where $\theta$ is the angle between the normal to the surface and the direction of propagation.

EXAMPLE 34.3 Solar Energy

The sun delivers about 1000 W/m² of electromagnetic flux to the earth's surface. (a) Calculate the total power that is incident on a roof of dimensions 8 m × 20 m.

*Solution* The Poynting vector has a magnitude of $S = 1000\ \text{W/m}^2$, which represents the power per unit area, or the light intensity. Assuming the radiation is incident *normal* to the roof (sun directly overhead), we get

$$\text{Power} = SA = (1000\ \text{W/m}^2)(8 \times 20\ \text{m}^2)$$
$$= 1.60 \times 10^5\ \text{W}$$

If this power could *all* be converted into electrical energy, it would provide more than enough power for the average home. However, solar energy is not easily harnessed, and the prospects for large-scale conversion are not as "bright" as they may appear from this simple calculation. For example, the conversion efficiency from solar to electrical energy is far less than 100% (typically, 10% for photovoltaic cells). Roof systems for converting solar energy to *thermal* energy have been built with efficiencies of around 50%; however, there are other practical problems with solar energy that must be considered, such as overcast days, geographic location, and energy storage.

(b) Determine the radiation pressure and radiation force on the roof assuming the roof covering is a perfect absorber.

*Solution* Using Equation 34.32 with $S = 1000\ \text{W/m}^2$, we find that the radiation pressure is

$$P = \frac{S}{c} = \frac{1000\ \text{W/m}^2}{3 \times 10^8\ \text{m/s}} = 3.33 \times 10^{-6}\ \text{N/m}^2$$

Because pressure equals force per unit area, this corresponds to a radiation force of

$$F = PA = (3.33 \times 10^{-6}\ \text{N/m}^2)(160\ \text{m}^2)$$
$$= 5.33 \times 10^{-4}\ \text{N}$$

Of course, this "load" is *far* less than the other loads one must contend with on roofs, such as the roof's own weight or a layer of snow.

Exercise 2 How much solar energy (in joules) is incident on the roof in 1 h?
Answer $5.76 \times 10^8$ J.

EXAMPLE 34.4 Poynting Vector for a Wire

A long, straight wire of resistance $R$, radius $a$, and length $\ell$ carries a constant current $I$ as in Figure 34.9. Calculate the Poynting vector for this wire.

*Solution* First, let us find the electric field **E** along the wire. If $V$ is the potential difference across the ends of the wire, then $V = IR$ and

$$E = V/\ell = IR/\ell$$

Recall that the magnetic field at the surface of the wire (Example 30.4) is given by

$$B = \mu_0 I/2\pi a$$

The vectors **E** and **B** are mutually *perpendicular*, as shown in Figure 34.9, and therefore $|\mathbf{E} \times \mathbf{B}| = EB$. Hence, the Poynting vector **S** is directed radially *inward* and has a magnitude

$$S = \frac{EB}{\mu} = \frac{1}{\mu}\frac{IR}{\ell}\frac{\mu_0 I}{2\pi a} = \frac{I^2R}{2\pi a\ell} = \frac{I^2R}{A}$$

where $A = 2\pi a\ell$ is the *surface* area of the wire, and the total area through which $S$ passes. From this result, we see that

$$SA = I^2R$$

where $SA$ has units of power (J/s = W). That is, *the rate at which electromagnetic energy flows into the wire, SA, equals the rate of energy (or power) dissipated as joule heat, $I^2R$.*

Exercise 3 A heater wire of radius 0.3 mm and resistance 5 Ω carries a current of 2 A. Determine the magnitude and direction of the Poynting vector for this wire.
Answer $1.06 \times 10^4\ \text{W/m}^2$ directed radially inward.

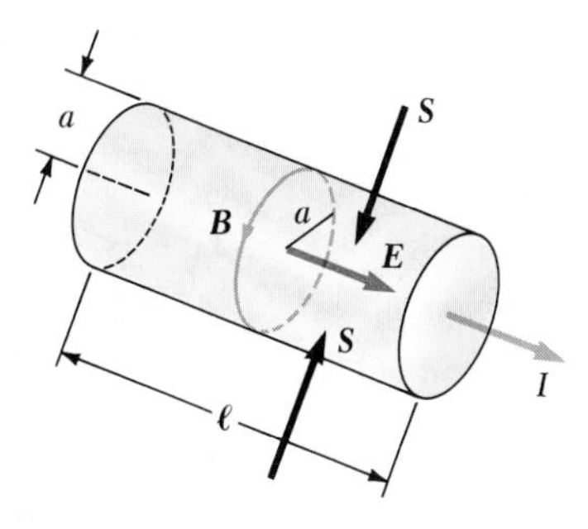


**Figure 34.9** (Example 34.4) A wire of length $\ell$, resistance $R$, and radius $a$ carrying a current $I$. The Poynting vector **S** is directed radially *inward*.

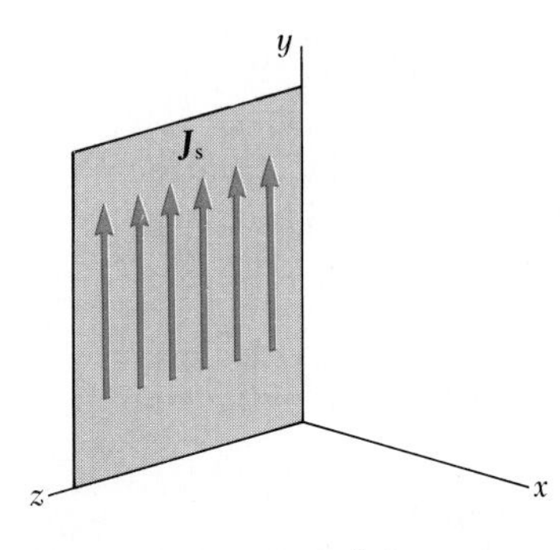


**Figure 34.10** An infinite current sheet lying in the $yz$ plane. The current density is sinusoidal and given by $J_s = J_0 \cos \omega t$. The magnetic field is everywhere parallel to the sheet and lies along $z$.

## *34.5 RADIATION FROM AN INFINITE CURRENT SHEET

In this section, we shall describe the fields radiated by a conductor carrying a time-varying current. The plane geometry we shall treat reduces the mathematical complexities one would encounter in a lower-symmetry situation, such as an oscillating electric dipole.

Consider an *infinite* conducting sheet lying in the $yz$ plane and carrying a *surface current per unit length* $\boldsymbol{J}_S$ in the $y$ direction, as in Figure 34.10. Let us assume that $J_S$ varies sinusoidally with time as

$$J_S = J_0 \cos \omega t$$

A similar problem for the case of a steady current was treated in Example 30.6, where we found that the magnetic field outside the sheet is everywhere parallel to the sheet and lies along the $z$ axis. The magnetic field was found to have a magnitude

Radiated magnetic field

$$B_z = -\mu_0 \frac{J_S}{2}$$

In the present situation, where $J_S$ varies with time, this equation for $B_z$ is valid only for distances *close* to the sheet. That is,

$$B_z = -\frac{\mu_0}{2} J_0 \cos \omega t \qquad \text{(for small values of } x\text{)}$$

To obtain the expression for $B_z$ for *arbitrary values* of $x$, we can investigate the following solution:[7]

$$B_z = -\frac{\mu_0 J_0}{2} \cos(kx - \omega t) \tag{34.35}$$

There are two things to note about this solution, which is unique to the geometry under consideration. First, it agrees with our original solution for small values of $x$. Second, it satisfies the wave equation as it is expressed in Equation 34.11. Hence, we conclude that the magnetic field lies along the $z$ axis and is characterized by a transverse traveling wave having an angular frequency $\omega$, wave number $k = 2\pi/\lambda$, and wave speed $c$.

We can obtain the radiated electric field that accompanies this varying magnetic field by using Equation 34.16:

Radiated electric field

$$E_y = cB_z = -\frac{\mu_0 J_0 c}{2} \cos(kx - \omega t) \tag{34.36}$$

That is, the electric field is in the $y$ direction, perpendicular to $\boldsymbol{B}$, and has the same space and time dependences.

These expressions for $B_z$ and $E_y$ show that the radiation field of an infinite current sheet carrying a sinusoidal current is a plane electromagnetic wave propagating with a speed $c$ along the $x$ axis, as shown in Figure 34.11.

We can calculate the Poynting vector for this wave by using Equation 34.24 together with Equations 34.35 and 34.36:

[7] Note that the solution could also be written in the form $\cos(\omega t - kx)$, which is equivalent to $\cos(kx - \omega t)$. That is, $\cos \theta$ is an even function, which means that $\cos(-\theta) = \cos \theta$.

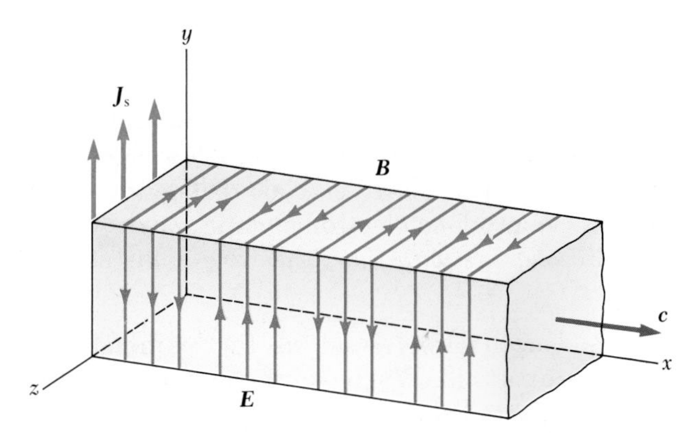


**Figure 34.11** Representation of the plane electromagnetic wave radiated by the infinite current sheet lying in the $yz$ plane. Note that $\mathbf{B}$ is in the $z$ direction, $\mathbf{E}$ is in the $y$ direction, and the direction of wave motion is along $x$. Both vectors have a $\cos(kx - \omega t)$ behavior.

$$S = \frac{EB}{\mu_0} = \frac{\mu_0 J_0^2 c}{4} \cos^2(kx - \omega t) \tag{34.37}$$

The intensity of the wave, which equals the average value of $S$, is

$$S_{av} = \frac{\mu_0 J_0^2 c}{8} \tag{34.38}$$

The intensity given by Equation 34.38 represents the average intensity of the outgoing wave on each side of the sheet. The total rate of energy emitted per unit area of the conductor is $2S_{av} = \mu_0 J_0^2 c/4$.

### EXAMPLE 34.5 An Infinite Sheet Carrying a Sinusoidal Current

An infinite current sheet lying in the $yz$ plane carries a sinusoidal current density that has a maximum value of 5 A/m. (a) Find the *maximum* values of the radiated magnetic field and electric field.

*Solution* From Equations 34.35 and 34.36, we see that the *maximum* values of $B_z$ and $E_y$ are given by

$$B_m = \frac{\mu_0 J_0}{2} \quad \text{and} \quad E_m = \frac{\mu_0 J_0 c}{2}$$

Using the values $\mu_0 = 4\pi \times 10^{-7}$ Wb/A · m, $J_0 = 5$ A/m, and $c = 3 \times 10^8$ m/s, we get

$$B_m = \frac{(4\pi \times 10^{-7}\ \text{Wb/A}\cdot\text{m})(5\ \text{A/m})}{2} = 3.14 \times 10^{-6}\ \text{T}$$

$$E_m = \frac{(4\pi \times 10^{-7}\ \text{Wb/A}\cdot\text{m})(5\ \text{A/m})(3 \times 10^8\ \text{m/s})}{2}$$

$$= 942\ \text{V/m}$$

(b) What is the average power incident on a second plane surface that is parallel to the sheet and has an area of 3 m²? (The length and width of the plate are both much larger than the wavelength of the light.)

*Solution* The power per unit area (the average value of the Poynting vector) radiated in each direction by the current sheet is given by Equation 34.38. Multiplying this by the area of the plane in question gives the incident power:

$$P = \left(\frac{\mu_0 J_0^2 c}{8}\right) A$$

$$= \frac{(4\pi \times 10^{-7}\ \text{Wb/A}\cdot\text{m})(5\ \text{A/m})^2(3 \times 10^8\ \text{m/s})}{8} (3\ \text{m}^2)$$

$$= 3.54 \times 10^3\ \text{W}$$

The result is *independent of the distance from the current sheet* since we are dealing with a plane wave.

## *34.6 THE PRODUCTION OF ELECTROMAGNETIC WAVES BY AN ANTENNA

Electromagnetic waves arise as a consequence of two effects: (1) a changing magnetic field produces an electric field and (2) a changing electric field produces a magnetic field. Therefore, it is clear that neither stationary charges nor steady currents can produce electromagnetic waves. Whenever the current through a wire *changes with time,* the wire emits electromagnetic radiation.

Accelerating charges produce *EM* radiation

The fundamental mechanism responsible for this radiation is the acceleration of a charged particle. Whenever a charged particle undergoes an acceleration, it must radiate energy.

An alternating voltage applied to the wires of an antenna forces an electric charge in the antenna to oscillate. This is a common technique for accelerating charged particles and is the source of the radio waves emitted by the antenna of a radio station.

Figure 34.12 illustrates the production of an electromagnetic wave by oscillating electric charges in an antenna. Two metal rods are connected to an ac generator, which causes charges to oscillate between the two rods. The output voltage of the generator is sinusoidal. At $t = 0$, the upper rod is given a maximum positive charge and the bottom rod an equal negative charge, as in Figure 34.12a. The electric field near the antenna at this instant is also shown in Figure 34.12a. As the charges oscillate, the rods become less charged, the field near the rods decreases in strength, and the downward-directed maximum electric field produced at $t = 0$ moves away from the rod. When the charges are neutralized, as in Figure 34.12b, the electric field has dropped to zero. This occurs at a time equal to one quarter of the period of oscillation.

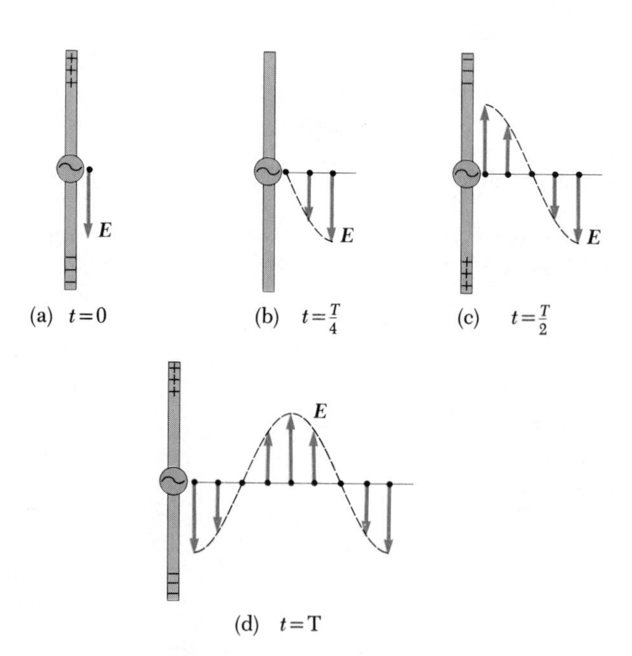


**Figure 34.12** The electric field set up by oscillating charges in an antenna. The field moves away from the antenna with the speed of light.

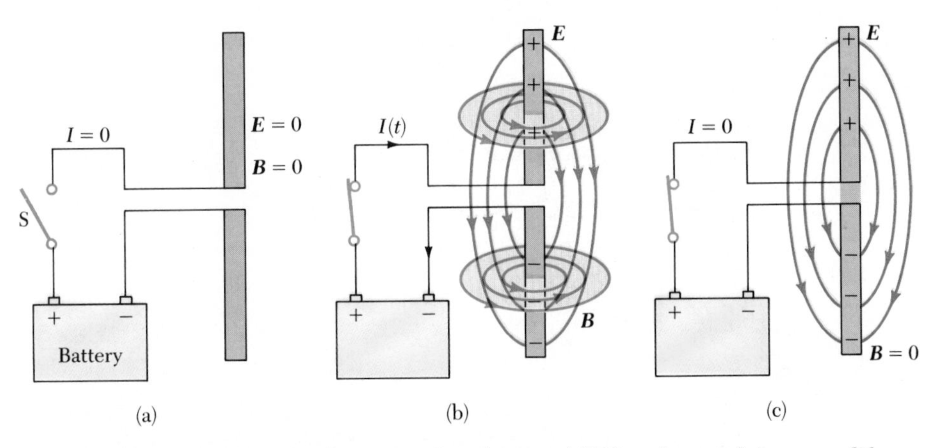


**Figure 34.13** A pair of metal rods connected to a battery. (a) When the switch is open and there is no current, the electric and magnetic fields are both zero. (b) After the switch is closed and the rods are being charged (so that a current exists), the rods generate changing electric and magnetic fields. (c) When the rods are fully charged, the current is zero, the electric field is a maximum, and the magnetic field is zero.

Continuing in this fashion, the upper rod soon obtains a maximum negative charge and the lower rod becomes positive, as in Figure 34.12c, resulting in an electric field directed upward. This occurs after a time equal to one half the period of oscillation. The oscillations continue as indicated in Figure 34.12d. A magnetic field oscillating perpendicular to the diagram in Figure 34.12 accompanies the oscillating electric field, but is not shown for clarity. The electric field near the antenna oscillates in phase with the charge distribution. That is, the field points down when the upper rod is positive and up when the upper rod is negative. Furthermore, the magnitude of the field at any instant depends on the amount of charge on the rods at that instant.

As the charges continue to oscillate (and accelerate) between the rods, the electric field set up by the charges moves away from the antenna at the speed of light. Figure 34.12 shows the electric field pattern at various times during the oscillation cycle. As you can see, one cycle of charge oscillation produces one full wavelength in the electric field pattern.

Next, consider what happens when two conducting rods are connected to the opposite ends of a battery (Fig. 34.13). Before the switch is closed, the current is zero, so there are no fields present (Fig. 34.13a). Just after the switch is closed, charge of opposite signs begins to build up on the rods (Fig. 34.13b), which corresponds to a time-varying current, $I(t)$. The changing charge causes the electric field to change, which in turn produces a magnetic field around the rods.[8] Finally, when the rods are fully charged, the current is zero and there is no magnetic field (Fig. 34.13c).

Now let us consider the production of electromagnetic waves by a *half-wave antenna.* In this arrangement, two conducting rods, each one quarter of a wavelength long, are connected to a source of alternating emf (such as an *LC* oscillator), as in Figure 34.14. The oscillator forces charges to accelerate back and forth between the two rods. Figure 34.14 shows the field configuration at some instant when the current is upward. The electric field lines resemble those of an electric dipole, that is, two equal and opposite charges. Since these

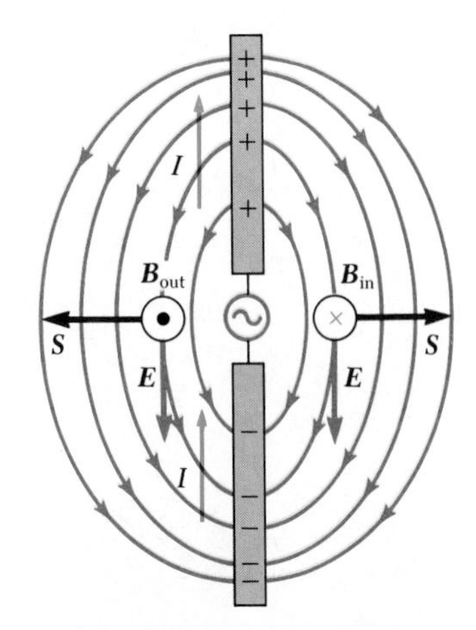


**Figure 34.14** A half-wave (dipole) antenna consists of two metal rods connected to an alternating voltage source. The diagram shows **E** and **B** at an instant when the current is upward. Note that the electric field lines resemble those of a dipole.

[8] We have neglected the field due to the wires leading to the rods. This is a good approximation if the circuit dimensions are small relative to the length of the rods.

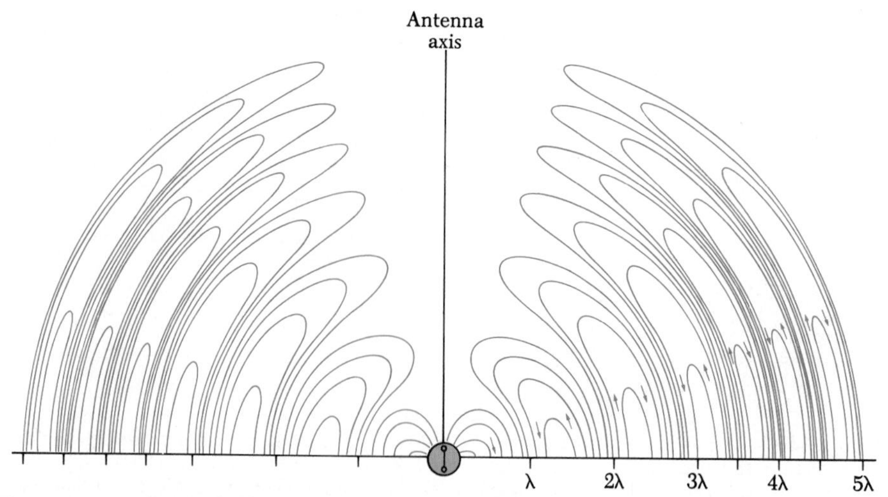


**Figure 34.15** Electric field lines surrounding an oscillating dipole at a given instant. The radiation fields propagate outward from the dipole with a speed $c$.

charges are continuously oscillating between the two rods, the antenna can be approximated by an oscillating electric dipole. The magnetic field lines form concentric circles about the antenna and are perpendicular to the electric field lines at all points. The magnetic field is zero at all points along the axis of the antenna. Furthermore, $\boldsymbol{E}$ and $\boldsymbol{B}$ are 90° out of phase in time, that is, $\boldsymbol{E}$ at some point reaches its maximum value when $\boldsymbol{B}$ is zero and vice versa. This is because when the charges at the ends of the rods are at a maximum, the current is zero.

At the two points shown in Figure 34.14, Poynting's vector $\boldsymbol{S}$ is radially outward. This indicates that energy is flowing away from the antenna at this instant. At later times, the fields and Poynting's vector change direction as the current alternates. Since $\boldsymbol{E}$ and $\boldsymbol{B}$ are 90° out of phase at points near the dipole, the net energy flow is zero. From this, we might conclude (incorrectly) that no energy is radiated by the dipole.

Since the dipole fields fall off as $1/r^3$ (as in the case of a static dipole), they are not important at large distances from the antenna. However, at these large distances, another effect produces the radiation field. The source of this radiation is the continuous induction of an electric field by a time-varying magnetic field and the induction of a magnetic field by a time-varying electric field. These are predicted by two of Maxwell's equations (Eqs. 34.3 and 34.4). The electric and magnetic fields produced in this manner are in phase with each other and vary as $1/r$. The result is an outward flow of energy at all times.

The electric field lines produced by an oscillating dipole at some instant are shown in Figure 34.15. Note that the intensity (and the power radiated) are a maximum in a plane that is perpendicular to the antenna and passing through its midpoint. Furthermore, the power radiated is zero along the antenna's axis. A mathematical solution to Maxwell's equations for the oscillating dipole shows that the intensity of the radiation field varies as $\sin^2 \theta/r^2$, where $\theta$ is measured from the axis of the antenna. The angular dependence of the radiation intensity (power per unit area) is sketched in Figure 34.16.

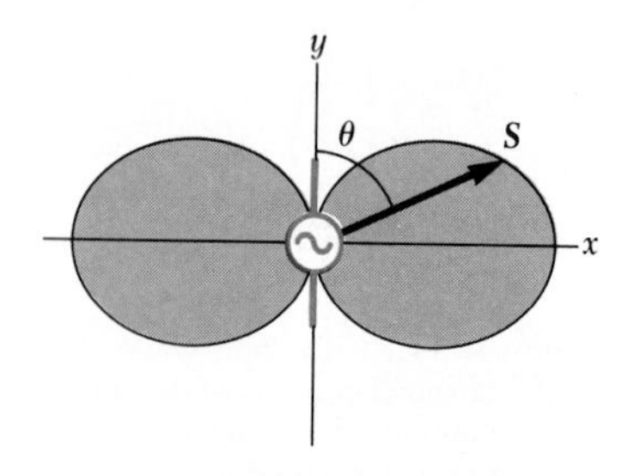


**Figure 34.16** Angular dependence of the intensity of radiation produced by an oscillating electric dipole.

Electromagnetic waves can also induce currents in a *receiving antenna.* The response of a dipole receiving antenna at a given position will be a maximum when its axis is parallel to the electric field at that point and zero when its axis is perpendicular to the electric field.

## 34.7 THE SPECTRUM OF ELECTROMAGNETIC WAVES

We have seen that all electromagnetic waves travel in a vacuum with the speed of light, $c$. These waves transport energy and momentum from some source to a receiver. In 1887, Hertz successfully generated and detected the radio-frequency electromagnetic waves predicted by Maxwell.[9] Maxwell himself had recognized as EM waves both visible light and the near infrared radiation discovered in 1800 by William Herschel. It is now known that other forms of electromagnetic waves exist which are distinguished by their frequency and wavelength.

Since all electromagnetic waves travel through vacuum with a speed $c$, their frequency $f$ and wavelength $\lambda$ are related by the important expression

$$c = f\lambda \tag{34.39}$$

The various types of electromagnetic waves are listed in Figure 34.17. Note the wide range of frequencies and wavelengths. For instance, a radio wave of frequency 5 MHz (a typical value) has a wavelength given by

Satellite television antennas are common for receiving television stations from satellites in orbit around the earth and are most widely used in rural locations. (© Hank Delespinasse/The IMAGE Bank)

[9] Following Hertz's discoveries, Marconi succeeded in developing a practical, long-range radio communication system. However, Hertz must be recognized as the true inventor of radio communication.

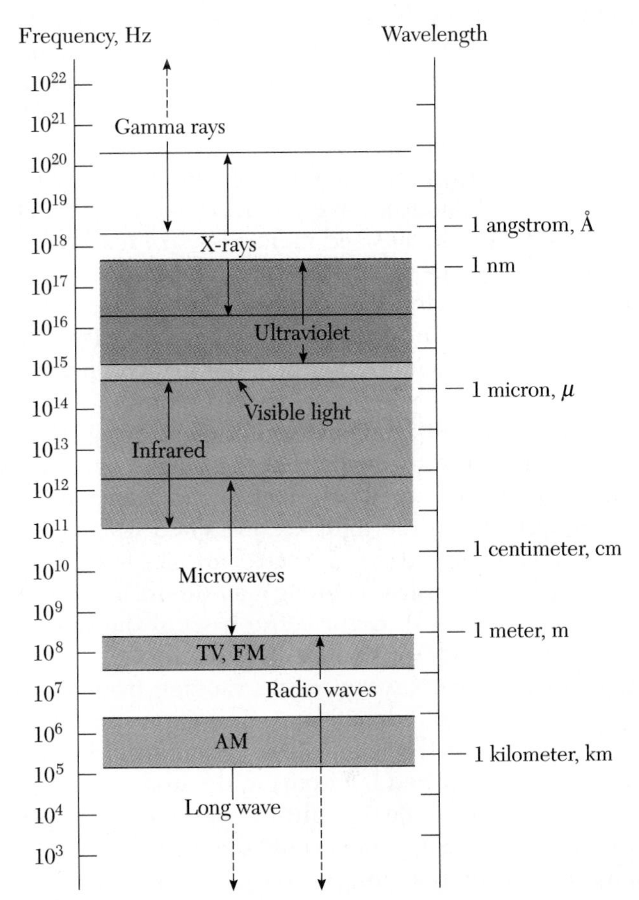


**Figure 34.17** The electromagnetic spectrum. Note the overlap between one type of wave and the next.

$$\lambda = \frac{c}{f} = \frac{3 \times 10^8 \text{ m/s}}{5 \times 10^6 \text{ s}^{-1}} = 60 \text{ m}$$

The following abbreviations are often used to designate short wavelengths and distances:

$$1 \text{ micrometer } (\mu\text{m}) = 10^{-6} \text{ m}$$
$$1 \text{ nanometer (nm)} = 10^{-9} \text{ m}$$
$$1 \text{ angstrom (Å)} = 10^{-10} \text{ m}$$

For example, the wavelengths of visible light range from 0.4 to 0.7 $\mu$m, or 400 to 700 nm, or 4000 to 7000 Å.

Let us give a brief description of these various waves in order of decreasing wavelength. There is no sharp dividing point between one kind of wave and the next. It should be noted that all forms of radiation are produced by accelerating charges.

Radio waves

**Radio waves,** which were discussed in the previous section, are the result of charges accelerating through conducting wires. They are generated by such electronic devices as *LC* oscillators and are used in radio and television communication systems.

Microwaves

**Microwaves** (short-wavelength radio waves) have wavelengths ranging between about 1 mm and 30 cm and are also generated by electronic devices. Because of their short wavelength, they are well suited for the radar systems used in aircraft navigation and for studying the atomic and molecular properties of matter. Microwave ovens represent an interesting domestic application of these waves. It has been suggested that solar energy could be harnessed by beaming microwaves down to earth from a solar collector in space.[10]

Infrared waves

**Infrared waves** (sometimes called *heat waves*) have wavelengths ranging from about 1 mm to the longest wavelength of visible light, $7 \times 10^{-7}$ m. These waves, produced by hot bodies and molecules, are readily absorbed by most materials. The infrared energy absorbed by a substance appears as heat since the energy agitates the atoms of the body, increasing their vibrational or translational motion, which results in a temperature rise. Infrared radiation has many practical and scientific applications, including physical therapy, infrared photography, and vibrational spectroscopy.

Visible waves

**Visible light,** the most familiar form of electromagnetic waves, may be defined as that part of the spectrum that the human eye can detect. Light is produced by the rearrangement of electrons in atoms and molecules. The various wavelengths of visible light are classified with colors ranging from violet ($\lambda \approx 4 \times 10^{-7}$ m) to red ($\lambda \approx 7 \times 10^{-7}$ m). The eye's sensitivity is a function of wavelength, the sensitivity being a maximum at a wavelength of about $5.6 \times 10^{-7}$ m (yellow-green). Light is the basis of the science of optics and optical instruments, which we shall deal with later.

Ultraviolet waves

**Ultraviolet light** covers wavelengths ranging from about $3.8 \times 10^{-7}$ m (380 nm) down to $6 \times 10^{-8}$ m (60 nm). The sun is an important source of ultraviolet light, which is the main cause of suntans. Most of the ultraviolet light from the sun is absorbed by atoms in the upper atmosphere, or stratosphere. This is fortunate since uv light in large quantities produces harmful effects on humans. One important constituent of the stratosphere is ozone ($O_3$), which results from reactions of oxygen with ultraviolet radiation. This

[10] P. Glaser, "Solar Power from Satellites," *Physics Today*, February, 1977, p. 30.

ozone shield converts lethal high-energy ultraviolet radiation into heat, which in turn warms the stratosphere. Recently, there has been a great deal of controversy concerning the possible depletion of the protective ozone layer as a result of the use of the freons used in aerosal spray cans and as refrigerants.

**X-rays** are electromagnetic waves with wavelengths in the range of about $10^{-8}$ m (10 nm) down to $10^{-13}$ m ($10^{-4}$ nm). The most common source of x-rays is the deceleration of high-energy electrons bombarding a metal target. X-rays are used as a diagnostic tool in medicine and as a treatment for certain forms of cancer. Since x-rays damage or destroy living tissues and organisms, care must be taken to avoid unnecessary exposure or overexposure. X-rays are also used in the study of crystal structure, since x-ray wavelengths are comparable to the atomic separation distances ($\approx 0.1$ nm) in solids.

X-rays

**Gamma rays** are electromagnetic waves emitted by radioactive nuclei (such as $^{60}$Co and $^{137}$Cs) and during certain nuclear reactions. They have wavelengths ranging from about $10^{-10}$ m to less than $10^{-14}$ m. They are highly penetrating and produce serious damage when absorbed by living tissues. Consequently, those working near such dangerous radiation must be protected with heavily absorbing materials, such as thick layers of lead.

Gamma rays

## SUMMARY

**Electromagnetic waves,** which are predicted by Maxwell's equations, have the following properties:

1. The electric and magnetic fields satisfy the following wave equations, which can be obtained from Maxwell's third and fourth equations:

Wave equations

$$\frac{\partial^2 E}{\partial x^2} = \mu_0 \epsilon_0 \frac{\partial^2 E}{\partial t^2} \tag{34.10}$$

$$\frac{\partial^2 B}{\partial x^2} = \mu_0 \epsilon_0 \frac{\partial^2 B}{\partial t^2} \tag{34.11}$$

2. Electromagnetic waves travel through a vacuum with the speed of light $c$, where

The speed of electromagnetic waves

$$c = \frac{1}{\sqrt{\mu_0 \epsilon_0}} = 3.00 \times 10^8 \text{ m/s} \tag{34.12}$$

3. The electric and magnetic fields of an electromagnetic wave are perpendicular to each other and perpendicular to the direction of wave propagation. (Hence, they are transverse waves.)
4. The instantaneous magnitudes of $|\boldsymbol{E}|$ and $|\boldsymbol{B}|$ in an electromagnetic wave are related by the expression

$$\frac{E}{B} = c \tag{34.16}$$

5. Electromagnetic waves carry energy. The rate of flow of energy crossing a unit area is described by the Poynting vector **S**, where

Poynting vector

$$\boldsymbol{S} = \frac{1}{\mu_0} \boldsymbol{E} \times \boldsymbol{B} \tag{34.23}$$

6. Electromagnetic waves carry momentum and hence can exert pressure on surfaces. If an electromagnetic wave whose Poynting vector is **S** is completely absorbed by a surface upon which it is normally incident, the radiation pressure on that surface is

$$P = \frac{S}{c} \quad \text{(complete absorption)} \tag{34.32}$$

If the surface totally reflects a normally incident wave, the pressure is doubled.

The electric and magnetic fields of a sinusoidal plane electromagnetic wave propagating in the positive $x$ direction can be written

Sinusoidal electric and magnetic fields

$$E = E_m \cos(kx - \omega t) \tag{34.14}$$

$$B = B_m \cos(kx - \omega t) \tag{34.15}$$

where $\omega$ is the angular frequency of the wave and $k$ is the wave number. These equations represent special solutions to the wave equations for $E$ and $B$. Since $\omega = 2\omega f$ and $k = 2\pi/\lambda$, where $f$ and $\lambda$ are the frequency and wavelength, respectively, one finds that

$$\frac{\omega}{k} = \lambda f = c$$

The average value of the Poynting vector for a plane electromagnetic wave has a magnitude given by

Wave intensity

$$S_{av} = \frac{E_m B_m}{2\mu_0} = \frac{E_m{}^2}{2\mu_0 c} = \frac{c}{2\mu_0} B_m{}^2 \tag{34.26}$$

The intensity of a sinusoidal plane electromagnetic wave equals the average value of the Poynting vector taken over one or more cycles.

The fundamental sources of electromagnetic waves are *accelerating electric charges*. For instance, radio waves emitted by an antenna arise from the continuous oscillation (and hence acceleration) of charges within the antenna structure.

The electromagnetic spectrum includes waves covering a broad range of frequencies and wavelengths. The frequency $f$ and wavelength $\lambda$ of a given wave are related by

$$c = f\lambda \tag{34.39}$$

## QUESTIONS

1. For a given incident energy of electromagnetic wave, why is the radiation pressure on a perfect reflecting surface twice as large as the pressure on a perfect absorbing surface?
2. In your own words, describe the physical significance of the Poynting vector.
3. Do all current-carrying conductors emit electromagnetic waves? Explain.
4. What is the fundamental source of electromagnetic radiation?
5. Electrical engineers often speak of the *radiation resistance* of an antenna. What do you suppose they mean by this phrase?
6. If a high-frequency current is passed through a solenoid containing a metallic core, the core will heat up by induction. This process also cooks foods in micro-

wave ovens. Explain why the materials heat up in these situations.

7. Certain orientations of the receiving antenna on a TV give better reception than others. Furthermore, the best orientation varies from station to station. Explain these observations.
8. Does a wire connected to a battery emit an electromagnetic wave? Explain.
9. If you charge a comb by running it through your hair and then hold the comb next to a bar magnet, do the electric and magnetic fields produced constitute an electromagnetic wave?
10. An empty plastic or glass dish removed from a microwave oven is cool to the touch. How can this be possible?
11. Often when you touch the indoor antenna on a television receiver, the reception instantly improves. Why?
12. Explain how the (dipole) VHF antenna of a television set works.
13. Explain how the (loop) UHF antenna of a television set works.
14. Explain why the voltage induced in a UHF antenna depends on the frequency of the signal, while the voltage in a VHF antenna does not.
15. List as many similarities and differences as you can between sound waves and light waves.
16. What does a radio wave do to the charges in the receiving antenna to provide a signal for your car radio?
17. What is meant by the terms "amplitude modulation" and "frequency modulation?"
18. What determines the height of an AM radio station's broadcast antenna?
19. Some radio transmitters use a "phased array" of antennas. What is their purpose?
20. What happens to the radio reception in an airplane as it flies over the (vertical) dipole antenna of the control tower?
21. When light (or other electromagnetic radiation) travels across a given region, what is it that moves?
22. Why should an infrared photograph of a person look different from a photograph taken with visible light?

## PROBLEMS

### Section 34.2 Plane Electromagnetic Waves

1. Verify that Equation 34.12 gives $c$ with dimensions of length per unit time.
2. An electromagnetic wave in vacuum has an electric field amplitude of 220 V/m. Calculate the amplitude of the corresponding magnetic field.
3. (a) Use the relationship $B = \mu_0 H$ described in Equation 30.37 (for free space where $\kappa_m = \mu_0$) together with the properties of $E$ and $B$ described in Section 34.2 to show that $E/H = \sqrt{\mu_0/\epsilon_0}$. (b) Calculate the numerical value of this ratio and show that it has SI units of ohms. (The ratio $E/H$ is referred to as the *impedance of free space*.)
4. Calculate the maximum value of the magnetic field in a region where the measured maximum value of the electric field is 5.3 mV/m.
5. The magnetic field amplitude of an electromagnetic wave is $5.4 \times 10^{-7}$ T. Calculate the electric field amplitude if the wave is traveling (a) in free space and (b) in a medium in which the speed of the wave is $0.8c$.
6. The speed of an electromagnetic wave traveling in a transparent substance is given by $v = 1/\sqrt{\kappa\mu_0\epsilon_0}$ where $\kappa$ is the dielectric constant of the substance. Determine the speed of light in water, which has a dielectric constant of 1.78 at optical frequencies.
7. Verify that the following pair of equations for $E$ and $B$ are solutions of Equations 34.7 and 34.8:

$$E = \frac{A}{\sqrt{\epsilon_0\mu_0}} e^{a(x-ct)} \quad \text{and} \quad B = Ae^{a(x-ct)}$$

8. Write down expressions for the electric and magnetic fields of a sinusoidal plane electromagnetic wave with a frequency of 3 GHz traveling in the positive $x$ direction. The amplitude of the electric field is 300 V/m.
9. Figure 34.3 shows a plane electromagnetic sinusoidal wave propagating in the $x$ direction. The wavelength is 50 m, and the electric field vibrates in the $xy$ plane with an amplitude of 22 V/m. Calculate (a) the sinusoidal frequency and (b) the magnitude and direction of $\mathbf{B}$ when the electric field has its maximum value in the negative $y$ direction. (c) Write an expression for $B$ in the form

$$B = B_m \cos(kx - \omega t)$$

with numerical values for $B_m$, $k$, and $\omega$.

10. Verify that the following equations are solutions to Equations 34.10 and 34.11, respectively:

$$E = E_m \cos(kx - \omega t)$$
$$B = B_m \cos(kx - \omega t).$$

11. The electric field in an electromagnetic wave in SI units is described by the equation

$$E_y = 100 \sin(10^7 x - \omega t)$$

Find (a) the amplitude of the corresponding magnetic wave, (b) the wavelength $\lambda$, and (c) the frequency $f$.

## Section 34.3 Energy Carried by Electromagnetic Waves

**12.** How much electromagnetic energy is contained per cubic meter near the earth's surface if the intensity of sunlight under clear skies is 1000 $W/m^2$?

13. At what distance from 100-W isotropic electromagnetic wave power source will $E_m = 15$ V/m?

14. A 10-mW laser has a beam diameter of 1.6 mm. (a) What is the intensity of the light, assuming it is uniform across the circular beam? (b) What is the average energy density of the laser beam?

**15.** What is the average magnitude of the Poynting vector at a distance of 5 miles from an isotropic radio transmitter, broadcasting with an average power of 250 kW?

**16.** The sun radiates electromagnetic energy at the rate of $P_{sun} = 3.85 \times 10^{26}$ W. (a) At what distance from the sun does the intensity of its radiation fall to 1000 $W/m^2$? (Compare this distance to the radius of the earth's orbit.) (b) At the distance you just found, what is the average energy density of the sun's radiation?

17. At a distance of 10 km from a radio transmitter, the amplitude of the **E** field is 0.20 V/m. What is the total power emitted by the radio transmitter?

**18.** The amplitude of the magnetic field in an electromagnetic wave is $B_m = 4.1 \times 10^{-8}$ T. What is the average intensity of this wave?

19. The filament of an incandescent lamp has a 150-Ω resistance, and carries a dc current of 1 A. The filament is 8 cm long and 0.9 mm in radius. (a) Calculate the Poynting vector at the surface of the filament. (b) Find the magnitude of the electric and magnetic fields at the surface of the filament.

20. A monochromatic light source emits 100 W of electromagnetic power uniformly in all directions. (a) Calculate the average electric-field energy density 1 m from the source. (b) Calculate the average magnetic-field energy density at the same distance from the source. (c) Find the wave intensity at this location.

21. A helium-neon laser intended for instructional use operates at a typical power of 5.0 mW. (a) Determine the maximum value of the electric field at a point where the cross section of the beam is 4 $mm^2$. (b) Calculate the electromagnetic energy in a 1-m length of the beam.

**22.** At a particular point, the electric field in an electromagnetic wave has the instantaneous value $E =$ 240 V/m. Compute the instantaneous magnitudes of the following quantities at the same point and time: (a) the magnetic field; (b) the Poynting vector; (c) the energy density of the electric field; (d) the energy density of the magnetic field.

23. At one location on the earth, the rms value of the magnetic field due to solar radiation is 1.8 $\mu T$. From this value calculate (a) the average electric field due to solar radiation, (b) the average energy density of the solar component of electromagnetic radiation at this location, and (c) the magnitude of the Poynting vector for the sun's radiation. (d) Compare the value found in (c) to the value of the solar flux given in Example 34.3.

24. A long wire has a radius of 1 mm and a resistance per unit length of 2 Ω/m. Determine the current required if the Poynting vector at the surface of the wire equals $2.68 \times 10^3$ $W/m^2$.

## Section 34.4 Momentum and Radiation Pressure

**25.** A radio wave transmits 25 $W/m^2$ of power per unit area. A plane surface of area $A$ is perpendicular to the direction of propagation of the wave. Calculate the radiation pressure on the surface if the surface is a perfect absorber.

**26.** A 100-mW laser beam is reflected back upon itself by a mirror. Calculate the force on the mirror.

27. A 15-mW helium-neon laser ($\lambda = 632.8$ nm) emits a beam of circular cross-section whose diameter is 2 mm. (a) Find the maximum electric field in the beam. (b) What total energy is contained in a 1-m length of the beam? (c) Find the momentum carried by a 1-m length of the beam.

28. A plane electromagnetic wave of intensity 6 $W/m^2$ strikes a small pocket mirror of area 40 $cm^2$ held perpendicular to the approaching wave. (a) What momentum does the wave transfer to the mirror each second? (b) Find the force that the wave exerts on the mirror.

29. A plane electromagnetic wave has an energy flux of 750 $W/m^2$. A flat, rectangular surface of dimensions 50 cm × 100 cm is placed perpendicular to the direction of the plane wave. If the surface absorbs half of the energy and reflects half (that is, it is a 50% reflecting surface), calculate (a) the total energy absorbed by the surface in a time of 1 min and (b) the momentum absorbed in this time.

30. (a) Assuming that the earth absorbs all the sunlight incident upon it, find the total force that the sun exerts on the earth due to radiation pressure. (b) Compare this value with the sun's gravitational attraction.

## Section 34.5 Radiation from an Infinite Current Sheet

**31.** A rectangular surface of dimensions 120 cm × 40 cm is parallel to and 4.4 m from a very large conducting sheet in which there is a sinusoidally varying surface current which has a maximum value of 10 A/m. (a) Calculate the average power incident on the smaller sheet. (b) What power per unit area is radiated by the current-carrying sheet?

**32.** A large current-carrying sheet is expected to radiate in each direction (normal to the plane of the sheet) at a rate equal to 570 W/m² (approximately one half of the solar constant). What maximum value of sinusoidal current density is required?

### Section 34.6 The Production of Electromagnetic Waves by an Antenna

33. What is the length of a half-wave antenna designed to broadcast 20 MHz radio waves?
34. An AM radio station broadcasts isotropically with an average power of 4 kW. A dipole receiving antenna, 65 cm long, is located 4 miles from the transmitter. Compute the emf induced by this signal between the ends of the receiving antenna.
35. A television set uses a dipole receiving antenna for VHF channels and a loop antenna for UHF channels. The UHF antenna produces a voltage from the changing *magnetic* flux through the loop. (a) Using Faraday's Law, derive an expression for the amplitude of the voltage that appears in a single-turn circular loop antenna with a radius $r$. The TV station broadcasts a signal with a frequency $f$, and the signal has an electric field amplitude $E_m$ and magnetic field amplitude $B_m$ at the receiving antenna's location. (b) If the electric field in the signal points vertically, what should be the orientation of the loop for best reception?
36. Figure 34.14 shows a Hertz antenna (also known as a half-wave antenna, since its length is $\lambda/2$). The Hertz antenna is located far enough from the ground that reflections do not significantly affect its radiation pattern. Most AM radio stations, however, use a Marconi antenna, which consists of the top half of a Hertz antenna. The lower end of this (quarter-wave) antenna is connected to earth ground, and the ground itself serves as the missing lower half. (a) What is the antenna height for a radio station broadcasting at the lower end of the AM band, at 560 kHz? (b) What is the height for a station broadcasting at 1600 kHz?
37. Two hand-held radio transceivers with dipole antennas are separated by a large fixed distance. Assuming a vertical transmitting antenna, what fraction of the maximum received power will occur in the receiving antenna when it is inclined from the vertical by (a) 15°; (b) 45°; (c) 90°?
38. Two radio-transmitting antennas are separated by half the broadcast wavelength and are driven in phase with each other. (a) In which direction is the strongest signal radiated? (b) In which direction is the weakest signal radiated?

### Section 34.7 The Spectrum of Electromagnetic Waves

39. What is the wavelength of an electromagnetic wave in free space that has a frequency of (a) $5 \times 10^{19}$ Hz and (b) $4 \times 10^{9}$ Hz?
40. The eye is most sensitive to light having a wavelength of $5.5 \times 10^{-7}$ m, which is in the green-yellow region of the electromagnetic spectrum. What is the frequency of this light?
41. Compute the frequency of the following electromagnetic waves: (a) microwaves ($\lambda = 1$ cm); (b) infrared radiation ($\lambda = 1$ $\mu$m); (c) yellow light ($\lambda = 580$ nm); (d) ultraviolet light ($\lambda = 100$ nm); and (e) x-rays ($\lambda = 1$ pm).
42. What are the wavelength ranges in (a) the AM radio band (540–1600 kHz), and (b) the FM radio band (88–108 MHz)?
43. There are 12 VHF channels (Channels 2–13) that lie in a frequency range from 54 MHz to 216 MHz. Each channel has a width of 6 MHz, with the two ranges 72–76 MHz and 88–174 MHz reserved for non-TV purposes. (Channel 2, for example, lies between 54 and 60 MHz.) Calculate the *wavelength* range for (a) Channel 4; (b) Channel 6; (c) Channel 8.

## ADDITIONAL PROBLEMS

44. Assume that the solar radiation incident on the earth is 1340 W/m². (This is the value of the solar flux above the earth's atmosphere.) (a) Calculate the total power radiated by the sun, taking the average earth-sun separation to be $1.49 \times 10^{11}$ m. (b) Determine the maximum values of the electric and magnetic fields at the earth's surface due to solar radiation.
45. A community plans to build a facility to convert solar radiation into electrical power. They require 1 MW of power ($10^6$ W), and the system to be installed has an efficiency of 30% (that is, 30% of the solar energy incident on the surface is converted to electrical energy). What must be the effective area of a perfectly absorbing surface used in such an installation, assuming a constant energy flux of 1000 W/m²?
46. A microwave source produces pulses of 20-GHz radiation, with each pulse lasting 1.0 ns. A parabolic reflector ($R = 6$ cm) is used to focus these into a parallel beam of radiation, as shown in Figure 34.18. The average power during each pulse is 25 kW. (a) What is the wavelength of these microwaves? (b) What is the total energy contained in each pulse? (c) Compute the average energy density inside each pulse. (d) Determine the amplitude of the electric and mag-

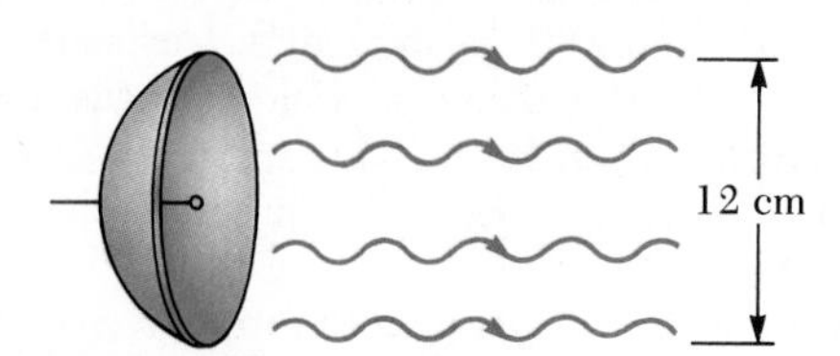


**Figure 34.18** (Problem 46).

netic fields in these microwaves. (e) If this pulsed beam strikes an absorbing surface, compute the force exerted on the surface during the 1-ns duration of each pulse.

47. A dish antenna with a diameter of 20 m receives (at normal incidence) a radio signal from a distant source, as shown in Figure 34.19. The radio signal is a continuous sinusoidal wave with amplitude $E_0 = 0.2\ \mu\text{V/m}$. Assume the antenna absorbs all the radiation that falls on the dish. (a) What is the amplitude of the magnetic field in this wave? (b) What is the intensity of the radiation received by this antenna? (c) What is the power received by the antenna? (d) What force is exerted on the antenna by the radio waves?

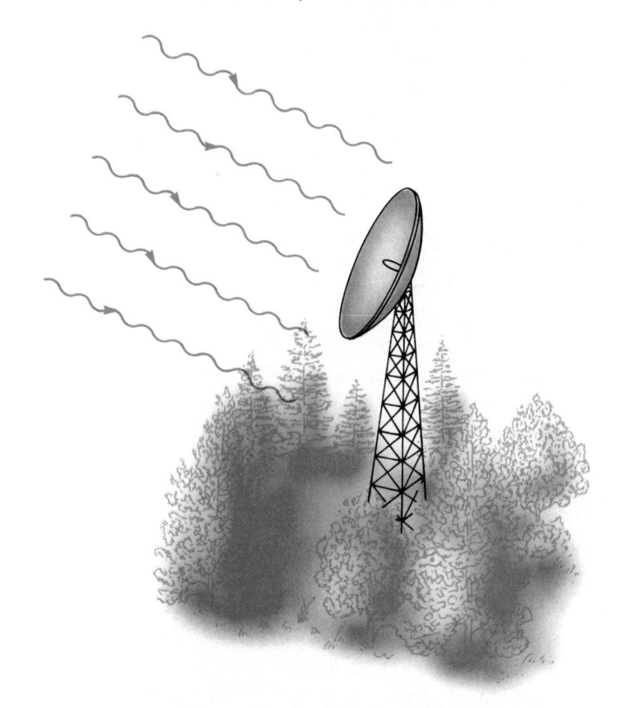

**Figure 34.19** (Problem 47).

48. Show that the instantaneous value of the Poynting vector has a magnitude given by

$$S = \frac{c}{2}(\epsilon_0 E^2 + \mu_0 H^2)$$

49. A thin tungsten filament of length 1 m radiates 60 W of energy in the form of electromagnetic waves. A perfectly absorbing surface in the form of a hollow cylinder of radius 5 cm and length 1 m is placed concentric with the filament. Calculate the radiation pressure acting on the cylinder. (Assume that the radiation is emitted in the radial direction, and neglect end effects.)

50. A group of astronauts plan to propel a spaceship by using a "sail" to reflect solar radiation. The sail is totally reflecting, oriented with its plane perpendicular to the direction to the sun, and 1 km × 1.5 km in size. What is the maximum acceleration that can be expected for a spaceship of 4 metric tons (4000 kg)? (Use the solar radiation data from Problem 44 and neglect gravitational forces.)

51. In 1965, Penzias and Wilson discovered the cosmic microwave radiation left over from the Big Bang expansion of the universe. The energy density of this radiation is $4 \times 10^{-14}\ \text{J/m}^3$. Determine the corresponding electric field amplitude.

52. The intensity of solar radiation at the top of the earth's atmosphere is $1350\ \text{W/m}^2$. Assuming that 60% of the arriving solar energy reaches the earth's surface and assuming that you absorb 50% of the incident energy, *estimate* the amount of solar energy you absorb in a 60-min sunbath.

53. An astronaut in a spacecraft moving with constant velocity wishes to increase the speed of the craft by using a laser beam attached to the spaceship. The laser beam emits 100 J of electromagnetic energy per pulse, and the laser is pulsed at the rate of 0.2 pulse/s. If the mass of the spaceship plus its contents is 5000 kg, for how long a time must the beam be on in order to increase the speed of the vehicle by 1 m/s in the direction of its initial motion? In what direction should the beam be pointed to achieve this?

54. Consider a small, spherical particle of radius $r$ located in space a distance $R$ from the sun. (a) Show that the ratio $F_{\text{rad}}/F_{\text{grav}} \propto 1/r$, where $F_{\text{rad}}$ = the force due to solar radiation and $F_{\text{grav}}$ = the force of gravitational attraction. (b) The result of (a) means that for a sufficiently small value of $r$ the force exerted on the particle due to solar radiation will exceed the force of gravitational attraction. Calculate the value of $r$ for which the particle will be in equilibrium under the two forces. (Assume that the particle has a perfectly absorbing surface and a mass density of $1.5\ \text{g/cm}^3$. Let the particle be located $3.75 \times 10^{11}$ m from the sun and use $214\ \text{W/m}^2$ as the value of the solar flux at that point.)

55. The torsion balance shown in Figure 34.8 is used in an experiment to measure radiation pressure. The torque constant (elastic restoring torque) of the suspension fiber is $1 \times 10^{-11}\ \text{N}\cdot\text{m/deg}$, and the length of the horizontal rod is 6 cm. The beam from a 3-mW helium-neon laser is incident on the black disk, and the mirror disk is completely shielded. Calculate the angle between the *equilibrium* positions of the horizontal bar when the beam is switched from "off" to "on."

56. Monoenergetic x-rays move through a material with a speed of 0.95$c$. The photon flux on a surface perpendicular to the x-ray beam is $10^{13}\ \text{photons/m}^2\cdot\text{s}$. (a) Calculate the density of photons (number per unit

volume) in the material. (b) If each photon has an energy of 8.88 keV, what is the energy density within the material?

57. A linearly polarized microwave of wavelength 1.5 cm is directed along the positive $x$ axis. The electric field vector has a maximum value of 175 V/m and vibrates in the $xy$ plane. (a) Assume that the magnetic field component of the wave can be written in the form $B = B_0 \sin(kx - \omega t)$ and give values for $B_0$, $k$, and $\omega$. Also, determine in which plane the magnetic field vector vibrates. (b) Calculate the Poynting vector for this wave. (c) What radiation pressure would this wave exert if directed at normal incidence onto a perfectly reflecting sheet? (d) What acceleration would be imparted to a 500-g sheet (perfectly reflecting and at normal incidence) with dimensions 1 m × 0.75 m?

58. A police radar unit transmits at a frequency of 10.525 GHz, better known as the "X-band." (a) Show that when this radar wave is reflected from the front of a moving car, the reflected wave is shifted in frequency by the amount $\Delta f \cong 2fv/c$, where $v$ is the speed of the car. (*Hint:* Treat the reflection of a wave as two separate processes: First, the motion of the car produces a "moving-observer" Doppler shift in the frequency of the waves striking the car, and then these Doppler-shifted waves are re-emitted by a moving source. Also, automobile speeds are low enough that you can use the binomial expansion $(1 + x)^n \cong 1 + nx$ in the acoustic Doppler formulas derived in Chapter 17.) (b) The unit is usually calibrated before and after an arrest for speeds of 35 mph and 80 mph. Calculate the frequency shift produced by these two speeds.

59. An astronaut, stranded in space "at rest" 10 m from his spacecraft, has a mass (including equipment) of 110 kg. He has a 100-W light source that forms a directed beam, so he decides to use the beam of light as a photon rocket to propel himself continuously toward the spacecraft. (a) Calculate how long it will take him to reach the spacecraft by this method. (b) Suppose, instead, he decides to throw the light source away in a direction opposite to the spacecraft. If the mass of the light source is 3 kg and, after being thrown, moves with a speed of 12 m/s *relative to the recoiling astronaut*, how long will the astronaut take to reach the spacecraft?

60. Assuming that the antenna of a 10-kW radio station radiates electromagnetic energy uniformly in all directions, compute the maximum value of the magnetic field at a distance of 5 km from the antenna, and compare this value with the magnetic field of the earth at the surface of the earth.

61. (a) For a parallel-plate capacitor having a small plate separation compared with the length or width of a plate, show that the displacement current is given by

$$I_d = C\frac{dV}{dt}$$

where $dV/dt$ is the time rate of change of potential difference across the plates. (b) Calculate the value of $dV/dt$ required to produce a displacement current of 1 A in a 1-$\mu$F capacitor.

62. Consider the situation shown in Figure 34.20. An electric field of 300 V/m is confined to a circular area 10 cm in diameter and directed outward from the plane of the figure. If the field is increasing at a rate of 20 V/m·s, what are the direction and magnitude of the magnetic field at the point $P$, 15 cm from the center of the circle?

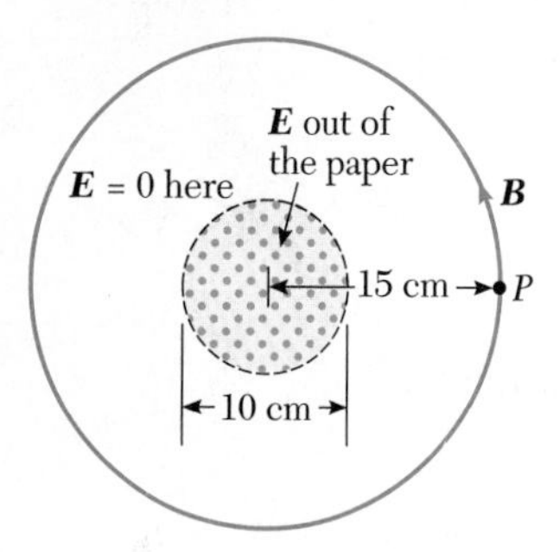


**Figure 34.20** (Problem 62).

63. A plane electromagnetic wave varies sinusoidally at 90 MHz as it travels along the $+x$ direction. The peak value of the electric field is 2 mV/m, and it is directed along the $\pm y$ direction. (a) Find the wavelength, the period, and the peak value $B_0$ of the magnetic field. (b) Write expressions in SI units for the space and time variations of the electric field and of the magnetic field. Include numerical values as well as subscripts to indicate coordinate directions. (c) Find the average power per unit area that this wave propagates through space. (d) Find the average energy density in the radiation (in units of J/m³). (e) What radiation pressure would this wave exert upon a perfectly reflecting surface at normal incidence?

*White light directed through a triangular glass prism creates a "Pheaenomena of Colours" on an old manuscript. (© Eric Lessing, Magnum Photos)*

# PART V
# Light and Optics

Scientists have long been intrigued by the nature of light, and philosophers have had endless arguments concerning the proper definition and perception of light. It is important to understand the nature of light because it is one of the basic ingredients of life on earth. Plants convert light energy from the sun to chemical energy through photosynthesis. Light is the principle means by which we are able to transmit and receive information from objects around us and throughout the universe.

The nature and properties of light have been a subject of great interest and speculation since ancient times. The Greeks believed that light consisted of tiny particles (corpuscles) that were emitted by a light source and then stimulated the perception of vision upon striking the observer's eye. Newton used this corpuscular theory to explain the reflection and refraction of light. In 1670, one of Newton's contemporaries, the Dutch scientist Christian Huygens, was able to explain many properties of light by proposing that light was wave-like in character. In 1803, Thomas Young showed that light beams can interfere with one another, giving strong support to the wave theory. In 1865, Maxwell developed a brilliant theory that electromagnetic waves travel with the speed of light (Chapter 34). By this time, the wave theory of light seemed to be on firm ground.

However, at the beginning of the 20th century, Max Planck returned to the corpuscular theory of light in order to explain the radiation emitted by hot objects. Albert Einstein used the same concept to explain the electrons emitted by a metal exposed to light (the photoelectric effect). We shall discuss these and other topics in the last part of this text, which is concerned with modern physics.

Today, scientists view light as having a dual nature. Sometimes light behaves as if it were particle-like, and sometimes it behaves as if it were wave-like.

In this part of the book, we shall concentrate on those aspects of light that are best understood through the wave model. First, we shall discuss the reflection of light at the boundary between two media and the refraction (bending) of light as it travels from one medium into another. We shall use these ideas to study the refraction of light as it passes through lenses and the reflection of light from surfaces. Then we shall describe how lenses and mirrors can be used to view objects with such instruments as cameras, telescopes, and microscopes. Finally, we shall study the phenomena of diffraction, polarization, and interference as they apply to light.

*"I procured me a Triangular glass-Prisme to try therewith the celebrated* Pheaenomena of Colours. . . . *I placed my Prisme at his entrance (the sunlight), that it might thereby be refracted to the opposite wall. It was a very pleasing divertisement to view the vivid and intense colours produced thereby; . . . I have often with admiration beheld that all the colours of the Prisme being made to converge, and thereby to be again mixed, as they were in the light before it was incident upon the Prisme, reproduced light, entirely and perfectly white, and not at all sensibly differing from the direct light of the sun. . . ."*

ISAAC NEWTON

# 35

# The Nature of Light and the Laws of Geometric Optics

*Rainbow over the Potala Palace, Lhasa, Tibet. (© Galen Rowell/Mountain Light)*

## 35.1 THE NATURE OF LIGHT

Before the beginning of the 19th century, light was considered to be a stream of particles that were emitted by a light source and stimulated the sense of sight upon entering the eye. The chief architect of this particle theory of light was, once again, Isaac Newton.[1] With this theory, Newton was able to provide a simple explanation of some known experimental facts concerning the nature of light, namely, the laws of reflection and refraction.

Most scientists accepted Newton's particle theory of light. However, during Newton's lifetime another theory was proposed — one that argued that light might be some sort of wave motion. In 1678, a Dutch physicist and astronomer, Christian Huygens (1629–1695), showed that a wave theory of light could explain the laws of reflection and refraction. The wave theory did not receive immediate acceptance for several reasons. All the waves known at the time (sound, water, etc.) traveled through some sort of medium. On the other hand, light could travel to us from the sun through the vacuum of space. Furthermore, it was argued that if light were some form of wave motion, the

[1] Isaac Newton, *Opticks*, 1704. [The fourth edition (1730) was reprinted by Dover Publications, New York, 1952.]

waves would be able to bend around obstacles; hence, we should be able to see around corners. It is now known that light does indeed bend around the edges of objects. This phenomenon, known as *diffraction,* is not easy to observe because light waves have such short wavelengths. Thus, although experimental evidence for the diffraction of light was discovered by Francesco Grimaldi (1618–1663) around 1660, most scientists rejected the wave theory and adhered to Newton's particle theory for more than a century. This was, for the most part, due to Newton's great reputation as a scientist.

The first clear demonstration of the wave nature of light was provided in 1801 by Thomas Young (1773–1829), who showed that, under appropriate conditions, light exhibits interference behavior. That is, at certain points in the vicinity of two sources, light waves can combine and cancel each other by destructive interference. Such behavior could not be explained at that time by a particle theory because there was no conceivable way by which two or more particles could come together and cancel one another. Several years later, a French physicist, Augustin Fresnel (1788–1829), performed a number of detailed experiments dealing with interference and diffraction phenomena. In 1850, Jean Foucault (1791–1868) provided further evidence of the inadequacy of the particle theory by showing that the speed of light in liquids is less than in air. According to the particle model of light, the speed of light would be higher in glasses and liquids than in air. Additional developments during the 19th century led to the general acceptance of the wave theory of light.

The most important development concerning the theory of light was the work of Maxwell, who in 1873 asserted that light was a form of high-frequency electromagnetic wave (Chapter 34). His theory predicted that these waves should have a speed of about $3 \times 10^8$ m/s. Within experimental error, this value is equal to the speed of light. As discussed in Chapter 34, Hertz provided experimental confirmation of Maxwell's theory in 1887 by producing and detecting electromagnetic waves. Furthermore, Hertz and other investigators showed that these *waves exhibited reflection, refraction, and all the other characteristic properties of waves.*

Although the classical theory of electricity and magnetism was able to explain most known properties of light, some subsequent experiments could not be explained by assuming that light was a wave. The most striking of these is the *photoelectric effect,* also discovered by Hertz. The photoelectric effect is the ejection of electrons from a metal whose surface is exposed to light. As one example of the difficulties that arose, experiments showed that the kinetic energy of an ejected electron is *independent* of the light intensity. This was in contradiction of the wave theory, which held that a more intense beam of light should add more energy to the electron. An explanation of this phenomenon was proposed by Einstein in 1905. Einstein's theory used the concept of quantization developed by Max Planck (1858–1947) in 1900. The quantization model assumes that the energy of a light wave is present in bundles of energy called *photons;* hence the energy is said to be *quantized.* (Any quantity that appears in discrete bundles is said to be quantized. For example, electric charge is quantized because it always appears in bundles equal to the elementary charge of $1.6 \times 10^{-19}$ C.) According to Einstein's theory, the energy of a photon is proportional to the frequency of the electromagnetic wave:

$$E = hf \tag{35.1}$$

Energy of a photon

where $h = 6.63 \times 10^{-34}$ J · s is **Planck's constant.** It is important to note that this theory retains some features of both the wave theory and the particle theory of light. As we shall discuss later, the photoelectric effect is the result of energy transfer from a single photon to an electron in the metal. That is, the electron interacts with one photon of light as if the electron had been struck by a particle. Yet this photon has wave-like characteristics because its energy is determined by the frequency (a wave-like quantity).

In view of these developments, one must regard light as having a *dual nature.* That is, *in some cases light acts like a wave and in others it acts like a particle.* Classical electromagnetic wave theory provides an adequate explanation of light propagation and of the effects of interference, whereas the photoelectric effect and other experiments involving the interaction of light with matter are best explained by assuming that light is a particle. Light is light, to be sure. However, the question, "Is light a wave or a particle?" is an inappropriate one. Sometimes it acts like a wave, sometimes like a particle. In the next few chapters, we shall investigate the wave nature of light.

## 35.2 MEASUREMENTS OF THE SPEED OF LIGHT

Light travels at such a high speed ($c \approx 3 \times 10^8$ m/s) that early attempts to measure its speed were unsuccessful. Galileo attempted to measure the speed of light by positioning two observers in towers separated by about 5 mi. Each observer carried a shuttered lantern. One observer would open his lantern first, and then the other would open his lantern at the moment he saw the light from the first lantern. The velocity could then be obtained, in principle, knowing the transit time of the light beams between lanterns. The results were inconclusive. Today, we realize (and as Galileo himself concluded) that it is impossible to measure the speed of light in this manner because the transit time of the light is very small compared with the reaction time of the observers.

We now describe two methods for determining the speed of light.

### Roemer's Method

The first successful estimate of the speed of light was made in 1675 by the Danish astronomer Ole Roemer (1644–1710). His technique involved astronomical observations of one of the moons of Jupiter, called Io. At that time, 4 of Jupiter's 14 moons had been discovered, and the periods of their orbits were known. Io, the innermost moon, has a period of about 42.5 h. This was measured by observing the eclipse of Io as it passed behind Jupiter (Fig. 35.1). The period of Jupiter is about 12 years, so as the earth moves through 180° about the sun, Jupiter revolves through only 15°.

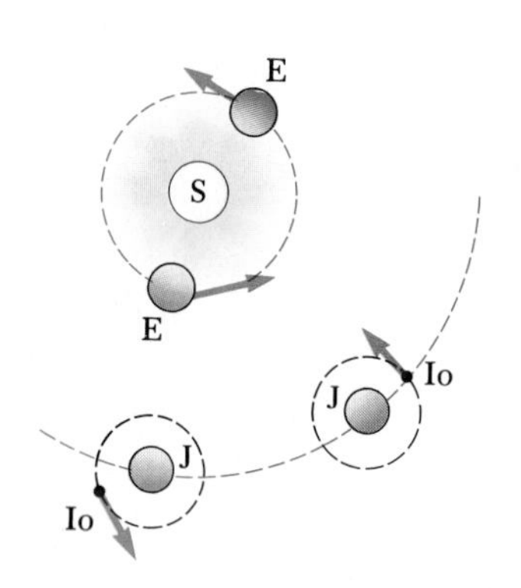


**Figure 35.1** Roemer's method for measuring the speed of light.

Using the orbital motion of Io as a clock, one would expect a constant period in its orbit over long time intervals. However, Roemer observed a systematic variation in Io's period during a year's time. He found that the periods were larger than average when the earth receded from Jupiter and smaller than average when the earth was approaching Jupiter. For example, if Io had a constant period, Roemer should have been able to see an eclipse occurring at a particular instant and be able to predict when an eclipse should begin at a later time in the year. However, when Roemer checked to see if the second eclipse did occur at the predicted time, he found that if the earth was receding from Jupiter, the eclipse was late. In fact, if the interval between observations was three months, the delay was approximately 600 s. Roemer attributed this variation in period to the fact that the distance between the earth and Jupiter was changing between the observations. In three months (one quarter of the period of the earth), the light from Jupiter has to travel an additional distance equal to the radius of the earth's orbit.

Using Roemer's data, Huygens estimated the lower limit for the speed of light to be about $2.3 \times 10^8$ m/s. This experiment is important historically because it demonstrated that light does have a finite speed and gave an estimate of this speed.

### Fizeau's Technique

The first successful method for measuring the speed of light using purely terrestrial techniques was developed in 1849 by Armand H. L. Fizeau (1819–1896). Figure 35.2 represents a simplified diagram of his apparatus.[2] The basic idea is to measure the total time it takes light to travel from some point to a distant mirror and back. If $d$ is the distance between the light source and the mirror and if the transit time for one round trip is $t$, then the speed of light is $c = 2d/t$. To measure the transit time, Fizeau used a rotating toothed wheel, which converts an otherwise continuous beam of light into a series of light pulses. Additionally, the rotation of the wheel controls what an observer at the light source sees. For example, if the light passing the opening at point $A$ in Figure 35.2 should return at the instant that tooth $B$ had rotated into position to cover the return path, the light would not reach the observer. At a faster rate of rotation, the opening at point $C$ could move into position to allow the reflected beam to pass and reach the observer. Knowing the distance $d$, the number of teeth in the wheel, and the angular velocity of the wheel, Fizeau arrived at a value of $c = 3.1 \times 10^8$ m/s. Similar measurements made by subsequent investigators yielded more precise values for $c$, approximately $2.9977 \times 10^8$ m/s.

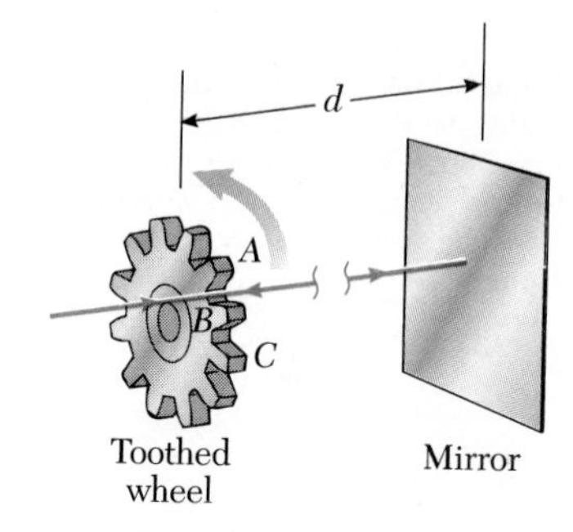


**Figure 35.2** Fizeau's method for measuring the speed of light using a rotating toothed wheel.

[2] The actual apparatus involved several lenses and mirrors which we have omitted for the sake of clarity. For more details, see F. W. Sears, *Optics*, 3rd ed., Reading, Mass., Addison-Wesley, 1958, Chapter 1.

**EXAMPLE 35.1 Measuring the Speed of Light with Fizeau's Toothed Wheel** □

Assume the toothed wheel of the Fizeau experiment has 360 teeth and is rotating with a speed of 27.5 rev/s when the light from the source is extinguished, that is, when a burst of light passing through opening $A$ in Figure 35.2 is blocked by tooth $B$ on return. If the distance to the mirror is 7500 m, find the speed of light.

*Solution* If the wheel has 360 teeth, it will turn through an angle of 1/720 rev in the time that passes while the light makes its round trip. From the definition of angular velocity, we see that the time is

$$t = \frac{\theta}{\omega} = \frac{(1/720)\text{ rev}}{27.5\text{ rev/s}} = 5.05 \times 10^{-5}\text{ s}$$

Hence, the speed of light is

$$c = \frac{2d}{t} = \frac{2\,(7500\text{ m})}{5.05 \times 10^{-5}\text{ s}} = 2.97 \times 10^8\text{ m/s}$$

## 35.3 THE RAY APPROXIMATION IN GEOMETRIC OPTICS

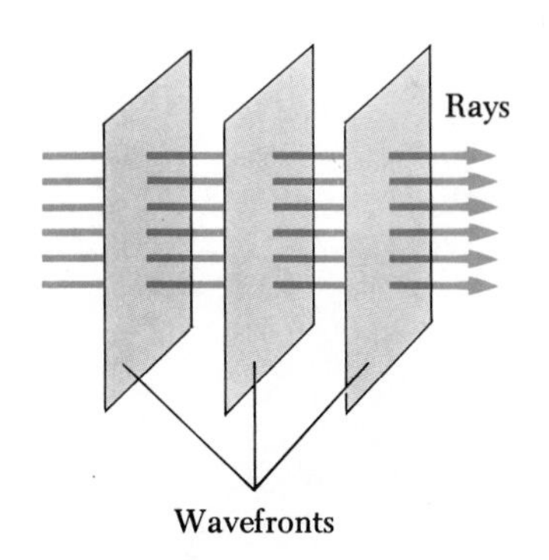


**Figure 35.3** A plane wave propagating to the right. Note that the rays, corresponding to the direction of wave motion, are straight lines perpendicular to the wavefronts.

In studying geometric optics here and in Chapter 36, we shall use what is called the *ray approximation*. To understand this approximation, first recall that the direction of energy flow of a wave, corresponding to the direction of wave propagation, is called a ray. The rays of a given wave are straight lines that are perpendicular to the wavefronts, as illustrated in Figure 35.3 for a plane wave. In the ray approximation, we assume that a wave moving through a medium travels in a straight line in the direction of its rays. That is, a ray is a line drawn in the direction in which the light is traveling. For example, a beam of sunlight passing through a darkened room traces out the path of a ray.

If the wave meets a barrier with a circular opening whose diameter is large relative to the wavelength, as in Figure 35.4a, the wave emerging from the opening continues to move in a straight line (apart from some small edge effects); hence, the ray approximation continues to be valid. On the other hand, if the diameter of the opening of the barrier is of the order of the wavelength, as in Figure 35.4b, the waves spread out from the opening in all directions. We say that the outgoing wave is noticeably *diffracted*. Finally, if the opening is small relative to the wavelength, the opening can be approximated as a point source of waves (Fig. 35.4c). Thus, the effect of diffraction is more pronounced as the ratio $d/\lambda$ approaches zero. Similar effects are seen when waves encounter an opaque circular object. In this case, when $\lambda \ll d$, the object casts a sharp shadow.

The ray approximation and the assumption that $\lambda \ll d$ will be used here and in Chapter 36, both dealing with geometric optics. This approximation is very good for the study of mirrors, lenses, prisms, and associated optical instruments, such as telescopes, cameras, and eyeglasses. We shall return to the subject of diffraction (where $\lambda \geq d$) in Chapter 38.

(a)

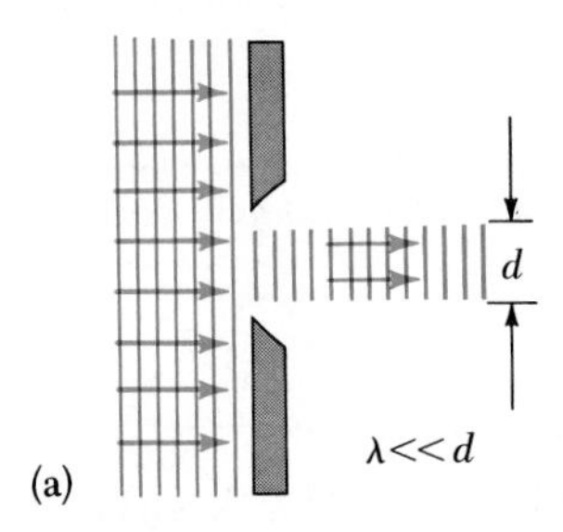


(b)

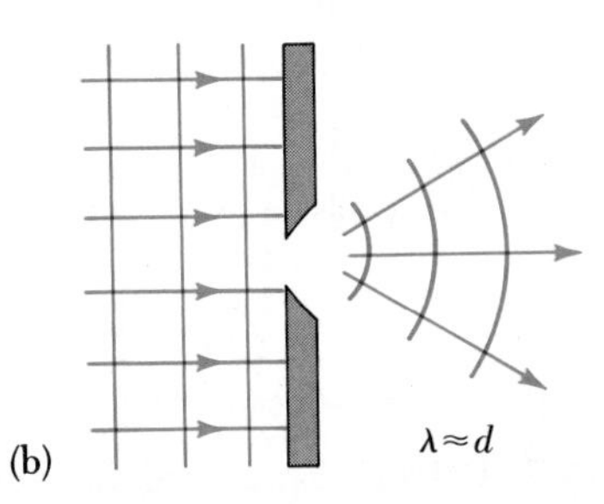


(c)

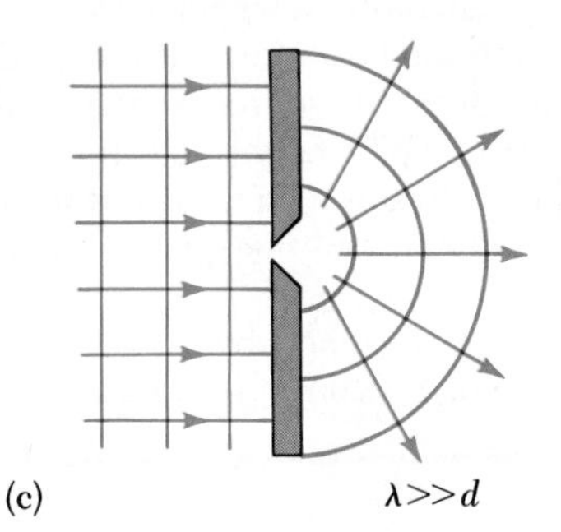


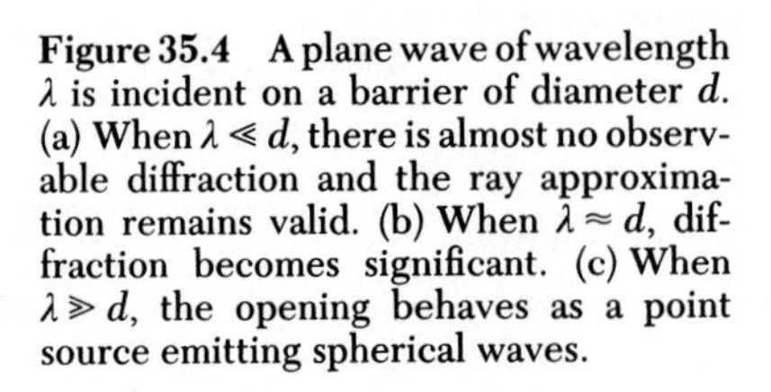

**Figure 35.4** A plane wave of wavelength $\lambda$ is incident on a barrier of diameter $d$. (a) When $\lambda \ll d$, there is almost no observable diffraction and the ray approximation remains valid. (b) When $\lambda \approx d$, diffraction becomes significant. (c) When $\lambda \gg d$, the opening behaves as a point source emitting spherical waves.

## 35.4 REFLECTION AND REFRACTION

### Reflection of Light

When a light ray traveling in a medium encounters a boundary leading into a second medium, part of the incident ray is reflected back into the first medium. Figure 35.5a shows several rays of a beam of light incident on a smooth, mirror-like, reflecting surface. The reflected rays are parallel to each other, as indicated in the figure. Reflection of light from such a smooth surface is called **specular reflection.** On the other hand, if the reflecting surface is rough, as in Figure 35.5b, the surface will reflect the rays in various directions. Reflection from any rough surface is known as **diffuse reflection.** A surface will behave as a smooth surface as long as the surface variations are small compared with the wavelength of the incident light. Photographs of specular reflection and diffuse reflection using laser light are shown in Figures 35.5c and 35.5d.

For instance, consider the two types of reflection one can observe from a road's surface while driving a car at night. When the road is dry, light from oncoming vehicles is scattered off the road in different directions (diffuse reflection) and the road is quite visible. On a rainy night, when the road is wet, the road irregularities are filled with water. Because the water surface is quite smooth, the light undergoes specular reflection. In this book, we shall concern ourselves only with specular reflection, and we shall use the term *reflection* to mean specular reflection.

Consider a light ray traveling in air and incident at an angle on a flat, smooth surface, as in Figure 35.6. The incident and reflected rays make angles

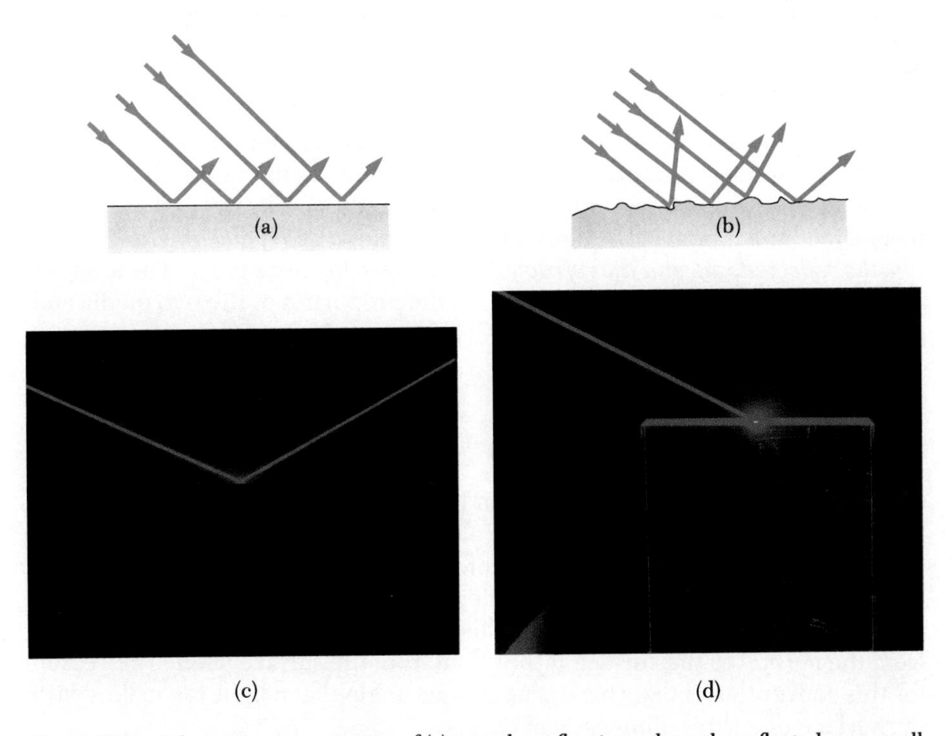


**Figure 35.5** Schematic representation of (a) specular reflection, where the reflected rays are all parallel to each other, and (b) diffuse reflection, where the reflected rays travel in random directions. (c) and (d) Photographs of specular and diffuse reflection using laser light. (Photographs courtesy of Henry Leap and Jim Lehman)

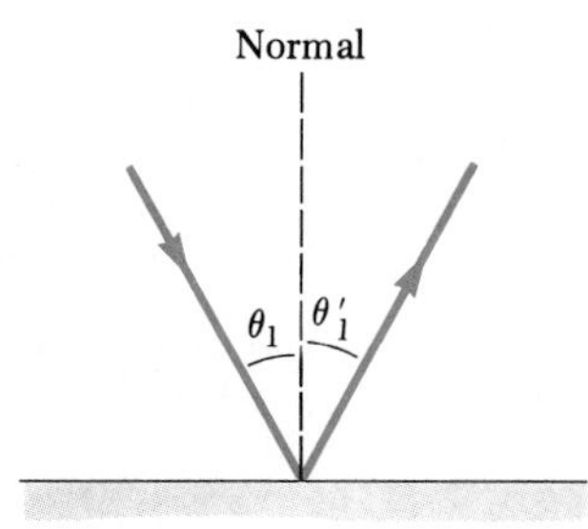


**Figure 35.6** According to the law of reflection, $\theta_1 = \theta'_1$. The incident ray, the reflected ray, and the normal all lie in the same plane.

$\theta_1$ and $\theta_1'$, respectively, with a line drawn perpendicular to the surface at the point where the incident ray strikes the surface. We shall call this line the *normal* to the surface. Experiments show that *the angle of reflection equals the angle of incidence,* that is,

Law of reflection

$$\theta_1' = \theta_1 \tag{35.2}$$

**EXAMPLE 35.2 The Double-Reflected Light Ray**

Two mirrors make an angle of 120° with each other, as in Figure 35.7. A ray is incident on mirror $M_1$ at an angle of 65° to the normal. Find the direction of the ray after it is reflected from mirror $M_2$.

*Solution* From the law of reflection, we see that the first ray also makes an angle of 65° with the normal. Thus, it follows that this same ray makes an angle of 90° − 65°, or 25°, with the horizontal. From the triangle made by the first reflected ray and the two mirrors, we see that the first reflected ray makes an angle of 35° with $M_2$ (since the sum of the interior angles of any triangle is 180°). This means that this ray makes an angle of 55° with the normal to $M_2$. Hence, from the law of reflection, it follows that the second reflected ray makes an angle of 55° with the normal to $M_2$. Finally, by comparing the direction of the ray incident on $M_1$ with its direction after reflecting from $M_2$, we see that the ray is rotated through 120°, which corresponds to the angle between the mirrors.

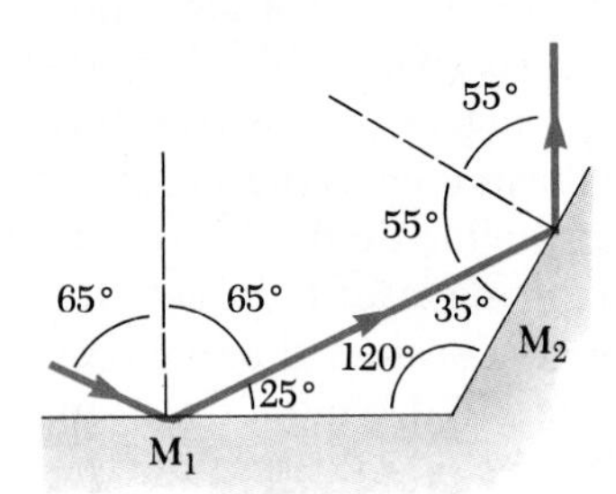


**Figure 35.7** (Example 35.2) Mirrors $M_1$ and $M_2$ make an angle of 120° with each other.

## Refraction of Light

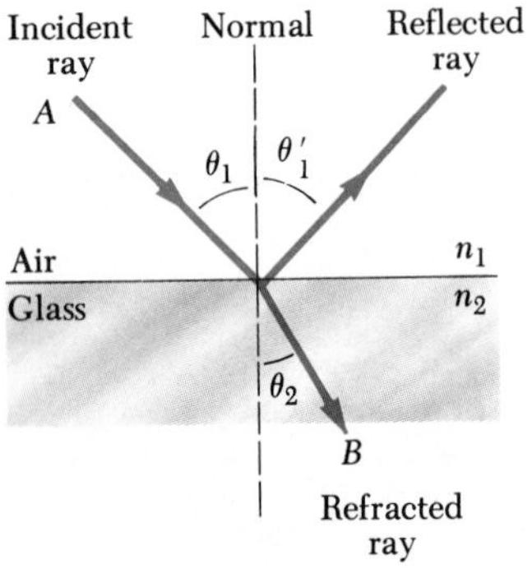

When a ray of light traveling through a transparent medium encounters a boundary leading into another transparent medium, as in Figure 35.8, part of the ray is reflected and part enters the second medium. The ray that enters the second medium is bent at the boundary and is said to be *refracted. The incident ray, the reflected ray, and the refracted ray all lie in the same plane.* The **angle of refraction,** $\theta_2$ in Figure 35.8, depends on the properties of the two media and on the angle of incidence through the relationship

$$\frac{\sin \theta_2}{\sin \theta_1} = \frac{v_2}{v_1} = \text{constant} \tag{35.3}$$

**Figure 35.8** A ray obliquely incident on an air-glass interface. The refracted ray is bent toward the normal since $n_2 > n_1$ and $v_2 < v_1$. All rays and the normal lie in the same plane.

where $v_1$ is the speed of light in medium 1 and $v_2$ is the speed of light in medium 2. The experimental discovery of this relationship is usually credited to Willebrord Snell (1591–1627) and is therefore known as **Snell's law.**[3] In Section 35.7, we shall verify the laws of reflection and refraction using Huygens' principle. The angles of incidence, reflection, and refraction are *all* measured from the *normal* to the surface rather than from the surface itself. The reason for this convention is that there is no unique angle that a light ray makes with the surface of a three-dimensional object.

[3] This law was also deduced from the particle theory of light by René Descartes (1596–1650) and hence is known as *Descartes' law* in France.

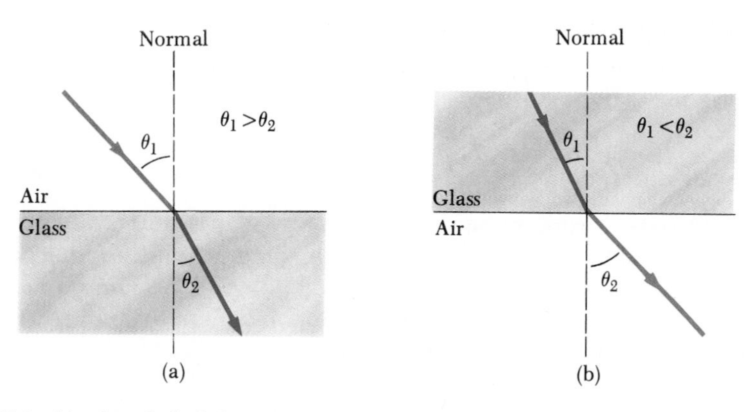


**Figure 35.9** (a) When the light beam moves from air into glass, its path is bent toward the normal. (b) When the beam moves from glass into air, its path is bent away from the normal.

The laser beam in this photograph is incident from the left on a container of water. Part of the beam is reflected, and part is transmitted. The refracted ray then reflects off a mirror at the bottom and emerges from the water at the right. Why is the transmitted ray more intense than the reflected ray? (Courtesy of Henry Leap and Jim Lehman)

It is found that *the path of a light ray through a refracting surface is reversible.* For example, the ray in Figure 35.8 travels from point $A$ to point $B$. If the ray originated at $B$, it would follow the same path to reach point $A$. In the latter case, however, the reflected ray would be in the glass.

When light moves from a material in which its speed is high to a material in which its speed is lower, the angle of refraction, $\theta_2$, is less than the angle of incidence, as shown in Figure 35.9a. If the ray moves from a material in which it moves slowly to a material in which it moves more rapidly, it is bent away from the normal, as shown in Figure 35.9b.

The behavior of light as it passes from air into another substance and then re-emerges into air is often a source of confusion to students. Let us take a look at what happens and see why this behavior is so different from other occurrences in our daily lives. When light travels in air, its speed is equal to $3 \times 10^8$ m/s, and its speed is reduced to about $2 \times 10^8$ m/s upon entering a block of glass. When the light re-emerges into air, its speed instantaneously increases to its original value of $3 \times 10^8$ m/s. This is far different from what happens, for example, when a bullet is fired through a block of wood. In this case, the speed of the bullet is reduced as it moves through the wood because some of its original energy is used to tear apart the fibers of the wood. When the bullet enters the air once again, it emerges at the speed it had just before leaving the block of wood.

In order to see why light behaves as it does, consider Figure 35.10, which represents a beam of light entering a piece of glass from the left. Once inside the glass, the light may encounter an electron bound to an atom, indicated as point $A$ in the figure. Let us assume that light is absorbed by the atom, which causes the electron to oscillate. The oscillating electron then acts as an antenna and radiates the beam of light toward an atom at point $B$, where the light is again absorbed by an atom at that point. The details of these absorptions and emissions are best explained in terms of quantum mechanics, a subject we shall study in the extended version of this text. For now, it is sufficient to think of the process as one in which the light passes from one atom to another through the glass. (The situation is somewhat analogous to a relay race in which a baton is passed between runners on the same team.) Although light travels from one atom to another with a speed of $3 \times 10^8$ m/s, the processes of absorption and emission of light by the atoms is very complex. However, the end result is that

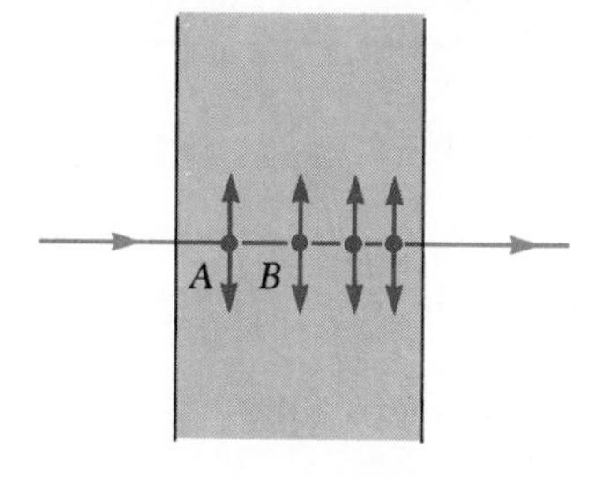


**Figure 35.10** Light passing from one atom to another in a medium.

the speed of light in one type of glass is lowered to $2 \times 10^8$ m/s. Once the light emerges into the air, the absorptions and emissions cease and its speed returns to the original value.

### The Law of Refraction

When light passes from one medium to another, it is refracted because the speed of light is different in the two media. In general, one finds that the speed of light in any material is less than the speed of light in vacuum. In fact, *light travels at its maximum speed in vacuum.* It is convenient to define the **index of refraction,** $n$, of a medium to be the ratio

Index of refraction

$$n = \frac{\text{speed of light in vacuum}}{\text{speed of light in a medium}} = \frac{c}{v} \qquad (35.4)$$

From this definition, we see that the index of refraction is a dimensionless number greater than unity since $v$ is always less than $c$. Furthermore, $n$ is equal to unity for vacuum. The indices of refraction for various substances measured with respect to vacuum are listed in Table 35.1.

As light travels from one medium to another, *the frequency of the light does not change.* To see why this is so, consider Figure 35.11. Wavefronts pass an observer at point $A$ in medium 1 with a certain frequency and are incident on the boundary between medium 1 and medium 2. The frequency with which the wavefronts pass an observer at point $B$ in medium 2 must equal the frequency at which they arrive at point $A$ in medium 1. If this were not the case, either wavefronts would be piling up at the boundary or they would be destroyed or created at the boundary. Since there is no mechanism for this to occur, the frequency must be a constant as a light ray passes from one medium into another.

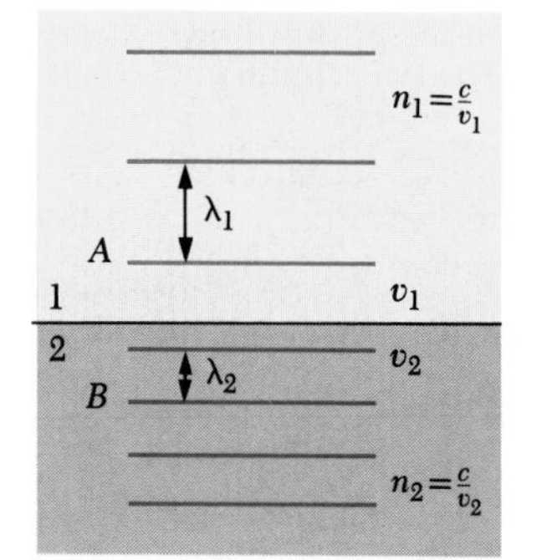


**Figure 35.11** As a wavefront moves from medium 1 to medium 2, its wavelength changes but its frequency remains constant.

Therefore, because the relation $v = f\lambda$ must be valid in both media and because $f_1 = f_2 = f$, we see that

$$v_1 = f\lambda_1 \qquad \text{and} \qquad v_2 = f\lambda_2$$

**TABLE 35.1 Index of Refraction for Various Substances Measured with Light of Vacuum Wavelength $\lambda_0 = 589$ nm**

| Substance | Index of Refraction | Substance | Index of Refraction |
|---|---|---|---|
| Solids at 20°C | | Liquids at 20°C | |
| Diamond (C) | 2.419 | Benzene | 1.501 |
| Fluorite ($CaF_2$) | 1.434 | Carbon disulfide | 1.628 |
| Fused quartz ($SiO_2$) | 1.458 | Carbon tetrachloride | 1.461 |
| Glass, crown | 1.52 | Ethyl alcohol | 1.361 |
| Glass, flint | 1.66 | Glycerine | 1.473 |
| Ice ($H_2O$) | 1.309 | Water | 1.333 |
| Polystyrene | 1.49 | Gases at 0°C, 1 atm | |
| Sodium chloride (NaCl) | 1.544 | Air | 1.000293 |
| Zircon | 1.923 | Carbon dioxide | 1.00045 |

where the subscripts refer to the two media. A relationship between index of refraction and wavelength can be obtained by dividing these two equations and making use of the definition of the index of refraction given by Equation 35.4:

$$\frac{\lambda_1}{\lambda_2} = \frac{v_1}{v_2} = \frac{c/n_1}{c/n_2} = \frac{n_2}{n_1} \qquad (35.5)$$

which gives

$$\lambda_1 n_1 = \lambda_2 n_2 \qquad (35.6)$$

If medium 1 is vacuum, or for all practical purposes air, then $n_1 = 1$. Hence, it follows from Equation 35.5 that the index of refraction of any medium can be expressed as the ratio

$$n = \frac{\lambda_0}{\lambda_n} \qquad (35.7)$$

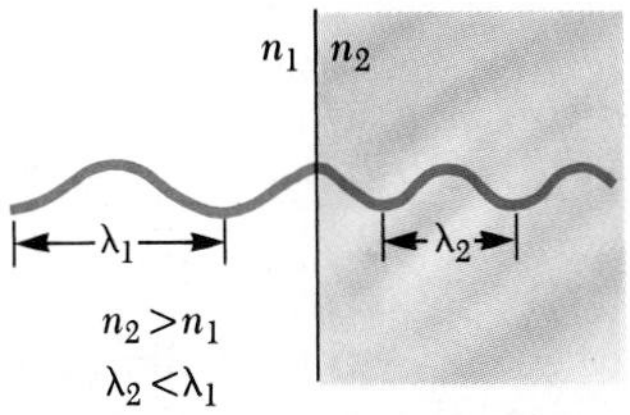

where $\lambda_0$ is the wavelength of light in vacuum and $\lambda_n$ is the wavelength in the medium whose index of refraction is $n$. A schematic representation of this reduction in wavelength is shown in Figure 35.12.

**Figure 35.12** Schematic diagram of the *reduction* in wavelength when light travels from a medium of low index of refraction to one of higher index of refraction.

We are now in a position to express Snell's law (Eq. 35.3) in an alternative form. If we substitute Equation 35.5 into Equation 35.3, we get

$$n_1 \sin\theta_1 = n_2 \sin\theta_2 \qquad (35.8)$$

Snell's law

This is the most widely used and practical form of **Snell's law.**

EXAMPLE 35.3 An Index of Refraction Measurement
A beam of light of wavelength 550 nm traveling in air is incident on a slab of transparent material. The incident beam makes an angle of 40° with the normal, and the refracted beam makes an angle of 26° with the normal. Find the index of refraction of the material.

*Solution* Snell's law of refraction (Eq. 35.8) with the given data, $\theta_1 = 40°$, $n_1 = 1.00$ for air, and $\theta_2 = 26°$, gives

$$n_1 \sin\theta_1 = n_2 \sin\theta_2$$

$$n_2 = \frac{n_1 \sin\theta_1}{\sin\theta_2} = (1.00)\frac{(\sin 40°)}{\sin 26°} = \frac{0.643}{0.438} = 1.47$$

If we compare this value with the data in Table 35.1, we see that the material is probably fused quartz.

Exercise 1 What is the wavelength of light in the material?
Answer 374 nm

EXAMPLE 35.4 Angle of Refraction for Glass
A light ray of wavelength 589 nm (produced by a sodium lamp) traveling through air is incident on a smooth, flat slab of crown glass at an angle of 30° to the normal, as sketched in Figure 35.13. Find the angle of refraction.

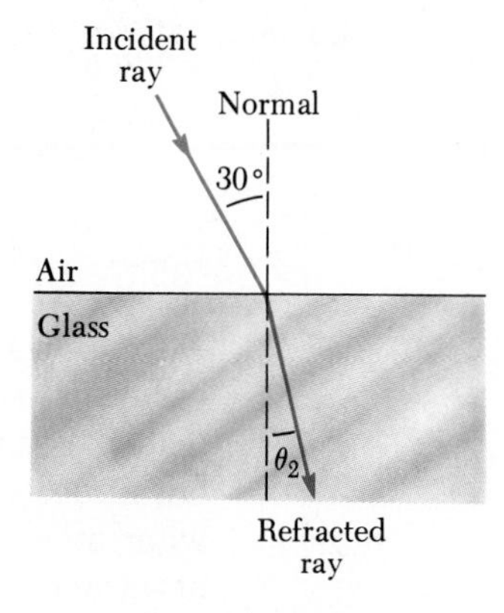


**Figure 35.13** (Example 35.4) Refraction of light by glass.

*Solution* Snell's law given by Equation 35.8 can be rearranged as

$$\sin \theta_2 = \frac{n_1}{n_2} \sin \theta_1$$

From Table 35.1, we find that $n_1 = 1.00$ for air and $n_2 = 1.52$ for glass. Therefore, the unknown refraction angle is

$$\sin \theta_2 = \left(\frac{1.00}{1.52}\right)(\sin 30°) = 0.329$$

$$\theta_2 = \sin^{-1}(0.329) = 19.2°$$

Thus we see that the ray is bent *toward* the normal, as expected.

Exercise 2 If the light ray moves from inside the glass toward the glass-air interface at an angle of 30° to the normal, determine the angle of refraction.
Answer 49.5° *away* from the normal.

**EXAMPLE 35.5 The Speed of Light in Quartz**
Light of wavelength 589 nm in vacuum passes through a piece of fused quartz ($n = 1.458$). (a) Find the speed of light in quartz.

*Solution* The speed of light in quartz can be easily obtained from Equation 35.4:

$$v = \frac{c}{n} = \frac{3 \times 10^8 \text{ m/s}}{1.458} = 2.058 \times 10^8 \text{ m/s}$$

(b) What is the wavelength of this light in quartz?

*Solution* We can use $\lambda_n = \lambda_0/n$ (Eq. 35.7) to calculate the wavelength in quartz, noting that we are given the wavelength in vacuum to be $\lambda_0 = 589$ nm:

$$\lambda_n = \frac{\lambda_0}{n} = \frac{589 \text{ nm}}{1.458} = 404 \text{ nm}$$

Exercise 3 Find the frequency of the light passing through the quartz.
Answer $5.09 \times 10^{14}$ Hz.

**EXAMPLE 35.6 Light Passing Through a Slab**
A light beam passes from medium 1 to medium 2 through a thick slab of material whose index of refraction is $n_2$ (Fig. 35.14). Show that the emerging beam is parallel to the incident beam.

*Solution* First, let us apply Snell's law to the upper surface:

$$(1) \qquad \sin \theta_2 = \frac{n_1}{n_2} \sin \theta_1$$

Applying Snell's law to the lower surface gives

$$(2) \qquad \sin \theta_3 = \frac{n_2}{n_1} \sin \theta_2$$

Substituting (1) into (2) gives

$$\sin \theta_3 = \frac{n_2}{n_1}\left(\frac{n_1}{n_2} \sin \theta_1\right) = \sin \theta_1$$

That is, $\theta_3 = \theta_1$, and so the layer does not alter the direction of the beam. It does, however, produce a displacement of the beam. The same result is obtained when light passes through multiple layers of materials.

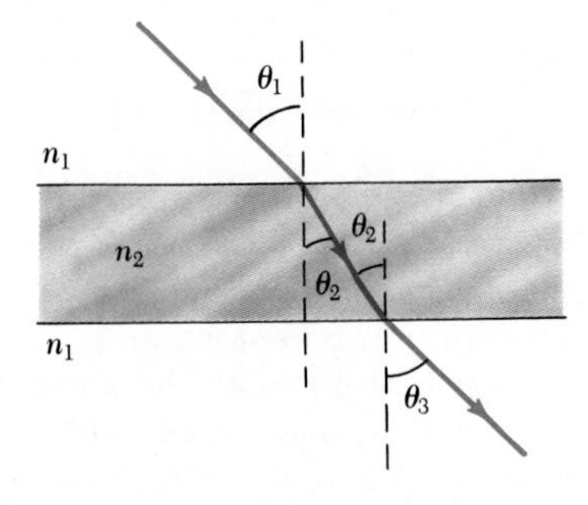


**Figure 35.14** (Example 35.6) When light passes through a flat slab of material, the emerging beam is parallel to the incident beam, and therefore $\theta_1 = \theta_3$.

## *35.5 DISPERSION AND PRISMS

An important property of the index of refraction is that it is different for different wavelengths of light. A graph of the index of refraction for three materials is shown in Figure 35.15. Since $n$ is a function of wavelength, Snell's law indicates that light of *different wavelengths* will be bent at *different angles* when incident on a refracting material. As we see from Figure 35.15, the index of refraction decreases with increasing wavelength. This means that blue light will bend more than red light when passing into a refracting material.

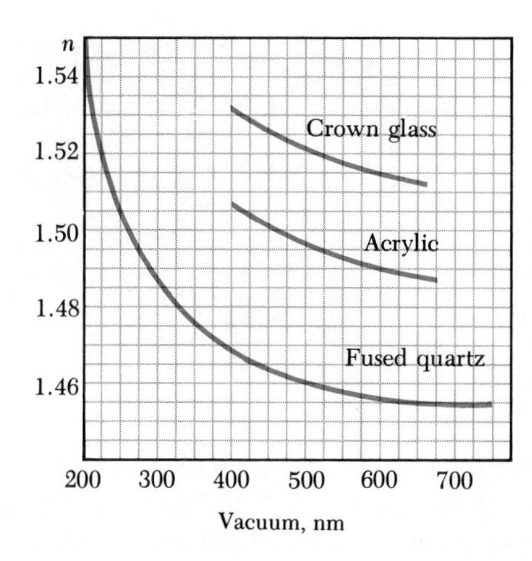


**Figure 35.15** Variation of index of refraction with vacuum wavelength for three materials.

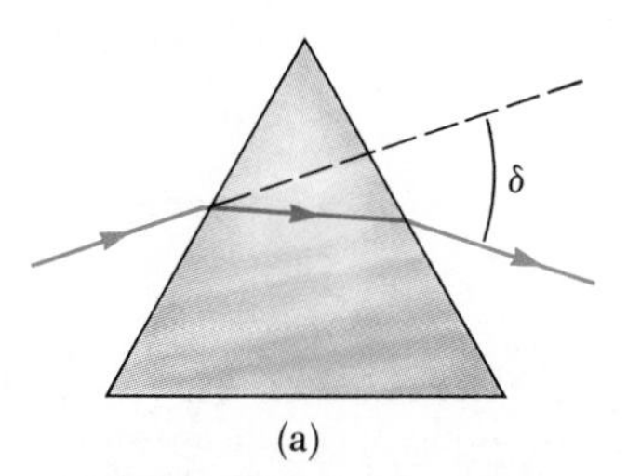


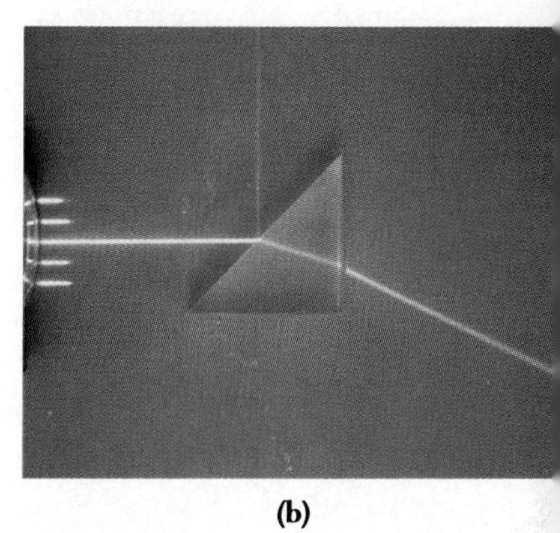

(b)

**Figure 35.16** (a) A prism refracts a light ray and deviates the light through an angle $\delta$. (b) The light is refracted as it passes through the prism. (Courtesy of Jim Lehman, James Madison University)

Any substance in which $n$ varies with wavelength is said to exhibit **dispersion.** To understand the effects that dispersion can have on light, let us consider what happens when light strikes a prism, as in Figure 35.16. A single ray of light incident on the prism from the left emerges bent away from its original direction of travel by an angle $\delta$, called the **angle of deviation.** Now suppose a beam of white light (a combination of all visible wavelengths) is incident on a prism, as in Figure 35.17. The rays that emerge from the second face spread out in a series of colors known as a **spectrum.** These colors, in order of decreasing wavelength, are red, orange, yellow, green, blue, indigo, and violet. Newton showed that each color had a particular angle of deviation, that the spectrum could not be further broken down, and that the colors could be recombined to form the original white light. Clearly, the angle of deviation, $\delta$, depends on the wavelength of a given color. Violet light deviates the most, red light deviates the least, and the remaining colors in the visible spectrum fall between these extremes. When light is spread out by a substance such as the prism, the light is said to be dispersed into a spectrum.

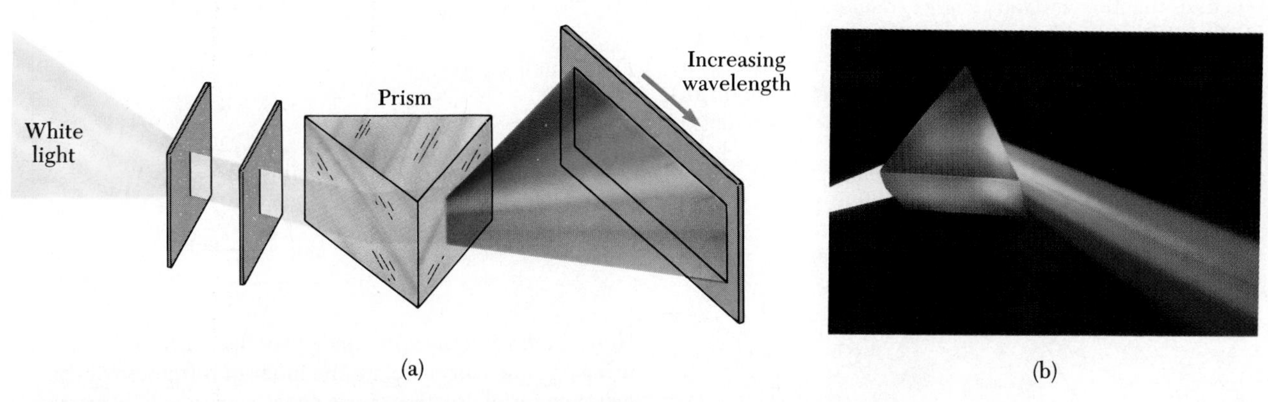


**Figure 35.17** (a) Dispersion of white light by a prism. Since $n$ varies with wavelength, the prism disperses the white light into its various spectral components. (b) The various colors are refracted at different angles because the index of refraction of the glass depends on wavelength. The blue light deviates the most, while red light deviates the least. (Photograph courtesy of Bausch and Lomb)

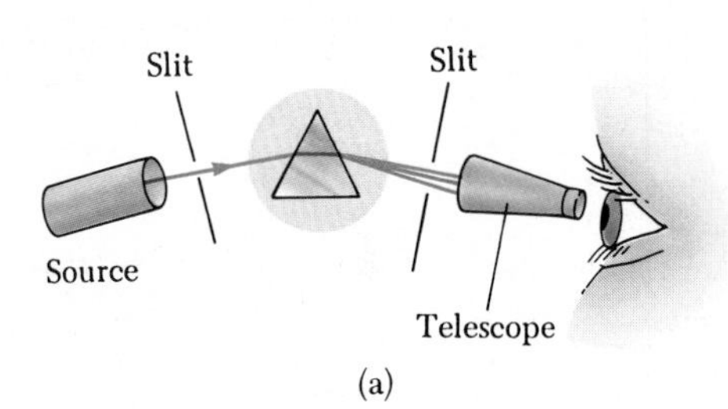


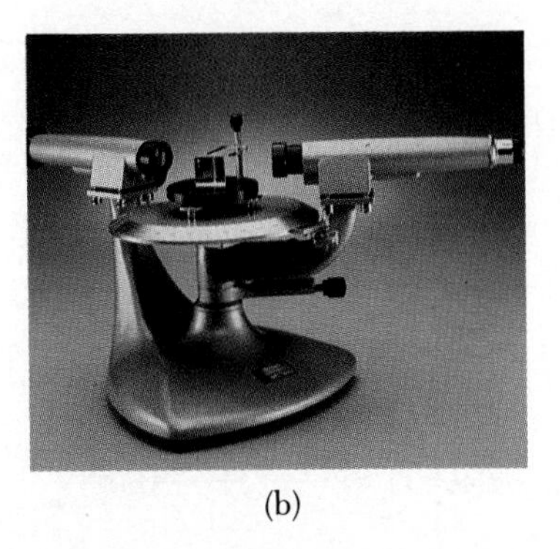


**Figure 35.18** (a) Diagram of a prism spectrometer. The various colors in the spectrum are viewed through a telescope. (b) Photograph of a prism spectrometer. (Courtesy of CENCO)

A prism is often used in an instrument known as a **prism spectrometer,** the essential elements of which are shown in Figure 35.18a. The instrument is commonly used to study the wavelengths emitted by a light source, such as a sodium vapor lamp. Light from the source is sent through a narrow, adjustable slit to produce a parallel, or collimated, beam. The light then passes through the prism and is dispersed into a spectrum. The refracted light is observed through a telescope. The experimenter sees an image of the slit through the eyepiece of the telescope. The telescope can be moved or the prism can be rotated in order to view the various images formed by different wavelengths at different angles of deviation.

All hot, low-pressure gases emit their own characteristic spectra. Thus, one use of a prism spectrometer is to identify gases. For example, sodium emits two wavelengths in the visible spectrum, which appear as two closely spaced yellow lines. Thus, a gas emitting these colors can be identified as having sodium as one of its constituents. Likewise, mercury vapor has its own characteristic spectrum, consisting of four prominent wavelengths—orange, green, blue, and violet lines—along with some wavelengths of lower intensity. The particular wavelengths emitted by a gas serve as "fingerprints" of that gas.

### EXAMPLE 35.7 Measuring $n$ Using a Prism

A prism is often used to measure the index of refraction of a transparent solid. Although we do not prove it here, one finds that the *minimum angle of deviation*, $\delta_m$, occurs at the angle of incidence $\theta_1$ where the refracted ray inside the prism makes the same angle $\alpha$ with the normal to the two prism faces,[4] as in Figure 35.19. Let us obtain an expression for the index of refraction of the prism material.

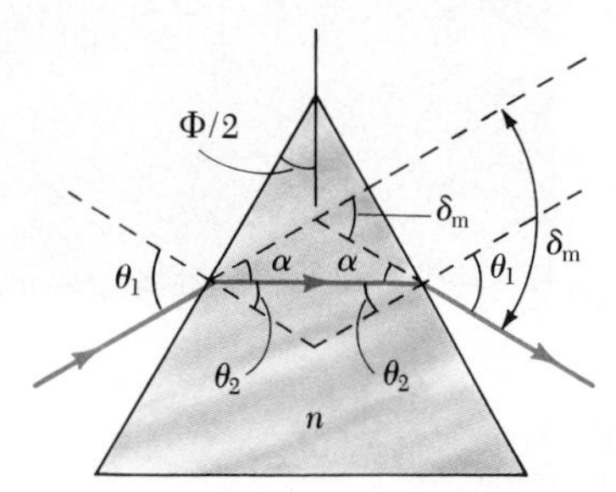


**Figure 35.19** (Example 35.7) A light ray passing through a prism at the minimum angle of deviation, $\delta_m$.

Using the geometry shown, one finds that $\theta_2 = \Phi/2$ and

$$\theta_1 = \theta_2 + \alpha = \frac{\Phi}{2} + \frac{\delta_m}{2} = \frac{\Phi + \delta_m}{2}$$

From Snell's law,

$$\sin \theta_1 = n \sin \theta_2$$

$$\sin\left(\frac{\Phi + \delta_m}{2}\right) = n \sin(\Phi/2)$$

$$n = \frac{\sin\left(\dfrac{\Phi + \delta_m}{2}\right)}{\sin(\Phi/2)} \qquad (35.9)$$

Hence, knowing the apex angle $\Phi$ of the prism and measuring $\delta_m$, one can calculate the index of refraction of the prism material. Furthermore, one can use a hollow prism to determine the values of $n$ for various liquids.

[4] For details, see F. A. Jenkins and H. E. White, *Fundamentals of Optics,* New York, McGraw-Hill, 1976, Chapter 2.

## 35.6 HUYGENS' PRINCIPLE

In this section, we shall develop the laws of reflection and refraction by using a geometric method proposed by Huygens in 1678. Huygens assumed that light is some form of wave motion rather than a stream of particles. He had no knowledge of the nature of light or of its electromagnetic character. Nevertheless, his simplified wave model is adequate for understanding many practical aspects of the propagation of light.

**Huygens' principle** is a geometric construction for determining the position of a new wavefront at some instant from the knowledge of an earlier wavefront. In Huygens' construction,

> all points on a given wavefront are taken as point sources for the production of spherical secondary waves, called wavelets, which propagate outward with speeds characteristic of waves in that medium. After some time has elapsed, the new position of the wavefront is the surface tangent to the wavelets.

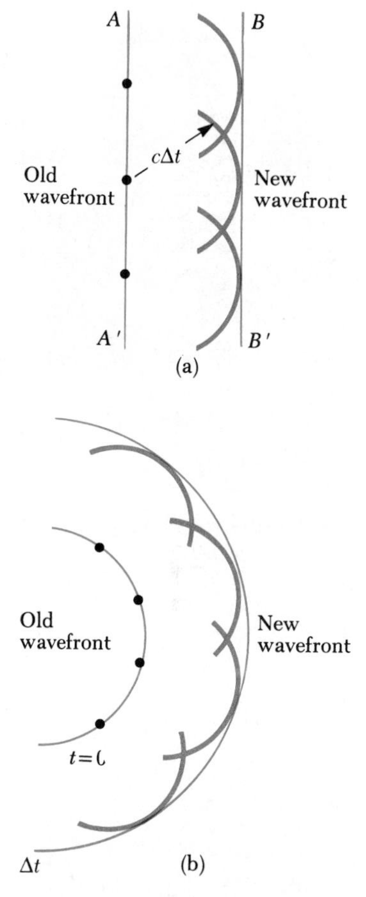


**Figure 35.20** Huygens' construction for (a) a plane wave propagating to the right and (b) a spherical wave.

Figure 35.20 illustrates two simple examples of Huygens' construction. First, consider a plane wave moving through free space, as in Figure 35.20a. At $t = 0$, the wavefront is indicated by the plane labeled $AA'$. (A wavefront is a surface passing through those points of a wave that are behaving identically. For instance, a wavefront could be a surface over which the wave is at a crest.) In Huygens' construction, each point on this wavefront is considered a point source. For clarity, only a few points on $AA'$ are shown. With these points as sources for the wavelets, we draw circles each of radius $c\ \Delta t$, where $c$ is the speed of light in free space and $\Delta t$ is the time of propagation from one wavefront to the next. The surface drawn tangent to these wavelets is the plane $BB'$, which is parallel to $AA'$. In a similar manner, Figure 35.20b shows Huygens' construction for an outgoing spherical wave.

A convincing demonstration of Huygens' principle is obtained with water waves in a shallow tank (called a ripple tank), as in Figure 35.21. Plane waves produced below the slit emerge above the slit as two-dimensional circular waves propagating outward.

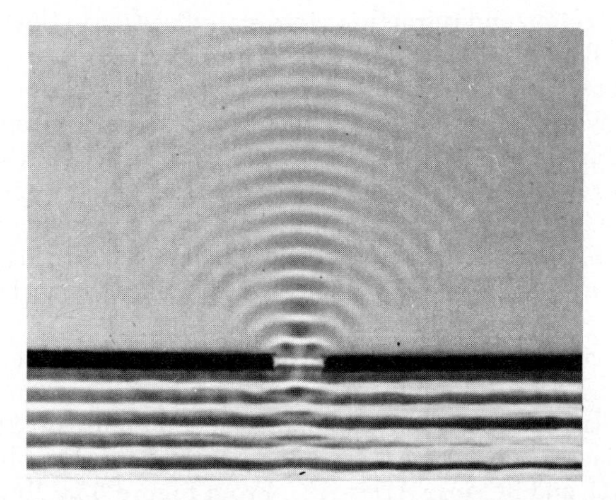

**Figure 35.21** Water waves in a ripple tank, which demonstrates Huygens' wavelets. A plane wave is incident on a barrier with a small opening. The opening acts as a source of circular wavelets. (Photography courtesy of Education Development Center, Newton, Mass.)

Biographical Sketch

**Christian Huygens**
*(1629–1695)*

Christian Huygens was a Dutch physicist and astronomer who is best known for his contributions to the fields of dynamics and optics. As a physicist, his accomplishments include invention of the pendulum clock and the first exposition of a wave theory of light. As an astronomer, Huygens was the first to recognize the rings around Saturn and to discover Titan, a satellite of Saturn.

Huygens was born in 1629 into a prominent family in The Hague. He was the son of Constantin Huygens, one of the most important figures of the Renaissance in Holland. Educated at the University of Leyden, Christian became a close friend of René Descartes, who was a frequent guest at the Huygens home. Huygens published his first paper in 1651 on the subject of the quadrature of various curves.

Huygens' reputation in optics and dynamics spread throughout Europe, and in 1663 he was elected a charter member of the Royal Society. Louis XIV lured Huygens to France in 1666, in accordance with the king's policy of collecting scholars for the glory of his regime. While in France, Huygens became one of the founders of the French Academy of Science.

In 1673, in Paris, Huygens published *Horologium Oscillatorium.* In this work he described a solution to the problem of the compound pendulum, for which he calculated the equivalent simple pendulum length. In the same publication he also derived a formula for computing the period of oscillation of a simple pendulum, and he explained his laws of centrifugal force for uniform motion in a circle.

Huygens returned to Holland in 1681, constructed some lenses of large focal lengths and invented the achromatic eyepiece for telescopes. Shortly after returning from a visit to England, where he met Isaac Newton, Huygens published his treatise on the wave theory of light. To Huygens, light was a vibratory motion in the ether, spreading out and producing the sensation of light when impinging on the eye. On the basis of this theory, he was able to deduce the laws of reflection and refraction and to explain the phenomenon of double refraction.

Huygens, second only to Newton among the greatest scientists in the second half of the 17th century, was the first to proceed in the field of dynamics beyond the point reached by Galileo and Descartes. It was Huygens who essentially solved the problem of centrifugal force. A solitary man, Huygens did not attract students or disciples and was very slow in publishing his findings. Huygens died in 1695 after a long illness.

(Rijksmuseum voor de Geschiedenis der Natuurwetenschappen. Courtesy AIP Niels Bohr Library)

### Huygens' Principle Applied to Reflection and Refraction

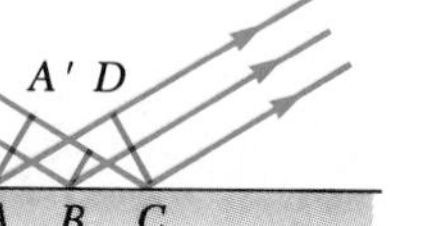


(b)

**Figure 35.22** (a) Huygens' construction for proving the law of reflection. (b) Triangle $ADC$ is identical to triangle $AA'C$.

The laws of reflection and refraction were stated earlier in this chapter without proof. We shall now derive these laws using Huygens' principle. Figure 35.22a will be used in our consideration of the law of reflection. The line $AA'$ represents a wavefront of the incident light. As ray 3 travels from $A'$ to $C$, ray 1 reflects from $A$ and produces a spherical wavelet of radius $AD$. (Recall that the radius of a Huygens wavelet is equal to $vt$.) Since the two wavelets having radii $A'C$ and $AD$ are in the same medium, they have the same velocity, $v$, and thus $AD = A'C$. Meanwhile, the spherical wavelet centered at $B$ has spread only half as far as the one centered at $A$ since ray 2 strikes the surface later than ray 1.

From Huygens' principle, we find that the reflected wavefront is $CD$, a line tangent to all the outgoing spherical wavelets. The remainder of the analysis depends upon geometry, as summarized in Figure 35.22b. Note that the right triangles $ADC$ and $AA'C$ are congruent because they have the same hypotenuse, $AC$, and because $AD = A'C$. From Figure 35.22b we have

$$\sin\theta_1 = \frac{A'C}{AC} \quad \text{and} \quad \sin\theta_1' = \frac{AD}{AC}$$

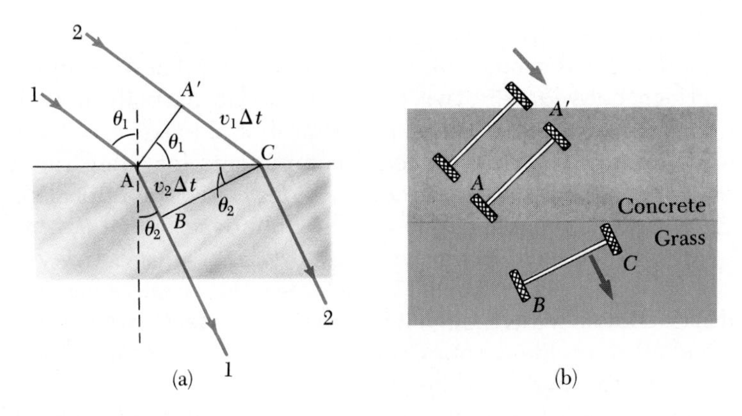


**Figure 35.23** (a) Huygens' construction for proving the law of refraction. (b) A mechanical analog of refraction.

Thus,

$$\sin \theta_1 = \sin \theta_1'$$

$$\theta_1 = \theta_1'$$

which is the law of reflection.

Now let us use Huygens' principle and Figure 35.23a to derive Snell's law of refraction. Note that in the time interval $\Delta t$, ray 1 moves from $A$ to $B$ and ray 2 moves from $A'$ to $C$. The radius of the outgoing spherical wavelet centered at $A$ is equal to $v_2 \, \Delta t$. The distance $A'C$ is equal to $v_1 \, \Delta t$. Geometric considerations show that angle $A'AC$ equals $\theta_1$ and angle $ACB$ equals $\theta_2$. From triangles $AA'C$ and $ACB$, we find that

$$\sin \theta_1 = \frac{v_1 \, \Delta t}{AC} \quad \text{and} \quad \sin \theta_2 = \frac{v_2 \, \Delta t}{AC}$$

If we divide these two equations, we get

$$\frac{\sin \theta_1}{\sin \theta_2} = \frac{v_1}{v_2}$$

But from Equation 35.4 we know that $v_1 = c/n_1$ and $v_2 = c/n_2$. Therefore,

$$\frac{\sin \theta_1}{\sin \theta_2} = \frac{c/n_1}{c/n_2} = \frac{n_2}{n_1}$$

$$n_1 \sin \theta_1 = n_2 \sin \theta_2$$

which is the law of refraction.

A mechanical analog of refraction is shown in Figure 35.23b. The wheels on a device such as a wagon change their direction as they move from a concrete surface to a grass surface.

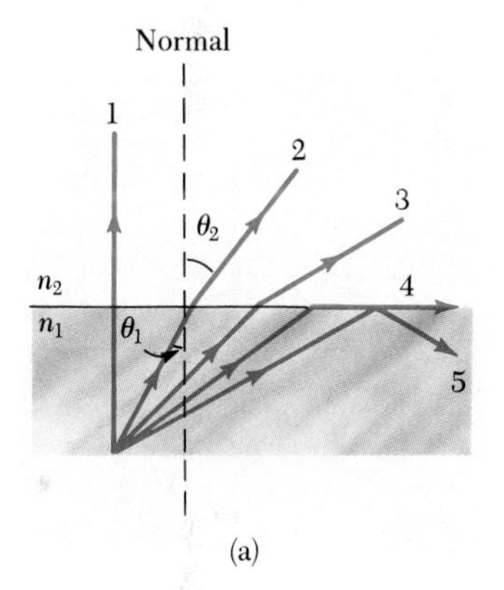


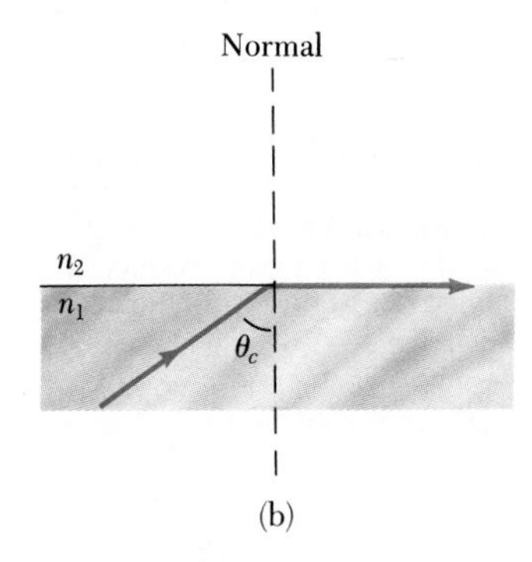


**Figure 35.24** (a) A ray from a medium of index of refraction $n_1$ to a medium of index of refraction $n_2$, where $n_1 > n_2$. As the angle of incidence increases, the angle of refraction increases until $\theta_2$ is 90° (ray 4). For even larger angles of incidence, total internal reflection occurs (ray 5). (b) The angle of incidence producing an angle of refraction equal to 90° is often called the *critical angle*, $\theta_c$.

## 35.7 TOTAL INTERNAL REFLECTION

An interesting effect called *total internal reflection* can occur when light attempts to move from a medium having a given index of refraction to one having a *lower* index of refraction. Consider a light beam traveling in medium 1 and meeting the boundary between medium 1 and medium 2, where $n_1$ is greater than $n_2$ (Fig. 35.24). Various possible directions of the beam are indicated by rays 1 through 5. The refracted rays are bent away from the normal because $n_1$ is greater than $n_2$. Furthermore, you should remember that when light refracts at the interface between the two media, it is also partially reflected. For example, the rays labeled 2, 3, and 4 in Figure 35.24 are *partially* reflected back into medium 1, but the reflected components for these rays are not shown. At some particular angle of incidence, $\theta_c$, called the **critical angle,** the refracted light ray will move parallel to the boundary so that $\theta_2 = 90°$ (Fig. 35.24b). *For angles of incidence greater than* $\theta_c$, the beam is entirely reflected at the boundary. Ray 5 in Figure 35.24a shows this occurrence. This ray is reflected at the boundary as though it had struck a perfectly reflecting surface. This ray, and all those like it, obey the law of reflection; that is, the angle of incidence equals the angle of reflection.

We can use Snell's law to find the critical angle. When $\theta_1 = \theta_c$, $\theta_2 = 90°$ and Snell's law (Eq. 35.8) gives

$$n_1 \sin \theta_c = n_2 \sin 90° = n_2$$

Critical angle

$$\sin \theta_c = \frac{n_2}{n_1} \qquad \text{(for } n_1 > n_2\text{)} \qquad (35.10)$$

This photograph shows nonparallel light rays entering a glass prism from the bottom. The two rays exiting at the right undergo total internal reflection at the longest side of the prism. The three rays exiting upward are refracted at this side as they leave the prism. (Courtesy of Henry Leap and Jim Lehman)

This equation can be used only when $n_1$ is greater than $n_2$. That is,

> total internal reflection occurs only when light attempts to move from a medium of given index of refraction to a medium of lower index of refraction.

If $n_1$ were less than $n_2$, Equation 35.10 would give $\sin \theta_c > 1$, which is an absurd result because the sine of an angle can never be greater than unity.

The critical angle for a substance in air is small for substances with a large index of refraction, such as diamond, where $n = 2.42$ and $\theta_c = 24°$. For crown glass, $n = 1.52$ and $\theta_c = 41°$. In fact, this property combined with proper faceting causes diamonds and crystal glass to sparkle.

One can use a prism and the phenomenon of total internal reflection to alter the direction of travel of a light beam. Two such possibilities are illustrated in Figure 35.25. In one case the light beam is deflected by 90° (Fig. 35.25a), and in the second case the path of the beam is reversed (Fig. 35.25b). A common application of total internal reflection is in a submarine periscope. In this device, two prisms are arranged as in Figure 35.25c so that an incident beam of light follows the path shown and one is able to "see around corners."

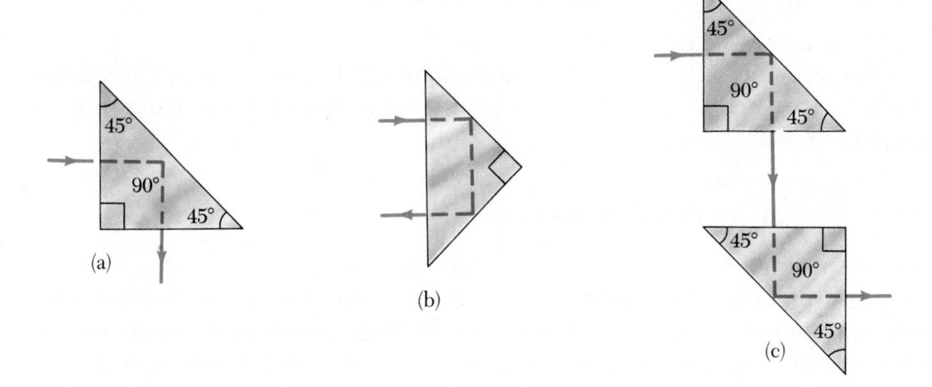


**Figure 35.25** Internal reflection in a prism. (a) The ray is deviated by 90°. (b) The direction of the ray is reversed. (c) Two prisms used as a periscope.

**EXAMPLE 35.8 A View from the Fish's Eye**
(a) Find the critical angle for a water-air boundary if the index of refraction of water is 1.33.

*Solution* Applying Equation 35.10, we find the critical angle to be

$$\sin\theta_c = \frac{n_2}{n_1} = \frac{1}{1.33} = 0.752$$

$$\theta_c = 48.8°$$

(b) Use the results of (a) to predict what a fish will see if it looks upward toward the water surface at an angle of 40°, 49°, and 60°.

*Solution* Because the path of a light ray is reversible, the fish can see out of the water if it looks toward the surface at an angle less than the critical angle. Thus, at 40°, the fish can see into the air above the water. At an angle of 49°, the critical angle for water, the light that reaches the fish has to skim along the water surface before being refracted to the fish's eye. At angles greater than the critical angle, the light reaching the fish comes via internal reflection at the surface. Thus, at 60°, the fish sees a reflection of some object on the bottom of the pool.

## Fiber Optics

Another interesting application of total internal reflection is the use of glass or transparent plastic rods to "pipe" light from one place to another. As indicated in Figure 35.26, light is confined to traveling within the rods, even around gentle curves, as the result of successive internal reflections. Such a "light pipe" will be flexible if thin fibers are used rather than thick rods. If a bundle of parallel fibers is used to construct an optical transmission line, images can be transferred from one point to another.

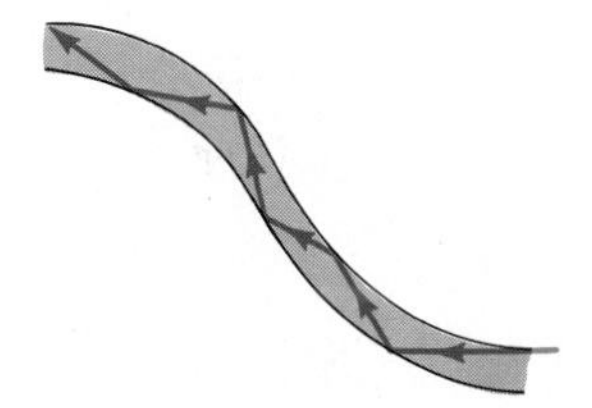

**Figure 35.26** Light travels in a curved transparent rod by multiple internal reflections.

This technique is used in a sizable industry known as *fiber optics* (see the essay at the end of this chapter). There is very little light intensity lost in these fibers as a result of reflections on the sides. Any loss in intensity is due essentially to reflections from the two ends and absorption by the fiber material. These devices are particularly useful when one wishes to view an image produced at inaccessible locations. For example, physicians often use this technique to examine internal organs of the body. The field of fiber optics is finding increasing use in telecommunications, since the fibers can carry a much higher volume of telephone calls or other forms of communication than electrical wires. The essay in this chapter discusses the use of fiber optics in the expanding field of telecommunications.

Spray of glass fiber optics consisting of 2000 individual strands, each measuring 60 μm thick. (Adam Hart, Davis/Science Photo Library/Photo Researchers, Inc.)

## *35.8 FERMAT'S PRINCIPLE

A general principle that can be used for determining the actual paths of light rays was developed by Pierre de Fermat (1601–1665). **Fermat's principle** states that

> when a light ray travels between any two points $P$ and $Q$, its actual path will be the one that requires the least time.

Fermat's principle is sometimes called the *principle of least time*. An obvious consequence of this principle is that when the rays travel in a single, homogeneous medium, the paths are straight lines because a straight line is the shortest distance between two points. Let us illustrate how to use Fermat's principle to derive the law of refraction.

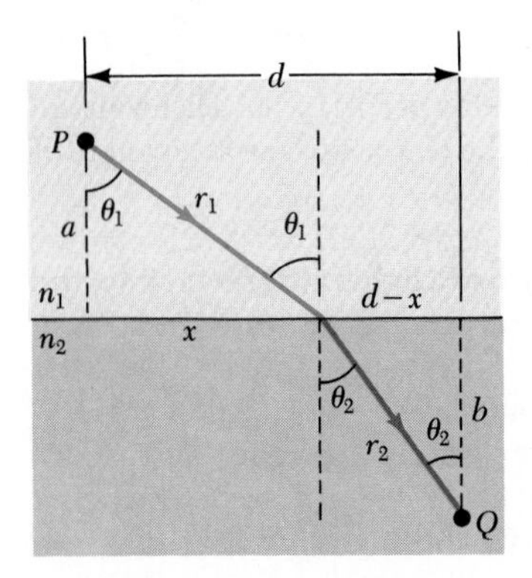


**Figure 35.27** Geometry for deriving the law of refraction using Fermat's principle.

Suppose a light ray is to travel from $P$ to $Q$, where $P$ is in medium 1 and $Q$ is in medium 2 (Fig. 35.27). The points $P$ and $Q$ are at perpendicular distances $a$ and $b$, respectively, from the interface. The speed of light is $c/n_1$ in medium 1 and $c/n_2$ in medium 2. Using the geometry of Figure 35.27, we see that the time it takes the ray to travel from $P$ to $Q$ is

$$t = \frac{r_1}{v_1} + \frac{r_2}{v_2} = \frac{\sqrt{a^2 + x^2}}{c/n_1} + \frac{\sqrt{b^2 + (d-x)^2}}{c/n_2}$$

We obtain the least time, or the minimum value of $t$, by taking the derivative of $t$ with respect to $x$ (the variable) and setting the derivative equal to zero. Using this procedure, we get

$$\frac{dt}{dx} = \frac{n_1}{c}\frac{d}{dx}(a^2 + x^2)^{1/2} + \frac{n^2}{c}\frac{d}{dx}[b^2 + (d-x)^2]^{1/2}$$

$$= \frac{n_1}{c}\left(\frac{1}{2}\right)\frac{2x}{(a^2 + x^2)^{1/2}} + \frac{n_2}{c}\left(\frac{1}{2}\right)\frac{2(d-x)(-1)}{[b^2 + (d-x)^2]^{1/2}}$$

$$\frac{dt}{dx} = \frac{n_1 x}{c(a^2 + x^2)^{1/2}} - \frac{n_2(d-x)}{c[b^2 + (d-x)^2]^{1/2}} = 0$$

From Figure 35.27 and recognizing $\sin\theta_1$ and $\sin\theta_2$ in this equation, we find that

$$n_1 \sin\theta_1 = n_2 \sin\theta_2$$

which is Snell's law of refraction.

It is a simple matter to use a similar procedure to derive the law of reflection. The calculation is left for you to carry out (Problem 46).

## SUMMARY

In geometric optics, we use the so-called **ray approximation,** in which we assume that a wave travels through a medium in straight lines in the direction of the rays. Furthermore, we neglect diffraction effects, which is a good approximation as long as the wavelength is short compared with any aperture dimensions.

The basic laws of geometric optics are the *laws of reflection and refraction* for light rays. The **law of reflection** states that the angle of reflection, $\theta_1'$, equals the angle of incidence, $\theta_1$. The **law of refraction, or Snell's Law,** states that

Snell's law of refraction

$$n_1 \sin\theta_1 = n_2 \sin\theta_2 \tag{35.8}$$

where $\theta_2$ is the angle of refraction and $n_1$ and $n_2$ are the indices of refraction in the two media. The incident ray, the reflected ray, the refracted ray, and the normal to the surface all lie in the same plane.

The **index of refraction** of a medium, $n$, is defined by the ratio

Index of refraction

$$n \equiv \frac{c}{v} \tag{35.4}$$

where $c$ is the speed of light in a vacuum and $v$ is the speed of light in the medium. In general, $n$ varies with wavelength and is given by

$$n = \frac{\lambda_0}{\lambda_n} \qquad (35.7)$$

Index of refraction and wavelength

where $\lambda_0$ is the vacuum wavelength and $\lambda_n$ is the wavelength in the medium.

**Huygens' principle** states that all points on a wavefront can be taken as point sources for the production of secondary wavelets. At some later time, the new position of the wavefront is the surface tangent to these secondary wavelets.

Huygens' principle

**Total internal reflection** can occur when light travels from a medium of high index of refraction to one of lower index of refraction. The minimum angle of incidence, $\theta_c$, for which total reflection occurs at an interface is given by

$$\sin \theta_c = \frac{n_2}{n_1} \qquad (\text{where } n_1 > n_2) \qquad (35.10)$$

Critical angle for total internal reflection

**Fermat's principle** states that when a light ray travels between two points, its path will be the one that requires the least time.

Fermat's principle

## QUESTIONS

1. Light of wavelength $\lambda$ is incident on a slit of width $d$. Under what conditions is the ray approximation valid? Under what circumstances will the slit produce significant diffraction?
2. Sound waves have much in common with light waves, including the properties of reflection and refraction. Give examples of such phenomena for sound waves.
3. Does a light ray traveling from one medium into another always bend toward the normal as in Figure 35.8? Explain.
4. As light travels from one medium to another, does its wavelength change? Does its frequency change? Does its velocity change? Explain.
5. A laser beam passing through a nonhomogeneous sugar solution is observed to follow a curved path. Explain.
6. A laser beam ($\lambda = 632.8$ nm) is incident on a piece of Lucite as in Figure 35.28. Part of the beam is reflected and part is refracted. What information can you get from this photograph?
7. Suppose blue light were used instead of red light in the experiment shown in Figure 35.28. Would the refracted beam be bent at a larger or smaller angle?
8. The level of water in a clear, colorless glass is easily observed with the naked eye. The level of liquid helium in a clear glass vessel is extremely difficult to see with the naked eye. Explain.
9. Describe an experiment in which internal reflection is used to determine the index of refraction of a medium.
10. Why does a diamond show flashes of color when observed under ordinary white light?
11. Explain why a diamond shows more "sparkle" than a glass crystal of the same shape and size.
12. Explain why an oar in the water appears bent.

**Figure 35.28** (Questions 6 and 7) Light from a helium-neon laser beam ($\lambda = 632.8$ nm) is incident on a block of Lucite. The photograph shows both reflected and refracted rays. Can you identify the incident, reflected, and refracted rays? From this photograph, estimate the index of refraction of Lucite at this wavelength. (Courtesy of Henry Leap and Jim Lehman)

13. Redesign the periscope of Figure 35.25c so that it can show you where you have been rather than where you are going.
14. Under certain circumstances, sound can be heard over extremely long distances. This frequently happens over a body of water, where the air near the water surface is cooler than the air higher up. Explain how the refraction of sound waves in such a situation could increase the distance over which the sound can be heard.
15. Why do astronomers looking at distant galaxies talk about looking backward in time?
16. A solar eclipse occurs when the moon gets between the earth and the sun. Use a diagram to show why some areas of the earth see a total eclipse, other areas see a partial eclipse, and most areas see no eclipse.
17. Some department stores have their windows slanted slightly inward at the bottom. This is to decrease the glare from streetlights or the sun, which would make it difficult for shoppers to see the display inside. Draw a sketch of a light ray reflecting off such a window to show how this technique works.
18. Suppose you are told only that two colors of light (X and Y) are sent through a glass prism and that X is bent more than Y. Which color travels more slowly in the prism?
19. Figure 35.29 represents sunlight striking a drop of water in the atmosphere. Use the laws of refraction and reflection and the fact that sunlight consists of a wide range of wavelengths to discuss the formation of rainbows.
20. Why does the arc of a rainbow appear with red colors on top and violet hues on the bottom?
21. How is it possible that a complete *circular* rainbow can sometimes be seen from an airplane?
22. You can make a corner reflector by placing three plane mirrors in the corner of a room where the ceiling meets the walls. Show that no matter where you are in the room, you can see yourself reflected in the mirrors—upside down.
23. Several corner reflectors were left on the moon's Sea of Tranquility by the astronauts of Apollo 11. How can scientists utilize a laser beam sent from Earth even today to determine the precise distance from the Earth to the moon?
24. What are the conditions for the production of a mirage? On a hot day, what is it that we are seeing when we observe "water on the road"?
25. The "professor in the box" shown in Figure 35.30 appears to be balancing himself on a few fingers with his feet elevated from the floor. How do you suppose this illusion was created? (The same fellow has also been known to "fly" at times.)

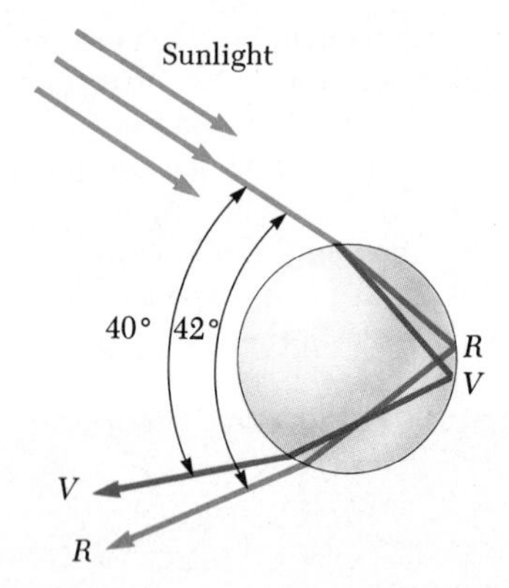


**Figure 35.29** (Question 19) Refraction of sunlight by a spherical raindrop.

**Figure 35.30** (Question 25). (Photograph courtesy of Henry Leap and Jim Lehman)

## PROBLEMS

### Section 35.2 Measurements of the Speed of Light

1. Experimenters at the National Bureau of Standards have made precise measurements of the speed of light using the property of electromagnetic waves that in vacuum the phase velocity of the waves is $c = \sqrt{1/\mu_0\epsilon_0}$, where $\mu_0$ (permeability constant) $= 4\pi \times 10^{-7}$ N · s²/C² and $\epsilon_0$ (permittivity constant) $= 8.854 \times 10^{-12}$ C²/N · m². What value (to four significant figures) does this give for the speed of light in vacuum?
2. As a result of his observations, Roemer concluded that the accumulated time interval between eclipses of the moon Io by the planet Jupiter increased by 22 min during a 6-month period as the earth moved from a point in its orbit, where its motion is *toward* Jupiter, to a diametrically opposite point where it moves *away*

from Jupiter. Using $1.5 \times 10^8$ km as the average radius of the earth's orbit about the sun, calculate the speed of light from these data.

3. Use Roemer's value of 22 min discussed in Problem 2 and the presently accepted value of the speed of light in vacuum to find an average value for the distance between the earth and the sun.
4. Michelson performed a very careful measurement of the speed of light using an improved version of the technique developed by Fizeau. In one of Michelson's experiments, the toothed wheel was replaced by a wheel with 32 identical mirrors mounted on its perimeter, with the plane of each mirror perpendicular to a radius of the wheel. The total light path was 8 miles in length (obtained by multiple reflections of a light beam within an evacuated tube 1 mile long). For what minimum angular velocity of the mirror would Michelson have calculated the speed of light to be $2.998 \times 10^8$ m/s?
5. In an experiment to measure the speed of light using the apparatus of Fizeau (Fig. 35.2), the distance between light source and mirror was 11.45 km and the wheel had 720 notches. The experimentally determined value of $c$ was $2.998 \times 10^8$ m/s. Calculate the minimum angular velocity of the wheel for this experiment.
6. If the Fizeau experiment is performed such that the round-trip distance for the light is 40 m, find the two lowest speeds of rotation that allow the light to pass through the notches. Assume that the wheel has 360 teeth and that the speed of light is $3 \times 10^8$ m/s. Repeat for a round-trip distance of 4000 m.
7. In Galileo's attempt to determine the speed of light, he and his companion were located on hilltops 9 km apart. What time interval between uncovering his lantern and seeing the light from his companion's lantern would Galileo have had to measure in order to obtain a value for the speed of light?

## Section 35.4 Reflection and Refraction

(*Note:* In this section if an index of refraction value is not given, refer to Table 35.1.)

8. A coin is on the bottom of a pool 1 m deep. What is the apparent depth of the coin, seen from above the water surface?
9. The wavelength of red helium-neon laser light in air is 632.8 nm. (a) What is its frequency? (b) What is its wavelength in glass of index of refraction 1.5? (c) What is its speed in glass?
10. A narrow beam of sodium yellow light is incident from air on a smooth surface of water at an angle $\theta_1 = 35°$. Determine the angle of refraction $\theta_2$ and the wavelength of the light in water.
11. An underwater scuba diver sees the sun at an apparent angle of 45° from the vertical. Where is the sun?
12. A light ray in air is incident on a water surface at an angle of 30° *with respect to the normal to the surface.* What is the angle of the refracted ray relative to the normal to the surface?
13. A ray of light in air is incident on a planar surface of fused quartz. The refracted ray makes an angle of 37° with the normal. Calculate the angle of incidence.
14. A light ray initially in water enters a transparent substance at an angle of incidence of 37°, and the transmitted ray is refracted at an angle of 25°. Calculate the speed of light in the transparent material.
15. A ray of light strikes a flat block of glass ($n = 1.50$) of thickness 2 cm at an angle of 30° with the normal. Trace the light beam through the glass, and find the angles of incidence and refraction at each surface.
16. Find the speed of light in (a) flint glass, (b) water, and (c) zircon.
17. Light of wavelength 436 nm in air enters a fishbowl filled with water, then exits through the crown glass wall of the container. What is the wavelength of the light (a) while in the water and (b) while in the glass?
18. Light is incident on the interface between air and polystyrene at an angle of 53°. The incident ray, initially traveling in air, is partially transmitted and partially reflected at the surface. What is the angle between the *refracted* and the *reflected* ray?
19. Light of wavelength $\lambda_0$ in vacuum has a wavelength of 438 nm in water and a wavelength of 390 nm in benzene. What is the index of refraction of water relative to benzene at the wavelength $\lambda_0$?
20. A cylindrical tank with an open top has a diameter of 3 m and is completely filled with water. When the setting sun reaches an angle of 28° above the horizon, sunlight ceases to illuminate the bottom of the tank. How deep is the tank?

## *Section 35.5 Dispersion and Prisms

21. A ray of light strikes the midpoint of one face of an equiangular glass prism ($n = 1.50$) at an angle of incidence of 30°. Trace the path of the light ray through the glass and find the angles of incidence and refraction at each surface.
22. Calculate the index of refraction of an equiangular prism for which the angle of minimum deviation is $\delta_m = 37°$.
23. A crown glass prism has an apex angle of 15°. What is the angle of minimum deviation of this prism for light of wavelength 525 nm? See Figure 35.15 for the value of $n$.
24. A certain kind of glass has an index of refraction of 1.6500 for blue light of wavelength 430 nm and an index of 1.6150 for red light of wavelength 680 nm. If a beam containing these two colors is incident at an angle of 30° on a piece of this glass, what is the angle between the two beams inside the glass?

25. Show that if the apex angle $\Phi$ of a prism is small, an approximate value for the angle of minimum deviation can be calculated from $\delta_m = (n-1)\Phi$.
26. For a particular prism and wavelength, $n = 1.62$. Compare the values found for the angle of minimum deviation in this prism when using the approximation of Problem 25 and the exact form given by Equation 35.9 when (a) $\Phi = 30°$ and (b) $\Phi = 10°$.
27. An experimental apparatus includes a prism made of sodium chloride. The angle of minimum deviation for light of wavelength 589 nm is to be 10°. What is the required apex angle of the prism?
28. Light of wavelength 700 nm is incident on the face of a fused quartz prism at an angle of 75° (with respect to the normal to the surface). The apex angle of the prism is 60°. Use the value of $n$ from Figure 35.15 and calculate the angle (a) of refraction at this (first) surface, (b) of incidence at the second surface, (c) of refraction at the second surface, and (d) between the incident and emerging rays.
29. A prism with apex angle 50° is made of cubic zirconia, with $n = 2.20$. What is the angle of minimum deviation $\delta_m$ for such a prism?
30. A triangular glass prism with apex angle 60° has an index of refraction $n = 1.5$. (a) What is the smallest angle of incidence $\theta_1$ for which a light ray can emerge from the other side? (See Figure 35.19.) (b) For what angle of incidence $\theta_1$ does the light ray leave at the same angle $\theta_1$?
31. The index of refraction for violet light in silica flint glass is 1.66, and the refractive index for red light is 1.62. What is the angular dispersion of visible light passing through a prism of apex angle 60° if the angle of incidence is 50°? (See Figure 35.31.)

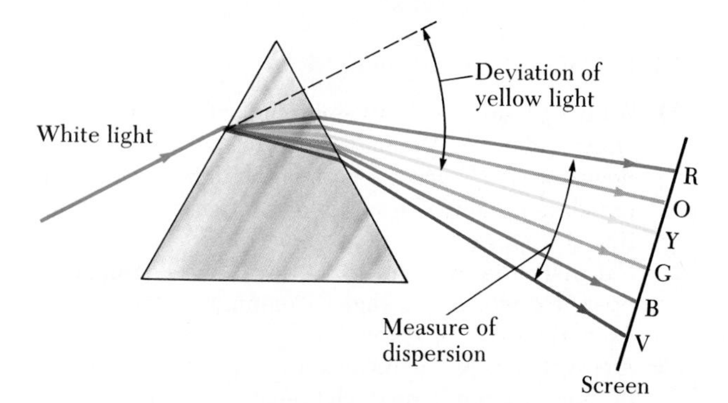


**Figure 35.31** (Problem 31).

### °Section 35.7 Total Internal Reflection

32. A large Lucite cube ($n = 1.59$) has a small air bubble (a defect in the casting process) below one surface. When a penny (diameter 1.9 cm) is placed directly over the bubble, the bubble cannot be seen by looking down into the cube at any angle. However, when a dime (radius 1.75 cm) is placed directly over it, the bubble can be seen by looking down into the cube. What is the range of the possible depths of the air bubble beneath the surface?
33. A fiber optic cable ($n = 1.50$) is submerged in water ($n = 1.33$). What is the critical angle for light to stay inside the cable?
34. A fish in a pond is located 15 m from shore. Below what depth would the fish be unable to see a small rock at the water's edge?
35. Calculate the critical angle for the following materials when surrounded by air: (a) diamond, (b) flint glass, (c) ice. (Assume that $\lambda = 589$ nm.)
36. Repeat Problem 35 when the materials are surrounded by water.
37. Consider a common mirage formed by super-heated air just above the roadway. If an observer viewing from 2 m above the road (where $n = 1.0003$) sees water up the road at $\theta_1 = 88.8°$, find the index of refraction of the air just above the road surface. (*Hint:* Treat this as a problem in total internal reflection.)
38. A light ray is incident on the interface between a diamond and air from within the diamond. What is the critical angle for total internal reflection? Use Table 35.1. (The smallness of $\theta_c$ for diamond means that light is easily "trapped" within a diamond and eventually emerges from the many cut faces; this makes a diamond more "brilliant" than stones with smaller $n$ and larger $\theta_c$.)
39. An optical fiber is made of a clear plastic with index of refraction $n = 1.50$. For what angles with the surface will light remain contained within the plastic "guide"?

### ADDITIONAL PROBLEMS

40. A narrow beam of light is incident from air onto a glass surface of index of refraction 1.56. Find the angle of incidence for which the corresponding angle of refraction will be one half the angle of incidence. (*Hint:* You might want to use the trigonometric identity $\sin 2\theta = 2 \sin\theta \cos\theta$.)
41. A specimen of glass has an index of refraction of 1.61 for the wavelength corresponding to the prominent bright line in the sodium spectrum. If an equiangular prism is made from this glass, what angle of incidence will result in minimum deviation of the sodium line?
42. A light beam is incident on a water surface from air. What is the maximum possible value for the angle of refraction?
43. A small underwater pool light is 1 m below the surface. What is the diameter of the circle of light on the surface, from which light emerges from the water?
44. When the sun is directly overhead, a narrow shaft of light enters a cathedral through a small hole in the

ceiling and forms a spot on the floor 10.0 m below. (a) At what speed (in cm/min) does the spot move across the (flat) floor? (b) If a mirror is placed on the floor to intercept the light, at what speed will the reflected spot move across the ceiling?

45. A drinking glass is 4 cm wide at the bottom, as shown in Figure 35.32. When an observer's eye is placed as shown, the observer sees the edge of the bottom of the glass. When this glass is filled with water, the observer sees the center of the bottom of the glass. Find the height of the glass.

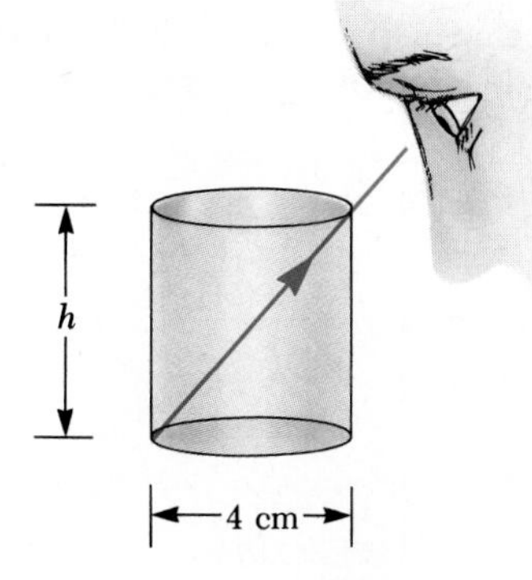


**Figure 35.32** (Problem 45).

46. Show that the time required for light to travel from a point source A in air a distance $h_1$ above a water surface to a point B at $h_2$ below the water surface is given by

$$t = \frac{h_1 \sec \theta_1}{c} + \frac{h_2 n \sec \theta_2}{c}$$

where $n$ is the index of refraction of water, $\theta_1$ is the angle of incidence, and $\theta_2$ is the angle of refraction.

47. Derive the law of reflection (Eq. 35.2) from Fermat's principle of least time. (See the procedure outlined in Section 35.9 for the derivation of the law of refraction from Fermat's principle.)

48. The angle between the two mirrors in Figure 35.33 is a right angle. The beam of light in the vertical plane $P$ strikes mirror 1 as shown. (a) Determine the distance the reflected light beam travels before striking mirror 2. (b) In what direction does the light beam travel after being reflected from mirror 2?

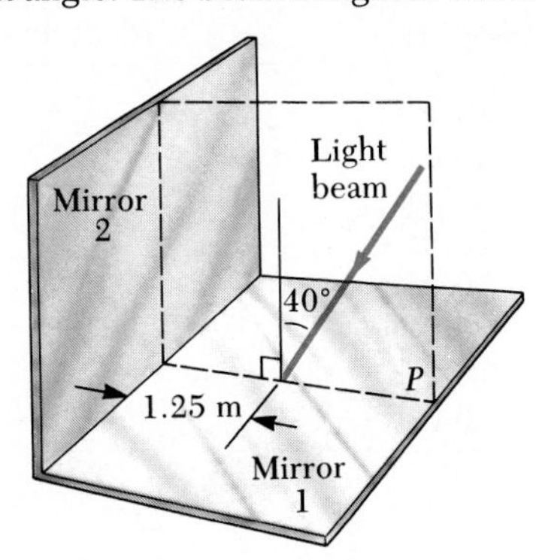


**Figure 35.33** (Problem 48).

49. A light ray of wavelength 589 nm is incident at an angle $\theta$ on the top surface of a block of polystyrene, as shown in Figure 35.34. (a) Find the maximum value of $\theta$ for which the refracted ray will undergo *total* internal reflection at the left vertical face of the block. (b) Repeat the calculation for the case in which the polystyrene block is immersed in water. (c) What happens if the block is immersed in carbon disulfide?

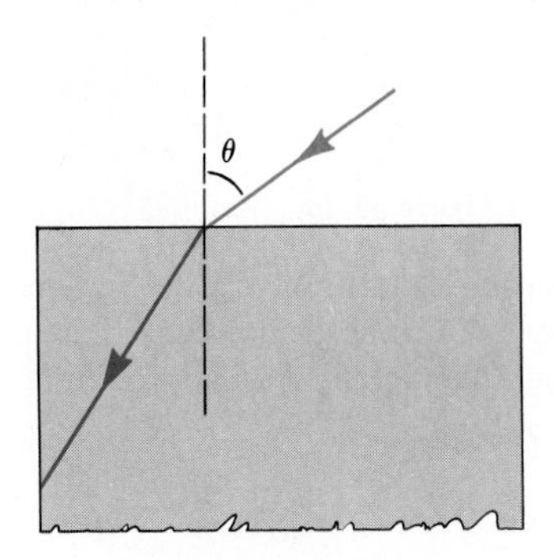


**Figure 35.34** (Problem 49).

50. One technique to measure the angle of a prism is shown in Figure 35.35. A parallel beam of light is directed on the angle such that the beam reflects from opposite sides. Show that the angular separation of the two beams is given by $B = 2A$.

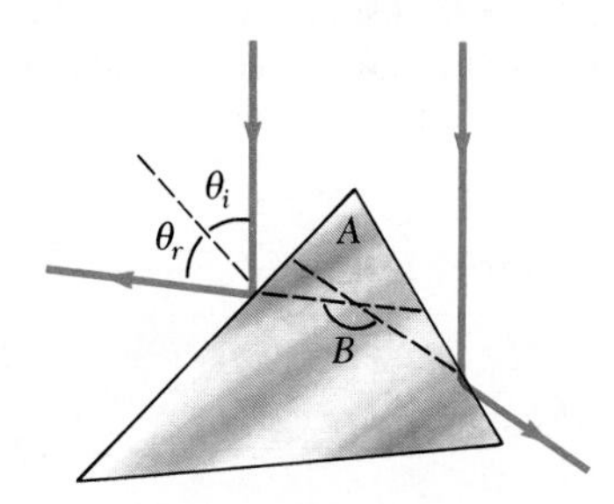


**Figure 35.35** (Problem 50).

51. A hiker stands on a mountain peak near sunset and observes a (primary) rainbow caused by water droplets in the air about 8 km away. The valley is 2 km below the mountain peak and is entirely flat. What fraction of the complete circular arc of the rainbow is visible to the hiker?

52. A fish is at a depth $d$ under water. Show that when viewed from an angle of incidence $\theta_1$, the *apparent depth* $z$ of the fish is

$$z = \frac{3d \cos \theta_1}{\sqrt{7 + 9 \cos^2 \theta_1}}$$

53. A light ray is incident on a prism and refracted at the first surface as shown in Figure 35.36. Let $\Phi$ represent the apex angle of the prism and $n$ its index of refraction. Find in terms of $n$ and $\Phi$ the smallest allowed value of the angle of incidence at the first surface for which the refracted ray will *not* undergo internal reflection at the second surface.

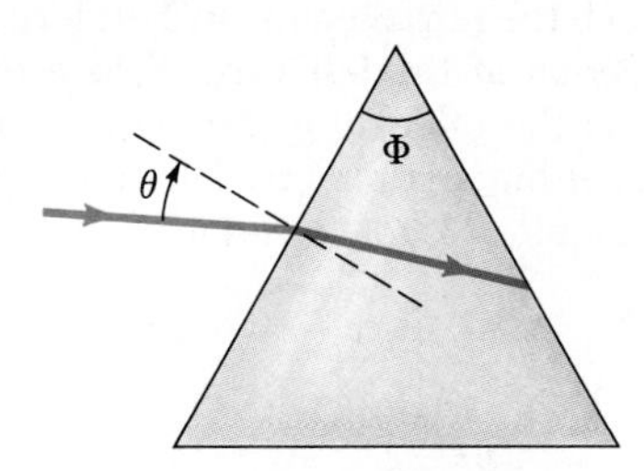


**Figure 35.36** (Problem 53).

54. The prism shown in Figure 35.37 has an index of refraction of 1.55. Light is incident at an angle of 20°. Determine the angle $\theta$ at which the light emerges.

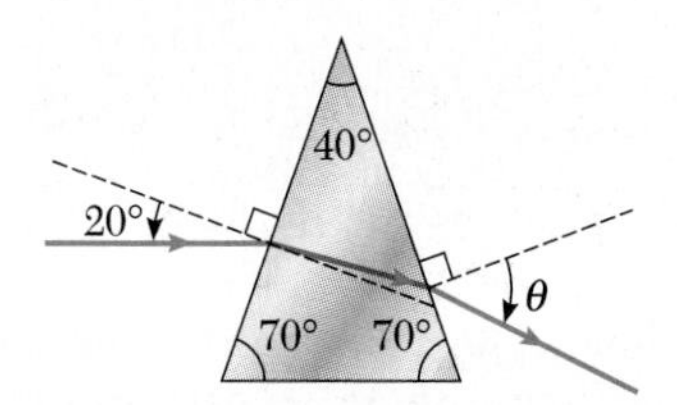


**Figure 35.37** (Problem 54).

55. A laser beam strikes one end of a slab of material, as shown in Figure 35.38. The index of refraction of the slab is 1.48. Determine the number of internal reflections of the beam before it emerges from the opposite end of the slab.

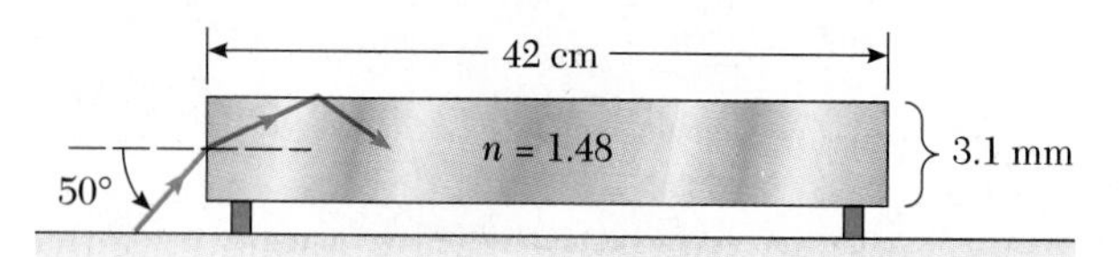


**Figure 35.38** (Problem 55).

56. Figure 35.39 shows a top view of a square enclosure. The inner surfaces are plane mirrors. A ray of light enters a small hole in the center of one mirror. (a) At what angle $\theta$ must the ray enter in order to exit through the hole after being reflected once by each of the other three mirrors? (b) Are there other values of $\theta$ for which the ray can exit after multiple reflections? If so, make a sketch of one of the ray's paths.

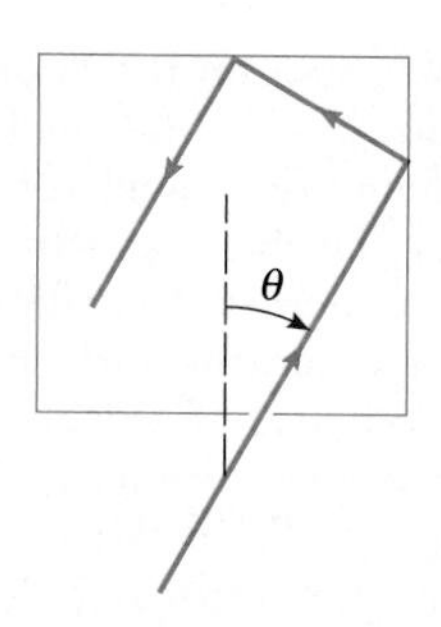


**Figure 35.39** (Problem 56).

57. The light beam in Figure 35.40 strikes surface 2 at the critical angle. Determine the angle of incidence $\theta_1$.

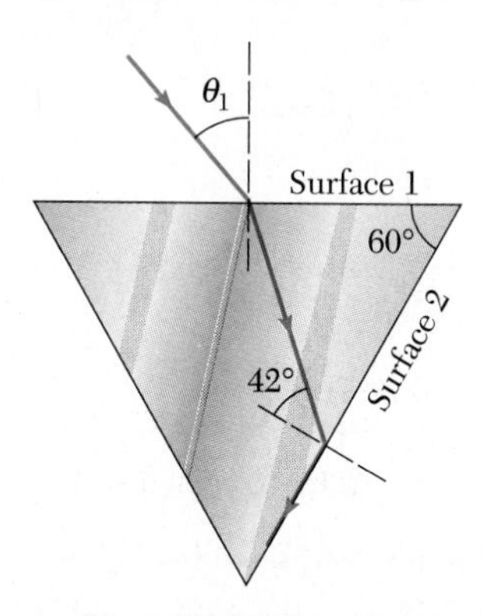


**Figure 35.40** (Problem 57).

Δ58. Students in an undergraduate laboratory allow a narrow beam of laser light to strike a water surface. They arrange to measure the angle of refraction for selected angles of incidence and record the data shown in the accompanying table. Use the data to verify Snell's law by plotting the sine of the angle of incidence versus the sine of the angle of refraction. Use the resulting plot to deduce the index of refraction of water.

| Angle of Incidence (degrees) | Angle of Refraction (degrees) |
|---|---|
| 10.0 | 7.5 |
| 20.0 | 15.1 |
| 30.0 | 22.3 |
| 40.0 | 28.7 |
| 50.0 | 35.2 |
| 60.0 | 40.3 |
| 70.0 | 45.3 |
| 80.0 | 47.7 |

# ESSAY

## Fiber Optics: The Ultimate in Telecommunications

**Edward A. Lacy**

In the 1980s light wave communications—photonics—came of age. The United States and other industrial nations went on a "high-fiber" diet, installing so much fiber-optic cable that the decade was called the "decade of glass." More than 3 million km of these cables were installed in the United States alone. In the process, thousands of miles of copper cables—both coaxial and twisted-pair—were made obsolete as far as long distance telecommunications were concerned. Copper cables have been displaced simply because they do not have the tremendous information-carrying capacity, called *bandwidth,* of fiber-optic cables.

With such bandwidth, fiber-optic systems can transmit thousands of telephone conversations, dozens of television programs, and numerous computer data signals over one or two flexible, hair-thin threads of ultrapure glass called *optical fibers.* Theoretically, the bandwidth of a fiber-optic cable is 50 THz (50 million MHz). For comparison, a television signal, the biggest user of bandwidth, requires 6 MHz of bandwidth, a telephone conversation just a few kilohertz.

Currently, such phenomenal bandwidths are not available because the input and output equipment to the cables are not capable of such speed. But even present systems, with their bandwidths of a few gigahertz, are able to transmit video conference signals and high definition television at economical prices.

These and other broadband services are not available to homes and businesses at present because of cost and government regulations. The long distance fiber-optic trunks across the nation are like superhighways. While most homes are within 50 miles of these super trunks, the connection between the trunks and homes and businesses is copper cable. In effect, the superhighways of telecommunications are connected to your home with dirt footpaths. To get the advantage of broadband services, the subscriber loop must also be replaced with fiber optics.

Running fiber all the way to a customer's home is presently too expensive. In addition, cable television companies and telephone companies are battling to see who will control the fiber to the home. Obviously, only one fiber would be needed per home. This fiber would carry telephone signals as well as cable TV and other signals. Government bodies will have to decide whether the phone company or the cable TV company will provide the service.

Meanwhile, local area networks (LANs) that interconnect computers at a common geographic site are being connected with optical fibers. In some cases these LANs do not use fiber optics because they have low bandwidth requirements. However, LANs at university campuses throughout the world are using fiber optics to

**Figure 1** Ultra-pure glass optical fibers carry voice, video and data signals in high-capacity telecommunications networks. (Photo courtesy of Corning Incorporated)

*(Continued on next page)*

**Figure 2** Optical fiber, the transmission medium of choice for telephone company interoffice trunks, proves in against copper for new and rebuild feeder systems. (Photo courtesy of Corning Incorporated)

connect host computers with terminals at dozens of buildings and hundreds of dormitory rooms.

In addition to giving extremely high bandwidth, fiber-optic systems have significantly smaller, lighter-weight cables than conventional telephone cables. An optical fiber with its protective jacket may be, typically, 0.635 cm in diameter, yet it can replace a 7.62-cm diameter bundle of copper wires now used to carry the same number of telephone conversations and other signals. The importance of this dramatic decrease in size is not obvious in uncongested rural areas; in major cities, however, where telephone cables must be placed underground, conduits are so crammed that they can scarcely accommodate a single additional copper cable.

Along with the size reduction, there is a corresponding weight reduction. For example, 94.5 kg of copper wire can be replaced with 3.6 kg of optical fiber. Weight reduction is important for the military services as it allows faster deployment of communication cables on battlefields. On the civilian side, it is important in huge jet aircraft, which use a surprising amount of copper cables between the various equipment and instrumentation on board. By replacing these cables with optical cables, up to 1 000 lb of weight may be saved, thereby giving better fuel consumption.

Perhaps of equal importance, fiber-optic systems are immune to electromagnetic interference such as lightning and arcs created in some factories. Because of this immunity, fiber-optic systems give accurate transmission of data—about 100 times better than transmissions over copper cables. For information transmitted in bits, the accuracy is typically one error in 100 million bits.

Still another advantage of fiber-optic systems is that it is difficult to tap them, as compared to telephone cables, for eavesdropping. In fact, fiber-optic systems are generally considered to be "secure," an important feature for the military. There are numerous other advantages that are important but discussion of them must be reserved for other texts.*

In a fiber-optic communication system, information is carried by lightwaves (photons) rather than by the movement of electrons as in metallic systems.

A fiber-optic communication system has three major components: a transmitter that converts electrical signals to light signals, an optical fiber for transmitting the signals, and a receiver that captures the signals at the other end of the fiber and converts them to electrical signals.

The key part of the transmitter is a light source—either a semiconductor laser diode or a light-emitting diode (LED). In general, laser diodes have greater capabili-

* E.A. Lacy, *Fiber Optics*, Englewood Cliffs, N.J., Prentice-Hall, 1982.

ties than LEDs but cost more, require more complex circuitry, and are less reliable. Laser diodes are generally used for long-haul routes, LEDs for short-haul loops. With either device, the light emitted is an invisible infrared signal with a wavelength of 1.31 μm (1310 nm). Wavelengths that are either higher or lower are significantly attenuated as they cannot pass through the "window" at 1.31 μm. Older systems had a wavelength of 0.85 μm; future systems may have a wavelength of 1.55 μm. Visible and ultraviolet light are impractical for fiber-optic systems because of high losses in the optical fiber. The laser diodes and LEDs used in this application are miniture units less than half the size of a thumbnail in order to couple the light into the tiny fibers effectively. Even so, a glass lens is frequently used to transfer the emitted light effectively to the fiber.

To transmit an audio, television, or computer data signal by light waves, it is necessary to change or *modulate* the light waves in accordance with the information in these signals. By varying the intensity of the light beam from the laser diode or the LED, *analog modulation* is achieved. By flashing the laser diode or LED on and off at an extremely fast rate, *digital modulation* is achieved.

In digital modulation, a pulse of light represents the number 1, and the absence of light represents 0. In a sense, instead of flashes of light traveling down the fiber, 1s and 0s are moving down the path. With computer-type equipment, any communication can be represented by a particular pattern or code of 1s and 0s. If the receiver is programmed to recognize such digital patterns, it can reconstruct the original signal from the 1s and 0s it receives.

Digital modulation is expressed in bits (short for "binary digit") per second, megabits (1 000 000 bits) per second, or gigabits (1 000 000 000 bits) per second. Engineers have demonstrated a fiber-optic system that can transmit 27 gigabits per second. At this rate, 400 000 phone conversations could be held simultaneously over this system.

As remarkable as these bandwidths are, engineers are pursuing other techniques to give even greater bandwidth. In *wavelength division multiplexing* (also called *color* multiplexing) the outputs of two or more lasers with different wavelengths are combined and sent through one optical fiber. In effect, this doubles the bandwidth. In *coherent modulation and detection* a laser sends a continuous beam whose frequency is varied by the messages that are being transmitted. At the receiver, this light beam is combined with a lightbeam that has been generated at the receiver. The frequencies are designed to be slightly different so that when they are combined there will be a new signal which is the difference of the two. The new signal can be processed, as in superheterodyne radio and television sets, much more easily than the original incoming lightbeam.

As you might suspect, the equipment used in digital modulation, such as encoders, is much more complicated than that used in analog modulation. Digital modulation also requires more bandwidth than analog modulation to send the same message. The former is, however, far more popular because it allows greater transmission distance with the same power and less expensive switching equipment. Thus, even though digital telecommunication is only a minority of present telecommunication now, it is rapidly replacing analog transmission which is used mainly for television signals.

Even though an optical fiber is usually made of glass, it is surprisingly tough; in fact, it can be bent and twisted just like wire. Splicing optical fiber, however, can be difficult: the ends of the fibers can be joined by fusion or with mechanical splices, but not by twisting and soldering as with copper cables. In splicing, great care must be taken to ensure precision mating of the tiny fibers. Otherwise the system will have enough losses to make it inoperable.

Optical fibers have very low transmission losses because of their ultra purity. If a window pane 1 km thick were to be made of such glass, it would be as transparent as an ordinary pane of glass. Despite this purity, the light waves eventually become dim *(Continued on next page)*

**Figure 3** Interference contrast light micrograph of a fiber optics wave guide of the type used for signal transmission in all aspects of communications. This particular type of fiber is known as a monomode. The faint impression of the innermost core of the fiber is visible as a darker shade of red (its refractive index is different from the surrounding material): it is along this inner core that the light signals travel. (Courtesy of Science Photo Library/Photo Researchers)

or attenuated because of absorption and scattering. *Absorption* occurs within the fiber when the light waves encounter impurities and are turned into heat. *Scattering* occurs primarily at splices or junctions in the fiber where light leaves the fiber because of imprecise connections.

Attenuation is measured in decibels per kilometer (dB/km). In most long haul circuits, the attenuation is less than 1 dB/km for signals being transmitted at a wavelength of 1310 nm. When equipment and fibers now being developed to transmit at a wavelength of 1550 nm are perfected, the losses are expected to be less than 0.25 dB/km.

Because of attenuation, the light signals must eventually be regenerated at intervals by devices called *repeaters.* A repeater is a combination of a fiber-optic receiver and a fiber-optic transmitter. The receiver decodes the signal and triggers the transmitter to produce an identical version, only now the signal has greater strength and purity. In the case of digital signals, it is also in better time synchronism. Repeaters are typically placed about 30 km apart, but in the newer systems they may be separated as much as 200 km or more. Whereas present repeaters convert the light signals to electrical signals and then back to light signals, *all-optical* repeaters being developed will convert weak signals directly to strong light signals, skipping the conversion from light signals to electric signals to light signals. By doing this, these repeaters will theoretically be much more sensitive and thus can be placed farther apart. In addition, they will be smaller, cost less, and have greater reliability.

Each fiber has three parts. At the center of the fiber is the *core,* which carries the light signal. A concentric layer of glass about 125 $\mu$m in diameter, called the *cladding,* surrounds the core. Because the cladding has a different index of refraction than the core, total internal reflection occurs in the core, keeping the light in the core. Surrounding the cladding is a polyurethane *jacket* that protects the fiber from abrasion, crushing, and chemicals. From one to several hundred fibers are grouped to form a cable.

In large-core fibers, typically with core diameter of 62.5 $\mu$m, light pulses can take numerous paths (called *modes*) as they bounce back and forth down the fiber. Because the different paths are not equal in length, some pulses will take longer to travel down the fiber, causing some of the pulses to overlap and thereby cause distortion. These *multimode* fibers, which are less expensive than other fibers, were widely used in the early days of fiber optics but now are used mainly for certain short-distance links such as local area networks.

In the newer fibers, the core is much smaller: only 8 $\mu$m in diameter, about one sixth the thickness of a sheet of paper. Because of the small diameter, only one light path is possible—straight down the core with no zigzagging. As there is only one path, there is much less distortion, giving these *single-mode* fibers a much higher bandwidth than multimode fibers.

At the end of the fiber, a photodiode converts the light signals to electric signals, which are then amplified and decoded, if necessary, to reform the signals originally transmitted.

While long-distance fiber-optic systems have received most of the attention, fiber-optic links are also useful for very short distances, such as *between* large computer mainframes and their peripheral terminals and printers. *Within* computers, fiber optics is being used to carry signals between circuit boards.

In an entirely different use of fiber optics, in a nontransmission application, optical fibers are being used as sensors to detect strain, pressure, temperature, and other stimulus. In this function, the fibers offer the advantages of compactness, sensitivity, and immunity to hostile environments, as compared to other sensors. For example, a fiber-optic sensor is being used to measure the temperature of a volcano. In another use, by embedding fiber-optic strain sensors in polymer composites, used to replace metal on some aircraft, it may be possible to give an aircraft a "smart skin" that would warn the pilot of dangerous strains in the wings or fuselage.

In one type of sensor, short lengths of optical fibers are made with intentional small lateral deformations called *microbends.* At these microbends some of the light radiates from the fiber. The behavior of this light can be influenced by temperature, acceleration, and other parameters to be sensed. In the optical interferometer type of sensor, light from a laser is transmitted down two paths: a reference path and a measuring path. The perturbation being measured will cause the light through the measuring path to be out of phase with the light through the reference path when they are recombined. This phase difference can then be converted to an amplitude change.

While most optical fibers are made of glass, *plastic* fibers serve a useful role in short-range data links, electronic billboards, automobile displays, and automobile electrical systems. Plastic fibers have excessive losses when used for distances greater than 1 km, but for short distances they have the advantages of being more flexible than glass fibers and less expensive.

## Suggested Readings

*Books*

Lacy, Edward A., *Fiber Optics,* Prentice-Hall, Inc. Englewood Cliffs, NJ, 1982 (an introductory text).

Basch, E.E. (Editor-in-chief), *Optical-Fiber Transmission;* Indianapolis, Howard W. Sams & Co., 1987 (for the more advanced student).

Chaffee, C. David, *The Rewiring of America: The Fiber Optics Revolution,* San Diego, Academic Press, CA, 1987.

*Trade Magazines*

*Lightwave,* The Journal of Fiber Optics
Laser focus/Electro-optics
Lasers and Applications

*Journals*

I.E.E.E. Spectrum
I.E.E.E. Journal of Lightwave Technology

## Essay Questions

1. How will business and society be affected if the cost of transferring information becomes insignificant?
2. Why is fiber optics useful as invasive medical probes?
3. Why would unauthorized "tapping" of optical fibers be difficult, if not impossible?
4. If economical superconductors are developed, will they displace optical fibers?
5. In addition to optical fibers, what devices had to be developed in order for lightwave communication to be practical?

## Essay Problems

1. What is the percentage of weight reduction of fiber-optic cables in comparison with copper cables?
2. If a telephone conversation requires 3 kHz, what is the theoretical maximum number of conversations that could be carried on a fiber-optic system without using special modulation techniques?
3. When fiber-optic systems start using the 1550 nm wavelength, how much further apart, relatively, can the repeaters be placed, as compared with spacing for the 1310-nm wavelength systems?

# 36

# Geometric Optics

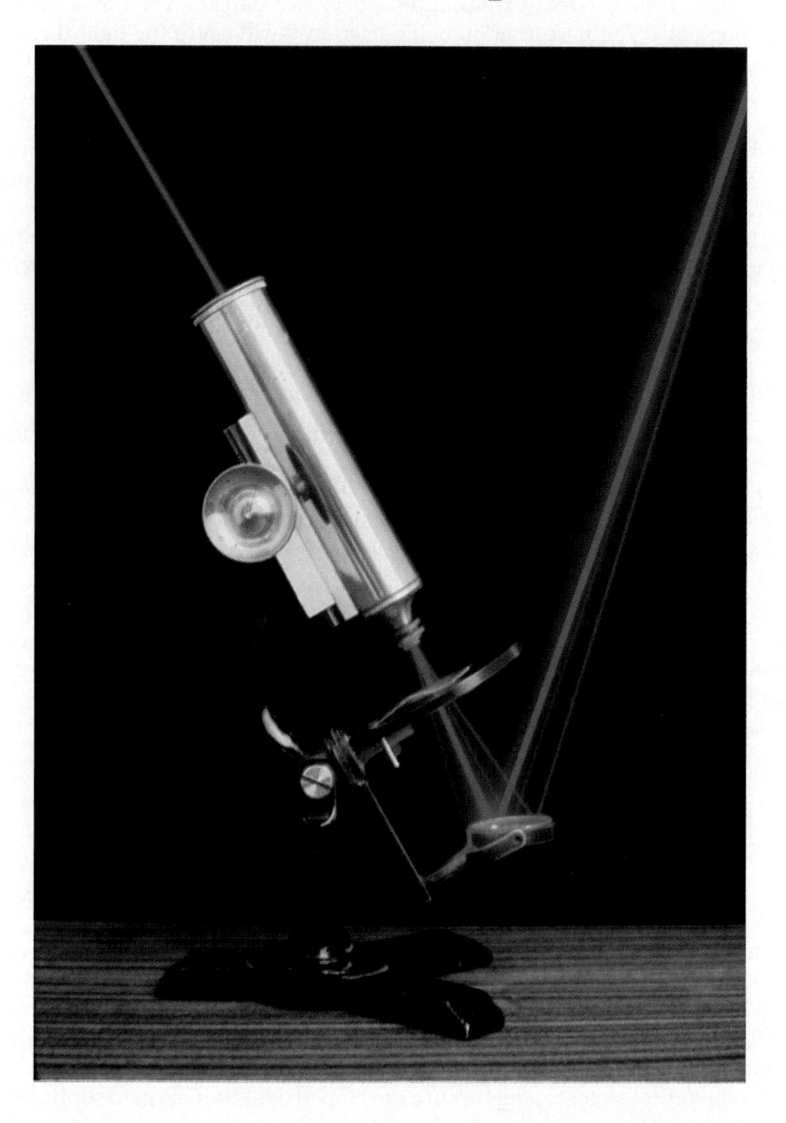

*A microscope is a common device used to view small objects with the assistance of a light source. This particular microscope is accompanied by a laser beam. (Garry Gay © THE IMAGE BANK)*

This chapter is concerned with the study of the formation of images when spherical waves fall on plane and spherical surfaces. We shall find that images can be formed by reflection or by refraction. From a practical viewpoint, mirrors and lenses are devices that work on the basis of image formation by reflection and refraction. Such devices, commonly used in optical instruments and systems, will be described in some detail. We shall continue to use the ray approximation and to assume that light travels in straight lines. This corresponds to the field of geometric optics. In subsequent chapters, we shall concern ourselves with interference and diffraction effects, or the field of wave optics.

## 36.1 IMAGES FORMED BY PLANE MIRRORS

One of the objectives of this chapter will be to discuss the manner in which optical elements such as lenses and mirrors form images. We shall begin this investigation by considering the simplest possible mirror, the plane mirror. Throughout our discussion of both mirrors and lenses, we shall use the ray model of light.

Consider a point source of light placed at $O$ in Figure 36.1, a distance $s$ in front of a plane mirror. The distance $s$ is called the **object distance.** Light rays leave the source and are reflected from the mirror. After reflection, the rays diverge (spread apart), but they appear to the viewer to come from a point $I$ located behind the mirror. Point $I$ is called the **image** of the object at $O$. Regardless of the system under study, images are always formed in the same way. *Images are formed at the point where rays of light actually intersect or at the point from which they appear to originate.* Since the rays in Figure 36.1 appear to originate at $I$, which is a distance $s'$ behind the mirror, this is the location of the image. The distance $s'$ is called the **image distance.**

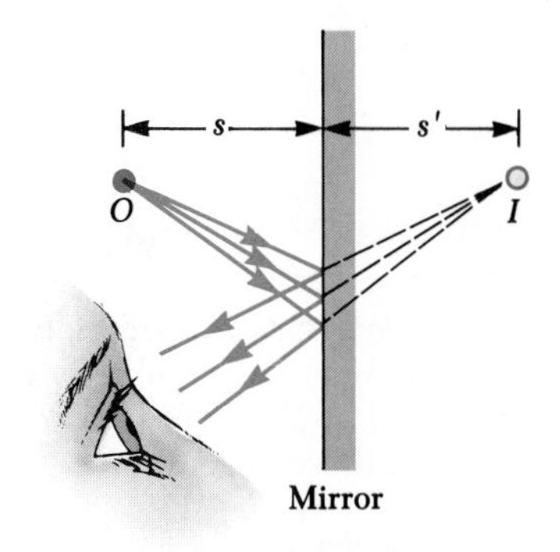


**Figure 36.1** An image formed by reflection from a plane mirror. The image point, $I$, is located behind the mirror at a distance $s'$, which is equal to the object distance, $s$.

Images are classified as real or virtual. A **real image** *is one in which light actually intersects, or passes through, the image point;* a **virtual image** *is one in which the light does not really pass through the image point but appears to diverge from that point.* The image formed by the plane mirror in Figure 36.1 is a virtual image. The images seen in plane mirrors *are always virtual* for real objects. Real images can usually be displayed on a screen (as at a movie), but virtual images cannot be displayed on a screen.

We shall examine some of the properties of the images formed by plane mirrors by using the simple geometric techniques shown in Figure 36.2. In order to find out where an image is formed, it is always necessary to follow at least two rays of light as they reflect from the mirror. One of those rays starts at $P$, follows a horizontal path to the mirror, and reflects back on itself. The second ray follows the oblique path $PR$ and reflects as shown. An observer to the left of the mirror would trace the two reflected rays back to the point from which they appear to have originated, that is, point $P'$. A continuation of this process for points on the object other than $P$ would result in a virtual image (drawn as a yellow arrow) to the right of the mirror. Since triangles $PQR$ and $P'QR$ are congruent, $PQ = P'Q$. Hence, we conclude that the *image formed by an object placed in front of a plane mirror is as far behind the mirror as the object is in front of the mirror.* Geometry also shows that the object height, $h$, equals the image height, $h'$. Let us define **lateral magnification, $M$,** as follows:

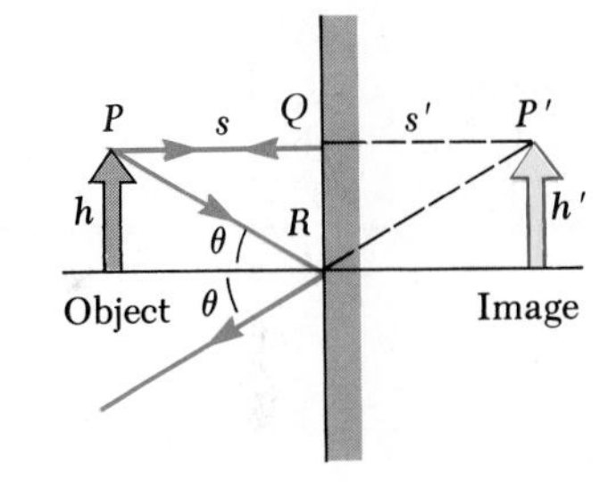


**Figure 36.2** Geometric construction used to locate the image of an object placed in front of a plane mirror. Because the triangles $PQR$ and $P'QR$ are congruent, $s = s'$ and $h = h'$.

$$M \equiv \frac{\text{image height}}{\text{object height}} = \frac{h'}{h} \qquad (36.1)$$

Magnification

This is a general definition of the lateral magnification of any type of mirror. $M = 1$ for a plane mirror because $h' = h$ in this case.

The image formed by a plane mirror has one more important property, that of right-left reversal between image and object. This reversal can be seen by standing in front of a mirror and raising your right hand. The image you see raises its left hand. Likewise, your hair appears to be parted on the opposite side and a mole on your right cheek appears to be on your left cheek.

Thus, we conclude that the image formed by a plane mirror has the following properties:

1. The image is as far behind the mirror as the object is in front.
2. The image is unmagnified, virtual, and erect. (By erect we mean that, if the object arrow points upward as in Figure 36.2, so does the image arrow.)
3. The image has right-left reversal.

**EXAMPLE 36.1 Multiple Images Formed by Two Mirrors**

Two plane mirrors are at right angles to each other, as in Figure 36.3, and an object is placed at point $O$. In this situation, multiple images are formed. Locate the positions of these images.

*Solution* The image of the object is at $I_1$, in mirror 1 and at $I_2$ in mirror 2. In addition, a third image is formed at $I_3$, which will be considered to be the image of $I_1$ in mirror 2 or, equivalently, the image of $I_2$ in mirror 1. That is, the image at $I_1$ (or $I_2$) serves as the object for $I_3$. When viewing $I_3$, note that the rays reflect twice after leaving the object at $O$.

Exercise 1 Sketch the rays corresponding to viewing the images at $I_1$ and $I_2$ and show that the light is reflected only once in these cases.

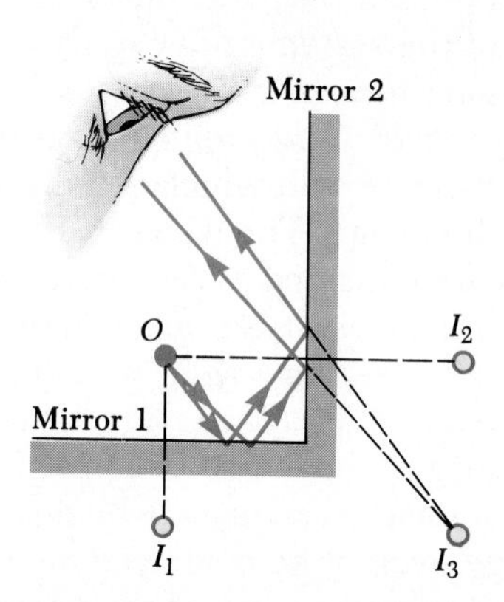


**Figure 36.3** (Example 36.1) When an object is placed in front of two mutually perpendicular mirrors as shown, three images are formed.

## 36.2 IMAGES FORMED BY SPHERICAL MIRRORS

### Concave Mirrors

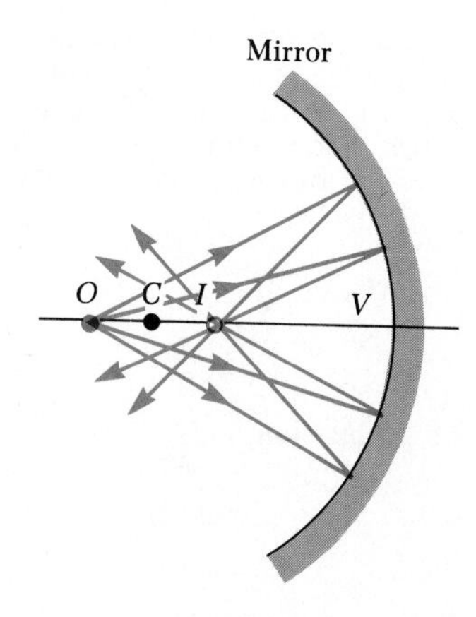


**Figure 36.4** A point object placed at $O$, outside the center of curvature of a concave spherical mirror, forms a real image at $I$. If the rays diverge from $O$ at small angles, they reflect through the same image point.

A **spherical mirror,** as its name implies, has the shape of a segment of a sphere. Figure 36.4 shows the cross section of a spherical mirror with light reflecting from its surface, represented by the solid curved line. Such a mirror, in which light is reflected from the inner, concave surface, is called a **concave mirror.** The mirror has a radius of curvature $R$, and its center of curvature is located at point $C$. Point $V$ is the center of the spherical segment, and a line drawn from $C$ to $V$ is called the **principal axis** of the mirror.

Now consider a point source of light placed at point $O$ in Figure 36.4, located on the principal axis and outside point $C$. Several diverging rays originating at $O$ are shown. After reflecting from the mirror, these rays converge and meet at $I$, called the **image point.** The rays then continue to diverge from $I$ as if there were an object there. As a result, we have a real image formed. *Real images are always formed at a point when reflected light actually passes through the point.*

In what follows, we shall assume that all rays that diverge from the object make a small angle with the principal axis. Such rays are called **paraxial rays.** All such rays reflect through the image point, as in Figure 36.4. Rays that are far from the principal axis, as in Figure 36.5, converge to other points on the principal axis, producing a blurred image. This effect, called **spherical aberra-**

**tion,** is present to some extent for any spherical mirror and will be discussed in Section 36.5.

We can use the geometry shown in Figure 36.6 to calculate the image distance, $s'$, from a knowledge of the object distance, $s$, and radius of curvature, $R$. By convention, these distances are measured from point $V$. Figure 36.6 shows two rays of light leaving the tip of the object. One of these rays passes through the center of curvature, $C$, of the mirror, hitting the mirror head on (perpendicular to the mirror surface) and reflecting back on itself. The second ray strikes the mirror at the center, point $V$, and reflects as shown, obeying the law of reflection. The image of the tip of the arrow is located at the point where these two rays intersect. From the largest triangle in Figure 36.6 we see that $\tan\theta = h/s$, while the blue-shaded triangle gives $\tan\theta = -h'/s'$. The negative sign signifies that the image is inverted, so $h'$ is negative. Thus, from Equation 36.1 and these results, we find that the magnification of the mirror is

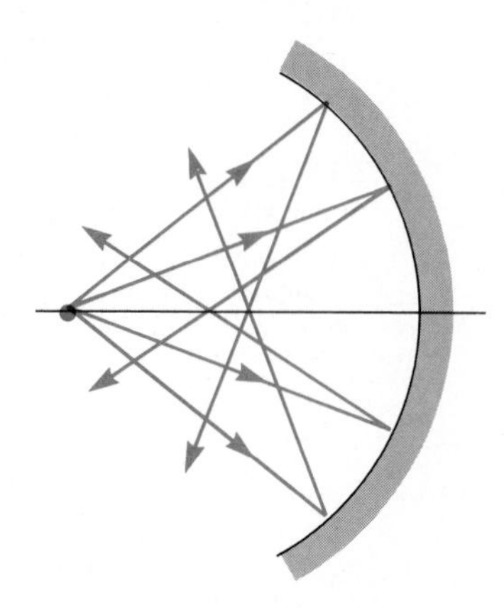

**Figure 36.5** Rays at large angles from the horizontal axis reflect from a spherical concave mirror to intersect the principal axis at different points, resulting in a blurred image. This is called *spherical aberration.*

$$M = \frac{h'}{h} = -\frac{s'}{s} \tag{36.2}$$

We also note from two other triangles in the figure that

$$\tan\alpha = \frac{h}{s-R} \quad \text{and} \quad \tan\alpha = -\frac{h'}{R-s'}$$

from which we find that

$$\frac{h'}{h} = -\frac{R-s'}{s-R} \tag{36.3}$$

If we compare Equations 36.2 and 36.3 we see that

$$\frac{R-s'}{s-R} = \frac{s'}{s}$$

Simple algebra reduces this to

$$\frac{1}{s} + \frac{1}{s'} = \frac{2}{R} \tag{36.4}$$

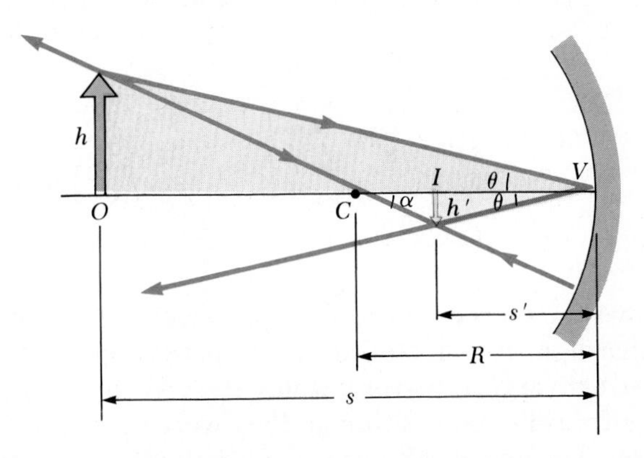


**Figure 36.6** The image formed by a spherical concave mirror where the object $O$ lies outside the center of curvature, $C$.

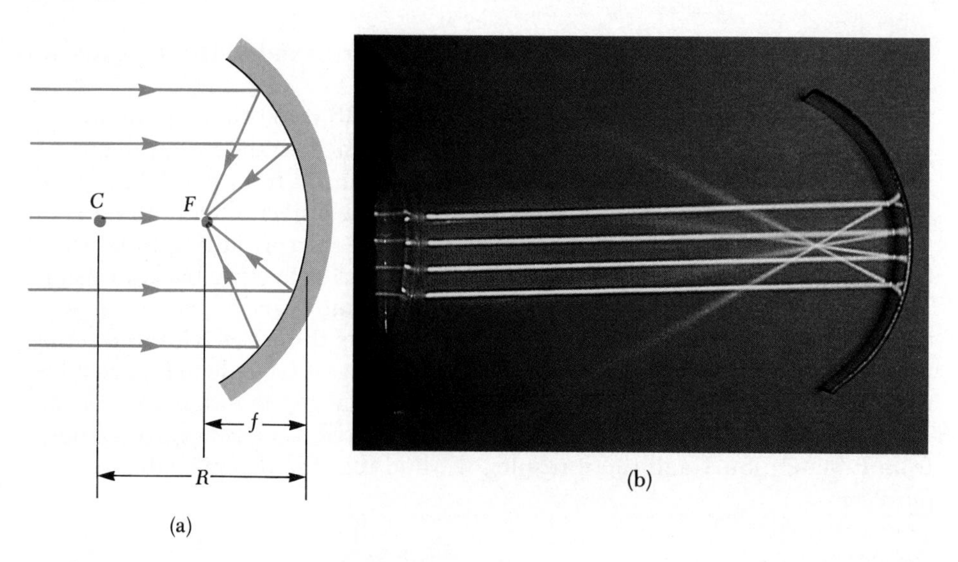


**Figure 36.7** (a) Light rays from a distant object ($s = \infty$) reflect from a concave mirror through the focal point, $F$. In this case, the image distance $s' = R/2 = f$, where $f$ is the focal length of the mirror. (b) Photograph of the reflection of parallel rays from a concave mirror. (Courtesy of Henry Leap and Jim Lehman)

This expression is called the **mirror equation.** This equation is applicable only to paraxial rays.

If the object is very far from the mirror, that is, if the object distance, $s$, is large enough compared with $R$ that $s$ can be said to approach infinity, then $1/s \approx 0$, and we see from Equation 36.4 that $s' \approx R/2$. That is, when the object is very far from the mirror, *the image point is halfway between the center of curvature and the center of the mirror*, as in Figure 36.7a. The rays are essentially parallel in this figure because the source is assumed to be very far from the mirror. We call the image point in this special case the **focal point,** $F$, and the image distance the **focal length,** $f$, where

Focal length

$$f = \frac{R}{2} \tag{36.5}$$

The mirror equation can therefore be expressed in terms of the focal length:

Mirror equation

$$\frac{1}{s} + \frac{1}{s'} = \frac{1}{f} \tag{36.6}$$

### Convex Mirrors

Figure 36.8 shows the formation of an image by a **convex mirror,** that is, one silvered such that light is reflected from the outer, convex surface. This is sometimes called a **diverging mirror** because the rays from any point on a real object diverge after reflection as though they were coming from some point behind the mirror. The image in Figure 36.8 is virtual rather than real because it lies behind the mirror at the location from which the reflected rays appear to

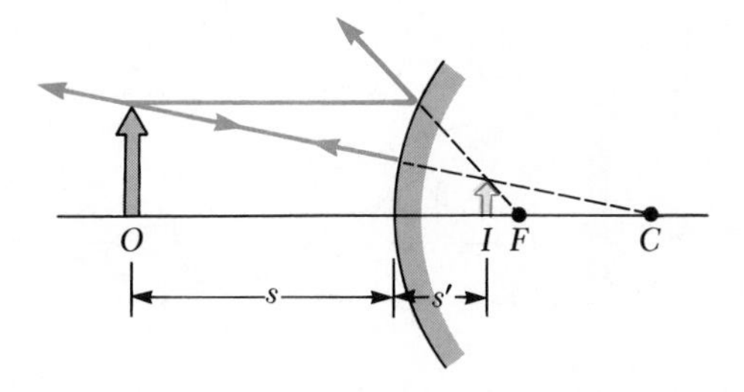


**Figure 36.8** Formation of an image by a spherical convex mirror. The image formed by the real object is virtual and erect.

originate. Furthermore, the image will always be erect, virtual, and smaller than the object, as shown in the figure.

We shall not derive any equations for convex spherical mirrors. Such derivations show that the equations developed for concave mirrors can be used if we adhere to a particular sign convention.

We can use Equations 36.2, 36.4, and 36.6 for either concave or convex mirrors if we adhere to the following procedure. Let us refer to the region in which light rays move as the *front side* of the mirror and the other side, where virtual images are formed, as the *back side.* For example, in Figures 36.6 and 36.8, the side to the left of the mirrors is the front side and the side to the right of the mirrors is the back side. Table 36.1 summarizes the sign conventions for all the necessary quantities.

### Ray Diagrams for Mirrors

The position and size of images formed by mirrors can be conveniently determined by using *ray diagrams.* These graphical constructions tell us the total nature of the image and can be used to check parameters calculated from the mirror and magnification equations. In these diagrams, one needs to know the position of the object and the location of the center of curvature. In order to locate the image, three rays are then constructed, as shown by the various examples in Figure 36.9. These rays all start from any object point (chosen to be the top) and are drawn as follows:

1. The first, labeled ray 1, is drawn from the top of the object parallel to the optical axis and is reflected back through the focal point, $F$.

**TABLE 36.1 Sign Convention for Mirrors**

| |
|---|
| $s$ is + if the object is in front of the mirror (real object).<br>$s$ is − if the object is in back of the mirror (virtual object). |
| $s'$ is + if the image is in front of the mirror (real image).<br>$s'$ is − if the image is in back of the mirror (virtual image). |
| Both $f$ and $R$ are + if the center of curvature is in front of the mirror (concave mirror).<br>Both $f$ and $R$ are − if the center of curvature is in back of the mirror (convex mirror). |
| If $M$ is positive, the image is erect.<br>If $M$ is negative, the image is inverted. |

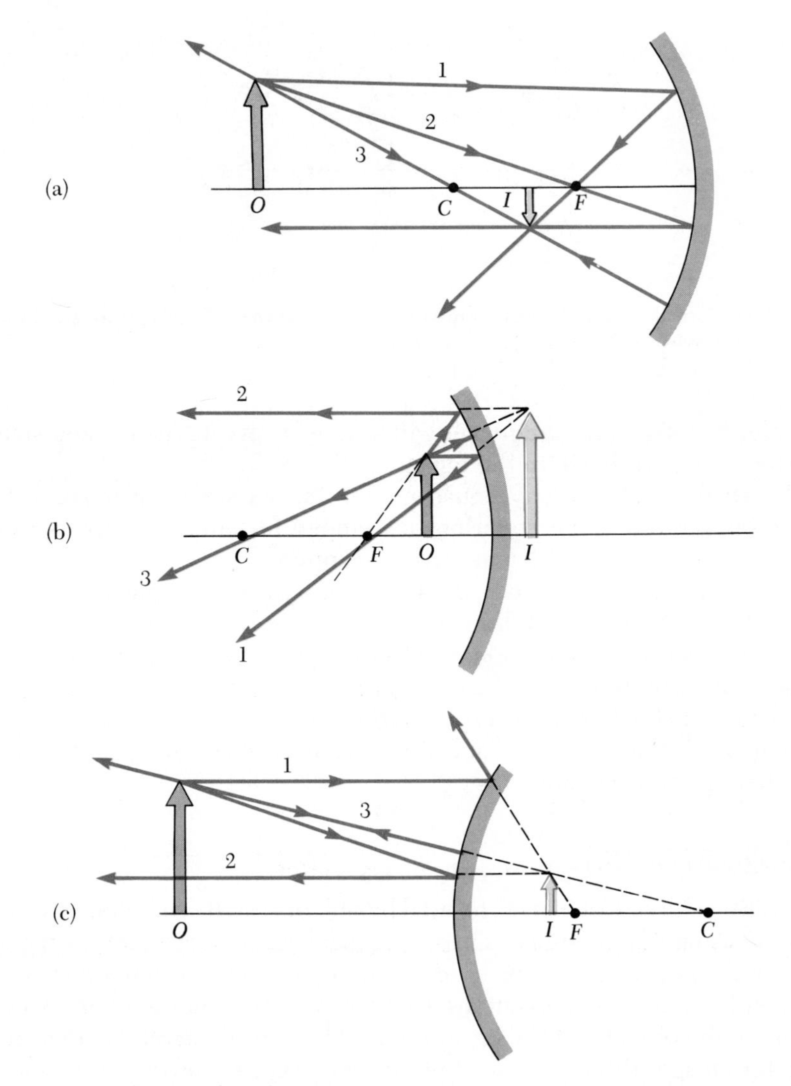


**Figure 36.9** Ray diagrams for spherical mirrors. (a) The object is located outside the center of curvature of a spherical concave mirror. (b) The object is located between the spherical concave mirror and the focal point, *F*. (c) The object is located in front of a spherical convex mirror.

2. The second, labeled ray 2, is drawn from the top of the object through the focal point *F* and is reflected back parallel to the optic axis.
3. The third, labeled ray 3, is drawn from the top of the object through the center of curvature, *C*, which is reflected back on itself.

The intersection of any *two* of these rays at a point locates the image. The third ray serves as a check on your construction. The image point obtained in this fashion must always agree with the value of $s'$ calculated from the mirror formula.

In the case of a concave mirror, note what happens as the object is moved closer to the mirror. The real, inverted image in Figure 36.9a moves to the left as the object approaches the focal point. When the object is at the focal point, the image is infinitely far to the left. However, when the object lies between the focal point and the vertex, as in Figure 36.9b, the image is virtual and erect. Finally, for the convex mirror shown in Figure 36.9c, the image of a real object is always virtual and erect. In this case, as the object distance increases, the virtual image decreases in size and approaches the focal point as $s$ approaches infinity. You should construct other diagrams to verify the variation of image position with object position.

**EXAMPLE 36.2 The Image for a Concave Mirror**
Assume that a certain concave spherical mirror has a focal length of 10 cm. Find the location of the image for object distances of (a) 25 cm, (b) 10 cm, and (c) 5 cm. Describe the image in each case.

*Solution* (a) For an object distance of 25 cm, we find the image distance using the mirror equation:

$$\frac{1}{s}+\frac{1}{s'}=\frac{1}{f}$$

$$\frac{1}{25\text{ cm}}+\frac{1}{s'}=\frac{1}{10\text{ cm}}$$

$$s' = 16.7\text{ cm}$$

The magnification is given by Equation 36.2:

$$M=-\frac{s'}{s}=-\frac{16.7\text{ cm}}{25\text{ cm}}=-0.668$$

Thus, the image is smaller than the object. Furthermore, the image is inverted because $M$ is negative. Finally, because $s'$ is positive, the image is located on the front side of the mirror and is real. This situation is pictured in Figure 36.9a.

(b) When the object distance is 10 cm, the object is located at the focal point. Substituting the values $s = 10$ cm and $f = 10$ cm into the mirror equation, we find

$$\frac{1}{10\text{ cm}}+\frac{1}{s'}=\frac{1}{10\text{ cm}}$$

$$s' = \infty$$

Thus, we see that rays of light originating from an object located at the focal point of a mirror are reflected such that the image is formed at an infinite distance from the mirror; that is, the rays travel parallel to one another after reflection.

(c) When the object is at the position $s = 5$ cm, it is inside the focal point of the mirror. In this case, the mirror equation gives

$$\frac{1}{5\text{ cm}}+\frac{1}{s'}=\frac{1}{10\text{ cm}}$$

$$s' = -10\text{ cm}$$

That is, the image is virtual since it is located behind the mirror. The magnification is

$$M=-\frac{s'}{s}=-\left(\frac{-10\text{ cm}}{5\text{ cm}}\right)=2$$

From this, we see that the image is magnified by a factor of 2, and the positive sign indicates that the image is erect (Fig. 36.9b).

Note the characteristics of the images formed by a concave spherical mirror. When the object is outside the focal point, the image is inverted and real; at the focal point, the image is formed at infinity; inside the focal point, the image is erect and virtual.

*Exercise 2* If the object distance is 20 cm, find the image distance and the magnification of the mirror.
*Answer* $s' = 20$ cm, $M = -1$.

**EXAMPLE 36.3 The Image for a Convex Mirror**
An object 3 cm high is placed 20 cm from a convex mirror having a focal length of 8 cm. Find (a) the position of the final image and (b) the magnification of the mirror.

*Solution* (a) Since the mirror is convex, its focal length is negative. To find the image position, we use the mirror equation:

$$\frac{1}{s}+\frac{1}{s'}=\frac{1}{f}=-\frac{1}{8\text{ cm}}$$

$$\frac{1}{s'}=-\frac{1}{8\text{ cm}}-\frac{1}{20\text{ cm}}$$

$$s' = -5.71\text{ cm}$$

The negative value of $s'$ indicates that the image is virtual, or behind the mirror, as in Figure 36.9c.

(b) The magnification of the mirror is

$$M=-\frac{s'}{s}=-\left(\frac{-5.71\text{ cm}}{20\text{ cm}}\right)=0.286$$

The image is erect because $M$ is positive.

*Exercise 3* Find the height of the image.
*Answer* 0.857 cm.

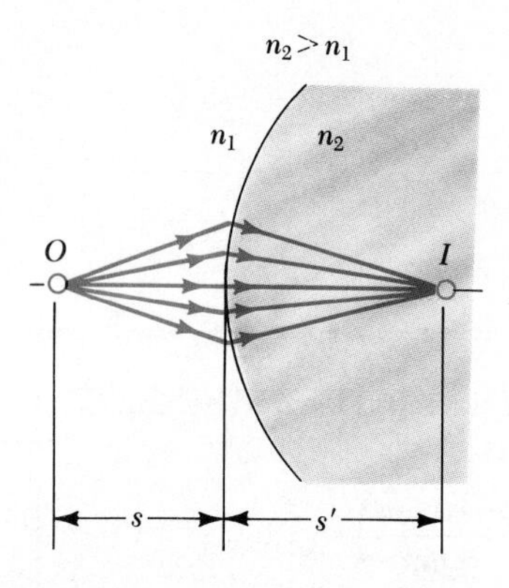


**Figure 36.10** Image formed by refraction at a spherical surface. Rays making small angles with the optic axis diverge from a point object at $O$ and pass through the image point, $I$.

## 36.3 IMAGES FORMED BY REFRACTION

In this section, we shall describe how images are formed by the refraction of rays at a spherical surface of a transparent material. Consider two transparent media with indices of refraction $n_1$ and $n_2$, where the boundary between the two media is a spherical surface of radius $R$ (Fig. 36.10). We shall assume that the object at point $O$ is in the medium whose index of refraction is $n_1$. Furthermore, let us consider paraxial rays leaving the point $O$ that make a small angle with the axis and with each other. As we shall see, all such rays originating at the object point will be refracted at the spherical surface and focus at a single point $I$, the image point.

Let us proceed by considering the geometric construction in Figure 36.11, which shows a single ray leaving point $O$ and focusing at point $I$. Snell's law applied to this refracted ray gives

$$n_1 \sin \theta_1 = n_2 \sin \theta_2$$

Because the angles $\theta_1$ and $\theta_2$ are assumed to be small, we can use the approximations $\sin \theta_1 \approx \theta_1$ and $\sin \theta_2 \approx \theta_2$ (angles in radians). Therefore, Snell's law becomes

$$n_1\theta_1 = n_2\theta_2$$

Now we make use of the fact that an exterior angle of any triangle equals the sum of the two opposite interior angles. Applying this to the triangles $OPC$ and $PIC$ in Figure 36.11 gives

$$\theta_1 = \alpha + \beta$$

$$\beta = \theta_2 + \gamma$$

If we combine the last three relations, and eliminate $\theta_1$ and $\theta_2$, we find

$$n_1\alpha + n_2\gamma = (n_2 - n_1)\beta \tag{36.7}$$

Again, in the small angle approximation, $\tan \theta \approx \theta$, so we can write the approximate relations

$$\alpha = \frac{d}{s}, \qquad \beta = \frac{d}{R}, \qquad \gamma = \frac{d}{s'}$$

where $d$ is the distance shown in Figure 36.11. We substitute these into Equation 36.7 and divide through by $d$ to give

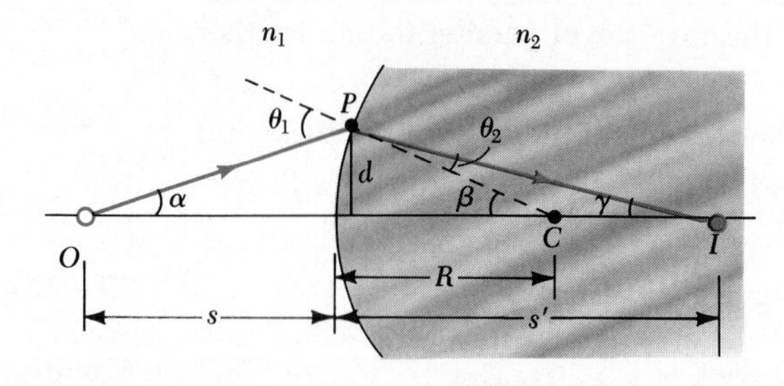


**Figure 36.11** Geometry used to derive Equation 36.8.

**TABLE 36.2 Sign Convention for Refracting Surfaces**

| |
|---|
| $s$ is + if the object is in front of the surface (real object). |
| $s$ is − if the object is in back of the surface (virtual object). |
| $s'$ is + if the image is in back of the surface (real image). |
| $s'$ is − if the image is in front of the surface (virtual image). |
| $R$ is + if the center of curvature is in back of the surface. |
| $R$ is − if the center of curvature is in front of the surface. |

$$\frac{n_1}{s} + \frac{n_2}{s'} = \frac{n_2 - n_1}{R} \tag{36.8}$$

For a fixed object distance $s$, the image distance $s'$ is independent of the angle that the ray makes with the axis. This tells us that all paraxial rays focus at the same point $I$.

As was the case for mirrors, we must use a sign convention if we are to apply this equation to a variety of circumstances. First note that real images are formed on the side of the surface that is *opposite* the side from which the light comes, in contrast to mirrors, where real images are formed on the *same* side of the reflecting surface. Therefore, *the sign convention for spherical refracting surfaces is similar to the convention for mirrors, recognizing the change in sides of the surface for real and virtual images.* For example, in Figure 36.11, $s$, $s'$, and $R$ are all positive.

The sign convention for spherical refracting surfaces is summarized in Table 36.2. The same sign convention will be used for thin lenses, which will be discussed in the next section. As with mirrors, we assume that the front of the refracting surface is the side from which the light approaches the surface.

### Plane Refracting Surfaces

If the refracting surface is a plane, then $R$ approaches infinity and Equation 36.8 reduces to

$$\frac{n_1}{s} = -\frac{n_2}{s'}$$

or

$$s' = -\frac{n_2}{n_1} s \tag{36.9}$$

The ratio $n_2/n_1$ represents the index of refraction of medium 2 *relative* to that of medium 1. From Equation 36.9 we see that the sign of $s'$ is opposite that of $s$. Thus, *the image formed by a plane refracting surface is on the same side of the surface as the object.* This is illustrated in Figure 36.12 for the situation in which $n_1$ is greater than $n_2$, where a virtual image is formed between the object and the surface. If $n_1$ is less than $n_2$, the image will still be virtual but will be formed to the left of the object.

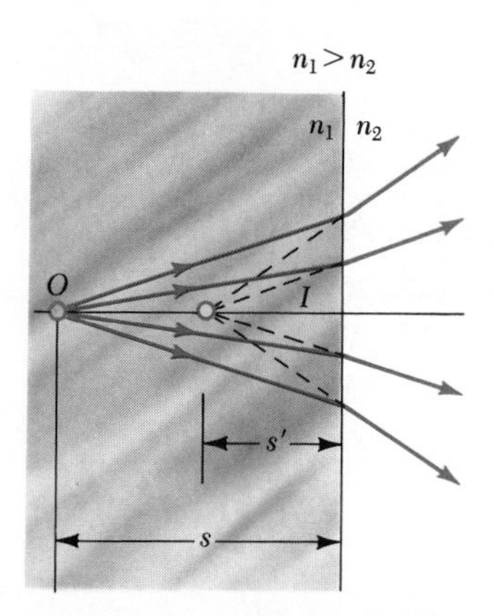


**Figure 36.12** The image formed by a plane refracting surface is virtual, that is, it forms to the left of the refracting surface. All rays are assumed to be paraxial.

**EXAMPLE 36.4 Gaze into the Crystal Ball**

A coin 2 cm in diameter is embedded in a solid glass ball of radius 30 cm (Fig. 36.13). The index of refraction of the ball is 1.5, and the coin is 20 cm from the surface. Find the position and height of the image.

*Solution* The rays originating from the object are refracted away from the normal at the surface and diverge outward. Hence, the image is formed in the glass and is virtual. Applying Equation 36.8 and taking $n_1 = 1.5$, $n_2 = 1$, $s = 20$ cm, and $R = -30$ cm, we get

$$\frac{n_1}{s} + \frac{n_2}{s'} = \frac{n_2 - n_1}{R}$$

$$\frac{1.5}{20 \text{ cm}} + \frac{1}{s'} = \frac{1 - 1.5}{-30 \text{ cm}}$$

$$s' = -17.1 \text{ cm}$$

The negative sign indicates that the image is in the same medium as the object (the side of incident light), in agreement with our ray diagram. Since the image is in the same medium as the object, it must be virtual.

Exercise 4 Find the diameter of the coin's image. Note that $M = -n_1 s'/n_2 s$.

Answer 2.58 cm.

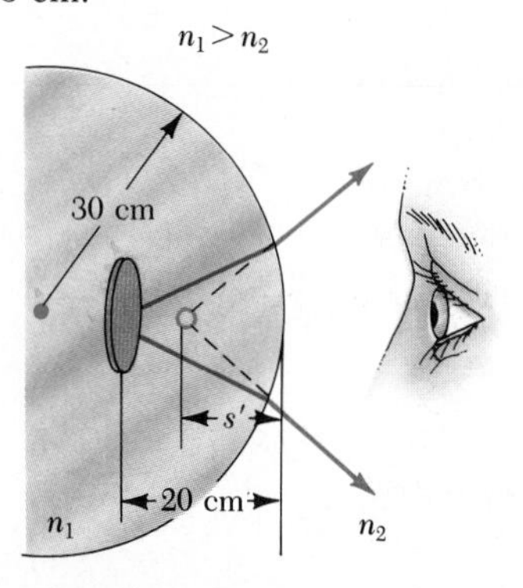


**Figure 36.13** (Example 36.4) A coin embedded in a glass ball forms a virtual image between the coin and the glass surface. All rays are assumed to be paraxial.

**EXAMPLE 36.5 The One That Got Away**

A small fish is swimming at a depth $d$ below the surface of a pond (Fig. 36.14). What is the *apparent depth* of the fish as viewed from directly overhead?

*Solution* In this example, the refracting surface is a plane, and so $R$ is infinite. Hence, we can use Equation 36.9 to determine the location of the image. Using the facts that $n_1 = 1.33$ for water and $s = d$ gives

$$s' = -\frac{n_2}{n_1} s = -\frac{1}{1.33} d = -0.750d$$

Again, since $s'$ is negative, the image is virtual, as indicated in Figure 36.14. The apparent depth is three fourths the actual depth. For instance, if $d = 4$ m, then $s' = -3$ m.

Exercise 5 If the fish is 10 cm long, how long is its image? Note that $M = -n_1 s'/n_2 s$.

Answer 10 cm.

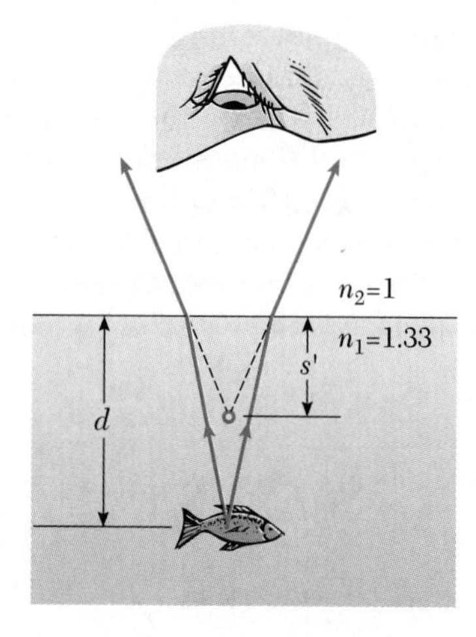


**Figure 36.14** (Example 36.5) The apparent depth, $s'$, of the fish is less than the true depth, $d$. All rays are assumed to be paraxial.

## 36.4 THIN LENSES

Lenses are commonly used to form images by refraction in optical instruments, such as cameras, telescopes, and microscopes. The methods discussed in the previous section will be used here to locate the image position. The essential idea in locating the final image of a lens is to *use the image formed by one refracting surface as the object for the second surface.*

Consider a lens having an index of refraction $n$ and two spherical surfaces of radii of curvature $R_1$ and $R_2$, as in Figure 36.15. An object is placed at point $O$ at a distance $s_1$ in front of the first refracting surface. For this example, $s_1$ has been chosen so as to produce a virtual image $I_1$, located to the left of the

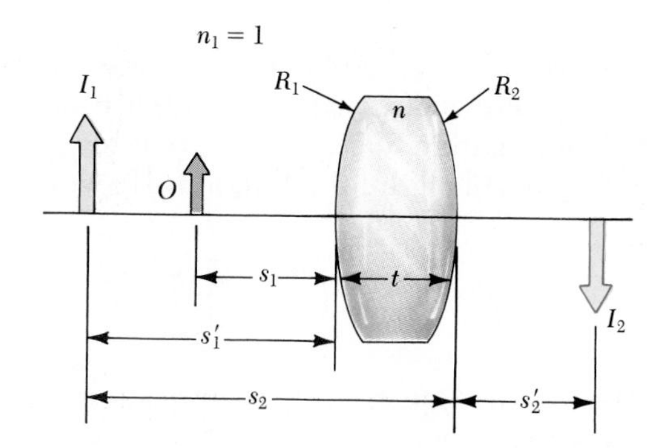


**Figure 36.15** To locate the image of a lens, the image at $I_1$ formed by the first surface is used as the object for the second surface. The final image is at $I_2$.

lens. This image is used as the object for the second surface, of radius $R_2$, which results in a real image $I_2$.

Using Equation 36.8 and assuming $n_1 = 1$, we find that the image formed by the first surface satisfies the equation

$$(1) \qquad \frac{1}{s_1} + \frac{n}{s_1'} = \frac{n-1}{R_1}$$

Now we apply Equation 36.8 to the second surface, taking $n_1 = n$ and $n_2 = 1$. That is, light approaches the second refracting surface *as if* it had come from the image, $I_1$, formed by the first refracting surface. Taking $s_2$ as the object distance and $s_2'$ as the image distance for the second surface gives

$$(2) \qquad \frac{n}{s_2} + \frac{1}{s_2'} = \frac{1-n}{R_2}$$

But $s_2 = -s_1' + t$, where $t$ is the thickness of the lens. (Remember $s_1'$ is a negative number and $s_2$ must be positive by our sign convention.) For a thin lens, we can neglect $t$. In this approximation and from Figure 36.15, we see that $s_2 = -s_1'$. Hence, (2) becomes

$$(3) \qquad -\frac{n}{s_1'} + \frac{1}{s_2'} = \frac{1-n}{R_2}$$

Adding (1) and (3), we find that

$$(4) \qquad \frac{1}{s_1} + \frac{1}{s_2'} = (n-1)\left(\frac{1}{R_1} - \frac{1}{R_2}\right)$$

For the thin lens, we can omit the subscripts on $s_1$ and $s_2'$ in (4) and call the object distance $s$ and the image distance $s'$, as in Figure 36.16. Hence, we can write (4) in the form

$$\frac{1}{s} + \frac{1}{s'} = (n-1)\left(\frac{1}{R_1} - \frac{1}{R_2}\right) \qquad (36.10)$$

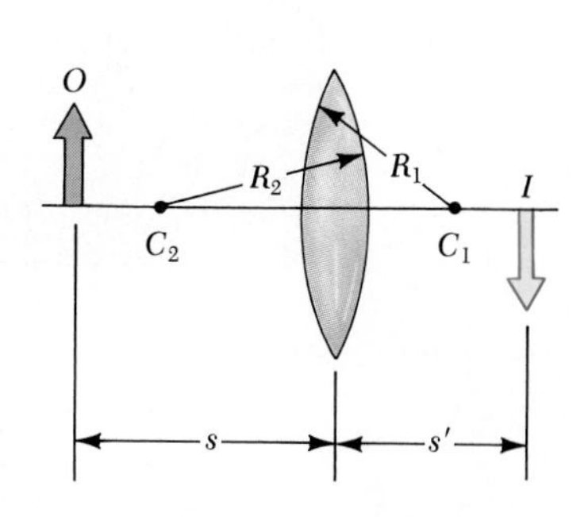


**Figure 36.16** The biconvex lens.

This expression relates the image distance $s'$ of a thin lens to the object distance $s$ and to the thin lens properties (index of refraction and radii of curvature). It is valid only for paraxial rays and only when the lens thickness is small relative to the radii $R_1$ and $R_2$.

We now define the focal length $f$ of a thin lens as the image distance that corresponds to an infinite object distance, as we did with mirrors. According to this definition and from Equation 36.10, we see that for $s \rightarrow \infty$, $f = s'$; therefore, the inverse of the focal length for a thin lens is given by

Lens makers' equation

$$\frac{1}{f} = (n - 1)\left(\frac{1}{R_1} - \frac{1}{R_2}\right) \tag{36.11}$$

Using this result, we can write Equation 36.10 in an alternate form identical to Equation 36.6 for mirrors:

$$\frac{1}{s} + \frac{1}{s'} = \frac{1}{f} \tag{36.12}$$

Equation 36.11 is called the **lens makers' equation,** since it enables one to calculate $f$ from the known properties of the lens. It can also be used to determine the values of $R_1$ and $R_2$ needed for a given index of refraction and desired focal length. A thin lens has *two* focal points, corresponding to incident parallel light rays traveling from the left or right. This is illustrated in Figure 36.17 for a biconvex lens (converging, positive $f$) and a biconcave lens (diverging, negative $f$). Focal point $F_1$ is sometimes called the *primary focal point,* and $F_2$ is called the *secondary focal point.*

Table 36.3 lists the signs of the quantities appearing in the thin lens equation. Note that the sign convention for thin lenses is the same as for refracting surfaces discussed in the previous section.

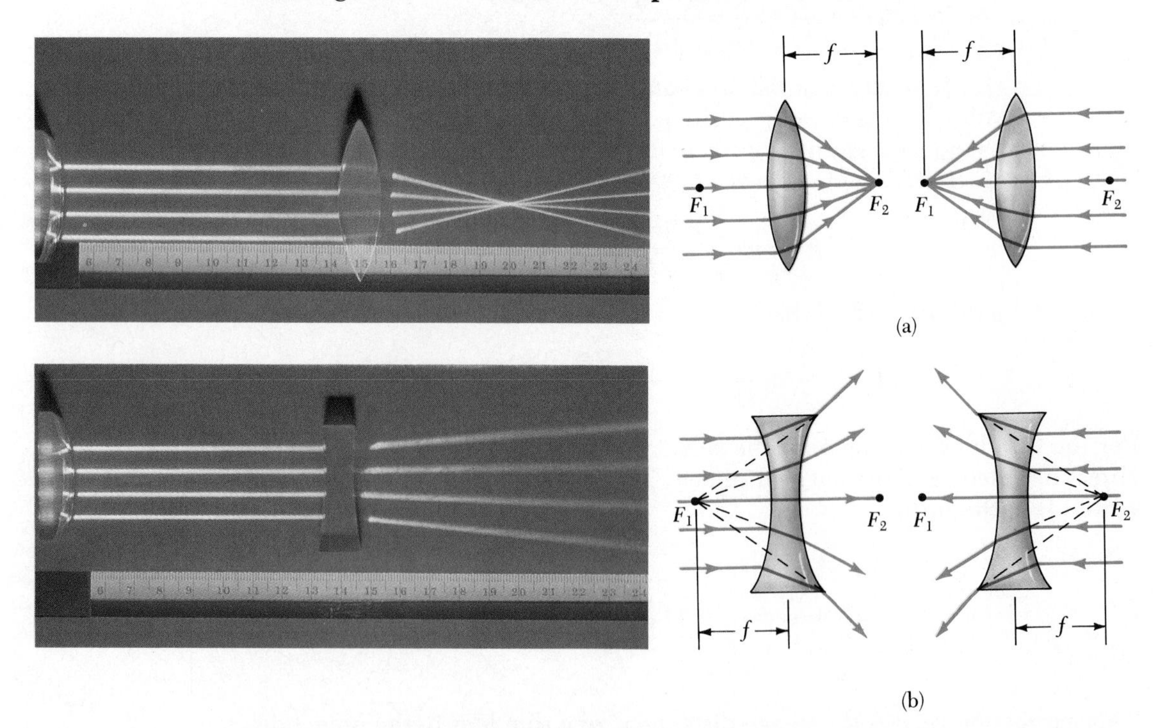


**Figure 36.17** (*Left*) Photographs of the effect of converging and diverging lenses on parallel rays. (Courtesy of Henry Leap and Jim Lehman) (*Right*) The principal focal points of (a) the biconvex lens and (b) the biconcave lens.

**TABLE 36.3 Sign Convention for Thin Lenses**

$s$ is + if the object is in front of the lens.
$s$ is − if the object is in back of the lens.

$s'$ is + if the image is in back of the lens.
$s'$ is − if the image is in front of the lens.

$R_1$ and $R_2$ are + if the center of curvature is in back of the lens.
$R_1$ and $R_2$ are − if the center of curvature is in front of the lens.

Applying these rules to the *converging* lens, we see that when $s > f$, the quantities $s$, $s'$, and $R_1$ are positive and $R_2$ is negative. Therefore, in the case of a converging lens, where a real object forms a real image, $s$, $s'$, and $f$ are all positive. Likewise, for a *diverging* lens, $s$ and $R_2$ are positive and $s'$ and $R_1$ are negative. Thus, $f$ is negative for a diverging lens.

Sketches of various lens shapes are shown in Figure 36.18. In general, note that a converging lens (positive $f$) is thicker at the center than at the edge, whereas a diverging lens (negative $f$) is thinner at the center than at the edge.

Consider a single thin lens illuminated by a *real* object, so that $s > 0$. As with mirrors, the *lateral magnification* of a thin lens is defined as the ratio of the image height $h'$ to the object height $h$. That is,

$$M = \frac{h'}{h} = -\frac{s'}{s}$$

From this expression, it follows that when $M$ is positive, the image is erect and on the same side of the lens as the object. When $M$ is negative, the image is inverted and on the side of the lens opposite the object.

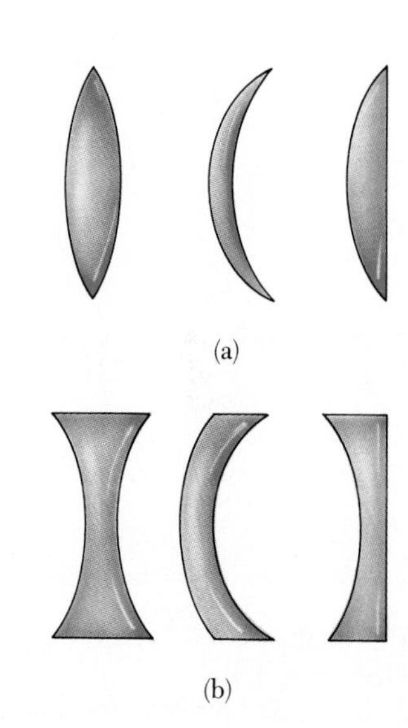


**Figure 36.18** Various lens shapes: (a) Converging lenses have a positive focal length and are thickest at the middle. (b) Diverging lenses have a negative focal length and are thickest at the edges.

### Ray Diagrams for Thin Lenses

Graphical methods, or ray diagrams, are very convenient for determining the image of a thin lens or a system of lenses. Such constructions should also help clarify the sign conventions that have been discussed. Figure 36.19 illustrates

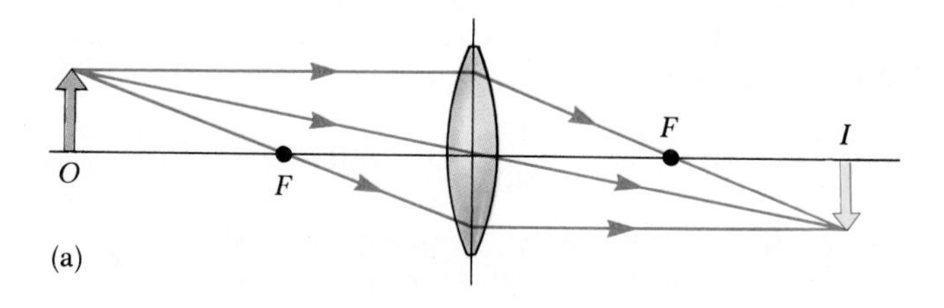


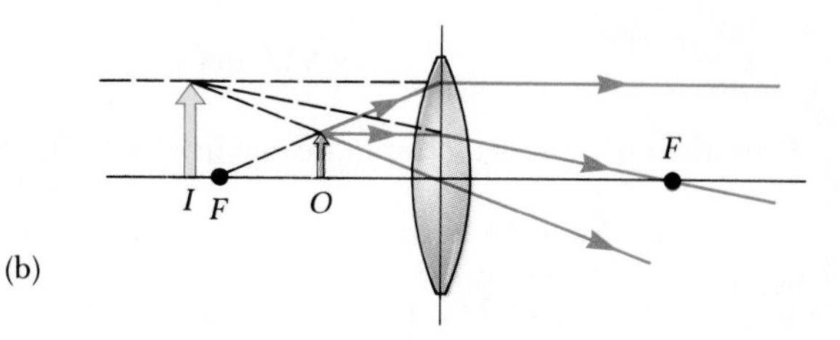


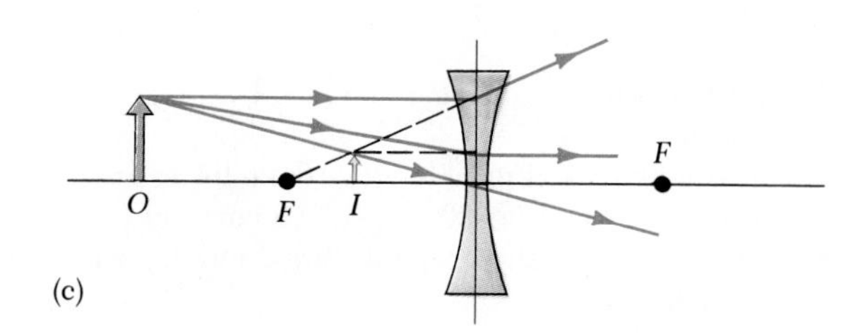


**Figure 36.19** Ray diagrams for locating the image of an object. (a) The object is located outside the focal point of a converging lens. (b) The object is located inside the focal point of a converging lens. (c) The object is located outside the focal point of a diverging lens.

this method for three different single-lens situations. To locate the image of a converging lens (Figs. 36.19a and 36.19b), the following three rays are drawn from the top of the object:

1. The first ray is drawn parallel to the optic axis. After being refracted by the lens, this ray passes through (or appears to come from) one of the focal points.
2. The second ray is drawn through the center of the lens. This ray continues in a straight line.
3. The third ray is drawn through the focal point $F$, and emerges from the lens parallel to the optic axis.

A similar construction is used to locate the image of a diverging lens, as shown in Figure 36.19c. The point of intersection of *any two* of the rays in these diagrams can be used to locate the image. The third ray serves as a check on your construction.

For the converging lens in Figure 36.19a, where the object is *outside* the front focal point ($s > f$), the image is real and inverted. On the other hand, when the real object is *inside* the front focal point ($s < f$), as in Figure 36.19b, the image is virtual, erect, and enlarged. Finally, for the diverging lens shown in Figure 36.19c, the image is always virtual and erect. These geometric constructions are reasonably accurate only if the distance between the rays and the principal axis is small compared to the radii of the lens surfaces.

**EXAMPLE 36.6 An Image Formed by a Diverging Lens**

A diverging lens has a focal length of $-20$ cm. An object 2 cm in height is placed 30 cm in front of the lens. Locate the position of the image.

*Solution* Using the thin lens equation with $s = 30$ cm and $f = -20$ cm, we get

$$\frac{1}{30\text{ cm}} + \frac{1}{s'} = -\frac{1}{20\text{ cm}}$$

$$s' = -12.0\text{ cm}$$

Thus, the image is virtual, as indicated in Figure 36.19c.

Exercise 6 Find the magnification of the lens and the height of the image.

Answer $M = 0.400$, $h' = 0.800$ cm.

**EXAMPLE 36.7 An Image Formed by a Converging Lens**

A converging lens of focal length 10 cm forms an image of an object placed (a) 30 cm, (b) 10 cm, and (c) 5 cm from the lens. Find the image distance and describe the image in each case.

*Solution* (a) The thin lens equation, Equation 36.12, can be used to find the image distance:

$$\frac{1}{s} + \frac{1}{s'} = \frac{1}{f}$$

$$\frac{1}{30\text{ cm}} + \frac{1}{s'} = \frac{1}{10\text{ cm}}$$

$$s' = 15.0\text{ cm}$$

The positive sign for the image distance tells us that the image is on the real side of the lens. The magnification of the lens is

$$M = -\frac{s'}{s} = -\frac{15\text{ cm}}{30\text{ cm}} = -0.500$$

Thus, the image is reduced in size by one half, and the negative sign for $M$ tells us that the image is inverted. The situation is like that pictured in Figure 36.19a.

(b) No calculation should be necessary for this case because we know that, when the object is placed at the focal point, the image will be formed at infinity. This is readily verified by substituting $s = 10$ cm into the lens equation.

(c) We now move inside the focal point, to an object distance of 5 cm. In this case, the lens equation gives

$$\frac{1}{5\text{ cm}} + \frac{1}{s'} = \frac{1}{10\text{ cm}}$$

$$s' = -10\text{ cm}$$

and

$$M = -\frac{s'}{s} = -\left(\frac{-10\text{ cm}}{5\text{ cm}}\right) = 2$$

The negative image distance tells us that the image is formed on the side of the lens from which the light is incident. The image is enlarged, and the positive sign for $M$ tells us that the image is erect, as shown in Figure 36.19b.

There are two general cases for a converging lens. When the real object is outside the focal point ($s > f$), the image is real, inverted, and smaller than the object. When the real object is inside the focal point ($s < f$), the image is virtual, erect, and enlarged.

**EXAMPLE 36.8 A Lens Under Water**

A converging glass lens ($n = 1.52$) has a focal length of 40 cm in air. Find its focal length when it is immersed in water, which has an index of refraction of 1.33.

*Solution* We can use the lens makers' formula (Eq. 36.11) in both cases, noting that $R_1$ and $R_2$ remain the same in air and water. In air, we have

$$\frac{1}{f_a} = (n-1)\left(\frac{1}{R_1} - \frac{1}{R_2}\right)$$

where $n = 1.52$. In water we get

$$\frac{1}{f_w} = (n'-1)\left(\frac{1}{R_1} - \frac{1}{R_2}\right)$$

where $n'$ is the index of refraction of glass *relative* to water. That is, $n' = 1.52/1.33 = 1.14$. Dividing the two equations gives

$$\frac{f_w}{f_a} = \frac{n-1}{n'-1} = \frac{1.52-1}{1.14-1} = 3.71$$

Since $f_a = 40$ cm, we find that

$$f_w = 3.71\, f_a = 3.71(40\text{ cm}) = 148\text{ cm}$$

In fact, the focal length of *any* glass lens is *increased* by the factor $(n-1)/(n'-1)$ when immersed in water.

## Combination of Thin Lenses

If two thin lenses are used to form an image, the system can be treated in the following manner. First, the image of the first lens is calculated as if the second lens were not present. The light then approaches the second lens *as if* it had come from the image formed by the first lens. Hence, the image of the first lens is treated as the object of the second lens. The image of the second lens is the final image of the system. If the image of the first lens lies to the right of the second lens, then the image is treated as a virtual object for the second lens (that is, $s$ negative). The same procedure can be extended to a system of three or more lenses. The overall magnification of a system of thin lenses equals the *product* of the magnification of the separate lenses.

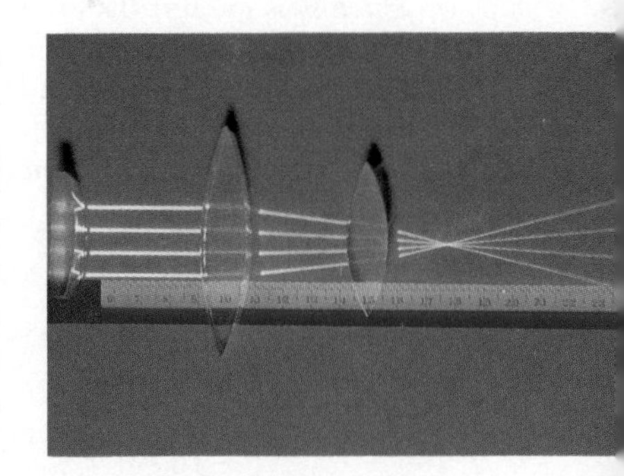

Light from a distant object brought into focus by two converging lenses. Can you estimate the overall focal length of this combination from the photo? (Courtesy of Henry Leap and Jim Lehman)

Now suppose two thin lenses of focal lengths $f_1$ and $f_2$ are placed in contact with each other. If $s$ is the object distance for the combination, then application of the thin lens equation to the first lens gives

$$\frac{1}{s} + \frac{1}{s_1'} = \frac{1}{f_1}$$

where $s_1'$ is the image distance for the first lens. Treating this image as the object for the second lens, we see that the object distance for the second lens must be $-s_1'$. Therefore, for the second lens

$$-\frac{1}{s_1'} + \frac{1}{s'} = \frac{1}{f_2}$$

where $s'$ is the final image distance from the second lens. Adding these equations eliminates $s_1'$ and gives

$$\frac{1}{s}+\frac{1}{s'}=\frac{1}{f_1}+\frac{1}{f_2}$$

Focal length of two thin lenses in contact

$$\frac{1}{f}=\frac{1}{f_1}+\frac{1}{f_2} \tag{36.13}$$

If the two *thin* lenses are in contact with one another, then $s'$ is also the distance of the final image from the first lens. Therefore, *two thin lenses in contact are equivalent to a single thin lens whose focal length is given by Equation 36.13.*

**EXAMPLE 36.9 Where is the Final Image?**
Two thin converging lenses of focal lengths 10 cm and 20 cm are separated by 20 cm, as in Figure 36.20. An object is placed 15 cm in front of the first lens. Find the position of the final image and the magnification of the system.

*Solution* First we find the image position for the first lens while neglecting the second lens:

$$\frac{1}{s_1}+\frac{1}{s_1'}=\frac{1}{15\text{ cm}}+\frac{1}{s_1'}=\frac{1}{10\text{ cm}}$$

$$s_1'=30.0\text{ cm}$$

where $s'_1$ is measured from the first lens.

Since $s_1'$ is greater than the separation between the two lenses, we see that the image of the first lens lies 10 cm to the *right* of the second lens. We take this as the object distance for the second lens. That is, we apply the thin lens equation to the second lens with $s_2=-10$ cm, where distances are now measured from the second lens, whose focal length is 20 cm:

$$\frac{1}{s_2}+\frac{1}{s_2'}=\frac{1}{f_2}$$

$$\frac{1}{-10\text{ cm}}+\frac{1}{s_2'}=\frac{1}{20\text{ cm}}$$

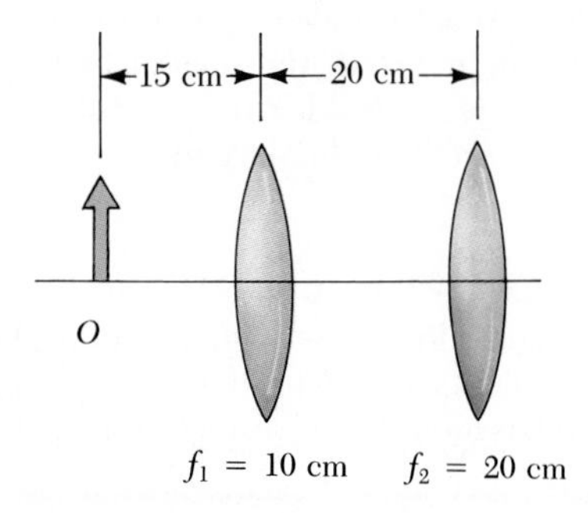


**Figure 36.20** (Example 36.9) A combination of two converging lenses.

Solving for $s_2'$ gives $s_2'=(20/3)$ cm. That is, the final image lies (20/3) cm to the *right* of the second lens.

The magnification of each lens separately is given by

$$M_1=\frac{-s_1'}{s_1}=-\frac{30\text{ cm}}{15\text{ cm}}=-2$$

$$M_2=\frac{-s_2'}{s_2}=-\frac{(20/3)\text{ cm}}{-10\text{ cm}}=\frac{2}{3}$$

The total magnification $M$ of the two lenses is the product $M_1M_2=(-2)(2/3)=-4/3$. Hence, the final image is real, inverted, and enlarged.

## *36.5 LENS ABERRATIONS

One of the basic problems of lenses and lens systems is the imperfect quality of the images. Such imperfect images are largely the result of defects in the shape and form of the lenses. The simple theory of mirrors and lenses assumes that rays make small angles with the optic axis. In this simple model, all rays leaving a point source focus at a single point, producing a sharp image. Clearly, this is not always true. For those cases where the approximations used in this theory do not hold, imperfect images are formed.

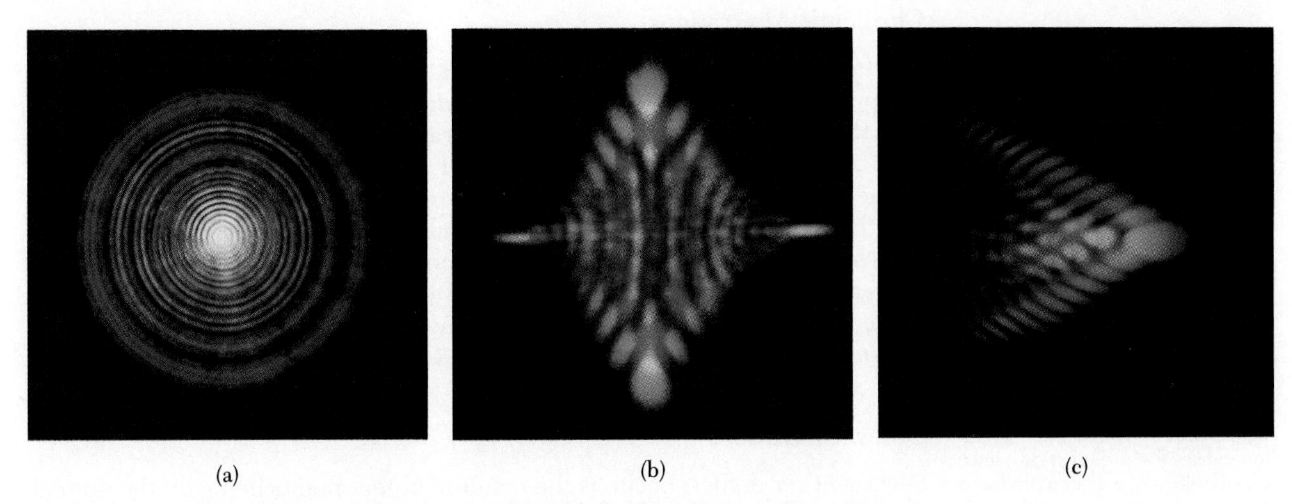


Lenses can produce various forms of aberrations, as shown by the blurred images of a point source in these photos. (a) Spherical aberration occurs when light passing through the lens at different distances from the optical axis is focused at different points. (b) Astigmatism is an aberration that occurs for objects that are not located on the optical axis of the lens. (c) Coma. This aberration occurs as light passing through the lens far from the optical axis focuses at a different part of the focal plane from light passing near the center of the lens. (Photos by Norman Goldberg)

If one wishes to perform a precise analysis of image formation, it is necessary to trace each ray using Snell's law at each refracting surface. This procedure shows that the rays from a point object do *not* focus at a single point. That is, there is no single point image; instead, the image is *blurred.* The departures of real (imperfect) images from the ideal image predicted by the simple theory are called **aberrations.** Two types of aberrations will now be described.

### Spherical Aberrations

Spherical aberrations result from the fact that the focal points of light rays far from the optic axis of a spherical lens (or mirror) are different from the focal points of rays of the same wavelength passing near the center. Figure 36.21 illustrates spherical aberration for parallel rays passing through a converging lens. Rays near the middle of the lens are imaged at greater distances from the lens than rays at the edges. Hence, there is no single focal length for a lens. Many cameras are equipped with an adjustable aperture to control the light intensity and reduce spherical aberration when possible. (An aperture is an opening that controls the amount of light transmitted through the lens.) Sharper images are produced as the aperture size is reduced, since only the central portion of the lens is exposed to the incident light. At the same time, however, less light is imaged. To compensate for this, a longer exposure time is used on the photographic film. A good example is the sharp image produced by a "pin-hole" camera, whose aperture size is approximately 1 mm.

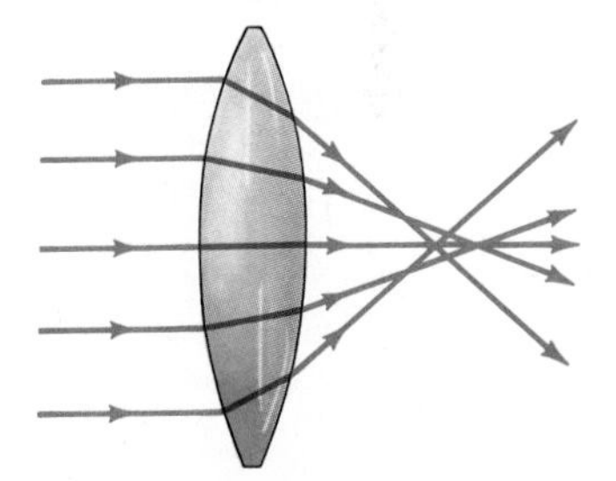

**Figure 36.21** Spherical aberration caused by a converging lens. Does a diverging lens cause spherical aberration?

In the case of mirrors used for very distant objects, one can eliminate, or at least minimize, spherical aberration by using a parabolic surface rather than a spherical surface. Parabolic surfaces are not often used, however, because those with high-quality optics are very expensive to make. Parallel light rays incident on such a surface focus at a common point. Parabolic reflecting surfaces are used in many astronomical telescopes in order to enhance the image quality. They are also used in flashlights, where a nearly parallel light beam is produced from a small lamp placed at the focus of the surface.

### Chromatic Aberrations

The fact that different wavelengths of light refracted by a lens focus at different points gives rise to *chromatic aberrations.* In Chapter 35, we described how the index of refraction of a material varies with wavelength. When white light passes through a lens, one finds, for example, that violet light rays are refracted more than red light rays (Fig. 36.22). From this we see that the focal length is larger for red light than for violet light. Other wavelengths (not shown in Fig. 36.22) would have intermediate focal points. The chromatic aberration for a diverging lens is opposite that for a converging lens. Chromatic aberration can be greatly reduced by using a combination of a converging and diverging lens made from two different types of glass.

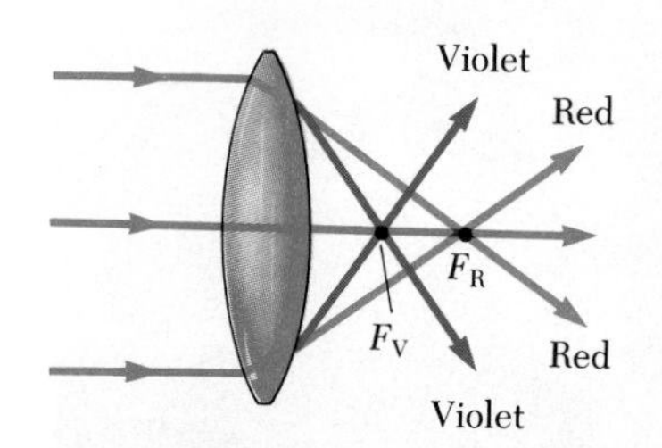


**Figure 36.22** Chromatic aberration caused by a converging lens. Rays of different wavelengths focus at different points.

### Other Aberrations

Several other defects occur as the result of object points being off the optical axis. *Astigmatism* results when a point object off the axis produces two line images at different points. A defect called *coma* is usually found in lenses with large spherical aberration. For this defect, an off-axis object produces a comet-shaped image. *Distortion* in an image exists for an extended object since magnification of off-axis points differs from magnification of those points near the axis. In high-quality optical systems, these defects are minimized by using properly designed, nonspherical surfaces or specific lens combinations.

## *36.6 THE CAMERA

The photographic **camera** is a simple optical instrument whose essential features are shown in Figure 36.23. It consists of a light-tight box, a converging lens that produces a real image, and a film behind the lens to receive the image. Focusing is accomplished by varying the distance between lens and film with an adjustable bellows in older style cameras or some other mechanical arrangement in modern cameras. For proper focusing, or sharp images, the lens-to-film distance will depend on the object distance as well as on the focal length of the lens. The shutter, located behind the lens, is a mechanical device that is opened for selected time intervals. With this arrangement, one can photograph moving objects by using short exposure times or dark scenes (low light levels) by using long exposure times. If this arrangement were not avail-

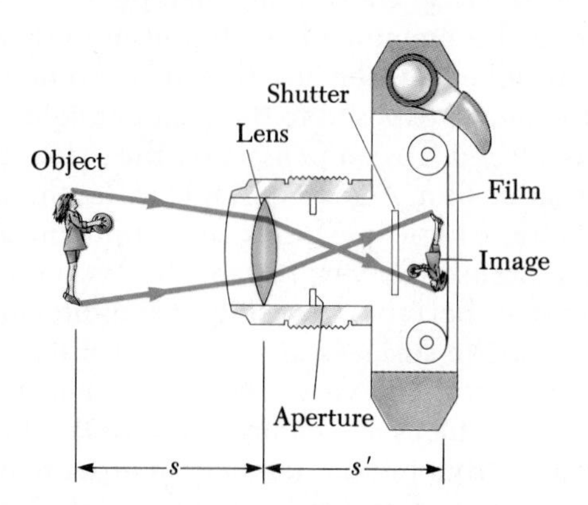


**Figure 36.23** Cross-sectional view of a simple camera. Note that in reality, $s \gg s'$.

able, it would be impossible to take stop-action photographs. For example, a rapidly moving vehicle could move enough in the time that the shutter was open to produce a blurred image. Another major cause of blurred image is the movement of the camera while the shutter is open. For this reason, you should use short exposure times, or use a tripod, even for stationary objects. Typical shutter speeds are 1/30, 1/60, 1/125, and 1/250 s. A stationary object is normally shot with a shutter speed of 1/60 s.

More expensive cameras also have an aperture of adjustable diameter either behind or in between the lenses to provide further control of the intensity of the light reaching the film. When an aperture of small diameter is used, only light from the central portion of the lens reaches the film and so the aberration is reduced somewhat.

The brightness (or energy flux) of the image focused on the film depends on the focal length of the lens and on the diameter $D$ of the lens. Clearly, the light intensity $I$ will be proportional to the area of the lens. Since the area is proportional to $D^2$, we conclude that $I \propto D^2$. Furthermore, the intensity is a measure of the energy received by the film per unit area of the image. Since the area of the image is proportional to $(s')^2$, and $s' \approx f$ (for objects with $s \gg f$), we conclude that the intensity is also proportional to $1/f^2$, so that $I \propto D^2/f^2$. The ratio $f/D$ is defined to be the *f-number* of a lens:

$$f\text{-number} \equiv \frac{f}{D} \qquad (36.14)$$

Hence, the intensity of light incident on the film can be expressed as

$$I \propto \frac{1}{(f/D)^2} \propto \frac{1}{(f\text{-number})^2} \qquad (36.15)$$

The $f$-number is a measure of the "light-concentrating" power and determines the "speed" of the lens. A "fast" lens has a small $f$-number. Fast lenses, with an $f$-number as low as about 1.4, are more expensive, because it is more difficult to keep aberrations acceptably small. Camera lenses are often marked with various $f$-numbers such as $f/2.8$, $f/4$, $f/5.6$, $f/8$, $f/11$, $f/16$. The various $f$-numbers are obtained by adjusting the aperture, which effectively changes $D$. The smallest $f$-number corresponds to the case where the aperture is wide open and the full lens area is in use. Simple cameras for routine snapshots usually have a fixed focal length and fixed aperture size, with an $f$-number of about $f/11$.

**EXAMPLE 36.10 Finding the Correct Exposure Time**

The lens of a certain 35-mm camera (where 35 mm is the width of the film strip) has a focal length of 55 mm and a speed of $f/1.8$. The correct exposure time for this speed under certain conditions is known to be (1/500) s. (a) Determine the diameter of the lens.

*Solution* From Equation 36.14, we find that

$$D = \frac{f}{f\text{-number}} = \frac{55 \text{ mm}}{1.8} = 30.6 \text{ mm}$$

(b) Calculate the correct exposure time if the $f$-number is changed to $f/4$ under the same lighting conditions.

*Solution* The total light energy received by each part of the image is proportional to the product of the flux and the exposure time. If $I$ is the light intensity reaching the film, then in a time $t$, the energy received by the film is $It$. Comparing the two situations, we require that $I_1t_1 = I_2t_2$, where $t_1$ is the correct exposure time for $f/1.8$ and $t_2$ is the correct exposure time for some other $f$-number. Using this result, together with Equation 36.15, we find that

$$\frac{t_1}{(f_1\text{-number})^2} = \frac{t_2}{(f_2\text{-number})^2}$$

$$t_2 = \left(\frac{f_2\text{-number}}{f_1\text{-number}}\right)^2 t_1 = \left(\frac{4}{1.8}\right)^2\left(\frac{1}{500}\text{ s}\right) \approx \frac{1}{100}\text{ s}$$

That is, as the aperture is reduced in size, the exposure time must increase.

## *36.7 THE EYE

The eye is an extremely complex part of the body, and because of its complexity, certain defects often arise that can cause the impairment of vision. In these cases, external aids, such as eyeglasses, are often used. In this section we shall describe the parts of the eye, their purpose, and some of the corrections that can be made when the eye does not function properly. You will find that the eye has much in common with the camera. Like the camera, a normal eye focuses light and produces a sharp image. However, the mechanisms by which the eye controls the amount of light admitted and adjusts itself to produce correctly focused images are far more complex, intricate, and effective than those in the most sophisticated camera. In all respects, the eye is an architectural wonder.

Figure 36.24 shows the essential parts of the eye. The front is covered by a transparent membrane called the *cornea.* This is followed by a clear liquid region (the *aqueous humor*), a variable aperture (the *iris* and *pupil*), and the *crystalline lens.* Most of the refraction occurs in the cornea because the liquid medium surrounding the lens has an average index of refraction close to that of the lens. The iris, which is the colored portion of the eye, is a muscular diaphragm that controls the size of the pupil. The iris regulates the amount of light entering the eye by dilating the pupil in light of low intensity and contracting the pupil in high-intensity light. The $f$-number range of the eye is from about $f/2.8$ to $f/16$.

Light entering the eye is focused by the cornea-lens system onto the back surface of the eye, called the *retina.* The surface of the retina consists of millions of sensitive structures called *rods* and *cones.* When stimulated by

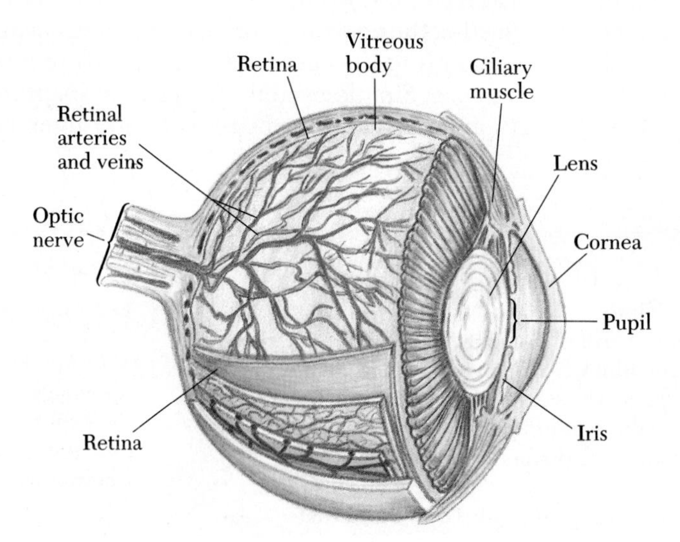


**Figure 36.24** Essential parts of the eye. Note the similarity between the eye and the simple camera. Can you correlate the parts of the eye with those of the camera?

light, these receptors send impulses via the optic nerve to the brain, where an image is perceived. By this process, a distinct image of an object is observed when the image falls on the retina.

The eye focuses on a given object by varying the shape of the pliable crystalline lens through an amazing process called **accommodation.** An important component in accommodation is the *ciliary muscle,* which is attached to the lens. When the eye is focused on distant objects, the ciliary muscle is relaxed. For an object distance of infinity, the focal length of the eye (the distance between the lens and the retina) is about 1.7 cm. The eye focuses on nearby objects by tensing the ciliary muscle. This action effectively decreases the focal length by slightly decreasing the radius of curvature of the lens, which allows the image to be focused on the retina. This lens adjustment takes place so swiftly that we are not even aware of the change. Again in this respect, even the finest electronic camera is a toy compared with the eye. It is evident that there is a limit to accommodation because objects that are very close to the eye produce blurred images.

The **near point** represents the closest distance for which the lens will produce a sharp image on the retina. This distance usually increases with age and has an average value of around 25 cm.

Typically, at age ten the near point of the eye is about 18 cm. This increases to about 25 cm at age 20, to 50 cm at age 40, and to 500 cm or greater at age 60.

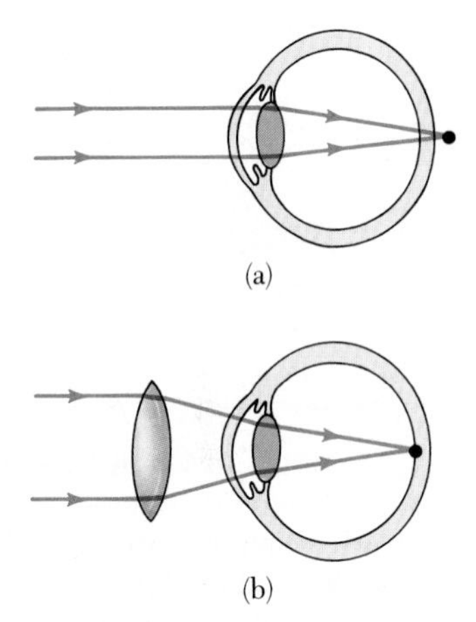

Figure 36.25 (a) A hyperopic eye (farsightedness) is slightly shorter than normal; hence the image of a distant object focuses behind the retina. (b) The condition can be corrected with a converging lens.

### Conditions of the Eye

Although the eye is one of the most remarkable organs in the body, it may have several abnormalities, which can often be corrected with eyeglasses, contact lenses, or surgery.

When the relaxed eye produces an image of a distant object *behind* the retina, as in Figure 36.25a, the condition is known as **hyperopia,** and the person is said to be farsighted. With this defect, nearby objects are blurred. Either the hyperopic eye is too short or the ciliary muscle is unable to change the shape of the lens enough to properly focus the image. The condition can be corrected with a converging lens, as shown in Figure 36.25b.

Another condition, known as **myopia,** or nearsightedness, occurs either when the eye is longer than normal or when the maximum focal length of the lens is insufficient to produce a clearly formed image on the retina. In this case, light from a distant object is focused in front of the retina (Fig. 36.26a). The distinguishing feature of this condition is that distant objects are not seen clearly. Nearsightedness can be corrected with a diverging lens, as in Figure 36.26b.

Beginning with middle age, most people lose some of their accommodation power as a result of a weakening of the ciliary muscle and a hardening of the lens. This causes an individual to become farsighted. Fortunately, the condition can be corrected with converging lenses.

A person may also have an eye defect known as **astigmatism,** in which light from a point source produces a line image on the retina. This condition arises either when the cornea or the crystalline lens or both are not perfectly spherical. Astigmatism can be corrected with lenses having different curvatures in two mutually perpendicular directions.

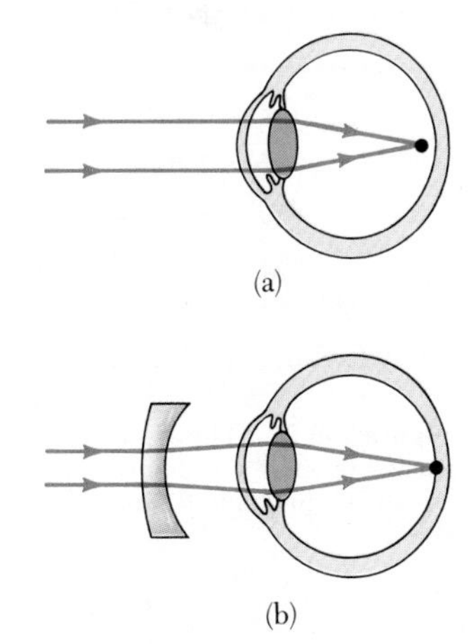

Figure 36.26 (a) A myopic eye (nearsightedness) is slightly longer than normal; hence the image of a distant object focuses in front of the retina. (b) The condition can be corrected with a diverging lens.

The eye is also subject to several diseases. One disease, which usually occurs in old age, is the formation of **cataracts,** where the lens becomes partially or totally opaque. One remedy for cataracts is surgical removal of the lens. Another disease, called **glaucoma,** arises from an abnormal increase in fluid pressure inside the eyeball. The pressure increase can lead to a swelling of the lens and to strong myopia. There is a chronic form of glaucoma in which the pressure increase causes a reduction in blood supply to the retina. This can eventually lead to blindness because the nerve fibers of the retina eventually die. If the disease is discovered early enough, it can be treated with medicine or surgery.

Optometrists and ophthalmologists usually prescribe lenses measured in **diopters.**

The **power,** $P$, of a lens in diopters equals the inverse of the focal length in meters, that is, $P = 1/f$.

For example, a converging lens whose focal length is $+20$ cm has a power of $+5$ diopters, and a diverging lens whose focal length is $-40$ cm has a power of $-2.5$ diopters.

EXAMPLE 36.11 A Case of Nearsightedness

A particular nearsighted person is unable to see objects clearly when they are beyond 50 cm (the far point of the eye). What should the focal length of the lens prescribed to correct this problem be?

*Solution* The purpose of the lens in this instance is to "move" an object from infinity to a distance where it can be seen clearly. This is accomplished by having the lens produce an image at the far point of the eye. From the thin lens equation, we have

$$\frac{1}{s} + \frac{1}{s'} = \frac{1}{\infty} - \frac{1}{50 \text{ cm}} = \frac{1}{f}$$

$$f = -50 \text{ cm}$$

Why did we use a negative sign for the image distance? As you should have suspected, the lens must be a diverging lens (negative focal length) to correct nearsightedness.

Exercise 6 What is the power of this lens?
Answer $-2$ diopters.

## *36.8 THE SIMPLE MAGNIFIER

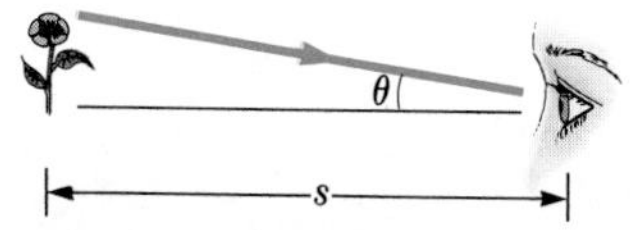

$\theta$

$s$

Figure 36.27 The size of the image formed on the retina depends on the angle $\theta$ subtended at the eye.

The simple magnifier is one of the simplest and most basic of all optical instruments because it consists of only a single converging lens. As the name implies, this device is used to increase the apparent size of an object. Suppose an object is viewed at some distance $s$ from the eye, as in Figure 36.27. Clearly, the size of the image formed at the retina depends on the angle $\theta$ subtended by the object at the eye. As the object moves closer to the eye, $\theta$ increases and a larger image is observed.[1] However, an average normal eye is unable to focus on an object closer than about 25 cm, the near point (Fig. 36.28a). Try it! Therefore, $\theta$ is maximum at the near point.

To further increase the apparent angular size of an object, a converging lens can be placed in front of the eye with the object located at point $O$, just inside the focal point of the lens, as in Figure 36.28b. At this location, the lens

[1] Regular eyeglasses give some magnification because the lenses are not located at the lens of the eye. On the other hand, contact lenses minimize this effect because of their close proximity to the lens of the eye.

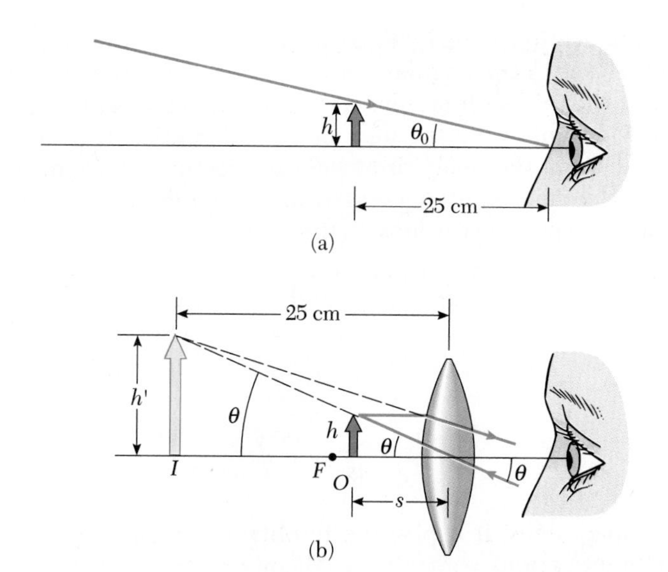


**Figure 36.28** (a) An object placed at the near point of the eye ($s = 25$ cm) subtends an angle $\theta_0$ at the eye, where $\theta_0 \approx h/25$. (b) An object placed near the focal point of a converging lens produces a magnified image, which subtends an angle $\theta \approx h'/25$ at the eye.

forms a virtual, erect, and enlarged image, as shown. Clearly, the lens increases the angular size of the object. We define the **angular magnification,** $m$, as the ratio of the angle subtended by an object with a lens in use (angle $\theta$ in Figure 36.28b) to that subtended by the object when it is placed at the near point with no lens (angle $\theta_0$ in Fig. 36.28a):

$$m \equiv \frac{\theta}{\theta_0} \tag{36.16}$$

Angular magnification

The angular magnification is a maximum when the image is at the near point of the eye, that is, when $s' = -25$ cm. The object distance corresponding to this image distance can be calculated from the thin lens formula:

$$\frac{1}{s} + \frac{1}{-25 \text{ cm}} = \frac{1}{f}$$

$$s = \frac{25f}{25 + f}$$

where $f$ is the focal length in centimeters. Let us now make the small angle approximation as follows:

$$\theta_0 \approx \frac{h}{25} \quad \text{and} \quad \theta \approx \frac{h}{s} \tag{36.17}$$

Thus, Equation 36.16 becomes

$$m = \frac{\theta}{\theta_0} = \frac{h/s}{h/25} = \frac{25}{s} = \frac{25}{25f/(25+f)}$$

$$m = 1 + \frac{25 \text{ cm}}{f} \tag{36.18}$$

The magnification given by Equation 36.18 is the ratio of the angular size seen with the lens to the angular size seen when the object is viewed at the near point of the eye with no lens. Actually, the eye can focus on an image formed anywhere between the near point and infinity. However, the eye is more relaxed when the image is at infinity (Section 36.7). In order for the image formed by the magnifying lens to appear at infinity, the object has to be placed at the focal point of the lens. In this case, the equations in 36.17 become

$$\theta_0 \approx \frac{h}{25} \quad \text{and} \quad \theta \approx \frac{h}{f}$$

and the magnification is

$$m = \frac{\theta}{\theta_0} = \frac{25 \text{ cm}}{f} \tag{36.19}$$

With a single lens, it is possible to obtain angular magnifications up to about 4 without serious aberrations. Magnifications up to about 20 can be achieved by using one or two additional lenses to correct for aberrations.

---

**EXAMPLE 36.12 Maximum Magnification of a Lens**
What is the maximum magnification of a lens having a focal length of 10 cm, and what is the magnification of this lens when the eye is relaxed?

*Solution* The maximum magnification occurs when the image formed by the lens is located at the near point of the eye. Under these circumstances, Equation 36.18 gives us the magnification as

$$m = 1 + \frac{25 \text{ cm}}{f} = 1 + \frac{25 \text{ cm}}{10 \text{ cm}} = 3.50$$

When the eye is relaxed, the image is at infinity. In this case, we use Equation 36.19:

$$m = \frac{25}{f} = \frac{25 \text{ cm}}{10 \text{ cm}} = 2.50$$

---

## *36.9 THE COMPOUND MICROSCOPE

A simple magnifier provides only limited assistance in inspecting the minute details of an object. Greater magnification can be achieved by combining two lenses in a device called a compound microscope, a schematic diagram of which is shown in Figure 36.29a. It consists of an objective lens with a very short focal length, $f_o$ (where $f_o < 1$ cm), and an ocular, or eyepiece lens, having a focal length, $f_e$, of a few centimeters. The two lenses are separated by a distance $L$, where $L$ is much greater than either $f_o$ or $f_e$. The object, which is placed just outside the focal length of the objective, forms a real, inverted image at $I_1$, which is at or close to the focal point of the eyepiece. The eyepiece, which serves as a simple magnifier, produces at $I_2$ an image of the image at $I_1$, and this image at $I_2$ is virtual and inverted. The lateral magnification, $M_1$, of the first image is $-s_1'/s_1$. Note from Figure 36.29a that $s_1'$ is approximately equal to $L$, and recall that the object is very close to the focal point of the objective; thus, $s_1 \approx f_o$. This gives a magnification for the objective of

$$M_1 \approx -\frac{L}{f_o}$$

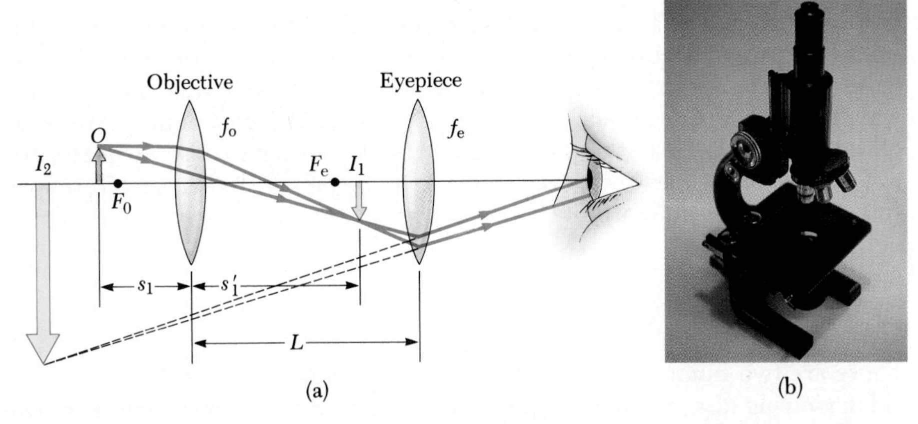


**Figure 36.29** (a) Diagram of a compound microscope, which consists of an objective lens and an eyepiece, or ocular, lens. (b) An old-fashioned compound microscope. The three-objective turret allows the user to switch to several different powers of magnification. Combinations of oculars with different focal lengths and different objectives can produce a wide range of magnifications. (Courtesy of Henry Leap and Jim Lehman)

The angular magnification of the eyepiece for an object (corresponding to the image at $I_1$) placed at the focal point is found from Equation 36.19 to be

$$m_e = \frac{25 \text{ cm}}{f_e}$$

The overall magnification of the compound microscope is defined as the product of the lateral and angular magnifications:

$$M = M_1 m_e = -\frac{L}{f_o}\left(\frac{25 \text{ cm}}{f_e}\right) \tag{36.20}$$

The negative sign indicates that the image is inverted.

The microscope has extended our vision to include the previously unknown details of incredibly small objects. The capabilities of this instrument have steadily increased with improved techniques in the precision grinding of lenses. A question that is often asked about microscopes is, "If you were extremely patient and careful, would it be possible to construct a microscope that would enable you to see an atom?" The answer to this question is no, as long as light is used to illuminate the object. The reason is that, in order to be seen, the object under a microscope must be at least as large as a wavelength of light. An atom is many times smaller than the wavelengths of visible light, and so its mysteries have to be probed using other types of "microscopes."

The wavelength dependence of the "seeing" ability of a wave can be illustrated by water waves set up in a bathtub in the following manner. Suppose you vibrate your hand in the water until waves having a wavelength of about 15 cm are moving along the surface. If you fix a small object, such as a toothpick, in the path of the waves, you will find that the waves are not disturbed appreciably by the toothpick but instead continue along their path, oblivious of the small object. Now suppose you fix a larger object, such as a toy sailboat, in the path of the waves. In this case, the waves are considerably "disturbed" by the object. In the first case, the toothpick is smaller than the wavelength of the waves, and as a result the waves do not "see" the toothpick. (The intensity of the scattered waves is low.) In the second case, the toy sailboat is about the same size as the wavelength of the waves and hence the sailboat creates a disturbance. That is, the object acts as the source of scattered waves that appear to come from it. Light waves behave in this same general way. The ability of an optical microscope to view an object depends on the size of the object relative to the wavelength of the light used to observe it. Hence, one will never be able to observe atoms or molecules with such a microscope, since their dimensions are small ($\approx 0.1$ nm) relative to the wavelength of the light ($\approx 500$ nm).

## *36.10 THE TELESCOPE

There are two fundamentally different types of **telescopes,** both designed to aid in viewing distant objects, such as the planets in our solar system. The two classifications are (1) the **refracting telescope,** which uses a combination of lenses to form an image, and (2) the **reflecting telescope,** which uses a curved mirror and a lens to form an image.

The telescope sketched in Figure 36.30a is a refracting telescope. The two lenses are arranged such that the objective forms a real, inverted image of the distant object very near the focal point of the eyepiece. Furthermore, the image at $I_1$ is formed at the focal point of the objective because the object is essentially at infinity. Hence, the two lenses are separated by a distance $f_o + f_e$, which corresponds to the length of the telescope's tube. The eyepiece finally forms, at $I_2$, an enlarged, inverted image of the image at $I_1$.

The angular magnification of the telescope is given by $\theta/\theta_0$, where $\theta_0$ is the angle subtended by the object at the objective and $\theta$ is the angle subtended by the final image. From the triangles in Figure 36.30a, and for small angles, we have

$$\theta \approx -\frac{h'}{f_e} \quad \text{and} \quad \theta_0 \approx \frac{h'}{f_o}$$

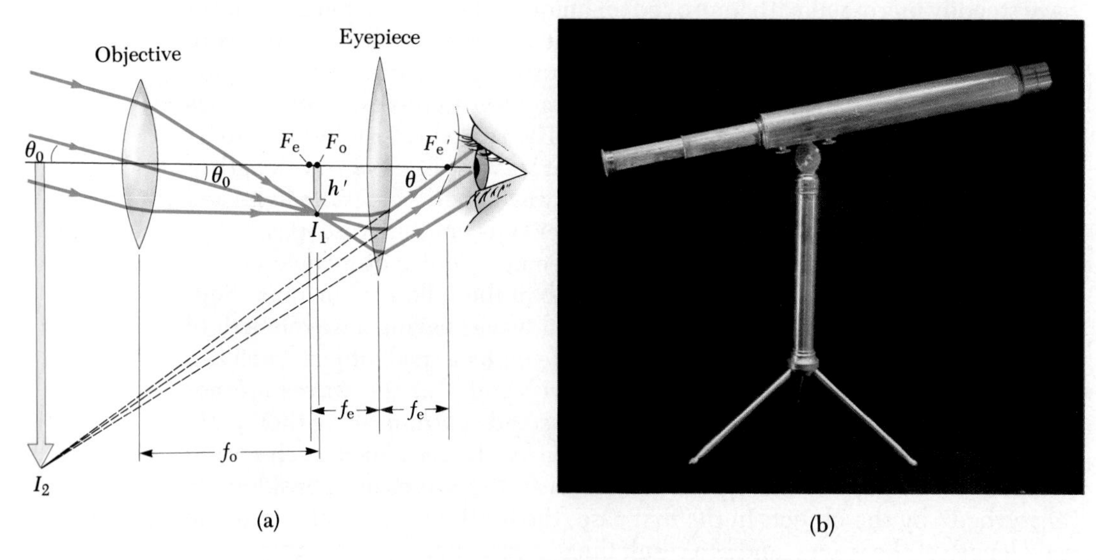


**Figure 36.30** (a) Diagram of a refracting telescope, with the object at infinity. (b) Photograph of a refracting telescope. (Photo courtesy of Henry Leap and Jim Lehman)

Hence, the angular magnification of the telescope can be expressed as

$$m = \frac{\theta}{\theta_0} = \frac{-h'/f_e}{h'/f_o} = -\frac{f_o}{f_e} \tag{36.21}$$

The minus sign indicates that the image is inverted. This says that the angular magnification of a telescope equals the ratio of the objective focal length to the eyepiece focal length. Here again, the magnification is the ratio of the angular size seen with the telescope to the angular size seen with the unaided eye.

In some applications, such as observing nearby objects like the sun, moon, or planets, magnification is important. However, stars are so far away that they always appear as small points of light regardless of how much magnification is used. Large research telescopes used to study very distant objects must have a large diameter in order to gather as much light as possible. It is difficult and expensive to manufacture large lenses for refracting telescopes. Another difficulty with large lenses is that their large weight leads to sagging, which is an additional source of aberration. These problems can be partially overcome by replacing the objective lens with a reflecting, concave mirror. Figure 36.31 shows the design for a typical reflecting telescope. Incoming light rays pass down the barrel of the telescope and are reflected by a parabolic mirror at the base. These rays converge toward point *A* in the figure, where an image would be formed. However, before this image is formed, a small flat mirror at point *M* reflects the light toward an opening in the side of the tube that passes into an eyepiece. This particular design is said to have a Newtonian focus because it was Newton who developed it. Note that the light never passes through glass in the reflecting telescope (except through the small eyepiece). As a result, problems associated with chromatic aberration are virtually eliminated.

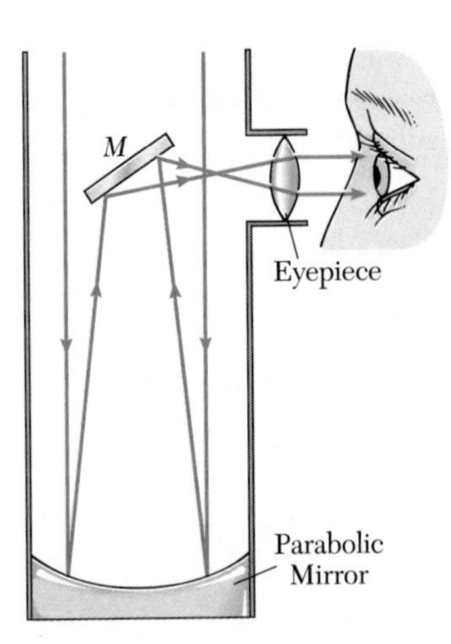


**Figure 36.31** A reflecting telescope with a Newtonian focus.

The largest telescope in the world is the 6-m-diameter reflecting telescope on Mount Pastukhov in the Caucasus, Soviet Union. The largest reflecting telescope in the United States is the 5-m-diameter instrument on Mount Palomar in California. In contrast, the largest refracting telescope in the world, which is located at the Yerkes Observatory in Williams Bay, Wisconsin, has a diameter of only 1 m.

## SUMMARY

The **magnification** *M* of a mirror or lens is defined as the ratio of the image height *h'* to the object height *h*:

Magnification of a mirror

$$M = \frac{h'}{h} = -\frac{s'}{s} \tag{36.2}$$

In the paraxial ray approximation, the object distance *s* and image distance *s'* for a spherical mirror of radius *R* are related by the **mirror equation**

Mirror equation

$$\frac{1}{s} + \frac{1}{s'} = \frac{2}{R} = \frac{1}{f} \tag{36.4, 36.6}$$

where $f = R/2$ is the **focal length** of the mirror.

An image can be formed by refraction from a spherical surface of radius $R$. The object and image distances for refraction from such a surface are related by

Formation of an image by refraction

$$\frac{n_1}{s} + \frac{n_2}{s'} = \frac{n_2 - n_1}{R} \tag{36.8}$$

where the light is incident in the medium of index of refraction $n_1$ and is refracted in the medium whose index of refraction is $n_2$.

The inverse of the **focal length** $f$ of a thin lens in air is given by

Lens makers' equation

$$\frac{1}{f} = (n - 1)\left(\frac{1}{R_1} - \frac{1}{R_2}\right) \tag{36.11}$$

**Converging lenses** have positive focal lengths, and **diverging lenses** have negative focal lengths.

For a thin lens, and in the paraxial ray approximation, the object and image distances are related by the **thin lens equation:**

Thin lens formula

$$\frac{1}{s} + \frac{1}{s'} = \frac{1}{f} \tag{36.12}$$

**Aberrations** are responsible for the formation of imperfect images by lenses and mirrors. **Spherical aberration** is due to the variation in focal points for parallel incident rays that strike the lens at various distances from the optical axis. **Chromatic aberration** arises from the fact that light of different wavelengths focuses at different points when refracted by a lens.

## QUESTIONS

1. When you look in a mirror, the image of your left and right sides is reversed, yet the image of your head and legs is not reversed. Explain.
2. Using a simple ray diagram, as in Figure 36.2, show that a plane mirror whose top is at eye level need not be as long as your height in order for you to see your entire body.
3. Consider a concave spherical mirror with a real object. Is the image always inverted? Is the image always real? Give conditions for your answers.
4. Repeat the previous question for a convex spherical mirror.
5. It is well known that distant objects viewed under water with the naked eye appear blurred and out of focus. On the other hand, the use of goggles provides the swimmer with a clear view of objects. Explain this, using the fact that the indices of refraction of the cornea, water, and air are 1.376, 1.333, and 1.00029, respectively.
6. Why does a clear stream always appear to be shallower than it actually is?
7. A person spearfishing in a boat sees a fish apparently located 3 m from the boat at a depth of 1 m. In order to hit the fish with his spear, should the person aim at the fish, above the fish, or below the fish? Explain.
8. Consider the image formed by a thin converging lens. Under what conditions will the image be (a) inverted, (b) erect, (c) real, (d) virtual, (e) larger than the object, and (f) smaller than the object?
9. Repeat Question 8 for a thin diverging lens.
10. If a cylinder of solid glass or clear plastic is placed above the words LEAD OXIDE and viewed from the side as shown in Figure 36.32, the word "LEAD" appears inverted but the word "OXIDE" does not. Explain.

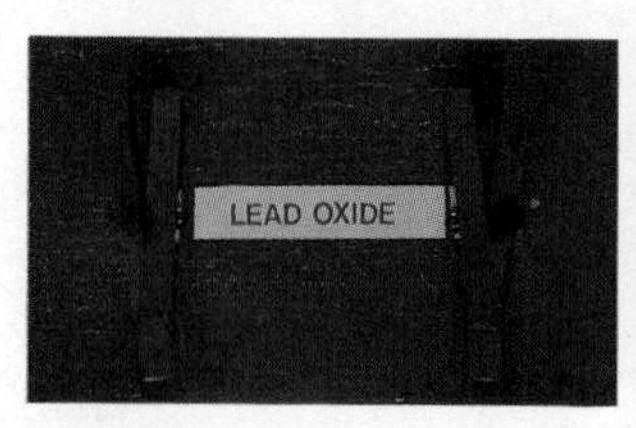

**Figure 36.32** (Question 10) (Courtesy of Henry Leap and Jim Lehman)

11. Describe two types of aberration common in a spherical lens.

12. Explain why a mirror cannot give rise to chromatic aberration.
13. What is the magnification of a plane mirror? What is its focal length?
14. Why do some emergency vehicles have the symbol ƎƆИA⅃UꓭMA written on the front?
15. Explain why a fish in a spherical goldfish bowl appears larger than it really is.
16. Lenses used in eyeglasses, whether converging or diverging, are always designed such that the middle of the lens curves away from the eye, like the center lenses of Figures 36.18a and 36.18b. Why?
17. A mirage is formed when the air gets gradually cooler as the height above the ground increases. What might happen if the air grows gradually warmer as the height is increased? This often happens over bodies of water or snow-covered ground: the effect is called looming.
18. Consider a spherical concave mirror, with the object located to the left of the mirror beyond the focal point. Using ray diagrams, show that the image of the object moves to the left as the object approaches the focal point.
19. In a Jules Verne novel, a piece of ice is shaped to form a magnifying lens to focus sunlight to start a fire. Is this possible?
20. The "$f$-number" of a camera is the focal length of the camera lens divided by its aperture (or diameter). How can the "$f$-number" of the lens be changed? How does this change the required exposure time?
21. A solar furnace can be constructed by using a concave mirror to reflect and focus sunlight into a furnace enclosure. What factors in the design of the reflecting mirror would guarantee that very high temperatures may be achieved?

## PROBLEMS

### Section 36.1 Images Formed by Plane Mirrors

1. In a physics laboratory experiment, a torque is applied to a small-diameter wire that is suspended vertically under tensile stress. It is necessary to measure accurately the small angle through which the wire turns as a consequence of the net torque. This is accomplished by attaching a small mirror to the wire and reflecting a beam of light off the mirror and onto a circular scale. Such an arrangement is known as an *optical lever* and is shown from a top view in Figure 36.33. Show that when the mirror turns through an angle $\theta$, the reflected beam is rotated by an angle $2\theta$.

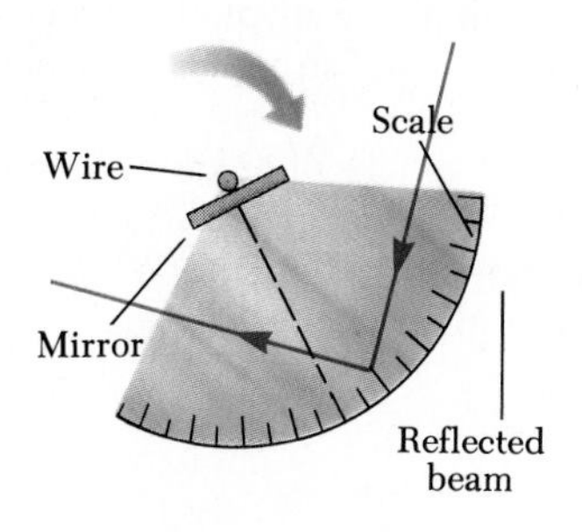


**Figure 36.33** (Problem 1).

2. Two plane mirrors, A and B, are in contact along one edge, and the planes of the two mirrors are at an angle of 45° with respect to each other (Fig. 36.34). A point object is placed at $P$ along the bisector of the angle between the two mirrors. Make a sketch similar to Figure 36.34 to a suitable scale and locate graphically (a) the image of $P$ in mirror A and the image of $P$ in mirror B. (b) Label the images found in (a) $P'_A$ and $P'_B$, respectively, and locate the image of $P'_A$ in mirror B and the image of $P'_B$ in mirror A. (c) Determine the total number of images for the arrangement described.

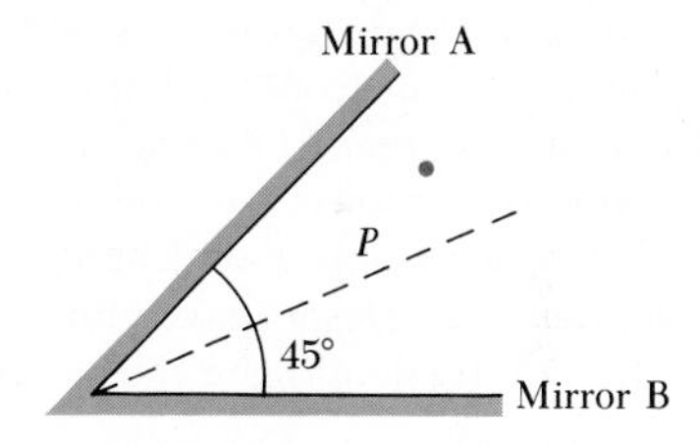


**Figure 36.34** (Problem 2).

3. Determine the minimum height of a vertical plane mirror in which a person 5′10″ in height can see his or her full image. (A ray diagram would be helpful.)

Two plane mirrors have their reflecting surfaces facing one another, with the edge of one mirror in contact with an edge of the other, so that the angle between the mirrors is $\alpha$. When an object is placed between the mirrors, a number of images are formed. In general, if the angle $\alpha$ between the two mirrors is such that $n\alpha = 360°$, where $n$ is an integer and the number of images formed is $n - 1$. Graphically, find all of the image positions for the case $n = 6$ when a point object is between the mirrors (but not on the angle bisector).

### Section 36.2 Images Formed by Spherical Mirrors

5. A concave mirror has a focal length of 40 cm. Determine the object position for which the resulting

image will be erect and four times the size of the object.

6. A convex mirror has a focal length of $-20$ cm. Determine an object location for which the image will be one half the size of the object.
7. A concave mirror has a radius of curvature of 60 cm. Calculate the image position and magnification of an object placed in front of the mirror at distances of (a) 90 cm and (b) 20 cm. (c) Draw ray diagrams to obtain the image in each case.
8. The real-image height of a concave mirror is observed to be four times larger than the object height when the object is 30 cm in front of the mirror. (a) What is the radius of curvature of the mirror? (b) Use a ray diagram to locate the image position corresponding to the given object position and the radius of curvature calculated in (a).
9. Calculate the image position and magnification for an object placed (a) 20 cm and (b) 60 cm in front of a convex mirror of focal length $-40$ cm. (c) Use ray diagrams to locate image positions corresponding to the object positions in (a) and (b).
10. A candle is 49 cm in front of a convex spherical mirror, of radius of curvature 70 cm. (a) Where is the image? (b) What is the magnification?
11. An object 2 cm high is placed 10 cm in front of a mirror. What type of mirror and what radius of curvature is needed for an upright image that is 4 cm high?
12. Use a ray diagram to demonstrate that the image of a real object placed in front of a spherical mirror is always virtual and erect when $s < |f|$.
13. A spherical mirror is to be used to form an image five times the size of an object on a screen located 5 m from the object. (a) Describe the type of mirror required. (b) Where should the mirror be positioned relative to the object?
14. A real object is located at the zero end of a meter stick. A concave mirror located at the 100-cm end of the meter stick forms an image of the object at the 70-cm position. A convex mirror placed at the 60-cm position forms a final image at the 10-cm point. What is the radius of curvature of the convex mirror?
15. A spherical convex mirror has a radius of 40 cm. Determine the position of the virtual image and magnification of the mirror for object distances of (a) 30 cm and (b) 60 cm. (c) Are the images erect or inverted?
16. An object is 15 cm from the surface of a reflective spherical Christmas-tree ornament 6 cm in diameter. What are the magnification and position of the image?

### Section 36.3 Images Formed by Refraction

17. A smooth block of ice ($n = 1.309$) rests on the floor with one face parallel to the floor. The block has a vertical thickness of 50 cm. Find the location of the image of a pattern in the floor covering as formed by rays that are nearly perpendicular to the block.
18. One end of a long glass rod ($n = 1.5$) is formed into the shape of a *convex* surface of radius 6 cm. An object is located in air along the axis of the rod. Find the image positions corresponding to the object at distances of (a) 20 cm, (b) 10 cm, and (c) 3 cm from the end of the rod.
19. Calculate the image positions corresponding to the object positions stated in Problem 18 if the end of the rod has the shape of a *concave* surface of radius 8 cm.
20. A lens made with a material of refractive index $n$ has a focal length $f$ in air. When immersed in a liquid with a refractive index $n_1$, the lens has a focal length $f'$. Derive the expression for $f'$ in terms of $f$, $n$, and $n_1$.
21. A glass sphere ($n = 1.50$) of radius 15 cm has a tiny air bubble located 5 cm from the center. The sphere is viewed along a direction parallel to the radius containing the bubble. What is the apparent depth of the bubble below the surface of the sphere?
22. A flint glass plate ($n = 1.66$) rests on the bottom of an aquarium tank. The plate is 8 cm thick (vertical dimension) and is covered with water ($n = 1.33$) to a depth of 12 cm. Calculate the apparent thickness of the plate as viewed from above the water. (Assume nearly normal incidence.)
23. A glass hemisphere is used as a paperweight with its flat face resting on a stack of papers. The radius of the circular cross section is 4 cm, and the index of refraction of the glass is 1.55. The center of the hemisphere is directly over a letter "O" that is 2.5 mm in diameter. What is the diameter of the image of the letter as seen looking along a vertical radius?
24. A goldfish is swimming in water inside a spherical plastic bowl of index of refraction 1.33. If the goldfish is 10 cm from the wall of the 15-cm-radius bowl, where does the goldfish appear to an observer outside the bowl?
25. A transparent sphere of unknown composition is observed to form an image of the sun on the surface of the sphere opposite the sun. What is the refractive index of the sphere material?

### Section 36.4 Thin Lenses

26. An object located 32 cm in front of a lens forms an image on a screen 8 cm behind the lens. (a) Find the focal length of the lens. (b) Determine the magnification. (c) Is the lens converging or diverging?
27. The left face of a biconvex lens has a radius of curvature of 12 cm, and the right face has a radius of curvature of 18 cm. The index of refraction of the glass is 1.44. (a) Calculate the focal length of the lens. (b) Calculate the focal length if the radii of curvature of the two faces are interchanged.

28. A microscope slide is placed in front of a converging lens with a focal length of 2.44 cm. The lens forms an image of the slide 12.9 cm from the slide. How far is the lens from the slide if the image is (a) real and (b) virtual?
29. What is the image distance of an object 1 m in front of a converging lens of focal length 20 cm? What is the magnification of the object? (See Figure 36.19a for a ray diagram.)
30. A person looks at a gem with a jeweler's microscope — a converging lens with a focal length of 12.5 cm. The microscope forms a virtual image 30.0 cm from the lens. Determine the magnification of the lens. Is the image upright or inverted?
31. Construct a ray diagram for the arrangement described in Problem 30.
32. A converging lens has a focal length of 40 cm. Calculate the size of the real image of an object 4 cm in height for the following object distances: (a) 50 cm, (b) 60 cm, (c) 80 cm, (d) 100 cm, (e) 200 cm, (f) $\infty$.
33. A real object is located 20 cm to the left of a diverging lens of focal length $f = -32$ cm. Determine (a) the location and (b) the magnification of the image.
34. Construct a ray diagram for the arrangement described in Problem 33.
35. A thin-walled, hollow convex lens is immersed in a tank of water. The hollow lens has $R_1 = 20$ cm and $R_2 = 30$ cm. Calculate the focal length of this "air lens" surrounded by water ($n = 1.33$). Use the derivation of Equation 36.11 as a guide.
36. A diverging lens is used to form a virtual image of a real object. The object is positioned 80 cm to the left of the lens, and the image is located 40 cm to the left of the lens. (a) Determine the focal length of the lens. (b) If the surfaces of the lens have radii of curvature of $R_1 = -40$ cm and $R_2 = -50$ cm, what is the value of the index of refraction of the lens?
37. The nickel's image in Figure 36.35 has twice the diameter of the nickel and is 2.84 cm from the lens. Determine the focal length of the magnifying lens.

Figure 36.35 (Problem 37).

### *Section 36.6 The Camera and *Section 36.7 The Eye

38. A camera is found to give proper film exposure when it is set at $f/16$ and the shutter is open for (1/32) s. Determine the correct exposure time if a setting of $f/8$ is used. (Assume the lighting conditions are unchanged.)
39. A camera is being used with correct exposure at $f/4$ and a shutter speed of (1/16) s. In order to "stop" a fast-moving subject, the shutter speed is changed to (1/128) s. Find the new $f$-number setting that should be used to maintain satisfactory exposure.
40. A 1.7-m tall woman stands 5 m in front of a camera equipped with a 50-mm focal length lens. What is the size of the image formed on the film?
41. A nearsighted person cannot see objects clearly beyond 25 cm (the far point). If the patient has no astigmatism and contact lenses are prescribed, what are the power and type of lens required to correct the patient's vision?
42. What is the unaided near point for a person required to wear lenses with a power of +1.5 diopters to read at 25 cm?
43. If the aqueous humor of the eye has an index of refraction of 1.34 and the distance from the vertex of the cornea to the retina is 2.2 cm, what is the radius of curvature of the cornea for which distant objects will be focused on the retina? (Assume all refraction occurs in the aqueous humor.)
44. Assume that the camera shown in Figure 36.23 has a fixed focal length of 6.5 cm and is adjusted to properly focus the image of a distant object. By how much and in what direction must the lens be moved in order to focus the image of an object at a distance of 2 m?
45. Figure 36.25a illustrates the case of a farsighted person who can focus clearly on objects that are more distant than 90 cm from the eye. Determine the power of lenses that will enable this person to read comfortably at a normal near point of 25 cm.
46. The eye of a nearsighted person is illustrated in Figure 36.26a. In this case, the person cannot focus clearly on objects that are more distant than 200 cm from the eye. Determine the power of lenses that will enable this person to see distant objects clearly.

### *Section 36.8 The Simple Magnifier, *Section 36.9 The Compound Microscope, and *Section 36.10 The Telescope

47. A philatelist examines the printing detail on a stamp using a convex lens of focal length 10 cm as a simple magnifier. The lens is held close to the eye, and the lens-to-object distance is adjusted so that the virtual image is formed at the normal near point (25 cm). Calculate the expected magnification.

48. A lens with focal length 5 cm is used as a magnifying glass. (a) To obtain maximum magnification, where should the object be placed? (b) What is the magnification?
49. The Yerkes refracting telescope has a 1-m diameter objective lens of focal length 20 m and an eyepiece of focal length 2.5 cm. (a) Determine the magnification of the planet Mars as seen through this telescope. (b) Are the polar caps right side up, or upside down?
50. The Palomar reflecting telescope has a parabolic mirror, with an 80-m focal length. Determine the magnification achieved when an eyepiece of 2.5 cm focal length is used.
51. An astronomical telescope has an objective with focal length 75 cm and an eyepiece with focal length 4 cm. What is the magnifying power of the instrument?
52. The distance between the eyepiece and the objective lens in a certain compound microscope is 23 cm. The focal length of the eyepiece is 2.5 cm, and the focal length of the objective is 0.4 cm. What is the overall magnification of the microscope?
53. The desired overall magnification of a compound microscope is 140×. The objective alone produces a lateral magnification of 12×. Determine the required focal length of the eyepiece lens.

## ADDITIONAL PROBLEMS

54. An object placed 10 cm from a concave spherical mirror produces a real image 8 cm from the mirror. If the object is moved to a new position 20 cm from the mirror, what is the position of the image? Is the final image real or virtual?
55. An object is located in a fixed position in front of a screen. A thin lens, placed between the object and the screen, produces a sharp image on the screen when it is in either of two positions that are 10 cm apart. The sizes of images in the two situations are in the ratio 3 : 2. (a) What is the focal length of the lens? (b) What is the distance from the screen to the object?
56. Show that, for a converging lens, unit magnification ($m = -1$) results when $s = 2f$. Under what condition is $m = +1$ approached?
57. Construct a graph of magnification $m$ as a function of $s$ for (a) a converging lens and (b) a diverging lens.
58. An object is placed 15 cm to the left of a convergent lens ($f_1 = 10$ cm). A divergent lens ($f_2 = -20$ cm) is placed 15 cm on the other side of the convergent lens. Locate and describe the final image formed by the two lenses.
59. A parallel beam of light enters a glass hemisphere perpendicular to the flat face, as shown in Figure 36.36. The radius is $R = 6$ cm and the index of refraction is $n = 1.560$. Determine the point at which the beam is focused. (Assume paraxial rays.)

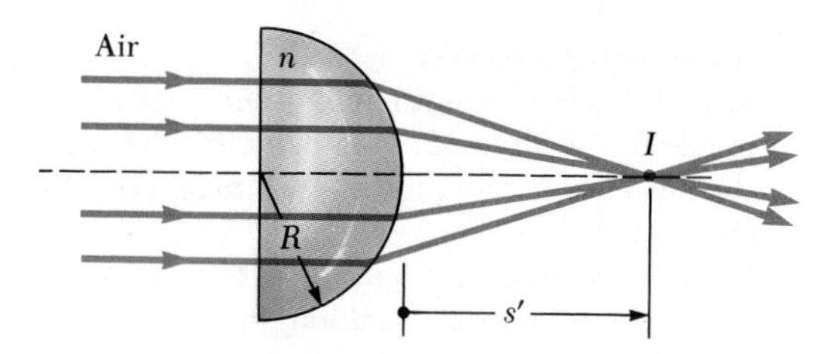


**Figure 36.36** (Problem 59).

60. A thin lens with refractive index $n$ is immersed in a liquid with index $n'$. Show that the focal length $f$ of the lens is given by

$$\frac{1}{f} = \left(\frac{n}{n'} - 1\right)\left(\frac{1}{R_1} - \frac{1}{R_2}\right)$$

61. A thin lens of focal length 20 cm lies on a horizontal front-surfaced mirror. How far above the lens should an object be held if its image is to coincide with the object?
62. Figure 36.37 shows a triangular enclosure whose inner walls are mirrors. A ray of light enters a small hole at the center of the short side. For each of the following, make a separate sketch showing the light path and find the angle $\theta$ for a ray that meets the stated conditions. (a) A light ray that is reflected once by each of the side mirrors and then exits through the hole. (b) A light ray that reflects only once and then exits. (c) Is there a path that reflects three times and then exits? If so, sketch the path and find $\theta$. (d) A ray that reflects four times and then exits.

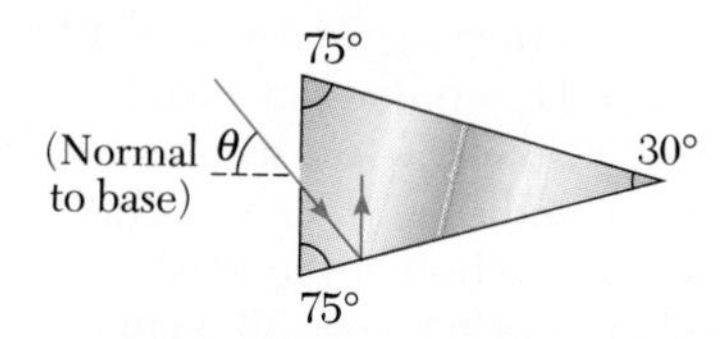


**Figure 36.37** (Problem 62).

63. A cataract-impaired lens may be surgically removed and replaced by a manufactured lens. The focal length of this "implant" is determined by the lens-to-retina distance, which is measured by a "sonar-like" device and by the requirement that the implant provide for correct distant vision (objects which are far from the eye). (a) If the distance from the lens to the retina is 22.4 mm, calculate the power of the implanted lens in diopters. (b) Since there is no accommodation and the implant allows for correct distant vision, a corrective lens for close work or reading must be used. Assume a reading distance of 33 cm and calculate the power of the lens in the reading glasses.
64. A concave mirror has a radius of 40 cm. (a) Calculate the image distance $s'$ for an arbitrary real object dis-

tance $s$. (b) Obtain values of $s'$ for object distances of 5 cm, 10 cm, 40 cm, and 60 cm. (c) Make a plot of $s'$ versus $s$ using the results of (b).

65. An object is placed 12 cm to the left of a diverging lens of focal length $-6$ cm. A converging lens of focal length 12 cm is placed a distance $d$ to the right of the diverging lens. Find the distance $d$ such that the final image is at infinity. Draw a ray diagram for this case.

66. A converging lens has a focal length of 20 cm. Find the position of the image for a real object at distances of (a) 50 cm, (b) 30 cm, (c) 10 cm. (d) Determine the magnification of the lens for these object distances and whether the image is erect or inverted. (e) Draw ray diagrams to locate the images for these object distances.

67. Find the object distances (in terms of $f$) of a thin converging lens of focal length $f$ if (a) the image is real and the image distance is four times the focal length and (b) the image is virtual and the image distance is three times the focal length. (c) Calculate the magnification of the lens for cases (a) and (b).

68. Consider light rays parallel to the principal axis approaching a concave *spherical* mirror. According to the mirror equation, rays close to the principal axis focus at $F$, a distance $R/2$ from the mirror. What about a ray farther from the principal axis? Will it be reflected to the axis at a point closer than $F$ or farther than $F$ from the mirror? Illustrate with an accurately drawn light ray that obeys the law of reflection.

69. A colored marble is dropped in a large tank filled with benzene ($n = 1.50$). (a) What is the depth of the tank if the apparent depth of the marble when viewed from directly above the tank is 35 cm? (b) If the marble has a diameter of 1.5 cm, what is its apparent diameter when viewed from directly above, outside the tank?

70. An object 1 cm in height is placed 4 cm to the left of a converging lens of focal length 8 cm. A diverging lens of focal length $-16$ cm is located 6 cm to the right of the converging lens. Find the position and size of the final image. Is the image inverted or erect? Real or virtual?

71. The disk of the sun subtends an angle of 0.5 degrees at the earth. What are the position and diameter of the solar image formed by a concave spherical mirror of radius 3 m?

72. A reflecting telescope with 2-m focal length and an eyepiece of 10-cm focal length is used to view the moon. Calculate the size of the image formed at the viewer's near point, 25 cm from the eye. (The diameter of the moon $= 3.5 \times 10^3$ km; the distance from the earth to the moon $= 3.84 \times 10^5$ km.)

73. The cornea of the eye has a radius of curvature of 0.80 cm. (a) What is the focal length of the reflecting surface of the eye? (b) If a \$20 gold piece 3.4 cm in diameter is held 25 cm from the cornea, what are the size and location of the reflected image?

74. Two converging lenses having focal lengths of 10.0 cm and 20.0 cm are located 50.0 cm apart as shown in Figure 36.38. The final image is to be located between the lenses at the position indicated. (a) How far to the left of the first lens should the object be positioned? (b) What is the overall magnification? (c) Is the final image erect or inverted?

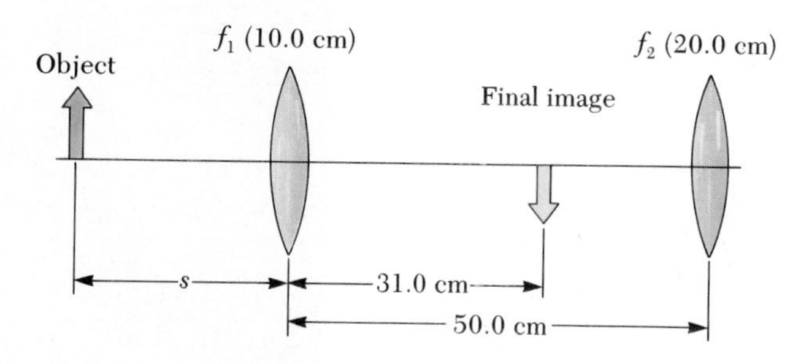


**Figure 36.38** (Problem 74).

75. In a darkened room a burning candle is placed 1.5 m from a white wall. A lens is placed between the candle and wall at a location that causes a larger, inverted image of the candle to form on the wall. When the lens is moved 90 cm toward the wall, another image of the candle is formed. Find (a) the two object distances that produce the images stated above and (b) the focal length of the lens. (c) Characterize the second image.

76. A lens and mirror have focal lengths of $+80$ cm and $-50$ cm, respectively, and the lens is located 1.0 m to the left of the mirror. An object is placed 1.0 m to the left of the lens. Locate the final image. State whether the image is erect or inverted and determine the overall magnification.

77. A "floating coin" illusion consists of two parabolic mirrors, each with a focal length 7.5 cm, facing each other so that their centers are 7.5 cm apart (Fig. 36.39). If a few coins are placed on the lower mirror, an image of the coins is formed at the small opening at the center of the top mirror. Show that the final image is formed at that location and describe its characteristics. (*Note:* a very startling effect is to shine a flashlight beam on these *images*. Even at a glancing angle, the incoming light beam is seemingly reflected off the *images* of the coins! Do you understand why?)

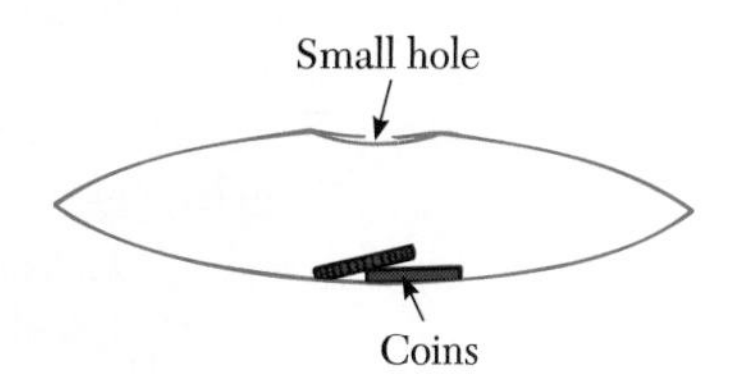


**Figure 36.39** (Problem 77).

# 37

# Interference of Light Waves

*Interference in soap bubbles. White light incident on soap bubbles (thin films of soap) forms a beautiful pattern of colors as a result of interference in the films, as described in Section 37.6. (Peter Aprahamian, Science Photo Library)*

In the previous chapter on geometric optics, we used the concept of light rays to examine what happens when light passes through a lens or reflects from a mirror. The next two chapters are concerned with the subject of *wave optics,* which deals with the phenomena of interference, diffraction, and polarization of light. These phenomena cannot be adequately explained with ray optics, but we shall describe how the wave theory of light leads to a satisfying description of such phenomena. This chapter is aimed at explaining various types of interference effects associated with light.

## 37.1 CONDITIONS FOR INTERFERENCE

In our discussion of wave interference in Chapter 18, we found that two waves could add together constructively or destructively. In constructive interference, the amplitude of the resultant wave is greater than that of either of the individual waves, while in destructive interference, the resultant amplitude is less than that of either of the individual waves. Light waves also undergo interference. Fundamentally, all interference associated with light waves arises as a result of combining the fields that constitute the individual waves.

Interference effects in light waves are not easy to observe because of the short wavelengths involved (about $4 \times 10^{-7}$ m to about $7 \times 10^{-7}$ m). In order to observe sustained interference in light waves, the following conditions should be met:

Conditions for interference

1. The sources must be **coherent,** that is, they must maintain a constant phase with respect to each other.
2. The sources must be **monochromatic,** that is, of a single wavelength.
3. The superposition principle must apply.

We shall now describe the characteristics of **coherent sources.** As we have said, two sources (producing two traveling waves) are needed to create interference. However, in order to produce a stable interference pattern, *the individual waves must maintain a constant phase relationship with one another.* When this situation prevails, the sources are said to be **coherent.** As an example, the sound waves emitted by two side-by-side loudspeakers driven by a single amplifier can produce interference because the two speakers respond to the amplifier in the same way at the same time.

Now, if two separate light sources are placed side by side, no interference effects are observed because in this case the light waves emitted by each of the sources are emitted independently of the other source; hence, their emissions do not maintain a constant phase relationship with each other over the time of observation. Light from an ordinary light source undergoes such random changes about once every $10^{-8}$ s. Therefore the conditions for constructive interference, destructive interference, or some intermediate state last for times of the order of $10^{-8}$ s. The result is that no interference effects are observed since the eye cannot follow such short-term changes. Such light sources are said to be **incoherent.**

A common method for producing two coherent light sources is to use one monochromatic source to illuminate a screen containing two small openings (usually in the shape of slits). The light emerging from the two slits is coherent because a single source produces the original light beam and the two slits serve only to separate the original beam into two parts (which, after all, is what was done to the sound signal discussed above). A random change in the light emitted by the source will occur in the two separate beams at the same time, and interference effects can still be observed.

## 37.2 YOUNG'S DOUBLE-SLIT EXPERIMENT

The phenomenon of interference in light waves from two sources was first demonstrated by Thomas Young in 1801. A schematic diagram of the apparatus used in this experiment is shown in Figure 37.1a. Light is incident on a screen, which is provided with a narrow slit $S_0$. The waves emerging from this slit arrive at a second screen, which contains two narrow, parallel slits, $S_1$ and $S_2$. These two slits serve as a pair of coherent light sources because waves emerging from them originate from the same wavefront and therefore maintain a constant phase relationship. The light from the two slits produces a visible pattern on screen C; the pattern consists of a series of bright and dark parallel bands called **fringes** (Fig. 37.1b). When the light from slits $S_1$ and $S_2$ arrives at a point on screen C such that constructive interference occurs at that location, a bright line appears. When the light from two slits combines de-

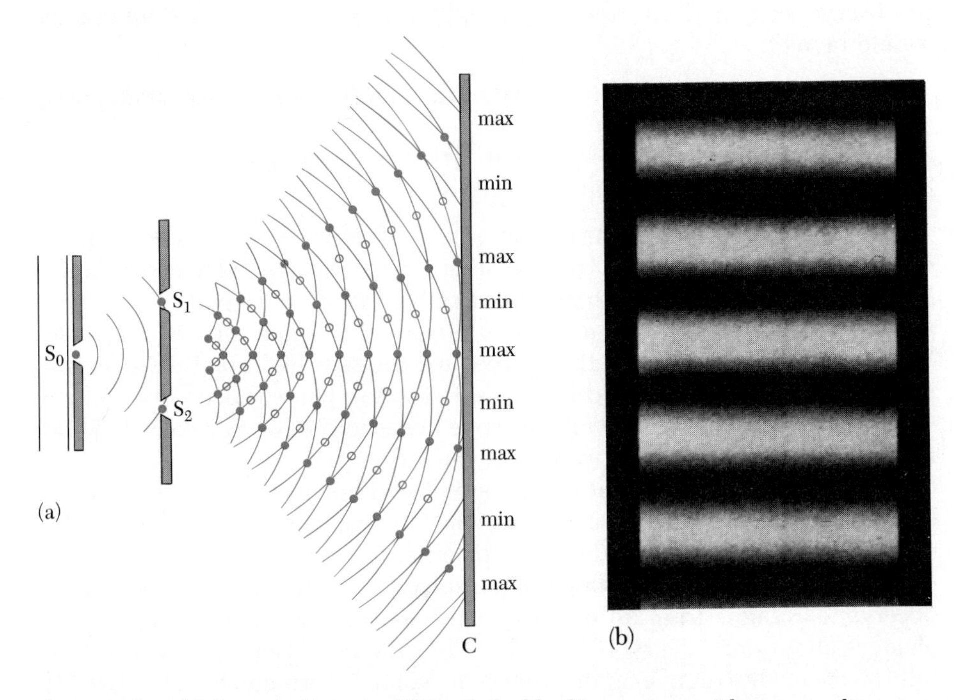


**Figure 37.1** (a) Schematic diagram of Young's double-slit experiment. The narrow slits act as sources of waves. Slits $S_1$ and $S_2$ behave as coherent sources which produce an interference pattern on screen C. (Note that this drawing is not to scale.) (b) The fringe pattern formed on screen C could look like this.

**Figure 37.2** Water waves set up by two vibrating sources produce an effect similar to Young's double-slit interference. The waves interfere constructively at X and destructively at Y.

structively at any location on the screen, a dark line results. Figure 37.2 is a photograph of an interference pattern produced by two coherent vibrating sources in a water tank.

Figure 37.3 is a schematic diagram of some of the ways the two waves can combine at the screen. In Figure 37.3a, the two waves, which leave the two slits in phase, strike the screen at the central point *P*. Since these waves travel an equal distance, they arrive in phase at *P*, and as a result constructive interference occurs at this location and a bright area is observed. In Figure 37.3b, the two light waves again start in phase, but the upper wave has to travel one wavelength farther to reach point *Q* on the screen. Since the upper wave falls behind the lower one by exactly one wavelength, they still arrive in phase at *Q*, and so a second bright light appears at this location. Now consider point *R*, midway between *P* and *Q* in Figure 37.3c. At this location, the upper wave has fallen half a wavelength behind the lower wave. This means that the trough from the bottom wave overlaps the crest from the upper wave, giving rise to destructive interference at *R*. For this reason, one observes a dark region at this location.

We can obtain a quantitative description of Young's experiment with the help of Figure 37.4. Consider a point *P* on the viewing screen; the screen is located a perpendicular distance *L* from the screen containing slits $S_1$ and $S_2$, which are separated by a distance *d*. Let us assume that the source is monochromatic. Under these conditions, the waves emerging from $S_1$ and $S_2$ have the same frequency and amplitude and are in phase. The light intensity on the screen at *P* is the resultant of the light coming from both slits. Note that a wave

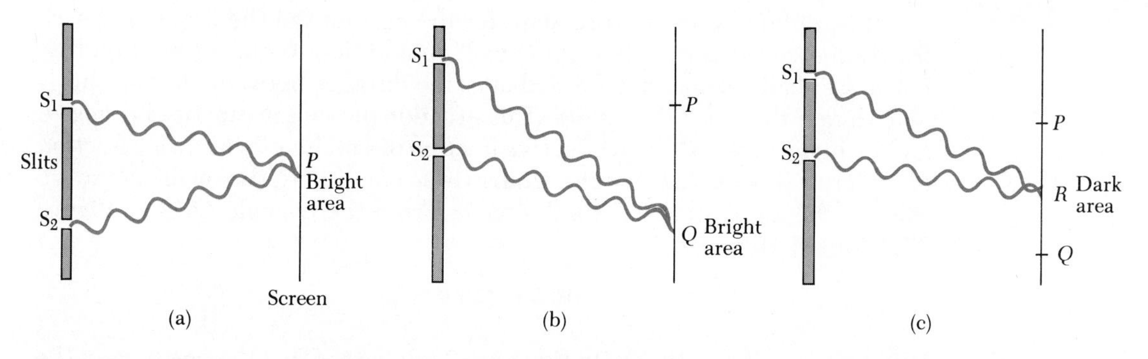


**Figure 37.3** (a) Constructive interference occurs at $P$ when the waves combine. (b) Constructive interference also occurs at $Q$. (c) Destructive interference occurs at $R$ because the wave from the upper slit falls half a wavelength behind the wave from the lower slit. (Note that these figures are not drawn to scale.)

from the lower slit travels farther than a wave from the upper slit by an amount equal to $d \sin \theta$. This distance is called the **path difference,** $\delta$, where

Path difference

$$\delta = r_2 - r_1 = d \sin \theta \qquad (37.1)$$

This equation assumes that $r_1$ and $r_2$ are parallel, which is approximately true because $L$ is much greater than $d$. As noted earlier, the value of this path difference will determine whether or not the two waves are in phase when they arrive at $P$. If the path difference is either zero or some integral multiple of the wavelength, the two waves are in phase at $P$ and constructive interference results. Therefore, the condition for bright fringes, or **constructive interference,** at $P$ is given by

Constructive interference

$$\delta = d \sin \theta = m\lambda \qquad (m = 0, \pm 1, \pm 2, \ldots) \qquad (37.2)$$

The number $m$ is called the **order number.** The central bright fringe at $\theta = 0$ ($m = 0$) is called the *zeroth-order maximum.* The first maximum on either side, when $m = \pm 1$, is called the *first-order maximum,* and so forth.

Similarly, when the path difference is an odd multiple of $\lambda/2$, the two waves arriving at $P$ will be 180° out of phase and will give rise to destructive interference. Therefore, the condition for dark fringes, or **destructive interference,** at $P$ is given by

Destructive interference

$$\delta = d \sin \theta = (m + \tfrac{1}{2})\lambda \qquad (m = 0, \pm 1, \pm 2, \ldots) \qquad (37.3)$$

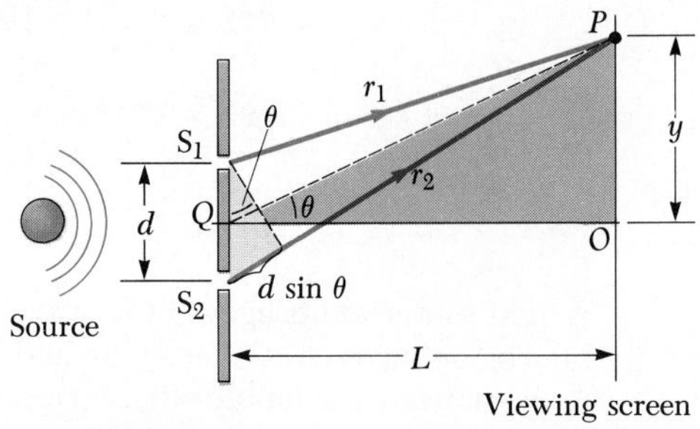

**Figure 37.4** Geometric construction for describing Young's double-slit experiment. The path difference between the two rays is $r_2 - r_1 = d \sin \theta$. For the approximation used in the text to be valid, it is essential that $L \gg d$. (Note that this figure is not drawn to scale.)

It is useful to obtain expressions for the positions of the bright and dark fringes measured vertically from $O$ to $P$. In addition to our assumption that $L \gg d$, we shall assume that $d \gg \lambda$; that is, the distance between the two slits is much larger than the wavelength. This situation prevails in practice because $L$ is often of the order of 1 m while $d$ is a fraction of a millimeter and $\lambda$ is a fraction of a micrometer for visible light. Under these conditions, $\theta$ is small, and so we can use the approximation $\sin\theta \approx \tan\theta$. From the triangle $OPQ$ in Figure 37.4, we see that

$$\sin\theta \approx \tan\theta = \frac{y}{L} \tag{37.4}$$

Using this result together with Equation 37.2, we see that the positions of the bright fringes measured from $O$ are given by

$$y_{\text{bright}} = \frac{\lambda L}{d}\, m \tag{37.5}$$

Similarly, using Equations 37.3 and 37.4, we find that the *dark fringes* are located at

$$y_{\text{dark}} = \frac{\lambda L}{d}\,(m + \tfrac{1}{2}) \tag{37.6}$$

As we shall demonstrate in Example 37.1, Young's double-slit experiment provides a method for measuring the wavelength of light. In fact, Young used this technique to make the first measurement of the wavelength of light. Additionally, the experiment gave the wave model of light a great deal of credibility. It was inconceivable that particles of light coming through the slits could cancel each other in a way that would explain the regions of darkness. Today we still use the phenomenon of interference to describe wave-like behavior in many observations.

**EXAMPLE 37.1 Measuring the Wavelength of a Light Source**

A screen is separated from a double-slit source by 1.2 m. The distance between the two slits is 0.03 mm. The second-order bright fringe ($m = 2$) is measured to be 4.5 cm from the center line. (a) Determine the wavelength of the light.

*Solution* We can use Equation 37.5, with $m = 2$, $y_2 = 4.5 \times 10^{-2}$ m, $L = 1.2$ m, and $d = 3 \times 10^{-5}$ m:

$$\lambda = \frac{dy_2}{mL} = \frac{(3 \times 10^{-5}\text{ m})(4.5 \times 10^{-2}\text{ m})}{2 \times 1.2\text{ m}}$$

$$= 5.62 \times 10^{-7}\text{ m} = 560\text{ nm}$$

(b) Calculate the distance between adjacent bright fringes.

*Solution* From Equation 37.5 and the results to (a), we get

$$y_{m+1} - y_m = \frac{\lambda L(m+1)}{d} - \frac{\lambda L m}{d}$$

$$= \frac{\lambda L}{d} = \frac{(5.62 \times 10^{-7}\text{ m})(1.2\text{ m})}{3 \times 10^{-5}\text{ m}}$$

$$= 2.25 \times 10^{-2}\text{ m} = 2.25\text{ cm}$$

**EXAMPLE 37.2 The Distance Between Bright Fringes**

A light source emits light of two wavelengths in the visible region, given by $\lambda = 430$ nm and $\lambda' = 510$ nm. The source is used in a double-slit interference experiment in which $L = 1.5$ m and $d = 0.025$ mm. Find the separation between the third-order bright fringes corresponding to these wavelengths.

*Solution* Using Equation 37.5, with $m = 3$ for the third-order bright fringes, we find that the values of the fringe positions corresponding to these two wavelengths are given by

$$y_3 = \frac{\lambda L}{d} m = 3\frac{\lambda L}{d} = 7.74 \times 10^{-2}\text{ m}$$

$$y'_3 = \frac{\lambda' L}{d} m = 3\frac{\lambda' L}{d} = 9.18 \times 10^{-2}\text{ m}$$

Hence, the separation between the two fringes is given by

$$\Delta y = y'_3 - y_3 = \frac{3(\lambda' - \lambda)}{d} L$$

$$= 1.44 \times 10^{-2}\text{ m} = 1.44\text{ cm}$$

## 37.3 INTENSITY DISTRIBUTION OF THE DOUBLE-SLIT INTERFERENCE PATTERN

We shall now calculate the distribution of light intensity associated with the double-slit interference pattern. Again, suppose that the two slits represent coherent sources of sinusoidal waves. Hence, they have the same angular frequency $\omega$ and a constant phase difference $\phi$. The total electric field intensity at the point $P$ on the screen in Figure 37.5 is the *vector superposition* of the two waves from slits $S_1$ and $S_2$. Assuming the two waves have the same amplitude $E_0$, we can write the electric field intensities at $P$ due to each wave separately as

$$E_1 = E_0 \sin \omega t \qquad \text{and} \qquad E_2 = E_0 \sin(\omega t + \phi) \tag{37.7}$$

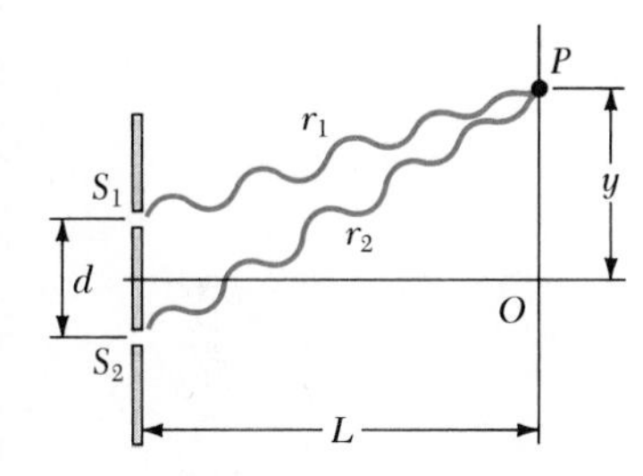


**Figure 37.5** Construction for analyzing the double-slit interference pattern. A bright region, or intensity maximum, is observed at $O$.

Although the waves have equal phase at the slits, *their phase difference $\phi$ at $P$ depends on the path difference $\delta = r_2 - r_1 = d \sin \theta$.* Since a path difference of $\lambda$ corresponds to a phase difference of $2\pi$ radians (constructive interference), while a path difference of $\lambda/2$ corresponds to a phase difference of $\pi$ radians (destructive interference), we obtain the ratio

$$\frac{\delta}{\phi} = \frac{\lambda}{2\pi}$$

Phase difference

$$\phi = \frac{2\pi}{\lambda}\delta = \frac{2\pi}{\lambda} d \sin \theta \tag{37.8}$$

This equation gives the precise dependence of the phase difference $\phi$ on the angle $\theta$.

Using the superposition principle and Equation 37.7, we can obtain the resultant electric field at the point $P$:

$$E_P = E_1 + E_2 = E_0[\sin \omega t + \sin(\omega t + \phi)] \tag{37.9}$$

To simplify this expression, we use the following trigonometric identity:

$$\sin A + \sin B = 2 \sin\left(\frac{A+B}{2}\right)\cos\left(\frac{A-B}{2}\right)$$

Taking $A = \omega t + \phi$ and $B = \omega t$, we can write Equation 37.9 in the form

$$E_P = 2E_0 \cos\left(\frac{\phi}{2}\right)\sin\left(\omega t + \frac{\phi}{2}\right) \tag{37.10}$$

Hence, the electric field at $P$ has the same frequency $\omega$, but its amplitude is multiplied by the factor $2 \cos(\phi/2)$. To check the consistency of this result, note that if $\phi = 0, 2\pi, 4\pi, \ldots$, the amplitude at $P$ is $2E_0$, corresponding to the condition for constructive interference. Referring to Equation 37.8, we find that our result is consistent with Equation 37.2. Likewise, if $\phi = \pi, 3\pi, 5\pi, \ldots$, the amplitude at $P$ is zero, which is consistent with Equation 37.3 for destructive interference.

Finally, to obtain an expression for the light intensity at $P$, recall that *the intensity of a wave is proportional to the square of the resultant electric field at that point* (Section 34.3). Using Equation 37.10, we can therefore express the intensity at $P$ as

$$I \propto E_P^2 = 4E_0^2 \cos^2(\phi/2) \sin^2\left(\omega t + \frac{\phi}{2}\right)$$

Since most light-detecting instruments measure the time average light intensity and the time average value of $\sin^2(\omega t + \phi/2)$ over one cycle is $1/2$, we can write the average intensity at $P$ as

$$I_{av} = I_0 \cos^2(\phi/2) \tag{37.11}$$

where $I_0$ is the *maximum* possible time average light intensity. [You should note that $I_0 = (E_0 + E_0)^2 = (2E_0)^2 = 4E_0^2$.] Substituting Equation 37.8 into Equation 37.11, we find that

$$I_{av} = I_0 \cos^2\left(\frac{\pi d \sin\theta}{\lambda}\right) \tag{37.12}$$

Alternatively, since $\sin\theta \approx y/L$ for small values of $\theta$, we can write Equation 37.12 in the form

$$I_{av} = I_0 \cos^2\left(\frac{\pi d}{\lambda L}\, y\right) \tag{37.13}$$

Constructive interference, which produces intensity maxima, occurs when the quantity $(\pi y d/\lambda L)$ is an integral multiple of $\pi$, corresponding to $y = (\lambda L/d)m$. This is consistent with Equation 37.5. A plot of the intensity distribution versus $\theta$ is given in Figure 37.6a. Note that the interference pattern consists of equally spaced fringes of equal intensity. However, the result is valid only if the slit-to-screen distance $L$ is large relative to the slit separation, and only for small values of $\theta$.

We have seen that the interference phenomena arising from two sources depend on the relative phase of the waves at a given point. Furthermore, the phase difference at a given point depends on the *path difference* between the two waves. The *resultant intensity at a point is proportional to the square of the resultant amplitude.* That is, the intensity is proportional to $(E_1 + E_2)^2$. It would be *incorrect* to calculate the resultant intensity by adding the intensities of the individual waves. This procedure would give a different quantity, namely, $E_1^2 + E_2^2$. Finally, $(E_1 + E_2)^2$ has the same *average* value as $E_1^2 + E_2^2$, when the time average is taken over all values of the phase difference between $E_1$ and $E_2$. Hence, there is no violation of energy conservation.

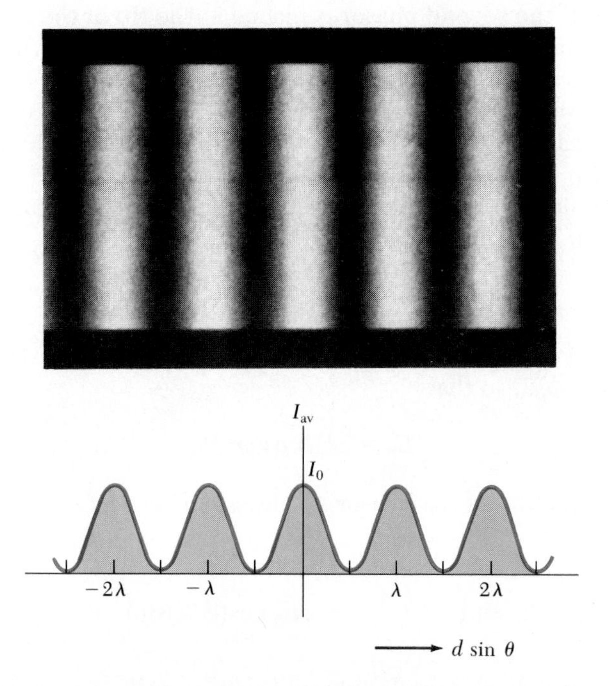


**Figure 37.6** Intensity distribution versus $d \sin \theta$ for the double-slit pattern when the screen is far from two slits ($L \gg d$). (Photo from M. Cagnet, M. Francon, and J.C. Thierr, *Atlas of Optical Phenomena*, Berlin, Springer-Verlag, 1962)

## 37.4 PHASOR ADDITION OF WAVES

In the previous section we combined two waves algebraically to obtain the resultant wave amplitude at some point on a screen. Unfortunately, this analytical procedure becomes rather cumbersome when several wave amplitudes have to be added. Since we shall eventually be interested in combining a large number of waves, we now describe a graphical procedure for this purpose.

Again, consider a sinusoidal wave whose electric field component is given by the expression

$$E_1 = E_0 \sin \omega t$$

where $E_0$ is the wave amplitude and $\omega$ is the angular frequency. This wave disturbance can be represented graphically with a vector of magnitude $E_0$, *rotating* about the origin in a counterclockwise direction with an angular frequency $\omega$, as in Figure 37.7a. Such a rotating vector is called a *phasor* and is commonly used in the field of electrical engineering (see Chapter 33). Note that the phasor makes an angle of $\omega t$ with the horizontal axis. The projection of the phasor on the vertical axis represents $E_1$, the magnitude of the wave disturbance at some time $t$. Hence, as the phasor rotates in a circle, the projection $E$ oscillates along the vertical axis about the origin.

Now consider a second sinusoidal wave whose electric field is given by

$$E_2 = E_0 \sin(\omega t + \phi)$$

That is, this wave has the same amplitude and frequency as $E_1$, but its phase is $\phi$ with respect to the wave $E_1$. The phasor representing the wave $E_2$ is shown in Figure 37.7b. The resultant wave, which is the sum of $E_1$ and $E_2$, can be obtained graphically by redrawing the phasors end to end, as in Figure 37.7c,

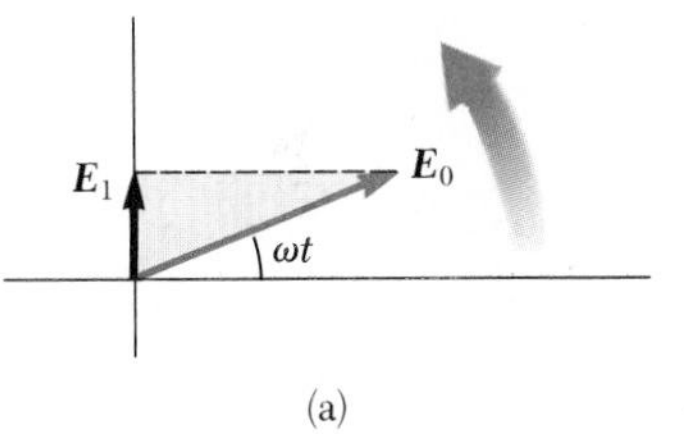


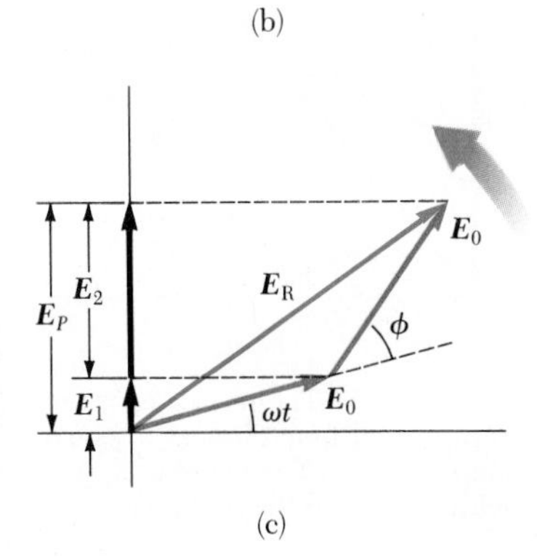


**Figure 37.7** (a) Phasor diagram for the wave disturbance $E_1 = E_0 \sin \omega t$. The phasor is a vector of length $E_0$ rotating counterclockwise. (b) Phasor diagram for the wave $E_2 = E_0 \sin(\omega t + \phi)$. (c) $E_R$ is the resultant phasor formed from the individual phasors shown in (a) and (b).

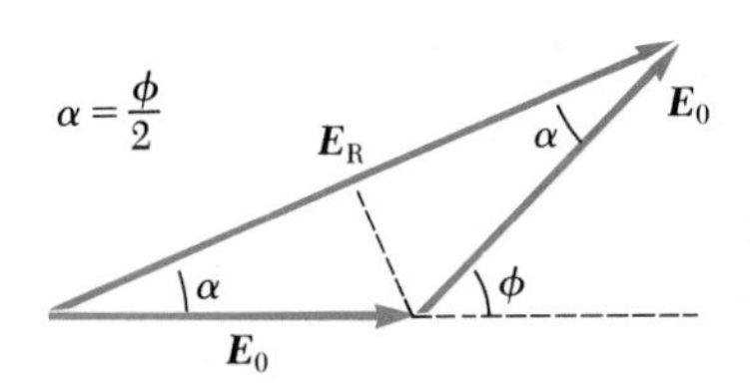

**Figure 37.8** Reconstruction of the resultant phasor $\boldsymbol{E_R}$. From the geometry, note that $\alpha = \phi/2$.

where the tail of the second phasor is placed at the tip of the first phasor. As with vector addition, the resultant phasor $\boldsymbol{E_R}$ runs from the tail of the first phasor to the tip of the second phasor. Furthermore, $\boldsymbol{E_R}$ rotates along with the two individual phasors at the same angular frequency $\omega$. The projection of $\boldsymbol{E_R}$ along the vertical axis equals the sum of the projections of the two phasors. That is, $E_P = E_1 + E_2$.

It is convenient to construct the phasors at $t = 0$ as in Figure 37.8. From the geometry of the triangle, we see that

$$E_R = E_0 \cos \alpha + E_0 \cos \alpha = 2E_0 \cos \alpha$$

Since the sum of the two opposite interior angles equals the exterior angle $\phi$, we see that $\alpha = \phi/2$, so that

$$E_R = 2E_0 \cos(\phi/2)$$

Hence, the projection of the phasor $\boldsymbol{E_R}$ along the *vertical axis* at any time $t$ is given by

$$E_P = E_R \sin\left(\omega t + \frac{\phi}{2}\right) = 2E_0 \cos(\phi/2) \sin\left(\omega t + \frac{\phi}{2}\right)$$

This is consistent with the result obtained algebraically, Equation 37.10. The resultant phasor has an amplitude $2E_0 \cos(\phi/2)$ and makes an angle of $\phi/2$ with the first phasor. Furthermore, the average intensity at $P$, which varies as $E_P{}^2$, is proportional to $\cos^2(\phi/2)$, as described previously in Equation 37.11.

We can now describe how to obtain the resultant of several waves which have the same frequency:

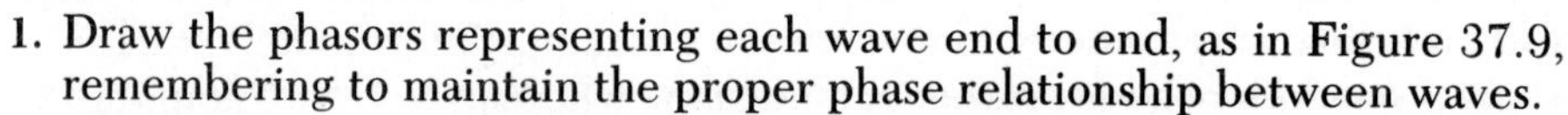

1. Draw the phasors representing each wave end to end, as in Figure 37.9, remembering to maintain the proper phase relationship between waves.
2. The resultant represented by the phasor $\boldsymbol{E_R}$ is the vector sum of the individual phasors. At each instant, the projection of $\boldsymbol{E_R}$ along the vertical axis represents the time variation of the resultant wave. The phase angle $\alpha$ of the resultant wave is the angle between $\boldsymbol{E_R}$ and the first phasor. From the construction in Figure 37.9, drawn for four phasors, we see that the phasor of the resultant wave is given by $E_P = E_R \sin(\omega t + \alpha)$.

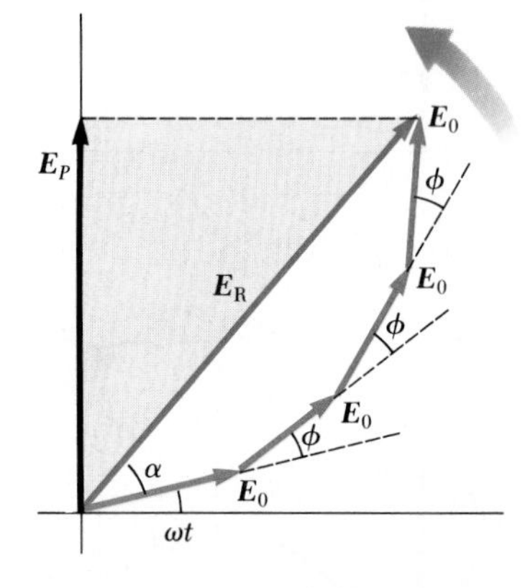

**Figure 37.9** The phasor $\boldsymbol{E_R}$ is the resultant of four phasors of equal amplitude $E_0$. The phase of $\boldsymbol{E_R}$ is $\alpha$ with respect to the first phasor.

### Phasor Diagrams for Two Coherent Sources

As an example of the phasor method, consider the interference pattern produced by two coherent sources, which was discussed in the previous section. Figure 37.10 represents the phasor diagrams for various values of the phase difference $\phi$, and the corresponding values of the path difference $\delta$, which are obtained using Equation 37.8.

From Figure 37.10, we see that the intensity at a point will be a maximum when $\boldsymbol{E_R}$ is a maximum. This occurs at values of $\phi$ equal to 0, $2\pi$, $4\pi$, etc. Likewise, we see that the intensity at some observation point will be zero when $\boldsymbol{E_R}$ is zero. The first zero-intensity point occurs at $\phi = 180°$, corresponding to $\delta = \lambda/2$, while the other zero points (not shown) occur at values of $\delta$ equal to $3\lambda/2$, $5\lambda/2$, etc. These results are in complete agreement with the analytical procedure described in the previous section.

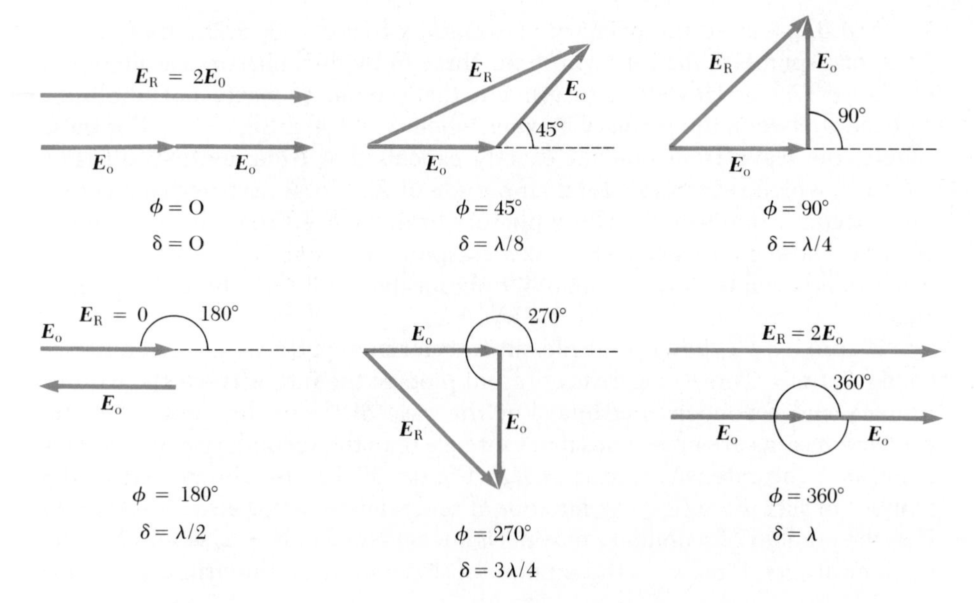


**Figure 37.10** Phasor diagrams for the double-slit interference pattern. The resultant phasor $E_R$ is a maximum when $\phi = 0, 2\pi, 4\pi, \ldots$ and is zero when $\phi = \pi, 3\pi, 5\pi, \ldots$.

### Three-Slit Interference Pattern

Using phasor diagrams, let us analyze the interference pattern caused by three equally spaced slits. The electric fields at a point $P$ on the screen due to waves from the individual slits can be expressed as

$$E_1 = E_0 \sin \omega t$$

$$E_2 = E_0 \sin(\omega t + \phi)$$

$$E_3 = E_0 \sin(\omega t + 2\phi)$$

where $\phi$ is the phase difference between waves from adjacent slits. Hence, the resultant field at $P$ can be obtained by using a phasor diagram as shown in Figure 37.11.

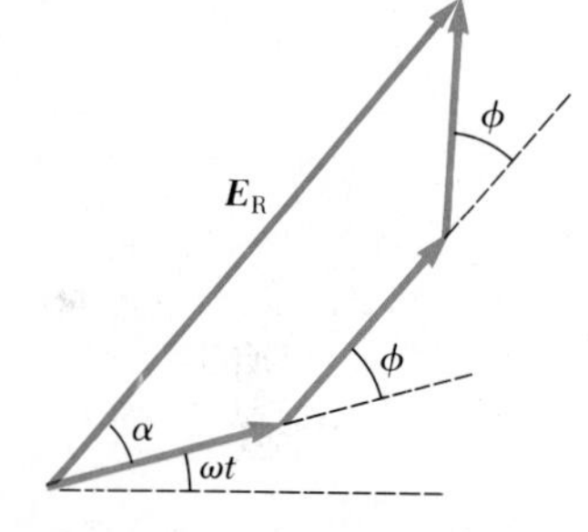


**Figure 37.11** Phasor diagram for three equally spaced slits.

The phasor diagrams for various specific values of $\phi$ for the three slits are shown in Figure 37.12. Note that the resultant amplitude at $P$ has a maximum

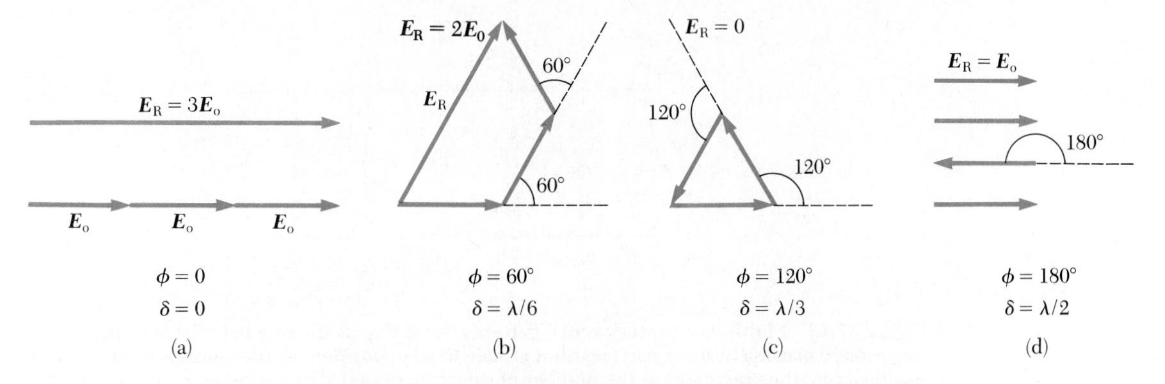


**Figure 37.12** Phasor diagrams for three equally spaced slits at various values of $\phi$. Note that there are primary maxima of amplitude $3E_0$ and secondary maxima of amplitude $E_0$.

value of $3E_0$ (called the primary maximum) when $\phi = 0, \pm 2\pi, \pm 4\pi, \ldots$. This corresponds to the case where the three individual phasors are aligned as in Figure 37.12a. However, we also find that secondary maxima of amplitude $E_0$ occur between the primary maxima when $\phi = \pm \pi, \pm 3\pi, \ldots$. For these points, the wave from one slit exactly cancels that from another slit (Fig. 37.12d), which results in a total amplitude of $E_0$. Total destructive interference occurs whenever the three phasors form a closed triangle as in Figure 37.12c. These points where $E_0 = 0$ correspond to $\phi = \pm 2\pi/3, \pm 4\pi/3, \ldots$. You should be able to construct other phasor diagrams for values of $\phi$ greater than $\pi$.

Figure 37.13 shows multiple-slit interference patterns for a number of configurations. These patterns represent plots of the intensity for the various primary and secondary maxima. For the case of three slits, note that the primary maxima are nine times more intense than the secondary maxima. This is because the intensity varies as $E_R^2$. Figure 37.13 also shows that as the number of slits increases, the number of secondary maxima also increases. In fact, the number of secondary maxima is always equal to N − 2, where N is the number of slits. Finally, as the number of slits increases, the primary maxima

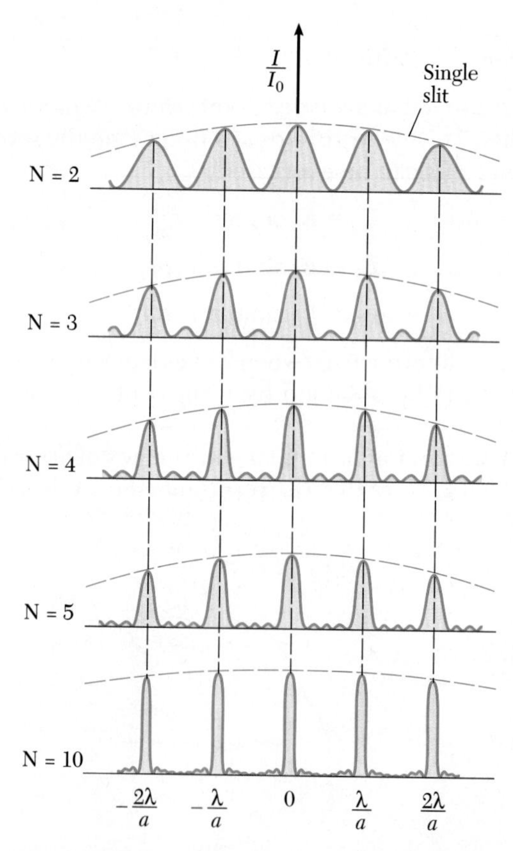


**Figure 37.13** Multiple-slit interference patterns. Note that as the number of slits is increased, the primary maxima become narrower but remain fixed in position. Furthermore, the number of secondary maxima increases as the number of slits is increased. The decrease in intensity of the maxima, as indicated by the dashed lines, is due to the phenomenon of diffraction, which is discussed in Chapter 38.

increase in intensity and become narrower, while the secondary maxima decrease in intensity relative to the primary maxima.

## 37.5 CHANGE OF PHASE DUE TO REFLECTION

We have described interference effects produced by two or more coherent light sources. Young's method for producing two coherent light sources involves illuminating a pair of slits with a single source. Another simple, yet ingenious, arrangement for producing an interference pattern with a single light source is known as *Lloyd's mirror.* A light source is placed at S close to a mirror, as illustrated in Figure 37.14. Waves can reach the viewing point $P$ either by the direct path $SP$ or by the path involving reflection from the mirror. The reflected ray can be treated as a ray originating from a source at $S'$, located behind the mirror. $S'$, which is the image of S, can be considered a virtual source. Hence, at observation points far from the source, one would expect an interference pattern due to waves from S and $S'$ just as is observed for two real coherent sources. An interference pattern is indeed observed. However, the positions of the dark and bright fringes are *reversed* relative to the pattern of two real coherent sources (Young's experiment). This is because the coherent sources at S and $S'$ differ in phase by 180°. This 180° phase change is produced upon reflection. To illustrate this further, consider the point $P'$, where the mirror meets the screen. This point is equidistant from S and $S'$. If path difference alone were responsible for the phase difference, one would expect to see a bright fringe at $P'$ (since the path difference is zero for this point), corresponding to the central fringe of the two-slit interference pattern. Instead, one observes a *dark* fringe at $P'$ because of the 180° phase change produced by reflection. In general,

> an electromagnetic wave undergoes a phase change of 180° upon reflection from a medium that is optically more dense than the one in which it was traveling. There is also a 180° phase change upon reflection from a conducting surface.

It is useful to draw an analogy between reflected light waves and the reflections of a transverse wave on a stretched string when the wave meets the boundary (Section 16.6). The reflected pulse on a string undergoes a phase change of 180° when it is reflected from a denser medium, such as a heavier string. On the other hand, there is no phase change if the pulse reflects from a less dense medium. Similarly, electromagnetic waves undergo a 180° phase change when reflected from a boundary leading to an optically denser medium. There is no phase change when the wave is reflected from a boundary leading to a less dense medium. In either case, the transmitted wave under-

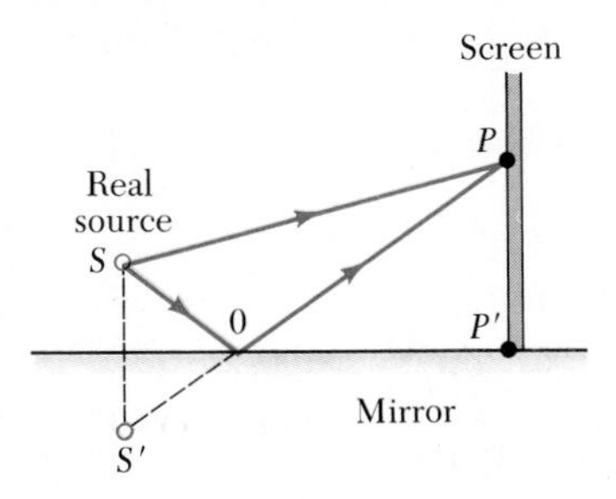


**Figure 37.14** Lloyd's mirror. An interference pattern is produced on a screen at $P$ as a result of the combination of the direct ray and the reflected ray. The reflected ray undergoes a phase change of 180°.

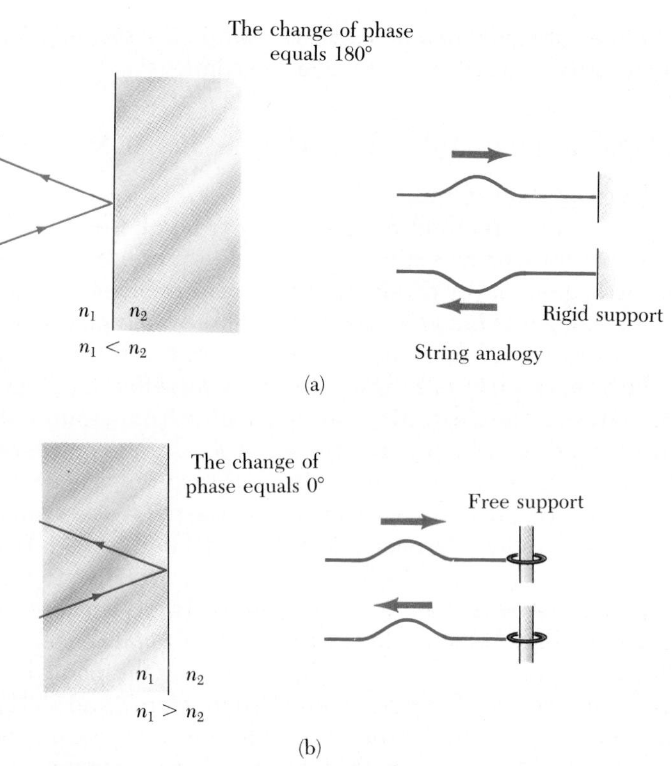


**Figure 37.15** (a) A ray reflecting from a medium of higher refractive index undergoes a 180° phase change. The right side shows the analogy with a reflected pulse on a string. (b) A ray reflecting from a medium of lower refractive index undergoes *no* phase change.

goes no phase change. These rules, summarized in Figure 37.15, can be deduced from Maxwell's equations, but the treatment is beyond the scope of this text.

## 37.6 INTERFERENCE IN THIN FILMS

Interference effects are commonly observed in thin films, such as thin layers of oil on water and soap bubbles. The various colors that are observed with ordinary white light result from the interference of waves reflected from the opposite surfaces of the film.

Consider a film of uniform thickness $t$ and index of refraction $n$, as in Figure 37.16. Let us assume that the light rays traveling in air are nearly normal to the surface. To determine whether the reflected rays interfere constructively or destructively, we must first note the following facts:

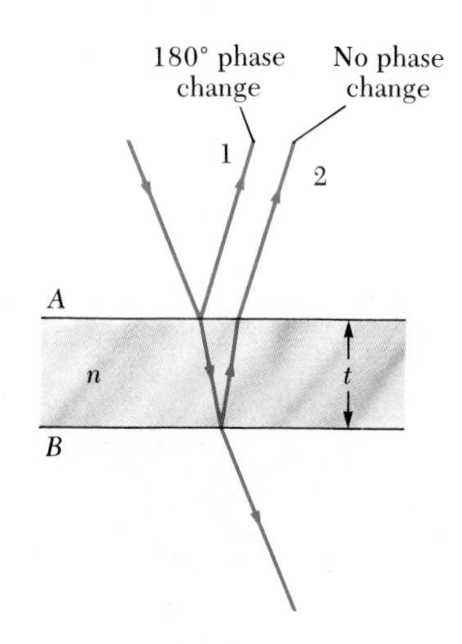


**Figure 37.16** Interference in light reflected from a thin film is due to a combination of rays reflected from the upper and lower surfaces.

1. *A wave traveling in a medium of low refractive index (air) undergoes a 180° phase change upon reflection from a medium of higher refractive index.* There is no phase change in the reflected wave if it reflects from a medium of lower refractive index.
2. The wavelength of light $\lambda_n$ in a medium whose refraction index is $n$ (Section 35.4) is given by

$$\lambda_n = \frac{\lambda}{n} \qquad (37.14)$$

where $\lambda$ is the wavelength of light in free space.

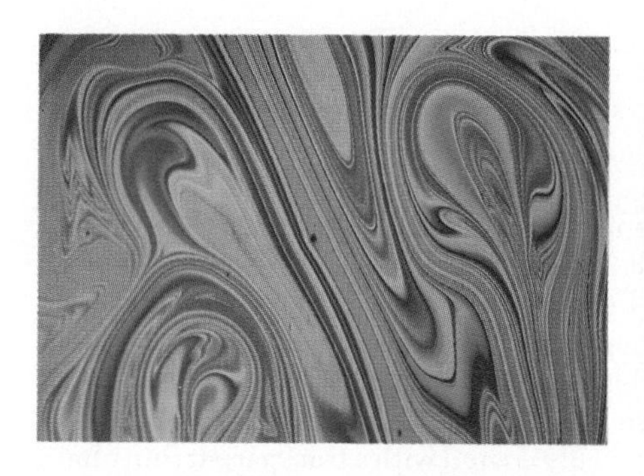

Thin film interference. A thin film of oil on water displays interference, as shown by the pattern of colors when white light is incident on the film. The film thickness varies, thereby producing the interesting color pattern. (© Tom Branch 1984, Photo Researchers)

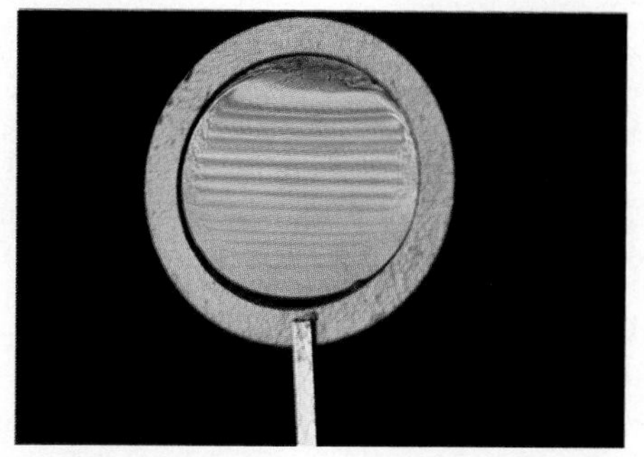

Interference in a vertical soap film of variable thickness. The top of the film appears darkest where the film is thinnest. (© 1983 Larry Mulvehill, Photo Researchers)

Let us apply these rules to the film described in Figure 37.16. According to the first rule, ray 1, which is reflected from the upper surface (A), undergoes a phase change of 180° with respect to the incident wave. On the other hand, ray 2, which is reflected from the lower surface (B), undergoes no phase change with respect to the incident wave. Therefore, ray 1 is 180° out of phase with respect to ray 2, which is equivalent to a path difference of $\lambda_n/2$. However, we must also consider that ray 2 travels an extra distance equal to $2t$ before the waves recombine. For example, if $2t = \lambda_n/2$, rays 1 and 2 will recombine in phase and constructive interference will result. In general, the condition for constructive interference can be expressed as

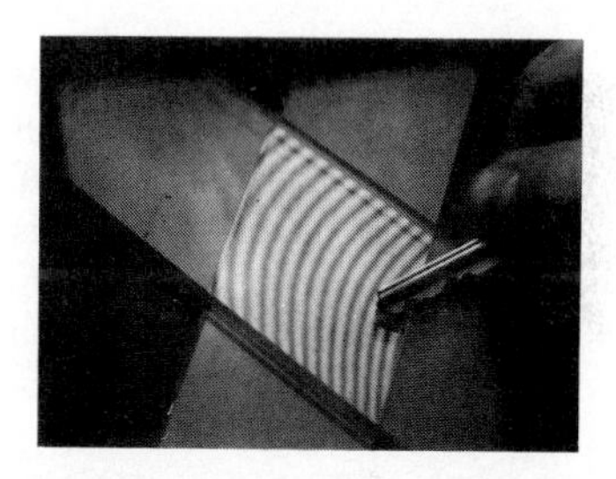

The thin film of air between two glass plates is responsible for the interference pattern. The lines are curved because pressure from the key bends the glass slightly, thus changing the thickness of the air film.

$$2t = (m + \tfrac{1}{2})\lambda_n \qquad (m = 0, 1, 2, \ldots) \qquad (37.15)$$

This condition takes into account two factors: (a) the difference in optical path length for the two rays (the term $m\lambda_n$) and (b) the 180° phase change upon reflection (the term $\lambda_n/2$). Since $\lambda_n = \lambda/n$, we can write Equation 37.15 in the form

$$2nt = (m + \tfrac{1}{2})\lambda \qquad (m = 0, 1, 2, \ldots) \qquad (37.16)$$

If the extra distance $2t$ traveled by ray 2 corresponds to a multiple of $\lambda_n$, the two waves will combine out of phase and destructive interference will result. The general equation for destructive interference is

$$2nt = m\lambda \qquad (m = 0, 1, 2, \ldots) \qquad (37.17)$$

These conditions for constructive and destructive interference are valid only when the film is surrounded by a common medium. The surrounding medium may have a refractive index less than or greater than that of the film. In either case, the rays reflected from the two surfaces will be out of phase by 180°. On the other hand, if the film is located between two *different* media, one of lower refractive index and one of higher refractive index, the conditions for constructive and destructive interference are *reversed*. In this case, either there is a phase change of 180° for both ray 1 reflecting from surface A and ray 2 reflecting from surface B or there is no phase change for either ray; hence, the net change in relative phase due to the reflections is *zero*.

### EXAMPLE 37.3 Interference in a Soap Film

Calculate the minimum thickness of a soap bubble film ($n = 1.33$) that will result in constructive interference in the reflected light if the film is illuminated with light whose wavelength in free space is 600 nm.

*Solution* The minimum film thickness for constructive interference in the reflected light corresponds to $m = 0$ in Equation 37.16. This gives $2nt = \lambda/2$, or

$$t = \frac{\lambda}{4n} = \frac{600 \text{ nm}}{4(1.33)} = 113 \text{ nm}$$

*Exercise 1* What other film thicknesses will produce constructive interference?

*Answer* 338 nm, 564 nm, 789 nm, and so on.

### EXAMPLE 37.4 Nonreflecting Coatings for Solar Cells

Solar cells are often coated with transparent thin film, such as silicon monoxide (SiO, $n = 1.45$), in order to minimize reflective losses from the surface (Fig. 37.17). A silicon solar cell ($n = 3.5$) is coated with a thin film of silicon monoxide for this purpose. Determine the minimum thickness of the film that will produce the least reflection at a wavelength of 550 nm, which is the center of the visible spectrum.

*Solution* The reflected light is a minimum when rays 1 and 2 meet the condition of destructive interference. Note that *both* rays undergo a 180° phase change upon reflection in this case, one from the upper and one from the lower surface. Hence, the net change in phase is zero due to reflection, and the condition for a reflection *minimum* requires a path difference of $\lambda_n/2$; hence $2t = \lambda/2n$ or

$$t = \frac{\lambda}{4n} = \frac{550 \text{ nm}}{4(1.45)} = 94.8 \text{ nm}$$

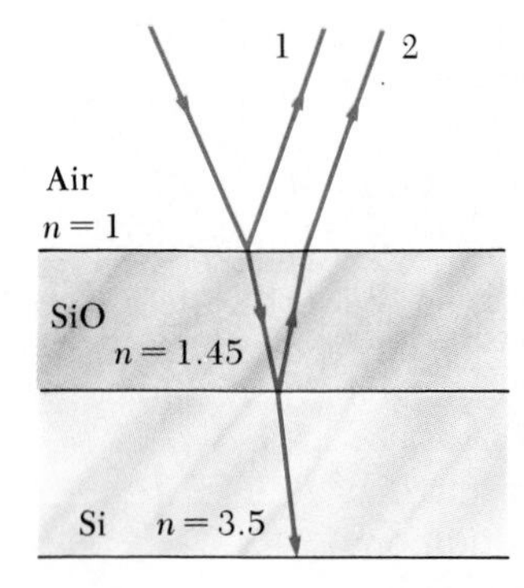


**Figure 37.17** (Example 37.4) Reflective losses from a silicon solar cell are minimized by coating it with a thin film of silicon monoxide.

Typically, such antireflecting coatings reduce the reflective loss from 30% (with no coating) to 10% (with coating). Such coatings therefore increase the cell's efficiency since more light is available to create charge carriers in the cell. In reality, the coating is never perfectly nonreflecting because the required thickness is wavelength-dependent and the incident light covers a wide range of wavelengths.

Glass lenses used in cameras and other optical instruments are usually coated with a transparent thin film, such as magnesium fluoride ($MgF_2$), to reduce or eliminate unwanted reflection. More important, such coatings enhance the transmission of light through the lenses.

### EXAMPLE 37.5 Interference in a Wedge-Shaped Film

A thin, wedge-shaped film of refractive index $n$ is illuminated with monochromatic light of wavelength $\lambda$, as illustrated in Figure 37.18. Describe the interference pattern observed for this case.

*Solution* The interference pattern is that of a thin film of variable thickness surrounded by air. Hence, the pattern will be a series of alternating bright and dark parallel bands. A dark band corresponding to destructive interference appears at point $O$, the apex, since the upper reflected ray undergoes a 180° phase change while the lower one does not. According to Equation 37.17, other dark bands appear when $2nt = m\lambda$, so that $t_1 = \lambda/2n$, $t_2 = \lambda/n$, $t_3 = 3\lambda/2n$, and so on. Similarly, bright bands will be observed when the thickness satisfies the condition $2nt = (m + \frac{1}{2})\lambda$, corresponding to thicknesses of $\lambda/4n$, $3\lambda/4n$, $5\lambda/4n$, and so on. If white light is used, bands of different colors will be observed at different points, corresponding to the different wavelengths of light.

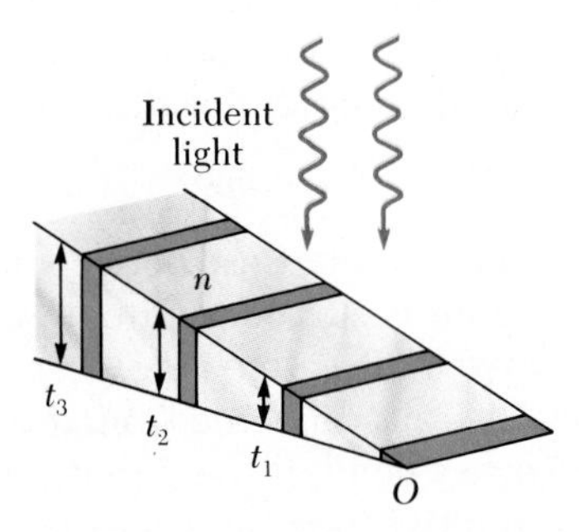


**Figure 37.18** (Example 37.5) Interference bands in reflected light can be observed by illuminating a wedge-shaped film with monochromatic light. The blue areas correspond to positions of destructive interference.

### Newton's Rings

Another method for observing interference of light waves is to place a plano-convex lens (one having one plane side and one convex side) on top of a plane glass surface as in Figure 37.19a. With this arrangement, the air film between the glass surfaces varies in thickness from zero at the point of contact to some value $t$ at $P$. If the radius of curvature of the lens $R$ is very large compared with the distance $r$, and if the system is viewed from above using light of wavelength $\lambda$, a pattern of light and dark rings is observed. A photograph of such a pattern is shown in Figure 37.19b. These circular fringes, discovered by Newton, are called **Newton's rings.** Newton's particle model of light could not explain the origin of the rings.

The interference effect is due to the combination of ray 1, reflected from the plane glass plate, with ray 2, reflected from the lower part of the lens. Ray 1 undergoes a phase change of 180° upon reflection, since it is reflected from a medium of higher refractive index, whereas ray 2 undergoes no phase change. Hence, the conditions for constructive and destructive interference are given by Equations 37.16 and 37.17, respectively, with $n = 1$ since the "film" is air. Here again, one might guess that the contact point $O$ would be bright, corresponding to constructive interference. Instead, the contact point is dark, as seen in Figure 37.19b, because ray 1, reflected from the plane surface, undergoes a 180° phase change with respect to ray 2. Using the geometry shown in Figure 37.19a, one can obtain expressions for the radii of the bright and dark bands in terms of the radius of curvature $R$ and wavelength $\lambda$. For example, the dark rings have radii given by $r \approx \sqrt{m\lambda R/n}$. The details are left as a problem for the reader (Problem 62). By measuring the radii of the rings, one can obtain the wavelength, provided $R$ is known. Conversely, if the wavelength is accurately known, one can use this effect to obtain $R$.

One of the important uses of Newton's rings is in the testing of optical lenses. A circular pattern like that pictured in Figure 37.19b is obtained only

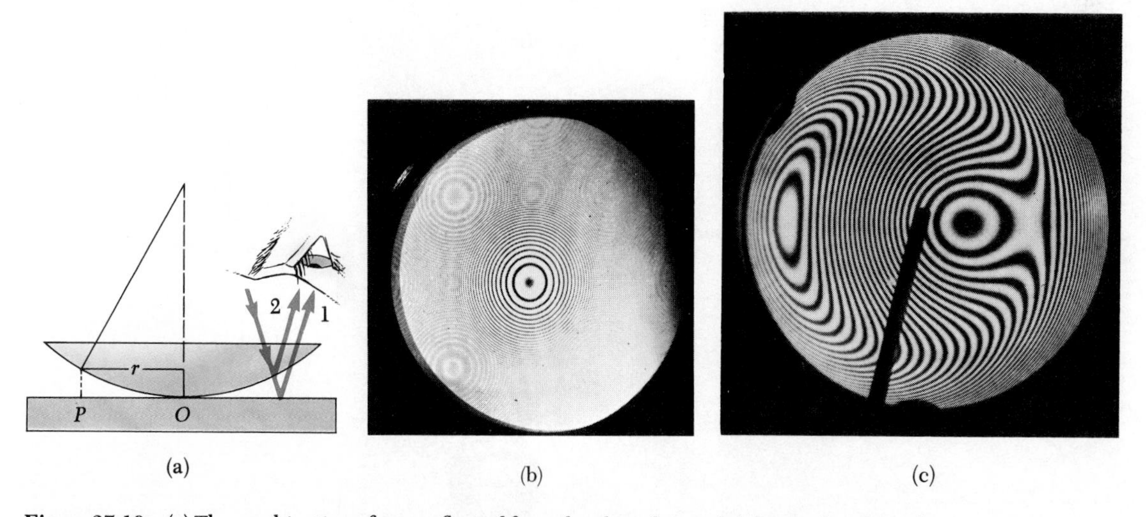


**Figure 37.19** (a) The combination of rays reflected from the glass plate and the curved surface of the lens give rise to an interference pattern known as Newton's rings. (b) Photograph of Newton's rings. (Courtesy of Bausch and Lomb Optical Co.) (c) This asymmetrical interference pattern indicates imperfections in the lens. (From Physical Science Study Committee, *College Physics*, Lexington, Mass., Heath, 1968)

when the lens is ground to a perfectly symmetric curvature. Variations from such symmetry might produce a pattern like that in Figure 37.19c. These variations give an indication of how the lens must be ground and polished in order to remove the imperfections.

## *37.7 THE MICHELSON INTERFEROMETER

The **interferometer**, invented by the American physicist A. A. Michelson (1852–1931), is an ingenious device that splits a light beam into two parts and then recombines them to form an interference pattern after they have traveled over different paths. The device can be used for obtaining accurate measurements of wavelengths or for precise length measurements.

A schematic diagram of the interferometer is shown in Figure 37.20. A beam of light provided by a monochromatic source is split into two rays by a partially silvered mirror M inclined at 45° relative to the incident light beam. One ray is reflected vertically upward toward mirror $M_1$ while the second ray is transmitted horizontally through M toward mirror $M_2$. Hence, the two rays travel separate paths $L_1$ and $L_2$. After reflecting from mirrors $M_1$ and $M_2$, the two rays eventually recombine to produce an interference pattern, which can be viewed through a telescope. The glass plate P, equal in thickness to mirror M, is placed in the path of the horizontal ray in order to insure that the two rays travel the same distance through glass.

The interference condition for the two rays is determined by the difference in their optical path lengths. When the two rays are viewed as shown, the

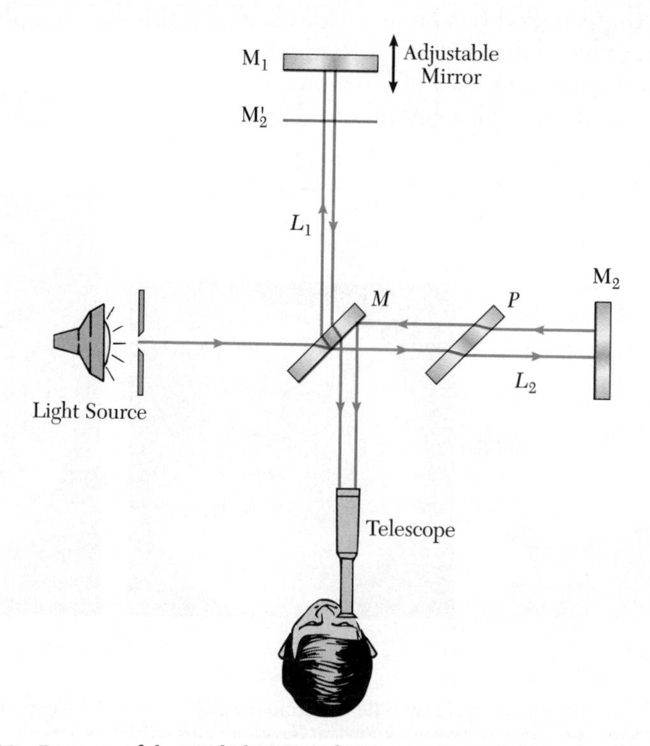


**Figure 37.20** Diagram of the Michelson interferometer. A single beam is split into two rays by the partially silvered mirror M. The path difference between the two rays is varied with the adjustable mirror $M_1$.

image of $M_2$ is at $M_2'$ parallel to $M_1$. Hence, $M_2'$ and $M_1$ form the equivalent of a parallel air film. The effective thickness of the air film is varied by moving mirror $M_1$ parallel to itself with a finely threaded screw. Under these conditions, the interference pattern is a series of bright and dark circular rings which resemble Newton's rings. If a dark circle appears at the center of the pattern, the two rays interfere destructively. If the mirror $M_1$ is then moved a distance of $\lambda/4$, the path difference changes by $\lambda/2$ (twice the separation between $M_1$ and $M_2'$). The two rays will now interfere constructively, giving a bright circle in the middle. As $M_1$ is moved an additional distance of $\lambda/4$, a dark circle will appear once again. Thus, we see that successive dark or bright circles are formed each time $M_1$ is moved a distance of $\lambda/4$. The wavelength of light is then measured by counting the number of fringe shifts for a given displacement of $M_1$. Conversely, if the wavelength is accurately known (as with a laser beam), mirror displacements can be measured to within a fraction of the wavelength.

Since the interferometer can accurately measure displacement, it is often used to make highly precise measurements of the length of mechanical components. The interferometer makes such precise wavelength measurements possible.

## SUMMARY

**Interference** of light waves is the result of the linear superposition of two or more waves at a given point. A sustained interference pattern is observed if (1) the sources are coherent (that is, they maintain a constant relative phase), (2) the sources are monochromatic (of a single wavelength), and (3) the linear superposition principle is applicable.

In Young's double-slit experiment, two slits separated by a distance $d$ are illuminated by a monochromatic light source. An interference pattern consisting of bright and dark fringes is observed on a screen a distance $L$ from the slits. The condition for bright fringes in the double-slit experiment (**constructive interference**) is given by

Conditions for constructive interference

$$d \sin \theta = m\lambda \qquad (m = 0, \pm 1, \pm 2, \ldots) \qquad (37.2)$$

The condition for dark fringes in the double-slit experiment (**destructive interference**) is

Conditions for destructive interference

$$d \sin \theta = (m + \tfrac{1}{2})\lambda \qquad (m = 0, \pm 1, \pm 2, \ldots) \qquad (37.3)$$

The index number $m$ is called the **order number** of the fringe.

The **average intensity** of the double-slit interference pattern is given by

$$I_{\text{av}} = I_0 \cos^2 \frac{\pi d \sin \theta}{\lambda} \qquad (37.12)$$

where $I_0$ is the maximum intensity on the screen.

When a series of $N$ slits is illuminated, a diffraction pattern is produced that can be viewed as interference arising from the superposition of a large number of waves. It is convenient to use phasor diagrams to simplify the analysis of interference from three or more equally spaced slits.

An electromagnetic wave undergoes a phase change of 180° upon reflection from an optically more dense medium or from any conducting surface.

The wavelength of light $\lambda_n$ in a medium whose refractive index is $n$ is given by

$$\lambda_n = \frac{\lambda}{n} \quad (37.14)$$

where $\lambda$ is the wavelength of light in free space.

The condition for constructive interference in a film of thickness $t$ and refractive index $n$ with a common medium on both sides of the film is given by

Constructive interference in thin films

$$2nt = (m + \tfrac{1}{2})\lambda \qquad (m = 0, 1, 2, \ldots) \quad (37.16)$$

Similarly, the condition for destructive interference in thin films is

Destructive interference in thin films

$$2nt = m\lambda \qquad (m = 0, 1, 2, \ldots) \quad (37.17)$$

## QUESTIONS

1. What is the necessary condition on the path length difference between two waves that interfere (a) constructively and (b) destructively?
2. Explain why two flashlights held close together will not produce an interference pattern on a distant screen.
3. If Young's double-slit experiment were performed under water, how would the observed interference pattern be affected?
4. What is the difference between interference and diffraction?
5. In Young's double-slit experiment, why do we use monochromatic light? If white light is used, how would the pattern change?
6. In the process of evaporation, a soap bubble appears black just before it breaks. Explain this phenomenon in terms of the phase changes that occur upon reflection from the two surfaces.
7. If an oil film is observed on water, the film appears brightest at the outer regions, where it is thinnest. From this information, what can you say about the index of refraction of the oil relative to that of water?
8. If a soap film on a wire loop is held in air, it appears black in the thinnest regions when observed by reflected light and shows a variety of colors in thicker regions, as in the photograph at the right. Explain.
9. A simple way of observing an interference pattern is to look at a distant light source through a stretched handkerchief or an opened umbrella. Explain how this works.
10. In order to observe interference in a thin film, why must the film not be very thick (on the order of a few wavelengths)?
11. A lens with outer radius of curvature $R$ and index of refraction $n$ rests on a flat glass plate. It is illuminated with white light from above. Is there a dark spot or a light spot at the center of the lens? What does it mean if the observed rings are noncircular?
12. Why is the lens on a good quality camera coated with a thin film?
13. Would it be possible to place a nonreflective coating on an airplane to cancel radar waves of wavelength 3 cm?
14. How is it that good quality pictures can be taken with a pinhole camera (if the pinhole is the right diameter)? Why isn't a lens required to get good focus?
15. Why is it so much easier to perform interference experiments with a laser than an ordinary light source?

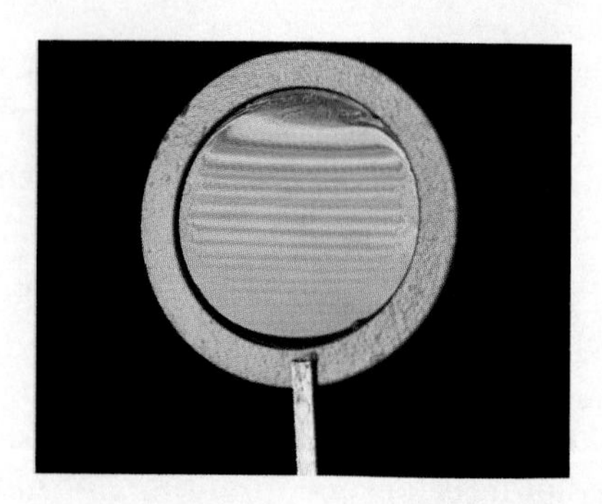

Question 8.

## PROBLEMS

### Section 37.2 Young's Double-Slit Experiment

1. A pair of narrow, parallel slits separated by 0.25 mm are illuminated by the green component from a mercury vapor lamp ($\lambda = 546.1$ nm). The interference pattern is observed on a screen located 1.2 m from the plane of the parallel slits. Calculate the distance (a) from the central maximum to the first bright region on either side of the central maximum and (b) between the first and second dark bands in the interference pattern.
2. A laser beam ($\lambda = 632.8$ nm) is incident upon two slits 0.2 mm apart. Approximately how far apart will the bright interference lines be on a screen 5 m from the double slits?
3. A Young's interference experiment is performed with blue-green argon laser light. The separation between the slits is 0.50 mm and the interference pattern on a screen 3.3 m away shows the first maximum at a distance of 3.4 mm from the center of the pattern. What is the wavelength of argon laser light?
4. Light from a helium-cadmium laser ($\lambda = 442$ nm) passes through a double-slit system with a slit separation $d = 0.40$ mm. Determine how far away a screen must be placed in order that a dark fringe appear directly opposite both slits.
5. The yellow component of light from a helium discharge tube ($\lambda = 587.5$ nm) is allowed to fall on a plane containing parallel slits that are 0.2 mm apart. A screen is located so that the second bright band in the interference pattern is at a distance equal to 10 slit spacings from the central maximum. What is the distance between the source plane and the screen?
6. A double slit with a spacing of 0.083 mm between the slits is 2.5 m from a screen. (a) If yellow light of wavelength 570 nm strikes the double slit, what is the separation between the zeroth- and first-order maxima on the screen? (b) If blue light of wavelength 410 nm strikes the double slit, what is the separation between the second- and fourth-order maxima? (c) Repeat Parts (a) and (b) for the minima.
7. A narrow slit is cut into each of two overlapping opaque squares. The slits are parallel, and the distance between them is adjustable. Monochromatic light of wavelength 600 nm illuminates the slits, and an interference pattern is formed on a screen 80 cm away. The third dark band is located 1.2 cm from the central bright band. What is the distance between the center of the central bright band and the center of the first bright band on either side of the central band?
8. Monochromatic light illuminates a double-slit system with a slit separation $d = 0.30$ mm. The second-order maximum occurs at $y = 4.0$ mm on a screen 1 m from the slits. Determine (a) the wavelength, (b) the position ($y$) of the third-order maximum, and (c) the angular position ($\theta$) of the $m = 1$ minimum.
9. In deriving Equation 37.5, it is assumed as stated in Equation 37.4 that $\sin\theta \approx \tan\theta$. The validity of this assumption depends on the requirement that $d \gg \lambda$. In the arrangement of Figure 37.4, let $L = 40$ cm, $d = 0.5$ mm, and $\lambda = 656.3$ nm (the red line in hydrogen). Under the assumption stated in Equation 37.4, the ninth-order dark band would be located 4.4631 mm on either side of the central maximum. What percent error is introduced by the assumption that $d \gg \lambda$?
10. Two radio antennas separated by 300 m as shown in Figure 37.21 simultaneously transmit identical signals (assume waves) on the same wavelength. A radio in a car traveling due north receives the signals. (a) If the car is at the position of the second maximum, what is the wavelength of the signals? (b) How much farther must the car travel to encounter the next minimum in reception? (*Caution:* Avoid small-angle approximations in this problem.)

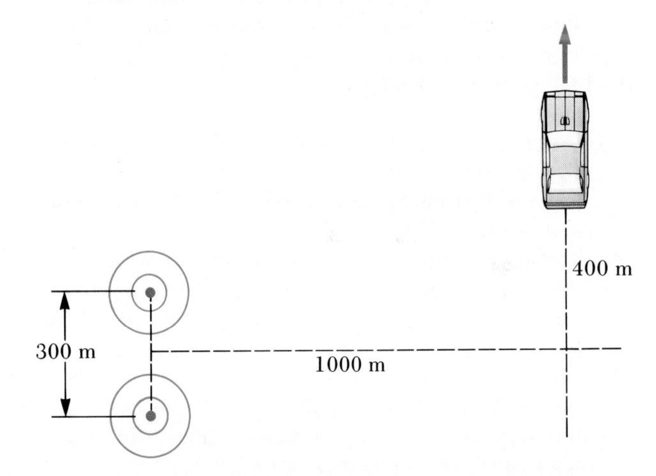


**Figure 37.21** (Problem 10).

11. Light of wavelength 546 nm (the intense green line from a mercury discharge tube) produces a Young's interference pattern in which the second-order minimum is along a direction that makes an angle of 18 minutes of arc relative to the direction to the central maximum. What is the distance between the parallel slits?
12. In a double-slit arrangement as illustrated in Figure 37.4, $d = 0.15$ mm, $L = 140$ cm, $\lambda = 643$ nm, and $y = 1.8$ cm. (a) What is the path difference $\delta$ for the two slits at the point $P$? (b) Express this path difference in terms of the wavelength. (c) Will point $P$ correspond to a maximum, a minimum, or an intermediate condition?

### Section 37.3 Intensity Distribution of the Double-Slit Interference Pattern

13. In the arrangement of Figure 37.4, let $L = 120$ cm and $d = 0.25$ cm. The slits are illuminated with light of wavelength 600 nm. Calculate the distance $y$ above the central maximum for which the average intensity on the screen will be 75% of the maximum.
14. Two slits are separated by a distance of 0.18 mm. An interference pattern is formed on a screen 80 cm away by the $H_\alpha$ line in hydrogen ($\lambda = 656.3$ nm). Calculate the fraction of the maximum intensity that would be measured at a point 0.6 cm above the central maximum.
15. In a double-slit interference experiment (Fig. 37.4), $d = 0.2$ mm, $L = 160$ cm, and $y = 1$ mm. What wavelength will result in an average intensity at $P$ that is 36% of the maximum?
16. In an arrangement similar to that illustrated in Figure 37.4, let $L = 140$ cm and $y = 8$ mm. Find the value of the ratio $d/\lambda$ for which the average intensity at point $P$ will be 60% of the maximum intensity.
17. Make a plot of $I/I_0$ as a function of $\theta$ (see Fig. 37.4) for the interference pattern produced by the arrangement described in Problem 1. Let $\theta$ range over the interval from 0 to 0.3°.
18. In Figure 37.4 let $L = 1.2$ m and $d = 0.12$ mm and assume that the slit system is illuminated with monochromatic light of wavelength 500 nm. Calculate the phase difference between the two wavefronts arriving at point $P$ from $S_1$ and $S_2$ when (a) $\theta = 0.5°$ and (b) $y = 5$ mm.
19. For the situation described in Problem 18, what is the value of $\theta$ for which (a) the phase difference will be equal to 0.333 rad and (b) the path difference will be $\lambda/4$?
20. The intensity on the screen at a certain point in a double-slit interference pattern is 64% of the maximum value. (a) What is the minimum phase difference (in radians) between sources that will produce this result? (b) Express the phase difference calculated in (a) as a path difference if the wavelength of the incident light is 486.1 nm ($H_\beta$ line).
21. At a particular location in a Young's interference pattern, the intensity on the screen is 6.4% of maximum. (a) Calculate the minimum phase difference in this case. (b) If the wavelength of light is 587.5 nm (from a helium discharge tube), determine the path difference.
22. Light from a helium-neon laser ($\lambda = 632.8$ nm) is incident on two parallel slits 0.2 mm apart. What is the distance to the first maximum and its intensity (relative to the central maximum) on a screen 2 m beyond the slits?
23. Two narrow parallel slits are separated by 0.85 mm and are illuminated by light with $\lambda = 6000$ Å. (a) What is the phase difference between the two interfering waves on a screen (2.8 m away) at a point 2.50 mm from the central bright fringe? (b) What is the ratio of the intensity at this point to the intensity at the center of a bright fringe?

### Section 37.4 Phasor Addition of Waves

24. The electric fields from three coherent sources are described by $E_1 = E_0 \sin \omega t$, $E_2 = E_0 \sin(\omega t + \phi)$, and $E_3 = E_0 \sin(\omega t + 2\phi)$. Let the resultant field be represented by $E_P = E_R \sin(\omega t + \alpha)$. Use the phasor method to find $E_R$ and $\alpha$ when (a) $\phi = 20°$, (b) $\phi = 60°$, (c) $\phi = 120°$.
25. Repeat Problem 24 when $\phi = (3\pi/2)$ radians.
26. Use the method of phasors to find the resultant (magnitude and phase angle) of two fields represented by $E_1 = 12 \sin \omega t$ and $E_2 = 18 \sin(\omega t + 60°)$. (Note that in this case the amplitudes of the two fields are unequal.)
27. Determine the resultant of the two waves $E_1 = 6 \sin(100\ \pi t)$ and $E_2 = 8 \sin(100\ \pi t + \pi/2)$.
28. Two coherent waves are described by

$$E_1 = E_0 \sin\left(\frac{2\pi x_1}{\lambda} - 2\pi f t + \frac{\pi}{6}\right)$$

$$E_2 = E_0 \sin\left(\frac{2\pi x_2}{\lambda} - 2\pi f t + \frac{\pi}{8}\right)$$

Determine the relationship between $x_1$ and $x_2$ that produces constructive interference when the two waves are superposed.

29. When illuminated, four equally spaced parallel slits act as multiple coherent sources, each differing in phase from the adjacent one by an angle $\phi$. Use a phasor diagram to determine the smallest value of $\phi$ for which the resultant of the four waves (assumed to be of equal amplitude) will be zero.
30. Sketch a phasor diagram to illustrate the resultant of $E_1 = E_{01} \sin \omega t$ and $E_2 = E_{02} \sin(\omega t + \phi)$, where $E_{02} = 1.5E_{01}$ and $\pi/6 \le \phi \le \pi/3$. Use the sketch and the law of cosines to show that, for two coherent waves, the resultant *intensity* can be written in the form $I_R = I_1 + I_2 + 2\sqrt{I_1 I_2} \cos \phi$.
31. Consider $N$ coherent sources described by $E_1 = E_0 \sin(\omega t + \phi)$, $E_2 = E_0 \sin(\omega t + 2\phi)$, $E_3 = E_0 \sin(\omega t + 3\phi)$, . . . , $E_N = E_0 \sin(\omega t + N\phi)$. Find the minimum value of $\phi$ for which $E_R = E_1 + E_2 + E_3 + \cdots E_N$ will be zero.

### Section 37.6 Interference in Thin Films

32. A material having an index of refraction of 1.30 is used to coat a piece of glass ($n = 1.50$). What should be the minimum thickness of this film in order to minimize reflected light at a wavelength of 500 nm?

33. A thin film of $MgF_2$ $10^{-5}$ cm thick ($n = 1.38$) is used to coat a camera lens. Will any wavelengths in the visible spectrum be intensified in the reflected light?

34. A soap bubble of index of refraction 1.33 strongly reflects both red and green colors in white light. What thickness of soap bubble allows this to happen? (In air, $\lambda$ red = 700 nm, $\lambda$ green = 500 nm.)

35. A soap bubble appears green when exposed to normally incident white light. Determine the minimum thickness of the soap film that can produce the observed effect. Take the wavelength of green light to be 500 nm. The index of refraction of the film is 1.33.

36. A thin layer of oil ($n = 1.25$) is floating on water. How thick is the oil in the region that reflects green light ($\lambda = 525$ nm)?

37. A thin layer of liquid methylene iodide ($n = 1.756$) is sandwiched between two flat parallel plates of glass. What must be the thickness of the liquid layer if normally incident light with $\lambda = 600$ nm is to be strongly reflected?

38. A beam of light of wavelength 580 nm passes through two closely spaced glass plates, as shown in Figure 37.22. For what minimum nonzero value of the plate separation, $d$, will the transmitted light be bright?

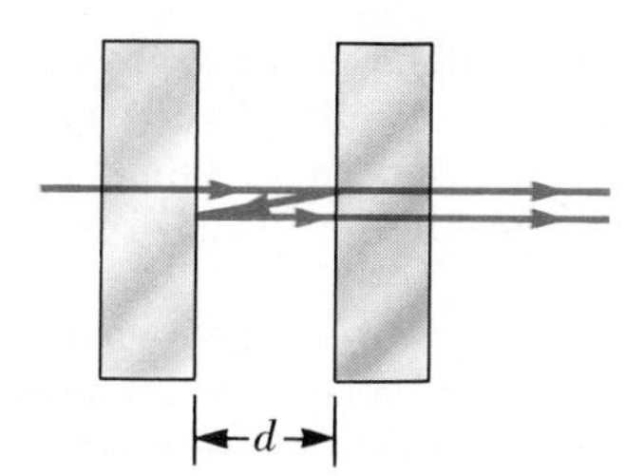


**Figure 37.22** (Problem 38).

39. An oil film ($n = 1.45$) floating on water is illuminated by white light at normal incidence. The film is 280 nm thick. Find (a) the dominant observed color in the reflected light and (b) the dominant color in the transmitted light. Explain your reasoning.

40. A possible means for making an airplane radar-invisible is to coat the plane with an antireflective polymer. If radar waves have a wavelength of 3 cm and the index of refraction of the polymer is $n = 1.5$, how thick would you make the coating?

41. A thin film of cryolite ($n = 1.35$) is applied to a camera lens ($n = 1.50$). The coating is designed to reflect wavelengths at the blue end of the spectrum and to transmit wavelengths in the near infrared. What minimum thickness will give high reflectivity at 450 nm and high transmission at 900 nm?

42. Two rectangular, optically flat glass plates ($n = 1.52$) are in contact along one end and are separated along the other end by a sheet of paper that is $4 \times 10^{-3}$ cm thick (Fig. 37.23). The top plate is illuminated by monochromatic light ($\lambda = 546.1$ nm). Calculate the number of dark parallel bands crossing the top plate (include the dark band at zero thickness along the edge of contact between the two plates).

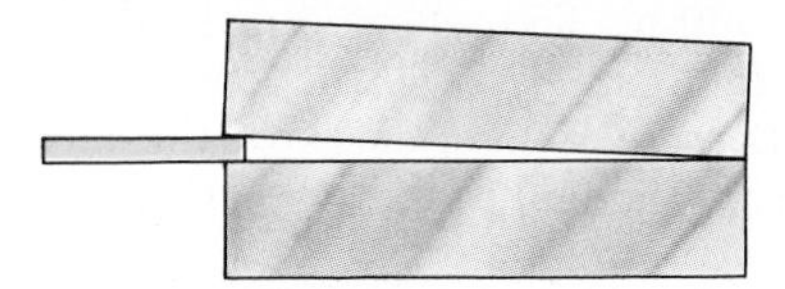

**Figure 37.23** (Problems 42 and 43).

43. An air wedge is formed between two glass plates separated at one edge by a very fine wire as in Figure 37.23. When the wedge is illuminated from above by light with a wavelength of 600 nm, 30 dark fringes are observed. Calculate the radius of the wire.

44. When a liquid is introduced into the air space between the lens and the plate in a Newton's-rings apparatus, the diameter of the tenth ring changes from 1.50 to 1.31 cm. Find the index of refraction of the liquid.

### *Section 37.7 The Michelson Interferometer

45. The mirror on one arm of a Michelson interferometer is displaced a distance $\Delta L$. During this displacement, 250 fringe shifts (formation of *successive dark or bright line fringes*) are counted. The light being used has a wavelength of 632.8 nm. Calculate the displacement $\Delta L$.

46. Light from a helium-cadmium laser is beamed into a Michelson interferometer. The movable mirror is displaced 0.382 mm, causing the interferometer pattern to reproduce itself 1700 times. Determine the wavelength of the laser light. What color is it?

47. Light of wavelength 550.5 nm is used to calibrate a Michelson interferometer. By use of a micrometer screw, the platform on which one mirror is mounted is moved 0.18 mm. How many *dark* fringe shifts are counted?

48. A microwave version of the Michelson interferometer uses waves whose wavelength is 4 cm. The detector receiving the two interfering microwave beams is sensitive to phase differences corresponding to path differences of one-tenth of a wavelength. What is the smallest displacement that can be detected with the interferometer?

## ADDITIONAL PROBLEMS

49. The calculation of Example 37.1 shows that the double-slit arrangement produced fringe separations of 2.2 cm for $\lambda = 560$ nm. Calculate the fringe separa-

tion for this same arrangement if the apparatus is submerged in a tank containing a 30% sugar solution ($n = 1.38$).

50. Interference effects are produced at point $P$ on a screen as a result of direct rays from a source of wavelength 500 nm and reflected rays off the mirror, as shown in Figure 37.24. If the source is 100 m to the left of the screen, and 1 cm above the mirror, find the distance $y$ (in mm) to the first dark band above the mirror.

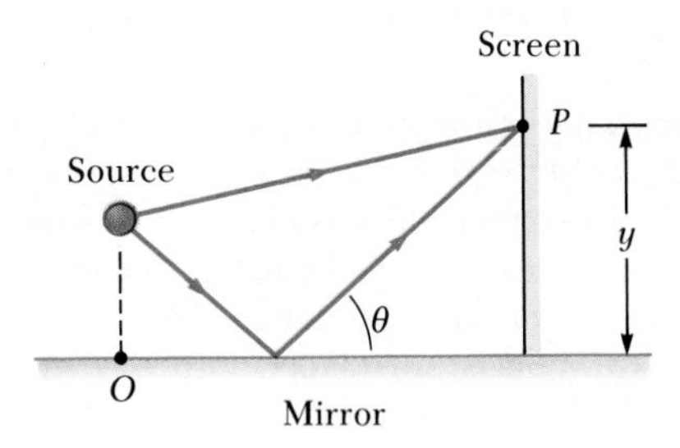


**Figure 37.24** (Problem 50).

51. In an application of interference effects to radioastronomy, Australian astronomers observed a 60-MHz radiosource both directly from the source and from its reflection from the sea. If the receiving dish is 20 m above sea level, what is the angle of the radiosource above the horizon at first maximum?

52. The waves from a radio station can reach a home receiver by two different paths. One is a straight line path from the transmitter to the home, a distance of 30 km. The second path is by reflection from the ionosphere (a layer of ionized air molecules near the top of the atmosphere). Assume this reflection takes place at a point midway between receiver and transmitter. If the wavelength broadcast by the radio station is 350 m, find the minimum height of the ionospheric layer that would produce destructive interference between the direct and reflected beams. (Assume no phase changes on reflection.)

53. Measurements are made of the intensity distribution in a Young's interference pattern (as illustrated in Figure 37.6). At a particular value of $y$ (distance from the center of the screen), it is found that $I/I_0 = 0.81$ when light of wavelength 600 nm is used. What wavelength of light should be used to reduce the relative intensity at the same location to 64%?

54. Waves broadcast by a 1500-kHz radio station arrive at a home receiver by two paths. One is a direct path, and the second is from reflection off an airplane directly above the home receiver. The airplane is approximately 100 m above the home receiver, and the direct distance from the station to the home is 20 km. What is the exact height of the airplane if destructive interference is occurring? (Assume no phase change occurs on reflection from the airplane.)

55. In a Young's interference experiment, the two slits are separated by 0.15 mm and the incident light includes light of wavelengths $\lambda_1 = 540$ nm and $\lambda_2 = 450$ nm. The overlapping interference patterns are formed on a screen 1.4 m from the slits. Calculate the minimum distance from the center of the screen to the point where a bright line of the $\lambda_1$ light coincides with a bright line of the $\lambda_2$ light.

56. Two sinusoidal phasors of the same amplitude $A$ and frequency $\omega$ have a phase difference $\phi$. Calculate the resultant amplitude of the two phasors both graphically and analytically if $\phi$ equals (a) 0, (b) 60°, (c) 90°.

**57.** Young's double-slit experiment is performed with sodium yellow light ($\lambda = 5890$ Å) and with a slits-to-screen distance of 2.0 m. The tenth interference minimum (dark fringe) is observed to be 7.26 mm from the central maximum. Determine the spacing of the slits.

58. A thin film of oil ($n = 1.38$) having a thickness $d$ floats on water. (a) Show that the equation

$$\lambda = \frac{2nd}{(m + \frac{1}{2})} \qquad m = 0, 1, 2, \ldots$$

where $d$ is the film thickness, gives the wavelengths that exhibit constructive interference for reflected light. (b) What color is the film at points where $d = 300$ nm? (The film is viewed from above in white light.)

59. A hair is placed at one edge between two flat glass plates 8 cm long. When this arrangement is illuminated with yellow light of wavelength 600 nm, a total of 121 dark bands are counted, starting at the point of contact of the two plates. How thick is the hair?

60. A glass plate ($n = 1.61$) is covered with a thin uniform layer of oil ($n = 1.2$). A nonmonochromatic light beam in air is incident normally on the oil surface. Observation of the reflected beam shows destructive interference at 500 nm and constructive interference at 750 nm with no intervening maxima or minima. Calculate the thickness of the oil film.

61. A piece of transparent material having an index of refraction $n$ is cut into the shape of a wedge as shown in Figure 37.25. The angle of the wedge is small, and monochromatic light of wavelength $\lambda$ is normally incident from above. If the height of the wedge is $h$ and the width is $\ell$, show that bright fringes occur at the positions $x = \lambda\ell(m + \frac{1}{2})/2hn$ and dark fringes occur at the positions $x = \lambda\ell m/2hn$, where $m = 0, 1, 2, \ldots$ and $x$ is measured as shown.

62. An air wedge is formed between two glass plates in contact along one edge and slightly separated at the opposite edge. When illuminated with monochromatic light from above, the reflected light reveals a

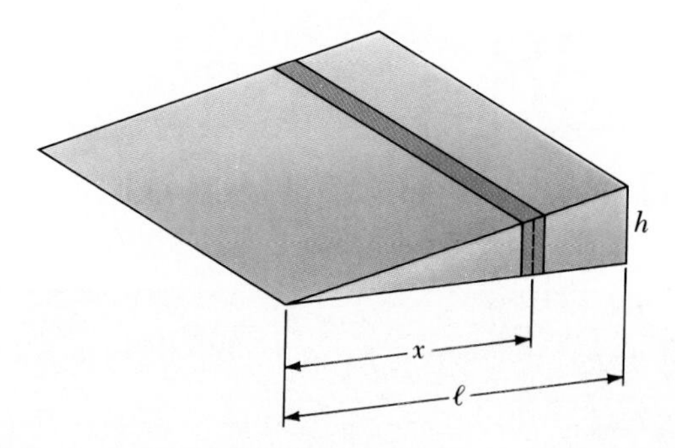


Figure 37.25 (Problem 61).

total of 85 dark fringes. Calculate the number of dark fringes that would appear if water ($n = 1.33$) were to replace the air between the plates.

63. The condition for constructive interference by reflection from a thin film in air as developed in Section 37.6 assumes nearly normal incidence. (a) Show that if the light is incident on the film at an angle $\phi_1 \gg 0$ (relative to the normal), then the condition for constructive interference is given by $2nt \cos \theta_2 = (m + \frac{1}{2})\lambda$, where $\theta_2$ is the angle of refraction. (b) Calculate the minimum thickness for constructive interference if sodium light ($\lambda = 5.9 \times 10^{-5}$ cm) is incident at an angle of 30° on a film with index of refraction 1.38.

**64.** Use the method of phasor addition to find the resultant amplitude and phase constant when the following three harmonic functions are combined: $E_1 = \sin(\omega t + \pi/6)$, $E_2 = 3 \sin(\omega t + 7\pi/2)$, $E_3 = 6 \sin(\omega t + 4\pi/3)$.

65. A fringe pattern is established in the field of view of a Michelson interferometer using light of wavelength 580 nm. A parallel-faced sheet of transparent material 2.5 μm thick is placed in front of one of the mirrors perpendicular to the incident and reflected light beams. An observer counts a fringe shift of six *dark* fringes. What is the index of refraction of the sheet?

66. A soap film ($n = 1.33$) is contained within a rectangular wire frame. The frame is held vertically so that the film drains downward due to gravity and becomes thicker at the bottom than at the top, where the thickness is essentially zero. The film is viewed in white light with near-normal incidence, and the first violet ($\lambda = 420$ nm) interference band is observed 3 cm from the top edge of the film. (a) Locate the first red ($\lambda = 680$ nm) interference band. (b) Determine the film thickness at the positions of the violet and red bands. (c) What is the wedge angle of the film?

67. A soap film such as that in Problem 66 is viewed in yellow light ($\lambda = 589$ nm) with near-normal incidence. Interference fringes with a uniform spacing of 4.5 mm are observed. What is the variation in thickness of the film per centimeter of vertical distance?

68. We can produce interference fringes using a *Lloyd's mirror* arrangement with a single monochromatic source with $\lambda = 606$ nm as in Figure 37.26. The image S′ of the source formed by the mirror acts as a second coherent source that interferes with $S_0$. If fringes spaced 1.2 mm apart are formed on a screen 2 m from the source $S_0$, find the vertical distance $h$ of the source above the plane of the reflecting surface.

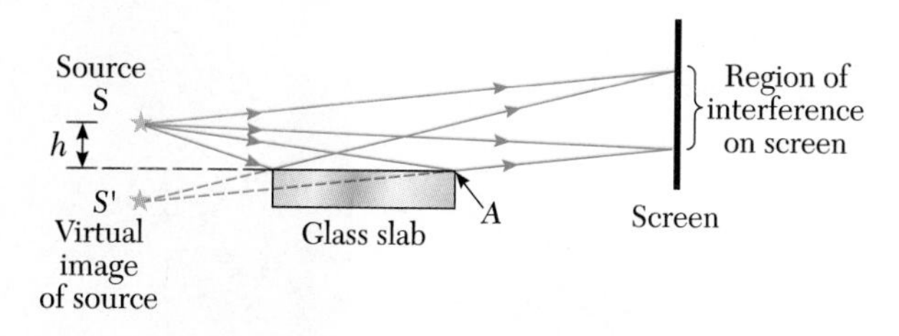


Figure 37.26 (Problem 68).

69. (a) Both sides of a uniform film of index of refraction $n$ and thickness $d$ are in contact with air. For normal incidence of light, an intensity minimum is observed in the reflected light at $\lambda_2$ and an intensity maximum is observed at $\lambda_1$, where $\lambda_1 > \lambda_2$. If there are no intensity minima observed between $\lambda_1$ and $\lambda_2$, show that the integer $m$ that appears in Equations 37.16 and 37.17 is given by $m = \lambda_1/2(\lambda_1 - \lambda_2)$. (b) Determine the thickness of the film if $n = 1.40$, $\lambda_1 = 500$ nm, and $\lambda_2 = 370$ nm.

70. Consider the double-slit arrangement shown in Figure 37.27, where the separation $d$ of the slits is 0.30 mm and the distance $L$ to the screen is 1 m. A thin sheet of transparent plastic of thickness 0.050 mm (about the thickness of this page) and a refractive index of $n = 1.50$, is placed over only the upper slit. As a result, the central maximum of the interference pattern moves upward a distance $y'$. Find this distance.

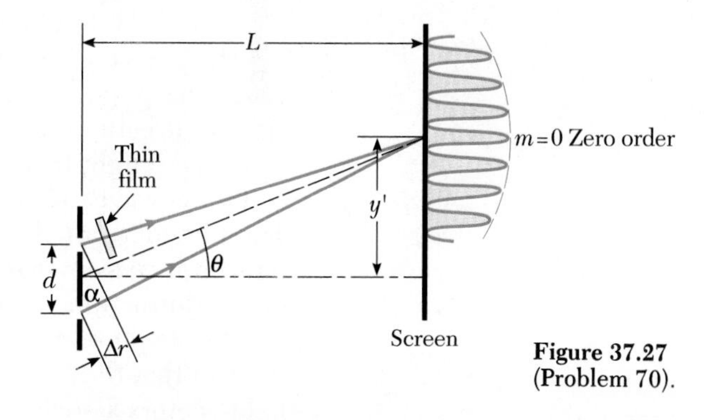


Figure 37.27 (Problem 70).

71. One slit of a double slit is wider than the other so that one slit emits light with three times greater amplitude than the other slit. Show that Equation 37.11 would then have the form $I = (4I_0/9)(1 + 3 \cos^2 \phi/2)$.

# 38

# Diffraction and Polarization

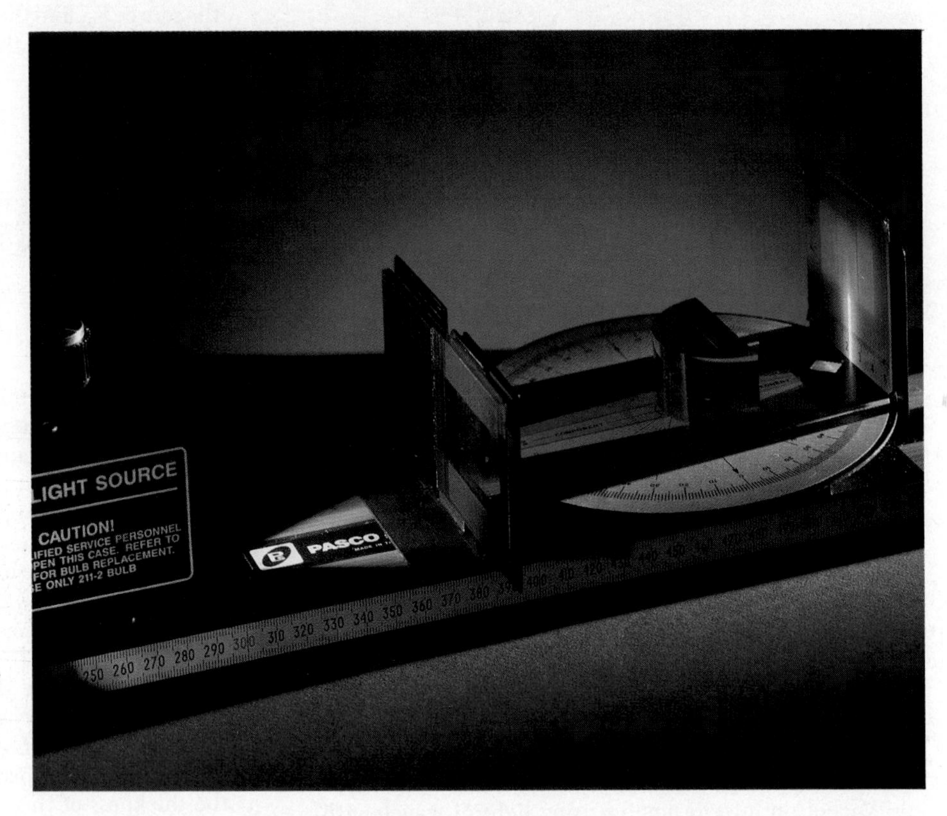


*White light is dispersed into its various spectral components using a diffraction grating, as shown in this photograph. (Courtesy of PASCO Scientific)*

In this chapter, we continue our treatment of wave optics with the discussion of diffraction and polarization phenomena. When light waves pass through a small aperture, an interference pattern is observed rather than a sharp spot of light cast by the aperture. This shows that light spreads in various directions beyond the aperture into regions where a shadow would be expected if light traveled in straight lines. Other waves, such as sound waves and water waves, also have this property of being able to bend around corners. This phenomenon, known as *diffraction,* can be regarded as a consequence of interference from many coherent wave sources. In other words, the phenomena of diffraction and interference are basically equivalent.

In Chapter 34, we discussed the properties of electromagnetic waves and the fact that they are transverse in nature. That is, the electric and magnetic field vectors associated with the wave are perpendicular to the direction of propagation. Under certain conditions, light waves can be plane-polarized. Although ordinary light is usually not polarized, we shall discuss various methods for producing polarized light, such as by using polarizing sheets.

## 38.1 INTRODUCTION TO DIFFRACTION

Suppose a light beam is incident on two slits, as in Young's double-slit experiment. If the light truly traveled in straight-line paths after passing through the slits, as in Figure 38.1a, the waves would not overlap and no interference pattern would be seen. Instead, Huygens' principle requires that the waves spread out from the slits as shown in Figure 38.1b. In other words, the light deviates from a straight-line path and enters the region that would otherwise be shadowed. This divergence of light from its initial line of travel is called **diffraction.**

In general, diffraction occurs when waves pass through small openings, around obstacles, or by relatively sharp edges. As an example of diffraction, consider the following. When an opaque object is placed between a point source of light and a screen, as in Figure 38.2a, the boundary between the shadowed and illuminated regions on the screen is not sharp. A careful inspection of the boundary shows that a small amount of light bends into the shadowed region. The region outside the shadow contains alternating light and dark bands, as in Figure 38.2b. The intensity in the first bright band is greater than the intensity in the region of uniform illumination. Effects of this type were first reported in the 17th century by Francesco Grimaldi.

Figure 38.3 shows the diffraction pattern and shadow of a penny. There is a bright spot at the center, circular fringes near the shadow's edge, and another set of fringes outside the shadow. This particular type of diffraction pattern was first observed in 1818 by Dominique Arago. The bright spot at the center of the shadow can be explained only through the use of the wave theory of light, which predicts constructive interference at this point. From the viewpoint of geometrical optics, the center of the pattern would be screened by the object, hence one would never expect to observe a central bright spot. It is interesting to point out a historical incident that occurred shortly before the central bright spot was observed. One of the supporters of geometrical optics, Simeon Poisson, argued that if Augustin Fresnel's wave theory of light were valid, then a central bright spot should be observed. Because the spot was

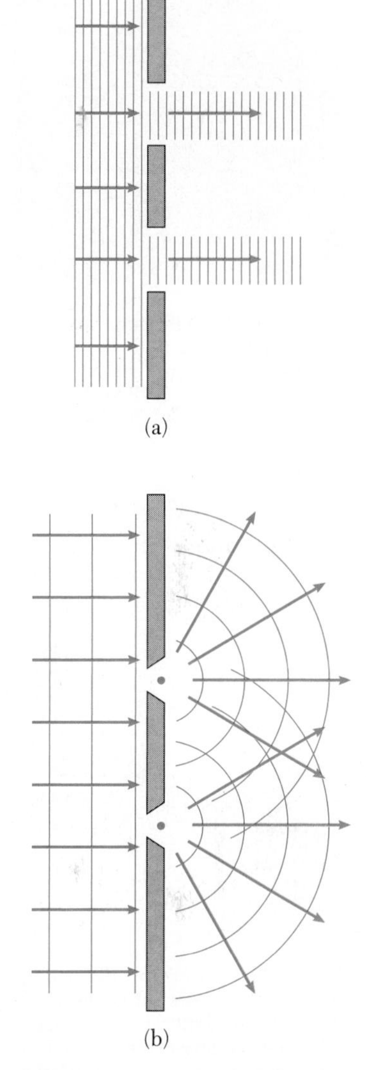


**Figure 38.1** (a) If light waves did not spread out after passing through the slits, no interference would occur. (b) The light waves from the two slits overlap as they spread out, producing interference fringes.

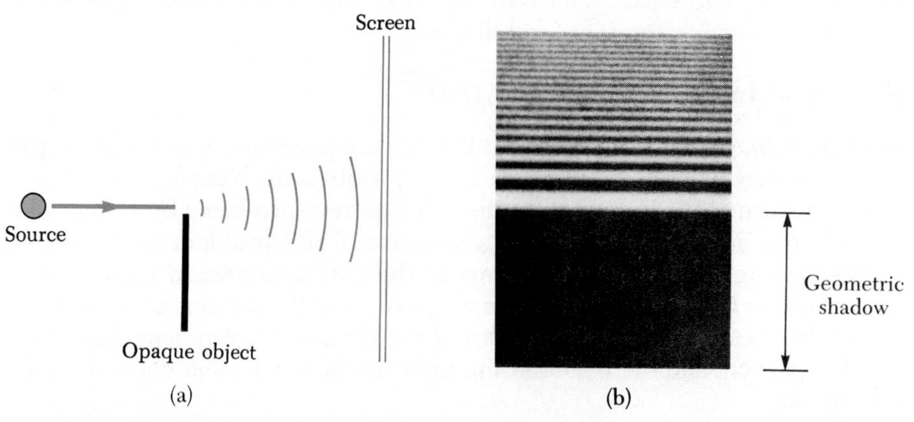


**Figure 38.2** (a) Light bends around the opaque object. (b) The result is a series of dark and light bands in the area that would be completely in the shadow if light did not bend around the object.

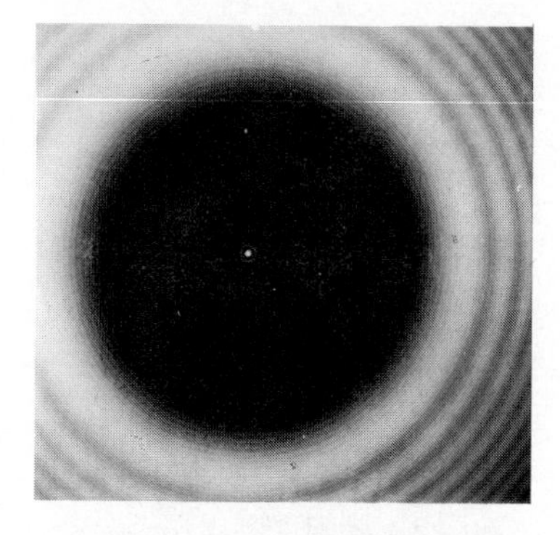

Figure 38.3 Diffraction pattern of a penny, taken with the penny midway between screen and source. (Courtesy of P.M. Rinard, from *Am. J. Phys.* 44:70, 1976)

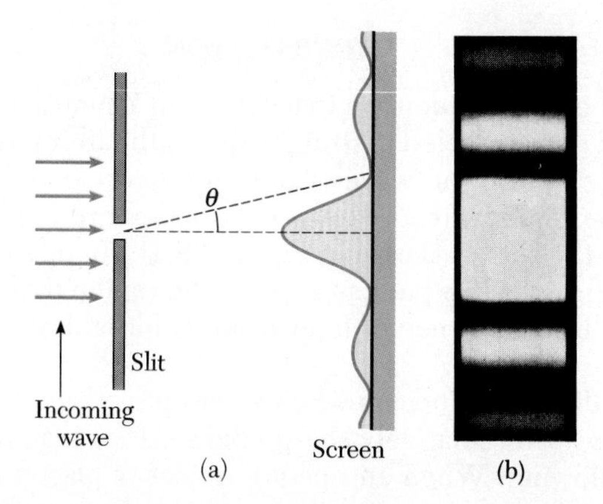


**Figure 38.4** (a) Fraunhofer diffraction pattern of a single slit. The pattern consists of a central bright region flanked by much weaker maxima. (Note that this is not to scale.) (b) Photograph of a single-slit Fraunhofer diffraction pattern. (From M. Cagnet, M. Francon, and J.C. Thierr, *Atlas of Optical Phenomena,* Berlin, Springer-Verlag, 1962, plate 18)

observed shortly thereafter, Poisson's prediction *reinforced* the wave theory rather than disproved it. This was certainly a most dramatic experimental proof of the wave nature of light.

Diffraction phenomena are usually classified as being of two types, which are named after the men who first explained them. The first type, called **Fraunhofer diffraction,** occurs when the rays reaching a point are approximately parallel. This can be achieved experimentally either by placing the observing screen far from the opening as in Figure 38.4a or by using a converging lens to focus the parallel rays on the screen. A bright fringe is observed along the axis at $\theta = 0$, with alternating dark and bright fringes on either side of the central bright fringe. Figure 38.4b is a photograph of a single-slit Fraunhofer diffraction pattern.

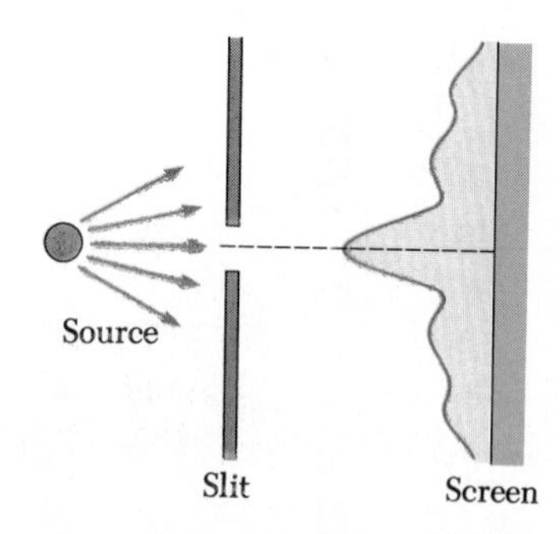


**Figure 38.5** A Fresnel diffraction pattern of a single slit is observed when the incident rays are not parallel and the observing screen is a finite distance from the slit. (Note that this is not to scale.)

When the observing screen is placed a finite distance from the slit and no lens is used to focus parallel rays, the observed pattern is called a **Fresnel diffraction** pattern (Fig. 38.5). The diffraction patterns shown in Figures 38.2b and 38.3 are examples of Fresnel diffraction. Fresnel diffraction is rather complex to treat quantitatively. Therefore, the following discussion will be restricted to Fraunhofer diffraction.

## 38.2 SINGLE-SLIT DIFFRACTION

Up until now, we have assumed that the slits are point sources of light. In this section, we shall determine how their finite width is the basis for understanding the nature of the Fraunhofer diffraction pattern produced by a single slit.

We can deduce some important features of this problem by examining waves coming from various portions of the slit, as shown in Figure 38.6. According to Huygens' principle, *each portion of the slit acts as a source of waves.* Hence, *light from one portion of the slit can interfere with light from another portion,* and the resultant intensity on the screen will depend on the direction $\theta$.

To analyze the diffraction pattern, it is convenient to divide the slit in two halves, as in Figure 38.6. All the waves that originate from the slit are in phase. Consider waves 1 and 3, which originate from the bottom and center of the slit, respectively. Wave 1 travels farther than wave 3 by an amount equal to the path difference $(a/2)\sin\theta$, where $a$ is the width of the slit. Similarly, the path difference between waves 2 and 4 is also $(a/2)\sin\theta$. If this path difference is exactly one half of a wavelength (corresponding to a phase difference of 180°), the two waves cancel each other and destructive interference results. This is true, in fact, for any two waves that originate at points separated by half the slit width because the phase difference between two such points is 180°. Therefore, waves from the upper half of the slit interfere *destructively* with waves from the lower half of the slit when

$$\frac{a}{2}\sin\theta = \frac{\lambda}{2}$$

or when

$$\sin\theta = \frac{\lambda}{a}$$

If we divide the slit into four parts rather than two and use similar reasoning, we find that the screen is also dark when

$$\sin\theta = \frac{2\lambda}{a}$$

Likewise, we can divide the slit into six parts and show that darkness occurs on the screen when

$$\sin\theta = \frac{3\lambda}{a}$$

Therefore, the general condition for **destructive interference** is

$$\sin\theta = m\frac{\lambda}{a} \qquad (m = \pm 1, \pm 2, \pm 3, \ldots) \qquad (38.1)$$

Condition for destructive interference

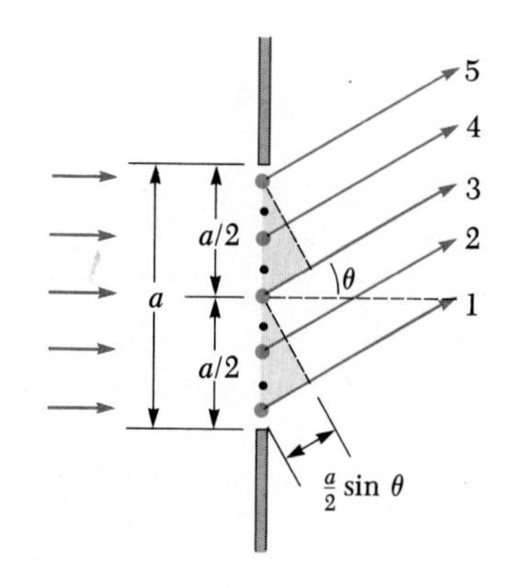


**Figure 38.6** Diffraction of light by a narrow slit of width $a$. Each portion of the slit acts as a point source of waves. The path difference between rays 1 and 3 or between rays 2 and 4 is equal to $(a/2)\sin\theta$. (Note that this is not to scale.)

Equation 38.1 gives the values of $\theta$ for which the diffraction pattern has zero intensity, that is, where a dark fringe is formed. However, it tells us nothing about the variation in intensity along the screen. The general features of the intensity distribution along the screen are shown in Figure 38.7. A broad central bright fringe is observed, flanked by much weaker, alternating

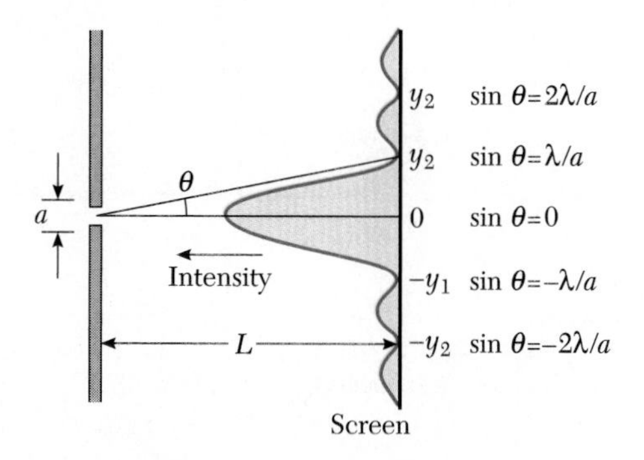


**Figure 38.7** Positions of the minima for the Fraunhofer diffraction pattern of a single slit of width $a$. The Fraunhofer diffraction pattern is obtained only if $L \gg a$. (Note that this is not to scale.)

bright fringes. The various dark fringes (points of zero intensity) occur at the values of $\theta$ that satisfy Equation 38.1. The position of the points of constructive interference lie approximately halfway between the dark fringes. Note that the central bright fringe is twice as wide as the weaker maxima.

**EXAMPLE 38.1 Where Are The Dark Fringes?**
Light of wavelength 580 nm is incident on a slit of width 0.30 mm. The observing screen is placed 2 m from the slit. Find the positions of the first dark fringes and the width of the central bright fringe.

*Solution* The first dark fringes that flank the central bright fringe correspond to $m = \pm 1$ in Equation 38.1. Hence, we find that

$$\sin \theta = \pm \frac{\lambda}{a} = \pm \frac{5.8 \times 10^{-7}\ \text{m}}{0.3 \times 10^{-3}\ \text{m}} = \pm 1.93 \times 10^{-3}$$

From the triangle in Figure 38.7, note that $\tan \theta = y_1/L$. Since $\theta$ is very small, we can use the approximation $\sin \theta \approx \tan \theta$, so that $\sin \theta \approx y_1/L$. Therefore, the positions of the first minima measured from the central axis are given by

$$y_1 \approx L \sin \theta = \pm L \frac{\lambda}{a} = \pm 3.87 \times 10^{-3}\ \text{m}$$

The positive and negative signs correspond to the dark fringes on either side of the central bright fringe. Hence, the width of the central bright fringe is equal to $2|y_1| = 7.73 \times 10^{-3}\ \text{m} = 7.73$ mm. Note that this value is *much larger* than the width of the slit. However, as the width of the slit is *increased*, the diffraction pattern will *narrow*, corresponding to smaller values of $\theta$. In fact, for large values of $a$, the various maxima and minima are so closely spaced that only a large central bright area is observed, which resembles the geometric image of the slit. This matter is of great importance in the design of lenses used in telescopes, microscopes, and other optical instruments.

**Exercise 1** Determine the width of the first order bright fringe.
**Answer** 3.87 mm.

### Intensity of the Single-Slit Diffraction Pattern

We can make use of phasors to determine the intensity distribution for the single-slit diffraction pattern. Imagine that a slit is divided into a large number of small zones, each of width $\Delta y$ as in Figure 38.8. Each zone acts as a source of coherent radiation, and each contributes an incremental electric field amplitude $\Delta E$ at some point $P$ on the screen. The total electric field amplitude $E$ at the point $P$ is obtained by summing the contributions from each zone. Note that the incremental electric field amplitudes between adjacent zones are out

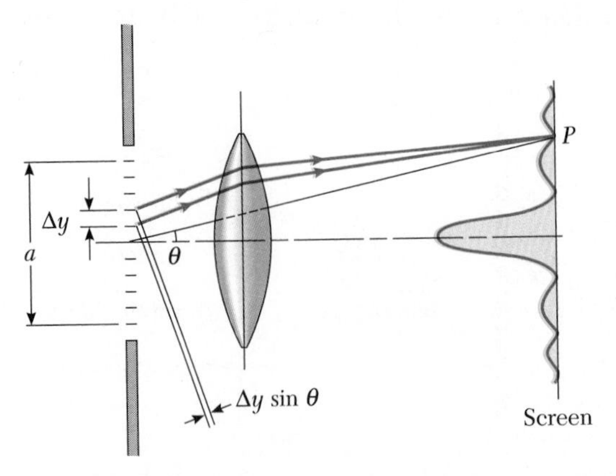


**Figure 38.8** Fraunhofer diffraction by a single slit. The intensity at the point $P$ on the screen is the resultant of all the incremental fields from zones of width $\Delta y$ leaving the slit.

of phase with one another by an amount $\Delta\beta$, corresponding to a path difference of $\Delta y \sin\theta$ (see Fig. 38.8). The ratio

$$\frac{\Delta\beta}{2\pi} = \frac{\Delta y \sin\theta}{\lambda}$$

can be written as

$$\Delta\beta = \frac{2\pi}{\lambda}\,\Delta y \sin\theta \qquad (38.2)$$

To find the total electric field amplitude on the screen at any angle $\theta$, we sum the incremental amplitudes due to each zone. For small values of $\theta$, we can assume that the amplitudes $\Delta E$ due to each zone are the same. It is convenient to use the phasor diagrams for various angles as in Figure 38.9. When $\theta = 0$, all phasors are aligned as in Figure 38.9a, since the waves from each zone are in phase. In this case, the total amplitude at the center of the screen is $E_0 = N\,\Delta E$, where $N$ is the number of zones. The amplitude $E_\theta$ at some small angle $\theta$ is shown in Figure 38.9b, where each phasor differs in phase from an adjacent one by an amount $\Delta\beta$. In this case, note that $E_\theta$ is the *vector sum* of the incremental amplitudes, and hence is given by the length of the chord. Therefore, $E_\theta < E_0$. The total phase difference between waves from the top and bottom portions of the slit is

$$\beta = N\,\Delta\beta = \frac{2\pi}{\lambda} N\,\Delta y \sin\theta = \frac{2\pi}{\lambda} a \sin\theta \qquad (38.3)$$

where $a = N\,\Delta y$ is the width of the slit.

As $\theta$ increases, the chain of phasors eventually forms a closed path as in Figure 38.9c. At this point, the vector sum is zero, so $E_\theta = 0$, corresponding to the first minimum on the screen. Noting that $\beta = N\,\Delta\beta = 2\pi$ in this situation, we see from Equation 38.3

$$2\pi = \frac{2\pi}{\lambda} a \sin\theta$$

or

$$\sin\theta = \frac{\lambda}{a}$$

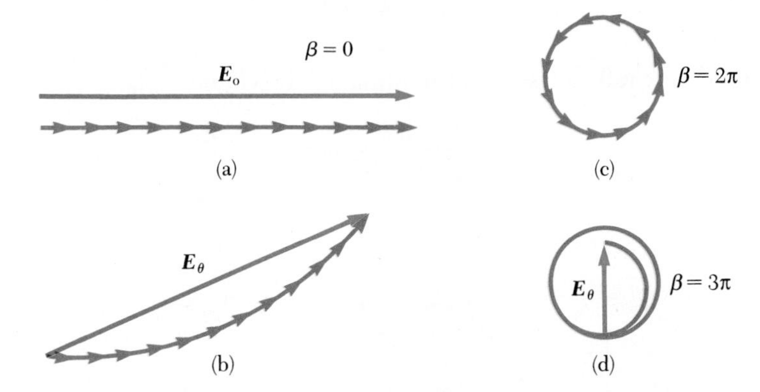


**Figure 38.9** Phasor diagrams for obtaining the various maxima and minima of the single slit diffraction pattern.

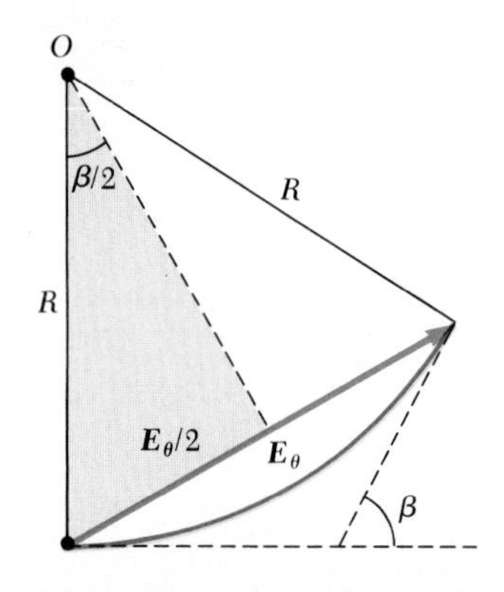


**Figure 38.10** Phasor diagram for a large number of coherent sources. Note that the ends of the phasors lie on a circular arc of radius $R$. The resultant amplitude $E_\theta$ equals the length of the chord.

That is, the first minimum in the diffraction pattern occurs when $\sin\theta = \lambda/a$, which agrees with Equation 38.1.

At even larger values of $\theta$, the spiral chain of phasors continues. For example, Figure 38.9d represents the situation corresponding to the second maximum, which occurs when $\beta \simeq 360° + 180° = 540°$ ($3\pi$ rad). The second minimum (two complete spirals not shown) corresponds to $\beta = 720°$ ($4\pi$ rad), which satisfies the condition $\sin\theta = 2\lambda/a$.

The total amplitude and intensity at any point on the screen can now be obtained by considering the limiting case where $\Delta y$ becomes infinitesimal ($dy$) and $N \rightarrow \infty$. In this limit, the phasor diagrams in Figure 38.9 become smooth curves, as in Figure 38.10. From this figure, we see that at some angle $\theta$, the wave amplitude on the screen, $E_\theta$, is equal to the chord length, while $E_0$ is the arc length. From the triangle whose angle is $\beta/2$, we see that

$$\sin\frac{\beta}{2} = \frac{E_\theta/2}{R}$$

where $R$ is the radius of curvature. But the arc length $E_0$ is equal to the product $R\beta$, where $\beta$ is in radians. Combining this with the expression above gives

$$E_\theta = 2R\sin\frac{\beta}{2} = 2\left(\frac{E_0}{\beta}\right)\sin\frac{\beta}{2}$$

or

$$E_\theta = E_0\left[\frac{\sin(\beta/2)}{\beta/2}\right]$$

Since the resultant intensity $I_\theta$ at $P$ is proportional to the square of the amplitude $E_\theta$, we find

$$I_\theta = I_0\left[\frac{\sin(\beta/2)}{\beta/2}\right]^2 \qquad (38.4)$$

where $I_0$ is the intensity at $\theta = 0$ (the central maximum), and $\beta = 2\pi a\sin\theta/\lambda$. Substitution of this expression for $\beta$ into Equation 38.4 gives

Intensity of a single-slit Fraunhofer diffraction pattern

$$I_\theta = I_0\left[\frac{\sin(\pi a\sin\theta/\lambda)}{\pi a\sin\theta/\lambda}\right]^2 \qquad (38.5)$$

From this result, we see that minima occur when

$$\frac{\pi a\sin\theta}{\lambda} = m\pi$$

or

Condition for intensity minima

$$\sin\theta = m\frac{\lambda}{a}$$

where $m = \pm1, \pm2, \pm3, \ldots$ . This is in agreement with our earlier result, given by Equation 38.1.

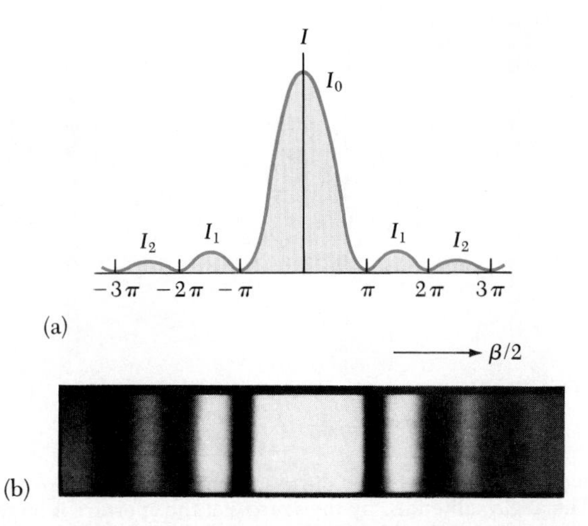


**Figure 38.11** (a) A plot of the intensity $I$ versus $\beta$ for the single-slit Fraunhofer diffraction pattern. (b) Photograph of a single-slit Fraunhofer diffraction pattern. (From M. Cagnet, M. Francon, and J.C. Thierr, *Atlas of Optical Phenomena,* Berlin, Springer-Verlag, 1962, plate 18)

Figure 38.11a represents a plot of Equation 38.5, and a photograph of a single-slit Fraunhofer diffraction pattern is shown in Figure 38.11b. Most of the light intensity is concentrated in the central bright fringe.

EXAMPLE 38.2 Relative Intensities of the Maxima

Find the ratio of intensities of the secondary maxima to the intensity of the central maximum for the single-slit Fraunhofer diffraction pattern.

*Solution* To a good approximation, we can assume that the secondary maxima lie midway between the zero points. From Figure 38.11a, we see that this corresponds to $\beta/2$ values of $3\pi/2$, $5\pi/2$, $7\pi/2$, . . . . Substituting these into Equation 38.4 gives for the first two ratios

$$\frac{I_1}{I_0} = \left[\frac{\sin(3\pi/2)}{(3\pi/2)}\right]^2 = \frac{1}{9\pi^2/4} = 0.045$$

$$\frac{I_2}{I_0} = \left[\frac{\sin(5\pi/2)}{5\pi/2}\right]^2 = \frac{1}{25\pi^2/4} = 0.016$$

That is, the secondary maximum that is adjacent to the central maximum has an intensity of 4.5% that of the central bright fringe, and the next secondary maximum has an intensity of 1.6% that of the central bright fringe.

Exercise 2 Determine the intensity of the secondary maximum corresponding to $m = 3$ relative to the central maximum.

Answer 0.0083.

## 38.3 RESOLUTION OF SINGLE-SLIT AND CIRCULAR APERTURES

The ability of optical systems such as microscopes and telescopes to distinguish between closely spaced objects is limited because of the wave nature of light. To understand this difficulty, consider Figure 38.12, which shows two light sources far from a narrow slit of width $a$. The sources can be considered as two point sources, $S_1$ and $S_2$, that are *not* coherent. For example, they could be two distant stars. If no diffraction occurred, one would observe two distinct bright spots (or images) on the screen at the right in the figure. However, because of diffraction, each source is imaged as a bright central region flanked by weaker bright and dark bands. What is observed on the screen is the sum of two diffraction patterns, one from $S_1$, and the other from $S_2$.

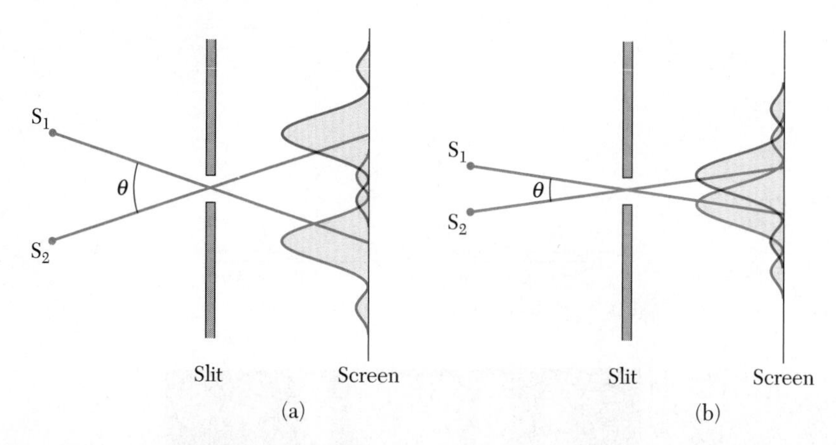


**Figure 38.12** Two point sources at some distance from a small aperture each produce a diffraction pattern. (a) The angle subtended by the sources at the aperture is large enough so that the diffraction patterns are distinguishable. (b) The angle subtended by the sources is so small that their diffraction patterns overlap and the images are not well resolved. (Note that the angles are greatly exaggerated.)

If the two sources are separated such that their central maxima do not overlap, as in Figure 38.12a, their images can be distinguished and are said to be *resolved.* If the sources are close together, however, as in Figure 38.12b, the two central maxima may overlap and the images are *not resolved.* To decide when two images are resolved, the following condition is often used:

When the central maximum of one image falls on the first minimum of another image, the images are said to be just resolved. This limiting condition of resolution is known as **Rayleigh's criterion.**

Figure 38.13 shows the diffraction patterns for three situations. When the objects are far apart, their images are well resolved (Fig. 38.13a). The images are just resolved when their angular separation satisfies Rayleigh's criterion (Fig. 38.13b). Finally, the images are not resolved in Figure 38.13c.

From Rayleigh's criterion, we can determine the minimum angular separation, $\theta_m$, subtended by the source at the slit such that their images will be just resolved. In Section 38.2, we found that the first minimum in a single-slit diffraction pattern occurs at the angle that satisfies the relationship

$$\sin\theta = \frac{\lambda}{a}$$

where $a$ is the width of the slit. According to Rayleigh's criterion, this expression gives the smallest angular separation for which the two images will be resolved. Because $\lambda \ll a$ in most situations, $\sin\theta$ is small and we can use the approximation $\sin\theta \approx \theta$. Therefore, the limiting angle of resolution for a slit of width $a$ is

Limiting angle of resolution for a slit

$$\theta_m = \frac{\lambda}{a} \qquad (38.6)$$

where $\theta_m$ is expressed in radians. Hence, the angle subtended by the two sources at the slit must be *greater* than $\lambda/a$ if the images are to be resolved.

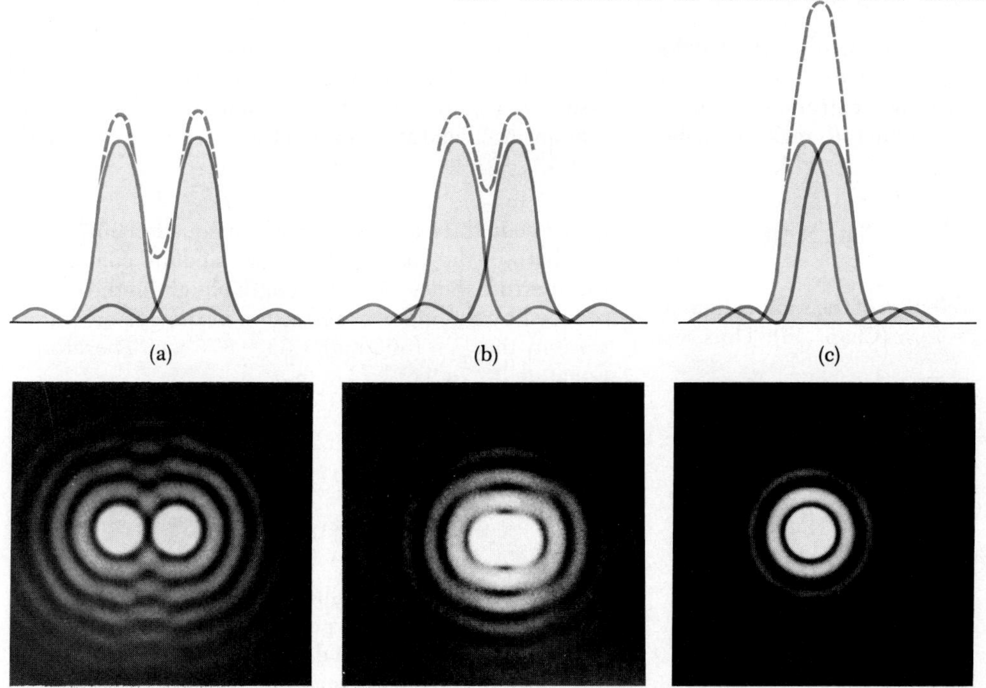


**Figure 38.13** The diffraction patterns of two point sources (solid curves) and the resultant pattern (dashed curves), for various angular separations of the sources. In each case, the dashed curve is the sum of the two solid curves. (a) The sources are far apart, and the patterns are well resolved. (b) The sources are closer together, and the patterns are just resolved. (c) The sources are so close together that the patterns are not resolved. (From M. Cagnet, M. Francon, and J.C. Thierr, *Atlas of Optical Phenomena,* Berlin, Springer-Verlag, 1962, plate 16)

Many optical systems use circular apertures rather than slits. The diffraction pattern of a circular aperture, illustrated in Figure 38.14, consists of a central circular bright disk surrounded by progressively fainter rings. Analysis shows that the limiting angle of resolution of the circular aperture is

$$\theta_m = 1.22 \frac{\lambda}{D} \qquad (38.7)$$

where $D$ is the diameter of the aperture. Note that Equation 38.7 is similar to Equation 38.6 except for the factor of 1.22, which arises from a complex mathematical analysis of diffraction from the circular aperture.

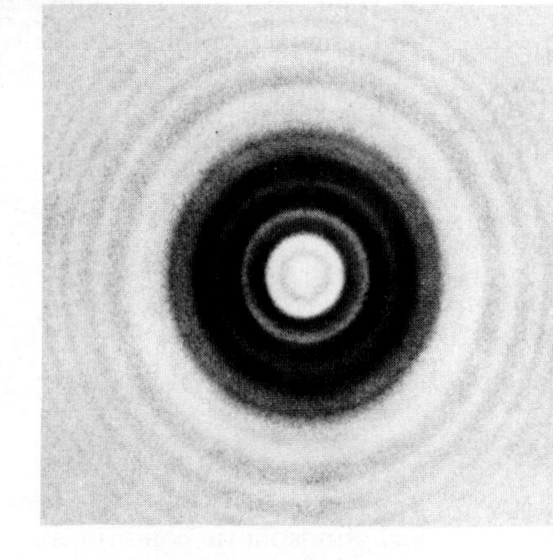

**Figure 38.14** The Fresnel diffraction pattern of a circular aperture consists of a central bright disk surrounded by concentric bright and dark rings. (From M. Cagnet, M. Francon, and J.C. Thierr, *Atlas of Optical Phenomena,* Berlin, Springer-Verlag, 1962, plate 34)

**EXAMPLE 38.3 Limiting Resolution of a Microscope**
Sodium light of wavelength 589 nm is used to view an object under a microscope. If the aperture of the objective has a diameter of 0.9 cm, (a) find the limiting angle of resolution. (b) Using visible light of any wavelength you desire, what is the maximum limit of resolution for this microscope? (c) Suppose water of index of refraction 1.33 fills the space between the object and the objective. What effect would this have on the resolving power of the microscope?

*Solution* (a) From Equation 38.7, we find the limiting angle of resolution to be

$$\theta_m = 1.22 \left( \frac{589 \times 10^{-9} \text{ m}}{0.9 \times 10^{-2} \text{ m}} \right) = 7.98 \times 10^{-5} \text{ rad}$$

This means that any two points on the object subtending an angle less than $8 \times 10^{-5}$ rad at the objective cannot be distinguished in the image.

(b) To obtain the smallest angle corresponding to the maximum limit of resolution, we have to use the shortest wavelength available in the visible spectrum. Violet light of wavelength 400 nm gives us a limiting angle of resolution of

$$\theta_m = 1.22\left(\frac{400 \times 10^{-9}\ \text{m}}{0.9 \times 10^{-2}\ \text{m}}\right) = 5.42 \times 10^{-5}\ \text{rad}$$

(c) In this case, the wavelength of the sodium light in the water is found by $\lambda_w = \lambda_a/n$ (Chap. 35). Thus, we have

$$\lambda_w = \frac{\lambda_a}{n} = \frac{589\ \text{nm}}{1.33} = 443\ \text{nm}$$

The limiting angle of resolution at this wavelength is

$$\theta_m = 1.22\left(\frac{443 \times 10^{-9}\ \text{m}}{0.9 \times 10^{-2}\ \text{m}}\right) = 6.00 \times 10^{-5}\ \text{rad}$$

### EXAMPLE 38.4 Resolution of a Telescope

The Hale telescope at Mount Palomar has a diameter of 200 in. What is its limiting angle of resolution at a wavelength of 600 nm?

*Solution* Because $D = 200$ in. $= 5.08$ m and the wavelength $\lambda = 6 \times 10^{-7}$ m, Equation 38.7 gives

$$\theta_m = 1.22\frac{\lambda}{D} = 1.22\left(\frac{6 \times 10^{-7}\ \text{m}}{5.08\ \text{m}}\right)$$

$$= 1.44 \times 10^{-7}\ \text{rad} = 0.03\ \text{s of arc}$$

Therefore, any two stars that subtend an angle greater than or equal to this value will be resolved (assuming ideal atmospheric conditions).

The Hale telescope can never reach its diffraction limit. Instead, its limiting angle of resolution is always set by atmospheric blurring. This seeing limit is usually about 1 s of arc and is never smaller than about 0.1 s of arc. (This is one of the reasons for the current interest in a large space telescope.)

*Exercise 3* The large radio telescope at Arecibo, Puerto Rico, has a diameter of 305 m, and is designed to detect radio waves at a wavelength of 0.75 m. Calculate the minimum angle of resolution for this telescope, and compare your answer with that of the Hale telescope.
*Answer* $3 \times 10^{-3}$ rad (10 min 19 s of arc), which is more than 10 000 times larger than the calculated minimum angle for the Hale telescope.

### EXAMPLE 38.5 Resolution of the Eye

Calculate the limiting angle of resolution for the eye, assuming a pupil diameter of 2 mm, a wavelength of 500 nm in air, and an index of refraction for the eye equal to 1.33.

*Solution* In this example, we can use Equation 38.7, noting that $\lambda$ is the wavelength in the medium containing the aperture. Since the wavelength of light in the eye is reduced by the index of refraction of the eye medium, we find that $\lambda = (500\ \text{nm})/1.33 = 376$ nm. Therefore, Equation 38.7 gives

$$\theta_m = 1.22\frac{\lambda}{D} = 1.22\left(\frac{3.76 \times 10^{-7}\ \text{m}}{2 \times 10^{-3}\ \text{m}}\right)$$

$$= 2.29 \times 10^{-4}\ \text{rad} = 0.0131°$$

We can use this result to calculate the minimum separation $d$ between two point sources that the eye can distinguish if they are at a distance $L$ from the observer (Fig. 38.15). Since $\theta_m$ is small, we see that

$$\sin\theta_m \approx \theta_m \approx d/L$$

$$d = L\theta_m$$

For example, if the objects are located at a distance of 25 cm from the eye (the near point), then

$$d = (25\ \text{cm})(2.29 \times 10^{-4}\ \text{rad}) = 5.73 \times 10^{-3}\ \text{cm}$$

This is approximately equal to the thickness of a human hair.

*Exercise 4* If the eye is dilated to a diameter of 5 mm, what is the minimum distance between two point sources that the eye can distinguish at a distance of 40 cm from the eye?
*Answer* $3.67 \times 10^{-3}$ cm

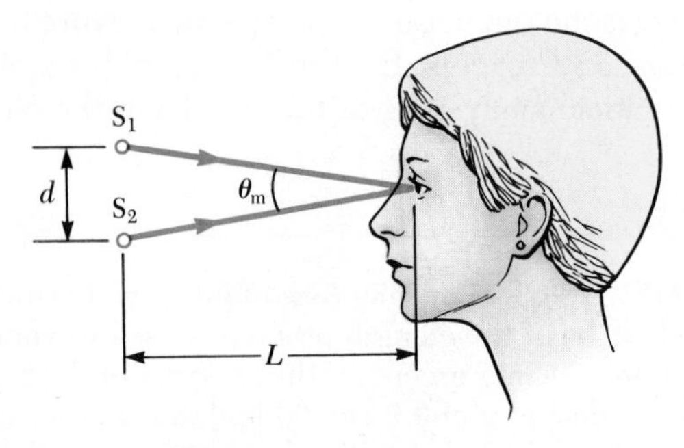


**Figure 38.15** (Example 38.5) Two point sources separated by a distance $d$ as observed by the eye.

## 38.4 THE DIFFRACTION GRATING

The diffraction grating, a very useful device for analyzing light sources, consists of a large number of equally spaced parallel slits. A grating can be made by engraving parallel lines on a glass plate with a precision machining technique. The spaces between each line are transparent to the light and hence act as separate slits. A typical grating contains several thousand lines per centimeter. For example, a grating ruled with 5000 lines/cm has a slit spacing $d$ equal to the inverse of this number, or $d = (1/5000)\ \text{cm} = 2 \times 10^{-4}\ \text{cm}$.

A schematic diagram of a section of plane diffraction grating is illustrated in Figure 38.16. A plane wave is incident from the left, normal to the plane of the grating. A converging lens can be used to bring the rays together at the point $P$. The intensity of the observed pattern on the screen is the result of the combined effects of interference and diffraction. Each slit produces diffraction, as was described in the previous section. The diffracted beams in turn interfere with each other to produce the final pattern. Moreover, each slit acts as a source of waves, where all waves start at the slits in phase. However, for some arbitrary direction $\theta$ measured from the horizontal, the waves must travel *different* path lengths before reaching a particular point $P$ on the screen. From Figure 38.16, note that the path difference between waves from any two adjacent slits is equal to $d \sin \theta$. If this path difference equals one wavelength or some integral multiple of a wavelength, waves from all slits will be in phase at $P$ and a bright line will be observed. Therefore, the condition for **maxima** in the interference pattern at the angle $\theta$ is

$$d \sin \theta = m\lambda \qquad (m = 0, 1, 2, 3, \ldots) \qquad (38.8)$$

Condition for interference maxima for a grating

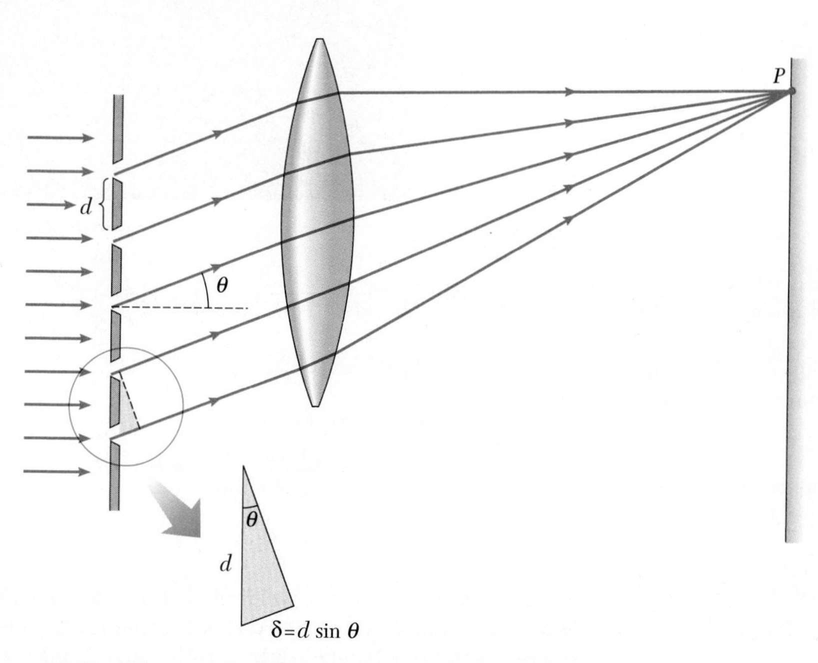


**Figure 38.16** Side view of a diffraction grating. The slit separation is $d$, and the path difference between adjacent slits is $d \sin \theta$.

This expression can be used to calculate the wavelength from a knowledge of the grating spacing and the angle of deviation $\theta$. The integer $m$ represents the *order number* of the diffraction pattern. If the incident radiation contains several wavelengths, the $m$th order maximum for each wavelength occurs at a specific angle. All wavelengths are seen at $\theta = 0$, corresponding to $m = 0$. This is called the *zeroth-order maximum*. The *first-order maximum*, corresponding to $m = 1$, is observed at an angle that satisfies the relationship $\sin\theta = \lambda/d$; the *second-order maximum*, corresponding to $m = 2$, is observed at a larger angle $\theta$, and so on.

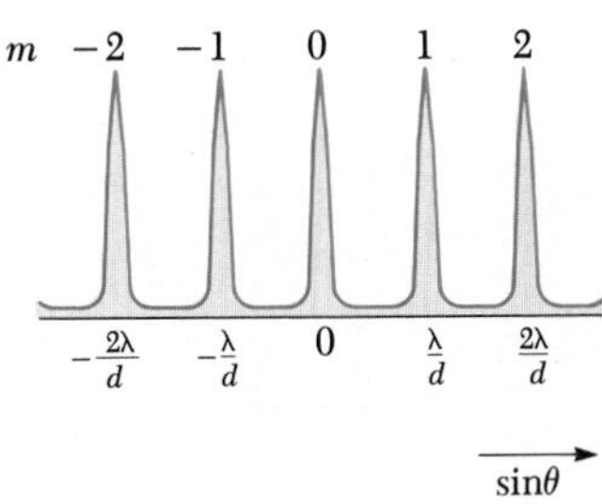


**Figure 38.17** Intensity versus $\sin\theta$ for a diffraction grating. The zeroth-, first-, and second-order maxima are shown.

A sketch of the intensity distribution for the diffraction grating is shown in Figure 38.17. If the source contains various wavelengths, a spectrum of lines will be observed at different positions for each order number. Note the sharpness of the principal maxima and the broad range of dark areas. This is in contrast to the broad bright fringes characteristic of the two-slit interference pattern (Fig. 37.1).

A simple arrangement that can be used to measure various orders of the diffraction pattern is shown in Figure 38.18. This is a form of a diffraction grating spectrometer. The light to be analyzed passes through a slit, and a parallel beam of light exits from the collimator, which is perpendicular to the grating. The diffracted light leaves the grating at angles that satisfy Equation 38.8. A telescope is used to view the image of the slit. The wavelength can be determined by measuring the precise angles at which the images of the slit appear for the various orders.

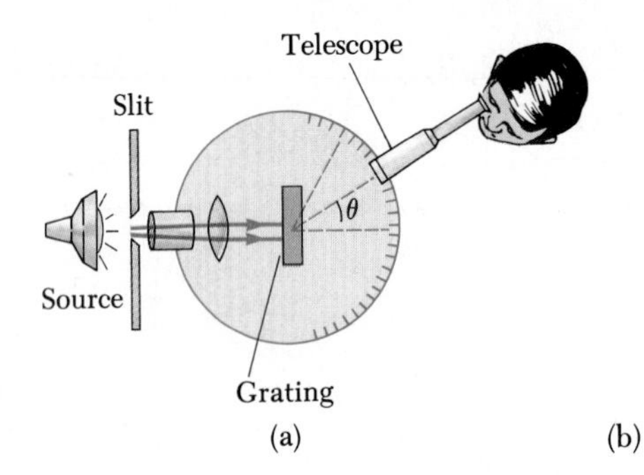

**Figure 38.18** (a) Diagram of a diffraction grating spectrometer. The collimated beam incident on the grating is diffracted into the various orders at the angles $\theta$ that satisfy the equation $d\sin\theta = m\lambda$, where $m = 0, 1, 2, \ldots$ (b) Photograph of a diffraction grating. (Courtesy of PASCO Scientific)

**EXAMPLE 38.6 The Orders of a Diffraction Grating**

Monochromatic light from a helium-neon laser ($\lambda = 632.8$ nm) is incident normally on a diffraction grating containing 6000 lines/cm. Find the angles at which one would observe the first-order maximum, the second-order maximum, and so forth.

*Solution* First, we must calculate the slit separation, which is equal to the inverse of the number of lines per cm:

$$d = (1/6000)\text{ cm} = 1.667 \times 10^{-4}\text{ cm} = 1667\text{ nm}$$

For the first-order maximum ($m = 1$), we get

$$\sin\theta_1 = \frac{\lambda}{d} = \frac{632.8\text{ nm}}{1667\text{ nm}} = 0.3797$$

$$\theta_1 = 22.31°$$

Likewise, for $m = 2$, we find that

$$\sin\theta_2 = \frac{2\lambda}{d} = \frac{2(632.8\text{ nm})}{1667\text{ nm}} = 0.7592$$

$$\theta_2 = 49.41°$$

However, for $m = 3$ we find that $\sin\theta_3 = 1.139$. Since $\sin\theta$ cannot exceed unity, this does not represent a realistic solution. Hence, only zeroth-, first-, and second-order maxima will be observed for this situation.

### Resolving Power of the Diffraction Grating

The diffraction grating is most useful for taking accurate wavelength measurements. Like the prism, the diffraction grating can be used to disperse a spectrum into its components. Of the two devices, the grating is more precise if one wants to distinguish between two closely spaced wavelengths. We say that the grating spectrometer has a "higher resolution" than a prism spectrometer. If $\lambda_1$ and $\lambda_2$ are the two nearly equal wavelengths between which the spectrometer can just barely distinguish, the **resolving power** $R$ of the grating is defined as

$$R \equiv \frac{\lambda}{\lambda_2 - \lambda_1} = \frac{\lambda}{\Delta\lambda} \qquad (38.9)$$

Resolving power

where $\lambda \approx \lambda_1 \approx \lambda_2$ and $\Delta\lambda = \lambda_2 - \lambda_1$. Thus, we see that a grating with a high resolving power can distinguish small differences in wavelength. Furthermore, if $N$ lines of the grating are illuminated, it can be shown that the resolving power in the $m$th order diffraction equals the product $Nm$:

$$R = Nm \qquad (38.10)$$

Resolving power of a grating

The derivation of Equation 38.10 is left as a problem (Problem 71). Thus, the resolving power increases with increasing order number. Furthermore, $R$ is large for a grating with a large number of illuminated slits. Note that for $m = 0$, $R = 0$, which signifies that *all wavelengths are indistinguishable* for the zeroth-order maximum. However, consider the second-order diffraction pattern ($m = 2$) of a grating that has 5000 rulings illuminated by the light source. The resolving power of such a grating in second-order is $R = 5000 \times 2 = 10\,000$. Therefore, the *minimum* wavelength separation between two spectral lines that can be just resolved, assuming a mean wavelength of 600 nm, is given by $\Delta\lambda = \lambda/R = 6 \times 10^{-2}$ nm. For the third-order principal maximum, we find $R = 15\,000$ and $\Delta\lambda = 4 \times 10^{-2}$ nm, and so on.

**EXAMPLE 38.7 Resolving the Sodium Spectral Lines**
Two strong lines in the spectrum of sodium have wavelengths of 589.00 nm and 589.59 nm. (a) What must the resolving power of the grating be in order to distinguish these wavelengths?

$$R = \frac{\lambda}{\Delta\lambda} = \frac{589.30 \text{ nm}}{589.59 \text{ nm} - 589.00 \text{ nm}} = \frac{589.30}{0.59} = 999$$

(b) In order to resolve these lines in the second-order spectrum ($m = 2$), how many lines of the grating must be illuminated?

From Equation 38.10 and the results to (a), we find that

$$N = \frac{R}{m} = \frac{999}{2} = 500 \text{ lines}$$

## *38.5 DIFFRACTION OF X-RAYS BY CRYSTALS

We have seen that the wavelength of light can be measured with a diffraction grating having a known number of rulings per unit length. In principle, the wavelength of any electromagnetic wave can be determined if a grating of the proper spacing (of the order of $\lambda$) is available. X-rays, discovered by W. Roentgen (1845–1923) in 1895, are electromagnetic waves with very short wavelengths (of the order of 1 Å = $10^{-10}$ m = 0.1 nm). Obviously, it would be impossible to construct a grating with such a small spacing. However, the atomic spacing in a solid is known to be about $10^{-10}$ m. In 1913, Max von Laue (1879–1960) suggested that the regular array of atoms in a crystal could act as a three-dimensional diffraction grating for x-rays. Subsequent experiments confirmed this prediction. The diffraction patterns that one observes are rather complicated because of the three-dimensional nature of the crystal. Nevertheless, x-ray diffraction has proved to be an invaluable technique for elucidating crystalline structures and for understanding the structure of matter.[1]

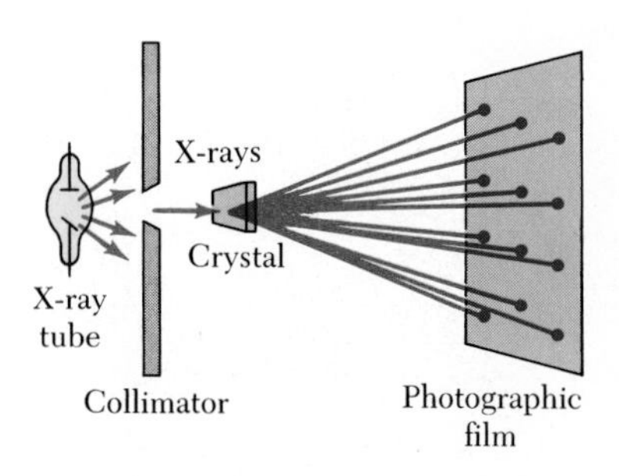


**Figure 38.19** Schematic diagram of the technique used to observe the diffraction of x-rays by a single crystal. The array of spots formed on the film by the strongly diffracted beams is called a Laue pattern.

Figure 38.19 is one experimental arrangement for observing x-ray diffraction from a crystal. A collimated beam of x-rays with a continuous range of wavelengths is incident on a crystal, such as one of sodium chloride, for example. The diffracted beams are very intense in certain directions, corresponding to constructive interference from waves reflected from layers of atoms in the crystal. The diffracted beams can be detected by a photographic film, and they form an array of spots known as a "Laue pattern." The crystalline structure is deduced by analyzing the positions and intensities of the various spots in the pattern.

The arrangement of atoms in a crystal of NaCl is shown in Figure 38.20. The smaller, dark spheres represent $Na^+$ ions and the larger, hollow spheres represent $Cl^-$ ions. Note that the ions are located at the corners of a cube; for this reason, the structure is said to have *cubic symmetry*.

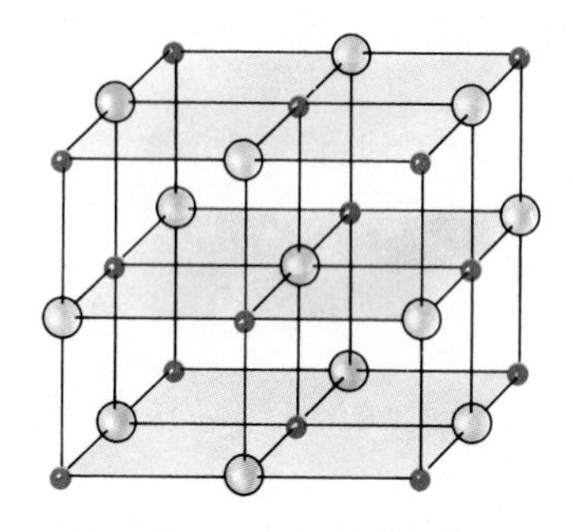

**Figure 38.20** A model of the cubic crystalline structure of sodium chloride. The larger blue spheres represent the $Cl^-$ ions, and the smaller red spheres represent the $Na^+$ ions. The length of the cube edge is $a = 0.562737$ nm.

A careful examination of the NaCl structure shows that the ions appear to lie in various planes. The shaded areas in Figure 38.20 represent one example in which the atoms lie in equally spaced planes. Now suppose an incident x-ray beam makes an angle $\theta$ with one of the planes, as in Figure 38.21. The beam can be reflected from both the upper and the lower plane of atoms. However, the geometric construction in Figure 38.21 shows that the beam reflected from the lower surface travels farther than the beam reflected from the upper surface. The effective path difference between the two beams is $2d \sin \theta$. The two beams will reinforce each other (constructive interference) when this path difference equals some integral multiple of the wavelength $\lambda$. The same is

[1] For more details on this subject, see Sir Lawrence Bragg, "X-Ray Crystallography," *Scientific American*, July 1968.

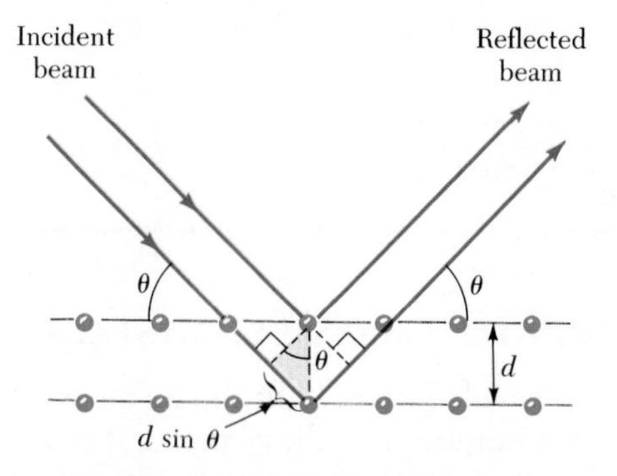


**Figure 38.21** A two-dimensional description of the reflection of an x-ray beam from two parallel crystalline planes separated by a distance $d$. The beam reflected from the lower plane travels farther than the one reflected from the upper plane by an amount equal to $2d \sin \theta$.

true for reflection from the entire family of parallel planes. Hence, the condition for constructive interference (maxima in the reflected wave) is given by

$$2d \sin \theta = m\lambda \qquad (m = 1, 2, 3, \ldots) \tag{38.11}$$

Bragg's law

This condition is known as **Bragg's law** after W. L. Bragg (1890–1971), who first derived the relationship. If the wavelength and diffraction angle are measured, Equation 38.11 can be used to calculate the spacing between atomic planes.

## 38.6 POLARIZATION OF LIGHT WAVES

The wave nature of light has been used to explain the phenomena of interference and diffraction. In Chapter 34 we described the transverse nature of light waves and, in fact, of all electromagnetic waves. Figure 38.22 shows that the electric and magnetic vectors associated with an electromagnetic wave are at right angles to each other and also to the direction of wave propagation. The phenomenon of polarization, which will be described in this section, is firm evidence of the transverse nature of electromagnetic waves.

An ordinary beam of light consists of a large number of waves emitted by the atoms or molecules of the light source. Each atom produces a wave with its own orientation of $\boldsymbol{E}$, as in Figure 38.22, corresponding to the direction of atomic vibration. The direction of polarization of the electromagnetic wave is defined to be the direction in which $\boldsymbol{E}$ is vibrating. However, since all directions of vibration are possible, the resultant electromagnetic wave is a superposition of waves produced by the individual atomic sources. The result is an **unpolarized** light wave, described in Figure 38.23a. The direction of wave propagation in this figure is perpendicular to the page. Note that *all* directions of the electric field vector are equally probable all lying in a plane perpendicular to the direction of propagation. At any given point and at some instant of time, there is only one resultant electric field, hence you should not be misled by the meaning of Figure 38.23a. A wave is said to be **linearly polarized** if $\boldsymbol{E}$ vibrates in the same direction *at all times* at a particular point, as in Figure 38.23b. (Sometimes such a wave is described as *plane-polarized,* or simply *polarized.*)

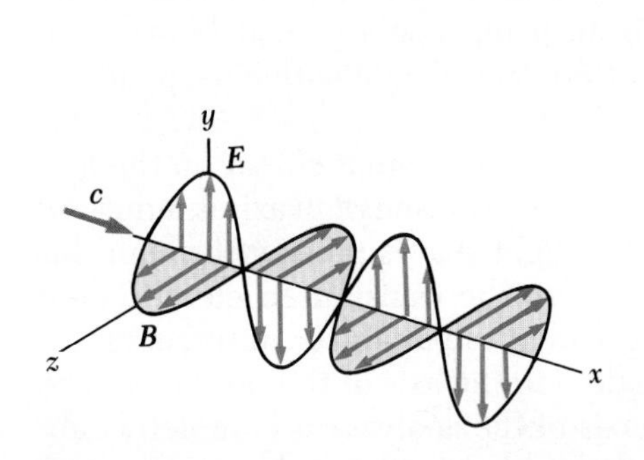


**Figure 38.22** Schematic diagram of an electromagnetic wave propagating in the $x$ direction. The electric field vector $\boldsymbol{E}$ vibrates in the $xy$ plane, and the magnetic field vector $\boldsymbol{B}$ vibrates in the $xz$ plane.

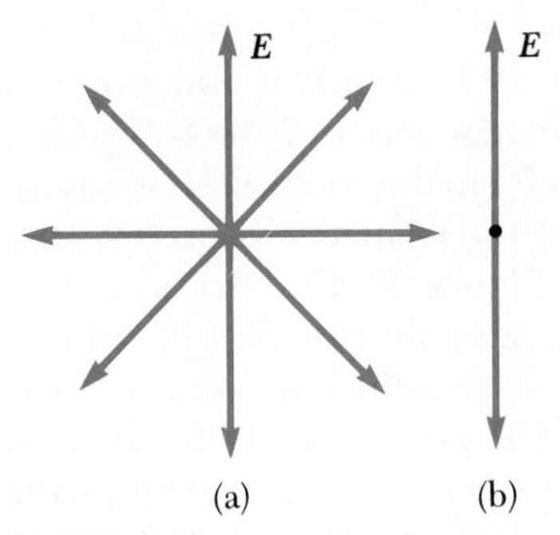


**Figure 38.23** (a) An unpolarized light beam viewed along the direction of propagation (perpendicular to the page). The transverse electric field vector can vibrate in any direction with equal probability. (b) A linearly polarized light beam with the electric field vector vibrating in the vertical direction.

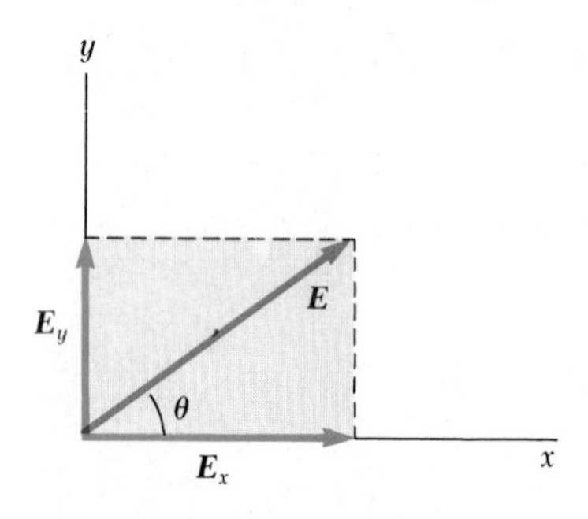


**Figure 38.24** A linearly polarized wave with $\boldsymbol{E}$ at an angle $\theta$ to $x$ has components $E_x = E \cos \theta$ and $E_y = E \sin \theta$.

Suppose a light beam traveling in the $z$ direction has an electric field vector that is at an angle $\theta$ with the $x$ axis at some instant, as in Figure 38.24. The vector has components $E_x$ and $E_y$ as shown. Obviously, the light is linearly polarized if one of these components is always zero or if the angle $\theta$ remains constant in time. However, if the tip of the vector $\boldsymbol{E}$ rotates in a circle with time, the wave is said to be **circularly polarized.** This occurs when the magnitudes of $E_x$ and $E_y$ are *equal,* but differ in phase by 90°. On the other hand, if the magnitudes of $E_x$ and $E_y$ are *not* equal, but differ in phase by 90°, the tip of $\boldsymbol{E}$ moves in an ellipse. Such a wave is said to be **elliptically polarized.** Finally, if $E_x$ and $E_y$ are, on the average, equal in magnitude, but have a randomly varying phase difference, the light beam is unpolarized.

It is possible to obtain a linearly polarized beam from an unpolarized beam by removing all waves from the beam except those whose electric field vectors oscillate in a single plane. We shall now discuss four different physical processes for producing polarized light from unpolarized light. These are (1) selective absorption, (2) reflection, (3) double refraction, and (4) scattering.

### Polarization by Selective Absorption

The most common technique for obtaining polarized light is to use a material that transmits waves whose electric field vectors vibrate in a plane parallel to a certain direction and absorbs those waves whose electric field vectors vibrate in other directions. Any substance that has the property of transmitting light with the electric field vector vibrating in only one direction is called a **dichroic substance.** In 1938, E. H. Land discovered a material, which he called **polaroid,** that polarizes light through selective absorption by oriented molecules. This material is fabricated in thin sheets of long-chain hydrocarbons, such as polyvinyl alcohol. The sheets are stretched during manufacture so that the long-chain molecules align.[2] After a sheet is dipped into a solution containing iodine, the molecules become conducting. However, the conduction takes place primarily along the hydrocarbon chains since the valence electrons of the molecules can move easily only along the chains (recall that valence electrons are "free" electrons that can readily move through the conductor). As a result, the molecules readily *absorb* light whose electric field vector is parallel to their length and *transmit* light whose electric field vector is perpendicular to their length. It is common to refer to the direction perpendicular to the molecular chains as the **transmission axis.** In an ideal polarizer, all light with $\boldsymbol{E}$ parallel to the transmission axis is transmitted, and all light with $\boldsymbol{E}$ perpendicular to the transmission axis is absorbed.

Figure 38.25 represents an unpolarized light beam incident on the first polarizing sheet, called the **polarizer,** where the transmission axis is indicated by the straight lines on the polarizer. The light that is passing through this sheet is polarized vertically as shown, where the transmitted electric field vector is $\boldsymbol{E}_0$. A second polarizing sheet, called the **analyzer,** intercepts this beam with its transmission axis at an angle $\theta$ to the axis of the polarizer. The component of $\boldsymbol{E}_0$ perpendicular to the axis of the analyzer is completely absorbed, and the component of $\boldsymbol{E}_0$ parallel to the axis of the analyzer is $E_0 \cos \theta$

[2] An earlier version of a Polaroid material developed by Land consisted of oriented dichroic crystals of quinine sulfate periodide imbedded in a plastic film.

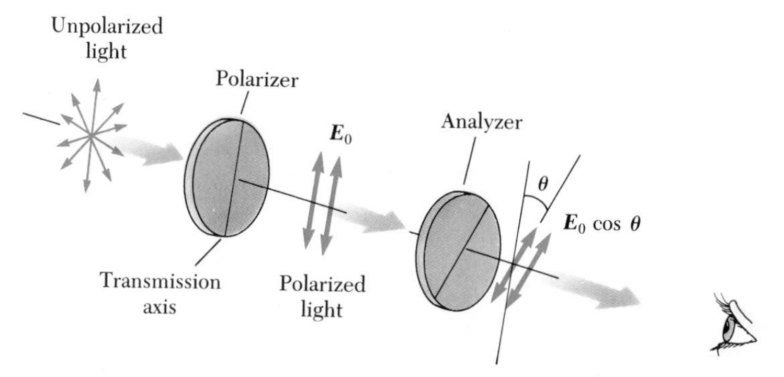


**Figure 38.25** Two polarizing sheets whose transmission axes make an angle $\theta$ with each other. Only a fraction of the polarized light incident on the analyzer is transmitted.

and since the transmitted intensity varies as the *square* of the transmitted amplitude, we conclude that the transmitted intensity varies as

$$I = I_0 \cos^2 \theta \tag{38.12}$$

where $I_0$ is the intensity of the polarized wave incident on the analyzer. This expression, known as **Malus's law,**[3] applies to any two polarizing materials whose transmission axes are at an angle $\theta$ to each other. From this expression, note that the transmitted intensity is a maximum when the transmission axes are parallel ($\theta = 0$ or $180°$). In addition, the transmitted intensity is zero (complete absorption by the analyzer) when the transmission axes are perpendicular to each other. This variation in transmitted intensity through a pair of polarizing sheets is illustrated in Figure 38.26.

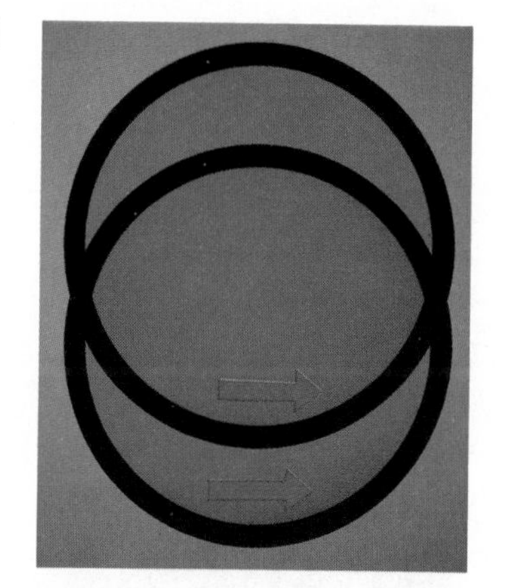

**Figure 38.26** *(Top)* Two polarizers whose transmission axes are aligned with each other. *(Bottom)* Two crossed polarizers, whose transmission axes are *perpendicular* to each other. (Courtesy of Henry Leap and Jim Lehman)

### Polarization by Reflection

Another method for obtaining polarized light is by reflection. When an unpolarized light beam is reflected from a surface, the reflected light is completely polarized, partially polarized, or unpolarized, depending on the angle of incidence. If the angle of incidence is either 0 or 90° (normal or grazing angles), the reflected beam is unpolarized. However, for intermediate angles of incidence, the reflected light is polarized to some extent. In fact, for one particular angle of incidence, the reflected light is completely polarized.

Suppose an unpolarized light beam is incident on a surface as in Figure 38.27a. The beam can be described by two electric field components, one parallel to the surface (the dots) and the other perpendicular to the first and to the direction of propagation (the arrows). It is found that the parallel component reflects more strongly than the other component, and this results in a partially polarized beam (Fig. 38.27a). Furthermore, the refracted ray is also partially polarized.

Now suppose the angle of incidence, $\theta_1$, is varied until the angle between the reflected and refracted beams is 90° (Fig. 38.27b). At this particular angle of incidence, experiments show that the reflected beam is completely polarized with its electric field vector parallel to the surface, while the refracted

[3] Named after its discoverer, E. L. Malus (1775–1812). Actually, Malus first discovered that reflected light was polarized by viewing it through a calcite ($CaCO_3$) crystal.

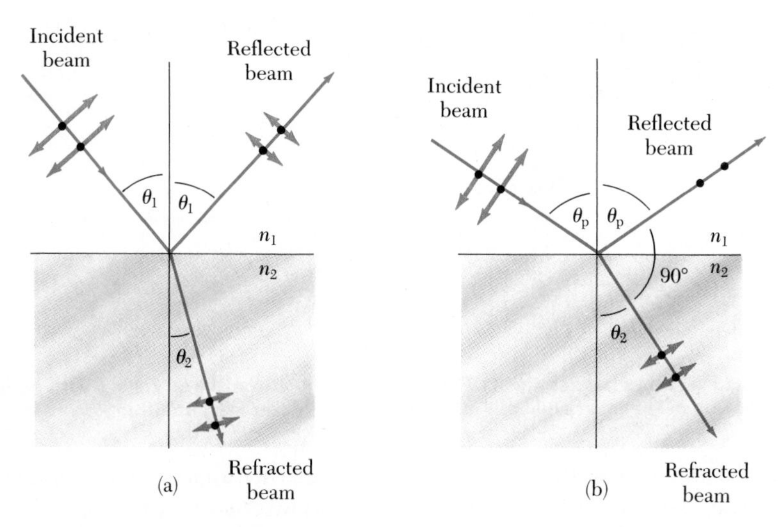


**Figure 38.27** (a) When unpolarized light is incident on a reflecting surface, the reflected and refracted beams are partially polarized. (b) The reflected beam is completely polarized when the angle of incidence equals the polarizing angle, $\theta_p$, which satisfies the equation $n = \tan \theta_p$.

The polarizing angle

beam is partially polarized. The angle of incidence at which this occurs is called the **polarizing angle,** $\theta_p$.

An expression relating the polarizing angle to the index of refraction, $n$, of the reflecting substance can be obtained by using Figure 38.27b. From this figure, we see that at the polarizing angle, $\theta_p + 90° + \theta_2 = 180°$, so that $\theta_2 = 90° - \theta_p$. Using Snell's law, we have

$$n = \frac{\sin \theta_1}{\sin \theta_2} = \frac{\sin \theta_p}{\sin \theta_2}$$

Because $\sin \theta_2 = \sin(90° - \theta_p) = \cos \theta_p$, the expression for $n$ can be written

$$n = \frac{\sin \theta_p}{\cos \theta_p} = \tan \theta_p \qquad (38.13)$$

Brewster's law

This expression is called **Brewster's law,** and the polarizing angle $\theta_p$ is sometimes called **Brewster's angle,** after its discoverer, Sir David Brewster (1781–1868). For example, the Brewster's angle for crown glass ($n = 1.52$) has the value $\theta_p = \tan^{-1}(1.52) = 56.7°$. Because $n$ varies with wavelength for a given substance, the Brewster's angle is also a function of the wavelength.

Polarization by reflection is a common phenomenon. Sunlight reflected from water, glass, and snow is partially polarized. If the surface is horizontal, the electric field vector of the reflected light will have a strong horizontal component. Sunglasses made of polarizing material reduce the glare of reflected light. The transmission axes of the lenses are oriented vertically so as to absorb the strong horizontal component of the reflected light.

### Polarization by Double Refraction

When light travels through an amorphous material, such as glass, it travels with a speed that is the same in all directions. That is, glass has a single index of refraction. However, in certain crystalline materials, such as calcite and quartz, the speed of light is *not* the same in all directions. Such materials are characterized by two indices of refraction. Hence, they are often referred to as **double-refracting** or **birefringent** materials.[4]

When an unpolarized beam of light enters a calcite crystal, the beam splits into two plane-polarized rays that travel with different velocities, corresponding to two different angles of refraction, as in Figure 38.28. The two rays are polarized in two mutually perpendicular directions, as indicated by the dots and arrows. One ray, called the **ordinary (O) ray,** is characterized by an index of refraction, $n_O$, that is the *same* in all directions. This means that if one could place a point source of light inside the crystal, as in Figure 38.29, the ordinary waves would spread out from the source as spheres.

The second plane-polarized ray, called the **extraordinary (E) ray,** travels with *different* speeds in different directions and hence is characterized by an index of refraction, $n_E$, that *varies* with the direction of propagation. A point source of light inside such a crystal would send out an extraordinary wave having wavefronts that are elliptical in cross section (Fig. 38.29). Note from Figure 38.29 that there is one direction, called the **optic axis,** along which the ordinary and extraordinary rays have the *same* velocity, corresponding to the direction for which $n_O = n_E$. The difference in velocity for the two rays is a maximum in the direction perpendicular to the optic axis. For example, in calcite $n_O = 1.658$ at a wavelength of 589.3 nm and $n_E$ varies from 1.658 along the optic axis to 1.486 perpendicular to the optic axis. Values for $n_O$ and $n_E$ for various double-refracting crystals are given in Table 38.1.

If one places a piece of calcite on a sheet of paper and then looks through the crystal at any writing on the paper, two images of the writing are seen, as shown in Figure 38.30. As can be seen from Figure 38.28, these two images correspond to one formed by the ordinary ray and the second formed by the extraordinary ray. If the two images are viewed through a sheet of rotating polarizing glass, they will alternately appear and disappear because the ordinary and extraordinary rays are plane-polarized along mutually perpendicular directions.

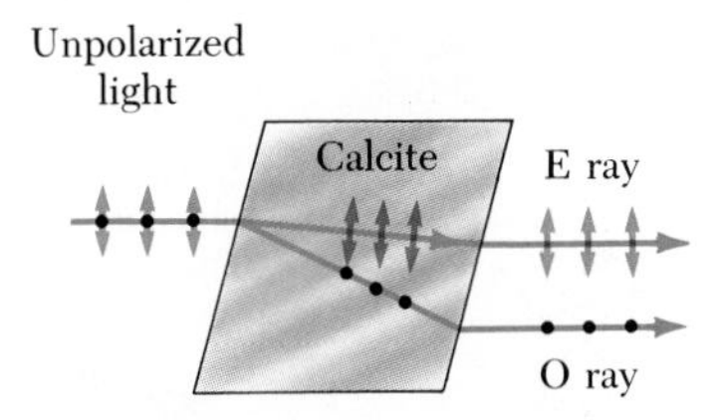


**Figure 38.28** An unpolarized light beam incident on a calcite crystal splits into an ordinary (O) ray and an extraordinary (E) ray. These two rays are polarized in mutually perpendicular directions. (Note that this is not to scale.)

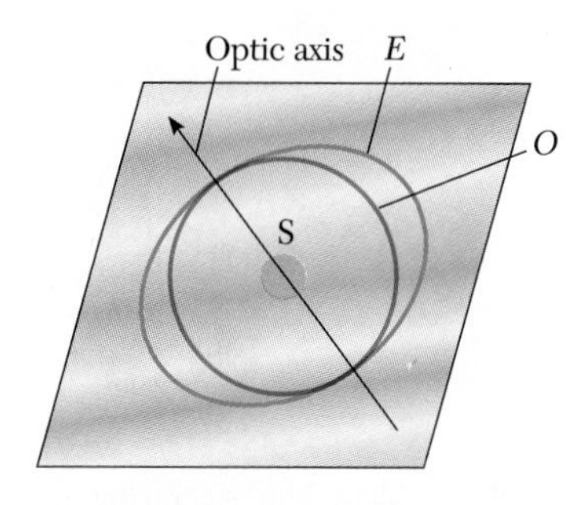


**Figure 38.29** A point source, S, inside a double refracting crystal produces a spherical wavefront corresponding to the ordinary ray (*O*) and an elliptical wavefront corresponding to the extraordinary ray (*E*). The two waves propagate with the same velocity along the optic axis.

## Polarization by Scattering

When light is incident on a system of particles, such as a gas, the electrons in the medium can absorb and reradiate part of the light. The absorption and reradiation of light by the medium, called **scattering,** is what causes sunlight

**Figure 38.30** A calcite crystal produces a double image because it is a birefringent (double-refracting) material. (Courtesy of Henry Leap and Jim Lehman)

[4] For a lucid treatment of this topic, see Elizabeth A. Wood, *Crystals and Light,* New York, Van Nostrand (Momentum), 1964, Chapter 12.

**TABLE 38.1 Indices of Refraction for Some Double-Refracting Crystals at a Wavelength of 589.3 nm**

| Crystal | $n_O$ | $n_E$ | $n_O/n_E$ |
|---|---|---|---|
| Calcite ($CaCO_3$) | 1.658 | 1.486 | 1.116 |
| Quartz ($SiO_2$) | 1.544 | 1.553 | 0.994 |
| Sodium nitrate ($NaNO_3$) | 1.587 | 1.336 | 1.188 |
| Sodium sulfite ($NaSO_3$) | 1.565 | 1.515 | 1.033 |
| Zinc chloride ($ZnCl_2$) | 1.687 | 1.713 | 0.985 |
| Zinc sulfide (ZnS) | 2.356 | 2.378 | 0.991 |

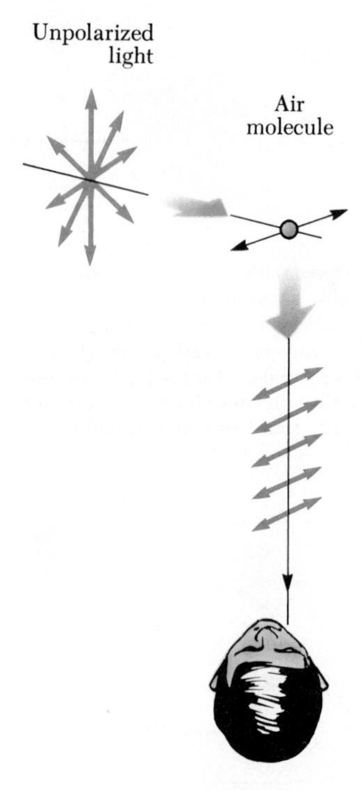


Figure 38.31 The scattering of unpolarized sunlight by air molecules. The light observed at right angles is plane-polarized because the vibrating molecule has a horizontal component of vibration.

reaching an observer on the earth from straight overhead to be partially polarized. You can observe this effect by looking directly up through a pair of sunglasses whose lenses are made of polarizing material. At certain orientations of the lenses, less light passes through than at others.

Figure 38.31 illustrates how the sunlight becomes partially polarized. The left side of the figure shows an incident unpolarized beam of sunlight traveling in the horizontal direction on the verge of striking an air molecule. When this beam strikes the air molecule, it sets the electrons of the molecule into vibration. These vibrating charges act like the vibrating charges in an antenna, except that these charges are vibrating in a complicated pattern. The horizontal part of the electric field vector in the incident wave causes the charges to vibrate horizontally, and the vertical part of the vector simultaneously causes them to vibrate vertically. A horizontally polarized wave is emitted by the electrons as a result of their horizontal motion, and a vertically polarized wave is emitted parallel to the earth as a result of their motion.

Some phenomena involving the scattering of light in the atmosphere can be understood as follows: When light of various wavelengths $\lambda$ is incident on air molecules of size $d$, where $d \ll \lambda$, the relative intensity of the scattered light varies as $1/\lambda^4$. The condition $d \ll \lambda$ is satisfied for scattering from $O_2$ and $N_2$ molecules in the atmosphere, whose diameters are about 0.2 nm. Hence shorter wavelengths (blue light) are scattered more efficiently than longer wavelengths (red light). Because sunlight contains wavelengths from the entire visible spectrum, the sky appears to be blue. (This also explains why the sky is black in outer space, where there is no atmosphere to scatter the sunlight!)

### Optical Activity

Many important practical applications of polarized light involve the use of certain materials which display the property of **optical activity.** A substance is said to be optically active if it rotates the plane of polarization of transmitted light. The angle through which the light is rotated by the material depends on the length of the sample and on the concentration if the substance is in solution. One optically active material is a solution of common sugar dextrose. A standard method for determining the concentration of sugar solutions is to measure the rotation produced by a fixed length of the solution.

Optical activity occurs in a material because of an asymmetry in the shape of its constituent molecules. For example, some proteins are optically active because of their spiral shape. Other materials, such as glass and plastic, become optically active when placed under stress. If polarized light is passed through an unstressed piece of plastic and then through an analyzer whose axis is perpendicular to that of the polarizer, none of the polarized light is transmitted. However, if the plastic is placed under stress, the regions of greatest stress produce the largest angles of rotation of polarized light. Hence, one observes a series of bright and dark bands in the transmitted light, with the bright bands corresponding to regions of greatest stress. Engineers often use this technique, called *optical stress analysis,* to assist in the design of various structures ranging from bridges to small tools and machine parts. A plastic model is built and analyzed under different load conditions to determine regions of potential weakness and failure under stress. Some examples of a plastic model under stress are shown in the photographs on the next page.

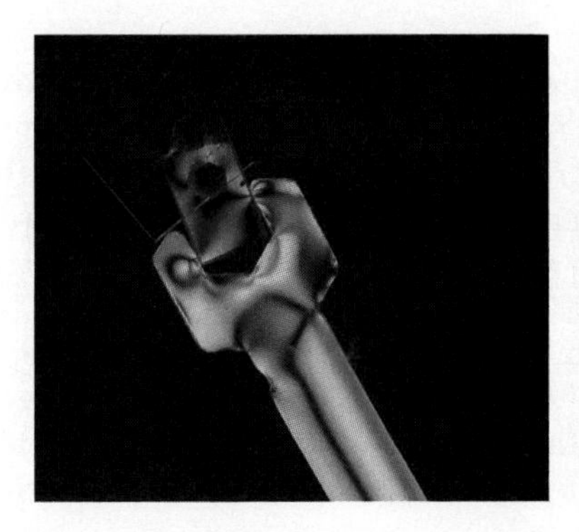

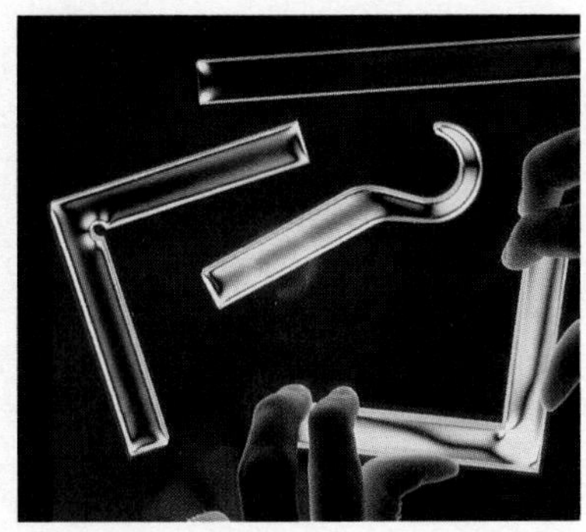

*(Left)* The strain distribution in a plastic wrench under stress. The test piece is viewed between two crossed polarizers. (Courtesy of Henry Leap and Jim Lehman) *(Right)* The photograph shows that the strain pattern disappears when the stress is released. (Courtesy of CENCO)

## SUMMARY

The phenomenon of **diffraction** arises from the interference of a large number or continuous distribution of coherent sources. Diffraction accounts for the deviation of light from a straight-line path when the light passes through an aperture or around obstacles.

The **Fraunhofer diffraction pattern** produced by a *single slit* of width $a$ on a distant screen consists of a central, bright maximum and alternating bright and dark regions of much lower intensities. The angles $\theta$ at which the diffraction pattern has *zero* intensity are given by

Condition for intensity minima in the single-slit diffraction pattern

$$\sin\theta = m\frac{\lambda}{a} \qquad (m = \pm 1, \pm 2, \pm 3, \ldots) \tag{38.1}$$

where $|m| \leq a/\lambda$.

The variation of intensity $I$ with angle $\theta$ is given by

Intensity of a single-slit Fraunhofer diffraction pattern

$$I_\theta = I_0\left[\frac{\sin(\beta/2)}{\beta/2}\right]^2 \tag{38.4}$$

where $\beta = 2\pi a \sin\theta/\lambda$ and $I_0$ is the intensity at $\theta = 0$.

Rayleigh's criterion

**Rayleigh's criterion**, which is a limiting condition of resolution, says that two images formed by an aperture are just distinguishable if the central maximum of the diffraction pattern for one image falls on the first minimum of the other image. The limiting angle of resolution for a slit of width $a$ is given by $\theta_m = \lambda/a$, and the limiting angle of resolution for a circular aperture of diameter $D$ is given by $\theta_m = 1.22\lambda/D$.

A **diffraction grating** consists of a large number of equally spaced, identical slits. The condition for intensity maxima in the interference pattern of a diffraction grating is given by

Condition for intensity maxima for a grating

$$d\sin\theta = m\lambda \qquad (m = 0, 1, 2, 3, \ldots) \tag{38.8}$$

where $d$ is the spacing between adjacent slits and $m$ is the *order number* of the diffraction pattern. The resolving power of a diffraction grating in the $m$th order of the diffraction pattern is given by $R = Nm$, where $N$ is the number of rulings in the grating.

Unpolarized light can be polarized by four processes: (1) selective absorption, (2) reflection, (3) double refraction, and (4) scattering.

When polarized light of intensity $I_0$ is incident on a polarizing film, the light transmitted through the film has an intensity equal to $I_0 \cos^2 \theta$, where $\theta$ is the angle between the transmission axis of the polarizer and the electric field vector of the incident light.

In general, light reflected from a dielectric material, such as glass, is partially polarized. However, the reflected light is completely polarized when the angle of incidence is such that the angle between the reflected and refracted beams is 90°. This angle of incidence, called the **polarizing angle** $\theta_p$, satisfies **Brewster's law**, given by

Brewster's law

$$n = \tan \theta_p \tag{38.13}$$

where $n$ is the index of refraction of the reflecting medium.

## QUESTIONS

1. If you place your thumb and index finger very close together and view light passing between them when they are a few cm in front of your eye, dark lines parallel to your thumb and finger will appear. Explain.
2. Observe the shadow of your book or some other straight edge when it is held a few inches above a table with a lamp several feet above the book. Why is the shadow of the book somewhat fuzzy at the edges?
3. What is the difference between Fraunhofer and Fresnel diffraction?
4. Although we can hear around corners, we cannot see around corners. How can you explain this in view of the fact that sound and light are both waves?
5. Describe the change in width of the central maximum of the single-slit diffraction pattern as the width of the slit is made smaller.
6. Assuming that the headlights of a car are point sources, estimate the maximum distance from an observer to the car at which the headlights are distinguishable from each other.
7. A laser beam is incident at a shallow angle on a machinist's ruler that has a finely calibrated scale. The rulings on the scale give rise to a diffraction pattern on a screen. Discuss how you can use this technique to obtain a measure of the wavelength of the laser light.
8. Certain sunglasses use a polarizing material to reduce the intensity of light reflected from shiny surfaces, such as water or the hood of a car. What orientation of polarization should the material have in order to be most effective?
9. Why is the sky black when viewed from the moon?
10. The diffraction grating effect is easily observed with everyday equipment. For example, a compact disk can be held so that light is reflected from it at a glancing angle. When held this way, various colors in the reflected light can be seen. Explain how this works.

(Question 10). Compact disks act as diffraction gratings when observed under white light. (© Bobbie Kingsley, Photo Researchers, Inc.)

11. The path of a light beam from a helium-neon laser can be made visible by placing chalk dust in the air (perhaps by shaking a blackboard eraser in the path of the light beam). Explain why the beam can be seen under these circumstances.
12. Is light from the sky polarized? Why is it that clouds seen through Polaroid glasses stand out in bold contrast to the sky?
13. If a coin is glued to a glass sheet and this arrangement is held in front of a helium-neon laser, the projected shadow shows diffraction rings around the coin and a bright spot in the center of the shadow. How is this possible?
14. If a fine wire is stretched across the path of a laser beam, is it possible to produce a diffraction pattern?
15. How would one determine the index of refraction of a flat piece of dark obsidian glass?

## PROBLEMS

### Section 38.2 Single-Slit Diffraction

1. In Figure 38.7, let the slit width $a = 0.5$ mm and assume incident monochromatic light of wavelength 460 nm. (a) Find the value of $\theta$ corresponding to the second dark fringe beyond the central maximum. (b) If the observing screen is located 120 cm in front of the slit, what is the distance $y$ from the center of the central maximum to the second dark fringe?
2. A Fraunhofer diffraction pattern is produced on a screen 140 cm from a single slit. The distance from the center of the central maximum to the first secondary maximum is $10^4\lambda$. Calculate the slit width.
3. The second bright fringe in a single-slit diffraction pattern is located 1.4 mm beyond the center of the central maximum. The screen is 80 cm from a slit opening of 0.8 mm. Assuming monochromatic incident light, calculate the wavelength.
4. Helium-neon laser light ($\lambda = 632.8$ nm) is sent through a 0.3-mm-wide single slit. What is the width of the central maximum on a screen 1 m in back of the slit?
5. The pupil of a cat's eye narrows to a slit of width 0.5 mm in daylight. What is the angular resolution? (Let the wavelength of light in the cat's eye be 500 nm.)
6. Light of wavelength 587.5 nm illuminates a single slit 0.75 mm in width. (a) At what distance from the slit should a screen be located if the first minimum in the diffraction pattern is to be 0.85 mm from the center of the screen? (b) What is the width of the central maximum?
7. A screen is placed 50 cm from a single slit, which is illuminated with light of wavelength 690 nm. If the distance between the first and third minima in the diffraction pattern is 3.0 mm, what is the width of the slit?
8. In equation 38.4, let $\beta/2 \equiv \phi$ and show that $I = 0.5I_0$ when $\sin\phi = \phi/\sqrt{2}$.
9. The equation $\sin\phi = \phi/\sqrt{2}$ found in Problem 8 is known as a *transcendental equation.* One method of solving such an equation is the graphical method. To illustrate this, let $\phi = \beta/2$, $y_1 = \sin\phi$, and $y_2 = \phi/\sqrt{2}$. Plot $y_1$ and $y_2$ on the same set of axes over a range from $\phi = 1$ rad to $\phi = \pi/2$ rad. Determine $\phi$ from the point of intersection of the two curves.
10. A beam of green light from a helium-cadmium laser is diffracted by a slit of width 0.55 mm. The diffraction pattern forms on a wall 2.06 m beyond the slit. The distance between the positions of zero intensity ($m = \pm 1$) is 4.1 mm. Estimate the wavelength of the laser light. The helium-cadmium laser can produce light at wavelengths of 441.2 nm or 537.8 nm.
11. A diffraction pattern is formed on a screen 120 cm away from a 0.4-mm-wide slit. Monochromatic light of 546.1 nm is used. Calculate the fractional intensity $I/I_0$ at a point on the screen 4.1 mm from the center of the principal maximum.
12. In the calculation of Example 38.2, we assumed that the first and second side maxima in the single-slit Fraunhofer diffraction pattern are at the locations $\beta/2 = 3\pi/2$ and $\beta/2 = 5\pi/2$. (a) Prove that these maxima are really at locations given by $\tan(\beta/2) = \beta/2$. (b) Find the first three positive solutions for $\beta/2$ to five digits.

### Section 38.3 Resolution of Single-Slit and Circular Apertures

13. A helium-neon laser emits light with a wavelength of 632.8 nm. The circular aperture through which the beam emerges has a diameter of 0.50 cm. Estimate the diameter of the beam at a distance of 10 km from the laser.
14. The moon is approximately 400 000 km from the earth. Can two lunar craters 50 km apart be resolved by a telescope on the earth whose mirror has a diameter of 15 cm? Can craters 1 km apart be resolved? Take the wavelength to be 700 nm and justify your answers with approximate calculations.
15. What is the minimum distance between two points that will permit them to be resolved at 1 km (a) using a terrestrial telescope with a 6.5-cm-diameter objective (assume $\lambda = 550$ nm) and (b) using the unaided eye (assume a pupil diameter of 2.5 mm)?
16. If we were to send a ruby laser beam ($\lambda = 694.3$ nm) outward from the barrel of a 2.7-m-diameter telescope, what would be the diameter of the big red spot when the beam hit the moon 384 000 km away? (Neglect atmospheric dispersion.)
17. The Hubble space telescope, now in earth orbit, has a mirror 2.4 m in diameter, ground to the wrong shape. If corrected, what is the best angular resolution it could achieve for visible light of wavelength 500 nm?
18. Find the radius of a star-image formed on the retina of the eye if the aperture diameter (the pupil) at night is 0.7 cm, and the length of the eye is 3 cm. Assume the wavelength of starlight in the eye is 500 nm.
19. At what distance could one theoretically distinguish two automobile headlights separated by 1.4 m? Assume a pupil diameter of 6 mm and yellow headlights ($\lambda = 580$ nm). The index of refraction in the eye is approximately 1.33.
20. A binary star system in the constellation Orion has an angular separation between the two stars of $10^{-5}$ radians. If $\lambda = 500$ nm, what is the smallest aperture

(diameter) telescope that can just resolve the two stars?

21. The Impressionist painter Georges Seurat created paintings with an enormous number of dots of pure pigment about 2 mm in diameter. The idea was to have colors such as red and green next to each other to form a scintillating canvas. Outside what distance would one be unable to discern individual dots on the canvas? (Assume $\lambda = 500$ nm within the eye and a pupil diameter of 4 mm.)
22. The angular resolution of a radio telescope is to be 0.1° when it operates at a wavelength of 3 mm. Approximately what minimum diameter is required for the telescope's receiving dish?
23. A circular radar antenna on a navy ship has a diameter of 2.1 m and radiates at a frequency of 15 GHz. Two small boats are located 9 km away from the ship. How close together could the boats be and still be detected as *two* objects?
24. When Mars is nearest the earth, the distance separating the two planets is $88.6 \times 10^6$ km. Mars is viewed through a telescope whose mirror has a diameter of 30 cm. (a) If the wavelength of the light is 590 nm, what is the angular resolution of the telescope? (b) What is the smallest distance that can be resolved between two points on Mars?

## Section 38.4 The Diffraction Grating

25. Collimated light from a hydrogen discharge tube is incident normally on a diffraction grating. The incident light includes four wavelength components: $\lambda_1 = 410.1$ nm, $\lambda_2 = 434.0$ nm, $\lambda_3 = 486.1$ nm, and $\lambda_4 = 656.3$ nm. There are 410 lines/mm in the grating. Calculate the angle between (a) $\lambda_1$ and $\lambda_4$ in the first-order spectrum and (b) $\lambda_1$ and $\lambda_3$ in the third-order spectrum.
26. Monochromatic light is incident on a grating that is 75 mm wide and ruled with 50 000 lines. The line is imaged at 32.5° in the second-order spectrum. Determine the wavelength of the incident light.
27. A grating with 250 lines/mm is used with an incandescent light source. Assume the visible spectrum to range in wavelength from 400 to 700 nm. In how many orders can one see (a) the entire visible spectrum and (b) the short-wavelength region?
28. A diffraction grating disperses white light so that the red wavelength $\lambda = 650$ nm appears in the second-order pattern at $\theta = 20°$. (a) Find the so-called *grating constant*—that is, the number of slits per centimeter. (b) Determine whether or not visible light of the third-order pattern appears at $\theta = 20°$.
29. The full width of a 3-cm-wide grating is illuminated by a sodium discharge tube. The lines in the grating are uniformly spaced at 775 nm. Calculate the angular separation in the first-order spectrum between the two wavelengths forming the sodium doublet ($\lambda_1 =$ 589.0 nm and $\lambda_2 = 589.6$ nm).
30. Light from an argon laser strikes a diffraction grating with 5310 lines per centimeter. The central and first-order principal maxima are separated by a distance of 0.488 m on a wall that is 1.72 m from the grating. Determine the wavelength of the laser light.
31. A source emits light with wavelengths of 531.62 nm and 531.81 nm. (a) What is the minimum number of lines required for a grating that resolves the two wavelengths in the first-order spectrum? (b) Determine the slit spacing for a grating 1.32 cm wide that has the required minimum number of lines.
32. A diffraction grating with 2500 rulings/cm is used to examine the sodium spectrum. Calculate the angular separation of the sodium yellow doublet lines (588.995 nm and 589.592 nm) in each of the first three orders.
33. The 501.5-nm line in helium is observed at an angle of 30° in the second-order spectrum of a diffraction grating. Calculate the angular deviation associated with the 667.8-nm line in helium in the first-order spectrum for the same grating.
34. A helium-neon laser ($\lambda = 632.8$ nm) is used to calibrate a diffraction grating. If the first-order maximum occurs at 20.5°, what is the line spacing, *d*?
35. White light is spread out into spectral hues by a diffraction grating. If the grating has 2000 lines per cm, at what angle will red light ($\lambda = 640$ nm) appear in first order?
36. Two spectral lines in a mixture of hydrogen ($H_2$) and deuterium ($D_2$) gas have wavelengths of 656.30 nm and 656.48 nm, respectively. What is the minimum number of lines a diffraction grating must have to resolve these two wavelengths in first order?
37. Monochromatic light from a He-Ne laser ($\lambda =$ 632.8 nm) is incident on a diffraction grating containing 4000 lines/cm. Determine the angle of the first-order maximum.

## *Section 38.5 Diffraction of X-Rays by Crystals

38. Potassium iodide (KI) has the same crystalline structure as that of NaCl, with $d = 0.353$ nm. A monochromatic x-ray beam shows a diffraction maximum when the grazing angle is 7.6°. Calculate the x-ray wavelength. (Assume first order.)
39. Monochromatic x-rays of the $K_\alpha$ line of potassium from a nickel target ($\lambda = 0.166$ nm) are incident on a KCl crystal surface. The interplanar distance in KCl is 0.314 nm. At what angle (relative to the surface) should the beam be directed in order that a second-order maximum be observed?
40. A monochromatic x-ray beam is incident on a NaCl crystal surface which has an interplanar spacing of 0.281 mm. The second-order maximum in the re-

flected beam is found when the angle between the incident beam and the surface is 20.5°. Determine the wavelength of the x-rays.

41. Show why the Bragg condition expressed by Equation 38.11 cannot be satisfied in cases where the wavelength is greater than $2d$ (the length of the unit cell).
42. The dimension labeled $a$ in Figure 38.20 is the edge length of the unit cell of NaCl. This is twice the distance between adjacent ions, which is the parameter $d$ in Equation 38.11. Calculate $d$ for the NaCl crystal from the following data: density $\rho = 2.164\ g/cm^3$, molecular mass $M = 58.45$ g/mol, and Avogadro's number $N_A = 6.022 \times 10^{23}$ atoms/mol.
43. X-rays of wavelength 0.14 nm are reflected from a NaCl crystal, and the first-order maximum occurs at an angle of 14.4°. (a) What value does this give for the interplane spacing of NaCl? (b) Compare the value found in (a) with the value calculated in Problem 42.
44. A wavelength of 0.129 nm characterizes $K_\beta$ x-rays from zinc. When a beam of these x-rays is incident on the surface of a crystal whose structure is similar to that of NaCl, a first-order maximum is observed at an angle of 8.15°. Calculate the interplanar spacing based on this information.
45. If the interplanar spacing of NaCl is 0.281 nm, what is the predicted angle at which x-rays of wavelength 0.14 nm will be diffracted in a first-order maximum?
46. In an x-ray diffraction experiment using x-rays of $\lambda = 0.5 \times 10^{-10}$ m, a first-order maximum occurred at 5°. Find the crystal plane spacing.

Section 38.6 Polarization of Light Waves

47. The angle of incidence of a light beam onto a reflecting surface is continuously variable. The reflected ray is found to be completely polarized when the angle of incidence is 48°. What is the index of refraction of the reflecting material?
48. Light is reflected from a smooth ice surface, and the reflected ray is completely polarized. Determine the angle of incidence. ($n = 1.309$ for ice.)
49. A light beam is incident on heavy flint glass ($n = 1.65$) at the polarizing angle. Calculate the angle of refraction for the transmitted ray.
50. How far above the horizon is the moon when its image reflected in calm water is completely polarized? ($n_{water} = 1.33$.)
51. Unpolarized light passes through two polaroid sheets. The axis of the first is vertical, the second is at 30° to the vertical. What fraction of the initial light is transmitted?
52. Vertically polarized light is passed through three successive polaroid filters. The transmission axes are at 30°, 60°, and 90° to the vertical. What percentage of the light gets through?
53. Plane-polarized light is incident on a single polarizing disk with the direction of $E_0$ parallel to the direction of the transmission axis. Through what angle should the disk be rotated so that the intensity in the transmitted beam will be reduced by a factor of (a) 3, (b) 5, (c) 10?
54. Three polarizing disks whose planes are parallel are centered on a common axis. The direction of the transmission axis in each case is shown, in Figure 38.32, relative to the common vertical direction. A plane-polarized beam of light with $E_0$ parallel to the vertical reference direction is incident from the left on the first disk with intensity $I_i = 10$ units (arbitrary). Calculate the transmitted intensity $I_f$ when (a) $\theta_1 = 20°$, $\theta_2 = 40°$, and $\theta_3 = 60°$; (b) $\theta_1 = 0°$, $\theta_2 = 30°$, and $\theta_3 = 60°$.

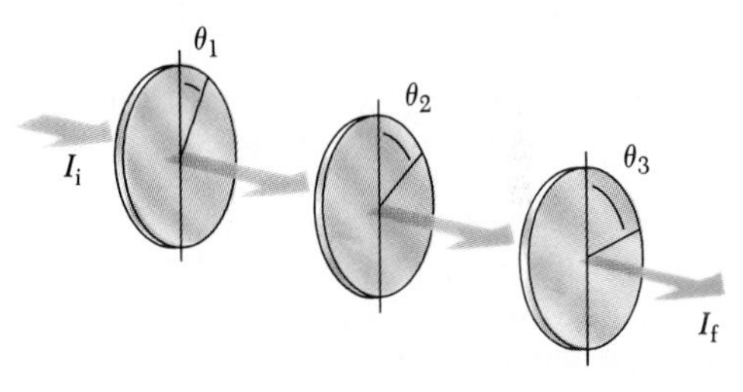


**Figure 38.32** (Problems 54 and 72).

55. The critical angle for sapphire surrounded by air is 34.4°. Calculate the polarizing angle for sapphire.
56. For a particular transparent medium surrounded by air, show that the critical angle for internal reflection and the polarizing angle are related by $\cot \theta_p = \sin \theta_c$.
57. If the polarizing angle for cubic zirconia ($ZrO_2$) is 65.6°, what is the index of refraction for this material?

## ADDITIONAL PROBLEMS

58. A solar eclipse is projected through a pinhole of diameter 0.5 mm and strikes a screen 2 meters away. (a) What is the diameter of the projected image? (b) What is the radius of the first diffraction minimum? Both the sun and moon have angular diameters of very nearly 0.5 degrees. (Assume $\lambda = 550$ nm.)
59. The hydrogen spectrum has a red line at 656 nm, and a blue line at 434 nm. What is the angular separation between two spectral lines obtained with a diffraction grating with 4500 lines/cm?
60. What are the approximate dimensions of the smallest object on earth that the astronauts can resolve by eye at 250 km height from the space shuttle? Assume $\lambda = 500$ nm light in the eye, and a pupil diameter of 0.005 m.

**61.** Grote Reber was a pioneer in radio astronomy. He constructed a radio telescope with a 10-m diameter receiving dish. What was the telescope's angular resolution for radio waves with a wavelength of 2 m?

**62.** An unpolarized beam of light is reflected from a glass surface. It is found that the reflected beam is linearly polarized when the light is incident from air at an angle of 58.6°. What is the refractive index of the glass?

63. Light consisting of two wavelength components is incident on a grating. The shorter-wavelength component has a wavelength of 440 nm. The third-order image of this component is coincident with the second-order image of the longer-wavelength component. Determine the value of the longer-wavelength component.

64. Sunlight is incident on a diffraction grating which has 2750 lines/cm. The second-order spectrum over the visible range (400–700 nm) is to be limited to 1.75 cm along a screen a distance $L$ from the grating. What is the required value of $L$?

65. A diffraction grating of length 4 cm contains 6000 rulings over a width of 2 cm. (a) What is the resolving power of this grating in the first three orders? (b) If two monochromatic waves incident on this grating have a mean wavelength of 400 nm, what is their wavelength separation if they are just resolved in the third order?

66. An American standard television picture is composed of about 485 horizontal lines of varying light intensity. Assume that your ability to resolve the lines is limited only by the Rayleigh criterion and that the pupils of your eyes are 5 mm in diameter. Calculate the ratio of minimum viewing distance to the vertical dimension of the picture such that you will not be able to resolve the lines. Assume that the average wavelength of the light coming from the screen is 550 nm.

67. Light of wavelength 500 nm is incident normally on a diffraction grating. If the third-order maximum of the diffraction pattern is observed at an angle of 32°, (a) what is the number of rulings per cm for the grating? (b) Determine the total number of primary maxima that can be observed in this situation.

68. Consider the case of a light beam containing two discrete wavelength components whose difference in wavelength $\Delta\lambda$ is small relative to the mean wavelength $\lambda$ incident on a diffraction grating. A useful measure of the angular separation of the maxima corresponding to the two wavelengths is the *dispersion D*, given by $D = d\theta/d\lambda$. (The dispersion of a grating should not be confused with its resolving power $R$, given by Equations 38.9 and 38.10.) (a) Starting with Equation 38.8 show that the dispersion can be written

$$D = \frac{\tan\theta}{\lambda}$$

(b) Calculate the dispersion in the third order for the grating described in Problem 65. Give the answer in units of deg/nm.

69. Two polarizing sheets are placed together with their transmission axes crossed so that no light is transmitted. A third sheet is inserted between them with its transmission axis at an angle of 45° with respect to each of the other axes. Find the fraction of incident unpolarized light intensity that will be transmitted by the combination of the three sheets. (Assume that each polarizing sheet is ideal.)

70. Figure 38.33a is a three-dimensional sketch of a birefringent crystal. The dotted lines illustrate how one could cut a thin parallel-faced slab of material from the larger specimen with the optic axis of the crystal parallel to the faces of the plate. A section cut from the crystal in this manner is known as a *retardation plate*. When a beam of light is incident on the plate perpendicular to the direction of the optic axis, as shown in Figure 38.33b, the O ray and the E ray travel along a single straight line, but with different speeds. (a) Let the thickness of the plate be $d$ and show that the phase difference between the O ray and the E ray in the transmitted beam is given by

$$\theta = \frac{2\pi d}{\lambda_0}|n_O - n_E|$$

where $\lambda_0$ is the wavelength in air. (Recall that the *optical path length* in a material is the product of the

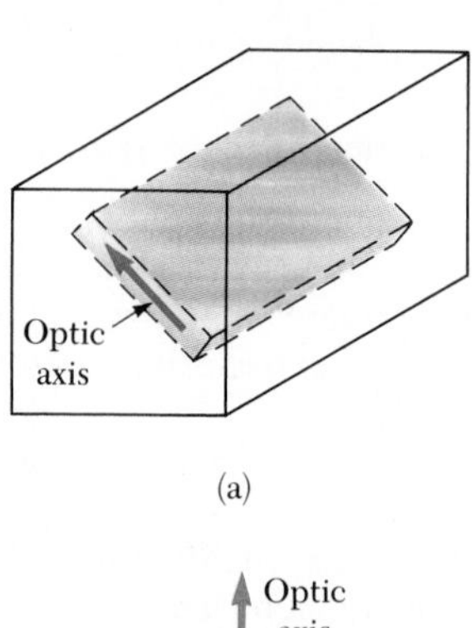


(a)

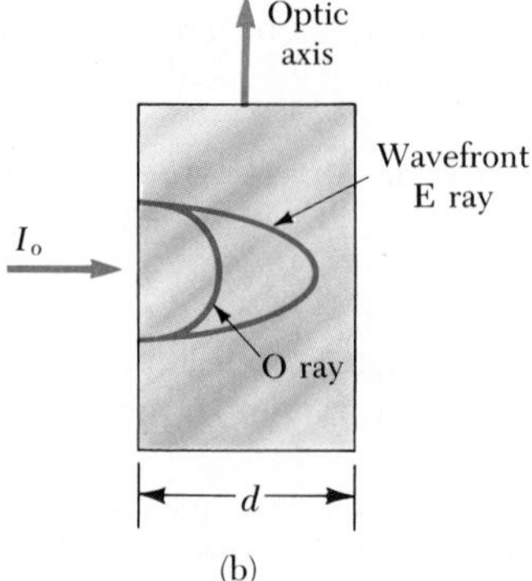


(b)

**Figure 38.33** (Problem 70).

geometric path and the index of refraction.) (b) If in a particular case the incident light has a wavelength of 550 nm, find the minimum value of $d$ for a quartz plate for which $\theta = \pi/2$. Such a plate is called a *quarter-wave plate.* (Use values of $n_O$ and $n_E$ from Table 38.1.)

71. Derive Equation 38.10 for the resolving power of a grating, $R = Nm$, where $N$ is the number of lines illuminated and $m$ is the order in the diffraction pattern. Remember that Rayleigh's criterion (Section 38.3) states that two wavelengths will be resolved when the principal maximum for one falls on the first minimum for the other.

72. In Figure 38.32, suppose that the left and right polarizing disks have their transmission axes perpendicular to each other. Also, let the center disk be rotated on the common axis with an angular velocity $\omega$. Show that if *unpolarized* light is incident on the left disk with an intensity $I_0$, the intensity of the beam emerging from the right disk will be

$$I = \frac{1}{16} I_0(1 - \cos 4\omega t)$$

This means that the intensity of the emerging beam will be modulated at a rate that is four times the rate of rotation of the center disk. (*Hint:* Use the trigonometric identities $\cos^2 \theta = (1 + \cos 2\theta)/2$ and $\sin^2 \theta = (1 - \cos 2\theta)/2$, and recall that $\theta = \omega t$.)

73. Suppose that the single slit opening in Figure 38.7 is 6 cm wide and is placed in front of a microwave source operating at a frequency of 7.5 GHz. (a) Calculate the angle subtended by the first minimum in the diffraction pattern. (b) What is the relative intensity $I/I_0$ at $\theta = 15°$? (c) Consider the case when there are *two* such sources, separated laterally by 20 cm, behind the single slit. What is the maximum distance between the plane of the sources and the slit if the diffraction patterns are to be resolved? (In this case, the approximation that $\sin \theta \approx \tan \theta$ may not be valid because of the relatively small value of the ratio $a/\lambda$.)

Δ74. Light from a helium-neon laser (wavelength = 632.8 nm) illuminates a single slit, and a diffraction pattern is formed on a screen 1.00 m from the slit. The following data record measurements of the relative intensity as a function of distance from the line formed by the undiffracted laser beam.

Make a plot of the relative intensity versus distance. Choose an appropriate value for the slit width $a$ and plot the theoretical expression for the relative intensity

$$\frac{I_\theta}{I_0} = \frac{\sin^2 (\beta/2)}{(\beta/2)^2}$$

on the same graph used for the experimental data. What value of $a$ gives the best fit of theory and experiment?

| Relative Intensity | Distance (mm) |
|---|---|
| 0.95 | 0.8 |
| 0.80 | 1.6 |
| 0.60 | 2.4 |
| 0.39 | 3.2 |
| 0.21 | 4.0 |
| 0.079 | 4.8 |
| 0.014 | 5.6 |
| 0.003 | 6.5 |
| 0.015 | 7.3 |
| 0.036 | 8.1 |
| 0.047 | 8.9 |
| 0.043 | 9.7 |
| 0.029 | 10.5 |
| 0.013 | 11.3 |
| 0.002 | 12.1 |
| 0.0003 | 12.9 |
| 0.005 | 13.7 |
| 0.012 | 14.5 |
| 0.016 | 15.3 |
| 0.015 | 16.1 |
| 0.010 | 16.9 |
| 0.0044 | 17.7 |
| 0.0006 | 18.5 |
| 0.0003 | 19.3 |
| 0.003 | 20.2 |

## CALCULATOR/COMPUTER PROBLEMS

75. Figure 38.11 shows the relative intensity of a single-slit Fraunhofer diffraction pattern as a function of the parameter $\beta/2 = (\pi a \sin \theta)/\lambda$. Make a plot of the relative intensity $I/I_0$ as a function of $\theta$, the angle subtended by a point on the screen at the slit, when (a) $\lambda = a$, (b) $\lambda = 0.5a$, (c) $\lambda = 0.1a$, (d) $\lambda = 0.05a$. Let $\theta$ range over the interval from 0 to 20° and choose a number of steps appropriate for each case.

76. From Equation 38.4 show that, in the Fraunhofer diffraction pattern of a single slit, the angular width of the central maximum at the point where $I = 0.5I_0$ is $\Delta\theta = 0.886\lambda/a$. (*Hint:* In Equation 38.4, let $\beta/2 = \phi$ and solve the resulting transcendental equation graphically; see Problem 9.)

77. Another method to solve the equation $\phi = \sqrt{2} \sin \phi$ in Problem 9 is to use a scientific calculator, guess a first value of $\phi$, see if it fits, and continue to update your estimate until the equation balances. How many steps (iterations) did this take? [Another approach is to apply the Newton-Raphson method to find the roots of $f(\phi) = \phi - \sqrt{2} \sin \phi$. In this approach, $\phi_2 = \phi_1 - f(\phi_1)/f'(\phi_1)$, where $f'$ is the first derivative of $f$.]

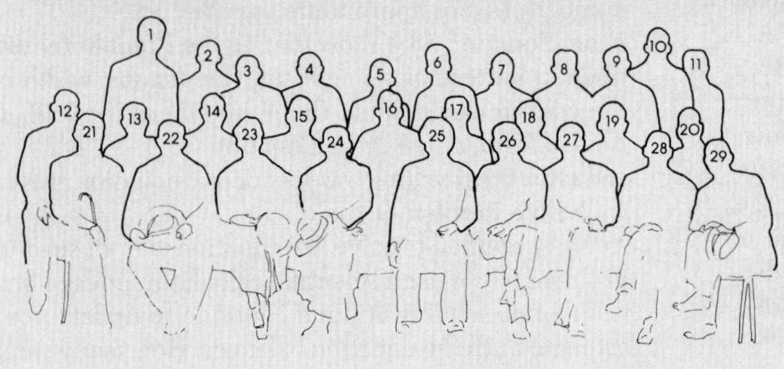

1. A. Piccard
2. E. Henriot
3. P. Ehrenfest
4. E. Herzen
5. Th. de Donder
6. E. Schroedinger
7. E. Verschaffelt
8. W. Pauli
9. W. Heisenberg
10. R.H. Fowler
11. L. Brillouin
12. P. Debye
13. M. Knudsen
14. W.L. Bragg
15. H.A. Kramers
16. P.A.M. Dirac
17. A.H. Compton
18. L.V. de Broglie
19. M. Born
20. N. Bohr
21. I. Langmuir
22. M. Planck
23. M. Curie
24. H.A. Lorentz
25. A. Einstein
26. P. Langevin
27. C.E. Guye
28. C.T.R. Wilson
29. O.W. Richardson

# PART VI
# Modern Physics

*"The new order of science makes possible, for the first time, a cooperative effort in which everyone is the gainer and no one the loser. The advent of modern science is the most important social event in history."*

KARL TAYLOR-COMPTON

At the end of the 19th century, scientists believed that they had learned most of what there was to know about physics. Newton's laws of motion and his universal theory of gravitation, Maxwell's theoretical work in unifying electricity and magnetism, and the laws of thermodynamics and kinetic theory were highly successful in explaining a wide variety of phenomena.

However, at the turn of the 20th century, a major revolution shook the world of physics. In 1900 Planck provided the basic ideas that led to the formulation of the quantum theory, and in 1905 Einstein formulated his brilliant special theory of relativity. The excitement of the times is captured in Einstein's own words: "It was a marvelous time to be alive." Both ideas were to have a profound effect on our understanding of nature. Within a few decades, these theories inspired new developments and theories in the fields of atomic physics, nuclear physics, and condensed matter physics.

The discussion of modern physics in this last part of the text will begin with a treatment of the special theory of relativity in Chapter 39. Although the concepts underlying this theory often violate our common sense, the theory provides us with a new and deeper view of physical laws. Next we shall discuss various developments in quantum theory (Chapter 40), which provides us with a successful model for understanding electrons, atoms, and molecules. The extended version of this text includes seven additional chapters on selected topics in modern physics. In these chapters, we cover the basic concepts of quantum mechanics, its application to atomic and molecular physics, and introductions to solid state physics, superconductivity, nuclear physics, particle physics, and cosmology.

You should keep in mind that, although modern physics has been developed during this century and has led to a multitude of important technological achievements, the story is still incomplete. Discoveries will continue to evolve during our lifetime, many of which will deepen or refine our understanding of nature and the world around us. It is still a "marvelous time to be alive."

*The "architects" of modern physics. This unique photograph shows many eminent scientists who participated in the fifth international congress of physics held in 1927 by the Solvay Institute in Brussels. At this and similar conferences, held regularly from 1911 on, scientists were able to discuss and share the many dramatic developments in atomic and nuclear physics. This elite company of scientists includes fifteen Nobel Prize winners in physics and three in chemistry. (Photograph courtesy of AIP Niels Bohr Library)*

# 39

# Relativity

*Albert Einstein, German physicist. Einstein is best known for his special theory of relativity. (Color added by John Austin.) (Photograph courtesy of Roger Richman Agency)*

Light waves and other forms of electromagnetic radiation travel through free space with a speed of $c = 3.00 \times 10^8$ m/s. As we shall see in this chapter, the speed of light is an upper limit for the speeds of particles and mechanical waves.

## 39.1 INTRODUCTION

Most of our everyday experiences and observations deal with objects that move at speeds much less than the speed of light. Newtonian mechanics, and the early ideas on space and time, were formulated to describe the motion of such objects. This formalism is very successful for describing a wide range of phenomena. Although Newtonian mechanics works very well at low speeds, it fails when applied to particles whose speeds approach that of light. Experimentally, one can test the predictions of the theory at large speeds by accelerating an electron through a large electric potential difference. For example, it is possible to accelerate an electron to a speed of $0.99c$ by using a potential difference of several million volts. According to Newtonian mechanics, if the potential difference (as well as the corresponding energy) is increased by a factor of 4, then the speed of the electron should be doubled to $1.98c$. However, experiments show that the speed of the electron always remains *less* than the speed of light, regardless of the size of the accelerating voltage. Since Newtonian mechanics places no upper limit on the speed that a particle can attain, it is contrary to modern experimental results and is clearly a limited theory.

In 1905, at the age of only 26, Einstein published his *special theory of relativity:*

> The relativity theory arose from necessity, from serious and deep contradictions in the old theory from which there seemed no escape. The strength of the new theory lies in the consistency and simplicity with which it solves all these difficulties, using only a few very convincing assumptions. . . .[1]

Although Einstein made many important contributions to science, the theory of relativity alone represents one of the greatest intellectual achievements of the 20th century. With this theory, one can correctly predict experimental observations over the range of speeds from $v = 0$ to velocities approaching the speed of light. Newtonian mechanics, which was accepted for over 200 years, is in fact a specialized case of Einstein's generalized theory. This chapter gives an introduction to the special theory of relativity, with emphasis on some of the consequences of the theory. A discussion of general relativity and some of its consequences and experimental tests is presented in the essay written by Clifford Will.

As we shall see, the special theory of relativity is based on two basic postulates:

The postulates of the special theory of relativity

1. The laws of physics are the same in all inertial reference frames.
2. The speed of light in a vacuum has the same value, $c = 3 \times 10^8$ m/s, in all inertial reference frames. (That is, the measured value of $c$ is independent of the motion of the observer or of the motion of the light source.)

Special relativity covers phenomena such as the slowing down of clocks and the contraction of lengths in moving reference frames as measured by a stationary observer. We shall also discuss the relativistic forms of momentum and energy, and some consequences of the famous mass-energy equivalence formula, $E = mc^2$. You may want to consult a number of excellent books on relativity for more details on the subject.[2]

Although it is well known that relativity plays an essential role in contemporary theoretical physics, it also has many practical applications, including the design of accelerators and other devices that utilize high-speed particles. We shall have occasion to use relativity in some subsequent chapters of this text, but often only the outcome of relativistic effects will be presented.

[1] A. Einstein and L. Infeld, *The Evolution of Physics,* New York, Simon and Schuster, 1961.

[2] The following books are recommended for more details on the theory of relativity at the introductory level: E. F. Taylor and J. A. Wheeler, *Spacetime Physics,* San Francisco, W. H. Freeman, 1963; R. Resnick, *Introduction to Special Relativity,* New York, Wiley, 1968; A. P. French, *Special Relativity,* New York, Norton, 1968; other suggested readings are selected reprints on "Special Relativity Theory" published by the American Institute of Physics.

## 39.2 THE PRINCIPLE OF RELATIVITY

In order to describe a physical event, it is necessary to establish a frame of reference, such as one that is fixed in the laboratory. You should recall from your studies in mechanics that Newton's laws are valid in *all* inertial frames of reference. Since an inertial frame of reference is defined as one in which Newton's first law is valid, one can say that *an inertial system is a system in which a free body exhibits no acceleration.* Furthermore, any system moving with constant velocity with respect to an inertial system is also an inertial system. There is no preferred frame. This means that the results of an experiment performed in a vehicle moving with uniform velocity will be identical to the results of the same experiment performed in the stationary laboratory.

Inertial frame of reference

According to the **principle of Newtonian relativity,** the laws of mechanics are the same in all inertial frames of reference.

For example, if you perform an experiment while at rest in a laboratory, and an observer in a passing car moving with constant velocity also observes your experiment, the laboratory coordinate system and the coordinate system of the moving car are both inertial reference frames. Therefore, if you find the laws of mechanics to be true in the lab, the person in the moving car must agree with your observation. This also implies that no mechanical experiment can detect any difference between the two inertial frames. The only thing that can be detected is the relative motion of one frame with respect to the other. That is, the notion of *absolute* motion through space is meaningless.

Suppose that some physical phenomenon, which we call an *event,* occurs in an inertial system. The event's location and time of occurrence can be specified by the coordinates $(x, y, z, t)$. We would like to be able to transform the space and time coordinates of the event from one inertial system to another moving with uniform relative velocity. This is accomplished by using a so-called *Galilean transformation.*

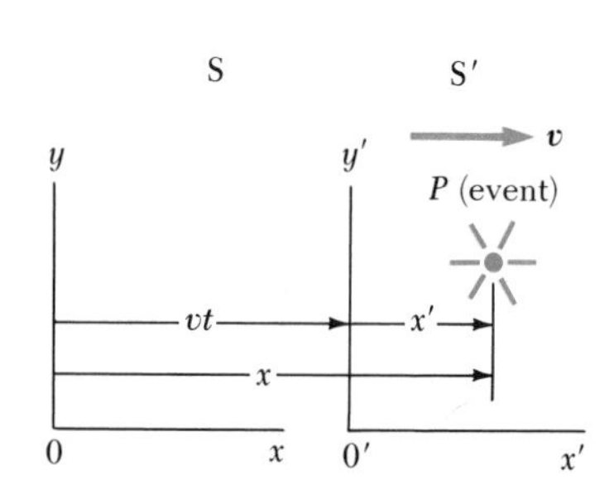


**Figure 39.1** An event occurs at a point $P$. The event is observed by two observers in inertial frames S and S′, where S′ moves with a velocity $v$ relative to S.

Consider two inertial systems S and S′, as in Figure 39.1. The system S′ moves with a constant velocity $v$ along the $xx'$ axes, where $v$ is measured relative to the system S. We assume that an event occurs at the point $P$ and that the origins of S and S′ coincide at $t = 0$. The event might be the "explosion" of a flashbulb or a heartbeat. An observer in system S would describe the event with space-time coordinates $(x, y, z, t)$, while an observer in system S′ would use $(x', y', z', t')$ to describe the same event. As we can see from Figure 39.1, these coordinates are related by the equations

Galilean transformation of coordinates

$$\begin{aligned} x' &= x - vt \\ y' &= y \\ z' &= z \\ t' &= t \end{aligned} \tag{39.1}$$

These equations constitute what is known as a **Galilean transformation of coordinates.** Note that the fourth coordinate, time, is *assumed* to be the same in both inertial systems. That is, within the framework of classical mechanics, clocks are universal, so that the time of an event for an observer in S is the same as the time for the same event in S′. Consequently, the time interval between two successive events should be the same for both observers. Although this assumption may seem obvious, it turns out to be *incorrect* when treating situations in which $v$ is comparable to the speed of light. In fact, this point represents one of the most profound differences between Newtonian concepts and the ideas contained in Einstein's theory of relativity.

Now suppose two events are separated by a distance $\Delta x$ and a time interval $\Delta t$ as measured by an observer in S. It follows from Equation 39.1 that the corresponding displacement $\Delta x'$ measured by an observer in S′ is given by $\Delta x' = \Delta x - v\,\Delta t$, where $\Delta x$ is the displacement measured by an observer in S. Since $\Delta t = \Delta t'$, we find that

$$\frac{\Delta x'}{\Delta t} = \frac{\Delta x}{\Delta t} - v$$

If the two events correspond to the passage of a moving object (or wave pulse) past two "milestones" in frame S, then $\Delta \boldsymbol{x}/\Delta t$ is the time-averaged velocity in S and $\Delta \boldsymbol{x}'/\Delta t$ is the average velocity in S′. In the time limit $\Delta t \rightarrow 0$, this becomes

$$\boldsymbol{u}'_x = \boldsymbol{u}_x - \boldsymbol{v} \tag{39.2}$$

Galilean addition law for velocities

where $\boldsymbol{u}_x$ and $\boldsymbol{u}'_x$ are the instantaneous velocities of the object relative to S and S′, respectively. The reader should note that the notation for $u$'s and $v$'s has been switched in going from Equation 4.25 to Equation 39.2.

This result, which is called the **Galilean addition law for velocities** (or Galilean velocity transformation), is used in everyday observations and is consistent with our intuitive notion of time and space. However, as we shall soon see, these results lead to serious contradictions when applied to electromagnetic waves.

## The Speed of Light

It is quite natural to ask whether the concept of Newtonian relativity in mechanics also applies to experiments in electricity, magnetism, optics, and other areas. For example, if one assumes that the laws of electricity and magnetism are the same in all inertial frames, a paradox concerning the velocity of light immediately arises. This can be understood by recalling that according to Maxwell's equations in electromagnetic theory, the velocity of light always has the fixed value of $(\mu_0\epsilon_0)^{-1/2} \approx 3.00 \times 10^8$ m/s. But this is in direct contradiction to what one would expect based on the Galilean addition law for velocities. According to this law, the velocity of light should *not* be the same in all inertial frames. For example, suppose a light pulse is sent out by an observer in a boxcar moving with a velocity $\boldsymbol{v}$ (Fig. 39.2). The light pulse has a velocity $\boldsymbol{c}$

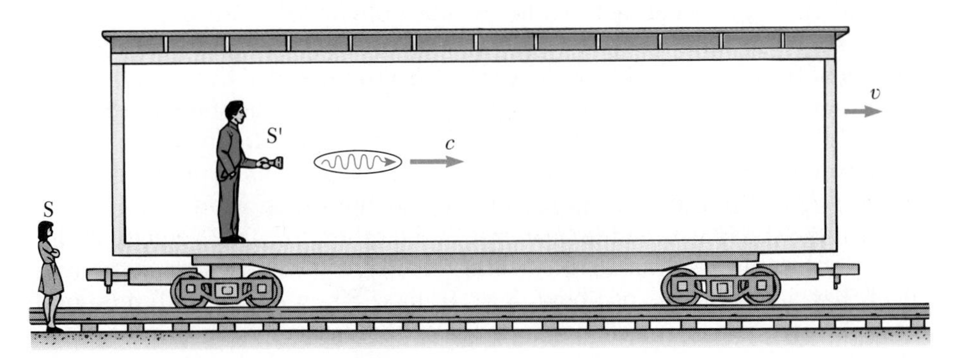


**Figure 39.2** A pulse of light is sent out by a person in a moving boxcar. According to Newtonian relativity, the speed of the pulse should be $c + v$ relative to a stationary observer.

relative to observer S′ in the boxcar. According to the ideas of Newtonian relativity, the velocity of the pulse relative to the stationary observer S outside the boxcar should be $\mathbf{c} + \mathbf{v}$. This is in obvious contradiction to Einstein's theory, which postulates that the velocity of the light pulse is the same for all observers.

In order to resolve this paradox, one must conclude that either (1) the Galilean addition law for velocities is incorrect, or else (2) the laws of electricity and magnetism are not the same in all inertial frames. If the Galilean addition law for velocities were incorrect, we would be forced to abandon the seemingly "obvious" notions of absolute time and absolute length that form the basis for the Galilean transformations.

If we assume that the second alternative is true, then a preferred reference frame must exist in which the speed of light has the value $c$, and that it is greater or less than this value in any other reference frame in accordance with the Galilean addition law for velocities. It is useful to draw an analogy with sound waves, which propagate through a medium such as air. The speed of sound in air is about 330 m/s when measured in a reference frame in which the air is stationary. However, the measured speed of sound is greater or less than this value when measured from a reference frame that is moving with respect to the source of sound.

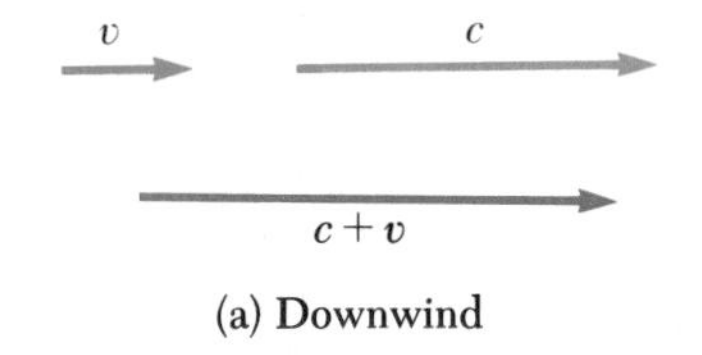


(a) Downwind

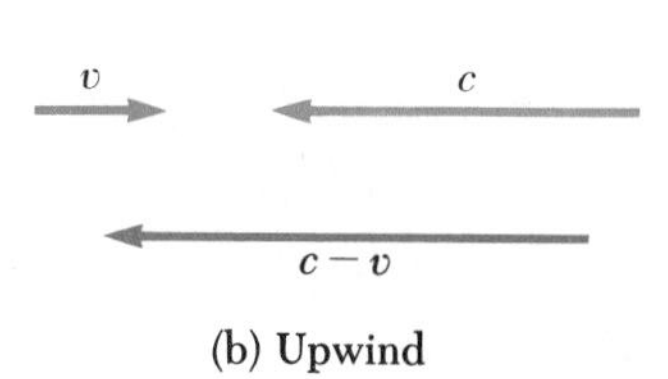


(b) Upwind

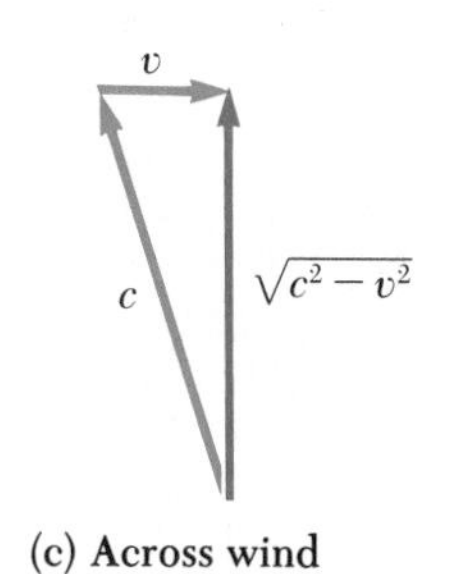


(c) Across wind

**Figure 39.3** If the velocity of the ether wind relative to the earth is $\mathbf{v}$, and $\mathbf{c}$ is the velocity of light relative to the ether, the speed of light relative to the earth is (a) $c + v$ in the downwind direction, (b) $c - v$ in the upwind direction, and (c) $(c^2 - v^2)^{1/2}$ in the direction perpendicular to the wind.

In the case of light signals (electromagnetic waves), recall that Maxwell's theory predicted that such waves must propagate through free space with a speed equal to the speed of light. However, Maxwell's theory does not require the presence of a medium for the wave propagation. This is in contrast to mechanical waves, such as water or sound waves, which do require a medium to support the disturbances. In the 19th century, physicists thought that electromagnetic waves also required a medium in order to propagate. They proposed that such a medium existed, and they gave it the name **luminiferous ether.** The ether was assumed to be present everywhere, even in empty space, and light waves were viewed as ether oscillations. Furthermore, the ether had to have the unusual properties of being a massless but rigid medium, and would have no effect on the motion of planets or other objects. Indeed, this is a strange concept. Additionally, it was found that the troublesome laws of electricity and magnetism would take on their simplest form in a frame of reference at *rest* with respect to this ether. This frame was called the *absolute frame.* The laws of electricity and magnetism would be valid in this absolute frame, but they would have to be modified in any reference frame moving with respect to the ether frame.

As a result of the importance attached to this absolute frame, it became of considerable interest in physics to prove by experiment that it existed. A direct method for detecting the ether wind would be to measure its influence on the speed of light relative to a frame of reference on earth. If $\mathbf{v}$ is the velocity of the ether relative to the earth, then the speed of light should have its maximum value, $c + v$, when propagating downwind as shown in Figure 39.3a. Likewise, the speed of light should have its minimum value, $c - v$, when propagating upwind as in Figure 39.3b, and some intermediate value, $(c^2 - v^2)^{1/2}$, in the direction perpendicular to the ether wind as in Figure 39.3c. If the sun is assumed to be at rest in the ether, then the velocity of the ether wind would be equal to the orbital velocity of the earth around the sun, which has a magnitude of about $3 \times 10^4$ m/s. Since $c = 3 \times 10^8$ m/s, one should be able to detect a change in speed of about 1 part in $10^4$ for measurements in the upwind or downwind directions. However, as we shall see in the next section, all attempts to detect such changes and establish the existence of the ether (and hence the absolute frame) proved futile!

## 39.3 THE MICHELSON-MORLEY EXPERIMENT

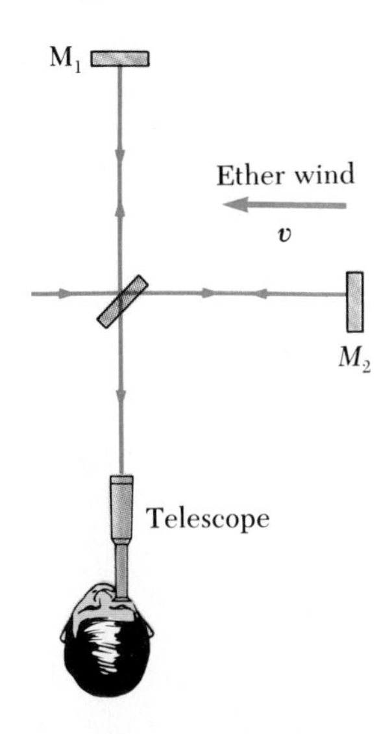


**Figure 39.4** According to the ether wind theory, the speed of light should be $c - v$ as the beam approaches mirror $M_2$ and $c + v$ after reflection.

The most famous experiment that was designed to detect small changes in the speed of light was performed in 1887 by A.A. Michelson (1852–1931) and E.W. Morley (1838–1923).[3] We should state at the outset that the outcome of the experiment was *negative,* thus contradicting the ether hypothesis. The experiment was designed to determine the velocity of the earth with respect to the hypothetical ether. The experimental tool used was the Michelson interferometer shown in Figure 39.4. Suppose one of the arms of the interferometer is aligned along the direction of the motion of the earth through space. The motion of the earth through the ether would be equivalent to the ether flowing past the earth in the opposite direction. This ether wind blowing in the opposite direction should cause the speed of light as measured in the earth's frame of reference to be $c - v$ as it approaches the mirror $M_2$ in Figure 39.4 and $c + v$ after reflection. The speed $v$ is the speed of the earth through space, and hence the speed of the ether wind, while $c$ is the speed of light in the ether frame. The two beams of light reflected from $M_1$ and $M_2$ recombine, and an interference pattern consisting of alternating dark and bright bands or fringes is formed. During the experiment, the interference pattern was observed while the interferometer was rotated through an angle of 90 degrees. This rotation would change the speed of the ether wind along the direction of the arms of the interferometer. The effect of this rotation should have been to cause the fringe pattern to shift slightly but measurably. Measurements failed to show any change in the interference pattern! The Michelson-Morley experiment was repeated by other researchers under various conditions and at different locations, but the results were always the same: *no fringe shift of the magnitude required was ever observed.*

The negative results of the Michelson-Morley experiment meant that it was impossible to measure the absolute velocity of the earth (the orbital velocity) with respect to the ether frame. However, as we shall see in the next section, Einstein developed a postulate for his theory of relativity that places quite a different interpretation on these null results. In later years, when more was known about the nature of light, the idea of an ether that permeates all of space was relegated to the ash heap of worn out concepts. Light is now understood to be *an electromagnetic wave, which requires no medium for its propagation.* As a result, the idea of having an ether in which these waves could travel became unnecessary.

Albert A. Michelson (1852–1931). (AIP Niels Bohr Library)

### Details of the Michelson-Morley Experiment

In order to understand the outcome of the Michelson-Morley experiment, let us assume that the interferometer shown in Figure 39.4 has two arms of equal length $L$. First consider the beam traveling parallel to the direction of the ether wind, which is taken to be horizontal in Figure 39.4. As the beam moves to the right, its speed is reduced by the wind and its speed with respect to the

[3] A. A. Michelson and E. W. Morley, *Am. J. Sci* **134**:333 (1887).

earth is $c - v$. On its return journey, as the light beam moves to the left downwind, its speed with respect to the earth is $c + v$. Thus, the time of travel to the right is $L/(c - v)$ and the time of travel to the left is $L/(c + v)$. The total time of travel for the round trip along the horizontal path is

$$t_1 = \frac{L}{c+v} + \frac{L}{c-v} = \frac{2Lc}{c^2 - v^2} = \frac{2L}{c}\left(1 - \frac{v^2}{c^2}\right)^{-1}$$

Now consider the light beam traveling perpendicular to the wind, which is the vertical direction in Figure 39.4. (The vertical direction here means vertical on the page.) Since the speed of the beam relative to the earth is $(c^2 - v^2)^{1/2}$ in this case (see Fig. 39.3), then the time of travel for each half of this trip is $L/(c^2 - v^2)^{1/2}$, and the total time of travel for the round trip is

$$t_2 = \frac{2L}{(c^2 - v^2)^{1/2}} = \frac{2L}{c}\left(1 - \frac{v^2}{c^2}\right)^{-1/2}$$

Thus, the time difference between the light beam traveling horizontally and the beam traveling vertically is

$$\Delta t = t_1 - t_2 = \frac{2L}{c}\left[\left(1 - \frac{v^2}{c^2}\right)^{-1} - \left(1 - \frac{v^2}{c^2}\right)^{-1/2}\right]$$

Since $v^2/c^2 \ll 1$, this expression can be simplified by using the following binomial expansion after dropping all terms higher than second order:

$$(1 - x)^n \approx 1 - nx \qquad (\text{for } x \ll 1)$$

In our case, $x = v^2/c^2$, and we find

$$t_1 - t_2 = \Delta t = \frac{Lv^2}{c^3} \tag{39.3}$$

Because the speed of the earth in its orbital path is approximately $3 \times 10^4$ m/s, the speed of the wind should be at least this great. The two light beams start out in phase and return to form an interference pattern. Let us assume that the interferometer is adjusted for parallel fringes and that a telescope is focused on one of these fringes. The time difference between the two light beams gives rise to a phase difference between the beams, producing an interference pattern when they combine at the position of the telescope. A difference in the pattern should be detected by rotating the interferometer through 90° in a horizontal plane, such that the two beams exchange roles. This results in a net time difference of twice that given by Equation 39.3. Thus, the path difference which corresponds to this time difference is

$$\Delta d = c(2\,\Delta t) = \frac{2Lv^2}{c^2}$$

The corresponding phase difference $\Delta\phi$ is equal to $2\pi/\lambda$ times this path difference, where $\lambda$ is the wavelength of light. That is,

$$\Delta\phi = \frac{4\pi L v^2}{\lambda c^2} \tag{39.4}$$

In the first experiments by Michelson and Morley, each light beam was reflected by mirrors many times to give an increased effective path length $L$ of about 11 meters. Using this value, and taking $v$ to be equal to $3 \times 10^4$ m/s, gives a path difference of

$$\Delta d = \frac{2(11 \text{ m})(3 \times 10^4 \text{ m/s})^2}{(3 \times 10^8 \text{ m/s})^2} = 2.2 \times 10^{-7} \text{ m}$$

This extra distance of travel should produce a noticeable shift in the fringe pattern. Specifically, calculations show that if one views the pattern while the interferometer is rotated through 90°, a shift of about 0.4 fringe should be observed. The instrument used by Michelson and Morley had the capability of detecting a shift in the fringe pattern as small as 0.01 fringe. However, *they detected no shift in the fringe pattern.* Since then, the experiment has been repeated many times by various scientists under various conditions, and no fringe shift has ever been detected. Thus, it was concluded that one cannot detect the motion of the earth with respect to the ether.

Many efforts were made to explain the null results of the Michelson-Morley experiment. For example, perhaps the earth drags the ether with it in its motion through space. To test this assumption, interferometer measurements were made at various altitudes, but again no fringe shift was detected. In the 1890s, G. F. Fitzgerald and H. A. Lorentz independently tried to explain the null results by making the following ad hoc assumption. They proposed that the length of an object moving along the direction of the ether wind would contract by a factor of $\sqrt{1 - v^2/c^2}$. The net result of this contraction would be a change in length of one of the arms of the interferometer such that no path difference would occur as it was rotated.

No experiment in the history of physics has received such valiant efforts to try to explain the absence of an expected result as did the Michelson-Morley experiment. The stage was set for the brilliant Albert Einstein, who solved the problem in 1905 with his special theory of relativity.

## 39.4 EINSTEIN'S PRINCIPLE OF RELATIVITY

In the previous section we noted the serious contradiction between the invariance of the speed of light and the Galilean addition law for velocities. In 1905, Albert Einstein proposed a theory that would resolve this contradiction but at the same time would completely alter our notion of space and time.[4] Einstein based his special theory of relativity on the following general hypothesis, which is called the **Principle of Relativity:**

All the laws of physics are the same in all inertial reference frames.

The postulates of special relativity

In Einstein's own words, ". . . the same laws of electrodynamics and optics will be valid for all frames of reference for which the equations of mechanics hold good." This simple statement is a generalization of Newton's principle of relativity, and is the foundation of the special theory of relativity.

An immediate consequence of the Principle of Relativity is as follows:

The speed of light in a vacuum has the same value in all inertial reference frames, $c = 2.99792458 \times 10^8$ m/s.

[4] A. Einstein, "On the Electrodynamics of Moving Bodies," *Ann. Physik* **17**: 891 (1905). For an English translation of this article and other publications by Einstein, see the book by H. Lorentz, A. Einstein, H. Minkowski, and H. Weyl, *The Principle of Relativity*, Dover, 1958.

Biographical Sketch

**Albert Einstein**
*(1879–1955)*

(Photograph Courtesy of AIP Niels Bohr Library)

**Albert Einstein,** one of the greatest physicists of all times, was born in Ulm, Germany. He showed little intellectual promise as a youngster, and left the highly disciplined German school system after one teacher stated "You will never amount to anything, Einstein." Following a vacation in Italy, he completed his education at the Swiss Federal Polytechnic School in 1901. Although Einstein attended very few lectures, he was able to pass the courses with the help of excellent lecture notes taken by a friend. Unable to find an academic position, Einstein accepted a position as a junior official in the Swiss Patent Office in Berne. In this setting, and during his "spare time," he continued his independent studies in theoretical physics. In 1905, at the age of 26, he published four scientific papers that revolutionized physics. (In that same year, he earned his Ph.D.) One of these papers, for which he was awarded the 1921 Nobel prize in physics, dealt with the photoelectric effect. Another was concerned with Brownian motion, the irregular motion of small particles suspended in a liquid. The remaining two papers were concerned with what is now considered his most important contribution of all, the special theory of relativity. In 1915, Einstein published his work on the general theory of relativity, which relates gravity to the structure of space and time. The most dramatic prediction of this theory is the degree to which light would be deflected by a gravitational field. Measurements made by astronomers on bright stars in the vicinity of the eclipsed sun in 1919 confirmed Einstein's prediction, and Einstein suddenly became a world celebrity. This and other predictions of the general theory of relativity are discussed in the delightful essay by Clifford Will that follows this chapter.

In 1913, following academic appointments in Switzerland and Czechoslovakia, Einstein accepted a special position created for him at the Kaiser Wilhelm Institute in Berlin. This made it possible for him to be able to devote all of his time to research, free of financial troubles and routine duties. Einstein left Germany in 1933, which was then under Hitler's power, thereby escaping the fate of millions of other European Jews. In the same year he accepted a special position at the Institute for Advanced Study in Princeton where he remained for the rest of his life. He became an American citizen in 1940. Although he was a pacifist, Einstein was persuaded by Leo Szilard to write a letter to President Franklin D. Roosevelt urging him to initiate a program to develop a nuclear bomb. The result was the successful six-year Manhattan project and two nuclear explosions in Japan which ended World War II in 1945.

Einstein made many important contributions to the development of modern physics, including the concept of the light quantum and the idea of stimulated emission of radiation, which led to the invention of the laser 40 years later. Einstein was deeply disturbed by the development of quantum mechanics in the 1920s despite his own role as a scientific revolutionary. In particular, he could never accept the probabilistic view of events in nature which is a central feature of the highly successful quantum theory. He once said, "God does not play dice with nature." The last few decades of his life were devoted to an unsuccessful search for a unified theory that would combine gravitation and electromagnetism into one picture.

In other words, anyone who measures the speed of light will get the same value, $c$. This implies that the ether simply does not exist. The Principle of Relativity, together with the above statement, are often referred to as the **two postulates of special relativity.**

Although the Michelson-Morley experiment was performed before Einstein published his work on relativity, it is not clear that Einstein was aware of the details of the experiment. Nonetheless, the null result of the experiment can be readily understood within the framework of Einstein's theory. According to his Principle of Relativity, the premises of the Michelson-Morley experiment were incorrect. In the process of trying to explain the expected results, we stated that when light traveled against the ether wind its speed was $c - v$, in accordance with the Galilean addition law for velocities. However, if the state of motion of the observer or of the source has no influence on the value found for the speed of light, one will always measure the value to be $c$. Likewise, the light makes the return trip after reflection from the mirror at a speed of $c$, and not with the speed $c + v$. Thus, the motion of the earth should not influence the fringe pattern observed in the Michelson-Morley experiment and a null result should be expected.

The Michelson-Morley experiment has been repeated many times, always giving the same null results. Modern versions of the experiment have compared the frequencies of resonant laser cavities of identical length oriented at right angles to each other. More recently, Doppler shift experiments using gamma rays emitted by a radioactive sample of $^{57}Fe$ placed an upper limit of about 5 cm/s on the ether wind velocity. These results have shown quite conclusively that the motion of the earth has no effect on the speed of light!

If we accept Einstein's theory of relativity, we must conclude that relative motion is unimportant when measuring the speed of light. At the same time, we must alter our common-sense notion of space and time and be prepared for some rather bizarre consequences.

## 39.5 DESCRIBING EVENTS IN RELATIVITY

Before we discuss the consequences of special relativity, we must first understand how an observer located in an inertial reference frame describes an event. As we mentioned earlier, an event is an occurrence which is described by three space coordinates and one time coordinate. Different observers in different inertial frames will, in general, describe the same event with different spacetime coordinates. The reference frame which is used to describe an event consists of a coordinate grid and a set of clocks located at the grid intersections as shown in Figure 39.5 in a two-dimensional representation. It is necessary that the array of clocks in the grid be synchronized. This can be accomplished with the help of light signals in many ways. For example, suppose the observer located at the origin of coordinates with his master clock sends out a pulse of light at $t = 0$. The light pulse takes a time $r/c$ to reach a second clock located a distance $r$ from the origin. Hence, the second clock will

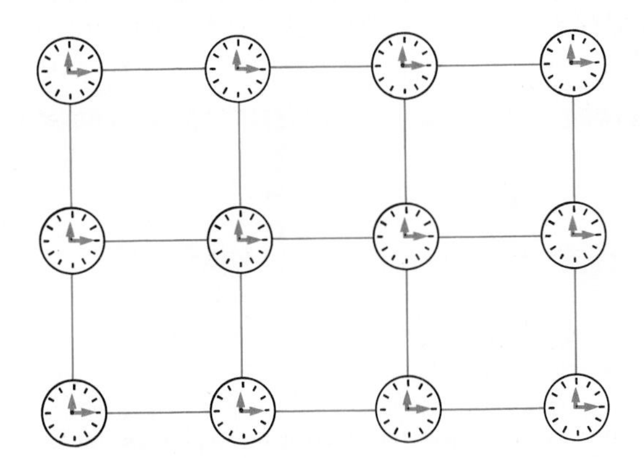

**Figure 39.5** In relativity, we use a reference frame consisting of a coordinate grid and a set of synchronized clocks.

be synchronized with the clock at the origin if the second clock reads a time $r/c$ at the instant the pulse reaches it. This procedure of synchronization assumes that the speed of light has the same value in all directions and in all inertial frames. Furthermore, the procedure concerns an event recorded by an observer in a specific inertial reference frame. An observer in some other inertial frame would assign different spacetime coordinates to events they observe using their own coordinate grid and array of clocks.

Almost everyone who has dabbled even superficially with science is aware of some of the startling predictions that arise because of Einstein's approach to relative motion. As we examine some of the consequences of relativity in the following three sections, we shall find that they conflict with our basic notions of space and time. We shall restrict our discussion to the concepts of length, time, and simultaneity, which are quite different in relativistic mechanics than in Newtonian mechanics. For example, we shall see that *the distance between two points and the time interval between two events depend on the frame of reference in which they are measured.* That is, *there is no such thing as absolute length or absolute time in relativity.* Furthermore, *events at different locations that occur simultaneously in one frame are not simultaneous in another frame.*

## 39.6 SIMULTANEITY

A basic premise of Newtonian mechanics is that a universal time scale exists which is the same for all observers. In fact, Newton wrote that "Absolute, true, and mathematical time, of itself, and from its own nature, flows equably without relation to anything external." Thus, Newton and his followers simply took simultaneity for granted. In his special theory of relativity, Einstein abandoned this assumption. According to Einstein, *time interval measurements depend on the reference frame in which the measurement is made.*

Einstein devised the following thought experiment to illustrate this point. A boxcar moves with uniform velocity, and two lightning bolts strike the ends of the boxcar, as in Figure 39.6a, leaving marks on the boxcar and ground. The marks left on the boxcar are labeled $A'$ and $B'$, while those on the ground are labeled $A$ and $B$. An observer at $O'$ moving with the boxcar is midway between $A'$ and $B'$, while a ground observer at $O$ is midway between $A$ and $B$. The events recorded by the observers are the light signals from the lightning bolts.

Let us assume that the two light signals reach the observer at $O$ at the same time, as indicated in Figure 39.6b. This observer realizes that the light signals

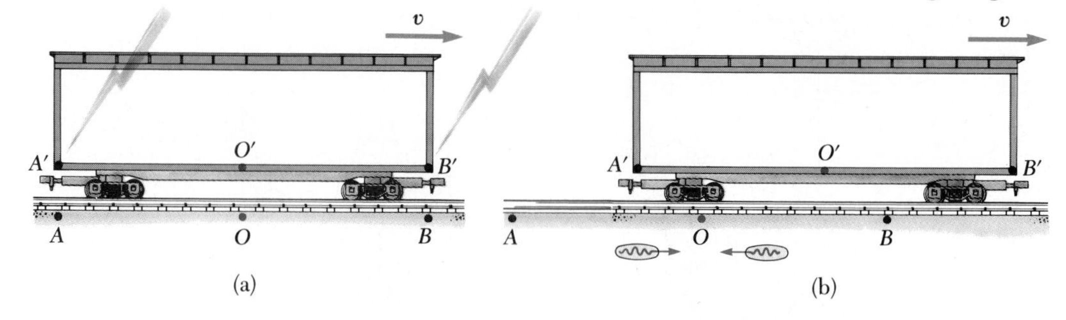


**Figure 39.6** Two lightning bolts strike the end of a moving boxcar. (a) The events appear to be simultaneous to the stationary observer at $O$, who is midway between $A$ and $B$. (b) The events do not appear to be simultaneous to the observer at $O'$, who claims that the front of the train is struck *before* the rear.

have traveled at the same speed over distances of equal length. Thus, the observer at $O$ rightly concludes that the events at $A$ and $B$ occurred simultaneously. Now consider the same events as viewed by the observer on the boxcar at $O'$. By the time the light has reached the observer at $O$, the observer at $O'$ has moved as indicated in Figure 39.6b. Thus, the light signal from $B'$ has already swept past $O'$, while the light from $A'$ has not yet reached $O'$. According to Einstein, the observer at $O'$ must find that light travels at the same speed as that measured by the observer at $O$. Therefore, the observer at $O'$ concludes that the lightning struck the front of the boxcar before it struck the back. This thought experiment clearly demonstrates that the two events, which appear to be simultaneous to the observer at $O$, do not appear to be simultaneous to the observer at $O'$. In other words,

> two events that are simultaneous in one reference frame are in general not simultaneous in a second frame moving with respect to the first. That is, simultaneity is not an absolute concept but one that depends upon the state of motion of the observer.

At this point, you might wonder which observer is right concerning the two events. The answer is that *both are correct,* because the Principle of Relativity states that *there is no preferred inertial frame of reference.* Although the two observers reach different conclusions, both are correct in their own reference frame because the concept of simultaneity is not absolute. This, in fact, is the central point of relativity—any uniformly moving frame of reference can be used to describe events and do physics. There is nothing wrong with the clocks and meter sticks used to perform measurements. It is simply that time intervals and length measurements depend on the state of motion of the observer. Observers in different inertial frames of reference will always measure different time intervals with their clocks and different distances with their meter sticks. However, they will both agree on the laws of physics in their respective frames, since they must be the same for all observers in uniform motion. It is the alteration of time and space that allows the laws of physics (including Maxwell's equations) to be the same for all observers in uniform motion.

## 39.7 THE RELATIVITY OF TIME

As we have seen, observers in different inertial frames will always measure different time intervals between a pair of events. This can be illustrated by considering a vehicle moving to the right with a speed $v$ as in Figure 39.7a. A mirror is fixed to the ceiling of the vehicle and an observer Liz at $O'$ at rest in this system holds a laser a distance $d$ below the mirror. At some instant, the laser emits a pulse of light directed towards the mirror (event 1), and at some later time after reflecting from the mirror, the pulse arrives back at the laser (event 2). Liz carries a clock $C'$ which she uses to measure the time interval $\Delta t'$ between these two events. Because the light pulse has a speed $c$, the time it takes the pulse to travel from Liz to the mirror and back to Liz can be found from the definition of speed:

$$\Delta t' = \frac{\text{distance traveled}}{\text{speed}} = \frac{2d}{c} \qquad (39.5)$$

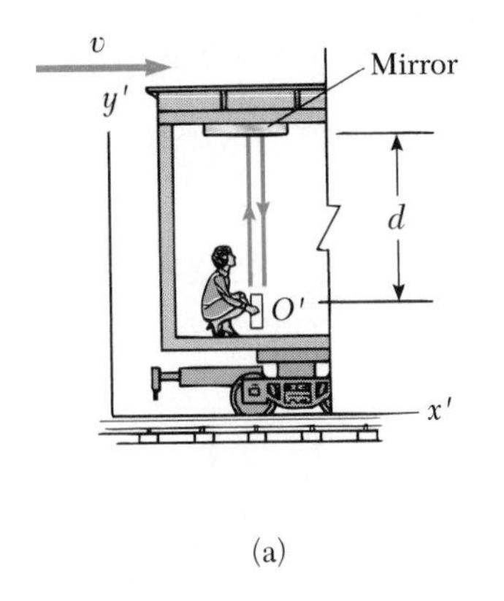


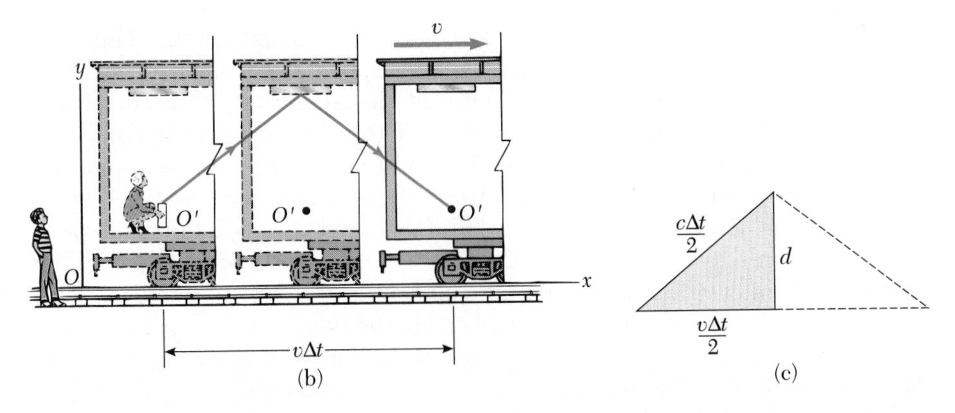


**Figure 39.7** (a) A mirror is fixed to a moving vehicle, and a light pulse leaves $O'$ at rest in the vehicle. (b) Relative to a stationary observer on earth, the mirror and $O'$ move with a speed $v$. Note that the distance the pulse travels is greater than $2d$ as measured by the stationary observer. (c) The right triangle for calculating the relationship between $\Delta t$ and $\Delta t'_{O'}$.

This time interval $\Delta t'$ measured by Liz who is at rest in the moving vehicle requires only a *single* clock $C'$ located at the *same place* in this frame.

Now consider the same pair of events as viewed by an observer named Mark at $O$ in a stationary frame as in Figure 39.7b. According to Mark, the mirror and laser are moving to the right with a speed $v$. By the time the light pulse reaches the mirror, the mirror moves a distance $v\,\Delta t/2$, where $\Delta t$ is the time it takes the light to travel from $O'$ to the mirror and back to $O'$ as measured by Mark. In other words, Mark concludes that, because of the motion of the vehicle, if the light is to hit the mirror, it must leave the laser at an angle with respect to the vertical direction. Comparing Figures 39.7a and 39.7b, we see that the light must travel farther for Mark than it does for Liz.

According to the second postulate of special relativity, the speed of light must be $c$ as measured by both observers. Since the light travels a farther distance for Mark, it follows that the time interval $\Delta t$ measured by Mark in the stationary frame is *longer* than the time interval $\Delta t'$ measured by Liz in the moving frame. To obtain a relationship between these two time intervals, it is convenient to use the right triangle shown in Figure 39.7c. The Pythagorean theorem applied to this triangle gives

$$\left(\frac{c\,\Delta t}{2}\right)^2 = \left(\frac{v\,\Delta t}{2}\right)^2 + d^2$$

Solving for $\Delta t$ gives

$$\Delta t = \frac{2d}{\sqrt{c^2 - v^2}} = \frac{2d}{c\sqrt{1 - \frac{v^2}{c^2}}} \tag{39.6}$$

Because $\Delta t' = 2d/c$, we can express Equation 39.6 as

Time dilation

$$\Delta t = \frac{\Delta t'}{\sqrt{1 - \frac{v^2}{c^2}}} = \gamma\,\Delta t' \tag{39.7}$$

where $\gamma = (1 - v^2/c^2)^{-1/2}$. The two events observed by Mark occur at *different* positions. In order to measure the time interval $\Delta t$, Mark must use *two synchronized clocks* located at *different places* in his reference frame. Thus, their situations are not symmetrical.

From Equation 39.7, we see that the time interval $\Delta t$ measured by Mark in the stationary frame is *longer* than the time interval $\Delta t'$ measured by Liz in the moving frame (because $\gamma$ is always greater than unity). That is, $\Delta t > \Delta t'$.

According to a stationary observer, a moving clock runs **slower** than an identical stationary clock. This effect is known as **time dilation.**

The time interval $\Delta t'$ in Equation 39.7 is called the **proper time.** In general, proper time is defined as *the time interval between two events as measured by an observer who sees the events occur at the same place.* In our case, the observer at $O'$ measures the proper time. That is, *proper time is always the time measured with a single clock at rest in that frame.*

We have seen that moving clocks run slow by a factor of $\gamma^{-1}$. This is true for ordinary mechanical clocks as well as for the light clock just described. In fact, we can generalize these results by stating that *all physical processes, including chemical reactions and biological processes, slow down relative to a stationary clock when they occur in a moving frame.* For example, the heartbeat of an astronaut moving through space would have to keep time with a clock inside the spaceship. Both the astronaut's clock and his heartbeat are slowed down relative to a stationary clock. The astronaut would not have any sensation of life slowing down in the spaceship.

Time dilation is a very real phenomenon that has been verified by various experiments. For example, muons are unstable elementary particles, which have a charge equal to that of an electron and a mass 207 times that of the electron. Muons can be produced by the absorption of cosmic radiation high in the atmosphere. These unstable particles have a lifetime of only 2.2 $\mu$s when measured in a reference frame at rest with respect to them. If we take 2.2 $\mu$s as the average lifetime of a muon and assume that their speed is close to the speed of light, we would find that these particles could travel a distance of only about 600 m before they decayed into something else (Fig. 39.8a). Hence, they could not reach the earth from the upper atmosphere where they are produced. However, experiments show that a large number of muons *do* reach the earth. The phenomenon of time dilation explains this effect. Relative to an observer on earth, the muons have a lifetime equal to $\gamma\tau$, where $\tau = 2.2$ $\mu$s is the lifetime in a frame of reference traveling with the muons. For example, for $v = 0.99c$, $\gamma \approx 7.1$ and $\gamma\tau \approx 16$ $\mu$s. Hence, the average distance traveled as measured by an observer on earth is $\gamma v\tau \approx 4800$ m, as indicated in Figure 39.8b.

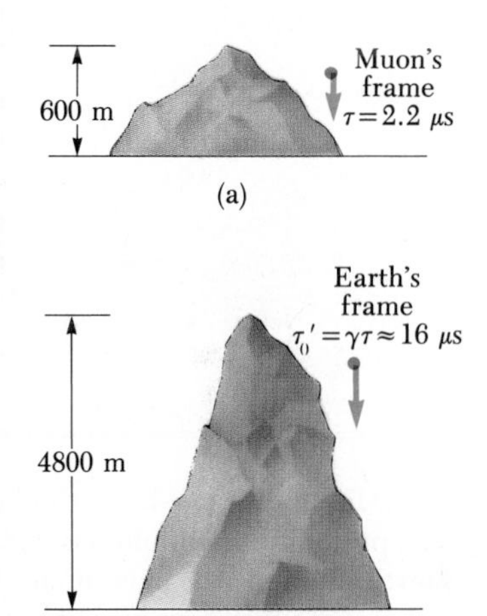


**Figure 39.8** (a) Muons traveling with a speed of $0.99c$ travel only about 600 m as measured in the muons' reference frame, where their lifetime is about 2.2 $\mu$s. (b) The muons travel about 4800 m as measured by an observer on earth. Because of time dilation, the muons' lifetime is longer as measured by the earth observer.

In 1976, experiments with muons were conducted at the laboratory of the European Council for Nuclear Research (CERN) in Geneva. Muons were injected into a large storage ring, reaching speeds of about $0.9994c$. Electrons produced by the decaying muons were detected by counters around the ring, enabling scientists to measure the decay rate, and hence the lifetime of the muons. The lifetime of the moving muons was measured to be about 30 times as long as that of the stationary muon (see Fig. 39.9), in agreement with the prediction of relativity to within two parts in a thousand.

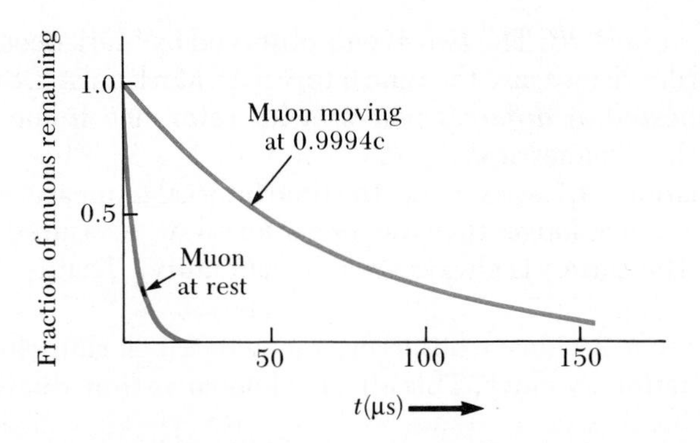


**Figure 39.9** Decay curves for muons traveling at a speed of $0.9994c$ and for muons at rest.

The results of an experiment reported by Hafele and Keating provided direct evidence for the phenomenon of time dilation.[5] The experiment involved the use of very stable cesium beam atomic clocks. Time intervals measured with four such clocks in jet flight were compared with time intervals measured by reference atomic clocks located at the U.S. Naval Observatory. (Because of the earth's rotation about its axis, a ground-based clock is not in a true inertial frame.) In order to compare these results with the theory, many factors had to be considered, including periods of acceleration and deceleration relative to the earth, variations in direction of travel, and the weaker gravitational field experienced by the flying clocks compared to the earth-based clock. Their results were in good agreement with the predictions of the special theory of relativity. In their paper, Hafele and Keating report the following: "Relative to the atomic time scale of the U.S. Naval Observatory, the flying clocks lost 59 ± 10 ns during the eastward trip and gained 273 ± 7 ns during the westward trip. . . . These results provide an unambiguous empirical resolution of the famous clock paradox with macroscopic clocks."

EXAMPLE 39.1 **What is the Period of the Pendulum?**
The period of a pendulum is measured to be 3.0 s in the inertial frame of the pendulum. What is the period of the pendulum when measured by an observer moving at a speed of $0.95c$ with respect to the pendulum?

*Solution* In this case, the proper time is equal to 3 s. We can use Equation 39.7 to calculate the period measured by the moving observer. This gives

$$T = \gamma T' = \frac{1}{\sqrt{1 - \frac{(0.95c)^2}{c^2}}} T' = (3.2)(3.0 \text{ s}) = 9.6 \text{ s}$$

That is, the observer moving at a speed of $0.95c$ observes the pendulum as oscillating more slowly.

### The Twin Paradox

An interesting consequence of time dilation is the so-called twin paradox. Consider a controlled experiment involving 20-year-old twin brothers Speedo and Goslo. Speedo, the more venturesome twin, sets out on a journey toward a star located 30 lightyears from earth. His spaceship is able to acceler-

[5] J. C. Hafele and R. E. Keating, "Around the World Atomic Clocks: Relativistic Time Gains Observed," *Science*, July 14, 1972, p. 168.

ate to a speed close to the speed of light. After reaching the star, Speedo becomes very homesick and immediately returns to earth at the same high speed. Upon his return, he is shocked to find that many things have changed. Old cities have expanded and new cities have appeared. Lifestyles, people's appearances, and transportation systems have changed dramatically. Speedo's twin brother, Goslo, has aged to about 80 years old and is now wiser, feeble, and somewhat hard of hearing. Speedo, on the other hand, has aged only about ten years. This is because his bodily processes slowed down during his travels in space.

It is quite natural to raise the question, "Which twin actually traveled at a speed close to the speed of light, and therefore would be the one who did not age?" Herein lies the paradox: From Goslo's frame of reference, he was at rest while his brother Speedo traveled at a high velocity. On the other hand, according to the space traveler Speedo, it was he who was at rest while his brother zoomed away from him on earth and then returned. This leads to contradictions as to which twin actually aged.

In order to resolve this paradox, it should be pointed out that the trip is not as symmetrical as we may have led you to believe. Speedo, the space traveler, had to experience a series of accelerations and decelerations during his journey to the star and back home, and therefore is not always in uniform motion. This means that Speedo has been in a noninertial frame during part of his trip, so that predictions based on special relativity are not valid in his frame. On the other hand, the brother on earth has been in an inertial frame, and he can make reliable predictions based on the special theory. The situation is not symmetrical since Speedo experiences forces when his spaceship turns around, whereas Goslo is not subject to such forces. Therefore, the space traveler will indeed be younger upon returning to earth.

## 39.8 THE RELATIVITY OF LENGTH

We have seen that measured time intervals are not absolute, that is, the time interval between two events depends on the frame of reference in which it is measured. Likewise, the measured distance between two points depends on the frame of reference. The **proper length** of an object is defined as *the length of the object measured in the reference frame in which the object is at rest.* You should note that proper length is defined similarly to proper time, in that proper time is the time measured by a clock at rest relative to both events. However, proper length is *not* the distance between two points measured at the same time. The length of an object measured in a reference frame in which the object is moving is always less than the proper length. This effect is known as **length contraction.**

To understand length contraction quantitatively, let us consider a spaceship traveling with a speed $v$ from one star to another, and two observers. An observer at rest on earth (and also assumed to be at rest with respect to the two stars) measures the distance between the stars to be $L'$, where $L'$ is the proper length. According to this observer, the time it takes the spaceship to complete the voyage is $\Delta t = L'v$. What does an observer in the moving spaceship measure for the distance between the stars? Because of time dilation, the space traveler measures a smaller time of travel: $\Delta t' = \Delta t/\gamma$. The space traveler claims to be at rest and sees the destination star as moving toward the spaceship with speed $v$. Since the space traveler reaches the star in the time $\Delta t'$, he or she concludes that the distance, $L$, between the stars is shorter than $L'$. This distance measured by the space traveler is given by

$$L = v\,\Delta t' = v\,\frac{\Delta t}{\gamma}$$

Since $L' = v\,\Delta t$, we see that $L = L'/\gamma$ or

Length contraction

$$L = L'\left(1 - \frac{v^2}{c^2}\right)^{1/2} \tag{39.8}$$

According to this result,

> if an observer at rest with respect to an object measures its length to be $L'$, an observer moving with a relative speed $v$ with respect to the object will find it to be shorter than its rest length by the factor $(1 - v^2/c^2)^{1/2}$.

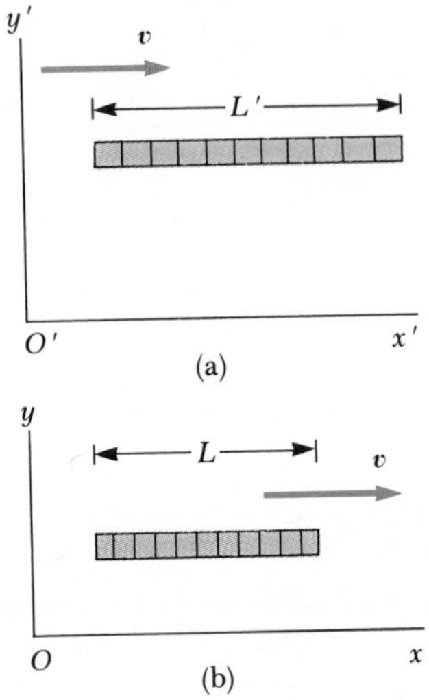

**Figure 39.10** (a) A stick as viewed by an observer in a frame attached to the stick (i.e., both have the same velocity). (b) The stick as seen by an observer in a frame in which the stick has a velocity $\boldsymbol{v}$ relative to the frame. The length is shorter than the proper length, $L'$, by a factor $(1 - v^2/c^2)^{1/2}$.

You should note that *the length contraction takes place only along the direction of motion.* For example, suppose a stick moves past a stationary earth observer with a speed $v$ as in Figure 39.10. The length of the stick as measured by an observer in the frame attached to it is the proper length $L'$, as illustrated in Figure 39.10a. The length of the stick, $L$, as measured by the earth observer in the stationary frame is shorter than $L'$ by the factor $(1 - v^2/c^2)^{1/2}$. Note that length contraction is a symmetrical effect: if the stick were at rest on earth, an observer in the moving frame would also measure its length to be shorter by the same factor $(1 - v^2/c^2)^{1/2}$.

It is important to emphasize that proper length and proper time are measured in *different* reference frames. As an example of this point, let us return to the example of the decaying muons moving at speeds close to the speed of light. An observer in the muon's reference frame would measure the proper lifetime, while an earth-based observer measures the proper height of the mountain in Figure 39.8. In the muon's reference frame, there is no time dilation, but the distance of travel is observed to be shorter when measured in this frame. Likewise, in the earth observer's reference frame, there is time dilation, but the distance of travel is measured to be the actual height of the mountain. Thus, when calculations on the muon are performed in both frames, one sees the effect of "offsetting penalties" and the outcome of the experiment is the same!

If an object in the shape of a box passing by could be photographed, its image would show length contraction, but its shape would also be distorted. This is illustrated in the computer-simulated drawings shown in Figure 39.11

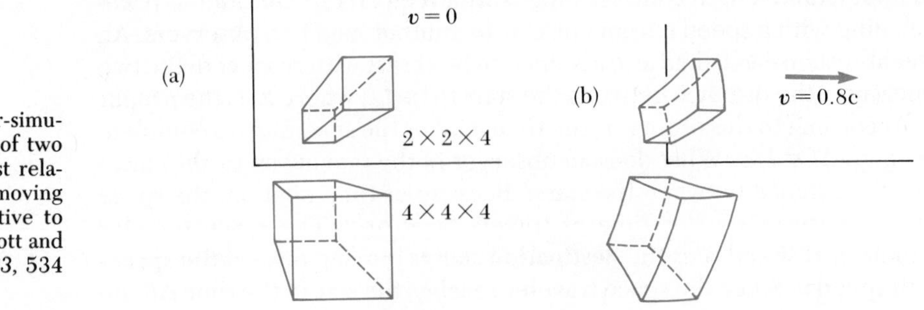


**Figure 39.11** Computer-simulated photographs of a set of two rectangular boxes (a) at rest relative to the camera and (b) moving at a velocity $\boldsymbol{v} = 0.8\boldsymbol{c}$ relative to the camera. [From G. D. Scott and M. R. Viner, *Am. J. Phys.* 33, 534 (1965).]

for a box moving past an observer with a speed $v = 0.8c$. When the shutter of the camera is opened, it records the shape of the object at the instant light from it began its journey to the camera. Since light from different parts of the object must arrive at the camera at the same time (when the photograph is taken), light from more distant parts of the object must start its journey earlier than light from closer parts. Hence, the photograph records different parts of the object at different times. This results in a highly distorted image.

**EXAMPLE 39.2 The Contraction of a Spaceship**
A spaceship is measured to be 100 m long while it is at rest with respect to an observer. If this spaceship now flies by the observer with a speed of $0.99c$, what length will the observer find for the spaceship?

*Solution* From Equation 39.8, the length measured by an observer in the spaceship is

$$L = L'\sqrt{1-\frac{v^2}{c^2}} = (100\text{ m})\sqrt{1-\frac{(0.99c)^2}{c^2}} = 14\text{ m}$$

Exercise 1 If the ship moves past the observer with a speed of $0.01c$, what length will the observer measure?
Answer 99.99 m.

**EXAMPLE 39.3 How High is the Spaceship?**
An observer on earth sees a spaceship at an altitude of 435 m moving downward toward the earth with a speed of $0.970c$. What is the altitude of the spaceship as measured by an observer in the spaceship?

*Solution* The moving observer in the ship finds the altitude to be

$$L = L'\sqrt{1-\frac{v^2}{c^2}} = (435\text{ m})\sqrt{1-\frac{(0.970c)^2}{c^2}}$$

$$= 106\text{ m}$$

**EXAMPLE 39.4 The Triangular Spaceship**
A spaceship in the form of a triangle flies by an observer with a speed of $0.950c$. When the ship is at rest (Fig. 39.12a), the distances $x$ and $y$ are found to be 50.0 m and 25 m, respectively. What is the shape of the ship as seen by an observer at rest when the ship is in motion along the direction shown in Figure 39.12b?

*Solution* The observer sees the horizontal length of the ship to be contracted to a length of

$$L = L'\sqrt{1-\frac{v^2}{c^2}} = (50.0\text{ m})\sqrt{1-\frac{(0.950c)^2}{c^2}}$$

$$= 15.6\text{ m}$$

The 25-m vertical height is unchanged because it is perpendicular to the direction of relative motion between the observer and the spaceship. Figure 39.12b represents the shape of the spaceship as seen by the observer at rest.

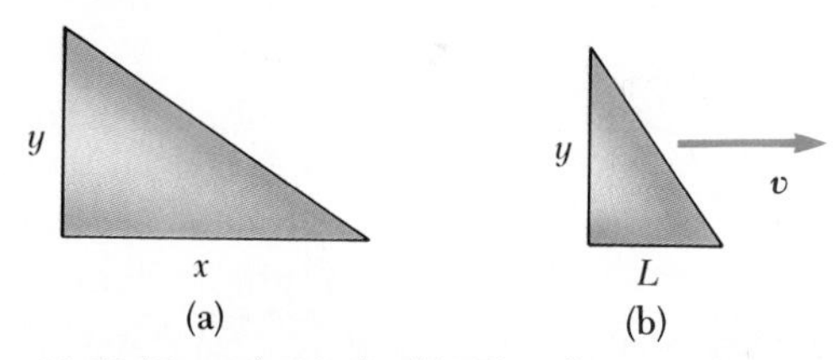


**Figure 39.12** (Example 39.4) (a) When the spaceship is at rest, its shape is as shown. (b) The spaceship appears to look like this when it moves to the right with a speed $v$. Note that only its $x$ dimension is contracted in this case.

## 39.9 THE LORENTZ TRANSFORMATION EQUATIONS

An event such as a flash of light is specified by three space coordinates and one time coordinate. Suppose that an event that occurs at some point $P$ is reported by two observers, one at rest in the S frame and another in the S′ frame moving to the right with a speed $v$ as in Figure 39.13. The observer in S reports the event with spacetime coordinates $(x, y, z, t)$, while the observer in S′ reports the same event using the coordinates $(x', y', z', t')$. We would like to find a relation between these coordinates that would be valid for all speeds. In Section 39.2, we found that the Galilean transformation of coordinates, given by Equation 39.1, does not agree with experiment at speeds comparable to the speed of light.

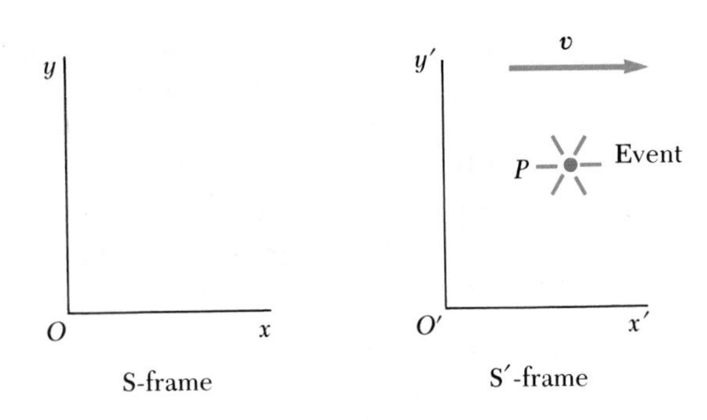


**Figure 39.13** Representation of an event that occurs at some point $P$ as observed by an observer at rest in the S frame and another in the S′ frame, which is moving to the right with a speed $v$.

The correct equations which are valid for speeds ranging from $v = 0$ to $v = c$, and enable us to transform some S to S′, are given by the **Lorentz transformation equations:**

Lorentz transformation for $S \rightarrow S'$

$$\begin{aligned} x' &= \gamma(x - vt) \\ y' &= y \\ z' &= z \\ t' &= \gamma\left(t - \frac{v}{c^2}x\right) \end{aligned} \tag{39.9}$$

where the symbol $\gamma$ (gamma) is defined as

$$\gamma \equiv \frac{1}{\sqrt{1 - \frac{v^2}{c^2}}} \tag{39.10}$$

The Lorentz transformation equations were originally derived by H. A. Lorentz (1853–1928) in 1890. However, it was Einstein who recognized their physical significance and took the bold step of interpreting them within the framework of the theory of relativity.

We see that the value for $t'$ assigned to an event by observer $O'$ depends both on the time $t$ and on the coordinate $x$ as measured by observer $O$. This is consistent with the notion that an event is characterized by four space-time coordinates $(x, y, z, t)$. In other words, space and time are not actually separate concepts, but are closely interwoven in relativity. This is unlike the case of the Galilean transformation in which $t = t'$.

If we wish to transform coordinates in the S′ frame to coordinates in the S frame, we simply replace $v$ by $-v$ and interchange the primed and unprimed coordinates in Equation 39.9. The resulting transformation is given by

Inverse Lorentz transformation for $S' \rightarrow S$

$$\begin{aligned} x &= \gamma(x' + vt') \\ y &= y' \\ z &= z' \\ t &= \gamma\left(t' + \frac{v}{c^2}x'\right) \end{aligned} \tag{39.11}$$

When $v \ll c$, the Lorentz transformation should reduce to the Galilean transformation. To check this, note that as $v \to 0$, $v/c^2 \ll 1$ and $v^2/c^2 \ll 1$, so that Equation 39.9 reduces in this limit to the Galilean coordinate transformation equations, given by

$$x' = x - vt \qquad y' = y \qquad z' = z \qquad t' = t$$

In many situations, we would like to know the *difference* in coordinates between two events or the time *interval* between two events as seen by observer $O$ and observer $O'$. This can be accomplished by writing the Lorentz equations in a form suitable for describing pairs of events. From Equations 39.9 and 39.11, we can express the differences between the four variables $x$, $x'$, $t$, and $t'$ in the form

$$\begin{aligned} \Delta x' &= \gamma(\Delta x - v\,\Delta t) \\ \Delta t' &= \gamma\left(\Delta t - \frac{v}{c^2}\,\Delta x\right) \end{aligned} \qquad S \to S' \qquad (39.12)$$

$$\begin{aligned} \Delta x &= \gamma(\Delta x' + v\Delta t') \\ \Delta t &= \gamma\left(\Delta t' + \frac{v}{c^2}\,\Delta x'\right) \end{aligned} \qquad S' \to S \qquad (39.13)$$

where $\Delta x' = x_2' - x_1'$ and $\Delta t' = t_2' - t_1$ are the differences measured by observer $O'$, while $\Delta x = x_2 - x_1$ and $\Delta t = t_2 - t_1$ are the differences measured by observer $O$. We have not included the expressions for relating the $y$ and $z$ coordinates since they are unaffected by motion along the $x$ direction.

EXAMPLE 39.5 Simultaneity and Time Dilation Revisited

Use the Lorentz transformation equations in difference form to show that (a) simultaneity is not an absolute concept and (b) moving clocks run slower than stationary clocks.

(a) **Simultaneity:** Suppose that two events are simultaneous according to the moving observer $O'$, so that $\Delta t' = 0$. From the expression for $\Delta t$ given in Equation 39.13, we see that in this case, $\Delta t = \gamma v\,\Delta x'/c^2$. That is, the time interval for the same two events as measured by observer $O$ is nonzero, so they will not appear to be simultaneous in $O$.

(b) **Time Dilation:** Suppose that observer $O'$ finds that two events occur at the same place ($\Delta x' = 0$), but at different times ($\Delta t' \neq 0$). In this situation, the expression for $\Delta t$ given in Equation 39.13 becomes $\Delta t = \gamma\,\Delta t'$. This is the equation for time dilation found earlier, Equation 39.7, where $\Delta t' = \Delta t$ is the proper time measured by the single clock in $O'$.

Exercise 2 Use the Lorentz transformation equations in difference form to confirm length contraction. That is, $L = L'/\gamma$.

## 39.10 LORENTZ VELOCITY TRANSFORMATION

Let us now derive the Lorentz velocity transformation, which is the relativistic counterpart of the Galilean velocity transformation. Suppose that an unaccelerated object is observed in the S′ frame at $x_1'$ at time $t_1'$ and at $x_2'$ at time $t_2'$. Its speed $u_x'$ measured in S′ is given by

$$u_x' = \frac{x_2' - x_1'}{t_2' - t_1'} = \frac{dx'}{dt'} \qquad (39.14)$$

Using Equation 39.9, we have

$$dx' = \gamma(dx - v\,dt)$$

$$dt' = \gamma\left(dt - \frac{v}{c^2}\,dx\right)$$

Substituting these into Equation 39.14 gives

$$u'_x = \frac{dx'}{dt'} = \frac{dx - v\,dt}{dt - \dfrac{v}{c^2}\,dx} = \frac{\dfrac{dx}{dt} - v}{1 - \dfrac{v}{c^2}\dfrac{dx}{dt}}$$

But $dx/dt$ is just the velocity component $u_x$ of the object measured in S, and so this expression becomes

Lorentz velocity transformation for $S \rightarrow S'$

$$u'_x = \frac{u_x - v}{1 - \dfrac{u_x v}{c^2}} \tag{39.15}$$

Similarly, if the object has velocity components along $y$ and $z$, the components in S′ are given by

$$u'_y = \frac{u_y}{\gamma\left(1 - \dfrac{u_x v}{c^2}\right)} \quad \text{and} \quad u'_z = \frac{u_z}{\gamma\left(1 - \dfrac{u_x v}{c^2}\right)} \tag{39.16}$$

When $u_x$ and $v$ are both much smaller than $c$ (the nonrelativistic case), the denominator of Equation 39.15 approaches unity, and $u'_x \approx u_x - v$. This corresponds to the Galilean velocity transformations. In the other extreme, when $u_x = c$, Equation 39.15 becomes

$$u'_x = \frac{c - v}{1 - \dfrac{cv}{c^2}} = \frac{c\left(1 - \dfrac{v}{c}\right)}{1 - \dfrac{v}{c}} = c$$

From this result, we see that an object moving with a speed $c$ relative to an observer in S also has a speed $c$ relative to an observer in S′—*independent* of the relative motion of S and S′. Note that this conclusion is consistent with Einstein's second postulate, namely, that the speed of light must be $c$ with respect to all inertial frames of reference. Furthermore, the speed of an object can never exceed $c$. That is, the speed of light is the "ultimate" speed. We shall return to this point later when we consider the energy of a particle.

To obtain $u_x$ in terms of $u'_x$, we replace $v$ by $-v$ in Equation 39.15 and interchange the roles of $u_x$ and $u'_x$. This gives

Inverse Lorentz velocity transformation for $S' \rightarrow S$

$$u_x = \frac{u'_x + v}{1 + \dfrac{u'_x v}{c^2}} \tag{39.17}$$

**EXAMPLE 39.6 Relative Velocity of Spaceships**
Two spaceships A and B are moving in *opposite* directions, as in Figure 39.14. An observer on the earth measures the speed of A to be $0.75c$ and the speed of B to be $0.85c$. Find the velocity of B with respect to A.

*Solution* This problem can be solved by taking the S′ frame as being attached to spacecraft A, so that $v = 0.75c$ relative to an observer on the earth (the S frame). Spacecraft B can be considered as an object moving with a velocity $u_x = -0.85c$ relative to the earth observer. Hence, the velocity of B with respect to A can be obtained using Equation 39.15:

$$u'_x = \frac{u_x - v}{1 - \dfrac{u_x v}{c^2}} = \frac{-0.85c - 0.75c}{1 - \dfrac{(-0.85c)(0.75c)}{c^2}}$$

$$= -0.98c$$

The negative sign for $u'_x$ indicates that spaceship B is moving in the negative $x$ direction as observed by A. Note that the result is less than $c$. That is, a body whose speed is less than $c$ in one frame of reference must have a speed less than $c$ in *any other* frame. If the Galilean velocity transformation were used in this example, we would find that $u'_x = u_x - v = -0.85c - 0.75c = -1.6c$, which is greater than $c$.

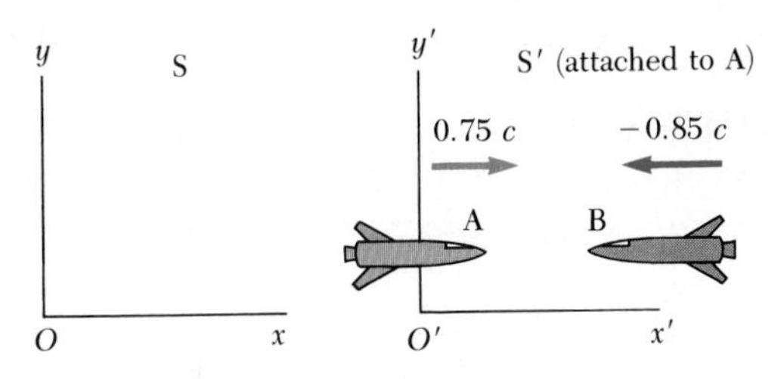


**Figure 39.14** (Example 39.6) Two spaceships *A* and *B* move in *opposite* directions. The speed of *B* relative to *A* is *less* than *c* and is obtained by using the relativistic velocity transformation.

**EXAMPLE 39.7 The Speeding Motorcycle**
Imagine a motorcycle rider moving with a speed of $0.8c$ past a stationary observer, as shown in Figure 39.15. If the rider tosses a ball in the forward direction with a speed of $0.7c$ relative to himself, what is the speed of the ball as seen by the stationary observer?

*Solution* In this situation, the velocity of the motorcycle with respect to the stationary observer is $v = 0.8c$. The velocity of the ball in the frame of reference of the motorcyclist is $0.7c$. Therefore, the velocity, $u_x$, of the ball relative to the stationary observer is

$$u_x = \frac{u'_x + v}{1 + \dfrac{u'_x v}{c^2}} = \frac{0.7c + 0.8c}{1 + \dfrac{(0.7c)(0.8c)}{c^2}} = 0.962c$$

**Exercise 3** Suppose that the motorcyclist moving with a speed $0.8c$ turns on a beam of light that moves away from him with a speed of $c$ in the same direction as the moving motorcycle. What would the stationary observer measure for the speed of the beam of light?
**Answer** $c$.

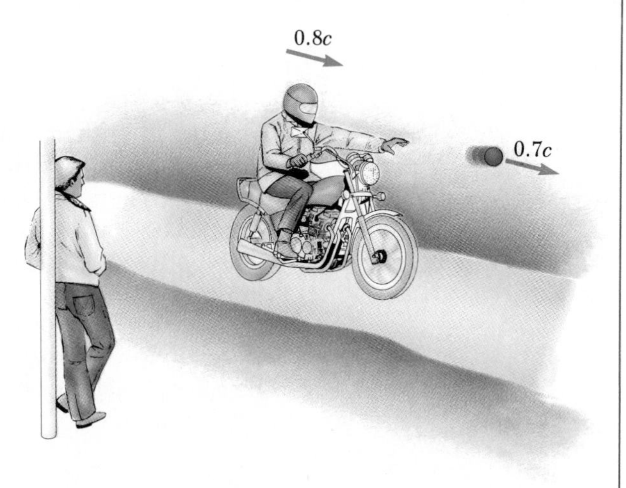


**Figure 39.15** (Example 39.7) A motorcyclist moves past a stationary observer with a speed of $0.8c$ and throws a ball in the direction of motion with a speed of $0.7c$ relative to himself.

## 39.11 RELATIVISTIC MOMENTUM

We have seen that the principle of relativity is satisfied if the Galilean transformation is replaced by the more general Lorentz transformation. Therefore, in order to properly describe the motion of material particles within the framework of special relativity, we must generalize Newton's laws and the definition of momentum and energy. These generalized definitions of momentum and energy will reduce to the classical (nonrelativistic) definitions for $v \ll c$.

First, recall that the conservation of momentum states that when two bodies collide, the total momentum remains constant, assuming the bodies are isolated (that is, they interact only with each other). Suppose the collision is described in a reference frame S in which the momentum is conserved. If the velocities in a second reference frame S′ are calculated using the Lorentz transformation and the classical definition of momentum, $\boldsymbol{p} = m\boldsymbol{u}$, one finds that momentum *is not* conserved in the second reference frame. However, since the laws of physics are the same in all inertial frames, the momentum must be conserved in all systems. In view of this condition and assuming the Lorentz transformation is correct, we must modifiy the definition of momentum.

Our definition of relativistic momentum $\boldsymbol{p}$ must satisfy the following conditions:

1. The relativistic momentum must be conserved in all collisions.
2. The relativistic momentum must approach the classical value $m\boldsymbol{u}$ as $\boldsymbol{u} \rightarrow 0$.

The correct relativistic equation for the momentum that satisfies these conditions is given by the expression

Definition of relativistic momentum

$$\boldsymbol{p} \equiv \frac{m\boldsymbol{u}}{\sqrt{1 - \frac{u^2}{c^2}}} = \gamma m\boldsymbol{u} \tag{39.18}$$

where $\boldsymbol{u}$ is the velocity of the particle. We use the symbol $\boldsymbol{u}$ for particle velocity rather than $\boldsymbol{v}$, which is used for the relative velocity of two reference frames. The proof of this generalized expression for $\boldsymbol{p}$ is beyond the scope of this text. When $u$ is much less than $c$, the denominator of Equation 39.18 approaches unity, so that $\boldsymbol{p}$ approaches $m\boldsymbol{u}$. Therefore, the relativistic equation for $\boldsymbol{p}$ reduces to the classical expression when $u$ is small compared with $c$.

In many situations, it is useful to interpret Equation 39.18 as the product of a relativistic mass, $\gamma m$, and the velocity of the object. Using this description, one can say that *the observed mass of an object increases with speed according to the relation*

$$\text{Relativistic mass} = \gamma m = \frac{m}{\sqrt{1 - \frac{v^2}{c^2}}} \tag{39.19}$$

where the relativistic mass is the mass of the object moving at a speed $v$ with respect to the observer and $m$ is the mass of the object as measured by an observer at rest with respect to the object. The quantity $m$ is often referred to as the **rest mass** of the object. (In future chapters, we shall sometimes use the symbol $m_0$ to represent rest mass.) Note that Equation 39.19 indicates that *the mass of an object increases as its speed increases.* This prediction has been borne out through many observations on elementary particles, such as energetic electrons.

The relativistic force $\boldsymbol{F}$ on a particle whose momentum is $\boldsymbol{p}$ is defined by the expression

$$\mathbf{F} \equiv \frac{d\mathbf{p}}{dt} \tag{39.20}$$

where $\mathbf{p}$ is given by Equation 39.18. This expression is identical to the classical statement of Newton's second law, which says that force equals the time rate of change of momentum.

**EXAMPLE 39.8 Momentum of an Electron**
An electron, which has a rest mass of $9.11 \times 10^{-31}$ kg, moves with a speed of $0.750c$. Find its relativistic momentum and compare this with the momentum calculated from the classical expression.

*Solution* Using Equation 39.18 with $u = 0.75c$, we have

$$p = \frac{mu}{\sqrt{1 - \frac{u^2}{c^2}}}$$

$$p = \frac{(9.11 \times 10^{-31}\text{ kg})(0.750 \times 3 \times 10^8\text{ m/s})}{\sqrt{1 - \frac{(0.750c)^2}{c^2}}}$$

$$= 3.10 \times 10^{-22}\text{ kg}\cdot\text{m/s}$$

The incorrect classical expression would give

$$\text{Momentum} = mu = 2.05 \times 10^{-22}\text{ kg}\cdot\text{m/s}$$

Hence, the correct relativistic result is 50% greater than the classical result!

It is left as a problem (Problem 55) to show that the acceleration $\mathbf{a}$ of a particle decreases under the action of a constant force, in which case $a \propto (1 - u^2/c^2)^{3/2}$. Furthermore, as the speed approaches $c$, the acceleration caused by any finite force approaches zero. Hence, it is impossible to accelerate a particle from rest to a speed equal to or greater than $c$.

## 39.12 RELATIVISTIC ENERGY

We have seen that the definition of momentum and the laws of motion required generalization to make them compatible with the principle of relativity. This implies that the relation between work and energy must also be modified.

In order to derive the relativistic form of the work-energy theorem, let us start with the definition of work done by a force $F$ and make use of the definition of relativistic force, Equation 39.20. That is,

$$W = \int_{x_1}^{x_2} F\,dx = \int_{x_1}^{x_2} \frac{dp}{dt}\,dx$$

where we have assumed that the force and motion are along the $x$ axis. In order to perform this integration, we make repeated use of the chain rule for derivatives in the following manner:

$$\left(\frac{dp}{dt}\right) dx = \left(\frac{dp}{du}\frac{du}{dt}\right) dx = \frac{dp}{du}\left(\frac{du}{dx}\frac{dx}{dt}\right) dx$$

$$= \frac{dp}{du}\,u\,\frac{du}{dx}\,dx = \frac{dp}{du}\,u\,du$$

Since $p$ depends on $u$ according to Equation 39.18, we have

$$\frac{dp}{du} = \frac{d}{du}\frac{mu}{\sqrt{1 - \frac{u^2}{c^2}}} = \frac{m}{\left(1 - \frac{u^2}{c^2}\right)^{3/2}}$$

Using these results, we can express the work as

$$W = \int_0^u \frac{dp}{du}\, u\, du = \int_0^u \frac{mu}{\left(1 - \frac{u^2}{c^2}\right)^{3/2}}\, du$$

where we have assumed that the particle is accelerated from rest to some final velocity $u$. Evaluating the integral, we find that

$$W = \frac{mc^2}{\sqrt{1 - \frac{u^2}{c^2}}} - mc^2 \qquad (39.21)$$

Recall that the work-energy theorem states the work done by a force acting on a particle equals the change in kinetic energy of the particle. Since the initial kinetic energy is zero, we conclude that the work $W$ is equivalent to the relativistic kinetic energy $K$, that is,

Relativistic kinetic energy

$$K = \frac{mc^2}{\sqrt{1 - \frac{u^2}{c^2}}} - mc^2 \qquad (39.22)$$

This equation has been confirmed by experiments using high-energy particle accelerators. At low speeds, where $u/c \ll 1$, Equation 39.22 should reduce to the classical expression $K = \frac{1}{2}mu^2$. We can check this by using the binomial expansion $(1 - x^2)^{-1/2} \approx 1 + \frac{1}{2}x^2 + \cdots$ for $x \ll 1$, where the higher-order powers of $x$ are neglected in the expansion. In our case, $x = u/c$, so that

$$\frac{1}{\sqrt{1 - \frac{u^2}{c^2}}} = \left(1 - \frac{u^2}{c^2}\right)^{-1/2} \approx 1 + \frac{1}{2}\frac{u^2}{c^2} + \cdots$$

Substituting this into Equation 39.22 gives

$$K \approx mc^2\left(1 + \frac{1}{2}\frac{u^2}{c^2} + \cdots\right) - mc^2 = \frac{1}{2}mu^2$$

which agrees with the classical result. A graph comparing the relativistic and nonrelativistic expressions is given in Figure 39.16. In the relativistic case, the particle speed never exceeds $c$, regardless of the kinetic energy. The two curves are in good agreement when $v \ll c$.

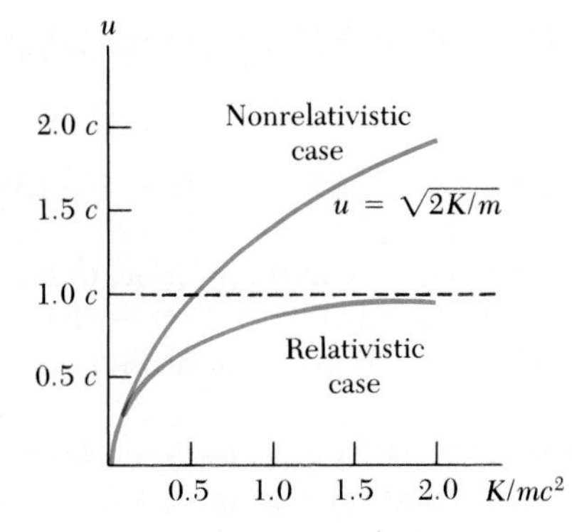


**Figure 39.16** A graph comparing relativistic and nonrelativistic kinetic energy. The energies are plotted versus speed. In the relativistic case, $u$ is always *less* than $c$.

It is useful to write the relativistic kinetic energy in the form

$$K = \gamma mc^2 - mc^2 \tag{39.23}$$

where

$$\gamma = \frac{1}{\sqrt{1 - \frac{u^2}{c^2}}}$$

The constant term $mc^2$, which is independent of the speed, is called the **rest energy** of the particle $E_0$. The term $\gamma mc^2$, which depends on the particle speed, is therefore the sum of the kinetic and rest energies. We define $\gamma mc^2$ to be the **total energy** $E$, that is,

$$E = \gamma mc^2 = K + mc^2 \tag{39.24}$$

Definition of total energy

or

$$E = \frac{mc^2}{\sqrt{1 - \frac{u^2}{c^2}}} \tag{39.25}$$

Energy-mass equivalence

This, of course, is Einstein's famous mass-energy equivalence equation. The relation $E = \gamma mc^2 = \gamma E_0$ shows that *mass is a form of energy.* Furthermore, this result shows that a small mass corresponds to an enormous amount of energy. This concept has revolutionized the field of nuclear physics.

In many situations, the momentum or energy of a particle is known rather than its speed. It is therefore useful to have an expression relating the total energy $E$ to the relativistic momentum $p$. This is accomplished by using the expressions $E = \gamma mc^2$ and $p = \gamma mu$. By squaring these equations and subtracting, we can eliminate $u$ (Problem 23). The result, after some algebra, is

$$E^2 = p^2c^2 + (mc^2)^2 \tag{39.26}$$

Energy-momentum relation

When the particle is at rest, $p = 0$, and so we see that $E = E_0 = mc^2$. That is, the total energy equals the rest energy. As we shall discuss in the next chapter, it is well established that there are particles that have zero mass, such as photons (quanta of electromagnetic radiation). If we set $m = 0$ in Equation 39.26, we see that

$$E = pc \tag{39.27}$$

This equation is an *exact* expression relating energy and momentum for photons and neutrinos, which always travel at the speed of light. In these cases, the particles have a characteristic wavelength $\lambda$, and their momentum is given by $p = h/\lambda$.

Finally, note that since the mass $m$ of a particle is independent of its motion, $m$ must have the same value in all reference frames. On the other hand, the total energy and momentum of a particle depend on the reference frame in which they are measured, since they both depend on velocity. Since $m$ is a constant, then according to Equation 39.26 the quantity $E^2 - p^2c^2$ must have the same value in all reference frames. That is, $E^2 - p^2c^2$ is *invariant* under a Lorentz transformation.

When dealing with electrons or other subatomic particles, it is convenient to express their energy in electron volts (eV), since the particles are usually given this energy by acceleration through a potential difference. The conversion factor is

$$1 \text{ eV} = 1.60 \times 10^{-19} \text{ J}$$

For example, the mass of an electron is $9.11 \times 10^{-31}$ kg. Hence, the rest energy of the electron is

$$mc^2 = (9.11 \times 10^{-31} \text{ kg})(3.00 \times 10^8 \text{ m/s})^2 = 8.20 \times 10^{-14} \text{ J}$$

Converting this to eV, we have

$$mc^2 = (8.20 \times 10^{-14} \text{ J})(1 \text{ eV}/1.60 \times 10^{-19} \text{ J}) = 0.511 \text{ MeV}$$

where 1 MeV $= 10^6$ eV.

**EXAMPLE 39.9 The Energy of a Speedy Electron**

An electron moves with a speed $u = 0.850c$. Find its total energy and kinetic energy in eV.

*Solution* Using the fact that the rest energy of the electron is 0.511 MeV together with Equation 39.25 gives

$$E = \frac{mc^2}{\sqrt{1 - \frac{u^2}{c^2}}} = \frac{0.511 \text{ MeV}}{\sqrt{1 - \frac{(0.85c)^2}{c^2}}}$$

$$= 1.90(0.511 \text{ MeV}) = 0.970 \text{ MeV}$$

The kinetic energy is obtained by subtracting the rest energy from the total energy:

$$K = E - mc^2 = 0.970 \text{ MeV} - 0.511 \text{ MeV}$$

$$= 0.459 \text{ MeV}$$

**EXAMPLE 39.10 The Energy of a Speedy Proton**

The total energy of a proton is three times its rest energy. (a) Find the proton's rest energy in eV.

*Solution*

$$\text{Rest energy} = mc^2 = (1.67 \times 10^{-27} \text{ kg})(3 \times 10^8 \text{ m/s})^2$$

$$= (1.50 \times 10^{-10} \text{ J})(1 \text{ eV}/1.60 \times 10^{-19} \text{ J})$$

$$= 939 \text{ MeV}$$

(b) With what speed is the proton moving?

*Solution* Since the total energy $E$ is three times the rest energy, Equation 39.25 gives

$$E = 3mc^2 = \frac{mc^2}{\sqrt{1 - \frac{u^2}{c^2}}}$$

$$3 = \frac{1}{\sqrt{1 - \frac{u^2}{c^2}}}$$

Solving for $u$ gives

$$\left(1 - \frac{u^2}{c^2}\right) = \frac{1}{9} \quad \text{or} \quad \frac{u^2}{c^2} = \frac{8}{9}$$

$$u = \frac{\sqrt{8}}{3}c = 2.83 \times 10^8 \text{ m/s}$$

(c) Determine the kinetic energy of the proton in eV.

*Solution*

$$K = E - mc^2 = 3mc^2 - mc^2 = 2mc^2$$

Since $mc^2 = 939$ MeV, $K = 1878$ MeV

(d) What is the proton's momentum?

*Solution* We can use Equation 39.26 to calculate the momentum with $E = 3mc^2$:

$$E^2 = p^2c^2 + (mc^2)^2 = (3mc^2)^2$$

$$p^2c^2 = 9(mc^2)^2 - (mc^2)^2 = 8(mc^2)^2$$

$$p = \sqrt{8}\,\frac{mc^2}{c} = \sqrt{8}\,\frac{(939 \text{ MeV})}{c} = 2656\,\frac{\text{MeV}}{c}$$

The unit of momentum is written MeV/$c$ for convenience.

## 39.13 CONFIRMATIONS AND CONSEQUENCES OF RELATIVITY THEORY

The special theory of relativity has been confirmed by a number of experiments. One important experiment concerned with the decay of muons, and time dilation in the muon's reference frame, was discussed in Section 39.7. This section describes further evidence of Einstein's special theory of relativity.

One of the first predictions of relativity that was experimentally confirmed is the variation of momentum with velocity. Experiments were performed as early as 1909 on electrons, which can easily be accelerated to speeds close to $c$ through the use of electric fields. When an energetic electron enters a magnetic field $\mathbf{B}$ with its velocity vector perpendicular to $\mathbf{B}$, a magnetic force is exerted on the electron, causing it to move in a circle of radius $r$. In this situation, the relativistic momentum of the electron is given by $p = eB/r$. From the relativistic equivalent of Newton's second law, $\mathbf{F} = d\mathbf{p}/dt$, the variation in momentum with kinetic energy can be checked experimentally. The results of such experiments on electrons and other charged particles support the relativistic expressions.

The release of enormous quantities of energy in nuclear fission and fusion processes is a manifestation of the equivalence of mass and energy. The conversion of mass into energy is, of course, the basis of atomic and hydrogen bombs, the most powerful and destructive weapons ever constructed. In fact, all reactions that release energy do so at the expense of mass (including chemical reactions). In a conventional nuclear reactor, the uranium nucleus undergoes fission, a reaction that results in several lighter fragments having considerable kinetic energy. In the case of $^{235}$U (the parent nucleus), which undergoes spontaneous fission, the fragments are two lighter nuclei and two neutrons. The total mass of the fragments is *less* than that of the parent nucleus by some amount $\Delta m$. The corresponding energy $\Delta mc^2$ associated with this mass difference is *exactly* equal to the total kinetic energy of the fragments. This kinetic energy is then used to produce heat and steam for the generation of electrical power.

Next, consider the basic fusion reaction in which two deuterium atoms combine to form one helium atom. This reaction is of major importance in current research and development of controlled-fusion reactors. The decrease in mass which results from the creation of one helium atom from two deuterium atoms is calculated to be $\Delta m = 4.25 \times 10^{-29}$ kg. Hence, the corresponding excess energy which results from one fusion reaction is given by the expression $\Delta mc^2 = 3.83 \times 10^{-12}$ J $= 23.9$ MeV. To appreciate the magnitude of this result, if 1 g of deuterium is converted into helium, the energy released is about $10^{12}$ J! At the 1988 cost of electrical energy, this would be worth about $50 000.

**EXAMPLE 39.11 Binding Energy of the Deuteron**

The mass of the deuteron, which is the nucleus of "heavy hydrogen," is not equal to the sum of the masses of its constituents, which are the proton and neutron. Calculate this mass difference and determine its energy equivalence.

*Solution* Using atomic mass units (u), we have

$$m_p = \text{mass of proton} = 1.007276 \text{ u}$$

$$m_n = \text{mass of neutron} = 1.008665 \text{ u}$$

$$m_p + m_n = 2.015941 \text{ u}$$

Since the mass of the deuteron is 2.013553 u, we see that the mass difference $\Delta m$ is 0.002388 u. By definition, 1 u $= 1.66 \times 10^{-27}$ kg, and therefore

$$\Delta m = 0.002388 \text{ u} = 3.96 \times 10^{-30} \text{ kg}$$

Using $E = \Delta mc^2$, we find that

$$E = \Delta mc^2 = (3.96 \times 10^{-30} \text{ kg})(3.00 \times 10^8 \text{ m/s})^2$$

$$= 3.56 \times 10^{-13} \text{ J}$$

$$= 2.23 \text{ MeV}$$

Therefore, the minimum energy required to separate the proton from the neutron of the deuterium nucleus (the binding energy) is 2.23 MeV.

## SUMMARY

The two basic postulates of the ***special theory of relativity*** are as follows:

1. All the laws of physics must be the same for all observers moving at constant velocity with respect to each other.
2. In particular, the speed of light must be the same for all inertial observers, independent of their relative motion.

In order to satisfy these postulates, the Galilean transformations must be replaced by the **Lorentz transformations** given by

$$\begin{aligned} x' &= \gamma(x - vt) \\ y' &= y \\ z' &= z \\ t' &= \gamma\left(t - \frac{v}{c^2}x\right) \end{aligned} \tag{39.9}$$

where

$$\gamma = \frac{1}{\sqrt{1 - \dfrac{v^2}{c^2}}} \tag{39.10}$$

In these equations, it is assumed that the primed system moves with a speed $v$ along the $xx'$ axes.

The relativistic form of the **velocity transformation** is

$$u'_x = \frac{u_x - v}{1 - \dfrac{u_x v}{c^2}} \tag{39.15}$$

where $u_x$ is the speed of an object as measured in the S frame and $u'_x$ is its speed measured in the S′ frame.

Some of the consequences of the special theory of relativity are as follows:

1. Clocks in motion relative to an observer appear to be slowed down by a factor $\gamma$. This is known as **time dilation.**
2. Lengths of objects in motion appear to be contracted in the direction of motion.
3. Events that are simultaneous for one observer are not simultaneous for another observer in motion relative to the first.

These three statements can be summarized by saying that duration, length, and simultaneity are not absolute concepts in relativity.

The relativistic expression for the **momentum** of a particle moving with a velocity $\boldsymbol{u}$ is

$$\boldsymbol{p} \equiv \frac{m\boldsymbol{u}}{\sqrt{1 - \dfrac{u^2}{c^2}}} = \gamma m\boldsymbol{u} \tag{39.18}$$

where

$$\gamma = \frac{1}{\sqrt{1 - \dfrac{u^2}{c^2}}}$$

The relativistic expression for the **kinetic energy** of a particle is

$$K = \gamma mc^2 - mc^2 \tag{39.23}$$

where $mc^2$ is called the **rest energy** of the particle.

The total energy $E$ of a particle is related to the mass through the famous **energy-mass equivalence** expression:

$$E = \gamma mc^2 = \frac{mc^2}{\sqrt{1 - \dfrac{u^2}{c^2}}} \tag{39.25}$$

Finally, the relativistic momentum is related to the total energy through the equation

$$E^2 = p^2c^2 + (mc^2)^2 \tag{39.26}$$

## QUESTIONS

1. What two speed measurements will two observers in relative motion *always* agree upon?
2. A spaceship in the shape of a sphere moves past an observer on earth with a speed $0.5c$. What shape will the observer see as the spaceship moves past?
3. An astronaut moves away from the earth at a speed close to the speed of light. If an observer on earth could make measurements of the astronaut's size and pulse rate, what changes (if any) would he or she measure? Would the astronaut measure any changes?
4. Two identically constructed clocks are synchronized. One is put in orbit around the earth while the other remains on earth. Which clock runs slower? When the moving clock returns to earth, will the two clocks still be synchronized?
5. Two lasers situated on a moving spacecraft are triggered simultaneously. An observer on the spacecraft claims to see the pulses of light simultaneously. What condition is necessary in order that a stationary observer agree that the two pulses are emitted simultaneously?
6. When we say that a moving clock runs slower than a stationary one, does this imply that there is something physically unusual about the moving clock?
7. When we speak of time dilation, do we mean that time passes more slowly in moving systems or that it simply appears to do so?
8. List some ways our day-to-day lives would change if the speed of light were only 50 m/s.
9. Give a physical argument which shows that it is impossible to accelerate an object of mass $m$ to the speed of light, even with a continuous force acting on it.
10. It is said that Einstein, in his teenage years, asked the question, "What would I see in a mirror if I carried it in my hands and ran at the speed of light?" How would you answer this question?
11. Since mass is a form of energy, can we conclude that a compressed spring has more mass than the same spring when it is not compressed? On the basis of your answer, which has more mass, a spinning planet or an otherwise identical but nonspinning planet?
12. Suppose astronauts were paid according to the time spent traveling in space. After a long voyage at a speed near that of light, a crew of astronauts return and open their pay envelopes. What will their reaction be?
13. What happens to the density of an object as its speed increases?
14. Consider the incorrect statement, "Matter can neither be created nor destroyed." How would you correct this statement in view of the special theory of relativity?
15. Some of the distant stars, called quasars, are receding from us at half the speed of light (or greater). What is the speed of light we receive from these quasars?
16. How is it possible that photons of light, with zero mass, have momentum?
17. The expression $E = mc^2$ is often given in popular descriptions of Einstein's work. Is this expression strictly correct? For example, does it accurately take into account the kinetic energy of a moving mass?
18. With regard to reference frames, how does general relativity differ from special relativity?
19. Imagine an astronaut on a trip to Sirius, which lies 8 lightyears from the earth. Upon arrival at Sirius, the astronaut finds that the trip lasted 6 years. If the trip was made at a constant speed of $0.8c$, how can the 8-lightyear distance be reconciled with the six-year duration?

## PROBLEMS

### Section 39.2 The Principle of Relativity

1. In a laboratory frame of reference, an observer notes that Newton's second law is valid. Show that it is also valid for an observer moving at a constant speed relative to the laboratory frame.
2. Show that Newton's second law is not valid in a reference frame moving past the laboratory frame of Problem 1 with a constant acceleration.
3. A 2000-kg car moving with a speed of 20 m/s collides with and sticks to a 1500-kg car at rest at a stop sign. Show that momentum is conserved in a reference frame moving with a speed of 10 m/s in the direction of the moving car.
4. A billiard ball of mass 0.3 kg moves with a speed of 5 m/s and collides elastically with a ball of mass 0.2 kg moving in the opposite direction with a speed of 3 m/s. Show that momentum is conserved in a frame of reference moving with a speed of 2 m/s in the direction of the second ball.
5. A ball is thrown at 20 m/s inside a boxcar moving along the tracks at 40 m/s. What is the speed of the ball with respect to the ground if it is thrown (a) forward, (b) backward, (c) out the side door?

### Section 39.7 The Relativity of Time

### Section 39.8 The Relativity of Length

6. With what speed will a clock have to be moving in order to run at a rate that is one half the rate of a clock at rest?
7. How fast must a meter stick be moving if its length is observed to shrink to 0.5 m?

8. A clock on a moving spacecraft runs 1 minute slower per day relative to an identical clock on earth. What is the speed of the spacecraft? (*Hint:* For $v/c \ll 1$, note that $\gamma \approx 1 + v^2/2c^2$.)
9. An atomic clock moves at 1000 km/h for one hour as measured by an identical clock on earth. How many nanoseconds slow will the moving clock be at the end of the one-hour interval?
10. Muons move in circular orbits at a speed of $0.9994c$ in a storage ring of radius 500 m. If a muon at rest decays into other particles after $T = 2.2\ \mu s$, how many trips around the storage ring do we expect the muons to make before they decay?
11. A spacecraft moves at a speed of $0.9c$. If its length is $L_0$ when measured from inside the spacecraft, what is its length measured by a ground observer?
12. The cosmic rays of highest energy are protons, with kinetic energy of $10^{13}$ MeV. (a) How long would it take a proton of this energy to travel across the Milky Way galaxy, of diameter $10^5$ lightyears, as measured in the proton's frame? (b) From the point of view of the proton, how many kilometers across is our galaxy?
13. A muon formed high in the earth's atmosphere travels at speed $v = 0.99c$ for a distance 4.6 km before it decays into an electron, a neutrino, and an antineutrino ($\mu^- \rightarrow e^- + \nu + \bar{\nu}$). (a) How long does the muon live as measured in its reference frame? (b) How far does the muon travel, as measured in its frame?
14. If astronauts could travel at $v = 0.95c$, we on earth would say it takes $(4.2/0.95) = 4.4$ years to reach Alpha Centauri, 4.2 lightyears away. The astronauts disagree. (a) How much time passes on the astronauts' clocks? (b) What is the distance to Alpha Centauri, as measured by the astronauts?

## Section 39.9 The Lorentz Transformation Equations

## Section 39.10 Lorentz Velocity Transformation

15. For what value of $v$ does $\gamma = 1.01$?
16. A certain quasar recedes from the earth with a speed $v = 0.87c$. A jet of material is ejected from the quasar toward the earth with a speed of $0.55c$ relative to the quasar. Find the speed of the ejected material relative to the earth.
17. Two jets of material fly away from the center of a radio galaxy in opposite directions. Both jets move at a speed of $0.75c$ relative to the galaxy. Determine the speed of one jet relative to the other.
18. A Klingon space ship moves away from the earth at a speed of $0.8c$ (Fig. 39.17). The Starship Enterprise pursues at a speed of $0.9c$ relative to the earth. Observers on earth see the Enterprise overtaking the Klingon ship at a relative speed of $0.1c$. With what speed is the Enterprise overtaking the Klingon ship as seen by the crew of the Enterprise?

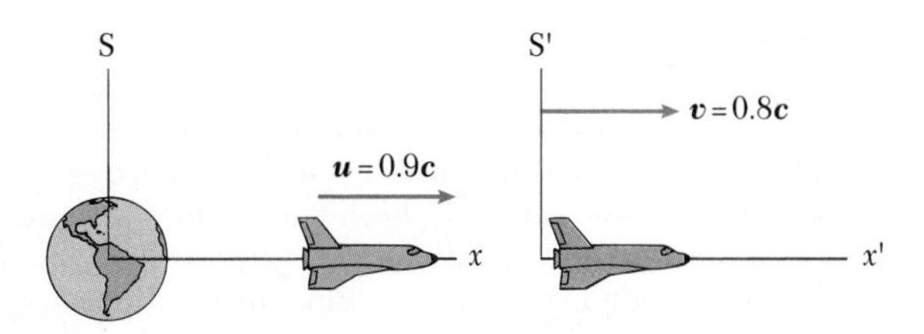


**Figure 39.17** The earth is frame S. The Klingon ship is frame S′. The Enterprise is the object whose motion is followed from S and S′. (Problem 18)

19. A cube of steel has a volume of 1 cm³ and a mass of 8 g when at rest on the earth. If this cube is now given a speed $v = 0.9c$, what is its density as measured by a stationary observer?

## Section 39.11 Relativistic Momentum

20. Calculate the momentum of a proton moving with a speed of (a) $0.01c$, (b) $0.5c$, (c) $0.9c$.
21. Find the momentum of a proton in MeV/$c$ units if its total energy is twice its rest energy.
22. A golf ball travels with a speed of 90 m/s. By what fraction does its relativistic momentum $p$ differ from its classical value $mv$? That is, find the ratio $(p - mv)/mv$.

## Section 39.12 Relativistic Energy

23. Show that the energy-momentum relationship given by $E^2 = p^2c^2 + (mc^2)^2$ follows from the expressions $E = \gamma mc^2$ and $p = \gamma mu$.
24. Show that the kinetic energy $K$, momentum $p$, and rest mass $m$ of a relativistic particle are related by $p^2c^2 = 2Kmc^2 + K^2$.
25. A proton moves with the speed of $0.95c$. Calculate its (a) rest energy, (b) total energy, and (c) kinetic energy.
26. Find the speed at which the relativistic kinetic energy equals two times the nonrelativistic value.
27. Find the speed of a particle whose total energy is twice that of its rest energy.
28. A proton in a high-energy accelerator is given a kinetic energy of 50 GeV. Determine the (a) momentum and (b) speed of the proton.
29. Determine the energy required to accelerate an electron from (a) $0.50c$ to $0.9c$ and (b) $0.90c$ to $0.99c$.
30. In a typical color television tube, the electrons are accelerated through a potential difference of 25 000 volts. (a) What speed do the electrons have when they strike the screen? (b) What is their kinetic energy (in Joules)?
31. Electrons are accelerated to an energy of $2 \times 10^{10}$ eV in the 3-km long Stanford Linear Accelerator. (a) What is the $\gamma$ factor for the electrons? (b) What is the speed of the 20-GeV electrons? (c) How long does the accelerator appear to a 20-GeV electron?

32. A spaceship of mass $10^6$ kg is to be accelerated to $0.6c$. (a) How much energy does this require? (b) How many kilograms of matter and antimatter (in equal proportions) will it take to provide this much energy?
33. A pion at rest ($m_\pi = 270m_e$) decays into a muon ($m_\mu = 206m_e$) and an antineutrino ($m_\nu = 0$) according to the following: $\pi^- \rightarrow \mu^- + \bar{\nu}$. Find the kinetic energy of the muon and the antineutrino in MeV. (*Hint:* Relativistic momentum is conserved.)

## Section 39.13 Confirmations and Consequences of Relativity Theory

34. A radium isotope decays by emitting an $\alpha$ particle to a radon isotope according to the scheme ${}^{226}_{88}\mathrm{Ra} \rightarrow {}^{222}_{86}\mathrm{Rn} + {}^{4}_{2}\mathrm{He}$. The masses of the atoms are 226.0254 (Ra), 222.0175 (Rn), and 4.0026 (He). How much energy is released as the result of this decay?
35. Consider the decay ${}^{55}_{24}\mathrm{Cr} \rightarrow {}^{55}_{25}\mathrm{Mn} + \mathrm{e}$, where e is an electron. The ${}^{55}\mathrm{Cr}$ nucleus has a mass of 54.9279 u, and the ${}^{55}\mathrm{Mn}$ nucleus has a mass of 54.9244 u. (a) Calculate the mass difference between the two nuclei in MeV. (b) What is the maximum kinetic energy of the emitted electron?
36. A ${}^{57}\mathrm{Fe}$ nucleus at rest emits a 14-keV photon. Use the conservation of energy and momentum to deduce the kinetic energy of the recoiling nucleus in eV. (Use $Mc^2 = 8.6 \times 10^{-9}$ J for the final state of the ${}^{57}\mathrm{Fe}$ nucleus.)
37. The power output of the sun is $3.8 \times 10^{26}$ W. How much matter is converted into energy in the sun each second?
38. A gamma ray (a very high energy photon of light) can produce an electron and an antielectron when it enters the electric field of a heavy nucleus: ($\gamma \rightarrow e^+ + e^-$). What is the minimum energy of the $\gamma$-ray to accomplish this task? (*Hint:* The masses of the electron and the antielectron are equal.)

## ADDITIONAL PROBLEMS

39. An astronaut is traveling in a space vehicle that has a speed of $0.50c$ relative to the earth. The astronaut measures his pulse rate to be 75 per minute. Signals generated by the astronaut's pulse are radioed to earth when the vehicle is moving perpendicular to a line that connects the vehicle with an earth observer. What pulse rate does the earth observer measure? What would be the pulse rate if the speed of the space vehicle were increased to $0.99c$?
40. An astronaut wishes to visit the Andromeda galaxy (2 million lightyears away) in a one-way trip that will take 30 years in the spaceship's frame of reference. Assuming that his speed is constant, how fast must he travel relative to the earth?
41. The net nuclear reaction inside the sun is $4\mathrm{p} \rightarrow \mathrm{He}^4 + \Delta E$. If the rest mass of each proton is 938.2 MeV and the rest mass of the $\mathrm{He}^4$ nucleus is 3727 MeV, calculate the percentage of the starting mass that is converted into energy.
42. The annual energy requirement for the United States is on the order of $10^{20}$ J. How many kilograms of matter would have to be completely converted to energy to meet this requirement?
43. An electron has a speed of $0.75c$. Find the speed of a proton which has (a) the same kinetic energy as the electron and (b) the same momentum as the electron.
44. How fast would a motorist have to be going to make a red light appear green? ($\lambda_{\text{red}} = 650$ nm, $\lambda_{\text{green}} =$ 550 nm.) In computing this, use the correct relativistic formula for Doppler shift:

$$\frac{\Delta\lambda}{\lambda} + 1 = \sqrt{\frac{c - v}{c + v}}$$

where $v$ is the approach velocity and $\lambda$ is the source wavelength.
45. Determine the velocity of recession of the quasar 3C9 if its redshift $\Delta\lambda/\lambda = 2$. Use the relativistic formula for Doppler shift:

$$\frac{\Delta\lambda}{\lambda} + 1 = \sqrt{\frac{c + u}{c - u}}$$

where $u =$ recessional velocity and $\Delta\lambda$ is the wavelength shift.
46. The average lifetime of a pi meson in its own frame of reference is $2.6 \times 10^{-8}$ s. If the meson moves with a speed of $0.95c$, what is (a) its mean lifetime as measured by an observer on earth and (b) the average distance it travels before decaying, as measured by an observer on earth?
47. If you travel on a jet plane from New York to Los Angeles (4000 km air distance), at an average speed of 1000 km/h, how much younger are you on arrival than you would have been had you remained in New York during the time it took the plane to make the journey? (*Hint:* Note that $T$, the time that would have been spent in New York, is extremely close to $T_0$, the time spent on the plane.)
48. A rod of length $L_0$ moves with a speed $v$ along the horizontal direction. The rod makes an angle of $\theta_0$ with respect to the $x'$ axis. (a) Show that the length of the rod as measured by a stationary observer is given by $L = L_0[1 - (v^2/c^2)\cos^2\theta_0]^{1/2}$. (b) Show that the angle that the rod makes with the $x$ axis is given by the expression $\tan\theta = \gamma \tan\theta_0$. These results show that the rod is both contracted and rotated. (Take the lower end of the rod to be at the origin of the primed coordinate system.)
49. An antiproton $\bar{\mathrm{p}}$ can be created in a large synchrotron by an energetic moving proton striking a stationary

proton: $p + p \rightarrow p + p + p + \bar{p}$. What minimum kinetic energy must the incoming proton have to produce an antiproton? The rest mass of the p and $\bar{p}$ is 938 MeV/$c^2$. (*Hint:* The incoming proton has speed $v$, the outgoing ensemble of 4 particles leaves with speed $u$.)

50. The muon is an unstable particle that spontaneously decays into an electron and two neutrinos. If the number of muons at $t = 0$ is $N_0$, the number at time $t$ is given by $N = N_0 e^{-t/\tau}$ where $\tau$ is the mean lifetime, equal to 2.2 $\mu$s. Suppose the muons move at a speed of $0.95c$ and there are $5 \times 10^4$ muons at $t = 0$. (a) What is the observed lifetime of the muons? (b) How many muons remain after traveling a distance of 3 km?

51. Imagine that the entire sun collapsed to a sphere of radius $R_g$ such that the work required to remove a small mass $m$ from the surface would be equal to its rest energy $mc^2$. This radius is called the *gravitational radius* for the sun. Find $R_g$. (It is believed that the ultimate fate of many stars is to collapse to their gravitational radii or smaller.)

52. Suppose that noted astronomers conclude that our sun is about to undergo a supernova explosion. In an effort to escape, we depart in a spaceship at $v = 0.8c$ and head toward the star Tau Ceti, 12 lightyears away. When we reach the midpoint (in space) of our journey, we see the supernova explosion of our sun and, unfortunately, at the same instant we see the explosion of Tau Ceti. (a) In the spaceship's frame of reference, should we conclude that the two explosions occurred simultaneously? If not, which occurred first? (b) In a frame of reference in which the sun and Tau Ceti are at rest, did they explode simultaneously? If not, which occurred first?

53. *Doppler effect for light.* If a light source moves with a speed $v$ relative to an observer, there is a shift in the observed frequency analogous to the Doppler effect for sound waves. Show that the observed frequency $f_0$ is related to the true frequency $f$ through the expression

$$f_0 = \sqrt{\frac{c \pm v_s}{c \mp v_s}}\, f$$

where the upper signs correspond to the source approaching the observer and the lower signs correspond to the source receding from the observer. [*Hint:* In the moving frame S′, the period is the proper time interval and is given by $T = 1/f$. Furthermore, the wavelength measured by the observer is $\lambda_0 = (c - v_s)T_0$, where $T_0$ is the period measured in s.]

54. *The red shift.* A light source recedes from an observer with a speed $v_s$, which is small compared with $c$. (a) Show that the fractional shift in the measured wavelength is given by the approximate expression

$$\frac{\Delta\lambda}{\lambda} \approx \frac{v_s}{c}$$

This result is known as the *red shift,* since the visible light is shifted toward the red. (Note that the proper period and measured period are approximately equal in this case.) (b) Spectroscopic measurements of light at $\lambda = 397$ nm coming from a galaxy in Ursa Major reveal a red shift of 20 nm. What is the recessional speed of the galaxy?

55. A charged particle moves along a straight line in a uniform electric field $\mathbf{E}$ with a speed $v$. If the motion and the electric field are both in the $x$ direction, (a) show that the acceleration of the charge $q$ in the $x$ direction is given by

$$a = \frac{dv}{dt} = \frac{qE}{m}\left(1 - \frac{v^2}{c^2}\right)^{3/2}$$

(b) Discuss the significance of the dependence of the acceleration on the speed. (c) If the particle starts from the rest at $x = 0$ at $t = 0$, how would you proceed to find the speed of the particle and its position after a time $t$ has elapsed?

# ESSAY

## The Renaissance of General Relativity

**Clifford M. Will**
*McDonnell Center for the Space Sciences, Washington University*

During the two decades from 1960 to 1980, the subject of general relativity experienced a rebirth. Despite its enormous influence on scientific thought in its early years, by the late 1950s general relativity had become a sterile, formalistic subject, cut off from the mainstream of physics. It was thought to have very little observational contact, and was believed to be an extremely difficult subject to learn and comprehend.

Yet by 1970, general relativity had become one of the most active and exciting branches of physics. It took on new roles both as a theoretical tool of the astrophysicist and as a playground for the elementary-particle physicist. New experimental tests verified its predictions in unheard-of-ways, and to remarkable levels of precision. Fields of study were created, such as "black-hole physics" and "gravitational-wave astronomy," that brought together the efforts of theorists and experimentalists. One of the most remarkable and important aspects of this renaissance of relativity was the degree to which experiment and observation motivated and complemented theoretical advances.

This was not always the case. In deriving general relativity during the final months of 1915, Einstein himself was not particularly motivated by a desire to account for observational results. Instead, he was driven by purely theoretical ideas of elegance and simplicity. His goal was to produce a theory of gravitation that incorporated in a natural way both the special theory of relativity that dealt with physics in inertial frames, and the principle of equivalence, the proposal that physics in a frame falling freely in a gravitational field was in some sense equivalent to physics in an inertial frame.

Once the theory was formulated, however, he did try to confront it with experiment by proposing three tests. One of these tests was an immediate success—the explanation of the anomalous advance in the perihelion of Mercury of 43 arcseconds per century, a problem that had bedeviled celestial mechanicians of the latter part of the 19th century. The next test, the deflection of light by the Sun, was such a success that it produced what today would be called a "media event." The measurements of the deflection, amounting to 1.75 arcseconds for a ray that grazes the Sun, by two teams of British astronomers in 1919, made Einstein an instant international celebrity. However, these measurements were not all that accurate, and subsequent measurements weren't much better. The third test, actually proposed by Einstein in 1907, was the gravitational redshift of light, but it was a test that remained unfulfilled until 1960, by which time it was no longer viewed as a true test of general relativity.

Cosmology was the other area where general relativity was believed to have observational relevance. Although the general relativistic picture of the expansion of the universe from a "big bang" was compatible with observations, there were problems. As late as the middle 1950s the measured values of the universal expansion rate implied that the universe was younger than the Earth! Even though this "age" problem had been resolved by 1960, cosmological observations were still in their infancy, and could not distinguish between various alternative models.

The turning point for general relativity came in the early 1960s, when discoveries of unusual astronomical objects such as quasars demonstrated that the theory would have important applications in astrophysical situations. Theorists found new ways to understand the theory and its observable consequences. Finally, the technological revolution of the last quarter century, together with the development of the interplanetary space program, provided new high-precision tools to carry out experimental tests of general relativity.

After 1960, the pace of research in general relativity and in an emerging field called "relativistic astrophysics" began to accelerate. New advances, both theoretical and observational, came at an ever increasing rate. They included the discovery of the cosmic background radiation; the analysis of the synthesis of helium from hydrogen in the big bang; observations of pulsars and of black-hole candidates; the development of the theory of relativistic stars and black holes; the theoretical study of gravitational radiation and the beginning of an experimental program to detect it; improved versions of old tests of general relativity, and brand new tests, discovered after 1960; the discovery of the binary pulsar, which provided evidence for gravity waves; the analysis of quantum effects outside black holes and of black hole evaporation; the discovery of a gravitational lens; and the beginnings of a unification of gravitation theory with the other interactions and with quantum mechanics.

During the two decades following 1960, general relativity rejoined the world of physics and astronomy. Research in relativity took on an increasingly interdisciplinary flavor, spanning such subjects as celestial mechanics, pure mathematics, experimental physics, quantum mechanics, observational astronomy, particle physics, and theoretical astrophysics.

## Einstein's Equivalence Principle

The foundations of general relativity are actually quite old, dating back to the equivalence principle of Galileo and Newton. In Newton's view, the principle of equivalence stated that all objects accelerate at the same rate in a gravitational field regardless of their mass or composition. This equality has been verified abundantly over the years, including classic experiments by the Hungarian physicist Baron Lorand von Eötvös around 100 years ago and recent experiments at Princeton and Moscow State Universities. The accuracy of these tests is better than one part in $10^{11}$. Einstein's insight was the recognition that, to an observer inside a freely falling laboratory, not only should objects float as if gravity were absent as a consequence of this equality, but also *all* laws of nongravitational physics, such as electromagnetism and quantum mechanics, should behave as if gravity were truly absent.

This idea is now known as the Einstein equivalence principle, and it was a key step, because it implied the converse: that in a reference frame where gravity is felt, such as in a laboratory on the Earth's surface, the effects of gravitation on physical laws can be obtained simply by mathematically transforming the laws from a freely falling frame to the laboratory frame. According to the branch of mathematics known as differential geometry, this is the same as saying that space-time is curved; in other words, that the effects of gravity are *indistinguishable* from the effects of being in curved space-time.

## Gravitational Redshift

An immediate consequence of Einstein's principle of equivalence is the gravitational redshift effect. It is not a consequence of general relativity itself, although Einstein believed that it was. One version of a gravitational redshift experiment measures the frequency shift between two identical clocks (meaning any device that produces a signal at a well-defined, steady frequency), placed at rest at different heights in a gravitational field. To derive the frequency shift from the equivalence principle, one imagines a freely falling frame that is momentarily at rest with respect to one clock at the moment the clock emits its signal. Because special relativity is valid in that frame, the frequency of the signal is unaffected by gravity as it travels from one clock to the other. However, by the time the signal reaches the second clock, the frame has picked up a downward velocity because it is in free fall in the gravitational field, and therefore, as seen by the falling frame, the second clock is moving upward. Thus the frequency seen by the second clock will appear to be shifted from the standard value by the Doppler effect. For small differences in height $h$ between clocks, the shift in the frequency $\Delta f$ is given by

$$\frac{\Delta f}{f} = \frac{v_{\text{fall}}}{c} = \frac{gh}{c^2}$$

*(continued on next page)*

where $g$ is the local gravitational acceleration and $c$ is the speed of light. If the receiver is at a lower height than the emitter, the received signal is shifted to higher frequencies ("blueshift"), while if the receiver is higher, the signal is shifted to lower frequencies ("redshift"). The generic name for the effect is "gravitational redshift."

The first and most famous high-precision redshift measurement was the Pound-Rebka experiment of 1960, which measured the frequency shift of gamma-ray photons from the decay of iron-57 ($^{57}Fe$) as they ascended or descended the Jefferson Physical Laboratory tower at Harvard University.

The most precise gravitational redshift experiment performed to date was a rocket experiment carried out in June 1976. A "hydrogen maser" atomic clock was flown on a Scout D rocket to an altitude of 10 000 km, and its frequency was compared to that of a similar clock on the ground by radio signals. After the effects of the rocket's motion were taken into account, the observations confirmed the gravitational redshift to 0.02 percent.

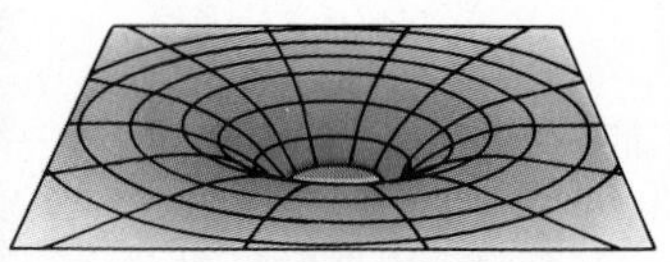

**Figure 1** The gravitation curvature of space. According to general relativity, gravity curves spacetime. This drawing shows how space is curved around a massive object such as the sun or a star. The red shaded region in the center indicates the location of the star. The greatest curvature is found immediately above the star's surface. Far from the star, where gravity is weak, spacetime is almost perfectly flat.

Because of these kinds of experiments, physicists are now convinced that spacetime *is* curved, and that the correct theory of gravity must be based on curved spacetime (see Fig. 1). That does not automatically imply general relativity, however, since it is not the only such theory. To test the specific predictions of general relativity for the amount and nature of space-time curvature, other experiments are needed.

## Deflection of Light

The first of these is a test that made Einstein's name a household word: the deflection of light. According to general relativity, a light ray that passes the Sun at a distance $d$ is deflected by an angle $\Delta\theta = 1''.75/d$, where $d$ is measured in units of the solar radius, and the notation $''$ denotes seconds of arc (see Fig. 2). Half of the deflection can be calculated by appealing only to the principle of equivalence. The idea is to determine what observers in a sequence of free-falling frames all along the trajectory of the photon would find for the deflection from one frame to the next. It turns out that the net deflection over all the frames is $0''.875/d$. The same result can be obtained by calculating the deflection of a particle by a massive body using ordinary Newtonian theory, and then taking the limit in which the particle's velocity becomes that of light. But the result of both versions is the deflection of light only relative to local straight lines, as defined for example by rigid rods laid end-to-end. However, because of the curvature of space, a local straight line defined by rods passing close to the Sun is itself bent relative to straight lines far from the Sun by an amount that yields the remaining half of the deflection.

The prediction of the bending of light by the Sun was one of the great successes of general relativity. Confirmation by the British astronomers Eddington and Crommelin of the bending of optical starlight observed during a total solar eclipse in the first

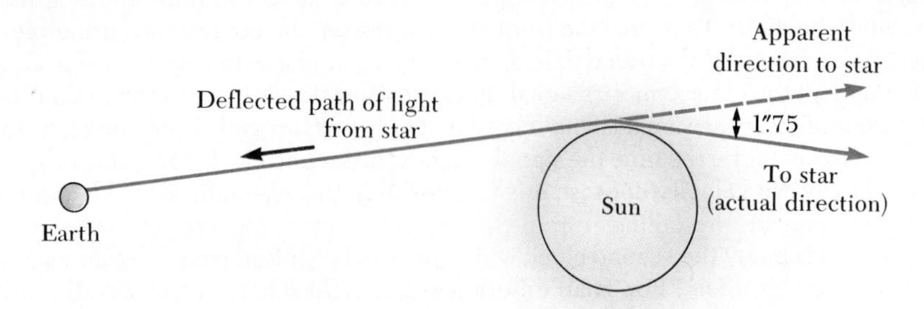


**Figure 2** Deflection of starlight passing near the sun. Because of this effect, the sun and other remote objects can act as a *gravitational lens*. In his general relativity theory, Einstein calculated that starlight just grazing the sun's surface should be deflected by an angle of 1″.75.

months following World War I helped make Einstein a celebrity. However, those measurements had only 30 percent accuracy, and succeeding eclipse experiments weren't much better; the results were scattered between one half and twice the Einstein value, and the accuracies were low.

However, the development of long-baseline radio interferometry produced a method for greatly improved determinations of the deflection of light. Radio interferometry is a technique of combining widely separated radio telescopes in such a way that the direction of a source of radio waves can be measured by determining the difference in phase of the signal received at the different telescopes. Modern interferometry has the ability to measure angular separations and changes in angles as small as $10^{-4}$ arcseconds. Coupled with this technological advance is a series of heavenly coincidences: each year groups of strong quasars pass near the Sun (as seen from the Earth). The idea is to measure the differential deflection of radio waves from one quasar relative to those from another as they pass near the Sun. A number of measurements of this kind were made almost annually over the period 1969–1975, yielding a confirmation of the predicted deflection to 1.5 percent.

The 1979 discovery of the "double" quasar Q0957 + 561 converted the deflection of light from a test of relativity to an important effect in astronomy and cosmology. The "double" quasar was found to be a multiple image of a single quasar caused by the gravitational lensing effect of a galaxy or a cluster of galaxies along the line of sight between us and the quasar. Several other such lenses have been found.

Closely related to light deflection is the "Shapiro time delay," a retardation of light signals that pass near the Sun. For instance, for a signal that grazes the Sun on a round trip from Earth to Mars at superior conjunction (when Mars is on the far side of the Sun), the round trip travel time is increased over what Newtonian theory would give by about 250 $\mu$s. The effect decreases with increasing distance of the signal from the Sun.

In the two decades following radio-astronomer Irwin Shapiro's 1964 discovery of this effect, several high-precision measurements have been made using the technique of radar-ranging to planets and spacecraft. Three types of targets were employed: planets such as Mercury and Venus, free-flying spacecraft such as Mariners 6 and 7, and combinations of planets and spacecraft, known as "anchored spacecraft," such as the Mariner 9 Mars orbiter and the 1976 Viking Mars landers and orbiters. The Viking experiments produced dramatic results, agreeing with the general relativistic prediction to one part in a thousand. This corresponded to a measurement accuracy in the Earth-Mars distance of 30 meters.

## Perihelion Shift

The explanation of the anomalous perihelion shift of Mercury's orbit was another of the triumphs of general relativity (see Figs. 3 and 4). This had been an unsolved problem in celestial mechanics for over half a century, since the announcement by Le Verrier in 1859 that, after the perturbing effects of the planets on Mercury's orbit had been accounted for, there remained in the data an unexplained advance in the perihelion of Mercury. The modern value for this discrepancy is 42.98 arcseconds per century. Many ad hoc proposals were made in an attempt to account for this excess, including the existence of a new planet Vulcan near the Sun, or a deviation from the inverse square law of gravitation. But each was doomed to failure. General relativity accounted for the anomalous shift in a natural way. Radar measurements of the orbit of Mercury since 1966 have led to improved accuracy, so that the relativistic advance is known to about 0.5 percent.

Other measurements carried out since 1960 have given further support to general relativity. Observations of the motion of the Moon using laser ranging to a collection of specially designed reflectors, deposited on the lunar surface during the

*(continued on next page)*

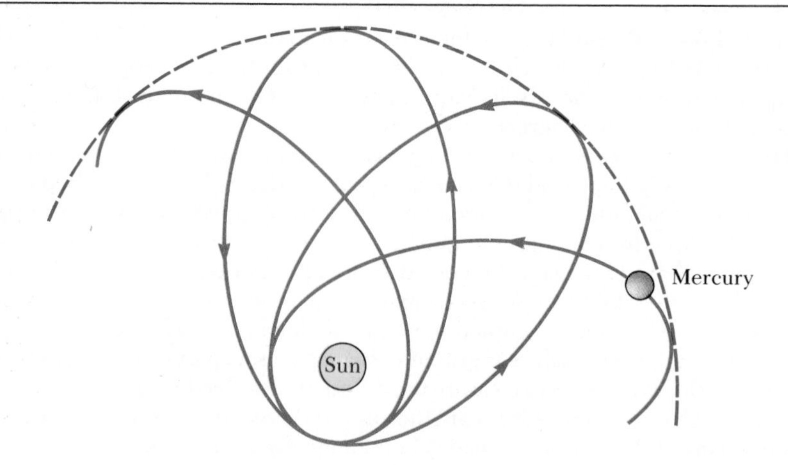


**Figure 3** Advance of perihelion of Mercury. The elliptical orbit of Mercury about the sun rotates very slowly relative to the system connected with the sun. General relativity successfully explains this small effect, which predicts that the direction of the perihelion should change by only 43 arcseconds per century.

Apollo and Luna missions, have shown that the Moon and the Earth accelerate toward the Sun equally to a part in $10^{11}$, an important confirmation of the equivalence principle applied to planetary bodies. Other measurements of planetary and lunar orbits have shown that the gravitational constant is a true constant of nature; it does not vary with time as the universe ages. Ambitious experimenters are currently designing an experiment using supercooled quartz gyroscopes to be placed in Earth orbit, to try to detect an important general relativistic effect called the "dragging of inertial frames," caused by the rotation of the Earth.

General relativity has passed every experimental test to which it has been put, and many alternative theories have fallen by the wayside. Most physicists now take the theory for granted, and look to see how it can be used as a practical tool in physics and astronomy.

## Gravitational Radiation

One of these new tools is gravitational radiation, a subject that is almost as old as general relativity itself. By 1916, Einstein had succeeded in showing that the field equations of general relativity admitted wavelike solutions analogous to those of electromagnetic theory. For example, a dumbbell rotating about an axis passing at right angles through its handle will emit gravitational waves that travel at the speed of light. They also carry energy away from the rotating dumbbell, just as electromagnetic waves carry energy away from a light source. That was about all that was heard on the subject for over 40 years, primarily because the effects associated with gravitational waves were extremely tiny, unlikely (it was thought) ever to be of experimental or observational interest.

But in 1968 the idea of gravitational radiation was resurrected by the stunning announcement by Joseph Weber that he had detected gravitational radiation of extraterrestrial origin by using massive aluminum bars as detectors. A passing gravitational wave acts as an oscillating gravitational force field that alternately compresses and extends the bar in the lengthwise direction. However, subsequent observations by other researchers using bars with sensitivity that was claimed to be better than Weber's failed to confirm Weber's results. His results are now generally regarded as a false alarm, although there is still no good explanation for the "events" that he recorded in his bars.

Nevertheless, Weber's experiments did initiate the program of gravitational-wave detection, and inspired other groups to build better detectors. Currently a dozen laboratories around the world are engaged in building and improving upon the

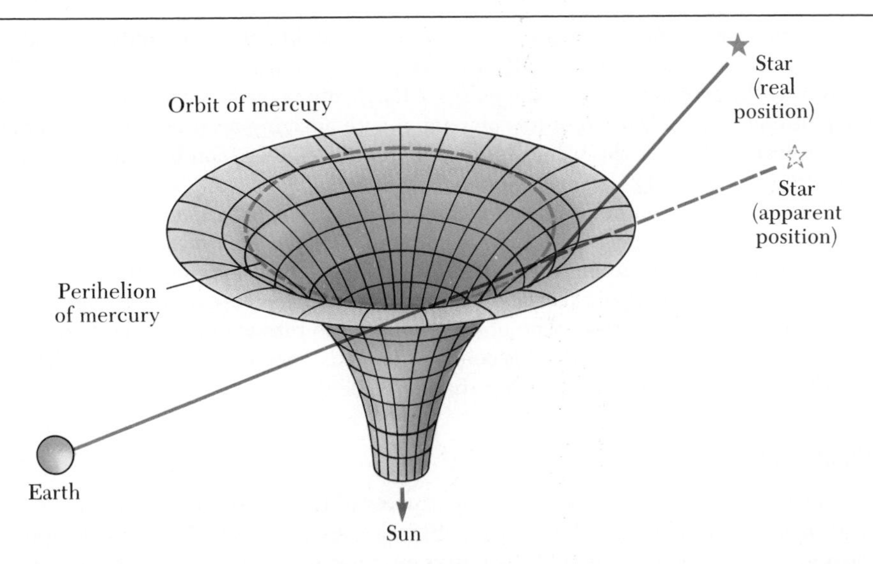


**Figure 4** Under the general theory of relativity, the presence of a massive body essentially warps the space nearby. This can account for both the bending of light near the sun and the advance of the perihelion point of Mercury by 43 arcseconds per century more than would otherwise be expected. The diagram shows how a two-dimensional surface warped into three dimensions can change the direction of a "straight" line that is constrained to its surface; the warping of space is analogous, although with a greater number of dimensions to consider. The effect is similar to the golfer putting on a warped green. Though the ball is hit in a straight line, we see it appear to curve. (Adapted from Pasachoff, Jay M., *Astronomy,* 3rd ed., Saunders College Publishing, 1987).

basic "Weber bar" detector, some using bigger bars, some using smaller bars, some working at room temperature, and some working near absolute zero. The goal in all cases is to reduce noise from thermal, electrical, and environmental sources as much as possible to have a hope of detecting the very weak oscillations produced by a gravitational wave. For a bar of one meter length, the challenge is to detect a variation in length smaller than $10^{-20}$ meters, or $10^{-5}$ of the radius of a proton.

Another important type of detector is the laser interferometer, currently under development at several laboratories. This device operates on the same principle as the interferometer used by Michelson and Morley, but uses a laser as the light source. A passing gravitational wave will change the length of one arm of the apparatus relative to the other, and cause the interference pattern to vary. One ambitious proposal is to build two independent interferometers, each with arm lengths of 4 km, and separated by 1000 km.

It is hoped that these detectors will eventually be sensitive enough to detect gravitational waves from many sources, both in our galaxy and in distant galaxies, such as collapsing stars, double star systems, colliding black holes, and possibly even gravitational waves left over from the big bang. When this happens, a new field of "gravitational-wave astronomy" will be born.

Although gravitational radiation has not been detected directly, we know that it exists, through a remarkable system known as the binary pulsar. Discovered in 1974 by radio astronomers Russell Hulse and Joseph Taylor, it consists of a pulsar (which is a rapidly spinning neutron star) and a companion star in orbit around each other. Although the companion has not been seen directly, it is also believed to be a neutron star. The two bodies are circling so closely—their orbit is about the size of the Sun, and the period of the orbit is eight hours—that the effects of general relativity are very large. For example, the periastron shift (the analogue of the perihelion shift for Mercury) is more than 4 degrees per *year*.

*(continued on next page)*

Furthermore, the pulsar acts as an extremely stable clock, its pulse period of approximately 59 milliseconds drifting by only a quarter of a nanosecond per year. By measuring the arrival times of radio pulses at Earth, observers were able to determine the motion of the pulsar about its companion with amazing accuracy. For example, the accurate value for the orbital period is 27906.98163 seconds, and the orbital eccentricity is 0.617127.

Like a rotating dumbbell, an orbiting binary system should emit gravitational radiation, and in the process lose some of its orbital energy. This energy loss will cause the pulsar and its companion to spiral in toward each other, and the orbital period to shorten. According to general relativity, the predicted decrease in the orbital period is 75 microseconds per year. The observed decrease rate is in agreement with the prediction to about 4 percent. This confirms the existence of gravitational radiation and the general relativistic equations that describe it.

### Black Holes

One of the most important and exciting aspects of the relativity renaissance is the study of and search for black holes. The subject began in the early 1960s as astrophysicists looked for explanations of the quasars, warmed up in the early 1970s following the possible detection of a black hole in an X-ray source, became white hot in the middle 1970s when black holes were found theoretically to evaporate, and continues today as one of the most active branches of general relativity.

However, the first glimmerings of the black hole idea date back to the 18th century, in the writings of a British amateur astronomer, the Reverend John Michell. Reasoning on the basis of the corpuscular theory that light would be attracted by gravity in the same way that ordinary matter is attracted, he noted that light emitted from the surface of a body such as the Earth or the Sun would be reduced in velocity by the time it reached great distances (Michell of course did not know special relativity). How large would a body of the same density as the Sun have to be in order that light emitted from it would be stopped and pulled back before reaching infinity? The answer he obtained was 500 times the diameter of the Sun. Light could therefore never escape from such a body. In today's language, such an object would be a supermassive black hole of about 100 million solar masses.

Although the general relativistic solution for a nonrotating black hole was discovered by Karl Schwarzschild in 1916, and a calculation of gravitational collapse to a black hole state was performed by J. Robert Oppenheimer and Hartland Snyder in 1939, black hole physics didn't really begin until the middle 1960s, when astronomers confronted the problem of the energy output of the quasars, and the mathematician Roy Kerr discovered the rotating black-hole solution.

A black hole is formed when a star has exhausted the thermonuclear fuel necessary to produce the heat and pressure that support it against gravity. The star begins to collapse, and if it is massive enough, it continues to collapse until the radius of the star approaches a value called the gravitational radius or Schwarzschild radius. In the nonrotating spherical case, this radius has a value given by $2GM/c^2$, where $M$ is the mass of the star. For a body of one solar mass, the gravitational radius is about 3 km; for a body of the mass of the Earth, it is about 9 mm. An observer sitting on the surface of the star sees the collapse continue to smaller and smaller radii, until both star and observer reach the origin $r = 0$, with consequences too horrible to describe in detail. On the other hand, an observer at great distances observes the collapse to slow down as the radius approaches the gravitional radius, a result of the gravitational redshift of the light signals sent outward. However, in this case, the redshifting or slowing down becomes so extreme that the star appears almost to stop just outside the gravitational radius. The distant observer never sees any signals emitted by the falling observer once the latter is inside the gravitational radius. Any signal emitted inside can never escape the sphere bounded by the gravitational radius, called the "event horizon."

Since the middle 1960s the laws of "black hole physics" have been established and codified. For example, it was discovered that the Kerr (rotating) and Schwarzschild (nonrotating) solutions are unique; there are no other black-hole solutions in general relativity. Inside every black hole resides a "singularity," a pathological region of spacetime where the gravitational forces become infinite, where time comes to an end for any observer unfortunate enough to hit the singularity, indeed where all the laws of physics break down. Fortunately the event horizon of the black hole prevents any bizarre phenomena that such singularities might cause from reaching the outside world (this notion has been dubbed "cosmic censorship"). In 1974, Stephen Hawking discovered that the laws of quantum mechanics applied to the physics outside a black hole required it to evaporate by the creation of particles with a thermal energy spectrum, and to have an associated temperature and entropy. The temperature of a Schwarzschild black hole is $T = hc^3/8\pi kGM$, where $h$ is Planck's constant and $k$ is Boltzmann's constant. This discovery demonstrated a remarkable connection between gravity, thermodynamics, and quantum mechanics, which helped renew the theoretical quest for a grand synthesis of all the fundamental interactions. For black holes of astronomical masses, however, the evaporation is completely negligible, since for a solar-mass black hole, $T \approx 10^{-6}$ K.

Although a great deal is known about black holes in theory, rather less is known about them observationally. There are several instances in which the evidence for the existence of black holes is impressive, but in all cases it is indirect. For instance, in the X-ray source Cygnus X1, the source of the X-rays is believed to be a collapsed object with a mass larger than about 6 solar masses in orbit around a giant star (see Figs. 5 and 6). The object cannot be a neutron star, because general relativity predicts that neutron stars must be lighter than about 3 solar masses. Thus the object must be a black hole. The X-rays are emitted by matter pulled from the surface of the companion star and sent into a spiralling orbit around the black hole. Similarly, there is evidence in the centers of certain galaxies, such as M87 and possibly even our own, of

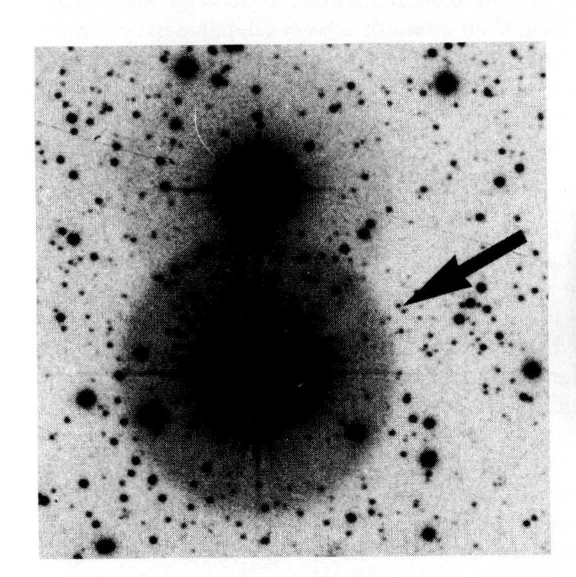

**Figure 5** The overexposed dark object in the center of this negative print is the blue supergiant star HDE 226868, which is thought to be the companion of the first black hole to be discovered, Cygnus X-1. The black hole in Cygnus X-1 that is thought to be orbiting the supergiant star is not visible. The image of the supergiant appears so large because it is overexposed on the film. (From Pasachoff, *Astronomy*.)

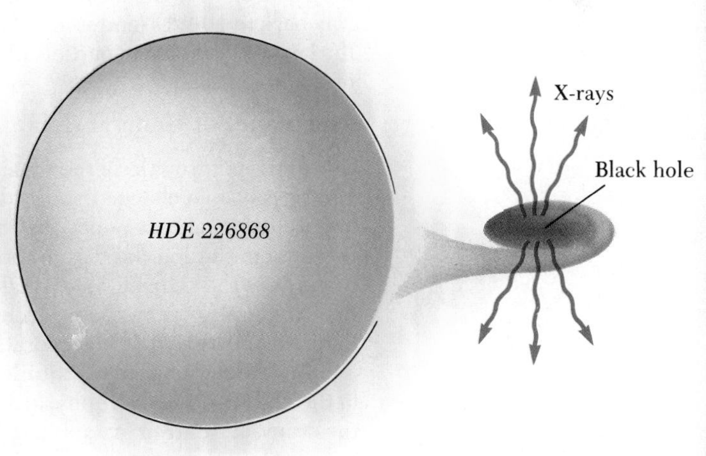


**Figure 6** The Cygnus X-1 black hole. The stellar wind from HDE 226868 pours matter onto a huge disk around its black hole companion. The infalling gases are heated to enormous temperatures as they spiral toward the black hole. The gases are so hot that they emit vast quantities of x-rays.

collapsed objects of between $10^2$ and $10^8$ solar masses. A black hole model is consistent with the observations, although it is not necessarily required. Accretion of matter onto supermassive rotating black holes may produce the jets of outflowing matter that are observed in many quasars and active galactic nuclei. These and other astrophysical processes that might aid in the detection of black holes are being studied by relativists and astrophysicists.

### Cosmology

The other area in which there has been a renaissance for general relativity is cosmology. Although Einstein in 1917 first used general relativity to calculate a model for the universe as a whole, the subject was not considered a serious branch of physics until the 1960s, when astronomical observations lent firm credence to the idea that the universe was expanding from a "big bang." For instance, observations of the rates at which galaxies are receding from us, coupled with improved determinations of their distances, implied that the universe was at least 10 billion years old, a value that was consistent with other observations such as the age of the Earth and of old star clusters.

In 1965 came the discovery of the cosmic background radiation by Arno Penzias and Robert Wilson. This radiation is the remains of the hot electromagnetic blackbody radiation that dominated the universe in its earlier phase, now cooled to 3 kelvins by the subsequent expansion of the universe. Next came calculations of the amount of helium that would be synthesized from hydrogen by thermonuclear fusion in the very early universe, around 1000 seconds after the big bang. The amount, approximately 25 percent by weight, was in agreement with the abundances of helium observed in stars and in interstellar space. This was an important confirmation of the hot big bang picture, because the amount of helium believed to be produced by fusion in the interiors of stars is woefully inadequate to explain the observed abundances.

Today, the general relativistic hot big-bang model of the universe has broad acceptance, and cosmologists now focus their attention on more detailed issues, such as how galaxies and other large-scale structures formed out of the hot primordial soup, and on what the universe might have been like earlier than 1000 seconds, all the way back to $10^{-36}$ s (and some brave cosmologists are going back even further) when the laws of elementary-particle physics may have played a major role in the evolution of the universe.

### Acceptance of General Relativity

One of the outgrowths of the renaissance of general relativity that has occurred since 1960 has been a change in attitude about the importance and use of the theory. Its importance as a fundamental theory of the nature of spacetime and gravitation has not been diminished in the least; if anything, it has been enhanced by the flowering of research in the subject that has taken place. Its importance as a foundation for other theories of physics has been strengthened by current searches for unified quantum theories of nature that incorporate gravity along with the other interactions.

But the real change in attitude about general relativity has been in its use as a tool in the real world. In astrophysical situations, general relativity plays a central role in the study of neutron stars, black holes, gravitational lenses, relativistic binary star systems, and the universe as a whole. Gravitational radiation may one day provide a completely new observational tool for exploring and examining the cosmos.

Relativity even plays a role in everyday life. For example, the gravitational redshift effect on clocks *must* be taken into account in satellite-based navigation systems, such as the US Air Force's Global Positioning System, in order to achieve the required positional accuracy of a few meters or time transfer accuracy of a few nanoseconds.

To general relativists, always eager to find practical consequences of their subject, these have been very welcome developments!

## Suggested Readings

Davies, P. C. W. *The Search for Gravity Waves,* Cambridge, England, Cambridge University Press, 1980.

Greenstein, G. *Frozen Star: Of Pulsars, Black Holes and the Fate of Stars,* New York, Freundlich Books, 1984.

Weinberg, S. *The First Three Minutes: A Modern View of the Origin of the Universe,* New York, Basic Books, 1977.

Will, C. M. *Was Einstein Right? Putting General Relativity to the Test,* New York, Basic Books, 1986.

## Essay Questions

1. Two stars are a certain distance from each other as seen in the night sky. Half a year later, the stars are now overhead during the day, and the Sun is now located midway between the two stars. Because of the deflection of light by the Sun, do the stars now appear closer together or farther apart?
2. When gravitational waves are emitted by the binary pulsar, the double-star system loses energy. As a consequence, its orbital velocity increases, and thus its kinetic energy increases. How is this possible?
3. The temperature of a black hole is inversely proportional to its mass, so that when it gains energy (which is equivalent to mass), its temperature goes down, and when it loses energy, its temperature goes up. What do you conclude about the sign of its heat capacity? Can a black hole ever be in thermal equilibrium with an infinite reservoir at a fixed temperature?

## Essay Problems

1. What is the gravitational frequency shift between two atomic clocks separated by 1000 km, both at sea level?
2. A pair of identical twin scientists set out to test the gravitational redshift effect in the following way. One climbs to the top of Mount Baldy and lives there for a year, while the other remains behind in the laboratory at the base of the mountain. When they are reunited after one year, one of the twins is slightly older than the other. Which one is older and why? If Mount Baldy is 5000 m high, what is the difference in age between the twins after one year?
3. Calculate the deflection of light for light rays that pass the Sun at distances of 2 solar radii, 10 solar radii, and 100 solar radii.
4. Gravitational waves travel with the same speed as light, $3 \times 10^{10}$ cm/s. Calculate the wavelength of gravitational waves emitted by the following sources: (a) a star collapsing to form a black hole, emitting waves with a frequency around 1000 Hz, and (b) the binary pulsar, emitting waves with a period of four hours (one half the orbital period).
5. What is the fractional decrease in the orbital period of the binary pulsar per year? Estimate how long the binary system will continue before gravitational radiation will reduce the orbital period to zero (in other words will bring the stars so close together that they will merge into one object).
6. Calculate the mass of a black hole (in solar mass units) whose Schwarzschild radius is equal to the radius of the Earth's orbit around the Sun ($1.5 \times 10^8$ km). One solar mass is about $2 \times 10^{33}$ g.
7. Show that a billion-ton black hole has a Schwarzschild radius comparable to the radius of a proton (1 ton $\approx 10^6$ g, proton radius $\approx 10^{-13}$ cm). What kind of object on Earth weighs a billion tons?
8. Determine the mass and radius of a black hole whose apparent density is equal to that of water, 1 g/cm$^3$. Assume that the volume of a black hole is given by $4\pi R^3/3$, and express the mass in solar mass units.
9. Calculate the mass of a black hole whose temperature is room temperature.

# 40

# Introduction to Quantum Physics

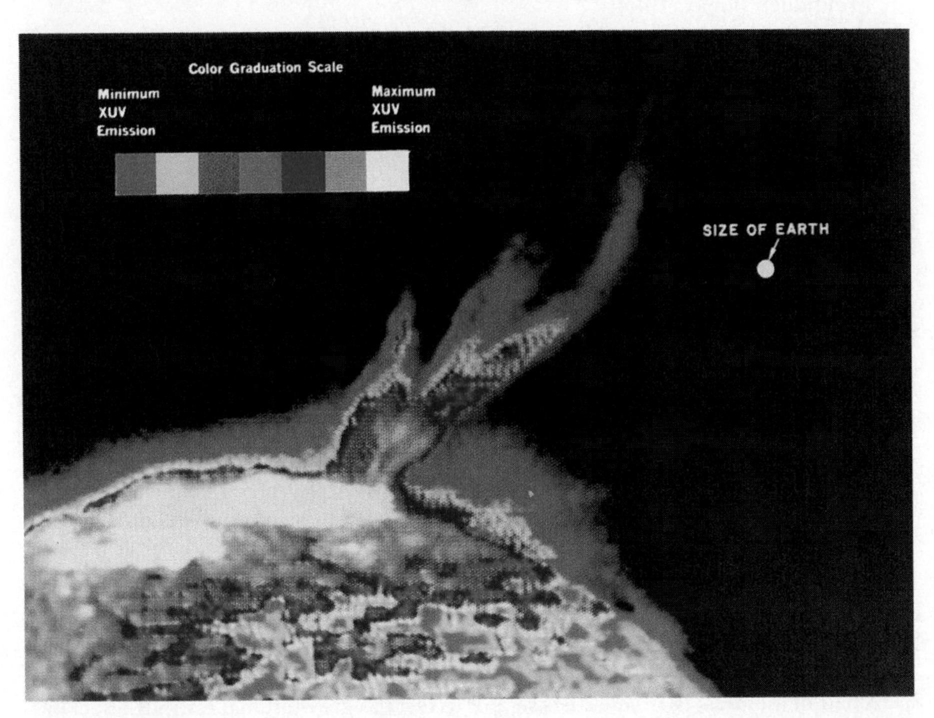


*A solar prominence recorded in wavelengths of extreme ultraviolet light. The temperatures of the churning gases at the sun's surface are about 4 000°C, but they increase rapidly to about 500 000°C in the chromosphere and up to some 3 000 000°C in the corona (black area of picture; the colors on the sun's surface represent various intensities of extreme ultraviolet radiation). When the energy of a solar explosion heads toward the earth, taking several days to reach us, it can produce so-called geomagnetic storms, which may disrupt broadcast communications and power transmission lines and cause navigational systems to go haywire. Such storms are also the cause of the beautiful aurora borealis. (Courtesy of NASA)*

In the previous chapter, we discussed the fact that Newtonian mechanics must be replaced by Einstein's special theory of relativity when we are dealing with particle velocities comparable to the speed of light. Although many problems were indeed resolved by the theory of relativity in the early part of the 20th century, many experimental and theoretical problems remained unanswered. Attempts to apply the laws of classical physics to explain the behavior of matter on the atomic scale were totally unsuccessful. Various phenomena, such as blackbody radiation, the photoelectric effect, and the emission of sharp spectral lines by atoms in a gas discharge, could not be understood within the framework of classical physics. We shall describe these phenomena because of their importance in subsequent developments.

Another revolution took place in physics between 1900 and 1930. This was the era of a new and more general scheme called *quantum mechanics.* This new approach was highly successful in explaining the behavior of atoms, molecules, and nuclei. Moreover, the quantum theory reduces to classical physics when applied to macroscopic systems. As with relativity, the quantum theory requires a modification of our ideas concerning the physical world.

The basic ideas of quantum theory were first introduced by Max Planck, but most of the subsequent mathematical developments and interpretations were made by a number of distinguished physicists, including Einstein, Bohr, Schrödinger, de Broglie, Heisenberg, Born, and Dirac. Despite the great success of the quantum theory, Einstein frequently played the role of critic, especially with regard to the manner in which the theory was interpreted. In

Biographical Sketch

**Max Planck**
*(1858–1947)*

(Courtesy of AIP Niels Bohr Library, W.F. Meggers Collection)

**Max Planck** was born in Kiel and received his college education in Munich and Berlin. He joined the faculty at Munich in 1880 and five years later received a professorship at Kiel University. He replaced Kirchhoff at the University of Berlin in 1889 and remained there until 1926. He introduced the concept of a "quantum of action" (Planck's constant $h$) in an attempt to explain the spectral distribution of blackbody radiation, which laid the foundations for quantum theory. He was awarded the Nobel prize in 1918 for this discovery of the quantized nature of energy. The work leading to the "lucky" blackbody radiation formula was described by Planck in his Nobel prize acceptance speech (1920): "But even if the radiation formula proved to be perfectly correct, it would after all have been only an interpolation formula found by lucky guesswork and thus, would have left us rather unsatisfied. I therefore strived from the day of its discovery, to give it a real physical interpretation and this led me to consider the relations between entropy and probability according to Boltzmann's ideas."

Planck's life was filled with personal tragedies. One of his sons was killed in action in World War I, and two daughters were lost in childbirth in the same period. His house was destroyed by bombs in World War II, and his son Erwin was executed by the Nazis in 1944, after being accused of plotting to assassinate Hitler.

Planck became president of the Kaiser Wilhelm Institute of Berlin in 1930. The institute was renamed the Max Planck Institute in his honor after World War II. Although Planck remained in Germany during the Hitler regime, he openly protested the Nazi treatment of his Jewish colleagues, and consequently was forced to resign his presidency in 1937. Following World War II, he was renamed the president of the Max Planck Institute. He spent the last two years of his life in Göttingen as an honored and respected scientist and humanitarian.

particular, Einstein did not accept Heisenberg's uncertainty principle, which says that it is impossible to obtain a precise simultaneous measurement of the position and the velocity of a particle. According to this principle, it is possible to predict only the *probability* of the future of a system, contrary to the deterministic view held by Einstein.[1]

An extensive study of quantum theory is certainly beyond the scope of this book. This chapter is simply an introduction to the underlying ideas of quantum theory. We shall also discuss some simple applications of quantum theory, including the photoelectric effect, the interpretation of atomic spectra, and the Compton effect.

## 40.1 BLACKBODY RADIATION AND PLANCK'S HYPOTHESIS

An object at any temperature is known to emit radiation sometimes referred to as **thermal radiation.** The characteristics of this radiation depend on the temperature and properties of the object. At low temperatures, the wavelengths of the thermal radiation are mainly in the infrared region and hence are not observed by the eye. As the temperature of the object is increased, it eventually begins to glow red. At sufficiently high temperatures, it appears to be white, as in the glow of the hot tungsten filament of a lightbulb. A careful study of thermal radiation shows that it consists of a continuous distribution of wavelengths from the infrared, visible, and ultraviolet portions of the spectrum.

[1] Einstein's views on the probabilistic nature of quantum theory are brought out in his statement, "God does not play dice with the universe."

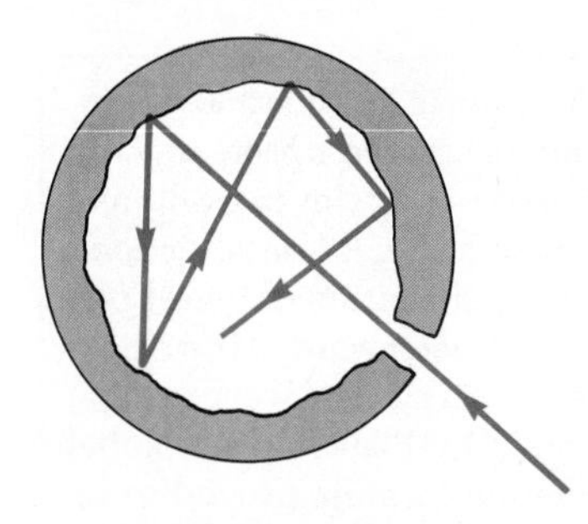

**Figure 40.1** The opening in the cavity of a body is a good approximation to a blackbody. As light enters the cavity through the small opening, part is reflected and part is absorbed on each reflection from the interior walls. After many reflections, essentially all of the incident energy is absorbed.

From a classical viewpoint, thermal radiation originates from accelerated charged particles near the surface of the object; those charges emit radiation much like small antennas. The thermally agitated charges can have a distribution of accelerations, which accounts for the continuous spectrum of radiation emitted by the object. By the end of the 19th century, it had become apparent that the classical theory of thermal radiation was inadequate. The basic problem was in understanding the observed distribution of wavelengths in the radiation emitted by a black body. By definition, a black body is an ideal system that absorbs all radiation incident on it. A good approximation to a black body is the inside of a hollow object, as shown in Figure 40.1. The nature of the radiation emitted through a small hole leading to the cavity depends only on the temperature of the cavity walls.

Experimental data for the distribution of energy for blackbody radiation at three different temperatures are shown in Figure 40.2. The radiated energy varies with wavelength and temperature. As the temperature of the black body increases, the total amount of energy it emits increases. Also, with increasing temperatures, the peak of the distribution shifts to shorter wavelengths. This shift was found to obey the following relationship, called **Wien's displacement law:**

$$\lambda_{\max} T = 0.2898 \times 10^{-2}\ \mathrm{m \cdot K} \tag{40.1}$$

where $\lambda_{\max}$ is the wavelength at which the curve peaks and $T$ is the absolute temperature of the object emitting the radiation.

Early attempts to explain these results based on classical theories failed. To describe the radiation spectrum, it is useful to define $I(\lambda, T)\, d\lambda$ to be the power per unit area emitted in the wavelength interval $d\lambda$. The result of a calculation based on a classical model of blackbody radiation known as the **Rayleigh-Jeans law** is

Rayleigh-Jeans law

$$I(\lambda, T) = \frac{2\pi ckT}{\lambda^4} \tag{40.2}$$

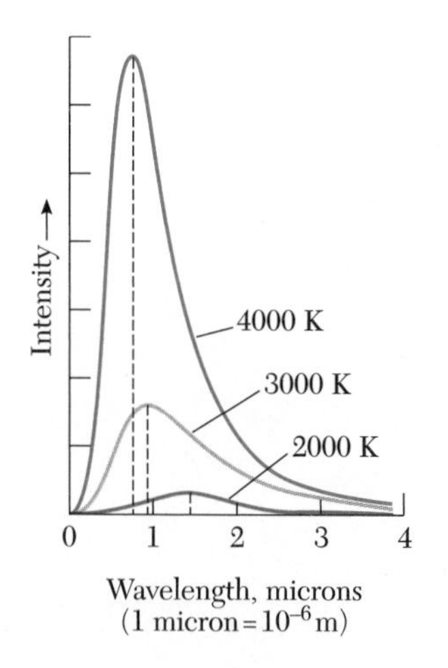


**Figure 40.2** Intensity of blackbody radiation versus wavelength at three temperatures. Note that the amount of radiation emitted (the area under a curve) increases with increasing temperature.

where $k$ is Boltzmann's constant. In this classical model of blackbody radiation, the atoms in the cavity walls are treated as a set of oscillators that emit electromagnetic waves at all wavelengths. This model leads to an average energy per oscillator that is proportional to $T$.

An experimental plot of the blackbody radiation spectrum is shown in Figure 40.3, together with the theoretical prediction of the Rayleigh-Jeans law. At long wavelengths, the Rayleigh-Jeans law is in reasonable agreement with experimental data. However, at short wavelengths there is major disagreement. This can be seen by noting that as $\lambda$ approaches zero, the function $I(\lambda, T)$ given by Equation 40.2 approaches infinity; that is, short-wavelength radiation should predominate. This is contrary to the experimental data plotted in Figure 40.3, which shows that as $\lambda$ approaches zero, $I(\lambda, T)$ also approaches zero. This contradiction is often called the **ultraviolet catastrophe.** Another major problem with classical theory is that it predicts an *infinite total energy density* since all wavelengths are possible.[2] Physically, an infinite energy in the electromagnetic field is an impossible situation.

[2] The total power per unit area $I = \int_0^\infty I(\lambda, T)\, d\lambda$ diverges to $\infty$ when all wavelengths are allowed.

In 1900 Max Planck discovered a formula for blackbody radiation that was in complete agreement with experiment at all wavelengths. Figure 40.3 shows that Planck's law fits the experimental results quite well. The empirical function proposed by Planck is given by

$$I(\lambda, T) = \frac{2\pi hc^2}{\lambda^5(e^{hc/\lambda kT} - 1)} \quad (40.3)$$

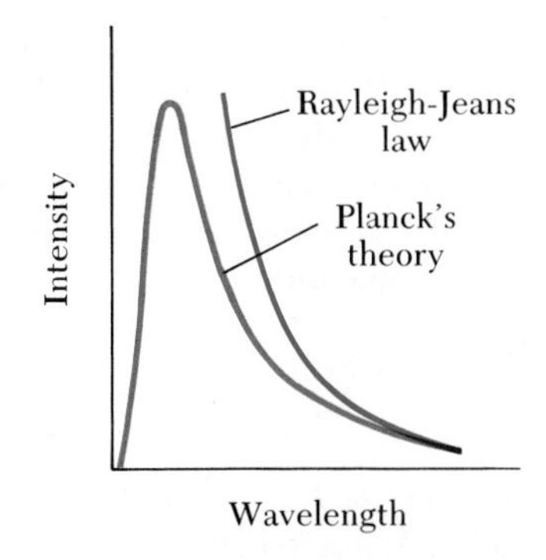


**Figure 40.3** Comparison between Planck's theory and classical theory for the distribution of blackbody radiation.

where $h$ is a constant that can be adjusted to fit the data. The current value of $h$, known as **Planck's constant,** is given by

$$h = 6.626 \times 10^{-34}\ \text{J}\cdot\text{s} \quad (40.4)$$

You should show that at long wavelengths, Planck's expression, Equation 40.3, reduces to the Rayleigh-Jeans expression given by Equation 40.2 . Furthermore, at short wavelengths, Planck's law predicts an exponential decrease in $I(\lambda, T)$ with decreasing wavelength, in agreement with experimental results.

In his theory, Planck made two bold and controversial assumptions concerning the nature of the oscillating molecules of the cavity walls:

1. The oscillating molecules that emit the radiation could have only *discrete* units of energy $E_n$ given by

Quantization of energy

$$E_n = nhf \quad (40.5)$$

   where $n$ is a positive integer called a **quantum number** and $f$ is the frequency of vibration of the molecules. The energies of the molecules are said to be *quantized,* and the allowed energy states are called *quantum states.*
2. The molecules emit or absorb energy in discrete units of light energy called **quanta** (or **photons,** as they are now called). They do so by "jumping" from one quantum state to another. If the quantum number $n$ changes by one unit, Equation 40.5 shows that the amount of energy radiated or absorbed by the molecule equals $hf$. Hence, the energy of a photon corresponding to the energy difference between two adjacent quantum states is given by

$$E = hf \quad (40.6)$$

   The molecule will radiate or absorb energy only when it changes quantum states. If it remains in one quantum state, no energy is absorbed or emitted. Figure 40.4 shows the quantized energy levels and allowed transitions proposed by Planck.

The key point in Planck's theory is the radical assumption of quantized energy states. This development marked the birth of the quantum theory. We should emphasize that Planck's work involved more than clever mathematical manipulation. In fact, Planck spent more than six years attempting to derive the blackbody distribution curve. In his own words, the emission problem "represents something absolute, and since I had always regarded the search for the absolute as the loftiest goal of all scientific activity, I eagerly set to work." This work was to occupy most of his life as he continued in his search for a physical interpretation of his formula and to reconcile the quantum concept with classical theory.

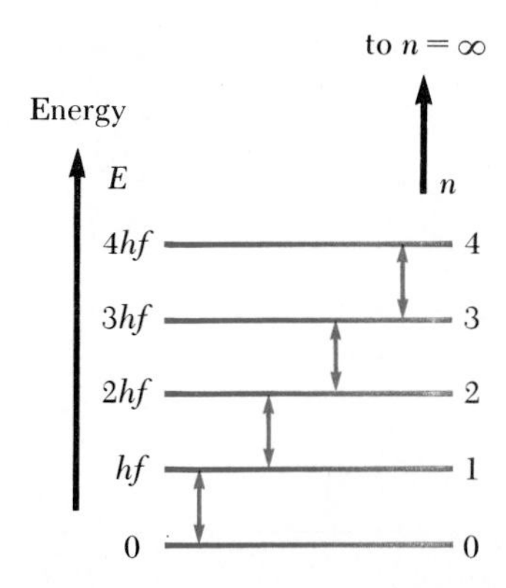


**Figure 40.4** Allowed energy levels for an oscillator of natural frequency $f$. Allowed transitions are indicated by vertical arrows.

**EXAMPLE 40.1 Thermal Radiation from the Human Body**

The temperature of the skin is approximately 35°C. What is the wavelength at which the peak occurs in the radiation emitted from the skin?

*Solution* From Wien's displacement law, we have

$$\lambda_{max} T = 0.2898 \times 10^{-2} \text{ m}\cdot\text{K}$$

Solving for $\lambda_{max}$, noting that 35°C corresponds to an absolute temperature of 308 K, we have

$$\lambda_{max} = \frac{0.2898 \times 10^{-2} \text{ m}\cdot\text{K}}{308 \text{ K}} = 9.40\ \mu\text{m}$$

This radiation is in the infrared region of the spectrum.

**EXAMPLE 40.2 The Quantized Oscillator**

A 2-kg mass is attached to a massless spring of force constant $k = 25$ N/m. The spring is stretched 0.4 m from its equilibrium position and released. (a) Find the total energy and frequency of oscillation according to classical calculations. (b) Assume that the energy is quantized and find the quantum number, $n$, for the system. (c) How much energy would be carried away in a one-quantum change?

*Solution* (a) The total energy of a simple harmonic oscillator having an amplitude $A$ is $\frac{1}{2}kA^2$. Therefore,

$$E = \tfrac{1}{2}kA^2 = \tfrac{1}{2}(25 \text{ N/m})(0.40 \text{ m})^2 = 2.0 \text{ J}$$

The frequency of oscillation is

$$f = \frac{1}{2\pi}\sqrt{\frac{k}{m}} = \frac{1}{2\pi}\sqrt{\frac{25 \text{ N/m}}{2.0 \text{ kg}}} = 0.56 \text{ Hz}$$

(b) If the energy is quantized, we have $E_n = nhf$, and from the result of (a) we have

$$E_n = nhf = n(6.626 \times 10^{-34} \text{ J}\cdot\text{s})(0.56 \text{ Hz}) = 2.0 \text{ J}$$

Therefore,

$$n = 5.4 \times 10^{33}$$

(c) The energy carried away in a one-quantum change of energy is

$$E = hf = (6.63 \times 10^{-34} \text{ J}\cdot\text{s})(0.56 \text{ Hz})$$

$$= 3.7 \times 10^{-34} \text{ J}$$

The energy carried away by a one-quantum change in energy is such a small fraction of the total energy of the oscillator that we could not expect to see such a small change in the system. Thus, even though the decrease in energy of a spring-mass system is quantized and does decrease by small quantum jumps, our senses perceive the decrease as continuous. Quantum effects become important and measurable only on the submicroscopic level of atoms and molecules.

**EXAMPLE 40.3 The Energy of a "Blue" Photon**

What is the energy carried by a quantum of light whose frequency equals $6.00 \times 10^{-14}$ Hz (blue light)?

*Solution* The energy carried by one quantum of light is given by Equation 40.6:

$$E = hf = (6.626 \times 10^{-34} \text{ J}\cdot\text{s})(6.00 \times 10^{14} \text{ Hz})$$

$$= 3.98 \times 10^{-19} \text{ J} = 2.45 \text{ eV}$$

Exercise 1 What is the wavelength of this light?
Answer 500 nm.

## 40.2 THE PHOTOELECTRIC EFFECT

In the latter part of the 19th century, experiments showed that, when light is incident on certain metallic surfaces, electrons are emitted from the surfaces. This phenomenon is known as the **photoelectric effect,** and the emitted electrons are called **photoelectrons.** The first discovery of this phenomenon was made by Hertz, who was also the first to produce the electromagnetic waves predicted by Maxwell.

Figure 40.5 is a schematic diagram of an apparatus in which the photoelectric effect can occur. An evacuated glass or quartz tube contains a metal plate, C, connected to the negative terminal of a battery. Another metal plate, A, is maintained at a positive potential by the battery. When the tube is kept in the dark, the ammeter reads zero, indicating that there is no current in the circuit. However, when monochromatic light of the appropriate wavelength shines on plate C, a current is detected by the ammeter, indicating a flow of charges across the gap between C and A. The current associated with this process arises from electrons emitted from the negative plate and collected at the positive plate.

A plot of the photoelectric current versus the potential difference, $V$, between A and C is shown in Figure 40.6 for two light intensities. Note that for large values of $V$, the current reaches a maximum value, corresponding to the case where all photoelectrons are collected at A. In addition, the current increases as the incident light intensity increases, as you might expect. Finally, when $V$ is negative, that is, when the battery in the circuit is reversed to make C positive and A negative, the photoelectrons are repelled by the negative plate A. Only those electrons having a kinetic energy greater than $eV$ will reach A, where $e$ is the charge on the electron. Furthermore, if $V$ is less than or equal to $V_s$, called the **stopping potential,** no electrons will reach A and the current will be zero. The stopping potential is *independent* of the radiation intensity. The maximum kinetic energy of the photoelectrons is related to the stopping potential through the relation

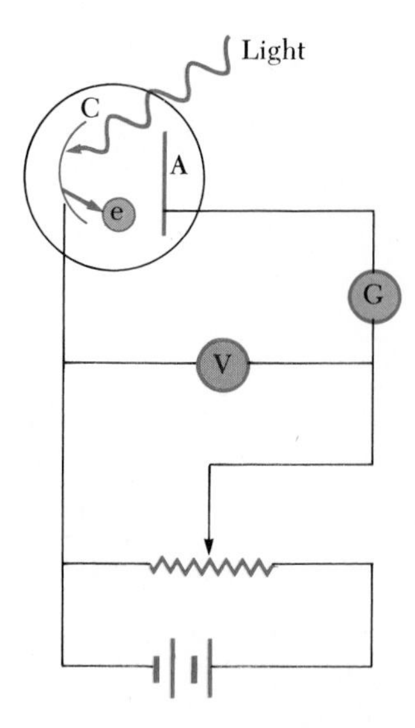


**Figure 40.5** Circuit diagram for observing the photoelectric effect. When light strikes plate C (the emitter), photoelectrons are ejected from the plate. The flow of electrons to plate A (the collector) constitutes a current in the circuit.

$$K_{max} = eV_s \tag{40.7}$$

Several features of the photoelectric effect could not be explained with classical physics or with the wave theory of light. The major observations that were not understood are as follows:

1. No electrons are emitted if the incident light frequency falls below some **cutoff frequency, $f_c$,** which is characteristic of the material being illuminated. For example, in the case of sodium, $f_c = 5.50 \times 10^{14}$ Hz. This is inconsistent with the wave theory, which predicts that the photoelectric effect should occur at any frequency, provided the light intensity is high enough.
2. If the light frequency exceeds the cutoff frequency, a photoelectric effect is observed and the number of photoelectrons emitted is proportional to the light intensity. However, the maximum kinetic energy of the photoelectrons is independent of light intensity, a fact that cannot be explained by the concepts of classical physics.
3. The maximum kinetic energy of the photoelectrons increases with increasing light frequency.
4. Electrons are emitted from the surface almost instantaneously (less than $10^{-9}$ s after the surface is illuminated), even at low light intensities. Classically, one would expect that the electrons would require some time to absorb the incident radiation before they acquired enough kinetic energy to escape from the metal.

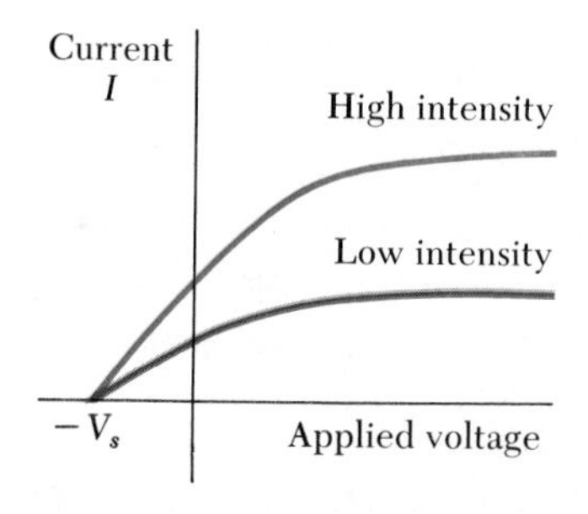


**Figure 40.6** Photoelectric current versus applied voltage for two light intensities. The current increases with intensity but reaches a saturation level for large values of $V$. At voltages equal to or less than $-V_s$, the current is zero.

A successful explanation of the photoelectric effect was given by Einstein in 1905, the same year he published his special theory of relativity. As part of a general paper on electromagnetic radiation, for which he received the Nobel prize in 1921, Einstein extended Planck's concept of quantization to electromagnetic waves. He assumed that light (or any electromagnetic wave) of frequency $f$ can be considered a stream of photons. Each photon has an energy $E$, given by

$$E = hf \tag{40.8}$$

Energy of a photon

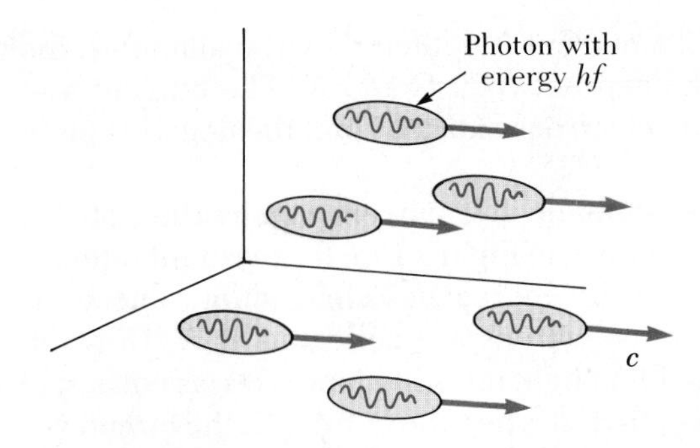


**Figure 40.7** Einstein's photon picture of "a traveling light wave."

where $h$ is Planck's constant. Einstein maintained that the energy of light is not distributed evenly over the classical wave front, but is concentrated in discrete regions (or in "bundles") called quanta or photons. A suggestive image of photons, not to be taken too literally, is shown in Figure 40.7. Einstein's simple view of the photoelectric effect was that a photon gives *all* its energy, $hf$, to a single electron in the metal. Electrons emitted from the surface of the metal possess the maximum kinetic energy, $K_{max}$. According to Einstein, the maximum kinetic energy for these liberated electrons is

Photoelectric effect equation

$$K_{max} = hf - \phi \tag{40.9}$$

where $\phi$ is called the **work function** of the metal. The work function represents the minimum energy with which an electron is bound in the metal, and is of the order of a few electron volts. Table 40.1 lists values of work functions measured for different metals.

With the photon theory of light, one can explain the features of the photoelectric effect that cannot be understood using classical concepts. These are briefly described in the order they were introduced earlier:

1. The fact that the photoelectric effect is not observed below a certain cutoff frequency follows from the fact that the energy of the photon must be greater than or equal to $\phi$. If the energy of the incoming photon is not equal to or greater than $\phi$, the electrons will never be ejected from the surface, regardless of the intensity of the light.
2. The fact that $K_{max}$ is independent of the light intensity can be understood with the following argument. If the light intensity is doubled, the number of photons is doubled, which doubles the number of photoelectrons emitted. However, their kinetic energy, which equals $hf - \phi$, depends only on the light frequency and the work function, not on the light intensity.
3. The fact that $K_{max}$ increases with increasing frequency is easily understood with Equation 40.9.
4. Finally, the fact that the electrons are emitted almost instantaneously is consistent with the particle theory of light, in which the incident energy appears in small packets and there is a one-to-one interaction between photons and electrons. This is in contrast to having the energy of the photons distributed uniformly over a large area.

**TABLE 40.1 Work Functions of Selected Metals**

| Metal | $\phi$ (eV) |
|---|---|
| Na | 2.28 |
| Al | 4.08 |
| Cu | 4.70 |
| Zn | 4.31 |
| Ag | 4.73 |
| Pt | 6.35 |
| Pb | 4.14 |
| Fe | 4.50 |

A final confirmation of Einstein's theory is a test of the prediction of a linear relationship between $f$ and $K_{max}$. Indeed, such a linear relationship is observed, as sketched in Figure 40.8. The slope of such a curve gives a value for $h$. The intercept on the horizontal axis gives the cutoff frequency, which is related to the work function through the relation $f_c = \phi/h$. This corresponds to a **cutoff wavelength** of

$$\lambda_c = \frac{c}{f_c} = \frac{c}{\phi/h} = \frac{hc}{\phi} \tag{40.10}$$

where $c$ is the speed of light ($3.00 \times 10^8$ m/s). Wavelengths *greater* than $\lambda_c$ incident on a material with a work function $\phi$ do not result in the emission of photoelectrons.

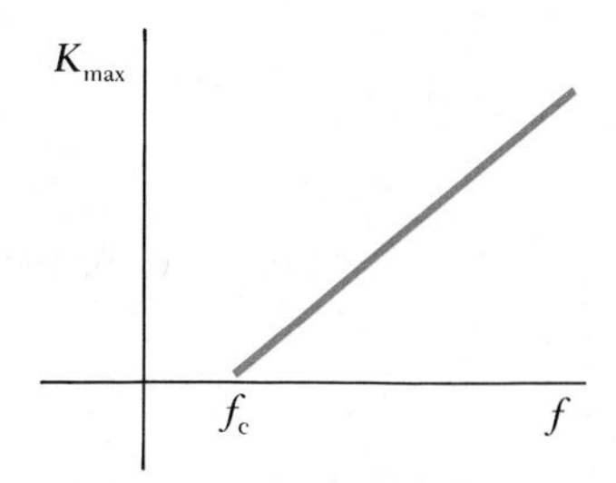


**Figure 40.8** A sketch of $K_{max}$ versus frequency of incident light for photoelectrons in a typical photoelectric effect experiment. Photons with frequency less than $f_c$ do not have sufficient energy to eject an electron from the metal.

**EXAMPLE 40.4 The Photoelectric Effect for Sodium**
A sodium surface is illuminated with light of wavelength 300 nm. The work function for sodium metal is 2.46 eV. Find (a) the kinetic energy of the ejected photoelectrons and (b) the cutoff wavelength for sodium.

*Solution* (a) The energy of the illuminating light beam is

$$E = hf = \frac{hc}{\lambda} = \frac{(6.626 \times 10^{-34}\ \text{J} \cdot \text{s})(3.00 \times 10^8\ \text{m/s})}{300 \times 10^{-9}\ \text{m}}$$

$$= 6.63 \times 10^{-19}\ \text{J} = \frac{6.63 \times 10^{-19}\ \text{J}}{1.60 \times 10^{-19}\ \text{J/eV}} = 4.14\ \text{eV}$$

where we have used the conversion of units $1\ \text{eV} = 1.60 \times 10^{-19}$ J. Using Equation 40.9 gives

$$K_{max} = hf - \phi = 4.14\ \text{eV} - 2.46\ \text{eV} = \boxed{1.68\ \text{eV}}$$

(b) The cutoff wavelength can be calculated from Equation 40.10 after we convert $\phi$ from electron volts to joules:

$$\phi = 2.46\ \text{eV} = (2.46\ \text{eV})(1.60 \times 10^{-19}\ \text{J/eV})$$
$$= 3.94 \times 10^{-19}\ \text{J}$$

Hence

$$\lambda_c = \frac{hc}{\phi} = \frac{(6.626 \times 10^{-34}\ \text{J} \cdot \text{s})(3.00 \times 10^8\ \text{m/s})}{3.94 \times 10^{-19}\ \text{J}}$$

$$= 5.05 \times 10^{-7}\ \text{m} = \boxed{505\ \text{nm}}$$

This wavelength is in the green region of the visible spectrum.

**Exercise 2** Calculate the maximum speed of the photoelectrons under the conditions described in this example.
**Answer** $7.68 \times 10^5$ m/s.

## 40.3 THE COMPTON EFFECT

In 1919, Einstein concluded that a photon of energy $E$ travels in a single direction (unlike a spherical wave) and carries a momentum equal to $E/c$ or $hf/c$. In his own words, "if a bundle of radiation causes a molecule to emit or absorb an energy packet $hf$, then momentum of quantity $hf/c$ is transferred to the molecule, directed along the line of the bundle for absorption and opposite the bundle for emission." In 1923, Arthur Holly Compton (1892–1962) and Peter Debye independently carried Einstein's idea of photon momentum farther. They realized that the scattering of x-ray photons from electrons could be explained by treating photons as point-like particles with energy $hf$ and momentum $hf/c$, and by conserving energy and momentum of the photon-electron pair in a collision.

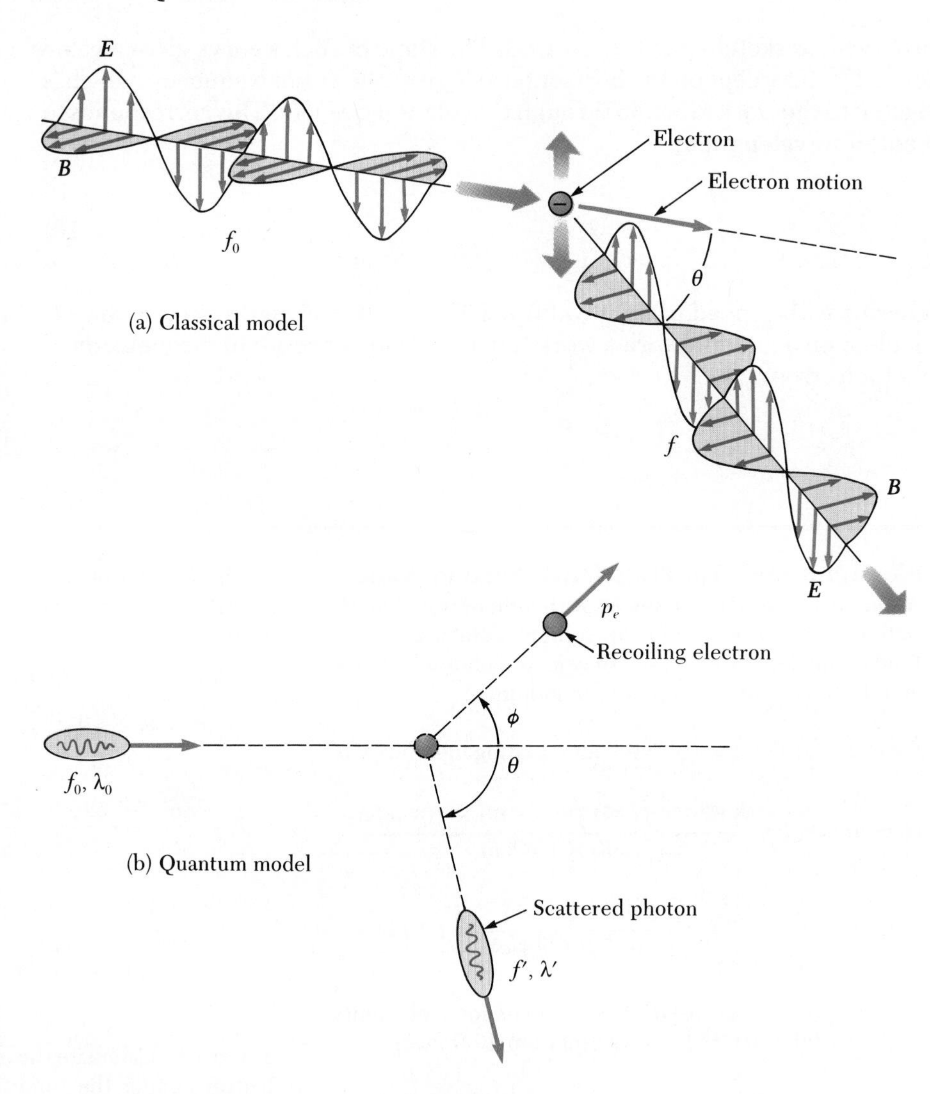


**Figure 40.9** X-ray scattering from an electron: (a) the classical model, (b) the quantum model.

Arthur Holly Compton (1892–1962), American physicist. (Courtesy of AIP Niels Bohr Library)

Prior to 1922, Compton and his coworkers had accumulated evidence that showed that classical wave theory failed to explain the scattering of x-rays from electrons. According to classical wave theory, incident electromagnetic waves of frequency $f_0$ should accelerate electrons, forcing them to oscillate and reradiate at a frequency $f < f_0$ as in Figure 40.9a. Furthermore, according to classical theory, the frequency or wavelength of the scattered radiation should depend on the time of exposure of the sample to the incident radiation as well as the intensity of the incident radiation. Contrary to these predictions, Compton's experimental results showed that the wavelength shift of x-rays scattered at a given angle depends *only on the scattering angle.* Figure 40.9b shows the quantum picture of the transfer of momentum and energy between an individual x-ray photon and an electron.

A schematic diagram of the apparatus used by Compton is shown in Figure 40.10a. In his original experiment Compton measured the dependence of scattered x-ray intensity on wavelength at three different scattering angles. The wavelength was measured with a rotating crystal spectrometer using carbon as the target, and the intensity was determined by an ioni-

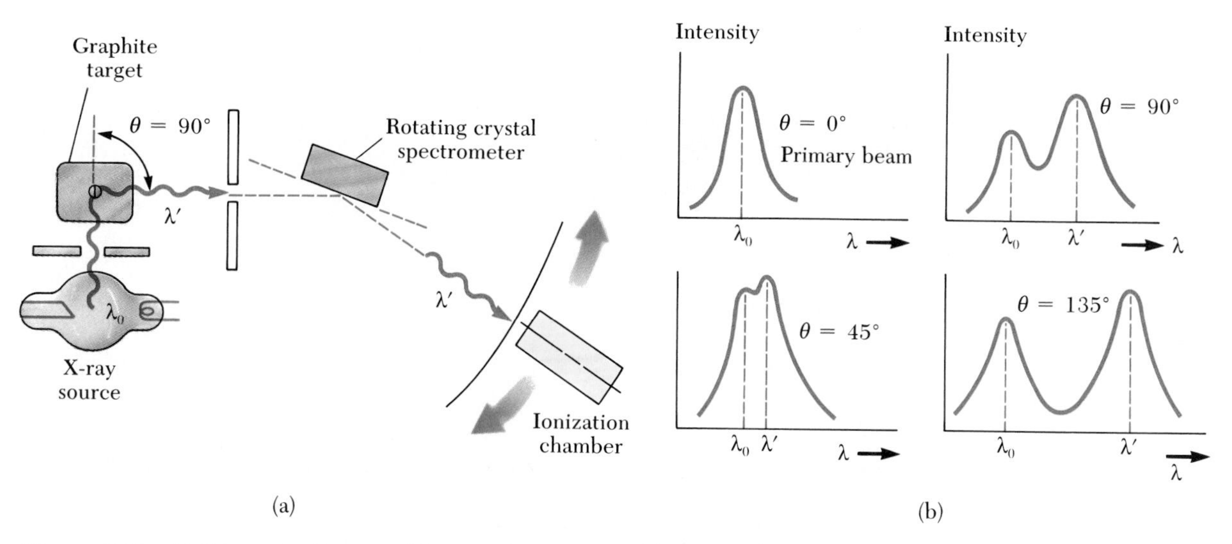


**Figure 40.10** (a) Schematic diagram of Compton's apparatus. (b) Scattered intensity versus wavelength of Compton scattering at $\theta = 0°$, $45°$, $90°$, and $135°$.

zation chamber that generated a current proportional to the x-ray intensity. The incident beam consisted of monochromatic x-rays of wavelength $\lambda_0 = 0.071$ nm. The experimental intensity versus wavelength plots observed by Compton for scattering angles of 0°, 45°, 90°, and 135° are shown in Figure 40.10b. They show two peaks, one at $\lambda_0$ and a shifted peak at $\lambda'$. The peak at $\lambda_0$ is caused by x-rays scattered from electrons that are tightly bound to the target atoms, while the shifted peak is caused by scattering of x-rays from free electrons in the target. In his analysis, Compton predicted that the shifted peak should depend on scattering angle $\theta$ as

$$\lambda' - \lambda_0 = \frac{h}{mc}(1 - \cos\theta) \qquad (40.11)$$

Compton shift equation

In this expression, known as the **Compton shift equation,** $m$ is the mass of the electron; $h/mc$ is called the **Compton wavelength** $\lambda_c$ of the electron; and has a currently accepted value of

$$\lambda_c = \frac{h}{mc} = 0.00243 \text{ nm}$$

Compton wavelength

Compton's measurements were in excellent agreement with the predictions of Equation 40.11. It is fair to say that these were the first experimental results to convince most physicists of the fundamental validity of the quantum theory!

### Derivation of the Compton Shift Equation

We can derive the Compton shift expression, Equation 40.11, by assuming that the photon exhibits particle-like behavior and collides elastically with a free electron initially at rest, as in Figure 40.11. In this model, the photon is treated as a particle of energy $E = hf = hc/\lambda$, with a rest mass of zero. In the scattering process, the total energy and total momentum of the system must be

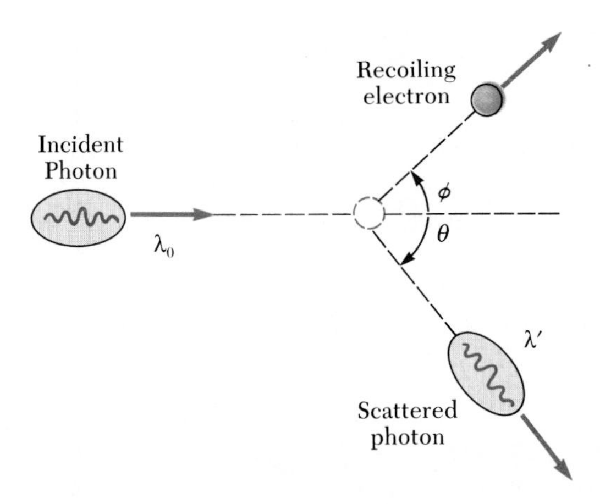


**Figure 40.11** Diagram representing Compton scattering of a photon by an electron. The scattered photon has less energy (or longer wavelength) than the incident photon.

conserved. Applying conservation of energy to this process gives

$$\frac{hc}{\lambda_0} = \frac{hc}{\lambda'} + K_e$$

where $hc/\lambda_0$ is the energy of the incident photon, $hc/\lambda'$ is the energy of the scattered photon, and $K_e$ is the kinetic energy of the recoiling electron. Since the electron may recoil at speeds comparable to the speed of light, we must use the relativistic expression for $K_e$, given by $K_e = \gamma mc^2 - mc^2$ (Equation 39.23). Therefore,

$$\frac{hc}{\lambda_0} = \frac{hc}{\lambda'} + \gamma mc^2 - mc^2 \tag{40.12}$$

where $\gamma = 1/\sqrt{1 - v^2/c^2}$.

Next, we can apply the law of conservation of momentum to this collision, noting that *both the x and y components of momentum are conserved.* Since the momentum of a photon has a magnitude given by $p = E/c = h/\lambda$, and since the relativistic expression for the momentum of the recoiling electron is $p_e = \gamma mv$ (Eq. 39.18), we obtain the following expressions for the $x$ and $y$ components of linear momentum:

$$x \text{ component:} \qquad \frac{h}{\lambda_0} = \frac{h}{\lambda'} \cos\theta + \gamma mv \cos\phi \tag{40.13}$$

$$y \text{ component:} \qquad 0 = \frac{h}{\lambda'} \sin\theta - \gamma mv \sin\phi \tag{40.14}$$

By eliminating $v$ and $\phi$ from Equations 40.12 to 40.14, we obtain a single expression that relates the remaining three variables ($\lambda'$, $\lambda_0$, and $\theta$). After some algebra (Problem 68), one obtains the Compton shift equation:

$$\Delta\lambda = \lambda' - \lambda_0 = \frac{h}{mc}(1 - \cos\theta)$$

**EXAMPLE 40.5 Compton Scattering at 45°**
X-rays of wavelength $\lambda_0 = 0.20$ nm are scattered from a block of material. The scattered x-rays are observed at an angle of 45° to the incident beam. Calculate the wavelength of the scattered x-rays at this angle.

*Solution* The shift in wavelength of the scattered x-rays is given by Equation 40.11. Taking $\theta = 45°$, we find that

$$\Delta\lambda = \frac{h}{mc}(1 - \cos\theta)$$

$$\Delta\lambda = \frac{6.626 \times 10^{-34}\ \text{J}\cdot\text{s}}{(9.11 \times 10^{-31}\ \text{kg})(3.00 \times 10^{8}\ \text{m/s})}(1 - \cos 45°)$$

$$= 7.10 \times 10^{-13}\ \text{m} = 0.000710\ \text{nm}$$

Hence, the wavelength of the scattered x-ray at this angle is

$$\lambda' = \Delta\lambda + \lambda_0 = \boxed{0.200710\ \text{nm}}$$

Exercise 3 Find the fraction of energy lost by the photon in this collision.
Answer Fraction $= \Delta\lambda/\lambda_0 = 0.00354$.

## 40.4 ATOMIC SPECTRA

As already pointed out in Section 40.1, all substances at some temperature emit thermal radiation which is characterized by a continuous distribution of wavelengths. The shape of the distribution depends on the temperature and properties of the substance. In sharp contrast to this continuous distribution spectrum is the discrete **line spectrum** emitted by a low-pressure gas subject to electric discharge. When the light from such a low-pressure gas discharge is examined with a spectroscope, it is found to consist of a few bright lines of pure color on a generally dark background. This contrasts sharply with the continuous rainbow of colors seen when a glowing solid is viewed through a spectroscope. Furthermore, as you can see from Figure 40.12a, the wavelengths

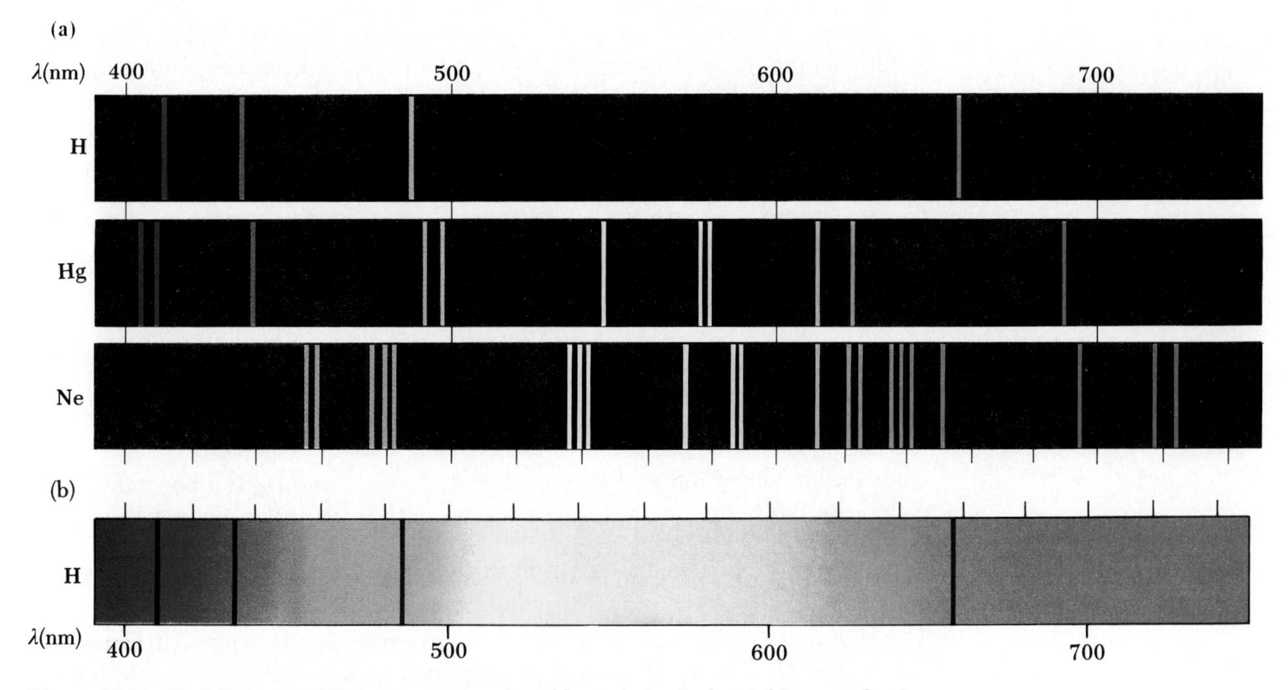


**Figure 40.12** Visible spectra (a) Line spectra produced by emission in the visible range for the elements hydrogen, helium, and neon. (b) The absorption spectrum for hydrogen. The dark absorption lines occur at the same wavelengths as the emission lines for hydrogen shown in (a). (Whitten, K.W., K.D. Gailey, and R.E. Davis, *General Chemistry*, 3rd ed., Saunders College Publishing, 1987).

contained in a given line spectrum are characteristic of the particular element emitting the light. The simplest line spectrum is observed for atomic hydrogen, and we shall describe this spectrum in detail. Other atoms such as mercury and neon emit completely different line spectra. Since no two elements emit the same line spectrum, this phenomenon represents a practical and sensitive technique for identifying the elements present in unknown samples.

Another form of spectroscopy which is very useful in analyzing substances is **absorption spectroscopy.** An absorption spectrum is obtained by passing light from a continuous source through a gas or dilute solution of the element being analyzed. The absorption spectrum consists of a series of dark lines superimposed on the otherwise continuous spectrum of that same element, as shown in Figure 40.12b for atomic hydrogen. In general, not all of the emission lines are present in the absorption spectrum.

The absorption spectrum of an element has many practical applications. For example, the continuous spectrum of radiation emitted by the sun must pass through the cooler gases of the solar atmosphere and through the earth's atmosphere. The various absorption lines observed in the solar spectrum have been used to identify elements in the solar atmosphere. In early studies of the solar spectrum, some lines were found that did not correspond to any known element. A new element had been discovered! The new element was named helium, after the Greek word for sun, *helios*. Scientists are able to examine the light from stars other than our sun in this fashion, but elements other than those present on earth have never been detected. Atomic absorption spectroscopy has also been a useful technique in analyzing heavy metal contamination of the food chain. For example, the first determination of high levels of mercury in tuna fish was made with atomic absorption.

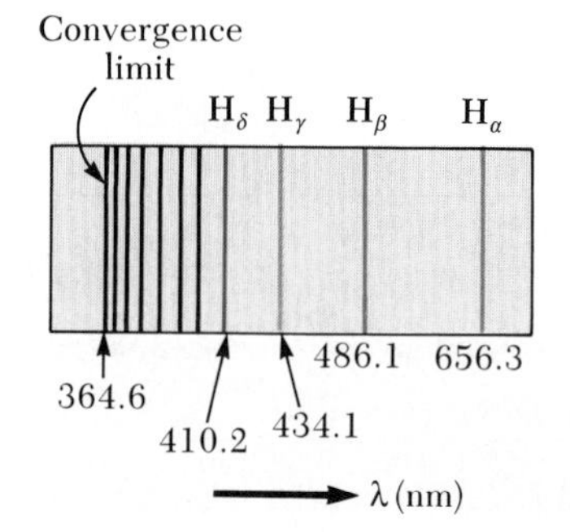


**Figure 40.13** The Balmer series of spectral lines for hydrogen (emission spectrum).

From 1860 to 1885, scientists had accumulated a great deal of data using spectroscopic measurements. In 1885, a Swiss school teacher, Johann Jacob Balmer (1825–1898) found a formula that correctly predicted the wavelengths of four visible emission lines of hydrogen: $H_\alpha$(red), $H_\beta$(green), $H_\gamma$(blue), and $H_\delta$(violet). Figure 40.13 shows these and other lines in the emission spectrum of hydrogen. The four visible lines occur at the wavelengths 656.3 nm, 486.1 nm, 434.1 nm, and 410.2 nm. The wavelengths of these lines (called the Balmer series) can be described by the empirical equation

Balmer series

$$\frac{1}{\lambda} = R_{\mathrm{H}}\left(\frac{1}{2^2} - \frac{1}{n^2}\right) \tag{40.15}$$

where $n$ may have integral values of 3, 4, 5, . . . and $R_{\mathrm{H}}$ is a constant, now called the **Rydberg constant.** If the wavelength is in meters, $R_{\mathrm{H}}$ has the value

Rydberg constant

$$R_{\mathrm{H}} = 1.0973732 \times 10^7 \text{ m}^{-1} \tag{40.16}$$

The $H_\alpha$ line in the Balmer series at 656.3 nm corresponds to $n = 3$ in Equation 40.15; the $H_\beta$ line at 486.1 nm corresponds to $n = 4$, and so on. Much to Balmer's delight, the measured spectral lines agreed with his empirical formula to within 0.1%!

Other line spectra for hydrogen were found following Balmer's discovery. These spectra are called the Lyman, Paschen, and Brackett series after their discoverers. The wavelengths of the lines in these series can be calculated by the following empirical formulas:

$$\frac{1}{\lambda} = R_{\mathrm{H}}\left(1 - \frac{1}{n^2}\right) \qquad n = 2, 3, 4, \ldots \qquad (40.17)$$

Lyman series

$$\frac{1}{\lambda} = R_{\mathrm{H}}\left(\frac{1}{3^2} - \frac{1}{n^2}\right) \qquad n = 4, 5, 6, \ldots \qquad (40.18)$$

Paschen series

$$\frac{1}{\lambda} = R_{\mathrm{H}}\left(\frac{1}{4^2} - \frac{1}{n^2}\right) \qquad n = 5, 6, 7, \ldots \qquad (40.19)$$

Brackett series

## 40.5 BOHR'S QUANTUM MODEL OF THE ATOM

At the beginning of the 20th century, scientists were perplexed by the failure of classical physics in explaining the characteristics of atomic spectra. Why did hydrogen emit only certain lines in the visible part of the spectrum? Furthermore, why did hydrogen absorb only those wavelengths which it emitted? In 1913, the Danish scientist Niels Bohr (1885–1963) provided an explanation of atomic spectra that included some features present contained in the currently accepted theory. Bohr's theory contained a combination of ideas from Planck's original quantum theory, Einstein's photon theory of light, and Rutherford's model of the atom. Bohr's model of the hydrogen atom contains some classical features as well as some revolutionary postulates that could not be justified within the framework of classical physics. The Bohr model can be applied quite successfully to such hydrogen-like ions as singly ionized helium and doubly ionized lithium. However, the theory does not properly describe the spectra of more complex atoms and ions.

The basic ideas of the Bohr theory as it applies to an atom of hydrogen are as follows:

Assumptions of the Bohr theory

1. The electron moves in circular orbits about the proton under the influence of the Coulomb force of attraction, as in Figure 40.14. So far nothing new!
2. Only certain orbits are stable. These stable orbits are ones in which the electron does not radiate. Hence the energy is fixed or stationary, and classical mechanics may be used to describe the electron's motion.
3. Radiation is emitted by the atom when the electron "jumps" from a more energic initial stationary state to a lower state. This "jump" cannot be visualized or treated classically. In particular, the frequency $f$ of the photon emitted in the jump *is independent of the frequency of the electron's orbital motion.* The frequency of the light emitted is related to the change in the atom's energy and is given by the Planck-Einstein formula

$$E_{\mathrm{i}} - E_{\mathrm{f}} = hf \qquad (40.20)$$

where $E_{\mathrm{i}}$ is the energy of the initial state, $E_{\mathrm{f}}$ is the energy of the final state, and $E_{\mathrm{i}} > E_{\mathrm{f}}$.
4. The size of the allowed electron orbits is determined by an additional quantum condition imposed on the electron's orbital angular momentum. Namely, the allowed orbits are those for which the electron's orbital angular momentum about the nucleus is an integral multiple of $\hbar = h/2\pi$,

$$mvr = n\hbar \qquad n = 1, 2, 3, \ldots \qquad (40.21)$$

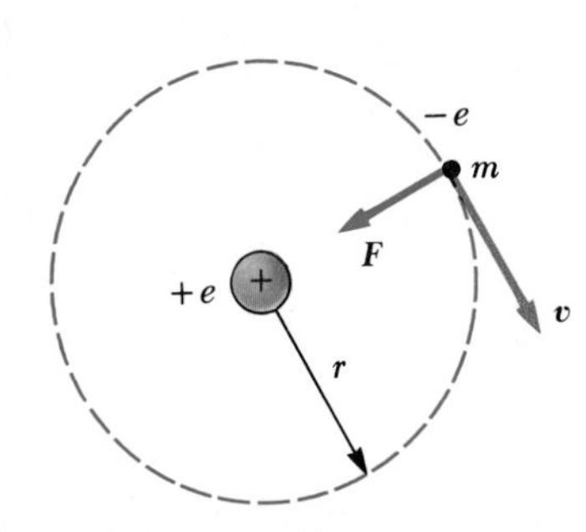


**Figure 40.14** Diagram representing Bohr's model of the hydrogen atom, in which the orbiting electron is allowed to be only in specific orbits of discrete radii.

Biographical Sketch

**Niels Bohr**
*(1885–1962)*

(Photograph courtesy of AIP Niels Bohr Library, Margarethe Bohr Collection)

**Niels Bohr, a Danish physicist, proposed the first quantum model of the atom. He was an active participant in the early development of quantum mechanics and provided much of its philosophical framework. He made many other important contributions to theoretical nuclear physics, including the development of the liquid-drop model of the nucleus and work in nuclear fission. He was awarded the 1922 Nobel prize in physics for his investigation of the structure of atoms and of the radiation emanating from them.***

**Bohr spent most of his life in Copenhagen, Denmark, and received his doctorate at the University of Copenhagen in 1911. The following year he traveled to England where he worked under J. J. Thomson at Cambridge, and then went to Manchester where we worked under Ernest Rutherford. He was married in 1912 and returned to the University of Copenhagen in 1916 as professor of physics. During the 1920s and 1930s, Bohr headed the Institute for Advanced Studies in Copenhagen, under the support of the Carlsberg brewery. (This was certainly the greatest benefit provided by beer to the field of theoretical physics.) The institute was a magnet for many of the world's best physicists and provided a forum for the exchange of ideas. Bohr, a firm believer in doing physics on a "man-to-man" basis, was always the initiator of probing questions, reflections, and discussions with his guests.**

**When Bohr visited the United States in 1939 to attend a scientific conference, he brought news that the fission of uranium had been discovered by Hahn and Strassman in Berlin. The results, confirmed by other scientists shortly thereafter, were the foundations of the atomic bomb developed in the United States during World War II. He returned to Denmark and was there during the German occupation in 1940. He escaped to Sweden in 1943 to avoid imprisonment, and helped to arrange the escape of many imperiled Danish citizens.**

**Although Bohr himself worked on the Manhattan project at Los Alamos until 1945, he strongly felt that openness between nations concerning nuclear weapons should be the first step in establishing their control. After the war, Bohr committed himself to many human issues including the development of peaceful uses of atomic energy. He was awarded the Atoms for Peace award in 1957. As John Archibald Wheeler summarizes, "Bohr was a great scientist. He was a great citizen of Denmark, of the World. He was a great human being."**

* For several interesting articles concerning Bohr, read the special Niels Bohr centennial issue of *Physics Today,* October 1985.

Using these four assumptions, we can now calculate the allowed energy levels and emission wavelengths of the hydrogen atom. Recall that the electrical potential energy of the system shown in Figure 40.14 is given by $U = qV = -ke^2/r$, where $k$ (the Coulomb constant) has the value $1/4\pi\epsilon_0$. Thus, the total energy of the atom, which contains both kinetic and potential energy terms, is given by

$$E = K + U = \tfrac{1}{2}mv^2 - k\frac{e^2}{r} \tag{40.22}$$

Applying Newton's second law to this system, we see that the Coulomb attractive force on the electron, $ke^2/r^2$, must equal the mass times the centripetal acceleration of the electron, or

$$\frac{ke^2}{r^2} = \frac{mv^2}{r}$$

From this expression, we immediately find the kinetic energy to be

$$K = \frac{mv^2}{2} = \frac{ke^2}{2r} \qquad (40.23)$$

Substituting this value of $K$ into Equation 40.22, we find that the total energy of the atom is

$$E = -\frac{ke^2}{2r} \qquad (40.24)$$

Note that the total energy is negative, indicating a bound electron-proton system. This means that energy in the amount of $ke^2/2r$ must be added to the atom to just remove the electron and make the total energy zero. An expression for $r$, the radius of the stationary orbits, may be obtained by solving Equations 40.21 and 40.23 for $v$ and equating the results:

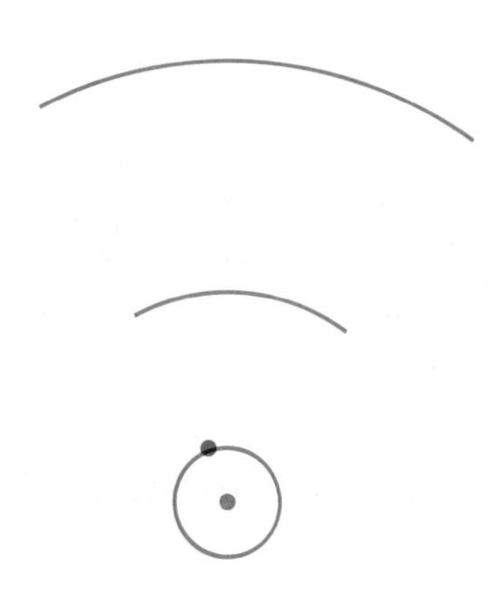

**Figure 40.15** The first three Bohr orbits for hydrogen.

**Radii of Bohr orbits in hydrogen**

$$r_n = \frac{n^2\hbar^2}{mke^2} \qquad n = 1, 2, 3, \ldots \qquad (40.25)$$

The orbit for which $n = 1$ has the smallest radius; it is called the **Bohr radius,** $a_0$, and has the value

$$a_0 = \frac{\hbar^2}{mke^2} = 0.529 \text{ Å} = 0.0529 \text{ nm} \qquad (40.26)$$

The fact that Bohr's theory gave an accurate value for the radius of hydrogen from first principles without any empirical calibration of orbit size was considered a striking triumph for this theory. The first three Bohr orbits are shown to scale in Figure 40.15.

The quantization of the orbit radii immediately leads to energy quantization. This can be seen by substituting $r_n = n^2a_0$ into Equation 40.24, giving for the allowed energy levels

$$E_n = -\frac{ke^2}{2a_0}\left(\frac{1}{n^2}\right) \qquad n = 1, 2, 3, \ldots \qquad (40.27)$$

Inserting numerical values into Equation 40.27 gives

$$E_n = -\frac{13.6}{n^2} \text{ eV} \qquad n = 1, 2, 3, \ldots \qquad (40.28)$$

The lowest stationary or non-radiating state is called the **ground state,** has $n = 1$, and has an energy $E_1 = -13.6$ eV. The next state or **first excited state** has $n = 2$, and an energy $E_2 = E_1/2^2 = -3.4$ eV. An energy level diagram showing the energies of these discrete energy states and the corresponding quantum numbers is shown in Figure 40.16. The uppermost level, corresponding to $n = \infty$ (or $r = \infty$) and $E = 0$, represents the state for which the electron is removed from the atom. The minimum energy required to ionize the atom (that is, to completely remove an electron in the ground state from the proton's influence) is called the **ionization energy.** As can be seen from Figure 40.16, the ionization energy for hydrogen based on Bohr's calculation is 13.6 eV. This constituted another major achievement for the Bohr theory, since the ionization energy for hydrogen had already been measured to be precisely 13.6 eV.

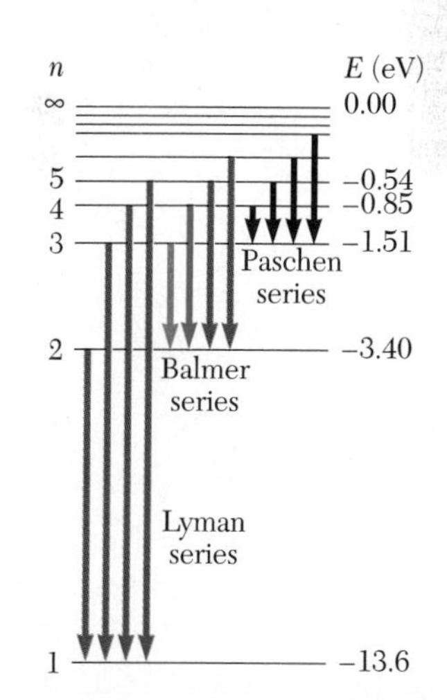


**Figure 40.16** An energy level diagram for hydrogen. In such diagrams the discrete allowed energies are plotted on the vertical axis. Nothing is plotted on the horizontal axis, but the horizontal extent of the diagram is made large enough to show allowed transitions. Note that the quantum numbers are given on the left.

Equation 40.27 together with Bohr's third postulate can be used to calculate the frequency of the photon emitted when the electron jumps from an outer orbit to an inner orbit:

$$f = \frac{E_i - E_f}{h} = \frac{ke^2}{2a_0 h}\left(\frac{1}{n_f^2} - \frac{1}{n_i^2}\right) \tag{40.29}$$

Since the quantity actually measured is wavelength, it is convenient to convert frequency to wavelength using $c = f\lambda$ to get

**Emission wavelengths of hydrogen**

$$\frac{1}{\lambda} = \frac{f}{c} = \frac{ke^2}{2a_0 hc}\left(\frac{1}{n_f^2} - \frac{1}{n_i^2}\right) \tag{40.30}$$

The remarkable fact is that the *theoretical* expression, Equation 40.30, is identical to a generalized form of the empirical relations discovered by Balmer and others, given by Equations 40.16 to 40.19,

$$\frac{1}{\lambda} = R_H\left(\frac{1}{n_f^2} - \frac{1}{n_i^2}\right) \tag{40.31}$$

provided that the combination of constants $ke^2/2a_0hc$ is equal to the experimentally determined Rydberg constant, $R_H = 1.0973732 \times 10^7\ m^{-1}$. After Bohr demonstrated the agreement of these two quantities to a precision of about 1%,it was soon recognized as the crowning achievement of his new theory of quantum mechanics. Furthermore, Bohr showed that all of the spectral series for hydrogen have a natural interpretation in his theory. These spectral series are shown as transitions between energy levels in Figure 40.16.

Bohr immediately extended his model for hydrogen to other elements in which all but one electron had been removed. Ionized elements such as $He^+$, $Li^{++}$, and $Be^{+++}$ were suspected to exist in hot stellar atmospheres, where frequent atomic collisions occurred with enough energy to completely remove one or more atomic electrons. Bohr showed that many mysterious lines observed in the sun and several stars could not be due to hydrogen, but were correctly predicted by his theory if attributed to singly ionized helium. In general, to describe a single electron orbiting a fixed nucleus of charge $+Ze$, Bohr's theory gives

$$r_n = (n^2)\frac{a_0}{Z} \tag{40.32}$$

and

$$E_n = -\frac{ke^2}{2a_0}\left(\frac{Z^2}{n^2}\right) \qquad n = 1, 2, 3, \ldots \tag{40.33}$$

**EXAMPLE 40.6 Spectral Lines from the Star $\zeta$-Puppis**
The mysterious lines observed by Pickering in 1896 in the spectrum of the star $\zeta$-Puppis fit the empirical formula

$$\frac{1}{\lambda} = R_H\left(\frac{1}{(n_f/2)^2} - \frac{1}{(n_i/2)^2}\right)$$

Show that these lines can be explained by the Bohr theory as originating from $He^+$.

*Solution* $He^+$ has $Z = 2$. Thus the allowed energy levels are given by Equation 40.33 as

$$E_n = -\frac{ke^2}{2a_0}\left(\frac{4}{n^2}\right)$$

Using $hf = E_i - E_f$ we find

$$f = \frac{E_i - E_f}{h} = \frac{ke^2}{2a_0 h}\left(\frac{4}{n_f^2} - \frac{4}{n_i^2}\right)$$

$$= \frac{ke^2}{2a_0 h}\left(\frac{1}{(n_f/2)^2} - \frac{1}{(n_i/2)^2}\right)$$

or

$$\frac{1}{\lambda} = \frac{f}{c} = \frac{ke^2}{2a_0 hc}\left(\frac{1}{(n_f/2)^2} - \frac{1}{(n_i/2)^2}\right)$$

This is the desired solution, since $R_H \equiv ke^2/2a_0hc$.

**EXAMPLE 40.7 An Electronic Transition in Hydrogen**

The electron in the hydrogen atom makes a transition from the $n = 2$ energy state to the ground state (corresponding to $n = 1$). Find the wavelength and frequency of the emitted photon.

*Solution* We can use Equation 40.31 directly to obtain $\lambda$, with $n_i = 2$ and $n_f = 1$:

$$\frac{1}{\lambda} = R_H\left(\frac{1}{n_f^2} - \frac{1}{n_i^2}\right)$$

$$\frac{1}{\lambda} = R_H\left(\frac{1}{1^2} - \frac{1}{2^2}\right) = \frac{3R_H}{4}$$

$$\lambda = \frac{4}{3R_H} = \frac{4}{3(1.097 \times 10^7\ \text{m}^{-1})}$$

$$= 1.215 \times 10^{-7}\ \text{m} = 121.5\ \text{nm}$$

This wavelength lies in the ultraviolet region.

Since $c = f\lambda$, the frequency of the photon is

$$f = \frac{c}{\lambda} = \frac{3.00 \times 10^8\ \text{m/s}}{1.215 \times 10^{-7}\ \text{m}} = 2.47 \times 10^{15}\ \text{Hz}$$

**Exercise 4** What is the wavelength of the photon emitted by hydrogen when the electron makes a transition from the $n = 3$ state to the $n = 1$ state?

**Answer** $\frac{9}{8R_H} = 102.6$ nm.

**EXAMPLE 40.8 The Balmer Series for Hydrogen**

The Balmer series for the hydrogen atom corresponds to electronic transitions that terminate in the state of quantum number $n = 2$, as shown in Figure 40.16. (a) Find the longest-wavelength photon emitted and determine its energy.

*Solution* The longest-wavelength photon in the Balmer series results from the transition from $n = 3$ to $n = 2$. Using Equation 40.31 gives

$$\frac{1}{\lambda} = R_H\left(\frac{1}{n_f^2} - \frac{1}{n_i^2}\right)$$

$$\frac{1}{\lambda_{max}} = R_H\left(\frac{1}{2^2} - \frac{1}{3^2}\right) = \frac{5}{36}R_H$$

$$\lambda_{max} = \frac{36}{5R_H} = \frac{36}{5(1.097 \times 10^7\ \text{m}^{-1})} = 656.3\ \text{nm}$$

This wavelength is in the red region of the visible spectrum.

The energy of this photon is

$$E_{photon} = hf = \frac{hc}{\lambda_{max}}$$

$$= \frac{(6.626 \times 10^{-34}\ \text{J}\cdot\text{s})(3.00 \times 10^8\ \text{m/s})}{656.3 \times 10^{-9}\ \text{m}}$$

$$= 3.03 \times 10^{-19}\ \text{J} = 1.89\ \text{eV}$$

We could also obtain the energy of the photon by using the expression $hf = E_3 - E_2$, where $E_2$ and $E_3$ are the energy levels of the hydrogen atom, which can be calculated from Equation 40.28. Note that this is the lowest-energy photon in this series since it involves the smallest energy change.

(b) Find the shortest-wavelength photon emitted in the Balmer series.

*Solution* The shortest-wavelength photon in the Balmer series is emitted when the electron makes a transition from $n = \infty$ to $n = 2$. Therefore

$$\frac{1}{\lambda_{min}} = R_H\left(\frac{1}{2^2} - \frac{1}{\infty}\right) = \frac{R_H}{4}$$

$$\lambda_{min} = \frac{4}{R_H} = \frac{4}{1.097 \times 10^7\ \text{m}^{-1}} = 364.6\ \text{nm}$$

This wavelength is in the ultraviolet region and corresponds to the series limit.

**Exercise 5** Find the energy of the shortest-wavelength photon emitted in the Balmer series for hydrogen.

**Answer** 3.40 eV.

Although the theoretical derivation of the line spectrum was a remarkable feat in itself, the scope, breadth, and impact of Bohr's monumental achievement is truly seen only when it is realized what else he treated in his three-part paper of 1913:

(a) He explained the limited number of lines seen in the absorption spectrum of hydrogen compared to the emission spectrum.
(b) He explained the emission of x-rays from atoms.
(c) He explained the chemical properties of atoms in terms of the electron shell model.
(d) He explained how atoms associate to form molecules.

### Bohr's Correspondence Principle

In our study of relativity, we found that newtonian mechanics cannot be used to describe phenomena that occur at speeds approaching the speed of light. Newtonian mechanics is a special case of relativistic mechanics and is usable only when $v$ is much less than $c$. Similarly, *quantum mechanics is in agreement with classical physics when the quantum numbers are very large.* This principle, first set forth by Bohr, is called the **correspondence principle.**

For example, consider an electron orbiting the hydrogen atom with $n >$ 10 000. For such large values of $n$, the energy differences between adjacent levels approach zero and the levels are nearly continuous. Consequently, the classical model is reasonably accurate in describing the system for large values of $n$. According to the classical picture, the frequency of the light emitted by the atom is equal to the frequency of revolution of the electron in its orbit about the nucleus. Calculations show that for $n >$ 10 000, this frequency is different from that predicted by quantum mechanics by less than 0.015%.

## SUMMARY

The characteristics of *blackbody radiation* cannot be explained using classical concepts. Planck first introduced the *quantum concept* when he assumed that the atomic oscillators responsible for this radiation existed only in discrete states.

The **photoelectric effect** is a process whereby electrons are ejected from a metallic surface when light is incident on that surface. Einstein provided a successful explanation of this effect by extending Planck's quantum hypothesis to the electromagnetic field. In this model, light is viewed as a stream of particles called *photons,* each with energy $E = hf$, where $f$ is the frequency and $h$ is Planck's constant. The kinetic energy of the ejected photoelectron is given by

Photoelectric effect equation

$$K_{\text{max}} = hf - \phi \tag{40.9}$$

where $\phi$ is the work function of the metal.

X-rays from an incident beam are scattered at various angles by electrons in a target such as carbon. In such a scattering event, a shift in wavelength is observed for the scattered x-rays, and the phenomenon is known as the **Compton effect.** Classical physics does not explain this effect. If the x-ray is treated as a photon, conservation of energy and momentum applied to the photon-electron collisions yields the following expression for the Compton shift:

Compton shift equation

$$\Delta\lambda = \frac{h}{mc}(1 - \cos\theta) \tag{40.11}$$

where $m$ is the mass of the electron, $c$ is the speed of light, and $\theta$ is the scattering angle.

The Bohr model of the atom is successful in describing the spectra of atomic hydrogen and hydrogen-like ions. One of the basic assumptions of the model is that the electron can exist only in discrete orbits such that the angular momentum $mvr$ is an integral multiple of $h/2\pi = \hbar$. Assuming circular orbits and a simple coulombic attraction between the electron and proton, the energies of the quantum states for hydrogen are calculated to be

$$E_n = -\frac{ke^2}{2a_0}\left(\frac{1}{n^2}\right) \quad (40.27)$$

Allowed energies of the hydrogen atom

where $k$ is the Coulomb constant, $e$ is electronic charge, $n$ is an integer called a *quantum number*, and $a_0 = 0.0529$ nm is the **Bohr radius.**

If the electron in the hydrogen atom makes a transition from an orbit whose quantum number is $n_i$ to one whose quantum number is $n_f$, where $n_f < n_i$, a photon is emitted by the atom whose frequency is given by

$$f = \frac{ke^2}{2a_0 h}\left(\frac{1}{n_f^2} - \frac{1}{n_i^2}\right) \quad (40.29)$$

Frequency of a photon emitted from hydrogen

Using $E = hf = hc/\lambda$, one can calculate the wavelengths of the photons for various transitions in which there is a change in quantum number, $n_i \rightarrow n_f$. The calculated wavelengths are in excellent agreement with observed line spectra.

## QUESTIONS

1. What assumptions were made by Planck in dealing with the problem of blackbody radiation? Discuss the consequences of these assumptions.
2. The classical model of blackbody radiation given by the Rayleigh-Jeans law has two major flaws. Identify them and explain how Planck's law deals with these two issues.
3. What is the difference between a "regular" electron and a photoelectron?
4. If the photoelectric effect is observed for one metal, can you conclude that the effect will also be observed for another metal under the same conditions? Explain.
5. In the photoelectric effect, explain why the photocurrent depends on the intensity of the light source but not on the frequency.
6. In the photoelectric effect, explain why the stopping potential depends on the frequency of light but not on the intensity.
7. Explain why the results of the photoelectric effect experiment may vary if an incandescent light with filters is used instead of using gas tubes.
8. Suppose the photoelectric effect occurs in a gaseous target rather than a solid. Will photoelectrons be produced at *all* frequencies of the incident photon? Explain.
9. How does the Compton effect differ from the photoelectric effect?
10. What assumptions were made by Compton in dealing with the scattering of a photon from an electron?
11. The Bohr theory of the hydrogen atom is based upon several assumptions. Discuss these assumptions and their significance. Do any of these assumptions contradict classical physics?
12. Suppose that the electron in the hydrogen atom obeyed classical mechanics rather than quantum mechanics. Why should such a "hypothetical" atom emit a continuous spectrum rather than the observed line spectrum?
13. Can the electron in the ground state of hydrogen absorb a photon of energy (a) *less* than 13.6 eV and (b) *greater* than 13.6 eV?
14. By studying the relative intensities of various spectral lines emitted by a gas, the temperature, pressure, and density of the gas can be determined. Explain how these three variables might affect the various shells the electrons could occupy about the nucleus of the atom.
15. Why aren't the relative intensities of the various spectral lines in a gas all the same?

16. Why would the spectral lines of the diatomic hydrogen be different from those of monatomic hydrogen?
17. Explain the significance behind the fact that the total energy of the atom in the Bohr model is negative.
18. What are the limitations on the Bohr model for the hydrogen atom?
19. An x-ray photon is scattered by an electron. What happens to the frequency of the scattered photon relative to that of the incident photon?
20. Why does the existence of a cutoff frequency in the photoelectric effect favor a particle theory for light rather than a wave theory?
21. Using Wien's law, calculate the wavelength of highest intensity given off by a human body. Using this information, explain why an infrared detector would be a useful alarm for security work.
22. All objects radiate energy. Why, then, are we not able to see all objects in a dark room?
23. Which has more energy, a photon of ultraviolet radiation or a photon of yellow light?
24. What effect, if any, would you expect the temperature of a material to have on the ease with which electrons can be ejected from it in the photoelectric effect?
25. Some stars are observed to be reddish, and some are blue. Which stars have the higher surface temperature? Explain.

## PROBLEMS

### Section 40.1 Blackbody Radiation and Planck's Hypothesis

1. Calculate the energy of a photon whose frequency is (a) $6.2 \times 10^{14}$ Hz, (b) 3.1 GHz, (c) 46 MHz. Express your answers in eV.
2. Determine the corresponding wavelengths for the photons described in Problem 1.
3. An FM radio transmitter has a power output of 150 kW and operates at a frequency of 99.7 MHz. How many photons per second does the transmitter emit?
4. The average power generated by the sun is equal to $3.74 \times 10^{26}$ W. Assuming the average wavelength of the sun's radiation to be 500 nm, find the number of photons emitted by the sun in 1 s.
5. Consider the mass-spring system described in Example 40.2. If the quantum number $n$ changes by unity, calculate the *fractional* change in energy of the oscillator.
6. A sodium-vapor lamp has a power output of 10 W. Using 589.3 nm as the average wavelength of the source, calculate the number of photons emitted per second.
7. Using Wien's displacement law, calculate the surface temperature of a red giant star that radiates with a peak wavelength of $\lambda_{max} = 650$ nm.
8. The radius of our sun is $6.96 \times 10^8$ m, and its total power output is $3.77 \times 10^{26}$ W. (a) Assuming that the sun's surface emits as an ideal black body, calculate its surface temperature. (b) Using the result of part (a), find the wavelength at the maximum of the spectral distribution of radiation from the sun.
9. What is the peak wavelength emitted by the human body? Assume a body temperature of 98.6°F and use the Wien displacement law. In what part of the electromagnetic spectrum does this wavelength lie?
10. A tungsten filament is heated to a temperature of 800°C. What is the wavelength of the most intense radiation?
11. The human eye is most sensitive to light with a wavelength $\lambda = 560$ nm. What temperature blackbody would radiate most intensely at this wavelength?
12. Show that at *long* wavelengths, Planck's radiation law (Eq. 40.3) reduces to the Rayleigh-Jeans law (Eq. 40.2).
13. Show that at *short* wavelengths or *low* temperatures, Planck's radiation law (Eq. 40.3) predicts an exponential decrease in $I(\lambda, T)$ given by *Wien's radiation law:*

$$I(\lambda, T) = \frac{2\pi hc^2}{\lambda^5} e^{-hc/\lambda kT}$$

### Section 40.2 The Photoelectric Effect

14. The photocurrent of a photocell is stopped by a retarding potential of 0.54 V for radiation of wavelength 750 nm. Find the work function for the material.
15. The work function for potassium is 2.24 eV. If potassium metal is illuminated with light of wavelength 480 nm, find (a) the maximum kinetic energy of the photoelectrons and (b) the cutoff wavelength.
16. Molybdenum has a work function of 4.2 eV. (a) Find the cutoff wavelength and threshold frequency for the photoelectric effect. (b) Calculate the stopping potential if the incident light has a wavelength of 180 nm.
17. When cesium metal is illuminated with light of wavelength 500 nm, the photoelectrons emitted have a maximum kinetic energy of 0.57 eV. Find (a) the work function of cesium and (b) the stopping potential if the incident light has a wavelength of 600 nm.
18. Lithium has a work function $\phi = 2.3$ eV. (a) Calculate the cutoff frequency for lithium. (b) Calculate the

stopping potential for 500-nm light falling on a clean lithium surface.

**19.** Two light sources are used in a photoelectric experiment to determine the work function for a particular metal surface. When the green light from a mercury lamp ($\lambda = 546.1$ nm) is used, a retarding potential of 1.70 V reduces the photocurrent to zero. Based on this measurement, what is the work function for this metal? (b) What stopping potential would be observed when using the yellow light from a helium discharge tube ($\lambda = 587.5$ nm)?

**20.** Electrons are ejected from a metallic surface with speeds ranging up to $4.6 \times 10^5$ m/s when light with a wavelength $\lambda = 625$ nm is used. (a) What is the work function of the surface? (b) What is the cutoff frequency for this surface?

**21.** Consider the metals lithium, beryllium, and mercury, which have work functions of 2.3 eV, 3.9 eV, and 4.5 eV, respectively. If light of wavelength 400 nm is incident on each of these metals, determine (a) which metals exhibit the photoelectric effect and (b) the maximum kinetic energy for the photoelectron in each case.

**22.** Light of wavelength 300 nm is incident on a metallic surface. If the stopping potential for the photoelectric effect is 1.2 V, find (a) the maximum energy of the emitted electrons, (b) the work function, and (c) the cutoff wavelength.

**23.** The active material in a photocell has a work function of 3.1 eV. Under reverse-bias conditions, the cutoff wavelength is found to be 270 nm. What is the value of the bias voltage?

**24.** Ultraviolet light is incident normally on the surface of a certain substance. The work function of the electrons in this substance is 3.44 eV. The incident light has an intensity of 0.055 W/m$^2$. The electrons are photoelectrically emitted with a maximum speed of $4.2 \times 10^5$ m/s. How fast can electrons be emitted from a square centimeter of the surface? Pretend that none of the photons are reflected or heat the surface.

### Section 40.3 The Compton Effect

**25.** Calculate the energy and momentum of a photon of wavelength 700 nm.

**26.** X-rays of wavelength 0.200 nm are scattered from a block of carbon. If the scattered radiation is detected at 60° to the incident beam, find (a) the Compton shift $\Delta\lambda$ and (b) the kinetic energy imparted to the recoiling electron.

**27.** What scattering angle will shift the wavelength of a 1-nm x-ray by 0.02%?

**28.** A gamma-ray photon with an energy equal to the rest energy of an electron (511 keV) collides with an electron that is initially at rest. Calculate the kinetic energy acquired by the electron if the photon is scattered 30° from its original line of approach.

**29.** A 0.03-nm x-ray photon is scattered from a free electron. (a) If the *shift* in the x-ray wavelength equals the Compton wavelength of the electron, what is the electron's kinetic energy after the collision? (b) What is its speed?

**30.** X-rays with an energy of 300 keV undergo Compton scattering from a target. If the scattered rays are detected at 37° relative to the incident rays, find (a) the Compton shift at this angle, (b) the energy of the scattered x-ray, and (c) the energy of the recoiling electron.

**31.** After a 0.80-nm x-ray photon scatters from a free electron, the electron recoils with a speed equal to $1.4 \times 10^6$ m/s. (a) What was the Compton shift in the photon's wavelength? (b) Through what angle was the photon scattered?

**32.** A metal target is placed in a beam of 662-keV gamma rays emitted by a radioactive isotope of cesium ($^{137}$Cs). Find the energy of those photons that are scattered through an angle of 90°. The electrons in the target may be considered as free electrons.

**33.** X-rays with a wavelength of 0.12 nm undergo Compton scattering. (a) Find the wavelength of photons scattered at angles of 30, 60, 90, 120, 150, and 180°. (b) Find the energy of the scattered electrons corresponding to these scattered x-rays. (c) Which one of the given scattering angles provides the electron with the greatest energy?

**34.** A 0.5-nm x-ray photon is deflected through a 134° angle in a Compton scattering event from a free electron. At what angle (with respect to the incident beam) is the recoiling electron found?

**35.** A 0.0016-nm photon scatters from a free electron. For what (photon) scattering angle will the recoiling electron and scattered photon have the same kinetic energy?

### Section 40.4 Atomic Spectra

**36.** Calculate the wavelengths of the first three lines in the Paschen series for hydrogen using Equation 40.18.

**37.** Calculate the wavelengths of the first three lines in the Lyman series for hydrogen using Equation 40.17.

**38.** (a) Compute the shortest wavelength in each of these hydrogen spectral series: Lyman, Balmer, Paschen, and Brackett. (b) Compute the energy (in eV) of the highest-energy photon produced in each of these series.

**39.** (a) What value of $n$ is associated with the Lyman series line in hydrogen whose wavelength is 94.96 nm? (b) Could this wavelength be associated with the Paschen or Brackett series?

**40.** Liquid oxygen has a bluish color. This means that the liquid preferentially absorbs light with wavelengths toward the red end of the visible spectrum. Although the oxygen molecule ($O_2$) does not strongly absorb radiation in the visible spectrum, it does absorb strongly at a wavelength of 1269 nm, which is in the infrared region of the spectrum. Research has shown that it is possible for *two* colliding $O_2$ molecules to absorb a *single* photon, sharing its energy equally. The transition that both molecules undergo is the same transition that results when they absorb radiation with a wavelength of 1269 nm. What is the wavelength of the single photon that causes this double transition? What is the corresponding color of this radiation?

### Section 40.5 Bohr's Quantum Model of the Atom

**41.** Use Equation 40.25 to calculate the radius of the first, second, and third Bohr orbits of hydrogen.

**42.** For a hydrogen atom in its ground state, use the Bohr model to compute (a) the orbital velocity of the electron, (b) the kinetic energy (in eV) of the electron, and (c) the electrical potential energy (in eV) of the atom.

**43.** (a) Construct an energy level diagram for the $He^+$ ion, for which $Z = 2$. (b) What is the ionization energy for $He^+$?

**44.** Construct an energy level diagram for the $Li^{2+}$ ion, for which $Z = 3$.

**45.** What is the radius of the first Bohr orbit in (a) $He^+$, (b) $Li^{2+}$, and (c) $Be^{3+}$?

**46.** Two hydrogen atoms collide head-on and end up with zero kinetic energy. Each then emits a photon with a wavelength of 121.6 nm ($n = 2$ to $n = 1$ transition). At what speed were the atoms moving before the collision?

**47.** A photon is emitted from a hydrogen atom which undergoes a transition from the state $n = 6$ to the state $n = 2$. Calculate (a) the energy, (b) the wavelength, and (c) the frequency of the emitted photon.

**48.** What is the energy of the photon that could cause (a) an electronic transition from the $n = 3$ state to the $n = 5$ state and (b) an electronic transition from the $n = 5$ state to the $n = 7$ state?

**49.** Four possible transitions for a hydrogen atom are listed below.

(A) $n_i = 2;\ n_f = 5$

(B) $n_i = 5;\ n_f = 3$

(C) $n_i = 7;\ n_f = 4$

(D) $n_i = 4;\ n_f = 7$

(a) Which transition will emit the shortest wavelength photon? (b) For which transition will the atom gain the most energy? (c) For which transition(s) does the atom lose energy?

**50.** (a) Using Equation 40.19 calculate the longest and shortest wavelengths for the Brackett series. (b) Determine the photon energies corresponding to these wavelengths.

**51.** Find the potential energy and kinetic energy of an electron in the first excited state of the hydrogen atom.

### ADDITIONAL PROBLEMS

**52.** A hydrogen atom is in its first excited state ($n = 2$). Using the Bohr theory of the atom, calculate (a) the radius of the orbit, (b) the linear momentum of the electron, (c) the angular momentum of the electron, (d) the kinetic energy, (e) the potential energy, and (f) the total energy.

**53.** *Positronium* is a hydrogen-like atom consisting of a positron (a positively charged electron) and an electron revolving around each other. Using the Bohr model, find the allowed radii (relative to the center of mass of the two particles) and the allowed energies of the system.

**54.** Figure 40.17 shows the stopping potential versus incident photon frequency for the photoelectric effect for sodium. Use these data points to find (a) the work function, (b) the ratio $h/e$, and (c) the cutoff wavelength. (Data taken from R. A. Millikan, *Phys. Rev.* 7:362 [1916].)

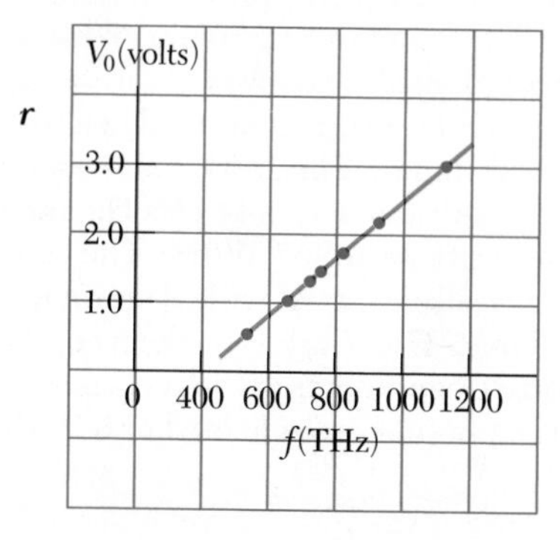


**Figure 40.17** (Problem 54).

**55.** *The Auger process.* An electron in chromium makes a transition from the $n = 2$ state to the $n = 1$ state without emitting a photon. Instead, the excess energy is transferred to an outer electron (in the $n = 4$ state), which is ejected by the atom. (This is called an *Auger process,* and the ejected electron is referred to as an *Auger electron.*) Use the Bohr theory to find the kinetic energy of the Auger electron.

**56.** Photons of wavelength 450 nm are incident on a metal. The most energetic electrons ejected from the metal are bent into a circular arc of radius 20 cm by a magnetic field whose strength is equal to $2 \times 10^{-5}$ T. What is the work function of the metal?

57. Gamma rays (high-energy photons) of energy 1.02 MeV are scattered from electrons that are initially at rest. If the scattering is *symmetric*, that is, if $\theta = \phi$, find (a) the scattering angle $\theta$ and (b) the energy of the scattered photons.
58. A 200-MeV photon is scattered at 40° by a free proton initially at rest. (a) Find the energy (in MeV) of the scattered photon. (b) What kinetic energy (in MeV) does the proton acquire?
59. A *muon* is a particle with a charge of $-e$ and a mass equal to 207 times the mass of an electron. Muonic lead is formed when a $^{208}$Pb nucleus captures a muon. According to the Bohr theory, what are the radius and energy of the ground state of muonic lead?
60. Use Bohr's model of the hydrogen atom to show that when the atom makes a transition from the state $n$ to the state $n-1$, the frequency of the emitted light is given by

$$f = \frac{2\pi^2 mk^2e^4}{h^3}\left[\frac{2n-1}{(n-1)^2n^2}\right]$$

Show that as $n \rightarrow \infty$, the expression above varies as $1/n^3$ and reduces to the classical frequency one would expect the atom to emit. (*Hint:* To calculate the classical frequency, note that the frequency of revolution is $v/2\pi r$, where $r$ is given by Eq. 40.25.) This is an example of the correspondence principle, which requires that the classical and quantum models agree for large values of $n$.

61. A muon (Problem 59) is captured by a deuteron to form a muonic atom. (a) Find the energy of the ground state and the first excited state. (b) What is the wavelength of the photon emitted when the atom makes a transition from the first excited state to the ground state?
62. Show that a photon cannot transfer all of its energy to a free electron. (*Hint:* Note that energy and momentum must be conserved.)
63. A photon of initial energy 0.1 MeV undergoes Compton scattering at an angle of 60°. Find (a) the energy of the scattered photon, (b) the recoil energy of the electron, and (c) the recoil angle of the electron.
64. The total power per unit area radiated by a blackbody at a temperature $T$ is given by the area under the $I(\lambda, T)$ versus $\lambda$ curve, as in Figure 40.2. (a) Show that this power per unit area is given by

$$\int_0^\infty I(\lambda, T)\, d\lambda = \sigma T^4$$

where $I(\lambda, T)$ is given by Planck's radiation law and $\sigma$ is a constant independent of $T$. This result is known as the *Stefan-Boltzmann law* (see Eq. 20.11). To carry out the integration, you should make the change of variable $x = hc/\lambda kT$ and use the fact that

$$\int_0^\infty \frac{x^3\, dx}{e^x - 1} = \frac{\pi^4}{15}$$

(b) Show that the Stefan-Boltzmann constant $\sigma$ has the value

$$\sigma = \frac{2\pi^5 k^4}{15c^2h^3} = 5.7 \times 10^{-8}\ \mathrm{W/m^2\cdot K^4}$$

65. Experiments indicate that a dark-adapted human eye can detect a single photon of visible light. Consider a point source that emits 2 W of light of wavelength 555 nm in all directions. How far away would this source have to be for, on the average, one photon per second to enter an eye whose pupil diameter is 6 mm?
66. A photon of initial energy $E_0$ undergoes a Compton scattering at an angle $\theta$ by a free electron (mass $m$) initially at rest. Using relativistic equations for energy and momentum conservation, derive the following relation for the final energy $E$ of the scattered photon: $E = E_0[1 - (E_0/mc^2)(1 - \cos\theta)]^{-1}$.
67. Show that the ratio of the Compton wavelength $\lambda_c$ to the de Broglie wavelength $\lambda$ for a *relativistic* electron is given by

$$\frac{\lambda_c}{\lambda} = \left[\left(\frac{E}{mc^2}\right)^2 - 1\right]^{1/2}$$

where $E$ is the total energy of the electron and $m$ is its mass.

68. Derive the formula for the Compton shift (Eq. 40.11) from Equations 40.12, 40.13, and 40.14.
69. An electron initially at rest recoils from a head-on collision with a photon. Show that the kinetic energy acquired by the electron is given by $2hf\alpha/(1+2\alpha)$, where $\alpha$ is the ratio of the photon's initial energy to the rest energy of the electron.

Δ70. The table below shows data obtained in a photoelectric experiment. (a) Using these data, make a graph similar to Figure 40.8 that plots as a straight line. From the graph, determine (b) an experimental value for Planck's constant (in joules second) and (c) the work function (in electron volts) for the surface. (Two significant figures for each answer are sufficient.)

| **Wavelength (nm)** | **Maximum Kinetic Energy of Photoelectrons (eV)** |
|---|---|
| 588 | 0.67 |
| 505 | 0.98 |
| 445 | 1.35 |
| 399 | 1.63 |

# 41

# Quantum Mechanics

*Viewing crystal surfaces. The surface of $TaSe_2$, "viewed" with a scanning tunneling microscope. The photograph is actually a charge density wave contour of the surface, where the various colors indicate regions of different charge densities. (Courtesy of Prof. R. V. Coleman, University of Virginia)*

In Chapter 40, we introduced the concept of quantization and the underlying ideas of quantum theory. In addition, we discussed Bohr's model of the hydrogen atom and emphasized that the model is an oversimplification of how one views the atom. Although the notion of quantization as assumed by Bohr turns out to be correct, his model has severe limitations. For example, the Bohr model does not explain the spectra of complex atoms, nor does it predict such details as variations in spectral line intensities and the splittings observed in certain spectral lines under controlled laboratory conditions. Finally, it does not enable us to understand how atoms interact with each other and how such interactions affect the observed physical and chemical properties of matter.

In this chapter, we shall discuss the approach to atomic phenomena called *quantum mechanics* or *wave mechanics*. This scheme, developed from 1925 to 1926 by Schrödinger, Heisenberg, and others, makes it possible to understand a host of phenomena involving atoms, molecules, nuclei, and solids. We shall begin by describing the idea of wave-particle duality. For example, a particle such as an electron can be viewed as having wave-like properties, and its wavelength can be calculated if its momentum is known. Next, we shall describe some of the basic features of the formalism of quantum mechanics and its application to simple, one-dimensional systems. For example, we shall treat the problem of a particle confined to a potential well with infinitely high barriers, the so-called particle in a box. In Chapter 42, we shall show how one can use such simple models to understand some of the characteristics of atomic structure. The chapter concludes with an essay on the scanning tunneling microscope, or STM, a remarkable device that uses quantum mechanical tunneling to make images of surfaces with resolution comparable to atomic dimensions.

## 41.1 PHOTONS AND ELECTROMAGNETIC WAVES

An explanation of phenomena such as the photoelectric effect and the Compton effect (Chapter 40) presents very convincing evidence in support of the photon (or particle) concept of light. These phenomena offer ironclad evidence that when light interacts with matter it behaves as if it were composed of particles with energy $hf$ and momentum $h/\lambda$. An obvious question that arises at this point is, "How can light be considered a photon when it exhibits wave-like properties?" On the one hand, we describe light in terms of photons having energy and momentum. On the other hand, we must also recognize that light and other electromagnetic waves exhibit interference and diffraction effects, which are consistent only with a wave interpretation. Which model is correct? Is light a wave or a particle? The answer depends on the specific phenomenon being observed. Some experiments can be better, or solely, explained on the basis of the photon concept, whereas others are best described, or can be described only, with a wave model. The end result is that *we must accept both models and admit that the true nature of light is not describable in terms of a single classical picture.* However, you should recognize that the same light beam that can eject photoelectrons from a metal can also be diffracted by a grating. In other words, *the photon theory and the wave theory of light complement each other.*

The success of the particle model of light in explaining the photoelectric effect and the Compton effect raises many other questions. If the photon is a particle, what is the meaning of the "frequency" and "wavelength" of the particle, and which determines its energy and momentum? Is light in some sense simultaneously a wave and a particle? Although photons have no rest mass, is there a simple expression for the mass of a "moving" photon? If a "moving" photon has mass, do photons experience gravitational attraction? What is the spatial extent of a photon, and how does an electron absorb or scatter a photon? Although answers to some of these questions are possible, some demand a view of atomic processes that is too pictorial and literal. Furthermore, many of these questions result from classical analogies such as colliding billiard balls and water waves breaking on a shore. Quantum mechanics gives light a more fluid and flexible nature by requiring that both the particle model and wave model of light are necessary and complementary. Neither model can be used exclusively to describe all properties of light. A complete understanding of the observed behavior of light is obtained only if the two models are combined in a complementary manner.

We can perhaps understand why photons are compatible with electromagnetic waves in the following manner. We may suspect that long-wavelength radio waves do not exhibit particle characteristics. Consider, for instance, radio waves at a frequency of 2.5 MHz. The energy of a photon having this frequency is only about $10^{-8}$ eV. From a practical viewpoint, this energy is too small to be detected as a single photon. A sensitive radio receiver might require as many as $10^{10}$ of these photons to produce a detectable signal. Such a large number of photons would appear, on the average, as a continuous wave. With such a large number of photons reaching the detector every second, it would be unlikely that any graininess would appear in the detected signal. That is, we would not be able to detect the individual photons striking the antenna.

Now consider what happens as we go to higher frequencies, or shorter wavelengths. In the visible region, it is possible to observe both the photon and the wave characteristics of light. As we mentioned earlier, a light beam shows interference phenomena and at the same time can produce photoelectrons, which can be understood best by using Einstein's photon concept. At even higher frequencies and correspondingly shorter wavelengths, the momentum and energy of the photon increase. Consequently, the photon nature of light becomes more evident than its wave nature. For example, absorption of an x-ray photon is easily detected as a single event. However, as the wavelength decreases, wave effects, such as interference and diffraction, become more difficult to observe. Very indirect methods are required to detect the wave nature of very-high-frequency radiation, such as gamma rays.

All forms of electromagnetic radiation can be described from two points of view. At one extreme, electromagnetic waves describe the overall interference pattern formed by a large number of photons. At the other extreme, the photon description is natural when we are dealing with a highly energetic photon of very short wavelength. Hence,

The dual nature of light

light has a dual nature: it exhibits both wave and photon characteristics.

## 41.2 THE WAVE PROPERTIES OF PARTICLES

Students first introduced to the dual nature of light often find the concept very difficult to accept. In the world around us, we are accustomed to regarding such things as a thrown baseball solely as particles and such things as sound waves solely as forms of wave motion. Every large-scale observation can be interpreted by considering either a wave explanation or a particle explanation, but in the world of photons and electrons, such distinctions are not as sharply drawn. Even more disconcerting is the fact that, under certain conditions, *particles such as electrons also exhibit wave characteristics.*

Louis de Broglie (1892–1987), a French physicist, was awarded the Nobel prize in 1929 for his discovery of the wave nature of electrons. "It would seem that the basic idea of quantum theory is the impossibility of imaging an isolated quantity of energy without associating with it a certain frequency." (AIP Niels Bohr Library)

The first bold step toward a new mechanics of atomic systems was taken by Louis Victor de Broglie in 1923. In his doctoral dissertation he postulated that *because photons have wave and particle characteristics, perhaps all forms of matter have wave as well as particle properties.* This was a highly revolutionary idea with no experimental confirmation at that time. According to de Broglie, electrons had a dual particle/wave nature. Accompanying every electron was a wave (not an electromagnetic wave!), which guided or "piloted" the electrons through space. He explained the source of this assertion in his 1929 Nobel Prize acceptance speech:

> On the one hand the quantum theory of light cannot be considered satisfactory since it defines the energy of a light corpuscle by the equation $E = hf$ containing the frequency $f$. Now a purely corpuscular theory contains nothing that enables us to define a frequency; for this reason alone, therefore, we are compelled, in the case of light, to introduce the idea of a corpuscle and that of periodicity simultaneously. On the other hand, determination of the stable motion of electrons in the atom introduces integers, and up to this point the only phenomena involving integers in physics were those of interference and of normal modes of

vibration. This fact suggested to me the idea that electrons too could not be considered simply as corpuscles, but that periodicity must be assigned to them also.

In Chapter 39, we found that the relationship between energy and momentum for a photon, which has a rest mass of zero, is $p = E/c$. We also know that the energy of a photon is

$$E = hf = \frac{hc}{\lambda} \tag{41.1}$$

Energy of a photon

Thus, the momentum of a photon can be expressed as

$$p = \frac{E}{c} = \frac{hc}{c\lambda} = \frac{h}{\lambda} \tag{41.2}$$

Momentum of a photon

From this equation we see that the photon wavelength can be specified by its momentum, or $\lambda = h/p$. De Broglie suggested that

material particles of momentum $p$ should also have wave properties and a corresponding wavelength.

Because the momentum of a particle of mass $m$ and velocity $v$ is $p = mv$, the **de Broglie wavelength** of a particle is

$$\lambda = \frac{h}{p} = \frac{h}{mv} \tag{41.3}$$

De Broglie wavelength

Furthermore, in analogy with photons, de Broglie postulated that the frequencies of matter waves (that is, waves associated with particles of nonzero rest mass) obey the Einstein relation $E = hf$, so that

$$f = \frac{E}{h} \tag{41.4}$$

Frequency of matter waves

The dual nature of matter is apparent in these two equations. That is, each equation contains both particle concepts ($mv$ and $E$) and wave concepts ($\lambda$ and $f$). The fact that these relationships are established experimentally for photons makes the de Broglie hypothesis that much easier to accept.

### Quantization of Angular Momentum in the Bohr Model

Bohr's model of the atom has many shortcomings and problems. For example, as the electrons revolve around the nucleus, how does one visualize the fact that only certain electronic energies are allowed? Why do all atoms of a given element have precisely the same physical properties regardless of the infinite variety of starting velocities and positions of the electrons in each atom?

De Broglie's great insight was to recognize that wave theories of matter handle these problems neatly by means of interference. Recall from Chapter 18 that a plucked guitar string, while initially subjected to a wide range of wavelengths, will support only standing wave patterns that have nodes at each end. Any free vibration of the string consists of a superposition of various amounts of many standing waves. This same reasoning can be applied to electron matter waves bent into a circle around the nucleus. All of the possible

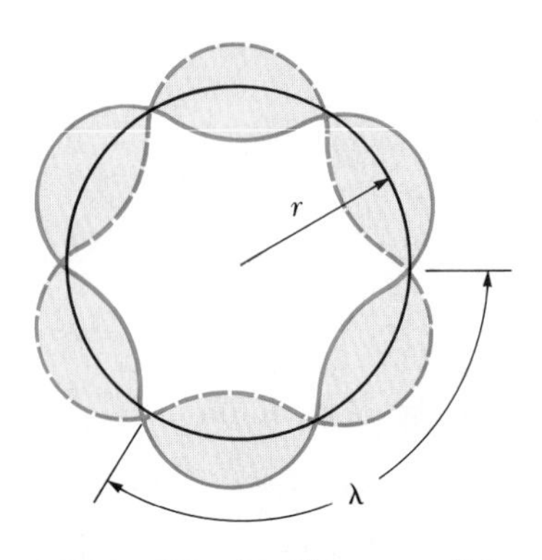


**Figure 41.1** Standing waves fit to a circular Bohr orbit. In this particular picture, three wavelengths are fit to the orbit, corresponding to the $n = 3$ energy state of the Bohr theory.

states of the electron are standing wave states, each with its own wavelength, speed, and energy. The residual standing wave patterns thus account for the identical nature of all atoms of a given element and show that atoms are more like vibrating drum heads with discrete modes of vibration than like miniature solar systems. This point of view is emphasized in Figure 41.1, which shows the standing wave pattern of the electron in the hydrogen atom corresponding to the $n = 3$ state of the Bohr theory.

Another aspect of the Bohr theory that is also easier to visualize physically by using de Broglie's hypothesis is the quantization of angular momentum. One simply needs to assume that the allowed Bohr orbits arise because the electron matter waves form standing waves when an integral number of wavelengths exactly fits into the circumference of a circular orbit. Thus

$$n\lambda = 2\pi r$$

where $r$ is the radius of the orbit. Since the de Broglie wavelength is given by $\lambda = h/mv$, we can write the above condition as $n(h/mv) = 2\pi r$, or

$$mvr = n\hbar$$

Note that this is precisely the Bohr condition of quantization of angular momentum.[1] Thus, we see that electron waves that fit the orbits are standing waves because of the boundary conditions imposed. These standing waves have discrete frequencies, corresponding to the allowed wavelengths. If $n\lambda \neq 2\pi r$, a standing-wave pattern can never form a closed circular orbit as in Figure 41.1. Furthermore, since the de Broglie waves have frequencies given by $f = E/h$, we see that the condition of standing waves implies quantized energies.

### The Davisson–Germer Experiment

De Broglie's proposal that any kind of particle exhibits both wave and particle properties was first regarded as pure speculation. If particles such as electrons had wave-like properties, then under the correct conditions they should exhibit interference phenomena. Three years later, in 1927, C.J. Davisson and L.H. Germer of the United States succeeded in measuring the wavelength of electrons. Their important discovery provided the first experimental confirmation of the matter waves proposed by de Broglie.

It is interesting to point out that the intent of the initial Davisson–Germer experiment was not to confirm the de Broglie hypothesis. In fact, their discovery was made by accident, as is often the case. The experiment involved the scattering of low energy electrons (about 54 eV) from a nickel target in a vacuum. During one experiment, the nickel surface was badly oxidized because of an accidental break in the vacuum system. After the nickel target was heated in a flowing stream of hydrogen to remove the oxide coating, subsequent experiments showed that the scattered electrons exhibited intensity maxima and minima at specific angles. The experimenters finally realized that the nickel had formed large crystal regions upon heating and that the regularly spaced planes of atoms in the crystalline regions served as a diffraction grating for electron matter waves.

[1] Note that de Broglie's analysis still failed to explain the fact that states having an orbital angular momentum quantum number of zero do exist.

Shortly thereafter, Davisson and Germer performed more extensive diffraction measurements on electrons scattered from single-crystal targets. Their results showed conclusively the wave nature of electrons and confirmed the de Broglie relation $p = h/\lambda$. In the same year, G.P. Thomson of Scotland also observed electron diffraction patterns by passing electrons through very thin gold foils. Diffraction patterns have since been observed for helium atoms, hydrogen atoms, and neutrons. Hence the universal nature of matter waves has been established in various ways.

The problem of understanding the dual nature of both matter and radiation is conceptually difficult because the two models seem to contradict each other. This problem as it applies to light was discussed earlier. Niels Bohr helped to resolve this problem in his principle of complementarity, which states that *the wave and particle models of either matter or radiation complement each other.* Neither model can be used exclusively to adequately describe matter or radiation. A complete understanding is obtained only if the two models are combined in a complementary manner.

The principle of complementarity

**EXAMPLE 41.1 The Wavelength of an Electron**
Calculate the de Broglie wavelength for an electron ($m = 9.11 \times 10^{-31}$ kg) moving with a speed of $10^7$ m/s.

*Solution* Equation 41.3 gives

$$\lambda = \frac{h}{mv} = \frac{6.63 \times 10^{-34}\ \text{J}\cdot\text{s}}{(9.11 \times 10^{-31}\ \text{kg})(10^7\ \text{m/s})}$$

$$= 7.28 \times 10^{-11}\ \text{m}$$

This wavelength corresponds to that of x-rays in the electromagnetic spectrum.

**Exercise 1** Find the de Broglie wavelength of a proton moving with a speed of $10^7$ m/s.
**Answer** $3.97 \times 10^{-14}$ m.

**EXAMPLE 41.2 The Wavelength of a Rock**
A rock of mass 50 g is thrown with a speed of 40 m/s. What is the de Broglie wavelength of the rock?

*Solution* From Equation 41.3, we have

$$\lambda = \frac{h}{mv} = \frac{6.63 \times 10^{-34}\ \text{J}\cdot\text{s}}{(50 \times 10^{-3}\ \text{kg})(40\ \text{m/s})} = 3.32 \times 10^{-34}\ \text{m}$$

Notice that this wavelength is much smaller than the size of any possible aperture through which the rock could pass. This means that we could not observe diffraction effects, and as a result the wave properties of large-scale objects cannot be observed.

**EXAMPLE 41.3 An Accelerated Charge**
A particle of charge $q$ and mass $m$ is accelerated from rest through a potential difference $V$. (a) Find its de Broglie wavelength.

*Solution* When a charge is accelerated from rest through a potential difference $V$, its gain in kinetic energy $\frac{1}{2}mv^2$ must equal its loss in potential energy $qV$, since energy is conserved. That is,

$$\tfrac{1}{2}mv^2 = qV$$

Since $p = mv$, we can express this in the form

$$\frac{p^2}{2m} = qV \qquad \text{or} \qquad p = \sqrt{2mqV}$$

Substituting this expression for $p$ into the de Broglie relation $\lambda = h/p$ gives

$$\lambda = \frac{h}{p} = \frac{h}{\sqrt{2mqV}}$$

(b) Calculate $\lambda$ if the particle is an electron and $V = 50$ V.

*Solution* The de Broglie wavelength of the electron accelerated through 50 V is

$$\lambda = \frac{h}{\sqrt{2mqV}}$$

$$= \frac{6.63 \times 10^{-34}\ \text{J}\cdot\text{s}}{\sqrt{2(9.11 \times 10^{-31}\ \text{kg})(1.6 \times 10^{-19}\ \text{C})(50\ \text{V})}}$$

$$= 1.74 \times 10^{-10}\ \text{m} = 0.174\ \text{nm}$$

This wavelength is of the order of atomic dimensions and the spacing between atoms in a solid. Such low-energy electrons are normally used in electron diffraction experiments.

## 41.3 THE DOUBLE-SLIT EXPERIMENT REVISITED

In the previous section and in Chapter 40, we saw evidence for both the wave properties and particle properties of electrons. One way to crystallize our ideas about the wave-particle duality is to consider a double-slit electron diffraction experiment. This experiment shows the impossibility of measuring *simultaneously* both wave and particle properties, and illustrates the use of the wave function in determining interference effects.

Consider a parallel beam of monoenergetic electrons incident on a double slit as in Figure 41.2. It shall be assumed that the individual slit openings are much smaller than the slit separation, $D$, so that single-slit diffraction effects are negligible. An electron detector, capable of detecting individual electrons, is located at a distance much greater than $D$. *If the detector collects electrons at different positions for a long enough time, one finds a typical wave interference pattern for the counts/min or probability of arrival of electrons.* This experiment is analogous to the interference of monochromatic light waves in Young's double-slit experiment described in Chapter 37. Such an interference pattern would not be expected if the electrons behaved as classical particles. If the experiment is carried out at lower beam intensities, the interference pattern is still observed if the time of the measurement is sufficiently long. This is illustrated in the computer-simulated patterns shown in Figure 41.3. Note that the interference pattern becomes clearer as the number of electrons reaching the screen increases.

If one imagines a single electron to produce in-phase "wavelets" at the slits, standard wave theory can be used to find the angular separation, $\theta$, of the central probability maximum from its neighboring minimum. The minimum

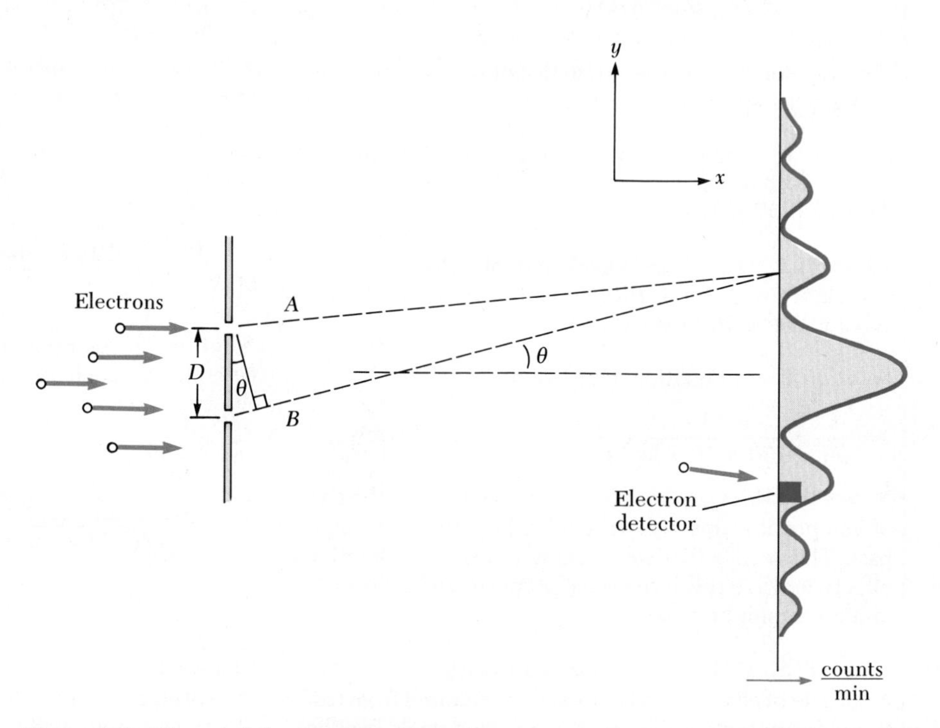


**Figure 41.2** Electron diffraction. The slit separation $D$ is much greater than the individual slit widths and much less than the distance between the slit and detector.

occurs when the path length difference between $A$ and $B$ is half a wavelength or

$$D \sin \theta = \frac{\lambda}{2}$$

As the electron's wavelength is given by $\lambda = h/p_x$, we see that

$$\sin \theta \approx \theta = \frac{h}{2p_x D}$$

for small $\theta$. Thus the dual nature of the electron is clearly shown in this experiment: *while the electrons are detected as particles at a localized spot at some instant of time, the probability of arrival at that spot is determined by finding the intensity of two interfering matter waves.*

But there is more. What happens if one slit is covered during the experiment? In this case one obtains a symmetric curve peaked around the center of the open slit, much like the pattern formed by bullets shot through a hole in armor plate. Plots of the counts per minute or probability of arrival of electrons with the lower or upper slit closed are shown in the central part of Figure 41.4. These are expressed as the appropriate square of the absolute value of some wave function, $|\psi_1|^2 = \psi_1^*\psi_1$ or $|\psi_2|^2 = \psi_2^*\psi_2$, where $\psi_1$ and $\psi_2$ represent the cases of the electron passing through slit 1 and slit 2, respectively. If an experiment is now performed with slit 2 blocked half of the time and then slit 1 blocked during the remaining time, the accumulated pattern of counts/min shown by the blue curve on the right side of Figure 41.4 is completely different from the case with both slits open. There is no longer a maximum probability of arrival of an electron at $\theta = 0$. In fact, *the interference pattern has been lost and the accumulated result is simply the sum of the individual results.* When only one slit is open at a time, we know the electron has the same localizability

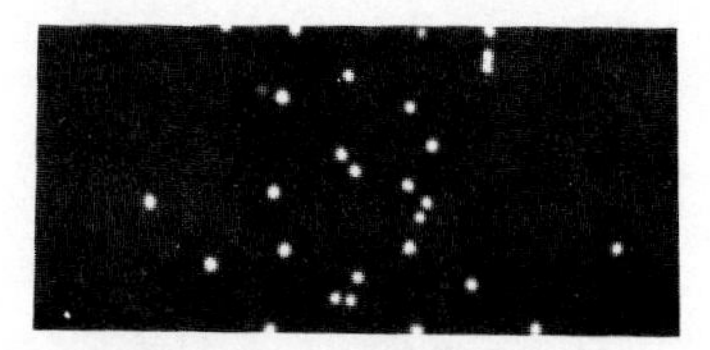

(a) After 28 electrons

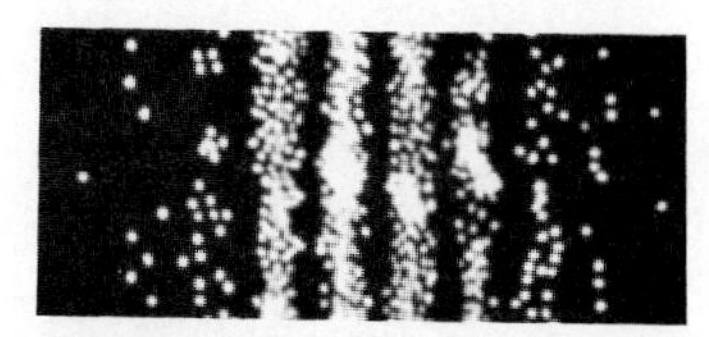

(b) After 1000 electrons

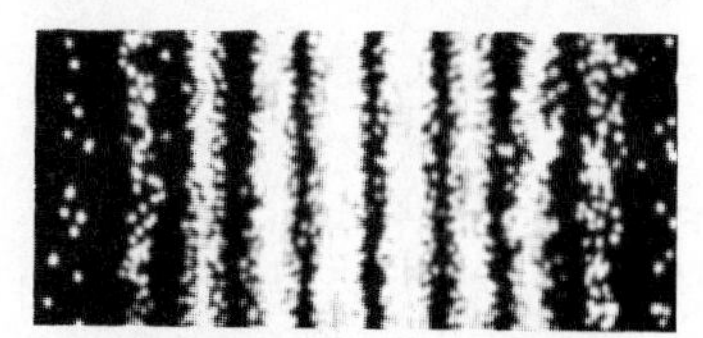

(c) After 10,000 electrons

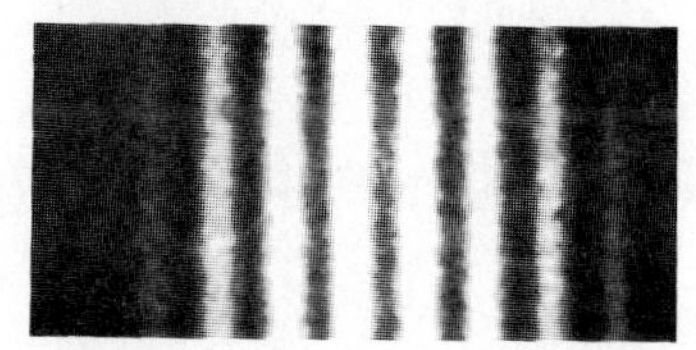

(d) Two slit electron pattern

**Figure 41.3** (a), (b), (c) Computer-simulated interference patterns for a beam of electrons incident on a double slit. (From E. R. Huggins, *Physics I*, New York, W. A. Benjamin, 1968). (d) Photograph of a double-slit interference pattern produced by electrons. (From C. Jönsson, *Zeitschrift fur Physik* **161**:454, 1961; used with permission)

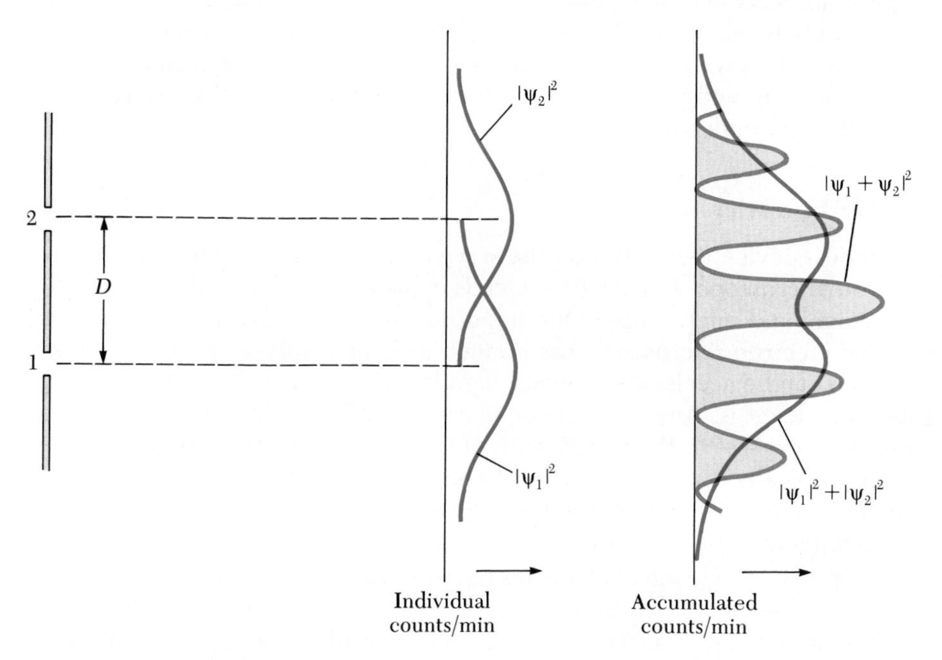


**Figure 41.4** Accumulated results from the two-slit electron diffraction experiment with each slit closed half the time. The result with both slits open is shown in red.

and indivisibility at the slits as we measure at the detector, since the electron clearly goes through slit 1 or slit 2. Thus the total must be analyzed as the sum of those electrons that come through slit 1, $|\psi_1|^2$, and those that come through slit 2, $|\psi_2|^2$.

When both slits are open, it is tempting to assume that the electron goes through either slit 1 or slit 2, and that the counts/min are again given by $|\psi_1|^2 + |\psi_2|^2$. We know, however, that the experimental results indicated by the red interference pattern in Figure 41.4 contradict this. Thus our assumption that the electron is localized and goes through only one slit when both slits are open must be wrong (a painful conclusion!). Somehow the electron must be simultaneously present at both slits to exhibit interference. In order to find the probability of detecting the electron at a particular point on the screen with both slits open, we may say that the electron is in a *superposition state* given by

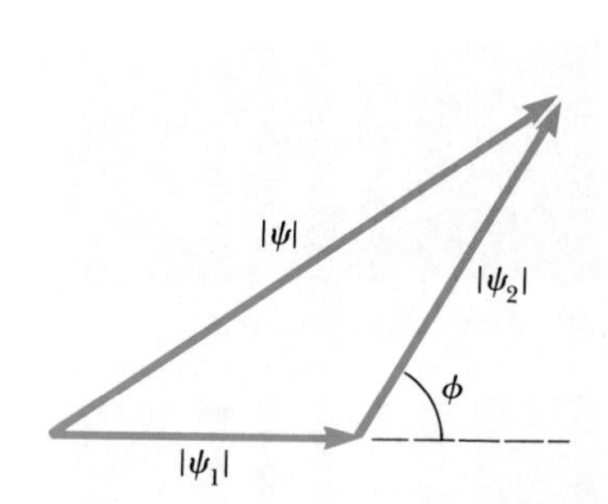


**Figure 41.5** Phasor diagram to represent the addition of two complex quantities, $\psi_1$ and $\psi_2$.

$$\psi = \psi_1 + \psi_2$$

Thus the probability of detecting the electron at the screen is equal to the quantity $|\psi_1 + \psi_2|^2$ and not $|\psi_1|^2 + |\psi_2|^2$. Since matter waves that start out in phase at the slits in general travel different distances to the screen (see Fig. 41.5), $\psi_1$ and $\psi_2$ will possess a relative phase difference $\phi$ at the screen. Using a phasor diagram (Fig. 41.5) to find $|\psi_1 + \psi_2|^2$ immediately yields

$$|\psi|^2 = |\psi_1 + \psi_2|^2 = |\psi_1|^2 + |\psi_2|^2 + 2|\psi_1||\psi_2| \cos \phi$$

where $|\psi_1|^2$ is the probability of detection if slit 1 is open and slit 2 is closed and $|\psi_2|^2$ is the probability of detection if slit 2 is open and slit 1 is closed. The last term $2|\psi_1||\psi_2| \cos \phi$ is the interference term, which arises from the relative phase $\phi$ of the waves in analogy with the phasor addition used in wave optics (Chapter 37).

In order to interpret these results, one is forced to conclude that an *electron interacts with both slits simultaneously*. If we attempt to determine experimentally which slit the electron goes through, the act of the measurement will destroy the interference pattern. It is impossible to determine which slit the electron will go through. In effect, *we can say only that the electron passes through both slits!*

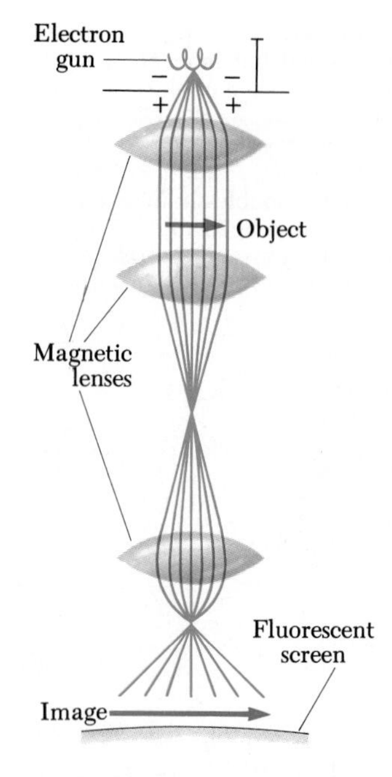


**Figure 41.6** Diagram of an electron microscope. The "lenses" that control the electron beam are magnetic deflection coils.

### The Electron Microscope

A practical device that relies on the wave characteristics of electrons is the **electron microscope** (Fig. 41.6), which is in many respects similar to an ordinary compound microscope. One important difference between the two is that the electron microscope has a much greater resolving power because electrons can be accelerated to very high kinetic energies, giving them very short wavelengths. Any microscope is capable of detecting details that are comparable in size to the wavelength of the radiation used to illuminate the object. Typically, the wavelengths of electrons are about 100 times shorter than those of the visible light used in optical microscopes. As a result, electron microscopes are able to distinguish details about 100 times smaller.

In operation, a beam of electrons falls on a thin slice of the material to be examined. The section to be examined must be very thin, typically a few hundred angstroms, in order to minimize undesirable effects, such as absorption or scattering of the electrons. The electron beam is controlled by electrostatic or magnetic deflection, which acts on the charges to focus the beam to an image. Rather than examining the image through an eyepiece as in an ordinary microscope, a magnetic lens forms an image on a fluorescent screen. The fluorescent screen is necessary because the image produced would not otherwise be visible. An example of a photograph taken by an electron microscope is shown in Figure 41.7.

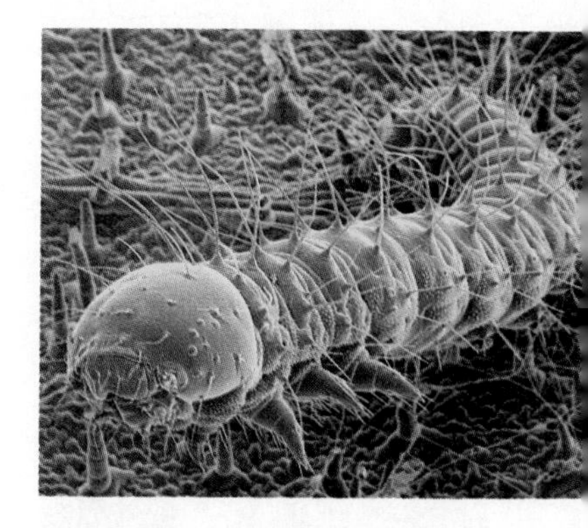

**Figure 41.7** Scanning electron microscope picture of a recently hatched caterpillar of a large white butterfly seen on the undersurface of a nasturtium leaf. (Courtesy of Dr. Jeremy Burgess, Science Photo Library)

## 41.4 THE UNCERTAINTY PRINCIPLE

If you were to measure the position and velocity of a particle at any instant, you would always be faced with experimental uncertainties in your measurements. According to classical mechanics, there is no fundamental barrier to an ultimate refinement of the apparatus and/or experimental procedures. That is, it would be possible, in principle, to make such measurements with arbitrarily small uncertainty or with infinite accuracy. Quantum theory predicts, however, that *it is fundamentally impossible to make simultaneous measurements of a particle's position and velocity with infinite accuracy.*

In 1927, Werner Heisenberg (1901–1976) first introduced the notion that it is impossible to determine simultaneously and with unlimited precision the position and momentum of a particle. In words, we may state the uncertainty principle as follows:

Uncertainty principle

If a measurement of position is made with precision $\Delta x$ and a simultaneous measurement of momentum is made with precision $\Delta p$, then the product of the two uncertainties can never be smaller than a number of the order of $\hbar$. That is,

$$\Delta x\, \Delta p \gtrsim \hbar \tag{41.5}$$

That is, *it is physically impossible to simultaneously measure the exact position and exact momentum of a particle.* If $\Delta x$ is very small, then $\Delta p$ will be large, and vice versa. In his paper of 1927, Heisenberg was careful to point out that the inescapable uncertainties $\Delta x$ and $\Delta p$ do not arise from imperfections in practical measuring instruments. Rather, they arise from the quantum structure of matter itself: from effects such as the *unpredictable* recoil of an electron when struck by an *indivisible* photon, or the diffraction of light or electrons passing through a small opening.

In order to understand the uncertainty principle, consider the following thought experiment introduced by Heisenberg. Suppose you wish to measure the position and momentum of an electron as accurately as possible. You might be able to do this by viewing the electron with a powerful light microscope. In order for you to see the electron, and thus determine its location, at least one photon of light must bound off the electron and pass through the microscope into your eye. This incident photon is shown moving toward the electron in Figure 41.8a. When the photon strikes the electron, as in Figure 41.8b, the photon transfers some of its energy and momentum to the electron. Thus, in the process of attempting to locate the electron very accurately (that is, by making $\Delta x$ very small), we have caused a rather large uncertainty in its momentum.

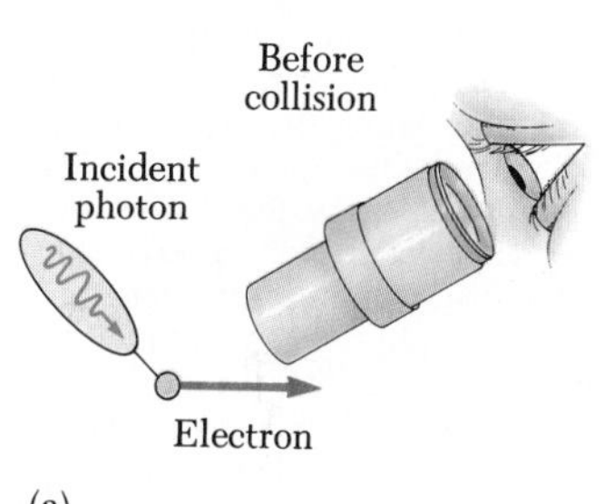


**Figure 41.8** A thought experiment for viewing an electron with a powerful microscope. (a) The electron is viewed before colliding with the photon. (b) The electron recoils (is disturbed) as the result of the collision with the photon.

Let us analyze the collision between the photon and the electron by first noting that the incoming photon has momentum $h/\lambda$. As a result of the collision, the photon transfers part or all of its momentum to the electron. Thus, the

Werner Heisenberg (1901–1976), a German theoretical physicist, obtained his PhD in 1923 at the University of Munich where he studied under Arnold Sommerfeld and became an enthusiastic mountain climber and skier. While physicists such as de Broglie and Schrödinger tried to develop physical models of the atom, Heisenberg developed an abstract mathematical model called matrix mechanics to explain the wavelengths of spectral lines. The more successful wave mechanics by Schrödinger announced a few months later was shown to be equivalent to Heisenberg's approach. Heisenberg made many other significant contributions to physics including his famous uncertainty principle, for which he received the Nobel Prize in 1932, the prediction of two forms of molecular hydrogen, and theoretical models of the nucleus.

uncertainty in the electron's momentum after the collision could be as large as the momentum of the incoming photon. That is, $\Delta p = h/\lambda$. Furthermore, since light also has wave properties, we would expect to be able to determine the position of the electron to within one wavelength of the light being used to view it, so that $\Delta x = \lambda$. Multiplying these two uncertainties gives

$$\Delta x \, \Delta p = \lambda \left(\frac{h}{\lambda}\right) = h$$

This represents the minimum in the products of the uncertainties. Since the uncertainty can always be greater than this minimum, we have

$$\Delta x \, \Delta p \gtrsim h$$

This agrees with Equation 41.5 (apart from a small numerical factor introduced by Heisenberg's more precise analysis).

Heisenberg's uncertainty principle enables us to better understand the dual wave–particle nature of light and matter. We have seen that the wave description is quite different from the particle description. Therefore, if an experiment is designed to reveal the particle character of an electron (such as the photoelectric effect), its wave character will become less apparent. Likewise, if the experiment is designed to accurately measure the electron's wave properties (such as diffraction from a crystal), its particle character will become less apparent.

Before concluding this section, we should point out that there is another uncertainty principle, which sets a limit on the accuracy with which the energy of a system, $\Delta E$, can be measured if a finite time interval, $\Delta t$, is allowed for the measurement. This energy-time uncertainty principle may be stated as

$$\Delta E \, \Delta t \gtrsim \hbar \qquad (41.6)$$

This uncertainty relation is plausible if one considers a frequency measurement of any wave. For example, consider the measurement of a 1000 Hz electrical wave. If our frequency measuring device has a fixed sensitivity of $\pm 1$ cycle, in one second we will measure a frequency of $(1000 \pm 1)$ cycles/1 s but in two seconds we will measure a frequency of $(2000 \pm 1)$ cycles/2 s. Thus the uncertainty in frequency, $\Delta f$, is inversely proportional to $\Delta t$, the time during which the measurement is made. This may be stated as

$$\Delta f \, \Delta t \approx 1$$

Since all quantum systems are "wavelike" and can be described by the relationship $E = hf$, we may substitute $\Delta f = \Delta E/h$ into the expression above to obtain

$$\Delta E \, \Delta t \approx h$$

in basic agreement with Equation 41.6.

We conclude this section with two examples of the types of calculations that can be done with the uncertainty principle. In the spirit of Fermi or Heisenberg, these "back of the envelope calculations" are surprising for their simplicity and essential description of quantum systems of which the details are unknown.

EXAMPLE 41.4 Locating an Electron

The speed of an electron is measured to have a value of $5.00 \times 10^3$ m/s to an accuracy of 0.003 percent. Find the uncertainty in determining the position of this electron.

*Solution* The momentum of the electron is

$$p = mv = (9.11 \times 10^{-31}\ \text{kg})(5.00 \times 10^3\ \text{m/s})$$
$$= 4.56 \times 10^{-27}\ \text{kg} \cdot \text{m/s}$$

Because the uncertainty in $p$ is 0.003% of this value, we get

$$\Delta p = 0.00003p = (0.00003)(4.56 \times 10^{-27}\ \text{kg} \cdot \text{m/s})$$
$$= 1.37 \times 10^{-31}\ \text{kg} \cdot \text{m/s}$$

The uncertainty in position can now be calculated by using this value of $\Delta p$ and Equation 41.5.

$$\Delta x\, \Delta p \gtrsim \frac{h}{2\pi}$$

$$\Delta x \gtrsim \frac{h}{2\pi \Delta p} = \frac{6.63 \times 10^{-34}\ \text{J} \cdot \text{s}}{2\pi(1.37 \times 10^{-31}\ \text{kg} \cdot \text{m/s})}$$

$$= 0.770 \times 10^{-3}\ \text{m} = \boxed{0.770\ \text{mm}}$$

EXAMPLE 41.5 The Width of Spectral Lines

Although an excited atom can radiate at any time from $t = 0$ to $t = \infty$, the average time after excitation at which a group of atoms radiates is called the **lifetime,** $\tau$. (a) If $\tau = 10^{-8}$ s, use the uncertainty principle to compute the line width $\Delta f$ produced by this finite lifetime.

*Solution* We use $\Delta E\, \Delta t \approx \hbar$, where $\Delta E = h\, \Delta f$, and where $\Delta t = 10^{-8}$ s is the average time available to measure the excited state. Thus,

$$\Delta f = \frac{1}{2\pi \times 10^{-8}\ \text{s}} = \boxed{1.6 \times 10^7\ \text{Hz}}$$

Note that $\Delta E$ is the uncertainty in energy of the excited state. It is also the uncertainty in the energy of the photon emitted by an atom in this state.

(b) If the wavelength of the spectral line involved in this process is 500 nm, find the fractional broadening $\Delta f/f$.

*Solution* First, we find the center frequency of this line as follows:

$$f_0 = \frac{c}{\lambda} = \frac{3 \times 10^8\ \text{m/s}}{500 \times 10^{-9}\ \text{m}} = 6.0 \times 10^{14}\ \text{Hz}$$

Hence,

$$\frac{\Delta f}{f_0} = \frac{1.6 \times 10^7\ \text{Hz}}{6.0 \times 10^{14}\ \text{Hz}} = \boxed{2.7 \times 10^{-8}}$$

This narrow natural linewidth can be seen with a sensitive interferometer. Usually, however, temperature and pressure effects overshadow the natural linewidth and broaden the line through mechanisms associated with the Doppler effect and collisions.

## 41.5 INTRODUCTION TO QUANTUM MECHANICS

There is a striking similarity between the behavior of light and the behavior of matter. Both have dualistic character in that they behave both as waves and as particles. We can carry the analogy even further. In the case of light waves, we described how a wave theory gives only the probability of finding a photon at a given point within a given time interval. Likewise, matter waves are described by a complex-valued wave function (usually denoted by $\psi$, the Greek letter psi) whose absolute square $|\psi|^2 = \psi^*\psi$ gives the probability of finding the particle at a given point at some instant, where $\psi^*$ is the complex conjugate of $\psi$. The wave function contains within it all the information that can be known about the particle.

This interpretation of matter waves was first suggested by Max Born (1882–1970) in 1928. In the same year, Erwin Schrödinger (1887–1961) proposed a wave equation that described the manner in which matter waves change in space and time. The *Schrödinger wave equation* represents a key element in the theory of quantum mechanics. Its role is as important in quantum mechnics as that played by Newton's laws of motion in classical mechanics. Schrödinger's wave equation has been successfully applied to the hydrogen atom and to many other microscopic systems. Its importance in most aspects of modern physics cannot be overemphasized.

The concepts of quantum mechanics, strange as they sometimes may seem, developed from older ideas in classical physics. In fact, if the techniques of quantum mechanics are applied to macroscopic systems rather than atomic systems, the results are essentially identical with those of classical physics. This blending of the two theories occurs when the de Broglie wavelength is small compared with the dimensions of the system. The situation is similar to the agreement between relativistic mechanics and classical mechanics when $v \ll c$. For these reasons, Newtonian physics can give an accurate description of a system only if the system's dimensions are large compared with atomic dimensions and only if the system is moving with a speed that is small relative to the speed of light.

A number of experiments have been conducted which show that matter has both a wave nature and a particle nature. A question which arises quite naturally in this regard is the following: If we are describing a particle such as an electron, how do we view what is waving? The answer to this question is quite clear in the case of waves on strings, water waves, and sound waves. These waves represent the propagation of a disturbance in a material medium. In each case, the wave is represented by some quantity that varies with time and position. For example, sound waves can be represented by a pressure variation $\Delta P$, and waves on strings can be represented by a transverse displacement $y$. In a similar manner, matter waves, or de Broglie waves, can be represented by a quantity $\psi$, called the **wave function.** In general, $\psi$ depends on both the position and the time of all the particles in a system and therefore is often written $\psi(x, y, z, t)$. The form of $\psi$ depends on the system being described and on the forces acting on the system. If $\psi$ is known for a particle, then the particular properties of that particle can be described. In fact, the fundamental problem of quantum mechanics is this: given the wave function at some instant, say $t = 0$, find the wave function at some later time $t$.

The wave function $\psi$

In Section 41.2, we found that the de Broglie equation relates the momentum of a particle to its wavelength through the relation $p = h/\lambda$. If a free particle has a precisely known momentum, its wave function is a sinusoidal wave of wavelength $\lambda = h/p$. The real part of the wave function for such a free particle moving along the $x$ axis can be written in the form

$$\psi(x) = A \sin\left(\frac{2\pi x}{\lambda}\right) = A \sin(kx) \tag{41.7}$$

where $k = 2\pi/\lambda$ is the wave number and $A$ is a constant.[2] As we mentioned earlier, the wave function is generally a function of both position and time. Equation 41.7 represents that part of the wave function dependent on position only. For this reason, one can view $\psi(x)$ as a "snapshot" of the wave at a given instant, as shown in Figure 41.9a. The wave function for a particle whose wavelength is not precisely defined is shown in Figure 41.9b. Since the wavelength is not precisely defined, it follows that the momentum is only approximately known. That is, if one were to measure the momentum of the particle, the result would have any value over some range, determined by the spread in wavelength.

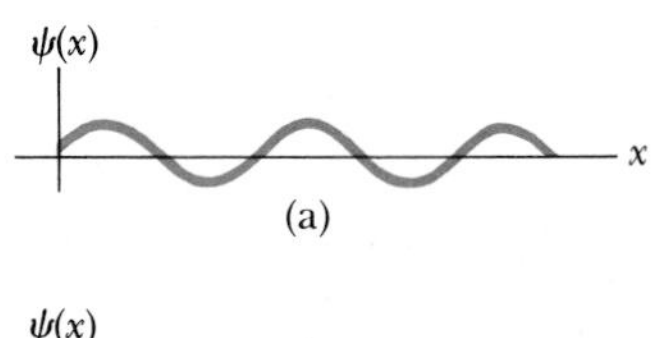


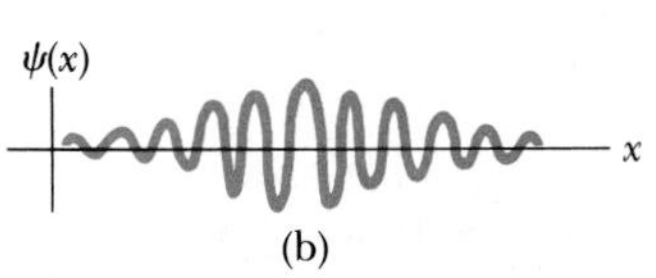


**Figure 41.9** (a) Wave function for a particle whose wavelength is precisely known. (b) Wave function for a particle whose wavelength is not precisely known and hence whose momentum is known only over some range of values.

[2] In general, we would write the wave function for the particle in the form $Ae^{ikx}$, where the imaginary part of the function describes the phase of the wave.

Although $\psi$ itself is not a quantity that one can measure, the quantity $|\psi|^2$ can be measured, where $|\psi|^2$ means the square of the absolute value of $\psi$. The quantity $|\psi|^2$ can be interpreted as follows. If $\psi$ represents a single particle, then $|\psi|^2$ is the probability per unit volume that the particle will be found at any given point. This interpretation, first suggested by Born in 1928, can also be stated in the following manner. If $dV$ is a small volume element surrounding some point, then the probability of finding the particle in that volume element is given by

$$\text{Probability} = |\psi|^2\, dV$$

In this chapter we shall be dealing with one-dimensional systems, where the particle must be located along the $x$ axis, thus we replace $dV$ by $dx$. In this case, the probability that the particle will be found in the infinitesimal interval $dx$ about the point $x$, denoted by $P(x)\, dx$, is

$$P(x)\, dx = |\psi|^2\, dx \tag{41.8}$$

Since the particle must be somewhere along the $x$ axis, the sum of the probabilities over all values of $x$ must be 1:

$$\int_{-\infty}^{\infty} |\psi|^2\, dx = 1 \tag{41.9}$$

Normalization condition on $\psi$

Any wave function satisfying Equation 41.9 is said to be *normalized*. The quantity $|\psi|^2$ is often called the **probability density.**[3] Note that normalization is simply a statement that the particle exists at some point at all times. If the probability were zero, the particle would not exist. Therefore, although it is not possible to specify the position of a particle with complete certainty, it is possible, through $|\psi|^2$, to specify the probability of observing it. Furthermore, *the probability of finding the particle in the interval $a \leq x \leq b$ is given by*

$|\psi|^2$ equals the probability density

$$P_{ab} = \int_a^b |\psi|^2\, dx \tag{41.10}$$

The probability $P_{ab}$ is the area under the curve of probability density versus $x$ between the points $x = a$ and $x = b$ as in Figure 41.10.

Experimentally, there is a finite probability of finding a particle at some point and at some instant. The value of the probability must lie between the limits 0 and 1. For example, if the probability is 0.3, this would signify a 30% chance of finding the body.

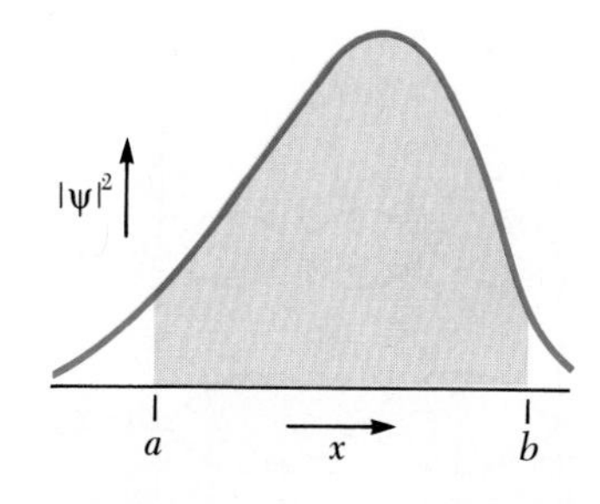


**Figure 41.10** The probability for a particle to be in the interval $a \leq x \leq b$ is the area under the curve from $a$ to $b$ of the probability density function $|\psi(x, t)|^2$.

The wave function $\psi$ satisfies a wave equation, just as the electric field associated with an electromagnetic wave satisfies a wave equation that follows from Maxwell's equations. The wave equation satisfied by $\psi$, called the *Schrödinger equation,* cannot be derived from any more fundamental laws. However, it is the basis upon which $\psi$ can be computed. Although $\psi$ itself is not a measurable quantity, all measurable quantities, such as the energy and momentum of a particle, can be derived from a knowledge of $\psi$. For example, once the wave function for a particle is known, it is possible to calculate the

[3] The wave function may be real or complex. In general, $\psi$ has a real and an imaginary component and its absolute value squared is given by $|\psi|^2 = \psi^*\psi$, where $\psi^*$ is the complex conjugate of $\psi$. For more details, see, for example, R.A. Serway, C.J. Moses, and C.A. Moyer, *Modern Physics*, Philadelphia, Saunders College Publishing, 1989, Chapter 5.

average position $x$ of the particle, after many experimental trials. This average position is called the **expectation value of $x$** and is defined by the equation

Expectation value of $x$

$$\langle x \rangle \equiv \int_{-\infty}^{\infty} x|\psi|^2\,dx \tag{41.11}$$

This expression implies that the particle is in a definite state, so that the probability density is time-independent. Note that the expectation value is equivalent to the average value of $x$ that one would obtain when dealing with a large number of particles in the same state. Furthermore, one can find the expectation value of *any* function $f(x)$ by using Equation 41.11 with $x$ replaced by $f(x)$.

## 41.6 A PARTICLE IN A BOX

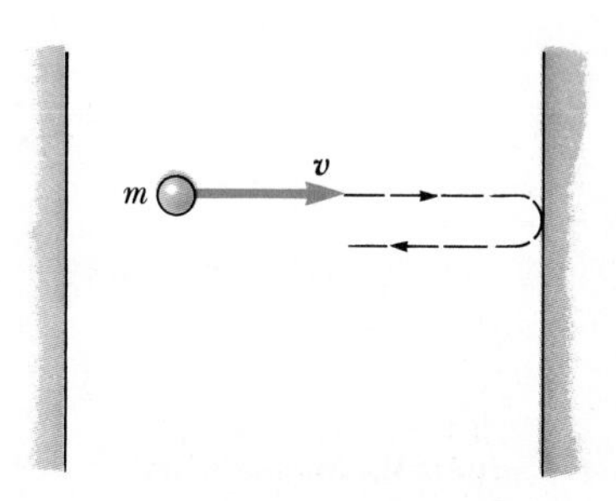


**Figure 41.11** A particle of mass $m$ and velocity $v$ confined to bouncing between two impenetrable walls.

From a classical viewpoint, if a particle is confined to moving along the $x$ axis and to bouncing back and forth between two impenetrable walls (Fig. 41.11), its motion is easy to describe. If the speed of the particle is $v$, then the magnitude of its momentum $mv$ remains constant, as does its kinetic energy. Furthermore, classical physics places no restrictions on the values of its momentum and energy. The wave mechanics approach to this problem is quite different and requires that we find the appropriate wave function consistent with the conditions of the situation.

Before we address this problem, it is instructive to review the classical situation of standing waves on a stretched string (Sections 18.2 and 18.3), which is analogous to the present particle-in-a-box problem. If a string of length $L$ is fixed at each end, we found that the standing waves set up in the string must have nodes at the ends, as in Figure 41.12. We require this condition for electrons, since the wave function must vanish at the boundaries. A resonance condition is achieved only when the length is some integral multiple of half-wavelengths. That is, we require that

$$L = n\frac{\lambda}{2}$$

or

$$\lambda = \frac{2L}{n} \qquad n = 1, 2, 3, \ldots \tag{41.12}$$

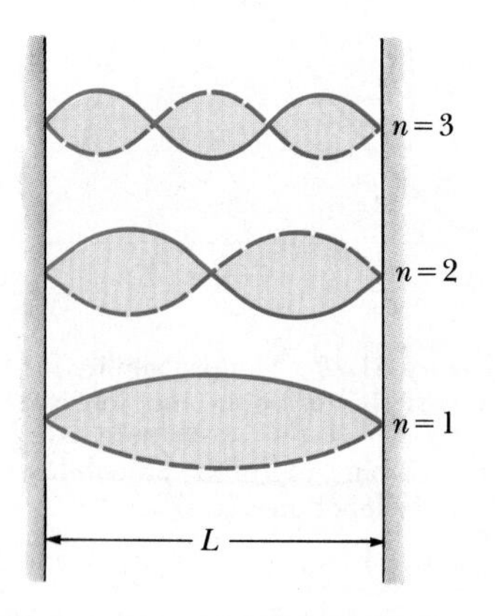


**Figure 41.12** Standing waves set up in a stretched string of length $L$.

This result shows that *the wavelength is quantized.*

As we saw in Section 18.2, each point on a standing wave oscillates with simple harmonic motion. Furthermore, each point oscillates with the same frequency, but the amplitude of oscillation, $y$, depends on the distance $x$ of the point measured from one end. We found that the position-dependent part of the wave function for a standing wave is given by

$$y(x) = A\sin(kx) \tag{41.13}$$

where $A$ is the maximum amplitude and $k = 2\pi/\lambda$ is the wave number. Since $\lambda = 2L/n$, we see that

$$k = \frac{2\pi}{\lambda} = \frac{2\pi}{2L/n} = n\,\frac{\pi}{L}$$

Substituting this into Equation 41.13 gives

$$y(x) = A \sin\left(\frac{n\pi x}{L}\right) \qquad (41.14)$$

From this expression, we see that the wave function meets the required boundary conditions, namely, $y = 0$ at $x = 0$ and at $x = L$ for all values of $n$. The wave functions for $n = 1$, 2, and 3 are plotted in Figure 41.12.

Now let us return to the wave mechanics description of a particle in a box. Assuming that the walls of the box are perfectly rigid, we must require that the probability of penetration is zero. This is equivalent to requiring that the wave function $\psi(x)$ be zero at the walls and outside the walls. That is, we must require that $\psi(x) = 0$ for $x = 0$ and for $x = L$. Only those wave functions that satisfy this condition are allowed. In analogy with standing waves on a string, the allowed wave functions are sinusoidal and given by

$$\psi(x) = A \sin\left(\frac{n\pi x}{L}\right) \qquad n = 1, 2, 3, \ldots \qquad (41.15)$$

Allowed wave functions for a particle in a box

where $A$ is the maximum value of the wave function. This shows that for a particle confined to a box and having a well-defined de Broglie wavelength, $\psi$ is represented by a sinusoidal wave. The allowed wavelengths are those for which the length $L$ is equal to an integral number of half-wavelengths, that is, $L = n\lambda/2$. These allowed states of the system are called **stationary states** since they represent standing waves, not traveling waves.

Figure 41.13 gives plots of the wave function $\psi$ versus $x$ and of the probability density $|\psi|^2$ versus $x$ for $n = 1$, 2, and 3. As we shall soon see, these states

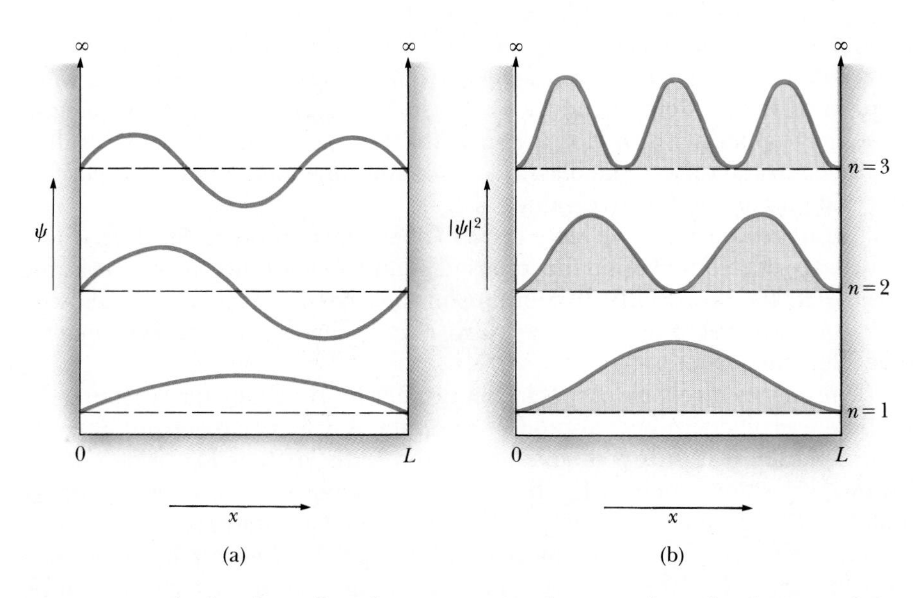


**Figure 41.13** The first three allowed stationary states for a particle confined to a one-dimensional box. (a) The wave functions for $n = 1$, 2, and 3. (b) The probability distributions for $n = 1$, 2, and 3.

correspond to the three lowest allowed energies for the particle. Note that although $\psi$ can be positive or negative, the probability density $|\psi|^2$ is always positive. From any viewpoint, a negative probability density would be meaningless.

Further inspection of Figure 41.13b shows that $|\psi|^2$ is always zero at the boundaries, indicating that it is impossible to find the particle at these points. In addition, $|\psi|^2$ is zero at other points, depending on the value of $n$. For $n = 2$, $|\psi|^2 = 0$ at the midpoint, $x = L/2$; for $n = 3$, $|\psi|^2 = 0$ at $x = L/3$ and at $x = 2L/3$. For $n = 1$, however, the probability of finding the particle is a maximum at $x = L/2$. For $n = 4$, $|\psi|^2$ has maxima also at $x = L/4$ and at $x = 3L/4$, and so on.

Since the wavelengths of the particle are restricted by the condition $\lambda = 2L/n$, the magnitude of the momentum is also restricted to specific values. We can obtain these values of the momentum using $p = h/\lambda$, so that

$$p = \frac{h}{\lambda} = \frac{h}{2L/n} = \frac{nh}{2L} \tag{41.16}$$

Using $p = mv$, we find that the allowed values of the kinetic energy are given by

$$E_n = \tfrac{1}{2}mv^2 = \frac{p^2}{2m} = \frac{(nh/2L)^2}{2m}$$

or

Allowed energies for a particle in a box

$$E_n = \left(\frac{h^2}{8mL^2}\right) n^2 \qquad n = 1, 2, 3, \ldots \tag{41.17}$$

As we see from this expression, *the energy of the particle is quantized*, as we would expect. The lowest allowed energy corresponds to $n = 1$, for which $E_1 = h^2/8mL^2$. Since $E_n = n^2E_1$, the excited states corresponding to $n = 2, 3, 4, \ldots$ have energies given by $4E_1, 9E_1, 16E_1, \ldots$. An energy level diagram describing the positions of the allowed states is given in Figure 41.14. Note that the state $n = 0$ is not allowed. This means that according to wave mechanics, the particle can never be at rest. The least energy the particle can have, corresponding to $n = 1$, is called the *zero-point energy*. This result is clearly contradictory to the classical viewpoint, in which $E = 0$ is an acceptable state, as are all positive values of $E$.

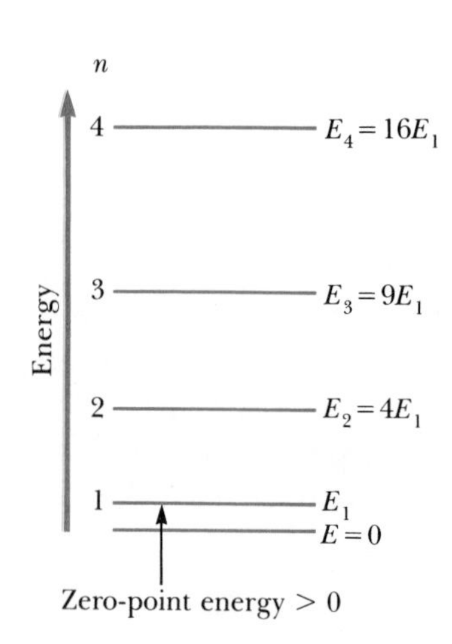


**Figure 41.14** Energy level diagram for a particle confined to a one-dimensional box of width $L$. The lowest allowed energy is $E_1$ and has the value $E_1 = h^2/8mL^2$.

Let us review what we have learned about this problem. Starting with a knowledge of $\psi$ (based upon the imposed boundary conditions), we were able to calculate the probability distribution for the particle for the various allowed states. As an added bonus, we were also able to calculate the allowed energy levels for the particle.

The energy levels are of special importance for the following reason. If the particle was electrically charged (for example, if it was an electron), it could emit a photon if it dropped from an excited state, such as $E_3$, to one of the lower-lying states, such as $E_2$. It could also absorb a photon whose energy matches the difference in energy between two allowed states. For example, if the photon frequency is $f$, the particle will jump from the state $E_1$ to the state $E_2$ if $hf = E_2 - E_1$. The processes of photon emission or absorption can be observed by spectroscopy, in which spectral wavelengths are a direct measure of such energy differences.

**EXAMPLE 41.6 A Bound Electron**
An electron is confined between two impenetrable walls which are 0.2 nm apart. Determine the energy levels for the states $n = 1$, 2, and 3.

*Solution* We can apply Equation 41.17, using $m = 9.11 \times 10^{-31}$ kg and $L = 0.2$ nm $= 2 \times 10^{-10}$ m. For the state $n = 1$, we get

$$E_1 = \frac{h^2}{8mL^2} = \frac{(6.63 \times 10^{-34}\ \text{J} \cdot \text{s})^2}{8(9.11 \times 10^{-31}\ \text{kg})(2 \times 10^{-10}\ \text{m})^2}$$

$$= 1.51 \times 10^{-18}\ \text{J} = 9.42\ \text{eV}$$

The energies of the $n = 2$ and $n = 3$ states are given by $E_2 = 4E_1 = 37.7$ eV and $E_3 = 9E_1 = 84.8$ eV. Although this is a rather primitive model, it can be used to describe an electron trapped in a vacant crystal site or to approximate the energy levels of a nucleon in a nucleus.

**EXAMPLE 41.7 Energy Quantization for a Macroscopic Object**
A small object of mass 1 mg is confined to moving between two rigid walls separated by 1 cm. (a) Calculate the minimum speed of the object.

*Solution* The minimum speed corresponds to the state for which $n = 1$. Using Equation 41.17 with $n = 1$ gives the zero-point energy:

$$E_1 = \frac{h^2}{8mL^2} = \frac{(6.63 \times 10^{-34}\ \text{J} \cdot \text{s})^2}{8(1 \times 10^{-6}\ \text{kg})(1 \times 10^{-2}\ \text{m})^2}$$

$$= 5.49 \times 10^{-58}\ \text{J}$$

Since $E = \frac{1}{2}mv^2$, we can find $v$ as follows:

$$\tfrac{1}{2}mv^2 = 5.49 \times 10^{-58}\ \text{J}$$

$$v = \left[\frac{2(5.49 \times 10^{-58}\ \text{J})}{1 \times 10^{-6}\ \text{kg}}\right]^{1/2} = 3.31 \times 10^{-26}\ \text{m/s}$$

This result is so small that the object can be considered to be at rest, which is what one would expect for a macroscopic object.

(b) If the speed of the object is $3 \times 10^{-2}$ m/s, find the corresponding value of $n$.

*Solution* The kinetic energy of the object is

$$E = \tfrac{1}{2}mv^2 = \tfrac{1}{2}(1 \times 10^{-6}\ \text{kg})(3 \times 10^{-2}\ \text{m/s})^2$$

$$= 4.5 \times 10^{-10}\ \text{J}$$

Since $E_n = n^2E_1$ and $E_1 = 5.49 \times 10^{-58}$ J, we find that

$$n^2E_1 = 4.5 \times 10^{-10}\ \text{J}$$

$$n = \left(\frac{4.5 \times 10^{-10}\ \text{J}}{E_1}\right)^{1/2} = \left(\frac{4.5 \times 10^{-10}\ \text{J}}{5.49 \times 10^{-58}\ \text{J}}\right)^{1/2}$$

$$\approx 9.05 \times 10^{23}$$

This value of $n$ is so large that we would never be able to distinguish the quantized nature of the energy levels. That is, the difference in energy between the two states $n_1 = 9.05 \times 10^{23}$ and $n_2 = (9.05 \times 10^{23}) + 1$ is too small to be detected experimentally. This is another example that illustrates the working of the correspondence principle, that is, as $n \rightarrow \infty$, the quantum description must agree with the classical result. In reality, the speed of the particle in this state cannot be measured since its position in the box cannot be specified.

**EXAMPLE 41.8 Model of an Atom**
An atom can be viewed as several electrons moving around a positively charged nucleus, where the electrons are subject mainly to the coulomb attraction of the nucleus (which is actually partially "screened" by the inner-core electrons). The potential well which each electron "sees" is sketched in Figure 41.15. Use the model of a particle in a box to estimate the energy (in eV) required to raise an electron from the state $n = 1$ to the state $n = 2$, assuming the atom has a radius of 0.1 nm.

*Solution* Using Equation 41.17 and taking the length $L$ of the box to be 0.2 nm (the diameter of the atom) and $m = 9.11 \times 10^{-31}$ kg, we find that

$$E_n = \left(\frac{h^2}{8mL^2}\right)n^2 = \frac{(6.63 \times 10^{-34}\ \text{J} \cdot \text{s})^2}{8(9.11 \times 10^{-31}\ \text{kg})(2 \times 10^{-10}\ \text{m})^2}\,n^2$$

$$= (1.51 \times 10^{-18})n^2\ \text{J} = 9.42n^2\ \text{eV}$$

Hence, the energy difference between the states $n = 1$ and $n = 2$ is

$$\Delta E = E_2 - E_1 = 9.42(2)^2 - 9.42(1)^2 = 28.3\ \text{eV}$$

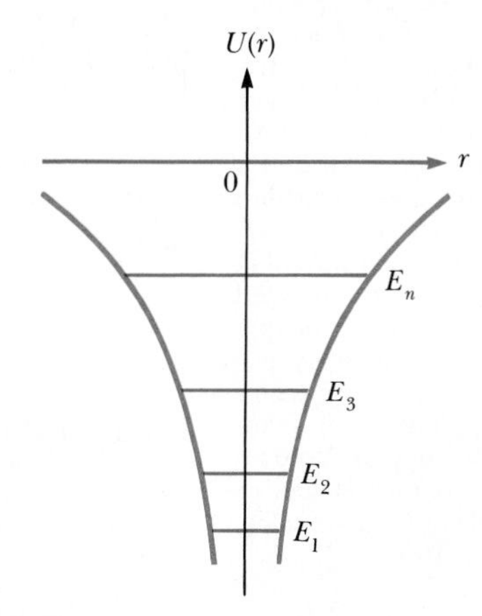


**Figure 41.15** (Example 41.8) Model of the potential energy versus $r$ for the one-electron atom.

We could also calculate the wavelength of the photon which would cause this transition, using the fact that $\Delta E = hc/\lambda$:

$$\lambda = \frac{hc}{\Delta E} = \frac{(6.63 \times 10^{-34}\ \text{J} \cdot \text{s})(3 \times 10^8\ \text{m/s})}{(28.3\ \text{eV} \times 1.6 \times 10^{-19}\ \text{J/eV})}$$

$$= 4.40 \times 10^{-8}\ \text{m} = 44.0\ \text{nm}$$

This wavelength is in the far ultraviolet region, and it is interesting to note that the result is roughly correct. Although this oversimplified model gives a good estimate for transitions between lowest-lying levels of the atom, the estimate gets progressively worse for higher-energy transitions.

## 41.7 THE SCHRÖDINGER EQUATION

Erwin Schrödinger (1887–1961) was an Austrian theoretical physicist best known as the creator of wave mechanics. He also produced important papers in the fields of statistical mechanics, color vision, and general relativity. Schrödinger did much to hasten the universal acceptance of quantum theory by demonstrating the mathematical equivalence between his wave mechanics and the more abstract matrix mechanics developed by Heisenberg. In 1927 Schrödinger accepted the chair of theoretical physics at the University of Berlin where he formed a close friendship with Max Planck. In 1933, he left Germany and eventually settled at the Dublin Institute of Advanced Study where he spent 17 happy, creative years working on problems in general relativity, cosmology, and the application of quantum physics to biology. In 1956 he returned home to Austria and to his beloved Tirolean mountains, where he died in 1961.

As we mentioned earlier, the wave function for de Broglie waves must satisfy an equation developed by Schrödinger in 1927. The basic problem in wave mechanics is to determine a solution to this equation, which in turn yields the allowed wave functions and energy levels of the system under consideration. Proper manipulation of the wave functions allows one to calculate *all* measurable features of the system.

In Chapter 16, we discussed the general form of the wave equation for waves traveling along the $x$ axis. This general form is

$$\frac{\partial^2 \psi}{\partial x^2} = \frac{1}{v^2}\frac{\partial^2 \psi}{\partial t^2} \tag{41.18}$$

where $v$ is the wave speed and where the wave function $\psi$ depends on $x$ and $t$. (We use $\psi$ here instead of $y$ because we are dealing with de Broglie waves.)

In describing de Broglie waves, let us confine our discussion to bound systems whose total energy $E$ remains constant. Since $E = hf$, the frequency of the de Broglie wave associated with the particle also remains constant. In this case, we can express the wave function $\psi(x, t)$ as the product of a term that depends only on $x$ and a term that depends only on $t$. That is,

$$\psi(x, t) = \psi(x)\cos(\omega t) \tag{41.19}$$

This is analogous to the case of standing waves on a string, where the wave function is represented by $y(x, t) = y(x)\cos \omega t$. The frequency-dependent part of the wave function is sinusoidal since the frequency is precisely known. Substituting Equation 41.19 into Equation 41.18 gives

$$\cos(\omega t)\frac{\partial^2 \psi}{\partial x^2} = -\left(\frac{\omega^2}{v^2}\right)\psi\cos(\omega t)$$

or

$$\frac{\partial^2 \psi}{\partial x^2} = -\left(\frac{\omega^2}{v^2}\right)\psi \tag{41.20}$$

Recall that $\omega = 2\pi f = 2\pi v/\lambda$ and, for de Broglie waves, $p = h/\lambda$. Therefore,

$$\frac{\omega^2}{v^2} = \left(\frac{2\pi}{\lambda}\right)^2 = \frac{4\pi^2}{h^2}p^2 = \frac{p^2}{\hbar^2}$$

Furthermore, we can express the total energy $E$ as the sum of the kinetic energy and the potential energy:

$$E = K + U = \frac{p^2}{2m} + U$$

so that

$$p^2 = 2m(E - U)$$

and

$$\frac{\omega^2}{v^2} = \frac{p^2}{\hbar^2} = \frac{2m}{\hbar^2}(E - U)$$

Substituting this result into Equation 41.20 gives

$$\frac{\partial^2 \psi}{\partial x^2} = -\frac{2m}{\hbar^2}(E - U)\psi \qquad (41.21)$$

Time-independent Schrödinger equation

This is the famous **Schrödinger equation** as it applies to a particle confined to moving along the $x$ axis. Since this equation is independent of time, it is commonly referred to as the *time-independent Schrödinger equation.* We shall not discuss the time-dependent Schrödinger equation in this text.

In principle, if the potential energy $U(x)$ is known for the system, one can solve Equation 41.21 and obtain the wave functions and energies for the allowed states. Since $U$ may vary with position, it is necessary to solve the equation in different regions of space. In the process, the wave functions for the different regions must join smoothly at the boundaries. In the language of mathematics, we require that $\psi(x)$ be *continuous.* Furthermore, in order that $\psi(x)$ obey the normalization condition, we require that $\psi(x)$ approach zero as $x$ approaches $\pm\infty$. Finally, $\psi(x)$ must be *single-valued* and $d\psi/dx$ must also be continuous for finite values of $U(x)$.

Required conditions for $\psi(x)$

It is important to recognize that the steps leading to Equation 41.21 do not represent a derivation of the Schrödinger equation. Rather, the procedure represents a plausibility argument based upon an analogy with other wave phenomena that are already familiar to us. The task of solving the Schrödinger equation may be very difficult, depending on the form of the potential energy function. As it turns out, the Schrödinger equation has been extremely successful in explaining the behavior of atomic and nuclear systems, whereas classical physics has failed to do so. Furthermore, when wave mechanics is applied to macroscopic objects, the results agree with classical physics, as required by the correspondence principle.

### The Particle in a Box

Let us solve the Schrödinger equation for the simple problem of a particle in a one-dimensional box of width $L$ (Fig. 41.16). The walls are infinitely high, corresponding to $U(x) = \infty$ for $x = 0$ and $x = L$. The potential energy is constant within the box, and it is convenient to choose $U = 0$ in this region. Hence, in the region $0 < x < L$, we can express the Schrödinger equation in the form

$$\frac{d^2\psi}{dx^2} = -\frac{2mE}{\hbar^2}\psi = -k^2\psi \qquad (41.22)$$

$$k = \frac{\sqrt{2mE}}{\hbar}$$

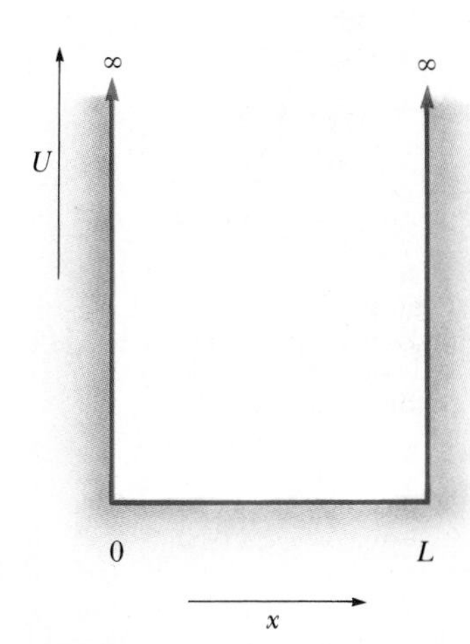


**Figure 41.16** Diagram of a one-dimensional box of width $L$ and infinitely high walls.

Since the walls are infinitely high, the particle cannot exist outside the box. Consequently, $\psi(x)$ must be zero outside the box and at the walls. The solution of Equation 41.22 that meets the boundary conditions $\psi(x) = 0$ at $x = 0$ and $x = L$ is

$$\psi(x) = A \sin(kx) \tag{41.23}$$

This can easily be verified by substitution into Equation 41.22. Note that the first boundary condition, $\psi(0) = 0$, is satisfied by Equation 41.23 since $\sin 0° = 0$. The second boundary condition, $\psi(L) = 0$, is satisfied only if $kL$ is an integral multiple of $\pi$, that is, if $kL = n\pi$, where $n$ is an integer. Since $k = \sqrt{2mE}/\hbar$, we get

$$kL = \frac{\sqrt{2mE}}{\hbar} L = n\pi$$

Solving for the allowed energies $E$ gives

$$E_n = \left(\frac{h^2}{8mL^2}\right) n^2 \tag{41.24}$$

Likewise, the allowed wave functions are given by

$$\psi_n(x) = A \sin\left(\frac{n\pi x}{L}\right) \tag{41.25}$$

These results agree with those obtained in the previous section. It is left as a problem (Problem 37) to show that the normalization constant $A$ for this solution is equal to $(2/L)^{1/2}$.

## *41.8 THE PARTICLE IN A WELL OF FINITE HEIGHT

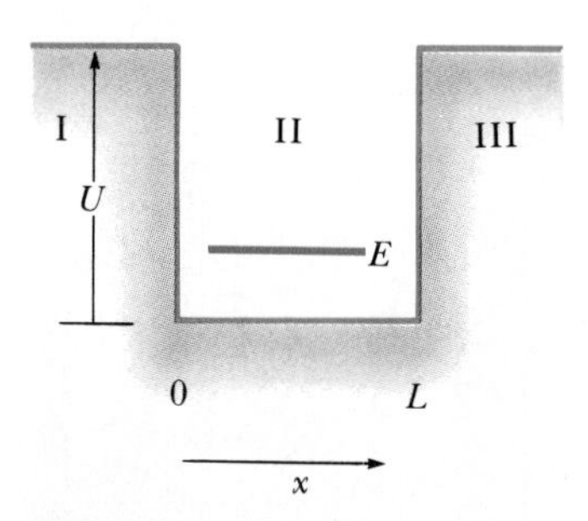


**Figure 41.17** Potential energy diagram of a well of finite height $U$ and width $L$. The energy $E$ of the particle is less than $U$.

Next, consider a particle located in a potential well of finite height $U$ and width $L$, as shown in Figure 41.17. The zero point of the energy is taken to be at the bottom of the well. If the energy $E$ of the particle is less than $U$, classically the particle is permanently bound in the region $0 < x < L$. However, according to quantum mechanics, there is a finite probability that the particle can be found outside this region. That is, the wave function is generally nonzero outside the well, in regions I and III, and so the probability density is also nonzero in these regions.

In region II, where $U = 0$, the allowed wave functions are again sinusoidal since they represent solutions of Equation 41.22. However, the boundary conditions no longer require that $\psi$ be zero at the walls, as was the case with infinitely high walls.

The Schrödinger equation for regions I and III may be written

$$\frac{d^2\psi}{dx^2} = \frac{2m(U - E)}{\hbar^2} \psi \tag{41.26}$$

Since $U > E$, the coefficient on the right-hand side of Equation 41.26 is necessarily positive. Therefore, we can express Equation 41.26 in the form

$$\frac{d^2\psi}{dx^2} = C^2\psi \tag{41.27}$$

where $C^2 = 2m(U - E)/\hbar^2$ is a positive constant in regions I and III. As you can verify by substitution, the general solution of Equation 41.27 is

$$\psi = Ae^{Cx} + Be^{-Cx}$$

where $A$ and $B$ are constants.

We can use this solution as a starting point for determining the appropriate form of the solutions for regions I and III. The function we choose for our solution must remain finite over the entire region under consideration. In region I, where $x < 0$, we must rule out the term $Be^{-Cx}$. In other words, we must require that $B = 0$ in region I in order to avoid an infinite value for $\psi$ for large values of $x$ measured in the negative direction. Likewise, in region III, where $x > L$, we must rule out the term $Ae^{Cx}$; this is accomplished by taking $A = 0$ in this region. This choice avoids an infinite value for $\psi$ for large positive $x$ values. Hence, the solutions in regions I and III are

$$\psi_{\text{I}} = Ae^{Cx} \qquad \text{for } x < 0$$

and

$$\psi_{\text{III}} = Be^{-Cx} \qquad \text{for } x > L$$

In region II the wave function is sinusoidal and has the general form

$$\psi_{\text{II}}(x) = F\sin(kx) + G\cos(kx)$$

where $F$ and $G$ are constants.

These results show that the wave functions in the exterior regions decay exponentially with distance. At large negative $x$ values $\psi_{\text{I}}$ approaches zero exponentially, and at large positive $x$ values $\psi_{\text{III}}$ approaches zero exponentially. These functions, together with the sinusoidal solution in region II, are shown in Figure 41.18a for the first three states. In evaluating the complete wave function, we require that

$$\psi_{\text{I}} = \psi_{\text{II}} \qquad \text{and} \qquad \frac{d\psi_{\text{I}}}{dx} = \frac{d\psi_{\text{II}}}{dx} \qquad \text{at } x = 0$$

and that

$$\psi_{\text{II}} = \psi_{\text{III}} \qquad \text{and} \qquad \frac{d\psi_{\text{II}}}{dx} = \frac{d\psi_{\text{III}}}{dx} \qquad \text{at } x = L$$

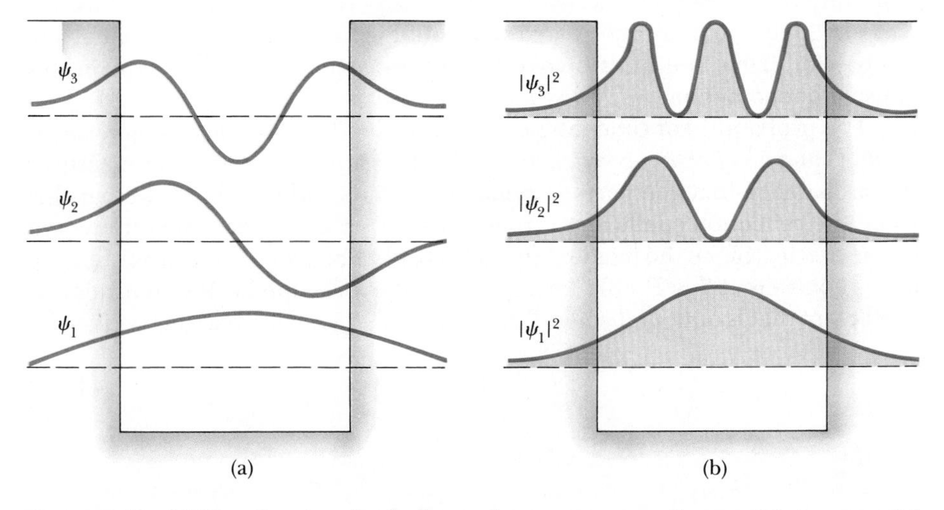


**Figure 41.18** (a) Wave functions for the lowest three energy states for a particle in a potential well of finite height. (b) Probability densities for the lowest three energy states for a particle in a potential well of finite height.

Figure 41.18b represents plots of the probability densities for these states. Note that in each case the wave functions join smoothly at the boundaries of the potential well. These boundary conditions and plots follow from the Schrödinger equation. Further inspection of Figure 41.18a shows that the wave functions are not equal to zero at the walls of the potential well and in the exterior regions. Therefore, the probability density is nonzero at these points. The fact that $\psi$ is nonzero at the walls *increases* the de Broglie wavelength in region II (compare the case of a particle in a potential well of infinite depth), and this in turn lowers the energy and momentum of the particle.

## *41.9 TUNNELING THROUGH A BARRIER

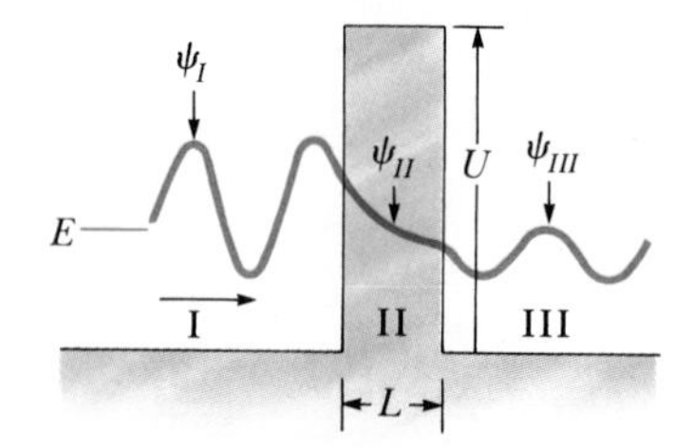


**Figure 41.19** Wave function for a particle incident from the left on a barrier of height $U$. Note that the wave function is sinusoidal in regions I and III but is exponentially decaying in region II.

A very interesting and peculiar phenomenon occurs when a particle strikes a barrier of finite height and width. Consider a particle of energy $E$ incident on a rectangular barrier of height $U$ and width $L$, where $E < U$ (Fig. 41.19). Classically, the particle is reflected by the barrier since it does not have sufficient energy to cross, or even penetrate it. Thus, regions II and III are classically *forbidden* to the particle. According to quantum mechanics, however, *all regions are accessible to the particle, regardless of its energy,* since the amplitude of the matter wave associated with the particle is nonzero everywhere. A typical waveform for this case, illustrated in Figure 41.19, shows the penetration of the wave into the barrier and beyond. The wave functions are sinusoidal to the left (region I) and right (region II) of the barrier and join smoothly with an exponentially decaying function within the barrier (region II). Since the probability of locating the particle is proportional to $|\psi|^2$, we conclude that the chance of finding the particle beyond the barrier in region III is nonzero. This barrier penetration is in complete disagreement with classical physics. The possibility of finding the particle on the far side of the barrier is called **tunneling** or **barrier penetration.** Although the particle *can never be observed inside the barrier* (because it would violate conservation of energy), it is able to tunnel through that region and be observed in region III. How can the particle penetrate the barrier? The answer is that it can't if we describe it as a classical particle. Only the de Broglie wave associated with the particle can penetrate the barrier. Thus, we must invoke the wave-particle duality to explain this unusual phenomenon.

The probability of tunneling can be described with a *transmission coefficient, T,* and a *reflection coefficient, R.* The transmission coefficient measures the probability that the particle penetrates to the other side of the barrier, while the reflection coefficient is the probability that the particle is reflected by the barrier. Since the incident particle is either reflected or transmitted, we must require that $T + R = 1$. An approximate expression for the transmission coefficient that is obtained when $T \ll 1$ (a very high or wide barrier) is given by

$$T \cong e^{-2KL} \tag{41.28}$$

where

$$K = \frac{\sqrt{2m(U - E)}}{\hbar} \tag{41.29}$$

**EXAMPLE 41.9 Transmission Coefficient for an Electron**

A 30-eV electron is incident on a square barrier of height 40 eV. What is the probability that the electron will tunnel through the barrier if its width is (a) 1 nm, and (b) 0.1 nm?

*Solution* (a) In this situation, the quantity $U - E$ has the value

$$U - E = (40 \text{ eV} - 30 \text{ eV}) = 10 \text{ eV} = 1.6 \times 10^{-18} \text{ J}$$

Using Equation 41.29, and given that $L = 1$ nm, the quantity $2KL$ is

$$2KL = 2\,\frac{\sqrt{2(9.11 \times 10^{-31}\text{ kg})(1.6 \times 10^{-18}\text{ J})}}{1.054 \times 10^{-34}\text{ J}\cdot\text{s}}\,(1 \times 10^{-9}\text{ m}) = 32.4$$

Thus, the probability of tunneling through the barrier is

$$T \cong e^{-2KL} = e^{-32.4} = \boxed{8.49 \times 10^{-15}}$$

That is, the electron has only about 1 chance in $10^{14}$ to tunnel through the 1-nm wide barrier.

(b) For $L = 0.1$ nm, we find $2KL = 3.24$, and

$$T \cong e^{-2KL} = e^{-3.24} = \boxed{0.0392}$$

This result shows that the electron has a high probability (4% chance) of penetrating the 0.1-nm barrier. Thus, reducing the width of the barrier by only one order of magnitude has increased the probability of tunneling by about twelve orders of magnitude!

### Applications of Tunneling

As we have seen, tunneling is a quantum phenomenon, and is a manifestation of the wave nature of matter. There are many examples in nature on the atomic and nuclear scales for which tunneling is very important. We shall briefly describe four such examples.

1. **Tunnel diode** The tunnel diode is a semiconductor device consisting of two oppositely charged regions separated by a very narrow neutral region. The current in this device is largely due to tunneling of electrons through the neutral region. The current, or rate of tunneling, can be controlled over a wide range by varying the bias voltage, which changes the height of the barrier.
2. **Josephson junction** The Josephson junction consists of two superconductors separated by a thin insulating oxide layer, 1 to 2 nm thick. Under appropriate conditions, electrons in the superconductors travel as pairs and tunnel from one superconductor to the other through the oxide layer. Several effects in this type of junction have been observed. For example, a dc current is observed across the junction *in the absence of electric or magnetic fields*. The current is proportional to $\sin\phi$, where $\phi$ is the phase difference between the wave functions in the two superconductors. When a bias voltage $V$ is applied across the junction, one observes oscillations in the current, with a frequency given by $f = 2eV/h$, where $e$ is the charge on the electron. We shall return to the discussion of Josephson junctions in Chapter 44.
3. **Alpha decay** One form of radioactive decay is the emission of alpha particles (the nuclei of helium atoms) by unstable, heavy nuclei. In order for the alpha particle to escape from the nucleus, it must penetrate a barrier due to a combination of the attractive nuclear force and the Coulomb repulsion between the alpha particle and the remaining part of the nucleus. Occasionally, an alpha particle tunnels through the barrier, which explains the basic mechanism for this type of decay and the large variations in the mean lifetimes of various radioactive nuclei. We shall return to this topic in Chapter 45.

4. **Scanning tunneling microscope** The scanning tunneling microscope, or STM, is a remarkable device which uses tunneling to create images of surfaces with resolution comparable to the size of a single atom. A small probe with a very fine tip is scanned very close to the surface of a specimen. A tunneling current is maintained between the probe and specimen; the current is very sensitive to the separation between the tip and specimen. By maintaining a constant tunneling current, a feedback signal is obtained which is used to raise and lower the probe as the surface is scanned. Since the vertical motion of the probe follows the contour of the specimen's surface, one obtains an image of the surface. A more detailed discussion of this important device is provided in the interesting essay that follows this chapter.

## *41.10 THE SIMPLE HARMONIC OSCILLATOR

Finally, let us consider the problem of a particle subject to a linear restoring force $F = -kx$, where $x$ is the displacement of the particle from equilibrium ($x = 0$) and $k$ is the force constant. The classical motion of a particle subject to such a force is simple harmonic motion, which was discussed in Chapter 13. The potential energy of such a system is given by

$$U = \tfrac{1}{2}kx^2 = \tfrac{1}{2}m\omega^2x^2$$

where the angular frequency of vibration is $\omega = \sqrt{k/m}$. Classically, if the particle is displaced from its equilibrium position and released, it will oscillate between the points $x = -A$ and $x = A$, where $A$ is the amplitude of motion. Furthermore, its total energy $E$ is given by

$$E = K + U = \tfrac{1}{2}kA^2 = \tfrac{1}{2}m\omega^2A^2$$

In the classical model, any value of $E$ is allowed and the total energy may be zero if the particle is at rest at $x = 0$.

The Schrödinger equation for this problem is obtained by substituting $U = \tfrac{1}{2}m\omega^2x^2$ into Equation 41.21. This gives

$$\frac{d^2\psi}{dx^2} = -\left[\left(\frac{2mE}{\hbar^2}\right) - \left(\frac{m\omega}{\hbar}\right)^2 x^2\right]\psi \tag{41.30}$$

The mathematical technique for solving this equation is beyond the level of this text. However, it is instructive to guess at a solution. We take as our guess the following wave function:

$$\psi = Be^{-Cx^2} \tag{41.31}$$

Substituting this function into Equation 41.30, we find that Equation 41.31 is a satisfactory solution to the Schrödinger equation provided that

$$C = \frac{m\omega}{2\hbar} \quad \text{and} \quad E = \tfrac{1}{2}\hbar\omega$$

It turns out that the solution we have guessed corresponds to the ground state of the system. This is the state that has the lowest energy, $\tfrac{1}{2}\hbar\omega$, which is the **zero-point energy** of the system. Since $C = m\omega/2\hbar$, it follows from Equation 41.31 that the **wave function** for this state is

Wave function for the ground state of a simple harmonic oscillator

$$\psi = Be^{-(m\omega/2\hbar)x^2} \tag{41.32}$$

Note that this is only one solution to Equation 41.30. The remaining solutions, which describe the excited states, are more complicated and will not be given here. However, all the solutions have the form of an exponential factor, $e^{-Cx^2}$, multiplied by a polynomial in $x$.

The energy levels of a harmonic oscillator are quantized, as we would expect. The energy of the state whose quantum number is $n$ is given by

Allowed energies for a simple harmonic oscillator

$$E_n = (n + \tfrac{1}{2})\hbar\omega \qquad n = 0, 1, 2, \ldots \tag{41.33}$$

The state $n = 0$ corresponds to the ground state, whose energy is $E_0 = \frac{1}{2}\hbar\omega$; the state $n = 1$ corresponds to the first excited state, whose energy is given by $E_1 = \frac{3}{2}\hbar\omega$; and so on. The energy level diagram for this system is shown in Figure 41.20. Note that the separations between consecutive levels are equal and given by

$$\Delta E = \hbar\omega \tag{41.34}$$

The probability densities for the first three states of a harmonic oscillator are indicated by the red curves in Figure 41.21. The blue lines represent the classical probability densities corresponding to the same energy, provided for comparison. Note that as $n$ increases, the agreement between the classical and quantum mechanics probabilities improves, as expected.

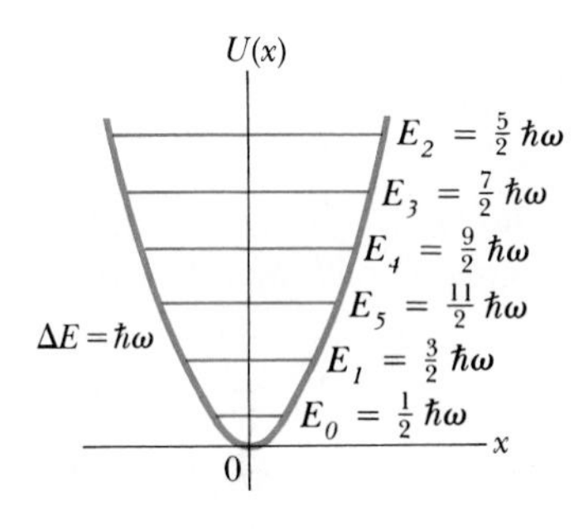


**Figure 41.20** Energy level diagram for a simple harmonic oscillator. Note that the levels are equally spaced, with a separation equal to $\hbar\omega$. The zero-point energy is equal to $E_0$.

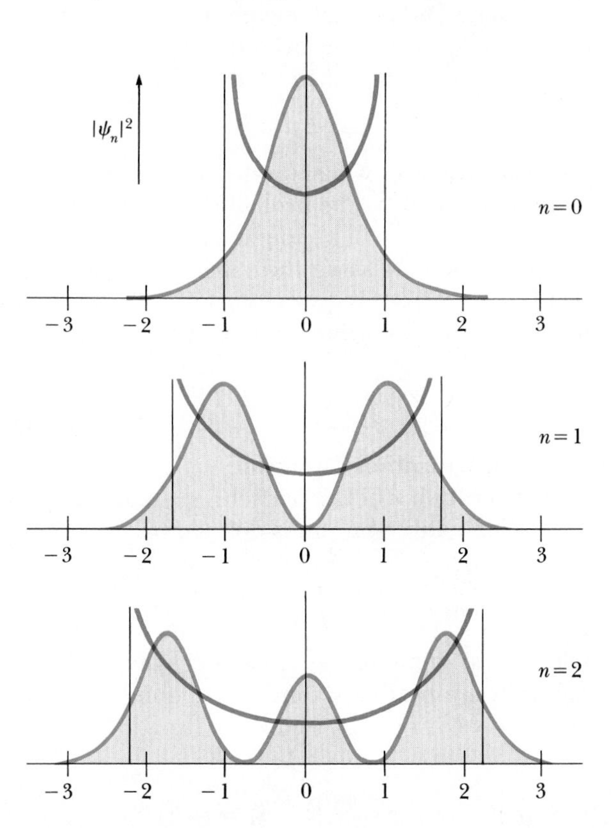


**Figure 41.21** Probability densities for a few states of a harmonic oscillator represented by the red curves. The blue curves represent the classical probabilities corresponding to the same energies. (From C.W. Sherwin, *Introduction to Quantum Mechanics*, New York, Holt, Rinehart and Winston, 1959; used with permission)

The quantum mechanics solution to the simple harmonic oscillator problem predicts a series of equally spaced energy levels with separations equal to $\hbar\omega$. This result represents a justification of Planck's quantum hypothesis, made 25 years before the Schrödinger equation was developed. Furthermore, the solution presented is useful for describing more complicated problems, such as molecular vibrations, which can be approximated by the idealized model of simple harmonic motion.

## SUMMARY

Every object of mass $m$ and momentum $p$ has wavelike properties with a wavelength given by the de Broglie relation

The de Broglie wavelength of an object with momentum $p$.

$$\lambda = \frac{h}{p} \tag{41.3}$$

By applying this wave theory of matter to electrons in atoms, de Broglie was able to explain the appearance of quantization in the Bohr model of hydrogen as a standing wave phenomenon.

The **uncertainty principle** states that if a measurement of position is made with precision $\Delta x$ and a *simultaneous* measurement of momentum is made with precision $\Delta p$, then the product of the two uncertainties can never be smaller than a number of the order of $\hbar$.

(The uncertainty principle)

$$\Delta x \, \Delta p \gtrsim \hbar \tag{41.5}$$

In quantum mechanics, matter waves (or de Broglie waves) are represented by a wave function $\psi(x, y, z, t)$. The probability per unit volume (or probability density) that a particle will be found at a point is equal to $|\psi|^2$. If the particle is confined to moving along the $x$ axis, then the probability that it will be located in an interval $dx$ is given by $|\psi|^2\, dx$. Furthermore, the sum of all these probabilities over all values of $x$ must be 1. That is,

Normalization condition on $\psi$

$$\int_{-\infty}^{\infty} |\psi|^2 \, dx = 1 \tag{41.9}$$

This is called the **normalization condition.**

The measured position $x$ of the particle, averaged over many trials, is called the **expectation value** of $x$ and is defined by

Expectation value of $x$

$$\langle x \rangle \equiv \int_{-\infty}^{\infty} x|\psi|^2 \, dx \tag{41.11}$$

If a particle of mass $m$ is confined to moving in a one-dimensional box of width $L$ whose walls are perfectly rigid, we require that $\psi$ be zero at the walls and outside the box. The **allowed wave functions** for the particle are given by

Allowed wave functions for a particle in a box

$$\psi(x) = A \sin\left(\frac{n\pi x}{L}\right) \qquad n = 1, 2, 3, \ldots \tag{41.15}$$

where $A$ is the maximum value of $\psi$. The particle has a well-defined wavelength $\lambda$ whose values are such that the width of the box $L$ is equal to an integral number of half wavelengths, that is, $L = n\lambda/2$. These allowed states are called **stationary states** of the system. The energies of a particle in a box are quantized and are given by

$$E_n = \left(\frac{h^2}{8mL^2}\right)n^2 \qquad n = 1, 2, 3, \ldots \tag{41.17}$$

Allowed energies for a particle in a box

The wave function must satisfy the **Schrödinger equation.** The time-independent Schrödinger equation for a particle confined to moving along the $x$ axis is

$$\frac{\partial^2 \psi}{\partial x^2} = -\frac{2m}{\hbar^2}(E - U)\psi \tag{41.21}$$

Time-independent Schrödinger equation

where $E$ is the total energy of the system and $U$ is the potential energy.

The approach of quantum mechanics is to solve Equation 41.21 for $\psi$ and $E$, given the potential energy $U(x)$ for the system. In doing so, we must place special restrictions on $\psi(x)$. We require (1) that $\psi(x)$ be continuous, (2) that $\psi(x)$ approach zero as $x$ approaches $\pm\infty$, (3) that $\psi(x)$ be single-valued, and (4) that $d\psi/dx$ be continuous for all finite values of $U(x)$.

Restrictions on $\psi(x)$

When a particle of energy $E$ meets a barrier of height $U$, where $E < U$, the particle has a finite probability of penetrating the barrier. Part of the incident wave is transmitted through the barrier, and part is reflected. This process, called **tunneling,** is the basic mechanism that explains the operation of the Josephson junction and the phenomenon of alpha decay in some radioactive nuclei.

When the simple harmonic oscillator system is solved using quantum mechanics, one finds that the energy of the system is quantized and has the values

$$E_n = (n + \tfrac{1}{2})\hbar\omega \qquad n = 0, 1, 2, \ldots \tag{41.33}$$

Allowed energies for a simple harmonic oscillator

The ground state of the system, corresponding to $n = 0$, is given by $E_0 = \frac{1}{2}\hbar\omega$. This is called the **zero-point energy.** The adjacent energy levels of the system have a constant separation equal to $\Delta E = \hbar\omega$.

## QUESTIONS

1. Is light a wave or a particle? Support your answer by citing specific experimental evidence.
2. Is an electron a particle or a wave? Support your answer by citing some experimental results.
3. An electron and a proton are accelerated from rest through the same potential difference. Which particle has the longer wavelength?
4. If matter has a wave nature, why is this wave-like character not observable in our daily experiences?
5. In what way does Bohr's model of the hydrogen atom violate the uncertainty principle?
6. Why is it impossible to simultaneously measure the position and velocity of a particle with infinite accuracy?
7. Suppose that a beam of electrons is incident on three or more slits. How would this influence the interference pattern? Would the state of an electron depend on the number of slits? Explain.
8. In describing the passage of electrons through a slit and arriving at a screen, Feynman said that "electrons arrive in lumps, like particles, but the probability of arrival of these lumps is determined as the intensity of the waves would be. It is in this sense that the electron behaves sometimes like a particle and sometimes like a wave." Elaborate on this point in your own words. (For a further discussion of this point, see R. Feynman, *The Character of Physical Law,* Cambridge, Mass., MIT Press, 1980, Chapter 6.)

9. The probability density at certain points for a particle in a box is zero, as seen in Figure 41.13b. Does this imply that the particle cannot move across these points? Explain.
10. Discuss the relation between the zero-point energy and the uncertainty principle.
11. As a particle of energy $E$ is reflected from a potential barrier of height $U$, where $E < U$, how does the amplitude of the reflected wave change as the barrier height is reduced?
12. A philosopher once said that "it is necessary for the very existence of science that the same conditions always produce the same results." In view of what has been discussed in this section, present an argument showing that this statement is false. How might the statement be reworded to make it acceptable?
13. In wave mechanics it is possible for the energy $E$ of a particle to be less than the potential energy, but classically this is not possible. Explain.
14. Consider two square wells of the same width, one with finite walls and the other with infinite walls. What can you say about the value of the energy and momentum of a particle trapped in the finite well as it compares to the energy and momentum of an identical particle in the infinite well?
15. Why cannot the lowest energy state of a harmonic oscillator be zero?
16. Why is an electron microscope more suitable than an optical microscope for "seeing" objects of an atomic size?
17. What is the Schrödinger equation? How is it useful in describing atomic phenomena?
18. Why was the Davisson-Germer diffraction of electrons an important experiment?
19. What is the significance of the wavefunction $\psi$?

## PROBLEMS

### Section 41.1 Photons and Electromagnetic Waves and Section 41.2 The Wave Properties of Particles

1. Calculate the de Broglie wavelength for a proton moving with a speed of $10^6$ m/s.
2. Calculate the de Broglie wavelength for an electron with kinetic energy (a) 50 eV and (b) 50 keV.
3. Calculate the de Broglie wavelength of a 75-kg person who is jogging at 5 m/s.
4. The "seeing" ability, or resolution, of radiation is determined by its wavelength. If the size of an atom is of the order of 0.1 nm, how fast must an electron travel to have a wavelength small enough to "see" an atom?
5. Find the de Broglie wavelength of a 0.15-kg ball moving with a speed of 20 m/s.
6. Calculate the de Broglie wavelength of a proton that is accelerated through a potential difference of 10 MV.
7. Show that the de Broglie wavelength of an electron accelerated from rest through a potential difference $V$ is given by $\lambda = 1.226/\sqrt{V}$ nm, where $V$ is in volts.
8. The distance between adjacent atoms in crystals is of the order of 1.0 Å. The use of electrons in diffraction studies of crystals requires that the de Broglie wavelength of the electrons is of the order of the distance between atoms of the crystals. What must be the minimum energy (in eV) of electrons to be used for this purpose?
9. An electron has a de Broglie wavelength equal to the diameter of the hydrogen atom. What is the kinetic energy of the electron? How does this energy compare with the ground-state energy of the hydrogen atom?
10. In order for an electron to be confined to a nucleus, its de Broglie wavelength would have to be less than $10^{-14}$ m. (a) What would be the kinetic energy of an electron confined to this region? (b) On the basis of this result, would you expect to find an electron in a nucleus? Explain.
11. In the Davisson–Germer experiment, 54-eV electrons were diffracted from a nickel lattice. If the first maximum in the diffraction pattern was observed at $\phi = 50°$ (Figure 41.22), what was the lattice spacing $d$ of nickel atoms?

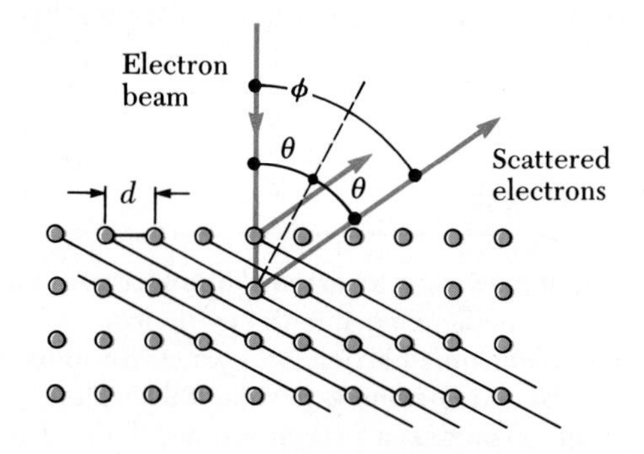


**Figure 41.22** (Problem 11).

12. Robert Hofstadter won the 1961 Nobel Prize in physics for his pioneering work in scattering 20-GeV electrons from nuclei. (a) What is the $\gamma$-factor for a 20-GeV electron, where $\gamma = (1 - v^2/c^2)^{-1/2}$? What is the momentum of the electron in kg·m/s? (b) What is the wavelength of a 20-GeV electron and how does it compare to the size of a nucleus?

13. Electrons are accelerated through 40 000 V in an electron microscope. What, theoretically, would be the smallest distance between objects that can be observed?

### Section 41.3 The Double-Slit Experiment Revisited

14. Through what potential difference would an electron have to be accelerated to give it a de Broglie wavelength of $10^{-10}$ m?
15. A monoenergetic beam of electrons is incident on a single slit of width 0.5 nm. A diffraction pattern is formed on a screen 20 cm from the slit. If the distance between successive minima of the diffraction pattern is 2.1 cm, what is the energy of the incident electrons?
16. A neutron beam with a selected speed of 0.4 m/s is directed through a double slit with a 1-mm separation. An array of detectors is placed 10 m from the slit. (a) What is the de Broglie wavelength of the neutrons? (b) How far off axis is the first zero-intensity point on the detector array? (c) Can we say which slit the neutron passed through? Explain.
17. The resolving power of a microscope is proportional to the wavelength used. If one wished to use a microscope to "see" an atom, a resolution of approximately $10^{-11}$ m (0.1 Å) would have to be obtained. (a) If electrons are used (electron microscope), what minimum kinetic energy is required for the electrons? (b) If photons are used what minimum photon energy is needed to obtain $10^{-11}$ m resolution?
18. An air rifle is used to shoot 1-g particles at a speed of 100 m/s through a hole of diameter 2 mm. How far from the rifle must an observer be in order to see the beam spread by 1 cm because of the uncertainty principle? Compare this answer to the diameter of the universe ($2 \times 10^{26}$ m).

### Section 41.4 The Uncertainty Principle

19. A light source is used to determine the location of an electron in an atom to a precision of 0.05 nm. What is the uncertainty in the velocity of the electron?
20. Suppose Fuzzy, the quantum-mechanical duck, lives in a world in which $h = 2\pi$ J·s. Fuzzy has a mass of 2.0 kg and is initially known to be within a region 1.0 m wide. (a) What is his minimum uncertainty in speed? (b) Assuming this uncertainty in speed to prevail for 5.0 s, determine the uncertainty in position after this time.
21. A proton has a kinetic energy of 1 MeV. If its momentum is measured with an uncertainty of 5%, what is the minimum uncertainty in its position?
22. An electron ($m = 9.11 \times 10^{-31}$ kg) and a bullet ($m = 0.02$ kg) each have a speed of 500 m/s, accurate to within 0.01%. Within what limits could we determine the position of the objects?
23. A small boy on a ladder drops small pellets toward a spot on the floor. (a) Show that according to the uncertainty principle the miss distance must be at least

$$\Delta x = \left(\frac{\hbar}{m}\right)^{1/2} \left(\frac{H}{2g}\right)^{1/4}$$

where $H$ is the initial vertical distance of each pellet above the floor and $m$ is the mass of each pellet. (b) If $H = 2$ m, and $m = \frac{1}{2}$ g, what is $\Delta x$?

### Section 41.5 Introduction to Quantum Mechanics

24. A free electron has a wave function

$$\psi(x) = A \sin(5 \times 10^{10} x)$$

where $x$ is measured in m. Find (a) the electron's de Broglie wavelength, (b) the electron's momentum, and (c) the electron's energy in eV.
25. An electron has a wavefunction

$$\psi(x) = \sqrt{\frac{2}{L}} \sin\left(\frac{2\pi x}{L}\right)$$

Find the probability of finding the electron between $x = 0$ and $x = \frac{L}{4}$.

### Section 41.6 A Particle in a Box

26. An electron is confined to a one-dimensional region in which its ground-state ($n = 1$) energy is 2 eV. (a) What is the width of the region? (b) How much energy is required to "promote" the electron to its first excited state?
27. Use the particle-in-a-box model to calculate the first three energy levels of a neutron trapped in a nucleus $2 \times 10^{-5}$ nm in diameter. Are the energy level differences realistic?
28. (a) Use the uncertainty principle to estimate the uncertainty in momentum for a particle in a box. (b) Estimate the ground state energy, and compare your result with the actual ground state energy.
29. A bead of mass 5 g slides freely on a wire 20 cm long. Treating this system as a particle in a one-dimensional box, calculate the value of $n$ corresponding to the state of the bead if it is moving at a speed of 0.1 nm per year (that is, apparently at rest).
30. An electron with an energy of approximately 6 eV moves between rigid walls exactly 1 nm apart. (a) Find the quantum number $n$ for the energy state that the electron occupies. (b) Find the exact value for the electron's energy.
31. An alpha particle in a nucleus can be modeled as a particle moving in a box of width $10^{-14}$ m (the approximate diameter of a nucleus). Using this model,

estimate the energy and momentum of an alpha particle in its lowest energy state. (The mass of an alpha particle is $4 \times 1.66 \times 10^{-27}$ kg.)

32. Find the wave functions for the three lowest energy states for a particle in a one-dimensional box. Note that the normalization constant for all three functions in equal to $\sqrt{2/L}$ (see Problem 37).

33. An electron is contained in a one-dimensional box of width 0.1 nm. (a) Draw an energy level diagram for the electron for levels up to $n = 4$. (b) Find the wavelengths of *all* photons that can be emitted by the electron in making transitions that would eventually get it from the $n = 4$ state to the $n = 1$ state.

34. Consider a particle moving in a one-dimensional box with walls at $x = -L/2$ and $x = L/2$. (a) Write the wave functions and probability densities for the states $n = 1, n = 2$, and $n = 3$. (b) Sketch the wave functions and probability densities. (*Hint:* Make an analogy to the case of a particle in a box with walls at $x = 0$ and $x = L$.)

35. A ruby laser emits light of wavelength 694.3 nm. If this light is due to transitions from the $n = 2$ state to the $n = 1$ state of an electron in a box, find the width of the box.

36. A proton is confined to moving in a one-dimensional box of width 0.2 nm. (a) Find the lowest possible energy of the proton. (b) What is the lowest possible energy of an electron confined to the same box? (c) How do you account for the large difference in your results for (a) and (b)?

## Section 41.7 The Schrödinger Equation

37. The wave function for a particle confined to moving in a one-dimensional box is given by

$$\psi(x) = A \sin\left(\frac{n\pi x}{L}\right)$$

Use the normalization condition on $\psi$ to show that the constant $A$ is given by

$$A = \sqrt{\frac{2}{L}}$$

*Hint:* Since the particle is confined to the box of width $L$, the normalization condition (Eq. 41.9) in this case becomes

$$\int_0^L |\psi|^2 \, dx = 1$$

38. A particle in the space $-a \leq x \leq a$ may be represented by either of the following wave functions:

$$\psi_1 = A \cos\left(\frac{\pi x}{2a}\right) \quad \text{or} \quad \psi_2 = B \sin\left(\frac{\pi x}{a}\right)$$

Using the normalization condition on $\psi$, find $A$ and $B$. (See Problem 37 for a helpful hint.)

39. The wave function of a particle is given by

$$\psi(x) = A \cos(kx) + B \sin(kx)$$

where $A$, $B$, and $k$ are constants. Show that $\psi$ is a solution of the Schrödinger equation (Eq. 41.21), assuming the particle is free ($U = 0$), and find the corresponding energy $E$ of the particle.

40. In a region of space, a particle with zero energy has a wave function given by

$$\psi(x) = Axe^{-x^2/L^2}$$

(a) Find the potential energy $U$ as a function of $x$. (b) Make a sketch of $U(x)$ versus $x$.

41. Show that the time dependent wavefunction

$$\psi = Ae^{i(kx-\omega t)}$$

where $k = 2\pi/\lambda$ is a solution to both the wave equation (Eq. 41.18) and the Schrödinger equation (Eq. 41.21).

## *Section 41.8 The Particle in a Well of Finite Height

42. Sketch the wave function $\psi(x)$ and the probability density $|\psi(x)|^2$ for the $n = 4$ state of a particle in a *finite* potential well. (See Fig. 41.18)

43. Suppose a particle is trapped in its *ground* state in a box with infinitely high walls (Fig. 41.13a). Now suppose the height of the left-hand barrier is suddenly lowered to a finite height. (a) Qualitatively sketch the wave function for the particle a short time later. (b) If the box has a width $L$, what is the wavelength of the wave that penetrates the barrier?

44. A particle with 7 eV of kinetic energy moves from a region where the potential is zero into one in which $U = 5$ eV (Fig. 41.23). Classically, one would expect the particle to continue on, although with less kinetic energy. According to quantum mechanics, the particle has a probability of being transmitted and a probability of being reflected. What are these probabilities?

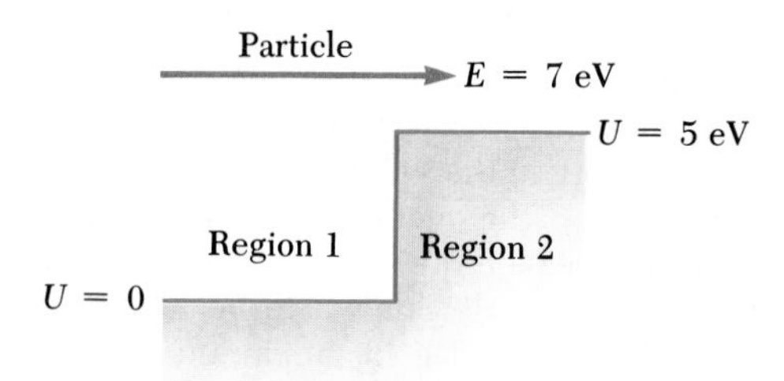


**Figure 41.23** (Problem 44).

## *Section 41.9 Tunneling Through a Barrier and *Section 41.10 The Simple Harmonic Oscillator

45. A 5-eV electron is incident on a barrier 0.2 nm thick and 10 eV high (Figure 41.24). (a) What is the probability that the electron will tunnel through the

barrier? (b) What is the probability that the electron will be reflected?

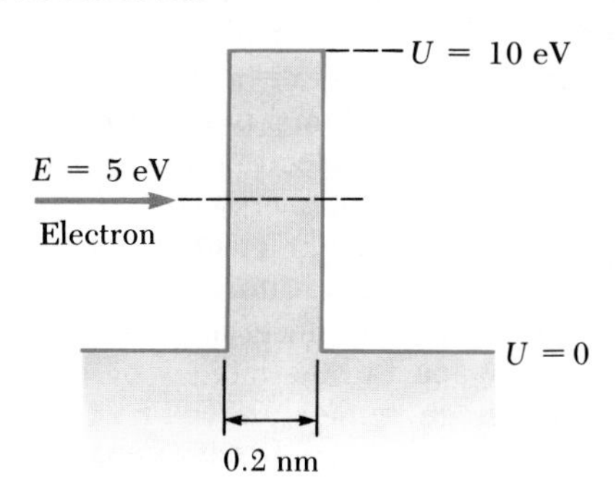


**Figure 41.24** (Problem 45).

46. A one-dimensional harmonic oscillator wavefunction is

$$\psi = Axe^{-bx^2}$$

(a) Show that $\psi$ satisfies Schrödinger's equation (Eq. 41.30). (b) Find $b$ and the total energy $E$. (c) Is this a ground state or a first excited state?

47. Show that the oscillator energies in Equation 41.34 correspond to the classical amplitudes

$$A_n = \sqrt{\frac{\hbar}{m\omega}(2n+1)}$$

48. The total energy of a particle moving with simple harmonic motion along the $x$ axis is

$$E = \frac{p_x^2}{2m} + \frac{Kx^2}{2}$$

where $p_x$ is the momentum of the particle and $K$ is the spring constant. (a) Using the uncertainty principle, show that this can be written as

$$E = \frac{p_x^2}{2m} + \frac{K\hbar^2}{2p_x^2}$$

(b) Show that the *minimum* kinetic energy of the harmonic oscillator is

$$K_{\text{min}} = \frac{p_x^2}{2m} = \frac{1}{2}\hbar\sqrt{\frac{K}{m}} = \frac{1}{2}\hbar\omega$$

## ADDITIONAL PROBLEMS

49. The nuclear potential that binds protons and neutrons in the nucleus of an atom is often approximated by a square well. Imagine a proton confined in an infinite square well of width $10^{-5}$ nm, a typical nuclear diameter. Calculate the wavelength and energy associated with the photon that is emitted when the proton undergoes a transition from the first excited state ($n = 2$) to the ground state ($n = 1$). In what region of the electromagnetic spectrum does this wavelength belong?

50. A two-slit electron diffraction experiment is done with slits of *unequal* widths. When only slit 1 is open, the number of electrons reaching the screen per second is 25 times the number of electrons reaching the screen per second when only slit 2 is open. When both slits are open, an interference pattern results in which the destructive interference is not complete. Find the ratio of the probability of an electron arriving at an interference maximum to the probability of an electron arriving at an adjacent interference minimum. (*Hint:* Use the superposition principle.)

51. An electron is trapped in an infinitely deep potential well 0.3 nm in width. (a) If the electron is in its ground state, what is the probability of finding it within 0.1 nm of the left-hand wall? (b) Repeat (a) for an electron in the 99th energy state above the ground state. (c) Are your answers consistent with the correspondence principle? (*Hint:* Use Eq. 41.10.)

52. An atom in an excited state 1.8 eV above the ground state remains in that excited state on the average of 2 $\mu$s before undergoing a transition to the ground state. Find (a) the frequency and (b) the wavelength of the emitted photon. (c) Find the approximate uncertainty in energy of the photon.

53. A $\pi^0$ meson is an unstable particle that is produced in high-energy particle collisions. It has a mass–energy equivalent of about 135 MeV, and it exists for an average lifetime of only $8.7 \times 10^{-17}$ s before decaying into two gamma rays. Using the uncertainty principle, estimate the fractional uncertainty $\Delta m/m$ in its mass determination.

54. The neutron has a mass of $1.67 \times 10^{-27}$ kg. Neutrons emitted in nuclear reactions can be slowed down via collisions with matter. They are referred to as thermal neutrons once they come into thermal equilibrium with their surroundings. The average kinetic energy ($3kT/2$) of a thermal neutron is approximately 0.04 eV. Calculate the de Broglie wavelength of a neutron with a kinetic energy of 0.04 eV. How does it compare with the characteristic atomic spacing in a crystal? Would you expect thermal neutrons to exhibit diffraction effects when scattered by a crystal?

**55.** A particle of mass $2 \times 10^{-28}$ kg is confined to a one-dimensional box of width $10^{-10}$ m (1 Å). For $n = 1$, (a) What is the particle wavelength? (b) Its momentum? (c) Its ground-state energy?

56. A particle is described by the wave function

$$\psi(x) = \begin{cases} A\cos\left(\dfrac{2\pi x}{L}\right) & \text{for } -\dfrac{L}{4} \le x \le \dfrac{L}{4} \\ 0 & \text{for other values of } x \end{cases}$$

(a) Determine the normalization constant $A$. (b) What is the probability that the particle will be found between $x = 0$ and $x = L/8$ if a measurement of its position is made? (*Hint:* Use Eq. 41.10.)

57. For a particle described by a wave function $\psi(x)$, the average value, or expectation value, of a physical quantity $f(x)$ associated with the particle is defined by

$$\langle f(x) \rangle = \int_{-\infty}^{\infty} f(x)|\psi|^2\,dx$$

(Brackets, $\langle \cdots \rangle$, are used to denote average values.) For a particle in a one-dimensional box extending from $x = 0$ to $x = L$, show that

$$\langle x^2 \rangle = \frac{L^2}{3} - \frac{L^2}{2n^2\pi^2}$$

58. (a) Show that the wavelength of a neutron is

$$\lambda = \frac{2.86 \times 10^{-11}}{\sqrt{K_n}}\ \text{m}$$

where $K_n$ is the kinetic energy of the neutron in electron volts. (b) What is the wavelength of a 1-keV neutron?

59. Particles incident from the left are confronted with a step potential as shown in Figure 41.25. The step has a height $U_0$, and the particles have energy $E > U_0$. Classically, all the particles would pass into the region of higher potential at the right. However, according to wave mechanics, one finds that a fraction of the particles are reflected at the barrier. The probability that a particle will be reflected, called the *reflection coefficient* $R$, is given by

$$R = \frac{(k_1 - k_2)^2}{(k_1 + k_2)^2}$$

where $k_1 = 2\pi/\lambda_1$ and $k_2 = 2\pi/\lambda_2$ are the wave numbers for the incident and transmitted particles, respectively. If $E = 2U_0$, what fraction of the incident particles are reflected? (This situation is analogous to the partial reflection and transmission of light striking an interface between two different media.)

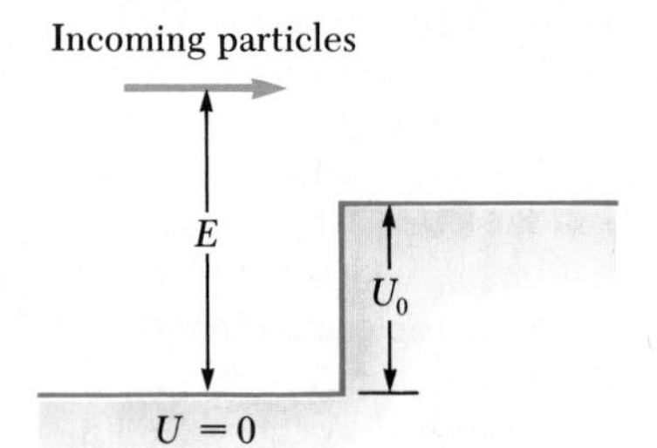


**Figure 41.25** (Problem 59).

60. A particle has a wave function given by

$$\psi(x) = \begin{cases} \sqrt{\dfrac{2}{a}}\, e^{-x/a} & \text{for } x > 0 \\ 0 & \text{for } x < 0 \end{cases}$$

(a) Find and sketch the probability density. (b) Find the probability that the particle will be found anywhere with $x < 0$. (c) Show that $\psi$ is normalized and then find the probability that the particle will be found between $x = 0$ and $x = a$.

61. An electron of momentum $p$ is at a distance $r$ from a stationary proton. The electron has a kinetic energy $K = p^2/2m$ and potential energy $U = -ke^2/r$. Its total energy is $E = K + U$. If the electron is bound to the proton to form a hydrogen atom, its average position is at the proton but the uncertainty in its position is approximately equal to the radius $r$ of its orbit. The electron's average momentum will be zero, but the uncertainty in its momentum will be given by the uncertainty principle. Treat the atom as a one-dimensional system in the following: (a) Estimate the uncertainty in the electron's momentum in terms of $r$. (b) Estimate the electron's kinetic, potential, and total energies in terms of $r$. (c) The actual value of $r$ is the one that *minimizes the total energy*, resulting in a stable atom. Find that value of $r$ and the resulting total energy. Compare your answer with the predictions of the Bohr theory.

62. An electron is confined to one-dimensional motion between two rigid walls separated by a distance $L$. (a) What is the probability of finding the electron within the interval $x = 0$ to $x = L/3$ from one wall if the electron is in its $n = 1$ state? (b) Compare this value with the classical probability.

63. A particle of mass $m_0$ is placed in a one-dimensional box of width $L$. The box is so small that the particle's motion is *relativistic*, so that $E = p^2/2m_0$ is *not valid*. (a) Derive an expression for the energy levels of the particle. (b) If the particle is an electron in a box of width $L = 10^{-12}$ m, find its lowest possible kinetic energy. By what percent is the nonrelativistic formula for the energy in error? (*Hint:* See Eq. 39.26.)

64. Consider a "crystal" consisting of two nuclei and two electrons as shown in Figure 41.26. (a) Taking into account all the pairs of interactions, find the potential energy of the system as a function of $d$. (b) Assuming the electrons to be restricted to a one-dimensional box of width $3d$, find the minimum kinetic energy of the two electrons. (c) Find the value of $d$ for which the total energy is a *minimum*. (d) Compare this value of $d$ with the spacing of atoms in lithium, which has a density of 0.53 g/cm³ and an atomic weight of 7. (This type of calculation can be used to estimate the density of crystals and certain stars.)

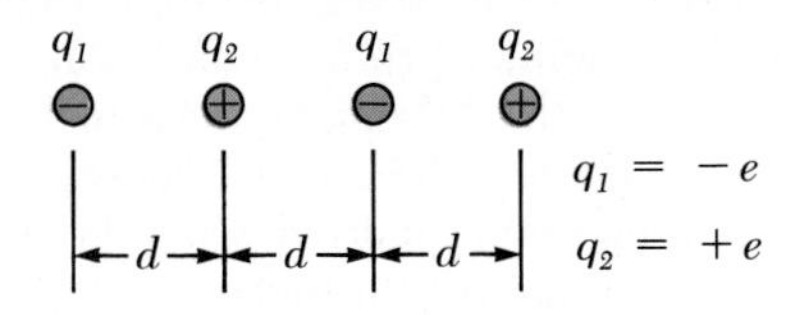


**Figure 41.26** (Problem 64).

65. An electron is represented by the time-independent wave function

$$\psi(x) = \begin{cases} Ae^{-\alpha x} & \text{for } x > 0 \\ Ae^{+\alpha x} & \text{for } x < 0 \end{cases}$$

(a) Sketch the wave function as a function of $x$. (b) Sketch the probability that the electron is found between $x$ and $x + dx$. (c) Why do you suppose this is a physically reasonable wave function? (d) Normalize the wave function. (e) Determine the probability of finding the electron somewhere in the range

$$x_1 = -\frac{1}{2\alpha} \text{ to } x_2 = \frac{1}{2\alpha}$$

66. The normalized ground-state wave function for the electron in the hydrogen atom is

$$\psi(r, \theta, \phi) = \frac{2}{\sqrt{4\pi}}\left(\frac{1}{a_0}\right)^{3/2} e^{-r/a_0}$$

where $r$ is the radial coordinate of the electron and $a_0$ is the Bohr radius. (a) Sketch the wave function versus $r$. (b) Show that the probability of finding the electron between $r$ and $r + dr$ is given by $|\psi(r)|^2 4\pi r^2\, dr$. (c) Sketch the probability versus $r$ and from your sketch find the radius at which the electron is most likely to be found. (d) Show that the wave function as given is normalized. (e) Find the probability of locating the electron between $x_1 = a_0/2$ and $x_2 = 3a_0/2$.

67. The *transmission coefficient* $T$ gives the probability that a particle of mass $m$ approaching the rectangular potential barrier of Figure 41.27 may "tunnel" through the barrier:

$$T = e^{-2KL} \qquad \text{where } K = \frac{\sqrt{2m(U - E)}}{\hbar}$$

Consider a barrier with $U = 5$ eV and having a width $L = 950$ pm. Suppose that an electron with energy

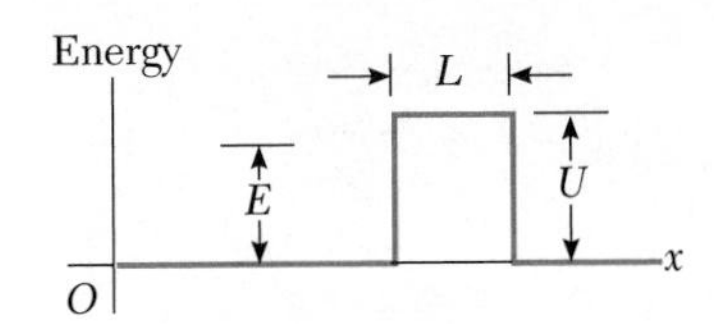


**Figure 41.27** (Problems 67 and 68).

$E = 4.5$ eV approaches the barrier. Classically, the electron could not pass through the barrier because $E < U$. However, quantum-mechanically there is a finite probability of tunneling. Calculate this probability.

68. In Problem 67, by how much would the width $L$ of the potential barrier have to be increased so that the chance of an incident 4.5-eV electron tunneling through the barrier is one in a million?

69. *The simple harmonic oscillator:* (a) Show that the wave function given by Equation 41.32 is a solution of the Schrödinger equation (Eq. 41.30) with energy given by $E = \frac{1}{2}\hbar\omega$. (b) The wave function given by

$$\psi(x) = Cxe^{-(m\omega/2\hbar)x^2}$$

is also a solution to the simple harmonic oscillator problem. Find the energy corresponding to this state. Can you identify this state?

70. An electron is trapped at a defect in a crystal. The defect may be modeled as a one-dimensional, rigid-walled box of width 1 nm. (a) Sketch the wave functions and probability densities for the $n = 1$ and $n = 2$ states. (b) For the $n = 1$ state, find the probability of finding the electron between $x_1 = 0.15$ nm and $x_2 = 0.35$ nm, where $x = 0$ is the left side of the box. (c) Repeat (b) for the $n = 2$ state. (d) Calculate the energies in eV of the $n = 1$ and $n = 2$ states. *Hint:* For (b) and (c), use Equation 41.10 and note that

$$\int \sin^2 ax\, dx = \tfrac{1}{2}x - \frac{1}{4a}\sin 2ax$$

71. *Normalization of wave functions:* (a) Find the normalization constant $A$ for a wave function made up of the two lowest states of a particle in a box. This wave function is given by

$$\psi(x) = A\left[\sin\left(\frac{\pi x}{L}\right) + 4\sin\left(\frac{2\pi x}{L}\right)\right]$$

(b) A particle is described in the space $-a \le x \le a$ by the wave function

$$\psi(x) = A\cos\left(\frac{\pi x}{2a}\right) + B\sin\left(\frac{\pi x}{a}\right)$$

Determine values for $A$ and $B$. (*Hint:* make use of the identity $\sin 2\theta = 2\sin\theta\cos\theta$.)

# ESSAY

## The Scanning Tunneling Microscope

**Roger A. Freedman and Paul K. Hansma**
*Department of Physics, University of California, Santa Barbara*

The basic idea of quantum mechanics, that particles have properties of waves and vice versa, is among the strangest found anywhere in science. Because of this strangeness, and because quantum mechanics mostly deals with the very small, it might seem to have little practical application. As we will show in this essay, however, one of the basic phenomena of quantum mechanics—the tunneling of a particle—is at the heart of a very practical device that is one of the most powerful microscopes ever built. This device, the *scanning tunneling microscope* or *STM,* enables physicists to make highly detailed images of surfaces with resolution comparable to the size of a *single atom.* Such images promise to revolutionize our understanding of structures and processes on the atomic scale.

Before discussing how the STM works, we first look at a sample of what the STM can do. An image made by a scanning tunneling microscope of the surface of a piece of gold is shown in Figure 1. You can easily see that the surface is not uniformly flat, but is a series of terraces separated by steps that are only one atom high. Gentle corrugations can be seen in the terraces, caused by subtle rearrangements of the gold atoms.

What makes the STM so remarkable is the fineness of the detail that can be seen in images such as Figure 1. The *resolution* in this image—that is, the size of the smallest detail that can be discerned—is about 2 Å ($2 \times 10^{-10}$ m). For an ordinary microscope, the resolution is limited by the wavelength of the waves used to make the image. Thus an optical microscope has a resolution of no better than 2000 Å, about half the wavelength of visible light, and so could never show the detail displayed in Figure 1. Electron microscopes can have a resolution of 2 Å by using electron waves of wavelength 4 Å or shorter. From the de Broglie formula $\lambda = h/p$, the electron momentum $p$ required to give this wavelength is 3100 eV/$c$, corresponding to an electron speed $v = p/m = 1.8 \times 10^6$ m/s. Electrons traveling at this high speed would penetrate into the interior of the piece of gold in Figure 1, and so would give no information about individual surface atoms.

The image in Figure 1 was made by Gerd Binnig, Heinrich Rohrer, and collaborators at the IBM Research Laboratory in Zurich, Switzerland. Binnig and Rohrer invented the STM and shared the 1986 Nobel Prize in Physics for their work. Such is the importance of this device that unlike most Nobel Prizes, which come decades after the original work, Binnig and Rohrer received their Nobel Prize just six years after their first experiments with an STM.

**Figure 1** Scanning tunneling microscope image of the surface of crystalline gold. The divisions on the scale are 5 Å. Successive scans are approximately 1.5 Å apart. The figure is from G. Binning, H. Rohrer, Ch. Gerber, and E. Stoll, *Surface Science* **144**:321, 1984.

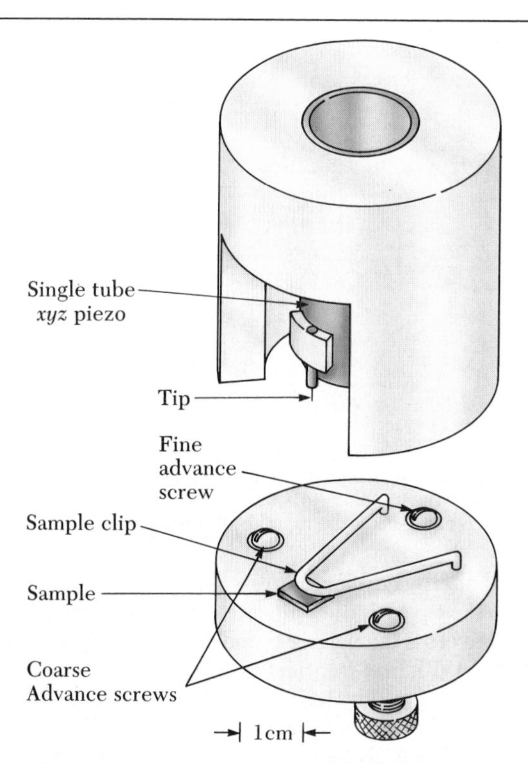


**Figure 2** Drawing of an actual STM head and base showing the essential components. Also depicted are the three screws used for controlling the mechanical approach of the tip to the sample. Three keys to a successful STM design are (1) A smooth mechanical approach mechanism, (2) rigidity, and (3) convenience in changing the sample and tip. (Based on a drawing from P.K. Hansma, V.B. Elings, O. Marti, and C.E. Bracker, *Science* **242**:209–16, 1988. Copyright 1988 by the AAAS.)

One design for an STM is shown in Figure 2. The basic idea behind its operation is very simple, as shown in Figure 3. A conducting probe with a very sharp tip is brought near the surface to be studied. Because it is attracted to the positive ions in the surface, an electron in the surface has a lower total energy than would an electron in the empty space between surface and tip. The same thing is true for an electron in the tip. In classical Newtonian mechanics, electrons could not move between the surface and tip because they would lack the energy to escape either material. But because the electrons obey quantum mechanics, they can "tunnel" across the barrier of empty space between the surface and the tip. Let us explore the operation of the STM in terms of the discussion of tunneling in Section 41.9.

For an electron in the apparatus of Figures 2 and 3, a plot of the energy as a function of position would look like Figure 4. The horizontal coordinate in this figure represents electron position. Now $L$ is to be interpreted as the distance between the surface and the tip, so that coordinates less than 0 refer to positions inside the surface material and coordinates greater than $L$ refer to positions inside the tip. The barrier height $U = q\phi$ is the potential energy difference between an electron outside the material and an electron in the material where $\phi$ is the work function of the material. That is, an electron in the surface or tip has potential energy $-U$ compared to one in vacuum. (We are assuming for the moment that the surface and tip are made of the same material. We will comment on this assumption shortly.) The kinetic energy of an electron in the surface is $E$ so that an amount of energy equal to $(U - E)$ must be given to an electron to remove it from the surface. Thus $(U - E)$ is the work function of an electron in the surface.

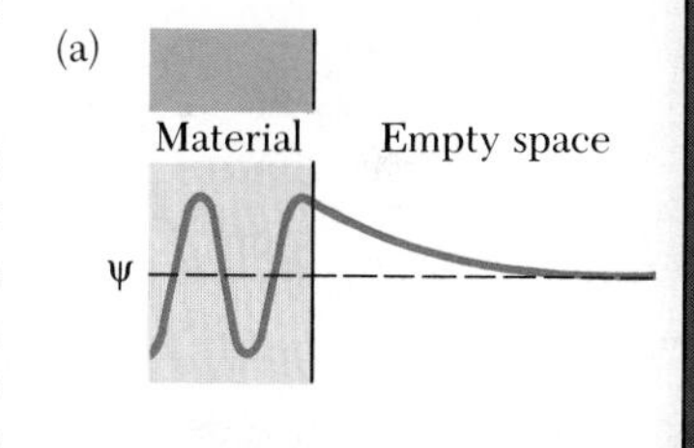


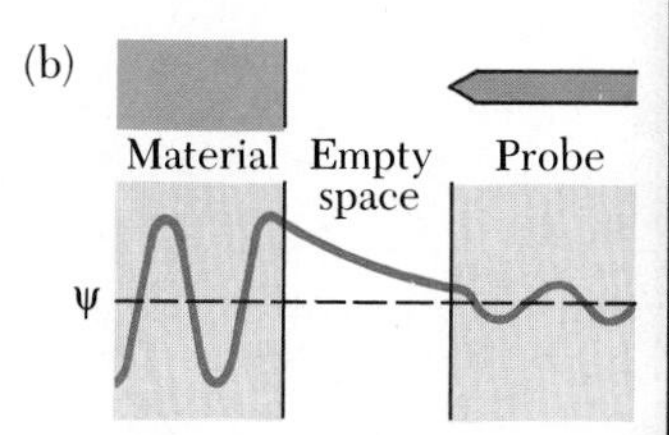


**Figure 3** (a) The wave function of an electron in the surface of the material to be studied. The wave function extends beyond the surface into the empty region. (b) The sharp tip of a conducting probe is brought close to the surface. The wave function of a surface electron penetrates into the tip, so that the electron can "tunnel" from surface to tip.

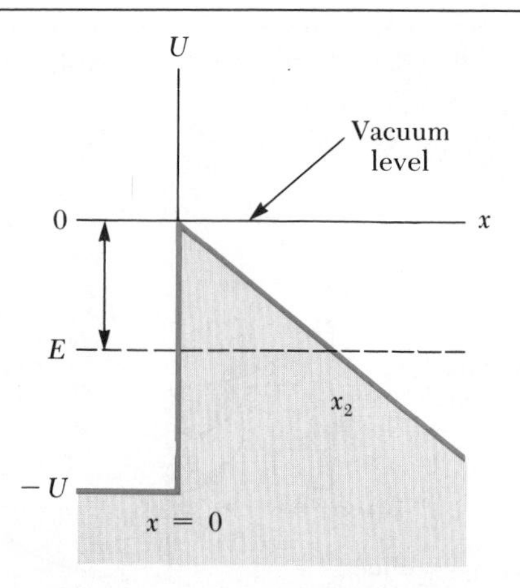


**Figure 4** The potential energy versus position for an electron in a metal. The potential energy is $-U$ when the electron is in the metal ($x < 0$) and is proportional to $x$ outside the metal ($x > 0$). An electron with energy $E$ can escape the metal by tunneling from $x = 0$ to the point $x_2$.

For the potential energy curve of Figure 4, one could expect as much tunneling from the surface into the tip as in the opposite direction. In an STM, the direction in which electrons tend to cross the barrier is controlled by applying a voltage between the surface and the tip. With preferential tunneling from the surface into the tip, the tip samples the distribution of electrons in and above the surface. Because of this "bias" voltage, the work functions of surface and tip are different, giving a preferred direction of tunneling. This is also automatically the case if the surface and tip are made of different materials. In addition, the top of the barrier in Figure 4 will not be flat but will be tilted to reflect the electric field between the surface and the tip. However, if the barrier energy $U$ is large compared to the difference between the surface and tip work functions, and if the bias voltage is small compared to $\phi = U/q$, we can ignore these complications in our calculations.

The characteristic scale of length for tunneling is set by the work function $(U - E)$. For a typical value $(U - E) = 4.0$ eV, this scale of length is

$$\delta = \frac{\hbar}{\sqrt{2m(U-E)}} = \frac{\hbar c}{\sqrt{2mc^2(U-E)}}$$
$$= \frac{1.973 \text{ keV}\cdot\text{Å}}{\sqrt{2(511 \text{ keV})(4.0 \times 10^{-3} \text{ keV})}} = 0.98 \text{ Å} \approx 1.0 \text{ Å}$$

The probability that a given electron will tunnel across the barrier is the transmission coefficient $T \cong e^{-2L/\delta}$, where $L$ is the barrier width. If the separation $L$ between surface and tip is not small compared to $\delta$, then the barrier is "wide" and we can use an approximate result for $T$. The current of electrons tunneling across the barrier is simply proportional to $T$. The tunneling current density can be shown to be

$$j = \frac{e^2 V}{4\pi^2 L\delta\hbar} e^{-2L/\delta}$$

In this expression $e$ is the charge of the electron and $V$ is the bias voltage between surface and tip.

We can see from this expression that the STM is very sensitive to the separation $L$ between tip and surface. This is because of the exponential dependence of the tunneling current on $L$ (this is much more important than the $1/L$ dependence). As we saw above, typically $\delta \approx 1.0$ Å. Hence increasing the distance $L$ by just 0.01 Å causes the tunneling current to be multiplied by a factor $e^{-2(0.01\text{ Å})/(1.0\text{ Å})} \approx 0.98$; i.e., the current decreases by 2%—a change that is measurable. For distances $L$ greater than 10 Å (that is, beyond a few atomic diameters), essentially no tunneling takes place. This sensitivity to $L$ is the basis of the operation of the STM: monitoring the tunneling current as the tip is scanned over the surface gives a sensitive measure of the topography of the surface. In this way the STM can measure the height of surface features to within 0.01 Å, or approximately one one-hundredth of an atomic diameter.

The STM also has excellent lateral resolution, that is, resolution of features in the plane of the surface. This is because the tips used are *very* sharp indeed, typically only an atom or two wide at their extreme end. Thus the tip samples the surface electrons only in a very tiny region approximately 2 Å wide, and so can "see" very fine detail. You might think that making such tips would be extremely difficult, but in fact it's relatively easy: sometimes just sharpening the tip on a fine grinding stone (or even with fine sandpaper) is enough to cause the tip atoms to rearrange by themselves into an atomically sharp configuration. (If you find this surprising, you're not alone. Binnig and Rohrer were no less surprised when they discovered this.)

There are two modes of operation for the STM, shown in Figure 5. In the *constant current mode* a convenient operating voltage (typically between 2 millivolts and 2 volts) is first established between surface and tip. The tip is then brought close enough to the surface to obtain measurable tunneling current. The tip is then scanned over the surface while the tunneling current $I$ is measured. A feedback network changes the vertical position of the tip, $z$, to keep the tunneling current constant, thereby keeping the separation $L$ between surface and tip constant. An image of the surface is made by plotting $z$ versus lateral position $(x, y)$. The simplest scheme for plotting the image is shown in the graph below the schematic view. The height $z$ is plotted versus the scan position $x$. An image consists of multiple scans displaced laterally from each other in the $y$ direction.

The constant current mode was historically the first to be used, and has the advantage that it can be used to track surfaces that are not atomically flat (as in Figure 1). However, the feedback network requires that the scanning be done relatively slowly. As a result, the sample being scanned must be held fixed in place for relatively long times to prevent image distortion.

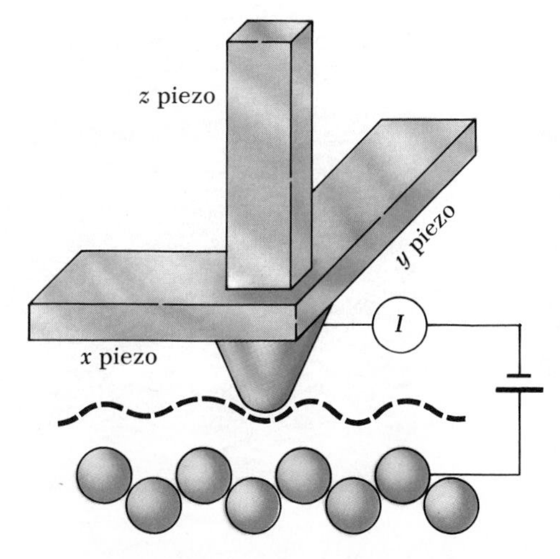


**Figure 5** A schematic view of an STM. The tip, shown as a rounded cone, is mounted on a piezoelectric $x, y, z$ scanner. A scan of the tip over the sample can reveal contours of the surface down to the atomic level. An STM image is composed of a series of scans displaced laterally from one another such as shown in Figure 6. (Based on a drawing from P.K. Hansma, V.B. Elings, O. Marti, and C.E. Bracker, *Science* **242**:209–16, 1988. Copyright 1988 by the AAAS.)

Alternatively, in the *constant height mode* the tip is scanned across the surface at constant voltage and nearly constant height while the current is monitored. In this case the feedback network responds only rapidly enough to keep the average current constant, which means that the tip maintains the same average separation from the surface. The image is then a plot of current $I$ versus lateral position $(x, y)$, as shown in the graph below the schematic. Again, multiple scans along $x$ are displayed laterally displaced in the $y$ direction. The image shows the substantial variation of tunneling current as the tip passes over surface features such as individual atoms.

The constant height mode allows much faster scanning of atomically flat surfaces (100 times faster than the constant current mode), since the tip does not have to be moved up and down over the surface "terrain." This fast scanning means that making an image of a surface requires only a short "exposure time." By making a sequence of such images, researchers may be able to study in real-time processes in which the surfaces rearrange themselves—in effect making an STM "movie."

Individual atoms have been imaged on a variety of surfaces, including those of so-called *layered materials* in which atoms are naturally arranged into two-dimensional layers. Figure 6 shows an example of atoms on one of these layered materials. In this image it is fascinating not only to see individual atoms, but also to note that some atoms are missing. Specifically, there are three atoms missing from Figure 6. Can you find the places where they belong?

Another remarkable aspect of the STM image in Figure 6 is that it was obtained with the surface and tip immersed in liquid nitrogen. While we assumed earlier in this essay that the space between the surface and tip must be empty, in fact electron tunneling can take place not just through vacuum but also through gases and liquids—even water. This seems very surprising since we think of water, especially water with salts dissolved in it, as a conductor. But water is only an *ionic* conductor. For electrons, water behaves as an insulator just as vacuum behaves as an insulator. Thus electrons can flow through water only by tunneling, which makes scanning tunneling microscopy possible "under water."

As an example, Figure 7 shows individual carbon atoms on a graphite surface. It was obtained for a surface immersed in a silver plating solution, which is highly

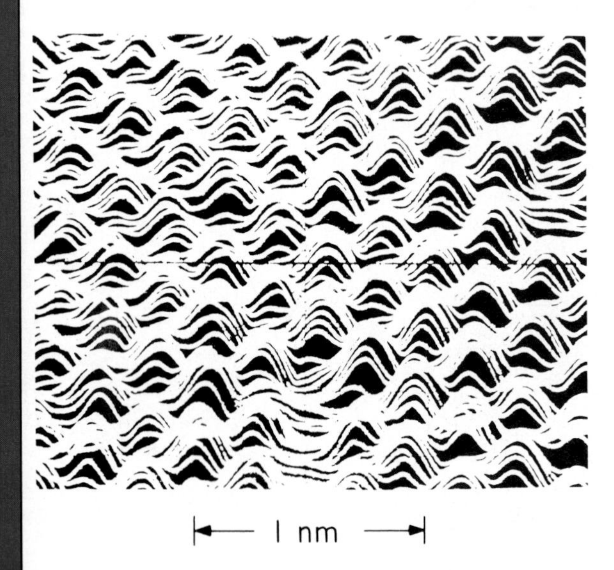


**Figure 6** Image of atoms on a surface of tantalum disulfide ($TaS_2$) immersed in liquid nitrogen. The figure is from C.G. Slough, W.W. McNairy, R.V. Coleman, B. Drake, and P.K. Hansma, *Physical Review B* **34**:994, 1986.

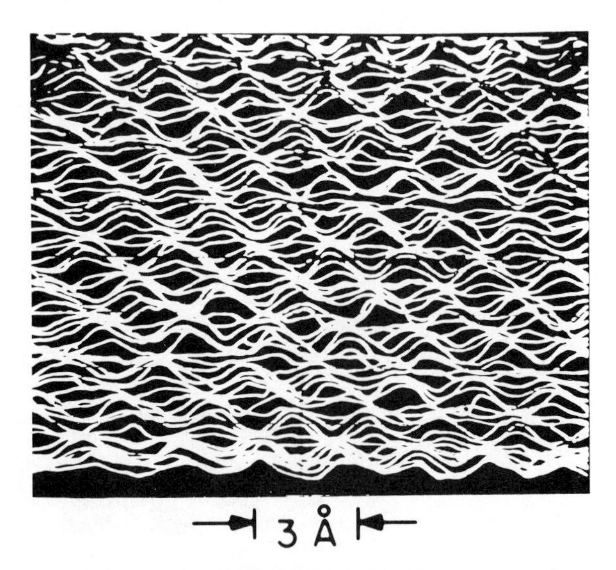


**Figure 7** Image of a graphite electrode in an electrolyte used for silver plating. The figure is from R. Sonnenfeld and B. Schardt, *Applied Physics Letters* **49**:1172, 1986.

conductive for ions but behaves as an insulator for electrons. (The sides of the conducting probe were sheathed with a nonconductor, so that the predominant current into the probe comes from electrons tunneling into the exposed tip. The design of STM used to make this particular image is the one shown in Figure 2.) Sonnenfeld and Schardt observed atoms on this graphite surface before plating it with silver, after "islands" of silver atoms were plated onto the surface, and after the silver was electrochemically stripped from the surface. Their work illustrates the promise of the scanning tunneling microscope for seeing processes that take place on an atomic scale.

While the original STMs were one-of-a-kind laboratory devices, commercial STMs have recently become available. Figure 8 is an image of a graphite surface in air made with such a commercial STM. Note the high quality of this image and the recognizable rings of carbon atoms. You may be able to see that three of the six carbon atoms in each ring *appear* lower than the other three. All six atoms are in fact at the same level, but the three that appear lower are bonded to carbon atoms lying directly beneath them in the underlying atomic layer. The atoms in the surface layer that appear higher do not lie directly over subsurface atoms, and hence are not bonded to carbon atoms beneath them. For the higher-appearing atoms, some of the electron density that would have been involved in bonding to atoms beneath the surface instead extends into the space above the surface. This extra electron density makes these atoms appear higher in Figure 8, since what the STM maps is the topography of the electron distribution.

The availability of commercial instruments should speed the use of scanning tunneling microscopy in a variety of applications (Figure 9). These include characterizing electrodes for electrochemistry (while the electrode is still in the electrolyte), characterizing the roughness of surfaces, measuring the quality of optical gratings, and even imaging replicas of biological structures.

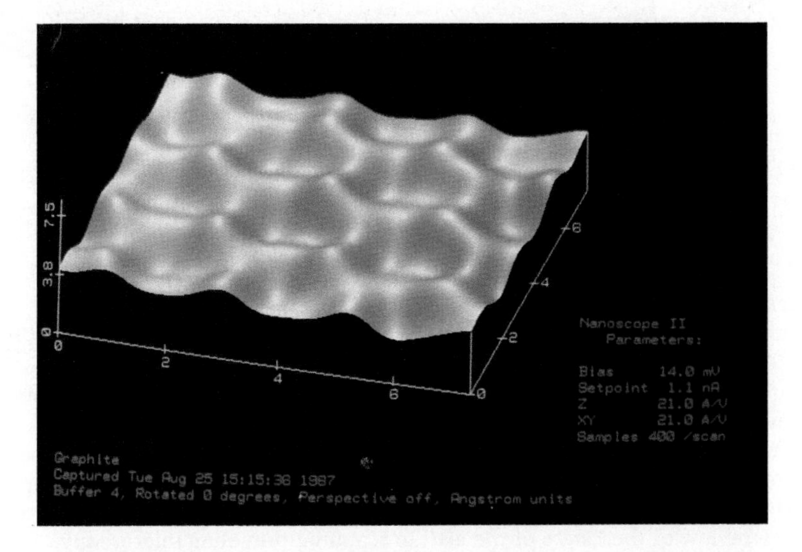


**Figure 8** The surface of graphite as "viewed" with a scanning tunneling microscope. This technique enables scientists to see small details on surfaces, with a resolution of about 2 Å. The contours seen here represent the arrangement of individual atoms on the crystal surface. (Obtained with a commercial STM: the Nanoscope II from Digital Instruments in Goleta, California)

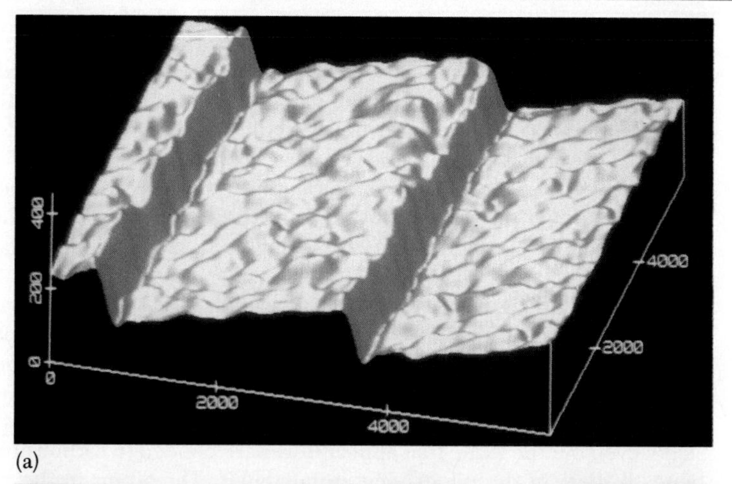


(a)

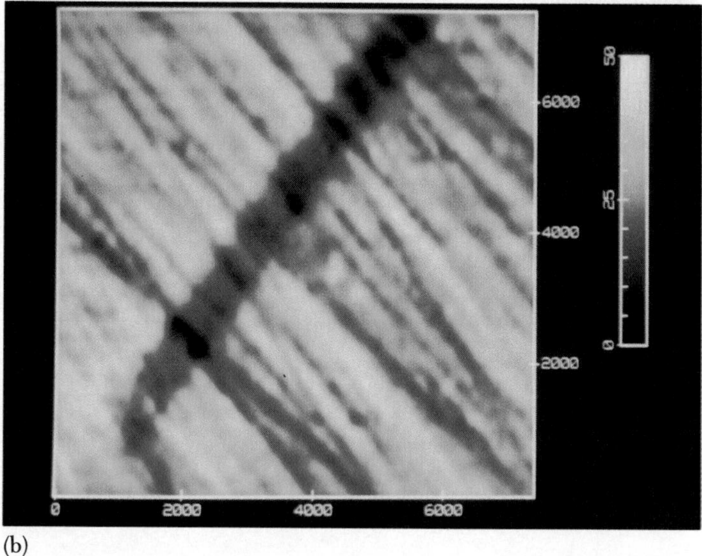


(b)

**Figure 9** Examples of technological applications of the STM. (a) A 6-$\mu$m by 6-$\mu$m scan of a gold diffraction-grating master. The grating has 375 lines per millimeter, and the steps are 120 nm high. (b) A thin-film magnetic recording head. The magnetic material is about 0.5 $\mu$m wide. Measurements show that the material is undercut about 20 nm below the surrounding material during lapping. (c) Bumps on a compact disk stamper. Five tracks are shown in this image. (d) The sharpened edge of a diamond-cutting tool. The diamond is a type 2B blue diamond which is conductive enough for the STM to image. The scales in all of these figures are in nanometers. These images were also obtained with the Nanoscope II. (Figure from P.K. Hansma, V.B. Elings, O. Marti, and C.E. Bracker, *Science* 242:209–16, 1988. Copyright 1988 by the AAAS.)

Perhaps the most remarkable thing about the scanning tunneling microscope is that its operation is based on a quantum mechanical phenomenon—tunneling—that was well understood in the 1920s, yet the STM itself wasn't built until the 1980s. What other applications of quantum mechanics may yet be waiting to be discovered?

## Suggested Readings


G. Binnig, H. Rohrer, Ch. Gerber, and E. Weibel, *Physical Review Letters* **49:** 57, 1982. The first description of the operation of a scanning tunneling microscope.

G. Binnig and H. Rohrer, *Sci. American,* August 1985, p. 50. A popular description of the STM and its applications.

C.F. Quate, *Physics Today,* August 1986, p. 26. An overview of the field of scanning tunneling microscopy, including insights into how it came to be developed.

P.K. Hansma and J. Tersoff, *Journal of Applied Physics* **61:** R1, 1987. A comprehensive review of the "state of the art" in scanning tunneling microscopy.

G. Binnig and H. Rohrer, *Reviews of Modern Physics* **59:** 615, 1987. The text of the lecture given on the occasion of the presentation of the 1986 Nobel Prize in Physics.

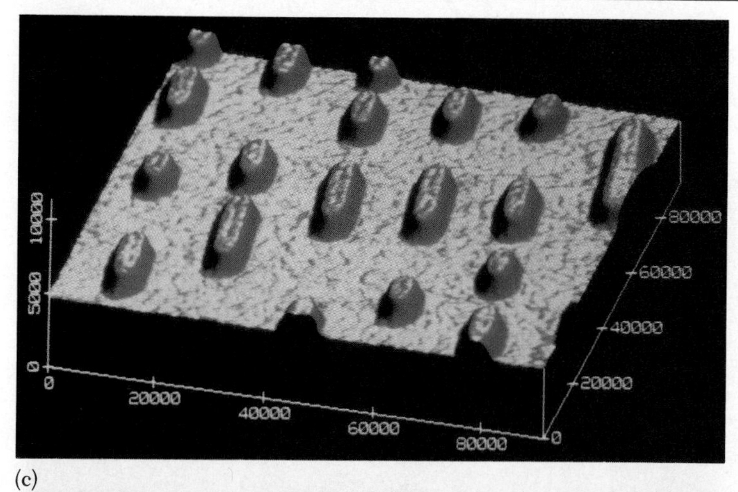

(c)

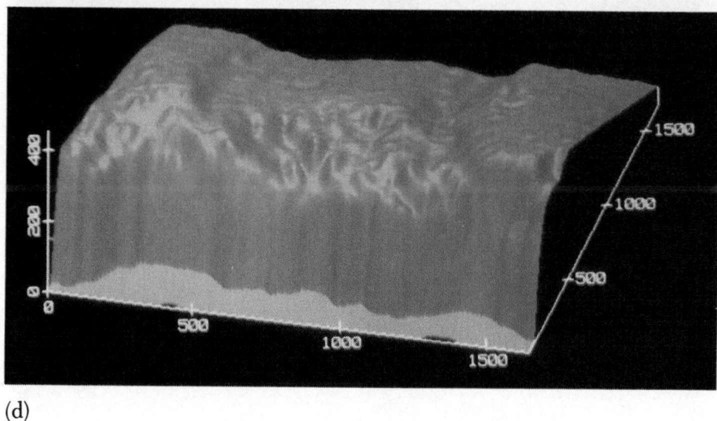

(d)

## Essay Questions and Problems

1. The density of the earth's atmosphere of air depends on altitude $z$ according to $\rho = \rho_0 e^{-z/z_0}$, where $z_0$, the "scale height" of the atmosphere, is 8430 meters. What is the scale height for the electron "atmosphere" above a conductor? Give both a formula and a numerical estimate.
2. Our discussion of STM operation was based on the assumption that the conducting probe has only a single tip. In fact a probe may have a number of protrusions on its end, each of which acts as a "tip." Hence it might be expected that such a probe would give multiple STM images of the surface, greatly complicating the analysis. Explain why there is in fact no problem with multiple images, provided that the multiple tips differ in proximity to the surface by 20 Å or more.
3. The STM, while using physical concepts that date from the 1920s, was not developed until the 1980s. Suggest some reasons why the STM was not invented half a century earlier.
4. It was stated in the essay that for conventional microscopes, the resolution is limited by the wavelength of the waves used to make the image. To see whether this guideline applies to the STM, estimate the wavelength of an electron in the surface of a conductor and compare your estimate to the vertical resolution (about 0.01 Å) and lateral resolution (about 2 Å). Does the guideline apply to the two resolutions? Why or why not?

# 42

# Atomic Physics

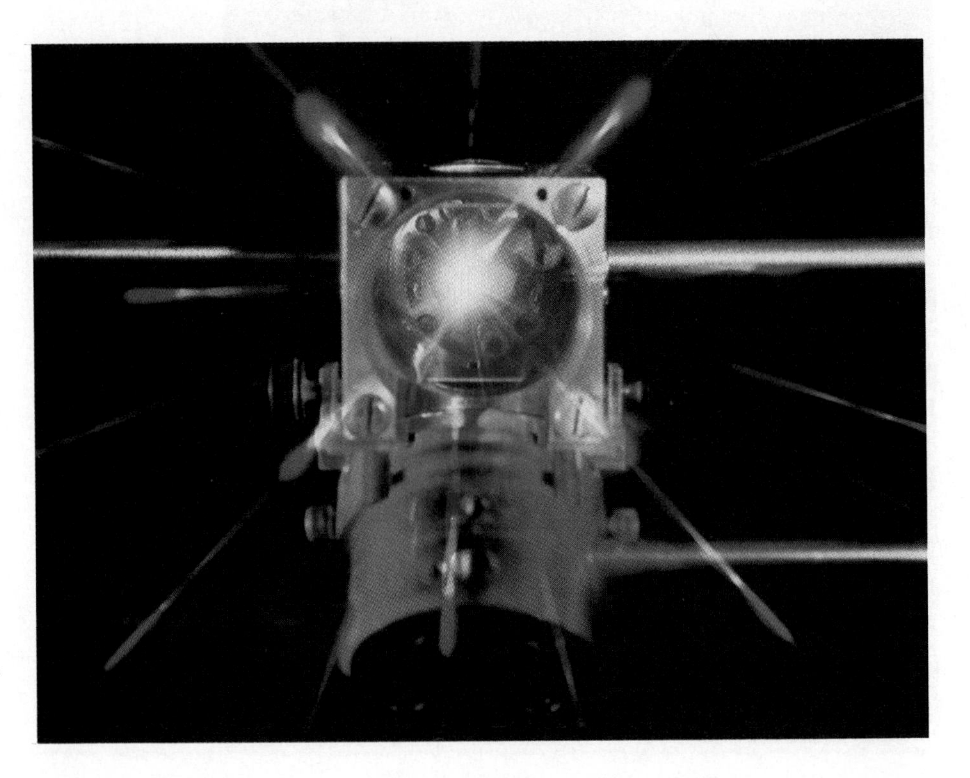

*Solar cell research. A "tandem" solar cell using two photovoltaic materials (silicon and gallium arsenide) is being tested in the laboratory. This design improves the efficiency of the cell, since the two semiconductors use different wavelengths of the sunlight in the process of converting electromagnetic energy into electricity. Part of the radiation that passes through the upper semiconducting layer of the solar cell is absorbed by the bottom layer.*

In Chapter 41, we introduced some of the basic concepts and techniques used in quantum mechanics along with their applications to various simple systems. This chapter deals with the application of quantum mechanics to the real world of atomic structure.

A large portion of this chapter is concerned with the study of the hydrogen atom from the viewpoint of quantum mechanics. Although the hydrogen atom is the simplest atomic system, it is an especially important system to understand for several reasons:

1. Much of what is learned about the hydrogen atom with its single electron can be extended to such single-electron ions as $He^+$ and $Li^{2+}$, which are hydrogen-like in their atomic structure.
2. The hydrogen atom is an ideal system for performing precise tests of theory against experiment and for improving our overall understanding of atomic structure.
3. The quantum numbers used to characterize the allowed states of hydrogen can be used to describe the allowed states of more complex atoms. This enables us to understand the periodic table of the elements, which is one of the greatest triumphs of quantum mechanics.
4. The basic ideas about atomic structure must be well understood before we attempt to deal with the complexities of molecular structures and the electronic structure of solids.

The full mathematical solution of the Schrödinger equation as applied to the hydrogen atom gives a complete and beautiful description of the various properties of the hydrogen atom. However, the mathematical procedures involved in this treatment are beyond the scope of this text, and so the details will be omitted. The solutions for some states of hydrogen will be discussed, together with the quantum numbers used to characterize various allowed stationary states. We shall also discuss the physical significance of the quantum numbers and the effect of a magnetic field on certain quantum states.

A new physical idea, known as the *exclusion principle,* is also presented in this chapter. This physical principle is extremely important in understanding the properties of multielectron atoms and the arrangement of elements in the periodic table. In fact, the implications of the exclusion principle are almost as far-reaching as those of the Schrödinger equation itself. Finally, we shall apply our knowledge of atomic structure to describe the mechanisms involved in the production of x-rays and in the operation of a laser.

Sir Joseph John Thomson (1856–1940), an English physicist and the recipient of the Nobel Prize in 1906. Thomson, usually considered the discoverer of the electron, opened up the field of subatomic particle physics with his extensive work dealing with the deflection of cathode rays (electrons) in an electric field. (The Bettmann Archive)

## 42.1 EARLY MODELS OF THE ATOM

The model of the atom in the days of Newton was that of a tiny, hard, indestructible sphere. This model provided a good basis for the kinetic theory of gases. However, new models had to be devised when later experiments revealed the electrical nature of atoms. J.J. Thomson suggested a model that described the atom as a volume of positive charge with electrons embedded throughout the volume, much like the seeds in a watermelon (Fig. 42.1). Thomson referred to this model as the "plum-pudding atom."

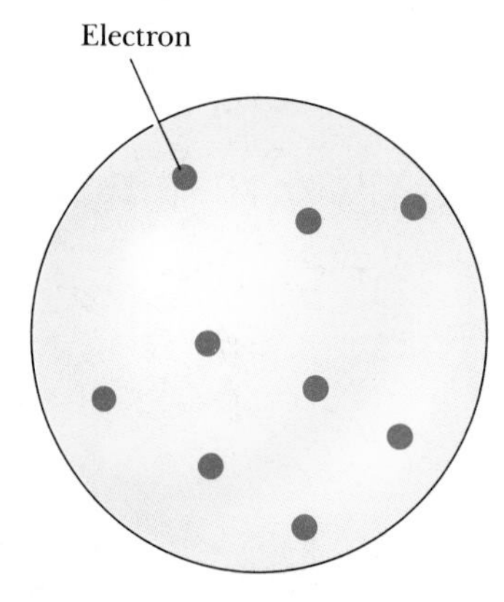


**Figure 42.1** Thomson's plum-pudding model of the atom.

In 1911, Ernest Rutherford and his students Hans Geiger and Ernst Marsden performed a critical experiment that showed that Thomson's plum-pudding model could not be correct. In this experiment, a beam of positively charged **alpha particles,** now known to be the nuclei of helium atoms, was projected into a thin metal foil, as in Figure 42.2a. The results of the experiment were astounding. It was found that most of the alpha particles passed through the foil as if it were empty space. Furthermore, many of the alpha particles that were deflected from their original direction of travel were scat-

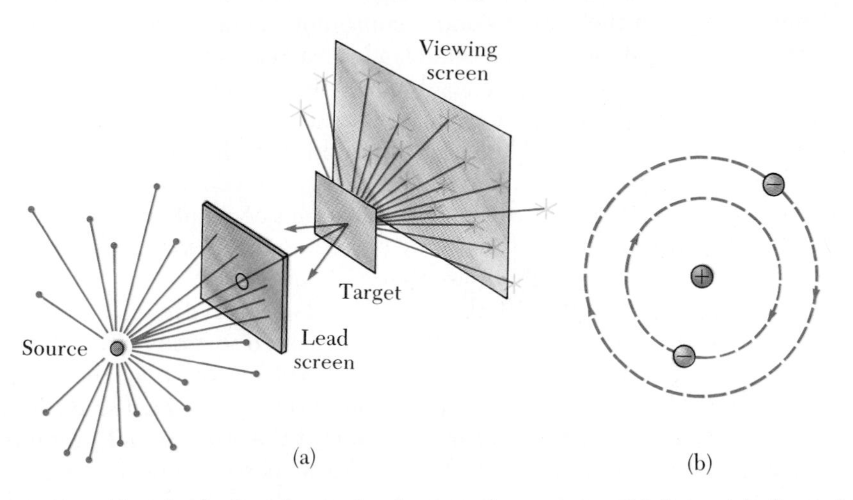


**Figure 42.2** (a) Rutherford's technique for observing the scattering of alpha particles from a thin foil target. The source is a naturally occurring radioactive substance, such as radium. (b) Rutherford's planetary model of the atom.

tered through very large angles. Some particles were even deflected backwards so as to reverse their direction of travel. When Geiger informed Rutherford that some alpha particles were scattered backwards, Rutherford wrote, "It was quite the most incredible event that has ever happened to me in my life. It was almost as incredible as if you fired a 15-inch shell at a piece of tissue paper and it came back and hit you."

Such large deflections were not expected on the basis of the plum-pudding model. According to this model, a positively charged alpha particle would never come close enough to a large enough volume of charge to cause any large-angle deflections. Rutherford explained these observations by assuming that the positive charge was concentrated in a region that was small relative to the size of the atom. He called this concentration of positive charge the **nucleus** of the atom. Any electrons belonging to the atom were assumed to be in the relatively large volume outside the nucleus. In order to explain why electrons in this outer region of the atom were not pulled into the nucleus, Rutherford developed a model similar to that of our solar system. He viewed the electrons as particles moving in orbits about the positively charged nucleus in the same manner as the planets orbit the sun, as in Figure 42.2b.

There are two basic difficulties with Rutherford's planetary model of the atom. As we saw in Chapter 40, an atom emits certain characteristic frequencies of electromagnetic radiation and no others. The Rutherford model is unable to explain this phenomenon. A second difficulty is that, according to classical theory, an accelerated charge must radiate energy in the form of electromagnetic waves. In the Rutherford model, the electrons undergo a centripetal acceleration because they move in circular paths about the nucleus. According to Maxwell's theory of electromagnetism, centripetally accelerated charges revolving with frequency $f$ should radiate electromagnetic waves of frequency $f$, as confirmed by Hertz in 1888. Unfortunately, this classical model leads to disaster. As the electron radiates energy, the radius of its orbit steadily decreases and its frequency of revolution increases. This leads to an ever increasing frequency of emitted radiation and an ultimate collapse of the atom as the electron plunges into the nucleus (Fig. 42.3).

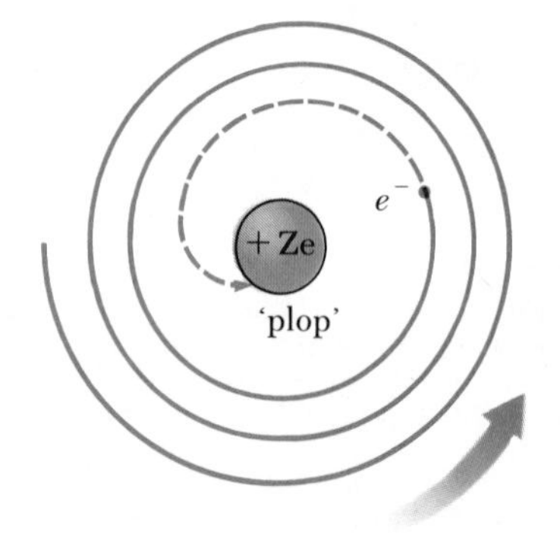


**Figure 42.3** The classical model of the nuclear atom.

Now the stage was set for Bohr! In order to circumvent the deductions of electrons falling into the nucleus and a continuous emission spectrum from elements, Bohr simply postulated that classical radiation theory did not hold for atomic sized systems. He overcame the problem of a classical electron that continually lost energy by applying Planck's ideas of quantized energy levels to orbiting atomic electrons. Thus he postulated that electrons in atoms are generally confined to stable, nonradiating energy levels and orbits called stationary states. Furthermore, he applied Einstein's concept of the photon to arrive at an expression for the frequency of light emitted when the electron jumps from one stationary state to another.

As we learned in Chapter 40, the Bohr model of the hydrogen atom was a tremendous success in certain areas in that it explained several features of the spectra of hydrogen that had previously defied explanation. It accounted for the Balmer and other series, it provided a numerical value for the Rydberg constant, it derived an expression for the radius of the atom, and it predicted the energy levels of hydrogen. Although these successes were important to scientists, it is perhaps even more important that the Bohr theory gave us a model of what the atom looks like and how it behaves. Once a basic model is constructed, refinements and modifications can be made to enlarge upon the concept and to explain finer details.

One of the first indications that there was a need for modification of the Bohr theory became apparent when improved spectroscopic techniques were used to examine the spectral lines of hydrogen. It was found that many of the lines in the Balmer and other series were not single lines at all. Instead, each line was actually a group of lines spaced very close together. An additional difficulty arose when it was observed that, in some situations, certain single spectral lines were split into three closely spaced lines when the atoms were placed in a strong magnetic field.

Efforts to explain these deviations from the Bohr model led to improvements in the theory. One of the changes introduced into the original theory was the concept that the electron could spin on its axis. Also, Arnold Sommerfeld improved the Bohr theory by introducing the theory of relativity into the analysis of the electron's motion. An electron in an elliptical orbit has a continuously changing velocity. This variation in the electron's velocity leads to variations in its energy.

Wolfgang Pauli and Niels Bohr watch a spinning top. (Courtesy of AIP Niels Bohr Library, Margarethe Bohr Collection.)

## 42.2 THE HYDROGEN ATOM

The hydrogen atom consists of one electron and one proton. In Chapter 40, we described how the Bohr model of atoms views the electron as a particle orbiting around the nucleus in nonradiating, quantized energy levels. The de Broglie model gave electrons a wave nature by viewing them as standing waves in allowed orbits. This standing wave description removed the objections to Bohr's postulates, which were quite arbitrary. However, the de Broglie model created other problems. The exact nature of the de Broglie wave was unspecified, and the model implied unlikely electron densities at large distances from the nucleus. These difficulties are fortunately removed when the methods of quantum mechanics are used to describe atoms.

The potential energy function for the hydrogen atom is given by

$$U(r) = -k\frac{e^2}{r} \tag{42.1}$$

where $k$ is the Coulomb constant and $r$ is the radial distance from the proton (situated at $r = 0$) to the electron. Figure 42.4 is a plot of this function versus $r/a_0$ where $a_0$ is the Bohr radius, equal to 0.0529 nm.

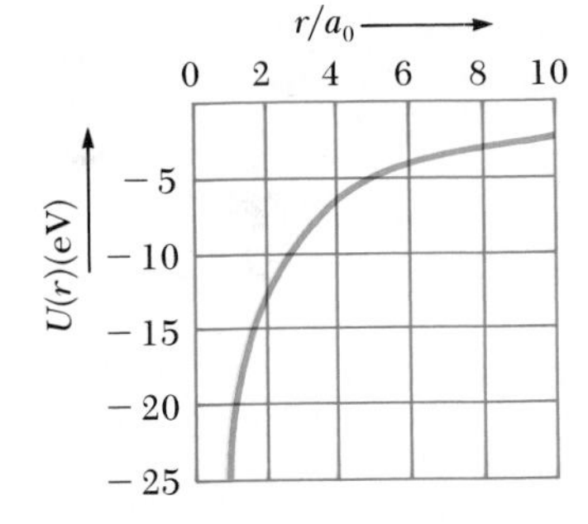


**Figure 42.4** The potential energy $U(r)$ versus the ratio $r/a_0$ for the hydrogen atom. The constant $a_0$ is the Bohr radius, and $r$ is the electron-proton separation.

The formal procedure for solving the problem of the hydrogen atom would be to substitute $U(r)$ into the Schrödinger equation and find appropriate solutions to the equation, as we did in Chapter 41 for other situations. The present problem is more complicated, however, because it is three-dimensional and because $U$ depends on the radial coordinate $r$. We shall not attempt to carry out these solutions. Rather, we shall simply describe the properties of these solutions and some of their implications with regard to atomic structure.

According to quantum mechanics, the energies of the allowed states for the hydrogen atom are given by

$$E_n = -\left(\frac{ke^2}{2a_0}\right)\frac{1}{n^2} = -\frac{13.6}{n^2}\text{ eV} \qquad n = 1, 2, 3, \ldots \tag{42.2}$$

Allowed energies for the hydrogen atom

This result is in exact agreement with that obtained in the Bohr theory. Note that the allowed energies depend only on the quantum number $n$.

In the case of one-dimensional problems, only one quantum number is needed to characterize a stationary state. In the three-dimensional problem of the hydrogen atom, three quantum numbers are needed for each stationary state, corresponding to three independent degrees of freedom for the electron. The three quantum numbers which emerge from the theory are represented by the symbols $n$, $\ell$, and $m_\ell$. The quantum number $n$ is called the **principal quantum number,** $\ell$ is called the **orbital quantum number,** and $m_\ell$ is called the **orbital magnetic quantum number.**

There are certain important relationships between these quantum numbers, as well as certain restrictions on their values. These restrictions are

Restrictions on the values of quantum numbers

The values of $n$ can range from 1 to $\infty$.
The values of $\ell$ can range from 0 to $n - 1$.
The values of $m_\ell$ can range from $-\ell$ to $\ell$.

For example, if $n = 1$, only $\ell = 0$ and $m_\ell = 0$ are permitted. If $n = 2$, the value of $\ell$ may be 0 or 1; if $\ell = 0$, then $m_\ell = 0$, but if $\ell = 1$, then $m_\ell$ may be 1, 0, or $-1$. Table 42.1 summarizes the rules for determining the allowed values of $\ell$ and $m_\ell$ for a given value of $n$.

For historical reasons, *all states with the same principal quantum number are said to form a* **shell.** These shells are identified by the letters K, L, M, . . . , which designate the states for which $n = 1, 2, 3, \ldots$ . Likewise, *the states having the same values of n and $\ell$ are said to form a* **subshell.** The letters *s, p, d, f, g, h,* . . . are used to designate the states for which $\ell = 0, 1, 2, 3, \ldots$ . These notations are summarized in Table 42.2. For example, the state designated by $3p$ has the quantum numbers $n = 3$ and $\ell = 1$; the $2s$ state has the quantum numbers $n = 2$ and $\ell = 0$.

States that violate the rules given in Table 42.1 cannot exist. For instance, one state that cannot exist is the $2d$ state, which would have $n = 2$ and $\ell = 2$. This state is not allowed because the highest allowed value of $\ell$ is $n - 1$, or 1 in this case. Thus, for $n = 2$, $2s$ and $2p$ are allowed states but $2d$, $2f$, . . . are not. For $n = 3$, the allowed states are $3s$, $3p$, and $3d$.

**EXAMPLE 42.1 The $n = 2$ Level of Hydrogen**
Determine the number of orbital states in the hydrogen atom corresponding to the principal quantum number $n = 2$ and calculate the energies of these states.

*Solution* When $n = 2$, $\ell$ can have the values 0 and 1. If $\ell = 0$, $m_\ell$ can only be 0; for $\ell = 1$, $m_\ell$ can be $-1$, 0, or 1. Hence, we have a state designated as the $2s$ state associated with the quantum numbers $n = 2$, $\ell = 0$, and $m_\ell = 0$, and three orbital states designated as $2p$ states for which the quantum numbers are $n = 2$, $\ell = 1$, $m_\ell = -1$; $n = 2$, $\ell = 1$, $m_\ell = 0$; and $n = 2$, $\ell = 1$, $m_\ell = 1$.

Because all of these orbital states have the same principal quantum number, $n = 2$, they also have the same energy, which can be calculated using $E_n = (-13.6/n^2)$ eV (Eq. 42.2). For $n = 2$, this gives

$$E_2 = -\frac{13.6}{2^2}\text{ eV} = -3.40\text{ eV}$$

**Exercise 1** How many possible states are there for the $n = 3$ level of hydrogen? For the $n = 4$ level?
**Answers** 9 states for $n = 3$, and 16 states for $n = 4$.

**TABLE 42.1 Three Quantum Numbers for the Hydrogen Atom**

| Quantum Number | Name | Allowed Values | Number of Allowed States |
|---|---|---|---|
| $n$ | Principal quantum number | $1, 2, 3, \ldots$ | Any number |
| $\ell$ | Orbital quantum number | $0, 1, 2, \ldots, n-1$ | $n$ |
| $m_\ell$ | Orbital magnetic quantum number | $-\ell, -\ell+1, \ldots, 0, \ldots, \ell-1, \ell$ | $2\ell+1$ |

## 42.3 THE SPIN MAGNETIC QUANTUM NUMBER

Example 42.1 was presented to give you some practice in manipulating quantum numbers. For example, we found that there are four orbital states corresponding to the principal quantum number $n = 2$. As we shall see in this section, there actually are *eight* electron states rather than four. This can be explained by requiring a fourth quantum number for each state, called the **spin magnetic quantum number, $m_s$.**

**TABLE 42.2 Atomic Shell and Subshell Notations**

| $n$ | Shell Symbol | $\ell$ | Subshell Symbol |
|---|---|---|---|
| 1 | K | 0 | $s$ |
| 2 | L | 1 | $p$ |
| 3 | M | 2 | $d$ |
| 4 | N | 3 | $f$ |
| 5 | O | 4 | $g$ |
| 6 | P | 5 | $h$ |
| . . . | | . . . | |

The need for this new quantum number first came about because of an unusual feature in the spectra of certain gases, such as sodium vapor. Close examination of one of the prominent lines of sodium shows that this line is, in fact, two very closely spaced lines. The wavelengths of these lines occur in the yellow region at 589.0 nm and 589.6 nm. In 1925, when this was first noticed, the theory of atomic structure had not been adequately developed to explain why there were two lines rather than one. To resolve this dilemma, Samuel Goudsmidt and George Uhlenbeck, following a suggestion by the Austrian physicist Wolfgang Pauli, proposed that a new quantum number, called the spin quantum number, must be added to the set of required quantum numbers in describing a quantum state.

In order to describe the spin quantum number, it is convenient (but incorrect) to think of the electron as spinning on its axis as it orbits the nucleus, just as the earth spins about its axis as it orbits the sun. There are only two ways that the electron can spin as it orbits the nucleus, as shown in Figure 42.5. If the direction of spin is as shown in Figure 42.5a, the electron is said to have "spin up." If the direction of spin is reversed, as in Figure 42.5b, the electron is said to have "spin down." The energy of the electron is slightly different for the two spin directions. As it turns out, this energy difference properly accounts for the observed splitting of the yellow sodium lines. The quantum numbers associated with the spin of the electron are $m_s = \frac{1}{2}$ for the spin-up state and $m_s = -\frac{1}{2}$ for the spin-down state. As we shall see in the following example, this added quantum number doubles the number of allowed states specified by the quantum numbers $n$, $\ell$, and $m_\ell$.

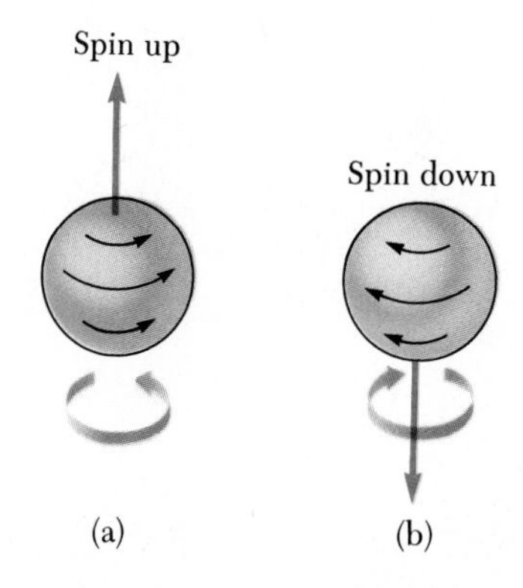


**Figure 42.5** The spin of an electron can be either (a) up or (b) down.

The classical description of electron spin given above is incorrect because quantum mechanics tells us that, since the electron cannot be precisely located in space, it cannot be considered to be spinning as pictured in Figure

42.5. In spite of this conceptual difficulty, all experimental evidence supports the fact that an electron does have some intrinsic property that can be described by the spin magnetic quantum number.

**EXAMPLE 42.2 Adding Electron Spin to Hydrogen**
Determine the quantum numbers associated with the possible states in the hydrogen atom corresponding to the principal quantum number $n = 2$.

*Solution* With the addition of the spin quantum number, we have the possibilities given in the table at the right.

| $n$ | $\ell$ | $m_\ell$ | $m_s$ | Subshell | Shell | Number of Electrons in Subshell |
|---|---|---|---|---|---|---|
| 2 | 0 | 0 | $\frac{1}{2}$ | 2s | L | 2 |
| 2 | 0 | 0 | $-\frac{1}{2}$ | | | |
| 2 | 1 | 1 | $\frac{1}{2}$ | 2p | L | 6 |
| 2 | 1 | 1 | $-\frac{1}{2}$ | | | |
| 2 | 1 | 0 | $\frac{1}{2}$ | | | |
| 2 | 1 | 0 | $-\frac{1}{2}$ | | | |
| 2 | 1 | $-1$ | $\frac{1}{2}$ | | | |
| 2 | 1 | $-1$ | $-\frac{1}{2}$ | | | |

Exercise 2 Show that for $n = 3$, there are 18 possible states. (This follows from the restrictions that the maximum number of electrons in the 3*s* state is 2, the maximum number in the 3*p* state is 6, and the maximum number in the 3*d* state is 10.)

## 42.4 THE WAVE FUNCTIONS FOR HYDROGEN

Since the potential energy of the hydrogen atom depends only on the radial distance *r*, we would expect that some of the allowed states can be represented by wave functions that depend only on *r*. This indeed is the case. The simplest wave function for hydrogen is the one that describes the 1*s* state and is designated $\psi_{1s}(r)$:

Wave function for hydrogen in its ground state

$$\psi_{1s}(r) = \frac{1}{\sqrt{\pi a_0^3}} e^{-r/a_0} \tag{42.3}$$

where $a_0$ is the Bohr radius, given by

Bohr radius

$$a_0 = \frac{\hbar^2}{mke^2} = 0.0529 \text{ nm} \tag{42.4}$$

In this expression, *k* is the Coulomb constant. Note that the wave function $\psi_{1s}(r)$ satisfies the condition that it approach zero as *r* approaches $\infty$ and is normalized as presented. Furthermore, since $\psi_{1s}$ depends only on *r*, it is *spherically symmetric*. This, in fact, is true for all *s* states. On the other hand, $\psi$ depends on other variables for states with $\ell$ values greater than zero.

Recall that the probability density (that is, the probability per unit volume) of finding the electron at any location is equal to $|\psi|^2$ if $\psi$ is normalized. The probability density for the 1*s* state is given by

$$|\psi_{1s}|^2 = \left(\frac{1}{\pi a_0^3}\right) e^{-2r/a_0} \tag{42.5}$$

Furthermore, the actual probability of finding the electron in a volume element $dV$ is given by $|\psi|^2 \, dV$. It is convenient to define the radial probability density function $P(r)$ as the probability of finding the electron in a spherical shell of radius $r$ and thickness $dr$. The volume of such a shell equals its surface area, $4\pi r^2$, multiplied by the shell thickness, $dr$ (Fig. 42.6). That is, the volume of the spherical shell is $dV = 4\pi r^2\,dr$, so that we get

$$P(r)\,dr = |\psi|^2\,dV = |\psi|^2 4\pi r^2\,dr$$

or

$$P(r) = 4\pi r^2|\psi|^2 \tag{42.6}$$

**Figure 42.6** A spherical shell of radius $r$ and thickness $dr$ has a volume equal to $4\pi r^2\,dr$.

Substituting Equation 42.5 into Equation 42.6 gives the radial probability density function for the hydrogen atom in its ground state:

$$P_{1s}(r) = \left(\frac{4r^2}{a_0^3}\right) e^{-2r/a_0} \tag{42.7}$$

Radial probability density for the 1s state of hydrogen

A plot of the function $P_{1s}(r)$ versus $r$ is presented in Figure 42.7a. The peak of the curve corresponds to the most probable value of $r$ for this particular state. The spherical symmetry of the distribution function is shown in Figure 42.7b.

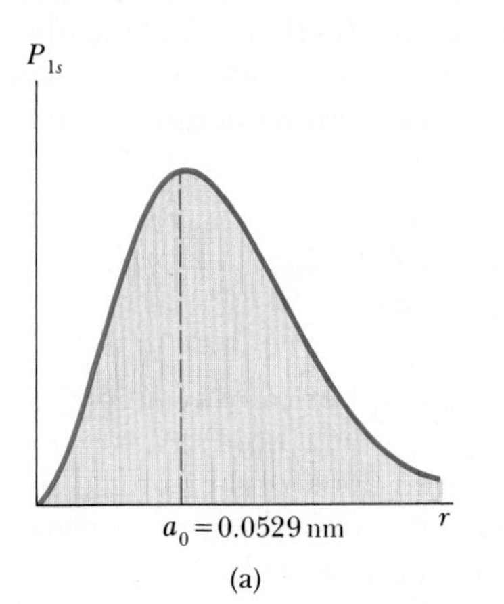


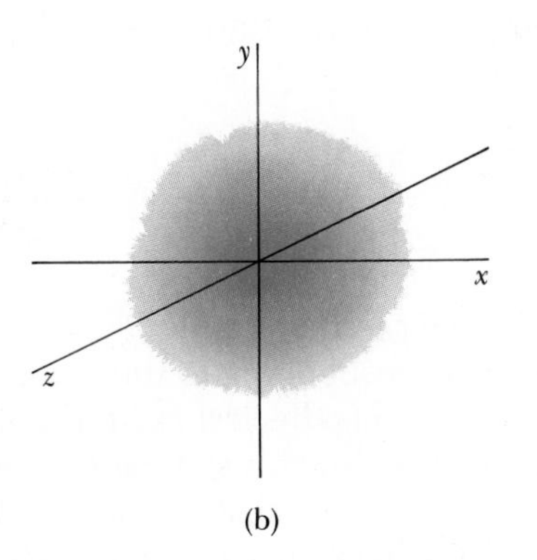


**Figure 42.7** (a) The probability of finding the electron as a function of distance from the nucleus for the hydrogen atom in the 1s (ground) state. Note that the probability has its maximum value when $r$ equals the first Bohr radius, $a_0$. (b) The spherical electron cloud for the hydrogen atom in its 1s state.

**EXAMPLE 42.3 The Ground State of Hydrogen** □
Calculate the most probable value of $r$ for an electron in the ground state of the hydrogen atom.

*Solution* The most probable value of $r$ corresponds to the peak of the plot of $P(r)$ versus $r$. The slope of the curve at this point is zero, and so we can evaluate the most probable value of $r$ by setting $dP/dr = 0$ and solving for $r$. Using Equation 42.7, we get

$$\frac{dP}{dr} = \frac{d}{dr}\left[\left(\frac{4r^2}{a_0^3}\right) e^{-2r/a_0}\right] = 0$$

Carrying out the derivative operation and simplifying the expression, we get

$$e^{-2r/a_0}\frac{d}{dr}(r^2) + r^2\frac{d}{dr}(e^{-2r/a_0}) = 0$$

$$2re^{-2r/a_0} + r^2(-2/a_0)e^{-2r/a_0} = 0$$

$$2r[1 - (r/a_0)]e^{-2r/a_0} = 0$$

This expression is satisfied if

$$1 - \frac{r}{a_0} = 0$$

or

$$r = a_0$$

**EXAMPLE 42.4 Probabilities for the Electron in Hydrogen**
Calculate the probability that the electron in the ground state of hydrogen will be found outside the first Bohr radius.

*Solution* The probability is found by integrating the radial probability density for this state, $P_{1s}(r)$, from the Bohr radius $a_0$ to $\infty$. Using Equation 42.7

$$P = \int_{a_0}^{\infty} P_{1s}(r)\, dr = \frac{4}{a_0^3} \int_{a_0}^{\infty} r^2 e^{-2r/a_0}\, dr$$

We can put the integral in dimensionless form by changing variables from $r$ to $z = 2r/a_0$. Noting that $z = 2$ when $r = a_0$, and that $dr = (a_0/2)\, dz$, we get

$$P = \frac{1}{2} \int_{2}^{\infty} z^2 e^{-z}\, dz = -\frac{1}{2} \{z^2 + 2z + 2\} e^{-z} \Big|_{2}^{\infty}$$

$$P = 5e^{-2} = \quad 0.677 \quad \text{or} \quad 67.7\%$$

Example 42.3 shows that, for the ground state of hydrogen, the most probable value of $r$ equals the first Bohr radius, $a_0$. It turns out that the average value of $r$ for the ground state of hydrogen is $\frac{3}{2}a_0$, which is 50% larger than the most probable value of $r$ (see Problem 40). The reason for this is the large asymmetry in the radial distribution function shown in Figure 42.7a, which has a greater area on the side to the right of the peak. According to wave mechanics, there is no sharply defined boundary to the atom. One could view the probability distribution for the electron as an effective "electron cloud."

The next-simplest wave function for the hydrogen atom is the one corresponding to the $2s$ state ($n = 2$, $\ell = 0$). The normalized wave function for this state is given by

Wave function for hydrogen in the $2s$ state

$$\psi_{2s}(r) = \frac{1}{4\sqrt{2\pi}} \left(\frac{1}{a_0}\right)^{3/2} \left[2 - \frac{r}{a_0}\right] e^{-r/2a_0} \tag{42.8}$$

Again, we see that $\psi_{2s}$ depends only on $r$ and is spherically symmetric. The energy corresponding to this state is given by $E_2 = -(13.6/4)\text{ eV} = -3.4\text{ eV}$. This represents the first excited state of hydrogen. Plots of the radial distribution function for this state and several other states of hydrogen are shown in Figure 42.8. Note that the plot for the $2s$ state has two peaks. In this case, the most probable value corresponds to that value of $r$ that has the highest value of $P$ ($\approx 5a_0$). An electron in the $2s$ state would be much farther from the nucleus (on the average) than an electron in the $1s$ state. Likewise, the average value of $r$ is even greater for the $3d$, $3p$, and $4d$ states.

As we have mentioned, all $s$ states have spherically symmetric wave functions. The other states are not spherically symmetric. For example, the three wave functions corresponding to the states for which $n = 2$ ($m_\ell = 1, 0$, or $-1$) can be expressed as appropriate linear combinations of the three $p$ states. Each state has a wave function that has distinct directional characteristics, indicated by the notations $p_x$, $p_y$, and $p_z$. Figure 42.9 shows views of the electron clouds for these states. Note that the three clouds have identical structure but differ in their orientation with respect to the $x$, $y$, and $z$ axes. The nonspherical wave functions for these states are

Wave functions for the $2p$ state

$$\begin{aligned} \psi_{2p_x} &= xF(r) \\ \psi_{2p_y} &= yF(r) \\ \psi_{2p_z} &= zF(r) \end{aligned} \tag{42.9}$$

where $F(r)$ is some exponential function of $r$. Wave functions with a highly directional character, such as these, play an important role in chemical bonding, the formation of molecules, and chemical properties.

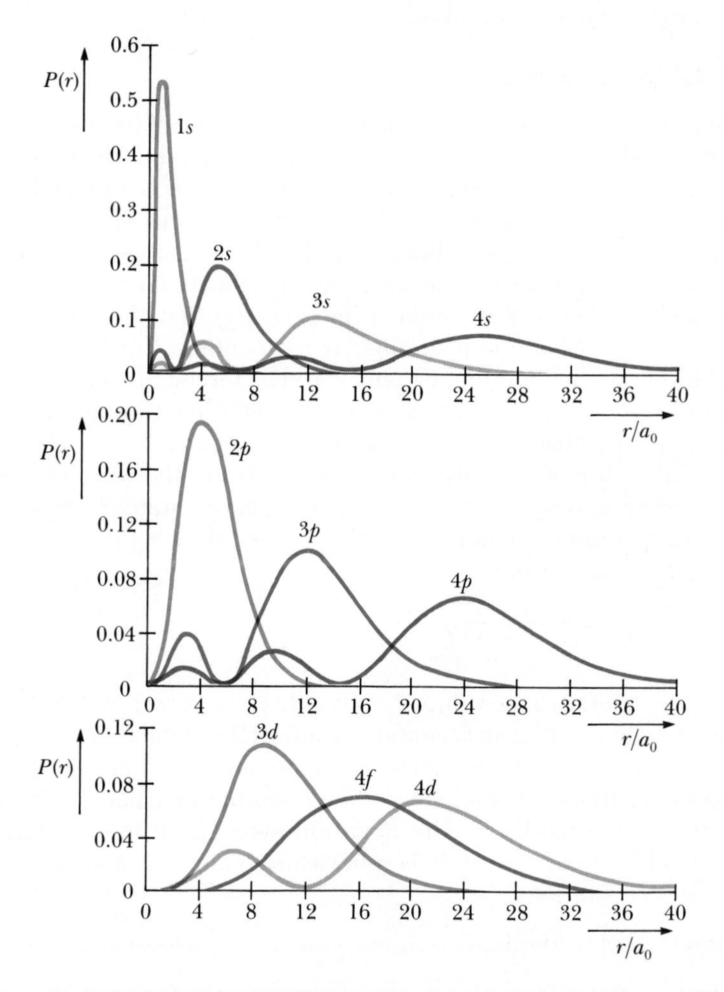


**Figure 42.8** The radial probability density function versus $r/a_0$ for several states of the hydrogen atom. (From E.U. Condon and G.H. Shortley, *The Theory of Atomic Spectra*, Cambridge, Cambridge University Press, 1953; used with permission)

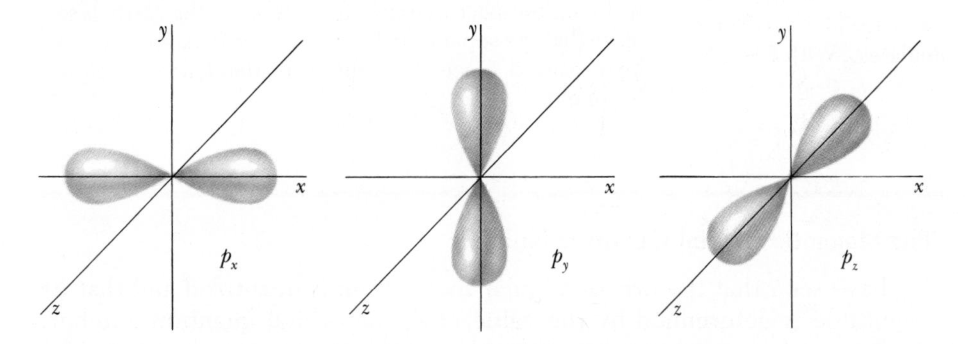


**Figure 42.9** Electron charge distribution for an electron in a $p$ state. The three charge distributions $p_x$, $p_y$, and $p_z$ have the same structure and differ only in their orientation in space.

## 42.5 THE QUANTUM NUMBERS

In the previous sections, we found that the energy of a particular state depends primarily on the principal quantum number. The purpose of this section is to describe the physical significance of the orbital quantum number and the magnetic orbital quantum number.

### The Orbital Quantum Number

If a particle moves in a circle of radius $r$, the magnitude of its angular momentum relative to the center of the circle is given by $L = mvr$. The direction of $\mathbf{L}$ is perpendicular to the plane of the circle. The sense of $\mathbf{L}$ is given by a right-hand rule.[1] From this result, we see that, according to classical physics, $L$ can have any value. However, the Bohr model of hydrogen postulates that the angular momentum is restricted to multiples of $\hbar$, that is, $mvr = n\hbar$. This model must be modified because it predicts (incorrectly) that the ground state of hydrogen ($n = 1$) has one unit of angular momentum. Furthermore, if $L$ is taken to be zero in the Bohr model, one would be forced to accept a picture of the electron as a *particle* oscillating along a straight line through the nucleus. This is a physically unacceptable situation.

These difficulties are resolved when one turns to the wave model of the atom. According to quantum mechanics, an atom in a state characterized by the principal quantum number $n$ can take on the following *discrete* values of orbital angular momentum:

Allowed value of $L$

$$L = \sqrt{\ell(\ell + 1)}\,\hbar \qquad \ell = 0, 1, 2, \ldots, n - 1 \qquad (42.10)$$

Because $\ell$ is restricted to the values $\ell = 0, 1, 2, \ldots, n - 1$, we see that $L = 0$ (corresponding to $\ell = 0$) is an acceptable value of the angular momentum. The fact that $L$ can be zero in this model serves to point out the inherent difficulties in attempting to describe results based on quantum mechanics in terms of a purely particle-like model. In the quantum mechanical interpretation, the electron cloud for the $L = 0$ state is spherically symmetric and has no fundamental axis of revolution.

[1] See Sections 11.2 and 11.3 for details on angular momentum if you have forgotten this material.

**EXAMPLE 42.5 Calculating $L$ for a $p$ State**

Calculate the orbital angular momentum of an electron in a $p$ state of hydrogen.

*Solution* Because we know that $\hbar = 1.054 \times 10^{-34}$ J·s, we can use Equation 42.10 to calculate $L$. With $\ell = 1$ for a $p$ state, we have

$$L = \sqrt{1(1 + 1)}\,\hbar = \sqrt{2}\,\hbar = 1.49 \times 10^{-34} \text{ J·s}$$

This number is extremely small relative to the orbital angular momentum of the earth orbiting the sun, which is about $2.7 \times 10^{40}$ J·s. The quantum number that describes $L$ for macroscopic objects, such as the earth, is so large that the separation between adjacent states cannot be measured. Once again, the correspondence principle is upheld.

### The Magnetic Orbital Quantum Number

We have seen that the orbital angular momentum is quantized and that its magnitude is determined by the value of $\ell$, the orbital quantum number. However, since angular momentum is a vector, its direction must also be specified. Recall from Chapter 30 that an orbiting electron can be considered an effective current loop with a corresponding magnetic moment. If such a moment is placed in a magnetic field **B**, it will interact with the field. Suppose a weak magnetic field is applied along the $z$ axis so that it defines a direction in space. According to quantum mechanics, $L^2$ and $L_z$, the projection of **L** along the $z$ axis, can have discrete values. The magnetic orbital quantum $m_\ell$ specifies the allowed values of $L_z$ according to the expression

$$L_z = m_\ell \hbar \qquad (42.11)$$

Allowed values of $L_z$

The fact that the direction of **L** is quantized with respect to an external magnetic field is often referred to as **space quantization.**

Let us look at the possible orientations of **L** for a given value of $\ell$. Recall that $m_\ell$ can have values ranging from $-\ell$ to $\ell$. If $\ell = 0$, then $m_\ell = 0$ and $L_z = 0$. In order for $L_z$ to be zero, **L** must be perpendicular to **B**. If $\ell = 1$, then the possible values of $m_\ell$ are $-1$, 0, and 1, so that $L_z$ may be $-\hbar$, 0, or $\hbar$. If $\ell = 2$, $m_\ell$ can be $-2$, $-1$, 0, 1, or 2, corresponding to $L_z$ values of $-2\hbar$, $-\hbar$, 0, $\hbar$, or $2\hbar$, and so on.

Space quantization

A vector model describing space quantization for $\ell = 2$ is shown in Figure 42.10a. Note that **L** can never be aligned parallel or antiparallel to **B** since $L_z$ must be smaller than the total angular momentum $L$. From a three-dimensional viewpoint, **L** must lie on the surface of a cone that makes an angle $\theta$ with the $z$ axis, as shown in Figure 42.10b. From the figure, we see that $\theta$ is also quantized and that its values are specified through the relation

$$\cos\theta = \frac{L_z}{|\mathbf{L}|} = \frac{m_\ell}{\sqrt{\ell(\ell+1)}} \qquad (42.12)$$

$\theta$ is quantized

Note that $m_\ell$ is never greater than $\ell$, and therefore $\theta$ can never be zero. (Classically, $\theta$ can have any value.)

Because of the uncertainty principle, the vector **L** does not point in a specific direction but rather traces out a cone in space. If **L** had a definite

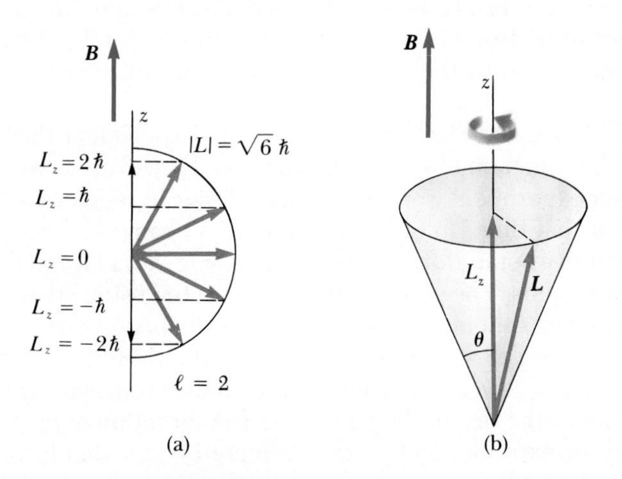


**Figure 42.10** (a) The allowed projections of the orbital angular momentum for the case $\ell = 2$. (b) The orbital angular momentum vector lies on the surface of a cone and precesses about the $z$ axis when a magnetic field **B** is applied in this direction.

value, then all three components $L_x$, $L_y$, and $L_z$ would be exactly specified. For the moment, let us assume that this is the case and let us suppose that the electron moves in the $xy$ plane, so that $\boldsymbol{L}$ is in the $z$ direction and $p_z = 0$. This would mean that $p_z$ is precisely known, which is in violation of the uncertainty principle, $\Delta p_z \Delta z \gtrsim \hbar$. In reality, only the magnitude of $\boldsymbol{L}$ and one component (say $L_z$) can have definite values. In other words, quantum mechanics allows us to specify $L$ and $L_z$ but not $L_x$ and $L_y$. Since the direction of $\boldsymbol{L}$ is constantly changing as it precesses about the $z$ axis, the average values of $L_x$ and $L_y$ are zero and $L_z$ maintains a fixed value of $m_\ell \hbar$.

**EXAMPLE 42.6 Space Quantization for Hydrogen**
Consider the hydrogen atom in the $\ell = 3$ state. Calculate the magnitude of the total angular momentum and the allowed values of $L_z$ and $\theta$.

*Solution* Since $\ell = 3$, we can calculate the total angular momentum using Equation 42.10:

$$L = \sqrt{\ell(\ell+1)}\hbar = \sqrt{3(3+1)}\hbar = 2\sqrt{3}\hbar$$

The allowed values of $L_z$ can be calculated using $L_z = m_\ell \hbar$, with $m_\ell = -3, -2, -1, 0, 1, 2$, and 3. This gives

$$L_z = -3\hbar, -2\hbar, -\hbar, 0, \hbar, 2\hbar, \text{ and } 3\hbar$$

Finally, we can calculate the allowed values of $\theta$ using Equation 42.12. Since $\ell = 3$, $\sqrt{\ell(\ell+1)} = 2\sqrt{3}$ and we have

$$\cos\theta = \frac{m_\ell}{2\sqrt{3}}$$

Substituting the allowed values of $m_\ell$ gives

$$\cos\theta = \pm 0.866, \pm 0.577, \pm 0.289, \text{ and } 0$$

or

$$\theta = 30.0°, 54.8°, 73.2°, 90.0°, 107°, 125°, \text{ and } 150°$$

## 42.6 ELECTRON SPIN

In Section 42.3 we introduced an inherent property of the electron called spin and the associated spin quantum number $m_s$. The theory of the atom in the absence of spin does not explain the fact that spectral lines often split into more components than predicted. Furthermore, without spin the theory does not explain why many spectral lines of atoms are actually two closely spaced lines. This splitting is referred to as *fine structure*. For example, the first line of the Balmer series of hydrogen shows a fine-structure splitting of 0.14 nm. Such effects are attributed to the existence of the spin magnetic moment of the electron.

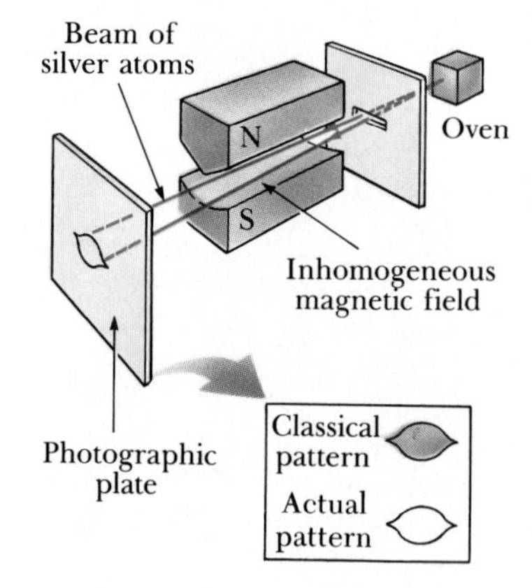


**Figure 42.11** The apparatus used by Stern and Gerlach to verify space quantization. A beam of neutral silver atoms is split into two components by a nonuniform magnetic field.

In 1921, Stern and Gerlach performed an experiment that first demonstrated the phenomenon of space quantization. However, as we shall see, the results of their experiment were not in quantitative agreement with the theory that existed at that time. In their experiment, a beam of neutral silver atoms was sent through a nonuniform magnetic field (Fig. 42.11), and this beam split into two components. The experiment was repeated using other atoms, and in each case the beam split into two or more components. The classical argument is as follows: If the $z$ direction is chosen to be the direction of the maximum inhomogeneity of $\boldsymbol{B}$, the net magnetic force on the atom is along the $z$ axis and is proportional to the magnetic moment in the direction of $\mu_z$. Classically, $\mu_z$ can have any orientation, and so the deflected beam should be spread out continuously. According to quantum mechanics, however, the deflected beam would have several components and the number of components would determine the various possible values of $\mu_z$. Hence, because the Stern-Gerlach experiment showed split beams, space quantization was at least qualitatively verified.

For the moment, let us assume that $\mu_z$ is due to the orbital angular momentum. Since $\mu_z$ is proportional to $m_\ell$, the number of possible values of $\mu_z$ is $2\ell + 1$. Furthermore, since $\ell$ is an integer, the number of values of $\mu_z$ is always odd. This prediction is clearly not consistent with the observations of Stern and Gerlach, who observed only two components in the deflected beam of silver atoms. Hence, one is forced to conclude that either quantum mechanics is incorrect or the model is in need of refinement.

In 1927, Phipps and Taylor repeated the Stern-Gerlach experiment using a beam of hydrogen atoms. This experiment is an important one because it deals with an atom with a single electron in its ground state, for which the theory makes reliable predictions. Recall that $\ell = 0$ for hydrogen in its ground state, and so $m_\ell = 0$. Hence, one would not expect the beam to be deflected by the field since $\mu_z$ would be zero. However, the experiment shows that the beam is again split into two components. On the basis of this result, one can only conclude that there is some contribution to the magnetic moment other than the orbital motion.

In 1925, Goudsmit and Uhlenbeck proposed that the electron has an intrinsic angular momentum apart from its orbital angular momentum. From a classical viewpoint, this intrinsic angular momentum is attributed to the charged electron spinning about its own axis and hence is called *electron spin*.[2] In other words, the total angular momentum of the electron in a particular electronic state contains both an orbital contribution $\boldsymbol{L}$ and a spin contribution **S**.

In 1929, Dirac solved the relativistic wave equation for the electron in a potential well using the relativistic form of the total energy. This theory confirmed the fundamental nature of electron spin. Furthermore, the theory showed that the electron spin could be described by a single quantum number $s$, whose value could be only $\frac{1}{2}$. The magnitude of the **spin angular momentum S** for the electron is given by

**Spin angular momentum of an electron**

$$S = \sqrt{s(s+1)}\hbar = \frac{\sqrt{3}}{2}\hbar \qquad (42.13)$$

Like orbital angular momentum, spin angular momentum is quantized in space, as described in Figure 42.12. Spin angular momentum can have two orientations, specified by the spin magnetic quantum number $m_s$, where $m_s$ has two possible values, $\pm\frac{1}{2}$. The $z$ component of spin angular momentum is given by

$$S_z = m_s\hbar = \pm\tfrac{1}{2}\hbar \qquad (42.14)$$

[2] Physicists often use the word *spin* when referring to *spin angular momentum*. For example, it is common to use the phrase *the electron has a spin of* $\frac{1}{2}$. The spin angular momentum of the electron *never changes*. This notion contradicts classical laws, where a rotating charge would slow down in the presence of an applied magnetic field because of the Faraday emf that accompanies the changing field. Further, if the electron is viewed as a spinning ball of charge subject to classical laws, parts of it near its surface would be rotating with velocities exceeding the speed of light. Thus, the classical picture must not be pressed too far; ultimately, the spinning electron is a quantum entity defying any simple classical description.

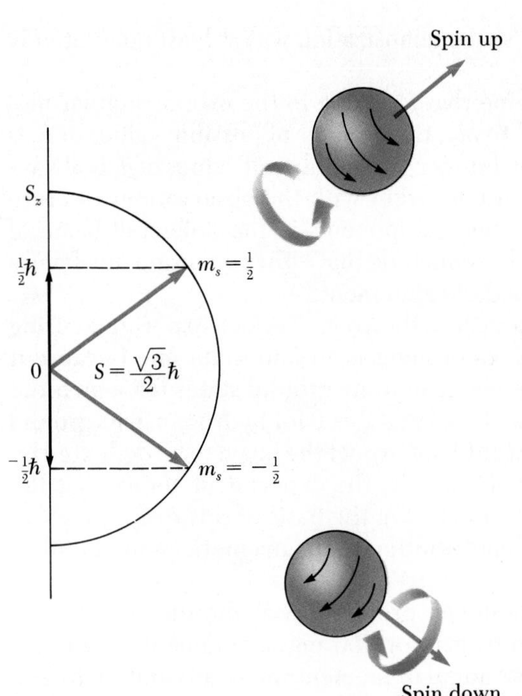


**Figure 42.12** The spin angular momentum also exhibits space quantization. This figure shows the two allowed orientations of the spin vector **S** for a spin ½ particle, such as the electron.

The two values $\pm\hbar/2$ for $S_z$ correspond to the two possible orientations for **S** shown in Figure 42.12. The value $m_s = +1/2$ refers to the up spin case, while the value $m_s = -1/2$ refers to the down spin case.

The spin magnetic moment of the electron, $\boldsymbol{\mu}_s$, is related to its spin angular momentum **S** by the expression

$$\boldsymbol{\mu}_s = -\frac{e}{m}\mathbf{S} \tag{42.15}$$

Since $S_z = \pm\frac{1}{2}\hbar$, the $z$ component of the spin magnetic moment can have the values

$$\mu_{sz} = \pm\frac{e\hbar}{2m} \tag{42.16}$$

The quantity $e\hbar/2m$, called the **Bohr magneton** $\mu_B$, has the numerical value of $9.27 \times 10^{-24}$ J/T (see Section 30.9). Note that the spin contribution to the angular momentum is *twice* the contribution of the orbital motion. The factor of 2 is explained in a relativistic treatment of the system first carried out by Dirac.[3]

One can now understand the outcome of the Stern-Gerlach experiment as follows. The observed moments for both silver and hydrogen are due to spin angular momentum and not to orbital angular momentum. A single-electron atom such as hydrogen has its electron quantized in the magnetic field in such a way that its $z$ component of spin angular momentum is either $\frac{1}{2}\hbar$ or $-\frac{1}{2}\hbar$, corresponding to $m_s = \pm\frac{1}{2}$.

[3] A more exact expression for $\mu_{sz}$, obtained from the Dirac theory of hydrogen, is $\mu_{sz} = -g_e eS/2m$, where $g_e = 2.00229$ for the "free" electron.

Electrons with spin $+\frac{1}{2}$ are deflected downward, and those with spin $-\frac{1}{2}$ are deflected upward. It is interesting to note that the Stern-Gerlach experiment provided two important results. First, it verified the concept of space quantization. Second, it showed that spin angular momentum existed even though this property was not recognized until long after the experiments were performed.

## 42.7 THE EXCLUSION PRINCIPLE AND THE PERIODIC TABLE

Earlier we found that the state of a hydrogen atom is specified by four quantum numbers: $n$, $\ell$, $m_\ell$ and $m_s$. For example, an electron in the ground state of hydrogen could have quantum numbers of $n = 1$, $\ell = 0$, $m_\ell = 0$, $m_s = \frac{1}{2}$. As it turns out, the state of an electron in any other atom may also be specified by this same set of quantum numbers. In fact, these four quantum numbers can be used to describe all the electronic states of an atom regardless of the number of electrons in its structure.

An obvious question that arises here is, "How many electrons can have a particular set of quantum numbers?" This important question was answered by Pauli in 1925 in a powerful statement known as the **exclusion principle:**

Exclusion principle

> No two electrons in an atom can ever be in the same quantum state; that is, no two electrons in the same atom can have the same set of quantum numbers $n$, $\ell$, $m_\ell$, *and* $m_s$.

It is interesting to note that if this principle were not valid, every electron would end up in the lowest energy state of the atom and the chemical behavior of the elements would be grossly modified. Nature as we know it would not exist! In reality, we can view the electronic structure of complex atoms as a succession of filled levels increasing in energy, where the outermost electrons are primarily responsible for the chemical properties of the element.

As a general rule, the order of filling of an atom's subshells with electrons is as follows. Once one subshell is filled, the next electron goes into the vacant subshell that is lowest in energy. One can understand this principle by recognizing that if the atom was not in the lowest energy state available to it, it would radiate energy until it reached this state.

Before we discuss the electronic configuration of various elements, it is convenient to define an *orbital* as the state of an electron characterized by the quantum numbers $n$, $\ell$, and $m_\ell$. From the exclusion principle, we see that *there can be only two electrons in any orbital.* One of these electrons has a spin quantum number $m_s = +\frac{1}{2}$, and the other has $m_s = -\frac{1}{2}$. Since each orbital is limited to two electrons, the number of electrons that can occupy the various levels is also limited.

Table 42.3 shows the number of allowed quantum states for an atom up to $n = 3$. The arrows pointing upward indicate $m_s = \frac{1}{2}$, and the arrows pointing

**TABLE 42.3 Allowed Quantum Numbers for an Atom up to $n = 3$**

| $n$ | 1 | 2 | | | | 3 | | | | | | | | |
|---|---|---|---|---|---|---|---|---|---|---|---|---|---|---|
| $\ell$ | 0 | 0 | 1 | | | 0 | 1 | | | 2 | | | | |
| $m_\ell$ | 0 | 0 | 1 | 0 | −1 | 0 | 1 | 0 | −1 | 2 | 1 | 0 | −1 | −2 |
| $m_s$ | ↑↓ | ↑↓ | ↑↓ | ↑↓ | ↑↓ | ↑↓ | ↑↓ | ↑↓ | ↑↓ | ↑↓ | ↑↓ | ↑↓ | ↑↓ | ↑↓ |

## Biographical Sketch

**Wolfgang Pauli**
*(1900–1958)*

(Photo taken by S. A. Goudsmit, AIP Niels Bohr Library)

**Wolfgang Pauli** was an extremely talented Austrian theoretical physicist who made important contributions in many areas of modern physics. At the age of 21, Pauli gained public recognition with a masterful review article on relativity, which is still considered to be one of the finest and most comprehensive introductions to the subject. Other major contributions were the discovery of the exclusion principle, the explanation of the connection between particle spin and statistics, theories of relativistic quantum electrodynamics, the neutrino hypothesis, and the hypothesis of nuclear spin. An article entitled "The Fundamental Principles of Quantum Mechanics," written by Pauli in 1933 for the *Handbuch der Physik*, is widely acknowledged to be one of the best treatments of quantum physics ever written. Pauli was a forceful and colorful character, well known for his witty and often caustic remarks directed at those who presented new theories in a less than perfectly clear manner. Pauli exerted great influence on his students and colleagues by forcing them with his sharp criticism to a deeper and clearer understanding. Victor Weisskopf, one of Pauli's famous students, has aptly described him as "the conscience of theoretical physics." Pauli's sharp sense of humor was also nicely captured by Weisskopf in the following anecdote:

"In a few weeks, Pauli asked me to come to Zurich. I came to the big door of his office, I knocked, and no answer. I knocked again and no answer. After about five minutes he said, rather roughly, "Who is it? Come in!" I opened the door, and here was Pauli — it was a very big office — at the other side of the room, at his desk, writing and writing. He said, "Who is this? First I must finish calculating." Again he let me wait for about five minutes and then: "Who is that?" "I am Weisskopf." "Uhh, Weisskopf, ja, you are my new assistant." Then he looked at me and said, "Now, you see I wanted to take Bethe, but Bethe works now on the solid state. Solid state I don't like, although I started it. This is why I took you." Then I said, "What can I do for you sir?" and he said "I shall give you right away a problem." He gave me a problem, some calculation, and then he said, "Go and work." So I went, and after 10 days or so, he came and said, "Well, show me what you have done." And I showed him. He looked at it and exclaimed: "I should have taken Bethe!"[1]

[1] From *Physics in the Twentieth Century: Selected Essays, My Life as a Physicist,* Victor F. Weisskopf, The MIT Press, 1972, p. 10.

downward indicate $m_s = -\frac{1}{2}$. The $n = 1$ shell can accommodate only two electrons since there is only one allowed orbital with $m_\ell = 0$. The $n = 2$ shell has two subshells, with $\ell = 0$ and $\ell = 1$. The $\ell = 0$ subshell is limited to only two electrons since $m_\ell = 0$. The $\ell = 1$ subshell has three allowed orbitals, corresponding to $m_\ell = 1$, 0, and $-1$. Since each orbital can accommodate two electrons, the $\ell = 1$ subshell can hold six electrons. Finally, the $n = 3$ shell has three subshells and nine orbitals and can accommodate up to 18 electrons. Note that each shell can accommodate up to $2n^2$ electrons.

The exclusion principle can be illustrated by an examination of the electronic arrangement in a few of the lighter atoms.

*Hydrogen* has only one electron, which, in its ground state, can be described by either of two sets of quantum numbers: 1, 0, 0, $\frac{1}{2}$ or 1, 0, 0, $-\frac{1}{2}$. The electronic configuration of this atom is often designated as $1s^1$. The notation $1s$ means that we are referring to a state for which $n = 1$ and $\ell = 0$, and the superscript indicates that one electron is present in this level.

Neutral *helium* has two electrons. In the ground state, the quantum numbers for these two electrons are 1, 0, 0, $\frac{1}{2}$ and 1, 0, 0, $-\frac{1}{2}$. There are no other possible combinations of quantum numbers for this level, and we say that the **K** shell is filled. Helium is designated by the notation $1s^2$.

Neutral *lithium* has three electrons. In the ground state, two of these are in the 1$s$ subshell and the third is in the 2$s$ subshell because this subshell is slightly lower in energy than the 2$p$ subshell. Hence, the electronic configuration for lithium is $1s^2\,2s^1$.

The electronic configurations of some successive elements are given in Figure 42.13. Note that the electronic configuration of *beryllium*, with its four electrons, is $1s^2 2s^2$, and *boron* has a configuration of $1s^2 2s^2 2p^1$. The 2$p$ electron in boron may be described by six sets of quantum numbers, each set corresponding to states of equal energy.

*Carbon* has six electrons, and a question arises concerning how to assign the two 2$p$ electrons. Do they go into the same orbital with paired spins (↑↓), or do they occupy different orbitals with unpaired spins (↑↑)? Experimental data show that the most stable configuration (that is, the one that is energetically preferred) is the latter, where the spins are unpaired. Hence, the two 2$p$ electrons in carbon and the three 2$p$ electrons in nitrogen have unpaired spins (Fig. 42.13). The general rule that governs such situations, called **Hund's rule,** states that

> when an atom has orbitals of equal energy, the order in which they are filled by electrons is such that a maximum number of electrons will have unpaired spins.

**Hund's rule**

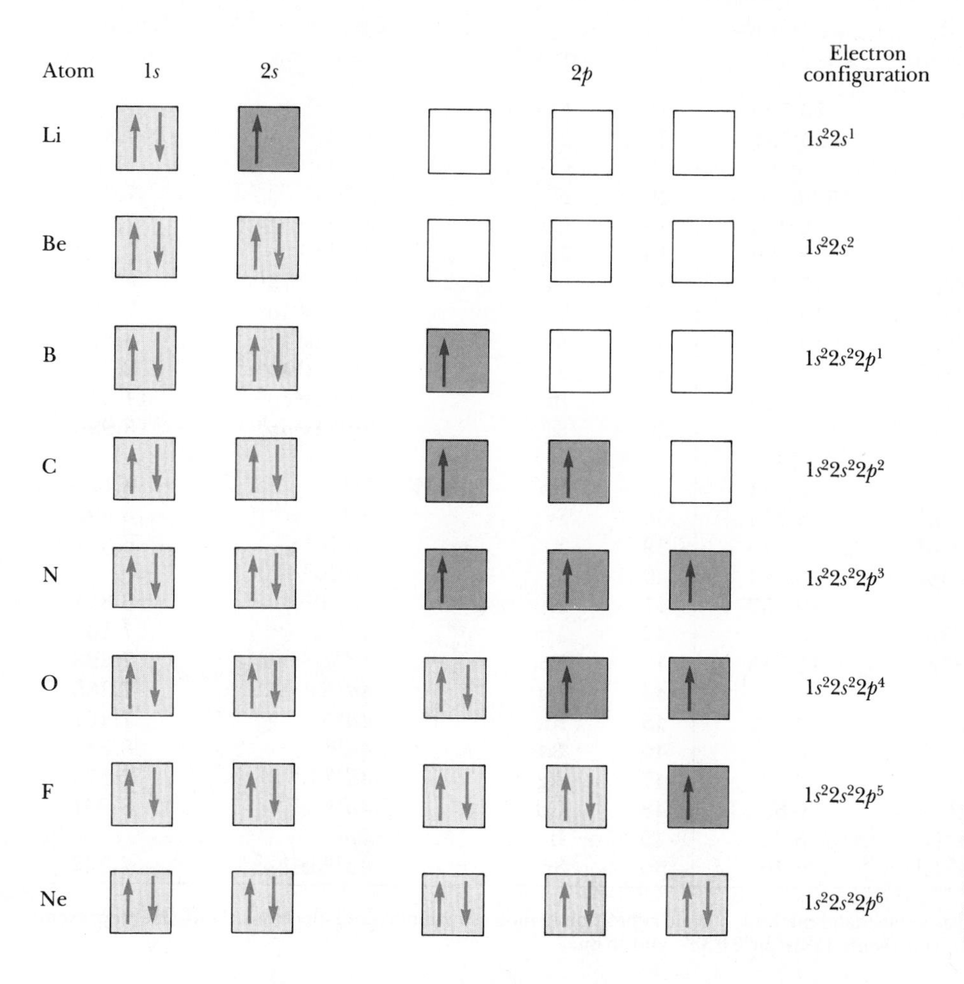


**Figure 42.13** The filling of electronic states must obey the Pauli exclusion principle and Hund's rule. The electronic configurations for these elements are given at the right.

Some exceptions to this rule occur in those elements which have sublevels that are close to being filled or half-filled. A more complete description of the electronic configuration of the elements can be found in most general chemistry texts.[4]

A complete list of electronic configurations is provided in Table 42.4. The first attempt at finding some order among the elements was made by a Russian chemist, Dmitri Mendeleev, in 1871. He arranged the atoms in a table similar to that shown in Appendix C according to their atomic weights and chemical similarities. The first table Mendeleev proposed contained many blank spaces, and he boldly stated that the gaps were there only because the elements had not yet been discovered. By noting the column in which these missing elements should be located, he was able to make rough predictions about their chemical properties. Within 20 years of this announcement, these elements were indeed discovered.

[4] For example, see W.L. Masterton, E.J. Slowinski, and C.L. Stanitski, *Chemical Principles*, 6th ed., Philadelphia, Saunders College Publishing, 1985, Chap. 7.

**TABLE 42.4 Electronic Configuration of the Elements**

| Z | Symbol | Ground Configuration | Ionization Energy (eV) | Z | Symbol | Ground Configuration | Ionization Energy (eV) |
|---|---|---|---|---|---|---|---|
| 1 | H | $1s^1$ | 13.595 | 25 | Mn | $3d^54s^2$ | 7.432 |
| 2 | He | $1s^2$ | 24.581 | 26 | Fe | $3d^64s^2$ | 7.87 |
| | | | | 27 | Co | $3d^74s^2$ | 7.86 |
| 3 | Li | [He] $2s^1$ | 5.39 | 28 | Ni | $3d^84s^2$ | 7.633 |
| 4 | Be | $2s^2$ | 9.320 | 29 | Cu | $3d^{10}4s^1$ | 7.724 |
| 5 | B | $2s^22p^1$ | 8.296 | 30 | Zn | $3d^{10}4s^2$ | 9.391 |
| 6 | C | $2s^22p^2$ | 11.256 | 31 | Ga | $3d^{10}4s^24p^1$ | 6.00 |
| 7 | N | $2s^22p^3$ | 14.545 | 32 | Ge | $3d^{10}4s^24p^2$ | 7.88 |
| 8 | O | $2s^22p^4$ | 13.614 | 33 | As | $3d^{10}4s^24p^3$ | 9.81 |
| 9 | F | $2s^22p^5$ | 17.418 | 34 | Se | $3d^{10}4s^24p^4$ | 9.75 |
| 10 | Ne | $2s^22p^6$ | 21.559 | 35 | Br | $3d^{10}4s^24p^5$ | 11.84 |
| | | | | 36 | Kr | $3d^{10}4s^24p^6$ | 13.996 |
| 11 | Na | [Ne] $3s^1$ | 5.138 | | | | |
| 12 | Mg | $3s^2$ | 7.644 | 37 | Rb | [Kr] $5s^1$ | 4.176 |
| 13 | Al | $3s^23p^1$ | 5.984 | 38 | Sr | $5s^2$ | 5.692 |
| 14 | Si | $3s^23p^2$ | 8.149 | 39 | Y | $4d^15s^2$ | 6.377 |
| 15 | P | $3s^23p^3$ | 10.484 | 40 | Zr | $4d^25s^2$ | |
| 16 | S | $3s^23p^4$ | 10.357 | 41 | Nb | $4d^45s^1$ | 6.881 |
| 17 | Cl | $3s^23p^5$ | 13.01 | 42 | Mo | $4d^55s^1$ | 7.10 |
| 18 | Ar | $3s^23p^6$ | 15.755 | 43 | Tc | $4d^55s^2$ | 7.228 |
| | | | | 44 | Ru | $4d^75s^1$ | 7.365 |
| 19 | K | [Ar] $4s^1$ | 4.339 | 45 | Rh | $4d^85s^1$ | 7.461 |
| 20 | Ca | $4s^2$ | 6.111 | 46 | Pd | $4d^{10}$ | 8.33 |
| 21 | Sc | $3d^14s^2$ | 6.54 | 47 | Ag | $4d^{10}5s^1$ | 7.574 |
| 22 | Ti | $3d^24s^2$ | 6.83 | 48 | Cd | $4d^{10}5s^2$ | 8.991 |
| 23 | V | $3d^34s^2$ | 6.74 | 49 | In | $5p^1$ | |
| 24 | Cr | $3d^54s^1$ | 6.76 | 50 | Sn | $4d^{10}5s^25p^2$ | 7.342 |

Note: The bracket notation is used as a shorthand method to avoid repetition in indicating inner-shell electrons. Thus, [He] represents $1s^2$, [Ne] represents $1s^22s^22p^6$, [Ar] represents $1s^22s^22p^63s^23p^6$, and so on.

The elements in the periodic table are arranged such that all those in a vertical column have similar chemical properties. For example, consider the elements in the last column: He (helium), Ne (neon), Ar (argon), Kr (krypton), Xe (xenon), and Rn (radon). The outstanding characteristic of these elements is that they do not normally take part in chemical reactions, that is, they do not join with other atoms to form molecules, and are therefore classified as being inert. Because of this aloofness, they are referred to as the noble gases. We can partially understand this behavior by looking at the electronic configurations shown in Table 42.4. This table also lists the ionization energies for various elements. The element helium is one in which the electronic configuration is $1s^2$. In other words, one shell is filled. Additionally, it is found that the electrons in this filled shell are considerably separated in energy from the next available level, the $2s$ level.

The electronic configuration for neon is $1s^22s^22p^6$. Again, the outermost shell is filled and there is a wide gap in energy between the $2p$ level and the $3s$ level. Argon has the configuration $1s^22s^22p^63s^23p^6$. Here, the $3p$ subshell is filled and there is a wide gap in energy between the $3p$ subshell and the $3d$

**TABLE 42.4 Electronic Configuration of the Elements *(Continued)***

| Z | Symbol | Ground Configuration | Ionization Energy (eV) | Z | Symbol | Ground Configuration | Ionization Energy (eV) |
|---|---|---|---|---|---|---|---|
| 51 | Sb | $4d^{10}5s^25p^3$ | 8.639 | 78 | Pt | $4f^{14}5d^86s^2$ | 8.88 |
| 52 | Te | $4d^{10}5s^25p^4$ | 9.01 | 79 | Au | [Xe, $4f^{14}5d^{10}$] $6s^1$ | 9.22 |
| 53 | I | $4d^{10}5s^25p^5$ | 10.454 | 80 | Hg | $6s^2$ | 10.434 |
| 54 | Xe | $4d^{10}5s^25p^6$ | 12.127 | 81 | Tl | $6s^26p^1$ | 6.106 |
| | | | | 82 | Pb | $6s^26p^2$ | 7.415 |
| 55 | Cs | [Xe] $6s^1$ | 3.893 | 83 | Bi | $6s^26p^3$ | 7.287 |
| 56 | Ba | $6s^2$ | 5.210 | 84 | Po | $6s^26p^4$ | 8.43 |
| 57 | La | $5d^16s^2$ | 5.61 | 85 | At | $6s^26p^5$ | |
| 58 | Ce | $4f^15d^16s^2$ | 6.54 | 86 | Rn | $6s^26p^6$ | 10.745 |
| 59 | Pr | $4f^36s^2$ | 5.48 | | | | |
| 60 | Nd | $4f^46s^2$ | 5.51 | 87 | Fr | [Rn] $7s^1$ | |
| 61 | Pm | $4f^56s^2$ | | 88 | Ra | $7s^2$ | 5.277 |
| 62 | Fm | $4f^66s^2$ | 5.6 | 89 | Ac | $6d^17s^2$ | 6.9 |
| 63 | Eu | $4f^76s^2$ | 5.67 | 90 | Th | $6d^27s^2$ | |
| 64 | Gd | $4f^75d^16s^2$ | 6.16 | 91 | Pa | $5f^26d^17s^2$ | |
| 65 | Tb | $4f^96s^2$ | 6.74 | 92 | U | $5f^36d^17s^2$ | 4.0 |
| 66 | Dy | $4f^{10}6s^2$ | | 93 | Np | $5f^46d^17s^2$ | |
| 67 | Ho | $4f^{11}6s^2$ | | 94 | Pu | $5f^67s^2$ | |
| 68 | Er | $4f^{12}6s^2$ | | 95 | Am | $5f^77s^2$ | |
| 69 | Tm | $4f^{13}6s^2$ | | 96 | Cm | $5f^76d^17s^2$ | |
| 70 | Yb | $4f^{14}6s^2$ | 6.22 | 97 | Bk | $5f^86d^17s^2$ | |
| 71 | Lu | $4f^{14}5d^16s^2$ | 6.15 | 98 | Cf | $5f^{10}7s^2$ | |
| 72 | Hf | $4f^{14}5d^26s^2$ | 7.0 | 99 | Es | $5f^{11}7s^2$ | |
| 73 | Ta | $4f^{14}5d^36s^2$ | 7.88 | 100 | Fm | $5f^{12}7s^2$ | |
| 74 | W | $4f^{14}5d^46s^2$ | 7.98 | 101 | Mv | $5f^{13}7s^2$ | |
| 75 | Re | $4f^{14}5d^56s^2$ | 7.87 | 102 | No | $5f^{14}7s^2$ | |
| 76 | Os | $4f^{14}5d^66s^2$ | 8.7 | 103 | Lw | $5f^{14}6d^17s^2$ | |
| 77 | Ir | $4f^{14}5d^76s^2$ | 9.2 | 104 | Ku | $5f^{14}6d^27s^2$ | |

subshell. We could continue this procedure through all the noble gases; the pattern remains the same. A noble gas is formed when either a shell or a subshell is filled and there is a large gap in energy before the next possible level is encountered.

## 42.8 ATOMIC SPECTRA: VISIBLE AND X-RAY

In Chapter 40 we briefly discussed the origin of the spectral lines for hydrogen and hydrogen-like ions. Recall that an atom will emit electromagnetic radiation if an electron in an excited state makes a transition to a lower energy state. The set of wavelengths observed for a specific species by such processes is called an **emission spectrum.** Likewise, atoms with electrons in the ground-state configuration can also absorb electromagnetic radiation at specific wavelengths, giving rise to an **absorption spectrum.** Such spectra can be used to identify the elements in gases.

The energy level diagram for hydrogen is shown in Figure 42.14. The states are labeled with the quantum numbers $n$ and $\ell$. The various diagonal lines drawn between the levels represent allowed transitions between stationary states. Whenever an electron makes a transition from a higher energy state to a lower one, a photon of light is emitted. The frequency of this photon is given by $f = \Delta E/h$, where $\Delta E$ is the energy difference between the two levels and $h$ is Planck's constant. The transitions that are allowed (that is, those corresponding to the diagonal lines in Fig. 42.14) are those for which $\ell$ changes by 1. That is, the **selection rule** ***for the allowed transitions*** is given by

Selection rule for allowed atomic transitions

$$\Delta\ell = \pm 1 \tag{42.17}$$

Transitions for which $\Delta\ell \neq \pm 1$ are said to be forbidden. Such transitions can actually occur, but their probability is negligible relative to the probability of the allowed transitions.

Since the orbital angular momentum of an atom changes when a photon is emitted or absorbed (that is, as a result of a transition) and since angular momentum must be conserved, we conclude that *the photon involved in the*

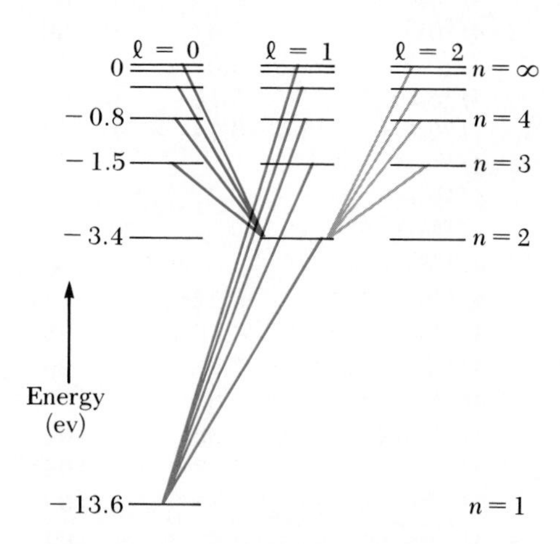


**Figure 42.14** Some allowed electronic transitions for hydrogen, represented by the colored lines. These transitions must obey the selection rule $\Delta\ell = \pm 1$.

*process must carry angular momentum.* In fact, the photon has an angular momentum equivalent to that of a particle having a spin of 1. It is also interesting to note that the angular momentum of the photon is consistent with the classical description of electromagnetic radiation. Hence, a photon has energy, linear momentum, and angular momentum.

The photon carries angular momentum

It is interesting to plot the ionization energy versus atomic number Z, as in Figure 42.15a. Note the pattern of 2, 8, 8, 18, 18, 32 for the ionization

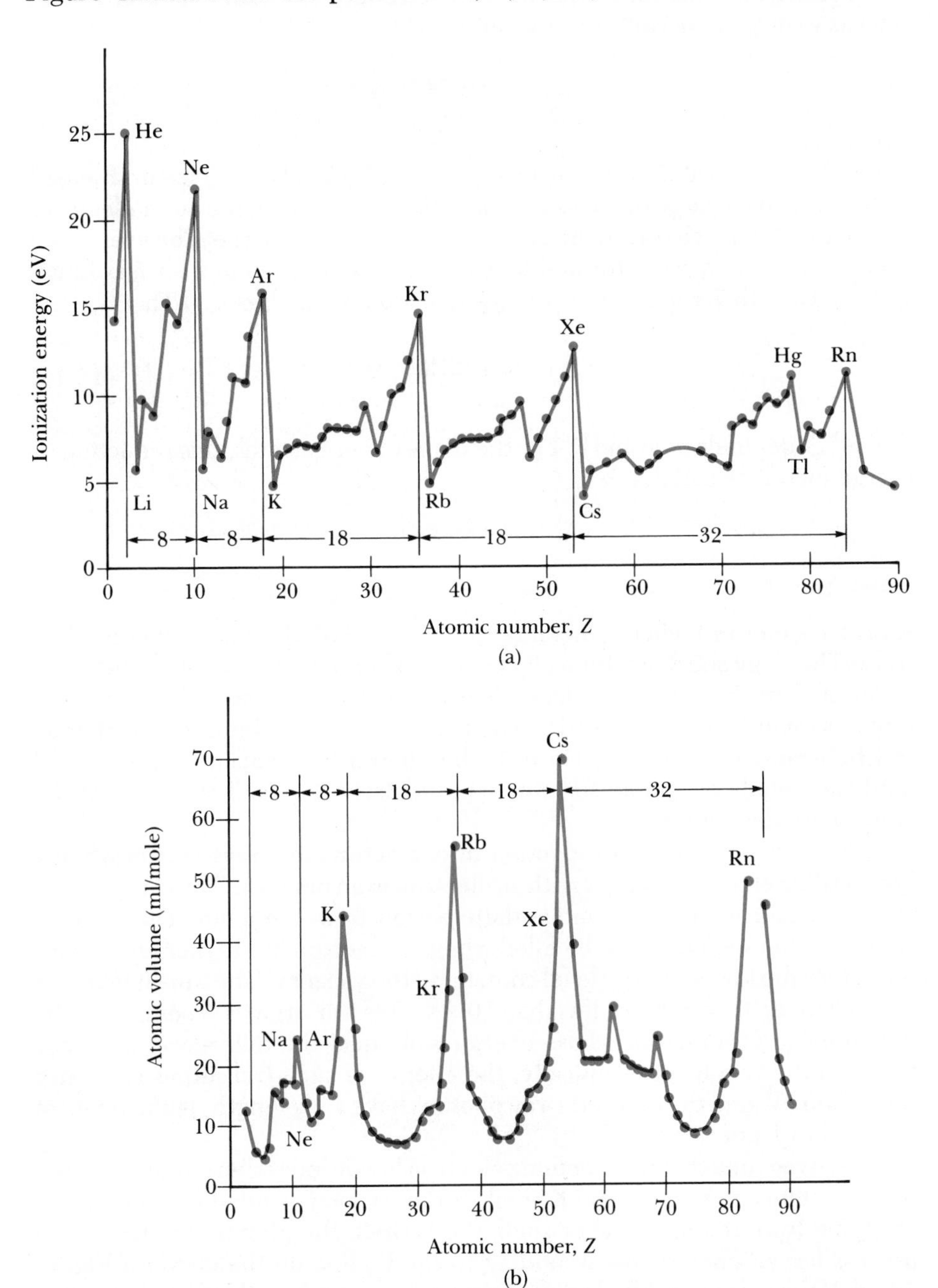


**Figure 42.15** (a) Ionization energy of the elements versus atomic number Z. (b) Atomic volume of the elements versus atomic number. (Adapted from J. Orear, *Physics*, New York, Macmillan, 1979.)

energies. This pattern follows from the Pauli exclusion principle and helps explain why the elements repeat their chemical properties in groups. For example, the peaks at $Z = 2$, 10, 18, and 36 correspond to the elements He, Ne, Ar, and Kr, which have filled shells. These elements have similar energies and chemical behavior. A similar repetitive pattern is observed in a plot of the atomic volume per atom versus atomic number (Fig. 42.15b).

Recall from Chapter 40 that the allowed energies for one-electron atoms, such as hydrogen or $He^+$, are given by

Allowed energies for one-electron atoms

$$E_n = -\frac{13.6Z^2}{n^2}\text{ eV} \tag{42.18}$$

For multielectron atoms, the nuclear charge $Ze$ is largely canceled or shielded by the negative charge of the inner-core electrons. Hence, the outer electrons interact with a net charge of the order of the electronic charge. The expression for the allowed energies for multielectron atoms has the same form as Equation 42.18 with Z replaced by an effective atomic number $Z_{eff}$. That is,

Allowed energies for multi-electron atoms

$$E_n = -\frac{13.6Z_{eff}^2}{n^2}\text{ eV} \tag{42.19}$$

where $Z_{eff}$ depends on $n$ and $\ell$. For the higher energy states, this reduction in charge increases and $Z_{eff} \rightarrow 1$.

## X-Ray Spectra

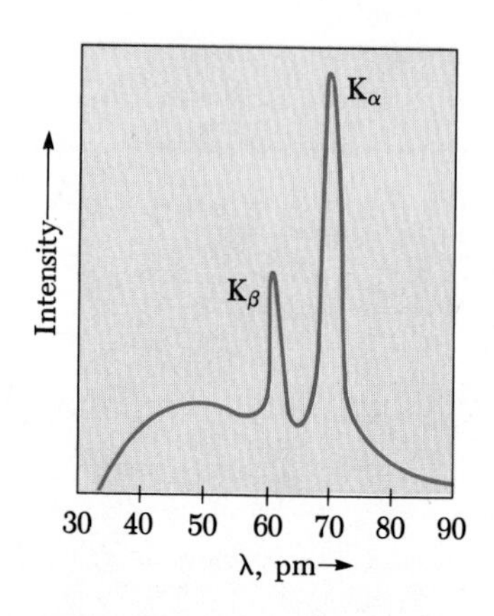


**Figure 42.16** The x-ray spectrum of a metal target consists of a broad continuous spectrum plus a number of sharp lines, which are due to *characteristic x-rays*. The data shown were obtained when 35-keV electrons bombarded a molybdenum target. Note that 1 pm = $10^{-12}$ m = $10^{-3}$ nm.

X-rays are emitted when a metal target is bombarded by high-energy electrons. The x-ray spectrum typically consists of a broad continuous band and a series of sharp lines that are dependent on the type of material used for the target, as shown in Figure 42.16. The presence of these lines, called **characteristic x-rays,** was discovered in 1908, but their origin remained unexplained until the details of atomic structure, particularly the shell structure of the atom, were developed.

The first step in the production of characteristic x-rays occurs when a bombarding electron collides with an electron in an inner shell of a target atom with sufficient energy to remove the electron from the atom. The vacancy created in the shell can now be filled when an electron in a higher level drops down into the lower energy level containing the vacancy. The time it takes for this to happen is very short, less than $10^{-9}$ s. This transition is accompanied by the emission of a photon whose energy will equal the difference in energy between the two levels. Typically, the energy of such transitions is greater than 1000 eV, and the emitted x-ray photons have wavelengths in the range of 0.01 nm to 1 nm.

Let us assume that the incoming electron has dislodged an atomic electron from the innermost shell, the K shell. If the vacancy is filled by an electron dropping from the next higher shell, the L shell, the photon emitted in the process has an energy corresponding to the $K_\alpha$ line on the curve of Figure 42.16. If the vacancy is filled by an electron dropping from the M shell, the line produced is called the $K_\beta$ line.

Other characteristic x-ray lines are formed when electrons drop from upper levels to vacancies other than those in the K shell. For example, L lines are produced when vacancies in the L shell are filled by electrons dropping from higher shells. An $L_\alpha$ line is produced as an electron drops from the M shell to the L shell, and an $L_\beta$ line is produced by a transition from the N shell to the L shell.

We can estimate the energy of the x-rays emitted by an atom as follows. Consider two electrons in the K shell of an atom whose atomic number is Z, where Z is the number of protons in the nucleus. Each electron partially shields the other from the charge of the nucleus, $Ze$, and so each electron is subject to an effective nuclear charge $Z_{\text{eff}} = (Z - 1)e$. We can now use Equation 42.19 to estimate the energy of either electron in the K shell (with $n = 1$). This gives

$$E_K = -(Z - 1)^2(13.6 \text{ eV}) \tag{42.20}$$

As we shall show in the following example, one can estimate the energy of an electron in an L or M shell in a similar fashion. Taking the energy difference between these two levels, one can then calculate the energy and wavelength of the emitted photon.

In 1914, Henry G.J. Moseley plotted the Z values for a number of elements versus $\sqrt{1/\lambda}$, where $\lambda$ is the wavelength of the $K_\alpha$ line for each element. He found that such a plot produced a straight line, as in Figure 42.17. This is consistent with rough calculations of the energy levels calculated using Equation 42.20. From this plot, Moseley was able to determine the Z values of other elements, which provided a periodic chart in excellent agreement with the known chemical properties of the elements.

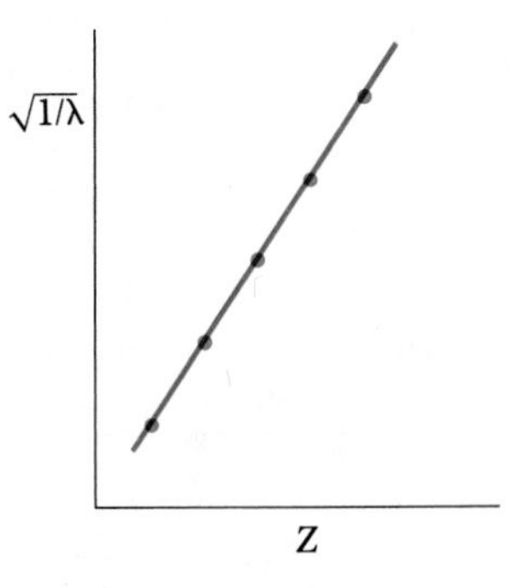


**Figure 42.17** A Moseley plot. A straight line is obtained when $\sqrt{1/\lambda}$ is plotted versus Z for the $K_\alpha$ x-ray lines of a number of elements.

**EXAMPLE 42.7 Estimating the Energy of an X-Ray**

Estimate the energy of the characteristic x-ray emitted from a tungsten target when an electron drops from an M shell ($n = 3$ state) to a vacancy in the K shell ($n = 1$ state).

*Solution* The atomic number for tungsten is $Z = 74$. Using Equation 42.20, we see that the energy of the electron in the K shell state is approximately

$$E_K = -(74 - 1)^2(13.6 \text{ eV}) = -72{,}500 \text{ eV}$$

The electron in the M shell ($n = 3$) is subject to an effective nuclear charge that depends on the number of electrons in the $n = 1$ and $n = 2$ states, which shield the nucleus. Because there are eight electrons in the $n = 2$ state and one electron in the $n = 1$ state, roughly nine electrons shield the nucleus, and so $Z_{\text{eff}} = Z - 9$. Hence, the energy of an electron in the M shell ($n = 3$), following Equation 42.19, is equal to

$$E_M = -Z_{\text{eff}}^2 E_3 = -(Z - 9)^2 \frac{E_0}{3^2}$$

$$= -(74 - 9)^2 \frac{(13.6 \text{ eV})}{9} = -6380 \text{ eV}$$

where $E_3$ is the energy of an electron in the $M = 3$ level of the hydrogen atom and $E_0$ is the ground state energy. Therefore, the emitted x-ray has an energy equal to $E_M - E_K = -6380 \text{ eV} - (-72\,500 \text{ eV}) = 66\,100 \text{ eV}$. Note that this energy difference is also equal to $hf = hc/\lambda$, where $\lambda$ is the wavelength of the emitted x-ray.

Exercise 3 Calculate the wavelength of the emitted x-ray for this transition.
Answer 0.0188 nm.

## 42.9 ATOMIC TRANSITIONS

In this section we shall look at some of the basic processes involved in an atomic system. It is necessary to understand these mechanisms before we can understand the operation of a laser, which we shall examine in the next section.

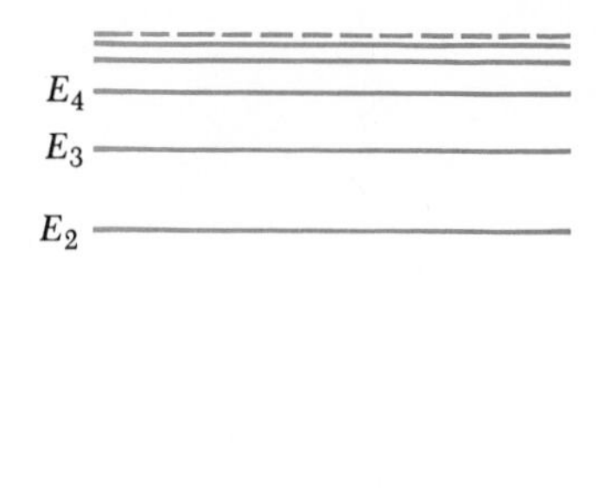


**Figure 42.18** Energy level diagram of an atom with various allowed states. The lowest energy state, $E_1$, is the ground state. All others are excited states.

We have seen that an atom will emit radiation only at certain frequencies, which correspond to the energy separation between the various allowed states. Consider an atom with many allowed energy states, labeled $E_1$, $E_2$, $E_3$, . . . , as in Figure 42.18. When light is incident on the atom, only those photons whose energy, $hf$, matches the energy separation $\Delta E$ between two levels can be absorbed by the atom. A schematic diagram representing this **stimulated absorption process** is shown in Figure 42.19. At ordinary temperatures, most of the atoms are in the ground state. If a vessel containing many atoms of a gaseous element is illuminated with a light beam containing all possible photon frequencies (that is, a continuous spectrum), only those photons of energies $E_2 - E_1$, $E_3 - E_1$, $E_4 - E_1$, $E_3 - E_2$, $E_4 - E_2$, and so on, can be absorbed. As a result of this absorption, some atoms are raised to various allowed higher energy levels called **excited states.**

Once an atom is in an excited state, there is a certain probability that it will jump back to a lower level by emitting a photon, as shown in Figure 42.20. This process is known as **spontaneous emission.** In typical cases, an atom will remain in an excited state for only about $10^{-8}$ s.

Finally, there is a third process, which is of importance in lasers, known as **stimulated emission.** Suppose an atom is in an excited state $E_2$, as in Figure 42.21, and a photon of energy $hf = E_2 - E_1$ is incident on it. The incoming photon will increase the probability that the electron will return to the ground state and thereby emit a second photon having the same energy, $hf$. This process of speeding up atomic transitions to lower levels is called stimulated emission. Note that there are two identical photons that result from this process, the incident photon and the emitted photon. The emitted photon will be exactly in phase with the incident photon. These photons can, in turn, stimulate other atoms to emit photons in a chain of similar processes. The many photons produced in this fashion are the source of the intense, coherent light in a laser.

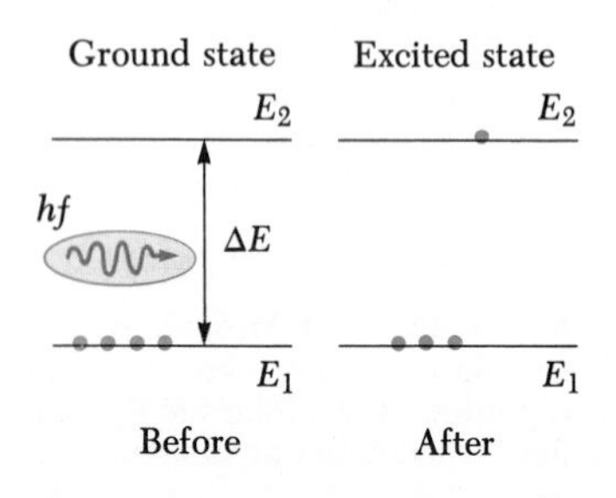


**Figure 42.19** Diagram representing the *stimulated absorption* of a photon by an atom. The dots represent electrons. One electron is transferred from the ground state to the excited state when the atom absorbs a photon whose energy $hf = E_2 - E_1$.

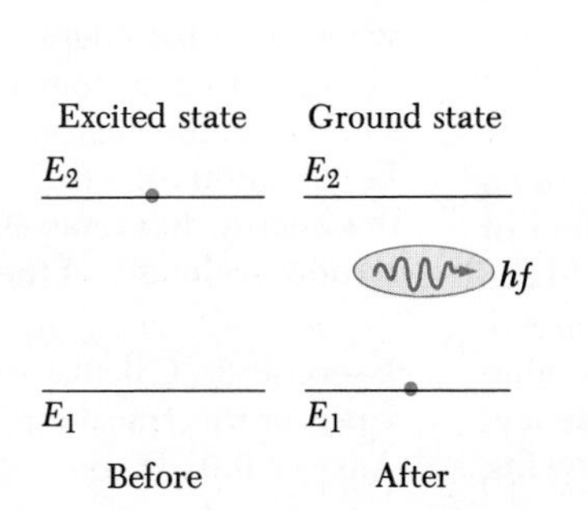


**Figure 42.20** Diagram representing the *spontaneous emission* of a photon by an atom that is initially in the excited state $E_2$. When the electron falls to the ground state, the atom emits a photon whose energy $hf = E_2 - E_1$.

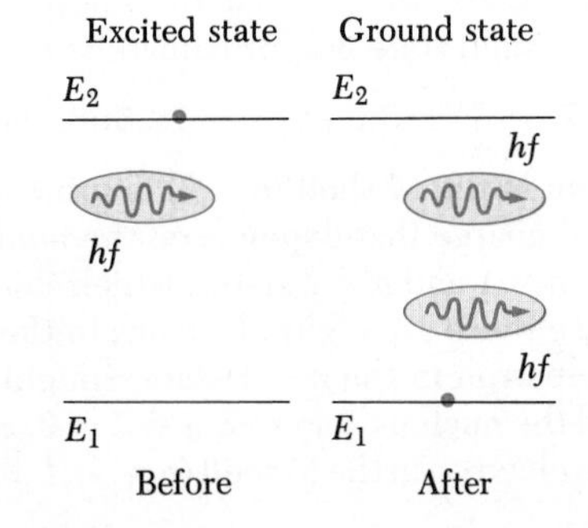


**Figure 42.21** Diagram representing the *stimulated emission* of a photon by an incoming photon of energy $hf$. Initially, the atom is in the excited state. The incoming photon stimulates the atom to emit a second photon of energy $hf = E_2 - E_1$.

## *42.10 LASERS AND HOLOGRAPHY

We have described how an incident photon can cause atomic transitions either upward (stimulated absorption) or downward (stimulated emission). Both processes are equally probable. When light is incident on a system of atoms, there is usually a net absorption of energy because there are many more atoms in the ground state than in excited states when the system is in thermal equilibrium. That is, in a normal situation, there are more atoms in state $E_1$ ready to absorb photons than there are atoms in states $E_2$, $E_3$, . . . ready to emit photons. However, if one can invert the situation so that there are more atoms in an excited state than in the ground state, a net emission of photons can result. Such a condition is called **population inversion.** This, in fact, is the fundamental principle involved in the operation of a **laser,** an acronym for *l*ight *a*mplification by *s*timulated *e*mission of *r*adiation. The amplification corresponds to a buildup of photons in the system as the result of a chain reaction of events.

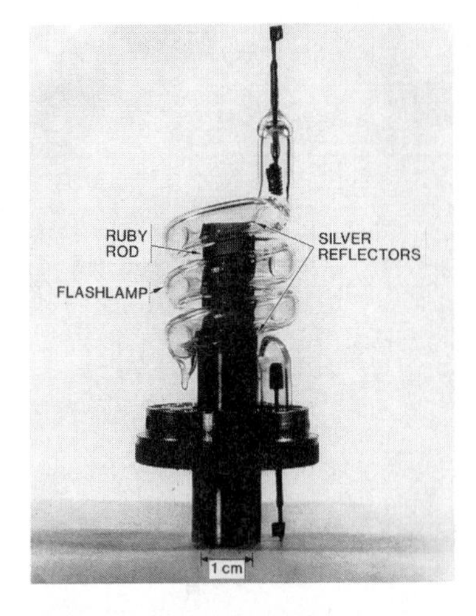

Photograph of the first ruby laser showing the flash lamp surrounding the ruby rod. (Courtesy of Hughes Aircraft Company)

The following three conditions must be satisfied in order to achieve laser action:

1. The system must be in a state of population inversion (that is, more atoms in an excited state than in the ground state).
2. The excited state of the system must be a *metastable state,* which means its lifetime must be long compared with the usually short lifetimes of excited states. When such is the case, stimulated emission will occur before spontaneous emission.
3. The emitted photons must be confined in the system long enough to allow them to stimulate further emission from other excited atoms. This is achieved by the use of reflecting mirrors at the ends of the system. One end is made totally reflecting, and the other is slightly transparent to allow the laser beam to escape.

One device that exhibits stimulated emission of radiation is the helium-neon gas laser. The energy level diagram for the neon atom in this system is shown in Figure 42.22. The mixture of helium and neon is confined to a glass tube sealed at the ends by mirrors. An oscillator connected to the tube causes electrons to sweep through the tube, colliding with the atoms of the gas and raising them into excited states. Neon atoms are excited to state $E_3$ through this process and also as a result of collisions with excited helium atoms. Stimulated emission occurs as the neon atoms make a transition to state $E_2$ and neighboring excited atoms are stimulated. This results in the production of coherent light at a wavelength of 632.8 nm.

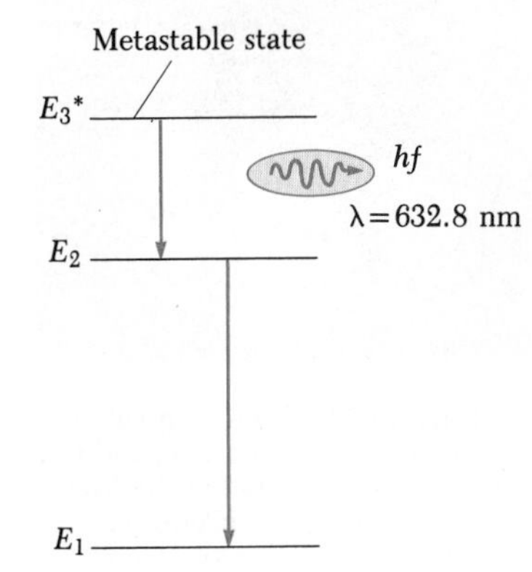

**Figure 42.22** Energy level diagram for the neon atom, which emits photons at a wavelength of 632.8 nm through stimulated emission. The photon at this wavelength arises from the transition $E_3^* \rightarrow E_2$. This is the source of coherent light in the helium-neon gas laser.

Since the development of the first laser in 1960, there has been a tremendous growth in laser technology. Lasers are now available that cover wavelengths in the infrared, visible, and ultraviolet regions. Applications include surgical "welding" of detached retinas, precision surveying and length measurement, a potential source for inducing nuclear fusion reactions, precision cutting of metals and other materials, and telephone communication along optical fibers. These and other applications are possible because of the unique characteristics of laser light. In addition to its being highly monochromatic and coherent, laser light is also highly directional and can be sharply focused to produce regions of extremely intense light energy. A more detailed description of lasers and some interesting applications are discussed in the essay at the end of this chapter.

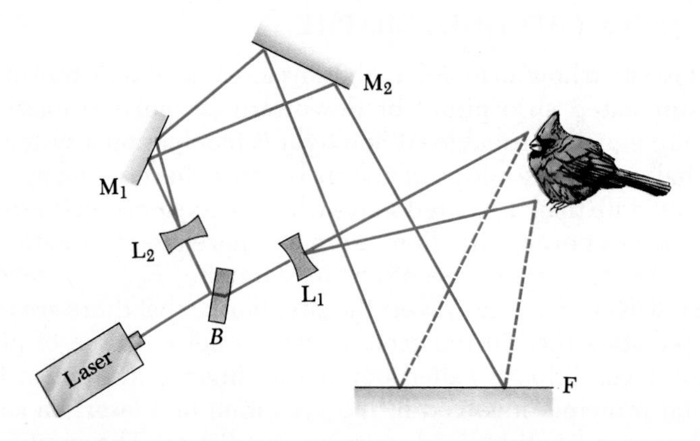


Figure 42.23 Experimental arrangement for producing a hologram.

## Holography

One of the most unusual and interesting applications of the laser is in the production of three-dimensional images of an object in a process called **holography.** Figure 42.23 shows how a hologram is made. Light from the laser is split into two parts by a half-silvered mirror at $B$. One part of the beam reflects off the object to be photographed and strikes an ordinary photographic film. The other half of the beam is diverged by lens $L_2$, reflects from mirrors $M_1$ and $M_2$, and finally strikes the film. The two beams overlap to form an extremely complicated interference pattern on the film. Such an interference pattern can be produced only if the phase relationship of the two waves is maintained constant throughout the exposure of the film. This condition is met if one uses light from a laser because such light is coherent (all of the photons in the beam have the same phase). The hologram records not only the intensity of the light scattered from the object (as in a conventional photograph), but also the phase difference between the reference beam and the beam scattered from the object. Because of this phase difference, an interference pattern is formed that produces an image with full three-dimensional perspective.

A hologram is best viewed by allowing coherent light to pass through the developed film as one looks back along the direction from which the beam comes. Fig. 42.24 is a photograph of a hologram made using a cylindrical film. One sees a three-dimensional image of the object such that as the viewer's head is moved, the perspective changes, as for an actual object. The applications of holography promise to be many and varied. For example, someday your television set may be replaced by one using holography.

Figure 42.24 Photograph of a hologram which uses a cylindrical film. Note the detail of the Volkswagen image. (Courtesy of CENCO.)

## *42.11 FLUORESCENCE AND PHOSPHORESCENCE

When an atom absorbs a photon of energy and ends up in an excited state, it can return to the ground state via some intermediate states, as shown in Figure 42.25. The photons emitted by the atom will have lower energy, and therefore lower frequency, than the absorbed photon. The process of converting ultraviolet light to visible light by this means is referred to as **fluorescence.**

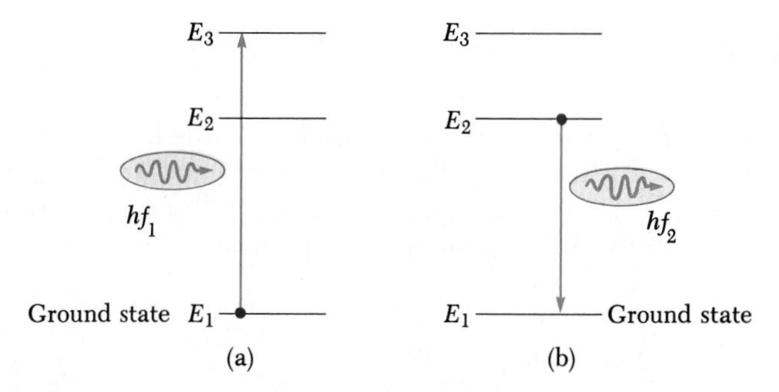


**Figure 42.25** The process of fluorescence. (a) An atom absorbs a photon of energy $hf_1$ and ends up in an excited state, $E_3$. (b) The atom emits a photon of energy $hf_2$ when the electron moves from an intermediate state, $E_2$, back to the ground state.

The common fluorescent light, which makes use of this principle, works as follows. Electrons are produced in the tube as a filament at the end is heated to sufficiently high temperatures. The electrons are accelerated by an applied voltage, and this causes them to collide with atoms of mercury vapor present in the tube. As a result of the collisions, many mercury atoms are raised to excited states. As the excited atoms drop to their normal levels, some ultraviolet photons are emitted, and these strike a phosphor-coated screen. The screen absorbs these photons and emits visible light by means of fluorescence. Different phosphors on the screen emit light of different colors. "Cool white" fluorescent lights emit nearly all the visible colors and hence the light is very white. "Warm white" fluorescent lights have a phosphor that emits more red light and thereby produces a "warm" glow. It is interesting to note that the fluorescent lights above the meat counter in a grocery store are usually "warm white" to give the meat a redder color.

Two common fluorescent materials that you may have in your medicine cabinet are Murine eye drops and Pearl Drops toothpaste. If you were to use these products and then stand under a "black light," your eyes and teeth would glow with a beautiful yellow color. (A black light is simply a lamp that emits ultraviolet light along with some visible violet-blue light.)

Fluorescence analysis is often used to identify compounds. This is made possible by the fact that every compound has a "fingerprint" associated with the specific wavelength at which it fluoresces.

As we have seen, a fluorescent material emits visible light only when ultraviolet radiation bombards it. Another class of materials, called **phosphorescent** materials, continue to glow long after the illumination has been removed. An excited atom in a fluorescent material drops to its normal level in about $10^{-8}$ s, but an excited atom of a phosphorescent material may remain in an excited metastable state for periods ranging from a few seconds to several hours. Eventually, the atom will drop to its normal state and emit a visible photon. For this reason, phosphorescent materials emit light long after being placed in the dark. Paints made from these substances are often used to decorate the hands of watches and clocks and to outline doors and stairways in large buildings so that these exits will be visible if there is a power failure.

## SUMMARY

The methods of wave mechanics can be applied to the hydrogen atom using the appropriate potential energy function $U(r) = -ke^2/r$ in the Schrödinger equation. The solution to this equation yields the wave functions for the allowed states and the allowed energies, given by

Allowed energies for the hydrogen atom

$$E_n = -\left(\frac{ke^2}{2a_0}\right)\frac{1}{n^2} = -\frac{13.6}{n^2}\text{ eV} \qquad n = 1, 2, 3, \ldots \qquad (42.2)$$

This is precisely the result obtained in the Bohr theory. The allowed energy depends only on the **principal quantum number** $n$. The allowed wave functions depend on three quantum numbers: $n$, $\ell$, and $m_\ell$, where $\ell$ is the **orbital quantum number** and $m_\ell$ is the **orbital magnetic quantum number.** The restrictions on the quantum numbers are as follows:

Allowed values for the quantum numbers

$$n = 1, 2, 3, \ldots$$
$$\ell = 0, 1, 2, \ldots, (n-1)$$
$$m_\ell = -\ell, -\ell + 1, \ldots, \ell - 1, \ell$$

All states with the same principal quantum number $n$ form a **shell,** identified by the letters K, L, M, . . . (corresponding to $n = 1, 2, 3, \ldots$). All states with the same values of both $n$ and $\ell$ form a **subshell,** designated by the letters $s, p, d, f, \ldots$ (corresponding to $\ell = 0, 1, 2, 3, \ldots$).

Spin magnetic quantum number

In order to completely describe a quantum state of the hydrogen atom, it is necessary to include a fourth quantum number, $m_s$, called the **spin magnetic quantum number.** This quantum number can have only two values, $\pm\frac{1}{2}$. In effect, this doubles the number of allowed states specified by the quantum numbers $n$, $\ell$, and $m_\ell$.

An atom in a state characterized by a specific $n$ can have the following values of **orbital angular momentum** $L$:

Allowed values of $L$

$$L = \sqrt{\ell(\ell + 1)}\hbar \qquad (42.10)$$

where $\ell$ is restricted to the values $\ell = 0, 1, 2, \ldots, n - 1$.

The allowed values of the projection of $\mathbf{L}$ along the $z$ axis are given by

Allowed values of $L_z$

$$L_z = m_\ell \hbar \qquad (42.11)$$

where $m_\ell$ is restricted to integer values lying between $-\ell$ and $\ell$. Note that only discrete values of $L_z$ are allowed, and these are determined by the restrictions on $m_\ell$. This quantization of $L_z$ is referred to as **space quantization.**

Space quantization was confirmed experimentally by Stern and Gerlach, who showed that a beam of neutral silver atoms is split into two components when passed through an inhomogeneous magnetic field. Subsequent experiments and analyses of such data showed that the electron has an intrinsic angular momentum called the **spin angular momentum.** That is, the total angular momentum of an electron in an atom can have two contributions, one arising from the spin of the electron ($\mathbf{S}$) and one arising from the orbital motion of the electron ($\mathbf{L}$).

Electronic spin can be described by a single quantum number $s = \frac{1}{2}$. The **magnitude of the spin angular momentum** is given by

Spin angular momentum of an electron

$$S = \frac{\sqrt{3}}{2}\hbar \tag{42.13}$$

and the $z$ component of **S** is given by

The $z$ component of spin angular momentum

$$S_z = m_s\hbar = \pm\tfrac{1}{2}\hbar \tag{42.14}$$

That is, the spin angular momentum is also quantized in space, as specified by the **spin magnetic quantum number** $m_s = \pm\frac{1}{2}$.

The magnetic moment $\mu_s$ associated with the spin angular momentum of an electron is given by

Relation between spin magnetic moment and spin angular momentum

$$\mu_s = -\frac{e}{m}\mathbf{S} \tag{42.15}$$

which is *twice* as large as the orbital magnetic moment. The $z$ component of $\mu_s$ can have the values

The $z$ component of spin magnetic moment

$$\mu_{sz} = \pm\frac{e\hbar}{2m} \tag{42.16}$$

Exclusion principle

The **exclusion principle** states that *no two electrons in an atom can ever be in the same quantum state.* In other words, no two electrons can have the same set of quantum numbers $n$, $\ell$, $m_\ell$, and $m_s$. Using this principle and the principle of minimum energy, one can determine the electronic configuration of the elements. This serves as a basis for understanding atomic structure and the chemical properties of the elements.

The allowed electronic transitions between any two levels in an atom are governed by the selection rule

Selection rule for allowed atomic transitions

$$\Delta\ell = \pm 1 \tag{42.17}$$

X-rays are emitted by atoms when an electron undergoes a transition from an outer shell into an electron vacancy in one of the inner shells. Transitions into a vacant state in the K shell give rise to the K series of spectral lines; transitions into a vacant state in the L shell create the L series of lines, and so on. The x-ray spectrum of a metal target consists of a set of sharp characteristic lines superimposed on a broad, continuous spectrum.

## QUESTIONS

1. Why are three quantum numbers needed to describe the state of a one-electron atom (neglecting spin)?
2. Compare the Bohr theory and the Schrödinger treatment of the hydrogen atom. Comment specifically on the total energy and orbital angular momentum.
3. Why is the direction of the orbital angular momentum of an electron opposite that of its magnetic moment?
4. Why is an inhomogeneous magnetic field used in the Stern-Gerlach experiment?
5. Could the Stern-Gerlach experiment be performed with ions rather than neutral atoms? Explain.
6. Describe some experiments that would support the conclusion that the spin quantum number for electrons can only have the values $\pm\frac{1}{2}$.
7. Discuss some of the consequences of the exclusion principle.
8. Why do lithium, potassium, and sodium exhibit similar chemical properties?
9. From Table 42–4, we find that the ionization energies for Li, Na, K, Rb, and Cs are 5.390, 5.138, 4.339, 4.176, and 3.893 eV, respectively. Explain why these values are to be expected in terms of the atomic structures.

10. Although electrons, protons, and neutrons obey the exclusion principle, some particles which have integral spin, such as photons (spin = 1), do not. Explain.
11. Explain why a photon must have a spin of 1.
12. An energy of about 21 eV is required to excite an electron in a helium atom from the $1s$ state to the $2s$ state. The same transition for the $He^+$ ion requires about twice as much energy. Explain why this is so.
13. Does the intensity of light from a laser fall off as $1/r^2$.
14. The absorption or emission spectrum of a gas consists of lines which broaden as the density of gas molecules increases. Why do you suppose this occurs?
15. How is it possible that electrons, with a probability distribution around a nucleus, can exist in states of definite energy (e.g., $1s$, $2p$, $3d$, . . .)?
16. It is easy to understand how two electrons (one spin up, one spin down) can fill the $1s$ shell for a helium atom. How is it possible that eight more electrons can fit into the $2s$, $2p$ level to complete the $1s^2 2s^2 2p^6$ shell for a neon atom?
17. In 1914, Henry Moseley was able to determine the atomic number of an element from its characteristic x-ray spectrum. How was this possible? (*Hint:* See Figs. 42.16 and 42.17.)
18. What are the advantages of using monochromatic light to view a holographic image?
19. Why is *stimulated emission* so important in the operation of a laser? (Interestingly, the concept of stimulated emission was first discussed by Albert Einstein 35 years before the first successful laser.)

## PROBLEMS

### Section 42.2 The Hydrogen Atom

1. (a) Determine the quantum numbers $\ell$ and $m_\ell$ for the $He^+$ ion in the state corresponding to $n = 3$. (b) What is the energy of this state?
2. (a) Determine the quantum numbers $\ell$ and $m_\ell$ for the $Li^{2+}$ ion in the states for which $n = 1$ and $n = 2$. (b) What are the energies of these states?
3. A general expression for the energy levels of one-electron atoms is

$$E_n = -\left(\frac{\mu k^2 q_1^2 q_2^2}{2\hbar^2}\right)\frac{1}{n^2}$$

where $k$ is the Coulomb constant, $q_1$ and $q_2$ are the charges of the two particles, and $\mu$ is the reduced mass given by $\mu = m_1 m_2/(m_1 + m_2)$. In Example 40.8, we found that the wavelength for the $n = 3$ to $n = 2$ transition of the hydrogen atom is 656.3 nm (visible red light). What are the wavelengths for this same transition in (a) positronium, which consists of an electron and a positron, and (b) singly ionized helium? (*Note:* A positron is a positively charged electron.)

4. The energy of an electron in a hydrogen atom is

$$E = \frac{p^2}{2m_e} - \frac{ke^2}{r}$$

According to the uncertainty principle, if the electron is localized within $r$, its momentum is $p$ must be at least $\hbar r$. Use this principle to find the *minimum* values of $E$ and $r$. Compare the results to those of Neils Bohr.

### Section 42.3 The Spin Magnetic Quantum Number and Section 42.4 The Wave Functions for Hydrogen

5. Show that the wave function $\psi_{1s}(r)$ given by Equation 42.3 is normalized.
6. Make plots of the wave function $\psi_{1s}(r)$ (Eq. 42.3) and the radial probability density function $P_{1s}(a)$ (Eq. 42.7) for hydrogen. Let $r$ range from 0 to $1.5a_0$, where $a_0$ is the Bohr radius.
7. The wavefunction for an electron in a $2p$-state in hydrogen is

$$\psi_{2p} = \frac{1}{\sqrt{3}(2a_0)^{3/2}}\frac{r}{a_0}e^{-r/2a_0}$$

What is the most likely distance from the H-nucleus to find an electron in the $2p$ state? (See Fig. 42.8.)

8. Show that the $1s$ wavefunction for an electron in hydrogen

$$\psi(r) = \frac{1}{\sqrt{\pi a_0^3}}e^{-r/a_0}$$

satisfies the radially symmetric Schrödinger equation

$$-\frac{\hbar^2}{2m}\left(\frac{d^2\psi}{dr^2} + \frac{2}{r}\frac{d\psi}{dr}\right) - \frac{ke^2}{r}\psi = E\psi$$

9. If a muon (a negatively charged particle with mass 206 times the electron's mass) is captured by a lead nucleus, $Z = 82$, the resulting system will behave like a one-electron atom. (a) What is the "Bohr radius" for a muon captured by a lead nucleus? (*Hint:* Use Equation 42.4.) (b) Using Equation 42.2 with $e$ replaced by $Ze$, calculate the ground state energy of a muon captured by a lead nucleus. (c) What is the transition energy for a muon descending from the $n = 2$ to the $n = 1$ level in a muonic lead atom?

## Section 42.5 The Quantum Numbers

10. Calculate the angular momentum for an electron in (a) the $4d$ state and (b) the $6f$ state.
11. If an electron has an angular orbital momentum of $4.714 \times 10^{-34}$ s, what is the orbital quantum number for this state of the electron?
12. List the possible sets of quantum numbers for electrons in (a) the $3d$ subshell and (b) the $3p$ subshell.
13. How many different sets of quantum numbers are possible for an electron for which (a) $n = 1$, (b) $n = 2$, (c) $n = 3$, (d) $n = 4$, and (e) $n = 5$? Check your results to show that they agree with the general rule that the number of different sets of quantum numbers is equal to $2n^2$.
14. (a) Write out the electronic configuration for oxygen ($Z = 8$). (b) Write out the values for the set of quantum numbers $n$, $\ell$, $m_\ell$, and $m_s$ for each of the electrons in oxygen.
15. Calculate the possible values of the $z$ component (the component along the direction of an external magnetic field) of angular momentum for an electron in a $d$ subshell.
16. Consider an atom whose $M$ shell is completely filled (with no additional electrons). (a) Identify the atom. (b) List the number of electrons in each of its subshells.
17. An electron is in the N shell. Determine the maximum value of the $z$ component of the angular momentum of the electron.
18. Determine the number of electrons that can occupy the $n = 3$ shell.
19. Find the possible values of $L$, $L_z$, and $\theta$ for an electron in a $3d$ state of hydrogen.
20. All objects, large and small, behave quantum-mechanically. (a) Estimate the quantum number $\ell$ for the earth in its orbit about the sun. (b) What energy change (in joules) would occur if the earth made a transition to an adjacent allowed state?

## Section 42.6 Electron Spin

21. The $z$-component of the electron's spin magnetic moment is given by the Bohr magneton, $\mu_B = e\hbar/2m$. Show that the Bohr magneton has the numerical value of $9.27 \times 10^{-24}$ J/T, or $5.79 \times 10^{-5}$ eV/T.
22. Like the electron, the nucleus of an atom has spin angular momentum and a corresponding magnetic moment. The $z$-component of the spin magnetic moment for a nucleus is characterized by the *nuclear magneton* $\mu_n = e\hbar/2m_p$, where $m_p$ is the proton mass. (a) Calculate the value of $\mu_n$ in J/T and in eV/T. (b) Determine the ratio $\mu_n/\mu_B$, and comment on your result.

## Section 42.7 The Exclusion Principle and The Periodic Table

23. Which electronic configuration has a lower energy: $[Ar]3d^4 4s^2$ or $[Ar]3d^5 4s^1$? Identify this element and discuss Hund's rule in this case.
24. Which electronic configuration has the lesser energy and the greater number of unpaired spins: $[Kr]4d^9 5s^1$ or $[Kr]4d^{10}$? Identify this element and discuss Hund's rule in this case. (*Note:* The notation [Kr] represents the filled configuration for Kr.)
25. Devise a table similar to that shown in Figure 42.13 for atoms with 11 through 19 electrons. Use Hund's rule and educated guesswork.
26. (a) Scanning through Table 42.4 in order of increasing atomic number, note that the electrons fill the subshells in such a way that those subshells with the lowest values of $n + \ell$ are filled first. If two subshells have the same value of $n + \ell$, the one with the lower value of $n$ is filled first. Using these two rules, write the order in which the subshells are filled through $n = 7$. (b) Predict the chemical valence for elements with atomic numbers 15, 47, and 86 and compare them with the actual valences.

## Section 42.8 Atomic Spectra: Visible and X-ray

27. If you wish to produce 1-Å x-rays in the laboratory, what is the minimum voltage you must use in accelerating the electrons?
28. What is the shortest x-ray wavelength that can be produced with an accelerating voltage of 10 kV?
29. A tungsten target is struck by electrons that have been accelerated from rest through a 40-kV potential difference. Find the shortest wavelength of the bremsstrahlung radiation emitted.
30. The $K_\alpha$ x-ray is the one emitted when an electron undergoes a transition from the L shell ($n = 2$) to the K shell ($n = 1$). Calculate the frequency of the $K_\alpha$ x-ray from a nickel target ($Z = 28$).
31. Find the wavelength of the $K_\alpha$ x-ray line that is emitted when electrons strike an iron target. Note that since the innermost shells are involved, $Z_{eff}$ is approximately $Z - 1$.
32. Use the method illustrated in Example 42.7 to calculate the wavelength of the x-ray emitted from a molybdenum target ($Z = 42$) when an electron undergoes a transition from the L shell ($n = 2$) to the K shell ($n = 1$).

## Section 42.9 Atomic Transitions

33. The familiar yellow light from a sodium vapor street lamp results from the $3p \rightarrow 3s$ transition in $^{11}Na$. Eval-

uate the wavelength of the light given that the energy difference $E_{3p} - E_{3s} = 2.1$ eV.

34. The wavelength of coherent ruby laser light is 694.3 nm. What is the energy difference (in eV) between the upper excited state and the lower unexcited energy state?

35. A ruby laser delivers a 10-ns pulse of 1 MW average power. If all the photons are of wavelength 694.3 nm, how many photons are contained in the pulse?

## ADDITIONAL PROBLEMS

36. Zirconium has two unpaired electrons in the $d$ subshell. (a) What are all possible values of $\ell$ and $s$ for each electron? (b) What are all possible values of $n$, $m_\ell$, and $m_s$? (c) What is the electron configuration in zirconium?

37. In the technique known as electron spin resonance (ESR), a sample containing unpaired electrons is placed in a magnetic field. Consider the simplest situation, that in which there is only one electron and therefore only two possible energy states, corresponding to $m_s = \pm\frac{1}{2}$. In ESR, the electron's spin magnetic moment is "flipped" from a lower energy state to a higher energy state by the absorption of a photon. (The lower energy state corresponds to the case where the magnetic moment $\mu_s$ is aligned against the magnetic field, and the higher energy state corresponds to the case where $\mu_s$ is aligned with the field.) What is the photon frequency required to excite an ESR transition in a magnetic field of 0.35 T?

38. A Nd:YAG laser used in eye surgery emits a 3-mJ pulse in 1 ns, focussed to a spot 30 $\mu$m in diameter on the retina. (a) Find (in SI units) the power per unit area at the retina. (This quantity is called the *irradiance.*) (b) What energy is delivered to an area of molecular size, say a circular area 0.6 nm in diameter?

39. A dimensionless number which often appears in atomic physics is the *fine-structure constant* $\alpha$, given by

$$\alpha = \frac{ke^2}{\hbar c}$$

where $k$ is the Coulomb constant. (a) Obtain a numerical value for $1/\alpha$. (b) In scattering experiments, the "size" of the electron is the *classical electron radius*, $r_e = ke^2/m_ec^2$. In terms of $\alpha$, what is the ratio of the Compton wavelength (Section 40.3), $\lambda_c = h/m_ec$, to the classical electron radius? (c) In terms of $\alpha$, what is the ratio of the Bohr radius, $a_0$, to the Compton wavelength? (d) In terms of $\alpha$, what is the ratio of the *Rydberg wavelength*, $1/R_H$, to the Bohr radius (Section 40.5)?

40. Show that the average value of $r$ for the 1s state of hydrogen has the value $3a_0/2$. (*Hint:* Use Eq. 42.7.)

41. Suppose that a hydrogen atom is in the 2s state. Taking $r = a_0$, calculate values for (a) $\psi_{2s}(a_0)$, (b) $|\psi_{2s}(a_0)|^2$, and (c) $P_{2s}(a_0)$. (*Hint:* Use Eq. 42.8.)

42. The carbon dioxide ($CO_2$) laser is one of the most powerful lasers developed. The energy difference between the two laser levels is 0.117 eV. Determine the frequency and wavelength of the radiation emitted by this laser. In what portion of the electromagnetic spectrum is this radiation?

43. Show that the wavefunction for an electron in the 2s-state in hydrogen

$$\psi(r) = \frac{1}{4\sqrt{2\pi}}\left(\frac{1}{a_0}\right)^{3/2}\left(2 - \frac{r}{a_0}\right)e^{-r/2a_0}$$

satisfies the radially symmetric Schrödinger equation

$$-\frac{\hbar^2}{2m}\left(\frac{d^2\psi}{dr^2} + \frac{2}{r}\frac{d\psi}{dr}\right) - \frac{ke^2}{r}\psi = E\psi$$

44. For the ground state of hydrogen, what is the probability of finding the electron closer to the nucleus than the Bohr radius corresponding to $n = 1$?

45. A pulsed ruby laser emits light at 694.4 nm. For a 14-ps pulse containing 3 J of energy, find (a) the physical length of the pulse as it travels through space and (b) the number of photons in the pulse. (c) If the beam has a circular cross section of 0.6 cm diameter, find the number of photons per cubic millimeter in the beam.

46. The number $N$ of atoms in a particular state is called the *population* of that state. This number depends on the energy of that state and the temperature. The population of atoms in a state of energy $E_n$ is given by a Boltzmann distribution expression:

$$N = N_0 e^{-E_n/kT}$$

where $N_0$ is the population of the state as $T \rightarrow \infty$. (a) Find the ratio of populations of the states $E_3^*$ to $E_2$ for the laser in Figure 42.22, assuming $T =$ 27°C. (b) Find the ratio of the populations of the two states in a ruby laser that produces a light beam of wavelength 694.3 nm at 4 K.

47. Consider a hydrogen atom in its ground state. (a) Treating the electron as a current loop of radius $a_0$, derive an expression for the magnetic field at the nucleus. (*Hint:* Use the Bohr theory of hydrogen and see Example 30.3.) (b) Find a numerical value for the magnetic field at the nucleus in this situation.

48. The force on a magnetic moment $\mu_z$ in a nonuniform magnetic field $B_z$ is given by

$$F_z = \mu_z \frac{dB_z}{dz}$$

If a beam of silver atoms travels a horizontal distance of 1 m through such a field and each atom has a speed of 100 m/s, how strong must the field gradient $dB_z/dz$ be in order to deflect the beam 1 mm?

49. (a) Calculate the most probable position for an electron in the 2*s* state of hydrogen. (*Hint:* Let $x = r/a_0$, find an equation for $x$, and show that $x = 5.236$ is a solution to this equation.) (b) Show that the wave function given by Equation 42.8 is normalized.

50. All atoms are roughly the same size. (a) To show this, estimate the diameters for aluminum, with molar atomic mass = 27 g/mol and density 2.70 g/cm³, and uranium, with molar atomic mass = 238 g/mol and density 18.9 g/cm³. (b) What do the results imply about the wave functions for inner-shell electrons as we progress to higher and higher atomic weight atoms? (*Hint:* The molar volume is roughly proportional to $D^3N_A$, where $D$ is the atomic diameter and $N_A$ is Avogadro's number.)

51. For hydrogen in the 1*s* state, what is the probability of finding the electron farther than $2.50a_0$ from the nucleus?

52. According to classical physics, an accelerated charge $e$ radiates at a rate

$$\frac{dE}{dt} = -\frac{1}{6\pi\epsilon_0}\frac{e^2a^2}{c^3}$$

(a) Show that an electron in a classical hydrogen atom (see Fig. 42.3) will spiral into the nucleus at a rate

$$\frac{dr}{dt} = -\frac{e^4}{12\pi^2\epsilon_0^2r^2m^2c^3}$$

(b) Find the time for the electron to reach $r = 0$, starting from $r_0 = 2 \times 10^{-10}$ m.

53. Light from a certain He-Ne laser has a power output of 1.0 mW and a cross-sectional area of 10 mm². The entire beam is incident on a metal target which requires 1.5 eV to remove an electron from its surface. (a) Perform a classical calculation to determine how long it would take one atom in the metal to absorb 1.5 eV from the incident beam. (*Hint:* Assume that the area of an atom is 1 Å² = $10^{-20}$ m² and first calculate the energy incident on each atom per second.) (b) Compare the (wrong) answer obtained in (a) to the actual response time for photoelectric emission ($\approx 10^{-9}$ s), and discuss the reasons for the large discrepancy.

54. In interstellar space, atomic hydrogen produces the sharp spectral line called the *21-cm radiation*, which astronomers find most helpful in detecting clouds of hydrogen between stars. This radiation is useful because interstellar dust that obscures visible wavelengths is transparent to these radio wavelengths. The radiation is not generated by an electron transition between energy states characterized by $n$. Instead, in the ground state ($n = 1$), the electron and proton spins may be *parallel* or *antiparallel*, with a resultant slight difference in these energy states. (a) Which condition has the higher energy? (b) The line is actually at 21.11 cm. What is the energy difference between the states? (c) The average lifetime in the excited state is about $10^7$ y. Calculate the associated uncertainty in energy of this excited energy level.

# ESSAY

## Lasers and Applications

**Isaac D. Abella**
*The University of Chicago*

### Laser Principles

The enormous growth of laser technology mentioned in Section 42.10 has stimulated a broad range of scientific and engineering applications that exploit some of the unique properties of laser light. These properties derive from the distinctive way laser light is produced in contrast to the generation of ordinary light. In an ordinary sodium vapor street lamp for example, the atoms *spontaneously* emit in random directions and at irregular times, over a broad spectrum, resulting in isotropic illumination of incoherent light. Laser light originates from atoms, ions, or molecules through a process of *stimulated emission* of radiation. The active laser medium is contained in an enclosure or *cavity* which organizes the normally random emission process into an intense directional, monochromatic and coherent wave. The end mirrors provide the essential optical feedback which selectively builds up the stimulating wave along the tube axis. Laser light has a well-defined phase, permitting a wide variety of applications based on interference or wave modulation.

Currently operating laser systems utilize a varietyof gases, solids, or liquids as the working laser substance. These devices are designed to emit either continuous or pulsed monochromatic beams, and operate over a broad range of the optical spectrum (ultraviolet, visible, infrared) with output powers ranging from microwatts to megawatts. The particular application determines the choice of laser system, wavelength, power level, or other relevant variables, since no one laser has all the desirable properties. As noted in Section 42.10, several conditions on populations of states in atoms must be satisfied for a laser to operate successfully. Population ratios of energy states of atoms in thermal equilibrium at temperature $T$ are normally described by the *Boltzmann* factor,

$$\frac{N_2}{N_1} = e^{-\Delta E/kT}$$

where $\Delta E$ is the energy difference between two given levels, and $k$ is Boltzmann's constant. Thus, the ratio of upper-state (2) to lower-state (1) populations for any reasonable laboratory temperature is always less than unity. This is the normal net absorptive case where stimulated emission does not play a dominant role. When the above ratio exceeds unity, an *inverted* population results, which permits net stimulated emission. This is sometimes described as a *negative temperature* condition, a non-physical temperature referring to a situation not in thermal equilbrium.

The requirement for population inversion, that is, more atoms in a particular excited state than in a lower state, essentially means that energy must be supplied from outside the system. Otherwise, atoms would eventually radiate, fall to the lowest energy state, and stop emitting altogether. Therefore, as required by conservation of energy, all laser systems must be connected to external energy sources, usually electrical, to maintain this non-thermal equilibrium situation as indicated in Figure 1. For example, atoms can be energized by electron impact in gaseous discharges (so called "electrical pumping"). We can also supply energy to lamps whose light populates excited states by photon absorption ("optical pumping") for those solids or liquids which do not conduct electric charge. These pumping mechanisms tend to have low efficiency (ratio of laser energy output to electric energy supplied), typically a few percent, with the balance discharged as heat into cooling water or circulating air.

Controlling the electrical input into the laser system provides a variable laser energy output, which may be important in many applications. Thus, the argon ion laser system can emit up to about 10 W in the green optical beam by adjustment of the electric current in the argon gas, which in turn controls the degree of population inversion. Chemical lasers, on the other hand, operate without direct electrical input. Several highly reactive gases are mixed in the laser chamber, with the energy released in the ensuing reaction populating the excited levels in the molecule. In this

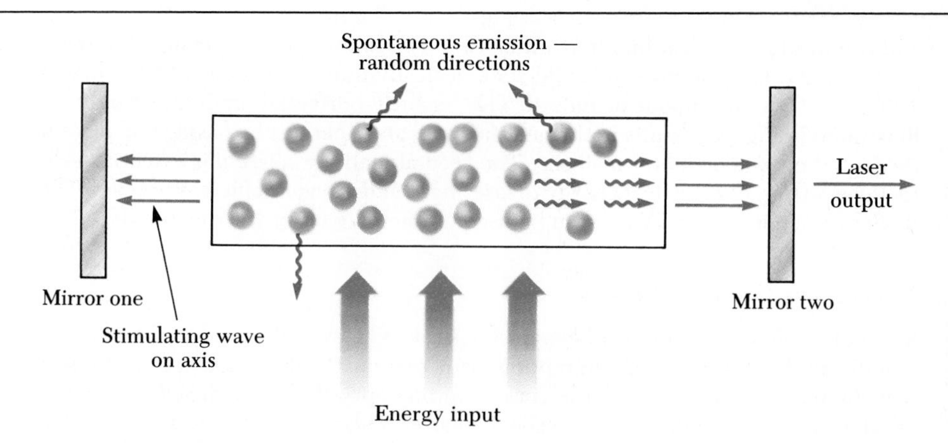


**Figure 1** A schematic of a laser design. The tube contains atoms, which represent the active medium. An external source of energy (optical, electrical, etc.) is needed to "pump" the atoms to excited energy states. The parallel end mirrors provide the feedback of the stimulating wave.

case, the reactants need to be resupplied for the laser to operate for any length of time.

Some laser systems have fluid media, containing dissolved dye molecules. The dye lasers are usually pumped to excited levels by an external laser. The advantage of this arrangement is that dye lasers can be continuously "tuned" over a wide range of wavelengths, using prisms or gratings, whereas the pump source has a fixed wavelength. Color variability is important for those cases where the laser is directed at materials whose absorption depends on wavelength. Thus, the laser can be tuned into exact coincidence with selected energy states. For example, blood does not absorb red light to any extent, which excludes red light use for most surgical applications on blood-rich tissue.

Recent developments in tunable solid state materials have permitted the design of tunable lasers without the need for unstable dye molecules; most notable are sapphire crystals containing titanium ions. These materials are optically pumped by flashlamps or fixed wavelength lasers acting to populate the upper state directly.

A variety of laser systems are in general use today. They include the 1 mW helium-neon laser (Fig. 2), usually operating in the red at 632.8 nm (although yellow and green beams are available); the argon ion laser, which operates in the green or blue up to 10 W; the carbon dioxide gas laser, which emits in the infrared at 10 $\mu$m

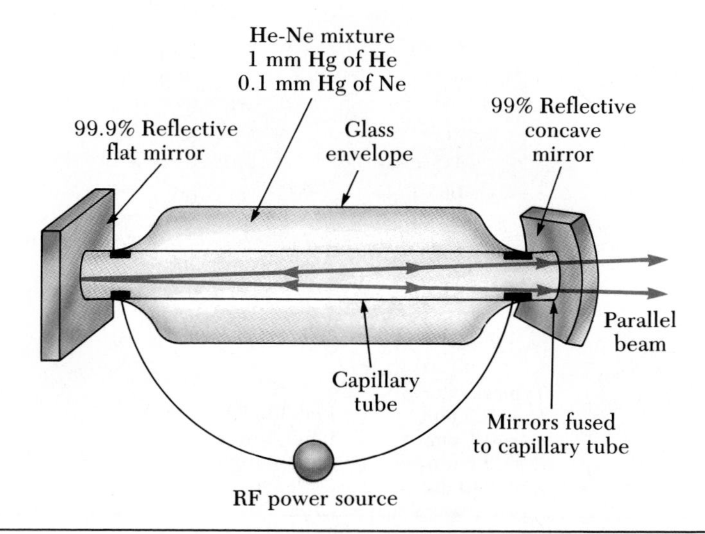


**Figure 2** A typical He-Ne gas laser.

and can produce several hundred watts; the neodymium doped yttrium aluminum garnet (YAG) laser, a powerful solid-state optically-pumped system which emits at 1.06 $\mu$m either continuous or pulsed. The recently perfected diode junction-laser illustrated in Figure 3 emits in the near infrared and operates by passage of current through the semiconductor material. The recombination radiation is essentially direct conversion of electrical energy to laser light and is a very efficient process. The diodes can emit up to 5 W and can be used to energize other laser materials.

## Non-Linear Optical Effects

Some of the first scientific applications for laser light were devoted to the study of non-linear effects. The usual assumption made prior to the invention of the laser was that the intensity of light, and the corresponding optical electric field $E$, was weak relative to electric fields already present in matter. This is equivalent to assuming that the effect of light on an atom is simply proportional to the optical electric field.

In the study of electricity and magnetism, we note that the property of matter is often included in the linear auxiliary equation $P = KE$ where the induced polarization $P$ in a dielectric medium is taken to be linear in the electric field, and $K$ is the dielectic constant or the polarizability of the matter. In general, this does not hold for arbitrary field strength and a more general equation, a power series in $E$, needs to written:

$$P = K_1E + K_2E^2 + K_3E^3 + \cdots$$

The consequence of including the second term in the series is significant. Suppose a strong red beam of laser light is introduced into a transparent crystal such as quartz. The effect of the non-linear term is to produce *blue* light at precisely the second harmonic of the incident light. To see why this should be so, consider the input light as a sinusoid, $E = E_0 \sin \omega t$, where $E_0$ is the amplitude of the light field in the crystal. An induced polarization now exists given by $P = K_2E_0^2 \sin^2\omega t$. This can written, using trigonometric methods, as

$$P = (\tfrac{1}{2})K_2E_0{}^2(1 - \cos 2\omega t)$$

Thus, a new polarization appears consisting of two terms: a constant (DC) field, and more importantly, a term oscillating at *twice* the input frequency, or half the wave-

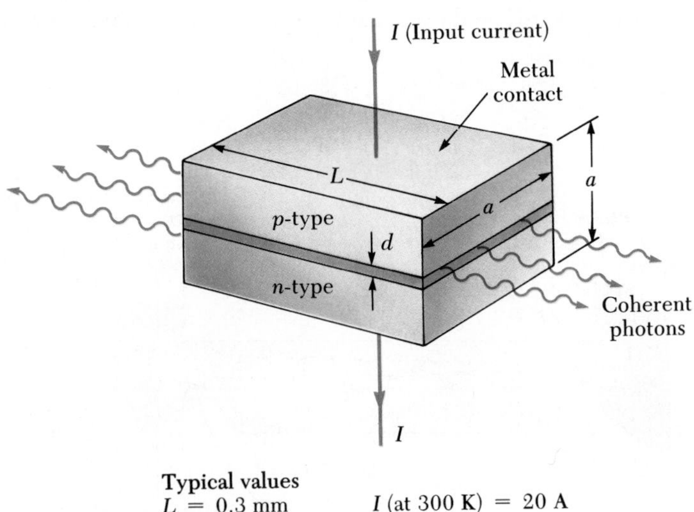


Typical values
$L = 0.3$ mm
$a = 0.1$ mm
$d = 2.0$ μm
$\lambda = 840$ nm

$I$ (at 300 K) = 20 A
Input power = 136 W
Light output power = 9 W

**Figure 3** A gallium arsenide *p-n* junction laser.

length. For laser light at 694 nm, the harmonic is in the ultraviolet at 347 nm. By way of analogy, harmonic distortion of sound often occurs in vibrating mechanical systems when non-linear responses are present. The optical Second Harmonic Generation (SHG) effect was observed very soon after the invention of the ruby laser and Second Harmonic Generation is now a standard method for increasing the frequency of laser light into the UV and soft x-ray region.

There are some subtle points that must be noted. For one thing, not all materials exhibit SHG. The effect is restricted to a class of crystals having a lack of *inversion symmetry* in their crystal structures, such as piezoelectric crystals. Materials that do not exhibit SHG are cubic crystals, water, or glass. Another complication is that the speed of light at different wavelengths is not the same (a consequence of dispersion), which leads to inefficient SHG. The driving field at the fundamental gets out of phase with the second harmonic field and some cancellation occurs. To solve this problem, a method known as *phase matching* is employed, where for special directions in some birefringent crystals, the wave speeds can be matched for orthogonal polarizations of light. When light is propagated along these phase matching directions, very efficient conversion to the second harmonic can be observed.

A second important application of non-linearity comes to atomic physics and quantum mechanics. In the Bohr theory of the atom, and its subsequent modern development, transitions between atomic energy levels occur with the emission or absorption of a single quantum of light, the photon. Thus for the Bohr atom, $\Delta E = hf$ implies a single photon transition. This holds for the case of weak light beams as described above, since the optical electric field is treated as a weak first-order perturbation on the fields already present in an atom.

A more detailed quantum mechanical treatment shows that non-linear absorption can take place with the *simultaneous* absorption of *two* quanta so long as the *sum* of the energies of the two photons carries the electron between two real atomic levels. This effect has been observed in atoms and molecules, for double or triple quantum absorption, and the effect is used to populate very highly energetic states of atoms with light that is not normally resonant with the state in question. Multiphoton effects usually require focussed laser beams where the optical electric fields are comparable to atomic fields.

## Holography

A method of lensless photography called holography has been developed with the advent of coherent laser sources, although the first holograms were made with ordinary monochromatic sodium yellow light. The basic ideas of holography can be understood on the basis of diffraction theory. In Section 38.4, we saw that a diffraction grating consisting of many narrow identical apertures produces various maxima in different orders; the grating equation was given as $m\lambda = d \sin \theta$. For a single laser beam incident on a grating of this sort we observe a number of diffracted beams, according to the order number $m$. These can be thought of as the multiple "images" of the original beam. Another view is that the beams represent a *spatial* Fourier analysis of the square aperture, which leads to many diffracted Fourier components.

Now consider a "grating" formed by the intersection of two plane waves incident on a photographic film. In such circumstances, one observes the stable bright and dark interference fringes. After the film is developed, a series of interference fringes is observed, which have a sinusoidal modulation of opacity, instead of the sharp-edge or "square" aperture of a ruled grating. If a laser beam is sent through a sinusoidal grating, only the first order beam is seen, corresponding to a single point image of the input beam. In Fourier terms, the spatial decomposition of a sinusoidal grating is a single Fourier component.

*(Continued on next page)*

Simply stated, holography consists of exposing a photographic film to laser light *scattered* from an object, together with a *reference* laser beam interfering with the scattered light on the plate. The "image" on the film is a sum of sinusoidal interference gratings each of which corresponds to some point on the object. The image is reconstructed by sending laser light through the film to produce the diffracted first-order image. Early holograms required laser reconstruction of the image.

Several conditions must be satisfied to generate successful holographic plates. If the laser is low power, say at the ten milliwatt level, exposures must be long enough to get reasonable film response. During this time, the coherent phases of the interfering waves must be stable. Vibrations and moving air currents could lead to a washout of the desired effect. Higher intensity lasers reduce the time constraints. For this reason, holograms tend to be made of stationary objects, although pulsed-laser holograms can also be made.

Many technological improvements on this scheme have been made, including thick holograms where the interference is distributed in three dimensions in the film. Ordinary white light can now be scattered in reflection from these thick films in analogy to the Bragg scattering that occurs in crystal x-ray diffraction (Section 38.5). We observe thick holograms in use on credit cards, viewed in ordinary white light, where they are intended to improve security.

## Other Applications

We shall describe several other applications that should serve to illustrate the wide variety of laser uses. First, lasers are used in precision long-range distance measurement (range-finding). It has become important, for astronomical and geophysical purposes, to measure as precisely as possible the distance from various points on the surface of the earth to a point on the moon's surface. To facilitate this, the Apollo astronauts set up a compact array, a 0.5 m square of reflector prisms on the moon, which allows laser pulses directed from an earth station to be retro-reflected to the same station. (See Section 35.7 on prism reflectors.) Using the known speed of light and the measured round-trip travel time of a 1 ns pulse, one can determine the earth-moon distance, 380 000 km, to a precision of better than 10 cm. Such information would be useful, for example, in making more reliable earthquake predictions and for learning more about the motions of the earth-moon system. This technique requires a high-power pulsed laser for its success, since a sufficient burst of photons must return to a collecting telescope on earth and be detected. Variations of this method are also used to measure the distance to inaccessible points on the earth.

The low-power helium-neon laser is the basis for a widespread technical innovation involving product labels. The laser beam can be focussed with a lens to a very small bright spot, which is then reflected from an oscillating mirror producing a swiftly moving dot image. If this spot is scanned over the product identification bar-code printed on supermarket products, the variation of the reflected light can be detected and decoded for speed and accuracy at the checkout counter. The spot of light must be small enough to resolve the different widths of the individual bars and bright enough to be "seen" in reflection by the optical detector below the counter. Since this laser is operated in public, the power must be low enough to be safe to the eye for any reasonable use of the system. This puts stringent limits on the type of laser used in this application.

Similarly, a laser (light-emitting diode) is used to decode the digital information on the compact audio laser-disc, the so-called CD. On the compact disc, the music has been digitized as pits and grooves embedded into a plastic-covered metal foil. The fluctuating reflection of the weak laser spot from the foil surface is detected by a photocell and decoded by digital to analog circuits to reproduce music with extremely high fidelity, without the noise or hiss associated with regular long-playing records or magnetic tape. There are also video versions of the laser disc. In all of these decoding applications, the essential laser properties that are exploited are: the accurate focusing of the beam, the monochromaticity (to be able to operate in the presence of background illumination), and enough power to observe the diffusely reflected light. High power lasers could lead to damage in these applications and are not employed.

The amount of information stored on a compact audio disc as digital data is estimated to be about 1 gigabyte ($10^9$ characters), which is enormously high density data storage. By way of comparison, the data storage on magnetic "floppy" discs in personal computers is typically 800 kilobytes ($8 \times 10^5$ characters). Developments are under way to transfer this optical storage technology to computer disc readers having both read and write capability. The CD disc as a read-only device with prerecorded computer data requires very little modification and is already in use to store encyclopedia or dictionary volumes. However, the optically erasable feature would require a combination of optical and magnetic methods. The reason the data density is so high on a CD has to do with the fine size of the optical spot that can be produced from a laser beam.

Novel medical applications utilize the fact that the different laser wavelengths can be absorbed in specific biological tissues. A widespread eye condition, glaucoma, is manifested by a high fluid pressure in the eye, which can lead to destruction of the optic nerve. A simple laser operation (iridectomy) can "burn" open a tiny hole in a clogged membrane, relieving the destructive pressure. Along the same lines, a serious side effect of diabetes is the formation of weak blood vessels (neovascularization), which often leak blood into extremities. When this occurs in the eye, vision deteriorates (diabetic retinopathy) leading to blindness in diabetic patients. It is now possible to direct the green light from the argon ion laser through the clear eye lens and eye fluid, focus on the retina edges, and photo-coagulate the leaky vessels. These procedures have greatly reduced cases of blindness due to glaucoma and diabetes.

Laser surgery is now a practical reality. Infrared light at 10 $\mu$m from a carbon dioxide laser can cut through muscle tissue, primarily by heating and evaporating the water contained in cellular material. Laser power of about 100 W is required in this

Scientist checking the performance of an experimental laser-cutting device mounted on a robot arm. The laser is being used to cut through a metal plate. (Courtesy of Philippe Plailly, Science Photo Library/Photo Researchers, Inc.)

An argon laser passing through a cornea and lens during eye surgery. (© Alexander Tsiaras, Science Source/Photo Researchers, Inc.)

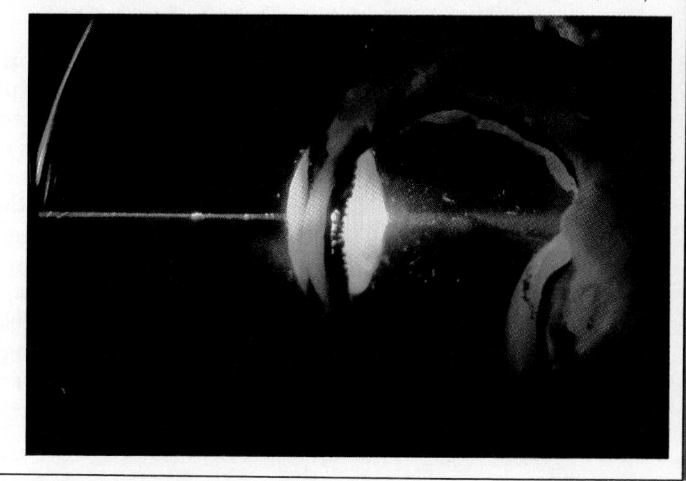

techique. The advantage of the "laser knife" over conventional methods is that laser radiation cuts and coagulates at the same time, leading to substantial reduction of blood loss. In addition, the technique virtually eliminates cell migration, which is very important in tumor removal. Furthermore, a laser beam can be trapped in fine glass-fiber light-guides (endoscopes) by means of total internal reflection (Sec. 35.7). The light fibers can be introduced through natural orifices, conducted around internal organs and directed to specific interior body locations, eliminating the need for massive surgery. For example, bleeding in the gastrointestinal tract can be optically cauterized by fiberoptic endoscopes inserted through the mouth.

Finally, we describe an application to biological and medical research. It is often important to isolate and collect unusual cells for study and growth. A laser cell-separator exploits the fact that specific cells can be tagged with fluorescent dyes. All cells are then dropped from a tiny charged nozzle and laser scanned for the dye tag. If triggered by the correct light-emitting tag, a small voltage applied to parallel plates deflects the falling electrically charged cell into a collection beaker. This is an efficient method for extracting the proverbial needles from the haystack.

For communications applications of lasers, see the essay on fiber optics in Chapter 35.

### Suggested Readings


Demtroder, W., *The Laser Spectroscopy,* New York, Springer, 1982.

O'Shea, D., R. Callen, and W.T. Rhodes, *Introduction to Lasers and Their Applications,* Reading, Mass., Addison-Wesley, 1977.

Shawlow, A.L., "Laser Light," *Sci. American,* September 1968, 120–126.

Siegman, A.E., *An Introduction to Lasers and Masers,* New York, McGraw-Hill, 1979.

*Physics Today,* Vol. 30, no. 5, May 1977, Special issue on Applications of Lasers in Research.


### Essay Questions

1. Discuss the main criteria that must be met to achieve laser action in a three-level system.
2. Why is it necessary to use a pumping light source in a laser?
3. Explain why the process of cooling the operating temperature of a laser increases its efficiency.
4. What are the main differences between a laser and an incandescent bulb as light sources?
5. Discuss some industrial and medical applications of lasers, and how they have modified our lifestyles.

### Essay Problems

1. (a) In the case of the earth-moon laser range-finder, what is the round-trip time for a laser pulse? (b) What precision in timing is required, that is, how small a time change needs to be detectable, to be able to measure the distance to an error of 10 cm? What effect does the earth's atmosphere have on this measurement?
2. A laser beam of wavelength $\lambda = 600$ nm is directed at the moon from a laser tube of 1-cm diameter. (a) Does the beam spread at all and what is the diameter of the "spot" on the moon's surface? (b) On the basis of your answer to (a), what strategy would you suggest for successful lunar-array illumination?
3. (a) Estimate how much chemical reagent is required to produce a chemical laser of 1 kW output, in mol/s. Pick a reasonable exothermic reaction rate in kcal/mol. (b) Estimate how many kilograms each of hydrogen and chlorine gas would be needed to make an HCl laser operate for an hour at 1-kW output.

# 43
# Molecules and Solids

*Natural quartz ($SIO_2$) crystals, one of the most common minerals on earth. Quartz crystals are used to make special lenses and prisms and in certain electronic applications. (Courtesy of Ward's Natural Science)*

The preceding chapter was concerned with the atomic and electronic structure of single atoms. Except for the inert gases, elements generally combine to form chemical compounds, that is, an aggregate of individual atoms joined by chemical bonds. The physical and chemical properties of molecules and solids depend fundamentally upon atomic and electronic structures.

In this chapter, we shall first describe the bonding mechanisms in molecules, the various modes of molecular excitation, and the radiation emitted or absorbed by molecules. We shall then take the next logical step and show how molecules combine to form solids. By examining their electronic distributions, we shall explain the differences between insulating, metallic, and semiconducting crystals. The chapter concludes with discussions of semiconducting junctions, the operation of several semiconductor devices, and an essay on the photovoltaic effect.

## 43.1 MOLECULAR BONDS

Two atoms combine to form a molecule because of a net attractive force between them when their separation is greater than their equilibrium separation in the molecule. Furthermore, the total energy of the stable bound molecule is *less* than the total energy of the separated atoms.

Fundamentally, the bonding mechanisms in a molecule are primarily due to electrostatic forces between atoms (or ions). When two atoms are separated by an infinite distance, the force between them is zero, as is the electrostatic potential energy of the system. As the atoms are brought closer together, both attractive and repulsive forces act. At very large separations, the dominant forces are attractive in nature. For small separations, repulsive forces between like charges begin to dominate. The potential energy of the system can be positive or negative, depending on the separation between the atoms.

The total potential energy of the system can be approximated by the expression

$$U = -\frac{A}{r^n} + \frac{B}{r^m} \qquad (43.1)$$

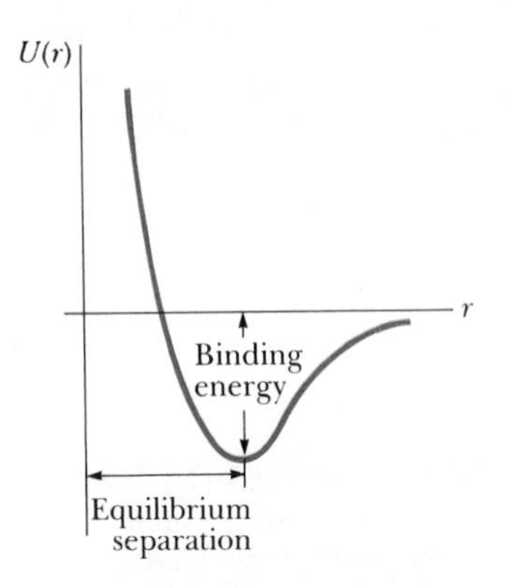


**Figure 43.1** The total potential energy as a function of the internuclear separation for a system of two atoms.

where $r$ is the internuclear separation, $A$ and $B$ are constants associated with the attractive and repulsive forces, and $n$ and $m$ are small integers. Figure 43.1 represents a sketch of the total potential energy versus internuclear separation. The potential energy for large separations is negative, corresponding to a net attractive force. At the equilibrium separation, the attractive and repulsive forces just balance and the potential energy has its minimum value.

A complete description of the binding mechanisms in molecules is a highly complex problem because it involves the mutual interactions of many particles. In this section, we shall discuss some simplified models in the following order of decreasing bond strength: the ionic bond, the covalent bond, the hydrogen bond, and the van der Waals bond. The metallic bond will be discussed in Section 43.3.

### Ionic Bonds

**Ionic bonds** are fundamentally due to the Coulomb attraction between oppositely charged ions. A familiar example of an ionically bonded molecule is sodium chloride, NaCl, which forms common table salt. Sodium, which has an electronic configuration $1s^22s^22p^63s$, is relatively easy to ionize, giving up its $3s$ valence electron to form a $Na^+$ ion. The energy required to ionize the atom to form $Na^+$ is 5.1 eV. Chlorine, which has an electronic configuration $1s^22s^22p^5$, is one electron short of the closed-shell structure of argon. Because closed-shell configurations are energetically more favorable, the $Cl^-$ ion is more stable than the neutral Cl atom. The energy released when an atom takes on an electron is called the **electron affinity.** For chlorine, the electron affinity is 3.6 eV. Therefore, an energy equal to $5.1 - 3.6 = 1.5$ eV must be provided to neutral Na and Cl atoms to change them into $Na^+$ and $Cl^-$ ions at infinite separation.

The total energy versus the internuclear separation for NaCl is shown in Figure 43.2. Note that the total energy of the molecule has a minimum value of $-4.2$ eV at the equilibrium separation of about 0.24 nm. The energy required to separate the NaCl molecule into neutral sodium and chlorine atoms, called the **dissociation energy,** is equal to 4.2 eV.

When the two ions are brought closer than 0.24 nm, the electrons in the closed shells begin to overlap, which results in a repulsion between the closed shells. When the ions are far apart, core electrons from one ion do not overlap with those of the other ion. However, as they are brought closer together, the core-electron wave functions begin to overlap. Because of the exclusion principle, some electrons must occupy a higher-energy state.

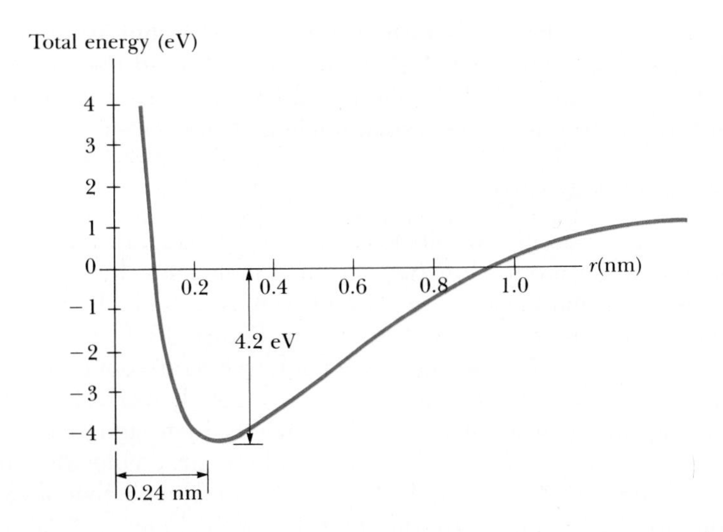


Figure 43.2 Total energy versus the internuclear separation for the NaCl molecule. Note that the dissociation energy is 4.2 eV.

## Covalent Bonds

**A covalent bond** between two atoms can be visualized as the sharing of electrons supplied by one or both atoms that form the molecule. Many diatomic molecules such as $H_2$, $F_2$, and CO owe their stability to covalent bonds. In the case of the $H_2$ molecule, the two electrons are equally shared between the nuclei, and form a so-called molecular orbital. The two electrons are more likely to be found between the two nuclei, hence the electron density is large in this region. The formation of the molecular orbital from the *s* orbitals of the two hydrogen atoms is shown in Figure 43.3. Because of the exclusion principle, the two electrons in the ground state of $H_2$ must have anti-parallel spins. If a third H atom is brought near the $H_2$ molecule, the third electron would have to occupy a higher energy quantum state because of the exclusion principle, which is an energetically unfavorable situation. Hence, the $H_3$ molecule is not stable and does not form.

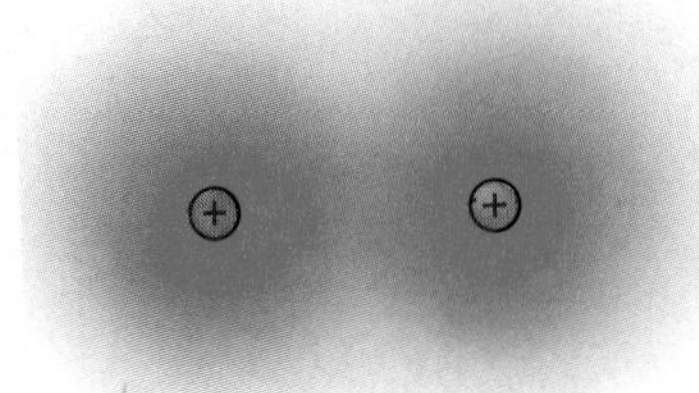


Figure 43.3 The covalent bond formed by the two 1*s* electrons of the $H_2$ molecule. The depth of blue color is proportional to the probability of finding an electron in that location.

More complex stable molecules such as $H_2O$, $CO_2$, and $CH_4$ are also formed by covalent bonds. Consider methane, $CH_4$, a typical organic molecule shown schematically in the electron sharing diagram of Figure 43.4a. In

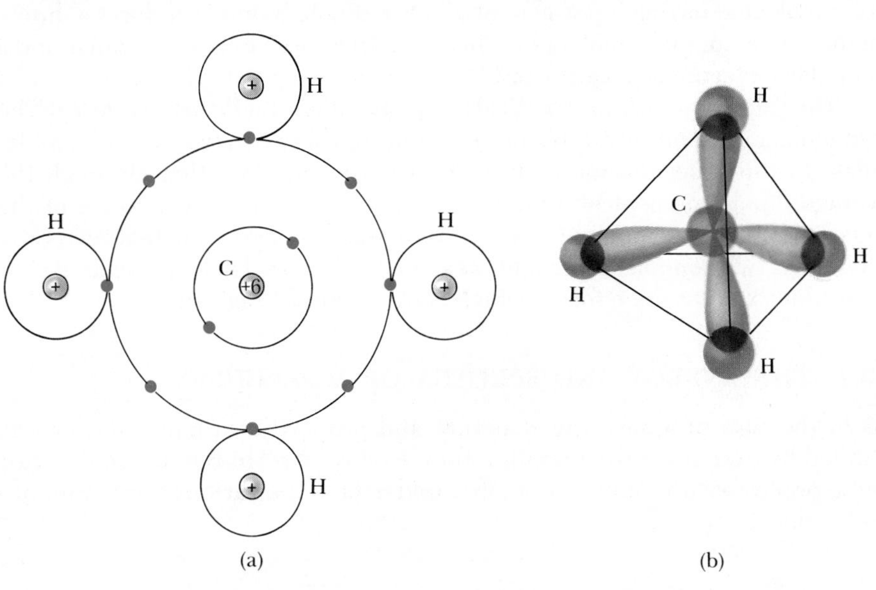


Figure 43.4 (a) Diagram of the four covalent bonds in the $CH_4$ molecule. (b) The spatial electron distribution of the four covalent bonds in the $CH_4$ molecule. Note that the carbon atom is at the center of a tetrahedron with hydrogen atoms at its corners.

this case four covalent bonds are formed between the carbon atom and each of the hydrogen atoms. The spatial electron distribution of the four covalent bonds is shown in Figure 43.4b. The four hydrogen nuclei are at the corners of a regular tetrahedron, with the carbon nucleus at the center.

### The Hydrogen Bond

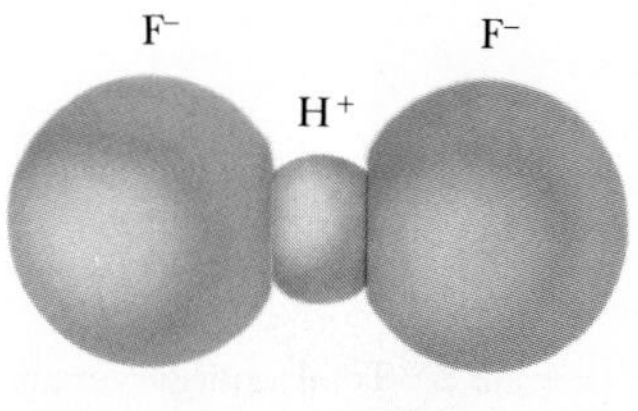

$F^-$ $F^-$ $H^+$

**Figure 43.5** Hydrogen bonding in the $(HF_2)^-$ molecule ion. Note that the two negative fluorine ions are bound by the positively charged proton between them.

Because hydrogen has only one electron, it is expected to form a covalent bond with only one other atom. However, in some molecules, hydrogen forms a different type of bond between two atoms or ions, called a **hydrogen bond.** One example of a hydrogen bond, shown in Figure 43.5, is the hydrogen difluoride ion, $(HF_2)^-$. The two negative fluorine ions are bound by the positively charged proton between them. This is a relatively weak chemical bond, with a binding energy of about 0.1 eV. Although the hydrogen bond is weak, it is the mechanism responsible for linking giant biological molecules and polymers. For example, in the case of the famous DNA molecule that has a double helix structure, hydrogen bonds link the turns of the helix.

### Van der Waals Bonds

If two molecules are some distance apart, they are attracted to each other by electrostatic forces. Likewise, atoms that do not form ionic or covalent bonds are attracted to each other by electrostatic forces. For this reason, at sufficiently low temperatures where thermal excitations are negligible, substances will condense into a liquid and then solidify (with the exception of helium, which does not solidify at atmospheric pressure). The weak electrostatic attractions between molecules are called **van der Waals forces.**

There are actually three types of van der Waals forces, which we shall briefly describe. The first type, called the *dipole-dipole force*, is an interaction between two molecules each having a permanent electric dipole moment. For example, polar molecules such as HCl and $H_2O$ have permanent electric dipole moments and attract other polar molecules. In effect, one molecule interacts with the electric field produced by another molecule. The attractive force turns out to be proportional to $1/r^7$.

The second type of van der Waals force is a *dipole-induced force* in which a polar molecule having a permanent electric dipole moment *induces* a dipole moment in a nonpolar molecule. This attractive force between a polar and a nonpolar molecule also varies as $1/r^7$.

The third type of van der Waals force is called the *dispersion force*. The dispersion force is an attractive force that occurs between two nonpolar molecules. In this case, the interaction results from the fact that although the average dipole moment of a nonpolar molecule is zero, the average of the square of the dipole moment is nonzero because of charge fluctuations. Consequently, two nonpolar molecules near each other tend to be correlated so as to produce an attractive force, which is the van der Waals force.

## 43.2 THE ENERGY AND SPECTRA OF MOLECULES

As in the case of atoms, the structure and properties of molecules can be studied by examining the radiation they emit or absorb. Before we describe these processes, it is important to first understand the various excitations of a molecule.

Consider a single molecule in the gaseous phase. The energy of the molecule can be divided into four categories: (1) electronic energy, due to the mutual interactions of the molecule's electrons and nuclei; (2) translational energy, due to the motion of the molecule's center of mass through space; (3) rotational energy, due to the rotation of the molecule about its center of mass; and (4) vibrational energy, due to the vibration of the molecule's constituent atoms. Thus, we can write the total energy of the molecule in the form

Excitations of a molecule

$$E = E_{\text{el}} + E_{\text{trans}} + E_{\text{rot}} + E_{\text{vib}}$$

The electronic energy of a molecule is very complex because it involves the interaction of many charged particles. Various techniques have been developed to obtain approximate solutions to such systems. In Chapter 21 we found that the average translational kinetic energy of a molecule for each degree of freedom ($x$, $y$, or $z$) is equal to $\frac{1}{2}kT$, where $k$ is Boltzmann's constant. Hence, on the average, the total translational kinetic energy of a molecule is $\frac{3}{2}kT$. From this result, we found that the average kinetic energy per mole of gas is $\frac{3}{2}RT$, where $R$ is the universal gas constant. The important point to understand here is the fact that each degree of freedom of a molecule contributes an energy of $\frac{1}{2}kT$ to the molecule. However, since the quantization of translational energy is unrelated to internal structure, the translational mode is unimportant for the interpretation of molecular spectra.

## Rotational Motion of a Molecule

Let us consider the rotation of a molecule about its center of mass. We shall confine our discussion to a diatomic molecule (Fig. 43.6a), although the same ideas can be extended to polyatomic molecules. The diatomic molecule has only two rotational degrees of freedom, corresponding to rotations about the $y$

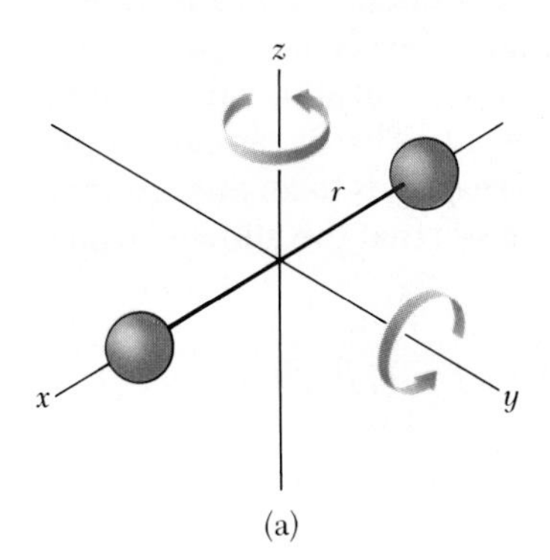


$$E_1 = \frac{\hbar^2}{2I}$$

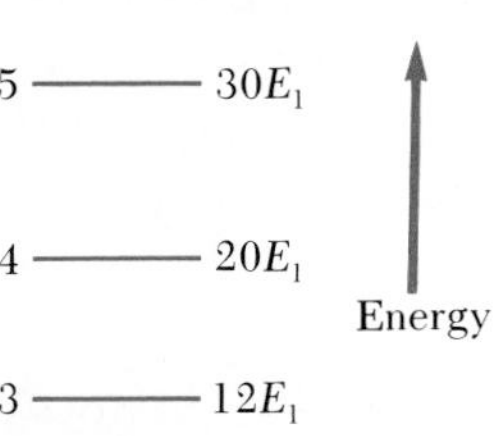


**Figure 43.6** (a) A diatomic molecule oriented along the $x$ axis has two rotational degrees of freedom, corresponding to rotation along the $y$ and $z$ axes. (b) Allowed rotational energies of a diatomic molecule as calculated using Equation 43.5.

and $z$ axes, that is, the axes perpendicular to the molecular axis.[1] If $\omega$ is the angular frequency of rotation about one of these axes, the rotational kinetic energy of the molecule can be expressed in the form

$$E_{\text{rot}} = \tfrac{1}{2} I\omega^2 \tag{43.2}$$

where $I$ is the moment of inertia of the molecule, which has the value

Moment of inertia for a diatomic molecule

$$I = \left(\frac{m_1 m_2}{m_1 + m_2}\right) r^2 = \mu r^2 \tag{43.3}$$

The parameter $\mu$ is the **reduced mass** of the molecule, and $r$ is the atomic separation (Fig. 43.6a).

The magnitude of the angular momentum of the molecule can be expressed as $I\omega$, which classically can have any value. Wave mechanics restricts the angular momentum values to multiples of $\hbar$. That is, the allowed values of rotational angular momentum are given by

Allowed values of rotational angular momentum

$$I\omega = \sqrt{J(J+1)}\,\hbar \qquad J = 0, 1, 2, \ldots \tag{43.4}$$

where $J$ is an integer called the **rotational quantum number.** Substituting Equation 43.4 into Equation 43.2, we get an expression for the allowed values of the rotational kinetic energy:

$$E_{\text{rot}} = \tfrac{1}{2} I\omega^2 = \frac{1}{2I}(I\omega)^2 = \frac{(\sqrt{J(J+1)}\,\hbar)^2}{2I}$$

or

Allowed values of rotational energy

$$E_{\text{rot}} = \frac{\hbar^2}{2I} J(J+1) \qquad J = 0, 1, 2, \ldots \tag{43.5}$$

Thus, we see that *the rotational energy of the molecule is quantized and depends on the moment of inertia of the molecule.* The allowed rotational energies of a diatomic molecule are plotted in Figure 43.6b. These results apply also to polyatomic molecules provided an appropriate generalization of $I$ is used.

The spacings between adjacent rotational energy levels for a molecule lie in the *microwave range* of frequencies ($f \approx 10^{11}$ Hz). The allowed rotational transitions correspond to the selection rule $\Delta J = \pm 1$. That is, an absorption line in the microwave spectrum of a molecule corresponds to an energy separation equal to $E_J - E_{J-1}$. From Equation 43.5, we see that the allowed transitions are given by the condition

Separation between adjacent rotational levels

$$\begin{aligned} \Delta E = E_J - E_{J-1} &= \frac{\hbar^2}{2I}[J(J+1) - (J-1)J] \\ &= \frac{\hbar^2}{I} J = \frac{h^2}{4\pi^2 I} J \end{aligned} \tag{43.6}$$

where $J$ is the quantum number of the higher energy state. Since $\Delta E = hf$, where $f$ is the frequency of the absorbed microwave photon, we see that the

[1] The excitation energy for rotations about the molecular axis is so large that such modes are not observable. This is because nearly all the molecular mass is concentrated within nuclear dimensions of the rotation axis, giving a negligibly small moment of inertia about the internuclear line.

**TABLE 43.1 Microwave Absorption Lines Corresponding to Several Rotational Transitions of the CO Molecule**

| Rotational Transition | Wavelength of the Absorption Line (m) | Frequency of the Absorption Line (Hz) |
|---|---|---|
| $J = 0 \longrightarrow J = 1$ | $2.60 \times 10^{-3}$ | $1.15 \times 10^{11}$ |
| $J = 1 \longrightarrow J = 2$ | $1.30 \times 10^{-3}$ | $2.30 \times 10^{11}$ |
| $J = 2 \longrightarrow J = 3$ | $8.77 \times 10^{-4}$ | $3.46 \times 10^{11}$ |
| $J = 3 \longrightarrow J = 4$ | $6.50 \times 10^{-4}$ | $4.61 \times 10^{11}$ |

From G. M. Barrows, *The Structure of Molecules*, New York, W. A. Benjamin, 1963.

allowed frequency for the transition $J = 0$ to $J = 1$ is given by $f_1 = h^2/4\pi^2 I$. Likewise, the frequency corresponding to the $J = 1$ to $J = 2$ transition is equal to $2f_1$, and so on. These predictions are in excellent agreement with the observed frequencies. The wavelengths and frequencies for the microwave absorption spectrum of the CO molecule are given in Table 43.1. From these data, one can evaluate the moment of inertia and the bond length of the molecule.

**EXAMPLE 43.1 Rotation of the CO Molecule**

The $J = 0$ to $J = 1$ rotational transition of the CO molecule occurs at a frequency of $1.15 \times 10^{11}$ Hz. (a) Use this information to calculate the moment of inertia of the molecule.

*Solution* From Equation 43.6, we see that the energy difference between the $J = 0$ and $J = 1$ rotational levels is $h^2/4\pi^2 I$. Equating this to the energy of the absorbed photon, we get

$$\frac{h^2}{4\pi^2 I} = hf$$

Solving for $I$ gives

$$I = \frac{h}{4\pi^2 f} = \frac{6.626 \times 10^{-34}\ \text{J} \cdot \text{s}}{4\pi^2(1.15 \times 10^{11}\ \text{s}^{-1})}$$

$$= \boxed{1.46 \times 10^{-46}\ \text{kg} \cdot \text{m}^2}$$

(b) Calculate the bond length of the molecule.

*Solution* Equation 43.3 can be used to calculate the bond length, but we first need to know the value for the reduced mass $\mu$ of the CO molecule. Since $m_1 = 12$ u and $m_2 = 16$ u, the reduced mass is given by

$$\mu = \frac{m_1 m_2}{m_1 + m_2} = \frac{(12\ \text{u})(16\ \text{u})}{12\ \text{u} + 16\ \text{u}} = 6.86\ \text{u}$$

$$= (6.86\ \text{u})\left(1.66 \times 10^{-27}\ \frac{\text{kg}}{\text{u}}\right) = 1.14 \times 10^{-26}\ \text{kg}$$

where we have used the fact that $1\ \text{u} = 1.66 \times 10^{-27}$ kg.

Substituting this value and the result of (a) into Equation 43.3, and solving for $r$, we get

$$r = \sqrt{\frac{I}{\mu}} = \sqrt{\frac{(1.46 \times 10^{-46}\ \text{kg} \cdot \text{m}^2)}{(1.14 \times 10^{-26}\ \text{kg})}}$$

$$= 1.13 \times 10^{-10}\ \text{m} = 1.13\ \text{Å} = \boxed{0.113\ \text{nm}}$$

This example illustrates the direct way one can calculate molecular dimensions and properties from spectroscopic measurements.

## Vibrational Motion of Molecules

As we mentioned earlier, another mode of excitation of a molecule is its vibrational motion. A molecule is a flexible structure whose atoms are bonded together by what can be considered "effective springs." If disturbed, the molecule can vibrate and acquire vibrational energy. This vibrational motion and corresponding vibrational energy can be altered if the molecule is exposed to radiation of the proper frequency.

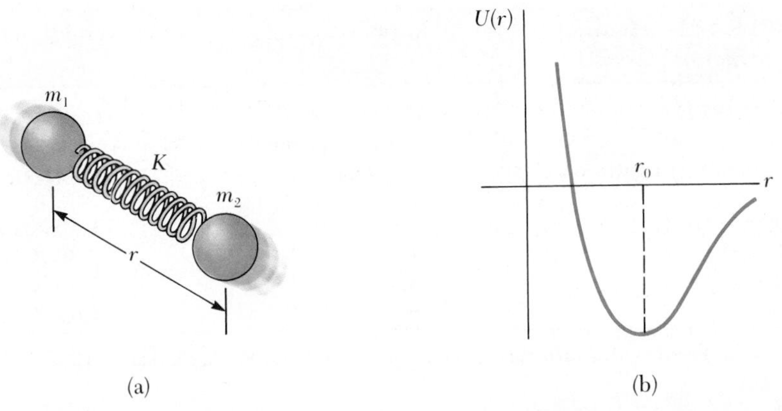


**Figure 43.7** (a) Model of a diatomic molecule whose atoms are bonded by effective spring of force constant $K$. The fundamental vibration is along the molecular axis. (b) A plot of the potential energy of a diatomic molecule versus atomic separation. The parameter $r_0$ is the equilibrium separation of the atoms.

Consider a diatomic molecule as shown in Figure 43.7a, where the effective spring has a force constant $K$. The potential energy versus atomic separation for such a molecule is sketched in Figure 43.7b, where $r_0$ is the equilibrium atomic separation. According to classical mechanics, the **frequency of vibration** for such a system is

Frequency of vibration of a diatomic molecule

$$f = \frac{1}{2\pi}\sqrt{\frac{K}{\mu}} \qquad (43.7)$$

where $\mu$ is the **reduced mass,** given by

Reduced mass

$$\mu = \frac{m_1 m_2}{m_1 + m_2} \qquad (43.8)$$

The quantum mechanics solution to this system shows that the energy is quantized. The allowed energies of vibration are given by

$$E_{\text{vib}} = (v + \tfrac{1}{2})hf \qquad v = 0, 1, 2, \ldots \qquad (43.9)$$

where $v$ is an integer called the **vibrational quantum number.** If the system is in the lowest vibrational state, for which $v = 0$, its energy is $\frac{1}{2}hf$ (the so-called **zero-point energy**). The accompanying vibration—the zero point motion—is *always present,* even if the molecule is not excited. In the first excited state, $v = 1$ and its energy is equal to $\frac{3}{2}hf$, and so on.

Substituting Equation 43.7 into Equation 43.9 gives the following expression for the vibrational energy:

$$E_{\text{vib}} = (v + \tfrac{1}{2})\frac{h}{2\pi}\sqrt{\frac{K}{\mu}} \qquad v = 0, 1, 2, \ldots \qquad (43.10)$$

The selection rule for the allowed vibrational transitions is given by $\Delta v = \pm 1$. From Equation 43.10, we see that the energy difference between any two *successive* vibrational levels is equal and given by

$$\Delta E_{\text{vib}} = \frac{h}{2\pi}\sqrt{\frac{K}{\mu}} = hf \qquad (43.11)$$

The vibrational energies of a diatomic molecule are plotted in Figure 43.8. At ordinary temperatures, most molecules have vibrational energies corresponding to the $v = 0$ state because the spacing between vibrational states is large compared to $kT$, where $k$ is Boltzmann's constant. Transitions between vibrational levels lie in the *infrared region* of the spectrum. The absorption frequencies corresponding to the $v = 0$ to $v = 1$ transition for several diatomic molecules are listed in Table 43.2, together with effective force constants calculated using Equation 43.11. The strength of a bond can be measured by the size of the effective force constant. For example, the CO molecule, which is bonded by several electrons, has a larger bonding strength than such single-bonded molecules as HCl.

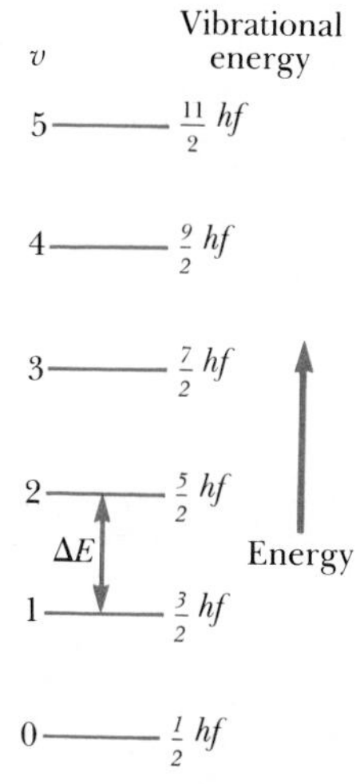

**Figure 43.8** Allowed vibrational energies of a diatomic molecule, where $f$ is the fundamental frequency of virbation, given by Equation 43.7. Note that the spacings between adjacent vibrational levels are equal.

**TABLE 43.2 Fundamental Vibrational Frequencies and Effective Force Constants for Some Diatomic Molecules**

| Molecule | Frequency (Hz), $v = 0$ to $v = 1$ | Force Constant (N/m) |
|---|---|---|
| HF | $8.72 \times 10^{13}$ | 970 |
| HCl | $8.66 \times 10^{13}$ | 480 |
| HBr | $7.68 \times 10^{13}$ | 410 |
| HI | $6.69 \times 10^{13}$ | 320 |
| CO | $6.42 \times 10^{13}$ | 1860 |
| NO | $5.63 \times 10^{13}$ | 1530 |

From G. M. Barrows, *The Structure of Molecules*, New York, W. A. Benjamin, 1963.

**EXAMPLE 43.2 Vibration of the CO Molecule**

The fundamental vibrational band for the CO molecule occurs at a frequency of $6.42 \times 10^{13}$ Hz. (a) Calculate the effective force constant for this molecule.

*Solution* This transition corresponds to the case where the electron moves from the $v = 0$ to the $v = 1$ vibrational state. From Equation 43.9, we see that the energy difference between these states is given by

$$\Delta E = \tfrac{3}{2}hf - \tfrac{1}{2}hf = hf$$

Using Equation 43.11 and the fact that the reduced mass is $\mu = 1.14 \times 10^{-26}$ kg for the CO molecule (Example 43.1), we get

$$\frac{h}{2\pi}\sqrt{\frac{K}{\mu}} = hf$$

$$K = 4\pi^2\mu f^2$$

$$= 4\pi^2(1.14 \times 10^{-26}\ \text{kg})(6.42 \times 10^{13}\ \text{s}^{-1})^2$$

$$= 1.85 \times 10^3\ \text{N/m}$$

(b) What is the maximum amplitude of vibration for this molecule in the $v = 0$ vibrational state?

*Solution* The maximum potential energy stored in the molecule is $\frac{1}{2}KA^2$, where $A$ is the amplitude of vibration. Equating this to the vibrational energy given by Equation 43.10, with $v = 0$, we get

$$\tfrac{1}{2}KA^2 = \frac{h}{4\pi}\sqrt{\frac{K}{\mu}}$$

Substituting the values $K = 1.86 \times 10^3$ N/m and $\mu = 1.14 \times 10^{-26}$ kg, we get

$$A^2 = \frac{h}{2\pi K}\sqrt{\frac{K}{\mu}} = \frac{h}{2\pi}\sqrt{\frac{1}{K\mu}}$$

$$= \frac{6.626 \times 10^{-34}}{2\pi}\sqrt{\frac{1}{(1.86 \times 10^3)(1.14 \times 10^{-26})}}$$

$$= 2.30 \times 10^{-23}\ \text{m}^2$$

so that

$$A = 4.79 \times 10^{-12}\ \text{m} = 0.0479\ \text{Å} = 4.79 \times 10^{-3}\ \text{nm}$$

Comparing this result with the bond length of 0.1123 nm, we see that the amplitude of vibration is about 4% of the bond length. Thus, we see that infrared spectroscopy provides useful information on the elastic properties (bond strengths) of molecules.

### Molecular Spectra

In general, an excited molecule will rotate and vibrate simultaneously. To a first approximation, these motions are independent, and the total energy of the molecule is given by the sum of Equations 43.5 and 43.9:

$$E = \frac{\hbar^2}{2I}[J(J+1)] + (v + \tfrac{1}{2})hf \qquad (43.12)$$

The energy levels can be calculated from this expression, and each level is indexed by two quantum numbers, $J$ and $v$. From these calculations, one can then construct an energy level diagram like that shown in Figure 43.9a. For each allowed value of the vibrational quantum number $v$, there is a complete set of rotational levels corresponding to $J = 0, 1, 2, \ldots$. Note that the energy separation between successive rotational levels is much smaller than the energy separation between successive vibrational levels. The allowed transitions between any two energy levels are subject to the following selection rules[2]:

$$\Delta J = \pm 1 \qquad \text{and} \qquad \Delta v = \pm 1$$

[2] The selection rule $\Delta J = \pm 1$ implies that the photon (which excites a transition) is a spin one particle with spin quantum number $s = 1$. Hence, this selection rule describes angular momentum conservation for the system molecule plus photon.

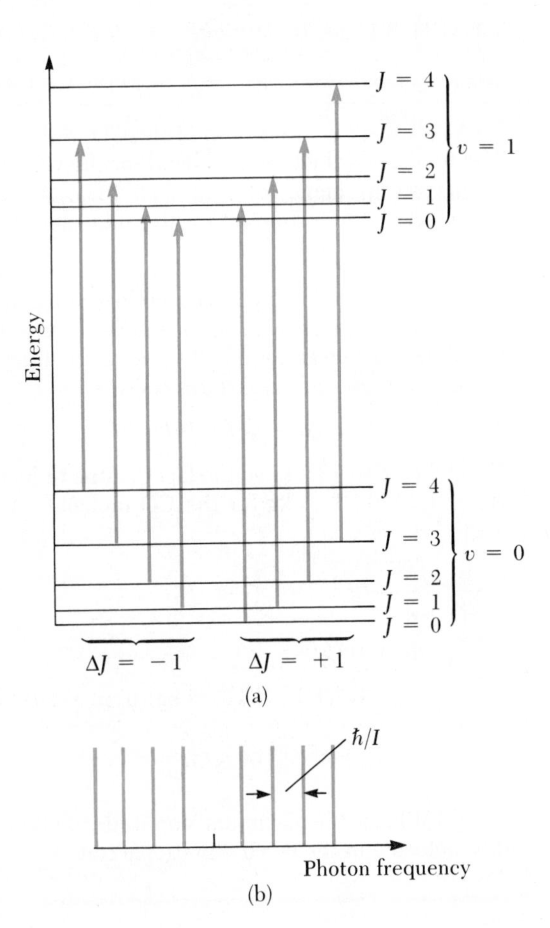


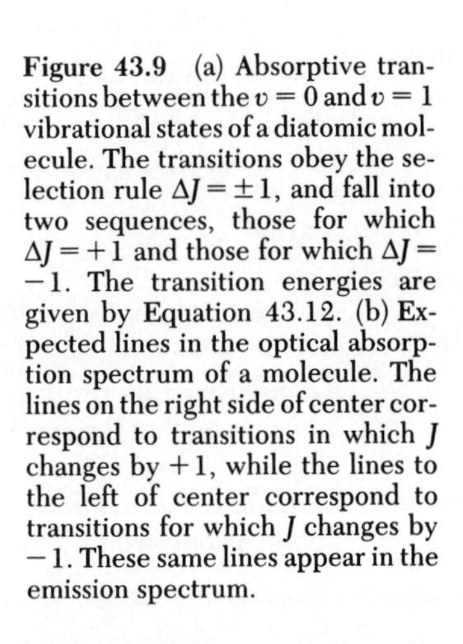

**Figure 43.9** (a) Absorptive transitions between the $v = 0$ and $v = 1$ vibrational states of a diatomic molecule. The transitions obey the selection rule $\Delta J = \pm 1$, and fall into two sequences, those for which $\Delta J = +1$ and those for which $\Delta J = -1$. The transition energies are given by Equation 43.12. (b) Expected lines in the optical absorption spectrum of a molecule. The lines on the right side of center correspond to transitions in which $J$ changes by $+1$, while the lines to the left of center correspond to transitions for which $J$ changes by $-1$. These same lines appear in the emission spectrum.

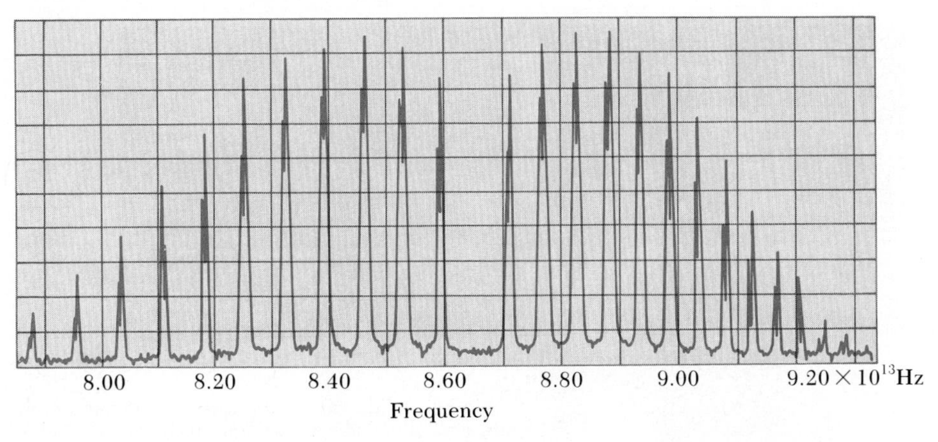


**Figure 43.10** The absorption spectrum of the HCl molecule. Each line is split into a doublet because chlorine has two isotopes, $^{35}$Cl and $^{37}$Cl, which have different nuclear masses.

These selection rules restrict the number of wavelengths observed in the spectra of molecules, since transitions are prohibited unless all rules are obeyed simultaneously. For example, one would not observe a pure rotational transition since this requires $\Delta v = 0$, in violation of the second selection rule. Likewise, a pure vibrational transition (corresponding to $\Delta J = 0$) is forbidden since it is in violation of the first selection rule. At ordinary temperatures, most of the molecules will be in the $v = 0$ vibrational state, but in various states of rotation. When the molecule absorbs a photon, $v$ increases by one unit while $J$ either increases or decreases by one unit as in Figure 43.9a. Thus, the molecular absorption spectrum consists of two groups of lines; the group at the right satisfy the selection rules $\Delta J = 1$ and $\Delta v = 1$, while the group at the left satisfy the selection rules $\Delta J = -1$ and $\Delta v = 1$. The energies of the absorbed photons can be calculated from Equation 43.12:

$$\Delta E = hf + \frac{\hbar^2}{I}(J+1) \qquad J = 0, 1, 2, \ldots \qquad (\Delta J = +1) \qquad (43.13a)$$

$$\Delta E = hf + \frac{\hbar^2}{I}J \qquad J = 1, 2, 3, \ldots \qquad (\Delta J = -1) \qquad (43.13b)$$

In these equations, $J$ is the rotational quantum number of the *initial* state. Equation 43.13a generates the series of equally spaced lines *above* the characteristic frequency $f$, while Equation 43.13b generates a series *below* this frequency. Adjacent lines are separated in frequency by the fundamental unit $\hbar/I$. Figure 43.9b shows the expected frequencies in the absorption spectrum of the molecule; these same frequencies appear in the emission spectrum.

The actual absorption spectrum of the HCl molecule shown in Figure 43.10 follows this pattern very well and reinforces our model. However, one peculiarity is apparent: each line is split into a doublet. This doubling occurs because of two chlorine isotopes ($^{35}$Cl and $^{37}$Cl) whose different masses give two distinct values for $I$.

## 43.3 BONDING IN SOLIDS

A crystalline solid consists of a large number of atoms arranged in a regular array, forming a periodic structure. The bonding schemes for molecules described in Section 43.1 are also appropriate in describing the bonding in

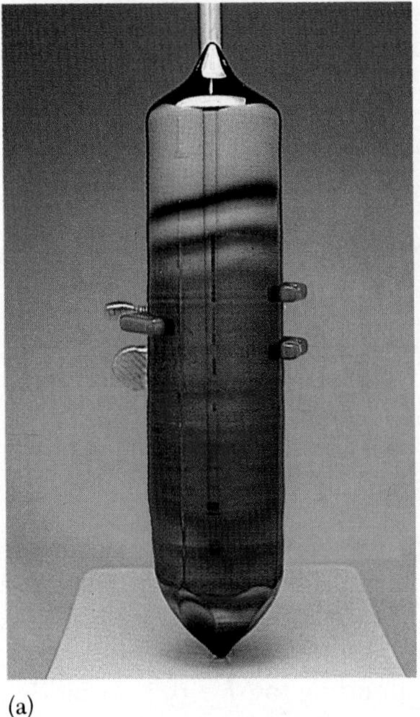

(a)

(b)

Crystalline solids. (a) A man-made cylinder of nearly pure crystalline silicon (Si), approximately 10 inches long. Such crystals are cut into wafer form and processed to make various semiconductor devices such as integrated circuits and solar cells. (© Charles D. Winters) (b) Amethyst crystals are an impure form of quartz crystals. Their color, which is due to iron impurities ($Fe^{3+}$), can range from pale lilac to deep purple, depending on the iron concentration. (Courtesy of Ward's Natural Science)

solids. For example, the ions in the NaCl crystal are ionically bonded, while the carbon atoms in the diamond structure form covalent bonds. Another type of bonding mechanism is the metallic bond, which is responsible for the cohesion of copper, silver, sodium, and other metals.

### Ionic Solids

Many crystals form by ionic bonding, where the dominant interaction is the Coulomb interaction between the ions. Consider the NaCl crystal shown in Figure 43.11, where each $Na^+$ ion has six nearest-neighbor $Cl^-$ ions, and each $Cl^-$ ion has six nearest-neighbor $Na^+$ ions. Each $Na^+$ ion is attracted to the six

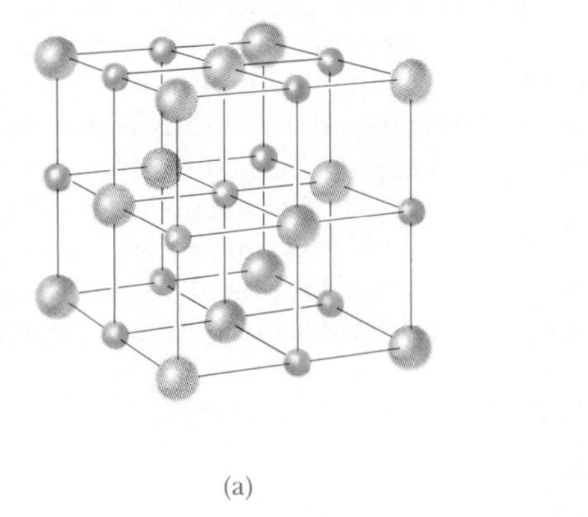

(a)

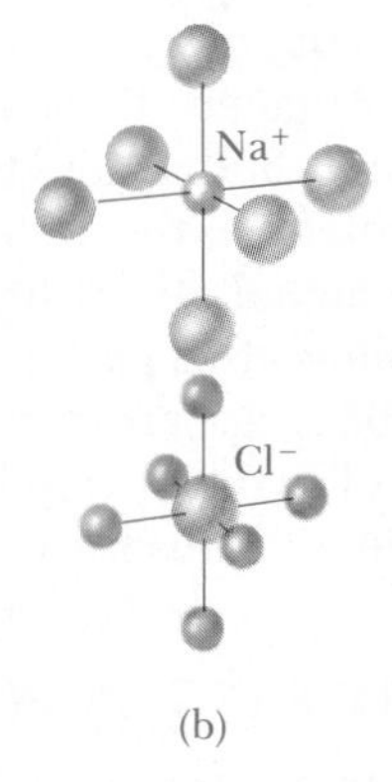


(b)

**Figure 43.11** (a) The crystal structure of NaCl. (b) In the NaCl structure, each positive sodium ion (red spheres) is surrounded by six negative chlorine ions (blue spheres), while each chlorine ion is surrounded by six sodium ions.

$Cl^-$ ions. The corresponding attractive potential energy is given by $-6ke^2/r$, where $r$ is the $Na^+$–$Cl^-$ separation. In addition, there are also 12 $Na^+$ ions at a distance of $\sqrt{2}\,r$ from the $Na^+$, which produce a weaker repulsive force on the $Na^+$ ion. Furthermore, beyond these 12 $Na^+$ ions one finds more $Cl^-$ ions that produce an attractive force, and so on. The net effect of all these interactions is a resultant negative electrostatic potential energy given by

$$U_{\text{attractive}} = -\alpha k \frac{e^2}{r} \tag{43.14}$$

where $\alpha$ is a constant called the **Madelung constant.** The value of $\alpha$ depends on the strength of the interaction and the specific crystal structure. For example, $\alpha = 1.7476$ for the NaCl structure. When the atoms are brought close together, the subshells tend not to overlap because of the exclusion principle, which introduces a repulsive potential energy term given by $B/r^m$ as discussed in Section 43.1. The total potential energy is therefore of the form

$$U_{\text{total}} = -\alpha k \frac{e^2}{r} + \frac{B}{r^m} \tag{43.15}$$

A plot of the total potential energy versus ion separation is shown in Figure 43.12. The potential energy has its minumum value $U_0$ at the equilibrium separation, when $r = r_0$. It is left as a problem (Problem 35) to show that the energy $U_0$ is given by

$$U_0 = -\alpha k \frac{e^2}{r_0}\left(1 - \frac{1}{m}\right) \tag{43.16}$$

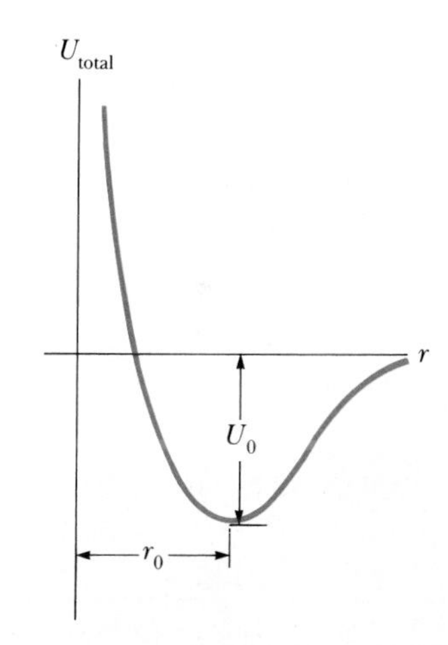


**Figure 43.12** The total potential energy versus ion separation for an ionic solid, where $U_0$ is the ionic cohesive energy and $r_0$ is the equilibrium separation between the ions.

The energy $U_0$ is called the **ionic cohesive energy** of the solid, and its absolute value represents the energy required to separate the solid into a collection of positive and negative ions. Its value for NaCl is about $-7.84$ eV. In order to calculate the **atomic cohesive energy,** which is the binding energy relative to the neutral atoms, one must add to this the ionization energy of Na (5.14 eV) and the energy released to form Cl ($-3.61$ eV). Hence, the atomic cohesive energy of NaCl is about $-6.31$ eV.

Properties of ionic solids

Ionic crystals have the following general properties:

1. They form relatively stable and hard crystals.
2. They are poor electrical conductors because there are no available free electrons.
3. They have high vaporization temperatures.
4. They are transparent to visible radiation but absorb strongly in the infrared region. This occurs because the electrons form such tightly bound shells in ionic solids that visible radiation does not contain sufficient energy to promote electrons to the next allowed shell. The strong infrared absorption (at 20 to 150 $\mu$m) occurs because the vibrations of the more massive ions have a lower natural frequency and experience resonant absorption in the low-energy infrared region.
5. They are usually quite soluble in polar liquids such as water. The water molecule, which has a permanent electric dipole moment, exerts an attractive force on the charged ions, which breaks the ionic bonds and dissolves the solid.

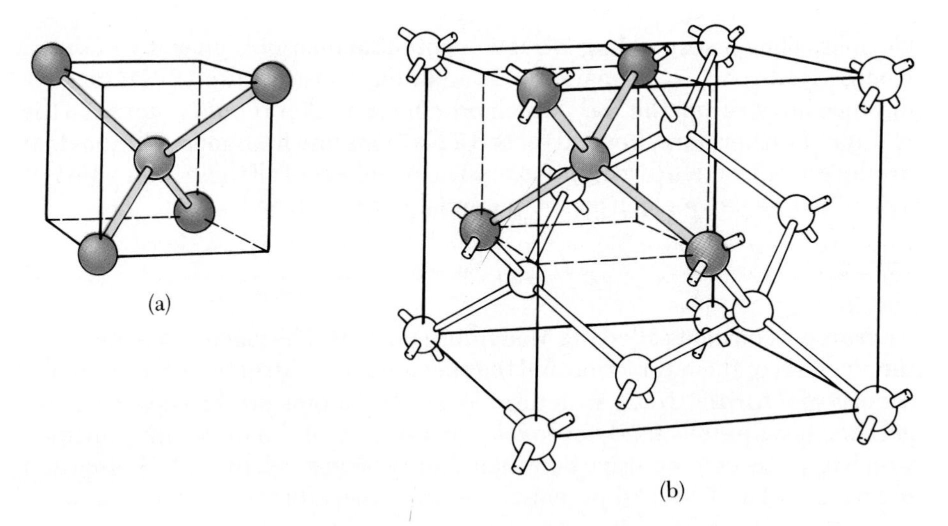


**Figure 43.13** (a) Each carbon atom in diamond is covalently bonded to four other carbons and forms a tetrahedral structure. (b) The crystal structure of diamond, showing the tetrahedral bond arrangement. (After W. Shockley, *Electrons and Holes in Semiconductors,* New York, Van Nostrand, 1950.)

## Covalent Crystals

**TABLE 43.3 The Cohesive Energies of Some Covalent Solids**

| Crystal | Cohesive Energy (eV) |
|---|---|
| C (diamond) | 7.37 |
| Si | 4.63 |
| Ge | 3.85 |
| InAs | 5.70 |
| SiC | 12.3 |
| ZnS | 6.32 |
| CuCl | 9.24 |

As we found in Section 43.1, the covalent bond is very strong and comparable to the ionic bond. Solid carbon, in the form of diamond, is a crystal whose atoms are covalently bonded. Because carbon has an electron configuration $1s^22s^22p^2$, it lacks four electrons with respect to a filled shell ($2p^6$). Hence, two carbon atoms have a strong attraction for each other, with a cohesive energy of 7.3 eV.

In the diamond structure, each atom is covalently bonded to four other carbon atoms located at four corners of a cube as in Figure 43.13a. In order to form such a configuration of bonds, the electron of each atom must be promoted to the configuration $1s^22s2p^3$, which requires an energy of about 4 eV. The crystal structure of diamond is shown in Figure 43.13b. Note that each carbon atom forms covalent bonds with four nearest-neighbor atoms. The basic structure of diamond is called *tetrahedral* (each carbon atom is at the center of a regular tetrahedron), and the angle between the bonds is 109.5°. Other crystals such as silicon and germanium have similar structures.

The properties of some covalent solids are given in Table 43.3. The cohesive energies are larger than for ionic solids, which accounts for the hardness of covalent solids. Diamond is particularly hard, and has an extremely high melting point (about 4000 K). In general, covalently bonded solids are very hard, have large bond energies and high melting points, are good insulators, and are transparent to visible light.

## Metallic Solids

Metallic bonds are generally weaker than ionic or covalent bonds. The valence electrons in a metal are relatively free to move throughout the material. There are a large number of such mobile electrons in a metal, typically one or two electrons per atom. The metal structure can be viewed as a "sea" or "gas" of nearly free electrons surrounded by a lattice of positive ions (Fig. 43.14). The binding mechanism in a metal is the attractive force between the positive ions and the electron gas.

Metals have a cohesive energy in the range of 1 to 3 eV, which is smaller than the cohesive energies of ionic or covalent solids. Since visible photons also have energies in this range, light interacts strongly with the free electrons in metals. Hence visible light is absorbed and re-emitted quite close to the surface of a metal, which accounts for the shiny nature of metallic surfaces. In addition to the high electrical conductivity of metals produced by the free electrons, the nondirectional nature of the metallic bond allows many different types of metallic atoms to be dissolved in a host metal in varying amounts. The resulting solid solutions or alloys may be designed to have particularly useful structural properties, such as high strength and low density, since these properties tend to change rather slowly and controllably with alloy composition.

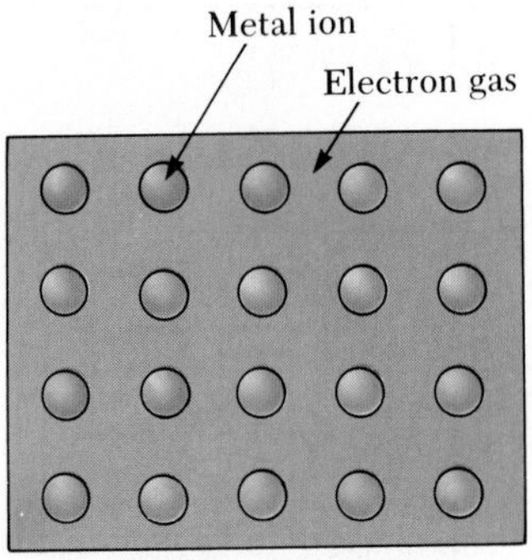

Metal ion

Electron gas

**Figure 43.14** Schematic diagram of a metal. The blue area represents the electron gas, while the red circles represent the positive metal ion cores.

## 43.4 BAND THEORY OF SOLIDS

If two identical atoms are very far apart, they do not interact and their electronic energy levels can be considered to be those of isolated atoms. Suppose the two atoms are sodium, each having a $3s$ electron with a specific, well-defined energy. As the two sodium atoms are brought closer together, their wave functions begin to overlap. When the interaction between them is strong enough, two different $3s$ levels will form as shown in Figure 43.15a.

When a large number of atoms are brought together to form a solid, a similar phenomenon occurs. As the atoms are brought close together, the various atomic energy levels begin to split. This splitting in levels for six atoms in close proximity is shown in Figure 43.15b. In this case, there are six energy levels and six overlapping wave functions for the system. Since the width of an energy band arising from a particular atomic energy level is independent of the number of atoms in a solid, the energy levels are more closely spaced in the case of six atoms than in the case of two atoms. If we extend this argument to a large number of atoms (of the order of $10^{23}$ atoms/cm$^3$), we obtain a large number of levels so closely spaced that they may be regarded as a *continuous band* of energy levels as in Figure 43.15c. In the case of sodium, it is common to refer to the continuous distribution of allowed energy levels as the $3s$ band because the band originates from the $3s$ levels of individual sodium atoms. In general, a crystalline solid has a large number of allowed energy bands that

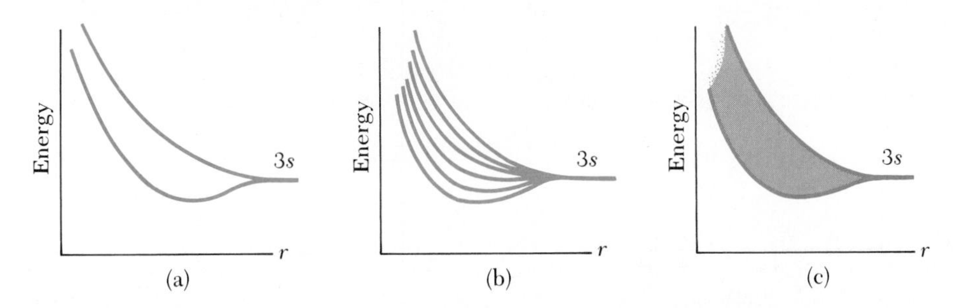


**Figure 43.15** (a) The splitting of the $3s$ levels when two sodium atoms are brought together. (b) The splitting of the $3s$ levels when six sodium atoms are brought together. (c) The formation of a $3s$ band when a large number of sodium atoms are assembled to form a solid.

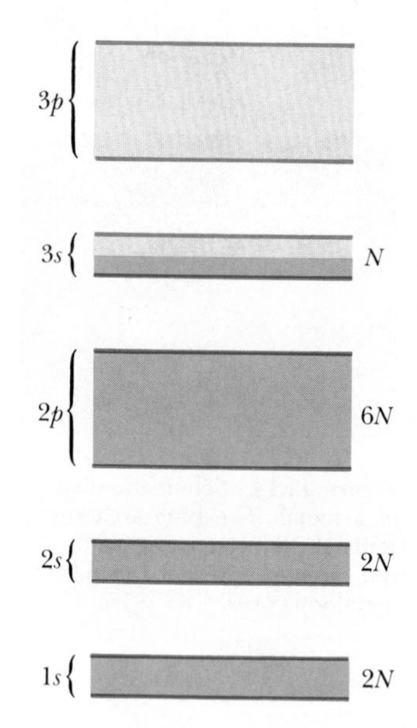


**Figure 43.16** The energy bands of sodium. The solid contains $N$ atoms. Note the energy gaps between the allowed bands.

arise from different atomic energy levels. Figure 43.16 shows the allowed energy bands of sodium. Note that energy gaps or forbidden energy bands for electrons occur between the allowed bands.

If the solid contains $N$ atoms, each energy band has $N$ energy levels. In the case of sodium, the $1s$, $2s$, and $2p$ bands are each full, as indicated by the gray-shaded areas in Figure 43.16. A level whose orbital angular momentum is $\ell$ can hold $2(2\ell + 1)$ electrons. The factor of 2 arises from the two possible electron spin orientations, while the factor $2\ell + 1$ corresponds to the number of possible orientations of the orbital angular momentum. The capacity of each band for a system of $N$ atoms is $2(2\ell + 1)N$ electrons. Hence, the $1s$ and $2s$ bands each contain $2N$ electrons ($\ell = 0$), while the $2p$ band contains $6N$ electrons ($\ell = 1$). Because sodium has only one $3s$ valence electron, and there are a total of $N$ atoms in the solid, the $3s$ band contains only $N$ electrons and is only half full. The $3p$ band, which is above the $3s$ band, is completely empty.

## 43.5 FREE-ELECTRON THEORY OF METALS

In this section, we shall discuss the free-electron theory of metals. In this model, one imagines that the valence electrons in the metal are not strongly bound to individual atoms but are free to move through the metal.

Statistical physics can be applied to a collection of particles in an effort to relate microscopic properties to its macroscopic properties. In the case of electrons, it is necessary to use quantum statistics, with the requirement that each state of the system can be occupied by only one electron. Each state is specified by a set of quantum numbers. All particles with half integral spin, called **Fermions,** must obey the Pauli exclusion principle. An electron is one example of a Fermion. The probability of finding an electron in a particular state of energy $E$ is

Fermi-Dirac distribution function

$$f(E) = \frac{1}{e^{(E-E_F)/kT} + 1} \tag{43.17}$$

where $E_F$ is called the **Fermi energy.** The function $f(E)$ is called the **Fermi-Dirac distribution function.** A plot of $f(E)$ versus $E$ at $T = 0$ K is shown in Figure 43.17a. Note that $f = 1$ for $E < E_F$, and $f = 0$ for $E > E_F$. That is, at 0 K,

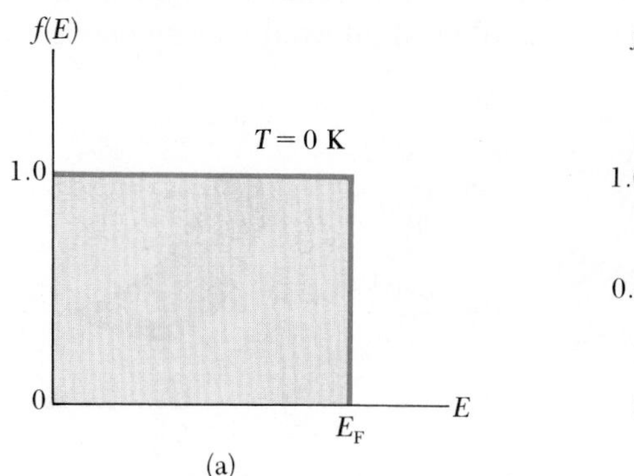


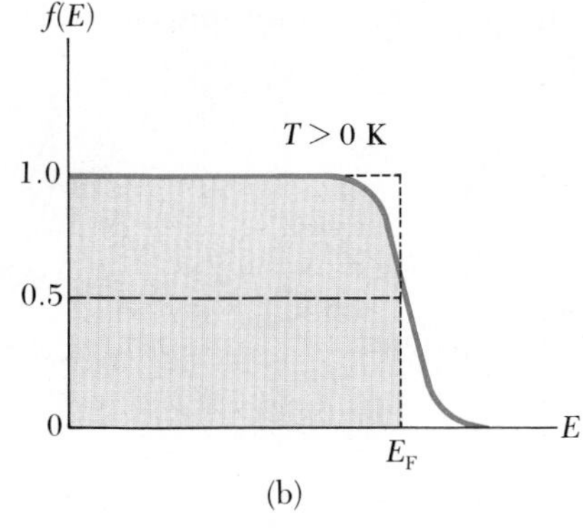


**Figure 43.17** A plot of the Fermi-Dirac distribution function at (a) $T = 0$ K, and (b) $T > 0$ K, where $E_F$ is the Fermi energy.

all states whose energies lie below the Fermi energy are occupied, while all states with energies greater than the Fermi energy are vacant. A plot of $f(E)$ versus $E$ at some temperature $T > 0$ is shown in Figure 43.17b. At $E = E_F$, the function $f(E)$ has the value $\frac{1}{2}$. The main variations in the Fermi-Dirac distribution function with temperature occur at the high energies in the vicinity of the Fermi energy. Note that only a small fraction of the levels with energies greater than the Fermi energy are occupied. Furthermore, a small fraction of the levels below the Fermi energy are empty.

In Chapter 41 we found that if a particle is confined to move in a one-dimensional box of length $L$, the allowed states have quantized energy levels given by

$$E = \frac{\hbar^2\pi^2}{2mL^2} n^2 \qquad n = 1, 2, 3, \ldots$$

The wave functions for these allowed states are standing waves given by $\psi = A \sin(n\pi x/L)$, which satisfy the boundary condition $\psi = 0$ at $x = 0$ and $x = L$.

Now imagine that an electron is moving in a three-dimensional box of edge length $L$ and volume $L^3$, where the walls of the box represent the surfaces of the metal. One could show (Problem 24) that the energy for such a particle is given by

$$E = \frac{\hbar^2\pi^2}{2mL^2}(n_x^2 + n_y^2 + n_z^2) \tag{43.18}$$

where $n_x$, $n_y$, and $n_z$ are quantum numbers. Again, the energy levels are quantized and each is characterized by this set of three quantum numbers (one for each degree of freedom) and the spin quantum number $m_s$. For example, the ground state corresponding to $n_x = n_y = n_z = 1$ has an energy equal to $3\hbar^2\pi^2/2mL^2$, and so on. In this model, we require that $\psi(x, y, z) = 0$ at the boundaries. This requirement results in solutions that are standing waves in three dimensions.

If the quantum numbers are treated as continuous variables, one finds that the number of allowed states per unit volume having energies between $E$ and $E + dE$ is

$$g(E)\, dE = CE^{1/2}\, dE \tag{43.19}$$

where

$$C = \frac{8\sqrt{2}\pi m^{3/2}}{h^3} \tag{43.20}$$

The function $g(E) = CE^{1/2}$ is called the **density of states function.**

In thermal equilibrium, the number of electrons per unit volume with energy between $E$ and $E + dE$ is equal to the product $f(E)g(E)\, dE$. That is,

$$N(E)\, dE = C \frac{E^{1/2}\, dE}{e^{(E-E_F)/kT} + 1} \tag{43.21}$$

A plot of $N(E)$ versus $E$ is given in Figure 43.18. If $n$ is the total number of electrons per unit volume, we require that

$$n = \int_0^\infty N(E)\, dE = C \int_0^\infty \frac{E^{1/2}\, dE}{e^{(E-E_F)/kT} + 1} \tag{43.22}$$

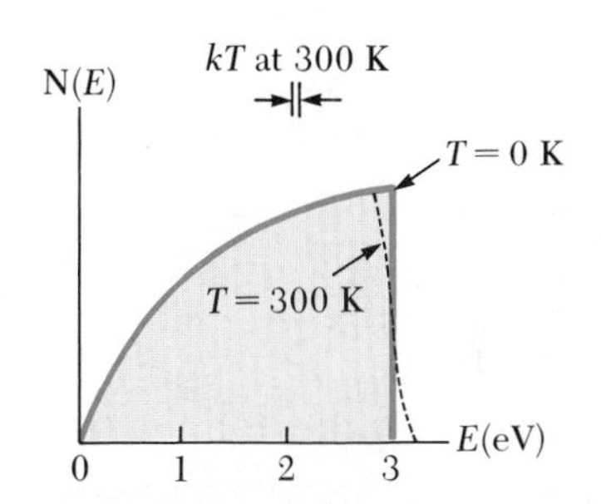


**Figure 43.18** A plot of the electron distribution versus energy in a metal at $T = 0$ K and $T = 300$ K. The Fermi energy is taken to be 3 eV.

This condition can be used to calculate the Fermi energy. At $T = 0$ K, the Fermi distribution function $f(E) = 1$ for $E < F_F$ and is zero for $E > E_F$. Therefore, at $T = 0$ K, Equation 43.22 becomes

$$n = C \int_0^{E_F} E^{1/2}\, dE = \tfrac{2}{3} C E_F^{3/2} \tag{43.23}$$

Substituting Equation 43.20 into Equation 43.23 and solving for $E_F$ gives

Fermi energy at $T = 0$ K

$$E_F = \frac{h^2}{2m}\left(\frac{3n}{8\pi}\right)^{2/3} \tag{43.24}$$

According to this result, $E_F$ shows a gradual increase with increasing electron concentration. This is expected, because the electrons fill the available energy states, two electrons per state, in accordance with the Pauli exclusion principle, up to the Fermi energy.

The order of magnitude of the Fermi energy for metals is about 5 eV. Values for the Fermi energy based on the free-electron theory are given in Table 43.4, together with values for the electron velocity at the Fermi level, defined by

$$\tfrac{1}{2} m v_F^2 = E_F \tag{43.25}$$

and the Fermi temperature, $T_F$, defined by

$$T_F = \frac{E_F}{k} \tag{43.26}$$

It is left as a problem (Problem 22) to show that the average energy of a conduction electron in a metal at 0 K is given by

$$E_{av} = \tfrac{3}{5} E_F \tag{43.27}$$

In summar,', we can consider a metal to be a system with a very large number of energy levels available to the valence electrons. These electrons fill these levels in accordance with the Pauli exclusion principle, beginning with $E = 0$ and ending with $E_F$. At $T = 0$ K, all levels below the Fermi energy are filled, while all levels above the Fermi energy are empty. Although the levels are discrete, they are so close together that the electrons have an almost continuous distribution of energy. At 300 K, a very small fraction of the valence electrons are excited above the Fermi energy. An estimate of this fraction is given in Example 43.4.

**TABLE 43.4 Calculated Values of Various Parameters for Metals at 300 K Based on the Free-Electron Theory**

| Metal | Electron Concentration ($m^{-3}$) | Fermi Energy (eV) | Fermi Velocity (m/s) | Fermi Temperature (K) |
|---|---|---|---|---|
| Li | $4.70 \times 10^{28}$ | 4.72 | $1.29 \times 10^{6}$ | $5.48 \times 10^{4}$ |
| Na | $2.65 \times 10^{28}$ | 3.23 | $1.07 \times 10^{6}$ | $3.75 \times 10^{4}$ |
| K | $1.40 \times 10^{28}$ | 2.12 | $0.86 \times 10^{6}$ | $2.46 \times 10^{4}$ |
| Cu | $8.49 \times 10^{28}$ | 7.05 | $1.57 \times 10^{6}$ | $8.12 \times 10^{4}$ |
| Ag | $5.85 \times 10^{28}$ | 5.48 | $1.39 \times 10^{6}$ | $6.36 \times 10^{4}$ |
| Au | $5.90 \times 10^{28}$ | 5.53 | $1.39 \times 10^{6}$ | $6.41 \times 10^{4}$ |

**EXAMPLE 43.3 The Fermi Energy of Gold**
Each atom of gold contributes one free electron to the metal. Compute (a) the Fermi energy, (b) the Fermi velocity, and (c) the Fermi temperature for gold.

*Solution* (a) The concentration of free electrons in gold is $5.90 \times 10^{28}\ \text{m}^{-3}$ (see Table 43.4). Substitution of this value into Equation 43.24 gives

$$E_F = \frac{h^2}{2m}\left(\frac{3n}{8\pi}\right)^{2/3}$$

$$= \frac{(6.626 \times 10^{-34}\ \text{J}\cdot\text{s})^2}{2(9.11 \times 10^{-31}\ \text{kg})}\left(\frac{3 \times 5.90 \times 10^{28}\ \text{m}^{-3}}{8\pi}\right)^{2/3}$$

$$= 8.85 \times 10^{-19}\ \text{J} = \boxed{5.53\ \text{eV}}$$

(b) The Fermi velocity is defined by the expression $\frac{1}{2}mv_F{}^2 = E_F$. Solving this for $v_F$ gives

$$v_F = \left(\frac{2E_F}{m}\right)^{1/2} = \left(\frac{2 \times 5.85 \times 10^{-19}\ \text{J}}{9.11 \times 10^{-31}\ \text{kg}}\right)^{1/2}$$

$$= \boxed{1.39 \times 10^6\ \text{m/s}}$$

(c) The Fermi temperature is defined by Equation 43.26:

$$T_F = \frac{E_F}{k} = \frac{8.85 \times 10^{-19}\ \text{J}}{1.38 \times 10^{-23}\ \text{J/K}} = \boxed{6.41 \times 10^4\ \text{K}}$$

Thus, a gas of classical particles would have to be heated to about 64 000 K to have an average energy per particle equal to the Fermi energy at 0 K!

**EXAMPLE 43.4 Excited Electrons in Copper**
Calculate an approximate value for the fraction of electrons that are excited from below $E_F$ to above $E_F$ when copper is heated from 0 K to 300 K.

*Solution* Only those electrons that are within a range of $kT \approx 0.025$ eV are affected by the change in temperature from 0 K to 300 K. Because the Fermi energy for copper is 7.0 eV, the fraction of electrons excited from below $E_F$ to above $E_F$ is of the order of $kT/E_F$. In this case, $kT/E_F = 0.025/7.0 = 0.0036$ or 0.36%. Thus, we see that only a very small fraction of the electrons are affected. A more precise analysis shows that the fraction excited is given by $9kT/16E_F$.

## 43.6 CONDUCTION IN METALS, INSULATORS, AND SEMICONDUCTORS

In Chapter 27 we found that good conductors contain a high density of charge carriers, whereas the density of charge carriers in insulators is nearly zero. Semiconductors are a class of technologically important materials with charge carrier densities intermediate between those of insulators and those of conductors. In this section, we provide a qualitative discussion of the mechanisms of conduction in these three classes of materials. The enormous variation in electrical conductivity of these materials may be explained in terms of energy bands.

### Metals

In Section 43.4, we described the energy band picture for the ground state of metallic sodium. If energy is added to the system (say in the form of heat), electrons will be able to move from filled states to one of many empty states. For example, electrons in the half-filled 3*s* band could absorb enough energy to move to the empty 3*p* band.

We can obtain a better understanding of the properties of metals by considering a half-filled band such as the 3*s* band of sodium. Figure 43.19 shows a half-filled band of a metal at $T = 0$ K, where the shaded region represents levels which are filled with electrons. Since electrons obey Fermi-Dirac statistics, all levels below the Fermi energy, $E_F$, are filled with electrons, while all levels above $E_F$ are empty. In the case of sodium, the Fermi energy lies in

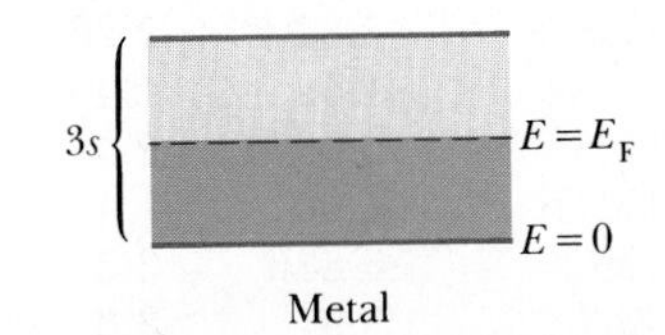


**Figure 43.19** A half-filled band of a conductor such as the 3*s* band of sodium. At $T = 0$ K, the Fermi energy lies in the middle of the band.

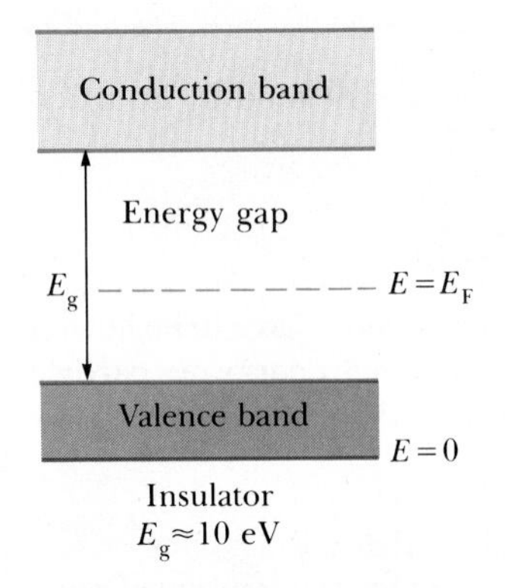


**Figure 43.20** An insulator at $T = 0$ K has a filled valence band and an empty conduction band. The Fermi level lies somewhere between these bands.

the middle of the band. At temperatures greater than 0 K, a few electrons are thermally excited to levels above $E_F$, but overall there is little change from the 0 K case. *However, if an electric field is applied to the metal, electrons with energies near the Fermi energy require only a small amount of additional energy from the field to reach nearby empty energy states.* Thus electrons are free to move with only a small applied field in a metal because there are many unoccupied states close to occupied energy states.

### Insulators

Now consider the two highest energy bands of a material having the lower band completely filled with electrons and the higher completely empty at 0 K (see Fig. 43.20). It is common to refer to the separation between the outermost filled and empty bands as the *energy gap*, $E_g$, of the material. The energy gap for an insulator is large, of the order of 10 eV. The lower band filled with electrons is called the *valence band*, and the upper empty band is the *conduction band*. The Fermi energy lies somewhere in the energy gap, as shown in Figure 43.20. Since the energy gap for an insulator is large ($\sim 10$ eV) compared to $kT$ at room temperature ($kT = 0.025$ eV at 300 K), the Fermi-Dirac distribution predicts that there will be very few electrons thermally excited into the upper band at normal temperatures. *Thus, although an insulator has many vacant states in the conduction band that can accept electrons, there are so few electrons actually occupying conduction band states that the overall contribution to electrical conductivity is very small, resulting in a high resistivity for insulators.*

### Semiconductors

Materials that have an energy gap of the order of 1 eV are called semiconductors. Table 43.5 shows the energy gaps for some representative materials. At $T = 0$ K, all electrons are in the valence band and there are no electrons in the conduction band. Thus semiconductors are *poor* conductors at low temperatures. At ordinary temperatures, however, the situation is quite different. The band structure of a semiconductor can be represented by the diagram shown in Figure 43.21. Since the Fermi level, $E_F$, is located at about the middle of the gap for a semiconductor, and since $E_g$ is small, appreciable numbers of electrons are thermally excited from the valence band to the conduction band. Since there are many empty nearby states in the conduction band, a small applied potential can easily raise the energy of the electrons in the conduction band, resulting in a moderate current. Because thermal excitation across the narrow gap is more probable at higher temperatures, the conductivity of semiconductors depends strongly on temperature and *increases* rapidly with temperature. This contrasts sharply with the conductivity of a metal, which *decreases slowly* with temperature.

**TABLE 43.5 Energy Gap Values for Some Semiconductors***

| Crystal | $E_g$ (eV) 0 K | $E_g$ (eV) 300 K |
|---|---|---|
| Si | 1.17 | 1.14 |
| Ge | 0.744 | 0.67 |
| InP | 1.42 | 1.35 |
| GaP | 2.32 | 2.26 |
| GaAs | 1.52 | 1.43 |
| CdS | 2.582 | 2.42 |
| CdTe | 1.607 | 1.45 |
| ZnO | 3.436 | 3.2 |
| ZnS | 3.91 | 3.6 |

* These data were taken from C. Kittel, *Introduction to Solid State Physics*, 5th ed., New York, John Wiley & Sons, 1976.

It is important to point out that there are both negative and positive charge carriers in a semiconductor. When an electron moves from the valence band into the conduction band, it leaves behind a vacant crystal site, or so-called **hole**, in the otherwise filled valence band. This hole (electron-deficient site) appears as a positive charge, $+e$. The hole acts as a charge carrier in the sense that a valence electron from a nearby bond can transfer into the hole, thereby filling it and leaving a hole behind in the electron's original place.

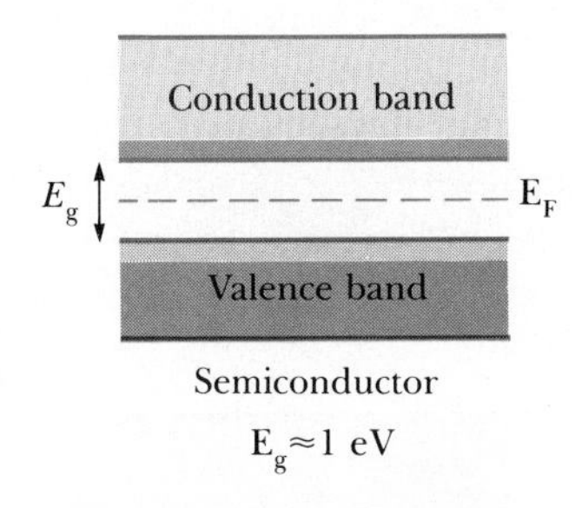


**Figure 43.21** The band structure of a semiconductor at ordinary temperatures ($T \approx 300$ K). Note that the energy gap is much smaller than in an insulator and that many electrons occupy states in the conduction band.

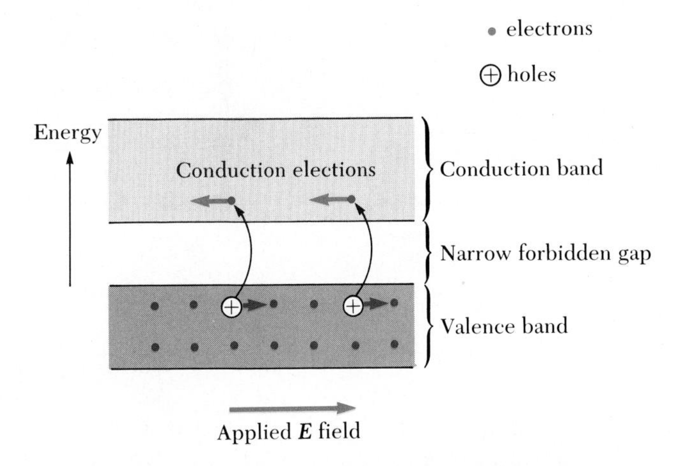


**Figure 43.22** An intrinsic semiconductor. Note that the electrons move toward the left and the holes move toward the right when the applied electric field is to the right as shown.

Thus the hole migrates through the material. In a pure crystal containing only one element or compound, there are equal numbers of conduction electrons and holes. Such combinations of charges are called electron-hole pairs, and a pure semiconductor that contains such pairs is called an *intrinsic semiconductor* (see Fig. 43.22). In the presence of an electric field, the holes move in the direction of the field and the conduction electrons move opposite the field.

## Doped Semiconductors

When impurities are added to semiconductors, their band structure and resistivities are modified. The process of adding impurities, called **doping,** is important in making devices and fabricating semiconductors with well-defined regions of different conductivity. For example, when an atom with five valence electrons, such as arsenic, is added to a semiconductor, four valence electrons participate in the covalent bonds and one electron is left over (Fig. 43.23a). This extra electron is nearly free and has an energy level in the band diagram that lies within the energy gap, just below the conduction band (Fig. 43.23b). Such a pentavalent atom in effect donates an electron to the structure and hence is referred to as a **donor atom.** Since the energy spacings between

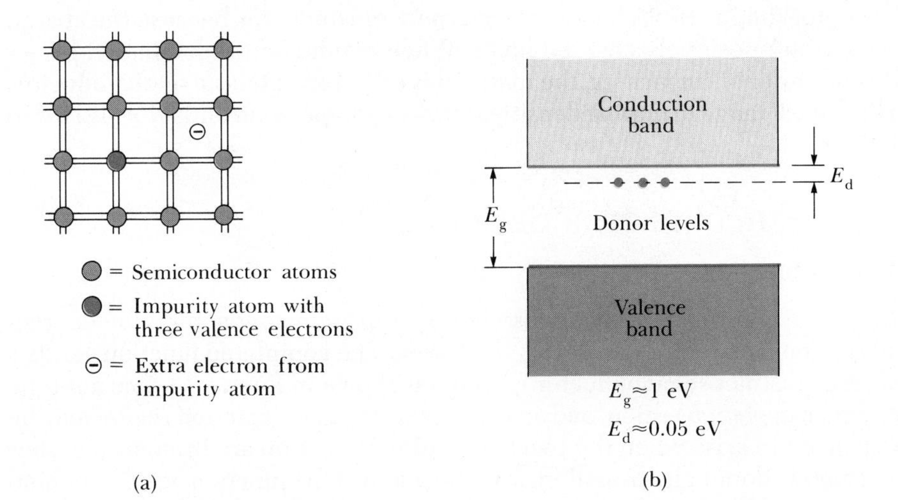


**Figure 43.23** (a) Two-dimensional representation of a semiconductor containing a donor atom (red spot). (b) Energy band diagram of a semiconductor in which the donor levels lie within the forbidden gap, just below the bottom of the conduction band.

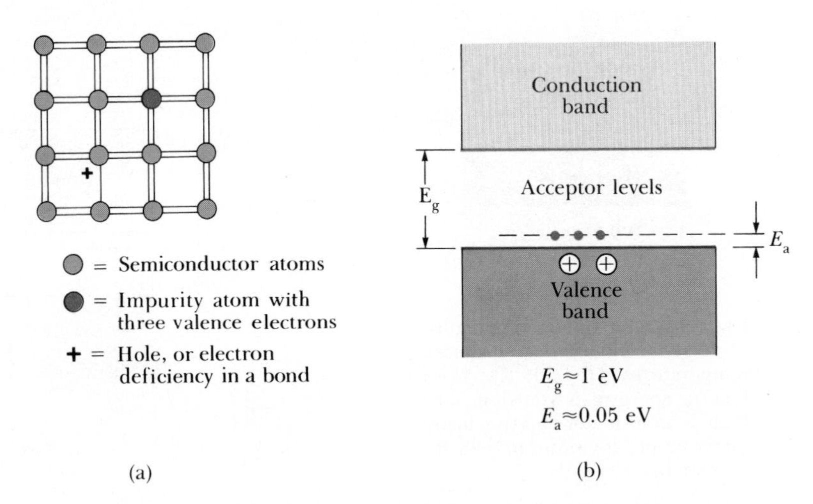


**Figure 43.24** (a) Two-dimensional representation of a semiconductor containing an acceptor atom (red spot). (b) Energy band diagram of a semiconductor in which the acceptor levels lie within the forbidden gap, just above the top of the valence band.

the donor levels and the bottom of the conduction band are very small (typically, about 0.05 eV), a small amount of thermal energy will cause an electron in these levels to move into the conduction band. (Recall that the average thermal energy of an electron at room temperature is about $kT \approx 0.026$ eV.) Semiconductors doped with donor atoms are called ***n*-type semiconductors** because the majority charge carriers are electrons, whose charge is negative.

If the semiconductor is doped with atoms with three valence electrons, such as indium and aluminum, the three electrons form covalent bonds with neighboring atoms, leaving an electron deficiency, or hole, in the fourth bond (Fig. 43.24a). The energy levels of such impurities also lie within the energy gap, just above the valence band, as indicated in Figure 43.24b. Electrons from the valence band have enough thermal energy at room temperature to fill these impurity levels, leaving behind a hole in the valence band. Because a trivalent atom in effect accepts an electron from the valence band, such impurities are referred to as **acceptors.** A semiconductor doped with trivalent (acceptor) impurities is known as a ***p*-type semiconductor** because the charge carriers are positively charged holes. When conduction is dominated by acceptor or donor impurities, the material is called an **extrinsic semiconductor.** The typical range of doping densities for *n*- or *p*-type semiconductors is $10^{13}$ to $10^{19}$ cm$^{-3}$.

## *43.7 SEMICONDUCTOR DEVICES

### The *p-n* Junction

Now let us consider what happens when a *p*-type semiconductor is joined to an *n*-type semiconductor to form a *p-n junction.* The completed junction consists of three distinct semiconductor regions, as shown in Figure 43.25a: a *p*-type region, a *depletion* region, and an *n*-type region. The depletion region may be visualized to arise when the two halves of the junction are brought together and mobile donor electrons diffuse to the *p* side of the junction, leaving behind

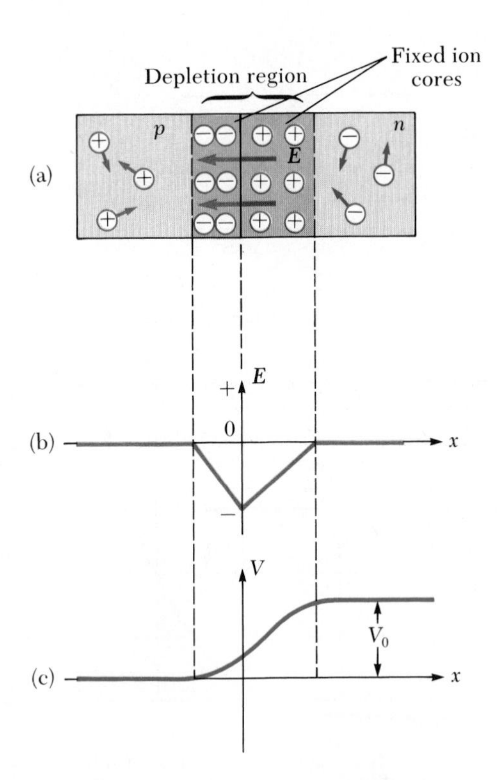


**Figure 43.25** (a) Physical arrangement of a *p-n* junction. (b) Built-in electric field versus *x* for the *p-n* junction. (c) Built-in potential versus *x* for the *p-n* junction.

immobile positive ion cores. (Conversely, holes diffuse to the *n* side and leave a region of fixed negative ion cores.) The region extending several microns from the junction is called the *depletion region* because it is depleted of *mobile* charge carriers. It also contains a built-in electric field on the order of $10^4$ to $10^6$ V/cm, which serves to sweep mobile charge out of this region and keep it truly depleted (Fig. 43.25b). This internal electric field creates a potential barrier $V_0$ (Fig. 43.25c) that prevents the further diffusion of holes and electrons across the junction and insures zero current through the junction when no external voltage is applied.

Perhaps the most notable feature of the *p-n* junction is its ability to pass current in only one direction. Such *diode* action is easiest to understand in terms of the potential diagram in Figure 43.25c. If a positive external voltage is applied to the *p* side of the junction, the overall barrier is decreased, resulting in a current that increases exponentially with increasing forward voltage or bias. For reverse bias (a positive external voltage applied to the *n* side of the junction), the potential barrier is increased, resulting in a very small reverse current that quickly reaches a saturation value, $I_0$, with increasing reverse bias. The current-voltage relation for an ideal diode is given by

$$I = I_0(e^{qV/kT} - 1) \tag{43.28}$$

where $q$ is the electronic charge, $k$ is Boltzmann's constant, and $T$ is the temperature in kelvins. Figure 43.26 shows a plot of an *I-V* characteristic for a real diode along with a schematic of a diode under forward bias.

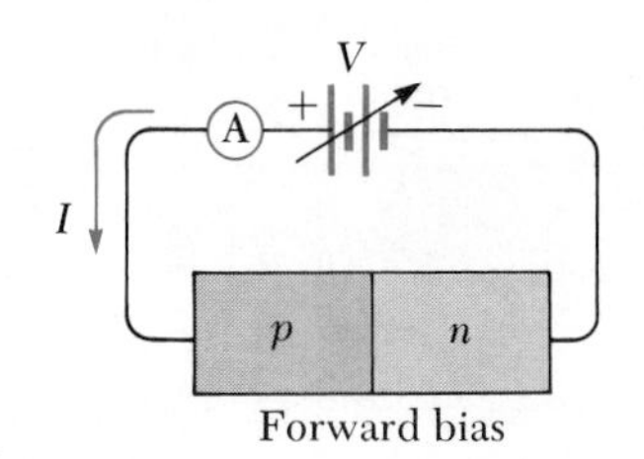


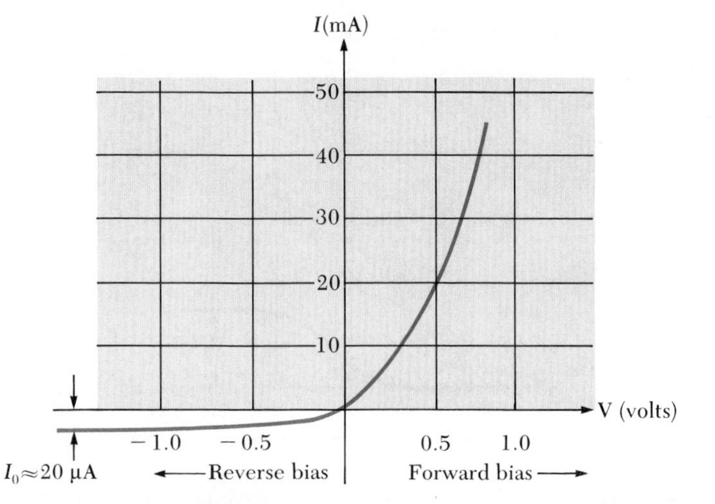


**Figure 43.26** The characteristic curve for a real diode.

(a)

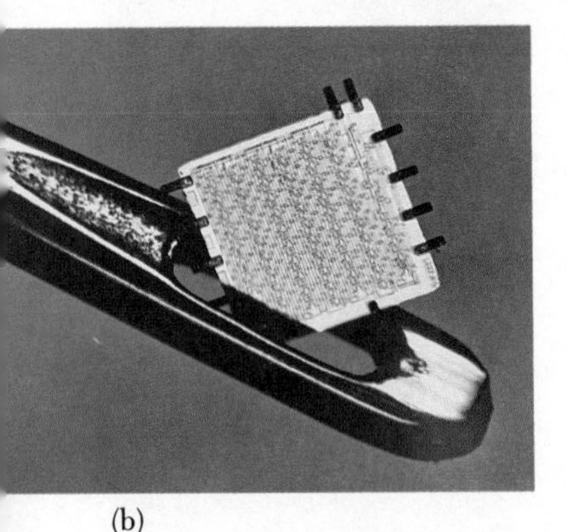

(b)

(a) A light-emitting diode (LED). (© C. Falco, Science Source/Photo Researchers) (b) The microprocessor chip shown here contains almost 150 000 transistors. Note that the chip is small enough to fit through the eye of a needle. (Courtesy of AT&T Archives)

### The Junction Transistor

The discovery of the transistor by John Bardeen, Walter Brattain, and William Shockley in 1948 totally revolutionized the world of electronics. For this work, these three men shared a Nobel prize in 1956. By 1960, the transistor had replaced the vacuum tube in many electronic applications. The advent of the transistor created a multibillion dollar industry that produces such popular devices as pocket radios, handheld calculators, computers, television receivers, and electronic games.

The junction transistor consists of a semiconducting material with a very narrow *n* region sandwiched between two *p* regions. This configuration is called the ***pnp*** **transistor.** Another configuration is the ***npn*** **transistor,** which consists of a *p* region sandwiched between two *n* regions. Because the operation of the two transistors is essentially the same, we shall describe only the *pnp* transistor.

The structure of the *pnp* transistor, together with its circuit symbol, is shown in Figure 43.27. The outer regions of the transistor are called the **emitter** and **collector,** and the narrow central region is called the **base.** The configuration contains two junctions: one junction is the interface between the emitter and the base, and the other is between the base and the collector.

Suppose a voltage is applied to the transistor such that the emitter is at a higher potential than the collector. (This is accomplished with battery $V_{ec}$ in Fig. 43.28.) If we think of the transistor as two diodes back to back, we see that the emitter-base junction is forward-biased and the base-collector junction is reverse-biased.

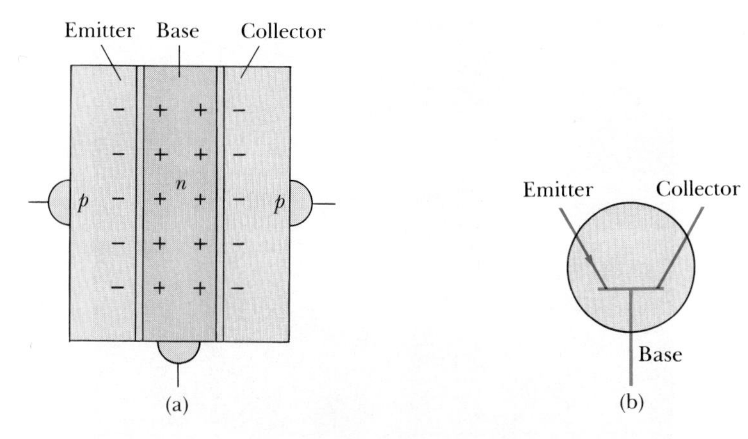


**Figure 43.27** (a) The *pnp* transistor consists of an *n* region (base) sandwiched between two *p* regions (the emitter and the collector). (b) The circuit symbol for the *pnp* transistor.

Because the *p*-type emitter is heavily doped relative to the base, nearly all of the current consists of holes moving across the emitter-base junction. Most of these holes do not recombine in the base because it is very narrow. The holes are finally accelerated across the reverse-biased base-collector junction, producing the current $I_e$ in Figure 43.28.

Although only a small percentage of holes recombine in the base, those that do limit the emitter current to a small value because positive charge carriers accumulate in the base and prevent holes from flowing into this region. In order to prevent this limitation of current, some of the positive charge on the base must be drawn off; this is accomplished by connecting the base to a second battery, $V_{eb}$ in Figure 43.28. Those positive charges that are not swept across the collector-base junction leave the base through this added pathway. This base current, $I_b$, is very small, but a small change in it can significantly change the collector current, $I_c$. If the transistor is properly biased, the collector (output) current will be directly proportional to the base (input) current and the transistor will act as a current amplifier. This condition may be written

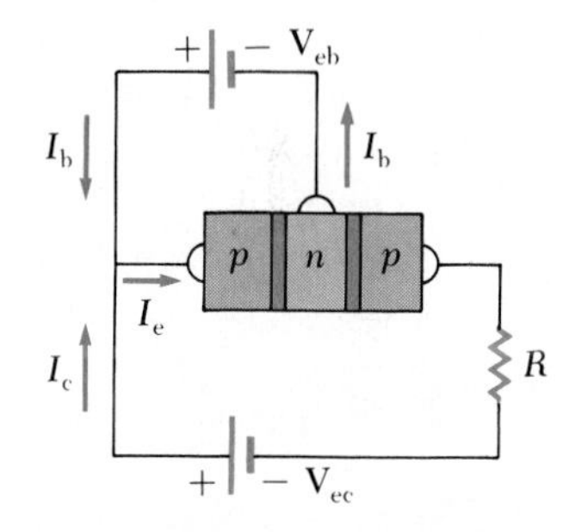


**Figure 43.28** A bias voltage $V_{eb}$ applied to the base as shown produces a small base current, $I_b$, which is used to control the collector current, $I_c$.

$$I_c = \beta I_b$$

where $\beta$, the current gain, is typically in the range from 10 to 100. Thus, the transistor may be used to amplify a small, time-varying signal. The small voltage to be amplified is placed in series with the battery $V_{eb}$ as shown in Figure 43.28. The time-dependent input signal produces a small variation in the base current. This results in a large change in collector current and hence a large change in voltage across the output resistor.

### The Integrated Circuit

The integrated circuit, invented independently by Jack Kilby at Texas Instruments in late 1958 and by Robert Noyce at Fairchild Camera and Instrument in early 1959, has been justly called "the most remarkable technology ever to hit mankind." Jack Kilby's first device is shown in Figure 43.29. ICs have indeed started a second industrial revolution and are found at the heart of computers, watches, cameras, automobiles, aircraft, robots, space vehicles, and all sorts of communication and switching networks. In simplest terms, an integrated circuit is a collection of interconnected transistors, diodes, resis-

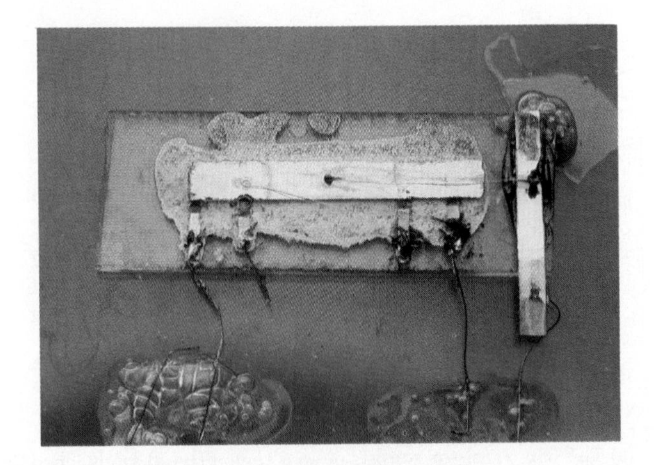

**Figure 43.29** Jack Kilby's first integrated circuit, tested on September 12, 1958. (Courtesy of Texas Instruments, Inc.)

tors, and capacitors fabricated on a single piece of silicon. State of the art chips easily contain several hundred thousand components in a 1 $cm^2$ area (Fig. 43.30).

It is interesting that integrated circuits were invented partly in an effort to achieve circuit miniaturization and partly in order to solve the interconnection problem spawned by the transistor. In the era of vacuum tubes, power and size considerations of individual components set modest limits on the number of components that could be interconnected in a given circuit. With the advent of the tiny, low-power, highly reliable transistor, design limits on the number of components disappeared and were replaced with the problem of wiring together hundreds of thousands of components. The magnitude of this problem can be appreciated when one considers that second generation computers (consisting of discrete transistors) contained several hundred thousand components requiring more than a million hand-soldered joints to be made and tested.

In addition to solving the interconnection problem, ICs possess the advantages of miniaturization and fast response, two attributes critical for high speed computers. The fast response is actually also a product of the miniaturization and close packing of components. This is so because the response time of a circuit depends on the time it takes for electrical signals traveling at about 1 ft/ns to pass from one component to another. This time is clearly reduced by the closer packing of components.

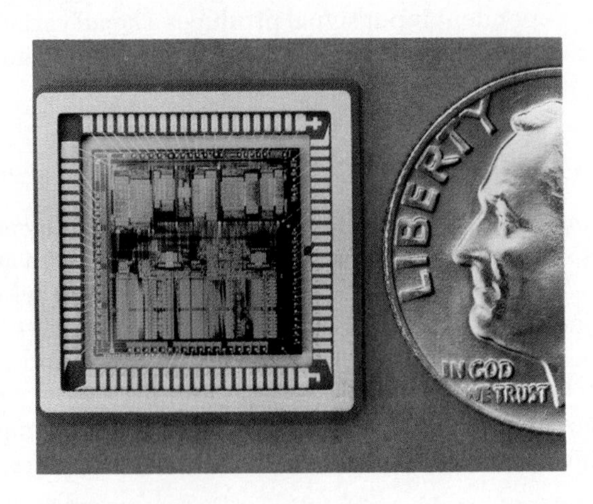


**Figure 43.30** A 32-bit microprocessor integrated circuit containing over 200 000 components is about the size of a dime. (Courtesy of A.T.&T. Bell Labs)

## SUMMARY

Two or more atoms combine to form molecules because of a net attractive force between the constituent atoms. The mechanisms responsible for the bonding in the molecule can be classified as follows:

Molecular bonds

1. **Ionic bonds** Certain molecules form ionic bonds primarily because of the Coulomb attraction between oppositely charged ions. Sodium chloride (NaCl) is one example of an ionically bonded molecule.
2. **Covalent bonds** The covalent bond in a molecule is formed by the sharing of valence electrons of its constituent atoms. For example, the two electrons of the $H_2$ molecule are equally shared between the nuclei.
3. **Hydrogen bonds** This type of bonding corresponds to the attraction of two negative ions by an intermediate hydrogen atom (a proton).
4. **Van der Waals bonds** This is a weak electrostatic bond between atoms that do not form ionic or covalent bonds. It is responsible for the condensation of inert gas atoms and nonpolar molecules into the liquid phase.

The energy of a molecule in a gas consists of contributions from the electrical interactions, the translation of the molecule, rotations, and vibrations. The **allowed values of the rotational energy** of a diatomic molecule are given by

Allowed values of rotational energy

$$E_{\text{rot}} = \frac{\hbar^2}{2I} J(J+1) \qquad J = 0, 1, 2, \ldots \tag{43.5}$$

where $I$ is the *moment of inertia* of the molecule and $J$ is an integer called the *rotational quantum number.* The selection rule for transitions between rotational levels (which lie in the microwave range of frequencies) is given by $\Delta J = \pm 1$.

The **allowed values of the vibrational energy** of a diatomic molecule are given by

Allowed values of vibrational energy

$$E_{\text{vib}} = (v + \tfrac{1}{2}) \frac{h}{2\pi} \sqrt{\frac{K}{\mu}} \qquad v = 0, 1, 2, \ldots \tag{43.10}$$

where $v$ is the *vibrational quantum number,* $K$ is the *force constant* of the effective spring bonding the molecule, and $\mu$ is the *reduced mass of the molecule.* The selection rule for allowed vibrational transitions (which lie in the infrared region) is $\Delta v = \pm 1$, and the separation between adjacent levels is equal.

Bonding mechanisms in solids

**Bonding mechanisms** in solids can be classified in a manner similar to the schemes already described for molecules. For example, the $Na^+$ and $Cl^-$ ions in NaCl form *ionic bonds,* while the carbon atoms in diamond form *covalent bonds.* Another type of bonding mechanism is the *metallic bond,* which is generally weaker than ionic or covalent bonds. The mechanism for the metallic bond in a metal is a net attractive force between the positive ion cores and the mobile valence electrons.

Band theory of solids

In a crystalline solid, the energy levels of the system form a set of bands. Electrons are allowed to occupy only certain bands. There are energy regions between the allowed bands corresponding to states which the electrons cannot occupy.

One can best understand the properties of metals, insulators, and semiconductors in terms of the **band theory of solids.** The valence band of a metal such as sodium is half-filled. Therefore, there are many electrons free to move throughout the metal and contribute to the conduction current. In an insulator at $T = 0$ K, the valence band is completely filled with electrons, while the conduction band is empty. The region between the valence band and the conduction band is called the **energy gap** of the material. The energy gap for an insulator is of the order of 10 eV. Because this gap is large compared to $kT$ at ordinary temperatures, very few electrons are thermally excited into the conduction band, which explains the small electrical conductivity of an insulator.

In the free-electron theory of metals, the valence electrons fill the quantized levels in accordance with the Pauli exclusion principle. The **number of states** per unit volume available to the conduction electrons with energies between $E$ and $E + dE$ is

Particle distribution in a metal

$$N(E)\,dE = C\frac{E^{1/2}\,dE}{e^{(E-E_F)/kT} + 1} \qquad (43.21)$$

where $C$ is a constant, and $E_F$ is the Fermi energy. At $T = 0$ K, all levels below $E_F$ are filled, while all levels above $E_F$ are empty.

The **Fermi energy** of a metal at $T = 0$ K is given by

Fermi energy at $T = 0$ K

$$E_F = \frac{h^2}{2m}\left(\frac{3n}{8\pi}\right)^{2/3} \qquad (43.24)$$

where $n$ is the total number of conduction electrons per unit volume. Only those electrons having energies near $E_F$ can contribute to the electrical conductivity of the metal.

Semiconductors

A **semiconductor** is a material with a small energy gap, of the order of 1 eV, and a valence band which is filled at $T = 0$ K. Because of their small energy gap, a significant number of electrons can be thermally excited from the valence band into the conduction band as the temperature increases. The band structures and electrical properties of a semiconductor can be modified by adding donor atoms with five valence electrons (such as arsenic), or by adding acceptor atoms with three valence electrons (such as indium). A semiconductor **doped** with donor impurity atoms is called an ***n*-type semiconductor,** while one doped with acceptor impurity atoms is called ***p*-type.** The energy levels of these impurity atoms fall within the energy gap of the material.

## QUESTIONS

1. Discuss the three major forms of excitation of a molecule (other than the translational motion) and the relative energies associated with the three excitations.
2. Explain the role of the Pauli exclusion principle in describing the electrical properties of metals.
3. Discuss the properties of a material that will determine whether it is a good electrical insulator or a good conductor.
4. Table 43.5 shows that the energy gaps for semiconductors decrease with increasing temperature. What do you suppose accounts for this behavior?
5. The resistivity of most metals increases with increasing temperature, whereas the resistivity of an intrinsic semiconductor decreases with increasing temperature. What explains this difference in behavior?

6. Discuss the differences in the band structures of metals, insulators, and semiconductors. How does the band structure model enable you to better understand the electrical properties of these materials?
7. Discuss the mechanisms responsible for the different types of bonds that can occur to form stable molecules.
8. Discuss the electrical, physical, and optical properties of ionically bonded solids.
9. Discuss the electrical and physical properties of covalently bonded solids.
10. Discuss the electrical and physical properties of metals.
11. When a photon is absorbed by a semiconductor, an electron-hole pair is said to be created. Give a physical explanation of this statement using the energy band model as the basis for your description.
12. Pentavalent atoms such as arsenic are donor atoms in a semiconductor such as silicon, while trivalent atoms such as indium are acceptors. Inspect the periodic table in Appendix C, and determine what other elements would be considered as either donors or acceptors.
13. Explain how a *p-n* junction diode operates as a rectifier.
14. What are the essential assumptions made in the free-electron theory of metals? How does the energy band model differ from the free-electron theory in describing the properties of metals?
15. How do the vibrational and rotational levels of heavy hydrogen ($D_2$) molecules compare with those of ordinary hydrogen ($H_2$) molecules?
16. Which is easier to excite in a diatomic molecule, rotational or vibrational motion?
17. The energy of visible light ranges between 1.8 eV and 3.2 eV. Does this explain why silicon with an energy gap of 1.1 eV (Table 43.5) appears black whereas diamond with an energy gap of 5.5 eV appears transparent?
18. Why is a *pnp* or *npn* "sandwich" (with the central region very thin) essential to the operation of a transistor?
19. How can the analysis of the rotational spectrum of a molecule lead to an estimate of the size of that molecule?

## PROBLEMS

### Section 43.1 Molecular Bonds

1. The separation between the $K^+$ and $Cl^-$ ions in a KCl molecule is $2.8 \times 10^{-10}$ m. Assuming the two ions act like point charges, (a) determine the force of attraction between them, and (b) determine the potential energy of attraction in eV.
2. A reasonable description of the potential energy between two atoms in a molecule is given by the *Lenard-Jones potential*

$$U = \frac{A}{r^{12}} - \frac{B}{r^6}$$

where $A$ and $B$ are constants. (a) Find the value $r_0$ at which the energy is a minimum in terms of $A$ and $B$. (b) How much energy $E$ is required to break up a diatomic molecule in terms of $A$ and $B$? (c) Evaluate $r_0$ in m and $E$ in eV for the $H_2$ molecule. In your calculations, let $A = 0.124 \times 10^{-120}$ eV·m$^{12}$ and $B = 1.488 \times 10^{-60}$ eV·m$^6$.

### Section 43.2 The Energy and Spectra of Molecules

3. Use the data in Table 43.2 to calculate the reduced mass of the NO molecule. Then compute a value for $\mu$ using Equation 43.8. Compare the two results.
4. The CO molecule makes a transition from the $J = 1$ to $J = 2$ rotational state when it absorbs a photon of frequency $2.30 \times 10^{11}$ Hz. Find the moment of inertia of the CO molecule.
5. Use the data in Table 43.2 to calculate the minimum amplitude of vibration for (a) the HI molecule and (b) the HF molecule. Which molecule has the weaker bond?
6. The nuclei of the diatomic oxygen molecule $O_2$ are separated by $1.2 \times 10^{-10}$ m. The mass of each oxygen atom in the molecule is $2.66 \times 10^{-26}$ kg. (a) Determine the rotational energies of an oxygen molecule in electron volts for the levels corresponding to $J = 0$, 1, and 2. (b) The effective force constant $K$ between the atoms in the oxygen molecule is 50 N/m. Determine the vibrational energies of the molecule in electron volts corresponding to $v = 0$, 1, and 2.
7. Figure 43.31 is a model of a benzene molecule. All atoms lie in a plane, and the carbon atoms form a regular hexagon, as do the hydrogen atoms. The carbon atoms are 0.110 nm apart center to center. Determine the allowed energies of rotation about an axis perpendicular to the plane of the paper through the center point $O$. Hydrogen and carbon atoms have masses of $1.67 \times 10^{-27}$ kg and $1.99 \times 10^{-26}$ kg, respectively.

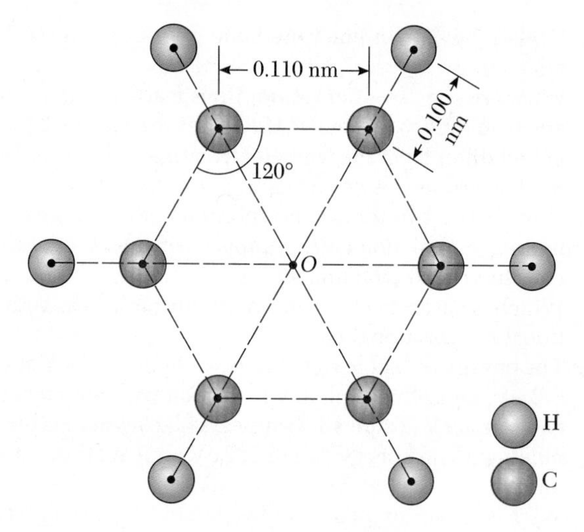


**Figure 43.31** (Problem 7).

8. The HF molecule has a bond length of 0.092 nm. (a) Calculate the reduced mass of the molecule. (b) Plot the potential energy versus internuclear separation in the vicinity of $r = 0.092$ nm.
9. The HCl molecule is excited to its first rotational energy level, corresponding to $J = 1$. If the distance between its nuclei is 0.1275 nm, what is the angular velocity of the molecule about its center of mass?
10. The distance between the protons in the $H_2$ molecule is $r = 0.75 \times 10^{-10}$ m. (a) Find the energy of the first rotational state $J = 1$, and (b) the wavelength of radiation emitted in the transition $J = 1$ to $J = 0$.
11. If the force constant of a vibrating HCl molecule is $K = 480$ N/m, estimate the energy difference between the ground state and the first vibrational level.
12. The rotational spectrum of the HCl molecule contains the following wavelengths: 0.0604 mm, 0.0690 mm, 0.0804 mm, 0.0964 mm, 0.1204 mm. What is the moment of inertia of the HCl molecule?
13. In the oxygen molecule $O_2$, the separation between the atoms is $1.2 \times 10^{-10}$ m. Treating the oxygen atoms as particles, determine the allowed values of the rotational energy for the $J = 1$ and $J = 2$ states.

## Section 43.3 Bonding in Solids

14. Calculate the ionic cohesive energy for NaCl using Equation 43.16. Take $\alpha = 1.7476$, $r_0 = 0.281$ nm, and $m = 8$.
15. Find the ionic cohesive energy for LiCl, which has the same crystal structure as NaCl. Take $r_0 = 0.257$ nm, and $m = 7$.
16. Consider a one-dimensional chain of alternating positive and negative ions. Show that the potential energy of an ion in this hypothetical crystal is

$$U(r) = -k\alpha \frac{e^2}{r}$$

where $\alpha = 2 \ln 2$ (the Madelung constant), and $r$ is the interionic spacing. [*Hint:* Make use of the series expansion for $\ln(1 + x)$.]

## Section 43.4 Band Theory of Solids and Section 43.5 Free-Electron Theory of Metals

17. Show that Equation 43.24 can be expressed as

$$E_F = (3.65 \times 10^{-19}) n^{2/3} \text{ eV}$$

where $E_F$ is in eV when $n$ is in electrons/m$^3$.
18. Calculate the probability that a conduction electron in copper at 300 K has an energy equal to 99% of the Fermi energy.
19. Find the probability that a conduction electron in a metal has an energy equal to the Fermi energy at the temperature 300 K.
20. Sodium is a monovalent metal having a density of 0.971 g/cm$^3$ and an atomic weight of 23.0 g/mol. Use this information to calculate (a) the density of charge carriers, (b) the Fermi energy, and (c) the Fermi velocity for sodium.
21. Calculate the energy of a conduction electron in silver at 800 K if the probability of finding the electron in that state is 0.95. Assume that the Fermi energy for silver is 5.48 eV at this temperature.
22. Show that the averge kinetic energy of a conduction electron in a metal at 0 K is given by

$$E_{av} = \tfrac{3}{5} E_F$$

(*Hint:* In general, the average kinetic energy is given by $E_{av} = \frac{1}{n} \int E\, N(E)\, dE$, where $n$ is the density of particles, and $N(E)dE$ is given by Equation 43.21.)
23. Consider a cube of gold 1 mm on an edge. Calculate the approximate number of conduction electrons in this cube whose energies lie in the range 4.000 eV to 4.025 eV.
24. An electron moves in a three-dimensional box of edge length $L$ and volume $L^3$. Show that the energy of the particle whose wave function is given by the expression $\psi = A \sin(k_x x) \sin(k_y y) \sin(k_z z)$ is

$$E = \frac{\hbar^2 \pi^2}{2mL^2} (n_x^2 + n_y^2 + n_z^2)$$

where the quantum numbers $(n_x, n_y, n_z)$ are $\geqq 1$. (*Hint:* The Schrödinger equation in three dimensions may be written

$$\frac{d^2\psi}{dx^2} + \frac{d^2\psi}{dy^2} + \frac{d^2\psi}{dz^2} = \frac{\hbar^2}{2m} (U - E)\psi$$

To confine the electron inside the box, we require $U = 0$ inside and $U = \infty$ outside the box).

25. (a) Consider a system of electrons confined to a three-dimensional box. Calculate the ratio of the number of allowed energy levels at 8.5 eV to the number of allowed energy levels at 7.0 eV. (b) Copper has a Fermi energy of 7.0 eV at 300 K. Calculate the ratio of the number of occupied levels at an energy of 8.5 eV to the number of occupied levels at the Fermi energy. Compare your answer with that obtained in part (a).

### Section 43.6 Conduction in Metals, Insulators, and Semiconductors

26. The energy gap for Si at 300 K is 1.14 eV. (a) Find the lowest frequency photon that will promote an electron from the valence band to the conduction band of silicon. (b) What is the wavelength of this photon?
27. From the optical absorption spectrum of a certain semiconductor, one finds that the longest wavelength radiation absorbed is 1.85 $\mu$m. Calculate the energy gap for this semiconductor.
28. A light-emitting diode (L.E.D.) made of the semiconductor GaAsP gives off red light ($\lambda = 650$ nm). Determine the band energy gap $E_g$ in the semiconductor.
29. Most solar radiation has a wavelength of $10^{-6}$ m or less. What is the energy gap that a solar cell material should have to be able to absorb this radiation? Is silicon appropriate (see Table 43.5)?

### *Section 43.7 Semiconductor Devices

30. (a) For what value of the bias voltage in Equation 43.28 does $I = 9I_0$? (b) For what value of $V$ does $I = -0.9I_0$? Assume $T = 300$ K.

## ADDITIONAL PROBLEMS

31. The hydrogen molecule comes apart (disassociates) when it is excited internally by 4.5 eV. Assuming that this molecule behaves like a harmonic oscillator with classical angular frequency $\omega = 8.28 \times 10^{14}$ rad/s, find the highest vibrational quantum number beneath the 4.5 eV dissociation energy.
32. The force holding the atoms together in a nitrogen molecule ($N_2$) has a spring constant of 52.0 N/m. The nitrogen atoms each have a mass of $2.32 \times 10^{-26}$ kg, and their nuclei are 0.12 nm apart. The molecule is excited to the second vibrational level and subsequently decays to the first vibrational level by cascading through the rotational states in between. How many photons are emitted?
33. The Fermi energy of copper at 300 K is 7.05 eV. (a) What is the average energy of a conduction electron in copper at 300 K? (b) At what temperature would the average energy of a molecule of an ideal gas equal the energy obtained in (a)?
34. Helium forms a solid as each helium atom forms a bond with four other atoms, and each of these bonds has an average energy of $1.74 \times 10^{-23}$ J. Find the latent heat of fusion for helium in J/g. The atomic weight of He is 4 g/mol and one mole contains $6.023 \times 10^{23}$ atoms.
35. Show that the ionic cohesive energy of an ionically bonded solid is given by Equation 43.16. (*Hint:* Start with Equation 43.15, and note that $dU/dr = 0$ at $r = r_0$.)
36. A particle of mass $m$ moves in one-dimensional motion in a region where its potential energy is given by

$$U(x) = \frac{A}{x^3} - \frac{B}{x} \qquad \text{(A and B are constants with appropriate units)}$$

The general shape of this function is shown in Figure 43.12, where $x$ replaces $r$. (a) Find the static equilibrium position $x_0$ of the particle in terms of $m$, $A$, and $B$. (b) Determine the depth $U_0$ of this potential well. (c) In moving along the $x$ axis, what maximum force toward the *negative x* direction does the particle experience?

37. The Madelung constant may be found by summing an infinite alternating series of terms giving the electrostatic potential energy between an $Na^+$ ion and its six nearest $Cl^-$ neighbors, its twelve next-nearest $Na^+$ neighbors, etc. (see Fig. 43.11a). (a) From this expression, show that the first *three* terms of the infinite series for the Madelung constant for the NaCl structure yield $\alpha = 2.13$. (b) Does this infinite series converge rapidly? Calculate the fourth term as a check.

## CALCULATOR/COMPUTER PROBLEMS

38. Write a program that will enable you to calculate the Fermi-Dirac distribution function, Equation 43.17, versus energy. Use your program to plot $f(E)$ versus $E$ for (a) $T = 0.2T_F$, and (b) $T = 0.5T_F$, where $T_F$ is the Fermi temperature, defined by Equation 43.26.
39. Copper has a Fermi energy of 7.05 eV and a conduction electron concentration of $8.49 \times 10^{28}$ m$^{-3}$ at 300 K. Write programs that will enable you to plot (a) the density of states, $g(E)$, versus $E$, (b) the particle distribution function, $N(E)$, as a function of energy at $T = 0$ K and (c) the particle distribution function versus $E$ at $T = 1000$ K. Your energy scales should range from $E = 0$ to $E = 10$ eV.

# ESSAY

## Photovoltaic Conversion

**John D. Meakin**
*Department of Mechanical Engineering & Institute of Energy Conversion, University of Delaware*

The photovoltaic effect occurs when light is absorbed by a semiconductor. The energy of the photons is transferred to electrons in the valence band of the semiconductor, promoting them into the conduction band and resulting in the formation of electron-hole pairs. Only photons with energies exceeding the band gap energy of the semiconductor can be effective in this process. If the semiconductor has a small band gap, a large fraction of the incident photons will be able to create electron-hole pairs giving a large current but the voltage generated across the solar cell will be low. With a large band gap semiconductor the voltage generated will be high but the current low. Analysis has shown that to convert sunlight, in which most of the photons have energies between 1 eV and 3 eV, the optimum band gap for maximum power conversion is about 1.5 eV.

In an isolated semiconductor the excited electron would eventually recombine with a hole in the valence band, emitting its excess energy as photons (photoemission) or phonons (heat), and no useful generation of electric energy would take place. To extract the energy of the photo-excited carriers as useable electricity requires the existence of a charge-separating junction such as a *p-n* homojunction or diode. The passage of the excess charge carriers, known as minority carriers (electrons in the *p*-region, holes in the *n*-region), across the junction prevents electron-hole recombination. Thus a photo-generated electron in *p*-type material has a limited lifetime in the presence of the stable population of holes but an unlimited lifetime after crossing the junction into the *n*-type material.

In the traditional crystalline silicon solar cell, the charge-separating region is formed by the diffusion of specific impurities, or dopants, into a wafer of silicon, creating regions of opposite conductivity type. Diffusion of carriers across the transition between the *n* and *p* regions, i.e., the junction, occurs until an equilibrium is established in which the electric field created prevents further diffusion of charges. This internal electric field constitutes the charge-separating barrier which is key to the operation of the solar cell. Electrons created by light in the *p*-type material migrate to the junction and are then swept into the *n*-type region. In an efficient cell this collection occurs before the electrons recombine with the stable population of holes in the *p* region. The energy absorbed from the light is thus converted into electrical energy which can be fed into an external circuit. (The analogous situation exists for holes created on the *n*-type side of the junction.)

Figure 1a illustrates a diode under illumination and shows the production of electron-hole pairs which are then separated by the built-in electric field. The *I-V* behavior of an ideal photo-diode is given by

$$I = I_O\,(e^{qV/kT} - 1) - I_{sc}$$

where $I$ and $V$ are the external current and voltage respectively, $q$ is the electronic charge and $k$ is Boltzmann's constant, $I_{sc}$ is the short circuit current and $I_O$ is the reverse saturation current of the diode. The corresponding *I-V* curve is given in Figure 1b and is seen to be that of a diode but displaced along the current axis by the "light-generated current."

If the outside circuit has zero resistance, a maximum current will flow as shown on the current-voltage plot. As the load resistance rises a voltage will be generated across the cell and the external current will eventually fall to zero at the open circuit voltage. In this situation, electrons accumulate on the *n*-side of the junction (holes on the *p*-side), biasing the junction in the opposite sense to the built-in field. The maximum voltage the cell can develop corresponds to the forward light-generated current exactly matching the reverse current. The maximum power generation occurs where the current-voltage product, *IV*, is largest. Under ideal conditions a single-junction solar cell in direct sunlight should convert about 22% of the incident solar energy into electricity. This figure ignores all the losses which inevitably occur in an actual cell, such as reflection from the front surface and various other electrical and optical

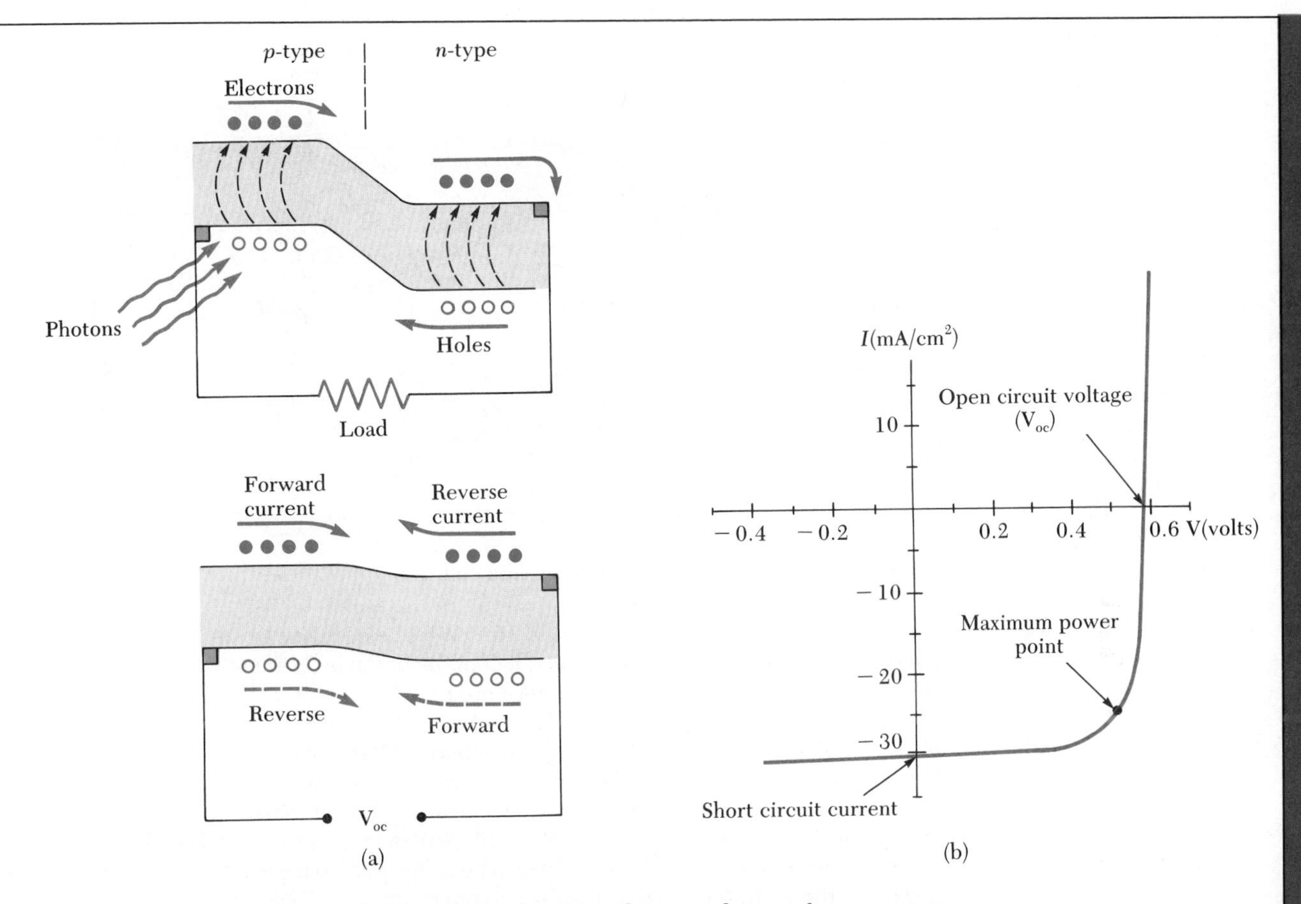


**Figure 1** (a) A *p-n* junction under illumination showing in the upper diagram the generation and collection of electrons and holes near the short circuit current point. As the external load increases, charges accumulate in the cell reducing the height of the charge-separating barrier until at open circuit there is no net current flow. (b) A typical current-voltage curve for a silicon solar cell under 1 kW/m² insolation.

losses. The theoretical efficiency limit is increased by operating the cell under concentrated illumination, which can be achieved using mirrors or lenses. Efficiencies over 30% are possible but the concentrating system must track the sun as it orbits across the sky.

In 1954 the first practical solar cells, with a structure similar to that shown in Figure 2, were made from single crystal wafers of semiconductor-grade silicon. These cells converted about 6% of the total incident sunlight into electrical power, i.e., they had an efficiency of 6%. Research Si cells have now reached over 20% efficiency under direct sunlight and almost 30% under sunlight concentrated by a factor of a few hundred times. Commercial cells are moving towards 20% efficiency yielding multicell modules of about 15% efficiency. Why then are solar cells not being used more extensively to generate electricity? To answer this question we must look more closely at the economics of electricity generation from solar cells.

The most significant costs are the initial capital outlays; operation and maintenance costs are small and there are no fuel costs. A simple analysis reveals the scale of investment that would be economically competitive. A module rated at 1 kW (i.e., able to produce 1 kW of electricity under sunlight of 1 kW/m²) will generate on

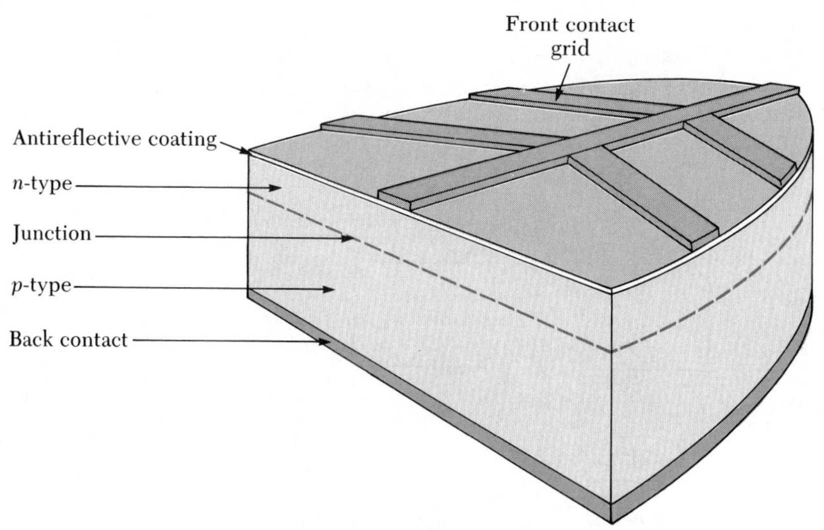


**Figure 2** A schematic drawing of a silicon solar cell.

average about 5 kWh/day or about $5 \times 10^4$ kWh of electricity worth \$1000–3000 over its lifetime of, say, 20 years. About half of the initial investment will be for costs other than the modules, which must therefore be available for a few hundred dollars per kilowatt. More sophisticated analyses taking into account inflation and carrying charges yield similar order of magnitude estimates. Although the first cells produced for space use cost about \$1 million/kW, extensive technical and manufacturing progress, coupled with much less stringent specifications for terrestrial use, has brought the present price down to about \$3000–5000/kW in large quantities. Is it likely that a further order of magnitude reduction can be achieved? To answer this question we must first return to the physics of solar cell operation.

The rate at which sunlight is absorbed by a semiconductor is given by

$$F = e^{-\alpha t}$$

where $F$ is the fraction of radiation absorbed by a thickness $t$ of semiconductor with an absorption coefficient $\alpha$. The absorption coefficient, $\alpha$, measured in $cm^{-1}$, varies with the wavelength of the light. It is very low for photon energies below the semiconductor band gap and rises at higher photon energies. Crystalline Si has a relatively low absorption coefficient, between $10^2$ and $10^4$ $cm^{-1}$, which means that a thickness of about 200 $\mu$m is necessary to absorb most of the sunlight. The carriers generated by the photons must then be able to travel about that far in order to reach the charge-separating junction before recombination occurs.

Efficient carrier collection in crystalline Si requires very pure and perfect material, and bringing the price of Si cells down by another factor of ten cannot be assured, although a number of ongoing developments may yield substantial reductions. These include methods to produce large-area thin sheets directly from a silicon melt to avoid the wafering step used with single crystals. Large-grained polycrystalline material, which is less expensive than single crystals, is also in use. Another promising approach is to texture the front and/or rear surface of the cell so that light undergoes multiple reflections within the cell. This lengthens the light path for a given cell thickness without increasing the distance that carriers must travel and may yield efficient cells as thin as 50 $\mu$m or less.

There are, however, many semiconducting compounds in addition to the semiconducting elements, Si, Ge, and grey Sn. The essential properties for a material to be a candidate for a low-cost solar cell are an appropriate band gap, say between 1.0 and

1.7 eV, and a very high absorption coefficient. The latter makes it possible to use very thin layers, which in turn relaxes the limitations on purity and perfection as short collection distances (short carrier lifetimes) now apply. Any compound is a potential semiconductor if the ratio of valence electrons to atoms, $e/a$, is 4, as is the case for the group IV elements.

Among the classes of semiconducting compounds are the II–VI (i.e., a compound AB where A is from group II and B from group VI), III–V, and I–III–$VI_2$ families. Presently, solar cells are being developed based on members of each of these groups with the expectation that an improved combination of higher efficiency and lower cost will result.

Representative of the III–V compounds are GaAs and InP, which in single crystal form have yielded solar cells of about 30% efficiency. As a general rule the III–V's have excellent carrier properties. There is a major effort to develop them for high speed integrated circuits, but their properties are seriously degraded in thin film form and single crystals may have to be used to maintain performance. If this should continue to be the case, a concentrating system may be essential in which an optical assembly is used to focus the energy from a large aperture onto a much smaller solar cell, thus reducing the total area of solar cell needed for a given output. (See Suggested Readings.)

The II–VI compounds are represented by CdTe, which has an ideal band gap of about 1.5 eV and which has yielded solar cells of about 12% efficiency. A number of techniques for forming large areas of thin film are being developed including electroplating, various types of vapor deposition, and a spray pyrolysis process.

The I–III–$VI_2$ material receiving most attention is $CuInSe_2$. This has a rather low band gap of 1.0 eV but, as we shall discuss shortly, that may be an advantage for two junction or tandem cells. A number of research groups have demonstrated efficiencies of well over 10% and recently 14% has been reported. All investigators have confirmed a very high degree of stability for these cells, which is particularly important for thin film cells that in the past have often shown unacceptably short useable lifetimes.

The reason for interest in all of the above compounds is their high absorptivity which means that very thin layers, as little as 1 $\mu$m, will completely absorb the useful solar spectrum. The carriers generated by the light need then only travel equally short distances to the junction and relatively impure and imperfect materials, such as polycrystalline thin films, can be used successfully.

In parallel with the development of polycrystalline materials, there has been an explosive growth in the investigation, and application, of amorphous materials based on an alloy of Si and H, conventionally designated a-Si:H. Thin films of this material are deposited by creating a plasma in a gas containing the two elements, most frequently $SiH_4$. The resulting solid contains tetrahedrally bonded Si but in a disordered, not a crystalline array. The hydrogen appears to heal any unsatisfied Si–Si bonds, eliminating energy states that would otherwise essentially eliminate the forbidden energy gap. The hydrogenated a-Si can therefore be doped in contrast to hydrogen-free amorphous silicon.

Progress in the science and technology of amorphous thin films has been remarkable and in about a decade a-Si:H solar cells have gone from a laboratory curiosity to a familiar component in solar-powered calculators. Panels with areas of 1 $ft^2$ and larger are now entering the market for battery charging and other uses.

Scientifically and technically, the field of photovoltaic conversion remains exciting and challenging. New materials are being developed and more efficient configurations reduced to practice. Rather than having one cell to harvest the entire solar spectrum, stacked or tandem structures are appearing. In these systems the top cell removes the high energy photons, allowing the longer wavelengths (lower energy photons) to be efficiently harvested in a second or even third cell, creating the

*(Continued)*

photovoltaic equivalent of a multistage turbine. The compound a-Si:H makes an ideal top cell as its band gap of about 1.6 eV efficiently uses the short wavelength light while transmitting the longer wavelengths into a bottom cell. Polycrystalline $CuInSe_2$ is presently the best available material for the bottom cell as its band gap, 1.0 eV, is close to the ideal value for a two-junction device. A total efficiency of about 16% has been achieved by stacking an a-Si:H cell on a $CuInSe_2$/CdS cell. Multijunction a-Si:H cells are also being developed using amorphous Si alloys containing C, Ge, or Sn to give various band gaps.

The present world market for solar cells, about 30 MW, is modest when compared to the installed electrical generating capacity, over 600 GW ($6 \times 10^{11}$ W) in the United States alone. However, production has doubled each year for some time and 10% market penetration by the year 2000 is quite possible. The economic incentive to develop alternative sources of electricity for use in the developed nations is controlled by the cost of nuclear energy and traditional fossil fuels, such as oil, gas, and coal. However, growing concern about the greenhouse effect and acid rain appear likely to also serve as an incentive for more rapid utilization of photovoltaics and other renewable energy resources.

Solar cells are ideally suited for remote locations and third world areas without grid systems, where the only alternatives are diesel generators or expendable batteries. Under such conditions it is already economical to use solar cells; for example the U.S. Coast Guard is converting all remote buoys to solar cell operation.

In spite of the uncertainty in the future level of solar electricity generation there can be no doubt that solar cells will make a very significant contribution and will continue to attract both scientific and commercial interest.

### Suggested Readings

Earl Cook, *Man, Energy, Society,* New York, W.H. Freeman, 1976.
Paul D. Maycock, and Edward N. Stirewalt, *Photovoltaics,* Andover, Mass., 1981.
Martin A. Green, *Solar Cells,* Englewood Cliffs, Prentice Hall, 1983.

### Essay Problems

1. Typically 1 $m^2$ of land will receive 2000 kWh of sunshine each year. Compare the total value of the energy generated by solar cells to the value of a typical crop such as corn. Assume that the cells have a conversion efficiency of 10% and the electricity is worth about \$0.05/kWh. A good crop yield is 150 bushels an acre worth about \$2/bushel.

You should find that the electricity is worth about two orders of magnitude more than the corn. However, one must also address the initial cost of the solar system and compare this to the annual costs of cultivating corn, typically about \$180/acre. It is hoped that the total cost of a solar system can be brought down to about \$100/m$^2$; what would be the initial cost of a one-acre solar system?

2. Amorphous silicon cells generally contain a-Si:H films about 1 $\mu$m thick. (a) How much Si would be needed to make enough cells to generate at the rate of 1 kW under a noon time illumination of 1 kW/m$^2$? Assume that the conversion efficiency is 10% and that a-Si:H has the same density as crystalline Si, i.e., 2.3 g/cm$^3$. (b) A peaking generator is normally rated at about 25 MW; how much ground area would be needed to set up an equivalent photovoltaic plant operating at noon? You should find that over 50 acres are needed for the solar system, much more than needed for the gas turbine. What factors would favor the solar system over the gas turbine? You should be able to think of environmental as well as resource-utilization considerations.
3. A car normally operates at a power rating of about 20 hp. What area of 10% efficiency cells would be needed to match this under 1 kW/m$^2$ insolation? Does simple conversion from gasoline to solar power seem feasible for a conventional solar car? What changes would have to be made and technological advances achieved in order for a solar car to be attractive to a suburban commuter?
4. The *IV* behavior of a solar cell can be represented by the expression

$$I = I_O[e^{qV/kT} - 1] - I_{sc}$$

where $I$ and $V$ are the current and voltage, respectively, $q$ is the electronic charge and $k$ is Boltzmann's constant, $I_{sc}$ is the short circuit current and $I_O$ is the reverse saturation current of the diode. (a) Develop an expression for the open circuit voltage of the cell, $V_{OC}$, i.e., the voltage generated when $I = 0$. ($I_{SC}$ is always very much greater than $I_O$.) (b) For Si cells $I_O$ is about $10^{-12}$ A/cm$^2$ and $I_{SC}$ about 40 mA/cm$^2$ under 1 kW/m$^2$ insolation. Show that the open circuit voltage at room temperature is about 0.6 V. ($q/kT$ at 300 K has the value of 40 V$^{-1}$.) (c) Using the above values draw the *I-V* curve for a 1 cm$^2$ cell and compute the maximum power output. You will need to calculate *IV* at various points on the *I-V* curve and find the maximum product value. At what efficiency do you conclude the cell is operating? It should be about 10%.

# 44

# Superconductivity

*Photograph of a small permanent magnet levitated above a pellet of the* $Y_1Ba_2Cu_3O_{7-\delta}$ (123) *superconductor cooled to liquid nitrogen temperature. (Courtesy of IBM Research)*

The phenomenon of superconductivity has always been very exciting both for fundamental scientific interest and because of its many technical applications. The recent discovery of high-temperature superconductivity in certain metallic oxides has sparked even greater excitement in the scientific and business communities. This major breakthrough is being hailed by many scientists to be as important as the invention of the transistor. For this reason, it is important that all students of science and engineering understand the basic electromagnetic properties of superconductors, and become aware of the scope of their current applications.

## 44.1 INTRODUCTION

Superconductors have many unusual electromagnetic properties, and most applications take advantage of such properties. For example, once a current is produced in a superconducting ring maintained at a sufficiently low temperature, it will persist with no measurable decay. The superconducting ring exhibits no electrical resistance to dc currents, no heating, and no losses. In addition to the property of zero resistance, certain superconductors can expel applied magnetic fields so that the field is always zero everywhere inside. As we shall see, classical physics cannot explain the behavior and properties of superconductors. In fact, the superconducting state is now known to be a special quantum condensation of electrons. This quantum behavior has been verified through such observations as the quantization of magnetic flux produced by a superconducting ring.

In this chapter, we first give a brief historical review of superconductivity, beginning with its discovery in 1911 and ending with the recent developments in high-temperature superconductivity. We shall attempt to develop an understanding of some of the electromagnetic properties displayed by superconductors. Whenever possible, simple physical arguments will be used to describe these properties. The essential features of the theory of superconductivity will be reviewed with the realization that a detailed study is beyond the scope of this text. We shall discuss many of the important applications of superconductivity, and speculate on potential future applications which could arise because of the recent discovery of high-temperature superconductors.

## 44.2 BRIEF HISTORICAL REVIEW

The era of low temperature physics began in 1908 when the Dutch physicist Heike Kamerlingh Onnes first liquified helium, which has a boiling temperature of only 4.2 K. Three years later, in 1911, Onnes and one of his assistants discovered the phenomenon of superconductivity while studying the resistivity of metals at low temperatures.[1] They first studied platinum, and found that its resistivity, when extrapolated to 0 K, depended on the purity of the sample. They then decided to study mercury since very pure samples could easily be prepared by distillation. Much to their surprise, the resistance of the Hg sample dropped sharply at 4.15 K to an unmeasurably small value. It was quite natural that Onnes would name this new phenomenon of perfect conductivity, *superconductivity.* Figure 44.1 shows the experimental results for Hg and Pt. Note that platinum does not exhibit superconducting behavior as indicated by its finite resistivity as $T$ approaches 0 K. In 1913, Onnes was awarded the Nobel prize in physics for the study of matter at low temperatures and the liquefaction of helium.

[1] H.K. Onnes, *Leiden Comm.*, **120b, 122b, 124c** (1911).

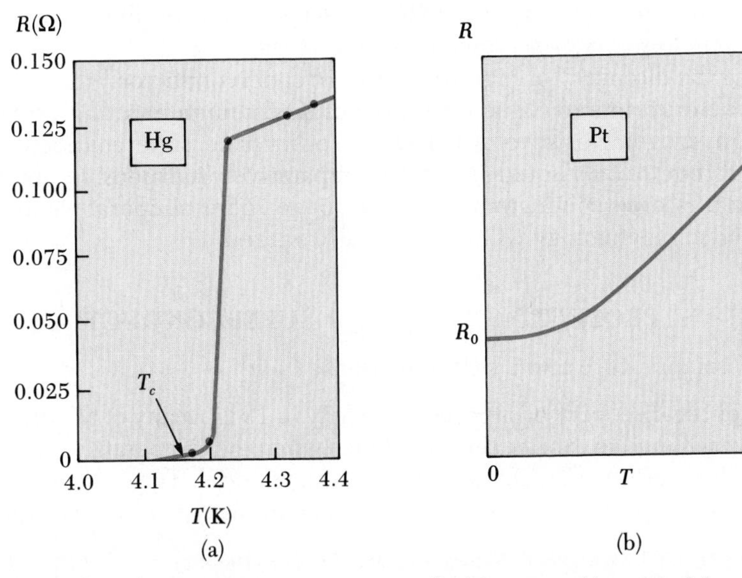


**Figure 44.1** Plots of resistance versus temperature for (a) mercury (the original data published by Onnes) and (b) platinum. Note that the resistance of mercury follows the path of a normal metal above the critical temperature $T_c$, and suddenly drops to zero at the critical temperature, which is 4.15 K for mercury. In contrast, the data for platinum show a finite resistance $R_0$ even at very low temperatures.

We now know that the resistivity of a superconductor is truly zero. Soon after the discovery by Onnes, many other elemental metals were found to exhibit zero resistance when the temperature was lowered below a certain characteristic temperature of the material called the **critical temperature, $T_c$**.

The magnetic properties of superconductors are as dramatic and as difficult to understand as their electrical properties. In 1933 W. Hans Meissner and Robert Ochsenfeld studied the magnetic behavior of superconductors and found that when certain superconductors in the presence of a magnetic field are cooled below their critical temperature, *the magnetic flux is expelled from the interior of the superconductor.*[2] Furthermore, it was found that these materials lost their superconductive behavior above a certain temperature-dependent **critical magnetic field, $B_c(T)$**. A phenomenological theory of superconductivity was developed in 1935 by Fritz and Heinz London,[3] but the actual nature and origin of the superconducting state were first explained by John Bardeen, Leon N. Cooper, and J. Robert Schrieffer in 1957.[4] A central feature of this theory, commonly referred to as the BCS theory, is the formation of bound two-electron states called Cooper pairs. In 1962, Brian D. Josephson predicted a tunneling current between two superconductors separated by a thin ($<2$ mm) insulating barrier, where the current is carried by these paired electrons.[5] Shortly thereafter, Josephson's predictions were verified, and today a whole field of device physics exists based on the Josephson effect. Early in 1986, J. Georg Bednorz and Karl Alex Müller reported evidence for superconductivity in an oxide of lanthanum, barium, and copper at a temperature of about 30 K.[6] This was a major breakthrough in superconductivity since the highest known value of $T_c$ at that time was about 23 K in a compound of niobium and germanium. This remarkable discovery, which marks the beginning of a new era of high-temperature superconductivity, has received worldwide attention in both the scientific community and the business world. Recently, researchers have reported critical temperatures as high as 125 K in more complex metallic oxides, but the mechanisms responsible for superconductivity in these materials remain unclear at this time.

Until the discovery of high-temperature superconductors, the use of superconductors required coolant baths of liquified helium (rare and expensive) or liquid hydrogen (very explosive). On the other hand, superconductors with $T_c > 77$ K require only liquid nitrogen (comparatively inexpensive, abundant, and inert). If superconductors with $T_c$'s above room temperature are ever found, human technology will be drastically altered.

## 44.3 SOME PROPERTIES OF TYPE I SUPERCONDUCTORS

### Critical Temperature and Critical Magnetic Field

Soon after the discovery of superconductivity in 1911, many of the elemental metals were found to exhibit zero resistance when the temperature was below a **critical temperature $T_c$**, characteristic of the material. The critical temperatures of some superconducting elements, classified as type I superconductors,

[2] W. Meissner and R. Ochsenfeld, *Naturwisschaften* **21**, 787 (1933).

[3] F. London and H. London, *Proc. Roy. Soc. (London)* **A149**, 71 (1935).

[4] J. Bardeen, L.N. Cooper, and J.R. Schrieffer, *Phys. Rev.* **108**, 1175 (1957).

[5] B.D. Josephson, *Phys. Letters* **1**, 251 (1962).

[6] J.G. Bednorz and K.A. Müller, *Z.Phys.* **B 64**, 189 (1986).

**TABLE 44.1 Critical Temperatures and Critical Magnetic Fields (measured at $T = 0$ K) of Some Elemental Superconductors**

| Superconductor | $T_c$ (K) | $B_c$ (0) in Tesla |
|---|---|---|
| Al | 1.196 | 0.0105 |
| Ga | 1.083 | 0.0058 |
| Hg | 4.153 | 0.0411 |
| In | 3.408 | 0.0281 |
| Nb | 9.26 | 0.1991 |
| Pb | 7.193 | 0.0803 |
| Sn | 3.722 | 0.0305 |
| Ta | 4.47 | 0.0829 |
| Ti | 0.39 | 0.010 |
| V | 5.30 | 0.1023 |
| W | 0.015 | 0.000115 |
| Zn | 0.85 | 0.0054 |

are given in Table 44.1. Copper, silver, and gold, which are excellent conductors, do not exhibit superconductivity.

When the critical temperature of a superconductor is measured in the presence of an applied magnetic field $\boldsymbol{B}$, the value of $T_c$ decreases with increasing magnetic field as indicated in Figure 44.2 for several type I superconductors. When the magnetic field exceeds a certain **critical magnetic field** $\boldsymbol{B}_c$, the superconducting state is destroyed, and the material behaves like a normal conductor with finite resistance.

It is found that the critical magnetic field varies with temperature according to the following approximate expression:

$$B_c(T) = B_c(0)\left[1 - \left(\frac{T}{T_c}\right)^2\right] \tag{44.1}$$

As you can see from this equation and Figure 44.2, the value of the critical field is a maximum at $T = 0$ K. The value of $B_c(0)$ is found by determining $B_c$ at some

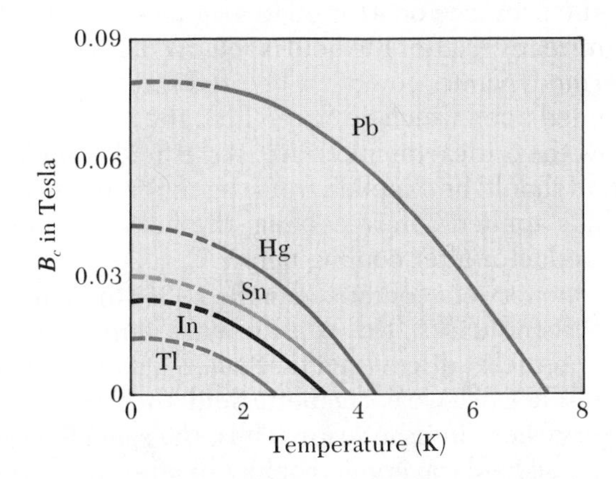


**Figure 44.2** The critical field $B_c(T)$ versus temperature $T$ for several type I superconductors. Extrapolations of these fields to 0 K give the critical fields listed in Table 44.1. For a given metal, the material is superconducting at fields and temperatures below its critical curve, and behaves like a normal conductor above that curve.

finite temperature, and extrapolating back to 0 K, which cannot be achieved. The maximum current that can be sustained in a type I superconductor is limited by the value of the critical field.

Note that $B_c(0)$ is the *maximum* magnetic field required to destroy superconductivity in a given material. If the applied field exceeds $B_c(0)$, the metal will never become superconducting at any temperature. Values for the critical field for type I superconductors are quite low, less than 0.2 T as shown in Table 44.1. For this reason, type I superconductors cannot be used to construct high-field magnets, called *superconducting magnets.* In the next section, we shall see that another class of superconductors, called type II superconductors, are ideally suited for this application.

### Magnetic Properties of Type I Superconductors

As we have seen, a superconductor has the important property of having zero dc resistance. We shall now describe some of its magnetic properties.

One can use simple arguments based on the laws of electricity and magnetism to show that the magnetic field inside a superconductor cannot change with time. According to Ohm's law, the electric field in a conductor is proportional to the resistance of the conductor. Thus, since $R = 0$ for a superconductor, *the electric field in its interior must be zero.* Now recall that Faraday's law of induction can be expressed as

$$\oint \boldsymbol{E} \cdot d\boldsymbol{s} = -\frac{d\Phi_m}{dt} \qquad (44.2)$$

That is, the line integral of the electric field $\boldsymbol{E}$ around any closed loop is equal to the negative rate of change in the magnetic flux $\Phi_m$ through the loop. Since $\boldsymbol{E}$ is zero everywhere inside the superconductor, the integral over any closed path in the superconductor is zero. Hence, $d\Phi_m/dt = 0$, which tells us that *the magnetic flux in the superconductor cannot change.* From this, we can conclude that $B$ $(=\Phi_m/A)$ *must remain constant inside the superconductor.*

Prior to 1933, it was assumed that superconductivity was a manifestation of perfect conductivity. If a perfect conductor is cooled below its critical temperature in the presence of an applied magnetic field the field should be trapped in its interior even after the field is removed. For a perfect conductor, equilibrium thermodynamics could not be used since its final state in a magnetic field depended upon which occurred first, the application of the field or the cooling below the critical temperature. If the field is applied after cooling below $T_c$, the field should be expelled from the superconductor. On the other hand, if the field is applied before cooling, the field should not be expelled from the superconductor after cooling below $T_c$.

When experiments were conducted in the 1930s to examine the magnetic behavior of superconductors, the results were quite different. In 1933, Meissner and Ochsenfeld[2] discovered that when a metal becomes superconducting in the presence of a weak magnetic field, the field is actually expelled so that $\boldsymbol{B} = 0$ everywhere in its interior. Thus, the same final state $\boldsymbol{B} = 0$ was achieved whether the field was applied before or after the material was cooled below its critical temperature. This effect is illustrated in Figure 44.3 for a material in the shape of a long cylinder. Note that the field penetrates the cylinder when its temperature is greater than $T_c$ as in Figure 44.3a. However,

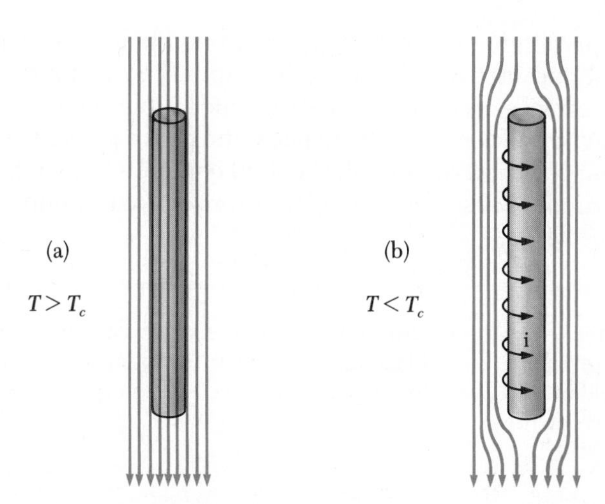


**Figure 44.3** A type I superconductor in the form of a long cylinder in the presence of an external magnetic field. (a) At temperatures above $T_c$, the field lines penetrate the sample since it is in its normal state. (b) When the rod is cooled to $T < T_c$, and the cylinder becomes superconducting, and flux is excluded from its interior by the induction of surface currents.

as the temperature is lowered below $T_c$, the field lines are spontaneously expelled from the interior of the superconductor as in Figure 44.3b. Thus, a type I superconductor is more than a perfect conductor (resistivity $\rho = 0$) but also a perfect diamagnet ($\boldsymbol{B} = 0$). The phenomenon of the expulsion of magnetic fields from the interior of a superconductor is known as the **Meissner effect.** The property that $\boldsymbol{B} = 0$ in the interior of a type I superconductor is as fundamental as the property of zero resistance. If the applied field is sufficiently large ($\boldsymbol{B} > \boldsymbol{B}_c$), the superconducting state will be destroyed and the field will penetrate the sample.

Because a superconductor is a perfect diamagnet, it will repel a permanent magnet. In fact, one can perform a most dazzling demonstration of the Meissner effect by floating a small permanent magnet above a superconductor and hence achieve magnetic levitation. A dramatic photograph of magnetic levitation is shown in Figure 44.4. The details of this demonstration are provided in Questions 29–35. A more detailed discussion of the magnetic properties of superconductors and magnetic levitation is given in the first essay that follows this chapter.

**Figure 44.4** Photograph of a small permanent magnet levitated above a disk of the superconductor $YBa_2Cu_3O_{7-\delta}$ which is at 77 K. (Courtesy of Argonne National Laboratory)

You should recall from the study of electricity that a good conductor expels static electric fields by moving charges to its surface. In effect, the surface charges produce an electric field which exactly cancels the externally applied field inside the conductor. In a similar manner, a superconductor expels magnetic fields by the formation of surface currents.

To illustrate this point, consider a type I superconductor in the shape of a long cylinder in an external magnetic field parallel to its long axis. Let us assume that the sample is initially at a temperature $T > T_c$ as in Figure 44.3a, so that the field penetrates the sample. As the sample is cooled to a temperature $T < T_c$, the field is expelled from the interior of the superconductor as in Figure 44.3b. In this case, surface currents are induced on the superconductor, producing a magnetic field which exactly cancels the externally applied field inside the superconductor. As you would expect, the surface currents disappear when the external magnetic field is removed.

In the superconducting state, and for applied fields less than the critical field, the field does not penetrate the bulk of a type I superconductor, and there are only surface currents. Hence, the type I superconductor behaves as a perfect diamagnet. When the applied magnetic field exceeds the critical field, the entire sample becomes normal, the field completely penetrates the sample, and the sample's resistance goes from zero to some value expected for a normal conductor.

**EXAMPLE 44.1 Critical Current in A Pb Wire**
A lead (Pb) wire has a radius of 3 mm and is at a temperature of 4.20 K. Find (a) the critical magnetic field in lead at this temperature, and (b) the maximum current the wire can carry at this temperature.

*Solution* (a) We can use Equation 44.1 to find the critical field at any temperature for a material if $B_c(0)$ and $T_c$ are known. From Table 44.1, we see that the critical magnetic field of lead at $T = 0$ K is 0.0803 T and its critical temperature is $T_c = 7.193$ K. Hence, Equation 44.1 gives

$$B_c(4.2\ \text{K}) = (0.0803\ \text{T})\left[1 - \left(\frac{4.20}{7.193}\right)^2\right]$$

$$= 0.0529\ \text{T}$$

(b) Recall from Ampere's law (Ch. 30) that the magnetic field generated by a wire carrying a steady current $I$ at an exterior point a distance $r$ from the wire is given by

$$B = \frac{\mu_0 I}{2\pi r}$$

When the current in the wire equals a certain critical current $I_c$, the magnetic field at the surface of the wire will equal the critical magnetic field $B_c$. (Note that $B = 0$ inside since all the current is on the wire's surface.) Using the expression above, and taking $r$ to be equal to the radius of the wire, we find

$$I = 2\pi r B/\mu_0 = 2\pi\,\frac{(3 \times 10^{-3}\ \text{m})(0.0529\ \text{T})}{4\pi \times 10^{-7}\ \text{N/A}^2}$$

$$= 794\ \text{A}$$

### Penetration Depth

As we have seen, magnetic fields are expelled from the interior of a type I superconductor by the formation of surface currents. In reality, these currents are not formed in an infinitessimally thin layer on the surface. Instead, they penetrate the surface to a small extent. Within this thin layer, which is about 100 nm in depth, the magnetic field **B** inside a type I superconductor decreases exponentially from its external value to zero according to the expression

$$B(x) = B_0 e^{-x/\lambda} \tag{44.3}$$

where it is assumed that the external magnetic field is parallel to the surface of the sample. In this equation, $B_0$ is the value of the magnetic field at the surface, $x$ is the distance from the surface to some interior point, and $\lambda$ is a parameter called the **penetration depth.** The variation of the magnetic field inside a semi-infinite type I superconductor with distance is plotted in Figure 44.5. The superconductor occupies the region on the positive side of the $x$ axis. As you can see, the magnetic field becomes very small at depths a few times $\lambda$ below the surface. Values for $\lambda$ are typically in the range of 10–100 nm.

The penetration depth varies with temperature according to the empirical expression

$$\lambda(T) = \lambda_0\left[1 - \left(\frac{T}{T_c}\right)^2\right]^{-1/2} \tag{44.4}$$

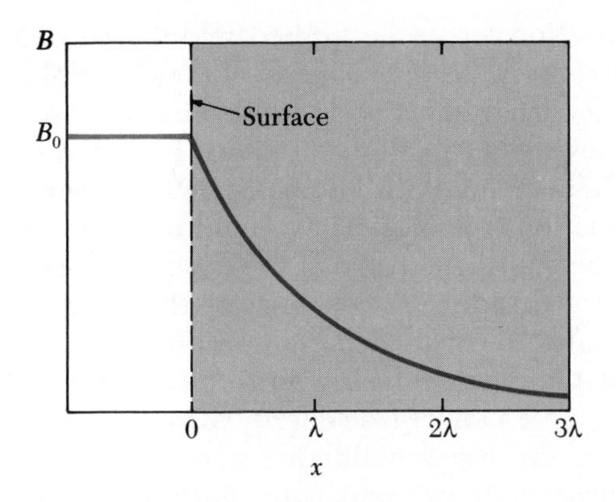


**Figure 44.5** The magnetic field $\boldsymbol{B}$ inside a semi-infinite superconductor versus distance $x$ from its surface. The field outside the superconductor (for $x < 0$) is $\boldsymbol{B}_0$, and the superconductor is to the right of the dashed line.

where $\lambda_0$ is the penetration depth at $T = 0$ K. From this we see that $\lambda$ becomes infinite as $T$ approaches $T_c$. Furthermore, as $T$ approaches $T_c$, while the sample is in the superconducting state, an applied magnetic field penetrates deeper into the sample. Ultimately, the field penetrates the entire sample ($\lambda$ becomes infinite), and the sample becomes normal.

The existence of field penetration is especially important when dealing with type I superconductors in the form of thin films or small particles. For example, if the thickness of the film is comparable to or less than $\lambda$, an applied field would penetrate the sample and flux expulsion would not be complete.

### Magnetization

When a bulk sample is placed in an external magnetic field $\boldsymbol{B}$, the sample acquires a magnetization $\boldsymbol{M}$ (see Section 30.9). The magnetic field $\boldsymbol{B}_{\text{in}}$ inside the sample is related to $\boldsymbol{B}$ and $\boldsymbol{M}$ through the relation $\boldsymbol{B}_{\text{in}} = \boldsymbol{B} + \mu_0\boldsymbol{M}$. When the sample is in the superconducting state, $\boldsymbol{B}_{\text{in}} = 0$ inside the superconductor; therefore it follows that the magnetization is given by

$$\boldsymbol{M} = -\frac{\boldsymbol{B}}{\mu_0} = \chi\boldsymbol{B} \tag{44.5}$$

where $\chi$ $(= -1/\mu_0)$ in this case is the **magnetic susceptibility.** That is, the magnetization of a superconductor *opposes* the external magnetic field and the magnetic susceptibility has its maximum negative value. This means that *a type I superconductor exhibits perfect diamagnetism, which is an essential property of the superconducting state.*

A plot of the magnetic field inside a type I superconductor versus the applied field (parallel to a long cylinder) at $T < T_c$ is shown in Figure 44.6a,

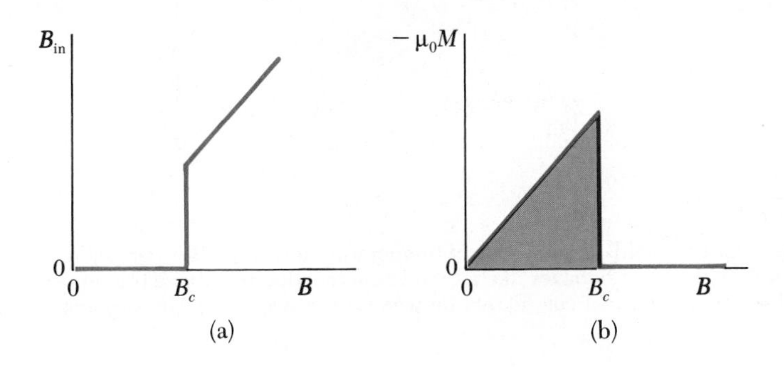


**Figure 44.6** The magnetic field dependent properties of a type I superconductor. (a) A plot of the internal field versus the applied field, where $\boldsymbol{B}_{\text{in}} = 0$ for $\boldsymbol{B} < \boldsymbol{B}_c$. (b) A plot of the magnetization versus the applied field. Note that $\boldsymbol{M} \cong 0$ for $\boldsymbol{B} > \boldsymbol{B}_c$.

while the magnetization versus the applied field at some constant temperature is plotted in Figure 44.6b. Note that when the applied field is greater than the critical field, $B_c$, the magnetization is approximately zero.

With the discovery of the Meissner effect, Fritz and Heinz London were able to develop phenomenological equations for type I superconductors based on equilibrium thermodynamics. They could explain the critical magnetic field in terms of the energy increase of the superconducting state due to the exclusion of flux from its interior, compared to the normal state, which allows the flux to penetrate. According to equilibrium thermodynamics, a system prefers to be in a state having the lowest free energy. Hence, the superconducting state must have a lower free energy than the normal state. If $E_s$ represents the energy of the superconducting state per unit volume, and $E_n$ the energy of the normal state per unit volume, then $E_s < E_n$ below $T_c$ and the material becomes superconducting. The exclusion of a field $B$ causes the total energy of the superconducting state to *increase* by an amount equal to $B^2/2\mu_0$ per unit volume. The critical field value is defined by the equation

$$E_s + \frac{B_c^{\,2}}{2\mu_0} = E_n \tag{44.6}$$

Since the London theory also gives the temperature dependence of $E_s$, an exact expression for $B_c(T)$ could be obtained. Note that the field exclusion energy $B_c^{\,2}/2\mu_0$ is just the area under the curve in Figure 44.6b.

## 44.4 TYPE II SUPERCONDUCTORS

By the 1950s it was established that there exists another class of superconductors known as type II superconductors. These materials are characterized by two critical magnetic fields, designated as $B_{c1}$ and $B_{c2}$ in Figure 44.7. When the applied field is less than the lower critical field $B_{c1}$, the material is entirely superconducting and there is no flux penetration, just as in the case of type I superconductors. When the applied field exceeds the upper critical field $B_{c2}$, the flux penetrates completely and the superconducting state is destroyed, just as for type I materials. However, for fields lying between $B_{c1}$ and $B_{c2}$, the material is in a *mixed* state, often referred to as the **vortex state.** While in the vortex state, the material can have zero resistance and has partial flux penetra-

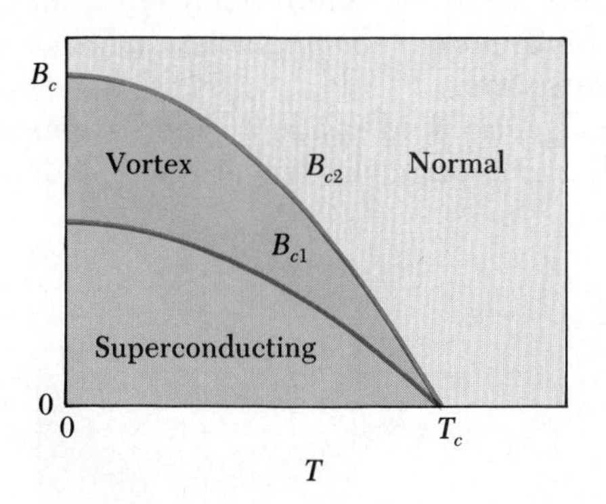


**Figure 44.7** Critical fields as a function of temperature for a type II superconductor. Below the lower critical field, $B_{c1}$, it behaves like a type I superconductor. Above the upper critical field, $B_{c2}$, it behaves like a normal conductor. Between these two fields, the superconductor is in a mixed state.

tion. The vortex regions are essentially filaments of normal material that run through the sample when the applied field exceeds the lower critical field as illustrated in Figure 44.8. As the applied field is increased in strength, the number of filaments increases until the field reaches the upper critical value, and the sample becomes normal.

One can view the vortex state as a cylindrical swirl of supercurrents surrounding a cylindrical normal metal core which allows some flux to penetrate the interior of the type II superconductor. Associated with each vortex filament is a magnetic field which is greatest at the core center and falls off exponentially outside the core with the characteristic penetration depth $\lambda$. The supercurrents are the "source" of $\boldsymbol{B}$ for each vortex. In type II superconductors, the radius of the normal metal core is smaller than the penetration depth.

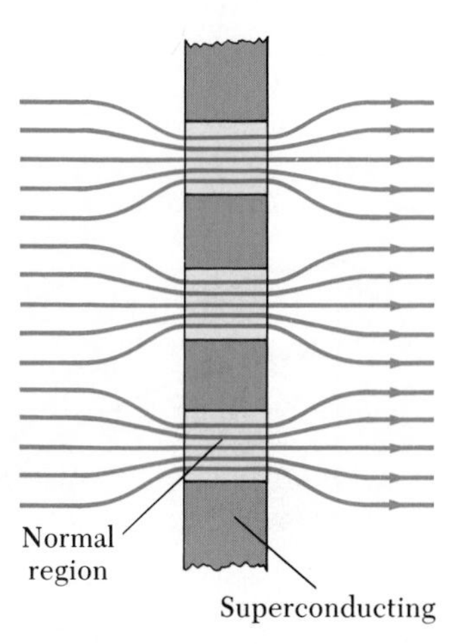


**Figure 44.8** A schematic diagram of a type II superconductor in the mixed state. The sample contains filaments of normal regions (shaded yellow) through which magnetic field lines can pass. The field lines are excluded from the superconducting regions (shaded in red).

Values for the critical temperature and upper critical magnetic field for several type II superconductors are given in Table 44.2. The values of $B_{c2}$ are very large when compared with the values of $B_c$ for type I superconductors (Table 44.1). Notice also that type II superconductors are compounds formed from elements of the transition and actinide series. Plots of the upper critical field variations with temperature for several type II superconductors are shown in Figure 44.9a. The three-dimensional plot in Figure 44.9b shows the variation of the critical temperature with both upper critical field and critical current density, $J_c$, for several type II superconductors. Note that the values of the critical fields are very large compared to the type I superconductors. For example, the upper critical field for the alloy $Nb_3(AlGe)$ is $B_{c2} = 44$ T, and its critical temperature is $T_c = 21$ K. For this reason, type II superconductors are well suited for constructing high-field superconducting magnets. For example, using the alloy NbTi, superconducting solenoids may be wound to produce and sustain magnetic fields in the range of 5 to 10 T *with no power consumption.* Iron core electromagnets rarely exceed 2 T with much higher power consumption.

Figure 44.10a represents a plot of the internal magnetic field versus the applied field for a type II superconductor, while Figure 44.10b is the corresponding magnetization versus the applied field. Again, note that the state of the material depends upon the magnitude of the applied field. That is, the material is in the flux exclusion superconducting state for $B < B_{c1}$, in the mixed state when $B$ lies between $B_{c1}$ and $B_{c2}$ and in the normal state when $B > B_{c2}$.

**TABLE 44.2 Values of the Critical Temperature and Upper Critical Magnetic Field (at $T = 0$ K) for Several Type II Superconductors**

| Superconductor | $T_c$ (K) | $B_{c2}$ (0) in Tesla |
|---|---|---|
| $Nb_3Al$ | 18.7 | 32.4 |
| $Nb_3Sn$ | 18.0 | 24.5 |
| $Nb_3Ge$ | 23 | 38 |
| NbN | 15.7 | 15.3 |
| NbTi | 9.3 | 15 |
| $Nb_3(AlGe)$ | 21 | 44 |
| $V_3Si$ | 16.9 | 23.5 |
| $V_3Ga$ | 14.8 | 20.8 |
| PbMoS | 14.4 | 60 |

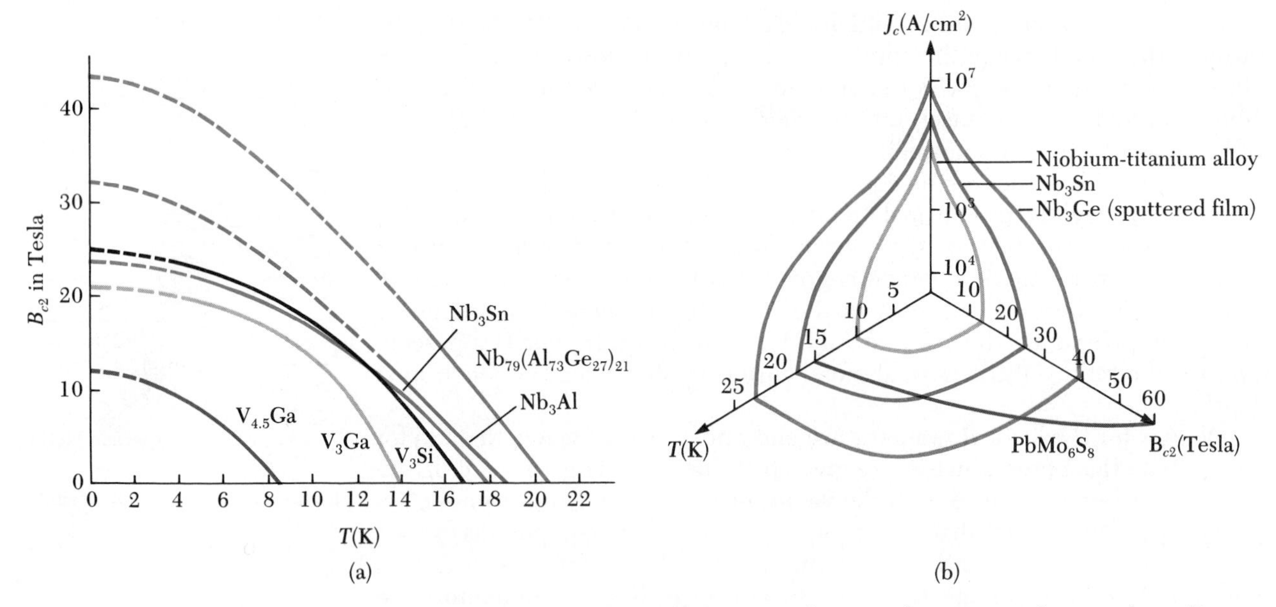


**Figure 44.9** (a) Upper critical field, $B_{c2}$, as a function of temperature for several type II superconductors. (From S. Foner, et al., *Physics Letters* **31A**:349, 1970) (b) A three-dimensional plot showing the variation of critical current density, $J_c$, with temperature, and the variation of the upper critical field with temperature for several type II superconductors.

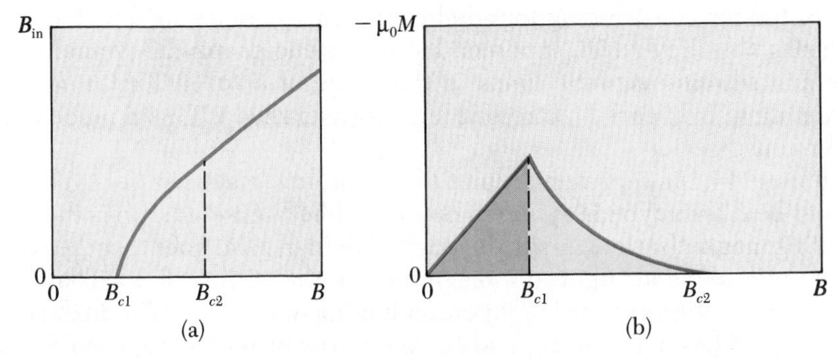


**Figure 44.10** The magnetic behavior of a type II superconductor. (a) Plot of the internal field versus the applied field. (b) Plot of the magnetization versus the applied field.

When a type II superconductor is in the mixed state, sufficiently large currents can lead to a motion of vortices perpendicular to the direction of the current. This vortex motion corresponds to a change in flux with time, and produces resistance in the material. By adding impurities or other special inclusions, one can effectively pin the vortices and prevent their motion, to produce zero resistance for the mixed state of the superconductor. The critical current for type II superconductors is determined by that value of the current, which when multiplied by the flux in the vortices, gives a Lorentz force which overcomes the strength of the pinning force.

**EXAMPLE 44.2 A Superconducting Solenoid**
A solenoid is to be constructed using wire made of the alloy $Nb_3Al$, which has an upper critical field of 32 T at $T = 0$ K and a critical temperature of 18 K. The wire has a radius of 1 mm, the solenoid is to be wound on a hollow cylinder of diameter 8 cm and length 90 cm, and there are to be 150 turns of wire per cm of length. How much current is required to produce a magnetic field of 5 T at the center of the solenoid?

*Solution* Recall from Chapter 30 that the magnetic field at the center of a tightly wound solenoid is given by $B = \mu_0 nI$, where $n$ is the number of turns per unit length along the solenoid, and $I$ is the current in the solenoid wire. Taking $n = 150$ turns/cm $= 1.5 \times 10^4$ turns/m $B = 5$ T, we find

$$I = B/\mu_0 n = \frac{5\ \text{T}}{(4\pi \times 10^{-7}\ \text{N/A}^2)(1.5 \times 10^4\ \text{m}^{-1})}$$

$$= 265\ \text{A}$$

Note that the length and diameter of the solenoid are not needed in this calculation.

**Exercise 1** What maximum current can the solenoid carry if its temperature is to be maintained at 15 K? (Note that $B$ near the solenoid windings is approximately equal to $B$ on its axis.)
**Answer** Using Equation 44.1, with $B_c(0) = 32$ T, we find $B_c = 9.78$ T at a temperature of 15 K. For this value of $B$, we find $I_{max} = 518$ A.

## 44.5 OTHER PROPERTIES OF SUPERCONDUCTORS

### Persistent Currents

Because the dc resistance of a superconductor is zero below its critical temperature, once a current is set up in the material, it will persist *without any applied voltage* (which follows from Ohm's law and the fact that $R = 0$ for the superconductor). These **persistent currents,** sometimes called *supercurrents,* have been observed to last for several years with no measurable losses. In one experiment conducted by S.S. Collins in Great Britain in 1956, a current was maintained in a superconducting ring for 2½ years. The current stopped only because a trucking strike delayed the delivery of liquid helium, which was necessary to maintain the ring below its critical temperature.[7]

To better understand the origin of these persistent currents, consider a loop of wire made of a superconducting material. Suppose the loop is placed in an external magnetic field $\boldsymbol{B}$ while it is in its normal state ($T > T_c$), and its temperature is then lowered below $T_c$ so it becomes superconducting as in Figure 44.11a. As in the previous example of the long cylinder, the flux is excluded from the interior of the wire because of the induced surface currents.

[7] This charming story was provided by Steve Van Wyk.

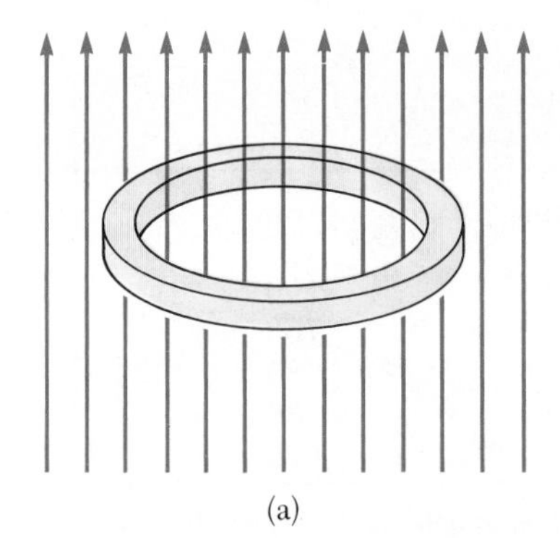


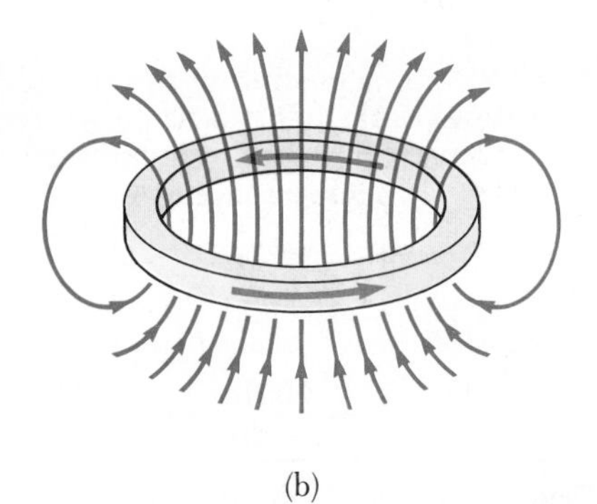


**Figure 44.11** (a) When a superconducting loop at $T < T_c$ is placed in an external magnetic field, flux passes through the hole in the loop even though it does not penetrate the interior of the superconductor. (b) After the external field is removed, the flux through the hole in the loop will remain trapped, and an induced current appears in the loop.

However, note that *flux lines still pass through the hole in the loop.* When the external field is turned off as in Figure 44.11b, *the flux that was inside the loop is trapped because the magnetic flux through the loop cannot change.*[8] The superconducting wire prevents the flux from going to zero through the appearance of a large spontaneous current induced by the collapsing external magnetic field. If the dc resistance of the superconducting wire is truly zero, this current should persist forever. Experimental results using a technique known as nuclear magnetic resonance indicate that such currents will persist for more than $10^5$ years! The resistivity of a superconductor based on such measurements has been shown to be less than $10^{-26}$ $\Omega \cdot m$. This reaffirms the fact that $R$ is zero for a superconductor. (See Problem 30 for a simple but convincing demonstration of zero resistance.)

Now consider what would happen if the loop is cooled to a temperature $T < T_c$ *before* the external field is turned on. When the field is turned on while the loop is maintained at this temperature, *flux must be excluded from the entire loop, including the hole* since the loop is in the superconducting state. Again, a current is induced in the loop to maintain zero flux through the loop and the interior of the wire. In this case, the current disappears when the external field is turned off.

### Coherence Length

Another important parameter associated with superconductivity is called the **coherence length,** $\xi$. One can think of the coherence length as the smallest dimension over which superconductivity can be established or destroyed. Alternatively, one can view the coherence length as the distance over which the electrons in a Cooper pair remain together. In the BCS theory, the coherence length is directly related to the distance over which the two electrons in a Cooper pair remain correlated. Typical values of $\lambda$ and $\xi$ at $T = 0$ K for selected superconductors are given in Table 44.3.

A superconductor will be type I if the coherence length is larger than the penetration depth, $\lambda$. Most pure metals fall into this category. On the other hand, an increase in the ratio $\lambda/\xi$ favors type II superconductivity. A detailed analysis shows that the coherence length and penetration depth depend on the mean free path of the electrons in the normal state. The mean free path of a

[8] Alternatively, one can apply Equation 44.2 taking the line integral of the $E$ field over the loop. Since $\rho = 0$ everywhere along a path on the superconductor, $E = 0$, the integral is zero, and $d\Phi/dt = 0$.

**TABLE 44.3 Penetration Depth and Coherence Length of Selected Superconductors at $T = 0$ K[a]**

| Superconductor | $\lambda$ (nm) | $\xi$ (nm) |
|---|---|---|
| Al | 16 | 160 |
| Cd | 110 | 760 |
| Pb | 37 | 83 |
| Nb | 39 | 38 |
| Sn | 34 | 23 |

[a] These are calculated values from C. Kittel, *Introduction to Solid State Physics,* New York, John Wiley, 1986.

metal can be reduced by adding impurities to the metal. As impurities are added to a metal, the penetration depth increases while the coherence length decreases. Thus, one can cause a metal to change from type I to type II by introducing another alloying element. For example, pure lead is a type I superconductor. However, it changes to a type II superconductor (with almost no change in $T_c$) by alloying it with 2% (by weight) of indium.

## 44.6 SPECIFIC HEAT

The thermal properties of superconductors have been extensively studied and compared with those of the same material in the normal state. One very important measurement is the specific heat of the material. When a small amount of heat is added to a normal metal, some of the heat energy is used to excite lattice vibrations, and the remainder is used to increase the kinetic energy of the conduction electrons. The *electronic specific heat C* is defined as the ratio of the heat absorbed by the system of electrons to the increase in temperature of the system. Figure 44.12 shows how the electronic specific heat varies with temperature for both the normal state and the superconducting state of gallium, a type I superconductor. The specific heat of the material in the normal state below the critical temperature is obtained by applying a sufficiently strong magnetic field to destroy superconductivity. This value can then be compared with that of the material in the superconducting state (in zero magnetic field). At low temperatures, the electronic specific heat of the material in the normal state, $C_n$, varies with temperature as $AT + BT^3$. The linear term arises from electronic excitations, while the cubic term is due to lattice vibrations. In comparison, the electronic specific heat of the material in the superconducting state, $C_s$, is substantially altered below the critical temperature. As the temperature is lowered starting from $T > T_c$, the specific heat first jumps to a very high value at $T_c$, and then falls well below the value for the normal state at very low temperatures. Analyses of such data show that at

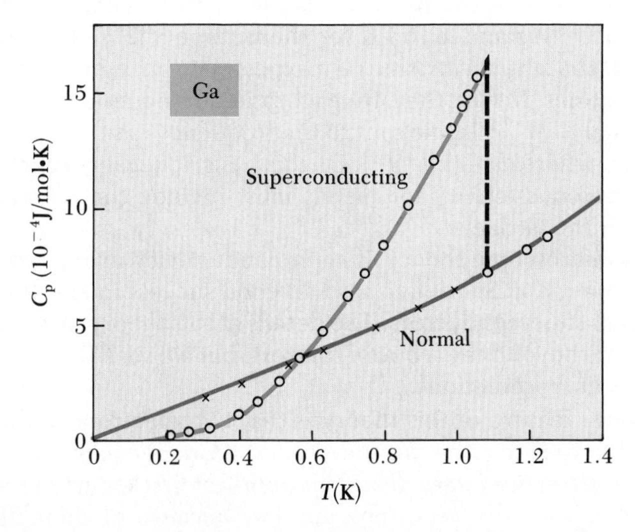


**Figure 44.12** Electronic specific heat versus temperature of gallium in the superconducting state (zero applied magnetic field), and in the normal state (in the presence of a 0.02 T magnetic field). For the superconducting state, note the discontinuity that occurs at $T_c$ and the exponential dependence on $1/T$ at low temperatures. (Taken from N. Phillips, *Phys. Rev.* **134,** 385, 1964)

temperatures well below $T_c$, the electronic part of the specific heat is dominated by a term that varies as $\exp(-\Delta/kT)$, where $\Delta$ is one-half of the energy gap. This result suggests the existence of an energy gap in the energy levels available to the electrons, where the energy gap is a measure of the thermal energy necessary to move electrons from the ground to the excited states. As we shall see in the next section, this energy gap and its temperature dependence are accounted for by the BCS theory of superconductivity.

## 44.7 THE BCS THEORY

Part of the resistivity of a normal metal is due to collisions between the free electrons and the thermally displaced ions of the metal lattice. The electrons can also encounter impurities and other defects in their journey through the metal. Soon after the discovery of superconductivity, scientists recognized that the superconducting state could never be explained with this classical model, since for real materials the electrons would always suffer some collisions. The phenomenon of superconductivity could not be understood by using a simple microscopic quantum-mechanical model, where one views an individual electron as a wave traveling through the material. Many phenomenological theories were proposed based on the known properties of superconductors. However, none of these theories was able to explain the basic mechanism of superconductivity (why do electrons enter the superconducting state), and why electrons in this state are not scattered by impurities and lattice vibrations.

Several important developments occurred in the 1950s that led to a better understanding of the superconducting state. In particular, many research groups reported that the transition temperatures of different isotopes of the same element decreased with increasing atomic mass. This observation, called the **isotope effect,** was early evidence that the motion of the lattice played an important role in the mechanism of superconductivity. For example, in the case of mercury, $T_c = 4.161$ K for the isotope $^{199}$Hg, as compared to 4.126 K for the isotope $^{204}$Hg and 4.153 K for the isotope $^{200}$Hg. The characteristic frequencies of the lattice vibrations are expected to change with the mass $M$ of the vibrating atoms. In fact, the lattice vibrational frequencies are expected to be proportional to $M^{-1/2}$ [analogous to the frequency $\omega$ of vibration of a mass-spring system, where $\omega = (k/M)^{1/2}$]. On this basis, it became apparent that any theory of superconductivity for metals must include the interaction of the electrons with the lattice.

The full microscopic theory of superconductivity presented in 1957 by Bardeen, Cooper, and Schrieffer has had good success in explaining the various features of superconductors. The details of this theory, now known as the BCS theory, are beyond the scope of this text, but we shall describe some of its main features and predictions.

The central feature of this theory is that two electrons in the superconductor are able to form a bound state called a **Cooper pair** if they somehow experience an *attractive interaction.* This notion at first seems counterintuitive since electrons normally repel one another because of their like charges. However, a net attraction could be obtained if the electrons interact with each other via the motion of the crystal lattice as the lattice structure is momentarily

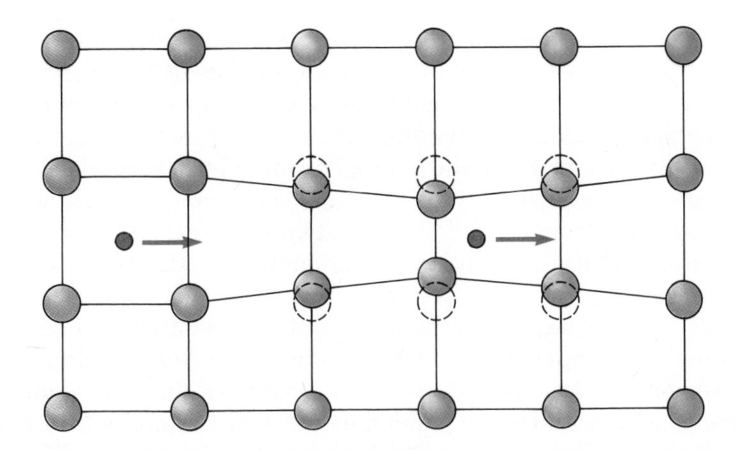

Figure 44.13 A schematic representation of the basis for the attractive interaction between two electrons via the lattice deformation. The first electron attracts the positive ions which move inwards from their equilibrium positions (dashed circles). This distorted region of the lattice has a net positive charge, and hence a nearby second electron is attracted to it.

deformed by a passing electron.[9] To illustrate this point, Figure 44.13 shows an instant of the first electron (red particle on the right) moving through the positive ions of the lattice. This causes the nearby ions (large blue spheres) to move inward towards the electron, resulting in a slight increase in the concentration of positive charge in this region. Another electron passing nearby later (the second electron of the Cooper pair) before the positive ions have had a chance to return to their equilibrium positions is therefore attracted to this distorted positively charged region. The net effect is a weak delayed attractive force between the two electrons via the motion of the positive ions. One researcher commented that "the following electron surfs on the virtual lattice wake of the leading electron." In more technical terms, one can say that the attractive force between the two electrons making up a Cooper pair is an *electron-lattice-electron interaction,* where the crystal lattice serves as the mediator of the attractive force. Some scientists refer to this as a *phonon mediated mechanism* since quantized lattice vibrations are called *phonons.*

A Cooper pair in a superconductor consists of two electrons with equal and opposite momenta and spin, as described schematically in Figure 44.14. Hence, in the superconducting state and in the absence of any supercurrents, *the Cooper pair forms a system with zero total momentum and zero spin.* Because Cooper pairs have zero spin, they behave as bosons and can all be in the same state. This is in contrast to electrons, which are Fermions with spin 1/2 that must obey the Pauli exclusion principle allowing only one electron in the same momentum and spin state. In the BCS theory, a ground state is constructed in which *all electrons form bound pairs.* In effect, all Cooper pairs are "locked" into the *same quantum state of zero momentum.* One can view this

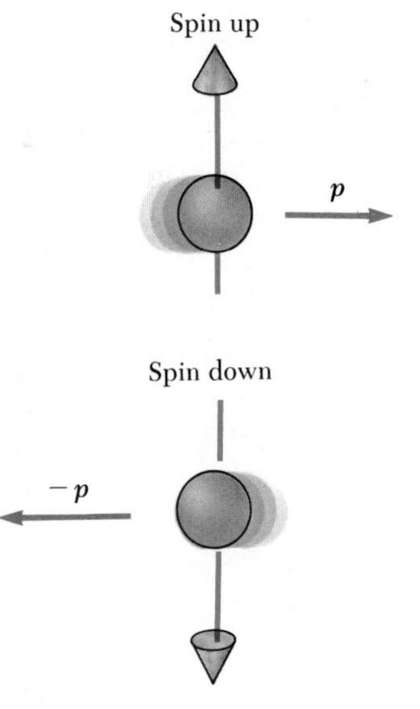


Figure 44.14 Schematic diagram of a Cooper pair. The electron moving to the right has a momentum $\boldsymbol{p}$ and its spin is up, while the electron moving to the left has a momentum $-\boldsymbol{p}$ and its spin is down. Hence the total momentum of the system is zero and the total spin is zero.

[9] For a lively description of this process, see *Electromagnetism: Path to Research,* Doris Teplitz, Ed., New York, Plenum Press, 1982. In particular, see Chapter 1, entitled "Electromagnetic Properties of Superconductors," by Brian B. Schwartz and Sonia Frota-Pessoa. You should note that the electron which causes the lattice to deform remains in that region for a very short time, $\approx 10^{-16}$ s, compared to the much longer time it takes the lattice to deform, $\approx 10^{-13}$ s. Thus, the sluggish ions continue to move inward for a time interval of about 1000 times longer than the response time of the electron, so the region is effectively positively charged between $10^{-16}$ s and $10^{-13}$ s.

state of affairs as a condensation of all electrons into the same state similar to the condensation of $^4$He molecules (which are also bosons) in superfluid liquid helium. It should also be noted that because the Cooper pairs have zero spin (and hence zero angular momentum), their wave functions are spherically symmetric (like the *s*-states of the hydrogen atom.) In a "semi-classical" sense they are always undergoing "head-on" collisions, and as such, are always moving in each others' "wake." Since the two electrons are in a bound state, their trajectories will always change directions to keep their separation within the coherence length, $\xi$, discussed in Section 44.5.

The BCS theory has been very successful in explaining the characteristic superconducting properties of zero resistance and flux expulsion. From a qualitative point of view, one can say that in order to reduce the momentum of any single pair by scattering, it would be necessary to simultaneously reduce the momentum of all the other pairs. It is an "all or nothing at all" situation. You cannot change the velocity of one Cooper pair without changing it for all of them.[10] Lattice imperfections and lattice vibrations, which effectively scatter electrons in normal metals, have no effect on Cooper pairs! In the absence of scattering, the resistivity is zero and the current persists forever. It is rather strange, and perhaps amazing, that the mechanism of lattice vibrations responsible (in part) for the resistivity of normal metals also provides the interaction giving rise to their superconductivity. Thus, Cu, Ag, and Au, which exhibit small lattice scattering at room temperature, are not superconductors, whereas Pb, Sn, Hg, and other modest conductors have strong lattice scattering at room temperature, and become superconductors at low temperatures.

As we mentioned earlier, the superconducting state is one in which the Cooper pairs act collectively, rather than independently. The condensation of all pairs into the same quantum state makes the system behave like a giant quantum mechanical system or "macromolecule." For this reason, one finds that superconductors exhibit quantum effects on a macroscopic scale, as compared to individual atoms and molecules, which exhibit quantum behavior on a microscopic scale. *The condensed state of the Cooper pairs is represented by a single coherent wave function* $\psi$. This wave function extends over the entire volume of the superconductor.

It is important to emphasize that the stability of the superconducting state is critically dependent upon strong correlation between Cooper pairs. In fact, the theory explains superconducting behavior in terms of the energy levels of the "macromolecule," and the existence of an **energy gap** between the ground and excited state of the system as in Figure 44.15a.[11] Note that there is no energy gap for a normal conductor as shown in Figure 44.15b. In a normal conductor, the Fermi energy, $E_F$, represents the largest kinetic energy the free electrons can have at $T = 0$ K.

The energy gap of a superconductor is very small, of the order of $kT_c$ ($\approx 10^{-3}$ eV) at 0 K, as compared to the energy gap in semiconductors, which is of the order of 1 eV, or the Fermi energy of a metal, which is of the

[10] Many authors choose to refer to this cooperative state of affairs as a **collective state.** As an analogy, one author wrote that the electrons in the paired state "move like mountain-climbers tied together by a rope: should one of them leave the ranks due to the irregularities of the terrain (caused by the thermal vibrations of the lattice atoms) his neighbors would pull him back."

[11] A Cooper pair is somewhat analogous to a helium atom, $^4_2$He, in that both are bosons with zero spin. It is well known that the superfluidity of liquid helium may be viewed as a condensation of bosons in the ground state. Likewise, superconductivity may be viewed as a superfluid state of Cooper pairs, all in the same quantum state.

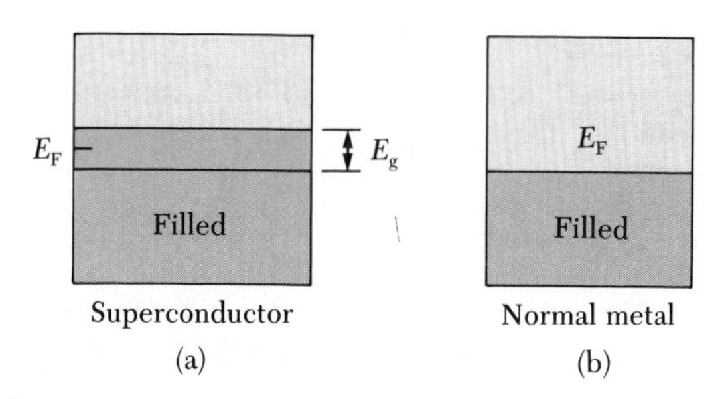


**Figure 44.15** (a) Simplified energy band structure for a superconductor. Note the energy gap between the lower filled states and the upper empty states. (b) The energy band structure for a normal conductor has no energy gap. At $T = 0$ K, all states below the Fermi energy, $E_F$, are filled, and all states above it are empty.

order of 5 eV. The energy gap of a superconductor, $E_g$, represents the energy needed to break up one of the Cooper pairs. The BCS theory predicts that the gap energy at $T = 0$ K is proportional to the critical temperature as follows:

$$E_g = 3.53\, kT_c \tag{44.7}$$

Thus, superconductors with large energy gaps have higher critical temperatures. The argument of the exponential function in the low temperature heat capacity discussed in the previous section contains a factor $\Delta$, which is found to be one-half of the energy gap; that is, $\Delta = E_g/2$. Furthermore, the energy gap values predicted by Equation 44.7 are in good agreement with the experimental values given in Table 44.4. The tunneling experiment used to obtain these values will be described later. As we noted earlier, the heat capacity in zero magnetic field undergoes a discontinuity at the critical temperature. (In thermodynamics, this is known as a second-order phase transition.) Furthermore, at finite temperatures, thermally excited individual electrons interact with the Cooper pairs and reduce the energy gap. The energy gap decreases continuously from its peak value at 0 K to zero at the critical temperature as shown in Figure 44.16 for several superconductors. The solid curve in Figure 44.16 represents the variation in the energy gap predicted by the BCS theory.

**TABLE 44.4 The Energy Gap for Several Superconductors at $T = 0$ K**

| Superconductor | $E_g$ (meV) |
|---|---|
| Al | 0.34 |
| Ga | 0.33 |
| Hg | 1.65 |
| In | 1.05 |
| Pb | 2.73 |
| Sn | 1.15 |
| Ta | 1.4 |
| Zn | 0.24 |
| La | 1.9 |
| Nb | 3.05 |

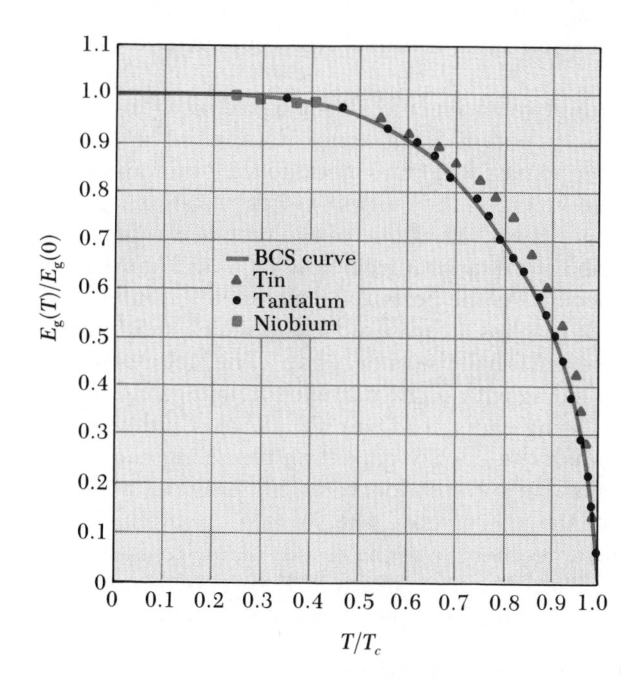


**Figure 44.16** The points on this graph represent reduced values of the observed energy gap $E_g(T)/E_g(0)$ as a function of the reduced temperature $T/T_c$ for tin, tantalum, and niobium. The solid curve gives the values predicted by the BCS theory. (The data are from electron tunneling measurements by P. Townsend and J. Sutton, *Phys. Rev.* 28:591, 1962)

**EXAMPLE 44.3 The Energy Gap for Lead**
Use Equation 44.7 to calculate the energy gap for lead and compare the answer with the experimental value in Table 44.4.

*Solution* Since $T_c = 7.193$ K for lead, Equation 44.7 gives

$$E_g = 3.53kT_c = (3.53)(1.38 \times 10^{-23}\ \text{J/K})(7.193\ \text{K})$$

$$= 3.50 \times 10^{-22}\ \text{J} = 0.00219\ \text{eV}$$

$$= \boxed{2.19 \times 10^{-3}\ \text{eV}}$$

The experimental value is $2.73 \times 10^{-3}$ eV, corresponding to a percentage *difference* of about 20%.

Because the two electrons of the Cooper pair have opposite spin angular momenta as indicated in Figure 44.14, an applied magnetic field will raise the energy of one electron and lower the energy of the other (see Problem 16). If the magnetic field is made strong enough, it will become more favorable for the pair to break up into a state where both spins are pointing in the same direction to lower their energy. This field corresponds to a critical field which compares the spin response of the superconducting state to that of the normal state. One must account for this effect when dealing with type II superconductors with very high values of the upper critical field.

## 44.8 ENERGY GAP MEASUREMENTS

### Single Particle Tunneling

The energy gaps in superconductors can be measured very precisely in *single particle tunneling* experiments (those involving normal electrons), first reported by Giaever in 1960.[12] Tunneling is a phenomenon in quantum mechanics that enables a particle to penetrate through a barrier even though classically it has insufficient energy to go over the barrier (see Section 41.9). If two metals are separated by an insulator, the insulator normally acts as a barrier to the motion of electrons between the two metals. However, if the insulator is made sufficiently thin (less than about 2 nm), there is a small probability that electrons will tunnel from one metal to the other across the barrier.

First, consider two normal metals separated by a thin insulating barrier as in Figure 44.17a. If a potential difference $V$ is applied between the two metals, electrons are able to pass from one metal to the other and a current is set up. For small applied voltages, the current-voltage relation is linear (the junction obeys Ohm's law.) However, if one of the metals is replaced by a superconductor maintained at a temperature below $T_c$, as in Figure 44.17b, something quite unusual occurs. As the potential difference $V$ is increased, no current is observed until $V$ reaches a threshold value which satisfies the relation $V_t = E_g/2e = \Delta/e$, where $\Delta$ is half the energy gap. (The factor of one half comes from the fact we are dealing with single particle tunneling, and the energy required is one half the binding energy of a pair, $2\Delta$.) That is, if the product $eV$ is at least as large as one-half the energy gap, then tunneling can occur between the normal metal and the superconductor. This provides a direct experimental measurement of the energy gap. The value of $\Delta$ obtained from such experiments is in good agreement with the results of low-temperature heat capacity measurements. The $I$-$V$ curve shown in Figure 44.17b shows the nonlinear relation for this junction. Note that as the temperature increases towards $T_c$,

[12] I. Giaever, *Phys. Rev Letters* **5**, 147 and 464 (1960).

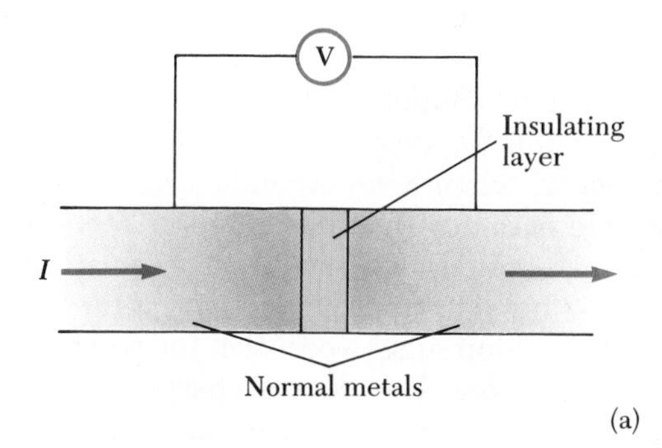


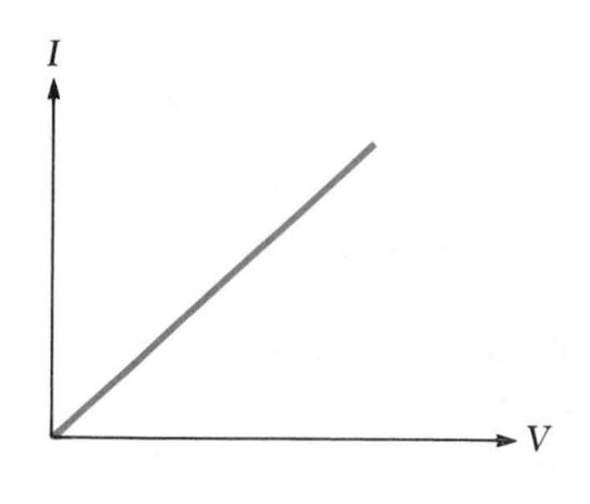


(a)

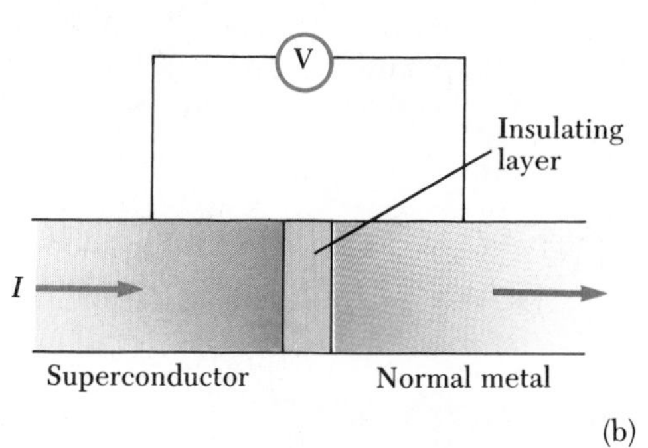


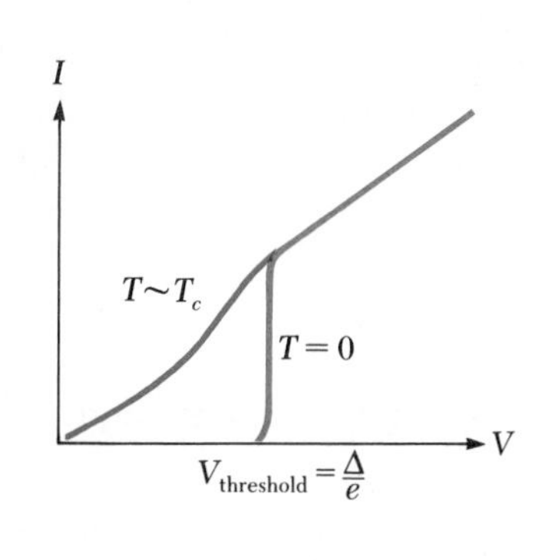


(b)

**Figure 44.17** (a) Current-voltage relation for electron tunneling through a thin insulator between two normal metals. The relation is linear for small currents and voltages. (b) Current-voltage relation of electron tunneling through a thin insulator between a superconductor and a normal metal. The relation is nonlinear and strongly temperature dependent. (Adapted from N.W. Ashcroft and N.D. Mermin, *Solid State Physics*, Philadelphia, Saunders College Publishing, 1975)

there is some tunneling current at voltages smaller than the energy gap threshold voltage. This is due to a combination of thermally excited single particle electrons and a decrease in the energy gap.

### Absorption of Electromagnetic Radiation

Another experiment that has been used to measure the energy gap of superconductors is the absorption of electromagnetic radiation. In the case of a semiconductor, you should recall that photons can be absorbed by the material when their energy is greater than the gap energy. Electrons in the valence band of the semiconductor absorb the incident photons, exciting the electrons across the gap into the conduction band. In a similar manner, superconductors absorb photons if their energy exceeds the gap energy, $2\Delta$. If the photon energy is less than $2\Delta$, no absorption occurs. When photons are absorbed by the superconductor, Cooper pairs are broken apart. Photon absorption in superconductors typically occurs in the range between microwave and infrared frequencies, as shown in the following example.

---

**EXAMPLE 44.4 The Absorption of Radiation by Lead**
Find the minimum frequency photon that will be absorbed by lead at $T = 0$ K.

*Solution* From Table 44.4, we see that the energy gap for lead is equal to $2.73 \times 10^{-3}$ eV. Setting this equal to the photon energy $hf$, and using the conversion 1 eV = $1.60 \times 10^{-19}$ J, we find

$$hf = 2\Delta = 2.73 \times 10^{-3}\ \text{eV} = 4.37 \times 10^{-22}\ \text{J}$$

or

$$f = \frac{4.37 \times 10^{-22}\ \text{J}}{6.626 \times 10^{-34}\ \text{J}\cdot\text{s}}$$

$$= 6.60 \times 10^{11}\ \text{Hz}$$

Exercise 2 What is the maximum wavelength of radiation that can be absorbed by lead at 0 K?
Answer $\lambda = c/f = 0.455$ mm. This lies between the microwave region and the far infrared.

---

## 44.9 FLUX QUANTIZATION

The phenomenon of flux exclusion by a superconductor described earlier in this chapter refers only to a simply connected object; that is, one with no holes or their topological equivalent. However, when a superconducting ring is placed in a magnetic field, and the field is removed as described in Section 44.5, flux lines are trapped and are maintained by a persistent circulating current as shown in Figure 44.18. Realizing that superconductivity is fundamentally a quantum phenomenon, Fritz London suggested[13] that the trapped magnetic flux should be quantized in units of $h/e$. The electric charge $e$ in the denominator arises because London assumed the supercurrent was carried by single electrons. Subsequent delicate measurements on very small superconducting hollow cylinders showed that the flux quantum is one-half the value postulated by London.[14] That is, the magnetic flux $\Phi$ is quantized not in units of $h/e$ but in units of $h/2e$, or

$$\Phi = \frac{nh}{2e} = n\Phi_0 \tag{44.8}$$

where $n$ is an integer and

$$\Phi_0 = \frac{h}{2e} = 2.0679 \times 10^{-15}\ \mathrm{T \cdot m^2} \tag{44.9}$$

is the **magnetic flux quantum.** This is in agreement with the BCS theory and the Cooper pair concept.

## 44.10 JOSEPHSON TUNNELING

In Section 44.8 we described the phenomenon of *single particle tunneling* from a normal metal through a thin insulating barrier into a superconductor. Now consider tunneling in the case of two superconductors separated by a thin insulator. In 1962, Brian Josephson proposed that in addition to single particle tunneling, Cooper pairs could also tunnel through such a junction. (See footnote 5.) Josephson predicted that the tunneling of pairs could occur without any resistance and would produce a dc current with zero applied voltage, and a

[13] F. London, *Superfluids,* Vol. I, New York, John Wiley, 1954.

[14] The effect was discovered by B.S. Deaver, Jr. and W.M. Fairbank, *Phys. Rev. Letters* 7; 43, 1961, and independently by R. Doll and M. Nabauer, *Phys. Rev. Letters* 7; 51, 1961.

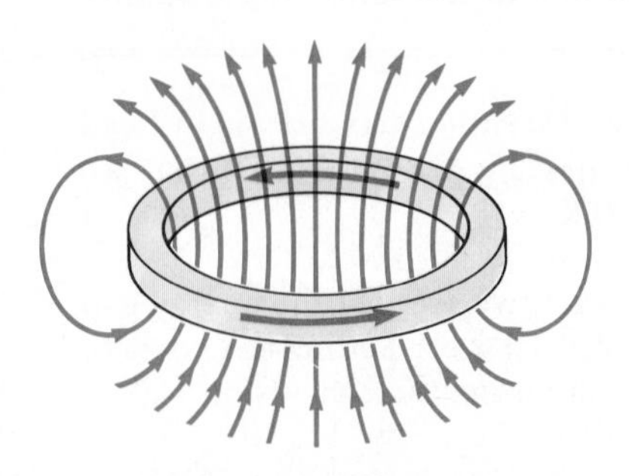

**Figure 44.18** The persistent current carried by a superconducting loop after it is removed from an external field. The trapped magnetic flux generated by the current is quantized in units of the magnetic flux quantum $\Phi_0$.

second effect in which an ac current would develop when a dc voltage is applied across the junction.

At first, physicists were very skeptical about Josephson's proposal since it was believed that single particle tunneling would mask the effects of (what seemed the less probable) pair tunneling. However, when the phase coherence of the pairs is taken into account, one finds that, under the appropriate conditions, the probability for tunneling of pairs across the junction is comparable to that of single particle tunneling. In fact, when the insulating barrier separating the two superconductors is made sufficiently thin, (say, $\approx 1$ nm) Josephson tunneling is as easy to observe as single particle tunneling.

In the remainder of this section, we shall describe three remarkable effects associated with the tunneling of Cooper pairs. These are commonly called the dc Josephson effect, the ac Josephson effect, and quantum interference.

### DC Josephson Effect

Consider two superconductors separated by a thin oxide layer, typically 1 to 2 nm thick, as in Figure 44.19a. Such a configuration is known as a Josephson junction. In a given superconductor, the pairs could be represented by a wave function $\Psi = \Psi_0 e^{i\phi}$, where $\phi$ is the phase and is the same for every pair. If one of the superconductors has a phase $\phi_1$ and the other a phase $\phi_2$, Josephson showed that at zero voltage a supercurrent would appear across the junction satisfying the relationship

$$I_s = I_m \sin(\phi_2 - \phi_1) = I_m \sin \delta \quad (44.10)$$

where $I_m$ is the maximum current across the junction under zero voltage conditions. The value of $I_m$ depends on the contact area between the two superconductors, and decreases exponentially with the thickness of the oxide separation between them.

The first confirmation of the dc Josephson effect was made in 1963 by Rowell and Anderson. Since then, all of Josephson's theoretical predictions have been verified. A plot of the current versus applied voltage for the Josephson junction is shown in Figure 44.19b.

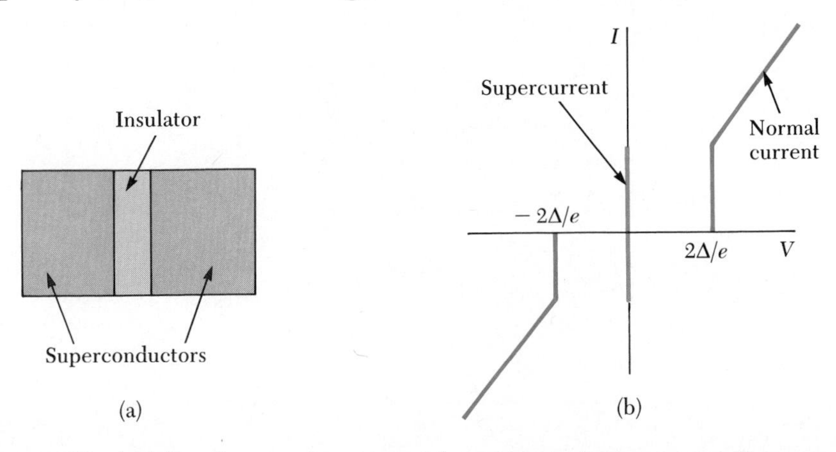


**Figure 44.19** (a) A Josephson junction consists of two superconductors separated by a very thin insulator. Cooper pairs are able to tunnel through this insulating barrier in the absence of an applied voltage, setting up a dc current. (b) The $I$-$V$ curve of a Josephson junction. When the bias current exceeds some threshold value, $I_m$, a voltage develops and the current undergoes oscillations with a frequency given by $f = 2eV/h$.

### AC Josephson Effect

When a dc voltage is applied across the Josephson junction, a most remarkable effect occurs. *The dc voltage generates an ac current* given by

$$I = I_m \sin(\delta - 2\pi ft) \tag{44.11}$$

where $\delta$ is a constant equal to the phase at $t = 0$, and $f$ is the frequency of the Josephson current, given by

$$f = \frac{2eV}{h} \tag{44.12}$$

A dc voltage of 1 $\mu$V results in a current frequency of 483.6 MHz. Precise measurements of the frequency and voltage have enabled physicists to obtain the ratio $e/h$ to unprecedented precision.

The ac Josephson effect can be demonstrated in various ways. One method is to apply a dc voltage and detect the electromagnetic radiation generated by the junction. Another method is to irradiate the junction with external radiation of frequency $f'$. With this method, one finds that a graph of the dc current versus voltage has steps when the voltages correspond to Josephson frequencies, $f$, that are integral multiples of the external frequency, $f'$; that is, when $V = hf/2e = nhf'/2e$ (see Fig. 44.20). Because the two sides of the junction are in different quantum states, the junction behaves like an atom undergoing transitions between these states as it absorbs or emits radiation. In effect, when a Cooper pair crosses the junction, a photon of frequency $f = 2eV/h$ is either emitted or absorbed by the system.

### Quantum Interference

The last effect we shall briefly describe deals with the behavior of the dc tunneling current in the presence of an external magnetic field. When a Josephson junction is subjected to a magnetic field, one finds that the maximum critical current in the junction depends on the magnetic flux through the junction. The tunneling current under these conditions is predicted to be

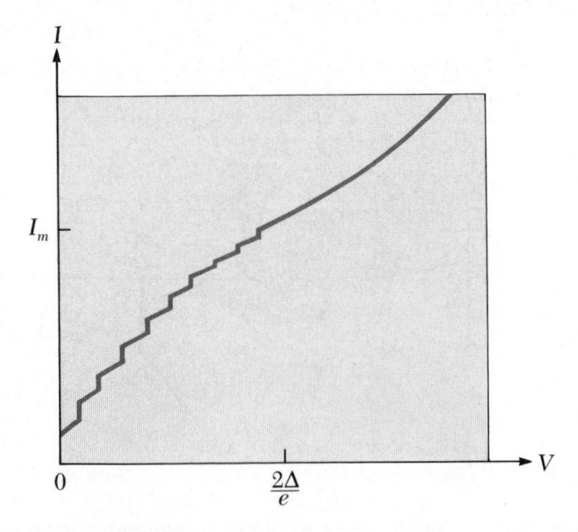


**Figure 44.20** A plot of the dc current as a function of the bias voltage for a Josephson junction placed in an electromagnetic field. At a frequency of 10 GHz, as many as 500 steps have been observed. (In the presence of an applied electromagnetic field, there is no clear onset of single particle tunneling.)

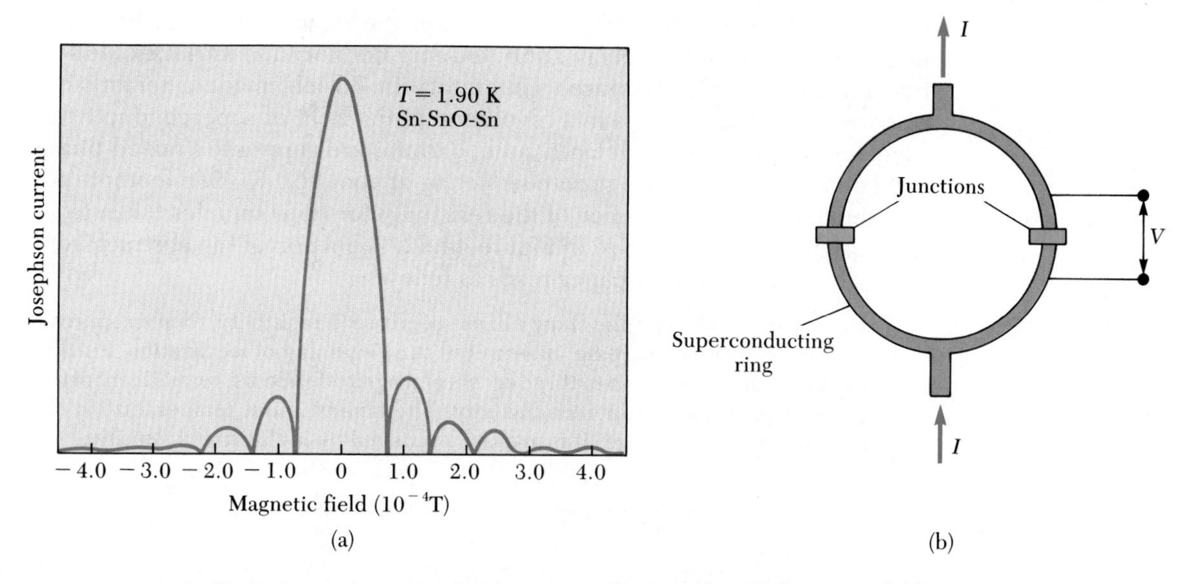


**Figure 44.21** (a) Maximum Josephson tunneling current as a function of applied magnetic field in an Sn-SnO-Sn junction. The zeros in the current are directly related to the number of flux quanta through the junction. (From *Superconductivity*, R.D. Parks, ed., Vol. 1, New York, Dekker, 1969). (b) Schematic diagram of a SQUID constructed from two Josephson junctions in parallel with each other.

periodic in the number of flux quanta through the junction. The maximum tunneling current as a function of magnetic field in an Sn-SnO-Sn junction at 1.90 K is shown in Figure 44.21a. Note that the current depends periodically on the magnetic flux. For typical junctions the field periodicity is about $10^{-4}$ T. This is not surprising in view of the quantum nature of the magnetic flux and the size of the flux quantum, as described in Section 44.9.

**If a superconducting circuit is constructed with two Josephson junctions in parallel with each other, as in the ring in Figure 44.21b, one can observe interference effects similar to the interference of light waves in Young's double slit experiment (see Ch. 37). In this case, the total current depends periodically on the flux inside the ring. Since the ring can have an area much greater than a single junction, the magnetic field sensitivity is greatly increased. A device that contains one or more Josephson junctions in a loop is called a SQUID, which is an acronym for *S*uperconducting *QU*antum *I*nterference *D*evice. SQUIDs are very useful for detecting very weak magnetic fields, of the order of $10^{-14}$ T, which is a very small fraction of the earth's field ($B_{earth} \approx 0.5 \times 10^{-4}$ T). Commercially available SQUIDs are able to detect a change in flux of about $10^{-5}\ \Phi_0 \approx 10^{-20}\ \mathrm{T \cdot m^2}$ in a bandwidth of 1 Hz. For example, they are being used to scan "brain-waves" corresponding to fields generated by current-carrying neurons. A more detailed discussion of SQUIDs and other superconducting devices is presented in the second essay following this chapter.**

## 44.11 HIGH-TEMPERATURE SUPERCONDUCTIVITY

**For many years, scientists have searched for new materials that would exhibit superconductivity at higher temperatures, and until recently, the alloy $Nb_3Ge$ had the highest known critical temperature, 23.2 K. There were various theoretical predictions that the maximum critical temperature for a superconduc-**

tor in which the electron-lattice interaction was important would be in the neighborhood of 30 K. Early in 1986, J. Georg Bednorz and Karl Alex Müller, two scientists at IBM Research Laboratory in Zurich, made a remarkable discovery that has resulted in a revolution in the field of superconductivity. They found that an oxide of lanthanum, barium, and copper in a mixed-phase form of a ceramic became superconducting at about 30 K. (See footnote 6.) The temperature dependence of the resistivity for their samples taken from their original paper is shown in Figure 44.22. A portion of the abstract from this rather unpretentious paper reads as follows:

> Upon cooling, the samples show a linear decrease in resistivity, then an approximately logarithmic increase, interpreted as a beginning of localization. Finally, an abrupt decrease by up to three orders of magnitude occurs, reminiscent of the onset of percolative superconductivity. The highest onset temperature is observed in the 30 K range. It is markedly reduced by high current densities.

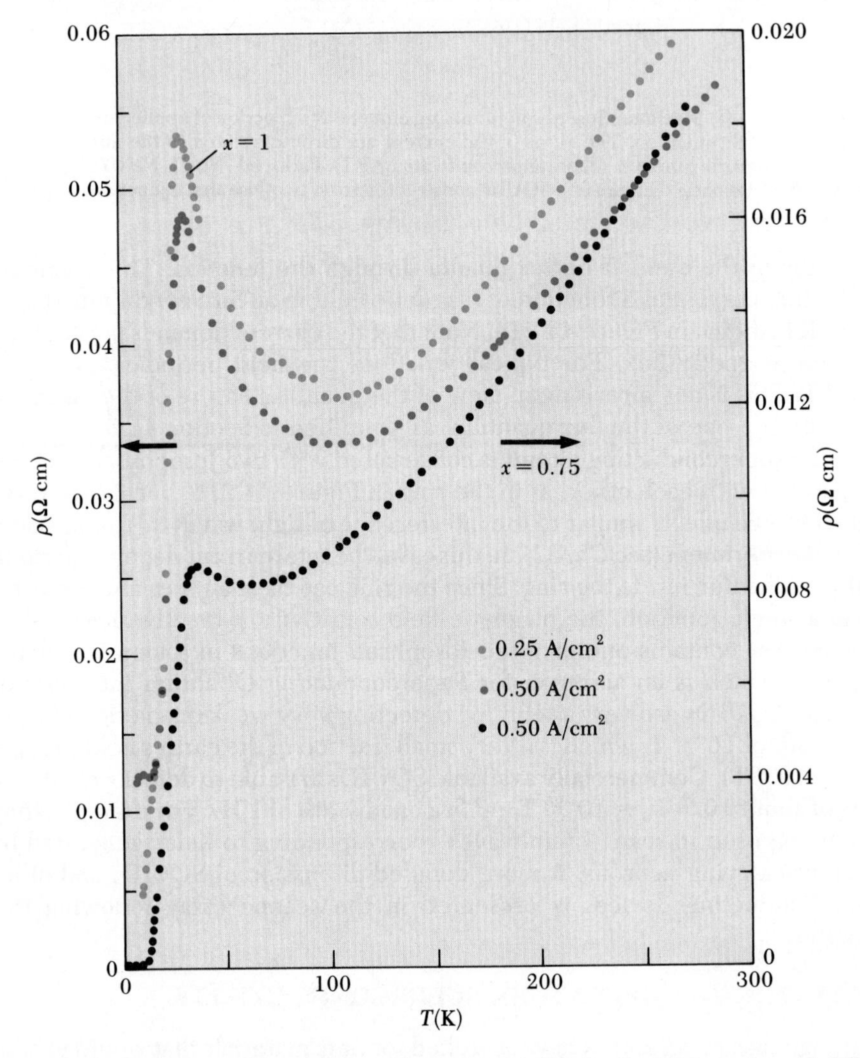


**Figure 44.22** Temperature dependence of the resistivity of Ba-La-Cu-O for samples with different concentrations of Ba and La. The left vertical scale is for the upper two curves, while the right scale is for the lower curve. The influence of current density is shown in the upper two curves. (Taken from J.G. Bednorz and K.A. Müller, *Z. Phys. B*, 64:189, 1986)

The superconducting phase was soon identified at other laboratories as the compound $La_{2-x}Ba_xCuO_4$, where $x \approx 0.2$. Announcements of higher temperature superconductivity were met with great skepticism, since such high critical temperatures were quite unexpected, especially in a metallic oxide. By replacing barium with strontium, researchers soon raised the value of $T_c$ to about 36 K. Inspired by these developments, scientists worldwide worked feverishly to discover materials with even higher $T_c$ values, and research in the superconducting behavior of metallic oxides accelerated at a tremendous pace. The year 1986 marked the beginning of a new era of high-temperature superconductivity. Bednorz and Müller were awarded the Nobel prize in 1987 (the fastest-ever recognition by the Nobel committee) for their most important discovery.

Early in 1987, research groups at the University of Alabama and the University of Houston announced the discovery of superconductivity near 92 K in a mixed-phase sample containing yttrium, barium, copper, and oxygen.[15] The discovery was confirmed in other laboratories around the world, and the superconducting phase was soon identified to be the compound $YBa_2Cu_3O_{7-\delta}$. A plot of the resistivity versus temperature for this compound is shown in Figure 44.23. This was an important milestone in high-temperature superconductivity, since the transition temperature of this compound is above the boiling point of liquid nitrogen (77 K), a coolant that is readily available, inexpensive, and simple to handle compared to liquid helium.

At the New York meeting of the American Physical Society held on March 18th, 1987, a special panel discussion on high-temperature superconductivity introduced the world to the newly discovered novel superconductors. This all-night session, which attracted a standing-room-only crowd of about 3000 persons, produced great excitement in the scientific community and has been referred to as the "Woodstock of Physics." Realizing the possibility of routinely operating superconducting devices at liquid nitrogen temperature and perhaps eventually at room temperature, thousands of scientists from various

[15] M.K. Wu, J.R. Ashburn, C.J. Torng, P.H. Hor, R.L. Meng, L. Gao, Z.J. Huang, Y.Q. Wang, and C.W. Chu, *Phys. Rev Letters* **58**; 908, 1987.

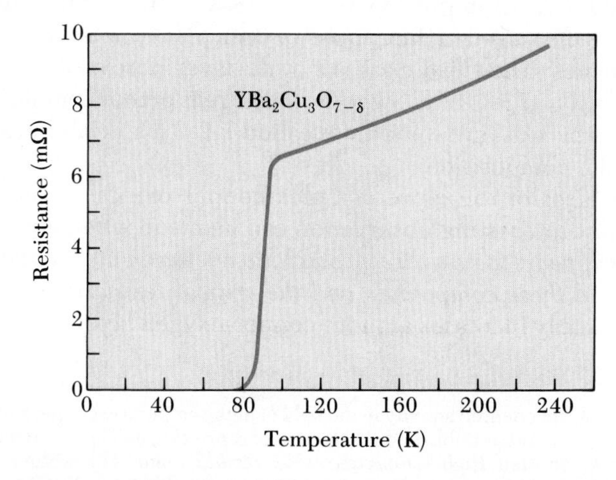


**Figure 44.23** Temperature dependence of the resistance of a sample of $YBa_2Cu_3O_{7-\delta}$ showing the transition temperature near 90 K. (Taken from M.K. Wu et al., *Phys. Rev. Letters* **58**:908, 1987)

disciplines entered the arena of superconductivity research. The exceptional interest in these novel materials is due to at least four factors:

1. The metallic oxides are relatively easy to fabricate and hence can be investigated at smaller laboratories and universities.
2. They have very high $T_c$ values, and very high upper critical magnetic fields, estimated to be greater than 100 T in several materials.
3. Their properties and the mechanisms responsible for their superconducting behavior represent a great challenge to theoreticians.
4. They may be of considerable technological importance for both existing applications and their potential use in nitrogen-temperature superconducting electronics and large-scale applications such as energy generation and transport, and magnetic levitation for high-speed transportation.

Thousands of articles have appeared in the literature in the last few years dealing with high-temperature superconductors.[16] It is beyond the scope of this text to attempt to review the literature. Instead, we shall present some of the highlights of recent discoveries, and describe some similarities and differences between the novel superconductors and the old (prior to 1986) materials.

Recently, several complex metallic oxides in the form of ceramics have been investigated, and critical temperatures above 100 K (triple-digit superconductivity) have been observed. Early in 1988, researchers reported the onset of superconductivity at about 120 K in a Bi-Sr-Ca-Cu-O compound and about 125 K in a Tl-Ba-Ca-Cu-O compound. The change in the values of $T_c$ since 1986 are highly dramatized if one plots the evolution of the transition temperature since the discovery of superconductivity as shown in Figure 44.24.

As you can see, the new high-$T_c$ materials are all copper oxides of one form or another. The various superconducting compounds that have been extensively studied to date can be classified in terms of the so-called perovskite crystal structures. The first class (Fig. 44.25a) is the cubic perovskite ($a = b = c$) such as $BaPb_{1-x}Bi_xO_3$, one of the original "high-$T_c$ materials," which has $T_c \approx 10$ K. The second class, known as the $KNiF_4$ structure (Fig. 44.25b) is a single-layer perovskite having a tetragonal distortion ($a = b \neq c$), of which $La_{1.85}Sr_{0.15}CuO_4$ is an example, having $T_c \approx 38$ K. Note that the lattice parameters $a$ and $b$ are measured in the copper-oxygen planes, while $c$ is perpendicular to these planes. The third class is a multi-layer perovskite (Fig. 44.25c), such as $YBa_2Cu_3O_7$ ($T_c \approx 92$ K), whose structure is orthorhombic ($a \neq b \neq c$). Compounds in this class are sometimes called 1-2-3 materials because of their relative metallic composition.

The structures of the more complex copper oxides are not illustrated here, but a most interesting observation can be made when examining these results. *There appears to be a direct correlation between the number of copper-oxygen layers in these compounds and the critical temperature.* The critical temperature clearly increases as more copper-oxygen layers are added to the

[16] Because of the unprecedented activity in the field of high-temperature superconductivity. The American Physical Society has published a collection of reprints from *Physical Review Letters* and *Physical Review B*, entitled *High-Temperature Superconductivity*. This volume contains 112 papers covering the period January–June, 1987, and is available from The American Physical Society, 335 East 45th Street, New York, NY 10017. Volumes II-A and II-B covering the second half of 1987 have recently become available as well.

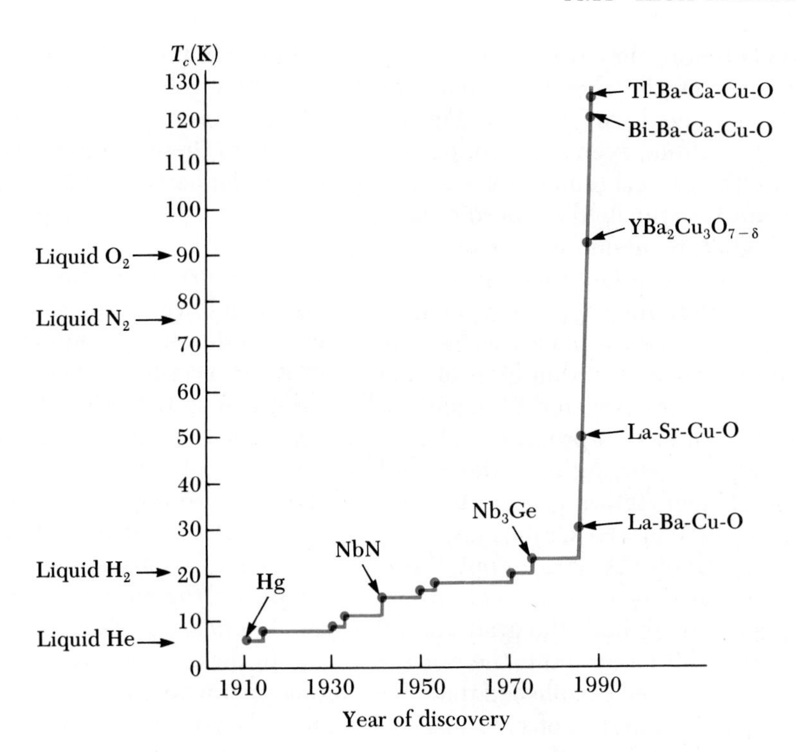


**Figure 44.24** Evolution of the superconductive transition temperature since the discovery of the phenomenon. Note the abrupt increase after the year 1986.

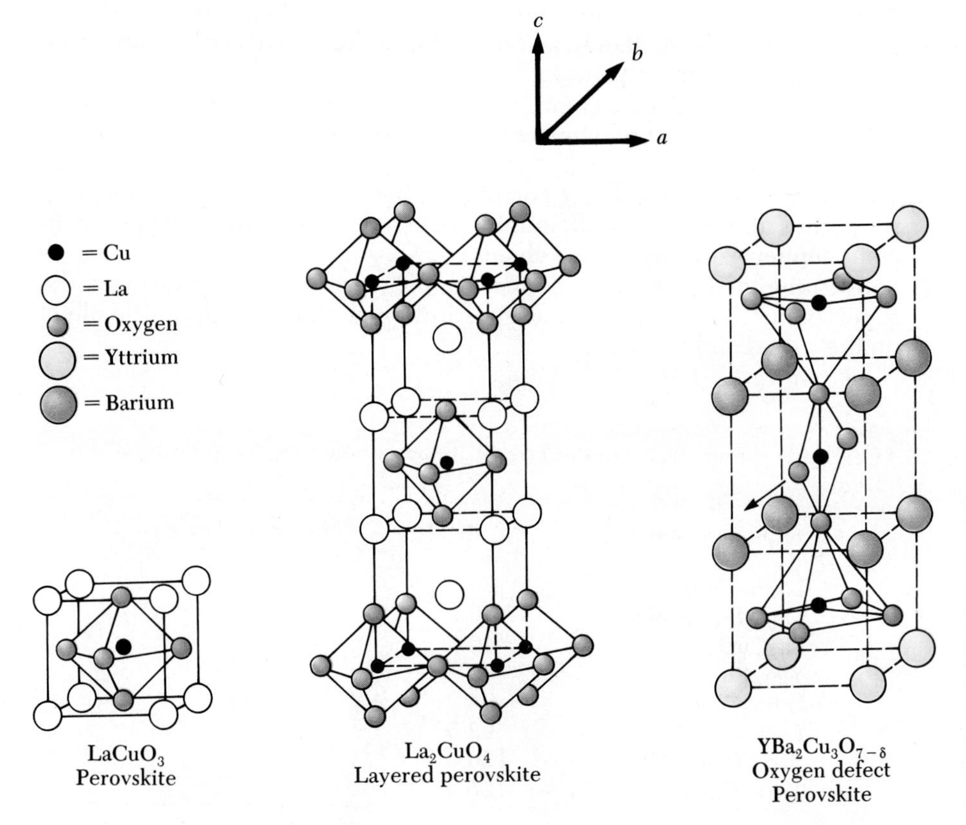


**Figure 44.25** Crystal structures for some of the new superconducting materials whose general class structure is perovskite. (a) The fundamental perovskite unit. (b) The single-layered perovskite having a tetragonal distortion ($a = b \neq c$). The first high-temperature superconductor discovered at 30 K falls in this class. (c) The double-layered perovskite with its orthorhombic structure ($a \neq b \neq c$). The compound $YBa_2Cu_3O_{7-\delta}$ has this structure. This structure is also related to the fundamental perovskite cube, but has missing oxygen atoms.

structures before they repeat. The valence of copper in the CuO and $CuO_2$ layers and how the chemical bonds are directed play roles that are being investigated (see these features in Fig. 44.25c). Hence, there is some reason to expect that adding even more copper-oxygen layers to these complex oxides will raise the critical temperature even higher. On the basis of such results, some experts in the field are predicting $T_c$ values above 200 K. Some properties of high-$T_c$ compounds are summarized in Table 44.5.

It is now well established that the maximum supercurrents in these structures are high in the copper-oxygen planes, and much lower in the direction perpendicular to these planes. In fact, critical current densities in the copper-oxygen ($a$, $b$) planes of thin films of about $10^{10}$ A/m$^2$ have been reported in $YBa_2Cu_3O_{7-\delta}$, as compared to much smaller values along the $c$ direction. In effect, the conduction process can be viewed as being two-dimensional. Unfortunately, the current densities in bulk ceramic samples are much lower because of such factors as grain boundary effects. For example, the critical current density in $YBa_2Cu_3O_{7-\delta}$ is in the range $10^5-10^7$ A/m$^2$ in bulk polycrystalline samples. As it turns out, this is much too low for most applications. Because these materials consist of tiny grains compacted together, the current must pass through both the grains and the grain boundaries. Most scientists believe that this factor limits the critical current in these materials.

It has been well established that these new copper oxides exhibit the two characteristic properties of superconductors, namely zero resistivity and diamagnetism. In addition, they are known to have the following properties:

1. They have been determined to be type II superconductors with very high upper critical fields, greater than 100 T.
2. They are very anisotropic in nature, as evidenced by their small resistivities in the copper-oxygen planes, and much higher resistivities in the direction perpendicular to these planes.
3. They have a granular or ceramic composition. Because of their ceramic nature, they have the unfortunate mechanical problem of being very brittle and inflexible.
4. There appears to be a direct correlation between their superconducting properties and their crystallographic structures that contain oxygen-deficient copper oxide layers and chains.
5. Substitution of atoms on the copper oxide layers degrades or destroys the superconductivity, while atomic substitution at other sites has little effect on the superconductivity.

**TABLE 44.5 Some Properties of High-$T_c$ Superconductors in the Form of Bulk Ceramics**

| Superconductor | $T_c$ (K) | $B_{c2}$ (0) in Tesla[a] |
|---|---|---|
| La-Ba-Cu-0 | 30 | |
| $La_{1.85}Sr_{0.15}CuO_5$ | 36.2 | >36 |
| $La_2CuO_4$ | 40 | |
| $YBa_2Cu_3O_{7-\delta}$ | 92 | ≈160 |
| $ErBa_2Cu_3O_{9-\delta}$ | 94 | >28 |
| $DyBa_2Cu_3O_7$ | 92.5 | |
| Bi-Sr-Ca-Cu-O | 120 | |
| Tl-Ba-Ca-Cu-O | 125 | |

[a] These are projected extrapolations based on data up to about 30 T.

6. Almost all the 1-2-3 materials have $T_c$ values close to 90 K, even though they differ in their bandgaps, high-temperature resistivity, critical current densities, critical magnetic fields, and so on.
7. The critical current densities in bulk polycrystalline samples are very low (of the order of $10^6$ A/m$^2$), but much higher in well-oriented thin films.

### Mechanisms of High-$T_c$ Superconductivity

It is important to note that although the framework of the conventional BCS theory has been quite successful in explaining the behavior and properties of the "old generation" superconductors, theoreticians are still trying to understand the nature of superconductivity in the "new generation" of high-$T_c$ metallic oxides. Various models and mechanisms have been proposed which are far too technical to describe here. However, it is interesting to note that many of the empirical observations related to the copper oxide superconductors are consistent with the predictions of the BCS theory. Evidence for this is as follows:

1. Many of the copper oxides have energy gaps which are in the range of the predicted BCS value of $3.53kT_c$, although there are wide discrepancies in results reported by various groups.
2. Pairs of charge carriers like Cooper pairs are involved in the process of superconductivity as shown by flux quantization experiments.
3. The discontinuity in the specific heat at $T_c$ is similar to that predicted by the BCS model.

Although the BCS model is consistent with such observations, the basic mechanisms to explain the behavior of the copper oxide superconductors remain to be identified conclusively. There is sufficient evidence to indicate that the carriers in the high-transition temperature oxides discussed here are hole pairs ("*p*-type" material) associated with missing oxygen atoms. Recent theories have proposed mechanisms that can be placed into three categories which can be used to obtain an effective attractive interaction between a pair of charge carriers. The first are those that retain some aspects of the electron-lattice-electron coupling of the BCS model. A much stronger coupling leading to a higher $T_c$ through this mechanism may be attributed to anharmonic lattice vibrations. The second class of mechanisms involves interactions with fluctuations of the electric charges, while the third class involves interactions with spin fluctuations and associated magnetic interactions. The recent discovery of another class of paired electron ("*n*-type") oxide superconductors has further complicated the puzzle. It suggests that either several mechanisms are possible, or it raises the requirement of a comprehensive theory that accounts for both electron and hole pairing. Clues as to how to achieve even higher critical temperatures and higher critical current densities may surface once mechanisms for high $T_c$ have been identified and a formal theory is developed.

## *44.12 APPLICATIONS

The discovery of high-temperature superconductors may very well introduce many important technological advances in the future, such as the potential of having superconducting devices in every household. However, many significant material science problems must be overcome before such applications become reality. Perhaps the most difficult technical problem is how to mold the brittle ceramic materials into useful shapes such as wires and ribbons for large scale applications, and thin films for small devices such as SQUIDs. Another major problem is the relatively low current densities measured in bulk ceramic compounds. Assuming that such problems are overcome, it is interesting to speculate on some of the future applications of these newly discovered materials.

An obvious application using the property of zero resistance to dc currents is the exploitation of low-loss electrical power transmission. A significant fraction of electrical power is lost as heat when current is passed through normal conductors such as copper. If power transmission lines could be made superconducting, these dc losses could be eliminated and there would be substantial savings in energy costs.

The new superconductors could have major impact in the field of electronics. Because of its switching properties, the Josephson junction can be used as a computer element. In addition, if one could use superconducting films to interconnect computer chips, chip size could be reduced and speeds would be enhanced because of this smaller size. Thus, information could be transmitted more rapidly and more chips could be contained on a circuit board with far less heat generation.

This prototype train, constructed in Japan, has superconducting magnets built into its base. A powerful magnetic field both levitates the train a few inches above the track, and propels it smoothly at speeds of 300 miles per hour or more. (© T. Matsumoto/SYGMA)

The phenomenon of magnetic levitation discussed earlier can be exploited in the field of transportation. In fact, a prototype train has already been constructed in Japan using superconducting magnets on the vehicle (with liquid helium as the coolant). The moving train levitates above a normal conducting metal track through eddy current repulsion. One can envision a future society of vehicles of all sorts gliding above a freeway making use of superconducting magnets. Some scientists are speculating that the first major market for levitating devices will be in the toy industry.

Another very important application of superconductivity, mentioned earlier, is the construction of superconducting magnets. These high-field magnets are crucial components in the operation of particle accelerators. Currently, all high-energy particle accelerators use liquid-helium-based superconducting technology. Again, there would be significant savings in cooling and operating costs if a liquid-nitrogen-based technology were developed.

An important application of superconducting magnets is a diagnostic tool called magnetic resonance imaging (MRI). This technique has played a prominent role in diagnostic medicine in the last few years because it uses relatively safe rf radiation to produce images of body sections, rather than x-rays. Because the technique relies on intense magnetic fields generated by superconducting magnets, the initial and operating costs of MRI systems are high. A liquid-nitrogen-cooled magnet could reduce such costs significantly. More details concerning MRI are presented in the essay by S. Marshall that concludes Chapter 45.

In the field of power generation, various companies and government laboratories have worked for years in developing superconducting motors and generators. In fact, a small-scale superconducting motor using the newly discovered ceramic superconductors has already been constructed at Argonne National Laboratory.

We have already mentioned some small-scale applications of superconductivity, namely SQUIDs and magnetometers which make use of the Josephson effect and quantum interference. Such devices are currently being used to measure and interpret the weak magnetic field generated by the brain. Other small-scale applications of Josephson junctions include their use as voltage standards and as infrared detectors. It should be noted that SQUIDs have been fabricated from films of $YBa_2Cu_3O_{7-\delta}$, and that Josephson junctions using this compound have been operated at liquid nitrogen temperature and above. It is quite likely that such small-scale applications will be influenced by the new generation of superconductors in the near future. Unfortunately, because of the brittle nature of these new materials, as well as their low critical current densities, one cannot at this time be optimistic about the future of large-scale applications.

## SUMMARY

Superconductors are materials whose dc resistance is zero below a certain temperature $T_c$, called the **critical temperature.** A second characteristic of a type I superconductor is that it behaves as a perfect diamagnet. At temperatures below $T_c$, any applied magnetic flux is expelled from the interior of a type I superconductor. This phenomenon of flux expulsion is known as the **Meissner effect.**

Applied magnetic fields are able to destroy the superconducting state of a material. The superconductivity of a type I superconductor is destroyed when the magnetic field exceeds the **critical magnetic field,** $B_c$, which is less than 0.2 T for the elemental superconductors.

A type II superconductor is characterized by two critical fields. When the applied field is less than the lower critical field, $B_{c1}$, the material is entirely superconducting and there is no flux penetration. When the applied field exceeds the upper critical field, $B_{c2}$, the superconducting state is destroyed and the flux penetrates the material completely. However, when the applied field lies between $B_{c1}$ and $B_{c2}$, the material is in a mixed state which is a combination of superconducting regions threaded by regions of normal resistance.

**Persistent currents** or supercurrents, once they are set up in a superconducting ring, have been shown to circulate for several years with no measurable losses and with zero applied voltage. This is a direct consequence of the fact that the dc resistance is truly zero in the superconducting state.

A central feature of the BCS theory of superconductivity for metals is the formation of a bound state called a **Cooper pair,** consisting of two electrons with equal and opposite momenta and opposite spins. The two electrons are able to form a bound state through a weak attractive interaction in which the crystal lattice serves as a mediator. In effect, one electron is weakly attracted to the other as it momentarily deforms the lattice as it moves. In the ground state of the superconducting system, all electrons form Cooper pairs, and all Cooper pairs are in the same quantum state of zero momentum. Thus, the superconducting state is represented by a single coherent wave function which extends over the entire volume of the sample. The BCS model predicts an energy gap given by $E_g = 3.53kT_c$, which is in contrast to a normal conductor, which has no energy gap. This energy gap represents the energy needed to break up one of the Cooper pairs, and is of the order of 1 meV for the elemental superconductors. The predictions of the BCS theory are in good agreement with most of the observed properties of conventional superconductors, such as specific heats, tunneling phenomena, absorption of electromagnetic radiation, and flux quantization.

The discovery of high-temperature superconductivity in 1986 by Bednorz and Müller has led to an uprecedented amount of research and excitement in the field of superconductivity in the last few years. Since this monumental discovery, many other milestones have been achieved by various researchers including the discovery of superconducting compounds with transition temperatures as high as 125 K. The high-$T_c$ compounds are all copper oxides whose critical temperatures appear to be linked to the number of copper-oxygen layers in the structures. The great interest in these discoveries is motivated by the potential of developing superconducting devices at high temperatures, with the ultimate goal of room-temperature operation. The new generation materials, which are known to be type II superconductors, have highly anisotropic resistivities and high upper critical fields. However, in the form of bulk ceramic samples, the materials have limited critical currents and are quite brittle. Although the BCS model appears to be consistent with most empirical observations, the basic mechanisms giving rise to the superconducting state remain undecided. A great deal of materials research will be necessary to overcome such problems before large-scale applications of high-$T_c$ superconductors are realized. Research in high-$T_c$ superconductivity and the ultimate goal of room-temperature superconductivity represent one of the most exciting challenges that scientists will face in the near future.

During a recent interview, Prof. M. Brian Maple, a research physicist at the University of California at San Diego, was asked what he found so fascinating about superconductivity.[17] His response was "For me the fascination of superconductivity is associated with the words perfect, infinite, and zero. A superconductor has the property of being a perfect conductor, or having infinite conductivity, or zero resistance. Whenever you see something that's perfect, infinite, or zero, truly zero, that's obviously a special state of affairs." (Of course, we can add to his response the property of perfect diamagnetism.) We hope that this introduction to the field of superconductivity will inspire the same fascination in many of you future scientists and engineers.

[17] *Supercurrents,* The Superconductivity Magazine, Vol. 2, February 1988, p. 13.

## QUESTIONS

1. Discuss the two major characteristics of a superconductor. Describe how you would measure these characteristics.
2. The properties of a superconductor are intimately associated with the words perfect, infinite, and zero, as noted in the closing comments. Discuss the meaning and significance of this statement.
3. Why is it not possible to explain the property of zero resistance using a classical model of charge transport through a solid?
4. Discuss the meanings of (a) the critical temperature of a superconductor, (b) the critical magnetic field, and (c) the critical current. (d) How are they related to each other?
5. Discuss the Meissner effect as it applies to type I and type II superconductors.
6. Discuss the differences between type I and type II superconductors. Discuss their similarities.
7. What are persistent currents, and how can they be set up in a superconductor?

8. A magnetic field applied to a superconductor decreases exponentially to zero inside the material with distance from the surface, which is characterized by the penetration depth. What happens to this penetration depth as $T$ approaches $T_c$?
9. The specific heat of a superconductor in the absence of a magnetic field undergoes an anomaly at the critical temperature, and shows an exponential decay towards zero below this temperature. What information does this behavior provide?
10. What is the isotope effect, and why does it play an important role in testing the validity of the BCS theory?
11. The concept of Cooper pairs is a central feature of the BCS theory of superconductivity. What are the Cooper pairs? Discuss their essential properties, such as their momentum, spin, binding energy, etc.
12. What is meant when it is said that Cooper pairs act collectively? How is it possible that all Cooper pairs can occupy the same quantum state?
13. How would you explain the fact that lattice imperfections and lattice vibrations can scatter electrons in normal metals, but have no effect on Cooper pairs?
14. Discuss the origin of the energy gap of a superconductor, and how the energy band structure of a superconductor differs from a normal conductor.
15. Why does the energy gap of a superconductor tend towards zero as the temperature approaches $T_c$ from below?
16. Describe two experiments that provide a direct measure of the energy gap of a superconductor.
17. Define single particle tunneling, and the conditions under which it can be observed. What information can one obtain from a tunneling experiment?
18. Define Josephson tunneling, and the conditions under which it can be observed. What is the difference between Josephson tunneling and single particle tunneling?
19. Describe the dc Josephson effect and the ac Josephson effect.
20. What are three useful devices that have been constructed which make use of the Josephson junction?
21. What is meant by magnetic flux quantization, and how is it related to the concept of Cooper pairs?
22. Discuss two milestones in the discovery of high-temperature superconductivity which occurred in 1986 and 1987.
23. The discovery of high-temperature superconductivity has been regarded by some as being as important to science and technology as the invention of the transistor. Discuss the rationale behind this statement.
24. The high critical temperatures of the new superconducting copper oxides make them very promising for developing devices that will operate at elevated temperatures. What are the present limiting features of these materials as far as possible applications are concerned?
25. How does one design a superconducting solenoid? What properties must you be concerned with in your design?
26. What results suggest that the superconductivity in the new high-$T_c$ copper oxides is two-dimensional in nature? In what manner do the crystal structures of these materials affect the superconductivity?
27. Because of the brittle nature of the new high-$T_c$ bulk ceramic oxides, it is not yet possible to form wires to construct superconducting magnets and other large-scale devices. On the other hand, much research is being conducted on thin-film devices using these materials. What kinds of problems associated with materials are likely to arise in developing this technology?
28. Discuss at least four features of high-$T_c$ superconductors that make them superior to the "old generation" superconductors. In what way are they inferior?

The following questions deal with the demonstration of the Meissner effect. A small permanent magnet is placed on top of a high-temperature superconductor (usually $YBa_2Cu_3O_{7-\delta}$), starting at room temperature. As the superconductor is cooled with liquid nitrogen (77 K), the permanent magnet is observed to levitate as shown in Figure 44.26—a most dazzling phenomenon. Assume for simplicity that the superconductor is type I.

**Figure 44.26** (Questions 29–35) Photograph of a small permanent magnet levitated above a disk of the superconductor $YBa_2Cu_3O_{7-\delta}$ which is at a temperature of 77 K. (Courtesy of Profs. J. Dinardo and Som Tyagi, Drexel University)

29. According to the Meissner effect, the magnetic field must be expelled from the superconductor below its critical temperature. How does this explain the levitation of the permanent magnet? What must happen to the superconductor in order to account for this behavior?
30. If the permanent magnet is set into rotation while levitated, it will continue to rotate for a long time on

its own. Can you think of an application that could make use of this frictionless magnetic bearing?

31. As soon as the permanent magnet has levitated, it gains potential energy. What accounts for this increase in mechanical energy? (This is a tricky one.)
32. When the levitated permanent magnet is pushed towards the edge of the superconductor, it will often fly off the edge. Why do you suppose this happens?
33. Why is it necessary to use a very small, but relatively strong, permanent magnet in this demonstration?
34. If the experiment is repeated by first cooling the superconductor below its critical temperature, and then placing the permanent magnet on top of it, will the permanent magnet still levitate? If so, would there be any difference in its elevation compared to the previous case?
35. If a calibrated thermocouple were attached to the superconductor to measure its temperature, describe how you could obtain the critical temperature of the superconductor through the levitation effect. (*Hint:* Start the observation below $T_c$ with a levitated permanent magnet.)

## PROBLEMS

### Section 44.3 Some Properties of Type I Superconductors and Section 44.4 Type II Superconductors

1. A wire made of $Nb_3Al$ has a radius of 2 mm and is maintained at 4.2 K. Using the data in Table 44.2, find (a) the upper critical field for this wire at this temperature, (b) the maximum current that can pass through the wire before its superconductivity is destroyed, and (c) the magnetic field at a distance of 6 mm from the surface of the wire when the current has its maximum value.
2. A superconducting solenoid is to be designed to generate a magnetic field of 10 T. (a) If the solenoid winding has 2000 turns/meter, what is the required current? (b) What force per meter length of wire is exerted on the solenoid's inner windings by the magnetic field?
3. Determine the current generated in a superconducting ring of niobium metal 2 cm in diameter if a magnetic field of 0.02 T directed *perpendicular* to the ring is suddenly decreased to zero. The inductance of the ring is $L = 3.1 \times 10^{-8}$ H.
4. Determine the amount of magnetic field energy in joules that needs to be added to destroy superconductivity in 1 cm$^3$ of lead near 0 K. Use the fact that $B_c(0)$ for lead is 0.08 T.
5. The penetration depth for lead (Pb) at $T = 0$ K is 39 nm. Find the penetration depth in lead at (a) $T =$ 1 K, (b) $T = 4.2$ K, and (c) $T = 7.0$ K.
6. Find the critical magnetic field in mercury (Hg) at (a) $T = 1$ K and (b) $T = 4$ K.

### Section 44.5 Other Properties of Superconductors

*Persistent Currents.* In an experiment carried out by S.C. Collins between 1955 and 1958, a current was maintained in a superconducting lead ring for 2½ years with no observed loss. If the inductance of the ring was $3.14 \times 10^{-8}$ H, and the sensitivity of the experiment was 1 part in $10^9$, determine the maximum resistance of the superconducting ring. (*Hint:* Treat this as a decaying current in an *RL* circuit, and recall that $e^{-x} \cong 1 - x$ for small $x$.)

8. *Velocity of Electron Flow.* Current is carried throughout the body of niobium-tin, a type II superconductor. If a niobium-tin wire of cross section 2 mm$^2$ can carry a maximum supercurrent of $10^5$ A, estimate the average speed of the superconducting electrons. (Assume the density of conducting electrons is equal to $n_s = 5 \times 10^{27}$ /m$^3$.)
9. *Diamagnetism.* When a superconducting material is placed in a magnetic field, surface currents are established that make the magnetic field inside the material truly zero. (That is, the material is perfectly diamagnetic.) Suppose that a circular disk, 2 cm in diameter, is placed in a magnetic field $B = 0.02$ T with the plane of the disk perpendicular to the field lines. Find the equivalent surface current if it all lies at the circumference of the disk.
10. *Surface Currents.* A rod of a superconducting material 2.5 cm long is placed in a uniform magnetic field of 0.54 T with its cylindrical axis along the magnetic field lines. (a) Sketch the directions of the applied field and the induced surface current, and (b) estimate the magnitude of the surface current.

### Section 44.7 The BCS Theory and Section 44.8 Energy Gap Measurements

11. Calculate the energy gap for each of the type I superconductors given in Table 44.1 as predicted by the BCS theory. Use the known values of their critical temperatures and compare your answers with the experimental values given in Table 44.4.
12. Calculate the energy gaps for each of the type II superconductors given in Table 44.2 as predicted by the BCS theory. How do your values compare with those found for the type I superconductors?
13. *High-Temperature Superconductor.* Estimate the energy gap $E_g$ for the high-temperature superconductor $YBa_2Cu_3O_{7-\delta}$, which has a transition temperature $T_c = 92$ K, assuming BCS theory holds.

14. *Isotope Effect.* According to the isotope effect, the critical temperature should be inversely proportional to the mass according to $T_c \propto M^{-\alpha}$. Use the data for mercury given below to determine the value of the constant $\alpha$ in the exponent. Is your result close to what you might expect based on a simple model?

| Isotope | $T_c$ (K) |
|---|---|
| $^{199}$Hg | 4.161 |
| $^{200}$Hg | 4.153 |
| $^{204}$Hg | 4.126 |

15. *Cooper Pairs.* A cooper pair of electrons in a type I superconductor has an average separation of about $10^{-4}$ cm. If these two electrons can interact within a volume of this diameter, how many other Cooper pairs have their centers within the volume occupied by one pair? Use the appropriate data for lead, which has $n_s = 2 \times 10^{22}$ electrons/cm$^3$.
16. *Dipole in a* ***B*** *Field.* The potential energy of a magnetic dipole of moment $\boldsymbol{\mu}$ in the presence of a magnetic field $\mathbf{B}$ is given by $U = -\boldsymbol{\mu} \cdot \mathbf{B}$. When an electron, which has spin of 1/2, is placed in a magnetic field, its magnetic moment can be aligned either with or against the field. Because its magnetic moment is negative, the higher energy state, having energy $E_2 = \mu B$, corresponds to the case where its moment is aligned with the field, while the lower energy state, with energy $E_1 = -\mu B$, corresponds to the case where its moment is aligned against the field. Thus, the energy separation between these two states is $\Delta E = 2\mu B$, where the magnetic moment of the electron is equal to $\mu = 5.79 \times 10^{-5}$ eV/T. (a) If a Cooper pair is subjected to a magnetic field of 38 T (the critical field for $Nb_3Ge$) calculate the energy separation between the "spin-up" electron and the "spin-down" electron. (b) Calculate the energy gap for $Nb_3Ge$ as predicted by the BCS theory at 0 K using the fact that $T_c = 23$ K. (c) How do your answers to (a) and (b) compare? What does this result suggest based on what you have learned about critical fields?

### Section 44.9 Flux Quantization and Section 44.10 Josephson Tunneling

17. Estimate the area of a ring that would fit one of your fingers, and calculate the magnetic flux through the ring due to the earth's magnetic field (take $B_{earth} = 5.8 \times 10^{-5}$ T). If this flux were quantized, how many fluxons would the ring enclose?
18. A Josephson junction (see Figure 44.19) is fabricated using indium for the superconducting layers. If the junction is biased with a dc voltage of 0.50 mV, find the frequency of the ac current generated. (For comparison, note that the energy gap of indium at $T = 0$ K is 1.05 meV.)
19. If a magnetic flux of $10^{-4}\Phi_0$ ($\frac{1}{10\,000}$ of the flux quantum) can be measured with a SQUID device (Figure 44.27), what is the smallest magnetic field change $\Delta B$ that can be detected with this device, if the ring has a radius of 2 mm?

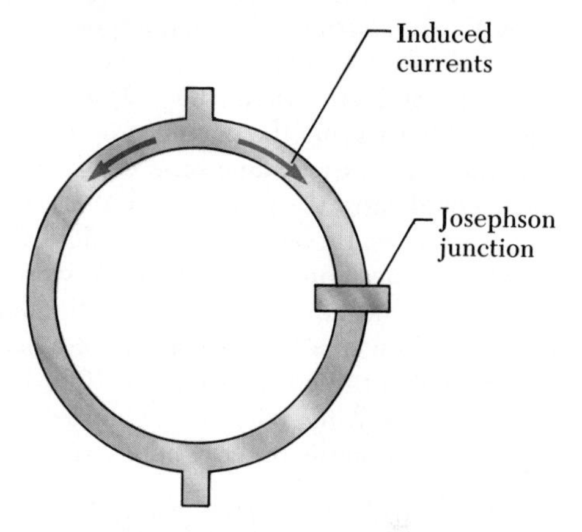


**Figure 44.27** Superconducting Quantum Interference Device (SQUID). (Problem 19).

20. A superconducting circular loop of very fine wire has a diameter of 2 mm and a self-inductance of 5 nH. The flux through the loop, $\Phi$, is the sum of the applied flux, $\Phi_{app}$, and the flux due to the supercurrent in the ring itself, $\Phi_{sc} = LI$, where $L$ is the self-inductance of the loop. Since the flux through the loop is quantized, we have

$$n\Phi_0 = \Phi_{app} + \Phi_{sc} = \Phi_{app} + LI$$

where $\Phi_0$ is the flux quantum. (a) If the applied flux is zero, what is the smallest current that can meet this quantization condition? (b) If the applied field is perpendicular to the plane of the loop and has a magnitude of $3 \times 10^{-9}$ T, find the smallest current that will circulate around the loop.

## ADDITIONAL PROBLEMS

21. A solenoid of diameter 8.0 cm and length 1 m is wound with 2000 turns of superconducting wire. If the solenoid carries a current of 150 A, find (a) the magnetic field at the center of the solenoid, (b) the magnetic flux through the center cross section of the solenoid, and (c) the number of flux quanta through the center.
22. *Energy Storage.* A novel method has been proposed to store electrical energy by fabricating a huge underground superconducting coil 1.0 km in diameter. This would carry a maximum current of 50 kA through each winding of a 150-turn $Nb_3Sn$ solenoid. (a) If the inductance of this huge coil is 50 H, what would be

the total energy stored? (b) What would be the compressive force acting per meter length between two adjacent windings 0.25 m apart?

23. *Superconducting Power Transmission.* Superconductors have also been proposed for power transmission lines. A single coaxial cable (Fig. 44.28) could carry $10^3$ MW (the output of a large power plant) at 200 kV, dc, over a distance of 1000 km without loss. The superconductor would be a 2-cm-radius inner $Nb_3Sn$ cable carrying the current $I$ in one direction, while the outer surrounding superconductor of 5-cm radius would carry the return current $I$. (a) In such a system, what would be the magnetic field at the surface of the inner conductor? (b) What would be the magnetic field at the inner surface of the outer conductor? (c) How much energy would be stored in the space between the conductors in a 1000-km superconducting line? (d) What is the force per meter length exerted on the outer conductor?

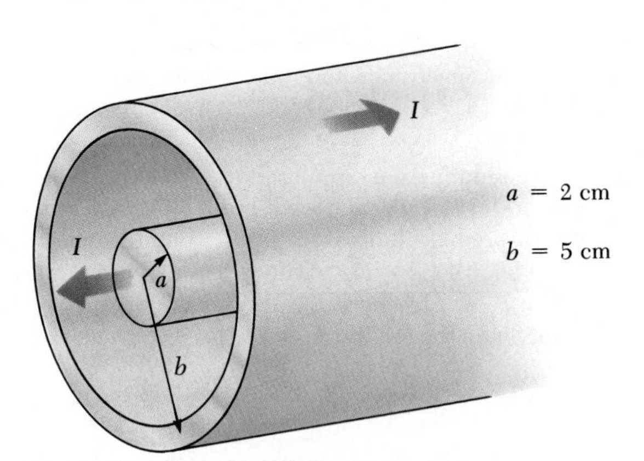


**Figure 44.28** (Problem 23.)

24. *Penetration Depth.* The $1/e$ penetration depth of a magnetic field into a superconductor is found from London's equations to be

$$\lambda = \sqrt{\frac{m_e}{\mu_0 n_s e^2}}$$

(a) Estimate the number of superconducting electrons per m$^3$ in lead at 0 K if the magnetic penetration depth near 0 K is given by $\lambda_0 = 4 \times 10^{-8}$ m. (b) Starting with Equation 44.4, determine the number of superconducting electrons per m$^3$ in lead as a function of temperature for $T \leq T_c$.

25. *"Floating" a Wire.* Is it possible to "float" a superconducting lead wire of radius 1 mm in the magnetic field of the earth? Assume the horizontal component of the earth's magnetic field is 0.5 gauss.

26. *Magnetic Field Inside a Wire.* A type II superconducting wire of radius $R$ carries current uniformly distributed through its cross section. If the total current carried by the wire is $I$, show that the magnetic energy per unit length *inside* the wire is $\frac{\mu_0 I^2}{16\pi}$.

27. *Magnetic Levitation.* If a very small but very strong permanent magnet is lowered toward a flat-bottomed type I superconducting dish, at some point it will float freely, or levitate, above the superconductor. This is because the superconductor is a perfect diamagnet, and it expels all magnetic flux. Therefore the superconductor acts like an identical magnet lying an equal distance below the surface. (See Figure 44.29.) At what height will the magnet "float" if its mass is 4 g and its magnetic moment $\mu = 0.25\ \text{A}\cdot\text{m}^2$? $\left(\textit{Hint:}\text{ The potential energy between two dipole magnets a distance } r \text{ apart is } \frac{\mu_0 \mu^2}{4\pi r^3}\right)$.

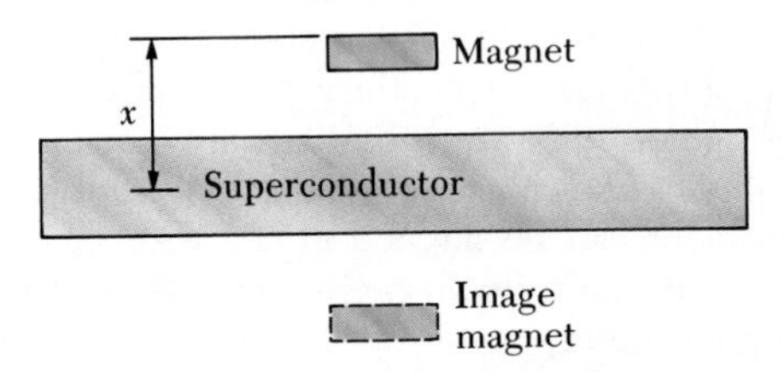


**Figure 44.29** Magnetic levitation experiment. (Problem 27.)

28. *Designing a Superconducting Solenoid.* A superconducting solenoid was made with $Nb_3Zr$ wire wound on a tube of diameter 10 cm. The solenoid winding consisted of a double layer of 0.5-mm diameter wire with 984 turns (corresponding to a coil length of 25 cm). (a) Calculate the inductance, $L$, of the solenoid, assuming its length is large compared to its diameter. (b) The magnitude of the persistent current in this solenoid in the superconducting state has been reported to decrease by one part in $10^9$ per hour. If the resistance of the solenoid is $R$, then the current in the circuit should decay (see Ch. 32) according to

$$I = I_0 e^{-Rt/L}$$

This resistance, although small, is due to magnetic flux migration in the type II superconductor. Determine an upper limit of the coil's resistance in the superconducting state. [The data were reported by J. File and R.G. Mills, *Phys. Rev. Letters* **10**, 93(1963).]

29. *Entropy Difference.* The entropy difference per unit volume between the normal and superconducting states is given by

$$\frac{\Delta S}{V} = -\frac{\partial}{\partial T}\left(\frac{B^2}{2\mu_0}\right)$$

where $\frac{B^2}{2\mu_0}$ is the magnetic energy per unit volume required to destroy superconductivity. Determine the entropy difference between the normal and superconducting states in 1 mol of lead at 4 K if the critical magnetic field $B_c(0) = 0.08$ T and $T_c = 7.2$ K.

30. *A Convincing Demonstration of Zero Resistance.* A direct and relatively simple demonstration of the property of zero dc resistance of a superconductor can be carried out using what is known as the *four-point probe method.* The probe that is used, shown in Figure 44.30, consists of a disc of $YBa_2Cu_3O_{7-\delta}$ (a high-$T_c$ superconductor) to which four wires are attached by indium solder (or some other suitable contact material). A constant current is maintained through the sample by applying a dc voltage between points *a* and *b*, and is measured with a dc ammeter. (The current can be varied with a variable resistance *R* as indicated.) The potential difference $V_{cd}$ between points *c* and *d* is measured with a dc digital voltmeter. When the probe is immersed in liquid nitrogen, the sample cools quickly to 77 K, which is below the critical temperature of the sample (92 K); the current remains approximately constant, but *the potential difference $V_{cd}$ drops abruptly to zero.* (a) How would you explain this observation on the basis of what you know about superconductors and your understanding of Ohm's law? (b) The data in Table 44.6 represent actual values of $V_{cd}$ for different values of $I$ taken on a sample at room temperature. Make an *I-V* plot of the data, and determine whether the sample behaves in a linear manner. From the data, obtain a value of the dc resistance of the sample at room temperature. (c) At room temperature, it is found that $V_{cd} = 2.234$ mV for $I = 100.3$ mA, but after the sample is cooled to 77 K, $V_{cd} = 0$ and $I = 98.1$ mA. What do you think might explain the slight decrease in the current upon cooling the sample to below the critical temperature?

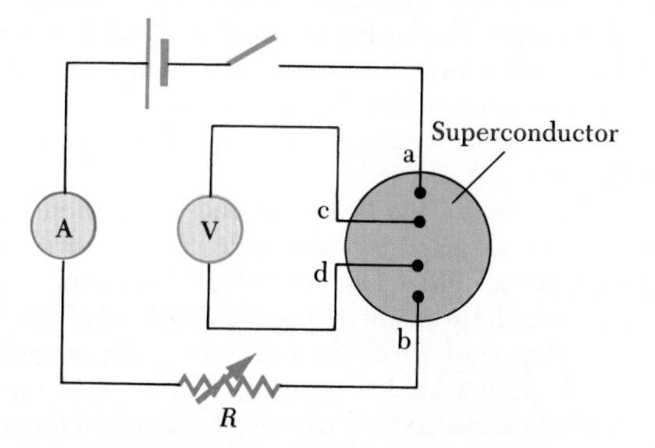


**Figure 44.30** (Problem 30) Circuit diagram used in the four-point probe measurement of the dc resistance of a sample. A dc digital ammeter is used to measure the current, while the potential difference between points *c* and *d* is measured with a dc digital voltmeter. Note that there is no voltage source in the inner loop circuit where $V_{cd}$ is measured. In our circuit, the current source was provided by a 6 V battery, and the values of the variable resistance *R* ranged from about 10 Ω to 100 Ω.

**TABLE 44.6 Current versus Potential Difference $V_{cd}$ Measured in a Bulk Ceramic Sample of $YBa_2Cu_3O_{7-\delta}$ at Room Temperature**[a]

| *I* (mA) | $V_{cd}$ (mV) |
|---|---|
| 57.8 | 1.356 |
| 61.5 | 1.441 |
| 68.3 | 1.602 |
| 76.8 | 1.802 |
| 87.5 | 2.053 |
| 102.2 | 2.398 |
| 123.7 | 2.904 |
| 155 | 3.61 |

[a] The current was supplied by a 6-V battery in series with a variable resistor *R*. The values of *R* ranged from 10 Ω to 100 Ω. The data are from the author's laboratory.

"If tachyons do exist, and if they do go faster than the speed of light, then I'm determined to find something that goes faster than tachyons."

# ESSAY

## Levitation and Suspension Effects in Superconductors

**Brian B. Schwartz**
*Department of Physics, Brooklyn College, and The American Physical Society*

### Introduction

The recent discovery of high-temperature superconductors has added a new image to our media-oriented visual society: that of a small magnet floating above a pellet of superconductor cooled to liquid nitrogen temperature, 77 K. A picture much like that in Figure 1 has appeared in scientific journals, newspapers, and popular as well as business-oriented magazines. There are two major hallmarks of superconductivity. First there exists a critical temperature $T_c$ below which the resistance goes to zero; second, below $T_c$, the magnetic field is excluded from the interior of the superconductor. This field exclusion effect is called the Meissner-Ochsenfeld effect. The levitation of a magnet above a superconductor, or vice versa, was well known before the discovery of high-$T_c$ superconductors; however, it was difficult to demonstrate. A special see-through glass-walled dewar operating at liquid helium temperature, 4 K, was required. The discovery of high-$T_c$ superconductors changed the ease with which the demonstration of levitation is possible. All one needs now is a Styrofoam cup and relatively cheap and easy-to-handle liquid nitrogen. Even more importantly, the demonstrator and the observers are now able to get a hands-on feel for the levitation effect. This ability to "tinker" is especially important because one can feel directly some of the special levitation properties of the high-$T_c$ materials. For example, the levitating superconductor exhibits lateral stability, and requires a force to move it sideways. This would not be the case if the superconductor simply excluded the magnetic field. Another easily observable effect is the fact that the levitation height can vary and thus is not unique. Even more surprising was the recent discovery of the suspension effect,[1] in which a chip of a high-temperature superconductor with strong pinning can be raised from a liquid nitrogen container and held suspended *below* the magnet in a "stable" position. (See Fig. 2.)

In this essay, we present a discussion of the levitation and suspension effects associated with superconductivity. In all cases, we will try to present a physical picture in terms of a simple understanding of the magnetic behavior of superconductors. We shall use the force law between the gradient of the magnetic field of a magnet and a magnetizable material. As a background, we define and describe the magnetization of type I superconductors, and another kind of superconductor called type II, which allows for some penetration of magnetic field in special quantized flux units called vortices. The flux vortices, produced by small circulating currents, can be effectively pinned and thus become unable to move sideways or into or out of the superconductor easily. The pinning effects are instrumental to understanding the

**Figure 1** Photograph of a small permanent magnet levitated above a pellet of the $Y_1Ba_2Cu_3O_{7-\delta}$ (123) superconductor cooled to liquid nitrogen temperature. (Courtesy of IBM Research)

suspension effect and the lateral stability as well as the possibility of many "stable" levitation and suspension heights.

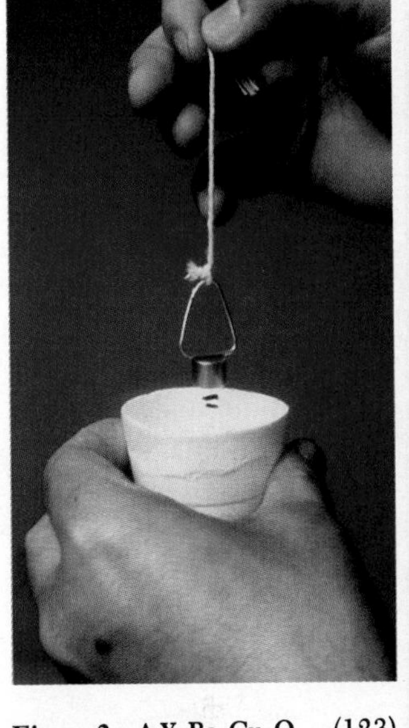

**Figure 2** A $Y_1Ba_2Cu_3O_{7-\delta}$(123) superconductor doped with AgO (the small black chip) is shown suspended below a magnet (the shiny cylinder handing from a string). The suspended superconductor has just been raised from the liquid nitrogen in the Styrofoam cup and is hence at 77 K.

## Perfect Diamagnetism and the Meissner Effect

If an electric field $\boldsymbol{E}$ in a metal acts on an electron of charge $e$ and mass $m$, Newton's force law gives $\boldsymbol{F} = e\boldsymbol{E} = m(d\boldsymbol{v}/dt)$. If one assumes perfect conductivity in which no resistive term scatters the electron velocity, and uses Maxwell's equations, one obtains the magnetic behavior for perfect conductors in which the magnetic field lines deep inside the perfect conductor do not change with time. As soon as one attempts to change the internal magnetic field, lossless currents are set up which maintain the internal field, $\boldsymbol{B}_{\text{in}}$, and thus within a perfect conductor $(d\boldsymbol{B}/dt) = 0$. In 1933, however, careful magnetic measurements by Meissner and Ochsenfeld showed that not only does $(d\boldsymbol{B}/dt) = 0$ hold within superconducting materials, but, in addition, the internal magnetic field itself is also zero, $\boldsymbol{B} = 0$. In other words, superconductors are not simply perfect conductors but also exhibit perfect diamagnetism. The magnetic field inside the superconductor is related to the external magnetic field $\boldsymbol{H}$ and the magnetization $\boldsymbol{M}$ of the superconductor through the equation

$$\boldsymbol{B} = \mu_0\boldsymbol{H} + \boldsymbol{M}$$

If, as Meissner observed for superconductors, $\boldsymbol{B} = 0$, then $\boldsymbol{M}$ is

$$\boldsymbol{M} = -(1/\mu_0)\boldsymbol{H} = \chi\boldsymbol{H}$$

where $\chi = (-1/\mu_0)$ is the magnetic susceptibility and corresponds to perfect diamagnetism. Superconductors which exhibit a complete Meissner effect with perfect diamagnetism are called type I superconductors. In Figure 3, we plot $\boldsymbol{B}$ and $\boldsymbol{M}$ versus the applied magnetic field $\boldsymbol{H}$. Since excluding magnetic field from the interior of a superconductor raises its free energy, there exists a critical field $\boldsymbol{H}_c$, above which it is energetically favorable for the superconductor to revert to the normal state where $\boldsymbol{B} = \mu_0\boldsymbol{H}$ and $\boldsymbol{M} = 0$.

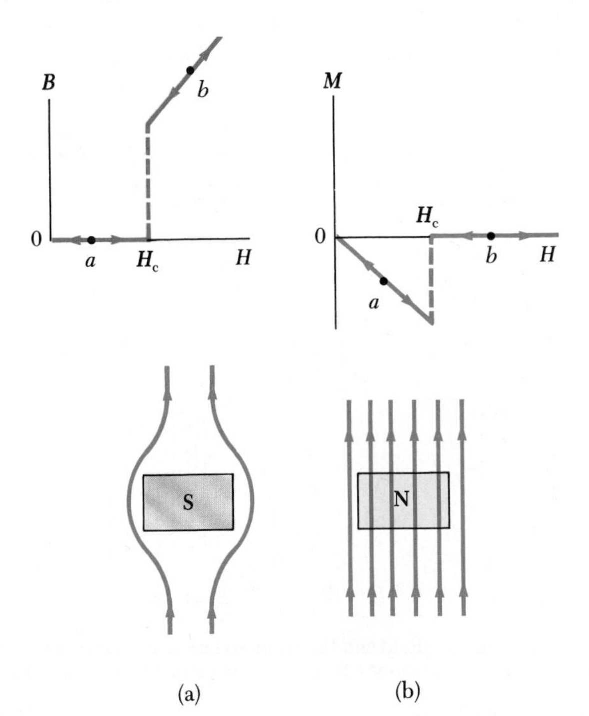


**Figure 3** The internal magnetic field and the magnetization vs. external field for a type I superconductor. The internal magnetic field profile sketched corresponds to parts (a) and (b).

## Incomplete Magnetic Flux Exclusion, Type II Superconductors

Almost all the elemental metals which exhibit superconductivity (except Nb) are type I superconductors. There is another class of magnetic behavior for some alloys (such as NbTi), compounds (such as $Nb_3Sn$), and the new high-$T_c$ superconductors, called type II superconductors. The magnetic field penetrates type II superconductors in quantized units of magnetic flux called **Abrikosov vortices,** corresponding to swirls of supercurrent producing a flux value of $2.07 \times 10^{-15}$ Wb. One must take into account the balance between the increase in energy required to create a vortex, and the lowering of the magnetic exclusion energy if flux can penetrate into the superconductor. Above a minimum field, it becomes energetically favorable to allow some flux to penetrate into the type II superconductor. The separation between type I and type II behavior is determined essentially by the ratio $\kappa$ of the penetration depth of the decaying magnetic field at the surface of a superconductor $\lambda$ to the superconducting correlation length, $\xi_0$, which corresponds to the length over which the electrons of the Cooper pair which are responsible for superconductivity are "bound." Type I superconductors have $\kappa < 1$ whereas type II superconductors have $\kappa > 1$. Figure 4 illustrates the magnetic behavior of an ideal type II superconductor. At a lower critical field $\boldsymbol{H}_{c1}$, it first becomes favorable for the flux vortices to penetrate. The material remains superconducting until the upper critical field value $\boldsymbol{H}_{c2}$ is reached. In the region $\boldsymbol{H}_{c1} < \boldsymbol{H} < \boldsymbol{H}_{c2}$, the field inside the superconductor, $\boldsymbol{B}$ is smaller than the external field $\boldsymbol{H}$ corresponding to a *negative* magnetization. At and above $\boldsymbol{H}_{c2}$, the superconductor returns to the normal state with $\boldsymbol{B} = \mu_0 \boldsymbol{H}$ and $\boldsymbol{M} = 0$.

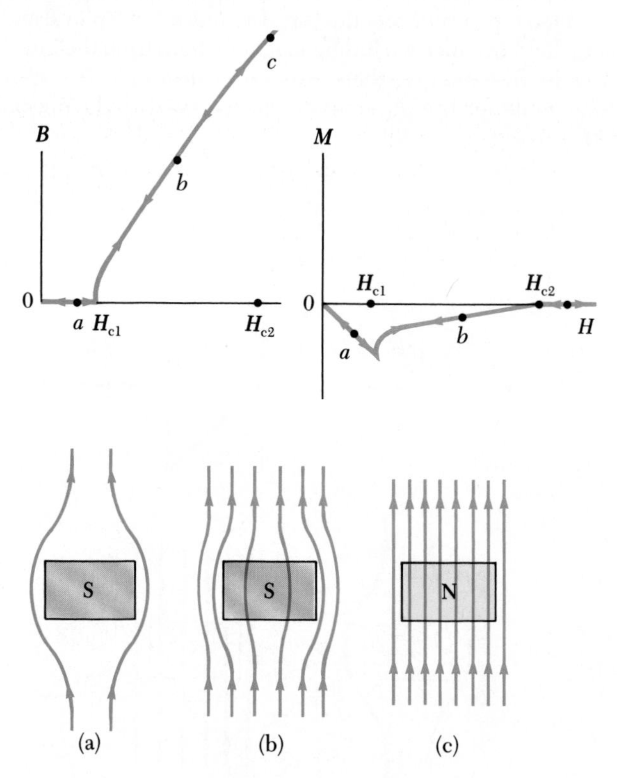


**Figure 4** The internal magnetic field and the magnetization vs. external field for an ideal type II superconductor. The internal magnetic field profile sketched corresponds to parts (a), (b), and (c).

## The Magnetic Behavior of Pinned Type II Superconductors

For ideal type II superconductors, the flux vortices can move into and out of the superconductor without any hindrance. The superconductor can always reach its thermodynamically most stable state in a magnetic field. The magnetization curves are reversible and each external field value gives a unique value of internal field $\mathbf{B}$. If however, the flux vortex or bundles of vortices which are responsible for the internal field are impeded in their movement (pinned), then a situation can arise in which the field inside the superconductor differs from that of an ideal type II superconductor.

A comparison can be made between the ideal type II behavior shown in Figure 4, and the pinned, irreversible behavior shown in Figure 5. Above $\mathbf{H}_{c1}$ for increasing magnetic fields, the flux for the pinned case does not penetrate freely into the superconductor. As a result, a screening current is set up such that the internal field is smaller than that in the thermodynamically ideal state. The change in the internal field and the magnetization at the lower critical field $\mathbf{H}_{c1}$ is not very sharp. The field is prevented from penetrating the superconductor until the pinning force is overcome

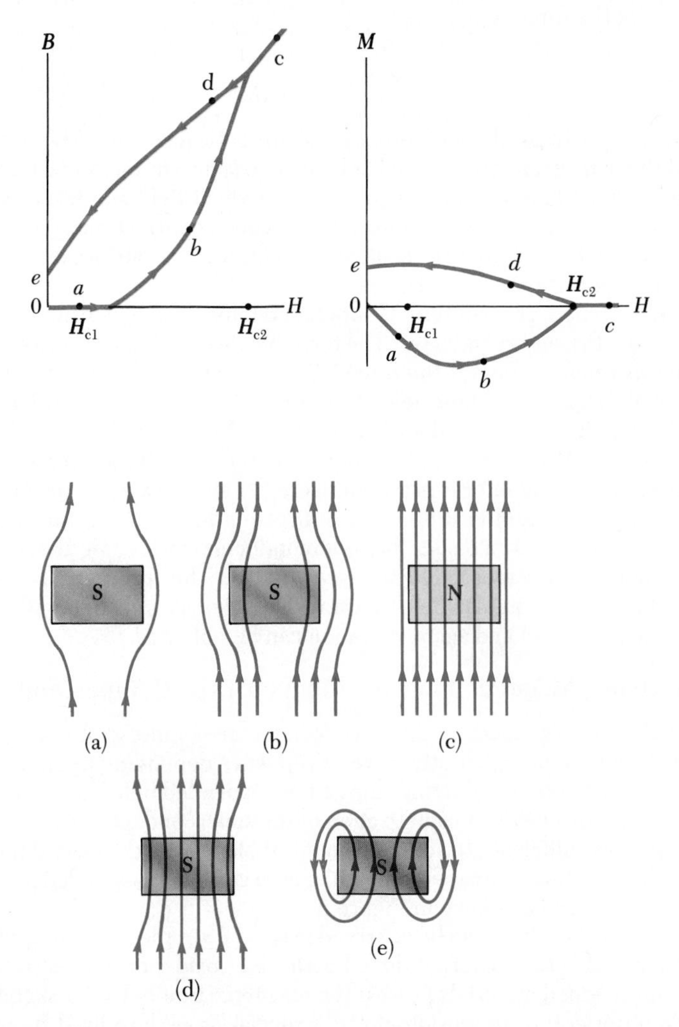


**Figure 5** The internal magnetic field and the magnetization vs. increasing and decreasing external field for a pinned, irreversible type II superconductor. The internal magnetic field sketched corresponds to parts (a), (b), (c), (d), and (e).

*(Continued)*

by the Lorentz force produced by the screening current acting on the quantized flux vortices. This model for flux motion is called the *Beam model* and produces the negative magnetization curve for increasing external field shown in Figure 5. The stronger the pinning force, the more difficult it is for flux to penetrate. The upper critical field $\boldsymbol{H}_{c2}$ remains the same since it is the magnetic field value for which the superconducting state returns to the normal state. If we now lower the external field from $\boldsymbol{H}_{c2}$, we get the opposite situation. The flux which is *inside* the superconductor at $\boldsymbol{H}_{c2}$ cannot move out freely. As a result, the internal field remains *higher* than the external field, leading to a *positive* magnetization. This positive magnetization remains even if the external field is reduced to zero. Figures 4b and 5b are sketches of the internal and external field behavior for an ideal and pinned type II superconductor in increasing and decreasing magnetic fields.

## The Force of a Magnetic Field on a Magnetizable Object

For simplicity, let us consider the case of the vertical force by a magnetic field $\boldsymbol{H}$ on a small (unit volume) specimen which is magnetized in the $z$ direction parallel to $\boldsymbol{B}$. The magnitude of the force is given by

$$F = \mu_0 M(H) \frac{dH}{dz}$$

where $(dH/dz)$ is the gradient of the external magnetic field and $M(H)$ is the magnetization of the material in the external field. Note that the force can be either positive or negative depending upon the signs of the gradient field and the magnetization. (This same force equation can be used to explain the attraction of a magnetizable material, like soft iron, to a permanent magnet, and the attraction and repulsion between two permanent magnets.)

For ideal type I or ideal type II superconductors, some external field is always excluded from the superconductor. The force between a magnet and a superconductor is *always repulsive,* and if the force is strong enough to balance the downward gravitational force, one obtains levitation. For example, if we consider a magnetic dipolar field along the $z$ axis, then one has $H \propto z^{-3}$ and $(dH/dz) \propto -z^{-4}$. For type I superconductors $M(H) = -(1/\mu_0)H$. Therefore, the force between the dipolar field and a type I superconductor is proportional to $z^{-7}$ and thus is in a positive direction for $z > 0$ (repulsive) and in the negative direction for $z < 0$ (again repulsive with respect to the dipole). As a result, the superconductor can levitate above a magnet or a magnet above the superconductor (see Fig. 1). The force on an ideal type II superconductor by a dipolar field is always negative as well since $M(H)$ is always diamagnetic (negative) and again levitation can be obtained.

## The Force of a Magnetic Field on a Pinned Type II Superconductor

The situation for a pinned type II superconductor is quite different. The vertical magnetic force can be either attractive *or* repulsive depending upon the *sign* of the magnetization. A superconducting chip can be suspended below a magnet if the field gradient times the positive magnetization of the superconductor gives rise to a magnetic force with sufficient attraction to counterbalance the downward force of gravity. In such a situation, the superconducting chip can hang suspended in the gradient of the magnetic field as shown in Figure 2.

The suspension effect can be observed in special samples of high-$T_c$ yttrium-barium-copper oxide (123 material) doped with silver oxide, for which strong pinning occurs. These doped materials possess the magnetization behavior sketched in Figure 6. To observe the suspension effect, a special procedure must be used, and is shown schematically in Figure 7. The permanent magnet is first lowered toward the superconducting chip, which is immersed in liquid nitrogen (Fig. 7a). We assume axial symmetry so only forces in the $z$ direction must be considered. The superconductor responds to the field of the magnet in such a way as to be repelled by the permanent magnet. This corresponds to the negative (diamagnetic) initial magnetization region in Figure 6 in which the diamagnetic superconductor and the permanent magnet repel one another. The chip is forced against the bottom of the cup by both gravity and the magnetic field. As the permanent magnet is lowered further, the

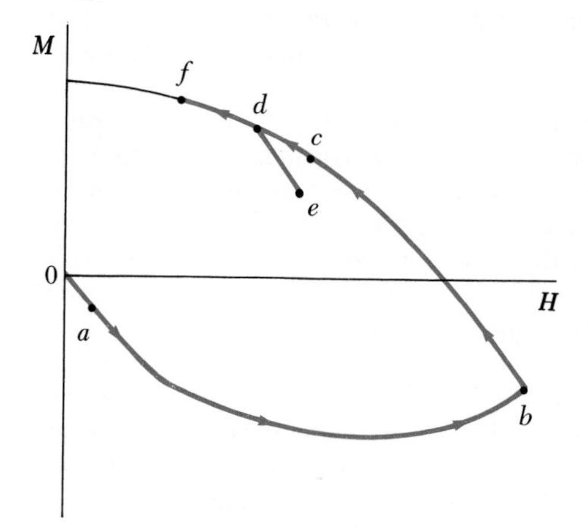


**Figure 6** A schematic representation of the magnetization curve corresponding to parts (a), (b), (c), (d), (e), and (f) in Figure 7. The suspension corresponds to point *d*. The curves *de* and *df* correspond to the situation in which the superconductor is disturbed such that it moves closer and further from the suspending magnet.

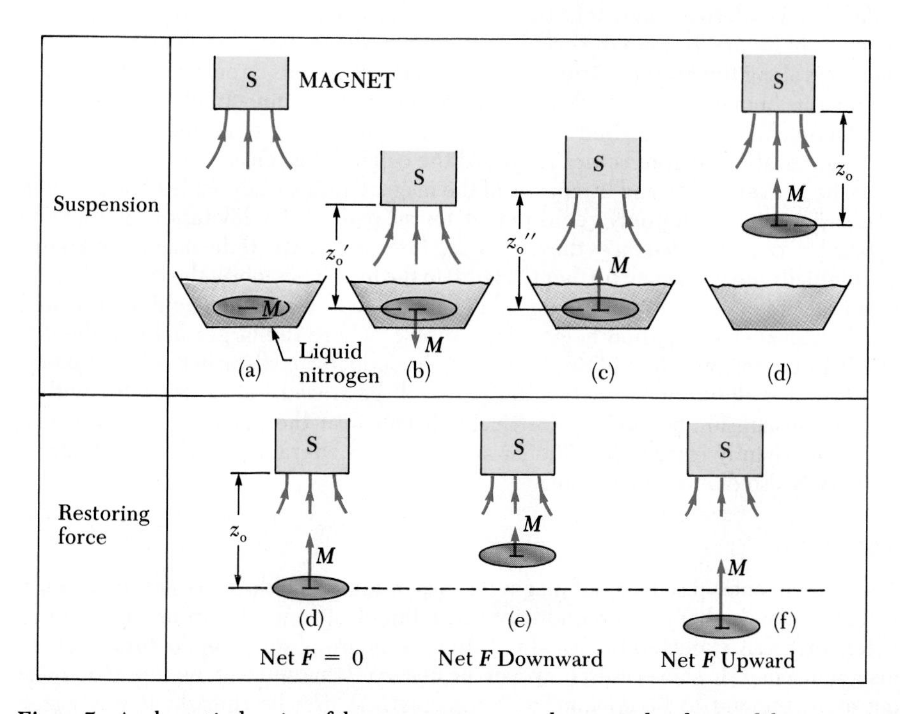


**Figure 7** A schematic drawing of the steps necessary to observe and understand the suspension effect: (a) Lowering the permanent magnet, negative magnetization. (b) Further lowering leads to stronger repulsion of the superconductor chip. (c) Pulling away of the magnet, with the reversal of the magnetization and an attraction to the magnet. (d) Suspension of the superconductor at a distance $z_0$. At this point, the upward magnetic force counterbalances the downward force of gravity. (e) The superconductor is disturbed such that it moves closer to the magnet. This results in a reduction of the positive magnetization. Superconducting screening currents flow without flux motion so the slope is steep corresponding to diamagnetic screening much like a perfect conductor. (f) The superconductor is disturbed such that it moves further from the magnet. This results in an increase of the positive magnetization.

*(Continued on next page)*

repulsive force can be even stronger (Fig. 7b) because the sample is more diamagnetic. At this point, the permanent magnet is moved away from the superconductor. This corresponds to reducing the external field, which in turn results in flux trapping and a positive magnetization (upper side of the loop in Fig. 6). This positive magnetization leads to an attraction of the superconductor to the permanent magnet (Fig. 7c). If the force of attraction is strong enough to counterbalance the force of gravity, the superconductor can be lifted out of the liquid nitrogen and will hang suspended in space below the permanent magnet (Fig. 7d and the photograph in Fig. 2), as long as it remains superconducting.

The attraction and ultimate suspension of the superconductor below the permanent magnet is unlike the case of the attraction due to a soft iron magnet. The force of attraction gets stronger as an iron specimen gets closer to the permanent magnet. In the case of the pinned type II superconductor, the attractive force gets stronger the *further* away the magnet is pulled from the superconductor. To understand this phenomenon, one must take into account the negative slope of the upper part of magnetization curve in Figure 6. If the suspended superconductor is disturbed slightly such that it moves closer to the permanent magnet (stronger magnetic field) than the original suspension point, as in Figure 7e, the flux remains constant in the superconductor and the screening current is reduced. This response is similar to what one expects for a perfect conductor. The positive magnetization of the superconductor is reduced and the attraction weakens. The gravitational force causes the superconductor to "fall back" towards the original suspension point. If the superconductor is disturbed such that it moves farther from the permanent magnet (weaker magnetic field), as in Figure 7f, than the original suspension point, the positive magnetization increases along the magnetization curve. The attraction gets stronger and the superconductor "moves up" towards the original suspension point. In effect, the superconductor acts as if it has a built-in servomechanism that acts to increase or decrease its magnetization so as to return it toward the original suspension point.

The irreversibility and hysteresis of the magnetization does not lead to a unique suspension point. The only requirement for suspension (or levitation) is that the upward force *counterbalances* the downward force of gravity. If the magnet in Figure 7b is initially lowered even further (larger $\boldsymbol{B}$ in the hysteresis loop), the reversal of the sign in magnetization upon pulling away would occur at a higher field. One would obtain a smaller suspension height, for a dipole, where the larger field gradient is closer to the magnet. In addition, if the superconductor is disturbed while it hangs suspended, a different magnetization path will be traced out, leading to another possible suspension height. The vertical behavior near the suspension point is discussed more fully elsewhere.[5] Similar arguments for lateral forces show that lateral stability is also due to flux pinning.

## Conclusion

The experimental discovery of magnetic suspension was a direct result of working with the new high-$T_c$ superconductors in a liquid nitrogen environment. Palmer Peters of NASA expected the usual repulsive response of superconductors to a permanent magnet. In some cases, however, he observed that some of the superconducting chips seemed to be sticking to his magnet. At first he thought it was water condensation or some direct contact surface-sticking effect. He noted, however, that the superconducting chip was truly hanging suspended in space below the magnet. Shortly after this observation, it was realized that pinning with flux tapping would give rise to a positive magnetization and could explain the suspension effect. Finally, one should note that the suspension of the superconductor at a fixed distance from a magnet does not depend upon a gravitational field. (For levitation, the magnetic force is always repulsive. In the absence of gravity, the superconductor and magnet move further and further away from each other.) In the procedure sketched in Figure 7, when the magnet direction is reversed, for zero gravity, the suspension would occur at the point where the magnetization $\boldsymbol{M} = 0$. At this point, there is zero net force on the superconductor and it will be pulled out of the liquid nitrogen and remain at a distance corresponding to the field value of the $\boldsymbol{M} = 0$ crossing.

## References


1. Y. Shapira, C.Y. Huang, E.J. McNiff, Jr., P.N. Peters, and B.B. Schwartz, Magnetization and Magnetic Suspension of YBaCuO–AgO Ceramic Superconductors, *Jour. of Mag. and Magnetic Materials,* **78**; 19, 1989. A review paper on the suspension effect including quantitative calculations. See Section 6 for a discussion of the vertical motion of the hanging superconductor when disturbed from the suspension point.
2. B.B. Schwartz and S. Frota-Pessoa, "Electromagnetic Properties of Superconductors," in *Electromagnetism: Paths to Research,* ed. Doris Teplitz, New York, Plenum Press, 1982. A good review of the properties of superconductors at the level of a college physics major. A good source of material for the advanced student.
3. M. Tinkham, *Introduction to Superconductivity,* New York, McGraw-Hill, Inc., 1975. A readable, clear presentation of the basic physics of superconductivity.
4. C.P. Bean, "Magnetization of High Field Superconductors," *Rev. Mod. Phys.* **36:** 31, 1964. A good discussion of the way the flux penetrates a pinned type II superconductor.
5. E.H. Brandt, "Levitation in Physics," *Science* **243**; 349, 1989. A comprehensive review of a host of levitation principles including acoustic, optical, electric, magnetic, and superconductive.


## Essay Questions

1. How would one use an electromagnet in order to suspend a ferromagnetically magnetizable material below it? Assume you are able to control the current in the electromagnet and could sense the height of suspended magnetic material. This very process is being used to develop a magnetically suspended high-speed train system in Germany.
2. When a current is set up in a high-field type II superconductor wire in a magnet, a Lorentz force acts on the flux lines. How does the pinning of the flux lines relate to the maximum value of the critical current? If there was no pinning; would a type II superconductor have resistance? (See reference 4.)
3. The levitation or suspension of a superconducting chip below or above a magnet is not the familiar situation of an object in a potential well since the restoring force is *not conservative.* Can you imagine other systems in which disturbances from a "stable" position return the object to a new "stable" position?

## Essay Problems

1. Calculate the gradient of the magnetic field along the cylindrical axis above a round cylindrical magnet of radius $R$, length $L$, and magnetization per unit volume $\boldsymbol{M}_0$.
2. How does the area between the magnetization curve and the $B_{in} = 0$ axis in Figures 3 and 4 relate to the increase in energy of the superconductor? At the critical field, what can one say about the energy of the superconductivity state and the normal state?
3. Using Newton's second law, the definition of current density, and Maxwell's equation, show that the equation for the change in magnetic field inside a perfect conductor is given by

$$\frac{d^2}{dx^2}\left(\frac{dB}{dt}\right) = \frac{1}{\lambda^2}\left(\frac{dB}{dt}\right) \quad \text{where} \quad \lambda = \left(\frac{m}{\mu_0 n e^2}\right)^{1/2}$$

# ESSAY

## Superconducting Devices*

**Clark A. Hamilton**
*National Institute of Standards and Technology, Electromagnetic Technology Division*

Superconductivity was discovered in 1911 by Dutch physicist Heike Kamerlingh Onnes. In experiments to measure the resistance of frozen mercury he made the startling discovery that the resistance vanished completely at a temperature a few degrees above absolute zero. Since that time superconductivity has been found in a large number of elemental metals and alloys. The first proposed application of this new phenomenon was to make powerful electromagnets. Magnets wound with copper wire are limited by the heat generated in the windings. Since superconductors can carry current with no heat generation it was assumed that extremely powerful magnets could easily be made. Unfortunately it was soon discovered that the superconductivity in materials such as mercury, tin, and lead is destroyed by only modest magnetic fields. In fact, the conditions under which superconductivity exists in any material can be described by a 3-dimensional diagram such as that shown in Figure 1a. The axes are temperature, magnetic field, and current density. The curved surface in this figure represents the transition between normal and superconducting behavior, and its intersections with the 3 axes are the critical current density ($J_c$), critical field ($H_c$), and critical temperature ($T_c$). The primary focus of superconducting materials research since 1911 has been to expand the volume of superconducting behavior in this diagram. This effort has generated a wide variety of what are now called conventional superconductors with critical temperatures up to 23 K and critical fields up to 40 T (Earth's magnetic field is 0.00005 T). The primary drawback of these conventional superconductors is the expense and inconvenience of their refrigeration systems. Nevertheless, the fabrication of superconducting magnets for magnetic resonance imaging, accelerator magnets, energy storage, fusion research, and other applications is a $200 million/year business.

In 1987, Karl Alex Müller and J. Georg Bednorz received the Nobel Prize in Physics for their discovery in 1986 of a new class of superconducting ceramics consisting of oxygen, copper, barium, and lanthanum. Expanding on their work, researchers at the University of Alabama and at the University of Houston developed the compound $Y_1Ba_2Cu_3O_{7-\delta}$, which has a transition temperature of about 92 K. For the first time, superconductivity became possible at liquid nitrogen temperatures. An explosion of research on the new materials has resulted in the discovery of many materials with transition temperatures near 100 K and hints of superconductivity at much higher temperatures. The potential of the new superconductors is illustrated in Figure 1b, which compares the critical parameters of the old and new materials. With their high critical temperatures and critical fields projected to hundreds of tesla, the parameter space where superconductivity exists has been greatly expanded. Unfor-



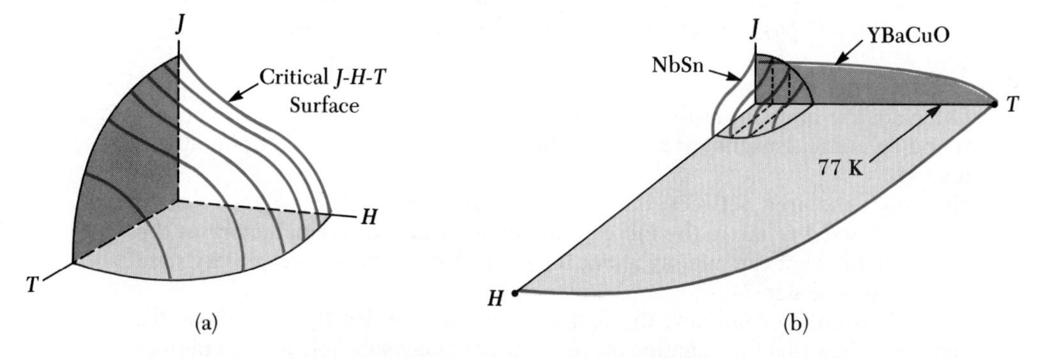


**Figure 1** (a) A three-dimensional diagram illustrating the surface separating normal and superconducting behavior as a function of temperature ($T$), magnetic field ($H$), and current density ($J$). (b) A comparison of the $J$-$H$-$T$ diagrams for $Nb_3Sn$ and $Y_1Ba_2Cu_3O_{7-\delta}$. (Courtesy of Alan Clark, National Bureau of Standards.)

tunately the critical current density of the new materials is low over most of the *H-T* plane. The ceramic nature of the new materials also poses difficulty in forming conductors that can be wound into coils. It is expected that these problems can be solved when a better understanding of the new materials evolves.

## Superconductivity

In 1957, John Bardeen, L.N. Cooper, and J.R. Schrieffer published a microscopic theory which shows that superconductivity results from an interaction between electrons and the lattice in which they flow [1,2,3]. Coulomb attraction between electrons and lattice ions produces a lattice distortion which can attract other electrons. At sufficiently low temperatures the attractive force overcomes thermal agitation and the electrons begin to condense into pairs. As the temperature is reduced further, more and more electrons condense into the paired state. The electron pairs flow through the lattice in a coherent manner and do not suffer the collisions which lead to resistance in normal conductors (see Section 27.5 for a discussion of conduction in normal metals).

We can model a superconductor as a material in which current is carried by two different electron populations. The first population consists of conduction electrons which behave as the electrons in any normal metal. The second population consists of electron pairs which behave coherently and do not exchange energy with the lattice. Since electron pair current has zero resistance, all direct current is transported by pairs. The condensation energy of the electron pairs produces an energy gap between normal and superconducting electrons which is $2\Delta \approx 2$ meV for conventional superconductors. This model is known as the two-fluid model.

## Josephson Junctions

In addition to the work on large-scale applications of superconductivity (primarily magnets), there has been considerable research on small-scale superconducting devices. Most of these devices take advantage of the unusual properties of a Josephson junction, which is simply two superconductors separated by an insulating barrier as shown in Figure 2. When the barrier is thin enough, electron pairs can tunnel from

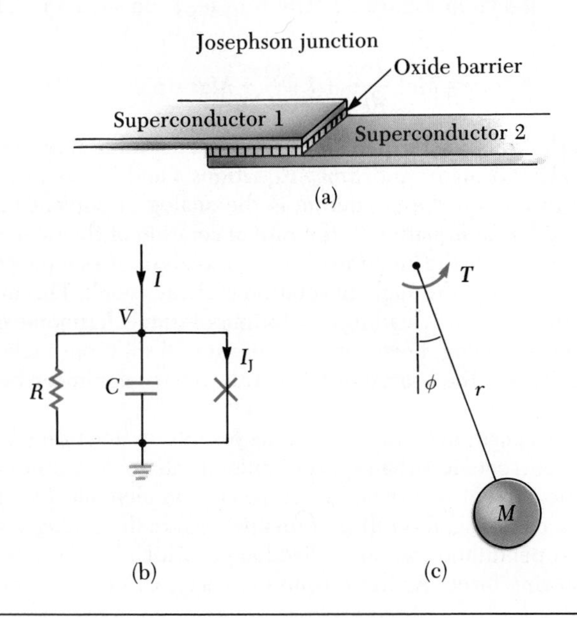


**Figure 2** (a) The geometry of a Josephson junction (b), its equivalent circuit (c), and a mechanical analog for the junction behavior.

one superconductor to the other. Brian Josephson shared the Nobel Prize in Physics in 1973 for his prediction of the behavior of such a device. The remainder of this essay will discuss the operation of Josephson devices and circuits.

Josephson solved the quantum mechanical equations for the pair current flow across an insulating barrier between two superconductors [1]. The result is the now famous Josephson equations

$$I_J = I_0 \sin \phi \tag{1}$$

$$\frac{d\phi}{dt} = \frac{2eV}{\hbar} \tag{2}$$

where $I_J$ is the electron pair current through the junction, $V$ is the voltage across the junction, $e$ is the electron charge, and $\hbar$ is Planck's constant divided by $2\pi$. The junction phase $\phi$ has a quantum mechanical interpretation, but for our purposes can be treated simply as an intermediate parameter. $I_0$ is a constant which decreases exponentially with the barrier thickness. The full behavior of the junction can be modeled by the equivalent circuit of Figure 2b. The resistance $R$ describes the flow of normal electrons, the $\times$ represents a pure Josephson element having a flow of electron pairs, $I_J$, given by Equations 1 and 2, and $C$ is the geometrical capacitance. Thus the total junction current is given by:

$$I = \frac{V}{R} + C\frac{dV}{dt} + I_0 \sin \phi \tag{3}$$

Substituting Equation 2 this becomes

$$I = C\frac{\hbar}{2e}\frac{d^2\phi}{dt^2} + \frac{\hbar}{2eR}\frac{d\phi}{dt} + I_0 \sin \phi \tag{4}$$

Under different drive and parameter conditions, Equation 4 leads to an almost infinite variety of behavior and is the subject of more than 100 papers.

## The Pendulum Analogy

In understanding Equation 4, it is useful to make the analogy between the current in a Josephson junction and the torque applied to a pendulum which is free to rotate in a viscous medium as shown in Figure 2c. The torque, $\tau$, on such a pendulum can be written

$$\tau = Mr^2\frac{d^2\phi}{dt^2} + k\frac{d\phi}{dt} + Mgr \sin \phi \tag{5}$$

where $k\, d\phi/dt$ is the damping torque, $Mr^2\, d^2\phi/dt^2$ is the inertial term and $Mgr \sin \phi$ is the gravitational restoring torque. Since Equations 4 and 5 have identical form we can say that the torque on the pendulum is the analog of current in a Josephson junction, and considering Equation 2, the rate of rotation of the pendulum $d\phi/dt$ is the analog of voltage across the junction. The usual analysis of a simple pendulum (see Section 13.4) assumes that the angle of rotation is always small. This makes possible the linearization of Equation 5, leading to solutions of simple harmonic motion. In the case of the Josephson analog, however, we consider all solutions including full 360 degree rotations. This is the source of the extraordinary nonlinear behavior of Josephson devices.

We are now in a position to understand the Josephson junction $I$-$V$ curve shown in Figure 3a. For currents less than $I_0$, the junction phase $\phi$ assumes a static value such that all of the current is carried by the Josephson element. Except for a small transient while $I$ is changing, the voltage remains zero. In the analog, a steady torque is applied and the pendulum assumes a fixed angle which balances the torque and gravitational restoring force. As the torque increases, $\phi$ increases. For an applied

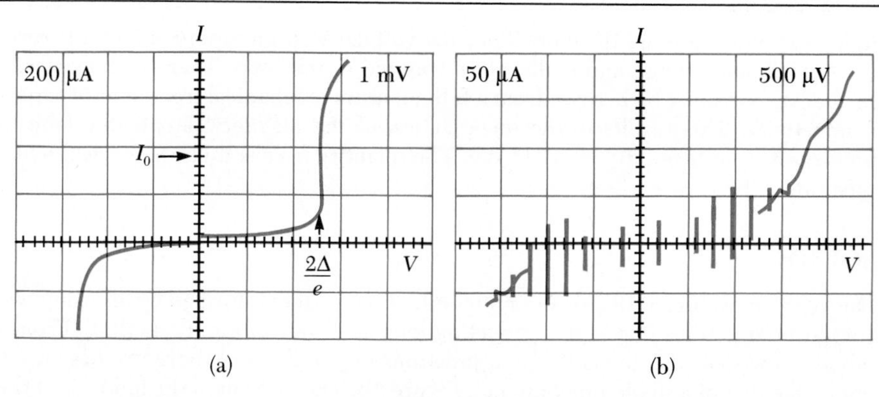


Figure 3 (a) The *I-V* curve of Josephson junction. (b) The *I*-curve when the junction is exposed to 96 GHz radiation. The quantized voltage levels occur at the values $V_n = nhf/2e$.

torque less than the critical value *Mgr*, Equation 5 has a static solution. When the torque exceeds the critical value, *Mgr*, the pendulum rotation accelerates until the viscous damping term $k\, d\phi/dt$ can absorb the applied torque. Applying this analog to the Josephson junction we see that when the current in the junction exceeds the critical value $I_0$, the junction switches to a voltage state in which the current oscillates and has an average value determined by the normal state resistance. As the current decreases, the voltage follows the resistive state curve until the current reaches a value $I_{\text{min}}$. Below $I_{\text{min}}$ the voltage falls abruptly to zero as the junction switches back into the superconducting state. In the analog, with decreasing torque, the pendulum rotation slows down to the point where the applied torque is insufficient to push it over the top. It then settles into a nonrotating state. The nonlinearity in the normal state curve for the junction can be explained as follows: For small voltages the resistance is high because there are comparatively few normal state electrons to tunnel across the barrier. At voltages greater than the superconducting energy gap voltage, $2\Delta/e$, there is sufficient energy from the bias source to break superconducting pairs and the number of normal carriers substantially increases, causing an abrupt increase in current.

## Josephson Voltage Standard

One of the most important applications of the Josephson effect comes from the fact that the Josephson current oscillates at a frequency proportional to the applied voltage, i.e., $f = 2eV/h$. Since this relation is independent of all other parameters and has no known corrections, it forms the basis of voltage standards throughout the world. In practice, these standards operate by applying a microwave current through the junction and allowing the Josephson oscillation to phase lock to some harmonic of the applied frequency. With the correct choice of design parameters the only stable operating points of the junction will be at one of the voltages $V = nhf/2e$ where $n$ is an integer. These voltages are very accurately known because frequency can be measured with great accuracy and $2e/h = 483.594$ GHz/mV is a defined value for the purpose of voltage standards. Figure 3b is the *I-V* curve of a junction which is driven at 96 GHz. About 10 quantized voltage levels span the range from $-1$ to 1 mV. Each of these levels represents a phase lock between a harmonic of the applied frequency at 96 GHz and the Josephson self oscillation. Practical voltage standards use up to 20 000 junctions in series to generate more than 150 000 quantized reference voltages spanning the range from $-10$ to $+10$ V.

## Flux Quantization

Another important aspect of superconducting electronics is the quantization of magnetic flux in a superconducting loop. Consider a closed loop surrounding magnetic flux $\Phi$. We know from Faraday's law (see Section 31.1) that any change in $\Phi$ must

*(Continued on next page)*

induce a voltage around the loop. Since the voltage across a superconductor is zero, the flux cannot change and is therefore trapped in the loop. There is, however, a further restriction which arises from a full quantum mechanical treatment of superconductivity. This results in the quantization of the magnetic flux in the loop in units of $\Phi_0 = h/2e = 2.07 \times 10^{-15}$ Wb. The quantity $h/2e$ is thus known as the flux quantum.

## SQUIDs

One of the most interesting of all superconducting devices is formed by inserting two Josephson junctions in a superconducting loop as shown in Figure 4a [2,3]. From a circuit viewpoint this looks like two junctions in parallel and therefore has an *I-V* curve like that of a single junction, i.e., Figure 4b. The current in the loop is just the sum of the currents in the two junctions. When a magnetic field is applied to the circuit it induces a circulating current just sufficient to exclude the flux from the loop. As long as the junction currents remain below the critical value all flux is excluded. However, if the critical current of either junction is exceeded, it switches momentarily to a resistive state and flux can enter the loop. In essence, the junctions act like gates which can allow flux to enter or leave the loop. A complete analysis of the circuit shows that the maximum current which can flow through the loop, $I_{max}$, is a periodic function of the applied magnetic flux $\Phi$. This periodic dependence is shown in Figure 4c. This pattern can be interpreted as the consequence of interference between the coherent superconducting pair states on either side of the junction. These devices are therefore called Superconducting QUantum Interference Devices or SQUIDs.

The most important application of SQUIDs is the measurement of small magnetic fields. This is accomplished by using a large pick-up loop to inductively couple flux

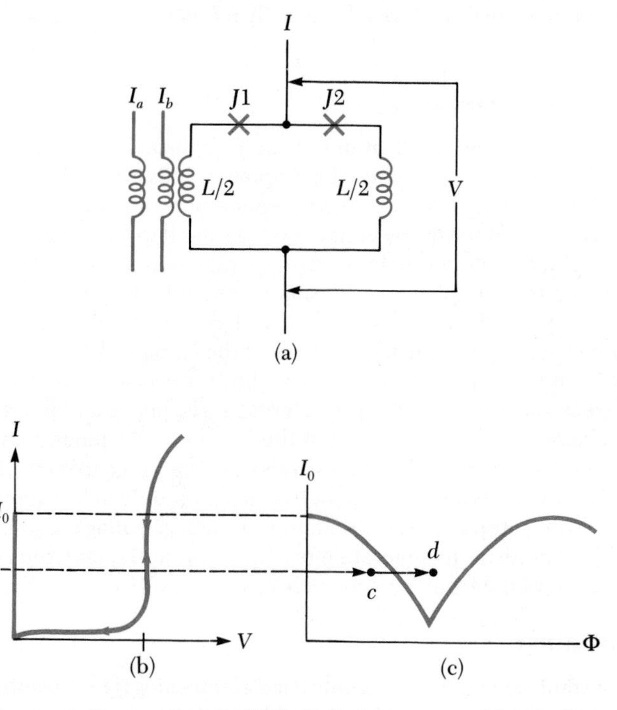


**Figure 4** (a) The circuit diagram for a SQUID. (b) Its *I-V* curve, and (c) the dependence of the critical current on applied magnetic flux.

into the SQUID loop. An electronic circuit senses the changes in critical current $I_{max}$ caused by changing flux in the SQUID loop. This arrangement can resolve flux changes of as little as $10^{-6}\Phi_0$. For a typical SQUID with an inductance $L = 100$ pH, this corresponds to an energy resolution of $\Delta E = \Delta\Phi^2/2L = 2 \times 10^{-32}$ J, which is approaching the theoretical limit set by Planck's constant.

A rapidly expanding application of SQUIDs is the measurement of magnetic fields caused by the small currents which flow in the heart and brain. Systems with as many as seven SQUIDs are being used to map these biomagnetic fields. This new noninvasive tool may one day be important in locating the source of epilepsy and other brain disorders. In such applications it is important to discriminate against distant sources of magnetic noise. This is accomplished by using two pick-up coils wound in opposite directions so that the SQUID is only sensitive to the *gradient* of the magnetic field.

## Digital Applications

The possibility of digital applications for Josephson junctions is suggested immediately by the bistable nature of the *I-V* curve in Figure 3a. At a fixed bias current, the junction can be in either the zero or finite voltage state representing a binary 0 or 1. In the mid-1970s a major effort was started to develop large scale computing systems using Josephson devices. [4]. The motivation for this is two-fold: first, Josephson junctions can switch states in only a few picoseconds, and second, the power dissipation of Josephson devices is nearly three orders of magnitude less than semiconductor devices. The low power dissipation results from the low logic levels (0 and 2 mV) of Josephson devices. Dissipation becomes increasingly important in ultra-high-speed computers because the need to limit propagation delay forces the computer into a small volume. For example, the propagation delay in typical transmission lines is about 1 ns for every 20 cm of length. Thus a computer with a 1 GHz clock rate must be squeezed into a volume in which the maximum signal path length is less than 20 cm. Since faster circuits generally dissipate more power and must be packaged with higher density, cooling in semiconductor computers becomes a limiting problem.

A variety of Josephson logic circuits have been proposed. We will consider just one, based on the SQUID circuit of Figure 4a. Suppose that the SQUID carries a current at the level of the lower dashed line in Figure 4. Current in the two input lines ($I_a$, $I_b$) moves the bias point horizontally along this line in Figure 4c. For example a current in either line will move the operating point to (c) while a current in both lines moves the operating point across the threshold curve to (d). When the threshold curve is crossed, the SQUID switches into the voltage state. The device thus has an output which is the AND function of the two inputs. Other logic combinations can be achieved by adjusting the coupling of the input lines. Memory can be added by using coupled SQUIDs where one SQUID senses the presence or absence of flux in a second SQUID. Figure 5 is a photograph of two coupled thin film SQUIDs which form a flip-flop circuit. These devices are made using photolithographic processes similar to those used in the semiconductor industry

Josephson logic and memory chips with thousands of junctions have been built and successfully operated. However, there are still significant problems to be overcome in developing a complete Josephson computer. One of the most important problems is the dependence of Josephson logic on threshold switching. This results in circuits which operate over only a small range of component parameters. Such circuits are said to have small *margins*. As process control improves, ever more complex digital circuits are being demonstrated. Recently a Josephson microprocessor with 24 000 junctions was successfully operated at a clock rate of 1.1 GHz. The development of a Josephson supercomputer will require solutions to a variety of memory density, interconnect, and packaging problems.

*(Continued on next page)*

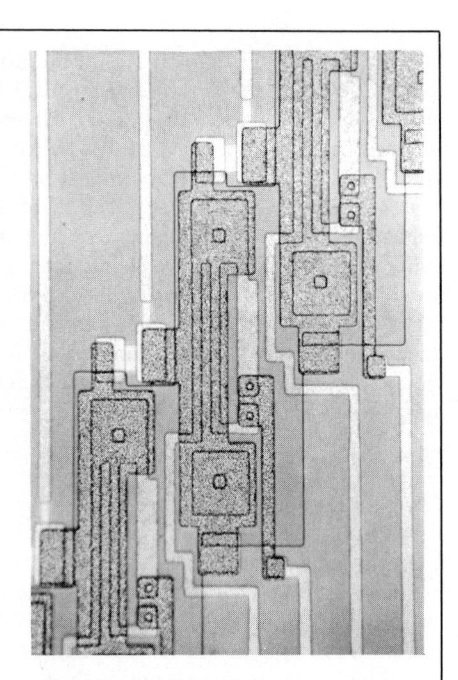

**Figure 5** A photograph of a Josephson integrated circuit consisting of two coupled SQUIDs. The minimum feature size is 4 microns.

## References


1. Feynmann, R., R.B. Leighton, and M. Sands, *The Feynmann Lectures on Physics,* Reading, Mass., Addison Wesley, Vol. 3, pp 21.1–21.18, 1965.
2. Langenberg, Donald N., Douglass J. Scalapino, and Barry N. Taylor, "The Josephson Effects," *Scientific American,* May 1966.
3. Mcdonald, Donald G., "Superconducting Electronics," *Physics Today,* February 1981.
4. Matisoo, Juri "The Superconducting Computer," *Scientific American,* Vol. 242 no. 5, pp. 50–65, 1980.


## Essay Questions

1. It is well known that superconductors carry dc current without loss. Using the 2-fluid model, describe a mechanism which would lead to loss for ac current. *Hint:* Consider the inertia of the electron pairs.
2. Starting with the pendulum analog of a Josephson junction, describe a similar mechanical analog for an inductor and a SQUID.
3. Using the concept of *gain* explain why transistor logic has much greater margins than Josephson logic.
4. The basic reason for the reduced power consumption of Josephson logic is the low operating voltage ($\approx 2$ mV). What fundamental barrier prevents room temperature logic from operating at such low voltage levels? *Hint:* Consider the fact that logic circuits are by nature nonlinear. How is nonlinearity related to the thermal energy $kT/e$?

## Essay Problems

1. Derive the Josephson frequency-voltage relation $f = 2eV/h$ from Equations 1 and 2.
2. For small values of $\phi$ Equations 1 and 2 can be seen to describe an inductance commonly called the Josephson inductance. The resonance of this inductance and the junction capacitance sets an upper limit on the frequency response of many Josephson devices. For a typical junction with $I_0 = 1$ mA and $C = 3$ pF what is this limit?
3. What is the approximate switching time of a Josephson junction with a critical current of 1 mA and a capacitance of 3 pF?

# 45

# Nuclear Structure

*This research scientist is studying a photograph of particle tracks made in a bubble chamber at Fermilab. The curved tracks are produced by charged particles moving through the chamber in the presence of an applied magnetic field. Negatively charged particles deflect in one direction, while positively charged particles deflect in the opposite direction. Various events such as delays and collisions can be identified by examining such tracks. (© Dan McCoy/Rainbow)*

In 1896, the year that marks the birth of nuclear physics, Becquerel discovered radioactivity in uranium compounds. A great deal of research followed this discovery in an attempt to understand and characterize the nature of the radiation emitted by radioactive nuclei. Pioneering work by Rutherford showed that the emitted radiation was of three types, which he called alpha, beta, and gamma rays. These are classified according to the nature of the electric charge they possess and according to their ability to penetrate matter and ionize air. Later experiments showed that alpha rays are actually helium nuclei, beta rays are electrons, and gamma rays are high-energy photons.

In 1911, Rutherford and his students Geiger and Marsden performed a number of important scattering experiments involving alpha particles. These experiments established that the nucleus of an atom can be regarded as essentially a point mass and point charge and that most of the atomic mass is contained in the nucleus. Furthermore, such studies demonstrated a totally new type of force, the *nuclear force,* which is predominant at distances of less than about $10^{-14}$ m and zero for large distances. That is, the nuclear force is a short-range force.

Other milestones in the development of nuclear physics include the following events:

1. The observation of nuclear reactions by Cockroft and Walton in 1930, using artificially accelerated particles
2. The discovery of the neutron by Chadwick in 1932
3. The discovery of artificial radioactivity by Joliot and Irene Curie in 1933
4. The discovery of nuclear fission in 1938 by Hahn and Strassmann
5. The development of the first controlled fission reactor in 1942 by Fermi and his collaborators

In this chapter we shall discuss the properties and structure of the atomic nucleus. We start by describing the basic properties of nuclei, followed by a discussion of nuclear forces and binding energy, nuclear models, and the phenomenon of radioactivity. We shall also discuss nuclear reactions and the various processes by which nuclei decay. The chapter concludes with an essay on Magnetic Resonance Imaging, which has many applications, especially in the field of medical diagnostics.

## 45.1 SOME PROPERTIES OF NUCLEI

All nuclei are composed of two types of particles: protons and neutrons. The only exception to this is the ordinary hydrogen nucleus, which is a single proton. In this section, we shall describe some of the properties of nuclei, such as their charge, mass, and radius. In doing so, we shall make use of the following quantities:

1. The **atomic number,** $Z$, which equals the number of protons in the nucleus
2. The **neutron number,** $N$, which equals the number of neutrons in the nucleus
3. The **mass number,** $A$, which equals the number of nucleons (neutrons plus protons) in the nucleus

It will be convenient for us to have a symbolic way of representing nuclei that will show how many protons and neutrons are present. The symbol to be used is ${}^{A}_{Z}\text{X}$, where X represents the chemical symbol for the element. For example, ${}^{56}_{26}\text{Fe}$ has a mass number of 56 and an atomic number of 26; therefore it contains 26 protons and 30 neutrons. When no confusion is likely to arise, we shall omit the subscript Z because the chemical symbol can always be used to determine Z.

The nuclei of all atoms of a particular element contain the same number of protons but often contain different numbers of neutrons. Nuclei that are related in this way are called **isotopes.**

Isotopes

The isotopes of an element have the same $Z$ value but different $N$ and $A$ values.

The natural abundances of isotopes can differ substantially. For example, ${}^{11}_{6}\text{C}$, ${}^{12}_{6}\text{C}$, ${}^{13}_{6}\text{C}$, and ${}^{14}_{6}\text{C}$ are four isotopes of carbon. The natural abundance of the ${}^{12}_{6}\text{C}$ isotope is about 98.9%, whereas that of the ${}^{13}_{6}\text{C}$ isotope is only about 1.1%. Some isotopes do not occur naturally but can be produced in the laboratory through nuclear reactions. Even the simplest element, hydrogen, has isotopes. They are ${}^{1}_{1}\text{H}$, the ordinary hydrogen nucleus; ${}^{2}_{1}\text{H}$, deuterium; and ${}^{3}_{1}\text{H}$, tritium.

### Charge and Mass

The proton carries a single positive charge, equal in magnitude to the charge $e$ on the electron (where $|e| = 1.6 \times 10^{-19}$ C). The neutron is electrically neutral, as its name implies. Because the neutron has no charge, it is difficult to detect.

Nuclear masses can be measured with great precision with the help of the mass spectrometer (Section 29.6) and the analysis of nuclear reactions. The proton is about 1836 times as massive as the electron, and the masses of the proton and the neutron are almost equal. It is convenient to define, for atomic masses, the **unified mass unit,** u, in such a way that the mass of the isotope $^{12}$C is exactly 12 u. That is, the mass of a nucleus (or atom) is measured relative to the mass of an atom of the neutral carbon-12 isotope (the nucleus plus six electrons). Thus, the mass of $^{12}$C is defined to be 12 u, where 1 u $= 1.660559 \times 10^{-27}$ kg. The proton and neutron each have a mass of about 1 u, and the electron has a mass that is only a small fraction of an atomic mass unit:

$$\text{Mass of proton} = 1.007276 \text{ u}$$

$$\text{Mass of neutron} = 1.008665 \text{ u}$$

$$\text{Mass of electron} = 0.0005486 \text{ u}$$

For reasons that will become apparent later, note that the mass of the neutron is greater than the combined masses of the proton and the electron.

Because the rest energy of a particle is given by $E_0 = mc^2$, it is often convenient to express the unified mass unit in terms of its energy equivalence. For a proton, we have

$$E_0 = mc^2 = (1.67 \times 10^{-27} \text{ kg})(3 \times 10^8 \text{ m/s})^2$$

$$= 1.50 \times 10^{-10} \text{ J} = 9.38 \times 10^8 \text{ eV} = 938 \text{ MeV}$$

Following this procedure, it is found that the rest energy of an electron is 0.511 MeV.

Nuclear physicists often express the unified mass unit in terms of the unit MeV/$c^2$, where

$$1 \text{ u} \equiv 1.660559 \times 10^{-27} \text{ kg} = 931.50 \text{ MeV}/c^2$$

The rest masses of the proton, neutron, and electron are given in Table 45.1. The masses and some other properties of selected isotopes are provided in Appendix A.

**TABLE 45.1 Rest Mass of the Proton, Neutron, and Electron in Various Units**

| | Mass | | |
|---|---|---|---|
| Particle | kg | u | MeV/$c^2$ |
| Proton | $1.6726 \times 10^{-27}$ | 1.007276 | 938.28 |
| Neutron | $1.6750 \times 10^{-27}$ | 1.008665 | 939.57 |
| Electron | $9.109 \times 10^{-31}$ | $5.486 \times 10^{-4}$ | 0.511 |

**EXAMPLE 45.1 The Unified Mass Unit**
Use Avogadro's number to show that the unified mass unit is $1\ u = 1.66 \times 10^{-27}$ kg.

*Solution* We know that 12 kg of $^{12}C$ contains Avogadro's number of atoms. Avogadro's number, $N_A$, has the value $6.02 \times 10^{23}$ atoms/g·mol $= 6.02 \times 10^{26}$ atoms/kg·mol.

Thus, the mass of one carbon atom is

$$\text{Mass of one } ^{12}\text{C atom} = \frac{12 \text{ kg}}{6.02 \times 10^{26} \text{ atoms}} = 1.99 \times 10^{-26} \text{ kg}$$

Since one atom of $^{12}C$ is defined to have a mass of 12 u, we find that

$$1\ u = \frac{1.99 \times 10^{-26} \text{ kg}}{12} = 1.66 \times 10^{-27} \text{ kg}$$

Ernest Rutherford (1871–1937), a New Zealander, was awarded the Nobel Prize in 1908 for discovering that atoms can be broken apart by alpha rays and for studying radioactivity. "On consideration, I realized that this scattering backward must be the result of a single collision, and when I made calculations I saw that it was impossible to get anything of that order of magnitude unless you took a system in which the greater part of the mass of the atom was concentrated in a minute nucleus. It was then that I had the idea of an atom with a minute massive center carrying a charge." (Photo courtesy of AIP Niels Bohr Library)

## The Size of Nuclei

The size and structure of nuclei were first investigated in the scattering experiments of Rutherford, discussed in Section 42.1. In these experiments, positively charged nuclei of helium atoms (alpha particles) were directed at a thin piece of metal foil. As the alpha particles moved through the foil, they often passed near a nucleus of the metal. Because of the positive charge on both the incident particles and the metal nuclei, particles were deflected from their straight-line paths by the Coulomb repulsive force. In fact, some particles were even deflected backwards, through an angle of 180° from the incident direction. These particles were apparently moving directly toward a nucleus, on a head-on collision course.

Rutherford applied the principle of conservation of energy to the alpha particles and nuclei and found an expression for the minimum distance, $d$, that the particle can approach the nucleus when moving directly toward it before it is turned around by Coulomb repulsion. In such a head-on collision, the kinetic energy of the incoming alpha particle must be converted completely to electrical potential energy when the particle stops at the point of closest approach and turns around (Fig. 45.1). If we equate the initial kinetic energy of the alpha particle to the electrical potential energy of the system (alpha particle plus target nucleus), we have

$$\tfrac{1}{2}mv^2 = k\frac{q_1 q_2}{r} = k\frac{(2e)(Ze)}{d}$$

Solving for $d$, the distance of closest approach, we get

$$d = \frac{4kZe^2}{mv^2}$$

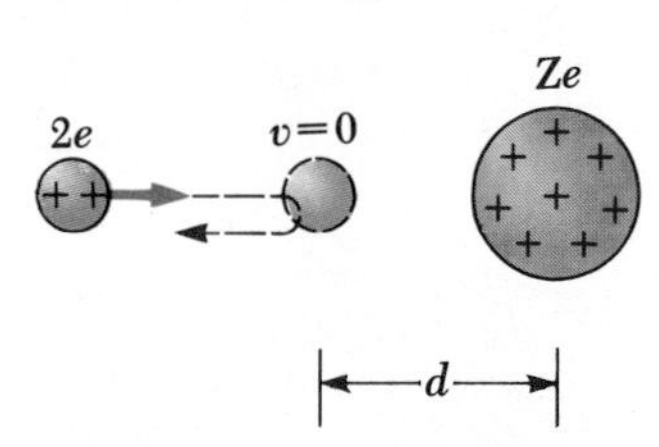


**Figure 45.1** An alpha particle on a head-on collision course with a nucleus of charge $Ze$. Because of the Coulomb repulsion between the like charges, the alpha particle will stop instantaneously at a distance $d$ from the target nucleus, called the distance of closest approach.

From this expression, Rutherford found that the alpha particles approached nuclei to within $3.2 \times 10^{-14}$ m when the foil was made of gold. Thus, the radius of the gold nucleus must be less than this value. For silver atoms, the distance of closest approach was found to be $2 \times 10^{-14}$ m. From these results, Rutherford concluded that the positive charge in an atom is concentrated in a small sphere, which he called the nucleus, whose radius is no greater than about $10^{-14}$ m. Because such small lengths are common in nuclear physics, a convenient unit of length that is used is the *femtometer* (fm), sometimes called the **fermi,** defined as

$$1 \text{ fm} \equiv 10^{-15} \text{ m}$$

Since the time of Rutherford's scattering experiments, a multitude of other experiments have shown that most nuclei are approximately spherical and have an average radius given by

$$r = r_0 A^{1/3} \qquad (45.1)$$

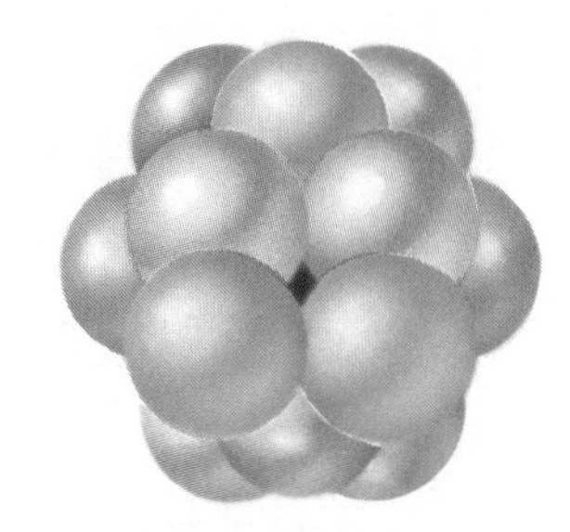

Figure 45.2 A nucleus can be visualized as a cluster of tightly packed spheres, where each sphere is a nucleon.

where $A$ is the mass number and $r_0$ is a constant equal to $1.2 \times 10^{-15}$ m. Because the volume of a sphere is proportional to the cube of its radius, $r$, it follows from Equation 45.1 that the volume of a nucleus (assumed to be spherical) is directly proportional to $A$, the total number of nucleons. This suggests that *all nuclei have nearly the same density.* When the nucleons combine to form a compound nucleus, they combine as though they were tightly packed spheres (Fig. 45.2). This fact has led to an analogy between the nucleus and a drop of liquid, in which the density of the drop is independent of its size. We shall discuss the liquid-drop model in Section 45.3.

**EXAMPLE 45.2 The Volume and Density of a Nucleus**
Find (a) an approximate expression for the mass of a nucleus of mass number $A$, (b) an expression for the volume of this nucleus in terms of the mass number, and (c) a numerical value for its density.

*Solution* (a) The mass of the proton is approximately equal to that of the neutron. Thus, if the mass of one of these particles is $m$, the mass of the nucleus is approximately $Am$.

(b) Assuming the nucleus is spherical and using Equation 45.1, we find that the volume is given by

$$V = \tfrac{4}{3}\pi r^3 = \tfrac{4}{3}\pi r_0{}^3 A$$

(c) The nuclear density can be found as follows:

$$\rho_n = \frac{\text{mass}}{\text{volume}} = \frac{Am}{\tfrac{4}{3}\pi r_0{}^3 A} = \frac{3m}{4\pi r_0{}^3}$$

Taking $r_0 = 1.2 \times 10^{-15}$ m and $m = 1.67 \times 10^{-27}$ kg, we find that

$$\rho_n = \frac{3(1.67 \times 10^{-27}\ \text{kg})}{4\pi(1.2 \times 10^{-15}\ \text{m})^3} = 2.3 \times 10^{17}\ \text{kg/m}^3$$

Because the density of water is only $10^3$ kg/m$^3$, the nuclear density is about $2.3 \times 10^{14}$ times as great as the density of water!

## Nuclear Stability

Since the nucleus consists of a closely packed collection of protons and neutrons, you might be surprised that it can exist. Because like charges (the protons) in close proximity exert very large repulsive electrostatic forces on each other, these forces should cause the nucleus to fly apart. However, nuclei are stable because of the presence of another force, which is called the **nuclear force.** This force, which has a very short range (about 2 fm), is an attractive force that acts between all nuclear particles. The protons attract each other via the nuclear force, and at the same time they repel each other through the Coulomb force. The nuclear force also acts between pairs of neutrons and between neutrons and protons.

There are about 400 stable nuclei; hundreds of others have been observed, but these are unstable. A plot of $N$ versus $Z$ for a number of stable nuclei is given in Figure 45.3. Note that light nuclei are most stable if they contain an equal number of protons and neutrons, that is, if $N = Z$. For example, the helium nucleus (two protons and two neutrons) is very stable. Also note that heavy nuclei are more stable if the number of neutrons exceeds the number of protons. This can be understood by recognizing that, as the number

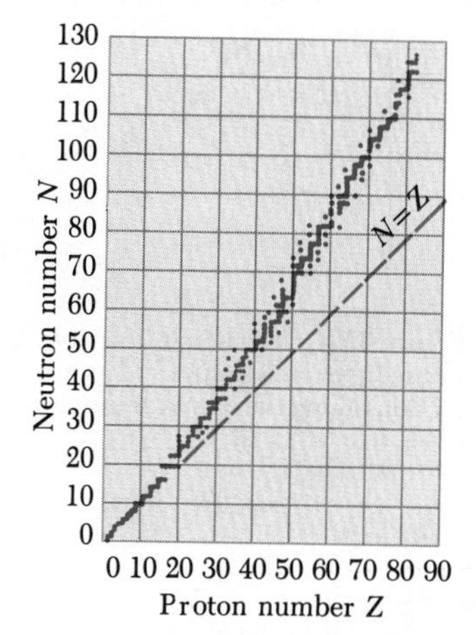


Figure 45.3 A plot of neutron number, $N$, versus atomic number, $Z$, for the stable nuclei. The dashed line, corresponding to the condition $N = Z$, is called the *line of stability*.

of protons increases, the strength of the Coulomb force increases, which tends to break the nucleus apart. As a result, more neutrons are needed to keep the nucleus stable since neutrons experience only the attractive nuclear forces. Eventually, the repulsive forces between protons cannot be compensated by the addition of more neutrons. This occurs when $Z = 83$. Elements that contain more than 83 protons do not have stable nuclei.

It is interesting that most stable nuclei have even values of $A$. Furthermore, only eight have $Z$ and $N$ numbers that are both odd. In fact, certain values of $Z$ and $N$ correspond to nuclei with unusually high stability. These values of $N$ and $Z$, called **magic numbers,** are given by

$$Z \text{ or } N = 2, 8, 20, 28, 50, 82, 126 \tag{45.2}$$

For example, the alpha particle (two protons and two neutrons), with $Z = 2$ and $N = 2$, is very stable.

## Nuclear Spin and Magnetic Moment

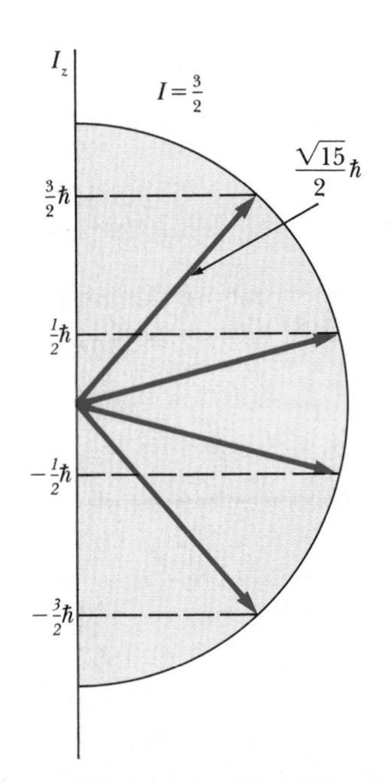


**Figure 45.4** The possible orientations of the nuclear spin and its projections along the $z$ axis for the case $I = \frac{3}{2}$.

In Chapter 42, we discussed the fact that an electron has an intrinsic angular momentum associated with its spin. Nuclei, like electrons, also have an intrinsic angular momentum. The magnitude of the *nuclear angular momentum* is $\sqrt{I(I+1)}\hbar$, where $I$ is a quantum number called the *nuclear spin* and may be an integer or half-integer. The maximum component of the angular momentum projected along any direction is $I\hbar$. Figure 45.4 illustrates the possible orientations of the nuclear spin and its projections along the $z$ axis for the case where $I = \frac{3}{2}$.

The nuclear angular momentum has a corresponding nuclear magnetic moment associated with it, similar to that of the electron. The *magnetic moment* of a nucleus is measured in terms of the **nuclear magneton** $\mu_n$, a unit of moment defined as

$$\mu_n \equiv \frac{e\hbar}{2m_p} = 5.05 \times 10^{-27} \text{ J/T} \tag{45.3}$$

This definition is analogous to that of the Bohr magneton, $\mu_B$, which corresponds to the spin magnetic moment of a free electron (Section 42.6). Note that $\mu_n$ is smaller than $\mu_B$ by a factor of about 2000, due to the large difference in masses of the proton and electron.

The magnetic moment of a free proton is not $\mu_n$ but $2.7928\mu_n$. Unfortunately, there is no general theory of nuclear magnetism that explains this value. Another surprising point is the fact that a neutron also has a magnetic moment, which has a value of $-1.9135\mu_n$. The minus sign indicates that its moment is opposite its spin angular momentum.

Larmor frequency

It is interesting that nuclear magnetic moments (as well as electronic magnetic moments) will precess when placed in an external magnetic field. The frequency at which they precess, called the **Larmor precessional frequency** $\omega_p$, is directly proportional to the magnetic field. This is described schematically in Figure 45.5a, where the magnetic field is along the $z$ axis. For example, the Larmor frequency of a proton in a magnetic field of 1 T is equal to 42.577 MHz. Recall that the potential energy of a magnetic dipole moment in an external magnetic field is given by $-\boldsymbol{\mu}\cdot\mathbf{B}$. When the projection of $\boldsymbol{\mu}$ is along

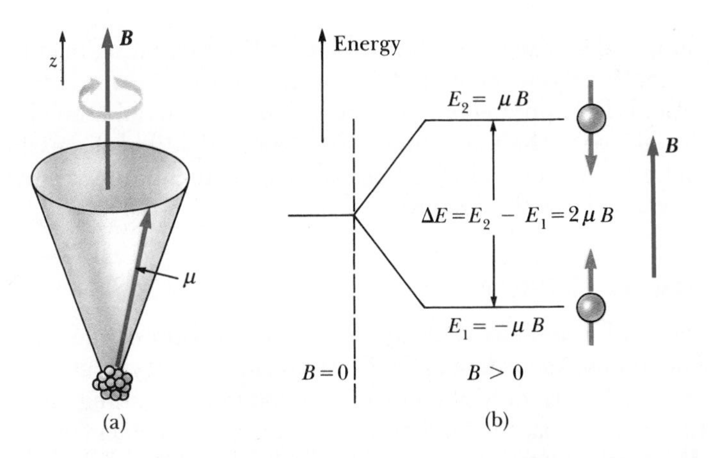


**Figure 45.5** (a) When a nucleus having a magnetic moment is placed in an external magnetic field, $B$, the magnetic moment precesses about the magnetic field with a frequency that is proportional to the field. (b) A proton, whose spin is ½, can occupy one of two energy states when placed in an external magnetic field. The lower energy state $E_1$ corresponds to the case where the spin is aligned with the field, and the higher energy state $E_2$ corresponds to the case where the spin is opposite the field. The reverse is true for electrons.

the field, the potential energy of the dipole moment is $-\mu B$, that is, it has its minimum value. When the projection of $\boldsymbol{\mu}$ is against the field, the potential energy is $\mu B$ and it has its maximum value. These two energy states for a nucleus with a spin of $\frac{1}{2}$ are shown in Figure 45.5b.

It is possible to observe transitions between these two spin states using a technique known as **nuclear magnetic resonance.** A dc magnetic field is introduced to align the magnetic moments (Fig. 45.5a), along with a second, weak, oscillating magnetic field oriented perpendicular to $\boldsymbol{B}$. When the frequency of the oscillating field is adjusted to match the Larmor precessional frequency, a torque acting on the precessing moments causes them to "flip" between the two spin states. These transitions result in a net absorption of energy by the spin system, which can be detected electronically. A diagram of the apparatus used in nuclear magnetic resonance is illustrated in Figure 45.6. The absorbed

**Nuclear magnetic resonance**

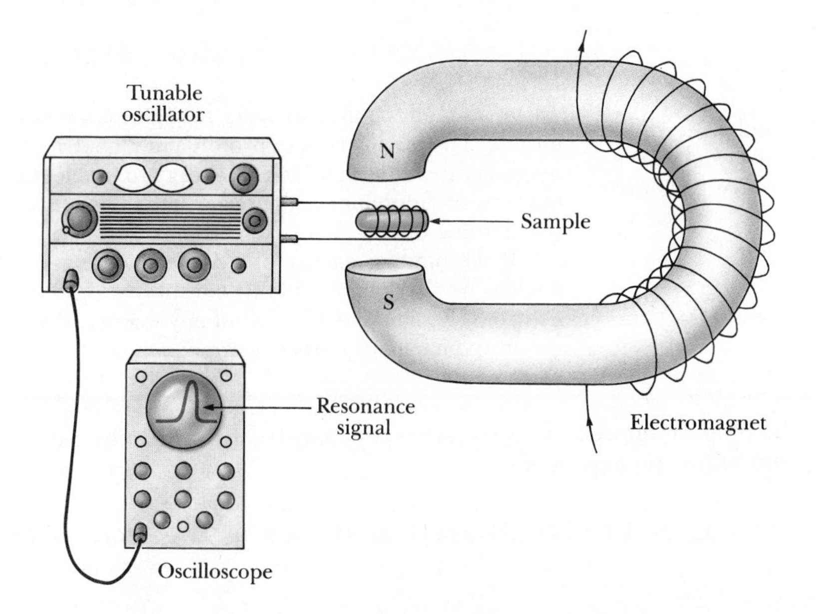


**Figure 45.6** An experimental arrangement for nuclear magnetic resonance. The rf magnetic field of the coil, provided by the variable-frequency oscillator, must be perpendicular to the dc magnetic field. When the nuclei in the sample meet the resonance condition, the spins absorb energy from the rf field of the coil, which changes the $Q$ of the circuit in which the coil is included.

energy is supplied by the generator producing the oscillating magnetic field. Nuclear magnetic resonance and electron spin resonance are extremely important methods for studying nuclear and atomic systems and the interactions of these systems with their surroundings. A more detailed discussion of nuclear magnetic resonance and an associated imaging technique are given in the essay that accompanies this chapter.

## 45.2 BINDING ENERGY

The total mass of a nucleus is always less than the sum of the masses of its individual nucleons. According to the Einstein mass-energy relationship, if the mass difference, $\Delta m$, is multiplied by $c^2$, we obtain the binding energy of the nucleus. In other words, *the total energy of the bound system (the nucleus) is less than the total energy of the separated nucleons.* Therefore, in order to separate a nucleus into protons and neutrons, energy must be delivered to the system.

In addition to simple radioactive decay, there are two important processes that result in energy release from the nucleus. In **nuclear fission,** a nucleus splits into two or more fragments. In **nuclear fusion,** two or more nucleons combine to form a heavier nucleus.

In any fission or fusion reaction, the total rest mass of the products is less than that of the reactants.

The decrease in rest mass is accompanied by a release of energy from the system. The following example illustrates this point for the deuteron.

**EXAMPLE 45.3 The Binding Energy of the Deuteron**
Calculate the binding energy of the deuteron, which consists of a proton and a neutron, given that the mass of the deuteron is 2.014102 u.

*Solution* We know that the proton and neutron masses are

$$m_p = 1.007825 \text{ u}$$

$$m_n = 1.008665 \text{ u}$$

Note that the masses used for the proton and deuteron in this example are actually those of the neutral atoms. We are able to use atomic masses for these calculations since the electron masses cancel. Therefore,

$$m_p + m_n = 2.016490 \text{ u}$$

To calculate the mass difference, we subtract the deuteron mass from this value:

$$\begin{aligned}\Delta m &= (m_p + m_n) - m_d \\ &= 2.016490 \text{ u} - 2.014102 \text{ u} \\ &= 0.002388 \text{ u}\end{aligned}$$

Since 1 u corresponds to an equivalent energy of 931.50 MeV (that is, $1 \text{ u} \cdot c^2 = 931.50$ MeV), the mass difference corresponds to the following binding energy:

$$E_b = (0.002388 \text{ u})(931.5 \text{ MeV/u}) = 2.224 \text{ MeV}$$

This result tells us that, in order to separate a deuteron into its constituent parts (a proton and a neutron), it is necessary to add 2.224 MeV of energy to the deuteron. One way of supplying the deuteron with this energy is by bombarding it with energetic particles.

If the binding energy of a nucleus were zero, the nucleus would separate into its constituent protons and neutrons without the addition of any energy; that is, it would spontaneously break apart.

The binding energy of any nucleus represented by ${}^A_Z\text{X}$ can be calculated using the following expression:

Binding energy of a nucleus

$$E_b \text{ (MeV)} = [Zm_H + Nm_n - M({}^A_Z\text{X})] \times 931.50 \text{ MeV/u} \tag{45.4}$$

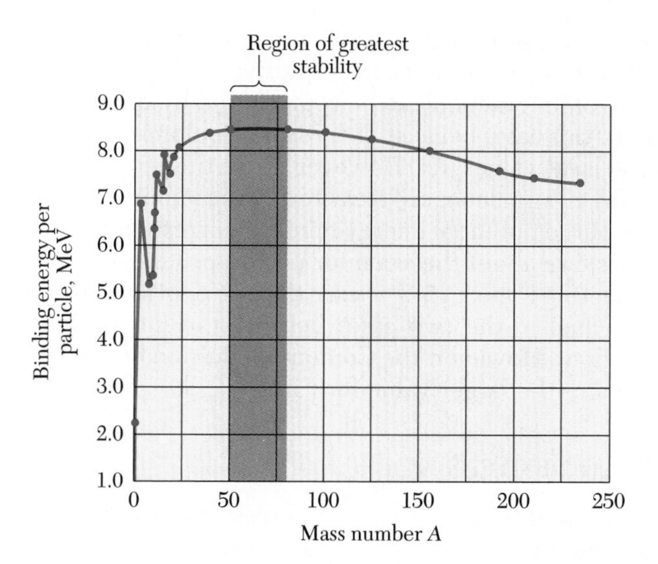


**Figure 45.7** A plot of binding energy per nucleon versus mass number for nuclei that lie along the line of stability in Figure 45.3.

where $m_H$ is the mass of the neutral hydrogen atom, $M(^A_Z X)$ represents the atomic mass of the associated compound nucleus, and the masses are all in unified mass units. Note that *atomic* masses are used in these calculations rather than *nuclear* masses since tables usually give atomic masses.[1]

It is interesting to examine a plot of the binding energy per nucleon, $E_b/A$, as a function of mass number for various stable nuclei (Fig. 45.7). Note that except for the lighter nuclei, the average binding energy per nucleon is about 8 MeV. For the deuteron, the average binding energy per nucleon is found to be $E_b/A = 2.224/2$ MeV $= 1.112$ MeV. Note that the curve in Figure 45.7 peaks in the vicinity of $A = 60$. That is, nuclei with mass numbers greater or less than 60 are not as strongly bound as those near the middle of the periodic table. As we shall see later, this fact allows energy to be released in fission and fusion reactions. Note that the curve in Figure 45.7 is slowly varying for $A > 40$, which suggests that the nuclear force saturates. In other words, a particular nucleon can interact with only a limited number of other nucleons, which can be viewed as being the "nearest neighbors" in the close-packed structure illustrated in Figure 45.2.

The strong nuclear force dominates the Coulomb repulsive force within the nucleus (at short ranges). If this were not the case, stable nuclei would not exist. Moreover, the strong nuclear force is nearly independent of charge. In other words, the nuclear forces associated with the proton-proton, proton-neutron, and neutron-neutron interactions are approximately the same, apart from the additional repulsive Coulomb force for the proton-proton interaction.

By performing scattering experiments, one can establish that the proton-proton interaction can be represented by the potential energy curve shown in Figure 45.8a. For large values of $r$, the Coulomb repulsive force is clearly dominant, corresponding to a positive potential energy. When $r$ is reduced to about 3 fm, a break occurs in the curve to negative values of potential energy.

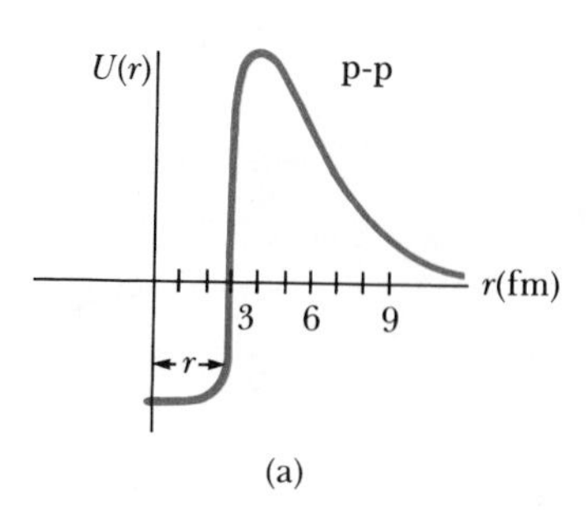


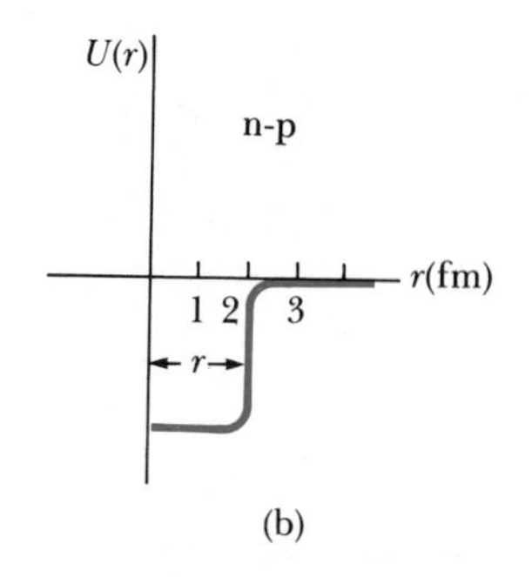


**Figure 45.8** (a) Potential energy versus separation for the proton-proton system. (b) Potential energy versus separation for the neutron-proton system. The difference in the two curves is due mainly to the large Coulomb repulsion in the case of the proton-proton interaction.

[1] It is possible to do this because electron masses cancel in such calculations. One exception to this is the $\beta^+$ decay process.

That is, as the proton-proton separation approaches the nuclear radius, the attractive nuclear force overcomes the repulsive Coulomb force. In contrast, the proton-neutron interaction can be represented by the diagram shown in Figure 45.8b. In this case, because there is no Coulomb force, the potential energy is zero at large values of $r$. However, when the proton-neutron separation is about 2 fm, the nucleons are attracted by the strong nuclear force. Note that the depths of the potential energy minima (corresponding to the nuclear binding energies) are about the same for proton-proton and proton-neutron curves. On inspecting Figure 45.8, we see that the depths of the potentials are approximately equal in the two cases because the interactions are nearly charge-independent. However, the Coulomb repulsion for the proton-proton interaction explains the larger value for $r$ in this case.

## 45.3 NUCLEAR MODELS

Although the detailed nature of nuclear forces is still not well understood, several phenomenological nuclear models have been proposed, and these are useful in understanding some features of nuclear experimental data and the mechanisms responsible for the binding energy. The models we shall discuss are (1) the liquid-drop model, which accounts for the nuclear binding energy, (2) the independent-particle model, which accounts for the existence of stable isotopes, and (3) the collective model.

### Liquid-Drop Model

The **liquid-drop model,** proposed by Bohr in 1936, treats the nucleons as if they were molecules in a drop of liquid. The nucleons interact strongly with each other and undergo frequent collisions as they "jiggle around" within the nucleus. This is analogous to the thermally agitated motion of molecules in a liquid.

The three major effects that influence the binding energy of the nucleus in the liquid-drop model are as follows:

1. **The volume effect** Earlier, we showed that the binding energy per nucleon is approximately constant, indicating that the nuclear force exhibits saturation (Fig. 45.7). Therefore, the binding energy is proportional to $A$ and to the nuclear volume. If a particular nucleon is adjacent to $n$ other nucleons and if the binding energy per pair of nucleons is $E_b$, the binding energy associated with the volume effect will be $C_1A$, where $C_1 = nE_b$ and where $E_b$ is approximately $\frac{1}{2}$ MeV for a typical interacting pair of nucleons.
2. **The surface effect** Since many of the nucleons will be on the surface of the drop, they will have fewer neighbors than those in the interior of the drop. Hence, these surface nucleons will reduce the binding energy by an amount proportional to $r^2$. Since $r^2 \propto A^{2/3}$ (Eq. 45.1), the reduction in energy can be expressed as $-C_2A^{2/3}$, where $C_2$ is a constant.
3. **The Coulomb repulsion effect** Each proton repels every other proton in the nucleus. The corresponding potential energy per pair of interacting particles is given by $ke^2/r$, where $k$ is the Coulomb constant. The total Coulomb energy represents the work required to assemble $Z$ protons from infinity to a sphere of volume $V$. This energy is proportional to the number of proton pairs $Z(Z-1)$ and is inversely proportional to the nuclear radius. Consequently, the reduction in energy that results from the Coulomb effect is $-C_3Z(Z-1)/A^{1/3}$.

Another small effect that contributes to the binding energy in this model is significant only for heavy nuclei having a large excess of neutrons. This effect gives rise to a binding energy term that is proportional to $(A - 2Z)^2/A$.

Adding these contributions, we get as the total binding energy

$$E_b = C_1A - C_2A^{2/3} - C_3\frac{Z(Z-1)}{A^{1/3}} - C_4\frac{(A-2Z)^2}{A} \tag{45.5}$$

Semiempirical binding energy formula

Equation 45.5 is often referred to as the **semiempirical binding energy formula.**[2]

The four constants in Equation 45.5 can be adjusted to fit this expression to experimental data. For nuclei with $A \geq 15$, the constants have the values

$$C_1 = 15.7 \text{ MeV} \qquad C_2 = 17.8 \text{ MeV}$$

$$C_3 = 0.71 \text{ MeV} \qquad C_4 = 23.6 \text{ MeV}$$

Equation 45.5, together with the constants given above, fits the known nuclear mass values very well. However, the liquid drop model does not account for some finer details of nuclear structure, such as stability rules and angular momentum. On the other hand, it does provide a qualitative description of the process of nuclear fission, shown schematically in Figure 45.9. If the drop vibrates with a large amplitude (which may be initiated by collision with another particle), it distorts and, under the right conditions, will break apart. We shall discuss the process of fission in Section 46.2.

### The Independent-Particle Model

We have mentioned the limitations of the liquid-drop model and its failure to explain some features of nuclear structure. The **independent-particle model,** often called the *shell model,* is based upon the assumption that *each nucleon moves in a well-defined orbit within the nucleus and in an averaged field produced by the other nucleons.* In this model, the nucleons exist in quantized energy states and there are few collisions between nucleons. Obviously, the assumptions of this model differ greatly from those made in the liquid-drop model.

[2] For more details on this model, see R.L. Sproull and W.A. Phillips, *Modern Physics,* New York, Wiley, 1980, Chap. 11.

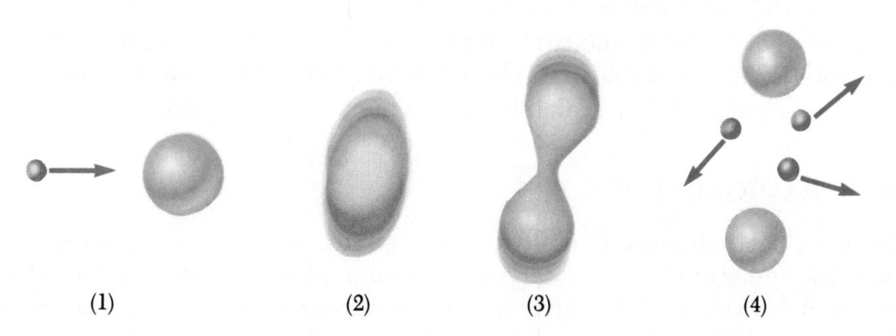


**Figure 45.9** Steps leading to fission according to the liquid-drop model of the nucleus.

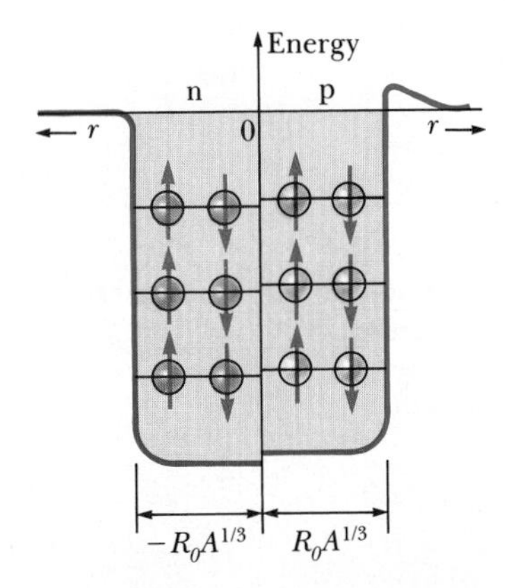


**Figure 45.10** A square-well potential containing 12 nucleons. The dark circles represent protons, and the open circles represent neutrons. The energy levels for the protons are slightly higher than those for the neutrons because of the coulomb potential in the protons. The difference in the levels increases as Z increases. Note that only two nucleons with opposite spin can occupy a given level, as required by the Pauli exclusion principle.

The quantized states occupied by the nucleons can be described by a set of quantum numbers. Since both the proton and the neutron have spin $\frac{1}{2}$, one can apply Pauli's exclusion principle to describe the allowed states (as we did for electrons in Chapter 42). That is, each state can contain only two protons (or two neutrons) with *opposite spin* (Fig. 45.10). The protons have a set of allowed states, and these states differ from those of the neutron since they are different particles. *The proton levels are higher in energy than the neutron levels as a result of the added Coulomb repulsion between protons.*

Using this model of the nucleus, one can begin to understand the stability of certain nuclear states. In particular, if two nucleons are to undergo a collision, the energy of each state after the collision must correspond to one of the allowed states. The collision will simply not occur if that allowed state is already filled by another nucleon. In other words, *a collision between two nucleons can occur only if the process does not violate the exclusion principle.* One can interpret this as an effective force that acts between nucleons to prevent a collision.

The independent-particle model is also quite successful in predicting the angular momentum of stable nuclei. In order to properly account for the stable nuclear states, it is necessary to include a large contribution from magnetic effects. In particular, the magnetic moment of each nucleon interacts with the magnetic field produced by its orbital motion.

Finally, it is possible to understand why nuclei containing an even number of protons and neutrons are more stable than others. In fact, there are 160 such stable isotopes (even-even nuclei). Any particular state is filled when it contains two protons (or two neutrons) with opposite spins. *An extra proton or neutron can be added to the nucleus only at the expense of increasing the energy of the nucleus.* This increase in energy leads to a nucleus that is less stable than the original nucleus. A careful inspection of the stable nuclei shows that *the majority have a special stability when their nucleons combine in pairs, which results in a total angular momentum of zero.* This accounts for the large number of high-stability nuclei (those with high binding energies) with the magic numbers given by Equation 45.2.

### The Collective Model

A third model of nuclear structure, known as the **collective model,** combines some features of both the liquid-drop model and the independent-particle model. The nucleus is considered to have some "extra" nucleons moving in quantized orbits, in addition to the filled core of nucleons. The extra nucleons are subject to the field produced by the core, as in the independent-particle scheme. Deformations can be set up in the core as a result of a strong interaction between the core and the extra nucleons, thereby setting up vibrational and rotational motions as in the liquid-drop model. The collective model has been very successful in explaining many nuclear phenomena.

## 45.4 RADIOACTIVITY

In 1896, Becquerel accidentally discovered that uranium salt crystals emit an invisible radiation that can darken a photographic plate even if the plate is covered to exclude light. After several observations of this type under controlled conditions, Becquerel concluded that the radiation emitted by the

**Marie Sklodowska Curie** was born in Poland shortly after the unsuccessful Polish revolt against the Russians in 1863. Following her high school education, she worked diligently to help meet the educational expenses of her older brother and sister who had left for Paris. At the same time, she managed to save enough money for her own trip to Paris and entered the Sorbonne in 1891. Although she lived under very frugal conditions during this period (fainting once from hunger in the classroom), she graduated at the top of her class.

In 1895 she married the French chemist Pierre Curie who was already known for the discovery of piezoelectricity. (A piezoelectric crystal exhibits a potential difference when it is under pressure.) Using piezoelectric materials to measure the activity of radioactive substances, she demonstrated the radioactive nature of the elements uranium and thorium. In 1898, she and her husband discovered a new radioactive element contained in uranium ore, which they called polonium, named after Madame Curie's native land. By the end of 1898, the Curies succeeded in isolating trace amounts of an even more radioactive element which they named radium. In an effort to produce weighable quantities of radium, they embarked on a painstaking effort of isolating radium from pitchblende, an ore rich in uranium. After four years of purifying and repurifying tons of ore, and using their own life savings to finance their work, the Curies succeeded in preparing about 0.1 g of radium. In 1903, Marie and Pierre Curie were awarded the Nobel prize in physics, which they shared with A. H. Becquerel, for their studies of radioactive substances.

After her husband's death in a tragic accident in 1906, Madame Curie took over his professorship at the Sorbonne. Unfortunately, she experienced prejudice in the scientific community because she was a woman. For example, after being nominated to the French Academy of Sciences, she was refused membership, losing by one vote.

In 1911, she was awarded a second Nobel prize in chemistry for the discovery of radium and polonium. She spent the last few decades of her life supervising the Paris Institute of Radium. Madame Curie died of leukemia caused by the years of exposure of radiation from radioactive substances.

Biographical Sketch

**Marie Curie**
*(1867–1934)*

crystals was of a new type, one that required no external stimulation. This process of spontaneous emission of radiation by uranium was soon to be called **radioactivity.** Subsequent experiments by other scientists showed that other substances were also radioactive. The most significant investigations of this type were conducted by Marie and Pierre Curie. After several years of careful and laborious chemical separation processes on tons of pitchblende, a radioactive ore, the Curies reported the discovery of two previously unknown elements, both of which were radioactive. These were named polonium and radium. Subsequent experiments, including Rutherford's famous work on alpha-particle scattering, suggested that radioactivity was the result of the decay, or disintegration, of unstable nuclei.

There are three types of radiation that can be emitted by a radioactive substance: alpha ($\alpha$) decay, where the emitted particles are $^4$He nuclei; beta ($\beta$) decay, in which the emitted particles are either electrons or positrons; and gamma ($\gamma$) decay, in which the emitted "rays" are high-energy photons. A positron is a particle similar to the electron in all respects except that it has a charge of $+e$ (the antimatter twin of the electron). The symbol $\beta^-$ is used to designate an electron, and $\beta^+$ designates a positron.

It is possible to distinguish these three forms of radiation using the scheme described in Figure 45.11. The radiation from a radioactive sample is directed

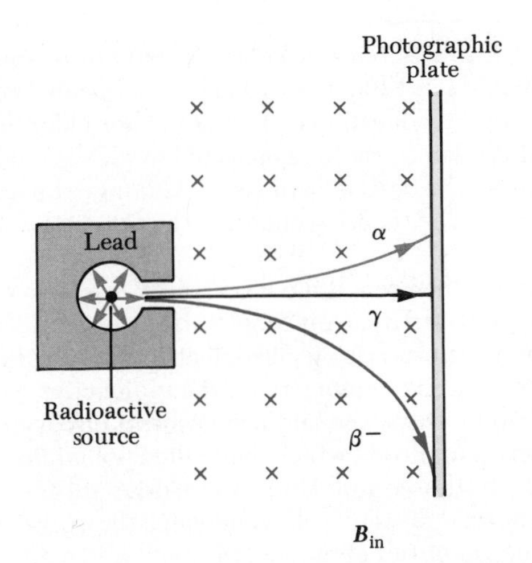


**Figure 45.11** The radiation from a radioactive source, such as radium, can be separated into three components by using a magnetic field to deflect the charged particles. The photographic plate at the right records the events.

into a region in which there is a magnetic field. Under these conditions, the beam splits into three components, two bending in opposite directions and the third experiencing no change in direction. From this simple observation, one can conclude that the radiation of the undeflected beam carries no charge (the gamma ray), the component deflected upward corresponds to positively charged particles (alpha particles), and the component deflected downward corresponds to negatively charged particles ($\beta^-$). If the beam includes a positron ($\beta^+$), it is deflected upward.

The three types of radiation have quite different penetrating powers. Alpha particles barely penetrate a sheet of paper, beta particles can penetrate a few millimeters of aluminum, and gamma rays can penetrate several centimeters of lead.

The rate at which a particular decay process occurs in a radioactive sample is proportional to the number of radioactive nuclei present (that is, those nuclei that have not yet decayed). If $N$ is the number of radioactive nuclei present at some instant, the rate of change of $N$ is

$$\frac{dN}{dt} = -\lambda N \tag{45.6}$$

where $\lambda$ is the **decay constant,** or **disintegration constant.** The minus sign indicates that $dN/dt$ is negative; that is, $N$ is decreasing in time.

If we write Equation 45.6 in the form

$$\frac{dN}{N} = -\lambda\, dt$$

we can integrate the expression to give

$$\int_{N_0}^{N} \frac{dN}{N} = -\lambda \int_0^t dt$$

$$\ln\left(\frac{N}{N_0}\right) = -\lambda t$$

or

Exponential decay

$$N = N_0 e^{-\lambda t} \tag{45.7}$$

The constant $N_0$ represents the number of radioactive nuclei at $t = 0$.

The **decay rate** $R$ $(= |dN/dt|)$ can be obtained by differentiating Equation 45.7 with respect to time. This gives

$$R = \left|\frac{dN}{dt}\right| = N_0\lambda e^{-\lambda t} = R_0 e^{-\lambda t} \tag{45.8}$$

where $R_0 = N_0\lambda$ is the decay rate at $t = 0$, and $R = \lambda N$. The decay rate of a sample is often referred to as its **activity.** Note that both $N$ and $R$ decrease exponentially with time. The plot of $N$ versus $t$ shown in Figure 45.12 illustrates the exponential decay law.

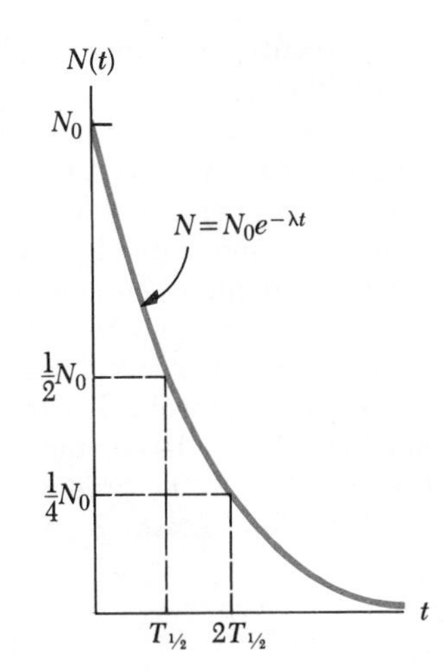


**Figure 45.12** Plot of the exponential decay law for radioactive nuclei. The vertical axis represents the number of radioactive nuclei present at any time $t$, and the horizontal axis is time. The time $T_{1/2}$ is the half-life of the sample.

Another useful parameter used to characterize the decay of a particular nucleus is the half-life, $T_{1/2}$.

The half-life of a radioactive substance is the time it takes half of a given number of radioactive nuclei to decay.

Setting $N = N_0/2$ and $t = T_{1/2}$ in Equation 45.7 gives

$$\frac{N_0}{2} = N_0 e^{-\lambda T_{1/2}}$$

Writing this in the form $e^{\lambda T_{1/2}} = 2$ and taking the natural logarithm of both sides, we get

$$T_{1/2} = \frac{\ln 2}{\lambda} = \frac{0.693}{\lambda} \tag{45.9}$$

Half-life equation

This is a convenient expression relating the half-life to the decay constant. Note that after an elapsed time of one half-life, there are $N_0/2$ radioactive nuclei remaining (by definition); after two half-lives, half of these will have decayed and $N_0/4$ radioactive nuclei will be left; after three half-lives, $N_0/8$ will be left, and so on. In general, after $n$ half-lives, the number of radioactive nuclei remaining is $N_0/2^n$. Thus, we see that the nuclear decay is independent of the past history of the sample.

The unit of activity is the **curie** (Ci), defined as

$$1 \text{ Ci} \equiv 3.7 \times 10^{10} \text{ decays/s}$$

The curie

This unit was selected as the original activity unit because it is the approximate activity of 1 g of radium. The SI unit of activity is called the **becquerel** (Bq):

$$1 \text{ Bq} \equiv 1 \text{ decay/s}$$

The becquerel

Therefore, $1 \text{ Ci} = 3.7 \times 10^{10}$ Bq. The most commonly used units of activity are the mCi ($10^{-3}$ Ci) and the $\mu$Ci ($10^{-6}$ Ci).

EXAMPLE 45.4 How Many Nuclei Are Left?

Carbon-14, $^{14}_{6}C$, is a radioactive isotope of carbon that has a half-life of 5730 years. If you start with a sample of 1000 carbon-14 nuclei, how many will still be around in 22 920 years?

*Solution* In 5730 years, half the sample will have decayed, leaving 500 carbon-14 nuclei remaining. In another 5730 years (for a total elapsed time of 11 460 years), the number will be reduced to 250 nuclei. After another 5730 years (total time 17 190 years), 125 remain. Finally, after four half-lives (22 920 years), only about 62 remain.

It should be noted here that these numbers represent ideal circumstances. Radioactive decay is an averaging process over a very large number of atoms, and the actual outcome depends on statistics. Our original sample in this example contained only 1000 nuclei, certainly not a very large number. Thus, if we were actually to count the number remaining after one half-life for this small sample, it probably would not be 500. If the initial number is very large, the probability of getting extremely close to the predicted number remaining after one half-life increases greatly.

**EXAMPLE 45.5 The Activity of Radium** □

The half-life of the radioactive nucleus $^{226}_{88}Ra$ is $1.6 \times 10^3$ years. If a sample contains $3 \times 10^{16}$ such nuclei, determine the activity at this time.

*Solution* First, let us calculate the decay constant $\lambda$ using Equation 45.9 and the fact that

$$T_{1/2} = 1.6 \times 10^3 \text{ years}$$
$$= (1.6 \times 10^3 \text{ years})(3.16 \times 10^7 \text{ s/year})$$
$$= 5.0 \times 10^{10} \text{ s}$$

Therefore,

$$\lambda = \frac{0.693}{T_{1/2}} = \frac{0.693}{5.0 \times 10^{10} \text{ s}} = 1.4 \times 10^{-11} \text{ s}^{-1}$$

We can calculate the activity of the sample at $t = 0$ using $R_0 = \lambda N_0$, where $R_0$ is the decay rate at $t = 0$ and $N_0$ is the number of radioactive nuclei present at $t = 0$. Since $N_0 = 3 \times 10^{16}$, we have

$$R_0 = \lambda N_0 = (1.4 \times 10^{-11} \text{ s}^{-1})(3 \times 10^{16})$$
$$= 4.1 \times 10^5 \text{ decays/s}$$

Since $1 \text{ Ci} = 3.7 \times 10^{10}$ decays/s, the activity, or decay rate, at $t = 0$ is

$$R_0 = 11.1 \ \mu\text{Ci}$$

**EXAMPLE 45.6 The Activity of Carbon**

A radioactive sample contains 3.50 $\mu$g of pure $^{11}_{6}C$, which has a half-life of 20.4 min. (a) Determine the number of nuclei present initially.

*Solution* The atomic mass of $^{11}_{6}C$ is approximately 11, and therefore 11 g will contain Avogadro's number $(6.02 \times 10^{23})$ of nuclei. Therefore, 3.50 $\mu$g of sample will contain $N$ nuclei, where

$$\frac{N}{6.02 \times 10^{23} \text{ nuclei/mol}} = \frac{3.50 \times 10^{-6} \text{ g}}{11 \text{ g/mol}}$$

or

$$N = 1.92 \times 10^{17} \text{ nuclei}$$

(b) What is the activity of the sample initially and after 8 h?

*Solution* Since $T_{1/2} = 20.4 \text{ min} = 1224 \text{ s}$, the decay constant is

$$\lambda = \frac{0.693}{T_{1/2}} = \frac{0.693}{1224 \text{ s}} = 5.66 \times 10^{-4} \text{ s}^{-1}$$

Therefore, the initial activity of the sample is

$$R_0 = \lambda N_0 = (5.66 \times 10^{-4} \text{ s}^{-1})(1.92 \times 10^{17})$$
$$= 1.08 \times 10^{14} \text{ decay/s}$$

We can use Equation 45.8 to find the activity at any time $t$. For $t = 8 \text{ h} = 2.88 \times 10^4$ s, we see that $\lambda t = 16.3$ and so

$$R = R_0 e^{-\lambda t} = (1.09 \times 10^{14} \text{ decays/s})e^{-16.3}$$

$$= 8.96 \times 10^6 \text{ decays/s}$$

A listing of activity versus time for this situation is given in Table 45.2.

Exercise 1 Calculate the number of radioactive nuclei remaining after 8 h.
Answer $N = 1.58 \times 10^{10}$ nuclei.

**TABLE 45.2 Activity Versus Time for the Sample Described in Example 45.6**

| $t$(h) | $R$(decays/s) |
|---|---|
| 0 | $1.08 \times 10^{14}$ |
| 1 | $1.41 \times 10^{13}$ |
| 2 | $1.84 \times 10^{12}$ |
| 3 | $2.39 \times 10^{11}$ |
| 4 | $3.12 \times 10^{10}$ |
| 5 | $4.06 \times 10^{9}$ |
| 6 | $5.28 \times 10^{8}$ |
| 7 | $6.88 \times 10^{7}$ |
| 8 | $8.96 \times 10^{6}$ |

**EXAMPLE 45.7 A Radioactive Isotope of Iodine**

A sample of the isotope $^{131}I$, which has a half-life of 8.04 days, has a measured activity of 5 mCi at the time of shipment. Upon receipt in a medical laboratory, the activity is measured to be 4.2 mCi. How much time has elapsed between the two measurements?

*Solution* We can make use of Equation 45.8:

$$R = R_0 e^{-\lambda t}$$

where $R_0$ is the initial activity and $R$ is the activity at time $t$. Writing this in the form

$$\frac{R}{R_0} = e^{-\lambda t}$$

and taking the natural logarithm of each side, we get

$$\ln\left(\frac{R}{R_0}\right) = -\lambda t$$

or

$$(1) \qquad t = -\frac{1}{\lambda}\ln\left(\frac{R}{R_0}\right)$$

To find $\lambda$, we can use Equation 45.9:

$$(2) \qquad \lambda = \frac{0.693}{T_{1/2}} = \frac{0.693}{8.04 \text{ days}}$$

Substituting (2) into (1) gives

$$t = -\left(\frac{8.04 \text{ days}}{0.693}\right)\ln\left(\frac{4.2 \text{ mCi}}{5.0 \text{ mCi}}\right) = 2.02 \text{ days}$$

## 45.5 THE DECAY PROCESSES

As we stated in the previous section, a radioactive nucleus spontaneously decays via three processes: alpha decay, beta decay, and gamma decay. Let us discuss these three processes in more detail.

### Alpha Decay

If a nucleus emits an alpha particle ($^4_2He$), it loses two protons and two neutrons. Therefore, $N$ decreases by 2, $Z$ decreases by 2, and $A$ decreases by 4. The decay can be written symbolically as

$$^A_Z X \longrightarrow {}^{A-4}_{Z-2}Y + {}^4_2He \qquad (45.10)$$

Alpha decay

where X is called the **parent nucleus** and Y the **daughter nucleus.** As examples, $^{238}U$ and $^{226}Ra$ are both alpha emitters and decay according to the schemes

$$^{238}_{92}U \longrightarrow {}^{234}_{90}Th + {}^4_2He \qquad (45.11)$$

$$^{226}_{88}Ra \longrightarrow {}^{222}_{86}Rn + {}^4_2He \qquad (45.12)$$

The half-life for $^{238}U$ decay is $4.47 \times 10^9$ years, and the half-life for $^{226}Ra$ decay is $1.60 \times 10^3$ years. In both cases, note that the $A$ of the daughter nucleus is 4 less than that of the parent nucleus. Likewise, $Z$ is reduced by 2. The differences are accounted for in the emitted alpha particle (the $^4He$ nucleus).

The decay of $^{226}Ra$ is shown in Figure 45.13. When one element changes into another, as in the process of alpha decay, the process is called *spontaneous decay*. As a general rule, (1) the sum of the mass numbers $A$ must be the same on both sides of the equation and (2) the sum of the charge numbers $Z$ must be the same on both sides of the equation. In addition, the total energy must be conserved. If we call $M_X$ the mass of the parent nucleus, $M_Y$ the mass of the daughter nucleus, and $M_\alpha$ the mass of the alpha particle, we can define the **disintegration energy $Q$:**

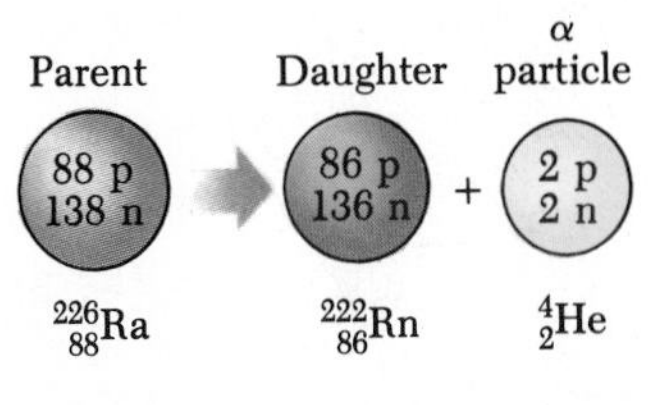

**Figure 45.13** Alpha decay of the $^{226}_{88}Ra$ nucleus.

$$Q = (M_X - M_Y - M_\alpha)c^2 \qquad (45.13)$$

The disintegration energy $Q$

Biographical Sketch

**Enrico Fermi**
*(1901–1954)*

**Enrico Fermi,** an Italian-American physicist, received his doctorate from the University of Pisa in 1922 and did postdoctorate work in Germany under Max Born. He returned to Italy in 1924 and became a professor of physics at the University of Rome in 1926. He received the Nobel Prize for physics in 1938 for his work dealing with the production of transuranic radioactive elements by neutron bombardment.

Fermi first became interested in physics at the age of 14 after reading an old physics book in Latin. He had an excellent scholastic record and was able to recite Dante's *Divine Comedy* and much of Aristotle from memory. His great ability to solve problems in theoretical physics and his skill for simplifying very complex situations made him somewhat of an oracle. He was also a gifted experimentalist and teacher of physics. During one of his early lecture trips to the United States in the depths of the Depression, a car which he had purchased became disabled and he pulled into a nearby gas station. After Fermi repaired the car with ease, the station owner immediately offered him a job.

Fermi and his family immigrated to the United States and he became a naturalized citizen in 1944. Once in America, Fermi accepted a position at Columbia University and later became a professor at the University of Chicago. After the Manhattan Project was established, Fermi was commissioned to design and build a structure (called an atomic pile) in which a self-sustained chain reaction might occur. The structure, built in the squash court of the University of Chicago, contained uranium in combination with graphite blocks to slow the neutrons to thermal velocities. Cadmium rods inserted in the "pile" were used to absorb neutrons and control the reaction rate. History was made at 3:45 P.M. on December 2, 1942 as the cadmium rods were slowly withdrawn, and a self-sustained chain reaction was observed. Fermi's earthshaking achievement of the world's first nuclear reactor marked the beginning of the atomic age.

Fermi died of cancer in 1954 at the age of 53. One year later, the one-hundredth element was discovered and named *fermium* in his honor.

Note that $Q$ will be in J if the masses are in kg and $c$ is the usual $3 \times 10^8$ m/s. However, when the nuclear masses are expressed in the more convenient unit u, the value of $Q$ can be calculated in MeV using the expression

$$Q = (M_X - M_Y - M_\alpha) \times 931.50 \text{ MeV/u} \qquad (45.14)$$

It is left as a problem (Problem 41) to show that 931.50 MeV/u is the correct conversion factor.

The residual energy $Q$ appears in the form of kinetic energy of the daughter nucleus and the alpha particle. The quantity given by Equation 45.13 is sometimes referred as the $Q$ value of the nuclear reaction. In the case of the $^{226}$Ra decay described in Figure 45.13, if the parent nucleus decays at rest, the residual kinetic energy of the products is 4.87 MeV. Most of the kinetic energy is associated with the alpha particle because this particle is much less massive than the recoiling daughter nucleus, $^{222}$Rn. That is, since momentum must be conserved, the lighter alpha particle recoils with a much higher velocity than the daughter nucleus. Generally, light particles carry off most of the energy in nuclear decays.

Finally, it is interesting to note that if one assumed that $^{238}$U (or other alpha emitters) decayed by emitting a proton or neutron, the mass of the decay products would exceed that of the parent nucleus, corresponding to negative $Q$ values. Therefore, such spontaneous decays do not occur.

**EXAMPLE 45.8 The Energy Liberated When Radium Decays**

The $^{226}$Ra nucleus undergoes alpha decay according to Equation 45.12. Calculate the $Q$ value for this process. Take the mass of $^{226}$Ra to be 226.025406 u, the mass of $^{222}$Rn to be 222.017574 u, and the mass of $^{4}_{2}$He to be 4.002603 u, as found in Table A.3. (Note that atomic mass units rather than nuclear mass units can be used here because the electron masses cancel in evaluating the mass differences.)

*Solution* Using Equation 45.14, we see that

$$Q = (M_X - M_Y - M_\alpha) \times 931.50 \text{ MeV/u}$$
$$= (226.025406 \text{ u} - 222.017574 \text{ u} - 4.002603 \text{ u}) \times 931.50 \text{ MeV/u}$$
$$= (0.005229 \text{ u}) \times \left(931.50 \frac{\text{MeV}}{\text{u}}\right) = 4.87 \text{ MeV}$$

It is left as a problem (Problem 75) to show that the kinetic energy of the alpha particle is about 4.8 MeV, whereas the recoiling daughter nucleus has only about 0.1 MeV of kinetic energy.

We now turn to the mechanism of alpha decay. Figure 45.14 is a plot of the potential energy versus distance $r$ from the nucleus for the alpha particle-nucleus system, where $R$ is the range of the nuclear force. The curve represents the combined effects of (1) the Coulomb repulsive energy, which gives the positive peak for $r > R$, and (2) the nuclear attractive force, which causes the curve to be negative for $r < R$. As we saw in Example 45.8, the disintegration energy is about 5 MeV, which is the approximate kinetic energy of the alpha particle, represented by the lower dotted line in Figure 45.14. According to classical physics, the alpha particle is trapped in the potential well. How, then, does it ever escape from the nucleus?

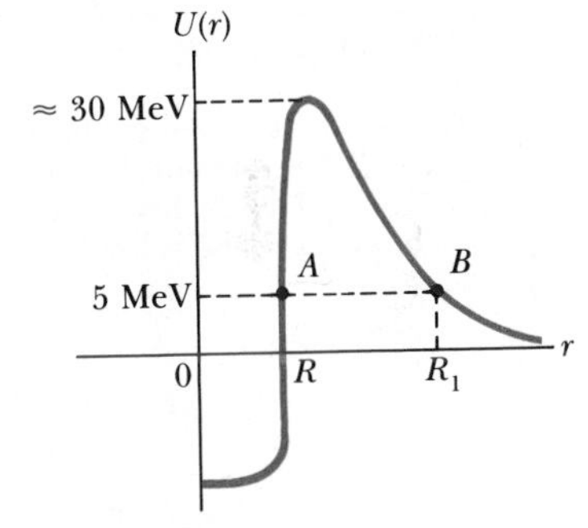


**Figure 45.14** Potential energy versus separation for the alpha particle–nucleus system. Classically, the energy of the alpha particle is not sufficiently large to overcome the barrier, and so the particle should not be able to escape the nucleus.

The answer to this question was first provided by Gamow and independently by Gurney and Condon in 1928, using quantum mechanics. Briefly, the view of quantum mechanics is that there is always some probability that the particle can penetrate (or tunnel) through the barrier (Section 41.9). Recall that the probability of locating the particle depends on its wave function $\psi$ and that the probability of tunneling is measured by $|\psi|^2$. Figure 45.15 is a sketch of the wave function for a particle of energy $E$ meeting a square barrier of finite height, which approximates the nuclear barrier. Note that the wave function is oscillating both inside and outside the barrier but is greatly reduced in amplitude because of the barrier. As the energy $E$ of the particle is increased, the probability of escaping also increases. Furthermore, the probability increases as the width of the barrier is decreased.

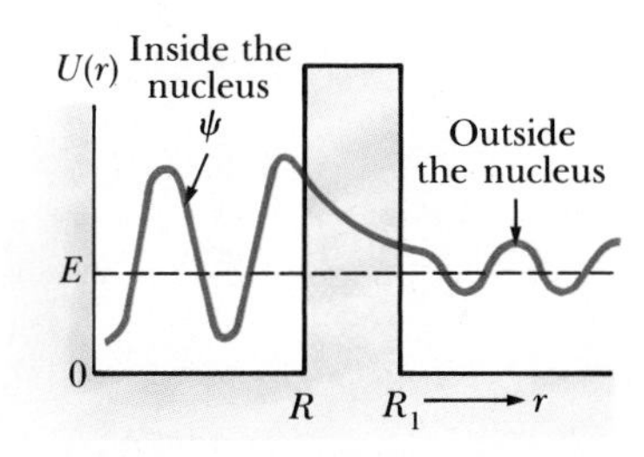


**Figure 45.15** The nuclear potential energy is modeled as a square barrier. The energy of the alpha particle is $E$, which is less than the height of the barrier. According to wave mechanics, the alpha particle has some chance of tunneling through the barrier, as indicated by the finite size of the wave function for $r > R_1$.

### Beta Decay

When a radioactive nucleus undergoes beta decay, the daughter nucleus has the same number of nucleons as the parent nucleus but the charge number is changed by 1. Symbolically, the two beta decay processes are

$$^{A}_{Z}\text{X} \longrightarrow {}^{A}_{Z+1}\text{Y} + \beta^- \quad (45.15)$$

$$^{A}_{Z}\text{X} \longrightarrow {}^{A}_{Z-1}\text{Y} + \beta^+ \quad (45.16)$$

Again, note that the nucleon number and total charge are both conserved in these decays. As we shall see later, *these processes are not described completely by these expressions.* We shall give reasons for this shortly.

Two typical beta decay processes are

Beta decay

$$^{14}_{6}C \longrightarrow {}^{14}_{7}N + \beta^-$$

$$^{12}_{7}N \longrightarrow {}^{12}_{6}C + \beta^+$$

It is important to note that the electron or positron involved in these decays is created within the nucleus as an initial step in the decay process. This is equivalent to saying that, during beta decay, a neutron in the nucleus is transformed into a proton. Indeed, the process $n \rightarrow p + \beta^-$ does occur.

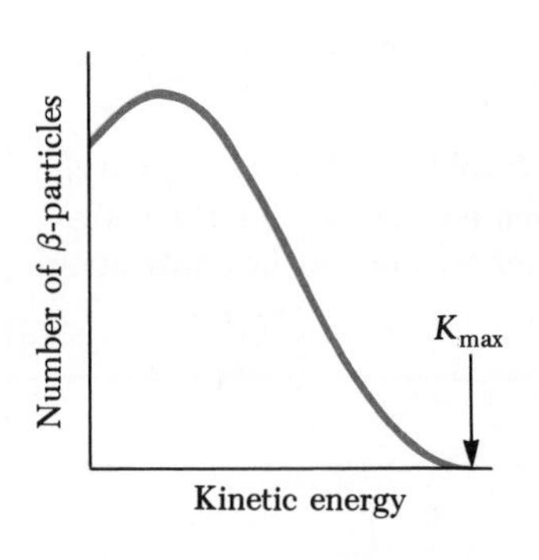


**Figure 45.16** A typical beta decay curve. The maximum kinetic energy observed for the beta particles corresponds to the value of $Q$ for the reaction.

Now consider the energy of the system before and after the decay. As with alpha decay, energy must be conserved. Experimentally, one finds that the beta particles are emitted over a continuous range of energies (Fig. 45.16). These results suggest that the decaying nuclei emit electrons with different energy. The kinetic energy of the electrons must be balanced by the decrease in mass of the system, that is, the $Q$ value. However, since all decaying nuclei have the same initial mass, *the Q value must be the same for each decay.* In view of this, why do the emitted electrons have different kinetic energies? The law of conservation of energy seems to be violated! Further analysis shows that, according to the decay processes given by Equations 45.15 and 45.16, the principles of conservation of both angular momentum (spin) and linear momentum are also violated!

After a great deal of experimental and theoretical study, Pauli in 1930 proposed that a third particle must be present to carry away the "missing" energy and momentum. Fermi later named this particle the **neutrino** (little neutral one) since it had to be electrically neutral and have little or no rest mass. Although it eluded detection for many years, the neutrino (symbol $\nu$) was finally detected experimentally in 1956.

The neutrino has the following properties:

Properties of the neutrino

1. It has zero electric charge.
2. It has a rest mass smaller than that of the electron, and in fact its mass may be zero (although recent experiments suggest that this may not be true).
3. It has a spin of $\frac{1}{2}$, which satisfies the law of conservation of angular momentum.
4. It interacts very weakly with matter and is therefore very difficult to detect.

We can now write the beta decay processes in their correct form:

$$^{14}_{6}C \longrightarrow {}^{14}_{7}N + \beta^- + \bar{\nu} \tag{45.17}$$

$$^{12}_{7}N \longrightarrow {}^{12}_{6}C + \beta^+ + \nu \tag{45.18}$$

$$n \longrightarrow p + \beta^- + \bar{\nu} \tag{45.19}$$

where the symbol $\bar{\nu}$ represents the **antineutrino.**

The positron $\beta^+$ emitted in the process given by Equation 45.18 is a particle that is identical to the electron except that it has a positive charge of $+e$. Because it is like the electron in all respects except charge, the positron is said to be an **antiparticle** to the electron. Likewise, the antineutrino is the antiparticle to the neutrino. We shall discuss antiparticles further in Chapter 47. For now, it suffices to say that *a neutrino is emitted in positron decay and an antineutrino is emitted in electron decay.*

### Carbon Dating

The beta decay of $^{14}C$ given by Equation 45.17 is commonly used to date organic samples. Cosmic rays (high-energy particles from outer space) in the upper atmosphere cause nuclear reactions that create $^{14}C$. In fact, the ratio of $^{14}C$ to $^{12}C$ isotopic abundance in the carbon dioxide molecules of our atmosphere has a constant value of about $1.3 \times 10^{-12}$. All living organisms have the same ratio of $^{14}C$ to $^{12}C$ because they continuously exchange carbon dioxide with their surroundings. When an organism dies, however, it no longer absorbs $^{14}C$ from the atmosphere, and so the ratio of $^{14}C$ to $^{12}C$ decreases as the result of the beta decay of $^{14}C$. It is therefore possible to measure the age of a material by measuring its activity per unit mass due to the decay of $^{14}C$. Using carbon dating, samples of wood, charcoal, bone, and shell have been identified as having lived from 1000 to 25 000 years ago. This knowledge has helped us to reconstruct the history of living organisms—including humans—during this time span.

A particularly interesting example is the dating of the Dead Sea Scrolls. This group of manuscripts was first discovered by a shepherd in 1947. Translation showed them to be religious documents, including most of the books of the Old Testament. Because of their historical and religious significance, scholars wanted to know their age. Carbon dating applied to fragments of the scrolls and to the material in which they were wrapped established their age at about 1950 years.

---

**EXAMPLE 45.9 Radioactive Dating**

A piece of charcoal of mass 25 g is found in some ruins of an ancient city. The sample shows a $^{14}C$ activity of 250 decays/min. How long has the tree that this charcoal came from been dead?

*Solution* First, let us calculate the decay constant for $^{14}C$, which has a half-life of 5730 years.

$$\lambda = \frac{0.693}{T_{1/2}} = \frac{0.693}{(5730\ \text{y})(3.16 \times 10^{7}\ \text{s/y})}$$

$$= 3.83 \times 10^{-12}\ \text{s}^{-1}$$

The number of $^{14}C$ nuclei can be calculated in two steps. First, the number of $^{12}C$ nuclei in 25 g of carbon is

$$N\,(^{12}\text{C}) = \frac{6.02 \times 10^{23}\ \text{nuclei/mol}}{12\ \text{g/mol}}\,(25\ \text{g})$$

$$= 1.25 \times 10^{24}\ \text{nuclei}$$

Assuming that the ratio of $^{14}C$ to $^{12}C$ is $1.3 \times 10^{-12}$, we see that the number of $^{14}C$ nuclei in 25 g *before* decay is

$$N_0\,(^{14}\text{C}) = (1.3 \times 10^{-12})(1.26 \times 10^{24})$$

$$= 1.63 \times 10^{12}\ \text{nuclei}$$

Hence, the initial activity of the sample is

$$R_0 = N_0\lambda = (1.63 \times 10^{12}\ \text{nuclei})(3.83 \times 10^{-12}\ \text{s}^{-1})$$

$$= 6.25\ \text{decays/s} = 375\ \text{decays/min}$$

We can now calculate the age of the charcoal using Equation 45.8, which relates the activity $R$ at any time $t$ to the initial activity $R_0$:

$$R = R_0 e^{-\lambda t} \quad \text{or} \quad e^{-\lambda t} = \frac{R}{R_0}$$

Since it is given that $R = 250$ decays/min and since we found that $R_0 = 375$ decays/min, we can calculate $t$ by taking the natural logarithm of both sides of the last equation:

$$-\lambda t = \ln\left(\frac{R}{R_0}\right) = \ln\left(\frac{250}{375}\right) = -0.405$$

$$t = \frac{0.405}{\lambda} = \frac{0.405}{3.84 \times 10^{-12}\ \text{s}^{-1}}$$

$$= 1.06 \times 10^{11}\ \text{s} = 3350\ \text{years}$$

This technique of radioactive-carbon dating has been successfully used to measure the age of many organic relics up to 25 000 years old.

A process that competes with $\beta^+$ decay is called **electron capture.** This occurs when a parent nucleus captures one of its own electrons and emits a neutrino. The final product after decay is a nucleus whose charge is $Z - 1$. In general, the electron-capture process can be expressed as

Electron capture

$$^A_ZX + ^{\ 0}_{-1}e \longrightarrow {}^{\ \ \ A}_{Z-1}X + \nu \tag{45.20}$$

In most cases, it is a K-shell electron that is captured, and one refers to this as **K capture.** One example of this process is the capture of an electron by $^7_4Be$ to become $^7_3Li$:

$$^7_4Be + ^{\ 0}_{-1}e \longrightarrow {}^7_3Li + \nu$$

## Gamma Decay

Very often, a nucleus that undergoes radioactive decay is left in an excited energy state. The nucleus can then undergo a second decay to a lower energy state, perhaps to the ground state, by emitting a photon. Photons emitted in such a de-excitation process are called gamma rays. Such photons have very high energy (in the range of 1 MeV to 1 GeV) relative to the energy of visible light (about 1 eV). Recall that the energy of photons emitted (or absorbed) by an atom equals the difference in energy between the two electronic states involved in the transition. Similarly, a gamma ray photon has an energy $hf$ that equals the energy difference $\Delta E$ between two nuclear energy levels. When a nucleus decays by emitting a gamma ray, it doesn't change, apart from the fact that it ends up in a lower energy state. We can represent a gamma decay process as

Gamma decay

$$^A_ZX^* \longrightarrow {}^A_ZX + \gamma \tag{45.21}$$

where $X^*$ indicates a nucleus in an excited state.

A nucleus may reach an excited state as the result of a violent collision with another particle. However, it is more common for a nucleus to be in an excited state after it has undergone a previous alpha or beta decay. The following sequence of events represents a typical situation in which gamma decay occurs:

$$^{12}_5B \longrightarrow {}^{12}_6C^* + ^{\ 0}_{-1}e + \bar{\nu} \tag{45.22}$$

$$^{12}_6C^* \longrightarrow {}^{12}_6C + \gamma \tag{45.23}$$

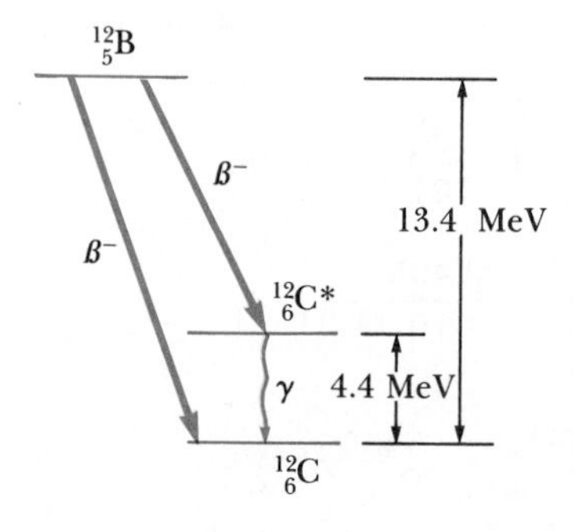

**Figure 45.17** The $^{12}B$ nucleus undergoes $\beta^-$ decay to two levels of $^{12}C$. The decay to the excited level, $^{12}C^*$, is followed by gamma decay to the ground state.

Figure 45.17 shows the decay scheme for $^{12}B$, which undergoes beta decay to either of two levels of $^{12}C$. It can either (1) decay directly to the ground state of $^{12}C$ by emitting a 13.4-MeV electron or (2) undergo $\beta^-$ decay to an excited state of $^{12}C^*$, followed by gamma decay to the ground state. This latter process results in the emission of a 9.0-MeV electron and a 4.4-MeV photon.

The various pathways by which a radioactive nucleus can undergo decay are summarized in Table 45.3.

**TABLE 45.3 Various Decay Pathways**

|  |  |
|---|---|
| Alpha decay | ${}^{A}_{Z}X \longrightarrow {}^{A-4}_{Z-2}X + {}^{4}_{2}He$ |
| Beta decay ($\beta^-$) | ${}^{A}_{Z}X \longrightarrow {}_{Z+1}^{A}X + \beta^- + \bar{\nu}$ |
| Beta decay ($\beta^+$) | ${}^{A}_{Z}X \longrightarrow {}_{Z-1}^{A}X + \beta^+ + \nu$ |
| Electron capture | ${}^{A}_{Z}X + {}_{-1}^{0}e \longrightarrow {}_{Z-1}^{A}X + \nu$ |
| Gamma decay | ${}^{A}_{Z}X^* \longrightarrow {}^{A}_{Z}X + \gamma$ |

## 45.6 NATURAL RADIOACTIVITY

Radioactive nuclei are generally classified into two groups: (1) unstable nuclei found in nature, which give rise to what is called **natural radioactivity,** and (2) nuclei produced in the laboratory through nuclear reactions, which exhibit **artificial radioactivity.**

There are three series of naturally occurring radioactive nuclei (Table 45.4). Each series starts with a specific long-lived radioactive isotope whose half-life exceeds that of any of its descendants. The three natural series begin with the isotopes $^{238}$U, $^{235}$U, and $^{232}$Th, and the corresponding stable end products are three isotopes of lead: $^{206}$Pb, $^{207}$Pb, and $^{208}$Pb. The fourth series in Table 45.4 begins with $^{237}$Np and has as its stable end product $^{209}$Bi. The element $^{237}$Np is a transuranic element (one having an atomic number greater than that of uranium) not found in nature. This element has a half-life of "only" $2.14 \times 10^6$ years.

Figure 45.18 shows the successive decays for the $^{232}$Th series. Note that $^{232}$Th first undergoes alpha decay to $^{228}$Ra. Next, $^{228}$Ra undergoes two successive $\beta$ decays to $^{228}$Th. The series continues and finally branches when it reaches $^{212}$Bi. At this point, there are two decay possibilities. The end of the decay series is the stable isotope $^{208}$Pb.

The two uranium series are somewhat more complex than the $^{232}$Th series. Also, there are several naturally occurring radioactive isotopes, such as $^{14}$C and $^{40}$K, that are not part of either decay series.

The existence of radioactive series in nature enables our environment to be constantly replenished with radioactive elements that would otherwise have disappeared long ago. For example, because the solar system is about $5 \times 10^9$ years old, the supply of $^{226}$Ra (whose half-life is only 1600 years) would have been depleted by radioactive decay long ago if it were not for the decay series that starts with $^{238}$U, whose half-life is equal to $4.47 \times 10^9$ years.

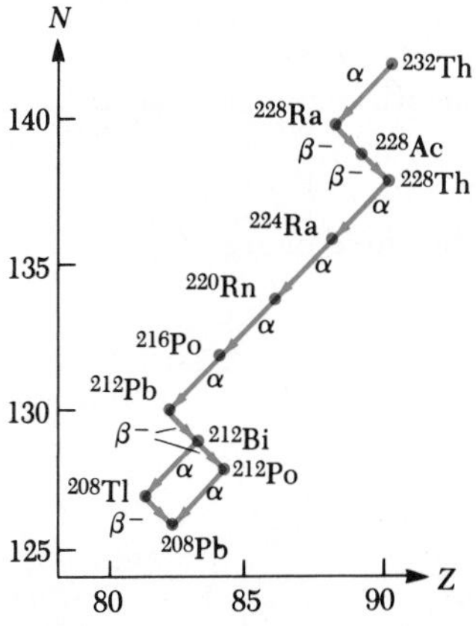

**TABLE 45.4 The Four Radioactive Series**

| Series | | Starting Isotope | Half-Life (years) | Stable End Product |
|---|---|---|---|---|
| Uranium | Natural | ${}^{238}_{92}U$ | $4.47 \times 10^9$ | ${}^{206}_{82}Pb$ |
| Actinium | Natural | ${}^{235}_{92}U$ | $7.04 \times 10^8$ | ${}^{207}_{82}Pb$ |
| Thorium | Natural | ${}^{232}_{90}Th$ | $1.41 \times 10^{10}$ | ${}^{208}_{82}Pb$ |
| Neptunium | | ${}^{237}_{93}Np$ | $2.14 \times 10^6$ | ${}^{209}_{83}Bi$ |

**Figure 45.18** Successive decays for the $^{232}$Th series.

## 45.7 NUCLEAR REACTIONS

It is possible to change the structure of nuclei by bombarding them with energetic particles. Such collisions, which change the identity or properties of the target nuclei, are called **nuclear reactions.** Rutherford was the first to observe nuclear reactions, in 1919, using naturally occurring radioactive sources for the bombarding particles. Since then, thousands of nuclear reactions have been observed following the development of charged-particle accelerators in the 1930s. With today's advanced technology in particle accelerators and particle detectors, it is possible to achieve particle energies of at least 1000 GeV = 1 TeV. These high-energy particles are used to create new particles whose properties are helping to solve the mystery of the nucleus.

Consider a reaction in which a target nucleus is bombarded by a particle a, resulting in a nucleus Y and a particle b:

Nuclear reaction

$$a + X \longrightarrow Y + b \tag{45.24}$$

Sometimes this reaction is written in the more compact form

$$X(a, b)Y$$

In the previous section, the $Q$ value, or disintegration energy, of a radioactive decay was defined as the energy released as the result of the decay process. Likewise, we define the **reaction energy $Q$** associated with a nuclear reaction as *the total energy released as the result of the reaction.* More specifically, $Q$ is defined as

Reaction energy $Q$

$$Q = (M_a + M_X - M_Y - M_b)c^2 \tag{45.25}$$

As an example, consider the reaction $^7$Li (p, $\alpha$)$^4$He, or

$$^1_1H + ^7_3Li \longrightarrow ^4_2He + ^4_2He$$

which has a $Q$ value of 17.3 MeV. This reaction was first observed by Cockroft and Walton in 1930 using accelerated charged particles. A reaction such as this, for which $Q$ is positive, is called **exothermic.** After the reaction, this residual energy appears as an increase in kinetic energy of (Y, b). That is, the loss in the mass of the system is balanced by an increase in the kinetic energy of the final particles. A reaction for which $Q$ is negative is called **endothermic.** An endothermic reaction will not occur unless the bombarding particle has a kinetic energy greater than $Q$. The minimum energy necessary for such a reaction to occur is called the **threshold energy.**

Exothermic reaction

Endothermic reaction

Threshold energy

The kinetics of nuclear reactions must obey the law of conservation of linear momentum. That is, the total linear momentum of the system of interacting particles must be the same before and after the reaction. This assumes that the only force acting on the interacting particles is their mutual force of interaction; that is, there are no external accelerating electric fields present near the colliding particles.

If a nuclear reaction occurs in which particles a and b are identical, so that X and Y are also necessarily identical, the reaction is called a *scattering event.* If kinetic energy is conserved as a result of the reaction (that is, if $Q = 0$), it is classified as *elastic scattering.* On the other hand, if $Q \neq 0$, kinetic energy is not

**TABLE 45.5** ***Q*** **Values for Nuclear Reactions Involving Light Nuclei**

| Reaction[a] | Measured $Q$-Value (MeV) |
|---|---|
| $^{2}$H(n, $\gamma$)$^{3}$H | $6.257 \pm 0.004$ |
| $^{2}$H(d, p)$^{3}$H | $4.032 \pm 0.004$ |
| $^{6}$Li(p, $\alpha$)$^{3}$H | $4.016 \pm 0.005$ |
| $^{6}$Li(d, p)$^{7}$Li | $5.020 \pm 0.006$ |
| $^{7}$Li(p, n)$^{7}$Be | $-1.645 \pm 0.001$ |
| $^{7}$Li(p, $\alpha$)$^{4}$He | $17.337 \pm 0.007$ |
| $^{9}$Be(n, $\gamma$)$^{10}$Be | $6.810 \pm 0.006$ |
| $^{9}$Be($\gamma$, n)$^{8}$Be | $-1.666 \pm 0.002$ |
| $^{9}$Be(d, p)$^{10}$Be | $4.585 \pm 0.005$ |
| $^{9}$Be(p, $\alpha$)$^{6}$Li | $2.132 \pm 0.006$ |
| $^{10}$B(n, $\alpha$)$^{7}$Li | $2.793 \pm 0.003$ |
| $^{10}$B(p, $\alpha$)$^{7}$Be | $1.148 \pm 0.003$ |
| $^{12}$C(n, $\gamma$)$^{13}$C | $4.948 \pm 0.004$ |
| $^{13}$C(p, n)$^{13}$N | $-3.003 \pm 0.002$ |
| $^{14}$N(n, p)$^{14}$C | $0.627 \pm 0.001$ |
| $^{14}$N(n, $\gamma$)$^{15}$N | $10.833 \pm 0.007$ |
| $^{18}$O(p, n)$^{18}$F | $-2.453 \pm 0.002$ |
| $^{19}$F(p, $\alpha$)$^{16}$O | $8.124 \pm 0.007$ |

From C.W. Li, W. Whaling, W.A. Fowler, and C.C. Lauritsen, *Physical Review* 83:512 (1951).
[a] The symbols n, p, d, $\alpha$, and $\gamma$ denote the neutron, proton, deuteron, alpha particle, and photon, respectively.

conserved and the reaction is called *inelastic scattering.* This terminology is identical to that used in dealing with the collision between macroscopic objects (Section 9.4).

A list of measured $Q$ values for a number of nuclear reactions involving light nuclei is given in Table 45.5.

In addition to energy and momentum, the total charge and total number of nucleons must be conserved in any nuclear reaction. For example, consider the reaction $^{19}$F(p, $\alpha$)$^{16}$O, which has a $Q$ value of 8.124 MeV. We can show this reaction more completely as

$$ {}^{1}_{1}\mathrm{H} + {}^{19}_{9}\mathrm{F} \longrightarrow {}^{16}_{8}\mathrm{O} + {}^{4}_{2}\mathrm{He} $$

We see that the total number of nucleons before the reaction $(1 + 19 = 20)$ is equal to the total number after the reaction $(16 + 4 = 20)$. Furthermore, the total charge $(Z = 10)$ is the same before and after the reaction.

## SUMMARY

A nuclear species can be represented by ${}^{A}_{Z}\mathrm{X}$, where $A$ is the **mass number,** which equals the total number of nucleons, and Z is the **atomic number,** which is the total number of protons. The total number of neutrons in a nucleus is the **neutron number** $N$, where $A = N + Z$. Elements with the same Z but different $A$ and $N$ values are called **isotopes.**

Radius of most nuclei

Assuming that nuclei are spherical, their radius is given by

$$r = r_0 A^{1/3} \tag{45.1}$$

where $r_0 = 1.2$ fm (1 fm $= 10^{-15}$ m).

Properties of stable nuclei

Light nuclei are most stable when the number of protons equals the number of neutrons. Heavy nuclei are most stable when the number of neutrons exceeds the number of protons. In addition, most stable nuclei have $Z$ and $N$ values that are both even. Nuclei with unusually high stability have $Z$ or $N$ values of 2, 8, 20, 28, 50, 82, and 126, called **magic numbers.**

Nuclei have an intrinsic spin angular momentum of magnitude $\sqrt{I(I+1)}\hbar$, where $I$ is the **nuclear spin.** The magnetic moment of a nucleus is measured in terms of the **nuclear magneton** $\mu_n$, where

Nuclear magneton

$$\mu_n = 5.05 \times 10^{-27} \text{ J/T} \tag{45.3}$$

The proton has a magnetic moment of $2.7928\mu_n$, and the neutron has a magnetic moment of $-1.9135\mu_n$. When a nuclear moment is placed in an external magnetic field, it precesses about the field with a frequency that is proportional to the field.

The difference in mass between the separate nucleons and that of the compound nucleus containing these nucleons, when multiplied by $c^2$, gives the **binding energy** $E_b$ of the nucleus: $E_b = \Delta mc^2$. We can calculate the binding energy of any nucleus ${}^A_Z\text{X}$ using the expression

Binding energy of a nucleus

$$E_b \text{ (MeV)} = [Zm_H + Nm_n - M({}^A_Z\text{X})] \times 931.50 \text{ MeV/u} \tag{45.4}$$

where all masses are atomic masses, $m_H$ is the mass of the hydrogen atom, and $m_n$ is the neutron mass.

Nuclei are stable because of the **strong nuclear force** between nucleons. This short-range force dominates the Coulomb repulsive force at distances of less than about 2 fm and is nearly independent of charge.

Liquid-drop model

The **liquid-drop model** of nuclear structure treats the nucleons as molecules in a drop of liquid. In this model, the three main contributions influencing the binding energy of the nucleus are the volume effect, the surface effect, and the Coulomb repulsion. Summing such contributions results in the **semiempirical binding energy formula** given by Equation 45.5.

Independent-particle model

The **independent-particle model** of nuclear structure, or shell model, assumes that each nucleon moves in a well-defined quantized orbit within the nucleus and in an averaged field produced by the other nucleons. The stability of certain nuclei can be explained with this model. Two nucleons will collide only if the energy of each state after the collision corresponds to one of the allowed nuclear states, that is, a state that does not violate the exclusion principle.

Collective model

The **collective model** of the nucleus combines some features from the liquid-drop model and some from the independent-particle model. It has been very successful in describing various nuclear phenomena.

Radioactive decay

A radioactive substance can undergo decay by three processes: alpha decay, beta decay, and gamma decay. An alpha particle is the ${}^4$He nucleus; a beta particle is either an electron ($\beta^-$) or a positron ($\beta^+$); a gamma particle is a high-energy photon.

If a radioactive material contains $N_0$ radioactive nuclei at $t = 0$, the number $N$ of nuclei remaining after a time $t$ has elapsed is given by

$$N = N_0 e^{-\lambda t} \tag{45.7}$$

Exponential decay

where $\lambda$ is the **decay constant, or disintegration constant.** The *decay rate,* or *activity,* of a radioactive substance is given by

$$R = \left|\frac{dN}{dt}\right| = R_0 e^{-\lambda t} \tag{45.8}$$

Decay rate

where $R_0 = N_0\lambda$ is the activity at $t = 0$. The **half-life** $T_{1/2}$ is defined as the time it takes half of a given number of radioactive nuclei to decay, where

$$T_{1/2} = \frac{0.693}{\lambda} \tag{45.9}$$

Half-life

The unit of activity is the **curie** (Ci), defined as 1 Ci $\equiv 3.7 \times 10^{10}$ decays/s. The SI unit of activity is the **becquerel** (Bq), defined as 1 decay/s.

Decay processes

Alpha decay can occur because, according to quantum mechanics, some nuclei have barriers that can be penetrated by the alpha particles (the tunneling process). This process is energetically more favorable for those nuclei having a large excess of neutrons. A nucleus can undergo beta decay in two ways. It can emit either an electron ($\beta^-$) and an antineutrino ($\overline{\nu}$) or a positron ($\beta^+$) and a neutrino ($\nu$). In the electron-capture process, the nucleus of an atom absorbs one of its own electrons (usually from the K shell) and emits a neutrino. In gamma decay, a nucleus in an excited state decays to its ground state and emits a gamma ray.

**Nuclear reactions** can occur when a target nucleus X is bombarded by a particle a, resulting in a nucleus Y and a particle b:

$$\mathrm{a} + \mathrm{X} \longrightarrow \mathrm{Y} + \mathrm{b} \quad \text{or} \quad \mathrm{X(a, b)Y} \tag{45.24}$$

Nuclear reaction

The energy released in such a reaction, called the **reaction energy** $Q$, is given by

$$Q = (M_a + M_X - M_Y - M_b)c^2 \tag{45.25}$$

Reaction energy $Q$

## QUESTIONS

1. Why are heavy nuclei unstable?
2. A proton precesses with a frequency $\omega_p$ in the presence of a magnetic field. If the magnetic field intensity is doubled, what happens to the precessional frequency?
3. Explain why nuclei that are well off the line of stability in Figure 45.3 tend to be unstable.
4. Why do nearly all the naturally occurring isotopes lie *above* the $N = Z$ line in Figure 45.3?
5. Consider two heavy nuclei X and Y with similar mass numbers. If X has the higher binding energy, which nucleus would tend to be more unstable?
6. Discuss the differences between the liquid-drop model and the independent-particle model of the nucleus.
7. How many values of $I_z$ are possible for $I = 5/2$? How many are possible for $I = 3$?
8. How can a neutron (which is electrically neutral) possess a magnetic moment?
9. In nuclear magnetic resonance (NMR), if the dc magnetic field is increased, how does this change the frequency of the ac field that excites a particular transition?
10. Would the liquid-drop or independent-particle nuclear model be more appropriate to predict the behavior of a nucleus in a fission reaction? Which would be more successful in predicting the magnetic moment of a given nucleus? Which could better explain the $\gamma$-ray spectrum of an excited nucleus?

11. If a nucleus has a half-life of 1 year, does this mean that it will be completely decayed after 2 years? Explain.
12. What fraction of a radioactive sample decays after two half-lives have elapsed?
13. Two samples of the same radioactive nuclide are prepared. Sample A has twice the initial activity of sample B. How does the half-life of A compare to the half-life of B? After each has passed through five half-lives, what is the ratio of their activities?
14. Explain why the half-lives for radioactive nuclei are essentially independent of temperature.
15. The radioactive nucleus $^{226}_{88}Ra$ has a half-life of about $1.6 \times 10^3$ years. Since the solar system is about 5 billion years old, how can you explain why we still can find this nucleus in nature?
16. Why is the electron involved in the reaction

$$^{14}_{6}C \rightarrow {}^{14}_{7}N + \beta^-$$

written as $\beta^-$, while the electron involved in the reaction

$$^{7}_{4}Be + {}^{0}_{-1}e \rightarrow {}^{7}_{3}Li + \nu$$

is written as $^{0}_{-1}e$?
17. A free neutron undergoes beta decays with a half-life of about 15 min. Can a free proton undergo a similar decay?
18. Explain in detail how you can determine the age of a sample using the technique of carbon dating.
19. What is the difference between a neutrino and a photon?
20. Does the $Q$ in Equation 45.25 represent the quantity (final mass − initial mass)$c^2$, or does it represent quantity (initial mass − final mass)$c^2$?
21. Use Equations 45.17 to 45.19 to explain why the neutrino must have a spin of $\frac{1}{2}$.
22. If a nucleus such as $^{226}Ra$ initially at rest undergoes alpha decay, which has more kinetic energy after the decay, the alpha particle or the daughter nucleus?
23. Can a nucleus emit alpha particles with different energies? Explain.
24. Explain why many heavy nuclei undergo alpha decay but do not spontaneously emit neutrons or protons.
25. If an alpha particle and an electron have the same kinetic energy, which undergoes the greater deflection when passed through a magnetic field?
26. If film is kept in a box, alpha particles from a radioactive source outside the box cannot expose the film but beta particles can. Explain.
27. Pick any beta decay process and show that the neutrino must have zero charge.
28. Suppose it could be shown that the cosmic ray intensity was much greater 10 000 years ago. How would this affect present values of the age of ancient samples of once-living matter?
29. Why is carbon dating unable to provide accurate estimates of very old material?
30. Element X has several isotopes. What do these isotopes have in common? How do they differ?
31. Explain the main differences between alpha, beta, and gamma rays.
32. How many protons are there in the nucleus $^{222}_{86}Rn$? How many neutrons? How many electrons are there in the neutral atom?

## PROBLEMS

Table 45.6 will be useful for many of these problems. A more complete list of atomic masses is given in Table A.3 in Appendix A.

### Section 45.1 Some Properties of Nuclei

1. Find the radius of (a) a nucleus of $^{4}_{2}He$ and (b) a nucleus of $^{238}_{92}U$. (c) What is the ratio of these radii?
2. In Example 45.2, the density of nuclear matter was calculated to be approximately $2.3 \times 10^{17}$ kg/m$^3$. What is this density in nucleons/fm$^3$? in g/cm$^3$?
3. The compressed core of a star formed in the wake of a supernova explosion can consist of pure nuclear material and is called a *pulsar* or *neutron star*. Calculate the mass of 10 cm$^3$ of a pulsar.
4. What would be the gravitational force between two golf balls (each of 4.3-cm diameter), one meter apart, if they were made of nuclear matter?
5. From Table A.3, identify the stable nuclei that correspond to the magic numbers given by Equation 45.2.
6. Consider the hydrogen atom to be a sphere of radius equal to the Bohr radius, $a_0$, and calculate the approximate value of the ratio of the nuclear density to the atomic density.
7. The unified mass unit u is exactly $\frac{1}{12}$ of the mass of an atom of $^{12}C$. This unit was adopted in 1961. Previous to that date, there were two different units in general use:

u (physical scale) = $\frac{1}{16}$ of the mass of $^{16}O$
u (chemical scale) = $\frac{1}{16}$ of the average mass of oxygen taking into account the relative isotopic abundances

Calculate the percent difference between the mass unit on the present scale and the older physical scale.

**TABLE 45.6 Some Atomic Masses**

| Element | Atomic Mass (u) | Element | Atomic Mass (u) |
|---|---|---|---|
| ${}^{4}_{2}He$ | 4.002603 | ${}^{27}_{13}Al$ | 26.981541 |
| ${}^{7}_{3}Li$ | 7.016004 | ${}^{30}_{15}P$ | 29.978310 |
| ${}^{9}_{4}Be$ | 9.012182 | ${}^{40}_{20}Ca$ | 39.962591 |
| ${}^{10}_{5}B$ | 10.012938 | ${}^{42}_{20}Ca$ | 41.95863 |
| ${}^{12}_{6}C$ | 12.000000 | ${}^{43}_{20}Ca$ | 42.958770 |
| ${}^{13}_{6}C$ | 13.003355 | ${}^{56}_{26}Fe$ | 55.934939 |
| ${}^{14}_{7}N$ | 14.003074 | ${}^{64}_{30}Zn$ | 63.929145 |
| ${}^{15}_{7}N$ | 15.000109 | ${}^{64}_{29}Cu$ | 63.929599 |
| ${}^{15}_{8}O$ | 15.003065 | ${}^{93}_{41}Nb$ | 92.906378 |
| ${}^{17}_{8}O$ | 16.999131 | ${}^{197}_{79}Au$ | 196.966560 |
| ${}^{18}_{8}O$ | 17.999159 | ${}^{202}_{80}Hg$ | 201.970632 |
| ${}^{18}_{9}F$ | 18.000937 | ${}^{216}_{84}Po$ | 216.001790 |
| ${}^{20}_{10}Ne$ | 19.992439 | ${}^{220}_{86}Rn$ | 220.011401 |
| ${}^{23}_{11}Na$ | 22.989770 | ${}^{234}_{90}Th$ | 234.043583 |
| ${}^{23}_{12}Mg$ | 22.994127 | ${}^{238}_{92}U$ | 238.050786 |

Use values of atomic masses from Table A.3 in Appendix A.

8. Refer to the list of selected isotopes in Table A.3 and consider only the stable nuclei listed. Identify the number of stable nuclei which are of the following four types: even Z, even N; even Z, odd N; odd Z, even N; and odd Z, odd N.
9. Certain stars at the end of their lives are thought to collapse, combining their protons and electrons to form a neutron star. Such a star could be thought of as a giant atomic nucleus. If a star of mass equal to that of the sun ($M = 1.99 \times 10^{30}$ kg) collapsed into neutrons ($m_n = 1.67 \times 10^{-27}$ kg), what would be the radius of such a star? (*Hint:* $r = r_0 A^{1/3}$.)
10. The Larmor precessional frequency is

$$f = \frac{\Delta E}{h} = \frac{2\mu B}{h}$$

Calculate the radio-wave frequency at which resonance absorption will occur for (a) free neutrons in a magnetic field of 1 T, (b) free protons in a magnetic field of 1 T, and (c) free protons in the earth's magnetic field at a location where the field strength is 50 $\mu$T.

11. Show that the quantity $e^2/4\pi\epsilon_0 = 1.440$ MeV·fm. (This is a convenient equivalence for computational purposes.)
12. How much kinetic energy must an alpha particle (charge $= 2 \times 1.6 \times 10^{-19}$ C) have to approach to within $10^{-14}$ m of a gold nucleus (charge $= 79 \times 1.6 \times 10^{-19}$ C)?
13. (a) Use energy methods to calculate the distance of closest approach for a head-on collision between an alpha particle with an initial energy of 0.5 MeV and a gold nucleus ($^{197}Au$) at rest. (Assume the gold nucleus remains at rest during the collision.) (b) What minimum initial speed must the alpha particle have in order to reach a distance of 300 fm?

### Section 45.2 Binding Energy

14. Calculate the binding energy per nucleon for (a) $^{2}H$, (b) $^{4}He$, (c) $^{56}Fe$, and (d) $^{238}U$.
15. In Example 45.3, the binding energy of the deuteron was calculated to be 2.224 MeV. This corresponds to a value of 1.112 MeV/nucleon. What is the binding energy per nucleon for the heaviest isotope of hydrogen, $^{3}H$ (called *tritium*)?
16. Using the value of the atomic mass of ${}^{56}_{26}Fe$ given in Table 45.6, find its binding energy. Then compute the binding energy per nucleon and compare your result with Figure 45.7.
17. ${}^{60}_{28}Ni$ has an atomic mass of 59.930789 u. (a) What is its *nuclear* mass? (b) What is the binding energy per nucleon? Is this a tightly bound or loosely bound nucleus?
18. (a) From Figure 45.8, estimate the electrostatic potential energy due to the Coulomb repulsion between two protons in contact. (b) Compare this value with the rest energy of an electron.
19. A pair of nuclei for which $Z_1 = N_2$ and $Z_2 = N_1$ are called *mirror isobars* (the atomic and neutron num-

bers are interchangeable). Binding energy measurements on these nuclei can be used to obtain evidence of the charge independence of nuclear forces (that is, proton-proton, proton-neutron, and neutron-neutron forces are approximately equal). Calculate the difference in binding energy for the two mirror nuclei $^{15}_{8}O$ and $^{15}_{7}N$.

20. Calculate the binding energy per nucleon for (a) $^{20}_{10}Ne$, (b) $^{40}_{20}Ca$, (c) $^{93}_{41}Nb$, and (d) $^{197}_{79}Au$.
21. Calculate the minimum energy required to remove a neutron from the $^{43}_{20}Ca$ nucleus.
22. Two isotopes having the same mass number are known as *isobars*. Calculate the difference in binding energy per nucleon for the isobars $^{23}_{11}Na$ and $^{23}_{12}Mg$. How do you account for the difference?
23. The $^{139}_{57}La$ isotope of lanthanum is stable. A radioactive, isobar (see Problem 22) of this lanthanum isotope, $^{139}_{59}Pr$, is located above the line of stable nuclei in Figure 45.3 and decays by $\beta^+$ emission. Another radioactive isobar of $^{139}La$, $^{139}_{55}Cs$, decays by $\beta^-$ emission and is located below the line of stable nuclei in Figure 45.3. (a) Which of these three isobars has the highest neutron-to-proton ratio? (b) Which has the greatest binding energy per nucleon? (c) Which of the two radioactive nuclei ($^{139}Pr$ or $^{139}Cs$) do you expect to be heavier?

### Section 45.3 Nuclear Models

24. (a) In the liquid-drop model of nuclear structure, why does the surface-effect term $-C_2A^{2/3}$ have a minus sign? (b) The binding energy of the nucleus increases as the volume-to-surface ratio increases. Calculate this ratio for both spherical and cubical shapes and explain which is more plausible for nuclei.
25. Using the graph in Figure 45.7, estimate how much energy is released when a nucleus of mass number 200 is split into two nuclei each of mass number 100.
26. Use Equation 45.5 and the given values for the constants $C_1$, $C_2$, $C_3$, and $C_4$ to calculate the binding energy per nucleon for the isobars $^{64}_{29}Cu$ and $^{64}_{30}Zn$.
27. (a) Use Equation 45.5 to compute the binding energy for $^{56}_{26}Fe$. (b) What percentage is contributed to the binding energy by each of the four terms?

### Section 45.4 Radioactivity

28. How much time elapses before 90.0% of the radioactivity of a sample of $^{72}_{33}As$ disappears as measured by the activity? The half-life of $^{72}_{33}As$ is 26 h.
29. A sample of radioactive material contains $10^{15}$ atoms and has an activity of $6.00 \times 10^{11}$ Bq. What is the half-life for this material?
30. The half-life of radioactive iodine-131 is eight days. Find the number of $^{131}I$ nuclei necessary to produce a sample of activity 1.0 $\mu$Ci.
31. A freshly prepared sample of a certain radioactive isotope has an activity of 10 mCi. After an elapsed time of 4 h, its activity is 8 mCi. (a) Find the decay constant and half-life of the isotope. (b) How many atoms of the isotope were contained in the freshly prepared sample? (c) What is the sample's activity 30 h after it is prepared?
32. A radioactive source with a half-life of 1.04 s has an initial activity of $10^4$ decays/s. Determine the number of unstable nuclei remaining in the sample after 1.7 s.
33. A sample of radioactive material is said to be *carrier-free* when no stable isotopes of the radioactive element are present. Calculate the mass of strontium in a 5-mCi sample of $^{90}Sr$, whose half-life is 28.8 years.
34. Determine the disintegration rate (activity) of 1 gram of $^{60}Co$. The half-life of $^{60}Co$ is 5.24 yr.
35. A laboratory stock solution is prepared with an initial activity due to $^{24}Na$ of 2.5 mCi/ml, and 10 ml of the stock solution is diluted (at $t_0 = 0$) to a working solution whose total volume is 250 ml. After 48 h, a 5-ml sample of the working solution is monitored with a counter. What is the measured activity? (*Note:* 1 ml = 1 milliliter.)
36. Start with Equation 45.8 and derive the following useful formulas for the decay constant and half-life:

$$\lambda = \frac{1}{t}\ln\left(\frac{R_0}{R}\right) \qquad T_{1/2} = \frac{(\ln 2)t}{\ln(R_0/R)}$$

37. The radioactive isotope $^{198}Au$ has a half-life of 64.8 h. A sample containing this isotope has an initial activity ($t = 0$) of 40 $\mu$Ci. Calculate the number of nuclei that will decay in the time interval between $t_1 = 10$ h and $t_2 = 12$ h.

### Section 45.5 The Decay Processes

38. Identify the missing nuclide (X) in each of the following reactions.

(a) $X \longrightarrow {}^{65}_{28}Ni + \gamma$

(b) $^{215}_{84}Po \longrightarrow X + \alpha$

(c) $X \longrightarrow {}^{55}_{26}Fe + \beta^+ + \nu$

(d) $^{109}_{48}Cd + X \longrightarrow {}^{109}_{47}Ag + \nu$

(e) $^{14}N(\alpha, X)^{17}O$

39. Find the energy released in the alpha decay of $^{238}_{92}U$:

$$^{238}_{92}U \longrightarrow {}^{234}_{90}Th + {}^{4}_{2}He$$

You will find the following mass values useful:

$$M(^{238}_{92}U) = 238.050786 \text{ u}$$

$$M(^{234}_{90}Th) = 234.043583 \text{ u}$$

$$M(^{4}_{2}He) = 4.002603 \text{ u}$$

40. A living specimen of organic material in equilibrium with the atmosphere contains one atom of $^{14}C$ (half-life = 5730 y) for every $7.7 \times 10^{11}$ stable carbon atoms. An archeological sample of wood (cellulose, $C_{12}H_{22}O_{11}$) contains 21.0 mg of carbon. When the sample is placed inside a beta counter whose counting efficiency is 88%, 837 counts are accumulated in one week. Assuming that the cosmic-ray flux and the earth's atmosphere have not changed appreciably since the sample was formed, find the age of the sample.
41. Perform a calculation to verify the accuracy of the conversion factor used in Equation 45.14. In addition to the definition value for u, you will need the following constants: $c = 2.99793 \times 10^8$ m/s and $e = 1.60219 \times 10^{-19}$ C.
42. Calculate the energy released in the alpha decay of $^{210}Po$.
43. A $^{239}Pu$ nucleus at rest undergoes alpha decay, leaving a $^{235}U$ nucleus in its ground state. Determine the kinetic energy of the alpha particle.
44. Determine which of the following suggested decays can occur spontaneously:

(a) $^{40}_{20}Ca \longrightarrow {}^{0}_{1}\beta + {}^{40}_{19}K$

(b) $^{98}_{44}Ru \longrightarrow {}^{4}_{2}He + {}^{94}_{42}Mo$

(c) $^{144}_{60}Nd \longrightarrow {}^{4}_{2}He + {}^{140}_{58}Ce$

45. Starting with $^{235}_{92}U$, the following sequence of decays is observed, ending with the stable isotope $^{207}_{82}Pb$ (Fig. 45.19). Enter the correct isotope symbol in each open square.

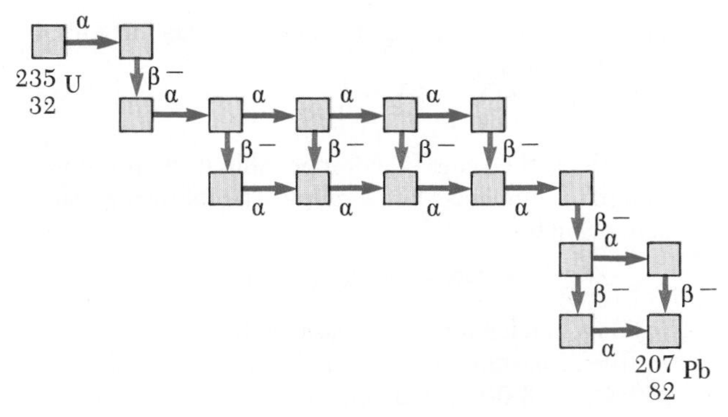


**Figure 45.19** (Problem 45).

46. Show that $\beta^-$ decay of $^{255}Md$ cannot occur.
47. The nucleus $^{15}_{8}O$ decays by electron capture. Write (a) the basic nuclear process and (b) the decay process referring to neutral atoms. (c) Determine the energy of the neutrino. Disregard the daughter's recoil.

Section 45.7 Nuclear Reactions

48. The following reaction, first observed in 1930, led to the discovery of the neutron by Chadwick:

$$^{9}_{4}Be(\alpha, n)^{12}_{6}C$$

Calculate the $Q$ value of this reaction.
49. The following is the first known reaction (achieved in 1934) in which the product nucleus is radioactive:

$$^{27}_{13}Al(\alpha, n)^{30}_{15}P$$

Calculate the $Q$ value of this reaction.
50. Determine the $Q$ associated with the spontaneous fission of $^{236}U$ into the fragments $^{90}Rb$ and $^{143}Cs$ with masses 89.914811 u and 142.927220 u, respectively. The masses of the other reacting particles are given in Appendix A.
51. Show that the following inverse reactions have the same *absolute value* of $Q$:

$$^{10}_{5}B(\alpha, p)^{13}_{6}C \quad \text{and} \quad ^{13}_{6}C(p, \alpha)^{10}_{5}B$$

52. Natural gold has only one isotope, $^{197}_{79}Au$. If natural gold is irradiated by a flux of slow neutrons, $\beta^-$ particles are emitted. (a) Write the appropriate reaction equations. (b) Calculate the maximum energy of the emitted beta particles. The mass of $^{198}_{80}Hg$ is 197.96675 u.
53. Using the appropriate reactions and $Q$ values from Table 45.5, calculate the mass of $^{8}Be$ and $^{10}Be$ in atomic mass units to four decimal places.

## ADDITIONAL PROBLEMS

54. (a) What fraction of the space in a tank of hydrogen ($H_2$) gas at STP is occupied by the hydrogen molecules themselves? Assume each hydrogen atom is a sphere with a 1-Å diameter, and a hydrogen molecule consists of two such spheres in contact. (b) What fraction of the space within a single hydrogen atom is occupied by its nucleus?
55. Copper as it occurs naturally consists of two stable isotopes, $^{63}Cu$ and $^{65}Cu$. What is the relative abundance of the two forms? In your calculations, take the atomic weight of copper to be 63.55 and take the masses of the two isotopes to be 62.95 u and 64.95 u.
56. (a) The first nuclear transmutation was achieved in 1919 by Rutherford, who bombarded nitrogen atoms with alpha particles emitted by the isotope $^{214}Bi$. The reaction is

$$^{4}_{2}He + {}^{14}_{7}N \longrightarrow {}^{17}_{8}O + {}^{1}_{1}H$$

What is the $Q$ value of the reaction? What is the threshold energy? (b) The first nuclear reaction utilizing particle accelerators was performed by Cockroft and Walton. In this case, accelerated protons were used to bombard lithium nuclei, producing the following reaction:

$$^{1}_{1}\text{H} + ^{7}_{3}\text{Li} \longrightarrow ^{4}_{2}\text{He} + ^{4}_{2}\text{He}$$

Since the masses of the particle involved in the reaction were well known, these results were used to obtain an early proof of the Einstein mass-energy relation. Calculate the $Q$ value of the reaction.

**57.** (a) One method of producing neutrons for experimental use is based on the bombardment of light nuclei by alpha particles. In one particular arrangement, alpha particles emitted by polonium are incident on beryllium nuclei and this results in the production of neutrons:

$$^{4}_{2}\text{He} + ^{9}_{4}\text{Be} \longrightarrow ^{12}_{6}\text{C} + ^{1}_{0}\text{n}$$

What is the $Q$ value for this reaction? (b) Neutrons are also often produced by small-particle accelerators. In one design, deuterons ($^{2}$H) that have been accelerated in a Van de Graaf generator are used to bombard other deuterium nuclei, resulting in the following reaction:

$$^{2}_{1}\text{H} + ^{2}_{1}\text{H} \longrightarrow ^{3}_{2}\text{He} + ^{1}_{0}\text{n}$$

Is this reaction exothermic or endothermic? Calculate its $Q$ value.

**58.** △ The activity of a sample of radioactive material was measured over 12 h, and the following *net* count rates were obtained at the times indicated:

| Time (h) | Counting Rate (counts/min) |
|---|---|
| 1 | 3100 |
| 2 | 2450 |
| 4 | 1480 |
| 6 | 910 |
| 8 | 545 |
| 10 | 330 |
| 12 | 200 |

(a) Plot the activity curve on semilog paper. (b) Determine the disintegration constant and half-life of the radioactive nuclei in the sample. (c) What counting rate would you expect for the sample at $t = 0$? (d) Assuming the efficiency of the counting instrument to be 10%, calculate the number of radioactive atoms in the sample at $t = 0$.

**59.** A by-product of some fission reactors is the isotope $^{239}_{94}$Pu, which is an alpha emitter with a half-life of 24 000 years:

$$^{239}_{94}\text{Pu} \longrightarrow ^{235}_{92}\text{U} + \alpha$$

Consider a sample of 1 kg of pure $^{239}_{94}$Pu at $t = 0$. Calculate (a) the number of $^{239}_{94}$Pu nuclei present at $t = 0$ and (b) the initial activity in the sample. (c) How long does the sample have to be stored if a "safe" activity level is 0.1 Bq?

**60.** A piece of charcoal has a mass of 25 g and is known to be about 25 000 years old. (a) Determine the number of decays per minute expected from this sample. (b) If the radioactive background in the counter without a sample is 20 counts/min and we assume 100% efficiency in counting, explain why 25 000 years is close to the limit of dating with this technique.

**61.** A large nuclear power reactor produces about 3000 MW of thermal power in its core. Three months after a reactor is shut down, the thermal power in the core is 10 MW, due to radioactive byproducts. Assuming that each emission delivers 1 MeV of energy to the thermal power, estimate the activity in bequerels three months after the reactor is shut down.

**62.** In a piece of rock from the moon the $^{87}$Rb content is assessed to be $1.82 \times 10^{10}$ atoms per gram of material. In a piece of the same rock, the $^{87}$Sr content is found to be $1.07 \times 10^{9}$ atoms per gram. (a) Determine the age of the rock. (b) Could the material in the rock actually be much older? What assumption is implicit in using the radioactive dating method? (The relevant decay is $^{87}\text{Rb} \rightarrow ^{87}\text{Sr} + e^{-}$. The half-life of the decay is $4.8 \times 10^{10}$ yr.)

**63.** In addition to the radioactive nuclei included in the natural decay series, there are several other radioactive nuclei that occur naturally. One of these is $^{147}$Sm, which is 15% naturally abundant and has a half-life of $1.3 \times 10^{10}$ years. Calculate the number of decays per second per gram (due to this isotope) in a sample of natural samarium. The atomic weight of samarium is 150.4. (Activity per unit mass is referred to as *specific activity.*)

**64.** (a) Why is the following inverse beta decay forbidden for a free proton:

$$\text{p} \longrightarrow \text{n} + \beta^{+} + \nu$$

(b) Why is the same reaction possible if the proton is bound in a nucleus? For example, the following reaction occurs:

$$^{13}_{7}\text{N} \longrightarrow ^{13}_{6}\text{C} + \beta^{+} + \nu$$

(c) How much energy is released in the reaction given in (b)? [Take the masses to be $m(\beta^{+}) = 0.000549$ u, $M(^{13}\text{C}) = 13.003355$ u, and $M(^{13}\text{N}) = 13.005739$ u.]

**65.** (a) Find the radius of the $^{12}_{6}$C nucleus. (b) Find the force of repulsion between a proton at the surface of a $^{12}_{6}$C nucleus and the remaining five protons. (c) How much work (in MeV) has to be done to overcome this electrostatic repulsion to put the last proton into the nucleus? (d) Repeat (a), (b), and (c) for $^{238}_{92}$U.

**66.** What activity in disintegrations/min·g would be expected for carbon samples from bones that are 2000 years old? (You should note from Example 45.9 that the ratio of $^{14}C$ to $^{12}C$ in living organisms is equal to $1.3 \times 10^{-12}$.)

67. Consider a hydrogen atom with the electron in the 1*s* state. The magnetic field at the nucleus produced by the orbiting electron has a value of 12.5 T. (See Problem 47, Chapter 42.) The proton can have its magnetic moment aligned in either of two directions perpendicular to the plane of the electron's orbit. Because of the interaction of the proton's magnetic moment with the electron's magnetic field, there will be a difference in energy between the states with the two different orientations of the proton's magnetic moment. Find that energy difference in eV.

68. Suppose that an electron resides in a nucleus with a diameter of $10^{-14}$ m. Determine the approximate energy (in MeV) that such an electron must have. Use the uncertainty principle, and interpret the uncertainty in position as the diameter of the nucleus and the uncertainty in momentum as the momentum. Your answer should justify using the relativistic approximation $E = pc$. (Since electrons emitted in beta decay seldom have energies greater than 1 MeV, the uncertainty principle supports the nonexistence of electrons in the nucleus.)

69. Carbon detonations are powerful nuclear reactions that temporarily tear apart the cores of massive stars late in their lives. These blasts are produced by carbon fusion, which requires a temperature of about $6 \times 10^8$ K to overcome the strong Coulomb repulsion between carbon nuclei. (a) Estimate the repulsive energy barrier to fusion, using the required ignition temperature for carbon fusion. (In other words, what is the kinetic energy for a carbon nucleus at $6 \times 10^8$ K?) (b) Calculate the energy (in MeV) released in each of these "carbon-burning" reactions:

$$^{12}C + ^{12}C \longrightarrow ^{20}Ne + ^{4}He$$

$$^{12}C + ^{12}C \longrightarrow ^{24}Mg + \gamma$$

(c) Calculate the energy (in kWh) given off when 2 kg of carbon completely fuses according to the first reaction.

70. When a material of interest is irradiated by neutrons, radioactive atoms are produced continually and some decay according to their given half-life. (a) If radioactive atoms are produced at a constant rate $R$ and their decay is governed by the conventional radioactive decay law, show that the number of radioactive atoms accumulated after an irradiation time $t$ is

$$N = \frac{R}{\lambda}(1 - e^{-\lambda t})$$

(b) What is the maximum number of radioactive atoms that can be produced?

71. Many radioisotopes have important industrial, medical, and research applications. One of these is $^{60}Co$, which has a half-life of 5.2 years and decays by the emission of a beta particle (energy 0.31 MeV) and two gamma photons (energies 1.17 MeV and 1.33 MeV). A scientist wishes to prepare a $^{60}Co$ sealed source that will have an activity of at least 10 Ci after 30 months of use. (a) What is the minimum initial mass of $^{60}Co$ required? (b) At what rate will the source emit energy after 30 months?

72. Consider a model of the nucleus in which the positive charge ($Ze$) is uniformly distributed throughout a sphere of radius $R$. By integrating the energy density, $\frac{1}{2}\epsilon_0 E^2$, over all space, show that the electrostatic energy may be written

$$U = \frac{3Z^2e^2}{20\pi\epsilon_0 R}$$

73. "Free neutrons" have a characteristic half-life of 12 min. What fraction of a group of free neutrons at thermal energy (0.04 eV) will decay before traveling a distance of 10 km?

74. When, after a reaction or disturbance of any kind, a nucleus is left in an excited state, it can return to its normal (ground) state by emission of a gamma-ray photon (or several photons). This process is illustrated by Equation 45.21. The emitting nucleus must recoil in order to conserve both energy and momentum. (a) Show that the recoil energy of the nucleus is given by

$$E_r = \frac{(\Delta E)^2}{2Mc^2}$$

where $\Delta E$ is the difference in energy between the excited and ground states of a nucleus of mass $M$. (b) Calculate the recoil energy of the $^{57}Fe$ nucleus when it decays by gamma emission from the 14.4-keV excited state. For this calculation, take the mass to be 57 u. (*Hint:* When writing the equation for conservation of energy, use $(Mv)^2/2M$ for the kinetic energy of the recoiling nucleus. Also, assume that $hf \ll Mc^2$ and use the binomial expansion.)

75. The decay of an unstable nucleus by alpha emission is represented by Equation 45.10. The disintegration energy $Q$ given by Equation 45.13 must be shared by the alpha particle and the daughter nucleus in order to conserve both energy and momentum in the decay process. (a) Show that $Q$ and $K_\alpha$, the kinetic energy of the alpha particle, are related by the expression

$$Q = K_\alpha\left(1 + \frac{M_\alpha}{M}\right)$$

where $M$ is the mass of the daughter nucleus. (b) Use the result of (a) to find the energy of the alpha particle emitted in the decay of $^{226}Ra$. (See Example 45.8 for the calculation of $Q$.)

76. Consider the deuterium-tritium fusion reaction with the tritium nucleus at rest:

$$^2_1H + ^3_1H \longrightarrow ^4_2He + ^1_0n$$

(a) From Equation 45.1, estimate the required distance of closest approach. (b) What is the Coulomb potential energy (in eV) at this distance? (c) If the deuteron has just enough energy to reach the distance of closest approach, what is the final velocity of the deuterium and tritium nuclei in terms of the initial deuteron velocity, $v_0$? (*Hint:* At this point, the two nuclei have a common velocity equal to the center-of-mass velocity.) (d) Use energy methods to find the minimum initial deuteron energy required to achieve fusion. (e) Why does the fusion reaction occur at much lower deuteron energies than that calculated in (d)?

77. The ground state of $^{93}_{43}Tc$ (atomic mass, 92.9102) decays by electron capture and $\beta^+$ to energy levels of the daughter (atomic mass in ground state, 92.9068) at 2.44 MeV, 2.03 MeV, 1.48 MeV, and 1.35 MeV. (a) For which of these levels are electron capture and $\beta^+$ decay allowed? (b) Identify the daughter and sketch the decay scheme, assuming all excited states de-excite by direct $\gamma$ decay to the ground state.

78. The isotope $^{25}_{11}Na$ decays by $\beta^-$ emission (disintegration energy = 4.83 MeV; $T_{1/2}$ = 60.3 s) to the ground state of $^{25}_{12}Mg$. The following $\gamma$ rays are also emitted: 0.40 MeV, 0.58 MeV, 0.98 MeV, 1.61 MeV. The 0.40-MeV $\gamma$ is the first $\gamma$ in a two-step cascade to the ground state of $^{25}_{12}Mg$. All other $\gamma$ rays decay from an excited state directly to the ground state. Construct the decay scheme.

79. Potassium as it occurs in nature includes a radioactive isotope, $^{40}K$, which has a half-life of $1.27 \times 10^9$ years and a relative abundance of 0.0012%. These nuclei decay by two different pathways: 89% by $\beta^-$ emission and 11% by $\beta^+$ emission. Calculate the total activity in Bq associated with 1 kg of KCl *due to $\beta^-$ emission.*

80. When the nuclear reaction represented by Equation 45.24 is endothermic, the disintegration energy $Q$ is negative. In order for the reaction to proceed, the incoming particle must have a minimum energy called the *threshold energy,* $E_{th}$. Some fraction of the energy of the incident particle is transferred to the compound nucleus in order to conserve momentum. Therefore, $E_{th}$ must be greater than $Q$.
(a) Show that

$$E_{th} = -Q\left(1 + \frac{M_a}{M_X}\right)$$

(b) Calculate the threshold energy of the incident alpha particle in the reaction

$$^4_2He + ^{14}_7N \longrightarrow ^{17}_8O + ^1_1H$$

81. During the manufacture of a steel engine component, radioactive iron ($^{59}Fe$) is included in the total mass of 0.2 kg. The component is placed in a test engine when the activity due to this isotope is 20 $\mu$Ci. After a 1000-h test period, oil is removed from the engine and found to contain enough $^{59}Fe$ to produce 800 disintegrations/min per liter of oil. The total volume of oil in the engine is 6.5 liters. Calculate the total mass worn from the engine component per hour of operation. (The half-life for $^{59}Fe$ is 45.1 days.)

Δ82. The rate of decay of a radioactive sample is measured at 10-s intervals, beginning at $t = 0$. The following data in counts per second are obtained: 1137, 861, 653, 495, 375, 284, 215, 163. (a) Plot these data on semilog graph paper and determine the best-fit straight line. (b) From the graph, determine the half-life of the sample.

Δ83. The rate of decay of a radioactive sample is measured at 1-min intervals starting at $t = 0$. The following data (in counts per second) are obtained: 260, 160, 101, 72, 35, 24, 13, 10, 6, 4. (a) Plot these data on semilog graph paer and sketch the best-fit straight line. Determine, to two significant figures, (b) the half-life for this sample and (c) the decay constant.

Δ84. *Student Determination of the Half-Life of* $^{137}Ba$. The radioactive barium isotope ($^{137}Ba$) has a relatively short half-life and can be easily extracted from a solution containing radioactive cesium ($^{137}Cs$). This barium isotope is commonly used in an undergraduate laboratory exercise for demonstrating the radioactive decay law. The data presented in Figure 45.20 were taken by undergraduate students using modest experimental equipment. Determine the half-life for the decay of $^{137}Ba$ using their data.

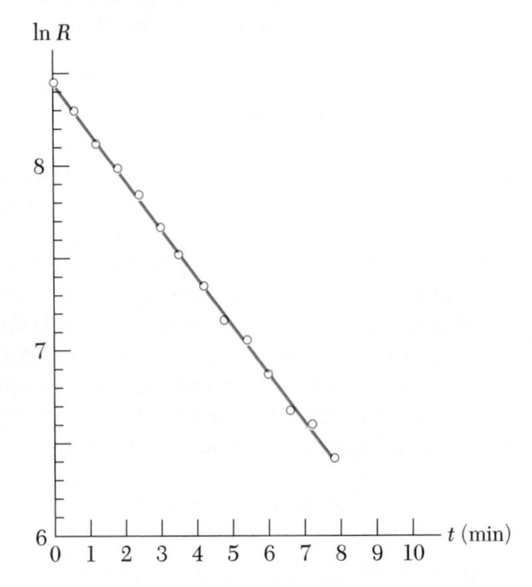


**Figure 45.20** (Problem 84).

# ESSAY

## Magnetic Resonance Imaging

**S.A. Marshall**
*Michigan Technological University*

### Introduction

Magnetic resonance imaging (MRI) is an outgrowth of various mapping procedures that have been applied to the reconstruction of images. Over the past several years these procedures have been greatly refined, the principal motivation for their refinements being their use in medicine as diagnostic tools. Examples of such mappings include medical diagnostics using x-rays and magnetic resonance, astrophysics, aids to navigation, and the detection and characterization of macroscopic defects in solids. Perhaps the most successful and highly developed application in the area of medicine has been computerized x-ray tomograph imaging, or CT. This is because CT is a diagnostic technique which is noninvasive and has minimal impact on the subject being imaged.

An image may be defined as one that allows for a faithful representation of an object and provides information on one or more aspects of the object. Consider an image as an object representation in, say, two dimensions whose image reconstruction data are sensitive to some property of the object. The object representation is the result of a one-to-one mapping operation which generates a point in image space for every point in object space. For practical reasons, every object point should generate a unique image point and *vice versa.* Furthermore, the corresponding object and image points should be unique and continuous, and the mapping should preserve angles between points.

An ordinary photograph possesses virtually all of the qualities that one might expect of an image. The photograph provides the viewer with a sense of the object and produces analog information that can be retrieved and subsequently subjected to manipulation. If the object is three-dimensional, the photograph can provide a perspective which can give the viewer a basis for interpreting the image. Another useful property of a photograph is its exposure or photographic density. Every point on a photograph's two-dimensional surface represents an exposure spot whose density will to some degree be faithful to the object it is meant to represent. Another aspect of the photograph is its set of emulsion grains and their locations. The surface of a photograph consists of a permanent two-dimensional record of the density of silver crystals at the various image points. This record represents analog information which can be converted to digital information by decomposing it into volume cells and assigning to each such cell a photographic density. Each cell may be given an appropriate set of identifying numbers, such as the $x$ and $y$ surface coordinates, and a photographic density number $z$. These numbers can then be assigned to each cell and processed to reconstruct an image.

Each cell of the image may be thought of as a picture element called a *pixel.* A set of numbers $(x, y, z)$ can be assigned to each pixel which refer to a column number, a row number, and a brightness number. From these data, and with the help of a computer and an imaging algorithm, one can reconstruct an image which can then be viewed on a computer's screen or presented as an $x$-$y$ graphic plot. Refinements may then be made to enhance the presentation of this raw image. For example, if the image were to be that of the sun's disk, the $z$-axis of the image could be color-enhanced with various shades to represent regions with different temperatures.

There are many applications for which the surface of an object as seen in an ordinary photograph is of limited value. If the regions of interest are located within an object's interior, then the surface photograph would not be very useful. A familiar example is the need for a medically noninvasive diagnostic tool for investigating regions of the body that lie beneath the skin, such as a broken bone, a foreign body, or a malformed internal organ. In many cases, x-rays could provide the medically required information. Other modes that provide images of internal regions of a body include ultrasound, $\gamma$-rays, magnetic resonance, and autography. These techniques produce a single sheet of information whose exposed area has pixel densities that are

*(Continued on next page)*

proportional to integrated radiation attenuation or photon counts for the cases of x-ray, $\gamma$-ray, or autograph images or to reflected sound wave intensities for ultrasound images.

Computerized tomography (CT) is a procedure which can make use of any of the radiations referred to above. However, the terminology is generally associated with x-ray imaging techniques. Tomography refers to a slice or sheet image of the object. If the formalisms of CT could be used with absorptions of radiations other than x-rays, it could be applied to the production of images constructed from such data.

Some very general ideas have been presented concerning the relationships between objects, representations, and images. These notions will now be incorporated into a discussion on the reconstruction of images using the techniques of CT and then applied to MRI.

### Computerized Tomography

Consider an object whose image is to be reconstructed by means of CT. This is accomplished by taking appropriate two-dimensional slice data of the object, accumulating these data in a computer's memory, and then constructing an image through the use of some imaging algorithm.

As an example, consider an object of some cross-sectional shape having a well-defined boundary whose body exhibits some distribution of x-ray absorbance. Let us assume that the object is located above the $x$-$y$ plane near the origin and that the x-rays uniformly illuminate the object from above. An x-ray detector with an adjustable iris is positioned beneath the object and the source as in Figure 1. The detector is scanned over the surface of the object, thereby generating a triplet of numbers $(x, y, z)$ which may be stored in the computer's memory. Of this triplet, $x$ and $y$ represent the horizontal and vertical positions of the detector, while $z$ represents the

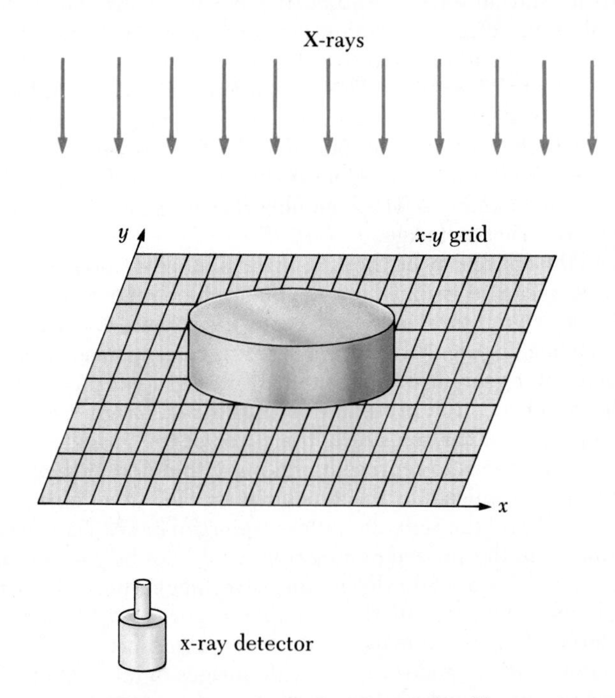


**Figure 1** One method of obtaining an image is by using a rotatable x-ray source that uniformly illuminates a stationary object. A x-ray detector placed beneath the object takes transmission intensity measurements at each of the indicated grid points.

intensity of the transmitted x-ray beam. Thus, at each location of the detector, the accumulated data can be converted to represent x-ray absorbance. The stored information is then subjected to an imaging algorithm which assigns to every picture element a brightness corresponding to the object's x-ray absorbance.

The resolution of the image will be a function of the iris's area, the collimation and intensity of the x-ray beam, the wavelength of the radiation, and the sensitivity of the detector. Within certain limits, these parameters may be selected so as to provide the desired image resolution. For example, higher resolution may be achieved by reducing the area of the detector's iris and choosing a finer scan of the detector. However, as the iris's area is reduced, the signal intensity at the detector is diminished. As the iris size is reduced further, the desired signals will eventually be overwhelmed by detector noise. Another parameter of limited variability is radiation wavelength. At shorter wavelengths (higher photon energies), x-rays tend to penetrate more deeply into body tissues and thereby yield more diagnostic information. However, this strategy is limited since the exposure of biological tissues to shorter x-ray wavelengths can result in cell damage.

Finally, computer memory places another practical limit on the degree of image resolution. Every picture element or pixel requires the storage of information such as its coordinates and x-ray absorbance number. Each image slice requires a large number of pixels, which may place heavy demand upon computer memory. As a result, the degree to which an image can be resolved will depend on the size of the computer's memory.

Another method of image production is reconstruction from rays. Consider an object in the $x$-$y$ plane whose center is located above the origin of the coordinate axes. An x-ray source located on a graduated circle is capable of rotating full circle about the center of coordinates. An x-ray detector is positioned opposite the x-ray source and follows the source as it is rotated about the object as in Figure 2. Such a system produces two pieces of image information. One is the orientation of the x-ray beam, and the second is the total x-ray absorbance. If a second detector samples the intensity of the x-ray beam before it enters the object, the difference in intensities of the x-rays can be determined without concern for either momentary fluctuations or gradual degradation of intensities in the x-ray source.

Another factor of x-ray absorbance which must be considered is the spectral character of the x-ray source. If different regions of an object have x-ray absorbances which depend upon the wavelength $\lambda$, one must use x-rays that are nearly monochromatic. Alternatively, corrections in the imaging algorithms must be made to account for the polychromatic nature of the x-ray source.

Successful procedures now exist for the reconstruction of images from x-ray absorption data. Currently, a number of highly developed computer-oriented mathe-

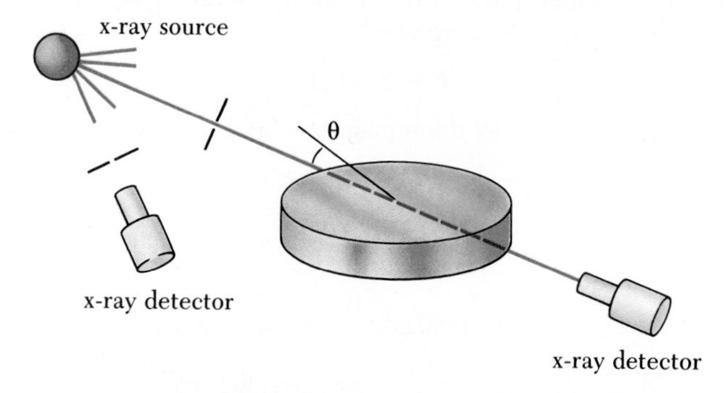


**Figure 2** Another method of obtaining images used in x-ray computerized tomography, CT. The x-ray source and detector scan the object by rotating over a full circle. The integrated x-ray intensity beam is obtained for each of a uniformly spaced set of discrete orientations.

*(Continued on next page)*

matical techniques are being used to reconstruct images from observational data. Although only a few primitive examples have been discussed, several techniques exist, each of which is designed to address a specific class of problems. The choice of one technique over another will most likely depend upon the nature of the imaging problem, its complexity, and the computer resources required to execute an image.

### Magnetic Resonance Spectroscopy

To understand magnetic resonance imaging, it is necessary first to become familiar with some of the concepts of magnetic resonance spectroscopy. Consider a system at the microscopic level characterized by the quantum mechanical angular momentum $L = J\hbar$, where $J$ is the angular momentum quantum number, and $\hbar$ is Planck's constant divided by $2\pi$. The absolute value squared of angular momentum is given by

$$L^2 = [J(J+1)]\hbar^2 \tag{1}$$

When a particle with angular momentum carries an electric charge, a magnetic dipole moment is generated whose magnitude is given by

$$\mu = \gamma\hbar\sqrt{J(J+1)} \tag{2}$$

where $\gamma$ is known as the magnetogyric ratio. The energy of a magnetic moment $\boldsymbol{\mu}$ in the presence of a magnetic field $\boldsymbol{H}$ is $E = -\boldsymbol{\mu}\cdot\boldsymbol{H}$, or its equivalent

$$E = -\mu H \cos\theta \tag{3}$$

where $\theta$ is the angle between the magnetic moment and the magnetic field. Classically, a magnetic moment can assume any orientation with respect to the magnetic field. That is, $\theta$ can take on any value from 0° to 180°. However, at the atomic or quantum level, the number of such orientations is restricted such that the projection of $\boldsymbol{\mu}$ upon the magnetic field direction is limited to a finite and well-prescribed set of values (see Section 42.5). Its projection upon the magnetic field is $\gamma m\hbar$, where $m$ is the "magnetic" quantum number whose values are restricted to the following:

$$m = J, (J-1), (J-2), \ldots, -(J-2), -(J-1), -J \tag{4}$$

The two most primitive examples of a magnetic moment are the proton and electron. Each of these particles, known as fermions, has an irreducible spin angular momentum whose quantum number is $S = \frac{1}{2}$. The magnetic moment associated with spin is given by

$$|\mu| = \gamma\hbar\sqrt{S(S+1)} \tag{5}$$

For the proton, the $\gamma$ factor is $\gamma_p = 2.675 \times 10^8\ \mathrm{s^{-1}\,T^{-1}}$ and for the electron it is $\gamma_e = -1.759 \times 10^{11}\ \mathrm{s^{-1}\,T^{-1}}$. Thus, a proton (or an electron) in a magnetic field has only two allowed states identified by the quantum numbers $m = \pm\frac{1}{2}$. The energies associated with these two states are given by

$$E = \pm\tfrac{1}{2}\gamma\hbar H \tag{6}$$

This linear dependence of energy upon magnetic field is shown in Figure 3.

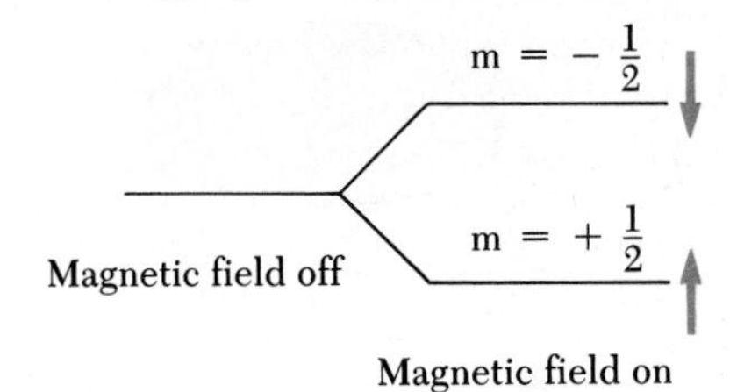


**Figure 3** A schematic representation of the energy dependence on magnetic field of a two-state quantum mechanical system corresponding to $S = \frac{1}{2}$.

A classical treatment shows that the time rate of change of angular momentum $\frac{d\boldsymbol{L}}{dt}$ is equal to the applied torque $\boldsymbol{\tau} = \boldsymbol{\mu} \times \boldsymbol{H}$. Hence, a magnetic dipole moment in a magnetic field will experience a time rate of change of angular momentum given by $\frac{d\boldsymbol{L}}{dt} = \boldsymbol{\mu} \times \boldsymbol{H}$, and since $\boldsymbol{\mu} = \gamma \boldsymbol{L}$ we find that

$$\frac{d\boldsymbol{\mu}}{dt} = \gamma(\boldsymbol{\mu} \times \boldsymbol{H}) \tag{7}$$

Since the torque acting on a system is perpendicular to both $\boldsymbol{\mu}$ and $\boldsymbol{H}$, the resulting motion will be a precession of $\boldsymbol{\mu}$ about the magnetic field. This is similar to the motion of a top spinning in a gravitational field. Furthermore, since the magnetic force acting upon a moving electrical charge is directed at right angles to its instantaneous velocity, no work is done by the magnetic force upon the dipole moment. Thus, once initiated, the precession will continue unattenuated until some dissipative mechanism is introduced to remove energy from the system.

If the system of nuclei is in thermal contact with a heat bath whose heat capacity is large compared to the system, what will be the outcome of total polarization? To answer this question, suppose that at some instant a system consisting of a very large number $N$ of protons were polarized so that all have spin quantum numbers $m = +\frac{1}{2}$. The total magnetization of the system would then be $M = \frac{1}{2}N\gamma\hbar$, and its energy would be $E = -\frac{1}{2}N\gamma\hbar H$. This would be a configuration of minimum spin energy and maximum spin order. If this system were completely isolated from the rest of the universe, it would be characterized by a spin temperature of absolute zero since it would have no spin energy to give up nor any means of increasing its spin order. Such a system could not be in thermal equilibrium with its heat bath. Eventually, a real system would arrive at the temperature of the bath and would have a magnetic polarization corresponding to that temperature. This limiting or equilibrium polarization is given by the Boltzmann relation $M_0 = \frac{1}{2}N\gamma\hbar e^{-\Delta E/kT}$, where $\Delta E = \gamma\hbar H$ is the difference in energy between the two spin states, $k$ is the Boltzmann constant, and $T$ is the absolute temperature. Further, the $z$-component of the magnetization $M_z$ will advance towards $M_0$, the equilibrium value of magnetization. The differential equation governing the return to equilibrium is

$$\frac{d}{dt} M_z(t) = -\frac{M_z - M_0}{T_1} \tag{8}$$

whose solution is given by

$$M_z(t) = M_0(1 - e^{t/T_1}) \tag{9}$$

where $t$ is time and $T_1$ is the longitudinal (spin-lattice) relaxation time, which characterizes the time it takes $M_z$ to return to thermal equilibrium.

Another parameter which characterizes the dynamics of a spin system moving towards thermal equilibrium is the transverse or spin-spin relaxation time $T_2$. To understand its nature, consider a system of nuclei in thermal equilibrium with its bath. Let the system be perturbed at some time $t = 0$ such that its spin order increases while its total energy remains unchanged. Although energy is conserved, the increase in spin order comes at the expense of an increase in transverse magnetic polarization, $M_{xy}$. As the system returns to thermal equilibrium, $M_{xy}$ goes to zero. The system returns to thermal equilibrium at a rate characterized by $T_2$. In general $T_2$ is shorter than or at best equal to $T_1$.

The details of the relaxation parameters $T_1$ and $T_2$ are known to be quite complicated and dependent upon several variables such as temperature, nuclear motion, the concentration of magnetic species, and the external magnetic field intensity. How-

*(Continued on next page)*

ever, the concept of a relaxation time has proven to be very useful in correlating specific interactions with the characteristics of drift towards equilibrium. Consequently, relaxation times are universally used in magnetic resonance spectroscopy.

Magnetic resonance can best be understood by using Equation 6 to find the energy difference between two states of a spin $\frac{1}{2}$ system of nuclei, say, protons, in the presence of a magnetic field $\boldsymbol{H}$. If the difference in energy $\Delta E = \gamma\hbar H$ is supplied by a radiation field whose photons have energy $h\omega = \gamma\hbar H$, then nuclei in the lower energy state will absorb photons and be raised to the upper energy state. The radiation field will also cause nuclei in the upper state to give up a photon of energy $\Delta E = \gamma\hbar H$ and thereby drop to the lower energy state. On average, the radiation field would lose no energy and the system of nuclei would remain unchanged. This assumes, however, that the upper and lower states are equally populated. In fact, at ordinary temperatures, that is not the case. Statistical mechanics shows that for a quantum system of spin $\frac{1}{2}$ particles, the relation between the populations $n_+$ and $n_-$ corresponding to the states $m = \pm\frac{1}{2}$ is given by

$$n_- = n_+ e^{-\Delta E/kT} \tag{10}$$

Thus, the lower energy state will be more populated to the extent determined by the exponential term known as the Boltzmann factor. If the energy difference $\Delta E$ is small compared to $kT$, then the following approximation holds:

$$n_- = n_+\left(1 - \frac{\Delta E}{kT}\right) \tag{11}$$

Since the probability of induced absorption is equal to the probability of induced emission, the result is a net absorption of photons because $n_- \leq n_+$.

If the radiation field is characterized by a single frequency $\omega_0$ or by a very narrow band of frequencies $\omega_0 \pm \delta\omega$, where $\delta\omega$ is small compared to $\omega_0$ but larger than the width of the resonance absorption, then one can observe an absorption line by choosing the magnetic field such that $\hbar\omega = \gamma\hbar H_0$. On the other hand, if the radiation frequency is held at $\omega_0$ while the magnetic field is scanned through $H_0$, then a resonance can be observed in the magnetic field domain. This is referred to as CW or continuous wave spectroscopy.

Another kind of spectroscopy that is more appropriate to magnetic resonance imaging is the pulsed variety. To understand pulse magnetic resonance spectroscopy, consider a system of protons in the presence of a steady magnetic field $\boldsymbol{H}_0$ along the $z$-axis. At ordinary temperatures, the protons will be slightly polarized and have a small but finite net magnetization $\boldsymbol{M}_z$ along the direction of the magnetic field, while the total nuclear magnetization $\boldsymbol{M}$ will precess about the $z$-axis. Because of this precession, the transverse magnetizations $\boldsymbol{M}_x$ and $\boldsymbol{M}_y$ will have average values that vanish. Now suppose a radiation field of frequency $\omega_0 = \gamma H_0$ is impressed upon this system of protons having an $x$-component whose time dependence is $H_x = H_x^{\,0}\cos(\omega_0 t)$. Such a linearly polarized field may be expressed as the sum of two counter-rotating circularly polarized fields $\boldsymbol{H}_r$ and $\boldsymbol{H}_\ell$, where $\boldsymbol{H}_r$ is a field that rotates about the $z$-axis in a counter clockwise manner, while $\boldsymbol{H}_\ell$ rotates about the $z$-axis in a clockwise manner. The component $\boldsymbol{H}_\ell$, which rotates in the sense opposite to that of precession, will on average, or over a rotation of $2\pi$ radians, receive as much torque from $\boldsymbol{M}$ as it imparts. Thus its influence on $\boldsymbol{M}$ may be ignored. The only component of $\boldsymbol{H}_x$ that affects the magnetization vector is $\boldsymbol{H}_r$, which rotates in the same sense as does the precession of the magnetization vector. In the rotating frame, $\boldsymbol{H}_r$ remains in phase with $\boldsymbol{M}$, thereby imparting a constant torque upon the latter. This added torque causes $\boldsymbol{M}$ to undergo a continuous increase in its polar angle $\theta$. The increase in $\theta$ will continue as long as the system is exposed to this oscillating field. This additional angular motion of $\boldsymbol{M}$ is referred to as *nutation*. As the tip of the vector $\boldsymbol{M}$ precesses about the direction of the applied field $\boldsymbol{H}_0$, it undergoes a nodding motion (see Figure 4). If the angular fre-

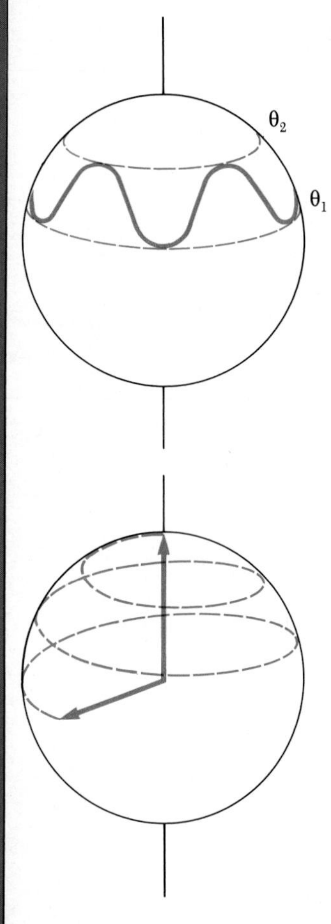


**Figure 4** The behavior of a magnetic moment in a uniform magnetic field. If no alternating magnetic field is applied to the system, a pure precession takes place. When an alternating field of angular velocity $\omega_0$ is applied, a nutation results in the motion of the moment. Eventually, the magnetization will tip into the $x'$-$y'$ plane.

quency of the radiation field is $\omega$, one finds that the effective magnetic field fixed in the rotating reference frame is given by

$$\boldsymbol{H}_{\text{eff}} = H_x\boldsymbol{i} + \left(H_0 - \frac{\omega}{\gamma}\right)\boldsymbol{k} \tag{12}$$

where $\boldsymbol{i}$ is a unit vector lying along the $x$-axis in the rotating frame and $\boldsymbol{k}$ is a unit vector along the $z$-axis fixed in the lab frame. When $\omega$ is equal to the Larmor angular velocity $\omega_0$, the term in brackets will vanish since $\omega_0 = \gamma H_0$. Furthermore, when $\omega = \omega_0$, the effective field $\boldsymbol{H}_{\text{eff}}$ is equal to $\boldsymbol{H}_r$.

Now suppose that an oscillating field $H_r$ of angular frequency $\omega_0$ is applied, for a short time $\delta t$, to a collection of protons in a magnetic field $H_0$. In the rotating frame, the vector $\boldsymbol{M}$ will dip through an angle given by $\theta = \gamma H_r\, \delta t$. If the strength of $H_r$ and the time of its duration $\delta t$ are such that $\gamma H_r\, \delta t = \pi$, then $\boldsymbol{M}$ will rotate into the negative $z$-axis. Such an orientation represents a state of maximum magnetic energy. Consequently, the system will return to thermal equilibrium in a manner characterized by the longitudinal relaxation time, $T_1$ (according to Eq. 9).

Next, suppose that the duration time of the pulse is halved so that $\gamma H_r\, \delta t = \pi/2$. At the end of this pulse, all of the nuclear magnetic moments will be aligned parallel to the $y'$-axis of the rotating frame. This is not a state of thermal equilibrium; hence the system will immediately move towards thermal equilibrium by giving up its $y'$ magnetization. It does so in a time characterized by $T_2$, the transverse relaxation time, assuming this time is short compared to $T_1$. The transverse magnetization of the system returns to its state of equilibrium according to the relation

$$M_{x'}(t) = M_{x'}{}^0 e^{-t/T_2} \tag{13}$$

The manner in which the system returns to thermal equilibrium is a randomization process caused by inhomogeneities in the external field and the effect of neighboring magnetic moments. The magnetization, initially polarized along $y'$, gradually randomizes in the $x'$-$y'$ plane, and "fans out" in this plane as in Figure 5. The time dependence of this process is given by Equation 13, and is referred to as free induction decay.

If the time lapsed is long compared to the relaxation time $T_2$, the transverse magnetization will vanish. However, if a second pulse is applied to the system, a 180° pulse, the magnetic moments will later refocus to yield the original transverse magnetization and a second magnetic resonance signal referred to as a *spin echo* will result. The spin echo signal can be viewed as one which is a synthesis of a free induction that decays as time flows in the negative direction and a second free induction that decays as time flows in the forward direction. The two are joined together at the zero of time at which their maxima coincide.

To recapitulate, a spin echo signal can be generated as the system of nuclei is exposed to a pair of radio frequency magnetic field pulses polarized along the rotating $x$-axis. The duration of these pulses must be such that $\gamma H_r\, \delta t$ is 90° for the first pulse and 180° for the second. The time between the two pulses should be several transverse relaxation times to insure that the transverse magnetization has time to partially randomize. Yet, it must be short compared to the longitudinal relaxation time to avoid the loss of magnetization as the system tends to return to thermal equilibrium. Figure 6 shows the consequences of a 90° pulse followed by a 180° pulse.

There is a fundamental difference between magnetic resonance absorption spectroscopy and the two forms of pulsed spectroscopy. In absorption spectroscopy, information is gathered in the frequency or magnetic field domain, while in pulsed spectroscopy, information is gathered in the time domain. Both techniques are equivalent in that the resonance absorption shape function can be obtained from the free induction shape function using a Fourier transformation. Pulsed spectroscopy has the

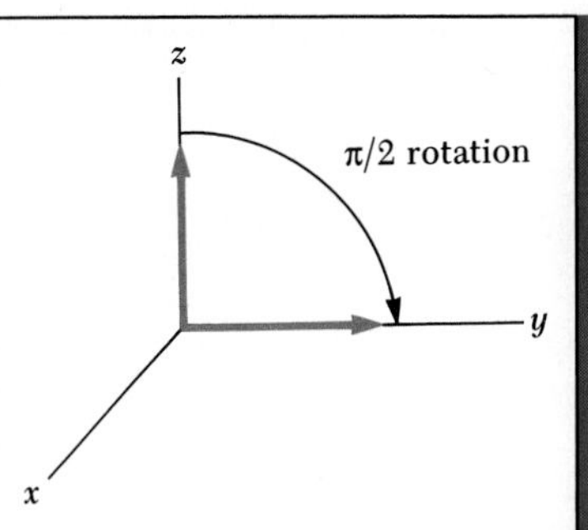


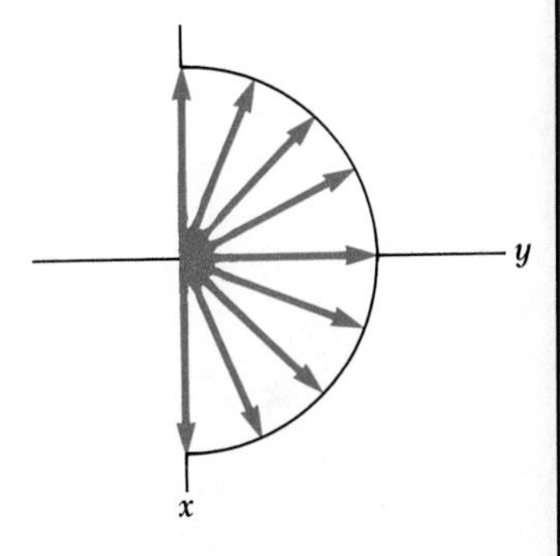


**Figure 5** A rotation of the magnetization vector from the laboratory $z$-axis into the rotating frame's $y'$-axis. After completing this rotation, all the magnetic moments in the system will be phase. As time evolves, this system of moments fans out in the manner discussed in the text.

*(Continued on next page)*

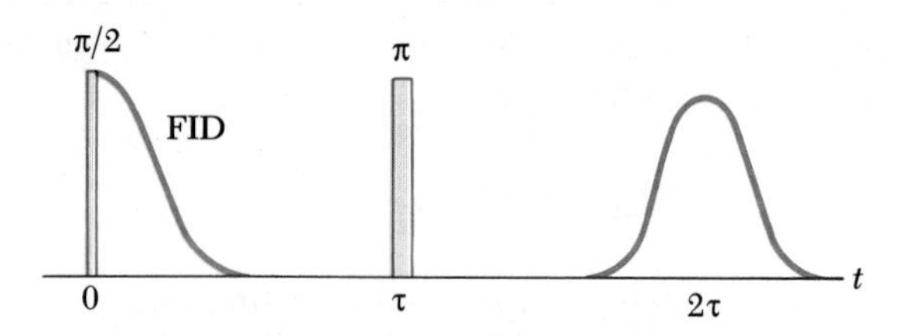


**Figure 6** A schematic representation sequence of two pulses, one of 90° duration followed by another of 180° duration. The consequences of this sequence of pulses is the generation of a free induction decay followed by a spin echo.

advantage of being very fast compared to absorption spectroscopy. For example, a single 90° pulse is sufficient to produce a signal which can be captured by a suitable signal processing device. This signal may then be subjected to an appropriate Fourier transformation to provide the absorption shape function. The entire process may be conducted very rapidly using a computer to first control the pulse sequencing and then to perform the various mathematical manipulations.

At the heart of magnetic resonance imaging is the magnetic resonance absorption shape function, or simply the absorption function, which is used to drive an imaging algorithm. In practice, an integrated density function is required for image reconstruction. This function is directly proportional to the integrated magnetic resonance absorption. Image reconstruction is obtained by establishing a set of procedures for encoding the coordinates of object volume elements with magnetic field strength values. One uses gradient magnetic fields (spatially varying fields) to produce this encoding. A procedure is established for encoding each point of a planar sheet with a unique magnetic field strength. Once this is accomplished, an image reconstruction algorithm may be used to produce an image.

To better understand the advantage of pulsed spectroscopy over frequency domain spectroscopy, consider the following experiment. Suppose the entire $x$-$y$ plane is subjected to a uniform magnetic field, and an *rf* pulse is applied to the system which rotates the magnetization by 90°. All magnetic moments will now be rotated from the $z$-axis to the rotating $y'$-axis, and free induction will proceed along with its decay. After some interval of time (short compared to $T_2$), let a gradient magnetic field be applied along the $x$-axis. Then each volume element of the object will have its magnetic moments precess at a frequency characterized by the local magnetic field identified with the space coordinates of the object. The result is a single complex signal which is the superposition of all free induction decay signals along the direction of the field gradient. This signal may then be subjected to a Fourier transformation which will yield the entire distribution of precession frequencies within the slice along with the signal intensity associated with each frequency. This information may then be used as input data to an imaging algorithm.

One of the earliest magnetic resonance images reported in the scientific literature is a view of two 1-mm diameter water-filled capillary tubes shown in Figure 7. By comparison, Figure 8 shows state-of-the-art magnetic resonance images of a human head. As mentioned earlier, MRI is a diagnostic technique that is considered to be noninvasive. Photons associated with the *rf* signals have energies of about $10^{-7}$ eV. Since molecular bond strengths are much larger (of the order of 1 eV), the *rf* photons cause little cellular damage. In comparison, x-rays or $\gamma$-rays have energies ranging from $10^4$ to $10^6$ eV and have the potential of causing considerable cellular damage. This is the main advantage that MRI has over other imaging techniques in medical diagnostics.

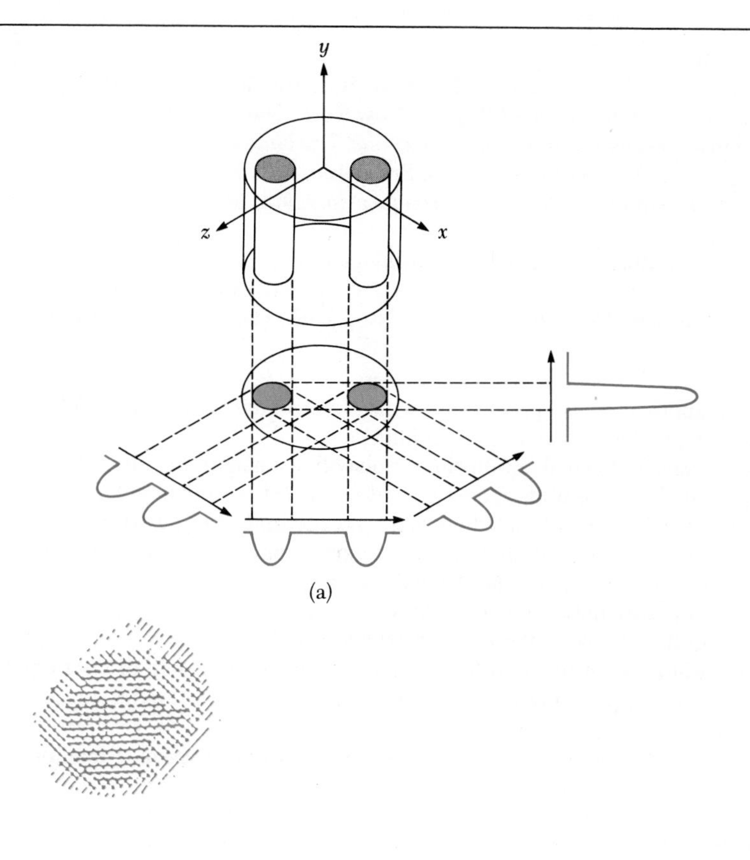


(a)

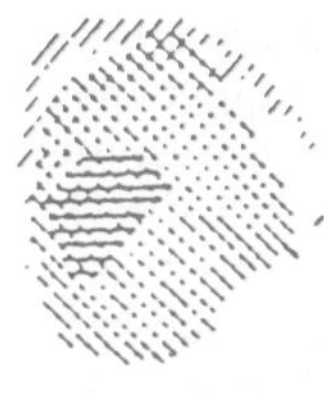

(b)

**Figure 7** An early magnetic resonance image of two water-filled capillary tubes 1 mm in diameter with their centers separated by about 3 mm. (a) This shows the method of projections used in producing the image. (b) The reconstructed image. (From P.C. Lauterbur, *Nature*, **242**:190, 1973.)

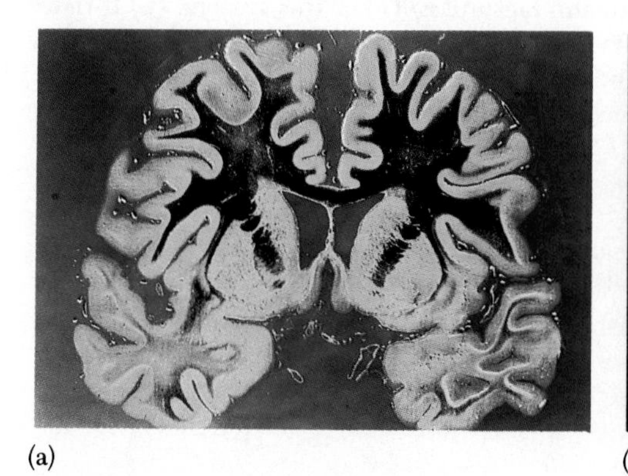

(a)

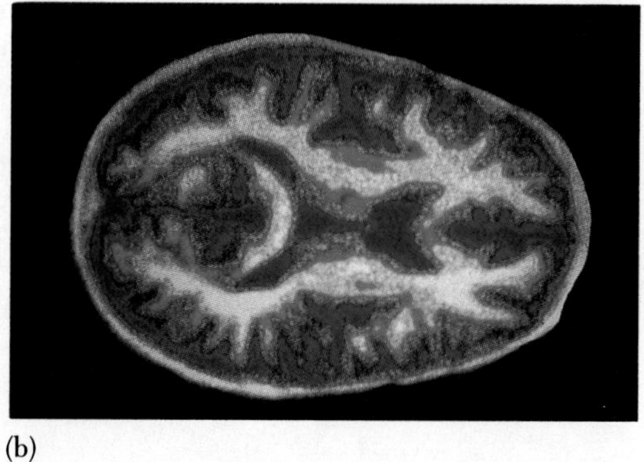

(b)

**Figure 8** State-of-the-art magnetic resonance images taken of the human head. (a) is a sagittal view and (b) is a coronal view. The slice images are of 1 cm thickness and the result of a $128 \times 128$ point reconstruction. (Photo Researchers, Inc.)

## Suggested Readings

Carrington, A., and A.D. McLachlan, *Introduction to Magnetic Resonance With Applications to Chemistry and Chemical Physics,* New York, Harper & Row, 1967.

Herman, G.T., *Image Reconstruction from Projections. The Fundamentals of Computerized Tomography,* New York, Academic Press, Inc., 1980.

Keller, P.J., *Basic Principles of Magnetic Resonance Imaging,* Milwaukee, General Electric Company, 1988.

Mansfield P., and P.G. Morris, *NMR Imaging in Biomedicine,* New York, Academic Press, Inc., 1982.

Slichter, C.P., *Principles of Magnetic Resonance,* New York, Harper & Row, 1963.

## Essay Questions

1. Why does a magnetic moment precess in the presence of a magnetic field? What factors determine its frequency of precession?
2. A system of protons in thermal equilibrium is placed in a magnetic field. Explain why the upper and lower spin states corresponding to $m = -\frac{1}{2}$ and $m = \frac{1}{2}$, respectively, are not equally populated. Which state has the higher population?
3. Discuss the concepts of longitudinal and transverse relaxation times, and why they are useful in characterizing the dynamics of a spin system.
4. In pulsed spectroscopy, how does one produce a spin echo signal? What are the requirements on the nature of the pulse durations?
5. Construct an energy level diagram for an electron in the presence of a steady magnetic field and compare it to Figure 3 for a proton. What are the differences, if any, in the two diagrams?
6. Why is pulsed spectroscopy more appropriate to magnetic resonance imaging than continuous wave spectroscopy?

## Essay Problems

1. A proton is placed in a steady magnetic field of 0.50 T. (a) What is the energy difference between the upper and lower spin states? (b) What frequency photon will excite a transition between these two states?
2. Repeat Problem 1 if the particle is an electron in a steady magnetic field of 0.50 T.
3. A sample of a particular substance containing both protons and electrons is placed in a steady magnetic field of 0.80 T. What is the Larmor precessional frequency for (a) the protons, and (b) the electrons?
4. A system of protons is placed in a steady magnetic field of 1.00 T at a temperature of 300 K. (a) Calculate the Boltzmann factor $\Delta E/kT$ for this system. (b) If there are $10^{23}$ protons in the upper energy state corresponding to $m = -\frac{1}{2}$, how many protons are there in the lower energy state corresponding to $m = \frac{1}{2}$?

# 46

# Nuclear Physics Applications

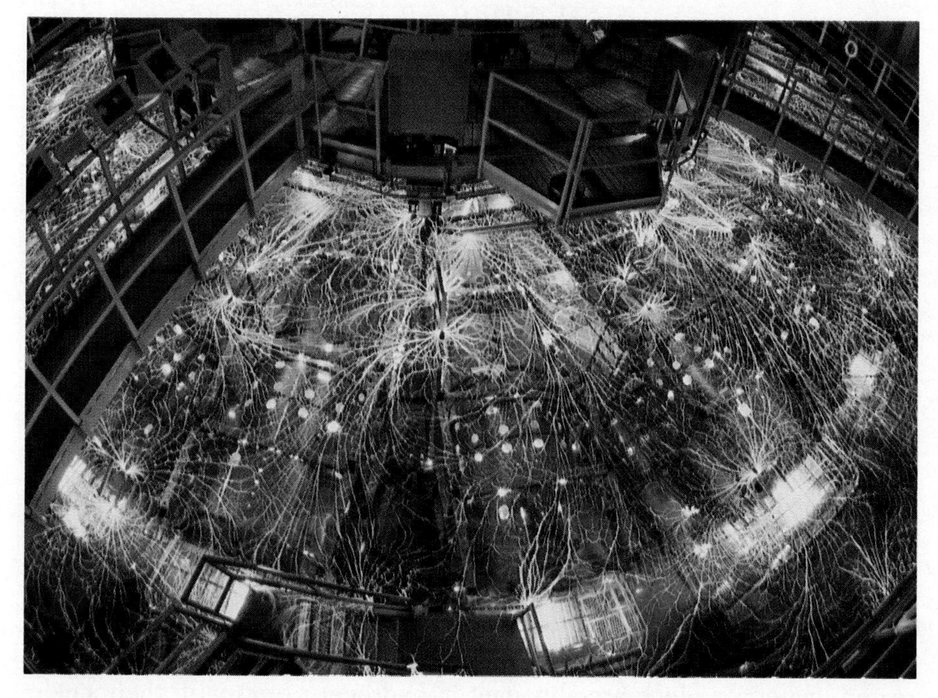

*Electrical discharge visible in the pulse-forming section of the Particle Beam Fusion Accelerator II, the nation's most powerful x-ray source, at Sandia National Laboratories. The discharges are due to air breakdown during firing at the surface of the water that covers this section. High-energy electrons are converted to x-rays at the center of the machine, which in turn are used to understand their effect on weapon systems and other components. (Courtesy of Sandia National Laboratories. Photo by Walter Dickenman)*

This chapter is concerned primarily with the two means by which energy can be derived from nuclear reactions. These two techniques are fission, in which a nucleus of large mass number splits, or fissions, into two smaller nuclei, and fusion, in which two light nuclei fuse to form a heavier nucleus. In either case, there is a release of energy, which can be used destructively through bombs or constructively through the production of electric power. We shall also examine several devices used to detect radiation. The chapter concludes with a discussion of some industrial and biological applications of radiation.

## 46.1 INTERACTIONS INVOLVING NEUTRONS

In order to understand the process of nuclear fission and the physics of the nuclear reactor, one must first understand the manner in which neutrons interact with nuclei. Because of their charge neutrality, neutrons are not subject to Coulomb forces. Nevertheless, very slow neutrons can wander through a material and cause nuclear reactions. Since neutrons interact very weakly with electrons, matter appears quite "open" to neutrons. In general, one finds that the rate of neutron-induced reactions increases as the neutron energy decreases. Free neutrons undergo beta decay with a mean lifetime of about 10 min. On the other hand, neutrons in matter are absorbed by nuclei before they decay.

When a fast neutron (energy greater than about 1 MeV) moves through matter, it undergoes many scattering events with the nuclei. In each such event, the neutron gives up some of its kinetic energy to a nucleus. The neutron continues to undergo collisions until its energy is of the order of the thermal energy $kT$, where $k$ is Boltzmann's constant and $T$ is the absolute temperature. A neutron with this energy is called a **thermal neutron.** At this low energy, there is a high probability that the neutron will be captured by a nucleus, accompanied by the emission of a gamma ray. This **neutron-capture process** can be written

Thermal neutron

Neutron capture

$$ {}^{1}_{0}\text{n} + {}^{A}_{Z}\text{X} \longrightarrow {}^{A+1}_{Z}\text{X} + \gamma \tag{46.1}$$

Although we did not indicate it here, the nucleus X is in an excited state X* for a very short time before it undergoes gamma decay. Also, the product nucleus ${}^{A+1}_{Z}\text{X}$ is usually radioactive and decays by beta emission.

The neutron-capture rate associated with the above process depends on the nature of the material with which the neutron interacts. For some materials, elastic collisions are dominant. Materials for which this occurs are called **moderators** since they slow down (or moderate) the originally energetic neutrons very effectively. Boron, graphite, and water are a few examples of moderator materials.

Moderator

As you might recall from Chapter 9, during an elastic collision between two particles, the maximum kinetic energy is transferred from one particle to the other when they have the same mass (Example 9.8). Consequently, a neutron loses all of its kinetic energy when it collides head-on with a proton, in analogy to the collision between a moving and a stationary billiard ball. If the collision is oblique, the neutron loses only part of its kinetic energy. For this reason, materials which are abundant in hydrogen atoms with their single-proton nuclei, such as paraffin and water, are good moderators for neutrons.

At some point, many of the neutrons in the moderator become thermal neutrons, which are neutrons in thermal equilibrium with the moderator material. Their average kinetic energy at room temperature is given by

$$K_{\text{av}} = \tfrac{3}{2}kT \approx 0.04 \text{ eV}$$

Thermal neutrons have a distribution of velocities, just as the molecules in a container of gas do (Chapter 21). A high-energy neutron, whose energy is several MeV, will thermalize (that is, reach $K_{\text{av}}$) in less than 1 ms when incident on a moderator. These thermal neutrons have a very high probability of undergoing neutron capture by the moderator nuclei.

## 46.2 NUCLEAR FISSION

Fission

**Nuclear fission** occurs when a heavy nucleus, such as $^{235}$U, splits, or fissions, into two smaller nuclei. In such a reaction, *the total rest mass of the products is less than the original rest mass.*

Nuclear fission was first observed in 1938 by Otto Hahn and Fritz Strassman, following some basic studies by Fermi. After bombarding uranium ($Z = 92$) with neutrons, Hahn and Strassman discovered among the reaction products two medium-mass elements, barium and lanthanum. Shortly thereafter, Lisa Meitner and Otto Frisch explained what had happened. The uranium nucleus had split into two nearly equal fragments after absorbing a neutron.

Such an occurrence was of considerable interest to physicists attempting to understand the nucleus, but it was to have even more far-reaching consequences. Measurements showed that about 200 MeV of energy is released in each fission event, and this fact was to affect the course of human history.

The fission of $^{235}$U by slow (low energy) neutrons can be represented by the equation

$$^{1}_{0}\text{n} + {}^{235}_{92}\text{U} \longrightarrow {}^{236}_{92}\text{U}^{*} \longrightarrow \text{X} + \text{Y} + \text{neutrons} \quad (46.2)$$

Fission of $^{235}$U

where $^{236}$U* is an intermediate state that lasts only for about $10^{-12}$ s before splitting into X and Y. The resulting nuclei, X and Y, are called **fission fragments.** There are many combinations of X and Y that satisfy the requirements of conservation of mass-energy and charge. In the fission of uranium, there are about 90 different daughter nuclei that can be formed. The process also results in the production of several neutrons, typically two or three. On the average 2.47 neutrons are released per event.

A typical reaction of this type is

$$^{1}_{0}\text{n} + {}^{235}_{92}\text{U} \longrightarrow {}^{141}_{56}\text{Ba} + {}^{92}_{36}\text{Kr} + 3\,({}^{1}_{0}\text{n}) \quad (46.3)$$

The fission fragments, barium and krypton, and the released neutrons have a great deal of kinetic energy following the fission event.

The breakup of the uranium nucleus can be compared to what happens to a drop of water when excess energy is added to it. All the atoms in the drop have energy, but this energy is not great enough to break up the drop. However, if enough energy is added to set the drop into vibration, it will undergo elongation and compression until the amplitude of vibration becomes large enough to cause the drop to break. In the uranium nucleus, a similar process occurs (Fig. 46.1). The sequence of events is as follows:

1. The $^{235}$U nucleus captures a thermal (slow-moving) neutron.
2. This capture results in the formation of $^{236}$U*, and the excess energy of this nucleus causes it to undergo violent oscillations.
3. The $^{236}$U* nucleus becomes highly distorted, and the force of repulsion between protons in the two halves of the dumbbell shape tends to increase the distortion.
4. The nucleus splits into two fragments, emitting several neutrons in the process.

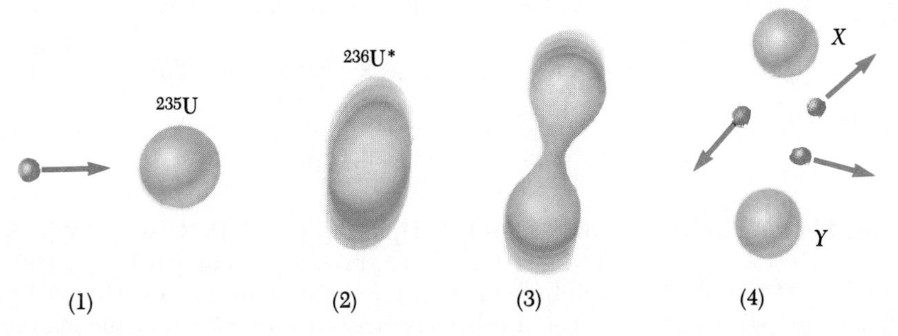


**Figure 46.1** The stages in a nuclear fission event as described by the liquid-drop model of the nucleus.

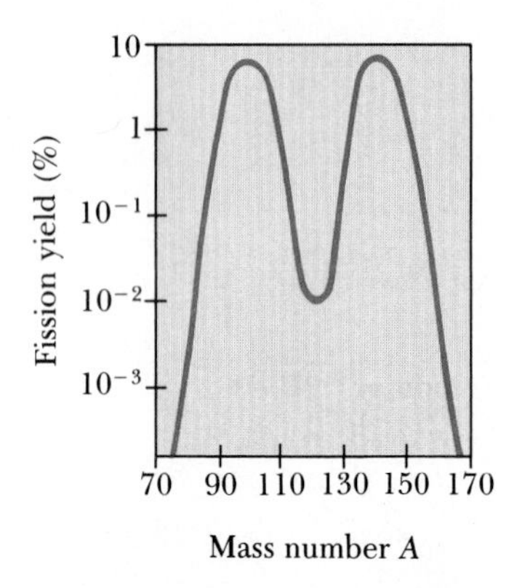


**Figure 46.2** The distribution of fission products versus mass number for the fission of $^{235}$U with slow neutrons. Note that the ordinate has a logarithmic scale.

Figure 46.2 is a graph of the distribution of fission products versus mass number $A$. The most probable fission events correspond to nonsymmetric events, that is, events for which the fission fragments have unequal masses. (Symmetric fission events corresponding to $A = 118$ are highly improbable, occurring in only about 0.01% of the cases.) The most probable events correspond to fission fragments with mass numbers of $A \approx 140$ and $A \approx 95$. These fragments, which share the protons and neutrons of the mother nucleus, both fall to the left of the stability line in Figure 45.3. In other words, *fragments that have a large excess of neutrons are unstable.* As a result, the neutron-rich fragments almost instantaneously release two or three neutrons. The remaining fragments are still rich in neutrons and proceed to decay to more stable nuclei through a succession of $\beta^-$ decays. In the process of such decay, gamma rays are also emitted by nuclei in excited states.

Let us estimate the disintegration energy $Q$ released in a typical fission process. From Figure 45.7 we see that the binding energy per nucleon is about 7.6 MeV for heavy nuclei (those having a mass number approximately equal to 240) and about 8.5 MeV for nuclei of intermediate mass. This means that the nucleons in the fission fragments are more tightly bound and therefore have less mass than the nucleons in the original heavy nucleus. This decrease in mass per nucleon appears as released energy when fission occurs. The amount of energy released is (8.5 − 7.6) MeV per nucleon. Assuming a total of 240 nucleons, we find that the energy released per fission event is

$$Q = (240 \text{ nucleons})\left(8.5 \frac{\text{MeV}}{\text{nucleon}} - 7.6 \frac{\text{MeV}}{\text{nucleon}}\right) = 220 \text{ MeV}$$

This is indeed a very large amount of energy relative to the amount of energy released in chemical processes. For example, the energy released in the combustion of one molecule of octane used in gasoline engines is about one millionth the energy released in a single fission event!

### EXAMPLE 46.1 The Fission of Uranium

Two other possible ways by which $^{235}$U can undergo fission when bombarded with a neutron are (1) by the release of $^{140}$Xe and $^{94}$Sr as fission fragments and (2) by the release of $^{132}$Sn and $^{101}$Mo as fission fragments. In each case, neutrons are also released. Find the number of neutrons released in each of these events.

*Solution* By balancing mass numbers and atomic numbers, we find that these reactions can be written

$${}^{1}_{0}\text{n} + {}^{235}_{92}\text{U} \longrightarrow {}^{140}_{54}\text{Xe} + {}^{94}_{38}\text{Sr} + 2\,({}^{1}_{0}\text{n})$$

$${}^{1}_{0}\text{n} + {}^{235}_{92}\text{U} \longrightarrow {}^{132}_{50}\text{Sn} + {}^{101}_{42}\text{Mo} + 3\,({}^{1}_{0}\text{n})$$

Thus, two neutrons are released in the first event and three in the second.

### EXAMPLE 46.2 The Energy Released in the Fission of $^{235}$U

Calculate the total energy released if 1 kg of $^{235}$U undergoes fission, taking the disintegration energy per event to be $Q = 208$ MeV (a more accurate value than the estimate given before).

*Solution* We need to know the number of nuclei in 1 kg of uranium. Since $A = 235$, the number of nuclei is

$$N = \left(\frac{6.02 \times 10^{23} \text{ nuclei/mol}}{235 \text{ g/mol}}\right)(10^3 \text{ g})$$

$$= 2.56 \times 10^{24} \text{ nuclei}$$

Hence the total disintegration energy is

$$E = NQ = (2.56 \times 10^{24} \text{ nuclei})\left(208 \frac{\text{MeV}}{\text{nucleus}}\right)$$

$$= 5.32 \times 10^{26} \text{ MeV}$$

Since 1 MeV is equivalent to $4.45 \times 10^{-20}$ kWh, $E = 2.37 \times 10^{7}$ kWh. This is enough energy to keep a 100-W lightbulb burning for about 30 000 years. Thus, 1 kg of $^{235}$U is a relatively large amount of fissionable material.

## 46.3 NUCLEAR REACTORS

We have seen that, when $^{235}$U undergoes fission, an average of about 2.5 neutrons are emitted per event. These neutrons can in turn trigger other nuclei to undergo fission, with the possibility of a chain reaction (Fig. 46.3). Calculations show that if the chain reaction is not controlled (that is, if it does not proceed slowly), it could result in a violent explosion, with the release of an enormous amount of energy, even from only 1 g of $^{235}$U. If the energy in 1 kg of $^{235}$U were released, it would be equivalent to detonating about 20 000 tons of TNT! This, of course, is the principle behind the first nuclear bomb, an uncontrolled fission reaction.

Chain reaction

A nuclear reactor is a system designed to maintain what is called a **self-sustained chain reaction.** This important process was first achieved in 1942 by Fermi at the University of Chicago, with natural uranium as the fuel (Fig. 46.4). Most reactors in operation today also use uranium as fuel. Natural uranium contains only about 0.7% of the $^{235}$U isotope, with the remaining 99.3% being the $^{238}$U isotope. This is important to the operation of a reactor because $^{238}$U almost never undergoes fission. Instead, it tends to absorb neutrons, producing neptunium and plutonium. For this reason, reactor fuels must be artificially enriched to contain a few percent of the $^{235}$U isotope.

Reproduction constant

Earlier, we mentioned that an average of about 2.5 neutrons are emitted in each fission event of $^{235}$U. In order to achieve a self-sustained chain reaction, one of these neutrons, on the average, must be captured by another $^{235}$U nucleus and cause it to undergo fission. A useful parameter for describing the level of reactor operation is the **reproduction constant** $K$ defined as *the average number of neutrons from each fission event that will cause another event.* As we have seen, $K$ can have a maximum value of 2.5 in the fission of uranium. However, in practice $K$ is less than this because of several factors, which we shall soon discuss.

A self-sustained chain reaction is achieved when $K = 1$. Under this condition, the reactor is said to be **critical.** When $K$ is less than unity, the reactor is subcritical and the reaction dies out. When $K$ is greater than unity, the reactor is said to be supercritical and a runaway reaction occurs. In a nuclear reactor

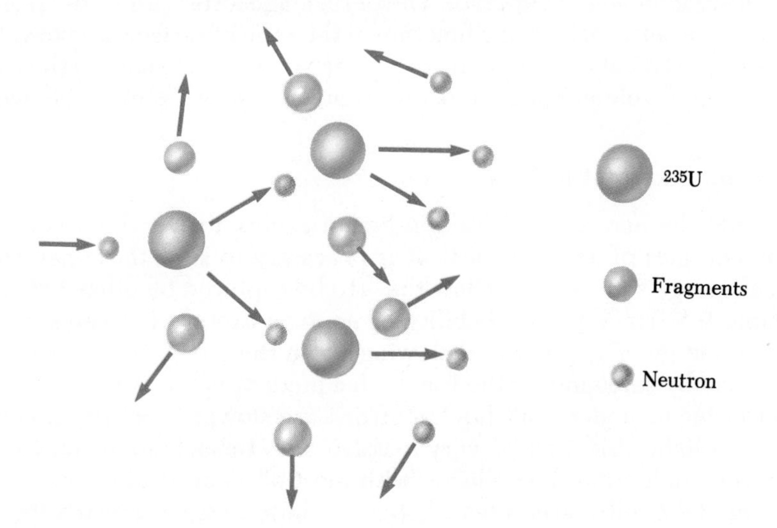


**Figure 46.3** A nuclear chain reaction.

**Figure 46.4** Sketch of the world's first reactor. Because of wartime secrecy, there are no photographs of the completed reactor. The reactor was composed of layers of graphite interspersed with uranium. A self-sustained chain reaction was first achieved on December 2, 1942. Word of the success was telephoned immediately to Washington with this message: "The Italian navigator has landed in the New World and found the natives very friendly." The historic event took place in an improvised laboratory in the racquet court under the west stands of the University of Chicago's Stagg Field and the Italian navigator was Fermi. (Courtesy of Chicago Historical Society)

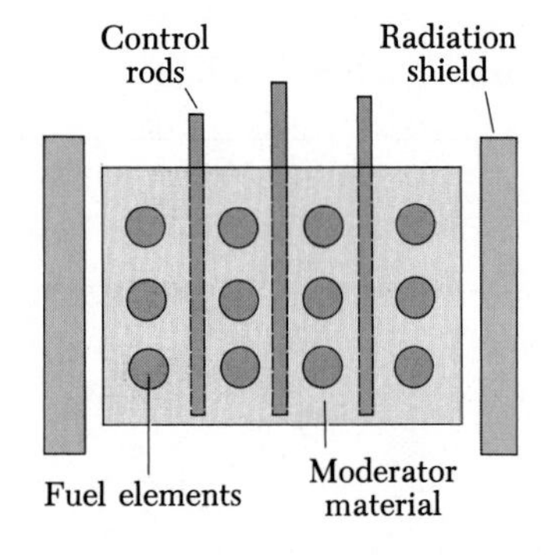


**Figure 46.5** Cross section of a reactor core surrounded by a radiation shield.

used to furnish power to a utility company, it is necessary to maintain a value of $K$ close to unity.

The basic design of a nuclear reactor is shown in Figure 46.5. The fuel elements consist of enriched uranium. The function of the remaining parts of the reactor and some aspects of its design will now be described.

### Neutron Leakage

In any reactor, a fraction of the neutrons produced in fission will leak out of the core before inducing other fission events. If the fraction leaking out is too large, the reactor will not operate. The percentage lost is large if the reactor is very small because leakage is a function of the ratio of surface area to volume. Therefore, a critical feature of the design of a reactor is to choose the correct surface area to volume ratio so that a sustained reaction can be achieved.

### Regulating Neutron Energies

Recall that the neutrons released in fission events are very energetic, with kinetic energies of about 2 MeV. It is necessary to slow these neutrons to thermal energies in order to allow them to be captured by other $^{235}U$ nuclei (Example 9.8) since the probability of neutron-capture increases with decreasing energy. The process of slowing down the energetic neutrons is accomplished by surrounding the fuel with a moderator substance.

In order to understand how neutrons are slowed, consider a collision between a light object and a very massive one. In such an event, the light object rebounds from the collision with most of its original kinetic energy. However, if the collision is between objects whose masses are nearly the same, the incoming projectile will transfer a large percentage of its kinetic energy to the target. In the first nuclear reactor ever constructed, Fermi placed bricks of graphite (carbon) between the fuel elements. Carbon nuclei are about 12 times more massive than neutrons, but after several collisions with carbon nuclei, a neutron is slowed sufficiently to increase its likelihood of fission with $^{235}$U. In this design carbon is the moderator; most modern reactors use water as the moderator.

### Neutron Capture

In the process of being slowed down, neutrons may be captured by nuclei that do not undergo fission. The most common event of this type is neutron capture by $^{238}$U. The probability of neutron capture by $^{238}$U is very high when the neutrons have high kinetic energies and very low when they have low kinetic energies. Thus the slowing down of the neutrons by the moderator serves the dual purpose of making them available for reaction with $^{235}$U and decreasing their chances of being captured by $^{238}$U.

### Control of Power Level

It is possible for a reactor to reach the critical stage ($K = 1$) after all the neutron losses described above are minimized. However, a method of control is needed to maintain a $K$ value near unity. If $K$ rises above this value, the heat produced in the runaway reaction would melt the reactor. To control the power level, control rods are inserted into the reactor core (Fig. 46.5). These rods are made of materials, such as cadmium, that are very efficient in absorbing neutrons. By adjusting the number and position of these control rods in the reactor core, the $K$ value can be varied and any power level within the design range of the reactor can be achieved.

A diagram of a pressurized-water reactor is shown in Figure 46.6. This type of reactor is commonly used in electric power plants in the United States. Fission events in the reactor core supply heat to the water contained in the primary (closed) loop and maintained at high pressure to keep it from boiling. This water also serves as the moderator. The hot water is pumped through a heat exchanger, and the heat is transferred to the water contained in the

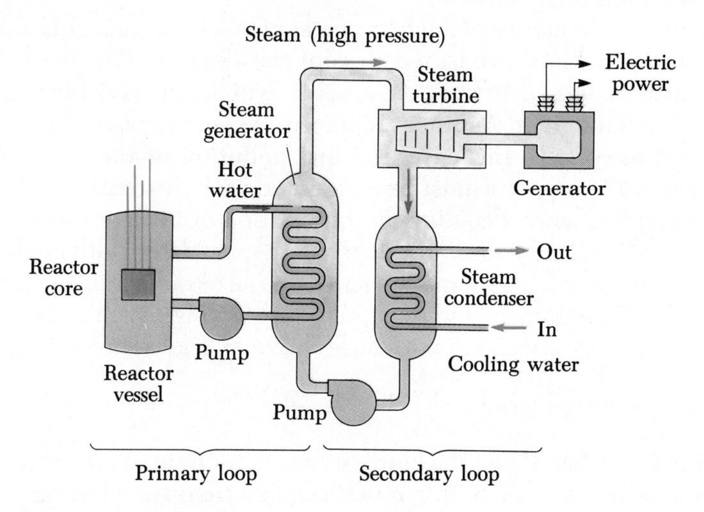


**Figure 46.6** Main components of a pressurized-water reactor.

secondary loop. The hot water in the secondary loop is converted to steam, which drives a turbine-generator system to create electric power. Note that the water in the secondary loop is isolated from the water in the primary loop in order to avoid contamination of the secondary water and steam by radioactive nuclei from the reactor core.

### Safety and Waste Disposal

There is justifiable concern about the safety of nuclear reactors and about their effect on the environment. This is particularly true in light of the 1979 near-disaster at Three Mile Island in Pennsylvania and the more recent tragedy at Chernobyl in the U.S.S.R. The problems of reactor safety are so vast and complex that we can only touch on a few of them here.

One of the inherent dangers in a nuclear reactor is the possibility that the water flow could be interrupted. If this should occur, it is conceivable that the temperature of the reactor could increase to the point where the fuel elements would melt, which would melt the bottom of the reactor and the ground below. This possibility has been referred to, appropriately enough, as the "China syndrome." Additionally, the large amounts of heat generated could lead to a high-pressure steam explosion (non-nuclear) that would spread radioactive material throughout the area surrounding the power plant. To decrease the chances of such an event as much as possible, all reactors are built with a backup cooling system that takes over if the regular cooling system fails.

Another problem of concern in nuclear fission reactors is the disposal of radioactive material when the reactor core is replaced. This waste material contains long-lived, highly radioactive isotopes and must be stored over long periods of time in such a way that there is no chance of environmental contamination. At present, sealing radioactive wastes in deep salt mines seems to be the most promising solution.

Other major concerns associated with the proliferation of nuclear power plants are the danger of sabotage at reactor sites and the danger that nuclear fuel (or waste) might be stolen during transport. In some instances, these stolen materials could be used to make an atomic bomb. The difficulties in handling and transporting such highly radioactive materials reduce the possibility of such terrorist activities.

Another consequence of nuclear power plants is thermal pollution. Water from a nearby river is often used to cool the reactor. This raises the river temperatures downstream from the reactor, which affects living organisms in and adjacent to the river. Another technique to cool the reactor water is to use evaporation towers. In this case, thermal pollution of the atmosphere is a consequence. Clearly, one must be concerned with the detrimental effects of thermal pollution when deciding on the location of nuclear power plants.

We have listed only a few of the problems associated with nuclear power reactors. Among these, the handling and disposal of radioactive wastes appear to be the chief problems. One must, of course, weigh such risks against the problems and risks associated with alternative energy sources.

## 46.4 NUCLEAR FUSION

In Chapter 45 we found that the binding energy for light nuclei (those having a mass number of less than 20) is much smaller than the binding energy for heavier nuclei (see Fig. 45.7). This suggests a possible process that is the reverse of fission. *When two light nuclei combine to form a heavier nucleus, the process is called* **nuclear fusion.** Because the mass of the final nucleus is less than the combined rest masses of the original nuclei, there is a loss of mass accompanied by a release of energy. The following are examples of such energy-liberating fusion reactions:

$$ {}^{1}_{1}\mathrm{H} + {}^{1}_{1}\mathrm{H} \longrightarrow {}^{2}_{1}\mathrm{H} + {}^{0}_{1}\mathrm{e} + \nu \qquad (46.4)$$

$$ {}^{1}_{1}\mathrm{H} + {}^{2}_{1}\mathrm{H} \longrightarrow {}^{3}_{2}\mathrm{He} + \gamma $$

This second reaction is followed by either

$$ {}^{1}_{1}\mathrm{H} + {}^{3}_{2}\mathrm{He} \longrightarrow {}^{4}_{2}\mathrm{He} + {}^{0}_{1}\mathrm{e} + \nu $$

or

$$ {}^{3}_{2}\mathrm{He} + {}^{3}_{2}\mathrm{He} \longrightarrow {}^{4}_{2}\mathrm{He} + {}^{1}_{1}\mathrm{H} + {}^{1}_{1}\mathrm{H} $$

These are the basic reactions in what is called the **proton-proton cycle,** believed to be one of the basic cycles by which energy is generated in the sun and other stars with an abundance of hydrogen. Most of the energy production takes place at the sun's interior, where the temperature is about $1.5 \times 10^7$ K. As we shall see later, such high temperatures are required in order to drive these reactions, and they are therefore called **thermonuclear fusion reactions.** The hydrogen (fusion) bomb, first exploded in 1952, is an example of an uncontrolled thermonuclear fusion reaction.

Thermonuclear reactions

All of the reactions in Equation 46.4 are exothermic, that is, there is a release of energy. An overall view of the proton-proton cycle is that four protons combine to form an alpha particle and two positrons, with the release of 25 MeV of energy in the process.

### Fusion Reactors

The enormous amount of energy released in fusion reactions suggests the possibility of harnessing this energy for useful purposes here on earth. A great deal of effort is currently under way to develop a sustained and controllable thermonuclear reactor—a fusion power reactor. Controlled fusion is often called the ultimate energy source because of the availability of its source of fuel: water. For example, if deuterium were used as the fuel, 0.12 g of it could be extracted from 1 gal of water at a cost of about four cents. Such rates would make the fuel costs of even an inefficient reactor almost insignificant. An additional advantage of fusion reactors is that comparatively few radioactive by-products are formed. As noted in Equation 46.4, the end product of the fusion of hydrogen nuclei is safe, nonradioactive helium. Unfortunately, a thermonuclear reactor that can deliver a net power output over a reasonable time interval is not yet a reality, and many difficulties must be resolved before a successful device is constructed.

We have seen that the sun's energy is based, in part, upon a set of reactions in which ordinary hydrogen is converted to helium. Unfortunately, the proton-proton interaction is not suitable for use in a fusion reactor because the event requires very high pressures and densities. The process works in the sun only because of the extremely high density of protons in the sun's interior.

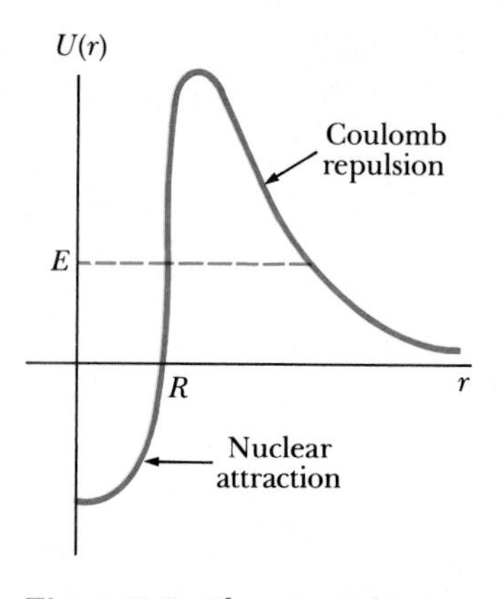


**Figure 46.7** The potential energy as a function of separation between two deuterons. The Coulomb repulsive force is dominant at long range, whereas the nuclear attractive force is dominant at short range, where $R$ is of the order of 1 fm. The two deuterons would require an energy $E$ greater than the height of the barrier to undergo fusion.

The fusion reactions that appear most promising in the construction of a fusion power reactor involve deuterium ($^2_1$H) and tritium ($^3_1$H), which are isotopes of hydrogen. These reactions are

$$\begin{aligned} {}^2_1\text{H} + {}^2_1\text{H} &\longrightarrow {}^3_2\text{He} + {}^1_0\text{n} \qquad Q = 3.27 \text{ MeV} \\ {}^2_1\text{H} + {}^2_1\text{H} &\longrightarrow {}^3_1\text{H} + {}^1_1\text{H} \qquad Q = 4.03 \text{ MeV} \\ {}^2_1\text{H} + {}^3_1\text{H} &\longrightarrow {}^4_2\text{He} + {}^1_0\text{n} \qquad Q = 17.59 \text{ MeV} \end{aligned} \tag{46.5}$$

where the $Q$ values represent the amount of energy released per reaction. As noted earlier, deuterium is available in almost unlimited quantities from our lakes and oceans and is very inexpensive to extract. Tritium, however, is radioactive ($T_{1/2} = 12.3$ years) and undergoes beta decay to $^3$He. For this reason, tritium does not occur naturally to any great extent and must be artificially produced.

One possible scheme for extracting energy from a fusion reactor is to surround the reactor core with a lithium blanket. Energetic particles, such as neutrons, would be absorbed by the molten lithium, and the thermal energy would then be transferred to a heat exchanger to generate steam. One advantage of this scheme is the formation of tritium from neutron-induced reactions in the lithium blanket; this tritium can be recirculated into the reactor for fuel. The two reactions involved in the breeding of tritium are as follows:

$$\begin{aligned} \text{n (fast)} + {}^7\text{Li} &\longrightarrow {}^3\text{H} + {}^4\text{He} + \text{n (slow)} \\ \text{n (slow)} + {}^6\text{Li} &\longrightarrow {}^3\text{H} + {}^4\text{He} + 4.8 \text{ MeV} \end{aligned} \tag{46.6}$$

One of the major problems in obtaining energy from nuclear fusion is the fact that the Coulomb repulsion force between two charged nuclei must be overcome before they can fuse. The potential energy as a function of particle separation for two deuterons (each with charge $+e$) is shown in Figure 46.7. The potential energy is positive in the region $r > R$, where the Coulomb repulsive force dominates, and negative in the region $r < R$, where the strong nuclear force dominates. The fundamental problem then is to give the two nuclei enough kinetic energy to overcome this repulsive force. This can be accomplished by heating the fuel to extremely high temperatures (to about $10^8$ K, far greater than the interior temperature of the sun). As you might expect, such high temperatures are not easy to obtain in the laboratory or a power plant. At these high temperatures, the atoms are ionized, and the system consists of a collection of electrons and nuclei, commonly referred to as a **plasma**. The following example illustrates how to estimate the temperature required to achieve the fusion of two deuterons.

High temperatures are required to overcome the large Coulomb barrier

**EXAMPLE 46.3 The Fusion of Two Deuterons**

The separation between two deuterons must be as little as about $10^{-14}$ m in order for the attractive nuclear force to overcome the repulsive Coulomb force. (a) Calculate the height of the potential barrier due to the repulsive force.

*Solution* The potential energy associated with two charges separated by a distance $r$ is given by

$$U = k\frac{q_1 q_2}{r}$$

where $k$ is the Coulomb constant. For the case of two deuterons, $q_1 = q_2 = +e$, so that

$$U = k\frac{e^2}{r} = \left(9 \times 10^9 \frac{\text{N}\cdot\text{m}^2}{\text{C}^2}\right)\frac{(1.6 \times 10^{-19}\text{ C})^2}{10^{-14}\text{ m}}$$

$$= 2.3 \times 10^{-14}\text{ J} = 0.14 \text{ MeV}$$

(b) Estimate the effective temperature required in order for a deuteron to overcome the potential barrier, assuming an energy of $\frac{3}{2}kT$ per deuteron (where in this case $k$ is Boltzmann's constant).

*Solution* Since the total Coulomb energy of the pair of deuterons is 0.14 MeV, the Coulomb energy per deuteron is 0.07 MeV = $1.2 \times 10^{-14}$ J. Setting this equal to the average thermal energy per deuteron gives

$$\tfrac{3}{2}kT = 1.2 \times 10^{-14}\ \mathrm{J}$$

where $k$ is equal to $1.38 \times 10^{-23}$ J/K. Solving for $T$ gives

$$T = \frac{2 \times (1.2 \times 10^{-14}\ \mathrm{J})}{3 \times (1.38 \times 10^{-23}\ \mathrm{J/K})} = 5.8 \times 10^{8}\ \mathrm{K}$$

Example 46.3 suggests that the deuterons must be heated to about $6 \times 10^{8}$ K in order to achieve fusion. This estimate of the required temperature is too high, however, because the particles in the plasma have a Maxwellian velocity distribution and therefore some fusion reactions will be caused by particles in the high-energy "tail" of this distribution. Furthermore, even those particles that do not have enough energy to overcome the barrier have some probability of penetrating the barrier through tunneling. When these effects are taken into account, a temperature of about $4 \times 10^{8}$ K appears adequate to fuse the two deuterons. As you might imagine, such high temperatures are not easy to attain in the laboratory.

Critical ignition temperature

The temperature at which the power generation rate exceeds the loss rate (due to mechanisms such as radiation losses) is called the **critical ignition temperature.** This temperature for the D-D (deuterium-deuterium) reaction is $4 \times 10^{8}$ K. According to $E \cong kT$, this temperature is equivalent to approximately 35 keV. It turns out that the critical ignition temperature for the D-T (deuterium-tritium) reaction is about $4.5 \times 10^{7}$ K, or only 4 keV. A plot of the power generated by fusion versus temperature for the two reactions is shown in Figure 46.8. The straight line in this plot represents the power loss due to the radiation mechanism known as **bremsstrahlung.** This is the principal mechanism of energy loss, in which radiation (primarily x-rays) is emitted as the result of electron-ion collisions within the plasma.[1]

Confinement time

In addition to the high temperature requirements, there are two other critical parameters that determine whether or not a thermonuclear reactor will be successful. These parameters are the **ion density,** $n$, and the **confinement time,** $\tau$. *The confinement time is the time the interacting ions are maintained at a temperature equal to or greater than the ignition temperature.* The density and confinement time must both be large enough to ensure that

[1] Cyclotron radiation is another loss mechanism; it is especially important in the case of the D-D reaction.

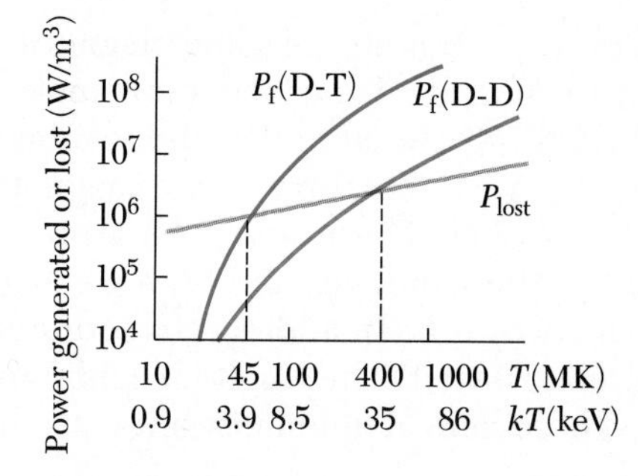


**Figure 46.8** The power generated or lost versus temperature for the D-D and D-T fusion reactions. When the generation rate $P_f$ exceeds the loss rate, ignition takes place.

more fusion energy will be released than is required to heat the plasma. In particular, **Lawson's criterion** states that a net energy output is possible under the following conditions:

Lawson's criterion

$$n\tau \geq 10^{14}\ \text{s/cm}^3 \qquad \text{(D-T)}$$
$$n\tau \geq 10^{16}\ \text{s/cm}^3 \qquad \text{(D-D)} \qquad (46.7)$$

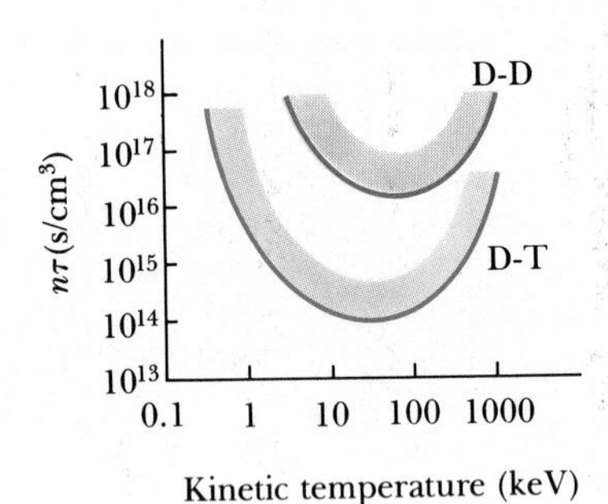


**Figure 46.9** The Lawson number $n\tau$ versus kinetic temperature for the D-T and D-D fusion reactions.

A graph of $n\tau$ versus kinetic temperature for the D-T and D-D reactions is given in Figure 46.9.

Lawson's criterion was arrived at by comparing the energy required to heat the plasma with the energy generated by the fusion process. The energy $E_h$ required to heat the plasma is proportional to the ion density $n$, whereas the energy *generated by the fusion process* is proportional to the product $n^2\tau$. Net energy is produced when the energy generated by the fusion process, $E_f$, exceeds $E_h$. This condition leads to Lawson's criterion.[2] The constants involved can be evaluated under suitable conditions. When $E_f = E_h$, a *break-even* point is reached. As of 1989, no laboratory has reported achieving the break-even condition.

In summary, the three basic requirements of a successful thermonuclear power reactor are as follows:

Requirements for a fusion power reactor

1. The plasma temperature must be very high—about $4.5 \times 10^7$ K for the D-T reaction and $4 \times 10^8$ K for the D-D reaction.
2. The ion density $n$ must be high. It is necessary to have a high density of interacting nuclei to increase the collision rate between particles.
3. The confinement time $\tau$ of the plasma must be long. In order to meet Lawson's criterion, the product $n\tau$ must be large. For a given value of $n$, the probability of fusion between two particles increases as the time of confinement increases.

Current efforts are aimed at meeting Lawson's criterion at temperatures exceeding the critical ignition temperature in one device. Although the minimum plasma densities have been achieved, the problem of confinement time has yet to be solved. How can one confine a plasma at a temperature of $10^8$ K for times of the order of 1 s? The two basic techniques under investigation to confine plasmas are magnetic field confinement and inertial confinement. Let us briefly describe these two techniques.

### Magnetic Field Confinement

Most fusion-related plasma experiments are using **magnetic field confinement** to contain the plasma. A toroidal device called a **tokamak,** first developed in the USSR, is shown in Figure 46.10. Note that the tokamak has a doughnut-shaped geometry (a toroid). A combination of two magnetic fields is used to confine and stabilize the plasma. These are (1) a strong toroidal field $\mathbf{B}_t$, produced by the current in the windings, and (2) a weaker "poloidal" field, produced by the toroidal current $I_t$. In addition to confining the plasma, the toroidal current is also used to heat it. The resultant confining field is helical, as shown in Figure 46.10. These helical field lines spiral around the plasma and

[2] The Lawson's criterion neglects the magnetic field energy, which is expected to be about 20 times greater than the plasma heat energy. For this reason, it is necessary to have a magnetic energy recovery system or to make use of superconducting magnets.

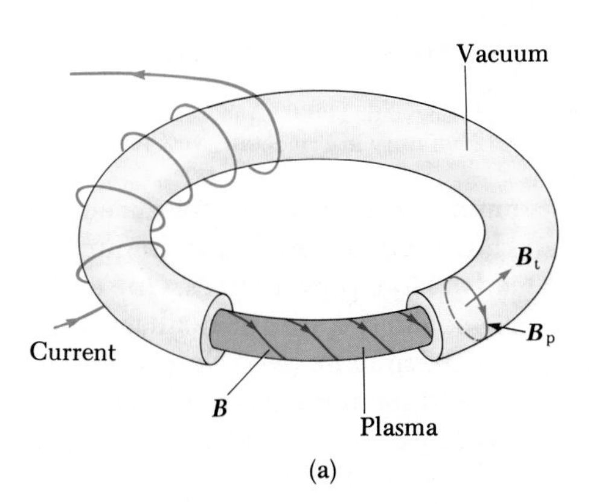

**Figure 46.10** (a) Schematic diagram of a tokamak used in the magnetic confinement scheme. Note that the total magnetic field $B$ is the superposition of the toroidal field $B_t$ and the poloidal field $B_p$. The plasma is trapped within the spiraling field lines as shown. (b) Photograph of the Princeton TFTR tokamak device, prior to the installation of the external structure components. (Courtesy of the Plasma Physics Laboratory at Princeton)

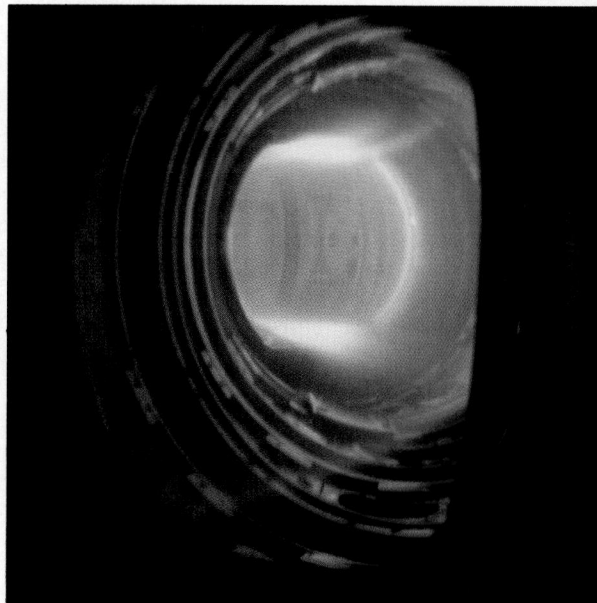

(Left) Alignment of the toroidal field coils of the TFTR machine. The coils are subject to large forces caused by their mutual magnetic interactions. (Right) The main limiter of the TFTR machine is brightly illuminated during an 800 kA, 1 s discharge within the vacuum vessel. (Courtesy of Princeton Plasma Physics Laboratory.)

**TABLE 46.1 Recent Parameters for Tokamaks[a]**

| | $n(\text{cm}^{-3})$ | $T$ (keV) | $\tau$(s) | $n\tau$ (s/cm³) |
|---|---|---|---|---|
| JET[b] | $3.5 \times 10^{13}$ | 25 | 1 | $3.5 \times 10^{13}$ |
| TFTR[b] | $2 \times 10^{13}$ | 34.6 | 1.4 | $3 \times 10^{13}$ |

[a] Cordey, Geoffrey, Robert Godston and Ronald Parker, "Progress Toward a Tokamak Fusion Reactor," *Physics Today*, January 1992.
[b] J.E.T. is the Joint European Torus located in Abington, England. TFTR is the Tokamak Fusion Test Reactor located at the Princeton Plasma Physics Laboratory in Princeton, New Jersey.

keep it from touching the walls of the vacuum chamber. If the plasma comes into contact with the walls, its temperature is reduced and heavy impurities sputtered from the walls "poison" it and lead to large power losses. One of the major breakthroughs in the last decade has been in the area of auxiliary heating to reach ignition temperatures. Recent experiments have shown that injecting a beam of energetic neutral particles into the plasma is a very efficient method of heating the plasma to ignition temperatures (5 to 10 keV). RF (radio frequency) heating will probably be needed for reactor-size plasmas.

In the last decade, dramatic progress has been made toward the development of a tokamak-based fusion reactor. Table 46.1 lists some recent characteristics of two major tokamaks. The TFTR reactor at Princeton has reported central ion temperatures of 34.6 keV (approximately $4 \times 10^8$ K), representing a fivefold increase since 1981. During this same period, plasma confinement times have increased from 0.02 s to about 1.4 s. Values for $n\tau$ for the D-T reaction are well above $10^{13}$ s/cm³ and close to the value required by Lawson's criterion (see Eq. 46.7). It is also important to note that in 1991, reaction rates of $6 \times 10^{17}$ D-T fusions per second have been reached in the JET tokamak at Abington, England, and $1 \times 10^{17}$ D-D fusions per second have been reported in the TFTR tokamak at Princeton. An international collaborative effort involving four major fusion programs is currently under way to design and build a fusion reactor called ITER (International Thermonuclear Experimental Reactor). This facility is designed to address the remaining technological and scientific issues that will establish the feasibility of fusion power.

A second magnetic confinement technique that has received much attention is the so-called magnetic mirror confinement scheme. The idea of this technique is to trap the plasma in a cylindrical tube by adding additional magnetic field coils at the ends of the tube (Fig. 46.11). The increased fields at the ends of the cylinder serve as magnetic "mirrors" (or a magnetic bottle) for the charged particles, which spiral around the field lines, thereby reducing the leakage problem and increasing the plasma density.

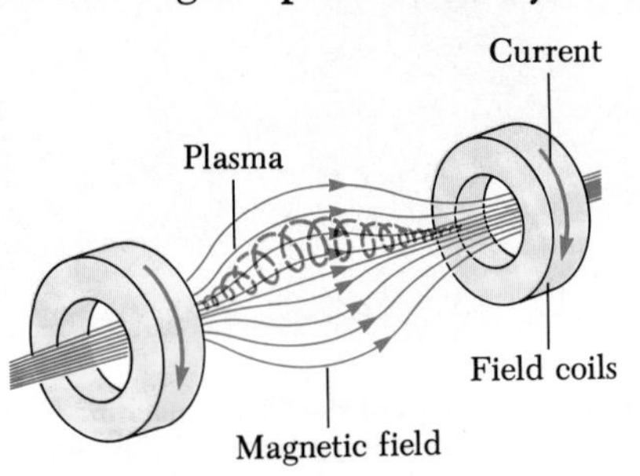


**Figure 46.11** Magnetic mirror confinement of a plasma.

### Inertial Confinement

The second technique for confining a plasma is called **inertial confinement.** This technique makes use of a target with a very high particle density. In this scheme, the confinement time is very short (typically $10^{-11}$ to $10^{-9}$ s), so that, because of their own inertia, the particles do not have a chance to move appreciably from their initial positions.

The laser fusion technique is the most common form of inertial confinement. A small D-T pellet, about 1 mm in diameter, is struck simultaneously by several focused, high-intensity laser beams, resulting in a large pulse of energy (Fig. 46.12). The energy from the laser pulse causes the surface of the fuel pellet to evaporate. The escaping particles produce a reaction force on the core of the pellet, resulting in a strong, inwardly moving, compressive shock wave. This shock wave increases the pressure and density of the core and produces a corresponding increase in temperature. When the temperature of the core reaches ignition temperature, fusion reactions cause the pellet to explode. The process can be viewed as a miniature hydrogen bomb. The SHIVA laser fusion project at the Lawrence Livermore Laboratory uses 20 synchronized laser pulses to deliver a total of 200 kJ of energy to a pellet in less than $10^{-9}$ s. This corresponds to a power of $2 \times 10^{14}$ W!

Advantages of fusion

If fusion power can be harnessed, it will offer several advantages: (1) the low cost and abundance of the fuel (deuterium), (2) the absence of weapons-grade material, (3) the impossibility of runaway accidents, and (4) a lesser radiation hazard than with fission. Some of the anticipated problem areas and disadvantages include (1) its as yet unestablished feasibility, (2) the very high proposed plant costs, (3) the possibility of the scarcity of lithium, (4) the

Problem areas and disadvantages of fusion

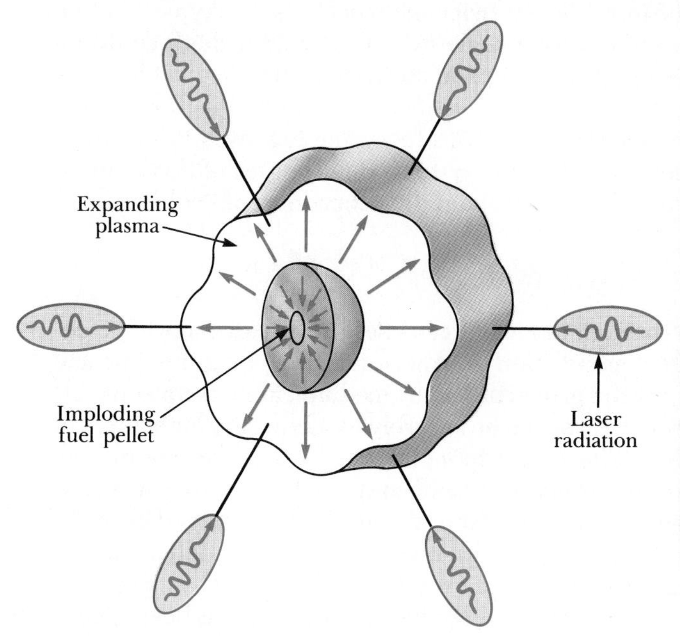


**Figure 46.12** In the inertial confinement scheme, a D-T fuel pellet undergoes fusion when struck by several high-intensity laser beams simultaneously.

This accelerator called Saturn, one of the most powerful x-ray sources in the world, has the potential of igniting a controlled laboratory fusion reaction. It uses up to 25 trillion watts of power to produce an electron beam current of $1.25 \times 10^{7}$ A. The energetic electrons are converted to x-rays at the center of the machine. The glow in this photograph is due to electrical discharge in underwater switches as the accelerator is fired. (Courtesy of Sandia National Laboratories.)

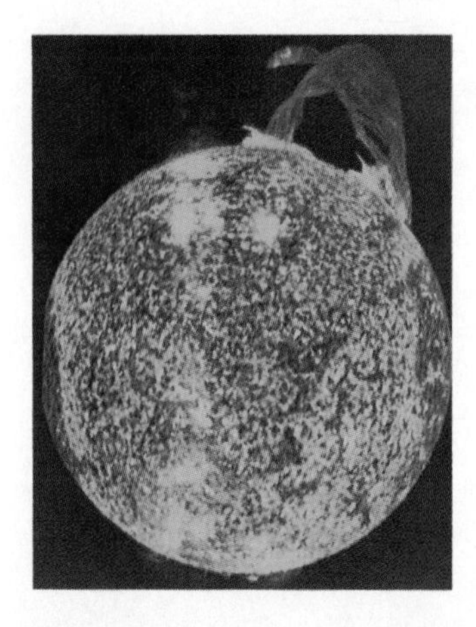

This photograph shows one of the most spectacular solar flares ever recorded; it spanned more than 588 000 km (367 000 mi) of the sun's surface. The flare gives the distinct impression of a twisted sheet of gas in the process of unwinding itself. During solar flares, which may occur several times a week, an enormous amount of energy is released, partly in the form of electromagnetic energy. (Courtesy of NASA)

limited supply of the helium and superconducting materials that are needed in the tokamak-type reactors, (5) structural damage and induced radioactivity due to neutron bombardment, and (6) the anticipated high degree of thermal pollution. If these basic problems and the engineering design factors can be resolved, nuclear fusion may become a feasible source of energy by the end of this century.

### Cold Fusion (?)

In March of 1989, chemists at the University of Utah reported experimental evidence of nuclear fusion in an electrolytic cell at room temperature. The cell consists of a palladium cathode surrounded by a platinum wire anode, immersed in a solution of lithium deuteroxide and 99.5% heavy water ($D_2O$). The researchers claimed that the cell produces heat at a rate which is four times greater than the input power. Since then, some groups have reported a net heat production in such cells, while others have not found any excess heat. However, net heat generation by itself does not prove that fusion has occurred. In order to confirm "cold fusion," one must be able to detect particles associated with the two most common reactions involving two deuterons (see Eq. 46.5). Physicists around the world have searched unsuccessfully for particles that would confirm "cold fusion." Neutron fluxes reported by a few groups appear to be far too low to explain the heat generated.[3]

In view of what has been discussed in this section regarding the conditions for fusion, how can "cold fusion" be possible? One possible explanation is that two closely separated deuterium nuclei trapped in the lattice could tunnel through the Coulomb barrier between them. The probability of tunneling could be significant if the internuclear spacing is much less than typical lattice separations. Other researchers have suggested that "cold fusion" could be caused by the acceleration of deuterons in strong internal electric fields at special lattice sites.

It is too early to come to any consensus regarding the mystery of events that occurs in the deuterium-palladium system. However, as of this writing, most physicists appear to be skeptical about the occurrence of "cold fusion."

## *46.5 RADIATION DAMAGE IN MATTER

When radiation passes through matter, it can cause severe damage. The degree and type of damage depend upon several factors, including the type and energy of the radiation and the properties of the medium. For example, metals used in nuclear reactor structures can be severely weakened by high fluxes of energetic neutrons, which often leads to metal fatigue. The damage in such situations is in the form of atomic displacements, often resulting in major alterations in the material properties. Materials can also be damaged by ionizing radiations, such as gamma rays and x-rays. For example, defects called *color centers* can be produced in inorganic crystals such as NaCl by irradiating the crystals with x-rays. One extensively studied color center has been identified as an electron trapped in a $Cl^-$ ion vacancy.

Radiation damage in biological organisms is primarily due to ionization effects in cells. The normal operation of a cell may be disrupted when highly

[3] *Physics Today*, June 1989, p. 17. This short article gives a review of the "current" status of "cold fusion."

reactive ions or radicals are formed as the result of ionizing radiation. For example, hydrogen and hydroxyl radicals produced from water molecules can induce chemical reactions that may break bonds in proteins and other vital molecules. Furthermore, the ionizing radiation may directly affect vital molecules by removing electrons from their structure. Large doses of radiation are especially dangerous, since damage to a great number of molecules in a cell may cause the cell to die. Although the death of a single cell is usually not a problem, the death of many cells may result in irreversible damage to the organism. Also, cells that do survive the radiation may become defective. These defective cells, upon dividing, can produce more defective cells and lead to cancer.

In biological systems, it is common to separate radiation damage into two categories, *somatic damage* and *genetic damage.* Somatic damage is radiation damage associated with all the body cells except the reproductive cells. Such damage can lead to cancer at high dose rates or seriously alter the characteristics of specific organisms. Genetic damage affects only the reproductive cells of the person exposed to the radiation. Damage to the genes in reproductive cells can lead to defective offspring. Clearly, one must be concerned about the effect of diagnostic treatments such as x-rays and other forms of radiation exposure.

There are several units used to quantify the amount, or *dose,* of any radiation that interacts with a substance.

The roentgen

The **roentgen** (R) is defined as that amount of ionizing radiation that will produce $\frac{1}{3} \times 10^{-9}$ C of electric charge in 1 $cm^3$ of air under standard conditions.

Equivalently, the roentgen is that amount of radiation that deposits an energy of $8.76 \times 10^{-3}$ J into 1 kg of air.

For most applications, the roentgen has been replaced by the *rad* (which is an acronym for *radiation absorbed dose*), defined as follows:

The rad

One **rad** is that amount of radiation that deposits $10^{-2}$ J of energy into 1 kg of absorbing material.

Although the rad is a perfectly good physical unit, it is not the best unit for measuring the degree of biological damage produced by radiation. This is because the degree of biological damage depends not only on the dose but also on the type of the radiation. For example, a given dose of alpha particles causes about ten times more biological damage than an equal dose of x-rays. The **RBE** (relative biological effectiveness) factor for a given type of radiation is defined as *the number of rad of x-radiation or gamma radiation that produces the same biological damage as 1 rad of the radiation being used.* The RBE factors for different types of radiation are given in Table 46.2. Note that the values are only approximate since they vary with particle energy and the form of the damage.

**TABLE 46.2 RBE[a] for Several Types of Radiation**

| Radiation | RBE Factor |
|---|---|
| X-rays and gamma rays | 1.0 |
| Beta particles | 1.0–1.7 |
| Alpha particles | 10–20 |
| Slow neutrons | 4–5 |
| Fast neutrons and protons | 10 |
| Heavy ions | 20 |

[a] RBE = relative biological effectiveness.

Finally, the **rem** (radiation equivalent in man) is defined as the product of the dose in rad and the RBE factor:

$$\text{dose in rem} \equiv \text{dose in rad} \times \text{RBE} \tag{46.8}$$

According to this definition, 1 rem of any two radiations will produce the same amount of biological damage. From Table 46.2, we see that a dose of 1 rad of fast neutrons represents an effective dose of 10 rem. On the other hand, 1 rad of gamma radiation is equivalent to a dose of 1 rem.

Low-level radiation from natural sources, such as cosmic rays and radioactive rocks and soil, delivers to each of us a dose of about 0.13 rem/year. The upper limit of radiation dose recommended by the U.S. government (apart from background radiation) is about 0.5 rem/year. Many occupations involve much higher radiation exposures, and so an upper limit of 5 rem/year has been set for combined whole-body exposure. Higher upper limits are permissible for certain parts of the body, such as the hands and forearms. A dose of 400 to 500 rem results in a mortality rate of about 50%. The most dangerous form of exposure is ingestion or inhalation of radioactive isotopes, especially those elements the body retains and concentrates, such as $^{90}$Sr. In some cases, a dose of 1000 rem can result from ingesting 1 mCi of radioactive material.

## *46.6 RADIATION DETECTORS

Various devices have been developed for detecting radiation. These devices are used for a variety of purposes, including medical diagnoses, radioactive dating measurements, and the measurement of background radiation.

Geiger counter

The **Geiger counter** (Fig. 46.13) is perhaps the most common device used to detect radiation. It can be considered the prototype of all counters that make use of ionization of a medium as the basic detection process. It consists of a cylindrical metal tube filled with gas at low pressure and a long wire along the axis of the tube. The wire is maintained at a high positive potential (about $10^3$ V) with respect to the tube. When a high-energy particle or photon enters the tube through a thin window at one end, some of the atoms of the gas can become ionized. The electrons removed from the atoms are attracted toward the positive wire, and in the process they ionize other atoms in their path. This results in an avalanche of electrons, which produces a current pulse at the output of the tube. After the pulse is amplified, it can be either used to trigger an electronic counter or delivered to a loudspeaker, which clicks each time a particle is detected.

Semiconductor diode detector

A **semiconductor diode detector** is essentially a reverse-bias *p-n* junction. Recall from Chapter 43 that a *p-n* junction diode passes current readily when forward-biased and prohibits the flow of current under reverse-bias condi-

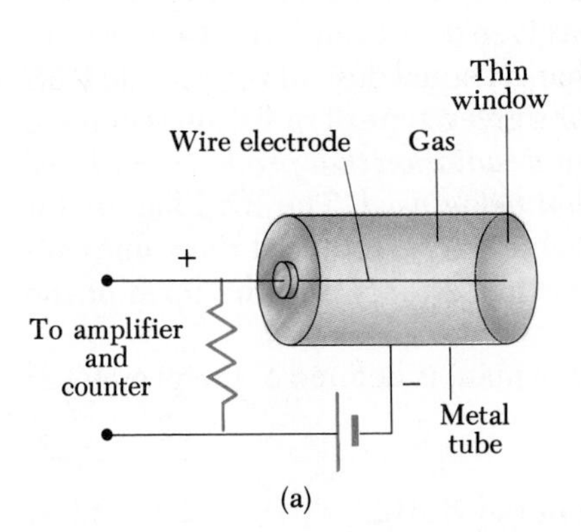


(a)

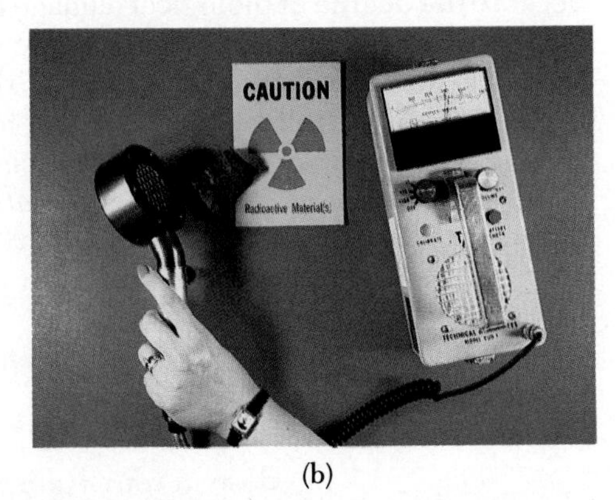


(b)

**Figure 46.13** (a) Diagram of a Geiger counter. The voltage between the central wire and the metal tube is usually about 1000 V. (b) Using a Geiger counter to measure the activity in a radioactive mineral. (Photo courtesy of Henry Leap and Jim Lehman.)

tions. As an energetic particle passes through the junction, electrons are excited into the conduction band and holes are formed in the valence band. The internal electric field sweeps the electrons toward the positive ($n$) side of the junction and the holes toward the negative ($p$) side. This creates a pulse of current, which can be measured with an electronic counter. In a typical device, the duration of the pulse is about $10^{-7}$ to $10^{-8}$ s.

Scintillation counter

**A scintillation counter** (Fig. 46.14) usually uses a solid or liquid material whose atoms are easily excited by the incoming radiation. These excited atoms emit visible light when they return to their ground state. Common materials used as scintillators are crystals of sodium iodide and certain plastics. If such a material is attached to one end of a device called a **photomultiplier** (PM) tube, the photons emitted by the scintillator can be converted to an electric signal. The PM tube consists of several electrodes, called *dynodes,* whose potentials are increased in succession along the length of the tube as shown in Figure

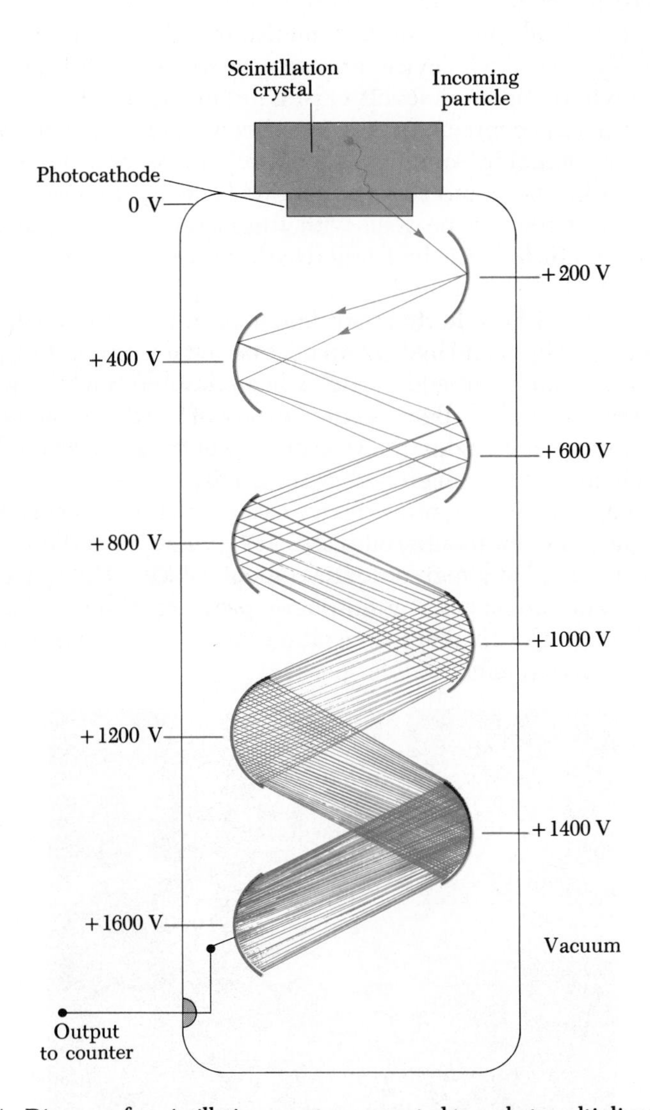


**Figure 46.14** Diagram of a scintillation counter connected to a photomultiplier tube.

46.14. The top of the tube contains a photocathode, which emits electrons by the photoelectric effect. As one of these emitted electrons strikes the first dynode, it has sufficient kinetic energy to eject several electrons. When these electrons are accelerated to the second dynode, many more electrons are ejected and an avalanche process occurs. The end result is 1 million or more electrons striking the last dynode. Hence, one particle striking the scintillator produces a sizable electric pulse at the output of the PM tube, and this pulse is in turn sent to an electronic counter. The scintillator device is much more sensitive than a Geiger counter, mainly because of the higher density of the detecting medium. It is especially sensitive to gamma rays, which interact more weakly with matter than do charged particles.

The devices described so far make use of ionization processes induced by energetic particles. Various devices can be used to view the tracks of charged particles directly. A **photographic emulsion** is the simplest example. A charged particle ionizes the atoms in an emulsion layer. The path of the particle corresponds to a family of points at which chemical changes have occurred in the emulsion. When the emulsion is developed, the particle's track becomes visible. Such devices are common in the film badges used in any environment where radiation levels must be monitored.

Cloud chamber

A **cloud chamber** contains a gas that has been supercooled to just below its usual condensation point. An energetic particle passing through ionizes the gas along its path. These ions serve as centers for condensation of the supercooled gas. The track can be seen with the naked eye and can be photographed. A magnetic field can be applied to determine the signs of the charges as they are deflected by the field.

Bubble chamber

A device called a **bubble chamber,** invented in 1952 by D. Glaser, makes use of a liquid (usually liquid hydrogen) maintained near its boiling point. Ions produced by incoming charged particles leave bubble tracks, which can be photographed (Fig. 46.15). Because the density of the detecting medium of a bubble chamber is much higher than the density of the gas in a cloud chamber, the bubble chamber has a much higher sensitivity.

A **spark chamber** is a counting device that consists of an array of conducting parallel plates. Even-numbered plates are grounded, and odd-numbered plates are maintained at a high potential (about 10 kV). The spaces between the plates contain a noble gas at atmospheric pressure. When a charged particle passes through the chamber, ionization occurs in the gas, resulting in a large surge of current and a visible spark.

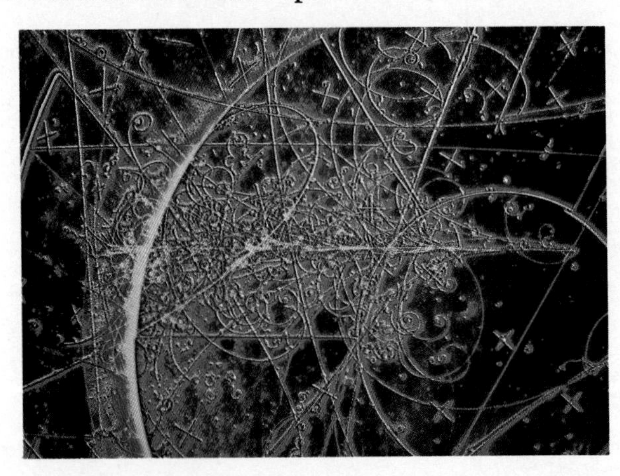

**Figure 46.15** Artificially colored bubble chamber photograph showing tracks of subatomic particles. (Courtesy of Photo Researchers, Inc./Science Photo Library)

## *46.7 USES OF RADIATION

Practical applications of nuclear physics are extremely widespread in manufacturing processes, medicine, and biology. Even a brief discussion of all the possibilities would fill an entire book, and to keep such a book up to date would require a number of revisions each year. In this section, we shall present a few of these applications and some of the underlying theories supporting them.

### Tracing

Radioactive particles can be used to trace chemicals participating in various reactions. One of the most valuable uses of radioactive tracers is in medicine. For example, $^{131}I$ is an artificially produced isotope of iodine (the natural, nonradioactive isotope is $^{127}I$). Iodine, which is a necessary nutrient for our bodies, is obtained largely through the intake of iodized salt and seafood. The thyroid gland plays a major role in the distribution of iodine throughout the body. In order to evaluate the performance of the thyroid, the patient drinks a very small amount of radioactive sodium iodide. Two hours later, the amount of iodine in the thyroid gland is determined by measuring the radiation intensity at the neck area.

A second medical application is indicated in Figure 46.16. Here a salt containing radioactive sodium is injected into a vein in the leg. The time at which the radioisotope arrives at another part of the body is detected with a radiation counter. The elapsed time is a good indication of the presence or absence of constrictions in the circulatory system.

The tracer technique is also useful in agricultural research. Suppose one wishes to determine the best method of fertilizing a plant. A certain element in a fertilizer, such as nitrogen, can be tagged with one of its radioactive isotopes. The fertilizer is then sprayed on one group of plants, sprinkled on the ground for a second group, and raked into the soil for a third. A Geiger counter is then used to track the nitrogen through the three types of plants.

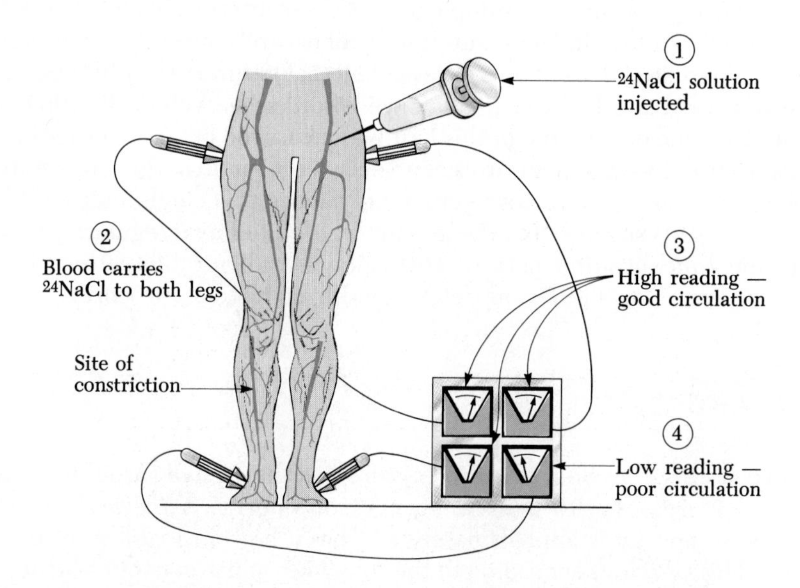


**Figure 46.16** A tracer technique for determining the condition of the human circulatory system.

Tracing techniques are as wide-ranging as human ingenuity can devise. Present applications range from checking the absorption of fluorine by teeth to checking contamination of food-processing equipment by cleansers to monitoring deterioration inside an automobile engine. In the latter case, a radioactive material is used in the manufacture of the pistons, and the oil is checked for radioactivity to determine the amount of wear on the pistons.

### Activation Analysis

For centuries, a standard method of identifying the elements in a sample of material has been chemical analysis, which involves testing a portion of the material for reactions with various chemicals. A second method is spectral analysis, which utilizes the fact that, when excited, each element emits its own characteristic set of electromagnetic wavelengths. These methods are now supplemented by a third technique, **neutron activation analysis.** Both chemical and spectral methods have the disadvantage that a fairly large sample of the material must be destroyed for the analysis. In addition, extremely small quantities of an element may go undetected by either method. Activation analysis has an advantage over the other two methods in both of these respects.

When the material under investigation is irradiated with neutrons, nuclei in the material will absorb the neutrons and be changed to different isotopes. Most of these isotopes will be radioactive. For example, $^{65}Cu$ absorbs a neutron to become $^{66}Cu$, which undergoes beta decay:

$$ {}_0^1n + {}_{29}^{65}Cu \longrightarrow {}_{29}^{66}Cu \longrightarrow {}_{30}^{66}Zn + {}_{-1}^{0}e \qquad (46.9)$$

The presence of the copper can be deduced because it is known that $^{66}Cu$ has a half-life of 5.1 min and decays with the emission of beta particles having maximum energies of 2.63 and 1.59 MeV. Also emitted in the decay of $^{66}Cu$ is a gamma ray having an energy of 1.04 MeV. Thus, by examining the radiation emitted by a substance after it has been exposed to neutron irradiation, one can detect extremely small traces of an element.

Neutron activation analysis is used routinely by a number of industries, but the following nonroutine example of its use is of interest. Napoleon died on the island of St. Helena in 1821, supposedly of natural causes. Over the years, suspicion has existed that his death was not all that natural. After his death, his head was shaved and locks of his hair were sold as souvenirs. In 1961, the amount of arsenic in a sample of this hair was measured by neutron activation analysis. Unusually large quantities of arsenic were found in the hair. (Activation analysis is so sensitive that very small pieces of a single hair could be analyzed.) Results showed that the arsenic was fed to him irregularly. In fact, the arsenic concentration pattern corresponded to the fluctuations in the severity of Napoleon's illness as determined from historical records.

## SUMMARY

The probability that neutrons are captured as they move through matter generally increases with decreasing neutron energy. A *thermal neutron* (one whose energy is approximately $kT$) has a high probability of being captured by a nucleus according to the following **neutron-capture process:**

Neutron capture

$$ {}_0^1n + {}_Z^AX \longrightarrow {}^{A+1}_{\;\;Z}X + \gamma \qquad (46.1)$$

Moderators

Energetic neutrons are slowed down readily in materials called **moderators.** In these materials, neutrons lose their energy mainly through elastic collisions.

Fission

The process of **nuclear fission** occurs when a very heavy nucleus, such as $^{235}U$, splits into two smaller fragments. Thermal neutrons can create fission in $^{235}U$ by the following process:

Fission of $^{235}U$

$$^{1}_{0}n + ^{235}_{92}U \longrightarrow ^{236}_{92}U^* \longrightarrow X + Y + \text{neutrons} \qquad (46.2)$$

where X and Y are the fission fragments and $^{236}U^*$ is a compound nucleus in an excited state. On the average, 2.47 neutrons are released per fission event. The fragments and neutrons have a great deal of kinetic energy following the fission event. The fragments then undergo a series of beta and gamma decays to various stable isotopes. The energy released per fission event is about 208 MeV.

Reproduction constant

The **reproduction constant** $K$ is defined as the average number of neutrons released from each fission event that will cause another event. In a power reactor, it is necessary to maintain a value of $K$ close to 1. The value of $K$ is affected by such factors as the reactor geometry, the mean energy of the neutrons, and the probability of neutron capture. Proper design of the reactor geometry is necessary to minimize neutron leakage from the reactor core. Neutron energies are regulated with a moderator material to slow down energetic neutrons and therefore increase the probability of neutron capture by other $^{235}U$ nuclei. The power level of the reactor is adjusted with control rods made of a material which is very efficient in absorbing neutrons. The value of $K$ can be adjusted by inserting the rods at various depths into the reactor core.

Fusion

**Nuclear fusion** is a process in which two light nuclei combine to form a heavier nucleus. A great deal of energy is released in such a process. The major obstacle in obtaining useful energy from fusion is the large Coulomb repulsive force between the charged nuclei at close separations. Sufficient energy must be supplied to the particles to overcome this Coulomb barrier and thereby enable the nuclear attractive force to take over. The temperature required to produce fusion is of the order of $10^8$ K. At such high temperatures, all matter is in the form of a **plasma,** which consists of positive ions and free electrons.

Critical ignition temperature

In a fusion reactor, the plasma temperature must reach at least the **critical ignition temperature,** which is the temperature at which the power generated by the fusion reactions *exceeds* the power lost in the system. The most promising fusion reaction is the D-T reaction, which has a critical ignition temperature of about $4.5 \times 10^7$ K. Two critical parameters involved in fusion reactor design are **ion density** $n$ **and confinement time** $\tau$. The confinement time is the time the interacting particles must be maintained at a temperature equal to or greater than the critical ignition temperature. **Lawson's criterion** states that for the D-T reaction, $n\tau \geq 10^{14}$ s/cm$^3$.

Confinement time

Lawson's criterion

## QUESTIONS

1. Explain the function of a moderator in a fission reactor.
2. Why is water a better shield against neutrons than lead or steel?
3. Why is molten sodium used as the primary coolant in a "breeder reactor"?
4. What is meant by "weapons-grade" fissile material?

5. Discuss the advantages and disadvantages of fission reactors from the point of view of safety, pollution, and resources. Make a comparison with power generated from the burning of fossil fuels.
6. In a fission reactor, nuclear reactions produce heat to drive a turbine generator. How is this heat produced?
7. Why would a fusion reactor produce less radioactive waste than a fission reactor?
8. Lawson's criterion states that the product of ion density and confinement time must exceed a certain number before a break-even fusion reaction can occur. Why should these two parameters determine the outcome?
9. Why is the temperature required for the D-T fusion less than that needed for the D-D fusion? Estimate the relative importance of Coulomb repulsion and nuclear attraction in each case.
10. What factors make a fusion reaction difficult to achieve?
11. Discuss the similarities and differences between fusion and fission.
12. Discuss the advantages and disadvantages of fusion power from the point of safety, pollution, and resources.
13. Discuss three major problems associated with the development of a controlled fusion reactor.
14. Describe two techniques that are being pursued in an effort to obtain power from nuclear fusion.
15. If two radioactive samples have the same activity measured in Ci, will they necessarily create the same damage to a medium? Explain.
16. Radiation is often used to sterilize such things as surgical equipment and packaged foods. Why do you suppose this works?
17. One method of treating cancer of the thyroid is to insert a small radioactive source directly into the tumor. The radiation emitted by the source can destroy cancerous cells. Very often, the radioactive isotope $^{131}_{53}I$ is injected into the bloodstream in this treatment. Why do you suppose iodine is used?
18. Why should a radiologist be extremely cautious about x-ray doses when treating pregnant women?
19. The design of a PM tube might suggest that any number of dynodes may be used to amplify a weak signal. What factors do you suppose would limit the amplification in this device?

## PROBLEMS

### Section 46.2 Nuclear Fission

1. Strontium-90 is a particularly dangerous fission product of $^{235}U$ because it is radioactive, and it substitutes for calcium in bones. What other direct fission products would accompany it in the neutron-induced fission of $^{235}U$? (*Note:* This reaction may release 2, 3, or 4 free neutrons.)
2. List the nuclear reactions required to "breed" fissile $^{239}Pu$ from nonfissile $^{238}U$ under fast neutron bombardment.
3. List the nuclear reactions required to "breed" fissile $^{233}U$ from nonfissile $^{232}Th$ under fast neutron bombardment.
4. (a) Find the energy released in the following fission reaction:

$$^{1}_{0}n + ^{235}_{92}U \longrightarrow ^{141}_{56}Ba + ^{92}_{36}Kr + 3(^{1}_{0}n)$$

The required masses are

$$M(^{1}_{0}n) = 1.008665 \text{ u}$$
$$M(^{235}_{92}U) = 235.043915 \text{ u}$$
$$M(^{141}_{56}Ba) = 140.9139 \text{ u}$$
$$M(^{92}_{36}Kr) = 91.8973 \text{ u}$$

(b) What fraction of the initial matter-energy of the system is lost?

5. Find the number of $^{6}Li$ and the number of $^{7}Li$ nuclei present in 2 kg of lithium. (The natural abundance of $^{6}Li$ is 7.5%; the remainder is $^{7}Li$.)
6. A typical nuclear fission power plant produces about 1000 MW of electrical power. Assume that the plant has an overall efficiency of 40% and that each fission produces 200 MeV of thermal energy. Calculate the mass of $^{235}U$ consumed each day.
7. Suppose enriched uranium containing 3.4% of the fissionable isotope $^{235}_{92}U$ is used as fuel for a ship. The water exerts an average frictional drag of $1.0 \times 10^5$ N on the ship. How far can the ship travel per kilogram of fuel? Assume that the energy released per fission event is 208 MeV and that the ship's engine has an efficiency of 20%.

### Section 46.3 Nuclear Reactors

8. In order to minimize neutron leakage from a reactor, the surface area-to-volume ratio should be a minimum for a given shape. For a given volume $V$, calculate this ratio for (a) a sphere, (b) a cube, and (c) a parallelepiped of dimensions $a \times a \times 2a$. (d) Which of these shapes would have the minimum leakage? Which would have the maximum leakage?
9. It has been estimated that there are $10^9$ tons of natural uranium available at concentrations exceeding 100 parts per million, of which 0.7% is $^{235}U$. If all the

world's energy needs ($7 \times 10^{12}$ J/s) were to be supplied by $^{235}U$ fission, how long would this supply last? (This estimate of uranium supply was taken from K. S. Deffeyes and I. D. MacGregor, *Scientific American*, January 1980, p. 66.)

## Section 46.4 Nuclear Fusion

**10.** Draw a vector diagram (showing $\boldsymbol{F}$, $\boldsymbol{v}$, and $\boldsymbol{B}$) for the Lorentz force acting on a positive ion spiralling toward one of the field coils in Figure 46.11. Note that the axial component of $\boldsymbol{F}$ reflects the ion back toward the center of the "magnetic bottle."

11. Inhomogeneities in the confining magnetic field make it difficult to contain plasmas. (a) Show that the magnetic field within a toroid is inversely proportional to $r$, where $r$ is the distance from the central axis of the toroid. (b) Show that the fractional inhomogeneity $(B_{max} - B_{min})/B_{avg}$ in a toroidal field with rectangular windings as shown in Figure 32.32 is approximately equal to $(b - a)/r$. (c) Evaluate the fractional inhomogeneity for a toroid with $b - a = 10$ cm, first with $r = 1$ m and then with $r = 10$ m.

**12.** Two nuclei with atomic numbers $Z_1$ and $Z_2$ approach each other with a total energy $E$. (a) If the minimum distance of approach for fusion to occur is $r = 10^{-14}$ m, find $E$ in terms of $Z_1$ and $Z_2$. (b) Calculate the minimum energy for fusion for the D-D and D-T reactions (the first and third reactions in Eq. 46.5).

13. To understand why containment of a plasma is necessary, consider the rate at which a plasma would be lost if it were not contained. (a) Estimate the rms speed of deuterons in a plasma at $4 \times 10^8$ K. (b) Estimate the time such a plasma would remain in a 10-cm cube if no steps were taken to contain it.

14. Of all the hydrogen nuclei in the ocean, 0.0156 of the mass is deuterium. The oceans have a volume of 317 million cubic miles. (a) If all the deuterium in the oceans were fused to $^4_2He$, how many joules of energy would be released? (b) Present world energy consumption is about $7 \times 10^{12}$ W. If consumption were 100 times greater, how many years would the energy calculated in (a) last?

15. It has been pointed out that fusion reactors are safe from explosion because there is never enough energy in the plasma to do much damage. (a) Using the data in Table 46.1, calculate the amount of energy stored in the plasma of the TFTR reactor. (b) How many kg of water could be boiled by this much energy? (The plasma volume of the TFTR reactor is about 50 m$^3$.)

16. In order to confine a stable plasma, the magnetic energy density in the magnetic field (Eq. 32.15) must exceed the pressure $2nkT$ of the plasma by a factor of at least 10. In the following, assume a confinement time $\tau = 1$ s. (a) Using Lawson's criterion, determine the required ion density. (b) From the ignition temperature criterion for the D-T reaction, determine the required plasma pressure. (c) Determine the magnitude of the magnetic field required to contain the plasma.

## *Section 46.5 Radiation Damage in Matter

**17.** A building has become accidentally contaminated with radioactivity. The longest-lived material in the building is strontium-90 ($^{90}_{38}Sr$ has an atomic mass 89.9077 and its half-life is 28.8 yr). If the building initially contained 1.0 kg of this substance and the safe level is less than 10.0 counts/min, how long will the building be unsafe?

**18.** A particular radioactive source produces 100 mrad of 2-MeV gamma rays per hour at a distance of 1.0 m. (a) How long could a person stand at this distance before accumulating an intolerable dose of 1 rem? (b) Assuming the gamma radiation is emitted isotropically, at what distance would a person receive a dose of 10 mrad/h from this source?

**19.** Assume that an x-ray technician takes an average of eight x-rays per day and receives a dose of 5 rem/year as a result. (a) Estimate the dose in rem per x-ray taken. (b) How does this result compare with low-level background radiation?

**20.** In terms of biological damage, how many rad of heavy ions is equivalent to 10 rad of x-rays?

21. A "clever" technician decides to heat some water for his coffee with an x-ray machine. If the machine produces 10 rad/s, how long will it take to raise the temperature of a cup of water by 50 C°?

**22.** Two workers using an industrial x-ray machine accidentally insert their hands in the x-ray beam for the same length of time. The first worker inserts one hand in the beam, and the second worker inserts both hands. Which worker received the larger dose in rad?

23. Technetium-99 is used in certain medical diagnostic procedures. If $\frac{1}{100}$ $\mu$g of $^{99}Tc$ is injected into a 60-kg patient and $\frac{1}{2}$ of the 0.14 MeV gamma rays are absorbed in the body, determine the total radiation dose received by the patient. (1 MeV $= 1.6 \times 10^{-13}$ J.)

**24.** A person whose mass is 75 kg is exposed to a dose of 20 rad. How many joules of energy are deposited in the person's body?

## *Section 46.6 Radiation Detectors

25. In a Geiger tube, the voltage between the electrodes is typically 1 kV and the current pulse discharges a 5-pF capacitor. (a) What is the energy amplification of this device for a 0.5-MeV beta ray? (b) How many electrons are avalanched by the initial electron?

26. In a PM tube, assume that there are seven dynodes with potentials of 100 V, 200 V, 300 V, . . . , 700 V. The average energy required to free an electron from

the dynode surface is 10 eV. For each incident electron, how many electrons are freed (a) at the first dynode and (b) at the last dynode? (c) What is the average energy supplied by the tube for each electron? (Assume an efficiency of 100%.)

## ADDITIONAL PROBLEMS

27. The probability of a given nuclear reaction increases dramatically above the "Coulomb barrier," which is the electrostatic potential energy of the two nuclei when their surfaces just touch. Compute the Coulomb barrier for the absorption of an alpha particle by a gold nucleus.
28. Compare the fractional mass-energy loss in a typical $^{235}U$ fission reaction with the fractional mass-energy loss in D-T fusion.
29. The half-life of tritium is 12 years. If the TFTR fusion reactor contains 50 $m^3$ of tritium at a density equal to $2.0 \times 10^{14}$ particles/$cm^3$, how many Ci of tritium are in the plasma? Compare this with a fission inventory of $4 \times 10^{10}$ Ci.
30. (a) Estimate the volume of space required to store the radioactive wastes that would be produced in one year if all the annual U.S. electricity production (which is about $2.2 \times 10^{12}$ kWh/year) came from uranium enriched to 3% $^{235}U$. (Assume that the conversion efficiency is 30% and that the waste is in the form of a liquid with a density of 1 g/$cm^3$.) (b) If the waste could be formed into a cube, what would be the length of the cube's sides?
31. A 2-MeV neutron is emitted in a fission reactor. If it loses one half of its kinetic energy in each collision with a moderator atom, how many collisions must it undergo in order to achieve thermal energy (0.039 eV)?
32. Consider a nucleus with $A = 176$, $R = 6.7$ fm, and $Z = 71$. (a) How much electric energy must be put in when the nucleus is split into its constituent neutrons and protons? (b) How much total energy input is required to effect the splitting? (*Hints:* Use the result of Chapter 45, Problem 72, and the semiempirical binding energy formula.)
33. About 1 of every 6500 water molecules contains one deuterium atom. (a) If all the deuterium atoms in 1 liter of water are fused in pairs according to the DD reaction $^2H + {}^2H \rightarrow {}^3He + n + 3.3$ MeV, how many joules of energy would be liberated? (b) Burning gasoline produces about $3.4 \times 10^7$ J/L. Compare the energy obtainable from the fusion of the deuterium in a liter of water with the energy liberated from the burning of a liter of gasoline.
34. The $\alpha$ emitter polonium-210 ($^{210}_{84}Po$) is used in a nuclear battery. Determine the initial power output of the battery if it contains 0.155 kg of $^{210}Po$. Assume that the efficiency for conversion of energy to electricity is 1.0%.
35. A particle cannot generally be localized to distances much smaller than its de Broglie wavelength. This means that a slow neutron appears to be larger to a target particle than does a fast neutron, in the sense that the slow neutron will probably be found over a large volume of space. For a thermal neutron at room temperature (300 K), find (a) the linear momentum and (b) the de Broglie wavelength. Compare this effective neutron size with both nuclear and atomic dimensions.
36. Consider a nucleus at rest, which then splits into two fragments, of masses $m_1$ and $m_2$. Show that the fraction of the total kinetic energy that is carried by $m_1$ is

$$\frac{K_1}{K_{tot}} = \frac{m_2}{m_1 + m_2}$$

and the fraction carried by $m_2$ is

$$\frac{K_2}{K_{tot}} = \frac{m_1}{m_1 + m_2}$$

assuming relativistic corrections can be ignored. (*Note:* If the parent nucleus was moving before the decay, then $m_1$ and $m_2$ still divide the kinetic energy as shown, as long as all velocities are measured in the center-of-mass frame of reference, in which the total momentum of the system is zero.)
37. Assuming that a deuteron and a triton are at rest when they fuse according to $^2H + {}^3H \rightarrow {}^4He + n + 17.6$ MeV, determine the kinetic energy acquired by the neutron.
38. A patient swallows a radiopharmaceutical tagged with phosphorus-32 ($^{32}_{15}P$), a $\beta^-$ emitter. The average kinetic energy of the electrons is 700 keV. If the initial activity is 5.22 MBq, determine the absorbed dose during a 10-day period. Assume the electrons are completely absorbed in 10.0 g of tissue. (*Hint:* Find the number of electrons emitted.)
39. (a) Calculate the energy (in kWh) released if 1 kg of $^{239}Pu$ undergoes complete fission and the energy released per fission event is 200 MeV. (b) Calculate the energy (in MeV) released in the D-T fusion:

$$^2_1H + {}^3_1H \longrightarrow {}^4_2He + {}^1_0n$$

(c) Calculate the energy (in kWh) released if 1 kg of deuterium undergoes fusion. (d) Calculate the energy (in kWh) released by the combustion of 1 kg of coal if each $C + O_2 \rightarrow CO_2$ reaction yields 4.2 eV. (e) List the advantages and disadvantages of each of these methods of energy generation.
40. The sun radiates energy at the rate of $4 \times 10^{23}$ kW. If the reaction

$$4({}^1_1H) \longrightarrow {}^4_2He + 2\beta^+ + 2\nu + \gamma$$

accounted for all the energy released, calculate (a) the number of protons fused per second and (b) the mass transformed into energy per second.

41. Suppose the target in a laser fusion reactor is a sphere of solid hydrogen, with a diameter of $1.5 \times 10^{-4}$ m and a density of 0.2 g/cm$^3$. Also assume that half of the nuclei are $^2$H and half are $^3$H. (a) If 1% of a 200-kJ laser pulse is delivered to this sphere, what temperature will the sphere reach? (b) If all of the hydrogen "burns" according to the D-T reaction, how many joules of energy will be released?

42. In a tokamak fusion reactor, suppose a 5-keV deuteron moves at an angle of 20° to the toroidal magnetic field. Assume that $B_t = 1.2$ T and $B_p = 0$. (a) Calculate the components of velocity parallel and perpendicular to $B_t$. (b) What is the radius of the spiral motion for the deuteron? (c) How far does the deuteron travel *along* the magnetic field before it completes one revolution around the magnetic field?

43. Consider the two nuclear reactions

$$\text{(I) A} + \text{B} \longrightarrow \text{C} + \text{E}$$

$$\text{(II) C} + \text{D} \longrightarrow \text{F} + \text{G}$$

(a) Show that the net $Q$ for these two reactions ($Q_{net} = Q_I + Q_{II}$) is identical to the $Q$ for the reaction

$$\text{A} + \text{B} + \text{D} \longrightarrow \text{E} + \text{F} + \text{G}$$

(b) One chain of reactions in the proton-proton cycle in the sun's interior is the following:

$${}^1_1\text{H} + {}^1_1\text{H} \longrightarrow {}^2_1\text{H} + {}^0_1\text{e} + \nu$$

$${}^1_1\text{H} + {}^2_1\text{H} \longrightarrow {}^3_2\text{He} + \gamma$$

$${}^1_1\text{H} + {}^3_2\text{He} \longrightarrow {}^4_2\text{He} + {}^0_1\text{e} + \nu$$

Based on part (a), what is $Q_{net}$ for this sequence of three reactions?

44. The carbon cycle, first proposed by Hans Bethe in 1939, is another cycle by which energy is released in stars and hydrogen is converted to helium. The carbon cycle requires higher temperatures than the proton-proton cycle. The series of reactions is

$${}^{12}\text{C} + {}^1\text{H} \longrightarrow {}^{13}\text{N} + \gamma$$

$${}^{13}\text{N} \longrightarrow {}^{13}\text{C} + \beta^+ + \nu$$

$${}^{13}\text{C} + {}^1\text{H} \longrightarrow {}^{14}\text{N} + \gamma$$

$${}^{14}\text{N} + {}^1\text{H} \longrightarrow {}^{15}\text{O} + \gamma$$

$${}^{15}\text{O} \longrightarrow {}^{15}\text{N} + \beta^+ + \nu$$

$${}^{15}\text{N} + {}^1\text{H} \longrightarrow {}^{12}\text{C} + {}^4\text{He}$$

(a) If the proton-proton cycle requires a temperature of $1.5 \times 10^7$ K, estimate the temperature required for the first step in the carbon cycle. (b) Calculate the $Q$ value for each step in the carbon cycle and the overall energy released. (c) Do you think the energy carried off by the neutrinos is deposited in the star? Explain.

45. When high-energy photons (x-rays or gamma rays) pass through matter, the intensity $I$ of the beam (measured in W/m$^2$) decreases exponentially according to

$$I = I_0 e^{-\mu x}$$

where $I_0$ is the intensity of the incident beam, and $I$ is the intensity of the beam that just passed through a thickness $x$ of material. The constant $\mu$ is known as the linear absorption coefficient and its value depends on the absorbing material and the wavelength of the photon beam. This wavelength (or energy) dependence allows us to filter out unwanted wavelengths from a broad-spectrum x-ray beam. (a) Two x-ray beams of wavelengths $\lambda_1$ and $\lambda_2$ and equal incident intensities pass through the same metal plate. Show that the ratio of the emergent beam intensities will be

$$\frac{I_2}{I_1} = e^{-[\mu_2 - \mu_1]x}$$

(b) Compute the ratio of intensities emerging from a 1-mm thick aluminum plate if the incident beam contains equal intensities of 50 pm and 100 pm x-rays. The values of $\mu$ for aluminum at these two wavelengths are given below:

$$\lambda_1 = 50 \text{ pm} \qquad \mu_1 = 5.4 \text{ cm}^{-1}$$

$$\lambda_2 = 100 \text{ pm} \qquad \mu_2 = 41.0 \text{ cm}^{-1}$$

(c) Repeat for a 10-mm aluminum plate.

# 47

# Particle Physics and Cosmology

*An artist's version of a high-energy particle colliding with a nucleus. The quark structure of the nucleus is indicated by the small colored spheres inside the nucleus. (Courtesy of Janie Martz/CEBAF)*

In this concluding chapter, we shall examine the properties and classifications of the various known subatomic particles and the fundamental interactions that govern their behavior. We shall also discuss the current theory of elementary particles in which all matter in nature is believed to be constructed from only two families of particles, quarks and leptons. Finally, we shall discuss how clarifications of such models might help scientists understand the evolution of the universe.

## 47.1 INTRODUCTION

The word "atom" is from the Greek word *atomos*, which means indivisible. At one time atoms were thought to be the indivisible constituents of matter; that is, they were regarded to be elementary particles. Discoveries in the early part of the 20th century revealed that the atom is not elementary, but has as its constituents protons, neutrons, and electrons. Until 1932, physicists viewed all matter as consisting of only three constituent particles: electrons, protons, and neutrons. With the exception of the free neutron, these particles are very stable. Beginning in 1945, many new particles were discovered in experiments involving high-energy collisions between known particles. These new particles are characteristically very unstable and have very short half-lives, ranging between $10^{-6}$ and $10^{-23}$ s. So far more than 300 of these unstable, temporary particles have been catalogued.

During the last 30 years, many powerful particle accelerators have been constructed throughout the world, making it possible to observe collisions of particles with greater violence under controlled laboratory conditions, so as to reveal the subatomic world in finer detail. Up until the 1960s, physicists were bewildered by the large number and variety of subatomic particles being discovered. They wondered if the particles were like animals in a zoo with no systematic relationship connecting them, or whether a pattern was emerging that would provide a better understanding of the elaborate structure in the subnuclear world. In the last two decades, physicists have made tremendous advances in our knowledge of the structure of matter by recognizing that all particles (with the exception of electrons, photons, and a few related particles) are made of smaller particles called quarks. Thus, protons and neutrons, for example, are not truly elementary but are systems of tightly bound quarks. The quark model has reduced the bewildering array of particles to a manageable number, and has been successful in predicting new quark combinations later found in many experiments.

## 47.2 THE FUNDAMENTAL FORCES IN NATURE

The key to understanding the properties of elementary particles is to be able to describe the forces between them. All particles in nature are subject to four fundamental forces: strong, electromagnetic, weak, and gravitational.

The **strong force** is very short-ranged and is responsible for the binding of neutrons and protons into nuclei. This force represents the "glue" that holds the nucleons together and is the strongest of all the fundamental forces. The strong force is very short-ranged and is negligible for separations greater than about $10^{-15}$ m (which is about the size of the nucleus). The **electromagnetic force,** which is about $10^{-2}$ times the strength of the strong force, is responsible

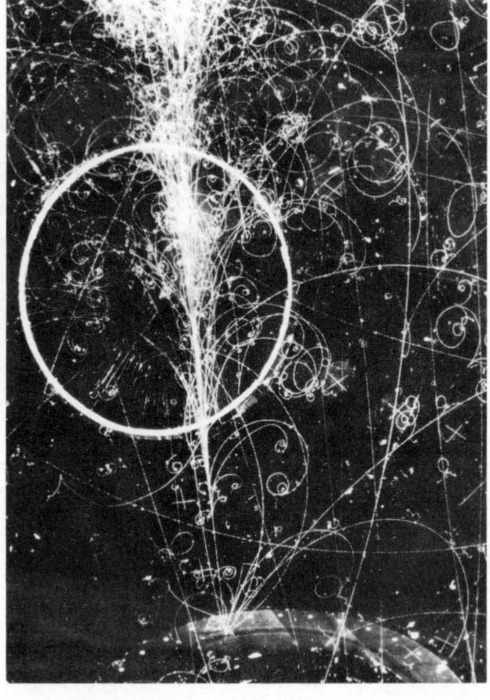

*(Left)* The main tunnel at Fermilab, now housing both the older ring of conventional magnets (red and blue, above) and the new ring of superconducting magnets (yellow, below). In its present mode of operation as the Tevatron, the superconducting ring receives injections of protons and antiprotons from the conventional ring. The two beams are accelerated in opposite directions around the ring to their final energy of 1 TeV and then collide together at interaction points. The Tevatron is the most powerful accelerator in the world today. ***(Right)*** Photograph of a particle interaction in the 15-ft bubble chamber at Fermilab. The photographs showing particle "tracks" are studied by scanners and experimenters. (Photographs courtesy of Fermi National Accelerator Laboratory)

**TABLE 47.1 Particle Interactions**

| Interaction (Force) | Relative Strength | Range of Force | Mediating Field Particle |
|---|---|---|---|
| Strong | 1 | Short ($\approx 1$ fm) | Gluon |
| Electromagnetic | $10^{-2}$ | Long ($\propto 1/r^2$) | Photon |
| Weak | $10^{-13}$ | Short ($\approx 10^{-3}$ fm) | $W^{\pm}$, Z |
| Gravitational | $10^{-38}$ | Long ($\propto 1/r^2$) | Graviton |

for the binding of atoms and molecules. It is a long-range force that decreases in strength as the inverse square of the separation between interacting particles. The **weak force** is a short-range nuclear force that tends to produce instability in certain nuclei. It is responsible for most radioactive decay processes such as beta decay, and its strength is only about $10^{-13}$ times that of the strong force. (As we shall discuss later, scientists now believe that the weak and electromagnetic forces are two manifestations of a single force called the electroweak force). Finally, the **gravitational force** is a long-range force that has a strength of only about $10^{-38}$ times that of the strong force. Although this familiar interaction is the force that holds the planets, stars, and galaxies together, its effect on elementary particles is negligible. Thus, the gravitational force is the weakest of all the fundamental forces.

In modern physics, one often describes the interactions between particles in terms of the exchange of field particles or quanta. In the case of the familiar electromagnetic interaction, the field particles are photons. In the language of modern physics, one can say that the electromagnetic force is *mediated* by photons, which are the quanta of the electromagnetic field. Likewise, the strong force is mediated by field particles called *gluons,* the weak force is mediated by particles called the W and Z *bosons,* and the gravitational force is mediated by quanta of the gravitational field called *gravitons.* These interactions, their ranges, and their relative strengths are summarized in Table 47.1.

## 47.3 POSITRONS AND OTHER ANTIPARTICLES

In the 1920s, the theoretical physicist Paul Adrien Maurice Dirac (1902–1984) developed a version of quantum mechanics that incorporated special relativity. Dirac's theory was successful in explaining the origin of the electron's spin and its magnetic moment. However, Dirac was faced with one major difficulty in this theory. His relativistic wave equation required solutions corresponding to negative energy states.[1] But if negative energy states existed, one would expect an electron in a state of positive energy to make a rapid transition to one of these states, emitting a photon in the process. Dirac was able to avoid this difficulty by postulating that all negative energy states were filled. Those electrons that occupy the negative energy states are called the "Dirac sea." Electrons in the Dirac sea are not directly observable because the Pauli exclusion principle does not allow them to react to external forces. However, if one of these negative energy states is vacant, leaving a hole in the sea of filled states, the hole can react to external forces and would be observ-

[1] P.A.M. Dirac, *The Principles of Quantum Mechanics,* 3rd ed., New York, Oxford University Press, Chapter 11, 1947.

able. (This is analogous to the behavior of a hole in the valence band of a semiconductor.) *The profound implication of this theory was that for every particle, there was also an antiparticle.* The antiparticle would have the same mass as the particle, but their charges would be opposite each other. For example, the electron's antiparticle (now called a *positron*) would have a mass of 0.511 MeV, and a positive charge of $1.6 \times 10^{-19}$ C. Usually we shall designate an antiparticle with a bar over the symbol for the particle. Thus, the positron is denoted by $\bar{e}$ (although sometimes the notation $e^+$ is preferred), the antiproton is denoted by $\bar{p}$, and the antineutrino is denoted by $\bar{\nu}$.

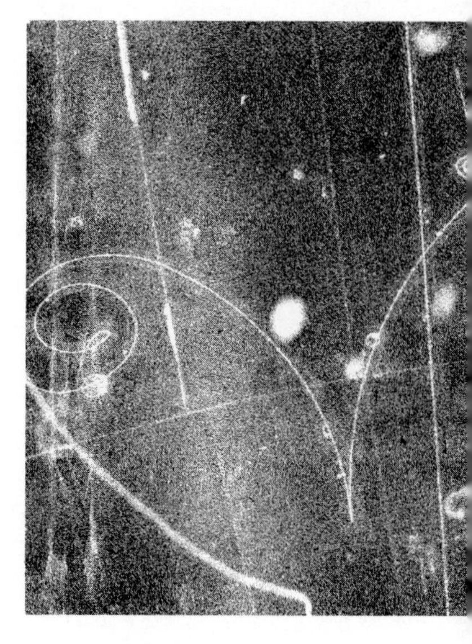

Pair creation in a strong magnetic field. (Fermilab Photo)

The positron was discovered by Carl Anderson in 1932 (the same year that the neutron was discovered), and in 1936 he was awarded the Nobel Prize for his discovery. Anderson made his discovery while examining tracks created by electronlike particles of positive charge in a cloud chamber. (These early experiments used cosmic rays—mostly energetic protons passing through interstellar space—to initiate high-energy reactions of the order of several GeV.) In order to discriminate between positive and negative charges, the cloud chamber was placed in a magnetic field, causing moving charges to follow curved paths. Anderson noted that some of the electronlike tracks deflected in a direction corresponding to a positively charged particle.

Since Anderson's initial discovery, the positron has been observed in a number of experiments. Perhaps the most common process for producing positrons is **pair production.** In this process, a gamma ray with sufficiently high energy collides with a nucleus and an electron-positron pair is created. Since the total rest energy of the electron-positron pair is $2mc^2 = 1.02$ MeV (where $m_0$ is the rest mass of the electron), the gamma ray must have at least this much energy to create an electron-positron pair. Thus, electromagnetic energy in the form of a gamma ray is transformed into mass in accordance with Einstein's famous relation $E = mc^2$. Figure 47.1 shows tracks of electron-positron pairs created by 300-MeV x-rays striking a lead sheet.

A process that is the reverse of pair production can also occur. Under the proper conditions, an electron and positron can annihilate and produce two

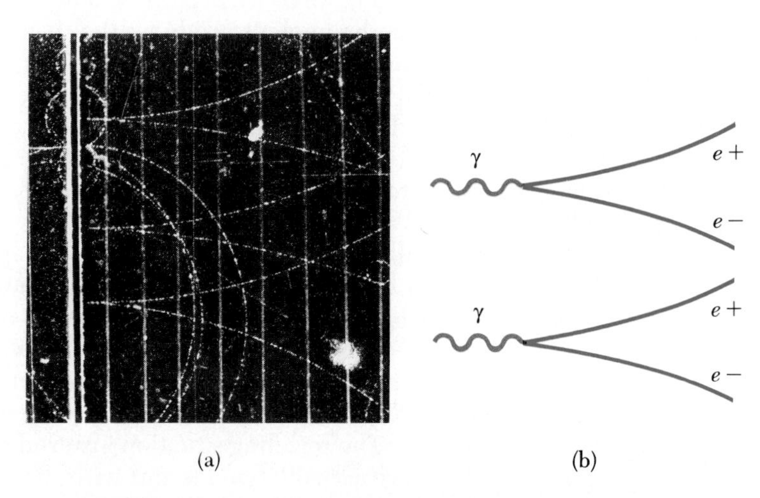



**Figure 47.1** (a) Bubble-chamber tracks of electron-positron pairs produced by 300-MeV $\gamma$-rays striking a lead sheet. (Courtesy of Lawrence Berkeley Laboratory, University of California) (b) Sketch of the pertinent pair-production events. Note that the positrons deflect upward while the electrons deflect downward in an applied magnetic field that is directed into the diagram.

photons that have a combined energy of at least 1.02 MeV. The reaction can be expressed as

$$e + \bar{e} \longrightarrow 2\gamma$$

Likewise, a proton-antiproton pair can annihilate to produce two gamma rays; however, this event is very rare.

Practically every known elementary particle has an antiparticle. Among the exceptions are the photon and the neutral pion ($\pi^0$). (Note that the $\pi^0$ and $\eta^0$ are their own antiparticles!) Following the construction of high-energy accelerators in the 1950s, many other antiparticles were discovered. These included the antiproton ($\bar{p}$) discovered by Emilio Segre and Owen Chamberlain in 1955, and the antineutron ($\bar{n}$) discovered shortly thereafter.

## 47.4 MESONS AND THE BEGINNING OF PARTICLE PHYSICS

The structure of matter as viewed by physicists in the mid-1930s was fairly simple. The building blocks of matter were considered to be the proton, the electron, and the neutron. Three other particles were known at the time: the gamma particle (photon), the neutrino, and the positron. These six particles were considered to be the fundamental constituents of matter. Although the accepted picture of the world was marvelously simple, no one was able to provide an answer to the following important question. If the nucleus of an atom contains many charged protons in close proximity which should strongly repel each other due to their like charges, what is the nature of the force that holds the nucleus together? Scientists recognized that this mysterious force must be much stronger than anything encountered in nature up to the time.

The first theory to explain the nature of the strong force was proposed in 1935 by the Japanese physicist Hideki Yukawa (1907–1981), an effort that later earned him the Nobel Prize. In order to understand Yukawa's theory, it is useful to first recall that *two atoms can form a covalent chemical bond by the exchange of electrons.* Similarly, in the modern views of electromagnetic interactions, *charged particles interact through the exchange of photons.* Yukawa used this same idea to explain the strong force by proposing a new particle whose exchange between nucleons in the nucleus produces the strong force with a range of about $10^{-15}$ m (the order of the nuclear diameter). Furthermore, he established that the range of the force is inversely proportional to the mass of this carrier particle, and predicted that the mass would be about 200 times the mass of the electron. Since the new particle would have a mass between that of the electron and proton, it was called a **meson** (from the Greek *meso,* meaning "middle").

In an effort to substantiate Yukawa's predictions, physicists began an experimental search for the meson by studying cosmic rays that enter the earth's atmosphere from interstellar space. In 1937, Carl Anderson and his collaborators discovered a particle whose mass was 106 MeV/$c^2$, which is about 207 times the mass of the electron. However, subsequent experiments showed that the particle interacted very weakly with matter, and hence could not be the carrier of the strong force. The puzzling situation inspired several theoreticians to propose that there are actually two mesons with slightly different masses. This idea was confirmed in 1947 with the discovery in cosmic rays of the pi meson ($\pi$), or simply *pion,* by Cecil Frank Powell (1903–1969) and Guiseppe P. S. Occhialini (1907–      ). The lighter meson discovered earlier by Anderson, now called a *muon* ($\mu$), has only weak and electromagnetic interaction and plays no role in the strong interaction.

The pion comes in three varieties, corresponding to three charge states: $\pi^+$, $\pi^-$, and $\pi^0$. The $\pi^+$ and $\pi^-$ particles have masses of 139.6 MeV/$c^2$, while the $\pi^0$ has a mass of 135.0 MeV/$c^2$. Pions and muons are very unstable particles. For example, the $\pi^-$ first decays into a muon and an antineutrino with a mean lifetime of about $2.6 \times 10^{-8}$ s. The muon then decays into an electron, a neutrino, and an antineutrino with a mean lifetime of $2.2 \times 10^{-6}$ s. The sequence of decays is

$$\pi^- \longrightarrow \mu^- + \bar{\nu} \qquad (47.1)$$

$$\mu^- \longrightarrow e + \nu + \bar{\nu}$$

The interaction between two particles can be represented in a simple diagram called a *Feynman diagram,* developed by Richard P. Feynman (1918–1988). Figure 47.2 is a Feynman diagram for the case of the electromagnetic interaction between two electrons. In this simple case, a single photon acts as the carrier of the electromagnetic force between the electrons. The photon transfers energy and momentum from one electron to the other in this interaction. These photons are called *virtual photons* because they can never be detected directly. This is because the photon is absorbed by the second electron very shortly after it is emitted by the first electron. The virtual photons violate the law of conservation of energy, but because of the uncertainty principle and the very short lifetime $\Delta t$, the photon's excess energy is less than the uncertainty in its energy, given by $\Delta E \approx \hbar/\Delta t$.

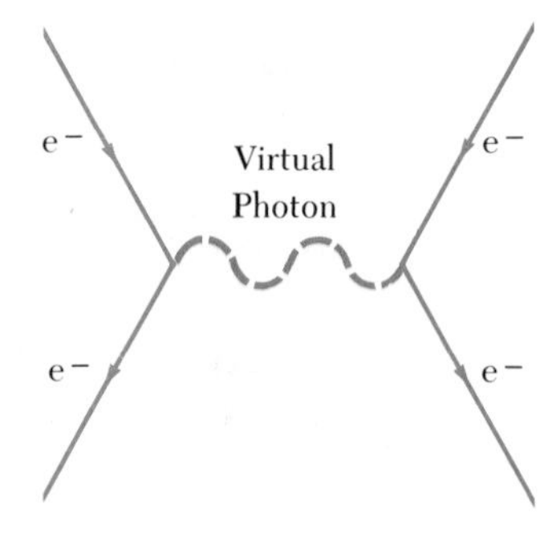

**Figure 47.2** Feynman diagram showing how a photon mediates the electromagnetic force between two interacting electrons.

Now consider the pion exchange between a proton and a neutron via the strong force. The Feynman diagram for this interaction is shown in Figure 47.3. One can reason that the energy $\Delta E$ needed to create a pion of mass $m_\pi$ is given by Einstein's equation $\Delta E = m_\pi c^2$. Again, the very existence of the pion violates conservation of energy by an amount, $\Delta E$, which is permitted by the uncertainty principle only if this energy is surrendered in a time, $\Delta t$, the time it takes the pion to transfer between nucleons. From the uncertainty principle, $\Delta E\, \Delta t \approx \hbar$, we get

$$\Delta t \approx \frac{\hbar}{\Delta E} = \frac{\hbar}{m_\pi c^2} \qquad (47.2)$$

Since the pion cannot travel faster than the speed of light, the maximum distance $d$ it can travel in a time $\Delta t$ is $c\,\Delta t$. Using Equation 47.2, and $d = c\,\Delta t$, we find this maximum distance to be

$$d \approx \frac{\hbar}{m_\pi c} \qquad (47.3)$$

From Chapter 45, we know that the range of the strong force has a value of about $1.5 \times 10^{-15}$ m. Using this value for $d$ in Equation 47.3, the rest energy of the pion is calculated to be

$$m_\pi c^2 \approx \frac{\hbar c}{d} = \frac{(1.05 \times 10^{-34}\ \text{J}\cdot\text{s})(3 \times 10^8\ \text{m/s})}{1.5 \times 10^{-15}\ \text{m}}$$

$$= 2.1 \times 10^{-11}\ \text{J} \cong 130\ \text{MeV}$$

This corresponds to a mass of 130 MeV/$c^2$ (about 250 times the mass of the electron), which is in good agreement with the observed mass of the pion.

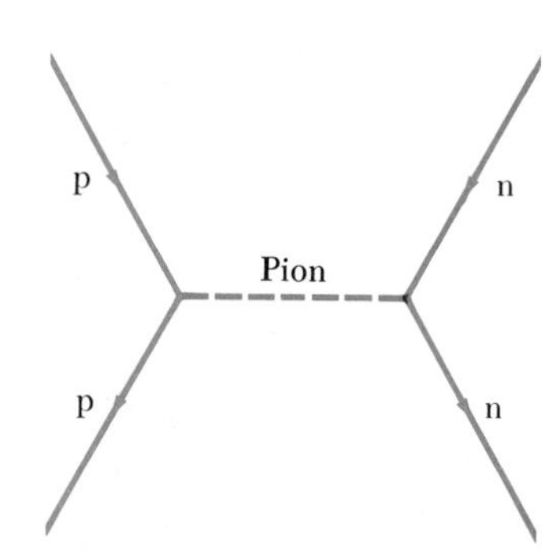

**Figure 47.3** Feynman diagram representing a proton interacting with a neutron via the strong force. In this case, the pion mediates the strong force.

# Biographical Sketch

## Richard P. Feynman
*(1918–1988)*

**Richard Phillips Feynman** was a brilliant theoretical physicist who together with Julian S. Schwinger and Shinichiro Tomonaga shared the 1965 Nobel Prize for Physics for their fundamental work in the principles of quantum electrodynamics. His many important contributions to physics include the invention of simple diagrams to represent particle interactions graphically, the theory of the weak interaction of subatomic particles, a reformulation of quantum mechanics, the theory of superfluid helium, and his contribution to physics education through the magnificent three-volume text, *The Feynman Lectures on Physics.*

Feynman did his undergraduate work at MIT and received his PhD in 1942 from Princeton University where he worked under John Archibald Wheeler. During World War II, he worked on the Manhattan project at Princeton and then at Los Alamos, New Mexico. He then joined the faculty at Cornell University in 1945 and was appointed professor of physics at California Institute of Technology in 1950 where he remained for the rest of his career.

It is well known that Feynman had a passion for finding new and better ways to formulate each problem, or, as he would say, "turning it around." In the early part of his career, he was fascinated with electrodynamics, and developed an intuitive view of quantum electrodynamics. Convinced that the electron could not interact with its own field, he said, "That was the beginning, and the idea seemed so obvious to me that I fell deeply in love with it . . . ." Often called the outstanding intuitionist of our age, he said in his Nobel acceptance speech, "Often, even in a physicist's sense, I did not have a demonstration of how to get all of these rules and equations, from conventional electrodynamics . . . I never really sat down, like Euclid did for the geometers of Greece, and made sure that you could get it all from a single set of axioms."

In 1986, Feynman was a member of the presidential commission to investigate the explosion of the space shuttle *Challenger*. In this capacity, he performed a simple experiment for the commission members which showed that one of the shuttle's O-ring seals was the likely cause of the disaster. After placing a seal in a pitcher of ice water, and squeezing it with a clamp, he demonstrated that the seal failed to spring back into shape once the clamp was removed.[1]

Feynman worked in physics with a style commensurate with his personality, that is, with energy, vitality, and humor. The following quotes from some of his colleagues are characteristic of the great impact he made on the scientific community.[2]

"A brilliant, vital, and amusing neighbor, Feynman was a stimulating (if sometimes exasperating) partner in discussions of profound issues—we would exchange ideas and silly jokes in between bouts of mathematical calculation—we struck sparks off each other, and it was exhilarating." *Murray Gell-Mann*

"Reading Feynman is a joy and a delight, for in his papers, as in his talks, Feynman communicated very directly, as though the reader were watching him derive the results at the blackboard." *David Pines*

"He loved puzzles and games. In fact, he saw all the world as a sort of game, whose progress of "behavior" follows certain rules, some known, some unknown . . . Find places or circumstances where the rules don't work, and invent new rules that do." *David L. Goodstein*

"Feynman was not a theorist's theorist, but a physicist's physicist and a teacher's teacher." *Valentine L. Telegdi*

Laurie M. Brown, one of his graduate students at Cornell, noted that Feynman, a playful showman, was "undervalued at first because of his rough manners, who in the end triumphs through native cleverness, psychological insight, common sense and the famous Feynman humor. . . . Whatever else Dick Feynman may have joked about, his love for physics approached reverence."

[1] Feynman's own account of this inquiry can be found in *Physics Today,* 4:26, February 1988.

[2] For more on Feynman's life and contributions, see the numerous articles in a special memorial issue of *Physics Today* 42, February 1989. For a personal account of Feynman, see his popular autobiographical books, *Surely You're Joking Mr. Feynman,* New York, Bantam Books, 1985, and *What Do You Care What Other People Think,* New York, W.W. Norton & Co., 1987.

The concept we have just described is quite revolutionary, and something we have not encountered as yet. In effect, it says that a proton can change into a proton plus a pion, as long as it returns to its original state in a very short time. High energy physicists often say that a nucleon undergoes "fluctuations" as it emits and absorbs pions. As we have seen, these fluctuations are a consequence of a combination of quantum mechanics (through the uncertainty principle) and special relativity (through Einstein's energy-mass relation $E = mc^2$).

This section has dealt with the particles which mediate the strong force, namely the pions, and the mediators of the electromagnetic force, photons. The graviton, which is the mediator of the gravitational force, has yet to be observed. The particles which mediate the weak nuclear force are referred to as $W^+$, $W^-$, and $Z^0$. The discovery of the $W^\pm$ and $Z^0$ particles at CERN was announced in 1983 by Carlo Rubbia (1934– ) and his associates using a proton-antiproton collider.[2] In this accelerator, protons and antiprotons that have a momentum of 270 GeV/$c$ undergo head-on collisions with each other. In some of the collisions, $W^\pm$ and $Z^0$ particles are produced, which are, in turn, identified by their decay products.

## 47.5 CLASSIFICATION OF PARTICLES

### Hadrons

All particles other than photons can be classified into two broad categories, hadrons and leptons, according to the interactions they experience. Particles that interact through the strong force are called *hadrons*. There are two classes of hadrons, known as *mesons* and *baryons*. These can be classified according to their masses and spins.

Mesons all have zero or integral spins (0 or 1), with masses that lie between the mass of the electron and the mass of the proton. All mesons are known to decay finally into electrons, positrons, neutrinos, and photons. The pion is the lightest of known mesons, with a mass of about 140 MeV/$c^2$ and a spin of 0. Another is the K meson, with a mass of about 500 MeV/$c^2$ and spin 0.

Baryons, which are the second class of hadrons, have a mass equal to or greater than the proton mass (hence the name *baryon*, which means *heavy* in Greek), and their spin is always a noninteger value (1/2 or 3/2). Protons and neutrons are included in the baryon family, as are many other particles. With the exception of the proton, all baryons decay in such a way that the end products include a proton. For example, the $\Xi$ hyperon first decays to a $\Lambda^0$ in about $10^{-10}$ s. The $\Lambda^0$ particle then decays to a proton and a $\pi^-$ in about $3 \times 10^{-10}$ s.

Today it is believed that hadrons are composed of more elemental units called *quarks*. Later we shall have more to say about the quark model. Some of the important properties of hadrons are listed in Table 47.2.

### Leptons

Leptons (from the Greek *leptos* meaning *small* or *light*) are a group of particles that participate in the weak interaction. All leptons have a spin of 1/2. In-

[2] Carlos Rubbia, an Italian physicist, and Simon van der Meer, a Dutch physicist, both at CERN, shared the 1984 Nobel Prize in Physics for the discovery of the $W^\pm$ and $Z^0$ particles and the development of the proton-antiproton collider.

**TABLE 47.2 A Table of Some Particles and Their Properties**

| Category | Particle Name | Symbol | Anti-particle | Rest Mass (MeV/$c^2$) | $B$ | $L_e$ | $L_\mu$ | $L_\tau$ | S | Lifetime (s) | Principal Decay Modes[a] |
|---|---|---|---|---|---|---|---|---|---|---|---|
| Photon | Photon | $\gamma$ | Self | 0 | 0 | 0 | 0 | 0 | 0 | Stable | |
| Leptons | Electron | $e^-$ | $e^+$ | 0.511 | 0 | +1 | 0 | 0 | 0 | Stable | |
| | Neutrino (e) | $\nu_e$ | $\bar{\nu}_e$ | 0(?) | 0 | +1 | 0 | 0 | 0 | Stable | |
| | Muon | $\mu^-$ | $\mu^+$ | 105.7 | 0 | 0 | +1 | 0 | 0 | $2.20 \times 10^{-6}$ | $e^-\bar{\nu}_e\nu_\mu$ |
| | Neutrino ($\mu$) | $\nu_\mu$ | $\bar{\nu}_\mu$ | 0(?) | 0 | 0 | +1 | 0 | 0 | Stable | |
| | Tau | $\tau^-$ | $\tau^+$ | 1784 | 0 | 0 | 0 | −1 | 0 | $<4 \times 10^{-13}$ | $\mu^-\bar{\nu}_\mu\nu_\tau$, $e^-\bar{\nu}_e\nu_t$, hadrons |
| | Neutrino ($\tau$) | $\nu_\tau$ | $\bar{\nu}_\tau$ | 0(?) | 0 | 0 | 0 | −1 | 0 | Stable | |
| Hadrons | | | | | | | | | | | |
| Mesons | Pion | $\pi^+$ | $\pi^-$ | 139.6 | 0 | 0 | 0 | 0 | 0 | $2.60 \times 10^{-8}$ | $\mu^+\nu_\mu$ |
| | | $\pi^0$ | Self | 135.0 | 0 | 0 | 0 | 0 | 0 | $0.83 \times 10^{-16}$ | $2\gamma$ |
| | Kaon | $K^+$ | $K^-$ | 493.7 | 0 | 0 | 0 | 0 | +1 | $1.24 \times 10^{-8}$ | $\mu^+\nu_\mu$, $\pi^+\pi^0$ |
| | | $K_S^0$ | $K_S^0$ | 497.7 | 0 | 0 | 0 | 0 | +1 | $0.89 \times 10^{-10}$ | $\pi^+\pi^-$, $2\pi^0$ |
| | | $K_L^0$ | $K_L^0$ | 497.7 | 0 | 0 | 0 | 0 | +1 | $5.2 \times 10^{-8}$ | $\pi^\pm e^\mp \overset{(-)}{\nu}_e$<br>$\pi^\pm \mu^\mp \overset{(-)}{\nu}_\mu$<br>$3\pi^0$ |
| | Eta | $\eta^0$ | Self | 548.8 | 0 | 0 | 0 | 0 | 0 | $<10^{-18}$ | $2\gamma$, $3\mu$ |
| Baryons | Proton | p | $\bar{p}$ | 938.3 | +1 | 0 | 0 | 0 | 0 | Stable | |
| | Neutron | n | $\bar{n}$ | 939.6 | +1 | 0 | 0 | 0 | 0 | 920 | $pe^-\bar{\nu}_e$ |
| | Lambda | $\Lambda^0$ | $\bar{\Lambda}^0$ | 1115.6 | +1 | 0 | 0 | 0 | −1 | $2.6 \times 10^{-10}$ | $p\pi^-$, $n\pi^0$ |
| | Sigma | $\Sigma^+$ | $\bar{\Sigma}^-$ | 1189.4 | +1 | 0 | 0 | 0 | −1 | $0.80 \times 10^{-10}$ | $p\pi^0$, $n\pi^+$ |
| | | $\Sigma^0$ | $\bar{\Sigma}^0$ | 1192.5 | +1 | 0 | 0 | 0 | −1 | $6 \times 10^{-20}$ | $\Lambda^0\gamma$ |
| | | $\Sigma^-$ | $\bar{\Sigma}^+$ | 1197.3 | +1 | 0 | 0 | 0 | −1 | $1.5 \times 10^{-10}$ | $n\pi^-$ |
| | Xi | $\Xi^0$ | $\bar{\Xi}^0$ | 1315 | +1 | 0 | 0 | 0 | −2 | $2.9 \times 10^{-10}$ | $\Lambda^0\pi^0$ |
| | | $\Xi^-$ | $\Xi^+$ | 1321 | +1 | 0 | 0 | 0 | −2 | $1.64 \times 10^{-10}$ | $\Lambda^0\pi^-$ |
| | Omega | $\Omega^-$ | $\Omega^+$ | 1672 | +1 | 0 | 0 | 0 | −3 | $0.82 \times 10^{-10}$ | $\Xi^0\pi^-$, $\Lambda^0K^-$ |

[a] A notation in this column such as $p\pi^-$, $n\pi^0$ means two possible decay modes. In this case, the two possible decays are $\Lambda^0 \rightarrow p + \pi^-$ or $\Lambda^0 \rightarrow n + \pi^0$.

cluded in this group are electrons, muons, and neutrinos, which are all less massive than the lightest hadron. Although hadrons have size and structure, leptons appear to be truly elementary particles with no structure (that is, pointlike).

Quite unlike the situation with hadrons, the number of known leptons is very limited. Currently, scientists believe there only are six leptons (each having an antiparticle): the electron, the muon, the tau, and a neutrino associated with each of these particles. (Note that the neutrino associated with the tau has not yet been observed in the laboratory.) The $\tau$ (tau) lepton, discovered in 1975, has a mass equal to about twice that of the proton. Associated with this heavy lepton is the $\tau$ neutrino ($\nu_\tau$). We now classify the six known leptons into three groups:

$$\begin{pmatrix} e^- \\ \nu_e \end{pmatrix} \quad \begin{pmatrix} \mu^- \\ \nu_\mu \end{pmatrix} \quad \begin{pmatrix} \tau^- \\ \nu_\tau \end{pmatrix}$$

Although neutrinos are thought to be massless, there is a possibility that they may have a small, nonzero mass. As we shall see later, a firm knowledge of the neutrino's mass could have great significance in cosmological models and the future of the universe.

## 47.6 CONSERVATION LAWS

In Chapter 45 we learned that conservation laws are important in understanding why certain decays or reactions occur and others do not. In general, the laws of conservation of energy, linear momentum, angular momentum, and electric charge provide us with a set of rules that all processes must follow. For example, conservation of electric charge requires that the total charge before a reaction must equal the total charge after the reaction.

A number of new conservation laws are important in the study of elementary particle decays and reactions. Two of these described in this section are the conservation of baryon number and the conservation of lepton number. Although these conservation laws have no theoretical foundation, they are supported by an abundance of empirical evidence.

### Baryon Number

The conservation of baryon number implies that whenever a baryon is created in a reaction or decay, an antibaryon is also created. This can be quantified by assigning a baryon number $B = +1$ to all baryons, $B = -1$ to all antibaryons, and $B = 0$ to all other particles. Thus, the **law of conservation of baryon number** can be stated as follows:

> Whenever a nuclear reaction or decay occurs, the sum of the baryon numbers before the process must equal the sum of the baryon numbers after the process.

An equivalent statement is that the net number of baryons remains constant in any process.

Note that if baryon number is absolutely conserved, the proton must be absolutely stable. If it were not for the law of conservation of baryon number, the proton could decay to a positron and a neutral pion. However, such a decay has never been observed. At the present, we can say only that the proton has a half-life of at least $10^{31}$ years (which could be compared with the estimated age of the universe, which is about $10^{10}$ years). In one recent version of the recent grand unified theory, or GUT, physicists have predicted that the proton is actually unstable. According to this theory, the baryon number (sometimes called the baryonic charge) cannot be absolutely conserved, whereas electric charge is always conserved.

**EXAMPLE 47.1 Checking Baryon Numbers**
Determine whether or not each of the following reactions can occur based on the law of conservation of baryon number.

$$p + n \longrightarrow p + p + n + \bar{p} \quad (1)$$

$$p + n \longrightarrow p + p + \bar{p} \quad (2)$$

*Solution* First, let us check reaction (1). Recall that $B = +1$ for baryons and $B = -1$ for antibaryons. Hence the left side of (1) gives a total baryon number of $1 + 1 = 2$. The right side of (1) gives a total baryon number of $1 + 1 + 1 + (-1) = 2$. Hence the reaction can occur provided the incoming proton has sufficient energy.

Now let us examine reaction (2). The left side of (2) again gives a total baryon number of $1 + 1 = 2$. However, the right side of (2) gives a total number of $1 + 1 + (-1) = 1$. Since the baryon number is not conserved, the reaction cannot occur.

### Lepton Number

Recall that there are three varieties of leptons: the electron, the muon, and the tau lepton. Each of these is accompanied by a neutrino. There are three separate conservation laws involving lepton numbers, one for each variety of lepton. The **law of conservation of electron-lepton number** states that

> the sum of the electron-lepton numbers before a reaction or decay must equal the sum of the electron-lepton numbers after the reaction or decay.

The electron ($e^-$) and the electron neutrino ($\nu_e$) are assigned a positive lepton number $L_e = +1$, the antileptons $e^+$ and $\bar{\nu}_e$ are assigned a negative lepton number $L_e = -1$, and all others have $L_e = 0$. For example, consider the decay of the neutron,

Neutron decay

$$n \longrightarrow p + e^- + \bar{\nu}_e$$

Before the decay, the electron-lepton number is $L_e = 0$, while after the decay the electron-lepton number is $0 + 1 + (-1) = 0$. Thus, the electron-lepton number is conserved. It is important to recognize that the baryon number must also be conserved. This can easily be seen by noting that before the decay $B = +1$, while after the decay the baryon number is $+1 + 0 + 0 = +1$.

Similarly, when a decay involves muons, the muon-lepton number, $L_\mu$, is conserved. The $\mu^-$ and the $\nu_\mu$ are assigned positive numbers, $L_\mu = +1$, the antimuons $\mu^+$ and $\bar{\nu}_\mu$ are assigned negative numbers, $L_\mu = -1$, while all others have $L_\mu = 0$. Finally, the tau-lepton number, $L_\tau$, is conserved, and similar assignments can be made for the $\tau$ lepton and its neutrino, $\nu_\tau$.

**EXAMPLE 47.2 Checking Lepton Numbers**

Determine which of the following decay schemes can occur on the basis of conservation of lepton-number.

$$\mu^- \longrightarrow e^- + \bar{\nu}_e + \nu_\mu \qquad (1)$$

$$\pi^+ \longrightarrow \mu^+ + \nu_\mu + \nu_e \qquad (2)$$

*Solution* First let us examine decay (1). Since this decay involves both a muon and an electron, $L_\mu$ and $L_e$ must both be conserved. Before the decay, $L_\mu = +1$ and $L_e = 0$. After the decay, $L_\mu = 0 + 0 + 1 = +1$, and $L_e = +1 - 1 + 0 = 0$. Thus, both numbers are conserved, and on this basis the decay mode is possible.

Now consider decay (2). Before the decay, $L_\mu = 0$ and $L_e = 0$. After the decay, $L_\mu = -1 + 1 + 0 = 0$, but $L_e = +1$. Thus, the decay is not possible because the electron-lepton number is not conserved.

Exercise 1 Determine whether the decay $\mu^- \rightarrow e^- + \bar{\nu}_e$ can occur.

Answer No. The muon-lepton number is $+1$ before the decay and is 0 after the decay. Thus, the muon-lepton number is not conserved.

## 47.7 STRANGE PARTICLES AND STRANGENESS

Many particles discovered in the 1950s were produced by the strong interaction of pions with protons and neutrons in the atmosphere. A group of these particles, namely the K, $\Lambda$, and $\Sigma$, were found to exhibit unusual properties in their production and decay, and hence were called *strange particles*. One unusual property is that they are always produced in pairs. For example, when a pion collides with a proton, two neutral strange particles are produced with high probability (see Fig. 47.4) following the reaction

$$\pi^- + p \longrightarrow K^0 + \Lambda^0$$

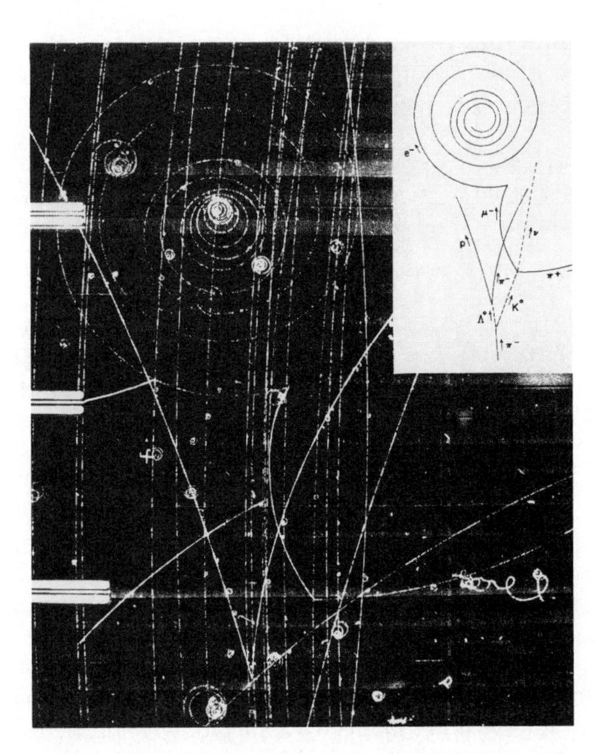


**Figure 47.4** The bubble-chamber photograph shows many events, and the inset represents a drawing of identified tracks. The strange particles $\Lambda^0$ and $K^0$ are formed (at the bottom) as the $\pi^-$ interacts with a proton according to $\pi^- + p \rightarrow \Lambda^0 + K^0$. (Note that the neutral particles leave no tracks, as indicated by the dashed lines.) The $\Lambda^0$ and $K^0$ then decay according to $\Lambda^0 \rightarrow \pi^- + p$ and $K^0 \rightarrow \pi + \mu^- + \nu_\mu$. (Courtesy of Lawrence Berkeley Laboratory, University of California, Photographic Services.)

On the other hand, the reaction $\pi^- + p \rightarrow K^0 + n$ never occurred, even though no known conservation laws were violated and the energy of the pion was sufficient to initiate the reaction. The second peculiar feature of strange particles is that although they are produced by the strong interaction at a high rate, they do not decay into strongly interacting particles at a very high rate as one might expect. Instead, they decay very slowly, which is characteristic of the weak interaction. Their half-lives are in the range $10^{-10}$ s to $10^{-8}$ s; strongly interacting particles have lifetimes of the order of $10^{-23}$ s.

In order to explain these unusual properties of strange particles, a new conservation law called conservation of strangeness was introduced, together with a new quantum number $S$ called the **strangeness.** The strangeness numbers for various particles are given in Table 47.2. The production of strange particles in pairs is explained by assigning $S = +1$ to one of the particles, and $S = -1$ to the other. The nonstrange particles such as the $\pi$ mesons, protons, and leptons are assigned strangeness $S = 0$. The **law of conservation of strangeness** can be stated as follows:

> Whenever a nuclear reaction or decay occurs, the sum of the strangeness numbers before the process must equal the sum of the strangeness numbers after the process.

One can explain the slow decay of strange particles by assuming that the strong and electromagnetic interactions obey the law of conservation of strangeness, while the weak interaction does not. Since the decay reaction involves the loss of one strange particle, it violates strangeness conservation, and hence proceeds slowly via the weak interaction.

**EXAMPLE 47.3 Is Strangeness Conserved?**
(a) Determine whether the following reaction occurs on the basis of conservation of strangeness.

$$\pi^0 + n \longrightarrow K^+ + \Sigma^-$$

*Solution* The initial state has a total strangeness $S = 0 + 0 = 0$. Since the strangeness of the $K^+$ is $S = +1$, and the strangeness of the $\Sigma^-$ is $S = -1$, the total strangeness of the final state is $+1 - 1 = 0$. Thus, we see that strangeness is conserved and the reaction is allowed.

(b) Show that the following reaction does not conserve strangeness, and hence cannot occur.

$$\pi^- + p \longrightarrow \pi^- + \Sigma^+$$

*Solution* In this case, the initial state has a total strangeness $S = 0 + 0 = 0$, while the final state has a total strangeness $S = 0 + (-1) = -1$. Thus strangeness is not conserved, so the reaction does not occur.

**Exercise 2** Show that the observed reaction

$$p + \pi^- \longrightarrow K^0 + \Lambda^0$$

obeys the law of conservation of strangeness.

## 47.8 THE EIGHTFOLD WAY

As we have seen, conserved quantities such as spin, baryon number, lepton number, and strangeness are labels we associate with particles. Many classification schemes have been proposed which group particles into families based on such labels. First, consider the first eight baryons listed in Table 47.2, all having a spin of one-half. The family consists of the proton, the neutron, and six other particles. If we plot their strangeness versus their charge using a sloping coordinate system, as in Figure 47.5, a fascinating pattern is observed. Six of the baryons form a hexagon, while the remaining two are at its center.

Now consider the family of mesons listed in Table 47.2 that have spin of zero. If we count both particles and antiparticles, there are a total of nine such mesons. A plot of strangeness versus charge for this family is shown in Figure 47.6. Again, a similar fascinating hexagonal pattern emerges. In this case, the

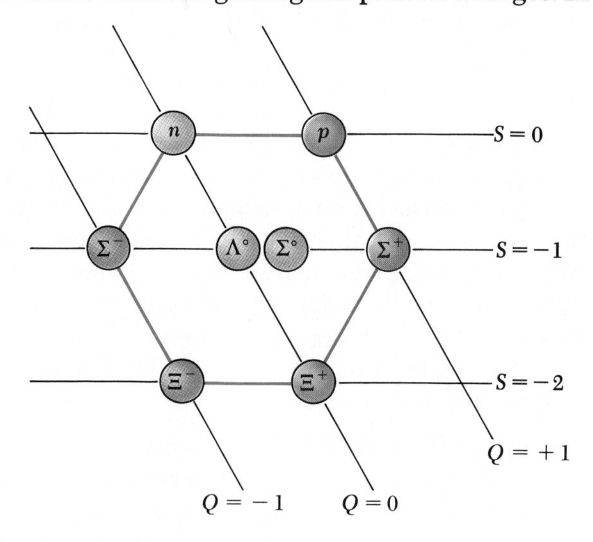


**Figure 47.5** The hexagonal Eightfold Way pattern for the eight spin-½ baryons. This strangeness versus charge plot uses a sloping axis for the charge number $Q$.

The American physicists Richard Feynman *(left)* and Murray Gell-Mann *(right)* won the Nobel prize in physics in 1965 and 1969, respectively, for their theoretical studies dealing with subatomic particles. (Photo courtesy of Michael R. Dressler)

particles on the perimeter of the hexagon lie opposite their antiparticles, and the remaining three (which form their own antiparticles) are at its center. These and related symmetric patterns, called the **Eightfold Way,** were proposed independently in 1961 by Murray Gell-Mann and Yuval Ne'eman.

The various groups of baryons and mesons can be displayed in many other symmetrical patterns within the framework of the Eightfold Way. For example, the family of spin-3/2 baryons contains ten particles arranged in a pattern like the tenpins in a bowling alley. After the pattern was proposed, one of the particles was missing—the particle had yet to be discovered. Gell-Mann predicted that the missing particle, which he called the omega minus ($\Omega^-$), should have a spin of 3/2, a charge of $-1$, a strangeness of $-3$, and a rest energy of about 1680 MeV. Shortly thereafter, in 1964, scientists at the Brookhaven National Laboratory found the missing particle through careful analyses of bubble chamber photographs, and confirmed all its predicted properties.

The patterns of the Eightfold Way in the field of particle physics have much in common with the periodic table in chemistry. Whenever a vacancy (a missing particle or element) occurs in the organized patterns, experimentalists have a guide for their investigations. Furthermore, the existence of the Eightfold Way patterns suggest that baryons and mesons have a more elemental substructure. In the next section, we shall describe that underlying structure, called the **quark model.**

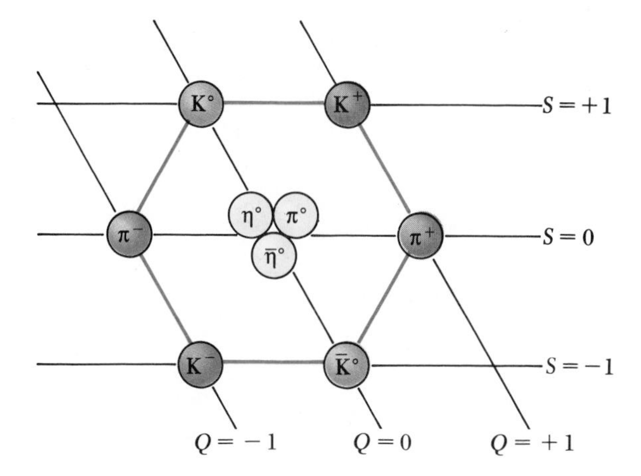


**Figure 47.6** The Eightfold Way pattern for the nine spin-zero mesons.

## 47.9 QUARKS—FINALLY

As we have noted, leptons appear to be truly elementary particles because they have no measurable size or internal structure, are limited in number, and do not seem to break down into smaller units. Hadrons, on the other hand, are complex particles having size and structure. Furthermore, we know that hadrons decay into other hadrons and are many in number. Table 47.2 lists only hadrons that are stable against hadronic decay. Hundreds of others have been discovered, and their properties have been determined. These facts strongly suggest that hadrons cannot be truly elementary, but have some substructure.

### The Original Quark Model

In 1963 Murray Gell-Mann and George Zweig independently proposed that the known hadrons (baryons and mesons) have a more elemental substructure. According to their model, all hadrons are composite systems of two or three fundamental constituents called **quarks.** Gell-Mann borrowed the word "quark" from the passage "Three quarks for Muster Mark" in James Joyce's book *Finnegan's Wake.* In the original quark model, there were three types of quarks designated by the symbols $u$, $d$, and $s$. These were given the arbitrary names *up, down,* and *sideways* (or now more commonly, *strange*). A most unusual property of quarks is that they have fractional electronic charges. The $u$, $d$, and $s$ quarks have charges of $+2e/3$, $-e/3$, and $-e/3$, respectively. Each quark has a baryon number of 1/3 and a spin of 1/2. Furthermore the $u$ and $d$ quarks have strangeness of 0, while the $s$ quark has strangeness of $-1$. Other properties of quarks and antiquarks are given in Table 47.3. Associated with each quark is an antiquark of opposite charge, baryon number, and strange-

**TABLE 47.3 Properties of Quarks and Antiquarks**

| Quarks | | | | | | | | |
|---|---|---|---|---|---|---|---|---|
| Name | Symbol | Spin | Charge | Baryon Number | Strangeness | Charm | Bottomness | Topness |
| Up | $u$ | $\frac{1}{2}$ | $+\frac{2}{3}e$ | $\frac{1}{3}$ | 0 | 0 | 0 | 0 |
| Down | $d$ | $\frac{1}{2}$ | $-\frac{1}{3}e$ | $\frac{1}{3}$ | 0 | 0 | 0 | 0 |
| Strange | $s$ | $\frac{1}{2}$ | $-\frac{1}{3}e$ | $\frac{1}{3}$ | $-1$ | 0 | 0 | 0 |
| Charmed | $c$ | $\frac{1}{2}$ | $+\frac{2}{3}e$ | $\frac{1}{3}$ | 0 | $+1$ | 0 | 0 |
| Bottom | $b$ | $\frac{1}{2}$ | $-\frac{1}{3}e$ | $\frac{1}{3}$ | 0 | 0 | $+1$ | 0 |
| Top (?) | $t$ | $\frac{1}{2}$ | $+\frac{2}{3}e$ | $\frac{1}{3}$ | 0 | 0 | 0 | $+1$ |

| Antiquarks | | | | | | | | |
|---|---|---|---|---|---|---|---|---|
| Name | Symbol | Spin | Charge | Baryon Number | Strangeness | Charm | Bottomness | Topness |
| Up | $\bar{u}$ | $\frac{1}{2}$ | $-\frac{2}{3}e$ | $-\frac{1}{3}$ | 0 | 0 | 0 | 0 |
| Down | $\bar{d}$ | $\frac{1}{2}$ | $+\frac{1}{3}e$ | $-\frac{1}{3}$ | 0 | 0 | 0 | 0 |
| Strange | $\bar{s}$ | $\frac{1}{2}$ | $+\frac{1}{3}e$ | $-\frac{1}{3}$ | $+1$ | 0 | 0 | 0 |
| Charmed | $\bar{c}$ | $\frac{1}{2}$ | $-\frac{2}{3}e$ | $-\frac{1}{3}$ | 0 | $-1$ | 0 | 0 |
| Bottom | $\bar{b}$ | $\frac{1}{2}$ | $+\frac{1}{3}e$ | $-\frac{1}{3}$ | 0 | 0 | $-1$ | 0 |
| Top (?) | $\bar{t}$ | $\frac{1}{2}$ | $-\frac{2}{3}e$ | $-\frac{1}{3}$ | 0 | 0 | 0 | $-1$ |

ness. The composition of all hadrons known at the time could be completely specified by three simple rules. (1) Mesons consist of one quark and one antiquark, giving them a baryon number of 0 as required. (2) Baryons consist of three quarks. (3) Antibaryons consist of three antiquarks. Table 47.4 lists the quark compositions of several mesons and baryons. For example, a $\pi^-$ meson contains one $\bar{u}$ and one $d$ quark (designated as $\bar{u}d$), with charge number $Q = -2e/3 - e/3 = -e$, $B = -1/3 + 1/3 = 0$, and $S = 0 + 0 = 0$. A proton, on the other hand, contains two $u$ quarks and one $d$ quark with $Q = +1$, $B = 1$, and $S = 0$. Note that just two of the quarks, $u$ and $d$, are contained in all hadrons encountered in ordinary matter (protons and neutrons). The third quark, $s$, is needed only to construct strange particles with a strangeness number of either $+1$ or $-1$. For example, the neutral particle $\Lambda^0$ has $uds$ as its quark composition, with $Q = 0$, $B = 1$, and $S = -1$. Figure 47.7 is a pictorial representation of the quark composition of several particles.

**TABLE 47.4 Quark Composition of Several Hadrons**

| Particle | Quark Composition |
|---|---|
| | Mesons |
| $\pi^+$ | $u\bar{d}$ |
| $\pi^-$ | $\bar{u}d$ |
| $K^+$ | $u\bar{s}$ |
| $K^-$ | $\bar{u}s$ |
| $K^0$ | $d\bar{s}$ |
| | Baryons |
| p | $uud$ |
| n | $udd$ |
| $\Lambda^0$ | $uds$ |
| $\Sigma^+$ | $uus$ |
| $\Sigma^0$ | $uds$ |
| $\Sigma^-$ | $dds$ |
| $\Xi^0$ | $uss$ |
| $\Xi^-$ | $dss$ |
| $\Omega^-$ | $sss$ |

### Charm and Other Recent Developments

Although the original quark model was highly successful in classifying particles into families, there were some discrepancies between predictions of the model and certain experimental decay rates. Consequently, a fourth quark was proposed by several physicists in 1967. They argued that if there are four leptons (as was thought at the time), then there should also be four quarks because of an underlying symmetry in nature. The fourth quark, designated by $c$, was given the new property, or quantum number, called **charm.** The *charmed* quark would have a charge $+2e/3$, but its new property of charm would distinguish it from the other three quarks. The new quark would have a charm of $C = +1$, while its antiquark would have a charm of $C = -1$ as indicated in Table 47.3. Charm, like strangeness, would be conserved in strong and electromagnetic interactions, but would not be conserved in weak interactions.

In 1974, a new heavy meson called the $J/\Psi$ particle (or simply $\Psi$) was discovered independently by a group led by Burton Richter at the Stanford

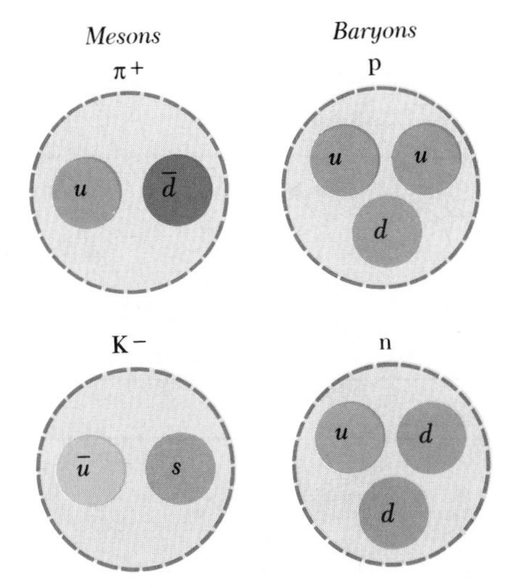


**Figure 47.7** Quark compositions of several particles. Note that the mesons on the left contain two quarks, while the baryons on the right contain three quarks.

Linear Accelerator (SLAC) and another group led by Samuel Ting at the Brookhaven National Laboratory. Richter and Ting were awarded the Nobel Prize in 1976 for this work. The $J/\Psi$ particle did not fit into the original three-quark model, but had the properties of a combination of a charmed quark and its antiquark ($c\bar{c}$). Its mass was much heavier than the other known mesons ($\sim 3100$ MeV/$c^2$) and its lifetime was much longer than other strongly decaying particles. Soon thereafter, other related charmed mesons including $D^+$ and $D^-$ were discovered corresponding to such quark combinations as $\bar{c}d$ and $c\bar{d}$, all of which have large masses and long lifetimes. In 1975, researchers at Stanford University reported strong evidence for the tau ($\tau$) lepton, with a mass of 1784 MeV/$c^2$. Such discoveries led to more elaborate quark models, and the proposal of two new quarks, named *top* ($t$), and *bottom* ($b$). (Some physicists prefer the whimsical names *truth* and *beauty*.) To distinguish these quarks from the old ones, quantum numbers called *topness* and *bottomness* were assigned to these new particles, and are included in Table 47.3. In 1977, researchers at the Fermi National Laboratory under the direction of Leon Lederman reported the discovery of a very massive new meson, Υ, whose composition is considered to be $b\bar{b}$.

At this point, you are probably wondering whether or not such discoveries will ever end. How many "building blocks" of matter really exist? At the present, physicists believe that the fundamental particles in nature include six quarks and six leptons (together with their antiparticles). Some of the properties of these particles are given in Table 47.5.

It should be noted that in spite of many extensive experimental efforts, no isolated quark has ever been observed. Physicists now believe that quarks are permanently confined inside ordinary particles because of an exceptionally strong force that prevents them from escaping. This force, called the "color" force, increases with separation distance (similar to the force of a spring); the properties of this force are discussed in the next section. The great strength of the force between quarks has been described by one author as follows:[3]

[3] Harald Fritzsch, *Quarks, The Stuff of Matter*, London, Allen Lane, 1983.

**TABLE 47.5 The Fundamental Particles and Some of Their Properties**

| Particle | Rest Energy | Charge |
|---|---|---|
| Quarks | | |
| $u$ | 360 MeV | $+\frac{2}{3}e$ |
| $d$ | 360 MeV | $-\frac{1}{3}e$ |
| $c$ | 1500 MeV | $+\frac{2}{3}e$ |
| $s$ | 540 MeV | $-\frac{1}{3}e$ |
| $t$ (?) | $\sim$100 GeV | $+\frac{2}{3}e$ |
| $b$ | 5 GeV | $-\frac{1}{3}e$ |
| Leptons | | |
| $e^-$ | 511 keV | $-e$ |
| $\mu^-$ | 107 MeV | $-e$ |
| $\tau^-$ | 1784 MeV | $-e$ |
| $\nu_e$ | $<$30 eV | 0 |
| $\nu_\mu$ | $<$0.5 MeV | 0 |
| $\nu_\tau$ | $<$250 MeV | 0 |

Quarks are slaves of their own color charge, . . . bound like prisoners of a chain gang. . . . Any locksmith can break the chain between two prisoners, but no locksmith is expert enough to break the gluon chains between quarks. Quarks remain slaves forever.

## 47.10 THE STANDARD MODEL

Shortly after the concept of quarks was proposed, scientists recognized that certain particles had quark compositions which were in violation of the Pauli exclusion principle. Recall that quarks are fermions with spins of 1/2, and hence are expected to follow the exclusion principle. One example is the $\Omega^-$ (*sss*) baryon predicted by Gell-Mann which contains three *s* quarks having parallel spins, giving it a total spin of 3/2. Other examples of baryons which have identical quarks with parallel spins are the $\Delta^{++}$ (*uuu*) and the $\Delta^-$ (*ddd*). To resolve this problem, it was suggested that quarks possess a new property called **color.** This property is similar in many respects to electric charge except that it occurs in three varieties (of color) called red, green, and blue. Another term that is used to distinguish among the six quarks is **flavor.** Thus, each flavor of quark can have three colors. (The whimsical names "color" and "flavor" should not be taken literally.) Of course, the antiquarks have the colors antired, antigreen, and antiblue. In order to satisfy the exclusion principle, all three quarks in a baryon must have different colors. A meson consists of a quark of one color and an antiquark of the corresponding anticolor. The result is that baryons and mesons are always colorless (or white). Furthermore, the new property of color increases the number of quarks by a factor of three.

Although the concept of color in the quark model was originally conceived to satisfy the exclusion principle, it also provided a better theory for explaining certain experimental results. For example, the modified theory correctly predicts the lifetime of the $\pi^0$ meson. The theory of how quarks interact with each other is called **quantum chromodynamics,** or QCD, because it is similar in its structure to quantum electrodynamics (the theory of interaction between electric charges). In QCD, the quark is said to carry a *color charge,* in analogy to electric charge. The strong force between quarks is often called the *color force.* As mentioned earlier, the strong interaction between hadrons is mediated by massless particles called **gluons** (analogous to photons for the electromagnetic force). According to the theory, there are eight gluons, six of which have color charge. Because of their color charge, quarks can attract each other and form composite particles. When a quark emits or absorbs a gluon, its color changes. For example, a blue quark that emits a gluon may become a red quark, while the red quark that absorbs this gluon becomes a blue quark. The color force between quarks is analogous to the electric force between charges—like colors repel and opposite colors attract. Therefore, two red quarks repel each other, but a red quark will be attracted to an antired quark. The attraction between quarks of opposite color to form a meson ($q\bar{q}$) is indicated in Figure 47.8a. Differently colored quarks also attract each other, but with less intensity than opposite colors of quark and antiquark. For example, a cluster of red, blue, and green quarks all attract each other to form baryons as indicated in Figure 47.8b. Thus, all baryons contain three quarks each of which has a different color.

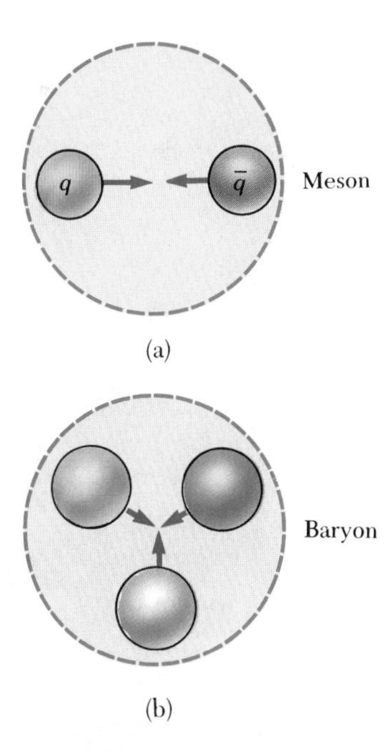


**Figure 47.8** (a) A red quark is attracted to an antired quark. This forms a meson whose quark structure is ($q\bar{q}$). (b) Three different colored quarks attract each other to form a baryon.

Recall that the weak force is believed to be mediated by the $W^+$, $W^-$, and $Z^0$ bosons (spin 1 particles). These particles are said to have *weak charge* just as a quark has color charge. Thus, each elementary particle can have mass, electric charge, color charge, and weak charge. Of course, one or more of these could be zero. Scientists now believe that the truly elementary particles are leptons and quarks, and the force mediators are the gluon, the photon, $W^{\pm}$, $Z^0$, and the graviton. (Note that quarks and leptons have spin ½ and hence are fermions, while the force mediators have spin 1 or higher, and are bosons.)

In 1979, Sheldon Glashow, Abdus Salam, and Steven Weinberg won a Nobel Prize for developing a theory that unified the electromagnetic and weak interactions. This so-called **electroweak theory** postulates that the weak and electromagnetic interactions have the same strength at very high particle energies. Thus, the two interactions are viewed as two different manifestations of a single unifying electroweak interaction. The photon and the three massive bosons ($W^{\pm}$ and $Z^0$) play a key role in the electroweak theory. The theory makes many concrete predictions, but perhaps the most spectacular is the prediction of the masses of the W and Z particles at about 82 GeV/$c^2$ and 93 GeV/$c^2$, respectively. The 1984 Nobel Prize was awarded to Carlo Rubbia and Simon van der Meer for their work leading to the discovery of these particles at just these energies at the CERN Laboratory in Geneva, Switzerland.

The combination of the electroweak theory and QCD for the strong interaction form what is referred to in high energy physics as the "Standard Model." Although the details of the Standard Model are complex, its essential ingredients can be summarized with the help of Figure 47.9. The strong force, mediated by gluons, holds quarks together to form composite particles such as protons, neutrons, and mesons. Leptons participate only in the electromagnetic and weak interactions. The electromagnetic force is mediated by photons, while the weak force is mediated by W and Z bosons. Note that all fundamental forces are mediated by spin 1 particles whose properties are given, to a large extent, by the symmetries involved in the theories. However, the Standard Model does not answer all questions. A major problem is why the photon has no mass while the W and Z bosons are massive particles. Because of their mass difference, the electromagnetic and weak forces are quite distinct

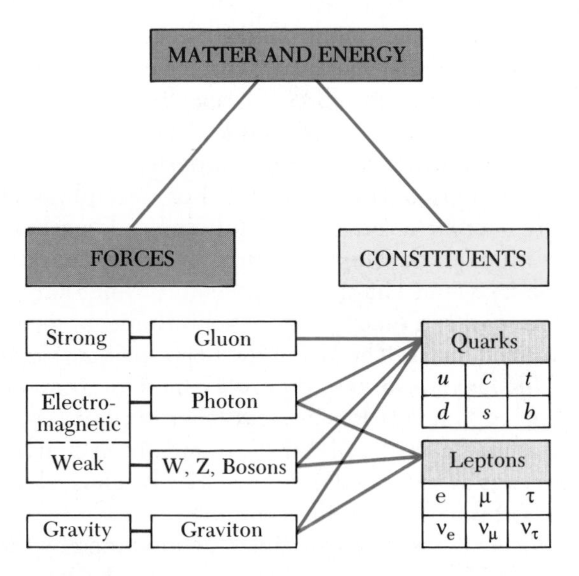


**Figure 47.9** The Standard Model of particle physics.

at low energies, but become similar in nature at very high energies. This behavior as one goes from low to high energies, called *symmetry breaking*, leaves open the question of the origin of particle masses. In order to resolve this problem, a hypothetical particle called the *Higgs boson* has been proposed, which provides a mechanism for breaking the electroweak symmetry. The Standard Model, including the Higgs mechanism, provides a logically consistent explanation of the massive nature of the W and Z bosons. Unfortunately, the Higgs boson has not yet been found, but physicists know that its mass should be less than 1 TeV ($10^{12}$ eV).

A computer reconstruction of the magnetic field distribution in the cross section of a prototype SSC magnet. In the green central area, the field is uniform to better than 0.02%. Other colors denote deviations of 0.04%, 0.06%, and so on. High quality field almost fills the whole beam chamber, ensuring that the proton beam will have excellent stability. (Courtesy Brookhaven National Laboratory)

In order to determine whether the Higgs boson exists, two quarks of at least 1 TeV of energy must collide, but calculations show that this requires injecting 40 TeV of energy within the volume of a proton. The excess energy is needed because of the quarks and gluons contained in the proton. Although no existing accelerator can provide this energy, scientists in the United States are currently planning to construct the world's largest and most powerful particle accelerator called the Superconducting Super Collider, or SSC, which will meet this need (see Fig. 47.10). The SSC will accelerate protons to 20 TeV (two protons traveling in opposite directions with this energy will give the required 40 TeV), it will have a circumference of 82.944 km (about 52 miles), and will cost about 4 billion dollars! The proton energies in the SSC will be twenty times greater than is currently available at Fermilab, and physicists expect to be able to explore distances down to about $10^{-18}$ m (one thousandth

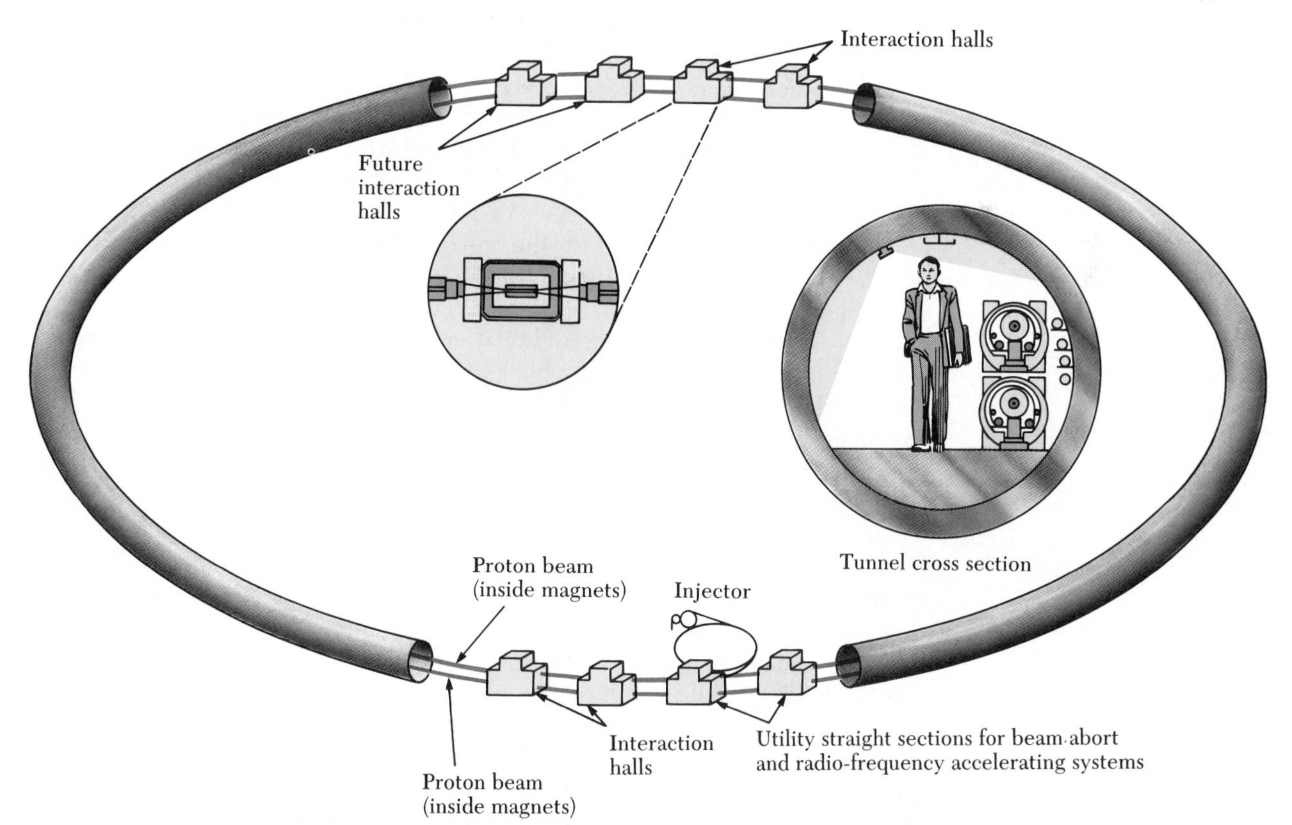


**Figure 47.10** The proposed layout of the SSC, which is 52 mi in circumference. The experimental halls and other components on the ring are not to scale. Access and service points are placed at regular intervals around the ring.

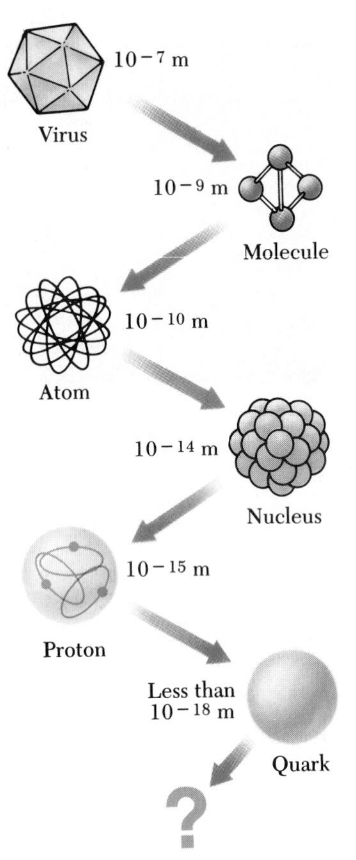


**Figure 47.11** Looking at matter with various microscopes reveals structures ranging in size from the smallest living thing, a virus, down to a quark, which has not yet been observed as an isolated particle.

the diameter of a proton). Thus, it will be a microscope of unparalleled power. Figure 47.11 shows the evolution of the various stages of matter that scientists have been able to investigate with various types of microscopes. The Department of Energy endorsed detailed research on the SSC in 1983, and federal funds have been allocated to support research and development on the superconducting magnets to be used in the accelerator. The SSC is expected to be completed in the mid-1990s if federal funding is provided.

Following the success of the electroweak theory, scientists attempted to combine it with QCD in a *grand unification theory* known as GUT. In this model, the electroweak force was merged with the strong color force to form a grand unified force. One version of the theory considers leptons and quarks as members of the same family that are able to change into each other by exchanging an appropriate messenger particle. Many GUT theories predict that protons are unstable, and will decay with a half-life of about $10^{31}$ years. This is far greater than the age of the universe; as yet proton decays have not been observed.

## 47.11 THE COSMIC CONNECTION

As we have seen, the world around us is dominated by protons, electrons, neutrons, and neutrinos. Some of the other more exotic particles can be seen in cosmic rays. However, most of the new particles are produced using large, expensive machines which accelerate protons and electrons to energies in the GeV and TeV range. These energies are enormous when compared to the thermal energy in today's universe. For example, kT is only about 1 keV at the center of the sun. On the other hand, the temperature of the early universe was high enough to reach energies of TeV and higher.

In this section we shall describe one of the most fascinating theories in all of science—the big bang theory of the creation of the universe—and the experimental evidence that supports it. This theory of cosmology states that the universe had a beginning, and further that the beginning was so cataclysmic that it is impossible to look back beyond the creation. According to this theory, the universe erupted from a point-like singularity about 15 to 20 billion years ago. The first few minutes after the Big Bang saw such extremes of energy that it is believed that all four interactions of physics were unified and that all matter melted down into an undifferentiated "quark soup."

The evolution of the four fundamental forces from the Big Bang to the present is shown in Figure 47.12. During the first $10^{-43}$ s (the ultra-hot epoch where $T \approx 10^{32}$ K), it is presumed that the strong, electroweak, and gravitational forces were joined to form a completely unified force. In the first $10^{-32}$ s following the Big Bang (the hot epoch where $T \approx 10^{29}$ K), gravity broke free of this unification while the strong and electroweak forces remained as one, described by a grand unification theory. This was a period when particle energies were so great ($>10^{16}$ GeV) that very massive particles as well as quarks, leptons, and their antiparticles existed. Then the universe rapidly expanded and cooled during the warm epoch when the temperatures ranged from $10^{29}$ to $10^{15}$ K (corresponding to energies ranging from $10^{16}$ to $10^{2}$ GeV). During this epoch, the strong and electroweak forces parted company, and the grand unification scheme was broken. As the universe continued to cool, about

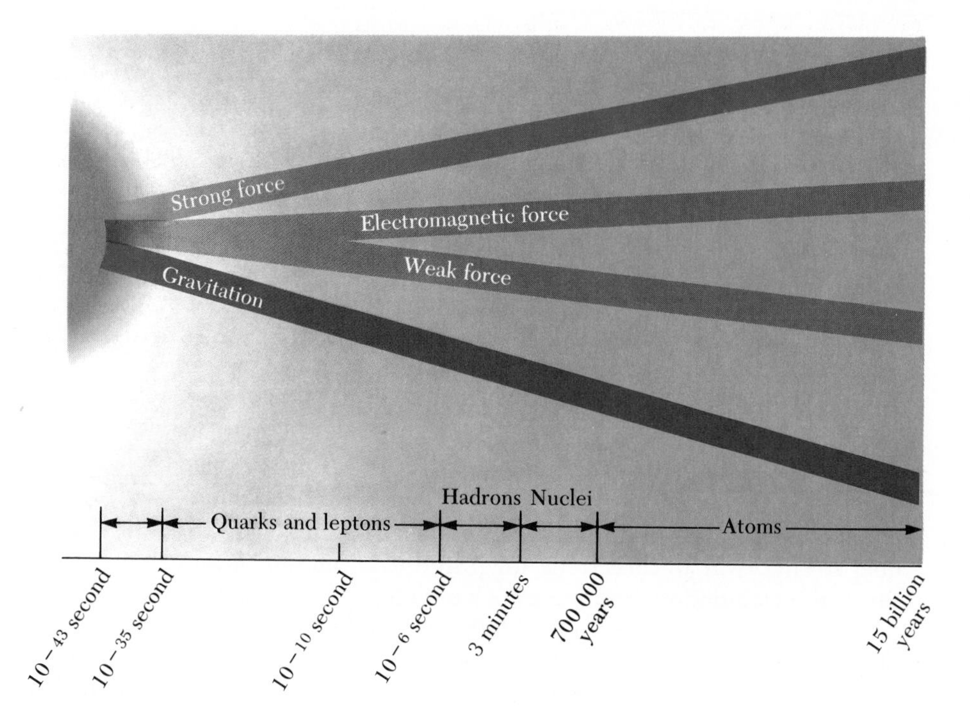


**Figure 47.12** A brief history of the universe from the Big Bang to the present. The four forces became distinguishable during the first microsecond. Following this, all the quarks combined to form the strongly interacting particles. However, the leptons remained separate, and exist as individually observable particles to this day.

$10^{-10}$ s after the Big Bang, the electroweak force split in two. This resulted in the four forces in the universe as we distinguish them today.

Until about 700 000 years after the Big Bang, the universe was "radiation dominated"; ions absorbed and reemitted photons, thereby ensuring thermal equilibrium of radiation and matter. Energetic radiation also prevented matter from forming clumps or even neutral hydrogen atoms. When the universe was about 700 000 years old, it had expanded and cooled to about 3000 K, and protons could bind to electrons to form neutral hydrogen atoms. Since neutral atoms do not appreciably scatter photons, the universe suddenly became transparent to photons. The photons were frozen in time with an energy distribution characteristic of a 3000-K black body. Radiation no longer dominated the universe and clumps of neutral matter steadily grew — first atoms, followed by molecules, gas clouds, stars, and finally galaxies.

## Observation of Radiation from the Primordial Fireball

**Figure 47.13** Robert W. Wilson (left) and Arno A. Penzias with Bell Telephone Laboratories horn-reflector antenna. (AT&T Bell Laboratories)

In 1965, Arno A. Penzias and Robert W. Wilson of Bell Labs were testing a sensitive microwave receiver and made an amazing discovery. A pesky signal producing a faint background "hiss" was interfering with their satellite communications experiments, and in spite of their valiant efforts, the signal remained. Ultimately, it would become clear that they were observing a microwave background radiation (at a wavelength of 7.35 cm) representing the left-over glow from the Big Bang.

The microwave horn that served as their receiving antenna is shown in Figure 47.13. The intensity of the detected signal remain unchanged as the

Biographical Sketch

**George Gamow**
*(1904–1968)*

George Gamow and his students, Ralph Alpher and Robert Herman, were the first to take the first half hour of the universe seriously. In a mostly overlooked paper published in 1948, they made truly remarkable cosmological predictions. They correctly calculated the abundances of hydrogen and helium after the first half hour (75% H and 25% He), and predicted that radiation from the Big Bang should still be present with an apparent temperature of about 5 K. In Gamow's own words, the universe's supply of hydrogen and helium was created very quickly, "in less time than it takes to cook a dish of duck and roast potatoes." The comment is characteristic of this interesting physicist, who is known as much for his explanation of alpha decay and theories of cosmology as for his delightful popular books, his cartoons, and his wonderful sense of humor. A classic Gamow story holds that having coauthored a paper with Alpher, he made Hans Bethe an honorary author so that the credits would read "Alpher, Bethe, and Gamow" (to be compared with the Greek letters $\alpha$, $\beta$, and $\gamma$).

antenna was pointed in different directions. The fact that the radiation had equal strengths in all directions suggested that the entire universe was the source of this radiation. Booting a flock of pigeons from the 20-foot horn (bird droppings in the horn were a problem) and cooling the microwave detector both failed to remove the "spurious" signal. Through a casual conversation, Penzias and Wilson discovered that a group at Princeton had predicted the residual radiation from the Big Bang and were planning an experiment seeking to confirm the theory. The excitement in the scientific community was high when Penzias and Wilson announced that they had already observed an excess microwave background compatible with a 3 K blackbody source.

Because the measurements of Penzias and Wilson were taken at a single wavelength, they did not completely confirm the radiation as 3 K blackbody radiation. Subsequent experiments by other groups added intensity data at different wavelengths as shown in Figure 47.14. The results confirm that the radiation is that of a black body at 2.9 K. This figure is, perhaps, the most clearcut evidence for the Big Bang theory. The 1978 Nobel prize in physics was awarded to Penzias and Wilson for their most important discovery.

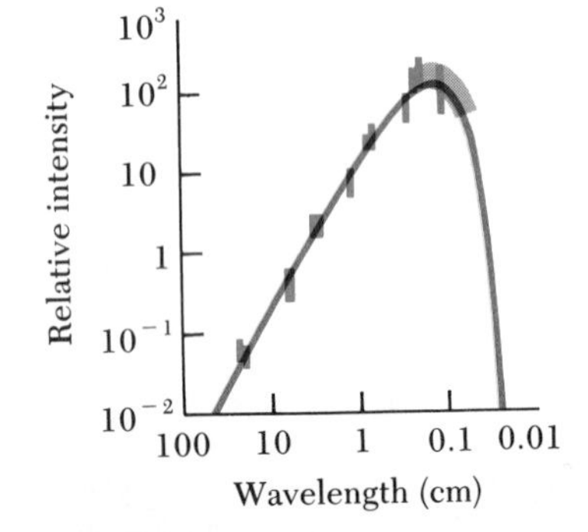


**Figure 47.14** Radiation spectrum of the Big Bang. The shaded areas are experimental results. The solid line is the spectrum calculated for a black body at 2.9 K.

### Other Evidence for the Expanding Universe

Most of the key discoveries supporting the theory of an expanding universe, and indirectly the Big Bang theory of cosmology, were made in the 20th century. Vesto Melvin Slipher, an American astronomer, reported that most nebulae were receding from the earth at speeds up to several million miles per hour. Slipher was one of the first to use the methods of Doppler shifts in spectral lines to measure velocities.

In the late 1920s, Edwin P. Hubble made the bold assertion that the universe as a whole was expanding. From 1928 to 1936 he and Milton Humason toiled at Mount Wilson to prove this assertion until they reached the limits of the 100-inch telescope. The results of this work and its continuation on the 200-inch telescope in the 1940s showed that the velocities of galaxies increase in direct proportion to their distance $R$ from us (see Fig. 47.15). This linear relation, known as Hubble's law, may be written

$$v = HR \tag{47.4}$$

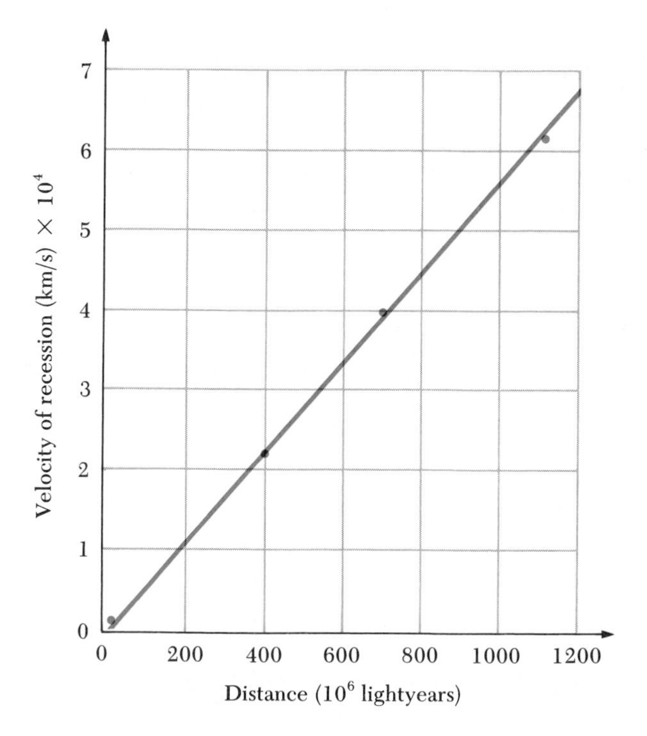


**Figure 47.15** Hubble's law: A plot of velocity of recession versus distance for four galaxies.

where $H$, called the *Hubble parameter*, has the approximate value

$$H = 17 \times 10^{-3} \text{ m/(s}\cdot\text{lightyear)}$$

EXAMPLE 47.4 Recession of a Quasar

A quasar is a very distant object whose speed can be measured from Doppler shift measurements in the light it emits. A certain quasar is measured to recede from the earth at a speed of $0.55c$. What is the distance from the earth to the quasar?

*Solution* We can find the distance from Hubble's law:

$$R = \frac{v}{H} = \frac{(0.55)(3 \times 10^8 \text{ m/s})}{17 \times 10^{-3} \text{ m/(s}\cdot\text{lightyear)}}$$

$$9.7 \times 10^9 \text{ lightyears}$$

Exercise 3 Assuming that the quasar has moved with the speed $0.55c$ ever since the Big Bang, estimate the age of the universe.

Answer $t = R/v = 1/H \approx 18$ billion years, which is in good agreement with other calculations.

## Will the Universe Expand Forever?

In the 1950s and 1960s Allan R. Sandage used the 200-inch telescope at Mount Palomar to measure the speeds of galaxies at distances of up to 6 billion lightyears. These measurements showed that these very distant galaxies were moving about 10 000 km/s faster than the Hubble law predicted. According to this result, the universe must have been expanding more rapidly 1 billion years ago and, consequently, the expansion is slowing.[4] (See Fig. 47.16.) The

[4] The data at large distances have large observational uncertainties and may be systematically in error from selection effects such as abnormal brightness in the most distant visible clusters.

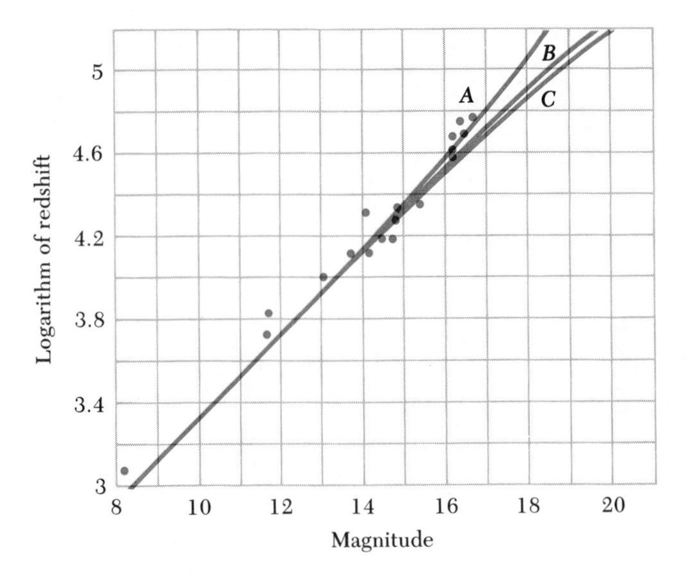


**Figure 47.16** Red shift or speed of recession versus apparent magnitude or distance of 18 faint clusters. Curve A is the trend suggested by the six faintest clusters of galaxies. Line C corresponds to a universe with a constant rate of expansion. If the data fall between B and C, the expansion slows but never stops. If the data fall to the left of B, expansion stops and contraction occurs.

question that remains concerns the rate at which the expansion is slowing. If the average mass density of the universe is less than the critical density (3 hydrogen atoms/m$^3$), the galaxies will slow in their outward rush but still escape to infinity. If the average density exceeds the critical value, the expansion will eventually stop and contraction will begin, possibly leading to a superdense state and another expansion or an oscillating universe. The critical mass density of the universe, $\rho_c$, can be estimated from energy considerations. Figure 47.17 shows a large section of the universe with radius $R$, containing galaxies with a total mass $M$. A galaxy of mass $m$ and velocity $v$ at $R$ will just escape to infinity with zero velocity if the sum of its kinetic energy and gravitational potential energy is zero. Thus,

$$E_{\text{total}} = 0 = K + U = \tfrac{1}{2}mv^2 - \frac{GmM}{R}$$

or

$$\tfrac{1}{2}mv^2 = \frac{Gm\frac{4}{3}\pi R^3\rho_c}{R}$$

or

$$v^2 = \frac{8\pi G}{3}R^2\rho_c \qquad (47.5)$$

Since the galaxy of mass $m$ obeys the Hubble law, $v = HR$, Equation 47.5 becomes

$$H^2 = \frac{8\pi G}{3}\rho_c$$

or

$$\rho_c = \frac{3H^2}{8\pi G} \qquad (47.6)$$

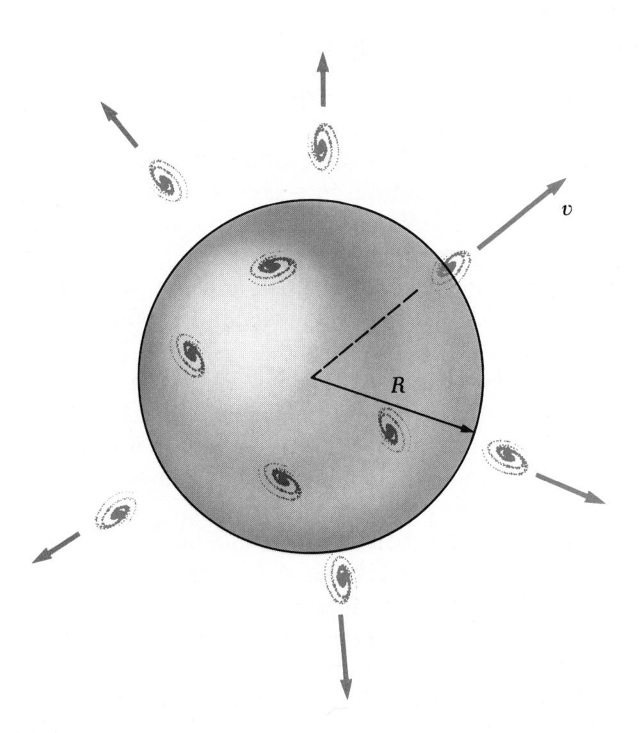


**Figure 47.17** A galaxy escaping from a large cluster contained within radius *R*. Only the mass within *R* slows the mass *m*.

Using $H = 17$ km/s/million lightyears, where 1 lightyear $= 9.46 \times 10^{12}$ km, and $G = 6.67 \times 10^{-8}$ cm$^3$/g·s$^2$ yields $\rho_c = 6 \times 10^{-30}$ g/cm$^3$. As the mass of a hydrogen atom is $1.67 \times 10^{-24}$ g, $\rho_c$ corresponds to $3 \times 10^{-6}$ hydrogen atoms per cm$^3$ or 3 atoms per m$^3$.

### Missing Mass in the Universe(?)

The visible matter in galaxies averages out to $5 \times 10^{-33}$ g/cm$^3$. The radiation in the universe has a mass equivalent of about 2% of the visible matter. Nonluminous matter (such as interstellar gas or black holes) may be estimated from the velocities of galaxies orbiting each other in a cluster. The higher the galaxy velocities, the more mass in the cluster. Results from measurements on the Coma cluster of galaxies, surprisingly, indicate that the amount of invisible matter is *20 to 30 times the amount* present in stars and luminous gas clouds. Yet even this large invisible component, if applied to the universe as a whole, leaves the observed mass density a factor of 10 less than $\rho_c$. This so-called *missing mass* (or *dark matter*) has been the subject of intense theoretical and experimental work, with exotic particles such as axions, photinos, and superstring particles suggested as candidates for the missing mass (see the essay by Virginia Trimble in Chapter 14). More mundane proposals have been that the missing mass is present in brown dwarfs, red dwarfs, or neutrinos. In fact, neutrinos are so abundant that a tiny neutrino rest mass on the order of 20 eV would furnish the missing mass and "close" the universe.

Although we are a bit more sure about the beginning of the universe, we are uncertain about its end. Will the universe expand forever? Will it collapse and repeat its expansion in an endless series of oscillations? Results and answers to these questions remain inconclusive and the exciting controversy continues.

## 47.12 PROBLEMS AND PERSPECTIVES

While particle physicists have been exploring the realm of the very small, cosmologists have been exploring cosmic history back to the first microsecond of the Big Bang. The observation of events that occur when two particles collide in an accelerator is essential in reconstructing the early moments in cosmic history. Perhaps the key to understanding the early universe is first to understand the world of elementary particles. Cosmologists and particle physicists now find they have many common goals and are joining hands to attempt to understand the physical world at its most fundamental level.

Our understanding of physics at short distances is far from complete. Particle physics is faced with many questions. Why is there so little antimatter in the universe? Do neutrinos have a small rest mass, and if so, how do they contribute to the "dark matter" of the universe? Is it possible to unify the strong and electroweak theories in a logical and consistent manner? Why do quarks and leptons form three similar but distinct families? Are muons the same as electrons (apart from their difference in mass), or do they have other subtle differences that have not been detected? Why are some particles charged and others neutral? Why do quarks carry a fractional charge? What determines the masses of the fundamental constituents? Can isolated quarks exist? The questions go on and on. Because of the rapid advances and new discoveries in the field of particle physics, by the time you read this book some of these questions will likely be resolved while others may emerge.

An important and obvious question that remains is whether leptons and quarks have a substructure. If they have a substructure, one could envision an infinite number of deeper structure levels. However, if leptons and quarks are indeed the ultimate constituents of matter, as physicists today tend to believe, we should be able to construct a final theory of the structure of matter as Einstein himself dreamed. In the view of many physicists, the end of the road is in sight, but it is anyone's guess as to how long it will take to reach that goal.

## SUMMARY

There are four fundamental forces in nature: strong (hadronic), electromagnetic, weak, and gravitational. The strong force is the force between nucleons that keeps the nucleus together. The weak force is responsible for beta decay. The electromagnetic and weak forces are now considered to be manifestations of a single force called the electroweak force.

An antiparticle and a particle have the same mass, but opposite charge. Furthermore, other properties may have opposite values such as lepton number and baryon number. It is possible to produce particle-antiparticle pairs in nuclear reactions if the available energy is greater than $2mc^2$, where $m$ is the rest mass of the particle (or antiparticle).

Particles other than photons are classified as being *hadrons* or *leptons*. Hadrons interact primarily through the strong force. We now know that hadrons have size and structure, and hence are not fundamental constituents of matter. There are two types of hadrons, called *baryons* and *mesons*. Mesons have a baryon number zero and have either zero or integral spin. Baryons, which generally are the most massive particles, have a nonzero baryon number and a spin of 1/2 or 3/2. The neutron and proton are examples of baryons.

Leptons have no structure or size, and are considered to be truly elementary particles. Leptons interact only through the weak and electromagnetic forces. These are six leptons, the electron $e^-$, the muon $\mu^-$, the tau $\tau^-$, and their neutrinos $\nu_e$, $\nu_\mu$, and $\nu_\tau$.

In all reactions and decays, quantities such as energy, linear momentum, angular momentum, electric charge, baryon number, and lepton number are strictly conserved. Certain particles have properties called *strangeness* and *charm.* These unusual properties are conserved only in those reactions and decays that occur via the strong force.

Recent theories in elementary particle physics have postulated that all hadrons are composed of smaller units known as *quarks.* Quarks have a fractional electric charge and a baryon number of 1/3. There are six flavors of quarks, up ($u$), down ($d$), strange ($s$), charmed ($c$), top ($t$), and bottom ($b$). All baryons contain three quarks, while all mesons contain one quark and one antiquark.

According to the theory of *quantum chromodynamics,* quarks have a property called *color,* and the strong force between quarks is referred to as the *color force.*

Every fundamental interaction is said to be mediated by the exchange of field particles. The electromagnetic interaction is mediated by the photon; the weak interaction is mediated by the $W^\pm$ and $Z^0$ bosons; the gravitational interaction is mediated by gravitons; the strong interaction is mediated by gluons.

The observation of background microwave radiation discovered by Penzias and Wilson strongly suggests that the universe started with a "Big Bang" about 15 billion years ago. The radiation is equivalent to that of a black body at a temperature of about 3 K.

Various astronomical measurements strongly suggest that the universe is expanding. According to *Hubble's law,* distant galaxies are receding from the earth at a speed given by

$$v = HR \tag{47.4}$$

where $R$ is the distance to the galaxy and $H$ is *Hubble's parameter,* which has the approximate value $17 \times 10^{-3}$ m/(s·lightyear).

## SUGGESTED READINGS

### Particle Physics


E.D. Bloom and G.J. Feldmann, "Quarkonium," *Sci. American,* May, 1982.

Frank Close, *The Cosmic Onion: Quarks and the Nature of the Universe,* The American Institute of Physics, 1986. A timely monograph on particle physics, including lively discussions of the Big Bang theory.

Harald Fritzsch, *Quarks, The Stuff of Matter,* London, Allen and Lane, 1983. An excellent introductory overview of elementary particle physics.

H. Harari, "The Structure of Quarks and Leptons," *Sci. American,* April, 1983.

J. David Jackson, Maury Tigner, and Stanley Wojcicki, "The Superconducting Supercollider," *Sci. American,* March, 1986.

Leon M. Lederman, "The Value of Fundamental Science," *Sci. American,* November, 1984.

N.B. Mistry, R.A. Poling, and E.H. Thorndike, "Particles with Naked Beauty," *Sci. American,* July, 1983.

Chris Quigg, "Elementary Particles and Forces," *Sci. American,* April, 1985.

James S. Trefil, *From Atoms to Quarks,* New York, Scribner, 1980. This is an excellent introduction to the world of particle physics.

Steven Weinberg, *The Discovery of Elementary Particles.* New York, Scientific American Library, W.H. Freeman and Company, 1983. This book emphasizes the important discoveries, experiments, and intellectual exercises which reshaped physics in the 20th century.

Steven Weinberg, "The Decay of the Proton," *Sci. American,* June, 1981.

### Cosmology and the Big Bang


John D. Barrow and Joseph Silk, "The Structure of the Early Universe," *Sci. American*, April, 1980.

George Gamow, "The Evolutionary Universe," *Sci. American*, September, 1956.

David Layzer, *Constructing the Universe*, Scientific American Library, New York, W.H. Freeman and Co., 1984, Chapters 7 and 8.

David L. Meier and Rashid A. Sunyaev, "Primeval Galaxies," *Sci. American*, November, 1979.

Richard A. Muller, "The Cosmic Background Radiation and the New Aether Drift," *Sci. American*, May, 1978.

Carl Sagan and Frank Drake, "The Search for Extraterrestrial Intelligence," *Sci. American*, May, 1975.

Allan R. Sandage, "The Red-Shift," *Sci. American*, September, 1956.


## QUESTIONS

1. Name the four fundamental interactions and the particles that meditate each interaction.
2. Discuss the quark model of hadrons, and describe the properties of quarks.
3. Discuss the differences between hadrons and leptons.
4. Describe the properties of baryons and mesons and the important differences between them.
5. Particles known as resonances have very short lifetimes, of the order of $10^{-23}$ s. From this information, would you guess they are hadrons or leptons? Explain.
6. The family of $K$ mesons all decay into final states that contain no protons or neutrons. What is the baryon number of the $K$ mesons?
7. The $\Xi^0$ particle decays by the weak interaction according to the decay mode $\Xi^0 \rightarrow \Lambda^0 + \pi^0$. Would you expect this decay to be fast or slow? Explain.
8. Identify the particle decays listed in Table 47.2 that occur by the weak interaction. Justify your answers.
9. Identify the particle decays listed in Table 47.2 that occur by the electromagnetic interaction. Justify your answers.
10. Two protons in a nucleus interact via the strong interaction. Are they also subject to the weak interaction?
11. Discuss the following conservation laws: energy, linear momentum, angular momentum, electric charge, baryon number, lepton number, and strangeness. Are all of these laws based on fundamental properties of nature? Explain.
12. An antibaryon interacts with a meson. Can a baryon be produced in such an interaction? Explain.
13. Discuss the essential features of the Standard Model of particle physics.
14. How many quarks are there in (a) a baryon, (b) an antibaryon, (c) a meson, and (d) an antimeson? How do you account for the fact that baryons have half-integral spins while mesons have spins of 0 or 1? (*Hint:* Quarks have spins of 1/2.)
15. In the theory of quantum chromodynamics, quarks come in three colors. How would you justify the statement that "all baryons and mesons are colorless"?
16. Which baryon was predicted to exist by Murray Gell-Mann in 1961? What is the supposed quark composition of this particle?
17. What is the quark composition of the $\Xi^-$ particle? (See Table 47.4.)
18. The W and Z bosons were first produced at CERN in 1983 (by having a beam of protons and a beam of antiprotons meet at high energy). Why was this an important discovery?
19. How did Edwin Hubble in 1928 determine that the universe is expanding?
20. How will the Hubble Space Telescope (scheduled to be launched into earth orbit by 1990) help determine the large-scale nature of the universe?

## PROBLEMS

### Section 47.3 Positrons and Other Antiparticles

1. Two photons are produced when a proton and antiproton annihilate each other. What is the minimum frequency and corresponding wavelength of each proton?
2. A photon produces a proton-antiproton pair according to the reaction $\gamma \rightarrow p + \bar{p}$. What is the frequency of the photon? What is its wavelength?

### Section 47.4 Mesons and the Beginning of Particle Physics

3. One of the mediators of the weak interaction is the $Z^0$ boson whose mass is 96 GeV/$c^2$. Use this information to find an approximate value for the range of the weak interaction.
4. Occasionally, high energy muons will collide with electrons and produce two neutrinos according to the

reaction $\mu^+ + e \rightarrow 2\nu$. What kind of neutrinos are these?

5. When a high-energy proton or pion traveling near the speed of light collides with a nucleus, it travels an average distance of $3 \times 10^{-15}$ m before interacting. From this information, estimate the time for the strong interaction to occur.
6. A neutral pi-meson at rest decays into 2 gamma rays according to

$$\pi^0 \longrightarrow \gamma + \gamma$$

Find the energy, momentum, and frequency of each gamma-ray photon.

### Section 47.5 The Classification of Particles

7. Name one possible decay mode (see Table 47.2) of each of the following particles: $\Omega^+$, $\overline{K}°$, $\overline{\Lambda}°$, $\overline{n}$

### Section 47.6 Conservation Laws

8. Each of the following reactions is forbidden. Determine a conservation law that is violated for each reaction.
(a) $p + \overline{p} \rightarrow \mu^+ + e$
(b) $\pi^- + p \rightarrow p + \pi^+$
(c) $p + p \rightarrow p + \pi^+$
(d) $p + p \rightarrow p + p + n$
(e) $\gamma + p \rightarrow n + \pi^0$
9. (a) Show that baryon number and charge are conserved in the following reactions of a pion with a proton.

$$\pi^+ + p \longrightarrow K^+ + \Sigma^+ \qquad (1)$$

$$\pi^+ + p \longrightarrow \pi^+ + \Sigma^+ \qquad (2)$$

(b) The first reaction is observed, but the second never occurs. Explain these observations.
10. The following reactions or decays involve one or more neutrinos. Supply the missing neutrinos ($\nu_e$, $\nu_\mu$, or $\nu_\tau$).
(a) $\pi^- \rightarrow \mu^- + ?$ (d) $? + n \rightarrow p + e$
(b) $K^+ \rightarrow \mu^+ + ?$ (e) $? + n \rightarrow p + \mu^-$
(c) $? + p \rightarrow n + e^+$ (f) $\mu^- \rightarrow e + ? + ?$
11. For the following two reactions, the first may occur but the second cannot. Explain.

$$K^0 \longrightarrow \pi^+ + \pi^- \qquad \text{(can occur)}$$

$$\Lambda^0 \longrightarrow \pi^+ + \pi^- \qquad \text{(cannot occur)}$$

12. Determine which of the reactions below can occur. For those that cannot occur, determine the conservation law (or laws) that each violates.
(a) $p \rightarrow \pi^+ + \pi^0$ (d) $\pi^+ \rightarrow \mu^+ + \nu_\mu$
(b) $p + p \rightarrow p + p + \pi^0$ (e) $n \rightarrow p + e + \overline{\nu}_e$
(c) $p + p \rightarrow p + \pi^+$ (f) $\pi^+ \rightarrow \mu^+ + n$

### Section 47.7 Strange Particles and Strangeness

13. Determine whether or not strangeness is conserved in the following decays and reactions.
(a) $\Lambda^0 \rightarrow p + \pi^-$ (d) $\pi^- + p \rightarrow \pi^- + \Sigma^+$
(b) $\pi^- + p \rightarrow \Lambda^0 + K^0$ (e) $\Xi^- \rightarrow \Lambda^0 + \pi^-$
(c) $\overline{p} + p \rightarrow \overline{\Lambda}^0 + \Lambda^0$ (f) $\Xi^0 \rightarrow p + \pi^-$
14. The neutral $\rho$ meson decays by the strong interaction into two pions according to $\rho^0 \rightarrow \pi^+ + \pi^-$ with a half-life of about $10^{-23}$ s. The neutral K meson also decays into two pions according to $K^0 \rightarrow \pi^+ + \pi^-$, but with a much longer half-life of about $10^{-10}$ s. How do you explain these observations?
15. Each of the following decays is forbidden. For each process, determine a conservation law that is violated.
(a) $\mu^- \rightarrow e + \gamma$ (d) $p \rightarrow e^+ + \pi^0$
(b) $n \rightarrow p + e + \nu_e$ (e) $\Xi^0 \rightarrow n + \pi^0$
(c) $\Lambda^0 \rightarrow p + \pi^0$

### Section 47.9 Quarks—Finally

16. The quark composition of the proton is *uud*, while that of the neutron is *udd*. Show that the charge, baryon number, and strangeness of these particles equal the sums of these numbers for their quark constituents.
17. The quark compositions of the $K^0$ and $\Lambda^0$ particles are $d\overline{s}$ and *uds*, respectively. Show that the charge, baryon number, and strangeness of these particles equal the sums of these numbers for the quark constituents.
18. What is the electrical charge of the baryons with the quark compositions (a) $\overline{u}\overline{u}\overline{d}$ and (b) $\overline{u}\overline{d}\overline{d}$. What are these baryons called?
19. Analyze each of the reactions in terms of their constituent quarks:
(a) $\pi^- + p \rightarrow K^0 + \Lambda^0$
(b) $\pi^+ + p \rightarrow K^+ + \Sigma^+$
(c) $K^- + p \rightarrow K^+ + K^0 + \Omega^-$
(d) $p + p \rightarrow K^0 + p + \pi^+ + ?$
In the last reaction, identify the mystery particle.

### Section 47.11 The Cosmic Connection

20. Using Hubble's Law (Equation 47.4), estimate the wavelength of the 590 nm sodium line emitted from galaxies (a) $2 \times 10^6$ lightyears away, (b) $2 \times 10^8$ lightyears away and (c) $2 \times 10^9$ lightyears away. *Hint:* Use the relativistic Doppler formula for wavelength $\lambda'$ of light emitted from a moving source:

$$\lambda' = \lambda \sqrt{\frac{1 + v/c}{1 - v/c}}$$

21. A distant quasar is moving away from the earth at such high speed that the blue 434-nm hydrogen line is observed at 650 nm, in the red portion of the spectrum. (a) How fast is the quasar receding? (See the hint in the preceding problem.) (b) Using Hubble's Law, determine the distance from earth to this quasar.

## ADDITIONAL PROBLEMS

22. The strong nuclear interaction has a range of approximately $1.4 \times 10^{-15}$ m. It is thought that an elementary particle is exchanged between the protons and neutrons in the nucleus, leading to an attractive force. (a) Utilize the uncertainty principle $\Delta E\,\Delta t \gtrsim \hbar$ to estimate the mass of the elementary particle if it moves at nearly the speed of light. (b) Using Table 47.2, identify the particle.
23. A gamma-ray photon strikes a stationary electron. Determine the minimum gamma-ray energy to make this reaction go:

$$\gamma + e^- \longrightarrow e^- + e^- + e^+$$

24. What are the kinetic energies of the proton and pi-meson resulting from the decay of a $\Lambda^0$ at rest?

$$\Lambda^0 \longrightarrow p + \pi^-$$

25. The energy flux of neutrinos from the sun is estimated to be on the order of 0.4 W/m$^2$, at the earth's surface. Estimate the fractional mass loss of the sun over $10^9$ years due to the radiation of neutrinos. (The mass of the sun is $2 \times 10^{30}$ kg. The distance of the earth from the sun is $1.5 \times 10^{11}$ m.)
26. A $\Sigma^0$ particle at rest decays according to

$$\Sigma^0 \longrightarrow \Lambda^0 + \gamma$$

Find the gamma-ray energy.

27. If a $K^0$ meson at rest decays in $0.9 \times 10^{-10}$ s, how far will a $K^0$ meson travel if it is moving at $0.96c$ through a bubble chamber?
28. A $\pi$-meson at rest decays according to $\pi^- \rightarrow \mu^- + \overline{\nu}_\mu$. What is the energy carried off by the neutrino? (Assume the neutrino moves off with the speed of light.) $m_\pi c^2 = 139.5$ MeV, $m_\mu c^2 = 105.7$ MeV, $m_\nu = 0$.
29. Two protons approach each other with equal and opposite velocities. What is the minimum kinetic energy of each of the protons if they are to produce a $\pi^+$ meson at rest in the reaction

$$p + p \longrightarrow p + n + \pi^+$$

30. What processes are described by the Feynman diagrams in Figure 47.18? What is the exchanged particle in each process?

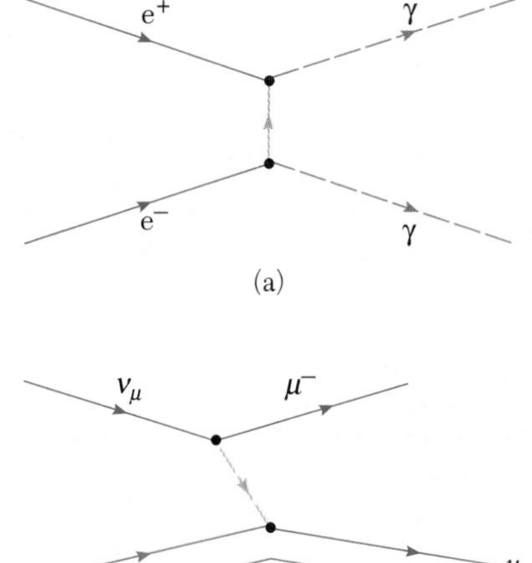


**Figure 47.18** (Problem 30).

### The Meaning of Success

To earn the respect of intelligent people and to win the affection of children;
To appreciate the beauty in nature and all that surrounds us;
To seek out and nurture the best in others;
To give the gift of yourself to others without the slightest thought of return, for it is in giving that we receive;
To have accomplished a task, whether it be saving a lost soul, healing a sick child, writing a book, or risking your life for a friend;
To have celebrated and laughed with great joy and enthusiasm and sung with exaltation;
To have hope even in times of despair, for as long as you have hope, you have life;
To love and be loved;
To be understood and to understand;
To know that even one life has breathed easier because you have lived;
This is the meaning of success.

ANONYMOUS and modified by
Ray Serway December, 1989

# Appendix A

**TABLE A.1 Conversion Factors**

**Length**

| | m | cm | km | in. | ft | mi |
|---|---|---|---|---|---|---|
| 1 meter | 1 | $10^2$ | $10^{-3}$ | 39.37 | 3.281 | $6.214 \times 10^{-4}$ |
| 1 centimeter | $10^{-2}$ | 1 | $10^{-5}$ | 0.3937 | $3.281 \times 10^{-2}$ | $6.214 \times 10^{-6}$ |
| 1 kilometer | $10^3$ | $10^5$ | 1 | $3.937 \times 10^4$ | $3.281 \times 10^3$ | 0.6214 |
| 1 inch | $2.540 \times 10^{-2}$ | 2.540 | $2.540 \times 10^{-5}$ | 1 | $8.333 \times 10^{-2}$ | $1.578 \times 10^{-5}$ |
| 1 foot | 0.3048 | 30.48 | $3.048 \times 10^{-4}$ | 12 | 1 | $1.894 \times 10^{-4}$ |
| 1 mile | 1609 | $1.609 \times 10^5$ | 1.609 | $6.336 \times 10^4$ | 5280 | 1 |

**Mass**

| | kg | g | slug | u |
|---|---|---|---|---|
| 1 kilogram | 1 | $10^3$ | $6.852 \times 10^{-2}$ | $6.024 \times 10^{26}$ |
| 1 gram | $10^{-3}$ | 1 | $6.852 \times 10^{-5}$ | $6.024 \times 10^{23}$ |
| 1 slug (lb/g) | 14.59 | $1.459 \times 10^4$ | 1 | $8.789 \times 10^{27}$ |
| 1 atomic mass unit | $1.660 \times 10^{-27}$ | $1.660 \times 10^{-24}$ | $1.137 \times 10^{-28}$ | 1 |

**Time**

| | s | min | h | day | year |
|---|---|---|---|---|---|
| 1 second | 1 | $1.667 \times 10^{-2}$ | $2.778 \times 10^{-4}$ | $1.157 \times 10^{-5}$ | $3.169 \times 10^{-8}$ |
| 1 minute | 60 | 1 | $1.667 \times 10^{-2}$ | $6.994 \times 10^{-4}$ | $1.901 \times 10^{-6}$ |
| 1 hour | 3600 | 60 | 1 | $4.167 \times 10^{-2}$ | $1.141 \times 10^{-4}$ |
| 1 day | $8.640 \times 10^4$ | 1440 | 24 | 1 | $2.738 \times 10^{-3}$ |
| 1 year | $3.156 \times 10^7$ | $5.259 \times 10^5$ | $8.766 \times 10^3$ | 365.2 | 1 |

**Speed**

| | m/s | cm/s | ft/s | mi/h |
|---|---|---|---|---|
| 1 meter/second | 1 | $10^2$ | 3.281 | 2.237 |
| 1 centimeter/second | $10^{-2}$ | 1 | $3.281 \times 10^{-2}$ | $2.237 \times 10^{-2}$ |
| 1 foot/second | 0.3048 | 30.48 | 1 | 0.6818 |
| 1 mile/hour | 0.4470 | 44.70 | 1.467 | 1 |

Note: 1 mi/min = 60 mi/h = 88 ft/s.

**Force**

| | N | dyn | lb |
|---|---|---|---|
| 1 newton | 1 | $10^5$ | 0.2248 |
| 1 dyne | $10^{-5}$ | 1 | $2.248 \times 10^{-6}$ |
| 1 pound | 4.448 | $4.448 \times 10^5$ | 1 |

**TABLE A.1 (Continued)**

**Work, Energy, Heat**

| | J | erg | ft · lb |
|---|---|---|---|
| 1 joule | 1 | $10^7$ | 0.7376 |
| 1 erg | $10^{-7}$ | 1 | $7.376 \times 10^{-8}$ |
| 1 ft · lb | 1.356 | $1.356 \times 10^7$ | 1 |
| 1 eV | $1.602 \times 10^{-19}$ | $1.602 \times 10^{-12}$ | $1.182 \times 10^{-19}$ |
| 1 cal | 4.186 | $4.186 \times 10^7$ | 3.087 |
| 1 Btu | $1.055 \times 10^3$ | $1.055 \times 10^{10}$ | $7.779 \times 10^2$ |
| 1 kWh | $3.600 \times 10^6$ | $3.600 \times 10^{13}$ | $2.655 \times 10^6$ |

| | eV | cal | Btu | kWh |
|---|---|---|---|---|
| 1 joule | $6.242 \times 10^{18}$ | 0.2389 | $9.481 \times 10^{-4}$ | $2.778 \times 10^{-7}$ |
| 1 erg | $6.242 \times 10^{11}$ | $2.389 \times 10^{-8}$ | $9.481 \times 10^{-11}$ | $2.778 \times 10^{-14}$ |
| 1 ft · lb | $8.464 \times 10^{18}$ | 0.3239 | $1.285 \times 10^{-3}$ | $3.766 \times 10^{-7}$ |
| 1 eV | 1 | $3.827 \times 10^{-20}$ | $1.519 \times 10^{-22}$ | $4.450 \times 10^{-26}$ |
| 1 cal | $2.613 \times 10^{19}$ | 1 | $3.968 \times 10^{-3}$ | $1.163 \times 10^{-6}$ |
| 1 Btu | $6.585 \times 10^{21}$ | $2.520 \times 10^2$ | 1 | $2.930 \times 10^{-4}$ |
| 1 kWh | $2.247 \times 10^{25}$ | $8.601 \times 10^5$ | $3.413 \times 10^2$ | 1 |

**Pressure**

| | Pa | dyn/cm² | atm |
|---|---|---|---|
| 1 pascal | 1 | 10 | $9.869 \times 10^{-6}$ |
| 1 dyne/centimeter² | $10^{-1}$ | 1 | $9.869 \times 10^{-7}$ |
| 1 atmosphere | $1.013 \times 10^5$ | $1.013 \times 10^6$ | 1 |
| 1 centimeter mercury* | $1.333 \times 10^3$ | $1.333 \times 10^4$ | $1.316 \times 10^{-2}$ |
| 1 pound/inch² | $6.895 \times 10^3$ | $6.895 \times 10^4$ | $6.805 \times 10^{-2}$ |
| 1 pound/foot² | 47.88 | $4.788 \times 10^2$ | $4.725 \times 10^{-4}$ |

| | cm Hg | lb/in.² | lb/ft² |
|---|---|---|---|
| 1 newton/meter² | $7.501 \times 10^{-4}$ | $1.450 \times 10^{-4}$ | $2.089 \times 10^{-2}$ |
| 1 dyne/centimeter² | $7.501 \times 10^{-5}$ | $1.450 \times 10^{-5}$ | $2.089 \times 10^{-3}$ |
| 1 atmosphere | 76 | 14.70 | $2.116 \times 10^3$ |
| 1 centimeter mercury* | 1 | 0.1943 | 27.85 |
| 1 pound/inch² | 5.171 | 1 | 144 |
| 1 pound/foot² | $3.591 \times 10^{-2}$ | $6.944 \times 10^{-3}$ | 1 |

* At 0°C and at a location where the acceleration due to gravity has its "standard" value, 9.80665 m/s².

**TABLE A.2 Symbols, Dimensions, and Units of Physical Quantities**

| Quantity | Common Symbol | Unit* | Dimensions† | Unit in Terms of Base SI Units |
|---|---|---|---|---|
| Acceleration | $a$ | m/s² | L/T² | m/s² |
| Amount of substance | $n$ | mole | | mol |
| Angle | $\theta, \phi$ | radian (rad) | 1 | |
| Angular acceleration | $\alpha$ | rad/s² | T⁻² | s⁻² |
| Angular frequency | $\omega$ | rad/s | T⁻¹ | s⁻¹ |
| Angular momentum | $L$ | kg · m²/s | ML²/T | kg · m²/s |
| Angular velocity | $\omega$ | rad/s | T⁻¹ | s⁻¹ |
| Area | $A$ | m² | L² | m² |

*(Table continues)*

TABLE A.2 (Continued)

| Quantity | Common Symbol | Unit* | Dimensions† | Unit in Terms of Base SI Units |
|---|---|---|---|---|
| Atomic number | $Z$ | | | |
| Capacitance | $C$ | farad (F)(=C/V) | $Q^2T^2/ML^2$ | $A^2 \cdot s^4/kg \cdot m^2$ |
| Charge | $q, Q, e$ | coulomb (C) | Q | $A \cdot s$ |
| Charge density | | | | |
| Line | $\lambda$ | C/m | Q/L | $A \cdot s/m$ |
| Surface | $\sigma$ | $C/m^2$ | $Q/L^2$ | $A \cdot s/m^2$ |
| Volume | $\rho$ | $C/m^3$ | $Q/L^3$ | $A \cdot s/m^3$ |
| Conductivity | $\sigma$ | $1/\Omega \cdot m$ | $Q^2T/ML^3$ | $A^2 \cdot s^3/kg \cdot m^3$ |
| Current | $I$ | AMPERE | Q/T | A |
| Current density | $J$ | $A/m^2$ | $Q/T^2$ | $A/m^2$ |
| Density | $\rho$ | $kg/m^3$ | $M/L^3$ | $kg/m^3$ |
| Dielectric constant | $\kappa$ | | | |
| Displacement | $s$ | METER | L | m |
| Distance | $d, h$ | | | |
| Length | $\ell, L$ | | | |
| Electric dipole moment | $p$ | $C \cdot m$ | QL | $A \cdot s \cdot m$ |
| Electric field | $E$ | V/m | $ML/QT^2$ | $kg \cdot m/A \cdot s^3$ |
| Electric flux | $\Phi$ | $V \cdot m$ | $ML^3/QT^2$ | $kg \cdot m^3/A \cdot s^3$ |
| Electromotive force | $\mathcal{E}$ | volt (V) | $ML^2/QT^2$ | $kg \cdot m^2/A \cdot s^3$ |
| Energy | $E, U, K$ | joule (J) | $ML^2/T^2$ | $kg \cdot m^2/s^2$ |
| Entropy | $S$ | J/K | $ML^2/T^2 {}^\circ K$ | $kg \cdot m^2/s^2 \cdot K$ |
| Force | $F$ | newton (N) | $ML/T^2$ | $kg \cdot m/s^2$ |
| Frequency | $f, \nu$ | hertz (Hz) | $T^{-1}$ | $s^{-1}$ |
| Heat | $Q$ | joule (J) | $ML^2/T^2$ | $kg \cdot m^2/s^2$ |
| Inductance | $L$ | henry (H) | $ML^2/Q^2$ | $kg \cdot m^2/A^2 \cdot s^2$ |
| Magnetic dipole moment | $\mu$ | $N \cdot m/T$ | $QL^2/T$ | $A \cdot m^2$ |
| Magnetic field | $B$ | tesla (T)(=$Wb/m^2$) | M/QT | $kg/A \cdot s^2$ |
| Magnetic flux | $\Phi_m$ | weber (Wb) | $ML^2/QT$ | $kg \cdot m^2/A \cdot s^2$ |
| Mass | $m, M$ | KILOGRAM | M | kg |
| Molar specific heat | $C$ | $J/mol \cdot K$ | | $kg \cdot m^2/s^2 \cdot kmol \cdot K$ |
| Moment of inertia | $I$ | $kg \cdot m^2$ | $ML^2$ | $kg \cdot m^2$ |
| Momentum | $p$ | $kg \cdot m/s$ | ML/T | $kg \cdot m/s$ |
| Period | $T$ | s | T | s |
| Permeability of space | $\mu_o$ | $N/A^2$ (=H/m) | $ML/Q^2T$ | $kg \cdot m/A^2 \cdot s^2$ |
| Permittivity of space | $\epsilon_o$ | $C^2/N \cdot m^2$ (= F/m) | $Q^2T^2/ML^3$ | $A^2 \cdot s^4/kg \cdot m^3$ |
| Potential (voltage) | $V$ | volt (V)(=J/C) | $ML^2/QT^2$ | $kg \cdot m^2/A \cdot s^3$ |
| Power | $P$ | watt (W)(=J/s) | $ML^2/T^3$ | $kg \cdot m^2/s^3$ |
| Pressure | $P, p$ | pascal (Pa) = ($N/m^2$) | $M/LT^2$ | $kg/m \cdot s^2$ |
| Resistance | $R$ | ohm ($\Omega$)(=V/A) | $ML^2/Q^2T$ | $kg \cdot m^2/A^2 \cdot s^3$ |
| Specific heat | $c$ | $J/kg \cdot K$ | $L^2/T^2 {}^\circ K$ | $m^2/s^2 \cdot K$ |
| Temperature | $T$ | KELVIN | °K | K |
| Time | $t$ | SECOND | T | s |
| Torque | $\tau$ | $N \cdot m$ | $ML^2/T^2$ | $kg \cdot m^2/s^2$ |
| Speed | $v$ | m/s | L/T | m/s |
| Volume | $V$ | $m^3$ | $L^3$ | $m^3$ |
| Wavelength | $\lambda$ | m | L | m |
| Work | $W$ | joule (J)(=$N \cdot m$) | $ML^2/T^2$ | $kg \cdot m^2/s^2$ |

* The base SI units are given in upper case letters.

† The symbols M, L, T, and Q denote mass, length, time, and charge, respectively.

**TABLE A.3 Table of Selected Atomic Masses***

| Atomic Number Z | Element | Symbol | Mass Number, A | Atomic Mass† | Percent Abundance, or Decay Mode (if radioactive)‡ | Half-Life (if radioactive) |
|---|---|---|---|---|---|---|
| 0 | (Neutron) | *n* | 1 | 1.008665 | $\beta^-$ | 10.6 min |
| 1 | Hydrogen | H | 1 | 1.007825 | 99.985 | |
| | Deuterium | D | 2 | 2.014102 | 0.015 | |
| | Tritium | T | 3 | 3.016049 | $\beta^-$ | 12.33 y |
| 2 | Helium | He | 3 | 3.016029 | 0.00014 | |
| | | | 4 | 4.002603 | ≈100 | |
| 3 | Lithium | Li | 6 | 6.015123 | 7.5 | |
| | | | 7 | 7.016005 | 92.5 | |
| 4 | Beryllium | Be | 7 | 7.016930 | EC, $\gamma$ | 53.3 days |
| | | | 8 | 8.005305 | $2\alpha$ | $6.7 \times 10^{-17}$ s |
| | | | 9 | 9.012183 | 100 | |
| 5 | Boron | B | 10 | 10.012938 | 19.8 | |
| | | | 11 | 11.009305 | 80.2 | |
| 6 | Carbon | C | 11 | 11.011433 | $\beta^+$, EC | 20.4 min |
| | | | 12 | 12.000000 | 98.89 | |
| | | | 13 | 13.003355 | 1.11 | |
| | | | 14 | 14.003242 | $\beta^-$ | 5730 y |
| 7 | Nitrogen | N | 13 | 13.005739 | $\beta^+$ | 9.96 min |
| | | | 14 | 14.003074 | 99.63 | |
| | | | 15 | 15.000109 | 0.37 | |
| 8 | Oxygen | O | 15 | 15.003065 | $\beta^+$, EC | 122 s |
| | | | 16 | 15.994915 | 99.759 | |
| | | | 18 | 17.999159 | 0.204 | |
| 9 | Fluorine | F | 19 | 18.998403 | 100 | |
| 10 | Neon | Ne | 20 | 19.992439 | 90.51 | |
| | | | 22 | 21.991384 | 9.22 | |
| 11 | Sodium | Na | 22 | 21.994435 | $\beta^+$, EC, $\gamma$ | 2.602 y |
| | | | 23 | 22.989770 | 100 | |
| | | | 24 | 23.990964 | $\beta^-$, $\gamma$ | 15.0 h |
| 12 | Magnesium | Mg | 24 | 23.985045 | 78.99 | |
| 13 | Aluminum | Al | 27 | 26.981541 | 100 | |
| 14 | Silicon | Si | 28 | 27.976928 | 92.23 | |
| | | | 31 | 30.975364 | $\beta^-$, $\gamma$ | 2.62 h |
| 15 | Phosphorus | P | 31 | 30.973763 | 100 | |
| | | | 32 | 31.973908 | $\beta^-$ | 14.28 days |
| 16 | Sulfur | S | 32 | 31.972072 | 95.0 | |
| | | | 35 | 34.969033 | $\beta^-$ | 87.4 days |
| 17 | Chlorine | Cl | 35 | 34.968853 | 75.77 | |
| | | | 37 | 36.965903 | 24.23 | |
| 18 | Argon | Ar | 40 | 39.962383 | 99.60 | |
| 19 | Potassium | K | 39 | 38.963708 | 93.26 | |
| | | | 40 | 39.964000 | $\beta^-$, EC, $\gamma$, $\beta^+$ | $1.28 \times 10^9$ y |
| 20 | Calcium | Ca | 40 | 39.962591 | 96.94 | |
| 21 | Scandium | Sc | 45 | 44.955914 | 100 | |
| 22 | Titanium | Ti | 48 | 47.947947 | 73.7 | |
| 23 | Vanadium | V | 51 | 50.943963 | 99.75 | |

*(Table continues)*

**TABLE A.3 (Continued)**

| Atomic Number Z | Element | Symbol | Mass Number, A | Atomic Mass† | Percent Abundance, or Decay Mode (if radioactive)‡ | Half-Life (if radioactive) |
|---|---|---|---|---|---|---|
| 24 | Chromium | Cr | 52 | 51.940510 | 83.79 | |
| 25 | Manganese | Mn | 55 | 54.938046 | 100 | |
| 26 | Iron | Fe | 56 | 55.934939 | 91.8 | |
| 27 | Cobalt | Co | 59 | 58.933198 | 100 | |
| | | | 60 | 59.933820 | $\beta^-, \gamma$ | 5.271 y |
| 28 | Nickel | Ni | 58 | 57.935347 | 68.3 | |
| | | | 60 | 59.930789 | 26.1 | |
| | | | 64 | 63.927968 | 0.91 | |
| 29 | Copper | Cu | 63 | 62.929599 | 69.2 | |
| | | | 64 | 63.929766 | $\beta^-, \beta^+$ | 12.7 h |
| | | | 65 | 64.927792 | 30.8 | |
| 30 | Zinc | Zn | 64 | 63.929145 | 48.6 | |
| | | | 66 | 65.926035 | 27.9 | |
| 31 | Gallium | Ga | 69 | 68.925581 | 60.1 | |
| 32 | Germanium | Ge | 72 | 71.922080 | 27.4 | |
| | | | 74 | 73.921179 | 36.5 | |
| 33 | Arsenic | As | 75 | 74.921596 | 100 | |
| 34 | Selenium | Se | 80 | 79.916521 | 49.8 | |
| 35 | Bromine | Br | 79 | 78.918336 | 50.69 | |
| 36 | Krypton | Kr | 84 | 83.911506 | 57.0 | |
| | | | 89 | 88.917563 | $\beta^-$ | 3.2 min |
| 37 | Rubidium | Rb | 85 | 84.911800 | 72.17 | |
| 38 | Strontium | Sr | 86 | 85.909273 | 9.8 | |
| | | | 88 | 87.905625 | 82.6 | |
| | | | 90 | 89.907746 | $\beta^-$ | 28.8 y |
| 39 | Yttrium | Y | 89 | 88.905856 | 100 | |
| 40 | Zirconium | Zr | 90 | 89.904708 | 51.5 | |
| 41 | Niobium | Nb | 93 | 92.906378 | 100 | |
| 42 | Molybdenum | Mo | 98 | 97.905405 | 24.1 | |
| 43 | Technetium | Tc | 98 | 97.907210 | $\beta^-, \gamma$ | $4.2 \times 10^6$ y |
| 44 | Ruthenium | Ru | 102 | 101.904348 | 31.6 | |
| 45 | Rhodium | Rh | 103 | 102.90550 | 100 | |
| 46 | Palladium | Pd | 106 | 105.90348 | 27.3 | |
| 47 | Silver | Ag | 107 | 106.905095 | 51.83 | |
| | | | 109 | 108.904754 | 48.17 | |
| 48 | Cadmium | Cd | 114 | 113.903361 | 28.7 | |
| 49 | Indium | In | 115 | 114.90388 | 95.7; $\beta^-$ | $5.1 \times 10^{14}$ y |
| 50 | Tin | Sn | 120 | 119.902199 | 32.4 | |
| 51 | Antimony | Sb | 121 | 120.903824 | 57.3 | |
| 52 | Tellurium | Te | 130 | 129.90623 | 34.5; $\beta^-$ | $2 \times 10^{21}$ y |
| 53 | Iodine | I | 127 | 126.904477 | 100 | |
| | | | 131 | 130.906118 | $\beta^-, \gamma$ | 8.04 days |

*(Table continues)*

° Data are taken from *Chart of the Nuclides,* 12th ed., General Electric, 1977, and from C. M. Lederer and V. S. Shirley, eds., *Table of Isotopes,* 7th ed., New York, John Wiley & Sons, Inc., 1978.

† The masses given in column (5) are those for the neutral atom, including the Z electrons.

‡ The process EC stands for "electron capture."

**TABLE A.3 (Continued)**

| Atomic Number Z | Element | Symbol | Mass Number, A | Atomic Mass† | Percent Abundance, or Decay Mode (if radioactive)‡ | Half-Life (if radioactive) |
|---|---|---|---|---|---|---|
| 54 | Xenon | Xe | 132 | 131.90415 | 26.9 | |
| | | | 136 | 135.90722 | 8.9 | |
| 55 | Cesium | Cs | 133 | 132.90543 | 100 | |
| 56 | Barium | Ba | 137 | 136.90582 | 11.2 | |
| | | | 138 | 137.90524 | 71.7 | |
| | | | 144 | 143.922673 | $\beta^-$ | 11.9 s |
| 57 | Lanthanum | La | 139 | 138.90636 | 99.911 | |
| 58 | Cerium | Ce | 140 | 139.90544 | 88.5 | |
| 59 | Praseodymium | Pr | 141 | 140.90766 | 100 | |
| 60 | Neodymium | Nd | 142 | 141.90773 | 27.2 | |
| 61 | Promethium | Pm | 145 | 144.91275 | EC, $\alpha$, $\gamma$ | 17.7 y |
| 62 | Samarium | Sm | 152 | 151.91974 | 26.6 | |
| 63 | Europium | Eu | 153 | 152.92124 | 52.1 | |
| 64 | Gadolinium | Gd | 158 | 157.92411 | 24.8 | |
| 65 | Terbium | Tb | 159 | 158.92535 | 100 | |
| 66 | Dysprosium | Dy | 164 | 163.92918 | 28.1 | |
| 67 | Holmium | Ho | 165 | 164.93033 | 100 | |
| 68 | Erbium | Er | 166 | 165.93031 | 33.4 | |
| 69 | Thulium | Tm | 169 | 168.93423 | 100 | |
| 70 | Ytterbium | Yb | 174 | 173.93887 | 31.6 | |
| 71 | Lutecium | Lu | 175 | 174.94079 | 97.39 | |
| 72 | Hafnium | Hf | 180 | 179.94656 | 35.2 | |
| 73 | Tantalum | Ta | 181 | 180.94801 | 99.988 | |
| 74 | Tungsten (wolfram) | W | 184 | 183.95095 | 30.7 | |
| 75 | Rhenium | Re | 187 | 186.95577 | 62.60, $\beta^-$ | $4 \times 10^{10}$ y |
| 76 | Osmium | Os | 191 | 190.96094 | $\beta^-$, $\gamma$ | 15.4 days |
| | | | 192 | 191.96149 | 41.0 | |
| 77 | Iridium | Ir | 191 | 190.96060 | 37.3 | |
| | | | 193 | 192.96294 | 62.7 | |
| 78 | Platinum | Pt | 195 | 194.96479 | 33.8 | |
| 79 | Gold | Au | 197 | 196.96656 | 100 | |
| 80 | Mercury | Hg | 202 | 201.97063 | 29.8 | |
| 81 | Thallium | Tl | 205 | 204.97441 | 70.5 | |
| | | | 208 | 207.981988 | $\beta^-$, $\gamma$ | 3.053 min |
| 82 | Lead | Pb | 204 | 203.973044 | $\beta^-$, 1.48 | $1.4 \times 10^{17}$ y |
| | | | 206 | 205.97446 | 24.1 | |
| | | | 207 | 206.97589 | 22.1 | |
| | | | 208 | 207.97664 | 52.3 | |
| | | | 210 | 209.98418 | $\alpha$, $\beta^-$, $\gamma$ | 22.3 y |
| | | | 211 | 210.98874 | $\beta^-$, $\gamma$ | 36.1 min |
| | | | 212 | 211.99188 | $\beta^-$, $\gamma$ | 10.64 h |
| | | | 214 | 213.99980 | $\beta^-$, $\gamma$ | 26.8 min |
| 83 | Bismuth | Bi | 209 | 208.98039 | 100 | |
| | | | 211 | 210.98726 | $\alpha$, $\beta^-$, $\gamma$ | 2.15 min |

*(Table continues)*

**TABLE A.3 (Continued)**

| Atomic Number Z | Element | Symbol | Mass Number, A | Atomic Mass† | Percent Abundance, or Decay Mode (if radioactive)‡ | Half-Life (if radioactive) |
|---|---|---|---|---|---|---|
| 84 | Polonium | Po | 210 | 209.98286 | $\alpha, \gamma$ | 138.38 days |
| | | | 214 | 213.99519 | $\alpha, \gamma$ | 164 $\mu s$ |
| 85 | Astatine | At | 218 | 218.00870 | $\alpha, \beta^-$ | $\approx 2$ s |
| 86 | Radon | Rn | 222 | 222.017574 | $\alpha, \gamma$ | 3.8235 days |
| 87 | Francium | Fr | 223 | 223.019734 | $\alpha, \beta^-, \gamma$ | 21.8 min |
| 88 | Radium | Ra | 226 | 226.025406 | $\alpha, \gamma$ | $1.60 \times 10^3$ y |
| | | | 228 | 228.031069 | $\beta^-$ | 5.76 y |
| 89 | Actinium | Ac | 227 | 227.027751 | $\alpha, \beta^-, \gamma$ | 21.773 y |
| 90 | Thorium | Th | 228 | 228.02873 | $\alpha, \gamma$ | 1.9131 y |
| | | | 232 | 232.038054 | 100, $\alpha, \gamma$ | $1.41 \times 10^{10}$ y |
| 91 | Protactinium | Pa | 231 | 231.035881 | $\alpha, \gamma$ | $3.28 \times 10^4$ y |
| 92 | Uranium | U | 232 | 232.03714 | $\alpha, \gamma$ | 72 y |
| | | | 233 | 233.039629 | $\alpha, \gamma$ | $1.592 \times 10^5$ y |
| | | | 235 | 235.043925 | 0.72; $\alpha, \gamma$ | $7.038 \times 10^8$ y |
| | | | 236 | 236.045563 | $\alpha, \gamma$ | $2.342 \times 10^7$ y |
| | | | 238 | 238.050786 | 99.275; $\alpha, \gamma$ | $4.468 \times 10^9$ y |
| | | | 239 | 239.054291 | $\beta^-, \gamma$ | 23.5 min |
| 93 | Neptunium | Np | 239 | 239.052932 | $\beta^-, \gamma$ | 2.35 days |
| 94 | Plutonium | Pu | 239 | 239.052158 | $\alpha, \gamma$ | $2.41 \times 10^4$ y |
| 95 | Americium | Am | 243 | 243.061374 | $\alpha, \gamma$ | $7.37 \times 10^3$ y |
| 96 | Curium | Cm | 245 | 245.065487 | $\alpha, \gamma$ | $8.5 \times 10^3$ y |
| 97 | Berkelium | Bk | 247 | 247.07003 | $\alpha, \gamma$ | $1.4 \times 10^3$ y |
| 98 | Californium | Cf | 249 | 249.074849 | $\alpha, \gamma$ | 351 y |
| 99 | Einsteinium | Es | 254 | 254.08802 | $\alpha, \gamma, \beta^-$ | 276 days |
| 100 | Fermium | Fm | 253 | 253.08518 | EC, $\alpha, \gamma$ | 3.0 days |
| 101 | Mendelevium | Md | 255 | 255.0911 | EC, $\alpha$ | 27 min |
| 102 | Nobelium | No | 255 | 255.0933 | EC, $\alpha$ | 3.1 min |
| 103 | Lawrencium | Lr | 257 | 257.0998 | $\alpha$ | $\approx 35$ s |
| 104 | Unnilquadium | Rf | 261 | 261.1087 | $\alpha$ | 1.1 min |
| 105 | Unnilpentium | Ha | 262 | 262.1138 | $\alpha$ | 0.7 min |
| 106 | Unnilhexium | | 263 | 263.1184 | $\alpha$ | 0.9 s |
| 107 | Unnilseptium | | 261 | 261 | $\alpha$ | 1–2 ms |

° Data are taken from *Chart of the Nuclides,* 12th ed., General Electric, 1977, and from C. M. Lederer and V. S. Shirley, eds., *Table of Isotopes,* 7th ed., New York, John Wiley & Sons, Inc., 1978.

† The masses given in column (5) are those for the neutral atom, including the Z electrons.

‡ The process EC stands for "electron capture."

# Appendix B

## Mathematics Review

These appendices in mathematics are intended as a brief review of operations and methods. Early in this course, you should be totally familiar with basic algebraic techniques, analytic geometry, and trigonometry. The appendices on differential and integral calculus are more detailed and are intended for those students who have difficulties in applying calculus concepts to physical situations.

### B.1 SCIENTIFIC NOTATION

Many quantities that scientists deal with often have very large or very small values. For example, the speed of light is about 300 000 000 m/s and the ink required to make the dot over an *i* in this textbook has a mass of about 0.000 000 001 kg. Obviously, it is very cumbersome to read, write, and keep track of numbers such as these. We avoid this problem by using a method dealing with powers of the number 10:

$$10^0 = 1$$

$$10^1 = 10$$

$$10^2 = 10 \times 10 = 100$$

$$10^3 = 10 \times 10 \times 10 = 1000$$

$$10^4 = 10 \times 10 \times 10 \times 10 = 10\,000$$

$$10^5 = 10 \times 10 \times 10 \times 10 \times 10 = 100\,000$$

and so on. The number of zeros corresponds to the power to which 10 is raised, called the **exponent** of 10. For example, the speed of light, 300 000 000 m/s, can be expressed as $3 \times 10^8$ m/s.

For numbers less than one, we note the following:

$$10^{-1} = \frac{1}{10} = 0.1$$

$$10^{-2} = \frac{1}{10 \times 10} = 0.01$$

$$10^{-3} = \frac{1}{10 \times 10 \times 10} = 0.001$$

$$10^{-4} = \frac{1}{10 \times 10 \times 10 \times 10} = 0.0001$$

$$10^{-5} = \frac{1}{10 \times 10 \times 10 \times 10 \times 10} = 0.00001$$

In these cases, the number of places the decimal point is to the left of the digit 1 equals the value of the (negative) exponent. Numbers that are expressed as some power of 10 multiplied by another number between 1 and 10 are said to be in **scientific notation.** For example, the scientific notation for 5 943 000 000 is $5.943 \times 10^9$ and that for 0.0000832 is $8.32 \times 10^{-5}$.

When numbers expressed in scientific notation are being multiplied, the following general rule is very useful:

$$10^n \times 10^m = 10^{n+m} \tag{B.1}$$

where $n$ and $m$ can be *any* numbers (not necessarily integers). For example, $10^2 \times 10^5 = 10^7$. The rule also applies if one of the exponents is negative. For example, $10^3 \times 10^{-8} = 10^{-5}$.

When dividing numbers expressed in scientific notation, note that

$$\frac{10^n}{10^m} = 10^n \times 10^{-m} = 10^{n-m} \tag{B.2}$$

### EXERCISES

With help from the above rules, verify the answers to the following:

1. $86{,}400 = 8.64 \times 10^4$
2. $9{,}816{,}762.5 = 9.8167625 \times 10^6$
3. $0.0000000398 = 3.98 \times 10^{-8}$
4. $(4 \times 10^8)(9 \times 10^9) = 3.6 \times 10^{18}$
5. $(3 \times 10^7)(6 \times 10^{-12}) = 1.8 \times 10^{-4}$
6. $\dfrac{75 \times 10^{-11}}{5 \times 10^{-3}} = 1.5 \times 10^{-7}$
7. $\dfrac{(3 \times 10^6)(8 \times 10^{-2})}{(2 \times 10^{17})(6 \times 10^5)} = 2 \times 10^{-18}$

## B.2 ALGEBRA

### Some Basic Rules

When algebraic operations are performed, the laws of arithmetic apply. Symbols such as $x$, $y$, and $z$ are usually used to represent quantities that are not specified, what are called the **unknowns.**

First, consider the equation

$$8x = 32$$

If we wish to solve for $x$, we can divide (or multiply) each side of the equation by the same factor without destroying the equality. In this case, if we divide both sides by 8, we have

$$\frac{8x}{8} = \frac{32}{8}$$

$$x = 4$$

Next consider the equation

$$x + 2 = 8$$

In this type of expression, we can add or subtract the same quantity from each side. If we subtract 2 from each side, we get

$$x + 2 - 2 = 8 - 2$$

$$x = 6$$

In general, if $x + a = b$, then $x = b - a$.

Now consider the equation

$$\frac{x}{5} = 9$$

If we multiply each side by 5, we are left with $x$ on the left by itself and 45 on the right:

$$\left(\frac{x}{5}\right)(5) = 9 \times 5$$

$$x = 45$$

In all cases, *whatever operation is performed on the left side of the equality must also be performed on the right side.*

The following rules for multiplying, dividing, adding, and subtracting fractions should be recalled, where $a$, $b$, and $c$ are three numbers:

| | Rule | Example |
|---|---|---|
| **Multiplying** | $\left(\frac{a}{b}\right)\left(\frac{c}{d}\right) = \frac{ac}{bd}$ | $\left(\frac{2}{3}\right)\left(\frac{4}{5}\right) = \frac{8}{15}$ |
| **Dividing** | $\frac{(a/b)}{(c/d)} = \frac{ad}{bc}$ | $\frac{2/3}{4/5} = \frac{(2)(5)}{(4)(3)} = \frac{10}{12}$ |
| **Adding** | $\frac{a}{b} \pm \frac{c}{d} = \frac{ad \pm bc}{bd}$ | $\frac{2}{3} - \frac{4}{5} = \frac{(2)(5) - (4)(3)}{(3)(5)} = -\frac{2}{15}$ |

## EXERCISES

In the following exercises, solve for $x$:

| | Answers |
|---|---|
| 1. $a = \frac{1}{1 + x}$ | $x = \frac{1 - a}{a}$ |
| 2. $3x - 5 = 13$ | $x = 6$ |
| 3. $ax - 5 = bx + 2$ | $x = \frac{7}{a - b}$ |
| 4. $\frac{5}{2x + 6} = \frac{3}{4x + 8}$ | $x = -\frac{11}{7}$ |

### Powers

When powers of a given quantity $x$ are multiplied, the following rule applies:

$$x^n x^m = x^{n+m} \tag{B.3}$$

For example, $x^2x^4 = x^{2+4} = x^6$.

When dividing the powers of a given quantity, the rule is

$$\frac{x^n}{x^m} = x^{n-m} \tag{B.4}$$

For example, $x^8/x^2 = x^{8-2} = x^6$.

A power that is a fraction, such as $\frac{1}{3}$, corresponds to a root as follows:

$$x^{1/n} = \sqrt[n]{x} \tag{B.5}$$

For example, $4^{1/3} = \sqrt[3]{4} = 1.5874$. (A scientific calculator is useful for such calculations.)

Finally, any quantity $x^n$ that is raised to the $m$th power is

$$(x^n)^m = x^{nm} \tag{B.6}$$

Table B.1 summarizes the rules of exponents.

**TABLE B.1 Rules of Exponents**

| |
|---|
| $x^0 = 1$ |
| $x^1 = x$ |
| $x^n x^m = x^{n+m}$ |
| $x^n/x^m = x^{n-m}$ |
| $x^{1/n} = \sqrt[n]{x}$ |
| $(x^n)^m = x^{nm}$ |

### EXERCISES

Verify the following:

1. $3^2 \times 3^3 = 243$
2. $x^5x^{-8} = x^{-3}$
3. $x^{10}/x^{-5} = x^{15}$
4. $5^{1/3} = 1.709975$ (Use your calculator.)
5. $60^{1/4} = 2.783158$ (Use your calculator.)
6. $(x^4)^3 = x^{12}$

### Factoring

Some useful formulas for factoring an equation are

$ax + ay + az = a(x + y + x)$ common factor

$a^2 + 2ab + b^2 = (a + b)^2$ perfect square

$a^2 - b^2 = (a + b)(a - b)$ differences of squares

### Quadratic Equations

The general form of a quadratic equation is

$$ax^2 + bx + c = 0 \tag{B.7}$$

where $x$ is the unknown quantity and $a$, $b$, and $c$ are numerical factors referred to as **coefficients** of the equation. This equation has two roots, given by

$$x = \frac{-b \pm \sqrt{b^2 - 4ac}}{2a} \quad \text{(B.8)}$$

If $b^2 \geq 4ac$, the roots will be real.

EXAMPLE 1

The equation $x^2 + 5x + 4 = 0$ has the following roots corresponding to the two signs of the square-root term:

$$x = \frac{-5 \pm \sqrt{5^2 - (4)(1)(4)}}{2(1)} = \frac{-5 \pm \sqrt{9}}{2} = \frac{-5 \pm 3}{2}$$

that is,

$$x_+ = \frac{-5 + 3}{2} = -1 \qquad x_- = \frac{-5 - 3}{2} = -4$$

where $x_+$ refers to the root corresponding to the positive sign and $x_-$ refers to the root corresponding to the negative sign.

## EXERCISES

Solve the following quadratic equations:

| | Answers | |
|---|---|---|
| 1. $x^2 + 2x - 3 = 0$ | $x_+ = 1$ | $x_- = -3$ |
| 2. $2x^2 - 5x + 2 = 0$ | $x_+ = 2$ | $x_- = \frac{1}{2}$ |
| 3. $2x^2 - 4x - 9 = 0$ | $x_+ = 1 + \sqrt{22}/2$ | $x_- = 1 - \sqrt{22}/2$ |

### Linear Equations

A linear equation has the general form

$$y = ax + b \quad \text{(B.9)}$$

where $a$ and $b$ are constants. This equation is referred to as being linear because the graph of $y$ versus $x$ is a straight line, as shown in Figure B.1. The constant $b$, called the **intercept,** represents the value of $y$ at which the straight line intersects the $y$ axis. The constant $a$ is equal to the **slope** of the straight line and is also equal to the tangent of the angle that the line makes with the $x$ axis. If any two points on the straight line are specified by the coordinates $(x_1, y_1)$ and $(x_2, y_2)$, as in Figure B.1, then the **slope** of the straight line can be expressed as

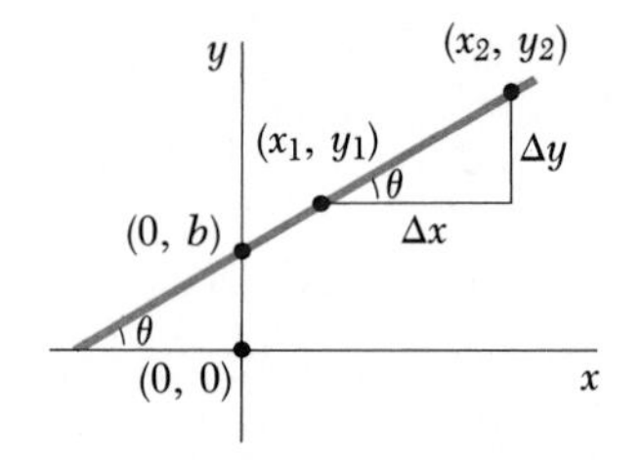

Figure B.1

$$\text{Slope} = \frac{y_2 - y_1}{x_2 - x_1} = \frac{\Delta y}{\Delta x} = \tan\theta \quad \text{(B.10)}$$

Note that $a$ and $b$ can have either positive or negative values. If $a > 0$, the straight line has a *positive* slope, as in Figure B.1. If $a < 0$, the straight line has a *negative* slope. In Figure B.1, both $a$ and $b$ are positive. Three other possible situations are shown in Figure B.2: $a > 0, b < 0$; $a < 0, b > 0$; and $a < 0, b < 0$.

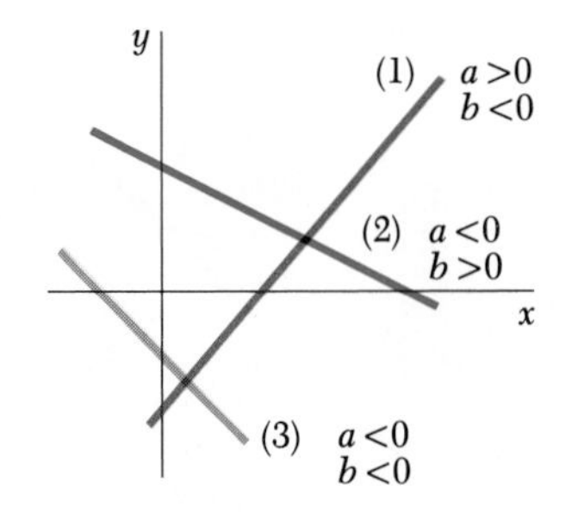


Figure B.2

## EXERCISES

**1.** Draw graphs of the following straight lines:
(a) $y = 5x + 3$ (b) $y = -2x + 4$ (c) $y = -3x - 6$
**2.** Find the slopes of the straight lines described in Exercise 1.
Answers (a) 5 (b) $-2$ (c) $-3$
**3.** Find the slopes of the straight lines that pass through the following sets of points:
(a) $(0, -4)$ and $(4, 2)$, (b) $(0, 0)$ and $(2, -5)$, and (c) $(-5, 2)$ and $(4, -2)$
Answers (a) 3/2 (b) $-5/2$ (c) $-4/9$

### Solving Simultaneous Linear Equations

Consider an equation such as $3x + 5y = 15$, which has two unknowns, $x$ and $y$. Such an equation does not have a unique solution. That is, $(x = 0, y = 3)$, $(x = 5, y = 0)$, and $(x = 2, y = 9/5)$ are all solutions to this equation.

If a problem has two unknowns, a unique solution is possible only if we have *two* equations. In general, if a problem has $n$ unknowns, its solution requires $n$ equations. In order to solve two simultaneous equations involving two unknowns, $x$ and $y$, we solve one of the equations for $x$ in terms of $y$ and substitute this expression into the other equation.

---

EXAMPLE 2
Solve the following two simultaneous equations:

$$(1)\ 5x + y = -8$$

$$(2)\ 2x - 2y = 4$$

*Solution* From (2), $x = y + 2$. Substitution of this into (1) gives

$$5(y + 2) + y = -8$$

$$6y = -18$$

$$y = -3$$

$$x = y + 2 = -1$$

*Alternate Solution* Multiply each term in (1) by the factor 2 and add the result to (2):

$$10x + 2y = -16$$

$$\underline{2x - 2y = 4}$$

$$12x = -12$$

$$x = -1$$

$$y = x - 2 = -3$$

---

Two linear equations with two unknowns can also be solved by a graphical method. If the straight lines corresponding to the two equations are plotted in

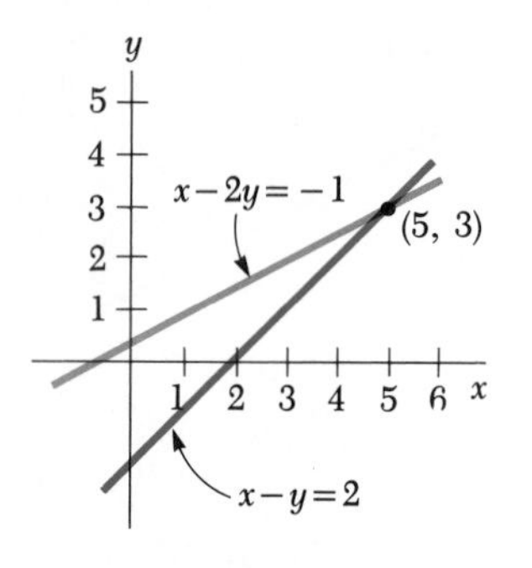


Figure B.3

a conventional coordinate system, the intersection of the two lines represents the solution. For example, consider the two equations

$$x - y = 2$$

$$x - 2y = -1$$

These are plotted in Figure B.3. The intersection of the two lines has the coordinates $x = 5$, $y = 3$. This represents the solution to the equations. You should check this solution by the analytical technique discussed above.

## EXERCISES

Solve the following pairs of simultaneous equations involving two unknowns:

| | Answers |
|---|---|
| 1. $x + y = 8$<br>$x - y = 2$ | $x = 5$, $y = 3$ |
| 2. $98 - T = 10a$<br>$T - 49 = 5a$ | $T = 65$, $a = 3.27$ |
| 3. $6x + 2y = 6$<br>$8x - 4y = 28$ | $x = 2$, $y = -3$ |

## Logarithms

Suppose that a quantity $x$ is expressed as a power of some quantity $a$:

$$x = a^y \tag{B.11}$$

The number $a$ is called the **base** number. The **logarithm** of $x$ with respect to the base $a$ is equal to the exponent to which the base must be raised in order to satisfy the expression $x = a^y$:

$$y = \log_a x \tag{B.12}$$

Conversely, the **antilogarithm** of $y$ is the number $x$:

$$x = \text{antilog}_a\, y \tag{B.13}$$

In practice, the two bases most often used are base 10, called the *common* logarithm base, and base $e = 2.718 \ldots$, called the *natural* logarithm base. When common logarithms are used,

$$y = \log_{10} x \qquad (\text{or } x = 10^y) \tag{B.14}$$

When natural logarithms are used,

$$y = \ln_e x \qquad (\text{or } x = e^y) \tag{B.15}$$

For example, $\log_{10} 52 = 1.716$, so that $\text{antilog}_{10}\, 1.716 = 10^{1.716} = 52$. Likewise, $\ln_e 52 = 3.951$, so $\text{antiln}_e\, 3.951 = e^{3.951} = 52$.

In general, note that you can convert between base 10 and base $e$ with the equality

$$\ln_e x = (2.302585) \log_{10} x \tag{B.16}$$

Finally, some useful properties of logarithms are as follows:

$$\log(ab) = \log a + \log b$$
$$\log(a/b) = \log a - \log b$$
$$\log(a^n) = n \log a$$
$$\ln e = 1$$
$$\ln e^a = a$$
$$\ln\left(\frac{1}{a}\right) = -\ln a$$

## B.3 GEOMETRY

The **distance** $d$ between two points whose coordinates are $(x_1, y_1)$ and $(x_2, y_2)$

$$d = \sqrt{(x_2 - x_1)^2 + (y_2 - y_1)^2} \tag{B.17}$$

The **radian measure:** the arc length $s$ of a circular arc (Fig. B.4) is proportional to the radius $r$ for a fixed value of $\theta$ (in radians)

$$s = r\theta$$
$$\theta = \frac{s}{r} \tag{B.18}$$

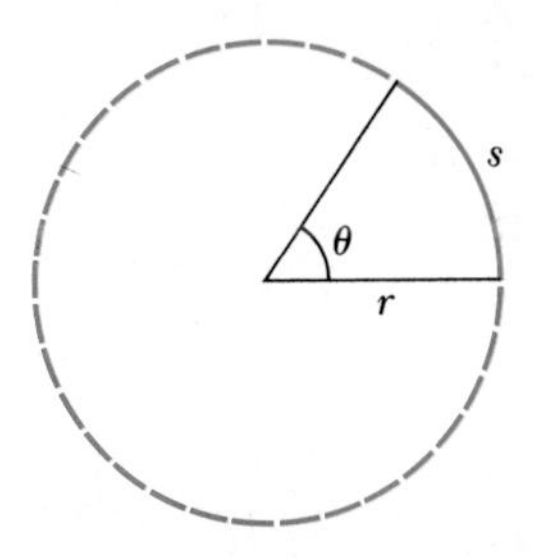

Figure B.4

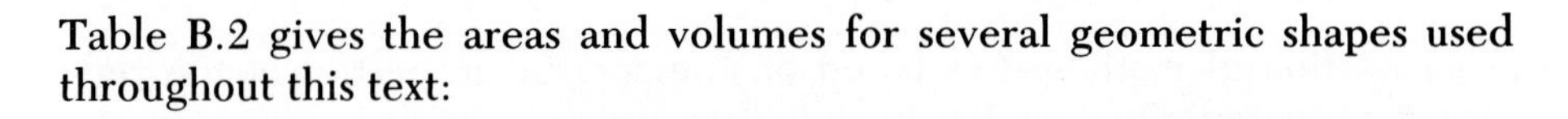

Table B.2 gives the areas and volumes for several geometric shapes used throughout this text:

**TABLE B.2 Useful Information for Geometry**

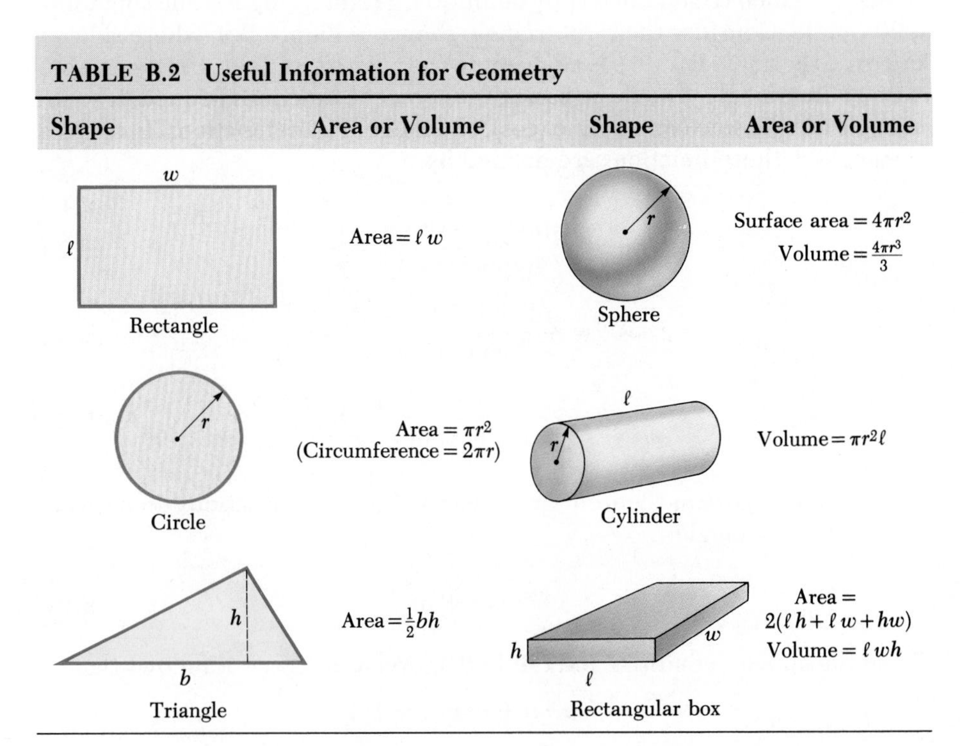

| Shape | Area or Volume | Shape | Area or Volume |
|---|---|---|---|
| Rectangle | Area $= \ell w$ | Sphere | Surface area $= 4\pi r^2$<br>Volume $= \frac{4\pi r^3}{3}$ |
| Circle | Area $= \pi r^2$<br>(Circumference $= 2\pi r$) | Cylinder | Volume $= \pi r^2 \ell$ |
| Triangle | Area $= \frac{1}{2}bh$ | Rectangular box | Area $=$<br>$2(\ell h + \ell w + hw)$<br>Volume $= \ell wh$ |

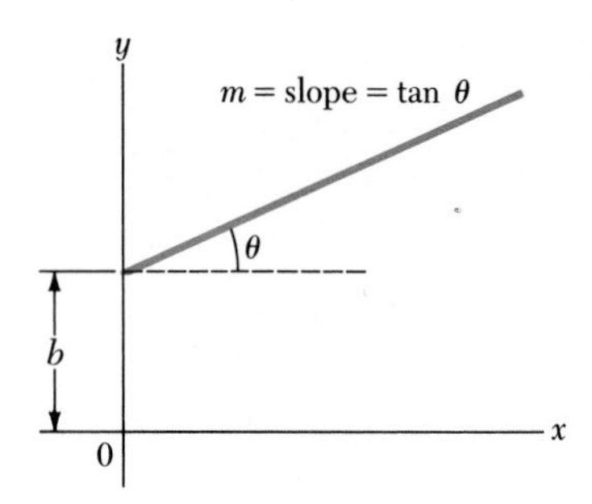

Figure B.5

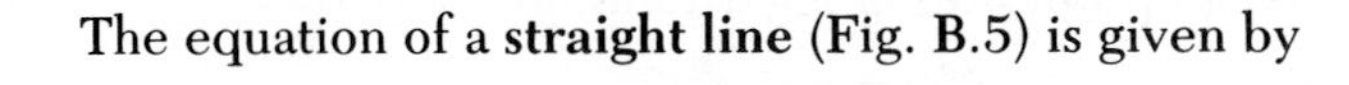

The equation of a **straight line** (Fig. B.5) is given by

$$y = mx + b \tag{B.19}$$

where $b$ is the $y$ intercept and $m$ is the slope of the line.

The equation of a **circle** of radius $R$ centered at the origin is

$$x^2 + y^2 = R^2 \tag{B.20}$$

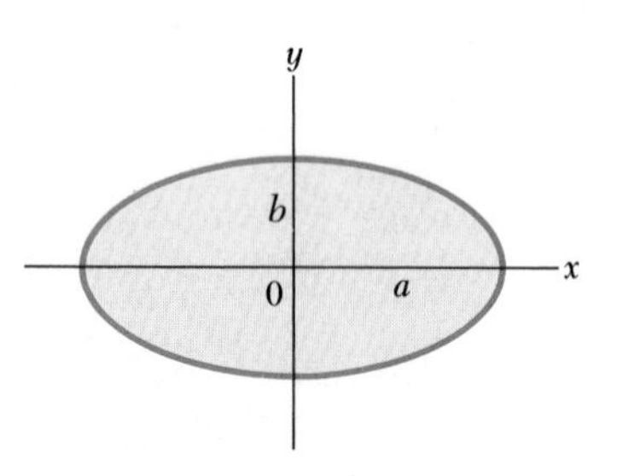

Figure B.6

The equation of an **ellipse** with the origin at its center (Fig. B.6) is

$$\frac{x^2}{a^2} + \frac{y^2}{b^2} = 1 \tag{B.21}$$

where $a$ is the length of the semi-major axis and $b$ is the length of the semi-minor axis.

The equation of a **parabola** whose vertex is at $y = b$ (Fig. B.7) is

$$y = ax^2 + b \tag{B.22}$$

The equation of a **rectangular hyperbola** (Fig. B.8) is

$$xy = \text{constant} \tag{B.23}$$

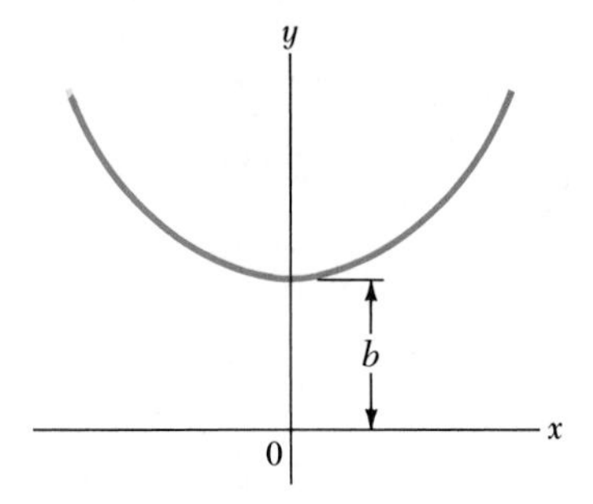

Figure B.7

## B.4 TRIGONOMETRY

That portion of mathematics based on the special properties of the right triangle is called trigonometry. By definition, a right triangle is one containing a 90° angle. Consider the right triangle shown in Figure B.9, where side $a$ is opposite the angle $\theta$, side $b$ is adjacent to the angle $\theta$, and side $c$ is the hypotenuse of the triangle. The three basic trigonometric functions defined by such a triangle are the sine (sin), cosine (cos), and tangent (tan) functions. In terms of the angle $\theta$, these functions are defined by

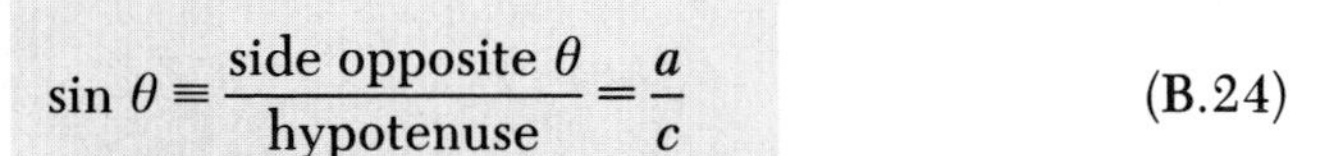

$$\sin\theta \equiv \frac{\text{side opposite } \theta}{\text{hypotenuse}} = \frac{a}{c} \tag{B.24}$$

$$\cos\theta \equiv \frac{\text{side adjacent to } \theta}{\text{hypotenuse}} = \frac{b}{c} \tag{B.25}$$

$$\tan\theta \equiv \frac{\text{side opposite } \theta}{\text{side adjacent to } \theta} = \frac{a}{b} \tag{B.26}$$

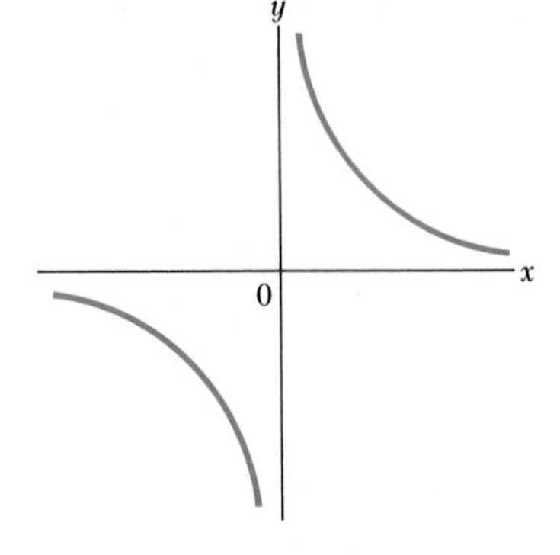

Figure B.8

The Pythagorean theorem provides the following relationship between the sides of a triangle:

$$c^2 = a^2 + b^2 \tag{B.27}$$

From the above definitions and the Pythagorean theorem, it follows that

$$\sin^2\theta + \cos^2\theta = 1$$

$$\tan \theta = \frac{\sin \theta}{\cos \theta}$$

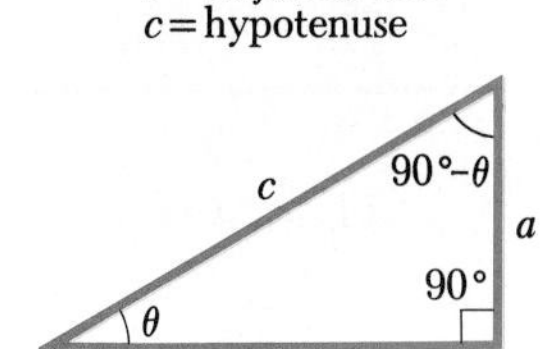


Figure B.9

The cosecant, secant, and cotangent functions are defined by

$$\csc \theta \equiv \frac{1}{\sin \theta} \qquad \sec \theta \equiv \frac{1}{\cos \theta} \qquad \cot \theta \equiv \frac{1}{\tan \theta}$$

The relations below follow directly from the right triangle shown in Figure B.9:

$$\begin{cases} \sin \theta = \cos(90^\circ - \theta) \\ \cos \theta = \sin(90^\circ - \theta) \\ \cot \theta = \tan(90^\circ - \theta) \end{cases}$$

Some properties of trigonometric functions are as follows:

$$\begin{cases} \sin(-\theta) = -\sin \theta \\ \cos(-\theta) = \cos \theta \\ \tan(-\theta) = -\tan \theta \end{cases}$$

The following relations apply to *any* triangle as shown in Figure B.10:

$$\alpha + \beta + \gamma = 180^\circ$$

Law of cosines
$$\begin{cases} a^2 = b^2 + c^2 - 2bc \cos \alpha \\ b^2 = a^2 + c^2 - 2ac \cos \beta \\ c^2 = a^2 + b^2 - 2ab \cos \gamma \end{cases}$$

Law of sines
$$\frac{a}{\sin \alpha} = \frac{b}{\sin \beta} = \frac{c}{\sin \gamma}$$

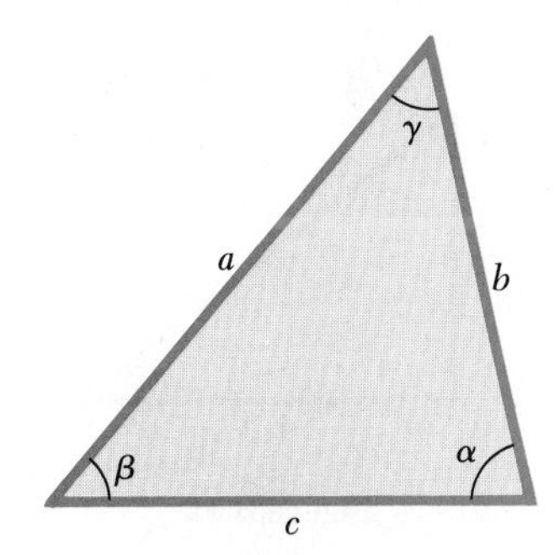


Figure B.10

Table B.3 lists a number of useful trigonometric identities.

**TABLE B.3 Some Trigonometric Identities**

| | |
|---|---|
| $\sin^2 \theta + \cos^2 \theta = 1$ | $\csc^2 \theta = 1 + \cot^2 \theta$ |
| $\sec^2 \theta = 1 + \tan^2 \theta$ | $\sin^2 \frac{\theta}{2} = \frac{1}{2}(1 - \cos \theta)$ |
| $\sin 2\theta = 2 \sin \theta \cos \theta$ | $\cos^2 \frac{\theta}{2} = \frac{1}{2}(1 + \cos \theta)$ |
| $\cos 2\theta = \cos^2 \theta - \sin^2 \theta$ | $1 - \cos \theta = 2 \sin^2 \frac{\theta}{2}$ |
| $\tan 2\theta = \frac{2 \tan \theta}{1 - \tan^2 \theta}$ | $\tan \frac{\theta}{2} = \sqrt{\frac{1 - \cos \theta}{1 + \cos \theta}}$ |
| $\sin(A \pm B) = \sin A \cos B \pm \cos A \sin B$ | |
| $\cos(A \pm B) = \cos A \cos B \mp \sin A \sin B$ | |
| $\sin A \pm \sin B = 2 \sin[\frac{1}{2}(A \pm B)]\cos[\frac{1}{2}(A \mp B)]$ | |
| $\cos A + \cos B = 2 \cos[\frac{1}{2}(A + B)]\cos[\frac{1}{2}(A - B)]$ | |
| $\cos A - \cos B = 2 \sin[\frac{1}{2}(A + B)]\sin[\frac{1}{2}(B - A)]$ | |

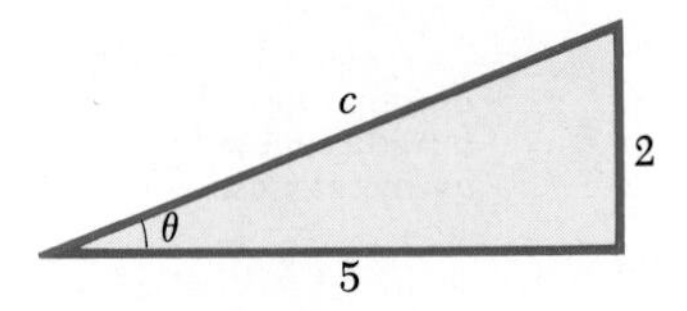


Figure B.11

**EXAMPLE 3**
Consider the right triangle in Figure B.11, in which $a = 2$, $b = 5$, and $c$ is unknown. From the Pythagorean theorem, we have

$$c^2 = a^2 + b^2 = 2^2 + 5^2 = 4 + 25 = 29$$

$$c = \sqrt{29} = \boxed{5.39}$$

To find the angle $\theta$, note that

$$\tan \theta = \frac{a}{b} = \frac{2}{5} = 0.400$$

From a table of functions or from a calculator, we have

$$\theta = \tan^{-1}(0.400) = \boxed{21.8^\circ}$$

where $\tan^{-1}(0.400)$ is the notation for "angle whose tangent is 0.400," sometimes written as arctan(0.400).

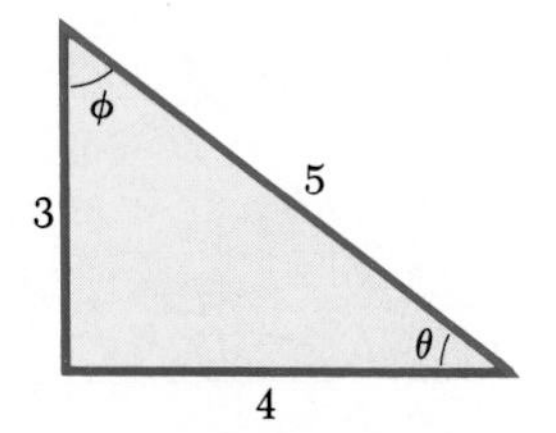


Figure B.12

## EXERCISES

**1.** In Figure B.12, find (a) the side opposite $\theta$, (b) the side adjacent to $\phi$, (c) $\cos\theta$, (d) $\sin\phi$, and (e) $\tan\phi$.
**Answers** (a) 3, (b) 3, (c) $\frac{4}{5}$, (d) $\frac{4}{5}$, and (e) $\frac{4}{3}$
**2.** In a certain right triangle, the two sides that are perpendicular to each other are 5 m and 7 m long. What is the length of the third side of the triangle?
**Answer** 8.60 m
**3.** A right triangle has a hypotenuse of length 3 m, and one of its angles is 30°. What is the length of (a) the side opposite the 30° angle and (b) the side adjacent to the 30° angle?
**Answers** (a) 1.5 m and (b) 2.60 m

## B.5 SERIES EXPANSIONS

$$(a + b)^n = a^n + \frac{n}{1!}a^{n-1}b + \frac{n(n-1)}{2!}a^{n-2}b^2 + \cdots$$

$$(1 + x)^n = 1 + nx + \frac{n(n-1)}{2!}x^2 + \cdots$$

$$e^x = 1 + x + \frac{x^2}{2!} + \frac{x^3}{3!} + \cdots$$

$$\ln(1 \pm x) = \pm x - \tfrac{1}{2}x^2 \pm \tfrac{1}{3}x^3 - \cdots$$

$$\left.\begin{aligned} \sin x &= x - \frac{x^3}{3!} + \frac{x^5}{5!} - \cdots \\ \cos x &= 1 - \frac{x^2}{2!} + \frac{x^4}{4!} - \cdots \\ \tan x &= x + \frac{x^3}{3} + \frac{2x^5}{15} + \cdots \quad |x| < \pi/2 \end{aligned}\right\} x \text{ in radians}$$

For $x \ll 1$, the following approximations can be used:

$$(1+x)^n \approx 1+nx \qquad \sin x \approx x$$

$$e^x \approx 1+x \qquad \cos x \approx 1$$

$$\ln(1 \pm x) \approx \pm x \qquad \tan x \approx x$$

## B.6 DIFFERENTIAL CALCULUS

In various branches of science, it is sometimes necessary to use the basic tools of calculus, first invented by Newton, to describe physical phenomena. The use of calculus is fundamental in the treatment of various problems in newtonian mechanics, electricity, and magnetism. In this section, we simply state some basic properties and "rules of thumb" that should be a useful review to the student.

First, a **function** must be specified that relates one variable to another (such as a coordinate as a function of time). Suppose one of the variables is called $y$ (the dependent variable), the other $x$ (the independent variable). We might have a function relation such as

$$y(x) = ax^3 + bx^2 + cx + d$$

If $a$, $b$, $c$, and $d$ are specified constants, then $y$ can be calculated for any value of $x$. We usually deal with continuous functions, that is, those for which $y$ varies "smoothly" with $x$.

The **derivative** of $y$ with respect to $x$ is defined as the limit of the slopes of chords drawn between two points on the $y$ versus $x$ curve as $\Delta x$ approaches zero. Mathematically, we write this definition as

$$\frac{dy}{dx} = \lim_{\Delta x \to 0} \frac{\Delta y}{\Delta x} = \lim_{\Delta x \to 0} \frac{y(x+\Delta x) - y(x)}{\Delta x} \tag{B.28}$$

where $\Delta y$ and $\Delta x$ are defined as $\Delta x = x_2 - x_1$ and $\Delta y = y_2 - y_1$ (see Fig. B.13).

A useful expression to remember when $y(x) = ax^n$, where $a$ is a *constant* and $n$ is *any* positive or negative number (integer or fraction), is

$$\frac{dy}{dx} = nax^{n-1} \tag{B.29}$$

If $y(x)$ is a polynomial or algebraic function of $x$, we apply Equation B.29 to *each* term in the polynomial and take $da/dx = 0$. It is important to note that

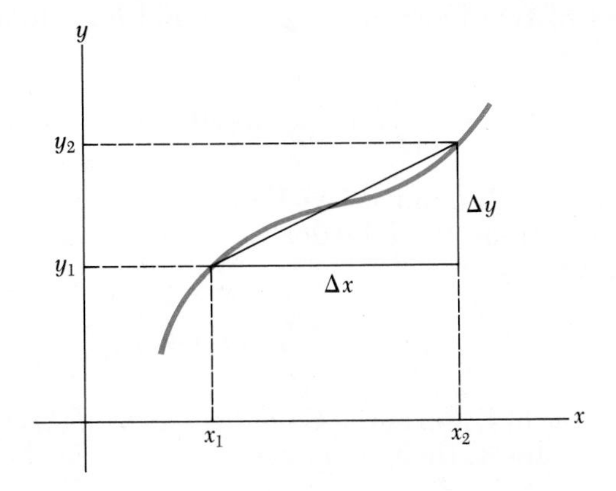


**Figure B.13**

$dy/dx$ *does not* mean $dy$ divided by $dx$, but is simply a notation of the limiting process of the derivative as defined by Equation B.28. In Examples 4 through 7, we evaluate the derivatives of several well-behaved functions.

---

**EXAMPLE 4**

Suppose $y(x)$ (that is, $y$ as a function of $x$) is given by

$$y(x) = ax^3 + bx + c$$

where $a$ and $b$ are constants. Then it follows that

$$y(x + \Delta x) = a(x + \Delta x)^3 + b(x + \Delta x) + c$$

$$y(x + \Delta x) = a(x^3 + 3x^2\,\Delta x + 3x\,\Delta x^2 + \Delta x^3) + b(x + \Delta x) + c$$

so

$$\Delta y = y(x + \Delta x) - y(x) = a(3x^2\,\Delta x + 3x\,\Delta x^2 + \Delta x^3) + b\,\Delta x$$

Substituting this into Equation B.28 gives

$$\frac{dy}{dx} = \lim_{\Delta x \to 0} \frac{\Delta y}{\Delta x} = \lim_{\Delta x \to 0} [3ax^2 + 3x\,\Delta x + \Delta x^2] + b$$

$$\frac{dy}{dx} = 3ax^2 + b$$

---

**EXAMPLE 5**

$$y(x) = 8x^5 + 4x^3 + 2x + 7$$

*Solution* Applying Equation B.29 to each term independently, and remembering that $d/dx$ (constant) $= 0$, we have

$$\frac{dy}{dx} = 8(5)x^4 + 4(3)x^2 + 2(1)x^0 + 0$$

$$\frac{dy}{dx} = 40x^4 + 12x^2 + 2$$

---

## Special Properties of the Derivative

A. **Derivative of the Product of Two Functions** If a function $y$ is given by the product of two functions, say, $g(x)$ and $h(x)$, then the derivative of $y$ is defined as

$$\frac{d}{dx} f(x) = \frac{d}{dx} [g(x)h(x)] = g\frac{dh}{dx} + h\frac{dg}{dx} \qquad \text{(B.30)}$$

B. **Derivative of the Sum of Two Functions** If a function $y$ is equal to the sum of two functions, then the derivative of the sum is equal to the sum of the derivatives:

$$\frac{d}{dx} f(x) = \frac{d}{dx} [g(x) + h(x)] = \frac{dg}{dx} + \frac{dh}{dx} \qquad \text{(B.31)}$$

C. **Chain Rule of Differential Calculus** If $y = f(x)$ and $x$ is a function of some other variable $z$, then $dy/dx$ can be written as the product of two derivatives:

$$\frac{dy}{dx} = \frac{dy}{dz}\frac{dz}{dx} \tag{B.32}$$

**D. The Second Derivative** The second derivative of $y$ with respect to $x$ is defined as the derivative of the function $dy/dx$ (or, the derivative of the derivative). It is usually written

$$\frac{d^2y}{dx^2} = \frac{d}{dx}\left(\frac{dy}{dx}\right) \tag{B.33}$$

---

**EXAMPLE 6**

Find the first derivative $y(x) = x^3/(x+1)^2$ with respect to $x$.

*Solution* We can rewrite this function as $y(x) = x^3(x+1)^{-2}$ and apply Equation B.30 directly:

$$\frac{dy}{dx} = (x+1)^{-2}\frac{d}{dx}(x^3) + x^3\frac{d}{dx}(x+1)^{-2}$$

$$= (x+1)^{-2}\,3x^2 + x^3(-2)(x+1)^{-3}$$

$$\frac{dy}{dx} = \frac{3x^2}{(x+1)^2} - \frac{2x^3}{(x+1)^3}$$

---

**EXAMPLE 7**

A useful formula that follows from Equation B.30 is the derivative of the quotient of two functions. Show that the expression is given by

$$\frac{d}{dx}\left[\frac{g(x)}{h(x)}\right] = \frac{h\dfrac{dg}{dx} - g\dfrac{dh}{dx}}{h^2}$$

*Solution* We can write the quotient as $gh^{-1}$ and then apply Equations B.29 and B.30:

$$\frac{d}{dx}\left(\frac{g}{h}\right) = \frac{d}{dx}(gh^{-1}) = g\frac{d}{dx}(h^{-1}) + h^{-1}\frac{d}{dx}(g)$$

$$= -gh^{-2}\frac{dh}{dx} + h^{-1}\frac{dg}{dx}$$

$$= \frac{h\dfrac{dg}{dx} - g\dfrac{dh}{dx}}{h^2}$$

---

Some of the more commonly used derivatives of functions are listed in Table B.4.

**TABLE B.4 Derivatives for Several Functions**

| |
|---|
| $\frac{d}{dx}(a) = 0$ |
| $\frac{d}{dx}(ax^n) = nax^{n-1}$ |
| $\frac{d}{dx}(e^{ax}) = ae^{ax}$ |
| $\frac{d}{dx}(\sin ax) = a\cos ax$ |
| $\frac{d}{dx}(\cos ax) = -a\sin ax$ |
| $\frac{d}{dx}(\tan ax) = a\sec^2 ax$ |
| $\frac{d}{dx}(\cot ax) = -a\csc^2 ax$ |
| $\frac{d}{dx}(\sec x) = \tan x\sec x$ |
| $\frac{d}{dx}(\csc x) = -\cot x\csc x$ |
| $\frac{d}{dx}(\ln ax) = \frac{1}{x}$ |

Note: The letters $a$ and $n$ are constants.

## B.7 INTEGRAL CALCULUS

We think of integration as the inverse of differentiation. As an example, consider the expression

$$f(x) = \frac{dy}{dx} = 3ax^2 + b$$

which was the result of differentiating the function

$$y(x) = ax^3 + bx + c$$

in Example 4. We can write the first expression $dy = f(x)dx = (3ax^2 + b)dx$ and obtain $y(x)$ by "summing" over all values of $x$. Mathematically, we write this inverse operation

$$y(x) = \int f(x)dx$$

For the function $f(x)$ given above,

$$y(x) = \int (3ax^2 + b)dx = ax^3 + bx + c$$

where $c$ is a constant of the integration. This type of integral is called an *indefinite integral* since its value depends on the choice of the constant $c$.

A general **indefinite integral** $I(x)$ is defined as

$$I(x) = \int f(x)dx \tag{B.34}$$

where $f(x)$ is called the *integrand* and $f(x) = \dfrac{dI(x)}{dx}$.

For a *general continuous* function $f(x)$, the integral can be described as the area under the curve bounded by $f(x)$ and the $x$ axis, between two specified values of $x$, say, $x_1$ and $x_2$, as in Figure B.14.

The area of the shaded element is approximately $f_i \Delta x_i$. If we sum all these area elements from $x_1$ to $x_2$ and take the limit of this sum as $\Delta x_i \rightarrow 0$, we obtain the *true* area under the curve bounded by $f(x)$ and $x$, between the limits $x_1$ and $x_2$:

$$\text{Area} = \lim_{\Delta x_i \rightarrow 0} \sum_i f(x_i)\Delta x_i = \int_{x_1}^{x_2} f(x)dx \tag{B.35}$$

Integrals of the type defined by Equation B.35 are called **definite integrals.**

One of the common types of integrals that arise in practical situations has the form

$$\int x^n \, dx = \frac{x^{n+1}}{n+1} + c \qquad (n \neq -1) \tag{B.36}$$

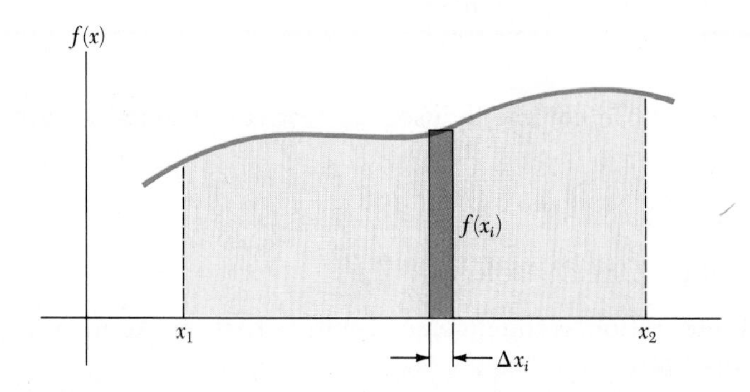


**Figure B.14**

This result is obvious since differentiation of the right-hand side with respect to $x$ gives $f(x) = x^n$ directly. If the limits of the integration are known, this integral becomes a *definite integral* and is written

$$\int_{x_1}^{x_2} x^n \, dx = \frac{x_2^{\,n+1} - x_1^{\,n+1}}{n+1} \qquad (n \neq -1) \tag{B.37}$$

**Examples**

1. $\displaystyle\int_0^a x^2 \, dx = \frac{x^3}{3}\bigg]_0^a = \frac{a^3}{3}$
2. $\displaystyle\int_0^b x^{3/2} \, dx = \frac{x^{5/2}}{5/2}\bigg]_0^b = \frac{2}{5} b^{5/2}$
3. $\displaystyle\int_3^5 x \, dx = \frac{x^2}{2}\bigg]_3^5 = \frac{5^2 - 3^2}{2} = 8$

### Partial Integration

Sometimes it is useful to apply the method of *partial integration* to evaluate certain integrals. The method uses the property that

$$\int u \, dv = uv - \int v \, du \tag{B.38}$$

where $u$ and $v$ are *carefully* chosen so as to reduce a complex integral to a simpler one. In many cases, several reductions have to be made. Consider the example

$$I(x) = \int x^2 e^x \, dx$$

This can be evaluated by integrating by parts twice. First, if we choose $u = x^2$, $v = e^x$, we get

$$\int x^2 e^x \, dx = \int x^2 \, d(e^x) = x^2 e^x - 2\int e^x x \, dx + c_1$$

Now, in the second term, choose $u = x$, $v = e^x$, which gives

$$\int x^2 e^x \, dx = x^2 e^x - 2x e^x + 2\int e^x \, dx + c_1$$

or

$$\int x^2 e^x \, dx = x^2 e^x - 2x e^x + 2e^x + c_2$$

### The Perfect Differential

Another useful method to remember is the use of the *perfect differential.* That is, we should sometimes look for a change of variable such that the differential of the function is the differential of the independent variable appearing in the integrand. For example, consider the integral

$$I(x) = \int \cos^2 x \sin x \, dx$$

This becomes easy to evaluate if we rewrite the differential as $d(\cos x) = -\sin x\, dx$. The integral then becomes

$$\int \cos^2 x \sin x\, dx = -\int \cos^2 x\, d(\cos x)$$

If we now change variables, letting $y = \cos x$, we get

$$\int \cos^2 x \sin x\, dx = -\int y^2 dy = -\frac{y^3}{3} + c = -\frac{\cos^3 x}{3} + c$$

Table B.5 lists some useful indefinite integrals. Table B.6 gives Gauss' probability integral and other definite integrals. A more complete list can be found in various handbooks, such as *The Handbook of Chemistry and Physics*, CRC Press.

**TABLE B.5 Some Indefinite Integrals (an arbitrary constant should be added to each of these integrals)**

$$\int x^n\, dx = \frac{x^{n+1}}{n+1} \qquad \text{(provided } n \neq -1\text{)}$$

$$\int \frac{dx}{x} = \int x^{-1}\, dx = \ln x$$

$$\int \frac{dx}{a+bx} = \frac{1}{b}\ln(a+bx)$$

$$\int \frac{dx}{(a+bx)^2} = -\frac{1}{b(a+bx)}$$

$$\int \frac{dx}{a^2+x^2} = \frac{1}{a}\tan^{-1}\frac{x}{a}$$

$$\int \frac{dx}{a^2-x^2} = \frac{1}{2a}\ln\frac{a+x}{a-x} \qquad (a^2 - x^2 > 0)$$

$$\int \frac{dx}{x^2-a^2} = \frac{1}{2a}\ln\frac{x-a}{x+a} \qquad (x^2 - a^2 > 0)$$

$$\int \frac{x\, dx}{a^2 \pm x^2} = \pm\tfrac{1}{2}\ln(a^2 \pm x^2)$$

$$\int \frac{dx}{\sqrt{a^2-x^2}} = \sin^{-1}\frac{x}{a} = -\cos^{-1}\frac{x}{a} \qquad (a^2 - x^2 > 0)$$

$$\int \frac{dx}{\sqrt{x^2 \pm a^2}} = \ln(x + \sqrt{x^2 \pm a^2})$$

$$\int \frac{x\, dx}{\sqrt{a^2-x^2}} = -\sqrt{a^2-x^2}$$

$$\int \frac{x\, dx}{\sqrt{x^2 \pm a^2}} = \sqrt{x^2 \pm a^2}$$

$$\int \sqrt{a^2-x^2}\, dx = \tfrac{1}{2}\left(x\sqrt{a^2-x^2} + a^2 \sin^{-1}\frac{x}{a}\right)$$

$$\int x\sqrt{a^2-x^2}\, dx = -\tfrac{1}{3}(a^2-x^2)^{3/2}$$

$$\int xe^{ax}\, dx = \frac{e^{ax}}{a^2}(ax-1)$$

$$\int \frac{dx}{a+be^{cx}} = \frac{x}{a} - \frac{1}{ac}\ln(a+be^{cx})$$

$$\int \sin ax\, dx = -\frac{1}{a}\cos ax$$

$$\int \cos ax\, dx = \frac{1}{a}\sin ax$$

$$\int \tan ax\, dx = -\frac{1}{a}\ln(\cos ax) = \frac{1}{a}\ln(\sec ax)$$

$$\int \cot ax\, dx = \frac{1}{a}\ln(\sin ax)$$

$$\int \sec ax\, dx = \frac{1}{a}\ln(\sec ax + \tan ax) = \frac{1}{a}\ln\left[\tan\left(\frac{ax}{2} + \frac{\pi}{4}\right)\right]$$

$$\int \csc ax\, dx = \frac{1}{a}\ln(\csc ax - \cot ax) = \frac{1}{a}\ln\left(\tan\frac{ax}{2}\right)$$

$$\int \sin^2 ax\, dx = \frac{x}{2} - \frac{\sin 2ax}{4a}$$

$$\int \cos^2 ax\, dx = \frac{x}{2} + \frac{\sin 2ax}{4a}$$

$$\int \frac{dx}{\sin^2 ax} = -\frac{1}{a}\cot ax$$

$$\int \frac{dx}{\cos^2 ax} = \frac{1}{a}\tan ax$$

$$\int \tan^2 ax\, dx = \frac{1}{a}(\tan ax) - x$$

$$\int \cot^2 ax\, dx = -\frac{1}{a}(\cot ax) - x$$

*(Table continues)*

**TABLE B.5 (Continued)**

$$\int \sqrt{x^2 \pm a^2}\, dx = \tfrac{1}{2}[x\sqrt{x^2 \pm a^2} \pm a^2 \ln(x + \sqrt{x^2 \pm a^2})]$$

$$\int x\,(\sqrt{x^2 \pm a^2})\, dx = \tfrac{1}{3}(x^2 \pm a^2)^{3/2}$$

$$\int e^{ax}\, dx = \frac{1}{a}\, e^{ax}$$

$$\int \ln ax\, dx = (x \ln ax) - x$$

$$\int \sin^{-1} ax\, dx = x(\sin^{-1} ax) + \frac{\sqrt{1 - a^2x^2}}{a}$$

$$\int \cos^{-1} ax\, dx = x(\cos^{-1} ax) - \frac{\sqrt{1 - a^2x^2}}{a}$$

$$\int \frac{dx}{(x^2 + a^2)^{3/2}} = -\frac{x}{a^2\sqrt{x^2 + a^2}}$$

$$\int \frac{x\, dx}{(x^2 + a^2)^{3/2}} = \frac{1}{\sqrt{x^2 + a^2}}$$

**TABLE B.6 Gauss' Probability Integral and Related Integrals**

$$I_0 = \int_0^\infty e^{-\alpha x^2}\, dx = \tfrac{1}{2}\sqrt{\frac{\pi}{\alpha}} \qquad \text{(Gauss' probability integral)}$$

$$I_1 = \int_0^\infty x e^{-\alpha x^2}\, dx = \frac{1}{2\alpha}$$

$$I_2 = \int_0^\infty x^2 e^{-\alpha x^2}\, dx = -\frac{dI_0}{d\alpha} = \tfrac{1}{4}\sqrt{\frac{\pi}{\alpha^3}}$$

$$I_3 = \int_0^\infty x^3 e^{-\alpha x^2}\, dx = -\frac{dI_1}{d\alpha} = \frac{1}{2\alpha^2}$$

$$I_4 = \int_0^\infty x^4 e^{-\alpha x^2}\, dx = \frac{d^2 I_0}{d\alpha^2} = \tfrac{3}{8}\sqrt{\frac{\pi}{\alpha^5}}$$

$$I_5 = \int_0^\infty x^5 e^{-\alpha x^2}\, dx = \frac{d^2 I_1}{d\alpha^2} = \frac{1}{\alpha^3}$$

$$\vdots$$

$$I_{2n} = (-1)^n \frac{d^n}{d\alpha^n} I_0$$

$$I_{2n+1} = (-1)^n \frac{d^n}{d\alpha^n} I_1$$

# Appendix C

## Periodic Table of the Elements*

Symbol — Ca 20 — Atomic number
Atomic mass† — 40.08
$4s^2$ — Electron configuration

| Group I | Group II | Transition elements | | | | | | |
|---|---|---|---|---|---|---|---|---|
| H 1<br>1.0080<br>$1s^1$ | | | | | | | | |
| Li 3<br>6.94<br>$2s^1$ | Be 4<br>9.012<br>$2s^2$ | | | | | | | |
| Na 11<br>22.99<br>$3s^1$ | Mg 12<br>24.31<br>$3s^2$ | | | | | | | |
| K 19<br>39.102<br>$4s^1$ | Ca 20<br>40.08<br>$4s^2$ | Sc 21<br>44.96<br>$3d^1 4s^2$ | Ti 22<br>47.90<br>$3d^2 4s^2$ | V 23<br>50.94<br>$3d^3 4s^2$ | Cr 24<br>51.996<br>$3d^5 4s^1$ | Mn 25<br>54.94<br>$3d^5 4s^2$ | Fe 26<br>55.85<br>$3d^6 4s^2$ | Co 27<br>58.93<br>$3d^7 4s^2$ |
| Rb 37<br>85.47<br>$5s^1$ | Sr 38<br>87.62<br>$5s^2$ | Y 39<br>88.906<br>$4d^1 5s^2$ | Zr 40<br>91.22<br>$4d^2 5s^2$ | Nb 41<br>92.91<br>$4d^4 5s^1$ | Mo 42<br>95.94<br>$4d^5 5s^1$ | Tc 43<br>(99)<br>$4d^5 5s^2$ | Ru 44<br>101.1<br>$4d^7 5s^1$ | Rh 45<br>102.91<br>$4d^8 5s^1$ |
| Cs 55<br>132.91<br>$6s^1$ | Ba 56<br>137.34<br>$6s^2$ | 57 - 71‡ | Hf 72<br>178.49<br>$5d^2 6s^2$ | Ta 73<br>180.95<br>$5d^3 6s^2$ | W 74<br>183.85<br>$5d^4 6s^2$ | Re 75<br>186.2<br>$5d^5 6s^2$ | Os 76<br>190.2<br>$5d^6 6s^2$ | Ir 77<br>192.2<br>$5d^7 6s^2$ |
| Fr 87<br>(223)<br>$7s^1$ | Ra 88<br>(226)<br>$7s^2$ | 89 - 103§ | Rf 104<br>(261)<br>$6d^2 7s^2$ | Ha 105<br>(262)<br>$6d^3 7s^2$ | Unh 106<br>(263) | Uns 107<br>(261) | | |

| Series | | | | | | |
|---|---|---|---|---|---|---|
| ‡Lanthanide series | La 57<br>138.91<br>$5d^1 6s^2$ | Ce 58<br>140.12<br>$5d^1 4f^1 6s^2$ | Pr 59<br>140.91<br>$4f^3 6s^2$ | Nd 60<br>144.24<br>$4f^4 6s^2$ | Pm 61<br>(147)<br>$4f^6 6s^2$ | Sm 62<br>150.4<br>$4f^6 6s^2$ |
| §Actinide series | Ac 89<br>(227)<br>$6d^1 7s^2$ | Th 90<br>(232)<br>$6d^2 7s^2$ | Pa 91<br>(231)<br>$5f^2 d^1 7s^2$ | U 92<br>(238)<br>$5f^3 6d^1 7s^2$ | Np 93<br>(239)<br>$5f^3 6d^0 7s^2$ | Pu 94<br>(239)<br>$5f^6 6d^0 7s^2$ |

° Atomic mass values given are averaged over isotopes in the percentages in which they exist in nature.

† For an unstable element, mass number of the most stable known isotope is given in parentheses.

| | | | Group III | Group IV | Group V | Group VI | Group VII | Group 0 |
|---|---|---|---|---|---|---|---|---|
| | | | | | | | H 1<br>1.0080<br>$2s^4$ | He 2<br>4.0026<br>$1s^2$ |
| | | | B 5<br>10.81<br>$2p^2$ | C 6<br>12.011<br>$2p^2$ | N 7<br>14.007<br>$2p^3$ | O 8<br>15.999<br>$2p^4$ | F 9<br>18.998<br>$2p^5$ | Ne 10<br>20.18<br>$2p^6$ |
| | | | Al 13<br>26.98<br>$3p^1$ | Si 14<br>28.09<br>$3p^2$ | P 15<br>30.97<br>$3p^3$ | S 16<br>32.06<br>$3p^4$ | Cl 17<br>35.453<br>$3p^5$ | Ar 18<br>39.948<br>$3p^6$ |
| Ni 28<br>58.71<br>$3d^8 4s^2$ | Cu 29<br>63.54<br>$3d^{10} 4s^1$ | Zn 30<br>65.37<br>$3d^{10} 4s^2$ | Ga 31<br>69.72<br>$4p^1$ | Ge 32<br>72.59<br>$4p^2$ | As 33<br>74.92<br>$4p^3$ | Se 34<br>78.96<br>$4p^4$ | Br 35<br>79.91<br>$4p^5$ | Kr 36<br>83.80<br>$4p^6$ |
| Pd 46<br>106.4<br>$4d^{10} 5s^6$ | Ag 47<br>107.87<br>$4d^{10} 5s^1$ | Cd 48<br>112.40<br>$4d^{10} 5s^2$ | In 49<br>114.82<br>$5p^1$ | Sn 50<br>118.69<br>$5p^2$ | Sb 51<br>121.75<br>$5p^3$ | Te 52<br>127.60<br>$5p^4$ | I 53<br>126.90<br>$5p^5$ | Xe 54<br>131.30<br>$5p^6$ |
| Pt 78<br>195.09<br>$5d^9 6s^1$ | Au 79<br>196.97<br>$5d^{10} 6s^1$ | Hg 80<br>200.59<br>$5d^{10} 6s^2$ | Tl 81<br>204.37<br>$6p^2$ | Pb 82<br>207.2<br>$6p^2$ | Bi 83<br>208.98<br>$6p^3$ | Po 84<br>(210)<br>$6p^4$ | At 85<br>(218)<br>$6p^5$ | Rn 86<br>(222)<br>$6p^6$ |

| | | | | | | | | |
|---|---|---|---|---|---|---|---|---|
| Eu 63<br>152.0<br>$4f^7 6s^2$ | Gd 64<br>157.25<br>$5d^1 4f^7 6s^2$ | Tb 65<br>158.92<br>$5d^1 4f^8 6s^2$ | Dy 66<br>162.50<br>$4f^{10} 6s^2$ | Ho 67<br>164.93<br>$4f^{11} 6s^2$ | Er 68<br>167.26<br>$4f^{12} 6s^2$ | Tm 69<br>168.93<br>$4f^{13} 6s^2$ | Yb 70<br>173.04<br>$4f^{14} 6s^2$ | Lu 71<br>174.97<br>$5d^1 4f^{14} 6s^2$ |
| Am 95<br>(243)<br>$5f^7 6d^0 7s^2$ | Cm 96<br>(245)<br>$5f^7 6d^1 7s^2$ | Bk 97<br>(247)<br>$5f^8 6d^1 7s^2$ | Cf 98<br>(249)<br>$5f^{10} 6d^0 7s^2$ | Es 99<br>(254)<br>$5s^{11} 6d^0 7s^2$ | Fm 100<br>(253)<br>$5f^{12} 6d^0 7s^2$ | Md 101<br>(255)<br>$5f^{13} 6d^0 7s^2$ | No 102<br>(255)<br>$6d^0 7s^2$ | Lr 103<br>(257)<br>$6d^1 7s^2$ |

# Appendix D

## SI Units

**TABLE D.1 SI Base Units**

| Base Quantity | SI Base Unit | |
|---|---|---|
| | Name | Symbol |
| Length | Meter | m |
| Mass | Kilogram | kg |
| Time | Second | s |
| Electric current | Ampere | A |
| Temperature | Kelvin | K |
| Amount of substance | Mole | mol |
| Luminous intensity | Candela | cd |

**TABLE D.2 Some Derived SI Units**

| Quantity | Name | Symbol | Expression in Terms of Base Units | Expression in Terms of Other SI Units |
|---|---|---|---|---|
| Plane angle | Radian | rad | m/m | |
| Frequency | Hertz | Hz | $s^{-1}$ | |
| Force | Newton | N | $kg \cdot m/s^2$ | J/m |
| Pressure | Pascal | Pa | $kg/m \cdot s^2$ | $N/m^2$ |
| Energy: work | Joule | J | $kg \cdot m^2/s^2$ | $N \cdot m$ |
| Power | Watt | W | $kg \cdot m^2/s^3$ | J/s |
| Electric charge | Coulomb | C | $A \cdot s$ | |
| Electric potential (emf) | Volt | V | $kg \cdot m^2/A \cdot s^3$ | W/A |
| Capacitance | Farad | F | $A^2 \cdot s^4/kg \cdot m^2$ | C/V |
| Electric resistance | Ohm | Ω | $kg \cdot m^2/A^2 \cdot s^3$ | V/A |
| Magnetic flux | Weber | Wb | $kg \cdot m^2/A \cdot s^2$ | $V \cdot s$ |
| Magnetic field intensity | Tesla | T | $kg/A \cdot s^2$ | $Wb/m^2$ |
| Inductance | Henry | H | $kg \cdot m^2/A^2 \cdot s^3$ | Wb/A |

# Appendix E

## Nobel Prizes

All Nobel Prizes in physics are listed (and marked with a P), as well as relevant Nobel Prizes in Chemistry (C). The key dates for some of the scientific work are supplied; they often antedate the prize considerably.

**1901** (P) *Wilhelm Roentgen* for discovering x-rays (1895).

**1902** (P) *Hendrik A. Lorentz* for predicting the Zeeman effect and *Pieter Zeeman* for discovering the Zeeman effect, the splitting of spectral lines in magnetic fields.

**1903** (P) *Antoine-Henri Becquerel* for discovering radioactivity (1896) and *Pierre* and *Marie Curie* for studying radioactivity.

**1904** (P) *Lord Rayleigh* for studying the density of gases and discovering argon.
(C) *William Ramsay* for discovering the inert gas elements helium, neon, xenon, and krypton, and placing them in the periodic table.

**1905** (P) *Philipp Lenard* for studying cathode rays, electrons (1898 – 1899).

**1906** (P) *J. J. Thomson* for studying electrical discharge through gases and discovering the electron (1897).

**1907** (P) *Albert A. Michelson* for inventing optical instruments and measuring the speed of light (1880s).

**1908** (P) *Gabriel Lippmann* for making the first color photographic plate, using interference methods (1891).
(C) *Ernest Rutherford* for discovering that atoms can be broken apart by alpha rays and for studying radioactivity.

**1909** (P) *Guglielmo Marconi* and *Carl Ferdinand Braun* for developing wireless telegraphy.

**1910** (P) *Johannes D. van der Waals* for studying the equation of state for gases and liquids (1881).

**1911** (P) *Wilhelm Wien* for discovering Wien's law giving the peak of a blackbody spectrum (1893).
(C) *Marie Curie* for discovering radium and polonium (1898) and isolating radium.

**1912** (P) *Nils Dalén* for inventing automatic gas regulators for lighthouses.

**1913** (P) *Heike Kamerlingh Onnes* for the discovery of superconductivity and liquefying helium (1908).

**1914** (P) *Max T. F. von Laue* for studying x-rays from their diffraction by crystals, showing that x-rays are electromagnetic waves (1912).
(C) *Theodore W. Richards* for determining the atomic weights of sixty elements, indicating the existence of isotopes.

**1915** (P) *William Henry Bragg* and *William Lawrence Bragg*, his son, for studying the diffraction of x-rays in crystals.

**1917** (P) *Charles Barkla* for studying atoms by x-ray scattering (1906).

**1918** (P) *Max Planck* for discovering energy quanta (1900).

**1919** (P) *Johannes Stark*, for discovering the Stark effect, the splitting of spectral lines in electric fields (1913).

**1920** (P) *Charles-Édouard Guillaume* for discovering invar, a nickel-steel alloy with low coefficient of expansion.
(C) *Walther Nernst* for studying heat changes in chemical reactions and formulating the third law of thermodynamics (1918).

**1921** (P) *Albert Einstein* for explaining the photoelectric effect and for his services to theoretical physics (1905).
(C) *Frederick Soddy* for studying the chemistry of radioactive substances and discovering isotopes (1912).

**1922** (P) *Niels Bohr* for his model of the atom and its radiation (1913).
(C) *Francis W. Aston* for using the mass spectrograph to study atomic weights, thus discovering 212 of the 287 naturally occurring isotopes.

**1923** (P) *Robert A. Millikan* for measuring the charge on an electron (1911) and for studying the photoelectric effect experimentally (1914).

**1924** (P) *Karl M. G. Siegbahn* for his work in x-ray spectroscopy.

**1925** (P) *James Franck* and *Gustav Hertz* for discovering the Franck-Hertz effect in electron-atom collisions.

**1926** (P) *Jean-Baptiste Perrin* for studying Brownian motion to validate the discontinuous structure of matter and measure the size of atoms.

**1927** (P) *Arthur Holly Compton* for discovering the Compton effect on x-rays, their change in wavelength when they collide with matter (1922), and *Charles T. R. Wilson* for inventing the cloud chamber, used to study charged particles (1906).

**1928** (P) *Owen W. Richardson* for studying the thermionic effect and electrons emitted by hot metals (1911).

**1929** (P) *Louis Victor de Broglie* for discovering the wave nature of electrons (1923).

**1930** (P) *Chandrasekhara Venkata Raman* for studying Raman scattering, the scattering of light by atoms and molecules with a change in wavelength (1928).

**1932** (P) *Werner Heisenberg* for creating quantum mechanics (1925).

**1933** (P) *Erwin Schrödinger* and *Paul A. M. Dirac* for developing wave mechanics (1925) and relativistic quantum mechanics (1927).
(C) *Harold Urey* for discovering heavy hydrogen, deuterium (1931).

**1935** (P) *James Chadwick* for discovering the neutron (1932).
(C) *Irène* and *Frédéric Joliot-Curie* for synthesizing new radioactive elements.

**1936** (P) *Carl D. Anderson* for discovering the positron in particular and antimatter in general (1932) and *Victor F. Hess* for discovering cosmic rays.
(C) *Peter J. W. Debye* for studying dipole moments and diffraction of x-rays and electrons in gases.

**1937** (P) *Clinton Davisson* and *George Thomson* for discovering the diffraction of electrons by crystals, confirming de Broglie's hypothesis (1927).

**1938** (P) *Enrico Fermi* for producing the transuranic radioactive elements by neutron irradiation (1934–1937).

**1939** (P) *Ernest O. Lawrence* for inventing the cyclotron.

**1943** (P) *Otto Stern* for developing molecular-beam studies (1923), and using them to discover the magnetic moment of the proton (1933).

**1944** (P) *Isidor I. Rabi* for discovering nuclear magnetic resonance in atomic and molecular beams.
(C) *Otto Hahn* for discovering nuclear fission (1938).

**1945** (P) *Wolfgang Pauli* for discovering the exclusion principle (1924).
**1946** (P) *Percy W. Bridgman* for studying physics at high pressures.
**1947** (P) *Edward V. Appleton* for studying the ionosphere.
**1948** (P) *Patrick M. S. Blackett* for studying nuclear physics with cloud-chamber photographs of cosmic-ray interactions.
**1949** (P) *Hideki Yukawa* for predicting the existence of mesons (1935).
**1950** (P) *Cecil F. Powell* for developing the method of studying cosmic rays with photographic emulsions and discovering new mesons.
**1951** (P) *John D. Cockcroft* and *Ernest T. S. Walton* for transmuting nuclei in an accelerator (1932).
(C) *Edwin M. McMillan* for producing neptunium (1940) and *Glenn T. Seaborg* for producing plutonium (1941) and further transuranic elements.
**1952** (P) *Felix Bloch* and *Edward Mills Purcell* for discovering nuclear magnetic resonance in liquids and gases (1946).
**1953** (P) *Frits Zernike* for inventing the phase-contrast microscope, which uses interference to provide high contrast.
**1954** (P) *Max Born* for interpreting the wave function as a probability (1926) and other quantum-mechanical discoveries and *Walther Bothe* for developing the coincidence method to study subatomic particles (1930–1931), producing, in particular, the particle interpreted by Chadwick as the neutron.
**1955** (P) *Willis E. Lamb, Jr.* for discovering the Lamb shift in the hydrogen spectrum (1947) and *Polykarp Kusch* for determining the magnetic moment of the electron (1947).
**1956** (P) *John Bardeen, Walter H. Brattain,* and *William Shockley* for inventing the transistor (1956).
**1957** (P) *T.-D. Lee* and *C.-N. Yang* for predicting that parity is not conserved in beta decay (1956).
**1958** (P) *Pavel A. Čerenkov* for discovering Čerenkov radiation (1935) and *Ilya M. Frank* and *Igor Tamm* for interpreting it (1937).
**1959** (P) *Emilio G. Segrè* and *Owen Chamberlain* for discovering the antiproton (1955).
**1960** (P) *Donald A. Glaser* for inventing the bubble chamber to study elementary particles (1952).
(C) *Willard Libby* for developing radiocarbon dating (1947).
**1961** (P) *Robert Hofstadter* for discovering internal structure in protons and neutrons and *Rudolf L. Mössbauer* for discovering the Mössbauer effect of recoilless gamma-ray emission (1957).
**1962** (P) *Lev Davidovich Landau* for studying liquid helium and other condensed matter theoretically.
**1963** (P) *Eugene P. Wigner* for applying symmetry principles to elementary-particle theory and *Maria Goeppert Mayer* and *J. Hans D. Jensen* for studying the shell model of nuclei (1947).
**1964** (P) *Charles H. Townes, Nikolai G. Basov,* and *Alexandr M. Prokhorov* for developing masers (1951–1952) and lasers.
**1965** (P) *Sin-itiro Tomonaga, Julian S. Schwinger,* and *Richard P. Feynman* for developing quantum electrodynamics (1948).
**1966** (P) *Alfred Kastler* for his optical methods of studying atomic energy levels.
**1967** (P) *Hans Albrecht Bethe* for discovering the routes of energy production in stars (1939).

**1968** (P) *Luis W. Alvarez* for discovering resonance states of elementary particles.

**1969** (P) *Murray Gell-Mann* for classifying elementary particles (1963).

**1970** (P) *Hannes Alfvén* for developing magnetohydrodynamic theory and *Louis Eugène Félix Néel* for discovering antiferromagnetism and ferrimagnetism (1930s).

**1971** (P) *Dennis Gabor* for developing holography (1947).
(C) *Gerhard Herzberg* for studying the structure of molecules spectroscopically.

**1972** (P) *John Bardeen, Leon N. Cooper,* and *John Robert Schrieffer* for explaining superconductivity (1957).

**1973** (P) *Leo Esaki* for discovering tunneling in semiconductors, *Ivar Giaever* for discovering tunneling in superconductors, and *Brian D. Josephson* for predicting the Josephson effect, which involves tunneling of paired electrons (1958–1962).

**1974** (P) *Anthony Hewish* for discovering pulsars and *Martin Ryle* for developing radio interferometry.

**1975** (P) *Aage N. Bohr, Ben R. Mottelson,* and *James Rainwater* for discovering why some nuclei take asymmetric shapes.

**1976** (P) *Burton Richter* and *Samuel C. C. Ting* for discovering the J/psi particle, the first charmed particle (1974).

**1977** (P) *John H. Van Vleck, Nevill F. Mott,* and *Philip W. Anderson* for studying solids quantum-mechanically.
(C) *Ilya Prigogine* for extending thermodynamics to show how life could arise in the face of the second law.

**1978** (P) *Arno A. Penzias* and *Robert W. Wilson* for discovering the cosmic background radiation (1965) and *Pyotr Kapitsa* for his studies of liquid helium.

**1979** (P) *Sheldon L. Glashow, Abdus Salam,* and *Steven Weinberg* for developing the theory that unified the weak and electromagnetic forces (1958–1971).

**1980** (P) *Val Fitch* and *James W. Cronin* for discovering CP (charge-parity) violation (1964), which possibly explains the cosmological dominance of matter over antimatter.

**1981** (P) *Nicolaas Bloembergen* and *Arthur L. Schawlow* for developing laser spectroscopy and *Kai M. Siegbahn* for developing high-resolution electron spectroscopy (1958).

**1982** (P) *Kenneth G. Wilson* for developing a method of constructing theories of phase transitions to analyze critical phenomena.

**1983** (P) *William A. Fowler* for theoretical studies of astrophysical nucleosynthesis and *Subramanyan Chandrasekhar* for studying physical processes of importance to stellar structure and evolution, including the prediction of white dwarf stars (1930).

**1984** (P) *Carlo Rubbia* for discovering the W and Z particles, verifying the electroweak unification, and *Simon van der Meer,* for developing the method of stochastic cooling of the CERN beam that allowed the discovery (1982–1983).

**1985** (P) *Klaus von Klitzing* for the quantized Hall effect, relating to conductivity in the presence of a magnetic field (1980).

**1986** (P) *Ernst Ruska* for inventing the electron microscope (1931), and *Gerd Binnig* and *Heinrich Rohrer* for inventing the scanning-tunneling electron microscope (1981).

**1987** (P) *J. Georg Bednorz* and *Karl Alex Müller* for the discovery of high temperature superconductivity (1986).

**1988** (P) *Leon M. Lederman, Melvin Schwartz,* and *Jack Steinberger* for a collaborative experiment that led to the development of a new tool for studying the weak nuclear force, which affects the radioactive decay of atoms.

**1989** (P) *Norman Ramsay* (U.S.), for various techniques in atomic physics; and *Hans Dehmelt* (U.S.) and *Wolfgang Paul* (Germany) for the development of techniques for trapping single charge particles.

**1990** (P) *Jerome Friedman, Henry Kendall* (both U.S.), and *Richard Taylor* (Canada) for experiments important to the development of the quark model.

**1991** (P) *Pierre-Gilles de Gennes* for discovering that methods developed for studying order phenomena in simple systems can be generalized to more complex forms of matter, in particular to liquid crystals and polymers.

# Appendix F

## Physics Problem Solving with a Spreadsheet

*D.A. Stetser, Miami University-Middletown*

### INTRODUCTION

Spreadsheet computer solutions of problems for a first course in physics provide the student with an effective means to increase understanding of complex phenomena. The year or semester of preparation to learn a high level language, such as BASIC or FORTRAN, reduces to a few hours of spreadsheet introduction. The spreadsheet with internal programming capability quickly models physical phenomena. Most spreadsheets allow immediate graphical representations of the variables. Spreadsheet modeling details the physical phenomena without the burden of writing cumbersome language code.

Spreadsheet modeling for this text is introduced by providing a floppy disk containing worksheet files to be retrieved into the Lotus 1-2-3 software program. While the worksheets provided require the Lotus 1-2-3 software, other spreadsheets, QUATTRO, TWIN, EXCEL, are equally applicable to problem solving. The spreadsheet programs are designed for use with IBM PC microcomputers (or IBM-compatible machines). The Lotus 1-2-3 software provides the blank row and column format. The rows and columns define cells on this electronic simulation of a sheet of paper with a grid system. Labels, numerical values and equations may be inserted into these cells. The equations acquire values in cells, algebraically manipulate these values, and compute results. The spreadsheet recalculates immediately after the keyboard insertion of a new parameter to facilitate a "what if" approach to the modeling of a phenomenon. Let us examine the details of one of the worksheet files on the disk.

### WORKSHEET EXAMPLE

The worksheet RAPGRO.wk1 models the solution to Question 3.15.

**Question 3.15** A rapidly growing plant doubles in height each week. At the end of the 25th day, the plant reaches the height of the building. At what time was the plant one-fourth the height of the building?

The *.wk1 files on the floppy disk represent models to examples and problems from the text. Additional problems extend the scope of the text material. The worksheets are used to model the problem with interactive input into the cells representing constant parameters. Suggestions are made for modifications to the worksheet in order to model other problems in the text. After examination of the provided spreadsheets, the student should modify the worksheets and create original worksheets. If you do not save a modified worksheet under the original name, it can be retrieved in its original form.

If you wish to save a modified worksheet, rename it **RAPGRO1**.wk1, for example.

With the Lotus 1-2-3 spreadsheet activated, the worksheet **RAPGRO** is retrieved by the keystrokes **/FR** and the selection of **METHOD** and then **RAPGRO** from the menu. The screen displays the worksheet.

```
RAPGRO  Q3.15  A Rapidly Growing Plant
DT=1 days                  DH=k*H*Dt          Hanal=Ho*EXP(k*t)
 k=0.1041 1/time            H=H+DH                  k=LN(2)/7
Ho=1 ft                                             k=0.099021
TAB ⟶ for Instructions
```

| t days | H ft | Hanal ft |
|---|---|---|
| 0 | 1.000 | 1 |
| 1 | 1.104 | 1.104089 |
| 2 | 1.219 | 1.219013 |
| 3 | 1.346 | 1.345900 |
| 4 | 1.486 | 1.485994 |
| 5 | 1.641 | 1.640670 |
| 6 | 1.812 | 1.811447 |
| 7 | 2.000 | 2 |
| 8 | 2.208 | 2.208179 |
| 9 | 2.438 | 2.438027 |
| 10 | 2.692 | 2.691800 |

**Instructions**

1. Enter the time interval, 1, into cell B3. By trial and error, adjust the constant $k$ until the height = 2 m at $t = 7$ days.
2. Enter $Dt = 2.5$ days, and determine the height of the plant at $t = 25$ days. This is the height of the building. Divide this value by 4, and find the time that corresponds to this height. Adjust the time interval to achieve an $H$ value that is equal to 1/4 the height of the building. This is a rough estimate.
3. The analytical solution is an exponential growth. The growth constant $k$ is determined analytically from:

```
H=Ho*EXP(k*t)
2=EXP(k*7)
LN(2)=k*7
k=LN(2)/7=0.099021
```

4. Notice that the approximate numerical solution is in best agreement with analytical solution when $k = 0.099021$ and small time intervals, $Dt = 0.1$ are entered.
5. A satisfactory numerical solution uses a time interval, $Dt = 1$. Larger values create errors. Copy the last row of the worksheet to include values for times extending to $t = 25$ days.

This example demonstrates the method of numerical analysis. The key difference between the analytical technique using calculus and numerical analysis is that the change in the independent variable time is infinitesimally small, using calculus, and finite, using numerical analysis. The dependent variable, *H*, assumes instantaneous values with calculus, and the *H* value is approximated by a constant value over the finite interval, *Dt*, using numerical analysis. The time rate of change in the height of the plant, *DH*/*Dt*, is proportional to the height. The change, *DH*, during this time interval equals k*H*Dt. The proportionality constant, *k*, is a growth constant. The worksheet calculates this change for each time interval and adds this change to the value of height for the previous interval. These *H* values are tabulated in a column which corresponds to time values in the first column. The analytical solution is tabulated in the last column with respect to times in the first column. The *k* value for the analytical equation

$$H = H_0 * e^{k*t}$$

is calculated automatically in a cell above the column containing the analytical solution values.

The graph of height versus time is viewed by pressing F10. The graph is constructed by selecting the graph type as XY and the horizontal and vertical variable ranges. In this example the height determined by numerical analysis and the height determined by analytical solution are chosen as two dependent

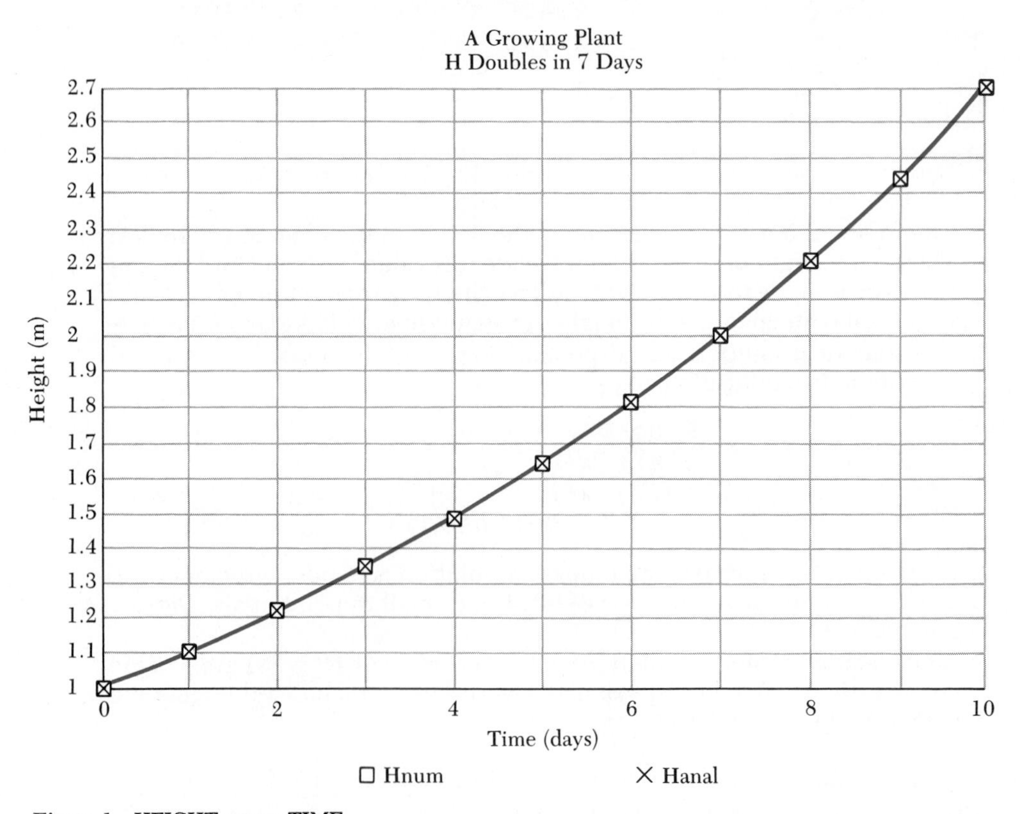


**Figure 1** HEIGHT versus TIME.

variables on the vertical ranges, and the independent variable time is placed on the horizontal range.

Modeling provides understanding of physical concepts without the constraints of simplifying assumptions, necessary for an analytical solution by the techniques of calculus.

## BASICS

A spreadsheet worksheet is similar in purpose to a program written in programming languages, such as BASIC, FORTRAN or Pascal. The spreadsheet software requires only a few hours of familiarization to begin creating models of physical phenomena. The graphic abilities of a spreadsheet are easily accessed compared to the code required for programming languages. The spreadsheet is ideally suited to satisfy the computational requirements for an introductory text in physics.

Your introduction to spreadsheet modeling is incorporated into text examples. Worksheets are provided on disk to expand upon the presentation in the text. You are requested to enter data or constants into the worksheet, observe and interpret the resulting computations and modify the spreadsheet to perform other calculations. Examine these text examples to determine the technique of their creation. As soon as possible, try to create original worksheets. Spreadsheet modeling is a tool to replace the electronic calculator, which replaced the slide rule that hung from my belt many years ago.

Problems from each chapter are suggested as appropriate for spreadsheet modeling. Many others are appropriate as well, and you should develop worksheets for any and all purposes. Worksheet solutions to these and other problems of a more difficult nature are provided on another disk offered to instructors of the course.

### The Worksheet

An electronic grid of rows and columns appears on your computer screen. The rows are labeled with numbers and the columns are labeled with letters. The specification of a column and a row indicates an individual cell. A1 is the cell address of the upper left corner. The cells are accessed by moving the cursor to the desired cell.

Entries to the cell are made from the keyboard.

1. A number is referred to as a value. This value is formatted as desired (Fixed, scientific, currency, etc.).

   0.238, 2.38E-1, $0.24

2. Text is referred to as a label. The labels may be right- or left-justified or centered. Values cannot be adjusted within the cell; if ′, ″, or ^ precedes a value, the computer recognizes the combination as a label.

   ′text — right justified label
   ″text — left justified label
   ^text — centered label

3. Formulas with values and cell addresses.

   The entry 0.238*A10 represents the value 0.238 multiplying the value contained in cell A10.

   If a letter is the first character, a symbol (+,−,() must precede the letter: +A10*0.238.

4. A spreadsheet function: mathematical, statistical, logic. The functions are always preceded by the @ symbol. A list of all functions is obtained from electronic HELP(F1) or from the software documentation.

   The function @SQRT(A10) produces the value 10, if A10 contains the value 100.

   The function @SIN(3.14/6) produces the value 0.5. The parentheses are required for function notation.

   The function @AVG(A10..A20) finds the average value of the values in the range of cells A10 to and including A20.

   The function @IF(A10 > 1, 1, 0) returns the value 1 to the cell if the value in cell A10 is greater than 1 and returns the value 0 to the cell if the value in the cell A10 is less than 1.

The spreadsheet command menu is accessed by the / key. Command choices appear across the top of the screen. The commands are in a tree structure that leads the user to the desired result. Experimentation is the best method for learning the commands. A useful command is the **COPY** command. Either moving the cursor to **COPY** and pressing **ENTER** or pressing the key **C** starts the copy sequence. The computer responds **FROM**, and the cell to copy from is highlighted with the cursor: you should then press **ENTER**. The computer responds TO, and the cell to which the first cell is to be copied is highlighted: press **ENTER.** A more useful copy sequence is for copying a formula in a cell to a range of cells. Cells may be copied in a relative or absolute fashion. A cell with an address (A10) is copied relatively. A cell with an address ($A$10) is copied in an absolute fashion. An example follows.

In the first row, the number in the first column is multiplied by the value 0.238. The result is shown in the second and third columns. The second column has the cell entry 0.238*E1. Copying this cell down over the four rows computes the values shown. This is described as relative copying. The third column has the cell entry 0.238*$E$1. Copying this cell down over the four rows computes the values in the third column. This is an absolute cell copy. The cell text entries are listed to the right of the numbers.

```
1   0.24   0.238        1  0.238*E1  0.238*$E$1
2   0.48   0.238        2  0.238*E2  0.238*$E$1
3   0.71   0.238        3  0.238*E3  0.238*$E$1
4   0.95   0.238        4  0.238*E4  0.238*$E$1
5   1.19   0.238        5  0.238*E5  0.238*$E$1
```

The simplest method to further introduce spreadsheet technique is to detail the construction of a worksheet. If possible, sit in front of a computer with the Lotus 1-2-3 spreadsheet installed. Insert a copy of the disk provided into the B floppy drive of the computer and enter the keystrokes:

**/FD** (sets the current directory)
**B:\method** (specifies the drive and the desired directory)
**/FR** (selects a group file for retrieval)

Move the cursor to EX1FF and press **ENTER.** The screen displays the worksheet.

```
EX1FF Free Fall                              Numerical Analysis
  A=g=         9.80 m/s/s                     Yex=Vo*t+0.5*A*t^2
   Vo=         0.00 m/s                       V=V+DV
   Dt=         1.00 s                         Y=Y+V*Dt
```

| t | V(t+0.5t) | Y(t) | Yex(t) |
|---|---|---|---|
| s | m/s | m | m |
| 0.00 | 4.90 | 0.00 | 0.00 |
| 1.00 | 14.70 | 4.90 | 4.90 |
| 2.00 | 24.50 | 19.60 | 19.60 |
| 3.00 | 34.30 | 44.10 | 44.10 |
| 4.00 | 44.10 | 78.40 | 78.40 |
| 5.00 | 53.90 | 122.50 | 122.50 |
| 6.00 | 63.70 | 176.40 | 176.40 |
| 7.00 | 73.50 | 240.10 | 240.10 |
| 8.00 | 83.30 | 313.60 | 313.60 |
| 9.00 | 93.10 | 396.90 | 396.90 |
| 10.00 | 102.90 | 490.00 | 490.00 |
| 11.00 | 112.70 | 592.90 | 592.90 |
| 12.00 | | 705.60 | 705.60 |

This worksheet computes the velocity at the middle of the time intervals defined in the first column. The acceleration, entered as a constant in cell B3, multiplied by the time interval *Dt*, entered as a constant in cell B5, calculates the change in velocity from the beginning of the interval to the midpoint of the time interval. Adding this velocity change to the initial value of velocity computes the velocity at the middle of the first time interval. Repeating the process for a whole time interval gives the velocity at the middle of the time intervals for the remainder of the column.

The distance fallen by the object is computed by adding the change in distance, DY = V*Dt, to the initial value, Y = Y + DY. The last column computes the exact result of an analytical solution, $Y = 0.5*A*t^2$. These equations are copied down over the range of the worksheet. The computational process approximates the phenomena of a freely falling body by assuming the velocity is a constant over the brief time interval *Dt*. The change in position over this interval is calculated by

$$Y_{final} - Y_{initial} = \text{Velocity*time interval}$$

The worksheet equations appear as follows:

```
EX1FF Free Fall                                Numerical Analysis
   A=g=          9.8 m/s/s                         Yex=Vo*t+0.5*A*t^2
    Vo=            0 m/s                           Y=Y+V*Dt
    Dt=            1 s

     t          V(t+0.5Dt)               Y(t)                  Yex(t)
    sec            m/s                    m                      m

              0+B$4+B$3*B$5/2                              00.5*(A8^2)*B$3
+A8+$B$5        +B8+B$3*B$5          +C8+B8*B$5             0.5*(A9^2)*B$3
+A9+$B$5        +B9+B$3*B$5          +C9+B9*B$5             0.5*(A10^2)*B$3
+A10+$B$5       +B10+B$3*B$5         +C10+B10*B$5           0.5*(A11^2)*B$3
+A11+$B$5       +B11+B$3*B$5         +C11+B11*B$5           0.5*(A12^2)*B$3
+A12+$B$5       +B12+B$3*B$5         +C12+B12*B$5           0.5*(A13^2)*B$3
+A13+$B$5       +B13+B$3*B$5         +C13+B13*B$5           0.5*(A14^2)*B$3
+A14+$B$5       +B14+B$3*B$5         +C14+B14*B$5           0.5*(A15^2)*B$3
+A15+$B$5       +B15+B$3*B$5         +C15+B15*B$5           0.5*(A16^2)*B$3
+A16+$B$5       +B16+B$3*B$5         +C16+B16*B$5           0.5*(A17^2)*B$3
+A17+$B$5       +B17+B$3*B$5         +C17+B17*B$5           0.5*(A18^2)*B$3
+A18+$B$5       +B18+B$3*B$5         +C18+B18*B$5           0.5*(A19^2)*B$3
+A19+$B$5                            +C19+B19*B$5           0.5*(A20^2)*B$3
```

**The Distance versus Time Graph**

1. Key **/GT** and select type with cursor at **XY** and press **ENTER.**
2. Key **X** to define the time range by moving the cursor to the top value in the time column, press period (.) to "pin" the top of the range and move the cursor to the bottom of the range and press **ENTER** to select this range.
3. Key **A** to define the distance range in the third column. Create the range just as you did for the time.
4. Titles and grids are created from the **OPTIONS** command.
5. View the graph with the **VIEW** command if you are in the graph command menu or **F10** from the **READY** mode.

The graph displays the quadratic dependence of distance on time for the falling object. The graph as well as the spreadsheet automatically adjust to new constants. Change the acceleration to that of the moon by placing the cursor on the cell with the acceleration value and press **F2** (the edit function). Edit this value by entering **/6** after the current value 9.8 and press **ENTER.** The worksheet recalculates for this new value of acceleration.

This worksheet is easily modified to include the effects of an initial velocity, Vo. The upward direction is now considered to be represented by a positive number. Downward directions are negative. Let the initial velocity equal +49 m/s and let the acceleration due to gravity equal −9.8 m/s/s. Enter these values into the cells B3 and B4. Your modified worksheet appears as follows:

```
EX2FFVO Free Fall with Initial Velocity Upward
 A=g=        -9.80 m/s/s               Yex=Vo*t+0.5*A*t^2
  Vo=        49.00 m/s                    V=V+DV
  Dt=         1.00 s                      Y=Y+V*Dt

   t         V(t+0.5Dt)                   Y(t)       Yex(t)
   s            m/s                        m           m

  0.00       44.10                         0.00        0.00
  1.00       34.30                        44.10       44.10
  2.00       24.50                        78.40       78.40
  3.00       14.70                       102.90      102.90
  4.00        4.90                       117.60      117.60
  5.00       -4.90                       122.50      122.50
  6.00      -14.70                       117.60      117.60
  7.00      -24.50                       102.90      102.90
  8.00      -34.30                        78.40       78.40
  9.00      -44.10                        44.10       44.10
 10.00                                     0.00        0.00
```

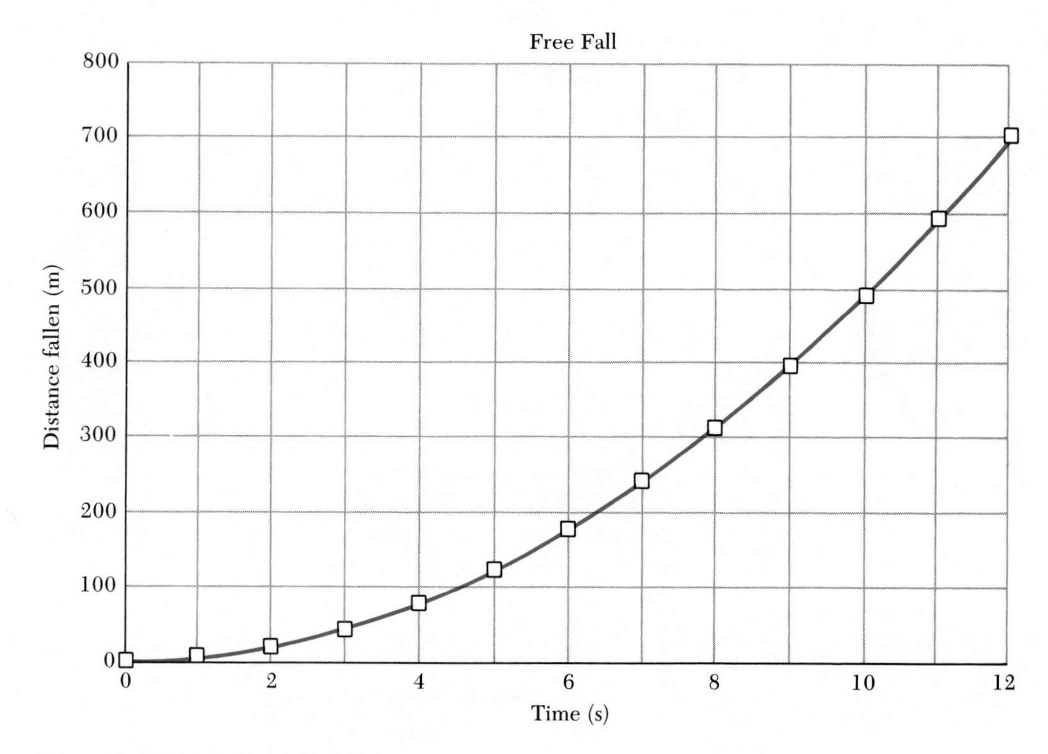


**Figure 2** DISTANCE versus TIME.

Your screen does not exactly look like this. You must clean up your new worksheet. Move the cursor to A1, type in the new file name **EX2FFVO** and press **ENTER**. Type in the new title "Free Fall with Initial Velocity Upward" in cell A2. Find the cell with "Numerical Analysis," and erase the contents by the keystrokes **/RE** and **ENTER** with the cursor specifying the cell. Erase the extraneous cells in the worksheet. The graph must be altered by adding a velocity range column, changing the titles and the legend. The ranges of the X and A variables must also be changed. Try to modify your worksheet to match the worksheet above.

The new graph appears as follows. The values for velocity represent the velocities at the times t + 0.5*Dt even though the velocity points appear graphically at t.

Notice that the velocity decreases, goes through zero, and becomes negative in a symmetric fashion. The object rises to a maximum height and returns to its starting height also in a symmetric fashion. The time up is equal to the time down. Experiment with different values of velocity and observe the results.

A further modification of the free fall worksheet describes the effect of air resistance on the object. Air resistance provides a force that is opposite in direction to the velocity, and the magnitude of this force is proportional to the velocity squared.

$$F=k*v^2$$

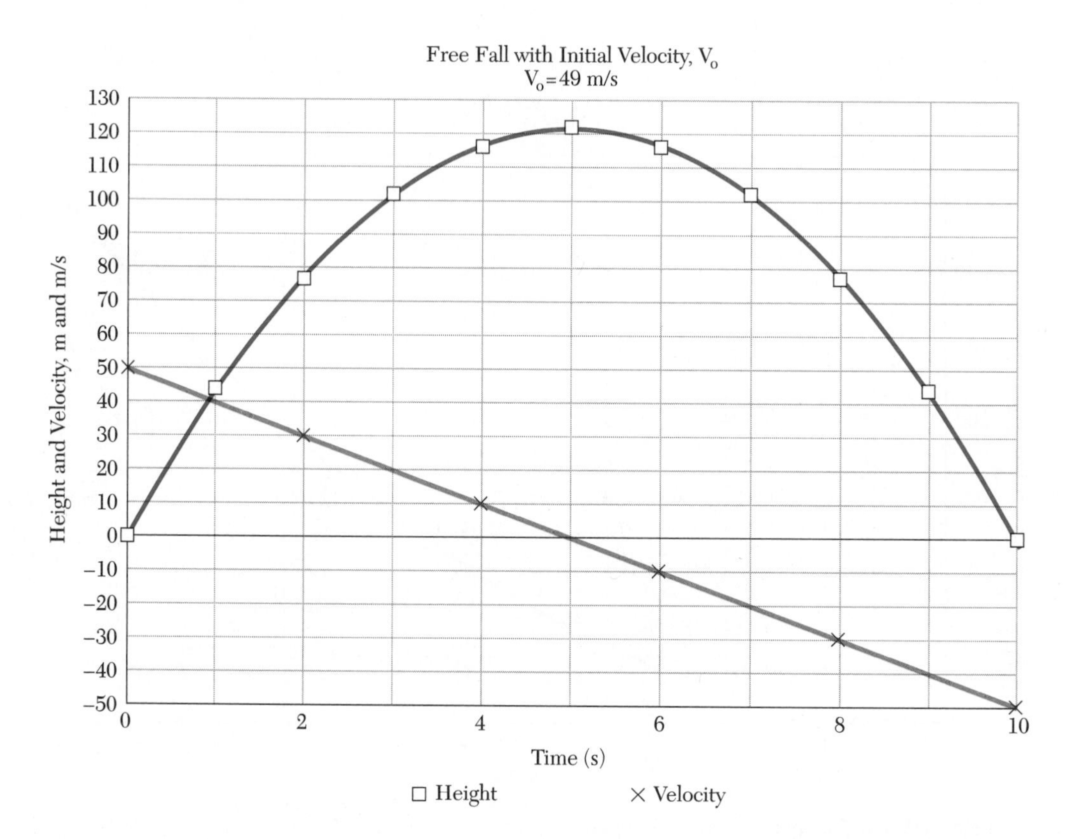


**Figure 3** Ht and V versus TIME.

The net acceleration for the rising object is:

$$a_r=-g-(k/m)*v^2$$

And the net acceleration for the falling object is:

$$a_f=-g+(k/m)*v^2$$

The acceleration is no longer a constant. It is a function of the velocity. Let us construct a worksheet that models an object thrown upward under the influence of gravity and air resistance. The key element in this worksheet is the calculation of a variable acceleration, dependent upon the velocity squared.

```
a=-9.8-v*@ABS(v)
```

The term −v*@ABS(v) gives a magnitude equal to $v^2$ and a direction opposite to the velocity. (@ABS(v) gives the absolute value.)

Retrieve this worksheet by the keystrokes **/FR**, select **METHOD** and press **ENTER**, select EX3VOAR and press **ENTER.** The following worksheet should appear on your screen.

```
EX3VOAR Free Fall with an Initial Velocity and
Air Resistance
 k=     0.15 m            DV=A*Dt                DY=Vav*Dt
 Vo=   20.00 m/s          V=V+DV                 Y=Y+DY
 Dt=    0.10 s            Vav=(VO+V)/2           A=-g-k*V*@ABS(V)
```

| t | A | DV | V | Vav | Y |
|---|---|---|---|---|---|
| s | m/s/s | m/s | m/s | m | m |
| 0.00 | -69.80 | | 20.00 | | 0.00 |
| 0.10 | -35.23 | -6.98 | 13.02 | 16.51 | 1.65 |
| 0.20 | -23.33 | -3.52 | 9.50 | 11.26 | 2.78 |
| 0.30 | -17.50 | -2.33 | 7.16 | 8.33 | 3.61 |
| 0.40 | -14.20 | -1.75 | 5.41 | 6.29 | 4.24 |
| 0.50 | -12.19 | -1.42 | 3.99 | 4.70 | 4.71 |
| 0.60 | -10.96 | -1.22 | 2.78 | 3.38 | 5.05 |
| 0.70 | -10.22 | -1.10 | 1.68 | 2.23 | 5.27 |
| 0.80 | -9.86 | -1.02 | 0.66 | 1.17 | 5.39 |
| 0.90 | -9.78 | -0.99 | -0.33 | 0.16 | 5.40 |
| 1.00 | -9.54 | -0.98 | -1.31 | -0.82 | 5.32 |
| 1.10 | -9.03 | -0.95 | -2.26 | -1.78 | 5.14 |
| 1.20 | -8.30 | -0.90 | -3.17 | -2.71 | 4.87 |
| 1.30 | -7.41 | -0.83 | -3.99 | -3.58 | 4.51 |
| 1.40 | -6.44 | -0.74 | -4.74 | -4.37 | 4.08 |
| 1.50 | -5.46 | -0.64 | -5.38 | -5.06 | 3.57 |
| 1.60 | -4.53 | -0.55 | -5.93 | -5.65 | 3.01 |
| 1.70 | -3.70 | -0.45 | -6.38 | -6.15 | 2.39 |
| 1.80 | -2.97 | -0.37 | -6.75 | -6.56 | 1.74 |
| 1.90 | -2.35 | -0.30 | -7.05 | -6.90 | 1.05 |
| 2.00 | -1.85 | -0.24 | -7.28 | -7.16 | 0.33 |

Let us get a fresh start and see if we can reconstruct this worksheet. I will lead you through the construction. First clear the screen. The keystrokes **/WEY** denote worksheet—erase—yes. You are not permanently erasing the file. Since it is saved on disk, you may retrieve it at any time by **/FR** and **ENTER** with the cursor on the filename.

Type the filename in cell A1, **EX3VOAR.** In cell B1 type the worksheet title, **Free Fall with an Initial Velocity and Air Resistance.** Cells, A4..A6, contain the labels for the constants, **"k=, "Vo=, "Dt=**. The values of the constants, **0.15, 20, 0.10** are entered in cells B4..B6. Type the units for the constants, **m, m/s, s** in cells C4..C6. The equations to the right of the constants represent the equations typed in the main body of the worksheet. Columns representing values of time, acceleration, change in velocity, velocity, average velocity and vertical position represent the main body of the worksheet. Rows 8 and 9 contain labels for the column headings and their units. Enter these as centered (^) labels.

The column for time starts with **0** entered in cell A10 for the initial time value. In cell A11 type the equation **+A10+b$6.** This will add the time interval DT to the value in cell A10. Copy this downward from A11 to A11..A30. This creates time intervals from 0 to 2.0 in increments, Dt = 0.10 s.

The acceleration is computed in the B column. Enter the equation **−9.8 − B$4*D10*@ABS(D10)** in cell B10. As yet there is nothing in cell D10; it will represent the initial velocity. Column C is for the change in velocity, Dv = a*Dt. Type the equation **+B10*B$6** in cell C11 to compute the change in velocity for the interval from t = 0 to t = 0.10 s. The **$** in **B$6** is to have that term copy absolutely (always as cell B6, which represents the con-

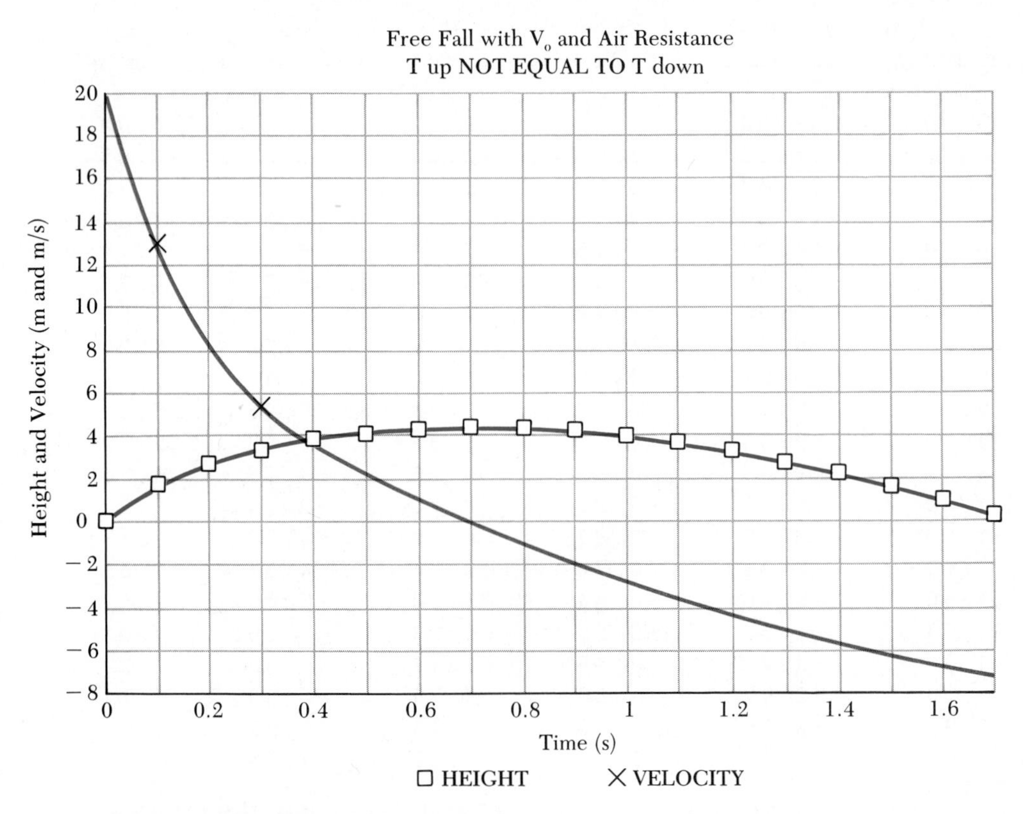


**Figure 4** V & Y versus TIME.

stant Dt). Column D starts out with the initial velocity value in cell B5; enter **+B$5** in cell D10. The equation **+D10+C11,** representing V + DV, is entered in cell D11. An average velocity is computed in column E by entering the equation **(D10 + D11)/2** in cell E11. These values are used to compute the vertical position in Column F. The vertical position is calculated from Y = Yo + $v_{ave}$*Dt. Type **0** in cell F10 and type **+F10 + E11*B$6** in cell F11.

The spreadsheet now takes form. Copy row 11 downward to row 30. Press **/C**, move the cursor to cell A11, press **.** to "pin" the from range and move the cursor to stretch the highlighted range A11..F11. Press **ENTER,** stretch the cursor over the total copy range A11..F30, at the command TO and again press **ENTER.** If you are very lucky, the worksheet takes form. The copy commands are like learning to ride a bicycle, only difficult until you do it once. If all else fails, retrieve the file from disk by **/FR,** selecting EX3VOAR and pressing **ENTER.**

Figure 4 uses an XY plot of velocity (A variable) and position (B variable) versus the independent variable time. Use the **/G** commands to create the graph. The graph is very informative in accessing the effects of air resistance. Notice the time up is not equal to the time down. Why not? Change the time interval to 0.20 s, and note what happens. Why does the velocity become constant? What happens with the acceleration as time progresses? Try to construct a qualitative picture of the quantitative results of the spreadsheet modeling.

These examples indicate the applicability of spreadsheet modeling to the solution of physics problems. With practice the spreadsheet software and the computer will enlarge your understanding of the physical laws describing natural phenomena. The simplifying assumptions, sometimes necessary for analytical solutions, are usually not needed. The computer leads us into more realistic descriptions of our world. The value of spreadsheet modeling is the ease of application, as compared to language programming, afforded to you to *create your own* worksheets.

# Index of Worksheets

| Chapter | Example | Filename | Problems | Filenames |
|---|---|---|---|---|
| 1 Measurement | | | P1.58,59 | MTHNOT.wk1 |
| 2 Vectors | 2.3 | VECTORS.wk1 | P2.17 | ROLCOS.wk1 |
| 3 Motion 1D | 3.2 | MOTION1D.wk1 | | GEOSER.wk1 |
| | | | P3.83 | VARACC.wk1 |
| 4 Motion 2D | 4.1 | MOTION2D.wk1 | | GOLFDR2.wk1 |
| | | | P4.66 | HOMER.wk1 |
| | | | P4.82 | SAFZON.wk1 |
| | | | P4.84 | COYOTE.wk1 |
| 5 Laws of Motion | 5.10 | LAWSMOT.wk1 | | |
| 6 Circular Motion | 6.5 | CIRCMOT.wk1 | P6.54 | ADRAG.wk1 |
| 7 Work & Energy | | | P7.84 | WVARF.wk1 |
| 8 PE & Cons. of E | 8.1 | BAFRFL.wk1 | | CONSME.wk1 |
| | | | P8.56 | PEG&PEND.wk1 |
| | | | P8.66 | FALROP.wk1 |
| 9 P & Collisions | 9.14 | ROCKET.wk1 | | |
| 10 Rotation | 10.6 | RBROT.wk1 | P10.6 | ROTWHL.wk1 |
| 11 Angular Momentum | | | Q11.15 | ROMORB.wk1 |
| 13 Oscillatory Mo. | 13.1 | OSCBOD.wk1 | | TORPEND.wk1 |
| | | | | DAMPOSC.wk1 |
| 14 Universal Grav. | 14.2 | GVERSH.wk1 | | CENFOR1.wk1 |
| | 14.8 | GFORCE.wk1 | | |
| 15 Solids & Fluids | 15.6 | FORDAM.wk1 | P15.76 | PVERSH.wk1 |
| 16 Wave Motion | 16.1 | PULSE.wk1 | P16.52 | TWVEL.wk1 |
| 17 Sound Waves | 17.4 | SNDWAV.wk1 | P17.9 | SNDSPD.wk1 |
| 18 Standing Waves | 18.2 | STAWAV.wk1 | P18.51 | BEATS.wk1 |
| 19 Temperature | 19.1 | CONVTEMP.wk1 | | COOLER.wk1 |
| 20 Heat & 1st Law | 20.9 | HTTRAN.wk1 | P20.79 | HTRATE.wk1 |
| 21 Kinetic Theory | 21.5 | MAXDIS.wk1 | | |
| 22 Entropy | | | | ENTROP.wk1 |
| | | | P23.28 | EFLDRD.wk1 |
| 23 Electric Fields | 23.10 | CHGDSK.wk1 | P23.79 | RODCHG.wk1 |
| | | | P23.80 | RNGCHG.wk1 |
| 24 Gauss' Law | 24.3 | SPSYCD.wk1 | | |

| Chapter | Example | Filename | Problems | Filenames |
|---|---|---|---|---|
| 25 Elec. Potential | 25.2 | MOPEFLD.wk1 | P25.40 | CHGLIN.wk1 |
| 26 Capacitance | 26.5 | REWCC.wk1 | | |
| 27 I & R | 27.4 | RESCOAX.wk1 | P27.70 | RESTOR.wk1 |
| 28 D. C. Circuits | 28.10 | DISCAP.wk1 | | |
| 29 Mag. Fields | 29.2 | FSEMIC.wk1 | | |
| 30 Sources Mag Fd | | | P30.86 | CURLOOP.wk1 |
| 31 Faraday's Law | 31.4 | BFORBAR.wk1 | | |
| 32 Inductance | 32.3 | RLCIR.wk1 | | |
| 33 AC Circuits | | | | LCRCIR.wk1 |
| | | | | LCRCIRD.wk1 |
| 34 E.M. Waves | | | P34.54 | FG&FR.wk1 |
| 35 Nature of Light | 35.1 | FIZEAU.wk1 | P35.5,6 | FIZEAU.wk1 |
| 36 Geo. Optics | | | | |
| 37 Interference | | | | |
| 38 Diff. & Polar. | | | | |
| 39 Relativity | | | | |
| 40 Intro. Q. Phys. | | | | |
| 41 Quant. Mech. | | | | |
| 42 Atomic Phy. | 42.3 | GSHYD.wk1 | | |
| 43 Mole. & Sol. | | | | |
| 44 Supercond. | | | P44.30 | ZERORES.wk1 |
| 45 Nuclear Str. | 45.5 | RADDATE.wk1 | P45.58 | RADMAT.wk1 |
| | | | | RADACT.wk1 |
| 46 Nuc. Phy. App. | | | | ALPSCAT.wk1 |
| 47 Part. Phy. & Cos. | | | | |

# Answers to Odd-numbered Problems

## CHAPTER 23

**1.** $5.14 \times 10^5$ N
**3.** $1.60 \times 10^{-9}$ N repelling one another
**5.** $(8.50 \times 10^{-2}$ N$)\boldsymbol{i}$
**7.** 0.873 N at 330°
**9.** 40.9 N at 263°
**11.** $2.51 \times 10^{-10}$
**13.** 3.60 MN down on the top and up on the bottom of the cloud
**15.** (a) $(-5.58 \times 10^{-11}$ N/C$)\boldsymbol{j}$ (b) $(1.02 \times 10^{-7}$ N/C$)\boldsymbol{j}$
**17.** (a) $(-5.20 \times 10^3$ N/C$)\boldsymbol{i}$ (b) $(2.93 \times 10^3$ N/C$)\boldsymbol{j}$ (c) $5.85 \times 10^3$ N/C at 225°
**19.** (a) $(1.29 \times 10^4$ N/C$)\boldsymbol{j}$ (b) $(-3.87 \times 10^{-2}$ N$)\boldsymbol{j}$
**21.** $\boldsymbol{E} = 0$
**23.** (a) at the center (b) $\left(\frac{\sqrt{3}\,kq}{a^2}\right)\boldsymbol{j}$
**25.** (a) $0.914\frac{kq}{a^2}$ at 225° (b) $0.914\frac{kq^2}{a^2}$ at 45°
**27.** 7.82 m to the left of the negative charge
**29.** $-\left(\frac{k\lambda_0}{x_0}\right)\boldsymbol{i}$
**31.** (a) $(6.65 \times 10^6$ N/C$)\boldsymbol{i}$ (b) $(2.42 \times 10^7$ N/C$)\boldsymbol{i}$ (c) $(6.40 \times 10^6$ N/C$)\boldsymbol{i}$ (d) $(6.65 \times 10^5$ N/C$)\boldsymbol{i}$
**33.** (a) 0.145 C/m³ (b) $1.94 \times 10^{-3}$ C/m²
**35.** (a) $9.35 \times 10^7$ N/C away from the center; $1.039 \times 10^8$ N/C is 10.0% larger (b) 515.1 kN/C away from the center; 519.3 kN/C is 0.8% larger
**37.** $7.20 \times 10^7$ N/C away from the center; $1.00 \times 10^8$ N/C axially away
**39.** $(-21.6$ MN/C$)\boldsymbol{i}$
**41.**

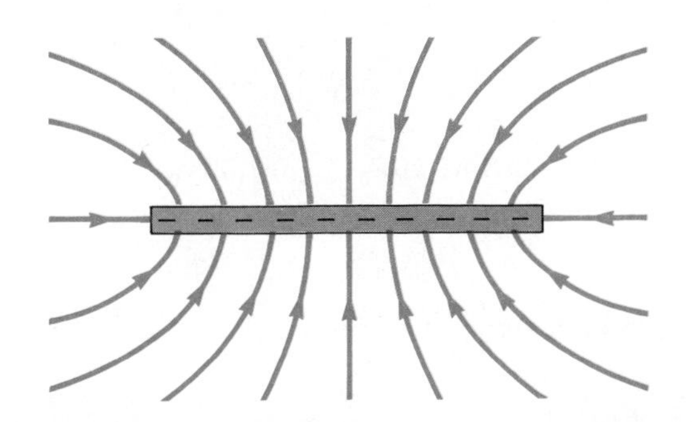

**43.** (a) $\frac{q_1}{q_2} = -1/3$ (b) $q_1$ is negative and $q_2$ is positive
**45.** (a) $6.14 \times 10^{10}$ m/s² (b) 19.5 μs (c) 11.7 m (d) 1.20 fJ
**47.** $1.00 \times 10^3$ N/C in the direction of the beam
**49.** (a) $(-5.75 \times 10^{13}$ m/s²$)\boldsymbol{i}$ (b) $2.84 \times 10^6$ m/s (c) 49.4 ns
**51.** (a) 111 ns (b) 5.67 mm (c) (450 km/s)$\boldsymbol{i}$ + (102 km/s)$\boldsymbol{j}$
**53.** (a) 36.9°, 53.1° (b) 167 ns, 221 ns
**55.** (a) 10.9 nC (b) $5.43 \times 10^{-3}$ N
**57.** (a) $\theta_1 = \theta_2$
**59.** 204 nC
**63.** (a) $-\left(\frac{4kq}{3a^2}\right)\boldsymbol{j}$ (b) (0, 2.00 m)
**67.** $5.27 \times 10^{17}$ m/s²; 0.854 mm
**71.** (a) $\boldsymbol{F} = \frac{kq^2}{s^2}(1.90)(\boldsymbol{i} + \boldsymbol{j} + \boldsymbol{k})$

(b) $F = 3.29\frac{kq^2}{s^2}$ in a direction away from the vertex diagonally opposite to it

## CHAPTER 24

**1.** (a) $1.98 \times 10^6$ Nm²/C (b) 0 (c) $1.92 \times 10^6$ Nm²/C
**3.** (a) aA (b) bA (c) 0
**5.** $4.14 \times 10^6$ N/C
**7.** $bhw^2/3$
**9.** $1.87 \times 10^3$ Nm²/C
**11.** $5.65 \times 10^5$ Nm²/C
**13.** (a) $1.36 \times 10^6$ Nm²/C (b) $6.78 \times 10^5$ Nm²/C (c) No, the same field lines go through spheres of all sizes.
**15.** $-6.89 \times 10^6$ Nm²/C. The number of lines entering exceeds the number leaving by 2.91 times or more.
**17.** 13.7 μC, No
**19.** 28.3 N·m²/C
**21.** (a) 761 nC (b) It may have any distribution. Any and all point and smeared-out charges, positive and negative, must add algebraically to +761 nC. (c) Total charge is −761 nC
**23.** (a) $\frac{Q}{2\epsilon_0}$ (out of the volume enclosed)

(b) $-\frac{Q}{2\epsilon_0}$ (into it).
**25.** (a) 0 (b) $7.20 \times 10^6$ N/C away from the center
**27.** (a) 0.713 μC (b) 5.7 μC
**29.** (a) 0 (b) $(3.66 \times 10^5$ N/C$)\hat{\boldsymbol{r}}$ (c) $(1.46 \times 10^6$ N/C$)\hat{\boldsymbol{r}}$ (d) $(6.50 \times 10^5$ N/C$)\hat{\boldsymbol{r}}$
**31.** $\boldsymbol{E} = (a/2\epsilon_0)\hat{\boldsymbol{r}}$
**33.** (a) $5.14 \times 10^4$ N/C outward (b) 646 Nm²/C
**35.** $\boldsymbol{E} = (\rho r/2\epsilon_0)\hat{\boldsymbol{r}}$

37. $5.08 \times 10^5$ N/C up.
39. (a) 0 (b) $5.40 \times 10^3$ N/C
(c) 540 N/C, both radially outward
41. (a) 80.0 nC/m² on each face (b) $(9.04 \times 10^3 \text{ N/C})\mathbf{k}$
(c) $(-9.04 \times 10^3 \text{ N/C})\mathbf{k}$
43. (a) 0 (b) $(125 \times 10^4 \text{ N/C})\hat{\mathbf{r}}$ (c) $(639 \text{ N/C})\hat{\mathbf{r}}$
(d) No change
45. (a) 0 (b) $(8.00 \times 10^7 \text{ N/C})\hat{\mathbf{r}}$ (c) 0
(d) $(7.35 \times 10^6 \text{ N/C})\hat{\mathbf{r}}$
47. (a) $-\lambda, +3\lambda$ (b) $\left(\frac{3\lambda}{2\pi\epsilon_0 r}\right)\hat{\mathbf{r}}$
49. (b) $\frac{Q}{2\epsilon_0}$ (c) $\frac{Q}{\epsilon_0}$
51. (a) $\mathbf{E} = \left(\frac{\rho r}{3\epsilon_0}\right)\hat{\mathbf{r}}$ for $r < a$; $\mathbf{E} = \left(\frac{kQ}{r^2}\right)\hat{\mathbf{r}}$ for $a < r < b$;
$\mathbf{E} = 0$ for $b < r < c$; $\mathbf{E} = \left(\frac{kQ}{r^2}\right)\hat{\mathbf{r}}$ for $r > c$
(b) $\sigma_1 = -\frac{Q}{4\pi b^2}$ inner; $\sigma_2 = +\frac{Q}{4\pi c^2}$ outer
53. (c) $f = \frac{1}{2\pi}\sqrt{\frac{ke^2}{mR^3}}$ (d) 102 pm
57. $\mathbf{g} = -\left(\frac{GMr}{R_e{}^3}\right)\hat{\mathbf{r}}$
59. (a) $\sigma/\epsilon_0$ to the left (b) zero (c) $\sigma/\epsilon_0$ to the right
63. $\mathbf{E} = \frac{\rho a}{3\epsilon_0}\mathbf{j}$

## CHAPTER 25

1. 0
3. (a) 152 km/s (b) $6.50 \times 10^6$ m/s
5. (a) 2.7 keV (b) 509 km/s
7. $6.41 \times 10^{-19}$ C
9. $2.10 \times 10^6$ m/s
11. 1.35 MJ
13. 432 V; 432 eV
15. $-38.9$ V; the origin
17. $1.56 \times 10^3$ N/C
19. 260 V, B
21. 2.00 m
23. 119 nC, 2.67 m
25. $-88.2$ kV
27. $-11.0$ MV
29. (a) $-386$ nJ. Positive binding energy would have to be put in to separate them. (b) 103 V
31. (a) $-27.3$ eV (b) $-6.81$ eV (c) 0
35. $W_{\text{ext}} = -20.1$ J
37. $-(0.553)\frac{kQ}{R}$
39. (a) C/m² (b) $k\alpha\left[L - d\ln\frac{d+L}{d}\right]$
41. $V = \frac{\sigma}{2\epsilon_0}(\sqrt{x^2+b^2} - \sqrt{x^2+a^2})$
43. $\mathbf{E} = (6xy - 5)\mathbf{i} + (3x^2 - 2z^2)\mathbf{j} - 4yz\mathbf{k}$; 7.07 N/C
45. (a) 10 V, $-11$ V, $-32$ V
(b) $(7 \text{ N/C})\mathbf{i}$, $(7 \text{ N/C})\mathbf{i}$, $(7 \text{ N/C})\mathbf{i}$
47. (a) 0 (b) $\left(\frac{kQ}{r^2}\right)\hat{\mathbf{r}}$
49. 3.38 MV; 3.00 MV is 11.1% smaller
51. $1.56 \times 10^{12}$ electrons removed
53. (a) 0, 1.67 MV (b) $5.85 \times 10^6$ N/C away, 1.17 MV (c) $1.19 \times 10^7$ N/C away, 1.67 MV
55. (a) $4.50 \times 10^7$ N/C outward, 30.0 MN/C outward
(b) 1.80 MV
57. (a) 450 kV (b) 7.50 μC
59. 5.00 μC
61. (a) 6.00 m (b) $-2.00$ μC
63. 1.73 MV
65. (a) 180 kV (b) 127 kV
67. 1.12 MV; 1.00 MV is smaller by 11.1%
69.

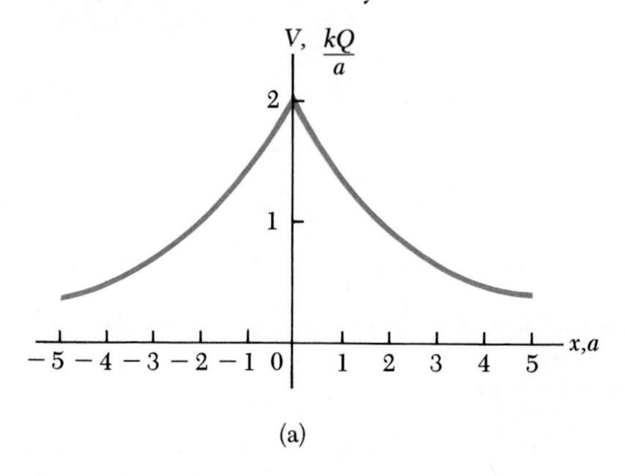


(a)

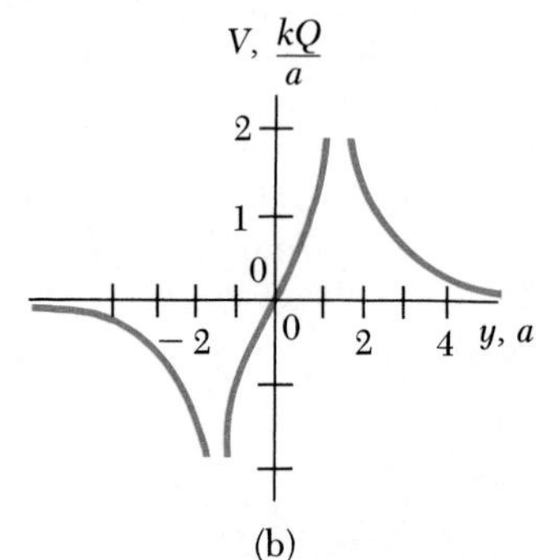


(b)

71. (a) $1.26\sigma_0$ (b) $1.26E_0$ (c) $1.59V_0$
73. $k\frac{Q^2}{2R}$
75. $V_2 - V_1 = -\frac{\lambda}{2\pi\epsilon_0}\ln\left(\frac{r_2}{r_1}\right)$
79. $E_y = k\frac{Q}{\ell y}\left[1 - \frac{y^2}{\ell^2 + y^2 + \ell\sqrt{\ell^2 + y^2}}\right]$
81. (a) $E_r = \frac{2kp\cos\theta}{r^3}$; $E_\theta = \frac{kp\sin\theta}{r^3}$; yes; no
(b) $\mathbf{E} = \frac{3kpxy\mathbf{i}}{(x^2+y^2)^{5/2}} + \frac{kp(2y^2 - x^2)\mathbf{j}}{(x^2+y^2)^{5/2}}$

83. $\frac{3}{5}\left(\frac{kQ^2}{R}\right)$

87. $x = 0.57735$ m and $x = 0.24859$ m

## CHAPTER 26

1. 13.3 kV
3. 1.40 pF
5. 684 $\mu$C
7. (a) 1.33 $\mu$C/m$^2$ (b) 13.3 pF
9. (a) $4\pi\epsilon_0(R_1 + R_2)$ (b) $\frac{Q_1}{Q_2} = \frac{R_1}{R_2}$
11. 1.52 mm 13. 4.42 $\mu$m
15. (a) $1.11 \times 10^4$ N/C toward the negative plate (b) 98.3 nC/m$^2$ (c) 3.74 pF (d) 74.8 pC
17. 109 mm
19. 3.54 nC
21. (a) 2.68 nF (b) 3.02 kV
23. (a) 15.6 pF (b) 256 kV
25. 66.7 nC
27. 18.0 $\mu$F
29. (a) 4.00 $\mu$F (b) 8.00 V, 4.00 V, 12.0 V, 24.0 $\mu$C, 24.0 $\mu$C, 24.0 $\mu$C
31. (a) 5.96 $\mu$F (b) 89.2 $\mu$C, 63.1 $\mu$C, 26.4 $\mu$C, 26.4 $\mu$C
33. 120 $\mu$C; 80.0 $\mu$C and 40.0 $\mu$C
35. (a) 12.0 $\mu$F (b) 24.0 $\mu$C, 14.4 $\mu$C, 19.2 $\mu$C, 19.2 $\mu$C
37. 10
39. 83.6 $\mu$C
41. 12.9 $\mu$F
43. $\frac{\epsilon_0 A}{(s-d)}$
45. 90.0 mJ
47. (a) 55.9 $\mu$C (b) 4.65 V
49. 800 pJ, 5.79 mJ/m$^3$
55. (a) 8.13 nF (b) 2.40 kV
57. 16.7 pF, 1.62 kV
59. 1.04 m
61. $\kappa = 8.00$
63. 10.1 V
65. 416 pF
67. 1.00 $\mu$F and 3.00 $\mu$F
69. 3.00 $\mu$F
71. 2.33
73. (a) 243 $\mu$J (b) 2.30 mJ
75. 4.29 $\mu$F
77. 480 V
79. 0.188 m$^2$
81. 3.00 $\mu$F
83. (b) $Q/Q_0 = \kappa$
85. (a) $\frac{\epsilon_0}{d}[\ell^2 + \ell x(\kappa - 1)]$ (b) $\frac{\epsilon_0 V^2}{2d}[\ell x(\kappa - 1) + \ell^2]$ (c) $\frac{\epsilon_0 V^2}{2d}\ell(\kappa - 1)$ to the right (d) $1.55 \times 10^{-3}$ N
89. 19.0 kV
91. 3.00 $\mu$F

## CHAPTER 27

1. 481 nA
3. 400 nA
5. 33.3 C
7. $5.90 \times 10^{28}$/m$^3$
9. (a) $1.50 \times 10^5$ A (b) $5.40 \times 10^8$ C
11. 13.3 $\mu$A/m$^2$
13. $1.32 \times 10^{11}$ A/m$^2$
15. 1.59 $\Omega$
17. $3.23 \times 10^6/\Omega\cdot$m
19. 2.57 mm
21. 1.56$R$
23. (a) 1.82 m (b) 280 $\mu$m
25. 6.43 A
27. (a) 3.75 k$\Omega$ (b) 536 m
29. (a) $3.15 \times 10^{-8}$ $\Omega\cdot$m (b) $6.35 \times 10^6$ A/m$^2$ (c) 49.9 mA (d) $6.59 \times 10^{-4}$ m/s (assume 1 conduction electron per atom) (e) 0.400 V
31. 0.125
33. 20.8 $\Omega$
35. 67.6°C
37. 26.2°C
39. $3.03 \times 10^7$ A/m$^2$
41. 21.2 nm
43. 0.833 W
45. 36.1%
47. 82.9 V
49. (a) 133 $\Omega$ (b) 9.42 m
51. 28.9 $\Omega$
53. 26.9 cents/day
55. (a) 184 W (b) 461°C
57. 2020°C
59. (a) 667 A (b) 50.0 km
63. (a) $R = \frac{\rho L}{\pi(r_b^2 - r_a^2)}$ (b) 37.4 M$\Omega$
69. (a) $\rho = 1.49 \times 10^{-6}$ $\Omega\cdot$m (b) $\rho = 1.70 \times 10^{-6}$ $\Omega\cdot$m

## CHAPTER 28

1. $R = 7.67$ $\Omega$
3. (a) 1.79 A (b) 10.4 V
5. 12.0 $\Omega$
7. (a) 6.73 $\Omega$ (b) 1.98 $\Omega$
9. (a) 4.59 $\Omega$ (b) 8.16%
11. $0.923\ \Omega \le R \le 9.0\ \Omega$
13. 1.00 k$\Omega$
15. 1.99 A, 1.17 A, 0.819 A
17. $\frac{6}{11}R$
19. 14.3 W, 28.5 W, 1.33 W, 4.00 W
21. (a) 0.227 A (b) 5.68 V
23. 470 $\Omega$; 220 $\Omega$
25. $-10.4$ V

27. $\frac{11}{13}$ A, $\frac{6}{13}$ A, $\frac{17}{13}$ A
29. Starter: 171 A    Battery: 0.283 A
31. 3.50 A, 2.50 A, 1.00 A
33. (a) $I_1 = \frac{5}{13}$ mA; $I_2 = \frac{40}{13}$ mA; $I_3 = \frac{35}{13}$ mA
(b) 69.2 V; c
35. (a) 0.667 A    (b) 50.0 $\mu$C
37. (a) 909 mA    (b) −1.82 V
39. 800 W, 450 W, 25.0 W, 25.0 W
41. 3.00 J
43. 4.06 $\mu$A
45. (a) 12.0 s    (b) $i(t) = (3.00\ \mu\text{A})e^{-t/12.1}$
$q(t) = (36.0\ \mu\text{C})[1 - e^{-t/12.1}]$
47. (a) 6.00 V    (b) 8.29 $\mu$s
49. (a) 6.50 mJ    (b) 6.50 mJ
51. 1.60 MΩ
53. 16.6 kΩ
55. 0.302 Ω
57. 49.9 kΩ
59. (b) 0.0501 Ω, 0.451 Ω
61. 0.588 A
63. 60.0
65. (a) 12.5 A, 6.25 A, 8.33 A    (b) 27.1 A; No, it would not be sufficient since the current drawn is greater than 25 A.
67. (a) 0.101 W    (b) 10.1 W
69. (a) 16.7 A    (b) 33.3 A    (c) The 120-V heater requires four times as much mass.
71. 6.00 Ω; 3.00 Ω
73. (a) 72.0 W    (b) 72.0 W
75. (a) 40 W    (b) 80 V, 40 V, 40 V
77. (a) $R \leq 1050\ \Omega$    (b) $R \geq 10.0\ \Omega$
79. (a) $R \to \infty$    (b) $R \to 0$    (c) $R \to r$
81. (a) 9.93 $\mu$C    (b) $3.37 \times 10^{-8}$ A
(c) $3.34 \times 10^{-7}$ W    (d) $3.37 \times 10^{-7}$ W
83. $T = (R_A + 2R_B)C \ln 2$
85. $R = 0.521\ \Omega$, 0.260 Ω, 0.260 Ω, assuming resistors are in series with the galvanometer
87. (a) 1/3 mA for $R_1$, $R_2$    (b) 50 $\mu$C
(c) $(0.278\ \text{mA})e^{-t/(0.18\text{s})}$    (d) 0.290 s
89. (a) 1.96 $\mu$C    (b) 53.3 Ω
91. (a) $\ln \frac{\mathcal{E}}{V} = (0.0118\ \text{s}^{-1})t + 0.0882$
(b) 85 s ± 6%; 8.5 $\mu$F ± 6%

## CHAPTER 29

1. (a) West    (b) zero deflection    (c) up    (d) down
3. $2.31 \times 10^{-13}$ N
5. 48.8° or 131°
7. $(12.3\boldsymbol{i} + 4.48\boldsymbol{j} - 1.60\boldsymbol{k}) \times 10^{-18}$ N
9. $2.34 \times 10^{-18}$ N
11. zero
13. $(-2.88\ \text{N})\boldsymbol{j}$
15. 0.245 T east
17. (a) 4.73 N    (b) 5.46 N    (c) 4.73 N
19. $(0.042\ \text{T})\boldsymbol{k}$
21. $F = 2\pi rIB \sin\theta$, up
23. 9.98 N·m, clockwise as seen looking in the negative $y$-direction
25. (a) 376 $\mu$A    (b) 1.67 $\mu$A
27. 58.9 mJ
29. 1.98 cm
31. 182 $\mu$T
33. $r_\alpha = r_d = \sqrt{2}r_p$
35. 7.88 pT
37. 2.99 u; ${}^3_1H^+$ or ${}^3_2He^+$
39. $5.93 \times 10^5$ N/C
41. $mg = 8.93 \times 10^{-30}$ N down, $qE = 1.60 \times 10^{-17}$ N up, $qvB = 4.74 \times 10^{-17}$ N down
43. 0.278 m
45. 31.2 cm
47. (a) $4.31 \times 10^7$ rad/s    (b) $5.17 \times 10^7$ m/s
49. 70.1 mT
51. $3.70 \times 10^{-9}$ m³/C
53. $4.32 \times 10^{-5}$ T
55. $7.37 \times 10^{28}$ electrons/m³
57. 128 mT at 78.7° below the horizon
59. (a) $(3.52\boldsymbol{i} - 1.60\boldsymbol{j}) \times 10^{-18}$ N    (b) 24.4°
61. 0.588 T
65. 3.13 cm
67. $3.82 \times 10^{-25}$ kg
69. $3.70 \times 10^{-24}$ N·m
71. (a) $(-8.00 \times 10^{-21}\ \text{kg}\cdot\text{m/s})\boldsymbol{j}$    (b) 8.91°

## CHAPTER 30

1. 200 nT
3. 31.4 cm
5. 28.3 $\mu$T
7. 24.6 $\mu$T
9. $\frac{\mu_0 I}{4\pi x}$ into the plane of the paper
11. 330 nT
13. (a) 189 $\mu$T    (b) 7.66 cm
15. 80.0 $\mu$N/m
17. 2000 N/m; attractive
19. $(-27.0\ \mu\text{N})\boldsymbol{i}$
21. 13.0 $\mu$T directed downward
23. (a) 3.98 kA    (b) ≈0
25. 5.03 T
27. (a) 3.60 T    (b) 1.94 T
29. 5.00 cm
31. (a) $6.40 \times 10^{-3}$ N/m, inward
(b) $\boldsymbol{F}$ is greatest at the outer surface
33. 26.7 mT, 37.7 mT
35. 4.74 mT
37. (a) 3.13 mWb    (b) zero
39. 2.27 $\mu$Wb
41. (a) $(8.00\ \mu\text{A})e^{-t/4}$    (b) 2.94 $\mu$A
43. (a) $11.3 \times 10^9$ V·m/s    (b) 100 mA
45. 1.0001
47. 191 mT

49. 277 mA
51. 150 $\mu$Wb
53. $M/H$
55. $1.27 \times 10^3$ turns
57. 2.02
59. (a) $1.88 \times 10^{46}$ (b) $8.74 \times 10^{20}$ kg
61. 675 A down
63. 81.7 A
65. $\boldsymbol{B} = \frac{\mu_0 I}{2\pi w} \ln\left(\frac{b+w}{b}\right) \boldsymbol{k}$
67. 594 A east
69. $1.43 \times 10^{-10}$ T directed away from the center
73. (a) $B = \frac{1}{3}\mu_0 b r_1^2$ (b) $B = \frac{\mu_0 b R^3}{3r_2}$
75. (a) 2.46 N (b) 107.3 m/s$^2$
77. (a) 12.0 kA-turns/m (b) $2.07 \times 10^{-5}$ T·m/A
79. 366 g/mol
83. $\frac{\mu_0 I}{4\pi}(1 - e^{-2\pi})$ perpendicularly out of the paper
85. $\frac{\mu_0 I}{\pi r}\left(\frac{2r^2 + a^2}{4r^2 + a^2}\right)$ up
87. $\frac{4}{3}\rho\mu_0\omega R^2$
89. $\frac{4}{15}\pi\rho\omega R^5$

## CHAPTER 31

1. 500 mV
3. 160 A
5. 2.67 T/s
7. 61.8 mV
9. $(200\ \mu\text{V})e^{-t/7}$
11. $Nn\pi R^2 \mu_0 I_0 \alpha e^{-\alpha t} = (68.2\ \text{mV})e^{-1.6t}$ counterclockwise
13. 272 m
15. $\frac{2}{3}\pi R^2 N B_0 \omega \sin \omega t$ counterclockwise looking to the left
17. 2.81 mT
19. (a) 3.00 N to the right (b) 6.00 W
21. 2.00 mV; the west end is positive in the northern hemisphere
23. 2.83 mV
25. (a) $\frac{1}{8} B\omega L^2$ (b) 360 mV
27. (a) to the right (b) to the right (c) to the right (d) into the plane of the paper
29. 0.742 T
31. 114 $\mu$V clockwise
33. $1.80 \times 10^{-3}$ N/C counterclockwise
35. (a) $(9.87 \times 10^{-3}\ \text{V/m})\cos(100\pi t)$ (b) clockwise
37. (a) 1.60 A counterclockwise (b) 20.1 $\mu$T (c) up
39. 12.6 mV
41. (a) 7.54 kV (b) $\boldsymbol{B}$ is parallel to the plane of the loop.
43. $(28.6\ \text{mV})\sin(4\pi t)$
45. (a) 0.640 N·m (b) 241 W
47. 0.513 T
49. (a) $F = \frac{N^2 B^2 w^2 v}{R}$ to the left (b) 0
(c) $F = \frac{N^2 B^2 w^2 v}{R}$ to the left
51. $(-2.87\boldsymbol{j} + 5.75\boldsymbol{k}) \times 10^9$ m/s$^2$
53. It is, with the top end in the picture positive.
57. (a) 36.0 V (b) 600 mWb/s (c) 35.9 V (d) 4.32 N·m
59. (a) $(1.19\ \text{V})\cos(120\pi t)$ (b) 88.5 mW
61. Moving east; 458 $\mu$V
63. It is, with the left end in the picture positive
65. 1.20 $\mu$C
67. 6.00 A
69. (a) 900 mA (b) 108 mN (c) b (d) No
71. (a) Counterclockwise (b) $\frac{K\pi r^2}{R}$
73. (a) $\frac{\mu_0 IL}{2\pi} \ln\left(\frac{h+w}{h}\right)$ (b) $-4.80\ \mu$V

## CHAPTER 32

1. 100 V
3. 4.69 mH
7. (a) 188 $\mu$T (b) 33.3 nWb (c) 375 $\mu$H (d) field and flux
9. 21.0 $\mu$Wb
11. $(18.8\ \text{V})\cos(377t)$
13. $\frac{1}{2}$
15. (a) 15.8 $\mu$H (b) 12.6 mH
19. 2.61 H
23. (a) 139 ms (b) 461 ms
25. (a) 5.66 ms (b) 1.22 A (c) 58.1 ms
27. (a) 113 mA (b) 600 mA
29. (a) 0.800 (b) 0
31. (a) 1.00 A (b) 12.0 V, 1.20 kV, 1.21 kV (c) 7.62 ms
33. 2.44 $\mu$J
35. (a) 20.0 W (b) 20.0 W (c) 0 (d) 20.0 J
37. (a) $8.06 \times 10^6$ J/m$^3$ (b) 6.32 kJ
39. 44.3 nJ/m$^3$ in the $\boldsymbol{E}$ field; 995 $\mu$J/m$^3$ in the $\boldsymbol{B}$ field
41. 30.7 $\mu$J and 72.2 $\mu$J
43. (a) 500 mJ (b) 17.0 W (c) 11.0 W
45. $(-0.168t + 0.112)$V
47. 80.0 mH
49. 553 $\mu$H
51. (a) 18.0 mH (b) 34.3 mH (c) $-9.00$ mV
53. 400 mA
55. (a) 5.19 mA (b) 1.72 MHz (c) 7.68 nJ
57. 2.60 pF
59. (a) 36.0 $\mu$F (b) 8.00 ms
61. (a) 6.03 J (b) 0.529 J (c) 6.56 J
63. (a) 2.51 kHz (b) 69.9 $\Omega$
65. (a) 4.47 krad/s (b) 4.36 krad/s (c) 2.53%
67. (a) True. (b) There is no ratio, for the *RLC* circuit is overdamped and has no frequency of oscillation.
69. 979 mH

73. 95.6 mH
79. (a) 72.0 V; b (c) 75.2 $\mu s$
81. $\frac{\mu_0}{\pi} \ln \left(\frac{d-a}{a}\right)$
83. 300 Ω
85. (a) 56.9 W (b) 36.0 W (c) 20.9 W

## CHAPTER 33

3. 2.95 A, 70.7 V
5. (a) −1.44 A (b) 1.44 A
7. 14.6 Hz
9. (a) 42.4 mH (b) 943 Hz
11. 7.03 H
13. 5.60 A
15. 76.8 mH
17. (a) $f > 41.3$ Hz (b) $X_C < 87.5$ Ω
19. 2.77 nC
21. 100 mA
23. 0.427 A
25. (a) 78.5 Ω (b) 1.59 kΩ (c) 1.52 kΩ (d) 138 mA (e) −84.3°
27. (a) 17.4° (b) voltage leads the current
29. 61.6 Hz
31. (a) 146 V (b) 213 V (c) 179 V (d) 33.4 V
33. (a) 194 V (b) current leads by 33.3°
35. 1.43 W
37. 8.00 W
39. $v(t) = (283 \text{ V}) \sin(628t)$
41. (a) 16.0 Ω (b) −12.0 Ω
43. 159 Hz
45. 1.82 pF
47. (a) 124 nF (b) 51.5 kV
49. (a) 5.00 A (b) 2.77 A (c) 2.77 A
51. (a) 613 $\mu$F (b) 0.756
53. (a) 0.9998 (b) 0.0199
55. (a) True, if $\omega_0 = (LC)^{-1/2}$ (b) 0.107, 0.999, 0.137
57. 0.317
59. 687 V
61. 2.38 A, 240 V
63. 25.3 A
65. 56.7 W
67. (a) 1.89 A (b) $V_R = 39.7$ V, $V_L = 30.1$ V, $V_C = 175$ V (c) $\cos\phi = 0.265$
(d)

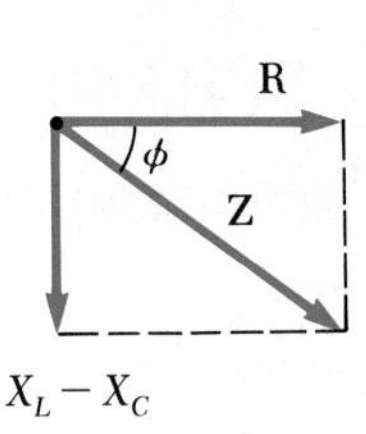


69. 99.6 mH
71. (a) 2.90 kW (b) $5.80 \times 10^{-3}$
(c) If the generator is limited to 4500 V, no more than 16 kW could be delivered to the load, never 5000 kW.
73. (a) Circuit (a) is a high-pass filter, (b) is a low-pass filter.
(b) $\frac{V_{out}}{V_{in}} = \frac{\sqrt{R_L^2 + X_L^2}}{\sqrt{R_L^2 + (X_L - X_C)^2}}$ for circuit (a);
$\frac{V_{out}}{V_{in}} = \frac{X_C}{\sqrt{R_L^2 + (X_L - X_C)^2}}$ for circuit (b)
75. 302 mH
77. (a) 200 mA; voltage leads by 36.8° (b) 40.0 V; $\phi = 0°$ (c) 20.0 V; $\phi = -90°$ (d) 50.0 V; $\phi = +90°$
83. (a) 173 Ω (b) 8.66 V
85. (a) 1.838 kHz
87.

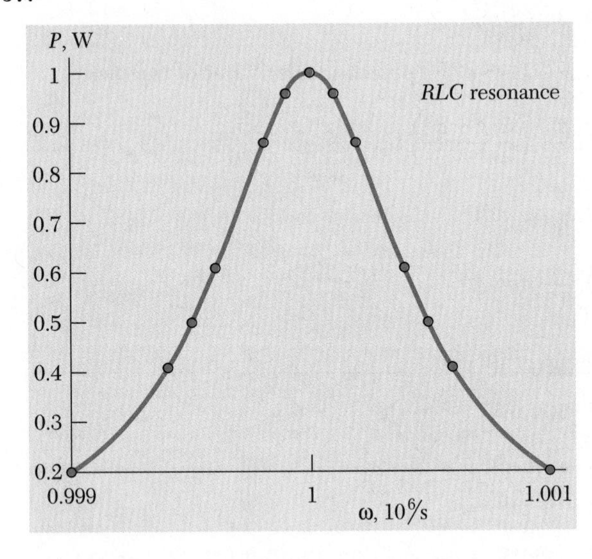


## CHAPTER 34

3. (b) 377 Ω
5. (a) 162 N/C (b) 130 N/C
9. (a) 6.00 MHz (b) (73.3 nT)($-\mathbf{k}$)
(c) $\mathbf{B} = (73.3 \text{ nT}) \cos(0.126x - 3.77 \times 10^7 t)(-\mathbf{k})$
11. (a) $B = 333$ nT (b) $\lambda = 628$ nm (c) $f = 4.77 \times 10^{14}$ Hz
13. 5.16 m
15. 307 $\mu W/m^2$
17. 66.7 kW
19. (a) 332 $kW/m^2$ (b) 1.88 kV/m, 222 $\mu$T
21. (a) 971 V/m (b) 16.7 pJ
23. (a) 540 N/C (b) 2.58 $\mu J/m^3$ (c) 774 $W/m^2$ (d) This is 77.4% of the flux mentioned in Example 34.3 and 57.8% of the 1340 $W/m^2$ intensity above the atmosphere. It may be cloudy at this location, or the sun may be setting.
25. 83.3 nPa
27. (a) 1.90 kV/m (b) 50.0 pJ (c) $1.67 \times 10^{-19}$ kg·m/s

29. (a) 11.3 kJ (b) $5.65 \times 10^{-5}$ kg·m/s
31. (a) 2.26 kW (b) 4.71 kW/m²
33. 7.50 m
35. (a) $\mathcal{E}_m = 2\pi^2 r^2 f B_m \cos\theta$, where $\theta$ is the angle between the magnetic field and the normal to the loop. (b) Vertical, with its plane pointing toward the broadcast antenna
37. (a) 0.933 (b) 0.500 (c) 0
39. (a) 6.00 pm (b) 7.50 cm
41. (a) 30.0 GHz (b) $3.00 \times 10^{14}$ Hz (c) $5.17 \times 10^{14}$ Hz (d) $3.00 \times 10^{15}$ Hz (e) $3.00 \times 10^{20}$ Hz
43. (a) 4.17 m to 4.55 m (b) 3.41 to 3.66 m (c) 1.61 m to 1.67 m
45. $3.33 \times 10^{3}$ m²
47. (a) $6.67 \times 10^{-16}$ T (b) $5.31 \times 10^{-17}$ W/m² (c) $1.67 \times 10^{-14}$ W (d) $5.56 \times 10^{-23}$ N
49. $6.37 \times 10^{-7}$ Pa
51. 95.1 mV/m
53. $7.50 \times 10^{10}$ s = 2370 y; out the back
55. $3.00 \times 10^{-2}$ degrees
57. (a) $B_0 = 583$ nT, $k = 419$ rad/m, $\omega = 1.26 \times 10^{11}$ rad/s; $xz$ (b) 40.6 W/m² in average value (c) 271 nPa (d) 406 nm/s²
59. (a) 22.6 h (b) 30.5 s
61. (b) 1.00 MV/s
63. (a) 3.33 m; 11.1 ns; 6.67 pT

(b) $\boldsymbol{E} = (2.00 \text{ mV/m}) \cos 2\pi \left(\dfrac{x}{3.33 \text{ m}} - \dfrac{t}{11.1 \text{ ns}}\right) \boldsymbol{j}$

$\boldsymbol{B} = (6.67 \text{ pT}) \cos 2\pi \left(\dfrac{x}{3.33 \text{ m}} - \dfrac{t}{11.1 \text{ ns}}\right) \boldsymbol{k}$

(c) $5.31 \times 10^{-9}$ W/m² (d) $1.77 \times 10^{-14}$ J/m³

(e) $3.54 \times 10^{-14}$ Pa

## CHAPTER 35

1. $2.998 \times 10^{8}$ m/s
3. $1.98 \times 10^{8}$ km
5. 114 rad/s
7. 20 μs, 60 μs
9. (a) $4.74 \times 10^{14}$ Hz (b) 422 nm (c) $2.00 \times 10^{8}$ m/s
11. 70.5° from the vertical
13. 61.3°
15. 19.5°, 19.5°, 30.0°.
17. (a) 327 nm (b) 287 nm
19. 0.890
21. 30.0°, 19.5° at entry; 40.5°, 77.1° at exit
23. 7.89°
27. 18.4°
29. 86.8°
31. 4.61°
33. 62.4°
35. (a) 24.4° (b) 37.0° (c) 49.8°
37. 1.00008
39. $\theta < 48.2°$
41. 53.6°
43. 2.27 m
45. 2.37 cm
49. 90°, 30°, No
51. 62.2%
53. $\sin^{-1}[(n^2 - 1)^{1/2} \sin\phi - \cos\phi]$ If $n \sin\phi \leq 1$, $\theta = 0$
55. 82
57. 28.7°

## CHAPTER 36

3. 2′11″
5. 30 cm
7. (a) $s' = 45.0$ cm, $M = -\frac{1}{2}$ (b) $s' = -60.0$ cm, $M = 3.00$ (c) Similar to Figures 36.6 and 36.9b
9. (a) $s' = -13.3$ cm, $M = 0.667$ (b) $s' = -24.0$ cm, $M = 0.400$ (c) See Figure 36.8
11. concave with radius 40.0 cm
13. (a) 2.08 m (concave) (b) 1.25 m in front of the object
15. (a) $s' = -12.0$ cm, $M = 0.400$ (b) $s' = -15.0$ cm, $M = 0.250$ (c) The images are erect
17. 11.8 cm above the floor
19. (a) −13.33 cm (b) −9.23 cm (c) −3.79 cm
21. 8.57 cm
23. 3.88 mm
25. 2
27. (a) 16.4 cm (b) 16.4 cm
29. 25.0 cm; −0.250
31.

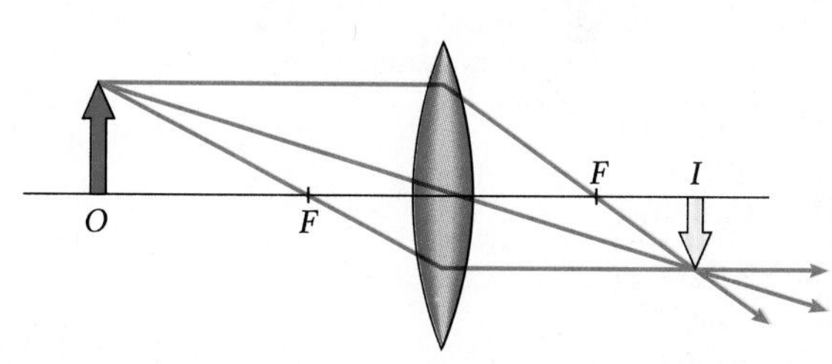


33. (a) −12.3 cm, to left of lens (b) −0.615
35. −48.4 cm
37. 2.84 cm
39. $\dfrac{f}{1.41}$
41. −4 diopters, a diverging lens
43. 0.558 cm
45. +2.89 diopters
47. 3.5
49. −800, image is inverted
51. −18.8
53. 2.14 cm
55. (a) 24.5 cm (b) 99.0 cm

57.

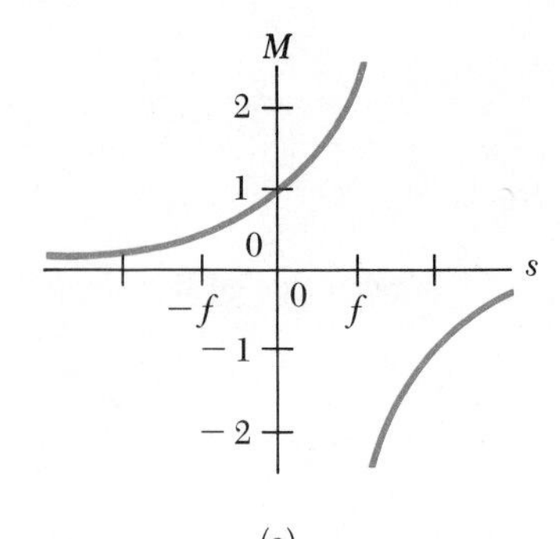


(a)

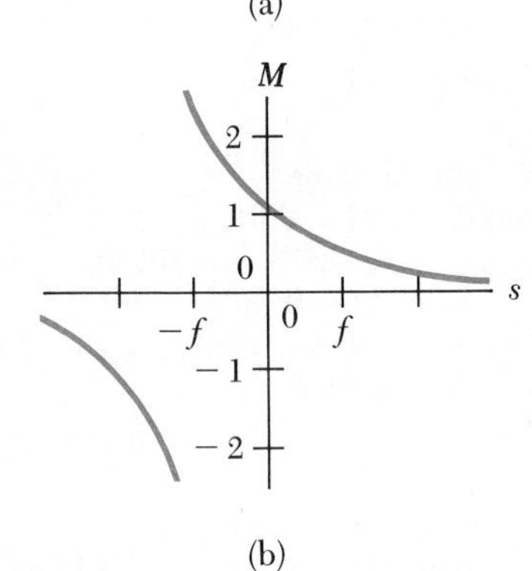


(b)

59. 10.7 cm beyond the curved surface
61. 20.0 cm
63. (a) 44.6 diopters (b) 3.03 diopters
65. $d = 8$ cm
67. (a) $1.33f$ (b) $0.750f$ (c) $-3.00$; $+4.00$
69. (a) 52.5 cm (b) 1.50 cm
71. $s' = 1.5$ m, $h' = -13.1$ mm
73. (a) $-0.400$ cm (b) $s' = -3.94$ mm, $h' = 535$ $\mu$m
75. (a) 30.0 cm and 120 cm (b) 24.0 cm
(c) real, inverted, diminished
77. real, inverted, actual size

## CHAPTER 37

1. (a) $2.62 \times 10^{-3}$ m (b) $2.62 \times 10^{-3}$ m
3. 515 nm
5. 0.340 m
7. $4.80 \times 10^{-3}$ m
9. $4.48 \times 10^{-3}$%
11. $1.56 \times 10^{-4}$ m
13. $4.80 \times 10^{-5}$ m
15. 423.5 nm
17.

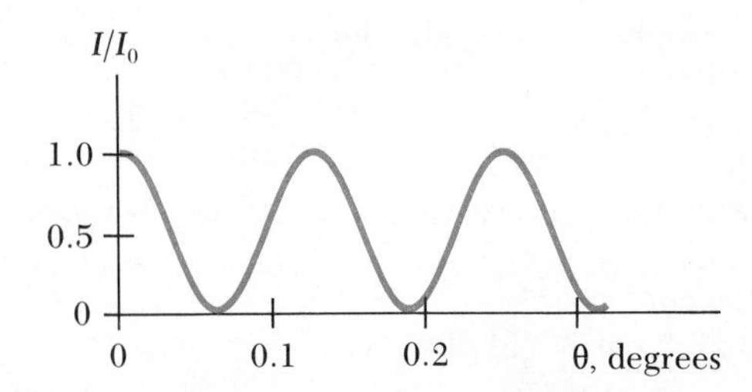


19. (a) $1.27 \times 10^{-2}$ degrees (b) $5.97 \times 10^{-2}$ degrees
21. (a) 2.63 rad (b) 246 nm
23. (a) 7.95 rad (b) 0.453
25. $E_R = 1.00E_0$, $\alpha = 3\pi/2$
27. $10 \sin(100\pi t + 0.93)$
29. $\pi/2$
31. 360°/N
33. No reflection maxima in the visible spectrum
35. 94.0 nm
37. 85.4 nm, or 256 nm, or 427 nm . . .
39. (a) Yellow (b) Violet
41. 167 nm
43. 4.35 $\mu$m
45. $3.96 \times 10^{-5}$ m
47. 654 dark fringes
49. 0.0162 m
51. 3.58°
53. 421 nm
55. 2.52 cm
57. 1.54 mm
59. $3.6 \times 10^{-5}$ m
61. $x_{\text{bright}} = \dfrac{\lambda \ell (m + \frac{1}{2})}{2hn}$, $x_{\text{dark}} = \dfrac{\lambda \ell m}{2hn}$
63. (b) 115 nm
65. 1.7
67. For each centimeter of vertical displacement, the thickness varies by 492 nm.
69. (b) 266 nm

## CHAPTER 38

1. (a) 0.105° (b) 2.20 mm
3. 560 nm
5. $\cong 10^{-3}$ rad
7. 0.230 mm
9. $\phi = \beta/2 = 1.392$ rad

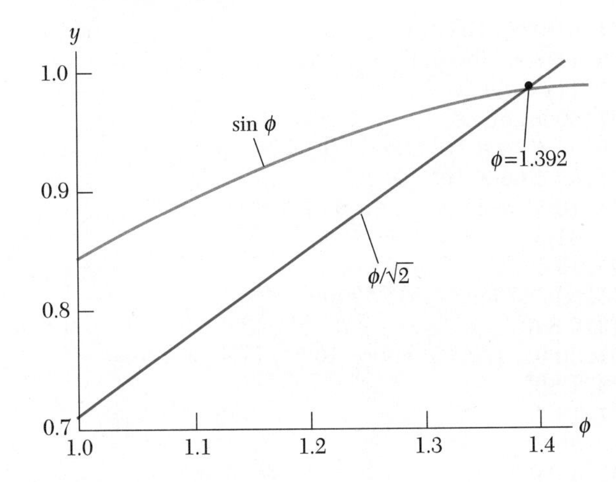


11. $1.62 \times 10^{-2}$
13. 3.09 m

15. (a) 1.03 cm (b) 20.6 cm
17. $2.54 \times 10^{-7}$ rad
19. 15.7 km
21. 13 m
23. 105 m
25. (a) 5.93° (b) 6.43°
27. (a) 5.7; 5 orders is the maximum
(b) 10; 10 orders to be in the short-wavelength region.
29. 0.0683°
31. (a) 2800 (b) 4.72 $\mu$m
33. 19.5°
35. 7.35°
37. 14.7°
39. 31.9°
41. It is not possible.
43. (a) 0.281 nm (b) 0.18%
45. 14.4°
47. 1.11
49. 31.2°
51. $\frac{3}{8} I_0$
53. (a) 54.7° (b) 63.4° (c) 71.6°
55. 60.5°
57. 2.2
59. 5.9°
61. 0.244 rad ≈ 14.0°
63. 660 nm
65. (a) 12 000, 24 000, 36 000 (b) 0.0111 nm
67. (a) $3.53 \times 10^3$ lines/cm
(b) Eleven maxima can be observed.
69. $\frac{1}{8}$
73. (a) 41.8° (b) 0.593 (c) 0.262 m

## CHAPTER 39

5. (a) 60 m/s (b) 20 m/s (c) 44.7 m/s
7. 0.866$c$
9. 1.54 ns
11. $0.436L_0$
13. (a) 2.18 $\mu$s (b) 649 m
15. 0.1404$c$
17. 0.960$c$
19. 42 g/cm$^3$
21. 1625 MeV/$c$
25. (a) 939.4 MeV (b) $3.008 \times 10^3$ MeV
(c) $2.069 \times 10^3$ MeV
27. 0.864$c$
29. (a) 0.582 MeV (b) 2.45 MeV
31. (a) $3.91 \times 10^4$ (b) 0.9999999997$c$ (c) 7.66 cm
33. 4 MeV and 29 MeV
35. (a) 3.29 MeV (b) 2.77 MeV
37. $4.2 \times 10^9$ kg/s
39. 65/min, 10.6/min.
41. 0.7%
43. (a) 0.0236$c$ (b) $6.18 \times 10^{-4}\, c$
45. 0.8$c$
47. 6.17 ns
49. 5600 MeV
51. 1.47 km
55. (b) For $v \ll c$, $a = qE/m$ as in the classical description. As $v \to c$, $a \to 0$, describing how the particle can never reach the speed of light.
(c) Perform $\int_0^v \left(1 - \frac{v^2}{c^2}\right)^{-3/2} dv = \int_0^t \frac{qE}{m}\, dt$ to obtain $v = \frac{qEct}{\sqrt{m^2c^2 + q^2E^2t^2}}$ and then

$$\int_0^x dx = \int_0^t \frac{qEtc}{\sqrt{m^2c^2 + q^2E^2t^2}}\, dt$$

## CHAPTER 40

1. (a) 2.57 eV (b) $1.28 \times 10^{-5}$ eV
(c) $1.91 \times 10^{-7}$ eV
3. $2.27 \times 10^{30}$ photons/s
5. $1.86 \times 10^{-34}$
7. $4.46 \times 10^3$ K
9. 9.35 $\mu$m; infrared
11. $5.18 \times 10^3$ K
15. (a) 0.350 eV (b) 555 nm
17. (a) 1.92 eV (b) 0.159 V
19. (a) 0.571 eV (b) 1.54 V
21. (a) only lithium (b) 0.808 eV
23. 1.50 V
25. 1.78 eV, $9.47 \times 10^{-28}$ kg·m/s
27. 23.4°
29. (a) 3.10 keV (b) $0.110c = 32.9 \times 10^6$ m/s
31. (a) $2.88 \times 10^{-12}$ m (b) 101°
33. (a) 0.1203 nm, 0.1212 nm, 0.1224 nm, 0.1236 nm, 0.1245 nm, 0.1248 nm (b) 28.0 eV, 104 eV, 205 eV, 305 eV, 377 eV, 403 eV (c) 180°
35. 70.1°
37. 121 nm, 103 nm, 97.2 nm
39. (a) 5 (b) No; No
41. 0.529 Å, 2.12 Å, 4.77 Å
43. (a) $E_n = -54.4 \frac{\text{eV}}{n^2}$, $n = 1, 2, 3, \ldots$
(b) $-54.4$ eV
45. (a) 0.265 Å (b) 0.177 Å (c) 0.132 Å
47. (a) 3.03 eV (b) 411 nm
(c) $7.32 \times 10^{14}$ Hz
49. (a) B (b) D (c) B and C
51. $-6.80$ eV, $+3.40$ eV
53. $r_n = (1.06\ \text{Å})n^2$, $E_n = -6.80\ \text{eV}/n^2$, $n = 1, 2, 3, \ldots$
55. 5.39 keV
57. (a) 41.4° (b) 680 keV
59. (a) 3.12 fm (b) $-18.9$ MeV
61. (a) $-6.67$ keV, $-668$ eV (b) 0.621 nm
63. (a) 91.1 keV (b) 8.90 keV
(c) 57.5°
65. $3.55 \times 10^6$ m

## CHAPTER 41

1. $3.97 \times 10^{-13}$ m
3. $1.77 \times 10^{-36}$ m
5. $2.21 \times 10^{-34}$ m
7. $1.23\sqrt{V}$ nm
9. 33.6 eV
11. 2.18 nm
13. 6.03 pm
15. 546 eV
17. (a) 15 keV (b) 124 keV
19. $2.3 \times 10^{6}$ m/s
21. $9.08 \times 10^{-14}$ m
23. $2.6 \times 10^{-16}$ m
25. 0.250
27. 0.513 MeV, 2.05 MeV, 4.62 MeV
29. $9.56 \times 10^{12}$
31. 0.516 MeV, $3.31 \times 10^{-20}$ kg·m/s
33. (a) 37.7 eV, 151 eV, 339 eV, 603 eV
    (b) 2.20 nm, 2.75 nm, 4.71 nm, 4.12 nm, 6.59 nm, 11.0 nm
35. 7.93 Å
37. $\pm\sqrt{\dfrac{2}{L}}$
39. $E = \dfrac{\hbar^2 k^2}{2m}$
43. (a)

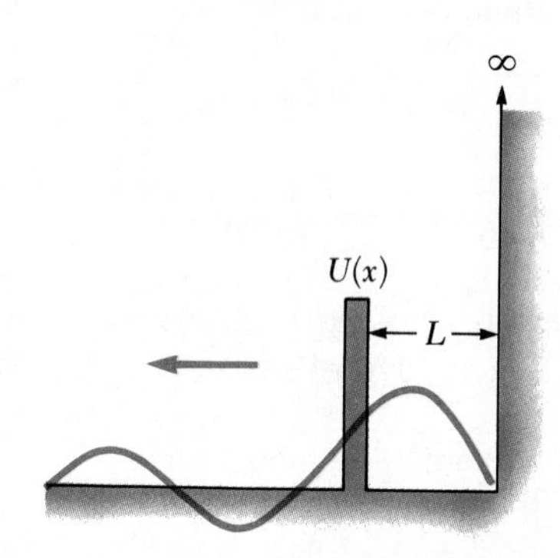


    (b) $2L$
45. 1% chance of transmission, 99% chance of reflection
49. $2.02 \times 10^{-4}$ nm; $\gamma$-ray
51. (a) 0.1955 (b) 0.3320
    (c) Yes. At large quantum numbers the probability approaches 1/3.
53. $5.62 \times 10^{-8}$
55. $\lambda = 2 \times 10^{-10}$ m, $p = 3.31 \times 10^{-24}$ kg·m/s, $E = 0.172$ eV
59. 0.0294
61. (a) $\Delta p \approx \dfrac{\hbar}{r}$ (b) $E = \dfrac{\hbar^2}{2mr^2} - \dfrac{ke^2}{r}$ (c) $-13.6$ eV
63. (a) $\sqrt{\left(\dfrac{nhc}{2L}\right)^2 + (m_0c^2)^2} - m_0c^2 \quad n = 1, 2, 3, \ldots$
    (b) $4.69 \times 10^{-14}$ J; 28.6%
65. (a)

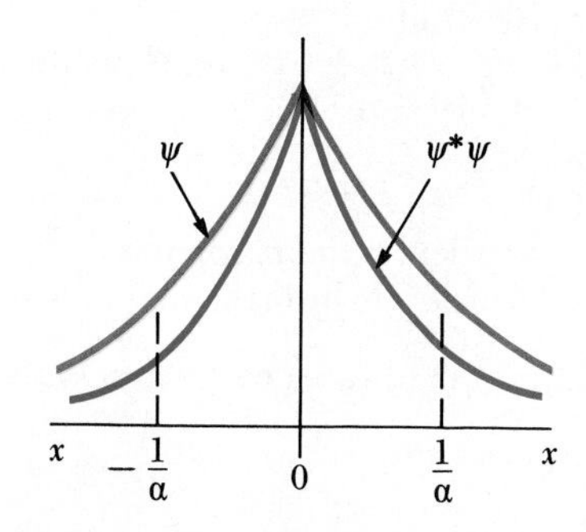


    (b) $P(x \to x + dx) = \psi^*\psi\, dx = \begin{cases} A^2 e^{-2\alpha x}\, dx \text{ for } x > 0 \\ A^2 e^{2\alpha x}\, dx \text{ for } x < 0 \end{cases}$
    (c) $\psi$ is continuous, $\psi \to 0$ as $x \to \pm\infty$, $\psi$ is finite at $-\infty < x < \infty$.
    (d) $A = \sqrt{\alpha}$ (e) 0.632
67. $1.03 \times 10^{-3}$
69. (b) $\frac{3}{2}\hbar\omega$; the first excited state
71. (a) $\sqrt{\dfrac{2}{17L}}$ (b) Any such that $|A|^2 + |B|^2 = \dfrac{1}{a}$

## CHAPTER 42

1. (a) $\ell = 0, m_\ell = 0$
    $\ell = 1, m_\ell = -1, 0, 1$
    $\ell = 2, m_\ell = -2, -1, 0, 1, 2$
    (b) All have energy $-6.05$ eV
3. 1312 nm (b) 164 nm
7. $r = 4a_0$
9. 0.000257 nm (b) $-18.8$ MeV (c) 14.1 MeV
11. 4
13. (a) 2 (b) 8 (c) 18 (d) 32 (e) 50
15. $\pm 2\hbar, \pm\hbar, 0$
17. $3\hbar$
19. $L = 2.58 \times 10^{-34}$ J·s; $L_z = -2\hbar, -\hbar, 0, \hbar$, and $2\hbar$
    $\theta = 145°, 114°, 90.0°, 65.9°$, and $35.3°$
21. $9.27 \times 10^{-24}$ J/T
25. $1s^2 2s^2 2p^6 3s^1$ — $^{11}$Na
    $1s^2 2s^2 2p^6 3s^2$ — $^{12}$Mg
    $1s^2 2s^2 2p^6 3s^2 3p^1$ — $^{13}$Al
    $1s^2 2s^2 2p^6 3s^2 3p^2$ — $^{14}$Si
    $1s^2 2s^2 2p^6 3s^2 3p^3$ — $^{15}$P
    $1s^2 2s^2 2p^6 3s^2 3p^4$ — $^{16}$S
    $1s^2 2s^2 2p^6 3s^2 3p^5$ — $^{17}$Cl
    $1s^2 2s^2 2p^6 3s^2 3p^6$ — $^{18}$Ar
    $1s^2 2s^2 2p^6 3s^2 3p^6 4s^1$ — $^{19}$K
27. $1.24 \times 10^{4}$ V
29. 0.0310 nm
31. 0.195 nm
33. 590 nm
35. $3.50 \times 10^{16}$ photons
37. 9.79 GHz
39. (a) 137.036 (b) $\dfrac{2\pi}{\alpha}$ (c) $\dfrac{1}{2\pi\alpha}$ (d) $\dfrac{4\pi}{\alpha}$

41. (a) $1.57 \times 10^{14}$ m$^{-3/2}$ (b) $2.47 \times 10^{28}$ m$^{-3}$
(c) $8.69 \times 10^{8}$ m$^{-1}$
45. (a) 4.20 mm (b) $1.05 \times 10^{19}$ photons
(c) $8.82 \times 10^{16}$/mm$^3$
47. (a) $\dfrac{\mu_0 \pi m^2 e^7}{8\epsilon_0{}^3 h^5}$ (b) 12.5 T
49. $5.24a_0$
51. 0.125
53. (a) 0.24 s (b) Light energy is quantized.

## CHAPTER 43

1. (a) $2.94 \times 10^{-9}$ N, (b) $-5.14$ eV
3. $12.4 \times 10^{-27}$ kg, $12.2 \times 10^{-27}$ kg
5. (a) 0.0118 nm (b) 0.00772 nm; HI is less stiff
7. 0, 36.9 $\mu$eV, 111 $\mu$eV, 221 $\mu$eV, 369 $\mu$eV, . . .
9. $5.69 \times 10^{12}$ rad/s
11. 0.358 eV
13. $3.63 \times 10^{-4}$ eV, $1.09 \times 10^{-3}$ eV
15. $-8.39$ eV
19. 0.500
21. 5.28 eV
23. $3.4 \times 10^{17}$ electrons
25. (a) 1.10 (b) $1.55 \times 10^{-25}$
27. 0.671 eV
29. 1.24 eV or less, yes
31. $v = 7$
33. (a) 4.23 eV (b) $3.27 \times 10^{4}$ K

## CHAPTER 44

1. (a) 30.76 T (b) $3.07 \times 10^{5}$ A (c) 7.69 T
3. 200 A
5. (a) 39.01 nm (b) 4.15 nm (c) 122 nm
7. $4 \times 10^{-25}$ Ω
9. 318 A
11.

| Superconductor | $E_g$ (meV) | Superconductor | $E_g$ (meV) |
|---|---|---|---|
| Al | 0.36 | Sn | 1.13 |
| Ga | 0.33 | Ta | 1.36 |
| Hg | 1.26 | Ti | 0.12 |
| In | 1.04 | V | 1.61 |
| Nb | 2.82 | W | 0.005 |
| Pb | 2.19 | Zn | 0.26 |

13. 0.028 eV
15. $5.2 \times 10^{9}$ pairs
17. $\sim 10^{7}$ fluxons
19. $1.65 \times 10^{-14}$ T
21. (a) 0.377 T (b) $1.89 \times 10^{-3}$ Wb
(c) $9.15 \times 10^{11}$
23. (a) 0.05 T (b) 0.02 T (c) $2.29 \times 10^{6}$ J
25. 1.4 T, No
27. 1.56 cm
29. $\Delta S = 9.83 \times 10^{-3}$ J/mol·K

## CHAPTER 45

1. (a) 1.9 fm (b) 7.44 fm (c) 3.92
3. $2.3 \times 10^{15}$ g
5. Even Z; He, O, Ca, Ni, Sn, Pb
Even N; T, He, N, O, Cl, K, Ca, V, Cr, Sr, Y, Zr, Xe, Ba, La, Ce, Pr, Nd, Pb, Bi, Po
7. $3.18 \times 10^{-2}$%
9. 12.7 km
13. (a) $4.55 \times 10^{-13}$ m (b) $6.03 \times 10^{6}$ m/s
15. 2.85 MeV/nucleon
17. (a) 59.915428 u (b) 8.781 MeV/nucleon, yes
19. 3.53 MeV
21. 7.93 MeV
23. (a) $\dfrac{A-Z}{Z}$ is greatest for $^{139}_{55}$Cs and is equal to 1.53.
(b) $^{139}$La (c) $^{139}$Cs
25. 160 MeV
27. (a) 491.3 MeV (b) 179%, 53%, 25%, 1%
29. 1155 s
31. (a) $1.55 \times 10^{-5}$ s$^{-1}$, 12.4 h
(b) $2.39 \times 10^{13}$ atoms (c) 1.87 mCi
33. 36.3 $\mu$g
35. 0.55 mCi
37. $9.46 \times 10^{9}$ nuclei
39. 4.28 MeV
43. 5.16 MeV
45.

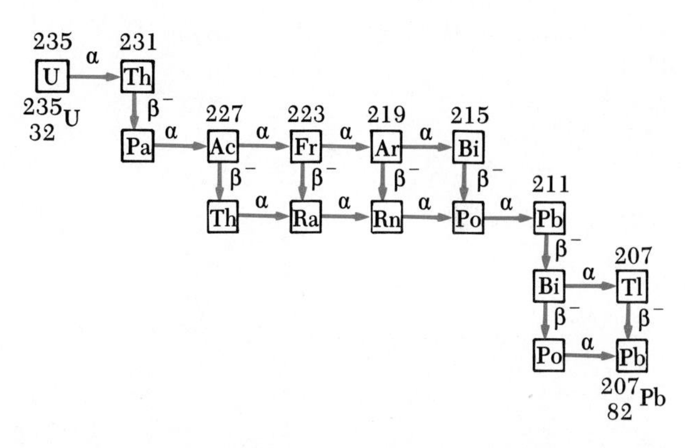


47. (b) $e^- + {}^{15}O \rightarrow {}^{15}N + \nu$ (c) $E_\nu = 2.75$ MeV
49. $-2.64$ MeV
53. 10.0135 u, 8.0053 u
55. 30% and 70%
57. (a) 5.70 MeV (b) 3.27 MeV; exothermic
59. (a) $6.40 \times 10^{24}$ (b) $58.6 \times 10^{11}$ decays/s
(c) 1.10 million years
61. $6 \times 10^{19}$ Bq
63. $1.02 \times 10^{3}$ Bq/g
65. (a) 2.75 fm (b) 152 N (c) 2.62 MeV
(d) $R = 7.44$ fm, $F = 379$ N, $U = 17.6$ MeV
67. $2.2 \times 10^{-6}$ eV
69. (a) $7.7 \times 10^{4}$ eV (b) 4.62 MeV, 13.9 MeV
(c) $10.3 \times 10^{6}$ kW·h
71. (a) 12.2 mg (b) 166 mW

73. 0.35%
75. (b) 4.79 MeV
77. (a) The 2.44 MeV state cannot be reached by $\beta^+$ decay. (b) ${}^{93}_{42}Mo$
79. $1.49 \times 10^3$ Bq
81. $4.45 \times 10^{-8}$ kg/h
83. (b) $T_{1/2} = 1.46$ min $\pm 6\%$ (c) $\lambda = 0.473$ min$^{-1} \pm 6\%$

## CHAPTER 46

1. ${}^{142}_{54}Xe$, ${}^{143}_{54}Xe$, ${}^{144}_{54}Xe$
3. ${}^{233}_{92}U + \beta^-$
5. $1.28 \times 10^{25}$, $1.61 \times 10^{26}$
7. 5800 km
9. 2664 y
11. (c) 10%, 1.0%
13. (a) $2.22 \times 10^{-6}$ m/s (b) $4.5 \times 10^{-8}$ s
15. (a) $2.4 \times 10^7$ J (b) 10.6 kg
17. 1570 y
19. (a) 0.0025 rem per x-ray (b) 38 times the background
21. About 24 days
23. 1.14 rad
25. (a) $3.1 \times 10^7$ (b) $3.1 \times 10^{10}$ electrons
27. 25.6 MeV
29. 488 Ci
31. 26 collisions
33. (a) 1.36 GJ (b) The fusion energy is 40 times greater than the gasoline energy.
35. (a) $4.55 \times 10^{-24}$ kg·m/s (b) 0.146 nm
37. 14.1 MeV
39. (a) $22 \times 10^6$ kW·h (b) 17.6 MeV for each D-T fusion (c) $2.34 \times 10^8$ kW·h (d) 9.37 kW·h
41. (a) $5.7 \times 10^8$ K (b) 120 kJ
43. (b) 25.7 MeV
45. (b) 35.2 (c) $2.89 \times 10^{15}$

## CHAPTER 47

1. $2.26 \times 10^{23}$ Hz, $1.32 \times 10^{-15}$ m
3. $2.06 \times 10^{-18}$ m
5. $10^{-23}$ s
7. $\Omega^+ \rightarrow \overline{\Lambda}^0 + K^+$
   $\overline{K}^0 \rightarrow \pi^+ + \pi^-$
   $\overline{\Lambda}^0 \rightarrow \overline{p} + \Pi^+$
   $\overline{n} \rightarrow \overline{p} + e^+ + \nu_e$
9. (b) Strangeness is not conserved.
11. Baryon number conservation
13. (a,b,c) conserved (d,e,f) not conserved
15. (a) lepton number (b) lepton number (c) strangeness and charge (d) baryon number (e) strangeness
19. (a) $d\overline{u} + uud \rightarrow d\overline{s} + uds$
   (b) $\overline{d}u + uud \rightarrow u\overline{s} + uus$
   (c) $\overline{u}s + uud \rightarrow \overline{u}s + d\overline{s} + sss$
   (d) $uud + uud \rightarrow d\overline{s} + uud + u\overline{d} + uds$
21. (a) $0.384c$, 38.4% of the speed of light (b) $6.7 \times 10^9$ lightyears
23. 2 MeV
25. One part in 50 million
27. 9.3 cm
29. 70.4 MeV for each proton

# Index

Page numbers in *italics* indicate illustrations; page numbers followed by "n" indicate footnotes; page numbers followed by "t" indicate tables.